# WINTER SKY



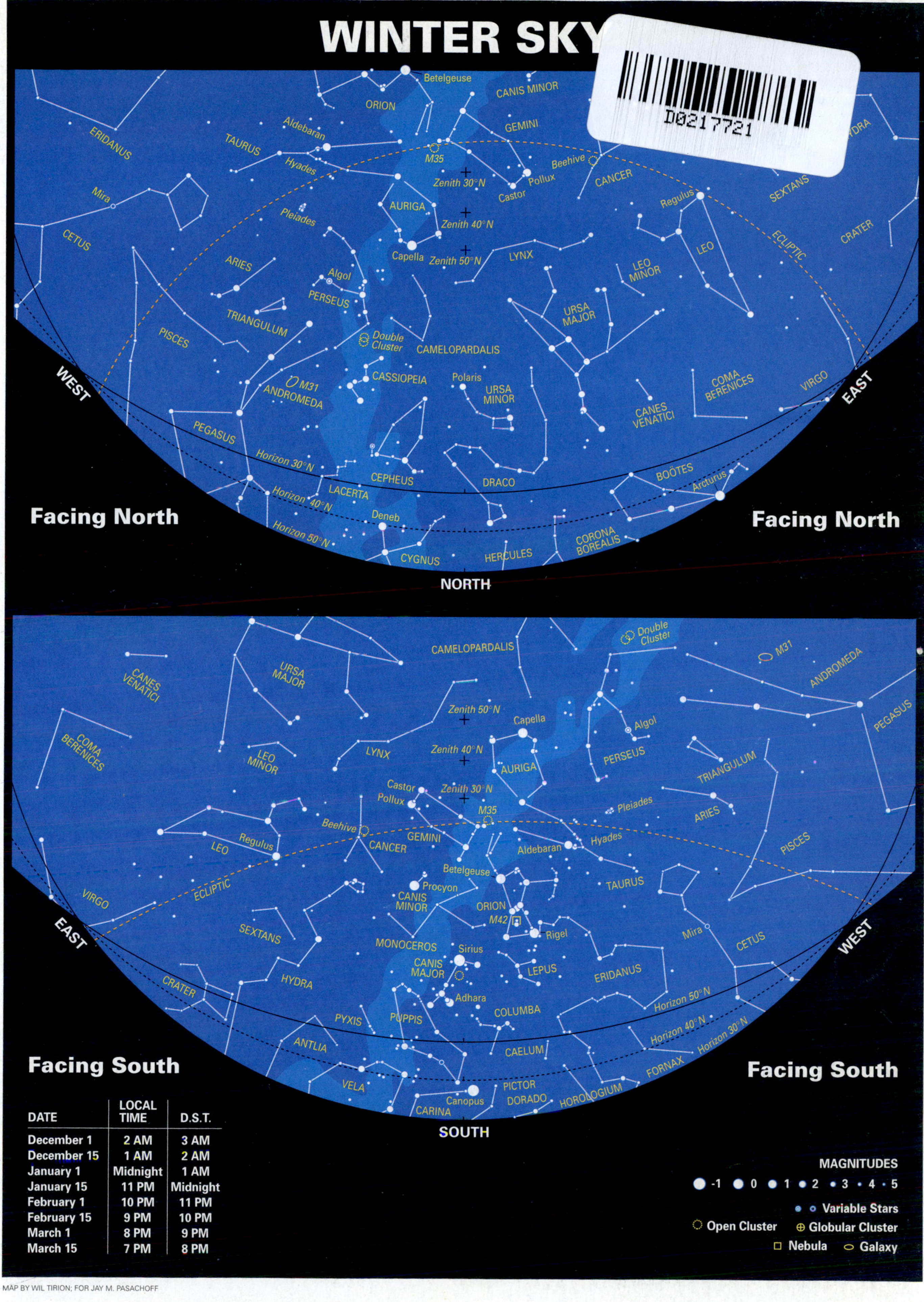


| DATE | LOCAL TIME | D.S.T. |
|---|---|---|
| December 1 | 2 AM | 3 AM |
| December 15 | 1 AM | 2 AM |
| January 1 | Midnight | 1 AM |
| January 15 | 11 PM | Midnight |
| February 1 | 10 PM | 11 PM |
| February 15 | 9 PM | 10 PM |
| March 1 | 8 PM | 9 PM |
| March 15 | 7 PM | 8 PM |

MAP BY WIL TIRION; FOR JAY M. PASACHOFF

# ASTRONOMY: FROM THE EARTH TO THE UNIVERSE

Sixth Edition

JAY M. PASACHOFF

Field Memorial Professor of Astronomy
Director of the Hopkins Observatory
Williams College
Williamstown, MA
http://www.williams.edu/astronomy/jay

Australia • Canada • Mexico • Singapore • Spain
United Kingdom • United States

Astronomy Editor: Kelley Tyner
Development Editor: Jennifer Pine
Project Editor: Robin Bonner
Marketing Manager: Kathleen McLellan
Production Manager: Alicia Jackson
Art Director: Jonel Sofian
Copy Editor: Sheryl Nelson
Illustrator: Rolin Graphics, Inc.
Cover Printer: Phoenix Color Corp.
Compositor: TechBooks, Inc.
Printer: Von Hoffmann Press



Printed in the United States of America
2 3 4 5 6 7 05 04 03 02

For more information about our products contact us at:
**Thomson Learning Academic Resource Center**
**1-800-423-0563**

For permission to use material from this text, contact us by:
**Phone:** 1-800-730-2214 **Fax:** 1-800-730-2215
**Web:** http://www.thomsonrights.com

**Cover:**
The total solar eclipse of the most recent maximum of the sunspot cycle; its streamers extend in all directions to reveal the corona's magnetic field. The image resulted from the author's eclipse expedition to Romania. *(Credit: © Jay M. Pasachoff, Williams College Expedition; photographs by Jay M. Pasachoff and Stephan Martin with image processing by Wendy Carlos (color photographs) and Daniel B. Seaton (monochromatic CCD images))*

**Back cover:**
The planetary nebula Menzel 3, the remnant of a dying sun-like star. This highly detailed image taken with the Hubble Space Telescope shows the complicated manner in which stars like the sun eject material as they die. The sun will go through a planetary-nebula stage in about 5 billion years. *(Credit: NASA, ESA & The Hubble Heritage Team (STScI/AURA))*

*Astronomy: From the Earth to the Universe* 6/e
ISBN: 0-03-033488-8
Library of Congress Control Number: 2001093157

**Asia**
Thomson Learning
60 Albert Street, #15-01
Albert Complex
Singapore 189969

**Australia**
Nelson Thomson Learning
102 Dodds Street
South Melbourne, Victoria 3205
Australia

**Canada**
Nelson Thomson Learning
1120 Birchmount Road
Toronto, Ontario MIK 5G4
Canada

**Europe/Middle East/Africa**
Thomson Learning
Berkshire House
168-173 High Holborn
London WCI V7AA
United Kingdom

**Latin America**
Thomson Learning
Seneca, 53
Colonia Polanco
11560 Mexico D.F.
Mexico

**Spain**
Paraninfo Thomson Learning
Calle/Magallanes, 25
28015 Madrid, Spain

# PREFACE

Astronomy enters the 21st century surfing not only the World Wide Web but also waves of increasing discoveries and capabilities. The giant telescopes of even a decade ago now seem like dwarfs alongside huge arrays of telescopes in Hawaii and Chile with mirrors of each separate unit as much as 10-m (400 inches) across. Astronomers using them see the giant stars and the dwarf stars in the sky in increasing detail. Indeed, coupled with electrons and computers new telescope capabilities are mapping and analyzing millions of galaxies to give us statistics and details in places where we only recently had a handful of examples.

It is astonishing to see how astronomy's spectacular successes keep increasing in number. The flood of observations from the Hubble Space Telescope and its partner the Chandra X-ray Observatory touch every field within astronomy. Jupiter and its moons from the Galileo spacecraft will soon be joined by images of Saturn and its moons from the Cassini spacecraft. Images of Mars show us individual rocks on the surface and signs that water flowed there, perhaps recently. The images of huge solar eruptions from the Solar and Heliospheric Observatory and the Transition Region and Coronal Explorer in space, and of the solar corona at total eclipses, reveal the levels of our Sun in their beauty. Scientists working with these data are advancing at a rapid pace in telling us about the Universe in which we live, and what it will be like in the future.

*Astronomy: From the Earth to the Universe* describes the current state of astronomy, both the fundamentals of astronomical knowledge that have been built up over decades and the exciting advances that are now taking place. I try to cover all

The spectral types of a wide variety of stars, from hottest to coolest, show different absorption lines, spectral lines cutting into the bands of color.

the branches of astronomy without slighting any of them; each teacher and each student may well find special interests that may be different from my own. One of my aims in writing this book is to show everyone that exciting forefront research is going on. One reason for this aim is the hope that people will support scientific research in general and astronomical research in particular. Another aim is that the discussions of how to think about the Universe and how astronomers deduce facts from meager observations will provide general training in how to approach a wide variety of matters that benefit from critical thinking.

This book is organized as a number of stories. In individual chapters, I often tell first what used to be known, then how space and other modern observations have transformed our understanding, and then what is scheduled for the future. The planets, for example, are organized in this way, with a separate chapter for each. Consequently, a professor can easily add photos (available as slides, overheads, CD-ROMs, and on the World Wide Web) and movies and keep a student's interest for a whole lecture on each planet, if desired. Comparative planetology is fit into the context, so that students have a good grounding in fundamentals of planetary science before they compare one planet to the next. This method helps keep different objects from melding in a student's mind. Not only planetary astronomy but also other parts of astronomy are organized as stories. For example, I tell the story of the discovery and understanding of pulsars and, later, the story of the discovery and explanation of quasars, with spectacular new Hubble images included. And within chapters, there are often smaller stories, such as how to observe with a telescope on Mauna Kea, what it is like to use the Hubble Space Telescope, what happens on a solar eclipse expedition, or how observing with the Very Large Array goes. (Each is singled out in a display called an "Astronomer's Notebook.") Thus the student learns about astronomy through a set of concrete examples, rather than merely being given overarching concepts without enough underpinning.

An artist's conception of the Next Generation Space Telescope.

## New to this Edition

It is exciting for me to add a wide variety of new images from the Hubble Space Telescope and from new telescopes on Earthbound mountaintops to the many other images previously presented. One of the most exciting developments in the last few years has been the discovery of planets orbiting other stars, and I am pleased to devote a whole chapter to that important topic.

All of science and much of the world have been transformed in the last few short years by the spread of the Internet and the World Wide Web. In the last few years, I have been using a Web site at http://www.williams.edu/astronomy/jay to provide updates for each chapter and a variety of links to astronomical information from scientists and sites around the globe. With this sixth edition, we are coupling this Web site with the official publisher's book Web site, http://www.harcourtcollege.com/astro/astronomy6e, integrating access to the textbook so that current information on a variety of topics that tend to change will be available to both students and faculty at my Web site. The site also incorporates review questions for students to use as self-tests and learning aids.

In the book and on the Web site, we are also providing references to and activities for one of the best astronomy CD-ROMs available, *RedShift College Edition*, in which Maris Multimedia has incorporated illustrated essays especially for Harcourt College Publishers. (We are also providing ways for students to inexpensively have their own copies, shrinkwrapped with the text itself.) Many of these activities will also be usable with other computer-planetarium-type programs.

I am especially proud that my texts are unique in that they have separate chapters covering branches of astronomy for which we newly have extensive information. Topics that receive here the separate chapters they deserve following contemporary research discoveries, in addition to Extrasolar Planets, are Neptune, Pluto, supernovae, pulsars/neutron stars, black holes, and quasars, topics that are especially interesting to students and faculty alike. We also provide special Photo Galleries of more wonderful images on several topics.

Of course, new scientific information has been added in every chapter. The speed with which astronomy information changes has led me often to single out by date new facts or theories. Hearing that some information was found "in 2000" or "in 2001" should help keep student interest high.

## Astronomy in the Next Decade

Each decade, a committee appointed by the U.S. National Academy of Sciences puts together an Astronomy and Astrophysics Survey, choosing priorities for the coming decade. The report issued in 2000 under the guidance of Joseph Taylor of Princeton and Christopher McKee of Berkeley has been widely acclaimed for the wisdom of its roadmap and for the hard choices that it has made in priorities. (I was privileged to be in graduate school at Harvard with Joe Taylor and to share an office as a postdoc at Caltech with Chris McKee.)

The committee recommended a number of initiatives, many of which are to address the following problems, each of which is discussed in this textbook:

- What is the structure of the Universe? What is it made of? How old is it? What is its fate?
- What occurred at the dawn of the modern universe, when the first stars and galaxies formed?
- How do black holes form and evolve, both those with the mass of a star and the huge ones in the centers of galaxies?
- How do stars and their planetary systems form? How do individual planets, ranging from Earth-like to Jupiter-like, form and evolve?
- What causes the astronomical environment—particularly the Sun and asteroids—to affect the Earth?

The Eagle Nebula, M16, imaged with the Hubble Space Telescope using special filters to bring out different kinds of gas in this false-color image. We see channels of gas and dust that will one day evaporate to reveal the stars now forming inside them.

Their highest priority major initiative is the Next Generation Space Telescope, which we hope for in 2008. The second priority is the Giant Segmented Mirror Telescope, a ground-based telescope 30-m across, with 10 times the collecting area of each of the current Keck telescopes. Their next priority is the Constellation X-ray Observatory, more powerful even than the Chandra X-ray Observatory and so better able to study the formation and evolution of black holes, among many other projects. They also recommend expansion of the Very Large Array set of radio telescopes, and construction of a large telescope with a wide field of view to carry out surveys of asteroids, supernovae, and other objects. For space projects, they recommend the Terrestrial Planet Finder that we discuss in Chapter 18, designed to find Earth-sized planets around nearby stars. And they recommend a space telescope to study long-wavelength infrared radiation. In addition to these major initiatives, they recommend a host of moderate initiatives including telescopes on the ground and in space to study a wide variety of objects in various parts of the spectrum. The only item on their list of small initiatives is a National Virtual Observatory. It is to integrate all the major astronomical archives into a digital database that can be used by anyone anywhere in the world.

## Educational Approach

I have paid special attention to making the books readable by students and the information accessible to them. The stories of the development of individual fields of study, as for the individual planets, are somewhat chronological to give students an opportunity to see how knowledge has grown. I include a lot of comparative planetology, but by covering each planet in a separate chapter provide the basic information that helps students keep all the solar-system objects straight in their minds. Further, the occasional mention of the names of the contemporary astronomers performing the research discussed personalizes astronomy and shows that it is a human science as well as giving credit where credit is due. Newton was not the last scientist worthy of being mentioned by name! Among those whose research has been so important to the development of astronomy are many women (starting as far back as Hypatia in ancient Alexandria).

In writing this book, I share the goals of a commission on the college curriculum of the Association of American Colleges, which reported that "a person who understands what science is recognizes that scientific concepts are created by acts of human intelligence and imagination; comprehends the distinction between observation and inference and between the occasional role of accidental discovery in scientific investigation and the deliberate strategy of forming and testing hypotheses; understands how theories are formed, tested, validated, and accorded provisional acceptance; and discriminates between conclusions that rest on unverified assertion and those that are developed from the application of scientific reasoning." The scientific method permeates the book.

What is science? The following statement was originally drafted by the Panel on Public Affairs of the American Physical Society, in an attempt to meet the perceived need for a very short statement that would differentiate science from pseudoscience. This statement has been endorsed as a proposal to other scientific societies by the Council of the American Physical Society, and was endorsed by the Executive Board of the American Association of Physics Teachers in 1999. The 2001 joint meeting of the American Association of Physics Teachers and the American Astronomical Society demonstrates the overlaps in interest between members of the two groups.

*Science is the systematic enterprise of gathering knowledge about the world and organizing and condensing that knowledge into testable laws and theories.*

*The success and credibility of science is anchored in the willingness of scientists to:*

1. *Expose their ideas and results to independent testing and replication by other scientists; this requires the complete and open exchange of data, procedures, and materials;*
2. *Abandon or modify accepted conclusions when confronted with more complete or reliable experimental evidence.*
3. *Adherence to these principles provides a mechanism for self-correction that is the foundation of the credibility of science.*

The first diagram showing the Sun as the center of what we now call the Solar System, from Copernicus's 1543 book.

I have tried to keep up with recent psychological research on the teaching of science and the understanding of concepts. My books, therefore, include many concrete examples that help students understand scientific situations more clearly. I continue to consult with my psychologist colleagues so as to be able to implement correctly some of the tenets of Piaget, while taking into account the results of post-Piagetan research.

Several recent projects on standards in education include the *National Standards in Science Education* of the National Academy of Sciences/National Research Council; *Project 2061* (the year of the next return of Halley's Comet) of the American Association for the Advancement of Science; and *Scope, Sequence, and Coordination of Secondary School Science,* a project of the National Science Teachers Association. I share with them the goals of providing understandable science for a wide variety of students, though I think that astronomy can be used as an example and a unifying theme for teaching more of science than these projects conclude. Some thoughts about the pitfalls of these standards appear in Chapter 1 of this book's homepage; I worry that the standards do not provide enough excitement for students and enough breadth to understand why science is important to understand. In my capacity as the Chair of the Astronomy Division of the American Association for the Advancement of Science, I adopted as one of my themes the role of astronomy in teaching science nationwide. In my current capacities as U.S. National Representative to and Vice-President of the International Astronomical Union's Commission on Education and Development, I have brought back many lessons about astronomy education from my colleagues all over the world and have worked to spread the word about developments here.

## Pedagogy

I have provided many aids to make this book easy to read and to study from. Many of the book's diagrams provide unique interpretive information; compare, for example, the diagrams that explain the Doppler effect or Hubble's law with those in other texts. Each section of my text is numbered to allow professors to include or omit parts of chapters. Some professors will choose to include discussions, for example, of the over four dozen new worlds photographed as moons of the giant planets in our solar system; others will choose to omit this material. Further, some sections and boxes appear with asterisks to show that these are especially easy to omit.

Important terminology is boldfaced in the text, listed in the key words at the end of each chapter, and defined in the glossary. The index provides further aid in finding explanations. Students should always be sure that they know all the points raised in the chapter summaries. End-of-chapter questions cover a range of material; some questions are straightforward and can be answered by merely reading the text, while others require more independent thought. Up-to-date appendices provide some standard and much recent information on planets, stars, constellations, and non-stellar objects. I am particularly pleased to have, in press here in advance of their official publication, the latest data about the nearest stars as measured with the Hipparcos spacecraft. (Even their order in distance from us has changed from what we had previously thought!)

A major feature first used by me in the previous edition of this book and expanded here is the providing of a set of common misconceptions for each chapter, which are presented along with the correct ideas. It has become clear in recent years that students' minds are not blank slates, but that they harbor misconceptions that must be tackled and unlearned before the correct material can be retained. My new chapter-by-chapter sections are designed to address this discovery. I also provide exercises for the World Wide Web; relevant sites are all available through this book's own Web site. Further, we present exercises for each chapter that can be done using the *RedShift College Edition* CD-ROM. This CD-ROM includes illustrated essays that cover some of the material in this book and that provide an additional interesting perspective. Other key features include:

The Hubble Space Telescope in orbit.

- The **Astronomer's Notebook** feature offers students an insider's view on observations and telescope use.
- The **Photo Gallery** includes artfully displayed images to help impress upon students the beauty and wonder of the universe. More than 1000 photos and illustrations bring astronomy to life on every page.
- **What's in a Name?** boxes explore the mythology and history of astronomical nomenclature.
- **Biography** boxes provide background and historical context for astronomy's most noted contributors.
- **In Review** boxes summarize and reinforce key topics discussed in the text.
- **A Deeper Discussion** boxes are optional and provide further discussion and insight to selected topics in the text.

## Ancillary Materials

The following ancillaries are available for the instructor:

**Instructor's Manual and Test Bank.** This manual contains course outlines, answers to all questions in the text, lists of audiovisual aids and organizations, laboratory experiments and exercises arranged for easy duplication, sample tests with answers (both quizzes and exams), and Test Bank objectives. The Test Bank portion contains over 1500 questions in multiple-choice and other formats.

**Overhead Transparencies.** This set contains 100 overhead transparency acetates.

**Slide Collection.** Available as an alternative to overhead transparencies.

**Computerized Test Bank.** All the questions from the printed test bank are presented in a computerized format that allows instructors to edit, add questions, and print assorted versions of the same test. Available for Windows™ and Macintosh® users.

**Instructor's Photo Resource Catalogue.** Many of this text's images are provided in an electronic file format for use in classroom presentations. This catalog is available both on a CD-ROM and as a downloadable Web file.

And for both the instructor and the student:

**RedShift College Edition™.** This CD-ROM (available shrink-wrapped with the text for students) is a special version of the powerful *RedShift* software that allows you to duplicate the look of the celestial sphere on your computer, to watch animated essays on various astronomical topics, to examine several astronomical phenomena, or to see the special features provided to accompany this text. A Workbook on the use of *RedShift College Edition* is also available (optionally also shrink-wrapped with the text for students and the CD-ROM).

**Online Student Study Guide.** Available only online, this site is a valuable reference and learning tool for students. Complete with practice questions, exercises and activities, definitions of terms, and other useful information, this guide is a student's roadmap to understanding astronomy.

The barred spiral galaxy NGC 1365.

**Astronomy Web site.** In addition to the book's official Web site, my astronomy Web site is accessible to all at http://www.williams.edu/astronomy/jay. Through it, I update chapters as new information becomes known and provide hotlinks to other sites.

For information about these ancillaries, contact your local Brooks/Cole sales representative. I can also often help in providing information or obtaining materials, so don't hesitate to call or e-mail me.

Brooks/Cole may provide complimentary instructional aids and supplements or supplement packages to those adopters qualified under our adoption policy. Please contact your sales representative for more information. If as an adopter or potential user, you receive supplements you do not need, please return them to your sales representative or send them to

Attn: Returns Department
Troy Warehouse
465 South Lincoln Drive
Troy, MO 63379

## Acknowledgments to Scientific and Educational Contributors

The publishers and I have always placed a heavy premium on accuracy in my books, and we have made certain that the manuscript and proof have been read not only by students for clarity and style and by astronomy teachers for pedagogical reasons but also by research astronomers for scientific accuracy.

I would particularly like to thank the reviewers of all or large sections of the manuscript of this edition, who include Victor Andersen (University of Houston), Nadine Barlow (University of Central Florida), Timothy Beers (Michigan State University), David Buckley (East Stroudsburg University), Geoff Clayton (Louisiana State University), Mark Devlin (University of Pennsylvania), Nelson Duller (Texas A&M University), Leonard Finegold (Drexel University), Pam Friese (Parkland Community College), Peter Hauschildt (University of Georgia), John Hoessel (University of Wisconsin-Madison), Richard Ignace (University of Iowa), Douglas Ingram (Texas Christian University), C. Renee James (Houston State University), Thomas Krause (Towson University), Mark Miksic (Queens College), Mark Slovak (Louisiana

State University), Curtis Struck (Iowa State University), Heidi VanTassell (Arizona State University) and George McClusky (Lehigh University).

We also thank reviewers of the manuscript of the previous edition: Jess T. Dowdy (Northeast Texas Community College), Raymond Wilson (Ohio Wesleyan University), Omer Blaes (University of California at Santa Barbara), Margaret Riedinger (University of Tennessee), Jill A. Marshall (Utah State University), Gary S. Weston (California State University, Hayward), Robert L. Zimmerman (University of Oregon), Solomon Gartenhaus (Purdue University), Alice L. Newman (California State University at Dominguez Hills), Timothy Beers (Michigan State University), John Laird (Bowling Green State University), John G. Hoessel (University of Wisconsin), Frederick M. Walter (State University of New York at Stony Brook), and David Kropp (Michigan State Teachers Association).

I also thank those astronomers who have commented on the latest versions of particular sections relevant to their fields of expertise, most recently Leo Blitz (University of California at Berkeley); James Head (Brown University), Moon and Mars, Milky Way; Daniel Green (Harvard-Smithsonian Center for Astrophysics), Kuiper belt and comets; S. Alan Stern (Southwest Research Institute, Boulder), Pluto and Kuiper belt; George Gatewood (University of Pittsburgh), astrometric search for planets; Geoff Marcy (University of California at Berkeley), extrasolar planets; Jeffrey McClintock (Harvard-Smithsonian Center for Astrophysics), black holes; Michael Garcia (Harvard-Smithsonian Center for Astrophysics), black holes; Gibor Basri (University of California at Berkeley), brown dwarfs; Bruce Carney (University of North Carolina), star clusters; Karen Kwitter (Williams College), planetary nebulae; Andrei Linde (Stanford), inflationary universe; Louise Prockter (Johns Hopkins University Applied Physics Lab), Ganymede; Joe Burns (Cornell), rings; Lew Snider (University of Illinois), interstellar molecules; and Marek Demianski (Williams College), cosmology. I appreciate earlier comments by G. Jeffrey Taylor, Hawaii Institute for Geophysics (University of Hawaii), Moon; Sean Solomon (DTM, Carnegie Institution of Washington), solar system; Arlin J. Krueger (NASA/GSFC), ozone hole; Robert Strom (LPL, University of Arizona), Mercury; Steve Saunders (JPL), Venus; Susan Stolovy (University of Arizona), galactic center; Dan Gezari (NASA/GSFC), galactic center; Carle Pieters (Brown University), Earth; Farhad Yusef-Zadeh (University of Illinois), galactic center; Paul Steinhardt (Princeton University), inflationary universe. I thank Jacques Beckersand and John Leibacher (National Solar Observatory) for their comments on solar projects.

The spiral galaxy M83, imaged with one of the Unit Telescopes of the Very Large Telescope in Chile.

I also thank A. W. Wolfendale (University of Durham, U.K.), cosmic rays; Bruce Partridge (Haverford College), cosmic background radiation; Ed Cheng (NASA/Goddard Space Flight Center), COBE; John Mather (NASA/Goddard Space Flight Center), COBE; Roman Juskiewicz (Copernicus Astronomical Center, Warsaw, and Institute for Advanced Study, Princeton), galaxy formation; John N. Bahcall (Institute for Advanced Study, Princeton), solar neutrinos; Roger Romani (Stanford University), millisecond pulsars; Daniel R. Stinebring (Oberlin College), pulsars; Joseph H. Taylor (Princeton University), pulsars; Stuart N. Vogel (University of Maryland), molecular studies; Noel Swerdlow (University of Chicago), Hipparchus; G. J. Toomer (Brown University), Hipparchus. I thank John P. Oliver (University of Florida) and Peter Katsaros for their detailed comments. I thank Brian Schmidt and Robert Kirshner (Harvard-Smithsonian Center for Astrophysics), Sandra Faber (Lick Observatory, University of California, Santa Cruz), Robin Ciardullo (Pennsylvania State University), and George Jacoby (National Optical Astronomy Observatories) for their detailed comments on the cosmic distance scale. I thank Wendy Freedman (Observatories of the Carnegie Institution of Washington) and Robert Kennicutt (University of Arizona) for the latest Hubble diagrams. I thank Norman Sperling for his comments.

Cheryl Gundy at the Space Telescope Science Institute has been wonderfully helpful in my use of Hubble Space Telescope images. Jurrie van der Woude and Ed McNevin of the Jet Propulsion Laboratory supplied some hard-to-find images. David Malin of the Anglo-Australian Observatory has been of special help in providing his magnificent color photographs, and we have even worked together to make Crab

Nebula color images for this book. Coral Cooksley at the AAO has helped with our obtaining the images. Andy Perala of the W. M. Keck Observatory has helped with Keck images. Peter Michaud of the Gemini Observatory has helped with Gemini materials. Brian Hadley of the Royal Observatory Edinburgh has provided nebular and galaxy images. Stephen P. Maran of the NASA/Goddard Space Flight Center, the Press Officer of the American Astronomical Society, has been especially helpful at meetings of the American Astronomical Society. Other providers of photographs are thanked in the illustration acknowledgments section at the end.

I remain grateful to the reviewers of earlier versions of my texts or sections in them, including Ronald J. Angione, Thomas T. Arny, James G. Baker, Michael Belton, Bruce E. Bohannon, Kenneth Brecher, Tom Bullock, Bernard Burke, Clark Chapman, Pamela Clark, Roy W. Clark, James W. Cristy, Martin Cohen, Leo Connolly, Peter Conti, Ernest R. Cowley, Lawrence Cram, Dale P. Cruikshank, Morris Davis, Raymond Davis, Jr., Gerard de Vaucouleurs, Dennis di Cicco, Richard B. Dunn, John J. Dykla, John A. Eddy, Farouk El-Baz, James L. Elliot, David S. Evans, J. Donald Fernie, William R. Forman, Peter V. Foukal, George D. Gatewood, John E. Gaustad, Tom Gehrels, Riccardo Giacconi, Owen Gingerich, Stephen T. Gottesman, Jonathan E. Grindlay, Alan R. Guth, Ian Halliday, U. O. Hermann, Darrel B. Hoff, James Houck, Robert F. Howard, W. N. Hubin, John Huchra, H. W. Ibser, Paul Johnson, Christine Jones, Bernard J. T. Jones, Agris Kalnajs, John Kielkopf, Robert Kirshner, David E. Koltenbah, Jerome Kristian, Edwin C. Krupp, Karl F. Kuhn, Marc L. Kutner, Karen B. Kwitter, John Lathrop, Stephen P. Lattanzio, Lawrence S. Lerner, Jeffrey L. Linsky, Sarah Lee Lippincott, Bruce Margon, Laurence A. Marschall, Brian Marsden, Janet Mattei, R. Newton Mayall, Everett Mendelsohn, George K. Miley, Freeman D. Miller, Alan T. Moffet, William R. Moomaw, David D. Morrison, R. Edward Nather, David Park, James Pierce, Carl B. Pilcher, James B. Pollack, B. E. Powell, James L. Regas, Edward L. Robertson, Thomas N. Robertson, Herbert Rood, Maarten Schmidt, David N. Schramm, Leon W. Schroeder, Richard L. Sears, P. Kenneth Seidelmann, Maurice Shapiro, Peter Shull, Jr., Joseph I. Silk, Lewis E. Snyder, Theodore Spickler, Alan Stockton, Robert G. Strom, Jean Pierre Swings, Eugene Tademaru, Laszlo Taksay, David L. Talent, Gustav Tammann, Joseph H. Taylor, Jr., Joe S. Tenn, David Theison, Laird Thompson, Aaron Todd, M. Nafi Toksöz, Juri Toomre, Kenneth D. Tucker, Brent Tully, Barry Turner, Peter van de Kamp, Gerald J. Wasserburg, Anthony Weitenbeck, Leonid Weliachew, Ray Weymann, John A. Wheeler, J. Craig Wheeler, Ewen A. Whitaker, Robert F. Wing, Reinhard A. Wobus, LeRoy A. Woodward, Susan Wyckoff, Anne C. Young, and Robert N. Zitter. Others reviewers to whom I am grateful are Yervant Terzian (Cornell University), Alexander G. Smith (University of Florida), Joseph Veverka (Cornell University), Bruce Margon (University of Washington), Paul Hodge (University of Washington), Hyron Spinrad (University of California, Berkeley), Gary Mechler (Pima Community College), Leonard Muldawer (Temple University), Paul B. Campbell (Western Kentucky University), Gordon E. Baird (University of Mississippi), Robert L. Mutel (University of Iowa), and Gordon B. Thompson (Rutgers University).

## Acknowledgments to Publishing Contributors

I appreciate the expert editorial work in Williamstown of Susan Kaufman. I am also grateful for the computer assistance of Stephan Martin, who also wonderfully maintains this book's Web site. I further thank Barbara Swanson for her continual aid. Rebecca Cover has carefully gone through the manuscript, making wise comments and helping me prepare for the new edition.

I thank Nancy P. Kutner for her excellent index.

I thank Steven Souza for his excellent work on the Instructor's Manual and Test Bank. Dr. Souza's expertise has helped bring the answers to questions and other useful information to a new level.

I thank many people at Saunders College Publishing for their efforts on behalf of my books. Kelley Tyner is my Acquisitions Editor at Saunders and Jennifer Pine is my Developmental Editor; they have helped in many ways, as has Kim Paschen, their Editorial Assistant. Robin Bonner and Jonel Sofian have done a wonderful job as Project Editor and Art Director. I continue to remain grateful to John J. Vondeling, Lloyd W. Black, Jennifer Bortel, and Sarah Fitz-Hugh, my editors on earlier editions. Alicia Jackson has been the Production Manager. Joanne Cassetti has been the Director of EDP. Caroline McGowan has been the Art Director on previous editions. George Kelvin and his studio, Science Graphics, provided many of the drawings for earlier versions.

I am grateful to John N. Bahcall and the Institute for Advanced Study, Princeton, and Harvey Tananbaum and the Harvard-Smithsonian Center for Astrophysics for their hospitality during my most recent sabbatical leaves. I have also enjoyed my work with Leon Golub of the Center for Astrophysics, and am glad to have written two recent books with him: *The Solar Corona* (Cambridge University Press, 1997) and *Nearest Star: The Surprising Science of Our Sun* (Harvard University Press, 2001).

Various members of my family have provided vital and valuable editorial services, in addition to their general support. My wife, Naomi Pasachoff, has taken time from her own projects to give expert readings of the various stages of proof. The participation of our daughters, Eloise and Deborah, now long out of college, in proofreading past editions has been a source of great satisfaction to me. Their growth has been monitored over the years alongside the Ahnighito meteorite. My father, Dr. Samuel S. Pasachoff, contributed so much to my books. I also appreciate the counsel of my mother, Anne T. Pasachoff.

The planetary nebula Menzel 3, imaged with the Hubble Space Telescope.

## To the Student

Astronomy is a very varied subject and you are sure to find some parts that interest you more than others. To get the most out of this text, you must read each chapter more than once. After you read the chapter the first time, you should go carefully through the list of key words, trying to identify or define each one. Whether or not you can do so, look up the word in the glossary and also find the definition that appears with the word the first time it is used in the chapter. The index will help you find these references. Next, read through the chapter again especially carefully. Read through the Summary, checking that you understand each of the major points. Check that you do not share any of the common misconceptions that are listed. Finally, answer a selection of the questions at the end of the chapter and test yourself with the questions for the chapter that are on this book's Web site. Also, be sure to examine the animations and explanations in *RedShift College Edition.*

I hope you not only enjoy this course but also learn things about astronomy and the way science is done that will keep you interested for decades to come.

## A Final Comment

I am extremely grateful to all the individuals named for their assistance. Of course, it is I who have put this all together, and I alone am responsible for it. I appreciate hearing from readers. I invite you to write me at Williams College—Hopkins Observatory, Williamstown, MA 01267, e-mail: jay.m.pasachoff@williams.edu.

Jay M. Pasachoff
Williamstown, MA 01267
June 2001

The spiral galaxy M51, with the red part of the image taken in the light of hydrogen gas to emphasize regions in which stars are forming.

# CONTENTS OVERVIEW

The Keyhole Nebula.

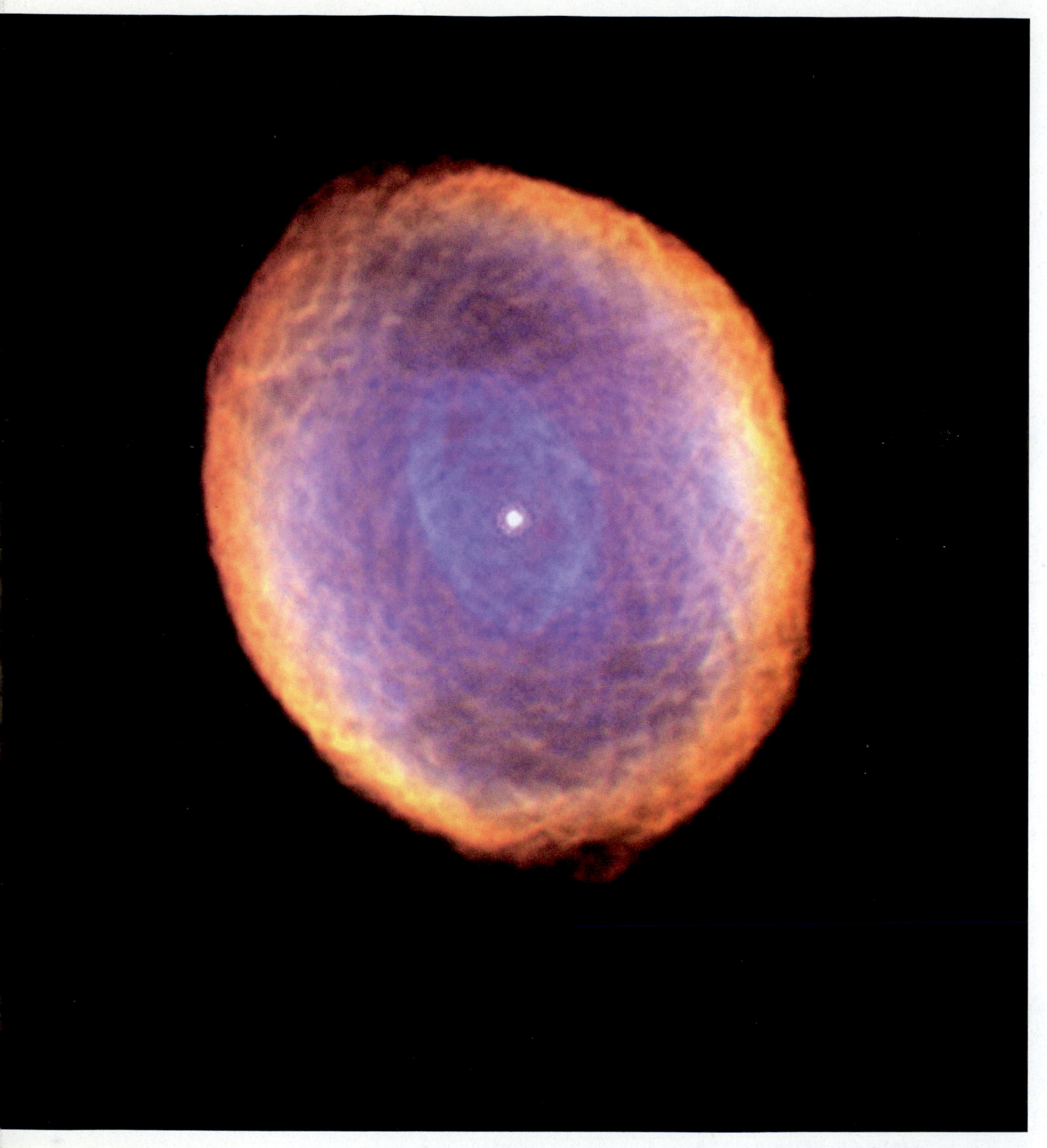

The Spirograph Nebula, IC 418, an example of gas ejected from a dying star to make what is called a planetary nebula.

# CONTENTS

NGC 2207, a pair of interacting galaxies.

The domes of the twin Keck Telescopes in Hawaii.

One spiral galaxy silhouetted in front of another, a rare circumstance, imaged with the Hubble Space Telescope.

Jupiter, imaged by the Cassini mission during its 2001–2002 Jupiter flyby.

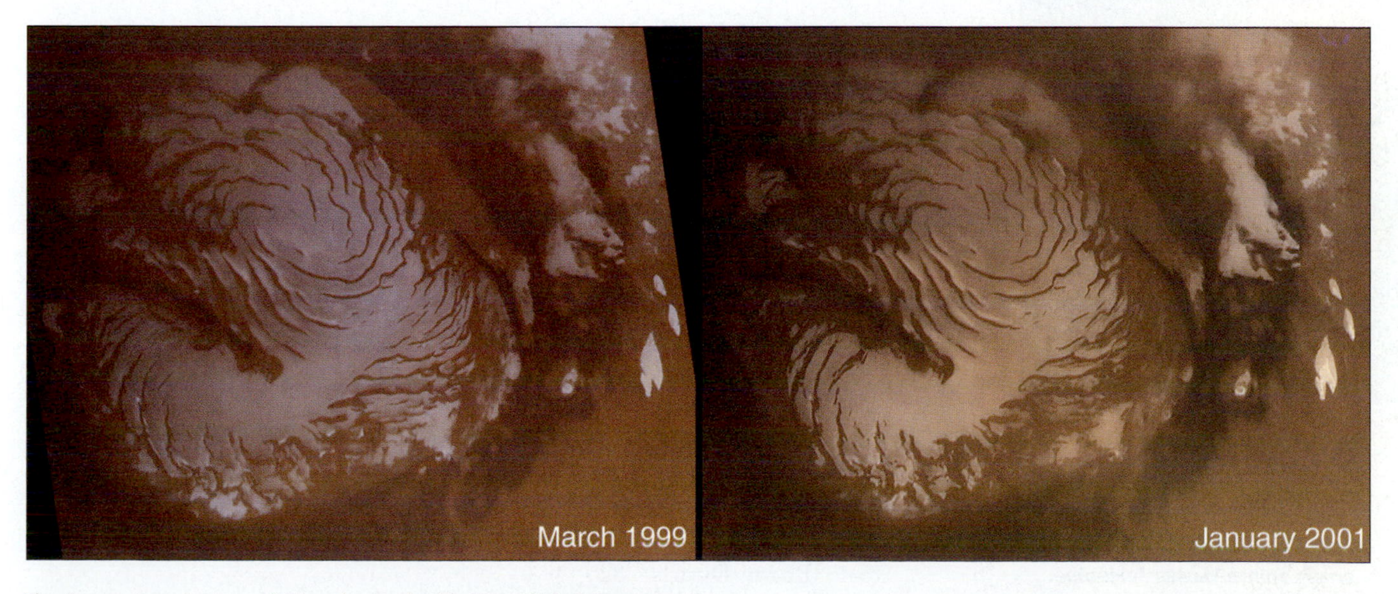

The icy north polar cap of Mars, imaged with Mars Global Surveyor.

Comet Hale-Bopp as seen in a wide-angle view.

The Veil Nebula, part of a supernova remnant—the debris from an exploded star. The image was taken with a huge electronic camera, with 8000 × 12,000 pixels, at the Canada-France-Hawaii Telescope.

The solar corona at the total solar eclipse of 1999, showing the circularly symmetric structure typical of the maximum of the solar-activity cycle.

The center of the Orion Nebula, imaged with the Subaru Telescope on Mauna Kea in Hawaii.

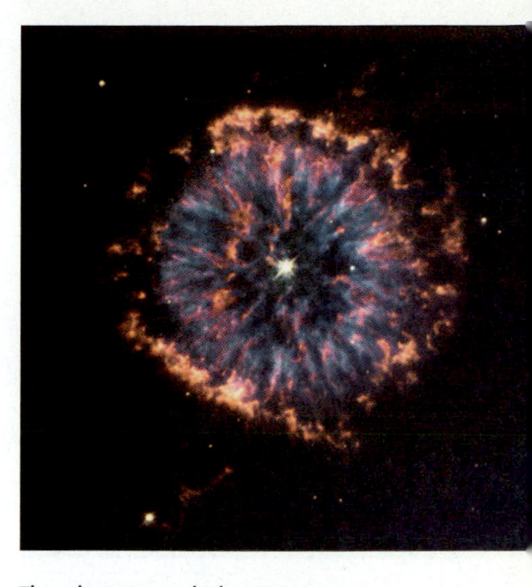

The planetary nebula NGC 6751, imaged with the Hubble Space Telescope. We see gas thrown out by a dying star that once resembled our Sun.

A group of galaxies, Hickson Compact Group 40. They are held together by their mutual gravity.

A quasar, PG 0052+251, imaged with the Hubble Space Telescope, reveals structure in the "fuzz" that surrounds the bright compact nucleus. It is one of the closest quasars, and is in the center of a normal spiral galaxy.



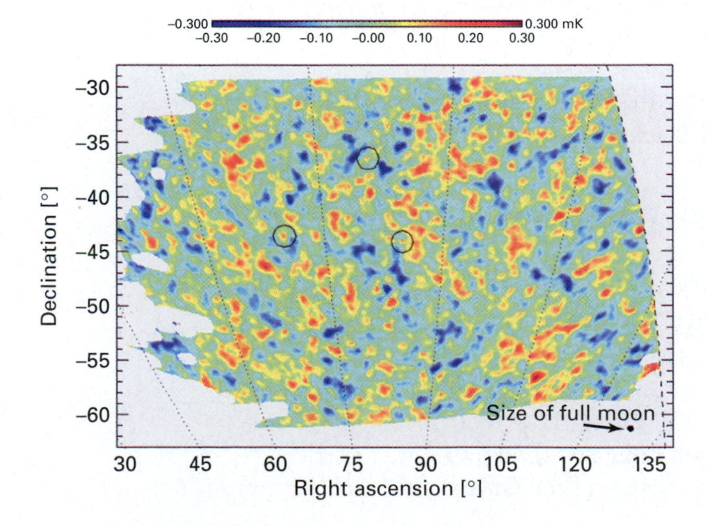


A map showing how the microwave intensity reveals slight ripples of temperature that grew from the Universe's earliest epoch into the large-scale structure we see today. (See Section 37.5f.)

The globular cluster M3, a grouping of tens of thousands of stars.

The Keyhole Nebula, imaged at much higher resolution with the Hubble Space Telescope than it was when Sir John Herschel discovered it in the 19th century. The circular Keyhole contains both bright filaments of hot gas and dark silhouetted clouds of cold molecules and dust. The light we see has been travelling toward us for 8000 years.

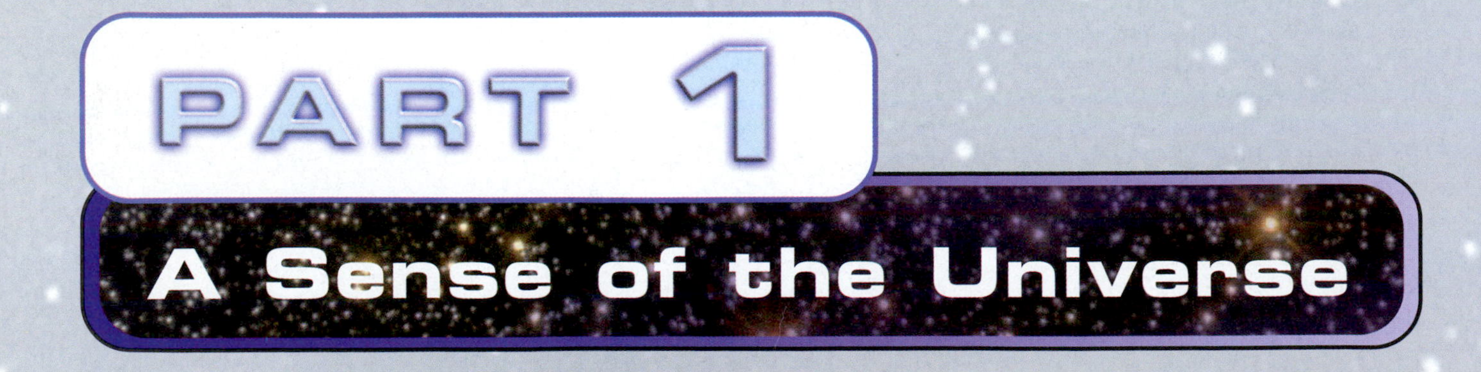

# PART 1

# A Sense of the Universe

Let us consider that the time between the origin of the Universe and the year 2000 was one day. Then it wasn't until nearly 5 p.m. that the Earth formed; the first fossils date from 7 p.m. The first humans appeared only 10 seconds ago, and it is only 1/300 second since Columbus reached America.

Still, the Sun should shine for another 9 hours; an astronomical time scale is much greater than the time scale of our daily lives. Astronomers use a wide range of technology and theories to find out about the Universe, what is in it, and what its future will be. This book surveys how we look out at the Universe around us and what we have found.

Throughout history, observations of the heavens have led to discoveries that have had major impacts on people. Even the dawn of mathematics may have followed ancient observations of the sky, made in order to keep track of seasons and seasonal floods in the fertile areas of the Earth. Because the Moon and planets are massive enough that gravity dominates their motions, and because their motions are free of complicated terrestrial forces such as friction, observations of these bodies' orbits led to an understanding of gravity and of the forces that govern all motion. We use our understanding of gravity to send spacecraft out to the planets (see figure to right).

In this first Part of the book, we discuss a variety of basic topics that are relevant to the study of astronomy. First, in Chapters 2 and 3, we talk about the main line of astronomical history that led us to today's understanding of the Universe. At the end of Chapter 2, we interpose some studies, known as archaeoastronomy, outside the main line. Then, in Chapters 4 and 5 we talk about the ways that we study the Universe, starting with a discussion of light and its properties. We go on to discuss types of telescopes and some of the major observatories that are now around the world and in space. At the end of this Part, in Chapter 6, we discuss how to observe the sky, how the positions of things in the sky are labelled, the causes of the seasons (which are not what most people think), and about time and calendars. It is easy to consider this material at some other point in the course, perhaps in conjunction with outdoor observations.

Many of the discoveries of tomorrow—perhaps new sources of energy, or perhaps something so revolutionary that it cannot now be predicted—will undoubtedly be based on concepts learned through such basic research as the study of astronomical systems. Considered in this sense, astronomy is an investment in our future.

Yet most of us study astronomy not for its technological and philosophical benefits but for its grandeur and inherent interest. We must stretch our minds to understand the strange objects and events that take place in the far reaches of space. The effort broadens us and continually fascinates.

Jupiter, imaged in 2000 from the Cassini spacecraft as it sped by en route to its arrival at Saturn in 2004.

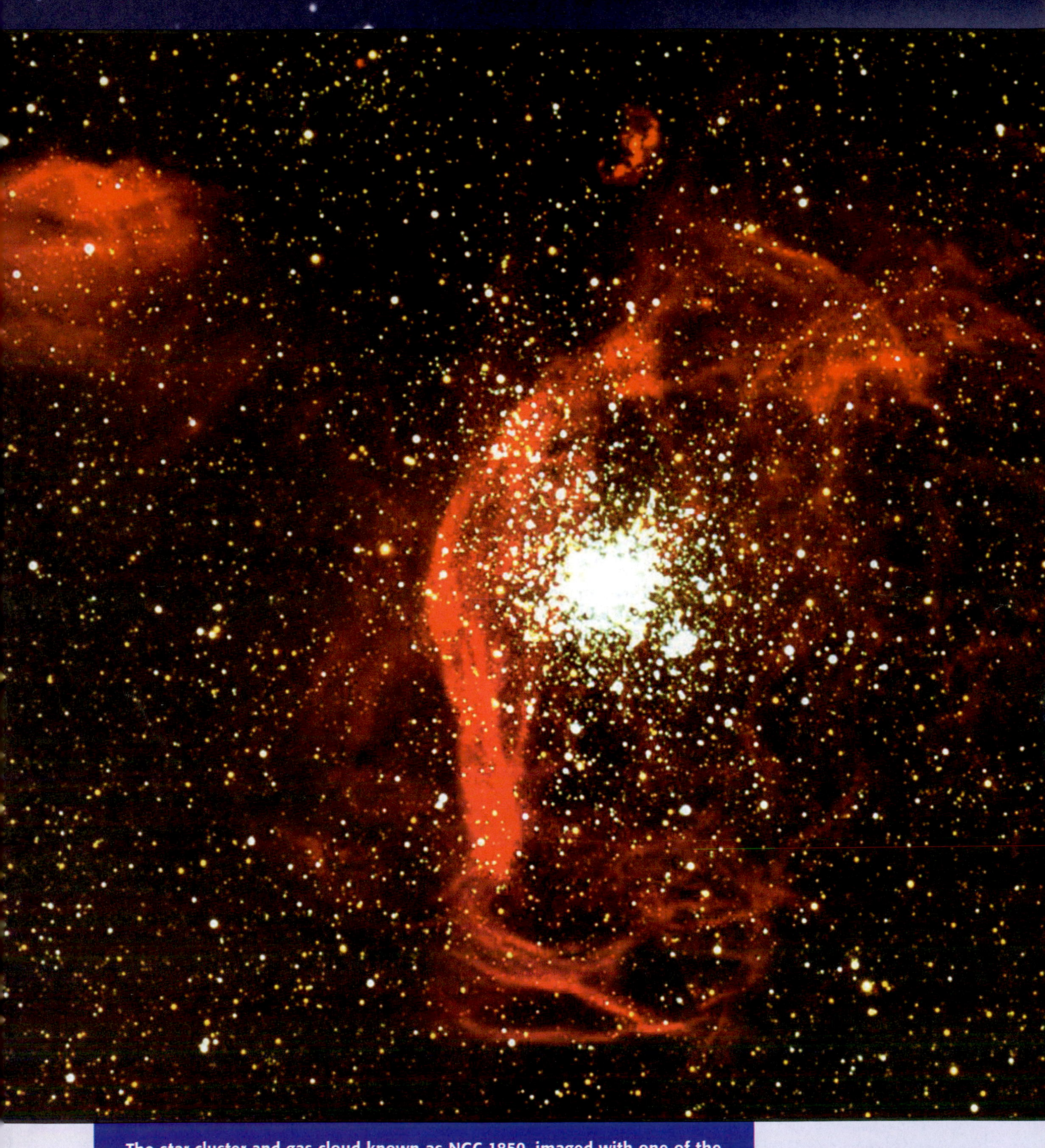

The star cluster and gas cloud known as NGC 1850, imaged with one of the four 8-meter-diameter telescopes of the European Southern Observatory's Very Large Telescope. We see a 40-million-year-old star cluster with a smaller, 4-million-year-old star cluster to its right. The red light from hydrogen gas shows clearly.

# Chapter 1

## Prologue: The Universe

**AIMS:** To get a feeling for the variety of objects in the Universe and a sense of scale

The Universe is a place of great variety—after all, it has everything in it! Some of the things astronomers study are of a size and scale that we humans can easily comprehend: the planets, for instance. Most astronomical objects, however, are so large and so far away that our minds have trouble grasping their sizes and distances.

Moreover, astronomers study the very small in addition to the very large. Most everything we know comes from the study of energy travelling through space in the form of "radiation"—of which light and radio waves are examples. The radiation we receive from distant bodies is emitted by atoms, which are much too small to see with the unaided eye. Also, the properties of the large astronomical objects are often determined by changes that take place on a minuscule scale—that of atoms or their nuclear cores. Further, the evolution of the Universe in its earliest stages depended on the still more fundamental particles within the nuclei. Thus the astronomer must understand the tiniest as well as the largest objects.

Such a variety of objects at very different distances from us or with very different properties often must be studied with widely differing techniques. Clearly, we use different methods to analyze the properties of solid particles like martian soil (using equipment in a spacecraft sitting on the martian surface) or the results of flying through the tail of Halley's Comet than we use to study the visible light, radio waves, or x-rays given off by a gaseous body like a quasar deep in space. Observations of the Sun made at solar eclipses (Fig. 1–1) can now be linked with observations of the Sun from space. The Hubble Space Telescope has given us especially clear views of a wide variety of objects.

**FIGURE 1–1** A scientist looks up at a total solar eclipse, seeing the solar corona surrounding the disk of the Moon.

The explosion of astronomical research in the last few decades has been fueled by our new ability to study radiation other than visible light—gamma rays, x-rays, ultraviolet radiation, infrared radiation, and radio waves. Astronomers' use of their new abilities to study such radiation is a major theme of this book. All the kinds of radiation together make up the **spectrum,** which we will discuss in Chapter 4. As we shall see there, we can think of radiation as waves, and all the types of radiation have similar properties except for the length of the waves. Still, although x-rays and visible light may be similar, our normal experiences tell us that very different techniques are necessary to study them.

The Earth's atmosphere shields us from most kinds of radiation, although light waves and radio waves do penetrate the atmosphere. For the last 40 years, we have been able to send satellites into orbit outside the Earth's atmosphere, and we are no longer limited to the study of radio and visible (that is, light) radiation. Many of the fascinating discoveries of recent years were made because of our extended senses. The first probable observations of giant black holes within our galaxy and other galaxies, for example, were made in the radio part of the spectrum, and we are increasingly observing stars in formation in the infrared, using telescopes on high mountains and in space. We will discuss how astronomers use all parts of the spectrum to help us understand the Universe.

We will see, also, how we get information from direct sampling of bodies in our solar system and from cosmic rays. These cosmic rays are atomic nuclei and subatomic particles whizzing through space; these particles are also known as "radiation," though they are different from the radiation we mentioned previously, which is akin to light. We have also detected subatomic particles called neutrinos both from the Sun and, spectacularly, from the explosion of a distant star. Neutrinos come in three types, and physicists didn't detect the third type until 2000. Just how the types may mix, and why we don't detect as many neutrinos from the Sun as predicted, are important questions with major consequences.

Further, we can now even observe the consequences of gravitational waves, whose existence was predicted by Einstein's general theory of relativity and confirmed from observations of a special pulsar. A pair of major new facilities to detect gravitational waves directly is starting to operate, with the possibility of exciting discoveries. We shall discuss all these methods of astronomical research in the book; remember that the index provides page references.

By the time we have finished the book, we will have covered a huge range of time, some 14 billion years from the formation of the Universe to the present (Fig. 1–2).

On this book's Web page, www.harcourtcollege.com/astro/etu6, we list some of the major discoveries that come along from week to week. A fun thing to do every day is to go to a popular Web page known as Astronomy Picture of the Day (see *Using Technology*, at the end of the chapter).

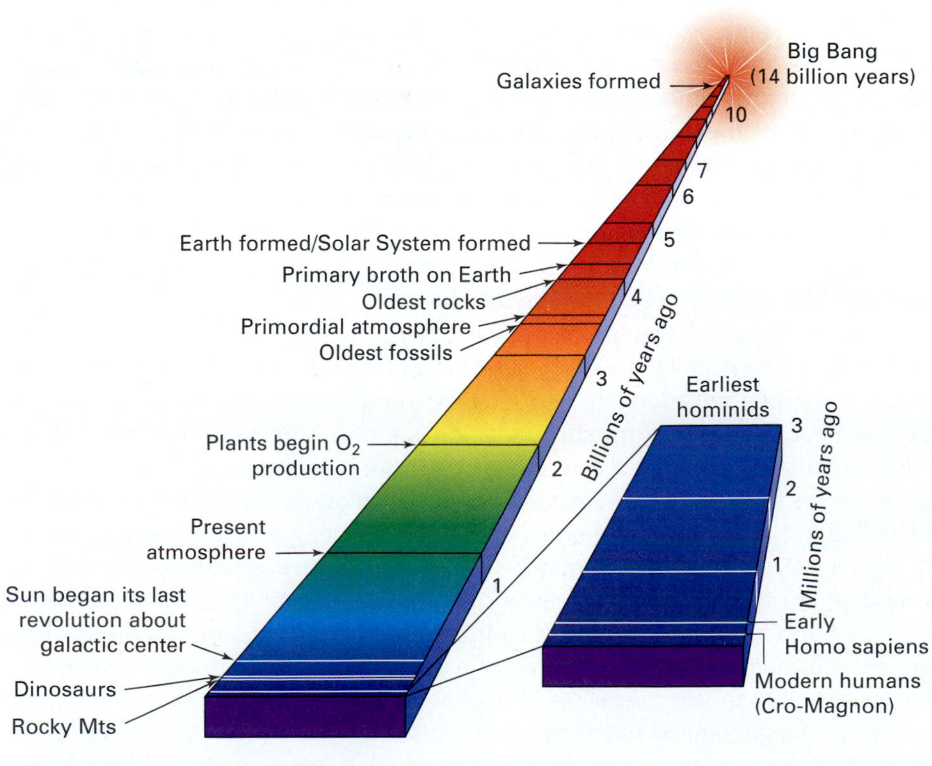


**FIGURE 1–2** A sense of time.

## 1.1 A SENSE OF SCALE

Let us get a sense of scale of the Universe, starting with sizes that are part of our experience. We can keep track of the size of our field of view as we expand in powers of 100: each diagram will show a square 100 times greater on a side. As we journey through space, every step we take will show a region 100 times larger in diameter than the previous picture. Let us begin with something 1 mm across (Fig. 1–3), a fly viewed through an electron microscope.

**FIGURE 1–3** 1 mm = 0.1 cm

10 cm = 100 mm

**FIGURE 1–4** A square 100 times larger on each side is 10 centimeters × 10 centimeters. (Since the area of a square is the length of a side squared, the area of a 10-cm square is 10,000 times the area of a 1-mm square.) The area encloses a flower.

10 m = 1000 cm

**FIGURE 1–5** Here we move far enough away to see an area 10 meters on a side. In the photo, Muhammed Ali is victorious.

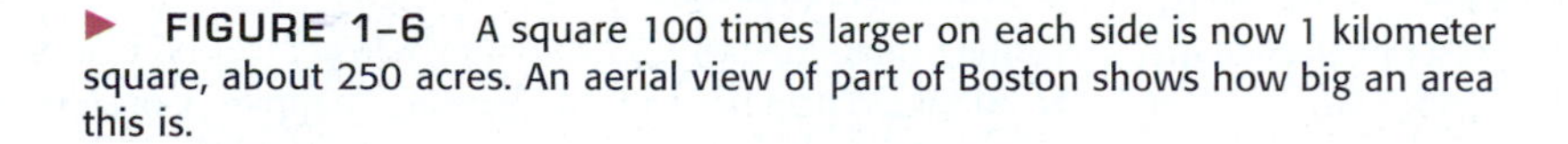

**FIGURE 1–6** A square 100 times larger on each side is now 1 kilometer square, about 250 acres. An aerial view of part of Boston shows how big an area this is.

1 km = $10^3$ m

*(text continues on page 7)*

## A DEEPER DISCUSSION

### BOX 1.1 Measuring Systems

In this book, we shall mostly use the International System of Units (Système international d'unités, or SI for short), which is commonly used in almost all countries around the world and by scientists in the United States. The basic unit of length is the meter (symbol: m), which is equivalent to 39.37 inches, slightly more than a yard. The basic unit of mass is the kilogram (symbol: kg), and the basic unit of time is the second (symbol: s; we use "sec" in this book for added clarity). Prefixes (Appendix 1) are used to define larger or smaller units of these base quantities. The most frequently used prefixes are "milli-" (symbol: m), meaning 1/1000; "kilo-" (symbol: k), meaning 1000 times; "micro-" (symbol: $\mu$), meaning 1/1,000,000; "mega-" (symbol: M), meaning 1,000,000 times; "nano-" (symbol: n), meaning one-billionth; and "giga-" (symbol: G), meaning one billion times. Thus 1 millimeter (1 mm) is 1/1000 of a meter, or about 0.04 inch, and a kilometer (1 km) is 1000 meters, or about 5/8 mile. For mass, the prefixes are used with "gram" (symbol: g), 1/1000 of the base unit "kilogram."

In table form, here are some examples of common prefixes. A complete list appears in Appendix 1.

$10^{-9}$ = 0.000 000 001, nano-
$10^{-6}$ = 0.000 001, micro-
$10^{-3}$ = 0.001, milli-
$10^{-2}$ = 0.01, centi-
$10^{0}$ = 1
$10^{1}$ = 10
$10^{2}$ = 100
$10^{3}$ = 1000, kilo-
$10^{6}$ = 1,000,000, mega-
$10^{9}$ = 1,000,000,000, giga-

Since Helen of Troy was said to have beauty enough to launch 1000 ships, a milliHelen, for example, would be a unit of beauty sufficient to launch one ship. Note that 10 (or anything else) to the zeroth power is 1 and to the first power is itself.

We will keep track of the powers of 10 by which we multiply 1 m by writing as an exponent the number of tens we multiply together; 1000 m, for example, is $10^3$ m, which is 1 km. We say more about this exponential notation in Box 1.2.

Astronomers still use various non-SI units, such as angstroms (symbol: Å) for wavelengths of light, where 10 Å = 1 nanometer. Though densities are sometimes seen in the SI unit of kg/m$^3$, people seem to relate better to g/cm$^3$ (read "grams per cubic centimeter"), a unit for which 1 g/cm$^3$ is the density of water. Light years and other common measures of distance used by astronomers are also non-SI.

The meter is now defined in terms of how far light travels in a certain time. The speed of light in a vacuum (a space in which matter is absent) is, according to the "special theory of relativity" that Albert Einstein advanced in 1905 (Box 31.1), the greatest speed that is physically attainable. Light travels at about 300,000 km/sec (186,000 miles/sec), fast enough to circle the Earth seven times in a single second. Even at that fantastic speed, we shall see that it would take years for us to reach the stars. Similarly, it has taken years for the light we see from stars to reach us, so we are really seeing the stars as they were years ago. In a sense, we are looking backward in time. The distance that light travels in a year is called a **light-year** (ly); note that the light-year is a unit of length rather than a unit of time even though the term "year" appears in it.

### BOX 1.2 Exponential Notation

In astronomy we often find ourselves writing numbers that have strings of zeros attached, so we use what is called either **scientific notation** or **exponential notation** to simplify our writing chores. Scientific notation helps prevent people from making mistakes when they copy long strings of numbers.

In scientific notation, which we use in Figures 1–3 to 1–17, we merely count the number of zeros and write the result as a superscript to the number 10. Thus the number 100,000,000, a 1 followed by 8 zeros, is written $10^8$. The superscript is called the **exponent.** We also say, "10 is raised to the eighth power." When a number is not a power of 10, we divide it into two parts: a number between 1 and 10, and a power of 10. Thus the number 3645 is written as $3.645 \times 10^3$. The exponent shows how many places the decimal point was moved to the left. A negative exponent shows how many places the decimal point was moved to the right.

When adding numbers in scientific notation, make sure to put each number into a form with the same exponent as the others'. For example, $2.5 \times 10^3$ could be written as $25 \times 10^2$.

#### *Discussion Question*

Adding quantities with different exponents can be compared to adding individual apples to bushels of apples. Consider the similarities between these concepts and how the difficulties may be resolved.

$100 \text{ km} = 10^5 \text{ m}$

▶ **FIGURE 1–7** The next square, 100 km on a side, encloses the cities of Boston (*top*) and Providence (*bottom left*). Note that though we are still bound to the limited area of the Earth, the area we can see is increasing rapidly.

$10{,}000 \text{ km} = 10^7 \text{ m}$

◀ **FIGURE 1–8** A square 10,000 km on a side (marked in red) covers nearly the entire Earth. The southwestern United States and Baja California are visible at center in the region clear of clouds.

$1{,}000{,}000 \text{ km} = 10^9 \text{ m} = 3 \text{ lt sec}$

◀ **FIGURE 1–9** When we have receded 100 times farther, we see a square 100 times larger in diameter: 1 million kilometers across. It encloses the orbit of the Moon around the Earth. We can measure with our wristwatches the amount of time that it takes light to travel this distance. If we were carrying on a conversation by radio with someone at this distance, there would be pauses of noticeable length after we finished speaking before we heard an answer. These pauses occur because radio waves, even at the speed of light, take that amount of time to travel. Laser pulses from Earth that are bounced off the Moon to find the distance to the Moon (Appendices 2 and 4) take a noticeable time to return. This photograph was taken by the Galileo spacecraft in 1990 when it passed by the Earth en route to Jupiter.

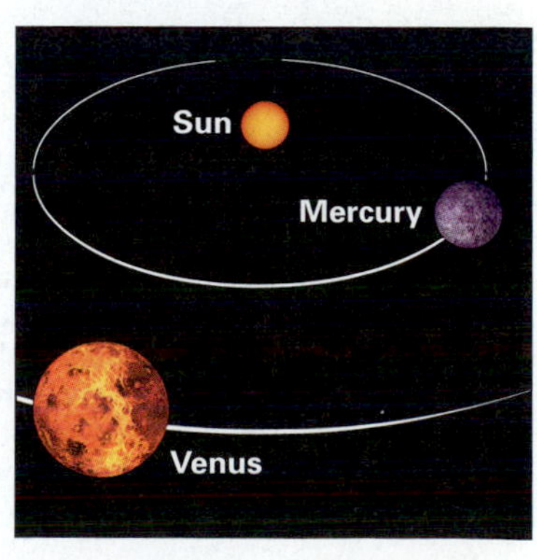


$10^{11} \text{ m} = 5 \text{ lt min}$

▶ **FIGURE 1–10** When we look on from 100 times farther away still, we see an area 100 million kilometers across, 2/3 the distance from the Earth to the Sun. We can now see the Sun and the two innermost planets in our field of view.

▶ **FIGURE 1–11** An area 10 billion kilometers across shows us the entire solar system in good perspective. It takes light 8 hours to travel this distance. The outer planets have become visible and are receding into the distance as our journey outward continues. Our spacecraft have now visited or passed the Moon, Mercury, Venus, Mars, Jupiter, Saturn, Uranus, Neptune, and their moons; the results will be discussed at length in Part 2. This artist's conception shows a Voyager spacecraft near Saturn.

$10^{13}$ m = 8 lt hr

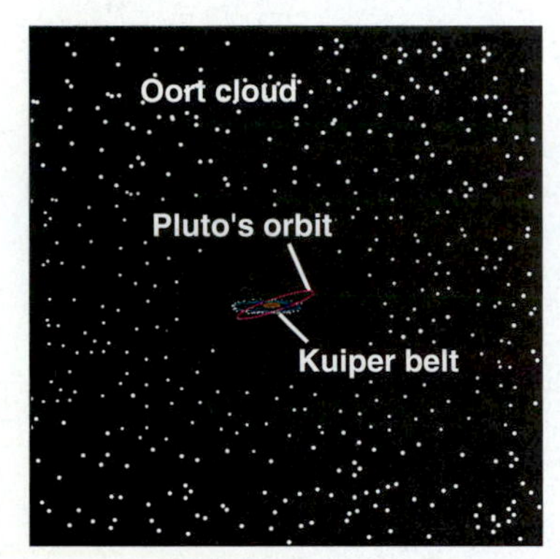


$10^{15}$ m = 38 lt days

◀ **FIGURE 1–12** From 100 times farther away, we see little that is new. The orbits of the outermost planets, surrounded by the Oort Cloud of incipient comets, mark the limits of our solar system. The solar system seems small from this distance, and we see the vastness of the empty space around us. A cube of this size has not yet reached the scale at which another star besides the Sun is visible.

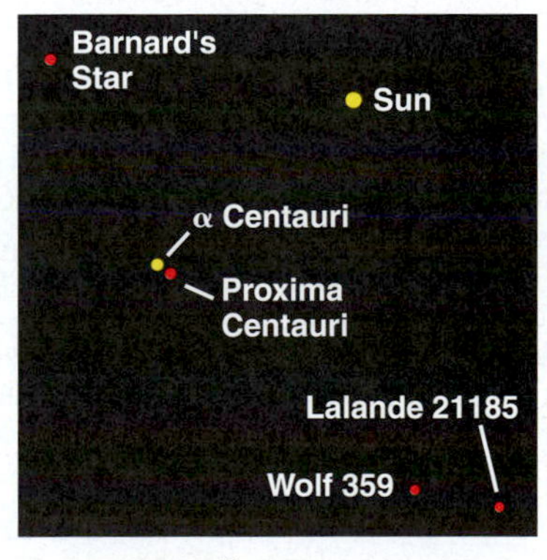


$10^{17}$ m = 10 ly

◀ **FIGURE 1–13** As we continue to recede from the solar system, the nearest stars finally come into view. We are seeing an area 10 light-years across, which contains only a few stars, most of whose names are unfamiliar (Appendix 7). Part 3 of this book discusses the properties of the stars, and Part 4 discusses how they live and die.

▶ **FIGURE 1–14** A faint band—the Milky Way—stretches across our sky when we look from Earth. By the time we are 100 times farther away from Earth than we were for the previous figure, we can see that this band is a fragment of our galaxy, the Milky Way Galaxy. We see not only many individual stars—this photo shows the stars of the constellation known as the Southern Cross—but also many clusters of stars and many areas of glowing, reflecting, or opaque gas or dust called nebulae. Between the stars, there is a lot of material (most of which is invisible to our eyes) that can be studied with radio telescopes on Earth or in infrared, ultraviolet, or x-rays with telescopes in space. Part 5 of this book is devoted to the study of our galaxy and its contents.

$10^{19}$ m = $10^{3}$ ly

▶ **FIGURE 1–15** In a field of view 100 times larger in diameter, we can now see an entire galaxy. The photograph shows the galaxy called NGC 7742, located in the direction of the constellation Pegasus. This galaxy, like all others, is far beyond the stars in that constellation. This galaxy shows arms wound in spiral form like a pinwheel. The galaxy in which we live also has spiral arms, though they are not wound tightly. NGC 7742 has a particularly active core (the yellow "yolk"), presumably surrounding a giant black hole.

$10^{21}$ m = $10^5$ ly

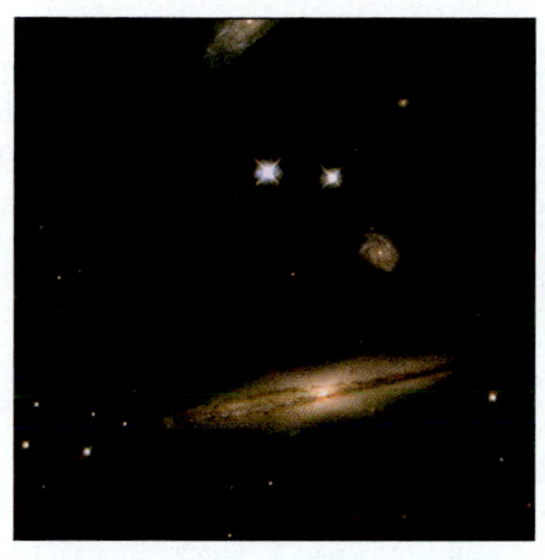

$10^{23}$ m = $10^7$ ly

◀ **FIGURE 1–16** Next we move sufficiently far away so that we can see an area 10 million light-years across. There are $10^{25}$ centimeters in 10 million light-years, about as many centimeters as there are grains of sand in all the beaches of the Earth. Our galaxy is in a cluster of galaxies, called the Local Group, that would take up only ⅓ of our angle of vision. In this group are all types of galaxies, which we will discuss in Chapter 34. The photograph shows part of a cluster of galaxies seen in the Hubble Deep Field (Section 38.7).

▶ **FIGURE 1–17** If we could see a field of view 1 billion light years across, our Local Group of galaxies would appear as but one of many clusters. It is difficult to observe on such a large scale, but new electronic methods used with telescopes have enabled astronomers to map galaxies on this large scale in recent years. We discuss this billion-light-year "slice of the Universe" in Chapter 35.

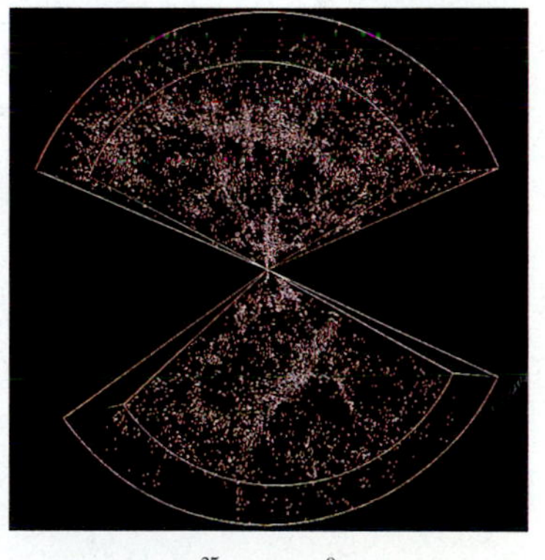

$10^{25}$ m = $10^9$ ly

At this last scale, we see superclusters—clusters of clusters of galaxies. We would be seeing almost to the distance of the quasars, the topic of Chapter 36. Quasars, the most distant objects known, seem to be explosive events related to giant black holes in the cores of galaxies. Light from the most distant quasars observed may have taken over 10 billion years to reach us on Earth. We are thus looking back to a time billions of years ago (Fig. 1–18). Since we think that the Universe began perhaps 14 billion years ago, we are looking back almost to the beginning of time.

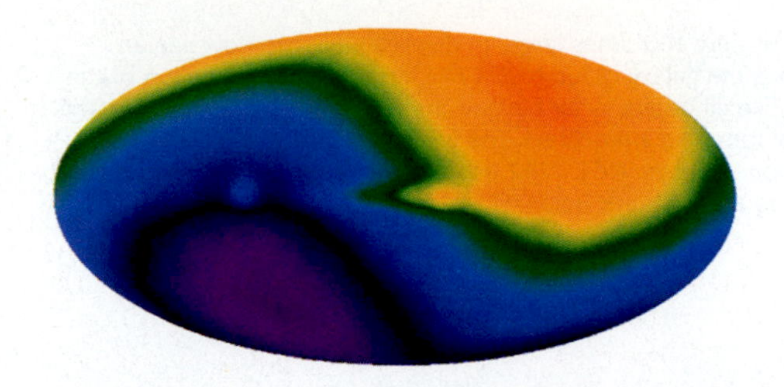

**FIGURE 1–18** We have even detected radiation from the Universe's earliest years. This map of the whole sky (Fig. 1–18) shows a slight difference in the Universe's temperature in opposite directions (Chapter 37), resulting from our Sun's motion. In more detailed maps, we are seeing the seeds from which today's clusters of galaxies grew. A combination of radio, ultraviolet, x-ray, and optical studies, together with theoretical work and experiments with giant atom smashers on Earth, is allowing us to explore the past and predict the future of the Universe.

It is impressive to realize that we have detected radiation from the Universe's earliest years. A combination of radio, ultraviolet, x-ray, and optical studies, together with theoretical work and experiments with giant atom smashers on Earth, is allowing us to explore the past and predict the future of the Universe. These wonderful ideas are discussed in Chapters 37 and 38.

## 1.2 A SENSE OF SPEED

Light travels so fast that, if it could bend, it would circle the Earth 7 times in a second. Its speed is 300,000 km/s. Light reaches the Moon in just over 1 second. If we could travel there at 100 km/hr, the speed of a car on a highway, since there are 3,600 seconds in an hour, we would be going (100 km/hr)/(3,600 hr/s) = 1/36 km/s. (Note that you can just cancel out units; the "/hr" in the first parenthesis cancels the "hr" in the second one.) In astronomy, we often give speeds "per second."

The distance travelled by a non-accelerating object is equal to the rate at which the object is travelling (its speed) times the time spent travelling ($d = vt$).

To reach the Moon (many basic facts about the solar system are shown in Appendix 2), which is about 400,000 km away, it would take time = distance/speed = (400,000 km)/(100 km/hr) = 4000 hours. To transform into days (note how we keep the units attached to the numbers), (4000 hours)/(24 hours/day) = 17 days.

### EXAMPLE 1.1 Distance from Speed and Time

***Question:*** How far does light travel in 1 hour?

***Answer:*** 1 hour = 3600 seconds. Then $d = vt = 300{,}000$ km/sec $\times$ 3600 sec = 1,080,000,000 km (just over 1 billion km).

### EXAMPLE 1.2 The Speed of Light

***Question:*** How fast does light travel, expressed in the unit light-years per year?

***Answer:*** Transform the equation to read "$v = d/t$." Then $v$ = 1 light-year/1 year = 1 light-year/year.

## 1.3 A SENSE OF TIME

A human sense of time, with our hearts beating about once every second, used to be much too short to sense astronomical timescales. But now that we can go beyond human timescales, we know about fascinating objects in the sky like pulsars, which send out pulses of radio waves up to several hundreds of times each second. Still, most objects in the sky change more slowly.

Many of our measures of time are linked, since ancient times, with astronomy. A day is the interval that it takes the Earth to rotate on its axis. To make a decision on whether to measure that rotation with respect to the Sun (a solar day) or the stars (a sidereal day) requires that you understand the astronomy involved (Chapter 6).

A month is based on a "moonth," the time it takes for the Moon to go through a cycle of phases. A year is based on the Earth–Sun system; it is the length of time it takes the Earth to revolve around the Sun.

The Sun itself has cycles, notably the 11-year sunspot cycle (Fig. 1–19). In that interval the Sun goes from active to quiet to active again; we have just completed the latest maximum of activity (Chapter 23).

Just as the Earth's period of revolution around the Sun is a year, a "galactic year" marks the period of time the Sun takes to revolve around the center of our Milky Way Galaxy. This galactic year is about 250 million of our Earth years.

Our galaxy may itself be orbiting other galaxies, but we cannot measure what the period might be. In the final chapters of this book, we will address the possibility—now out of favor—that the Universe as a whole has a period of variation, which might be on the order of 50 billion years, 10 times older than our Sun itself.

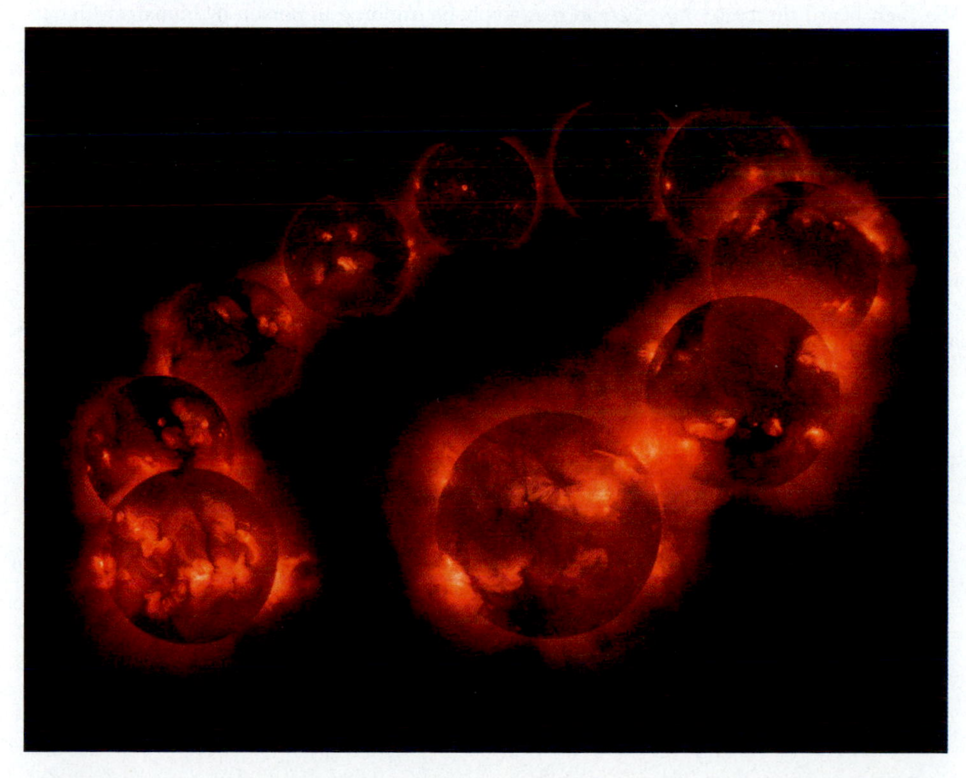

**FIGURE 1–19** A montage of x-ray images of the Sun over the 11-year sunspot cycle, from hyper-activity in 1990 to low activity and back to high activity in 2001. (See Figure 23–4.)

## CORRECTING MISCONCEPTIONS

| ✖ *Incorrect* | ✔ *Correct* |
| --- | --- |
| Space is full of material. | Space is relatively empty. |
| You cannot add $(5 \times 10^7) + (6 \times 10^8)$. | Each part is a number, so the two parts can certainly be added. To do so, write each term so that all terms have the same exponent. |
| We understand most of what there is to know about the Universe. | We have a lot to learn. |

## KEY WORDS

spectrum  light-year  scientific notation  exponential notation  exponent

## QUESTIONS

1. Why do we say that our senses have been expanded in recent years?

†2. The speed of light is $3 \times 10^5$ km/sec. Express this number in m/sec and in cm/sec.

†3. During the Apollo explorations of the Moon, we had a direct demonstration of the finite speed of light when ground controllers spoke to the astronauts. The sound from the astronauts' earpieces was sometimes picked up by the astronauts' microphones and retransmitted to Earth as radio signals, which travel at the speed of light. We then heard the controllers' words repeated. What is the time delay between the original and the "echo," assuming that no other delays were introduced in the signal? (The distance to the Moon and the speed of light are given in Appendix 2.)

†4. The time delay in sending commands to the Voyager 2 spacecraft when it was near Uranus was 2 hours 45 minutes. What was the distance in km from the Earth to Uranus at that time? What was the distance in Astronomical Units (A.U.), where one Astronomical Unit is the average radius of the Earth's orbit? (See Appendix 2.)

†5. The distance to the Andromeda Galaxy is $2 \times 10^6$ light years. If we could travel at one-tenth the speed of light, how long would a round trip take?

†6. How long would it take to travel to Andromeda, which is $2 \times 10^6$ light years away, at 1000 km/hr, the speed of a jet plane?

7. List the following in order of increasing size: (a) light-year, (b) distance from Earth to Sun, (c) size of Local Group, (d) size of football stadium, (e) size of our galaxy, (f) distance to a quasar.

8. Of the examples of scale in this chapter, which would you characterize as part of "everyday" experience? What range of scale does this encompass? How does this range compare with the total range covered in the chapter?

9. What is the largest of the scales discussed in this chapter that could reasonably be explored in person by humans with current technology?

†10. (a) Write the following in scientific notation: 4642; 70,000; 34.7. (b) Write the following in scientific notation: 0.254; 0.0046; 0.10243. (c) Write out the following in an ordinary string of digits: $2.54 \times 10^6$; $2.004 \times 10^2$.

†11. What is (a) $(2 \times 10^5) + (4.5 \times 10^5)$; (b) $(5 \times 10^7) + (6 \times 10^8)$; (c) $(5 \times 10^3)(2.5 \times 10^7)$; (d) $(7 \times 10^6)(8 \times 10^4)$? Write the answers in scientific notation, with only one digit to the left of the decimal point.

†12. What percentage of the age of the Universe has elapsed since the appearance of *Homo sapiens* on the Earth? (*Hint:* refer to the material opening Part 1.)

†13. How many galactic years old is our Universe?

† This question requires a numerical solution.

## USING TECHNOLOGY

### W World Wide Web

1. Use this book's Web site (http://www.harcourtcollege.com/astro/etu6) to explore general ideas about astronomy and astronomy teaching.
2. Sign on to the Space Telescope Science Institute's Web site to see some of the Hubble Space Telescope's latest observations (http://oposite.stsci.edu/pubinfo/pictures.html).
3. Use this book's Web site to link to sunspot counts, and note where we are in the current sunspot cycle. Look at an image of today's sunspots.
4. Check today's Astronomy Picture of the Day (http://antwrp.gsfc.nasa.gov/apod).

### REDSHIFT

1. Use a few of the RedShift College Edition features called "Guided Tours" to find out about the Universe.
2. Look out in the direction of Mars tonight. Find the names of some of the stars beyond Mars. How many times more distant are these stars than Mars is from Earth? What is the galaxy in the region generally behind Mars? How far away is it?
3. For general orientation, look at the Guided Tour called "From Big Bang to Galaxies."

**Copernicus's original heliocentric diagram, from the first edition of his 1543 book, *De Revolutionibus (Concerning the Revolutions).***

# Chapter 2

## The Early History of Astronomy

AIMS: To follow the theories of the Universe developed and held during the millennia of recorded history before the invention of the telescope

Though much of modern astronomy deals with explaining the Universe, throughout history astronomy has dealt with such practical things as keeping time, marking the arrival of the seasons, and predicting eclipses of the Sun and the Moon. In this chapter we shall discuss the origins of astronomy and trace its development, from the Earth-centered view of the solar system that dominated thought from the time of the Greeks through the Sun-centered view developed by Copernicus in the 16th century and the then-unusual idea of Tycho Brahe that observations of the highest possible quality were valuable.

Figures in the sky—constellations—were recorded by the Sumerians as far back as 2000 B.C., and maybe earlier. The bull, the lion, and the scorpion are constellations that date from this time. Other constellations that we have in the present time were recorded by a Greek, Thales of Miletus, in about 600 B.C., at the dawn of Greek astronomy. Certain constellations were common to several civilizations. The Chinese also had a lion, a scorpion, a hunter, and a dipper, for example (Fig. 2–1).

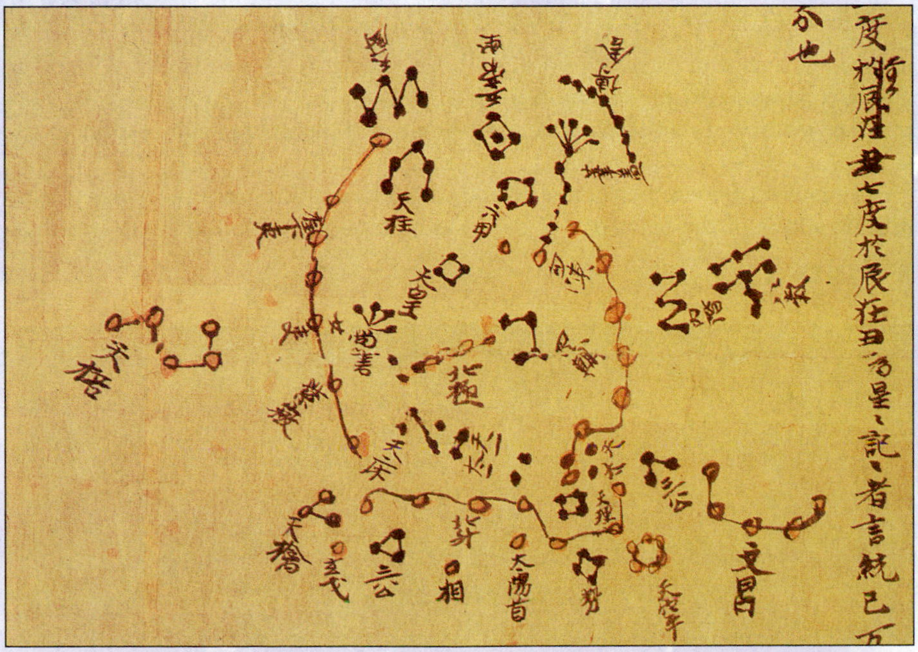

FIGURE 2–1 A Chinese manuscript from the 10th century, the oldest existing portable star map.

## 2.1 EGYPTIAN AND OTHER EARLY ASTRONOMY

In these days of digital watches, electronic calculators, and accurate time on radio and television, it is hard to imagine what life was like in 1000 B.C. or earlier. No mechanical clocks existed and even the idea of a calendar was vague. Yet each summer, the Nile River flooded the neighboring farmlands. In a land where changes in temperature do not clearly mark the seasons, Egyptian farmers wanted to know how to predict when the flooding would occur. Farmers everywhere, even those not dependent on annual flooding, wanted to know when to expect the seasonal rains so that they could plant their crops in ample time.

Eventually, some people noticed that they could base a calendar on Sirius, the brightest star in the sky. It became visible in the eastern sky before sunrise in June or early July, just the time of year when the crops should be planted and when the Nile was about to flood. The first visibility of an object in the pre-dawn sky, after months in which it has been up only in the daytime and thus invisible, is the object's **heliacal rising** (pronounced "he-lye'e-kl"). The heliacal rising of Sirius served as a marker for Egyptian astronomers and priests, enabling them to let everybody know that the floods would soon come. We still have traces of this history in the term "the Dog Days" for the sultry period of summer marked by Sirius, the Dog Star, being near the Sun. (According to the myth of that time, Sirius adds to the Sun's heat.)

## 2.2 GREEK ASTRONOMY: THE EARTH AT THE CENTER

In ancient times, five planets were known—Mercury, Venus, Mars, Jupiter, and Saturn. When we observe these planets in the sky, we notice that their positions vary from night to night with respect to each other and with respect to the stars. The planets and Moon appear to move on or near the **ecliptic,** the apparent path of the Sun across the sky. The stars, on the other hand, are so far away that their positions are relatively fixed with respect to each other. The fact that the planets appear as "wandering stars" was known to the ancients; our word "planet" comes from the Greek word for "wanderer." Of course, both stars and planets move together across our sky essentially once every 24 hours; by the "wandering of the planets" we mean that the planets appear to move at a slightly different rate, so that over a period of weeks or months they change position with respect to the fixed stars.

The planets do not always move in the sky in the same direction with respect to the stars. Most of the time the planets appear to drift eastward with respect to the background stars. But sometimes they drift backwards, that is, westward. We call the backward motion **retrograde motion** (Fig. 2–2). It contrasts with "prograde" motion, the ordinary direction in which the Sun and Moon drift with respect to the stars.

The ancient Greeks began to explain the motions of the planets by making theoretical models of the geometry of the solar system. For example, by determining

**FIGURE 2–2** Retrograde motion. A planetarium simulation of the path of Mars over 8 months, showing the retrograde loop. Mars passes first near the V-shaped Hyades star cluster, with its reddish star Aldebaran, reverses direction, and then reverses direction again near the Pleiades star cluster.

**FIGURE 2–3** Aristotle (*pointing down*) and Plato (*pointing up*) in Raphael's (1483–1520) *The School of Athens.*

which of the known planets had the longest periods of retrograde motion, they were able to discover the order of distance of the planets.

One of the earliest and greatest philosophers, Aristotle, lived in Greece about 350 B.C. (Fig. 2–3). He summarized the astronomical knowledge of his day into a qualitative cosmology that remained dominant for 1800 years. On the basis of what seemed to be very good evidence—what he saw—Aristotle thought, and actually believed that he knew, that the Earth was at the center of the Universe and that the planets, the Sun, and the stars revolved around it (Fig. 2–4). The Universe was made up of a set of 55 celestial spheres that fit around each other; each had rotation as its natural motion. Each of the heavenly bodies was carried around the heavens by one of the spheres. The motions of the spheres affected each other and combined to account for the various observed motions of the planets, including revolution around the Earth, retrograde motion, and motion above and below the ecliptic. The outermost sphere was that of the fixed stars, beyond which lay the prime mover, ***primum mobile,*** that caused the general rotation of the stars overhead.

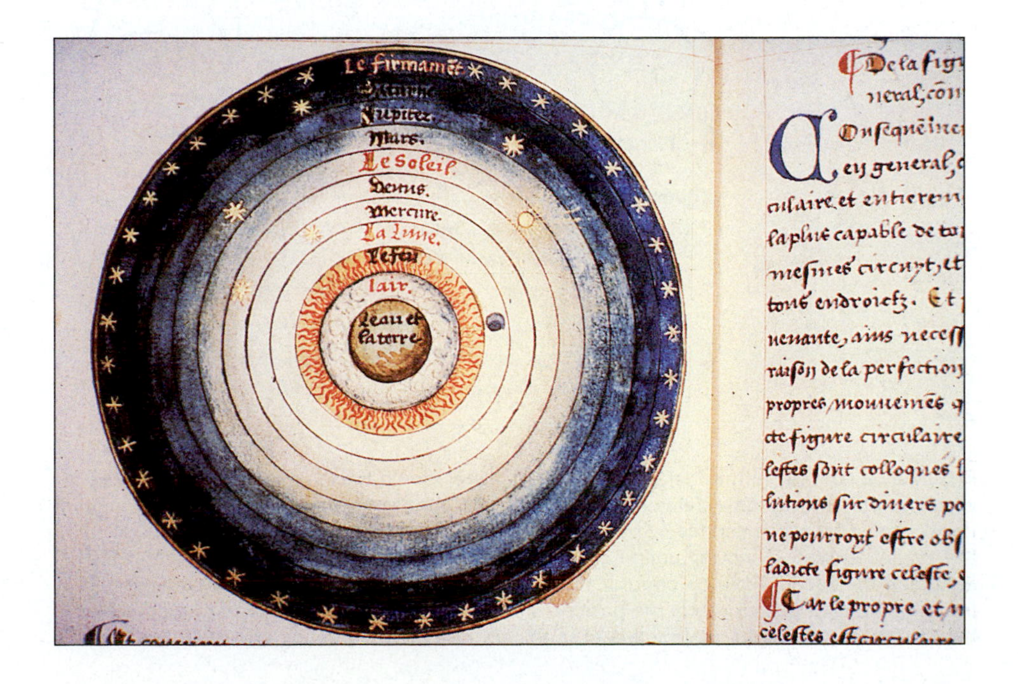


**FIGURE 2–4** Aristotle's cosmological system, with water and earth at the center, surrounded by air, fire, the Moon, Mercury, Venus, the Sun, Mars, Jupiter, Saturn, and the firmament of fixed stars.

FIGURE 2–5 Ptolemy, in a 15th-century drawing.

Aristotle's theories ranged through much of science. He held that below the sphere of the Moon everything was made of four basic "elements": earth, air, fire, and water. The fifth "essence"—the quintessence—was a perfect, unchanging, transparent element of which the celestial spheres were thought to be formed. As for what we now call physics, Aristotle thought that a force has to be continuously applied to cause an object to keep moving. (For example, the air displaced from the tip of a moving arrow was thought to come around to the back of the arrow to push it forward.) We now know the opposite—that moving objects tend to keep moving. He also distinguished between linear and circular motion, thinking that both were natural. It took Newton to set us straight on these points, as we shall see in the following chapter.

Aristotle's theories dominated scientific thinking for almost two thousand years, until the Renaissance. Unfortunately, most of his theories were far from what we now consider to be correct, so we tend to think that the widespread acceptance of Aristotelian physics impeded the development of science.

In about A.D. 140, almost 500 years after Aristotle, the Greek astronomer Claudius Ptolemy (Fig. 2–5), in Alexandria, presented a detailed theory of the Universe that explained the retrograde motion. Ptolemy's model was Earth-centered, as was Aristotle's. To account for the retrograde motion of the planets, the planets had to be moving not simply on large circles around the Earth but rather on smaller circles, called **epicycles,** whose centers moved around the Earth on larger circles, called **deferents** (Fig. 2–6). (The notion of epicycles and deferents had been advanced earlier by such astronomers as Hipparchus, whose work on star catalogues we will discuss in Section 25.1.) It seemed natural that the planets should follow circles in their motion, since circles were thought to be "perfect" figures.

Sometimes the center of the deferent was not centered at the Earth; such an off-center circle is known as an **eccentric,** short for "eccentric circle." The epicycles moved at a constant rate of angular motion (that is, the angle through which they moved was the same for each identical period of time). However, another complication was that the point around which the epicycle moved at this constant rate was neither at the center of the Earth nor at the center of the deferent. The

*(text continues on page 20)*

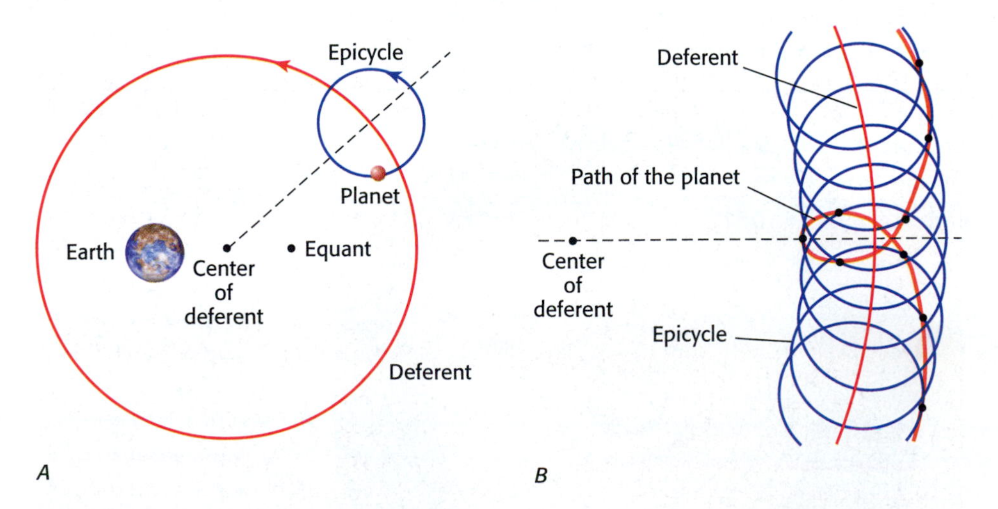


FIGURE 2–6 (*A*) In the Ptolemaic system, a planet would move around on an epicycle, which, in turn, moved on a deferent. Variations in the apparent speed of the planet's motion in the sky could be accounted for by having the epicycle move uniformly around a point called the equant, instead of moving uniformly around the Earth or the center of the deferent. The Earth and the equant were on opposite sides of the center of the deferent and equally spaced from it. When the planet was in the position shown, it would be in retrograde motion. (*B*) In the Ptolemaic system, the projected path in the sky of a planet in retrograde motion is shown.

IN REVIEW

## BOX 2.1 Ptolemaic Terms

*deferent*—a large circle centered approximately, but not exactly, on the Earth

*epicycle*—a small circle whose center moves along a deferent. The planets move on the epicycles.

*eccentric*—a circle, such as a deferent, that is not exactly centered at the Earth

*equant*—the point around which the epicycle moves at a uniform rate

*Don't forget the key point: that the Ptolemaic system is based on circles.*

### Discussion Question

Why aren't an epicycle and a deferent enough? That is, why do we need an eccentric and an equant?

A DEEPER DISCUSSION

## BOX 2.2 Eratosthenes and the Size of the Earth

Eratosthenes (pronounced eh-ra-tos'then-ees) was born in about 273 B.C. He received his education in Athens and then spent the latter half of his life in the city of Alexandria (which is in Egypt but which was then under Greek rule).

The Greek mathematician Pythagoras and his disciples had earlier recognized that the Earth is round, but this idea was not generally accepted. Eratosthenes set out to do no less than to measure the size of the whole Earth.

He made the measurement by looking in a deep well or by using a **gnomon** (pronounced noh'mon), basically an upright stick that is allowed to cast a shadow in sunlight. (Gnomons, simple as they now seem to us when we see them on sundials, were among the prime astronomical instruments available at that time.) Eratosthenes used observations made at two cities that were on the same line of longitude; that is, one was due north of the other. The length of the shadow cast by the gnomon (or, perhaps, the shadow inside a well) at noon varied from day to day. At one of his cities (Syene, near what is now Aswan, Egypt) on one day of the year (which we now call the summer solstice), that shadow vanished. This vanishing indicated that the Sun was directly overhead in Syene at that moment. From the length of the shadow at the same date and time at his second city, Alexandria, when there was no shadow at Syene, he concluded that Alexandria was one-fiftieth of a circle around the Earth from Syene (see figure).

Further, Eratosthenes knew the distance on the Earth's surface between Syene and Alexandria: it was 5000 stadia, where a stadium was a unit of distance equal to the size of an athletic stadium, about 160 meters. (The distance had actually been paced out by someone trained for the evenness in the size of his step.)

Eratosthenes concluded that if one-fiftieth of the Earth's circumference was 5000 stadia, then the Earth must be 250,000 stadia around. The value is in reasonable agreement with the value we now know to be the actual size of the Earth. But even more important than the particular number is the idea that, with the application of logic, the size of the Earth could be measured and fathomed by mere humans even with primitive technology.

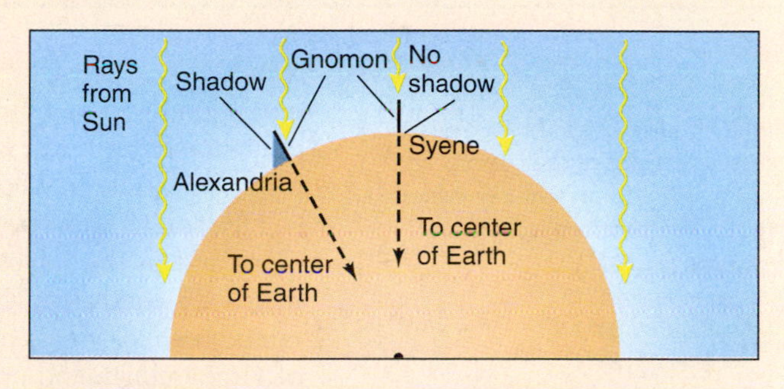


When the Sun is directly overhead at Syene, a gnomon (an upright stick) casts no shadow. At the same time, the Sun is not directly overhead at Alexandria, which is due north of Syene. From this angle and the distance between the cities, Eratosthenes calculated the size of the Earth. Note that the Sun is so far away from the Earth that sunlight hitting the Earth at one city is parallel to sunlight hitting the Earth at the other. The angular scale is greatly exaggerated here. Tradition has it that the observation may have been made from a well at Syene; when the Sun was directly overhead at Syene, it illuminated the entire bottom of the well.

### Discussion Question

What would you have to do yourself to measure the size of the Earth using Eratosthenes's method? Where would you have to travel? Why can we now be more accurate than Eratosthenes himself?

A DEEPER DISCUSSION

## BOX 2.3 Stellar Parallaxes and the Motion of the Earth

Most Greek astronomers held that the Earth was fixed at the center of the Universe. They reasoned (correctly) that if the Earth itself moved, during the course of the year the stars would be at slightly different heights in the sky because of our changing vantage point. Such an apparent movement resulting from a change in vantage point is called **parallax** (see Section 25.2). Parallax is similar to the apparent motion of your thumb when you view your thumb at the end of your outstretched arm, looking first with one eye and then the other. The thumb appears to jump left and right across the background. The closer you hold your thumb to your face, the greater its displacement appears.

Parallaxes of stars are not visible to the naked eye, nor could they be observed with the instruments available to the ancient Greeks. Actually, the nearest stars do have parallaxes. The closer stars are like your thumb, and the distant stars are the background. The two points from which we observe are the positions of the Earth on different parts of its orbit around the Sun. The parallaxes of even the nearby stars are too small to have been observed before the invention of the telescope (and were not published until Friedrich Bessel first measured stellar parallax in 1838). In 1997, fantastically accurate parallaxes for over 100,000 stars, measured from the Hipparcos satellite, were published.

The Greek astronomers deserve credit for the fact that, at least in this respect, their theories agreed better with observations than did the heliocentric (Sun-centered) theory. It was only much later that accurate observations changed the weight of the evidence on this point.

***Discussion Question***

How does parallax affect the speed an analog speedometer shows in a car when seen by (a) the driver, and (b) the passenger?

Ptolemy tried to account for the fact that the lengths of the seasons are not equal. The present-day values for the northern hemisphere are

| | |
|---|---|
| spring | $92^d19^h$ |
| summer | $93^d15^h$ |
| autumn | $89^d20^h$ |
| winter | $89^d0^h$ |

Given the season lengths, the Sun's orbit around the Earth cannot be both round and centered at the Earth. Ptolemy and Hipparchus invoked an eccentric, that is, an off-center circle, or an equivalent epicycle.

epicycle moved at a uniform angular rate about still another point, the **equant.** The equant and the Earth were equally spaced on opposite sides of the center of the deferent, as the figure shows.

Ptolemy's views were very influential in the study of astronomy; versions of his ideas and of the tables of planetary motions that he computed were accepted for nearly 15 centuries. His major work, which became known as the *Almagest,* contained both his ideas and a summary of the ideas of his predecessors (especially those of Hipparchus), and provides most of our knowledge of Greek astronomy.

Several hundred years later, in Alexandria around A.D. 400, Hypatia came to symbolize learning and science. Only the titles of her books are still known, but from them we see that she wrote about astronomy and about mathematics. We also know from letters that she knew how to construct an astrolabe, which is used to find positions in the sky. Her murder coincided with the beginning of the end of Alexandria as an academic site.

## 2.3 OTHER ANCIENT ASTRONOMY*

At the same time that Greek astronomers were developing their ideas, Babylonian priests were tabulating positions of the Moon and planets. The major work was carried out during the period from about 700 B.C. until about A.D. 50. This work is sometimes referred to as Chaldean (pronounced something like "kal-dee'an" in English). Chaldea was the southern part of Babylon.

The Babylonian tablets that have survived show lists and tables of planetary positions and eclipses. The tablets also show predictions of such quantities as the times when planets would be closest to (in **conjunction** with—technically, at the same longitude along the Sun's path) and opposite to (in **opposition** to—technically, 180° away in longitude along the Sun's path) the Sun in the sky and when objects would be visible for the first or last time in a year. Babylonian methods were communicated to the Greeks, and it is through the Greeks that Babylonian astronomy influenced Western thought.

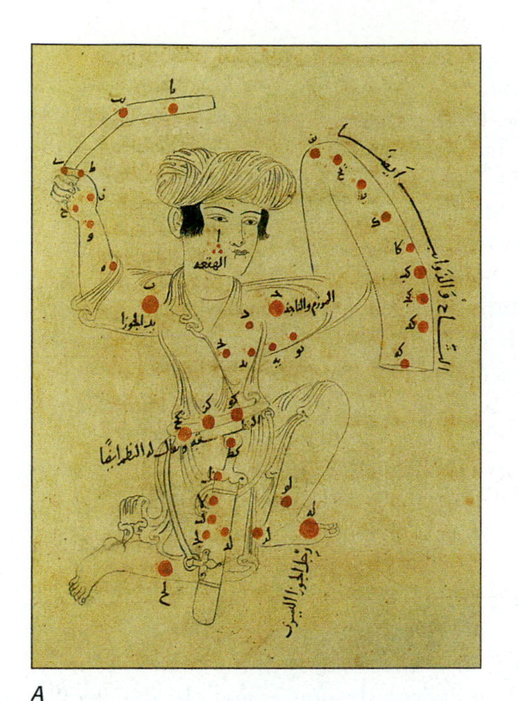

A

B

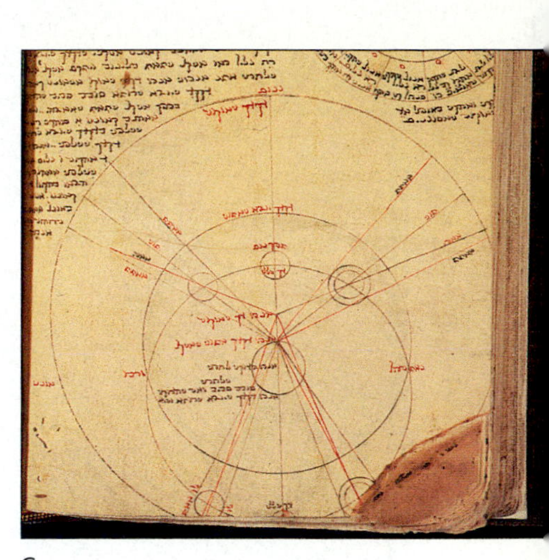

C

**FIGURE 2–7** (*A*) The constellation Orion, from the *Book of the Stars* by Al-Sufi, a 10th-century Islamic astronomer who carried out the principal revisions of Ptolemy's work during the Middle Ages. Note the three stars that make Orion's belt, the star in his sword, and the bright star in his right shoulder. In this pre-telescopic age, astronomers did not know of the Orion Nebula, which is in his sword. (*B*) An astronomer is shown holding an astrolabe and discussing the laws of nature with his fellow philosophers under a sky full of stars in this page from a manuscript copy of *The Guide to the Perplexed* by the 12th-century Jewish philosopher Moses Maimonides. (*C*) Astronomical studies from a 15th-century manuscript by the Jewish astronomer Abraham bar Hiyya ha-Nasi.

## 2.4 ASTRONOMY OF THE MIDDLE AGES*

The Middle Ages, which we might loosely define as the thousand years up to about 1500, were not marked by epochal astronomical discoveries or significant models, such as those advanced by Ptolemy and Aristotle. Nonetheless, the period was not as dead as is commonly presented.

Aristotle's ideas provided the framework for most of the medieval discussion of astronomy. Earth, air, fire, and water remained the four known "elements," and his false notions of motion persisted.

Arab astronomers were active (Fig. 2–7*A*), and translations of Ptolemy's major work, the *Almagest,* were made from the Greek into Arabic in the 8th and 9th centuries; only through such Arabic translations do we now know of several Greek works. (*Almagest* is, in fact, a translation of the Arabic word for "the greatest.") Major observatories were built and careful records were kept of planetary positions, though the telescope had not yet been invented. Jewish astronomers also wrote about celestial objects (Fig. 2–7*B* and *C*).

In the late 13th century, King Alfonso X collected a group of scholars at Toledo, Spain. The astronomical tables they computed, known as the *Alphonsine Tables,* remained in general use for centuries. Such tables provided basic information that could be used to calculate the positions of the planets at any time in the past or future.

One of the most famous medieval observatories was built in Samarkand (Fig. 2–8), Uzbekistan, by Ulugh Begh (a grandson of the more famous Tartar chief Tamerlane). Ulugh Begh's star catalogue shows the high quality of astronomy there. Using

**FIGURE 2–8** The monument built over the altitude quadrant in the center of Ulugh Begh's observatory at Samarkand.

FIGURE 2–9 A comet from the *Nuremberg Chronicle* (1493), one of four woodblocks used to reproduce 13 comets.

the measuring instruments available in about 1420, his results were only slightly better than those found in Ptolemy's *Almagest.*

In the mid-15th century, in Europe, Regiomontanus (the Latinized city name by which John Müller was known) and his teacher discovered how inaccurate the *Alphonsine Tables* could be. For example, an eclipse of the Moon occurred an hour later than the *Tables* had predicted, and Mars appeared two degrees (four times the Moon's diameter) from its forecast position. Regiomontanus ultimately settled in Nuremberg, in what is now Germany, where he started a printing press and published much of the astronomical literature of that period. He printed theoretical works and also the first **ephemeris**—a book of the planetary positions (the name is based on the Latin word for "changing," as in "ephemeral"). The positions were based on the *Alphonsine Tables.* Regiomontanus also completed a summary of Ptolemy's *Almagest* that had been begun by his teacher. Nuremberg remained a center of scientific printing; the *Nuremberg Chronicle* (1493) included reports of many comets and other astronomical events (Fig. 2–9).

## 2.5 NICOLAUS COPERNICUS: THE SUN AT THE CENTER

Helios was the Sun god in Greek mythology.

The major credit for the breakthrough in our understanding of the solar system belongs to Nicolaus Copernicus (Fig. 2–10), a Polish astronomer and cleric. Over 400 years ago, Copernicus advanced a **heliocentric**—Sun-centered—theory (Figs. 2–11 and 2–12). He suggested that the retrograde motion of the planets could be readily explained if the Sun, rather than the Earth, were at the center of the Universe; that the Earth is a planet; and that the planets move around the Sun in circles.

Aristarchus of Samos, a Greek scientist, had suggested a heliocentric theory 18 centuries earlier, though only fragments of his work are known and we do not know how detailed a picture of planetary motions he presented. His heliocentric suggestion required the then apparently ridiculous notion that the Earth itself moved, in contradiction to our senses and to the theories of Aristotle. If the Earth is rotating, for example, why aren't birds and clouds left behind the moving Earth? Only the 17th-century discovery, first by Galileo and then by Isaac Newton, of laws of motion substantially different from Aristotle's solved this dilemma. By the time of Copernicus, Aristarchus's heliocentric idea had long been overwhelmed by the **geocentric**—Earth-centered—theories of Aristotle and Ptolemy.

FIGURE 2–10 Copernicus in a painting hanging in the Torun Museum.

Copernicus's theory, although it put the Sun instead of the Earth at the center of the solar system (equivalent, at that time, to the Universe), still assumed that the orbits of celestial objects were circles. The notion that circles were perfect, and that celestial bodies had to follow such perfect orbits, shows that Copernicus had not needed to break away entirely from the old ideas, since observations at the time did not show a discrepancy. The so-called Copernican revolution, in the sense that our modern approach to science involves a comparison of nature and understanding, really occurred later.

Since Copernicus's theory contained only circular orbits, he still invoked the presence of some epicycles in order to improve agreement between theory and observation. Copernicus was proud, though, that he had eliminated the equant. In any case, the detailed predictions that Copernicus himself computed on the basis of his theory were not in much better agreement with the existing observations than tables based on Ptolemy's model, because in many cases Copernicus still used Ptolemy's

FIGURE 2–11 The pages of Copernicus's original manuscript in which he drew his heliocentric system. The manuscript was later set into type and published in 1543. The Sun (sol) is at the center surrounded by Mercury (Merc), Venus (Veneris), Earth (Telluris), Mars (Martis), Jupiter (Jovis), Saturn (Saturnus), and the fixed stars. Each planet is assigned the space between the circles, rather than the circles themselves, as is commonly believed. The manuscript is in the University library in Cracow.

net, in quo terram cum orbe lunari tanquam epicyclo contineri
diximus. Quinto loco Venus nono menſe reducitur. Sextum
deniq; locum Mercurius tenet, octuaginta dierum ſpacio circũ
currens. In medio uero omnium reſidet Sol. Quis enim in hoc



lcherimo templo lampadem hanc in alio uel meliori loco
et, quàm unde totum ſimul poſsit illuminare? Siquidem n
pte quidam lucernam mundi, alij mentem, alij rectorem u

A

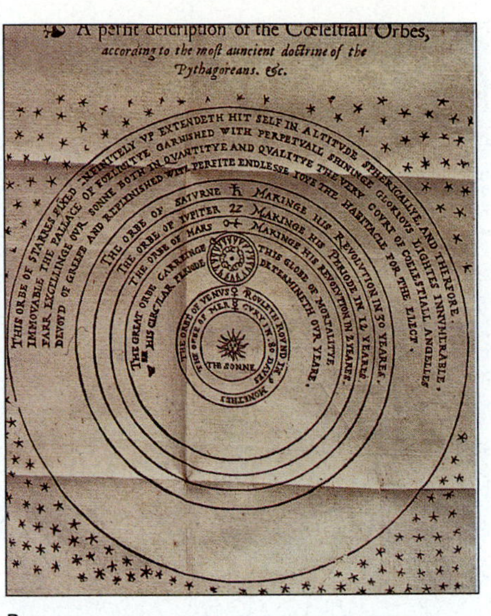


B

**FIGURE 2–12** (*A*) Copernicus's heliocentric diagram as printed in *De Revolutionibus* (1543). (*B*) The first English diagram of the Copernican system, which appeared in 1568 as an appendix by Thomas Digges to a book by his father.

observations. The heliocentric theory appealed to Copernicus and to many of his contemporaries on philosophical grounds, rather than because direct comparison of observations with theory showed the new theory to be better. In fact, there was at that time no observational way to distinguish between the heliocentric and geocentric models.

Copernicus's heliocentric theory was published in 1543 (Fig. 2–13) in the book he called *De Revolutionibus (Concerning the Revolutions)* (Fig. 2–14). The theory explained the retrograde motion of the planets as follows:

Let us consider, first, an outer planet like Mars as seen from the Earth. As the Earth approaches the part of its orbit that is closest to Mars (which is orbiting the Sun much more slowly than is the Earth), the projection of the Earth-Mars line outward to the stars (which are essentially infinitely far away compared to the planets) moves slightly against the stellar background. As the Earth comes to the point in its orbit closest to Mars, and then passes it, the projection of the line joining the two planets can actually seem to go backward, since the Earth is going at a greater speed than Mars. Then, as the Earth continues around its orbit, Mars appears to go forward again. A similar explanation can be demonstrated for the retrograde loops of the inner planets.

The idea that the Sun was at the center of the solar system led Copernicus to two additional important results. First, he was able to work out the distances to the planets, based on observation of their motion across the sky with respect to the changing position of the Sun. Second, he was able to derive the periods of the planets, the length of time they take to revolve around the Sun, based on the length of time they took to return to the position in the sky opposite to the direction of the Sun. Third, the heliocentric theory explained why Mercury and Venus never strayed far from the Sun.

His ability to derive these results played a large role in persuading Copernicus of the superiority of his heliocentric system.

Contrary to popular belief, Copernicus didn't elevate the Sun in its status by putting it at the center of what we now call the Solar System. Actually—as Dennis Danielson, editor of *The Book of the Cosmos,* has pointed out—gross, heavy things settle at a low point, and the Earth, by being at the center in the Ptolemaic system, was at such a low point. That is the reason why Dante, in his *Divine Comedy* (written more than two centuries before Copernicus's book), placed hell at the center of the Earth, far from the beauty of the celestial spheres.

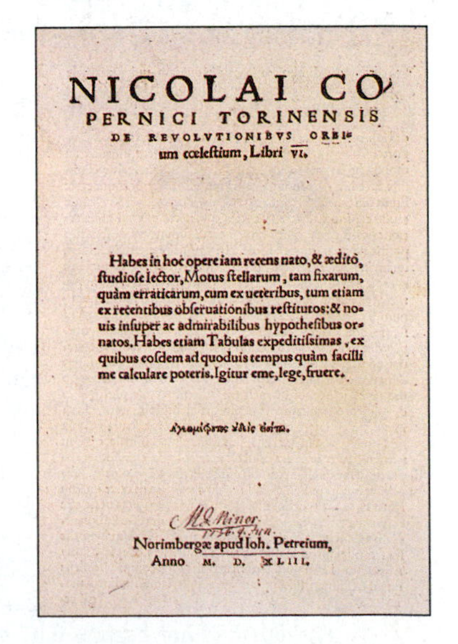
NICOLAI CO-
PERNICI TORINENSIS
DE REVOLVTIONIBVS ORBI-
um cœleſtium, Libri VI.

Habes in hoc opere iam recens nato, & ædito, ſtudioſe lector, Motus ſtellarum, tam fixarum, quàm erraticarum, cum ex ueteribus, tum etiam ex recentibus obſeruationibus reſtitutos: & nouis inſuper ac admirabilibus hypotheſibus ornatos. Habes etiam Tabulas expeditiſsimas, ex quibus eoſdem ad quoduis tempus quàm facillime calculare poteris. Igitur eme, lege, fruere.

Norimbergæ apud Ioh. Petreium,
Anno M. D. XLIII.

**FIGURE 2–13** The title page of the first edition (1543) of *De Revolutionibus*.

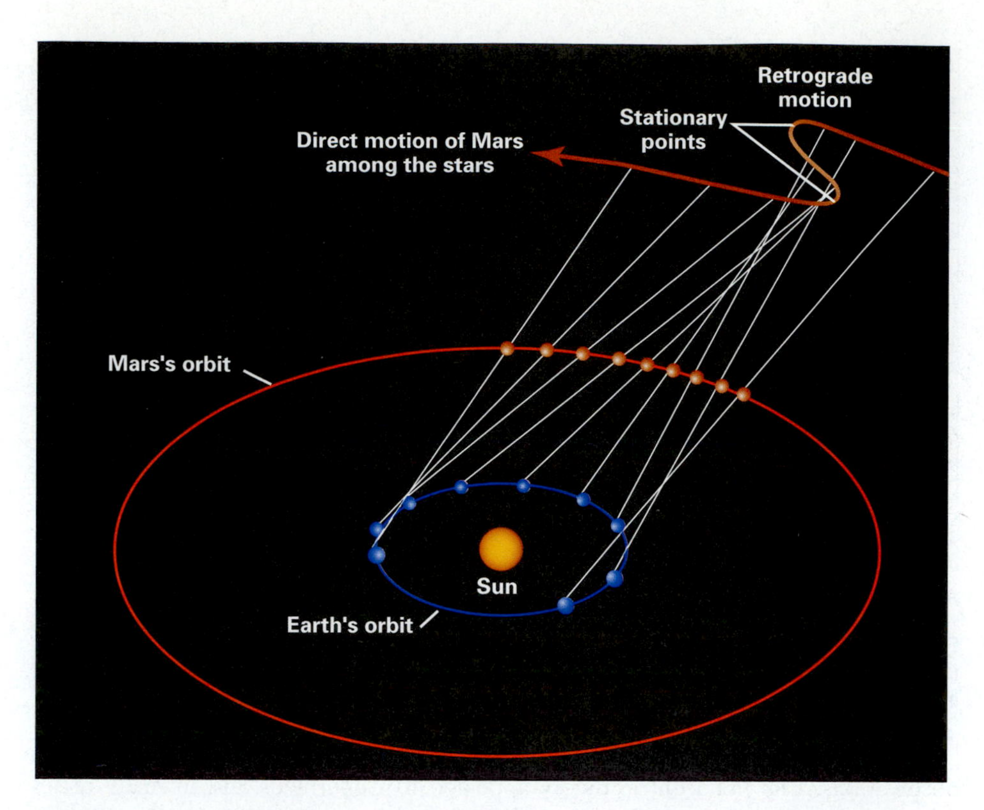


**FIGURE 2–14** The Copernican theory explains retrograde motion as an effect of projection. For each of the nine positions of Mars shown from right to left, follow the line from Earth's position through Mars's position to see the projection of Mars against the sky. Mars's forward motion appears to slow down as the Earth overtakes it. Between the two stationary points, Mars appears in retrograde motion; that is, it appears to move backward with respect to the stars. The drawing shows the explanation of retrograde motion for Mars or for the other superior planets; that is, planets whose orbits lie outside that of the Earth. Similar drawings can explain retrograde motion for inferior planets, that is, planets whose orbits lie inside that of the Earth (namely, Mercury and Venus).

## *BOX 2.4 Copernicus and His Contemporaries

Nicolaus Copernicus was born in 1473, and through family connections was destined from his youth for an ecclesiastical career. At the University of Cracow, which he entered in 1490, his studies included mathematics and astronomy, the latter including the books of Regiomontanus. Though his official career was as a religious administrator, he spent much time travelling abroad, and it is clear that mathematics and astronomy were always among his greatest interests. Copernicus probably wrote a draft of his heliocentric ideas around 1510, a few years after he returned from Italy. From 1512 on, he lived in Frauenburg (now Frombork, Poland). He wrote a lengthy manuscript about his ideas and continued to develop his ideas and manuscript even while carrying on official duties of various kinds. Indeed, he had even studied medicine in Italy and acted as medical advisor to his uncle, the bishop, though he probably never practiced as a physician.

Though it is not certain why he did not publish his book earlier, his reasons for withholding the manuscript from publication probably included a general desire to perfect the manuscript and also probably involved his realization that to publish the book would involve him in controversy.

Though Copernicus had not published his work, it appears that he was well-known as a mathematician and astronomer among experts in those fields. His idea that the Earth is moving and the Sun and stars are not was sufficiently unusual that it became known even to non-astronomers.

Copernicus received the final printed pages of his book on the day of his death in 1543. Although some people, including Martin Luther, noticed a contradiction between the Copernican theory and the Bible, little controversy occurred at that time. In particular, the Pope—to whom the book had been dedicated—did not object.

### *Discussion Question*

Would you be as secretive as Copernicus if you made such a great discovery? Why?

A DEEPER DISCUSSION

### BOX 2.5 Copernicus's Early Editions

Owen Gingerich, of the Harvard-Smithsonian Center for Astrophysics, is soon to publish his census of the two 16th-century editions of Copernicus's *De Revolutionibus* (1543 and 1566). By studying which copies were censored and what notes were handwritten into the various copies by readers of the time, he has been able to study the spread of belief in Copernicanism.

After decades of valiant searching, by 2001 he has found 276 copies of the first edition and 322 of the second edition. Although no one knows precisely how many were printed, it is likely that more than half of the original copies of each edition survive, largely because people understood that the book was very important even if most of it was too technical for all but the specialists. Dozens of astronomers read the book carefully in the 1500s, often adding their own notes in the margins or copying notes from their teachers. But in the absence of empirical proof of the Earth's motion, most of these astronomers suspended judgment on the reality of the heliocentric arrangement.

By 1620 a number of astronomers began to argue for the physical reality of the Copernican system, so the Roman Inquisition issued ten specific "corrections" to make the argument appear only hypothetical. The census shows that about 2/3 of the books in Italy were censored in that way, but almost none outside of Italy were altered. The other Catholic countries must have considered the "Galileo affair" a local Italian imbroglio!

## 2.6 TYCHO BRAHE

In the last part of the 16th century, not long after Copernicus's death, Tycho Brahe began a series of observations of Mars and other planets. Tycho, a Danish nobleman, set up an observatory on an island off the mainland of Denmark. The building was called Uraniborg (after Urania, the muse of astronomy). The telescope had not yet been invented, but Tycho used giant instruments to make observations of unprecedented accuracy. In 1597, Tycho lost his financial support in Denmark and moved to Prague, arriving two years later. A young assistant, Johannes Kepler, came there to work with him. At Tycho's death, in 1601, Kepler—who had worked with Tycho for only 10 months—was left to analyze all the observations that Tycho and his assistants had made (though getting access to the data from Tycho's family proved troublesome for Kepler). In Section 3.1, we will see how valuable Kepler's theory, and thus Tycho's observations, turned out to be.

Proof that the Earth rotates on its axis—more evidence that Copernicus was correct—was not forthcoming until the 19th century, when the French physicist Foucault suspended a long pendulum from the dome of the Pantheon in Paris. The inertia of the heavy bob kept the pendulum swinging back and forth in the same plane even while the Earth rotated below it. The pendulum appeared to rotate slowly. Observers were rotating with the Earth, while the plane of the pendulum was fixed with respect to the stars. Such a **Foucault pendulum** (Fig. 2–15) at the North or South Pole would rotate once a day, for example. The rate of apparent rotation of a particular pendulum depends on its latitude.

## 2.7 FEATURE: ARCHAEOASTRONOMY

Before we continue with the linear development of today's astronomy, let us look at an interesting side set of developments. In the next chapter, we return to the successors to Copernicus and Tycho.

The main line of astronomy that has led to today's conception of the Universe was largely European. There is currently increasing interest in discovering the extent of astronomical understanding in other parts of the world. Astronomical development in Asia was substantial, and many written documents were left. From still other parts of the world, surviving documents are fewer and understanding is often demonstrated in astronomical alignments of buildings and other objects. These investigations link archaeology and astronomy, and so they are called **archaeoastronomy.** We shall deal here with only four examples: Stonehenge in ancient England from four and five thousand years ago, Neolithic sites in Ireland also from four

*(text continues on page 27)*

**FIGURE 2–15** The Foucault pendulum at the Griffith Observatory in Los Angeles. As the Earth rotates, the swinging ball knocks over successive sticks. A pendulum at the North Pole would appear to rotate in 24 hours as the Earth turns beneath it; at the equator, a pendulum would not appear to rotate at all. At the latitude of the pendulum shown, a Foucault pendulum rotates in about 42 hours.

BIOGRAPHY

## BOX 2.6 Tycho Brahe

Tycho Brahe was born in 1546 to a Danish noble family. As a child, he was taken away and raised by a wealthy uncle. In 1560, a total eclipse was visible in Portugal, and the young Tycho witnessed the partial phases in Denmark. Though the event itself was not spectacular—partial eclipses never are—Tycho, at the age of 14, was so struck by the ability of astronomers to predict the event that he devoted his life from then on to making an accurate body of observations.

Three years later, he witnessed a conjunction of Saturn and Jupiter—a time when the two planets came very close together in the sky. But the tables based on the work of Copernicus were a few days off in predicting the event (and the older *Alphonsine Tables* were a month off). Tycho's desire to improve such tables was renewed and strengthened. Though his family wanted him to devote himself to the law instead of wasting time on astronomy, he finally won out.

When Tycho was 20, he dueled with swords with a fellow student. During the duel the bridge of his nose was cut off. For the rest of his life he wore a gold and silver replacement and was forever rubbing the remainder with ointment. Portraits made during his life and the relief on his tomb show a line across his nose, though just how much was actually cut off is not now definitely known (see figure).

In 1572, Tycho was astounded to discover a new star in the sky, so bright that it outshone Venus. It was what we now call a "supernova"; indeed, we now call it "Tycho's supernova." It was the explosion of a star, and remained visible in the sky for 18 months. Tycho had been building bigger instruments for measuring positions in the sky better than any of his contemporaries had, and he was able to observe very precisely that the supernova was not changing in position. It was thus a real star rather than a nearby object.

He published a book about the supernova, and his fame spread. In 1576, the king of Denmark offered to set Tycho up on the island of Hveen with funds to build a major observatory, as well as various other grants. The following 20 years saw the construction on Hveen of Uraniborg. Later, he built a second observatory on Hveen. Both had huge instruments able to measure angles and positions in the sky to unprecedented accuracy for that pre-telescopic time (see figure).

Unfortunately for Tycho, a new king came into power in Denmark in 1588, and Tycho's influence waned. Tycho had always been an argumentative and egotistical fellow, and he fell out of favor in the countryside and in the court. Finally, in 1597, his financial support cut, he left Denmark. Two years later he settled in Prague, at the invitation of the Holy Roman Emperor, Rudolph II.

In 1601, Tycho attended a dinner where etiquette prevented anyone from leaving the table before the baron (or at least Tycho thought so). The guests drank freely, and Tycho wound up with a urinary infection. Within two weeks he was dead of it, perhaps weakened by mercury poisoning from his alchemical investigations.

### *Discussion Question*

What would you do if you tried to observe an eclipse and the prediction was a week off? Contrast your actions with Tycho's.

The effigy of Tycho Brahe on his tomb in Prague. The dividing line on his nose shows.

Tycho's observatory at Uraniborg, Denmark. Here Tycho is seen showing the mural quadrant (a marked quarter-circle on a wall) that he used to measure the altitudes at which stars and planets crossed the meridian.

thousand or more years ago, Mayan sites in Central America from about a thousand years, and North American sites also from the early part of the last millennium. In all these cases, the recent work suggests that astronomical knowledge was greater than we might have expected.

### 2.7a Stonehenge

On Salisbury Plain in southern England, a set of massive stones set on end marks one of the strangest monuments known to humanity. Nowadays the stones stand next to a busy motorway and are a popular tourist site, but once they formed part of a religious site.

Stonehenge (Fig. 2–16) contains a set of giant standing stones weighing 25 tons each. They form a circle surrounding two horseshoe-shaped patterns. The circle is more than 30 meters wide and 4 meters high, and some of the pairs of stones have massive stone crosspieces (lintels) raised 4 meters above the ground. An incredible amount of work went into making this monument. Why did the site merit so much effort?

The realization that the earliest parts of Stonehenge date from 5000 years ago, and the discovery that many of the stones were brought from hundreds of miles away to this particular spot, make the matter even more interesting. "Why?" is not the only question. How?

Actually, Stonehenge turns out to have been built over a period of many centuries and has distinct phases. The oldest period was the date at which certain holes, called "post holes," were made at the site. With revised dating reported at a 1996 conference, these holes may date from as long ago as 3000 B.C. The latest period, in which the large stones were brought, is now dated at 2450 B.C., centuries earlier than previously thought.

It has been realized since 1771 (after having been forgotten for perhaps thousands of years) that at the summer solstice—the day in the year (currently June 21st) when the Sun is the farthest north and the day is the longest—the Sun rises directly over a particular stone, called the Heelstone, as seen from the center of Stonehenge (Fig. 2–17). The Heelstone is located 60 meters outside the outer circle of stones. We think that the weather was better in that region of the world 4000 years ago than it is now, so, fortunately, more days would have been clear enough for astronomical objects to be seen.

A

B

**FIGURE 2–16** (*A*) Stonehenge, with its giant stones standing on Salisbury Plain in southern England. It may have been used for astronomical prediction. (*B*) Carhenge, in Alliance, Nebraska, a recreation of Stonehenge with modern building blocks (that is, cars).

FIGURE 2–17 The Sun rises over the Heelstone on the day of the summer solstice when seen from the center of the ring of stones.

A companion stone to the Heelstone, helping define a corridor for the sunlight on midsummer mornings, was discovered only two decades ago. Note, however, that the Sun's position at sunrise on the solstice is indistinguishable from its sunrise position for several days earlier and later. So Stonehenge may be more of a ceremonial center with astronomical aspects than a scientific observatory.

Alignments between less prominent pairs of stones point to sunrise at other significant times of the year, such as the equinoxes, which fall midway between the solstices. Some alignments with the Moon are also suspected. All of this evidence points to the fact that the people who built the first stages of Stonehenge must have had substantial astronomical knowledge, including ideas of the year and the calendar (Fig. 2–18).

A still more elaborate theory was put forward in the 1950s by Gerald Hawkins, an American astronomer, and elaborated upon by Fred Hoyle, a British astronomer, that Stonehenge could be used to predict eclipses. (Having astronomers step in dismayed some professional archaeologists, and it is still difficult to assess the validity of Stonehenge theories only on their merits.) A half-century later, the idea that Stonehenge was an eclipse-predictor is not accepted.

Are we reading too much into these silent stones? It has been said that every age has the Stonehenge it deserves, or desires. Probably Stonehenge was not an observatory but was a sacred site in which astronomical alignments contributed to the symbolism. Still, its astronomical associations seem real.

### 2.7b Newgrange and Knowth

FIGURE 2–18 This tiny sundial models and mocks Stonehenge as a pocket watch: "Stonehenge: 5,000 years old and still ticking! At last, you can predict an eclipse and tell the local apparent time with this beautiful pocket time piece. Delight your more erudite friends and amaze your Druid neighbors," runs the advertisement.

There are several other neolithic sites whose astronomical significance seems clear. One of the most prominent is Newgrange, in the middle of Ireland. In that region are several huge mounds, and the most prominent has been excavated and reconstructed (Fig. 2–19). Built between 3500 and 2700 B.C., it is situated on the summit of a low hill with a commanding view. Such objects are known as "passage graves." They represent the stone age, prior to the bronze age. Newgrange has a huge central chamber 50 feet high, with a long tunnel as an opening. Over the entrance to the tunnel, a slit is aligned so that at the winter solstice, the rays from the rising Sun illuminate the tunnel and the chamber for about 15 minutes. The solar orientation shows that they were used for at least ceremonial astronomical purposes. Some of the large stones used to line the tunnel and elsewhere are richly carved. Newgrange was already fading in importance when Stonehenge was built.

Not far away are a series of other mounds, including Knowth (Fig. 2–20). On many stones of both Newgrange and Knowth, intricate patterns have been incised. At Knowth, too, tunnels are oriented toward astronomically significant directions.

### 2.7c Mayan Astronomy

About 1000 years ago, the Mayan people had a rich culture in what is now the Yucatán peninsula of Mexico and Guatemala, as can be seen from such sites as Chichen Itzá in Mexico and Tikal in Guatemala. Many features of their cities and buildings are aligned to astronomical phenomena (Fig. 2–21). Information from this period comes not only from archaeological measurement and excavation of buildings but also from the three remaining Mayan books, or codices. (All other Mayan books, incredibly, were burned by the Europeans who came later on.)

Besides setting their calendar by the Sun and the Moon, the Maya also based their calendar on the rising and setting of Venus. Though they could not observe the phases directly (after all, the phases were not discovered until 1610, when Galileo had a telescope available), they could tell from the positions of Venus in the sky—and by the dates of its heliacal rising—that it had a periodic cycle. The Venus table is accurate to 1 day in 500 years. It seems from the astronomical alignments of some of their architecture that they were concerned with trying to configure Venus's

**FIGURE 2–19** The 20th-century reconstruction of Newgrange, originally constructed in the Boyne Valley in Ireland between 3500 and 2700 B.C.

**FIGURE 2–20** One of the Neolithic tombs at Knowth, near Newgrange in Ireland. It, too, has astronomical alignments.

movements with their agricultural calendar, clearly showing how practical astronomy can be.

These cycles are set out in detail in one of the surviving Mayan books, the *Dresden Codex*, from at least 500 years ago. We have only recently learned how to read the text, but we have been able to read the numbers for a longer time. Two large sections that contain eclipse tables and Venus tables have been deciphered. They show the Maya's high level of sophistication in mathematics and astronomy. Even though they may have been studying astronomy for different reasons from our own, they were still also striving for precise knowledge.

**FIGURE 2–21** The Caracol at Chichen Itzá in the Yucatán. Alignments involving windows, walls, and the horizontal shafts at the top of the building seem to point to the locations of the Sun and Venus on significant days.

### 2.7d Native American Astronomy

In recent years, the American west has been searched for signs of Native American interest in astronomy.

In the Great Plains that lie on the eastern edge of the Rocky Mountains, certain Native American monuments can be found that, like Stonehenge and Newgrange, show important astronomical alignments. For example, in Wyoming there exists a "medicine wheel" (Fig. 2–22), the term "medicine" having connotations of magic. This site probably dates from A.D. 1400 to 1700. John A. Eddy of the High Altitude Observatory in Boulder, Colorado, has investigated the astronomical significance of the Bighorn Medicine Wheel. His analysis seemed to indicate that lines drawn between significant markings point not only to the sunrise and sunset on the day of the summer solstice but also toward the rising points of the three brightest stars that rise shortly before the Sun (have their heliacal risings) in the summer. Edward Krupp of the Griffith Observatory, Anthony Aveni of Colgate University, and others have continued to consider the site. Krupp builds on the fact that the line of sunrise at the summer solstice, which continues to the only outlying cairn, bisects the design. Considering also what he knows about Cheyenne ritual, he concludes that it is at least consistent that the site was used in ceremonial contexts.

About 50 medicine wheels are known, ranging in size from a few meters to a hundred meters across. Some are known to be several thousand years old, contemporaneous with the pyramids and with Stonehenge. And some of the older medicine wheels show the same alignment with the solstice and rising brightest stars as the Bighorn site. Krupp thinks that the medicine wheels are not observatories, though they did incorporate astronomical symbolism, perhaps as part of shamans' acquiring power.

**FIGURE 2–22** The Bighorn Medicine Wheel in Wyoming. Lines drawn between pairs of the major points extend toward such astronomical directions as those of the solstice sunrise and sunset and the rising positions of the stars Aldebaran, Rigel, and Sirius.

**FIGURE 2–23** Sunset on midsummer night at a ruin of a 700-year-old Anasazi building, in New Mexico.

As with other archaeoastronomical sites, there is always the question of whether these alignments were made intentionally or by accident. Statistical arguments have been made that indicate that it is unlikely for so many coincidences to occur. Not all the medicine wheels show astronomical alignments, though, and we want to be able to conclude that we are not fooling ourselves by choosing only ones that agree with our ideas.

Aveni points out that a site more clearly astronomical in importance is Ohio's Hopewell Octagon Mounds. So astronomical orientations are widespread in the Americas.

In Chaco Canyon, New Mexico, various Anasazi ruins from the late 13th century are aligned to key astronomical directions, such as the location on the horizon of sunset on midsummer's night (Fig. 2–23).

The pitfalls of interpreting stones and drawings from hundreds of years ago have been shown in the interpretation of a rock drawing that has long been known in Chaco Canyon. It was suggested several decades ago that Native Americans observed and drew the explosion of a star—a supernova—in A.D. 1054, an event that was recorded in China and the Middle East but for which, surprisingly, no European sightings are known. But the Native American evidence consists of a drawing (see Figure 29–10*A*), and is ambiguous. Did the drawing indeed show the supernova near the crescent moon? This interpretation was thrown into doubt by a photograph (see Figure 29–10*B*) of the crescent moon near Venus. Maybe the supposed supernova was actually Venus. We must beware of reading our own thoughts into the minds of people who lived long ago.

## CORRECTING MISCONCEPTIONS

| ✖ *Incorrect* | ✔ *Correct* |
|---|---|
| The equant is at the center of the deferent (see Chapter 3). | The equant is part of a Ptolemaic Universe, and is on the other side of the center of the deferent (a circle) from the Earth. |
| Copernicus knew about ellipses. | Everything in Copernicus's ideas was circular. |
| Copernicus's explanation for retrograde motion works only for the outer planets. | It works for the inner planets too. |

## SUMMARY AND OUTLINE

Egyptian astronomy (Section 2.1)
- Astronomy served practical purposes of farmers

Greek astronomy (Section 2.2)
- Aristotle and Ptolemy had geocentric theories
- They needed epicycles, deferents, equants to explain planetary motions

Chaldean, Chinese astronomy (Section 2.3)

Medieval astronomy (Section 2.4)
- Interpreting Aristotle
- Arab translation of Ptolemy's *Almagest*

Heliocentric theory (Section 2.5)
- Copernicus, *De Revolutionibus* published in 1543
- Retrograde motion explained as a projection effect
- Copernican system still used circular orbits
- Copernicus preferred his system on philosophical grounds rather than superiority of his predictions, because ancient observations were still used and Copernican system used circular orbits that didn't always fit the data

Tycho Brahe (Section 2.6)
- Amassed the best observations obtained up to then

Archaeoastronomy (Section 2.7)
- Alignments based on astronomical events

## KEY WORDS

heliacal rising
ecliptic
retrograde motion
*primum mobile*
epicycles
deferents
eccentric
equant
gnomon
parallax
conjunction
opposition
ephemeris
heliocentric
geocentric
Foucault pendulum
archaeoastronomy

## QUESTIONS

1. Discuss the velocity that a planet must have around its epicycle in the Ptolemaic theory with respect to the velocity that the epicycle has around the deferent if we are to observe retrograde motion.
2. Discuss the differences between the theories of Aristotle and Ptolemy.
3. (a) Imagining you lived on Saturn, describe the phases that the Earth would seem to go through on the Ptolemaic system. (b) Now describe the phases that the Earth would seem to go through on the Copernican system.
4. (a) Imagining you lived on Saturn, describe the phases that Pluto would seem to go through on the Ptolemaic system. (Hint: Check the relative distances of Saturn and Pluto from the Sun in Appendix 3.) (b) Now describe the phases that Pluto would seem to go through on the Copernican system.
5. Using the arguments we have given for retrograde motion of Mars, explain the retrograde motion of Venus as seen from Earth.
6. Define (a) epicycle; (b) deferent; (c) eccentric.
7. Draw a diagram showing the positions of the Earth, the Sun, and Mars to show (a) Mars in conjunction and (b) Mars in opposition.
8. We are launching spacecraft to Mars every 26 months, a practice that started in 1996. Explain why we have chosen this interval in terms of Mars's orbit.
9. Could Eratosthenes have carried out his measurement using a gnomon at your latitude? Why or why not?
10. Relate Eratosthenes's measurements to the myth that it was not known at the time of Columbus that the Earth is round.
11. Discuss the following statement: "With the addition of epicycles, the geocentric theory of the Solar System could be made to agree with observations. Since the geocentric theory was around first, and therefore better known, it should have been kept."
12. Discuss the significance of a Foucault pendulum. Describe how and why its behavior would appear to change at the equator and at the North Pole.
13. Do the astronomical alignments of Stonehenge, if established, show that the people who constructed Stonehenge III, the large stones, were astronomically sophisticated? Explain.
14. Discuss two signs of astronomy in the Americas prior to A.D. 1500.

## USING TECHNOLOGY

### W World Wide Web

1. Check the history of astronomy sites.
2. Learn the latest on Stonehenge dating at http://www.eng-h.gov.uk/stoneh.
3. Check out the site of the Center for Archaeoastronomy and its many links at http://www.wam.umd.edu/~tlaloc/archastro.

### *REDSHIFT*

1. Duplicate the retrograde motion of Mars as seen from the Earth.
2. Put yourself on Mars and see if the Earth goes through retrograde motion.

**The frontispiece to Galileo's *Dialogue Concerning the Two World Systems* (1632), which led to his being taken before the Inquisition since he had supposedly been warned in 1616 not to teach the Copernican theory. (Historians still argue about just what he had been warned.) According to the labels, Copernicus is to the right, with Aristotle and Ptolemy at the left; Copernicus was drawn with Galileo's face, however.**

# Chapter 3

# The Origin of Modern Astronomy

> **AIMS:** To follow how modern astronomy developed during the 17th century through the ideas of Kepler, Galileo, and Newton

The 17th century was a fertile time in the history of ideas. The Renaissance was well under way in Italy. The timeline at the end of the chapter gives some idea of the overlap between science and the humanities. To give just one specific correspondence, Shakespeare wrote *Hamlet* in England in 1601, a crucial time for the scientific matters that we will discuss in this chapter. In what was to become the United States, Jamestown was settled in Virginia in 1607, and the Pilgrims landed on Cape Cod and Plymouth Rock in 1620.

The intellectual ferment of the 17th century set the tone for science up to our time. Aristotle's ideas about the nature of our physical world were overturned. Kepler's discovery that the planets do not orbit the Sun in circles shattered ancient ideas and led to what has long been called (perhaps misleadingly) "the Copernican revolution." Galileo peered farther into the Universe than had ever been done and started methods of inquiry we still use. And Newton (spurred on by his colleague Halley) developed the laws of physics we still treat as fundamental. (The 20th century, with theories of relativity and quantum mechanics, saw the addition of other fundamental laws of physics.)

In this chapter, we consider the period from Kepler's starting work on the orbit of Mars in 1600, through Galileo's use of the telescope in 1609, through Newton's publication of his major work in 1687 with the aid of Halley. This period marks the origin of what we might call the modern era of astronomy; historical developments in astronomy since that time are treated in this text as we discuss the various separate topics.

## 3.1 JOHANNES KEPLER

Johannes Kepler, as a young mathematician in the late 16th century, had a mystical view of the universe. He had studied with a professor who was one of the first to believe in the Copernican view of the Universe, and he also came to believe in this heliocentric view. (Throughout this text, you will find names omitted in many places, such as that of Kepler's professor in the preceding sentence, so that it doesn't look as though there are too many names to remember.) Kepler wanted to discover the cause of the regularity in the orbits of the planets. While he was teaching a high-school class in 1595, it occurred to him that the ancient Greek discovery that there are five known perfect geometrical solid figures might be the key to explaining the orbits of the planets. Each of the five "regular figures"—the tetrahedron (pyramid), the cube, the

## BIOGRAPHY

### BOX 3.1 Johannes Kepler

Shortly after Tycho Brahe arrived in Prague, he was joined by an assistant, Johannes Kepler. Kepler had been born in 1571 in Weil der Stadt, near what is now Stuttgart, Germany. He was of a poor family and had gone to school on scholarships.

At the university, he had been fortunate to study under one of the few professors in the world who taught the new Copernicanism. Kepler was officially enrolled in religion, but a position opened to teach mathematics, and Kepler thus became a teacher. Four years later all Lutheran teachers were expelled from Graz, Kepler among them, and in 1600 Kepler joined Tycho Brahe in Prague.

Kepler was eager to work with the great man's data, though he had trouble prying them out of Tycho. Tycho, on his part, recognized Kepler's promise and wanted his assistance. This common need overcame their mutual friction, though their relationship was stormy.

At Tycho's death, Kepler succeeded him in official positions. Most important, after a battle over Tycho's scientific collections, he took over Tycho's data.

In 1604, Kepler observed a supernova and wrote a book about it. It was the second supernova to appear in 32 years, and was also seen by Galileo. But Kepler's study has led us to call it "Kepler's supernova." We have not seen another supernova in our galaxy since (though we shall be discussing a supernova in a satellite galaxy of our own).

By the time of the supernova, through analysis of Tycho's data, Kepler had discovered what we now call his second law. He was soon to find his first law. It took four additional years, though, to arrange and finance publication. Much later on, he deduced his third law.

Ptolemy, Copernicus, Kepler, and Tycho.

In 1627, Kepler published a set of tables of planetary positions, called the *Rudolphine Tables*, after his patron, the Holy Roman Emperor, Rudolph II. These, of course, were based on the knowledge of his three laws and were thus much more accurate than previous tables.

The abdication of Kepler's patron, the Emperor, led Kepler to leave Prague and move to Linz. Much of his time had to be devoted to financing the printing of the tables and his other expenses—and to defending his mother against the charge that she was a witch. He died in 1630, while travelling to collect an old debt.

#### Discussion Question

Kepler had a supernova named after him. What is an example of some important thing you might do to have such an honor happen to you?

octahedron (8 faces), the dodecahedron (12 faces), and the icosahedron (20 faces)—could be enclosed in a sphere. Motivated by a search for celestial harmony, Kepler suggested that spaces in the shapes of each of these regular solids came between each pair of the spheres that carried Copernicus's six planets (Fig. 3–1).

Though Tycho did not accept Kepler's idea, he was impressed with Kepler's mathematical and astronomical skill. When Tycho went to Prague, he invited Kepler to join him. Tycho's observational data showed that the tables then in use did not adequately predict planetary positions. Kepler started to study Tycho's planetary positions in 1600, and he carried out detailed numerical calculations by hand (Fig. 3–2).

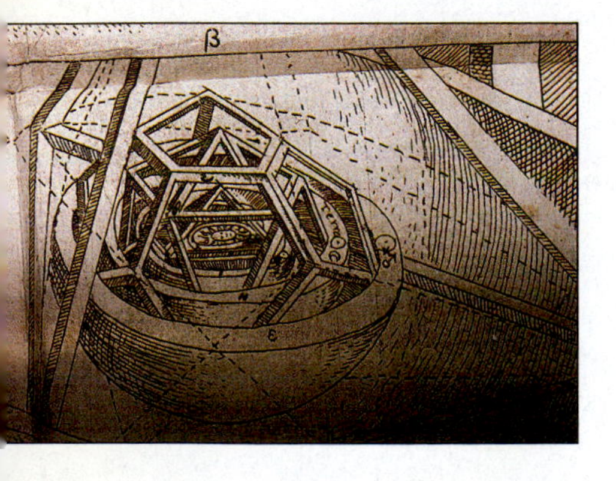

**FIGURE 3–1** Kepler, in his *Mysterium Cosmographicum* (1596), suggested that since there were six known planets and five regular solids known from Greek geometry, the planets could be moving on unseen spheres separated by distances corresponding to these nested solids. These solids are regular in that all their sides are the same: the pyramid (4 triangles), the cube (6 squares), the octahedron (8 triangles), the dodecahedron (12 pentagons), and the icosahedron (20 triangles). No other regular solids exist. The solids were centered on the Sun; Kepler was already a Copernican. Ultimately, this stage of his search for celestial harmony proved unsatisfactory.

**FIGURE 3–2** The plaque over Johannes Kepler's house in Prague.

(Nowadays we could use a computer to calculate in seconds results that took Kepler weeks to work out.) Superseding his earlier ideas on regular solids and celestial harmony, he based his work on his theory of the physical causes of the planets' motion. Although in retrospect we recognize that Kepler's celestial physics was wrong, he nonetheless succeeded in deriving two fundamental regularities in planetary motion that were later successfully described by Newton's ideas about mechanics and universal gravitation (which we will shortly discuss). Kepler's calculations made sense out of the observations of Mars that Tycho had made, and thus they cleared up the discrepancies between the predictions and the observations of Mars's changing position in the sky. In 1609, shortly before Galileo was to first turn a telescope on the sky, Kepler published in his book *Astronomia Nova (The New Astronomy)* the first two of his three laws. His third law followed in 1618 in his book *Harmonices Mundi (Harmonies of the World)*. In it, he expressed the relationship between the planets' distances from the Sun and their periods, which he also attempted to explain physically but less satisfactorily. Newtonian mechanics also triumphed in its explanation of this phenomenon.

Kepler and Galileo were contemporaries, and they even corresponded (though Galileo did not answer all of Kepler's letters). But their influence on the development of each other's scientific thought was small.

### 3.1a Kepler's First Law (1609)

Kepler's first law, published in 1609, says that *the planets orbit the Sun in ellipses, with the Sun at one focus*. Even Copernicus had assumed that the planets followed "perfect" orbits—namely, circles—though he had included the admittedly imperfect Earth as a planet. The discovery by Kepler that the orbits were in fact ellipses greatly improved the fit of his calculations to the observations. Kepler also concluded that the planets were like the Earth in being imperfect and made of matter.

An ellipse is a curve defined in the following way: Choose any two points on a plane; these points are called the **foci** (fo'si) (each is a **focus**). From any point on the ellipse, we can draw two lines, one to each focus. The sum of the lengths of these two lines is the same for each point on the ellipse (Fig. 3–3).

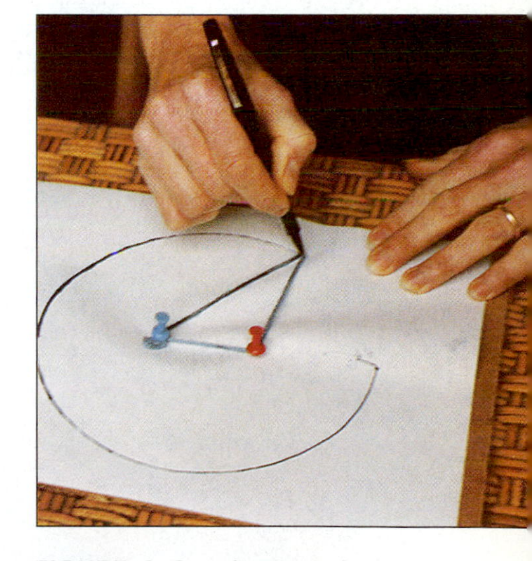

**FIGURE 3–3** It is easy to draw an ellipse. Put two nails or thumbtacks in a piece of paper, and link them with a piece of string that has some slack in it. If you pull a pen around while the pen keeps the string taut, the pen will necessarily trace out an ellipse. It does so since the string doesn't change in length and is equal to the sum of the lengths of the lines from the point to the two foci. That this sum remains constant is the defining property of an ellipse.

The **major axis** of an ellipse is the line within the ellipse that passes through the two foci, or the length of that line (Fig. 3–4). We often speak of the **semimajor axis,** which is just half the length of the major axis ("semi-" is a prefix from the Greek that means "half"). The **minor axis** is the part of the line lying within the ellipse that is drawn perpendicular to the major axis and passing through its midpoint, or the length of that line. When one of the foci lies on top of the other, the major and minor axes are the same length, and the ellipse is the special case that we call a **circle.**

We often describe how far an ellipse is "out of round" by giving its **eccentricity** (ek-sen-tri'-city) (Fig. 3–5), the distance between the foci divided by the length of the major axis. (We will sometimes use the term "ellipticity" to mean eccentricity.) For a circle, the distance between the foci is zero, so its eccentricity is zero.

Ellipses (and therefore circles), parabolas, and hyperbolas more generally fall into the class of **conic sections,** that is, sections of a cone (Fig. 3–6). If you slice off

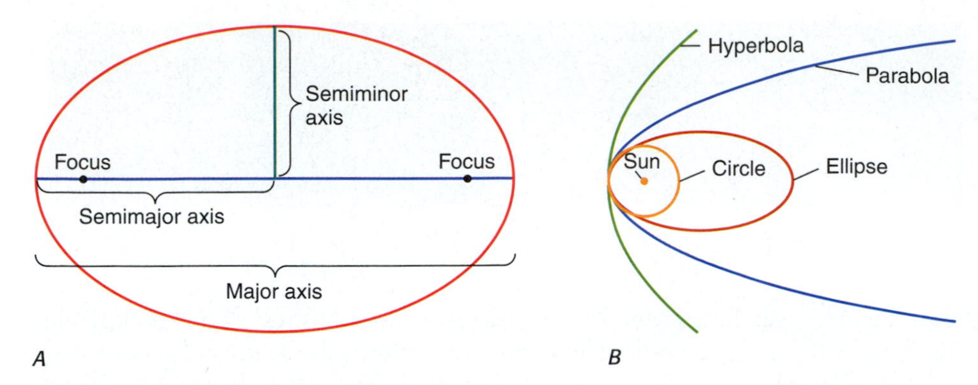


**FIGURE 3–4** (*A*) The parts of an ellipse. (*B*) Conic sections with a common focus. The ellipse shown has the same **perihelion** distance (closest approach to the Sun) as does the circle. Its *eccentricity,* the distance between its foci divided by its major axis, is 0.8. If the perihelion distance is kept constant but the eccentricity is allowed to reach 1, then we have a parabola. For eccentricities greater than 1, we have hyperbolas.

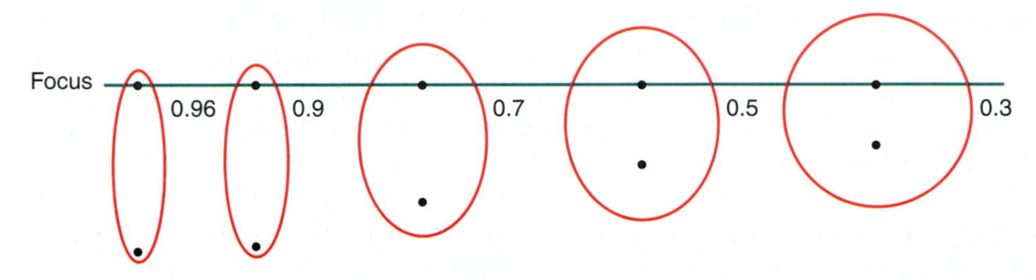


**FIGURE 3–5** A series of ellipses of the same major axis but different eccentricities. The *foci* are marked; these are the two points inside with the property that the sum of the distances from any point on the circumference to the foci is constant. As the eccentricity—distance between the foci divided by the major axis—approaches 1, the ellipse approaches a parabola. As the eccentricity approaches zero, the foci come closer and closer together. A circle is an ellipse of zero eccentricity. (Note that "focus" is the singular form; one can never have "one foci.")

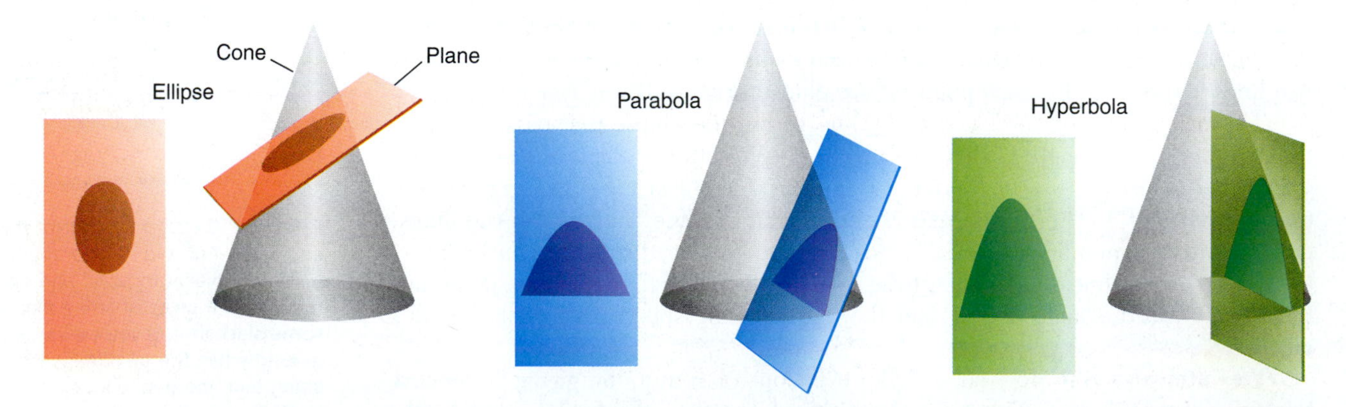


**FIGURE 3–6** An ellipse is a conic section, in that the intersection of a cone and a plane that passes through the sides of a cone (and not the bottom) is an ellipse. If the plane is parallel to an edge of the cone, a parabola results. If the plane is tipped further over so that it is neither parallel to the edge nor intersects the side, then a hyperbola results. So parabolas and hyperbolas are conic sections as well.

the top of a cone, the shape of the slice is an ellipse. (If, when you slice off the top, you cut perpendicularly to the axis, then the ellipse is a circle.) If your cut is parallel to the side of the cone, a parabola results. If your cut is tipped still further over, a hyperbola results.

For a planet orbiting the Sun, the Sun is at one focus of the elliptical orbit; nothing special marks the other focus (we say that it is "empty").

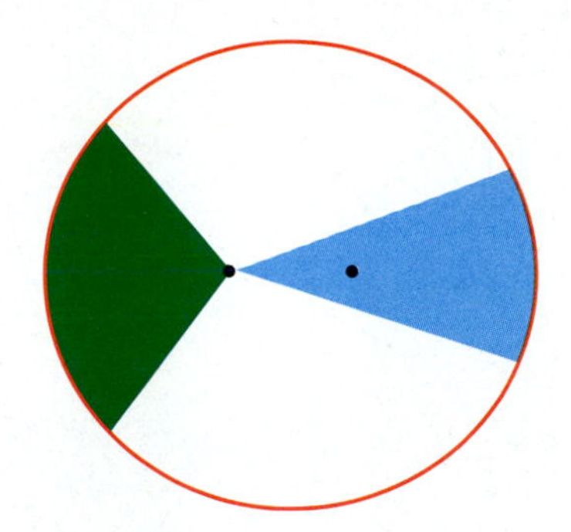

**FIGURE 3–7** Kepler's second law states that the two shaded sectors, which represent the areas covered by a line drawn from a focus of the ellipse to an orbiting planet in a given length of time, are equal in area. The Sun is at this focus; nothing is at the other focus.

## 3.1b Kepler's Second Law (1609)

The second law, also known as the **law of equal areas,** describes the speed with which the planets travel in their orbits. Kepler's second law says that *the line joining the Sun and a planet sweeps through equal areas in equal times***.** When a planet is at its greatest distance from the Sun in its elliptical orbit, the line joining it with the Sun sweeps out a long, skinny sector. (A **sector** is the area bounded by two straight lines from a focus of an ellipse and the part of the ellipse joining their outer ends.) The planet travels relatively slowly in this part of its orbit. By Kepler's second law, the long, skinny sector must have the same area as the short, fat sector formed in the same period of time when the planet is closer to the Sun (Fig. 3–7). The planet thus travels faster in its orbit when it is closer to the Sun. (Note that for a circle, the speed is the same at every point.)

The uniformity of motion of a planet in its orbit had been an important philosophical concept since the time of the Greeks. Kepler's second law replaced the uniform rate of revolution with a uniform rate of sweeping out area, maintaining the concept that something was uniform.

Kepler's second law is especially noticeable for objects with very elliptical orbits, such as comets. It describes why Halley's Comet sweeps so quickly, within months, through the inner part of the solar system, and takes the rest of its 76-year period moving slowly through the outer parts.

## 3.1c Kepler's Third Law (1618)

Kepler's third law deals with the length of time a planet takes to orbit the Sun, called its **period** of revolution. It relates the period to a measure of the planet's distance from the Sun.

To remember that it is the period of time that is squared rather than cubed, think of Times Square in New York.

Kepler's third law says that *the square of the period of revolution is proportional to the cube of the semimajor axis of the ellipse*. That is, if the cube of the semimajor axis goes up, the square of the period goes up by the same factor. It is often easiest to express Kepler's third law by relating values for a planet to values for the Earth. If $P$ is the period of revolution of a planet and $R$ is the semimajor axis (where $R$ stands for radius; this term often appears as $a$ but $R$ seems simpler to remember),

$$\frac{P^2_{\text{Planet}}}{P^2_{\text{Earth}}} = \frac{R^3_{\text{Planet}}}{R^3_{\text{Earth}}}.$$

We can choose to work in units that are convenient for us on the Earth. We call the semimajor axis of the Earth's orbit—the average distance from the Sun to the Earth—1 **Astronomical Unit** (1 A.U.). Similarly, the unit of time that the Earth takes to revolve around the Sun is defined as **one year.** Using these values, Kepler's third law appears in a simple form, since the numbers on the bottom of the equation are just 1.

### EXAMPLE 3.1 Kepler's Third Law

***Question:*** If we know from observation that Jupiter takes 11.86 years to revolve around the Sun, how do we find Jupiter's distance from the Sun?

***Answer:*** $\frac{(11.86)^2}{P^2_{\text{Earth}}} = \frac{R^3_{\text{Jupiter}}}{R^3_{\text{Earth}}}$

If we put all periods in years and all distances in A.U., we find

$$\frac{(11.86 \text{ years})^2_{\text{Jupiter}}}{1} = \frac{R^3_{\text{Jupiter}}}{1}$$

Nowadays, we can calculate $R$ easily with a pocket calculator, but that doesn't give us a feeling for the wanted magnitude of the answer. Astronomers often prefer approximate values that can be calculated in your head. For example, $(11.86)^2$ can be rounded off to $12^2 = 144$. We can then easily find the cube root of 144 by trial-and-error. It is not hard to calculate that $4^3 = 64$ (which is too small), $5^3 = 125$ (which is too small), and $6^3 = 216$ (which is too large). So in your head you can easily work out that the semimajor axis of Jupiter's orbit around the Sun is between 5 and 6, and it is a little over 5 A.U. (Since Jupiter's orbit is almost round, we can think of this value as the radius of Jupiter's orbit.) The actual value is 5.2 A.U.

**FIGURE 3–8** The frontispiece of Kepler's *Rudolphine Tables* shows Hipparchus, Ptolemy, Tycho Brahe, and Copernicus as pillars of astronomy.

We have used Kepler's third law to determine how the period of a planet revolving around the Sun is related to the size of its orbit. But how do we find the constant of proportionality between $P^2$ and $R^3$, that is, the number by which $R^3$ must be multiplied to get $P^2$ if we don't do the simplification of comparing with the values for the Earth? In the next section, we shall see that Isaac Newton derived mathematically that the constant depends on the mass of the central body, which is the Sun in this case.

Kepler used his laws to compute a set of tables of positions of the planets, the Moon, and the stars, named the *Rudolphine Tables* (Fig. 3–8) after the Emperor Rudolph of the Holy Roman Empire, its sponsor. The *Rudolphine Tables* were printed in 1627.

The period of revolution of satellites around other bodies follows Kepler's laws as well (Fig. 3–9). The laws, with different constants of proportionality, apply to artificial satellites in orbit around the Earth (Fig. 3–10) and to the moons of other planets.

## IN REVIEW

### BOX 3.2 Summary of Kepler's Laws of Planetary Motion

To this day, we consider Kepler's laws to be the basic description of the motions of solar system objects.

1. The planets orbit the Sun in ellipses, with the Sun at one focus.
2. The line joining the Sun and a planet sweeps through equal areas in equal times.
3. The square of the period of a planet is proportional to the cube of the semimajor axis of its orbit—half the longest dimension of the ellipse. (Loosely: The square of the period is proportional to the cube of the planets' distance from the Sun.)

***Discussion Question***

What would a comet's orbit be like if the Sun could be at the center of its orbit instead of at a focus? Sketch the two different cases and specify which is correct.

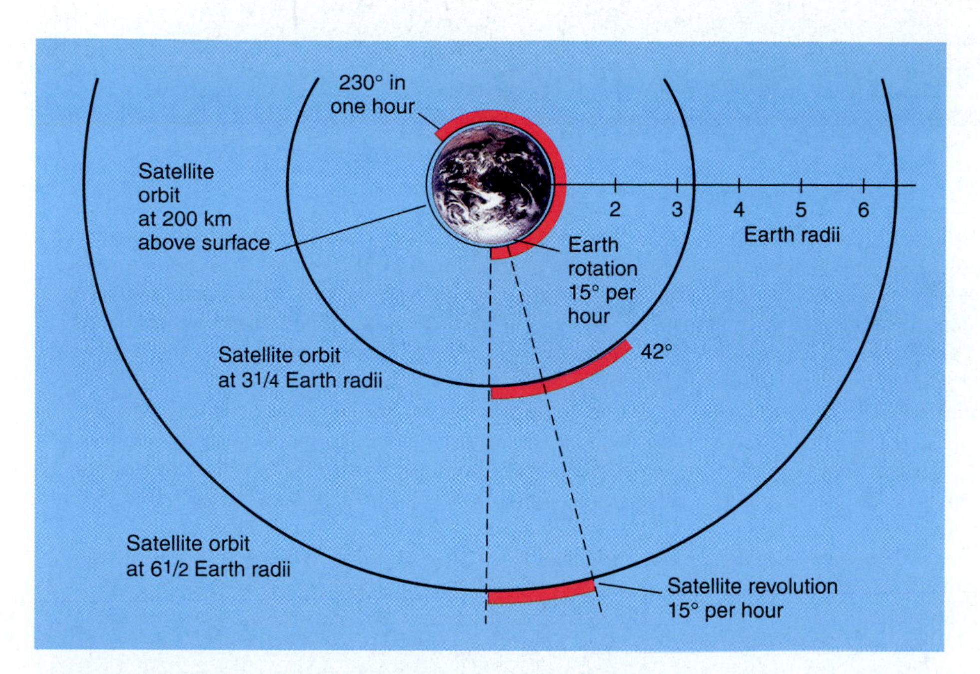


**FIGURE 3–9** Most satellites, including the space shuttles, are only 200 km or so above the Earth's surface and orbit the Earth in about 90 minutes. From Kepler's third law we can see that the period of a satellite's orbit gets longer as the satellite gets higher. It does so both because the satellite travels more slowly and because the path of the orbit is longer. At about 6½ Earth radii, the satellite orbits at the same velocity as that at which the Earth's surface rotates underneath. Thus the satellite is in synchronous revolution, and (if its orbit is not inclined to the equator) always remains over the same location on Earth ("geos," in Greek). This property makes such **geosynchronous satellites** useful for monitoring weather or for relaying communications.

To launch a geosynchronous satellite, one first puts it into low Earth orbit, where its speed is 28,000 km/hr. Increasing its speed to 36,900 km/hr then puts it into an elliptical "transfer orbit," whose low point is its old orbit and whose high point is the synchronous orbit. When it reaches its high point, apogee, it is moving at 5,800 km/hr, and increasing its speed to 11,100 km/hr puts it into a circular orbit at that height.

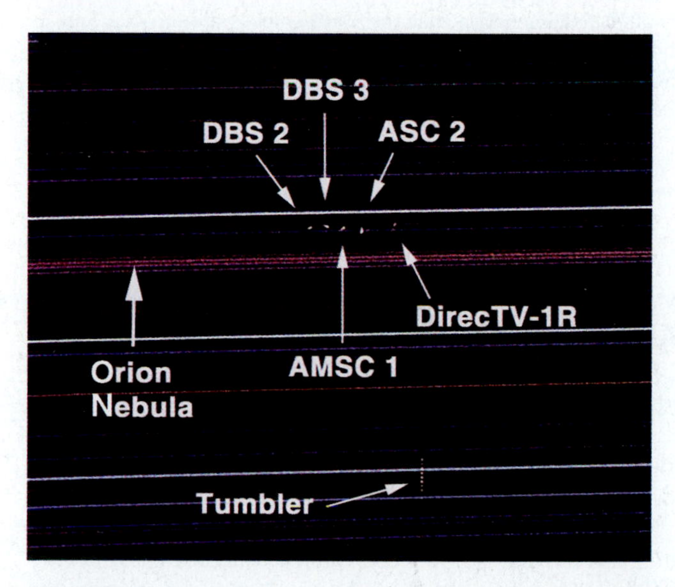


**FIGURE 3–10** In this time exposure, the stars trail but the series of geosynchronous satellites hovering over the Earth's equator, many delivering TV or telephone signals, appear as points.

BIOGRAPHY

## BOX 3.3 Galileo

Galileo Galilei was born in Pisa, in northern Italy, in 1564. Although he studied medicine, his aptitudes were greater in other areas of science and mathematics. At the age of 25 he began to teach mathematics (and astronomy) at the University of Pisa and to undertake experimental investigations. This was an unusual thing to do, because most scientific investigation of the time involved the interpretation of old texts, such as those of Aristotle.

In the 1590s, Galileo became a professor at Padua, in the Venetian Republic. At some point in this decade, he adopted Copernicus's heliocentric theory. In 1604, Galileo observed and studied a new star in the sky, a "nova" (though we now know that this event was actually a supernova). Its existence showed that, contrary to Aristotle's theories, things did change in the distant parts of the Universe. (What was distant for Galileo no longer seems so extremely distant for us.)

In late 1609 or early 1610, Galileo was the first to use a telescope for astronomy. We shall discuss his amazing and rather varied discoveries in the following pages. It is hard to believe that one person could have discovered so many significant things.

In 1613, Galileo published his acceptance of the Copernican theory; the book in which he did so was received and acknowledged with thanks by the Cardinal who, significantly, was Pope at the time of Galileo's trial. Just how Galileo got out of favor with the Church has been the subject of much study. His bitter quarrel with a Jesuit astronomer, Christopher Scheiner, over who deserved the priority for the discovery of sunspots, did not put him in good stead with the Jesuits. Still, the disagreement on Copernicanism was more fundamental than personal, and the Church's evolving views against the Copernican system were undoubtedly more important than personal feelings.

In 1616, Galileo was warned against holding or defending the Copernican theory. There has been much discussion of how serious the warning was and how seriously Galileo took it. Soon thereafter, Copernicus's *De Revolutionibus* was placed on the Index of Prohibited Books, though it became permitted in 1620, when certain "corrections," making its principles appear as methods of calculation rather than as physical truths, were made in about a dozen places. (See Box 2.5 for more on Copernicus's early editions.)

Over the following years, Galileo was certainly not outspoken in his Copernican beliefs, though he also did not cease to hold them. In 1629, he completed his major manuscript, *Dialogue on the Two Great World Systems,* these systems being the Ptolemaic and Copernican (see the photograph opening the chapter). With some difficulty, clearance from the censors was obtained, and the book was published in 1632. It was written in contemporary Italian, instead of the traditional Latin, and could reach a wide audience. Galileo was called before the Inquisition, and the book was condemned.

Galileo's trial provided high drama, and has indeed been transformed into drama on the stage. He was convicted of heresy for holding "that the Sun is the center of the Earth['s orbit] and does not move from east to west, and that the Earth moves and is not the center of the world." Because he had disobeyed orders and had effectively defended the Copernican system, he was sentenced to house arrest. He lived nine more years, until 1642, his eyesight and his health failing. Though during this time he went blind, he continued his writing (including an important book on mechanics) until the end of his life.

***Discussion Question***

When Galileo was warned against teaching Copernicanism, he was shown the instruments of torture. The story is that he recanted his conclusion that the Earth moves around the Sun but, when he went outside, whispered to himself the equivalent of "but still it moves." (This traditional story has no solid backing.) Should Galileo have recanted?

**FIGURE 3–11** Galileo.

## 3.2 GALILEO GALILEI

As art, music, and architecture began to flourish after the Middle Ages, in the era known as the Renaissance, astronomy also developed. Early in the Renaissance, in 1500, Leonardo da Vinci deduced that when we are able to see the dark part of the Moon faintly lighted, we do so because of **earthshine,** the reflection of sunlight off the Earth and over to the Moon.

One hundred years later, also in what is now Italy, Galileo Galilei (Fig. 3–11) began his scientific work. Galileo began to believe in the Copernican system in the 1590s and later provided important observational confirmation of the theory. In late 1609 or early 1610, simultaneously with the first settlements in the American colonies, Galileo became among the first to use a telescope for astronomical observation.

A

B

**FIGURE 3–12** (*A*) One of Galileo's original drawings of the Moon from his *Sidereus Nuncius,* made with his small, unmounted (and therefore shaky) telescope, compared with (*B*) a modern photograph. Notice how well Galileo did.

Thomas Harriot in England had observed the Moon with a telescope a few months before Galileo did, but he did not report the features. It seems reasonable that Galileo, unlike Harriot, had been sensitized to interpreting surfaces and shadows by the Italian Renaissance and by his related training in drawing.

In his book *Sidereus Nuncius* (*The Starry Message,* though traditionally called *The Starry Messenger*), published in 1610, he reported that with this telescope he could see many more stars than he could with the naked eye and could see that the Milky Way and certain other hazy-appearing regions of the sky actually contained individual stars. He described views of the Moon (Fig. 3–12), including the discovery of mountains, craters, and the relatively dark regions he called (and that we still call) **maria** (pronounced mar'ee-a), meaning seas. And, perhaps most importantly, he discovered that small bodies revolved around Jupiter (Figs. 3–13 and 3–14). This discovery proved that all bodies did not revolve around the Earth, and also, by displaying something that Aristotle and Ptolemy obviously had not known about, it showed that Aristotle and Ptolemy were not omniscient. The notion that Aristotle and Ptolemy had been omniscient had kept people from trying to discover new things.

Subsequently, Galileo found that Saturn had a more complex shape than that of a sphere (though it took better telescopes to actually show the rings). He was one of

A

me crederentur, non nullam tamen intulerunt admirationem, eo quod ſecundum exactam lineam rectam, atque Eclypticæ pararellam diſpoſitæ videbantur: ac ceteris magnitudine paribus ſplendidiores: eratque illarum inter ſe & ad Iouem talis conſtitutio.

Ori. * * O * Occ.

E ex parte,

B

**FIGURE 3–13** (*A*) Jupiter and its four Galilean satellites, observed from the Earth. (*B*) Asterisks indicate the moons of Jupiter on this page of Galileo's *Sidereus Nuncius* (1610). Directions are east ("Ori.," for orient) and west ("Occ.," for occident).

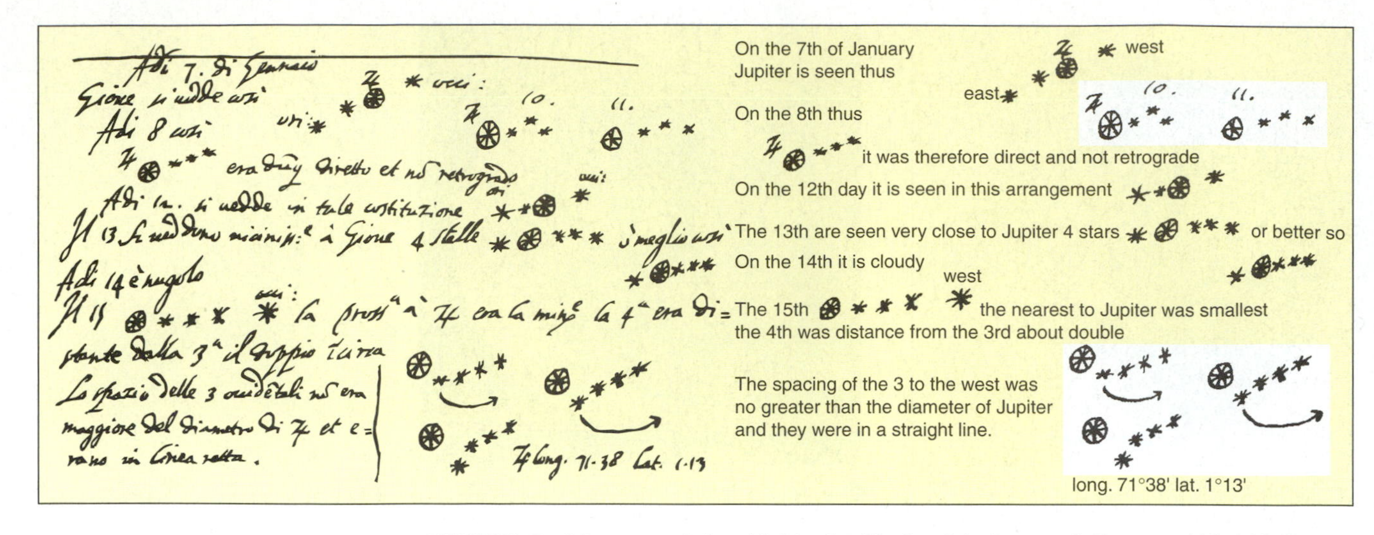


**FIGURE 3–14** A translation (*right*) of Galileo's original notes (*left*) summarizing his first observations of Jupiter's moons in January 1610. The highlighted areas were probably added later. It had not yet occurred to Galileo that the objects were moons in revolution around Jupiter.

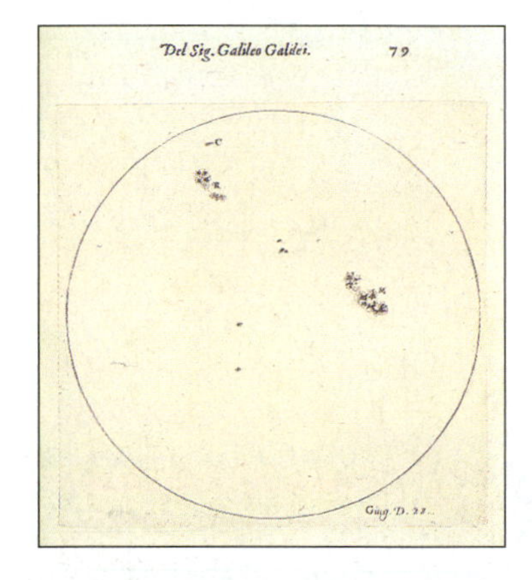


**FIGURE 3–15** Sunspots observed by Galileo in 1612, from his book of the following year.

several people who almost simultaneously discovered sunspots, and he studied their motion on the surface of the Sun (Fig. 3–15). In this way, he discovered that the Sun rotates. Galileo also discovered that Venus went through an entire series of phases (Figs. 3–16 and 3–17). This full set of phases could not be explained on the basis of Ptolemaic theory (Fig. 3–18), because if Venus travelled on an epicycle located between the Earth and the Sun, Venus should always appear as a crescent; since it would be almost between us and the Sun, we would see only the small bit of its lighted side that came around the edge from our point of view. (The observations matched both Copernicus's heliocentric theory and also a hybrid theory of Tycho Brahe, but Copernicanism is the survivor.) Also, more generally, Galileo's observations showed that Venus was a body similar to the Earth and the Moon in that it received light from the Sun rather than generating its own.

Although the Roman Catholic Church was not concerned about models made to explain observations, it was concerned about assertions that those models represented physical truth. It is difficult to say how much the Church feared that Galileo would generalize his ideas to a philosophical and religious level. In his old age, Galileo was forced by the Inquisition to recant his belief in the Copernican theory. The controversy has continued to the present day. Even recently, the Vatican has had a group

**FIGURE 3–16** The phases of Venus. Actual photographs of the phases of Venus at eight-day intervals. Note how Venus's size changes with its phase.

**FIGURE 3–17** These computer calculations, for which the Sun is at a constant location under the horizon, show how Venus's size and phase vary as its position changes with respect to the Sun. You can see that Venus is a crescent only when it is on the part of its orbit that is closer to the Earth than the Sun, so it appears relatively large in that phase.

studying how Galileo was treated. Pope John Paul II set up the group in 1979 "in loyal recognition of wrongs from whatever side they come . . . [to] dispel the mistrust that still opposes, in many minds, a fruitful concord between science and faith." In 1983, he reported to a group celebrating the 350th anniversary of the publication of Galileo's *Dialogue* that the research on the treatment of Galileo was "progressing very encouragingly" and that the Church upheld the freedom of scientific research.

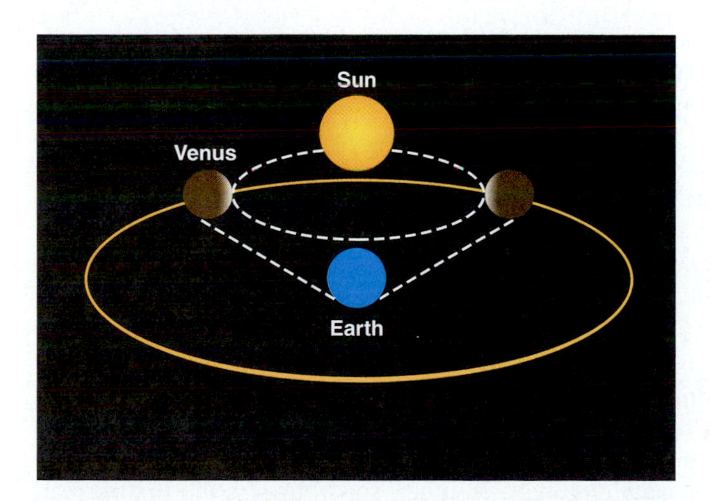


*A*

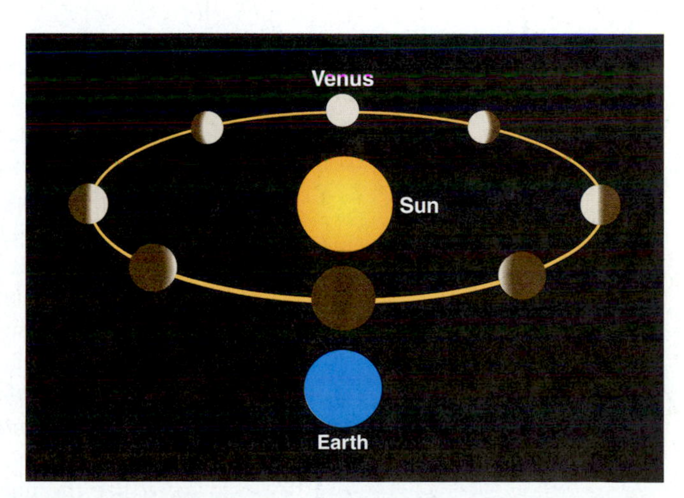


*B*

**FIGURE 3–18** In the Ptolemaic theory (*A*), Venus and the Sun both orbited the Earth, but because it is known that Venus never gets far from the Sun in the sky, the center of Venus's epicycle was restricted to always fall on the line joining the center of the deferent and the Sun. In this diagram Venus could never get farther from the Sun than the region restricted by the dotted lines. Thus, Venus would always appear as a crescent, though before the telescope was invented, this could not be verified. In the heliocentric theory (*B*), Venus is sometimes on the near side of the Sun, where it appears as a crescent, and it is sometimes on the far side, where we can see half or more of Venus illuminated. This agrees with Galileo's observations, a modern version of which appears in previous figures.

**FIGURE 3–19** Isaac Newton in 1702. Three years later, he was knighted.

**FIGURE 3–20** Newton's house at Woolsthorpe.

Studies of Galileo and his troubles with the Church continue. In 1992, the Pope formally closed the investigation, concluding that both sides acted in good faith but that the Inquisition had been overly harsh when it condemned Galileo in 1633. The Pope also stated that Galileo had been a better theologian than those opposing him with a highly literal reading of the Scriptures, and he commended Galileo's statement that "The Bible teaches how to go to heaven, not how the heavens go." The Pope's statement, however, was not a formal pardon.

## 3.3 ISAAC NEWTON

It was not until many years after Kepler had discovered his three laws of planetary orbits by laborious calculation that the laws were derived mathematically from basic physical principles. The credit for doing so belongs to Isaac Newton (Fig. 3–19). Newton was born in England in 1642 (Fig. 3–20), the year of Galileo's death. He was to become the greatest scientist of his time and one of the greatest of all time. Later in this book (Section 4.4) we will describe his discovery that visible light can be broken down into a spectrum and his invention of the reflecting telescope. Here we shall speak only of his ideas about motion and gravitation.

PHILOSOPHIÆ
NATURALIS
PRINCIPIA
MATHEMATICA.

Autore *J. S. NEWTON*, *Trin. Coll. Cantab. Soc.* Matheseos Professore *Lucasiano*, & Societatis Regalis Sodali.

IMPRIMATUR.
S. PEPYS, *Reg. Soc.* PRÆSES.
*Julii* 5. 1686.

LONDINI,
Jussu *Societatis Regiæ* ac Typis *Josephi Streater*. Prostant Venales apud *Sam. Smith* ad insignia Principis *Walliæ* in Cœmiterio D. *Pauli*, aliosq; nonnullos Bibliopolas. *Anno* MDCLXXXVII.

**FIGURE 3–21** The title page of the first edition of Isaac Newton's *Principia Mathematica*, published in 1687.

Newton set modern physics on its feet by deriving laws showing how objects move on the Earth and in space and by finding the law that describes gravity. In order to work out the law of gravity, Newton had to invent calculus! The *Philosophiae Naturalis Principia Mathematica (Mathematical Principles of Natural Philosophy)*, known as *The Principia* for short, appeared in 1687 (Fig. 3–21).

Newton—whether or not you believe that an apple fell on his head—was the first to realize the universality of gravity. In other words, he was the first to realize that the same law that describes how objects fall on Earth describes how objects move far out in space. In particular, he realized that the Moon is always falling toward the Earth. However, at the same time, the Moon moves forward in space in its orbit. Think of the Moon moving above the Earth's surface from left to right. As the Moon falls toward Earth, the Moon approaches the Earth closer on the right. But you can see that if the Earth's surface also curves downward toward the right at the same angle, the Moon would never get any closer to the Earth (Fig. 3–22). For this reason, the Moon remains in orbit, always falling but constantly missing the Earth.

Newton formulated the law describing the force of gravity ($F$) between two bodies. It always tends to pull the two bodies together, and its strength varies directly with the product of the masses of the two bodies ($m_1$ and $m_2$) and inversely with the square of the distance ($d$) between them:

$$\text{force of gravity} \propto \frac{m_1 m_2}{d^2}.$$

Since "proportional to" ($\propto$) simply stands for multiplying by a constant, we can write

$$F = \frac{G m_1 m_2}{d^2},$$

where the constant $G$ (given, as are many constants, to more accuracy in Appendix 2), is about $6.67 \times 10^{-11}$ $m^3/kg{\cdot}s$.

This relation of Newton's is known as the **law of universal gravitation;** it is a universal law in that it works all over the Universe rather than being limited to local applicability (only on Earth or even only in the solar system).

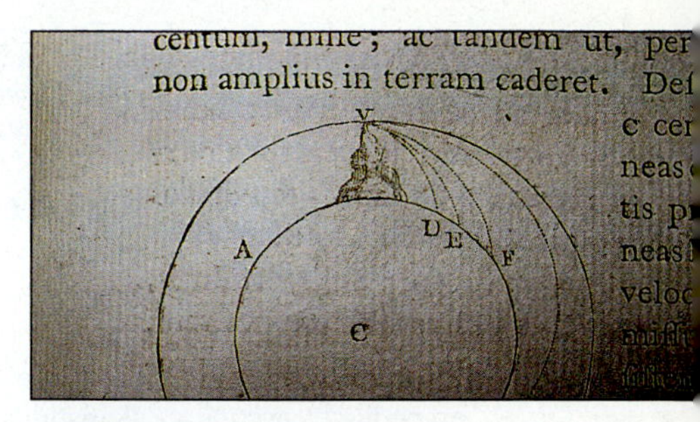


FIGURE 3–22 Newton's thought demonstration (that is, a demonstration in his mind) of the effect of increasing the velocity of a projectile, something ejected and then subject only to the force of gravity. If the projectile is moving fast enough, the Earth curves away to match the projectile's falling, and the projectile is in orbit. It took about 300 years before Newton's theoretical idea became reality.

$\propto$ means "is proportional to"; it means that the left-hand side is equal to a constant times the right-hand side. The constant, whose value is given in Appendix 2 (rather than here, so that you won't be tempted to memorize it), is called $G$ or the **universal gravitational constant** because to the best of our knowledge it is constant throughout the Universe.

## 3.3a Newton's Laws of Motion

In his *Principia* (Fig. 3–23), Newton advanced not only the law of universal gravitation but also three laws of motion. The laws govern the motion of objects in the normal circumstances of our everyday world. Whole courses are given on Newton's laws, and here we have space to discuss them only briefly.

Newton's first law was a development of ideas that had earlier been advanced by Galileo and others. The law states that every object tends to remain either at rest or in uniform motion in a fixed direction unless some outside force changes its state. This was in direct conflict with the Aristotelian idea that forces have to be continually applied to keep a body in motion. The property of mass that resists change in its motion is called **inertia,** and the first law is often called the **law of inertia.**

Newton's second law of motion states that the strength of a force is equal to the amount of mass involved times the acceleration it undergoes: $F = ma$. We actually define mass (technically, "inertial mass") from this formula: A larger mass requires a larger force to give it the same acceleration obtained for a smaller mass and by a smaller force.

Newton's third law is usually stated as follows: "For every action there is an equal and opposite reaction." To every force corresponds another force of equal magnitude but opposite direction. It is this principle, for example, that makes a jet plane go—the force expelling gas backwards from the engine is the action, and the force moving the plane forward is the reaction. It is not significant which force is called the action and which the reaction; the important point is that the force of the plane on the gas and that of the gas on the plane are equal and opposite. The same principle causes a balloon to fly when you blow it up without tying and let go. And it is Newton's third law that explains why the engines of space shuttles and other rockets can propel them even when they are in empty space. (A propeller airplane also flies forward because of Newton's third law; the force of the propeller pushing air back and that of the air pushing the propeller forward are action and reaction.)

FIGURE 3–23 The beginning of the handwritten "fair copy" for the publisher of Newton's manuscript for the *Principia.* The manuscript is in the Royal Society in London.

A DEEPER DISCUSSION

## BOX 3.4 Kepler's Third Law, Newton, and Planetary Masses*

Astronomers nowadays often make rough calculations to test whether physical processes under consideration could conceivably be valid. Astronomy has also had a long tradition of exceedingly accurate calculations. Pushing accuracy to one more decimal place sometimes leads to important results.

For example, Kepler's third law, in its original form—the period of a planet squared is proportional to its distance from the Sun cubed ($P^2$ = constant × $R^3$)—holds to a reasonably high degree of accuracy and seemed completely accurate when Kepler did his work. But now we have more accurate observations. If we consider each of the planets in turn, and if a term involving the sum of the masses of the Sun and the planet under consideration is included in the equation

$$P^2 = \frac{constant}{m_{\text{Sun}} + m_{\text{Planet}}} \times R^3,$$

the agreement with observation is improved in the fourth decimal place for the planet Jupiter and to a lesser extent for Saturn. The masses of the other planets are too small to have a detectable effect; even the effects caused by Jupiter and Saturn are tiny. Newton derived the equation in its general form:

$$P^2 = \frac{4\pi^2}{G(m_1 + m_2)} \times a^3,$$

for a body with mass $m_2$ revolving in an elliptical orbit with semimajor axis $a$ around a body with mass $m_1$. The constant $G$ is the **universal gravitational constant.** Newton's formula shows that the planet's mass contributes to the value of the proportionality constant, but its effect is very small.

Kepler's third law, and its subsequent generalization by Newton, applies not only to planets orbiting the Sun but also to any bodies orbiting other bodies under the control of gravity. Thus it also applies to satellites orbiting planets. We determine the mass of the Earth by studying the orbit of our Moon, and we determine the mass of Jupiter by studying the orbits of its moons. Until recently, we were unable to determine reliably the mass of Pluto because we could not observe a moon in orbit around it. The discovery of a moon of Pluto in 1978 finally allowed us to determine Pluto's mass, and we discovered that our previously best estimates (based on Pluto's gravitational effects on Uranus) were way off—much too high. The same formula can be applied to binary stars (pairs of stars orbiting each other, as we discuss in Chapter 26) to find their masses.

### Discussion Question

Explain why the form of Kepler's law discussed here is more useful for understanding the satellites of Jupiter than the simpler form giving the proportionality.

BIOGRAPHY

## BOX 3.5 Isaac Newton

Isaac Newton (Fig. 3–22) was born in Woolsthorpe, Lincolnshire, England, on December 25, 1642. Even now, almost 350 years later, scientists still refer regularly to "Newton's laws of motion" and to "Newton's laws of gravitation." In optics, we refer to "Newtonian telescopes" and to "Newton's rings," a series of concentric rings that appear often when two pieces of glass are placed together.

Newton's most intellectually fertile years were those right after his graduation from college when he returned home to the country (Fig. 3–23) because fear of the plague shut down many cities. Many of his basic theories stemmed from that period.

For many years he developed his ideas about the nature of motion and about gravitation. In order to derive them mathematically, he invented calculus. (A rival on the European continent independently invented calculus, and the two fought long and hard as to who should get credit.) Newton long withheld publishing his results, possibly out of shyness. Finally Edmond Halley—whose name we associate with the famous comet—persuaded him to publish his work. A few years later, in 1687, the *Philosophiae Naturalis Principia Mathematica* (*Mathematical Principles of Natural Philosophy*), known as *The Principia* (prin-ci'pe-a), was published with Halley's aid. In it, Newton showed that the motions of the planets and comets could all be explained by the same law of gravitation that governed bodies on Earth. In fact, he derived Kepler's laws on theoretical grounds.

Newton was professor of mathematics at Cambridge University and later in life went into government service as Master of the Mint in London. He lived until 1727. His tomb in Westminster Abbey bears the epitaph: "Mortals, congratulate yourselves that so great a man has lived for the honor of the human race."

### Discussion Question

Why is Newton's law of gravitation called "universal"?

## CORRECTING MISCONCEPTIONS

**✖ Incorrect**

Kepler's law for moons around a planet has the same constant as for planets around the Sun.

The relation between period and distance for a planet's moons doesn't tell us anything special.

**✔ Correct**

Newton derived the form of Kepler's law that includes the mass of the central object; the constant relating period squared and distance cubed includes this mass, so it depends on what the central object is. (See Box 3.4.)

The relation between period and distance for a planet's moons gives us the planet's mass.

## SUMMARY AND OUTLINE

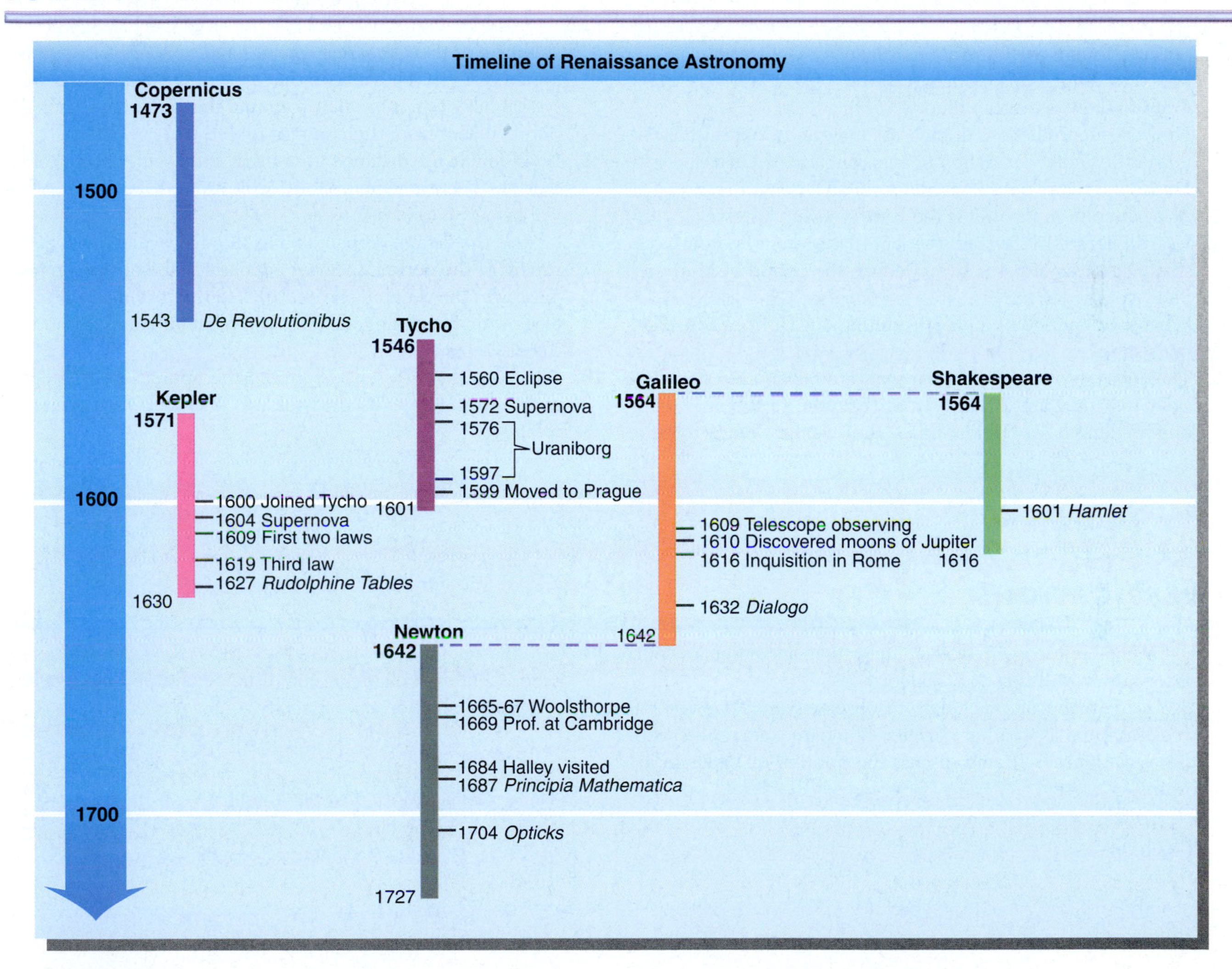


Kepler (Section 3.1)
- Studied Tycho's observations and discovered three laws of planetary motion
- Kepler's laws: (1) orbits are ellipses; (2) equal areas are swept out in equal times; (3) period squared is proportional to distance cubed (period in years squared equals distance in A.U. cubed).

Galileo (Section 3.2)
- First to use a telescope for astronomical observations
- His observations supported the heliocentric theory and further showed that Aristotle and Ptolemy had not known everything.

Newton (Section 3.3)
- Worked out laws of gravity and motion
- Gravity: $F = Gm_1m_2/d^2$
- First law of motion: inertia
- Second law of motion: $F = ma$
- Third law of motion: For every action there is an equal and opposite reaction
- Was able to derive Kepler's laws and find the masses of the planets

## KEY WORDS

foci
focus
major axis
semimajor axis
minor axis
circle
eccentricity
conic sections
law of equal areas
sector
period
Astronomical Unit
geosynchronous satellites
one year
earthshine
maria
law of universal gravitation
universal gravitational constant
inertia
law of inertia

## QUESTIONS

1. Discuss four new observations made by Galileo, and show how they tended to support, tended to oppose, or were irrelevant to the Copernican theory.
2. How do the predictions of the Ptolemaic and Copernican systems differ for the variation in apparent *size* of Venus?
3. Do Kepler's laws permit circular orbits? Discuss.
4. At what point in its orbit is the Earth moving fastest?
5. Use Kepler's third law and the fact that Mercury's orbit has a semimajor axis of 0.4 A.U. to deduce the period of Mercury. Show your work.
6. What is the period of a planet orbiting 1 A.U. from a 9-solar-mass star?
7. The denizens of a planet far, far away around another star consider that they are 1 flibbit from their sun (a flibbit is their unit of length—a trillion times the average length of the antennae of their king and queen), and that they orbit with a period of 1 pip. They notice that a purple planet in their solar system takes 8 pips to revolve around their sun. How far is the purple planet from their sun (in flibbits)?
8. If we double the distance from the Earth's center of a satellite that has been in synchronous orbit around the Earth, what will its new period be?
9. (a) Use the data in Appendix 3 to show for which planets the square of the period does not equal the cube of the semimajor axis to the accuracy given. (b) Use the formula in Box 3.4 to show that including the effects of planetary masses removes the discrepancy.
10. What was the difference between the approaches of Kepler and Newton to the discovery of laws that controlled planetary orbits?

## INVESTIGATIONS

1. Use a small telescope to observe the Moon, and compare your observations with those of Galileo.
2. Use a telescope or binoculars to observe Venus for several weeks as often as weather permits. Compare your results with those of Figure 3-16 and discuss the position of Venus in its orbit.
3. Use the method of Figure 3-3 to draw several ellipses. Experiment with changing the distance between the foci and observe its effect on the ellipticity.

## USING TECHNOLOGY

### W World Wide Web

1. Search the Web for information about the lives of Galileo, Kepler, and Newton.
2. Where is Galileo's finger kept?

### REDSHIFT

1. Show the relation between the phases of Venus as seen from Earth and the relative positions of the Earth, Venus, and Sun.
2. Show the view of the Earth that you would have from Venus for each of the positions for which you considered phases in the previous question.
3. Duplicate Galileo's observations of January 7, 1610, when he discovered the moons of Jupiter.
4. Compare the Moon's surface with Galileo's drawing, and try to identify all the major features shown on each.
5. Ponder *The Science of Astronomy: Revolutionaries: Copernicus and Galileo*

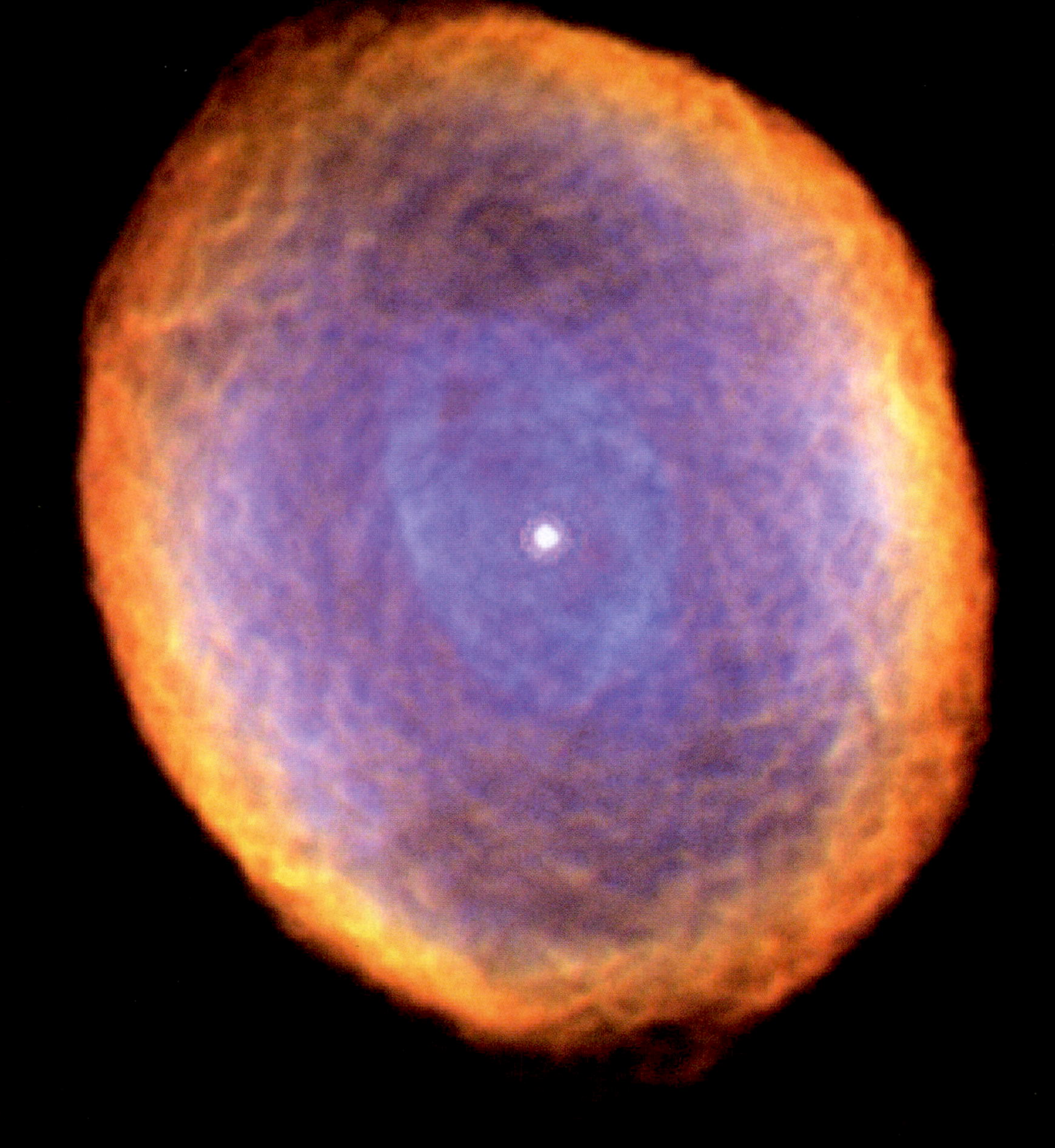

The planetary nebula IC 418 in the constellation Lepus. The central star, whose outer layers were ejected to form the planetary nebula, is visible at center. This Hubble Space Telescope image clearly shows structure lacing the nebula. Our Sun will die in this way in about 5 billion years.

Telescopes allow us to see more spatial detail than we can see with the eye. In general, the use of telescopes allows the collection of more light than the eye can see, and thus permits us to observe fainter objects. The Earth's atmosphere blurs light passing through it; but by placing the Hubble Space Telescope above the atmosphere, we have been able to achieve higher spatial resolution.

# Chapter 4

# Light and Telescopes

**AIMS:** To understand the spectrum and the fundamental principles of telescopes

Although the word "telescope" makes most of us think of objects with lenses and long tubes, that type of telescope is used only to observe "light." Studying ordinary "visible" light, the kind we see with our eyes, is not the only way we can study the Universe. We shall see in this chapter that light is only one type of **electromagnetic radiation**—a certain way in which energy moves through space. Other types of radiation are gamma rays, x-rays, ultraviolet light, infrared light, and radio waves. These other types of radiation are all fundamentally identical to ordinary light, also known as "visible light" (which we will often call, simply, "light," using common usage), though in practice we must usually use different methods to observe them. (Some kinds of "radiation" are not "electromagnetic radiation" but are rather tiny particles moving through space; we do not discuss them in this chapter. In astronomy, we usually use the word "radiation" to mean "electromagnetic radiation.")

In this chapter, we discuss the properties of radiation and how studying it enables scientists to study the Universe; after all, we cannot touch a star, and though we have brought bits of the Moon back to Earth for study and a few rocks from Mars are lying around, we are not yet able to bring back samples from even the nearest planets.

## 4.1 THE SPECTRUM

It was discovered over 300 years ago that when ordinary light is passed through a prism, a band of color like the rainbow comes out the other side. Thus "white light" is composed of all the colors of the rainbow (Fig. 4–1).

FIGURE 4–1 When white light passes through a prism, a full optical spectrum results. It ranges through ROYGBIV.

These colors are always spread out in a specific order, which has traditionally been remembered by the initials of the friendly fellow **ROY G BIV,** which stand for ***R***ed, ***O***range, ***Y***ellow, ***G***reen, ***B***lue, ***I***ndigo, ***V***iolet. No matter what rainbow you watch, or what prism you use, the order of the colors never changes.

We can understand why light contains the different colors if we think of light as waves of radiation. These waves are all travelling at the same speed, $3 \times 10^8$ m/sec (that is, meters per second). (In American "customary units," this speed is equivalent to 186,000 miles/sec.) Light travels at a speed the equivalent of seven times around the Earth each second. This speed is normally called the **speed of light.** (The phrase "the speed of light" usually really means "the speed of light in a vacuum"; light normally travels at a slightly lower speed because the light is not usually in a perfect **vacuum,** empty space where no matter is present.)

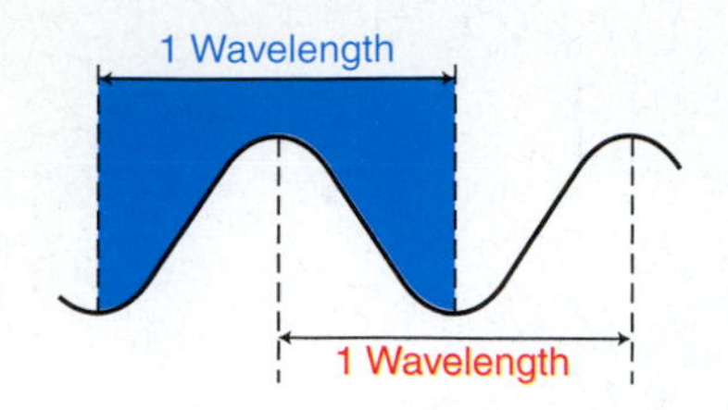


**FIGURE 4–2** The wavelength is the length over which a wave repeats.

The distance between one crest of a wave to the next or one trough to the next, or in fact between any point on a wave and the similar point on the next wave, is called the **wavelength** (Fig. 4–2). Light of different wavelengths appears as different colors. Red light has approximately $1\frac{1}{2}$ times the wavelength of blue light. Yellow light has a wavelength in between the two. Actually, there is a continuous distribution of wavelengths, and one color blends subtly into the next.

The wavelengths of light are very short: just a few ten-millionths of a meter. Astronomers use a unit of length called an **angstrom,** named after the 19th-century Swedish physicist A. J. Ångstrom. One angstrom (1 Å) is $10^{-10}$ m. (Remember that this number is 1 divided by $10^{10}$, which is 0.000 000 000 1 m.) The wavelength of violet light is approximately 4000 angstroms (4000 Å). Orange light is approximately 6000 Å, and red light is approximately 6500 Å in wavelength. (10 Å make 1 nanometer, abbreviated 1 nm, the standard International System (SI) unit, so violet light is 400 nm and red light is 650 nm in wavelength.)

The human eye is not sensitive to radiation of wavelengths much shorter than 4000 Å or much longer than 6600 Å, so the "colors" with which we are familiar are all in this range. But other devices exist that can measure light at shorter and longer wavelengths. At wavelengths shorter than violet, the radiation is called **ultraviolet;** at wavelengths longer than red, the radiation is called **infrared.** (Note that this last word is not "infared," a common misspelling; "infra" is a Latin prefix that means "below.")

In Chapter 24, just prior to the discussion of stellar spectra, we discuss some laws that show how much light of each color is given off for a given temperature. The cooler the star, the redder the light; the hotter the star, the bluer the light. Since all planets are cooler than all stars, most of the light they emit by themselves is in the infrared. They also reflect sunlight, whose spectrum is most intense in the visible. So a graph of the amount of light we get from a planet has two peaks in the intensity: one in the visible from reflected sunlight and one in the infrared from emitted infrared.

Intensity is technically the amount of energy per second in a unit band of wavelength that passes into a unit of area in a unit solid angle. Basically, you can think of it as the power in a given angstrom of wavelength or as a sort of brightness at a given color.

All the types of radiation—visible or invisible—have much in common. Many of us are familiar with the fields of force produced by a magnet—a magnetic field. Though that field is invisible, if we put a compass in it, the compass needle shows which way the magnetic field is pointing. Further, the field of force produced by a charged object like a hair comb after you use it on a dry day is an electric field. It has been known for over a century that fields that vary rapidly behave in a way that no one would guess from the study of magnets and combs. In fact, light, x-rays, and radio waves are all examples of rapidly varying electric fields and magnetic fields that have "broken away" from their sources and move rapidly through space. For this reason, these types of radiation are referred to as "electromagnetic radiation."

We can draw the entire **electromagnetic spectrum,** often simply called the "spectrum" (plural: **spectra**), ranging from radiation of wavelength shorter than 1 Å to radiation of wavelength many meters long (Fig. 4–3).

Note that from a scientific point of view there is no qualitative difference between types of radiation at different wavelengths. They can all be thought of as electromagnetic waves, waves of varying electric and magnetic fields. Light waves comprise but one limited range of wavelengths. When an electromagnetic wave has a wavelength of 1 Å, we call it an x-ray. When it has a wavelength of 5000 Å, we call it light. When it has a wavelength of 1 cm, we call it a radio wave. Of course, there are obvious practical differences in the methods by which we detect x-rays, light, and radio waves, but the principles that govern their existence are the same.

Note also that light occupies only a very small portion of the entire electromagnetic spectrum. It is obvious that the new ability that astronomers have to study parts of the electromagnetic spectrum other than light waves enables us to increase our knowledge of celestial objects manifold.

Only certain parts of the electromagnetic spectrum can penetrate the Earth's atmosphere. We say that the Earth's atmosphere has "windows" for the parts of the

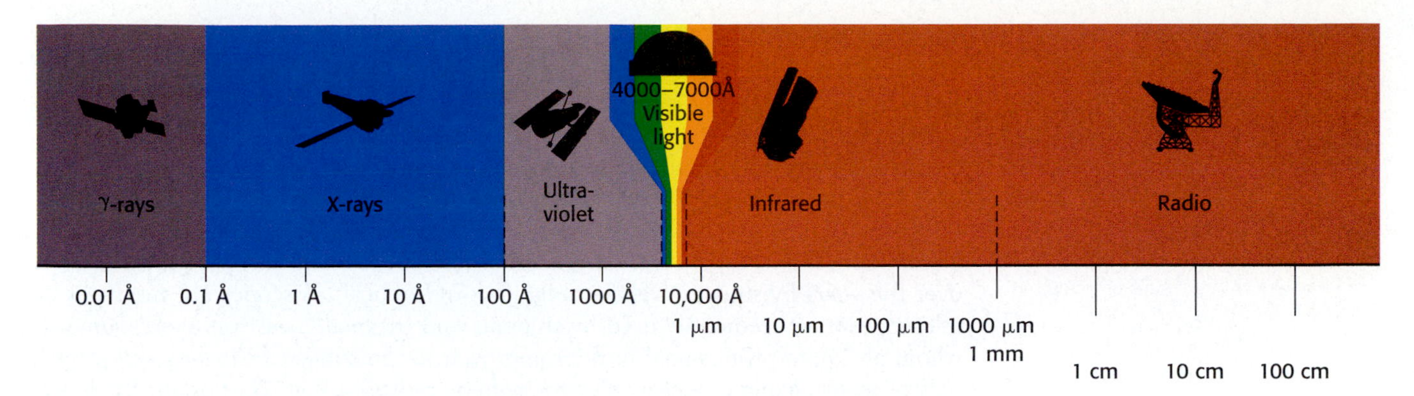


**FIGURE 4–3** The electromagnetic spectrum. The silhouettes shown represent telescopes or spacecraft used for observing that part of the spectrum: the Compton Gamma-Ray Observatory, the Chandra X-ray Observatory, the Hubble Space Telescope, the dome of a ground-based telescope, the Space Infrared Telescope Facility, and the Green Bank Telescope.

spectrum that can pass through it. The atmosphere is transparent at these windows and opaque at other parts of the spectrum. One window passes what we call "light," which astronomers technically call "visible light," or "the visible," or "the optical part of the spectrum." Another window falls in the radio part of the spectrum, and modern astronomy uses "radio telescopes" to detect that radiation (Fig. 4–4).

But we of the Earth are no longer bound to our planet's surface; balloons, rockets, and satellites carry telescopes above the atmosphere to observe in parts of the electromagnetic spectrum that do not pass through the Earth's atmosphere. In recent years, we have been able to make observations all across the spectrum. It may seem strange, in view of the long-time identification of astronomy with visible observations, to realize that optical studies no longer dominate astronomy.

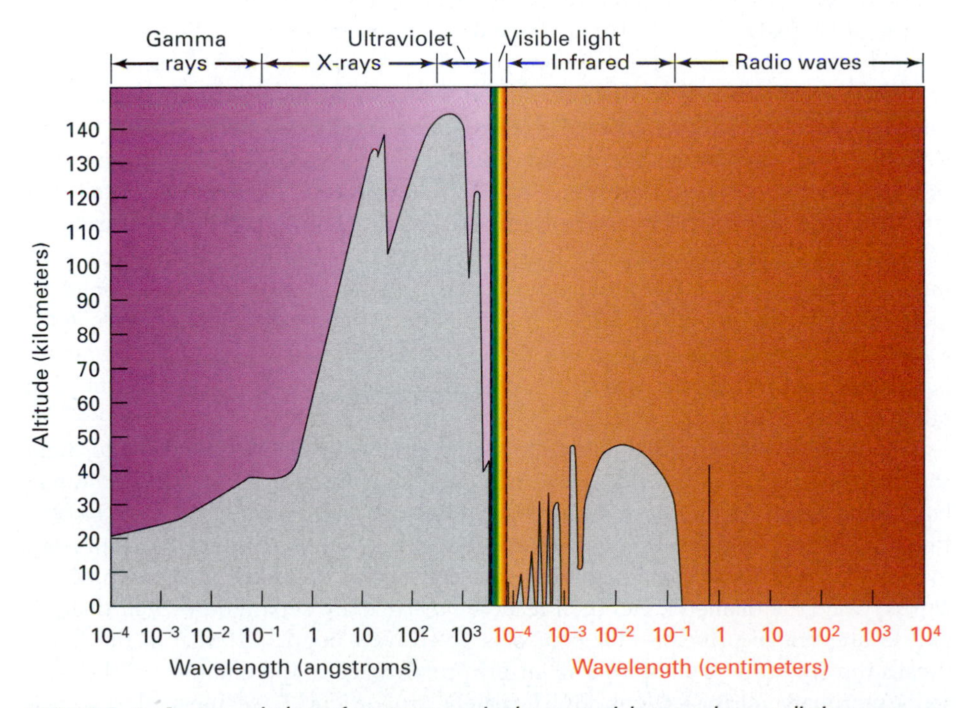


**FIGURE 4–4** In a window of transparency in the terrestrial atmosphere, radiation can penetrate to the Earth's surface. The curve specifies the altitude at which the intensity of arriving radiation is reduced to half its original value. When this happens high in the atmosphere, little or no radiation of that wavelength reaches the ground.

## 4.2 WHAT A TELESCOPE IS

How do you define "telescope"? The question has no simple answer, since a "telescope" to observe gamma rays may be a package of electronic sensors launched above the atmosphere, while a radio telescope may be a large number of small aerials strung over acres of landscape. We will begin, nevertheless, by discussing telescopes of the traditional types, which observe the radiation in the visible part of the spectrum. These "optical telescopes" are important because of the many things we have learned over the years by studying visible radiation. And optical telescopes are the types of telescopes most frequently used by students and by amateur astronomers (some of whom are quite professional in their approach to the subject). Further, the principles of focusing and detecting electromagnetic radiation that were originally developed through optical observations are widely used throughout the spectrum.

Contrary to popular belief, the most important purpose for which most optical telescopes are used is to gather light. (We will drop the qualifying word "optical.") True, telescopes can be used to magnify as well, but for the most part astronomers are interested in observing fainter and fainter objects and so must collect more light to make these objects detectable. Sometimes magnification is important—such as for many types of observations of the Sun or planets—but, with a handful of exceptions, stars are so far away that they appear as mere points of light no matter how much magnification is applied.

When we look at the sky with our naked eyes, several limitations come into play. First of all, our eyes see in only one part of the spectrum (the visible). Another limitation is that we can see only the light that passes through an opening (aperture) of a certain diameter—the pupils of our eyes. In the dark, our pupils dilate so that as much light as possible can enter, but the apertures are still only a few millimeters across (up to 8 mm for young people and decreasing to perhaps 5 mm for older people). An additional limitation is that of time. Our brains distinguish a new image about 30 times a second, and so we are unable to store faint images over a long time in order to accumulate a brighter image. Astronomers overcome these additional limitations by using a telescope to gather light and a recording device, such as a photographic plate or an electronic detector, to store the light.

A further advantage of a telescope over the eye, or of a large telescope over a smaller telescope, is that of **resolution,** the ability to distinguish finer details in an image or to distinguish two adjacent objects (Fig. 4–5). A telescope is capable of considerably better resolution than the eye, which is limited to a resolution of 1 arc min (the diameter of a dime at a distance of 60 m), that is, the ability to see detail that is 1 arc min across or distinguish between two stars that are separated by 1 arc min. (The larger the aperture accepting light, the better the resolution, and the eye's maximum opening is only a few millimeters across. The retina actually sets a resolution limit somewhat worse, about 3 arc min.) The aperture of each lens of binoculars is a few times larger, so binoculars can reveal details a few times finer. Thus binoculars or a telescope can distinguish the two components of a double star from each other in many cases where the unaided eye is unable to do so.

Actually, resolution depends not only on the aperture but also on the wavelength of radiation being observed; the example we consider at right is green light of approximately 5000 Å.

As the wavelength of radiation doubles, resolution is halved. For example, a telescope that can resolve two sources emitting 5000 Å (green) light that are 1 arc second apart could only resolve two 10,000 Å (infrared) sources if they were 2 arc sec apart. The limit of resolution for visible light, called Dawes's limit, is approximately $2 \times 10^{-3}\ \lambda/d$ arc sec, when the wavelength, $\lambda$, is in Å and the diameter of a telescope, $d$, is in cm.

The Earth's atmosphere limits us to seeing detail on celestial objects larger than roughly ½ arc sec across (the diameter of a dime at a distance of 7 km or of a human hair two football fields away). Consider a telescope with a collecting area 10 centimeters across. It can resolve double stars (that is, tell that there are two stars present instead of one blob) that are separated by 1 arc second. A telescope 20 centimeters across, twice the diameter, can then resolve stars that are separated by half that angle, ½ arc sec. In principle, for light of a given wavelength, the size of the finest details resolved (the resolution) is inversely proportional to the diameter of the telescope's primary mirror or lens. By "inversely proportional," we mean that as one quantity goes up, the other goes down by the same factor.

Even for large telescopes, the best resolution that can be achieved on the Earth's surface on an average good night is limited by turbulence in the Earth's atmosphere

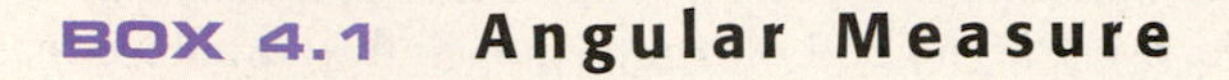

# A DEEPER DISCUSSION

## BOX 4.1 Angular Measure

Since Babylonian times, the circle has been divided into parts based on a system that uses multiples of 60. If we are at the center of a circle, we consider it to be divided into 360 parts as it extends all around us. Each part is called 1 degree (1°). Each degree, in turn, is divided into 60 minutes of arc (also written 60 arc min), often simply called 60 minutes (60'). And each minute is divided into 60 seconds of arc (also written 60 arc sec), often simply called 60 seconds (60"). Thus in 1°, we have $60 \times 60 = 3600$ seconds of arc.

In this book we usually mention angular measure only to give you a sense of appearance. It is useful to realize that your fist at the end of your outstretched arm covers about 10° (we say "subtends 10°") as seen from your eye, and your thumb's width subtends about 2°. The Moon covers about ½° of the sky; try covering the full Moon with your thumb one clear night.

### *Discussion Question*

If you were calculating the angular size of an object, would you find it easier to work in hundredths of a degree or in seconds of arc? Why? Compare the sizes of these units.

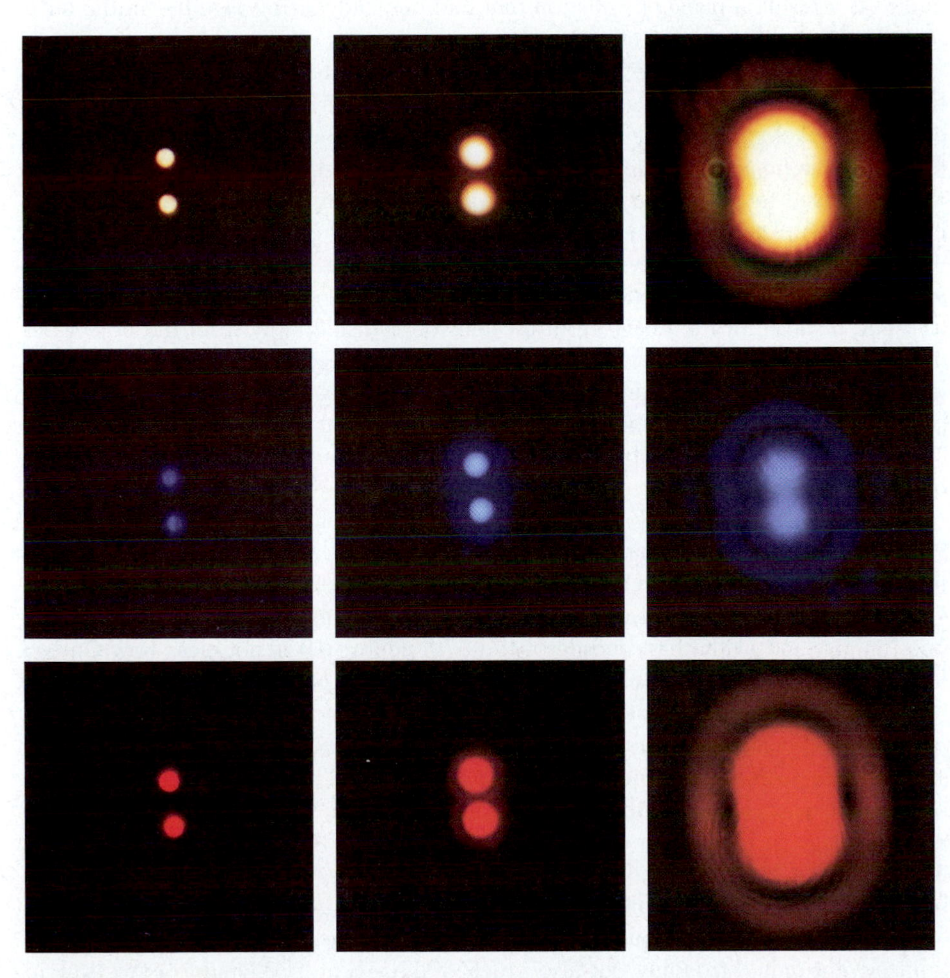

**FIGURE 4–5** In addition to their primary use for gathering light from faint objects, telescopes are often used to increase the resolution, our ability to distinguish details in an image. We see a pair of point sources (sources so small or far away that they appear as points) in the right-hand column that are not quite resolved in (*top row*) white light, (*middle row*) violet light (about 4000 Å), and (*bottom row*) red light (about 6500 Å). Note that the rightmost violet image is substantially better resolved than Dawes's limit, the condition in which two sources can barely be resolved, while the resolution is just at Dawes's limit for the longer, red wavelength. In this series, the images differ by different aperture size in the lab; from left to right, the aperture used to observe the image is smaller by a factor of 2 in successive images. In astronomy, as long as we are not limited by "seeing" (the fineness of detail we can detect), a larger image disk would come from a smaller telescope or from observing at a longer wavelength.

**FIGURE 4–6** The moon rock is visible straight on, but because it is mounted in a plastic box it is also visible as a greatly refracted image through the right side. Further, a third image is visible as a reflection in the far right corner of the box.

to about ½ arc sec. Thus increasing the size of a telescope above a certain limit no longer improves its resolution, even on the best of nights, though the advantage that a larger telescope gathers more light remains. However, techniques have now been developed and are rapidly being improved to compensate for atmospheric turbulence, thus drastically improving images. We will discuss such "active optics" later on.

## 4.3 LENSES AND TELESCOPES

Light (or other electromagnetic radiation) bends when it passes from one medium into another. This bending is called **refraction** (Fig. 4–6). A "medium" is simply any material (that is, a transparent object, air, interstellar space) through which the radiation travels. Electromagnetic radiation can even travel through a vacuum.

Consider parallel waves of light hitting a block of glass at an angle (Fig. 4–7). Some of the waves hit the glass before others. Since light travels 50 per cent more slowly in glass than in air, these waves travel more slowly than the waves that are still in air. As a result, a plane of radiation that had been advancing together in the air is travelling in a different direction in the glass.

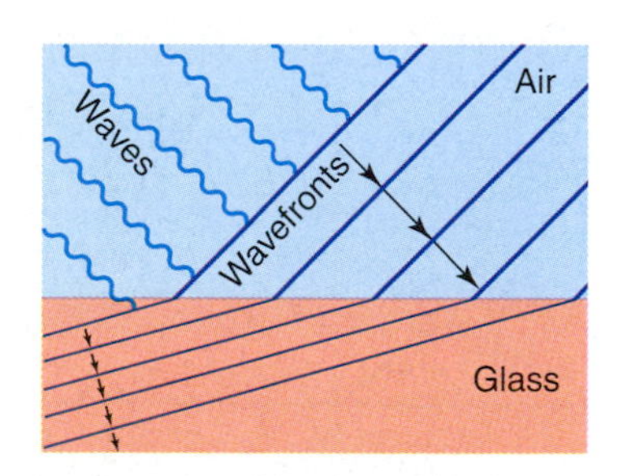


**FIGURE 4–7** Light travels at a speed in glass that is different from the speed that it has in air. This change leads to refraction.

A suitable curved piece of glass can be made so that all the light that travels through it is bent, bringing all the rays of each given wavelength to a focus. Such a curved piece of glass is called a **lens.** The lens and cornea of your eye do a similar thing to make images of objects of the world on your retina—each point of the object is focused to a point on the image. An eyeglass lens helps your eye's lens and cornea accomplish this task.

A lens uses the property of refraction to focus light. A telescope that has a lens as its major element is called a **refracting telescope.**

The distance of the image from the lens on the side away from the object being observed depends in part on the distance of the object from the front side of the lens. Astronomical objects are all so far away that we can treat them, for the purpose of forming images, as though they were infinitely far away. We say that they are "at infinity." The images of objects at infinity fall at a distance called the **focal length** behind the lens (Fig. 4–8).

A particular telescope forms an image of an area of the sky called its **field of view.** Objects that are at angles far from the center of the field of view may not be focused as well as objects in the center.

The technique of using a lens to focus on faraway objects was, we think, developed in Holland in the first decade of the 17th century. It is not clear how Galileo heard of the process, but it is known that Galileo quickly bought or ground a lens and made a simple telescope that he demonstrated in 1610 to the Senate in Venice. He put this small lens and another smaller lens at opposite ends of a tube (Fig. 4–9). The second lens, called the **eyepiece,** is used to examine and to magnify the image made by the first lens, called the **objective.** Galileo's telescope was very small, but it magnified enough to impress the nobles of Venice, who had assembled to see the new invention. The telescope had obvious commercial value, enabling the group to see boats coming to Venice before the boats were visible to the naked eye.

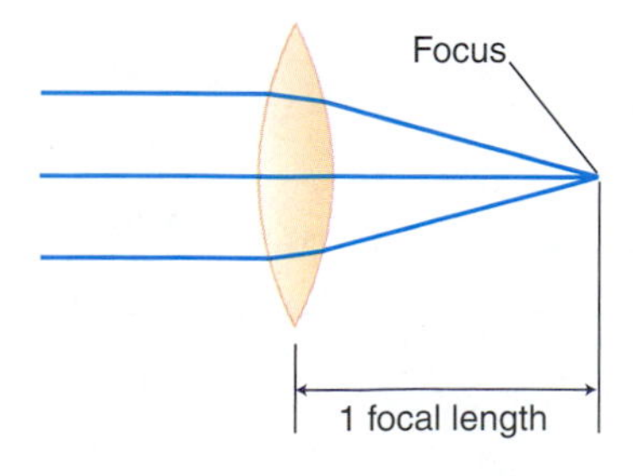


**FIGURE 4–8** The focal length is the distance behind a lens to the point at which objects at infinity are focused. The focal length of the human eye is about 2.5 cm (1 inch).

Galileo turned his simple telescope on the heavens, and what he saw revolutionized not only astronomy but also much of 17th-century thought (as we saw in Chapter 3). He discovered, for example, that the Milky Way includes many stars (Fig. 4–10), that there are mountains on the Moon, that Jupiter has satellites of its own, and that Venus has phases.

Refracting telescopes suffer from several problems. For one thing, lenses suffer from **chromatic aberration,** the effect whereby different colors are focused at different points (Fig. 4–11). The speed of light in a substance—in glass, for example—depends on the wavelength of the light. As a result, light composed of different colors bends by different amounts, and not all wavelengths can be brought to a focus at the same point. Ingenious methods of making "compound" lenses of different glasses that have slightly differing properties have succeeded in reducing this problem. But even after these methods of reducing chromatic aberration have been used, enough chromatic aberration remains to present a fundamental problem for the use of refracting telescopes. It is commonly said that the opaqueness of the glass in a very large lens, and the lens's sagging under gravity, have also limited the use of large refractors. There is some evidence, though, that larger refractors could have been made if any observatory had demanded one.

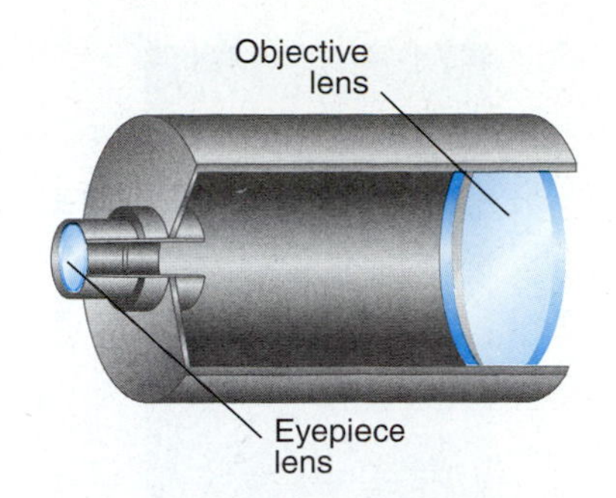


**FIGURE 4–9** A simple refracting telescope consists of an objective lens and an eyepiece. The eyepiece shown here is a more modern type than the one used by Galileo. (Galileo's was concave—narrower at the center than at the edges—and this one is convex—wider at the center.) Modern eyepieces usually contain several pieces of carefully chosen shapes and materials. The telescope shown here, with a double convex eyepiece lens, gives an inverted image (that is, objects appear upside down).

## 4.4 REFLECTING TELESCOPES

**Reflecting telescopes** are based on the principle of **reflection,** with which we are so familiar from ordinary household mirrors. A mirror reflects the light that hits it so that the light bounces off at the same angle at which it approached (Fig. 4–12). A flat mirror gives an image the same size as the object being reflected, but funhouse mirrors, which are not flat, cause images to be distorted.

We can construct a mirror in a shape so that it reflects incoming light to a focus. Let us consider a spherical mirror. If you were inside and at the center of a giant spherical mirror, in whatever direction you looked you would see your own image. Whatever is at a spherical mirror's center is imaged at the same point, its center (Fig. 4–13). If we were to use just a portion of a sphere, it would still image whatever was at the center of its curvature back at that same point.

But the stars are far away, and so would not be at the center of curvature if we used a spherical mirror. Thus we can use a mirror that takes advantage of the fact that a parabola focuses **parallel light** to a point (Fig. 4–14). We say that the light from the stars and planets is "parallel light" because the individual light rays are diverging by such an imperceptible amount by the time they reach us on Earth that they are practically parallel.

A parabola is a two-dimensional curve that has the property of focusing parallel rays to a point. In many cases, a telescope mirror is actually a **paraboloid,** which is the three-dimensional curve generated when a parabola is rotated around its axis of symmetry. Over a small area, a paraboloid differs from a sphere only very slightly, and we can ordinarily make a parabolic mirror by first making a spherical mirror and then deepening the center slightly.

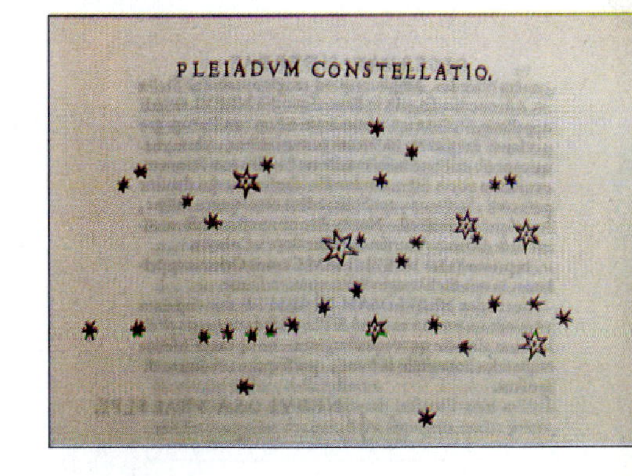


**FIGURE 4–10** Galileo, in his book of 1610, did not draw the Milky Way, but he did draw many stars in the Pleiades, in the Orion Nebula, and in the star cluster known as Praesepe.

Reflecting telescopes have several advantages over refracting telescopes. The angle at which light is reflected does not depend on the color of the light, so chromatic aberration is not a problem for reflectors. Telescope makers normally deposit a thin coat of a highly reflecting material on the front of a suitable ground and polished surface. Since light does not penetrate the surface, what is inside the telescope mirror does not matter too much. Further, a network of supports can be placed all across the back of the telescope mirror, to prevent the mirror from sagging under the force of gravity. All these advantages allow reflecting telescopes to be made much larger than refracting telescopes. (Only reflecting telescopes can be used at wavelengths shorter than those of visible light, because most ultraviolet radiation does not pass through glass.)

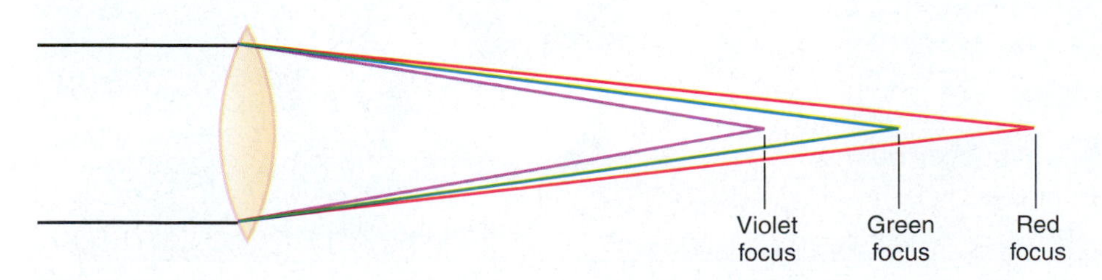


**FIGURE 4–11** The focal length of a lens is different for different wavelengths, leading to chromatic aberration.

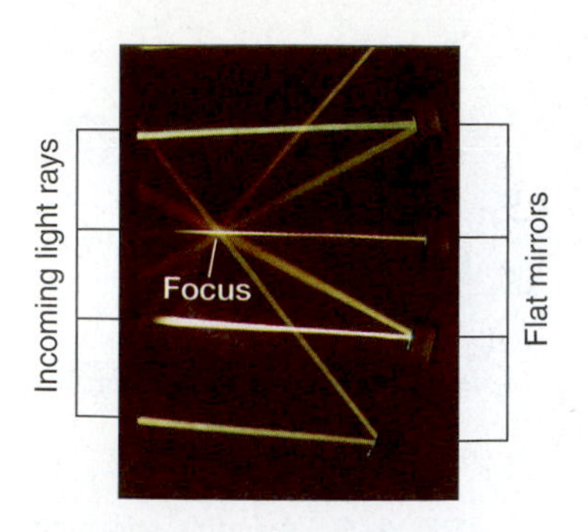


FIGURE 4–12 An actual photo to prove that a curved mirror is equivalent to a set of small flat mirrors spaced around the curve. Each individual beam of light bounces off a flat mirror at the same angle at which it hit the mirror. By arranging several flat mirrors on a curve, as shown here, incoming rays of light can be bent to a single point. A telescope mirror is made so that it is curved by the exact amount necessary to reflect all incoming parallel rays of light (Fig. 4–14) to a single point.

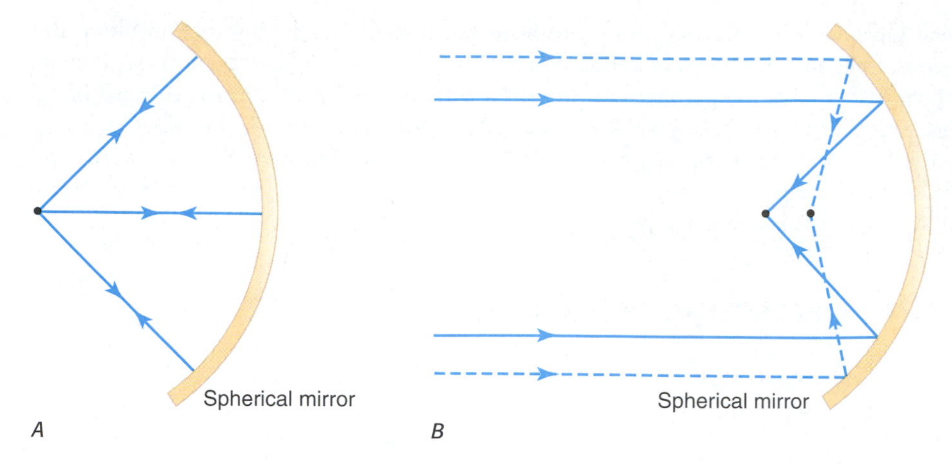


FIGURE 4–13 (*A*) A spherical mirror focuses light that originates at its center of curvature back on itself. (*B*) A spherical mirror, though, suffers from spherical aberration in that it does not perfectly focus light from very large distances, for which the incoming rays are essentially parallel.

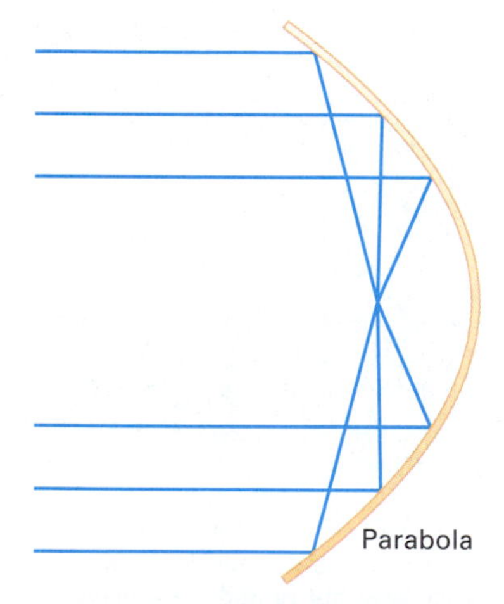


FIGURE 4–14 A parabola focuses parallel light to a point.

A potential problem with small reflecting telescopes is the need to gather the reflected light without impeding the incoming light. If you put your head at the focal point of a small telescope mirror, the back of your head would block the incoming light and you would see nothing at all!

Isaac Newton got around this problem 300 years ago by putting a small diagonal mirror a short distance in front of the focal point to reflect the light to the side of the telescope and out (Fig. 4–15). This design is a **Newtonian** telescope (Fig. 4–16) and is in widespread use. Newton actually invented the reflecting telescope because he mistakenly thought that chromatic aberration would always be too serious a problem for refracting telescopes. He was wrong—using compound instead of

FIGURE 4–15 A model of Newton's original telescope. It is housed in the Royal Society, London. Though often called Newton's original, it was probably made later.

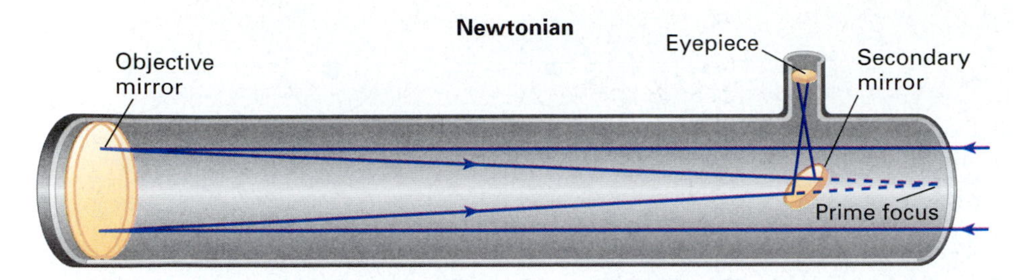


**FIGURE 4–16** A Newtonian reflector has a flat secondary mirror tilted at a 45° angle.

single lenses can adequately correct for chromatic aberration—but we have benefitted from Newton's mistake in many ways.

Two contemporaries of Newton invented alternative types of reflecting telescopes. G. Cassegrain, a French optician, and James Gregory, a Scottish astronomer, invented telescopes in which the light was reflected by the secondary mirror through a small hole in the center of the primary mirror. These designs are called **Cassegrain** and **Gregorian** telescopes (Fig. 4–17). The latest huge telescopes use a **Nasmyth focus,** in which the light comes out the side of the telescope's tube in a constant direction. The main mirror reflects the light up to a secondary mirror, which reflects it out the axis around which the telescope can rock back and forth to provide its adjustments in altitude.

## 4.5 LIGHT-GATHERING POWER AND SEEING

The principal function of large telescopes is to gather light. The amount of light collected by a telescope mirror is proportional to the area of the mirror, $\pi r^2$. [It is often convenient to work with the diameter $d$, which is twice the radius $r$, since the diameter is the number usually mentioned. Since $r = d/2$, the formula for the area of a circle can be written as $\pi(d/2)^2 = \pi(d^2/4)$.]

The ratio of the areas of two telescopes is thus the ratio of the square of their diameters $(d_1^2/d_2^2)$, which is usually more easily calculated as the square of the ratio of their diameters $(d_1/d_2)^2$. For example, each of the 10-m telescopes that are now the world's largest (Fig. 4–18) has an area four times greater than the area of the 5-meter telescope that was long the largest. Thus with exposures of equal duration, a 10-m collects four times more light than a 5-m. For the same exposure time, it therefore records fainter stars. Or for two stars of equal brightness, it can take the image four times faster.

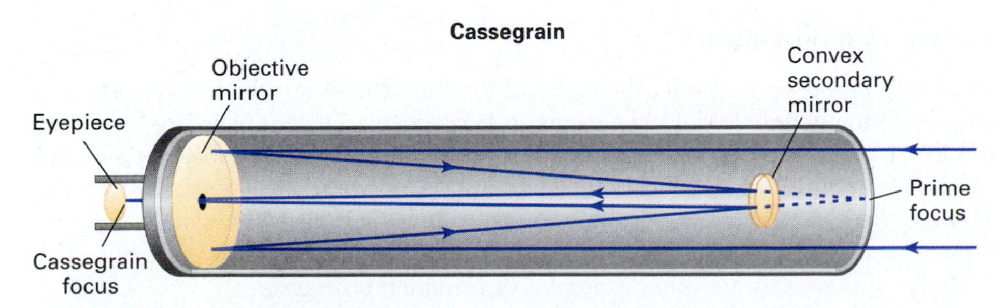


**FIGURE 4–17** A Cassegrain (Cassegrainian) telescope has a convex secondary mirror and normally has a short tube relative to its mirror diameters. Most large telescopes have a Cassegrain focus. A Gregorian telescope has a concave secondary mirror that is located beyond the focus.

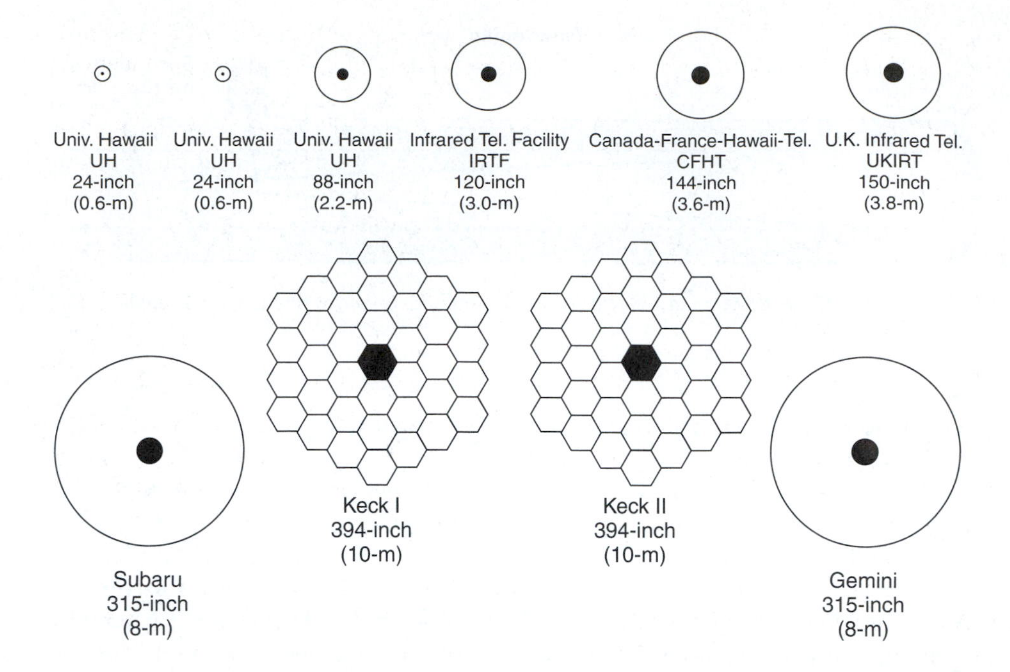


**FIGURE 4–18** Sizes of telescopes on Mauna Kea.

## 4.5a Seeing, Transparency, and Light Pollution

A telescope's light-gathering power (and its ability to concentrate light in as small an area as possible) is almost always all that is important for the study of individual stars, since the stars are so far away that no details can be discerned no matter how much magnification is used. The main limitation on the smallest image size that can be detected by ground-based telescopes is caused not in the telescope itself, but rather by turbulence in the Earth's atmosphere. The limitations are connected with the twinkling effect of stars. The steadiness of the Earth's atmosphere is called, technically, the **seeing.** We say, when the atmosphere is steady, "The seeing is good tonight." "How was the seeing?" is a polite question to ask astronomers about their most recent observing runs. Another factor of importance is the **transparency,** or how clear the sky is. It is quite possible for the transparency to be good but the seeing to be very bad, or for the transparency to be bad (a hazy sky, for example) but the seeing excellent.

A problem of modern civilization is **light pollution** (Fig. 4–19). Astronomers try to work with states, cities, towns, electric utilities, and architects, among others, to limit the amount of light shining into the sky. Sometimes, improvement is simply a matter of putting shields on streetlights, to reflect the light down at the ground where it is wanted. Often, shutting off outdoor lighting late at night or avoiding internally illuminated signs helps.

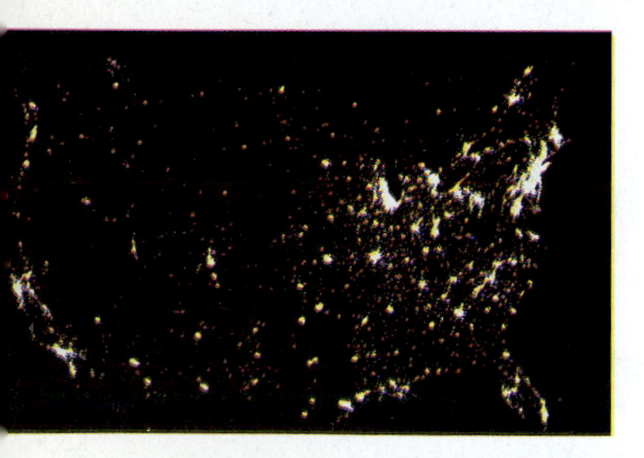

**FIGURE 4–19** The continental United States photographed from a satellite at night, showing how much light is escaping into the sky.

## 4.5b Magnification*

For some purposes, including the study of the Sun, the observation of planets, and the resolution of double stars, the magnification provided by the telescope does play a role. Magnification can be important for the study of **extended objects** as distinguished from **point objects**—like stars—almost all of whose images are points. (The Hubble Space Telescope and some arrays of telescopes can measure the apparent sizes of a few dozen stars, but all the other billions of stars remain points to us.) Nebulae and galaxies are other examples of extended objects.

**Magnification** is the number of times an object's angular size is enlarged; the angular size, which is commonly expressed in degrees, minutes, or seconds of arc, gives the angle across the sky that an object appears to cover in any one direction. Magnification works out to be equal to the focal length of the telescope objective (the primary mirror or lens) divided by the focal length of the eyepiece. By simply substituting eyepieces of different focal lengths, you can get different magnifications. However, there comes a point at which the atmosphere limits you from getting more detail, even with a shorter eyepiece.

A telescope of 1-meter (or 1000 mm) focal length will give a magnification of 25 when used with a 40-millimeter eyepiece, that is, (1 m)/(40 mm) = 25, an ordinary combination for amateur observing. With a 10-millimeter eyepiece, the same telescope has a magnification of 100. We would say that we were observing with "100 power."

The field of view of a telescope depends on the eyepiece used. The higher the magnification, the smaller the field of view. The field of view is usually indicated by degrees or minutes of arc from side to side in the sky. For example, with an eyepiece that allows the Moon to fill the image, the field of view is ½°. An eyepiece of half the focal length will have twice the magnification and a field of view of ¼°.

## 4.6 SPECTROSCOPY

FIGURE 4–20 A compact disc or laser disc contains so many lines close together that it acts like a diffraction grating and makes a spectrum.

We have discussed the collection of quantities of light by the use of large mirrors or lenses, as well as the focusing of this light to a point or suitable small area. Often we want to break up the light into a spectrum instead of merely photographing the area of sky at which the telescope is pointed.

A prism breaks up light into its spectrum for the same reasons that light focused by a lens has chromatic aberration. A wave front is refracted as it hits the side of the prism, but it is refracted by different amounts at different wavelengths.

Today, astronomical spectra are no longer usually made by means of prisms. Light may be broken up into a spectrum, without need for a prism, by lines that are ruled on a surface very close together. And I do mean close together: often 10,000 parallel lines are ruled in each centimeter (25,000 lines in a given inch)! Such a ruled surface is called a **diffraction grating,** since light is said to be "diffracted" when it bends to pass around an obstacle or through an opening. You can see the diffraction on a grating when you look at the surface of a compact disc (Fig. 4–20).

When a beam of parallel light falls on a prism or grating (Fig. 4–21), the colors that result are often indistinct because the spectrum of the light in one place in the

*A*

*B*

*C*

FIGURE 4–21 Both (*A*) prisms and (*B*) diffraction gratings make spectra. (*C*) Photography through a large prism over the telescope's objective—an "objective prism"—gives a spectrum for each star in the field of view.

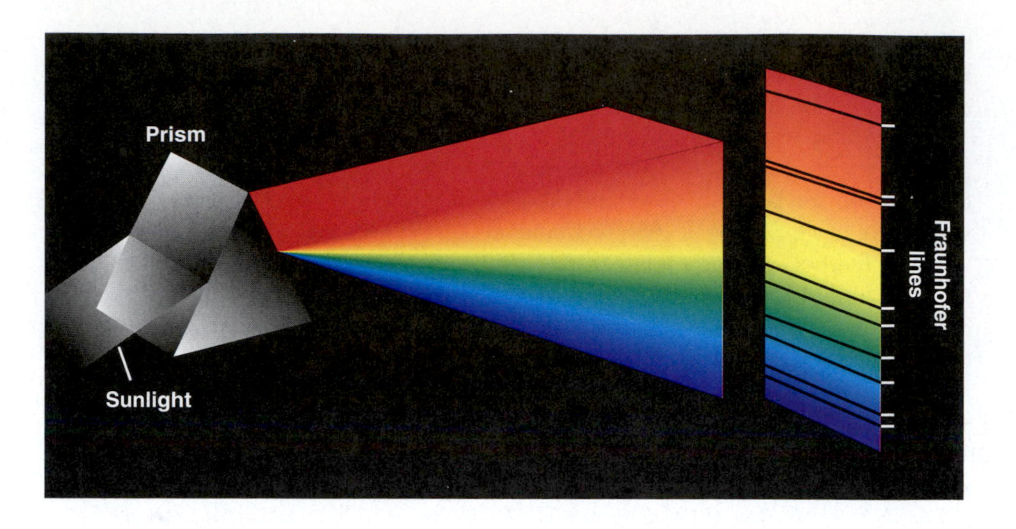


**FIGURE 4–22** When a narrow beam of sunlight (the sunlight that has passed through a slit, for example) is dispersed by a prism, we see not only a continuous spectrum but also dark Fraunhofer lines.

beam overlaps the spectrum of an adjacent spot in the beam. We can limit the blurring that results by allowing only the light that passes through a long, thin opening—called a **slit**—to fall on the prism or grating.

Scientists use optics resembling a small telescope to examine the spectrum that results. A spectrum-making device, usually including everything between the slit at one end and the viewing eyepiece at the other, is called a **spectroscope.** When the resultant spectrum is not viewed with the eye but is rather recorded either with a photographic plate or electronically, the device is called a **spectrograph.**

## 4.7 SPECTRAL LINES

Stations on the radio, places in the spectrum where there is energy, represent spectral lines. In particular, they are emission lines.

Note that "absorb" ends with a "b," but "absorption" is spelled with a "p."

As early as 1666, Isaac Newton showed that sunlight is composed of all the colors of the rainbow. William Wollaston in 1804 and Joseph Fraunhofer in 1811 also studied sunlight as it was dispersed (spread out) into its rainbow of component colors (Fig. 4–22). Wollaston was able to see that at certain colors there were a few gaps that looked like dark lines across the spectrum and Fraunhofer was the first to map many such dark lines with reasonable accuracy (Fig. 4–23). These gaps are thus called **spectral lines;** the continuous radiation in which the gaps appear is called the **continuum** (plural: **continua**). The dark lines in the spectrum of the Sun and in the spectra of stars are gaps that represent the lessening of electromagnetic radiation at those particular wavelengths. These dark lines are called **absorption lines;** they are

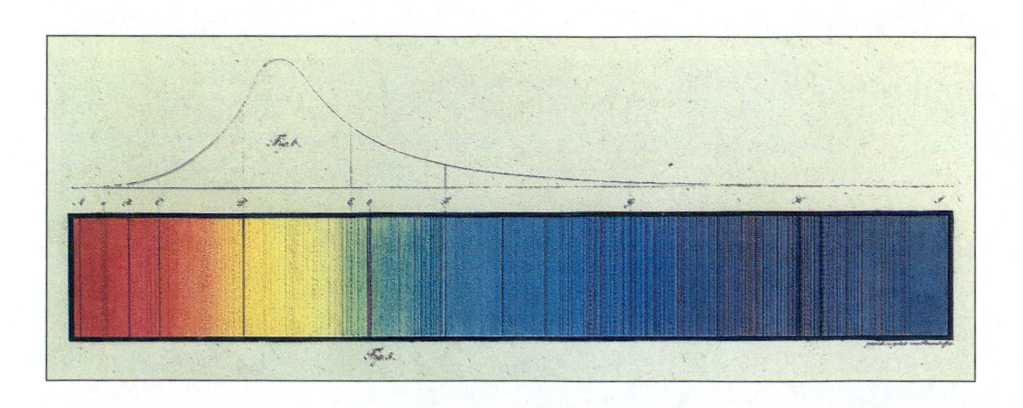

**FIGURE 4–23** Fraunhofer's original spectrum. Only the D and H lines retain their original notation from that time. We now know that the C line is from hydrogen, the D line is from sodium, and the H line is from ionized calcium.

also known (for the Sun, originally, but now also for other objects) as **Fraunhofer lines.** In contrast, an ordinary light bulb (an "incandescent" bulb) gives off only a continuum; it has no spectral lines.

It is also possible to have wavelengths at which there is somewhat more radiation than at neighboring wavelengths; lines at these wavelengths are called **emission lines.** We shall see that the nature of a spectrum, and whether we see emission or absorption lines, can provide considerable information about the nature of the body that was the source of light. We say that the lines are "in emission" or "in absorption."

The explanation of the spectral lines was discovered during the 19th century in laboratories on Earth. Patterns of spectral lines can be explained as the absorption or emission of energy at particular wavelengths by atoms of chemical elements in gaseous form. If the gaseous form of any specific element is heated, it re-emits the energy as a characteristic set of emission lines. That element, and only that element, has that specific set of spectral lines. If, on the other hand, a continuous spectrum radiated by a source of energy at a high temperature is permitted to pass through cooler gas of any specific element, a set of absorption lines appears in the continuous spectrum at the same characteristic wavelengths as those of the emission lines of that element. In this case, the gas has subtracted energy from the continuous spectrum passing through. It subtracted the energy at the set of wavelengths that is characteristic of that element. This rule was discovered by the German chemist Gustav Kirchhoff in 1859. If a continuous spectrum is directed first through the vapor of one absorbing element and then through the vapor of a second absorbing element, or through a mixture of the two gases, then the absorption spectrum that results will show the characteristic spectral absorption lines of both elements.

Though the exact German pronunciation is difficult for native speakers of English, the name Kirchhoff is normally pronounced in English with a hard "ch" (that is, like "k"): Kirk'hoff. Note the double "h."

Most solids, and gases under some conditions, can give off continuous spectra. Continuous spectra represent energy spread over a wide range of wavelengths instead of being concentrated at just a few.

The same characteristic patterns of spectral lines that we detect on Earth are observed in the spectra of stars, so we conclude that the same chemical elements are in the outermost layers of the stars. They absorb radiation from a continuous spectrum generated below them in the star, and thus cause the formation of absorption lines. These absorption lines form even though the same gas, if it could be seen from the side, would be observed giving off emission lines (Fig. 4–24). But we don't ordinarily see enough gas from angles that show emission lines to allow emission lines to be seen in the spectra of stars. *Stars show absorption spectra* and almost no stars show emission lines.

We make use of the fact that each element has its own characteristic pattern of lines when it is in a certain range of conditions of temperature and density (a measure of how closely the star's matter is packed). Thus we can study the absorption spectrum of a star (1) to identify the chemical constituents of the star's atmosphere (in other words, the types of atoms that make up the gaseous outer layers of the star), (2) to find the temperature of the surface of the star, and (3) to find the density of the radiating matter.

Not only individual atoms but also molecules—groups of linked atoms—exist in cooler bodies. (Such cooler bodies include the coolest stars, some of the gas between the stars, and the atmospheres of planets.) Molecules also have characteristic absorption—often in wide regions of spectrum (known as bands, to distinguish these broad regions from spectral lines)—and what we say for identifying elements goes for molecules as well.

When we look at a solid body, like a planet or the Moon, we see reflected sunlight, so we see the absorption lines from the Sun's spectrum. If the planet has an atmosphere, we also see absorption lines formed by gases in that atmosphere.

Where does the radiation that is absorbed go? A basic physical law called the law of conservation of energy says that the radiation cannot simply disappear. The energy of the radiation may be taken up in a collision of the absorbing atom with another atom. Alternatively, the radiation may be emitted again, sometimes at the

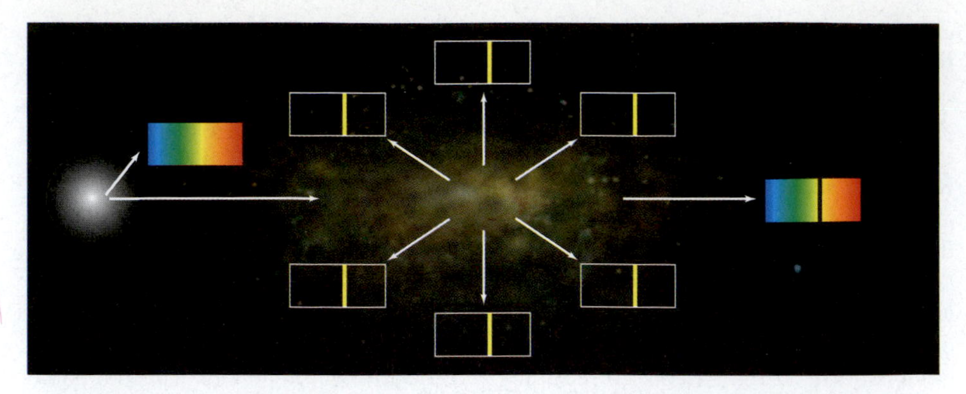

**FIGURE 4–24** The fact that stars have absorption spectra shows that the atmospheres of stars are cooler farther out. The diagram shows how a cooler gas seen in front of a hotter gas causes absorption lines. As part of the explanation, the diagram also shows how it can be that the same gas can be seen either as having emission lines or as causing absorption. Which case it is depends on the angle at which we look. When we view a hotter source that emits a continuum through a cooler gas, we may see absorption lines. But if you look at that same cooler gas from a different angle, you may see it giving off emission lines at the same wavelengths as the absorption lines.

Each of the inset boxes shows the spectrum that is the one you would measure if you were at the tip of the arrow looking back along the arrow. Note that the view from the right, in which you look past the cooler gas toward the hotter gas, shows an absorption line (which, in this simplified diagram, just appears as the dark vertical line in the middle of the yellow). Note further that from any other angle, the gas in the center appears to be giving off an emission line (which, in this simplified version, is the bright yellow vertical line). Only when you look through one source silhouetted against a hotter source do you see any absorption lines.

same wavelength, but in random directions. Thus fewer bits of energy proceed straight ahead at that particular wavelength than were originally heading in that direction. This subtraction of energy leads to the appearance of an absorption line when we look from that direction.

If one element or molecule is present in relatively great abundance, then its characteristic spectral lines will be especially strong. By observing the spectrum of a star or planet, we can tell not only which kinds of atoms or molecules are present but also their relative abundances.

The method of spectral analysis is a powerful tool that can be used to explore the universe from our vantage point on Earth. It tells us about the planets, the stars, and the other things in the Universe. It also has many uses outside of astronomy. For example, by analyzing the spectrum, we can determine the presence of impurities in an alloy deep inside a blast furnace in a steel plant on Earth. We can measure from afar the constituents of lava erupting from a volcano; indeed, colleagues and I have done so using the same spectrometer we used to study solar eclipses. Sensitive methods developed by astronomers trying to advance our knowledge of the Universe often are put to practical uses in fields unrelated to astronomy.

## 4.8 THE DOPPLER EFFECT

The **Doppler effect** is one of the most important tools that astronomers can use to understand the Universe. Without having to measure the distance to an object, they can use the Doppler effect to determine how fast it is moving toward or away from us. The method can be used whenever we have a spectral line to study.

The Doppler effect in sound is familiar to most of us, and its analog in electromagnetic radiation, including light, is very similar. It is from the Doppler effect that a train whistle or a car's motor changes in pitch as it first approaches you and then passes you and recedes. First consider a wave given off by a stationary object. It has the same pitch when heard from any direction (Fig. 4–25). As an object that emits sound waves approaches, it has moved closer to you by the time it emits a subsequent part of a wave than it had been when it started emitting the wave. Thus the wavelengths seem compressed, and the pitch is higher. After the emitting source has passed you, the wavelengths are stretched, and the pitch of the sound is lower (Fig. 4–26).

With light waves, the effect is similar. As a body emitting light, or other electromagnetic radiation, approaches you, the wavelengths become slightly shorter than they would be if the body were at rest. Visible radiation is thus shifted slightly in the direction of the blue. (It doesn't actually have to become blue, only be shifted in that direction.) We say that the radiation is **blueshifted.** Conversely, when the emitting object is receding, the radiation is said to be **redshifted.** We generalize these terms to types of radiation other than light and say that radiation is blueshifted whenever it changes to shorter wavelengths and redshifted whenever it changes to longer wavelengths.

The point of all this is that we can measure a Doppler shift for any object we see for which we can detect at least one spectral line. We can then compare the line's position with that from some unmoving object; the amount it is shifted to the left or right tells how fast the object is moving toward or away from us. In Chapter 25, in the context of stars, we will return to the Doppler effect. There we will discuss a simple equation that allows us to compute the speed of an object toward or away from us, given our measurement of the amount of redshifting or blueshifting.

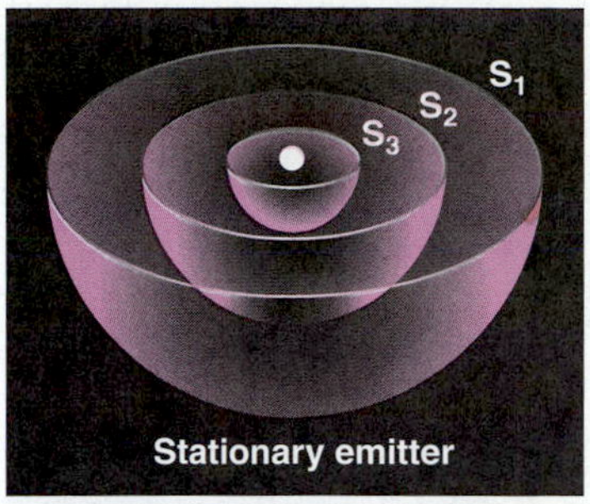

Stationary emitter

**FIGURE 4–25** An object emits waves of radiation that can be represented by spheres at the peaks of the wave, each centered on the object and expanding.

"Doppler radar" uses the Doppler effect to analyze radio signals bounced off rainstorms or other weather systems. The Doppler shifts detected reveal the systems' motions.

## 4.9 OBSERVING AT SHORT WAVELENGTHS

From antiquity until 1930, observations in a tiny fraction of the electromagnetic spectrum, the visible part, were the only way that observational astronomers could study the Universe. Most of the beautiful images that we have in our minds of astronomical objects are based on optical studies, since most of us depend on our eyes to discover what is around us. But it is now over 100 years since Wilhelm Roentgen discovered x-rays, and x-ray astronomy is now a standard part of the astronomer's arsenal.

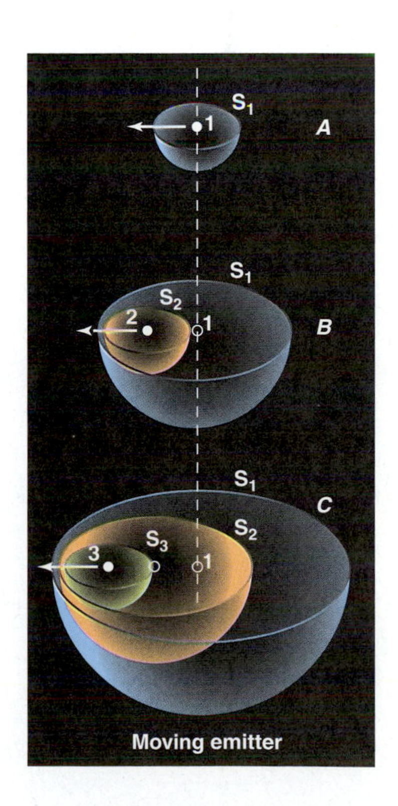


**FIGURE 4–26** In the drawing, the emitter is moving in the direction of the arrow. In part *A,* we see that the peak of the wave emitted when the emitter was at point 1 becomes a sphere (labelled $S_1$) around point 1, though the emitter has since moved toward the left. In part *B*, some time later, sphere S has continued to grow around point 1 (even though the emitter is no longer there), and we also see sphere $S_2$, which shows the position of the peaks emitted when the emitter had moved to point 2. Sphere $S_2$ is thus centered on point 2, even though the emitter has continued to move on. In part *C*, still later, yet a third peak of the wave, $S_3$, has been emitted, this time centered on point 3, while spheres $S_1$ and $S_2$ have continued to expand.

Observers who are being approached by the emitting source (those on the left side of the emitter, as shown on the left side of part *C*) see the three peaks drawn coming past them relatively bunched together (that is, at shorter intervals of time). The wavelength is shorter, since we measure the wavelength from one peak to the next. Thus they have a higher frequency, since more of the shorter waves pass the observer in a given interval of time.

This shorter wavelength is seen as a color farther toward the blue than the original, a **blueshift.** Observers from whom the emitter is receding (those on the right side of the emitter) see the three peaks coming past them with decreased frequency (at increased intervals of time), as though the wavelength were longer. This corresponds to a color farther toward the red, a **redshift.**

Once a light wave is emitted, it travels at a constant speed ("the speed of light," $3 \times 10^{10}$ cm/sec, equivalent to seven times around the earth in a second), so a shorter (lower) wavelength corresponds to a higher frequency, and a longer (higher) wavelength corresponds to a lower frequency.

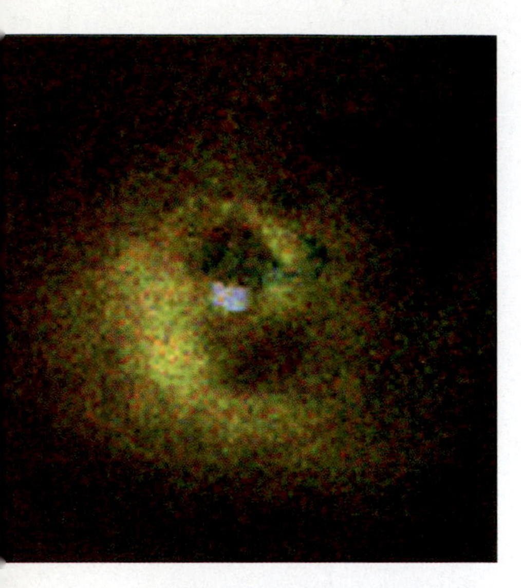

**FIGURE 4–27** A Chandra X-ray Observatory image of the Perseus cluster of galaxies, showing the hot gas present.

Thus far, we have talked of the spectrum in terms of the wavelengths of the radiation, which implies that radiation is a wave. But when light is absorbed, the energy is always transferred in discrete amounts that depend on the wavelength of the light. The radiation acts as though it were composed of particles rather than continuous waves. So it is often useful to think of light as particles of energy called **photons** instead of as waves. Photons that correspond to light of different wavelengths have different energies. We will come back to this idea in Section 24.3. But you will find it useful to have a picture in your mind of photons of high energy, and thus high penetrating power, corresponding to relatively short wavelengths (like x-rays). Similarly, photons corresponding to long wavelengths (radio waves or infrared) don't contain much energy, often not even enough to affect a photographic plate.

The wave theory and the particle theory are but different ways of understanding light. Each provides a "model"—a framework for thinking. In actuality, both the wave theory and the particle theory accurately explain how light acts at different times. The fact that we make scientific models to help us comprehend reality does not change the fact that reality may be more complicated.

Ultraviolet and x-ray photons have shorter wavelengths and greater energies than photons of visible light. Thus they too have enough energy to interact with the silver grains on film. Photographic methods can be used, therefore, throughout the x-ray, ultraviolet, and visible parts of the spectrum. Electronic devices like the ones that work in the visible work in the x-ray and ultraviolet, too (Fig. 4–27).

At this shorter end of the spectrum, gamma rays, x-rays, and ultraviolet light do not come through the Earth's atmosphere, as was illustrated in Figure 4-4. Ozone, a molecule of three atoms of oxygen ($O_3$), is located in a broad layer between about 20 and 40 km in altitude. The ozone prevents all the radiation at wavelengths less than approximately 3000 Å from penetrating. To get above the ozone layer, astronomers have used telescopes and other detectors that were launched in rockets that were aloft for brief times and in orbiting satellites.

## 4.10 OBSERVING AT LONG WAVELENGTHS

Infrared and radio photons have longer wavelengths and thus lower energies than visible photons. They do not have enough energy to interact with ordinary photographic plates. Some special films can be used at the shortest infrared wavelengths, but for the most part astronomers have to employ methods of detection involving electronic devices in these regions of the spectrum. As we saw in Figure 4-4, some of the infrared radiation reaches Earth, especially at some of the mountaintop observatories, but most of the infrared radiation is blocked. The Figure also showed that the whole radio spectrum comes through a window of atmospheric transparency.

### 4.10a Infrared Astronomy

Infrared wavelengths range from about 10,000 Å = 1 micrometer ($\mu$m) to about 1000 micrometers = 1 millimeter. Astronomers still mostly use the old name "micron" ($\mu$) for the SI unit "micrometer."

Another major limitation in the infrared, besides the insensitivity of film, is that the Earth's atmosphere has few windows of transparency. Most of the atmospheric absorption of infrared wavelengths is caused by water vapor, which is located at relatively low levels in our atmosphere compared to the ozone that causes the absorption in the ultraviolet. Thus we do not have to go as high to observe infrared as we do to observe ultraviolet. It is sometimes sufficient to send up instruments attached to huge balloons (Fig. 4–28), though spacecraft are also important.

**FIGURE 4–28** A balloon carries aloft the BOOMERanG telescope used to study the cosmic microwave background (Chapter 37) on the borderline of wavelength between infrared and radio.

## 4.10b Radio Astronomy

The radio region of the spectrum, at even longer wavelengths, is a major window of transparency. In the following chapter, we will discuss some of the world's best radio telescopes. We discuss radio astronomy results later in this text, especially when we discuss galaxies and cosmology.

We cannot always pinpoint the discovery of a whole field of research as decisively as that of radio astronomy. In 1931, Karl Jansky of what was then the Bell Telephone Laboratories in New Jersey was experimenting with a radio antenna to track down all the sources of noise that might limit the performance of short-wave radiotelephone systems (Fig. 4–29). After a time, he noticed that a certain static

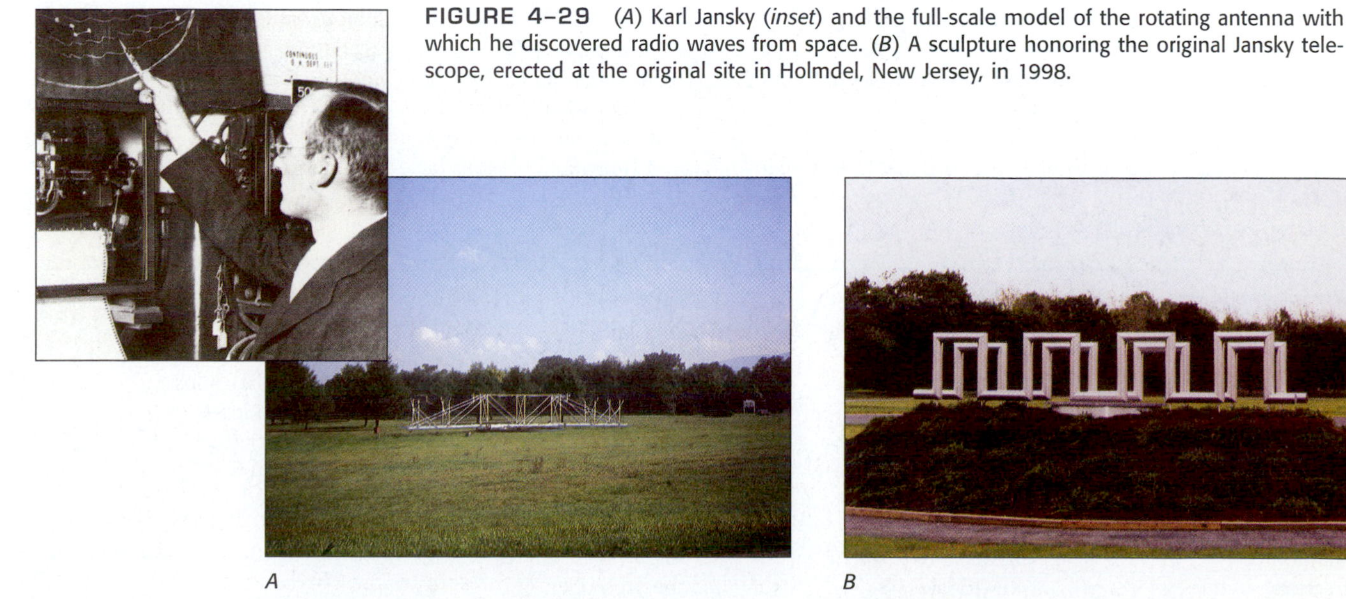

**FIGURE 4–29** (*A*) Karl Jansky (*inset*) and the full-scale model of the rotating antenna with which he discovered radio waves from space. (*B*) A sculpture honoring the original Jansky telescope, erected at the original site in Holmdel, New Jersey, in 1998.

*A* *B*

appeared at approximately the same time every day. Then he made the key observation: the static was actually appearing four minutes earlier each day. This was the link between the static that he was observing and the rest of the Universe. Jansky's static kept sidereal time, just as the stars do (see Section 6.3), and thus was coming from outside our solar system! Jansky was actually receiving radiation from the center of our galaxy.

We hope that radio astronomy from the ground survives the launch of multiple satellites for worldwide telephones and the widespread radio noise that results. Maybe we will have to do our radio astronomy one day from the far side of the Moon.

## MISCONCEPTIONS

| ✖ **Incorrect** | ✔ **Correct** |
| --- | --- |
| All radiation is harmful. | Radiation includes ordinary light. |
| Spelling: infared | Spelling: infrared |
| Spelling: Kirchoff | Spelling: Kirchhoff |
| Spelling: absorbtion | Spelling: absorption |
| Spelling: lense | Spelling: lens |
| Telescopes are mostly used to magnify things. | Telescopes are mostly used to collect light, so we can view fainter things. |

## SUMMARY AND OUTLINE

The spectrum (Section 4.1)
- From short wavelengths to long wavelengths, the various types of radiation are known as gamma rays, x-rays, ultraviolet, visible, infrared, and radio.
- Parts of the visible spectrum are ROY G BIV.
- All radiation travels at the "speed of light"; in a vacuum this is $3 \times 10^8$ m/sec.
- Windows of transparency of the Earth's atmosphere exist for visible light and radio waves (plus some parts of the infrared).
- Continuous radiation is called the continuum.

What a telescope is (Section 4.2)
- Resolution depends on the diameter of the telescope.

Lenses and telescopes (Section 4.3)
- Refracting telescopes use lenses to focus light.

Reflecting telescopes (Section 4.4)
- Reflecting telescopes use parabolic mirrors to focus parallel light.

Light-gathering power and seeing (Section 4.5)
- Light-gathering power is directly proportional to diameter squared.
- "Seeing" is the steadiness of the image.

Spectroscopy and spectral lines (Sections 4.6 and 4.7)
- Absorption lines (Fraunhofer lines) are gaps in the continuum shown as narrow wavelength regions where the intensity is diminished from that of the neighboring continuum.
- Emission lines are narrow wavelength regions where the intensity is relatively greater than neighboring wavelengths (which are either continuum or zero in intensity).
- A given element has a characteristic set of spectral lines under specific conditions of temperature and pressure.

The Doppler effect (Section 4.8)
- In sound or light, we measure motion toward or away from us.
- In light, blueshifts are motion toward us and redshifts are motion away.

Ultraviolet and x-ray astronomy (Section 4.9)
- Individual photons have more energy than photons of light.

Infrared and radio astronomy (Section 4.10)
- Individual photons have less energy than photons of light.

## KEY WORDS

electromagnetic radiation
speed of light
vacuum
wavelength
angstrom
ultraviolet
infrared
electromagnetic spectrum (spectra)
resolution
refraction
lens
refracting telescope
focal length
field of view
eyepiece
objective
chromatic aberration
reflecting telescope
reflection
parallel light
paraboloid
Newtonian
Cassegrain
Gregorian
Nasmyth focus
seeing
transparency
light pollution
extended objects
point objects
magnification
diffraction grating
slit
spectroscope
spectrograph
spectral lines
continuum (continua)
absorption lines
Fraunhofer lines
emission lines
Doppler effect
blueshifted
redshifted
photons

## QUESTIONS

1. What advantage does a reflecting telescope have over a refracting telescope?
2. What limits the resolving power of the 5-meter telescope?
3. Why might some stars appear double in blue light though they could not be resolved in red light?
†4. Use Dawes's limit to calculate how far apart two double stars have to be in order to be seen as separate with a 20-cm telescope in yellow light.
5. For each of the following, identify whether it is a characteristic of a reflecting telescope, a refracting telescope, both, or neither. Give any limitations on the applicability of your answer.
   (a) Is free of chromatic aberration
   (b) Has more severe spherical aberration
   (c) Can be used for photography
   (d) Has an objective supported only by its rim
   (e) Can be made in larger sizes
6. Do stars have, in general, absorption lines or emission lines? Explain how such lines are formed.
7. Use a diagram to explain whether we can detect a redshift or blueshift of a star moving from side to side across the sky.
8. Why can't we use ordinary photographic plates to record infrared images?
9. Why must we observe ultraviolet radiation using rockets or satellites, whereas balloons are sufficient for infrared observations?
10. Why can radio astronomers observe during the day, whereas optical astronomers are (for the most part) limited to nighttime observing?

† This question requires a numerical solution.

## INVESTIGATIONS

1. Use eyeglass lenses or other lenses with a cardboard tube to make a refracting telescope. Try it out on the Moon.
2. Get diffraction grating and use it to study the spectra of incandescent bulbs, fluorescent lights, streetlights, and the Moon.

## USING TECHNOLOGY

### W World Wide Web

1. Look up telescopes that operate in different parts of the spectrum and their observatories or spacecraft.
2. Find out the latest discoveries from the Hubble Space Telescope.

### REDSHIFT

1. Zoom in on the double star Alcor and Mizar in the Big Dipper, and see how far they are separated from each other. Examine which of them is double. Compare with the 1 arc second separation that is typically the minimum from ground-based telescopes.

The Chandra X-ray Observatory.

# Chapter 5

# Observatories and Space Missions

If you forced a group of astronomers to choose one and only one instrument to be marooned with on a desert island, most would probably choose . . . a computer. The notion that astronomers spend most of their time at telescopes is far from the case in modern times. Still, telescopes have been and continue to be very important to the development of astronomy, and in this chapter we will discuss how we study the stars with telescopes and the observatories where they are located.

**AIMS:** To understand the instruments used in astronomy and to learn where they are located

## 5.1 OPTICAL OBSERVATORIES

Once upon a time, over a hundred years ago, telescopes were put up wherever astronomers happened to be located. Now all major observatories are built at remote sites chosen for the quality of their observing conditions. Let us discuss some of the most prominent sites. In the first subsection we discuss telescopes in use through the mid-1990s. In the following subsection we discuss the new generation of even larger telescopes.

### 5.1a Observatory Sites

The largest refracting telescope in the world has a lens 1 meter (40 inches) across. It is at the Yerkes Observatory in Williams Bay, Wisconsin, and went into use during the 1890s.

On Mount Wilson, near Los Angeles, George Ellery Hale built a 2.5-m telescope, with a mirror made of plate glass. It began operation in 1917 and was the largest telescope in the world for 30 years. Though it was shut down in 1985, it has since been reopened.

One of the limitations of any telescope, reflecting or refracting, is the length of time that the mirror or lens takes to reach its equilibrium shape when exposed to the temperature of the cold night air. In the 1930s, the Corning Glass Works invented Pyrex, a type of glass that is less sensitive to temperature variations than ordinary glass. Workers there cast, with great difficulty, a mirror blank (the disk from which the reflecting surface is shaped) a full 5.08 meters (200 inches) across (Fig. 5–1).

However, the construction of the 200-inch telescope was held up by many factors, including World War II. Only in 1949 did active observing begin with the telescope. It stands on Palomar Mountain in southern California, and, for many years, it was the largest in the world. It is named the Hale telescope, after George Ellery Hale.

**FIGURE 5–1** The Palomar 5-m mirror blank was very difficult to cast. Here we see the unsuccessful first try, which is on display at the Corning Glass Works in Corning, New York. The mirror's back is ribbed to reduce its weight.

**FIGURE 5–2** The 3.5-m Wisconsin–Indiana–Yale–National Optical Astronomy Observatories (WIYN) telescope on Kitt Peak. Its efficient design cut costs, and improvements in air flow from new discoveries about telescope domes and sites allow the telescope to produce images of high resolution.

After many years of being run entirely by Caltech, half the operation of this 5-m telescope is now shared by Cornell University and by the Jet Propulsion Laboratory.

The National Optical Astronomy Observatories (NOAO) include the Kitt Peak National Observatory, on a sacred mountain of the Tohono O'odham, about 80 km (50 miles) west of Tucson, Arizona (Fig. 5–2). Also part of NOAO is the National Solar Observatory, with telescopes on Kitt Peak and on Sacramento Peak, New Mexico. The new Apache Point Observatory of a consortium of universities has its 3.5-m telescope near the latter.

Part of the sky is never visible from sites in the northern hemisphere, and many of the most interesting astronomical objects in the sky are only visible or best visible from the southern hemisphere. Therefore, in the last decades, special emphasis has been placed on constructing telescopes at southern sites. Some of these sites are on coastal mountains west of the Andes range in Chile. The Cerro Tololo Inter-American Observatory there has long had a 4-m (158-inch) twin to the National Optical Astronomy Observatory's 4-m telescope on Kitt Peak.

The Observatories of the Carnegie Institution of Washington boast the Las Campanas Observatory, which includes a 2.5-m (101-inch) telescope on Cerro las Campanas, another Chilean peak. The European Southern Observatory, an intergovernmental group from several European nations, has a dozen telescopes on La Silla, still another Chilean mountaintop. It is quite common in the astronomical community to go off to Chile for a while "to observe."

One of the most important observatory sites is in Hawaii on Mauna Kea on the island of Hawaii. Many large telescopes are there, including national telescopes of the United Kingdom, Canada, France, and Japan in addition to those of American organizations (Fig. 5–3). This site is at such a high altitude, 4200 m (13,800 ft), that the air above is particularly dry. Since water vapor, which blocks infrared radiation, is minimal, these telescopes are especially suitable for observations at infrared and even longer wavelengths known as "submillimeter." Also, the sky is very dark, and

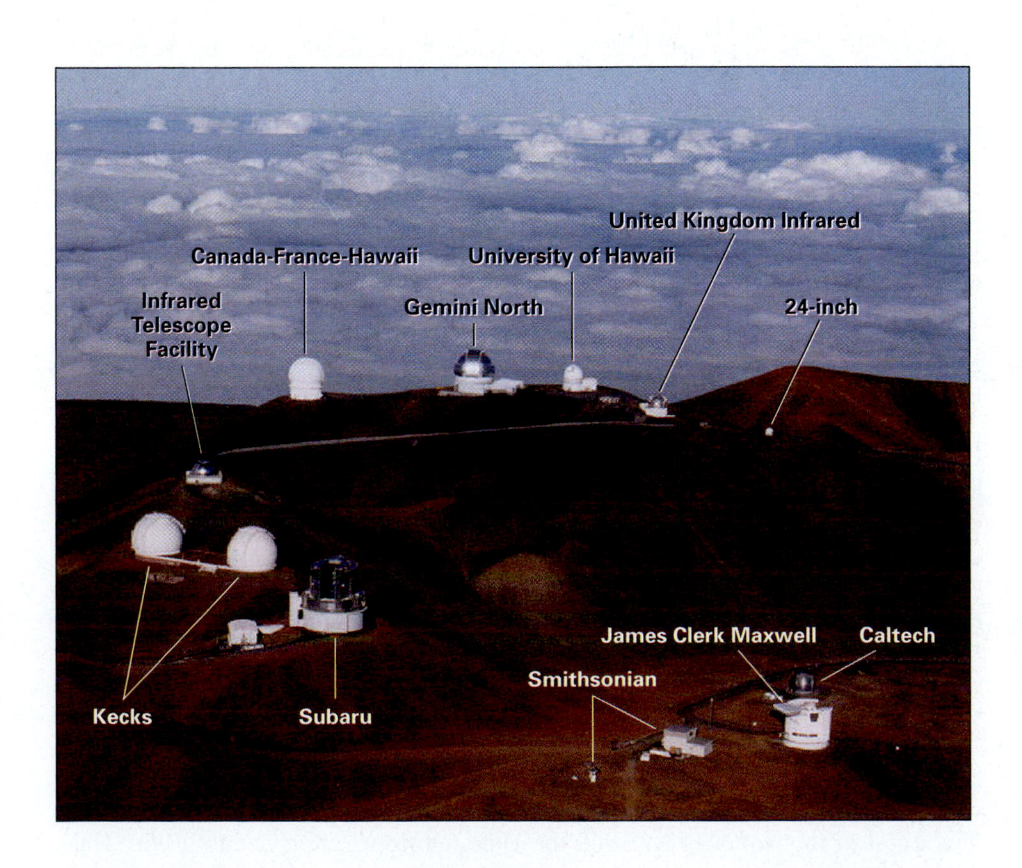


**FIGURE 5–3** Mauna Kea, with its many huge telescopes. The Keck twin domes with the Subaru dome near them are at mid-left; the NASA Infrared Telescope Facility is above them; and the Canada-France-Hawaii, Gemini North, University of Hawaii 88-inch, United Kingdom Infrared, and University of Hawaii 24-inch telescopes are on a ridge at top. In the "millimeter valley" at lower right, we see the Caltech 10-m submillimeter and James Clerk Maxwell 15-m telescopes. (Maxwell, perhaps the third greatest physicist of all time—after Newton and Einstein—unified electricity and magnetism theoretically.) We also see the first to be installed of eight 6-m telescopes and the assembly building of the Smithsonian Astrophysical Observatory's submillimeter array.

the atmosphere is often exceptionally steady. Thus Mauna Kea now includes huge telescopes, including the 10-m W. M. Keck Telescopes (Fig. 5–4).

"Adaptive optics" is a new technology in which the shape of a mirror can be modified from moment to moment to cancel the blurring effects of the atmosphere (Fig. 5–5). It is increasingly used on many of the largest telescopes to provide the highest resolution images.

**FIGURE 5–4** The domes of the 10-m W. M. Keck Telescopes (*center and right*). The mirror of each is made of 36 contiguous hexagonal segments, which are continually adjustable. The Subaru Telescope is adjacent (*left*).

## 5.1b The New Generation of Optical Telescopes

Telescopes in space solve some scientific problems, but they also suggest many new kinds of observations that can best or most efficiently be made from the ground. The need for more large ground-based telescopes continues.

Several giant telescopes are employing new technologies to provide large light-gathering power at relatively low cost compared to older standard designs (Table 5–1). The first two are the twin Keck Telescopes on Mauna Kea in Hawaii, a joint project of the California Institute of Technology and the University of California. The mirror of each is 10 meters across, twice the diameter and four times the surface area of Palomar's. Since we cannot construct a single 10-m mirror, U.C. scientists worked out a plan to make the mirror of many smaller segments.

Several other giant telescopes rely on a breakthrough by Roger Angel of the University of Arizona in making relatively inexpensive, lightweight mirrors. The Astrophysical Research Consortium (ARC) of Chicago–New Mexico State–Princeton–Washington–Washington State–Johns Hopkins Universities and the Wisconsin–Indiana–Yale–NOAO (WIYN) group (the latter was pictured above) is each using one of his mirrors in a 3.5-m size. Elsewhere in Arizona, a single Angel mirror 6.5 m in diameter has replaced the several mirrors of the Multiple Mirror Telescope. A Johns Hopkins University–Carnegie Institution–University of Arizona–Harvard University group plans a pair of 6.5-m telescopes called Magellan in Chile. The University of Arizona; the Arcetri Observatory of Florence, Italy; Ohio State University; a consortium of German institutions; and the Research Corporation of Tucson, Arizona, are constructing a pair of Angel 8-m mirrors on a single mounting in a dome on Mt. Graham, Arizona, to make the Large Binocular Telescope for 2004.

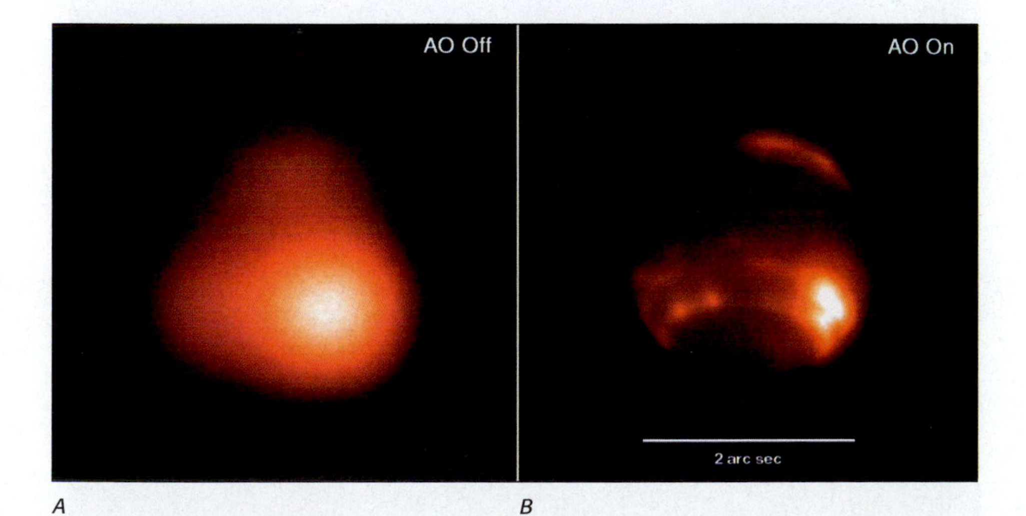


**FIGURE 5–5** Adaptive optics. (*A*) Rapidly changing the shape of a mirror can compensate for changes in the shape of incoming light caused by the Earth's atmosphere. (*B*) The double star ξ (xi) Ursae Majoris with the adaptive-optics (AO) system turned off (*left*) and on (*right*). Obviously, the adaptive-optics system improves the image dramatically.

## TABLE 5.1 The World's Largest Optical Telescopes

| Name | Description | Owner | Site |
|---|---|---|---|
| **Fully Steerable** | | | |
| Very Large Telescope | four 8-m telescopes + outliers | European Southern Obs. | Chile |
| Keck Telescopes | two 10-m telescopes, outliers planned | Caltech/U. Cal | Hawaii |
| Large Binocular Tel | two 8-m telescopes | U. Arizona and others | Arizona |
| Gran Telescopio Canarias | 10-m telescope | Spain (2002) | Canary Islands |
| Subaru Telescope | 8-m telescope | Japanese Natl. Obs. | Hawaii |
| Gemini North | 8-m telescope | U.S./international | Hawaii |
| Gemini South | 8-m telescope | U.S./international | Chile |
| Magellan | two 6.5-m telescopes | Carnegie, Harvard, U. Mich., U. Az., MIT | Chile |
| MMT | 6-m telescope | Smithsonian Astrophys. Obs. | Arizona |
| **Limited-Pointing Mounting** | | | |
| Hobby–Eberly Tel | 9-m (11-m × 10-m) | Penn State/U. Texas + | Texas |
| South African Large Tel | 9-m (11-m × 10-m) | South Africa + | South Africa |

The European Southern Observatory has built the Very Large Telescope (VLT), a set of four 8-m telescopes close together in Chile (Fig. 5–6). They are usable separately, and eventually, when techniques are further developed, will be used together. The mirrors are very thin, and thus their shape must be continuously monitored and controlled. Three 1.8-m auxiliary telescopes are planned.

The individual telescopes came into use one at a time, until all were available in 2000. They were named in the Mapuche language of indigenous people of a re-

**FIGURE 5–6** The European Southern Observatory's Very Large Telescope in Chile. The individual Unit Telescopes have been named in the Mapuche language of indigenous people from southern Chile. The names are Antu (the Sun), Kueyen (The Moon), Melipal (the Southern Cross), and Yepun (Venus, as the evening star).

**FIGURE 5–7** The Gemini North telescope on Mauna Kea. Its building and dome are wide open to the outside air, to equalize temperature quickly in order to limit an outflow of turbulent air.

gion of Chile. With their translations, they are: Antu, the Sun; Kueyen, the Moon; Melipal, The Southern Cross; and Yepun, Venus (the evening star).

A Japanese 8.2-m telescope, also with a "thin meniscus mirror," has been constructed on Mauna Kea. It is called "Subaru," the Japanese word for the Pleiades.

Other huge telescopes are under way. The U.S. National Optical Astronomy Observatories has managed the U.S. end of the Gemini Project of two 8-m mirrors, one in the northern hemisphere, opened in 2000, and another in the southern hemisphere (Fig. 5–7). The project is 50 percent American; the responsibility for the remainder is divided among Britain, Canada, Chile, Argentina, and Brazil.

The Hobby–Eberly Telescope, a huge telescope with a segmented mirror, was designed for spectroscopy. The areas of the segments cover 11-m × 10-m, but a huge focus set of instruments makes it equivalent to a still-huge 9-m telescope. Its innovative design saves money by limiting its ability to point at random positions in the sky. It is fixed to point only at an altitude of 55°, and swivels in azimuth. It is in Texas, operated by a collaboration of Pennsylvania State University, the University of Texas, Stanford University, and the German universities of Munich and Göttingen. A similar telescope, the Southern African Large Telescope (SALT) is being built in South Africa, funded by that country and universities in the United States (Rutgers, Wisconsin, and Carnegie Mellon), Poland, and New Zealand. Spain is building a 10.4-m telescope on La Palma in the Canary Islands, already a major observatory site.

With segmented mirrors, like the Keck and Hobby–Eberly Telescopes, there is no inherent limit to how large a telescope can be. The California Institute of Technology and the University of California are considering building the California Extremely Large Telescope (CELT), 30-m across. It would be built on Mauna Kea or in Chile in the next ten to fifteen years. A U.S. National Facility also 30-m across, the Giant Segmented Mirror Telescope (GSMT), is also under consideration, to be paid for by the National Science Foundation. The European Southern Observatory is planning the Overwhelmingly Large Telescope (OWL), 100-m across, for even further in the future. The present plans call for it to be made of 2,000 individual segments.

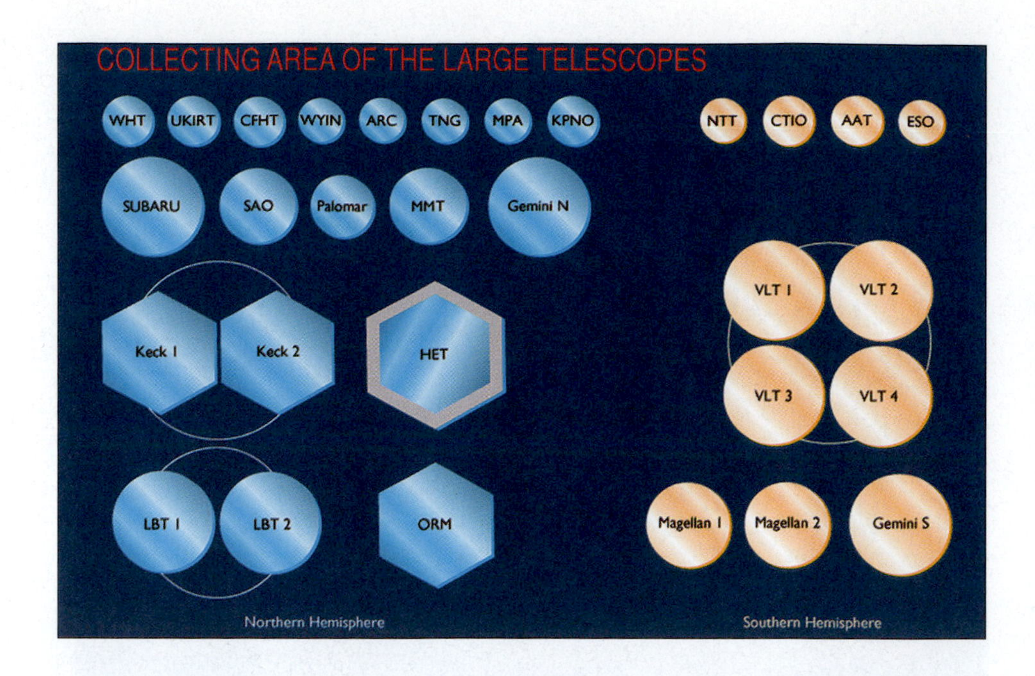


**FIGURE 5–8** The relative sizes of the largest telescopes in the world, most of them very new.

## 5.2 WIDE-FIELD TELESCOPES

### 5.2a Schmidt Telescopes

Bernhard Schmidt, working in Germany in about 1930, invented a type of telescope that combines some of the best features of both refractors and reflectors. The result allows photography of a much wider field of view than do other telescopes. In such a **Schmidt telescope,** the main optical element is a large mirror, but it is spherical instead of being paraboloidal. Before reaching the mirror, the light passes through a thin lens, called a **correcting plate,** that distorts the incoming light in just the way that is necessary to have the spherical mirror focus it over a wide field (Fig. 5–9). The correcting plate is too thin to introduce much chromatic aberration over the visible part of the spectrum.

One of the largest Schmidt cameras in the world is on Palomar Mountain, in a dome near that of the 5-m telescope. Renamed the Oschin Schmidt camera, it has a correcting plate 1.2 meters (49 inches) across (though it is traditionally known as the "48-inch Schmidt") and a spherical mirror 1.8 meters (72 inches) in diameter. Its field of view is about 7° across, which is much broader than the 5-m reflector's field of view (only about 2 minutes of arc across).

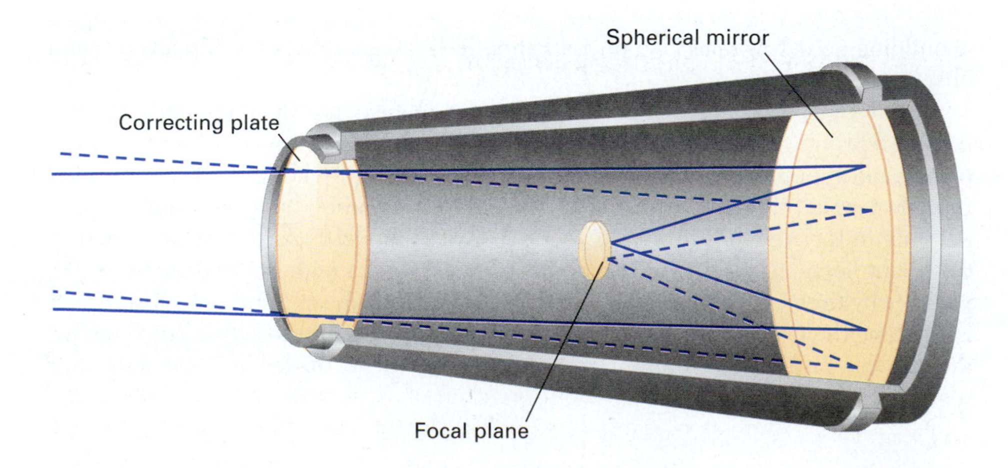


**FIGURE 5–9** By having a non-spherical thin lens called a correcting plate, a Schmidt camera is able to focus a wide angle of sky onto a curved piece of film located at the focal plane.

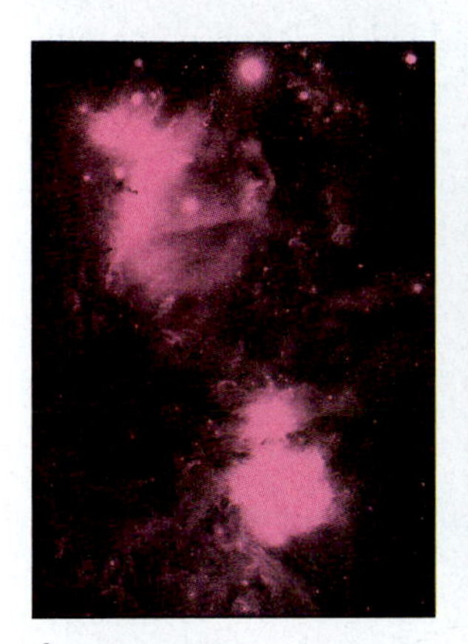

**FIGURE 5–10** (*A*) Three individual images, each taken with the U.K. Schmidt telescope through a filter in one color, are printed together to make the full-color image. (*B*) This color view of the Horsehead Nebula in Orion is from the top left quarter of the three monochromatic images.

*A* *B*

The field of view of the 5-m telescope is so small that it would be hopeless to try to map the whole sky with it because it would take too long. But the 1.2-meter Schmidt has been used to map the entire sky that is visible from Palomar. The survey was carried out through both red and blue filters; 900 pairs of plates were taken over seven years, from 1949 to 1956. Plates from this Palomar Observatory–National Geographic Society Sky Survey are widely used.

Newer Schmidt cameras—the United Kingdom Schmidt Telescope (Fig. 5–10), at the Anglo-Australian Observatory in Siding Spring, Australia, and another at the European Southern Observatory—have photographed the southern sky in the blue, red, and near infrared including the one-quarter of the sky that cannot be seen from Palomar.

A color-independent correcting plate was installed on the Palomar 1.2-m Schmidt. A second-generation sky survey was then made, with blue, red, and infrared ("near infrared," about 8300 Å) being exposed of the whole sky north of the celestial equator. The red and blue photographs are enabling astronomers to compare the skies of the 1980s and 1990s with those of 30 years earlier. The new survey is revealing objects four times fainter than those in the original survey, primarily because of Kodak's progress in providing more sensitive and finer-grain emulsions, as well as providing an infrared view. It takes 894 triplets of photographic plates to cover the northern sky.

The Space Telescope Science Institute, the University of Minnesota, and others have electronically scanned not only both Palomar Observatory Sky Surveys but also special partial surveys. Many of these scans are available on-line, and the original Palomar Sky Survey is available at lower resolution on a set of CD-ROMs.

### 5.2b The Forthcoming Large Survey Telescope

One of the major initiatives in the 2000 Astronomy and Astrophysics Survey's report is the Large-aperture Synoptic Survey Telescope (LSST). (Synoptic, to an astronomer, means following something as it changes over time.) It is to be much larger than any previous large-field telescope, with a mirror 6.5-m across. From a ground-based site yet to be selected, it will survey the sky each week, detecting objects as faint as 24th magnitude. The data from the LSST is to be available on the Internet to everyone, professional and general public alike.

In Chapter 18, we will discuss near-earth asteroids and the possibility of their doing great damage when one hits the Earth. The LSST should be able to discover 90 per cent of these objects larger than 300 meters across within a decade, and so help our assessing what the risk is. In Chapter 29, we will discuss supernovae, explosions of stars. The LSST should discover thousands of faint supernovae. Since supernovae are bright beacons that we can see quite far out in space, the study of these faint ones should help us understand the structure of the Universe and our Universe's early epochs.

On a long term, the weekly maps of the sky can be added together, potentially allowing the exceedingly faint structure of the Universe to be discerned. The dark matter (Chapter 37) may be mapped in this way.

No schedule has been approved for the LSST, but astronomers have a good track record in getting approval for the highest priority items on the Astronomy and Astrophysics Survey's lists.

## 5.3 THE HUBBLE SPACE TELESCOPE

In 1990, NASA launched the Hubble Space Telescope (HST), the largest telescope in space to date. It is a reflecting telescope with a mirror 2.4 m (94 inches) in diameter (Fig. 5–11). The HST was designed to have three major advantages over ground-based telescopes, even though it is much smaller than the major ones. First, since it is above the Earth's atmosphere, the pinpoint structure of the images is limited only by the mirror's size. (Telescopes on Earth, on the other hand, are limited by the atmospheric "seeing.") The HST was designed to give images only one-tenth of an arc second across, ten times finer than normal good seeing at major ground-based observatories and three times better than the best moments of seeing. Second, starlight is concentrated into smaller images compared with images seen by ground-based telescopes, and the sky above the atmosphere is extremely dark, so the HST can detect fainter objects than can be seen from Earth. Third, with no atmosphere above the telescope to absorb the ultraviolet and infrared, these parts of the spectrum are detectable.

FIGURE 5–11 The 2.4-m (94-inch) mirror of the Hubble Space Telescope, after it was ground and polished and covered with a reflective coating. This single telescope has already made numerous discoveries but is raising as many questions as it will answer. So we need still more ground-based telescopes as partners in the enterprise. We discuss Edwin Hubble's work on galaxies and his discoveries about the Universe in Chapters 34 and 35.

This orbiting observatory was named the Edwin P. Hubble Space Telescope after the scientist who, in the 1920s, discovered that there are other galaxies comparable in scale to our own. Hubble also provided the observations that were interpreted to show that the Universe is expanding.

The project cost 2 billion dollars; most of the funds are from NASA. The European Space Agency provided the solar array that supplies power, one of the instruments, and some of the staff. The telescope is controlled from the Space Telescope Science Institute (STScI) on the campus of the Johns Hopkins University in Baltimore. The HST is designed to work for at least 15 years and to be visited by astronauts every three years or so for maintenance and for upgrading of instruments.

The Hubble Space Telescope has proved to be a great success by actually providing images 10 times clearer than images available from Earth. For the first three years after launch, the images were not as good as they should have been because HST's mirror was made in slightly the wrong shape. Apparently, while it was being shaped back in 1980, an optical system used to test it was made slightly the wrong size, and it indicated that the mirror was in the right shape when it wasn't. The result was spherical aberration, which blurred the images. Instead of 80 percent of the light going into the central tenth of an arc second—the size of resolution expected—only about 10 per cent went into that region. For some purposes, and especially for bright objects, the images could be processed with computers on the ground to yield the desired resolution. Many observations were made that were superior to any that could be made from the ground. Still the exposures were longer than originally planned, and fainter objects could not be properly studied.

Knowing that the problem was a straightforward spherical aberration of the main mirror allowed a fix to be made, much as eyeglasses improve a person's vision (Fig. 5–12). The main camera, the Wide Field and Planetary Camera, which makes about half the observations of Hubble, was to be replaced in any case in a servicing mission after about three years. A small mirror in it was reground to correct the spherical aberration of the main mirror. For the other instruments, an ingenious system known as COSTAR (*C*orrective *O*ptics *S*pace *T*elescope *A*xial *R*eplacement) replaced the least used of HST's instruments. By erecting tiny mirrors on movable stalks in front of the other instruments, COSTAR corrected them all. (The High Speed Photometer, the prospectively least-used instrument, had to be removed to make space for COSTAR.) The replacement camera, now known as Wide Field and Planetary Camera 2, and the COSTAR were installed by astronauts aboard a space shuttle in 1993. The astronauts also replaced the solar panels, gyroscopes, and other items that needed repair. The mission was a spectacular success, and the Hubble Space Telescope went into full working order.

**FIGURE 5–12** The Hubble Space Telescope, seen as a space shuttle approached.

Images taken with the Wide Field and Planetary Camera 2 are of much higher quality than the pre-repair images (Fig. 5–13). Still, HST is only one 2.4-m telescope, and it is raising as many questions as it answers. It has taken hundreds of thousands of observations. A selection is available for anyone to see on the Web; a Hubble Heritage Program releases a new image on the first Thursday of each month, with the images chosen for their popular appeal.

A mission in 1997 installed an infrared camera known as NICMOS (*N*ear *I*nfrared *C*amera/*M*ulti *O*bject *S*pectrograph), extending Hubble's capabilities to longer wavelengths. It is sensitive from 1 to 2.5 $\mu$m. Unfortunately, its lifetime was shortened because of a problem that led to increased heat transfer. STIS (*S*pace *T*elescope *I*maging *S*pectrograph) was also installed. It is sensitive from the ultraviolet through 1 $\mu$m, the visible/infrared boundary. STIS can simultaneously take spectra of each point along a whole narrow straight line drawn onto an object, so it is much more efficient than the earlier spectrographs, which could only do one of those points at a time.

A 1999 upgrade mission restored various telescope systems to good health and reboosted the spacecraft into a higher orbit (Fig. 5–14). A subsequent upgrade mission in 2002 is to replace the Faint Object Camera with the Advanced Camera for Surveys and to provide an electronic cooler that will restore NICMOS to functioning. The new camera will not only be more sensitive but also will have a wider field of view than the older camera, improving Hubble's observing efficiency for surveys by a factor of 10. The Cosmic Origins Spectrograph and the Wide Field Camera 3 are slated for a subsequent mission.The current plan is to maintain the telescope through at least 2008, when its successor may be ready.

### 5.3a The Next Generation Space Telescope

Astronomers, engineers, and NASA administrators are actively working on the Next Generation Space Telescope, NGST (Fig. 5–15). From basing its mission on studying normal galaxies in the early Universe, on searching for distant stars with planets around them, and on probing a variety of other scientific points, the idea developed

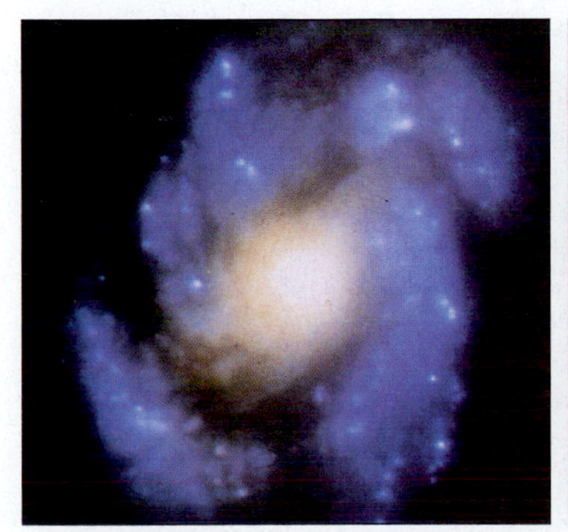

Wide Field and Planetary Camera 1

Wide Field and Planetary Camera 2

**FIGURE 5–13** This comparison image of the core of the spiral galaxy M100 in the Virgo Cluster shows the dramatic improvement in HST's view of the Universe. The picture clearly illustrates that the corrective optics incorporated within the Wide Field and Planetary Camera 2 compensate fully for the optical aberration in Hubble's primary mirror.

**FIGURE 5–14** An astronaut works on the Hubble Space Telescope during its 1999 servicing. The gyroscopes that are used to stabilize and point the telescope were replaced.

to launch a telescope with an 8-m mirror. It was "downsized" to 6-m in 2001. Following the precept of NASA Administrator Dan Goldin, who served into 2001, that projects should be "better, cheaper, faster," with the "faster" referring to the length of time it takes to get them done, the mirror will be excellent for work in the infrared but not of the same quality for visible-light observations. (Since it is most useful to measure the roughness and accuracy of a mirror's surface in units that correspond to the wavelengths being reflected, the longer wavelengths of infrared radiation make the necessary accuracy of the mirror less in absolute units—like millimeters or micrometers—than it would be for visible light.)

Detailed plans are now being made for launch of this Next Generation Space Telescope sometime in or after the year 2008. It would go to one of the Lagrangian points far from Earth, where the gravity of the Earth and Sun, as well as the effect of the Earth's orbital motion, balance. There, it would not go through Hubble's day-night cycles and so would be able to observe for a higher fraction of the time.

## 5.4 RECORDING THE DATA

**FIGURE 5–15** An artist's conception of the Next Generation Space Telescope, due for launch around 2008.

Astronomers want a more permanent record than is afforded by merely observing the image and more accuracy than can be guaranteed in a sketch. Formerly, we put a photographic film or "plate" (short for photographic plate, a layer of light-sensitive material on a sheet of glass) instead of an eyepiece at the observer's end of the telescope, so as to get a permanent record of the image or spectrum.

Basically, a photographic plate (or film) consists of a glass or plastic backing covered with grains of a silver compound. When these grains are struck by enough light, they undergo a chemical change. The result, after development, is a "negative" image, with the darkest areas corresponding to the brightest parts of the incident image. (The words "negative" and "positive" were first applied to photography by the astronomer John Herschel in the 19th century.) Inspection of a photographic plate under high magnification shows the grainy structure of the image (Fig. 5–16).

Other methods of recording data have been developed. The day of film is past, even though new emulsions now available are more sensitive and have finer grain than older emulsions. The future lies in electronic devices.

Electronic devices are more sensitive than photographic film to faint signals and can also be made to be sensitive in a wider region of the spectrum. Measurements of intensity, usually made with such electronic devices, are called **photometry.** Often the intensity of an object is measured through each of several colored filters in turn, to find its color, that is, how its continuous spectrum differs at different wavelengths.

Astronomers now use electronic devices built on "chips," which are related to the technology that has brought electronic calculators and digital watches to their present versatility and inexpensive price. These devices are used in place of photographic plates to make either direct images of astronomical objects or spectra. They come in several types, especially charge-coupled devices, known as **CCDs** (Fig. 5–17). When light hits the surface of the chip, electrons are released. Each discrete area on the chip is known as a "picture element," or **pixel** (Fig. 5–18). A CCD separately accumulates these electrons (each of which has an electric charge) on each pixel of its surface, and then scans the resulting charge off that surface with internal circuitry. The name "CCD" refers to the scanning method, which involves shifting the electrical signal from each point to its neighbor one space at a time, analogous to what happens with water in a fire-fighting bucket brigade when buckets of water are passed from one individual to the next. (In this way, the *charge* on different pixels is *coupled*.) A CCD's picture is created by scanning the picture elements line by line. CCDs are almost 100 times more sensitive than film. Also, adjacent objects that differ greatly in brightness can be accurately compared on a CCD image, whereas film would not permit such a comparison. You will see video cameras with relatively cheap CCDs advertised in your local newspaper and available at the mall.

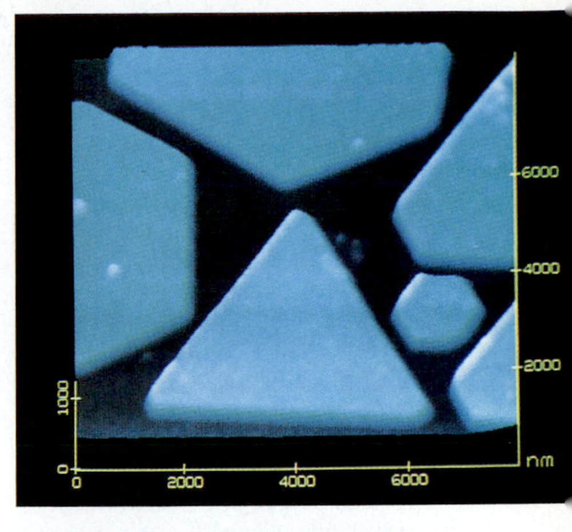


**FIGURE 5–16** Individual photographic grains on a microscopic view of an astronomical emulsion (103a-O, where the "a" stands for an astronomical emulsion that remains sensitive for long exposures and the "O" stands for blue sensitivity) enlarged 2000 times.

CCDs have become more commonplace at observatories than film. The imaging system on the Hubble Space Telescope is a composite of four arrays, each of which is 800 × 800 pixels; larger arrays have since become available. Other types of detectors are used in the infrared, where CCDs are not sensitive. The data from these detectors are fed into computers, which are used by astronomers to manipulate the data to analyze them.

Not all CCDs are alike; they differ not only in size but also in sensitivity. One difference is whether the light falls on the front or back of the chip, compared with where the electrodes are attached. Back illuminated CCDs with the chip's thickness thinned are the most sensitive.

CCD signals come off in digital form and are stored on computers or on digital storage devices like digital tape or CD-ROM. So much CCD data is around that nobody knows where it all is and certainly not what it all shows. One of the projects suggested in the year 2000's decade review, sponsored by the National Academy of Sciences, is the National Virtual Observatory. It would index all kinds of observations already made and available at a wide variety of sites.

**FIGURE 5–17** The University of Hawaii and the Canada-France-Hawaii Telescope have built this huge mosaic of 12 individual CCD detectors from MIT's Lincoln Labs, totalling 12,288 × 8,192 pixels. CCDs, charge-coupled devices, are being used as detectors in many telescopes. They are normally buried in electronics and cooled by liquid nitrogen. Here it is demonstrated by its builder, Gerry Lupino of the University of Hawaii.

### 5.4a Fiber Optics*

Many long-distance telephone lines are now optical fibers, over which signals are sent as flashes from tiny lasers. The laser beams travel basically straight along the fibers; if they should deviate by a small angle, they are reflected off the sides so as to keep them travelling forward, even when the fiber is bent (Fig. 5–19).

In astronomy, optical fibers are beginning to be used to make telescopes more efficient. For example, a hundred or more optical fibers can be arranged so that each fiber accepts the light from a different star or galaxy. The other ends of the fibers can be arranged so that they form a line, making the light coming out cover the whole slit of a spectrograph. Each part of the length of the spectrograph slit thus represents light from a different star or galaxy. The procedure allows a great gain in the efficiency of the spectrograph, for it now gathers hundreds of spectra at the same time instead of just one spectra.

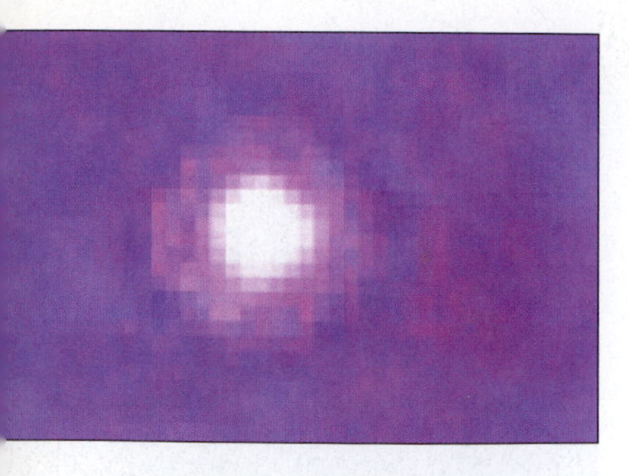

FIGURE 5–18 A CCD image; the individual picture elements (pixels) can be seen when it is sufficiently enlarged.

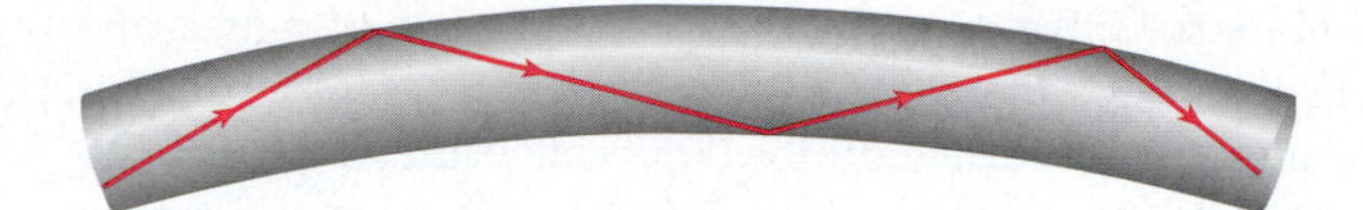

▲ FIGURE 5–19 As laser light bounces off the walls of optical fibers at small angles, it keeps travelling forward along the fiber, even when the fiber is bent.

## 5.5 THE SLOAN DIGITAL SKY SURVEY, 2dF, AND 6dF

The twentieth century saw mapping of the sky object by object. In the 21st, several projects are taking advantage of the CCD and fiber-optics technologies just described to switch to wholesale mapping.

The Sloan Digital Sky Survey uses a new 2.5-m telescope at Apache Point Observatory in New Mexico, but its main innovations are in the data recording and analysis rather than in the telescope itself. It has sets of CCDs behind color filters (Fig. 5–20), and records the stars and galaxies as they revolve overhead, passing over first one set of CCDs and then another. Over a five-year period, it is to image one-quarter of the sky in five colors. The clearest nights are devoted to imaging, while the less clear (but not cloudy!) nights will be used to take spectra of, and so measure distances, to a million galaxies, 100,000 quasars, and multitudes of brown dwarfs, objects we discuss later on. So many spectra can be taken because of the fiber optics used (Fig. 5–21). The three-dimensional picture of the Universe that will result should lead to important advances in our understanding of the overall structure. Even in its early runs, it has discovered several of the most distant quasars and thus the most distant objects in the Universe. We include some of its galaxy images in Figure 34–11.

The survey will reach 40 times farther into space than the Palomar Observatory Sky Surveys did and so should supersede it. So much data will be downloaded—10 trillion bytes, an amount of information comparable to all that of the materials in the Library of Congress—that astronomers have teamed up with the Fermi National Accelerator Laboratory, which is experienced from their own atom-smashing projects in handling huge amounts of data. Parts of the data reduction may be parcelled out to individuals' home computers, much as the computing of SETI@home is distributed (Chapter 21). The Sloan Digital Sky Survey is a joint project of The University of Chicago, Fermilab, the Institute for Advanced Study, the Japan Participation Group, The Johns Hopkins University, the Max Planck Institute for Astronomy,

FIGURE 5–20 The back end of the Sloan Digital Sky Survey. There are five colors of filters; the CCDs are behind them. The color bands are known as u′ (ultraviolet, with a 3500 Å peak), g′ (blue-green, with a 4800 Å peak), r′ (red, with a 6250 Å peak), i′ (far red/near infrared, with a 7700 Å peak), and z′ (near infrared, with a 9100 Å peak).

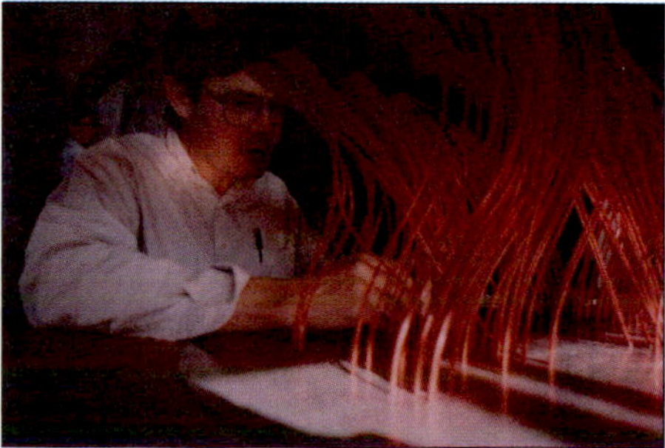

FIGURE 5–21 The Sloan Digital Sky Survey's 150 fiber-optics strands that take galaxy light from the focus to the CCDs.

Princeton University, the U.S. Naval Observatory, and the University of Washington. It uses a special telescope in New Mexico (Fig. 5–22).

At the Anglo-Australian Observatory in Australia, the 4-m telescope used to be considered large, but now it is so much smaller than the new, huge telescopes that it has been largely turned over to survey projects. Their project to use fiber optics to take spectra of 250,000 galaxies is known as **2dF,** which stands for two-degree field of view. They will use the spectra to make a three-dimensional map of the southern sky. A 6dF project is also under way using the wide field of the 1.2-m UK Schmidt Telescope, also on the same mountain in Australia, to cover a larger region of the sky. In 6dF, 150 fibers will allow 150 galaxy spectra to be obtained simultaneously, and 100,000 spectra total are planned. Another advantage of 6dF over 2dF is that the objects to be observed will be chosen from the 2MASS catalogue (Section 5.7a). Such infrared sources may be more likely to be extremely distant; 6dF could be complete by 2005. In part of the sky, Sloan, 2dF, and 6dF will overlap. It is hoped that a clone of 6dF could operate in the northern sky.

**FIGURE 5–22** The Sloan Digital Sky Survey's telescope in New Mexico, with its unusual dome-less design.

Three-dimensional maps of the Universe from Sloan and 2dF appear as Figs. 35–18 and 35–19. Each includes distance measurements of tens of thousands of galaxies.

## 5.6 OBSERVING AT SHORT WAVELENGTHS

The visible extends, at its shortest wavelength, down to the violet and to the edge of what we call "ultraviolet." Ordinary film is sensitive down to the shortest wavelengths that come through the Earth's atmosphere. Indeed, film is especially sensitive in this ultraviolet, violet, and blue, so older observations were principally made in this part of the spectrum. The electronic detectors we now use are more sensitive in the red than in the blue or violet.

Nowadays, when we talk of "ultraviolet astronomy," we mean observations of the part of the spectrum shortward of the limit at which the atmosphere stops transmitting radiation. Thus ultraviolet astronomy must be carried out above the Earth's atmosphere from spacecraft.

### 5.6a Ultraviolet and X-Ray Astronomy

The Hubble Space Telescope is the largest telescope ever in space for observing the ultraviolet part of the spectrum. Earlier, the International Ultraviolet Explorer sent back valuable ultraviolet spectra for over 20 years, though it has been superseded. Some ultraviolet telescopes were carried into space for shorter missions on space shuttles. The Far Ultraviolet Spectroscopic Explorer (FUSE), launched in 1999, is taking high-resolution spectra of a wide variety of objects (Fig. 5–23). In Chapter 37, we will see how it is being used to investigate the origin of the lightest elements.

Beyond even what we call the ultraviolet, we have the "extreme ultraviolet," at still shorter wavelengths. The Extreme Ultraviolet Explorer (EUVE) spacecraft mapped the sky in the spectral region from 10 Å to 1000 Å between 1992 and 2000. Because the gas between the stars was expected to block the extreme ultraviolet, this window of the spectrum was the last to be opened. Fortunately, the interstellar gas is inhomogeneous, so the EUVE spacecraft has been able to look through the spaces to observe a wide variety of sources. Hundreds of sources have been catalogued, including not only hot stars but also a few objects from outside our galaxy.

**FIGURE 5–23** The Far Ultraviolet Spectrographic Explorer (FUSE).

### 5.6b X-Ray Telescopes

In the shortest wavelength regions, x-rays and gamma-rays, we cannot merely use mirrors to image the incident radiation in ordinary fashion, since the x-rays will pass right through the mirrors! Fortunately, x-rays can still be bounced off a surface if they strike the surface at a very low angle. This phenomenon is called **grazing incidence.** This principle is similar to that of skipping stones across the water. If you throw a stone straight down at a lake surface, the stone will sink immediately. But if

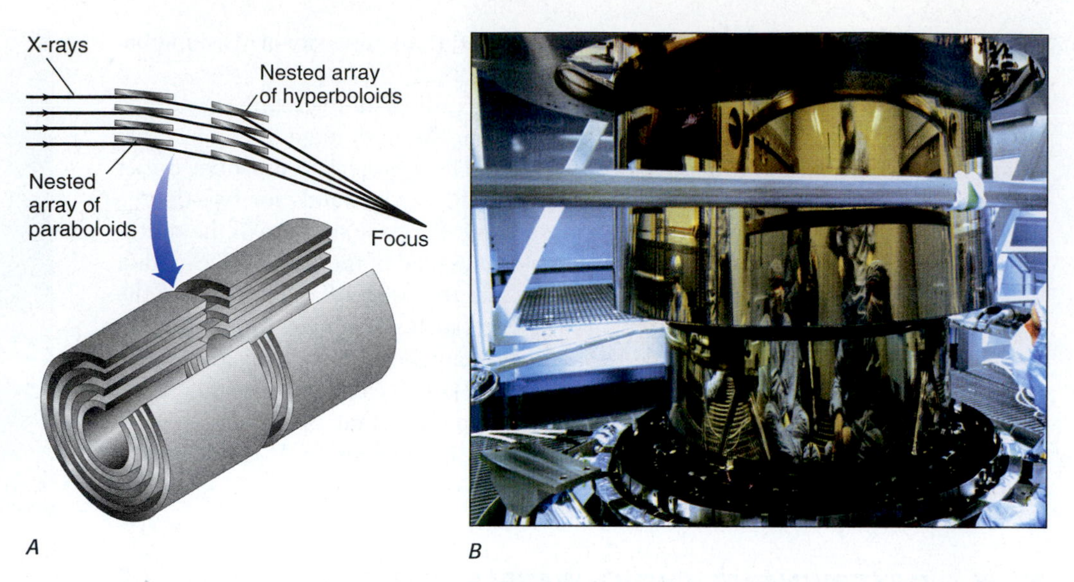


**FIGURE 5–24** (*A*) Similar paraboloid/hyperboloid mirror arrangements, used at grazing incidence and all sharing the same focus, are nested within each other to increase the area of telescope surface that intercepts x-rays. (*B*) The outermost mirror set of the Chandra X-ray Observatory.

you throw a stone out at the surface some distance in front of you, the stone could bounce up and skip along a few times.

By carefully choosing a variety of curved surfaces that suitably allow x-radiation to "skip" along, astronomers can now make telescopes that actually make x-ray images. But the telescopes appear very different from optical telescopes (Fig. 5–24A).

In order to form such high-energy photons, processes must be going on out in space that involve energies much higher than most ordinary processes that go on at the surface layers of stars. The study of the processes that bring photons or particles of matter to high energies is called **high-energy astrophysics.** In the late 1970s, NASA launched a series of three High-Energy Astronomy Observatories (HEAOs, pronounced "hee-ohs") to study high-energy astrophysics. Other x-ray observatories were launched by the European Space Agency, the Japanese, and the Soviets.

The best of the previous generation of x-ray satellites included the U.S.–German–British ROSAT (**Ro**entgen **Sat**ellite, named for Wilhelm Roentgen, the German scientist who discovered x-rays a hundred years ago), launched in 1990. ROSAT provided a survey of the entire sky at a sensitivity 1000 times better than the last full-sky map before the mission ended in 1999. The Rossi X-ray Timing Explorer, named for the American space scientist Bruno Rossi, gives observations of x-rays with high-time resolution.

The most recent of NASA's series of Great Observatories is the Chandra X-ray Observatory (Fig. 5–24B), launched in 1999. The Chandra X-ray Center at the Smithsonian Astrophysical Observatory in Cambridge, Massachusetts, helps scientists study the data. Chandra uses four nested pairs of cylindrical mirrors with the inner surface of each cylinder shaped into a paraboloid and a hyperboloid. It makes high-resolution x-ray images and uses gratings to take spectra.

The XMM-Newton mission, one of the European Space Agency's "cornerstone" missions, was launched in 1999. Though its spatial resolution is worse than Chandra's, it has more collecting area and so is more sensitive. It carries x-ray spectrometers and carries out simultaneous imaging in the ultraviolet and visible spectral regions.

Eventually, the Constellation-X Observatory will replace Chandra with a suite of four x-ray telescopes to concentrate on the formation and evolution of black holes.

At still shorter wavelengths, the Compton Gamma-Ray Observatory was launched in 1991, and it will be described subsequently (Section 32.4), when we discuss its observations of objects in our Milky Way Galaxy and beyond. It was purposely crashed into the Pacific Ocean in 1999, when some of its control systems failed. An Italian–Dutch satellite named BeppoSAX, launched in 1996, observes both x-rays and gamma rays.

The High-Energy Transient Explorer spacecraft was launched in 2000 to study the rapid bursts of gamma-rays that occur about once a day, as we will learn in Chapter 32.

The successor to Compton, with 30 times the sensitivity, is to be the Gamma-ray Large Area Space Telescope (GLAST), a joint project of NASA and the Department of Energy. We hope for its launch in 2006.

The techniques of interferometry that we will discuss in Section 34.8 show promise of making x-ray telescopes fantastically more capable around 2015. Once the techniques are mastered, the resolution (the detail on the image) would be about a thousand times finer than even Chandra's capabilities. Then we may really be able to image black holes or the coronas of some of the nearer stars. But to make this work, we have to find a way of sending up a fleet of dozens of spacecraft orbiting in formation and maintaining relative distances to 20 nanometers!

## 5.7 OBSERVING AT LONG WAVELENGTHS

Many useful observations can be made in the infrared from high mountaintops through "windows of transparency" in the atmosphere. Most of the infrared does not penetrate to Earth. The whole radio spectrum is a "window of transparency," though.

### 5.7a Infrared Astronomy

The existence of infrared windows has led to the construction of infrared telescopes at such high, dry sites as Mauna Kea in Hawaii. (The higher and drier, the wider and the more transparent the windows are, since water vapor absorbs many wavelengths of infrared radiation.) The Astronomer's Notebook describes an observing run with the Infrared Telescope Facility on Mauna Kea.

The infrared covers a factor of a thousand in wavelength, so it is almost misleading to discuss it all together. At the short infrared wavelength of two micrometers, one can observe from Earth's surface through a window of transparency. At present, the Two Micron All Sky Survey (2MASS) is mapping the entire sky in three wavelength bands: the so-called J band at 1.25 micrometers, the so-called H-band at 1.65 micrometers, and the so-called K-band at 2.17 micrometers. The project, operated by the University of Massachusetts at Amherst with participants from a dozen institutions, uses telescopes in Arizona and in Chile. Image processing is at the Infrared Processing and Analysis Center of Caltech's Jet Propulsion Laboratory, known as IPAC. Many parts of the sky are already available on-line. Ultimately, 2MASS will detect and catalogue over 300 million stars and over 1 million galaxies.

Still higher in altitude, and so above more of the absorbing water vapor, NASA's Kuiper Airborne Observatory, an instrumented airplane, carried a 0.9-m telescope aloft. Such infrared observations are especially useful to study gas and dust clouds in space, which cannot be observed as well from the ground. NASA is planning a more capable successor airplane named SOFIA: Stratospheric Observatory for Infrared Astronomy. It will carry a 2.5-m telescope (larger than Hubble's) starting in 2004.

The Infrared Astronomical Satellite (IRAS) mapped the infrared sky in 1983. It was able to study the infrared at wavelengths longer than those observable from the ground, since its position above the atmosphere placed it above the infrared radiation from the air in our atmosphere. Its telescope was cooled to only 2°C above absolute zero (that is, to 2 K) to prevent infrared radiation from the telescope from swamping the signal from celestial objects. When the liquid helium used for cooling ran out after 10 months, the spacecraft could no longer be used.

*(text continues on page 88)*

# ASTRONOMER'S NOTEBOOK

## A Night at Mauna Kea

An observing run with one of the world's largest telescopes highlights the work of many astronomers. The construction of the world's largest telescopes at the top of Mauna Kea in Hawaii has made that mountain a beehive of activity.

*Mauna Kea is so high*
*That you are halfway to the sky.*
*It's a site where, yes, you may*
*Use an infrared array.*
*(If the object's faint for you,*
*Use the Keck, now I or II.)*

Hawaii's Mauna Kea was chosen because of its outstanding observing conditions. Many of these conditions stem from the fact that the summit is so high—4200 m (13,800 ft) above sea level. The top of Mauna Kea is above so much of the atmosphere's water vapor, all but 10 per cent, that observations can be made in many parts of the infrared inaccessible from other terrestrial observatories.

Mauna Kea has other general advantages as an observatory site. It is so isolated that the sky above is particularly dark, allowing especially faint objects to be seen. And the flow of air across the mountaintop is often particularly smooth, leading to exceptional "seeing" for both visible and infrared observations. Moreover, the top of the mountain extends above most clouds that may cover the island below it. Mauna Kea is perhaps the world's best site in that these advantages are linked with a large number of clear nights—about 75 per cent. Though there are taller mountains in the world, none has such favorable conditions for astronomy (Fig. A).

The 3-m Infrared Telescope Facility (IRTF) is sponsored by NASA and operated by the University of Hawaii (Fig. B). Since it is a national facility, astronomers from all over the United States can propose observations. Every six months, a committee of infrared astronomers from all over the United States chooses the best proposals and makes out the observing schedule for the next half year.

During the months before your observing run, you make detailed lists of objects to observe, prepare charts of the objects' positions in the sky based on existing star

**FIGURE A** The summit of Mauna Kea in a wide-angle view.

**FIGURE B** The Infrared Telescope Facility on Mauna Kea, a 3-m telescope optimized for infrared observations.

maps and photographs, and plan the details of your observing procedure. The air is very thin at the telescope. You may not think as clearly or rapidly as you do at lower altitudes. Therefore, it is wise and necessary to plan each detail in advance.

The time for your observing run comes, and you fly off to Honolulu. You may spend a day or two there at the Institute for Astronomy of the University of Hawaii, consulting with the resident scientists, checking last-minute details, catching up with the time change, and perhaps giving a colloquium about your work. Then comes the brief plane trip to the island of Hawaii, the largest of the islands in the State of Hawaii. It is usually called simply "the Big Island."

Your plane may be met by one of the Telescope Operators, who will be assisting you with your observations. The Telescope Operators know the telescope and its systems very well and are responsible for the telescope, its operation, and its safety. You drive together up the mountain, as the scenery changes from tropical to relatively barren, crossing dark lava flows. First stop is the Ellison Onizuka House at Hale Pohaku, the mid-level facility at an altitude of 2750 m (9000 ft), now named after the Hawaiian-born astronaut who died aboard the Space Shuttle Challenger. The astronomers and technicians sleep and eat there. Since the top of Mauna Kea is too high for people to work comfortably there for too long, everyone spends much time at the mid-level.

The rules say that you must spend a full 24 hours acclimatizing to the high altitude before you can begin your own telescope run. So you overlap with the last night of the run of the people using the telescope before you. This familiarizes you with the telescope and its operation. A combination of time and altitude adjustment sends you to bed early this first night.

The next night is yours. In the afternoon, you and the Telescope Operator may make a special trip up the mountain to install the systems you will be using to record data at the Cassegrain focus of the telescope. Since objects at normal outdoor or room temperatures give off enough radiation in the infrared to bother your observations, much of the instrumentation you install is cooled. Observations at the shorter of the infrared wavelengths are now made with infrared arrays. Technological advances are making arrays available at longer and longer wavelengths. Infrared detectors are cooled to reduce their own infrared radiation so that it does not swamp the faint infrared celestial radiation. To keep the liquid helium from boiling away too fast, liquid nitrogen at a temperature of 77 K (−196°C) surrounds the helium.

After dinner, you return to the summit to start your observing. You dress warmly, since the nighttime temperature approaches freezing at this altitude, even in the summertime. (Though you are mainly in a warm room for observing, you have to get from the car to the dome, and you sometimes go outside to check the sky or into the dome to check the telescope.) Since the sky is fairly dark at infrared wavelengths, you can start observing even before sunset. The telescope is operated by computer and points at the coordinates you type in for the object you want to observe. A video screen displays an image of the object. You have already prepared your list of objects to observe and the exposure time necessary to get the quality of observation you need. You also have to take calibration exposures, making sure that you can account for any differences in sensitivity of individual pixels of your detector. At the end of your run, you almost always wish you could have observed more objects, used more filters, or made additional exposures.

After your night's observing you drive down the mountain in the early morning sun. The clouds you see may be a kilometer below you. The tiny pimples on the landscape that you first see turn out, as you get closer, to be giant cinder cones that tower above you. They are from past volcanic eruptions. The mountain around you is barren, littered with huge volcanic rocks, resembling surface views from the Viking landers on Mars more than any terrestrial view you have ever seen. Then some vegetation appears, and you reach Hale Pohaku, ready for breakfast and a day's sleep. The next night you start over again.

When you leave Mauna Kea, it is a shock to leave the pristine air above the clouds. But, happily, you take home with you data about the objects you have observed. The data are in the form of graphs, computer printouts, floppy disks, and computer tape. Some data may be transferred over the Internet directly to your computer back home. It often takes you months to study the data you gathered in a few brief days at the top of the world.

As Internet capabilities progress, fewer astronomers go up to the mountain. Some use computers to direct the observations from the base stations at lower altitudes on Hawaii. In some cases, you might even observe from your office at your home institution. The images as well as information about the telescope itself simply appears on your computer screen. That may be more efficient—but it sure is less fun.

FIGURE 5–25 An image from the 2MASS (Two Micron All Sky Survey) survey. The wispy structures in the picture are "infrared cirrus," a discovery of the 1980s Infrared Astronomical Observatory (IRAS). The data are displayed as false-color images, each color corresponding to a different infrared wavelength, which corresponds roughly to a different temperature.

In its brief but spectacular lifetime, IRAS made many discoveries. For example, it discovered a patchy network of infrared radiation; by analogy with cirrus clouds in the terrestrial sky, this network is called "infrared cirrus." (Fig. 5–25)

The European Infrared Space Observatory (ISO) was operating in the mid-1990s. It had spectrometers for long-wavelength and short-wavelength infrared spectra, a camera, and an imaging device for photometry and polarization measurements. NASA's Space Infrared Telescope Facility (SIRTF)—the final Great Observatory to accompany the Compton Gamma-Ray Observatory, the Hubble Space Telescope, and the Chandra X-ray Observatory—is being planned for launch in 2002 (Fig. 5–26).

To return to the Earth's surface, a huge, 10-m telescope is being discussed for the South Pole, where conditions are exceptionally dry, allowing more of the infrared to come through than at average locations on Earth.

Perhaps the most important recent instrumental development in astronomy is the "infrared array." While infrared images used to be laboriously built up point by point, infrared-sensitive elements can now be put onto silicon chips in configurations of $256 \times 256$, imaging 65,536 pixels at the same time. Further, the spacing between the elements is unchanging, and the atmospheric transmission is the same for all elements at a given time, both improvements on the former method of raster scanning (that is, back-and-forth while descending line-by-line). The center of our galaxy and star-forming regions like the Orion Nebula are among the types of objects for which infrared-array images are particularly valuable and for which examples are included later in this book. An infrared array was put on the Hubble Space Telescope in 1997 and is to be reactivated in 2002.

## 5.7b Radio Astronomy

The principles of radio astronomy are exactly the same as those of optical astronomy: radio waves and light waves are alike—only the wavelengths differ. Both are simply forms of electromagnetic radiation. But different technologies are necessary to detect the signals. Radio waves cause electrical changes in antennas, and these faint electrical signals can be detected with instruments that we call radio receivers.

If we want to collect and focus radio waves just as we collect and focus light waves, we must find a means to concentrate the radio waves at a point at which we can place an aerial. Lenses to focus radio waves are impractically heavy, so refracting radio telescopes are not used. However, radio waves will bounce off metal surfaces. Thus we can make reflecting radio telescopes that work on the same principle as reflecting optical telescopes (of which the 10-m optical telescopes on Mauna Kea are examples). The reflecting surface, called a "dish" rather than a mirror because it is of less-than-optical quality, is usually made of metal. The dish need not look shiny to our eyes, as long as it looks shiny to incoming radio waves.

FIGURE 5–26 NASA's Space Infrared Telescope Facility (SIRTF), scheduled for launch in 2002, should be able to observe objects in space at the long infrared wavelengths. Here we see its lightweight beryllium mirror.

We want dishes as big as possible for two reasons: First, a larger surface area means that the telescope will be that much more sensitive. Second, a larger dish has better resolution. Since radio wavelengths are much longer than optical wavelengths, any single-dish radio telescope has many fewer radio wavelengths that fit across it than an optical telescope has optical wavelengths. The resolution of any single radio telescope is thus far inferior to that of any single optical telescope (Section 4.2). The basic point is that measurements of the size of a telescope are most meaningful when they are in units of the wavelength of the radiation being observed (Fig. 5–27). Thus an optical telescope one meter across used to observe optical light, which is about one two-millionth of a meter in wavelength, is two million wavelengths across. The largest fully steerable radio telescope used to be at the Max Planck Institute for Radio Astronomy near Bonn, Germany. It is 100 meters (330 feet) in diameter. Though the Bonn Telescope is 100 m in diameter, larger than a football field, it is only 1000 wavelengths across when used to observe radio waves 10 centimeters in wavelength (100 m/10 cm = 1000). The 100-m Green Bank Telescope in Green Bank, West Virginia, has an unusual design with a view of the sky unobstructed by its own re-

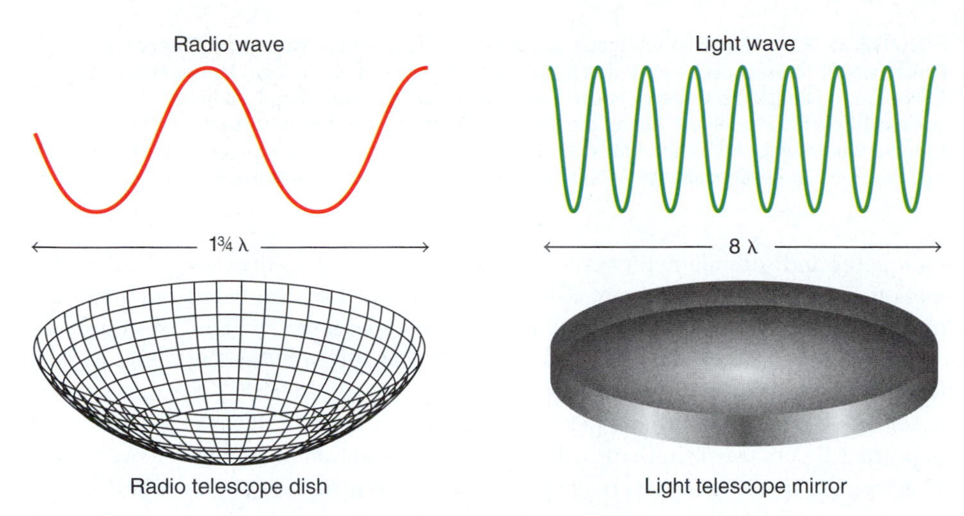


**FIGURE 5–27** It is more meaningful to measure the diameter of telescope mirrors in terms of the wavelength of the radiation that is being observed than it is to measure it in terms of units like centimeters that have no particularly relevant significance. In the diagram the radio dish at left is only 1³/4 wavelengths across (the Greek letter lambda stands for wavelength), whereas the mirror at right is 8 wavelengths across, making it effectively much bigger. To make it easier to see the differences in the ratios, the wavelengths are greatly exaggerated in this diagram relative to the size of any actual reflectors. For the Byrd Green Bank Telescope used to observe 1-cm radio waves, 100 m diameter ÷ 0.1 m per wave = 1000 wavelengths. For the 10-m Keck telescope used to observe 5000 Å light in the middle of the visible spectrum, 10 m ÷ 0.5 $\mu$m = 20,000,000 light waves. So large optical telescopes are millions of wavelengths of light across.

ceiving antenna (Fig. 5–28). For this and other reasons, it is especially sensitive. It was named after Senator Robert Byrd just after it was opened in 2000.

Radio telescopes used to study millimeter-length radio waves do not have to be as physically large as telescopes meant to study longer wavelengths. For example, to have the same relative size, a telescope to study waves of wavelength 2 mm need be only 2 meters across. Thus a 14-m (45-foot) radio telescope (Fig. 5–29) has 7 times better resolution at its wavelength than the physically larger 100-m telescope has for the longer-wavelength radiation. The Five College Radio Astronomy Group from Massachusetts is in a partnership with Mexico to build an even larger millimeter-wave telescope there.

A radio telescope dish receives radiation only from within a narrow cone called the **beam.** The larger the telescope, measured with respect to the wavelength of radiation observed, the narrower the width of the beam. Single-dish radio telescopes have broad beams and cannot resolve any spatial details smaller than the size of their beams. Astronomers now use the technique known as interferometry to improve the resolution, using a technique to be discussed in Section 34.8, close to the places in this book where we see results from this method. The technique has led, for example, to the giant array of 27 radio telescopes in New Mexico called, prosaically, the VLA (Very Large Array), and to the Very Long Baseline Array (Fig. 5–30). An Expanded VLA (EVLA) is under way, with eight additional telescopes spaced throughout New Mexico. Arrays of millimeter-wavelength radio telescopes are now operated at the Hat Creek Radio Observatory in California (Fig. 5–31) and at IRAM's site in France (Fig. 5–32). The international Atacama Large Millimeter Array (ALMA), with 64 transposable 12-m antennas, is to be built on a high, dry plateau in Chile, with American, European, and Japanese participation (Fig. 5–33). It should receive

**FIGURE 5–28** The Byrd Green Bank Telescope in Green Bank, West Virginia, opened for scientific operation in 2001. Its off-axis configuration gives cleaner images than the older type of radio telescope.

**FIGURE 5–29** The radio telescope of the Five College Radio Astronomy Observatory (University of Massachusetts at Amherst, Amherst College, Smith College, Mount Holyoke College, and Hampshire College) is used to study millimeter-wavelength radiation. It is enclosed in a radome, which keeps the Sun and snow off it but is almost entirely transparent to radio waves. A joint U. Mass.–Mexico 50-m telescope is being constructed in Mexico. It will be made of 126 hexagonal segments, a design pioneered by the Keck telescopes.

millimeter and submillimeter waves that penetrate the dust that keeps stars in formation from our view. A Square Kilometer Array, with a huge collecting area composed of hundreds of widely distributed smaller telescopes, is being seriously discussed, to be constructed perhaps starting in 2015, with even the continent on which it would be located still undecided (Fig. 5–34). It is to have 100 times the sensitivity of today's best radio telescopes. It is, further, to be able to make images at the important 21-cm wavelength of hydrogen with a resolution of 0.1 arc sec over a field 1° across. Its specifications were chosen to allow it to map the structures that formed

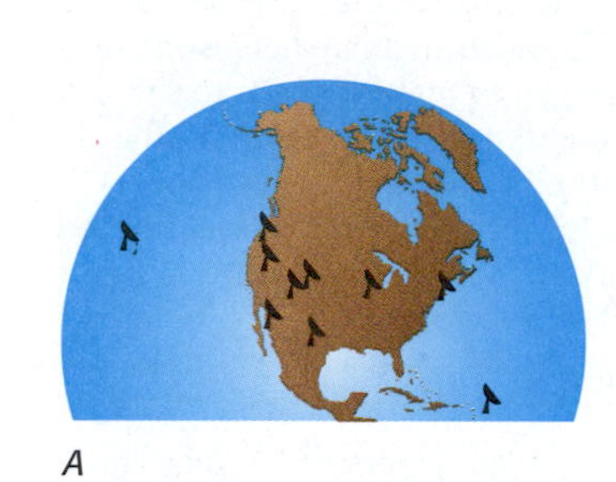

A

B

**FIGURE 5–30** The technique of interferometry works not only for the Very Large Array, whose telescopes are adjacent, but also for the Very-Long-Baseline Array, whose telescopes are spread out across the world's western hemisphere.

**FIGURE 5–31** The millimeter interferometer known as BIMA (U. Cal. at Berkeley–U. Illinois–U. Maryland Array).

**FIGURE 5–32** Three of the four 15-m dishes of the millimeter-wavelength interferometer of IRAM (Institut de Radio Astronomie Millimétrique), a French–German–Spanish consortium. IRAM's interferometer is in France. They also have a 30-m millimeter-wavelength single dish in Spain. Most of the observing time is used to study interstellar molecules.

**FIGURE 5–33** A prototype ALMA dish was set up at the Nobeyama Radio Observatory in Japan.

**FIGURE 5–34** An artist's conception of the Square Kilometer Array, perhaps to be built in some configuration after 2015.

in the early Universe. The consortium planning it includes members from the United States, Europe, Canada, Australia, China, and India.

The submillimeter region of the spectrum, between long infrared and short radio waves, is now being opened, especially by telescopes on Mauna Kea. Following the Submillimeter Wave Astronomy Satellite (SWAS), an American–German satellite, the European Space Agency's larger Herschel Space Observatory, with a 3.5-m telescope and instruments to observe the far infrared and submillimeter spectrum, is being scheduled for launch in 2007. (William Herschel, who discovered the planet Uranus in 1781, subsequently discovered infrared radiation.)

## CORRECTING MISCONCEPTIONS

| ✖ ***Incorrect*** | ✔ ***Correct*** |
| --- | --- |
| The Hubble Space Telescope takes images only in visible light. | Hubble is sensitive in the ultraviolet and infrared. |
| X-rays and radioactivity have always been known. | Roentgen discovered x-rays. Becquerel discovered radioactivity. Marie Curie isolated new, strongly radioactive elements. |

## SUMMARY AND OUTLINE

Observatory sites (Sections 5.1 and 5.2)
- Sites across the world
- New technologies to make larger telescopes, up to the 10-m Keck telescopes
- Schmidt telescopes have wide fields.

The Hubble Space Telescope (Section 5.3)
- Increased resolution; ultraviolet, visible, and infrared sensitivity
- It is working very well since its repair.

Recording the data (Section 5.4)
- Film and electronic devices (especially CCDs) for gathering photons
- Fiber optics are making observing more efficient.

Ultraviolet and x-ray astronomy (Section 5.5)
- Grazing incidence techniques used.
- Chandra launched in 1999.

Infrared and radio astronomy (Section 5.6)
- Infrared: High altitude sites and electronic devices are necessary.
- Space Infrared Telescope Facility 2002 launch
- Radio: Single-dish radio telescopes have lower resolution than interferometers.

## KEY WORDS

Schmidt telescope
correcting plate
photometry
CCD
pixel
2dF
grazing incidence
high-energy astrophysics
beam

## QUESTIONS

1. The Gemini project plans to have twin telescopes, one in the northern and one in the southern hemispheres. Why does it matter which hemisphere a telescope is in, north or south?
2. Why doesn't it matter whether a telescope is in the eastern or western hemispheres?
3. List the important criteria in choosing a site for an optical observatory meant to study stars and galaxies.

†4. How many times more light is gathered by the Keck Telescope, whose mirror is equivalent in area to that of a circle 10 m in diameter, than by the Palomar Hale Telescope, whose mirror is 5 m in diameter, in any given interval of time?

†5. A 4-m telescope has a Cassegrain hole 1 m in diameter. What percentage of the total area is taken up by the hole?

†6. The Keck Telescope is made of 36 hexagons each 1.8 m in diagonal diameter, with a central hexagon omitted. By breaking down each hexagon into triangles, and using the formula for the area of a triangle, derive the area of the Keck mirror in square meters and the size of the equivalent round mirror if its central hole were 1 m in diameter.

7. Describe the problems that occurred with the Hubble Space Telescope and what the current situation is.
8. What are two advantages of the Hubble Space Telescope over the Keck Telescope? What are two advantages of the Keck Telescope over the Hubble Space Telescope?

†9. The Palomar–National Geographic Sky Survey made with the Oschin Schmidt telescope covers the sky in about 900 pairs of plates. Using the field of view given in the text, and comparing with a 30-arc-min field of view for the Kitt Peak 4-m reflector, calculate how many red/blue pairs of photographs would be needed to cover the sky with the latter.

10. What are two ways of making telescope mirrors larger than previous ones?
11. What are the similarities and differences between making radio observations and using a reflector for optical observations? Compare the radiation path, the detection of signals, and limiting factors.
12. Why is it sometimes better to use a small telescope in orbit around the Earth than it is to use a large telescope on a mountaintop?
13. Why is it better for some purposes to use a medium-size telescope on a mountain instead of a telescope in space?
14. What are two reasons why the Hubble Space Telescope is able to observe fainter objects than we can now study from the ground?
15. What were the results of the first repair mission to the Hubble Space Telescope?
16. What were the results of the second repair mission to the Hubble Space Telescope?

† This question requires a numerical solution.

## INVESTIGATIONS

1. Cut out hexagons, and see how they fit together to make the mirror of the Keck telescope. Try to do the same with pentagons and octagons.
2. Coat the inside of an umbrella with aluminum foil, and see if you can use that surface to concentrate signals from a radio station onto a portable radio.
3. Make two holes in a piece of cardboard, the first a centimeter across and the second two centimeters across. Fit pieces of uncooked spaghetti through the holes. Verify that the area of the hole increases with the square of its diameter.

## USING TECHNOLOGY

### W World Wide Web

1. Look for the latest discoveries of the Hubble Space Telescope on its homepage.
2. Check for images of astronomical objects in the homepages of major observatories such as the National Optical Astronomy Observatories, the European Southern Observatory, and the Anglo-Australian Observatory.
3. Look at the professional journals *Astrophysical Journal* and the *Astrophysical Journal Letters* on the homepage of the American Astronomical Society. Check out some of the types of discoveries being reported.
4. Look up the history of our knowledge of the infrared at http://www.ipac.caltech.edu/Outreach/Edu and http://sirtf.caltech.edu/SSC_EPO.html.

### REDSHIFT

1. Use the Map of the Earth to find the sites of Observatories. Locate the nearest of the observatories shown to where you are. Appreciate the variation in latitude of some of the world's major observatories: try Mauna Kea, Kitt Peak, and Cerro Tololo, for example.
2. The Gemini telescopes are at Mauna Kea and Cerro Tololo. Use RedShift to check the range of the sky visible from each tonight. Can the Large Magellanic Cloud be seen from either? Can the galaxy M31 in Andromeda be seen from either?

This long exposure shows stars as well as Comet C/1996 B2 (Hyakutake) circling the north celestial pole.

# Chapter 6

# The Sky and the Calendar

*This chapter can be treated at any point in the course.*

**AIMS:** To understand how astronomers locate objects in the sky, how objects appear to move in the sky, and time zones and calendars

When we look up at the sky on a dark night from a location outside a city, we see a fantastic sight. If the Moon is up, its splendor can steal the show; not only does its pearly white appearance draw our attention, but also the light it gives off makes the sky so bright that many faint objects cannot be seen. But when the Moon is down, we see bright jewels in the inky sky. Generally, a few of the brightest shine steadily, which tells us that they are planets. The rest twinkle—sometimes gently, sometimes fiercely—which reveals them to be stars.

The Milky Way arches across the sky, and if we are in a good location on a dark night, it is quite obvious to the naked eye. (City dwellers may never see the Milky Way at all.) If you know where to look, from the northern hemisphere you can see a hazy spot that is actually a galaxy, rather than individual stars. It is known as the Andromeda Galaxy from its location in the constellation Andromeda. Along with one other galaxy, it is the farthest in the Universe we can see with the naked eye.

If we are lucky, a bright comet may be in the sky, but this happens rarely. More often, meteors—"shooting stars"—dart across the sky above. And we may see the steady light of a spacecraft cross the sky in a few minutes. Some spacecraft even seem to flare up dramatically as their solar panels reflect sunlight toward us.

## 6.1 THE CONSTELLATIONS

Long, long ago, when Egyptian and other ancient astronomers were beginning to study and understand the sky, they divided the sky into regions containing fairly distinct groups of stars. The groups, called **constellations,** were given names, and stories were associated with them, perhaps to make them easier to remember.

Actually, the constellations are merely areas in the sky that happen to have stars in particular directions as we see them from the Earth. There is no physical significance to the apparent groupings, nor are the stars in a given constellation necessarily associated with each other in any direct manner (Fig. 6–1).

Many of the stories we now associate with the constellations come from Greek mythology (Fig. 6–2*A*), though the names may have been associated with particular constellations more to honor Greek heroes than because the constellations actually looked like these people. Other civilizations (American Indians, for example) attached their own names, pictures, and stories to the stars and constellations (Fig. 6–2*B*). For

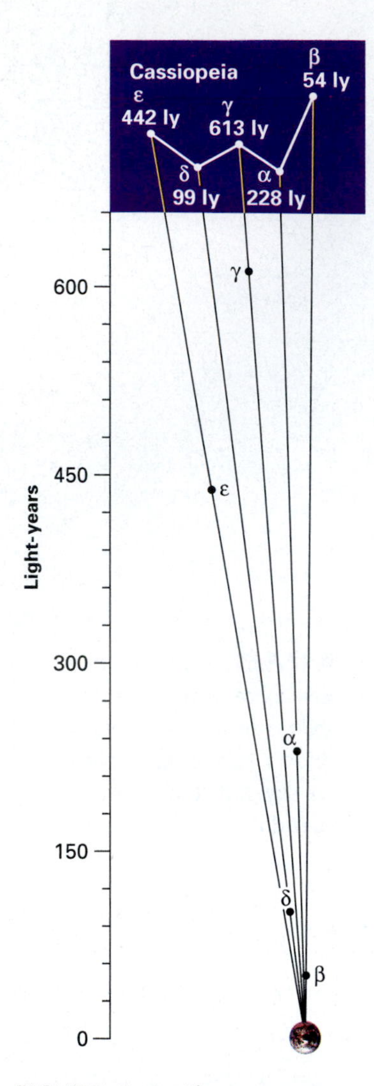


FIGURE 6–1 The stars we see as a constellation are actually differing distances from us. In this case, in the lower part of the figure we see the true distances of the brightest stars in Cassiopeia. Their appearance projected on the sky is shown in the upper part.

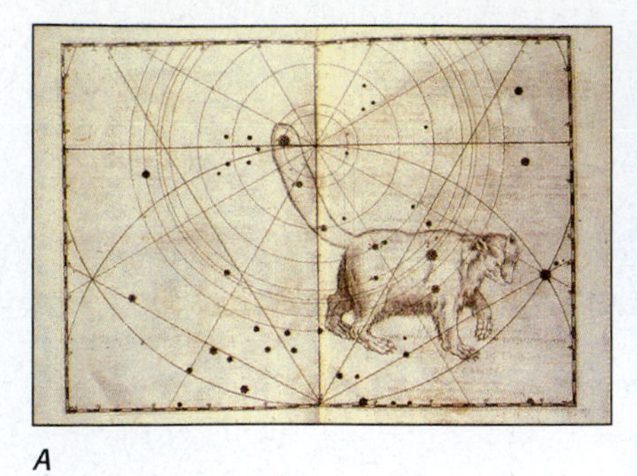

A

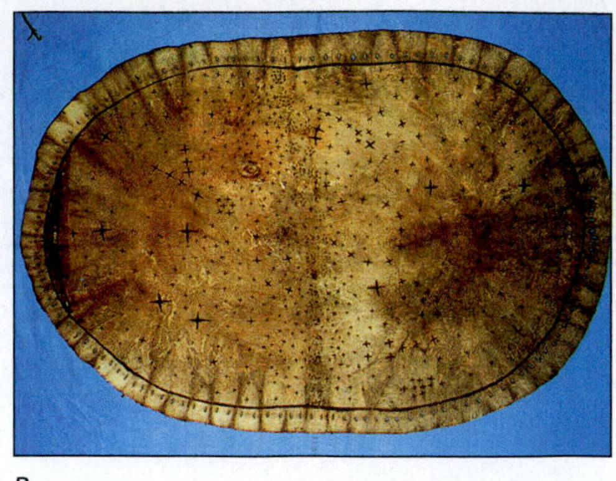

B

C

FIGURE 6–2 (*A*) Bayer, in 1603, used Greek letters to mark the brightest stars in constellations; he also used lower-case Latin letters. Here we see Ursa Minor, the Little Bear. (*B*) A sky chart from the Pawnee Indians who lived along the Platte River in what is now Nebraska. (*C*) Twelve constellations through which the Sun, Moon, and planets pass make up the zodiac. (*Drawing by Handelsman; © The New Yorker Collection 1978 Handelsman from cartoonbank.com. All rights reserved.*)

The International Astronomical Union put the scheme of constellations on a definite system in 1930. The sky was officially divided into 88 constellations (see Appendix 9) with definite boundaries, and every star is now associated with one and only one constellation.

example, the star that is nearest to stationary in the sky, with all the other stars seemingly revolving around it, we call Polaris; it is the North Star. (It isn't particularly bright; it just happens to be well located.) The Norse called it the Jeweled Nailhead, the Mongols called it the Golden Peg, the Chinese called it Emperor of Heaven, and the Skidi band of Pawnee Indians called it the Chief Star. In our Draco, the Dragon, the Egyptians saw the Hippopotamus. Our Orion, the Hunter, was the White Tiger to the Chinese of three thousand years ago.

Sometimes familiar groupings in the sky do not make up a constellation. Such groupings are called **asterisms;** the Big Dipper is an example, because it is only part of the constellation Ursa Major (the Big Bear).

In the star atlas he published in 1603, Johann Bayer assigned Greek letters (Appendix 6) in alphabetical order to the stars in each constellation, usually roughly in order of brightness. Thus $\alpha$ (alpha) is usually the brightest star in a constellation, $\beta$ (beta) is the second brightest, and so on. (In the Big Dipper, the letters go around the bowl rather than by brightness.) The Greek letters are used with the genitival form ("of . . .") of the constellation name (Appendix 9), as in $\alpha$ Orionis, meaning "alpha of Orion," which is Betelgeuse. For fainter stars, we sometimes use the Latin letters Bayer used when he ran out of Greek letters. Or we may use the numbers assigned to stars from side to side in a constellation in order of position (right ascension, Section 6.3) by John Flamsteed in England a century later: 61 Cygni, for example.

## 6.2 TWINKLING

What about the twinkling? It is not a property of the stars themselves, but merely an effect of our Earth's atmosphere. The starlight is always being bent by moving volumes of air in our atmosphere. These volumes of air bend light, primarily because they are different in temperature from their surroundings and also because they are different in density and water-vapor content. The effect makes the images of the stars appear to be larger than points, to dance around slightly (such image motion is called "seeing," as discussed in Section 4.5a), and to change rapidly in intensity (a property called scintillation). The change in intensity—scintillation—is what we non-technically call "twinkling" (Fig. 6–3).

Planets, unlike stars, do not usually seem to twinkle. They seem steadier because they are close enough to Earth so that they appear as tiny disks large enough to be seen through telescopes. Though each point of the disk of light representing a planet may change slightly in intensity, the disk is made of so many points of light that the total intensity doesn't change. To the naked eye the planets thus appear to shine more steadily than the stars. But when the air is especially turbulent, or when a planet is so low in the sky that we see it through a long column of air, even a planet may twinkle.

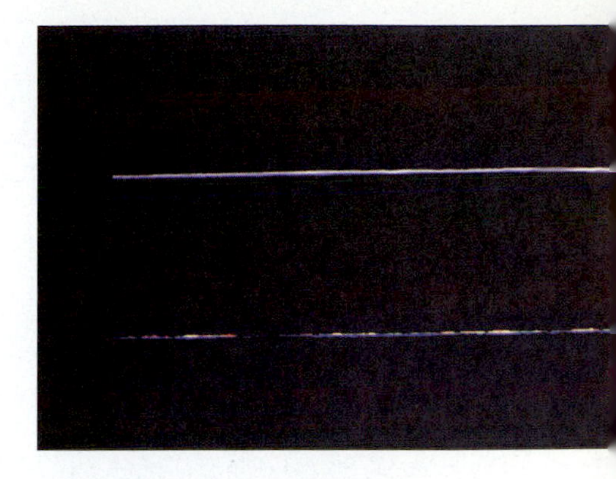

**FIGURE 6–3** A camera was moved steadily, from left to right, with the bright star Sirius in view. The star trails show twinkling. Also, the camera was moved from left to right when pointed at Jupiter, which left a solid, non-twinkling trail.

## 6.3 COORDINATE SYSTEMS*

The stars and other astronomical objects that we study are at a wide range of distances from the Earth. Though we may have to know the distance to an object to understand how much energy it is giving off, we need to know only an object's direction to observe it. It is thus often useful to think of the astronomical objects as being at a common distance, all hung on the inside of a large (imaginary) **celestial sphere.** In this section, we see how to describe the positions of objects on the celestial sphere.

Both astronomers and geographers have established systems of coordinates—**coordinate systems**—to designate the positions of places in the sky or on the Earth. The geographers' system is familiar to most of us: longitude and latitude. Lines (actually half-circles) of longitude on Earth called **meridians** run from the north pole to the south pole. The zero circle of longitude has been adopted, by international convention, to run through the former site of the Royal Observatory at Greenwich in England (Fig. 6–4). We measure longitude by the number of degrees east or west an object is from the meridian that passes through Greenwich, the Prime Meridian. Latitudes are defined by parallel circles that run around the Earth, all parallel to the equator. Latitude 0° corresponds to the equator; latitude ±90° corresponds to the poles. Los Angeles, for example, is at longitude 118° W and latitude 34°N.

**FIGURE 6–4** The international zero circle of longitude in Greenwich, England.

The astronomers' system for the sky is similar except that astronomers use the names **right ascension** for the celestial analogue of longitude (both measure east-west) and **declination** for the celestial analogue of latitude (both measure north-south). Right ascension and declination are measured with respect to a **celestial equator,** which lies above the Earth's equator, and **celestial poles,** which are on the extensions of the Earth's axis of spin (Fig. 6–5). The declination of the celestial equator is 0°; that of the north celestial pole +90° and that of the south celestial pole –90°. A faint star happens to be within 1° of the north celestial pole. We call it Polaris. It is easy to find because the end stars of the bowl of the Big Dipper point at it.

Right ascension and declination form a coordinate system fixed to the stars. To observers on Earth, the stars appear to revolve every 23 hours 56 minutes. The coordinate system thus appears to revolve at the same rate. Actually, of course, the Earth is rotating while the stars and celestial coordinate system remain fixed.

Although the stars are fixed in their positions in the sky, the Sun's position varies through the whole range of right ascension each year. The path of the Sun in the sky with respect to the stars is the **ecliptic.** The ecliptic is inclined by 23½° with respect to the celestial equator, since the Earth's axis is tipped by that amount. The ecliptic and the celestial equator cross at two points. The Sun crosses one of these points,

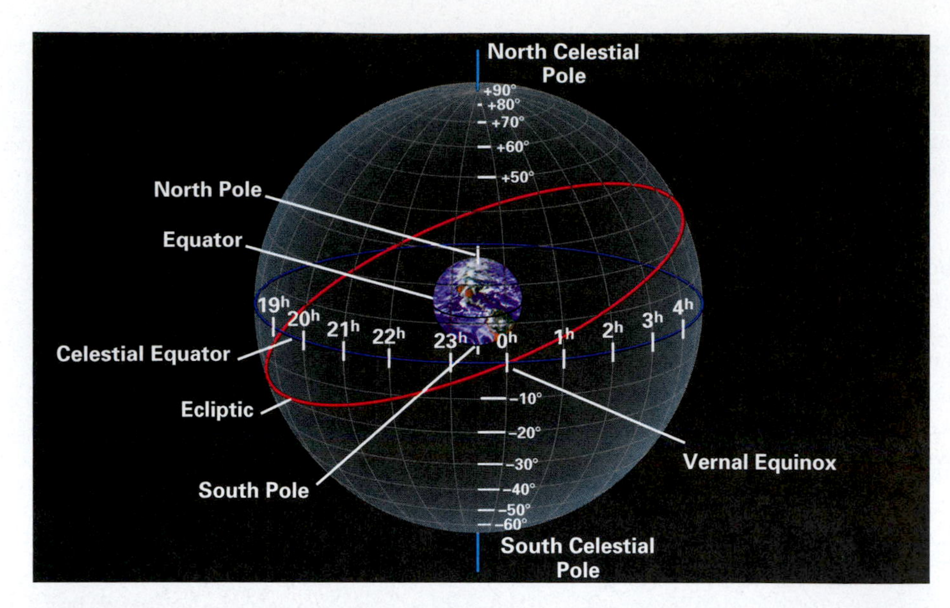


**FIGURE 6–5** The celestial equator is the projection of the Earth's equator onto the sky, and the ecliptic is the Sun's apparent path through the stars in the course of a year. The vernal equinox is one of the intersections of the ecliptic with the celestial equator and is the zero-point of right ascension. From a given location at the latitude of the United States, the stars nearest the north celestial pole never set and the stars nearest the south celestial pole never rise above the horizon.

Right ascension is measured along the celestial equator. Each hour of right ascension equals 15°. Declination is measured perpendicularly to right ascension (–10°, –20°, etc.).

the **vernal equinox,** on the first day of northern-hemisphere spring. The Sun crosses the other intersection, the **autumnal equinox,** on the first day of autumn. The vernal equinox is the zero-point of right ascension.

Right ascension is kept in units of time; since the sky passes once around each day, 360° = 24 hours, and we can divide both sides by 24 to see that 15° = 1 hour of right ascension. One degree (°) is subdivided into 60 minutes of arc (60′), each of which in turn is subdivided into 60 seconds of arc (60″); similarly an hour is divided into 60 minutes ($^{m}$), each of which in turn is subdivided into 60 seconds ($^{s}$). For example, the position of Sirius is right ascension $6^{h}45^{m}$ and declination –16°43′.

## EXAMPLE 6.1 Right Ascension and Declination

On September 22 each year, the day of the autumnal equinox, the Sun has declination 0°. In the yearly *Astronomical Almanac,* you can look up the right ascension and declination of the Sun and other objects for each day. At some time on this day, the Sun has right ascension 12 hours 00 minutes 00 seconds.

In Appendix 5 of this text, you can look up that Sirius has right ascension 6 hours 45 minutes and declination –16°43′, as stated above.

*Questions:*

(a) What are the right ascension and declination of the Sun on the vernal equinox, approximately March 21?

(b) What are the right ascension and declination of the Sun on the summer solstice, approximately June 21?

(c) What are the right ascension and declination of Sirius on the summer solstice?

*Answers:*

(a) On the equinoxes, the declination of the Sun is 0°. But the Earth will have moved halfway around the Sun in the six months between the autumnal and vernal equinoxes, so from the Earth looking at the Sun, the Sun is halfway around in the sky with respect to the stars. So its right ascension must be 0°, halfway around from 12°.

(b) At the summer solstice, the Sun's declination is +23½°, the highest it ever gets. Its right ascension advances 24 hours every 12 months. On September 21, its right ascension is 12 hours, as given; on the winter solstice, December 21, it is 18 hours; on the vernal equinox, it is 24 hours; and on the summer solstice, June 21, it is 6 hours.

(c) The stars are fixed in the sky on the system of right ascension and declination, so Sirius's right ascension and declination remain the same as they were given for the autumnal equinox.

Technically, we measure right ascension and declination with the use of **hour circles,** great circles running through the celestial poles. (A "great circle" on a sphere is a circle that is also on a plane that goes through the center of the sphere; it is the largest possible circle that can be drawn on the sphere's surface. On a sphere, the shortest distance between two points is on a great circle.) The hour circles cross the celestial equator perpendicularly. Right ascension is the angle to a body's hour circle, measured eastward along the celestial equator from the vernal equinox. Declination is the angle of an object north (+) or south (−) of the celestial equator along an hour circle.

Astronomers have set up a timekeeping system called **sidereal time** (sidereal means "by the stars"). Each location on Earth has a **meridian,** the great circle linking the north and south poles and passing through the **zenith,** the point directly overhead. The sidereal time at any location is the length of time since the vernal equinox has crossed the local meridian. As a result, the sidereal time is equal to the right ascension of any star on that place's meridian.

### EXAMPLE 6.2 Sidereal Time by the Stars

***Question:*** At what sidereal time does Barnard's star, one of the closest stars to us, cross due south of you?

***Answer:*** Barnard's star, from Appendix 7, has r.a. $17^{h}57.9^{m}$ and dec. $+04°41'$. From the right ascension, it crosses your meridian facing south at sidereal time $17^{h}57.9^{m}$.

Note that Appendix 5 shows the brightest stars and Appendix 7 is for the nearest stars.

Each north-south line on the Earth has the same sidereal time at each instant. Astronomers find sidereal time convenient because a star whose right ascension matches the current sidereal time is as high in the sky as it ever gets. So astronomers have clocks that run on sidereal time, to help them tell easily whether a star is favorably placed for observing.

Sidereal time and solar time differ, though both are caused by the rotation of the Earth on its axis. A **sidereal day** (a day by the stars) is the length of time that it takes the vernal equinox to return to the celestial meridian. A **solar day** is the length of time that the Sun takes to return to your meridian. Since the Earth revolves around the Sun once a year, by the time a day has passed, the Earth has moved 1⁄365 of the way around the Sun. Thus after the Earth has turned far enough for the stars to return to the same apparent positions in the sky, the Earth must still turn an additional 1⁄365 of 24 hours (24 hours/365 = 4 minutes) for the Sun to return to your meridian. A solar day is thus approximately 4 minutes (actually 3 minutes 56 seconds of time) longer than a sidereal day (Fig. 6–6). Thus the stars rise about 4 minutes

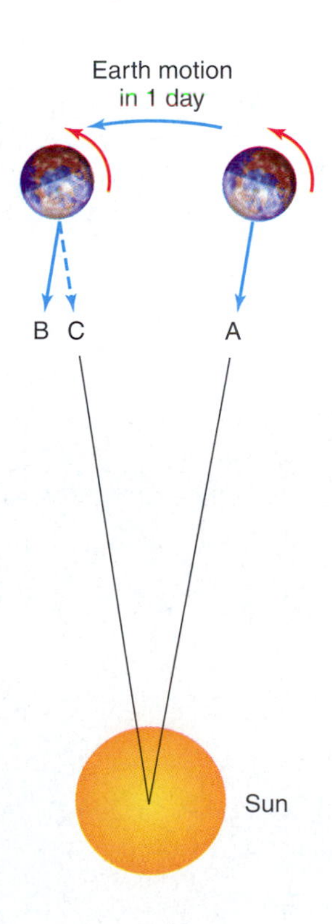


**FIGURE 6–6** While the Earth rotates once on its own axis with respect to the stars (one sidereal day), it also moves slightly in its orbit around the Sun. Thus after one sidereal day, arrow A becomes arrow B. But one solar day has passed only when arrow B rotates a little farther and becomes arrow C. This takes an additional 4 minutes, making a solar day 4 minutes longer than a sidereal day.

per day earlier according to the solar time on our normal clocks. After 30 days, a given star rises 30 days × 4 min/day = 120 minutes or 2 hours earlier. Because of this effect, which constellations are up at night changes from season to season.

Every observatory has both sidereal clocks, for the astronomers to tell when to observe their stars, and solar clocks, for the astronomers to gauge when sunrise will come and to know when to go to dinner. A solar clock and a sidereal clock show the same time (on a 24-hour system) only one instant each year, the autumnal equinox. (An equinox is both a point in the sky and the time when the Sun passes that point.) The next day the sidereal clock is 4 minutes ahead, the second day afterward it is 8 minutes ahead, and so on. Six months later the two clocks differ by 12 hours, and the stars that were formerly at their highest at midnight are then at their highest at noon, when the Sun is out. As a result, they may not be visible at all at that season.

## EXAMPLE 6.3 Sidereal and Solar Time

***Question:*** At what solar time does Barnard's star cross due south of you on September 22nd? On September 26th?

***Answer:*** From Example 6.1, Barnard's star crosses your meridian facing south at sidereal time $17^h57.9^m$. This sidereal time corresponds to solar time 17:58, which is 5:58 p.m. on September 21st. Four days later, the sidereal clock at midnight reads 16 minutes later than the solar clock, so the solar time for Barnard's star will be 16 minutes earlier, or 5:42 p.m. (ignoring the partial shift over the course of a day). After all, the stars rise 4 minutes earlier each day by solar time.

Though the coordinates are basically fixed to the stars, a small effect called **precession** causes a slow drift of the coordinate system with respect to the stars with a 26,000-year period (Fig. 6–7). Precession takes place because the spinning Earth is pulled by the Sun's gravity (and to a lesser extent by the gravitational pull of other planets). The effect of precession is that Earth's axis doesn't always point exactly at the same spot in the sky; the axis rather traces out a small circle. (The effect is like the wobbling of a spinning top.) The axis takes approximately 26,000 years to return to the same orientation. About halfway through the cycle from now—in, say, A.D. 15,000 (our A.D. 2000 plus half of 26,000 years)—the north star will be Vega. But don't worry—Polaris will be our north star again in about 26,000 years.

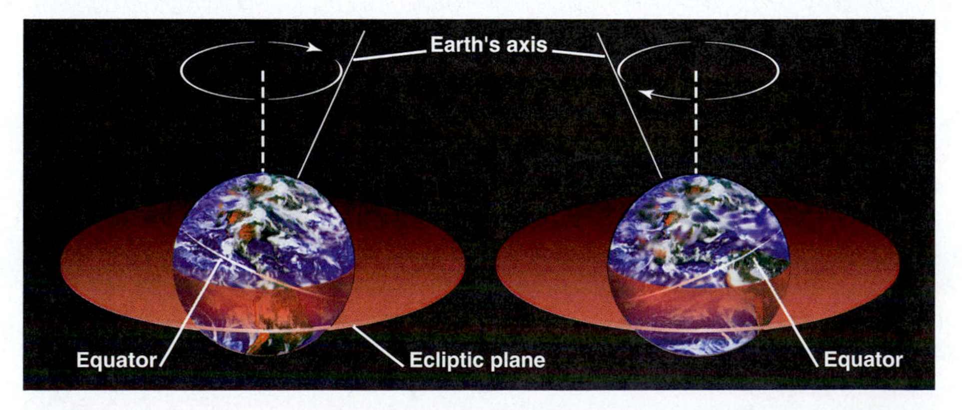


**FIGURE 6–7** The Earth's axis precesses with a period of 26,000 years. The two positions shown are separated by 13,000 years.

As the Earth's pole precesses, the equator moves with it (since the Earth is a rigid body). The celestial equator and the ecliptic will always maintain the $23\frac{1}{2}°$ angle between them, but the points of intersection, the equinoxes, will change. Thus, over the 26,000-year precession cycle, the vernal equinox will move through all the signs of the zodiac. It is now in the constellation Pisces and approaching Aquarius (and thus the celebration in the musical *Hair* of the "Age of Aquarius").

Because the pole moves, the celestial equator—which is always 90° from the pole—moves. Hence the **equinoxes,** which are the intersections of the celestial equator and the ecliptic, precess—that is, they apparently move slowly along the ecliptic. Because of this **precession of the equinoxes,** the right ascension and declination of objects in the sky change slowly. (The formulas for computing these changes are given in Appendix 2.)

As a result of precession, one has to make small corrections in any catalogue of celestial positions to update them to the present time. Precession is a small effect—the change in celestial coordinates is less than one minute of arc (1/60 of a degree) per year—and is much less for some parts of the sky. Thus it need not be taken into account for casual observing, though star maps and catalogues are now drawn for "epoch 2000.0."

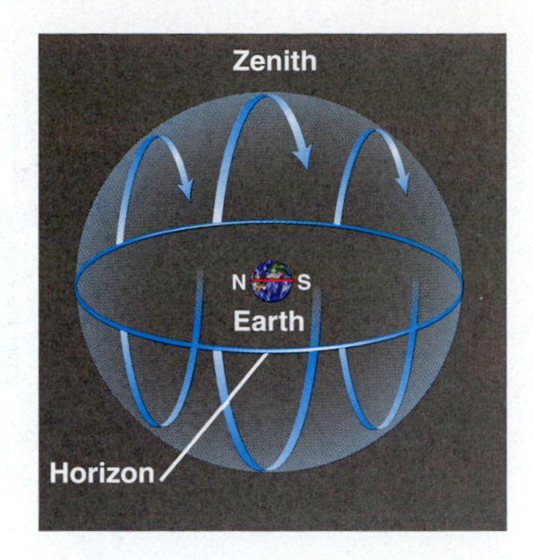


**FIGURE 6–8** From the equator, the stars rise straight up, pass right across the sky, and set straight down.

## 6.4 MOTIONS IN THE SKY*

At the latitudes of the United States, which range from +25° for the tip of the Florida Keys up to +49° for the Canadian border, and down to +19° in Hawaii and up to +67° in Alaska, the stars rise and set at angles to the horizon. In order to understand the situation, it is best first to visualize simpler cases. These concepts involve spherical geometry, which is difficult to visualize without practical experience with a telescope or in a planetarium.

If we were standing on the equator, the stars would rise perpendicularly to the horizon (Fig. 6–8). The north celestial pole would lie exactly on the horizon in the north, and the south celestial pole would lie exactly on the horizon in the south. Each star would rise somewhere on the eastern half of the horizon; each would remain "up" for twelve hours, and then would set. By waiting long enough, we would be able to see all the stars, no matter what their declination. The Sun, no matter what its declination, would also rise, be up for twelve hours, and then set, so day and night would each last twelve hours.

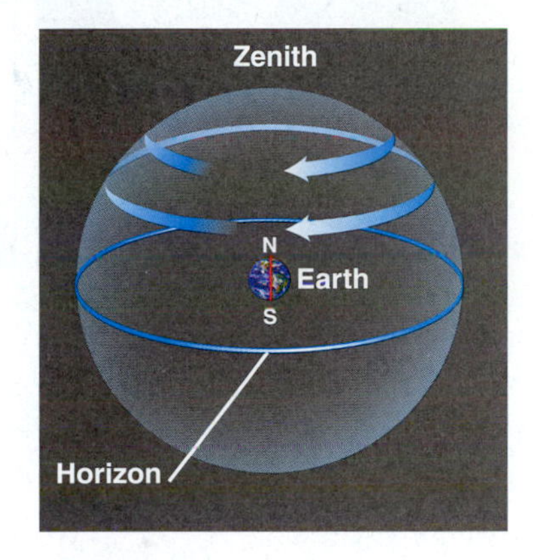


**FIGURE 6–9** From the pole, the stars move around the sky in circles parallel to the horizon, never rising or setting.

If, on the other hand, we were standing on the north pole, the north celestial pole would be directly overhead, and the celestial equator would be on the horizon (Fig. 6–9). All the stars would move around the sky in circles parallel to the horizon. Since the celestial equator would be on the horizon, we could see only the stars with northern declinations. The stars with southern declinations would never be visible.

Let us consider a latitude between the equator and the north pole, say +40°. There the stars seem to rise out of the horizon at oblique angles. The north celestial pole (with the north star nearby it) is always visible in the northern sky, and is at an **altitude** of 40° above the horizon, where "altitude" is the angle measured upward perpendicularly to the horizon. The star Polaris happens to be located within one degree of the north celestial pole, and so is called the **pole star.** (It is only a 2nd-magnitude star, not so very bright, and is famous only because of its location rather than because of its brightness.) If you can see the north celestial pole, then the south celestial pole must be hidden. The celestial pole you can see is the only fixed point in the sky (Fig. 6–10).

### EXAMPLE 6.4 The Altitude of a Star

*Question:* What is the altitude of Barnard's star as it crosses due south of you?

*Answer:* If you are at +40° latitude, the celestial equator is 40° south of your zenith. Since Barnard's star is slightly north of the celestial equator, it is $40° - (+04°41') = 35°19'$ south of your zenith. Its altitude is therefore $90° - 35°19' = 54°41'$.

If you point a camera at the sky and leave its lens open for a long time—many minutes or hours—the stars appear as trails. Those near celestial poles (Fig. 6–11

*A*

**FIGURE 6–10** (*A*) In the real world, there is no East Star or West Star, since the Earth rotates on its axis. Only the positions of the north and south celestial poles are steady. (*Cartoon by Charles Schulz. ©1970 United Feature Syndicate, Inc.*)

*B*

**FIGURE 6–10** (*B*) Lucy is still making things up. Unless you are right on the equator, you can never see the north and south celestial poles in the sky at the same time, since they are 180° from each other in the sky. Further, there is no South Star; that is, there is no obvious star near the south celestial pole. (*Cartoon by Charles Schulz. ©1970 United Feature Syndicate, Inc.*)

and chapter opening photograph) move in relatively small and obvious circles around the poles. The circles followed by stars farther from the celestial poles are so large that sections of them seem straight (Fig. 6–12).

When astronomers want to know if a star is favorably placed for observing, they must know both its right ascension and declination (Section 6.3). By seeing if the right ascension is reasonably close to the sidereal time, they can tell if it is the best time of year at which to observe it. But they must also know the star's declination to know how long it will be above the horizon each day.

Telescopes are often mounted at an angle such that one axis—the **polar axis**—points directly at the north celestial pole. Since all stars move across the sky in circles centered at the pole, the telescope must merely turn about that axis to keep up with stellar motions. The other axis of the telescope is used to point the telescope in declination. Since motion around only one axis is necessary to track the stars, one need have only a single motor set to rotate once every 24 sidereal hours. Arrangements of this type are called **equatorial mounts.**

**FIGURE 6–11** A montage of images of Comet C/1996 B2 (Hyakutake) taken every hour, showing its rotation around the north celestial pole. Multiple images of the stars also show as they circle the pole.

The alternative to this system is to mount a telescope such that one axis goes up-down (that is, changes in altitude) and the other goes around (that is, changes in **azimuth**). Computers make the necessary calculations that allow many new large and small telescopes to be mounted using such an **alt-azimuth** system.

**FIGURE 6–12** Near the celestial equator, the star circles are so large that they appear almost straight. Here we are looking east past the twin Keck telescopes on Mauna Kea.

## 6.5 POSITIONS OF THE SUN, MOON, AND PLANETS*

As the Sun moves along the ecliptic each year, it crosses the vernal equinox on approximately March 21st and the autumnal equinox on approximately September 22nd. On these days, the Sun's declination is 0°; the Sun's declination varies over the year from +23½° to −23½° (Fig. 6–13).

These points are called **equinoxes** ("equal nights"; *nox* is Latin for night) because the daytime and the nighttime are supposedly equal on these days. Actually, because refraction (bending) of light by the Earth's atmosphere makes the Sun appear to rise a little early and set a little late, because the top of the Sun rises ahead of the Sun's midpoint, and because the Sun's apparent rate of travel across the sky varies over the year, daytime exceeds nighttime at U.S. latitudes by about 10 minutes on the days of the equinoxes. The dates of equal daytime and nighttime precede the vernal equinox and follow the autumnal equinox by a few days.

If we were at the north pole, whenever the Sun had a northern declination, it would be above our horizon and we would have daytime. The Sun would move in a circle all around us, moving essentially parallel to the horizon. From day to day it would appear slightly higher in the sky for 3 months, and then move lower. The date when it is highest in the sky is the summer **solstice.** It occurs on approximately June 21st each year. On that date the Sun is 23½° above the horizon because its declina-

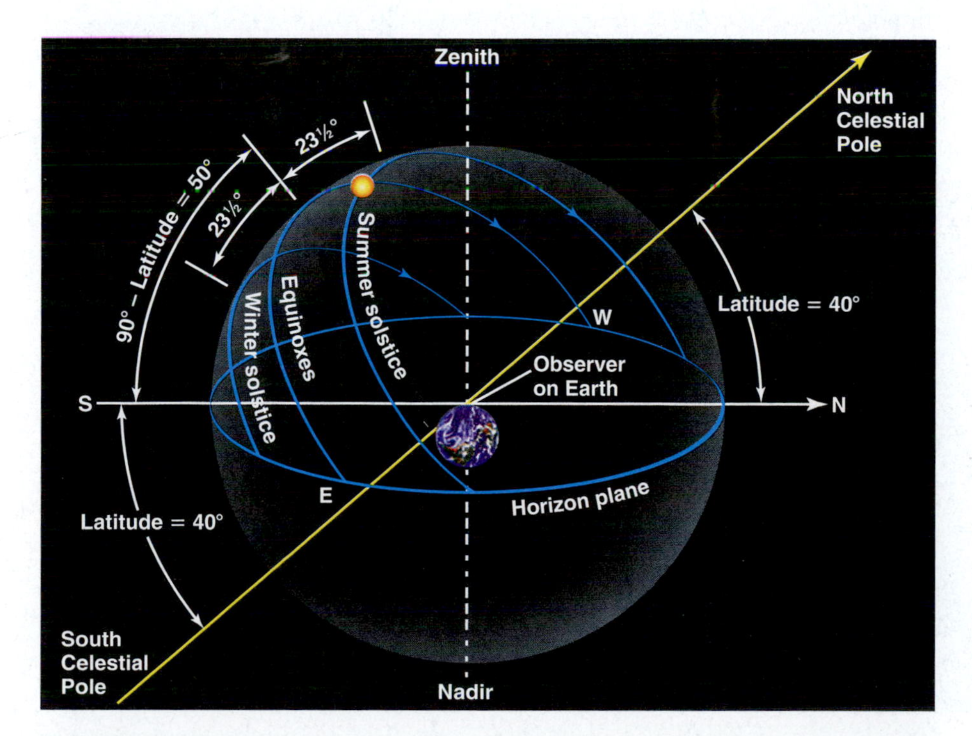


**FIGURE 6–13** The blue arcs from the equator up and over the top of the sphere shown are the path of the Sun at different times of the year. Around the summer solstice, June 21st, the Sun (shown in yellow) is at its highest declination, rises highest in the sky, stays up longer (because, as shown, more of its path is above the horizon), and rises and sets farthest to the north. The opposite is true near the winter solstice, December 21st. The diagram is drawn for latitude 40°.

**FIGURE 6–14** In this series taken in June from northern Norway, above the Arctic Circle, one photograph was taken each hour for an entire day. The Sun never set, a phenomenon known as the **midnight sun.** Since the site was not at the north pole, the Sun and stars moved somewhat higher and lower in the sky in the course of a day.

tion is $+23\frac{1}{2}°$. The time of the year when the Sun never sets is known as the time of the **midnight sun** (Fig. 6–14), which lasts six months at the poles and shorter times at other locations within $23\frac{1}{2}°$ of the poles.

Since the Sun goes $23\frac{1}{2}°$ above the celestial equator, the midnight sun is visible at some time anywhere within $23\frac{1}{2}°$ of the north pole, a boundary at $66\frac{1}{2}°$ latitude known as the Arctic Circle. The midnight sun seen from within $23\frac{1}{2}°$ of the south pole, within the Antarctic Circle, is six months out of phase with that near the north pole.

When the Sun is at the summer solstice, it is at its greatest northern declination and is above the horizon of all northern hemisphere observers for the longest time

**FIGURE 6–15** The seasons occur because the Earth's axis is tipped with respect to the plane or the orbit in which it revolves around the Sun. The dotted line is drawn perpendicularly to the plane of the Earth's orbit. When the northern hemisphere is tilted toward the Sun, it has its summertime; at the same time, the southern hemisphere is having its winter. At both locations of the Earth shown, the Earth rotates through many 24-hour day-night cycles before its motion around the Sun moves it appreciably. The diagram is not to scale.

(Figure 6–14 continued)

each day. Thus daytimes in the summer are longer than daytimes in the winter, when the Sun is at its lowest declinations. In the winter, the Sun not only is above the horizon for a shorter period each day but also never rises very high in the sky. As a result, the weather is colder. The instant of the Sun's lowest declination is the winter solstice. It is winter in the northern hemisphere when it is summer in the southern.

The seasons (Fig. 6–15), thus, are caused by the variation of declination of the Sun, which in turn is caused by the fact that the Earth's axis of spin is tipped by 23½°. If the Earth weren't tipped at all, then the Sun would always shine directly down on the Earth's equator; as the Earth rotated under the Sun once a day, people on the equator would always see the Sun overhead at noon and people at north latitude 45° would always see the Sun 45° down from the zenith at noon. But because the Earth's axis of spin is tipped, sometimes your latitude is tipped toward the Sun at noon. Then the Sun is high in the sky at noon and it is your summertime. About six months later, your latitude will be tipped away from the Sun at noon. Then the Sun is low in the sky at noon and it is your winter. The effect of the tipping is much greater than the minor effect of the differing distances between the Earth and the Sun caused by the Earth's orbit being slightly elliptical.

Many if not most people misunderstand the cause of the seasons. It is important to realize that the seasons are not caused by variation of the distance between the Earth and the Sun. Indeed, the Earth is closest to the Sun each year on or about January 4th, which falls in the northern-hemisphere winter. The variation of the Earth's distance from the Sun is very small, since the Earth's orbit—though elliptical—is very close to round. Thus only the tilt causes the seasons.

The Moon goes around the Earth once each month, and so the Moon's right ascension changes through the entire range of ascension once each month. Since the Moon's orbit is inclined to the celestial equator (though only by about 5°), the Moon's declination also varies.

The planets' motions in the sky are less easy to categorize, but they also change their right ascension and declination from day to day. Because their orbits are not very inclined to the ecliptic plane, they do not stray too far from the ecliptic.

## 6.6 TIME AND THE INTERNATIONAL DATE LINE*

Every city and town on Earth used to have its own time system, based on the Sun, until widespread railroad travel made this diversity inconvenient. In 1884, an international conference agreed on a series of longitudinal time zones. Now all localities in the same zone have a standard time (Fig. 6–16). Since there are 24 hours in a day, the 360° of longitude around the Earth are divided into 24 standard time zones, each 15° wide. Each time zone is centered on a meridian of longitude exactly divisible by 15. Because the time is the same throughout each zone, the Sun is not directly overhead at noon at each point in a given time zone but in principle is less than about a half-hour off. Standard time is based on a **mean solar day,** the average length of a solar day.

As the Sun seems to move in the sky from east to west, the time in any one place gets later. We can visualize noon, and each hour, moving around the world from east to west, minute by minute. We get a particular time back 24 hours later, but if the hours circled the world continuously, the date would not have changed. So we specify a north-south line and have the date change there. We call it the **international date line.** England won for Greenwich, then the site of the Royal Observatory, the distinction of having the basic line of longitude (the Prime Meridian), 0°. Since the

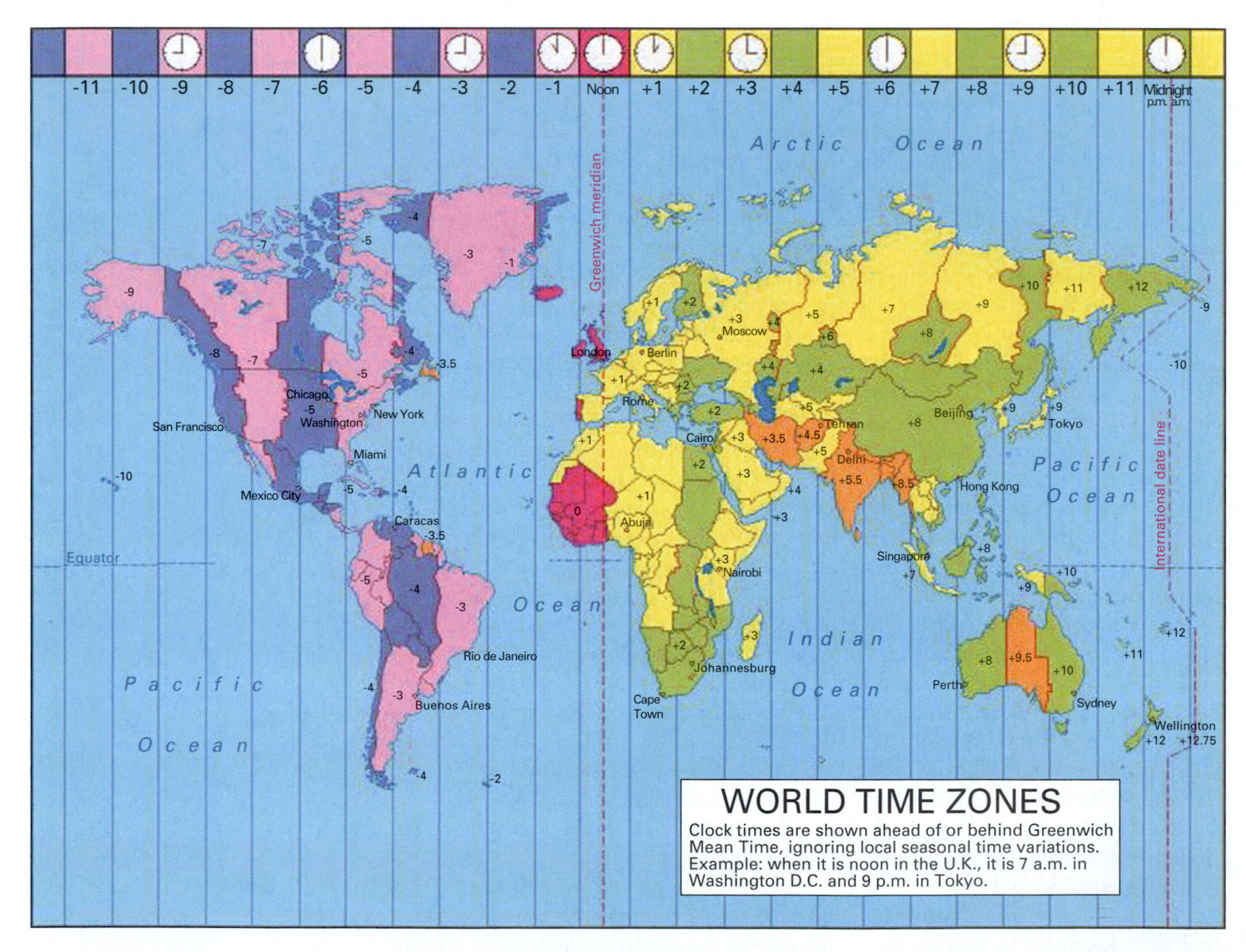


**FIGURE 6–16** Although in principle the Earth is neatly divided into 24 time zones, in practice, political and geographic boundaries have made the system much less regular. The existence of daylight-saving time in some places and not in others further confuses the time zone system. Other countries, shown in purple, have time zones that differ by one-half hour from a neighboring zone. India and Nepal actually differ from each other by 15 minutes! At the international date line, the date changes. In the United States, most states have daylight-saving time for seven months a year—from the first Sunday in April until the last Sunday in October. But Arizona, Hawaii, and parts of Indiana have standard time year-round. The United States has nine time zones: Atlantic (for the U.S. Virgin Islands and Puerto Rico), Eastern, Central, Mountain, Western, Alaska, Hawaii, Samoa, and Chamorro. The Chamorro time zone was named in 2001 after the original inhabitants of Guam and the Northern Marianas.

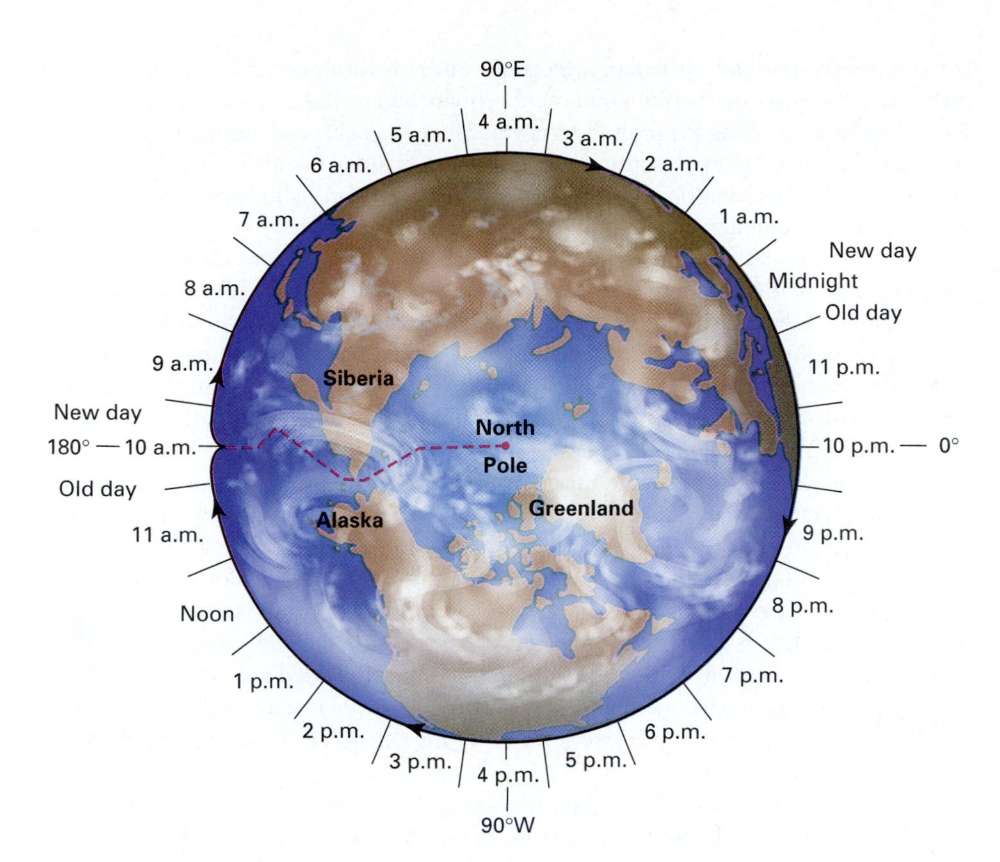


**FIGURE 6–17** This top view of the Earth, looking down from above the north pole, shows how days are born and die at the international date line. When you cross the international date line, your calendar changes by one day.

international date line would disrupt the calendars of those who crossed it, that line was put as far away from the populated areas of Europe as possible—near or along the 180° longitude line. When one flies from Hawaii to Japan across the international date line, the day will change from, say, Tuesday to Wednesday. The international date line passes from north to south through the Pacific Ocean and actually bends to avoid cutting through continents or groups of islands, thus providing them with the same date as their nearest neighbor, as shown in Figure 6–17. Some Pacific islands even switched from one side of the date line to the other by moving the date line in order to attract more tourists on New Year's Eve 2000.

In the summer, in order to make the daylight last into later hours, many countries have adopted daylight-saving time. Clocks are set ahead 1 hour on a certain date in the spring. Thus if darkness falls at 6 p.m. E.S.T. (Eastern Standard Time), that time is called 7 p.m. E.D.T. (Eastern Daylight Time), and most people have an extra hour of daylight after work. In most places, that hour is taken away in the fall, though some places have adopted daylight-saving time all year. The phrase to remember to help you set your clocks is "fall back, spring ahead." Of course, daylight-saving time is just a bookkeeping change in how we name the hours, and doesn't result from any astronomical changes.

Daylight-saving time can be controversial. Hawaii, Alaska, and parts of Indiana don't follow it, though Indiana is considering unifying its practices. Some of the election reforms being proposed involve setting up uniform times across the country for poll closings and delaying the end of daylight-saving time in presidential election years until after election day.

## 6.7 CALENDARS*

The period of time that the Earth takes to revolve once around the Sun is called, of course, a **year.** This period is about 365¼ mean solar days. A **sidereal year** is the interval of time that it takes the Sun to return to a given position with respect to the stars. A **solar year** (in particular, a **tropical year**) is the interval between passages of the Sun through the vernal equinox, the point where the ecliptic crosses the celestial equator travelling from south to north. Since tropical years are growing shorter by about half a second per century, we refer to a standard tropical year: 2000. The speed of the Earth's rotation varies slightly and cannot be predicted accurately enough for long-term predictions of the highest precision.

Roman calendars had, at different times, different numbers of days in a year, so the dates rapidly drifted out of synchronization with the seasons (which follow solar years). Julius Caesar decreed that 46 B.C. would be a 445-day year in order to catch up, and he defined a calendar, the **Julian calendar,** that would be more accurate. This calendar had years that were normally 365 days in length, with an extra day inserted every fourth year to bring the average year to 365¼ days in length. The fourth years were, and are, called **leap years.**

The Julian calendar was much more accurate than its predecessors, but still imprecise; the actual solar year 1997 was 365 days 5 hours 48 minutes 45.2 seconds long, some 11 minutes 14.8 seconds shorter than 365¼ days (365 days 6 hours). By 1582, the calendar was about 10 days out of phase with the date on which Easter had occurred at the time of a religious council 1250 years earlier, and Pope Gregory XIII issued a bull—a proclamation—to correct the situation. He dropped 10 days from 1582. Many citizens of that time objected to the supposed loss of the time from their lives and to the commercial complications. Does one pay a full month's rent for the month in which the days were omitted, for example? "Give us back our fortnight," they cried.

In the **Gregorian calendar,** the calendar that we now use, years that are evenly divisible by four are leap years, except that three out of every four century years, the ones not divisible evenly by 400, have only 365 days. Thus 1600 was a leap year; 1700, 1800, and 1900 were not; and 2000 was again a leap year.

Although many countries adopted the Gregorian calendar as soon as it was promulgated, Great Britain (and its American colonies) did not adopt it until 1752, when 11 days were skipped. As a result, we say that George Washington's birthday was on February 22nd, even though a calendar when he was born read February 11th. Also, the beginning of the year was changed from March to January. Washington was born in February 1731 (about a month before the end of 1731), often then written February 1731/2 (Fig. 6–18), but we now refer to his date of birth as February 22nd, 1732. It will be over 3000 years before our Gregorian calendar is as much as one day out of step.

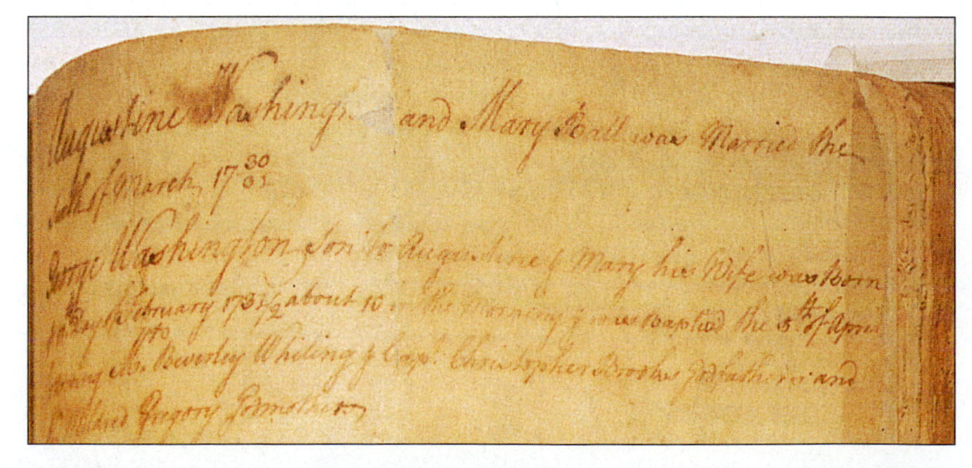
and Mary Ball was Married the
March 17 30/31
George Washington Son to Augustine & Mary his Wife was Born
February 1731/2 about 10 in the Morning & was baptis'd the
Mr. Beverley Whiting & Capt. Christopher Brooks Godfathers and
Mildred Gregory Godmother

**FIGURE 6–18** In George Washington's family Bible, his date of birth is given as "11th day of February 1731/2." Some contemporaries would have said 1731; we now say 1732.

## CORRECTING MISCONCEPTIONS

| ✖ Incorrect | ✔ Correct |
|---|---|
| Twinkling results from stars changing in brightness. | Twinkling is caused in the Earth's atmosphere. |
| Polaris is the brightest star in the sky. | Polaris is not a particularly bright star; it is merely easy to find and at a key location. |
| Changes in the Earth's distance from the Sun causes the seasons. | The seasons are caused by the tilt of the Earth's axis, and the Earth's distance from the Sun is irrelevant. |
| A set of stars is a constellation. | Constellations are regions of the sky, including space between and past the stars we see. |
| The Sun is a convenient timekeeper to help us observe the stars. | Our watches use solar time, which differs by 4 minutes a day from sidereal time. |
| The Moon is only up at night. | The Moon is often up in the daytime. |

## SUMMARY AND OUTLINE

The constellations (Section 6.1)
- Stars in them are not necessarily physically grouped.

Twinkling (Section 6.2)
- Stars appear to twinkle; planets usually do not.
- Twinkling is caused in the Earth's atmosphere.

Coordinate systems (Section 6.3)
- Celestial longitude and latitude are right ascension and declination.
- The celestial equator and celestial poles are the points in the sky that are on the extensions of the Earth's equator and poles into space. The ecliptic and the celestial equator cross at the equinoxes.
- Sidereal time is time by the stars.
- Sidereal day: a given right ascension returns to your meridian
  - Solar day: the Sun returns to your meridian
- Precession is the slow drift in the coordinate system; 26,000-year period.

Motions in the sky (Section 6.4)
- Stars rise and set; at the Earth's equator we would see them do so perpendicularly to the horizon.
- At the Earth's poles, we would see the stars move around parallel to the horizon.

Motions of the Sun, Moon, and planets (Section 6.5)
- At or near the poles, the Sun and Moon rise and set only when they change sufficiently in declination; the Sun goes through this sequence once a year—thus the "midnight sun."
- The seasons are caused by the Sun's variations in declination.

Time and the International Date Line (Section 6.6)
- Standard time is based on a mean solar day.
- The date changes at the international date line.
- Daylight-saving time: "fall back, spring ahead"

Calendars (Section 6.7)
- Leap years are needed to keep up with the 365¼-day year.
- The Julian calendar, introduced by Julius Caesar, was a reasonably good calendar, but, over the centuries, the days drifted.
- We now use the Gregorian calendar, set up in 1582.

## KEY WORDS

| | | | |
|---|---|---|---|
| constellations | ecliptic* | precession of the equinoxes* | midnight sun* |
| asterisms | vernal equinox* | equinoxes* | mean solar day* |
| celestial sphere* | autumnal equinox* | altitude* | international date line* |
| coordinate systems* | hour circles* | pole star* | year* |
| meridians* | sidereal time* | polar axis* | sidereal year* |
| right ascension* | zenith* | equatorial mounts* | solar year (tropical year)* |
| declination* | sidereal day* | azimuth* | Julian calendar* |
| celestial equator* | solar day* | alt-azimuth* | leap years* |
| celestial poles* | precession* | equinoxes* | Gregorian calendar* |

*These terms appear in an optional section.

## QUESTIONS

1. Explain why we cannot tell by merely looking in the sky that stars in a given constellation are at different distances, whereas in a room we can easily tell that objects are at different distances from us. What is the difference between the two situations?
2. Why is the Big Dipper only an asterism while the Big Bear is a constellation?
3. Can a star in the sky not be part of a constellation? Explain.
4. Can you reason that since all the stars in the constellation Pegasus appear close together in the sky, they must have formed at about the same time? Explain.
5. What is the difference between "seeing" and "twinkling"?
6. If you look toward the horizon, are the stars you see likely to be twinkling more or less than the stars overhead? Explain.
7. Would the planet Venus seem to twinkle by the same amount, more, or less if it were at the outskirts of the solar system?
8. Explain how it is that some stars never rise in our sky, while others never set.
9. By comparing their right ascensions and declinations (Appendix 5, the brightest stars), describe whether Sirius and Canopus, the two brightest stars, are close together or far apart in the sky. Explain.
10. †When Arcturus (Appendix 5) is due south of you, what is the sidereal time where you are?
11. †Between the vernal equinox, March 21st, and the autumnal equinox, about 6 months later, ignoring precession,
    (a) by how much does the right ascension of the Sun change?
    (b) by how much does the declination of the Sun change?
    (c) by how much does the right ascension of Sirius change?
    (d) by how much does the declination of Sirius change?
12. (a) Describe how the declination of the Sun varies over the year.
    (b) Does its right ascension increase or decrease from day to day? Justify your answer.
13. We normally express longitude on the Earth in degrees. Why would it make sense to express longitude in units of time?
14. †By how many hours do sidereal clocks and solar clocks differ at the summer solstice?
15. †Divide the number of minutes in a day by 365 to find out by how many minutes each day the sidereal day drifts with respect to the solar day, as the Earth goes 1/365 of the way around the Sun. Show your work.
16. †If a planet always keeps the same side toward the Sun, how many sidereal days are there in a year on that planet?
17. †When it is 6 p.m. on October 1st in New York City, what time of day and what date is it in Tokyo?
18. †When it is noon on April 1st in Los Angeles, describe how to use Figure 6-16 to find the date and time in China. Follow through both going westward across the international date line and going eastward across the Atlantic, and describe why you get the same answer.
19. What is the advantage of an equatorial mount?
20. What new development has led large telescopes to be placed on alt-azimuth mounts? Be explicit about the change and what made it possible.
21. A pair of friends walked south 2 kilometers, east 2 kilometers, and then north 2 kilometers, only to wind up back where they started, at which point they were eaten by a bear. What color was the bear? Explain.

† This question requires a numerical solution.

## TOPICS FOR DISCUSSION

1. Is it better to see the stars through the eyepiece of a telescope or to use an electronic detector and view on a TV screen? Discuss pros and cons of each.
2. When did the current millennium begin? Why do you think so?

## USING TECHNOLOGY

### W World Wide Web

1. Locate an old star atlas with beautiful constellation diagrams.
2. For time zones see http://www.time.gov or http://www.timezoneconverter.com.

### REDSHIFT

1. Show the path of the Sun in the sky for the solstices, the equinoxes, and halfway in between. Notice the maximum altitude of the Sun and the length of the day.
2. Find the time at which the Sun rises, the time at which the Sun sets, and the length of the day for each day during the week before and after the summer solstice. Explain why the longest day is not exactly at the solstice.
3. Duplicate the midnight sun effect at latitudes 70°, 80°, and 90°.
4. The analemma is the path of the Sun in the sky at the same time of day (say, 10 a.m.) each year. Duplicate the analemma and compare it with the image you find on this book's Web site.
5. Examine this evening's sky, and note the Sky Reference Markers: the zenith, the celestial equator, the ecliptic, and the north and south celestial poles.
6. Learn about celestial coordinates from The Story of the Universe: The Celestial Sphere; the Guided Tour: RA/Dec coordinates; and the Tutorial: 9 The Complete Sky: The Celestial Sphere.
7. Learn about the constellations from Guided Tours: Finding Your Way Around the Real Sky; and Science of Astronomy: Night on the Mountain: Constellations and the Deep Sky.
8. Learn about the seasons from Astronomy Lab: Ancient Astronomers: The Seasons; Guided Tours: Time and Seasons; and Tutorials: 2 The Seasons. Make sure you understand how the inclination of the Earth's axis rather than the distance of the Earth from the Sun causes the seasons.

Four of the planets were visible at one time in this image from the Solar and Heliospheric Observatory (SOHO) satellite, in which a disk hides (occults) the bright image of the Sun (the size of the central circle). Overexposure makes columns of the CCD bright to either side of the planets. Mercury is at left, below the Pleiades star cluster. Venus, Jupiter, and Saturn, from brightest to faintest, are at right.

# PART 2

# The Solar System

The Earth and the rest of the Solar System may be important to us, but they are only minor companions to the stars. In "Captain Stormfield's Visit to Heaven," by Mark Twain, the Captain races with a comet and gets off course. He comes into heaven by a wrong gate, and finds that nobody there has heard of "the world." ("*The* world, there's billions of them!" says a gatekeeper.) Finally, the gatekeepers send someone up in a balloon to try to detect "the world" on a huge map. The balloonist has to travel so far that he rises into clouds. After a day or two of searching, he comes back to report that he has found it: an unimportant planet named "the Wart."

We too must learn humility as we ponder the other objects in space. And while it is no doubt the case that the heavens are filled with a vast assortment of suns and planets more spectacular than our own, still, the Solar System is our own local environment. We would like to understand it and come to terms with it as best we can. Besides, in understanding our own Solar System, we may even find some keys to understanding the rest of the Universe.

To get an idea of its scale, imagine that the Solar System is scaled down and placed on a map of the United States. Let us say that the Sun is a hot ball of gas taking up all of Rockefeller Center, about a kilometer across, in the center of New York City.

Mercury is then a ball 4 meters across at the distance of mid–Long Island, and Venus is a 10-meter ball one and a half times farther away. The Earth, only slightly bigger, is located at the distance of Trenton, New Jersey. Mars, half Earth's diameter, 5 meters across, is located past Philadelphia.

Only beyond Mars are the planets much different in size from the Earth, and the separations become much greater. Jupiter is 100 meters across, the size of a baseball stadium, past Pittsburgh at the Ohio line. Saturn without its rings is a little smaller than Jupiter (including the rings it is a little larger), and is past Cincinnati toward the Indiana line. Uranus and Neptune are each about 30 meters across, about the size of a baseball infield, and are at the distance of Topeka and Santa Fe, respectively. And Pluto, a 2-meter ball, travels on an elliptical track that extends as far away as Los Angeles, 40 times farther away from the Sun than the Earth is. (At present, Pluto is in Montana, slightly farther out than Neptune.) Occasionally a comet sweeps in from Oregon or even from much farther away, passes around the Sun, and returns in the general direction from which it came.

The planets fall naturally into two groups. The first group, the **terrestrial planets,** consists of Mercury, Venus, Earth, and Mars. All are rocky in nature. The terrestrial planets are not very large and have densities about five times that of water.

The second group, the **giant planets,** consists of Jupiter, Saturn, Uranus, and Neptune. All these planets are much larger than the terrestrial planets, and also much less dense, ranging down to slightly below the density of water. (Density is mass per unit volume; things with lower density than water float in water while rocks, iron objects, and other things with higher density than water sink.) Jupiter and Saturn are largely gaseous in nature, similar to the Sun in composition. Uranus and Neptune also have thick gaseous atmospheres; Voyager 2, the Hubble Space Telescope, and ground-based telescopes using active optics have shown Neptune's atmosphere to be unexpectedly active. The giant planets all have fascinating moons, some the size of the inner planets. Pluto, the planet with the largest orbit, is anomalous in several of its properties, and so it may have had a very different history from the other planets. Some think it shouldn't be called a planet at all.

From time to time, beautiful comets come from the outer part of the Solar System into our own. The Sun heats them, and they give off gas and dust to make huge tails. The comets bring us material preserved from the early years of the Solar System for us to study.

Between the realms of the terrestrial planets and the giant planets are the orbits of thousands of chunks of small "minor planets." These **asteroids** range up to 1000 kilometers across. Some similar objects form a belt of material, called the Kuiper belt, beyond the orbit of Neptune. Sometimes much smaller chunks of interplanetary rock penetrate the Earth's atmosphere and hit the ground. We shall discuss these **meteorites** (and where they came from) together with asteroids.

Many people are interested in the planets in order to study their history—how they formed, how they have evolved since, and how they will change in the future. Others are more interested in what the planets are like today. Still others are interested in the planets mainly to consider whether they are harboring life forms, and recent analyses of meteorites from and water on Mars have added to the interest in this topic.

Because the Moon and some other objects in the Solar System have undergone less erosion on their surfaces than has the Earth, we now see them as they appeared eons ago. In this way, the study of the planets and other objects in the Solar System as they are now provides information about the Solar System's early stages.

Now that spacecraft have explored so many planets close up—all but Pluto—we can discuss many general properties of planets or of groups of planets. This study of **comparative planetology** gives us important insights into our own planet. But studies too restricted to comparative planetology can make us lose sight of the interesting individual nature of each planet and of the stages by which our knowledge has jumped during the past decades. In the following chapters, we thus treat each planet individually and, for each, tell a chronological story of how we have learned about it, while also stressing common features. We can then better appreciate not only the planets themselves, but also the projects and individual researchers who have been uncovering their mysteries.

The relative sizes of the planets from Mercury through Neptune.

# Chapter 7

# The Structure and Origin of the Solar System

The basic motions of the objects in the Solar System can be understood by the application of Kepler's laws, which we discussed in Section 3.1. As we saw, Newton then put Kepler's laws on a theoretical foundation. Since Newton's laws of motion and gravitation provide the current framework for our analysis of planetary motions, calculations can now be carried out to great precision (and need to be in order to send spacecraft to the planets). Indeed, though Einstein's work on what we call "general relativity" is in some sense a generalization of Newton's work, for our Solar System it is usually satisfactory to use the Newtonian theory.

In this chapter, we describe the Solar System and some aspects of its structure. First we briefly describe the central object of our Solar System: the Sun. Later on, in Chapter 22, we shall treat the Sun as a star; here we merely describe its most basic structure. Next, we describe such commonly observed phenomena as phases and eclipses, and then we discuss the general arrangement of the planets and their orbits. We conclude by discussing theories that try to explain how the Solar System began, although, as we shall see, this topic is not yet satisfactorily understood.

**AIMS:** To discuss the Sun as the center of our Solar System, the scale and structure of our Solar System, and our Solar System's origin

## 7.1 THE SUN: OUR STAR

The Sun is the most important celestial object to us on Earth. It provides the energy that allows life to flourish. Though the Sun is just an average star, similar to thousands of pinpoints of light we perceive dimly in the nighttime sky, the Sun is so close to us that we feel its energy strongly. It is only 8 light minutes away from us, compared with over 4 light years for the next nearest star.

If we project an image of the Sun onto a piece of paper, we see that the Sun has dark regions on it—sunspots (Fig. 7–1). (We could observe the sunspots by looking at them directly through a telescope instead of using the telescope to project the image so that the telescope is at our backs, but the telescope has to be very heavily filtered to cut out all but about one-millionth of the sunlight, or else we would burn our eyes. One should never stare at the Sun; it is rare for eye damage to occur but it does happen when people do not take proper precautions.) The sunspots appear embedded in the surface of the Sun from which we receive the solar energy. This everyday surface (that is, the surface we see every day) is called the **photosphere,** the sphere from which we get the Sun's light (from the Greek *photos,* meaning "light"). The photosphere is about 1.4 million kilometers (1 million miles) in diameter, about 1 percent of the size of the Earth's orbit.

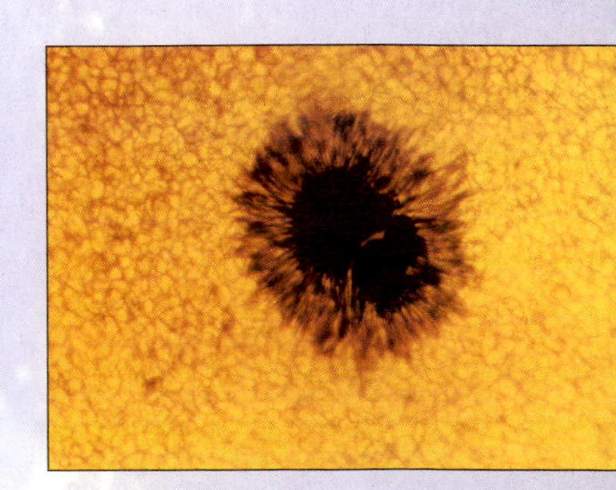

FIGURE 7–1 A sunspot, several times larger than the Earth.

A DEEPER DISCUSSION

### BOX 7.1 Why Is the Sky Blue?

The Sun is so bright that it lights up our entire sky. Sunlight bounces around among (astronomers say "scatters off") air molecules, making the sky blue. This scattering of sunlight by air molecules brightens the sky so much that the other stars are hidden from us in the daytime. The shorter the wavelength, the more efficiently light bounces around. As a result, blue light, with its shorter wavelength, bounces more than red light and fills the sky. The red light from the Sun tends to go straight ahead without being scattered; at sunset, we see this reddish light preferentially, since the blue light has been scattered away from our line of vision toward the setting Sun.

The energy from the photosphere travels outward through the Solar System and beyond in the form of electromagnetic radiation. Most of this solar radiation is in the visible and ultraviolet parts of the spectrum. We see the planets because they reflect the solar radiation toward us; only the Sun in the Solar System forms all its own energy inside itself. (Jupiter and the other giant planets radiate a bit of their own energy but overwhelmingly the light we get from them is reflected sunlight.)

Mercury and Venus are closer to the Sun than the Earth and so receive more solar radiation on each square centimeter of their surfaces. Jupiter and its moons, on the other hand, are farther from the Sun than the Earth and so receive less solar radiation. Thus the outer objects of the Solar System are colder than the inner ones.

The Sun doesn't end at the photosphere. It has layers of gas higher up, but these layers are much less dense than the photosphere and are usually invisible to us. From the ground they can be observed mainly during solar eclipses, as we shall see in Section 7.3. The uppermost level of the Sun is called the **corona** (the Latin word for "crown"), since it looks like a crown around the Sun (Fig. 7–2). The solar corona is continually expanding into space, forming a **solar wind** of particles. The solar wind

**FIGURE 7–2** The solar corona surrounds the dark disk of the Moon in this composite photograph of the 1998 total solar eclipse. (It is interesting to note that during a solar eclipse, the "dark side" of the Moon is the side nearest to us; do not confuse "dark side" with "far side.")

particles hit the Earth's upper atmosphere, causing magnetic storms on Earth. Thus we must understand the Sun's atmosphere to understand the environment of the planets in space.

## 7.2 THE PHASES OF THE MOON AND PLANETS

From the simple observation that the apparent shapes of the Moon and planets change, we can draw conclusions that are important for our understanding of the mechanics of the Solar System. In this section, we shall see how the positions of the Sun, Earth, and other solar-system objects determine the appearance of these objects.

The **phases** of moons or planets are the shapes of the sunlighted areas as seen from a given vantage point. The fact that the Moon goes through a set of such phases approximately once every month is perhaps the most familiar everyday astronomical observation (Fig. 7–3). In fact, the name "month" comes from the word "moon." The actual period of the phases, the interval between a particular phase of the Moon and its next repetition, is approximately $29\frac{1}{2}$ Earth days. This period can vary by as much as 13 hours.

The explanation of the phases is quite simple: The Moon is a sphere, and at all times the side that faces the Sun is illuminated and the side that faces away from the Sun is dark (Fig. 7–4). The phase of the Moon that we see from the Earth, as the Moon revolves around us, depends on the relative orientation of the three bodies: Sun, Moon, and Earth. The situation is relatively simple because of the fact that the plane of the Moon's revolution around the Earth is nearly, although not quite, the same as the plane of the Earth's revolution around the Sun.

"The Moon" is often capitalized to distinguish it from moons of other planets; in general writing, it is usually written with a small "m." In our discussion of the solar system and the history of astronomy, we are capitalizing "Earth" to put it on a par with the other planets and the Moon. For consistency, we are also capitalizing "Sun" and "Universe" there.

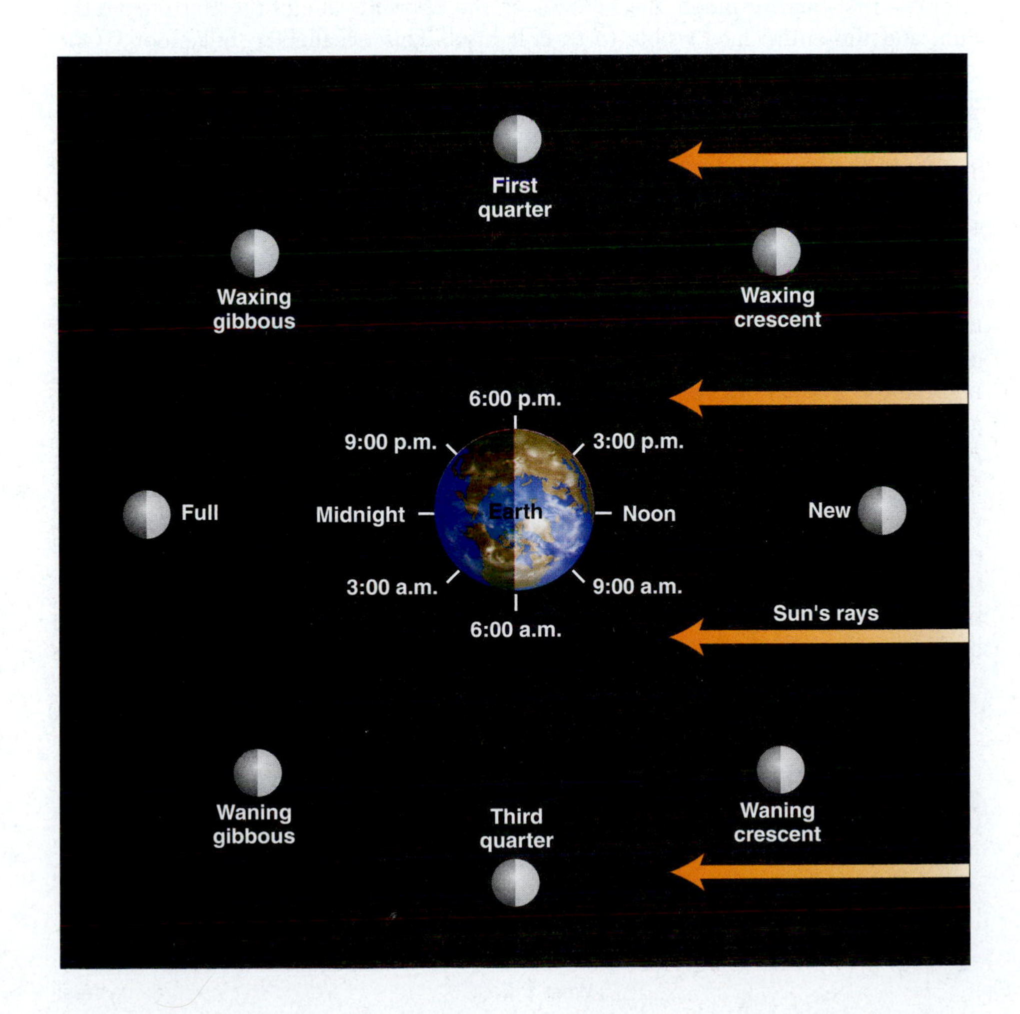


**FIGURE 7–3** The phases of the Moon depend on the Moon's position in its orbit around the Earth. Here we visualize the situation as if we could be high above the Earth's orbit, looking down. The times given show when the Moon of the phase drawn above each one is highest in the sky, that is, when it crosses the north-south meridian.

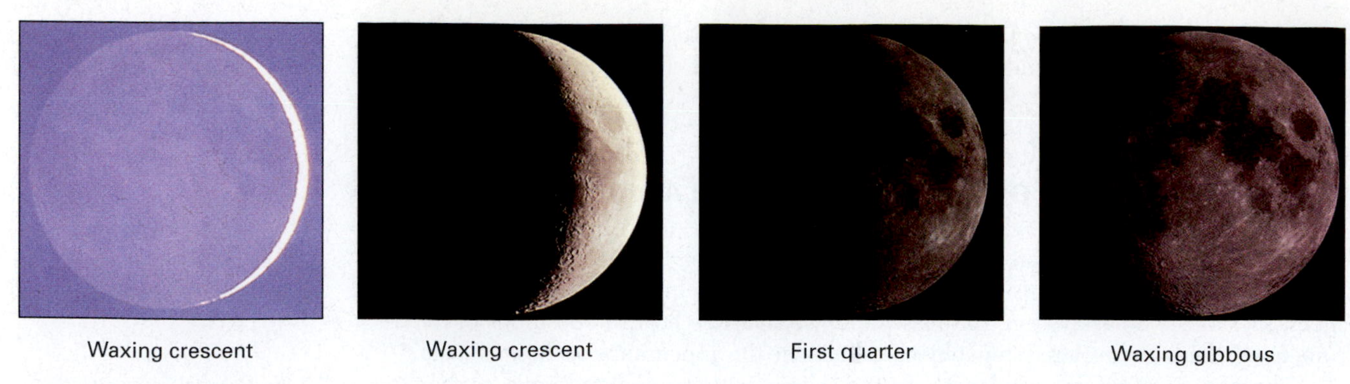


**FIGURE 7–4** The phases of the Moon.

Basically, when the Moon is almost exactly between the Earth and the Sun, the dark side of the Moon faces us. We call this view a "new moon." A few days earlier or later we see a sliver of the lighted side of the Moon, and call this view a "crescent." As the month wears on, the crescent gets bigger, and about 7 days after new moon, half the side of the Moon that faces us is lighted. We sometimes call this view a "half moon." Since this situation occurs one-fourth of the way through the phases, it is also called a "first-quarter moon." (Instead of apologizing for the fact that the same phase is called both "quarter" and "half," I'll just continue with a straight face and try to pretend that there is nothing strange about it.) When the lighted part of the Moon is growing, we say that the Moon is "waxing."

When over half the Moon's disk is visible, we have a "gibbous" moon. One week after the first-quarter moon, the Moon is on the opposite side of the Earth from the Sun, and the entire face visible to us is lighted. This is called a "full moon." One week later, when we see a half moon again, we have a "third-quarter moon." As the illuminated fraction of the Moon diminishes, we say that the Moon is "waning." When the Moon reaches "new moon" again, it repeats the cycle of phases.

Note that since the phase of the Moon is related to the position of the Moon with respect to the Sun, if you know the phase, you can tell when the Moon will rise. For example, since the Moon is 180° across the sky from the Sun when it is full, a full moon is always rising just as the Sun sets (Fig. 7–5). Each day thereafter, the Moon rises about 50 minutes later (24 hours divided by 28.8, the period of revolution of the Moon around the Earth), although the exact time difference also depends on the Moon's changing declination. The third-quarter moon, then, rises near midnight and is high in the sky at sunrise. The new moon rises with the Sun in the east at dawn. The first-quarter moon rises near noon and is high in the sky at sunset. It is often visible in the late afternoon.

The Moon is not the only object in the Solar System that is seen to go through phases. Mercury and Venus both orbit inside the Earth's orbit, and so sometimes we see the side that faces away from the Sun and sometimes we see the side that faces toward the Sun. Thus at times Mercury and Venus are seen as crescents, though it takes a telescope to observe their shapes. Spacecraft to the outer planets have looked back and seen the Earth and the other planets and their moons in all their phases as crescents (Fig. 7–6) as well.

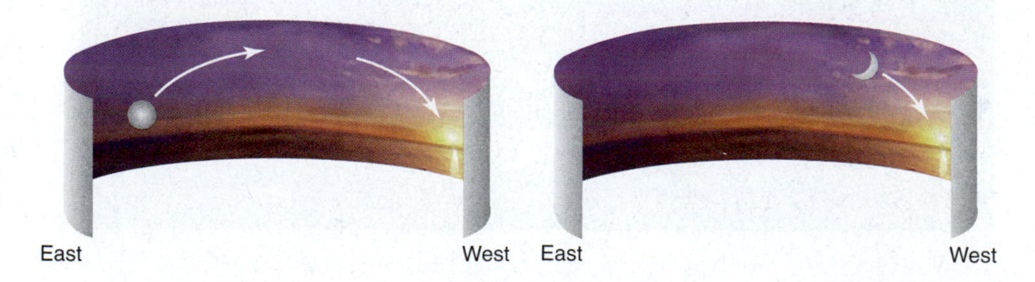


**FIGURE 7–5** Because the phase of the Moon depends on its position in the sky with respect to the Sun, we can see why a full moon is always rising at sunset while a crescent moon is either setting at sunset, as shown here, or rising at sunrise.

Full moon

Waning gibbous

Third quarter

Waning crescent

## 7.3 ECLIPSES

Because the Moon's orbit around the Earth and the Earth's orbit around the Sun are not precisely in the same plane (Fig. 7–7), the Moon usually passes slightly above or below the Earth's shadow at full moon, and the Earth usually passes slightly above or below the Moon's shadow at new moon. But every once in a while, up to seven times a year, the Moon is at the part of its orbit that crosses the Earth's orbital plane at full moon or new moon. When that happens, we have a lunar eclipse or a solar eclipse (Fig. 7–8).

When the Moon comes directly between the Earth and the Sun, we have an eclipse of the Sun. A total solar eclipse—when the Moon covers the whole surface of the Sun that is normally visible—is a relatively rare phenomenon, occurring only every 1½ years or so. The Moon must be fairly precisely aligned between the Sun and the Earth to have an eclipse; only those people in a narrow band on the surface of the Earth see the eclipse.

Many more people see a total lunar eclipse than a total solar eclipse when one occurs. At a total lunar eclipse, the Moon lies entirely in the Earth's shadow, and sunlight is entirely cut off from it. So anywhere on the Earth that the Moon has risen—basically, half the world—the eclipse is visible. The precise celestial alignment necessary to see a total solar eclipse isn't necessary to see a total lunar eclipse.

**FIGURE 7–6** Earth and Moon, from the Galileo spacecraft en route to Jupiter.

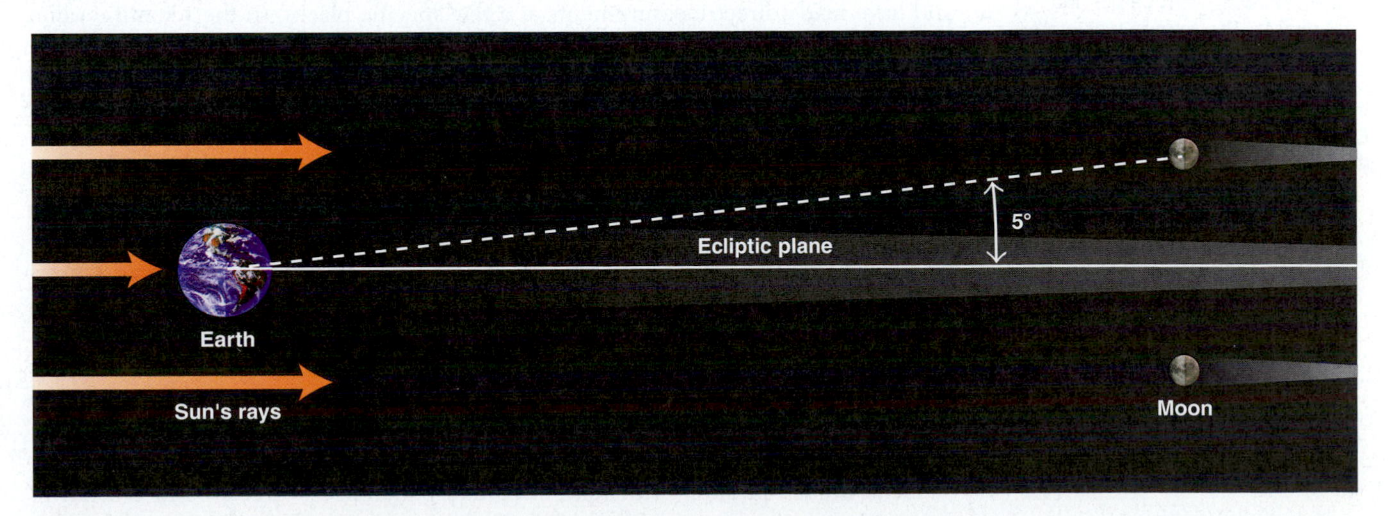

**FIGURE 7–7** The plane of the Moon's orbit is tipped with respect to the plane of the Earth's orbit, so the Moon usually passes above or below the Earth's shadow. Therefore, we don't have lunar eclipses most months.

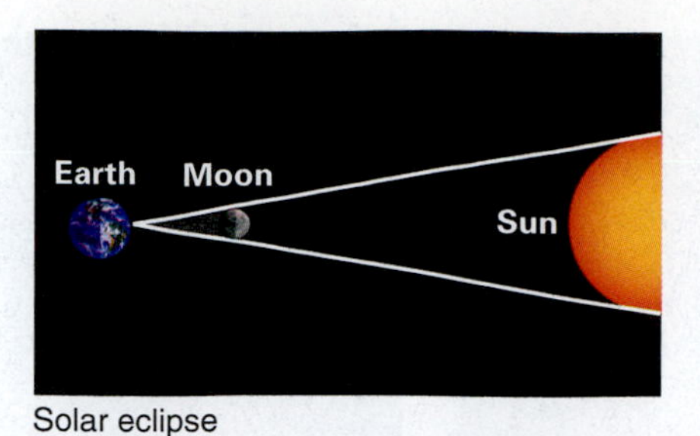


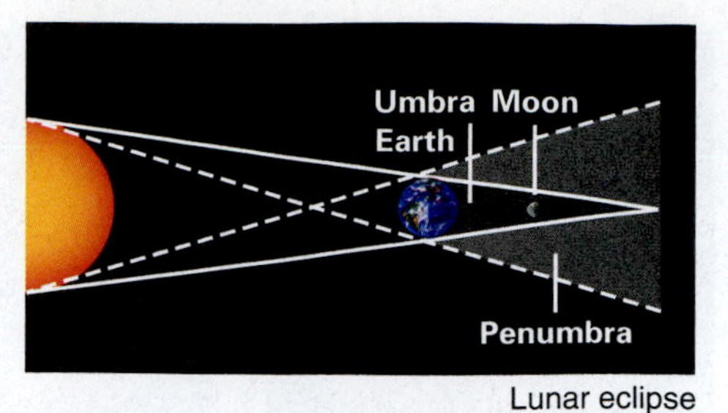


Solar eclipse

Lunar eclipse

**FIGURE 7–8** When the Moon is between the Earth and the Sun, we observe an eclipse of the Sun. When the Moon is on the far side of the Earth from the Sun, we see a lunar eclipse. The part of the Earth's shadow that is only partially shielded from the Sun's view is called the **penumbra;** the part of the Earth's shadow that is entirely shielded from the Sun is called the **umbra.** The distances in this diagram are not to scale.

## 7.3a Lunar Eclipses

A total lunar eclipse is a much more leisurely event to watch than a total solar eclipse. The partial phase, when the Earth's shadow gradually covers the Moon, lasts for hours. And then the total phase, when the Moon is entirely within the Earth's shadow, can last for over an hour. During this time, the sunlight is not entirely shut off from the Moon. A small amount is refracted around the edge of the Earth by our atmosphere. Most of the blue light is taken out during the sunlight's passage through our atmosphere (as we saw in Box 7.1). Since blue light scatters off molecules in the atmosphere better than redder light does, it is sent off in all directions; this effect explains how blue skies are made for the people part way around the globe from the point at which the Sun is overhead. The remaining light is reddish; this is the light that falls on the Moon (Fig. 7–9). Thus, the eclipsed Moon appears reddish. Some lunar eclipses are very dark with very little red showing; these occur when there is a lot of volcanic dust in the Earth's atmosphere.

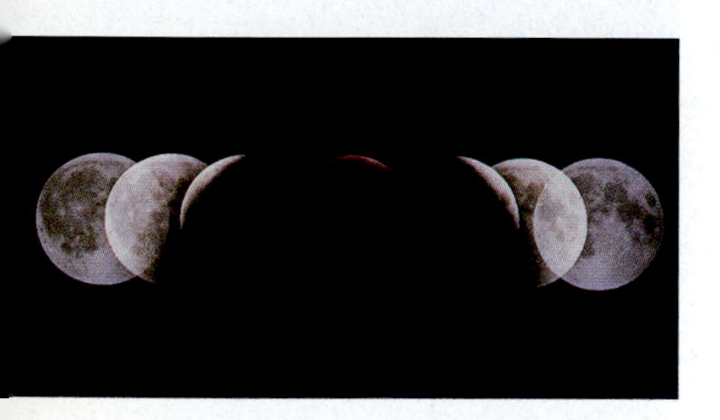

**FIGURE 7–9** A total lunar eclipse, showing both partial and total phases. The camera was guided on the position of the Earth's umbral shadow. Note that the photograph shows that the Earth's shadow is round, and thus that the Earth is round.

The next total lunar eclipses visible in the United States (eastern states) and Europe will be on May 16, 2003; November 9, 2003; and October 28, 2004. The United States's west coast, Hawaii, and Alaska won't have a total lunar eclipse until August 28, 2007.

In principle, scientists can make observations at lunar eclipses to study the properties of the lunar surface. For example, we can monitor the rate of cooling of areas on the lunar surface with infrared radiometers. But now we have landed on the Moon and have made direct measurements at a few specific places, so the scientific value of lunar eclipses has been greatly diminished. There remains little of scientific value to do during a lunar eclipse, which is quite different from the situation at a total solar eclipse.

## 7.3b Solar Eclipses

The outer layers of the Sun are fainter than the blue sky and so can't normally be seen during daylight. This blue sky, which we see every day, is given its brightness by the scattering of sunlight. To study the outer layers of the Sun, we need a way to remove the blue sky while the Sun is up. A total solar eclipse does just that for us.

Solar eclipses arise because of a happy coincidence: Though the Moon is 400 times smaller in diameter than the solar photosphere, it is also 400 times closer to the Earth. Because of this, the Sun and the Moon subtend almost exactly the same angle in the sky—about ½° (Fig. 7–10).

The Moon's position in the sky, at certain points in its orbit around the Earth, comes close to the position of the Sun. This happens approximately once a month at the time of the new moon. Since the lunar orbit is inclined with respect to the

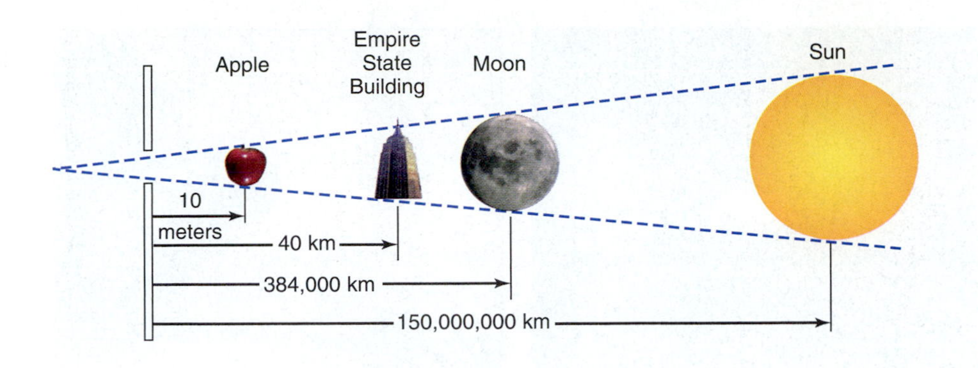


**FIGURE 7-10** An apple, the Empire State Building, the Moon, and the Sun are very different from each other in size, but here they subtend the same angle because they are at different distances from us.

Earth's orbit, the Moon usually passes above or below the line joining the Earth and the Sun. But occasionally the Moon passes close enough to the Earth–Sun line that the Moon's shadow falls upon the surface of the Earth.

At a total solar eclipse, the lunar shadow barely reaches the Earth's surface (Fig. 7–11). As the Moon moves through space on its orbit, and as the Earth rotates, this lunar shadow sweeps across the Earth's surface in a band up to 300 km wide. Only observers stationed within this narrow band can see the total eclipse.

From a wide area outside the band of totality, one sees only a partial eclipse. Sometimes the Moon, Sun, and Earth are not precisely aligned and the darkest part of the shadow—called the **umbra**—never hits the Earth. We are then in the intermediate part of the shadow, which is called the **penumbra.** Only a partial eclipse is visible on Earth under these circumstances. As long as the slightest bit of photosphere is visible, even as little as 1 percent, one cannot see the important eclipse phenomena—the faint outer layers of the Sun—that are both beautiful and the subject of scientific study. Thus, partial eclipses are of little value for most scientific purposes. After all, the photosphere is 1,000,000 times brighter than the outermost layer, the corona; if 1 percent of the photosphere is showing, then we still have 10,000 times more light from the photosphere than from the corona, which is enough to ruin our opportunity to see the corona.

If you are fortunate enough to be standing in the zone of a total eclipse, you will sense excitement all around you as the approaching eclipse is anticipated. An hour or so before totality begins, the partial phase of the eclipse starts. Nothing is visible to the naked eye immediately, but if you were to look at the Sun through a special filter, you would see that the Moon was encroaching upon the Sun (Fig. 7–12). At this stage of the eclipse, it is necessary to look at the Sun through a filter or to project the image of the Sun with a telescope or a pinhole camera onto a surface. You need the filter to protect your eyes because the photosphere is visible throughout the partial phases before and after totality. Its direct image on your retina for an extended time could cause burning and blindness, though such cases are very rare.

The eclipse progresses gradually, and by 15 minutes before totality the sky grows dark as if a storm were gathering. During the minute or two before totality begins, bands of light and dark race across the ground. These **shadow bands** are caused in the Earth's atmosphere, and thus they tell us about the terrestrial rather than the solar atmosphere.

**FIGURE 7-11** A solar eclipse with the Earth, the Moon, and the distance between them shown to actual scale.

*A* *B*

**FIGURE 7–12** The partially eclipsed Sun, settling behind trees, seen from Seattle during the eclipse of July 30, 2000. (*A*) A narrow-angle view. (*B*) A wide-angle view.

You still need the special filter to watch the final seconds of the partial phases. As the total phase—totality—begins, the bright light of the solar photosphere passing through valleys on the edge of the Moon glistens like a series of bright beads; these bright points of light are called **Baily's beads.** The last bead seems so bright that it looks like the diamond on a ring—the **diamond-ring effect** (Fig. 7–13). For a few seconds, the **chromosphere** (the layer of the Sun's atmosphere between the photosphere and the corona) is visible as a pinkish band around the leading edge of the Moon.

**FIGURE 7–13** The diamond-ring effect at the 1999 eclipse in Romania.

Then the corona comes into view in all its glory. You see streamers of gas near the Sun's equator and finer plumes near its poles. At this stage, the photosphere is totally hidden, so it is perfectly safe to look at the corona with the naked eye and without filters. The corona is about as bright as the full moon and is equally safe to look at.

The total phase may last a few seconds, or it may last as long as about 7 minutes. Cameras click away, computers hum, spectrographs are operated, photographs are taken through special filters, rockets photograph the spectrum from above the Earth's atmosphere, and tons of equipment are used to study the corona during this brief time of totality.

At the end of the eclipse, the diamond ring appears on the other side of the Sun, followed by Baily's beads and then the partial phases. All too soon, the eclipse is over.

### 7.3c Observing Solar Eclipses

On the average, a solar eclipse occurs somewhere in the world every year and a half. The band of totality, though, usually does not cross populated areas of the Earth, and astronomers often have to travel great distances to carry out their observations. Figure 7–14 shows the paths and dates of future total solar eclipses.

Sometimes the Moon subtends a slightly smaller angle in the sky than the Sun because the Moon is on the part of its orbit that is relatively far from the Earth. When a well-aligned eclipse occurs in such a circumstance, the Moon doesn't quite cover the Sun. An annulus—a ring—of photospheric light remains visible, so we call this type of solar eclipse an **annular eclipse** (Fig. 7–15). Such an annular eclipse crossed a narrow band across the United States on May 10, 1994. The rest of the United States saw a partial eclipse. Only 90 percent of the Sun's diameter was covered. Annularity lasted about 6 minutes for observers on the center line.

The last total eclipse to cross the continental United States and Canada occurred in 1979. The last total eclipse in the United States crossed Hawaii before reaching Mexico and parts of Central and South America in 1991. The next total eclipse in

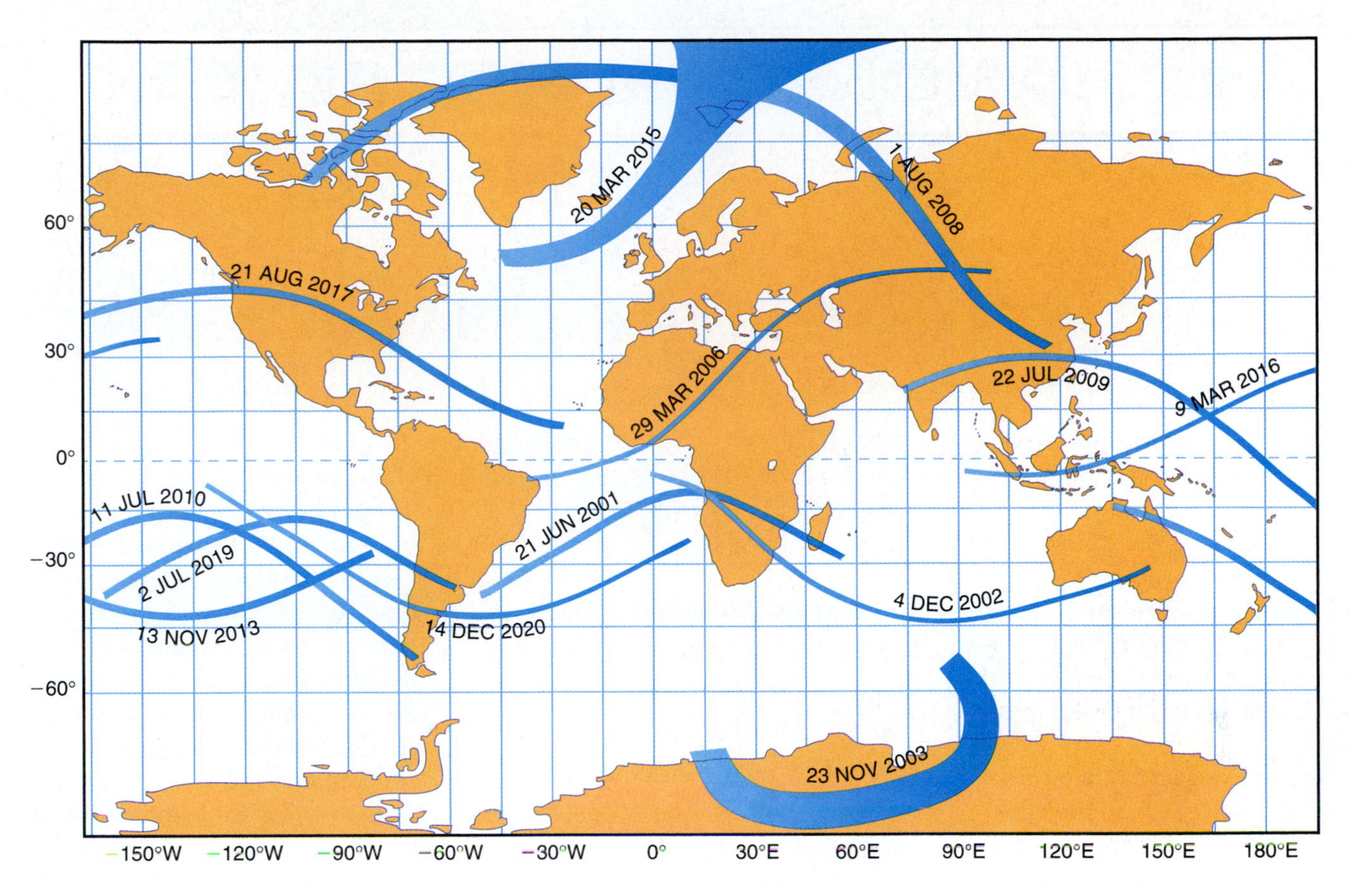


**FIGURE 7–14** The paths of the Moon's shadow during total solar eclipses occurring between 2000 and 2020. Anyone standing along such a path on the dates indicated will see a total solar eclipse.

the continental United States won't be until 2017. Accessible parts of Canada won't see another total eclipse until 2024.

Though a total solar eclipse is visible from a random location on Earth only every 350 years, on average, travelling allows anyone to see a total solar eclipse about every 18 months. The 1999 total solar eclipse that crossed Europe, Turkey, the Middle

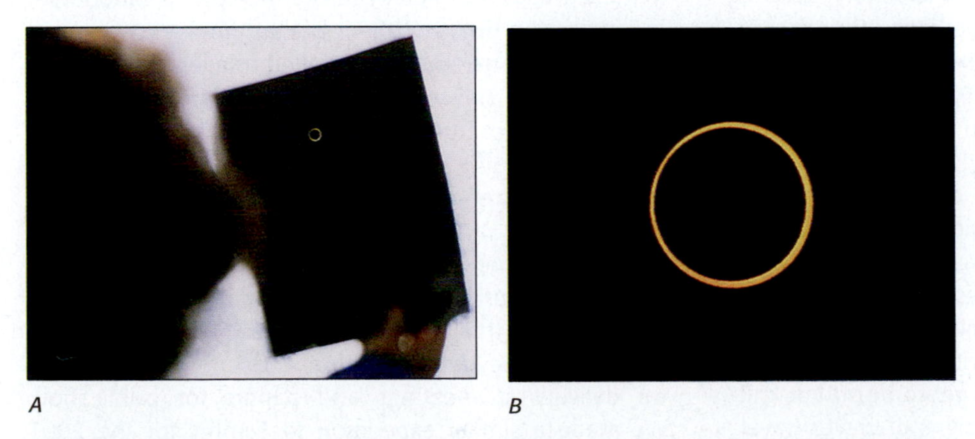

*A* *B*

**FIGURE 7–15** *(A)* Using a dense filter to observe an annular eclipse. In an annular eclipse, the Moon appears smaller than the Sun and a ring (annulus) of the bright everyday Sun remains visible. *(B)* During an annular eclipse, an annulus (ring) of photosphere remains visible. Here it is broken by the mountains of the Moon during the annular eclipse viewed from Australia in 1999.

ASTRONOMER'S NOTEBOOK

# A Solar Eclipse Expedition

Eclipses are like Olympic track events. Just as athletes prepare themselves thoroughly and for a long time to go to distant places every four years to be able to run the 100 meters, eclipse astronomers prepare carefully and often travel great distances to study total eclipses. During total eclipses, the corona and other solar phenomena outside the solar limb are best visible.

*Just as athletes prepare themselves thoroughly and for a long time to go to distant places every four years to be able to run the 100 meters, eclipse astronomers prepare carefully and often travel great distances to study total eclipses.*

Eclipses repeat with an 18-year 11 1/3-day interval. After that time, another solar eclipse crosses basically the same track, except 1/3 of the way around the Earth because of the 1/3 day. Within each of these 18-year 11 1/3-day intervals, a dozen total eclipses occur. One of those eclipses is especially long, approximately 7 minutes in duration. The most recent of these long eclipses occurred on July 11, 1991. We await its successor, which will cross China in 2009. The total solar eclipse of August 11, 1999, crossed Europe and the path extended as far as India. It may have been viewed by more people than any other total eclipse ever. Sometimes we have the opposite situation—eclipses with poor weather prospects and perhaps other circumstances like extreme cold. Few people will travel to Antarctica for the November 23, 2003, total solar eclipse.

For the 1999 eclipse, a dozen undergraduate students joined several colleagues and me at the very center of the eclipse path, in a small city in Romania. We took about a ton and a half of equipment, and were on site about 10 days in advance to set up, test, and align our equipment. We had telescopes, filters, computers, and other optical and electronic equipment.

Two of my experiments dealt with the heating of the solar corona. How does the corona get to be so hot, millions of degrees? We searched for signs of waves travelling along loops of gas held in place in the corona by magnetic fields. We did so by looking for oscillations of intensity in small regions of the corona, observing through a special filter that passed only the radiation from these loops. A second experiment was to map the coronal temperature. Though it is known that the temperature is millions of degrees, the temperature you measure differs depending on how you measure it and what aspect of the Sun's corona you are studying. We were mapping the temperature of the coronal electrons. By prior arrangement made using a satellite phone, we were able to point our telescope at the same region that the TRACE satellite was observing, and used the SOHO satellite's observations of the corona on preceding days to select that location. (We discuss these two satellites and the wonderful images and other data they take in Chapters 22 and 23.)

A third experiment was linked with two telescopes on SOHO. The coronagraph experiment on SOHO can see the outer corona and an experiment using filters in the extreme ultraviolet can see the corona silhouetted in front of the disk. But there is a gap in the coverage in the lower corona, and we supplied images to fill in that gap. There are still some things that are possible only at eclipses and not from space! Our oscillation experiment, mentioned above, also did something that could not be done from space—observe at a cadence as fast as a tenth of a second.

The weather at our site was very clear, and all our instruments worked nicely throughout the 2 minutes and 23 seconds of totality. My students and I have been hard at work on the data reduction and on preparing talks for delivery at meetings and papers for publication. We made a similar expedition to Zambia for the 2001 eclipse.

East, and extended as far as India was seen by millions (Fig. 7–16). The 2001 total solar eclipse that crossed land only at southern Africa was seen by many fewer people (Fig. 7–17). The 2002 total solar eclipse, which will cross southern Africa and wind up low in the sky in southern Australia, will also be viewed by relatively few people.

In these days of orbiting satellites, is it worth travelling to observe eclipses? There is much to be said for the benefits of eclipse observing. Eclipse observations are a relatively inexpensive way, compared to space research, of observing the chromosphere and corona to find out such quantities as the temperature, density, and magnetic field structure. And eclipse views are much better, though only for the minutes of totality, than views from mountain-based telescopes known as coronagraphs. Even realizing that some eclipse experiments may not work out because of the pressure of time or because of bad weather, eclipse observations are still very cost-effective. It is much less expensive to study the corona during an eclipse than as part of a space experiment. And for some kinds of observations, those at the highest resolutions, space techniques have not yet matched ground-based eclipse capabilities. In any case, the data collected at eclipses can be used alongside data collected from space satellites studying the Sun.

A

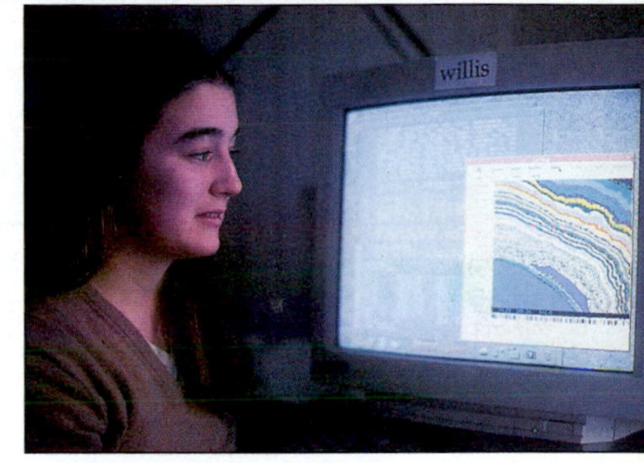


B

**FIGURE 7–16** (*A*) The corona in the sky over our instruments during the 1999 total eclipse in Romania. (*B*) A student ponders some of the data about the solar corona, as observed at the 1999 eclipse in Romania.

## 7.4 THE REVOLUTION AND ROTATION OF THE PLANETS

The motion of the planets around the Sun in their orbits is called **revolution.** The spinning of a planet is called **rotation.** The Earth, for example, revolves around the Sun in 1 year and rotates on its axis in 1 day (thus defining these terms).

The orbits of all the planets lie in approximately the same plane. The planets thus pass through only a disk, rather than a full sphere. Little is known of the parts of the Solar System away from this disk, although many of the comets originate in an extended spherical cloud around the Sun.

The plane of the Earth's orbit around the Sun is called the **ecliptic plane** (Fig. 7–18). So if we had a vantage point from afar, as in the figure, we would see the Earth's orbit traced out on the ecliptic plane. Of course, we are on the Earth rather than outside the Solar System looking in. As a result, for us the Sun appears to move across the sky with respect to the stars. The path the Sun takes among the stars, as seen from the Earth, is called the ecliptic (see also Section 6.3). So the ecliptic is the Sun's path in the sky, and the ecliptic plane is the plane in which the Earth's orbit lies.

Because they orbit close to the ecliptic, the inner planets go close to the Sun as seen from Earth. When they are at their closest, we say they are at "inferior conjunction." When the alignment is perfect, the inner planets go in front of the Sun as seen from Earth: at that position, they are **in transit.** Transits of Venus come in pairs separated by 8 years; the pairs are at intervals of over 100 years. The last transits were in the 19th century; no human now alive has seen a transit of Venus! The next pair will be in 2004 and 2012. Transits of Mercury (Fig. 7–19) are more frequent; there were 11 in the 20th century.

The **inclination** of the orbit of a planet is the angle between the plane of its orbit and the plane of the Earth's orbit. Thus the inclination of the Earth's orbit is 0°. The inclinations of the orbits of the other planets with respect to the ecliptic are small, with the exception of Pluto's 17° (Fig. 7–20). Mercury's inclination is 7°; the rest of the planets have inclinations of less than 4°.

**FIGURE 7–17** Skies were clear across southern Africa for the June 21, 2001, total solar eclipse. This eclipse digital image was taken from my expedition to Lusaka, Zambia, and has a central image inserted showing the sun from the Solar and Heliospheric Observatory in space.

The fact that the planets all orbit the Sun in essentially the same plane is one of the most important facts that we know about the Solar System. Its explanation is at the base of most models of the formation of the Solar System. Central to that explanation is a property that astronomers and physicists use in analyzing spinning or revolving objects: **angular momentum.** The amount of angular momentum of a small body revolving around a large central body is the distance between the two bodies' centers times the momentum of the small body. The momentum (sometimes

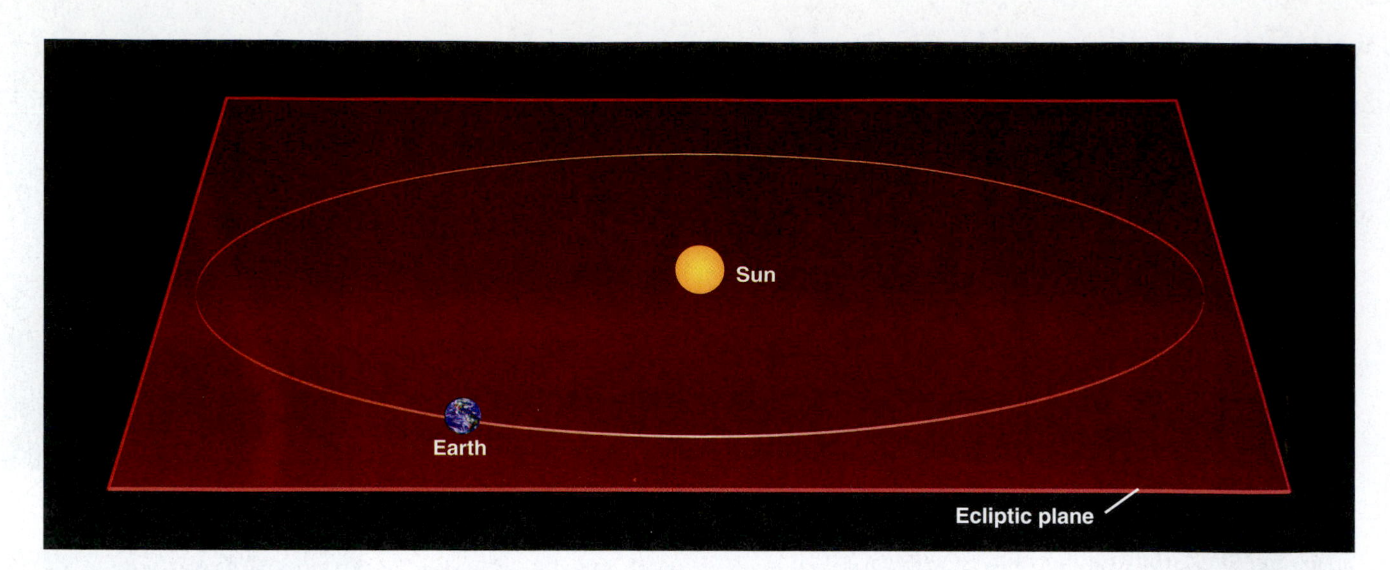


**FIGURE 7–18** The ecliptic plane is the plane of the Earth's orbit around the Sun.

called "linear momentum" to distinguish it from angular momentum) is the body's mass times its velocity. Therefore, the angular momentum of an orbiting body is the distance-from-the-center × mass × velocity. The distance from the center can be called the radius $r$. For velocity $v$, and mass $m$, we have

$$\text{angular momentum} = rmv.$$

The importance of angular momentum lies in the fact that it is "conserved"; that is, the total angular momentum of the system (the sum of the angular momenta of the different parts of the system) doesn't change even though the distribution of angular momentum among the parts may change. The total angular momentum of the Solar System should thus be the same as it was in the past, unless some mechanism is carrying angular momentum away. (We will learn about the cause of the tides in the chapter on the Earth; tidal friction is one way that angular momentum can be lost.)

Simple and familiar examples of the conservation of angular momentum are ice skaters (Fig. 7–21). They may start themselves spinning by exerting force on the ice with their skates, thus giving themselves a certain angular momentum. When they want to spin faster, they draw their arms in. This changes the distribution of mass so that the mass is effectively closer to their centers (the axes around which they are spinning), and to compensate for this they rotate more quickly so that their angular momentum remains the same (that is, is conserved).

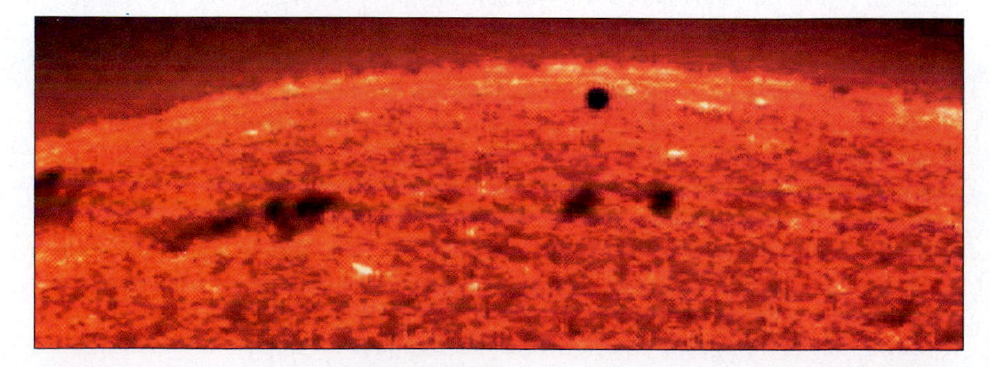

**FIGURE 7–19** The transit of Mercury on November 17, 1999. Mercury is visible in silhouette against this ultraviolet image of the solar corona from the TRACE satellite (which we discuss in Chapters 22 and 23).

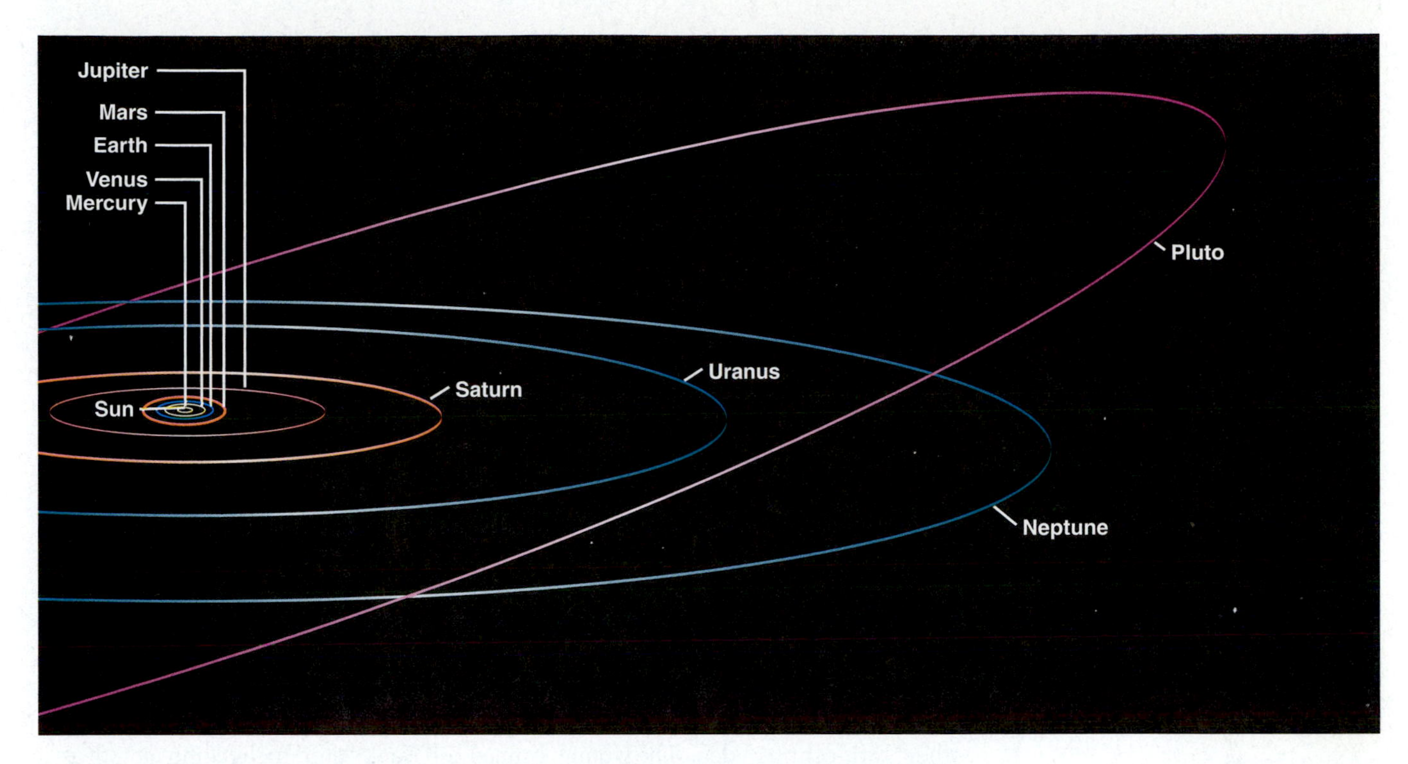


**FIGURE 7–20** The orbits of the planets, with the exception of Pluto, have only small inclinations to the ecliptic plane.

**FIGURE 7–21** An ice skater spins faster when she draws her arms in. Since angular momentum is conserved, concentrating her mass closer to her spin axis than when her arms are outstretched leads to a compensating increase in spin. Here we see Tara Lipinski, gold medalist at the 1998 Winter Olympics in Nagano, Japan, holding her arms in to spin fast. The spin causes her skirt to flare out. The skaters at the 2002 Winter Olympics in Salt Lake City will similarly use the conservation of angular momentum to spin fast.

Thus from the fact that every planet is revolving in the same plane and in the same direction, we deduce that the Solar System was formed out of material that was rotating in that direction. We would be very surprised to find a planet revolving around the Sun in a different direction—and we do not—or, to a lesser extent, to find a planet rotating in a direction opposite to that of the other planets. There are, however, three examples of such backward rotation—Uranus, Venus, and Pluto—which must each be carefully considered to find the reason. (We will see that collisions with other objects are invoked to cause these nonstandard directions of rotation.)

## 7.5 THEORIES OF COSMOGONY

The Solar System exhibits many regularities, and theories of its formation must account for them. The orbits of the planets are almost, but not quite, circular, and all lie in essentially the same plane. All the planets revolve around the Sun in the same direction, which is the same direction in which the Sun rotates. Moreover, the Sun and almost all the planets and planetary satellites rotate in that same direction. Some planets have families of satellites that revolve around them in a manner similar to the way that the planets revolve around the Sun. And cosmogonical theories must explain the spacing of the planetary orbits and the distribution of planetary sizes and compositions.

The study of the origin of the solar system is called cosmogony (pronounced with a hard "g").

René Descartes, the French philosopher, was one of the first to consider the origin of the Solar System in what we would call a scientific manner. In his theory, proposed in 1644, circular eddies—called vortices—of all sizes were formed at the beginning of the Solar System in a primordial gas and eventually settled down to become the various celestial bodies.

After Newton proved that Descartes's vortex theory was invalid, it was 60 years before the next major developments. The Comte de Buffon suggested in France in 1745 that the planets were formed by material ejected from the Sun when what he called a "comet" hit it. (At that time, the size of comets was unknown, and it was thought that comets were objects as massive as the Sun itself.) Later versions of Buffon's theory, called **catastrophe theories,** followed a similar line of reasoning, although they spoke explicitly of collision with another star. A variation postulated that the material for the planets was drawn out of the Sun by the gravitational attraction of a passing star. This latter possibility is called a **tidal theory.**

Theoretical calculations show that gas drawn out of a star in collision or by a tidal force would not condense into planets, but would rather disperse. Hence, catastrophe and tidal theories are currently out of favor. Further, they predict that only very few planetary systems would exist, since only very few stellar collisions or near-collisions would have taken place in the lifetime of the galaxy. The new observational evidence that many stars have planets (Chapter 18) requires methods of formation that can account for more planetary systems.

The theories of cosmogony that astronomers now tend to accept stem from another 18th-century idea. The young Immanuel Kant, later to become noted as a philosopher, suggested in 1755 that the Sun and the planets were formed by the same type of process. In 1796, the Marquis Pierre Simon de Laplace (pronounced La Plahce), the French mathematician, independently advanced a type of theory similar to Kant's when he postulated that the Sun and the planets all formed from a spinning cloud of gas called a **nebula.** Laplace called his idea the **nebular hypothesis,** using the word "hypothesis" because he had no proof that it was correct. The spinning gas supposedly threw off rings that eventually condensed to become the planets. But though these beginnings of modern cosmogony were laid down in the time of Benjamin Franklin and George Washington, the theory still has not been completely understood or quantified, for not all the stages that the primordial gas would have had to follow are clear.

Further, some of the details of Laplace's theory were thought for a time to be impossible. For example, it was calculated in the last century, by the great British physicist James Clerk Maxwell, that the planets would not have been able to condense out of the rings of gas and also that they could not have started out rotating as fast as they do.

So for a time, the nebular hypothesis was not accepted. But the current theories of cosmogony again follow Kant and Laplace (Fig. 7–22). In these **nebular theories,** the Sun and the planets condensed out of what is called a **primeval solar nebula.** Laplace's ring formation mechanism is no longer thought to be applicable; only the concept of joint formation of the Sun and planets from a single, rotationally flattened cloud has survived the centuries. We now think that some five billion years ago, billions of years after the galaxies began to form, smaller clouds of gas and dust began to contract out of interstellar space. Our Solar System originated from such a cloud. Similar interstellar clouds of gas and dust can now be detected at many locations in our galaxy.

Before they had data about gamma rays, scientists believed that the collapse of gas to form the Solar System was set off by shock waves (Fig. 7–23) from a nearby supernova. The evidence concerns the unusual abundances in meteorites (Section 20.1) of a certain radioactive isotope of aluminum. It had been thought that the isotopes could only have been formed in such supernova explosions. Their lifetimes are too short for them to have been formed very long before our solar nebula's collapse. But studies of gamma-ray data from a series of satellites in the late 1900s revealed

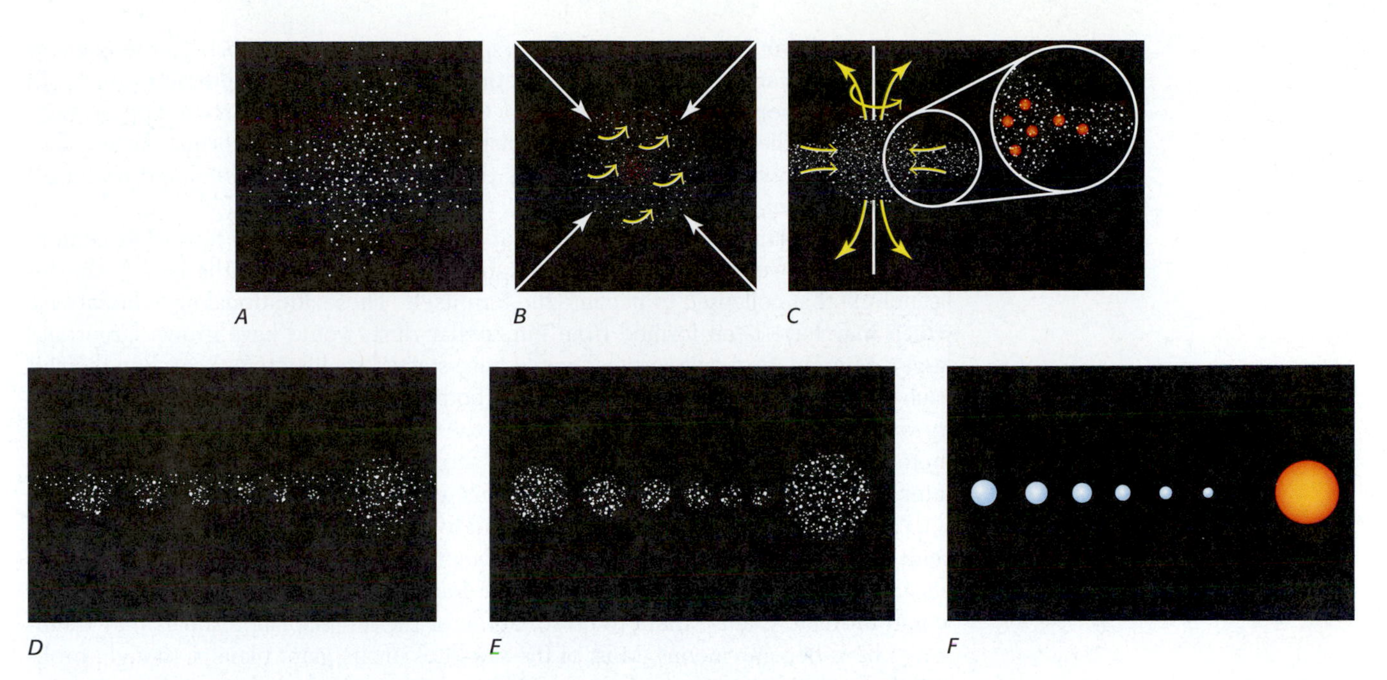


**FIGURE 7–22** (*A*) The leading model for the formation of the solar system. The protosolar nebula condenses because of the force of its own gravity (*B*). Between stages C and D, it contracts to form the protosun and a large number of small bodies called planetesimals. The planetesimals clump together to form protoplanets (*D*), which, in turn, contract to become planets (*E* and *F*). Some of the planetesimals may have become moons or asteroids (which are discussed in Section 20.2).

that this radioactive isotope of aluminum is widespread throughout interstellar space. So no local supernova acting as a trigger was needed to account for the isotope in our Solar System.

However the cloud collapse started, random fluctuations in the gas and dust would have given the primeval solar nebula a small net spin from the beginning. As the nebula contracted, it would have begun to spin faster because of the conservation of angular momentum (the same reason that the ice skater spins faster; we will later learn about pulsars, collapsed stars some of which spin hundreds of times a sec-

**FIGURE 7–23** Larger and smaller shock waves formed by an F/A-18 Hornet plane as it broke the sound barrier are visible. The clouds of condensation formed as the shock wave cooled the air.

ond). Gravity would have contracted the spinning nebula into a disk: In the plane of the nebula's rotation, the spin had the effect of a force that counteracts gravity. In the directions perpendicular to this plane, though, there was no force to oppose gravity's pull, and the nebula collapsed in that direction. Planetary formation in a flattened, rotating nebula naturally explains the regularity of the orbital and rotational characteristics of the Solar System.

How did planets form out of the solar nebula as it contracted toward its center? Perhaps there were additional agglomerations to the **protosun,** the part of the solar nebula that collapsed to become the Sun itself. These additional agglomerations, which may have been formed from interstellar dust, would have grown larger and larger. Micrometer-sized particles would have started sticking together when they hit each other, eventually clumping into larger bodies centimeters or meters across; gravity would then have increased the size of the clumps until there were bodies kilometers and even hundreds of kilometers across within about 10,000 years. These intermediate bodies, hundreds of kilometers across, are called **planetesimals.**

The planetesimals combined under the force of gravity over tens of millions of years to form **protoplanets.** These protoplanets may have been larger than the planets that resulted from them because they had not yet contracted, though gravity would ultimately cause them to do so. Some of the larger planetesimals may themselves have become moons. Most of the satellites of the giant planets, though, probably formed in mini-solar nebulae around each protoplanet.

As the Sun condensed, some of the energy it gained from its contraction heated it until its center reached the temperature at which nuclear-fusion reactions began taking place. The planets, on the other hand, were simply not massive enough to heat up sufficiently to have nuclear reactions start.

Billions of planetesimals may have become comets (Chapter 18) and have been far beyond the planets in the deep freeze of outer space for billions of years. The extreme interest in the most recent passage of Halley's Comet and the approval of our current space missions to comets stem in large part from the chance to study such pristine material.

Planetesimals may also have played additional roles in the early years of the Solar System. We shall see later on how collisions with planetesimals may have fragmented Mercury and stripped off its outer layers, may have stopped Venus from spinning as fast as other planets, may have tipped Uranus onto its side, and may have stripped off material from the Earth that then became our Moon.

Several modifications of the basic theory of the formation of the planets have been worked out to accommodate particular observational facts. For example, we have to explain why the inner planets are small, rocky, and dense, while the next group of planets out are large and primarily made of light elements. Since the protosun would have been made primarily out of hydrogen and helium, with just traces of the heavier elements formed in earlier cyclings of the material through stars and supernovae, we need a mechanism for ridding the inner regions of the Solar System of this lighter material. One possible way is to say that the Sun flared fiercely and/or often in its younger days (similar to a T Tauri star, as described in Section 27.1) and that the lighter elements were blown out of the inner part of the Solar System. But proto-Jupiter and the other outer protoplanets, which were much farther away from the Sun, would have retained thick atmospheres of hydrogen and helium because of their high gravity.

Scientists can calculate, based on their theories, how the temperature decreased with distance from the center of the solar nebula, ranging from almost 2000 K close to the protosun down to only about 20 K at the distance of Pluto. Various elements were able to condense—solidify—at different locations because of their differing temperatures. For example, in the positions occupied by the terrestrial planets, it was too hot for icy substances to form, though such ices are present in the giant planets.

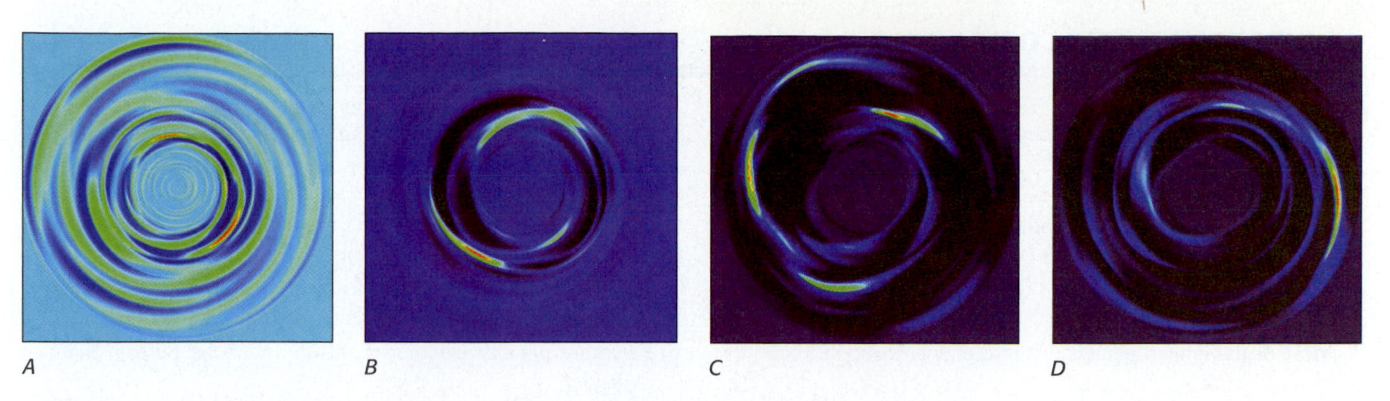


**FIGURE 7–24** A false-color representation of density clumps in the inner 20 A.U. of a model of the early solar nebula. A cross-section through the midplane of the nebula is shown. Orange corresponds to the highest density, and black to the lowest density. The four pictures cover about 300 years.

Alan Boss of the Carnegie Institution of Washington models in a computer what the solar nebula was like (Fig. 7–24). His calculations show that the equilibrium temperatures in a low mass solar nebula (0.02 times the mass of the Sun) are expected to be on the order of 1000 K at the asteroid belt (2.5 A.U.) and even higher at the Earth's location, during the phase when the nebula is still accreting gas and dust from the infalling molecular cloud. Temperatures fall to the point at which ice condenses (150 K) around 4 A.U., however, allowing these ices to remain solid there. They can then accumulate into bodies that can attract hydrogen gas and form the giant planets Jupiter and Saturn.

A new model of solar-system formation was advanced in the late 1990s by Frank Shu of the University of California at Berkeley. He considers that the disk of gas that collapsed from the solar nebula brought massive amounts of dust toward its center. The temperature there would have been 2000 K and most materials would have been vaporized. Shu theorized that jets of gas formed and ejected the vaporized dust outward at high speed. Later in the book (Chapter 20), we will learn that we find meteorites (rocks from space found on Earth) that contain round inclusions known as "chondrules." They had been thought to form far from the Sun, where it was cooler, but Shu's model has them forming close in. In the model, the molten particles solidified into the chondrules as they travelled outward. Some went to form the planets while some landed in the asteroid belt between Mars and Jupiter. In 2001, scientists reported their analysis of two meteorites in which metal grains of iron and nickel were found alongside the chondrules. The state of these iron-nickel grains shows that they formed in a hot location in only a few days (long ago), and were then quickly transported to a much colder region. Further, they have been stored in this colder region (the asteroid belt), since they would have changed state if they had been warmed. The observations match Shu's model, for they seem to verify that some of the Solar System's materials moved quickly from hot regions to cool ones.

All the modeling until the last few years was directed to explaining our own Solar System, with its arrangement of rocky planets close in and giant planets farther out. Now that dozens of giant planets have been discovered around other stars (Chapter 18), our ideas of the formation of planetary systems in general and our own Solar System in particular will certainly be modified. Many of these giant planets are closer to their parent stars than Mercury is to our Sun even though they are more massive than Jupiter! We weren't expecting that, for we had worked out explanations of why Jupiter and the other giant planets are farther out than the rocky planets. Did we foolishly convince ourselves that our Solar System was typical? Are the calculations right that seem to show that the planets in other Solar Systems could have formed far out and then been sent farther in through gravitational interactions?

## CORRECTING MISCONCEPTIONS

| ✖ *Incorrect* | ✔ *Correct* |
| --- | --- |
| The dark side of the Moon faces away from us. | The dark side of the Moon is the side away from the Sun, and it faces us at new moon. Do not confuse "dark side" with "far side." |
| Small rocky planets are found close to the main star of a solar system and gas giants beyond. | The discovery of giant planets close in to distant stars has forced us to rethink our ideas. |
| All planets rotate in the same direction. | Three planets rotate backward with respect to the other planets and to the sense of their orbital motion. |
| Most of the known planets are in our Solar System. | Far more planets are known outside our Solar System than inside. |

## SUMMARY AND OUTLINE

The Sun, our star (Section 7.1)
The phases of the Moon and planets (Section 7.2)
Eclipse phenomena (Section 7.3)
  Partial phases, Baily's beads, diamond ring, totality
Planetary rotation and revolution (Section 7.4)
  The ecliptic plane; inclinations of the orbits of the planets; conservation of angular momentum
  All planets revolve and most rotate in the same direction.
Theories of cosmogony (Section 7.5)
  Catastrophe theories
  Nebular theories of Kant and Laplace have newer versions: protosun, planetesimals, and protoplanets.
  The discovery of planets around other stars gives us other types of solar systems to study.

## KEY WORDS

photosphere
corona
solar wind
phases
umbra
penumbra
shadow bands
Baily's beads
diamond-ring effect
chromosphere
annular eclipse
revolution
rotation
ecliptic plane
in transit
inclination
angular momentum
catastrophe theories
tidal theory
nebula
nebular hypothesis
nebular theories
primeval solar nebula
protosun
planetesimals
protoplanets

## QUESTIONS

1. Sketch the Sun, labelling the corona and the solar wind.
2. Suppose that you live on the Moon. Sketch the phases of the Earth that you would observe for various times during the Earth's month.
3. If you lived on the Moon, would the motion of the planets appear to the eye any different from the motion we see from Earth?
4. If you lived on the Moon, how would the position of the Earth change in your sky over time?
5. Discuss the circumstances under which Uranus and Neptune are seen as crescents.
6. (a)Why isn't there a solar eclipse once a month? (b) Whenever there is a total solar eclipse, a lunar eclipse occurs either two weeks before or two weeks later. Explain.
7. If you lived on the Moon, what would you observe during an eclipse of the Moon? How would an eclipse of the Sun by the Earth differ from an eclipse of the Sun by the Moon that we observe from Earth?
8. Sketch what you would see if you were on Mars and the Earth passed between you and the Sun. Would you see an eclipse? Why?
9. Why can't we observe the corona every day from any location on Earth?
10. Describe why we cannot see the solar corona during the partial phases of an eclipse.
11. Where and when is the next total solar eclipse?
12. Describe relative advantages of ground-based eclipse studies and satellite studies of the corona.
13. Explain how Pluto can have a longer period than Neptune, even though Pluto was closer to the Sun until the year 1999.
14. Explain how conservation of angular momentum applies to a diver doing somersaults or twists. What can divers do to make sure they are vertical when they hit the water?
15. If two planets are of the same mass but different distances from the Sun, which will have the higher angular momentum around the Sun? (Hint: Use Kepler's third law.)
16. If the planets condensed out of the same primeval nebula as the Sun, why didn't they become stars?
17. Which planets are likely to have their original atmospheres? Explain.

## INVESTIGATIONS

1. On ice skates or Rollerblade in-line skates, or on a turntable, investigate the conservation of your own angular momentum.

## USING TECHNOLOGY

### W World Wide Web

1. Use a search engine or look at http://www.totalsolareclipse.net to find out about the locations from which the next two solar eclipses can be observed and about the predicted weather in those places during the eclipse.
2. Use the link on this book's Web site to look at the Web site known as The Nine Planets about the topics in this chapter.

### REDSHIFT

1. Position yourself high above the Sun and look down at the Solar System with the inner planets orbiting.
2. Position yourself higher and view all the planets orbiting.
3. Investigate the period in which Pluto is closer to the Sun than Neptune.
4. Put yourself in the plane of the ecliptic far beyond Pluto and investigate the angles of the planes of the planets' orbits.
5. Recreate the phases of the Moon. Ponder Tutorials: The phases of the Moon: 4 Moon's orbit around the Earth.
6. Check the phases of Venus from different locations in the Solar System, inside and outside its orbit. Study the phases from Guided Tours: Revolutionaries: The phases of Venus.
7. Duplicate the next solar eclipse, with vantage points in sequence from Earth, Moon, and Sun. Animate the view from each location.
8. Duplicate the next lunar eclipse, with vantage points in sequence from Earth, Moon, and Sun.
9. Consider Story of the Universe: Orbits in the Solar System; Tutorials: The Solar System; and Science of Astronomy: Night on the Mountain: The Planet Dossier.
10. On the second CD of RedShift College Edition, look at Movies: History of the Solar System: Formation of the Planets.

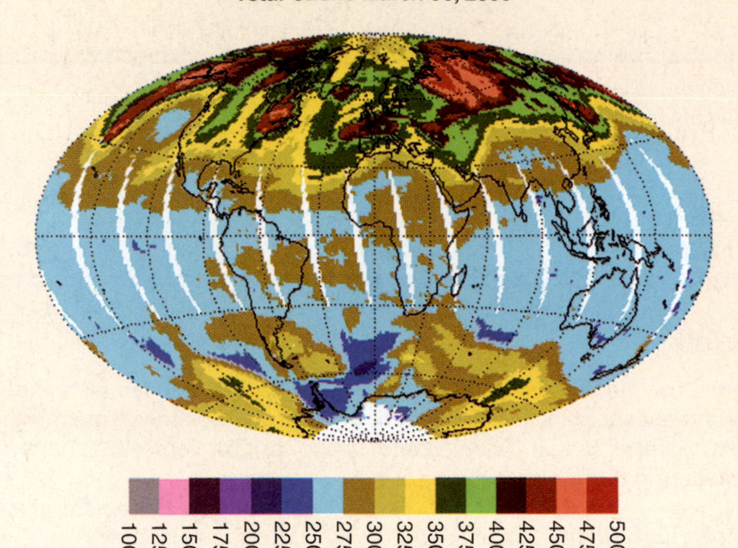


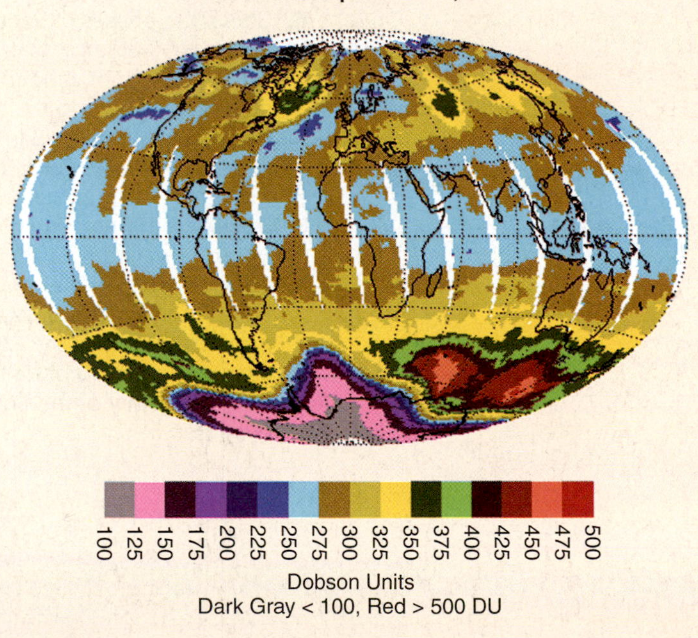


The largest ozone hole ever (purple and gray region surrounding the South Pole in the lower image), in late September 2000, contrasted with the ozone percentages six months earlier, when no ozone hole was present on Earth.

# Chapter 8

# Our Earth

We know the Earth intimately; a mere wrinkle on its surface is a mountain to us. Weather satellites now give us a constant global view and show us large atmospheric systems linking all parts of the Earth. More rarely, earthquakes or volcanoes bring us messages from below the Earth's surface. But no longer do we have to treat the Earth as one of a kind; by studying the other planets we can make comparisons that allow us to understand both the other planets and our own (Fig. 8–1). In this chapter, we will summarize some of our knowledge of the Earth's structure and history. In later chapters, we will compare the Earth with the other planets.

**AIMS:** To describe and understand the Earth on a planetary scale and to assess such current problems as the greenhouse effect and the ozone hole

Seven of the planets in the solar system have moons—that is, smaller bodies orbiting them. Most planets have more moons than the Earth. Our Earth–Moon system and the Pluto–Charon system are the only ones, though, in which the satellite contains as much as one per cent of the mass of the parent. The moons around other planets contain from 800 to more than a billion times less mass than their parent planets.

The study of the Earth's interior and solid surface is called **geology.** We can study the Earth so intricately that studies of other parts of the Earth have their own names, such as oceanography and meteorology. ("Meteorology," the study of weather, comes from the Greek root meaning astronomical phenomenon; the astronomical word "meteor" came later on from the same root.)

## 8.1 THE EARTH'S INTERIOR

Geologists study how the Earth vibrates as a result of large shocks, such as earthquakes. Much of our knowledge of the structure of the Earth's interior comes from **seismology,** the study of these vibrations. The vibrations travel through different types of material at different speeds. Also, when the **seismic waves** strike the boundary between different types of material, they are reflected and refracted, just as light is when it strikes a glass lens. From piecing together seismological and other geological evidence, geologists have been able to develop a picture of the Earth's interior (Fig. 8–2). The terms and divisions apply to many other solar-system bodies as well.

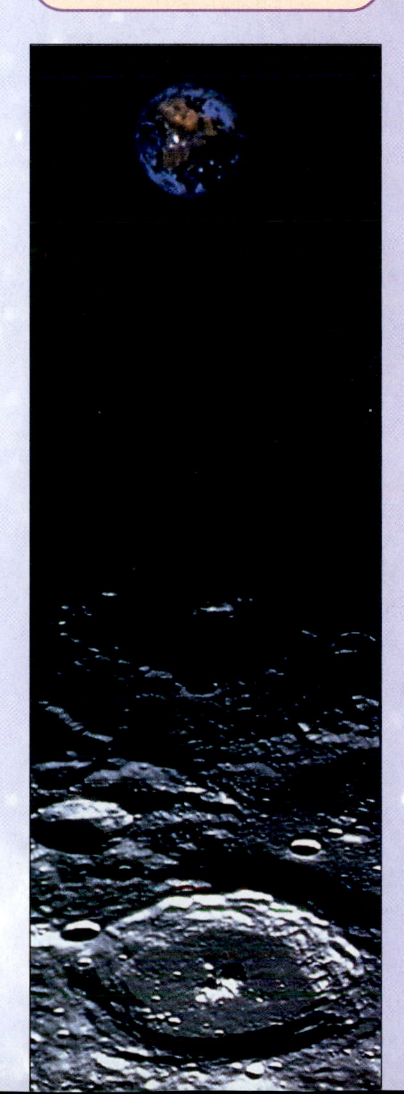

**FIGURE 8–1** The Earth, with a lunar crater in the foreground, imaged in 1994 by the Clementine spacecraft. We see the full Earth over the lunar north pole. It is clear over Africa. The angular distance between the images of the Earth and Moon was electronically reduced in this image.

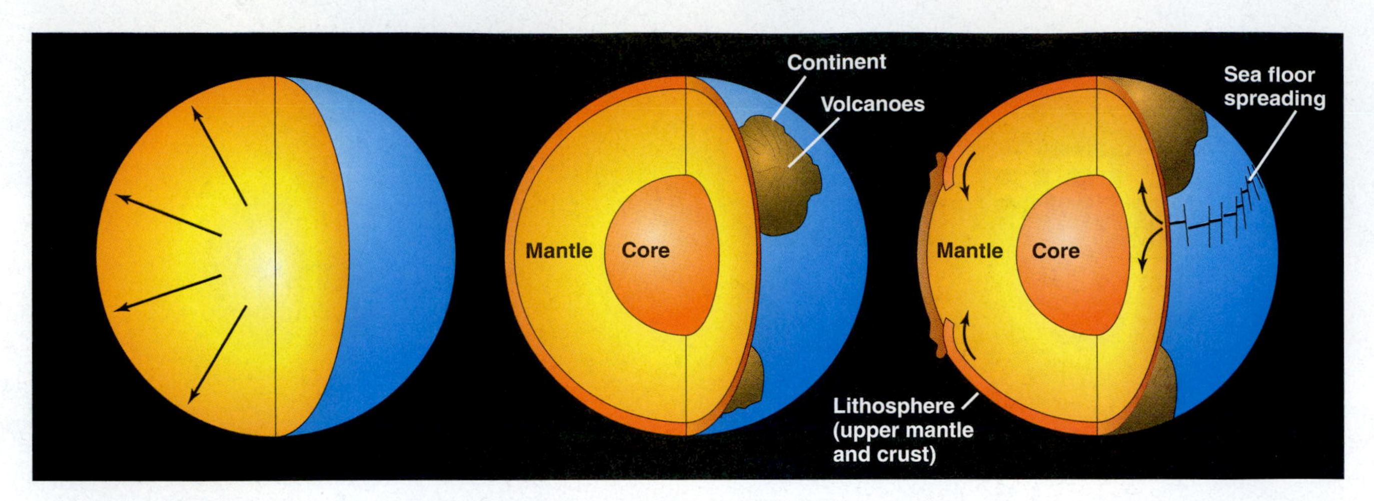


**FIGURE 8–2** The structure of the Earth and stages in its evolution. (*left*) Tens of millions of years after its formation, radioactive elements along with gravitational compression and impact of debris produced melting and differentiation. Heavy materials sank inward and light materials floated outward. During this time, the original atmosphere (consisting mostly of hydrogen) was blown away by the solar wind. The atmosphere that replaced it contained methane, ammonia, and water. (*center*) The heaviest materials formed the core, and the lightest materials formed the crust. (*right*) The crust broke into rigid plates that carried the continents and moved very slowly away from areas of sea-floor spreading.

The innermost region is called the **core.** It consists primarily of iron and nickel. The central part of the core—about the size of the Moon—may be solid, but the outer part is probably a very dense liquid. In a surprising discovery, it was measured in 1996 that the inner core rotates each day about ⅔ second quicker than the outer core. Over 100 years, this effect has given the inner, solid core an extra quarter turn than the outer, liquid core. The discovery may lead to new insights on changes in the Earth's magnetic field, which is generated by motions in the core.

Outside the core is the **mantle,** and on top of the mantle is the thin outer layer called the **crust.** The crust and the outer part of the mantle together are called the **lithosphere.** The lithosphere, a rigid layer, surrounds a zone, the **aesthenosphere,** that is partially melted (Fig. 8–3).

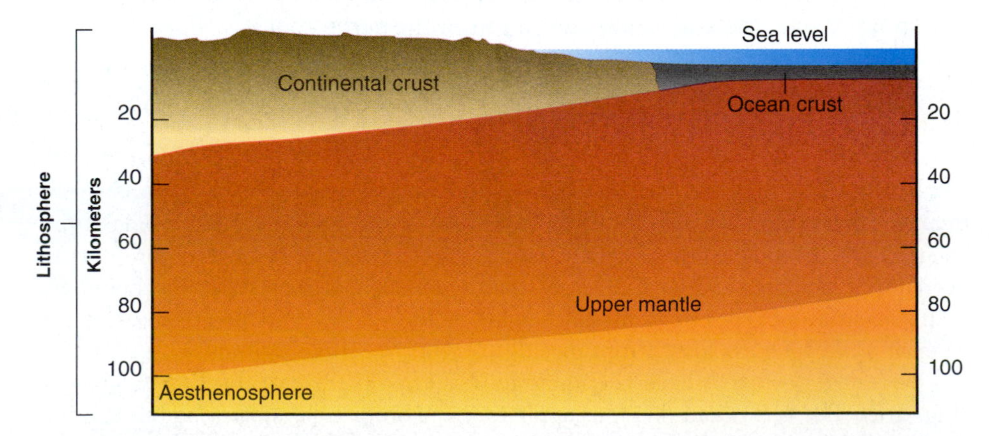


**FIGURE 8–3** Continental crust is thicker than ocean crust and has a lower density. The lithosphere, composed of the crust and the outer upper mantle, is a rigid layer. The crust is different in composition from the mantle. Below the lithosphere is a partially melted layer known as the aesthenosphere.

How did such a layered structure develop? The Earth formed surrounded by a lot of debris in the form of dust and rocks. The young Earth was probably subject to constant bombardment from this debris. This bombardment heated the surface to the point where it began to melt, producing lava. However, this process could not have been responsible for heating the interior of the Earth, because the rock out of which the Earth was made conducts heat very poorly.

Much of the original heat for the Earth's interior came from gravitational energy released as particles accreted to form the Earth. But the major source of heat for the interior is the natural radioactivity within the Earth. The different forms of individual atoms are called **isotopes;** an isotope undergoes chemical reactions in exactly the same way as another isotope of the same element, but has a different mass (Section 28.3). Certain isotopes are unstable—they spontaneously undergo nuclear reactions. In these reactions, particles such as electrons or alpha particles (helium nuclei) are given off with high energy. These energetic particles collide with the atoms in the rock and give some of their energy to these atoms. The rock becomes hotter. When the Earth was forming, there was a sufficient amount of radioactive material in its interior to cause a great deal of heating. And since the material conducted energy poorly, the heat generated was trapped. As a result, the interior got hotter and hotter.

After about a billion years, the upper parts of the Earth's interior had become so hot from radioactivity that the iron melted and sank to the center, forming the core. The process by which the differences occur at different distances from the center is called **differentiation.** This primary round of differentiation as well as later secondary and tertiary rounds is responsible for the present layered structure of the Earth.

Energy released from gravity as the iron fell toward the center provided enough energy to heat the Earth's interior considerably. Eventually other materials also melted. Since about $^1/_3$ the Earth's mass is iron, the formation of a liquid iron core was a major step in the Earth's history. Lighter materials or those with lower melting points floated up to form the crust. The material between remained as the mantle. As the Earth cooled, various materials, because of their different densities and freezing points (the temperature at which they change from liquid to solid), solidified at different distances from the center. Thus the Earth is different at different distances from its center.

We will see later on that planetary scientists try to find out which of the various planets, moons, and asteroids in the solar system are differentiated. Many are, indicating that internal activity has taken place in them.

## 8.2 PLATE TECTONICS

After various rounds of differentiation, most of the radioactive elements ended up in the outer layers of the Earth. Thus these elements provide a heat source not far below the ground. Their presence leads to a general **heat flow** outward through the upper mantle and crust (the lithosphere). The power flowing outward through each square centimeter of the lithosphere is small—10 million times less than that necessary to light an average light bulb and thousands of times less than that reaching us from the Sun. However, the terrestrial heat flow does have important geological consequences.

In some geologically active areas, the heat-flow rate is much higher than average, indicating that the source of heat is close to the surface. In a few places, the outflowing **geothermal energy** in these regions is being tapped as an energy source.

The lithosphere is a relatively cool, rigid layer; it is segmented into **plates** thousands of kilometers in extent but only about 50 km thick. Because of the internal heating, the lithosphere sits on top of a hot layer where the rock is soft, though it is

A DEEPER DISCUSSION

## BOX 8.1 The Oldest Rocks on Earth

Plate tectonics has transformed much of the Earth's surface, so few rocks remain from the Earth's earliest era. The oldest known rocks on Earth are about 4 billion years old. Two tiny crystals of zircon, found in Australia, are the only known objects that are older. One was dated at 4.28 billion years old and, in 2001, a second one was dated at 4.4 billion years old. Our Earth was only 150 million years old at the time. The crystals were dated by comparing the relative amounts of uranium and lead in them.

The scientists involved found oxygen atoms in the crystals. Their analysis of these atoms indicates that the Earth's surface was already cool enough for liquid water to form at the time the zircon formed. The observation places important limits on theories of how the Earth formed, and on when the Moon's material might have been ripped out of the Earth. Since life on Earth is thought to have formed in oceans, the observations may push to earlier times the date when we think life on Earth originated.

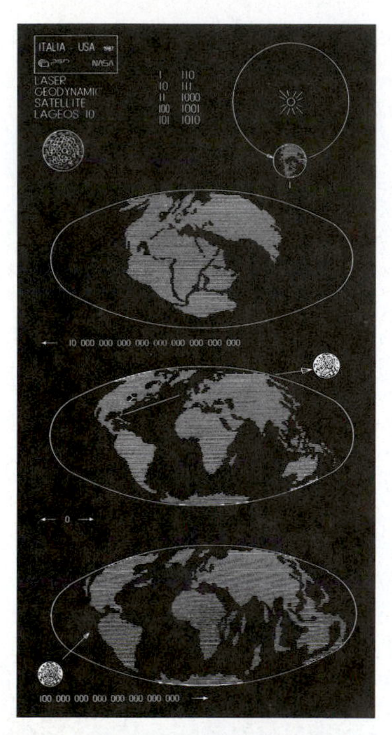


**FIGURE 8–4** The satellites LAGEOS (Laser Geodynamic Satellite) I and II provide information about the Earth's rotation and the crust's movements. They contain 426 corner-reflecting mirrors, which bounce laser beams back in exactly the direction from which they came. Scientists on Earth time the interval between their sending out a laser pulse and receiving the reflected signal, then use the results to figure out exactly where on Earth they are. LAGEOS I and LAGEOS II bear plaques showing the continents in their positions 270 million years in the past (*top*), at present (*middle*), and 8 million years in the future (*bottom*), according to our knowledge of continental drift. (The dates are given in the binary system.) Data on continental drift are gathered by reflecting laser beams from telescopes on Earth off the 426 retroreflectors that cover the satellite's surface. (Each retroreflector is a cube whose interior reflects incident light back in the direction from which it came.) Measurements of the time until the beams return can be interpreted to show the accurate position of the ground station. We can now measure the position to an accuracy of about 1 cm. LAGEOS II was launched in 1992, jointly by NASA and the Italian Space Agency.

not hot enough to melt completely. The lithosphere actually floats on top of this soft layer. The hot material beneath the lithosphere's rigid plates churns slowly in convective motions (circulation involving rising hot material; boiling is an example), carrying the plates around over the surface of the Earth. This theory, called **plate tectonics,** explains the observed **continental drift**—the slow drifting of the continents from their original positions. ("Tectonics" is from the Greek word meaning "to build.") "Continental drift" was the original observation that is now explained by the theory of plate tectonics.

That the continents have moved on the surface of the Earth, which once seemed unreasonable, is generally accepted now that we know that it is but a symptom of plate tectonics. The continents were once connected as two supercontinents, one called Gondwanaland (after a province of India of geological interest) and a northern supercontinent called Laurasia. (These may have, in turn, separated from a single supercontinent called Pangaea, which means "all lands.") Over two hundred million years or so, the continents have moved apart as plates have separated. We can see from their shapes how they originally fit together (Fig. 8–4). Finding similar fossils and rock types along two opposite coastlines, which were once adjacent but are now widely separated, provides proof. In the future, we expect California to separate from the rest of the United States, Australia to be linked to Asia, and many other changes to occur.

**FIGURE 8–5** The San Andreas fault marks the boundary between the California and Pacific plates.

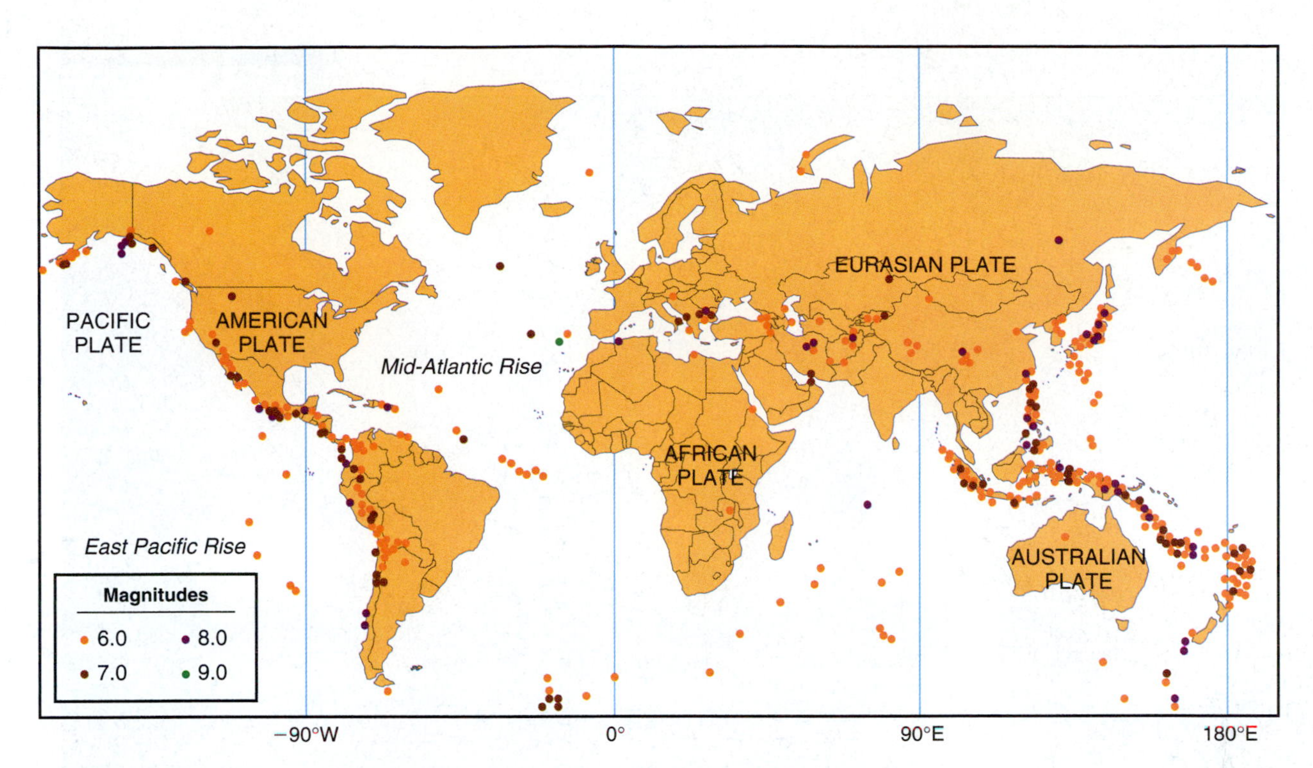


**FIGURE 8–6** This plot of earthquakes greater than 6.7 on the Richter scale over 15 years shows that earthquakes occur preferentially at plate boundaries. The principal tectonic plates are labelled.

The boundaries between the plates (Fig. 8–5) are geologically active areas. Therefore, these boundaries are traced out by the regions where earthquakes (Fig. 8–6) and most of the volcanoes (Fig. 8–7) occur. The boundaries where two plates are moving apart mark regions where molten material is being pushed up from the hotter interior to the surface, such as the **mid-Atlantic ridge** (Fig. 8–8). Molten

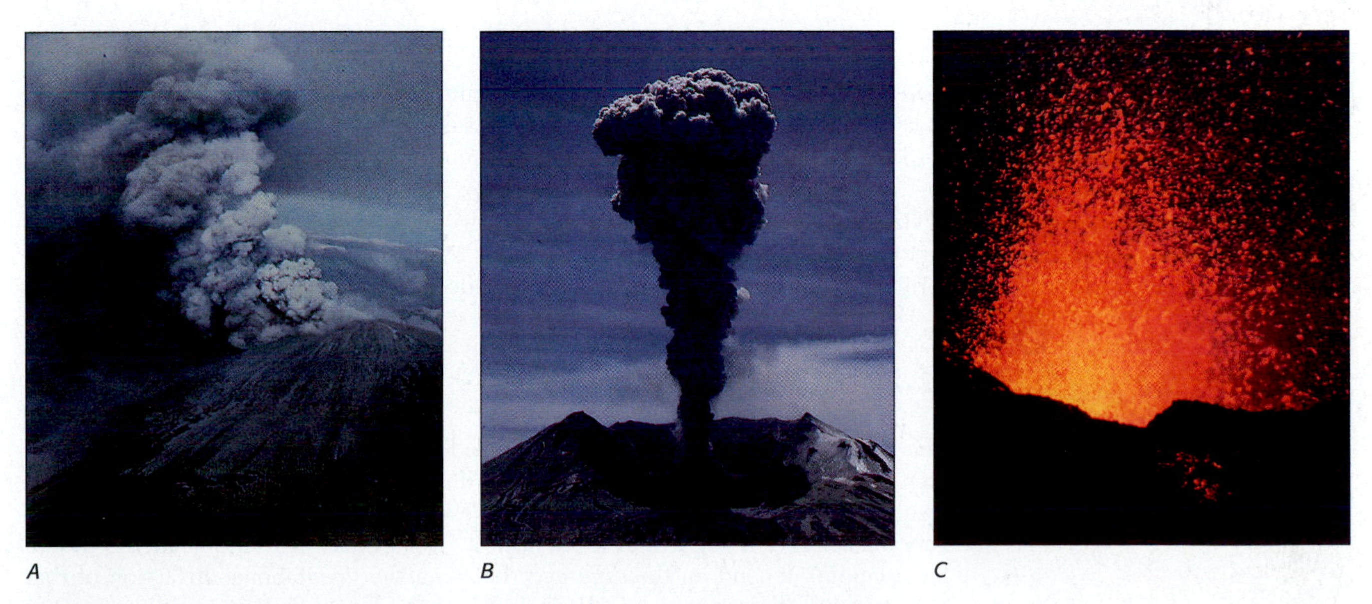


**FIGURE 8–7** (*A*) Mount St. Helens, shown here during its May 18, 1980, eruption, lies on a plate boundary. (*B*) After its devastating eruption. (*C*) Kilauea volcano, on the island of Hawaii, erupts less explosively.

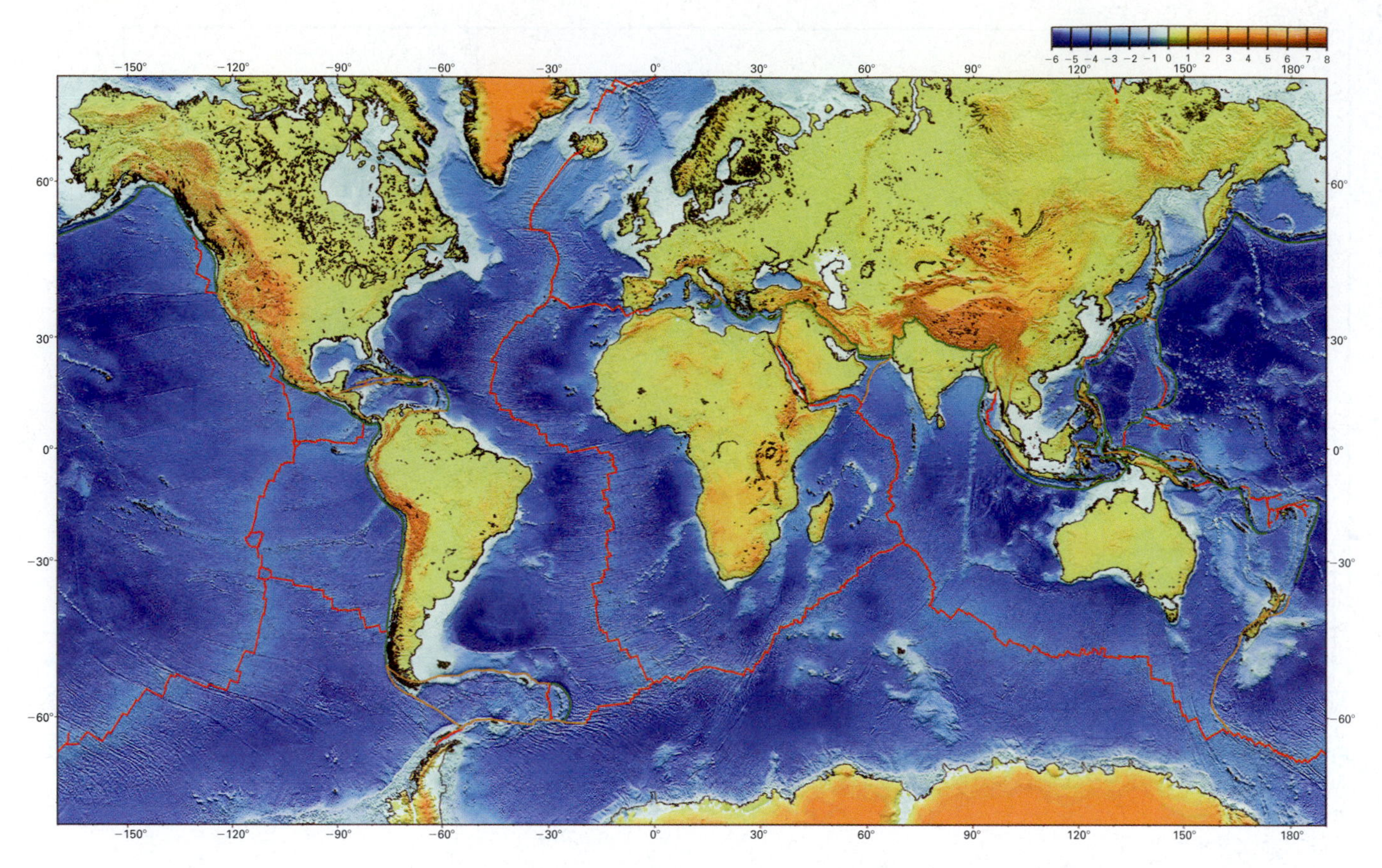


**FIGURE 8–8** The Earth's ocean floor, mapped with microwaves bounced off the ocean's surface by the U.S. Navy's Geosat satellite and declassified in 1995. Features on the sea floor with excess gravity attract the water enough to cause structure on the ocean's surface that can be mapped, though a 2000-meter-high seamount produces a bump on the surface only 2 meters high. High gravity is shown in yellow, orange, and red; low gravity is shown in blue and violet. The feature running from north to south in the middle of the ocean is the **mid-Atlantic ridge.** It marks the boundary between plates that are moving apart. Continents drift at about the rate that fingernails grow.

material is being forced up through the center of the ridge and is being deposited as lava flows on either side, producing new sea floor. The motion of the plates is also responsible for the formation of the great mountain ranges. When two plates come together, one may be forced under the other, and the other may be raised. The "ring of fire" volcanoes around the Pacific Ocean (including Mount St. Helens in Washington) were formed in this way. Another example is the Himalayas, which were formed when the Indian subcontinent collided with the Asian continent.

## 8.3 TIDES

The tides—like the Moon's passing overhead—occur about an hour later each day. So it has long been accepted that tides are directly associated with the Moon. Tides result from the fact that the force of gravity exerted by the Moon (or any other body) gets weaker as you get farther away from it. Tides caused by an object (the Moon, for example) depend on the *difference* between the gravitational attraction of that object at different points on another object, so the forces that cause tides are often called **differential forces.** Let us first, in this section, discuss the astronomical causes of tides. Toward the end of the section, we will mention the local effects of the shape of the ocean floor on the actual tides at an individual location.

Low tide (as at point B)    High tide (as at points A and C)

**FIGURE 8–9** (*top*) A schematic representation of the tidal effects caused by the Moon. The arrows represent the acceleration of each point that results from the gravitational pull of the Moon (exaggerated in the drawing). The water at point A has greater acceleration toward the Moon than does point B; since the Earth is solid, the whole Earth moves with point B. (Tides in the solid Earth exist, but are much smaller than tides in the oceans.) Similarly, the solid Earth is pulled away from the water at point C. Note that although the tidal force results from gravity, it is the *difference* between the gravitational forces at two places and is not the same as the "gravitational force." Tidal forces are important in many astrophysical situations, including rings around planets and accretion disks around neutron stars and black holes. (*bottom*) The tidal range at Kennebunkport, Maine, where the shape of the ocean floor leads to an especially high tidal range. Note that the Moon and the Earth are drawn to a different scale from the separation.

To explain the tides in Earth's oceans, consider, for simplicity, that the Earth is completely covered with water. We might first say that the water closest to the Moon is attracted toward the Moon with the greatest force and so is the location of high tide as the Earth rotates. If this were all there were to the case, high tides would occur about once a day. However, two high tides occur daily, separated by roughly 12.5 hours. Low tides come in between.

To see how we get two high tides a day, consider three points, A, B, and C, where B represents the solid Earth, A is the ocean nearest the Moon, and C is the ocean farthest from the Moon (Fig. 8–9 *top*). Since the Moon's gravity weakens with distance, it is greater at point A than at B, and greater at point B than at C. If the Earth and Moon were not in orbit about each other, all these points would fall toward the Moon, moving apart as they fell. (They would actually move apart only slightly, since the pull of the Earth's gravity on the oceans is stronger than that of the Moon.)

Thus the high tide on the side of the Earth that is near the Moon is a result of the water's being pulled toward the Moon more than the solid Earth is being pulled. Unlike the water, the land moves as a solid body; so on the opposite side of the Earth, the ground moves toward the Moon more than does the water in the same location, causing another high tide. In between the locations of the high tides, the water has rushed elsewhere so we have low tides. As the Earth rotates once a day, a point on its surface moves through two cycles of tides: high-low-high-low. Since the Moon is moving in its orbit around the Earth, a point on the Earth's surface has to rotate longer than 12 hours to return to a spot nearest to the Moon, and so to begin a new cycle. Thus the tides repeat about every 12.5 hours.

Note that the tidal force is not the same as the gravitational force. The tidal force is, rather, the difference between the gravitational forces computed at two different places—such as the side of the Earth near the Moon and the side of the Earth farthest from the Moon. While the gravitational force of one body on another varies inversely with the square of their distance, the tidal force of one body on another varies with the cube of their distance.

The Sun's effect on the Earth's tides is only about half as much as the Moon's. Though the Sun exerts a greater gravitational force on the Earth than does the Moon, the Sun is so far away that its force does not change very much from one side of the Earth to the other. And it is only the change in force from one place to another that accounts for tides.

At a time of new or full moon, the tidal effects of the Sun and Moon work in the same direction, and we have especially high tides, called **spring tides** (related to the German word "springen" meaning "to rise up"; the word has nothing to do with the season spring). At the time of the first- and last-quarter moons, the effects of the Sun and Moon do not add, and we have **neap tides,** when high and low tides differ least.

Tides vary as the distance of the Moon from the Earth changes. When the Moon is relatively close to the Earth at new moon or full moon, the tidal range is relatively great. Thus people who live near the ocean worry especially about storms that occur at new moon or full moon (including the times of eclipses).

Many people mistakenly think that there is some special effect on the tides at the days of the equinoxes. Actually, since the equinox merely occurs when the Sun crosses the equator, there is no major effect at the equinoxes. If the Moon is relatively close to the Earth, then the new moon or full moon closest to the equinox will have a higher tidal range, but such events are not usually on the day of an equinox.

The existence of continents that block the flow of water around the Earth, and the shape of the ocean floor, have major effects on the tides (Fig. 8-9 *bottom*). The movement of water is not immediate as the Moon passes overhead, and water may undergo reflection from distant shores. In some places, only one tide a day occurs because of such local circumstances and because of the Moon's changing declination above or below the celestial equator. The height of the tide varies too. But all such cases can be understood as modifications of the basic tidal theory discussed previously.

Though tides show up more dramatically in the Earth's oceans than in the solid Earth, the latter does exhibit a response to tidal forces. A tidal bulge in the solid Earth some 30 cm high circles the Earth under and across from the Moon. It was measured in 1996, when satellite measurements of ocean tides allowed them to be subtracted out accurately enough, that this solid tidal bulge is delayed 20 seconds from the Moon's passage. Further, deformations of the Earth's crust caused by lunar tides cause the giant atom smasher at the European Center for Particle Physics (CERN) to change slightly in size with the period of the Moon's orbit. The giant circle in the Large Electron Positron Collider (LEP), 26.7 km in circumference, stretches and shrinks by nearly 1 mm. These deviations change the energy of the circulating beam of particles by enough to make this tidal effect the dominant source of error in evaluating the mass of the exotic particles (Z-particles; see Chapter 38) being studied.

## 8.4 THE EARTH'S ATMOSPHERE

The Earth's atmosphere presses down on us all the time, though we are used to it. The force pressing down per unit of area is called **pressure.** We define the average pressure at the surface of the Earth to be **one atmosphere.** (In SI units, 1 atmosphere is 101,300 pascals, where 1 pascal is the pressure of a force of 1 newton pressing on each square meter.) When we express the pressure in atmospheres, we are really taking the ratio of the pressure at any level to that at the surface. The pressure in the atmosphere falls off sharply with increasing height. The temperature (Fig. 8–10) varies less regularly with altitude.

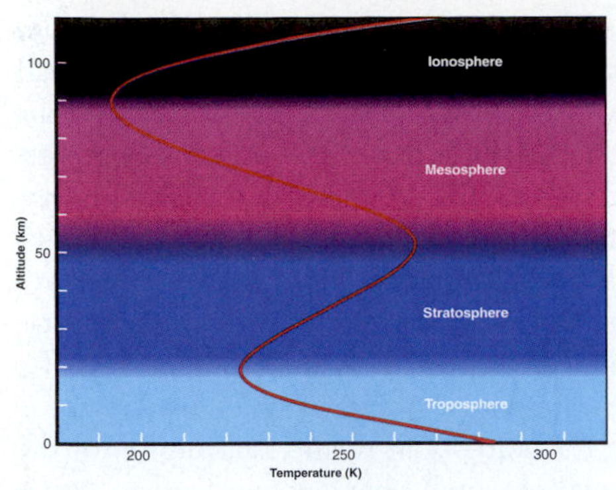


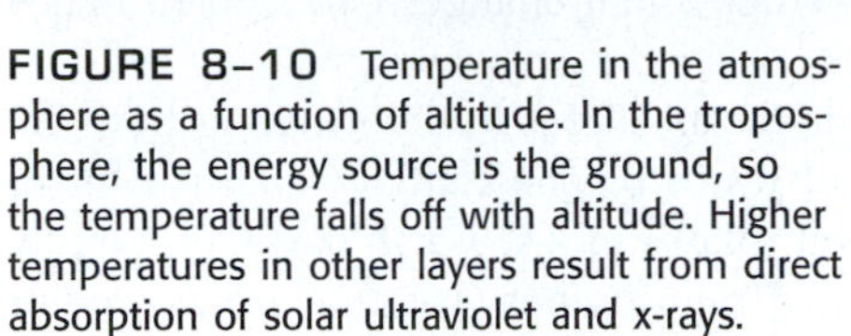

**FIGURE 8–10** Temperature in the atmosphere as a function of altitude. In the troposphere, the energy source is the ground, so the temperature falls off with altitude. Higher temperatures in other layers result from direct absorption of solar ultraviolet and x-rays.

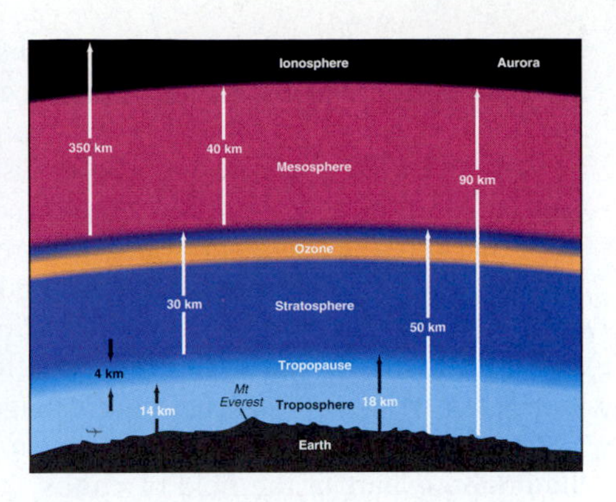


**FIGURE 8–11** The layers of the Earth's atmosphere showing the location of the part of the stratosphere where ozone is formed. Once formed, the ozone is transported to lower stratospheric layers.

It is convenient to divide the atmosphere into layers (Fig. 8–11), according to the composition and the physical processes that determine the temperature.

The Earth's weather is confined to the very thin **troposphere.** At the top of the troposphere, the pressure is only about 10 per cent of its value at the ground. A major source of heat for the troposphere is infrared radiation from the ground, so the temperature of the troposphere decreases with altitude.

It isn't true that water going down the drain curves to the right in the northern hemisphere and to the left in the southern hemisphere, since the shape of the sink and the injection fluctuations of the faucet are even greater effects. However, the germ of truth is there, since huge masses of air in the Earth's atmosphere tend to curve in one direction in one hemisphere (northern or southern) and the opposite direction in the other hemisphere as they move from equatorial regions toward the poles. As a result, the hurricanes—low-pressure regions—we see on our weather maps are always curving clockwise in the northern hemisphere. Also in our troposphere, lesser low-pressure storms curve in that direction as well. We will save the discussion of why—an application of Newton's laws that is known as the Coriolis force—for the Jupiter chapter, when we discuss huger storms than we have on Earth.

Basically, warm air rises upward in the troposphere near the equator and travels toward the poles. When it cools enough to become denser than the air below, it sinks. In the Earth's troposphere, giant cells of circulation known as "Hadley Cells" extend to about 30° north and south of the equator. Less prominent Hadley Cells go from 30° to 60° in latitude, and still another set goes from 60° toward the poles.

Above the troposphere are the **stratosphere** and the **mesosphere.** The upper stratosphere and lower mesosphere contain the **ozone layer.** (Ozone is a molecule, consisting of three oxygen atoms: $O_3$; it absorbs ultraviolet radiation from the Sun. The oxygen our bodies use is $O_2$.) The absorption of ultraviolet radiation by the ozone means that these layers get much of their energy directly from the Sun. The temperature is thus higher there than it is at the top of the troposphere.

Above the mesosphere is the **ionosphere,** where many of the atoms are ionized. The most energetic photons from the Sun (such as x-rays) are absorbed here. Thus the temperature gets quite high. Because of this rising temperature, this layer is also called the **thermosphere.** Since many of the atoms in this layer are ionized, the ionosphere contains many free electrons. These electrons reflect very long wavelength radio signals. When the conditions are right, radio waves bounce off the ionosphere, which allows us to tune in distant radio stations. When solar activity is high, there may be a lot of this "skip."

**FIGURE 8–12** The Earth from space, in an image from the Clementine spacecraft (described in the next chapter). The image clearly shows weather patterns.

Our knowledge of our atmosphere has been greatly enhanced by observations from satellites above it and from spacecraft to the Moon (Fig. 8–12). Scientists carry out calculations using the most powerful supercomputers to interpret the global data and to predict how the atmosphere will behave. The equations are essentially the same as those for the internal temperature and structure of stars, except that the sources of energy are different. Our ability to predict weather (which is short term) and climate (which is long term) is improving. However, we now realize that there is a fundamental limitation in how well we will ever begin to predict the weather—the butterfly-wing effect. Even so tiny a change as a flap of a butterfly's wing can cause adjacent changes that cause adjacent changes and so on until some major weather change—even a tornado—can result. Discussion of the limitation is part of the field of **chaos,** in which small changes can lead to major effects.

The rotation of the Earth has a very important effect in determining how the winds blow. Comparison of the circulation of winds on the Earth (which rotates in 1 Earth day); on slowly rotating Venus (which rotates in 243 Earth days); on rapidly rotating Jupiter, Saturn, and Neptune; and on Uranus (whose rotation axis can point nearly at the Sun) helps us understand the weather on Earth.

## 8.5 OF WATER, GLOBAL WARMING, THE OZONE HOLE, AND LIFE

Why is there is so much water on Earth—in our atmosphere and in our oceans? We care especially because water seems to be necessary for life. How have the absence of features that have made Earth a hospitable planet for life as we know it made the other planets so unattractive? Can we make sure we are not changing the Earth's atmosphere so that we will wind up with an atmosphere like that of, say, Venus, which has clouds of sulfuric acid drops?

There are many theories as to how the Earth got its water in the first place. One interesting idea worthy of note is the theory that the water came in collisions with many comets as the Earth formed. Some of the Earth's early atmosphere was belched out by volcanoes. In any case, massive amounts of water are present. Over the eons, the ultraviolet radiation hitting Earth affected the atmosphere greatly—though less so since a protective layer of ozone formed, blocking most of the ultraviolet.

The water on the Earth has spurred the incorporation of carbon dioxide into rocks. The carbon dioxide dissolves in the water, and the carbon is then incorporated into the shells of sea animals; these shells eventually form rocks. On Venus the carbon dioxide is still in the atmosphere—in fact, it makes up over 95 per cent of the atmosphere. The carbon dioxide is the major cause of trapping solar energy in Venus's atmosphere, thus keeping the surface of Venus very hot, too hot for life. This trapping is known as the "greenhouse effect," and we shall study it further in the chapter on Venus. Apparently water and plant life have saved the Earth from suffering the same fate as Venus. Indeed, the amount of warming from the greenhouse effect that occurs on Earth is enough to transform our atmosphere from a temperature too cold for us to be comfortable by about 33°C to the temperature we actually have. Perhaps if life had arisen on Venus at the same time that it arose on Earth, it would have modified Venus's atmosphere as it did the terrestrial one.

Our Earth's atmosphere currently provides about 33°C of greenhouse warming, heating our air enough to make the Earth livable. But we are now adding so much carbon dioxide to our air that we risk additional warming, with dire consequences.

Extrapolations of recent trends (Fig. 8–13) for the Earth's atmosphere project that the greenhouse effect will cause a "global warming" by an average of 2°C over the next century. This amount of warming predicted could cause substantial climate change, with an increase in severe storms and a change in the location of agricultural belts as only some of the consequences. Representatives of the world's governments have been holding occasional meetings to set goals for reducing the emission of "greenhouse gases"—principally carbon dioxide—that will reduce greenhouse warming

*(continued on p. 146)*

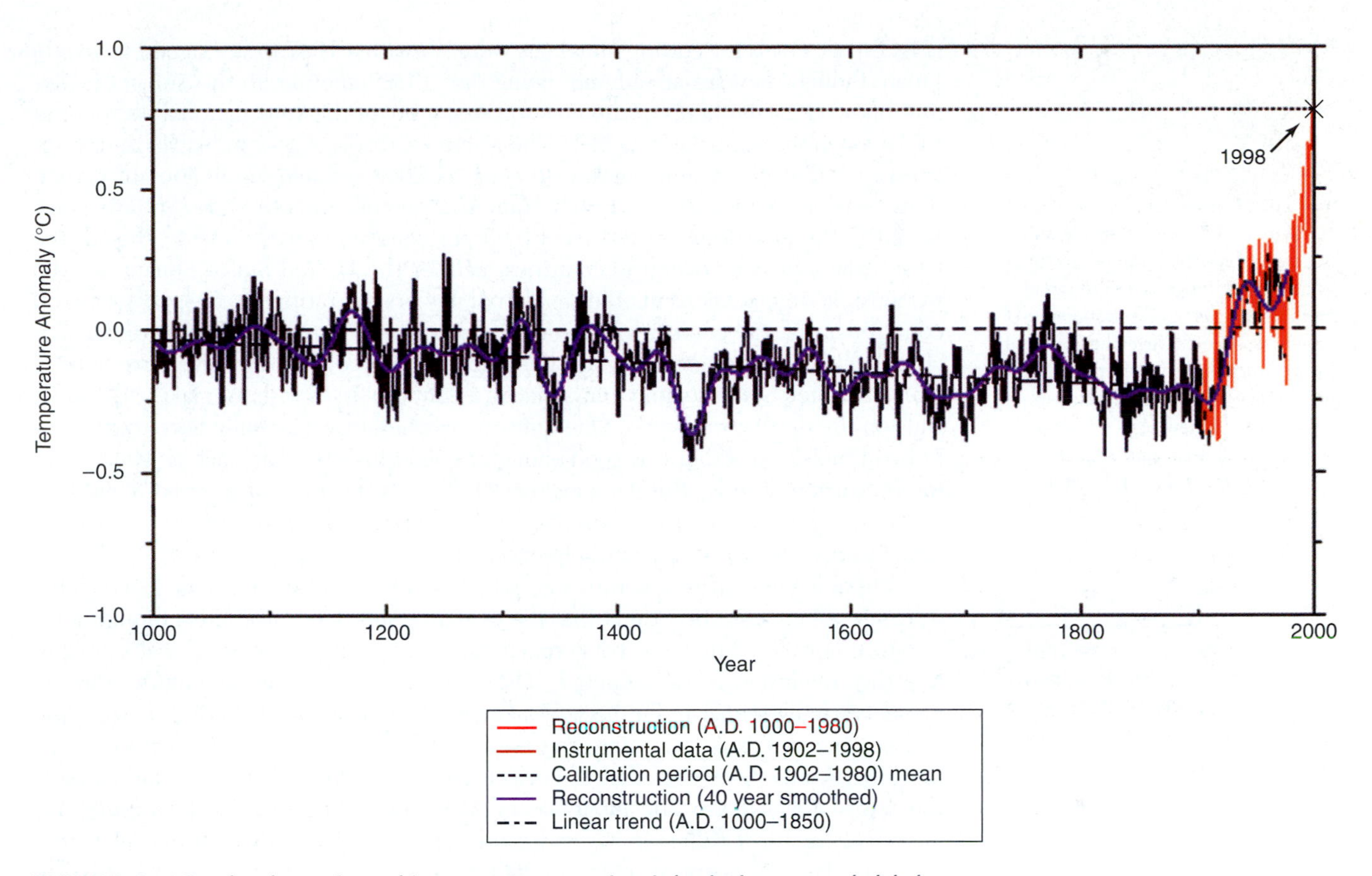


**FIGURE 8–13** The change in Earth's temperature over time is beginning to reveal global warming. Here we see the Earth's overall temperature for the last thousand years. The purple solid line is the best average, smoothed over 40-year periods. The black curve shows a reconstruction from 1000-1980, the red shows instrumental data since 1902, and the horizontal dashed line shows the mean from 1902 to 1980; the 95% confidence limits are shown in yellow. A linear cooling through 1850, shown with long and short dashes, has been reversed.

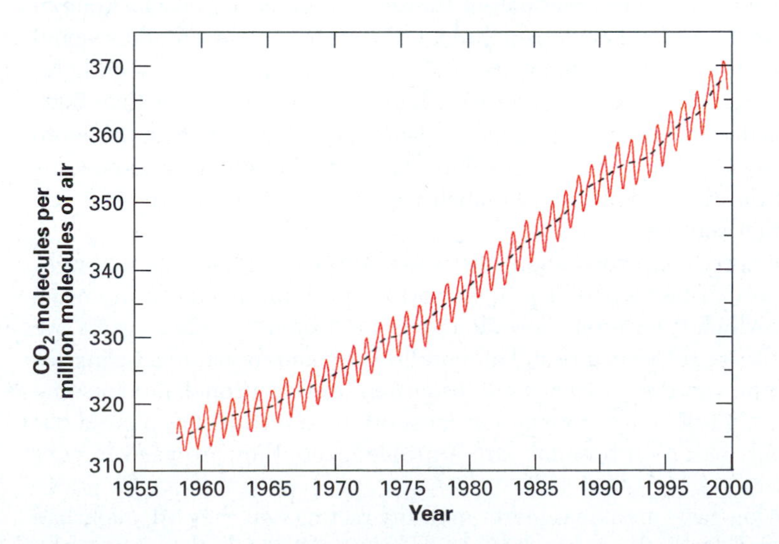


**FIGURE 8–14** The concentration of atmospheric carbon dioxide in parts per million (ppm) of dry air as it changed with time, observed at the Mauna Loa Observatory in Hawaii. In addition to the yearly cycle, there is an overall increase.

Paleoclimate analyses have looked at the interplay between the evolution of a star (brighter with age), and the evolution of carbon sequestering biology. The compensation among these processes has resulted in a relatively stable climate for much of the history of life on Earth. Concern about the continuing stability of a climate to which we have tuned our social systems (agriculture, sources of fresh and clean water, coastal land use, etc.) is with the rate as well as the magnitude of change. This is why changes in greenhouse gases documented over the past century are of concern.

The UN Framework Convention on Climate Change (http://www.unfccc.de), which commits us to, among other things "stabilization of greenhouse gas concentrations in the atmosphere at a level that would prevent dangerous anthropogenic interference with the climate system" was signed by President George H. W. Bush and ratified by the US Congress in 1992. At last count it has been ratified by 186 nations.

The updated statement in the 2001 IPCC document is "There is new and stronger evidence that most of the warming observed over the last 50 years is attributable to human activities."

(Fig. 8–14). Goals for cutting emissions were set with a 1997 treaty signed in Kyoto, Japan, though they are already not being met. (The meeting on this subject in Kyoto followed on the heels of the General Assembly of the International Astronomical Union in the same meeting hall.) The Kyoto Protocol calls for industrialized countries, including the United States, to start by 2008 reducing emissions of carbon dioxide, which comes largely from burning coal and oil, and other greenhouse gases. By 2012, the emissions are to be cut by 5 per cent below 1990 levels. Though the treaty was signed by over 150 countries, neither the United States Senate nor the responsible organizations in other industrial powers has ratified it. Several follow-up conferences of 150 countries have taken place. The question of what to do has become politically charged, especially because the Protocol calls for developed nations like the United States to limit emission of greenhouse gases without making the same calls on the developing world. One question is whether we actually have to cut uses of fossil fuels or whether it is good enough to plant forests, which act as "sinks," taking up carbon dioxide. But it seems that old forests (because of their soils and tree roots) are better than newly planted ones for sopping up carbon dioxide, so making new forest plantations may not help enough.

There is some current controversy whether we have already detected the global warming. Many scientists think that the evidence shows that we have, though temperature normally fluctuates by so much, and it is so hard to make global averages, that the conclusion is not definitive. There are a few scientific dissenters from the global-warming scenario, but very few. It seems to most scientists that the possible consequences of global warming are so severe that we must take action without waiting for definitive evidence that the effect has begun. Most of the ice and snow at the top of Kilimanjaro, the mountain in Africa described in Ernest Hemingway's novel *The Snows of Kilimanjaro*, has already disappeared and the rest may go within 20 years. The discovery by a ship in 2000 that there was open water at the North Pole made for headlines. It turns out that the Arctic Ocean indeed has open regions that can include the North Pole from time to time, but that the ice en route was only half as thick as it was a decade or more ago. So there is apparently a real danger that the Arctic ice pack may melt. The discovery was a graphic reminder of how humans are affecting our planet.

Another current worry for our future is the "ozone hole" that appears each September over Antarctica. Ozone ($O_3$) plays a vital role on Earth: ozone in the upper atmosphere keeps out the shortest-wavelength ultraviolet radiation from the Sun. The part of the ultraviolet that has wavelengths slightly longer than 3000 Å, where the Earth's atmosphere becomes transparent, causes suntans at the beach. None of the wavelengths shorter than 3000 Å pass through the ozone. If they did, they would kill us. Most scientists agree that amounts of ultraviolet radiation even slightly increased over the amount we already receive, whether longer or shorter than 3000 Å, would increase the incidence of skin cancer. Plants and animals might suffer more severe consequences. (The problem of too little ozone in the upper atmosphere is quite different from the problem of too much ozone at ground level as pollution from cars and other sources.)

Scientists are very concerned about the way that our technology is threatening the ozone layer of our Earth. The most serious problem is chlorofluorocarbons (CFCs), of which the Freons (a trade name) are examples. These gases are used as refrigerants in refrigerators and air conditioners and in manufacturing silicon chips and other products. They were also often used as propellants in some aerosol cans; the United States (though not most other countries) has phased out this usage. Halons, based on bromine, are a problem too; they are used to control fires.

CFCs and halon gases are not as inert—non-interacting—as they originally had been thought to be. Once they are released into the atmosphere, they rise to the stratosphere where they eventually break down. The individual chlorine and bromine

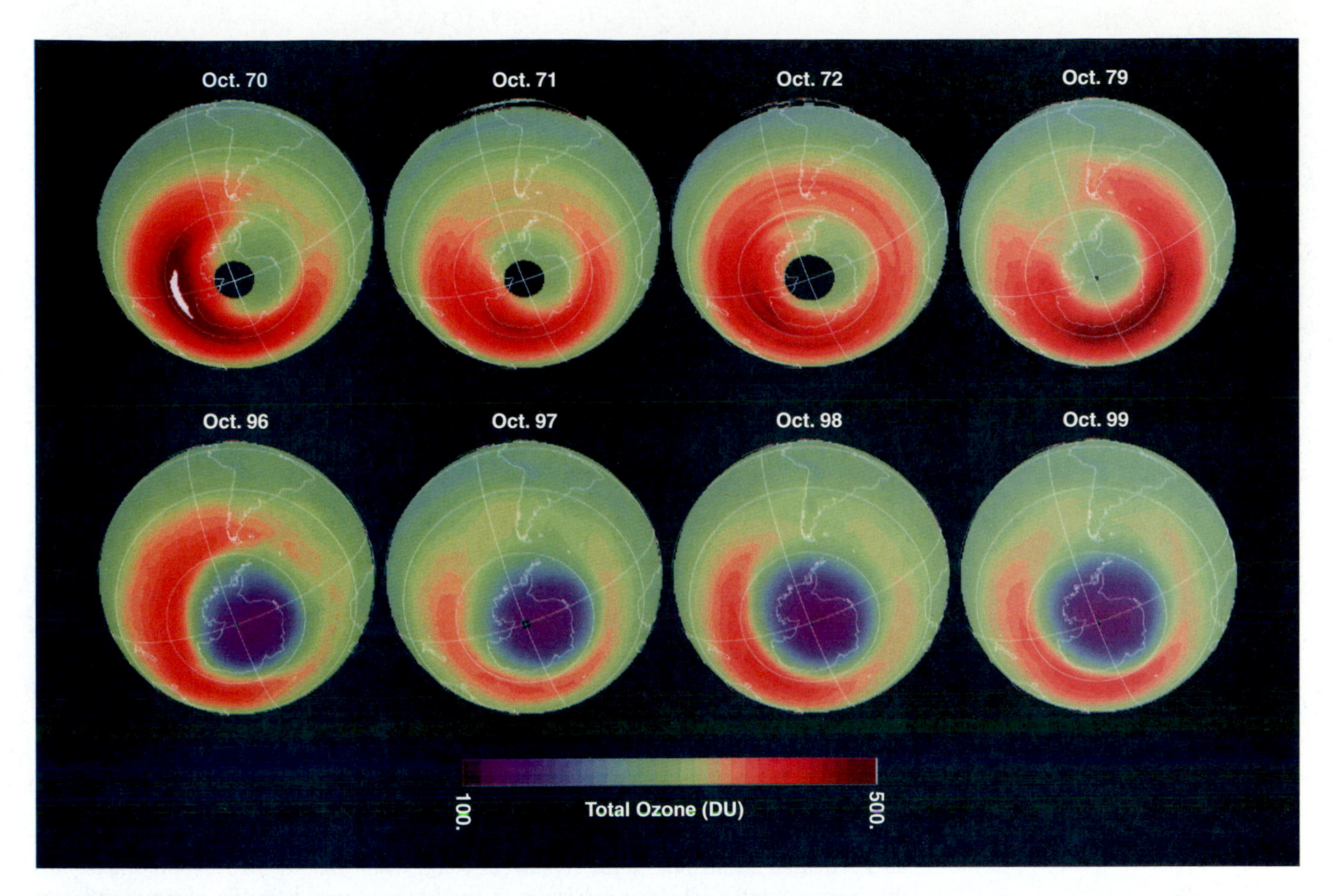


**FIGURE 8–15** The ozone hole (purple and blue) that appears over Antarctica each Antarctic spring; it is a result of the addition of CFCs to the atmosphere. The black spot represents an absence of data.

atoms transform the ozone ($O_3$) into other molecules (such as $O_2$ and chlorine monoxide, ClO).

Predictions of the global depletion of ozone have ranged from 3 to 18 per cent in the next century. A region of diminished ozone, an "ozone hole," first developed in the early 1980s, grew in intensity throughout the mid-1980s and has since grown yearly, in both size and intensity (Fig. 8–15). It is centered over Antarctica and extends up to the tip of South America. In it, the concentration of ozone drops drastically during each Antarctic springtime (Fig. 8–16). The ozone hole in the year 2000 became larger than ever, over three times larger than the size of the United States. A good word for the effect is "scary."

Is the ozone hole a forewarning of a significant worldwide change? Most scientists are quite worried about it. Ozone depletion appears in the northern hemisphere spring, too, bringing the resulting increase of ultraviolet radiation penetrating to the Earth's surface to more populated regions of the globe, but the extreme low ozone levels of the widespread, long-lasting southern hole have not appeared in the north due to differences in circulation.

Why is the Antarctic region most vulnerable? A vortex of winds at the pole, unimpeded over the frozen continent, collects photochemical products from the CFCs, such as HCl, that had been formed in the summertime stratosphere. The supercold Antarctic night allows ice crystals to form high in the stratosphere. The CFC product molecules attach themselves to the ice crystals, which helps them break down

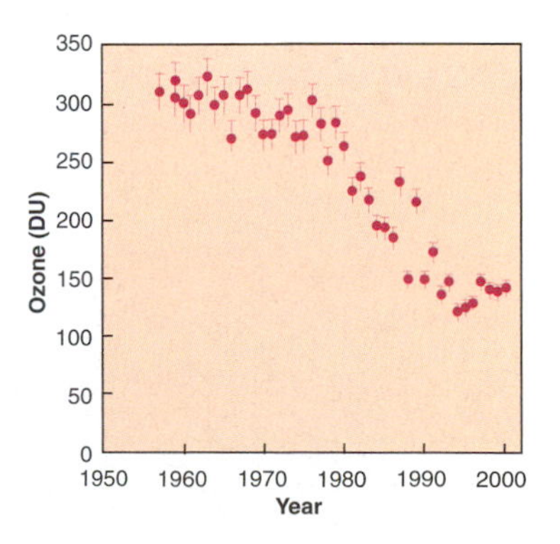


**FIGURE 8–16** The amount of ozone present over Antarctica each Antarctic spring diminished severely over the last few decades, making a region at those times that we call the ozone hole.

IN REVIEW

## BOX 8.2 Global Warming vs. Ozone Hole

Do not confuse:

*Global warming,* which comes from the greenhouse effect. It is caused by carbon dioxide and other gases in the Earth's atmosphere trapping solar energy. The Earth's average temperature is rising as a result.

*Ozone hole,* which is caused largely by complicated molecules, known as chlorofluorocarbons because they include fluorine and carbon, acting as catalysts to help ozone (the $O_3$) molecule high above us to break down. The breakdown products no longer keep ultraviolet radiation, which is hazardous to life, from coming through.

further to $Cl_2$. When the Sun rises at these high latitudes after months of darkness, the sunlight breaks up the $Cl_2$ and starts the reactions in which the chlorine monoxide breaks up the ozone. The chlorine monoxide, after it has destroyed an ozone molecule, undergoes a reaction with an oxygen atom that releases a chlorine atom. This chlorine atom destroys still more ozone, so a single CFC molecule can result in the destruction of thousands of ozone molecules. Sherwood Roland, Mario Molina, and Paul Crutzen received a Nobel Prize in Chemistry for their discovery and analysis of the ozone-loss problem.

In the northern hemisphere, sulfate aerosols from erupting volcanoes have taken the place of the ice crystals at times, but some especially cold northern winters have also provided stratospheric ice crystals in years when the vortex remains stable into the spring months. The ozone has diminished in northern latitudes at times, but not by enough to make it another "ozone hole."

The amount of ozone is usually measured in Dobson units, the amount of atmospheric ozone in a column of cross-section 1 square centimeter and named after the British atmospheric physicist G. M. B. Dobson. The ozone hole at its deepest yearly depth descends to less than ⅓ the normal 350 Dobson units.

We have already put enough gases into the atmosphere to make a permanent change in the ozone layer. Spacecraft are measuring the chlorine content of the upper atmosphere to improve our assessment of the situation, radio astronomers are measuring molecules in the atmosphere, and other astronomical methods are also being applied. Major Antarctic surveys are being carried out on the ground and from airplanes. The effect seems to be even more serious than we had first realized. The use of Freons in aerosol cans is non-essential and has been eliminated. "Essential uses"—such as refrigeration, industrial processing of metals, and the making of computer chips—are being met with less destructive substitutes that are being developed. Worldwide meetings have led to agreements in phasing out the use of CFCs. These air masses, now at ground level, should be migrating to the stratosphere, though we do not expect to be able to measure the recovery of the ozone layer until about 2005 or 2010.

One lesson to learn from all these considerations is that basic scientific research can expose possible side effects of ordinary things we otherwise take for granted. The dangers from refrigerants and aerosol propellants were understood in part because analyses were going on to understand chlorine compounds in Venus's atmosphere. Mars now provides a simpler system than the Earth to study ozone because only water vapor destroys it there. Studies of the atmospheres of other planets make us appreciate and help us understand the fundamental benefits of our own atmosphere. We must make certain that basic scientific research goes forward on all fronts

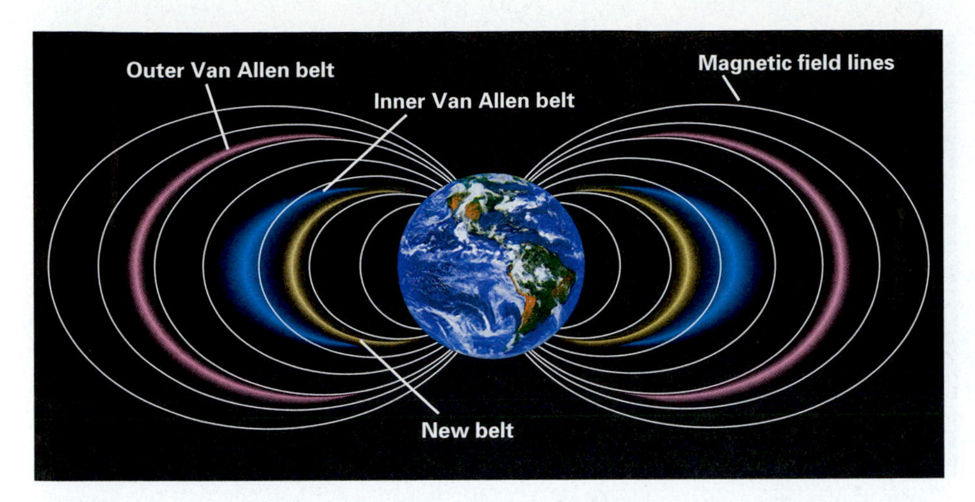


FIGURE 8–17 The outer Van Allen belt contains mainly electrons; the traditional "inner belt" contains mainly protons; and the newly discovered belt, which is within the other two belts, contains mainly ions of oxygen, nitrogen, and neon from interstellar space.

so that we will be prepared in the future for whatever problems might arise. Research limited to applied problems won't be enough.

## 8.6 THE VAN ALLEN BELTS

In 1958, the first American space satellite carried aloft, among other things, a device to search for cosmic rays, high-energy particles from space. Instead, this device, under the direction of James A. Van Allen of the University of Iowa, detected a high-altitude region filled with charged particles of high energies. We now know that there are actually three such regions—the **Van Allen belts**—surrounding the Earth, like small, medium, and large doughnuts (Fig. 8–17). The third radiation belt, discovered in 1993, contains particles from interstellar space that have penetrated the solar system.

The particles in the Van Allen belts are trapped by the Earth's magnetic field. The force on charged particles is perpendicular to the direction of magnetic-field lines, but the particles are kept by the magnetic-field lines from escaping completely. So as the charged particles move forward, the force that is pushing them perpendicularly makes them spiral around the magnetic-field lines.

As a particle moves from above the equator toward one of the magnetic poles, the field lines get closer together. At some point, the particle can travel no farther. The magnetic force stops the particle's forward motion and makes it go back in the other direction as it continues its spiral. Scientists are now making such **magnetic mirrors** in terrestrial laboratories as part of one of the leading methods to keep a hot plasma (a gas of charged particles) trapped in one place long enough to start fusion. (We will learn in Chapter 27 about fusion processes in the stars.) In the fusion process, hydrogen nuclei are combined to make heavier nuclei and much energy is released. Controlled fusion may one day provide much of the energy needed to support our civilization, though major technical problems still remain to be solved.

## 8.7 THE AURORA AND THE MAGNETOSPHERE

Charged particles, often from solar storms, are guided by the Earth's magnetic field toward the Earth's magnetic poles. When they hit our atmosphere, molecules in our atmosphere glow. We see this glow as the beautiful northern and southern

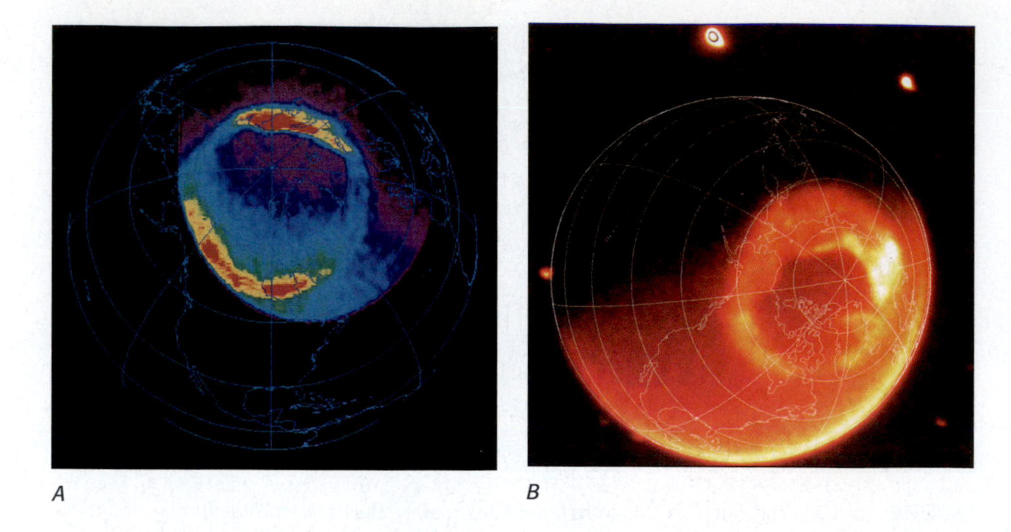

**FIGURE 8–18** (*A*) The aurora forms an oval around the south pole, which is seen in this view on April 6, 1996, from the NASA spacecraft Polar. Its UltraViolet Imager simultaneously observes the aurora on both the day and the night sides of the Earth for the first time by using special narrowband filters. (*B*) The auroral oval around the north pole from the IMAGE (Imager for Magnetopause to Aurora Global Exploration) spacecraft.

lights—the **aurora borealis** and **aurora australis**, respectively (Figs. 8–18 and 8–19). Magnetic waves from the Sun drive the aurora even when there is no solar storm.

The Earth is alone among the inner planets in having a respectable magnetic field; Mercury's magnetic field is very weak, and any magnetic fields that Venus or Mars might have are too weak to be detected. Motions in the Earth's core related to the Earth's fairly rapid spin and its internal heat source lead to the presence of our magnetic field. The region around the Earth in which the magnetic field keeps out the charged particles that are flowing out from the Sun (the "solar wind") is known as the **magnetosphere.** The magnetosphere is the subject of observation by a variety of spacecraft (Fig. 8–20).

**FIGURE 8–19** The aurora with a Perseid meteor, seen in Texas. Only rare, powerful auroras are seen that far south.

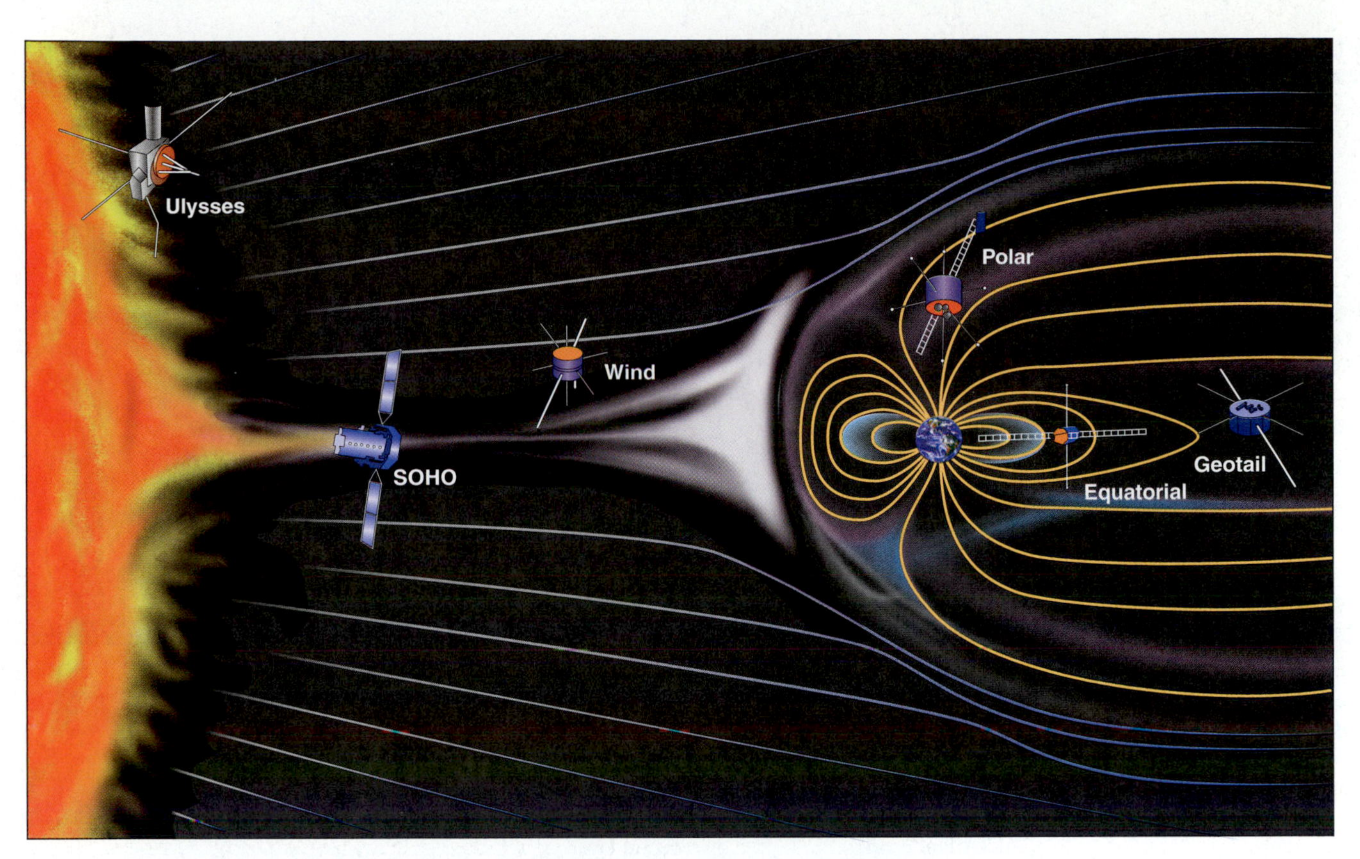


**FIGURE 8–20** A coordinated set of spacecraft studying the relation of the Earth and the Sun includes Ulysses, which flew over the solar poles; SOHO, the Solar and Heliospheric Observatory, which moves in a small circle 1,000,000 km toward the Sun from the Earth; and the spacecraft Wind, Polar, Equatorial, and Geotail nearer the Earth. The drawing also shows lines of force in interplanetary space and in the Earth's magnetosphere. Cluster II, a cluster of four spacecraft orbiting the Earth in close formation to monitor the solar wind, was launched in 2000. It replaced an earlier set of four satellites, which were destroyed at their launch.

## CORRECTING MISCONCEPTIONS

| ✖ *Incorrect* | ✔ *Correct* |
| --- | --- |
| The tidal force is the gravitational force. | The tidal force is the difference between the gravitational force at one place and the gravitational force at another place. |
| The tidal force varies with the square of the distance. | The tidal force varies with the cube of the distance. |
| The high tide on Earth is below the Moon. | High tides occur both on the side of Earth facing the Moon and on the opposite side. |
| Spring tides occur in the spring. | "Spring" means to spring up, and spring tides occur each month. |
| Only the Moon causes tides. | The Sun also causes tides on Earth, but they are weaker than tides caused by the Moon. |
| The ozone hole is permanent. | The ozone hole forms over the south pole in its springtime, lasts a couple of months, and disappears. |

## SUMMARY AND OUTLINE

Earth's layers (Section 8.1)
- Core, mantle, and crust; natural radioactivity is source of heat for interior.

Continental drift (Section 8.2)
- Geothermal energy comes from heat flow at active areas.
- Continents are on moving plates; the theory of plate tectonics explains the motion.

Tides (Section 8.3)
- From differential forces
- Spring tides when lunar and solar tides add; neap tides when lunar and solar tides are perpendicular.

Atmosphere (Section 8.4)
- All weather confined to thin troposphere.
- Ozone layer between stratosphere and mesosphere
- Ionosphere (also called thermosphere) reflects radio waves.

Water allows life on Earth (Section 8.5).
- We may be endangering life by modifying our atmosphere.

Van Allen belts (Section 8.6)
- Discovered in data from first American satellite
- Composed of charged particles trapped by Earth's magnetic field

Aurora (Section 8.7)
- Interaction of solar wind and Earth's atmosphere

## KEY WORDS

geology
seismology
seismic waves
core
mantle
crust
lithosphere
aesthenosphere
isotopes
differentiation
heat flow
geothermal energy
plates
plate tectonics
continental drift
mid-Atlantic ridge
differential forces
spring tides
neap tides
pressure
one atmosphere
troposphere
stratosphere
mesosphere
ozone layer
ionosphere
thermosphere
chaos
Van Allen belts
magnetic mirrors
aurora borealis
aurora australis
magnetosphere

## QUESTIONS

1. Of the following in the Earth, which is the densest: (a) crust; (b) mantle; (c) lithosphere; (d) core?
2. (a)Explain the origins of tides. (b) If the Moon were twice as far away from the Earth as it actually is, how would tides be affected?
3. [†] Using the inverse-square law of gravity, calculate the difference (that is, the tidal force) between the Moon's gravitational force on a 1-gram mass located on one side of the Earth and the gravitational force on the same mass located on the other side of the Earth, given the radius of the Earth (Appendix 3) and the Earth–Moon distance (Appendix 4). Compare your result for this difference with the Moon's gravitational force on the 1-gram mass; it is sufficient to use the value of the force at the Earth's center. In particular, what percentage of the gravitational force is the tidal force?
4. Draw a diagram showing the positions of the Earth, Moon, and Sun at a time when there is the least difference between high and low tides.
5. What is the source of most of the radiation that heats the troposphere?
6. Look at a globe and make a list or sketches of which pieces of the various continents would fit like a jigsaw puzzle and so probably lined up with each other before the continents drifted apart.
7. At what date is the ozone hole over Antarctica a maximum? Explain why in terms of heating from the Sun.
8. How does the temperature vary with height in the troposphere? Does it increase, decrease, or have some different form of relation? How does it vary in the stratosphere?
9. Space-shuttle astronauts orbit 175 km above the Earth. Communications satellites are $5\frac{1}{2}$ Earth radii above the surface. Compare their locations with the Van Allen belts, which can cause false readings on instruments.

[†] This question requires a numerical solution.

## INVESTIGATIONS

1. If you are near an ocean, find the intervals between successive high tides. If the interval differs from 12.6 hours, investigate the effect of local topology.
2. Investigate current uses of CFCs and possible alternatives.
3. Watch the International Space Station go overhead, using predicted times available on the Web. It is best visible just after sunset or before sunrise, when it is in sunlight but we are in darkness.

## USING TECHNOLOGY

### W World Wide Web

1. Find out about recent volcanic eruptions.
2. Find out about the distribution of earthquakes.
3. Scan for recent information on global warming or the ozone layer.

### REDSHIFT

1. Examine the Map of the Earth. Find your location on it. Turn on the list of Observatories, and locate the observatories nearest to you.
2. View the Earth from different planets, and notice its phases over time.
3. Duplicate the time of moonrise for a month, in sequence, and relate to the spacing between successive high tides. Choose a time when the Moon is not changing rapidly in declination.
4. From the *RedShift* main program, see the Earth orbit the Sun. Locate yourself high above the solar system and look down over time.

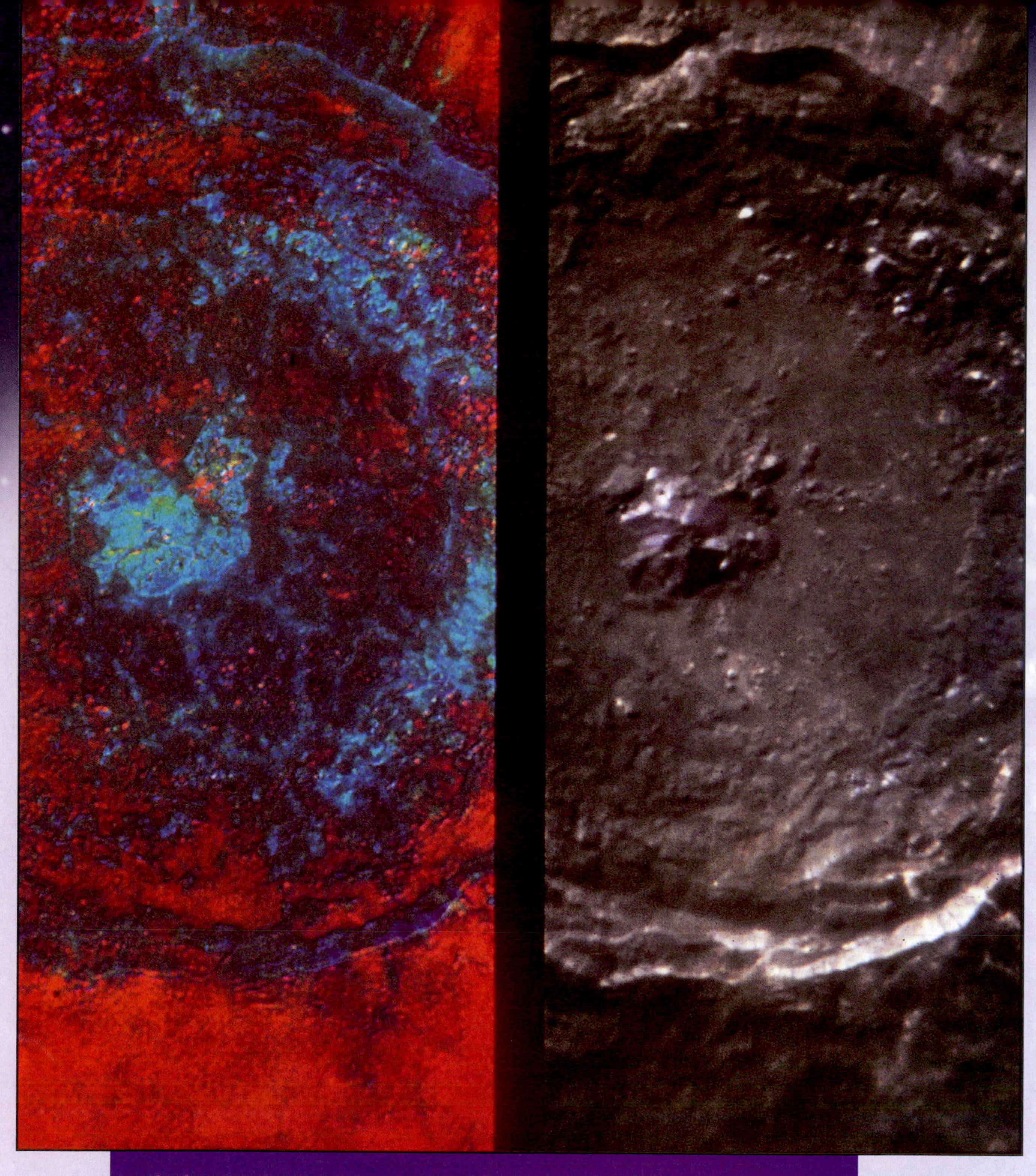

The lunar crater Tycho, imaged from the Clementine spacecraft. It is about 100 km in diameter; only a part is shown here. Its central peak shows clearly. The color composite uses the ratios of intensity as measured through different filters in the visible and near infrared. Such ratios reveal different types of material on the Moon's surface.

# Chapter 9

# Our Moon

The Earth's nearest celestial neighbor—the Moon—is only 380,000 km (238,000 miles) away from us on average, so close that it appears sufficiently large and bright to dominate our nighttime sky. The Moon's stark beauty has called our attention since the beginning of history, and studies of the Moon's position and motion led to the earliest consideration of the Solar System, to the prediction of tides, and to the establishment of the calendar.

**AIMS:** To see how direct exploration of the Moon has increased our knowledge, although such fundamental questions as how the Moon was formed remain the subject of current research

## 9.1 THE APPEARANCE OF THE MOON

The fact that the Moon's surface has different kinds of areas on it is obvious to the naked eye. Even a small telescope—Galileo's, for example—reveals a surface pockmarked with craters. The **highlands** are heavily cratered. Other areas, called **maria** (pronounced mar′ee-ah; singular, **mare,** pronounced mar′ay), are relatively smooth, and indeed the name comes from the Latin word for "sea" (Fig. 9–1). But, to wax poetical, there are no ships sailing on the lunar seas and no water in them; the Moon is a dry, airless, barren place. The Moon's mass is only 1/81 that of the Earth, and the gravity at its surface is only 1/6 that of the Earth. Essentially all of any atmosphere and any water that may once have been present would long since have escaped into space. The Moon is about 1/4 the diameter of the Earth—most solar system moons are much smaller fractions of the sizes of their parents. The Moon is a particularly valuable source of information about early epochs of the Solar System because its surface is relatively unchanged, whereas Earth's surface has been transformed by plate tectonics.

Besides the smooth maria and the cratered highlands, other types of structures visible on the Moon include **mountain ranges** and **valleys.** The mountains are formed by debris (we will learn later about collisions of meteoroids with the Moon, which eject such debris), though, unlike mountains on Earth, which are formed from plate tectonics. Lunar **rilles** are cracks that can extend for hundreds of kilometers along the surface. Some are relatively straight while others are sinuous. Raised **ridges** also occur. The craters themselves come in all sizes, ranging from as much as 295 km across for Bailly down to tiny fractions of a millimeter. Crater **rims** can be as much as several kilometers high over the crater floors, much higher than the Grand Canyon's rim stands above the Colorado River.

The Moon rotates on its axis at the same rate as it revolves around the Earth, always keeping the same face in our direction. The Earth's gravity has locked the Moon in this pattern, first causing a bulge in the distribution of the lunar mass and

**FIGURE 9–1** This Earth-based photograph shows a mare (a lunar sea, so called because it looks so smooth) at top and a highland with craters below.

then interacting with the bulge to prevent the Moon from rotating freely. As a result of this interlock, we always see essentially the same side of the Moon from our vantage point on Earth. Because of **librations** of the Moon—the apparent slight turning of the Moon—we see, at one time or another, 5/8 of the whole surface. Librations arise from several causes, including the varying speed of the Moon in its elliptical orbit around the Earth. (The varying speed allows us to see around the sides, for example, and sometimes we can see a bit over the top or around under the bottom.)

When the Moon is full (Fig. 9–2), it is bright enough to cast shadows or even to read by. But full moon is a bad time to try to observe lunar surface structure, for any shadows we see on the surface of the Moon are short. When the Moon is a crescent or even a half moon, however, the part of the Moon facing us is covered with long shadows. The lunar features then stand out in bold relief. Shadows are longest near the **terminator,** the line separating day from night.

The Moon makes an orbit of the Earth, as seen from far away from the Earth–Moon system, in 27⅓ days, the **sidereal revolution period** of the Moon. But since the Earth is moving in its orbit around the Sun, the phases repeat with a different period (Box 9.1). The Moon makes an orbit of the Earth with respect to the position of the Sun in 29½ days; the interval between successive new moons is the **synodic revolution period** of the Moon (Fig. 9–3). Because the same side of the Moon always faces the Earth, different regions face the Sun as the Moon orbits. As a result, the terminator moves completely around the Moon with this 29½-day synodic period. Most locations on the Moon are thus in sunlight for about 15 days, during which time they become very hot—130°C (265°F)—and then in darkness for about 15 days, during which time their temperature drops to as low as –110°C (–170°F).

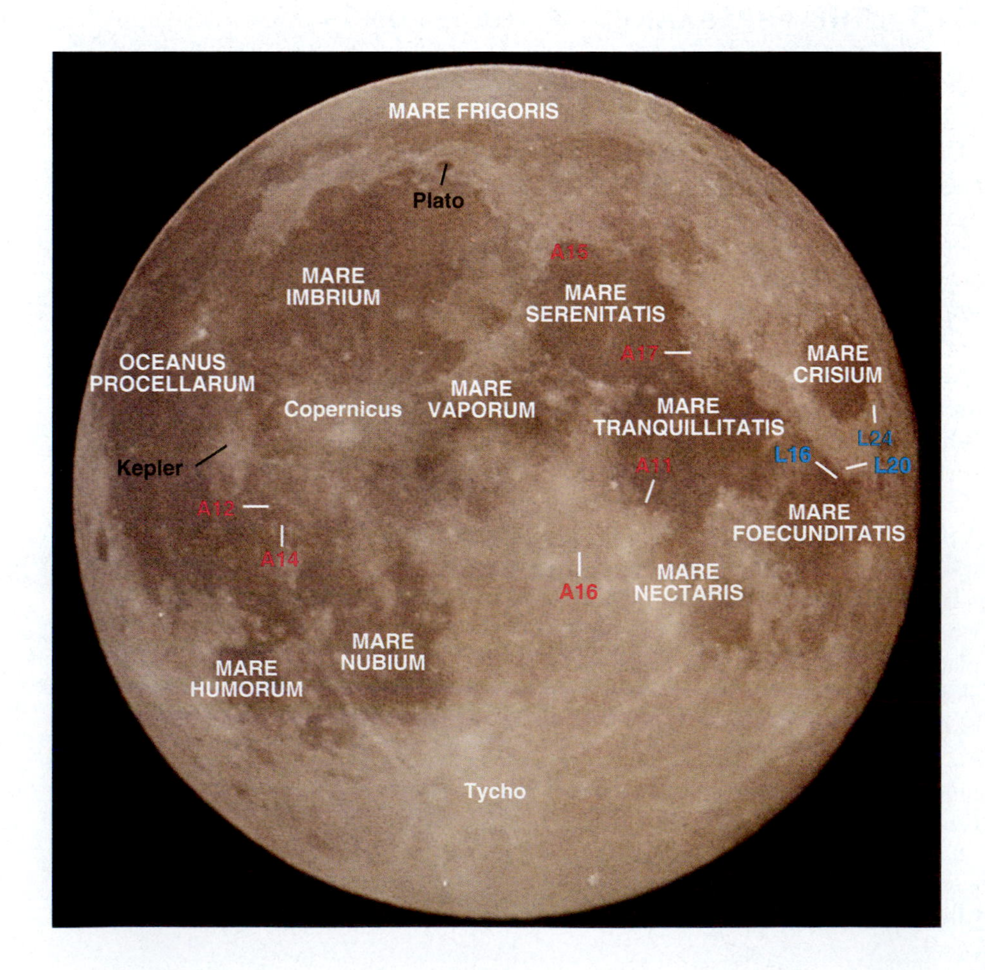


**FIGURE 9–2** The full moon. Note the dark maria and the lighter, heavily cratered highlands. The positions of the 6 American Apollo (A) and 3 Soviet Luna (L) missions from which material was returned to Earth for analysis, are marked. This photograph is oriented with north up, as we see the Moon with our naked eyes or through binoculars; a telescope normally inverts the image.

A DEEPER DISCUSSION

**BOX 9.1 Sidereal and Synodic Periods**

*Sidereal:* with respect to the stars. For example, the Moon revolves around the Earth in about 27⅓ days with respect to the stars.

*Synodic:* with respect to some other body, usually the Sun (as seen from the Earth). For example, while the Moon makes one sidereal revolution around the Earth, the Earth moves about 1/13 of the way around the Sun (about 27°). Thus the Moon must travel a little farther to catch up. The synodic period of the Moon, the synodic month, is about 29½ days.

***Discussion Question***

Is the interval between full moons determined by the sidereal or the synodic month?

The cycle of phases that we see from Earth also repeats with this 29½-day period —a **synodic month.**

In some sense, before the period of exploration by the Apollo program, as we shall discuss next, we knew more about almost any star than we did about the Moon. As a solid body, the Moon reflects the solar spectrum rather than emitting one of its own, so we were hard pressed to determine even the composition or the physical properties of the Moon's surface. Some information about surface mineralogy could be gleaned from how the surface transformed the solar spectrum while reflecting it, but not as much as we would have liked.

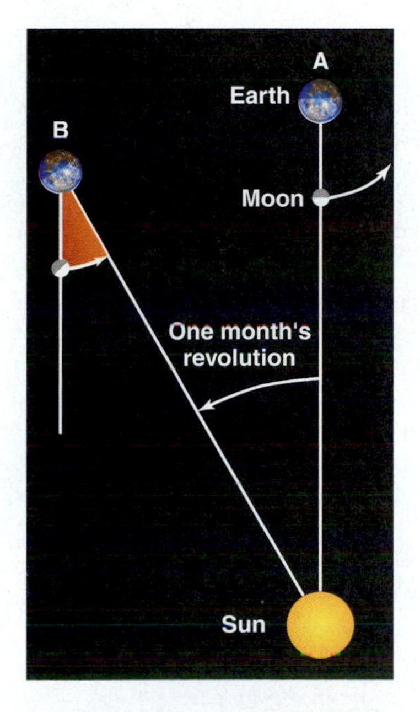


FIGURE 9–3 After the Moon has completed one revolution around the Earth with respect to the stars, which it does in 27⅓ days, it has moved from A to B. The Moon has still not swung far enough around to be in the same position again with respect to the Sun because the Earth has revolved one month's worth around the Sun. It takes about an extra two days for the Moon to complete its revolution with respect to the Sun, which gives a synodic rotation period of about 29½ days. In the extra two days, both Earth and Moon have continued to move around the Sun (toward the left in the diagram).

## 9.2 LUNAR EXPLORATION

The space age began on October 4, 1957, when the U.S.S.R. launched its first Sputnik (the Russian word for "travelling companion") into orbit. The shock of this event galvanized the American space program, and within months American spacecraft were also in Earth orbit.

In 1959, the Soviet Union sent its Luna 3 spacecraft around the Moon; Luna 3 radioed back the first murky photographs of the Moon's far side. Now that we have high-resolution maps, it is easy to forget how big an advance that was.

In 1961, President John F. Kennedy proclaimed that it would be a U.S. national goal to put a man on the Moon, and bring him safely back to Earth, by 1970. This grandiose goal led to the largest and most expensive coordinated program of any kind in world history. Though the Soviets had firsts in robotic exploration of the Moon, once the United States landed people on the Moon the Soviets claimed that they hadn't been trying to do so. Only in 1989 did the Soviets admit that they too had had a program to land people on the Moon.

The American lunar program, under the direction of the National Aeronautics and Space Administration (NASA), proceeded in gentle stages, starting with Alan Shepard rocketing into space for a short time and continuing with John Glenn as the first American to orbit the Earth.

### 9.2a The Apollo Missions

After a series of one-person and two-person spacecraft in Earth orbit and parallel robotic exploration of the Moon, the three-person Apollo missions began. In Apollo 8, three astronauts circled the Moon on Christmas Eve 1968 and returned to Earth.

About six months later, Apollo 11 brought humans to land on the Moon for the first time (Table 9–1). It went into orbit around the Moon after a three-day journey from Earth, and a small spacecraft called the Lunar Module separated from the larger Command Module (Fig. 9–4). On July 20, 1969—a date that from the long-range

Over 300 people have travelled in space.

**TABLE 9-1 Missions to the Lunar Surface and Back to Earth**

| | | | |
|---|---|---|---|
| Apollo 11 | U.S.A. | 1969 | robotic |
| Apollo 12 | U.S.A. | 1969 | robotic |
| Luna 16 | U.S.S.R. | 1970 | unrobotic |
| * | | | |
| Apollo 14 | U.S.A. | 1971 | robotic |
| Apollo 15 | U.S.A. | 1971 | robotic |
| Luna 20 | U.S.S.R. | 1972 | unrobotic |
| Apollo 16 | U.S.A. | 1972 | robotic |
| Apollo 17 | U.S.A. | 1972 | robotic |
| Luna 24 | U.S.S.R. | 1976 | unrobotic |

*An explosion on Apollo 13 while en route to the Moon led to an emergency return of the astronauts to Earth without landing on the Moon, though they had to circle the Moon to reach Earth.

**FIGURE 9–4** The Apollo 11 Lunar Module returning to the Command Module after the first landing of humans on the Moon. Smith's Sea is in the background with the Earth rising behind it.

**FIGURE 9–5** Neil Armstrong, the first person to set foot on the Moon, took this photograph of his fellow astronaut Buzz Aldrin climbing down from the Lunar Module of Apollo 11 on July 20, 1969. The site is called Tranquillity Base, as it is in the Sea of Tranquillity (Mare Tranquillitatis).

standard of history may be the most significant of the millennium—Neil Armstrong and Buzz Aldrin left Michael Collins orbiting in the Command Module and landed on the Moon (Fig. 9–5). In the preceding days there had been much discussion of what Armstrong's historic first words should be, and millions listened as he said, "One small step for man, one giant leap for mankind." (Though a controversial issue, the most authoritative commentators conclude that he meant to say "for a man." When, in 1984, a space shuttle astronaut became the first human to fly freely in orbit, untethered, he joked, "That may have been one small step for Neil, but it's a heck of a big leap for me." On hearing that he won the 1987 Nobel Prize in Literature, Soviet-émigré/American-citizen Joseph Brodsky said, "A big step for me, a small one for mankind.")

A

B

**FIGURE 9–6** (*A*) Eugene Cernan riding on the Lunar Rover during the Apollo 17 mission. The mountain in the background is the east end of the South Massif. (*B*) The Apollo astronauts left their rovers behind. (*Drawing by Alan Dunn; the New Yorker Collection 1971 Alan Dunn, from cartoonbank.com. All Rights Reserved.*)

The Lunar Module carried many experiments, including devices to test the soil, a camera to take stereo photos of lunar soil, a sheet of aluminum with which to capture particles from the solar wind, and a seismometer. Later Lunar Modules carried additional experiments, some even including a vehicle (Fig. 9–6). Six Apollo missions in all (Fig. 9–7), ending with Apollo 17 in 1972, carried people to the Moon.

The Soviet Union sent three robotic spacecraft to land on the lunar surface, collect lunar soil, and return it to Earth—Lunas 16, 20, and 24. They collected 300 grams of lunar soil in 1970, 1972, and 1976, only 10 per cent the weight of this book. They drilled to about 50 cm below the lunar surface and brought long, thin cylinders of material back to Earth. Two other missions carried rovers, Lunokhod 1 and Lunokhod 2, that travelled many kilometers across the lunar surface over periods of months.

## 9.3 THE RESULTS FROM APOLLO

The kilometers of film exposed by the astronauts, the 382 kg of rock brought back to Earth, the lunar seismograph data recorded on tape, and other data have been studied by hundreds of scientists from all over the world. The data have led to new views of several basic questions and have raised many new questions about the Moon and the Solar System. The next subsections will describe what we learned and some of the questions we are still asking.

### 9.3a The Composition of the Lunar Surface

The rocks astronauts found on the Moon are types that are familiar to terrestrial geologists. Almost all the rocks are **igneous,** which means that they were formed by the cooling of lava. The Moon has few **sedimentary** rocks, which are formed by settling, most commonly on Earth out of water and also from deposits carried by wind. (Earth's sedimentary rocks include limestone and shale and cover ⅔ of Earth's surface. On the Moon, only matter ejected when craters were formed is available to be

**FIGURE 9–7** The Apollo 16 landing site imaged in 1994 by the Clementine spacecraft from its orbit around the Moon.

**FIGURE 9–8** A 1.5-kg basalt from Apollo 15. Note the many vesicles, spherical cavities that arise in volcanic rock from gas trapped when the rock formed.

**FIGURE 9–9** An anorthosite. The dark part is dust covering the breccia to which the anorthosite was attached.

**FIGURE 9–10** A breccia, dark gray and white, returned to Earth by Apollo 15. It weighs 4.5 kg. A 1-cm block shows the scale.

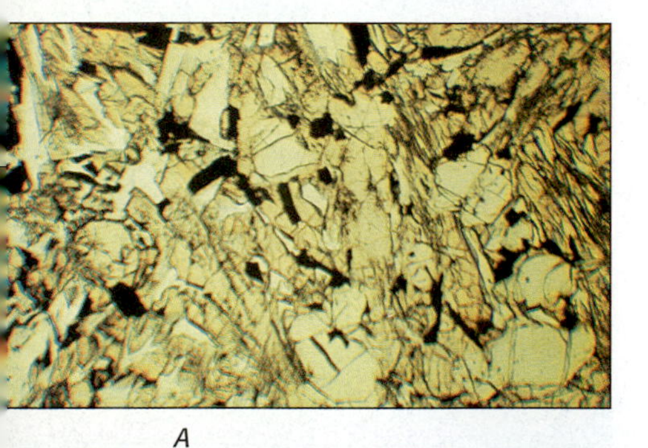

*A*

*B*

**FIGURE 9–11** A thin section of a lunar rock, seen through a microscope (*A*) in unpolarized light, and (*B*) in polarized light. In the latter, polarized light has passed through a section of lunar rock so thin as to be transparent. Different minerals in the rock rotate the plane of polarization differently, leading to the color effect when the resulting light is passed through a second polarizer.

made into sedimentary rocks.) The Moon's crust is thick—12 per cent of its volume; the Earth's continental crust is only 0.5 per cent of the Earth's volume.

In the maria, the rocks are mainly **basalts,** a certain kind of rock formed from molten lava (Fig. 9–8). The highland rocks are **anorthosites** (Fig. 9–9). (Anorthosites—defined by their particular combination of minerals—are rare on Earth, though they are abundant in the Adirondack Mountains, for example.) Anorthosites, though they have also cooled from molten material, have done so under different conditions from basalts and have taken longer to cool.

In both the maria and the highlands, some of the rocks are **breccias** (Fig. 9–10), mixtures of fragments of several different types of rock that have been compacted and welded together. (Some geologists thus consider breccias to be sedimentary.) The ratio of breccias to crystalline rocks is much higher in the highlands than in the maria because there have been many more impacts in the visible highland surface, as shown by the greater number of craters.

Polarizing microscopes, also used by geologists, are useful for studying lunar rocks (Fig. 9–11). When lunar and terrestrial rocks are compared, it is easy to see the difference. The lunar rocks contain no water, and so they never underwent the reactions that terrestrial rocks undergo.

The astronauts also collected some **lunar soils,** bits of dust and larger fragments from the Moon's surface. This material makes the **regolith,** which was built up by bombardment of the lunar surface by meteorites of all sizes over billions of years; the material is thrown up and then falls back. Microscopes showed that some of these soils contain small glassy globules (Fig. 9–12), which are not common on Earth. Some of these glassy globules and glassy coatings were undoubtedly formed when rock melted as it was ejected from the site of a meteorite impact and then cooled. It was useful to have a trained geologist on the Moon, Harrison Schmitt (Fig. 9–13), though a pity that he was the only geologist to reach there.

Almost all of these rocks and soils have lower proportions of elements with low melting points (**volatile** elements) than do rocks and soils on the Earth. On the other hand, they contain relatively high proportions of elements with high melting points (**refractory** elements) like calcium, aluminum, and titanium. Elements that are even rarer on Earth, such as uranium, thorium, and the rare-earth elements, are also found in greater abundances. (Will we be mining on the Moon one day?) We will come back to this topic in Section 9.4.

Since none of the lunar rocks contains any trace of water bound inside its minerals, clearly water never existed on the Moon. This eliminates the possibility that life evolved there.

## 9.3b Lunar Chronology

One way of dating the surface of a moon or planet is to count the number of craters in a given area, a method that was used before Apollo. Even from the Earth we can count the larger lunar craters. If we assume that the events that cause craters—whether they are impacts of meteoroids or the eruptions of volcanoes—continue over a long period of time and that no strong erosion occurs, surely those locations with the greatest number of craters must be the oldest. Relatively smooth areas—like maria—must have been covered over with volcanic material at some relatively recent time (which is still billions of years ago). When one crater is superimposed on another, we can be certain that the superimposed crater is the younger one.

A few craters on the Moon, notably Tycho and Copernicus (Fig. 9–14), have thrown out obvious rays of lighter-colored matter, best seen near the time of full moon. The rays are secondary craters formed by ejected material when the primary crater was formed. Since these rays extend over other craters, the craters with rays must have formed later. The youngest rayed craters may be very young indeed—perhaps only a few hundred million years. The rays darken with time, so rays that may have once existed near other craters are now indistinguishable from the rest of the surface. Rays are visible for about a billion years.

Crater counts and the superimposition of one crater on another give only relative ages. We could find the absolute ages only when rocks were physically returned to Earth. Scientists worked out the dates by comparing the current ratio of radioactive isotopes to nonradioactive isotopes present in the rocks with the ratios that all rocks have when they form, as determined from studies of the same isotopes in rocks on Earth. From the known rate of radioactive decay (Box 9.2), they could work out the ages. The oldest rocks that were found at the locations sampled on the Moon were formed 4.44 billion years ago. The youngest rocks were formed 3.1 billion years ago. (As is commonly and properly done in science, ages are given to the accuracy to which they can be measured. For example, giving the extra decimal place in 4.44 billion means that the accuracy is to within plus or

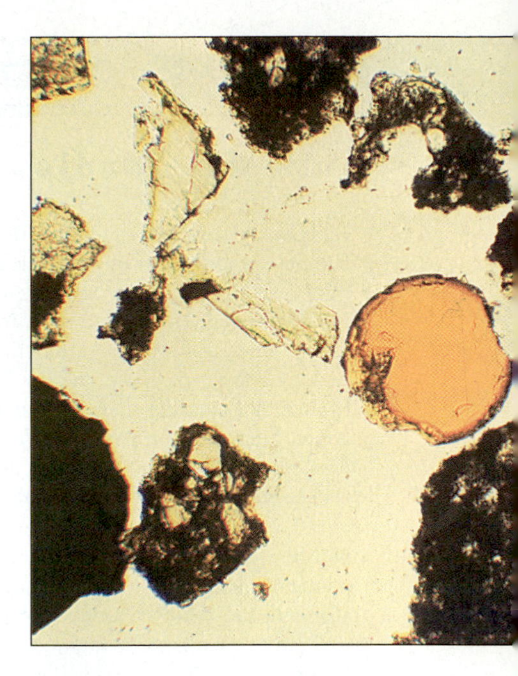

**FIGURE 9–12** An enlargement of a thin section including a glassy spherule from the lunar dust collected by the Apollo 11 mission. The shape and composition of the spherule indicate that it was created when a meteoroid crashed into the Moon, melting lunar material and splashing it long distances.

**FIGURE 9–13** Scientist-astronaut Harrison H. Schmitt, the only Ph.D. to have stood on the Moon, during the Apollo 17 mission to the Taurus-Littrow region.

**FIGURE 9–14** The new moon in earthshine with rayed crater Tycho, imaged in 1994 by the Clementine spacecraft.

A DEEPER DISCUSSION

## BOX 9.2 Radioactivity

Radioactive substances are those that decay spontaneously; that is, one isotope (form of a chemical element) decays into another; we shall say more about isotopes in Chapter 27. Stable isotopes remain unchanged over time. For certain pairs of isotopes—one radioactive and the other stable—we can calculate how long the radioactive one has been decaying from the following: We measure their ratio now and from other considerations know the proportion of the two when the rock was formed. Since we know the rate at which the radioactive one is decaying, we can tell how long the process has been going on.

For example, let us say we examine an isotope of thorium and how it decays relative to another element. What if when a lunar rock cooled, it had 400 atoms of thorium for every 400 atoms of the other element. The half-life of the thorium isotope often used is 1,000,000 years. Thus after 1,000,000 years, there are only 200 atoms of thorium compared with 400 atoms of the other element. After an additional 1,000,000 years has passed, the number of atoms of thorium falls in half again, to 100. So if we examine a lunar rock and find that there is $1/4$ the number of atoms of thorium compared to the other type, then we can say that the rock is two half-lives old, or—for the sample numbers given—2,000,000 years.

One of the first Nobel Prizes in Physics was given to Antoine Becquerel for discovering radioactivity and to Marie Curie and Pierre Curie for their work on radioactivity. A few years later, Marie Curie received a Nobel Prize in Chemistry, for discovering radium and polonium and isolating radium. The Curies' daughter Irène Joliot-Curie and her husband Frédéric Joliot-Curie later also shared a Nobel Prize, in their case for discovering artificial radioactivity.

### *Discussion Question*

Give a numerical example showing how to determine age using carbon dating in a partially radioactive system where ceramic pots have been found. Start with 1000 atoms and a lifetime of 640 years for the radioactive isotope of carbon used. Follow the numbers through four lifetimes.

minus 0.01 billion—or 10 million—years, while 3.1 billion implies that the accuracy is to within plus or minus 0.1 billion—or 100 million years.) The 4.44-billion-year age was especially precise because many lunar rocks apparently formed from a single catastrophic event on the Moon, and we therefore have many separate measurements pointing to this age. Many ages seem to be 4.35 billion years, probably when the magma ocean finally crystallized.

The ages of highland and maria rocks are significantly different. The original differentiation was finished by 4.4 billion years ago and other highland rocks were formed until about 3.9 billion years ago. The maria were formed between 3.8 and 3.1 billion years ago.

All the observations can be explained on the basis of the following general picture (Fig. 9–15): The Moon formed 4.45 billion years ago. We know that the top 500 km or so (or, some say, the whole Moon) was molten. The surface could have been entirely melted by the original heat from the Moon's formation or by an intense bombardment of meteoroids (or debris from the period of formation of the Moon and the Earth). The decay of radioactive elements also provided heat. Then the surface cooled, within about 200 million years. From 4.2 to 3.9 billion years ago, bombardment (perhaps by planetesimals) caused most of the craters we see today. Some lunar scientists believe that the impact rate was not constant during that interval, and that it increased dramatically around 3.9 billion years ago, producing many of the large-impact basins we see on the Moon. Most of the lunar rocks brought back by the Apollo astronauts were from that era. And, analysis of meteorites once blasted off the Moon and now found on Earth provided evidence for the cataclysm 3.9 billion years ago. Further, if asteroids pounded the Moon then, asteroids must have simultaneously pounded the Earth, which is a much bigger target. The bombardment would have had a big effect on the evolution of life on Earth.

About 3.8 billion years ago, the interior of the Moon heated up sufficiently (from radioactive elements inside) that volcanism began; lava flowed onto the lunar sur-

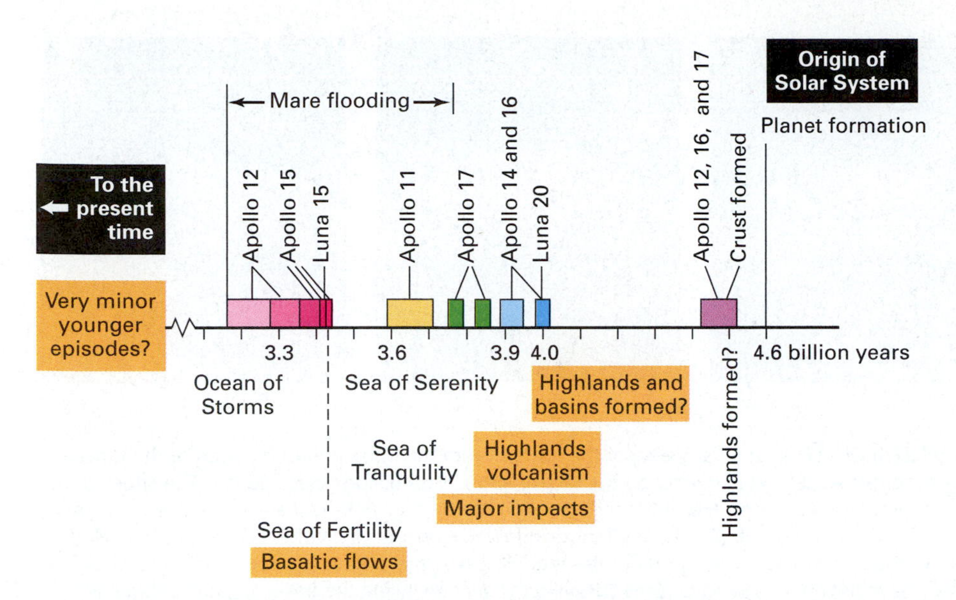


**FIGURE 9–15** The chronology of the lunar surface, based on work carried out at the Lunatic Asylum, a Caltech laboratory. (The common meaning of the word "lunatic" in modern speech comes from an old superstition about the Moon.) The ages of rocks found in 8 missions are shown in boxes. The names and descriptions below the line took place at the times indicated by the positions of the words. Note how many missions it took to get a sampling of many different ages on the lunar surface.

face and filled the largest basins that resulted from the earlier bombardment, thus forming the maria (Fig. 9–16). By 3.1 billion years ago, the era of volcanism was almost over, with only a final flow of lava between 3.0 and 2.5 billion years ago. The Moon has been geologically pretty quiet since then.

Up to this time, the Earth and the Moon had shared similar histories. But active lunar history stopped about 3 billion years ago, while the Earth continued to be geologically active. Because the Earth's interior continued to send gas into the atmosphere and because the Earth's higher gravity retained that atmosphere, the Earth developed conditions in which life evolved. The Moon, because it is smaller than the Earth, presumably lost its heat more quickly and also generated a thicker crust.

Almost all the rocks on the Earth are younger than 3 billion years of age; erosion and the remolding of the continents as they move slowly over the Earth's surface, according to the theory of plate tectonics, have taken their toll. Though the oldest single rock ever discovered on Earth has an age of 4.1 or 4.2 billion years, few rocks are older than 3 billion years. So we must look to extraterrestrial bodies—the Moon, meteorites, or comets—that have not suffered the effects of plate tectonics or erosion (which occurs in the presence of water or an atmosphere) to study the first billion years of the Solar System. In particular, we can deduce that the Earth, larger and more massive than the Moon, would have been hit by large objects hundreds of times about 4 billion years ago, with terrestrial craters hundreds of kilometers across forming.

## 9.3c The Origin of the Craters

The debate over whether the lunar craters were formed by meteoritic impact (Fig. 9–17) or by volcanic action began in pre-Apollo times. The results from Apollo indicate that most craters resulted from meteoritic impact, though some small fraction may have come from volcanism. Though only a few craters may have resulted from

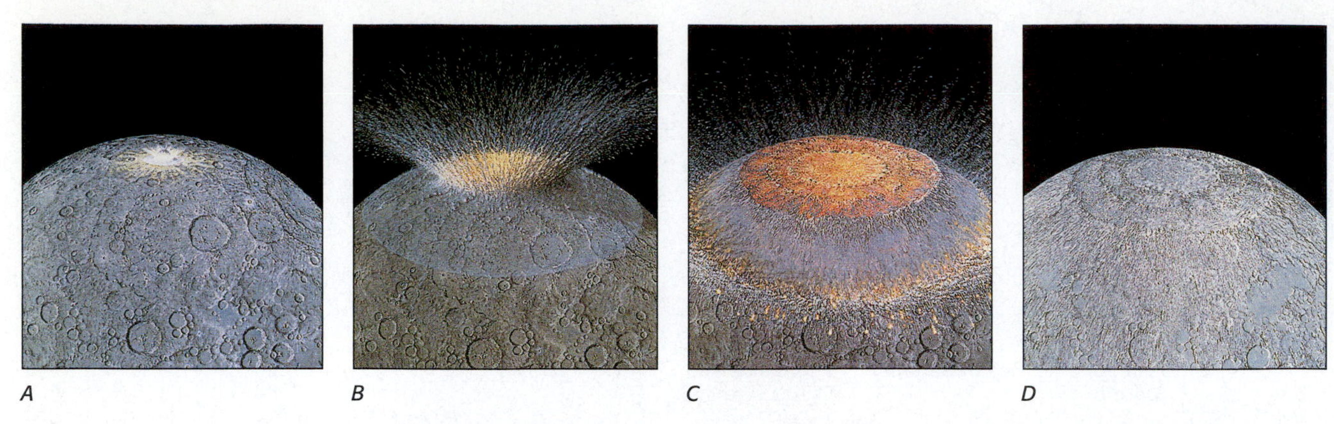


**FIGURE 9–16** An artist's view of the formation of the Mare Imbrium region of the lunar surface. (*A*) A meteoroid impact on the Moon, about 3.85 billion years ago. (*B*) The shock of the meteoroid impact began the Imbrium crater. (*C*) As the dust and heat subsided, the 1300-km Imbrium crater was left. (*D*) The molten rock flowed over outlying craters and cooled, leaving lunar mountains as the remainder of the rim. (*E*) Lavas that welled up from inside the Moon may have begun to flow as early as 3.8 billion years ago, filling the basin, as they did in Mare Tranquillitatis. (*F*) By 3.3 billion years ago, the lava flooding was nearly complete. (*G*) The final flow of thick lava came 2.5 to 3.0 billion years ago. (*H*) Subsequent cratering has left the Mare Imbrium on today's Moon.

**FIGURE 9–17** This photograph, made with a very short-exposure "strobe" light, shows the result of a falling milk drop. The formation of a lunar crater by a meteoroid is similar, because the energy of the impact makes the surface material flow like a liquid. A second milk drop is falling in the same location.

volcanism, there are many other signs of volcanic activity, including the lava flows that filled the maria. The Marius Hills in Oceanus Procellarum, for example, have many domes and rilles that apparently resulted from repeated volcanism. Mare lavas may have flowed from such areas.

The photographs of the far side of the Moon (Fig. 9–18) have shown us that the near and far hemispheres are quite different in overall appearance. The maria, which are so conspicuous on the near side, seem almost absent from the far side, which is

**FIGURE 9–18** The far side of the Moon, which takes up the right side of this image, looks very different from the near side in that there are few maria. Since the far-side crust is about 100-km thick, in contrast to the thinner near-side crust, less of the dark basalt can well up from the mantle to form maria. The Clementine spacecraft found near-side crust as thin as 10 km under near-side maria. Mare Crisium, which appears to us at the Moon's edge (called the "limb"), is in the top center of this Apollo 8 view.

**FIGURE 9–19** An oblique view of part of the lunar far side, from Clementine, shows how rough it looks.

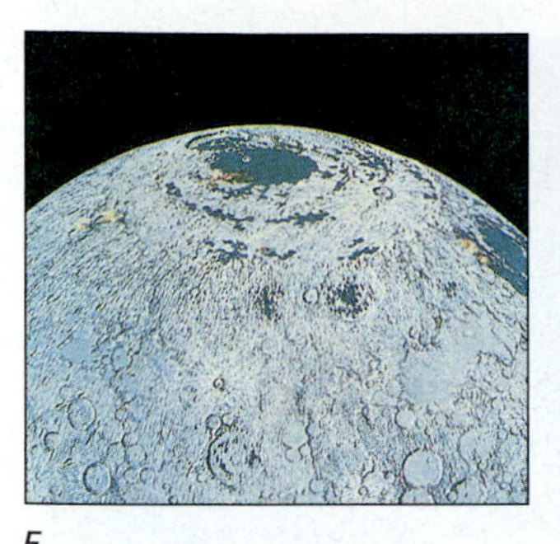
E

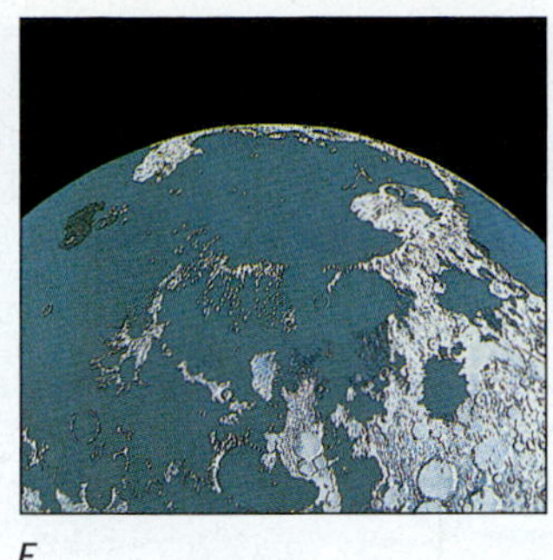
F

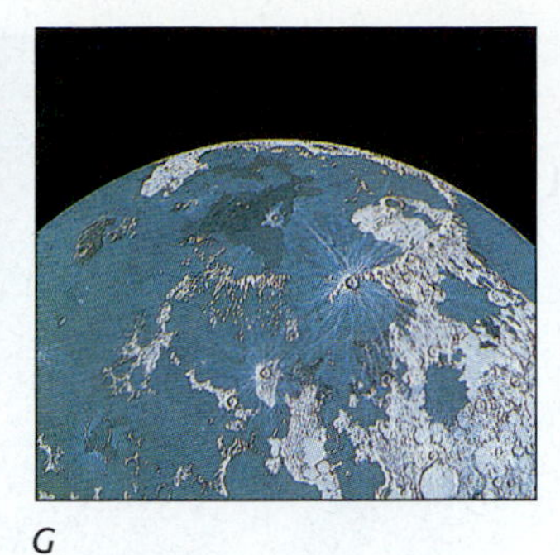
G

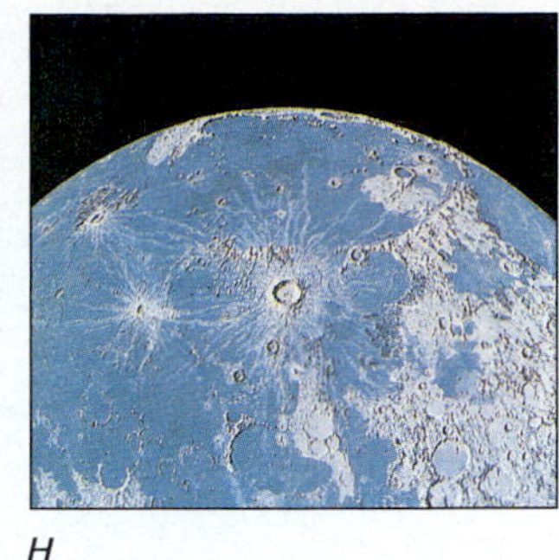
H

cratered all over. Actually, a similar number of basins are present on the far side, but the crust is thicker there, so the basins were less often filled with lava. The lunar crust is perhaps 65 km thick on the near side and twice as thick on the far side. This asymmetry may explain the different appearances of the sides, because lava would be less likely to flow through the far side's thicker crust.

Further, the difference in cratering between the two sides may have been a secondary result of an uneven distribution of the Moon's mass. Once any asymmetry was set up, the Earth's gravity locked one side toward us. The side that faces the Earth is somewhat shielded by the Earth and its gravity from collisions. The other side can plow into interplanetary rocks directly. Thus it can be understood in this way why the far side of the Moon shows many impacts and is very rough (Fig. 9–19).

In later chapters, we shall meet other cratered objects in our Solar System. They include not only the planets Mercury and Mars but also most of the moons of the outer planets. Some of the craters are quite old, since they were caused by impacts with planetesimals left over from the era in which the planets formed. Others are from more recent impacts from asteroids and comets in the Solar System and, primarily, comets in the outer Solar System.

## 9.3d The Lunar Interior

Before the Moon landings, it was widely thought that the Moon was a simple body, with the same composition throughout. But we now know it to be differentiated (Fig. 9–20), like the planets. Below its crust of relatively light material, a silica-rich mantle makes up most if not all of the interior. It may also have a metallic (iron-rich) core, though the Moon's relatively low average density shows that any such core would take up a much smaller fraction of the lunar interior than the Earth's core does of the Earth's interior.

The Moon was measured to be seismically quiet compared with the Earth. Perhaps three moonquakes per year reach 4 on the Richter scale used to measure earthquakes on Earth. This strength on Earth could be felt but would not cause damage. Our lunar stations worked for only 8 years, however, so we do not know how often giant moonquakes occur. Unfortunately, the seismometers and other instruments on the Moon were shut off by NASA in 1977 as an economy measure.

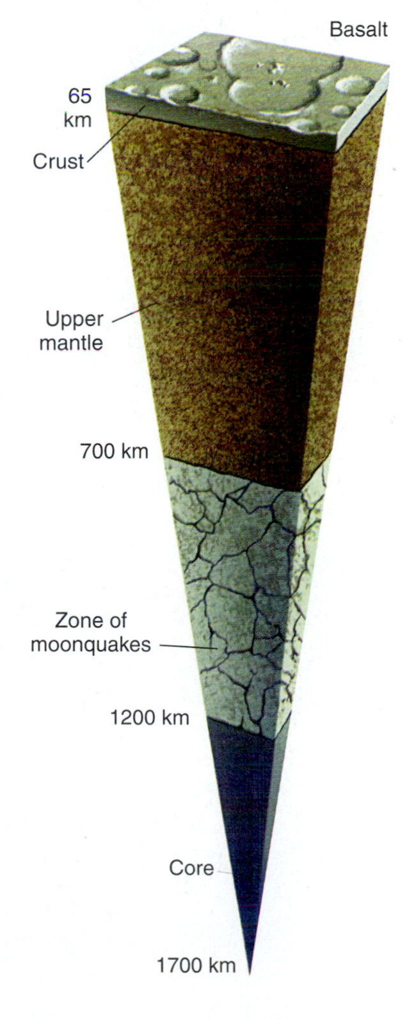


**FIGURE 9–20** The Moon's interior. The depth of basalts is greater under maria, which are largely on the side of the Moon nearest the Earth. Almost all the 10,000 moonquakes observed originated in a zone halfway down toward the center of the Moon, a distance ten times deeper than most terrestrial earthquakes. This fact can be used to interpret conditions in the lunar interior. If too much of the interior of the Moon were molten, the zone of moonquakes probably would have sunk instead of remaining suspended there. The deep moonquakes came from about 80 locations, and were triggered at each location twice a month by tidal forces resulting from the variation in the Earth–Moon distance.

**FIGURE 9–21** Buzz Aldrin and the experiments he had deployed. In the foreground, we see the seismic experiment. The laser-ranging retroreflector is behind it.

The seismometers worked long enough to capture the single important event when a meteorite hit the far side of the Moon on July 21, 1972. The impact generated seismic waves strong enough that we would have expected them to travel through to the near side, where the seismometers were all located. But one type of seismic wave—so-called shear waves, in which the matter carrying the wave has to vibrate from side to side—apparently did not travel through the core while the other types did. Since shear waves don't travel through molten material, most scientists deduce that at least part of the core is molten.

But if the core of the Moon were fully molten, as is the Earth's core, we would have expected to find a more intense magnetic field than the one we have detected. (The Earth's magnetic field is caused by motions in the metallic, liquid outer core of the Earth.) Yet the Moon has no global magnetic field. We do detect a weak magnetic field frozen into lunar rocks. It might be left over from a core that was molten long ago but has since cooled. We do not have a good understanding, either, of why the magnetic fields of other planets are as we find them.

Apollo astronauts made direct measurements of the rate at which heat flows upward through the top of the lunar crust. The rate is one-third that of the heat flow on Earth. The value is important for checking theories of the lunar interior, since it tells us about the distribution of radioactive elements under the surface.

Tracking the orbits of the Apollo Command Modules and other satellites that orbited the Moon also told us about the lunar interior. If the Moon were a perfect, uniform sphere, the spacecraft orbits would have been perfect ellipses. We interpret the deviation of the orbits from an ellipse as an effect of an asymmetric distribution of lunar mass.

One of the major surprises of the lunar missions was the discovery in this way of **mascons,** regions of **mas**s **con**centrations near and under most maria. (A few large mascons are known on Earth.) The mascons led to anomalies in the gravitational field, that is, deviations of the gravitational field from the orientation expected from a purely spherical and uniform mass. The mascons may be lava that is denser than the surrounding matter. The existence of the mascons is evidence that the whole lunar interior is not molten, for if it were, then these mascons could not remain near the surface. However, the mascons could be supported by an upper mantle or crust of sufficient thickness and stiffness.

One source of knowledge about the lunar interior continues to be studied. Sets of retroreflectors were left on the Moon (Fig. 9–21) by the astronauts of Apollo 11, 14, and 15 and on the Lunokhod 2. Each corner cube in these retroreflectors has the property that a beam of light that hits it is reflected back into exactly the same direction from which it came. When illuminated by the faint pulse received from a strong laser focused by a telescope on Earth (Fig. 9–22), the retroreflectors send back enough light to telescopes on Earth to be barely detectable—only about one of the $10^{18}$ photons sent out in a pulse is detected when it returns about 2.5 sec later. The studies, carried out mostly by telescopes at the McDonald Observatory in Texas and in France, allow the Earth–Moon distance to be accurately determined by measuring the time interval between sending out a brief pulse of laser light and its return. The distance at any time (it is constantly changing, of course) is now known to an accuracy of less than 10 cm. The Earth's rotation, the Moon's orbital motion, and the Moon's variations from constant rotation are also measured. One example of deductions from lunar ranging concerns the Moon's interior. The deviations from constant lunar rotation are caused by the gravitational forces from the Sun and planets acting on the non-spherical Moon and the mass distribution within it. By monitoring these deviations very precisely, one can infer various properties of the Moon's interior. Only the existence of a fluid layer at the interface between a fluid lunar core of radius about 330 km and the mantle can explain the laser-ranging observations. This conclusion that a fluid core does exist is consistent with similar conclusions based on the waves from the far-side meteorite impact, which had been challenged as statistically uncertain.

**FIGURE 9–22** A laser beam sent to the Moon at the McDonald Observatory of the University of Texas.

## 9.4 THE ORIGIN OF THE MOON

The leading models for the origin of the Moon that were considered at the time of the Apollo missions were the following:

- **Capture:** The Moon was formed far from the Earth in another part of the Solar System and was later captured by the Earth's gravity.
- **Fission:** The Moon was separated from the material that formed the Earth; the Earth spun up (that is, increased its rate of spin) and the Moon somehow spun off.
- **Co-accretion:** The Moon was formed near to and simultaneously with the Earth in the Solar System—a "double planet" hypothesis.

But studies since Apollo have ruled out the first two of these. The capture mechanism was long thought unlikely anyway, because a capture at just the right angle and speed was very improbable. Just this fact alone couldn't rule it out, since it only had to happen once. But it was finally ruled out because lunar samples show that the Earth and the Moon have similar relative abundances of oxygen isotopes, so the Moon couldn't have been formed too far away, given that the relative abundances of oxygen isotopes varied with position in the solar nebula. The fission hypothesis was ruled out in part because the Earth would not have been spinning fast enough to eject the material. Further, though the oxygen isotopes may be similar, the Apollo samples showed that other abundances of chemical elements are too different. Recent studies have also made the co-accretion model less likely than it seemed. It cannot explain why the Moon is enriched in "refractory" elements, those that boil at high temperatures (Section 9.3a). Refractory elements such as aluminum, calcium, thorium, and rare earths have abundances on the Moon about 50 per cent higher than on Earth. The Moon also has lower abundances of "volatile" elements, those that boil away easily. Further, the co-accretion model cannot explain how the Earth's and Moon's rotation and revolution periods, respectively, came out as they now are, given current models of the angular momentum of the Earth–Moon system.

It was originally hoped that new minerals discovered on the Moon might help solve the question of the Moon's origin. Though some minerals (Fig. 9–23) that do not exist on the Earth have been found on the Moon, we now think that they result from different conditions of formation rather than from abundance differences.

**FIGURE 9–23** A crystal of armalcolite (*Arm*strong-*Al*drin-*Col*lins, the crew of Apollo 11), examined under a polarization microscope. This mineral has been found only on the Moon.

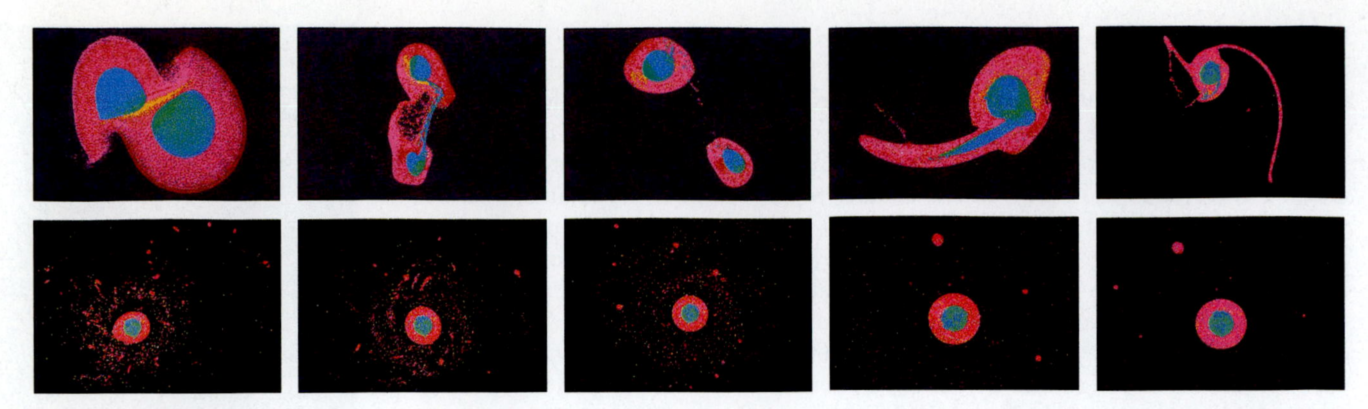

**FIGURE 9–24** A computer simulation of a collision between the proto-Earth and an impactor. Each is composed of an iron core and a rock mantle. The internal energy in each increases with both temperature and pressure. Increasing internal energy is shown as dark red, light red, brown, and yellow for rock, and dark blue, light blue, dark green, and light green for iron. In the first images, the impactor hits the proto-Earth, separates, and then falls in again, pulling out a tail of matter. At the end of this sequence, which takes six days, several large clumps and many smaller clumps are left. This material then clumps together to form the Moon.

The model that is now favored is:

- **Giant impact:** A Mars-sized planetesimal hit the proto-Earth off-center, ejecting matter.

One version of such a model has the matter ejected in gaseous form. This matter ordinarily would have fallen back onto the Earth, but since it was gaseous, pressure differences could exist. These presumably caused some of the matter to start moving rapidly enough to go into orbit, and some asymmetry established an orbiting direction; the other matter fell back. Another version has clumps of matter ejected; the clumps were later broken up by tidal forces and formed a disk around the Earth. In both versions, the disk of orbiting material eventually coalesced into the Moon. The model explains, among other things, why the Moon's iron core is so much smaller than Earth's: the core of the impactor remained with the Earth. Differences in abundances come largely from the inclusion of so much material from the impactor in what is now the Moon. The volatiles were driven off from the material that became the Moon in the heating from the collision. And the angular momentum of the impactor gave the Earth a large amount of angular momentum, making it spin rapidly. (It has since lost some of that angular momentum, as we see in the next paragraph.) Computer simulations (Fig. 9–24) are endorsing the main points of the giant-impact model. The iron in the Moon's core came from the impactor, since only the Earth's mantle went into the Moon's formation. The fact that the Moon's density is so much less than Earth's fits with the giant-impact model, though it is not a unique proof of it. Though consensus grows that the giant impact theory is correct, not all scientists in the field agree.

We on Earth are lucky that the Moon formed, since it may have played an important role in the origin of life. Without the tides caused by the Moon, perhaps life would not have arisen or would not have been able to leave the ocean. The Earth's rotation throws the high tides ahead of the Moon orbiting overhead. The Moon pulls back on the water with greater force than average in the region of high tide, and that water, hitting the continents, slows the Earth's rotation. Billions of years ago, we used to rotate in about 6 hours instead of our current 24. The length of our day is getting longer by about 2 milliseconds per century. To conserve angular momentum in the Earth–Moon system, the Moon is gradually spiralling away from the Earth by about 5 cm/year.

Geological studies released in 2000 of thin layers in sedimentary terrestrial rocks from South Africa show what tides on Earth were like 3.2 billion years ago. The uniformity of the layers apparently shows that the Moon's orbit was not very elliptical, and was rather more similar to today's orbit than had been thought. The discovery is in line with the giant-impact model for lunar formation, since a captured Moon would more likely have been in a very elliptical orbit.

## 9.5 POST-APOLLO LUNAR STUDIES

Ground-based studies of the Moon still bring surprises. Following the spectroscopic discovery of sodium in Mercury's atmosphere, a small amount of sodium was similarly discovered on the Moon, forming a very thin lunar atmosphere (Fig. 9–25). The sodium is probably ejected from the Moon's surface by incoming sunlight.

One surprise post-Apollo addition to lunar studies came in 1982, when it was realized that a meteorite found in Antarctica probably came from the Moon (Fig. 9–26). The rock was apparently ejected from the Moon when a crater was formed. The dozens of lunar meteorites now known, and all the lunar samples that have not been distributed or destroyed in analysis, are kept free of contamination in a Lunar Curatorial Facility in Houston for future study. They are kept in an atmosphere of nitrogen, to eliminate oxidation. The idea that these rocks came from the Moon is generally accepted, as is the presence on Earth of a few rocks from Mars. In Section 21.1b, we discuss the claim, not widely accepted, that fossils from ancient life forms were found in one of the meteorites from Mars.

For a long interval after the Apollo missions, little lunar exploration went on, with only one Soviet craft reaching the Moon. In 1990, Japan sent a small spacecraft into lunar orbit. NASA's Galileo mission went past the Moon in 1991, en route to Jupiter. The Galileo spacecraft (Fig. 9–27) tested some of the mapping methods it would later use at Jupiter.

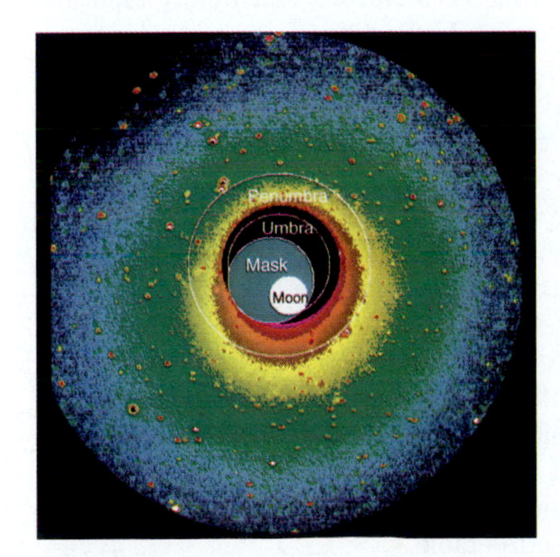


**FIGURE 9–25** The Moon's tenuous sodium atmosphere, seen during a lunar eclipse. The glow extends to nine times the lunar radius. Boston University researchers conclude that the sodium is ejected by sunlight striking the Moon's surface.

**FIGURE 9–26** This rock is one of many meteorites found on Earth in Antarctica. Under a microscope and in mineralogical and isotopic analyses, it seems like a sample from the lunar highlands (including anorthosites, for example) and quite unlike any terrestrial rock or any other meteorite. Several dozen meteorites—all breccias—are thought to have come from the Moon, and a dozen meteorites are thought to come from Mars, as we shall discuss later on.

**FIGURE 9–27** The Moon, from the Galileo spacecraft. Colors, from the ratios of brightness in different filters, show variations in the Moon's composition. The Orientale basin and portions of the far-side highlands (*red, at left*) that we do not see from Earth are similar to soils found at the Apollo 16 site. Highland regions shown in yellow are enhanced in iron. Some of the mare basalts have high (*blue*) titanium dioxide while others (*orange*) have lower titanium dioxide.

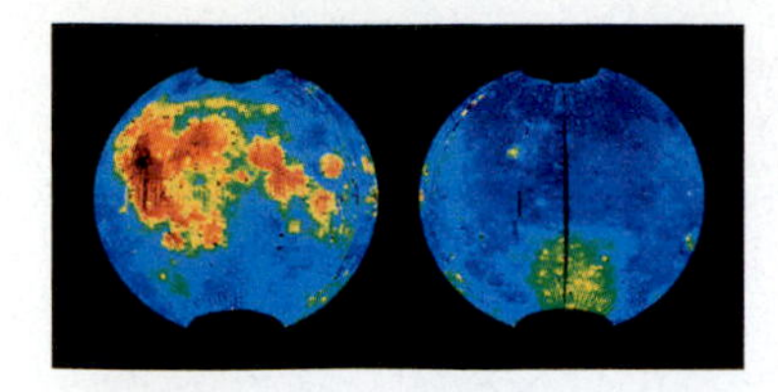

**FIGURE 9–28** A map of the iron distribution on the Moon from the Clementine spacecraft. The data were taken through two filters in the near infrared. Iron levels turn out to be high both in maria on the near side and in a region near the south pole on the far side. This south-pole region is the Aiken Basin, mapped by a laser altimeter on Clementine. It is about 2500 km across and over 12 km deep, the largest and deepest basin known in the Solar System.

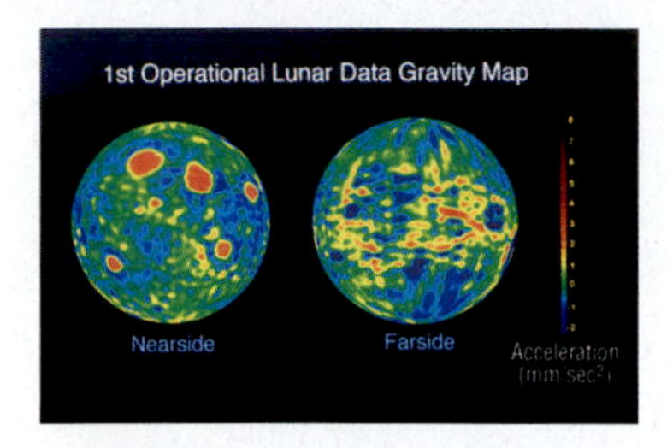


**FIGURE 9–29** A Lunar Prospector map of the distribution of gravity on the Moon showing hemispheres of the near side and the far side.

Though NASA hadn't had any lunar missions since the Apollo missions ended in 1972, in 1994 a U.S. Department of Defense spacecraft called Clementine went into space to test imaging devices and orbited the Moon for 71 days. It surveyed the Moon in 11 wavelength bands from the ultraviolet end of the visible through the near infrared with high resolution, mapping the Moon in more detail than any other celestial body. The spacecraft was named after the prospector's daughter in the old song; after all, both were related to a search for minerals and are "lost and gone forever."

Since the Clementine spacecraft was in a polar orbit, Clementine mapped all the lunar latitudes, unlike earlier spacecraft, which had been in orbits more nearly parallel to the Moon's equator and so never went near the poles. A laser-ranging system gave data to a resolution of 40 m, an improvement on earlier mapping. The altitude of the lunar surface from some smooth average shape varies up and down by 16 km, a $\frac{1}{3}$ greater distance than Apollo had found because of Clementine's greater coverage. The data are being used to catalogue rock types and the minerals the rocks contain (Fig. 9–28). Clementine also found a number of ancient multi-ringed basins in addition to the newer ones already known. Its detailed studies of craters have revealed how material was mixed as the craters formed, information of widespread applicability in the Solar System. Clementine's studies of mascons from their gravity have found variations in the strengths of the lunar lithosphere, probably showing that the Moon has been heated and cooled in a complex way.

In 1998, NASA launched its Lunar Prospector mission (Fig. 9–29). Its name showed that it had the same theme as Clementine: to find out what minerals are on the lunar surface. Lunar Prospector mapped the lunar surface's chemical composition and the Moon's magnetic and gravity fields. It looked for water ice in the permanent shadows that Clementine discovered near the Moon's poles. It reported, controversially, that it verified the interpretation of radar results advanced by Clementine scientists that ice chunks are indeed present there. Finding water on the Moon would make it much easier for humans to explore or even settle there. But Lunar Prospector didn't detect water directly; it detected merely neutrons that theory indicated would be given off when particles in the solar wind interact with the hydrogen in lunar ice. Maybe the neutrons were themselves from the solar wind; Apollo 17 astronaut Harrison Schmitt, the only geologist ever to have walked on the Moon, told me, in 1999, that he thinks such neutrons from the solar wind were the source of the Lunar Prospector measurements. The Lunar Prospector spacecraft had been crashed into the lunar surface earlier in 1999, purposely picking a location in a north-polar crater in hope that a plume of water vapor might rise and be detected by telescopes on Earth. But none of the telescopes detected any sign of the crash. Scientists using the giant radio telescope at the Arecibo Observatory on Earth have also reported that their observations of the same regions do not show the presence of ice.

Still, it would be wonderful to find water on the Moon. Schmitt points out that the discovery of water would make it possible for an eventual inhabited lunar base to supply itself, which is much easier and cheaper than bringing all their supplies from Earth. He thinks that such a base could be set up in 10 years, given the approval of funding. But no such plans are on the table.

## 9.6 FUTURE LUNAR MISSIONS

The European Space Agency plans a SMART-1 (Small Mission for Advanced Research and Technology) mission to the Moon, partly to test their solar electric propulsion system. The spacecraft is to go into geosynchronous orbit, and then to spiral slowly out to the Moon's distance with a low but steady amount of thrust from xenon gas ionized and expelled at high speed. Still, in a lunar polar orbit, it will use its 15 kg of scientific instruments to study the Moon's surface geology, topography, and mineralogy. They hope for a 2002 launch.

Japan plans two missions to the Moon. (Their Hiten mission, basketball-sized, went into lunar orbit in 1992.) In 2003, they will launch Lunar-A, which will include two probes that will penetrate the Moon's surface. One will impact on the side nearest Earth and the other on the lunar far side. They will measure the flow of heat through the surface and will carry seismometers. The seismometers will be five times more sensitive than those on Apollo were. The measurements of moonquakes they expect should reveal the size of the Moon's core. Knowledge of the core, in turn, will help us decide on models of the Moon's formation. The orbiter that relays the penetrators' signals back to Earth will itself make black-and-white images. In 2004, they will launch their *Sel*enographic and *En*gineering *E*xplorer (SELENE, matching the name of the Greek goddess of the Moon). This more substantial mission is to contain a low-altitude orbiter for high-resolution mapping, a lander, and a high-altitude satellite to relay the data back to Earth.

## CORRECTING MISCONCEPTIONS

| ✖ **Incorrect** | ✔ **Correct** |
| --- | --- |
| The Moon is too far away to analyze its composition. | The Moon is made out of rock, with composition discovered by the Apollo missions. |
| We can see all parts of the Moon from Earth. | Only about 5/8 of the Moon ever faces us. |
| The Moon's far side is always dark. | The Moon's "far side" faces away from us but is sometimes dark and sometimes light. The dark side is merely the face that is not being hit by sunlight. |

## SUMMARY AND OUTLINE

Lunar features (Section 9.1)
- Maria, highlands, craters, mountains, valleys, rilles, ridges
- Visibility dependent on Sun's angle; terminator

Revolution and rotation (Box 9.1)
- Sidereal: with respect to the stars
- Synodic: with respect to another body

Lunar exploration (Section 9.2)
- Robotic and crewed lines of development merge in Apollo.
- Six Apollo landings: 1969 to 1972
- Three Soviet robotic spacecraft brought samples back.
- Two Soviet Lunokhods roved many kilometers.
- Several spacecraft, especially Clementine, in the 1990s
- Composition of the lunar surface (Section 9.3a)
  - Mare basalts and highland anorthosites from cooled lava
  - More breccias—broken up and re-formed—in highlands
- Chronology (Section 9.3b)
  - Relative dating by crater counting
  - Radioactive dating gives absolute ages.
- Craters (Section 9.3c)
  - Almost all formed in meteoritic impact.
  - Signs of volcanism also present on surface.
  - Far side has few maria, quite different from near side.
- Interior (Section 9.3d)
  - Moon is differentiated into a core, a mantle, and a crust.
  - Seismographs revealed that interior is molten, though lack of strong lunar magnetic field indicates the opposite.
  - Weak moonquakes occur regularly.
  - Mascons discovered from their gravitational effects.

Origin (Section 9.4)
- Capture, fission, and co-accretion theories were standard.
- Newer theories are interaction of Earth-orbiting and Sun-orbiting planetesimals and ejection of a gaseous ring.
- Ejection of a gaseous ring seems most viable.

Post-Apollo studies (Section 9.5)
- Lunar meteorites found in Antarctica.
- A thin lunar sodium atmosphere
- Clementine satellite mapped the whole Moon at many wavelengths in 1994.
- Lunar Prospector may have confirmed the discovery of water.

## KEY WORDS

highlands
maria
mare
mountain ranges
valleys
rilles
ridges
rims
librations
terminator
sidereal revolution period
synodic revolution period
synodic month
igneous
sedimentary
basalts
anorthosites
breccias
lunar soils
regolith
volatile
refractory
mascons

## QUESTIONS

†1. About how many Moons would fit inside one Earth?
2. Compare the lengths of sidereal and synodic months. Which is longer? Why?
3. If the Moon's mass is 1/81 Earth's, why is gravity at the Moon's surface as great as 1/6 that at the Earth's surface?
4. Describe the Earth's terminator and its position relative to the Earth's surface, comparing it to the Moon's terminator and how it changes position.
5. Why is the heat flow rate related to the radioactive material content in the lunar surface?
6. What does cratering tell you about the age of the surface of the Moon, compared to that of the Earth's surface?
7. Why is it not surprising that the rocks in the lunar highlands are older than those in the maria?
8. Why are we more likely to learn about the early history of the Earth by studying the rocks from the Moon rather than those on the Earth?
9. Using the lunar-chronology chart (Fig. 9-15), describe the relative ages of the lava flows in the Seas of Fertility and Serenity. Which Apollo missions went there?
10. What do mascons tell us about the lunar interior?
11. Choose one of the proposed theories to describe the origin of the Moon, and discuss the evidence pro and con.
12. (a) Describe the American and Soviet lunar explorations. (b) What can you say about future plans?
13. Why was Clementine's mapping more complete than Apollo's?
14. Why was it reasonable to compare Clementine with a lunar prospector?

† This question requires a numerical solution.

## TOPICS FOR DISCUSSION

1. Discuss the scientific, political, and financial arguments for resuming (a) robotic and (b) crewed exploration of the Moon.
2. Discuss the scientific, political, and financial arguments for building the NASA Space Station.

## INVESTIGATIONS

1. Use three small, rectangular mirrors to make a corner-cube retroreflector and investigate its properties.
2. Put two large flat mirrors together perpendicularly or find a place where two such mirrors meet and try shaking hands with your image. How does the result differ from your trying to shake hands with your image in a regular, single mirror?

## USING TECHNOLOGY

### W World Wide Web

1. View images of the Moon from the Clementine spacecraft at http://www.nrl.navy.mil/clementine or http://nssdc.gsfc.nasa.gov/planetary/clementine.html. Follow the next mission at http://sci.esa.int/smart-1.
2. Read about the history of the Apollo program at the National Air and Space Museum's site at http://www.nasm.edu/apollo or at NASA's Kennedy Space Center's site at http://www.ksc.nasa.gov/history/apollo/apollo.html.
3. Check for recent research at the Hawaii Institute of Geophysics and Planetology's site at http://www.psrd.hawaii.edu.
   It covers "Planetary Science Research Discoveries" ranging over a wide variety of topics.
4. Learn about the Earth–Moon interaction, and the Moon's gradual recession from the Earth, at http://www.aspsky.org/education/tnl/33/33.html.

### REDSHIFT

1. Locate in time and duplicate the circumstances of the next lunar eclipse, both from Earth and from the Moon.
2. Examine the Moon and locate craters named after famous people of your choice.
3. View the video clips about the Apollo astronauts on the Moon.
4. Show the Moon's librations, and list some of the features that are only sometimes facing Earth.
5. From the *RedShift* main program, see the Moon orbit the Earth. Locate yourself high above the Solar System and look down over time. Then locate yourself in the ecliptic and notice the tilt of the lunar orbit.
6. Pick a crater on the Moon. What is its diameter? What is the basis of the fame of the person it is named for? How does it appear when you zoom in on it and watch it for a month, allowing the terminator to sweep across it?

A photomosaic of Mercury from Mariner 10, showing its cratered surface, surrounding an artist's conception of Mercury's thick core (*yellow*) and thin lithosphere (*red*). The symbol for Mercury appears at the top of the opposite page.

# Chapter 10

## Mercury

Mercury is the innermost planet and until very recently has been one of the least understood. Except for that of distant Pluto, Mercury's elliptical orbit around the Sun is the most out-of-round of any of the planets in our Solar System. (The difference between the maximum and minimum distances of Mercury from the Sun is as much as 40 per cent of the average distance, compared with less than 4 per cent for the Earth.) Its average distance from the Sun is 58 million kilometers, which is $^{4}/_{10}$ of the Earth's average distance. Thus Mercury is 0.4 A.U. from the Sun. It is also, except for Pluto, the least massive planet in our Solar System; it has only $5^{1}/_{2}$ per cent the mass of the Earth.

**AIMS:** To discuss the difficulties in studying Mercury from the Earth, the results of space observations, and exciting new ground-based observations

Since we on the Earth are outside Mercury's orbit looking in at it, Mercury always appears close to the Sun in the sky (Fig. 10–1). At times it rises just before sunrise, and at times it sets just after sunset, but it is never up when the sky is really dark. The maximum angle from the Sun at which we can see it is 28°, which means (since the sky moves 15° every hour) that the Sun always rises or sets within about two hours of Mercury's rising or setting. Of course, the difference in time is usually even less than this maximum. Consequently, whenever Mercury is visible, its light has to pass obliquely through the Earth's atmosphere to reach us.

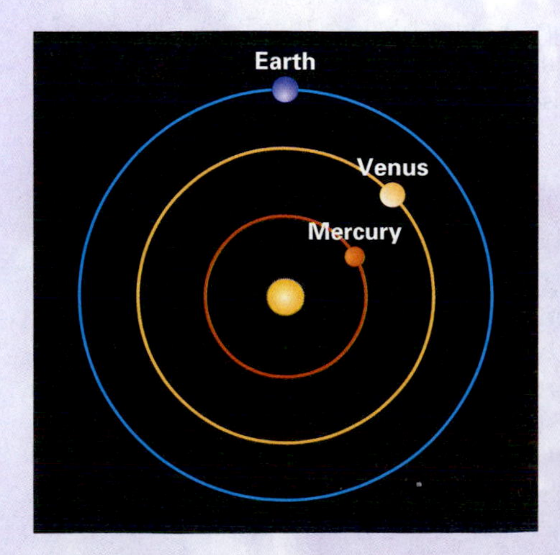


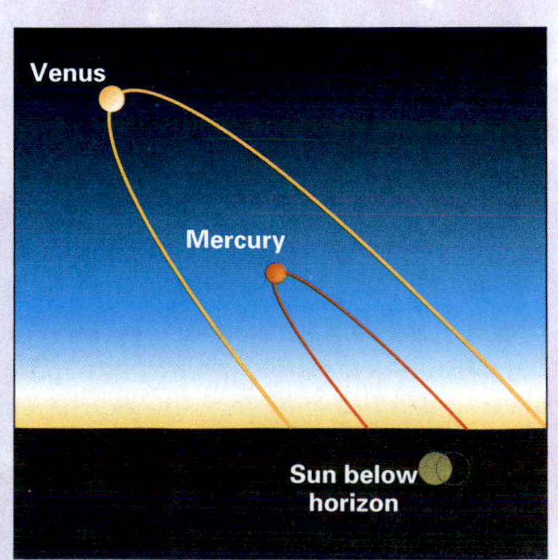


**FIGURE 10–1** Since Mercury's orbit is inside that of the Earth and relatively close to the Sun (*left*), Mercury is never seen against a really dark sky. A view from the Earth appears at right, showing Mercury and Venus at their greatest respective distances from the Sun.

**FIGURE 10–2** On rare occasions, Mercury goes into transit across the Sun; that is, we see it as a black dot crossing the Sun. Here we see an image of Mercury crossing near the edge of the Sun at the transit of November 17, 1999. A photograph taken with the TRACE satellite, which is studying the Sun, was Figure 7-19. The next transits of Mercury will be on May 7, 2003, which won't be visible from the United States, and November 8, 2006, which will. Mercury takes about 5 hours to cross the Sun's disk.

Since we are looking close to the horizon, along a long path through turbulent air, our view is blurred. Thus astronomers have never gotten a really good view of Mercury from the Earth, even with the largest telescopes. Many people have never seen it at all. (Copernicus's deathbed regret, the story has it, was that he had never seen Mercury.) Even the best photographs taken from the Earth show Mercury as only a fuzzy ball with faint, indistinct markings. We will show the best of these images at the end of the chapter.

An unusual time to see Mercury is when it is in transit, that is, when it passes between the Earth and Sun so that we see it silhouetted, as happens once every few years (Fig. 10–2).

## 10.1 THE ROTATION OF MERCURY

From studying drawings and photographs, astronomers of the first half of the twentieth century did as well as they could to describe Mercury's surface. A few features could, it seemed, barely be distinguished, and the astronomers watched to see how long those features took to rotate around the planet. From these observations they decided that Mercury rotated in the same amount of time that it took to revolve around the Sun. Thus they thought that one side always faced the Sun and the other side always faced away from the Sun. This apparent conclusion led to the fascinating further conclusion that Mercury could be both the hottest planet and the coldest planet in the Solar System.

It seemed reasonable that there could be a bulge in the distribution of mass of Mercury. The side that was bulging would be attracted to the Sun by gravity, locking the rotation to the revolution, just as the Moon is locked to the Earth. Such a match of periods is called **synchronous rotation.** Synchronous rotation implies that the periods of rotation and revolution are equal, and so the less massive body would always keep the same face toward the more massive body. Our Moon and all of the largest moons of the outer planets are in synchronous rotation.

But early radio astronomy studies of Mercury indicated that the dark side of Mercury was too hot for a surface that was always in the shade. (Simply, for an object of a given size at a given distance, the stronger the radio signal it gives off, the higher its temperature.) Perhaps the regions being studied weren't always in the shade after all.

Later, we became able not only to receive radio signals emitted by Mercury, but also to transmit radio signals from Earth and detect the echo. This technique is called **radar.** "Radar" is an acronym for *ra*dio *d*etection *a*nd *r*anging. Since Mercury is rotating, one side of the planet is always receding relative to the other. Such motion can be measured from Earth with the Doppler effect. The results were a surprise: Scientists had been wrong about the period of Mercury's rotation. It actually rotates in 58.6441 days. Mercury's 59-day period of rotation is exactly ⅔ of the 88-day period of its revolution that scientists had originally equated with its rotation period; thus the planet rotates three times for each two times it revolves around the Sun.

Although its tendency to continue rotating is too strong to be overcome by the gravitational grip of the Sun, the Sun's steadying pull, trying to slow the rotation, is strongest every 1½ rotations. At those times, Mercury is in its perihelion position, so the gravitational bulge on Mercury is in a line with Mercury's center and the Sun. Mercury's spin was probably once much faster and was slowed down by the fact that the Sun's gravity attracted the bulge more than it attracted the rest of the planet. A **gravitational interlock** is at work, also called a **tidal lock.** But the relation of rotation to revolution is 2:3 instead of the 1:1 relation of our Moon. Mercury is not in synchronous rotation after all, unlike the celestial objects we discussed three paragraphs earlier.

The rotation period is measured with respect to the stars; that is, the period is one mercurian sidereal day, the interval between successive returns of the stars to the same position in the sky. Mercury's rotation and revolution combine to give a value for the rotation of Mercury relative to the Sun (that is, a mercurian solar day) that is neither the 59-day **sidereal rotation period** nor the 88-day period of revolution. As can be seen from careful analysis of the combination (Fig. 10–3), if we lived on Mercury we would measure each day and each night to be 88 Earth days long. We would alternately be fried and frozen for 88 Earth days at a time. Mercury's **solar rotation period** is thus 176 days long, twice the period of Mercury's revolution.

Since we now know that different sides of Mercury face the Sun at different times, the temperature at the point where the Sun is overhead doesn't get as hot as

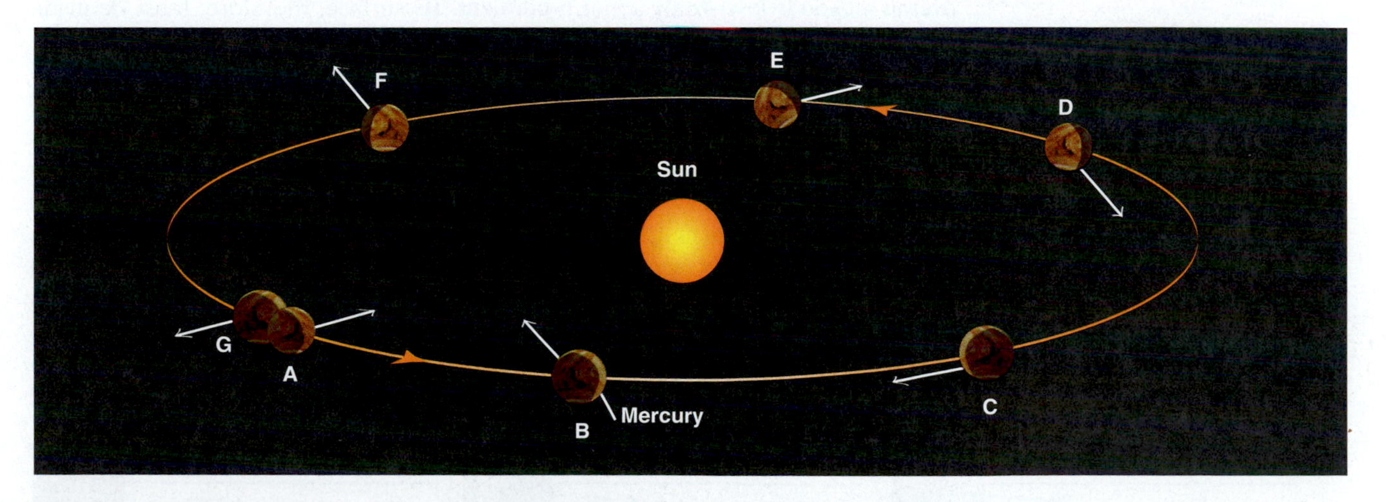


**FIGURE 10–3** Follow the arrow that starts facing rightward toward the Sun in the image of Mercury at the left of the figure (*A*), as Mercury revolves along the line. Mercury, and thus the arrow, rotates once with respect to the stars in 59 days, when Mercury has moved only ⅔ of the way around the Sun (*E*). Note that after one full revolution of Mercury around the Sun, the arrow faces away from the Sun (*G*). It takes another full revolution, a second 88 days, for the arrow to face the Sun again. Thus the rotation period with respect to the Sun is twice 88, or 176, days.

it would if Mercury were in synchronous rotation. The temperature at this point is about 425°C (800°F).

We know from Kepler's second law that Mercury travels around the Sun at different speeds at different times in its eccentric orbit. This effect, coupled with Mercury's slow rotation on its axis, would lead to an interesting effect if we could stand on its surface. From some locations we would see the Sun rise for an Earth day or two, and then retreat below the horizon from which it had just come, when the speed of Mercury's revolution around the Sun dropped below the speed of Mercury's rotation on its own axis. Later, the Sun would rise again and then continue across the sky.

No harm was done by the scientists' original misconception of Mercury's rotational period, but the story teaches all of us a lesson: we should not be too sure of so-called facts. Don't believe everything you read here, either.

## 10.2 GROUND-BASED VISUAL OBSERVATIONS

The albedo (from the Latin for whiteness) is the ratio of light reflected from a body to light received by it.

Even though the details of the surface of Mercury can't be studied very well from the Earth, there are other properties of the planet that can be better studied. For example, we can measure Mercury's **albedo,** the fraction of sunlight hitting Mercury that is reflected (Fig. 10–4). We can measure the albedo because we know how much sunlight hits Mercury (we know the brightness of the Sun and the distance of Mercury from the Sun). Then we can easily calculate at any given time how much light Mercury reflects, from both (1) how bright Mercury looks to us and (2) its distance from the Earth. Once we have a measure of the albedo, we can compare it with the albedo of materials on the Earth and on the Moon and thus learn something of what the surface of Mercury is like.

Let us consider some examples of albedo. An ideal mirror reflects all the light that hits it; its albedo is thus 100 per cent. (The very best real mirrors have albedoes of as much as 96 per cent.) A black cloth reflects essentially none of the light; its albedo (in the visible part of the spectrum, anyway) is almost 0 per cent. Mercury's overall albedo is low—only about 6 per cent. Its surface, therefore, must be made of a dark—that is, poorly reflecting—material. The albedo of the Moon is similar. In fact, Mercury (or the Moon) appears bright to us only because it is contrasted against a relatively dark sky; if it were silhouetted against something white, it would look relatively dark.

From Mercury's apparent angular size and its distance from the Earth—which can be determined from knowledge of its orbit—we have determined that Mercury is less than half the diameter of the Earth. Because Mercury has no moon, we can determine its mass only from its gravitational effects on bodies that pass near it, such as occasional asteroids or comets. The most accurate value we now have for Mercury's

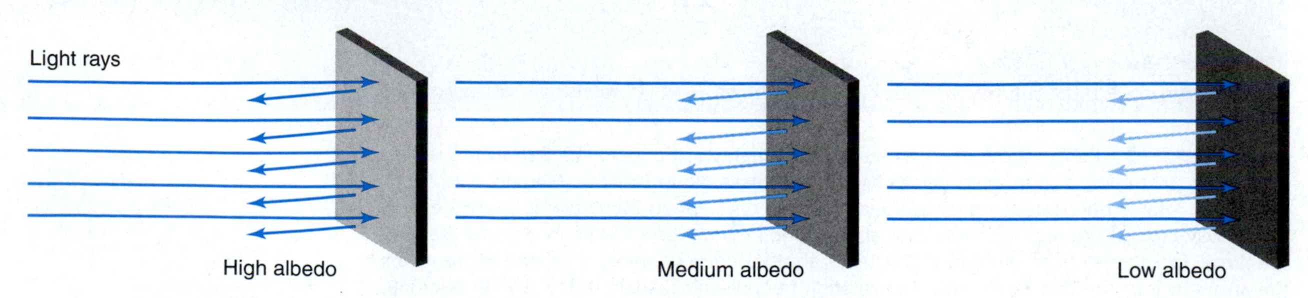


**FIGURE 10–4** **Albedo** is the fraction of radiation reflected. A surface of low albedo looks dark.

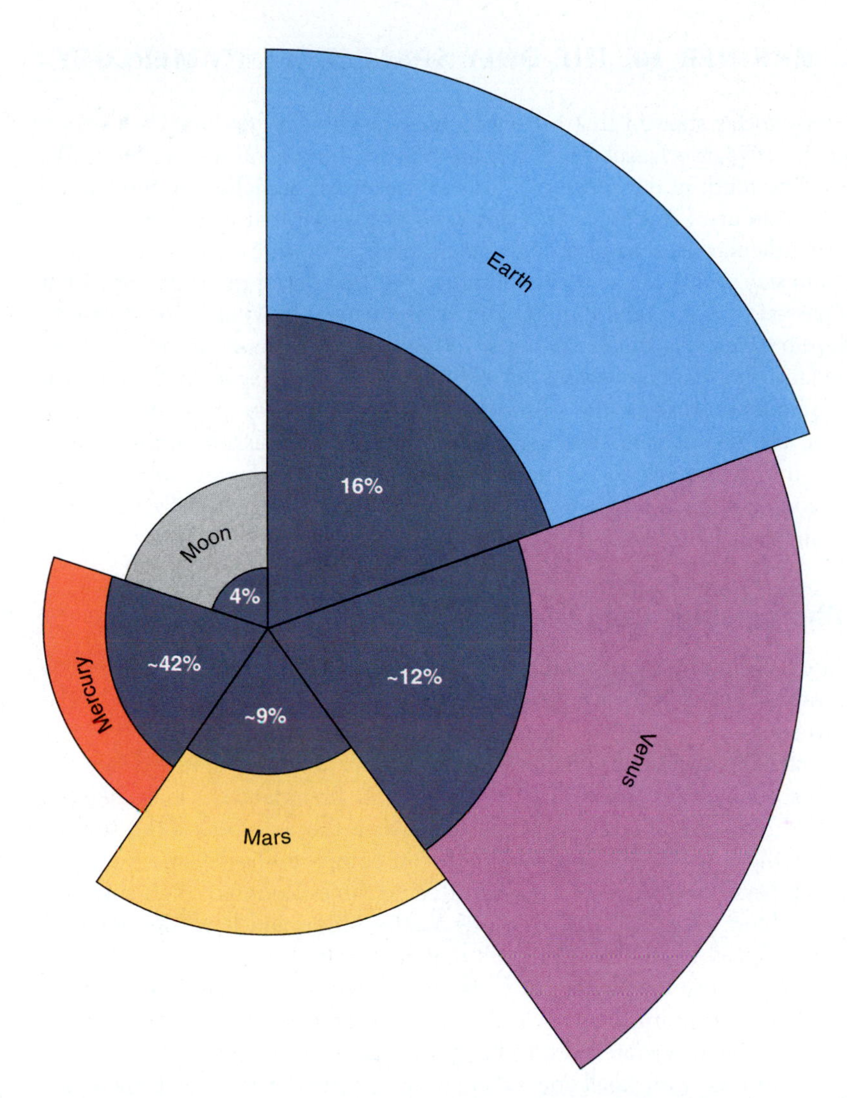


**FIGURE 10–5** The core of Mercury takes up a large fraction of its volume.

mass came from tracking the Mariner 10 spacecraft flyby. We find that Mercury's mass is five times greater than that of our Moon and 5½ per cent that of the Earth.

Mercury's density (its mass divided by its volume) can thus be calculated; it turns out to be 5.5 grams/centimeter$^3$, roughly the same density as that of Venus and the Earth. So Mercury's core (Fig. 10–5), like Venus's and the Earth's, must be dense; it too must be made of iron. Since Mercury has less mass than the Earth and is therefore less compressed, you might normally expect it to have lower density than Earth. But since it has the same density, it must have some portion that is of especially high density. Thus Mercury's core must have even more iron in proportion to its rock than the Earth, since the iron has relatively high density compared with the rock. Mercury's core takes up perhaps 42 per cent of the volume, or 70 per cent of the mass. (The Earth's core, by comparison, takes up only 16 per cent of the Earth's volume, and the Moon's core only 4 per cent of the Moon's volume.)

Mercury's surface undergoes the greatest variation in surface temperature, though Venus's atmosphere, as we shall see in the following chapter, makes that planet's surface hotter than Mercury's. Mercury's surface, during the long solar day under the nearby Sun, reaches 427°C, hot enough to melt zinc. At night, without an atmosphere to retain energy, the surface cools to below –183°C (–300°F).

## 10.3 MARINER 10, THE ONLY SPACECRAFT TO MERCURY

FIGURE 10–6 Mercury, photographed from a distance of 35,000 km as Mariner 10 approached the planet for the first time, shows a heavily cratered surface with many low hills. The valley at the bottom is 7 km wide and over 100 km long. The large, flat-floored crater is about 80 km in diameter; craters over 20 km wide have flat floors, while small craters are bowl-shaped like bullet craters.

The discussions so far showed that astronomers, typically, had deduced a lot from limited data. In 1974, we learned much more about Mercury in a brief time. We flew right by. The tenth in the series of Mariner spacecraft launched by the United States went to Mercury. (The Mariner series were spacecraft that were stabilized on all three perpendicular axes, as opposed to spacecraft that were stabilized by spinning, so it could stay pointed in a specific direction, getting better photographs.) First Mariner 10 passed by Venus and then had its orbit changed by Venus's gravity to direct it to Mercury. Tracking the orbits improved our measurements of the gravity of these planets and thus of their masses. In addition, the 475-kg spacecraft had a variety of instruments on board. One was a device to measure the magnetic fields in space and near the two planets. Another measured infrared emission of the planets and from that information deduced their temperatures. Two other instruments—a pair of television cameras—provided not only the greatest popular interest but also many important data.

### 10.3a Photographic Results

When Mariner 10 flew by Mercury the first time (yes, it went back again), it took 1800 photographs that were transmitted to Earth. It came as close as 750 km to Mercury's surface.

The most striking overall impression is that Mercury is heavily cratered (see the photograph that opened this chapter and Fig. 10–6). At first glance, it looks like the Moon! But there are several basic differences between the features on the surface of Mercury and those on the lunar surface. We can compare how the mass and location in the solar system of these two bodies affected the evolution of their surfaces.

Mercury's craters seem flatter than those on the Moon and have thinner rims (Fig. 10–7), perhaps as a result of Mercury's higher gravity.

There are three major types of surfaces on Mercury. One type, the "smooth plains," corresponds to maria on the Moon. The other two types, the "intercrater plains" and the "heavily cratered terrain," cover the rest of Mercury. The intercrater plains are covered with small craters and are distributed within the heavily cratered terrain. This situation is unlike that of the Moon, whose regions of heavily cratered terrain do not contain large expanses of more lightly cratered plains.

*A*

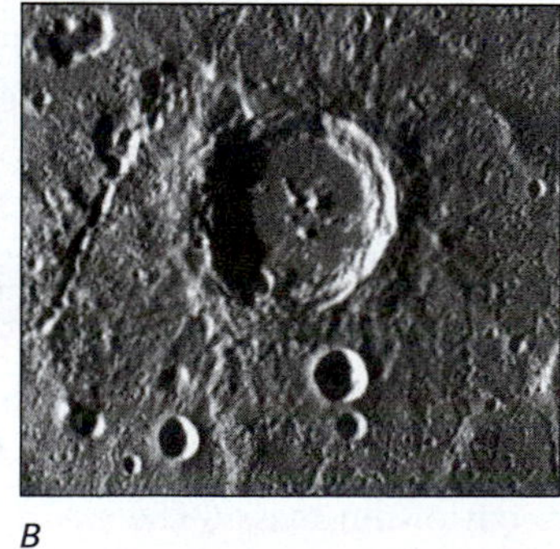

*B*

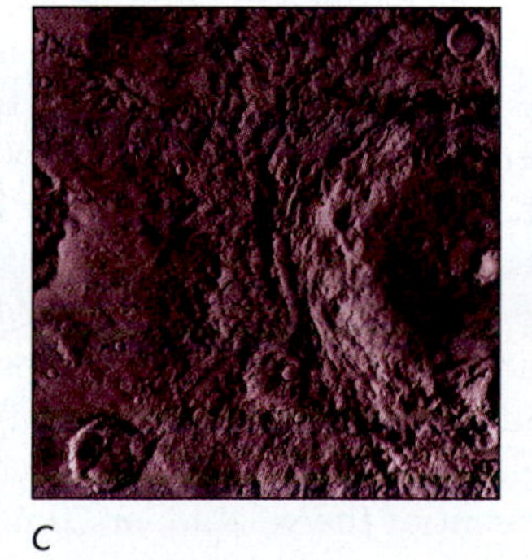

*C*

*D*

FIGURE 10–7 A comparison of (*A*) lunar craters, (*B*) craters on Mercury, (*C*) craters on Mars, and (*D*) craters on Venus. Though it is not noticeable on these images, material has been continuously thrown out less far on Mercury than on the Moon because of Mercury's higher gravity. The martian crater shows flow across its surface, possibly resulting from the melting of a permafrost layer under the surface. The photographs are dominated by Copernicus on the Moon, 93 km in diameter; a similar-sized flat-bottomed crater on Mercury; Cerulli on Mars, 115 km in diameter; and craters on Venus that are 37 to 50 km in diameter.

A DEEPER DISCUSSION

## BOX 10.1 Naming the Features of Mercury

The mapping of the surface of Mercury led to a need for names. The scarps were named for historical ships of discovery and exploration, such as *Endeavour* (Captain Cook's ship), *Santa Maria* (Columbus's ship), and *Victoria* (the first ship to sail around the world, which it did in 1519–22 under Magellan and his successors). Most plains were given the name of Mercury in different languages, such as Tir (in ancient Persian), Odin (an ancient Norse god), and Suisei (Japanese). Craters are being named for nonscientific authors, composers, and artists, in order to complement the lunar naming system, which honors scientists. So we find Mozart, Beethoven, Brontë, Michelangelo, and Matisse on Mercury.

### Discussion Question

Who would be your top 20 candidates for names of surface features on Mercury from 20th-century artists and musicians? Would Ringo be on your list? Sir Elton? (He was knighted in 1998.)

Most of the craters themselves seem to have been formed by impacts of meteorites. The secondary craters, caused by material ejected as primary craters were formed, are closer to the primaries than on the Moon, presumably because of Mercury's higher surface gravity. The smooth plains have relatively few craters compared with the intercrater plains. The smooth plains are sufficiently extensive that they are probably volcanic. Further evidence for their volcanism is the fact that they sometimes overlap the intercrater plains. Smaller, brighter craters are sometimes superimposed on the larger craters and thus must have been made afterward.

Some craters have rays of higher albedo emanating from them (Fig. 10–8), at least superficially resembling rays around some lunar craters. As on the Moon, these are the freshest craters.

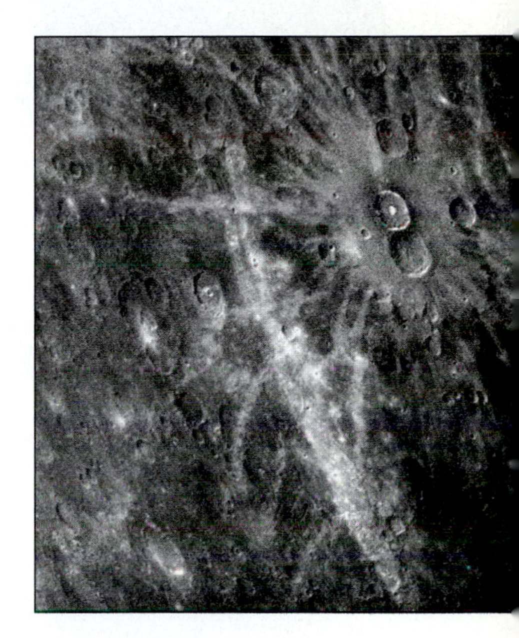

**FIGURE 10–8** A field of rays radiating from a pair of craters, each about 40 km across.

Mercury has no obvious fault systems such as the San Andreas fault in California on the Earth. There appear to be some signs of small-scale geologic tensions like troughs, widely distributed around the planet. They are very old. From their orientations, they are believed to have resulted from stresses in the planet's crust as Mercury changed shape while tides from the Sun's gravity (we discussed the cause of tides in Chapter 8) were slowing down its rotation.

One interesting kind of feature that is visible on Mercury is lines of cliffs hundreds of kilometers long; on Mercury, as on Earth, such lines of cliffs are called **scarps.** The scarps are particularly apparent in the region of Mercury's south pole (Fig. 10–9). These scarps are global in scale, not just isolated, and are relatively youthful. The scarps are wrinkles in Mercury's crust. Probably Mercury's core was once totally molten, and shrank as it solidified. This shrinking combined with shrinking of Mercury's rocky mantle as it cooled would have caused the crust to buckle, creating the scarps in at least the quantity that we now observe. A fairly recent insight is that Mercury may be the fragment of a giant early collision that nearly stripped it to its core, thus accounting for its large proportion of iron.

One of the largest flat regions on Mercury is the Caloris Basin (Fig. 10–10). It was the result of a gigantic impact. One part of the mercurian landscape seems particularly different from the rest. It seems to be grooved, with relatively smooth areas between the hills and valleys. It is called the "hilly and lineated terrain" (Fig. 10–11). No other areas like this are known on Mercury, and just a couple have been found on the Moon. This terrain is 180° around Mercury from the Caloris Basin, the site of a major meteorite impact. Shock waves from that impact apparently were focused halfway around the planet (Fig. 10–12).

The Mariner 10 mission was a navigational coup not only because it used the gravity of Venus to get the spacecraft to Mercury, but also because scientists and engineers were able to find an orbit around the Sun that brought the spacecraft back

FIGURE 10–9 A prominent scarp, Santa Maria Rupes. The existence and commonness of these faults indicate that Mercury compressed by 1 to 2 kilometers after the surface solidified and craters formed. This image is 200 km across. A view from the side would show that one side is higher than the other.

FIGURE 10–10 The fractured and ridged planes of the Caloris Basin. It is 1300 km in diameter and is bounded by mountains 2 km high. It is similar in size and appearance to Mare Imbrium on the Earth's Moon and so resulted from the impact of a body tens of kilometers in diameter. It is named Caloris from the Greek word for "hot" because it is almost directly under the Sun when Mercury is at perihelion. (As we saw, it is actually directly facing the Sun at only alternate perihelions.) This image represents an area about 100 km across.

to Mercury several times over. Every six months Mariner 10 and Mercury returned to the same place at the same time. As long as the gas jets for adjusting and positioning Mariner functioned, it was able to make additional measurements and to send back additional pictures in order to increase the photographic coverage. On its second visit, for example, in September 1974, Mariner 10 was able to study the south pole and the region around it for the first time. This pass was devoted to photographic studies. The spacecraft came within 48,000 km of Mercury, farther away than the 750-km minimum of the first pass, but the data were still very valuable. On its third visit, in March 1975, it had the closest encounter ever—only 300 km above the surface. Thus it was able to photograph part of the surface with a high resolution of only 50 meters. Then the spacecraft ran out of fuel for the small jets that controlled the direction in which it was pointing.

In total, Mariner 10 sent back images of 45 per cent of Mercury's surface, with an average resolution of 1 km. We have little idea what the rest of the surface is like.

## 10.3b Infrared Results

Mariner's infrared radiometer gave data that indicate that the surface of Mercury is covered with fine dust, as is the surface of the Moon, to a depth of at least several centimeters. Astronauts sent to Mercury, though the harsh conditions means that they won't go in the foreseeable future, will leave footprints behind them.

The Mariner 10 mission gave more accurate measurements of the temperature changes across Mercury than had been determined from the Earth. Within a few hundred kilometers of the terminator, the line between Mercury's day and night, the temperature falls from about 525°C to about 150°C (700 K to about 425 K) and then drops even lower—to –185°C (90 K)—farther across into the dark side of the planet.

FIGURE 10–11 "Weird terrain," also known as hilly and lineated terrain. We see the region diametrically across Mercury from the Caloris Basin, Caloris's antipodal point. The shock wave produced by the impact that formed Caloris was reflected and focused here, breaking up the crust. The photograph shows a region about 100 km across.

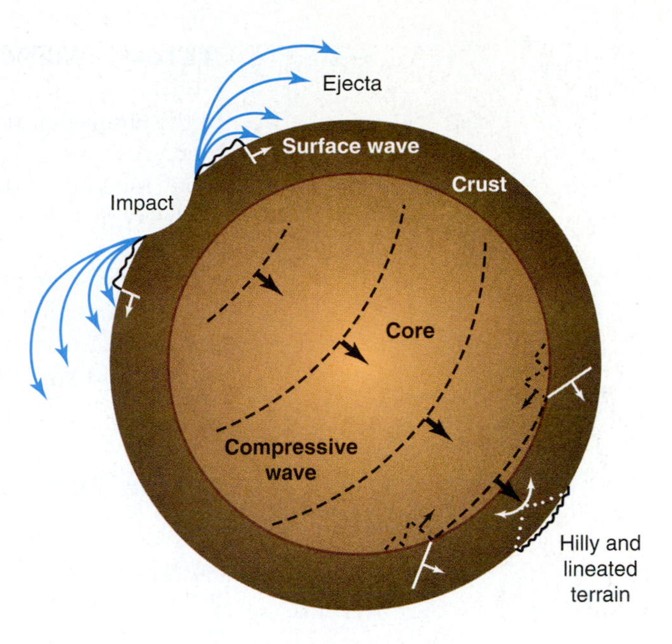


FIGURE 10–12 The formation of the hilly and lineated terrain.

### 10.3c Results from Other Types of Observations

Close-up spectral measurements showed that Mercury even has an atmosphere, although an all-but-negligible one. It is only a few billionths as dense as the Earth's, so slight that even someone standing on Mercury would need special instruments to detect it. Traces of hydrogen, helium, oxygen, and argon were detected with a spectrometer that operated in the ultraviolet. The presence of helium was a particular surprise, because helium is a light element and had been expected to escape from Mercury's weak gravity within a few hours. So it must be constantly replaced. All these gases probably come from the solar wind. Mercury's thin atmosphere, thus, isn't a stable atmosphere like the one we are used to on Earth.

One more surprise—perhaps the biggest of the mission—was that a magnetic field was detected in space near Mercury. It was discovered on Mariner 10's first pass and then confirmed on the third pass. The field is weak; extrapolated down to the surface it is about 1 per cent of the Earth's. It may be that the field is due to an active dynamo (liquid that produces an electromagnetic field using the energy from its own flow through a magnetic field) in a still-molten shell of sulfur-rich iron surrounding the now frozen inner core. The magnetic field can persist only with a partially molten core, and the core would have solidified if it contained less than 0.2 per cent sulfur. This large an abundance of sulfur, a volatile material, is unexpected so close to the Sun, and would indicate that Mercury is composed of material that originally condensed farther from the Sun. An alternate model, that the magnetic field has been frozen into Mercury for billions of years, has not been ruled out, though. In the latter case, Mercury's core could have solidified long ago. We need more close-up observations to decide which model is correct.

Mariner 10 detected lots of electrons near Mercury. Perhaps they are trapped in some sort of belt by the magnetic field, similar to the Van Allen belts around the Earth. But perhaps they are bound to Mercury for shorter times than electrons trapped by the Earth's magnetic field.

## 10.4 MERCURY RESEARCH REJUVENATED

In the absence of return spacecraft visits, progress in interpreting Mercury slowed. Mercury's geologic processes continued to be interpreted, and new thinking arose about Mercury's origin, composition, and thermal evolution. Then the astonishing discovery of sodium in Mercury's thin atmosphere showed that there was lots to do with telescopes on Earth.

Finally, some of the Mariner 10 images are being reprocessed to show features better and to take out the dividing lines in the mosaics. Mark Robinson of Northwestern University is a specialist in this work, and many reprocessed images can be found on his Web site.

### 10.4a Mercury's Atmosphere

The sodium in Mercury's atmosphere (Fig. 10–13) was discovered with the ground-based telescope of the University of Texas, used for daytime observations of the planet at times when it was farthest from the Sun. The spectra of planetary surfaces show mainly the continuum and absorption lines of the solar spectrum reflected to us. Surprisingly, spectra of Mercury showed not only the sodium absorption lines (the yellow "D lines" that make the yellow color you see in many streetlights or when you throw salt into a flame) but also bright, narrow emission lines. The emission lines were shown to be from Mercury itself since they match the changing Doppler shift of Mercury as it orbits the Sun. Some potassium has been discovered in a similar manner. The effects are so strong that we could have found it all along.

Mercury's atmosphere isn't like atmospheres with which we are familiar, like the Earth's stable atmosphere or the more extensive atmospheres of the giant planets. Mercury's high temperature and low gravity means that atmospheric particles continually escape, so what we see has a recent replenishment.

Mercury's atmosphere is not only transitory but also very tenuous. Still, we know that there is more sodium in it than any other element—150,000 atoms per cubic centimeter compared with 4500 of helium and smaller amounts of oxygen, potassium, and hydrogen. The spectrographs aboard Mariner 10 had worked only in the ultraviolet, while the sodium lines are in the visible part of the spectrum. It had been thought that the sodium was ejected into Mercury's atmosphere as the result of impact on Mercury's surface of either solar-wind particles (a process called "sputtering") or, on a larger scale, meteorites—between 6 and 60 tons per day. We shall see in Chapter 13 that Jupiter's moon Io has a cloud of sodium particles near it; these

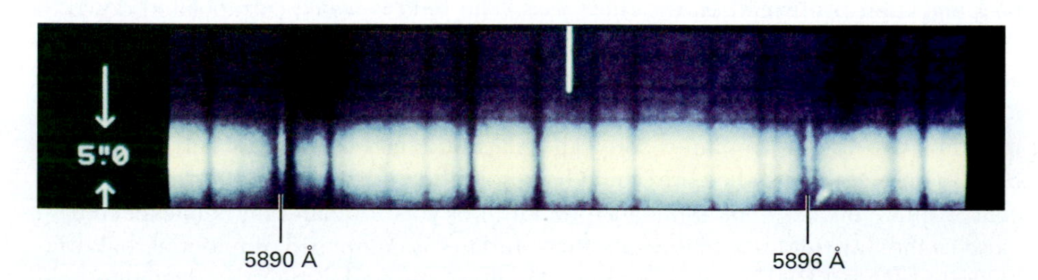


**FIGURE 10–13** A photographic spectrum of the planet Mercury showing an enlargement of the small region of spectrum including the pair of sodium spectral lines known as the D lines. Overall, Mercury's spectrum is that of the Sun, since we see reflected sunlight; thus we see a continuous spectrum with several dark absorption lines. The surprise was that just to the side of each dark absorption line of sodium is a bright emission line marked with the D-line wavelengths of 5890 Å and 5896 Å, respectively. The emission lines show the presence of Mercury's atmosphere.

particles have presumably been ejected from Io by a process similar to the one in force at Mercury, though the particles at Io causing the sputtering would come from Jupiter rather than directly from the solar wind. Calculations show that the sodium particles at Mercury would be continually ejected and so would have to be continually replenished. The situation apparently resembles more the "coma" of a comet (Chapter 19) than a planetary atmosphere like Earth's. Since potassium and sodium are enhanced when Caloris is in view, these elements have likely diffused up through Mercury's crust. As we said previously, the lighter elements found in the atmosphere come from the solar wind, and are constantly being replenished and escaping.

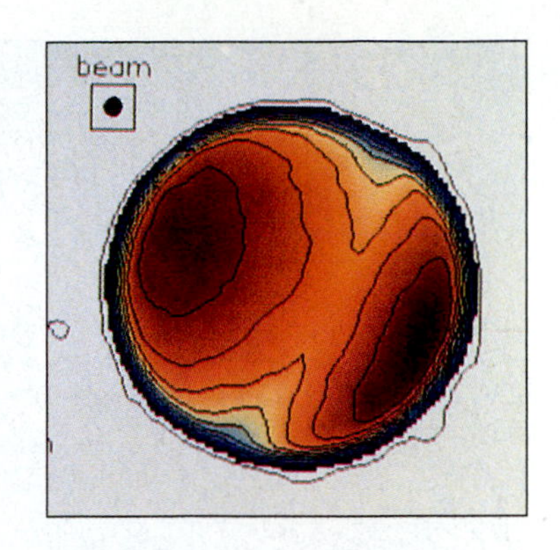


**FIGURE 10–14** Mercury in radio waves, made with the Very Large Array. At the frequency of observation, the surface is translucent, allowing us to study the soil to a depth that depends on frequency. The maps show the temperature distribution. Red is hotter. This map, at a wavelength of 3.6 cm, shows the level 70 cm below the surface.

### 10.4b Ground-Based Radio and Radar Observations

Scientists use the Very Large Array set of radio telescopes to map Mercury's temperature structure slightly under its surface (Fig. 10–14). The insulating properties of Mercury's surface make the temperature constant, as the observations verified.

During the 1970s and 1980s, extensive radar observations were made from the Goldstone and Arecibo antennas on Earth of the band extending 15° north and south of Mercury's equator. Surface resolution was 12 to 20 km around the equator and 50 to 100 km perpendicularly. The radar observations show especially the height structure across the surface, with a resolution of 150 m, and the roughness of the surface. The radar features have been compared with the features observed during the Mariner 10 flyby, which imaged only about half the planet. The craters, with their floors flatter than the Moon's craters, and the scarps show clearly on the radar maps. The radar data that show the side of Mercury not imaged by Mariner 10 reveal that it too is dominated by intercrater plains, though it appears different enough to show that Mercury is basically asymmetric.

### 10.4c Mercury's History

A generally accepted scenario for Mercury's early years holds that a core like that of the Earth developed, and the planet's crust expanded. After the expansion opened paths for magma to flow outward from the interior, the volcanic flooding caused the intercrater plains. As the core cooled, the crust contracted, and the scarps resulted. At about the same time, less extensive volcanic flooding formed the smooth plains. Solar tides on Mercury slowed its original rotation and led to the formation of the large-scale linear features that crisscross perpendicularly.

The sequence is quite different from that of the Moon's early years. The tentative conclusion reached soon after Mariner 10 flew by that Mercury resembled the Moon has now been replaced in the minds of researchers by the idea that Mercury is unique. (Similarly, the first images returned from Mars led researchers to think it looked like the Moon, but later observations and study revealed otherwise.) The distribution of major types of surface features on Mercury (Fig. 10–15) and of most of the tectonic features is very different from that of the Moon. For example, most of the hemisphere of Mercury that we have pictures of is a certain kind of terrain—the intercrater plains. These intercrater plains are intermediate in albedo and roughness; most are of an average albedo and roughness but a fraction are lighter and an equal fraction darker, while a fraction are rougher and an equal fraction relatively smooth. On the Moon, the bright, rugged lunar highlands dominate. Many tectonic features—features resulting from deformations of the planet's crust—have been detected in Mercury's smooth plains; many fewer such features are visible in the analogous lunar maria. (The Moon has wrinkle ridges and rilles, but they seem more local in origin and extent. There is no comparable scarp system.)

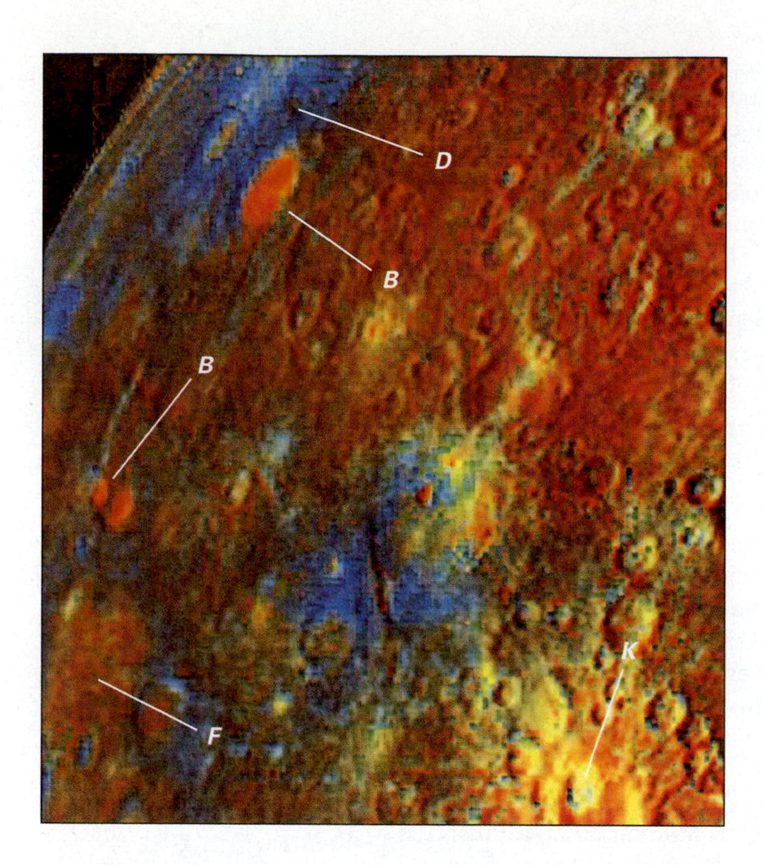


**FIGURE 10–15** An enhanced, false-color map of part of Mercury, observed from the Mariner 10 mission. Colors show regions in which different minerals are dominant. *K* is the crater Kuiper, which may be showing material excavated from an underground unit of unusual composition; *D* is a unit that seems to have its titanium content enhanced; *B* marks regions that may represent primitive crustal material; and *F* marks material that follows the boundaries of plains and that is therefore interpreted to be lava flows.

The dominant view is that the oldest features on Mercury are 4.2 billion years old. The Caloris Basin formed from a giant impact, and the scarps formed as Mercury cooled and shrank. The smooth plains then formed by 3.8 billion years ago as lava erupted from Caloris and other large impact basins.

Radar observations from Earth in 1991 revealed strange echoes that may come from an ice cap (Fig. 10–16). The radar penetrates perhaps 1 or 2 meters; we can-

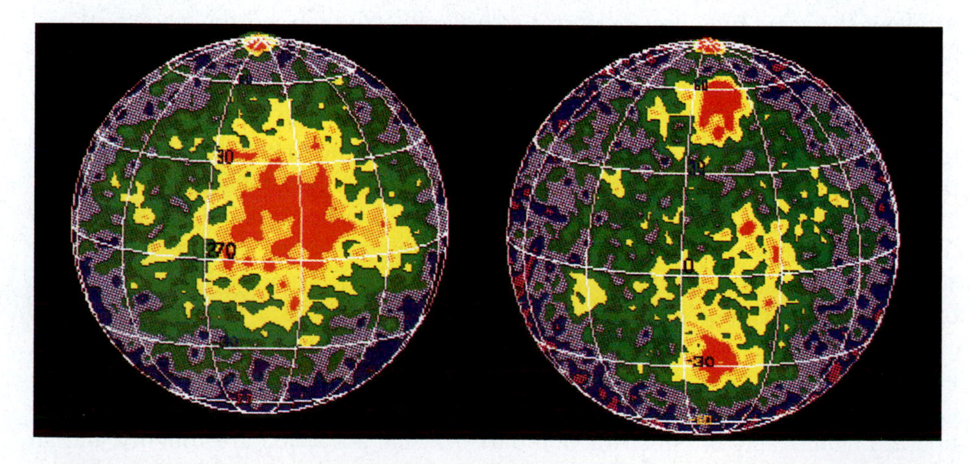

**FIGURE 10–16** On each of these two radar images of Mercury, the brightest echo is precisely at the north pole. Further, the polarization of the radio radiation being returned is unique. Scientists have concluded that the reflecting medium is ice, and probably water ice. It would lie just under the surface. The two newly discovered "basins" in the right-hand image have different polarization properties. They lie at lower (warmer) latitudes, where ice would have less chance of persisting. On these radar maps, made at a wavelength of 3.5 cm with the Very Large Array (VLA) and Goldstone antennas, the left image shows the entire hemisphere not seen by Mariner 10 and the right image is 90° farther around in longitude.

not tell whether 10 cm or more of dirt is shielding the ice. Since Mercury's axis of rotation is not inclined to its orbit, some regions at the poles could be perpetually shielded from sunlight and very cold. We can see them because the tilt of the Earth's orbit relative to Mercury's takes us high enough to look on at an 11° angle. We may also see ice-filled craters (Fig. 10–17).

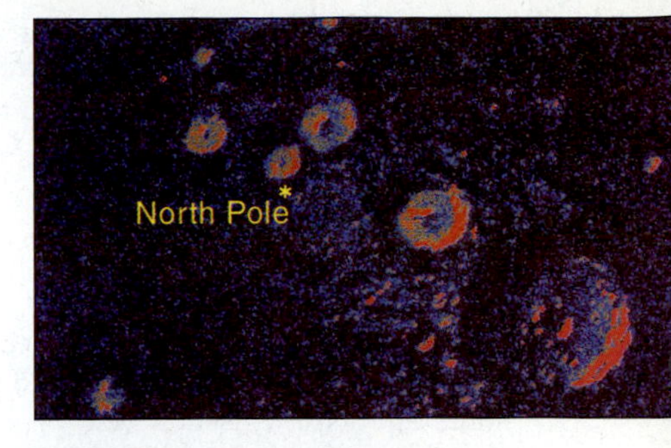


**FIGURE 10–17** Mercury's north pole, imaged in 1999 using the upgraded Arecibo radar. The image measures 450 km on a side; its resolution is 1.5 km, a factor of 10 better than the pre-upgrade radar. The bright features are most likely water ice deposits located in the permanently shaded floors of craters. These regions have been shielded from sunlight for so long that ice may have persisted.

### 10.4d Ground-Based Imaging

Even through Earth's turbulent atmosphere, images of Mercury have been obtained that show surface features. Thousands of short (1/60-second) exposures were taken over an hour and a half, and then the 50 or so best were selected and added together (Fig. 10–18). Some broad detail is visible even in areas that were not imaged from Mariner 10.

**FIGURE 10–18** The best visible-light image of Mercury ever taken from Earth, released in 2000. It is a combination of several dozen very short exposures, the result of an effort to find moments when the atmospheric seeing was best.

## 10.5 MORE MERCURY MISSIONS AT LAST

NASA's *ME*rcury *S*urface *S*pace *EN*vironment, *GE*ochemistry and *R*anging (or MESSENGER) spacecraft is being planned for launch in 2004 to take advantage of the best gravity-assist alignments, which allow a spacecraft to reach its objective either quicker or with less expenditure of fuel. After gravity assists from flying close by Earth, Venus, and Mercury, it would go into orbit around Mercury in 2009. From a polar orbit, the spacecraft will map the surface and the magnetic field for about a year. Observations of Mercury's crust should tell us its geologic history and reveal whether volcanism has occurred. The question of whether polar craters are filled with ice should be answered definitively. Also, the source of Mercury's tenuous atmosphere will be better understood. Studies of whether Mercury's rotation rate varies slightly should reveal if Mercury has any liquid in its core, presumably its outer core. Though Mercury's magnetic field is about 1 per cent the strength of Earth's, that is stronger than was expected on the basis of its slow rotation, even with a partially liquid core. So the magnetic-field studies should give further insights into the nature of Mercury's core. Theory has it that such magnetic fields are usually generated by a dynamo involving the continual rise and fall of molten material. But we could be seeing a remanent field, frozen in place for eons and not now being generated.

The European Space Agency is planning a major mission to Mercury to be launched in 2009, in collaboration with the Japanese Space Agency. It will be known as Bepi Colombo, after the Italian scientist who worked out how to use gravity assists to send spacecraft to planets. It is to use a new system of solar propulsion, ejecting ions of xenon at very high speed, allowing a travel time to Mercury of less than 4 years. Bepi Colombo will have a lander and an orbiter that will study Mercury's surface and magnetosphere. Within a decade we may have photographs taken by the lander on Mercury's surface as well as direct measurements of the composition of the surface rocks. A year later, the European Space Agency plans to launch a Mercury Planetary Orbiter, which will go into a polar orbit, mapping features as small as 10 m across.

### CORRECTING MISCONCEPTIONS

| ✖ *Incorrect* | ✔ *Correct* |
|---|---|
| Mercury is all very hot. | Some parts of Mercury are permanently shaded and are cold. Water ice may even be present. |
| We know a lot about Mercury. | We don't even know what half of it looks like. |

## SUMMARY AND OUTLINE

Mercury is difficult to observe from the Earth; it is never far from the Sun in the sky.
Radio astronomy (Section 10.1)
- Surface temperatures
- Rotation period linked to orbit: 1 day to 2 years

Albedo (Section 10.2)
- Low albedo means a poor reflector.
- Mercury has a low albedo, only 6 per cent.

Mass and density (Section 10.2)
- Mass is 5½ per cent that of Earth.
- Density is similar to Earth's.

Mariner 10 observations (Section 10.3)
- Photographic results
- Types of objects: craters, maria, scarps, hilly and lineated terrain
- Mechanisms: impacts, volcanism, shrinkage of the crust
- Results from infrared observations
  - Dust on the surface
  - Temperature measurements
- Results from other types of observations
  - A small atmosphere
    - Where does the helium come from?
- A big surprise: a magnetic field, which tells us about the histories of Mercury and of the Earth

New ground-based observations (Section 10.4)
- Sodium is the major constituent of Mercury's atmosphere.
- Radar studies show that Mercury is very different from the Moon.
- Radar shows Mercury may have an ice cap and ice-filled craters.

MESSENGER mission planned for arrival at Mercury in 2008 (Section 10.5)

## KEY WORDS

synchronous rotation
radar
gravitational interlock
tidal lock
sidereal rotation period
solar rotation period
albedo
scarps

## QUESTIONS

1. Assume that on a given day, Mercury sets after the Sun. Draw a diagram, or a few diagrams, to show that the height of Mercury above the horizon depends on the angle that the Sun's path in the sky makes with the horizon as the Sun sets. Discuss how this depends on the latitude or longitude of the observer.
2. If Mercury did always keep the same side toward the Sun, does that mean that the night side would always face the same stars? Draw a diagram to illustrate your answer.
3. Explain why a day/night cycle on Mercury is 176 Earth day/night cycles long.
4. What did radar tell us about Mercury? How did it do so?
5. Given Mercury's measured albedo, if about 1 erg hits a square meter of Mercury's surface per second, how much energy is reflected from that area? (An erg is a unit of energy.)
6. If ice has an albedo of 70 to 80 per cent and basalt has an albedo of 5 to 20 per cent, what can you say about the surface of Mercury based on its measured albedo?

7. If you increased the albedo of Mercury, would its temperature increase or decrease? Explain.
8. List those properties of Mercury that could be better measured by spacecraft observations than by Earth-based observations.
9. How would you distinguish an old crater from a new one?
10. What evidence is there for erosion on Mercury? Does this mean there must have been water on the surface?
11. List three major findings of Mariner 10.

## DISCUSSION QUESTION

1. Think of how you would design a system that transmits still photographs taken by a spacecraft back to Earth and then produces the picture. The system can be simpler than your TV, since still pictures are involved.

## USING TECHNOLOGY

### W World Wide Web

1. Examine the images of Mercury visible on The Nine Planets and also from the USGS, and contrast them with the images published in this book.
2. Check on the status and intentions of the MESSENGER mission: http://sd-www.jhuapl.edu/MESSENGER

### *REDSHIFT*

1. Is Mercury in the morning sky or the evening sky tonight? When would it best be visible?
2. Travel around Mercury using the main *RedShift* engine.
3. Explore the surface of Mercury, locating the features listed previously.

A global view of the surface of Venus as imaged with radar from NASA's Magellan spacecraft. Increasing height is shown by color changing from blue to red. The high volcano Maxwell appears near the top. The symbol for Venus appears at the top of the opposite page.

# Chapter 11

# Venus

**AIMS:** To discuss Venus's atmosphere and surface and to use this knowledge to improve our understanding of the Earth's structure and atmosphere

Venus and the Earth are sister planets: their sizes, masses, and densities are about the same. Both presumably formed from the same kinds of planetesimals, so they have similar overall compositions. But they are as different from each other as the wicked sisters were from Cinderella. The Earth is lush: it has oceans and rainstorms of water; an atmosphere containing oxygen; and creatures swimming in the sea, flying in the air, and walking on the ground. On the other hand, Venus is a hot, dry, forbidding planet with temperatures constantly over 475°C (900°F), a planet so unpleasant that it gives a sense of foreboding. Conditions are so harsh that it seems unlikely that life could have developed there. Why is Venus like that? How did these harsh conditions come about? Can it happen to us here on Earth?

Venus orbits the Sun at a distance of 0.7 A.U. Although it comes closer to us than any other planet—as close as 45 million kilometers—we cannot see its surface from Earth because it is always shrouded in heavy clouds (Fig. 11–1). Pre-spacecraft observers in the past saw faint hints of structure in the clouds, which seemed to indicate that these clouds might circle the planet in about 4 days, rotating in the opposite sense from Venus's orbital revolution. The clouds, though, never part to allow us to see the surface.

Radar penetrates the clouds and over the last decades has enabled astronomers to make very detailed maps of Venus's surface.

A

B

**FIGURE 11–1** (*A*) Venus is the bright object over the Earth's atmosphere in this photograph from a space shuttle. (*B*) A crescent Venus observed with the 2.5-m Mt. Wilson telescope. We see only a layer of clouds.

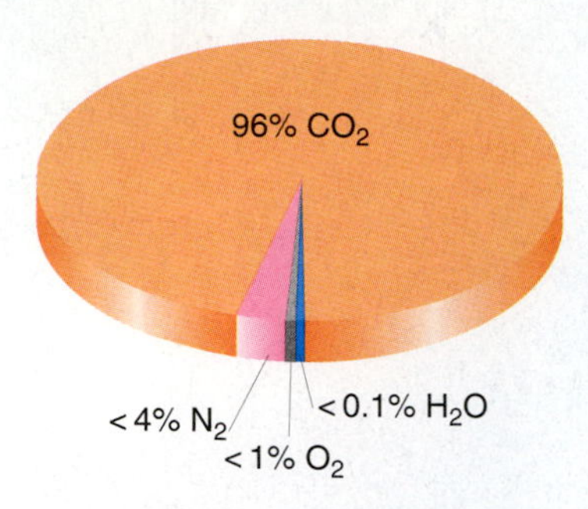


**FIGURE 11–2** The composition of Venus's atmosphere.

> Astronomers often use the adjective **cytherean** to describe Venus. Cythera was the island home of Venus, the Roman goddess of love. The adjectives "venusian" and "venerean" are now in increasing use.

## 11.1 THE ATMOSPHERE OF VENUS

Spectra of Venus taken with ground-based telescopes on Earth show a high concentration of carbon dioxide in the atmosphere of Venus. In fact, carbon dioxide makes up 96 per cent of Venus's atmosphere; nitrogen makes up almost all the rest (Fig. 11–2). The Earth's atmosphere, by comparison, is mainly nitrogen, with a fair amount of oxygen as well. Carbon dioxide makes up less than 0.1 per cent of the terrestrial atmosphere.

The surface pressure of Venus's atmosphere is 90 times higher than the pressure of Earth's atmosphere, as a result of the large amount of carbon dioxide in the former. Carbon dioxide on Earth mixes with rain to dissolve rocks. The dissolved rock and carbon dioxide eventually flow into the oceans, where they precipitate to form some terrestrial rocks, often with the help of life forms. (Many limestones, for example, formed from marine life under the Earth's oceans.) If this carbon dioxide were released from the Earth's rocks, along with other carbon dioxide trapped in sea water, our atmosphere would become as dense and have as high a pressure as that of Venus. Venus, slightly closer to the Sun than Earth and thus hotter, had no oceans in which the carbon dioxide could dissolve nor life to help take up the carbon. Thus the carbon dioxide remains in Venus's atmosphere.

Also, Venus has probably lost almost all the water it ever had. Since Venus is closer to the Sun than the Earth is, its lower atmosphere was hotter even early on. The result was that more water vapor went into its upper atmosphere, where solar ultraviolet radiation broke it up into hydrogen and oxygen. The hydrogen, a light gas, escaped easily; the oxygen has combined with other gases or with iron on Venus's surface.

Studies from the Earth show that the clouds on Venus are primarily composed of droplets of sulfuric acid, $H_2SO_4$, with a relatively small quantity of water in the form of droplets mixed in. Sulfuric acid may sound strange as a cloud constituent, but the Earth too has a significant layer of sulfuric acid droplets in its stratosphere. However, the water in the lower layers of the Earth's atmosphere, circulating because of weather, washes the sulfur compounds out of these layers, whereas Venus has sulfur compounds in the lower layers of its atmosphere in addition to those in its clouds. (On Earth, we are troubled by such "acid rain," which affects many of our lakes and kills life in them.)

Could life have arisen on Venus in an early, hot ocean? Even if it had, it would have been destroyed when the oceans disappeared and Venus heated up so.

## 11.2 TRANSITS OF VENUS

In Section 7.4, we saw that transits of Venus across the Sun's disk are very rare, and that none has occurred since the pair in 1874 and 1882 (Figure 11–3). But we have waited long enough: the transit of 2004 won't be long now. Venus will take 6 hours to pass in front of the Sun. Historically, transits of Venus were important for determining the absolute scale of the Solar System, though now we can do that with radar or by simply sending spacecraft. (Before then, we had only the relative distances to the planets, from Kepler's third law.) Still, the June 8, 2004, and the June 5–6, 2012, events will be intellectually worth watching. Since Venus is both larger than Mercury and, at transit, closer to the Earth, Venus's silhouette appears larger than Mercury's (which we saw in Figure 10–2) and therefore easier to see (through special filters, of course, since we are looking toward the Sun).

The atmosphere of Venus wound up confounding the many observers at the transits of 1761 and 1769, when many expeditions tried to determine the scale of the Solar System. The method depended on accurately measuring the times when Venus went onto the solar disk and when it exited. But Venus turned out not to appear simply as a black spot. When it entered the disk, a dark join appeared between Venus and the Sun's limb. This "black drop effect" made the timing inaccurate. Similarly, a black drop distorted the timing of Venus's exit. Still, the size of the Solar System was measured more accurately than ever before. And Captain Cook, after he observed the 1769 transit in Tahiti, found and mapped New Zealand, which we can therefore consider a spin-off of astronomy and of transits of Venus.

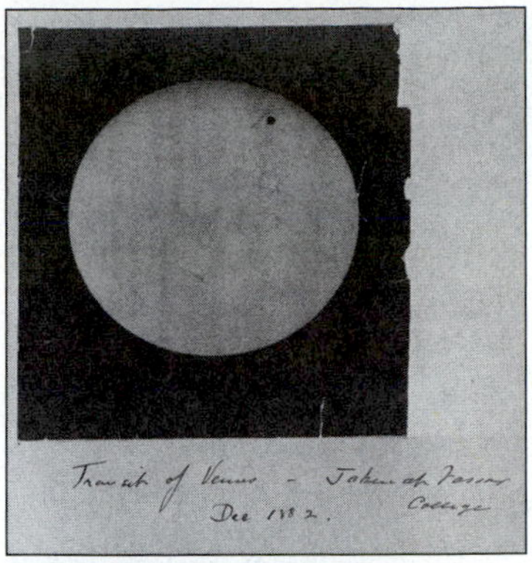

**FIGURE 11–3** The transit of Venus of 1882. It was photographed at Vassar College by Prof. Maria Mitchell and her students. Mitchell was one of the most famous women in science in the 19th century, following her discovery of a comet for which she won a gold medal from the King of Denmark.

## 11.3 THE ROTATION OF VENUS

In 1961, radar—radio waves sent out and the echo received—penetrated Venus's clouds, allowing us to determine an accurate rotation period for the planet's surface. Venus, because it comes closer to Earth, is an easier target for radar than Mercury (Section 10.1). Venus rotates in 243 days with respect to the stars in the direction opposite its orbital revolution around the Sun; such backward motion is called **retrograde rotation,** to distinguish it from forward (direct) rotation. Venus revolves around the Sun in 225 Earth days. Venus's periods of rotation and revolution combine, in a way similar to that in which Mercury's sidereal day and year combine, so that a solar day on Venus corresponds to 117 Earth days; that is, the planet's rotation brings the Sun back to the same position in the sky every 117 days.

The notion that Venus is in retrograde rotation seems very strange to astronomers, since all the known planets revolve around the Sun in the same direction, and most of the planets (Uranus and Pluto are other exceptions) and most satellites also rotate in that same direction. Because of the conservation of angular momentum (Section 7.4), and because the original material from which the planets coalesced was undoubtedly rotating, we expect all the planets to revolve and rotate in the same sense.

Nobody knows definitely why Venus rotates "the wrong way." One possibility is that when Venus was forming, the planetesimals formed clumps of different sizes. Perhaps the second largest clump struck the largest clump at an angle that caused the result to rotate backwards. Scientists do not like *ad hoc* ("for this special purpose") explanations like this one, since ad hoc explanations often do not apply in general. But we have not been able to think of any better explanation in this case.

## 11.4 THE TEMPERATURE OF VENUS

Though we can't see through Venus's clouds with our eyes, radio waves emitted by the surface penetrate the clouds. The intensity of radiation of a surface depends on its temperature, so long before we landed spacecraft on Venus scientists deduced the temperature of Venus's surface by studying Venus's radio emission. The surface is very hot, about 750 K (900°F). Infrared radiation penetrates outward through the clouds somewhat, though looking in the infrared does not allow us to see all the way to Venus's surface (Fig. 11–4).

In addition to measuring the temperature on Venus, scientists theoretically calculate approximately what it would be if Venus's atmosphere allowed all radiation coming back out from the planet's surface to pass through it. This value—less than 375 K (215°F)—is much lower than the measured values. The high temperatures derived from radio measurements indicate that Venus traps much of the solar energy that hits it.

The process by which this happens on Venus is similar to the process that is generally—though incorrectly—thought to occur in greenhouses here on the Earth. The

Now that spacecraft have enabled us to understand the runaway greenhouse effect on Venus, we can much better understand the effect of adding carbon dioxide from burning fossil fuels to Earth's climate. We have already increased the amount of carbon dioxide in Earth's atmosphere by 30 per cent; some scientists have predicted that it may even double in the next 50 years, which could result in a worldwide temperature rise of 2°C. The effects on people, climate, land use, agriculture, and so on, would be major (Sections 8.5 and 11.4).

**FIGURE 11–4** A false-color, infrared image of Venus, taken with an infrared-array camera at wavelengths that penetrate the upper visible cloud level of Venus's atmosphere. We see a crescent of reflected sunlight and thermal emission from deep in the atmosphere or from the surface. The dark features probably arise within the middle cloud level of Venus, 50 km above the surface and 20 km beneath the visible clouds. They rotate in the retrograde direction in about six days, which implies that the wind is blowing at 70 m/sec. These infrared features are lower and rotate more slowly than do the ultraviolet clouds. A camera on the Galileo spacecraft also imaged Venus in the infrared. Several wavelengths penetrated nearly to the surface. Results show that the lowlands are about 100°C hotter than the highlands.

process is thus called the **greenhouse effect** (Fig. 11–5); it was suggested for Venus by Carl Sagan, later of Cornell, and James B. Pollack of NASA/Ames.

Sunlight diffuses through Venus's atmosphere in the form of radiation in the visible part of the spectrum. The sunlight is absorbed by, and so heats up, the surface of Venus. At the temperatures that result, the radiation that the surface gives off is mostly in the infrared. But the carbon dioxide and other constituents of Venus's atmosphere are together opaque to infrared radiation, so the energy is trapped. Thus Venus heats up above the temperature it would have if the atmosphere were transparent; the surface radiates more and more energy as the planet heats up until a balance is struck between the rate at which energy flows in from the Sun and the rate at which it trickles out (as infrared) through the atmosphere (Fig. 11–6). The situation is so extreme on Venus that we say a "runaway greenhouse effect" is taking place there. ("Runaway" implies that the process is out of control and could continue indefinitely.) Understanding such processes involving the transfer of energy is but one of the practical results of the study of astronomy.

Greenhouses on Earth don't work quite this way. The closed glass of greenhouses on Earth prevents convection and the mixing in of cold outside air. The trapping of solar energy by the "greenhouse effect" is a less important process in an actual greenhouse. Try not to be bothered by the fact that the greenhouse in your backyard is not heated by the "greenhouse effect."

The Earth's atmosphere is now gaining carbon dioxide, and it is predicted that global warming of 2°C will occur over the next century (Section 8.5). We are currently increasing the amount of carbon dioxide in the Earth's atmosphere by burning fossil fuels; studies of Venus and its greenhouse effect have shown us how important it is to be careful about the consequences of our energy use. Even a rise in Earth's temperature of 2°C could have catastrophic consequences on agriculture, sea level (leading to coastal flooding), and climate. Bands for agriculture and for disease-carrying insects would change. We must very soon limit use worldwide of coal, oil, and gas to prevent disastrous consequences from this global warming. Renewable energy sources like solar power, wind power, and geothermal energy or methods that don't emit carbon dioxide, principally nuclear power (which comes from nuclear fission processes, in which elements like uranium are broken into smaller parts), are

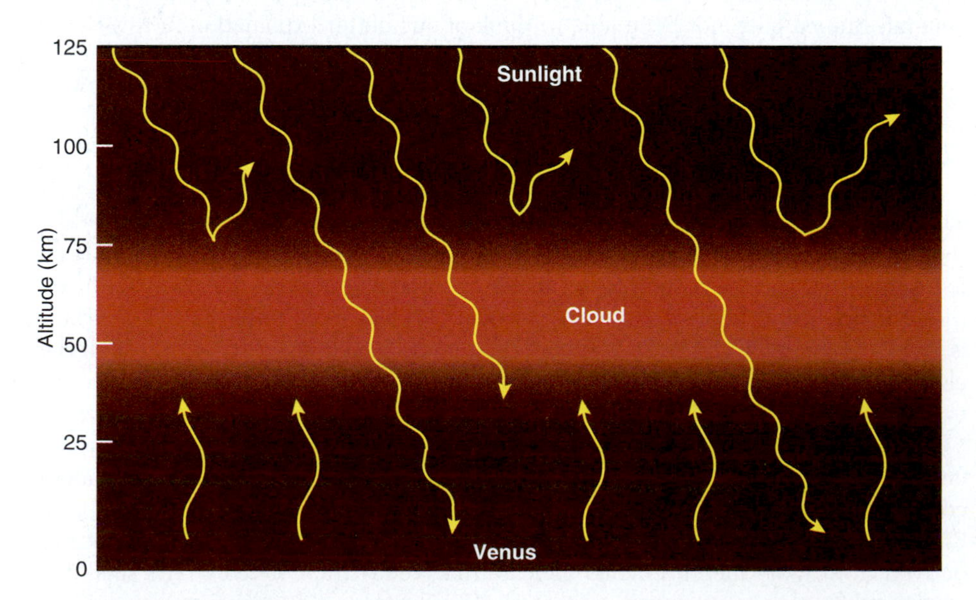


**FIGURE 11–5** Sunlight can penetrate Venus's clouds, so the surface is illuminated with radiation in the visible part of the spectrum. Venus's own radiation is mostly in the infrared. This infrared radiation is trapped, causing a phenomenon called the **greenhouse effect.**

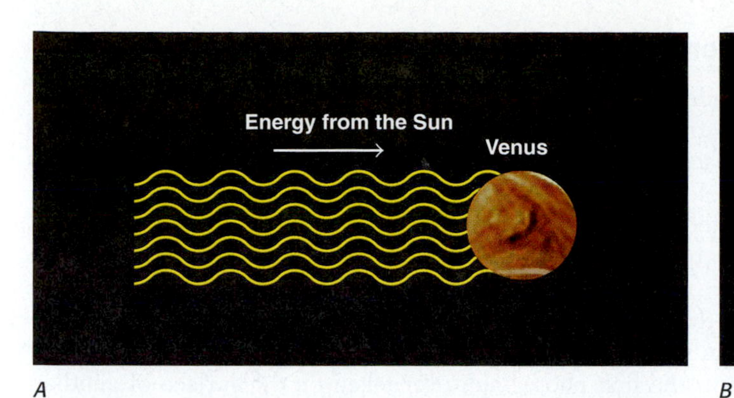


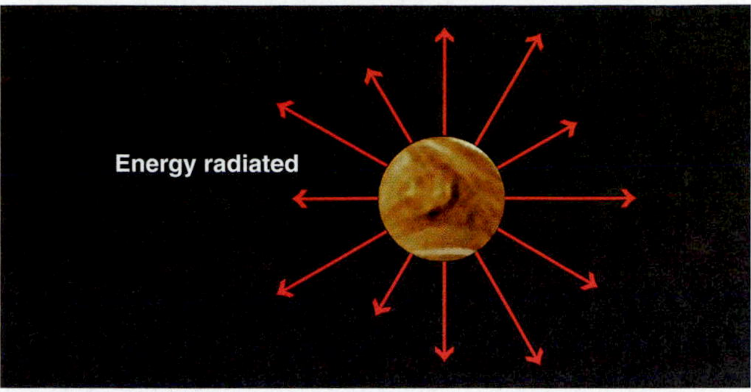


A B

**FIGURE 11–6** Venus receives energy from only one direction, but it heats up and radiates energy (mostly in the infrared) in all directions. From balancing the energy input and output, astronomers can calculate what Venus's temperature would be if Venus had no atmosphere. This type of calculation—balancing quantities theoretically to make a prediction to be compared with observations—is typical of those made by astronomers.

available. We hope that power from nuclear fusion (in which light elements like hydrogen or its deuterium isotope are fused, becoming helium) may become available in the next half century or so.

## 11.5 THE SURFACE OF VENUS

We can study the surface of Venus by using radar to penetrate Venus's clouds. Scientists have used radar with huge Earth-based radio telescopes, such as the giant 1000-ft (304-m) dish at Arecibo, Puerto Rico, to map as much as a quarter of Venus's surface (the part that faces Earth when Venus comes closest) with a resolution of up to 1.5 to 2 km. Regions that reflect a large percentage of the radar beam back at Earth show up as bright, and other regions as dark. But radars on Earth, though they gave us our earliest views of Venus's surfaces, cannot match the global mapping and amount of detail from radars sent on spacecraft orbiting Venus.

Still, Earth-based radars told us many important things about Venus's surface. (None of the radars tell us about the clouds, since the radio waves penetrate them.) For example, a bright area near a large plain was called Maxwell; we will see later on that Maxwell turns out to be a huge mountain (Fig. 11–7). (The 19th-century Scottish scientist James Clerk Maxwell, who ranks with Newton and Einstein as one of the three greatest physicists of all time, discovered the laws that link electricity and magnetism. Every time we use a radio or television, or a computer, we can thank Maxwell for his work. Maxwell's work is also at the basis of radar, with which his mountain was discovered.)

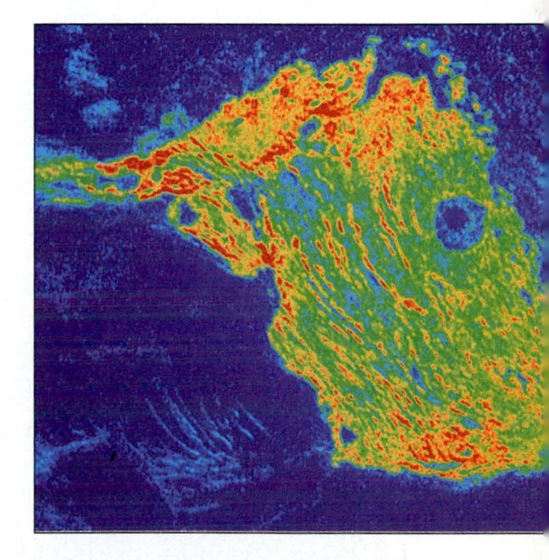

**FIGURE 11–7** The Arecibo Observatory's high-resolution radar image of Maxwell Montes, with a resolution of about 2 km. The summit caldera (a crater caused by the collapse of a volcanic cone) is named Cleopatra. Brighter areas show regions where more power bounces back to us from Venus; an increase in reflected power usually corresponds to rougher terrain than the lesser power from darker areas.

## 11.6 SPACE STUDIES OF VENUS

Venus was an early and frequent target of both American and Soviet space missions. We can divide the results into four categories: (1) observations from the surface of Venus, sent back by landers; (2) observations of the clouds of Venus, from spacecraft outside them and, later, penetrating them; (3) observations of the topography and radio reflectance properties of Venus's surface, from spacecraft bearing radar; and (4) observations of the gravitational field.

### 11.6a Venus, from Landers on Its Surface

The problem with sending a lander to Venus is that it not only must get there, but also must survive the extreme temperatures and pressures at Venus's surface. In 1970, the Soviet Venera 7 spacecraft radioed 23 minutes of data (but no photos) back from the surface of Venus before it succumbed. Two years later, the lander from the Soviet Venera 8 survived on the surface of Venus for 50 minutes. Both missions confirmed the Earth-based results of high temperatures, high pressures, and high carbon dioxide content.

Soviet spacecraft that landed on Venus in 1975 survived long enough to send back photographs. The single photograph that the Venera 9 lander took in the 53 minutes it survived was the first photograph ever taken on the surface of another planet. It showed a clear image of sharp-edged, angular rocks. This came as a surprise to some scientists, who had thought erosion would be rapid in Venus's dense atmosphere and that the rocks should therefore have become smooth or disintegrated into sand. Three days later, the Venera 10 lander sent data, including photographs for 65 minutes. The rocks at its site were not sharp; they resembled huge pancakes, and between them were sections of cooled lava or debris of weathered rock.

The Veneras measured the surface wind velocity to be only 1 to 4 km/hr. The low wind speed makes it seem likely that erosion is caused not by sand blasting but by temperature changes, chemical changes (which can be very efficient at the high temperature of Venus), and other mechanisms.

Both American and Soviet scientists, interpreting radio signals they detected, reported that lightning occurred at a frequent rate, still a controversial result. The Soviet landers detected up to 25 pulses of energy per second between 5 and 11 km above the surface, later confirmed by the Pioneer Venus Orbiter. The lightning appears to be concentrated over Beta Regio and the eastern part of Aphrodite Terra, two regions of Venus we will later see look like volcanoes. But no signs of active volcanism have been found, even by the latest high-resolution imagery that we will later discuss. Much later, in 1998 and 1999, the Cassini spacecraft flew by Venus en route to Saturn. It did not detect the types of radio signals typical of lightning. We know the radio receivers worked because they detected such signals when Cassini flew by Earth. The Cassini scientists concluded that if lightning exists in Venus's atmosphere, it is either very rare or very different from lightning in Earth's atmosphere.

The remaining Soviet landers lasted between one and slightly over two hours on Venus's surface. The Soviet Veneras 11 and 12 measured the chemical composition of the atmosphere as they landed. They didn't send back photos because their lens caps wouldn't come off! Veneras 13 and 14 carried cameras that showed a variety of sizes and shapes of rocks (Fig. 11–8). They showed that the light that filters through the clouds makes Venus's sky a dull orange. The landing sites were chosen in consultation with American scientists on the basis of Pioneer Venus data, representing the first example of such Soviet–American cooperation on planetary exploration.

The landers carried devices that measured the composition of the surface of the planet. Samples were drilled and brought into a low-pressure container at an Earth-like temperature. X-rays emitted when the samples were irradiated (by radioactive sources) revealed which elements were present. (A "fluorescent" process occurs when radiation at one wavelength is given off in response to another type or wavelength of radiation, so the devices are known as x-ray fluorescence spectrometers. Note that "fluorescent," which has the same root as the word "influence," is spelled with a "uo," and not an "ou" as in "flour" that is used for bread.) The gamma-ray spectrometers on earlier Venera spacecraft had been able to measure only naturally radioactive isotopes of uranium, thorium, and potassium. The chemical makeup of the regions of Venus where the new spacecraft landed is similar to that of basalt, a volcanic rock. Since they had landed in regions near Beta Regio that had the overall topography of a volcanic region, the result indicates that the same chemical processes are at work on both Venus and the Earth.

**FIGURE 11–8** (*A*) The view from Venera 13, showing a variety of sizes and textures of rocks. The spacecraft landed in the foothills of a mountainous region and survived for 127 minutes. (*B*) The view from Venera 14, showing flat structures without the smaller rocks of the other site. The flat structures may be large flat rocks or may be fine material held together. Venera 13 found basalts more typical of the Earth's continental crust and the Moon's highlands. Venera 14 found basalts typical of the Earth's ocean floor and the Moon's maria. The cameras first looked off to the side, then scanned downward as though looking at its feet, and then scanned up to the other side. As a result, the horizons appear as slanted boundaries at upper left and right. Note the lens caps; the one from Venera 14 rolled until it blocked a soil probe.

### 11.6b Venus's Clouds from Above

Studies of Venus's atmosphere have great practical value. The better we understand the interaction of solar heating, planetary rotation, and chemical composition in setting up an atmospheric circulation, the better we will understand our Earth's atmosphere. We then may be better able to predict the weather and discover jet routes that would aid air travel, for example. The potential financial return from this knowledge is enormous: it would be many times the investment we have made in planetary exploration.

The United States spacecraft Mariner 10 took thousands of photographs of Venus in 1974 as it passed by en route to Mercury. It went sufficiently close and its imaging optics were good enough that it was able to study the structure in the clouds with resolution as fine as 100 meters (Fig. 11–9). We could observe much finer details from the spacecraft than from Earth. The cloud structure shows only when we view it in ultraviolet light; the clouds look bland in the visible part of the spectrum.

In the ultraviolet, the clouds appear as long, delicate streaks, like terrestrial cirrus clouds. In Venus's tropics the clouds also show a mottling, which suggests that convection, the boiling phenomenon, is going on.

The United States and the Soviet Union both sent spacecraft to Venus in 1978. NASA's Pioneer Venus 1 was the first spacecraft to go into orbit around Venus; it survived until it entered the Venus atmosphere in 1992. Its elongated elliptical orbit sometimes brought the spacecraft as low as 150 km in altitude; it extended higher than 65,000 km. Its dozen experiments included cameras to study Venus's weather by photographing the planet regularly in ultraviolet light.

**FIGURE 11–9** The circulation of the clouds of Venus, photographed from Mariner 10 in the ultraviolet near 3550 Å. The contrast has been electronically enhanced. The clouds on Venus are produced by reflections from small particles of unknown composition that are embedded within a thick layer of sulfuric acid particles. They are thus different from terrestrial clouds, in spite of their similar appearance.

Venus's wind patterns undergo long-term patterns of change. The jet-stream pattern seen at mid-latitudes when Mariner 10 flew by in 1974 had given way to a pattern of cloud and wind acting like a solid body by the time the Pioneer Venus Orbiter arrived in 1979. Still further changes in the atmosphere occurred by the time NASA's Galileo spacecraft flew by in 1990 en route to Jupiter (Fig. 11–10). The upper atmosphere moves westward around the planet at tremendous speed, circling

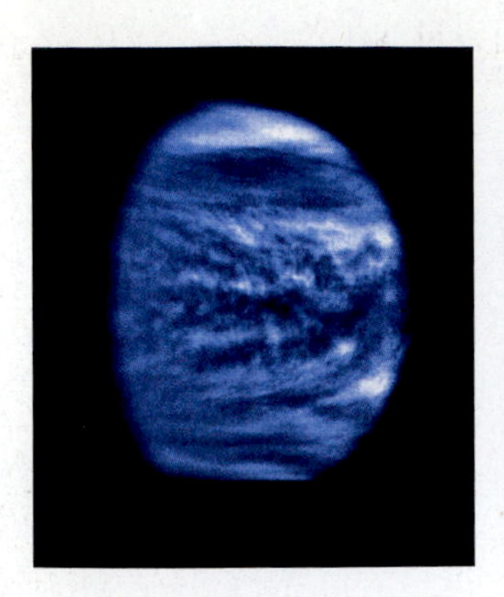
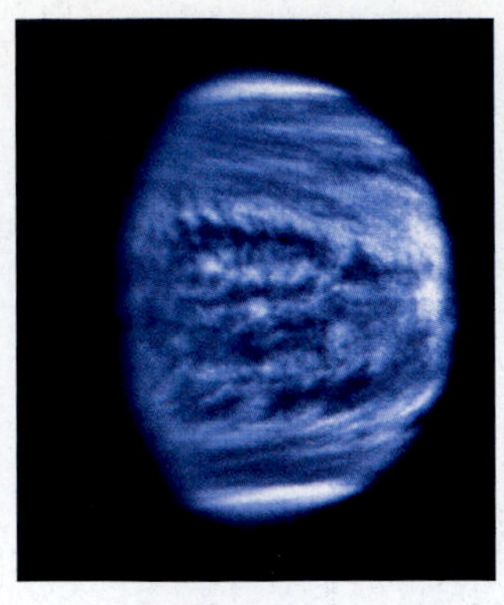

**FIGURE 11–10** Changes in Venus's clouds over several hours, observed by the Galileo spacecraft when it flew by in 1990.

the planet every 4 days. A high-altitude haze layer (over the main cloud layer) appears and disappears over a period of years. It contains tiny droplets of sulfuric acid.

The Pioneer Orbiter also provided evidence that volcanoes are active on Venus, still a controversial conclusion. The abundance of sulfur dioxide it found on arrival was far higher than the upper limits of previous observations, but it then dropped by a factor of 10 in the next five years. Over the same five-year period, the haze dropped drastically, as it had about 20 years earlier. Some have speculated that the effect comes from huge eruptions that pour sulfur dioxide and other gases into the atmosphere. The spacecraft lasted in orbit around Venus until 1992.

Pioneer Venus 2 arrived at Venus at about the same time as Pioneer Venus 1 and carried probes. Before they crashed, the probes found that high-speed winds at the upper levels are coupled to other high-speed winds at lower altitudes. The lowest part of Venus's atmosphere, however, is relatively stagnant. Winds in this lower half of the atmosphere do not exceed about 18 km/hr. This lower half is much denser than the upper half, however, so the amount of momentum in the lower- and upper-level winds may be the same.

The probes verified that carbon dioxide makes up 96 per cent of the atmosphere, as we had already known from ground-based observations. They found less water vapor but more forms of sulfur than had been expected in Venus's lower atmosphere. Knowledge of these gases is particularly important since the presence of carbon dioxide alone is not sufficient to cause the greenhouse effect. Carbon dioxide's spectrum has gaps in its blockage of infrared that would let out too much energy at some infrared wavelengths. Thus Venus's carbon dioxide absorbs only about half the infrared radiated from the ground. But comparison of computer models with the Pioneer Venus results tells us that the gaps are plugged by water vapor, sulfuric acid droplets with other types of cloud and haze particles, and sulfur dioxide.

Though we knew from the ground that Venus's clouds were largely made of sulfuric acid droplets, Pioneer Venus gave us details. Venus's clouds start 48 km above its surface and extend upward about 30 km in three distinct layers. They extend much higher than terrestrial clouds, which are made of water droplets and rarely go above 10 km. (The upper layers of the Earth's atmosphere, above the clouds, also contain sulfuric acid droplets but are thin enough to be transparent, unlike Venus's layers of sulfuric acid droplets. Even without the sulfuric acid, Venus's thick carbon dioxide atmosphere would make the atmosphere opaque.) Below the clouds, the atmosphere is fairly free of dust and cloud particles.

In 1985, a pair of joint Soviet–French probes was carried to Venus by Soviet rockets en route to Halley's Comet. Since the Soviets pronounced Halley as "Galley," the probes were called Vega (for **Ve**nus **Ga**lley). One of the probes floated in Venus's atmosphere on a balloon.

Nowadays, we can map Venus's clouds with the Hubble Space Telescope (Fig. 11–11). Thus we could, observing time permitting, follow changes over long periods of time. In reality, Hubble hadn't looked at Venus between January 1995 and the end of 2000.

No magnetic field was detected for Venus from any of the spacecraft. Possible reasons for the absence could be that Venus's core is solid everywhere, preventing the circulation thought necessary to create a magnetic field, or that the material's conductivity is too low. Though Venus rotates slowly, some measurable magnetic field might have been expected. One new line of argument holds that the absence may indicate that Venus's core is entirely fluid, if the circulating fluid that forms the magnetic field gets its energy in the same way as for the Earth. (In the case of the Earth, the fluid is thought to get its energy from energy released by the gradual solidifying of the Earth's inner core. So if Venus's core is solidifying, we would expect a magnetic field there.) The lack of a magnetic field means that Venus has no magnetosphere to prevent the solar wind from hitting the planet.

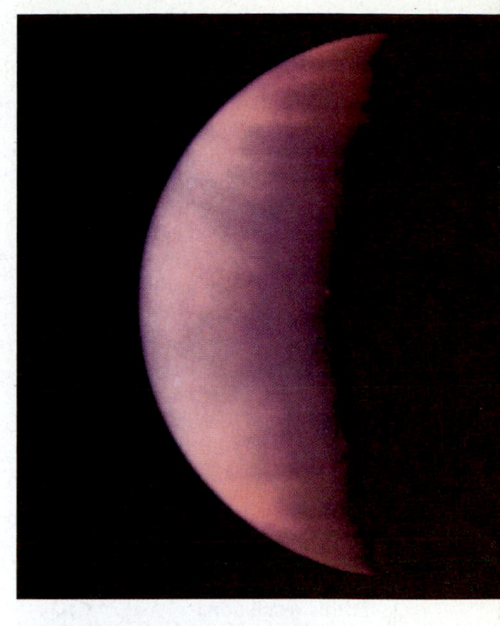

**FIGURE 11–11** A Hubble Space Telescope view of Venus's sulfuric acid clouds. The polar regions are bright, possibly because of small particles that overlie the main clouds. False color tints this ultraviolet view.

### 11.6c Radar Mapping of the Surface of Venus

The Pioneer Venus Orbiter carried a small radar altimeter and imager to study the topography of Venus's surface, both measuring altitudes and making images. The image resolution was not as good as that of the best Earth-based radar studies of Venus, but the orbiting radar mapped a much wider area. It mapped over 90 per cent of the surface with a resolution typically better than 30 km for images and 100–200 km for altimetry.

From the Pioneer radar altimeter map (Fig. 11–12), we now know that 60 per cent of Venus's surface is covered by a rolling plain, flat to within $\pm 1$ km. Only about 16 per cent of Venus's surface lies below this plain, a much smaller fraction than the two-thirds of the Earth covered by ocean floor. As we have seen in Chapter 8, the Earth's current surface is geologically young, formed by moving "plates" as part of continental drift or "plate tectonics." No such plate tectonics show on Venus.

Two large features, the size of small Earth continents, extend several kilometers above the mean elevation. A northern continent, Ishtar Terra, is about the size of the continental United States. A giant mountain on it known as Maxwell Montes, 11 km high (Fig. 11–13), is 2 km taller than Earth's Mt. Everest (measured from terrestrial sea level). Maxwell had formerly been known only as a bright spot on Earth-based radar images. Ishtar's western part is a broad plateau, about as high as the highest plateau on Earth (the Tibetan plateau) but twice as large. A southern continent, Aphrodite Terra, is about twice as large as the northern continent and is much rougher.

## A DEEPER DISCUSSION

### BOX 11.1 Naming the Features of Venus

The equatorial continent on Venus was named Aphrodite Terra and the northern continent Ishtar Terra. Aphrodite, the Greek goddess of love, was the equivalent of the Roman goddess Venus. Ishtar was the Babylonian goddess of love and war, daughter of the Moon and sister of the Sun.

Major features of Venus have been named after mythical goddesses, minor features after other mythical female figures, and still smaller circular features after famous women, like the physicist Lise Meitner, who identified nuclear fission.

***Discussion Question***

Who would be your top 20 candidates for names of surface features on Venus from 20th-century artists and musicians?

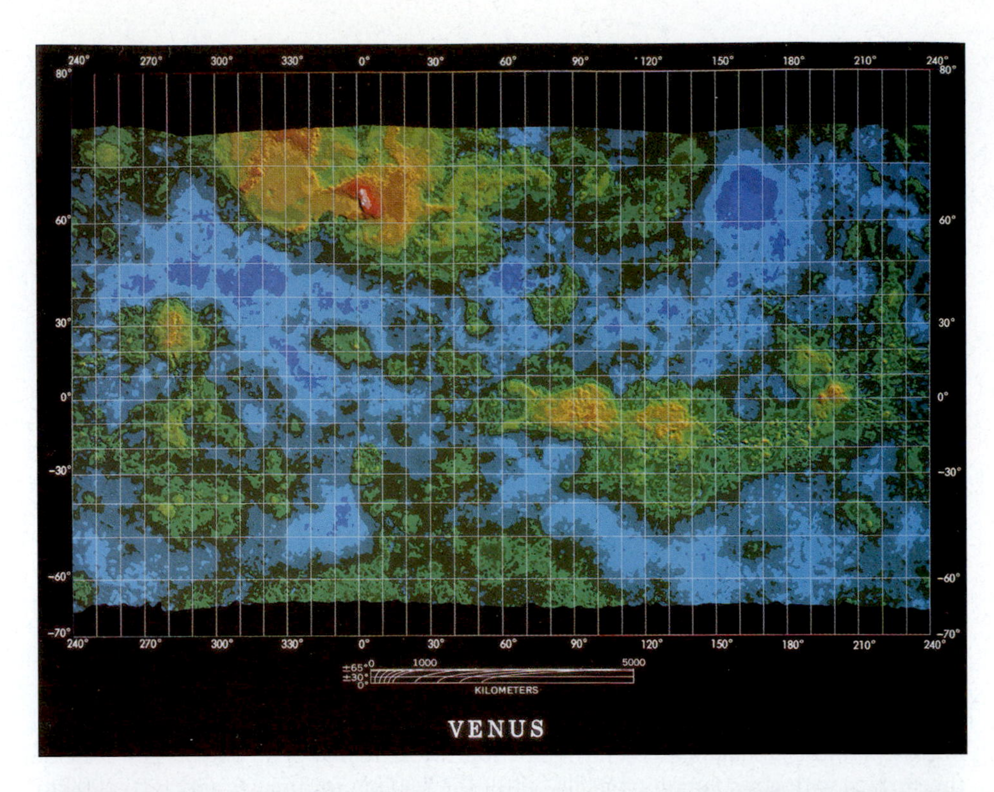


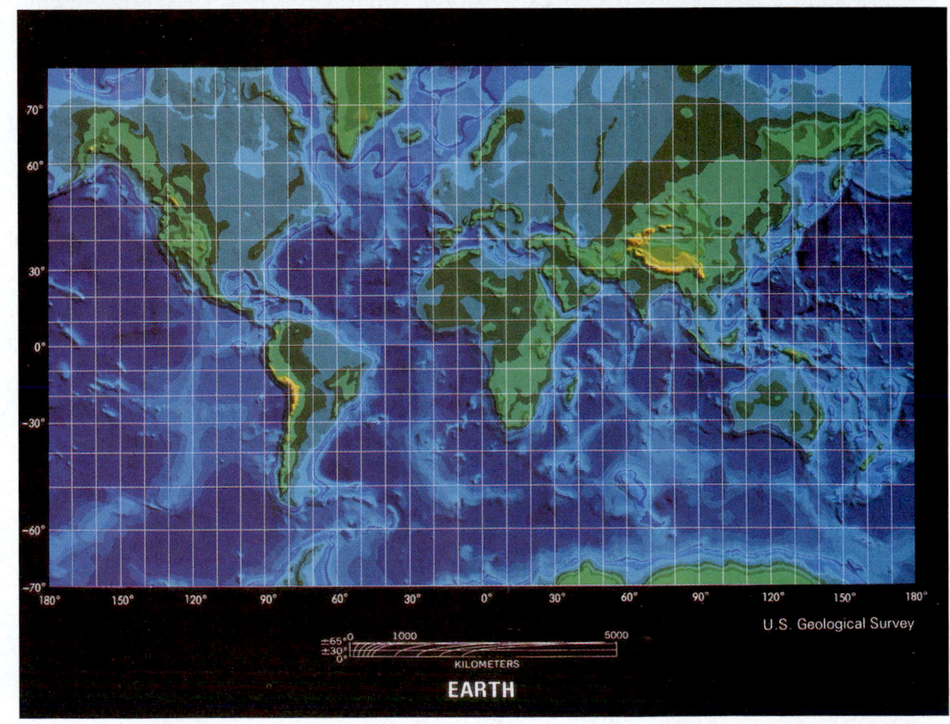


**FIGURE 11–12** Venus's surface, based on the Pioneer Venus radar map, compared with Earth's surface at the same resolution (50–100 km). Two continents exist on Venus: Aphrodite Terra (equatorial), which is comparable in scale with Africa, and Ishtar Terra (northern), which is comparable in scale with the continental United States or Australia. Sixty per cent of Venus's surface is covered with a huge, rolling plain. (The term *Planitia* is Latin for "plain.") Only about 16 per cent of Venus's surface is covered with lowlands; in comparison, over two-thirds of Earth's surface is covered with oceans. Mid-ocean ridges, a sign of spreading plates, are not visible on Venus, though this comparison shows that they would be detectable on Earth at this resolution.

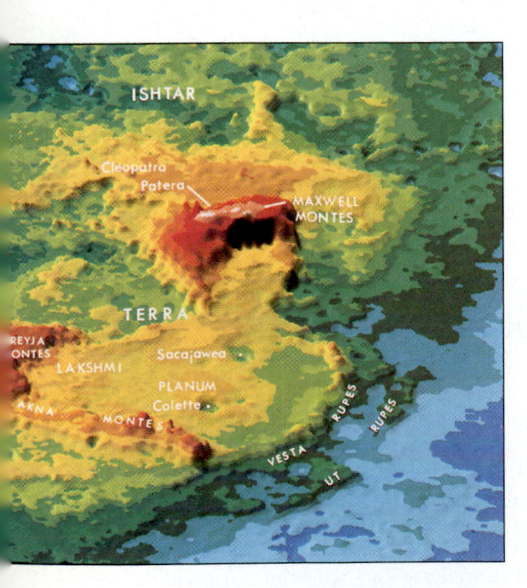


**FIGURE 11–13** Part of Ishtar Terra, with the giant Maxwell Montes, from the Pioneer Venus Orbiter radar.

The Soviet Veneras 15 and 16 reached Venus in 1983. They carried higher-resolution radars. They mapped about 25 per cent of the planet, chiefly around the north pole, at a resolution of 1 to 2 km.

NASA's Magellan spacecraft (Fig. 11–14) reached Venus in 1990. It went into an elliptical orbit that extended nearly up to Venus's poles. Magellan carried a **synthetic-aperture radar,** that is, a radar that allows scientists to combine (synthesize) data from a sequence of positions as the spacecraft flies along its trajectory.

In this way, it simulates results that would otherwise have required an antenna (aperture) as large as the length of trajectory traced out during the observations (which was typically of the order of 1 or 2 km). As a result, it achieved higher resolution. The resolution that resulted is about ten times better than the resolution of Venera 15 and 16 and over a hundred times more detailed than those from Pioneer Venus. Magellan mapped about 99 per cent of Venus's surface (Fig. 11–15) with a resolution of about 200 m. No signs of global, connected continental plate boundaries—which on Earth show up as global-scale ocean rifts and ridges, for example—appear even under Magellan's improved resolution. So though many volcanic features are visible on Venus, signs that Venus was at some time internally hot, its surface was confirmed by Magellan to be a single plate (Fig. 11–16).

Computer processing can transform the radar data into perspective images, making it seem as though we were looking down at a slant at the surface (Fig. 11–17). Apparent lava flows can be seen on many of the mountains and plains (Fig. 11–18). We must remember, though, that these are radar images and not visual ones, so what appears bright means only that it is a strong reflector of radio waves of a certain wavelength. Bright regions, thus, may have rocky, rough surfaces with a typical rock size of about the wavelength of the radio waves—12.6 cm.

Craters (Fig. 11–19) range in size from 2 km to 275 km. Sometimes craters have ejecta asymmetrically around them, showing the effect of Venus's atmosphere as well as of the angle of impact. Other types of surface features have been found on Venus that aren't known elsewhere, such as giant circular features that are sometimes hundreds of kilometers across. Each of these "coronas," with concentric rings, may have resulted from a plume of rising magma that pushed up the crust, which eventually collapsed like a fallen soufflé. Other circular features, pancake volcanoes, probably came from very thick lava.

**FIGURE 11–14** The Magellan spacecraft as it was launched from a space shuttle in 1989. It was in orbit around Venus through 1994.

**FIGURE 11–15** A pole-to-pole radar map of surface roughness of Venus from the Magellan spacecraft. Maxwell Montes is at top, left of center, and Aphrodite Terra, a highland, is right of center. Halos surrounding some of the younger impact craters appear as dark patches. The number of craters matches an average surface age of 400 million years, less than 10% the age of the planet.

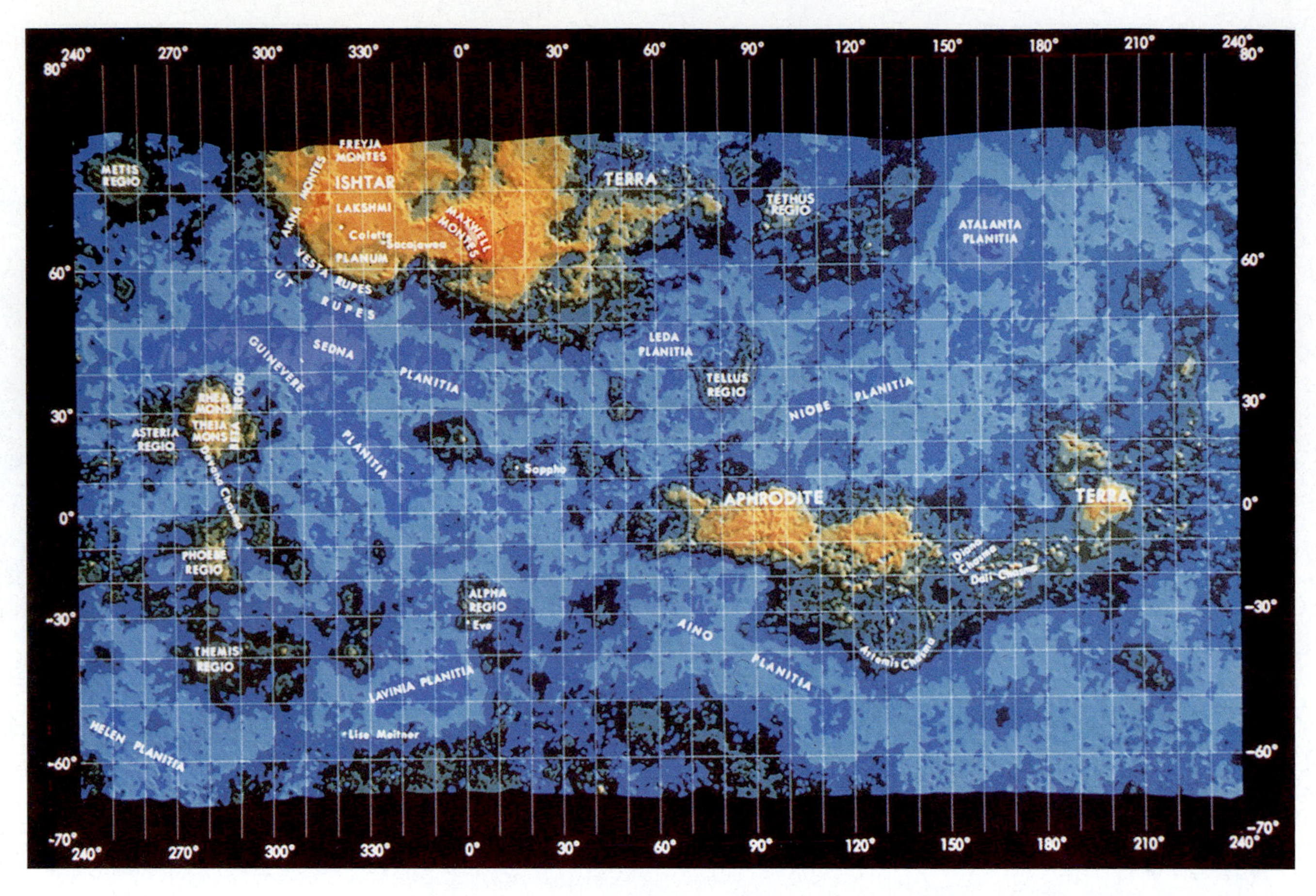


**FIGURE 11–16** A radar map of Venus from the Magellan spacecraft, with color changing from blue to red indicating increasing height. The single rolling plain that dominates Venus's surface, colored blue, shows no sign of plate tectonics.

The Magellan images show few craters, so something must have resurfaced the planet at some time. There are almost no very small craters; Venus has one-thousandth the crater density of our Moon. This observation can be explained if the smallest meteoroids burned up in Venus's atmosphere before they could hit the surface. Slightly larger meteoroids break up as they pass through the atmosphere, creating clusters of 5- to 15-km-wide craters on the surface.

Based on the number of larger craters, the surface of Venus is about half a billion years old. It isn't known, though, whether there was a widespread episode of volcanism then, or whether the resurfacing took place more gradually, at different

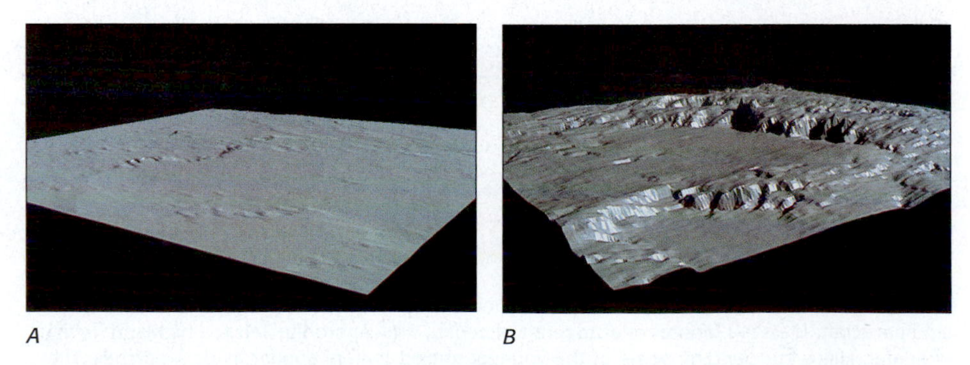

*A* *B*

**FIGURE 11–17** On the Magellan computer-generated images like these, the vertical heights, seen first (*A*) at real scale, are usually (*B*) exaggerated by a factor of about 20.

**FIGURE 11–18** Maat Mons, in a simulated perspective view based on radar data from Magellan and arbitrarily given the orange color that comes through Venus's clouds. Our viewpoint is 560 km north of Maat Mons at an elevation of 1.7 km. We see lava flows extending hundreds of kilometers across the foreground fractured plains. Maat Mons is 8 km high; its height here is exaggerated by a factor of about 20.

**FIGURE 11–19** Three large impact craters in a simulated perspective view. Howe Crater, at center, is 37 km across. Danilova, at upper left, is 48 km across, and Aglaonice, at upper right, is 63 km across. Their ejecta are bright to the radar and are therefore rough. The crater floors are dark and therefore smooth, presumably from flowing lava. As is common with meteorite impact craters, we see terraced inner walls and central peaks. The heights are exaggerated by a factor of about 20.

times in different places. Evidence for the first possibility is that few craters are seen partially covered with lava or destroyed by faults, which is also further evidence that Venus is not very geologically active. But some young craters are covered, so we know some volcanism has occurred in geologically recent times, within hundreds of millions of years.

Though Venus's surface is so obscured by its atmosphere that we cannot see it, radar imaging, height measurements, and other data have been put together to give a view of what Venus would look like if we could see it from an altitude of 6.5 km (Fig. 11–20). The image looks east toward Maxwell Montes.

Venus's gravity field was measured by following Magellan's path around Venus very carefully. After the radar program was completed, Magellan was put into a

**FIGURE 11–20** What Venus might look like from an altitude of 6.5 km, looking east toward Maxwell Montes. The image shows a region the size of West Virginia, with the horizon 350 km away. David Anderson of Southern Methodist University put the image together based on radar, altimetry, and other data, though he minimized the foreground haze. The surface was made brown because basalt is known to be present; a computer model was used to mock up clouds.

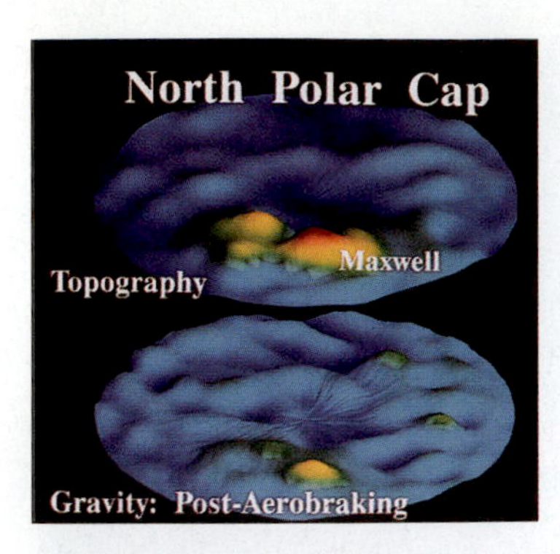


**FIGURE 11–21** The orbit of the Magellan spacecraft was changed in 1993, using the atmosphere to slow it up ("aerobraking") to a near-circular orbit about 250 km above the surface. The change greatly improved the gravity data resolution for high-latitude regions, including Maxwell Montes. Gravity variations are much smaller there than were expected.

circular orbit so that it could better map Venus's gravity field (Fig. 11–21), especially at high latitudes. Such a map reveals what Venus is like under its surface and leads to the conclusion that the lithosphere is strong enough to support the mountains and other surface relief. Further, distortions in the lithosphere show that convection is going on below it, in the mantle.

The Magellan spacecraft, which was crashed into Venus's surface in 1994, has gone down in the list of NASA's all-time big successes.

Our space results, coupled with our ground-based knowledge, show us that Venus is even more different from the Earth than had previously been imagined. Among the differences are Venus's slow rotation, its one-plate surface, the absence of a satellite, the extreme weakness or absence of a magnetic field, the lack of water in its atmosphere, and its high surface temperature.

## 11.7 VENUS, AGAIN, FROM EARTH

The mighty Keck I telescope studied Venus in 1999. It found a faint glow from atomic oxygen in Venus's atmosphere on Venus's night side. Studies of this green emission line in the Earth's atmosphere have helped trace terrestrial upper-atmospheric winds and temperatures. Perhaps we will now have similar success at Venus.

## CORRECTING MISCONCEPTIONS

| ✖ *Incorrect* | ✔ *Correct* |
|---|---|
| Venus's atmosphere is like Earth's. | Venus is hot enough to melt lead and has 100 times Earth's surface pressure. |
| Spacecraft discovered that Venus is very hot. | We knew from radio waves received on Earth that Venus is very hot. |
| Spacecraft discovered that Venus's atmosphere is 96 per cent carbon dioxide. | We knew the composition of Venus's atmosphere from Earth. |
| Radar from spacecraft can study the clouds. | Radar at the wavelengths used at Venus penetrates the clouds and studies only the surface, unlike terrestrial weather radar. |

## SUMMARY AND OUTLINE

Venus's atmosphere (Section 11.1)
- Mainly composed of carbon dioxide
- Clouds primarily composed of sulfuric acid droplets.

Transits of Venus (Section 11.2)
- Those of 2004 and 2012 will be first since 1882.
- Historically used for measuring scale of Solar System

A slow rotation period, in retrograde (Section 11.3)

The temperature of Venus (Section 11.4)
- A high temperature, 750 K, caused by the greenhouse effect.
- Greenhouse effect: visible light comes in, turns to infrared; infrared can't escape.

Radar mapping from Earth (Section 11.5)
- Maxwell Montes (named after James Clerk Maxwell) discovered.

Space studies of Venus (Section 11.6)
- Landers on the surface (Section 11.6a)
  - A Soviet specialty; survival for 1 to 2 hours, photographs showing rocks
  - Venera 13 and 14: landers, photographs showed sky is orange; soil is basaltic.
- Venus's clouds (Section 11.6b)
  - U.S. Mariner 10 observations of clouds and atmosphere
  - U.S. Pioneer Venus 1 orbited Venus, followed evolution of clouds.
  - Pioneer Venus 2 contained probes.
  - Measured composition of atmosphere and temperatures
  - Observed lightning
- The surface of Venus (Section 11.6c)
  - Radar maps show volcanoes and mountains.
  - NASA Magellan, in orbit 1990–1994, high resolution, 99 per cent coverage
  - No plate tectonics

Venus's atmosphere (Section 11.7)
- Oxygen emission line observed; may trace atmospheric structure.

## KEY WORDS

cytherean
retrograde rotation
greenhouse effect
synthetic-aperture radar

## QUESTIONS

1. Make a table displaying the major similarities and differences between the Earth and Venus.
2. Why does Venus have more carbon dioxide in its atmosphere than does the Earth?
3. Why do we think that there have been significant external effects on the rotation of Venus?
4. If observers from another planet tried to gauge the rotation of the Earth by watching the clouds, what would they find?
5. Suppose a planet had an atmosphere that was opaque in the visible but transparent in the infrared. Describe how the effect of this type of atmosphere on the planet's temperature differs from the greenhouse effect.
6. Why do radar observations of Venus provide more data about the surface structure than a flyby with close-up cameras?
†7. If one removed all the $CO_2$ from the atmosphere of Venus, the pressure of the remaining constituents would be how many times the pressure of the Earth's atmosphere?
8. Some scientists have argued that if the Earth had been slightly closer to the Sun, it would have turned out like Venus; outline the logic behind this conclusion.
9. Outline what you think Venus would be like if we sent an expedition to Venus that resulted in the loss of almost all of the $CO_2$ from its atmosphere.
10. Why do we say that Venus is the Earth's "sister planet"?
11. Compare and contrast the observations of Venus's surface made from the Venera landers.
12. Do radar observations of Venus study the surface or the clouds? Explain.
13. Describe the major radar results from Venus spacecraft.
14. What is the advantage of synthetic-aperture radar over ordinary radar?
15. What signs of volcanism are there on Venus?
16. Look up additional results from the Magellan spacecraft and describe them.

† This question requires a numerical solution.

## USING TECHNOLOGY

### W World Wide Web

1. Check out the Magellan homepage.
2. Read about Venus on Nine Planets (http://www.nineplanets.org) and on the JPL site (http://pds.jpl.nasa.gov/planets/).
3. Find a variety of photographs of Venus's clouds.
4. Locate sites describing transits of Venus and of Mercury.

### REDSHIFT

1. When can you see Venus in your sky? Tonight or tomorrow morning? How bright will it be, compared with the brightest star?
2. On disk 2, view the Venus movies from Magellan.
3. Using RedShift to put yourself high above the Solar System, demonstrate the relation between Venus's orbital period and rotational period that gives a day 127 Earth days long.
4. Use RedShift to compute a transit of Venus and a conjunction of Jupiter and Venus. When will the next of each occur?
5. Look at Venus in Science of Astronomy: Night on the Mountain: The Planet Dossier.
6. Look at Astronomy: Worlds of Space: The Greenhouse Effect.
7. On the map of Venus, find Maxwell Montes, Cleopatra, and other features.

A high-resolution image of a small portion of Mars, showing the geological detail studied by the Mars Global Surveyor that is still in orbit around Mars. We see gullies eroded into the wall of a meteor impact crater in Noachis Terra. Scientists attribute the apparent channels and associated aprons of debris to groundwater seepage, surface runoff, and debris flow. The gullies may indicate that liquid water is close to the surface of Mars even today. The symbol for Mars appears at the top of the opposite page.

# Chapter 12

## Mars

AIMS: To discuss the atmosphere and surface features of Mars, to see the results of spacecraft in orbit and on the surface, and to assess the chances of finding life there

Mars has long been the planet of greatest interest to scientists and nonscientists alike. Its interesting appearance as a reddish object in the night sky and some past scientific studies have made Mars the prime object of speculation as to whether or not extraterrestrial life exists.

In 1877, the Italian astronomer Giovanni Schiaparelli published the results of a long series of telescopic observations he had made of Mars. He reported that he had seen ***canali*** on the surface. When this Italian word for "channels" was improperly translated into "canals," which seemed to connote that they were dug by intelligent life, public interest in Mars increased.

In the United States, Percival Lowell grew interested in the problem and in 1894 established an observatory in Flagstaff, Arizona, to study Mars. Over the next decades, there were endless debates over just what had been seen. We now know that the channels or canals Schiaparelli and other observers reported (Fig. 12–1) are not present on Mars—the positions of the *canali* do not even always overlap the spots and markings that are actually on the martian surface. But hope of finding life in the Solar System springs eternal, and the latest studies have indicated the presence of considerable quantities of liquid water in Mars's past (Fig. 12–2), a fact that leads many astronomers to hope that life could have formed during those periods. Still, as we shall describe, the Viking spacecraft that landed on Mars found no signs of life. The shocking 1996 announcement that signs of fossilized early life might have been found in a meteorite found on Earth but thought to be from Mars could, if confirmed, revolutionize our thought. But, though reputable scientists made the report, their interpretation that signs of past life were found is not widely accepted. We discuss the evidence briefly at the end of this chapter and more fully in Section 21.1, where we discuss the larger picture of the search for life on planets other than our own.

FIGURE 12–1 An old drawing of Mars that showed "canals."

### 12.1 CHARACTERISTICS OF MARS

Mars is a small planet, 6800 km across, which is only about half the diameter and one-eighth the volume of Earth or Venus, although somewhat larger than Mercury. Mars's atmosphere is thin—at the surface its pressure is only 1 per cent of the surface pressure of Earth's atmosphere—but it might be sufficient for certain kinds of life.

From Earth, we have relatively little trouble in seeing through the martian atmosphere to inspect that planet's surface (Fig. 12–3), except when a martian dust

**FIGURE 12–2** An orientation map showing the location of the closeup of the Chapter Opener.

storm is raging. At other times, we are limited mainly by the turbulence in our own thicker atmosphere, which causes unsteadiness in the images and limits our resolution to 60 km at best on Mars. We can, however, follow features on the surface of Mars and measure the rotation period quite accurately. It turns out to be 24 hours 37 minutes 22.6 seconds from one martian noontime to the next, so a day on Mars lasts nearly the same length of time as a day on Earth.

Unlike the orbits of Mercury or Venus, the orbit of Mars is outside the Earth's, so it is much easier to observe in the night sky (Fig. 12–4). Because of Mars's elliptical orbit around the Sun, at some oppositions it is closer to Earth than at others. At such times, high-resolution views can be captured in the brief moments of the best seeing.

Mars revolves around the Sun in 23 Earth months. The axis of its rotation is tipped at a 25° angle from the plane of its orbit, nearly the same as the Earth's 23½° tilt. Because the tilt of the axis causes the seasons, we know that Mars goes through a year with four seasons just as the Earth does. Note that, given Mars's orbital period, each season is about twice as long as a season on Earth.

We have watched the effect of the seasons on Mars over the last century. In the martian winter, in a given hemisphere, there is a polar cap. As the martian spring comes to the northern hemisphere, the north polar cap shrinks and material at more

**FIGURE 12–3** A photograph of Mars taken from Earth at an extremely close opposition, when Mars was as close to Earth as it ever gets. (When three celestial bodies are in a line, the alignment is called a **syzygy,** and the particular case when the planet is on the opposite side of the Earth from the Sun is called an **opposition.** Oppositions of Mars occur at intervals of 26 Earth months on the average.) The darker regions spread in the martian springtime. One of the polar caps is clearly visible. At opposition, Mars's disk can be 8 times larger than at **conjunction,** when a planet passes the Sun's right ascension in the sky (which is approximately when the planet is closest to the Sun in the sky). The next oppositions will occur on August 28, 2003, and October 30, 2005. Mars ranges in brightness from that of Sirius, the brightest star, to three times brighter.

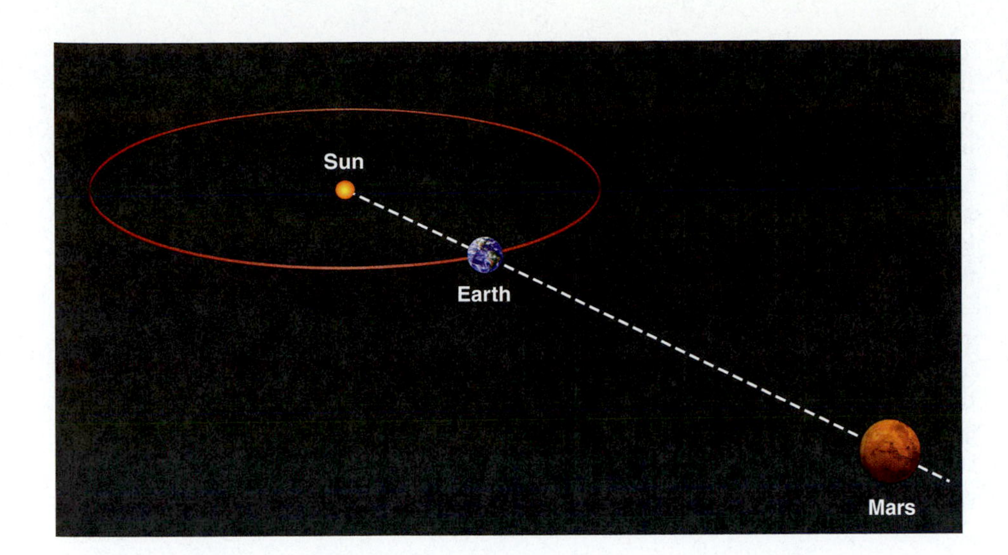


**FIGURE 12–4** When Mars is at opposition, it is high in our nighttime sky and thus easy to observe. We can then simply observe in the direction opposite from that of the Sun. We are then looking at the lighted face.

temperate zones darkens. The surface of Mars is always mainly reddish, with darker gray areas that appear blue-green. (They aren't really blue-green, but for physiological reasons of color contrast they appear so next to the reddish dust.) In the spring, the darker regions spread. Half a martian year later, the same thing happens in the southern hemisphere.

One possible explanation that long ago seemed obvious for these changes is biological: martian vegetation could be blooming or spreading in the spring. But there are other explanations, too. The theory that now seems most reasonable is that each year at the end of southern-hemisphere springtime, a global dust storm starts, covering Mars's entire surface with light dust. Then the winds reach velocities as high as hundreds of kilometers per hour and blow fine, light-colored dust off some of the slopes. This exposes the dark areas underneath. Gradually, as Mars passes through its seasons over the next year, the location where the dust is stripped away changes, mimicking the color change we would expect from vegetation. Finally, a global dust storm starts up again to renew the cycle. Astronomers still debate why the dust is reddish; the presence of iron oxide (rust) would explain the color in general, but might not satisfy the detailed measurements.

Mars has ⅒ the mass of the Earth; from the mass and radius one can easily calculate that its average density is slightly less than 4 grams/cm$^3$, not much higher than that of the Moon. This is substantially less than the density of 5½ grams/cm$^3$ shared by the three innermost terrestrial planets and indicates that Mars's overall composition must be fundamentally different from that of these other planets. Mars probably has a smaller core and a thicker crust than the Earth.

## 12.2 SPACE OBSERVATIONS

Mars has been the target of series of spacecraft launched by both the United States and the Soviet Union. The earliest of these spacecraft were flybys, which were in good position for only a few hours as they passed by. The first flyby, NASA's Mariner 4, sent back photos in 1965 of a section of Mars showing only a cratered surface. This seemed to indicate that Mars resembles the Moon more than it does the Earth. Mariners 6 and 7 confirmed the cratering four years later, and also showed some signs of erosion.

In 1971, the United States sent out a spacecraft not just to fly by Mars for only a few hours, but actually to orbit the planet and send back data for a year or more.

FIGURE 12–5 A Viking image showing several volcanoes on the Tharsis Ridge of Mars. Olympus Mons is nearest the top.

This spacecraft, Mariner 9, went into orbit around Mars after a five-month voyage from Earth. But when it reached Mars, our first reaction was as much disappointment as elation. While Mariner 9 was en route, a tremendous dust storm had come up, almost completely obscuring the entire surface of the planet. Only the south polar cap and four dark spots were visible. The storm began to settle after a few weeks, with the polar caps best visible through the thinning dust. Finally, three months after the spacecraft arrived, the surface of Mars was completely visible, and Mariner 9 could proceed with its mapping mission.

The data taken after the dust storm had ended showed four major "geological" areas on Mars: volcanic regions, canyon areas, expanses of craters, and terraced areas near the poles. These features were subsequently imaged with higher resolution from the Viking missions, NASA missions that reached Mars in 1976 and carried on studies for several years. Since then, Mars Pathfinder in 1996 and Mars Global Surveyor, still sending back photos, have given us images of even higher quality. The photographs we show here are from those later missions.

## 12.3 MAIN FEATURES OF MARS'S SURFACE AND ATMOSPHERE

A chief surprise of the Mariner 9 mission was the discovery of extensive areas of volcanism. Four dark spots on the martian surface as seen from Earth proved to be huge volcanoes. They are on a 2000-km-long ridge known as Tharsis (Fig. 12–5). The largest, which corresponds to the surface marking long known as Nix Olympica, "the snow of Olympus," is named Olympus Mons, "Mount Olympus." It is a huge volcano—600 km at its base and about 25 km high (Fig. 12–6). It is crowned with a crater 65 km wide; Manhattan Island could be easily dropped inside (Fig. 12–7). (The tallest volcano on Earth is Mauna Kea, in the Hawaiian islands, if we measure its height from its base deep below the ocean. Mauna Kea is taller than Everest, though still only 9 km high. The average level of the Tharsis ridge itself is 10 km.) Olympus Mons and the other martian volcanoes are "shield volcanoes," which have gradually sloping sides (Fig. 12–8); Mauna Kea (Fig. 12–9) is a terrestrial example.

Another surprise on Mars was the discovery of systems of canyons. One tremendous set of canyons—about 5000 kilometers long—is as long as the United States is wide and comparable in size to the Rift Valley in Africa, the longest geological fault on Earth. The main canyon is 200 km wide and up to 7 km deep. Here again the size of the canyon with respect to Mars is proportionately larger than any geologic formations on the Earth. Venus, the Earth, and Mars each seems to have a large rift. The canyons were named after the Mariner 9 spacecraft from which they were discovered (Fig. 12–10).

Perhaps the most amazing discovery on Mars was the presence of sinuous channels. These are on a smaller scale than the *canali* that Schiaparelli had seen and are entirely different phenomena. Some of the channels seemed to show tributaries (Fig. 12–11), and the bottoms of some of the channels seemed to show the same characteristic features as streambeds on Earth. Mars Global Surveyor observed such channels in even higher detail (Fig. 12–12). Even though water cannot exist on the surface of Mars under today's conditions, because the low pressure would allow the water to immediately vaporize, it is difficult to think of ways to explain the channels satisfactorily other than to say that they were cut by running water in the past. Later

FIGURE 12–6 Olympus Mons, from Mars Global Surveyor. We see a region about 800 km across.

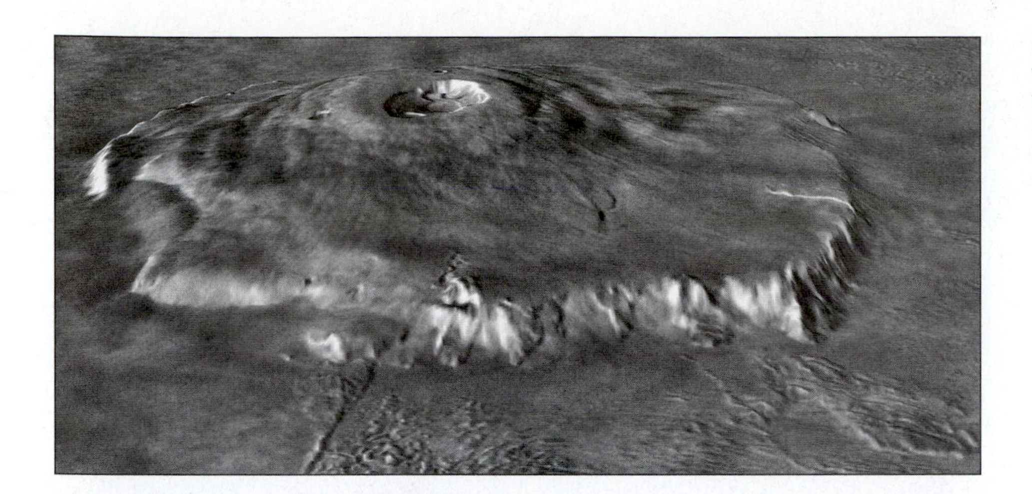

**FIGURE 12-7** Olympus Mons, with altitude measurements from the Mars Orbiter Laser Altimeter on Mars Global Surveyor draped over a mosaic of Viking images. Height is exaggerated by a factor of 10.

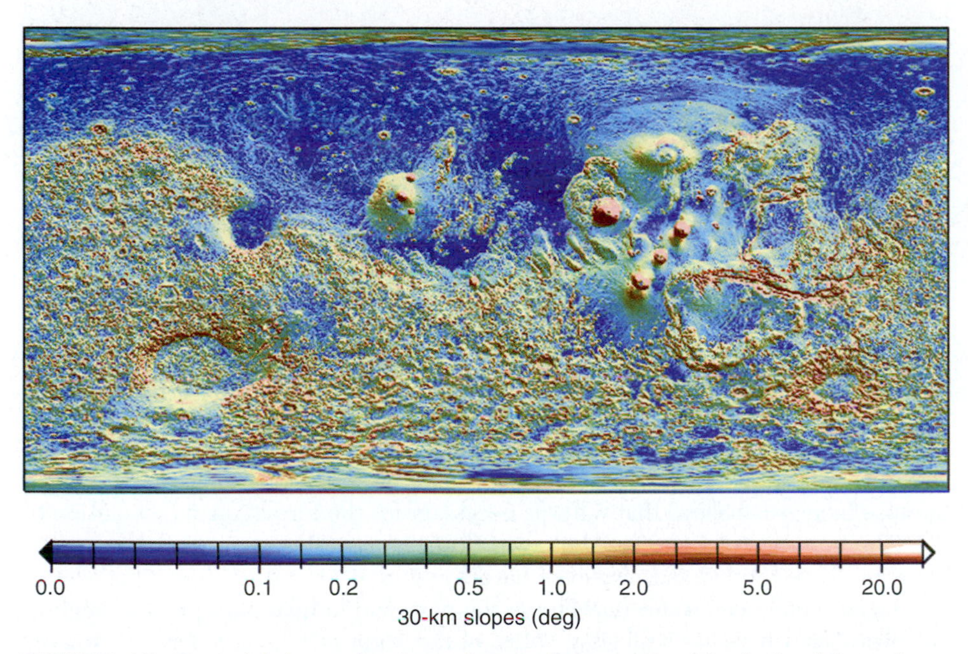


**FIGURE 12-8** A global map showing surface gradients on Mars. The flanks of the major volcanoes slope only 2.5°–5°, comparable to the shield volcanoes in Hawaii. The northern hemisphere is flatter than the southern hemisphere. The slopes were measured with the Mars Orbiter Laser Altimeter on Mars Global Surveyor.

**FIGURE 12-9** Olympus Mons on Mars is similar to shield volcanoes on Earth, volcanoes with very gradual slopes that resemble warriors' shields lying on the ground. Mauna Kea in Hawaii, with its gradual rise, is typical of shield volcanoes.

**FIGURE 12–10** The giant system of canyons, Valles Marineris, and several giant volcanoes (*left*) show clearly in this mosaic of 102 images from one of the Viking Orbiters. The volcanoes are Arsia Mons (*below*), Pavonis Mons (*center*), and Ascraeus Mons (*above*).

in this chapter, we will see images taken at even higher resolutions that also are generally interpreted as showing the past presence of water on Mars.

This indication that water most likely flowed on Mars is particularly interesting because biologists believe that water is necessary for the formation and evolution of life. The presence of water on Mars, therefore, even in the past, may indicate that life could have formed and may even have survived.

If there had been water on Mars in the past, and in quantities great enough to cut riverbeds, where has it all gone? Most of the water may be in a permafrost layer beneath middle latitudes and polar regions. A small amount of the water is bound in the polar caps (Fig. 12–13). Up to the time of Mariner 9, we thought that the polar caps were mostly frozen carbon dioxide—"dry ice"; the presence of water was controversial. We now know that the large polar caps that extend to latitude 50° during the winter are carbon dioxide (Fig. 12–14). But when a cap shrinks during its hemisphere's summer, a residual polar cap of water ice remains (Fig 12–15), especially in the northern hemisphere.

The presence of water is thought to be necessary for the origin of life as we know it, so there is much interest in searching for water on Mars. It may not be flowing in Percival Lowell's canals, now known to be nonexistent, but signs of flowing

**FIGURE 12–11** Stream channels with tributaries indicate to most scientists that water flowed on Mars in the past. This Viking view shows the boundary scarp between the ancient cratered highlands and the northern plains. In the site shown, Mangala Valles, the ancient river channels run north through the ancient cratered highlands, but north of the scarp are covered by young lava flows. The flow-like lobes of the large impact crater may reveal the presence of subsurface water or ice.

**FIGURE 12–12** A Mars Global Surveyor closeup of the Gorgonum Chaos region of the martian southern hemisphere. The feature looks as though it was cut by a flow of water for an extended period of time. Gullies that may have been formed by seeping ground water emanate from a specific layer near the tops of the troughs. The presence of so many gullies in the same layer in each mesa suggests that the layer is an aquifer, a region particularly effective in storing and conducting water. This aquifer seems to be only a few hundred meters below the surface. The region shown is only 3 km wide.

water are being found from the Mars Global Surveyor spacecraft in orbit. One relevant image appeared to open this chapter, and we discuss the evidence for flowing water later in this chapter.

The martian atmosphere is composed of 90 per cent carbon dioxide with small amounts of carbon monoxide, oxygen, and water. Scientists measured the density of the atmosphere by monitoring the changes of radio signals from spacecraft as they went behind Mars. As the spacecraft were occulted by the atmosphere of the planet, we could tell the rate at which the density drops. The surface pressure is less than 1 per cent of that near Earth's surface and decreases with altitude slightly more than it does on Earth. The atmosphere is too thin to affect the surface temperature, in contrast to the huge effects that the atmospheres of Venus and the Earth have on climate.

**FIGURE 12–13** The martian north polar cap, imaged from Mars Global Surveyor during Mars's early spring in that hemisphere. We see the north polar layered deposits, a terrain that scientists believe is composed of ice and dust deposited over millions of years. The swirls are channels eroded into this deposit.

◀ **FIGURE 12–14** The martian south polar cap, observed from Mars Global Surveyor in 1999 just before the start of southern spring on Mars. We see the south polar cap as it retreats; just above it, we can see wisps of a dust storm, caused by colder air blowing off the cap into warmer regions. The volcano Arsia Mons is at upper left.

▶ **FIGURE 12–15** A mosaic of Viking images of the south polar cap during the martian southern summer when the cap has retreated the most. The south residual cap is about 400 km across, and at least the exposed surface is thought to be mainly frozen carbon dioxide. Water ice may be present in layers underneath. By contrast, the north polar residual cap is made of water ice with no frozen carbon dioxide.

## 12.4 VIKING ORBITERS

In the summer of 1976, two elaborate U.S. spacecraft named Viking reached Mars after 10-month flights. Each contained two parts: an orbiter and a lander. The orbiter served two roles: it mapped and analyzed the martian surface using its cameras and other instruments and relayed the lander's radio signals to Earth. The lander served two purposes as well: it studied the rocks and weather near the surface of Mars and sampled the surface in order to decide whether there was life on Mars!

Viking 1 reached Mars in June of 1976 after a flight that led to spectacular large-scale views of the martian surface. The first month of its orbit around Mars was devoted to radioing back pictures to scientists on Earth who were trying to determine a relatively safe spot for its lander to touch down.

**FIGURE 12–16** Candor Chasm in Valles Marineris, imaged by one of the Viking Orbiters. The landscape has been shaped by tectonics and wind, and perhaps also by water and volcanism.

The orbiter observed the volcanoes and the huge canyon (Fig. 12–16) at higher resolution than did Mariner 9. The landslides that can be seen in the walls have more recently been imaged at even higher resolutions.

These detailed views of Mars allow us to interpret better the similarities and differences that this planet—with its huge canyons and gigantic volcanoes—has with respect to the Earth. For example, Mars has exceedingly large, gently sloping volcanoes but no signs of the long mountain ranges or deep mid-ocean ridges that on Earth tell us that plate tectonics has been and is taking place. Many of the large volcanoes on Mars are "shield volcanoes"—a type that has gently sloping sides formed by the rapid spread of lava, as we saw in Figure 12–9. On Earth, we also have steep-sided volcanoes, which occur where the continental plates are overlapping, as in the case of Mt. St. Helens, the Aleutian Islands, or Mount Fujiyama. No chains of such volcanoes exist on Mars.

Perhaps the volcanic features on Mars can get so huge because plate tectonics is absent there. If molten rock flowing upward causes volcanoes to form, then on Mars the features just get bigger and bigger for hundreds of millions of years, since the volcanoes stay over the sources and do not drift away.

**FIGURE 12–17** This oblique view shows layers in the atmosphere of Mars.

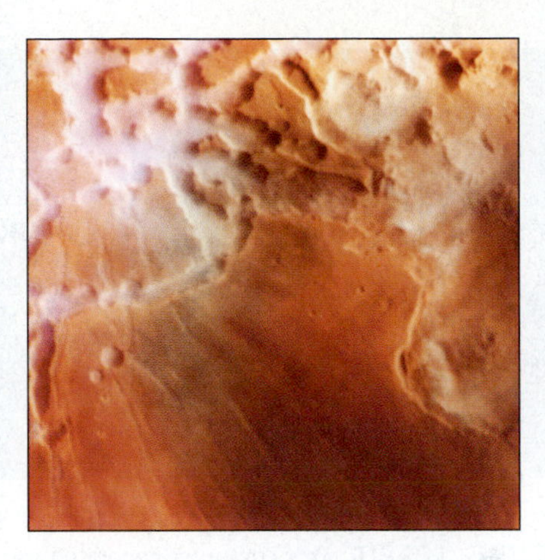

**FIGURE 12–18** Clouds of water ice form over Labyrinthus Noctis at sunrise.

**FIGURE 12–19** A martian dust storm over 300 km across inside the Argyre Basin. This martian storm resembles satellite pictures of storms on Earth. The temperature is too warm for the clouds to be carbon dioxide; they therefore must be made of water ice. In this view, taken at the end of southern winter, the polar cap is retreating and covers only the southern half of Argyre.

The presence of two orbiters circling Mars allowed a division of duties. One continued to relay data from the two Viking landers to Earth, while the orbit of the other was changed so that it passed over Mars's north pole. Summertime heat in the northern hemisphere caused the main polar cap (thought to be made of dry ice—frozen carbon dioxide) to retreat, leaving only a residual ice cap. From the amount of water vapor detected with the orbiter's spectrograph, and from the fact that the temperature was too high for this residual ice cap to be frozen carbon dioxide, the conclusion has been reached that it is made of water ice. This important result indicates the probable presence of lots of water on the martian surface in the past—in accordance with the existence of the streambeds—and seems to raise the possibility that life has existed or does exist on Mars. The layered look of the polar regions indicates that the general atmospheric conditions of Mars have fluctuated over eons, as do Earth's. This record of climate change episodes resulted from long-term variations in the eccentricity and tilt of Mars's orbit and in the tilt of Mars's axis of rotation.

Dating of the stream channels became possible late in the Viking mission when exceptionally high-resolution photos showed craters in the streambeds. The higher the density of craters, the older the underlying flow. So the earlier idea that all streambeds were formed at one stage early in Mars's history when the atmosphere was denser was not quite correct; flows must have occurred over a long period of time. Note that these dates are all relative; finding absolute dates will have to await the return of samples from Mars to Earth.

Observations from the Viking orbiters showed Mars's atmosphere (Fig. 12–17). The lengthy period of observation led to the discovery of weather patterns on Mars. With its rotation period similar to that of Earth, many features of Mars's weather are similar to our own. Surface temperatures measured from the landers ranged from a low of –125°C (–190°F) at the northern site of Lander 1 to over 25°C (80°F) at Lander 2. Temperature varied each day by 35–50°C (60–90°F), helping clouds to form (Fig. 12–18).

Both Viking landers and orbiters lasted long enough to show seasonal changes on Mars. The yearly temperature range spans the –111°C (–190°F) freezing point of carbon dioxide, which thus sometimes forms frost.

The orbiters showed dust storms (Fig. 12–19), which generally form at a time that is in the early summer in the southern hemisphere and in winter in the northern

**FIGURE 12–20** This computer-generated mosaic of Viking 2 images uses false color to show variations in the surface chemistry of Mars. Red shows rocks that contain a high proportion of iron oxides; dark blue shows fresher, less oxidized materials, probably volcanic basalts; bright turquoise shows surface frosts and fogs made from carbon dioxide or water ice; and brown, orange, and yellow show deposits of sand and dust.

hemisphere. The storms are seasonal partly because Mars is closest to the Sun at this time and the increased solar heating causes the atmosphere to circulate. The overall climates of the northern and southern hemispheres are different, probably because Mars's orbit is more elliptical than Earth's. As the result of careful study of the martian images, we now can categorize the surface into geological regions (Fig. 12–20).

As for magnetic field, Viking didn't find any around Mars. If there is any, it is too weak for it to have been measured with the instrumental capabilities of that time. A conclusion from the Mars Global Surveyor now in orbit around Mars is that the magnetic field switched off some 4 billion years ago, only half a billion years after Mars formed. The "solar wind" of particles from the Sun would then have been able to reach Mars's atmosphere and blow almost all of it away.

## 12.5 VIKING LANDERS

On July 20, 1976, exactly seven years after the first landing of people on the Moon, Viking 1's lander descended safely onto a plain called Chryse. The views showed rocks of several kinds, covered with yellowish-brown material that is probably an iron oxide (rust) compound (Fig. 12–21). Sand dunes were also visible. The sky on Mars turns out to be yellowish-brown, almost pinkish, from dust suspended in the air as a result of one of Mars's frequent dust storms. Craters up to 600 m across were visible near the horizon; they were probably deposited from the flow that formed the channel in which the spacecraft landed.

A series of experiments aboard the lander was designed to search for signs of life. A long arm was deployed, and a shovel at its end dug up a bit of the martian surface. The soil was dumped into three experiments that searched for such signs of life as respiration and metabolism. The results were astonishing at first: the experiments sent back signals that seemed similar to those that would be caused on Earth by biological rather than by mere chemical processes. But later results were less spectacular, and non-biological explanations seem more likely. It is probable that some strange chemical process mimicked life in these experiments. In particular, the soil contained strongly oxidized materials, and heating these rocks in the tests caused oxygen to be given off that was at first mistaken for signs of life forms. The presence of this oxidized soil and the lack of a protective ozone layer in Mars's atmosphere make Mars a very inhospitable environment for life as we know it.

**FIGURE 12–21** A Viking 1 view shows the red rocks and pink sky of Mars. The sampler arm at right has dug a trench at far left.

**Viking 1** (dates at or on Mars)
*Orbiter:*
June 19, 1976–Aug. 7, 1980
*Lander:*
July 20, 1976–Nov. 13, 1982

One experiment gave much more negative results for the probability that there is life on Mars. It analyzed the soil and looked for traces of organic compounds. On Earth, many organic compounds left over from dead forms of life remain in the soil; the life forms themselves are only a tiny fraction of the organic material. Yet these experiments found no trace of organic material. On Mars, who knows? Perhaps life forms evolved that efficiently used up their predecessors. The absence of organic material in martian soil is to many people a strong argument against the presence of life on Mars. Still, a Jet Propulsion Laboratory team of scientists proposed in 2000 that a chemical agent, superoxide radical ions, could be formed on Mars and that it could decompose organic material that was there. (An oxide is a single oxygen ion with a charge of –2; a peroxide is an ionized oxygen molecule with a charge of –2, and a superoxide is an ionized oxygen molecule with a charge of –1.) In their laboratory, they exposed mineral grains that resembled those in martian soil to a simulated martian atmosphere. Next, they duplicated the sunlight on Mars by irradiating the grains with ultraviolet light. They found superoxide radical ions on the grains in their test tubes. If they are correct, not finding organic material now doesn't mean that it wasn't there in the past.

The Viking 2 lander descended on September 3, 1976, on a site that was thought to be much more favorable to possible life forms than the Viking 1 site because it was closer to the north pole with its prospective water supply. It too found yellowish-brown dust and a pinkish sky (Fig. 12–22). Compared with Viking 1, Viking 2 found a smaller variety of types of rocks, but more large rocks and more pitted rocks. Such a distribution would result if the surface there had been ejected from the crater Mie. The atmosphere at the second site, Utopia Planitia, contains three or four times more water vapor than had been observed near Chryse (the first landing site), 7500 km away. Viking 2's experiments in the search for life sent back data similar to Viking 1's.

Both Viking landers survived on the surface of Mars to send back meteorological information over several seasons and to observe seasonal changes on Mars's surface.

Even if the life signs detected by Viking come from chemical rather than biological processes, as seems likely, we have still learned of fascinating new chemistry going on. When life arose on Earth, it probably took up chemical processes that had previously existed. Similarly, if life began on Mars in the past or will begin there in the future (assuming our visits didn't contaminate Mars and ruin the chances for indigenous life), we might expect the life forms to use chemical processes that already

**FIGURE 12–22** A Viking 2 view of Mars. The spacecraft landed on a rock and so was at an angle. The boom that supports Viking's weather station cuts through the center of the picture. ("Chance of precipitation," the local newscaster would say, "is 0%.")

**Viking 2** (dates at or on Mars)
*Orbiter:*
Aug. 7, 1976–July 25, 1978
*Lander:*
Sept. 3, 1976–Apr. 12, 1980

existed. So even if we haven't detected life itself, we may well have learned important things about its origin.

## 12.6 SATELLITES OF MARS

In Greek mythology, Phobos (Fear) and Deimos (Terror) were companions of Ares, the equivalent of the Roman war god, Mars.

Mars has two moons, Phobos and Deimos. In a sense, these are the first moons we have met in our study of the Solar System, for Mercury and Venus have no moons at all, and the Earth's Moon is so large relative to its parent that we may consider the Earth and Moon as a double-planet system. The moons of Mars are mere chunks of rock, only 27 and 15 km across, respectively.

Phobos and Deimos are very minor satellites. They revolve very rapidly; Phobos completes an orbit of Mars in only 7 hours 40 minutes, less than a martian day. Deimos orbits in about 30 hours, slower than Mars's rotation period. Because Phobos orbits more rapidly than Mars rotates, an observer on the surface of Mars would see Phobos moving conspicuously backwards compared to Deimos, the other planets, and the stars. Because Mars's gravity continually pulls back on Phobos, it is slowly spiralling downward. It should hit Mars in only 30 million years!

It is amusing to note that in 1727, long before the moons were discovered, Jonathan Swift invented two moons of Mars for *Gulliver's Travels.* This is widely considered to have been a lucky guess, but it seemed reasonable at that time for numerological purposes. It may even have dated back to Kepler, who had considered in 1610 that the planets interior to the Earth—Mercury and Venus—had no known moons, the Earth had 1, while Jupiter, the next planet out from Mars, had 4 known moons. He reasoned that Mars should have the intermediate value of 2. Both Swift's moons and the real ones are very small and revolve very quickly. The two non-fictional moons were discovered in 1877 at a favorable opposition of Mars to Earth.

Mariner 9 and the Vikings made close-up photographs and studies of Phobos and Deimos. Each turns out to be not at all like our Moon, which is a spherical, planet-like body. Phobos and Deimos are just cratered chunks of rock; they may be fragments of a former single martian moon that collided with a large meteoroid. Alternatively, they could be the final remains of an early swarm of small moons that have since collided with Mars. Or they could be captured asteroids.

Phobos and Deimos are too small to have enough gravity to have made them round (Fig. 12–23). If you were standing on them, you could throw a rock fast enough so that it would escape into space!

**FIGURE 12–23** Mars's moons Deimos (*lower left*) and Phobos (*lower right*), with the asteroid Gaspra (*top*) for comparison, shown to the same scale. Deimos measures about 15 km × 12 km × 11 km. Its surface is heavily cratered and thus presumably very old. The surface seems smoother than that of Phobos, though, because a regolith 10 m thick covers Deimos. Phobos is about 27 km across. The spectra of Deimos and Phobos are similar to those of one type of dark asteroid. The Deimos and Phobos images were from Viking; the Gaspra image was from Galileo.

Phobos, which is only about 27 km in its longest dimension and 19 km in its shortest, has a crater on it that is 8 km across, a large fraction of Phobos's circumference. Deimos is even smaller than Phobos, only around 15 km in its longest dimension and 11 km in its shortest. From their sizes, we can calculate their albedoes; Phobos and Deimos turn out to be about as dark as the darkest maria on the Moon, with albedoes of only 7 or 8 per cent. From more detailed but similar comparisons of how the albedoes vary with wavelength, we deduce that they may be made of dark carbon-rich ("carbonaceous") rock, similar to some of the asteroids. They may even have originated in the asteroid belt.

We can conclude that the moons are fairly old because they have been around long enough for their surfaces to be extensively cratered. The rotation of both moons is gravitationally linked to Mars, with the same side always facing the planet, just as our Moon is linked to Earth.

## 12.7 FAILED MISSIONS TO MARS

A pair of Soviet spacecraft launched in 1988 was supposed to rendezvous with Phobos while orbiting Mars, and so they were named Phobos 1 and Phobos 2. The orbits were to be so similar in size and shape to that of Phobos that the spacecraft would essentially hover only 50 meters above Phobos's surface! They were even to shoot powerful laser beams and ion beams at Phobos's surface in order to analyze the spectrum of the debris. Part of each spacecraft was to land on Phobos. Unfortunately, an incorrect signal was sent to the first spacecraft while en route, and contact with the spacecraft was lost. The second spacecraft survived until it was in orbit around Mars, but it too failed just as it was about to go into orbit around Phobos. Still, it sent back 40 images of Phobos and mapped over 80 per cent of its surface. It measured a spectrum from the ultraviolet through the infrared that shows Phobos's low albedo, similar to that of the carbonaceous chondritic meteorites (Chapter 20), which is the same as that of some types of asteroids. The irregular nature of Phobos supports the idea that it is an asteroid captured by Mars. Since many asteroids are thought to have dry surfaces covering icy interiors, so may Phobos and Deimos. The interiors could provide water, oxygen, and hydrogen for expeditions

FIGURE 12–24 A half-billion-dollar photograph, one of two images of Mars taken by the Mars Observer before it exploded. The two images were taken when the spacecraft was 5.8 million km from Mars, 28 days before the encounter. The resolution in this image is about 20 km per pixel. Syrtis Major, a region of volcanic plains and dark sand dunes, is at center. The bright impact basin Hellas is at bottom.

around Mars. Accurate tracking of the spacecraft allowed scientists to refine their measurement of Phobos's density to 1.95 $g/cm^3$, a reduction of 10 per cent from the Viking measurements that makes it closer to the density of carbonaceous chondrites.

Though the Phobos 2 mission was truncated, some of the non-photographic instruments, including an infrared scanner for study of Mars itself, sent back most of the data for which they were programmed. The temperature maps were sharp and showed the physical characteristics of different regions of martian soil. Further, a French infrared spectrometer and a gamma-ray spectrometer aboard made a mineralogical map of Mars's surface. They showed that the water content of minerals on the slopes of the Tharsis volcanoes is as much as 20 per cent higher than the surrounding plains. Other instruments mapped water vapor and other gases in the martian atmosphere.

NASA launched its Mars Observer spacecraft in 1992 to map surface chemistry, study atmospheric content and seasonal weather patterns, and make a radar map (Fig. 12–24). Mars Observer failed just as it approached Mars, probably from a small explosion while adjustments were being made in its fuel lines.

The Russians have also been planning to send a series of spacecraft to study the surface of Mars. Unfortunately, their mission launched in late 1996 did not escape Earth's gravity and was destroyed. The Russian space program is now so underfunded that no replacement or successor is foreseen.

In the next section, we will discuss Mars Pathfinder, which arrived on Mars in 1997. It was done on a low budget, less than a quarter the budget of the failed Mars Observer. The idea was to have many inexpensive spacecraft for planetary exploration instead of a few very expensive and elaborate ones, so that even if one failed, others would work. Mars Pathfinder's success apparently made NASA too cocky, and NASA's "better, faster, cheaper" mantra apparently got too cheap to have adequate quality control. When both of 1999's NASA spacecraft failed, the old engineering adage "better, faster, cheaper: choose two" seemed more appropriate. But it was too late to save Mars Polar Lander and Mars Climate Orbiter.

## 12.8 MARS PATHFINDER

**Mars Pathfinder**
July 4, 1997–March 10, 1998
(dates on Mars)

In 1996, NASA launched two small, lightweight missions that arrived at Mars in 1997. One spacecraft, Mars Pathfinder, landed on July 4, 1997, in an old streambed. Images confirm that water has massively flowed here in the past, since all the rocks are

FIGURE 12–25 A 360° panorama from Mars Pathfinder. The Mars Pathfinder's rover, named Sojourner, landed on July 4, 1997, on the flood plain of Ares Vallis. The team running the observations named features seen after cartoon characters.

tilted in the same direction. A wide variety of shapes and colors of rocks is present, washed down by ancient floods to this site.

Thousands of photographs from both the Lander, known as the Carl Sagan Memorial Station, and from the Sojourner rover (Fig. 12–25) have been radioed back to Earth, showing Mars's many types of rocks, distant hills, and pinkish sky (caused by blowing reddish dust).

Mars Pathfinder carried a small rover, named Sojourner after the 1850s abolitionist Sojourner Truth. The Sojourner rover carried a device that when placed directly against a rock would then send out nuclear particles and record the backscattered radiation. From that information, scientists on Earth can tell what elements are in the rocks, and can use that information to deduce what minerals are present.

## 12.9 MARS GLOBAL SURVEYOR

**Mars Global Surveyor**
Sept. 12, 1997–present
(dates orbiting Mars)

The other spacecraft launched in 1997, the Mars Global Surveyor, is orbiting Mars to study weather and geology. Its pixels (individual picture elements), at its altitude above the surface, show details as small as 1.5 m across. It carries backup copies of several of the instruments aboard the failed Mars Observer. It is working so well that we can almost forget that Mars Climate Orbiter failed. In fact, we can almost forget that Mars Global Surveyor didn't work perfectly, and that it took an extra year to get it into its proper orbit around Mars.

One of its astounding results, reported in 2000, is that liquid water may exist near, or even on, Mars's surface. A few of the extreme closeups shown by Malin Space Science Systems' Mars Orbiter Camera—less than 1 per cent—show erosional features. They look like gullies or like deposits that on Earth are produced by water seeping from springs (Fig. 12–26). They show erosion at the top, a discrete channel flowing downward, and deposition of material at the bottom. Geologists report that these features can be accounted for by seepage from a spring. The water carves channels in the slope, and the surfaces above collapse (Fig. 12–27).

The resolution is so high that sometimes even individual boulders show, as well as sometimes the paths the boulders took in rolling down. The features, further, appear to be young in a geological sense (which could be a few years or a few million years). Scientists conclude that water must be closer to Mars's crust than had been assumed—hundreds instead of thousands of meters. It had been thought that the top layers of Mars's crust were too cold for liquid water to be present. Models have

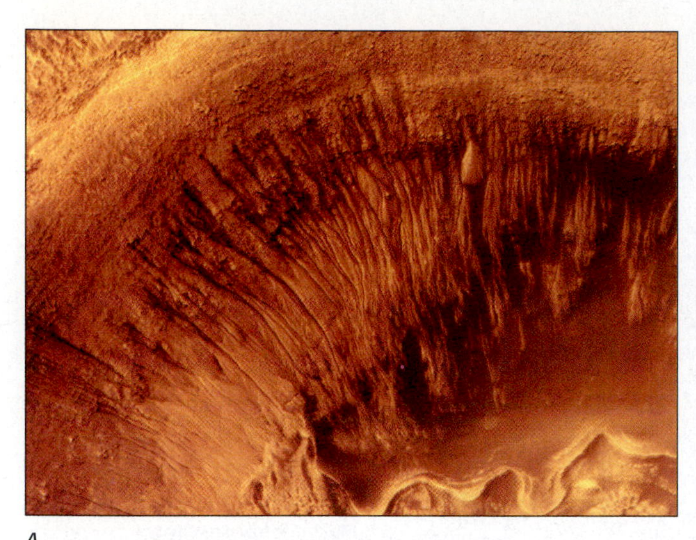

A

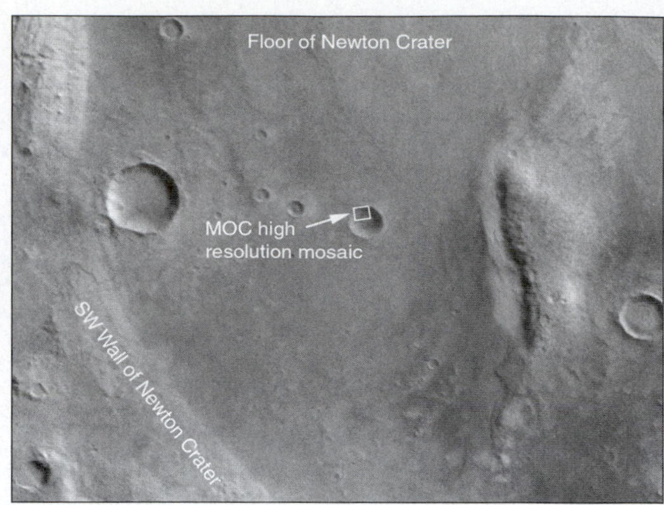


B

**FIGURE 12–26** Newton Crater on Mars, a large basin formed by an asteroid impact that probably occurred over 3 billion years ago. (*A*) A Mars Global Surveyor high-resolution view of the north wall of a small subsidiary crater, also formed by an impact. It is about 7 km across, seven times bigger than Meteor Crater in Arizona. The Mars Orbiter Camera reveals many narrow gullies eroded into the crater. They apparently formed by flowing water and debris. The debris created lobed and finger-like deposits at the base of the crater. Many of the finger-like deposits have small channels, which indicates that a liquid—most likely water—flowed in these areas. Each of hundreds of individual flow events would have constituted a competition among evaporation, freezing, and gravity. (*B*) A wider view of Newton Crater, obtained from the Viking 1 orbiter, with the location and size of the high-resolution image marked.

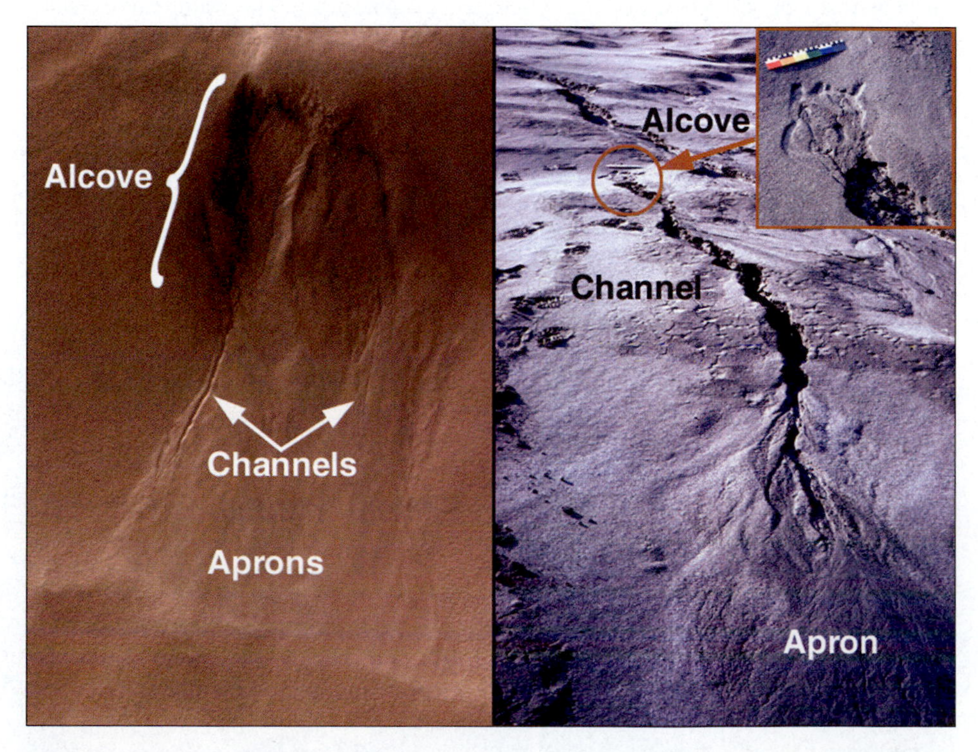


**FIGURE 12–27** Signs of flowing water, imaged with Mars Global Surveyor. As is currently understood, water seeps from between layers of rock on the cliff or crater wall. An alcove forms above the site of seepage, as water comes out of the ground and undermines the material from which it is seeping. The alcove forms from eroded material that collapsed and slid downhill.

A

B

**FIGURE 12–28** Layered rock, apparently sedimentary, imaged with Mars Global Surveyor.

to explain, also, how the water survives long enough on the surface before it evaporates in Mars's extremely low atmospheric pressure. If the water flowed a few million years ago, as cannot be ruled out, then the different tilt of Mars's axis may have played a role, making that part of the surface warmer. There are many doubters that the flow of water has been proven, so further observations and further theoretical analysis are necessary.

Some scientists have suggested that the gullies are formed by breakouts of liquid carbon dioxide rather than water.

The presence of water, as we have seen, is thought to be necessary for the origin and survival of life. So the apparent discovery that water on Mars is closer to the surface than expected is of great interest to those searching for extraterrestrial life. If further studies confirm the tentative conclusion about flowing water, the search for life will go on at even a greater interest level.

Mars Global Surveyor also determined that Mars's crust is significantly magnetized, which is consistent with the idea that a dynamo once existed in molten material in Mars's core. Though Mars would have had a global magnetic field long ago, it now has none.

Overall, Mars Global Surveyor has mapped Mars sufficiently to greatly improve models of the crust and lithosphere. It has found globally distinctive zones in which Mars's crust has evolved, with basalts in the south and andesites in the north. If the extensive layering visible in Valles Marineris and elsewhere is possibly volcanic in nature, it would indicate that volcanism played a larger role in Mars's history than had been thought.

But Michael Malin, whose camera has taken all the images, has concluded that he sees sedimentary rocks on Mars (Fig. 12–28). The layers of rock he sees would have been laid down in lakes and shallow seas. They come from an era 4.3 to 3.5 billion years ago, and indicate that Mars then had a warmer climate. Such a climate might have been more favorable to life originating there.

## 12.10 FUTURE MARS MISSIONS

The Japanese launched their Nozomi (formerly Planet-B) spacecraft to Mars in 1998. It won't reach Mars until 2004. It carries a dozen instruments, including a camera to take global images of Mars and closeup images of Phobos and Deimos. It will study also Mars's upper atmosphere and its interaction with the solar wind.

NASA's Mars Odyssey, launched in April 2001, as of this writing is to arrive at Mars in October 2001. It is an orbiter, and is to study the types of minerals on Mars's surface. By measuring how much hydrogen can be found in the shallow subsurfaces of the planet, it may reveal the past history and present status of water on Mars. Also, it will measure how much radiation reaches Mars's surface, important to know before we send humans there. It also carries an infrared imager but no optical camera.

Because the Earth and Mars are relatively close together in their orbits every 26 months, it requires less rocket power and time for missions then. Thus launches are spaced at that interval.

In 2003, NASA is to launch twin rovers to arrive on Mars in January 2004. The spacecraft are to investigate the geology of the surface with special attention to signs of subsurface water. Cornell scientist Steve Squyres, who is Principal Investigator of the Athena science laboratory on board, says of the rovers' scientific package, "It has everything a human field geologist has, and then much more. It has 20/20 vision, the ability to get inside rocks using the rock abrasion tool; it has spectrometers to tell us what rocks are made of; and it has a microscopic imager to tell us what things look like at a fine scale. It can use all these tools together to read the geologic record at the landing site and to tell us what conditions were once like, how much water was there, and, in particular, how habitable the site was—how suitable it would have been as an abode for life."

Further off, and thus less certain, are a series of NASA missions that will culminate in a return to Earth of martian soil. In 2005, a Mars Reconnaissance Orbiter is to continue the work of Mars Global Surveyor. This powerful spacecraft is to measure thousands of martian landscapes at 20- to 30-cm resolution. In 2007, a lander is to be sent. It will carry a hazard-avoidance system to improve its chances of land-

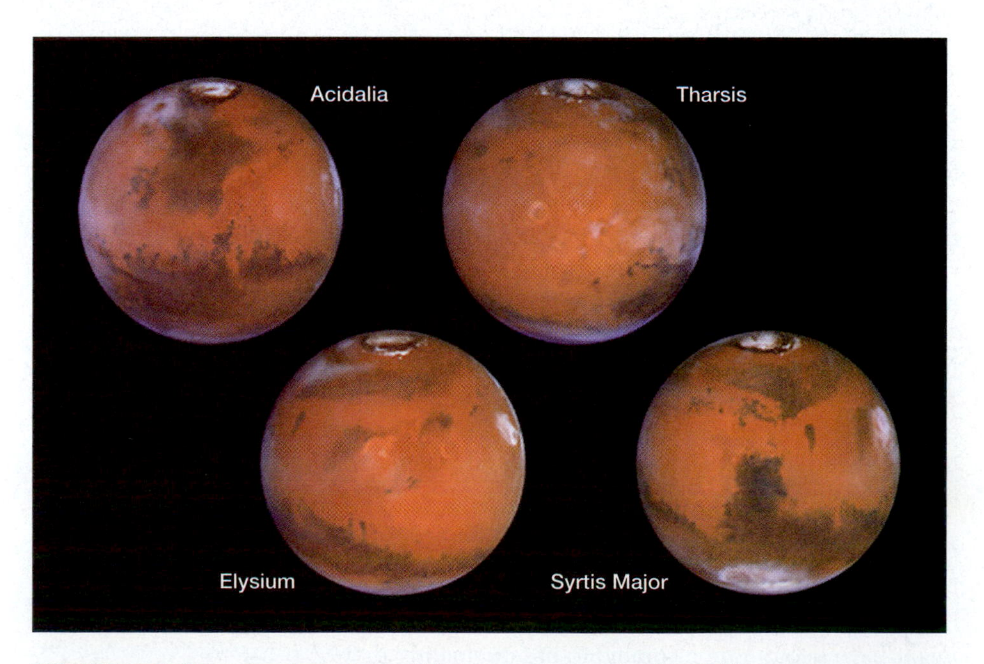


**FIGURE 12–29** Mars from the Hubble Space Telescope's Wide Field and Planetary Camera 2 in 1999, during its closest approach to Earth in 8 years. Details as small as 20 km show. There have been few major dust storms recently, perhaps the explanation of the measured cooling by 20°C since the time of our Viking visits. The cooling shows up as high cirrus clouds that show especially on the limb. Though the larger bright and dark markings remain stable, some of the bright smaller markings seen from Viking 25 years ago are now dark and vice versa as they are covered and uncovered by sand and dust.

ing safely. NASA hopes to have a robotic mobile science laboratory aboard. Also in 2007, NASA will launch a low-budget orbiter, the first of a series called Scout; it might carry balloons or a small airplane, either of which would carry scientific instruments. In that same year, NASA and the Italian Space Agency may send a satellite whose primary purpose in orbiting Mars would be to relay communications from other spacecraft. NASA and the French Research Ministry are planning to collaborate on small landers. In 2009, NASA and the Italian Space Agency may send a spacecraft that would prospect for water from orbit with a radar; as we saw for Mercury, such radar signals penetrate the ground.

When will the sample-return mission go? It may be launched in 2014, with material returned to Earth two years later. No crewed missions are currently on NASA's drawing board. Mars is so far away that, although Apollo astronauts got to the Moon in four days, astronauts would have to travel a couple of years to get to Mars. (Uncrewed rockets can be launched at higher speeds.) Human physiology joins cost in limiting this venture.

Images of Mars from Earth orbit with the Hubble Space Telescope (Fig. 12–29) continue to be made every few months.

## 12.11 ROCKS FROM MARS

Though no samples have been returned from Mars to Earth by spacecraft, 14 meteorites (Chapter 20) have been identified as coming from Mars. The ratio of isotopes of oxygen they carry does not appear on the Earth, the Moon, or in other meteorites. At first, scientists were skeptical that we had actual martian rocks on Earth, but theoreticians showed how the impact of a large meteorite on Mars could blast off material without heating it excessively, putting it on a gentle path toward Earth.

Three of the rocks, which are basalts, are 1.3 billion years old, presumably the age of the volcanic flows on Mars that formed them. They are called nakhlites, after the prototype, the Nakhla meteorite. Another group is only 165 million years old, only 4 per cent the age of Mars. They are the shergottites, after the prototype, Shergotty. The most recent to be put in this group, named Los Angeles, was identified in the year 2000, though it had been found a few years earlier in the desert near Los Angeles. We know the Mars meteorites were ejected from Mars less than 12 million years ago, from studying the tracks of cosmic rays in them. The minerals in them are similar to those found at hot springs on Earth. Might hot springs exist on Mars and be sites where life has formed?

Within one of the 1.3 billion-year-old meteorites, a rustlike material has been found that is only half that age. Since the rust is a sign that water was present, and since that age is too old for the result to be a contaminant picked up on Earth, the finding indicates that water flowed on Mars within the last 15 per cent of martian history. Some minerals found may indicate even more recent water flows. And we have already discussed the Mars Global Surveyor evidence that would fit with very recent water flows.

Twelve of the Mars meteorites are thought to come from the region of Olympus Mons and the Tharsis volcanoes. After all, that is a young volcanic region thought to be of the appropriate age. The spectra of the rocks seem to match. But why more meteorites have not been identified from other parts of Mars is not understood.

One meteorite thought to be from Mars, dated at 4.5 billion years old instead of the younger ages of the others, has been reported to contain organic compounds that could even be fossilized remains of former life. Named ALH84001, it is so old that it must be from Mars's earliest crust, as soon as it solidified from magma. A team of scientists has claimed that this and some other tantalizing evidence indicate that life may have existed on Mars over 3.6 billion years ago. But there are more doubters than believers. Chains of magnetic crystals found in the meteorite are claimed to be the latest evidence in favor of life having formed—if they are not a terrestrial con-

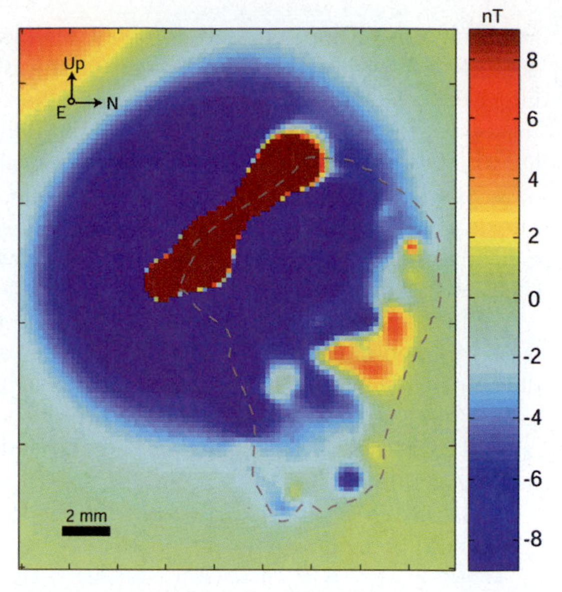


**FIGURE 12–30** ALH84001, a meteorite from Mars found on Earth. (*left*) A small piece of the fusion crust (marked with arrows) on the surface. This crust formed during the meteorite's passage through the Earth's atmosphere. (*right*) The magnetic field measured for this piece. Colors show the field strength in nanoteslas (nT). The crust was remagnetized by Earth's field. But the original varied pattern of magnetization survived inside the meteorite, showing that it wasn't heated enough to destroy the pattern.

tamination. Some research shows that meteorites could have been ejected from Mars without being heated greatly (Fig. 12–30), which could even allow life to survive the ejection and passage to Earth. We discuss a range of this evidence in the discussion of the search for life in the Universe (Section 21.1).

Studies of how the meteorite loses magnetism when heated have revealed that it was never heated above 40°C (104°F), not far from human body temperature and therefore a temperature at which bacteria could surely survive.

## CORRECTING MISCONCEPTIONS

**✖ *Incorrect***

Mars is not much farther than the Moon, so astronauts can go there without much more difficulty than the Apollo program.

We are certain that life never arose on Mars.

**✔ *Correct***

Mars is, even at its closest, 200 times farther than the Moon, so it is much more difficult to get to.

Most scientists agree that no signs of life have been found, but the evidence for liquid water on Mars leaves many hopeful.

## SUMMARY AND OUTLINE

Observations from the Earth (Section 12.1)
- Rotation; surface; dust storms; seasons

Space observations (Section 12.2)
- Many space probes sent from the United States and the U.S.S.R.
- Mariner 9 orbiter mapped all of Mars

General surface and atmosphere (Section 12.3)
- Volcanic regions indicating geological activity; canyons larger than Earth counterparts; craters and terraced areas near the poles; signs of water

Viking Orbiters (Section 12.4)
- Mapped surface in greater detail

Viking Landers (Section 12.5)
- Close-up of photographs of surface rocks; yellowish-brown rocks, dust, and sky; analysis of soil
- Search for life

Satellites of Mars (Section 12.6)
- Phobos and Deimos are both irregular chunks of rock.

Failed missions (Section 12.7)
- Russian Phobos, NASA Mars Orbiter

1999's smaller missions: Mars Climate Orbiter, Mars Polar Lander
Mars Pathfinder (Section 12.8)
Lander with rover
Mars Global Surveyor (Section 12.9)
Future missions (Section 12.10)
26-month intervals, sample return hoped for in 2011–2014
Rocks from Mars (Section 12.11)
Over a dozen meteorites from Mars found on Earth.
They didn't heat excessively, so life could have been preserved.
Controversial suggestion that fossils found is widely doubted.

## KEY WORDS

canali　syzygy　opposition　conjunction

## QUESTIONS

1. Outline the features of Mars that made scientists think that it was a good place to search for life.
2. If Mars were closer to the Sun, would you expect its atmosphere to be more or less dense than it is now? Explain.
†3. From the relative masses and radii (see Appendix 3), verify that the density of Mars is about 70 per cent that of Earth.
†4. Approximately how much more solar radiation strikes a square meter of the Earth than a square meter of Mars?
5. Compare the tallest volcanoes on Earth and Mars relative to the diameters of the planets.
6. What evidence is there that there is, or has been, water on Mars?
7. Why is Mars's sky pinkish?
8. Describe the composition of Mars's polar caps. Explain the evidence.
9. List the various techniques for determining the composition of the atmosphere of Mars.
10. Compare the temperature ranges on Mercury, Venus, Earth, and Mars.
11. List the evidence from Viking for and against the existence of life on Mars.
12. (a) Aside from the biology experiments, list three types of observations made from the Viking landers. (b) What are two types of observations made from the Viking orbiters?
13. Why aren't Phobos and Deimos regular, round objects like the Earth's Moon?
14. Describe the accomplishments and failures of the Soviet Phobos mission and the U.S. Mars Observer mission.
15. Compare the capability of the Hubble Space Telescope with that of spacecraft for imaging of Mars.
16. What was the capability of Mars Pathfinder and what did it find?
17. What is the capability of Mars Global Surveyor and what is it finding?

† This question requires a numerical solution.

## TOPICS FOR DISCUSSION

1. Discuss the problems of sterilizing spacecraft, and whether we should risk contaminating Mars by landing spacecraft. Now that we know that rocks from Mars reached Earth without being heated very much, does that change your conclusion?
2. Discuss the additional problems of contamination that crewed exploration would bring. Analyze whether, in this context, we should proceed with crewed exploration in the next decade or two.
3. Should a crewed mission to Mars be a major goal for the United States?
4. Should missions to Mars be international?

## USING TECHNOLOGY

### W World Wide Web

1. Examine the images of Mars and its moons visible on The Nine Planets, and contrast them with the images published in this book.
2. Explore the Mars Pathfinder and Mars Global Surveyor homepages at JPL (http://mars.sgi.com) as well as the Mars Global Surveyor homepage at Malin Space Science Systems (http://www.msss.com) to find out the latest about these Mars missions. Check http://mars.jpl.nasa.gov/odyssey for Mars Odyssey and http://athena.cornell.edu for the 2004 rovers.

### REDSHIFT

1. When is Mars visible tonight?
2. Travel through and around Mars and its system of moons.
3. Explore the surface of Mars, locating Valles Marineris, Olympus Mons, and several craters of your choice.
4. Observe Mars from Phobos.

Jupiter's Great Red Spot and the four Galilean satellites—Io, Europa, Ganymede, and Callisto. The symbol for Jupiter appears at the top of the facing page.

# Chapter 13

# Jupiter ♃

The planets beyond the asteroid belt—Jupiter, Saturn, Uranus, and Neptune—are very different from the four "terrestrial" planets. These **giant planets,** also called **jovian planets,** not only are much bigger and more massive but also are less dense. These facts suggest that the internal structure of these giant planets is entirely different from that of the four terrestrial planets.

**AIMS:** To describe Jupiter, the dominant planet of the Solar System; to discuss it as an example of a class of planets; to meet a new major set of bodies—Jupiter's moons; and to see how the Galileo spacecraft is now exploring the Jupiter system

## 13.1 FUNDAMENTAL PROPERTIES

The largest planet, Jupiter, dominates the Sun's planetary system. Jupiter is 5 A.U. from the Sun and revolves around it once every 12 years. It alone contains two-thirds of the mass in the Solar System outside of the Sun, 318 times as much mass as the Earth. Jupiter has at least 16 moons of its own and so is a miniature planetary system in itself. It is often seen as a bright object in our night sky, and observations with even a small telescope reveal bands of clouds across its surface and show four of its moons.

Jupiter, also called Jove, was the chief Roman deity. "Jovian" is the adjectival form of "Jupiter."

Jupiter is more than 11 times greater in diameter than the Earth. From its volume and mass, we calculate its density to be 1.3 g/cm$^3$, not much greater than the 1 g/cm$^3$ density of water. The density tells us that any core of heavy elements (such as iron) that Jupiter may have takes up a smaller fraction of Jupiter's mass than the cores of the inner planets make up of their planets' masses. Jupiter, rather, is composed mainly of the lighter elements, hydrogen and helium. Jupiter's chemical composition—almost entirely hydrogen and helium—is closer to that of the Sun and stars than it is to that of the Earth (Fig. 13–1). In Jupiter's outer parts, most of the hydrogen is in the form of hydrogen molecules ($H_2$). We can make direct measurements only for Jupiter's atmosphere. In fact, 86.1 per cent of Jupiter's atmosphere is hydrogen and 13.8 per cent is helium, so that those two gases comprise 99.9 per cent of the atmosphere. Because of the presence of so much hydrogen and the high pressures and temperatures in the planet's deeper layers, other molecules involving hydrogen besides $H_2$ are not uncommon: We find ammonia ($NH_3$) instead of the nitrogen molecules ($N_2$) of our atmosphere and methane ($CH_4$) instead of the carbon dioxide of the atmospheres of the inner planets. Still, though they are more visible than the hydrogen and helium for technical reasons having to do with their spectra, the ammonia and methane total only about 0.1 per cent of the atmosphere.

Jupiter, though 318 times more massive than Earth, is still only one tenth of 1 per cent the mass of the Sun.

Jupiter isn't solid; it has no crustal surface at all. At deeper and deeper levels, its gas just gets denser and denser, eventually liquefying. We know from Jupiter's average density that it must have heavier elements at its core, but we know little about them.

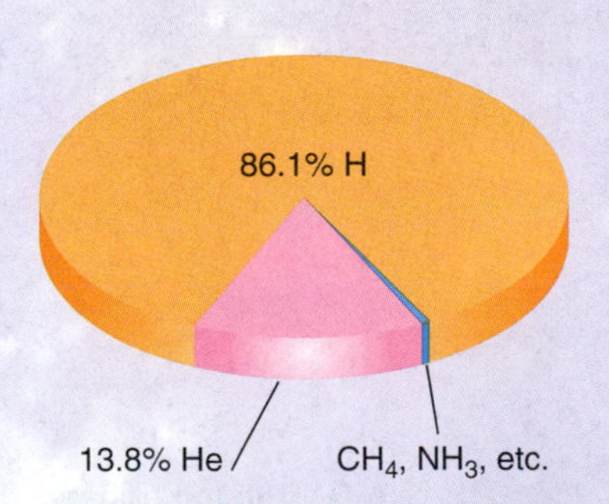


**FIGURE 13–1** The composition of Jupiter.

FIGURE 13–2 Jupiter, photographed from the Earth, shows belts and zones of different shades and colors. By convention (that is, the set of terms we arbitrarily choose to use), the bright horizontal bands are called **zones** and the dark horizontal bands are called **belts.** Adjacent belts and zones rotate at different speeds.

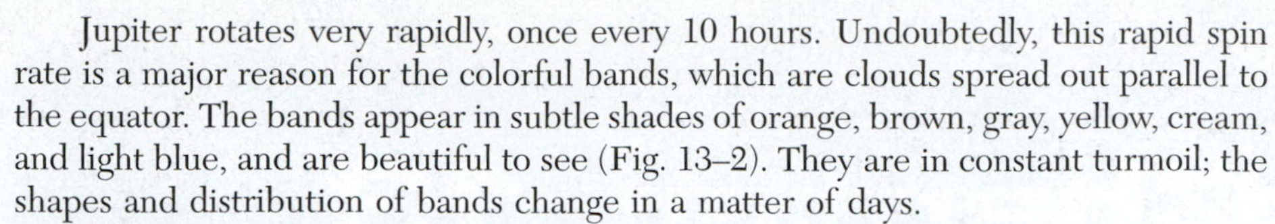

Jupiter rotates very rapidly, once every 10 hours. Undoubtedly, this rapid spin rate is a major reason for the colorful bands, which are clouds spread out parallel to the equator. The bands appear in subtle shades of orange, brown, gray, yellow, cream, and light blue, and are beautiful to see (Fig. 13–2). They are in constant turmoil; the shapes and distribution of bands change in a matter of days.

The most prominent feature of the visible cloud surface of Jupiter is a large reddish oval known as the **Great Red Spot.** It is about 13,000 km × 26,000 km, larger than the Earth, and drifts about slowly with respect to the clouds as the planet rotates. The Great Red Spot is a relatively stable feature, for it has been visible for over 350 years. Sometimes it is relatively prominent, and at other times the color may even disappear for a few years. Other, smaller spots are also present. In particular, three white spots formed about 60 years ago. Two merged in 1998 and the third joined the merger in 2000, forming a giant storm 12,000 kilometers long.

Jupiter's rapid rotation makes the planet bulge at the equator. The distance from pole to pole is 7 per cent less than the equatorial diameter; we say that Jupiter is **oblate** (Fig. 13–3). Jupiter's axis of rotation is only 3° from the axis of Jupiter's orbit around the Sun, so that, unlike the Earth and Mars, Jupiter has no seasons.

In 1955, intense bursts of radio radiation were discovered to be coming from Jupiter. Though it was quite a surprise to detect any radio signals at all, several kinds of signals were detected. Some bursts are correlated with the positions of Jupiter's innermost moon, Io, with respect to Jupiter and to Earth. At shorter radio wavelengths, Jupiter emits continuous radiation.

The fact that Jupiter emits radio waves indicates that Jupiter, even more so than the Earth, has a strong magnetic field and strong **radiation belts** (actually, belts filled with magnetic fields in which particles are trapped, large-scale versions of the Van Allen belts of Earth). We can account for the presence of this radio radiation only with mechanisms that involve such magnetic fields and belts. High-energy particles passing through space interact with Jupiter's magnetic field to produce the radio emission (Fig. 13–4). The Cassini spacecraft imaged Jupiter's magnetosphere when it passed by in 2000–2001 en route to Saturn.

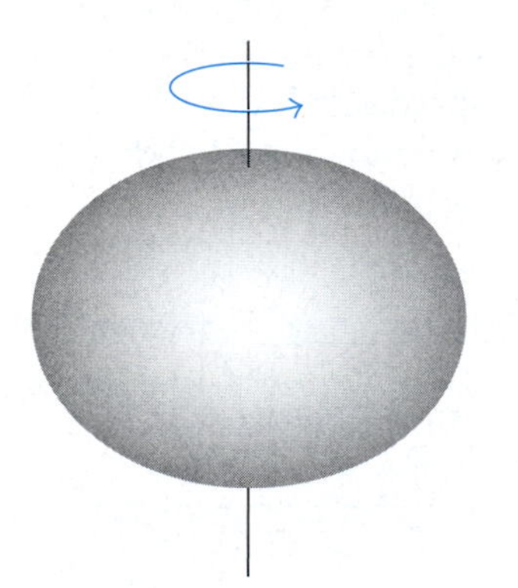

FIGURE 13–3 The top ellipsoid is **oblate;** the bottom ellipsoid is prolate.

## 13.2 JUPITER'S MOONS: EARLY VIEWS

Jupiter has at least 17 satellites. Four of the innermost satellites were discovered by Galileo in 1610 when he first looked at Jupiter with his telescope. These four moons are called the **Galilean satellites.** One of these moons, Ganymede, at 5276 km in diameter, is the largest satellite in the Solar System and is larger than the planet Mer-

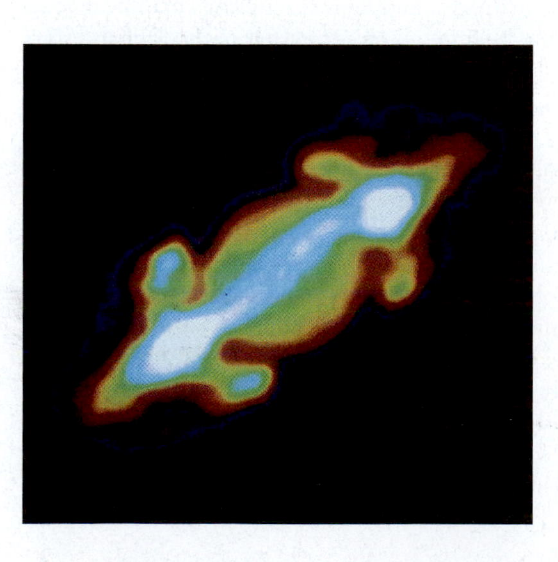

FIGURE 13–4 The radio emission from Jupiter, imaged with the Very Large Array (VLA), the giant set of radio telescopes in New Mexico, shows that planet's radiation belts. Observations were taken at a wavelength of 21 cm. The emission comes from electrons circling around the magnetic field in the Van Allen belts and extends beyond 4 Jupiter radii. White shows the strongest emission.

WHAT'S IN A NAME?

## BOX 13.1 Jupiter's Satellites in Mythology

All the moons except Amalthea are named after lovers of Zeus, the Greek equivalent of Jupiter. Amalthea, a goat-nymph, was Zeus's nurse, and out of gratitude he made her into the constellation Capricorn. Zeus changed Io into a heifer to hide her from Hera's jealousy; in honor of Io, the crescent moon has horns.

Ganymede was a Trojan youth carried off by an eagle to be Jupiter's cupbearer (the constellation Aquarius). Callisto was punished for her affair with Zeus by being changed into a bear. She was then slain by mistake, and rescued by Zeus by being transformed into the Great Bear in the sky. Jealous Hera persuaded the sea god to forbid Callisto to ever bathe in the sea, which is why Ursa Major never sinks below the horizon.

Europa was carried off to Crete on the back of Zeus, who took the form of a white bull. She became Minos's mother. Pasiphae (also the name of one of Jupiter's moons) was the wife of Minos and the mother of the Minotaur.

***Discussion Questions***

**1.** What are some other myths that have been or could be invented that involve the positions or motions of constellations in the sky? For example, can the colors of leaves in autumn come from color spilling out of the Big Dipper, as a Native American myth holds?

**2.** Suggest names that the International Astronomical Union might consider for the newly discovered satellites of Jupiter. The names must be associated with classical myths involving Jupiter.

cury. The moons have been observed from a variety of telescopes on the ground and in space.

The Galilean satellites have played a very important role in the history of astronomy. The fact that these particular satellites were noticed to be going around another planet, like a solar system in miniature, supported Copernicus's heliocentric model of our Solar System. Not everything revolved around the Earth! A half-century later, Ole Rømer's study of the times that the moons eclipse each other led him to deduce—correctly—that light travels at a finite speed. When Jupiter and its moons are moving away from Earth, the times between eclipses of the moons by Jupiter are longer than when Jupiter and its moons are approaching. (Question 4 at the end of this chapter deals further with this important result.)

The dozen or so moons of Jupiter that were discovered before the most recently found set of small moons fall into three groups. The first group includes the five innermost satellites then known (Amalthea and the Galilean satellites), which are also the largest. Their orbits are all less than 2 million km in radius. The orbits of the four moons in the second group are all about six times farther out. The orbits of the four outermost moons in the third group are another five to six times farther out than those of the first group. These outermost moons revolve in the direction opposite from that of Jupiter's rotation (that is, retrograde), while the other moons revolve prograde. Moons in the outer two groups may be bodies captured by Jupiter's gravity after their formation elsewhere, for they have high inclinations and/or eccentricities.

Additional moons have been discovered in the last few decades from the ground or from space.

## 13.3 SPACECRAFT OBSERVATIONS

The Pioneer 10 and Pioneer 11 spacecraft gave us our first close-up views of Jupiter in 1973 and 1974 as they flew by. A second revolution in our understanding of Jupiter occurred in 1979, when Voyager 1 and Voyager 2 also flew by Jupiter. The Ulysses spacecraft sent back a small amount of Jupiter data in the early 1990s when it used Jupiter's gravitational pull to put it into orbit high above the Sun. And a third

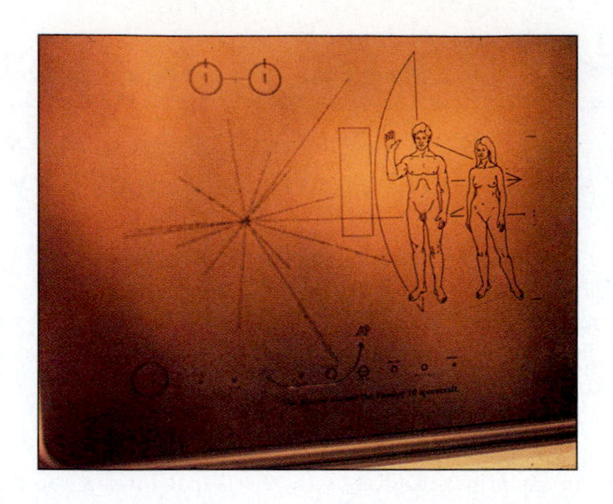

**FIGURE 13–5** The plaques borne by Pioneers 10 and 11. A man and a woman are shown standing in front of an outline of the spacecraft, for scale. The spin-flip of the hydrogen atom (Section 33.4) is shown at top left. It also provides a scale, because even travellers from another solar system would know that the wavelength that results from the spin-flip is a certain length, which is 21 cm in our units. The Sun and planets (including distinctively ringed Saturn) and the spacecraft's trajectory from Earth are shown at bottom. (Will the visitors to our Solar System realize that we had not yet discovered the rings around Jupiter, Uranus, and Neptune?) The directions and periods of several pulsars are shown in the rays extending from the midpoint at left. Numbers are given in the binary system.

Project Galileo travelled by a VEEGA orbit: Venus-Earth-Earth Gravity Assist.

| | |
|---|---|
| Launch | Oct. 18, 1989 |
| Venus flyby | Feb. 9, 1990 |
| Earth flyby | Dec. 8, 1990 |
| Asteroid (Gaspra) | Oct. 29, 1991 |
| Earth flyby | Dec. 8, 1992 |
| Asteroid (Ida) | Aug. 28, 1993 |
| Jupiter arrival | Dec. 7, 1995 |

revolution occurred, from the Galileo spacecraft that was in orbit around Jupiter starting in 1995. An orbital spacecraft is obviously far superior to a one-time "flyby" spacecraft in gathering useful data. The Cassini spacecraft's 2000–2001 flyby, though, provided useful information in tandem with Galileo.

Each spacecraft carried many types of instruments to measure properties of Jupiter, its satellites, and the space around them. The observations made with the imaging equipment were of the greatest popular interest. The resolution of the Voyager images was five times better than that of the best images we can obtain from Earth, and some of the Galileo images were another 70 times better. The image transmission was remarkable for a vehicle so far away that its signals travel about an hour to reach us. The energy in the signal would have to be collected for billions of years to light a Christmas tree bulb for just one second!

The Pioneer and Voyager spacecraft are travelling fast enough to escape the Solar System. As they depart, the spacecraft are radioing back valuable information about other planets and about interplanetary conditions. In about 80,000 years, the spacecraft will be about three light-years away from the Solar System. The Pioneers bear plaques that give information about their origin and about life on Earth, in case some interstellar traveller from another solar system happens to pick up one of the spacecraft (Fig. 13–5). The Voyagers bear more elaborate mementos, as we shall describe in Chapter 21.

The most recent Jupiter spacecraft, NASA's Galileo, carried both an orbiter and an atmospheric probe. It took a slow route to Jupiter because of safety concerns ruling out the use of liquid-fueled rockets from crewed spacecraft like the space shuttle, from which Galileo was launched. Galileo received a gravity assist as it swung around Venus (!) and then received two gravity assists in two passes by the Earth

## A DEEPER DISCUSSION

### BOX 13.2 Gravity Assists

Jupiter's large mass makes it a handy source of energy to use to send probes to more distant planets. If you bounce a ball off a wall, it comes back to you at about the same velocity at which you threw it. But if you bounce the ball off a train that is coming right at you, the train adds energy to the ball and the ball comes back at you rapidly: the ball's speed as you threw it plus twice the train's speed toward you. Similarly, some of Jupiter's energy can be transferred to a spacecraft through a gravitational interaction even without the spacecraft physically hitting Jupiter. This **gravity-assist** method has sent Pioneer 11 and the Voyagers to Saturn, Voyager 2 beyond to Uranus and Neptune, and Galileo to Jupiter. It has been used in other planetary missions as well.

**FIGURE 13–6** Jupiter, photographed from the Cassini spacecraft in 2000 as it passed by en route to Saturn.

separated by two years. It arrived at Jupiter on December 7, 1995. Unfortunately, its main antenna did not open fully. The smaller antenna that did work allowed only a much slower data rate, so some of the expected results had to be sacrificed, but most of the important data were obtained.

The probe transmitted data for 57 minutes as it fell hundreds of kilometers through the jovian atmosphere. It gave us accurate measurements of Jupiter's composition; the heights of the cloud layers; and the variations of temperature, density, and pressure.

The orbiter has come so close to many of Jupiter's moons that pictures have much better resolutions even than those from the Voyagers. The Cassini spacecraft, on its path to Saturn, passed Jupiter in 2000–2001 (Fig. 13–6), travelling much more slowly than Voyager and was able to study Jupiter for about 5 months. The sections that follow will describe Jupiter as we now know it, illustrated with images from the Voyagers and from Galileo.

### 13.3a The Great Red Spot

The Great Red Spot shows very clearly in many of the images (Fig. 13–7 ). It is a gaseous island (or isolated storm) many times larger than the Earth. Its top lies about

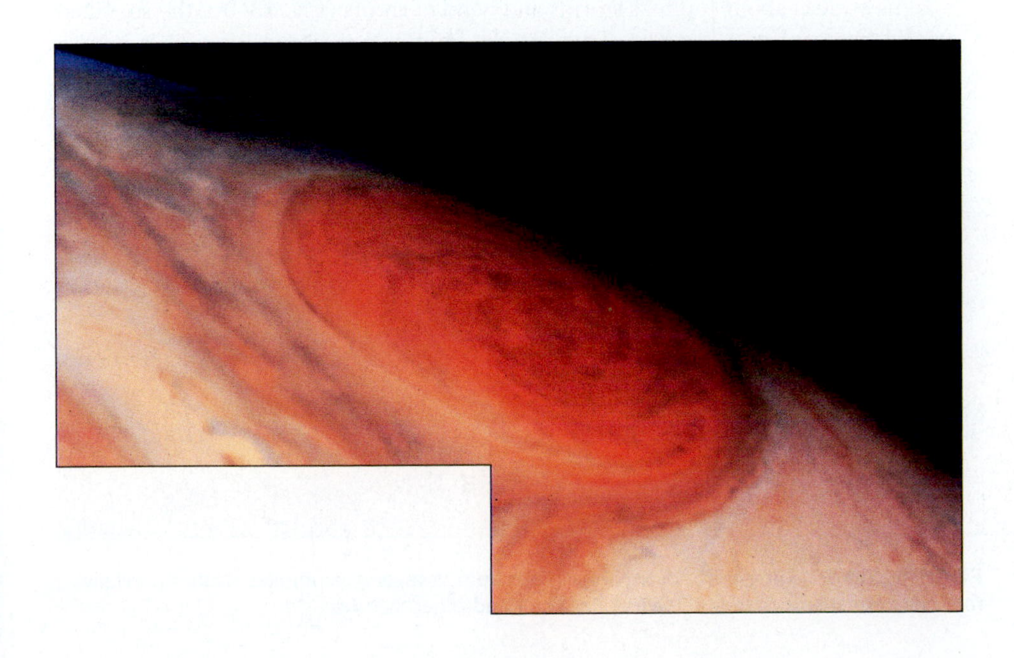

**FIGURE 13–7** The Great Red Spot, in an image from the Galileo spacecraft. It rotates in the anticyclonic sense (counterclockwise in that hemisphere). The Great Red Spot is now about 26,000 km by 13,000 km and has been shrinking for 50 years. It is probably mostly ammonia gas and ice clouds, and it goes 20 to 40 km deep.

**FIGURE 13–8** Jupiter from (*A*) Voyager 1, and (*B*) 4 ½ months later, from Voyager 2. Note how the white ovals have drifted around the Great Red Spot. Winds blow at different speeds at different latitudes, so clouds pass each other.

8 km above the neighboring cloud tops. Storms on Earth and other rotating planets usually rotate. Storms are called cyclones when their centers have low pressure and anticyclones when their centers have high pressure. The Great Red Spot is an anticyclonic storm lying in the southern hemisphere; it has counterclockwise winds (Fig. 13–8). We still have many things to learn about Jupiter's Great Red Spot. No upwelling has been detected in its center, though some would be expected in such a huge storm. We don't even know why it is red, though traces of phosphorus have been suggested as the coloring agent.

Time-lapse photographs made from Voyager data have revealed the flow within and around the Spot. Clouds within the Spot rotate counterclockwise with a period of about 6 days. The photographs also show how the Great Red Spot interacts with surrounding clouds and smaller spots. Similar spots, the so-called white ovals, occur at other latitudes and have been shown by Voyager to have similar internal flows, but none is as large or as colorful as the Great Red Spot.

In new ideas about Jupiter's storms and wind currents (Fig. 13–9 ), the so-called **Coriolis force** plays a key role. The Coriolis force is not a true force; it is merely an illusion caused by our motion on a rotating object. People are most familiar with

**FIGURE 13–9** The surface of Jupiter as seen from Voyager 1 is unrolled. Note the relative motions from side to side of the Great Red Spot and other features.

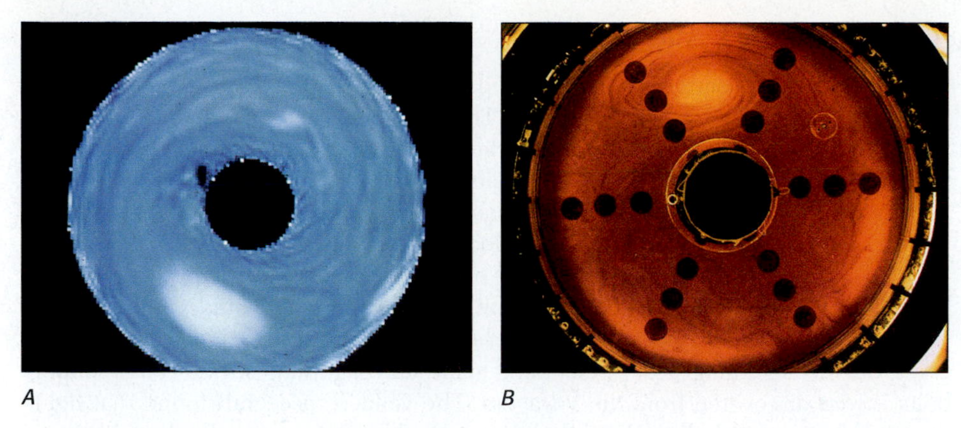

**FIGURE 13–10** (*A*) A University of California, Berkeley, computer model of the Great Red Spot, the result of merging many weaker spots. (*B*) A laboratory simulation of the Great Red Spot in a water-filled vat at the University of Texas. Researchers spin the vat to simulate Jupiter's rotation while pumping water in through the six valves nearest the center and out through the valves in the middle ring (the outer ring was not used). Vortices form, small vortices merge into larger ones, and a large vortex like Jupiter's Great Red Spot results.

it on Earth, using the Coriolis-force concept in weather forecasting to explain the motion of air with respect to the system of rotating latitude and longitude coordinates we fix to the Earth. Consider the following to explain the Coriolis force: Imagine you are looking down on the north pole of the Earth as it rotates from west to east. A point on the Earth's equator rotates with a greater velocity than a point at a latitude above (or below) the equator. If you were to shoot an object northward from the equator, the object has not only the northward velocity you gave it but also a relatively large eastward velocity because of the Earth's rotation. Since the higher latitude is not moving eastward as fast in km/hour, the object will land eastward of its original longitude. If we were at the higher latitude watching the projectile, we would think it was curving eastward. Similarly, an object shot southward from the equator would also appear to curve eastward. Though only the Earth's rotation and no actual change in direction is involved, if you were underneath looking up it would look as though the object has swerved eastward. Since we know that objects swerve because of forces, we make up a fictional force to explain this apparent swerve. This fictional force is known as the Coriolis force. Using the idea of a Coriolis force often simplifies calculations.

Jupiter rotates very rapidly, so the effect that we call the Coriolis force is particularly strong. Computer models show that under such conditions, large vortices like the Great Red Spot are produced. When many unequal spots are produced, the stronger ones absorb the weaker ones until only a single vortex is left. Such vortices last a long time. A terrestrial laboratory experiment has matched computer calculations and developed such a devouring and persisting vortex in a water tank (Fig. 13–10).

Alternative models, also using supercomputer calculations, indicate that the condition with one giant spot is a natural state, since it minimizes the energy present in the large-scale flow on Jupiter's surface. Often, in physics, the natural state is one of minimum energy, just as a ball runs down a hill and remains in equilibrium in a valley.

Though giant vortices aren't as obvious on Earth as they are on Jupiter, some large-scale (continental) weather systems on Earth—with high-pressure regions sitting in the same place for weeks—may correspond. Circulating rings that break off from the Gulf Stream in the Atlantic Ocean are other possible parallels. So our studies of Jupiter may have applications on Earth in understanding weather and ocean currents.

A

B

**FIGURE 13–11** (*A*) On ground-based infrared observations, the belts show the radiation characteristic of the temperature of a relatively deep layer of the atmosphere. The Great Red Spot and the zones appear relatively dark, because high clouds block some of the radiation from reaching us. This infrared view was taken with the NASA Infrared Telescope Facility on Mauna Kea. A 256 × 256 pixel infrared array allowed imaging all of Jupiter at one time. (*B*) An earlier view from Voyager shows belts and zones in the visible part of the spectrum.

### 13.3b Jupiter's Atmosphere

Jupiter has bright bands called **zones** and dark bands called **belts.** The zones are rising gas, while the belts are falling gas. The tops of these dark belts are somewhat lower (about 20 km) than the tops of the zones and so are about 10 K warmer (Fig. 13–11). The colors we see in the belts may have to do with sulfur compounds in the middle clouds, or perhaps with organic molecules.

Voyager measurements of wind velocities showed that each hemisphere of Jupiter has a half-dozen currents blowing eastward or westward. The Earth, in contrast, has only one westward current at low latitudes (the trade winds) and one eastward current at middle latitudes (the prevailing westerlies).

Extensive lightning storms, including giant-sized lightning strikes called "superbolts," were discovered from the Voyagers. The Galileo spacecraft found that lightning is associated with thunderhead systems, which contain water clouds and storms. The giant aurorae first found from the Voyagers are now monitored with the Hubble Space Telescope.

Pioneer 11 gave scientists their first look at Jupiter's poles, which never show well from Earth. At high latitudes the band pattern we see at the equator is destroyed. The bands break up into eddies. The Voyagers allowed detailed mapping of both poles. With such information, we can now study "comparative atmospheres": those of the Earth, Venus, Mars, Jupiter, Saturn, Uranus, and Neptune.

The probe (Fig. 13–12) from the Galileo spacecraft took direct samples of Jupiter's atmosphere. The helium concentration, for example, is somewhat more than expected but closely matches the Sun's, in line with the idea that Jupiter and the Sun coalesced out of the same primordial cloud. The biggest surprise was that the water-vapor content was much less than expected. The probe did not detect the thick clouds of water, ammonia, hydrogen sulfide, and their compounds that were expected. Months later, the Galileo orbiter sent back pictures showing towering clouds, like thunderclouds, on Jupiter, endorsing the possibility, now widely accepted, that the probe happened to go into an unusually hot and dry region. Such regions make up only about 1 per cent of Jupiter's surface. They reflect relatively little sunlight but emit infrared radiation. Dry, downwelling air in these hot spots stripped the regions of various condensible gases.

In all, the Galileo probe traversed about 600 km of Jupiter's atmosphere, only about 1 per cent of its radius. The probe found that Jupiter's winds were stronger

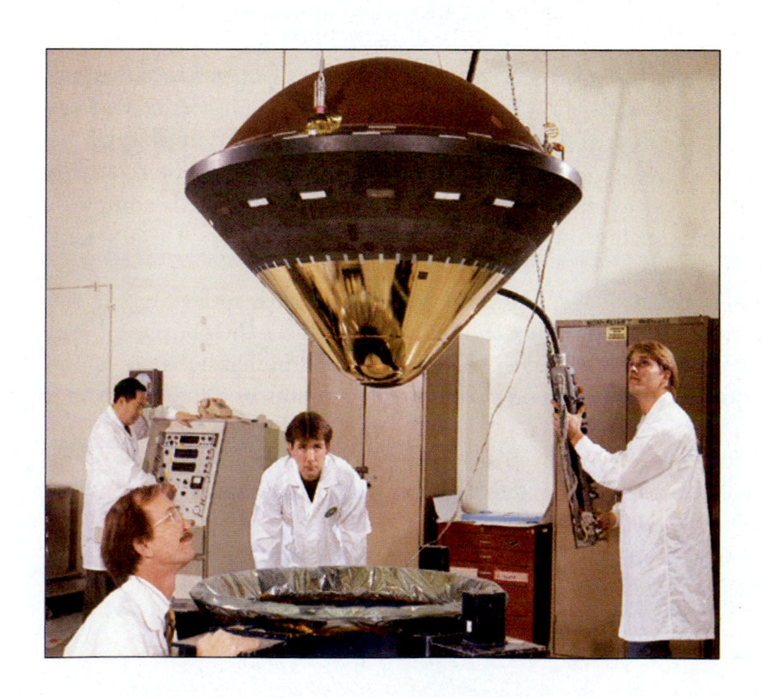

**FIGURE 13–12** The probe portion of Project Galileo. Six instruments inside the probe measured composition of Jupiter's atmosphere, located clouds, studied lightning and radio emission, found where energy is absorbed, and measured precisely the ratio of hydrogen to helium.

A DEEPER DISCUSSION

## BOX 13.3 Comet Shoemaker-Levy 9 Crashes into Jupiter

In what many observers called the most exciting week ever in astronomy, pieces of Periodic Comet Shoemaker-Levy 9 (shown in Chapter 19 on comets), D/1993 F2, crashed into Jupiter in July 1994. The comet had been in orbit around Jupiter and in perhaps 1992 had been broken into a chain of about 20 objects by Jupiter's tidal forces. In an unprecedented manner, almost every telescope in the world and in space was trained on Jupiter to see what would happen.

The results exceeded everyone's expectations. The impacts left huge Earth-sized spots on Jupiter (Figs. *A* and *B*) and often raised plumes that could be seen rising off Jupiter's edge. The Hubble Space Telescope saw the spots the most clearly, but they could be seen by telescopes of almost every size, including small backyard ones. The impacts heated Jupiter's gas to tens of thousands of degrees and so appeared bright in the infrared (Fig. *C*). Spectra showed sulfur and other elements, presumably dredged up from lower levels of Jupiter's atmosphere than we normally see.

Had any of these fragments hit Earth, they would have caused a crater as large as the state of Rhode Island, with dust thrown up to much greater distances.

### Discussion Question

Imagine that we hadn't known in advance about Comet Shoemaker-Levy 9. Would there have been a panic on Earth if amateur astronomers started picking up spots on Jupiter, one by one day after day?

(*A*) Jupiter, with scars from the July 1994 crash of Periodic Comet Shoemaker-Levy 9. The image was taken with the Hubble Space Telescope. (*B*) The shape of the ring left on Jupiter's surface showed that the fragment of Periodic Comet Shoemaker-Levy 9 that caused it entered at an angle. The image was taken with the Hubble Space Telescope. (*C*) This infrared image shows bright plumes of hot gas from the collisions of comet fragments with Jupiter. This image was taken with the Keck Telescope.

than expected, increasing with depth until the atmospheric pressure equalled several times Earth's. The wind speed levelled off at about 200 m/sec until the probe stopped sending data at a pressure of 24 times that of the Earth's atmosphere. These data show that energy to drive the winds comes from below, a different situation from the Earth's winds being driven by solar heating.

No lightning flashes were seen by the probe, though radio signals of tens of thousands of lightning strikes from farther away were recorded in the hour the probe lasted in Jupiter's atmosphere. This rate is a tenth that of lightning on Earth. Since lightning strikes make organic molecules, the organic component of Jupiter's atmosphere may be less than we had thought.

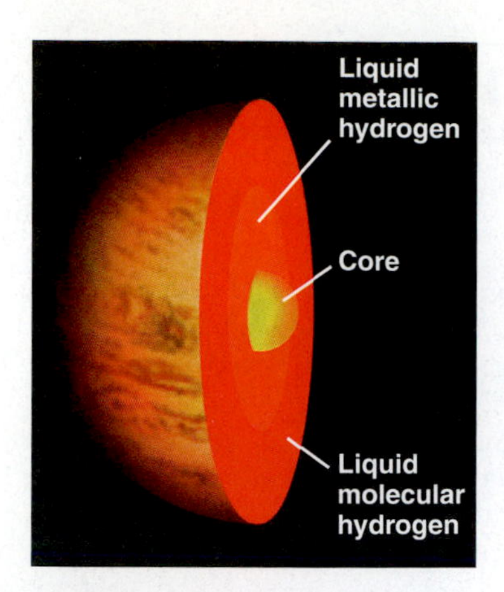


**FIGURE 13–13** A model of Jupiter's interior.

The probe found two to three times the solar values for various elements, including sulfur, nitrogen, carbon, and the rare gases argon, krypton, and xenon. These discoveries will be important for theories of how Jupiter formed.

The Cassini spacecraft's 5-month-long encounter during 2000–2001 enabled the spacecraft to make time-lapse movies of Jupiter's atmosphere.

### 13.3c Jupiter's Interior

Spacecraft data increased our understanding not only of the atmosphere but also of the interior of Jupiter (Fig. 13–13). Most of the interior is in liquid form. Jupiter's central temperature may be between 13,000 and 35,000 K. The central pressure is 100 million times the pressure of the Earth's atmosphere measured at our sea level because of Jupiter's great mass pressing in. Because of this high pressure, Jupiter's interior is composed of ultracompressed hydrogen surrounding a rocky core consisting of 20 Earth masses of iron and silicates. The heavy elements should have sunk to the center, but we have no direct evidence that they are there. If Jupiter were gaseous or fluid throughout, it would appear more flattened, more oblate, than it is. This fact is evidence that the core is solid and rocky.

The lower levels of the liquid hydrogen are in the form of molecules that have lost their outer electrons. In this state, the hydrogen conducts heat and electricity. Since these properties are basic to our normal terrestrial definition of "metal," we say that the hydrogen is in a "metallic" state. This liquid metallic region makes up 75 per cent of Jupiter's mass. It is probably where Jupiter's magnetic field is generated by the interaction of mass, motion, and an electric field. (This process also takes place in "dynamos" on Earth, which convert mechanical energy into electricity.)

A major limitation on our ability to model Jupiter's interior is our lack of understanding of hydrogen under extreme pressure. Scientists on Earth have been experimenting with ultrahigh pressures by using diamond anvils—bits of diamond with small ends that are forced together. Since pressure is force divided by area, the ability to have very small areas leads to very high pressures. It was discovered, for example, that at 2.5 million times the pressure of our atmosphere, hydrogen solidifies. Experiments continue. Among future steps is the measurement of the electrical conductivity of the solid hydrogen.

Jupiter radiates 1.6 times as much heat as it receives from the Sun, leading to Jupiter's complex and beautiful cloud circulation pattern. There must be some internal energy source—perhaps the energy remaining from Jupiter's collapse from a primordial gas cloud 20 million km across to a protoplanet 700,000 km across, five times the present size of Jupiter. Less energy has been released since, though Jupiter is undoubtedly still contracting. It lacks the mass necessary by a factor of about 75, however, to have heated up enough for nuclear reactions to begin, which would have made it into a star.

### 13.3d Jupiter's Magnetic Field

The Pioneer missions showed that Jupiter's tremendous magnetic field is even more intense than many scientists had expected, a result confirmed by the Voyagers (Fig. 13–14). At the height of Jupiter's clouds, the magnetic field is ten times that of the Earth, which itself has a strong field.

The inner field is "toroidal"—that is, shaped like a doughnut—containing several shells like giant versions of the Earth's Van Allen belts. High-energy protons and electrons are trapped there. The satellites Amalthea, Io, Europa, and Ganymede travel through this region. The Galileo spacecraft even discovered a mini-magnetosphere around Ganymede, in the midst of Jupiter's main magnetosphere. The middle region of Jupiter's magnetosphere, with charged particles being whirled around rapidly by the rotation of Jupiter's magnetic field, does not have a terrestrial counterpart. The outer region of Jupiter's magnetic field interacts with the solar wind

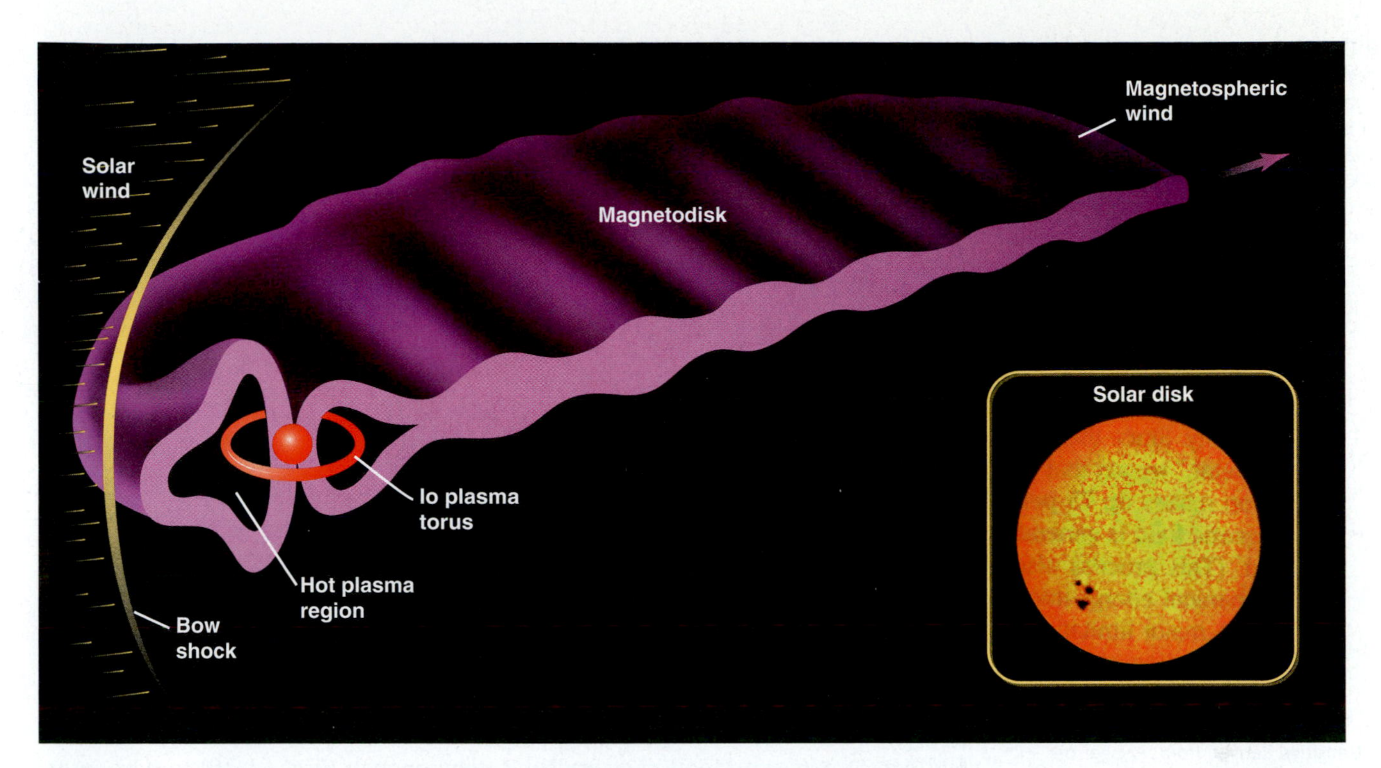


**FIGURE 13–14** An artist's view of Jupiter's magnetosphere, the region of space occupied by the planet's magnetic field. It rotates at hundreds of thousands of kilometers per hour around the planet, like a big wheel with Jupiter at the hub. The inner magnetosphere is shaped like a doughnut with the planet in the hole. The highly unstable outer magnetosphere is shaped as though the outer part of the doughnut had been squashed. The outer magnetosphere is "spongy" in that it pulses in the solar wind like a huge jellyfish. It often shrinks to one-third of its largest size. Volcanic emissions from Io form a torus of plasma; the plasma then escapes into the rest of the magnetosphere. Jupiter's rapid rotation heats some of the plasma to a few hundred million degrees. Eventually this hot plasma flows away from the planet at high speeds to form a magnetospheric wind. The size of the solar disk is shown for scale.

as far as 7 million km from the planet and forms a shock wave, as does the bow of a ship plowing through the ocean. When the solar wind is strong, Jupiter's outer magnetic field (flattened like a pancake) is pushed in.

Pioneer 11 observed a 10-hour fluctuation in the magnetic field. This fluctuation occurs because Jupiter's magnetic axis is tilted from the axis of rotation, and the center of the field is not exactly at the center of the planet. The magnetic field thus resembles a wobbly disk, with the 10-hour fluctuation resulting from a 10-hour period of rotation for Jupiter. Also, Jupiter's magnetic poles are reversed compared with the Earth's: our Earth's magnetic north pole is on the same side of the plane of the planetary orbits as is Jupiter's magnetic south pole.

The Voyagers discovered fascinating things about the magnetic fields of each of the giant planets, which we shall summarize in the Uranus and Neptune chapters. Galileo, in orbit around Jupiter, sampled Jupiter's magnetosphere extensively. The magnetic field causes auroral ovals (Fig. 13–15), like those on Earth.

Galileo and Cassini were able to simultaneously sample Jupiter's magnetosphere at different points during the latter's encounter with Jupiter in 2000–2001.

## 13.4 JUPITER'S MOONS CLOSE UP

Through Voyager close-ups, the satellites of Jupiter have become known to us as worlds with personalities of their own. The four Galilean satellites, in particular, were formerly known only as dots of light. Since these satellites range from 0.9 to 1.5 times

**FIGURE 13–15** Jupiter's magnetic field revealed by the northern and southern auroral ovals, imaged in the ultraviolet from the Hubble Space Telescope. They are about 500 km above the level in Jupiter's atmosphere where the pressure is the same as Earth's surface pressure.

Densities derived from Voyager observations (g/cm$^3$; remember that water has a density of 1.0 in these units)

| | |
|---|---|
| Io | 3.5 |
| Europa | 3.0 |
| Ganymede | 1.9 |
| Callisto | 1.8 |

The low densities indicate that Ganymede and Callisto contain large amounts of water ice. The high heat of Jupiter's first 100 million years drove off lighter elements from these two moons.

the size of our own Moon, they are substantial enough (Fig. 13–16) to have interesting surfaces and histories. Galileo images are now giving us even closer views. The Galileo data have led to new models of the satellites' interiors (Fig. 13–17).

Io provided the biggest surprises. Scientists knew that Io gave off particles as it went around Jupiter, but it was discovered with Voyager 1 that these particles resulted from active volcanos on the satellite (Fig. 13–18), which we can monitor with the Hubble Space Telescope in Earth orbit every few months (Fig. 13–19). Eight volcanos were seen actually erupting (Fig. 13–20), many more than erupt on the Earth at any one time. When Voyager 2 went by a few months later, most of the same volcanos were still erupting. (Incidentally, astronomers do not agree whether to say "eye′oh" or "ee′oh.") Galileo sees some of the same volcanos and some different ones (see the *Photo Gallery* on pp. 242–243).

**FIGURE 13–16** Jupiter's Galilean moons to scale. We see (from left to right) Io, Europa, Ganymede, and Callisto. Each is an individual, very different from the other moons. The International Astronomical Union names features on the moons of the outer planets after the people who discovered them or other relevant associations. For example, names on Europa have to do with southern Europe.

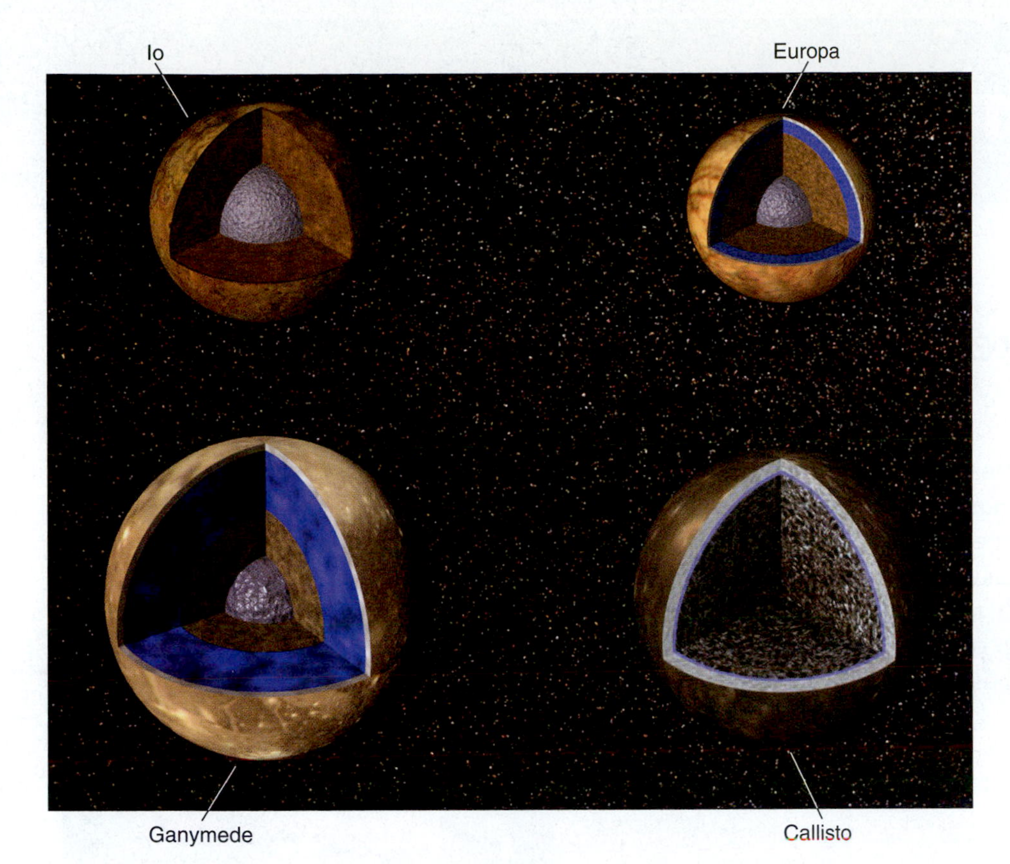


FIGURE 13–17 Models of the interiors of Jupiter's Galilean satellites. Data from the Galileo spacecraft indicate that Ganymede is separated into a metallic core, rock mantle, and ice-rich outer shell, while Io has a metallic core and rock mantle but no ice. Europa seems to have a metallic core surrounded by a rock mantle and a water ice-liquid outer shell perhaps 100 km thick.

Io's surface has been transformed by the volcanos and overall is the youngest surface we have observed in our Solar System. Gravitational forces from the other Galilean satellites and from Jupiter itself pull Io slightly inward and outward as it orbits around Jupiter. Tidal forces from these objects (as we saw, the difference between gravity on one side of Io and the gravity pulling on the other side) flex the satellite. Its surface moves in and out by as much as 100 meters. This squeezing and unsqueezing creates heat from friction, which presumably heats the interior and leads

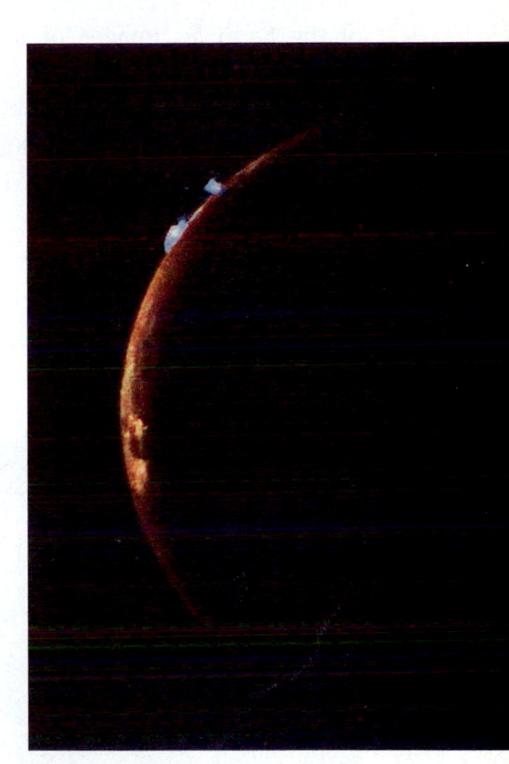

FIGURE 13–18 The volcanos of Io can be seen erupting on this photograph. Linda Morabito, a JPL engineer, discovered the first volcano. At the time, she was studying images that had been purposely overexposed to bring out the stars so that they could be used for navigation. She saw the large volcanic cloud barely visible on the right limb. The plume extends upward for about 250 km and is scattering sunlight toward us. Names on Io are from mythological characters that have to do with fire, including the Greek Prometheus and the Hawaiian volcano god Pele.

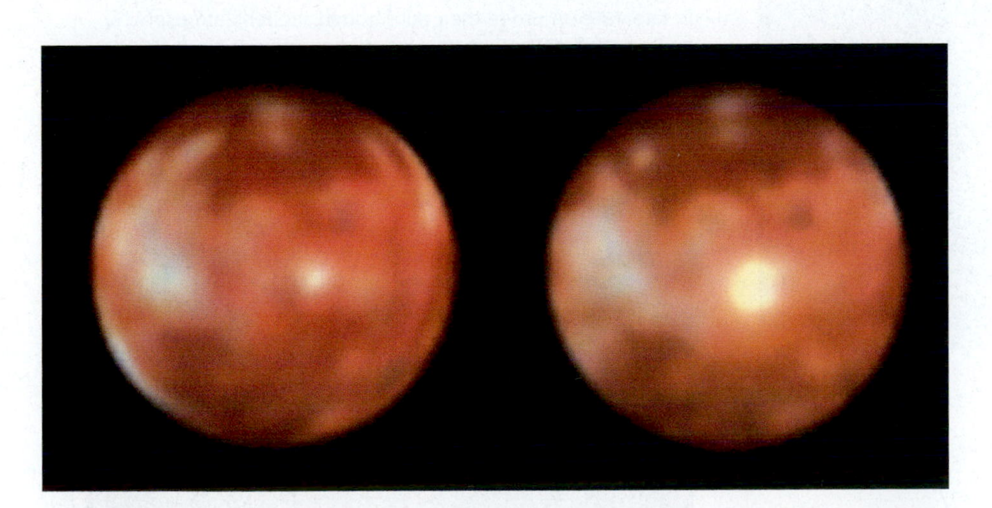

FIGURE 13–19 Monitoring Io with the Hubble Space Telescope reveals changes. Here a 300-km-diameter spot has emerged. The volcano Ra Patera has ejected frozen gas, possibly in a lava flow. Though the temperature on Io's surface averages about −150°C (−240°F), the volcanic hot spots can be 1000°C (1800°F). The images were taken with the Wide Field and Planetary Camera 2, through filters in the near infrared, violet, and yellow.

*(text continues on p. 244)*

# PHOTO GALLERY

## The Jupiter System

Though the Hubble Space Telescope can take images ten times finer than others from the distance of the Earth, its images of Jupiter's moons do not match those from Voyager or Galileo. Infrared images of Jupiter and Io from the Earth are of considerable value. The images now being obtained from the Galileo spacecraft will be unsurpassed for decades.

▲ The deployment of the Galileo spacecraft and its upper-stage rocket from a space shuttle in 1989. It arrived at Jupiter in late 1995. A probe then fell through Jupiter's atmosphere, and an orbiter went into a path from which we are viewing Jupiter and its moons.

▲ Io in front of Jupiter, imaged from Cassini in 2001.

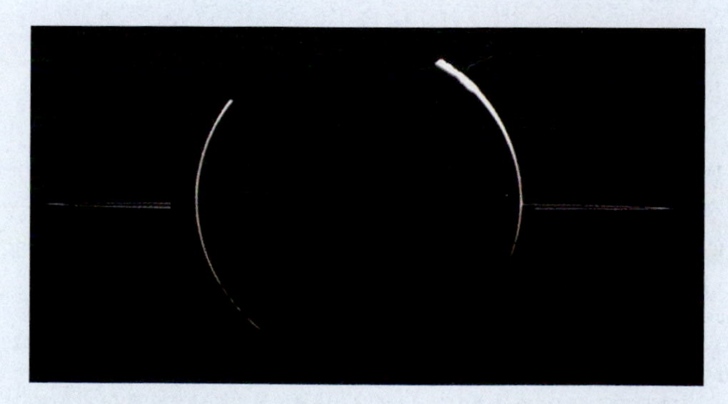

▲ Jupiter's main ring, from Galileo.

Io, from Galileo in 2000, showing extreme close-ups of irregular depressions, each one known as a patera. They correspond to active volcanic centers. Sometimes, lava erupts from the straight edges that seem to correspond to fractures in Io's crust.

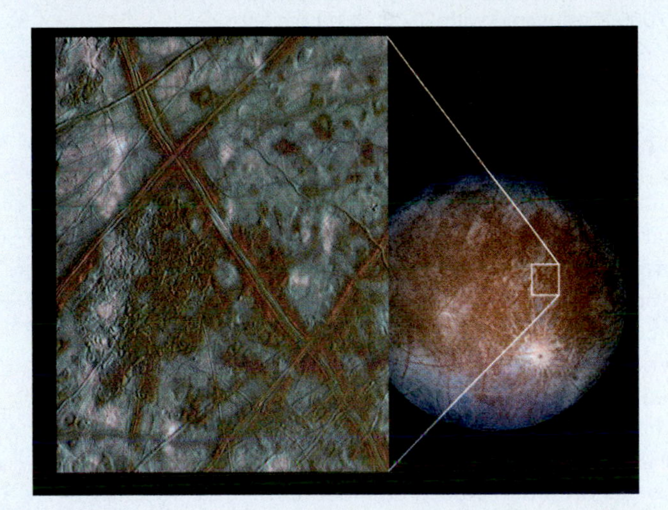

Part of Europa's crust apparently made up of blocks of ice that broke apart and "rafted" into new positions. In this false-color view, icy plains are shown in blue tones. Reddish-brown areas are non-icy material that erupted.

Ganymede's trailing hemisphere, in enhanced color. Craters, frosty polar caps; bright, grooved terrain; and older, dark furrowed areas show.

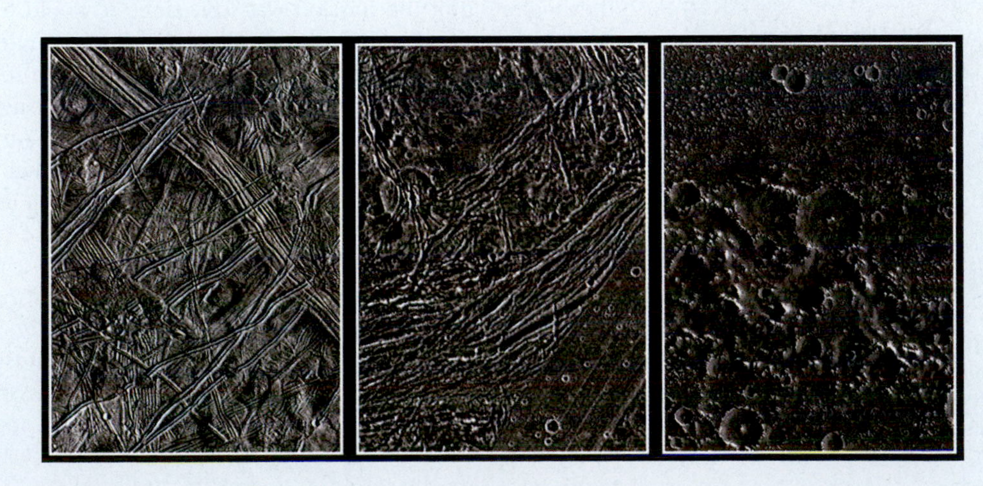

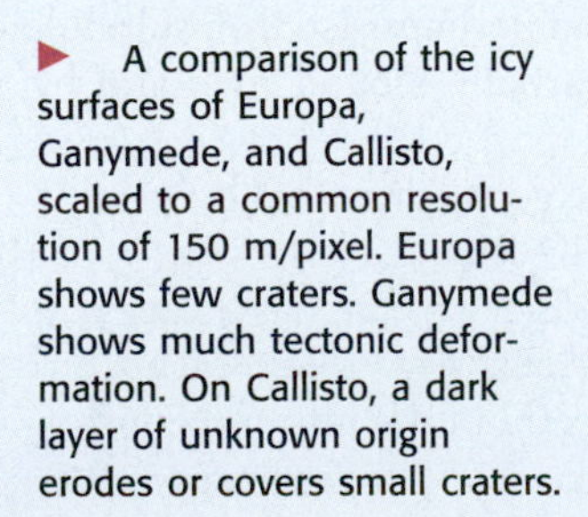

A comparison of the icy surfaces of Europa, Ganymede, and Callisto, scaled to a common resolution of 150 m/pixel. Europa shows few craters. Ganymede shows much tectonic deformation. On Callisto, a dark layer of unknown origin erodes or covers small craters.

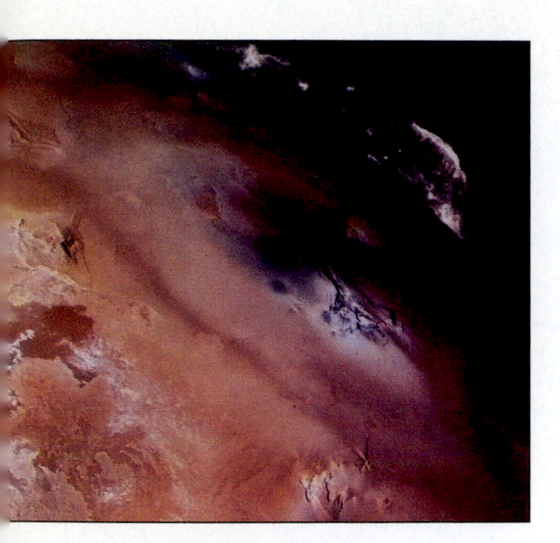

**FIGURE 13–20** Io's surface has been transformed by the sulfur erupted from volcanos like the one on the limb.

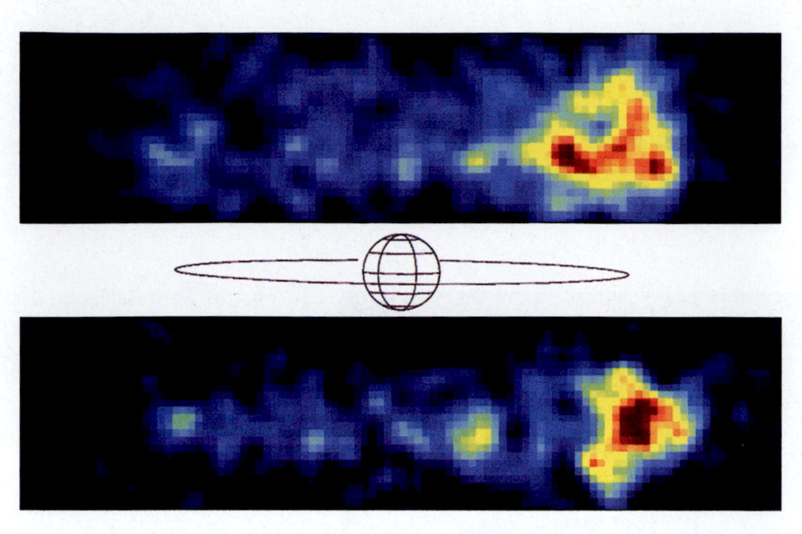

**FIGURE 13–21** An ultraviolet image of the Io torus, doubly ionized sulfur *(top)* and ionized oxygen *(bottom)* trapped in Jupiter's magnetic field at the same distance as Io, made with the Extreme Ultraviolet Explorer (EUVE). The more highly ionized the gases are, the hotter they must be, since electrons are stripped off increasingly at higher temperatures. The false color shows that the right side is 80,000 K compared with 70,000 K on the left side.

to the volcanism. To experience an analogous source of energy, bend a paper clip back and forth a few times, and then touch it to your lower lip. You should feel the warmth generated by the flexing.

Because of the internal heating, Io may have an underground ocean of liquid sulfur. The surface of Io is covered with sulfur and sulfur compounds, including frozen sulfur dioxide, and the atmosphere is full of sulfur dioxide. Sulfur can take on many colors, depending on how it cools. Given the smell and poisonous nature of sulfur, Io certainly wouldn't be a pleasant place to visit.

The Voyagers found that Io has, besides the volcanos, mountains up to 15 km tall, higher than Earth's Mt. Everest. Sulfur mountains that high couldn't support themselves, so perhaps these are silicate mountains with merely a coating of sulfur. A small amount of water ice has been detected. The mountains, it has recently been concluded, are filled blocks of crust.

Once we knew about them from the Voyagers, we could follow the volcanos and hot spots with infrared detectors on telescopes on Earth. The Loki lava lake remains the major radiator, even after many years.

Because of the problems with the antenna and thus the rate that data could be sent back to Earth by the Galileo spacecraft, the data from its closest Io passage at the beginning of the mission was lost, since it would have conflicted with the data being relayed from the Probe. Because the intense magnetic field at Io might damage the spacecraft, scientists and engineers waited to retarget Io for a close-up view until late in the mission (see the Photo Gallery). During 2000, Galileo came close to Io several times. Tracking Io's gravitational attraction on Galileo pinpointed Io's density at 3.53 (compared with 1 for water) and shows that Io has a two-layer structure. Its core, undoubtedly made of metal (presumably iron and perhaps also iron sulfide), is 1800 km across and is surrounded by a mantle of partially molten rock and by crust.

Gases—including sulfur, sodium, potassium, and oxygen—given off by Io fill a doughnut-shaped region that surrounds Io's orbit (Fig. 13–21).

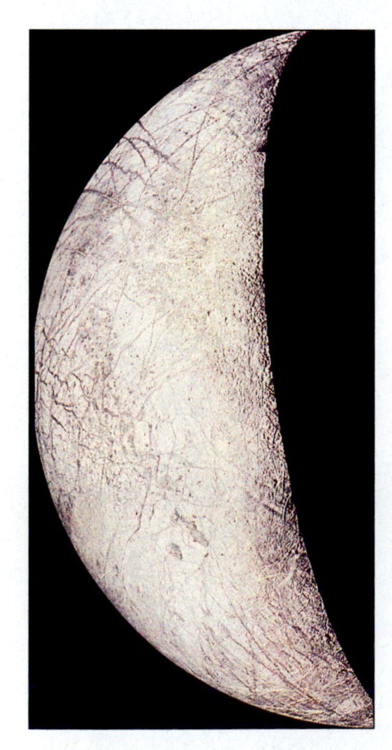

**FIGURE 13–22** The surface of Europa, which is very smooth, is covered by this complex array of streaks. Few impact craters are visible in this Voyager image.

Europa, the brightest of Jupiter's Galilean satellites (Fig. 13–22), has a very flat, uncratered surface and is covered with narrow dark stripes (Fig. 13–23). This appearance suggests that the surface we see is ice. And spectra in the infrared confirm this guess. The markings may be fracture systems in the ice, like fractures in the

**FIGURE 13–23** Close-ups from Galileo of icy regions on Europa, undoubtedly the crust of a global ocean.

large fields of sea ice near the Earth's north pole. Few craters are visible, suggesting that the ice was soft enough below the crust to close in the craters. Either internal radioactivity or a gravitational heating like that inside Io may have provided the heat to soften the ice. Because it possibly has liquid water and some extra heating, some scientists consider it a worthy location to check for signs of life. Spectra taken with the Hubble Space Telescope have even found traces of oxygen in Europa's atmosphere. Images from the Galileo spacecraft have revealed detail about the cracks, and found what looks like shifted blocks of ice, making it seem even more likely that there is an ocean underneath. The ice may be 20 km or so thick, and the ocean below could be 100 km or so deep.

Another instrument that indicates the presence of a Europan ocean is Galileo's magnetometer. It measures a magnetic field on Europa that oscillates in a way that matches an object with a conducting shell. This shell could well be a global layer of water with a salty composition similar to Earth's seawater, and no other possibilities are obvious. Indeed, some infrared spectroscopy has suggested the presence of salt on Europa's surface, though that salt could come from the ice itself rather than from an eruption through the ice from an underlying ocean.

An orbiting spacecraft around Europa could measure the altitude and gravitational field accurately enough to see the effect of tides. The information would let us know definitively whether there is an ocean and how thick the ice is. Such a Europa Orbiter is being planned, though a cutback is risked as NASA puts more money into its Mars exploration program.

The largest satellite, Ganymede (Figs. 13–24), shows many craters alongside weird, grooved terrain (Fig. 13–25). On the whole, the icy surface has been darkened by the dust it has collected over eons. Occasionally, an impact has caused a crater and revealed fresh, whiter ice. Ganymede is larger than Mercury but less dense. From Ganymede's density, we deduce that it contains large amounts of water ice surrounding a rocky core.

About ⅓ of Ganymede has a dark, ancient surface, with lots of craters. This dark terrain, made of a rock-ice mixture, is crossed by swaths of bright, younger terrain of water ice. The rest of Ganymede is also bright and younger. A comparison of Voyager and the higher-resolution Galileo images has allowed 3D maps to be made that show the heights of features. The results, released in 2001, indicate that the bright terrain is at many different heights. The smoothest, youngest swaths are as much as 200 m lower in elevation, and may be the result of flooding by water ice that seeped up from below. The ice would have melted inside as the mutual gravity of Jupiter, Io, and Europa deformed the moon this way and that and so heated its interior. The rougher, more heavily fractured bright terrain stands higher at a similar elevation to the dark terrain. The scientists conclude that the dark terrain has been sliced up to form the higher bright terrain.

**FIGURE 13–24** Ganymede, Jupiter's largest satellite. Many impact craters, some with systems of bright rays, show. The large dark region has been named Galileo Regio. Low-albedo features on satellites are named for astronomers.

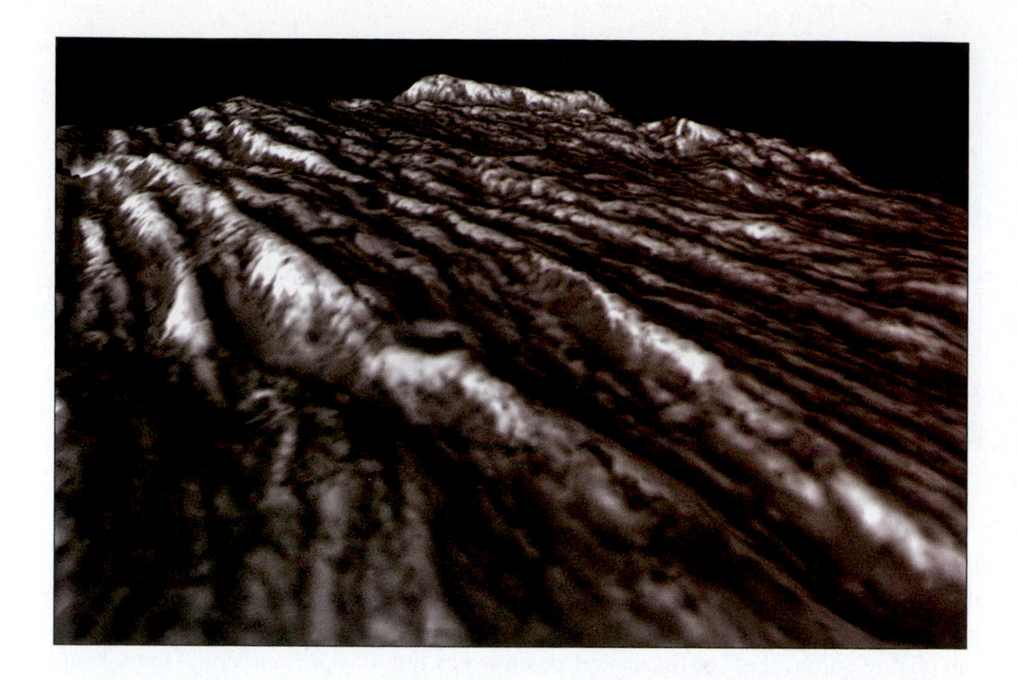

**FIGURE 13–25** A computer-generated view, with exaggerated perspective, of ridges on Ganymede. Ganymede's icy surface has been fractured and broken into many parallel ridges and troughs. Such bright grooved terrain covers over half of Ganymede's surface.

Other scientists, in scientific disagreement, suggest that the smooth, low swaths are actually regions of Ganymede that have been completely pulled apart. New material, possibly ice warmed enough to flow, would rise to fill the gaps.

Galileo's magnetometer detected a magnetic field on Ganymede—part permanent and part induced by Jupiter. The induced magnetic field indicates that Ganymede may have an ocean under its ice, since salty water would create the effects observed. But, Ganymede's ocean is apparently under 170 km of ice, much farther down than Europa's, whose top may be only 20 km from its surface.

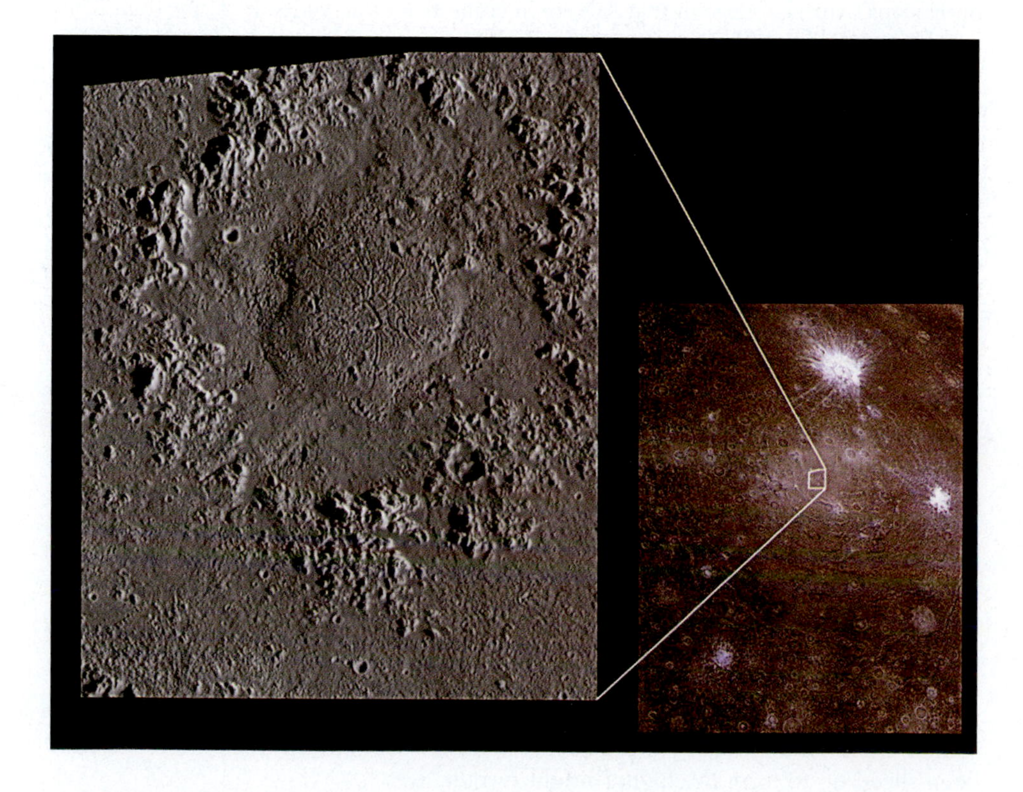

**FIGURE 13–26** The Asgard multi-ring structure, about 1700 km across. It is one of the largest impact features on Callisto. The high-resolution Galileo image on the left has a resolution of only 90 m/pixel. Callisto is cratered all over, very uniformly. Several craters have rays or concentric rings. Few craters larger than 50 km are visible, from which we deduce that Callisto's crust cannot be very firm. The fact that the limb is so smooth indicates that there is no high relief. Because there are so many craters, the surface must be very old, probably over 4 billion years.

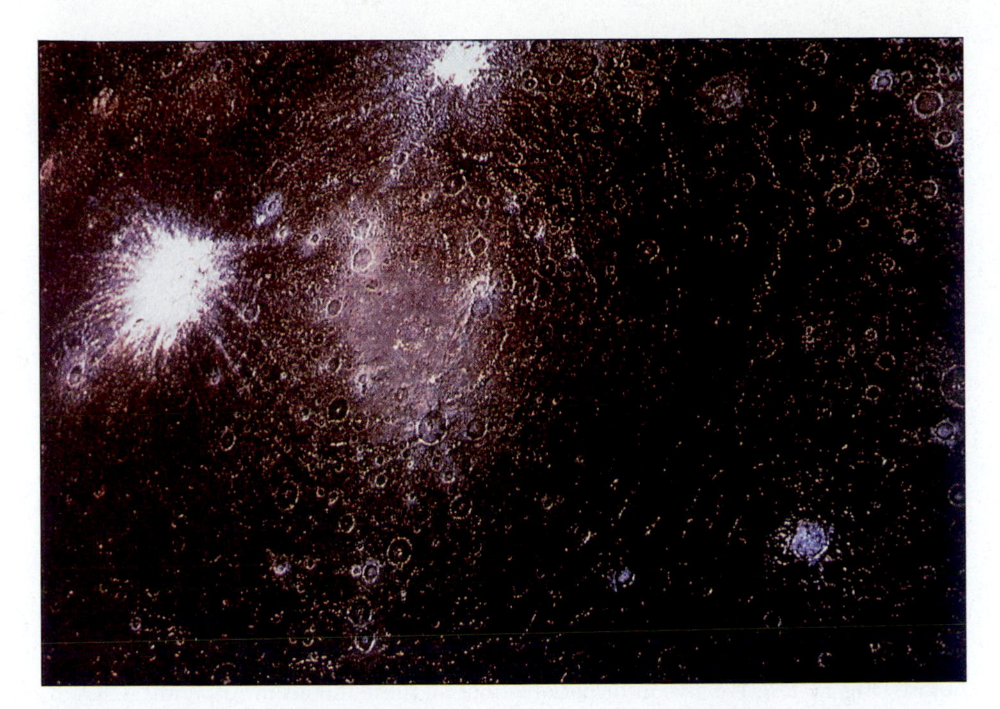

**FIGURE 13–27** A pair of ancient multi-ringed impact basins on Callisto, imaged from the Galileo spacecraft in a combination of low-resolution color data and higher resolution infrared data. The impact region Asgard *(center)* is surrounded by concentric rings up to 1700 km in diameter. The bright icy materials excavated by the younger craters contrast with the darker and redder coatings on older surfaces. Galileo images show, surprisingly, that Callisto has very few craters smaller than 100 m in diameter. They may be disintegrating on their own, perhaps from electrostatic charges.

Some parts of Ganymede also show lateral displacements, where grooves have slid sideways, like those that occur in some places on Earth (for example, the San Andreas fault in California). Thus further studies of Ganymede may enhance our understanding of terrestrial earthquakes. Other signs of modification of Ganymede's surface by tectonic processes in its early years have also been seen.

The surprise discovery from Galileo that Ganymede has its own magnetosphere means that Ganymede must have its own magnetic field. This discovery makes Ganymede the only satellite in the Solar System known to have this property.

Callisto has so many craters (Fig. 13–26) that its surface must be the oldest of Jupiter's Galilean satellites. Spectra indicate that Callisto is probably also covered with ice. No small craters are found; perhaps landslides or volcanism erased them or there were few small impactors. A huge bull's-eye formation, Valhalla, contains about 10 concentric rings, no doubt resulting from a huge impact. Ripples spreading from the impact froze into the ice to make Valhalla. Despite its general appearance, Valhalla differs greatly in detail from ringed features on the Moon and Mercury; its rings are asymmetric, for example. The Galileo spacecraft has sent back spectacular close-up images of Callisto (Fig. 13–27).

Voyager also observed Amalthea, formerly thought to be Jupiter's innermost satellite. This small chunk of rock is irregular and oblong in shape, looking like a dark red potato complete with potato "eyes." It is comparable in size to many asteroids, ten times larger than the moons of Mars and ten times smaller than the Galilean satellites. Its reddish color may be caused by material from Io.

The Voyagers discovered three previously unknown satellites of Jupiter. When the Galileo spacecraft, in its final months, was allowed to plunge into the inner jovian system, it imaged Adrastea and these other inner satellites better than ever

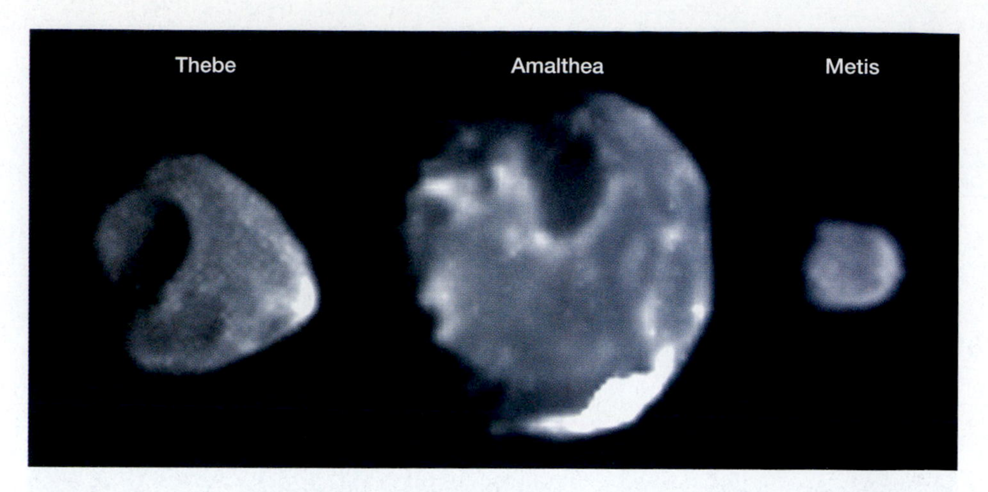


**FIGURE 13–28** Thebe, Amalthea, and Metis, in images taken in 2000 from the Galileo spacecraft. Resolutions vary from 2 to 4 km. Note how rough Amalthea's terminator is. Overexposed white material near Amalthea's crater Gaea shows. Amalthea is about the size of New York's Long Island.

before (Fig 13–28). We give information about all these moons in Appendix 4. There may well be more to come.

Studies of Jupiter's moons tell us about the formation of the Jupiter system and help us better understand the early stages of the entire Solar System.

## 13.5 JUPITER'S RING

Though Jupiter wasn't expected to have a ring, Voyager 1 was programmed to look for one just in case; Saturn's ring, of course, was well known, and Uranus's ring had been discovered only a few years earlier. The Voyager 1 photograph indeed showed a wispy ring of material around Jupiter at about 1.8 times Jupiter's radius, inside its innermost moon. As a result, Voyager 2 was targeted to take a series of photographs of the ring (Fig. 13–29). From the far side looking back, the ring appeared unexpectedly bright. This brightness results from very small—micron-size—particles in the ring that scatter the light toward us (Fig. 13–30). Within the main ring, fainter material appears to extend down to Jupiter's surface. A diffuse "gossamer ring" is outside the main ring.

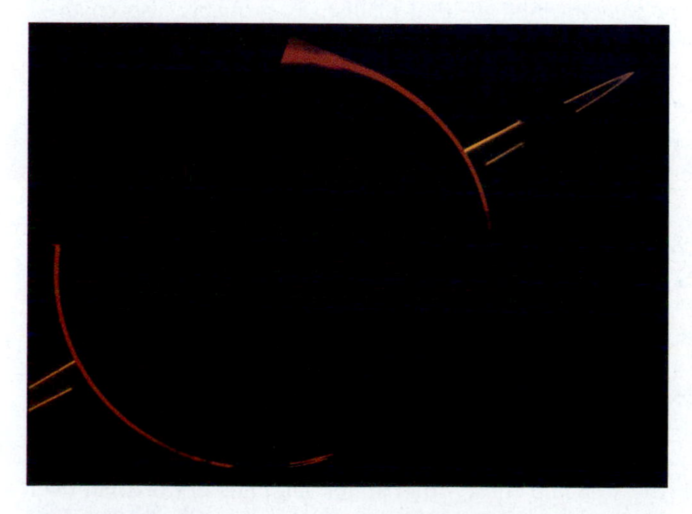

**FIGURE 13–29** Jupiter's ring, in a backlighted view, seen from Voyager 2.

The ring particles may come from Io, or else they may come from comet and meteor debris or from material knocked off the innermost moons by meteorites. The individual particles probably remain in the ring only temporarily. Perhaps Jupiter's newly discovered innermost moon causes the rings to end where they do.

In the next chapter, we shall discuss the fundamental ideas about how gravity leads to the formation of planetary rings, in the context of the most famous rings of all: the rings of Saturn.

**FIGURE 13–30** Jupiter's ring (overexposed), with Jupiter's moon Europa behind it. The Galileo spacecraft was looking at light scattered forward through the ring, and at the side of Europa away from the Sun, illuminated by reflected light off Jupiter itself.

## 13.6 FUTURE STUDY OF JUPITER

Voyager 1, Voyager 2, and Galileo were spectacular successes. The data they provided about the planet Jupiter, its satellites, and its ring dazzled the eye and allowed us to make new theories that revolutionized our understanding.

The resolution of images of Jupiter taken with the Hubble Space Telescope in Earth orbit is approximately equal to those from the Voyager and Galileo approach phases (Fig. 13–31), as were Cassini images as it passed in 2000–2001 en route to Saturn, so we can maintain our tracking of Jupiter's clouds.

As part of NASA's Fire and Ice program, a Europa Orbiter is being planned. It would be part of the search for life, with a first goal of determining whether Europa has a global ocean under its ice.

New discoveries about the Jupiter system can still be made from the ground (Fig. 13–32). Astronomers at the Spacewatch project at the University of Arizona discovered, in 1999, the first new moon of Jupiter found in over 25 years. This 5-km satellite, S/1999 J1, joined a group of other outer satellites that make eccentric orbits around Jupiter about every two years. The floodgate was opened, and observers in Hawaii discovered a dozen additional satellites (Fig. 13–32). We look forward to future ground-based discoveries with the new generation of telescopes and their adaptive optics.

**FIGURE 13–31** Jupiter, photographed with the Wide Field and Planetary Camera 2 on the Hubble Space Telescope.

**FIGURE 13–32** The orbits of Jupiter's newly discovered moons (green), plotted with a selection of the orbits of previously known moons (white). Only the innermost (Io) and the outermost (Callisto) of the four Galilean satellites are plotted.

## CORRECTING MISCONCEPTIONS

| ✖ Incorrect | ✔ Correct |
| --- | --- |
| Jupiter's moons are hard to see. | Even binoculars or a very small telescope can show moons orbiting Jupiter. |
| Jupiter is much like Earth. | Jupiter dominates the Solar System, and its size and mass give it many properties very different from Earth's. |
| Only Saturn has rings. | Jupiter and, as we shall see, Uranus and Neptune also have rings. |

## SUMMARY AND OUTLINE

Fundamental properties (Section 13.1)
- Highest mass and largest diameter of any planet
- Low density: 1.3 grams/cm$^3$
- Composition primarily of hydrogen and helium
- Rapid rotation, causing oblateness of disk
- Colored bands on surface; Great Red Spot
- Intense bursts of radio radiation
- Strong magnetic field and radiation belts

Jupiter's moons (Section 13.2)
- Historic role in Copernican theory

Spacecraft observations (Section 13.3)
- Pioneer 10 and 11 flybys in 1973 and 1974
- Voyager 1 and 2 flybys in 1979; increased resolution
- Galileo went into orbit in 1995; probe plus lengthy series of observations
- Great Red Spot is a giant, anticyclonic storm.
- Gaseous atmosphere and liquid interior
- Jupiter radiates 1.6 times the heat it receives.
- Tremendous magnetic field detected.

Spacecraft observations of moons (Section 13.4)
- Io is transformed by volcanos.
- Europa has a flat surface, crisscrossed with dark markings. A subsurface ocean causes blocks of ice on the surface to move around.
- Ganymede has many craters and grooved terrain; Galileo found a magnetic field.
- Callisto is covered with craters, including a bull's-eye.
- Amalthea is irregular and oblong.

Jupiter's ring (Section 13.5)
- A narrow ring, composed of small particles

Future study of Jupiter (Section 13.6)
- Europa Orbiter

## KEY WORDS

giant planets
jovian planets
Great Red Spot
oblate
radiation belts
Galilean satellites
Coriolis force
zones
belts
gravity assist

## QUESTIONS

1. Why does Jupiter appear brighter than Mars despite its greater distance from the Earth and Sun?
2. Even though Jupiter's atmosphere is very active, the Great Red Spot has persisted for a long time. How is this possible?
3. (a)How did we first know that Jupiter has a magnetic field? (b) What did spacecraft studies show about the magnetic field?
4. [†] From Jupiter's distance from the Earth (Appendix 3), the size of the orbits of the Galilean satellites (Appendix 4), and the speed of light (Appendix 2), calculate how "late" the moons of Jupiter appear to arrive in their predicted positions when Jupiter is at its farthest point from the Earth compared to when Jupiter is closest to the Earth. Rømer used the opposite of this calculation in 1675 to test whether light travels at finite or infinite speed.
5. What advantages over the 5-m Palomar telescope on Earth did Voyagers 1 and 2 have for making images of Jupiter?
6. How did the Galileo probe improve our knowledge about Jupiter's interior?
7. Describe applications of computer simulations to analyzing structure on Jupiter.
8. How does the interior of Jupiter differ from the interior of the Earth? What experiments on Earth can help interpret the Jupiter interior?
9. Which moons of Jupiter are icy? How do we know?
10. Contrast the volcanos of Io with those of Earth.
11. Compare the surfaces of Callisto, Io, and the Earth's Moon. Show what this comparison tells us about the ages of features on the surfaces.

[†] This question requires a numerical solution.

12. Using the information given in the text and in Appendices 3 and 4, sketch the Jupiter system, including all the moons and the ring. Mark the groups of moons.
13. (a)Describe the gravity-assist method. (b) What does it help us do?
14. Explain how the Hubble Space Telescope can match Voyager's resolution even though it is farther from Jupiter.
†15. Given the resolution of the Hubble Space Telescope, calculate the size of features visible on Jupiter. Provide sample distances between cities on Earth for comparison.
†16. Repeat the calculation of Question 15 for Io, and comment on what kinds of structures might be seen.
17. Contrast the highest resolution images from Voyager with those from Galileo.

## PROJECTS

1. Describe the origins in Greek mythology of the names of each of Jupiter's moons.
2. From your own observations made over a four-hour interval one night, plot the positions of the Galilean satellites with respect to Jupiter itself at half-hour intervals.
3. Over a one-week interval, plot the positions of the Galilean satellites from night to night with respect to Jupiter itself. Using the observations in Projects 2 and 3 together, deduce the projection on the sky of the orbits of the individual satellites.
4. Learn how to look up the predicted positions of Jupiter's Galilean satellites using computer programs like *RedShift*. Compare with your own observations.
5. Find out the latest results from the Galileo spacecraft. Discuss the effect of the failure of the main antenna to open fully.

## TOPICS FOR DISCUSSION

1. Although many of the most recent results about Jupiter came from spacecraft, prior ground-based studies had told us many things. Discuss the status of our pre-1973 knowledge of Jupiter, and specify both some things about which space research did not add appreciably to our knowledge and some things about which space research led to a major revision of our knowledge.
2. If we could somehow suspend a space station in Jupiter's clouds, the astronauts would find a huge gravitational force on them. What are some of the effects that this would have? (The novels *Slapstick* by Kurt Vonnegut and *Einstein's Dreams* by Alan Lightman deal with some of the problems that would arise if gravity were different in strength from what we are used to.)

## USING TECHNOLOGY

### W World Wide Web

1. Examine the latest images and animations of Jupiter and its moons, and contrast them with the images published in this book. See http://photojournal.jpl.nasa.gov, http://pirl.lpl.arizona.edu/pirl/Galileo/Releases, and the history at http://www.nineplanets.org.
2. Study the posted interpretation of data from the Galileo probe at http://www.jpl.nasa.gov/galileo.
3. View movies of Ganymede and other satellites rotating.
4. Check out the images of Jupiter taken in 2000 and 2001 when Cassini went by at http://www.jpl.nasa.gov/jupiterflyby and http://ciclops.lpl.arizona.edu.

### *REDSHIFT*

1. When is Jupiter visible tonight?
2. Where are Jupiter's Galilean satellites with respect to Jupiter in the sky at 9 p.m. tonight or, if they can't be seen then, at the peak of Jupiter's visibility?
3. Use the main *RedShift* engine to observe a conjunction of Jupiter and Venus and to observe Jupiter and its moons from Ganymede. When is the next such conjunction?
4. Look at *Principia Mathematica*: The Mass of Jupiter; Night on the Mountain: The Planet Dossier; Science of Astronomy: Worlds of Space: Moons, Asteroids, and Comets.
5. Look at the *RedShift* movies on disk 2: Jupiter, Great Red Spot.

Saturn and some of its moons, photographed from the Voyager 1 spacecraft. Enceladus is in the foreground, Mimas is above it in front of Saturn's disk, Dione is at lower left, Rhea is at middle left, Titan is at top left, Iapetus is at top center, and Tethys is at upper right. The symbol for Saturn appears at the top of the opposite page.

# Chapter 14

## Saturn ♄

Saturn is the most beautiful object in our Solar System. The glory of its system of rings makes it stand out even in small telescopes.

Saturn, like Jupiter, Uranus, and Neptune, is a giant planet. Its diameter, without its rings, is 9 times greater than Earth's; its mass is 95 times greater. Saturn is 9.5 A.U. from the Sun and has a lengthy year, equivalent to 30 Earth years.

The giant planets have low densities. Saturn's is only 0.7 g/cm$^3$, 70 per cent the density of water (Fig. 14–1) and only about half of Jupiter's, which is itself only about a quarter of Earth's. The bulk of Saturn is hydrogen molecules and helium. Deep in the interior where the pressure is sufficiently high, the hydrogen molecules are converted into metallic hydrogen in liquid form. Saturn could have a core of heavy elements, including rocky material, making up 20 per cent of its mass.

Saturn's atmosphere (Fig. 14–2) has relatively less helium compared with hydrogen than Jupiter's—92.4 per cent hydrogen molecules and 7.4 per cent helium for Saturn compared with 86.1 and 13.8 per cent, respectively, for Jupiter. (The remaining 0.2 per cent is almost all methane, plus some ammonia and traces of other molecules.) We think that Saturn's internal temperature is sufficiently lower than Jupiter's to prevent helium from dissolving in the liquid hydrogen. Helium droplets thus fall under the force of gravity toward Saturn's center. Energy is made available that makes the ratio of energy radiated to energy received higher for Saturn than for Jupiter.

**AIMS:** To describe Saturn, its rings, and its moons, and to see how scientists using the Voyager spacecraft multiplied our knowledge of them

**FIGURE 14–1** Since Saturn's density is lower than that of water, it would float, like Ivory Soap, if we could find a big enough bathtub. But it would leave a ring.

**FIGURE 14–2** A storm that developed on Saturn was photographed with the Hubble Space Telescope.

## A DEEPER DISCUSSION

### BOX 14.1 The Roche Limit

E. A. Roche defined the limit in 1849 as the shortest distance from a spherical body within which a liquid body does not break up because the body's own gravitation pulling itself together is stronger than the tidal forces pulling it apart. Similar limits can be defined for solid bodies and non-spherical bodies and for the case where the orbiting body is taking on individual molecules. The limits depend on both the mass of the planet and the density of the ring particles. Thus the existence of rings at a given radius can help distinguish whether the rings are made of ice or of stony material. Saturn's rings, to be within the Roche limit, must be made primarily of ice.

***Discussion Question***

Why isn't the Hubble Space Telescope torn apart by tidal forces, given that it orbits inside the Earth's Roche limit?

## 14.1 SATURN FROM THE EARTH

The rings of Saturn are material that was torn apart by Saturn's gravity. These bits of matter spread out in concentric rings around Saturn. They are particles of ice, with sizes ranging from tiny to those of cars and larger. We also know that they cannot be leftover material from the time Saturn formed, as we will discuss below (Section 14.2a). Thus there must still be collisions of moons or other objects near Saturn that replenish the rings with dust and rocks, or else a moon is pulled apart every once in a while by the tidal effect we will now discuss.

Artificial satellites that we send up to orbit around the Earth are constructed of sufficiently rigid materials that they do not break up even though they are within the Earth's Roche limit.

Every massive object has a sphere, called the **Roche limit** (Box 14.1), inside of which blobs of matter cannot be held together by their mutual gravity. The forces that tend to tear the blobs apart from each other are tidal forces (Section 8.3); they arise because some blobs are closer to the planet than others and are thus subject to greater forces of gravity.

The radius of the Roche limit varies with the amount of mass in the parent body and the nature of the bodies but is usually about $2\frac{1}{2}$ times the radius of the larger body. The Sun also has a Roche limit, but all the planets lie outside it. The natural moons of the various planets lie outside the planets' respective Roche limits. Saturn's rings lie inside Saturn's Roche limit, so it is not surprising that the material in the rings is spread out rather than being collected into a single orbiting satellite.

FIGURE 14–3 Saturn, from the Hubble Space Telescope, during the Earth's crossing of its ring plane.

FIGURE 14–4 An infrared view shows the rings very well.

A DEEPER DISCUSSION

### BOX 14.2 Cassini's Division

In 1610, Galileo discovered with a new invention, the telescope, that Saturn was not round; it seemed to have "ears." In 1656, the Dutch astronomer Christiaan Huygens published an anagram—a coded message made by scrambling letters (a procedure then often followed to establish priority for a discovery, since years later one could point back to the message)—explaining that Saturn has a ring. Huygens's open publication of his result in a book followed three years later. Giovanni Cassini, who worked at the Paris Observatory some years later, observed the largest gap in the ring in 1675, and this gap is named after him. Cassini has also been honored with the spacecraft, Cassini, that is en route to Saturn.

***Discussion Question***

Paralleling the anagram with which Huygens disguised his announcement of his discovery of Saturn's rings, try to make an anagram for "[Your name] discovered that it has rings."

The rings extend far out in Saturn's equatorial plane and, like it, are inclined to the planet's orbit. Over a 30-year period, we sometimes see them from above their northern side, sometimes from below their southern side, and at intermediate angles in between. When seen edge on, they are all but invisible (Fig. 14–3). At such times, new moons might be found, though the ones tentatively reported from the most recent edge-on views turned out to be merely clumps of ring material.

Saturn has several concentric major rings visible from Earth. The brightest ring (the "B-ring") is separated from a fainter broad outer ring (the "A-ring") by an apparent gap called **Cassini's division** (Box 14.2). Another ring (the "C-ring," or "crepe ring") is inside the brightest ring. It has long been thought that the gaps between the major rings result from gravitational effects from Saturn's satellites, though this is uncertain. Finer structure in the rings has been studied from spacecraft, as we shall see. Infrared views (Fig. 14–4) show the radiation from the rings rather than their reflection of solar light.

The rotation of Saturn's rings can be measured directly from Earth with a spectrograph, for different parts of the rings have different Doppler shifts. We know that the rings are not solid objects because they rotate at different angular speeds at different radii. These speeds correspond to Kepler's laws, with Saturn itself as the massive object at a focus of the orbits. (Later in the book, in Chapter 33, we will look at rotation graphs for galaxies, and find that the curves do not correspond to Kepler's law. We will be forced, therefore, to conclude that no specific central object provides all the mass of those systems.)

The rings are relatively much flatter than a compact disc (CD). Radar waves were first bounced off the rings in 1973. The result of the radar experiments showed that the particles in the rings are probably rough chunks of ice at least a few centimeters and possibly a meter across.

Like Jupiter, Saturn rotates quickly on its axis, also in about 10 hours. As a result, it is oblate—bulges at the equator—by 10 per cent. Saturn's delicately colored bands of clouds rotate 10 per cent more slowly at high latitudes than at the equator. Methane, ammonia, and molecular hydrogen have been detected spectrographically.

Saturn gives off radio signals, as does Jupiter, an indication to earthbound astronomers that Saturn also has a magnetic field. And, like Jupiter, Saturn has a source of internal heating.

## 14.2 SATURN FROM SPACE

Pioneer 11, which had passed Jupiter in 1974, reached Saturn in 1979. But Pioneer 11 had the handicap of having to travel across the diameter of the Solar System from Jupiter to reach Saturn, while the Voyagers travelled a much shorter route. So Voyager 1 reached Saturn in 1980, only a year after Pioneer 11. Voyager 2 arrived in 1981. Between the time of the launch of Pioneer 11 and the launches of the Voyagers, experimental and electronic capabilities had improved so much that the Voyagers could transmit data at 100 times the rate of Pioneer 11.

### 14.2a Saturn's Rings

From Pioneer 11, the rings were visible for the first time from a vantage point different from the one we have on Earth. In the backlighted view obtained by Pioneer 11, Cassini's division, visible as a dark (and thus apparently empty) band from Earth, appeared bright, with its own dark line of material. The outermost major ring, the A-ring, showed structure as well. The brightest ring observed from Earth, the B-ring, appeared dark on the Pioneer 11 view, showing that it is too opaque to allow light to pass through it, presumably because its particles are packed so closely together. The rings appear to have low mass and therefore low density, endorsing the idea that they are made up largely of ice. An icy composition would also explain why atomic hydrogen was found around the rings; it could result from dissociation of water ice (water molecules' separation into their component atoms).

The passage of Voyager 1 by Saturn was one of the most glorious events of the space program. The resolutions of the Voyager's cameras were much higher than the resolution of Pioneer 11's instrument, which had been designed for measuring polarization rather than for photography. Voyager 2's cameras turned out to be twice as sensitive as Voyager 1's and had higher resolution. Though the subdued colors in Saturn's clouds made the Voyagers' photographs of the surface less spectacular than those of Jupiter, the structure and beauty of the rings dazzled everyone (Fig. 14–5).

The closer the spacecraft got to the rings, the more rings became apparent. Before Voyager, scientists discussed whether there were three, four, five, or six rings. But as Voyager 1 approached, the photographs increasingly showed that each of the known rings was actually divided into many narrower rings. By the time Voyager 1 had passed Saturn, we knew of a thousand rings (Figs. 14–6 to 14–7). Voyager 2 saw still more. Further, a device on Voyager 2 was able to track the change in brightness of a star as it was seen through the rings and found even finer divisions. The number of these rings (sometimes called "ringlets") is in the hundreds of thousands. The

**FIGURE 14–5** Saturn and its rings in an image taken by Voyager 2. Saturn's moons Tethys, Dione, and Rhea appear below it. The shadow of Tethys appears on Saturn's disk. *(All Voyager photos courtesy of Jet Propulsion Laboratory/Caltech/NASA)*

**FIGURE 14–6** The complexity of Saturn's rings. About 60 can be seen on this enhanced image, taken by Voyager 1 when it was 8 million km from Saturn. The color difference indicates different surface composition of the rocks making up the rings. The main "C-ring" appears blue except for three ringlets that appear the same yellow color as the "B-ring." The C-ring material is the color of dirty ice. The original photographs from which this print was made were taken through ultraviolet, blue, and green filters.

words "rings" or "ringlets" for these features, though, perhaps imply that they persist in their current form, which may not be the case. The total amount of mass in the rings is about the mass of a middle-sized moon, such as Mimas.

Everyone had expected that collisions between particles in Saturn's rings would make the rings perfectly uniform. But there was a big surprise. As Voyager 1 approached Saturn, we saw that there was changing structure in the rings aligned in the radial direction. "Spokes" can be seen in the rings when they are seen at the proper angle (look back at Figure 14-5). Differential rotation must make the spokes dissipate soon after they form. The spokes look dark from the side illuminated by the Sun (Fig. 14–8), but look bright from behind (Fig. 14–9). This information showed that the particles in the spokes were very small, about 1 micron in size, since

**FIGURE 14–7** Saturn's rings and their shadow in natural color.

FIGURE 14–8 Dark, radial spokes became visible in the rings as the Voyagers approached. They formed and dispersed within hours.

FIGURE 14–9 When sunlight scatters forward, the spokes appear bright.

only small particles—like terrestrial dust in a sunbeam—reflect light in this way. And it seems that the spoke material is elevated above the plane of Saturn's rings. Since gravity wouldn't cause this, electrostatic forces may be repelling the spoke particles. The idea gained backing, from laboratory tests published in 2000, in which particles in a vacuum were bombarded with ultraviolet light.

It also came as a surprise that some of the rings are not circular and uniform in appearance, since ring particles would be expected to spread out uniformly. As you follow around the rings, at least some of them change in brightness or in radius from one location to another.

Before the flybys, it had been thought that the structure of the rings and the gaps between them were determined by gravitational effects of Saturn's several known moons on the orbiting ring particles. If a particle in the rings has a period that is a simple fraction of a moon's period, then the particle and the moon would come back in the same configuration regularly and be relatively close together in their elliptical orbits. This pattern reinforces the gravitational pull of the moon and sweeps away particles at those locations. But this explanation had been worked out with fewer than a half-dozen rings known. The ring structure the Voyagers discovered is too complex to be explained in this way alone. Consideration of the sixteen or so satellites cannot explain thousands of rings. And the gaps turned out to be in the wrong place for this model to work.

Enya, a popular singer/songwriter, named an album "Shepherd Moons" after this phenomenon.

At least some of the rings are kept narrow by "shepherding" satellites. From Kepler's laws, we can deduce that the outer satellite is moving slightly slower than the ring particles, and the inner one slightly faster. If a ring particle should move outward, it would be pulled backward by the outer satellite, diminishing its angular momentum. This would make it sink back toward the ring. Conversely, if a ring particle should move inward, it would be pulled ahead by the inner satellite, increasing its angular momentum. This would push it back toward the ring. Thus material tends to stay in the narrow ring.

Scientists were astonished to find on the Voyager 1 images that the outer ring, the narrow F-ring discovered by Pioneer 11, seems to be made of three braids (Fig. 14–10). Scientists explained the braided strands as resulting from the gravity of small moons located on either side of the ring. These moons are thought to "shepherd" the ring particles to keep them from straying. Then, compounding the series of surprises, Voyager 2 images showed that the rings were no longer intertwined.

A post-Voyager-1 theory said that many of the narrowest gaps may be swept clean by a variety of small moons—"ringmoons"—embedded in them. These objects would

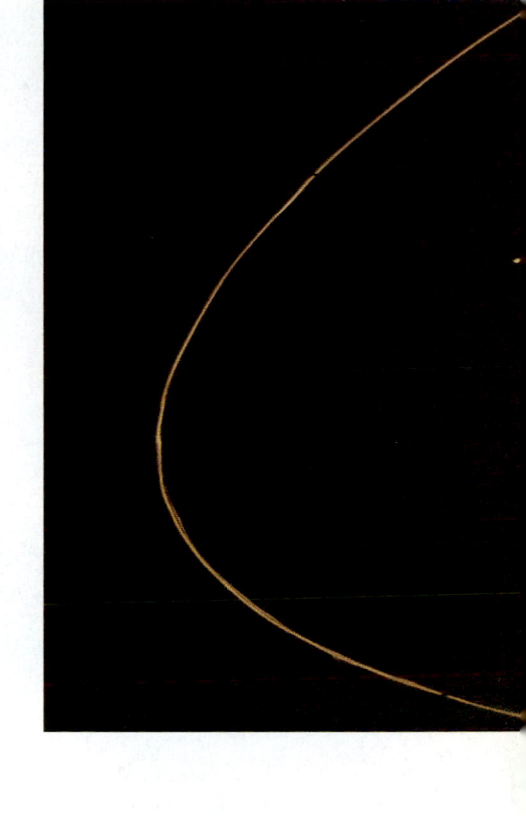

**FIGURE 14–10** Two narrow, braided, 10-km-wide rings are visible, as is a broader diffuse component about 35 km wide, on this Voyager 1 image of the F-ring. The Voyager 2 images of the F-ring showed no signs of braiding. Scientists from London concluded in 1997 that the F-ring's unusual and changing appearance can be explained by a ring of at least four strands with similar but slightly different orbits. The rings, made of 1-cm ice particles, are shepherded by the moons Pandora and Prometheus. The Hubble Space Telescope imaged the F-ring in 1995, the only sighting since those from the Voyagers.

be present in addition to the smaller icy snowballs that make up the bulk of the ring material. Unfortunately for the theoreticians, Voyager 2 did not find these larger objects, but no better reason for the narrow rings and gaps has been found.

Spiral waves were observed in the rings, and are thought to be generated by the gravity of the moons. These waves make the orbits of the individual ring particles collapse. Because the waves are tiny, they don't disturb the system very much, and the rings lose their particles on a timescale of hundreds of millions of years. Since our Solar System is billions of years old, we deduce that the rings are replenished. The new particles presumably come from the surfaces of the moons or even from the occasional collision between moons and the breakup of moons.

### 14.2b Saturn Below the Rings

The structure in Saturn's clouds is of much lower contrast than that in Jupiter's clouds. After all, Saturn is colder, so it has different chemical reactions. Even so, cloud structure was revealed to the Voyagers at their closest approaches. A haze layer present for Voyager 1 cleared up by the time Voyager 2 arrived, so more details could then be seen. Turbulence in the belts and zones showed up clearly. A few circulating ovals similar to Jupiter's Great Red Spot and ovals were detected (Fig. 14–11). Saturn's ovals are also storms in high-pressure regions, circulating oppositely to the low-pressure storms on Earth. Study of cloud features gave the first direct view of the rotation of Saturn's clouds and allowed detailed measurements of the velocities of rotation to be made.

Extremely high winds, 450 km/hr (four times higher than the winds on Jupiter (Fig. 14–12)), were measured. On Saturn, the variations in wind speed do not seem to correlate with the positions of belts and zones (Fig. 14–13), unlike on Jupiter. As

A

B

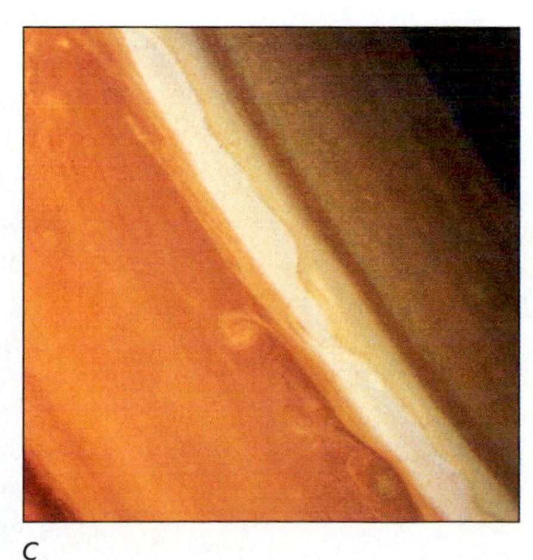

C

**FIGURE 14–11** (*A*) A false-color image brings out the divisions between belts and zones. The red oval below center is about the diameter of Earth. The shadow of Dione appears. (*B*) The red oval shows more clearly. (*C*) This curled cloud lasted for many months.

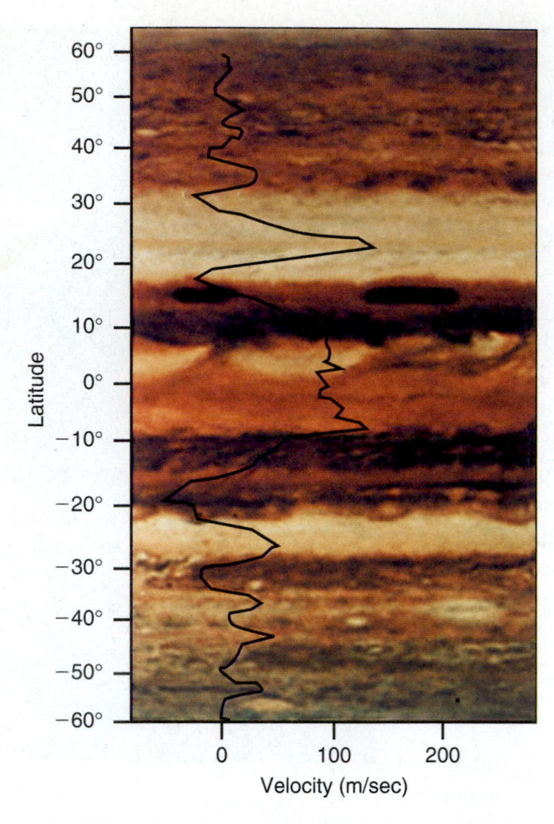


**FIGURE 14–12** Jupiter's winds, to compare with Saturn's winds in the adjacent photo. A graph of the winds, defined with respect to a fixed rotation rate, is superimposed on a Voyager photo of the corresponding regions of Jupiter. Negative velocities correspond to currents travelling from east to west. The east–west currents, especially an eastward equatorial jet of 100 m/sec (360 km/hr), are apparent. Latitudes range from +60° at top to −60° at bottom, and velocities from −50 m/sec at left to 150 m/sec at the highest peak toward the right.

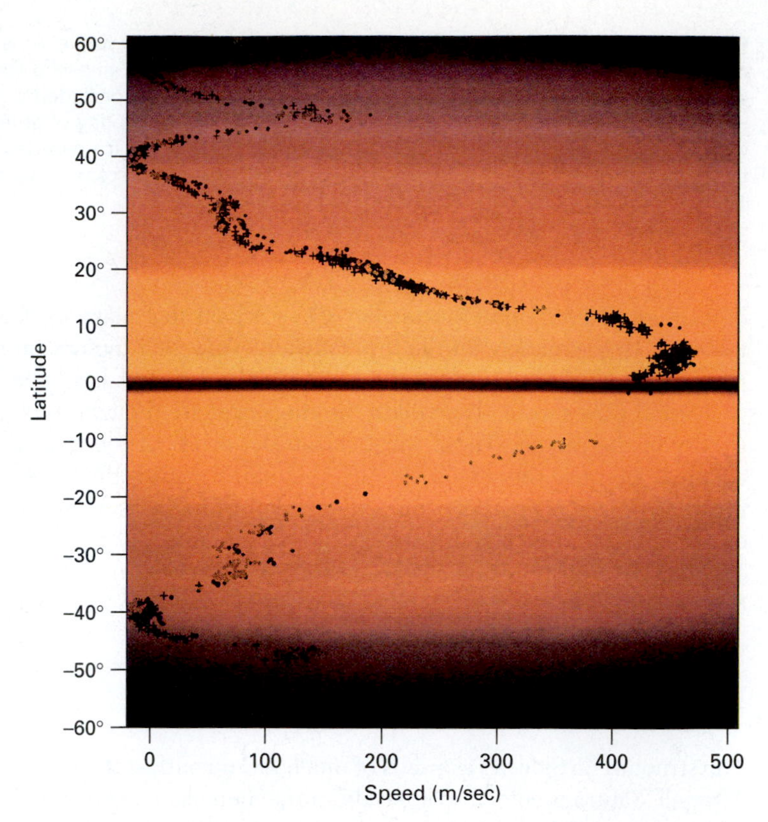


**FIGURE 14–13** Saturn's winds, to compare with Jupiter's winds in the adjacent photo. We see a graph of the wind velocities at each latitude, superimposed on a Voyager photo. The winds on Saturn are much higher at the equator and do not correlate with the cloud belts, in contrast with Jupiter. Thus the circulations of the atmospheres of Jupiter and Saturn are very different. Latitudes range from +60° at top to −60° at bottom, and velocities from 0 at left to 450 m/sec at right.

on Jupiter but unlike the case for Earth, the winds seem to be driven by rotating eddies, which in turn get their energy from the planet's interior. Infrared scans showed that the temperature decreases 10 K from equator to pole.

Saturn radiates about 2.5 times more energy than it absorbs from the Sun. One interpretation is that only ⅓ of Saturn's heat is energy remaining from its formation and from continuing gravitational contraction. The rest would be generated by the gravitational energy released by helium sinking through the liquid hydrogen in Saturn's interior. Saturn, unlike Jupiter, is cold enough for helium to condense, allowing gravitational energy to be generated in this way.

### 14.2c Saturn's Magnetic Field

Pioneer 11 had revealed the magnetic field at Saturn's equator to be somewhat weaker than had been anticipated, only ⅔ of the field present at the Earth's equator. Remember, though, that Saturn is much larger than the Earth, and so its equator is much farther from its center. Saturn's magnetic axis appears to be aligned with its rotational axis, unlike Jupiter's, the Earth's, and Mercury's. This observation forces us to formulate new theories of how magnetic fields form inside planets, since it had been thought that the relative tilt of magnetic and rotational axes led to the generation of the magnetic field. The total strength of Saturn's magnetic field (which depends on both the field's value and the planet's volume) is 1000 times stronger than Earth's and 20 times weaker than Jupiter's (Fig. 14–14). The magnetic field leads to radio emission, which is given off by the charged particles it traps.

Saturn has belts of charged particles (Van Allen belts), which are larger than Earth's but smaller than Jupiter's. Though Saturn's magnetic field passes through the rings, the charged particles can't, so the Van Allen belts begin only beyond the rings. Even there, the number of charged particles is depleted by the satellites. Thus the belts are weaker than Jupiter's. The size of the belts fluctuates as the pressure of the solar wind varies, allowing them to expand or forcing them to contract.

Since the rotational and magnetic axes are aligned, scientists were hard pressed to explain why a radio pulse—a blip of radio emission—occurred every rotation. By comparing Voyager 1 and 2 results, scientists pinpointed the area from which the pulse comes. It may be the result of some unusual structure fixed in place in Saturn's interior. The existence of the pulse allowed Saturn's rotation period to be accurately measured.

Titan, Saturn's largest moon, is sometimes within the belts and sometimes outside. Many of the particles in Saturn's magnetosphere probably come from Titan's atmosphere. A huge torus—doughnut—of neutral hydrogen from this source fills the region from Titan inward to Rhea.

A variety of instruments aboard the Voyagers studied the ionized gas around Saturn and the magnetic field. The region between the moons Dione and Rhea is filled with a "plasma torus." (A "plasma" is a gas containing both ions and electrons. It is electrically neutral but is affected by electric and magnetic fields.) A cloud of this plasma is at the tremendously high temperature of 600 million K, making it the hottest known place in the Solar System. The high temperature means that the velocities of individual particles are high. Since the density is extremely low, though, the total energy in the cloud is not large.

The three passes through the magnetic field and plasma around Saturn, by Pioneer 11 and the two Voyagers, surely did not allow us to understand this complex object as well as we would like. We must await the arrival of the Cassini spacecraft at Saturn in 2004 (Section 14.4).

**FIGURE 14–14** Saturn's magnetic field leads to aurorae around Saturn's poles.

### 14.2d Saturn's Family of Moons

Farther out than the rings, we find the satellites of Saturn (though a few small satellites are mixed with the rings). Like those of Jupiter, we now know them to have the personalities of independent worlds and are no longer merely dots of light in a telescope.

The largest of Saturn's satellites, Titan, is larger than the planet Mercury and has an atmosphere (Fig. 14–15) with several layers of haze. We had known something about Titan's atmosphere from ground-based studies. Spectra had showed signs of methane, polarization measurements had found signs of haze, and infrared measurements had indicated that hydrocarbons might be present. But a better understanding of Titan's atmosphere had to wait for the Voyagers. Studies of how the radio signals faded when Voyager 1 went behind Titan showed that Titan's atmosphere is 5 times denser than Earth's. Surface pressure on Titan is $1^1/_2$ times that on Earth. Though we had thought that Titan was the largest moon in our Solar System, Titan's atmosphere is so thick that its diameter measured at its surface is actually slightly smaller than the diameter of the surface of Jupiter's Ganymede.

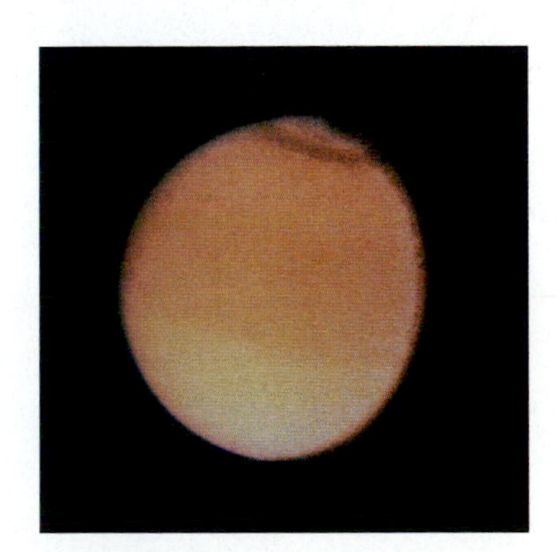

**FIGURE 14–15** Titan was disappointingly featureless even to Voyager 1's cameras because of its thick, smoggy atmosphere. Its northern polar region was relatively dark. The southern hemisphere was lighter than the northern.

The Voyagers' ultraviolet spectrometers detected nitrogen ($N_2$), which makes up the bulk of Titan's atmosphere. The methane is only a minor constituent, perhaps 1 per cent. Titan's huge doughnut-shaped hydrogen cloud presumably results from the breakdown of its methane ($CH_4$). The atmosphere may resemble the oxygen-free primordial atmosphere in which organisms on Earth first breathed.

The temperature near the surface, deduced from measurements made with the Voyagers' infrared radiometers, is only about –180°C, somewhat warmed by the

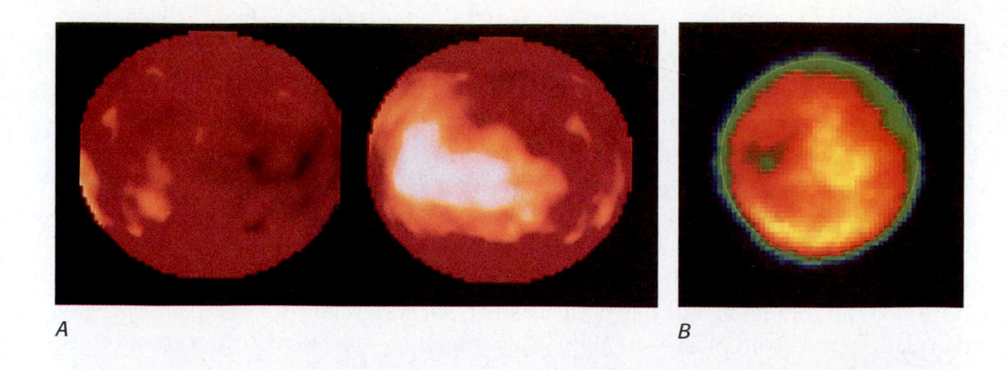


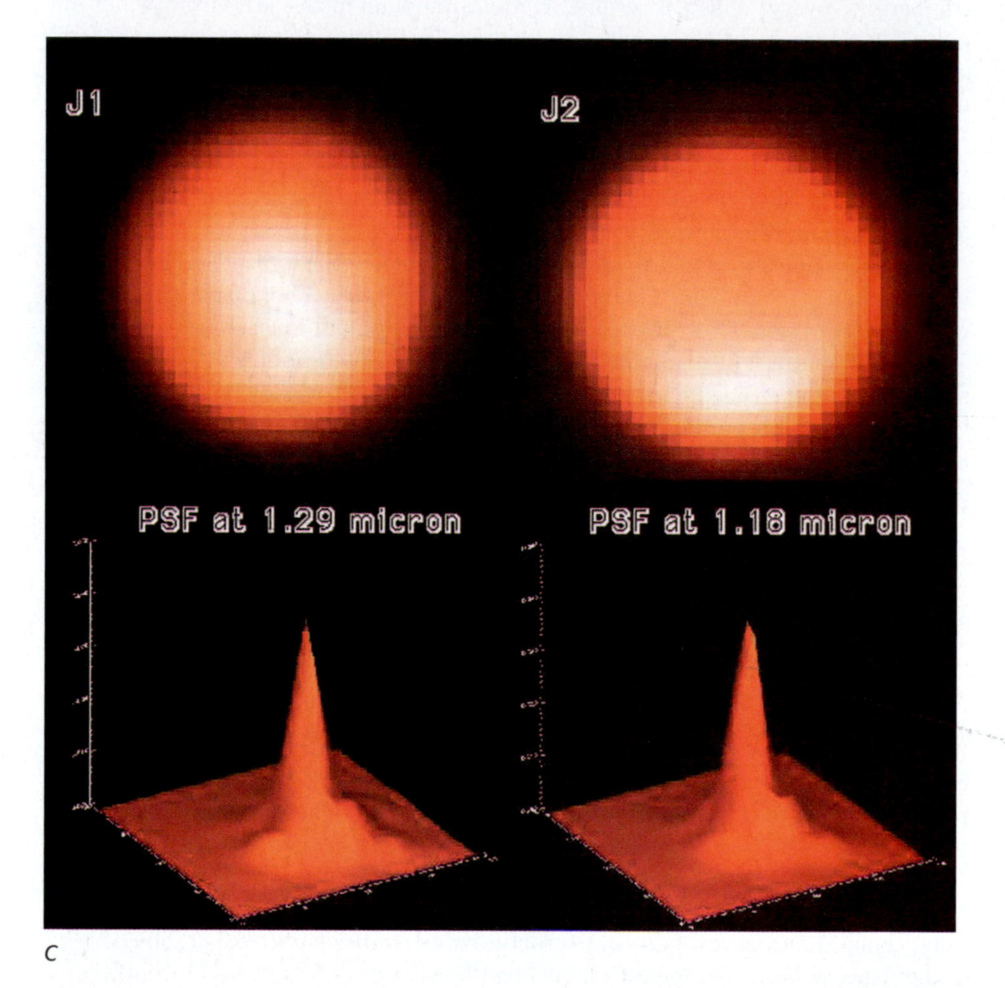


**FIGURE 14–16** (*A*) The Hubble Space Telescope has glimpsed Titan's surface through its atmospheric haze by observing at near infrared wavelengths. The large, bright region—about the size of Australia—may be a continent, a huge cloud, or the mark of a meteor impact. (*B*) A Keck image made at a longer infrared wavelength using a special technique of combining many short exposures gives a view of Titan's surface with higher contrast and resolution twice as good. The infrared dark areas may be basins filled with liquid methane or ethane, and the bright regions may be a mixture of water-ice and rock. (*C*) A pair of images taken in the near infrared with adaptive optics on the Canada–France–Hawaii Telescope. The 3-D diagrams at bottom show the "point-spread function" (PSF) for each wavelength, that is, how spread out the image of a point would appear, with intensity on the vertical axis and position on the horizontal axes.

greenhouse effect but still extremely cold. This temperature is near that of methane's "triple point," at which it can be in any of the physical states—solid, liquid, or gas. So methane may play the role on Titan that water does on Earth, probably with ethane ($C_2H_2$) mixed in to make the mixture stable. Most of Titan may be covered with methane-ethane lakes or oceans, and other parts may be covered with methane ice or snow.

Infrared observations (Fig. 14–16) show that Titan's surface is not uniform. Radar data suggest the presence of water ice. A bright region the size of Australia could be a very large range of ice mountains near Titan's equator. The ice could be continually eroding under methane rain that forms from a surrounding methane ocean. Only with such an ocean could as much methane as is present in the atmosphere survive more than 10 thousand years or so.

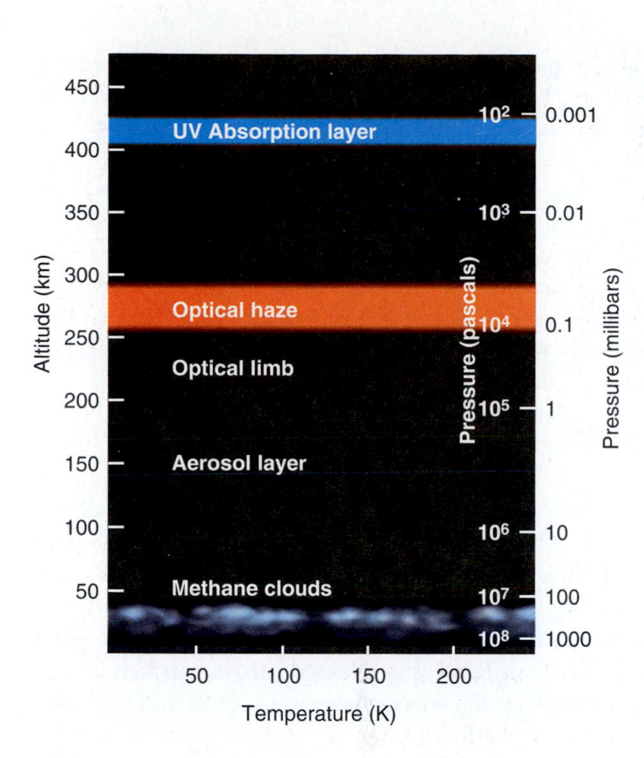


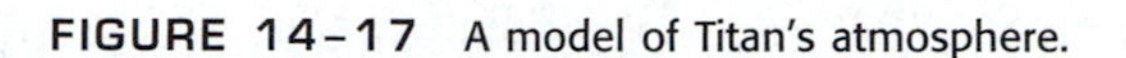

**FIGURE 14–17** A model of Titan's atmosphere.

Titan's clouds are opaque to visible light and probably contain droplets of liquid nitrogen and of liquid methane (Fig. 14–17). It seems that Titan's atmosphere is optically opaque because of the action of sunlight on methane and other molecules in it, forming a sort of smog and giving it its reddish tint. Smog on Earth forms in a similar way. Some of the color may result from the bombardment of Titan's atmosphere by protons and electrons trapped in Saturn's magnetic field. Other energy to run the chemical reactions may come from the solar wind, since Titan's orbit is large enough to go in part outside Saturn's protective magnetic field.

A view looking back from a Voyager showed the extent of the atmosphere (Fig. 14–18). Spectra taken in 1999 with the United Kingdom Infrared Telescope on Mauna Kea by scientists from Northern Arizona University and the Gemini Observatory have indicated not only that large clouds are present below the haze but also that small clouds, covering less than 1 per cent of Titan, change from hour to hour. The smaller clouds are at altitudes where convection in the atmosphere should cause cumulus clouds of methane droplets to form. Since Titan (and Saturn) receive only 1 per cent the solar energy that our Earth receives per cubic meter, the clouds would have to get their energy from below rather than from sunlight coming from above. From these clouds methane rain would fall.

Some of the organic molecules formed in Titan's atmosphere may rain downward as a tar-like substance. If the drops are large enough, they could reach Titan's surface. Thus the surface, hidden from our view, may be covered with an organic crust about a kilometer thick, perhaps partly dissolved in liquid methane. These chemicals are similar to those from which we think life evolved on the primitive Earth. The discovery of traces of carbon monoxide provides further evidence for an atmosphere resembling the atmosphere that some scientists think existed on the early Earth. But it is probably too cold on Titan for life to begin. Indeed, this cold may explain why Titan has an atmosphere while Jupiter's Ganymede and Callisto do not. Titan was sufficiently cold that its ice could trap large quantities of nitrogen and methane, from which other gases formed. Ganymede and Callisto were slightly too warm to do so. At their distance from the Sun, closer in than Titan, and possibly with additional warming from Jupiter itself, their gases escaped.

Titan

**FIGURE 14–18** Looking back at Titan's dark side, Voyager 2 recorded how extensive Titan's atmosphere is by photographing sunlight scattered by small particles hundreds of kilometers high.

The surfaces of Saturn's other moons, all icy, are so cold that the ice acts as rigidly as rock and can retain craters. The Voyagers found that the moons' mean densities are all 1.0 to 1.5 (on the common scale of g/cm$^3$, with water density = 1.0), which means that they are probably mostly water ice throughout with some rocky material included. Unexpectedly, the densities do not decrease with distance from the planet, as do the densities of the planets with distance from the Sun and of Jupiter's moons with distance from that planet.

In addition to Titan, four of Saturn's moons are over 1000 km across, so we are dealing with major bodies (see Appendix 4), about $\frac{1}{3}$ the size of Earth's Moon. For Saturn as for Jupiter, most satellites perpetually have the same side facing their planet. Let us consider the major ones in order from the inside out.

Mimas is saturated with craters, including many a few kilometers across. It boasts a huge impact structure, named Herschel, that is $\frac{1}{4}$ the diameter of the entire moon (Fig. 14–19). The crater has a raised rim and a central peak, typical of large impact craters on the Earth's Moon and on the terrestrial planets. The canyon visible halfway around Mimas (Fig. 14–20) may be a result of the impact. The energy of the impact may have partly shattered the satellite. The energy propagated through Mimas's surface and interior, and was apparently focused halfway around, making the canyon.

Enceladus's surface (Fig. 14–21) has both smooth regions and regions covered with impact craters. But the spectrum of the light reflected off the surface is uniform over the whole moon, indicating that the surface layer is uniform and probably relatively young. Also, its albedo is very high. Further, the existence of smooth regions suggests that Enceladus's surface was melted comparatively recently and so may be geologically active today. The internal heating may be the result of a gravitational tug of war with Saturn and the other moons, as for Jupiter's Io, though the situation is not as clear, and no explanation is widely accepted. Linear sets of grooves on Enceladus, tens of kilometers long, are probably geologic faults in the crust, resembling those on Jupiter's Ganymede.

Tethys also has a large circular feature (Fig. 14–22), called Odysseus, that is $\frac{1}{2}$ the diameter of the entire moon. The difference between heavily and lightly cratered regions may indicate that internal activity took place early in Tethys's history. The side of Tethys that faces Saturn (Fig. 14–23) shows not only many craters (Fig. 14–24) but also a large canyon, Ithaca Chasma, that goes $\frac{2}{3}$ the way around the satellite. The canyon, or trench, which is several kilometers deep, could have resulted from

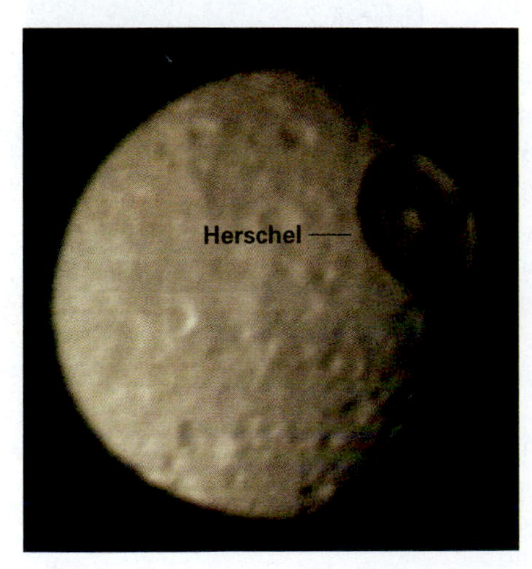


Mimas

**FIGURE 14–19** The impact feature on Mimas, named Herschel after the discoverer of Uranus, is about 130 km in diameter.

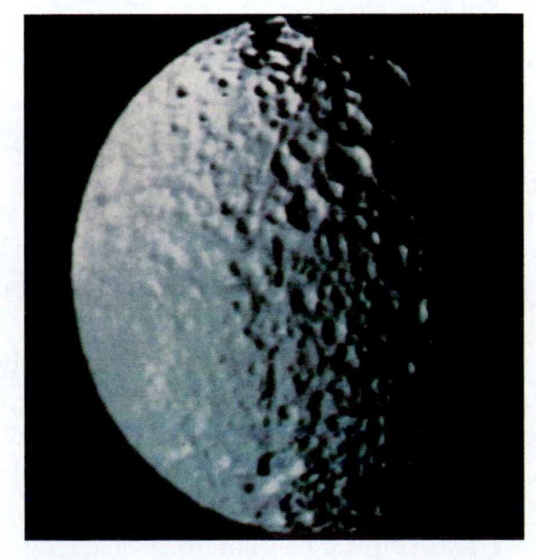

Mimas

**FIGURE 14–20** The other side of Mimas, showing the trough crossing the center that may be the result of the impact.

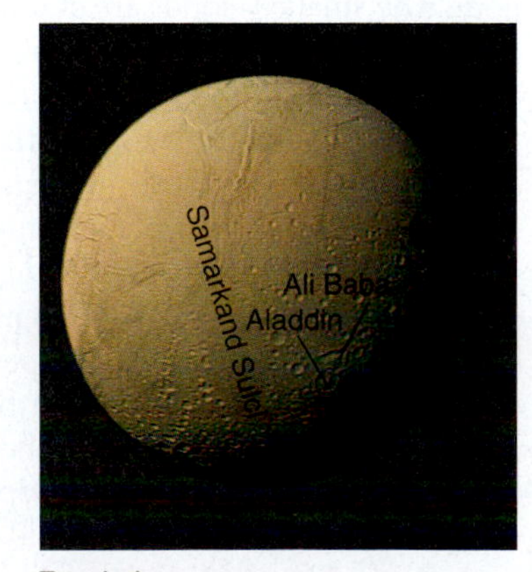


Enceladus

**FIGURE 14–21** Enceladus from Voyager 2. Enceladus resembles Jupiter's Ganymede in spite of being 10 times smaller. Resolution is 2 km on this Voyager 2 mosaic.

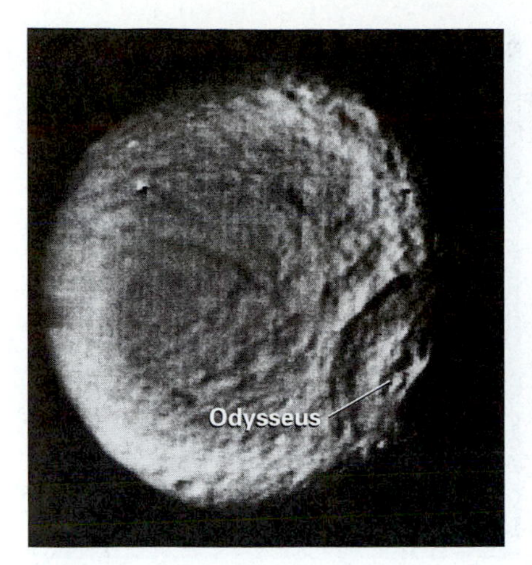


Tethys

**FIGURE 14-22** A computer-enhanced view of Tethys, showing a large, bright, circular feature 180 km across (named Odysseus). The crater has been flattened by the flow of softer ice and probably formed early in Tethys's history when its interior was still relatively warm and soft.

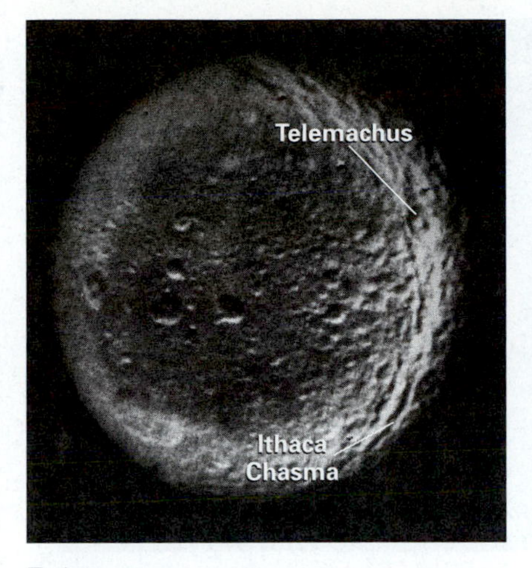


Tethys

**FIGURE 14-23** Craters and a 750-km-long canyon or trench on Tethys. The large crater at the upper right sits on the canyon or trench. Both this and the preceding image are from Voyager 2, with resolutions of 9 and 5 km.

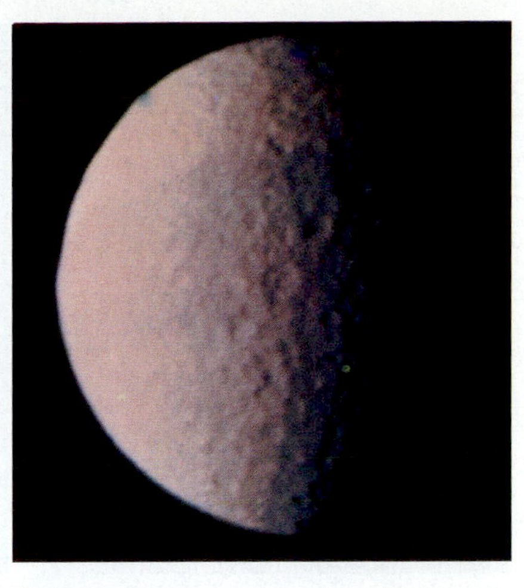

Tethys

**FIGURE 14-24** The best color view of Tethys shows a cratered terminator.

the expansion of Tethys as its warm interior froze. Or it could be a fault through the entire planet, a sign of the brittleness of ice under these extremely cold conditions.

Dione, too, shows many impact craters, some of them with rays of debris (Fig. 14–25). Many valleys are visible in Dione's icy crust (Fig. 14–26). Scientists wonder why Dione's two sides are so different. The wisps of material on the side of Dione that trails as Dione orbits Saturn may have erupted from inside.

Dione

**FIGURE 14-25** Dione, showing impact craters, debris, ridges or valleys, and wispy rays.

Two associated satellites share Tethys's orbit, always slightly ahead of Tethys itself. A small moon also shares Dione's orbit. No moons were known to share orbits in any planetary system before spacecraft studied the Saturn system from close up.

Rhea has impact craters (Fig. 14–27). Some craters, with sharp rims, must be fresh; others, with subdued rims, must be ancient. Rhea's other side, the side that is trailing as Rhea moves in its orbit and so collides with fewer objects, has wispy, light markings. Slush and mud could have flowed long ago out of Rhea's interior to cover all the large craters in some areas. The debris in the Saturn system continued to make small craters.

Next out is Titan, which we have already discussed.

Hyperion, when photographed by Voyager 2, turned out to look like a hamburger or a hockey puck (Fig. 14–28). The rotation of Hyperion is what is technically called "chaotic," with abrupt and unpredictable changes in orientation occurring. The effects of Hyperion's eccentric orbit, the strong gravity from Titan, and other gravitational forces on it result in this chaotic rotation. The field of "chaos" is a new one for science. Hyperion and Mimas are essentially the same size but are vastly different in appearance. Surely they had different histories.

Strangely, the side of Iapetus (Fig. 14–29) that precedes in its orbit is over five times darker than the side that trails. The large circular feature, Cassini Regio, is probably an impact feature outlined by dark material. Iapetus's density indicates that it may be the sole moon to contain an interior partly made of methane ice in addition to water ice. Water ice covers the bright part of the surface, giving it its albedo of about 50 per cent. The ice is covered with craters. Perhaps the dark material, whose albedo is less than 5 per cent, is a hydrocarbon formed by sunlight reacting

Dione

**FIGURE 14-26** The largest crater on Dione is about 100 km in diameter. The sinuous valleys were probably formed by geologic faults.

Rhea

**FIGURE 14–27** Rhea has craters as large as 300 km across, many with central peaks. This Voyager 1 close-up of Rhea has a resolution of 2.5 km. The area shows an ancient, heavily cratered region. White areas on the edges of several craters in the upper right are probably fresh ice either on steep slopes or deposited by leaks from inside.

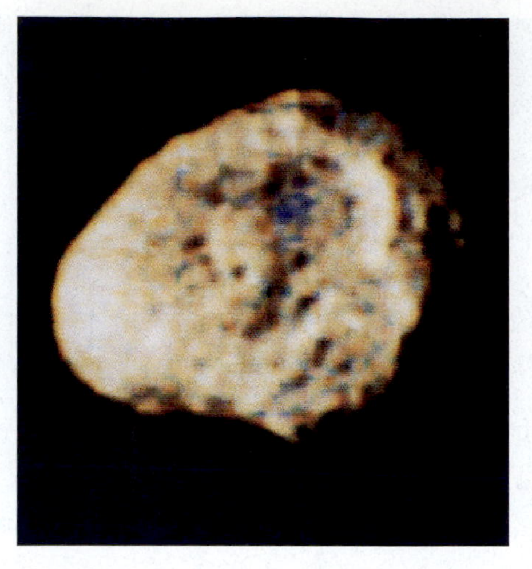

Hyperion

**FIGURE 14–28** Hyperion, an irregularly shaped satellite roughly 410 km × 220 km.

Iapetus

**FIGURE 14–29** Iapetus, whose trailing side is five times brighter than its leading one. The dark side may have accumulated dust spiralling in toward Saturn or may be covered with hydrocarbons. It could be as black as pitch because it is pitch! A large circular feature, Cassini Regio, is also visible.

on some of the methane that wells up. The spectrum of the dark material resembles that of some asteroids and meteorites.

Saturn's outermost satellite, Phoebe, is probably a captured asteroid and will be discussed in Section 20.2 on asteroids.

In addition to these nine moons long known, several others have been detected from the ground in the last decades, and several have been confirmed or discovered by the Pioneer and Voyager spacecraft. Voyager even provided close-ups of one of the recent ground-based discoveries (Fig. 14–30). This satellite is too small for gravity to have pulled it into a spherical shape. Oddly, it apparently very closely shares an orbit with another newly discovered satellite. The orbits are not absolutely identical, so the two approach each other every few years. Their gravitational fields are strong enough to affect each other only when they are very close. When this happens, they twirl around each other, interchange orbits, and are off on their merry ways again, separating from each other. No other case of such a gravitational interaction is known in the Solar System. They may be halves of a moon that broke apart.

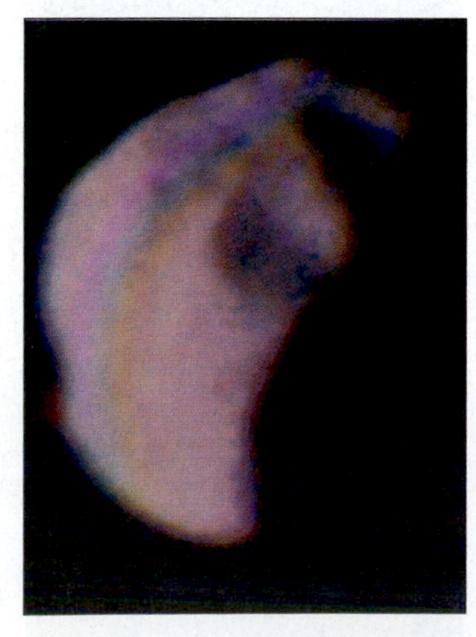

Janus

**FIGURE 14–30** Saturn's tooth-shaped moon Janus shared an orbit with Epimetheus. It is only about 220 km × 200 km × 160 km. The shadow of Saturn's F-ring crossed the satellite and was photographed in a series of six images superimposed here. The shadow appears as colored stripes, since the images were taken through different colored filters.

When Saturn was turned so that we were in the plane of its rings in 1995 and 1996, the brightness of the rings diminished greatly, and we could search to see whether there were additional moons to be discovered that were previously hidden in the rings' glare. Several bright points were found in Hubble Space Telescope images. (The rings disappeared almost entirely, indicating that they are no more than 20 meters thick.) Rather than moons, however, they turned out to be more likely clouds of orbiting ring debris. But several moons, irregularly shaped and between 10 and 50 km across, were reported in 2000 and 2001 from data taken with ground-based telescopes in Chile and Hawaii. The discoveries give Saturn at least 30 known moons. The orbits of these newly discovered moons are so large and so eccentric that the moons were no doubt captured by Saturn, unlike the large regular moons that we think formed from a disk of dust and gas that surrounded the planet as it formed.

So Saturn has a wide variety of moons, each with its own character (Fig. 14–31). The discoveries at Saturn's moons from Voyager reinforced the idea that there were many more interesting and different bodies in the Solar System than we had thought.

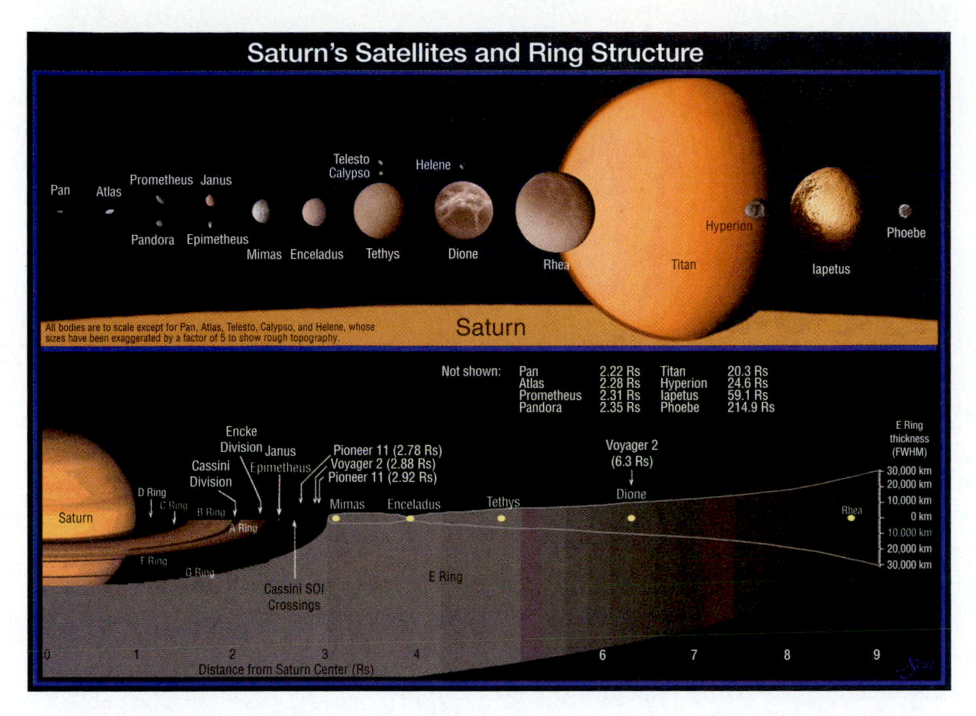


FIGURE 14–31 Saturn's moons and rings, to scale.

## 14.3 BEYOND SATURN

Yet another superlative achievement for Voyager 1 was the view it gave us when it looked back on the crescent Saturn. The spacecraft is now travelling up and out of the Solar System.

The Voyagers' passages through the Saturn system exchanged a lot of unknowns for a lot of knowledge plus many specific new mysteries. Monitoring Saturn with the Hubble Space Telescope has since provided much new information. Storms on Saturn have been studied, for example. And, as we discussed, surface structure has been seen on Titan.

Studies of the changes in the radio signals from the spacecraft when it went behind the rings showed that the rings are only about 20 m thick, equivalent to the thinness of a CD-ROM that is 30 km across.

## 14.4 CASSINI

A joint NASA/European Space Agency/Italian Space Agency mission called Cassini (after the discoverer of the gap in Saturn's rings) was launched in 1997. It has taken gravity assists from Venus, the Earth, and Jupiter, and has passed near an asteroid. As we have seen, it has tested out its instruments on the Jupiter system; its cameras are working very well. Cassini will arrive at Saturn on July 1, 2004 (Fig. 14–32).

Cassini will tour Saturn and its moons for four years. It will extensively sample Saturn's magnetosphere. The orbiter is to make over 30 close flybys of Titan and at least four of other major satellites. For example, it will try to detect small, geyser-like volcanoes on Enceladus. Cassini will also try to determine whether the dark matter on Iapetus comes from the interior or from an external source.

On January 14, 2005, Cassini is to drop a probe named Huygens into Titan's atmosphere. As it falls, it is to make dozens of panoramas, and to measure the infrared absorption spectrum to detect and measure organic molecules. Because Christiaan Huygens discovered Titan in March 1655, his name was chosen for the probe, which is the contribution of the European Space Agency. In 2000, a flaw was discovered in the communications system that the Huygens probe will use to send its

**FIGURE 14–32** Cassini, which will arrive at the Saturn system in 2004, with the Huygens probe, which will plummet into the atmosphere of Saturn's moon Titan.

data back to the Cassini orbiter, which will relay it to Earth. Apparently, the probe will be going at varying velocities that will Doppler-shift its frequencies largely out of the passband of the orbiter's receivers, whose bandwidth is narrower than expected. Engineers are working hard on the problem, to save as much as possible of the Titan data from Huygens.

## CORRECTING MISCONCEPTIONS

| ✖ *Incorrect* | ✔ *Correct* |
| --- | --- |
| The probe on Cassini will go into Saturn's atmosphere. | The probe on Cassini will go into the atmosphere of Saturn's moon Titan. |
| Saturn's rings are solid. | Saturn's rings are composed of individual chunks of rock and ice. |

## SUMMARY AND OUTLINE

Giant planet with system of rings; low density: 0.7 g/cm$^3$
Ring system (Section 14.1)
- Inclined to orbit; chunks of rock and ice orbiting inside Roche limit
- Cassini's division apparently dark
- Radar shows the rings are very thin.
- Rapid rotation makes planet oblate.
- Magnetic field
- Spacecraft
- Pioneer 11 (1979); Voyagers 1 and 2 (1980 and 1981)

Rings (Section 14.2a)
- Divided into thousands of ringlets
- Radial spokes made out of small particles held up by electrostatic force
- Ring structure affected by shepherding satellites

Saturn's surface and interior (Section 14.2b)
- High-pressure storms found; winds four times higher than those on Jupiter and do not match belts and zones

Magnetic field (Section 14.2c)
- Surface field only ⅔ that of Earth, but total field is 1000 times stronger
- Rotational and magnetic axes aligned
- Plasma torus of extremely high temperature

Moons (Section 14.2d)
- Titan: reddish smog, atmosphere mostly nitrogen; surface pressure is 1½ Earth's; methane near its triple point; 1 apparent continent, methane lakes or oceans may be present
- Other moons are icy.
- Mimas: huge crater and giant canyon
- Enceladus: recent activity and internal heating
- Tethys: large crater and giant canyon
- Dione: craters, rays, wisps
- Rhea: craters and wisps; no large craters
- Hyperion: hamburger-shaped; same size as Mimas
- Iapetus: one side very dark, the other bright
- Phoebe: captured asteroid
- Small moons newly discovered

Cassini mission (Section 14.4)
- Huygens Probe for Titan, 4-year orbiter

## KEY WORDS

Roche limit

Cassini's division

## QUESTIONS

1. What are the similarities between Jupiter and Saturn?
†2. How many times smaller is the angular size of the Sun as viewed from Saturn than the apparent angular size of the Sun from Earth? Compare with the 3-arc-min resolution of the naked eye.
†3. When Jupiter and Saturn are closest to each other, what is the angular size of Jupiter as viewed from Saturn?
4. What is the Roche limit, and how does it apply to Saturn's rings?
†5. Calculate the relative strength of the tidal force from Saturn itself on a moon the size of Titan (a) at Saturn's B-ring and (b) at Titan's orbit.
6. Sketch Saturn's major rings and the orbits of the largest moons, to scale.
7. Describe the major developments in our understanding of Saturn, from ground-based observations to Pioneer 11 to the Voyagers.
8. Why are the moons of the giant planets more appealing for direct exploration by humans than the planets themselves?
9. Why does Saturn have so many rings? What holds the material in a narrow ring?
10. Explain how parts of the rings can look dark from one side but bright from the other.
11. What did the Voyagers reveal about Cassini's division?
12. What are "spokes" in Saturn's rings, and how might they be caused?
13. What have we learned about Titan?
14. Explain how methane on Titan may act similarly to water on Earth.
15. Describe two of Saturn's moons other than Titan.
16. What is the Cassini mission?

† This question requires a numerical solution.

## USING TECHNOLOGY

### W World Wide Web

1. Examine the images of Saturn and its moons visible on The Nine Planets (http://www.nineplanets.org) and also from the USGS (at http://wwwflag.wr.usgs.gov/USGSFlag/Space/Jupiter/Mosaics), and contrast them with the set of images published in this book.
2. Look at the sites that report new satellites of Saturn, including http://www.obs-nice.fr/saturn and http://pinks.physics.mcmaster.ca/Saturn, as well as the JPL orbits at http://ssd.jpl.nasa.gov/sat_elem.html and http://ssd.jpl.nasa.gov/sat_props.html.
3. Follow the Cassini mission at http://ciclops.lpl.arizona.edu.

### REDSHIFT

1. When is Saturn visible tonight?
2. Where is Titan with respect to Saturn in the sky at 9 p.m. tonight or, if it can't be seen then, at the peak of Saturn's visibility? Will it reach naked-eye brightness?
3. Travel through and around Saturn and its system of moons.
4. Look at The Planet Dossier; and Science of Astronomy: Worlds of Space: Moons, Asteroids, and Comets.
5. Look at the relevant *RedShift* movies on disk 2.

A montage of Uranus and its moons. Ariel is at lower right, Umbriel is small to its left, Titania is at top left, Oberon is to the left of Uranus, and Miranda is at the top of Uranus. The symbol for Uranus appears at the top of the facing page.

# Chapter 15

# Uranus

**AIMS:** To study the third giant planet—Uranus—and its rings, magnetic field, and moons, seeing, especially, what the Voyager 2 spacecraft revealed

Uranus (pronounced U'ranus) and Neptune, the two giant planets beyond Saturn, are each about 50,000 km across and about 15 times more massive than the Earth. Uranus and Neptune are like Jupiter and Saturn in having no solid surfaces. However, compared with Jupiter and Saturn, they have higher proportions of heavier elements mixed in with the hydrogen and helium of which they are almost entirely formed. Uranus's and Neptune's mass is approximately equal to the mass of the cores of Jupiter or Saturn, and most of these planets are made of rock and ice. This mass and composition suggest that the cores of all four giant planets formed similarly but that Uranus and Neptune accumulated less atmosphere.

The densities of Uranus and Neptune are low (1.2 and 1.6 grams/cm$^3$, respectively). Thus we know that they are made out of hydrogen and helium. Their albedoes are high, so we have long known that they are covered with clouds.

Until Voyager 2's flyby of Uranus in 1986, little was known about this distant planet, though its rings had been discovered and thoroughly analyzed from the ground during the preceding decade. In this chapter, we shall describe our new knowledge. As a result of Voyager 2's data about Uranus and, subsequently, Neptune, we have found that the interior structure of those two planets is considerably different from that of Jupiter and Saturn. As a result, we now consider Jupiter and Saturn to be "gas giants," while Uranus and Neptune are "ice giants."

Uranus, in Greek mythology, was the personification of Heaven and ruler of the world, the son and husband of Gaea, the Earth.

## 15.1 URANUS AS SEEN FROM EARTH

Uranus was the first planet to be discovered that had not been known to the ancients. The English astronomer and musician William Herschel reported the discovery in 1781. Actually, Uranus had been plotted as a star on several sky maps during the hundred years prior to Herschel's discovery but had not been noticed as anything special.

Uranus revolves around the Sun in 84 years, at an average distance of more than 19 A.U. from the Sun. Uranus never appears larger than 4.1 seconds of arc, so studying its surface from the Earth is difficult (Fig. 15–1).

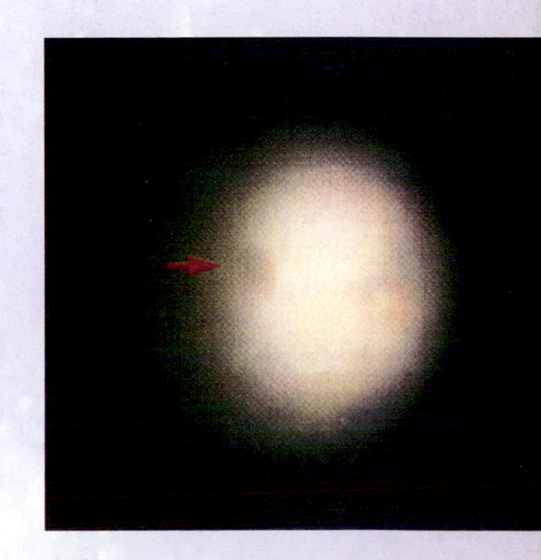

**FIGURE 15–1** Uranus, seen in three infrared bands using an adaptive-optics system to provide resolution of 0.5 arc sec. Uranus's disk was 3.7 arc sec across. The south pole is 1 arc sec to the left center. The dark spot is at latitude −35°.

**FIGURE 15–2** Uranus and its moons, observed from the Earth. The moons were only points of light.

Uranus is so far from the Sun that its outer layers are very cold. Studies of its infrared radiation give a temperature of 58 K. There is no evidence for an internal heat source, unlike the case for Jupiter, Saturn, and Neptune. The discovery of an aurora using the International Ultraviolet Explorer (IUE) spacecraft was our first indication that Uranus has a magnetic field.

From Earth, we had discovered five moons of Uranus. They range from 320 to 1690 km across. They orbit in the plane of Uranus's equator. Little was known about them based on their images from ground-based telescopes (Fig. 15–2), but spectra revealed water ice.

## 15.2 THE ROTATION OF URANUS

The other planets rotate such that their axes of rotation are very roughly parallel to their axes of revolution around the Sun. Uranus is different, for its axis of rotation is roughly perpendicular to the other planetary axes, lying only 8° from the plane of its orbit (Fig. 15–3). Its rotation is considered retrograde because its axis is tilted more than 90°. Sometimes one of Uranus's poles faces the Earth, 21 years later its equator crosses our field of view, and then another 21 years later the other pole faces the Earth. Polar regions remain alternately in sunlight and in darkness for decades. Thus there could be strange seasonal effects on Uranus. When we understand just how the seasonal changes in heating affect the clouds, we will be closer to understanding our own Earth's weather systems.

It is important to know about the period of rotation because models for the structure of Uranus's interior depend on the rotation period. Values measured from the Earth differed wildly—ranging from 12 to 24 hours.

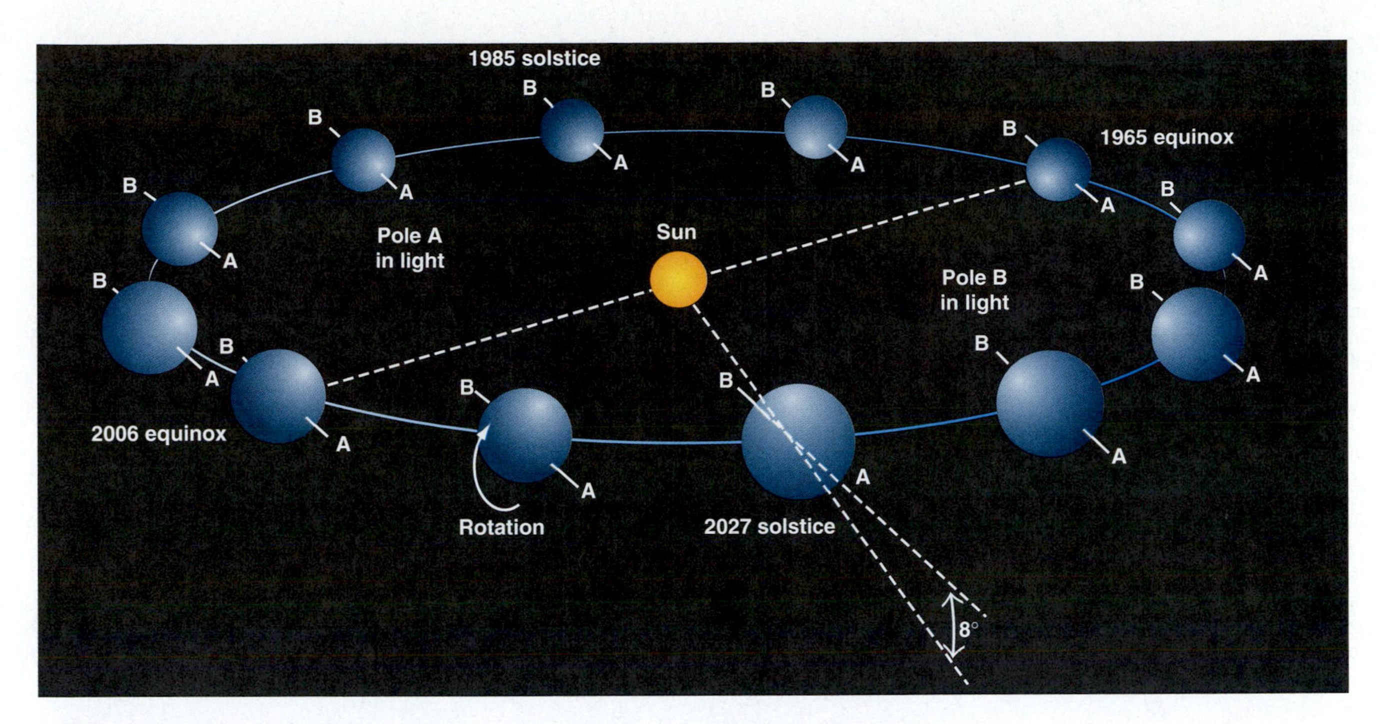


**FIGURE 15–3** Uranus's axis of rotation lies in the plane of its orbit. Notice how the planet's poles come within 8° of pointing toward the Sun, while ¼ of an orbit before or afterward, the Sun is almost over the equator.

Presumably Uranus was knocked on its side in a collision with a giant planetesimal or asteroid. The crash had to have happened early on, or else Uranus's moons wouldn't be orbiting in Uranus's equatorial plane.

## 15.3 URANUS'S RINGS

In 1977, Uranus occulted (passed in front of) a faint star. Predictions showed that the occultation would be visible only from the Indian Ocean southwest of Australia. James Elliot, then of Cornell and now of MIT, led a team that observed the occultation from an instrumented airplane. Surprisingly, about half an hour before the predicted time of occultation, they detected a few slight dips in the star's brightness (Fig. 15–4). (The equipment had been turned on early because the exact time of the occultation was uncertain.) They recorded similar dips, in the reverse order, about half an hour after the occultation. The dips indicated that Uranus is surrounded by at least five rings. More occultations showed a total of 9.

The rings have radii between 42,000 and 52,000 km, 1.7 to 2.1 times the radius of the planet. (Uranus's innermost moon is 130,000 km out.) Though the rings at first could be detected only when they occulted stars, now we can image them using techniques in the near-infrared. In particular, the Hubble Space Telescope can photograph them directly (Fig. 15–5).

The uranian rings are very narrow from side to side; some are only a few kilometers wide. No material has been detected between them. How can narrow rings exist, when the tendency of colliding particles would be to spread out? The discovery of Uranus's narrow rings led to the suggestion that gravity from a small satellite—a "ringmoon"—in each ring keeps the particles together. This model, set up for Uranus, turned out to be applicable to the narrow ringlets of Saturn discovered by the Voyagers.

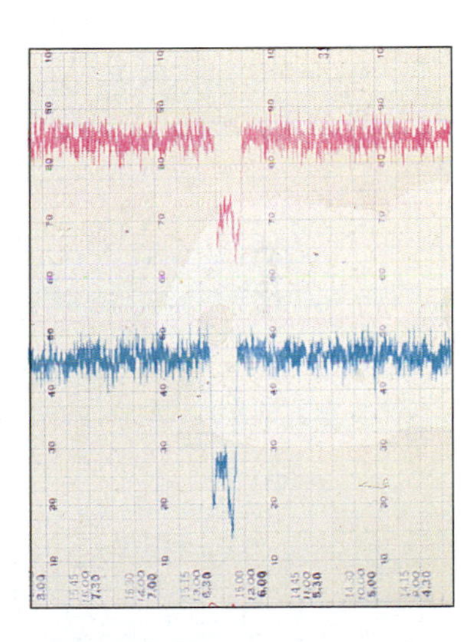

**FIGURE 15–4** The Uranus occultation by the E-ring; different channels are recorded in different colors. Time goes from right to left. This occultation started minutes after the scientists started recording data.

**FIGURE 15–5** Uranus and its rings, imaged in the near-infrared with the Hubble Space Telescope. Using a filter at the wavelength of the methane absorption from Uranus's disk diminishes the intensity of the disk. Since we see the rings by reflected sunlight, which does not have a diminished intensity at the methane wavelength, the rings appear relatively bright. We also see three levels of haze (the red level is highest) around the planet's disk.

## 15.4 VOYAGER 2 AT URANUS

Voyager 2 visited the Uranus system on January 24, 1986. The spacecraft rushed through quickly, since Uranus and its moons are oriented like a bull's-eye in the sky. At its passage, it was so far from Earth that the radio signals took 2 hours 45 minutes to reach us.

Sunlight was only 1/370 as bright at Uranus as at Earth, so exposures had to be long. Fortunately, engineers at the Jet Propulsion Laboratory had developed a technique of turning the spacecraft during the exposure to minimize blurring.

### 15.4a Uranus's Surface

Voyager 2 came as close as 80,000 km—a fifth of the distance from the Earth to the Moon—from Uranus's surface. Even from that close up, the surface of Uranus appeared bland (Fig. 15–6). This observation verified scientists' expectations; after all, Uranus is farther from the Sun than Jupiter or Saturn, and thus colder, limiting

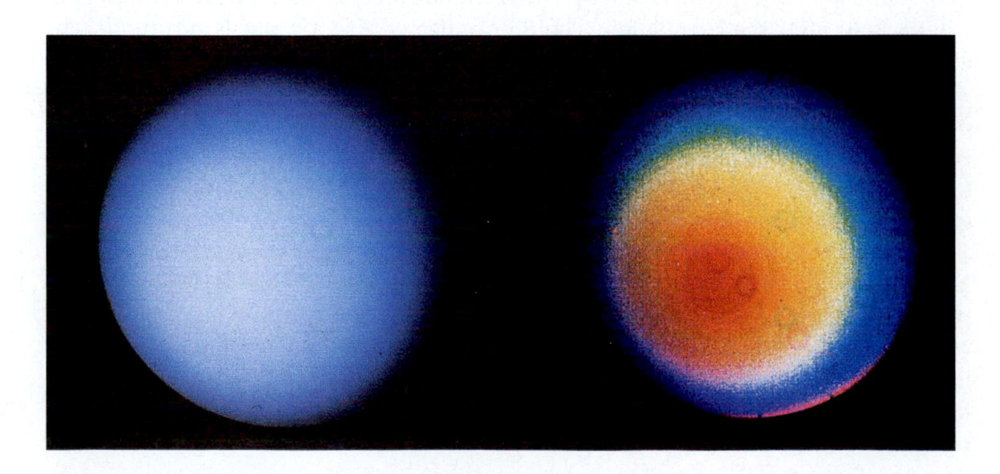

**FIGURE 15–6** Uranus, as the Voyager 2 spacecraft approached. The picture on the left shows Uranus as the human eye would see it. On the right, false colors bring out slight differences in contrast.

chemical reactions. The clouds form only relatively deep in the atmosphere. From measurements of the amount of radiation given off, scientists find that Uranus has no internal heat source.

The Voyager observations revealed a dark polar cap in Uranus's clouds. The observations are best interpreted in terms of high-level photochemical haze added to the effect of sunlight scattered by hydrogen molecules and helium atoms. At lower levels, the abundance of methane gas ($CH_4$) increases. It is this methane gas that absorbs the orange and red wavelengths from the sunlight that hits Uranus, leaving mostly blue-green in the light scattered back toward us. The relative amounts of hydrogen and helium are somewhat larger than for Jupiter and Saturn, but still similar to the Sun's atmosphere. The density of Uranus, though, is too high for Uranus to be entirely of solar composition, and $CH_4$ clouds are present, so the percentage of carbon may be 20 times higher in Uranus than in the Sun. The theory that Uranus formed from a core of rocky and icy planetesimals explains this abundance.

The elongated bright features (Fig. 15–7) may be places where convection carries the clouds, believed to be of methane ice, higher than usual. They are above more of the haze, so they appear brighter. It may be that material is carried upward to form these tails at higher altitudes. These clouds could be observed for over a dozen rotations each, with periods of about $16^1/_2$ hours.

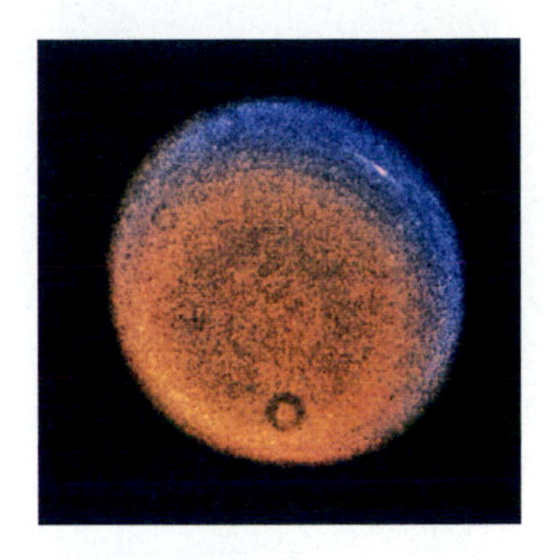

**FIGURE 15–7** An exaggerated false-color view. (The doughnut shapes are out-of-focus dust spots.) The rotation periods were: larger cloud (35° latitude), 16.3 hours; smaller, fainter clouds (27° latitude), 16.9 hours. Note the elongated feature near top.

A few low-contrast bands were seen on Uranus, especially in mid-latitudes. The patterns in the clouds and the motions of the clouds show that the winds circulate east-west rather than north-south. Winds on Jupiter and Saturn act like these, and winds on other planets do so to a lesser extent. The fact that clouds are drawn out in latitude shows how important rotation is for weather on a planet, more important than whether the Sun is overhead, heating the pole more than the equator. It was a surprise to find that both poles, even the one out of sunlight, are about the same temperature. The equator is nearly as warm. Comparing such a strange weather system with Earth's will help us understand our own weather better. We continue to monitor Uranus's clouds, with telescopes on Earth and Hubble in space. The Sun is now illuminating more of Uranus's northern hemisphere each year. Bright clouds form along the edge of the sunlight, but don't last. The longer-lived southern clouds may form in a different way.

Several experiments gathered information about Uranus's upper atmosphere. For example, an infrared radiometer measured the temperature of gases and how it varies with depth. The variation of a star's ultraviolet light as the planet apparently passed between it and the spacecraft also tells us about Uranus's atmosphere, as does the variation of Voyager's own radio beacon as detected on Earth.

A strange ultraviolet emission was discovered at Uranus. Called the **electroglow,** it occurs on the sunlighted side 1500 km above the top of the clouds. Apparently sunlight splits hydrogen molecules ($H_2$) into protons and electrons. The electrons somehow acquire extra energy and collide with remaining hydrogen molecules, and the electroglow results. The extra energy may involve Uranus's magnetic field. The electroglow is too strong to allow any auroras to be seen on the daylight side. On the night side, where there is no electroglow, a faint ultraviolet aurora was detected. We now realize that a faint electroglow also exists on Jupiter, Saturn, and Titan.

## 15.4b Uranus's Rings

Images from Voyager 2 showed the 9 rings known from ground-based occultation observations, plus a faint 10th ring and a broad, faint 11th ring. All but one are not quite circular, and most are inclined relative to the plane of Uranus's equator. They are basically gray but differ slightly in color and therefore composition.

The detailed Voyager observations showed that all the rings vary in width and apparent darkness with position around the ring. The ring known as the epsilon ring has the widest variation, from 20 km wide closest to Uranus up to 96 km farthest away. Other rings are closer to 10 km across, as had been known from the occultations.

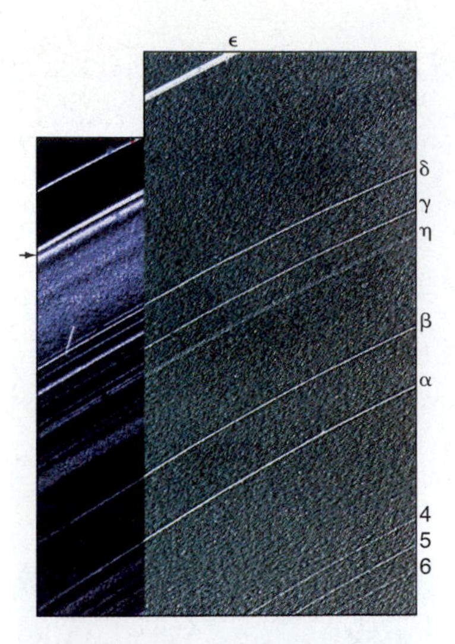


FIGURE 15–8 When the Voyager spacecraft viewed through the rings back toward the Sun, the backlighted view *(left)* revealed dust lanes between the known rings, which also show in this view. The streaks are stars. The forward-lighted view *(right)* is provided for comparison. One of the rings discovered from Voyager is marked with an arrow.

The ground-based results that the rings were only about 2 per cent reflective enabled Voyager scientists to set proper exposures. The epsilon ring was resolved into two bright inner and outer regions with a darker region between them. The low albedoes and relative absence of color make a sharp contrast with the brighter, reddish surfaces of the ring particles and moons around Jupiter and Saturn.

Following the unexpected brightness of Jupiter's ring when seen backlighted—looking through it toward the Sun—Voyager scientists arranged many exposures looking toward the Sun through Uranus's rings. But, as they watched in the control room, these exposures were all turning out blank. Apparently, there is less small dust in Uranus's rings than expected, since it is this size of dust that would cause such forward scattering of sunlight. Some process, perhaps an extended upper atmosphere of Uranus, must be removing these micrometer-sized particles from Uranus's rings.

After many blank exposures, when a photograph clearly showing ring structure appeared on the monitoring screen, the image was so good that scientists at first thought that it might be a joke some colleague was playing. But it turned out to be the single long exposure that was taken (Fig. 15–8). At the time, Uranus was hiding both Earth and Sun from the spacecraft, allowing the 96-second exposure of the backlighted rings to be taken. All 9 main rings are identifiable. The bright feature next to the wide band corresponds to the 10th ring. The dark lane just inside this ring corresponds to the orbit of a newly discovered moon, perhaps a shepherding satellite. There is dust in Uranus's rings after all! About 50 dust bands also appeared; Uranus's upper atmosphere extends into the rings and should draw the dust out of the rings in less than 1000 years.

The rings were also studied by observing how their apparent brightness changed as they occulted a star (Fig. 15–9). As for Saturn, this method gives the most precise information on the distribution of material in the ring system. Ring particles beyond the detected rings showed up when the "plasma-wave experiment" (one which measures the ionized gas) registered direct hits as the spacecraft passed through the plane of the ring system. The fact that all the previously known rings affected the reception of the Voyager's radio signal on Earth when the rings occulted the spacecraft as seen from Earth means that they must contain particles about the same size as the radio wavelengths used. Since they were detected at both 3.6-cm and 13-cm wavelengths, ring particles must be at least many centimeters in size.

Two of the new satellites discovered—Cordelia and Ophelia—are apparently the shepherds for the epsilon ring (Fig. 15–10). Shepherds for the other rings have not been found, though. Perhaps the shepherds are too small to have been seen by Voyager.

The rings of Uranus are apparently younger than 100 million years of age; one reason for this conclusion is that the masses of the shepherds are too low to have held the ring material in place longer. Therefore, there must have been a source of material to form them, other than the material from which Uranus formed. That source may well have been an icy moon destroyed by a giant impact on it by a meteoroid or comet. Particles in the dust rings seen only in the backlighted image may come from a different source—perhaps from the surfaces of the current moons.

The method of occultation used to discover Uranus' rings is a powerful one with many uses. In the next chapters we will see how it has been used not only for finding rings around Neptune but also for investigating the size and atmosphere of Pluto.

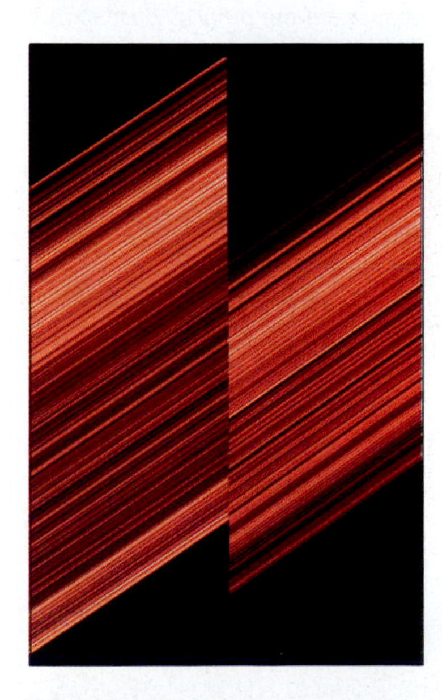

FIGURE 15–9 The epsilon ring, about 100 km wide here, showed a wide bright band, a wide fainter band, and a narrow bright band when Voyager viewed it occulting the star sigma Sagittarii. False color shows regions with less ring material (reddish) and more ring material (yellow). The ring varies in width. We see here slices 31 and 22 km wide, respectively.

### 15.4c Uranus's Interior and Magnetic Field

Though the presence of a magnetic field had been suspected, its direct detection by instruments on Voyager 2 was a major result of the mission. Voyager detected radio waves from Uranus only about two days before arrival; electrons in Uranus's ionosphere apparently block radio noise. Less than 8 hours before arrival at Uranus, about 500,000 km away, the spacecraft reached Uranus's magnetosphere. It is intrinsically about 50 times stronger than Earth's. Surprisingly, it is tipped 60° with respect to

Uranus's axis of rotation. And even more surprisingly, it is centered around a point offset from Uranus's center by 8000 km (Fig. 15–11). Since the field is so tilted, it winds up like a corkscrew as Uranus rotates. Uranus's magnetosphere contains belts of protons and electrons, similar to the Earth's Van Allen belts.

Two possibilities for why the axis of Uranus's magnetic field is so tipped are that (a) the magnetic field is in the midst of changing direction, as the Earth's does every 200,000 years, or (b) an Earth-sized object rammed into Uranus early on and tipped it on its side. But these ideas seem *ad hoc,* which, we have seen, is undesirable for scientists. Less *ad hoc,* and therefore perhaps more plausible, is (c) the idea that the dynamo is in a thin, electrically conducting shell outside the core of the planet, rather than in a deeply seated core as for the Earth, which would lead to a centrally based magnetic field. Jupiter's shell, though not as displaced, is thought to arise in the metallic hydrogen shell that surrounds the core. The discovery, as we shall see in the following chapter, that Neptune's magnetic field is also very inclined and offset makes the last idea more appealing, because it can be applied elsewhere. There may well be a layer of slush-ice inside both Uranus and Neptune, with their magnetic fields generated in it.

Studies of radio bursts, whose emission is linked to magnetic field lines anchored deep inside the planet, repeat with a period of 17.24 hours. Thus Uranus's interior rotates more slowly than its atmosphere.

All the rings and most of the moons of Uranus lie within the magnetosphere. They are thus protected from the solar wind.

Using lasers to form high pressures in a terrestrial laboratory, Berkeley scientists have matched the high pressures of the deep interiors of Uranus and Neptune. They reached temperatures of 2000 to 3000 K with pressures of 500,000 atmospheres. They found that the methane molecules break apart into carbon and hydrogen atoms. The carbon atoms then form diamonds, as they verified in the lab and reported in 1999. The theory is that the diamonds should rain downward toward the

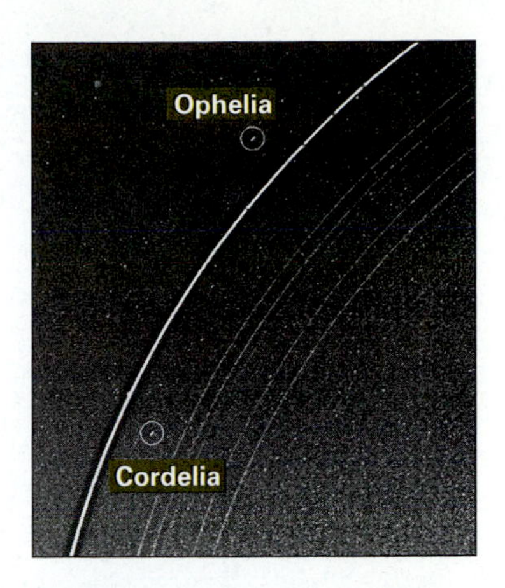


**FIGURE 15–10** Three days before closest approach, Voyager 2 discovered the two "shepherd satellites" for the epsilon ring. Based on their brightness and their likely albedo, they are 20 and 30 km in diameter, respectively. All 9 of the previously known rings are also visible. The image was processed to enhance narrow features, giving the epsilon ring an artificial dark halo. The dark blips on the ring are also artifacts.

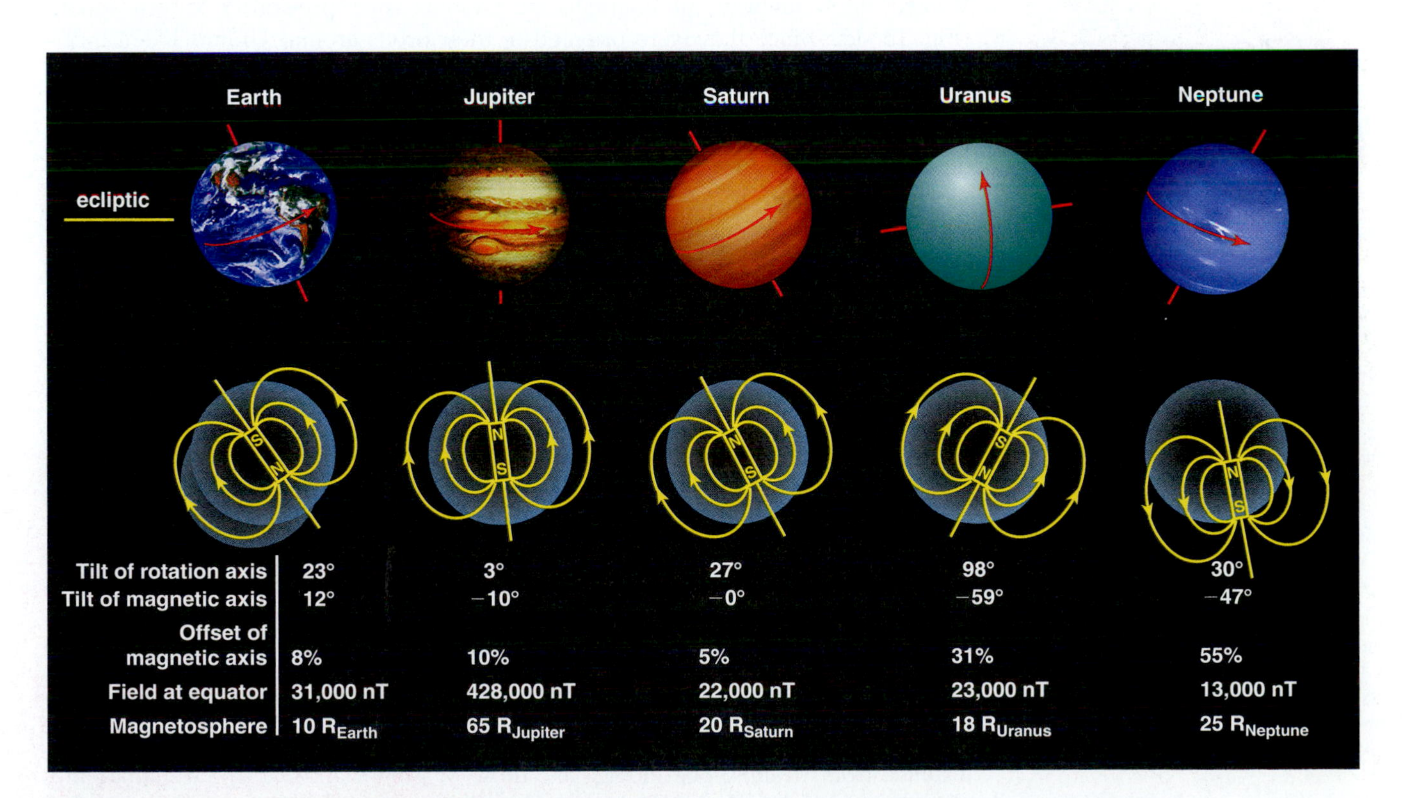


**FIGURE 15–11** Uranus's magnetic field is tipped and offset from Uranus's center; see also Table 16-1.

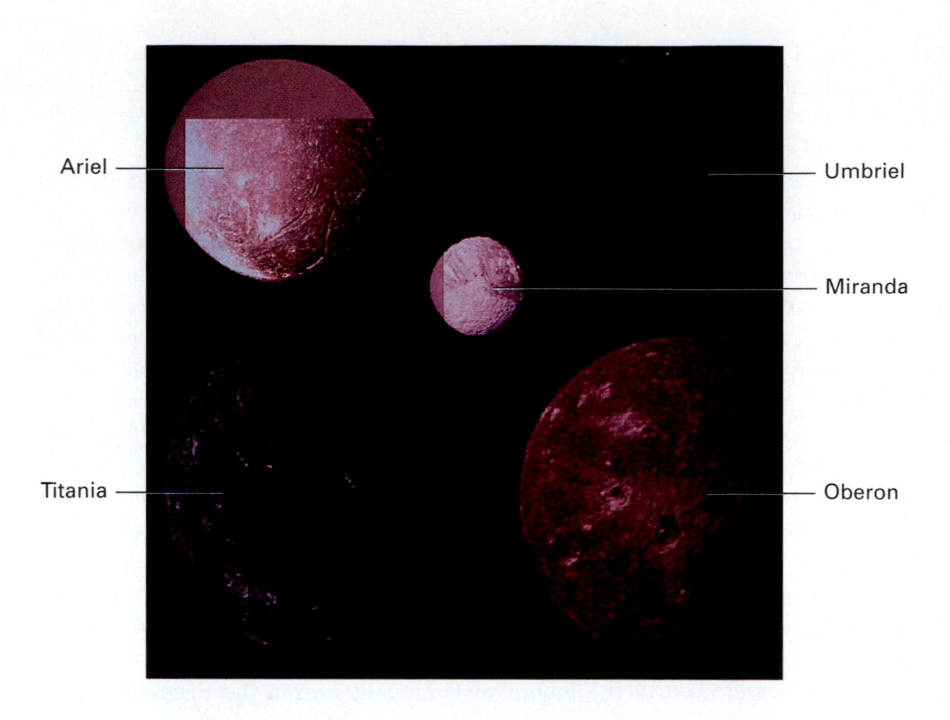


**FIGURE 15–12** The five largest moons of Uranus, reproduced in their relative sizes and albedoes.

centers of the planets. Perhaps the rain of diamonds contributes to heating these planets. How big are the diamonds? Perhaps dust or pebble size. But they sure would be hard to retrieve.

### 15.4d Uranus's Moons

Voyager took fantastic photographs of Uranus's five previously known moons (Fig. 15–12). Since they were oriented in their orbits around Uranus like a bull's-eye, the spacecraft sped by them all in a few hours. Nonetheless, the images we have show us their distinctive individual characters. I find it helpful to remember Miranda, Ariel, Umbriel, Titania, and Oberon in order outward by their initials: MAUTO. Titania and Oberon are the largest, about the same size as each other and Saturn's moons Iapetus and Rhea. Uranus's moons all rotate synchronously, once per orbit, leaving the same side always facing Uranus. Since the Sun shone almost directly down on their south poles, we have imaged only their southern hemispheres; the northern hemispheres remain unknown to us.

In general, as had been discovered from ground-based measurements, the moons of Uranus are darker (have lower albedoes) than the moons of Saturn, except for Phoebe and the dark regions of Iapetus. There is no obvious trend in albedo with distance from Uranus. The dark material seems evenly dark across the spectrum, that is, gray rather than colored. Carbon in the form of soot or graphite would explain this grayness; iron-bearing silicates and sulfur are ruled out because of their reddish color. The dark material in the rings is similarly gray.

Only for Miranda did the spacecraft fly sufficiently close to allow us to measure the moon's mass from its gravitational effect. The masses of the other major satellites have been found from detailed calculations of their gravitational effects on each other, measured on Voyager data as small variations of their positions from their average orbit. The measured masses plus the observed sizes of the moons allow the densities to be calculated. They turn out to be about 1.5 times the density of water, higher than those of Saturn's icy satellites, and do not vary in any consistent way with distance from Uranus. The moons, like those of Jupiter and Saturn, are apparently mixtures with varying proportions of rock and ices of water, ammonia, methane, and other chemicals.

A DEEPER DISCUSSION

## BOX 15.1 Naming the Moons of Uranus

Six years after he discovered Uranus, William Herschel reported the discovery of several moons around it. In 1851, William Lassell discovered two additional moons of Uranus and showed that only two of Herschel's discoveries were real. Lassell assigned names from the fairy kingdom. At his request, William Herschel's astronomer son John Herschel selected the names Oberon and Titania for the moons his father had discovered. Oberon is King of the Fairies and Titania is Queen of the Fairies in Shakespeare's *A Midsummer Night's Dream.* Umbriel and Ariel are the names of guardian spirits in Alexander Pope's *The Rape of the Lock,* and Ariel is also a character in Shakespeare's *The Tempest*. Miranda wasn't discovered until 1948, when it was named for the heroine of *The Tempest.*

The first rings to be discovered, by James Elliot and colleagues in 1977, were assigned Greek letters. The rings first noted by others were assigned numbers.

Permanent names for the moons were assigned by the International Astronomical Union. The first moon to be discovered by Voyager 2 was named Puck, after the spirit in Shakespeare's *A Midsummer Night's Dream*. Most other moons were named after Shakespearean heroines. Belinda's name comes from Pope's *The Rape of the Lock*. This moon's name fulfills the poet's prophecy:

> "This lock, the muse shall consecrate to fame,
> And midst the stars inscribe Belinda's name."

The International Astronomical Union continued the Shakespearean naming scheme at their 2000 General Assembly with Caliban and Sycorax, also from *The Tempest.*

***Discussion Question***

What names might you assign to Uranus's rings? Why do they make a suitable set and fit with the normal naming criteria?

All the satellites have craters on them. The largest craters probably formed in collisions with planetesimals or other debris of the formation of the planets during the early years of the solar system, more than 4 billion years ago. Many of the smaller craters were probably caused by secondary debris of collisions between satellites and of impacts with satellites. We think that whatever craters are still forming result from impact with short-period comets captured into orbits near Uranus's.

Ariel and Umbriel are similar in size and density, but they differ greatly in the way they look. Their geological histories must thus be very different. Ariel is about four times farther out than the rings and Umbriel is about five times farther out than the rings.

Ariel (Fig. 15–13) has the youngest surface of Uranus's moons and is the brightest. It shows signs of geological activity on a large scale: many of the fractures of Ariel's crust are global, extending over the entire surface we can see. Ariel has fault valleys but few large craters, the presence of the former and the absence of the latter both being signs of geological activity. Apparently material has flowed from under Ariel's surface; in one location, a half-buried crater is seen. Sinuous valleys similar to those on the Moon are found, some with parallel ridges alongside. These features appear similar to lava channels and ridges known on the Earth and other inner planets. What material is flowing? It is too cold for water to melt. A mixture of ammonia and water ice has been suggested. And perhaps tidal effects from the other moons could have heated Ariel in the past.

Umbriel (Fig. 15–14), though similar physically, is much darker. Its ground-based spectrum shows weaker signs of water ice than do spectra of Uranus's other major moons. Umbriel's surface is strikingly uniform, also in contrast to Uranus's other major moons. It is covered with craters; Umbriel has the oldest surfaces, similar to the ancient cratered highlands of the inner planets. Its surface may date back 4.5 billion years to the formation of the solar system. The bright ring is a high-albedo deposit that covers the floor of a 40-km-diameter crater, Wunda. It must be a relatively young feature. Why is Umbriel so uniform? Perhaps a uniform blanket of dark material has coated the surface, covering earlier markings. Since only Umbriel is coated, the material would have to have come from Umbriel itself, perhaps from a large impact that

**FIGURE 15–13** Ariel, 1200 km across, is covered with small craters and shows many valleys and scarps. At this resolution, the highest obtained, we see linear grooves that presumably resulted from tectonic activity and smooth patches that show where material was deposited.

**FIGURE 15–14** Umbriel, also 1200 km across, is the darkest of Uranus's moons and shows the least geological activity. The nature of the bright ring is unknown, though it might be a frost deposit around an impact crater.

sent up lots of material. The bright ring would have been deposited more recently or formed in a region where the blanket was thin. A similar model has been advanced for Saturn's Enceladus. The major objection to this model is that such an impact has a low probability. An alternate model is that something in Umbriel erupted explosively; methane has been suggested. In this model, the bright ring might be connected with the eruption. The major objection to this model is that we see no signs on the rest of the surface that volcanism occurred. Remember, though, we cannot see Umbriel's northern hemisphere—perhaps huge volcanoes are there. There is a lot we don't know about Umbriel, including why no craters with bright rays are seen.

Just as Ariel and Umbriel are similar in size, so are Titania and Oberon, each about 1600 km across; they are twice as far out as Ariel and Umbriel. The albedoes of the latter pair are similar to each other, 20 to 30 per cent. Both are gray. The visible surface of Titania shows many craters 10 to 50 km across, but few large ones (Fig. 15–15). The large ones would have been formed earlier, at the same time as others on other moons, since only in the early years of the Solar System, when planetesimals were still around, did a full range of size of impactors exist. (After the planets and their moons formed, the other bodies left over were all fairly small.) The large craters that would have once been on Titania were perhaps erased by material that flowed out of Titania's interior. Titania's surface shows many scarps from 2 to 5 km high. Halves of craters split by the scarps are displaced, showing that the scarps are relatively young. A few bright deposits are visible on some of the scarps, probably showing fresh material from underground.

Oberon has many large craters greater than 100 km across (Fig. 15–16). As with Umbriel, parts resemble the ancient cratered highlands of the inner planets. Such craters resulted from impacts of the debris left over from the formation of the plan-

Titania

**FIGURE 15-15** Titania, 1600 km across, shows so many impact craters that it must have an old surface. The walls of some of the fault valleys may appear bright because of frost deposits.

Oberon

**FIGURE 15-16** Oberon, also 1600 km across, was not imaged as well as the other moons. The bright patches may be patterns caused by the formation of an impact crater in an icy surface.

ets and are over 4 billion years old. The presence of large circular regions shows that Oberon has been geologically active. Some very dark material occurs in the floors of a few large craters. The dark patches are similar in some ways to dark patches in craters on Iapetus's trailing hemisphere and to volcanic deposits in several lunar craters. It seems that the dark patches were a fluid from inside Oberon that either was originally dark or turned dark after it reached the surface.

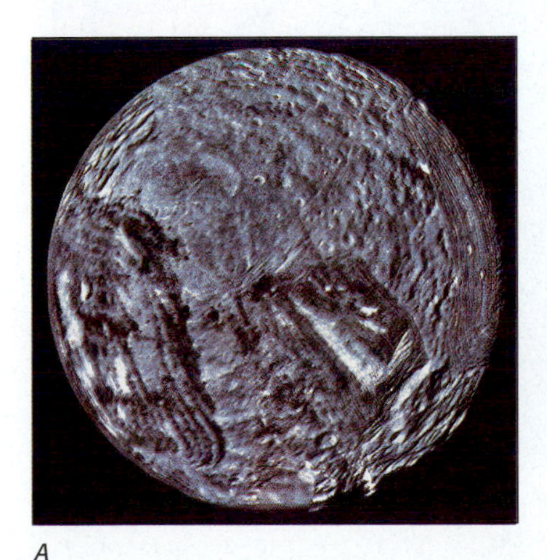

*A*

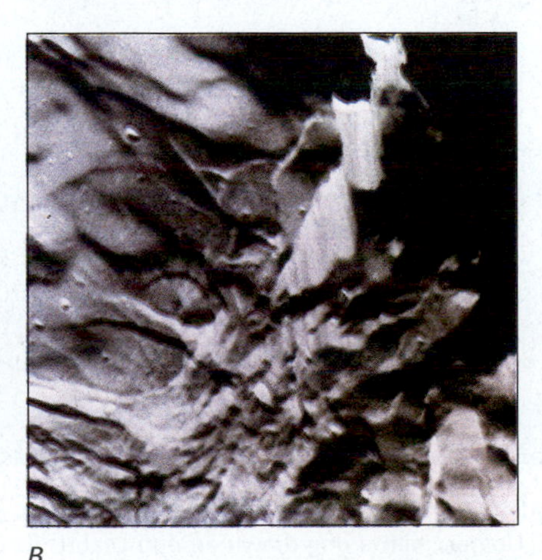

*B*

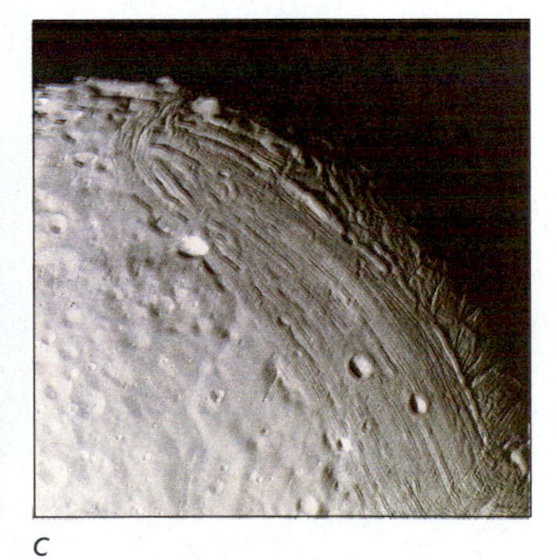

*C*

**FIGURE 15-17** (*A*) Miranda, only 500 km in diameter, shows several different types of geological regions on its surface, in spite of its small size. The south pole is almost exactly in the center of this image. (*B*) The giant canyon on the edge of Miranda, at a resolution of 0.6 km. The grooves and troughs can be a few kilometers deep. (*C*) At left, an ancient, cratered terrain of rolling hills and degraded craters; at center, a younger, grooved terrain; and near the terminator, a still younger, complex terrain that ends abruptly at the grooves.

Inside the orbits of these four moons lies the orbit of Miranda, half the size of Ariel and Umbriel and a third the size of Titania and Oberon. Many people thought it was too small to have much activity. But they were wrong—Miranda turns out to be as varied a body as exists in the Solar System. As Voyager 2 approached Miranda, a bright region shaped like a chevron—a V-shape—was visible (Fig. 15–17). Only an internal process could have caused such a sharp angle to appear. Some of the ridges resemble the grooved terrain found on Jupiter's Ganymede. In the highest-resolution images, the dark bands proved to be scarps. In some locations, erupted material has apparently flowed over and partly buried the grooves and ridges. A huge valley showed on Miranda's edge (Fig. 15–18).

How did Miranda get hot enough inside for material to be molten and so to erupt? Perhaps a collision tore it apart long ago. When the pieces came together, they gave Miranda its jumbled surface. Energy might have been released from gravity as denser blocks sank. Tidal heating, as for Io, is another source of energy that may have contributed. Though the satellites of Uranus do not interact in that manner now, they might have long ago.

Voyager 2 discovered 10 new satellites (Fig. 15–19), all in nearly circular orbits and all but one between the epsilon ring and the orbit of Miranda. They can be photographed with the Hubble Space Telescope (Fig. 15–20), which also follows weather on Uranus. Additional satellites of Uranus are discovered on old Voyager photographs from time to time, and they are listed in Appendix 4.

**FIGURE 15–18** A montage of photographs of the giant valley on the edge of Miranda and Uranus, with a ring drawn around Uranus for scale.

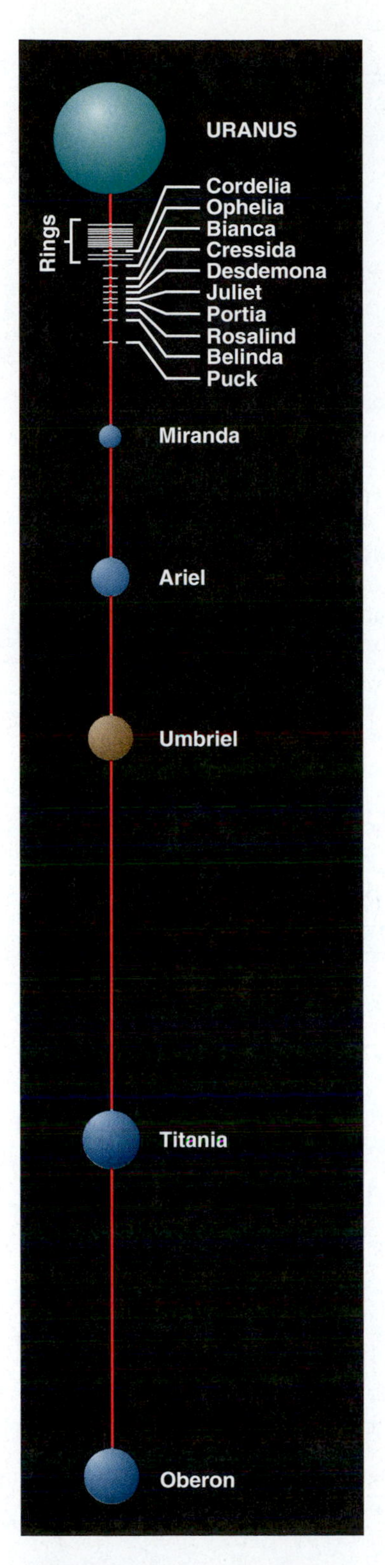


**FIGURE 15–19** The rings and moons of Uranus. The moons are in correct scale with respect to each other but are drawn with disks at a larger scale than Uranus and the positions of the rings and moons.

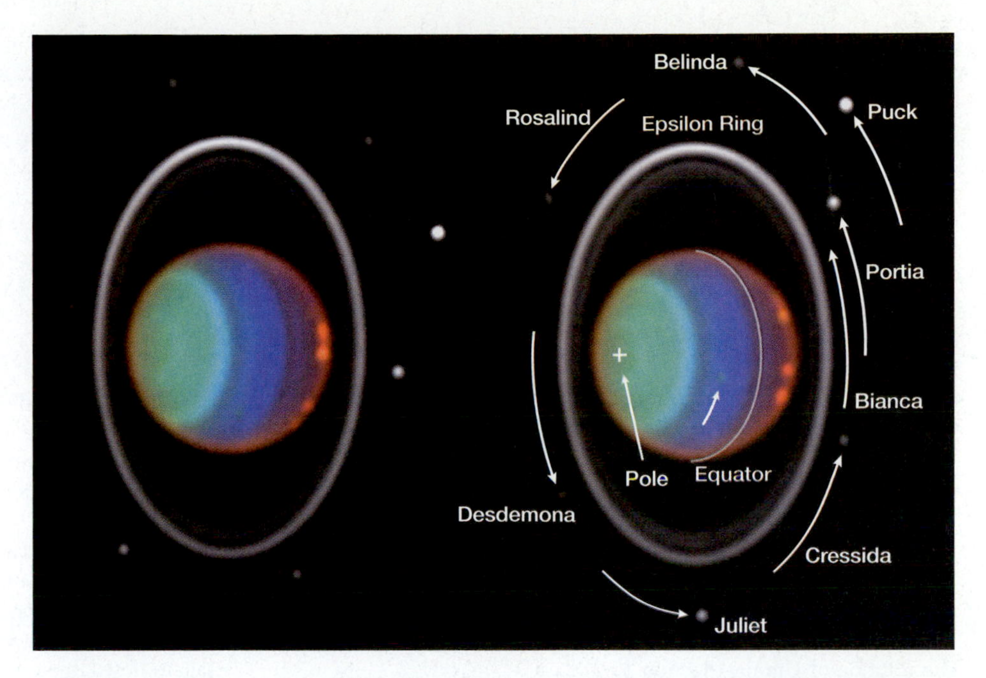


**FIGURE 15–20** Uranus and its rings and moons imaged with the Hubble Space Telescope's Near Infrared Camera. Eight of the 10 small satellites discovered by Voyager 2 show; their orbital periods are less than 1 day. Their motion in the 90 minutes between exposures is shown by the arrows. The images of the satellites have been enhanced relative to the rings and disk.

The albedoes of all Uranus's smaller satellites are low, 5 to 7 per cent, in contrast to the higher albedoes of the planet's larger satellites and the even higher albedoes of Saturn's smaller satellites. Thus the composition of these small moons, or perhaps the conditions under which these small moons have formed and evolved, must have been different for the Uranus system than for the Saturn system.

## CORRECTING MISCONCEPTIONS

**✖ *Incorrect***

Uranus's rings have always been known.

We have a lot of material on Uranus and its moons.

**✔ *Correct***

Uranus's rings were discovered only two decades ago.

We had only one flyby: Voyager 2. We couldn't image even half of each moon.

## SUMMARY AND OUTLINE

Uranus and Neptune are each 50,000 km in diameter and 15 times more massive than Earth.

Uranus as seen from Earth (Section 15.1)
- First planet discovered that was not known to ancients
- No detail visible on surface; perhaps surrounded by thick methane clouds
- No internal heat source
- Aurora indicates magnetic field.

The rotation of Uranus (Section 15.2)
- Axis lies in plane of solar system.

Uranus's rings (Section 15.3)
- Nine narrow rings discovered in 1977 at stellar occultation.
- Can now be imaged in infrared from Hubble and from Earth.

Voyager 2 at Uranus (Section 15.4)
- Dark polar cap; mid-latitude brightening
- A few methane clouds probably above the haze; rotation period $16^1/_2$ hours
- Both poles the same temperature
- Electroglow—ultraviolet emission
- Few small dust particles in rings; particles at least several cm across
- Magnetic field intrinsically 50 times stronger than Earth's, tipped 60° and offset from center
- Five largest moons are mixtures of rock and ices.
- Ariel: youngest surface; brightest; global fractures, fault valleys, few large craters
- Umbriel: much darker, uniform surface covered with craters, perhaps dark coating of material; unexplained bright ring
- Titania: few large craters, many young scarps, some fresh material
- Oberon: many large craters, scarps, dark patches in crater floors
- Miranda: part with undulating cratered plains; part with scarps that show as dark bands

## KEY WORD

electroglow

## QUESTIONS

1. What is strange about the direction in which Uranus rotates?
2. Using Appendix 4, compare the sizes of the moons of Uranus with other objects in the solar system.
3. Explain how the occultation of a star can help us learn the diameter of a planet.
4. How were the rings of Uranus discovered?
5. What did the Voyager 2 encounter tell us about the rings that we didn't already know?
6. How do we learn whether a planet's rotation or the orientation of its axis determines its weather?
7. Why does Uranus appear blue-green?
8. What causes the electroglow on Uranus?
9. Compare frontlighted and backlighted views of Uranus's rings. What does the comparison tell us about their composition?
10. Compare Uranus's magnetosphere with Earth's.
11. Why didn't we observe the northern hemispheres of Uranus's moons from Voyager 2?
12. Describe the physical properties of two of Uranus's five major moons.
13. Which of Uranus's moons has the youngest surface? Why do we think so?
14. Which of Uranus's moons have the oldest surfaces? Why do we think so?
15. Why can we now image Uranus's moons, while we couldn't earlier?

## USING TECHNOLOGY

### W World Wide Web

1. Examine the images of Uranus, its moons, and its rings at http://photojournal.jpl.nasa.gov and www.nineplanets.org. See images and animations of all the planets at http://www.solarviews.com.
2. Find the Hubble images of Uranus and consider Uranus's changing atmosphere.
3. Use this book's Web site to see if any more moons of Uranus have been discovered. Look at the latest listings for new moons at http://ssd.jpl.nasa.gov/sat_props.html.

### REDSHIFT

1. Examine the Uranus system. Note the orientation of Uranus's rotation and of the orbits of its satellites.
2. Where is Uranus in the sky tonight? How bright is it?

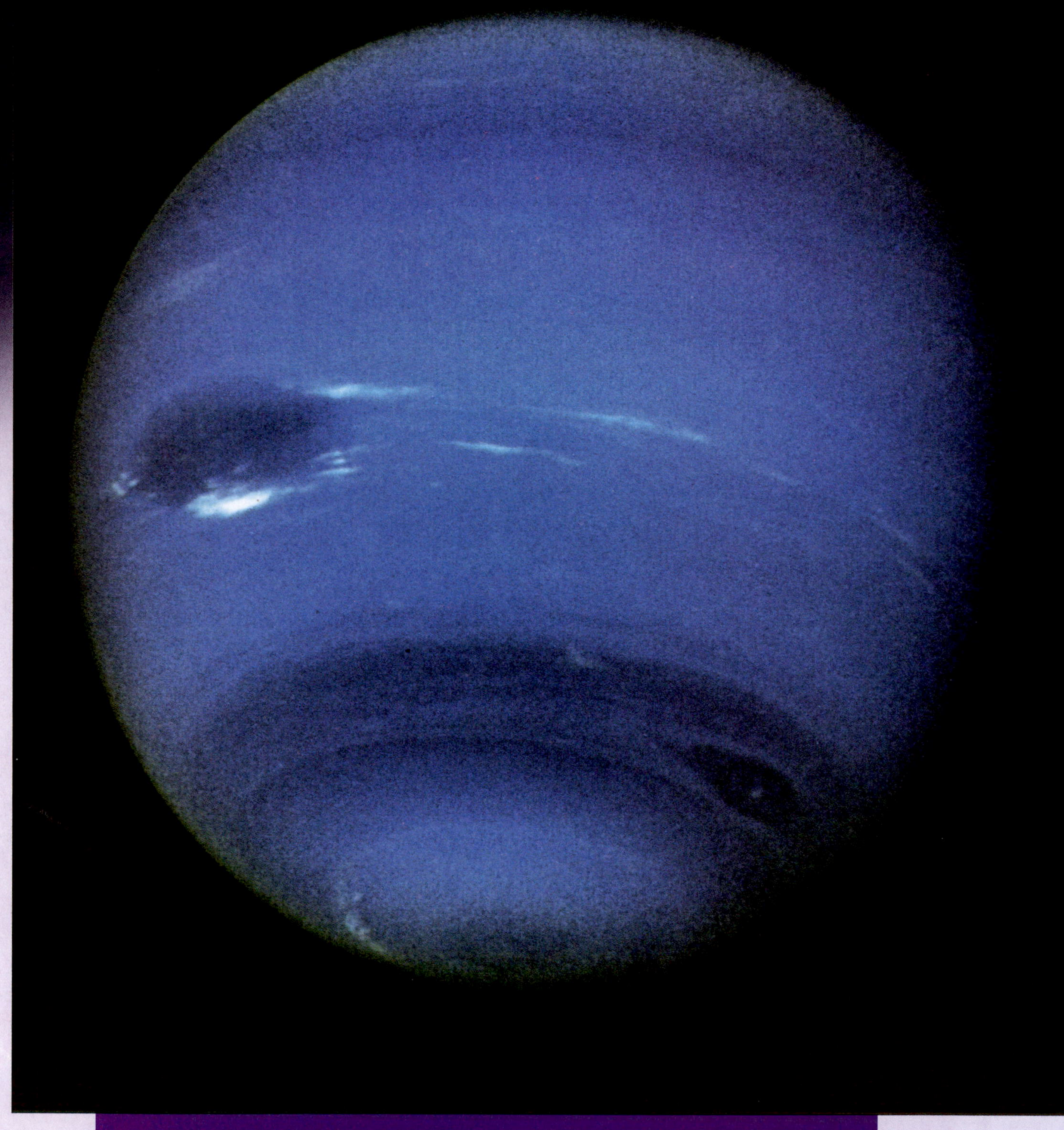

Neptune, with the Great Dark Spot present at the time of the Voyager 2 flyby. Neptune's blue-green color comes from absorption of the red part of the solar spectrum by traces of methane mixed into Neptune's atmosphere. The symbol for Neptune appears at the top of the facing page.

# Chapter 16

# Neptune

Neptune is like Uranus in many ways. Neptune is also about 50,000 km across and is about 17 times as massive as the Earth. Like Uranus, Neptune has a high albedo and a low density. Both probably have rocky cores of 10 to 15 times the Earth's mass. Both are probably 10 to 20 per cent hydrogen and helium. But Neptune and Uranus are more like cousins than like siblings. Uranus's surface is bland, while Neptune has many clouds. Uranus has no internal heat source, while Neptune has a strong one.

Though Uranus and Neptune are giant planets, they are fairly different from Jupiter and Saturn. Uranus and Neptune's hydrogen/helium content is lower than that of Jupiter or Saturn. A higher proportion of Uranus and Neptune probably formed from icy and rocky planetesimals. The rocky cores of Jupiter and Saturn, though probably the same absolute size as those of Uranus and Neptune, comprise a much smaller fraction of their masses.

Another similarity that Neptune shares with Uranus is that its secrets were bared to us by NASA's Voyager 2 spacecraft. Voyager 2 swept by Neptune and its rings and moons in August 1989 (Fig. 16–1). The visit was the last "first" of solar system exploration, aside from a visit to tiny Pluto. Over the last three decades, each of the other planets has been visited close up by spacecraft for the first time.

Scientists are also nostalgic about the Neptune visit for another reason—the knowledge we gained will have to last us for a long time. It is difficult to foresee that any further visit to Neptune will take place for many decades.

**AIMS:** To study the farthest giant planet—Neptune—and its rings, magnetic field, and moons, seeing, especially, what the Voyager 2 spacecraft revealed

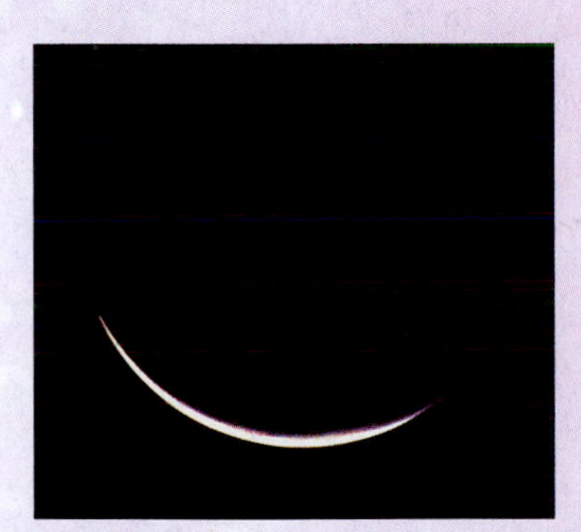

**FIGURE 16–1** The crescent Triton, just after Voyager 2's closest approach.

## 16.1 NEPTUNE FROM EARTH

Neptune is even farther from the Sun than Uranus, 30 A.U. compared to about 19 A.U. Neptune takes 165 years to orbit the Sun. Its discovery was a triumph of the modern era of Newtonian astronomy. Mathematicians analyzed the deviations of Uranus (then the outermost known planet) from the orbit it would follow if gravity from only the Sun and the other known planets were acting on it. Scientists suspected that the small deviations were caused by gravitational interaction with another planet.

Neptune, in Roman mythology, was the god of the sea. The planet Neptune's trident symbol reflects that origin.

John C. Adams, in England, predicted positions for the yet-to-be-observed planet in 1845 (Fig. 16–2). Unfortunately, the astronomy professor at Cambridge did not bother to try to check this prediction with a telescope; after all, Adams was to him merely a recent college graduate. Adams went to London to meet the Astronomer Royal (the chief British astronomer), but the A.R.'s butler did not permit dinner to

Memoranda.
1841. July 3. Formed a design, in the beginning of this week, of investigating, as soon as possible after taking my degree, the irregularities in the motion of Uranus, wh. are yet unaccounted for; in order to find whether they may be attributed to the action of an undiscovered planet beyond it; and if possible thence to determine the elements of its orbit, &c approximately, wh. wd. probably lead to its discovery.

**FIGURE 16–2** From Adams's diary, kept while he was in college: "1841. July 3. Formed a design in the beginning of this week, of investigating, as soon as possible after taking my degree, the irregularities in the motion of Uranus which are yet unaccounted for."

be interrupted, and Adams went away. Though Adams left a copy of his calculations, when the Astronomer Royal requested further information, partly to test Adams's abilities, Adams did not take the request seriously and did not respond at first. The very "proper" Astronomer Royal took offense and did not choose to have further dealings with Adams. The story then continues in France, where a year later Urbain Leverrier was independently working on predicting the position of the undetected planet.

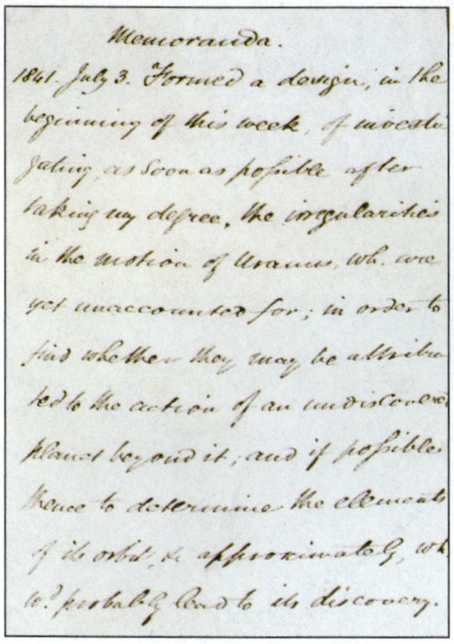

When the Astronomer Royal saw in the scientific journals that Leverrier's work was progressing well (Adams's work had not been made public), for nationalistic reasons he began to be more responsive to Adams's calculations. But the search for the new planet, though begun in Cambridge, was carried out halfheartedly. Nor did French observers take up the search. Leverrier sent his predictions to an acquaintance at the observatory in Berlin, where a star atlas had recently been completed. The Berlin observer, Johann Galle, enthusiastically began observing and discovered Neptune within hours by comparing the sky against the new atlas.

Years of acrimonious, nationalistic debate followed over who (and which country) should receive the credit for the prediction. (Adams and Leverrier themselves became friends.) We now credit both Adams and Leverrier. With hindsight, we see that they each assumed a radius for the new planet's orbit based on a numerological scheme for the distances of planets. (The scheme is known as Bode's law; no justification for it has ever been found and, in any case, it fails for Pluto.) The value Adams and Leverrier assumed was incorrect, but luckily it gave the right position anyway at that time.

Neptune has not yet made a full orbit since it was located in 1846. But it now seems that Galileo actually observed Neptune in late 1612 and early 1613, which would more than double the period of time over which it has been observed. Charles Kowal, then a Palomar astronomer, and Stillman Drake, a historian of science, tracked

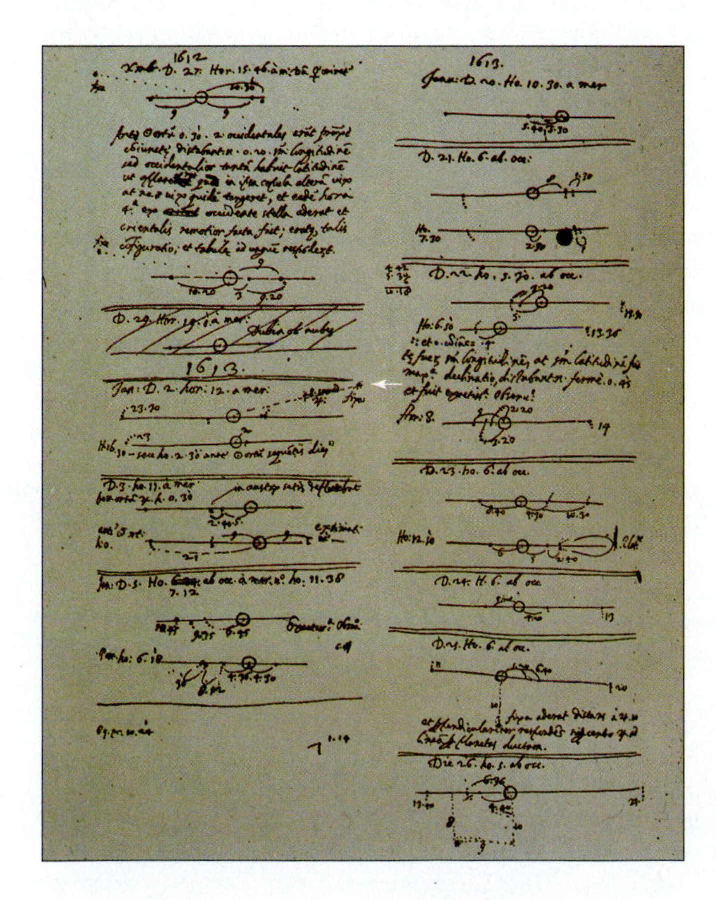

**FIGURE 16–3** Galileo's notebook from late December 1612, showing a * at the end of a dotted line, marking an apparently fixed star to the side of Jupiter and its moons. The "star," at extreme left, was apparently Neptune. The horizontal line through Jupiter extends 24 Jupiter radii to each side. Jupiter's moons appear as dots on the line.

The episode shows the importance of keeping good lab notebooks, a lesson we should all take to heart.

down Galileo's observing records from 1612 and 1613, when their calculations showed that Neptune passed close to Jupiter. Galileo, in fact, twice recorded in his notebooks stars that were very close to Jupiter (Fig. 16–3), stars that modern catalogues do not show. Galileo even once noted that one of the "stars" actually seemed to have moved from night to night, as a planet would. The objects that Galileo saw were very close to but not quite exactly where our calculations of Neptune's orbit show that Neptune would have been at that time. Presumably, Galileo saw Neptune. These observations make the span over which we have observed Neptune about 385 years instead of only 150 years, more than doubling it. It had been thought that the positions he measured would improve our knowledge of Neptune's orbit, but some of the sightings may not have been drawn to scale in his notebook. One sighting does seem to match today's best set of positions JPL (Jet Propulsion Lab) scientists calculate for Neptune. It matches closely enough that it rules out the presence of an unseen massive planet in the outer solar system.

Neptune's angular size in our sky is so small that it is always very difficult to study. Even measuring its diameter accurately from Earth is hard and is best done when it occults stars. From observations of the rate at which the stars dim, astronomers deduce information about Neptune's upper atmosphere, including its temperature and pressure structure. From the length of time that the star is hidden by the disk of Neptune, they measure Neptune's diameter. An accurate value for the diameter is important for calculating the planet's density, thus leading to deductions about its composition.

Neptune's density is 1.6 grams/cm$^3$, higher than Uranus's and Jupiter's 1.3 grams/cm$^3$ and much higher than Saturn's 0.7 grams/cm$^3$. Thus Neptune must have a higher proportion of heavy elements. Models predict the presence of cores of about the same mass—10 to 15 Earth masses—at the centers of all the giant planets, which makes these cores take up a larger fraction of Uranus and Neptune than they do of Jupiter and Saturn.

Neptune, like Uranus, appears bluish in a telescope because of its atmospheric methane (Fig. 16–4). In both cases, the atmosphere is mostly hydrogen, with helium

If the Sun emitted strongly in the ultraviolet soon after the formation of the Solar System, Uranus and Neptune were too far out for the Sun's gravity to keep most of the hydrogen and helium from being driven off. This model, from Berkeley-NASA/Ames scientists, explains why Uranus and Neptune have so little hydrogen and helium relative to Jupiter and Saturn.

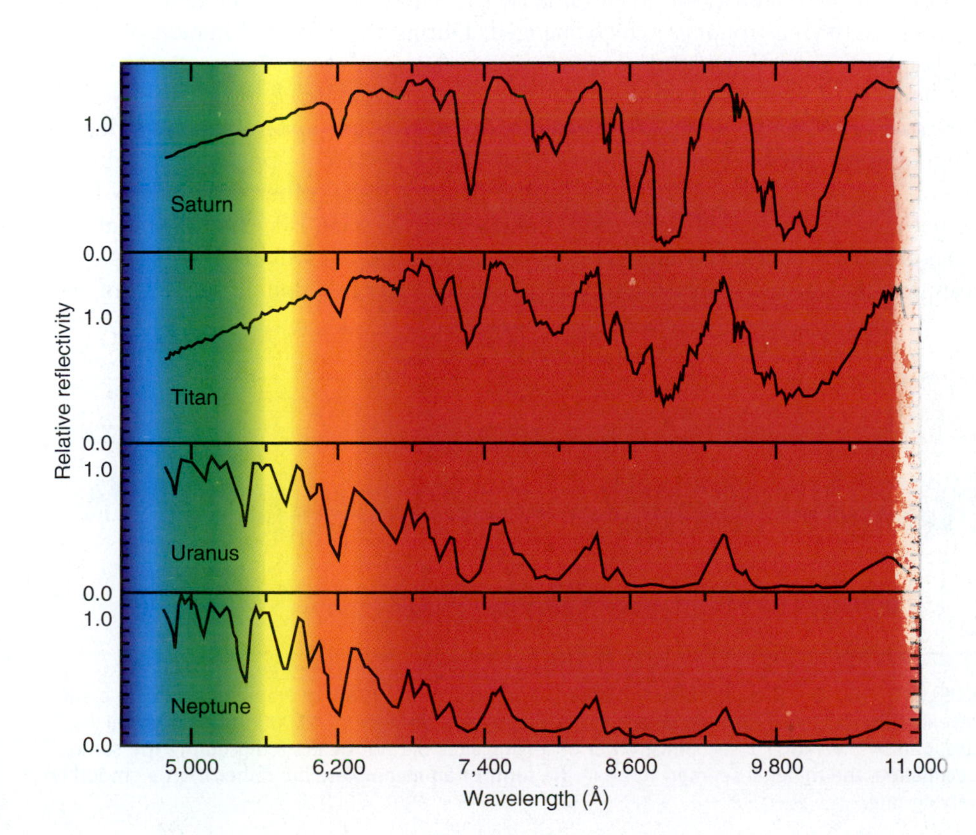


**FIGURE 16–4** The ground-based spectra of the giant planets and Saturn's moon Titan, divided by the solar spectrum to show the effect of the atmosphere of the planets and of Titan. The absorption bands (which appear as dips) are from methane. Note how strong the methane absorption is in the spectra of Uranus and Neptune. A trace of ammonia is also detectable in the spectrum of Saturn. The graphs extend from blue into infrared (which begins at about 7000 Å).

A

B

**FIGURE 16–5** *(A)* Neptune, imaged in the infrared light (8900 Å) that is strongly absorbed by methane gas in Neptune's atmosphere, making most of the planet appear dark. The bright regions are clouds of methane ice crystals above the methane gas, reflecting sunlight. The northern hemisphere appeared brighter than the southern, a reversal of the earlier situation. Neptune's disk was 2.3 arc sec across; the seeing was less than 1 arc sec. Note how the individual pixels show. *(B)* Neptune and Triton in the infrared from 2MASS.

next in abundance, but the trace of methane absorbs most of the red light instead of reflecting it. Like Uranus's, Neptune's rotational period is difficult to determine from the Earth. Pre-Voyager assessments were very uncertain, unknown even to within a few hours.

Like Jupiter and Saturn, Neptune radiates substantially more heat (2.7 times more) than it receives from the Sun. Uranus, in contrast, seems not to have an internal heat source, even after the more precise measurements made from Voyager 2. A recent idea for Neptune's internal heating is that diamond dust forms from methane at a depth where the pressure is sufficiently high. The model is based on laboratory experiments on methane under high pressure. Methane is a carbon atom surrounded by 4 hydrogen atoms. Under high pressure, the bonds holding the hydrogen atom dissolve and the carbons bond to each other, forming diamonds. If the diamond dust rained downward toward Neptune's core, friction could release energy.

Markings on Neptune have been detected from the ground (Fig. 16–5) on images taken electronically. The images show bright regions in the northern and southern hemispheres that are probably discrete clouds separated by a dark equatorial band. Motion caused by Neptune's rotation can also be seen.

Two of Neptune's moons were known from Earth-based studies. Triton is large, apparently a little larger than our Moon, and was discovered only a month after the first observation of Neptune. (It was named after a mythical sea god, son of Poseidon, the Greek equivalent of Neptune.) Triton is massive enough to have an atmosphere and a melted interior. Triton orbits Neptune in the retrograde direction, the only reasonably large moon in the solar system to do so, and is thus probably a captured object. Spectroscopy seemed to show that Triton has methane ice or methane frost, and possibly water ice and liquid or gaseous nitrogen as well. A second moon, Nereid (the Greek word meaning "sea nymph"), is much smaller, 340 kilometers across. It revolves in a very elliptical orbit with an average radius 15 times greater than Triton's. Its orbit is so elliptical that it is sometimes as close as 1.3 million kilometers and sometimes as far as 9.6 million kilometers from Neptune.

Does Neptune have rings, like the other giant planets? There was no obvious reason why it shouldn't, but no rings had been found at several occultations. Then, starting in 1983, astronomers' luck changed. During the 1983 occultation of a star by Neptune, good observations were obtained. (My own site, in Indonesia, at which students and I planned to observe the occultation, was clouded out.) Two observatories reported the same dip at the occultation, so Neptune apparently had a ring or rings after all, since a dip in brightness means that something—presumably a ring—is going in front of the star (Fig. 16–6). Since that time, other occultations have sometimes shown signs of rings and sometimes not, so scientists concluded that any rings present would not completely surround Neptune but would rather be partial arcs. Indeed, on one occasion only one of the two stars in a double-star system was

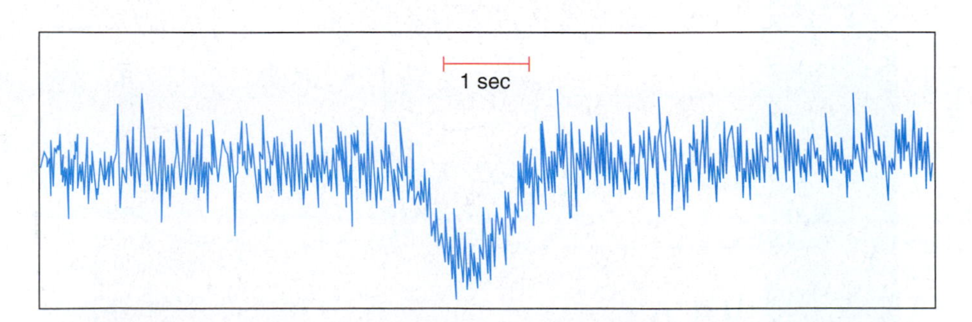


**FIGURE 16–6** This photoelectric observation on July 22, 1984, of Neptune occulting a star revealed the presence of material orbiting Neptune at a distance of 67,000 km. This signal dropped by 32% for 1.2 sec. Since other observing sites did not detect a dip during the same occultation, the material seemed to be in the form of an incomplete arc rather than a smoothly varying ring.

occulted, so the arcs apparently could end abruptly. Neptune's rings are thus apparently very different from the rings around Jupiter, Saturn, or Uranus, and much theoretical and observational work is concentrating on understanding this fourth solar-system ring system. It may be telling us about the positions and masses of otherwise unknown satellites.

## 16.2 VOYAGER 2 AT NEPTUNE

Voyager 2 reached the Neptune system on August 24, 1989. Neptune and Voyager were then 4 light hours 6 light minutes away from Earth when Neptune, Triton, and its other moons were transformed from mere points of light to actual worlds. Since it was Voyager 2's last destination, JPL scientists and engineers were able to direct the spacecraft as close to Triton as they pleased. They were thus able to get images of Triton of extremely high resolution. The spacecraft provided views that dazzled both the eye and the mind.

"I used to think that the idea of space probes was to answer questions. That's hopeless. You invariably raise more questions than you answer."—Andrew Ingersoll, Caltech, member of the Voyager 2 imaging team

As Voyager approached, even from far away it was obvious that Neptune had very active weather systems (Fig. 16–7). Why should Neptune be more active than Uranus, when Neptune is farther out from the Sun than Uranus and thus should be colder? It apparently has to do with the existence of a substantial extra internal heat source in Neptune besides the heat source in Uranus.

**FIGURE 16–7** A false-color image of Neptune from Voyager 2. The areas that appear red are a semitransparent haze surrounding the planet. Weather systems show.

As Voyager neared Neptune, an Earth-sized region soon called the Great Dark Spot (Fig. 16–8) became noticeable in Neptune's southern hemisphere. The Great Dark Spot, though colorless, seems analogous to Jupiter's Great Red Spot in several ways, including relative scale and position in the planet's southern hemisphere. It was very difficult for scientists to tell which way the Great Dark Spot was rotating, though, and thus whether it is cyclonic or anticyclonic, because successive exposures would show a new set of cloud tracers rather than movement of the clouds that had been previously seen. It was finally decided that the Great Dark Spot rotates counterclockwise, in the sense that is anticyclonic in the southern hemisphere, and has a period of 17 days. (In the northern hemisphere, the senses of cyclonic and anticyclonic are opposite from those senses in the southern hemisphere.) The Great Dark Spot is therefore a high-pressure region. The cirrus-like clouds form at the edge of the Great Dark Spot as the spot's high pressure forces upward methane-rich gas.

Several other cloud systems were also seen on Neptune's disk (Fig. 16–9). The "Scooter" moved quite rapidly around, to the south of the Great Dark Spot. A second major dark spot had a bright core. Since each of the features moves around Neptune at a different speed, they are not usually as close to each other as they were for the photograph.

In movies of the Great Dark Spot, it looked floppy, changing shape as time progressed. So it wasn't a complete surprise when Hubble Space Telescope images taken several years later showed that the Great Dark Spot had disappeared. A newer dark spot later developed in the northern hemisphere. Both spots might be holes in Neptune's methane clouds that allow us to observe to lower levels.

Neptune's high clouds are made of methane ice. Uniquely in the Solar System, the shadows of high clouds were seen (Fig. 16–10). Thus there must be two cloud decks separated by a clear region about 50 km in depth. The lower, more opaque cloud deck may be made of hydrogen sulfide ice particles. The Scooter seems to be an upwelling from this lower cloud deck and may lie over a long-lived hot spot. The cloud shadows appeared less distinct at short wavelengths (through the violet filter) than at longer wavelengths (through the orange filter). The effect apparently results from scattering by molecules in the extensive atmosphere between the cloud decks, which sends light into the shadows. The molecules scatter blue light much more efficiently than red light, so the shadows are least illuminated at the longest wavelengths. Since more blue light than red light arrives at the shadows, they appear blue, even though they are being illuminated with white light from the Sun.

**FIGURE 16–8** Neptune's Great Dark Spot, a giant storm in Neptune's atmosphere. It took up about the same span of latitude and longitude as the Great Red Spot does on Jupiter. Neptune's Great Dark Spot was about the size of the Earth.

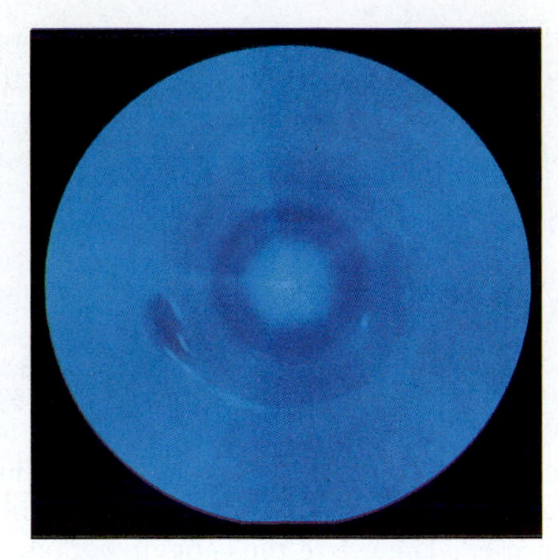

**FIGURE 16–9** Cloud systems in Neptune's southern hemisphere. Neptune's south pole is at the center of this polar projection. The outer edge of the disk corresponds to 15° north latitude. The oval storm drifts in latitude at up to 7000 km/hr.

**FIGURE 16–10** High-altitude cloud systems casting shadows, through a clear region, 50 km down to a lower cloud deck. The clouds are elongated at constant latitude, near 29° north, and the Sun is at lower left. The clouds are 50 to 200 km long.

Neptune's equatorial radius was measured to be 24,760 km. The amount of radiation Neptune gives off corresponds to an average temperature of 59.3 K (–353°F). (As we have seen, Neptune radiates about 2.7 times as much energy as it absorbs from the Sun.) Neptune receives the most solar energy at its equator, yet its equator and poles are about the same temperature. Mid-latitudes are several degrees cooler.

Radio bursts from Neptune were detected from Voyager at intervals of 16.11 hours, which must therefore be the rotation period of Neptune's interior. Upper atmosphere wind speeds varied greatly, from 560 m/sec westward to 20 m/sec eastward with respect to this underlying rotation rate. On Neptune, the Earth, and Uranus, the equatorial winds blow more slowly than the interior rotates. By contrast, equatorial winds on Venus, Jupiter, Saturn, and the Sun blow more rapidly than the interior rotates. We now have quite a variety of planetary atmospheres to help us understand the basic causes of circulation.

The magnetometer on board detected an extensive magnetic field around Neptune, starting with Voyager's crossing a bow shock 35 Neptune radii out. The field, as for Uranus, is both greatly tipped and offset from Neptune's center (Table 16–1). Neptune's field is inclined 47°, and its center is offset 55 per cent of Neptune's radius from the center of the planet. Neptune's magnetic field was the first to be penetrated at the pole by Voyager and was the first case observed in which

**TABLE 16–1** Planetary Magnetic Fields

| | Tilted Field Character | | | Field Strength | | |
|---|---|---|---|---|---|---|
| | Tilt | Moment (tesla-meter$^3$) | Offset ($R_{planet}$) | Equator | Minimum (microteslas)* | Maximum |
| Mercury | 14.7° | $0.005 \times 10^{15}$ | ? | 0.3300 | 0.3 | 0.6 |
| Earth | 11.7° | $8.0 \times 10^{15}$ | 0.08 | 31 | 24 | 62 |
| Jupiter | −9.6° | $160{,}000 \times 10^{15}$ | 0.10 | 428 | 300 | 1440 |
| Saturn | −0.0° | $4{,}700 \times 10^{15}$ | 0.05 | 22 | 18 | 84 |
| Uranus | −58.6° | $380 \times 10^{15}$ | 0.31 | 23 | 10 | 100 |
| Neptune | −46.8° | $280 \times 10^{15}$ | 0.55 | 13 | 10 | 100 |

*1 microtesla = $10^{-2}$ gauss

the solar wind hits near the magnetic pole rather than near the magnetic equator. Faint auroras were detected on both Neptune and Triton.

Table 16–1 compares the magnetic fields on the planets out to Neptune. We give first the value of the field on the planet's surface at the magnetic equator, which is related both to the total field in the planet and to the planet's size. We also give the total field (technically, the "magnetic moment," which is the product of the field at the planet's magnetic equator and the planet's volume). The tilt given is positive if the north magnetic pole is above the ecliptic (that is, for Mercury and the Earth) and negative if it is below the ecliptic (that is, for the jovian planets). Much of the data were shown graphically in the previous chapter (Figure 15–11), since we first ran into a weirdly offset magnetic field at Uranus, but it seems appropriate to go over this summary table while we discuss Neptune both for the sake of Neptune's own strange magnetic field and because it is the final chapter on the giant planets.

The polarity of Neptune's magnetic field is the same as that of Jupiter's, Saturn's, and Uranus's, all of which are opposite to the direction of the Earth's field with respect to the direction that the planets rotate. If not for the offset, the strength of magnetic field at Neptune's equator would be 13 microteslas, about one-third the Earth's equatorial field and about two-thirds the equatorial fields of Saturn and Uranus. The fact that Neptune's field is so tipped and offset indicates that there must be some fundamental reason for Neptune's and Uranus's fields being so, rather than mere chance collisions with planetesimals or chance observation during a period of magnetic reversal. The presence of an electromagnetic dynamo in an electrically conducting shell outside the planets' cores rather than deep in the core as for Earth and Jupiter could account for the phenomenon, since it would provide a reason for a field not symmetric about the center.

## 16.3 NEPTUNE'S RINGS

Based on the ground-based studies that implied the existence of ring arcs, which are hard to account for theoretically, there was great interest in the Voyager observations of Neptune's rings. At a great distance, it was thought that the ring arcs had actually been seen. But as Voyager grew closer to Neptune, it became clear that the rings went all the way around Neptune (Fig. 16–11), at its equator. Two distinct rings were

All the material—rings and satellites—within Neptune's Roche limit would form a single body 260 km across; the corresponding diameter would be 390 km for Saturn and 150 km for Uranus.

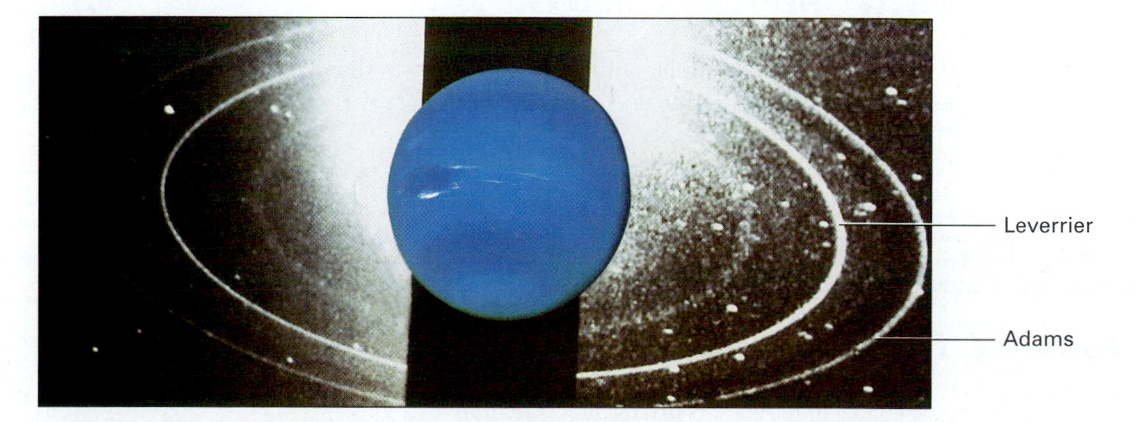


**FIGURE 16–11** Neptune's rings, a composite of views of the two sides of the planet. Each exposure lasted nearly 10 minutes, with 1 hour 27 minutes between exposures, and was taken looking back toward the Sun from a distance of 280,000 km. Unfortunately, the bright ring clumps that had been thought to be arcs were on the opposite side of the planet during each exposure and do not show. The faint, broad inner ring at about 42,000 km from Neptune's *(superimposed)* center and the "plateau" that extends smoothly from the 53,000-km ring to about halfway between it and the other bright ring are also visible. Numerous stars are visible in the background.

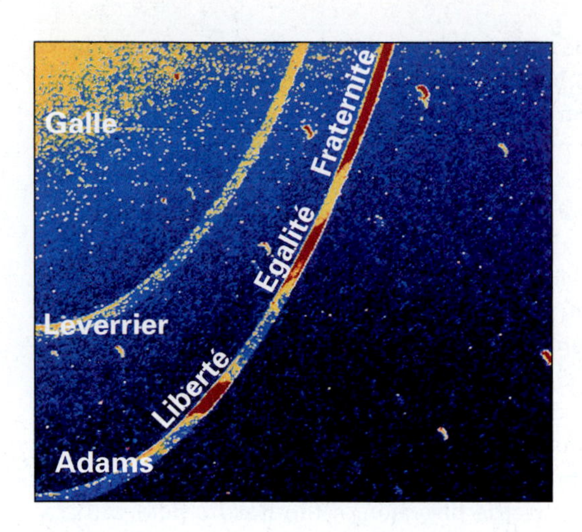


**FIGURE 16–12** Two of Neptune's rings, showing the clumpy structure in one of them that led ground-based scientists to conclude from occultation data that Neptune had ring arcs. The rings are 53,000 and 63,000 km, respectively, from Neptune's center. These photographs were taken as Voyager 2 left Neptune, 1.1 million km farther out in the Solar System and so are backlighted. The main clumps are 6° to 8° long.

detected at 53,000 and 63,000 km from Neptune's center, as well as a faint inner ring at about 42,000 km and an unprecedentedly broad "plateau" ring that extends from the 53,000-km bright ring halfway out to the other bright ring. The rings showed up in backlighted images taken after Voyager had passed Neptune. The material rotates in the same direction as Neptune itself, and the rings are very circular and lie very close to Neptune's equator.

The material in one of Neptune's rings is very clumpy (Fig. 16–12), a characteristic that had led to the incorrect pre-Voyager deduction that ring arcs existed. The term "ring arcs" is expected to fade until it is a mere historical curiosity.

The fact that Neptune's rings are so much brighter when backlighted tells us about the sizes of particles in them. Two of the rings and the clumps in the third have a hundred times more dust-size grains than most of the rings of Uranus and Saturn. Since dust particles settle out of the rings, new sources—such as collisions of moonlets or bombardment of Neptune's icy moons by meteoroids—must continually occur. Though much less dusty, Saturn's F ring and Encke Gap ringlet and one of Uranus's rings are similarly narrow.

## 16.4 NEPTUNE'S MOONS

As Voyager 2 approached Neptune, a new moon was discovered fairly early on, and then five more closer to the encounter. This number was smaller than some had ex-

## A DEEPER DISCUSSION

### BOX 16.1 Naming the Rings of Neptune

The rings of Neptune were named by a committee of the International Astronomical Union. There are three major rings, and they were named after the two theoretical predictors and the actual observational discoverer. The three subdivisions of the outer ring, on the suggestion of the French member of the committee, were named *Liberté, Egalité,* and *Fraternité* after the slogan of the French Revolution (1789).

***Discussion Question***

Given that the heavens are not supposed to be politicized, was the choice of names for the subdivisions of the outer ring justified? Do you agree with the decision that the names were basic rights and not political terms?

**FIGURE 16–13** Triton's southern region, including its south polar ice cap. Most of the cap may be nitrogen ice. The pinkish color may come from the action of ultraviolet light and magnetospheric radiation upon methane in the atmosphere and surface. The dark streaks represent material spread downwind from eruptions. The long, thin, dark streak marked with arrows is an erupting plume. Though the streaks are dark relative to other features on Triton, they are still ten times more reflective than the surface of our own Moon. The most detailed information on the photograph was taken in black and white through a clear filter; color information was added from lower-resolution images.

pected, given the number of moons discovered by Voyager around Jupiter, Saturn, and Uranus.

Everyone was waiting for the encounter with Triton, and nobody was disappointed with the results (Fig. 16–13). For one thing, as had been hoped, Triton's atmosphere is sufficiently thin that we could see through to the surface. Nitrogen gas is its dominant component. Triton turned out to have an incredibly varied surface, one of the weirdest in the entire solar system. (Io and Miranda have their fans. What is your favorite for weirdest moon?) Triton's density is 2.07 grams per cubic centimeter, so it is probably about 70 per cent rock and 30 per cent water ice. It is more dense than any jovian-planet satellite except Io and Europa. Its density is a major datum used to construct models of Triton's interior.

Much of the region imaged was near the south polar cap. This region is apparently covered with seasonal ice, probably nitrogen. The temperature during this late-spring season was 37K (−236°C = −393°F). The ice appeared slightly reddish, probably from organic material formed by the action of solar ultraviolet and magnetospheric particles upon methane in the atmosphere and surface. Some bluish nitrogen frost appeared at equatorial latitudes. Methane and nitrogen were the only materials directly detected on its surface, but craters and cliffs would slump if they were made of methane, so water ice must be the major component either by itself or with ammonia ice mixed in. Most of the craters are thought to have formed from impacts by comets over the past 3.5 billion years; Neptune's gravity captures comets into orbits that cross its own orbit with fairly short periods.

Giant faults cross Triton's surface. About 50 dark streaks were seen to be aligned parallel to each other. They are apparently dark material vented from below as ice volcanoes and spread out across Triton's surface by winds in the thin atmosphere. One of the plumes is seen to rise about 8 km before blowing downwind, presumably showing a change in Triton's atmosphere at that altitude. A leading model is that sunlight penetrates nitrogen ice, which is transparent, for a few meters. Some of the nitrogen ice sublimes (that is, changes directly from its solid to a gaseous state) and escapes through vents in the surface to make the observed geysers. This method

**FIGURE 16–14** A flat "lake" region on Triton, about 300 km across. The impact crater is one of the largest seen anywhere on Triton.

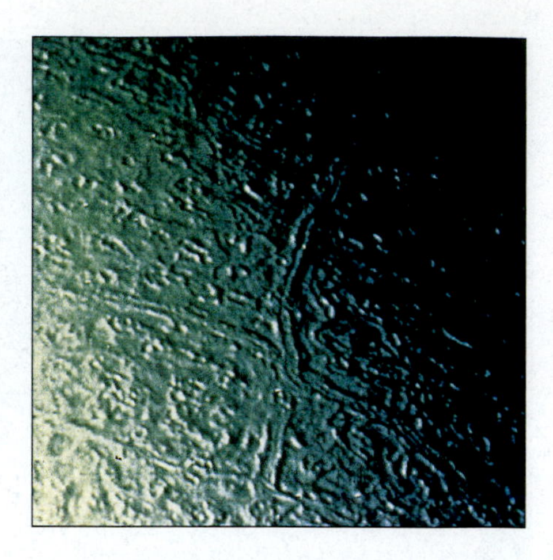

**FIGURE 16–15** Triton from 120,000 km. The long feature is probably a narrow down-dropped fault block.

involves a different type of energy source than the internal energy that powers the Earth's volcanoes or the flexing that powers Io's. In any case, on Triton, dark material that had been on the surface is carried by the plume as much as 150 km downwind. Since the streaks are on top of seasonal ice, they are probably less than 100 years old (Neptune's year is 164 Earth years long).

Much of the trailing (western) hemisphere of Triton is so puckered that it is called the cantaloupe terrain. The cantaloupe terrain contains 30-km-diameter depressions crisscrossed by ridges. Other terrains seem to have been more recently covered with material, presumably material that flowed from under Triton's surface. Triton has few impact craters. Some regions of Triton are broad, flat basins (Fig. 16–14), possibly old impact basins, though this is still debated. The ones shown have been extensively modified by flooding, melting, faulting, and collapse. The basins may have been filled and partially emptied several times. The most recent eruption is apparently in the middle of the central depression. Since water on Triton acts like rock on Earth, liquid water (perhaps with ammonia mixed in) acts like lava, and we see multiple layers like those of terrestrial calderas.

Triton has obviously been geologically active (Fig. 16–15). Some of the faults are quite long. We have not seen such surfaces sculpted in ice in this way on other solar-system bodies. These depressions look as though the icy surface has melted and collapsed (Fig. 16–16). These frozen lakes of water ice appear mainly in the eastern hemisphere, which leads as Triton orbits. A thin coating of methane and nitrogen ices and resulting materials apparently covers most or all of Triton. A computer-generated view shows a calculated perspective of one of the depressions on Triton (Fig. 16–17).

Since Triton is in a retrograde orbit around Neptune, it was probably born elsewhere in the Solar System and later captured by Neptune. Tidal forces from Neptune would then have kept Triton molten until its orbit became circular. While it was molten, the heavier rocky material would have settled to form a 2000-km-diameter core.

After Voyager 2's flyby at an altitude of only 3000 km above Triton's surface, the spacecraft could look back to see Triton as a crescent, as we saw in Figure 16–1. The diffuse haze it saw was probably photochemical smog, not unlike that over many of our cities on Earth.

Triton orbits with the same side always facing Neptune. Its orbit is inclined 157°. Nereid (Fig. 16–18), whose orbit is inclined 29°, was not well imaged by Voyager. Its orbit is so elliptical that it seems unlikely that its rotation is locked to Neptune.

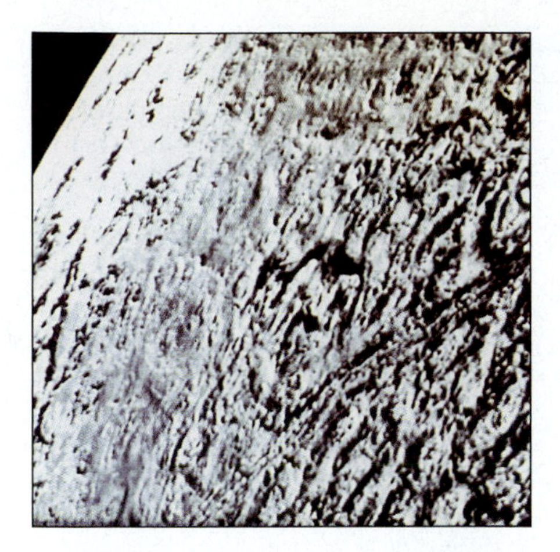

**FIGURE 16–16** Triton from 40,000 km. The depressions may have been caused by melting and collapsing of the icy surface.

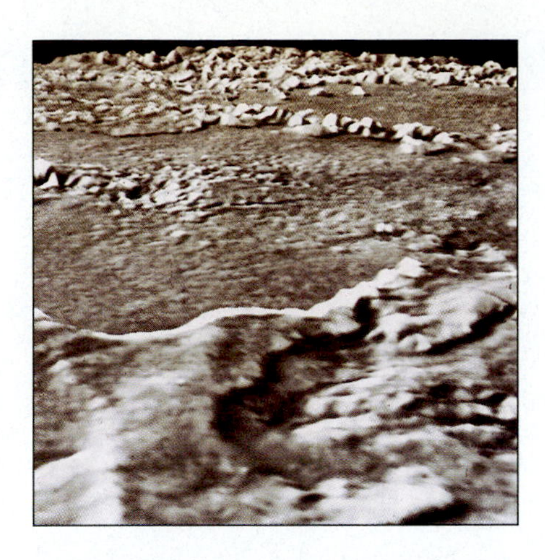

**FIGURE 16–17** A computer-generated perspective view of one of Triton's caldera-like depressions.

However, Voyager did not detect variations over time, which would give its rotation period.

All six of the newly discovered moons orbit Neptune in circular orbits in the same direction as Neptune's orbital motion, near Neptune's equatorial plane. Surprisingly, one of these newly discovered moons (Fig. 16–19) turned out to be larger than Nereid (Table 16–2). Named Proteus, it had not been discovered from Earth because of its closeness to Neptune, making it lost in Neptune's glare.

Most of the newly discovered moons are closer to Neptune than Neptune's rings. Since objects cannot grow by accretion within the Roche limit, these moons must have formed elsewhere and moved into their current orbits. Two of the moons orbit just inside two of the rings and may serve as inner shepherds that keep ring material from spiralling inward. No outer shepherds have been detected, though they would not have been detectable if they were smaller than 12 km across. And no moons were found in the positions that would have explained the observed clumpiness in one of the rings. Perhaps the ring material was accreted so recently that it has not yet spread out evenly around the ring.

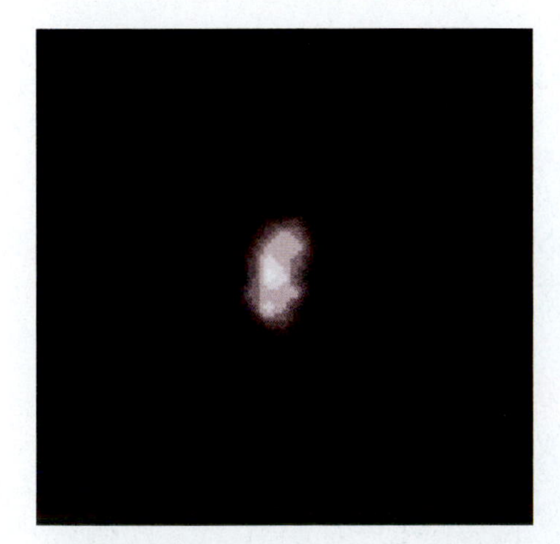

**FIGURE 16–18** Nereid. No brightness variations were detected, and not even the position of its pole is known. We see a poorly resolved crescent, concave at the right.

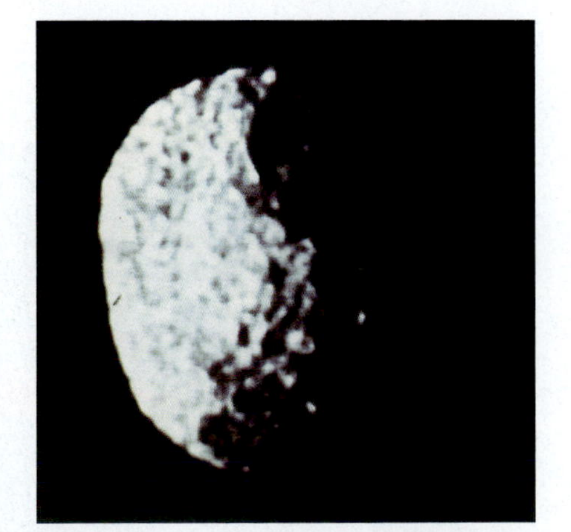

**FIGURE 16–19** Proteus, the second-largest moon in the neptunian system, approximately 210 km in radius. Its large crater shows that it was nearly broken apart.

**TABLE 16-2** The Moons and Rings of Neptune

| Name | Radius of Object (km) | Orbit (km) | Radius of Period (hr) |
|---|---|---|---|
| **Neptune's atmosphere** | | **24,760** | |
| *Inner ring (Galle)* | | *41,900* | |
| Naiad | 27 ± 8 | 48,230 | 7.1 |
| Thalassa | 40 ± 8 | 50,070 | 7.5 |
| Despina | 75 ± 15 | 52,530 | 8.0 |
| *Bright ring (Leverrier)* | | *53,200* | |
| *Lassell ring (Plateau ring)* | | *53,200–57,500* | |
| *Arago ring* | | *57,500* | |
| Galatea | 90 ± 10 | 61,950 | 10.3 |
| *Outer ring (Adams)* | | | |
| *Courage, Liberté, Egalité, Fraternité* | | *62,900* | |
| Larissa | 95 ± 10 | 73,550 | 13.3 |
| Proteus | 200 ± 10 | 117,640 | 26.9 |
| Triton | 1,356 × 1,362 × 1,352 ± 5 | 354,800 | 141.0 |
| Nereid | 170 ± 25 | 5,513,400 | 8,643.1 |

## 16.5 VOYAGER DEPARTS

All too soon, Voyager completed its tour of the Solar System and departed. Its look back at Neptune gave us a new perspective on that planet and its atmosphere. Voyager 2 was an unqualified success, sending back fantastic data over a much longer period than had been thought possible. It made the so-called "grand tour" of the outer planets a reality.

Where is the end of the solar system? The **heliopause** is the boundary between the hot solar wind (100,000 K) flowing outward from the Sun and the cooler (10,000 K) gas that is typical between the stars. Voyager 1, now over 50 A.U. from the Sun, and Voyager 2, now over 40 A.U. out, started picking up powerful radio signals in mid-1992 (Fig. 16–20). Scientists think that these signals are from bursts that formed at the heliopause when the region was struck by solar-wind gas enhanced by powerful solar flares. If so, the heliopause is somewhere between 115 and 180 A.U.

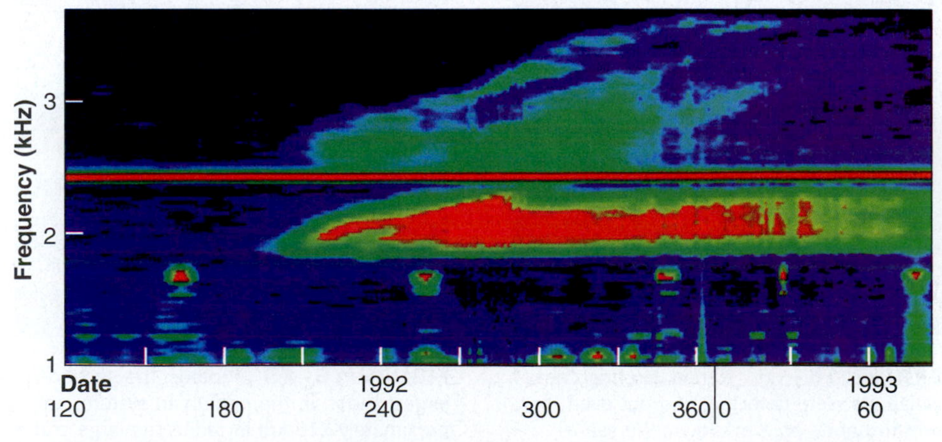


FIGURE 16–20 The Voyager 2 plot of the radio emission thought to come from the heliopause, marking the edge of our Solar System. Voyager 1's plot is similar. Color shows the intensity of the electric field (red is most intense).

## 16.6 CURRENT OBSERVATIONS OF NEPTUNE

The rings of Uranus and Neptune were discovered from the Earth and not from spacecraft. The method of occultations is very powerful, and it even applies at times to Neptune's giant moon Triton. I was fortunate to participate in a series of Earth-based expeditions to observe occultations of stars by Triton. If the star is bright enough, we can observe how its light is affected by Triton's atmosphere. Only rarely does Triton go in front of a star bright enough to use, given that we must take exposures only 1 sec long or shorter because of the changing situation. Also, the light from the star is so parallel that Triton's shadow on the Earth is the same size as Triton itself, only about 2700 km across, meaning that only part of the Earth is in the shadow. Further, the position of Triton is not well enough known, given the long lever arm at its distance, to allow for the predictions of the positions of the edges of the shadow to be known to within a few hundred km.

In spite of all these difficulties, James Elliot of MIT, the discoverer of Uranus's rings—who has since applied the technique to Neptune's rings and other objects—has organized a few such expeditions. In 1997, several collaborating expeditions of astronomers went to Australia and to Hawaii to try to get in Triton's shadow as cast by the star. It turns out that on the second occasion, Elliot got the best observations with the Hubble Space Telescope, which happened to pass the right place at the right time. When Triton blocked out the star, which was a few times brighter than Triton itself, the brightness of the merged image dropped measurably. From the way in which it dropped, he and other collaborating scientists could analyze what Triton's atmosphere was like. Triton's temperature turned out to be a few degrees warmer than it was when Voyager flew by, so our joint publication, in the journal *Nature*, was called "Global Warming on Triton." The applications to global warming on Earth are obvious. Our other expedition was described in a joint publication in the journal *Icarus*, the major journal of planetary science.

The giant telescopes on Mauna Kea and in Chile are able to image with higher resolution than ever before, thanks not only to general improvements like increased ventilation in telescope domes but also to jumps in quality provided by adaptive optics. Images of Neptune, in particular, are improved enough to clearly see clouds even from these earthbound telescopes (Fig. 16–21). Even smaller scale structure including waves within narrow bands of brightness encircling Neptune can be seen. As on Jupiter, such structures may be related to twirling regions on the planet. Using adaptive optics techniques, even the ring arcs can be seen (Fig. 16–22).

The Hubble Space Telescope is able to observe Neptune, its rings, and some of its moons whenever time is found in its schedule (Fig. 16–23).

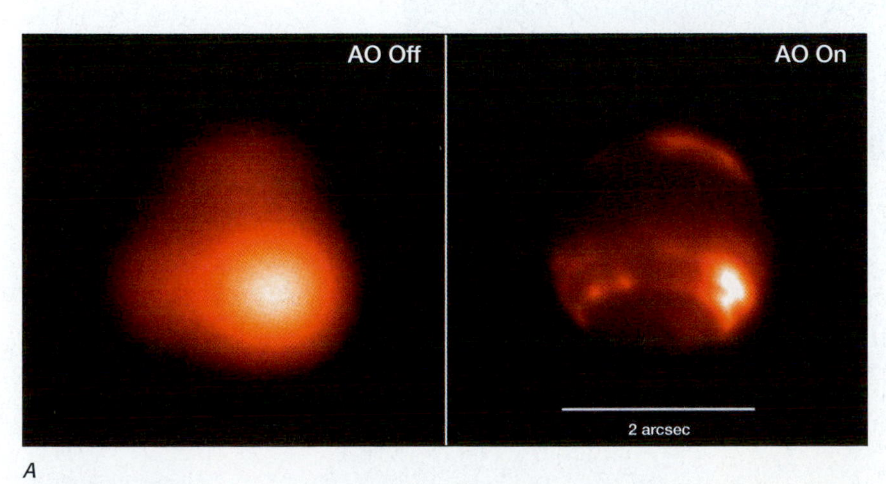


A

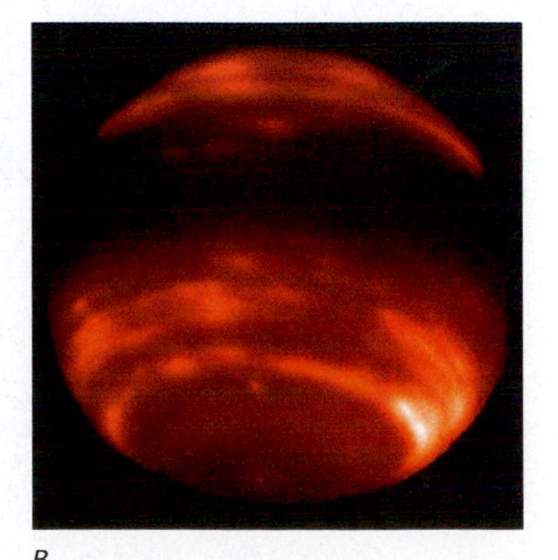

B

**FIGURE 16–21** Neptune, (*A*) observed with the 5-m Palomar telescope and with the adaptive optics (AO) turned off *(left)* and on *(right)*. (*B*) Neptune, observed with adaptive optics on the Keck II Telescope. A storm system is prominent at lower right.

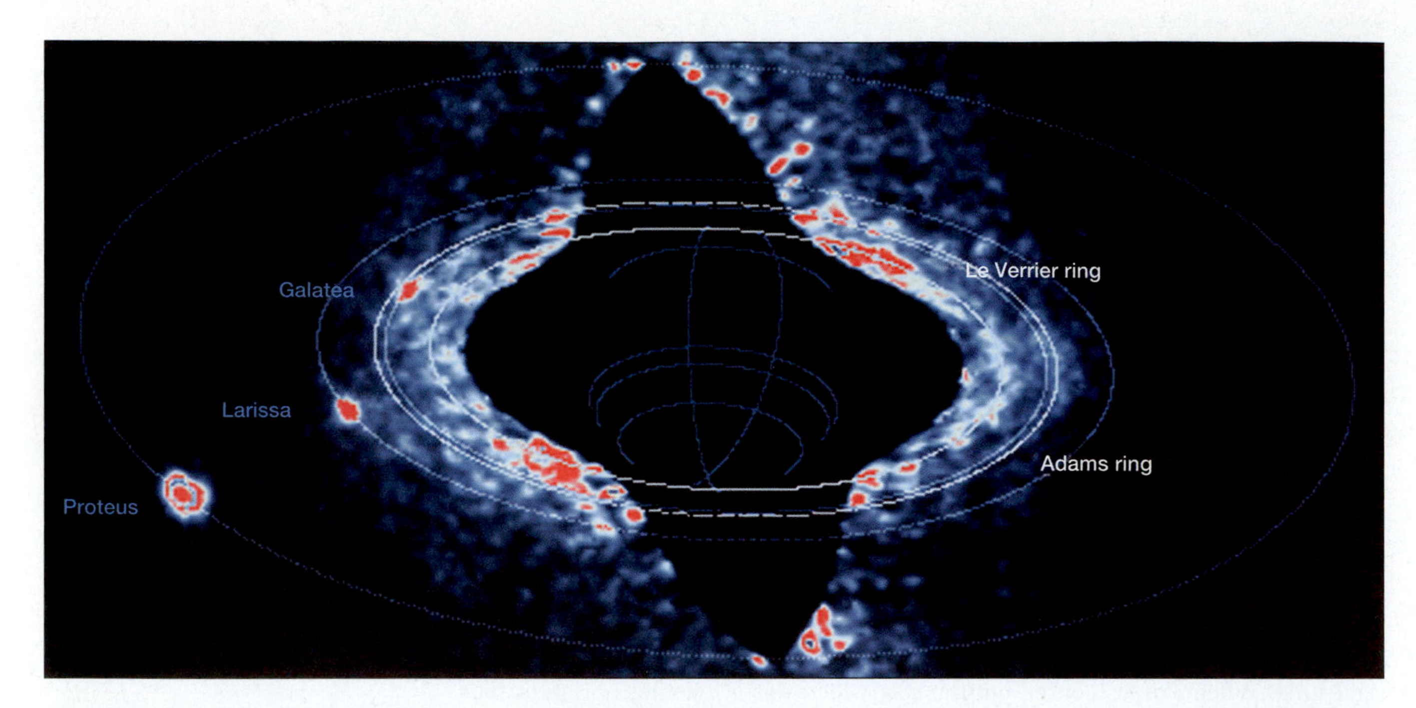


**FIGURE 16–22** An infrared adaptive-optics false-color image taken at the Canada–France–Hawaii Telescope through a narrow-band filter centered at the wavelength of methane. At such wavelengths, we see deeper into Neptune's atmosphere than otherwise. Neptune itself has been masked out. The satellites Galatea, Larissa, and Proteus appear elongated because of their orbital motion during the 10-minute-long exposure. The gravity of Galatea seems to maintain the ring arcs, which would otherwise each come apart because of their different velocities at different positions.

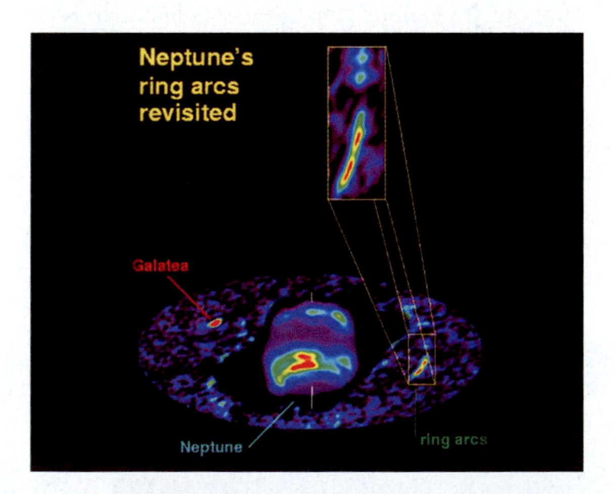


**FIGURE 16–23** Neptune, its moon Galatea, and some of its ring arcs, imaged with the Hubble Space Telescope.

## CORRECTING MISCONCEPTIONS

| ✖ *Incorrect* | ✔ *Correct* |
| --- | --- |
| Neptune's Great Dark Spot is a permanent feature. | Neptune's Great Dark Spot, discovered by Voyager 2, disappeared a few years later. |
| Neptune is so cold that it is uninteresting. | Neptune has more weather than Uranus; Triton is a fascinating moon. |

## SUMMARY AND OUTLINE

Neptune from Earth (Section 16.1)
- Large planet with low density and high albedo
- Predictions by Adams and Leverrier—part science and part luck
- Atmosphere of methane and molecular hydrogen
- Two known moons: Triton and Nereid
- Internal heating present
- A ring may have been discovered at an occultation.

Neptune from Voyager 2 (Section 16.2)
- Great Dark Spot discovered
- Cloud shadows indicate a 50-km clear region in atmosphere.

Neptune's Rings (Section 16.3)
- They are complete, after all, though clumpy.
- Two thin rings plus two broad rings discovered.
- They show up best in forward scattering, indicating small particle size.

Neptune's Moons (Section 16.4)
- Triton has the most varied surface in the Solar System.
- Triton may have ice volcanoes that emitted streaks.
- Giant faults, cantaloupe terrain, pinkish ice visible on Triton
- One of six newly discovered moons is bigger than Nereid.

Beyond Neptune (Section 16.5)
- Heliopause: end of the Solar System

Neptune Today (Section 16.6)
- Occultation observations of Triton's atmosphere
- Observations with adaptive optics from terrestrial telescopes
- Observations from the Hubble Space Telescope

## KEY WORD

heliopause

## QUESTIONS

1. Which planets are known to have significant internal heat sources? Compare and contrast these sources and the evidence for their existence.
2. †What fraction of its orbit has Neptune traversed since it was discovered? Since it was first seen?
3. Why do we have clearer images of Triton than we have of the moons of Jupiter?
4. Who deserves credit for the discovery of Neptune? Compare the relative claims of Galileo, Adams, Leverrier, and Galle.
5. Describe the relative roles of methane and of hydrogen gas in the atmosphere of Neptune.
6. Compare the Great Dark Spot on Neptune with the Great Red Spot on Jupiter.
7. Describe the history of ring discoveries and of "ring arcs" on Neptune.
8. Compare Triton in size and surface with other major moons of the Solar System, like Ganymede and Titan.
9. Describe some terrains in Triton's surface.
10. Discuss the location of Neptune's moons and their relation with the Roche limit.

† This question requires a numerical solution.

## USING TECHNOLOGY

### W World Wide Web

1. Examine the Photojournal, The Nine Planets, and Views of the Solar System images and descriptions of Neptune and its moons. See http://photojournal.jpl.nasa.gov, http://www.nineplanets.org, http://www.solarviews.com, respectively.
2. Find the Hubble images of Neptune and consider Neptune's changing atmosphere.

### REDSHIFT

1. Duplicate the Neptune system.
2. Where in the sky is Neptune tonight?

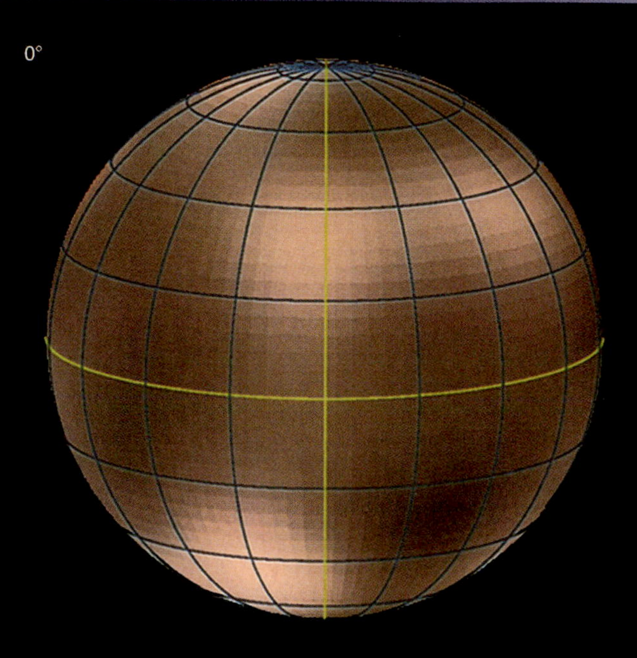

Hubble Space Telescope views of Pluto, at the highest resolution ever obtained on this object. The HST image of Pluto is like standing in Los Angeles and looking at the spots on a soccer ball in San Francisco. The symbol for Pluto appears at the top of the facing page.

# Chapter 17

# Pluto and the Kuiper Belt

♇

Pluto, the planet with the largest known orbit, is a deviant. Its orbit is the most eccentric and has the greatest inclination with respect to the ecliptic plane, near which all the other planets revolve.

Pluto reached perihelion, its closest distance from the Sun, in 1989. Its 249-year orbit is so eccentric that part lies inside the orbit of Neptune. It was on that part of its orbit until 1999. Thus in a sense, Pluto was the eighth planet for a while, though the significant point is really that the semimajor axis of Pluto's orbit, 40 A.U., is greater than that of Neptune's. Pluto is so far away and so small that even with today's best telescopes on Earth, we cannot see any detail on its surface.

The discovery of Pluto was the result of a long search for an additional planet that, together with Neptune, was causing perturbations in the orbit of Uranus. (The orbit of Neptune itself was not well enough known to rely on.) A young amateur astronomer, Clyde Tombaugh, was hired by the Lowell Observatory to conduct the search, and in 1930, he found the dot of light that is Pluto after a year of diligent study of photographic plates (Fig. 17–1). From its slow motion with respect to the stars from night to night (Fig. 17–2), Pluto was identified as a new planet and

**AIMS:** To show how studies of tiny Pluto and its moon Charon have told us about the masses, sizes, and compositions of these distant solar-system objects

**FIGURE 17–1** Clyde Tombaugh, at the equipment with which he discovered Pluto.

**FIGURE 17–2** Small sections of the plates on which Tombaugh discovered Pluto. On February 18, 1930, Tombaugh noticed that one dot among many had moved between January 23, 1930 (*left*), and January 29, 1930 (*right*). He didn't have the arrows on the plates, of course!

named for the Roman god of the underworld. As we shall see later on, though, Pluto does not have enough mass to have caused noticeable effects on Uranus's orbit. Tombaugh's discovery resulted, rather, from his hard work and skill in searching the sky instead of from the predictions. Tombaugh's death in 1997 ended people's delight in having someone alive who had actually discovered a planet in our Solar System.

If we were standing on Pluto, the Sun would appear a thousand times fainter than it does to us on Earth. We would need a telescope to see the solar disk, which would be about the same size that Jupiter appears from Earth.

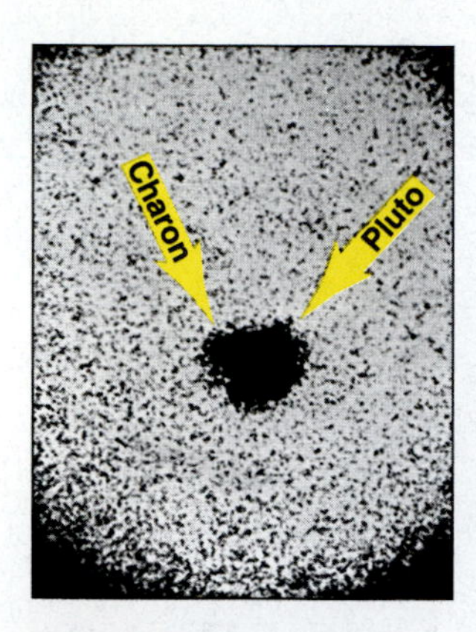


**FIGURE 17–3** The discovery image of Charon and Pluto, taken on July 2, 1978. The bump at upper left was interpreted as the satellite. Analysis shows that Charon is 1/2 the size of Pluto and is separated from Pluto by only about 8 Pluto diameters (compared with the 30 Earth diameters that separate the Earth and the Moon). So Pluto/Charon are almost a double-planet system. The fact that few ground-based photographs taken since Charon's discovery show it more clearly than this picture illustrates why it had not been previously discovered.

## 17.1 PLUTO'S MASS

Pluto is so far away that even such basics as the mass and diameter of Pluto have been very difficult to determine. It had been hard to deduce the mass of Pluto because to do so required measuring Pluto's gravitational effect on Uranus, a much more massive body. (The orbit of Neptune is still too poorly known to be of much use.) Moreover, Pluto has made less than one revolution around the Sun since its discovery, thus providing little of its path for detailed study. As recently as 1968, it was calculated that Pluto had 91 per cent the mass of the Earth, though soon thereafter it was realized that the data were actually not reliable enough to make any conclusions about the value of Pluto's mass. Later studies of the orbit of Uranus seemed (wrongly) to indicate that Pluto's mass was 11 per cent that of Earth, but these observations were also very uncertain.

The situation changed drastically in 1978 with the surprise discovery that Pluto has a satellite. The presence of a satellite allowed us to deduce the mass of the planet by applying Newton's form of Kepler's third law.

A U.S. Naval Observatory scientist, James W. Christy, was studying plates taken in a new series of observations to refine our knowledge of Pluto's orbit. He noticed that some of the photographs seemed to show a bump on the side of Pluto's image (Fig. 17–3). The bump was much too large to be a mountain and turned out to be a moon orbiting with the period that we had previously measured for the variation of Pluto's brightness—6 days 9 hours 17 minutes. The moon has been named Charon, after the boatman who rowed passengers across the River Styx to the realm of Hades (the Greek version of Pluto) in Greek mythology (and informally pronounced "Shar'on" by those in the know, with an "sh" sound to match that in the name of the discoverer's wife, Charlene, instead of with the "k" sound favored by experts in Latin). We had to wait for the Hubble Space Telescope to see Pluto and Charon resolved from each other (Fig. 17–4). Advances in resolution at big telescopes over the last decades, though, have enabled these telescopes to show the pair better as two objects instead of one blur.

From even the discovery photograph, the photograph on which Christy first located Charon, we can measure the separation of the two objects, and from the separation and the period we can get an approximate idea of the distance between Pluto and Charon. This measurement allows us to calculate the sum of the masses of Pluto and Charon. By measuring their relative sizes and estimating their relative densities, we can get approximate values for their individual masses.

Pluto's mass turns out to be only 1/400 the mass of Earth, a far cry from the 91 per cent it was once thought to be. As we shall discuss later on, Pluto's tiny mass reduces its credibility as a true major planet.

It is ironic that, now that we know Pluto's mass, we realize that it is not massive enough to cause the perturbations in Uranus's orbit that originally led to Pluto's discovery. Thus the pre-discovery prediction was actually wrong, and the discovery of Pluto was purely the reward for Tombaugh's capable search in a zone of the sky near the ecliptic.

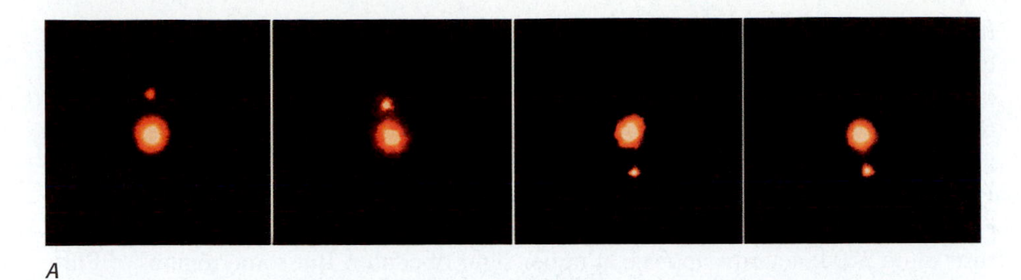

A

B

**FIGURE 17–4** (*A*) Ground-based views of Pluto and Charon, showing them orbiting around each other. This series of images was taken in 1999 at the highest possible resolution at near-infrared wavelengths, with adaptive optics (mirror-shape distortion in real time, to compensate for atmospheric turbulence) using the new Gemini North telescope on Mauna Kea. (*B*) Pluto and Charon imaged with the Hubble Space Telescope's Faint Object Camera.

## 17.2 PLUTO'S SIZE AND SURFACE

The diameter of Pluto was measured directly for the first time in 1979. The technique of speckle interferometry, which involves studying a series of images taken rapidly enough to freeze out the effect of the Earth's turbulent atmosphere, was used. The method gave 3000 to 3600 km for Pluto's diameter, roughly the coast-to-coast size of the United States. Pluto is thus much smaller than Mars (6800 km) or Mercury (4800 km) and is approximately the same size as our Moon (3500 km).

Between 1985 and 1990, Pluto and Charon passed in front of each other as they orbited each other every 6.4 days. These mutual occultations lasted a few months each year. Since they depend on the alignment of Charon's orbit around Pluto, we won't see that effect again from Earth for almost another 120 years (half Pluto's orbit around the Sun), so we were indeed lucky to catch it. When we measure the apparent brightness of Pluto, we are really receiving light from both Pluto and Charon together. Their blocking each other (and an additional contribution from the shadow of one object on the other) leads to dips in the total brightness we measure (Fig. 17–5). These accurate measurements showed that Pluto is 2300 ± 14 km in

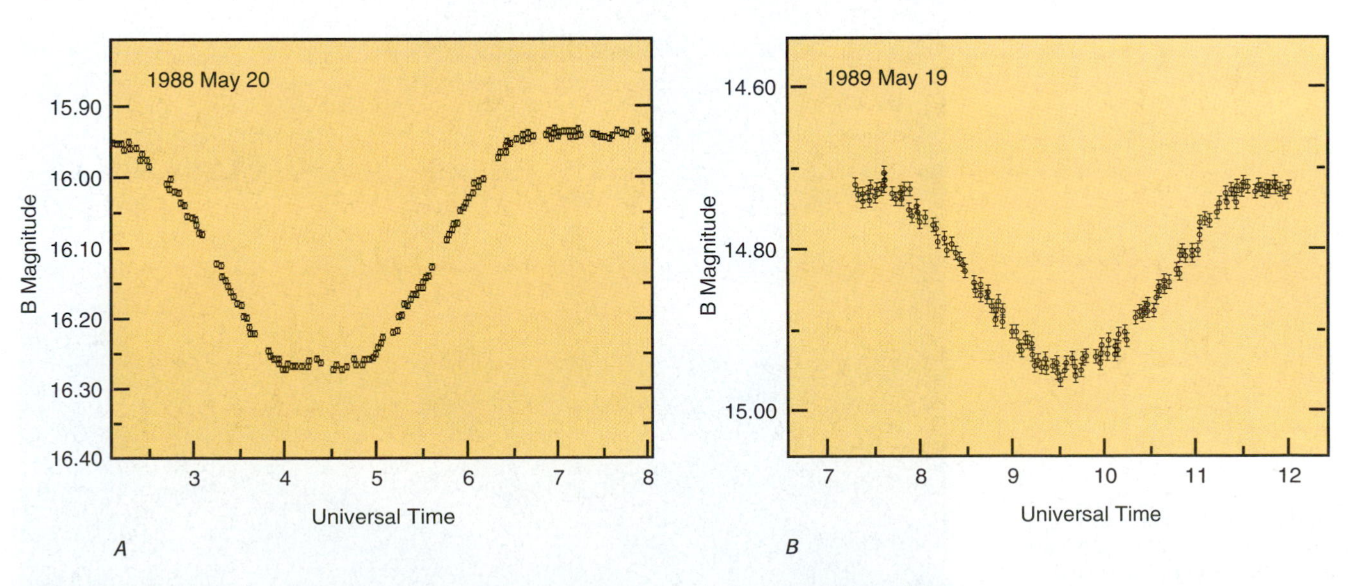


**FIGURE 17–5** (*A*) A 1988 occultation of Pluto by Charon. Note the flat bottom of the light curve, as Charon spent some time entirely in front of Pluto. (*B*) By 1989, the light curve no longer had a flat bottom, as Charon only partly occulted Pluto.

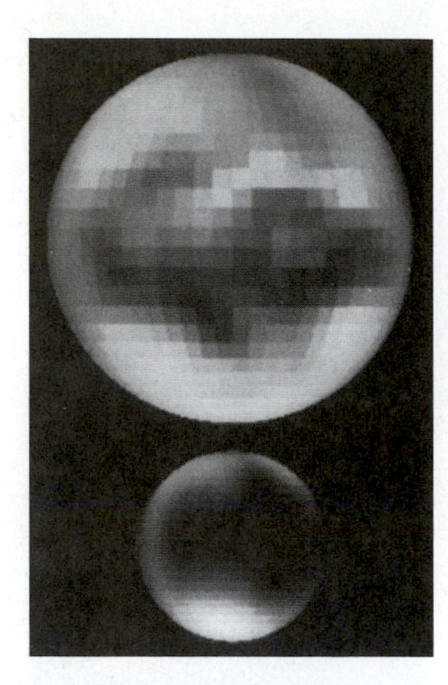

**FIGURE 17–6** Computer-generated renderings of the surface brightness of Pluto (*top*) and Charon (*bottom*) based on their mutual occultations. Here we see an equatorial view.

diameter, smaller than expected, and Charon is 1186 ± 20 km. Modelling the light curve can even show what albedoes different parts of the side of Pluto's surface facing Charon are likely to have (Fig. 17–6). For example, Pluto's south pole is brighter than the north pole, so has bright deposits of frost. The depth of the drop of the light curve is twice as great when Charon passes in front of Pluto compared with Pluto passing in front of Charon, so Pluto must be brighter than Charon. Pluto's average albedo is about 50 per cent, and Charon's is about 37 per cent. A bright feature some 250 km across appears at upper right of center. But some regions of Pluto are very dark, with one region 500 × 300 km having an albedo less than 0.1.

The mean density of Pluto and Charon together is 2 g/cm$^3$. Since Pluto contributes $\frac{7}{8}$ of the mass, Pluto's density cannot be far from 2 g/cm$^3$, which is about four times too great for Pluto to be made entirely of methane ice. It must be largely rock, probably with about 25 per cent water ice and a bit of methane ice forming a mantle around a rocky core. The current estimates for the densities are 2.0 g/cm$^3$ for Pluto and 1.2 to 1.3 g/cm$^3$ for Charon.

The complete hiding of Charon behind Pluto as seen from Earth in 1987 allowed the spectrum of Pluto alone to be recorded. This was then subtracted from the Pluto + Charon spectrum to give Charon's spectrum. Charon's infrared spectrum shows that water ice is present, but not methane or ammonia frost. Lower mass apparently did not allow it to retain the methane that Pluto did.

Observations of Pluto and Charon from a satellite in the infrared have been interpreted to indicate that Pluto has a dark equatorial band and bright polar caps of methane ice that vary over a Pluto year. Infrared observations in the 1990s confirmed an earlier report that nitrogen ice ($N_2$) also exists on Pluto's surface. The solid nitrogen is apparently mixed with small amounts of frozen methane and carbon dioxide.

The surface temperature of Pluto has been measured with the James Clerk Maxwell Telescope on Mauna Kea in the submillimeter region of the spectrum. David Jewitt of the University of Hawaii and his collaborators found lower temperatures than the infrared spacecraft had, showing that Pluto's surface is not uniform in appearance or color. He suggests that the surface may be part cold regions coated by nitrogen ice (close to 34 K) and part warmer regions (closer to 50 K) without nitrogen ice that match the dark patches on Pluto's surface. (Recall that 0°C = 273 K.)

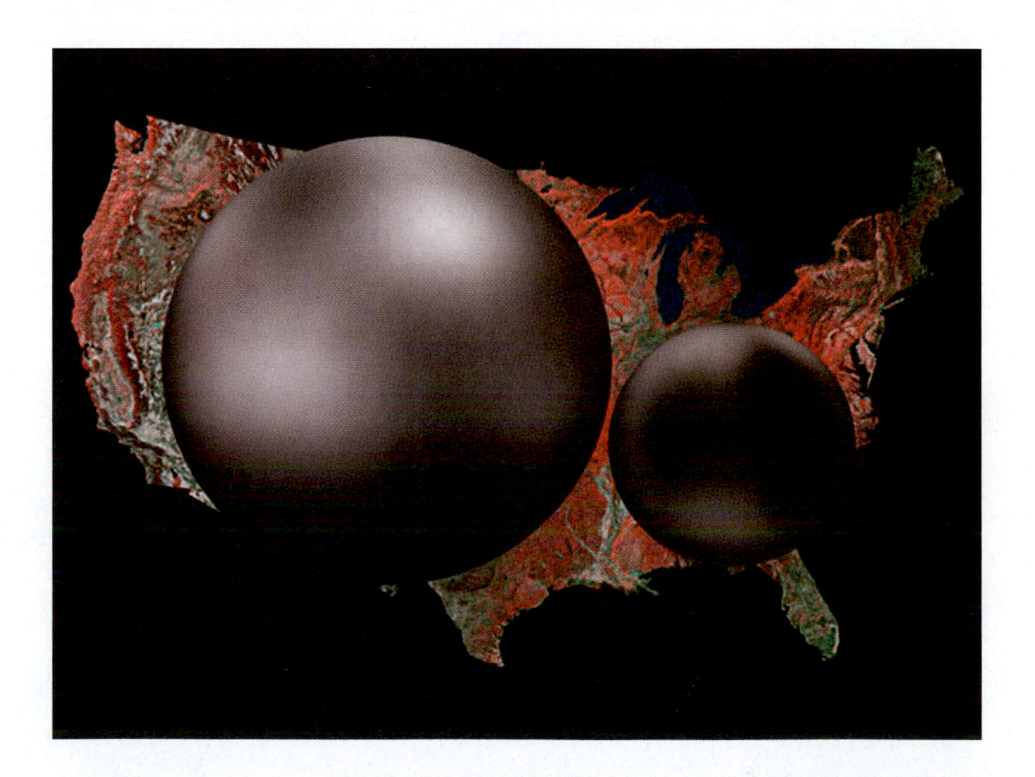

**FIGURE 17–7** Pluto and Charon, showing their sizes relative to each other and to the United States. The surface features correspond to Hubble Space Telescope observations.

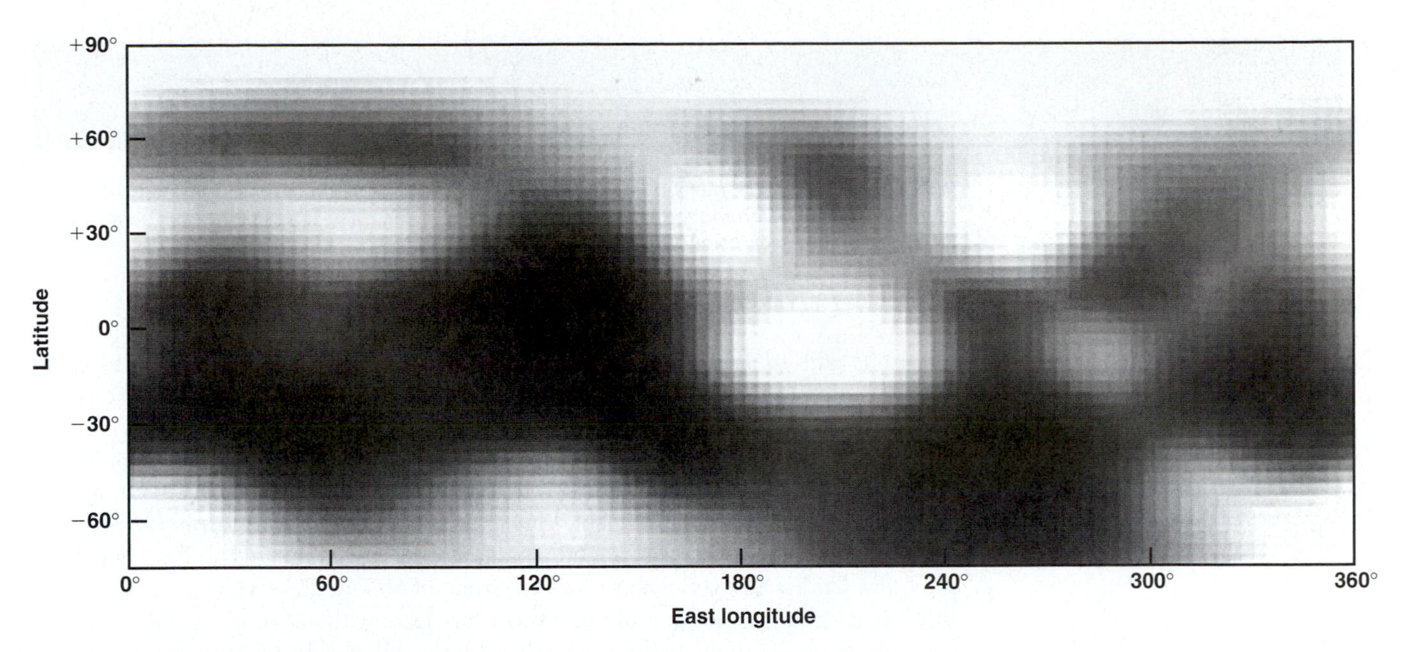


**FIGURE 17–8** A Mercator projection of the surface of Pluto, based on Hubble observations, shows a dozen areas of light and dark.

Hubble Space Telescope images of Pluto (Fig. 17–7) provide our present peak of knowledge of Pluto's surface. The resulting map (Fig. 17–8) shows a dozen areas of bright and dark on Pluto. The only problem is that we don't know what the bright and dark areas are! Are the bright areas bright because they are high clouds near mountains or because they are low haze and frost? We don't know. We merely know that the contrast is about as extreme as we have found in the solar system.

## 17.3 PLUTO'S ATMOSPHERE

Pluto is not massive enough to retain much of an atmosphere. But a tenuous (thin) atmosphere of methane was detected in the 1980s from infrared spectral observations. From the confirmation more recently that nitrogen ice is also present on Pluto's surface, coupled with the knowledge that molecular nitrogen is much more volatile than methane, scientists deduced that most of Pluto's atmosphere is nitrogen and that methane is a trace contaminant.

We found out a lot about Pluto's atmosphere when Pluto occulted (hid) a 12th-magnitude star—over 2 million times fainter than the faintest stars that can be seen with the naked eye, yet relatively bright for such an occultation and bright enough to be seen easily through a telescope. For this event—which took place on June 9, 1988—several observers travelled to Australia and New Zealand to observe the event from the ground and from the Kuiper Airborne Observatory, a NASA airplane instrumented for infrared observations. The star and Pluto were observed approaching each other in the telescope field of view, and then the total amount of light dimmed (Fig. 17–9). The light curves (graphs of intensity vs. time) showed the light dimming gradually rather than abruptly, so Pluto has a substantial atmosphere. The curves indicate either that a layer of smog or haze absorbs all the light from the star, or that a temperature inversion—where the temperature decreases and then increases as you move out from the planet's surface—in Pluto's atmosphere bends the light away from us; a changing atmospheric density, as results from a temperature change, would bend light. The smog or haze layer would have to extend at least 46 km from the surface. It would be so opaque that the starlight would be absorbed

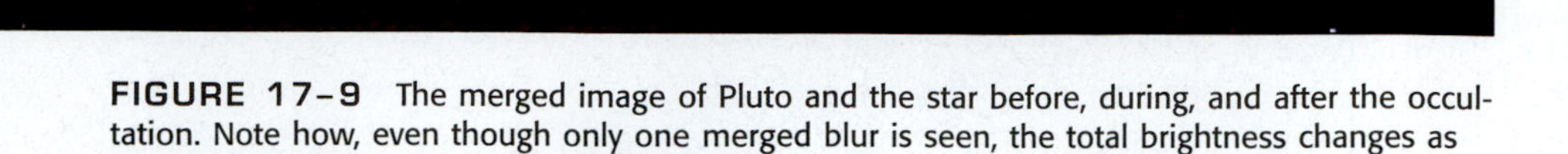

**FIGURE 17–9** The merged image of Pluto and the star before, during, and after the occultation. Note how, even though only one merged blur is seen, the total brightness changes as Pluto comes between us and the star.

before its path to us passed next to Pluto's surface. Calculations show that the point where the temperature would stop decreasing and begin increasing would be closer to the surface.

In either case, we are no longer sure that the value that we have for Pluto's diameter from Pluto and Charon's occultations of each other is the true surface diameter; it may include some of Pluto's atmosphere. The atmospheric pressure deduced for Pluto's surface is, nonetheless, only 1/100,000 that of the Earth's surface pressure, like Earth's atmosphere at an altitude of 80 km. Some calculations indicate that sharp changes in the temperature variation with altitude can explain the occultation observations without need for the haze layer. In addition to the overall variations viewed, structure in Pluto's atmosphere made "spikes" in the light curve (Fig. 17–10). It has also been concluded from the light curve that a gas heavier than

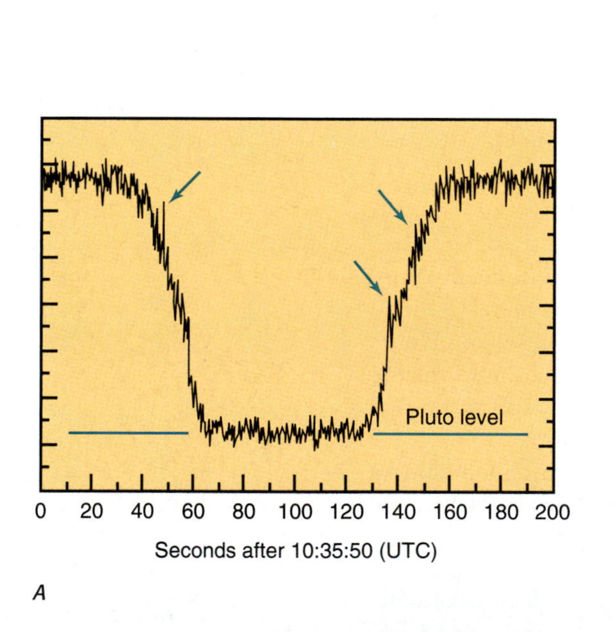


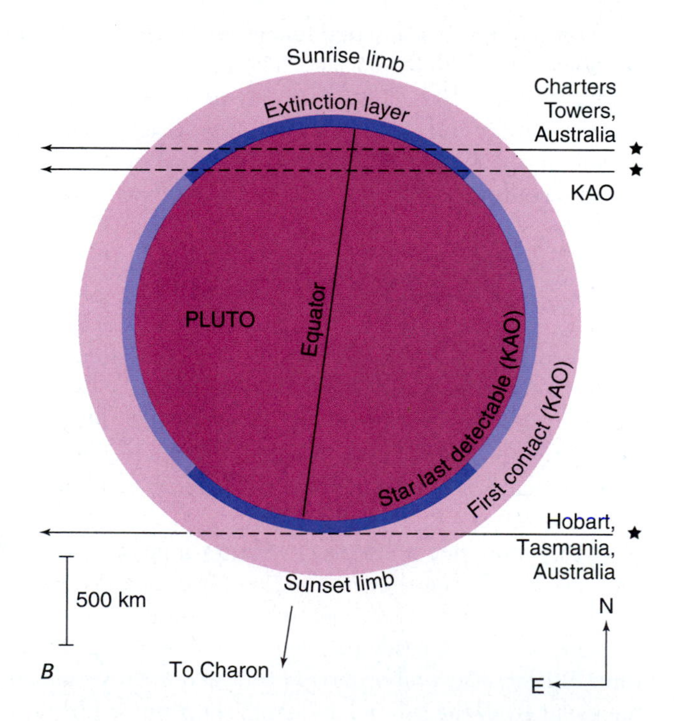


**FIGURE 17–10** (*A*) The light curve, intensity vs. time (Universal Coordinated Time), measured by MIT scientists from the Kuiper Airborne Observatory when Pluto occulted a star in 1988. The slope is steepest near the base, showing a change in the atmosphere. Arrows mark spikes resulting from the starlight passing through structure in Pluto's atmosphere where the temperature changes by a few per cent. The change in the slope of the sides of the dip reveals layers in Pluto's atmosphere. (*B*) The path of the star behind Pluto as visible from an airplane (KAO, the Kuiper Airborne Observatory) and two ground-based observatories. The horizontal lines with arrows shows the apparent path of Pluto across the Earth's surface above the observing sites; the lines are separated by the north-south distances on Earth between observing sites, as shown by the scale bar at lower left. The lines are dashed where the observations showed the occultation taking place. The circles showing Pluto and its atmosphere were drawn on to correspond with the dashed parts of the lines.

methane, perhaps carbon monoxide or nitrogen, must be present in Pluto's atmosphere.

It is thought that Pluto's atmosphere will snow out or freeze out over the next decade or so, because Pluto is moving farther from the Sun. If we want to get there to inspect it, we had better go soon. NASA's project for this purpose is discussed at the end of this chapter.

## 17.4 WHAT IS PLUTO?

Since we finally have accurate values of Pluto's mass and size, we know Pluto's density, which turns out to be too low to be from solid rock. Since only ices have such low density, Pluto must be made of frozen materials, probably around a rocky core. Its composition is thus more similar to that of the satellites of the giant planets than to that of the terrestrial planets. Spectral evidence indicates that non-icy materials (possibly silicates) are probably present in addition to methane ice.

Pluto remains a strange object in that it is so small next to the giants, and in that its orbit is so eccentric and so highly inclined to the ecliptic. The orbits of Neptune and Pluto are affected at present by their mutual gravities in such a way that their orbital positions relative to each other repeat in a cycle every 20,000 years. The two planets can now never come closer to each other than 18 A.U. Though Pluto may still have broken away from Neptune when another object passed nearby or even hit it, breaking off the outer layers and leaving the denser rocky core, this idea is not in favor. In fact, Pluto's orbit is chaotic, and for this reason we cannot reliably calculate it too far into the past or future. There is much research now that studies Pluto as the largest of a set of objects that we discuss in the next section.

Other lines of argument also favor the idea that Pluto originated independently of Neptune. Calculations of how the outer planets and their moons formed show that plentiful carbon monoxide in the protoplanets would have become methane ($CH_4$). The oxygen would then have been free to combine with other hydrogen to form water ($H_2O$). Methane-rich planets with icy moons would have resulted, which is just what has been found at Saturn, Uranus, and Neptune. Pluto, however, has too little water ice to fit the scenario, indicating that it did not originate as a moon.

## A DEEPER DISCUSSION

### BOX 17.1 Is Pluto a Planet?

Is Pluto a planet? It isn't possible to make a firm definition of what a planet is. Perhaps it is merely a matter of semantics—the meanings of words—rather than a scientific point. Still, perhaps if Pluto were to be found today, now that a couple of dozen objects in the Kuiper belt have been found, we would be classifying it as one of them. But historically, that is not what happened. And historically, as we shall see in the following chapters, there has been a distinction in our minds between the nine major planets and the thousands of minor planets, which are also known as asteroids. Pluto has been considered one of the Solar System's major planets for over 70 years, and it remains so in my mind.

The question as to whether Pluto is a full-fledged planet or not has proved to be of widespread public interest. A debate on the subject was held at the International Astronomical Union's General Assembly in August 2000, with a majority holding that it is a planet. The Planetary Sciences Division of the American Astronomical Society put out a statement in 1999, also holding that Pluto's planethood is not in doubt. The Rose Center for Earth and Space, the location of the Hayden Planetarium in New York City, omitted Pluto from its set of models of planets when it opened in 2000, and the controversy over its doing so erupted onto the front page of *The New York Times* in 2001.

The debate about Pluto's status has dwelled on the low-mass end of the category known as planets. In the next chapter, we will see how the discoveries of many planets and perhaps other objects in orbit around stars other than our Sun have led to a debate about the meaning of planet on the high-mass end as well.

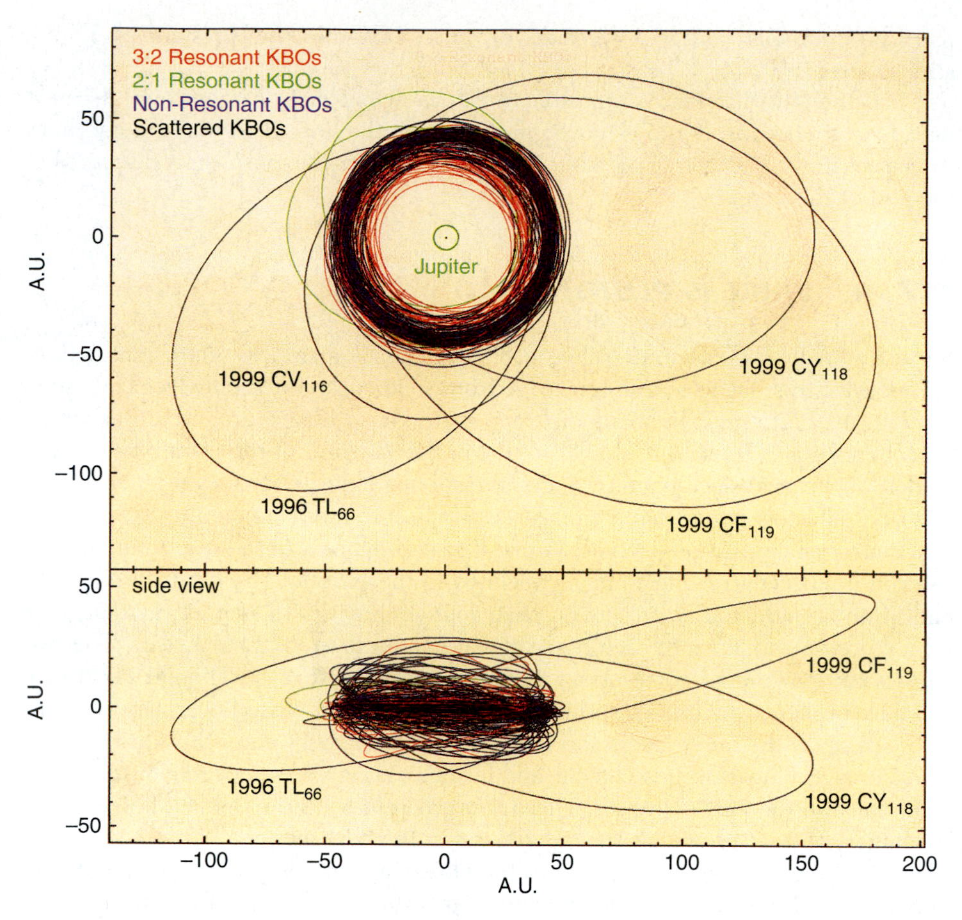


**FIGURE 17–11** Orbits of Jupiter and the Kuiper belt objects.

## 17.5 PLUTO AND THE KUIPER BELT

Over the last few years, over a hundred icy objects have been found in Pluto's part of the Solar System, with orbital radii between 30 and 50 A.U. (Fig. 17–11). David Jewitt of the University of Hawaii and Jane Luu, now at the University of Leiden in the Netherlands, were the discoverers of most of the known objects. The objects match predictions of a band of incipient comets in that part of the Solar System. The band was named the **Kuiper belt,** after the planetary scientist Gerard Kuiper (pronounced koy'per). It is perhaps 10 A.U. thick and extends from about the orbit of Neptune about twice as far out. Members are known as Kuiper belt objects or as trans-neptunian objects. A full search may turn up tens of thousands of these objects larger than 100 km across.

The Kuiper belt objects may be leftover planetesimals, and Pluto may be merely the largest of these objects (Fig. 17–12). The idea would explain why we find a small, rocky body like Pluto out beyond the giant planets. Though Pluto's 60 per cent albedo contrasts with the approximately 4 per cent albedoes of the Kuiper belt objects, one could say that because Pluto is the largest such object it is big enough to develop an atmosphere with frost. At this writing, research is going on to characterize better the new discovery of the second brightest known object in the outer Solar System. Found by the Spacewatch Project at the University of Arizona, its orbit is about the same size as Pluto's and, depending on its albedo, it may be about ⅓ Pluto's diameter.

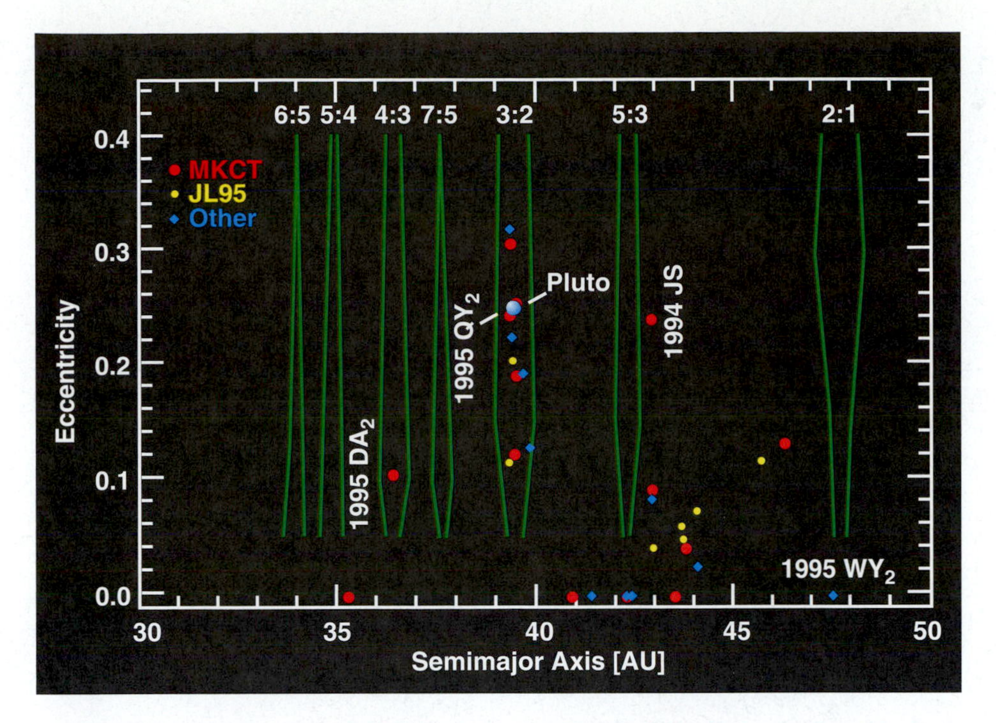


**FIGURE 17–12** The eccentricities (*vertical axis*) of Kuiper belt objects are plotted vs. their semimajor axes (*horizontal axis*). Green lines mark "resonance regions," where the periods of the objects are related to the periods of Neptune by ratios of small integers. For example, Pluto is in the 3:2 resonance, meaning that it completes 2 orbits for every 3 orbits completed by Neptune.

## 17.6 FURTHER PLANETS IN OUR SOLAR SYSTEM?

Are there still further planets beyond Pluto? Tombaugh continued his search (Fig. 17–13), and found none. Other optical searches have been made since, also without success. But space satellites have detected thousands of objects in the infrared part of the spectrum. Further, the sky has been recently mapped from the ground in the infrared by 2MASS. A 10th planet that was sufficiently large or close or had an internal energy source might have been recorded. It may one day turn up in the analysis of the massive amounts of data that have accumulated.

The data for Uranus's orbit have been reanalyzed using only the twentieth-century measurements for the positions of Uranus and the Voyager value for the mass of Neptune. This reanalysis does not find the deviation in Uranus's actual orbit from its expected orbit that is found when older data are used together with the new data. By that standard, there is no indication that a 10th planet exists.

Also, tracking data on the Pioneer 10 spacecraft, now out of the Solar System past the orbits of Neptune and Pluto, does not show any gravitational effect from a 10th planet. The limit set for the mass that any 10th planet could have is now very low.

**FIGURE 17–13** One of the original plates on which Tombaugh discovered Pluto. This plate is still at the Lowell Observatory; the other half of the pair is in the Smithsonian Institution. Imagine how many star images are packed on a plate of this large size. A small section was shown in Figure 17–2.

## 17.7 A MISSION TO PLUTO

A NASA mission to Pluto is being considered. A large part of the point is to get there before Pluto's tenuous atmosphere condenses to the ground as snow, as Pluto in its orbit recedes from the Sun. We won't have another chance to examine Pluto's

**FIGURE 17–14** An artist's conception of the Pluto–Kuiper Express.

atmosphere for 200 years. Given cost constraints and advances in electronics, the mission will weigh less than 10 per cent of a Voyager.

A long-term conception of the mission, known as Pluto–Kuiper Express (Fig. 17–14) was to map Pluto with a resolution of about 1 km in black-and-white and 3 to 10 km in color and infrared. An ultraviolet spectrometer was to examine Pluto's atmosphere, assuming the spacecraft gets there before it freezes out. Tracking the Doppler shift from the spacecraft should tell about Pluto's interior, based on its gravity field. Did our water on Earth come from comets that started in the Kuiper belt? Pluto–Kuiper Express was to help us with the answer.

Pluto–Kuiper Express was supposed to be launched in 2004 with arrival at Pluto at 2012, thought to be well before the time when the atmosphere freezes out. But in 2000, ostensibly because of all the money NASA moved into the Mars exploration program, the planned Pluto–Kuiper Express was delayed. The delay would have meant that the chance for a gravity assist was lost, and arrival at Jupiter wouldn't be until 2020. That could well mean that the atmosphere would be no longer detectable. The delay was so strongly protested by planetary scientists and others that NASA opened a competition for cheaper, alternative missions to Pluto that could get there sooner. Look on the Web for the latest.

## CORRECTING MISCONCEPTIONS

**✖ *Incorrect***

Pluto is the 9th planet because it always orbits outside Neptune.

Pluto is a big planet, like the other outer planets.

Pluto was discovered because of its effect on the orbit of Neptune, then the outermost planet known.

**✔ *Correct***

Pluto is the 9th planet because the semimajor axis of Pluto's orbit is larger than that of Neptune.

Pluto is much smaller than Earth; so small its planethood has been challenged.

Tombaugh's search was based on Uranus, since Neptune's orbit was not sufficiently well known. In any case, Pluto has so little mass that its effect on either of those bodies is not measurable.

## SUMMARY AND OUTLINE

Basic properties
- Most eccentric orbit; greatest inclination to ecliptic
- Rotation accurately determined by varying albedo
- Discovered in 1930 by diligent search near ecliptic

Determining radius, mass, and thus density (Sections 17.1 and 17.2)
- Discovery of a moon (Charon) allows mass to be deduced; it is only 0.2 per cent that of Earth; density thus is similar to that of satellites of outer planets.
- Radius measured from mutual occultations of Pluto and Charon.
- Separation of Pluto and Charon measured with Hubble Space Telescope.

Surface (Section 17.2)
- Imaged with Hubble Space Telescope

Atmosphere measured in an occultation (Section 17.3)

Is Pluto a planet? (Section 17.4)
- Historical point of view says yes.

Pluto and the Kuiper belt (Section 17.5)
- Pluto could be merely largest Kuiper belt object.

No evidence for further solar-system planets (Section 17.6)

Pluto Express to map Pluto/Charon (Section 17.7)

## KEY WORD

Kuiper belt

## QUESTIONS

[†]1. What fraction of its orbit has Pluto traversed since it was discovered?
2. What evidence suggests that Pluto is not a "normal" planet?
[†]3. Using Chapter 24, determine at what wavelength a cold body like a 10th planet would give off the most radiation. What part of the spectrum is this and how does it affect the chance of discovering such a planet? Use a reasonable estimate for the temperature, such as that given for Pluto in this chapter, and Wien's law (Section 24.1).
4. Aside from its size and mass, how does Pluto differ from the giant planets that are its neighbors in the solar system?
5. From the separation of Pluto and Charon, show how to calculate the combined mass of Pluto and comment on how to deduce Pluto's own mass.
6. Describe what mutual occultations of Pluto and Charon are and what they have told us.
[†]7. During what years have there been mutual occultations of Pluto and Charon? During this period, how many days and hours separated the dips in the light curve? Explain.
8. Describe the occultation observations of Pluto's atmosphere.
9. How did the Hubble Space Telescope improve our understanding of Pluto/Charon?
10. Why does Planet X, a prospective 10th solar-system planet, now seem less likely to exist?

[†]This question requires a numerical solution.

## TOPICS FOR DISCUSSION

1. Can you write a definitive definition of a planet? Try.
2. Is Pluto a planet?

## USING TECHNOLOGY

### W World Wide Web

1. Examine JPL's Fire and Ice page to find out the status of Pluto missions: http://www.jpl.nasa.gov/ice_fire.
2. Examine the Hubble Space Telescope's still and movie images of Pluto.
3. See the International Astronomical Union's statement at http://www.iau.org/IAV/FAQ/PlutoPR.html.

### REDSHIFT

1. Move through the Pluto/Charon system.
2. When will Pluto next be well placed for observing in the evening sky? If you can see to 15th magnitude with a medium-sized telescope, will you be able to see it?

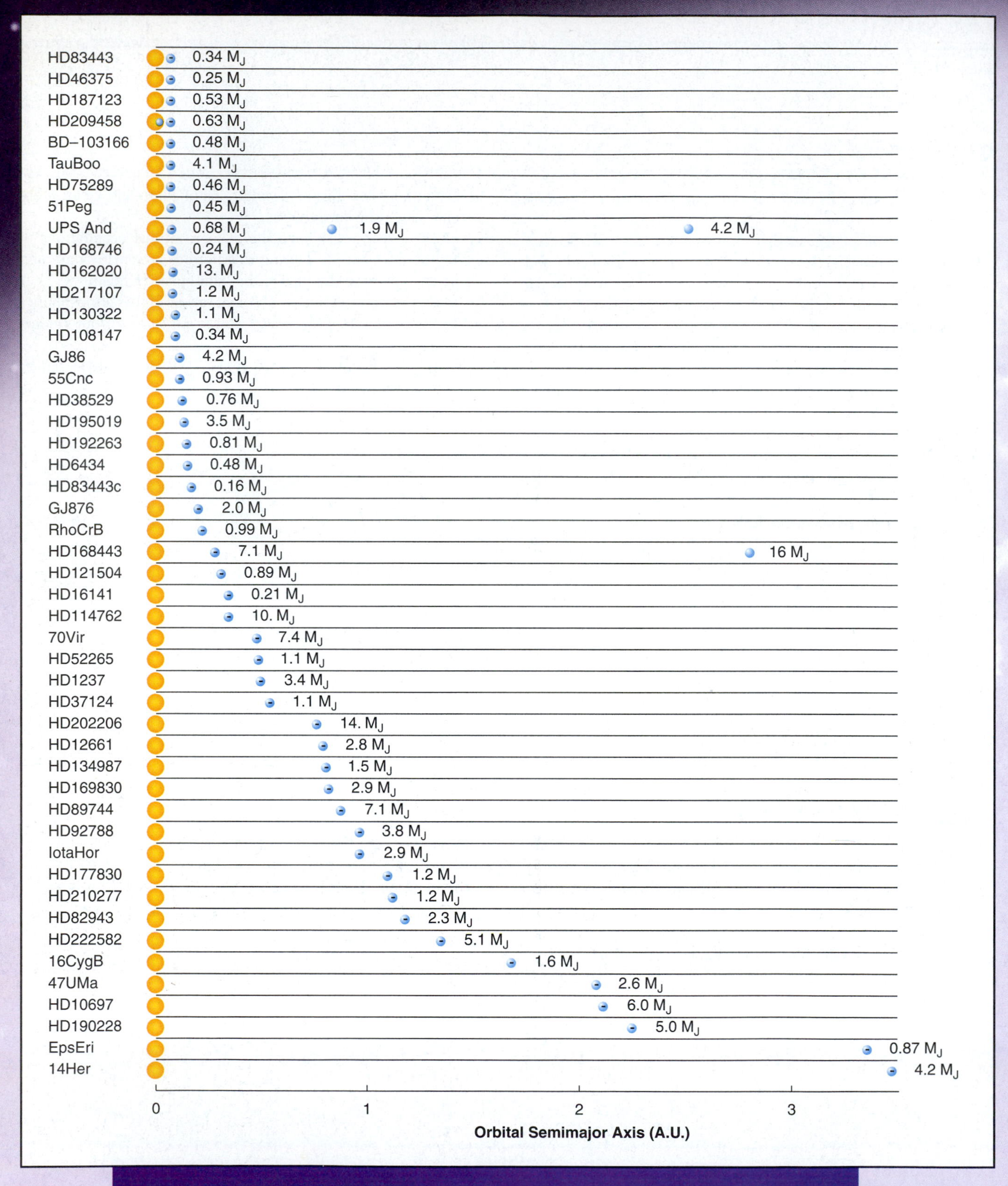


**The characteristics of the first planets discovered around nearby sunlike stars.**

# Chapter 18

# Extrasolar Planets: Other Solar Systems?

We have seen how difficult it was to detect even the outermost planets in our own Solar System. At the four-light-year distance of Proxima Centauri, the nearest star to the Sun, any planets around a star would be too faint for us to observe directly. We have a better chance of detecting such a planet by observing its gravitational effect on the star itself as a changing Doppler shift. And in the 1990s, this promise was fulfilled. We now know about more planets outside our Solar System than inside.

**AIMS:** To learn about the planets recently discovered orbiting around stars other than the Sun

Since we cannot see extra-solar-system planets directly, we search for planets around stars by their gravitational effects. Imagine looking onto a ballroom dance floor where couples are waltzing. Each member of each couple moves alternately closer to you and farther from you. Their center of mass moves smoothly while the partners revolve around that center of mass. Now imagine that one of the partners is invisible. You could still know that this unseen individual is there by watching the visible partner move to and fro. We use the same technique to infer objects invisible to us—planets—by watching the motions of the stars they accompany.

The science of **extrasolar planets,** or **exoplanets,** has progressed mightily in the last decade (Fig. 18–1). The tremendous success has led to a concentration of effort. Observing time has been allotted on the largest telescopes in the world and several relevant space missions have been planned. As we discover hundreds of planets orbiting other stars, the science turns from mere discovery to investigation of the statistics, which should lead to an understanding of the formation of solar systems in general, including our own.

FIGURE 18–1 Geoff Marcy and Paul Butler, discoverers of most of the extrasolar planets, with the 3-m telescope at the Lick Observatory in California that they used for many of their early observations. With their successes, they now have lots of telescope time at the 10-m Keck telescope.

## 18.1 OLDER SEARCHES FOR PLANETS

Older than the technique of searching for Doppler effects from planets is the study of the "proper motion of a star"—its apparent motion across the sky with respect to the other stars. We want to see whether the star follows a straight line or appears to wobble. Any wobble would have to result from an object too faint to see that is orbiting the visible star. (We will discuss such methods in general in Section 25.7.)

The easiest stars to observe in a search for planets are the stars with the largest proper motion, for these are likely to be the closest to us, and thus any wobble would cover the greatest angular distance that we could hope for. The star with the largest proper motion, Barnard's star, is only 6 light-years away from the Sun (Fig. 18–2). It is the nearest star to the Sun except for the Alpha Centauri star

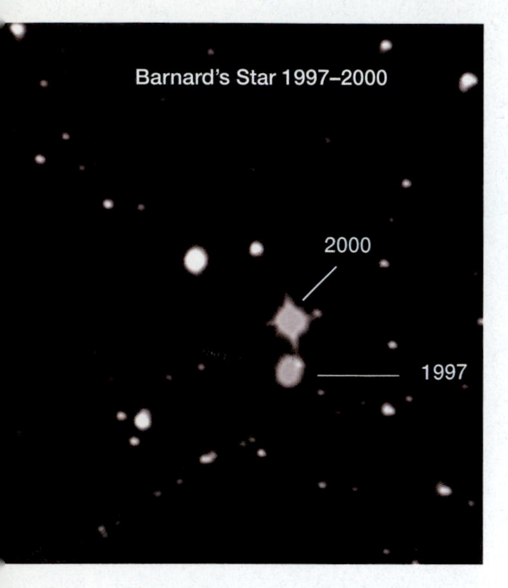


**FIGURE 18–2** The motion of Barnard's star across the sky, noticeable even over a short time interval.

system (Appendix 7). It moves across the sky by the same angle as the diameter of the Moon in only 180 years.

Studies of decades of photographic images of Barnard's star seemed to show slight wobbles in its proper motion. In the 1960s, it was reported that these wobbles showed the existence of a planet. Further analysis seemed to show that two planets were present in the system. The planets would have to be very massive; the Earth could not have made a measurable difference in the proper motion. A possible planet around epsilon Eridani, another very nearby star, was also reported. Epsilon Eridani is the tenth-nearest star to Earth, at 10.5 light-years away. But it turns out that the wobbles did not really exist; the "wobbles" resulted from uncertainties in the measurements and from a change in the magnification of the images on the photographic plates that resulted from a one-time change in the position of the lens in the telescope used for the observations.

George Gatewood of the University of Pittsburgh has recently reevaluated the continuing long-term astrometric observations of Barnard's star and epsilon Eridani and used his own newer instrumentation. He can rule out, for example, the existence of planets between 2 and 10 Jupiter masses in orbits from a few tenths of an A.U. out to 2 A.U. Gatewood finds, from his 12 years of astrometry with his current instrument, an object with about Jupiter's mass about epsilon Eridani. But he does not yet call it a planet, and is continuing his study.

## 18.2 PULSAR PLANETS

The first widely accepted detections of planetary-mass objects around other stars came only in 1991, and then in a surprising way. They involved a **pulsar** (Chapter 30), a dead star that gives off sharp pulses of radio waves at very regular intervals on a very rapid time scale. One pulsar was found whose pulses, instead of coming regularly every hundredth of a second, seemed to sometimes come a bit early or a bit late (Fig. 18–3). This variation had a six-month period. Then a second pulsar with similar properties was found, but with a more complicated apparent variation. From these apparent variations, scientists deduced that planets were going around the pulsars. The speeding up of the pulses corresponds to the pulsar approaching us, and the slowing down of the pulses corresponds to the pulsar receding. Since a basic law of physics holds that the center of mass of the system moves in a constant direction unless affected by an outside force, we can deduce that an invisible object and the pulsar are each orbiting the center of mass. They stay on opposite sides of it.

Embarrassingly, the first report of a detection, the one with the six-month period, was a mistake. Something in the computer program used for analysis had failed to account for the eccentricity of the Earth's orbit around the Sun. But the second reported detection, of two massive planets around a pulsar pulsing 162 times each second, has held up. Made by Alex Wolszczan, now of Penn State, it was confirmed in 1994. At that time, Wolszczan confirmed his prediction that the two planets would pass so close to each other that their gravity would change their orbits noticeably. When he detected the gravitational interaction of the two planets with each other, his result was widely accepted. Another planet, with a mass like that of our Moon, has since been found in the system (Fig. 18–4, on p. 318).

An additional pulsar planet was found in 1999. It is in an eccentric orbit around a pulsar in the core of the globular cluster M4. (Several pulsars have been found in such clusters, which are good places to search because they contain 100,000 stars in a small region of the sky.) Its mass is estimated to be about ten times that of Jupiter.

Many people are particularly interested in planets around stars other than our own as part of a hope that life may be found in the Universe outside our own Solar System. These pulsar planets, though, probably do not bring us closer to finding life outside our Solar System. The neighborhood of a pulsar is so raked with powerful radiation that it seems most unlikely that life could exist there. And if the planet were

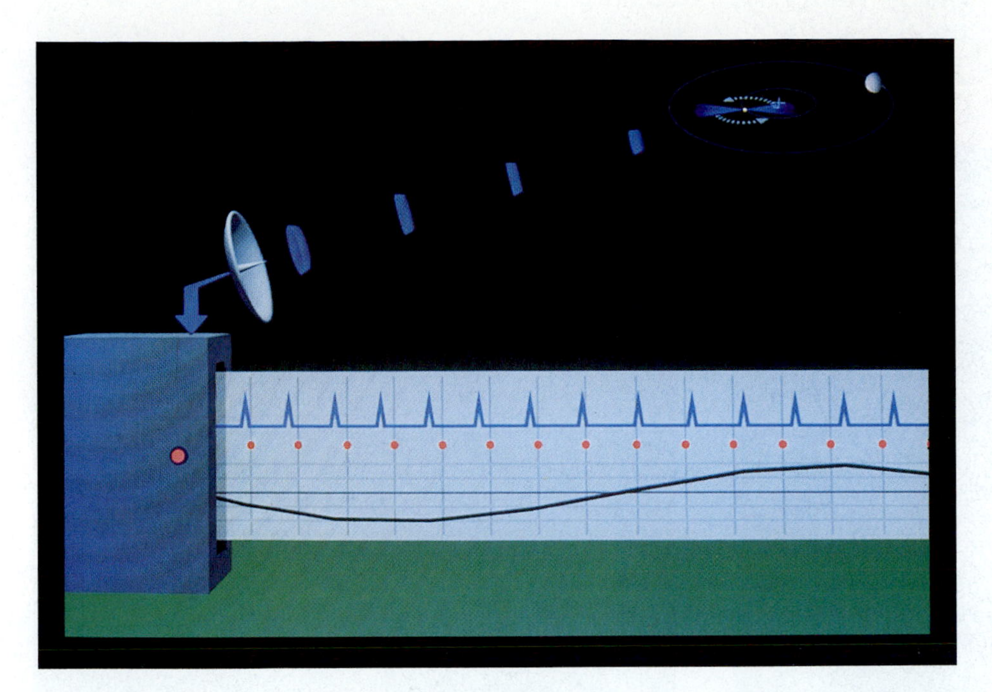

FIGURE 18–3 The pulsar and its planet, as shown at top right, are always at opposite sides of the system's center of mass, which is shown with a + sign. The pulsar, a star of about as much mass as our Sun, is only about 20 km across, much smaller than the planet, which is about 100,000 km across. Curved, dotted arrows indicate that the pulsar is rotating on its axis, sweeping the beams of radio waves (also shown in blue) around the sky. Each time one of the beams passes us, we detect a pulse of radio waves. (Pulsars got their name from those pulses of radio waves.) The pulsar's orbit around the center of mass is much, much smaller than the planet's orbit, because the pulsar is so much more massive than the planet. The pulsar's pulses of radio waves are shown in blue moving from upper right to lower left, where they are intercepted by a radio telescope and fed into the receiver. The average pulse period is only 6.2 milliseconds, meaning the star spins entirely around in that time interval (assuming only one side of the beam ever points at us). On the chart shown at bottom, we see a graph of the intensity of radio waves over time. The pulses show as blue spikes; they sometimes arrive slightly before or after regularly spaced pulses (*red dots*) would. The difference in arrival time is graphed at the bottom. It reveals that there must be planets orbiting the pulsar, since if the pulsar did not move to and fro, that black curve at the bottom would be a horizontal straight line.

a survivor of a supernova—the explosion of a star that is thought to explain how pulsars form—there would have been even more radiation.

## 18.3 EXTRASOLAR PLANETS IN SUNLIKE SYSTEMS

### 18.3a Finding Extrasolar Planets

Starting in 1995, planets more like those we already know in our own Solar System have been discovered around other stars. The technique involves using a spectrograph to monitor the visible-light wavelengths of spectral lines from some nearby stars (Fig. 18–5). If Doppler shifts show that the star is going away and then toward us with a certain period, we can deduce that a planet is going toward and away from us, the opposite directions, with the same period. A Swiss scientist and his student, Michel Mayor and Didier Queloz, used the technique to find a planet revolving about the sunlike star 51 Pegasi (Fig. 18–6A). (This numbering system for stars, invented by the Astronomer Royal John Flamsteed two hundred years ago, is still used for stars of intermediate brightness.) The planet had an astonishingly short period: only 4.2 days, which would have placed it far inside Mercury's orbit if it had been in our own Solar System.

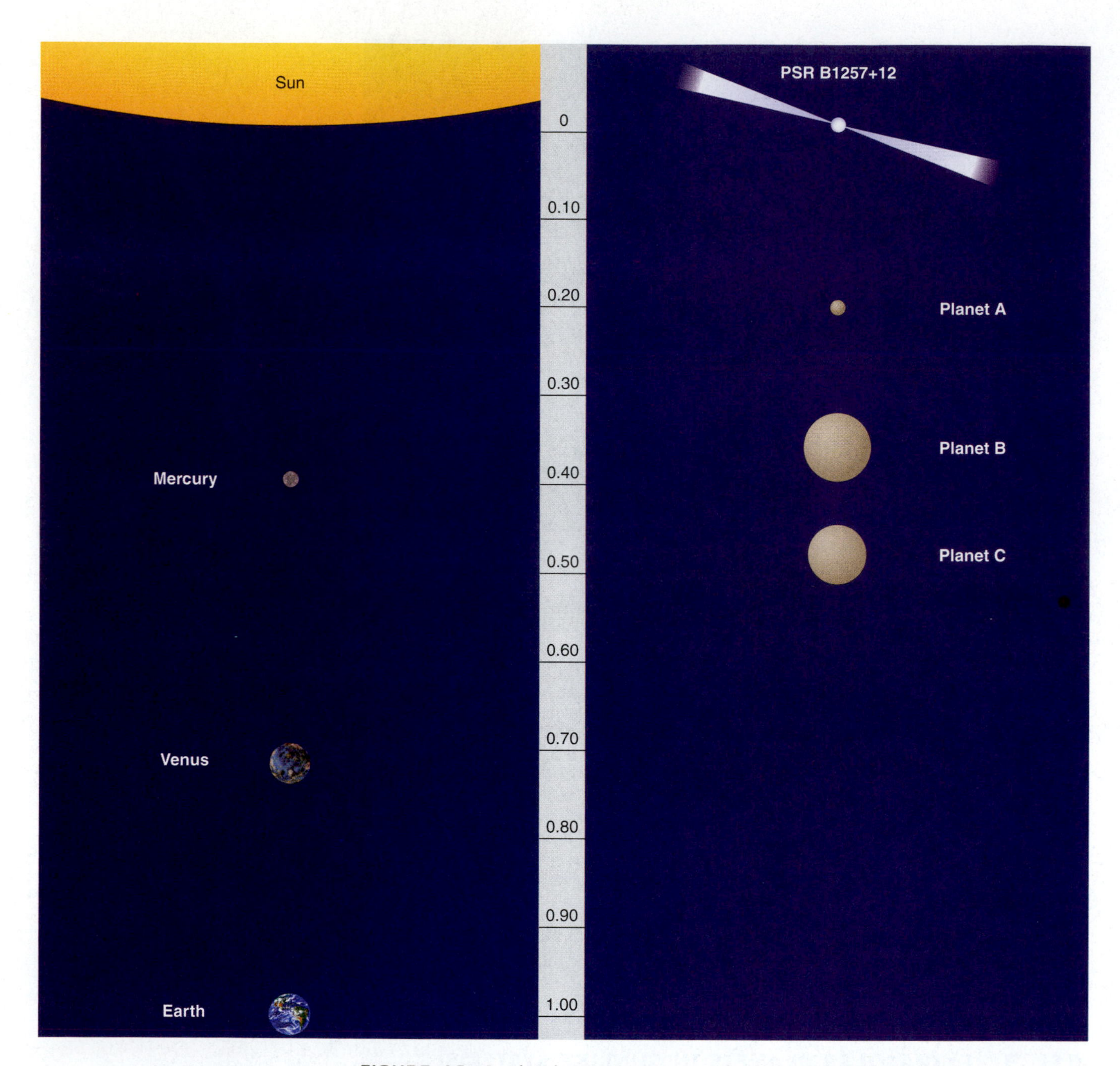


**FIGURE 18–4** The planetary system around pulsar PSR B1257+12, with the inner part of our Solar System on the same scale. The planetary sizes are on a different scale from the planetary spacings.

Geoff Marcy and Paul Butler, a pair of American astronomers then at San Francisco State University and the University of California at Berkeley, had been collecting similar data for some years on a variety of stars, but they had assumed that the effect would be so difficult to detect that they had not yet run their computer analyses of the data. They were, rather, improving their detection sensitivity by improving the accuracy of their wavelength measurements. Marcy and Butler had been hoping to find a Jupiter-like planet in a Jupiter-like orbit, perhaps 10 years in duration. But the 4.2-day period of the new planet didn't require lengthy data sequences. Given the 51 Peg discovery, Marcy and Butler used their next observing run a few days later to confirm the planet around 51 Peg (Fig. 18–6*B*), and immediately found

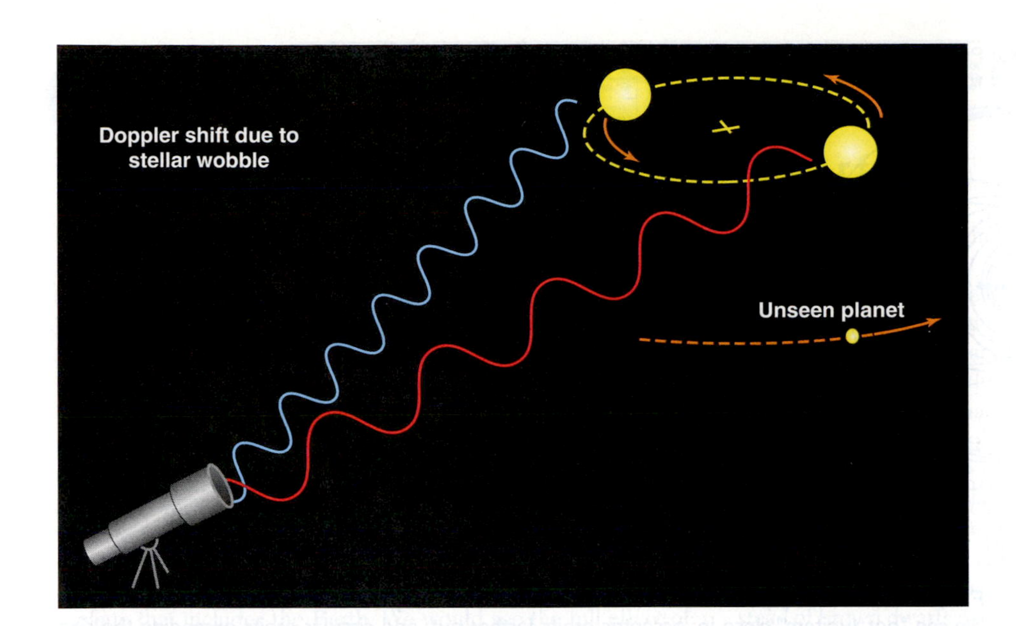


**FIGURE 18–5** The technique used by Geoff Marcy and Paul Butler to detect planets around sunlike stars by searching for Doppler effects in the stars' spectra.

planets around two additional sunlike stars in their data. Since then, both teams, and others as well, have detected dozens of planets around still other stars (Fig. 18–7). Marcy is now at the University of California at Berkeley and Butler is at the Department of Terrestrial Magnetism of the Carnegie Institution of Washington, in Washington, D.C.

Returning to the planet search through study of changing Doppler shifts, the planets discovered by this technique are more massive and are in weirder parts of their star systems than we had expected on the basis of our own Solar System. For

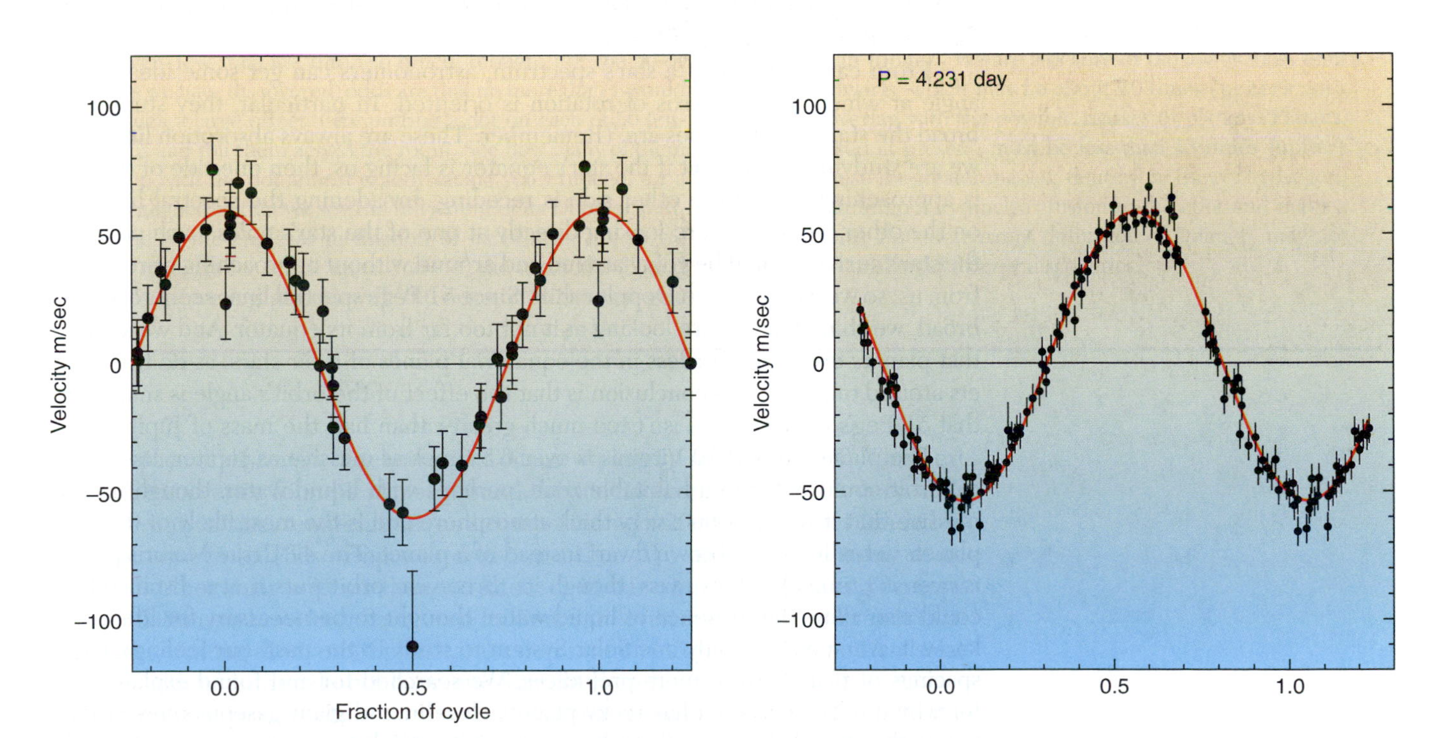


**FIGURE 18–6** *(A)* The Doppler shift of 51 Peg from Michel Mayor and Didier Queloz that revealed the presence of a planet in the system. The vertical error bars show the uncertainties of individual measurements, but they were able anyway to figure out how the star's motion varied. *(B)* Geoff Marcy and Paul Butler quickly confirmed the Mayor and Queloz discovery in 51 Peg, and refined the error bars. Both teams went on to discover many new planets.

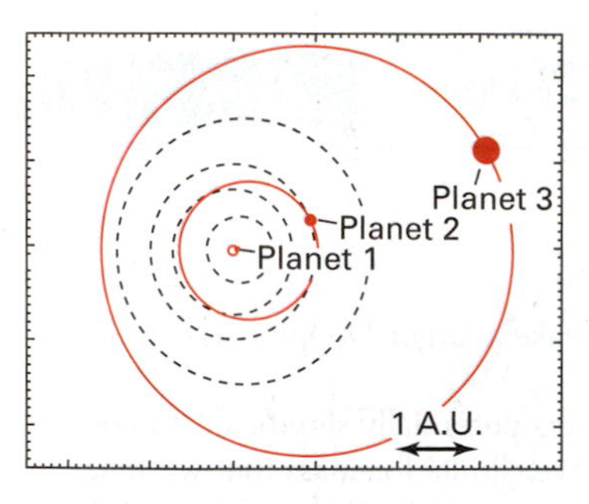


**FIGURE 18-8** A diagram of the orbits of the three planets (*red dots*) detected around the star $\upsilon$ (upsilon) Andromedae. The star, 44 light-years from us, has a mass about 30% greater than our Sun's and is about halfway through its 6-billion-year lifetime. The dashed circles show the orbits of Mercury, Venus, Earth, and Mars overplotted to give the scale of the orbits. Jupiter, at 5 A.U., would be outside the boundaries of this plot. The innermost planet has at least 3/4 Jupiter's mass and a period of only 4.6 days. The middle planet has at least twice Jupiter's mass and a very elliptical orbit with a period of 242 days. The outer planet has a mass of at least four times Jupiter's mass and orbits every 4 years. Because both short-period and long-period/elliptical-orbit planets exist in the same system, perhaps we have found the equivalent of a Rosetta Stone for understanding why these two types of planets exist.

Jupiter's mass in an eccentric orbit well inside the size of Earth's orbit is accompanied by a more massive object of over 17 Jupiters, which is located beyond Earth's orbital diameter. We know it has less than 40 Jupiter masses, or else Hipparcos would have picked up its star moving in position. By mass alone, it would be a brown dwarf, but it sure looks like it is in a planetary system. The definition we have of "planet" may have to change to accommodate objects like this one.

The statistics are beginning to emerge from a study of the first fifty extrasolar planets. About 3/4 of 1 per cent of nearby solar-type stars have jovian planets in circular orbits that take between 3 and 5 Earth days. These are sometimes called "hot Jupiters," since they are so close to their parent stars (within 1/10 A.U.) that temperatures are very high, over 1000 degrees. Another 7 per cent of these nearby solar-type stars have jovian planets whose orbits are very eccentric. The ones we have found so far are within 3.5 A.U. of their stars; planets with larger orbits would have longer periods, so we must observe for longer times before we can find out whether they exist. Our best searches at present can detect motions as small as 3 m/sec, dozens of times better than the previous standard, but we must sustain these searches for over a decade to detect planets more like those in our own Solar System. (The shift in the Sun caused by Jupiter's orbiting that we would search for if we were on a distant star is 13 m/sec, which is about 28 miles per hour.) At present, we are sensitive to Saturn-mass planets only within 1 A.U. and Neptune-mass planets within 0.1 A.U. So we expect the bars at the left side of Figure 18-7 to rise even more rapidly than they do with the present measurements.

How do these planets form, given that they are in places or orbits so different from those of planets in our own Solar System? Most theories at present have giant planets indeed forming at great distances from their stars, like those in our own Solar System. After all, there isn't enough matter in the inner Solar System to form giant planets. Some of these giant planets then later migrate into the smaller orbits, an idea that had been suggested for our own Solar System even before extrasolar planets were discovered. Perhaps several massive planets are present in these systems and their mutual gravity slingshots one or more into close orbits. Perhaps a star—a passing star or an orbiting companion star—intrudes into this system, perturbing the orbits of the planets. One or more of the planets may be put into very eccentric orbits and others may be ejected from the system entirely! (In Chapter 26, we discuss the OGLE observations that could potentially detect such lone planets wandering through space far from their parent stars. One can debate whether isolated, planetary-mass objects we might find are actually planets, which depends on the definition of planet we adopt.) Some planets, when they are close to the star, are brought into circular orbits by tidal forces. Others may spiral into the star and be consumed. So these models can explain both the close-in planets and the planets with larger orbits of great ellipticity. Whether these ideas are valid or whether they are remnants of our centuries of thinking about the scale and structure of our own Solar System remains to be seen.

Why have these searches finally succeeded after decades of searching for planets? A combination of observational technique at the telescope and computer calibration of the wavelengths over small intervals, in comparison with known wavelengths from a source on Earth, allows wavelengths to be measured with unprecedented accuracy. The Doppler technique has the advantage that as long as the stars are bright enough to have their spectra measured with high statistical accuracy, it doesn't matter how far away they are. The planets so far reported by this method are around stars that range from 15 to 50 light-years from the Sun, reasonably close but not among the hundred stars closest to us. The method grows increasingly sensitive, extending now to planets even less massive than Saturn. But it is still nowhere near sensitive enough to detect planets of the mass of the Earth. The largest telescopes in the world, including the Keck telescope in Hawaii, are currently used in the search, to provide enough light to allow the spectra of fainter and fainter stars to be observed with sufficient accuracy.

Why aren't the systems we are finding more like our own Solar System? For one thing, the Doppler method is sensitive especially to giant planets close in to their stars. As we observe for longer intervals, planets with larger orbits and thus longer periods—corresponding more closely to Jupiter's in Jupiter-like orbits—will be discovered. But in studying our Solar System we were formerly working from a sample of one, and now we have a wide variety of planetary systems to study. Clearly, the statistics of the orbits and planetary masses will give us a more complete (and correct) understanding of how planetary systems form.

### 18.3b Seeing Extrasolar Planets

Note that none of these planets can be seen directly, even with the Hubble Space Telescope. The planets are too faint and would be lost in the glare of their adjacent stars. Many people would be happier if we could actually see one of these planets. The dramatic discovery of a planet going in front of its star provided a confirmation that removed any doubt of the reality of the discoveries (Fig. 18–9A). The discovery was made in 1999 by two teams, one headed by Gregory Henry of Tennessee State University working with Marcy and Butler and the other by Dave Charbonneau working with Robert Noyes and David Latham, all of the Harvard-Smithsonian Center for Astrophysics, and Tim Brown of the National Center for

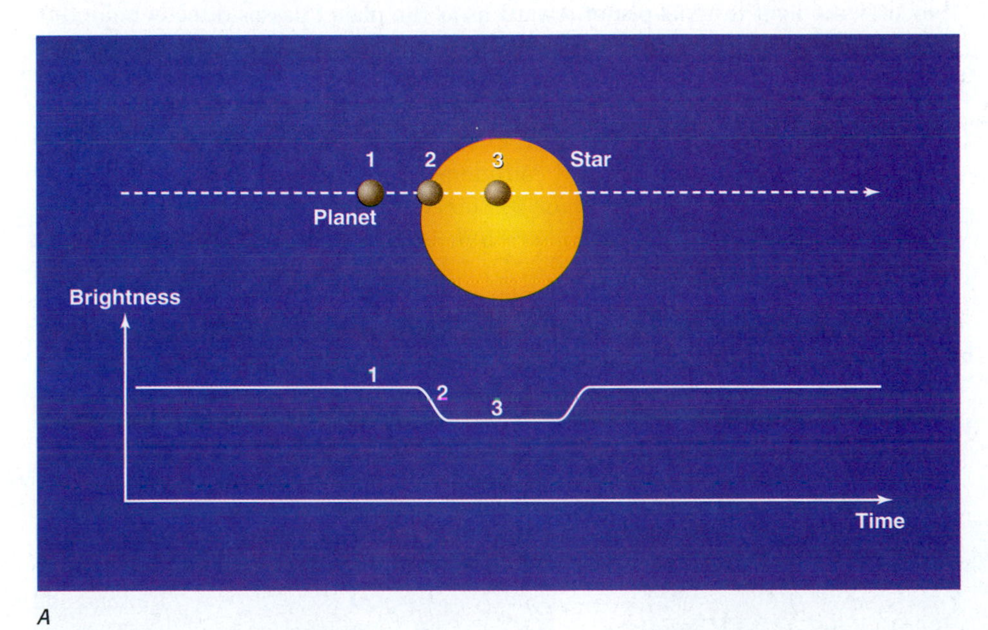


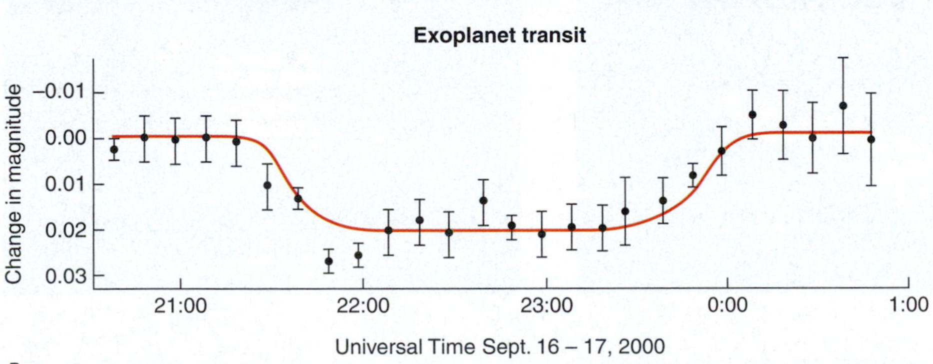


**FIGURE 18–9** (*A*) Schematic showing the dips that would result from an extrasolar planet. (*B*) Even amateur astronomers can detect the transits of certain extrasolar planets through careful photometry, as we see here.

Atmospheric Research in Boulder. The dip in the star's light, only about a hundredth of a magnitude, is nonetheless measurable with careful photometry (Fig. 18–9*B*). Such dips are known as transits, as they are for Mercury or Venus. Since these dips can be measured, and correspond exactly to the times deduced from the Doppler observations, these small, dark objects are surely out there. For this star, HD 209458 in Pegasus, the planet's mass is 0.63 Jupiter's. Its orbit is so small that it passes in front of its star every 3.52 days. Timing the dips shows that its radius is 30 per cent larger than Jupiter's, which proves that it is a gas giant, perhaps puffed up even more than normal by the star's energy.

The French Space Agency's Corot Mission, scheduled for launch in 2004, is to examine light curves of stars (that is, graphs of the brightnesses of stars over time) in two colors. One of the two major goals of the spacecraft is to monitor planetary transits. If terrestrial planets in orbits about the size of Earth's are common, it should find a few dozen extrasolar planets around the 25,000 stars that it is to monitor. Any transit should dim the star equally in all wavelengths, so having observations at two colors will be a good check. An Earth-size planet should dim its star by only 0.01 per cent. Note that the transit method is independent of the Doppler method that has been successful so far.

Up to now, no planet has been discovered around a nearby, sunlike star through its transit. But an effect known as "gravitational lensing," which we will discuss later in this book (in Chapter 36), should occasionally cause a star to brighten slightly as it focuses the light from its planet toward us as the planet passes directly behind it. Several surveys are under way to search for such signs of planets; they survey distant clusters of galaxies so that they can observe many, many stars at the same time. A handful of detections of possible planets has been reported.

Techniques known as interferometry, using several widely spaced optical telescopes together, must be developed considerably, perhaps for decades, before we succeed in imaging the planets whose existence we have now inferred from their gravitational effects. Ground-based interferometers are being developed for the twin Keck Telescopes in Hawaii and the four Unit Telescopes of the European Southern Observatory's Very Large Telescope in Chile. Using these large telescopes should improve the sensitivity of the Doppler measurements by a factor of 10 to 100.

NASA's Space Interferometry Mission (SIM) is meant to detect extrasolar planets and is to be launched within a half-dozen years (Figure 18–10). It is to carry several

*A*

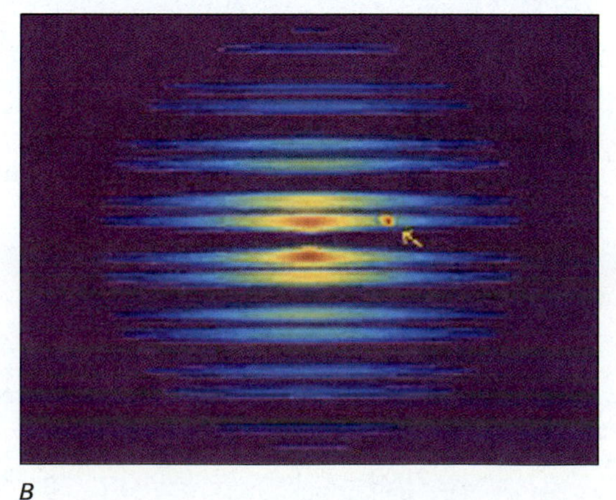

*B*

**FIGURE 18–10** (*A*) An artist's conception of the Space Interferometry Mission (SIM). (*B*) In the technique of nulling, the interferometer makes images in which part is sensitive and part is insensitive. The star itself is always placed on an insensitive part. Then by rotating the interferometer, sometimes a planet is in the sensitive part and shows up and sometimes is in the insensitive part and doesn't. This technique makes the existence of a planet more obvious.

small mirrors separated by 10 meters. In its five-year approved mission, it should be able to find planets as small as 10 Earth masses with periods of revolution less than about five years, which corresponds to semi-major axes of about 3 A.U. If it could last a dozen years, it should be able to detect accurately, for stars within about 30 light-years, planets having Neptune-like masses orbiting 5 A.U. from their parent stars. NASA is pointing to an even more sensitive mission, Terrestrial Planet Finder (TPF), to be able to detect planets of masses more nearly comparable to our own Earth. It will have an interferometer with mirrors separated by 100 m, that is, football-field size. Merged with a European Space Agency project, TPF may be launched in 2012. These interferometers are to work in the infrared at wavelengths of about 10 micrometers. They will use a technique called "nulling," which reduces the sensitivity so much at the position of the star itself that the star's brightness is zeroed out, or "nulled." With no bright star image present, the planet becomes easier to detect.

New planets are being discovered at such a rate that we can best keep track of them on the World Wide Web with the Extrasolar Planets Encyclopedia (http://www.obspm.fr/planets, also available through this book's Web site). After many years of searching for planets and searching for brown dwarfs, we have recently found examples of both. We are in a golden age for the discovery of such sub-stellar objects.

### 18.3c Looking for Life on Extrasolar Planets

What is the chance of finding life forms on these planets? (See also Chapter 20 for the search for life in the Universe.) The planets we find so far seem to be gas giants and wouldn't have solid surfaces, so any life would be very different from what we are used to. Further, the planets in elliptical orbits seem to swing through temperature zones so drastically different that life might not arise. Gravity from these massive planets in elliptical orbit may also eject small, Earth-like planets from the planetary systems.

At first, people pointed out that we were finding these extrasolar planets in sun-like stars, so planets might form easily. It turns out on further investigation that the stars with giant planets seem to have abundances of heavy elements (those heavier than hydrogen and helium) two or three times greater than those abundances in the Sun. Thus for some reason, the interstellar clouds of gas that are enriched in these elements might form giant planets more easily. Since collisions of planetesimals to form planets should occur more often at higher densities, which would result from higher proportions of heavy elements, perhaps the correlation of planets with higher abundances is expected. Still, it would make the fraction of stars with planets smaller than otherwise. We have a lot to learn.

## 18.4 BROWN DWARFS

What is the latest on brown dwarfs, objects that didn't quite become stars but that are too massive to be planets? Theoretical calculations indicate that objects that have less than 80 times the mass of Jupiter cannot be stars. These brown dwarfs all have about the same radius, about that of Jupiter, which is about 1/10 that of the Sun. Since there has been some controversy whether some or all of the objects just discussed were really brown dwarfs—a type of star—instead of planets, let us talk about brown dwarfs here.

For many years, brown dwarfs were a type of joke. Their discovery had been reported a few times and then withdrawn. We joshed that reporters at news conferences at meetings of the American Astronomical Society had a button on their word processors that they could press to give, "Astronomers today announced the discovery of brown dwarfs, a type of almost star that was not quite massive enough to start nuclear fusion."

> Where did the name "brown dwarf" come from? People didn't agree, before they were discovered, how these objects would look. "Brown" is a mixture of many colors, and Jill Tarter (whose subsequent searching for extraterrestrial intelligence we will discuss in Chapter 21) suggested the name to continue the ambiguity.

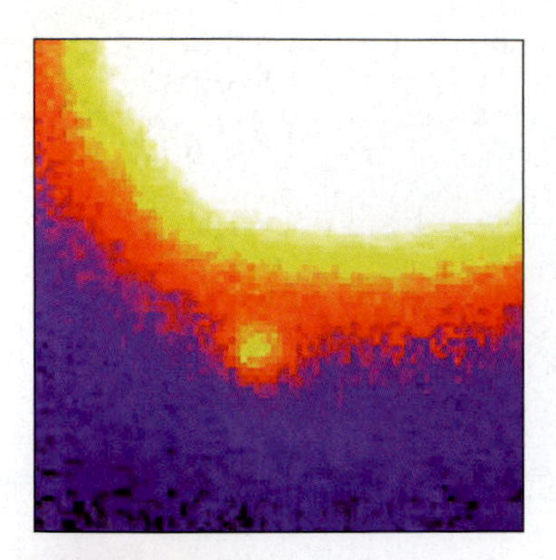

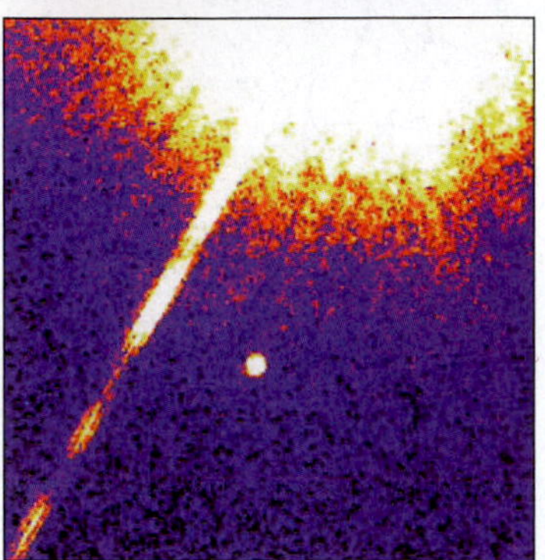

FIGURE 18–11 The first brown dwarf to be unambiguously discovered, Gliese 229B. A ground-based image (*top*) barely shows the brown dwarf orbiting the larger star, because the atmosphere blurs the larger star's image, but the Hubble Space Telescope (*bottom*) shows the brown dwarf clearly. (The straight streak is an artifact in the imaging.)

All objects result from a balance between gravity pushing in and something pushing back. Solid objects have electromagnetic forces between the atoms to prevent them from collapsing. Stars, which are gaseous, are a continual balance between gravity pulling inward and some pressure pushing outward. In normal stars, that pressure comes from energy generated by nuclear fusion, most usually with hydrogen fusing into helium, a process that gives off energy. But it takes a lot of mass to generate enough gravity to start the fusion process by providing a high density and temperature. Only that high density and temperature overcome the electric repulsion of protons for each other, which prevents them from fusing. In brown dwarfs, normal fusion of protons doesn't occur. But it is slightly easier for deuterons, nuclei with one proton and one neutron, to fuse, so deuteron fusion can go on for a while in brown dwarfs. Still, most of their energy comes from their contraction. They are considered not-quite stars.

The controversy comes in the question "what is a massive planet and what is a brown dwarf?" The overlap may come in the mass region about ten times greater than that of Jupiter. If all the supposed planets newly discovered were in that range, then doubters would claim that we haven't discovered planets after all but merely a lot of brown dwarfs. As a type of star, albeit a failed star, a brown dwarf doesn't engender as much excitement as does a planet. But now that we have planets with insufficient mass to become brown dwarfs, the controversy has lessened.

The first object to be generally accepted as a brown dwarf was GC 165B, the faint companion of the white dwarf GC 165. It has only 1/10,000 the luminosity of a star, which puts it on the boundary between stars and brown dwarfs. The star Gliese 229 has a companion, Gliese 229B, with a luminosity only six-millionths that of the Sun and a surface temperature below 1200 K (Fig. 18–11). These conditions are below those of which stars are capable, so we think it is the first unambiguous brown dwarf. Further, the star's infrared spectrum clearly shows methane and water vapor, other signs expected of brown dwarfs since they would be destroyed in the hot atmospheres of ordinary stars. Models indicate that the object has between 30 and 55 times the mass of Jupiter. We will say more about brown dwarfs when we discuss stellar spectra in Chapter 24.

## 18.5 FREE-FLOATING OBJECTS

The question "what is a planet" became even more complicated with the reports of "free-floating planets" observed in images of several nebulae in our galaxy. These objects have 5 to 10 times Jupiter's mass, so are not massive enough to be brown dwarfs. Since they have less than 13 Jupiter masses, they do not fuse deuterium, which makes them closer to planets than to brown dwarfs. But to most people, the word "planet" invokes the idea of a central star that is orbited, so many people call these things "free-floating objects."

After all, these objects are likely to have formed by the collapse of an interstellar cloud of gas and dust, rather than around a planet. Calling them "planets" might imply falsely that they were stripped off parent stars.

The terminology is still in flux. "Grey dwarf" has been suggested.

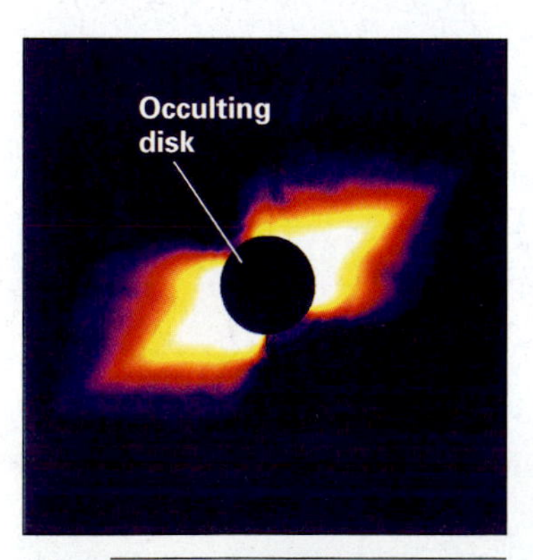


FIGURE 18–12 This picture of what may be another solar system in formation was taken with a ground-based telescope. A circular "occulting disk" blocked out the star $\beta$ (beta) Pictoris; shielding its brightness and observing in the infrared revealed the material that surrounds it.

## 18.6 SOLAR SYSTEMS IN FORMATION

The indications from infrared satellites that solar systems may be forming around stars, as deduced from the excess infrared measured (over that expected for stars of that temperature), are not as satisfactory as direct sightings of planets (which we have not yet made) or detections of their gravitational effects (which we have just discussed). The ground-based observations that we now have of clouds of material orbiting a star (Fig. 18–12) are also more persuasive than mere infrared excesses. The

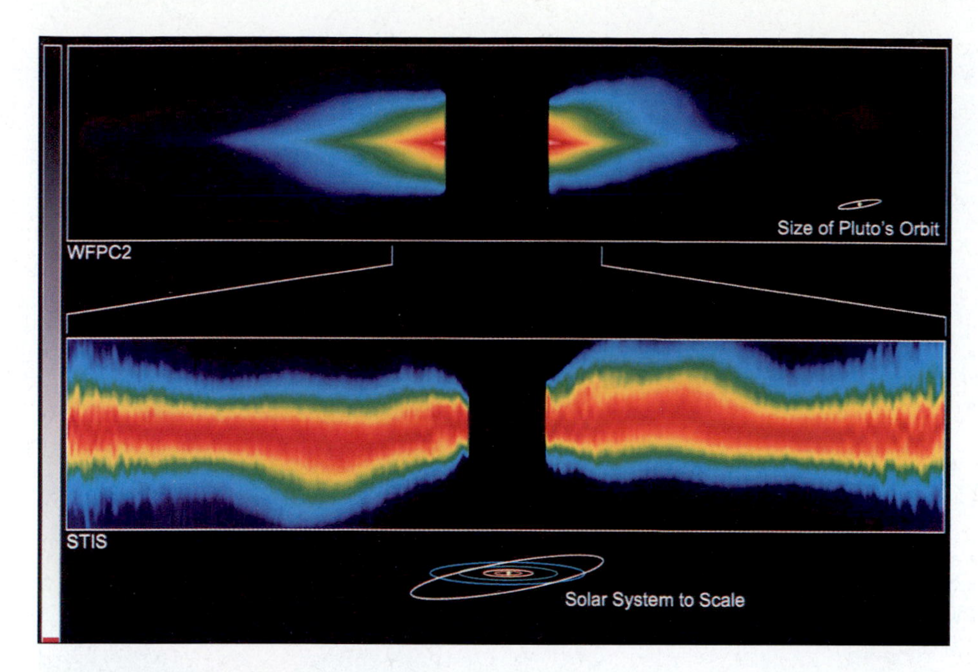


**FIGURE 18–13** The region around the southern star beta Pictoris, with the star itself blocked out in the telescope. This Hubble Space Telescope image with its Space Telescope Imaging Spectrograph clearly shows a side view of a disk of dust, which astronomers think is a planetary system in formation. The dips in the disk may indicate the positions of orbiting planets.

best evidence comes from the star $\beta$ (beta) Pictoris. The material present in a disk around the star apparently consists of ices, carbon-rich substances, and silicates. Our Earth formed from such materials, so planets may be forming at beta Pic. The density of the material indicates that planets may indeed be present. Little material is present in the inside portion of the disk; it may have been swept away by the planets. And its shape, revealed in detail by the Hubble Space Telescope, may indicate a planet within (Fig. 18–13).

Hubble has also revealed small, circular regions that have the characteristics of protoplanetary disks (Fig. 18–14). These "proplyds" appear to be thin disks with central stars and are the sizes expected—about 100 A.U. across. Now that infrared imaging devices are on board, they will be further studied.

The infrared observations of the stars epsilon Eridani, Fomalhaut, Vega, and beta Pictoris are best fit by a model in which the dust clouds around them have central regions in which dust is depleted (Fig. 18–15). Large, perhaps Earth-sized, objects could be sweeping the dust away. These stars have not been under observation by the Marcy/Butler group since A stars like these have too few spectral lines for accurate Doppler measurements.

Observations from the Infrared Space Observatory directly showed the size of Vega's dust disk for the first time (Fig. 18–16). It appears to be about 300 A.U. across. Ground-based infrared images reveal warm clouds of dust and gas surrounding newly formed stars. Optical and infrared studies by a UCLA–U. Arizona–Cerro Tololo–U. Toronto team have found not only a dust distribution surrounding the star HR 4796A but also a hole in that distribution with an inner radius somewhere between the size of Pluto's orbit and a few times larger than that. Further, they find that the dust grains are larger than interstellar grains and thus have grown from their original state, so perhaps still larger objects have also coalesced. They conclude that a likely explanation is the presence of a low-mass object, possibly a planet, at half that radius, whose gravity is sweeping out the hole.

In sum, in addition to seeing extraterrestrial planets directly, we now have several cases in which we seem to be seeing the dusty material of solar systems in for-

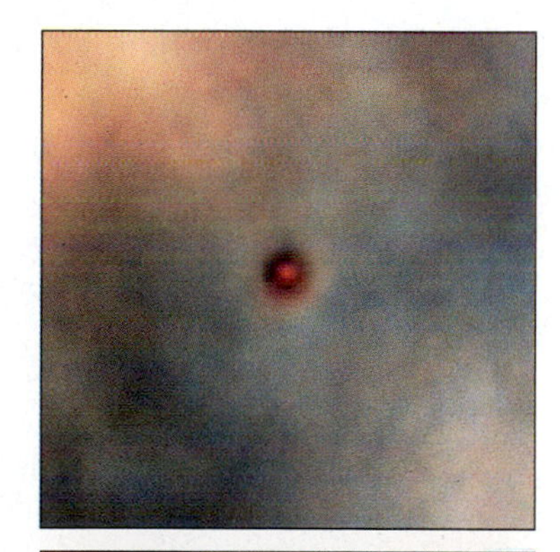

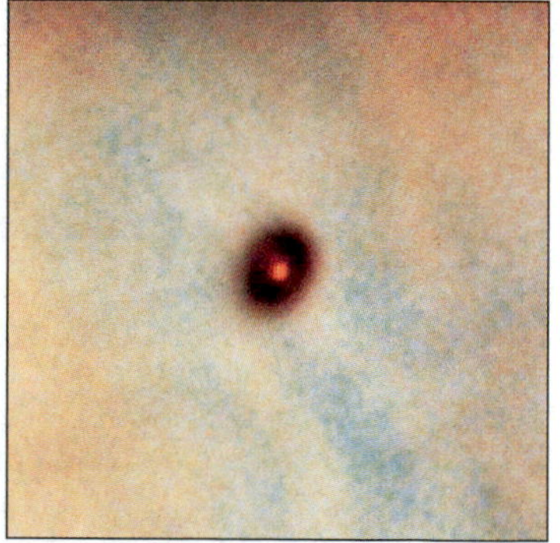

**FIGURE 18–14** The Hubble Space Telescope has revealed protoplanetary disks around several stars.

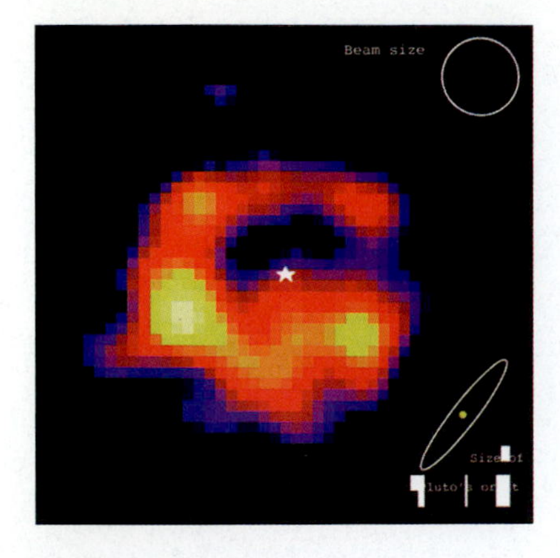


◄ **FIGURE 18–15** A dust ring around the star epsilon Eridani, imaged at 0.85 mm in the submillimeter part of the spectrum from a telescope in Hawaii. The peak of the dust ring is 60 A.U. from the star. The total amount of mass in it is about the same as that in comets orbiting in our Solar System. The ring is about the same size as and may be analogous to the Kuiper belt (see Chapter 17) around our own Sun, with the inner region cleared by planetesimals. The bright spot at the left in the ring may be dust trapped or otherwise affected by a planet there.

► **FIGURE 18–16** The disk of Vega as resolved by the Infrared Space Observatory (ISO), the European infrared observatory in orbit in the mid-1990s. The disk was imaged at the relatively long infrared wavelengths of 60 and 90 micrometers; we see the former.

mation and even some smaller nebulae out of which individual stars are forming. We are newly able to follow all the stages of solar-system formation, and look forward to applying this knowledge to the birth of our own Solar System.

## CORRECTING MISCONCEPTIONS

**✖ *Incorrect***

The planets discovered around other stars can be photographed.

Other solar systems also have small, rocky planets close in and Jupiter-like planets in Jupiter-like orbits.

**✔ *Correct***

It will be decades before we can image these planets.

Other solar systems often have Jupiter-like planets very close in.

## SUMMARY AND OUTLINE

Astrometric searches (Section 18.1)
- Search for wobbles in proper motions has been ambiguous.
- Barnard's star planetary report not correct

Pulsar planets (Section 18.2)
- Planets were found around a pulsar from variations in pulse arrival time.

Extrasolar planets around normal stars (Section 18.3)
- Planets have been found around several nearby sunlike stars from Doppler effects.
- We are now sensitive only to massive planets.
- We are detecting some planets in 3-to-5-day orbits and others in larger orbits that are very elliptical.

Brown dwarfs (Section 18.4)
- Need good definition that separates brown dwarfs and planets
- Free-floating objects are found (Section 18.5)

Solar systems in formation (Section 18.6)
- Beta Pictoris shows a disk of dust.
- Excess infrared radiation may reveal a planet.

## KEY WORDS

extrasolar planets
exoplanets
pulsar

## QUESTIONS

1. Discuss the evidence for and against the idea that Barnard's star has planets around it.
2. How can the Doppler effect reveal the presence of a planet?
3. Compare the effect by which pulsar planets were discovered with the effect by which planets around sunlike stars have been discovered. What is the underlying similarity?
4. Relate the Doppler-effect discoveries of planets around other stars to Kepler's and Newton's forms of Kepler's third law. Why wouldn't Kepler's form have been sufficient? Why, unlike Newton's form for planets orbiting the Sun, can we get an indication of the mass of the orbiting rather than only the central object? (Look back at Newton's form of Kepler's third law to answer this last question.)
5. Why don't the Doppler discoveries give a planet's mass exactly?
6. Does a star with a large proper motion necessarily have planets? Explain.
7. What is one advantage of the radial-velocity method over the proper-motion method for detecting distant planets?
8. What is the difference between a massive planet and a brown dwarf?
9. How might the idea that giant planets are formed at Jupiter-like distances from their stars be saved, even in the face of observations of "hot Jupiters" close in?
10. What is the evidence that the beta Pictoris system is a solar system in formation?

## USING TECHNOLOGY

### W World Wide Web

1. Check on the latest extrasolar planets and brown dwarfs at the Extrasolar Planets Encyclopedia site, http://www.obspm.fr/planets.
2. Check on extrasolar planets at Geoff Marcy's site, http://www.exoplanets.com, and Michel Mayor's site, http://obswww.unige.ch/~udry/planet.
3. See what the VLT and the Hubble Space Telescope have found about brown dwarfs.

### REDSHIFT

1. Is $\beta$ Pictoris ever visible from your location? Explore the effect of latitude.
2. How long is the newly discovered planet in Ursa Major visible tonight?
3. Find the locations in the sky, and rising and setting times, for a half dozen of the stars around which planets are known or suspected.

Comet Hale-Bopp in the night sky.

# Chapter 19

# Comets

Besides the planets and their moons, many other objects are in the family of the Sun. The most spectacular, as seen from Earth, are comets. Bright comets have been noted throughout history, instilling in observers great awe of the heavens.

It has been realized since the time of Tycho Brahe, who studied the comet of 1577, that comets are not in the Earth's atmosphere. From the fact that the comet did not show a parallax when observed from different locations on Earth, Tycho deduced that the comet was at least three times farther away from the Earth than the Moon.

Astronomers' current interest in comets comes because we think comets are made of material that has been safely tucked away in the deep freeze of the outer Solar System for billions of years. Thus by studying comets, we are learning about pristine material left over from the origin of our Solar System. We hope to use what we learn to understand our origins.

**AIMS:** To discuss how comets bring us information about the origin of the Solar System, and what recent studies of Halley's Comet and other comets have taught us

## 19.1 OBSERVING COMETS

Every few years, a bright **comet** is visible in our sky. From a small, bright area called the **head,** a **tail** may extend gracefully over one-sixth (30°) or more of the sky (Fig. 19–1). The tail of a comet is always directed away from the Sun.

In the last few years, several bright comets have appeared in our sky, bringing pleasure to people around the world. In 1996, comet C/1996 B2 (Hyakutake) was bright enough that all you had to do was step outside and look up, even in or near a city. In 1997, comet C/1995 O1 (Hale-Bopp) was even brighter (Fig. 19–2). Both provided opportunities for professional astronomers to use the wide range of equipment now available to learn in detail about the working of comets. Hyakutake was even observed by a spacecraft when it was too close to the Sun to be seen from Earth (Fig. 19–3).

Although a comet's tail may give an impression of motion because it extends out only to one side, the comet does not move noticeably across the sky as we watch. With binoculars or a telescope, however, an observer can accurately note the position of the head with respect to nearby stars and detect that the comet is moving at a slightly different rate from the stars as comet and stars rise and set together. Within days or weeks a bright comet will have faded to below naked-eye brightness, though it can be followed for additional weeks with binoculars and then for additional months with telescopes.

**FIGURE 19–1** Comet Hale-Bopp was so bright that it was even visible over Central Park in the middle of New York City.

Comets have long been seen as omens:

"When beggars die, there are no comets seen;

The heavens themselves blaze forth the death of princes."

—Shakespeare, *Julius Caesar*

The "long hair" that is the tail led to the name "comet," which comes from the Greek for "long-haired star," *aster kometes*.

**FIGURE 19–2** Comet C/1995 O1 (Hale-Bopp) was the brightest comet in decades at its peak in March and April 1997. Its 40-km-across nucleus gave off large quantities of gas and dust that made its long tails. The broad, yellowish dust tail curved behind the comet's head. We saw it by the sunlight that it reflected toward us. The much fainter, blue tail is made of ions given off by the comet's nucleus and is blown in the direction opposite to the Sun by the solar wind. We detected it by its own emission, though it was much less visible to the eye than it was to the camera. Hale-Bopp was so bright that scientists discovered a third type of tail: a faint one from neutral sodium.

Most comets are much fainter than the ones we have just described. About a dozen new comets are discovered each year, and most become known only to astronomers. Many of these are making their first passage near the Earth and Sun in recorded history. An additional few are "rediscovered" each year—that is, from the orbits derived from past occurrences, it can be predicted when and approximately

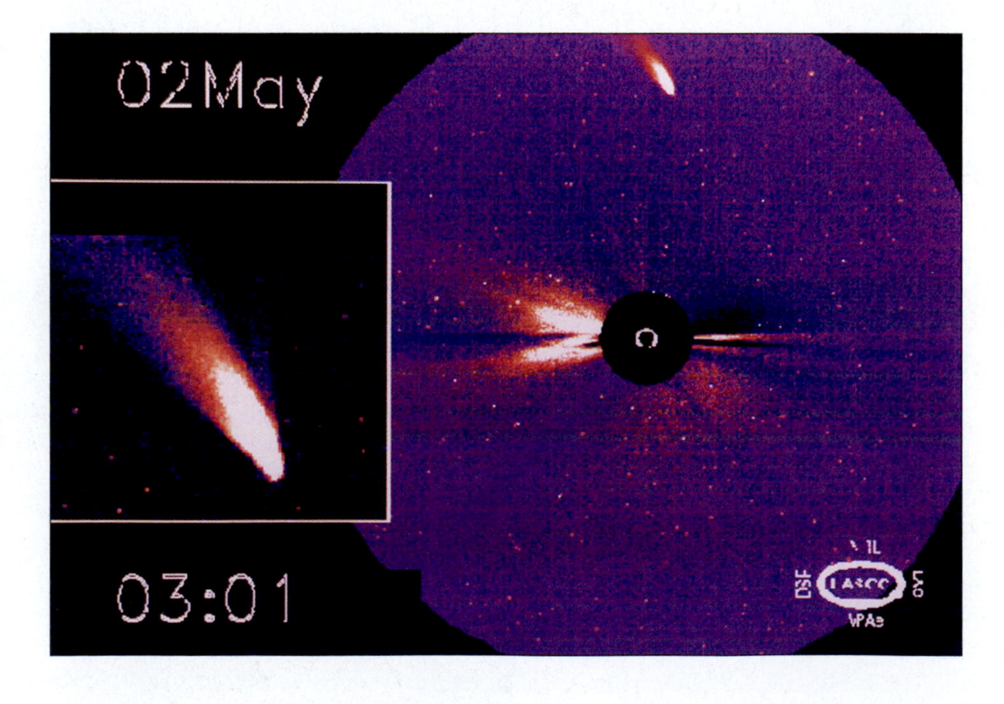


**FIGURE 19–3** Comet 1996 B2 (Hyakutake) from the Solar and Heliospheric Observatory (SOHO), when it was at perihelion and could not be seen from Earth. A telescope in orbit made this image with a coronagraph, an instrument that masks the Sun (whose outline is drawn in at center). The comet appears at the top.

where in the sky a comet will again become visible. Halley's Comet (Section 19.4) belongs to the latter group. Up to the present time, about 1000 comets have been discovered.

## 19.2 THE COMPOSITION OF COMETS

At the center of a comet's head is its **nucleus,** which is a few kilometers across. The most widely accepted theory of the composition of comets, advanced in 1950 by Fred L. Whipple of what is now the Harvard-Smithsonian Center for Astrophysics, is that the nucleus is like a **dirty snowball** a few kilometers across. The nucleus is apparently made of ices of such molecules as water ($H_2O$), carbon dioxide ($CO_2$), ammonia ($NH_3$), and methane ($CH_4$), with dust mixed in.

This dirty-snowball model explains many observed features of comets, including why the orbits of comets do not appear to follow the laws of gravity accurately. When sunlight evaporates the ices, molecules are expelled from the nucleus. This action generates an equal and opposite reaction, like a jet engine. Since the comet nucleus is rotating, the force is not always directly away from the Sun, even though the evaporation is triggered on the sunny side. So comets show the effects of "non-gravitational forces" in addition to the effect of solar gravity.

The nucleus itself is so small that we cannot observe it directly from Earth, but we have determined the nucleus's size with radar observations. The rest of the head is the **coma** (pronounced coh'ma), which may grow to be over a million km across. The coma shines partly because its gas and dust are reflecting sunlight toward us and partly because gases liberated from the nucleus are excited enough by sunlight that they radiate. (We discussed such emission spectra in Section 4.7.)

The Hubble Space Telescope has viewed the inner parts of the heads of recent nearby comets (though not all the way to the nuclei), shown us detail, and enabled us to study their rotation (Fig. 19–4). With rockets, we became able to observe from above the Earth's atmosphere, which blocks ultraviolet light. As a result, we could detect in comets the Lyman-alpha emission line of hydrogen, which is in the ultraviolet. This emission from hydrogen gas shows that a huge hydrogen cloud a million km in diameter (Fig. 19–5) surrounds a comet's head. The hydrogen cloud probably results from the breakup of water molecules by ultraviolet light from the Sun. We have long been able to detect spectral lines from simple molecules in the visible, and rockets and spacecraft enable us to take spectra in the ultraviolet (Fig. 19–6) that confirm that comets contain lots of water and carbon dioxide. Comet C/1995 01

*A*

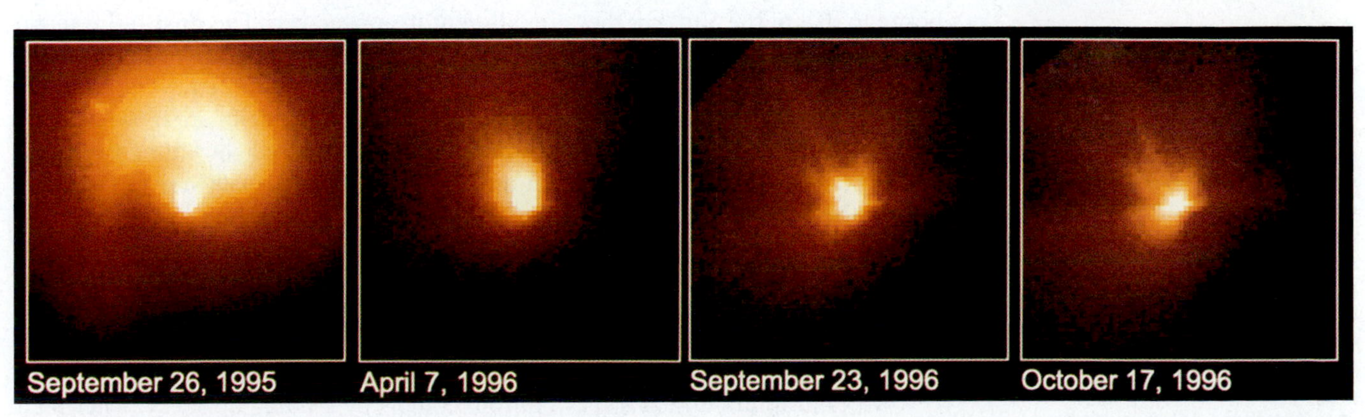


*B*

**FIGURE 19–4** The heads of (*A*) C/1996 B2 (Hyakutake) and of (*B*) C/1995 O1 (Hale-Bopp), imaged with the Hubble Space Telescope.

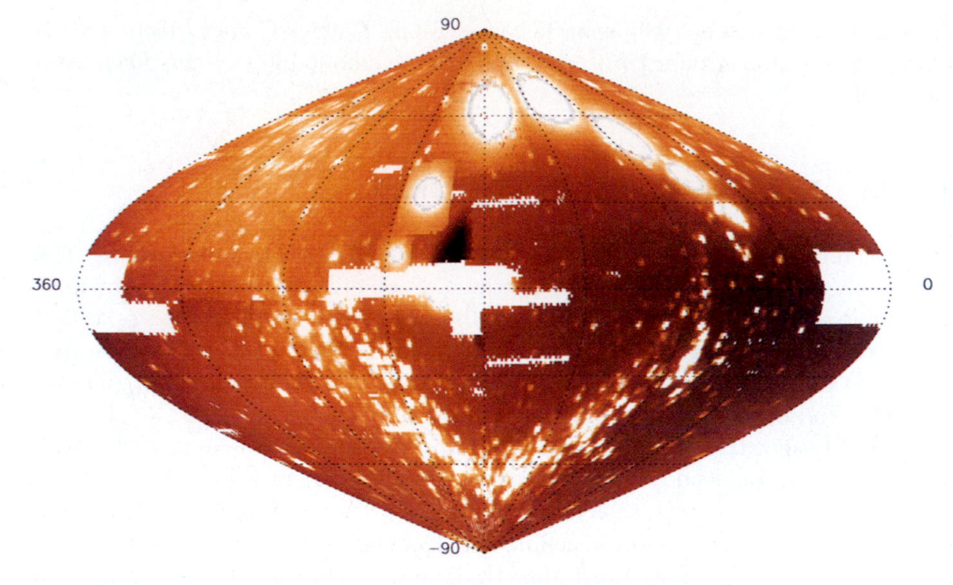


**FIGURE 19–5** The oval blotches near the top show the huge hydrogen cloud around Comet C/1996 B2 (Hyakutake). The images were taken in the Lyman-alpha line of hydrogen at 1216 Å in the ultraviolet. The whole-sky map shown here (unrolling the sky the way a map of the whole Earth enrolls the spherical Earth) was taken with the SWAN instrument on the Solar and Heliospheric Observatory (SOHO) in space. The ecliptic—the path of the Sun—is along the equator of the map. When the comet was only 0.1 A.U. from the Earth, the hydrogen cloud subtended 60°. That's 1/3 of the way across the sky from horizon to horizon! The oblong images are the comet's hydrogen cloud on different days; the white rectangular patches along the centerline are regions where no data are available because the Sun, the opposite direction to the Sun, and the spacecraft itself are at those locations.

(Hale-Bopp) was so bright that many molecules were detected from their spectral lines in the radio part of the spectrum. It turned out that the rate at which many trace molecules were emitted differed from the rate at which water vapor was given off, so these molecules had not been incorporated in the water ice as had been thought.

A comet's tail can extend over 1 A.U. (150 million km), so comets can be the largest objects in the Solar System. The tail of Comet Hyakutake was discovered to have affected a spacecraft 550 million kilometers away, meaning that the tail must be at least that long. But the amount of matter in the tail is very small—the tail is a much better vacuum than we can make in laboratories on Earth.

Many comets actually have two tails. Both extend generally in the direction opposite to that of the Sun, but they are different in appearance. The **dust tail** is made of dust particles released from the ices of the nucleus when they are vaporized. The dust particles are left behind in the comet's orbit. In addition to following their orbital motion around the Sun, following Kepler's laws, the particles are blown slightly away from the Sun by the pressure caused by photons of sunlight hitting the particles. As a result of the comet's orbital motion, the dust tail usually curves smoothly behind the comet.

The **gas tail** (also called the **ion tail**) is composed of ions blown out more or less straight behind the comet by the solar wind. As puffs of ionized gas are blown out and as the solar wind varies, the ion tail takes on a structured appearance. Each puff of matter can be seen. Magnetic fields in the solar wind carry only ionized matter along with them; the neutral atoms are left behind in the coma.

We can capture dust that may be from comets by sending up sticky plates on a NASA U-2 aircraft sent high above terrestrial pollution. The abundances of the elements in the dust are similar to those of meteorites rather than to terrestrial material. The particles may date back to the origin of the Solar System.

*A*

O C

*B*

**FIGURE 19–6** (*A*) Comet 1P/Halley in 1986. Its straight, narrow bluish gas tail extends diagonally up to the right at the right side of the visible object. Its gently curving dust tail extends upward toward its left. The color difference does not show well on this image. (*B*) The ultraviolet spectrum of Halley's Comet, showing strong spectral lines of oxygen at 1304 Å, and carbon at 1561 Å and 1657 Å. The oxygen comes from the dissociation of water. These observations have shown that the carbon most likely comes from the dissociation of carbon monoxide.

A comet—head and tail together—contains less than a billionth of the mass of the Earth. It has been said that a comet is as close as something can come to being nothing.

## 19.3 THE ORIGIN AND EVOLUTION OF COMETS

### 19.3a The Oort Cloud and the Kuiper Belt

It is now generally accepted that trillions of comet cores surround the Solar System in a sphere perhaps 50,000 A.U. (almost 1 light-year) in radius, a thousand times as far as Pluto's orbit. This sphere is known as the **Oort Comet Cloud** after Jan H. Oort, the Dutch astronomer who advanced the theory in 1950. The total mass of matter in the cloud is perhaps three times that of Earth. Occasionally some of the comets' cores leave the Oort Cloud, probably because the gravity of a star that penetrates the cloud tugs them out of place. That kind of event might occur a few times every million years, according to Hipparcos data. Some of these comets are ejected from the Solar System, while others come closer to the Sun, approaching it in long ellipses. The comets' orbits may be altered if they pass near a jovian planet. Comets are not limited to the plane of the ecliptic and come in randomly from all angles; from that observation, it was deduced that the Oort Cloud is almost spherical. The Oort Cloud is huge; an Oort Cloud the size of ours around $\alpha$ (alpha) Centauri, the nearest star to the Sun, would span 20° of sky when viewed from Earth, about the size of the halo seen around the Moon on some nights.

The objects in the Oort Cloud were once planetesimals as far from the Sun as the region in which Jupiter, Saturn, Uranus, and Neptune now orbit, according to current theory. Through gravitational interactions, the giant planets moved outward by about 30 per cent from where they were formed, and the much smaller objects were sent much farther out to form the Oort Cloud. New calculations indicate that collisions among these objects pulverized them. This process reduced the mass in the Oort Cloud compared with the mass we previously thought was there. The latest value for the Oort Cloud, as of 2001, is that it contains 3.5 solar masses of material.

Though most long-period comets come from the Oort Cloud, calculations indicate that short-period comets (those with periods less than 200 years) come directly from a reservoir just beyond the orbit of Neptune, known as the **Kuiper belt.** One piece of evidence for the Kuiper belt is that these short-period comets tend to be in the plane of the Solar System. Some of these Kuiper belt objects can be thrown by the planets into highly eccentric orbits that extend out 50,000 A.U., into the Oort Cloud. By Kepler's second law, these objects spend much more of their time in the Oort Cloud than in the inner parts of their orbits.

David Jewitt of the University of Hawaii and Jane Luu, now of the Leiden Observatory, have led the way in discovering Kuiper belt objects since they found the first in 1992. The sightings have turned these objects into observational facts, after many years of being merely theoretical constructs (Fig. 19–7). From the more than 300 found thus far, they project that there may be 140,000 such objects over 100 km in diameter outside Neptune's orbit, between 30 and 50 A.U. A separate class of objects in the outer Solar System, descended from and distinguished from the Kuiper belt objects by their planet-crossing orbits, are known as Centaurs. We shall discuss them in Section 19.5, along with the first such object to be discovered. We have already discussed the relation of Pluto to the Kuiper belt in Section 17.5.

SIRTF, the *S*pace *I*nfrared *T*elescope *F*acility, will be able to detect the energy emitted (thermally) by Kuiper belt objects instead of merely the sunlight they reflect. The measurements will lead to accurate sizes.

Since comets come close to us, bringing icy material from far out in the Solar System, we are eager to study them in detail. Since this material has largely been

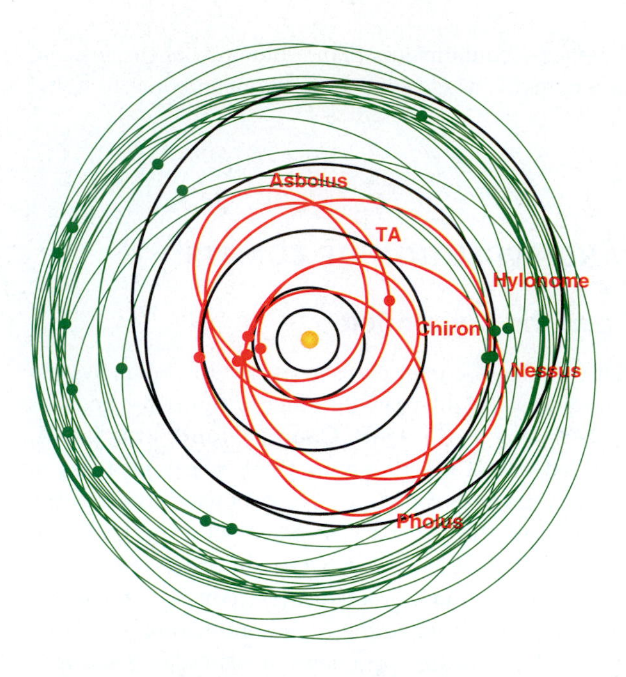


**FIGURE 19–7** Orbits of the outer planets (*black*), several Kuiper belt objects (*green*), and six of the Centaurs (*red*). The dots indicate the current positions of the objects.

out in the deep freeze of space in the outskirts of our Solar System, we hope that studying it will give us information about what the early years of our Solar System were like.

But not all comets may be suitable for such revelations. Alan Stern of the Southwest Research Institute, Boulder, has deduced on the basis of numerical simulations that most of the comets with periods of less than 200 years were created fairly recently in the Kuiper belt from collisions between larger, icy objects. These short-period comets are, therefore, not as ancient as had been thought. Their surfaces may have been melted or gained additional material in the collisions, so we cannot assume that their surfaces are pristine and so reveal the earliest Solar System. Further, calculations show that, given the sparse amount of material available at their locations, the observed Kuiper belt objects could not have grown to be as they are in the age of the Solar System. From his calculations of how the collisions led to the distribution of short-period comets that now exist, Stern concludes that the Kuiper belt was once 100 times more massive than it is now.

Further, Stern and a group of colleagues have detected argon in the tail of Hale-Bopp. Argon evaporates at a temperature of 40 K, so he deduces that Hale-Bopp never spent much time in a region warmer than that. Earlier, other astronomers had found that neon was absent. This fact means that the temperature was more than 25 K, the point at which neon evaporates. So Hale-Bopp must have formed in a region where the temperature was between 25 K and 40 K, matching the region between the orbits of Uranus and Neptune.

## 19.3b Comet Chemistry

As a comet gets close enough to the Sun, the solar radiation begins to vaporize the icy nucleus. Over the last decade or so, we have finally been able to detect the molecules in the nucleus—the **parent molecules**—directly, through observations with radio telescopes. They include water ($H_2O$), carbon dioxide ($CO_2$), carbon monoxide (CO), hydrogen cyanide (HCN), formaldehyde ($H_2CO$), methanol ($CH_3OH$, also known as wood alcohol), and isocyanoacetylene ($HC_2CN$). Further, the distribution of the isotopes of hydrogen, carbon, oxygen, and nitrogen in them have been studied, and may contain clues to the origin of the comets. Until recently we had mostly detected in the head only the simpler molecules into which the parent molecules break down—the **daughter molecules,** such as H, OH, and O, which might be re-

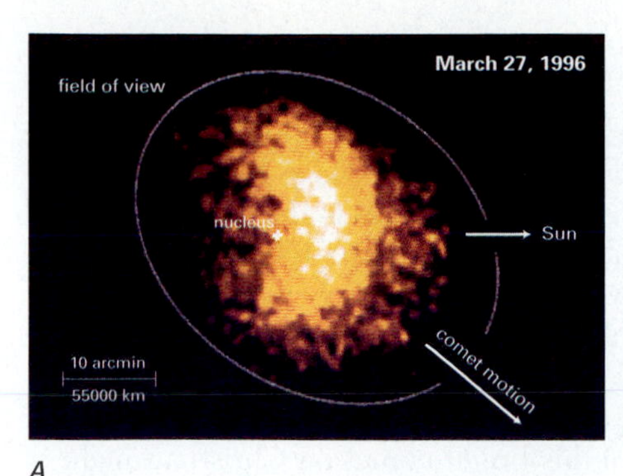


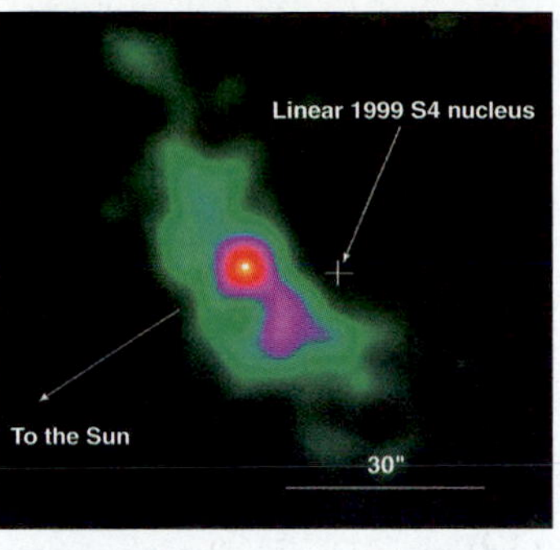


*A* *B*

**FIGURE 19–8** (*A*) C/1996 B2 (Hyakutake), seen as a surprisingly strong source in x-rays and, as here, extreme ultraviolet (XUV) radiation from a spacecraft. (*B*) X-rays from comet LINEAR seen from the Chandra X-ray Observatory.

sults of the breakdown of the parent $H_2O$ (water). Similarly, daughters NH and $NH_2$ can result from the parent $NH_3$. The close passages of Comets Hyakutake and Hale-Bopp have given us the opportunity of observing additional molecules. The concentrations of methane and ethane in the former were much higher than expected, leading to the idea that this comet may be of a different type from 1P/Halley.

One surprise from Hyakutake was the brightness of x-rays from it. When the Roentgen Satellite (ROSAT, named after the discoverer of x-rays) was turned to it, scientists thought that a long exposure might just eke out a detection. But the source was strong in both x-rays and the neighboring extreme ultraviolet region of the spectrum (Fig. 19–8). Where did so much energy come from? The answer was found when the Chandra X-ray Observatory observed Comet LINEAR (C/1999 S4). (Its strange name is an acronym for the search instrument at MIT's Lincoln Labs: Lincoln Near Earth Asteroid Research.) When ionized oxygen and nitrogen in the solar wind capture electrons, the electrons are at a high energy level. They emit x-rays as they drop to lower energy levels. (This Comet LINEAR itself was an interesting story: its nucleus broke apart into bits over a period of hours in August 2000, and the bits then disappeared.)

Continuing our story of a comet approaching the Sun, the tail forms, and it grows longer as more of the nucleus is vaporized. Even though the tail can be millions of kilometers long, it is still so tenuous that only 1/500 of the mass of the nucleus may be lost. Thus a comet may last for many passages around the Sun. But some comets hit the Sun and are destroyed (Fig. 19–9).

The comet is brightest and its tail is generally longest (in millions of kilometers) at perihelion. However, because of the angle at which we view the tail from the Earth, it may not appear the longest (in degrees across the sky) at this stage. Following perihelion, as the comet recedes from the Sun, its tail fades; the head and nucleus receive less solar energy and fade as well. The comet may be lost until its next return, which could be as short as 3.3 years (P2, Encke's Comet) or as long as 100,000 years or more. With each reappearance, a comet loses a little mass and eventually disappears. We shall see in Section 20.1 that some of the meteoroids are left in its orbit. Some of the asteroids, particularly those that cross the Earth's orbit, may be dead comet nuclei.

Both theoretical and laboratory studies in recent years have given us an improved understanding of the chemistry of the ice and dust floating between the stars. Some of the material that we now find in comets may have originated even before the Solar System was formed. In some cases, icy coats form on mineral grains, and the strong ultraviolet light from the Sun can transform the ice into substances including carbon compounds. After these compounds were brought to Earth's surface in collisions with comets, they may have played a role in the origin of life on Earth.

**FIGURE 19–9** The outer solar corona extends outward near the solar equator in this view from the Solar and Heliospheric Observatory (SOHO) spacecraft. A comet plunging into the Sun is also seen. A disk that blocks the bright everyday Sun and the finger that holds it in place appear in silhouette; the size of the solar photosphere is drawn on. In the background, we see stars and the Milky Way. SOHO is discovering one comet a month plunging into the Sun, over 200 so far. All these sungrazing comets belong to a single "family" in a common orbit. They are from a single comet that broke up.

FIGURE 19–10 Edmond Halley.

## 19.4 HALLEY'S COMET

In 1705, the English astronomer Edmond Halley (Fig. 19–10) applied the new theory of gravity developed by his friend and senior colleague Isaac Newton to determine the orbits of comets from observations of their positions in the sky. He reported that the orbits of the bright comets that had appeared in 1531, 1607, and 1682 were about the same. He was troubled, though, that the intervals between appearances were not quite equal. When he resolved this difficulty by analyzing the effect on the comet's orbit by the gravity of Jupiter and Saturn, Halley suggested that all the appearances corresponded to a single comet orbiting the Sun. He predicted that it would return in 1758. The reappearance of this bright comet on Christmas night of that year, 16 years after Halley's death, was the proof of Halley's hypothesis; the comet has since been known as Halley's Comet. It seems probable that the bright comets reported every 74 to 79 years since 240 B.C. were earlier appearances. The fact that Halley's Comet has been observed dozens of times (Fig. 19–11) endorses the calculations that show that less than 1 per cent of a cometary nucleus's mass is lost at each perihelion passage.

Halley's Comet has a long, elliptical, retrograde orbit that extends far out of the ecliptic plane (Fig. 19–12). But where did it come from originally, before gravity of the giant planets captured it into its current orbit? Harold Levison from Southwest Research Institute, Boulder, and colleagues calculated the motions of 27,700 simulated comets, taking account of the gravity of the Sun and the four giant planets as well as of the rest of the Milky Way Galaxy and the stars that are predicted to come into our neighborhood during the next billion years. They deduced that Halley's Comet (and presumably a bunch of similar comets) was originally in a flattened, inner part of the Oort cloud, about 20,000 A.U. out. If the inner part of the Oort cloud is in fact flattened, that would affect the distribution of comets that could one day crash into the Earth as the result of a passing star.

### 19.4a Halley's Comet Comes

At each of its apparitions, the world waits eagerly for Halley's Comet to return (Fig. 19–13). It was picked up three years in advance of its closest approach, the 1986 date that was widely known as the date of its return, by astronomers using a

FIGURE 19–11 The Bayeux Tapestry (which hangs in Bayeux, France) showed people looking up at the comet before King Harold was defeated by William the Conqueror in A.D. 1066. *(By special permission of the City of Bayeux.)*

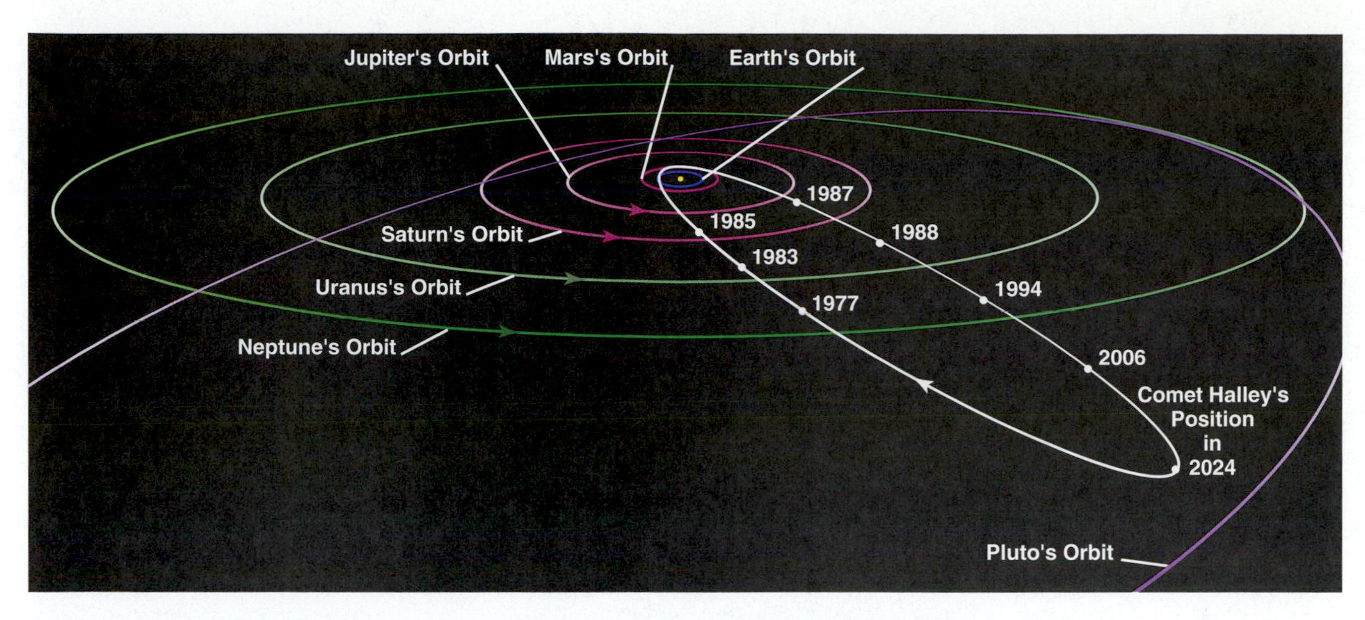


**FIGURE 19–12** Halley's Comet moves on an elliptical orbit in the opposite direction from the revolution of the planets.

sensitive CCD. It appeared then merely as a fuzzy spot in the sky. It was then followed with large telescopes to see how it changed. In particular, discovering how far from the Sun it was when it became active helped tell what its surface is made from, since the different ices that were thought to be on the surface sublime (change from solid to gas form) at different temperatures.

Astronomers know the orbit of Halley's Comet very well, but predicting its brightness is more difficult, since its brightness depends not only on how far the comet is from the Sun and Earth but also on how fast it gives off gas and dust. Still, it was certainly known that Halley's Comet would never become very bright on this apparition because of the Earth's position when the comet passed closest to the Sun—perihelion—on February 9, 1986. The Earth would then be on the opposite side of the Sun. The best chance to see the comet from Earth seemed to be two months later, when the comet passed closest to Earth and we could see it broadside. Nevertheless, the predictions showed that it would be the faintest as seen from Earth in any appearance in the last 2000 years.

Further, during the post-perihelion period the comet would be far south of the celestial equator. The curve of the Earth would prevent people at temperate northern latitudes from seeing the comet well.

There is a lore about Halley's Comet, and, even with the warnings, astronomers and the public alike wanted to see it. Observations started early, years before the tail was long. The comet came sufficiently close that radio techniques could be used to observe it. Optical observatories were also in full swing on the project (Fig. 19–14).

When Halley's Comet emerged from perihelion, it had indeed grown a beautiful tail. Alas, the tail was not very bright. It could be seen, barely, with binoculars; at its peak in March it could also be seen with the naked eye, though it was never spectacular. From a professional point of view, the observations were a great success, since it was a relatively bright object for large telescopes, and high-quality images and spectra could be taken. The comet was so well watched for such a long time that a series of "tail-disconnection events" were discovered to happen very regularly (Fig. 19–15). All or part of the tail would come loose, undoubtedly when the solar wind changed its magnetic polarity (that is, the direction toward north and south magnetic poles switched). It would then drift behind the comet, and the comet would grow a new tail. We shall continue with this aspect of the story in Section 19.4c.

**FIGURE 19–13** A cartoon about awaiting the 1910 apparition of Halley's Comet, William Heath Robinson's "Searching for Halley's Comet at Greenwich Observatory."

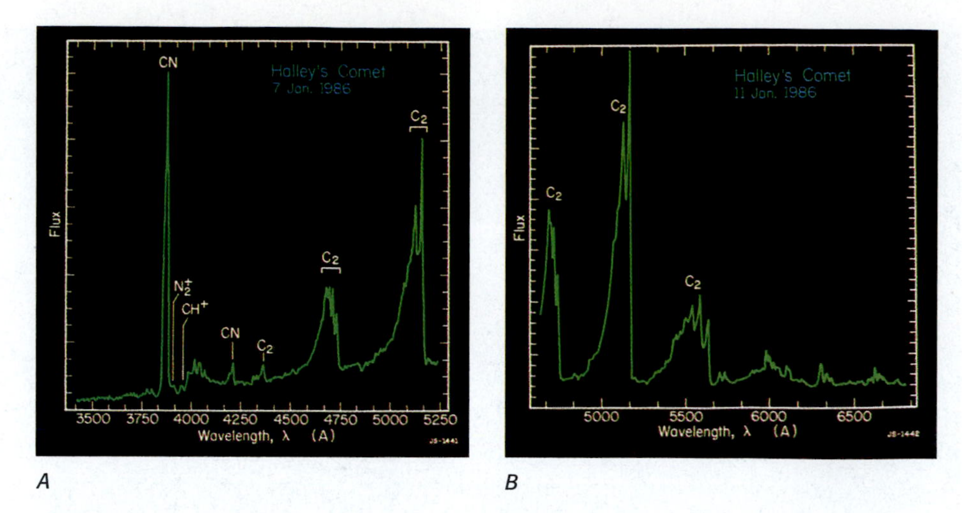


**FIGURE 19–14** Spectra of Halley's Comet in January 1986.

### 19.4b Halley Close Up

A month after perihelion, an international armada of spacecraft reached Halley's vicinity, from the U.S.S.R., Japan, and the European Space Agency. Since it is expensive in terms of fuel to go out of the Earth's orbital plane, all the spacecraft intersected the comet when Halley passed through the Earth's orbital plane.

The Soviet Vega 1 (*Ve*nus-*Ga*lley, where Galley is the Russian word for Halley) spacecraft arrived first. It passed, eventually, within 9000 km from Halley's nucleus. It had no high-resolution imaging device, so the images were difficult to interpret, but the nucleus was surely pinpointed. Other devices on board included dust detectors, charged particle detectors, a spectrometer, and a magnetometer. It was immediately obvious that there were more small dust particles than expected.

Two days later, the Japanese Suisei (meaning "Comet") spacecraft flew 151,000 km from the comet. Even at that relatively great distance, it encountered dust from

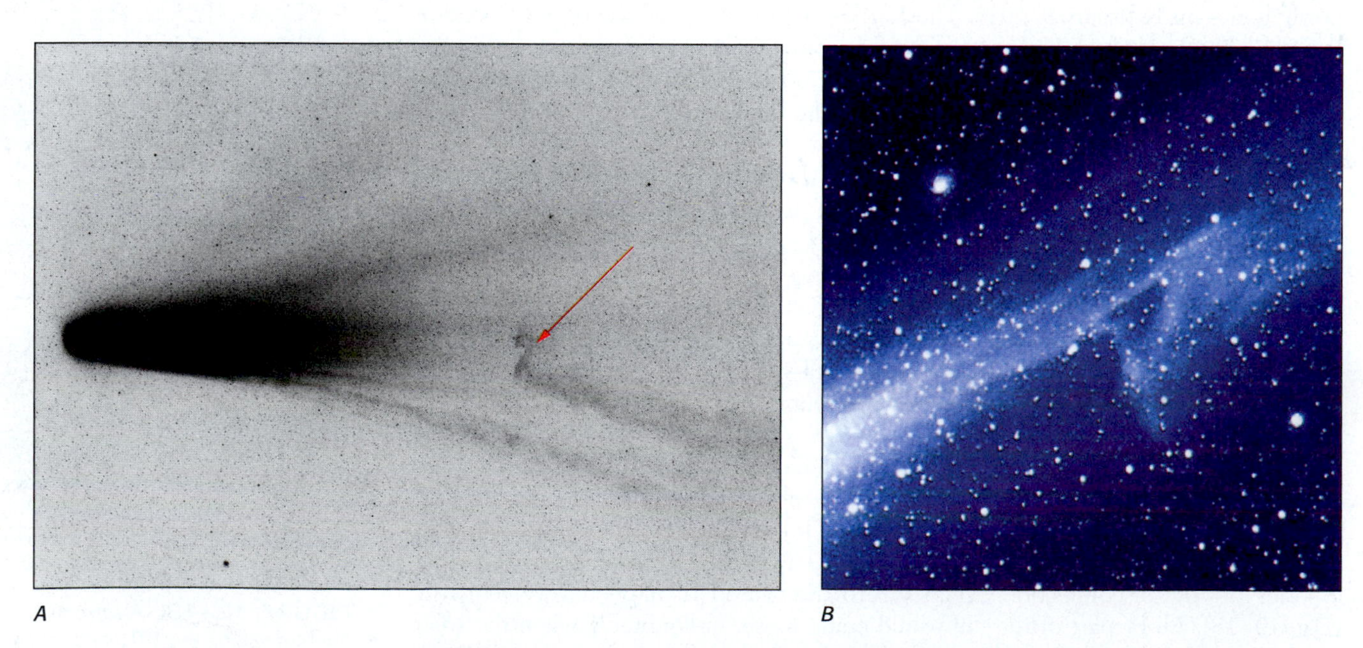


**FIGURE 19–15** (*A*) Halley's Comet in March 1986, showing a tail disconnection event. (*B*) A tail disconnection event on Comet C/1996 B2 (Hyakutake).

Halley. It found the hydrogen cloud around the comet changing in brightness from day to day. From its measurements, we could estimate the total amount of water evaporated from Halley's nucleus.

The next day, Vega 2 flew 8000 km from Halley's nucleus with its battery of instruments (Fig. 19–16). It was apparently facing a less active side of Halley, since it encountered only one-third the dust that Vega 1 had. A world network of radio telescopes had tracked the spacecraft carefully. This information was then hurriedly used to redirect the European Space Agency's Giotto mission so that it could pass about 600 km from the nucleus. Formerly, the position of the nucleus had not been known to that accuracy.

Two days later, the Japanese Sakigake (meaning "Forerunner") spacecraft flew 7 million km from Halley's nucleus, sampling the solar wind there.

After two more days, the European Space Agency's Giotto spacecraft flew close to Halley. Giotto got its name from the Italian painter who had included an image of Halley's Comet, which had been a bright object in the sky in A.D. 1301, as the star of Bethlehem in a fresco he did soon thereafter. It had always been known that the spacecraft would be travelling so fast relative to the comet that an impact by a dust particle could destroy the spacecraft or knock it askew. Indeed, because of that possibility, the data were sent back to Earth in "real time" rather than recorded. Excitement mounted as Giotto came closer and closer to the nucleus, sending back photos and other data all the while. The spacecraft was 8 light minutes from Earth at the time, too far for controllers on Earth to receive signals about problems and to send back corrections, so the spacecraft had automatic systems to follow the brightest part of the image. This brightest part turned out to be the jets rather than the nucleus, so within 14,400 km the nucleus moved out of the field of view. The best images came from processing several images together, including some from a longer view and some from closer up (Fig. 19–17). At least seven jets seemed to come together to form a fan-shaped coma.

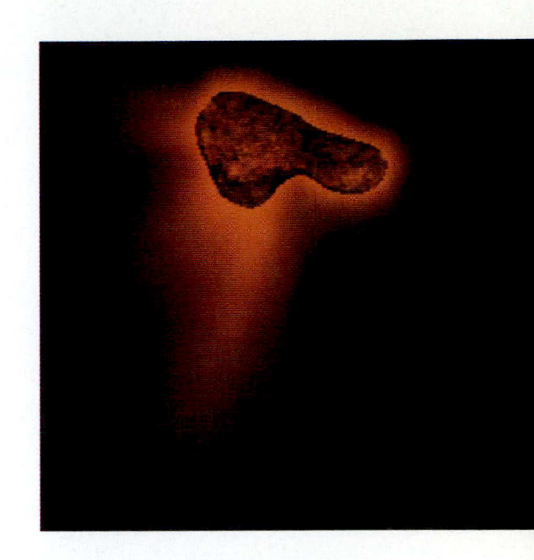

**FIGURE 19–16** A computer-processed Vega-2 image of the nucleus of Halley's Comet, taken on March 9, 1986.

The nucleus became visible, though partly masked by the dust jets. These jets are regions of the coma in which there are three to ten times as many dust particles (1 to 10 micrometers in size) as usual. The nucleus turned out to be potato-shaped (Fig. 19–18), with a surface area of 100 square km. The surface has depressions and hills. The long diameter is 16 km, and the smaller diameters of the cigar-shaped ellipse are 8 km. These dimensions are larger than expected and mean that the nucleus must be darker than expected. After all, if it were larger than expected but at the predicted albedo, it would be brighter than it is observed to be. The albedo of the nucleus is only 2 to 4 per cent, making the nucleus as dark as velvet, equally dark as the other dark objects in the Solar System such as the rings of Uranus and the dark side of Iapetus. The jets do not seem to be active on the night side of the nucleus.

Dramatically, 14 seconds before closest approach, Giotto was hit by a relatively large dust particle. The impact tilted the spacecraft's axis and made it wobble by up to 1.8°, enough to cause Giotto to lose steady communication with Earth for half an hour. Observations from the other side of the comet, including images of the sunlit side that would have allowed an exact determination of the comet's size and a mapping of the locations of the bases of the jets on its surface, were thus lost. Still, Giotto was certainly a spectacular success.

Giotto carried 10 instruments in addition to the camera. Among them were mass spectrometers to measure the types of particles present, detectors for dust, equipment to listen for radio signals that revealed the densities of gas and dust in the coma, detectors for ions, and a magnetometer to measure the magnetic field. As expected, water turned out to be the dominant parent molecule in Halley's coma—80 per cent by volume. Rockets sent up from Earth at about the same time made ultraviolet images (Fig. 19–19).

Some of Halley's dust particles are made only of hydrogen, carbon, nitrogen, and oxygen (Fig. 19–20A); this composition resembles that of the oldest type of

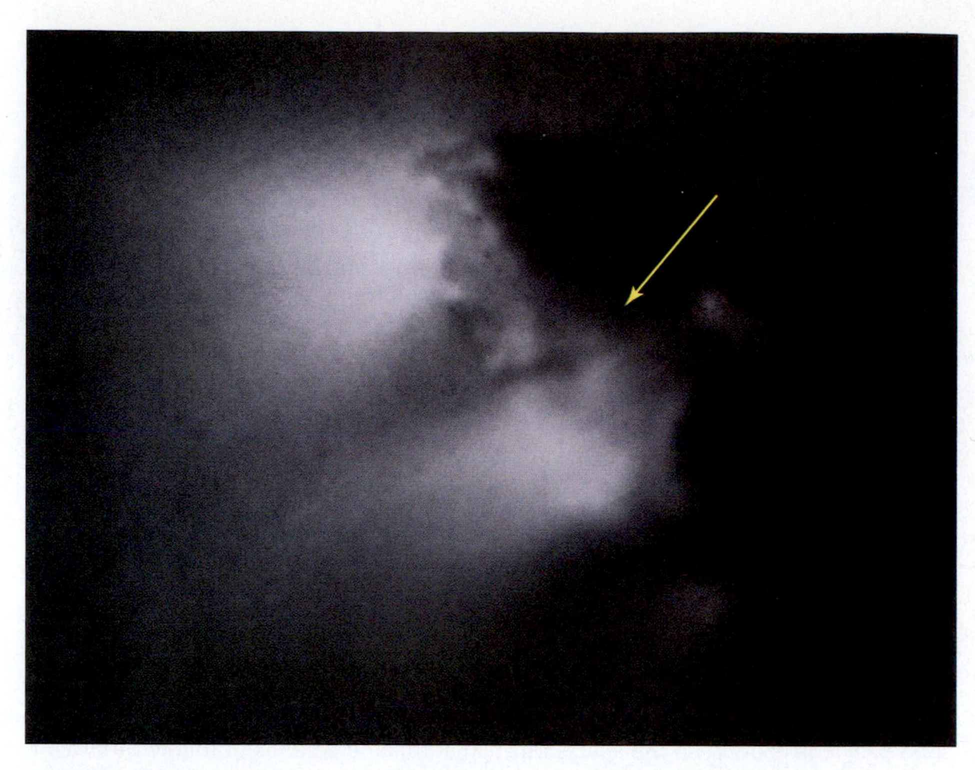

**FIGURE 19–17** The nucleus of Halley's Comet from the Giotto spacecraft. The dark, potato-shaped nucleus is visible in silhouette. It is about 16 km by 8 km; the frame is 30 km across. Two bright jets of dust reflecting sunlight are visible to the left of Halley's nucleus. The small bright region in the center of the nucleus is probably a raised region of its surface that is struck by sunlight.

**FIGURE 19–18** This potato looked remarkably like the nucleus of Halley's Comet, except 10,000 times smaller.

meteorite (Section 20.1b). It thus indicates that these particles may be from the earliest years of the Solar System. Other particles are rich in these same elements, though with other constituents present as well (Fig. 19–20*B*); the distribution is similar to but not quite the same as the composition of carbonaceous chondrites, the type of meteorite we had thought close to being cometary material. Most dust particles are a mixture of the two types. The many tiny dust grains detected, each a million times smaller than a particle in cigarette smoke, are smaller than dust particles known in meteorites. But interstellar particles are thought to be of such sizes, endorsing the idea that the material in comets comes from the earliest period of the Solar System.

Giotto detected 12,000 impacts of dust particles, which ranged from $10^{-17}$ (the lower limit of detectability) to $10^{-4}$ gram. Similarly to the Vega results, there were more dust particles of relatively low mass than had been expected. Perhaps 0.1 to 1 gram of matter was encountered in the form of dust.

As a result of all the flybys, Whipple's model of a comet as a dirty snowball was upheld. But though the existence of gas jets had been known from ground-based observations, nobody had anticipated that the emission of gas and dust would be so localized. The rest of the surface is apparently coated with a black crust. Since no bright regions were discovered at all, the ice is "dirty" even in the active centers. About 30 tons per hour of water and 5 tons per hour of dust were lost at the time the spacecraft flew by; the comet will survive for a few dozen more passes before it

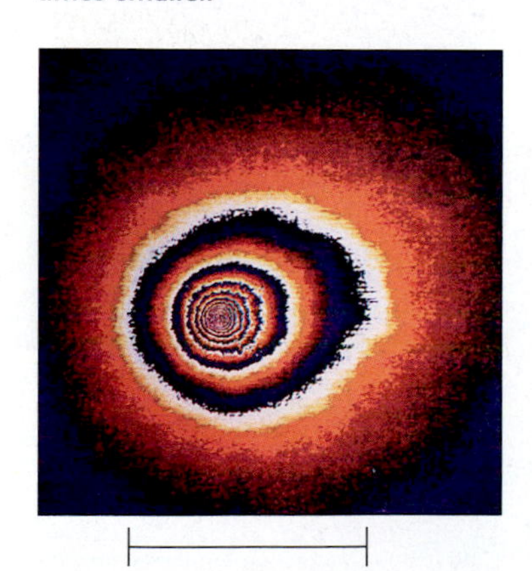

10 million kilometers

◀ **FIGURE 19–19** An image of Halley's Comet in the Lyman-$\alpha$ spectral line of hydrogen.

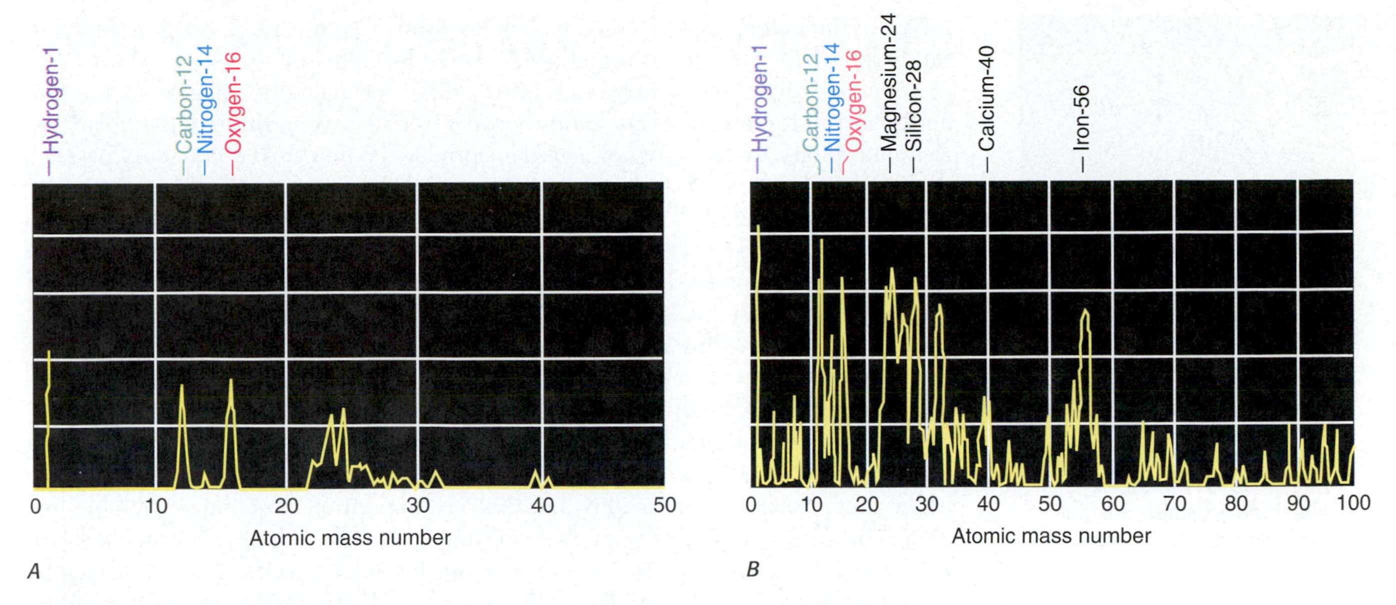


**FIGURE 19–20** The mass spectrometer on the Vega spacecraft showed the atomic mass number of each element in dust particles. The horizontal axis shows mass number (hydrogen = 1, carbon = 12) and the vertical axis shows the abundance. (*A*) A CHON dust particle, composed almost entirely of carbon, hydrogen, oxygen, and nitrogen. (*B*) Other dust particles contain other elements as well.

eventually sublimes away, though perhaps its crust will stop its activity before then, leaving a black hulk in orbit.

The observation of the mountains and valleys indicates that they are probably not made of the volatile materials (ices). Thus there may be a dusty substructure to Halley's Comet; it might be called an icy dustball rather than a dusty (dirty) snowball. Indeed, the nucleus may be made up of a number of 1-km cometesimals.

### 19.4c Halley from the Ground

Following the Halley space encounters, the comet was seen at its best from the ground. Excellent spectra were obtained in the optical (Fig. 19–21) and in the

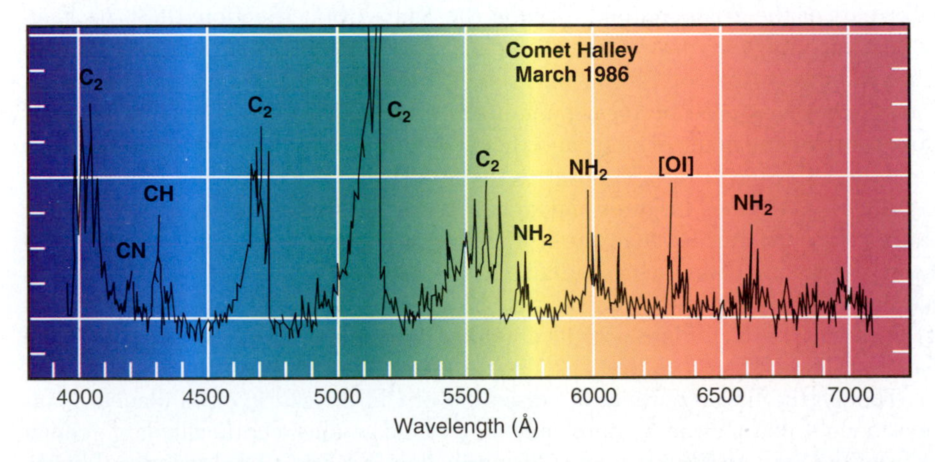


**FIGURE 19–21** The visible spectrum of Halley's Comet on March 14, 1986, when it was at its brightest. We see several spectral lines and bands from molecules. Their relative strengths are approximately the same as they are in other comets, except for $NH_2$, which was slightly stronger in Halley.

**FIGURE 19–22** Comet Hale-Bopp (C/1995 O1) in the sky, showing both a whitish dust tail and a bluish gas tail.

infrared. Radio telescopes studied molecules. Water vapor is the most prevalent gas, but carbon monoxide and carbon dioxide were also detected.

The solar wind has sectors, each with a different magnetic polarity. As the Sun and solar wind rotate, the boundaries between sectors sweep through the Solar System in a spiral, like the spiral of a garden sprinkler. When such a boundary passes a comet, the plasma tail can disconnect. Apparently such an event took place in late March, for when the full moon waned and the sky became dark enough to see the comet again, the comet had little tail. At that time, long before the comet had been supposed to become so dim, the comet's head was only barely visible to the naked eye, and both head and a short tail were only faintly visible with binoculars. The hoped-for tail stretching across a good fraction of the sky never materialized. We have done better with our recent comets C/1996 B2 (Hyakutake) and C/1995 O1 (Hale-Bopp) (Fig. 19–22).

Halley now continues to recede from the Sun and will do so until 2024. With the Hubble Space Telescope, Halley may never be "lost" again. Halley will be back in the inner Solar System in 2061, but this next apparition won't be a favorable one either for Earth-based observers. Only in May 2134 will we have a spectacular view from the Earth's surface—though we may not be limited to the Earth's surface by that time. But don't wait until then. When you hear that a bright, spectacular comet is almost here, don't miss it.

## 19.5 CHIRON AND THE CENTAURS

Chiron (ki′ron), a small object in our outer Solar System, was discovered in 1977. (In Greek mythology, Chiron was the wisest of the centaurs—each of whom was half man and half horse—and the teacher of Achilles.) Calculations of its orbit permitted its image to be found on photographs as far back as 1895, so its orbit is now well determined. It has a 51-year period and an eccentricity of 0.38. Its distance from the Sun ranges from 8.5 A.U., within the orbit of Saturn, to almost as far out as the orbit of Uranus. 2060 Chiron (it has been numbered in the asteroid scheme) is so faint that it must be small, and an occultation has shown that it is between about 150 and 210 km across. Occultations are very useful for making measurements since just timing how long the star is hidden by the asteroid tells us how big the asteroid is at that line across it. A few such lines across it (known as "chords") allow us to deduce the object's size and shape.

Now that we have detected hundreds of other subplanetary objects in the outer part of the Solar System, forming the Kuiper belt (Section 19.3), we realize that probably Chiron was once a member of this belt. Dozens of objects with orbits similar to that of Chiron have been found and are known as **Centaurs.** Though they may well be former members of the Kuiper belt, now their orbits cross those of the giant planets, and their distances from the Sun range between 5 and 30 A.U. Since they are closer to the Sun than the Kuiper belt objects, they are at least 100 times brighter and so can be better studied. Further, from solar heating they are warmer than the Kuiper belt objects so are more likely to show gases. With their diameters of hundreds of kilometers, they are intermediate in size between comets, whose nuclei are 1 to 20 km in diameter, and small icy planets or moons, such as Pluto with its 2300-km diameter or Triton with its 2700-km diameter.

Though the first reports of the discovery of Chiron said it was a planet, it was soon thought that Chiron was probably an asteroid or an exceptionally large comet far from the Sun. As Chiron reached its perihelion in 1995, it brightened and began to show a coma (Fig. 19–23, on p. 346). As an asteroid-sized object with a coma, it is intermediate between comets and asteroids. Many people just classify it as a comet. Its orbit is chaotic, in that gravitational interactions would not have allowed it to be in its current orbit since the origin of the Solar System.

A DEEPER DISCUSSION

## BOX 19.1 Discovering a Comet

Though Edmond Halley did not discover the comet named after him, nowadays a comet is generally named after its discoverer—the first person (or spacecraft or observing group) to detect it (or the first two or three if they independently find it within an interval of days). Until 1995, comets were also temporarily assigned letters in their order of discovery in a given year: for example, Halley's Comet was 1982i. Then, a year or two later when all the comets that had passed near the Sun in that earlier year were likely to be known, Roman numerals were assigned in order of their perihelion passage: for example, 1986 III.

Now we have a newer system of naming comets. The numbering system is primary, and the powers that be are trying to minimize the names. The notation starts with P/ for a periodic comet with a period less than 200 years, C/ for an ordinary comet (with period more than 200 years), D/ for a defunct comet, or X/ for a questionable detection. Periodic comets have been numbered consecutively, from 1P for Halley's Comet through the present limit in the low hundreds. Other comets, and even some apparitions of periodic comets, are given letters to signify the half-month (the first half of January is A, the second half of January is B, the first half of February is C, and so on) and numbers in that half-month. The name can follow in parentheses. So C/1996 B2 (Hyakutake) signifies that a comet whose period is longer than 200 years was discovered during the second half of January 1996, that it was the second comet discovered in that half-month, and that the discoverer was Hyakutake. 1P/1682 Q1 was the appearance viewed by Newton and Halley in 1682 of what we now call Halley's Comet. Just plain 1P is Halley's Comet in general.

Many discoverers of comets are amateur astronomers, including some who examine the sky each night with large binoculars in hope of finding a comet. To discover a comet in this way, one must know the sky very well, so that one can tell if a faint, fuzzy object is a new comet or a well-known nebula. It is for that reason that Charles Messier made a famous eighteenth-century list of nebulae (which we discuss in the introduction to Part VII of this book and list in Appendix 8). Messier's list of things not to confuse with comets turned out to be a list of interesting things in the sky. Thus though these objects were not of interest to Messier, they are many of the objects that amateur astronomers now like best to observe. Once you could win a gold medal from the King of Denmark for discovering a comet, as Maria Mitchell did when she discovered one from Nantucket in 1847. She went on to be professor of astronomy at Vassar.

If you think you have discovered a comet, send an e-mail message on the Internet with your report to the Central Bureau for Astronomical Telegrams, cbat@cfa.harvard.edu. (You might want to check first with your local observatory, which may know if a new comet is in the sky.) Faxing (617) 495-7231 is a much less desirable method. In any case, you should identify the direction of the comet's motion, its position, and its brightness. Don't forget to identify yourself by giving your name, address, and telephone number. If you are the first (or maybe even the second or third) to find the comet, it will be named after you. The Bureau's homepage on the World Wide Web is http://cfa-www.harvard.edu/cfa/ps/cbat.html, and instructions are given there on how to report a discovery.

Comet LINEAR broke apart into small bits while under intensive observation during 2000.

One of the most famous comet discoverers in history was Caroline Herschel, who found eight comets. She worked with her brother William, the discoverer of Uranus, on many projects but did most of the comet work herself. The most comets have been found by Carolyn and Eugene Shoemaker. They have discovered more than thirty. David Levy, who worked with the Shoemakers in taking images, discovered 7 on his own visually.

Many comets, particularly the ones discovered by professional astronomers, are discovered not by eye but by examination of photographs or electronic images taken with telescopes. The photographs usually have been taken for other purposes, so the discovery of a comet exemplifies serendipity —a fortuitous extra discovery. The Solar and Heliospheric Observatory has found hundreds of comets so near the Sun that they have not been seen in any other way. Comet discovery in the last few years has been taken over by automatic programs. The LINEAR program, as of this writing, has discovered 56 comets. The Spacewatch program at the University of Arizona is also finding many comets.

### *Discussion Question*

How much effort would it be worth to you to discover a comet? How many hours would you be willing to spend if a single discovery were guaranteed? Comet searchers usually sweep their binoculars back and forth along the eastern horizon, hoping to catch the comet soon after it rises and before someone farther west discovers it. Allow an average of 1 hour per night for the viewing that is often necessary.

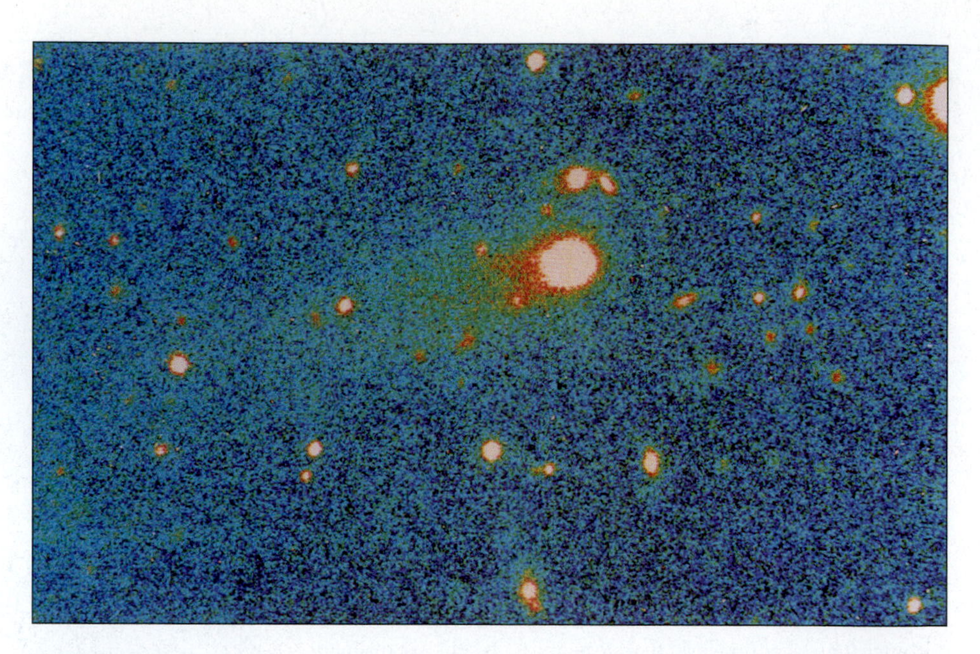

**FIGURE 19–23** Chiron: A coma shows; that is, the image is more diffuse than the images of the stars.

Chiron may really be telling us that there is no sharp distinction between asteroids and comets. Traditionally, comets give off gases because they contain ices, while asteroids are rocky; but other evidence already indicates that asteroids can contain ices, and comets can contain rock. We see that distinctions between comets and asteroids may be slight, particularly now that we know of Centaurs and Kuiper-belt asteroid-like objects closer in than the Oort Cloud though beyond the main asteroid belt.

Other Centaurs include Pholus, whose surface is much redder than Chiron's. All the Centaurs except Chiron were discovered since Pholus was found in 1992. The pace of discovery in the outer Solar System is rapid and increasing.

## 19.6 A COMET CRASHES ON JUPITER

Eugene Shoemaker was such an important contributor to our understanding of the Solar System that a spacecraft sent to the asteroid Eros was renamed after him: NEAR Shoemaker. Don't confuse the defunct comet Shoemaker-Levy 9 with the spacecraft NEAR Shoemaker.

Though most comets are in orbit around the Sun, Jupiter can affect their orbits greatly and sometimes even capture a comet into orbit around itself. Periodic Comet Shoemaker-Levy 9 was broken into many parts (Fig. 19–24) by Jupiter in 1992 and sent into an orbit around Jupiter. The various bits hit Jupiter in 1994 with astoundingly visible results (Box 13.3). The rings it left on Jupiter show us the hydrocarbon and other constituents of the comet.

## 19.7 FUTURE COMET STUDIES

New comets appear all the time and sometimes become reasonably bright. We can revel in the wonderful recent views and observations of C/1996 B2 (Hyakutake) and C/1995 O1 (Hale-Bopp) and hope for more bright comets to study. Astronomers remain enthusiastic not only about the beauty of comets but about their ability to tell us about the earliest epoch of our Solar System.

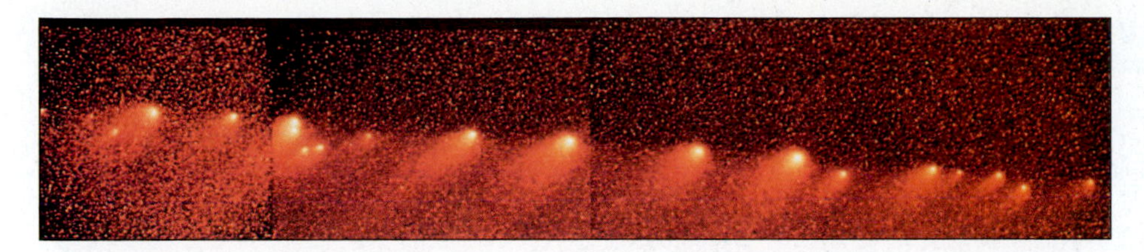

**FIGURE 19–24** Comet Shoemaker-Levy 9 viewed from the Hubble Space Telescope. Over 20 fragments, each with its own tail, appear in this image taken six months before their collisions with Jupiter. On the International Astronomical Union Circular No. 5800 announcing his calculations, Brian Marsden wrote, "the comet's minimum distance from the center of Jupiter will be only 0.0003 AU (Jupiter's radius being 0.0005 AU) on 1994 July 25.4." In his understated way, he was announcing the impending collision.

A NASA technology testing spacecraft, Deep Space 1, visited asteroid Braille in 1999 using its ion-propulsion system. It reached Comet Borrelly in 2001 and sent back close-up images that revealed the shape and rugged terrain of its 10-km-long nucleus.

NASA launched its spacecraft Stardust in 1999, with Donald Brownlee of the University of Washington at Seattle as Principle Investigator. It will fly through the extended coma of the active periodic comet 81P/Wild-2 in 2004, not only photographing the comet but also bringing back some of its dust for analysis on Earth. It is to capture the dust in an aerogel (a material of ultra-low density) and drop it to the Utah desert by parachute in 2006 (Fig. 19–25). A NASA spacecraft to be launched in 2002, CONTOUR (Comet Nucleus Tour) is to visit at least 3 comets over a half-dozen years. NASA's Deep Impact mission, to be launched in 2004, is to blast a projectile into Comet Tempel 1 on July 4, 2005. The impactor is to make a football-field-sized crater; the flyby spacecraft will image the explosion and measure the composition of the outflowing gas.

**FIGURE 19–25** Aerogel, a frothy, extremely light substance. Even a large amount weighs very little, so a rocket can carry more of it into space than it could of a denser material.

The European Space Agency is building its Rosetta mission for launch in 2003. Rosetta is to reach comet 46P/Wirtanen at its 2011 aphelion (its farthest point from the Sun) and travel with it through its perihelion in 2013. It will also land a probe on the comet's nucleus. Just as the Rosetta Stone in the British Museum—with its Greek, Egyptian hieroglyphic, and Demotic writing of the same text—enabled us to read past languages, it is hoped that the Rosetta mission will enable us to understand the early period of our Solar System through study of pristine objects like comets.

## CORRECTING MISCONCEPTIONS

| ✖ Incorrect | ✔ Correct |
|---|---|
| The orbit of Halley's Comet extends outward equally to either side of the Sun. | Because the elliptical orbit has the Sun at one focus, and because the focus is close to one narrow end of the ellipse, a comet's orbit extends much farther to one side. |
| A comet's tail trails behind the comet. | Comet tails point away from the Sun, even if they go ahead of the comet. |

## SUMMARY AND OBSERVATIONS

Observing comets (Section 19.1)
- The head (nucleus and coma together); the tail
- Recent bright comets C/1996 B2 (Hyakutake) in 1996 and C/1995 O1 (Hale-Bopp) in 1997

The composition of comets (Section 19.2)
- Dirty-snowball theory: nucleus of ice with dust mixed in
- Coma is gases vaporized from nucleus; molecules present.
- Ultraviolet space observations revealed huge hydrogen cloud.
- Dust tail: sunlight reflected off particles; ion tail (gas tail): sunlight reemitted by ions blown back by solar wind

The origin and evolution of comets (Section 19.3)
- Origin of long-period comets: Oort Comet Cloud
- Origin of short-period comets: Kuiper belt

Halley's Comet (Section 19.4)
- Earth badly placed for us to see the 1986 apparition, but observatories could obtain data.
- An armada of spacecraft in March 1986
- European Space Agency's Giotto imaged nucleus from 605 km.
- Brightest features are dust jets.
- Nucleus is dark as velvet; 15 km × 10 km; covered with crust except for jet regions
- Water is dominant molecule; many other molecules detected.
- Bow wave, small magnetic field detected.
- Whipple's dirty-snowball model upheld.
- Rotation of solar wind sectors causes tail disconnection events.

Chiron (Section 19.5)
- First thought to be a distant asteroid; now known to be a comet or at least an intermediate object between comets and asteroids.
- Former member of Kuiper belt

Comet Shoemaker-Levy 9 (Section 19.6)
- Jupiter broke it into bits and captured it.
- It hit Jupiter and left its material for analysis.

Future spacecraft (Section 19.7)
- Stardust to capture comet dust.
- Rosetta to orbit with a comet and land a probe.

## KEY WORDS

comet
head
tail
nucleus
dirty snowball
coma
dust tail
gas tail
ion tail
Oort Comet Cloud
Kuiper belt
parent molecules
daughter molecules
Centaurs

## QUESTIONS

1. In what part of its orbit does a comet travel head first? Explain.
2. Why is the Messier catalogue important for comet hunters?
3. Would you expect comets to follow the ecliptic? Explain.
4. †How far is the Oort Comet Cloud from the Sun, relative to the distance from the Sun to Pluto?
5. Sketch the relation of the Oort Comet Cloud and the Kuiper belt, and describe their different roles in providing comets.
6. The energy that we see as light from a comet tail comes from what source or sources?
7. What part of a comet has the most mass?
8. Why was the large cloud of hydrogen that surrounds the head of a comet not detected until 1970?
9. Explain why Halley's Comet showed delicate structure in its tail. Which of its tails shows that structure?
10. †Use Kepler's third law and the known period of Halley's Comet to deduce its semimajor axis. Sketch the orbit's shape on the Solar System to scale, using Kepler's first law to position the Sun.
11. †Use Kepler's third law and the 3000-year period of Comet C/1995 O1 Hale-Bopp to deduce its semimajor axis. Perihelion is 0.9 A.U. Relate the result to the size of planetary orbits and sketch the orbit.
12. Why wasn't the 1986 passage of Halley's Comet a good one for earthbound viewers?
13. Which spacecraft imaged Halley's jets?
14. Describe the nucleus of Halley's Comet.
15. Describe the dust discovered near the nucleus of Halley's Comet.
16. Compare Whipple's model for comets with the discoveries of the spacecraft to Halley's Comet.
17. How steady is Halley's output of gas? Explain.
18. Why do photographs of Halley's Comet look more spectacular than the comet looked to the eye?
19. How could Chiron, which shows a coma like a comet, be confused for so long with an asteroid?
20. What were the effects of Comet Shoemaker-Levy 9 on Jupiter at its 1994 collision? Investigate.

†This question requires a numerical solution.

## DISCUSSION QUESTION

1. Two families of dust are gathered by upper atmosphere missions. One family has a relatively low velocity (little more than Earth's escape velocity), and the other a relatively high velocity. Which do you think comes from asteroidal dust and which from comet dust, and why?

## USING TECHNOLOGY

### W World Wide Web

1. Examine comet homepages available through this book's homepage and find views of dust tails and gas tails.
2. Examine the current and expected comets on the homepage of Gary Kronk at http://amsmeteors.org and http://SkyPub.com.
3. Find out about the spacecraft en route to or planned for comet exploration, such as http://stardust.jpl.nasa.gov, http://www.contour2002.org, and http://sse.jpl.nasa.gov/missions/comet_missns/comet-di.html. Deep Space 1 is at http://nmp.jpl.nasa.gov/ds1.
4. Follow the pace of discovery of Centaurs, and compare their names with Greek mythology's names for centaurs at the list of the International Astronomical Union's Central Bureau for Astronomical Telegrams, http://cfawww.harvard.edu/iau/lists/Centaurs.html.

### REDSHIFT

1. Follow the orbit of Halley's Comet, looking both down from high above the Solar System and in the plane of the Solar System. How high out of the plane does it go?
2. Measuring on your screen, verify Kepler's second law for Halley's Comet, using triangles at perihelion, aphelion, and in between.
3. Compare the distance of Halley's Comet from Earth at closest approach for its 1910, 1986, 2061, and 2136 apparitions, and relate to its observed brightness.
4. Consider the content of Guided Tours: Revolutionaries: Halley's Comet; Astronomy Lab: Worlds of Space: Comets; and Science of Astronomy: Worlds of Space: Moons, Asteroids, and Comets.
5. On disc 2, look at the movie *Comet Hale-Bopp*.

Leonids (also see Fig. 20–9). A time exposure of the 1998 Leonid meteor shower, showing fireballs that appeared over a 45-minute period. The Big Dipper is standing on its handle to the right, and the radiant in Leo is off the upper-right of the frame.

# Chapter 20

# Meteorites and Asteroids

Asteroids and meteoroids are members of our Solar System that are solid, like planets, but are much smaller. We shall see how they, like the comets, are storehouses of information about the Solar System's origin. The meteoroids, which we discuss first, are probably chunks broken off asteroids. As for many things in astronomy, we are learning that the distinction between two things with different names is more a matter of degree than a fundamental difference. We will see this convergence between asteroids and meteoroids.

In the following chapter, we shall discuss the exciting (but not widely accepted) ideas that a few meteorites found on Earth not only are from Mars but also reveal that life arose there independently of life on Earth.

**AIMS:** To describe meteorites and asteroids, and to see how their history may provide us with information about the origin of the Solar System and even perhaps the origin of life

## 20.1 METEOROIDS AND METEORITES

There are many small chunks of matter in interplanetary space, ranging up to tens of meters across. When these chunks are in space, they are called **meteoroids.** When one hits the Earth's atmosphere, friction slows it down and heats it up—usually at a height of about 100 km—until all or most of it is vaporized. Such events result in streaks of light in the sky, which we call **meteors** (popularly known as **shooting stars**). Most meteors are the burning up of dust specks; technically tiny meteoroids, these particles are probably given off by comets.

The brightest meteors can be brighter than the full moon. We call such bright objects **fireballs** (Fig. 20–1). We can sometimes even hear the sounds of their passage and of their breaking up into smaller bits. A fireball that breaks up in this way is a **bolide.** When a fragment of a meteoroid survives its passage through the Earth's atmosphere, it is called a **meteorite.** We now also refer to objects that hit the surface of the Moon or of other planets as meteorites.

FIGURE 20–1 A fireball observed on August 10, 1972, from Grand Teton National Park in Wyoming. The fireball was visible for 101 seconds and might have reached magnitude −19. Thus it was about 80 m across and weighed about one million tons. It skipped on the atmosphere, like a rock skipped across a lake (or x-rays at grazing incidence from Chandra's mirrors), coming 58 km from hitting the Earth's surface.

### 20.1a Types and Sizes of Meteorites and Meteors

Tiny meteorites less than a millimeter across, **micrometeorites,** are the major cause of small-scale erosion on the Moon. Micrometeorites also hit the Earth's upper atmosphere all the time, and remnants can be collected for analysis from balloons or airplanes. The micrometeorites are thought to be debris from comet tails and asteroids. They are known as micrometeorites rather than micrometeoroids, even though they don't usually reach the ground. They may have been only the size of a grain of sand, and they are often sufficiently slowed down that they are not vaporized before they reach the ground. The resulting dust—100 tons a day of it—can be sampled by collecting ice from the Arctic or Antarctic, or from mountaintops.

Tektites, small, rounded glassy objects that are found at several locations on Earth, may have splashed out from meteorite impacts. Tektites were once thought to come from the Moon, though it is now accepted that they come from large impacts on Earth.

Space near the Earth is full of meteoroids of all sizes, with the smallest being most abundant. Most of the small particles, less than 1 mm across, probably come from comets. Most of the large particles, more than 1 cm across, may come from collisions of asteroids in the asteroid belt between Mars and Jupiter. It is generally these larger meteoroids that become meteorites. From matching the spectrum of reflected sunlight, some meteorite types can even be assigned to certain asteroids.

There are several kinds of meteorites. Most of the meteorites that are found (as opposed to most of those that exist) have a very high iron content—about 90 per cent; the rest is nickel. These **iron meteorites** (**irons,** for short) are thus very dense—that is, they weigh quite a lot for a given volume (Fig. 20–2).

Most meteorites that hit the Earth are stony in nature and are often referred to simply as **stones** (Fig. 20–3). Because stony meteorites resemble ordinary rocks and disintegrate with weathering, they are not usually discovered (that is, identified as meteorites) unless their fall is observed. That explains why most meteorites discovered at random are irons. But when a fall is observed, most meteorites recovered (known as "finds") are stones. The stony meteorites have a high content of silicates; only about 10 per cent of their mass is nickel and iron. Most stony meteorites are of a type called **chondrites** because they contain rounded particles of a certain mineral composition, which are called "chondrules." Astronomers think the chondrules condensed from the original solar nebula. (Indeed, there is evidence that half the

A

B

FIGURE 20–2 (*A*) A 15-ton iron meteorite, the Willamette meteorite, at the Rose Center for Earth and Space, which also contains the Hayden Planetarium, at the American Museum of Natural History in New York. Note how dense the meteorite has to be to have this high mass, given its small size. (*B*) The largest mass of the Cape York meteorite, a 31-metric-ton piece known as "Ahnighito" or "the Tent." This stony meteorite was discovered by Inuits in Greenland in the early 1800s and was brought to New York by William Peary and Matthew Henson in 1897. It is the largest on display in a museum (second in size only to a meteorite still in the ground in Namibia) and is in the American Museum of Natural History in New York City.

**TABLE 20–1** Meteorites

| Composition | Seen Falling | Finds; Not Seen Falling |
|---|---|---|
| Irons | 6% | 66% |
| Stony-irons | 2% | 8% |
| Stones | 92% | 26% |

**FIGURE 20–3** Most stony meteorites are known as "chondrites," from the roundish inclusions known as "chondrules" in them.

ordinary chondrites are chips off a single asteroid: Hebe.) The stony meteorites without chondrules are called **achondrites;** they may have once been chondrites but have melted and cooled.

A third kind of meteorite is a **stony-iron,** which contains roughly equal quantities of silicates and iron-nickel alloy. Most stony-irons are of a type called "pallasites"; they contain beautiful green crystals of the mineral olivine.

A large terrestrial crater that is meteoritic in origin is the Barringer Meteor Crater in Arizona (Fig. 20–4). It resulted from what was perhaps the most recent large meteorite to hit the Earth, for it was formed only 40,000 years ago. Dozens of other impact craters are known on Earth. Some buckyballs, arrangements of 60 or 70 atoms of carbon in a soccer-ball shape, have been found at the giant crater in Ontario, making us think they are associated with the impact. (Buckyballs are a third form of pure carbon, in addition to graphite and diamond.) Finding these organic materials confirms the idea that organics can be delivered to Earth by comets and meteors. It has even been suggested that the world's largest diamonds, black stones called carbonadoes that can be the size of baseballs, arrive on Earth in asteroids. They would have been formed in exploding giant stars.

Every few years a meteorite is discovered on Earth immediately after its fall. The chance of a meteorite's landing on someone's house or car is very small, but it happens every three years or so (Fig. 20–5)! It has recently been calculated on the basis of the number of detectable falls, a person's size, the population and size of North America, etc., that on the average a meteorite should hit a person every 180 years in North America. The only such documented case occurred on November 30, 1954, when a nine-pound stony meteorite crashed into a house in Sylacauga,

**FIGURE 20–4** The Barringer "meteor crater" (actually a meteorite crater) in Arizona. It is 1.2 km in diameter. Dozens of other terrestrial craters are now known, many from aerial or space photographs. The largest may be a depression over 400 km across under the Antarctic ice pack, comparable with lunar craters. Another very large crater, in Hudson Bay, Canada, is filled with water. Most either are disguised in such ways or have eroded away.

A

**WIZARD OF ID** *Brant Parker & Johnny Hart*

B

*By permission of Johnny Hart and Creators Syndicate, Inc.*

**FIGURE 20–5** (*A*) In 1996, a Minnesota man found a meteorite had hit his Geo Metro. A 23-g fragment was in the windshield wiper well and a 65-g fragment was on the ground. Here he is with the meteorite fit into the shattered part of the windshield. It is an ordinary chondrite. (*B*) Meteorites can be quite dark and have a high carbon content. *© 2000 Creators Syndicate, Inc. "Wizard of Id" cartoon by Brant Parker and Johnny Hart.*

Alabama, and hit Mrs. Hewlett Hodges on a bounce. She was bruised, but not seriously injured. The one-ton meteorite that crashed into Turkmenistan in 1999, leaving a crater 5 m across and 3.5 m deep, was much more impressive.

Often the positions of fireballs in the sky are tracked in the hope of finding fresh meteorite falls. The newly discovered meteorites are rushed to laboratories, since studies of the isotopes in them can reveal how long they have been in space. Many meteorites have recently been found in the Antarctic, where they have been well preserved as they accumulated over the years (Fig. 20–6). They are being kept free of human contamination. Since meteors are well preserved when we find them in Antarctica, the fraction of stony meteorites we find there compared with iron meteorites is higher than the fraction of stones found elsewhere.

A few meteorites that are found seem to have come from the Moon or even from Mars (Fig. 20–7). In appearance and composition, the rock shown in the image resembles certain basaltic rocks, which is unlikely in asteroids. Yet the relative abundances of oxygen isotopes in this meteorite are unlike those from Earth or from any other meteorite. Further, this meteorite is 1.3 billion years old, one of a dozen similar objects known; Mars was still forming crust from molten rock at that time. Most other meteorites are 4.5 billion years old. The separate discovery of a meteorite that probably came from the Moon makes it likely that other meteorites hitting the Moon or Mars can blast rocks off them without pulverizing the rocks. (Any

## A DEEPER DISCUSSION

### BOX 20.1 The Tunguska Explosion

Many theories have been advanced to explain an explosion near the Tunguska River in Siberia in 1908. Most scientists have long thought that it was a fragment of a meteoroid or comet entering our atmosphere. No remnant has been found, even though trees were blown outward for kilometers around. Signs of dust from the explosion have been found at widely separated regions of the Earth.

Computer simulations that explained the pattern of trees blown down found that the object was a stony asteroid about 40 meters across and travelling at 15 km/sec. The explosion reached 15,000°C and consumed the fragments of the object.

***Discussion Question***

Should we plan to have atomic bombs and powerful rockets ready to head off meteoroids and asteroids we find heading toward Earth? Why or why not?

**FIGURE 20–6** Ursula Marvin of the Harvard-Smithsonian Center for Astrophysics (prone) examining a typical-size Antarctic meteorite. The meteorites have probably been long buried in ice and are made visible when wind erodes the ice. Thousands of meteorites were found on this expedition.

reddish dust on Mars's surface would have been blown off.) Analysis has indicated that the martian meteorites were ejected from Mars millions of years ago. It certainly is cheaper to pick up rocks from Mars in this way instead of sending a spacecraft there. One of the supposed Mars meteorites has been found to carry organic (carbon-containing) compounds, again raising the possibility of organic compounds' existing on Mars. A controversy is going on as to whether the rock contains signs that primitive life forms existed on Mars; most astronomers are skeptical.

As we mentioned in the chapter on Mars, the most recent meteor from Mars was found near Los Angeles. Such meteorites apparently take from hundreds of thousands to hundreds of millions of years in their transfer to the Earth. Studies of how often they were hit by particles in interstellar space give us this information, as well as indicating that they arrived on Earth tens of thousands of years ago.

In recent years, scientists have realized that some of the meteorites that have never been heated contain grains of interstellar dust, which were given off by stars. These are presumably left over from before the formation of the Solar System. These interstellar grains can be recognized by their unusual ratios of abundances of different isotopes of carbon, nitrogen, and other elements. So dust from the stars lands at our feet for us to study.

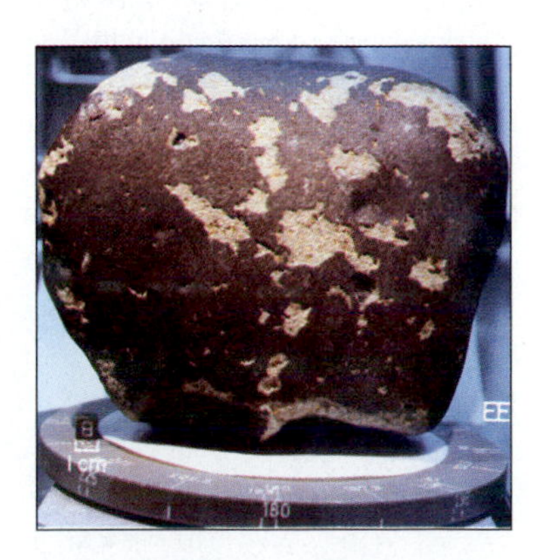


**FIGURE 20–7** This meteorite found in Antarctica is made of material that makes us think it is a rock from Mars.

Some small fraction of meteorites are going so fast that they are probably not from our Solar System. Radar studies show that some 15 per cent of meteors travel faster than 270,000 km/hr, the escape velocity for our Solar System, and aren't coming from the direction of planets that could have accelerated them. An even smaller fraction is coming much faster. So these meteorites are coming from nearby stars or from dust in interstellar space.

This interstellar matter makes up most of the material beyond the asteroid belt; some of it was detected by the Ulysses spacecraft when it went by Jupiter. Most of the dust material in space near the Earth, as we have said, is long-term solar-system material, dust from comets or chips off asteroids.

### 20.1b Carbonaceous Chondrites*

**FIGURE 20–8** Close-up of a part of the Murchison meteorite.

Objects that contain large quantities of carbon, often in organic compounds, are **carbonaceous. Carbonaceous chondrites**—most types of which are stony meteorites with chondrules and a high carbon content—are rare. One such carbonaceous chondrite—the Murchison meteorite (Fig. 20–8)—contains simple amino acids, building blocks of life. It fell near Murchison, Victoria, Australia, in 1969. Analysis of these amino acids showed that they were truly extraterrestrial and had not simply contaminated the sample on Earth. The formation of such complicated organic chemicals in cold, isolated places like meteoroids is one of several indications that the precursors of life develop naturally and gradually. We shall be developing more of this evidence in the following chapter.

Only five of the nearly 1000 meteorites recovered from falls that were observed over the last two centuries were the most primitive type of carbonaceous chondrites. (This type of carbonaceous chondrite is the richest of volatile elements—those that evaporate most easily. By "primitive," we mean that the relative abundances of its elements are indistinguishable from those measured in spectra of the solar surface. They both presumably show the values from the solar nebula from which the Sun and other objects condensed.) Finding these primitive carbonaceous chondrites from fallen meteorites is even rarer—only two bits totalling less than 15 grams were found out of the tens of thousands of meteorites. The most recent primitive carbonaceous chondrite seen to fall was at Tagish Lake, British Columbia, on January 18, 2000, and was recovered a week later. The finder saw the black stones on the icy lake, and knew that it would be useful not to handle them and to keep them frozen. The opportunity to have the never-thawed specimens is unequaled. When warmed to room temperature, the volatile compounds evaporate and a sulfuric odor is smelled. Scientists had hoped that the studies that are being carried out on this object should expand our knowledge of how organic matter that fell to Earth may have played a role in the origin of life. Strangely, contrary to expectation, the meteorite turned out to have almost no amino acids or other organic acids in it. Perhaps it came from a different part of the asteroid belt than the locations from which we get most meteorites. Scientists have suggested that it is our first example of a meteorite from the outer part of the asteroid belt, which is farther from the Sun and thus contains materials less transformed over time. Clays found in the Tagish Lake meteorite show that liquid water was present at one time, presumably melted from ice by radioactivity. Perhaps the water washed away the amino acids.

> The Tagish Lake meteorite, found in the Yukon after a fireball was seen in the sky, contains more carbon and nitrogen than any other. It is very rich in interstellar diamond grains, which shows that it formed in the outermost part of the protosolar nebula. It is perhaps the most primitive material ever found on Earth.

We measure the ages of meteorites by studying the ratios of stable isotopes in them, some of which are the daughter products of radioactive isotopes. (We also date lunar and terrestrial rocks in this way. We discussed radioactivity in Box 9.2.) The measurements show that most solid materials were formed up to 4.6 billion years ago, the beginning of the Solar System. The chondrites may date from the formation of the Solar System itself; some may be pieces of planetesimals. The abundances of the elements in meteorites thus tell us about the solar nebula from which the Solar System formed, or even about conditions before that. In fact, up to the time of the Moon landings, meteorites were the only extraterrestrial material we could get our hands on.

Most of our knowledge of the abundances of the elements in the Solar System has come either from analysis of the spectra of the Sun, planets, and their moons; or from analysis of meteorites. (We have on-site measurements, as well, on the Moon, Mars, Jupiter, and some comets.)

Buckyballs, those combinations of 60 atoms of carbon in the shape of a soccer ball, have been discovered in a few meteorites. (The 1996 Nobel Prize for Chemistry was awarded for the discovery of buckyballs, which are easy to form on Earth though had not previously been identified.) Some small fraction of these buckyballs—one in a million—has a helium atom trapped inside its closed structure. To get this fraction, the buckyballs had to have formed in a place with 500 times the helium pressure of Earth's atmosphere, so we think these buckyballs were formed in the neighborhood of red giant stars. Did enough carbon get delivered to Earth's early atmosphere in this way to affect the creation of life? Discussion on this point continues, with the survival in the meteorite impact of the carbon in the buckyballs favoring the possibility.

The first Kuiper belt object to be numbered is 20,000 Varuna. This "Trans Neptunian Object" has been named after a Hindu deity, lord of the cosmos. Varuna's diameter is estimated to be 1000 km. A Kuiper belt object discovered even more recently, 2001 KX76, has a diameter of 1270 km, assuming a 4% albedo. This size would make it larger than Ceres, the largest asteroid, or Pluto's moon Charon. The Lowell Observatory, MIT, and Large Binocular Telescope Observatory scientists who found it, as part of their Deep Ecliptic Survey, suggest that even larger objects remain undiscovered in the Kuiper belt.

We have discussed the Kuiper belt objects and the Centaurs in Sections 19.3 and 19.5. Both can be considered types of asteroids. The Kuiper belt objects orbit beyond Neptune, while the Centaurs' orbits have been modified to send them into elliptical orbits between the orbits of Jupiter and Neptune.

### 20.1c Meteor Showers

On any clear night, a naked-eye observer may see a few **sporadic** meteors an hour, that is, meteors that are not part of a meteor shower. (Just try going out to a field in the country on a night when the Moon is not up and watching the sky for an hour.) Meteors often occur in **showers,** when meteors are seen at a rate far above average. During a shower several meteors may be visible to the naked eye each minute, though a rate this high is rare. Meteor showers occur at the same time each year (Table 20–2) and represent the Earth's passing through the orbits of defunct comets and hitting the meteoroids left behind. The duration of a shower depends on how spread out the meteoroids are in the former comet's orbit. Meteorites do not usually result from showers, so presumably the meteoroids that cause a shower are very small.

The rate at which meteors are seen usually increases after midnight on the night of a shower (or otherwise), because that side of the Earth is then turned so that it plows through the oncoming interplanetary debris. If the Moon is gibbous or full, then the sky is too bright to see the shower well. The meteors in a shower are seen in all parts of the sky, but their trajectories all seem to emanate from a single point

**TABLE 20–2** Meteor Showers*

| | Date of Maximum | Duration Above 25% of Maximum | Approximate Limits | Number Per Hour at Maximum | Parent |
|---|---|---|---|---|---|
| Quadrantids | Jan. 4 | 1 day | Jan. 1–6 | 110 | — |
| Lyrids | April 22 | 2 days | April 20–24 | 12 | C/1861 G1 |
| $\eta$ Aquarids | May 5 | 3 days | May 1–8 | 20 | 1P/Halley |
| $\Delta$ Aquarids | July 27–28 | 7 days | July 15–Aug. 15 | 35 | — |
| Perseids | Aug. 12 | 5 days | July 25–Aug. 18 | 68 | 107P/Swift-Tuttle |
| Orionids | Oct. 21 | 2 days | Oct. 16–26 | 30 | 1P/Halley |
| Taurids | Nov. 8 | Spread out | Oct. 20–Nov. 30 | 12 | 2P/Encke |
| Leonids | Nov. 17 | Spread out | Nov. 15–19 | 10 | 55P/Tempel-Tuttle |
| Geminids | Dec. 14 | 3 days | Dec. 7–15 | 58 | 3200 Phaethon |

*The number of sporadic meteors per hour is 7 under perfect conditions. The visibility of showers depends mostly on how bright the Moon is on the date of the shower, which depends on its phase. Meteors are best seen with the naked eye; using a telescope or binoculars merely restricts your field of view.

*A*

*B*

**FIGURE 20–9** The 1999 Leonid meteor shower. Thousands of fainter meteors were seen in the peak hour. (*A*) Exposures totalling 20 minutes. (*B*) A fireball—an extremely bright meteor—streaking away from the backward-question-mark shape of the constellation Leo.

in the sky, the **radiant.** A shower is usually named after the constellation that contains its radiant. The existence of a radiant is just an optical illusion; perspective makes the set of parallel paths approaching us appear to emanate from a point, like railroad tracks apparently converging in the distance. Some meteor showers, like the Perseids we can expect each year on August 12, are fairly steady year after year. The Perseids are the most widely seen, in large part because of the warm weather in the northern hemisphere that makes outdoor observing pleasant. Other showers, like November 17's Leonids, are occasionally spectacular, as that shower was in 1999, when the 33-year period of the peak of this shower brought thousands of meteors in the sky per hour (Fig. 20–9)! Theoretical work completed shortly before the 1999 peak enables scientists to predict accurately which strand of dust from Comet Tempel-Tuttle will hit the Earth in a given year, and so predict how intense the meteor shower will be and what time the peak will occur. Their prediction was dead on for the 1999 peak so their theory is widely accepted. The 2000 peak was not quite as high, though still with hundreds of meteors per hour, and the 2001 peak is predicted to perhaps match the highest level.

The average meteor we see is the death of a meteoroid about 100 km high and moving about 30 km/sec. The small amount of sodium, calcium, silicon, iron, and other heavy elements that is left when the meteoroid vaporizes results in a radio-reflecting variable layer of the Earth's upper atmosphere. The electrons in the trails formed by the meteoroids also reflect radio waves and can be used to reflect radio messages for up to 15 seconds each. Some trucking companies are using these free reflections to keep in touch with their fleets.

## 20.2 ASTEROIDS

The nine known planets were not the only bodies to result from the agglomeration of planetesimals 4.6 billion years ago. Hundreds of thousands of **minor planets,** called **asteroids,** also resulted (Fig. 20–10).

Most of the asteroids have elliptical orbits between the orbits of Mars and Jupiter, in a zone called the **asteroid belt** (Fig. 20–11). A million or so larger than 1 km apparently exist there. Indeed, observers were not completely surprised when the first asteroid was discovered, on January 1, 1801, because Bode's law, a numerological

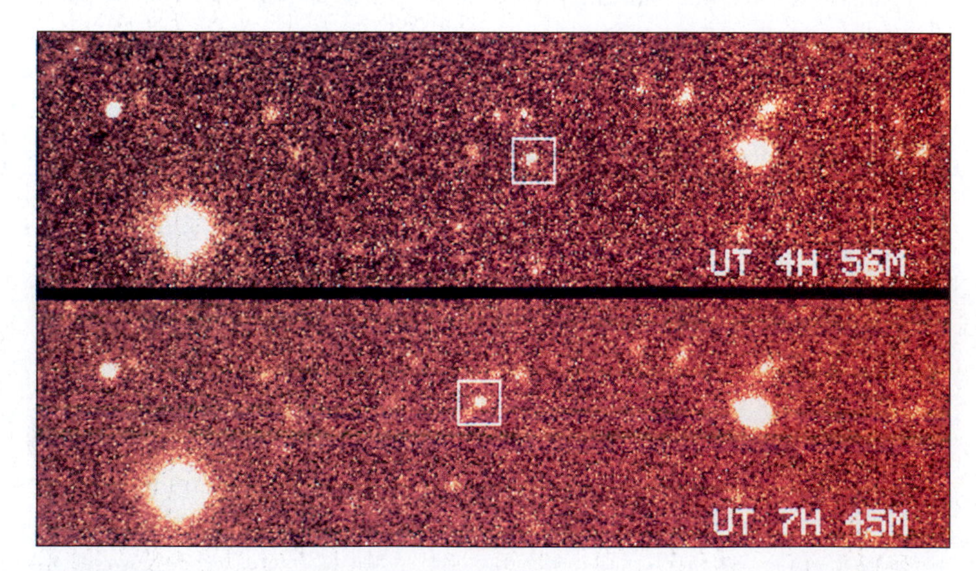


**FIGURE 20–10** A distant asteroid, 1995 QY9, one of the Kuiper belt objects, imaged with one of the Keck 10-m telescopes. We see its motion relative to the background stars after an interval of less than 3 hours.

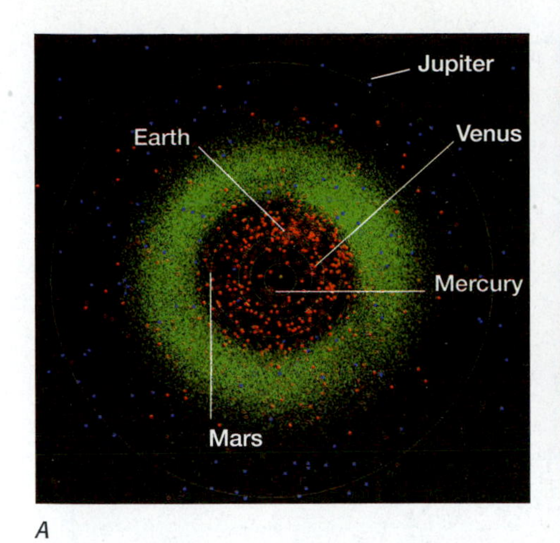


A

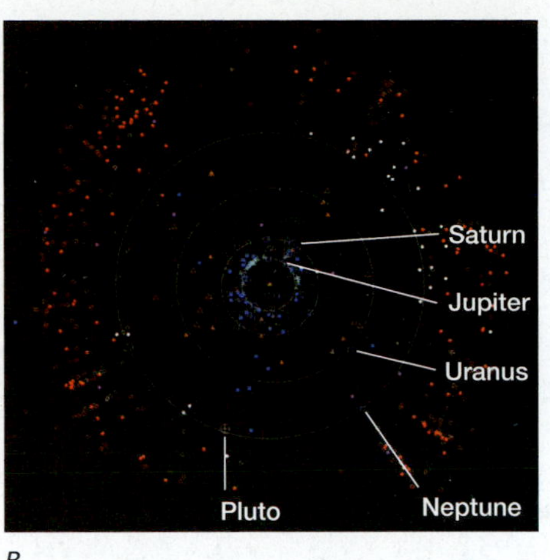


B

**FIGURE 20–11** The current positions and orbits of major and minor planets. The orbits of the major planets are shown in light blue, with their current location marked by large colored dots. Comets are shown as squares; the filled squares are for numbered periodic comets. (A) The inner Solar System. The locations of the asteroids (minor planets) are shown as green circles. Objects whose perihelia are within 1.3 A.U. are shown by red circles. Jupiter's Trojan asteroids are shown in deep blue. (B) The outer Solar System. Centaurs are shown as orange triangles, Plutinos (in 2:3 resonance wtih Neptune) as white circles, and run-of-the-mill asteroids as red circles.

formula for the sizes of planetary orbits (still without known justification), predicted the existence of a new planet in approximately that orbit.

The first asteroid was discovered by a Sicilian clergyman/astronomer, Giuseppe Piazzi, and was named Ceres after the Roman goddess of harvests. Though Piazzi followed Ceres's motion in the sky for over a month, he became ill, and before he could observe again, the object moved into evening twilight where it could no longer be seen. Ceres was lost. But the problem led to a great advance in mathematics: the young mathematician Carl Friedrich Gauss developed an important way of plotting the orbits of objects given only a few observations. The orbit Gauss worked out led to the rediscovery of Ceres. Modern versions of Gauss's methods are still in use today, though calculations are now done by computer rather than by hand.

Though the discovery of Ceres was a welcome surprise, even more surprising was the subsequent discovery in the next few years of three more asteroids. The new asteroids were named Pallas, Juno, and Vesta, also after goddesses, beginning the generally observed tradition of assigning female names to asteroids. The next asteroid wasn't discovered until 1845, and over 10,000 are now named and numbered (Fig. 20–12). They are assigned numbers when their orbits become well determined and can then be named by the discoverer. The number and name of an asteroid are often listed together: 1 Ceres, 16 Psyche, and 433 Eros, for example. Asteroids rarely come within a million kilometers of each other, though occasionally collisions do occur, producing meteoroids.

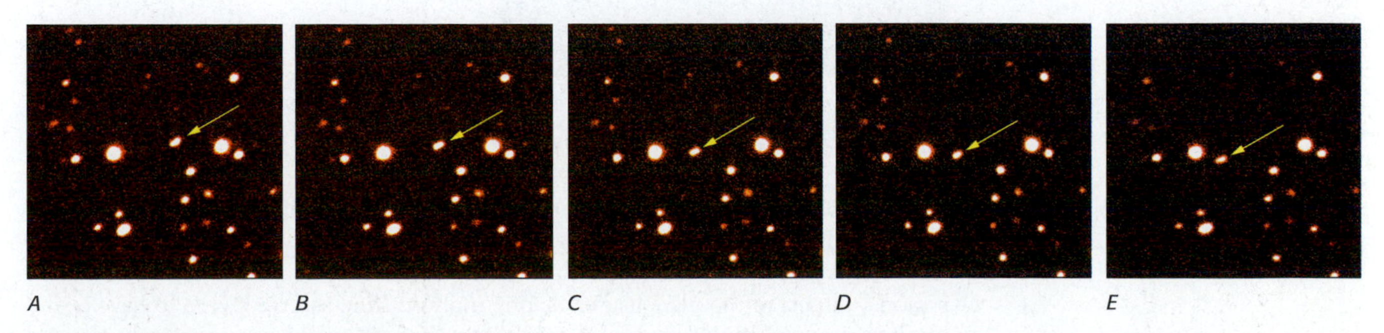

A B C D E

**FIGURE 20–12** Asteroid 5100 Pasachoff moves significantly against the background of stars in this series of exposures spanning six minutes. The asteroid appears elongated only because of its motion during the individual exposures.

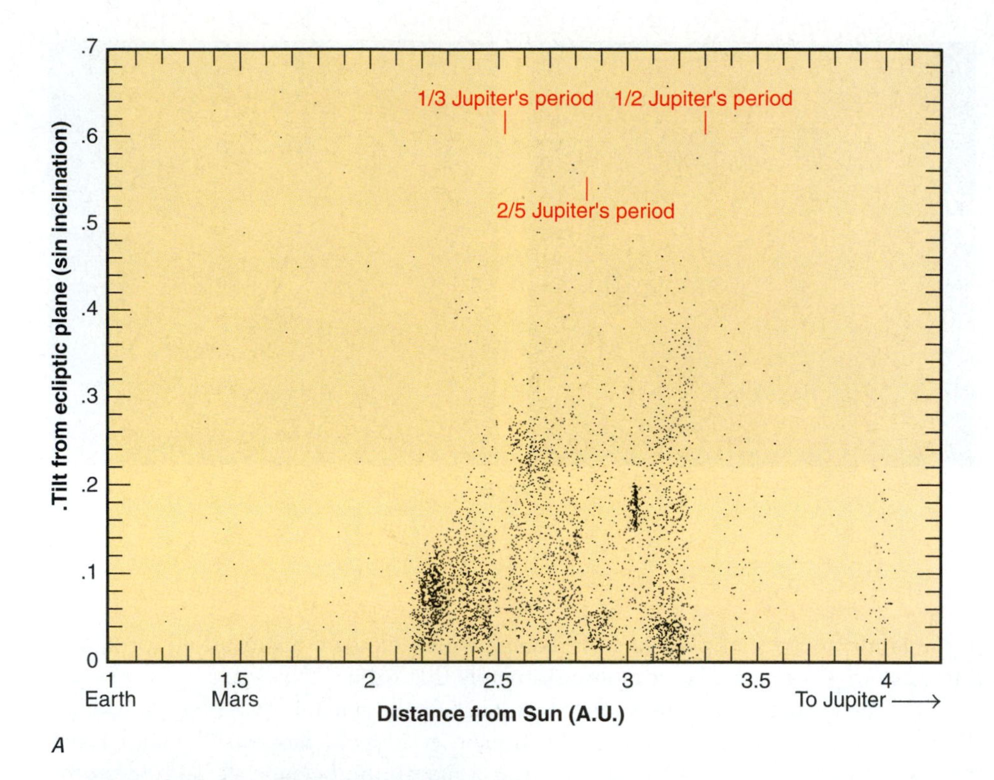


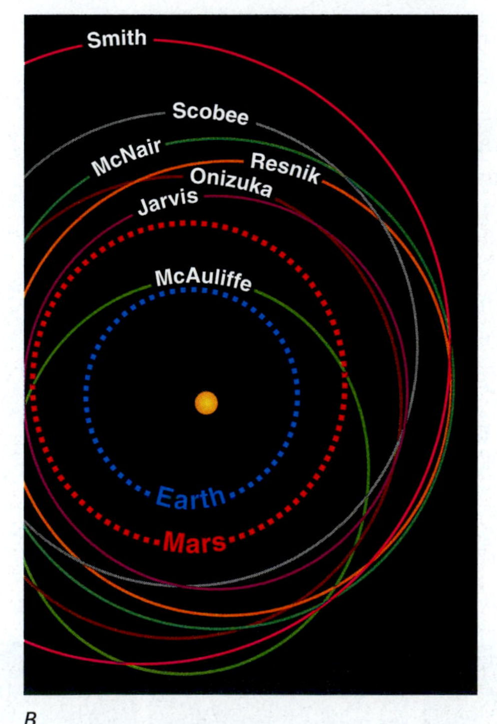


**FIGURE 20–13** (*A*) A graph of the positions of many asteroids shows gaps at certain values of semimajor axis (*horizontal axis*). These empty orbits are known as Kirkwood gaps. The vertical axis shows the sine of the angle of inclination of the orbits to the ecliptic plane. We show it here merely to spread out the graph vertically. (*B*) Asteroids have been named for each of the seven astronauts who died in the explosion of the space shuttle *Challenger* in 1986.

A graph of the sizes of the orbits of the asteroids reveals gaps in their distribution, found in the last century. Gravitational effects of the major planets cause these "Kirkwood gaps" in the asteroid belt (Fig. 20–13*A*). These gaps result from the gravitational effects of the planets and chaotic motion. If an asteroid were to have a simple fraction (½, ⅓ and so on) of Jupiter's period, it would pass relatively close to Jupiter while at the same part of its orbit each time it went by. Thus it would be tugged out of its orbit very effectively by Jupiter's gravity. So those orbits are empty.

Scientists at the Lowell Observatory in Flagstaff, Arizona, have discovered a large number of asteroids. They named seven of them after the seven astronauts who died in the explosion of the space shuttle *Challenger* (Fig. 20–13*B*).

Only half a dozen known asteroids are larger than 300 km across, and over 200 more are larger than 100 km across. The largest asteroids known (Fig. 20–14) are the size of some of the moons of the planets. Small asteroids may be only 1 km or less across. All the asteroids together contain less than 0.1 per cent of the mass of the Earth. More than 100,000 asteroids could be detected with Earth-based telescopes. The spins of almost 500 asteroids are now known; all but a dozen rotate with periods between 3 and 30 hours, like most major planets. It does not seem, however, that the spins result from primordial processes; the spins are apparently telling us something about angular-momentum transfer.

All together then, the asteroids in the asteroid belt total only 2% of the mass of the Moon. A third of that mass is in the largest asteroid, Ceres.

The asteroids at the inner edge of the asteroid belt are mostly stony in nature, and the ones at the outer edge are darker (as a result of being more carbonaceous). They affect sunlight reflected off them enough that scientists have been able to make several classes of asteroids based on their spectra. The spectra, in turn, reveal the asteroids' compositions.

Spacecraft to Jupiter and beyond travelled through the asteroid belt for many months and showed that the amount of dust among the asteroids is not much increased over the amount of interplanetary dust in the vicinity of the Earth. Particles the size of dust grains are only about three times more plentiful in the asteroid belt, and smaller particles are less plentiful. So the asteroid belt will not be a hazard for space travel to the outer parts of the Solar System.

Some asteroids turn out to have their own satellites (Fig. 20–15). In Figure 20–19, we will see another such pair.

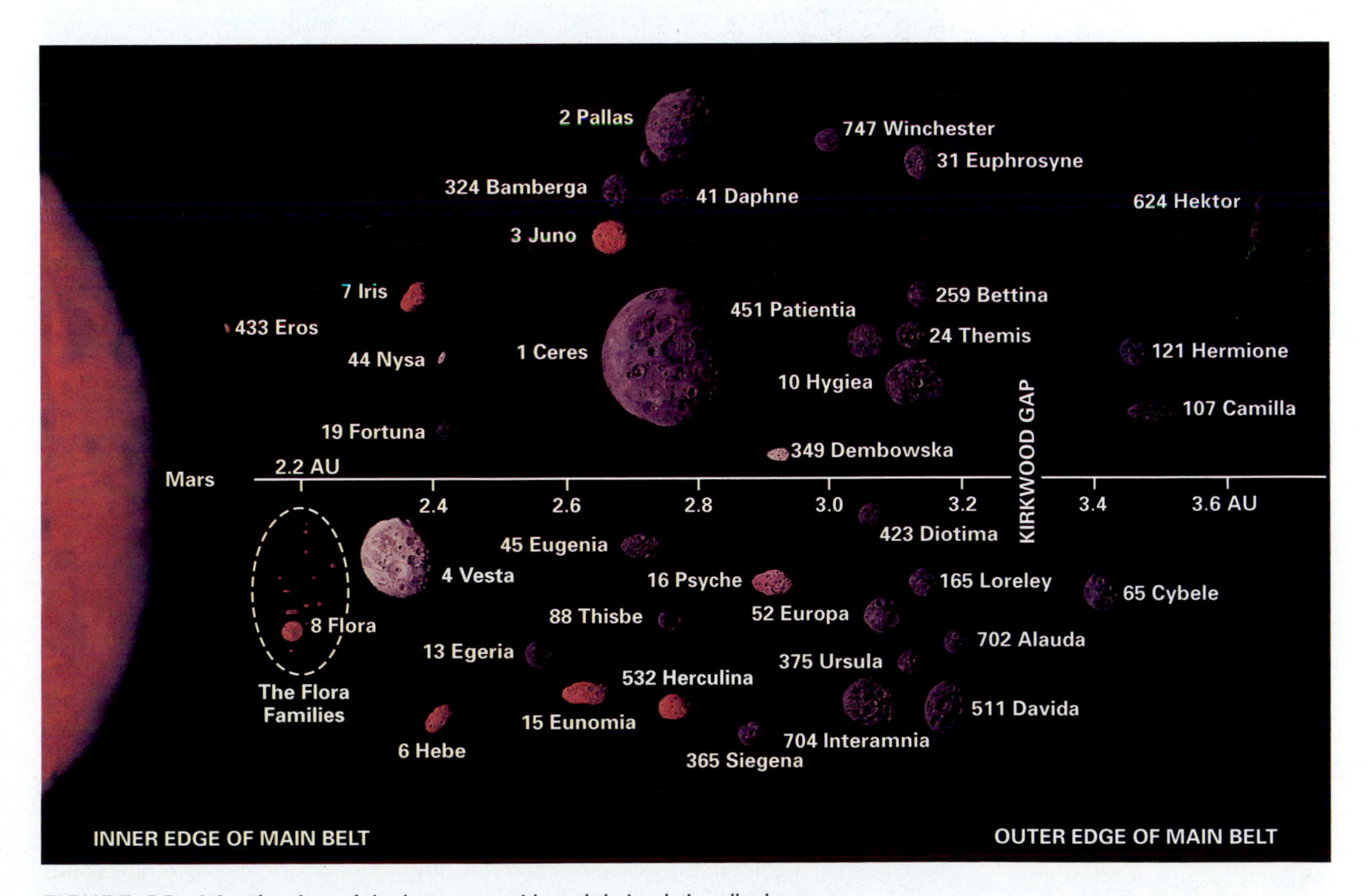


**FIGURE 20–14** The sizes of the larger asteroids and their relative albedoes.

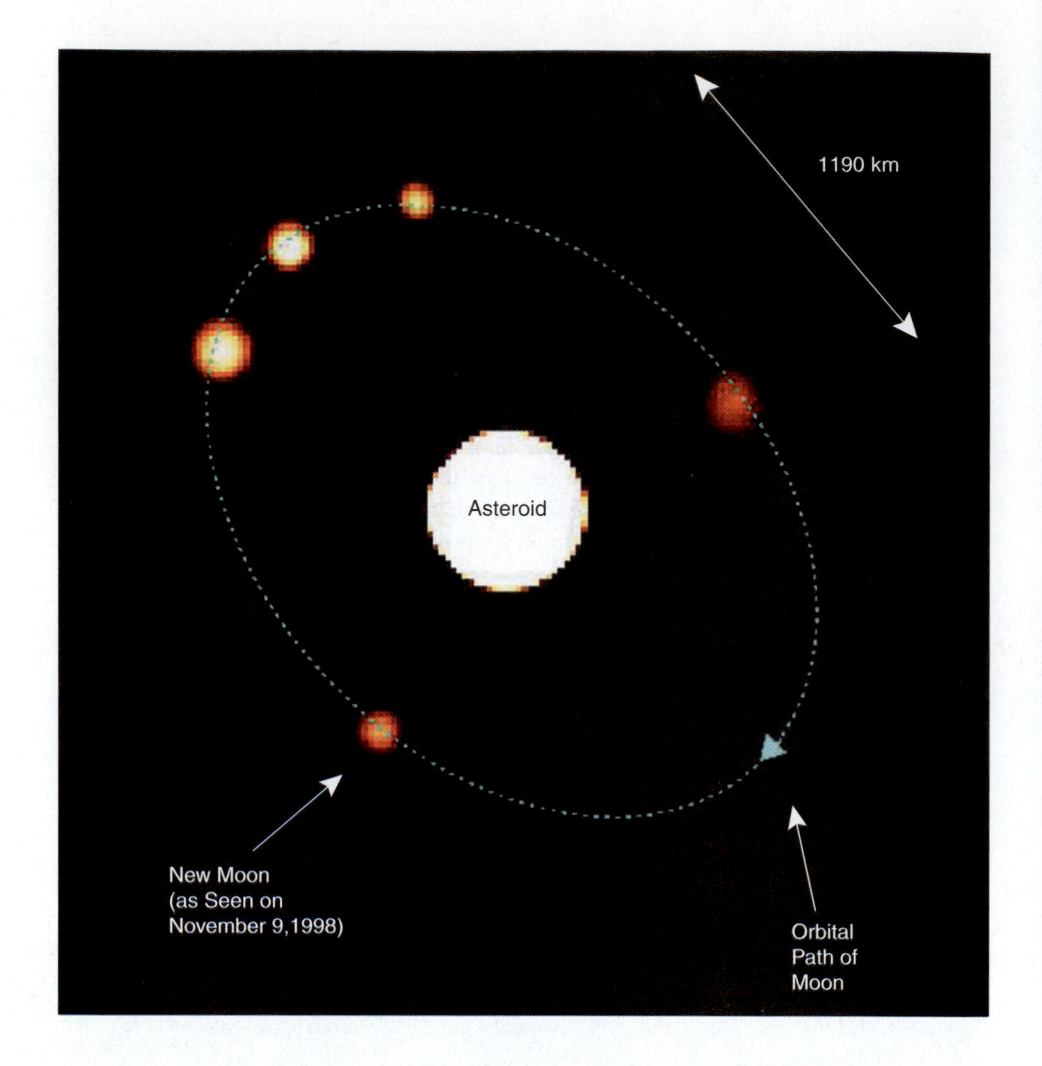


**FIGURE 20–15** A ground-based series of images, superimposed on each other, showing asteroid 45 Eugenia with its companion "moon" circling it clockwise in 4.7 days. The orbit is shown in green; the four spikes are artifacts in the telescope. Eugenia is 215 km across and its moon only 13 km across. The orbital period and the 1190 km distance between Eugenia and its moon give, by Kepler's third law, a low density for Eugenia. At only 1.2 g/cm$^3$, it must be a flying rubble pile.

How do pieces of asteroid reach Earth, where we find them as meteorites? Italian and Czech scientists, in 2000, explained that bits of asteroids, set free after an asteroid collision, can slowly drift into one of the narrow orbital bands from which the gravity of Jupiter or Saturn can eject them. They drift because one side gets more sunlight than the other, giving off more energy, with the fragment going in the opposite direction according to Newton's third law (action–reaction). It can take millions of years for them to drift far enough in space that they happen into these bands, which accounts for the long timescale measured for them in space through studying their exposure to cosmic rays. Once they are in the bands, the gravitational tugs that cause the Kirkwood gaps can send them Earthward.

Project LINEAR (*Li*ncoln *N*ear *E*arth *A*steroid *R*esearch, based at MIT's Lincoln Labs) has made, as of the beginning of 2001, over 400,000 asteroid sightings, representing over 60,000 asteroids. The University of Arizona's Spacewatch Camera, on Kitt Peak, is finding many new asteroids down to tens of meters across. Project LONEOS (*L*owell *O*bservatory *N*ear-*E*arth *O*bject *S*earch), observing from Flagstaff, Arizona, and Caltech's NEAT (*N*ear-*E*arth *A*steroid *T*racking), observing from Maui, Hawaii, Project Catalina, observing from Arizona and Australia, and a Japanese Spaceguard program are also in progress. They are all using CCD detectors and repetitive scanning to find objects that move from frame to frame. Caltech scientists are also using the Schmidt telescopes on Palomar Mountain. So the number of known asteroids is skyrocketing. The 10,000th asteroid was numbered in 1999 and already the 20,000th was numbered in 2001. As we will see later on in this chapter, the stakes have become higher in the asteroid-search business as we realize that asteroids could collide with Earth, killing billions of people.

20000 Varuna, with an estimated diameter of 1000 km, is the largest Kuiper belt object yet discovered (Pluto excepted). Varuna is lord of death and lord of the cosmos for Hindus.

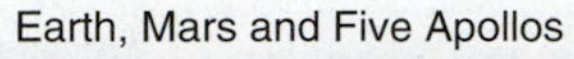


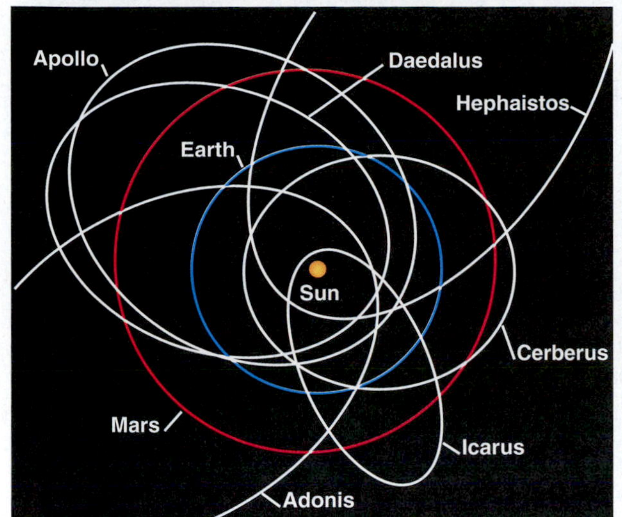


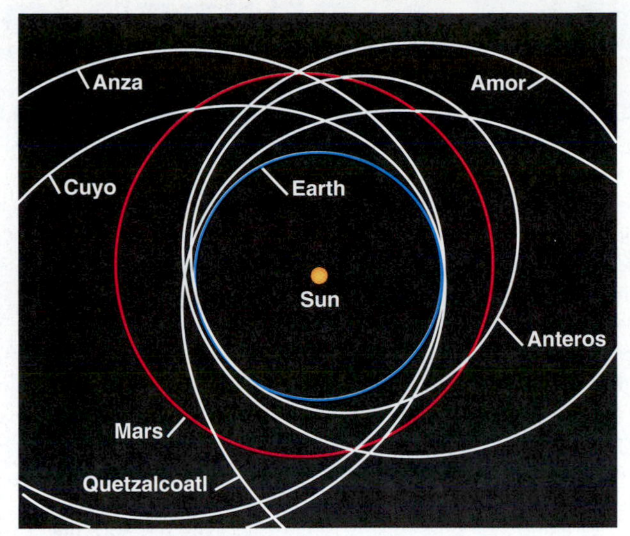


**FIGURE 20–16** The orbits of Apollo and Amor asteroids, whose orbits approach or cross that of the Earth.

## 20.2a Near-Earth Asteroids*

**Aten asteroids, Apollo asteroids,** and 10 per cent of the **Amor asteroids** have orbits that cross the orbit of the Earth (Fig. 20–16). The Aten asteroids, further, have orbits whose major axes are smaller than Earth's. (Though both Apollos and Amors have orbits whose major axes are larger than Earth's, the perihelions of the Amors are slightly farther from the Sun than those of Apollos. Technical definitions are in the glossary.) Until recently, we had observed only dozens of them, but Project LINEAR by the beginning of 2001 had observed about 100 Atens, 600 Apollos, and 600 Amors. Many of them may be dead comets.

Two asteroids even go inside the orbit of Mercury. The first to be found was named Icarus, after the inventor's son in Greek mythology who flew too near the Sun. On June 14, 1968, Icarus came within 6 million km of the Earth. One day in 1996, we were surprised to learn that an asteroid would pass only 453,000 km from Earth, about the distance to the Moon, in five days. This object, called 1996 JA1, is about 220 m across. If it had hit the Earth, it would have made a 2000-megaton explosion and left a crater kilometers across. Every few months, we are now discovering smaller objects, about 10 m across, coming even closer, with the nearest having passed only about 100,000 km from us. The French writer Antoine de Saint-Exupéry drew his Little Prince on just such a small asteroid (Fig. 20–17).

**FIGURE 20–17** Antoine de Saint-Exupéry's Little Prince tending his asteroid.

Most of the small asteroids that pass near the Earth—Icarus, Eros, Toro, Geographos, and Alinda—are among the stony asteroids. Three of the largest asteroids —Ceres, Hygiea, and Davida—are, rather, among the members of the carbonaceous group. A third type of asteroid may be mostly composed of iron and nickel. The differences may be a direct result of the variation of the solar nebula with distance from the protosun. Differences in chemical composition provide another reason not to believe the old theory that the asteroids represent the breakup of a planet that once existed between Mars and Jupiter. (There isn't enough mass present in asteroids anyway.)

Near-Earth asteroids, asteroids whose orbits bring them close to Earth, may well be the source of most meteorites, which could be the debris of collisions occurring when these asteroids visit the asteroid belt as they traverse their elliptical orbits. Eventually, most near-Earth asteroids will probably collide with the Earth. Measurements released in 2000 indicate that 1000 to 2000 of them are greater

A DEEPER DISCUSSION

## BOX 20.2 The Extinction of the Dinosaurs

On Earth 65 million years ago, dinosaurs and many other species were extinguished. Evidence has recently been accumulating that these extinctions were sudden and were caused by climate changes resulting from a 10-km-diameter asteroid hitting the Earth. Among the signs is the element iridium released from the asteroid in the impact, found worldwide in a narrow layer of rock. Subsequently, shocked quartz was found, something that must have come from an impact.

The impact would have raised so much dust into the atmosphere that sunlight could have been shut out for months; many species of plants and animals would not have been able to survive the cold. In particular, large animals like dinosaurs (Fig. A) could not have taken refuge in caves the way the smaller ones like mammals may have. Further evidence in favor of this theory is a worldwide enhancement of soot in a layer of rock 65 million years old and an ocean core that shows an adjacent layer suddenly devoid of life. The idea is very relevant to the theory that thermonuclear war could lead to a devastating "nuclear winter"—when sunlight would be prevented from hitting the Earth's surface by dust thrown up and by smoke in the atmosphere caused by the fires following a nuclear exchange—or at least, as the latest computer models seem to show, a significant but less drastic "nuclear autumn." Such calculations are also tied up with studies of the greenhouse effect. An impact could have released a lot of carbon dioxide into the atmosphere, raising the temperature by 20°C for years. Calculations show that chemical interactions with the carbon dioxide also would have been disastrous. They would have led to nitrogen-oxide smog, water poisoned with trace metals leached from soil and rock, and heavily corrosive acid rain.

As for the relation of an asteroid impact to the disappearance of the dinosaurs, evidence contrary to the above argument also exists. In particular, dinosaurs may have disappeared more gradually, which would mean that a catastrophic event was, therefore, not the cause.

Some scientists continue to search for terrestrial explanations for the mass extinctions. One such theory involves instabilities that would have arisen deep inside the Earth, just above the core. Blobs of material would have floated upward, carrying the iridium that is common within the Earth as well as other materials. In the model, the material would then have been ejected in volcanic eruptions. Research, and controversy, continues.

Maps of the Earth's gravitational field have revealed a buried crater at the northern end of Mexico's Yucatán peninsula, since the high-density impacted rock has especially strong gravity (Fig. B). It is the largest crater on Earth, hundreds of kilometers across. It is increasingly widely accepted that this crater, Chicxulub, is the site of the meteor impact 65 million years ago that killed off many species, including the dinosaurs.

***Discussion Question***

Should we build a large telescope for perhaps $50 million to spend all its time looking for meteoroids, asteroids, or comets that might hit the Earth? Why or why not?

*A*

**FIGURE A** The dinosaurs and many other species disappeared suddenly 65 million years ago, marking the end of the Cretaceous era. In the movie *Jurassic Park,* dinosaurs were cloned from their DNA, something we can't do—yet.

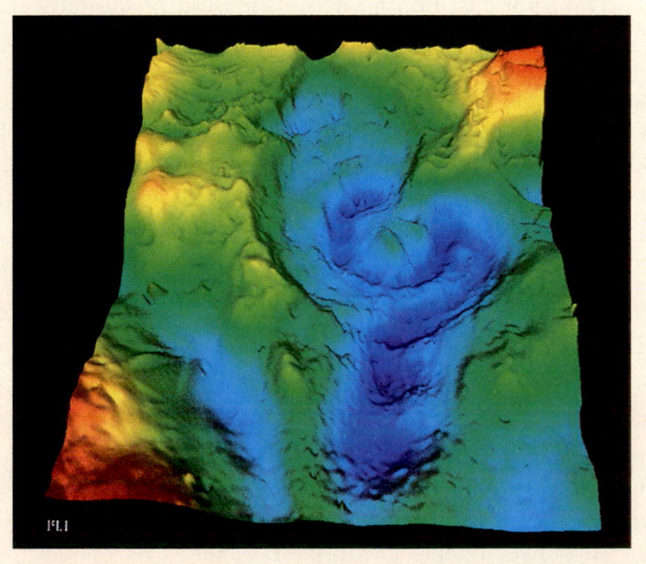

*B*

**FIGURE B** The impact structure known as Chicxulub, at the north end of the Yucatán. The blue areas are low-density rocks broken up by the impact. The green mound at the center is denser and probably represents a rebound at the point of impact. It is the largest crater known on Earth.

than 1 km in diameter. None have been found larger than 10 km across. Statistics show that a collision of the tremendous magnitude of a 1-km asteroid should take place every few hundred million years and could have drastic consequences for life on Earth. Calculations indicate a 1 per cent chance of a collision per thousand years. There are a hundred times more smaller objects, with a 1 per cent chance that an asteroid greater than 300 m in diameter would hit the Earth in the next century. Such a collision could kill hundreds of thousands or millions of people (Table 20–3).

Near-Earth asteroids have sparked a current debate. At the current rate, we could discover 90 per cent of them in the next decade (Figure 20–18). A major source of discovery is Project LINEAR. It has found 70 per cent of the near-Earth asteroids discovered in the last couple of years. As we discussed, LINEAR is also discovering many comets.

We have the capability of discovering more such asteroids and discovering them sooner if we devote more resources to the task, perhaps even building a telescope dedicated to the purpose. It has been suggested that since nuclear bombs might be used to change the course of an asteroid, nudging its track a few orbits before it intersected with Earth, the discussion might be encouraged by people wanting to continue developing nuclear weapons. The matter is very controversial. Still, current statistics indicate that the chance of being killed as the result of an asteroid collision with Earth is about the same as the chance of dying in a crash each time you take an airplane flight (Table 20–4). The former would kill billions of people but happen only every few hundred million years. Thankfully, the chances of both are very low.

**Even more devastating than the extinction event 65 million years ago was one that took place 250 million years ago. It killed 90 per cent of marine species and 70 per cent of backboned land animals from the Permian period. The Triassic period, which followed, had more active animals.**

### 20.2b Distant Asteroids

Another asteroid group is the Trojans, which always oscillate about the points 60° ahead of or behind Jupiter in their solar orbits (two of Jupiter's "Lagrangian points"). At the Lagrangian points, the effects of the orbital velocity and the gravitational pulls

**TABLE 20–3** Consequences of an Asteroid Collision

| Asteroid Diameter (meters) | Average Interval Between Impacts (years) | Consequences of Asteroid Strike | Possible Area of Direct Destruction for Land Impacts |
|---|---|---|---|
| Less than 40 | | Asteroids detonate in upper atmosphere. | |
| 75 | 1000 | Iron asteroids leave craters on the ground; stony asteroids explode in the lower atmosphere. | Large city such as Washington, London or Moscow |
| 160 | 4000 | Asteroids explode on impact with the Earth's surface; ocean impacts raise significant tsunamis. | Large urban area such as New York |
| 350 | 16,000 | Impacts on land produce craters; ocean impacts cause ocean-wide tsunamis. | Small state such as Delaware |
| 700 | 63,000 | Land impacts produce large craters; ocean impacts can cause tsunamis reaching the scale of the hemisphere. | Medium-sized state such as Virginia |
| 1700 | 250,000 | Land and ocean impacts throw enough dust into the atmosphere to affect climate and agriculture; ocean impacts cause global tsunamis. | Large state such as California |
| 3000 | 1,000,000 | Global climate change; global ejecta from the impact sparks widespread fires. | Large nation such as Mexico or India |
| 7000 | 10,000,000 | Global fires and probable mass extinction. | Area approaching continental scale, such as the U.S. |
| 16,000 | 100,000,000 | Large mass extinction that threatens the survival of all advanced life forms. | An entire continent |

*Source:* Report of the Task Force on Potentially Hazardous Near Earth Objects.

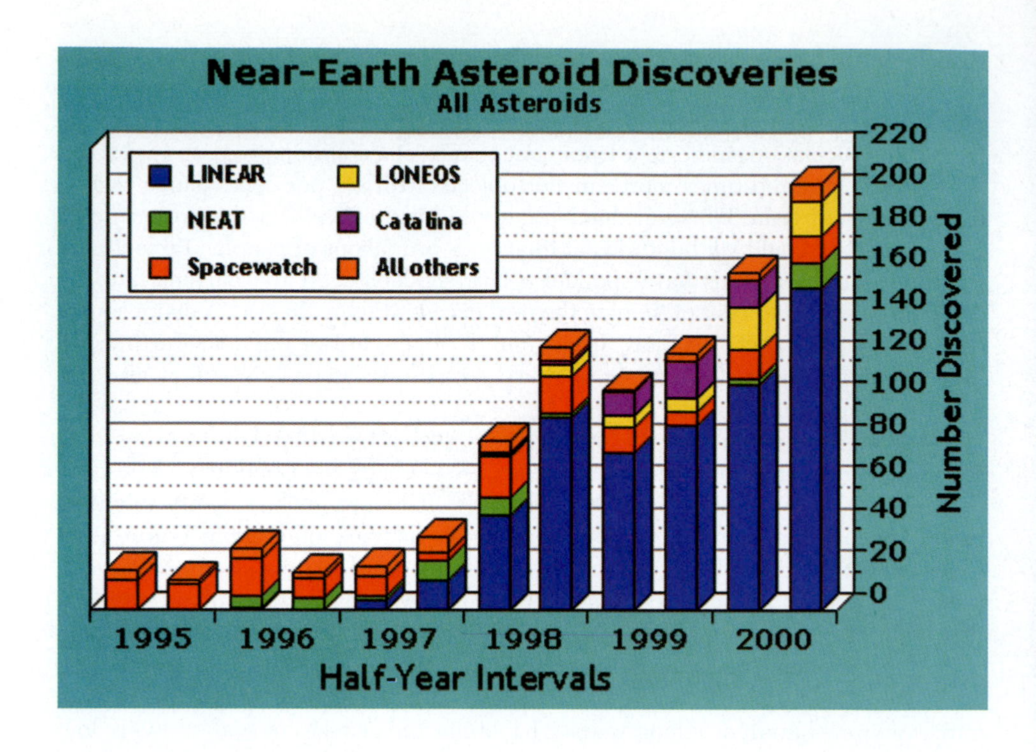


**FIGURE 20–18** The results of current search programs for Near-Earth Asteroids.

The first Trojan asteroid to be discovered was named Achilles. Later it was decided to name the asteroids in the Lagrangian point ahead of Jupiter after Greek heroes and those in the point trailing Jupiter after Trojan heroes. But the first Trojan asteroids to be named don't follow this rule. 617 Patroclus, a Greek, is in the Trojan camp. Similarly, there is one Trojan, 624 Hektor, in the Greek camp. In the *Iliad*, Homer's epic poem of the Trojan War, Achilles, the bravest of the Greeks, slew Hektor, the greatest Trojan warrior, in revenge for Hektor's killing his friend Patroclus, who had been wearing Achilles's armor.

of the Sun and of Jupiter balance, so material does not move toward one or the other. Recent observations have more than tripled the number of Trojan asteroids known to 200.

We have discussed the Kuiper belt objects and the Centaurs in Sections 19.3 and 19.5. Both can be considered types of asteroids. The Kuiper belt objects orbit beyond Neptune, while the Centaurs' orbits have been modified to send them into elliptical orbits between the orbits of Jupiter and Neptune.

### 20.2b Radar Observations

As ground-based radars become stronger, scientists have been able to reflect radar off a few asteroids. Some have proved to be irregular in shape, perhaps even in the form of a dumbbell (Fig. 20–19).

**TABLE 20–4** Asteroid Relative Liabilities

| Cause of Death | Probability |
|---|---|
| Traffic accident | 1 in 100 |
| Murder | 1 in 300 |
| Fire | 1 in 800 |
| Firearms accident | 1 in 2500 |
| Electrocution | 1 in 5000 |
| Passenger aircraft crash | 1 in 20,000 |
| **Asteroid or comet impact** | **1 in 20,000** |
| Flood | 1 in 30,000 |
| Tornado | 1 in 60,000 |
| Venomous bite or sting | 1 in 100,000 |
| Firework accident | 1 in 1 million |
| Food poisoning by botulism | 1 in 3 million |

*Source:* Nature (*vol 367, p 33*)

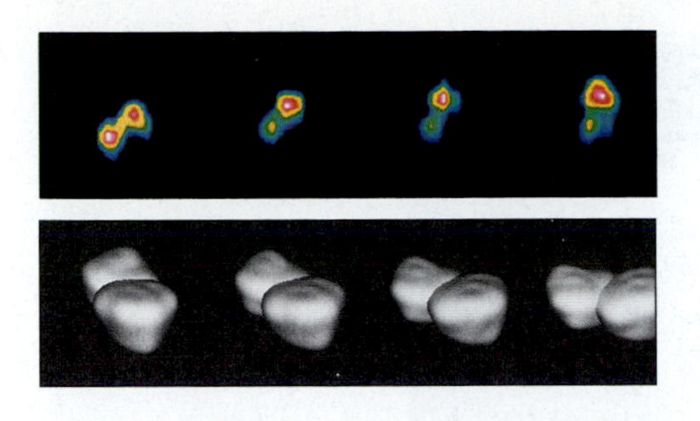

**FIGURE 20–19** (*Top*) False-color images of the radar signals from asteroid 4769 Castalia. (*Bottom*) Computer reconstructions based on a 3-dimensional model calculated from the radar data.

Are asteroids solid or are they clumps of rubble flying together through space? Studies of the rotation speeds of asteroids show that almost none rotate more rapidly than once every $2^1/4$ hours, the speed at which their rotation would throw off matter. Thus asteroids could be clumps of rock held together by their light mutual gravity. We will see later in this chapter that going up close to asteroids with spacecraft is the way to resolve the question, or at least to determine which asteroids fit in which category.

## 20.3 ASTEROIDS OBSERVED FROM SPACE

The Galileo spacecraft en route to Jupiter passed close to asteroid Gaspra on October 29, 1991 (Fig. 20–20*A*). The illuminated part is 16 km × 12 km. It rotates every 7 hours. One surprise so far is that the spacecraft detected a magnetic field, so Gaspra is probably made of metal and is magnetized. Galileo passed the asteroid Ida (Fig. 20–20*B*) on August 28, 1993. Ida has a small satellite, probably formed at the same time as Ida when a larger asteroid broke up. Asteroid satellites have since been found to be common, with more pairs discovered in 1999 and 2000. Theoreticians had no trouble describing how Gaspra and Ida may have both formed as fragments of a collision of a larger ancestral asteroid. But the additional discoveries don't always

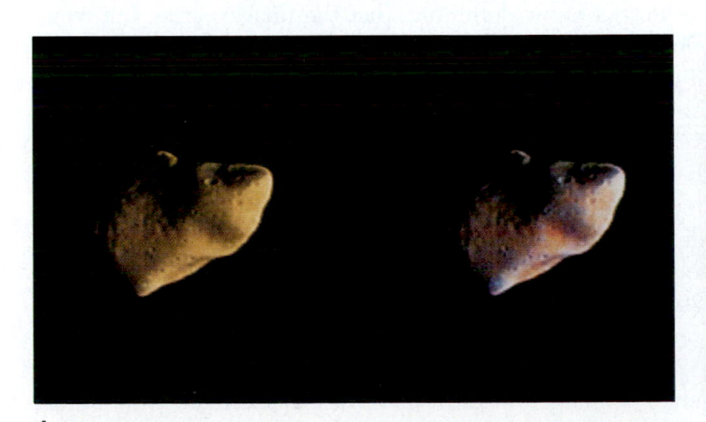

*A*

*B*

**FIGURE 20–20** (*A*) Gaspra, in this high-resolution image taken by the Galileo spacecraft, looks like a fragment of a larger body. The illuminated part seen here is 16 km by 12 km. Galileo discovered that Gaspra rotates every 7 hours. We see a natural color image of Gaspra at left. At right, the color-enhanced image brings out variations in the composition of Gaspra's surface. (*B*) Ida and its moon (or satellite) Dactyl photographed from the Galileo spacecraft at a range of 10,500 km. Filters passing light at 4100 Å (violet), 7560 Å (infrared), and 9860 Å (infrared) are used, enhancing the color since the CCD is most sensitive in the infrared. Subtle color variations show differences in the physical state and composition of the soil. The bluish areas around some of the craters suggest a difference in the abundance or composition of iron-bearing minerals there. Dactyl differs from all areas of Ida in both infrared and violet filters.

**FIGURE 20–21** 253 Mathilde imaged from the Near Earth Asteroid Rendezvous mission (later renamed NEAR Shoemaker) in 1997.

fit that model. The adaptive-optics discovery, in particular, that the asteroid Antiope is really made of two equal-size bodies orbiting each other is harder to understand. One suggestion is that an ancestral Antiope was reduced to a collection of rubble by earlier collisions, and then a glancing blow started the rubble pile spinning, at which it flew apart into two equal-size piles.

The Cassini mission, when it passed through the asteroid belt en route to Saturn, imaged asteroid Masursky, though did not come as close as Galileo did to its asteroids so didn't take photos as detailed.

The Hubble Space Telescope has imaged Vesta, which had already been known to be made of basalt. We do not know definitely what the images show, but some features look like lava flows. Apparently, Vesta is large enough for its radioactive elements to have heated up its inside. Since meteorites on Earth have been identified as coming from Vesta, Vesta joins the Moon and Mars as the only celestial objects from which we have samples.

NASA's Near Earth Asteroid Rendezvous (NEAR) mission was launched in 1996. First, it passed asteroid 253 Mathilde (Fig. 20–21) in 1997. From its low density, 66-km-long Mathilde seems to be a rubble pile, made of rocks that are gravitationally bound only loosely with spaces between them. Later, NEAR went into orbit around 433 Eros (Fig. 20–22) on February 14, 2000, (suitably, given the asteroid's name, Valentine's Day) resolving features as small as 1 km across on the 32-km-long elongated asteroid (Fig. 20–23). It was then renamed NEAR Shoemaker, after the planetary geologist Eugene Shoemaker. Most of its surface is saturated with craters smaller than 1 km in diameter. The largest crater is over 5 km across. Grooves and ridges are also seen, and perhaps show planes of weakness that resulted from frequent collisions. The International Astronomical Union named features on Eros with names of great lovers in history and literature, such as Orpheus and Eurydice, Casanova, Cupid, Lolita, and Don Quixote.

Tracking the spacecraft's radio signals has shown that Eros's density is 2.7 g/cm$^3$. The acceleration of gravity at various points on Eros's surface ranges from 2.1 to 5.5 mm/sec$^2$, compared with 10 m/sec$^2$, over 1000 times greater, on Earth's surface. By contrast, NEAR Shoemaker measured Mathilde's density to be only 1.3. The conclusion is that Eros is a solid object, in contrast with Mathilde as a rubble pile. Eros's

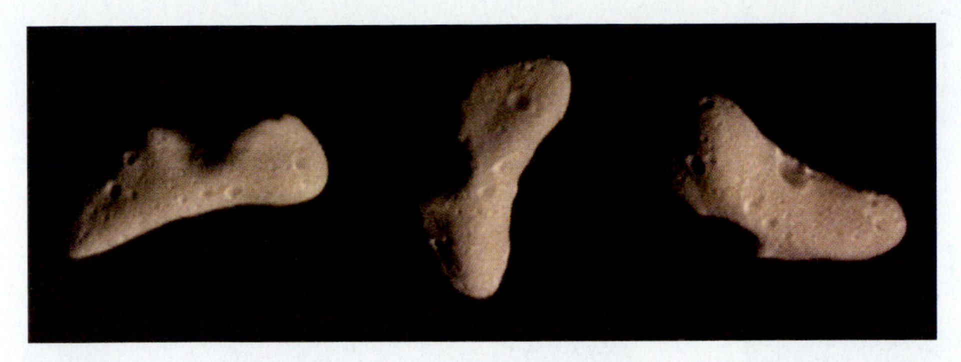

**FIGURE 20–22** The rotation of asteroid 433 Eros, imaged from the Near Earth Asteroid Rendezvous mission (renamed NEAR Shoemaker) in orbit around it. Eros's surface looks a uniform butterscotch color in visible light but shows more variation in the infrared.

albedo of 27 per cent, in contrast with Mathilde's 4 per cent albedo, coupled with the carbon-based mixtures detected in spectra of Mathilde's surface but not Eros's, show further tremendous differences between the two asteroids. Mathilde's structure may have allowed it to survive the collisions that made at least 5 craters between 19 and 33 km in diameter while not even showing surface cracks.

Mathilde, with its black color and spectrum, typifies a broad category of asteroid-belt object known as C-type asteroids. They are found especially in the outer regions of the asteroid belt. Eros, with its gray color and spectrum, typifies S-type asteroids, a broad category that predominates in the inner asteroid belt. It may typify the composition of the primordial matter from which the Earth and other objects of the inner Solar System formed.

*(text continues on page 373)*

**FIGURE 20–23** Eros, from NEAR Shoemaker in orbit only 220 km above its surface. Surface detail down to 35 m is visible. Note how different its opposite hemispheres appear.

# PHOTO GALLERY

## Eros: An Asteroid Orbited Up Close

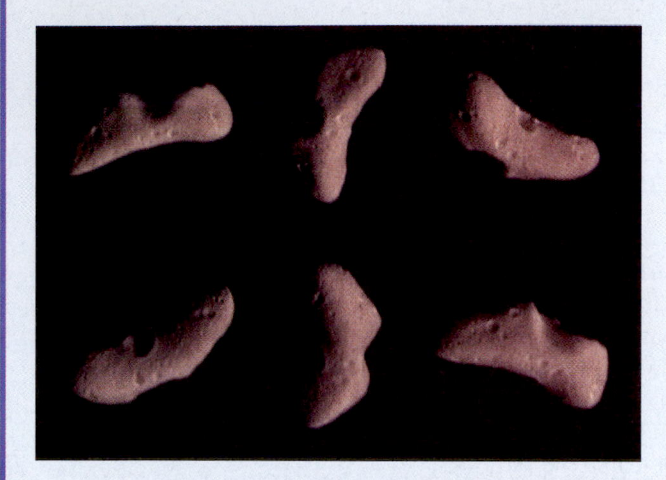

▲ Asteroid 433 Eros, tumbling over 5½ hours as the Near Earth Asteroid Rendezvous spacecraft approached.

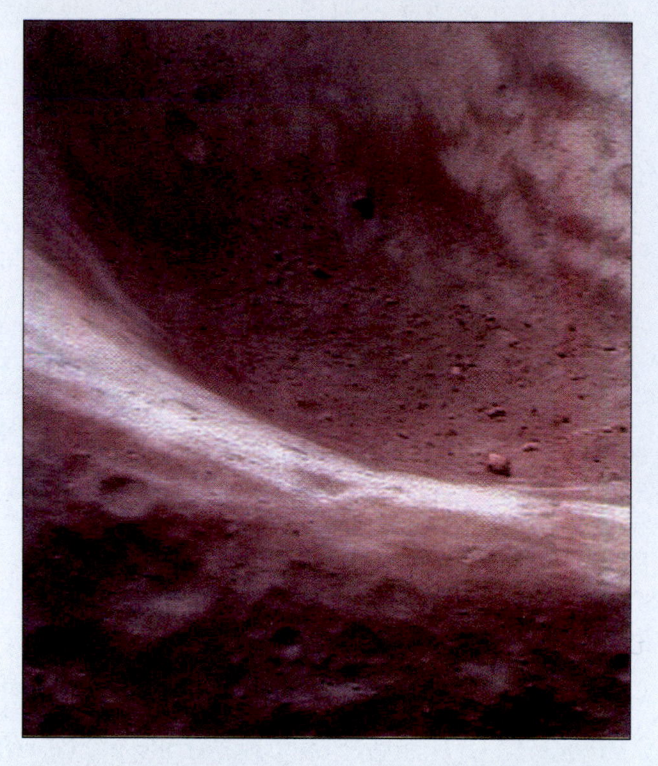

▲ Inside Eros's 5.3-km-diameter crater, showing rocks and regolith. Reddish regions have been subjected to space weathering.

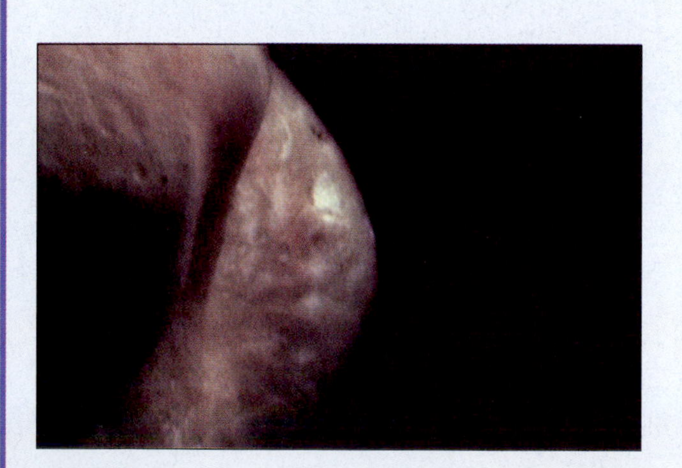

▲ This false color view, a combination of infrared and blue images, shows color differences between parts of the asteroid. The bright and greenish-gray appearing regions near the rim of the saddle are thought to represent relatively fresh exposures of soil from under the surface. The pinkish regions have been modified by space weathering from small impacts and the solar wind.

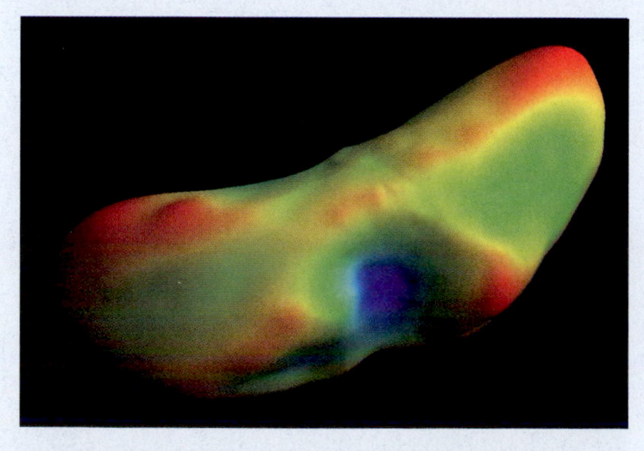

▲ A color display of "uphill" (*red*) and "downhill" (*blue*) sloping regions, to illustrate gravitational topography painted onto Eros's shape. Tracking of the radio signals from NEAR Shoemaker allowed Eros's gravitational pull on it to be measured. A ball dropped onto a red spot would try to roll across the nearest green area to the nearest blue area.

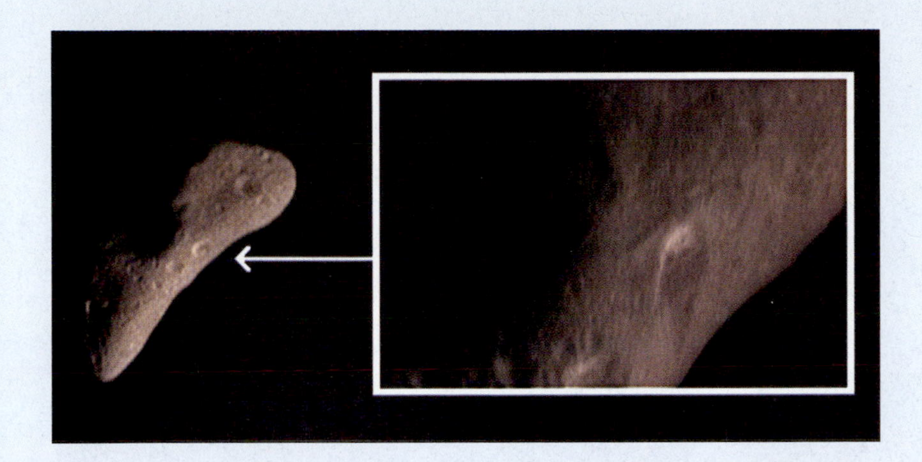

Surface details showed better once NEAR Shoemaker went into orbit around Eros.

A mosaic from a low-altitude flyover of Eros from an altitude of only 6.4 km, showing rocks as small as 1.4 meters across.

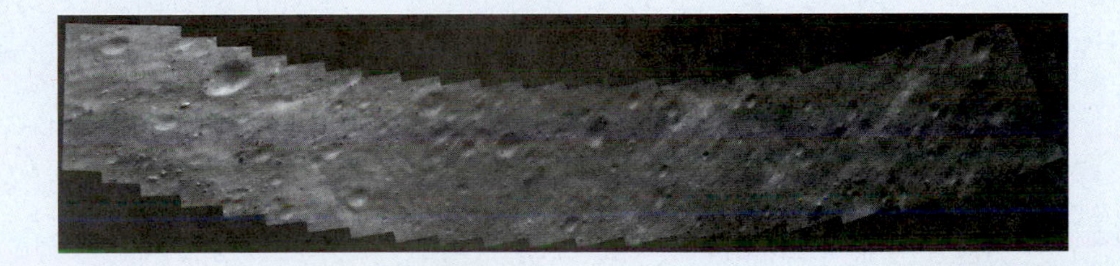

Rocks on Eros; the boulder in lower center is about 15-m across.

*(Photo Gallery continues next page)*

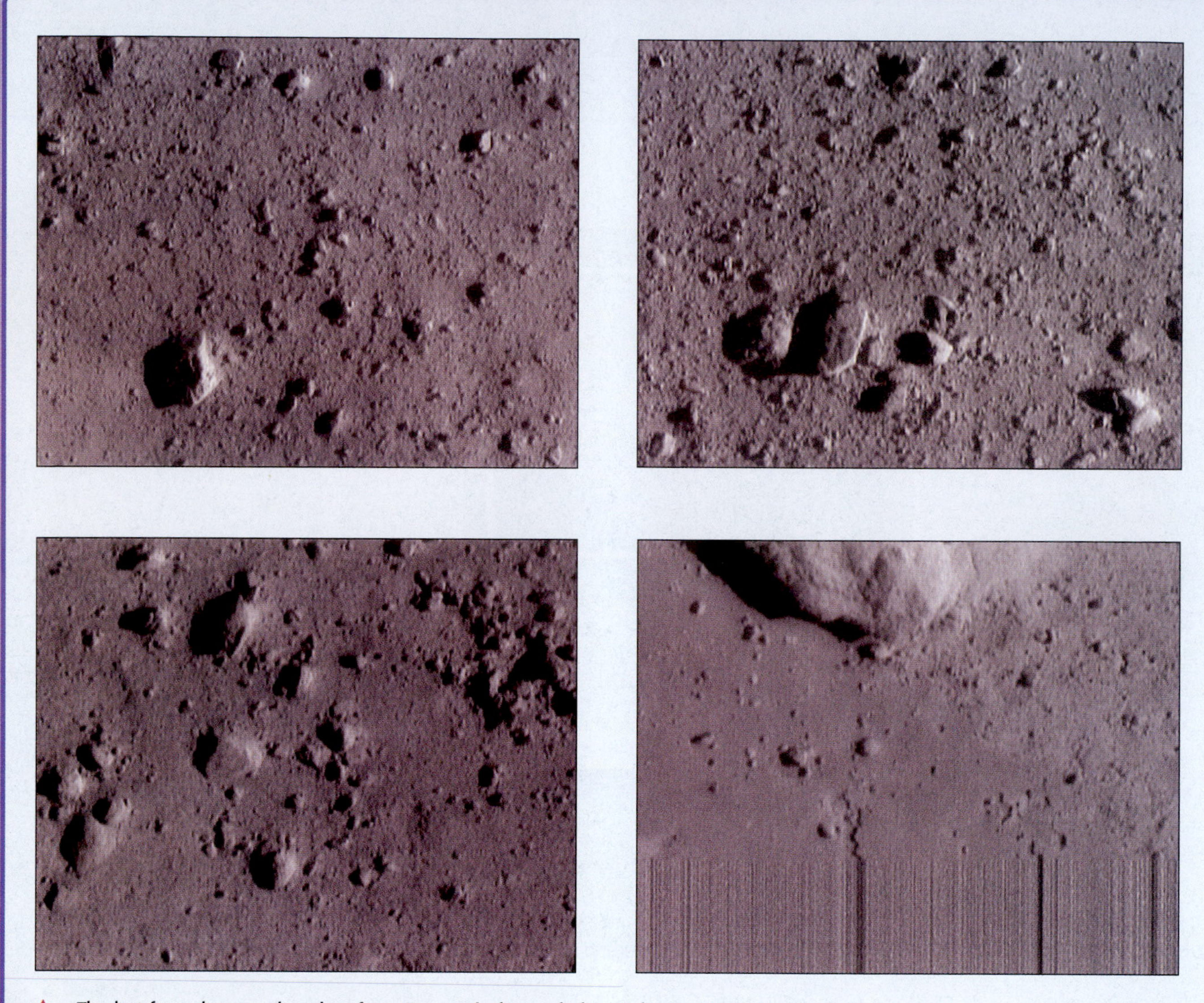

▲ The last four photographs taken from Eros as it descended on February 12, 2001. The last photograph shows features as small as 1 cm across! The transmission of the photo was cut when the spacecraft impacted Eros.

NEAR Shoemaker sent back data about Eros's surface structure, composition, topography, and internal structure. Eros is probably a fragment of a larger asteroid released by a collision. By seeing what x-rays and gamma-rays were given off when the asteroid was zapped by solar flares, the spacecraft was able to measure the abundances of the elements on Eros's surface, linking the properties of this asteroid to ordinary chondrites found as meteorites on Earth. The joint studies seem to be showing how "space weathering" can make asteroids appear redder over time. The studies are so detailed that they link astronomy with geology even more. In its year of orbiting Eros at altitudes measured in tens of kilometers, NEAR Shoemaker took 160,000 photos and made the most detailed map of any celestial body other than the Moon. Movies are available on the Web.

On February 12, 2001, NEAR Shoemaker was gradually lowered onto Eros's surface, hitting at the speed of a brisk walk. After all, Eros's surface gravity is so slight that a 200-pound person would weigh only 2 ounces there. The spacecraft was taking photographs all the way down, seeing finer and finer detail. The scientists had seen features not seen on other solar-system bodies, such as rocks sticking up out of the dust for no apparent reason, and they wanted the closest views.

The surface of Eros they saw close-up was unlike that of any other known object. Fine dust has filled in low regions, forming flat surfaces that are being referred to as "ponds." The "ponds" are surrounded by rougher transition zones being referred to as "beaches." Boulders stick up, with perhaps a million of them larger than 8 meters across existing on Eros's surface. Perhaps the boulders rise to the top in the same way that larger nuts in a can of mixed nuts wind up at the top as smaller ones settle. In the final image, some irregularly shaped regions tens of centimeters across have settled or been compressed by a few centimeters, and are being called "footprints." Scientists are far from figuring out why Eros's surface looks this way.

Incredibly, the spacecraft survived the landing. Its gamma-ray spectrometer sent back data about the surface composition for several days, though the camera was no longer able to make images of Eros's surface.

Eros is a Near-Earth Object. Should we find one of them threatening to collide with Earth, we must know whether the object is solid or a rubble pile to make an informed decision on how to deal with the threat. Further, Near-Earth Objects may turn out to be rich resources of minerals that we can mine. Studies of the mineralogy of Eros will always be at the basis of our knowledge relevant to selecting the appropriate objects.

Another particularly interesting Near-Earth asteroid is 3753 Cruithne. It all but shares the Earth's orbit! However, its highly inclined version of Earth's orbit keeps it from ever colliding with our planet. It never gets closer than 0.1 A.U., or about 40 times the distance from the Earth to the Moon, and won't be any closer than three times that for the next 28 million years. Its strange motion is described at http://aries.phys.yorku/ca/ ~wiegert/3753.html. Cruithne got its name from the first Celtic group to go to the British Isles (a group also known as the Picts), going from the European continent between about 800 and 500 B.C. The European Rosetta mission, following its launch in 2003, is to pass asteroids Otawara and Siwa in its 8-year journey to Comet 46P/Wirtanen.

## CORRECTING MISCONCEPTIONS

| ✖ Incorrect | ✔ Correct |
|---|---|
| Shooting stars are stars. | Shooting stars are interplanetary dust burning up in our atmosphere. |
| The asteroid belt is densely packed. | Distances between asteroids are very great, unlike in the movies. |
| The Kirkwood gaps have spacings that are fractions of a planet's orbital radius. | The Kirkwood gaps arise at spacings that correspond to orbits that are simple fractions (1/2, 1/3) of the period of Jupiter. |
| Asteroids are not a hazard for us on Earth. | The odds are that asteroids hit the Earth with some long-term (though rare) regularity and do great damage. |

## SUMMARY AND OUTLINE

Meteoroids (Section 20.1)
- *Meteoroids:* in space; *meteors:* in the air; *meteorites:* on the ground
- Most meteorites that hit the Earth are chondrites, which resemble stones and are therefore often undetected; iron meteorites are more often found.
- Meteorites may be chips of asteroids and thus give us material with which to investigate the origin of the Solar System.
- Some meteors arrive in showers; others are sporadic.

Asteroids, also called minor planets (Section 20.2)
- Most are in asteroid belt, 2.2 to 3.2 solar radii (between orbits of Mars and Jupiter).
- Apollo and Aten asteroids come within Earth's orbit.
- Asteroids can be hazards to life on Earth.
- Trojan asteroids in the Lagrangian points (Jupiter ± 60°)
- Radar images can be obtained.

Space observations (Section 20.3)
- Galileo viewed Gaspra and Ida/Dactyl.
- Near Earth Asteroid Rendezvous (NEAR) Shoemaker orbited and landed on Eros.
- Rosetta mission: asteroids and comet

## KEY WORDS

meteoroids
meteors
shooting stars
fireballs
bolide
meteorite
micrometeorites
iron meteorites
irons
stones
tektites
chondrites
achondrites
stony-iron
carbonaceous*
carbonaceous chondrites*
sporadic
showers
radiant
minor planets
asteroids
asteroid belt
Aten asteroids*
Apollo asteroids*
Amor asteroids*

*This term appears in an optional section.

## QUESTIONS

1. What is the relationship between meteorites and asteroids?
2. Why don't most meteoroids reach the Earth's surface to become meteorites?
3. Why do some meteor showers last only a day while others can last several weeks?
4. Why are meteorites important in our study of the Solar System?
5. Compare asteroids with the moons of the planets.
6. How might the occultation of a star by an asteroid tell us whether the asteroid has an atmosphere?
7. Using Appendix 4, compare the sizes of the largest asteroids with those of the smallest planetary satellites.
8. What are three locations in the Solar System from which meteorites come?
9. Why do many people think that an asteroid had to do with the extinction of the dinosaurs? Argue both pro and con.
10. What has the Galileo spacecraft found out about asteroids?
11. Why do we expect to find out a lot more about asteroids in the next few years?
12. What is the origin of each part of the name of the NEAR Shoemaker mission?
13. What did NEAR Shoemaker study and why was the mission unusual?
14. What are three major results from the NEAR Shoemaker mission?

## OBSERVING PROJECT

1. Observe the next meteor shower. No instruments will be necessary; the naked eye is best. Try to trace back the trails you see to find the radiant. What constellation is it in? If you observe for a long time, check whether you see more meteors after midnight, though this effect also depends on when the peak of the comet dust distribution crosses the Earth's orbit.

## DISCUSSION QUESTION

1. Consider why the percentage of iron meteors is different in falls vs. finds. Extrapolate those findings to discuss bias and systematic errors in observations. How might such errors skew your measurements of stars in distant galaxies?

## USING TECHNOLOGY

### W World Wide Web

1. From the mission homepages, follow the status of space missions to asteroids.
2. Check on Hubble views of asteroids.
3. Follow the latest numbers of asteroids of different types discovered in the Project LINEAR homepage (http://www.ll.mit.edu/LINEAR/), the Spacewatch homepage (http://www.lpl.arizona.edu/spacewatch), the NEAR homepage (http://near.jhuapl.edu) and the Central Bureau for Astronomical Telegrams homepage (http:// cfa-www.harvard.edu/iau/lists/Unusual.html and /OuterPlot.html).
4. On the Spacewatch site (http://www.lpl.arizona.edu/spacewatch), read about the history of asteroid research.
5. Follow the Near Earth Object programs at http://neo.jpl.nasa.gov.

### REDSHIFT

1. From high above the Solar System, view the asteroid belt. Turn off the stars and deep-sky objects to enhance the view. Evaluate the distribution of asteroids.
2. Follow the orbits of the 3 asteroids visited by spacecraft. See if they are in the evening sky tonight.
3. Distinguish among orbits of Apollo, Amor, and Aten asteroids.
4. Explore how the distribution of asteroids changes with the size range you display.
5. Consider the content of Science of Astronomy: Worlds of Space: Moons, Asteroids, and Comets.

**Yoda.** **(From Star Wars: Episode I – The Phantom Menace,** © Lucasfilm, Ltd. & courtesy of Lucasfilm, Ltd. All rights reserved. Used under authorization.)

# Chapter 21

# Life in the Universe

We have discussed the nine planets and the dozens of moons in the Solar System and have found most of them to be places that seem hostile to terrestrial life forms. Yet some locations besides the Earth—Mars, with its signs of ancient running water, and perhaps Titan or Europa—have characteristics that allow us to convince ourselves that life may have existed there in the past or might even be present now or develop in the future. **Exobiology,** or **astrobiology,** is the study of life elsewhere than on Earth. **Astrobiology** is the study of life anywhere in the Universe, including past, present, and future life on Earth.

**AIMS:** To discuss the possibility that intelligent life exists elsewhere in the Universe besides on the Earth and our chances of communicating with it, and to assess the reports of life discovered on Mars

In our first real attempt to search for life on another planet, the Viking landers carried out biological and chemical experiments with martian soil. The results seemed to show that there is probably no life on Mars. But the idea of life on Mars was brought to the fore in 1996 by the interpretation by some scientists that they have found signs that life on Mars existed in its early years. This idea has not been generally accepted, but it is still causing much discussion. We shall discuss this research in Section 21.1b.

Since it seems reasonable that life as we know it would be on planetary bodies, let us first discuss the chances of life arising elsewhere in our Solar System. Next, we consider the chances that life has arisen in more distant parts of our Galaxy or elsewhere in the Universe. In Chapter 18, we saw that planets have been discovered orbiting dozens of sunlike stars. Perhaps these planets, or smaller planets or moons in those planetary systems, would be suitable abodes for life.

In any case, we conclude this chapter with a discussion of ongoing searches for communications from intelligent life elsewhere in the Universe.

## 21.1 THE ORIGIN OF LIFE

It would be very helpful if we could state a clear, concise definition of life, but unfortunately that is not possible. Biologists state several criteria that are ordinarily satisfied by life forms—reproduction, or the capacity to undergo mutation and evolution, for example. Still, there exist forms on the fringes of life—viruses, for example, which need a host organism in order to reproduce—and scientists cannot always agree whether some of these things are "alive" or not.

In science fiction, authors sometimes conceive of beings that show such signs of life as the capability of intelligent thought, even though the being may share few of the other criteria that we ordinarily recognize. In Fred Hoyle's classic novel *The Black*

FIGURE 21–1 "I'll tell you something else I think. I think there are other bowls somewhere out there with intelligent life just like ours." *(Drawing by Frank Modell; © The New Yorker Collection 1987, Frank Modell, from http://cartoonbank.com. All Rights Reserved.)*

*Cloud,* for example, an interstellar cloud of gas and dust is alive—and very smart. But we can make no concrete deductions if we allow such wild possibilities, and exobiologists prefer to limit the definition of life to forms that are more like "life as we know it." See Fig. 21–1.

This rationale implies, for example, that extraterrestrial life is based on complicated chains of molecules that involve carbon atoms. Carbon is one of the most common elements on Earth. (Cosmically, for every million atoms of hydrogen there are about 100,000 atoms of helium, 850 atoms of oxygen, and 400 atoms of carbon, but the Earth has lost most of its hydrogen and helium.) Life on Earth is governed by deoxyribonucleic acid (DNA) and ribonucleic acid (RNA), two carbon-containing molecules that control the mechanisms of heredity. Chemically, carbon is able to form "bonds" with several other atoms simultaneously, making these long carbon-bearing chains possible. In fact, we speak of compounds that contain carbon atoms as **organic.** In addition, a few scientists consider that silicon-based life could exist, since silicon is almost as cosmically abundant as carbon and can also form several simultaneous bonds (though its chemistry is less rich).

## 21.1a Creating Organic Compounds

How hard is it to build up long organic chains? To the surprise of many, an experiment performed in the 1950s showed that it was much easier than had been supposed to make organic molecules. At the University of Chicago, at the suggestion of Harold Urey, Stanley Miller put quantities of several simple molecules in a glass jar from which air had been removed. Water vapor ($H_2O$), methane ($CH_4$), and ammonia ($NH_3$) were included along with hydrogen gas. Miller exposed the mixture to electric sparks, simulating the lightning that may exist in the early stages of the formation of a planetary atmosphere. After a few days, molecules containing several carbon atoms had formed in the jar; these organic molecules were even complex enough to include simple amino acids, the building blocks of life. Modern versions of these experiments have created even more complex organic molecules from a wide variety of simple actions on simple molecules. Such molecules may have mixed in the oceans to become a "primordial soup" of organic molecules (Fig. 21–2).

FIGURE 21–2 The simple solution of organic material in the oceans, from which life may have arisen, is informally known as "primordial soup."

Since the original experiment of Urey and Miller, most scientists have come to think that the Earth's primitive atmosphere was not made of methane and ammonia, which would have disappeared soon after the Earth's formation. Such an atmosphere, with molecules rich in hydrogen atoms, is known as "reducing." Rather, the gases in today's oceans and atmosphere were released from the Earth's interior by volcanos. When life formed, the atmosphere might have consisted of carbon dioxide, water, and carbon monoxide, but no methane and ammonia. The atmosphere, in short, lacked the free oxygen of our present-day atmosphere. It could, however, supply the oxygen of its molecules. We can call it a "neutral" atmosphere compared with today's atmosphere with free oxygen, which is known as an "oxidizing" atmosphere. Miller's methods would not make organic materials in such a neutral atmosphere. As of now, Miller states that nobody knows whether the Earth's early atmosphere was reducing or oxidizing, and that his early work remains valid.

Laboratories on Earth can work on the problem of the formation of complicated molecules (Fig. 21–3). At the University of Maryland, Cyril Ponnamperuma and colleagues synthesized all the chemical bases of human genes in a single experiment that simulated the Earth's primitive atmosphere. (The more complicated nucleotides, which have not been synthesized in this way, are composed of bases, the sugars ribose or deoxyribose, and phosphate.) They also detected them in a well-studied meteorite known as the Murchison meteorite (Section 20.1b). Scientists have also found extraterrestrial amino acids in two meteorites that had been long frozen in Antarctic ice (Section 20.1a).

However, mere amino acids or even DNA or RNA molecules are not life itself. A jar containing a mixture of all the atoms that are in a human being is not the same

A

B

**FIGURE 21-3** (*A*) Bishun Khare and Carl Sagan, of Cornell University, in their laboratory several decades ago. The apparatus was used to simulate conditions thought to be present on the primitive Earth, to see whether complex compounds such as amino acids form easily from the gases of which Earth's atmosphere consisted at that time. (*B*) Louis J. Allamandola with equipment used for similar purposes in his Astrochemistry Laboratory at the NASA/Ames Research Center.

as a human being. This is a vital gap in the chain; astronomers certainly are not qualified to say what supplies the "spark" of life.

Sidney Altman of Yale and Thomas Cech of the University of Colorado won the 1989 Nobel Prize in Chemistry for studies that indicate that RNA was a key molecule at the origin of life. They discovered that RNA can act as an enzyme, a class of molecules that accelerate chemical reactions. (For example, an enzyme in human saliva helps change starch into the glucose sugar molecule.) RNA accelerates reactions by a million times and so makes it possible for life to exist. Altman and Cech discovered that not only could RNA rather than proteins be enzymes but also that RNA can copy itself in a limited way. Thus though we depend on DNA to carry the genetic code, early life may have used RNA. In 1996, MIT scientists were able to make RNA act as a template in a test tube, making the base molecules ("nucleotides") join into new RNA.

Five top chemists involved in studies of the origin of life (including Stanley Miller), all in the San Diego area, have set up an institute to further their work on the early chemistry of life forms, funded by NASA's exobiology program. Did organic molecules become more and more complicated, leading to RNA and then DNA? Just how this might have happened is a subject of study at this institute.

In any case, many astronomers think that since it is not difficult to form complex molecules, life may well have arisen not only on the Earth but also in other locations. Even if life is not found in our Solar System, there are so many other stars in space, and so many with planets around them, that life could well have arisen there independently of life on Earth.

## 21.1b Life Discovered on Mars?

The world was electrified in 1996 when a group of distinguished scientists reported that they had found signs that life arose on Mars independently of life on Earth. (Since we know that meteorites are exchanged between Earth and Mars, it is possible that one planet "seeded" the other with life and thus life exists on Mars but

A

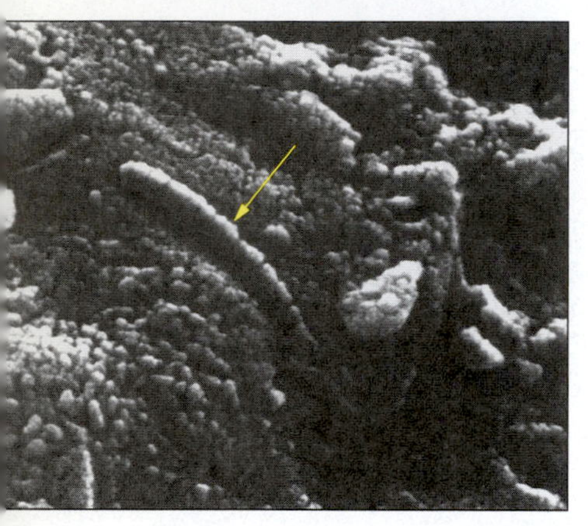

B

**FIGURE 21–4** (*A*) The martian meteorite suspected of harboring signs of early life on Mars, ALH84001 (where the ALH stands for Allan Hills in Antarctica). (*B*) Structures in the meteorite, seen with a microscope. It has been proposed that they are remnants of primitive life forms, but this conclusion is not widely accepted.

is related to Earth life.) The scientists, from NASA's Johnson Space Center and from Stanford University, were examining a meteorite found in Antarctica a dozen years earlier (Fig. 21–4). Though it was 4.5 billion years old, much older than the other dozen meteorites thought to be from Mars, the ratio of isotopes of oxygen in it matches results found for the Mars atmosphere by the Viking lander, indicating that this meteor was also martian. It would have been blasted off Mars's surface 15 million years ago and landed on Earth about 13,000 years ago.

Deep inside, the scientists found certain organic compounds. Though similar compounds have been found in interstellar space, where they clearly did not originate from living things, the distribution of the compounds seems very different. Further, the quantity of these compounds increased as you looked farther in the rock, a different situation from other meteorites. And in close proximity to these compounds, some tubelike structures were seen (Fig. 21–4*B*) that could be the result of living processes. A group of British scientists have similarly found material that they think resulted from living organisms in both this meteorite and another one. This second meteorite, also found in Antarctica, was formed 165 million years ago and was blown off Mars 600,000 years ago. The original group of scientists later found the tubelike structures in one of the 1.3 billion-year-old meteorites, so skeptics can argue that it is a mere natural and common formation. These types of Mars meteorites were discussed in Chapter 12.11.

All the evidence that they have found life that originated on Mars 4.5 billion years ago is indirect, yet these careful scientists conclude that the totality of the evidence is best interpreted as in terms of formation by a living organism. Still, most other scientists have not been convinced. A major question is how hot the vapors around the supposed fossils got before they crystallized, and whether that temperature was too hot to have allowed biological processes. Some scientists have concluded that it was too hot. Further, another set of scientists has concluded that the supposed fossils are actually merely crystals of iron oxide and that they have features inconsistent with their once having been alive. Still other scientists have found similar organic compounds in Antarctic ice samples that are known not to have any biological presence. At this writing, more scientists seem to be skeptical rather than accepting of the conclusion that life on Mars has been discovered.

Clearly the results will be carefully debated, and it may take years before the matter is further resolved. The Mars Global Surveyor spacecraft now at Mars does not contain any instruments that bear directly on the question, though its highly detailed images revealed signs that water has flowed on Mars, exciting those who think that where water was, life arose. The next generation of robotic exploration of Mars will further the search for life on Mars, leading up to a return of martian soil to Earth in perhaps 15 years. Certainly, the report has given new impetus to the exploration of Mars and shown the need for further governmental and other support for the investigation.

## 21.2 THE STATISTICS OF EXTRATERRESTRIAL LIFE

Instead of phrasing one all-or-nothing question about life in the Universe, we can break down the problem into a chain of simpler questions. This procedure was developed by Frank Drake, a Cornell astronomer now at the University of California at Santa Cruz, and extended by, among others, Carl Sagan of Cornell and Joseph Shklovskii, a Soviet astronomer.

### 21.2a The Probability of Finding Life

First we consider the probability that stars at the centers of solar systems are suitable to allow intelligent life to evolve. For example, the most massive stars evolve relatively quickly, and they probably stay in the stable state that characterizes the bulk of their lifetimes for too short a time to allow intelligent life to evolve.

Second, we ask what the chances are that a suitable star has planets. With the new detections of planets around other stars, most scientists think that the chances are probably pretty high.

Third, we need planets with suitable conditions for the origin of life. A planet like Jupiter might be ruled out, for example, because it lacks a solid surface and because its surface gravity is high. (Alternatively, though, one could consider a liquid region, if it were at a suitable temperature, to be as advantageous as were the oceans on Earth to the development of life here. Or a satellite could provide the solid surface.) Also, planets probably must be in orbits in which the temperature does not fluctuate too much. Even if the planets are in multiple-star systems, as most stars seem to be, such orbits exist. (These orbits are either far from a central pair of stars or, if the stars are far apart, close to one of the stars.) Calculations are made from time to time of the range of distances from the Sun over which planets are sufficiently warm for life to begin, treating the carbon-dioxide balance and the greenhouse effect. The models for the size of this "habitable zone" must explain the "Goldilocks effect": why Venus is too hot, Mars is too cold, but the Earth is just right. (A habitable zone in our Galaxy has also been defined; the Sun and all the nearby stars are in it.)

Fourth, we have to consider the fraction of the suitable planets on which life actually begins. This is the biggest uncertainty, for if this fraction is zero (with the Earth being a unique exception), then we get nowhere with this entire line of reasoning. Still, the discovery that amino acids can be formed in laboratory simulations of primitive atmospheres, and the discovery by radio astronomers of complex molecules in interstellar space (as complicated as ammonia, formic acid, and vinegar, as we will discuss in Chapter 33), indicate to many astronomers that it is relatively easy to form complicated molecules. Amino acids, much less complicated than DNA but also basic to life as we know it, have even been found in meteorites. The discovery from radio observations that molecules are associated with sites of star formation strengthens the idea that these molecules may eventually become part of planets that form, even if most are destroyed by the high temperature in the planet-forming process. Many astronomers choose to think that the fraction of planetary systems on which life begins may be high.

A previously unknown kind of single-celled organisms called **archaea** was discovered in undersea thermal vents in 1982. (Methanogens and halophiles, other types of archaea, had previously been known.) Though at first these organisms were thought to be rare, we now realize that they are one of three major types of life, along with eucarya (which include plants and animals as well as smaller objects such as ciliates and slime molds) and bacteria. Indeed, we have found archaea all around the extreme environments where they were first found, and we now think that more than half the Earth's biomass may be archaea! The sequencing of the entire genome of one type of archaeon revealed 1738 genes, of which only 44 per cent resemble genes from other organisms. (The human genome, as reported in 2001, has about 30,000 genes.) Studying archaea indicates that life may have formed very early in the Earth's history. Some of these living creatures—found in oxygen-free places like Yellowstone's hot springs—take in carbon dioxide and hydrogen and give off methane. Their RNA (genetic material) is different from that of other bacteria or of plants or animals in sequence though not in basic chemical structure. The discovery supports the idea that life evolved before oxygen appeared in the Earth's primeval atmosphere. These and some other living bacteria do not survive in the presence of oxygen. The earliest organisms on Earth for which there is fossil evidence are around 3.5 billion years old, and look very similar to modern cyanobacteria, a type of bacteria (not archaeon) that can carry out photosynthesis.

Further, underwater exploration has discovered organisms that thrive on sulfur from geothermal sources instead of on solar energy. We were wrong when we thought that life needed sunlight! Some of these organisms are bacteria that metabolize hydrogen sulfide to reproduce and grow, while others are animals that eat the

A very different point of view, that the chemistry that takes place on the surface of minerals is important for the origin of life, has been espoused. Such reactions could occur in hot environments like undersea volcanos. The idea that the prebiotic chemicals were formed on surfaces open to water differs from the dominant primordial soup idea.

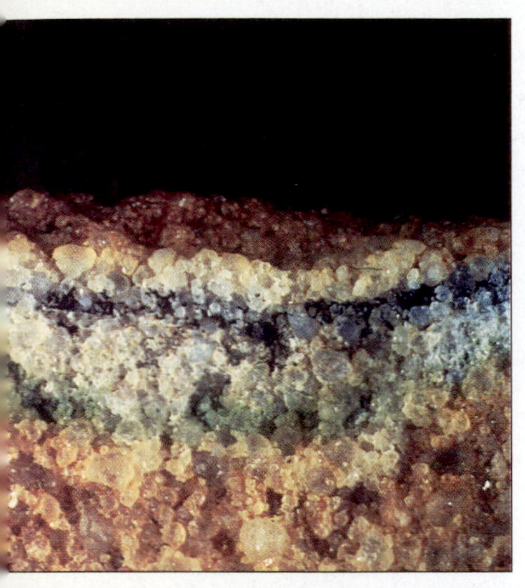

A

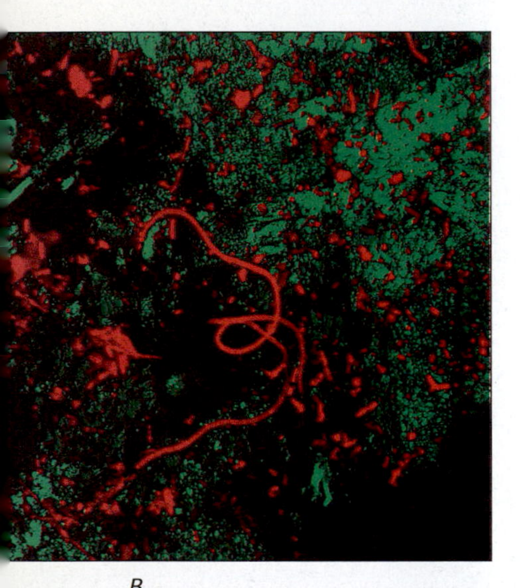

B

C

**FIGURE 21–5** (*A*) We see the inside of an Antarctic rock, with a lichen growing safely insulated from the external cold. The image shows the outer 1 cm of the rock. (*B*) Extremophile bacteria that live over 1 km below the Earth's surface, which fluoresce as red patches on this microscope image. The bacteria, discovered in deep wells, survive on hydrogen generated by a chemical reaction between water and ferrous silicates in the surrounding basaltic rock. Thus they do not need sunlight (or oxygen). Do similar types of microbes exist on Mars? (*C*) A mat of extremophile bacteria found deep underground. The colony shown is 10 mm across.

bacteria. So though they couldn't support the types of animal and plant life with which we are most familiar, environments on other planets may not be as hostile to life as we had thought. Even in the most apparently hostile places on Earth, we have now found living things (Fig. 21–5). We may even find practical use for some of these discoveries: the archaea reported that survive temperatures over 135°C may find use in biotechnology or in household or industrial detergents for removing stains. It is these hardy examples of life on Earth—sometimes called "extremophiles"—that seem to raise the odds that life will be found on Mars one day.

If we want to have meaningful conversations with aliens, we must have a situation where not merely life but intelligent life has evolved. We cannot converse with algae or paramecia, and certainly not with the organic compounds reported in the Mars meteorite. Furthermore, the life must have developed a technological civilization capable of interstellar communication. These considerations reduce the probabilities somewhat, but it has still been calculated that there may be technologically advanced civilizations within a few hundred light-years of the Sun.

Not everyone agrees that complex and, eventually, intelligent life commonly evolves. Geologist Peter Ward and astronomer Donald Brownlee (whose work on the Stardust mission to a comet we have read about) conclude that even if simple life is widespread, complex life is much rarer. They conclude in a 1999 book, *Rare Earth: Why Complex Life Is Uncommon in the Universe*, that so many unusual circumstances helped life on Earth evolve to complex forms, and so many extinction events have occurred, that intelligent life would be rare in our Galaxy. This conclusion is shared by most, but not all, biologists as well.

One aspect of this problem was expressed in the last term of the Drake-Sagan equation. It deals with the important question of the lifetime of the technological civilization itself. We now have the capability of destroying our civilization either dramatically in a flurry of hydrogen bombs or more slowly by, for example, altering our climate, lessening our ozone shield, or increasing the level of atmospheric pollution. It is a sobering question to ask whether the lifetime of a technological civilization is measured in decades, or whether all the problems that we have—political, environmental, and otherwise—can be overcome, leaving our civilization to last for millions or billions of years.

## 21.2b Evaluating the Probability of Life

At this point we can try to estimate (to guess, really, in some cases) fractions for each of these simpler questions within our chain of reasoning. We can then multiply these fractions together to get an answer to the larger question of the probability of extraterrestrial life. Reasonable assumptions lead to the conclusion that there may be millions or billions of planets in this galaxy on which life may have evolved. The nearest one may be within dozens of light-years of the Sun. On this basis, many if not most professional astronomers have come to believe that intelligent life probably exists in many places in the Universe. Carl Sagan has estimated that a million stars in the Milky Way Galaxy may be supporting technological civilizations.

Though this point of view has been gaining much popularity, a skeptical reaction exists. Evaluating the Drake-Sagan Equation (Box 21.1) with a more pessimistic set of numbers can lead to the conclusion that we earthlings are alone in our Galaxy.

## A DEEPER DISCUSSION

### BOX 21.1 The Probability of Life in the Universe

Our discussion follows the lines of an equation written out to estimate the number of civilizations in our galaxy that would be able to contact each other, designated by the letter $N$. In 1961, Frank Drake, then at Cornell, wrote

$$N = R^* f_p n_e f_l f_i f_c L,$$

where $R^*$ is the rate at which stars form in our galaxy, $f_p$ is the fraction of these stars that have planets, $n_e$ is the number of planets per solar system that are suitable for life to survive (for example, those that have Earth-like atmospheres), $f_l$ is the fraction of these planets on which life actually arises, $f_i$ is the fraction of these life forms that develop intelligence, $f_c$ is the fraction of the intelligent species that choose to communicate with other civilizations and develop adequate technology, and $L$ is the lifetime of such a civilization. Now that we have seen that satellites of the outer planets are worlds with personalities of their own, perhaps we should replace $f_p$ with a new factor $f_{mp}$ (moons and planets) that takes account of the suitable moons in a Solar System.

We can study habitable zones, for example, by analyzing the limits below which water would freeze because carbon monoxide clouds would reflect too much sunlight and above which a runaway greenhouse effect would occur. One analysis shows this zone to be from 0.8 to 1.4 A.U. for a star like the Sun and 1.7 to 2.8 A.U. for stars half again as massive. The position of the habitable zone will change as the star evolves. In the planetary systems with "hot Jupiters" on very elliptical orbits, these giant planets would disrupt the orbits of smaller, Earthlike planets, and perhaps prevent planets from orbiting consistently in the habitable zone for as long as is necessary for life to arise.

One of the largest uncertainties in this "Drake Equation" is $f_l$, which could be essentially zero or could be close to 1. The result is also very sensitive to the value one chooses for $L$—is it 1 century or a billion years? Do civilizations destroy themselves? Or perhaps they just go off the airwaves. Depending on the alternatives one chooses, one can predict that there are dozens of communicating civilizations within 100 light-years or that the Earth is unique in the galaxy in having one.

#### *Discussion Question*

Is breaking down the big question into smaller questions, as is done here, really a help in evaluating the probability?

One reason for doubting that the Universe is teeming with life is that extraterrestrials have not established contact with us. Where are they all? To complicate things further, is it necessarily true that if intelligent life evolved, they would choose to explore space or to send out messages?

Still, if the statements that life has been discovered in an ancient meteorite from Mars are confirmed, then we would be two for two in discovering life on hospitable planets, or three for three if we turn up life in Europa's hidden ocean. We would then certainly be justified in concluding that life is ubiquitous in the cosmos. But we are far from that point.

"Extraordinary claims require extraordinary proof." Carl Sagan

## 21.3 INTERSTELLAR COMMUNICATION

What are the chances of our visiting or being visited by representatives of these civilizations? Pioneers 10 and 11 and Voyagers 1 and 2 are even now carrying messages out of the Solar System in case an alien interstellar traveller should happen to encounter these spacecraft (Fig. 21–6). But it seems unlikely that humans themselves can travel the great distances to the stars, unless someday we develop spaceships to carry whole families and cities on indefinitely long voyages into space.

Yet there is plenty of time in the future for interstellar space travel to develop. The Sun has another 5 billion years to go in its current stable state. The solar luminosity will increase eventually, so perhaps we have about 1 billion years on Earth before our oceans boil away. In comparison, recorded history on Earth has existed for only about 5000 years, one-millionth of the age of the Solar System.

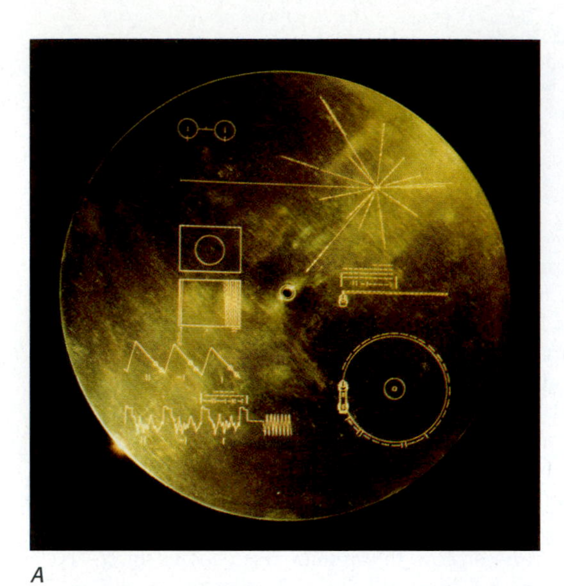

A

B

**FIGURE 21–6** (*A*) The goldplated copper record bearing two hours' worth of Earth sounds and a coded form of photographs, carried by Voyagers 1 and 2. The sounds include a car, a steamboat, a train, a rainstorm, a rocket blastoff, a baby crying, animals in the jungle, and greetings in various languages. Musical selections include Bach, Beethoven, rock, jazz, and folk music. (*B*) The record includes 116 photographs. One of them is this view, which I took when in Australia for an eclipse. It shows Heron Island on the Great Barrier Reef in Australia, in order to illustrate an island, an ocean, waves, a beach, and signs of life.

We can hope even now to communicate over interstellar distances by means of radio signals. We have known the basic principles of radio for only a hundred years, and powerful radio, television, and radar transmitters have existed for less than 70 years. Even now, though, we are sending out signals into space on the normal broadcast channels. Waves bearing the reports from Lindbergh's 1927 pioneering crossing of the Atlantic Ocean are expanding into space and at present are about 75 light-years from Earth. And once a week a new episode of *Friends* is carried into the depths of the Universe. Military radars are even stronger than our commercial broadcast stations, though the waves travel no faster.

From another star, the Sun would appear to be a variable radio source, because the Earth and Sun would be unresolved and the Earth's radio signals would therefore appear to be coming from the Sun. The signal would vary with a period of 24 hours, since it would peak each day as specific concentrations of transmitters rotated to the side of the Earth facing our listener.

In 1974, the giant radio telescope at Arecibo, Puerto Rico (Fig. 21–7), run by Cornell University, was upgraded. (It was upgraded again in the late 1990s.) At the rededication ceremony, a powerful signal was sent out into space, bearing a message from the people on Earth (Fig. 21–8).

The signal from Arecibo was directed at the globular cluster M13 (Fig. 21–9) in the constellation Hercules, on the theory that the presence of 190,000 closely packed stars in that location would increase the chances of our signal being received

**FIGURE 21–7** The Arecibo telescope in Puerto Rico, used once to send a message into space. The telescope is 305 meters across, the largest telescope on Earth; you can see it in the James Bond film *Goldeneye* and in *Contact.* Since its 1996 upgrade, it is sensitive enough to pick up a cellular telephone on Venus. But because of a fuss made over broadcasting the message at its earlier upgrade, allowing aliens (if any) to know where we are, no broadcast marked the latest upgrade. With its new focus device high above the dish, it can now also observe at wavelengths down to 3 cm, about three times shorter than previously.

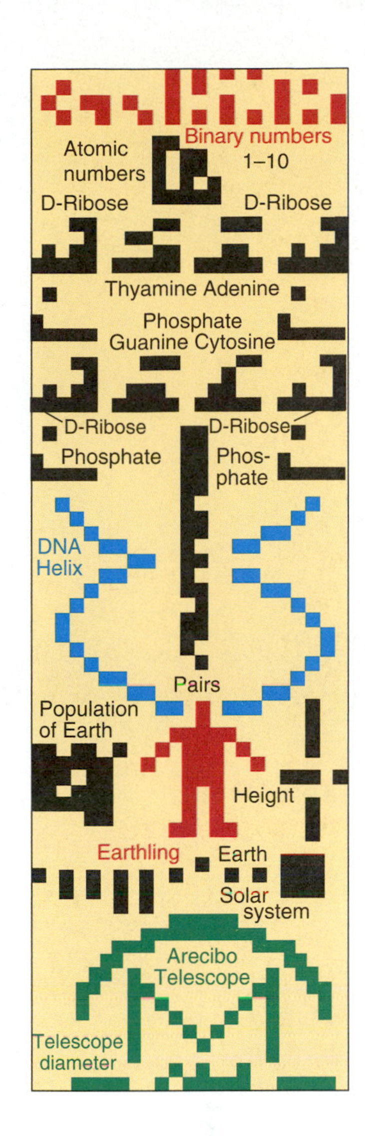


**FIGURE 21–8** The message sent to the globular cluster M13, plotted out and with a translation into English added. The basic binary-system count at upper right is provided with a position-marking square below each number. The message was sent as a string of 1679 consecutive characters, in 73 groups of 23 characters each. (The numbers 73 and 23 are each prime, and we figure that a civilization advanced enough to detect this message would realize that 1679 is a product of two primes.) There were two kinds of characters, each represented by a frequency; the two kinds of characters are reproduced here as 0s and 1s. Data rates and studies of prime numbers have advanced so much since this message was sent that much higher-resolution pictures could be sent nowadays.

by a civilization on one of them. But the travel time of the message (at the speed of light) is 24,000 years to M13, so we certainly could not expect to have an answer before twice 24,000, or 48,000 years, have passed. If anybody (or any*thing*) is observing our Sun when the signal arrives, the radio brightness of the Sun will increase by 10 million times for a 3-minute period. A similar signal, if received from a distant star, could be the giveaway that there is intelligent life there. But our Arecibo message was broadcast only briefly, so the aliens in M13 could try in vain to verify that the signal was really from afar.

For a long time, this Arecibo message was the only major signal that we have purposely sent out as our contribution to the possible interstellar dialogue. In the 1990s, a Canadian group of scientists used a radio telescope in Ukraine to send a more elaborate message using a language they invented. They used a radio wavelength of 13 cm and targeted 5 nearby sunlike stars, so if any response were to come, we would have it in a decade or so.

## 21.4 THE SEARCH FOR INTELLIGENT LIFE

For planets in our Solar System, we can search for life by direct exploration, as we did with the Viking landers on Mars, but electromagnetic waves, which travel at the speed of light, seem a much more sensible way to search for extraterrestrial life outside our Solar System. How would you go about trying to find out if there were life

**FIGURE 21–9** The globular cluster M13, toward which a 3-minute message was sent with the Arecibo radio telescope in 1974. The cluster contains perhaps 100,000 stars, so beings on a planet around any of them can potentially detect the message in about 24,000 years.

on a distant planet? If we could observe the presence of abundant oxygen molecules or discover from the spectrum of a distant star that the molecules there are out of their normal balance, that could tell us that life exists there. For example, methane ($CH_4$) and oxygen ($O_2$) are not chemically stable together in a planetary atmosphere. They can coexist only if they are being constantly generated—presumably by life.

The Galileo spacecraft, when it flew close by Earth in 1992, detected a strong radio signal, which was perhaps a navigation beacon. Its images of the Earth's southern hemisphere show only a few signs of life, chiefly agricultural boundaries in Australia. It did detect oxygen spectroscopically (in the form of ozone) as well as water vapor, possible but not definitive signs of life.

The most promising way to detect the presence of not merely life but actually intelligent life at great distances appears to be a search in the radio part of the spectrum. But it would be too overwhelming a task to listen for signals at all frequencies in all directions at all times. One must make some reasonable guesses on how to proceed.

A few frequencies in the radio spectrum seem especially fundamental, such as the spin-flip line of neutral hydrogen—the simplest element—at 21 cm (as described in Chapter 33). This wavelength corresponds to 1420 MHz, a frequency over ten times higher than stations at the high end of the normal FM band. We might conclude that creatures on a far-off planet would decide that we would be most likely to listen near this frequency because it is so fundamental.

On the other hand, perhaps the great abundance of radiation from hydrogen itself would clog this frequency, and one or more of the other frequencies that correspond to strong natural radiation should be preferred. The "water hole," the wavelength range between the radio spectral lines of H and OH, has a minimum of radio noise from background celestial sources, the telescope's receiver, and the Earth's atmosphere, and so it is another favored possibility. If we ever detect radiation at any frequency, we would immediately start to wonder why we had not realized that this frequency was the obvious choice.

In 1960, Frank Drake used a telescope at the National Radio Astronomy Observatory for a few months to listen for signals from two of the nearest stars—tau Ceti and epsilon Eridani. He was searching for any abnormal kind of signal, a sharp burst of energy, for example. He called this investigation Project Ozma after the queen of the land of Oz (Fig. 21–10) in L. Frank Baum's later stories.

The observations were later extended in a more methodical search called Ozma II. Starting in 1973, Ben Zuckerman of the University of Maryland (now at UCLA) and Patrick Palmer of the University of Chicago systematically monitored over 600 of the nearest stars of supposedly suitable spectral type. They used computers to search the data they recorded for any sign of special signals; Ozma II obtained data at 10 million times the rate of the original Ozma. Over the course of several days, they studied each star for a total of about a half hour. Nothing turned up.

**FIGURE 21–10** Dorothy and Ozma climb the magic stairway. *(From* Glinda of Oz, *by L. Frank Baum, illustrated by Roy Neill, copyright 1920)*

Over a dozen separate searches have since been undertaken in the radio spectrum, some concentrating on all-sky coverage and others on coverage of individual stars over time. Many astronomers believe that carrying out these brief searches was worthwhile, since we had no idea of what we would find. But now many feel that the early promise has not been justified, and the chance that there is extraterrestrial life seems much lower.

Still, more and more astronomers are interested in the detection of a signal in the "*s*earch for *e*xtra*t*errestrial *i*ntelligence" (SETI). The programs include scanning in all directions in the sky and listening for a long time in certain promising directions for artificial signals.

Paul Horowitz of Harvard, with support from the Planetary Society (a public-interest group) as well as a private donation from the moviemaker Steven Spielberg, developed a relatively inexpensive way of scanning 8 million ultra-narrow frequency bands simultaneously. He set up his Project META (*M*egachannel *E*xtra*T*errestrial *A*ssay) on a 26-m Harvard radio telescope for long-term scanning of certain "magic"

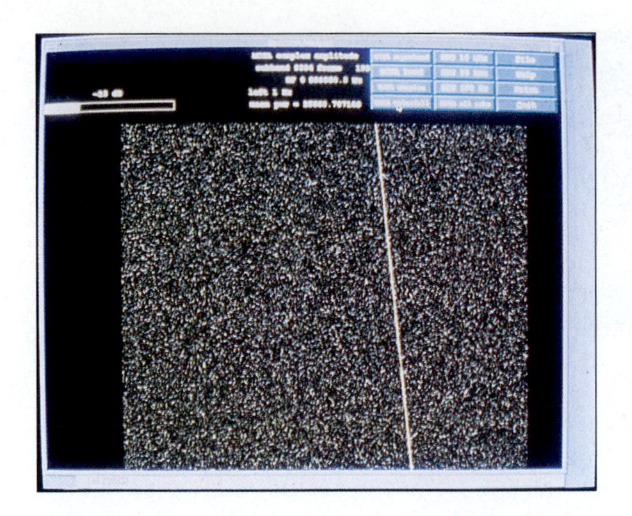

**FIGURE 21–11** This "waterfall" display from Project Phoenix shows the instantaneous power in 462 narrow channels, each only 2 Hz wide. Frequency varies from left to right, and time varies from top to bottom. Such a display is known in the SETI trade as a "waterfall." Each dot represents a channel, and the size of the dot is proportional to the strength of the signal. As a new spectrum is plotted in the top row across the screen, the older spectra move downward. Here, the signal changes steadily in frequency, so it slants on the screen. It is readily distinguishable from the background noise.

frequencies that seem most likely (21 cm, OH, $1/2 \times 21$ cm, etc.), covering a wide range of sky. They did a megaOzma per minute and then upgraded to Project BETA, with a billion channels. After a dozen years of searching, no clear extraterrestrial signals have been distinguished. A clone of META in Argentina is sampling the southern sky.

An ambitious effort was begun by a NASA-sponsored group on October 12, 1992, the 500th anniversary of Columbus's landing in the New World. It made use of new sophisticated signal-processing capabilities of powerful computers to search millions of radio channels simultaneously in the microwave region of the spectrum (Fig. 21–11). This High Resolution Microwave Survey, which used both a 34-m antenna in California and the 305-m antenna at Arecibo, was both a sky survey and a targeted study of individual objects. In its first fraction of a second, this NASA search surpassed the entire Project Ozma. But Congress cut off funds anyway, after about a year. The targeted search of the project, known as Project Phoenix, is continuing, backed by funds contributed by private individuals to an organization known as the SETI Institute in California. Led by Drake and astrophysicist Jill Tarter (Fig. 21–12), Project Phoenix first used the Parkes radio telescope in Australia, then used a telescope at the National Radio Astronomy Observatory in Green Bank, West Virginia, and now is using the refurbished Arecibo dish. (Dr. Tarter was the model for the scientist in Carl Sagan's novel *Contact,* played by Jodie Foster in the movie version.)

One might think that the scientists sit by the computer, waiting for a SETI signal to show up. But, as SETI Institute scientist Seth Shostak explains, such signals show up all the time, once every 5 minutes, on average! The next step is then to check whether that signal is extraterrestrial and otherwise unexplained. The SETI Institute scientists use a second large telescope, in England, to check whether such signals are truly extraterrestrial. Of course, none of the signals has turned out to be an extraterrestrial message.

One SETI effort, SETI@home, has caught the attention of the general public. A team of scientists at the University of California at Berkeley are using the giant radio telescope at Arecibo to listen for radio signals from intelligent life. It started with a 4-million-channel receiver piggybacking on other observations by using a different receiver than the one other scientists are using. Instead of processing all the signals in their own computers, they are sending a small fraction of these data out to people all over the world for processing on personal computers at people's homes and offices (Fig. 21–13). After all, at night, most computers sit unused. SETI@home is a screensaver program, which uses your computer to check SETI data when you aren't using the computer. A chunk of data is sent out over the Internet to your computer, and when it is analyzed, your computer sends it back. But you don't get a re-

**FIGURE 21–12** Jill Tarter, Director of Project Phoenix of the SETI Institute.

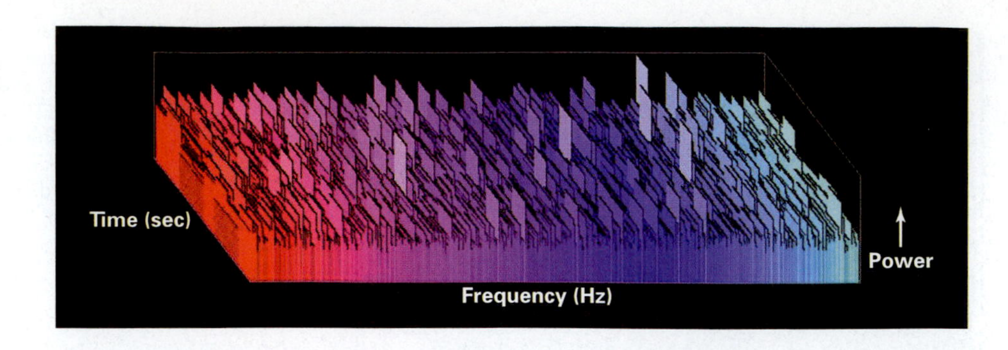


**FIGURE 21–13** Output on the screen of a computer that has been used to analyze some of the radio data collected by the "seti@home" project. There have been no clear signs of intelligent life among the data processed so far.

SETI@home is an enlargement of Project SERENDIP (*S*earch for *E*xtraterrestrial *R*adio *E*missions from *N*earby *D*eveloped *I*ntelligent *P*opulations). A Southern SERENDIP is being carried out with the Parkes radio telescope in Australia.

port on your computer's calculations, since it might have found only terrestrial interference. The telescope they are using points all over the sky, so it will be many months before the chunk of sky for which your computer calculated is again in view, allowing any suspicious signals to be tested. The Planetary Society, a public-interest group based in Pasadena, California, has become the lead sponsor of SETI@home, which has already signed up over two million users from all over the world. The technique of "distributed computing" pioneered by SETI@home has since been applied to the genome project in biology and is finding still other uses, including commercial ones.

Our improved abilities to observe in the infrared might pick up evidence of advanced civilizations in another way. Freeman Dyson, at the Institute for Advanced Study in Princeton, has suggested that a civilization might redistribute some of its planet's mass around its star to soak up stellar energy that would otherwise escape. Infrared spacecraft can pick up the infrared glow caused by material that blocks even 1 per cent of the sunlight of a star like the Sun at a distance of about 200 light-years. This distance would include 100,000 stars. Our infrared observational capabilities are growing rapidly, with the 2MASS ground-based program mapping the whole sky at short infrared wavelengths and with the *S*pace *I*nf*r*ared *T*elescope *F*acility (SIRTF) about to be launched.

Some people, including Paul Horowitz at Harvard, have turned to "optical SETI," looking for signals in visible light. They calculate that extraterrestrials could decide on various scientific grounds to send out signals using powerful lasers (which stands for "*l*ight *a*mplification through *s*timulated *e*mission of *r*adiation"; note the word "light" in its name), perhaps using pulses of nanosecond durations. One can argue whether extraterrestrials would be able to aim narrow optical beams over tens and hundreds of light-years. Interestingly, some of the scientists involved have calculated that the stars would become so bright while the signal is being sent that they could be seen in the daytime. That idea raises the possibility that the giant astronomical telescopes and the many smaller amateur telescopes that are now used only at night could be used for optical SETI in the rest of each 24-hour period. A 1.8-m telescope, sponsored by the Planetary Society, is being built under Horowitz's guidance at Harvard, Massachusetts, for optical SETI purposes. It should be able to detect optical pulses as brief as 1 billionth of a second. Today's lasers, pointed outward, could outshine our Sun by a factor of 5000 for that interval.

We mustn't be too quick to close out other possibilities than radio waves in searching for extraterrestrial intelligence. For example, neutrinos (subatomic particles that scarcely interact with matter) travel at or close to the speed of light. Just because we find neutrinos hard to detect doesn't prove that other civilizations haven't chosen them to carry messages.

I once was coauthor of an article that suggested that electromagnetic radiation itself—whether radio or optical—might not be the right way to look or listen for signals, and that beams of neutrinos could bring messages from an advanced civilization. Neutrino beams are created by atom smashers, and can be put out in very narrow beams. In order to test for "neutrino oscillations," whether the three known types of neutrinos change among themselves (as we will discuss in Chapter 27.6), neutrino beams are being sent hundreds of kilometers through the Earth (which does not absorb neutrinos) to special detectors. So we on Earth are already in the rudimentary phases of neutrino communication.

**FIGURE 21–14** An artist's conception of the Allen Telescope Array.

Returning to radio SETI, a major SETI telescope is being built. This Allen Telescope Array received major funding from Paul Allen, founder of Microsoft along with Bill Gates (and also owner of the Portland Trail Blazers basketball team and the Seattle Seahawks football team). When completed, the Allen Telescope Array will consist of 500 to 1000 small radio telescopes linked together (Fig. 21–14). It will be used simultaneously for radio-astronomy research by Berkeley scientists and for SETI by SETI Institute scientists. Because of the savings allowed by mass production of these dishes resulting from the mass proliferation of satellite television, the total price should be much less than it would have been previously. A benefit of using many small telescopes is that the array becomes useful even when it is only partially completed.

A still larger radio telescope is also being developed, though it is not as near to being built. National representatives from the United States and other countries have agreed to build the Square Kilometer Array. It is eventually to have at least ten times the collecting area of the Arecibo radio telescope. Whether it is better to use many small telescopes or 10 Arecibos is an example of the type of question still to be resolved. The instrument's final sensitivity will be incredible, both for radio astronomy and for SETI.

The list of searches has grown too long to include completely. Someday we may be scanning from radio telescopes on the far side of the Moon, shielded by the Moon's bulk from radio interference from Earth, with a world-wide network of optical telescopes, or with giant tanks of pure water to detect neutrinos.

## 21.5 UFOs AND THE SCIENTIFIC METHOD

But why, you may ask, if most astronomers accept the probability that life exists elsewhere in the Universe, do they not accept the idea that unidentified flying objects (UFOs) represent visitation from these other civilizations (Fig. 21–15)? The answer to this question leads us not only to explore the nature of UFOs but also to consider the nature of knowledge and truth. The discussion that follows is a personal view, but one that is shared by many scientists.

**FIGURE 21–15** The Far Side *(by Gary Larson © 1982 Farworks, Inc. All rights reserved. Used by permission. )*

## 21.5a UFOs

First, most of the sightings of UFOs that are reported can be explained in terms of natural phenomena. Astronomers are experts on strange effects that the Earth's atmosphere can display, and many UFOs can be explained by such effects. When Venus shines brightly near the horizon, for example, in my capacity as local astronomer I sometimes get telephone calls from people asking me about the UFO. It is not well known that a planet or star low on the horizon can seem to flash red and green because of atmospheric refraction. Atmospheric effects can affect radar waves as well as visible light.

Sometimes other natural phenomena—flocks of birds, for example—are reported as UFOs. One should not accept explanations that UFOs are flying saucers from other planets before more mundane explanations—including hoaxes, exaggeration, and fraud—are exhausted.

For many of the effects that have been reported, the UFOs would have been defying well-established laws of physics. Where are the sonic booms, for example, from rapidly moving UFOs? Scientists treat challenges to laws of physics very seriously, since our science and technology are based on these laws.

Most professional astronomers feel that UFOs can be so completely explained by natural phenomena that they are not worthy of more of our time. Furthermore, some individuals may ask why we reject the identification of UFOs with flying saucers, when—they may say—that explanation is "just as good an explanation as any other." Let us go on to discover what scientists mean by "truth" and how that applies to the above question.

## 21.5b Of Truth and Theories

At every instant, we can explain what is happening in a variety of ways. When we flip a light switch, for example, we assume that the switch closes an electric circuit in the wall and allows the electricity to flow. But it is certainly possible, although not very likely, that the switch activates a relay that turns on a radio that broadcasts a message to an alien on Mars. The Martian then might send back a telepathic message to the electricity to flow, making the light go on. The latter explanation sounds so unlikely that we don't seriously consider it. We would even call the former explanation "true" without qualification.

We usually regard as "true" the simplest explanation that satisfies all the data we have about any given thing. This principle is known as **Occam's Razor;** it is named after the 14th-century British philosopher who originally proposed it. Without this rule, we would always be subject to such complicated doubts that we would accept nothing as known.

> Occam's Razor, sometimes called the Principle of Simplicity, is a razor in the sense that it is a cutting edge that allows distinctions to be made among theories.

Science is based on Occam's Razor, though we don't usually bother to think about it. Sometimes, something we call "true" might be more accurately described as a theory. In principle, the scientific method is based on hypotheses and theories. (The actual working of science is more complicated.) A **hypothesis** (plural: hypotheses) is an explanation that is advanced to explain certain facts. When it is shown that the hypothesis actually explains most or all of the facts known, then we may call it a **theory.** We often test a theory by seeing whether it can predict things that were not previously observed, and then by trying to confirm whether the predictions are valid.

An example of a theory is the Newtonian theory of gravitation, which for many years explained almost all the planetary motions. Only a small discrepancy in the orbit of Mercury, as we describe in Section 31.1, remained unexplained. In 1916, Albert Einstein presented a general theory of relativity as a better explanation of gravitation. The theory explained the discrepancy in Mercury's orbit. It also made certain predictions about the positions of stars during total solar eclipses. When his predictions were verified, his theory was widely accepted.

The philosopher of science Karl Popper advanced the idea decades ago that a scientific theory is "falsifiable," in that it must be possible to prove it wrong. Newton's theory of gravitation was clearly falsifiable in that observations of the orbit of Mercury and at eclipses showed that it did not give correct predictions. Einstein's theory is falsifiable, since presumably there are experiments that could contradict its predictions, and such experiments are carried on regularly. As of now, though, no disagreements have been found. In any case, the idea of "falsification" is key to many modern philosophers of science.

Is Newton's theory "true"? Yes, in most regions of space. Is Einstein's theory "true"? We say so, although we may also think that one day a new theory will come along that is more general than Einstein's in the same way that Einstein's is more general than Newton's.

How does this view of truth tie in with the discussion of UFOs? Scientists have assessed the probability of UFOs being flying saucers from other worlds, and most have decided that the probability is so low that the possibility is not even worth considering. We have better things to do with our time and with our national resources. We have so many other, simpler explanations of the phenomena that are reported as UFOs that when we apply Occam's Razor, we call the identifications of UFOs with extraterrestrial visitation "false." UFOs may be unidentified, but they are probably not flying, nor for the most part are they objects.

An increasingly common discussion in some humanities circles in recent years concerns post-modernist views that science is a construct of humans, and that truth is all relative. We scientists, on the other hand, are largely convinced that we are actually approaching truth about many things in astronomy, even if we haven't cleared up all the uncertainties. In 1996, an NYU physicist submitted a hoax paper to a major humanities journal that was having a post-modernist discussion of science. In the paper, the physicist made what he thought were ridiculous assertions. The paper was nonetheless published, and the hoax was revealed with the statement that it proved the lack of value of the post-modernist work, given that its practitioners couldn't even screen out such ridiculousness. The fight reached even the front page of *The New York Times*. The episode left many scientists feeling good and some of the post-modernists hissing with rage.

In an age when we legitimately send astronauts permanently into space on an International Space Station (Figure 21–16), and when our robotic spacecraft explore even the outer Solar System, it is too bad that so many people spend their time thinking about aliens in UFOs instead of investigating the real exploration of space.

**FIGURE 21–16** The International Space Station. We may never again have a time when no human is in orbit.

## CORRECTING MISCONCEPTIONS

✖ ***Incorrect***

We know the best frequencies at which to listen for messages from extraterrestrials.

✔ ***Correct***

It is unclear that we can correctly guess frequencies that unknown extraterrestrials might choose.

## SUMMARY AND OUTLINE

The origin of life (Section 21.1)
- Earth life is based on complex chains of carbon-bearing (organic) molecules; carbon (and silicon) form such chains easily; organic molecules are easily formed under laboratory conditions that simulate primitive atmospheres.

Statistical chances for extraterrestrial life (Section 21.2)
- Problem broken down into stages; major uncertainties include what the chance is that life will form given the component parts, and what the lifetime of a civilization is likely to be

Signals we are sending from Earth (Section 21.3)
- Leakage of radio, television, and radar
- Signal beamed from Arecibo toward globular cluster M13
- Current signal beamed from Ukraine

The search for life (Section 21.4)
- Experiments on Viking
- Radio SETI: Projects Ozma, Ozma II, BETA, SETI@home
- Optical SETI

UFOs and the scientific method (Section 21.5)
- Definition of "truth"; Occam's Razor

## KEY WORDS

exobiology
astrobiology
organic
archaea
Occam's Razor
hypothesis
theory

## QUESTIONS

1. What is the significance of the Miller-Urey experiments?
2. How have the Miller-Urey experiments been improved upon?
3. [†] If 1/10 of all stars are of suitable type for life to develop, 1 per cent of all stars have planets, and 10 per cent of planetary systems have a planet at a suitable distance from the star, what fraction of stars have a planet suitable for life? How many such stars would there be in our galaxy?
4. Supply your own numbers in the Drake equation, and calculate the result. Justify your choices.
5. [†] On the message sent to M13, work out the binary value given for the population of Earth and compare it with the actual value. (This question assumes a knowledge of the binary system.)
6. [†] On the message sent to M13, work out the binary value given for the size of the Arecibo telescope, and compare it with the actual value. (This question assumes a knowledge of the binary system.)
7. Why do we think that intelligent life wouldn't evolve in planets around stars much more massive than the Sun?
8. List three means by which we might detect extraterrestrial intelligence.
9. Why can't we hope to carry on a conversation at a normal rate with extraterrestrials on a distant star?
10. Describe how Einstein's general theory of relativity serves as an example of the scientific method.

[†] This question requires a numerical solution.

## TOPICS FOR DISCUSSION

1. Comment on the possibility of another solar system's scientists' observing the Sun, suspecting it to be a likely star to have populated planets, and beaming a signal in our direction.
2. Using a grid 37 × 41 (two prime numbers), work out a message you might send. Show the string of 0s and 1s in this binary-system notation.
3. Find out what the largest known factorable product of two primes is (or at least some order of magnitude of this number), and assess how much data could be included as images in terms of numbers of TV screens. Studies of factoring large primes are very important in contemporary cryptograph, even providing security codes for ordinary banking transactions.
4. Can scientific investigation find truth?

## USING TECHNOLOGY

### W World Wide Web

1. See an interview with Stanley Miller about how life began at http://www.accessexcellence.com/WN/NM/miller.html.
2. Examine the homepage of the SETI Institute: http://www.seti.org and find out about the status of the Allen Telescope Array.
3. Check out SETI@home and perhaps even join the search: http://setiathome.berkeley.edu.
4. Find about the current plans for the Square Kilometer Array at http://www.ras.ucalgary.ca/SKA and linked sites.
5. Learn about optical SETI at http://www.oseti.org.
6. Read and consider the Declaration of Principles Concerning Activities Following the Detection of Extraterrestrial Intelligence at http://www.seti.org/post-detection.html.
7. Use search engines to look up Science Studies, Alan Sokal, and Norman J. Levitt to find their published papers. Examine the discussion that followed on the controversy about whether science is a human construction.
8. Learn about the idea that fossils from life forms have been found in Mars meteorite ALH 84001 at http://www.jsc. nasa.gov/pao/flash/marslife/video.htm and at http://earthsky.com/Features/Articles/life-on-mars.html.

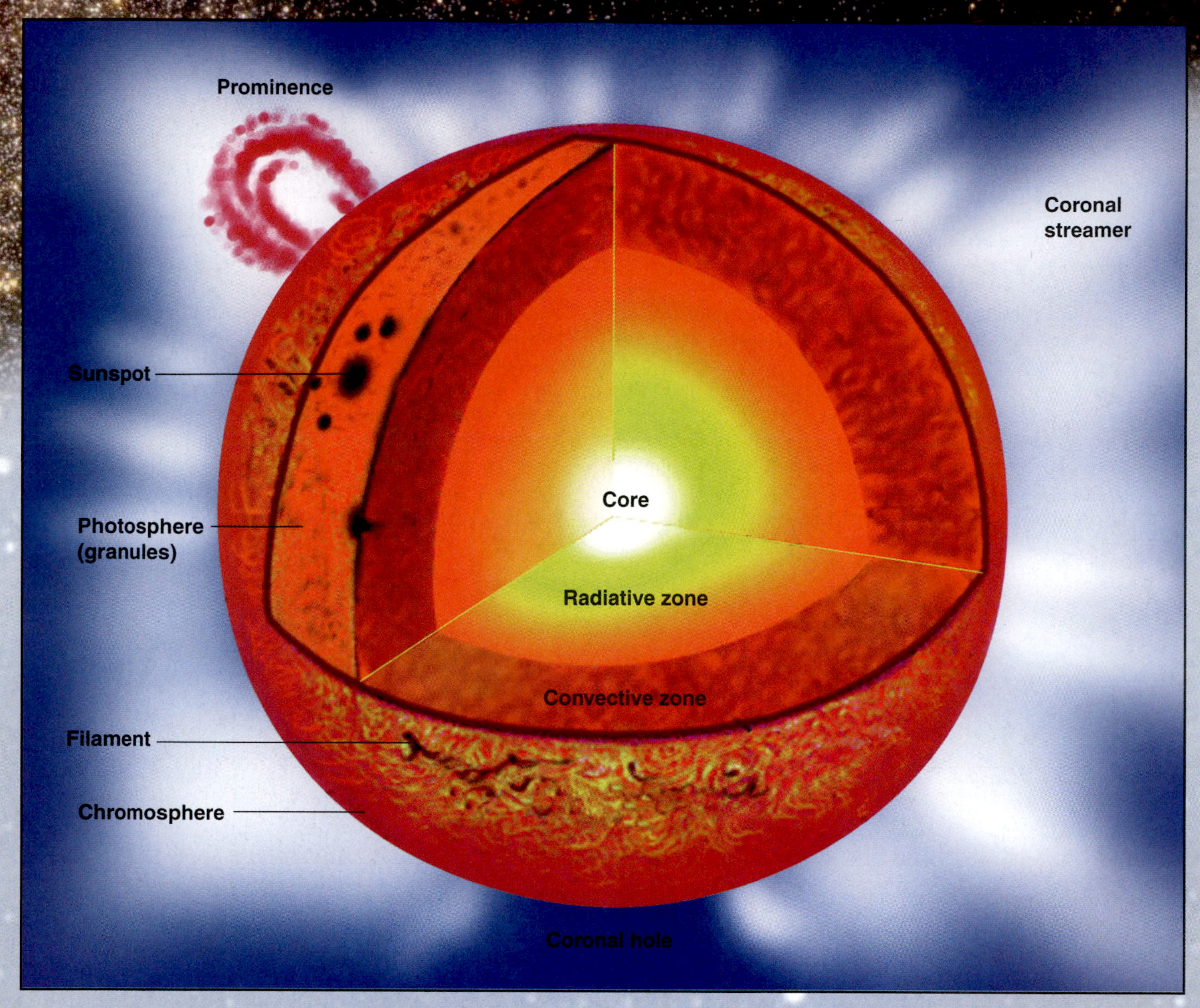


The parts of the solar atmosphere and interior. The solar surface is depicted as it appears through a hydrogen filter.

The interior has a core, where the energy is generated, surrounded by a zone in which radiation dominates and a zone in which convection dominates. On the surface—the photosphere—we find sunspots, loops of gas that are dark filaments when seen in projection against the disk and that are bright prominences when seen off the Sun's edge. Granulation, composed of small regions known as granules, is best seen in ordinary light instead of the hydrogen-light shown. The chromosphere, a thin layer above the photosphere, shows best in hydrogen-light and is composed of small spikes known as spicules. The white corona above it extends far into space. Its shape, including its huge streamers, is determined by the Sun's magnetic field. Regions where the corona is relatively absent are known as coronal holes. Inside the Sun we see traces of waves studied with helioseismology. Most waves curve as the pressure changes. Scientists are searching for waves affected by gravity.

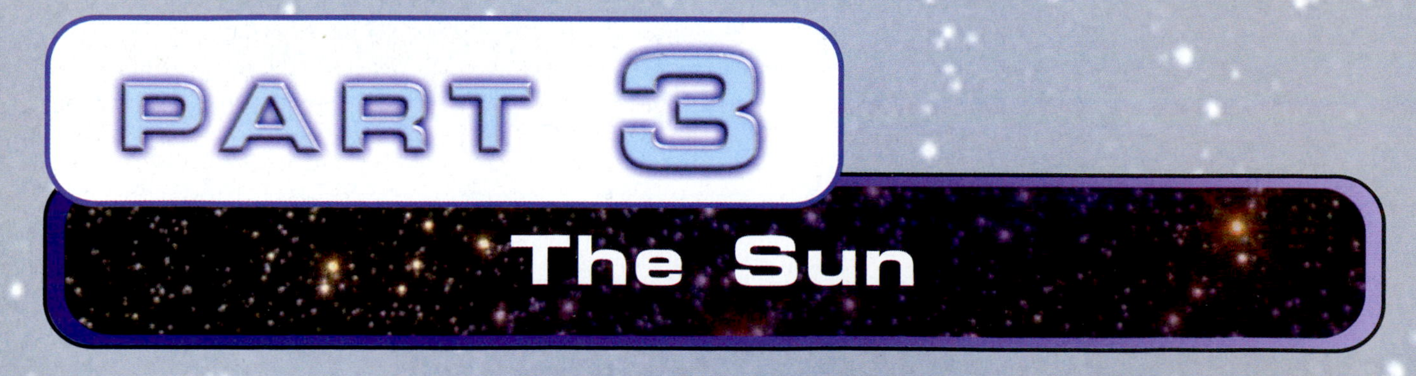

The center of our Solar System is, of course, the Sun. The Sun, a million times larger in volume than our Earth, contains 99.9 per cent of the mass of the Solar System. It provides the light and other energy that we need to live. Further, by studying the Sun, we not only learn about the properties of a particular star but also can study the details of processes that undoubtedly take place in more distant stars as well. In the following chapter, we discuss the **quiet sun,** the solar phenomena that appear every day. In the next chapter, we will discuss the **active sun,** solar phenomena that appear non-uniformly on the Sun and vary over time. The influence of these active-sun phenomena spreads throughout interplanetary space and has a strong effect on the Earth and on our environment in space.

We often think of the Sun as the bright ball of gas that appears to travel across our sky every day. We are then seeing only one layer of the Sun, part of its atmosphere; the properties of the solar interior below that layer and of the rest of the solar atmosphere above that layer are very different. The outermost parts of the solar atmosphere even extend through interplanetary space beyond the orbit of the Earth.

The layer of the Sun that we see on a normal day is called the **photosphere,** which simply means the sphere from which the light comes (from the Greek *photos,* meaning "light"). As is typical of many stars, about 94 per cent of the atoms and nuclei in the outer parts are hydrogen, about 5.9 per cent are helium, and a mixture of all the other elements makes up the remaining one-tenth of one per cent. The overall composition of the interior is not very different, though deep in the core helium is being formed out of hydrogen nuclei.

The Sun is an average star, since stars exist that are much hotter and much cooler, and that are intrinsically much brighter and much fainter. Radiation from the photosphere peaks (is strongest) in the middle of the visible spectrum—the yellow and the green; after all, our eyes evolved over time to be sensitive to that region of the spectrum because the greatest amount of the solar radiation occurred there (given that it penetrated our Earth's atmosphere). It is fun to speculate that if we lived on a planet orbiting an object that emitted mostly x-rays (and if the planet's atmosphere did not absorb x-rays), we, like Superman (if he could detect radiation at the same wavelengths at which he could emit radiation), would have x-ray vision.

Beneath the photosphere is the solar **interior.** All the solar energy is generated there at the solar **core,** which is about 10 per cent of the solar diameter at this stage of the Sun's life. The temperature there is about 15 million kelvins. (At such high temperatures, the 273°C difference between the kelvin and Celsius temperature scales is negligible; this temperature is also about 15 million °C.) Energy is generated in the Sun by nuclear fusion in the Sun's core. (All the other stars generate their energy similarly, also by nuclear fusion, as we will study in Chapter 27.) In these deep parts of the Sun, energy is carried by radiation. Above that region, in the upper part of the solar interior, energy is carried by convection, the way that matter moves when it is heated below and carries energy as it moves upward against gravity. We cannot see the solar interior but are developing ways of learning about it by studying its effects on the solar surface.

The photosphere is the lowest level of the **solar atmosphere.** Though the Sun is gaseous through and through, with no solid parts, we still use the term "atmosphere" for the upper part of the solar material. The parts of the atmosphere above the photosphere are very tenuous and contribute only a small fraction to the total mass of the Sun. In the visible part of the spectrum, these upper layers, very much fainter than the photosphere, cannot be seen with the naked eye except during a solar eclipse, when the Moon blocks the photospheric radiation from reaching our eyes directly. Now we can also study these upper layers with special instruments on the ground and in orbit around the Earth.

Just above the photosphere is a jagged, spiky layer about 10,000 km thick, only about 1.5 per cent of the solar radius. This layer glows colorfully pinkish when seen at an eclipse and is thus called the **chromosphere** (from the Greek *chromos,* meaning "color"). Above the chromosphere, a ghostly white halo called the **corona** (from the Latin, meaning "crown") extends tens of millions of kilometers into space. The corona is continually expanding into interplanetary space. What we call the **solar wind** is based on this expansion, but the *So*lar and *H*eliospheric *O*bservatory (SOHO) now aloft and other evidence have shown us more completely how the solar wind is formed. We shall discuss all these things in the next two chapters.

Loops of gas in the solar corona, imaged at high spatial resolution in the ultraviolet with the Transition Region and Coronal Explorer (TRACE) spacecraft. The hot gas is held in these loops by the Sun's magnetic field.

# Chapter 22

# Our Star, the Sun

The solar photosphere is about 1.4 million km (1 million miles) across. The disk of the Sun (the apparent surface of the photosphere) takes up about one-half a degree across the sky; we say that it **subtends** one-half degree. This angle is large enough for us to be able to make images of detailed structure on the solar surface.

In this chapter, we discuss the basic structure of the Sun. First we discuss the photosphere, and then we discuss higher levels of the solar atmosphere. In the next chapter, we discuss sunspots and other solar activity. In Chapter 27, we discuss the interiors of the Sun and of the other stars, and what makes them shine.

**AIMS:** To study the Sun, which is both the central part of our Solar System and the nearest star, an example of all the other stars whose surfaces we cannot observe in such detail, and to see how the Solar and Heliospheric Observatory is telling us about it

## 22.1 THE PHOTOSPHERE

The Sun is a normal star, with properties about in the middle of the range of brightness and temperature possible for stars. Its surface temperature is about 5800 K. The Sun is the only star close enough to allow us to study its surface in detail. In recent years, we have become able to verify directly from the ground and from space that many of its features indeed also appear on other stars.

Stars like the Sun live about 10 billion years. The Sun is about halfway through its lifetime, with about 5 billion years to go.

### 22.1a High-Resolution Observations of the Photosphere

One major limitation in observing the Sun is the turbulence in the Earth's atmosphere, which also causes the twinkling of stars. The problem is even more serious for studies of the Sun, since the Sun is up in the daytime when the atmosphere is heated by the solar radiation and so is more turbulent than it is at night.

Only at the very best observing sites, specially chosen for their steady solar observing characteristics, can one see detail on the Sun subtending an angle as small as 1 second of arc. This corresponds to about 700 km on the solar surface, the distance from Boston to Washington, D.C. Occasionally, objects 1/2 arc second across can be seen. Since atmospheric turbulence causes bad "seeing" that limits our ability to observe small-scale detail, it had been thought that it would do little good to build solar telescopes larger than about 50 cm (20 inches) in order to increase resolution, even though larger telescopes are inherently capable of resolving finer detail. But advances in stabilizing seeing, and the need to collect more light even from the Sun to allow detailed spectroscopic analysis, have led to ongoing plans at the U.S. National Solar Observatory to build a 4-m solar telescope, as we will discuss at the end of this chapter.

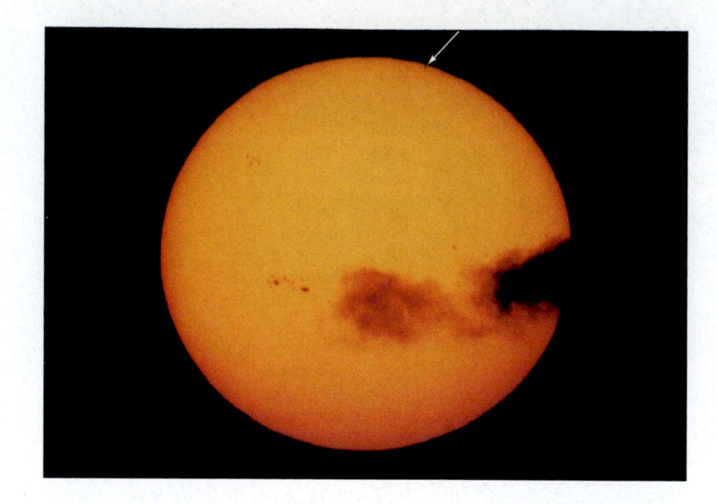

**FIGURE 22–1** The solar photosphere, including sunspots and silhouetted clouds. Barely visible at top is the planet Mercury in silhouette during its November 17, 1999, transit.

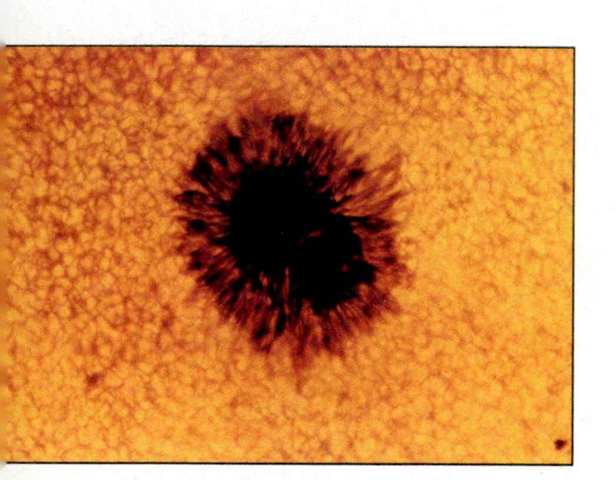

**FIGURE 22–2** A small region of the solar surface, showing the salt-and-pepper granulation and a sunspot.

Sometimes we observe the Sun in **white light**—all the visible radiation taken together (Fig. 22–1). When we study the solar surface in white light with 1 arc second resolution, we see a salt-and-pepper texture called **granulation** (Fig. 22–2). The effect is similar to that seen in boiling liquids on Earth, which are undergoing **convection.** Convection, the transport of energy as hot matter moves upward and cooler matter falls, is one of the basic ways in which energy can be moved. Convection carries energy to your boiling eggs, for example. Conduction and radiation are the other major methods of energy transport.

Granulation on the Sun is an effect of convection. Each granule is only about 1000 km across and represents a volume of gas that is rising from and falling to a shell of convection, called the **convection zone,** located below the photosphere. The granules are convectively carrying energy from the hot solar interior to the base of the photosphere. But the granules are about the same size as the limit of our resolution, so they are difficult to study.

### 22.1b The Photospheric Spectrum

**FIGURE 22–3** Fraunhofer's original spectrum, shown on a German postage stamp. We see a continuous spectrum, from red on the left to violet on the right, crossed from top to bottom by the dark lines that Fraunhofer discovered, now known as "Fraunhofer lines" or "absorption lines."

The spectrum of the solar photosphere, like that of just about every star, is a continuous spectrum crossed by dark lines known as absorption lines (Fig. 22–3). Hundreds of thousands of these absorption lines, which are also called Fraunhofer lines, have been photographed (Fig. 22–4) and catalogued. They come from most of the chemical elements, although some of the elements have many lines in their spectra and some have very few. Iron has many lines in the spectrum. The hydrogen Balmer lines are strong but few in number. At the relatively low temperatures of the photosphere for a star, the spectral lines of helium do not appear. We discussed the formation of spectral lines in Section 4.7, which material you may want to review.

Fraunhofer, in 1814, labelled the strongest of the absorption lines in the solar spectrum with letters from A through H. His C line, in the red, is now known to be the first line in a series of hydrogen lines and is called H$\alpha$ (H alpha). We still use Fraunhofer's notation for some of the strong lines: The D lines, a pair of lines close together in the yellow part of the spectrum, are caused by neutral sodium (Na I; read "sodium one"). The H line and the K line (which was named later), both in the part of the violet spectrum that is barely visible to the eye, are caused by calcium atoms that have been stripped of one electron each (Ca II; read "calcium two"). We will say more of such "ionized" states of matter in Section 24.3.

The spectral lines are formed as continuous sunlight travels through the outer layers of the star. The elements in those outer layers absorb at various wavelengths to form the Fraunhofer lines. But where does the continuous spectrum (called the **continuum**) come from? It is formed for the most part at somewhat lower levels of

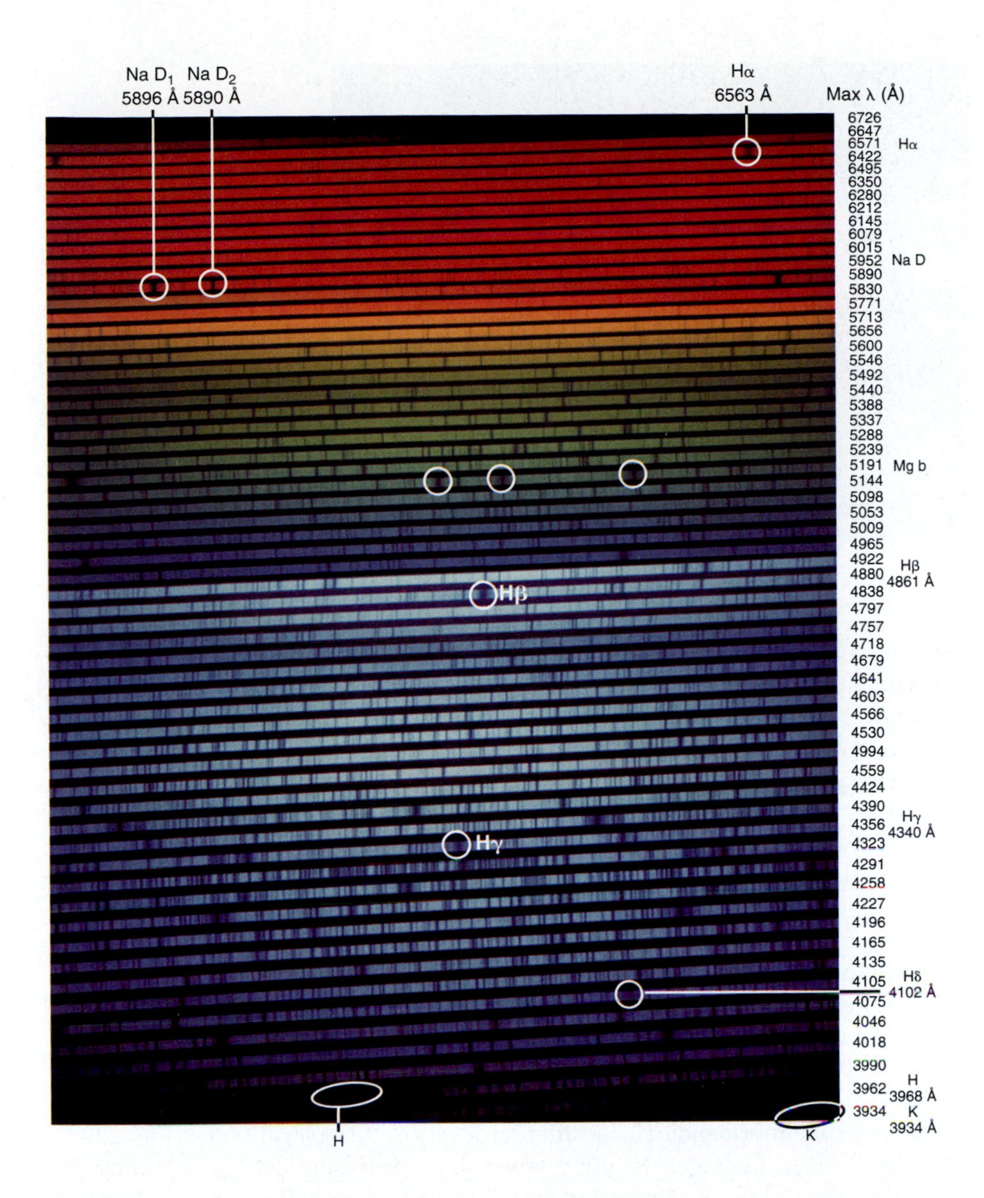


**FIGURE 22-4** A highly spread-out solar spectrum, from ultraviolet through red. The strips that we see from top to bottom should be placed next to each other. A few of the strongest absorption lines are marked and their wavelengths given. The numbers in a column at the right give the maximum wavelengths for each of the strips.

the solar atmosphere, though the levels at which the continuum and most of the absorption lines are formed are mixed together throughout the photospheric layers. The energy we receive on Earth in the form of solar radiation has been transported upward to these layers from the solar interior.

In the photosphere, the continuum we see in the visible part of the spectrum comes from hydrogen. Though hydrogen normally has one electron, the continuous spectrum is formed in the interaction of this hydrogen atom with a second electron. Sometimes the second electron briefly joins the hydrogen atom, and sometimes the second electron is merely deflected by the hydrogen atom. This situation, in which an atom gains an extra electron, is rarer than the case in which an atom loses an electron. Since an atom that has lost an electron has a positive charge, this more common case is called a "positive ion" (and the word "positive" is usually dropped). In the case of the solar photosphere, the hydrogen has, instead, an extra electron, so is called a "negative ion." Thus we say that the continuous emission from the Sun in the visible part of the spectrum is caused by the negative hydrogen ion.

## 22.1c Solar Seismology

The whole Sun's surface is oscillating with periods ranging from minutes up to hours. Basically, it is ringing like a bell. However, since the waves that cause the ringing

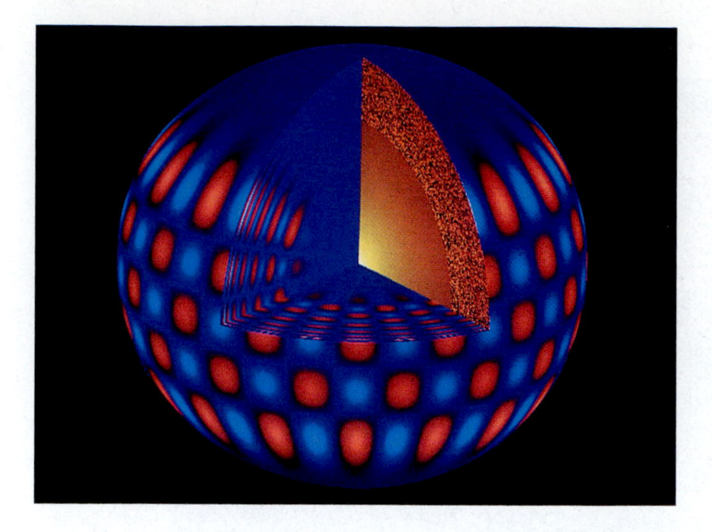

**FIGURE 22–5** One of the thousands of possible modes of the Sun's oscillations. Red represents zones of expansion and blue represents zones of contraction.

come randomly and continually, the Sun is ringing like a bell that is continually hit by particles in a sandstorm rather than a bell hit once. We think that motions in the solar convection zone make the waves that cause the ringing.

The Sun has many individual "modes of vibration," unique combinations of some parts moving inward while others move outward; each of these modes vibrates at a different frequency. We see the superposition (or simultaneous existence) of millions of these modes of vibration of the Sun (Fig. 22–5). We study the oscillations from their Doppler shifts. Interpreting these oscillations in the photosphere is telling us what the inside of the Sun is like and is similar to the way that geologists find out about the inside of the earth by interpreting seismic waves. The studies are thus called **solar seismology** (or **helioseismology**). The waves travel downward into the Sun until the rapid increase of temperature (which increases their speed) makes them bend back up. They then travel upward until they reflect off the bottom of the photosphere. The waves are trapped between upper and lower levels, and their wavelengths depend on how deep the lower level is. Thus studying different wavelengths gives us information about different levels in the solar interior (Fig. 22–6). Deep in the Sun, hydrogen is fusing into helium, releasing energy, following the "proton-proton chain," a sequence of fusion that we discuss for stars in general in Chapter 27.

Solar seismology, also called helioseismology, is the study of the inside of the Sun through the observation of waves on its surface.

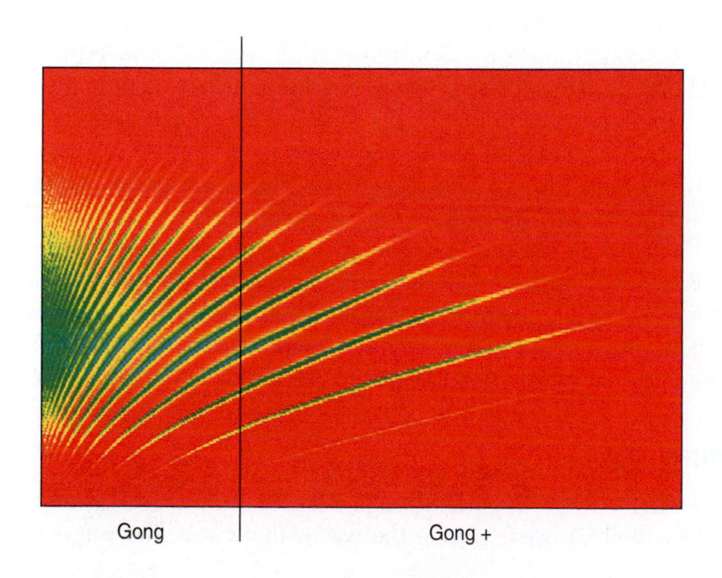


**FIGURE 22–6** Helioseismology data from the updated version of the Global Oscillation Network Group (GONG+). Plotted is the number of waves around the equator (as in Fig. 22–5) along the *y*-axis vs. the number of waves perpendicular to the equator along the *x*-axis. The agreement between theory and observation is excellent. Since the theoretical predictions depend on the temperature and density inside the Sun, we can adjust our model for temperature and density to match the observations as well as possible.

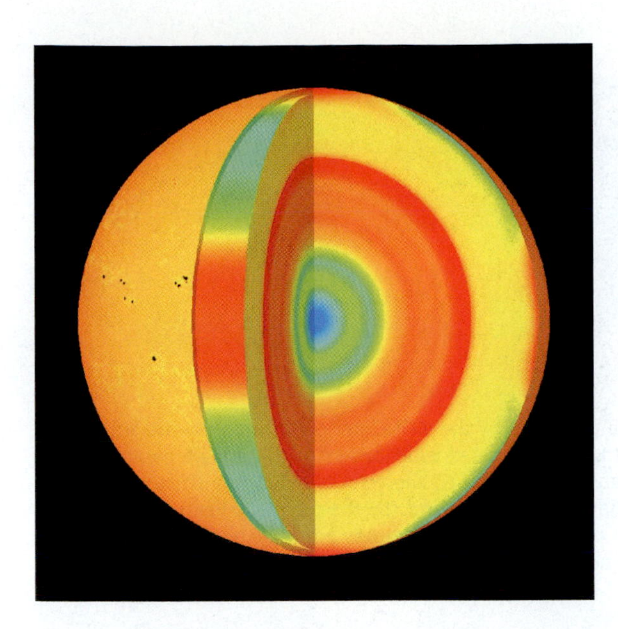

**FIGURE 22–7** The rotation periods of different layers and regions of the Sun, based on helioseismology results. Red is the shortest rotation period (26 days), green corresponds to 28 days, and violet corresponds to 36 days. Thus the measurements show that the surface rotation periods, which vary from 25 days at the equator to 36 days at the poles, persist inward through the solar convection zone. Further toward the center, the data show that the Sun rotates as a solid body with a 27-day period.

Theoretical astronomers can predict the periods of waves that will exist, but their predictions depend on how temperature, density, pressure, chemical composition, and even rotation speed vary under the solar surface. Matching predictions with the observations has led to the conclusion that the Sun's convection zone is deeper than had been thought. It takes up the outer 30 per cent of the solar radius. The results also show how fast the inside of the Sun can be rotating relative to the photosphere (Fig. 22–7). It had been thought that the center of the Sun might rotate much more rapidly than the surface, but solar seismology has taught us otherwise. Thus we can deduce that the center of the Sun has lost a lot of angular momentum in some way that we do not understand. This loss of angular momentum has consequences for theories of how the Solar System formed.

Since many of the modes of vibration have nearly but not quite identical periods, astronomers must observe for many days without a break to sort them out. Some of the first tries at continuous time-series, eliminating nighttime gaps, were made by coordinating existing observatories scattered around the world at different longitudes, such as one in Hawaii and one in the Canary Islands. Further, the waves have been studied with a telescope at the Earth's South Pole, where the Sun is up for months without setting.

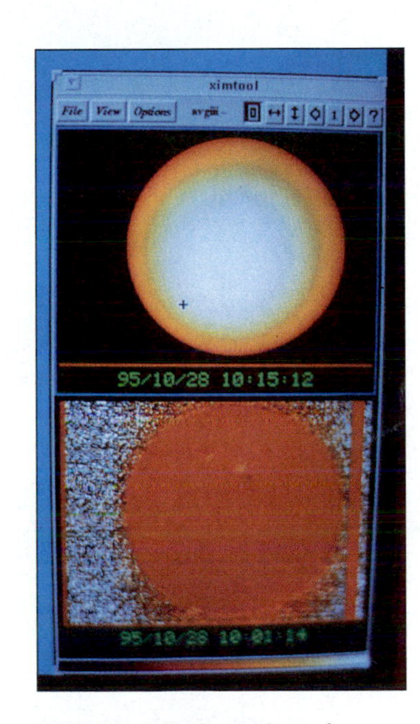


**FIGURE 22–8** A view of a computer screen at the GONG site in India. It shows images of the Sun used to study the solar interior through the effects of waves on the surface.

To provide continuous coverage of the Sun, specially designed telescopes were installed at a network of sites around the world (Fig. 22–8). Since the Sun is ringing like a bell, the project has the acronym GONG—*G*lobal *O*scillation *N*etwork *G*roup (Fig. 22–9); it is the major current project in ground-based solar astronomy. The six sites, spaced around the globe so that at least one sees the Sun all the time, are at Big Bear, California; Mauna Loa, Hawaii; Learmonth, Australia; Udaipur, India; the Canary Islands; and Chile. (Weather and instrument breakdowns leave gaps only a few per cent of the time.) They are now being upgraded to provide higher spatial resolution, reaching the limit set by the Earth's atmospheric seeing, in a project known as GONG+. The resulting improvements, which also include a new data system, will allow study of local regions of the Sun, especially the upper convection zone. Also, magnetograms will be made continuously, instead of hourly. Because significant variations of the solar interior have been discovered and appear to be linked to the Sun's 11-year cycle of magnetic activity (which we discuss in the following chapter), the project will continue for at least a cycle.

Continuous helioseismology observations are also being obtained by three of the instruments on the SOHO (Solar and Heliospheric Observatory) spacecraft. Other ground-based networks also exist.

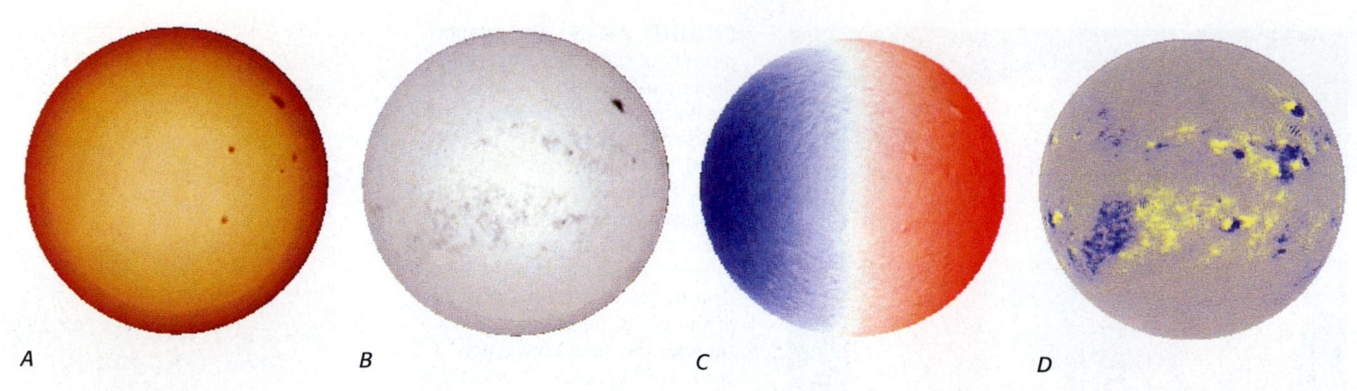


**FIGURE 22–9** The types of data produced by the GONG project include (*A*) the solar intensity; (*B*) the modulation of brightness after overall variations of brightness are subtracted, measured from the darkness of an absorption line; (*C*) the velocity field, here showing the overall redshift and blueshift from the Sun's rotation; and (*D*) the magnetic field, here with purple and yellow showing opposite polarities. GONG measures intensity, brightness modulation, and velocity once each minute and magnetic field every 20 minutes.

"Stellar seismology" applies to other stars the same techniques as helioseismology, and will one day allow astronomers to explore the interiors of distant stars. Theoretically, as stars evolve, their voices should deepen because of changes in their interiors. This prediction was verified in 2001 when pulsations with a 17-minute period, corresponding to the Sun's 5-minute oscillations, were discovered in the star beta Hydri, 24 light-years from us. Beta Hydri is 7 billion years old, compared with our Sun's 4.5 billion-year age, though it has the same mass and temperature as the Sun.

An Australian-built telescope called MONS (*M*easuring *O*scillations in *N*earby *S*tars) will be launched aboard a Danish satellite in 2004. To study stellar seismology, MONS will observe stars for about one month each over the course of its two-year mission.

## 22.2 THE CHROMOSPHERE

The chromosphere is intermediate in temperature and density between the material below it (the photosphere) and the material above it (the corona). Thus the chromosphere contains gas at temperatures about 10,000 K. It is thousands of times less dense than the photosphere but millions of times denser than the corona. When even a little energy is injected into the chromosphere from below, it raises the temperature substantially because it is spread among relatively few particles. Thus a "temperature minimum" is reached at the top of the photosphere. Temperatures at positions either up or down in the solar atmosphere are higher.

### 22.2a The Appearance of the Chromosphere

Under high resolution, we see that the chromosphere is not a spherical shell around the Sun but rather is composed of small spikes called **spicules.** The spicules rise and fall, and their appearance has been compared to that of blades of grass or burning prairies. This chromospheric gas is at a temperature of approximately 10,000 K, slightly hotter than the photosphere below it. The chromosphere emits radiation especially strongly at the red color of a spectral line of hydrogen, so is especially well observed with a filter that passes only that color (Fig. 22–10).

Spicules are more-or-less 1 arc second in diameter and perhaps ten times that in height, which corresponds to about 700 km across and 7000 km tall. They seem to have lifetimes of about 5 to 15 minutes, and there may be approximately half a million of them on the surface of the Sun at any given moment.

Studies of velocities on the solar surface showed the existence of large, organized convection cells, called **supergranulation.** Supergranulation cells look somewhat like polygons of approximately 30,000 km diameter. Supergranulation is an entirely different phenomenon from granulation. Each supergranulation cell may contain hundreds of individual granules.

Matter wells up in the middle of a supergranule and then slowly moves horizontally across the solar surface to the supergranule boundaries. The matter then sinks back down at the boundaries. This slow circulation of matter seems to be a basic process of the lower part of the solar atmosphere. The network of supergranulation boundaries is especially visible in the radiation of the "H and K lines," spectral lines of ionized calcium (Fig. 22–11). In both cases, the solar gas is so absorbing that our line of sight stops at the chromospheric level from seeing farther downward into the Sun. Thus even in the center of the solar disk, we see the chromosphere when looking in hydrogen or calcium radiation.

Ultraviolet spectra of distant stars have shown unmistakable signs of chromospheres in stars of spectral types like the Sun. Thus by studying the solar chromosphere we are also learning what the chromospheres of other stars are like.

**FIGURE 22–10** The Sun, photographed through a filter passing only the Hα line of hydrogen, shows the chromosphere. We see an image from August 11, 1999, the day of a total eclipse. Dark filaments snake across the Sun, and light plages are also visible around active regions.

## 22.2b The Chromospheric Spectrum at the Limb

During the few seconds at an eclipse of the Sun that the chromosphere is visible, its spectrum can be taken. This type of observation has been performed ever since the first spectroscopes were taken to eclipses in 1868. Since the chromosphere appears during that brief moment at an eclipse as hot gas silhouetted against dark sky, the chromospheric spectrum consists of emission lines.

The chromospheric emission lines appear to flash into view at the beginning and at the end of totality, so the visible spectrum of the chromosphere is known as the **flash spectrum** (Fig. 22–12).

Astronomers have been able to study the chromospheric spectrum in the ultraviolet using telescopes on spacecraft; the spectrum in the ultraviolet contains emission lines, at which there is more intensity than at neighboring wavelengths. These emission lines appear even when we look at the center of the solar disk. So the situation is quite different from the solar photosphere, which shows only absorption lines at the center of the solar disk.

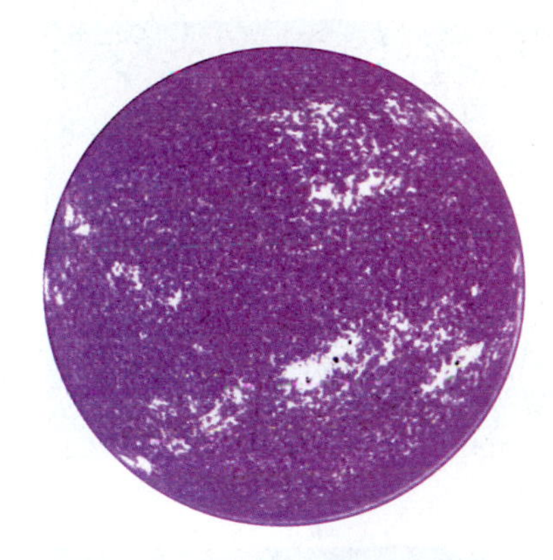

**FIGURE 22–11** A filter passing only the H or K line of ionized calcium also shows the chromosphere. The color is harder to see when looking through an eyepiece at a telescope, and the Sun is fainter than in Hα, but the supergranulation shows up particularly well.

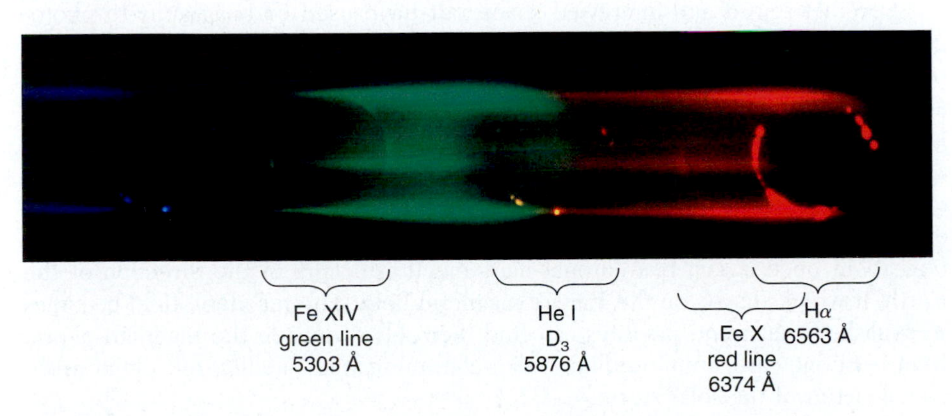


**FIGURE 22–12** A spectrum of the prominences and corona at the 1999 total solar eclipse. The last traces of photospheric spectrum show as the band of color. The bright points are the emission lines from the chromosphere and prominences. The yellow helium $D_3$ line was the emission line from which helium was first identified over a hundred years ago. The coronal "green line," radiation from iron in the corona heated to about 2,000,000 kelvins and thus highly ionized, shows as a faint but complete circle. The spectrum was taken without using a slit in the spectrograph, so the shape of the gas at the edge of the Sun is reproduced in the spectrum.

**FIGURE 22–13** The corona at the total solar eclipse of 1998, photographed from Aruba, with a superimposed picture in the center taken in the ultraviolet from the SOHO spacecraft. The ultraviolet image also shows the hot coronal gas.

## 22.3 THE CORONA

During total solar eclipses, when first the photosphere and then the chromosphere are completely hidden from view, a faint white halo around the Sun becomes visible. This **corona** (Fig. 22–13) is the outermost part of the solar atmosphere and extends throughout the Solar System. Close to the solar limb, the corona's temperature is about 2 million K.

### 22.3a The Structure and Temperature of the Corona

Even though the temperature of the corona is so high, the actual amount of energy in the solar corona is not large. The temperature quoted is actually a measure of how fast individual particles (electrons, in particular) are moving. There aren't very many coronal particles, even though each particle has a high velocity. The corona has less than one-billionth the density of the Earth's atmosphere, and would be considered to be a very good vacuum in a laboratory on Earth. For this reason, the corona serves as a unique and valuable celestial laboratory in which we can study gaseous plasmas (ionized gases) in a near vacuum.

The corona shows broad streamers extending outward. The streamers' shapes arise from the Sun's magnetic field. There are always streamers at lower solar latitudes. During the peak years of the 11-year sunspot cycle, which we discuss in the next chapter, there are also streamers at high latitudes. Near the minimum of the solar activity cycle, we see, rather, polar plumes.

### 22.3b Coronal Mass Ejections

The corona is normally too faint to be seen from the Earth's surface except at an eclipse of the Sun (Section 7.3b) because it is fainter than the everyday blue sky. But at certain locations on mountain peaks on the surface of the Earth, the sky is especially clear and dust-free, and the innermost part of the corona can be seen (Fig. 22–14*A*). Special telescopes called **coronagraphs** block out the solar photosphere so that they can study the corona from such sites; again, only the innermost part is detectable. Coronagraphs are built with special attention to low scattering of light, since their object is not to gather a lot of light but to prevent the strong photospheric radiation from being scattered about within the telescope. From the moon, we can observe the corona without atmospheric scattering (Fig. 22–14*B*).

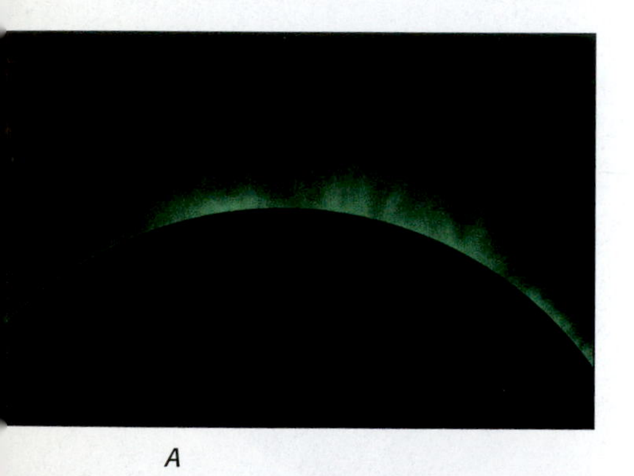

*A*

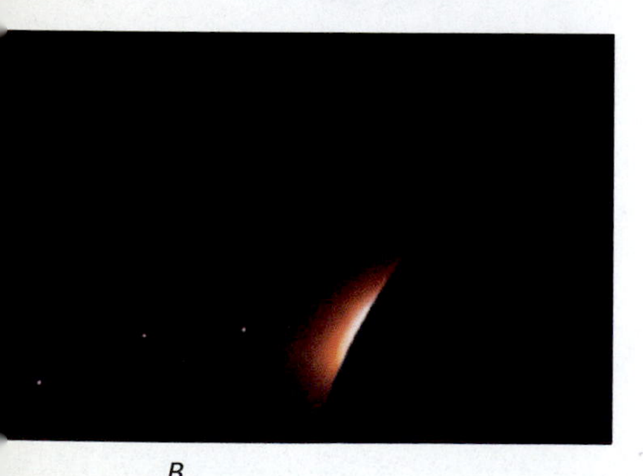

*B*

**FIGURE 22–14** (*A*) From a few mountain sites, the innermost corona can be photographed without need for an eclipse. The corona shows up best in its green emission line from 13-times-ionized iron at 5303 Å. (*B*) The Clementine satellite, when in orbit around the Moon in 1994, observed the solar corona rising before the solar photosphere came into view. Mercury, Mars, and Saturn appear to the left.

Several crewed and uncrewed spacecraft have used coronagraphs to photograph the corona, hour by hour, in visible light. These satellites studied the corona to much greater distances from the solar surface than can be studied with coronagraphs on Earth. Among the major conclusions of the research is that the corona is much more dynamic than we had thought. For example, many blobs of matter are ejected from the corona into interplanetary space (Fig. 22–15). These **coronal mass ejections** occur often. Scientists using the SOHO coronagraphs discovered that even at the minimum of the sunspot cycle, there is a coronal mass ejection about once a day. If a coronal mass ejection occurs in the direction of the Earth, it wreaks havoc on the Earth's magnetic field. The magnetic field becomes so squashed that some satellites can find themselves outside the magnetosphere. In at least one case, a multimillion-dollar communication satellite fell silent, probably a victim of the solar storm.

Coronagraphs in space hide not only the solar photosphere but also at least an extra fraction of a solar radius, because the inner corona is so bright that its light scatters in the telescope, so the innermost corona remains hidden. Thus solar eclipses remain the best way to study the inner corona. Coordinating eclipse observations with satellite observations is necessary for the most complete picture.

**FIGURE 22–15** A coronal mass ejection photographed by the Solar and Heliospheric Observatory (SOHO). The image is 30 solar radii across.

The "occulting disk," a disk that blocks (occults) the inner corona, and the pylon holding it in place are visible coming from lower left. The occulting disk covers the solar photosphere and several solar radii of inner corona. The size of the solar photosphere has been drawn on for scale.

## 22.3c The Heating of the Corona

The visible region of the coronal spectrum, when observed at eclipses, shows a continuum and also both absorption lines and emission lines. The emission lines do not correspond to any normal spectral lines known in laboratories on Earth or on other stars, and for many years their identification was one of the major problems in solar astronomy. The lines were even given the name of an element: coronium. In the late 1930s, it was discovered that they arose in atoms that were multiply ionized. This was the major indication that the corona was very hot (and that coronium doesn't exist). In the photosphere we find atoms that are neutral, singly ionized (missing one electron), or doubly ionized (missing two electrons)—Ca I, Ca II, and Ca III, for example. In the corona, on the other hand, we find ions that are ionized approximately a dozen times (Fig. 22–16). (The coronal "green line" seen in the visible spectrum during eclipses, for example, is iron that has lost 13 of its normal quota of 26 electrons.) The corona must be very hot indeed—millions of degrees—to have enough energy to strip that many electrons off atoms.

The gas in the corona is so hot that it emits mainly x-rays, photons of high energy. The photosphere, on the other hand, is too cool to emit x-rays. As a result, when photographs of the Sun are taken in the x-ray region of the spectrum, they show the corona and its structure (Fig. 22–17). A solar x-ray telescope is now aloft on the Japanese Yohkoh ("Sunbeam") spacecraft. The telescope, an American/Japanese collaboration, is being used to follow changes in the corona.

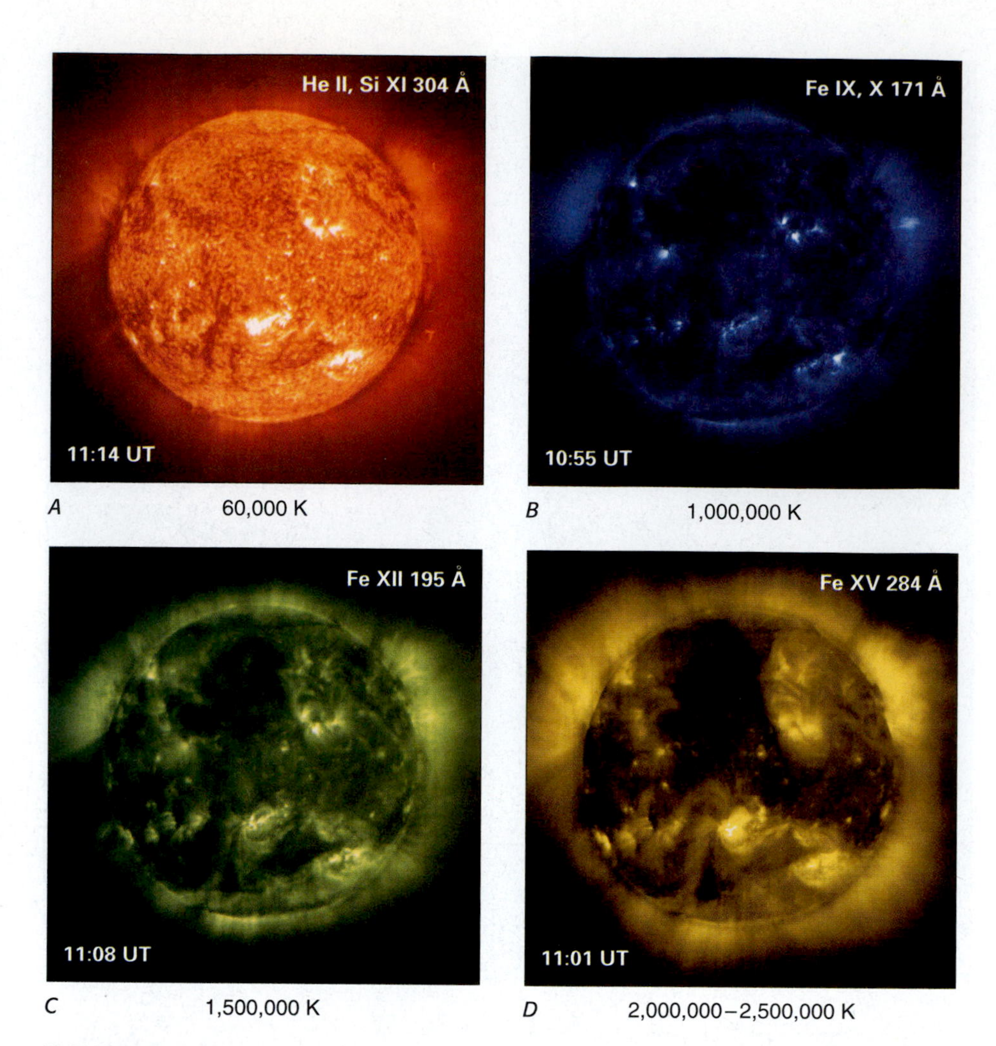


**FIGURE 22–16** A set of images taken August 11, 1999, the day of a total eclipse, through the four filters of the Extreme Ultraviolet Imaging Telescope (EIT) on the SOHO spacecraft. (*A*) Helium gas at a temperature of about 60,000 K. (*B*) Ionized iron at a temperature of about 1,000,000 K. (*C*) Ionized iron at a temperature of about 1,500,000 K. (*D*) Ionized iron at a temperature of about 2,000,000 to 2,500,000 K. Note that the spacecraft is deep in space beyond the Moon, so does not see an eclipse.

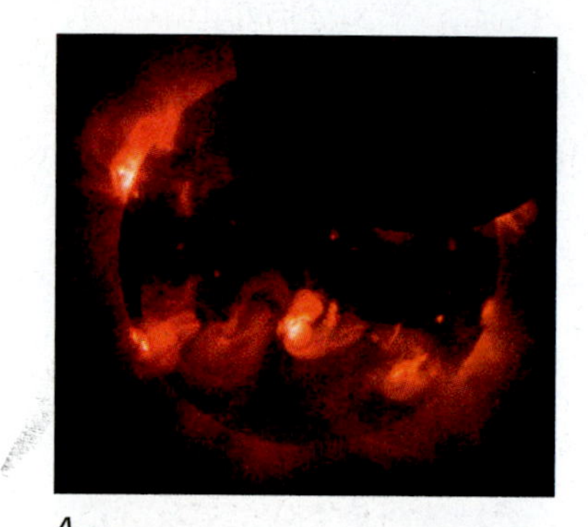


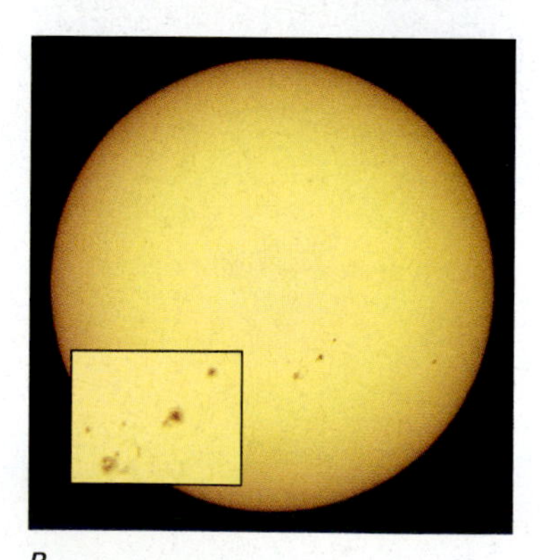


**FIGURE 22–17** (*A*) The corona in soft x-rays (*top*) corresponding to a white light image (*bottom*), observed on August 11, 1999, from the Yohkoh spacecraft during the eclipse. Yohkoh saw only a partial eclipse. (*B*) The solar photosphere, in an image also taken on the day of the 1999 total eclipse. The inset shows detail in the sunspots.

The x-ray pictures show structure that varies with the 11-year cycle of solar activity, which we shall discuss in Chapter 23 (see also Fig. 1–17). The brightest regions visible in x-rays are part of "active regions" that can also be seen in white light. (In white light, these active regions include sunspots.) X-ray images show a higher layer of the solar atmosphere than the white-light images show.

Detailed examination of the x-ray images shows that most, if not all, of the radiation of the inner corona is emitted by loops of gas joining points separated from each other on the solar surface. The corona seems to be composed entirely of these loops. One important line of thought that follows from this idea is that we must understand the physics of coronal loops in order to understand how the corona is heated. It is not sufficient to think in terms of a uniform corona, since the corona is obviously so non-uniform. The loops show the shape of the magnetic field that permeates the corona, though we are unable to measure the strength of the field directly.

The Transition Region and Explorer (TRACE) spacecraft has shown the coronal loops in even greater detail. Its high resolution, seen in the image opening this chapter, confirms the idea that the corona is entirely made of loops. Detailed looks should reveal just where on loops the gas is heated—at the base, at the top, or all along, for example. Such information should help us understand coronal heating. But results released so far on the position of the heating have been controversial, and no result yet is generally accepted.

The x-ray image also shows a very dark area at the Sun's north pole and extending downward across the center of the solar disk. (As the Sun rotates from side to side, obviously, the part extending across the center of the disk will not usually be facing us.) These dark locations are **coronal holes,** regions of the corona that are particularly cool and quiet. The density of gas in those areas is lower than the density in adjacent areas. Observations from rockets, when available, provide x-ray and far ultraviolet images in higher detail (Fig. 22–18) than is available from Yohkoh. A NASA/Smithsonian rocket flight in New Mexico set to coincide in time as closely as possible with the 2001 total solar eclipse in Africa radioed back about 20 high-resolution far-ultraviolet observations, including one showing 3-million-kelvin gas from sixteen-times-ionized nickel, though the payload and more data were then lost. The TRACE spacecraft sends back high-resolution observations in the far ultraviolet (Fig. 22–19) continuously.

NASA's Extreme Ultraviolet Explorer spacecraft, EUVE, catalogued hundreds of stellar coronas and took spectra until its lifetime ended in 2001. Some show the star's rotation; others show flares or coronal mass ejections. Many of the stars thus seem similar to our Sun. The x-ray satellites ROSAT and Chandra have also detected many coronas.

There is usually a coronal hole at one or both of the solar poles. Less often, we find additional coronal holes at lower solar latitudes. The regions of the coronal holes seem very different from other parts of the Sun.

Since the corona is so hot, the peak of its radiation (according to its Planck curve, Section 24.2) is in the x-ray region of the spectrum. X-rays don't come through the Earth's atmosphere, so they can only be studied from rockets and spacecraft. The TRACE spacecraft, launched in 1998, doesn't view x-rays but observes the corona in the ultraviolet close to the x-ray/UV dividing line with much higher resolution than previous solar spacecraft.

**FIGURE 22–18** A technique of x-ray optics allows imaging the Sun with reflections at normal rather than grazing incidence. This rocket photograph from 1989 shows the Sun at the 63.5-Å radiation from 15-times-ionized iron. The resolution was 3/4 arc sec. We see coronal loops.

Since energy doesn't travel from cooler regions to hotter ones, it seems odd that the corona should be so much hotter than the solar photosphere and chromosphere below it. Thus we deduce that some mechanism must be carrying energy upward from below the solar surface and depositing that energy in the corona.

Solar physicists used to think that the corona was heated to millions of degrees by shock waves that carried energy upward from underneath the photosphere. Shock waves, a kind of sound wave, are abrupt changes in pressure that occur when matter moves faster than the speed of sound at a location. The theory of shock-wave heating was discarded when a set of observations of the chromosphere from a spacecraft showed that the shock waves were carrying, as they passed through, 1000 times too little energy to heat the corona. It was also discovered from the Einstein Observatory x-ray spacecraft that even some stars give off enough x-rays that they must have hot coronas, even though their internal structure is such that they cannot generate shock waves. The x-ray observations show that different mechanisms are necessary

**FIGURE 22–19** An ultraviolet image, taken from space on the day of the 1999 total solar eclipse, shows small coronal loops. The image was made with the TRACE spacecraft.

FIGURE 22–20 The solar corona at the total eclipse of June 21, 2001, photographed from Zambia.

to heat the coronas of other stars. Some of these different mechanisms probably heat the solar corona, too.

Though the old ideas about coronal heating have been discarded, there is no agreement about what the real mechanism is for heating the corona. Apparently, the energy is generated in connection with the solar magnetic field. This idea explains the x-ray pictures, which show that the corona is hotter above active regions. A half-dozen different models of how waves in the magnetic field could heat the corona are now being considered.

## 22.4 FUTURE SOLAR ECLIPSES

Total eclipses of the Sun, which occur somewhere in the world every 18 months or so, are a particularly productive way of studying the outer layers of the Sun. In Section 7.3, we discussed them and surveyed several past and future eclipses. The next total solar eclipse will be on December 4, 2002, in southern Africa and south-central Australia. The November 23, 2003, eclipse will be seen by very few people, since it will be visible only from Antarctica and adjacent ocean. Then, aside from brief totality over the mid-Pacific on April 8, 2005, we won't have a total solar eclipse until March 29, 2006.

My own recent eclipse experiments have been devoted to testing predictions of theories that involve magnetic fields to heat the corona and to fostering the liaison between space-based coverage from SOHO and TRACE and ground-based eclipse images.

## 22.5 FUTURE SOLAR SPACE MISSIONS

The Japanese space agency, with major American participation, is planning for the launch of Solar-B in 2005, with a projected lifetime of 7 years, through the next solar maximum. (Yohkoh had been Solar-A until it was launched; Solar-B will also eventually have a new name.) It is to carry a 50-cm-diameter solar optical telescope, much larger than previous solar telescopes in space. In addition to making these high-resolution visible-light images, it will also have an extreme-ultraviolet imaging spectrometer and a grazing-incidence telescope to study soft x-rays. The telescope should have resolution of only 1/4 arc sec, substantially better than any previous spacecraft and several times better than ground-based resolution of about 1 arc sec. So it should be able to resolve and study spicules and other fine structure in the solar chromosphere as well as studying flares and other active phenomena.

NASA's planned Solar Probe was cancelled in 2001 for budgetary reasons. It was to travel deep into the solar corona, coming as close as 3 solar radii from the photosphere. A special design would have allowed the spacecraft to survive at the temperatures there. It was to be launched in 2008 and to arrive at the Sun via Jupiter in 2011. Among its major tasks was to investigate how the solar corona is heated.

The European Space Agency's Solar Orbiter is being investigated for launch in the coming decade. It would be in an orbit that brings it close to the Sun at a speed that makes it hover over one solar region for a time. It will also send images of the Sun's far side.

## 22.6 ADVANCED TECHNOLOGY SOLAR TELESCOPE

Observational solar research has continually shown that interesting and important activity takes place on smaller and smaller scales. Many lines of observation have revealed, further, that the magnetic fields are the dominant drivers of almost everything that happens in the upper solar photosphere, chromosphere, and corona. The activity seems to be driven by the same convective motions in the lower photosphere that we see as granulation.

Until recently, the observationalists were way ahead of the theoreticians in modelling the solar atmosphere. The theoreticians, even those with the fastest and most powerful computers, had to smear out the fine structure in the solar atmosphere. But in recent years, the theoreticians—including both those who use numerical models on computers and those who advance concepts—have caught up. Theoretical models now exist that may even exceed the spatial resolutions of the best observations in some cases.

Thus for reasons of theoretical modelling, we want observations of still higher resolutions. Fortunately, technologies have been advancing that should make this advance possible. Many have been developed for nighttime telescopes, such as new ways of making optics and techniques for limiting the degrading of seeing from thermal causes in the telescope or telescope dome. Further, adaptive optics is increasingly allowing higher-resolution observations than the Earth's atmospheric turbulence would normally allow. Adaptive optics has already been successful in improving solar observations at telescopes on Sacramento Peak in New Mexico and on La Palma in the Canary Islands.

Thus after many years of relatively small solar telescopes, plans are advancing for the U.S. National Solar Observatory to build an Advanced Technology Solar Telescope. It is to have a main mirror 4 m in diameter, and have a very low level of light scattered by the optics so that observations can be made of the coronal magnetic field. Its spatial resolution is to be only 0.25 arc sec at visible wavelengths, which corresponds to 200 km on the Sun, and a third that at an infrared spectral line that is especially useful for measuring magnetic fields. It will work in spectral regions ranging from the entire visible through about 30 micrometers in the infrared. Molecular spectral lines from sunspots, in particular, are found in the infrared. The telescope's location has yet to be chosen. We hope it will be available by about 2010.

## CORRECTING MISCONCEPTIONS

| ✖ *Incorrect* | ✔ *Correct* |
|---|---|
| All parts of the Sun, as seen from Earth, have the same spectrum. | The photospheric spectrum is in absorption and the chromospheric and coronal spectra are in emission. |
| The Sun has parts that are solid. | The Sun is gaseous throughout. |
| The Sun emits only yellow light. | The Sun emits radiation all across the spectrum, with yellow-green as the strongest color. |
| The Sun shines because gas is burning there. | Burning is a chemical process, while the Sun and stars shine by nuclear processes (fusion). |

## SUMMARY AND OUTLINE

Parts of the Sun (Part 3)
- Interior (15 million K), photosphere including granulation (5800 K), chromosphere and spicules (15,000 K), supergranulation, corona (2,000,000 K), coronal holes (perhaps 1,500,000 K)

The photosphere (Section 22.1)
- High-resolution observations: seeing limited; convection causes granulation (Section 22.1a)
- Sun ringing like a bell; interpretation shows conditions in solar interior; GONG and SOHO now in operation with continuous observations of the solar surface (Section 22.1b)
- Spectra (Section 22.1b)
- Photospheric spectrum: continuum with Fraunhofer lines; abundances of the elements (Section 22.1b)
- Solar seismology (Section 22.1c)

The chromosphere (Section 22.2)
- Chromospheric spectrum: emission at the limb during eclipses in the visible part of the spectrum (flash spectrum) (Section 22.2b)
- Coronal spectrum: identification of coronium with highly ionized ions shows the high temperature (Section 22.3c)
- Spicules; supergranulation; chromospheric network

The corona (Section 22.3)
- Streamers; coronal mass ejections; eclipses; coronagraphs on the ground and in space
- Heating of the corona (Section 22.3c)
- New theories involving solar magnetism stem from space observations; x-ray observations show link with magnetic fields.

Solar eclipses (Section 22.4)
- Good way to study chromosphere and corona

Future space missions (Section 22.5)
- Solar-B (2005) and Solar Orbiter (this decade)

The Advanced Technology Solar Telescope (Section 22.6)

## KEY WORDS

quiet sun
active sun
photosphere
interior
core
solar atmosphere
chromosphere
corona
solar wind
subtends
white light
granulation
convection
convection zone
continuum
solar seismology
helioseismology
spicules
supergranulation
flash spectrum
corona
coronagraphs
coronal mass ejection
coronal holes

## QUESTIONS

1. Sketch the Sun, labelling the interior, the photosphere, the chromosphere, and the corona. Give the approximate temperature of each.
2. What elements make most of the lines in the Fraunhofer spectrum? What elements make the strongest lines? Why?
3. Explain why the photospheric spectrum is an absorption spectrum and the chromospheric spectrum seen at an eclipse is an emission spectrum.
4. Sketch the solar spectrum with the colors from red to blue across a page in a line, including the strongest lines: H$\alpha$, H$\beta$, H$\gamma$, H and K, and the D lines. Your sketch can be either a drawing showing what a photographic spectrum would look like, or a graph of intensity vs. wavelength.
5. How can we learn about the interior of the Sun by studying its surface?
6. What are the advantages of GONG and SOHO for solar seismology over observing from a single ground-based telescope?
7. Graph the temperature of the interior and atmosphere of the Sun as a function of distance from the center.
8. How often are coronal mass ejections seen?
9. How do we know that the corona is hot?
10. Describe relative advantages of ground-based eclipse studies and satellite studies of the corona.

## TOPICS FOR DISCUSSION

1. Discuss the relative importance of solar observations (a) from the ground, (b) at eclipses, (c) from uncrewed satellites, and (d) from crewed satellites.
2. Is the discussion of the Sun in this text most closely related to the part of this book on stars or to the part on the Solar System, or is it equally linked to both?

## INVESTIGATIONS

1. Find out when the next total solar eclipse occurs and study its path.
2. Use the World Wide Web, *RedShift*, the *Canon of Solar Eclipses*, or the *Field Guide to the Stars and Planets* to find out when the next partial solar eclipse will be visible from your location.

## USING TECHNOLOGY

### W World Wide Web

1. Access the NASA Reference Publications, available through the Web site of the International Astronomical Union's Working Group on Solar Eclipses (http://www.williams.edu/astronomy/IAU_eclipses). Assess the path and the weather prospects for the next total solar eclipse.
2. Check today's images of the Sun through the links on this book's Web site. Describe what the Sun looks like when seen through different filters.
3. Plan a virtual eclipse expedition and learn about the historic work of Maria Mitchell at http://depts.vassar.edu/~physastr/mariamitchell.
4. Learn the latest about Solar-B (http://www.ssl.msfc.nasa.gov/ssl/pad/solar), the cancelled Solar Probe (http://www.jpl.nasa.gov/ice_fire/sprobe.htm), and Solar Orbiter (http://sci.esa.int/home/solarorbiter).
5. Check on the status of the Advanced Technology Solar Telescope (http://www.sunspot.noao.edu/ATST).

### REDSHIFT

1. Use the Eclipse Search menu item in the main *RedShift* engine to locate the next total solar eclipse and to view the passage of the shadow across the Earth's surface. Note the difference between the umbra and penumbra.
2. Use the Eclipse Search menu item in the main *RedShift* engine to find out when the next total solar eclipse will be visible from the United States and when the next total solar eclipse will be visible from Canada. View the passages of the eclipses across the Earth's surface.
3. Learn from "Story of the Universe": The Sun.
4. Watch the movies "The Sun in X-rays" and "SOHO Sees Coronal Mass Ejection."
5. How far over is the north pole of the Sun tipped as seen from Earth, and how does it change over the year?

The Sun on August 11, 1999, the day of a total solar eclipse. Images from the Solar and Heliospheric Observatory show million-degree coronal gas on the solar disk and in the outer corona. The middle region is covered by a compound image made from several CCD exposures taken during the total eclipse as observed from Romania by the author's expedition.

The shape of the corona is typical of the maximum of the solar activity cycle, with streamers coming off the Sun in all directions. At minimum, streamers are seen only near the Sun's equator.

# Chapter 23

# Solar Activity and the Earth

A host of time-varying phenomena are superimposed on the basic structure of the Sun. Many of them, notably the sunspots, vary with an 11-year cycle, which is called the **solar-activity cycle.** Particles and radiation from the Sun reach the Earth, especially in solar flares and in coronal mass ejections, and these phenomena are more frequent at the maximum of the solar-activity cycle. In this chapter, we discuss the solar-activity cycle and its consequences for Earth. We also discuss the almost steady amount of energy we on Earth receive from the Sun.

**AIMS:** To learn about the time-varying structures of the Sun, and how these structures are linked with the Sun's magnetic field, and to learn about the relation of the Sun and the Earth through an outflow of solar particles and through solar radiation

## 23.1 SUNSPOTS

**Sunspots** (Fig. 23–1) are the most obvious sign of solar activity. They are areas of the Sun that appear relatively dark when seen in white light. Sunspots appear dark because they are giving off less radiation than the photosphere that surrounds them. This implies that they are cooler areas of the solar surface, since cooler gas radiates less than hotter gas. Actually, if we could somehow remove a sunspot from the solar surface and put it off in space, it would appear bright against the dark sky; a large one would give off as much light as the full moon.

A sunspot includes a very dark central region called the **umbra,** from the Latin for "shadow" (plural: *umbrae*). The umbra is surrounded by a **penumbra** (plural: *penumbrae*), which is not as dark (just as during an eclipse the umbra of the shadow is the darkest part and the penumbra is less dark).

To understand sunspots, we must understand magnetic fields. When you file tiny bits of iron off a nail, and put the filings near a simple bar magnet on Earth, the filings show a pattern (Fig. 23–2). The magnet is said to have a north pole and a south pole; the magnetic field linking them is characterized by what we call **magnetic lines of force,** or **magnetic field lines** (after all, the iron filings are spread out in what look like lines). The Earth (as well as some other planets) has a magnetic field that has many characteristics in common with that of a bar magnet. The structure seen in the solar corona, as shown on the facing page, including huge streamers, results from matter being constrained by the solar magnetic field.

We can measure magnetic fields on the Sun by using a spectroscopic method. In the presence of a magnetic field, certain spectral lines are split into a number of components, and the amount of the splitting depends on the strength of the magnetic field (Fig. 23–3). The scientific process by which the splitting is made is known as the **Zeeman effect,** after a Dutch physicist of about 100 years ago.

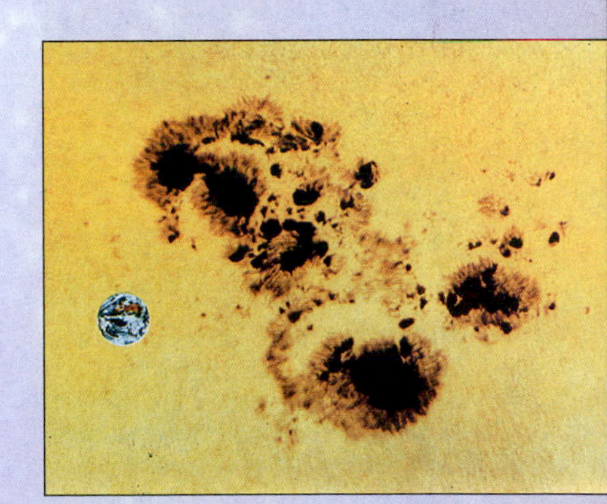

FIGURE 23–1 A sunspot, showing the dark **umbra** surrounded by the lighter **penumbra.** Granulation is visible in the surrounding photosphere. A photo of the Earth is superimposed to show its relative size.

## A DEEPER DISCUSSION

### BOX 23.1 The Origin of Sunspots

Although the details of the formation of sunspots are not yet understood, a general picture was suggested in 1961. In this model, just under the solar photosphere the magnetic field lines are bunched in tubes of magnetic field that wind around the Sun.

The Sun rotates approximately once each Earth month. Different latitudes on the Sun rotate at different angular speeds—**differential rotation.** Gas at the equator rotates in 25 days and gas at 40° latitude rotates in about 28 days. The differential rotation shows in the measured Doppler shifts.

A line of force that may have started out north–south on the solar surface is wrapped around the Sun by the differential rotation. These lines collect as tubes located not far beneath the solar photosphere. Sometimes buoyant forces carry part of a tube upward until the tube sticks up through the solar photosphere. Where the tube emerges we see a sunspot of one magnetic polarity, and where the tube returns through the surface we see a sunspot of the other polarity.

Because of the differential rotation, the spiral winding of the magnetic lines of force is tighter at higher latitudes on the Sun than at lower latitudes. Thus the instability that allows part of a tube to be carried to the surface arises first at higher solar latitudes. As the solar cycle wears on, the differential rotation continues, and the tubes rise to the surface at lower and lower latitudes. This explains why spots form at higher latitudes earlier in the sunspot cycle than they do later in the cycle.

Sunspots arise from the interaction of the solar differential rotation with turbulence and motions in the convective zone. Dynamos in factories on Earth also depend on the interaction of rotation and magnetic fields; thus these theories for the solar activity are called **dynamo theories,** and one says that the sunspots are generated by the **solar dynamo.**

#### *Discussion Question*

Try twisting rubber bands to make kinks. How do the kinks compare to sunspots, and what corresponds to the north and south polarities?

**FIGURE 23–2** Lines of force from a bar magnet are outlined by iron filings. One end of the magnet is called a north pole and the other is called a south pole. Similar poles ("like poles")—a pair of norths or a pair of souths—repel each other, and unlike poles (one north and one south) attract each other. Lines of force go between opposite poles.

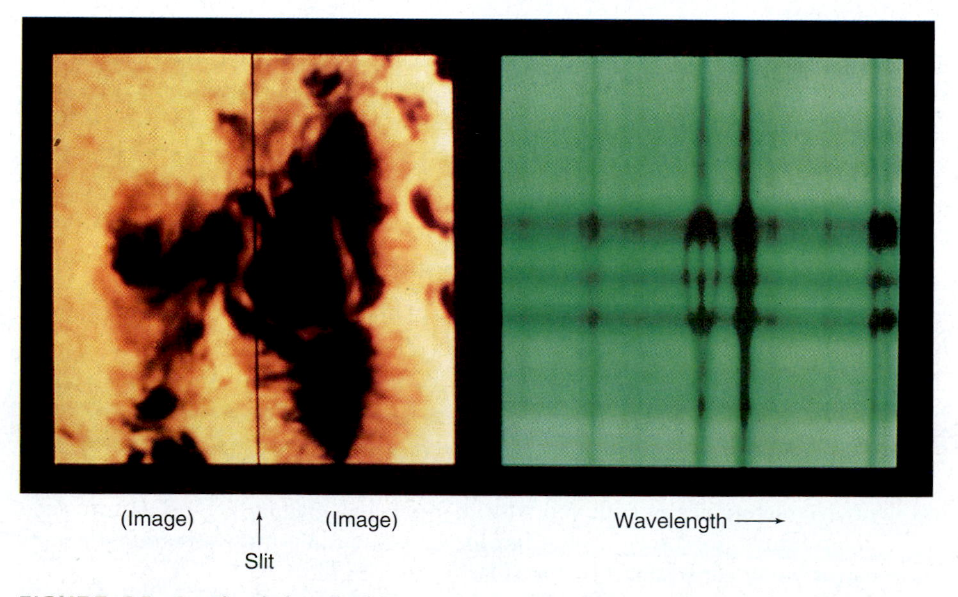


**FIGURE 23–3** The dark vertical line across the sunspot is the slit of the spectrograph onto which the image was projected; the light that we are not seeing in this image is that light that is being analyzed by the spectrograph. At right is a small, very spread-out portion of the spectrum of the sunspot region. Note that some of the spectral lines are split into several parts at the sunspot umbra. The split results because of the magnetic field, by the Zeeman effect. The stronger the magnetic field, the more the lines are split. By measuring the splitting, scientists build up maps of the solar magnetic field, which is highest in sunspot umbrae.

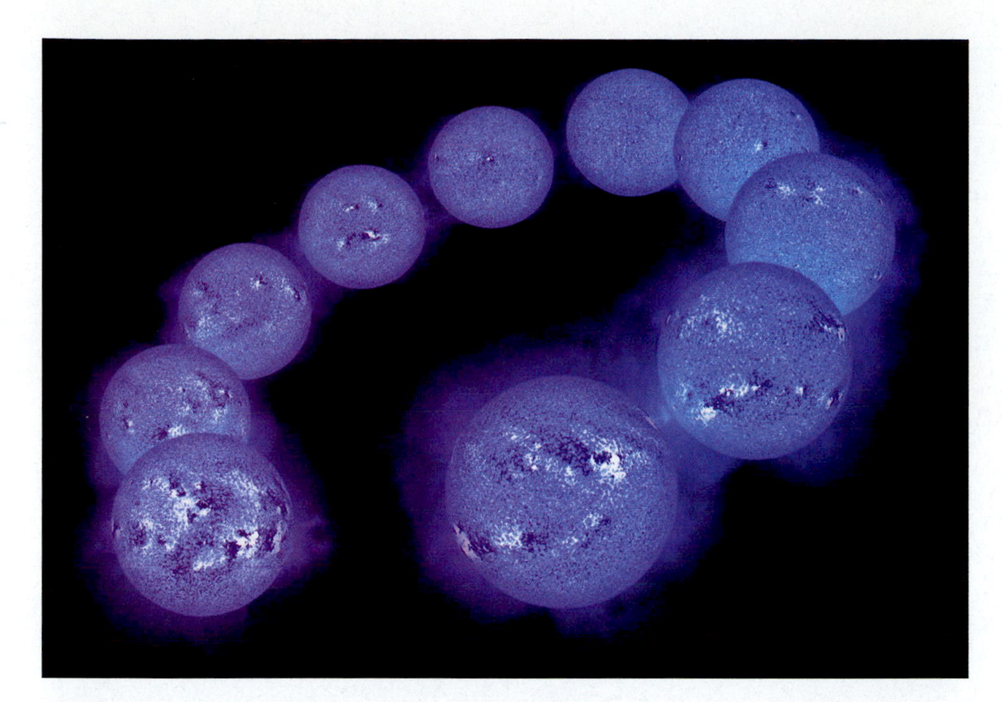

**FIGURE 23–4** The evolution of the magnetic field on the Sun over a sunspot cycle. These photographs correspond with the series of x-ray images that appeared in Figure 1-19. The regions of strong fields correspond to the positions of sunspots. Notice how the two members of a pair of sunspots have polarity opposite to each other, and how as the Sun rotates the polarity that comes first is different above and below the equator. Near the minimum of the solar-activity cycle, no large regions of concentrated magnetic field are visible, and, as a result, no sunspots are present on the Sun.

Measurements of the solar magnetic field were first made by George Ellery Hale in the United States. (Hale later went on beyond solar astronomy to found important observatories, and the giant 5-m Palomar Observatory telescope is named after him.) He showed, in 1908, that the sunspots are regions of very high magnetic field strength on the Sun, thousands of times more powerful than the Earth's magnetic field or than the average solar magnetic field. Sunspots usually occur in pairs, and often these pairs are part of larger groups. In each pair, one sunspot will have a polarity typical of a north magnetic pole and the other will have a polarity typical of a south magnetic pole (Fig. 23–4).

Magnetic fields are able to restrain matter (though only if it has an electric charge, such as ionized rather than neutral atoms). (This restraint of matter is the property some scientists are trying to exploit on Earth to contain superheated matter sufficiently long to allow nuclear fusion for energy production to take place. We wish we could make the process work on Earth as well as it works on the Sun.) The strongest magnetic fields in the Sun occur in sunspots. The magnetic fields in sunspots restrain the motions of the matter there, and in particular they keep convection from carrying energy to photospheric heights from lower, hotter levels. This results in sunspots being cooler and darker, though exactly why they remain so for weeks is not known. Many observatories around the world, including the National Solar Observatory (Fig. 23–5), monitor sunspots and their magnetic fields.

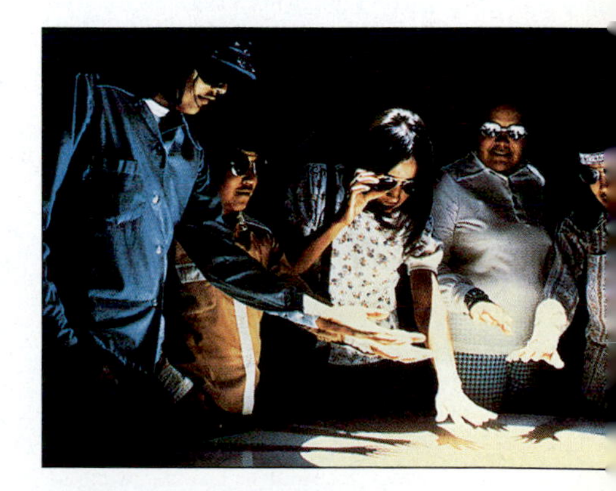

**FIGURE 23–5** The National Solar Observatory's telescopes on Kitt Peak are on the Tohono O'odham Reservation. Here Navajo students view the giant solar image projected onto an observing table from the largest of the solar telescopes.

The parts of the corona above sunspots are hotter and denser than the normal corona. Presumably the charged matter is guided upward by magnetic fields carrying its energy. These locations are prominent in radio and x-ray maps of the Sun.

Sunspots were discovered in 1610, independently by Galileo in Italy, Fabricius and Christopher Scheiner in Germany, and Thomas Harriot in England. In about

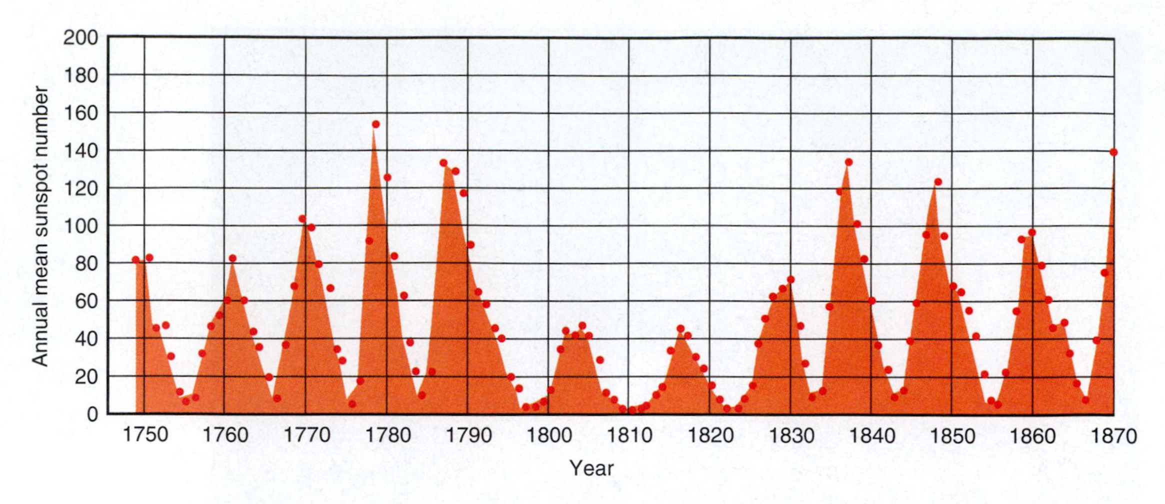


**FIGURE 23–6** The 11-year sunspot cycle is but one manifestation of the solar-activity cycle. As the graph shows, it is a very significant effect, not one of those small effects that has to be pulled out of the data by careful analysis.

1850, it was realized that the number of sunspots varies with an 11-year cycle, as is shown in Fig. 23–6. This variation is called the **sunspot cycle.** (Individual cycles can be as short as 8 years or as long as 12 years.) Since many related signs of solar activity (such as sunspots) vary with the same period, the variations are part of the solar-activity cycle.

Besides the specific magnetic fields in sunspots, the Sun seems to have a weak overall magnetic field with a north magnetic pole and a south magnetic pole, which may entirely result from the sum of the weak magnetic fields left over at locations of vanished sunspots. Every 11-year cycle, the north magnetic pole and south magnetic pole on the Sun reverse polarity; what had been a north magnetic pole is then a south magnetic pole and vice versa. This changeover occurs a year or two after the number of sunspots has reached its maximum. For a time during the changeover, the Sun may even have two north magnetic poles or two south magnetic poles! But the Sun is not a simple bar magnet, so this strange-sounding occurrence is not prohibited. Because of the changeover, it is 22 years before the Sun returns to its original configuration, so the real period of the solar-activity cycle is 22 years.

Sunspots are thought to be a result of differential rotation (Fig. 23–7) winding up the Sun's magnetic field (Fig. 23–8). Box 23.1 goes into the theoretical explanation. But the details of why the solar-activity cycle occurs, and why it is about 11 years long, are not understood.

At the beginning of a sunspot cycle, sunspots of that cycle appear at high solar latitudes. (Sunspots of the older cycle may still be visible at lower latitudes.) As a solar cycle advances year by year, sunspots appear closer to the equator (Fig. 23–9). The graph is called, for obvious reasons, the **butterfly diagram.**

The solar-activity cycle is demonstrated in a variety of ways, including the shape of the corona as seen at eclipses (Fig. 23–10). At sunspot minimum, only a few streamers are present, so the corona appears elongated. At sunspot maximum, so many streamers are present that the corona appears almost round.

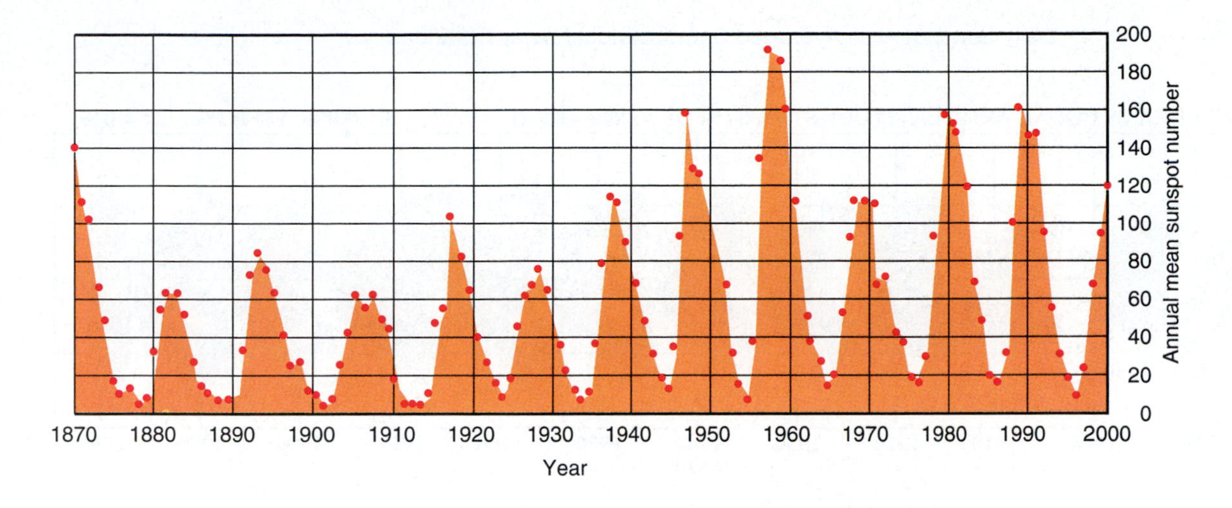


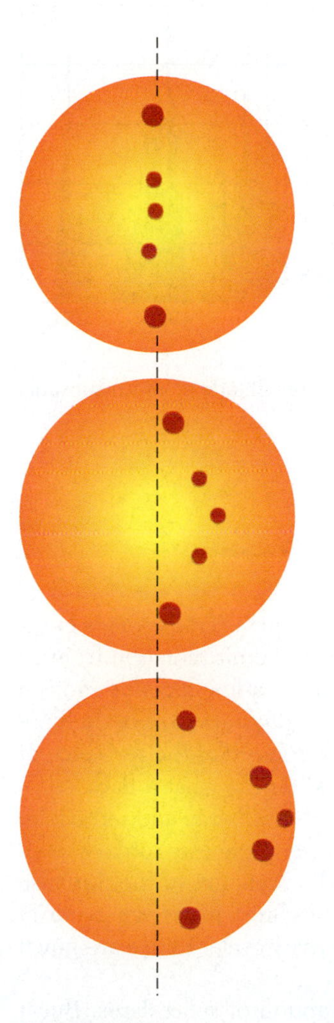

**FIGURE 23–7** The notion that the Sun rotates differentially is illustrated by this series, which shows the progress of a schematic line of sunspots month by month. The equator rotates faster than the poles by about 4 days per month.

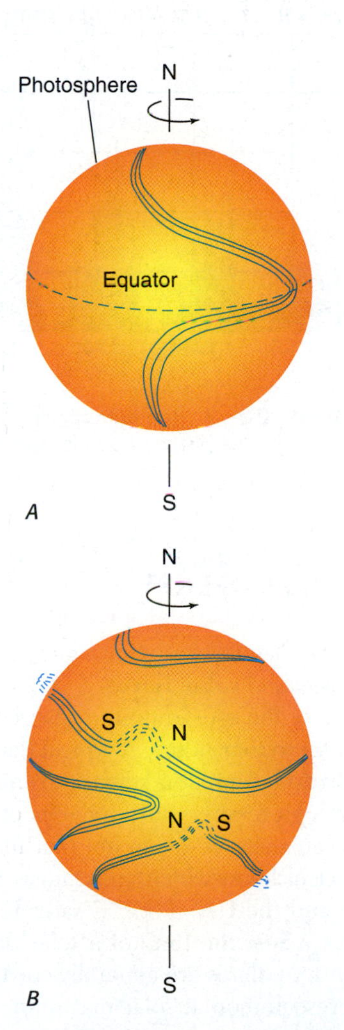


**FIGURE 23–8** (*A*) A leading model to explain sunspots suggests that the solar differential rotation winds tubes of magnetic flux around and around the Sun. (*B*) When the tubes kink and penetrate the solar surface, we see the sunspots that occur in the areas of strong magnetic field.

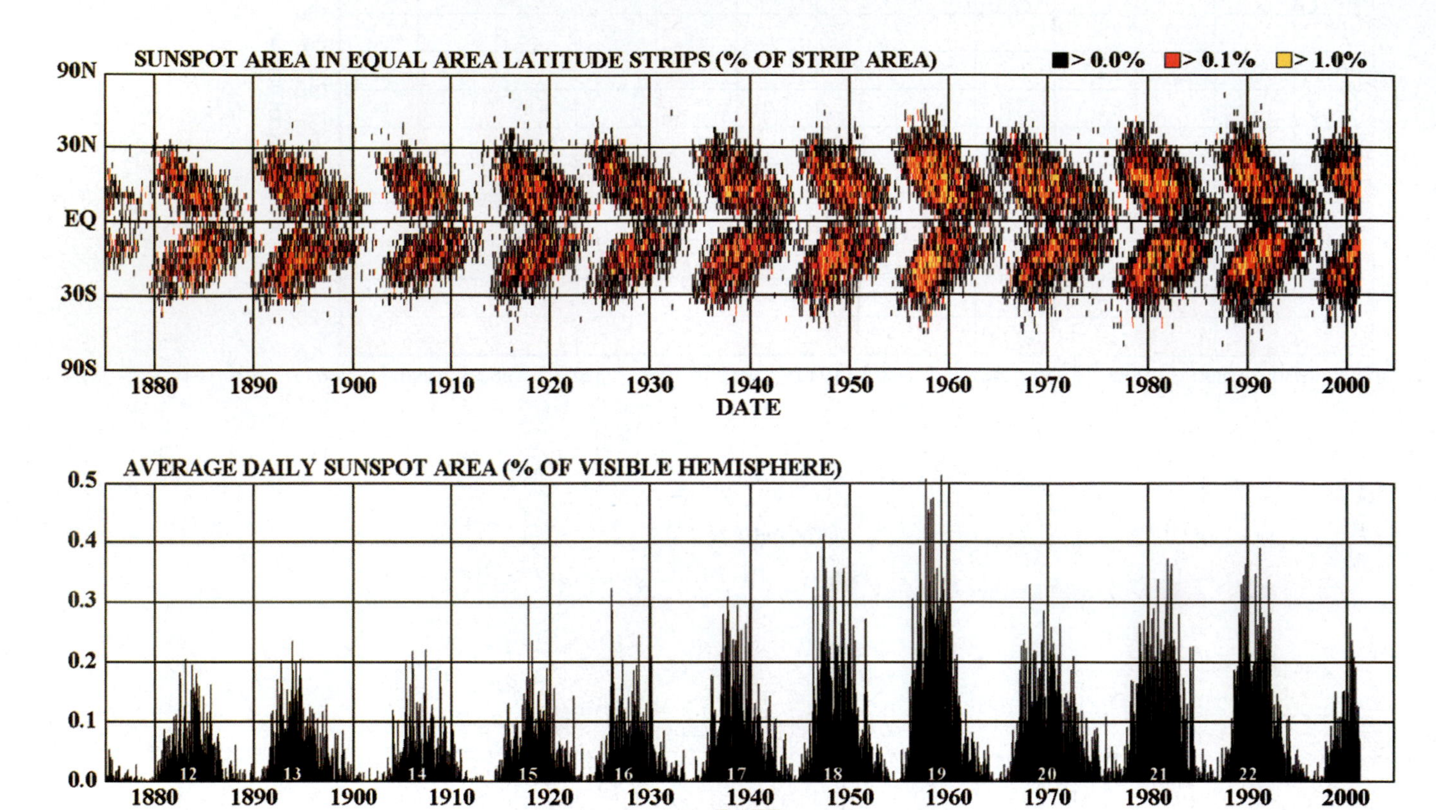


FIGURE 23–9 The butterfly diagram shows the dependence on latitude at which sunspots appear with the phase of the sunspot cycle.

## 23.2 FLARES

Violent activity sometimes occurs in the regions around sunspots. Tremendous eruptions called **solar flares** (Fig. 23–11) can eject particles and emit radiation from all parts of the spectrum into space. These solar storms begin in a few seconds and can last up to four hours. A typical flare lifetime is 20 minutes. Temperatures in the flare can reach 5 million kelvins, even hotter than the quiet corona. It is the presence of this high temperature, millions of degrees, that we use to define what a flare is.

Solar flares are so hot that they emit continuum x-rays and x-ray and ultraviolet spectral lines, which are studied from satellites such as the Japanese Yohkoh satellite and the U.S. TRACE satellite (Fig. 23–12). The radio emission of the Sun also increases at the time of a solar flare. Though the very brightest flares can occur at any time, flares are generally correlated with the solar-activity cycle. Flares are much more common at solar maximum.

No specific model is accepted as explaining the eruption of solar flares. But it is clear that a tremendous amount of energy is stored in the solar magnetic fields in sunspot regions. Something unknown triggers the release of the energy from the twisted or stressed magnetic field. An important recent discovery is that many flares are caused by coronal mass ejections. They go through the region of twisted or stressed magnetic field, causing a disruption that starts the flare. Much progress has been made in the last few years in understanding the site of the eruption within a flare; tracing the exact times at which ultraviolet, x-ray, and radio bursts of radiation are detected allows solar astronomers to follow the flare's evolution.

*A*

*B*

**FIGURE 23–10** The shape of the corona varies with the solar-activity cycle. At solar maximum, there are so many coronal streamers that the coronal shape is very round. Nearer solar minimum, only a few streamers are present, making the coronal shape elongated. (*A*) The eclipse of February 26, 1998, viewed from Aruba, when the Sun was still at a relatively low point of its cycle. (*B*) The solar-maximum eclipse of August 11, 1999, viewed from Romania, in which there are streamers going off in all directions, making the overall shape relatively round. The corona at the June 21, 2001, eclipse was also close enough to solar maximum to have streamers in all directions.

**FIGURE 23–11** The brightening of regions of the Sun, seen in the hydrogen-alpha line, marking the position of a solar flare. To make a flare, magnetic field lines change the way their north and south magnetic poles are connected. This "reconnection" releases high-energy particles, which follow the magnetic field lines back down to lower levels of the solar atmosphere. We see these regions brightening drastically on this image of the February 5, 2000, solar flare.

Energetic flare particles that are ejected from the Sun reach the Earth in a few hours or days and can cause disruptions in radio transmission. It has recently been realized that it is often the coronal mass ejections (Section 22.3b) rather than particles

*(text continues on page 423)*

# PHOTO GALLERY

## SOHO, the Solar and Heliospheric Observatory, and TRACE, the Transition Region and Coronal Explorer

Since 1996, an important solar observatory in space has been studying the Sun. Since it is studying not only the Sun itself but also the space between the Sun and the Earth, it is called the Solar and Heliospheric Observatory, or SOHO for short. It is mainly a project of the European Space Agency, but it has major participation by NASA and other U.S. institutions. It was joined in 1998 by the Transition Region and Coronal Explorer, TRACE, a smaller and less versatile satellite but whose telescope has higher spatial resolution. This high resolution and its spectral coverage allow it to do a superb job of studying not only the solar corona but also the transition region between the solar chromosphere and corona.

Most space observatories, including the Hubble Space Telescope, are in orbit only a few hundred kilometers above the Earth's surface and go through day and night on a 90-minute cycle. But to make long, continuous observations of the Sun over a period of years, SOHO was sent a million miles (1.5 million kilometers) toward the Sun, to a location from which the

He II 304 Å 60,000 kelvins (upper chromosphere/lower transition region)

Fe IX/X 171 Å 1 million kelvins (corona)

Fe XII 195 Å 1.5 million kelvins (corona)

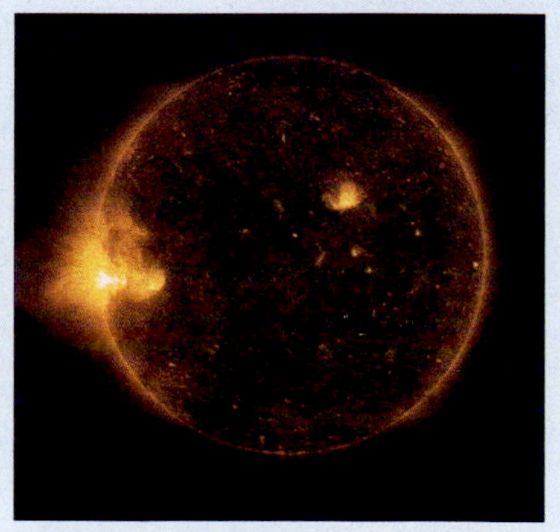

Fe XV 284 Å 2 to 2.5 million kelvins (corona)

A set of images for a given day for the Extreme Ultraviolet Imaging Telescope (EIT) on SOHO.

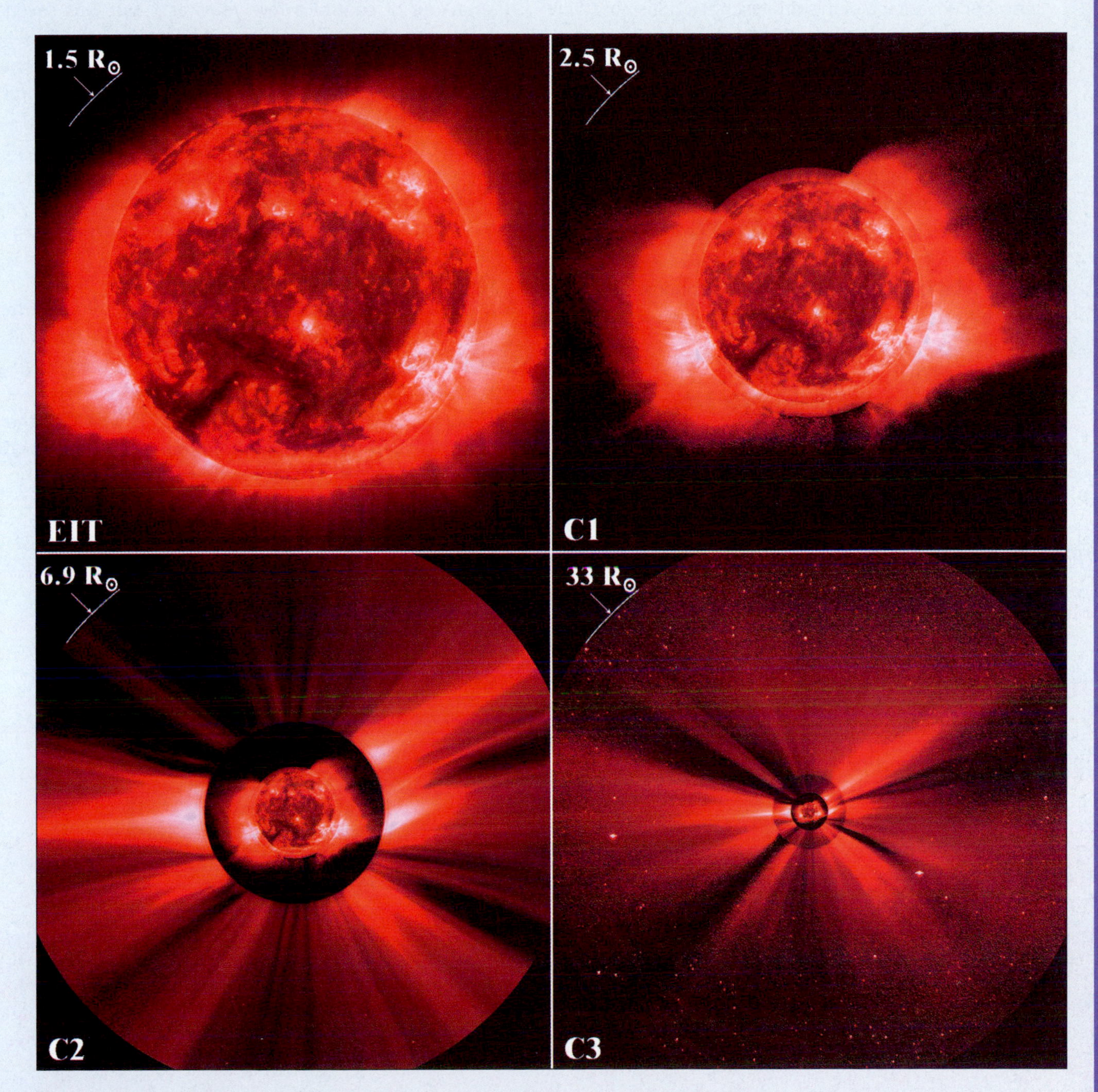


A set of images showing the EIT and the three LASCO coronagraphs, increasingly compounded into a joint image. The inner LASCO coronagraph is defunct, and the gap can be filled only at the times of total solar eclipses with ground-based coronal images.

Sun is always visible. It reached a point, known as a Lagrangian point, where effects of the Sun's gravity, the Earth's gravity, and the spacecraft's motion cancel out. As seen from Earth, SOHO travels in a small circle around this point, so we say it has a "halo orbit." Another advantage of this orbit is that it does not swing toward and away from the Sun as do satellites in Earth orbit, so there are no Doppler effects to remove from the data.

The TRACE spacecraft also has an unusual orbit. Its orbit goes near the Earth's poles but in a way that it always passes directly under the Sun. As the Earth goes around the Sun in the course of a year, TRACE's orbit moves around the Earth at the same rate, so that the plane of the satellite's orbit always faces the Sun. So its orbit is called "polar sun-synchronous." The orbit is tilted enough that the Sun is always in view, never blocked by the Earth.

One of SOHO's telescopes, the *E*xtreme Ultraviolet *I*maging *T*elescope (EIT), makes pictures of the Sun through four filters, each showing gases at a different temperature. The lowest temperature, characteristic of helium gas, is about 60,000 K. The other temperatures are characteristic of different states of gaseous iron in the corona. The hotter the temperature, the more energy has been available to drive electrons off atoms, and the more electrons have been driven off. We use Roman numerals to say how many electrons an ion has lost: I for the first state with no electrons lost, II for one electron lost, etc. Characteristic temperatures are 1 million K for iron-X (9 electrons lost); 2 million K for iron-XIV (13 electrons lost), and 4 million K for iron-XVI (15 electrons lost). Within the corona, these high-temperature gases are slightly differently distributed, though they are all bound by the Sun's magnetic field.

Another imaging device is the Large Angle Spectroscopic Coronagraph (LASCO), which has disks that block out the bright solar photosphere and some region around it so as to be able to see the corona. (Technology does not yet allow us to block only the photosphere and image all the corona; the inner part is always still lost.) LASCO actually contains three coronagraphs, each observing a different part of the corona, with the highest resolution farthest in. SOHO spun out of control a few years ago, but was saved by talented NASA engineers and brought back to life—except for the innermost of the LASCO coronagraphs. The set of coronagraphs has found that the corona ejects some of its mass—"coronal mass ejections"—more often than daily, even during the minimum of the sunspot cycle, a higher rate than expected.

SOHO has an additional 10 instruments, some of which study helioseismology and some of which measure the solar-wind particles directly when they pass the spacecraft.

We have already seen TRACE's high-resolution images of coronal loops. It has the same three filters as SOHO that pass spectral lines emitted by high-temperature gas. It also carries additional filters that show the lower-temperature gas that is typical of the chromosphere (Lyman-alpha at about 10,000 K) and the transition region (carbon-IV at 100,000 K).

A TRACE image at 171 Å (Fe IX and Fe X) showing 1-million-degree gas held by the corona's magnetic field into narrow loops.

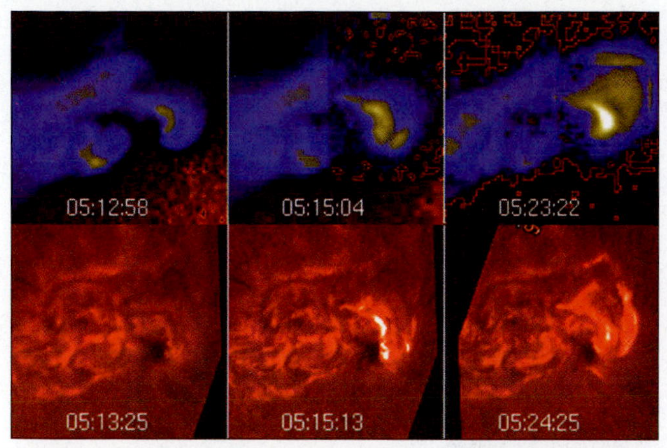

A

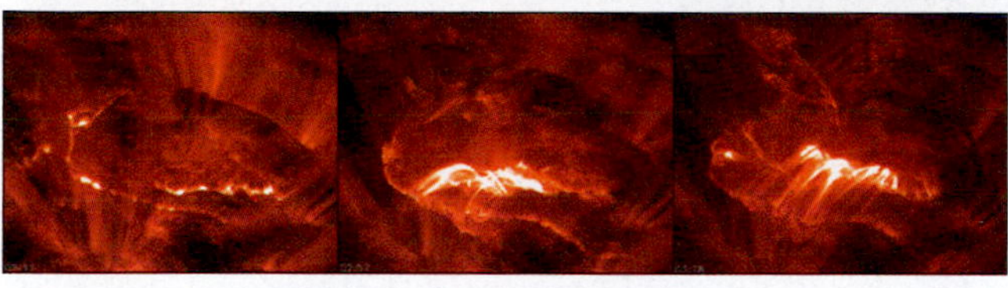

B

**FIGURE 23–12** Solar flares show as abrupt brightenings on the solar disk. (*A*) A time-sequence over a 12-minute interval showing a solar flare seen in both x-rays by Yohkoh (*top*, with x-ray intensity represented in false color) and in H-alpha (reproduced in its actual red color) with a ground-based telescope. To prevent saturation, the flare's image is a shorter exposure in parts B and C of the Yohkoh images. (*B*) A time-sequence over a 46-minute interval showing, in false color representing intensity in an ultraviolet emission line emitted by million-degree gas, a solar flare from the TRACE satellite.

ejected in flares that cause these disruptions. Both flares and coronal mass ejections also cause the aurorae—the **aurora borealis** is the "northern lights," and the **aurora australis** is the "southern lights" (Section 8.7). A buildup of electrons from the Sun even knocked out both Canadian weather satellites in 1993. A coronal mass ejection knocked out a communication satellite in 1997 and another one killed an x-ray astronomy satellite in 2000. A solar flare in 2001 was so strong that it was above the range sensors could measure. Pilots were kept on the ground for a while at northern latitudes, and an aurora was seen as far south as Texas. Because of these solar-terrestrial relationships, high priority is placed on understanding solar activity and being able to predict it. The U.S. government even has a bureau for **space weather** to forecast solar storms that might have consequences for Earth, just as it has a terrestrial weather bureau.

## 23.3 PLAGES, FILAMENTS, AND PROMINENCES

Studies of the solar atmosphere in H$\alpha$ radiation (Fig. 23–13) also reveal other types of solar activity. Bright areas called **plages** (pronounced "plah'jes," the French word for beaches) surround the entire sunspot region. Dark **filaments** are seen threading their way across the Sun in the vicinity of sunspots. The longest filaments can

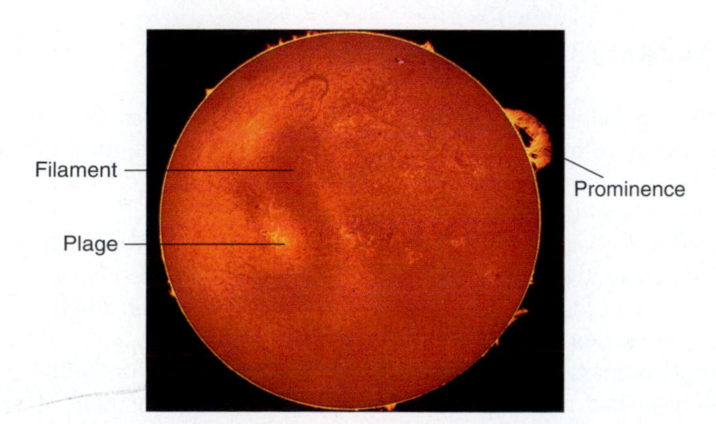


**FIGURE 23–13** The Sun, photographed in hydrogen radiation at solar maximum, shows bright areas called plages and dark filaments around the many sunspot regions.

FIGURE 23–14 Prominences, especially at left and lower right, as seen in a photograph of the Sun during the 1999 total eclipse. The chromosphere, as well as the last bit of the bright white diamond-ring effect, shows at upper left. The prominences and chromosphere appear pinkish because they radiate mainly the hydrogen-alpha emission line.

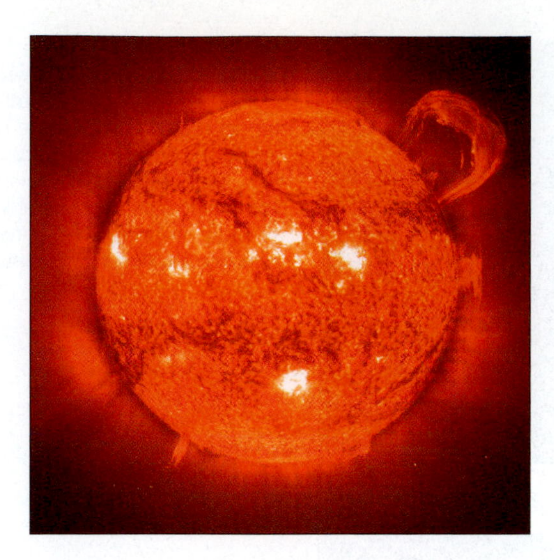

FIGURE 23–15 A full-disk image of the Sun, showing a huge prominence erupting at upper right and smaller prominences distributed around the limb, notably at the 3 o'clock and 7 o'clock positions. The image, from SOHO in space, was made in the ultraviolet spectral line from helium that, like H-alpha in the visible part of the spectrum, corresponds to chromospheric and prominence temperatures.

extend for 100,000 km. They mark the locations of zero magnetic fields that separate regions of positive and negative magnetic polarities. When filaments happen to be on the limb of the Sun, they can be seen to project into space, often in beautiful shapes. Then they are called **prominences** (Figs. 23–14 and 23–15).

Prominences can be seen with the eye at solar eclipses, when their pinkish glow, resulting from their emission in H$\alpha$ and a few other spectral lines, is visible to the eye. They can be observed from the ground even without an eclipse, if an H$\alpha$ filter is used. They have also been observed in the ultraviolet from Skylab and now SOHO and TRACE. Prominences appear to be composed of matter in a condition of temperature and density similar to matter in the quiet chromosphere, somewhat hotter and less dense than the photosphere. Thus they are at temperatures of about 10,000 kelvins, far lower than temperatures of solar flares.

Sometimes prominences can hover above the Sun, supported by magnetic fields, for weeks or months. They are then called **quiescent prominences,** which can extend tens of thousands of kilometers above the limb. Other prominences can seem to undergo rapid changes. But do not confuse these eruptive prominences with flares; the temperatures of eruptive prominences are only a few tens of thousands of degrees, whereas the temperatures within flares are millions of degrees.

In sum, the regions around sunspots show many kinds of solar activity, and so they are called **active regions.** Solar flares are explosive events with temperatures that reach millions of degrees. Solar prominences, which are filaments seen in projection against the sky when they rotate to the limb, are much cooler and change much more slowly. Commonly, photographs in general magazines or books that are labelled "flares" actually show prominences.

## 23.4 SOLAR-TERRESTRIAL RELATIONS

Careful studies of the solar activity cycle are now increasing our understanding of how the Sun affects the Earth. Although for many years scientists were skeptical of the idea that solar activity could have a direct effect on the Earth's weather, scientists currently seem to be accepting more and more the possibility of such a relationship. It was wondered in the early years of the twentieth century how the Sun might be related to the magnetic disturbances measured on Earth. Only with space observations did we discover that the link is through solar coronal holes (Fig. 23–16) and coronal mass ejections (Section 22.3b).

An extreme test of the interaction may be provided by the interesting probability that there were no sunspots at all on the Sun from 1645 to 1715! The sunspot

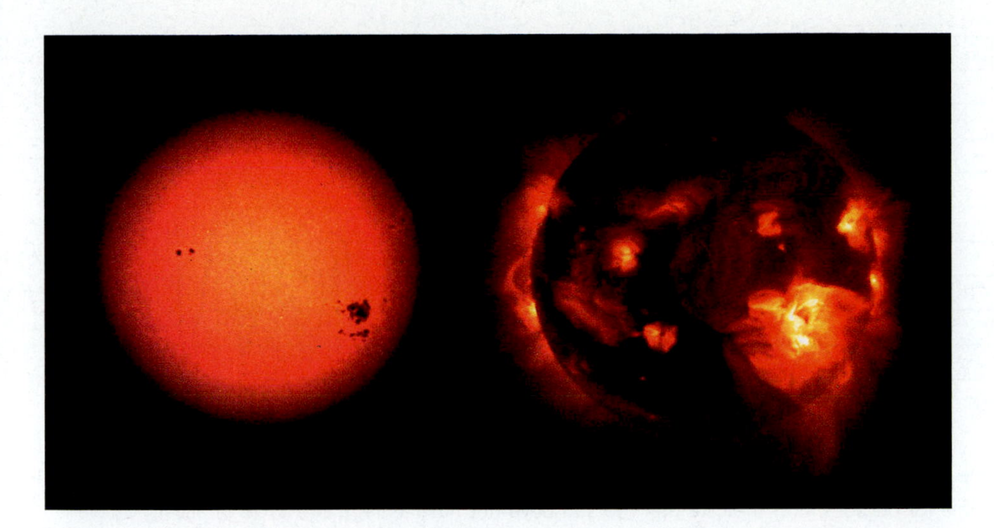

**FIGURE 23–16** A coronal hole rotates across the Sun, as shown in a Yohkoh x-ray image. This coronal hole is visible just left of center, connected with a coronal hole at the lower and left limbs. Gas escaping from coronal holes, the solar wind, interacts with the Earth. The white-light image on the left corresponds to the x-ray image at right. You can see, for example, how the brightest x-ray emission corresponds to the active region marked in the white-light image by a sunspot group. Figure 1–19 showed the evolution of the x-ray corona in a series of Yohkoh images taken over the last solar-activity cycle.

cycle may not have been operating for that 70-year period (Fig. 23–17). This period, called the **Maunder minimum,** was known to the British astronomer E. Walter Maunder and others in the early years of the twentieth century but was largely forgotten until its importance was noted and stressed by John A. Eddy, then of the High Altitude Observatory. (Maunder and his astronomer wife, Elizabeth Maunder, had also discovered the butterfly diagram.) Although no counts of sunspots exist for most of that period, there is evidence that people were looking for sunspots; it seems reasonable that there were no counts of sunspots because they were not there and not merely because nobody was observing. Indeed, the discovery of a sunspot would lead to a scientific article! A variety of indirect evidence has also been brought to bear on the question. For example, the solar corona appeared very weak when observed at eclipses during that period.

It may be significant that the anomalous sunspotless period coincided with a "Little Ice Age" in Europe and with a drought in the southwestern United States. An important conclusion from the existence of this sunspotless period is that the solar-activity cycle may be much less regular than we had thought. Further, the sunspots may not be connected to processes deep inside the Sun, which we think don't change on such time scales.

The evidence for the Maunder minimum is indirect and has been challenged. For example, the Little Ice Age could have resulted from volcanic dust in the air or from changing patterns of land use rather than from a lack of sunspots. It should come as no particular surprise that several mechanisms could affect the Earth's climate on this time scale.

## 23.5 THE SOLAR WIND*

At about the time of the launch of the first Earth-orbiting satellites, in 1957, it was realized that the corona must be expanding into space. This phenomenon is the basis of the **solar wind.** The expansion causes comet tails always to point away from the Sun, an observation that had previously led to suspicions that the solar wind existed.

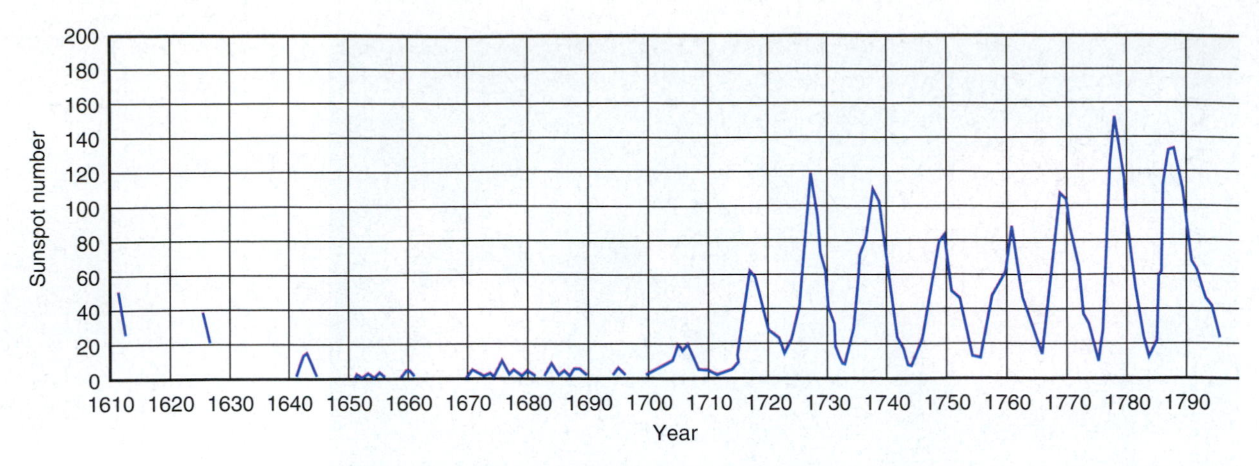


**FIGURE 23–17** The Maunder minimum (1645–1715), when sunspot activity was negligible for decades, may indicate that the Sun does not have as regular a cycle of activity as we had thought. Activity has been extrapolated into the oldest dates (pre-1650) with measurements made by indirect means (such as the frequency of auroras).

If the temperature of the solar corona is so high, why doesn't the Earth burn up in the solar wind? Remember that high temperature means merely that particles are randomly moving very fast. There are very few coronal particles in a given volume, though, so the total energy there is low. Similarly, you would burn yourself much worse in a scalding bath than from a match's flame, even though the flame is at a higher temperature.

We now realize that there are two types of solar wind. The "fast solar wind" has a velocity of about 800 km/sec. This part of the solar wind apparently emanates from coronal holes. The gas takes about 10 days to reach the Earth. The "slow solar wind" goes about half as fast and comes from some source near the Sun's equator. It may come simply from an expansion of the solar corona because of its high temperature. Far enough from the Sun, as the solar gravity weakens, the pressure of the gas simply makes the solar wind accelerate to supersonic speeds.

The density of particles in the solar wind, always low, decreases with distance from the Sun. There are only about 5 particles in each cubic centimeter at the distance of the Earth's orbit. The particles are a mixture of ions and electrons. The solar wind extends into space far beyond the orbit of the Earth, possibly even beyond the orbit of Pluto. The Pioneer and Voyager spacecraft are now exploring the region beyond Neptune, and they have not yet come to the end of the solar wind.

The Earth's outer atmosphere is bathed in the solar wind. Thus research on the nature of the solar wind and on the structure of the corona is necessary to understand our environment in the Solar System. Ulysses, a European spacecraft, flew over the solar south pole in 1994 and the solar north pole in 1995, to give information about the solar wind above the plane in which the planets all lie (Fig. 23–18). It proved that the fast solar wind comes from coronal holes. Ulysses was sent to the Sun with a gravity assist from Jupiter. It later returned to Jupiter so that it could travel over and around the Sun again during the sunspot maximum that has recently ended.

Several satellites to study the solar wind are part of NASA's Sun-Earth Connection Program. "Wind" was launched in 1994; "Polar" was launched in 1996. The European Space Agency's set of satellites known as "Cluster," meant to orbit in formation to study details of the propagation of the solar wind near the Earth, were on an experimental rocket to save money on the launch. (The launch was free because of the experimental nature of the rocket.) The gamble was lost, as these satellites were destroyed along with the rocket when it veered off course. Replacements—named Samba, Salsa, Rumba, and Tango—were launched in 2000. These identical spacecraft, each bearing 11 instruments, fly in a tight pyramid formation to allow scientists to compile three-dimensional maps of the solar wind that passes the spacecraft. Their eccentric orbits, about 20,000 km by 120,000 km, take them 1/3 of the way to the Moon, which brings them down the Earth's magnetotail and even beyond the magnetosphere. They have monitored particles from the Sun compressing the magnetosphere to half its usual size. The improved understanding of the magnetosphere

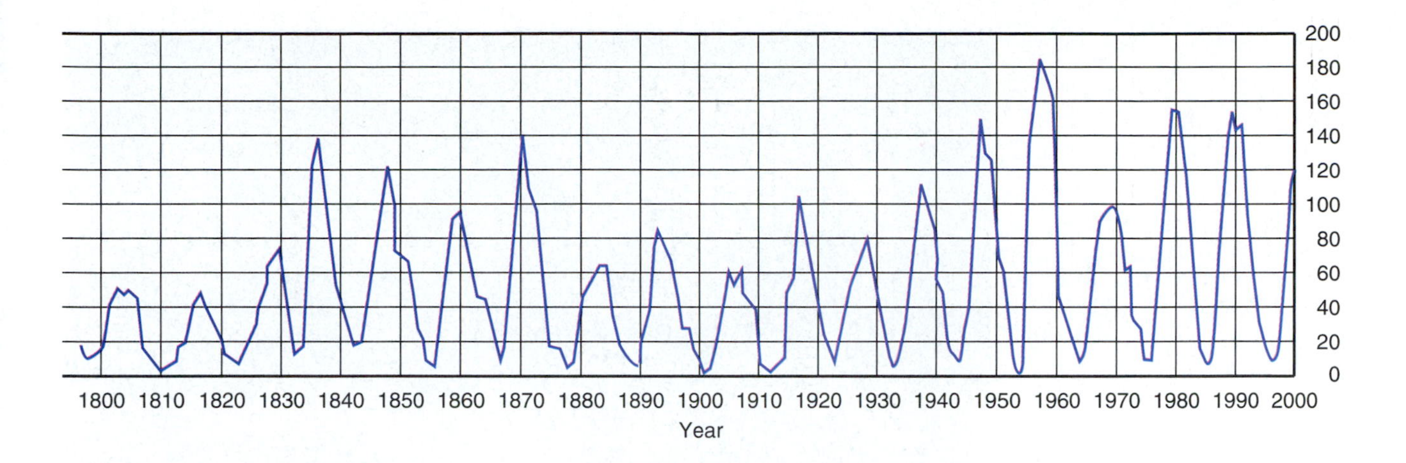


Cluster is bringing could help satellite operators develop early warning systems for when they should protect their spacecraft.

Since its launch in 2000, NASA's "Image" spacecraft (*I*mager for *M*agnetopause-to-*A*urora *G*lobal *E*xploration) has been sending back images that reveal the effects of the solar wind by observing the Earth's changing magnetosphere (Fig. 23–19).

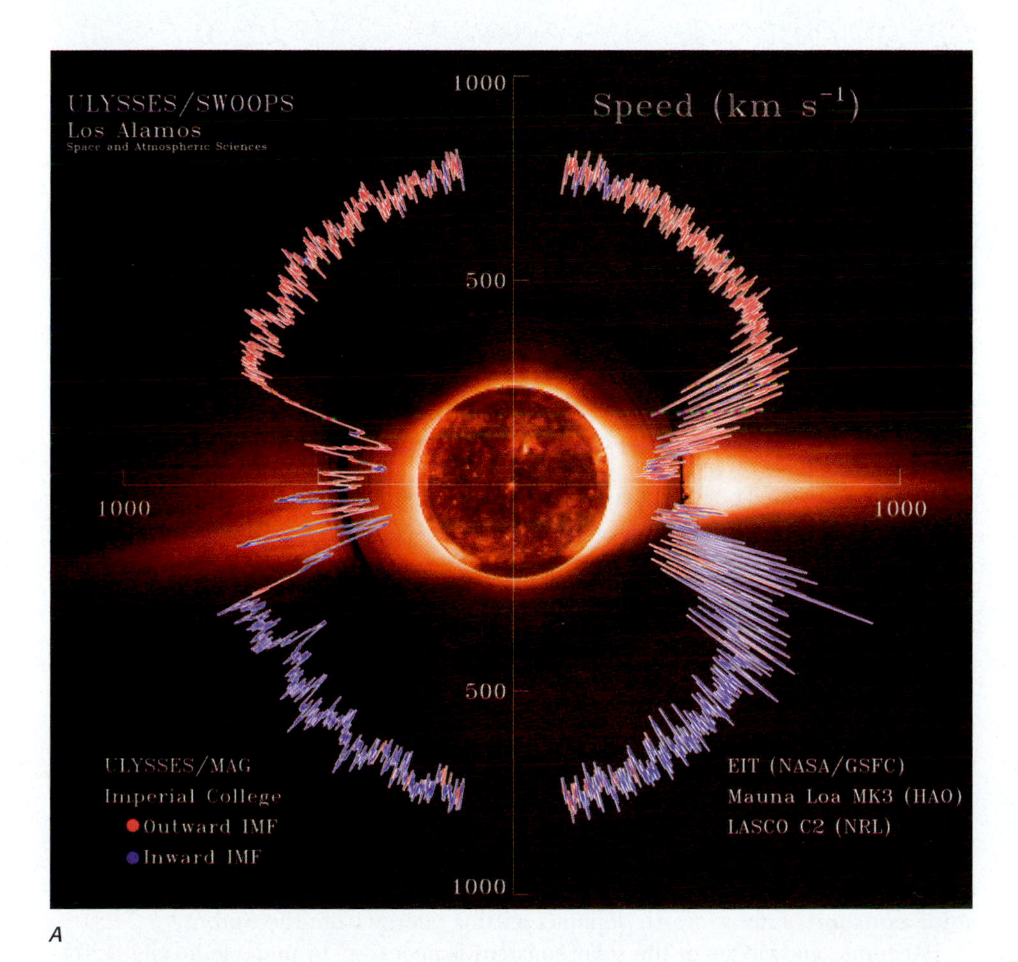


*A*

*B*

**FIGURE 23–18** (*A*) The speed of the solar wind at different angles from the Sun, superimposed on an image of the Sun compounded from observations of two instruments on SOHO and from the Mark IV coronagraph on Mauna Kea in Hawaii. Note the lower speeds of the solar wind above the Sun's equator compared with the higher speeds above higher solar latitudes. (*B*) The Ulysses spacecraft, which has passed over the Sun's poles and now is going around again.

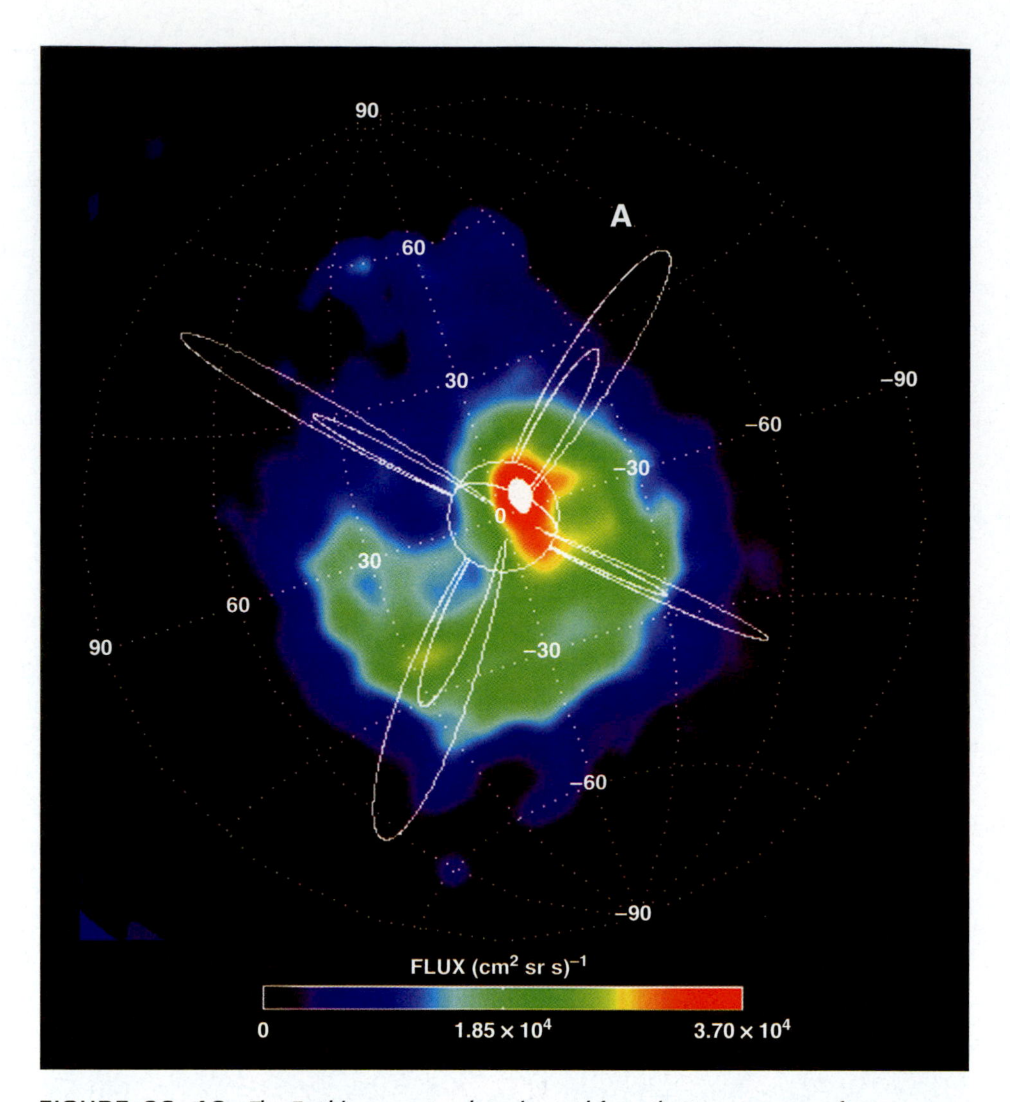


**FIGURE 23-19** The Earth's magnetosphere imaged from the Image spacecraft.

The Solar and Heliospheric Observatory (SOHO) has three instruments that monitor the solar wind as it flows past. One of the instruments, for example, picked up a high-energy proton only 43 minutes after the maximum of a solar flare. SOHO's instruments can often find the origin of changes in the speed or density of the solar wind.

NASA's Living With a Star Program, enlarged for 2001, stresses the effect of solar activity on life here on Earth.

## 23.6 THE SOLAR CONSTANT*

The solar constant is the amount of energy per second that would hit each square meter of the Earth at its average distance from the Sun if the Earth had no atmosphere. 99% of solar energy is in the range from 2760 Å to 49,000 Å (nearly 5 microns), and 99.9% is between 217 Å and 10.94 microns.

Every second a certain amount of solar energy passes through each square meter of space at the average distance of the Earth from the Sun. This quantity is called the **solar constant.** Life on Earth depends on this energy from the Sun.

Accurate knowledge of the solar constant is necessary to understand the terrestrial atmosphere, for to interpret our atmosphere completely we must know all the ways in which it can gain and lose energy. Further, knowledge of the solar constant enables us to calculate the amount of energy that the Sun itself is giving off, and thus it gives us an accurate measurement on which to base our quantitative understanding of the radiation of all the stars.

The solar constant (which is measured above the Earth's atmosphere) is about 1368 watts/m$^2$ (about 2 calories/cm$^2$/min; watts are the SI unit of power, and calories are a non-SI unit of energy in which 1 calorie will raise 1 gram of water by 1°C). Measurements from the ground gave an accuracy of 1.5 per cent. An instrument on the Solar Maximum Mission gave much higher accuracy—0.1 per cent—between 1980 and 1989. Short-term variations of about 0.2 per cent appeared. These dips in the Sun's overall intensity turn out to be correlated with large sunspots crossing the Sun. Thus sunspots actually lower the amount of energy we on Earth receive from the Sun, at least temporarily. A successor instrument is aloft and others are planned. In fact, since the solar panels always have to point toward the Sun even for weather satellites whose cameras point down at Earth, NASA is putting solar x-ray telescopes as well as monitors to measure the solar constant on the solar-panel structures of GOES satellites.

One interesting aim is to determine if there are long-term changes in the solar constant, perhaps arising from changes in the luminosity of the Sun. The Solar Maximum Mission instrument showed a slight but steady decline between 1980 and 1985, but the solar constant then increased as the Sun went toward the maximum of the sunspot cycle. We continue to monitor the solar constant, especially with the VIRGO instrument on the SOHO spacecraft (Fig. 23–20). Certainly, our climate would be profoundly affected by even small long-term changes. Of course, since the solar constant changes with time, it isn't really a constant at all.

Makers of solar-energy systems for heating and generating electricity start their calculations with the solar constant. But much of the Sun's energy is lost in passing through the Earth's atmosphere and in conversion to a usable form, not to mention the effect of cloudy days. For all these reasons, less solar energy is available to us than the solar constant itself.

Note that energy from the Sun comes at a very steady rate and should last for billions of years. By contrast, the energy we are now burning as fossil fuels represents solar energy received in the past and stored up. Once used, it cannot quickly be replaced.

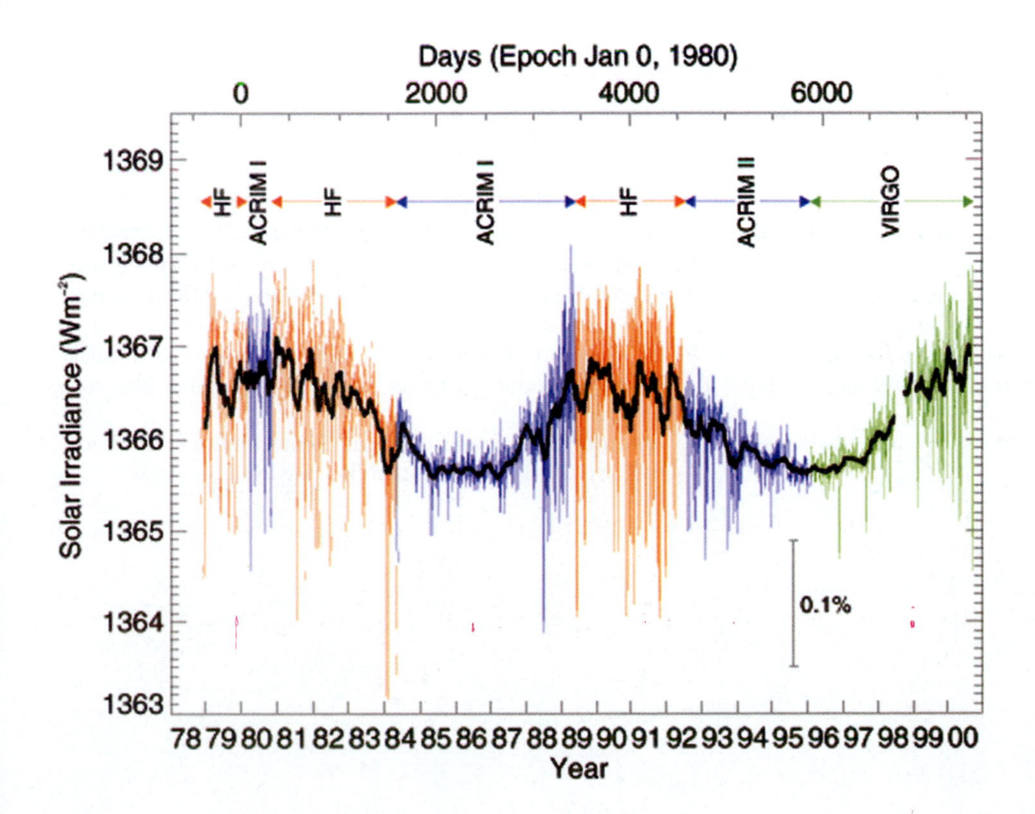


**FIGURE 23–20** Measurements of the solar constant from space have been made from the Nimbus-7 satellite, from the Solar Maximum Mission and the successor instrument on the Upper Atmospheric Research Satellite, and (with less accuracy) from the Earth Radiation Budget Satellite. All radiation—x-rays to radio waves—that passes through a small hole was measured. The day-to-day variations correlate nicely with the passage of sunspots across the solar disk; the fluctuations are greater near solar maximum. A long-term variation with the solar-activity cycle also shows. Current observations are made with VIRGO aboard SOHO.

## CORRECTING MISCONCEPTIONS

| ✖ *Incorrect* | ✔ *Correct* |
| --- | --- |
| There are always sunspots on the Sun. | At solar minimum, there may be weeks without sunspots. During the Maunder minimum, there were decades without sunspots. |
| Red spots on the edge of the Sun are flares. | Often these objects are prominences, which are much cooler than flares. |

## SUMMARY AND OUTLINE

Solar-activity cycle (Sections 23.1–3)
  Sunspots, flares, plages, filaments, and prominences all linked to magnetic field structure
Solar-terrestrial relations (Section 23.4)
  Possible effect of solar activity on terrestrial weather
  May not have been any solar activity during the Maunder minimum
Solar wind flows from coronal holes (Section 23.5).
Solar constant and its variability (Section 23.6)

## KEY WORDS

solar-activity cycle
sunspots
umbra
penumbra
magnetic lines of force
magnetic field lines
Zeeman effect
sunspot cycle
differential rotation
dynamo theories
solar dynamo
butterfly diagram
solar flares
aurora borealis
aurora australis
space weather
plages
filaments
prominences
quiescent prominences
active regions
Maunder minimum
solar wind*
solar constant*

*This term appears in an optional section.

## QUESTIONS

1. Why does a sunspot appear dark?
2. Define and contrast a prominence and a filament.
3. List three phenomena that vary with the solar activity cycle.
4. Describe the sunspot cycle and a mechanism that explains it.
5. What is a key difference between a flare and an erupting prominence?
6. What are the differences between the solar wind and the normal "wind" on Earth?
7. From what structures on the Sun does the solar wind come?
8. (a) What is the solar constant? (b) Why is it difficult to measure precisely?
9. Of what part of the spectrum is it most important to make accurate measurements in order to determine the solar constant?
10. Referring to Appendix 3 for data about Mars and the Earth, what would the solar constant be if we lived on Mars?

## USING TECHNOLOGY

### W World Wide Web

1. Sign onto the Solar Indices Data Center to find the latest sunspot numbers at http://www.oma.be/KSB-ORB/SIDC/index.html.
2. Sign onto the Marshall Space Flight Center site not only to find the latest butterfly diagram but also to see images and learn about all aspects of the Sun: http://science.msfc.nasa.gov/ssl/pad/solar/sunspots.htm.
3. View today's solar images, using the links on this book's Web site, and assess the solar activity.
4. Monitor the Sun and its activity at http://www.spaceweather.com

### REDSHIFT

1. Consider Astronomy Lab: Sunspot Cycle I and Sunspot Cycle II.
2. View the movie "Eruption of a Solar Prominence."

The Rosette Nebula in the constellation Monoceros, a nebula with an open cluster of stars (NGC 2244) at its center.

# PART 4
# The Stars

When we look up at the sky at night, most of the objects we see are stars. In the daytime, we see the sun, which is itself a star. The moon, the planets, even a comet may give a beautiful show, but, however spectacular, they are only minor actors on the stage of observational astronomy.

From the center of a city, we may not be able to see very many stars, because city light scattered by the earth's atmosphere makes the light level of the sky brighter than most stars. All together, about 6000 stars are bright enough to be seen with the unaided eye under good observing conditions.

The stars we see defining a constellation may be at very different distances from us. One star may appear bright because it is relatively close to us even though it is intrinsically faint; another star may appear bright even though it is far away because it is intrinsically very luminous. When we look at the constellations, we are observing only the directions of the stars and not whether the stars are physically close together. Nor do the constellations tell us anything about the nature of the stars. Astronomers tend not to be interested in the constellations because the constellations do not give useful information about how the universe works. After all, the constellations merely tell the directions in which stars lie. Most astronomers want to know *why* and *how*. Why is there a star? How does it shine? Such studies of the workings of the universe are called **astrophysics.** Almost all modern-day astronomers are astrophysicists as well, for they not only make observations but also think about their meaning.

Some properties of the stars that can be seen with the naked eye tell us about the natures of the stars themselves. For example, some stars seem blue-white in the sky, while others appear slightly reddish. From information of this nature, astronomers are able to determine the temperatures of the stars. Chapter 24 is devoted to the basic properties of stars and some of the ways we find out such information. Chapter 25 describes some of the ingenious ways we use basic information about stars. We can sort stars into categories and tell how far away they are and how they move in space.

Most stars occur in pairs or in larger groups, and we shall discuss types of groupings and what we learn from their study in Chapter 26. Binary stars are important, for example, in finding the masses of stars. Some stars vary in brightness, a property that astronomers use to tell the scale of the universe itself. The study of star clusters leads to our understanding of the ages of stars and how they evolve.

In Chapters 22 and 23, we studied an average star in detail. It is the only star we can see close up—the sun. The phenomena we observe on the sun take place on other stars as well, as has recently been verified by telescopes in space.

In recent decades, the heavens have been studied in other ways besides observing the ordinary light that is given off by many astronomical objects. Other forms of electromagnetic radiation—radio waves, x-rays, gamma rays, ultraviolet, and infrared—and interstellar particles called cosmic rays are increasingly studied from the earth's surface or from space. Many contemporary astronomers—even many who consider themselves observers (who mainly carry out observations) rather than theoreticians (who do not make observations, but rather construct theories)—have never looked through an optical telescope. Still, astronomy began with optical studies, and our story begins there. This part of the book will tell us how we study stars and what they teach us. Part 5 will take up the life stories of stars.

Channels of gas and dust, which will one day evaporate to reveal the stars now forming inside them. This false-color Hubble Space Telescope image of the Eagle Nebula, M16, reveals these channels. The image was taken through filters passing light of singly ionized sulfur (reproduced as red), hydrogen (reproduced as green), and doubly ionized oxygen (reproduced as blue).

# Chapter 24

# Stars and Their Spectra

**AIMS:** To study the colors and spectra of ordinary stars, and to see how they tell us about the temperatures of the stars' surfaces

What are these stars that we see as twinkling dots of light in the nighttime sky? The stars are luminous balls of gas scattered throughout space. All the stars we see are among the approximately 100 billion stellar members of a collection of stars and other matter called the Milky Way Galaxy. The sun, which is the star that gives most of the energy to our planet, is so close to us that we can see detail on its surface, but it is just an ordinary star like the rest.

Stars are balls of gas held together by the force of gravity, the same force that keeps us on the ground no matter where we are on the earth. Long ago, gravity compressed large amounts of gas and dust (Fig. 24–1) into dense spheres that became stars. The dust vaporized, and balls of gas remained.

Stars generate their own energy and light. It has been only about 70 years since people realized that stars shine by nuclear fusion. The discovery that this process operates in the stars explained how the stars could live so long—10 billion years or so for the sun—given that the methods previously considered gave lifetimes of only hundreds of thousands of years. Now we are using the nearest

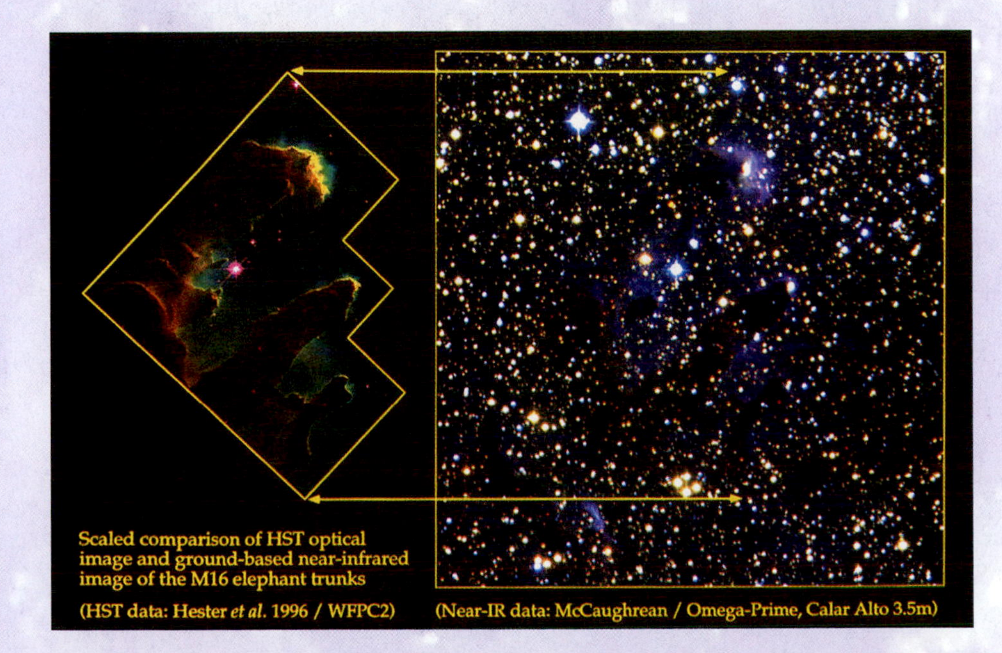


FIGURE 24–1 The orientation of the Hubble Space Telescope visible-light image of the Eagle Nebula (*opposite*) on a near-infrared image taken with a ground-based telescope.

From here on in the book, sun, solar system, and universe are written with lowercase initial letters.

star, our sun, as a direct source of energy in the form of its light and as a laboratory to study the physics of hot gases in magnetic fields. The latter study may tell us how to tame the fusion process and re-create it on earth.

We see only the outer layers of the stars; the interiors are hidden from our view. The stars are gaseous through and through; they have no solid parts. The outer layers do not generate energy by themselves but glow from energy transported outward from the stellar interiors. So when we study light from the stars, we are not observing the processes of energy generation directly. We will study these processes in Chapter 27, where we start to study the life cycles of stars. In this chapter, we shall concentrate on how we study the stars themselves.

## 24.1 THE COLORS OF THE STARS

In Section 4.1, we saw that the different colors we perceive result from radiation in certain ranges of wavelength. For example, an object appears blue because most of its visible radiation is at wavelengths between roughly 4300 and 5000 Å (angstroms). We will now see that how a star's radiation is distributed in wavelength depends on the temperature of the surface of that star. By measuring the color of a star, we can determine the wavelength distribution and thus deduce the star's temperature.

To determine a star's color, we can pass the light through a spectrograph and measure the intensity at each wavelength. More simply, we can measure the light that emerges from filters, each of which allows only a certain group of wavelengths to pass. (Note that even if a star appears to be one color to the eye, it is still giving off radiation at every wavelength.)

When we examine a set of these graphs, we can see that, spectral lines aside, the radiation follows a fairly smooth curve (Fig. 24–2). The radiation "peaks" in intensity at a certain wavelength (that is, has a maximum in its intensity at that wavelength) and decreases in amount more slowly on the long wavelength side of the peak than it does on the short wavelength side. Note very carefully that each curve represents a graph of intensity versus wavelength.

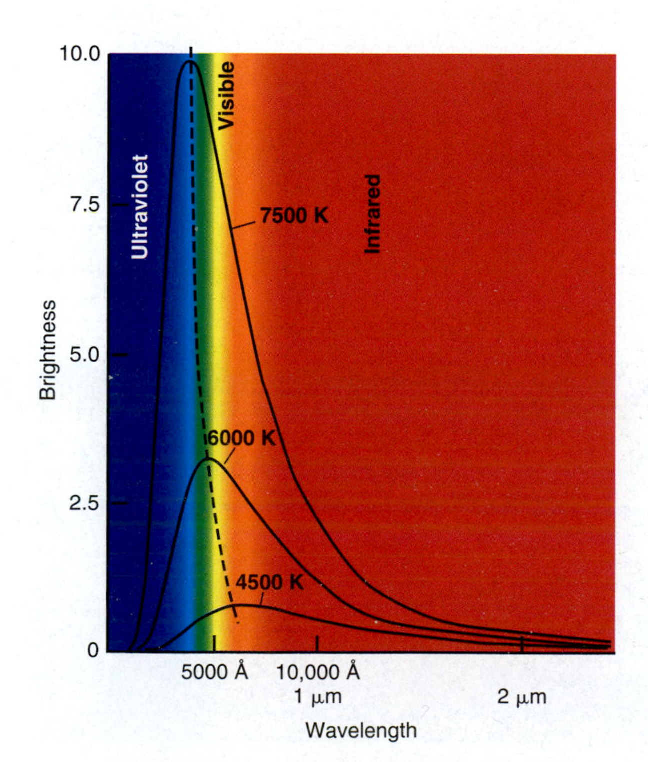


**FIGURE 24–2** The intensity of radiation for different stellar temperatures, according to Planck's law (Section 24.2). The wavelength scale is linear; that is, equal spaces signify equal wavelength intervals. The dotted line shows how the peak of the curve shifts to shorter wavelengths as temperature increases; this is known as Wien's displacement law. The total energy emitted at a given temperature—the area under the curve—is given by the Stefan-Boltzmann law.

A DEEPER DISCUSSION

## BOX 24.1 Wien's Displacement Law

$$\lambda_{max}T = \text{constant},$$

where $\lambda_{max}$ is the wavelength (the lowercase Greek letter lambda) at which the energy given off is at a maximum, and $T$ is the temperature. The numerical value of the constant is given in Appendix 2. Note: Wien's displacement law has nothing to say about the height (intensity) of the peak.

***Discussion Question***

What is $\lambda_{max}$ for the sun? Express it in angstroms and in meters.

The curves in the graph can be best understood by first considering the radiation that represents different temperatures. When you first put an iron poker in the fire, it glows faintly red. Then as it gets hotter it becomes redder. If the poker could be even hotter without melting, it would turn yellow, white, and then blue-white. White-hot is hotter than red-hot.

A set of physical laws governs the sequence of events when material is heated. As the material gets hotter, the peak of radiation shifts toward shorter wavelengths (the blue). This shift of wavelength of the peak is known as **Wien's displacement law** (Box 24.1 and Table 24–1).

Note that, for lines 2–5 of Table 24–1, we keep doubling the temperature and so wind up halving the peak wavelength.

Also, as the material gets hotter, the total energy of the radiation grows quickly. Notice in the figure how the total energy, which is the area under each curve, is much greater for higher temperatures. The energy follows the **Stefan-Boltzmann law** (Box 24.2). This law says that the total energy emitted from each square centimeter of a source in each second grows as the fourth power of the temperature ($T^4 = T \times T \times T \times T$). If we compare the radiation from two same-sized regions of a surface at different temperatures, the Stefan-Boltzmann law tells us that the ratio of the energies emitted is the fourth power of the ratio of the temperatures. That is, take the ratio of the temperatures and raise that ratio to the fourth power to find the ratio of the energies emitted. For example, consider the same gas at 10,000 K and at 5000 K. Because 10,000 K is twice 5000 K, the 10,000 K gas gives off $2^4 = 2 \times 2 \times 2 \times 2 = 16$ times more energy than does the same amount of gas at 5000 K. (We must measure from absolute zero, the minimum temperature possible. The Kelvin temperature scale begins there.) Note how even a small increase in temperature makes a star much brighter.

The Stefan-Boltzmann law relates the amount of energy a gas gives off to its temperature if we consider emission from a constant surface area of gas. It is further true that a larger volume of gas at the same temperature gives off more energy. So the actual amount of energy a star emits depends both on how hot its surface is and on how large its surface is.

**TABLE 24–1** Wien's Displacement Law

| Temperature | Wavelength of Peak | Spectral Region of Wavelength of Peak |
|---|---|---|
| 3 K | 9,660,000 Å = 0.97 mm | infrared–radio |
| 3,000 K | 9660 Å | infrared |
| 6,000 K | 4830 Å | blue |
| 12,000 K | 2415 Å | ultraviolet |
| 24,000 K | 1207 Å | ultraviolet |

## A DEEPER DISCUSSION

### BOX 24.2 Stefan-Boltzmann Law

$$Energy = \sigma T^4.$$

Energy ($E$) is the "strength" of the radiation per unit area. The numerical value of $\sigma$, the Stefan-Boltzmann constant, is given in Appendix 2.

***Discussion Question***

If you raise the temperature of your oven for making the Thanksgiving turkey from 325 to 375 degrees, by how much are you raising the energy per second? Pay close attention to the units of the temperature scale.

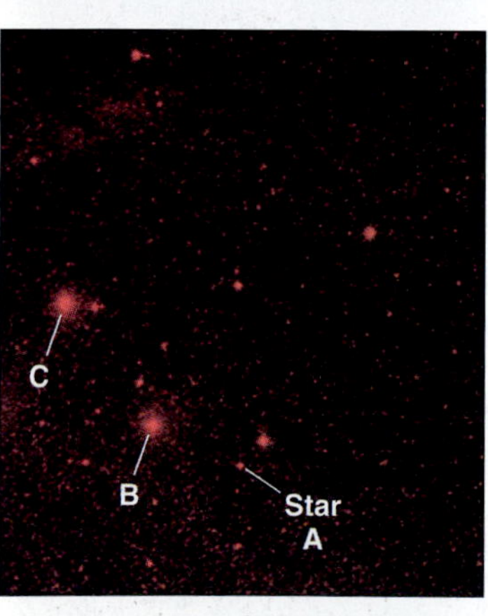


**FIGURE 24–3** The top photograph is a view of several stars in blue light, and the bottom photograph is a view of the same stars in red light. Star A appears about the same brightness in both. Star B, a relatively hot star, appears brighter in the blue photograph. Star C, a relatively cool star, appears brighter in the red photograph.

Note: Boltzmann has 2 n's and Wien is spelled with "ie," not "ei." Note also the "c" in "Planck."

The overall color that a star appears in the sky depends on the color of the peak of a star's energy distribution. Thus from Wien's displacement law, we see that the colors of stars in the sky tell us the stars' temperatures. The reddish star Betelgeuse, for example, is a comparatively cool star, while the blue-white star Sirius is very hot. By simply measuring the intensity of a star with a set of filters at different wavelengths (colors), we can find out the temperature of the star (Fig. 24–3). In Section 25.3b, we discuss the "color index" we measure from comparing the brightness as seen in different filters.

## 24.2 PLANCK'S LAW AND BLACK BODIES

Wien's displacement law tells you the wavelength at which the intensity reaches its peak, and the Stefan-Boltzmann law tells you the area under a graph of intensity versus wavelength, but neither tells you the shape of the curve on the graph. The law that gives the shape is much more general than its application to the stars. In principle, if any gas is heated, the graph showing the intensity at each wavelength for its continuous emission follows a certain peaked distribution called **Planck's law,** suggested in 1900 by the German physicist Max Planck (Fig. 24–4). Wien's dis-

**FIGURE 24–4** Max Planck (*left*) presented an equation in 1900 that explained the distribution of radiation with wavelength. Five years later, Albert Einstein (*right*), in suggesting that light acts as though it is made up of particles, found an important connection between his new theory and the earlier work of Planck. The photograph was taken years later.

IN REVIEW

## BOX 24.3 Do Not Confuse: Curves Showing Black-Body Radiation

Don't let the fact that Figure 24–2 simultaneously shows similarly shaped graphs for different temperatures confuse you; each of the individual curves represents gas at a different temperature and shows the intensity of radiation on the vertical axis and the wavelength on the horizontal axis. Remember: for each temperature—say, 6000 K—the curve shows the intensity of light given off for each wavelength.

Many students are confused by the set of curves for temperature, wavelength, and intensity. All too often, students mistakenly put "temperature" on one of the axes when asked on an exam to draw these radiation curves, which are so fundamental to astronomy.

A similar example is a set of data for gender, height, and weight. The graph showing height and weight is different for males and females. But you would never try to put "female" on one of the axes. Similarly, the diagram shows a set of curves for temperature, wavelength, and intensity. But it is only wavelength and intensity that you graph against each other.

placement law and the Stefan-Boltzmann law, which had been discovered earlier from study of experimental data, can be derived from Planck's law. Actual gases may not, and usually don't, follow Planck's law exactly. For example, Planck's law governs only the continuous spectrum; any emission or absorption in the form of spectral lines does not follow Planck's law.

Because of these difficulties, scientists consider an idealized case: a fictional object called a **black body,** which is defined as something that absorbs all radiation that falls on it. Note that the "black" in a black body is an idealized black that absorbs 100 per cent of the light that hits it. Ordinary black paint, on the other hand, does not absorb 100 per cent of the light that hits it. Further, something that looks black in visible light might not seem very black at all in the infrared; if so, it would not be a real black body.

An important physical law that governs the radiation from a black body says that anything that is a good absorber is a good emitter too. The radiation from a black body follows Planck's law. As a black body is heated, the peak of the radiation shifts toward the low-wavelength end of the spectrum (Wien's displacement law; note: only the wavelength of the peak is described by Wien's law, which says nothing about the height of the peak), and the total amount of radiation grows rapidly (Stefan-Boltzmann law). Note that the curves of intensities emitted by same-sized black bodies at different temperatures in Figure 24-2 do not cross. In particular, you should realize that if you heat a gas that gives off continuous radiation, it emits more continuous radiation at every wavelength. If you trace vertical lines upward on the figure, no matter which wavelength you are investigating, you cross the curve for the cooler gas before you reach upward high enough to reach the curve for the hotter gas.

To a certain extent, the atmospheres of stars appear as black bodies: over a broad region of the spectrum that includes the visible, the continuous radiation from stars follows Planck's law fairly well.

## 24.3 THE FORMATION OF SPECTRAL LINES

The spectrum of the sun, as we saw in Section 4.7, is a continuous rainbow of color crossed by a set of dark lines. All stars have similar spectra—a continuous rainbow crossed by dark lines, which are known as "absorption lines" or "Fraunhofer lines." Most of what we know about stars comes from studying these spectral lines. In this section, we shall discuss spectral lines themselves; we will go on to their appearance in stars in a following section.

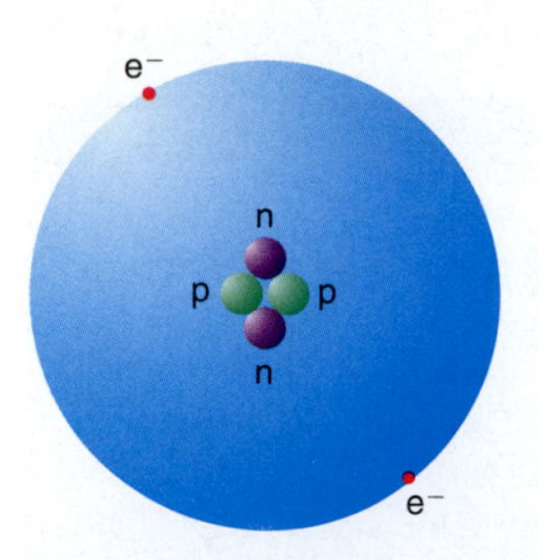


**FIGURE 24–5** A helium atom contains two protons and two neutrons in its nucleus and two orbiting electrons.

Spectral lines arise when there is a change in the amount of energy present in any given atom. We can think of an atom (Fig. 24–5) as consisting of particles in its core, called the **nucleus,** surrounded by orbiting particles, called **electrons.** Examples of nuclear particles are the **proton,** which carries one unit of electric charge, and the **neutron,** which has no electric charge. (Protons and neutrons are, in turn, made up of particles called **quarks,** as we discuss further in Chapter 37. The quark structure, though fundamental for understanding the makeup of our world and our universe, does not affect spectra.) An electron, which has one unit of negative electric charge, is only 1/1800 as massive as a proton or a neutron. Normally, an atom has the same number of electrons as it has protons. The negative and positive charges balance, leaving the atom electrically neutral.

The amount of energy that an atom can have is governed by the laws of **quantum mechanics,** a field of physics that was developed in the 1920s and whose applications to spectra were further worked out in the 1930s. According to the laws of quantum mechanics, light (and other electromagnetic radiation) has the properties of waves in some circumstances and the properties of particles under other circumstances. Don't worry about understanding this dual set of properties of light intuitively; some of the greatest physicists of the time had difficulty adjusting to the idea. Still, quantum mechanics has been worked out to explain experimental observations of spectra, and it is now thoroughly verified and accepted. Its development was one of the major scientific achievements of the twentieth century. For the purpose of this book, you need know only the following simple aspect of quantum mechanics:

According to the laws of quantum mechanics, an atom can exist only in a specific set of energy states, as opposed to a whole continuum (continuous range) of energy states. An atom can have only discrete values of energy; that is, the energy cannot vary continuously. For example, the energy corresponding to one of the states might be 10.2 electron volts (eV). The next energy state allowable by the quantum mechanical rules might be 12.0 eV. We say that these are **allowed states.** The atom *simply cannot have* an energy between 10.2 and 12.0 eV. *Never does this atom ever have, say, 10.5 eV of energy.* The energy states are discrete (separate and distinct); we say that they are "quantized," and hence the name "quantum mechanics." The discrete energy states are called **energy levels.** An analogy is that I can stand on one step of a staircase or on another step, but I cannot hover at a height in between.

When an atom drops from a higher energy state to a lower energy state without colliding with another atom, the difference in energy is sent off as a bundle of radiation. This bundle of energy, a **quantum,** is also called a **photon,** which may be thought of as a particle of electromagnetic radiation (Fig. 24–6), an energy bullet. Photons always travel at the speed of light. Each photon has a specific energy.

A link between the particle version of light and the wave version is seen in the equation that relates the energy of the photon with the wavelength it has: $E = hc/\lambda$. In this equation, $E$ is the energy, $h$ is a constant named after Planck, $c$ is the speed of light (as it would be in a vacuum), and $\lambda$ (the Greek lower-case letter lambda) is the wavelength. Since $\lambda$ is in the denominator (the bottom term) on the right-hand side of the equation, a small $\lambda$ corresponds to a photon of great energy. (When we divide by a small number, we get a large result.) When, on the other hand, $\lambda$ is larger, the photon has less energy. For example, an x-ray photon of wavelength 1 Å has much more energy than does a photon of visible light of wavelength 5000 Å. We can see that from the equation, since $hc/(1\text{ Å})$ is much greater than $hc/(5000\text{ Å})$.

### 24.3a Emission and Absorption Lines

Experimentally, we find that each element (hydrogen, say) has a set of spectral lines. It is the same set, at the same wavelengths, no matter if we are looking at absorption lines or emission lines. In this subsection, we will discuss why the emission lines and absorption lines of a given element are at the same wavelengths. We will soon

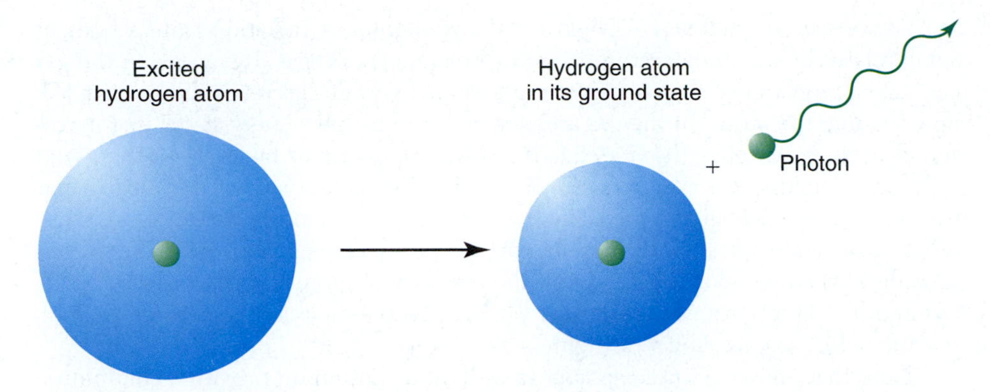


**FIGURE 24-6** Since the hydrogen atom has only a single electron, it is a particularly simple case to study. The lowest possible energy state of an atom is called its ground state. All other energy states are called excited states: the second energy level, immediately above the ground state, is the "first excited state," the next is the "second excited state," etc. When an atom in an excited state gives off a photon, it drops back to a lower energy state, perhaps even to the ground state. We see the photons for a given transition as an emission line.

see, though, that stars have essentially only absorption lines. We will meet emission lines in later chapters, when we discuss clouds of gas between the stars.

In a cold gas, most of the atoms are on the lowest possible energy level. When a gas is heated, many of its atoms are raised from this lowest energy level to higher energy states. From the higher energy levels, they then spontaneously drop back to their lowest energy levels, emitting photons as they do so. Each new photon carries away an amount of energy that corresponds to the difference in energy between the higher and lower level. Each value of energy difference corresponds to a particular wavelength (Fig. 24–7*A* and *B*). The new photons provide a spectrum with some wavelengths brighter than neighboring wavelengths. These brighter wavelengths are the *emission* lines.

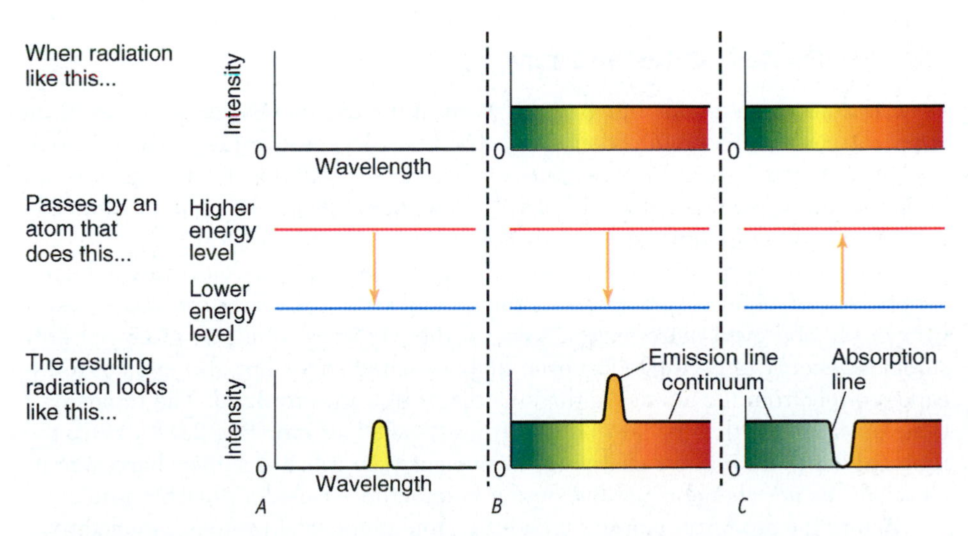


**FIGURE 24-7** When photons are emitted, we see an emission line; a continuum may or may not be present (Cases A and B). An absorption line must absorb radiation from something. Hence an absorption line necessarily appears in a continuum (Case C). The top and bottom horizontal rows are graphs of the spectrum before and after the radiation passes by an atom; a schematic diagram of the atom's energy levels (only two levels are shown to simplify the situation) is shown in the center row.

As we saw in Section 4.7, when we allow continuous radiation from a body at some relatively high temperature to pass through a cooler gas, the atoms in the gas can take energy out of the continuous radiation at certain discrete wavelengths. We now see that the atoms in the gas are changed into higher energy states that correspond to these wavelengths (Fig. 24–7*C*). Because energy is taken up in the gas at these wavelengths, less radiation remains to travel straight ahead at these particular wavelengths to reach an observer. These wavelengths of decreased radiation are those of the *absorption* lines. In the visible part of the spectrum, these wavelengths appear dark when we look back through the gas. (The energy absorbed is soon emitted in other directions, which explains why we can see emission lines when we look from the side, as was shown in Figure 4-24.)

Note that emission lines can appear without a continuum or with a continuum, since emission lines are merely energy added to the radiation field at certain wavelengths. But absorption lines must be absorbed *from* something, namely, the continuum, which is just the name for a part of the spectrum where there is some radiation over a continuous range of wavelengths.

A normal stellar spectrum is a continuum with absorption lines. We interpret this simple fact together with the idea that outer layers of gas in a star absorb energy coming from deeper layers. This combination leads us to the important conclusion that the temperature in the outer layers of the star is decreasing with distance from the center of the star. That is, layers are cooler as you go out. The presence of absorption lines in the spectrum we see tells us that we have been looking through cooler gas at hotter gas. Because the temperature declines as you go outward in a star, stellar spectra rarely show emission lines, which result from the presence of gas hotter than any background continuum. In the rare cases in which emission lines appear, the stars are surrounded by hot shells of gas.

So what do we see when we take the spectrum of the sun or of some other star? We see a band of color crossed by dark lines at which specific colors are absent. You can consider how strong the absorption is at the wavelengths of dark lines, and what those wavelengths are. Then you can smooth over those dark lines, and separately consider the continuous spectrum. We have already seen that the continuous spectrum, by itself, can tell you the star's temperature. Later in this chapter, we will see how, separately, the spectral lines also tell you the star's temperature.

### 24.3b Excited States and Ions

We can think of an atom's energy states as resulting from a change in energy of an atom's electrons. (The production of spectral lines does not involve changes in the nucleus.) If a gas were at a temperature of absolute zero, all the electrons in it would be in the lowest possible energy levels. The lowest energy level that an electron can be in is called the **ground state** of that atom.

As we add energy to the atom, we can "excite" one or more electrons to higher energy levels. The level directly above the ground state is the **first excited state.** If we were to add even more energy, some of the electrons would be given not only sufficient energy to be excited to even higher excited states but also enough to escape entirely from the atom. We then say that the atom is **ionized.** The remnant of the atom with less than its quota of electrons is called an **ion** (Fig. 24–8). Since the remnant has more positive charges in its nucleus than it has negative charges on its electrons, its net charge is positive, and it is sometimes called a **positive ion.**

Before the atom was ionized, it had the same number of protons and electrons, and so it was electrically neutral. Such an atom is called a neutral atom and is denoted with the Roman numeral I. For example, neutral helium is denoted He I. When the atom has lost one electron (is singly ionized), we use the Roman numeral II. Thus when helium has lost one electron, it is called He II (read "helium two"). If an atom is doubly ionized, that is, has lost two electrons, it is in state III, and so on. The Roman numeral is the number of electrons lost +1.

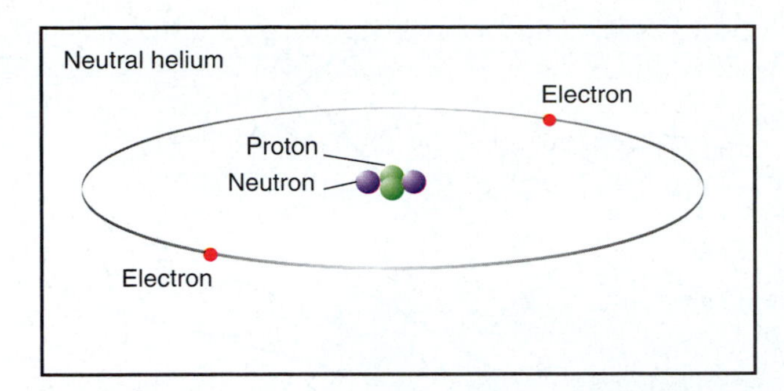


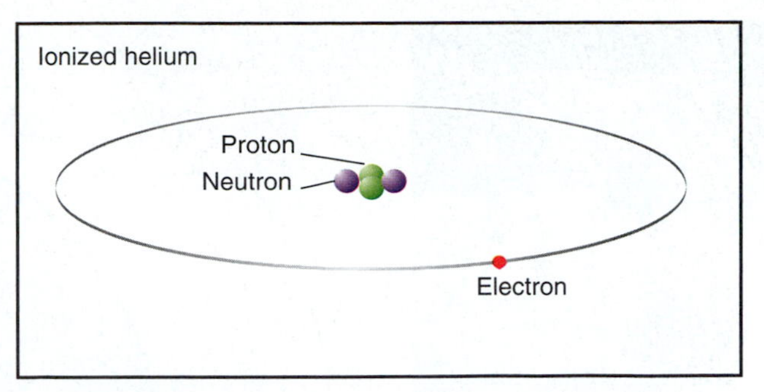


**FIGURE 24–8** An atom missing one or more electrons is ionized. Neutral helium is He I, and singly ionized helium is He II.

Atoms can be excited or ionized either by collisions (an atom's colliding with another atom or part of an atom) or by radiation (an atom's absorbing energy in the form of a photon). Collisions are the most important when the density is sufficiently high, since then atoms collide frequently with electrons. Collisions can also de-excite atoms. Certain spectral lines come only from regions of low density, since only there can atoms excited to the necessary states survive collisions long enough. Such lines, once thought to be from "nebulium" or "coronium," are now known to be from ordinary elements in unusual states that survive at low densities.

## 24.4 THE HYDROGEN SPECTRUM

Hydrogen gas emits, or absorbs, a set of photons that correspond to spectral lines that fall across the visible spectrum in a distinctive pattern (Fig. 24–9). This set of hydrogen lines is called the **Balmer series.** The strongest line is in the red, the second strongest is in the blue, and the other lines continue through the blue and ultraviolet, with the spacing between the lines getting smaller and smaller. Interestingly, the lines were first seen (in 1881) as a series in astronomical photographs rather than in a terrestrial laboratory. Johann Jakob Balmer (1825–1898), a Swiss schoolteacher, noticed the regular sequence when he was teaching a class. He described the converging series of hydrogen lines as a mathematical formula, and the series of lines was named after him. The converging nature of the series is particularly important, because it is very distinctive and allows you at a glance to pick out the presence of hydrogen.

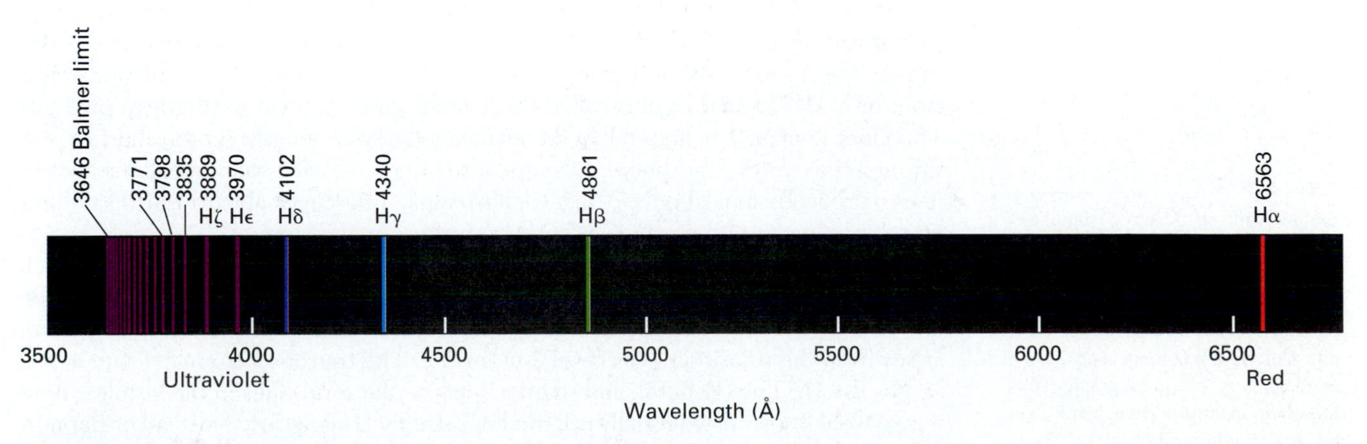


**FIGURE 24–9** The Balmer series, representing transitions down to or up from the second energy state of hydrogen. The strongest line in this series, Hα (H-alpha), is in the red.

A

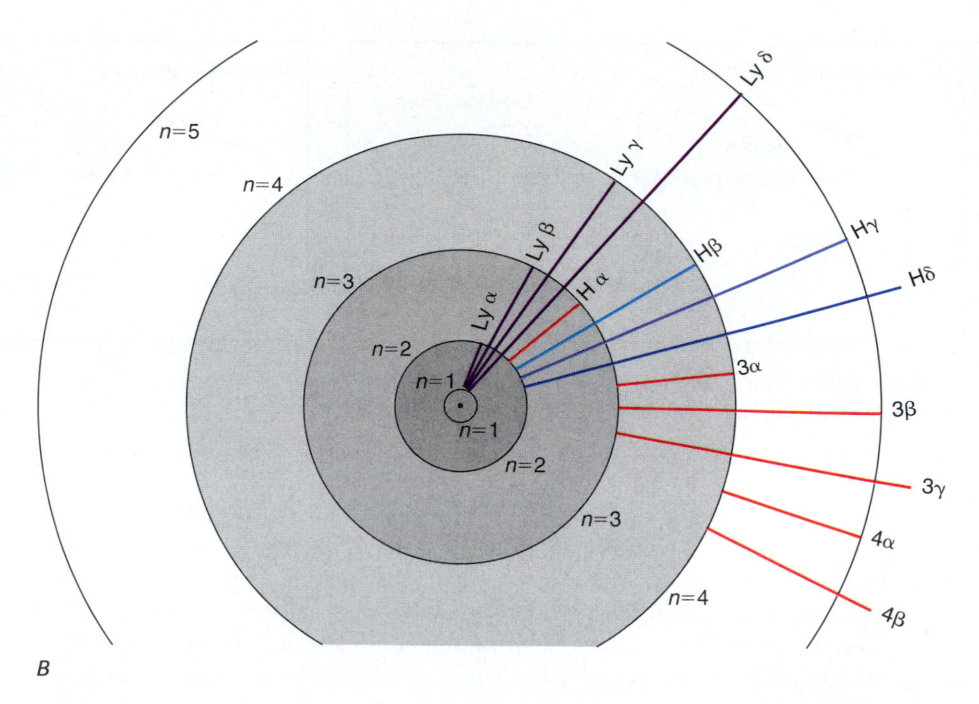


B

**FIGURE 24–10** (*A*) Niels Bohr and his wife, Margrethe. They are major characters in Michael Frayn's play *Copenhagen,* which has been playing very successfully on the stage in both London and New York. (Bohr worked in Copenhagen and in physics that city is very associated with him.) (*B*) The representation of hydrogen energy levels known as the Bohr atom.

It is easy to visualize how the hydrogen spectrum is formed by using the mental picture that Niels Bohr (Fig. 24–10*A*), the Danish physicist, laid out in 1913 to explain the mathematical formula for the visible hydrogen lines that Balmer had figured out decades earlier. In the **Bohr atom** (Fig. 24–10*B*), electrons can have orbits of different sizes that correspond to different energy levels. Only certain orbits are allowable, which is the same as saying that the energy levels are quantized. Bohr didn't know why only certain levels were allowable; he guessed that it was true because he needed that assumption to explain the observed spectrum. (Though the Bohr atom is a model of the hydrogen atoms, other atoms similarly have sets of energy levels, though they don't follow such a simple pattern.) This simple picture of the atom, part of the older "quantum theory," was superseded a dozen years later by the development of quantum mechanics, but the notion that the energy levels are quantized remains.

We use the letter $n$ to label the energy levels. It is called the **principal quantum number.** We call the energy level for $n = 1$ the **ground level** or **ground state,** as it is the lowest possible energy state. The hydrogen atom's series of transitions from or to the ground level is called the Lyman series, after the American physicist Theodore Lyman. The lines fall in the ultraviolet, at wavelengths far too short to pass through the earth's atmosphere. Lyman's observations of what we now call the Lyman lines were made in a laboratory in a vacuum tank. Telescopes aboard Earth satellites now enable us to observe Lyman lines from the sun and stars.

Lyman went so far as to climb a mountain in 1915, hoping that from its greater altitude there might happen to be a window of transparency that would pass Lyman alpha. Unfortunately, there is no such window, and his photographs of the solar spectrum were blank in the region where ultraviolet radiation would have appeared.

The series of transitions with $n = 2$ as the lowest level is the Balmer series, which we have already seen is in the visible part of the spectrum. The bright red emission line in the visible part of the hydrogen spectrum, known as the H$\alpha$ line (H-alpha), arises from the transition from level 3 to level 2. The transition from $n = 4$ to $n = 2$ causes the H$\beta$ line (H-beta), and so on. (Because the series falls in the visible where it is so well observed, we usually call the lines simply H-alpha, etc., instead of Balmer-alpha, etc.) There are many other series of hydrogen lines, each corresponding to transitions to a given lower level (Fig. 24–11).

IN REVIEW

## BOX 24.4 Do Not Confuse: Balmer Series and Bohr Atom

*Balmer series*: a series of spectral lines from hydrogen; the Balmer series of lines starts in the red and converges to the violet.

*Bohr atom*: the model of the hydrogen atom with discrete energy levels spaced in a certain way so that jumps of electrons between the levels lead to the observed spectral lines.

***Discussion Question***

Relate the Balmer series and the Bohr atom.

***Answer***

The Balmer series is an observational property, a set of spectral lines that are observed. The Bohr atom explains it.

Since the higher energy levels have greater energy, a spectral line caused by a transition from a higher level to a lower level yields an emission line. When, on the other hand, continuous radiation falls on cool hydrogen gas, some of the atoms in the gas can be raised to higher energy levels. Absorption lines result. The Balmer series is seen in stars in absorption, but when we look just at glowing hydrogen gas, we see the Balmer series in emission.

The hydrogen atom, with its lone electron, is a very simple case. More complicated atoms have more complicated sets of energy levels, and thus their spectral lines have less simple patterns.

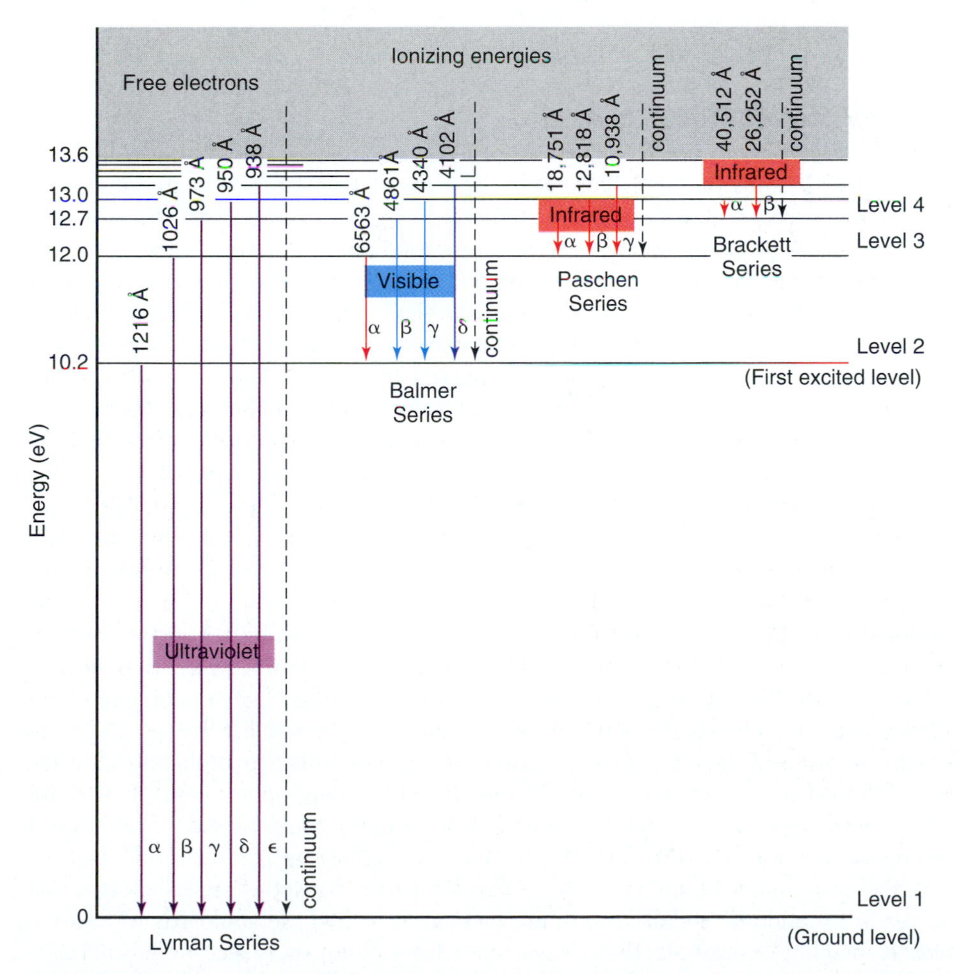


**FIGURE 24–11** The energy levels of hydrogen and the series of transitions among the lowest of these levels.

**FIGURE 24–12** Part of the computing staff of the Harvard College Observatory in about 1917. (The people were the "computers" in those days.) Annie Jump Cannon, fifth from the right, classified over 500,000 spectra in the decades following 1896. Her catalogue, called the *Henry Draper Catalogue* after the scientist whose widow was the benefactor who made the investigation possible, is still in use today. Many stars are still known by their HD (Henry Draper catalogue) numbers. Also in this photograph is Henrietta S. Leavitt, fifth from the left, whose work on variable stars we shall discuss in Section 25.4b.

**FIGURE 24–13** Annie Jump Cannon classifying spectra.

Each of the chemical elements has its own sets of energy levels, one set for each stage of ionization. Each has its own sets of spectral lines. (The Lyman and Balmer series are only for neutral hydrogen; the lines of most other elements do not form such a regular pattern.)

## 24.5 SPECTRAL TYPES

In the last decades of the 19th century, spectra of thousands of stars were photographed. Ways of classifying the different spectra were developed (Fig. 24–12). The early work on photographic spectra was carried out at the Harvard College Observatory, where Willamina Fleming classified over 10,000 spectra as well as discovering hundreds of variable stars. The most famous worker at the vital task of classifying spectra was one of her successors at Harvard, Annie Jump Cannon (Fig. 24–13). She classified hundreds of thousands of spectra. Photographic plates with low dispersion (that is, with the wavelengths not spread out very much) were used.

At first, the stellar spectra were classified only by the strength of certain absorption lines from hydrogen, and were lettered alphabetically: A for stars with the strongest hydrogen lines, B for stars with slightly weaker lines, and so on. These categories are called **spectral types** (Fig. 24–14). It was later realized that the types of spectra varied primarily because of differing temperatures of the **stellar atmospheres**—the stars' outer layers. The hydrogen lines were strongest in stars in the middle range of stellar temperatures and were weaker at both higher and lower temperatures. In the hottest stars, too much hydrogen is ionized for the lines to be strong. In stars cooler than spectral type A, too few hydrogen atoms have enough energy for electrons to be on levels above the ground state. When we now list the spectral types of stars in order of decreasing temperature, they are no longer in alphabetical order. From hottest to coolest, the spectral types used have long been O B A F G K M. The spectral types are divided into tenths. For example, the sequence, from hottest to coolest, includes F8, F9, G0, G1, G2, and so on. The sun is a G2 star.

The assignment of spectral types follows a particular set of stellar spectra that serves as a standard. Stellar spectra are compared with these standards to find the closest match. For example, there is one star whose spectrum is recorded for all time as the prototypical G2 star. If you take the spectrum of any star, you see which spectral type (such as G2) it most closely matches.

O stars (that is, stars of type O) are the hottest, and the temperatures of M stars are more than 10 times lower. Generations of American students and teachers have remembered the spectral types by the mnemonic: Oh, Be A Fine Girl [now also Guy], Kiss Me. The lettered spectral types have been subdivided into 10 subcategories each. For example, the hottest B stars are B0, followed by B1, B2, B3, and so on. Spectral type B9 is followed by spectral type A0. It is fairly easy to tell the spectral type of a star by mere inspection of its spectrum.

Historically, the hotter stars are called "early types" and the cooler stars are called "late types," though the words "early" and "late" are misleading since individual stars do not actually change spectral types during the long stable period that takes up most of their lives.

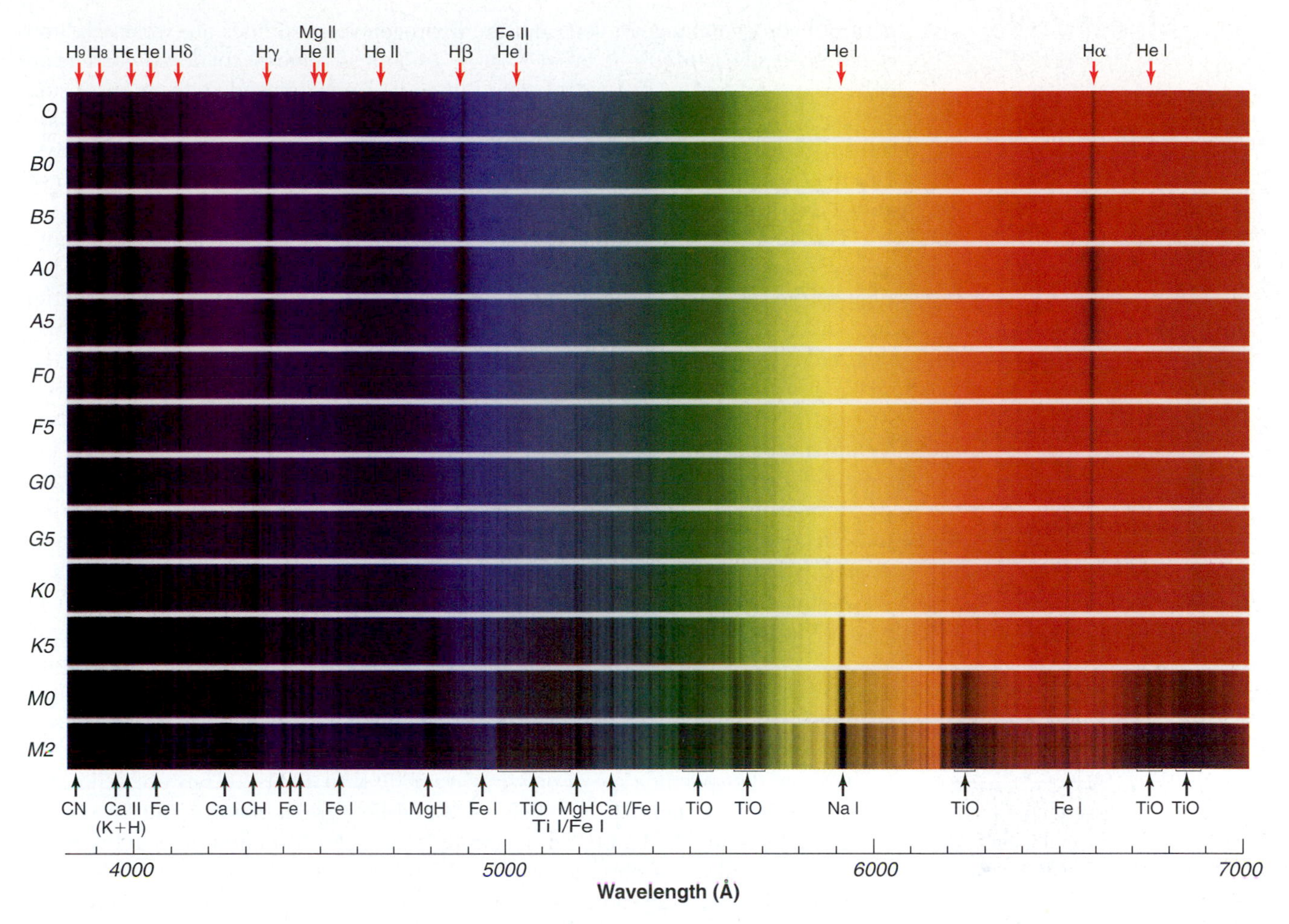


**FIGURE 24–14** Computer simulation of the spectra of stars from a wide range of spectral types. Note that all the stellar spectra shown have absorption lines. The numbers following the letters in the left-hand column represent a subdivision of the spectral types into tenths; the step from B8 to B9 is equal to the adjacent step from B9 to A0. The abbreviations of the elements forming the lines and the individual wavelengths (in angstroms) of the most prominent lines appear. Notice how the presence or absence of certain spectral lines, like the H and K lines of ionized calcium, can quickly indicate the spectral type. Notice also how the hydrogen lines are strongest in A stars and how the series of lines (*red*) can easily be spotted by the way it converges. M stars show not only numerous spectral lines from individual atoms but also wide "molecular bands."

With infrared studies and other advances in astronomy, types of stars known as brown dwarfs were discovered starting in the 1990s. They are cooler and less massive than even normal M stars. As I will elaborate below, new spectral types have been named to include them.

## 24.5a Stellar Spectra

When we observe the light from a star, we are seeing only the radiation from thin outer layers of that star's atmosphere. The continuum and the absorption lines are formed in these layers. (Again, ordinary stars have only absorption lines, not emission lines.) The interior is hidden within.

As we consider stars of different spectral types, we can observe the effect of temperature. For example, the hottest stars, those of type O, have such high temperatures that there is enough energy to remove the outermost electron or electrons from most of the atoms. Since most of the hydrogen is ionized, relatively few neutral hydrogen atoms are left, and the hydrogen spectral lines are weak. Helium is not so easily ionized—it takes a large amount of energy to do so—yet even helium appears not in its neutral state but in its singly ionized state in an O star. Temperatures range from 30,000 K to 60,000 K in O stars. O stars are relatively rare, since they have short lifetimes; none of the nearest stars to us are O stars. The system really begins with spectral types O3 and O4, of which only a handful are known at any distance. No O0, O1, or O2 stars apparently exist.

The development of infrared arrays has led to the discovery of ionized helium in some stars heavily shrouded by a cocoon of dust. The sensitivity of the arrays has allowed the spectrum of these objects to be observed at the wavelength of spectral lines of ionized helium. The presence of ionized helium proves that an O star is inside; no other type of star is hot enough.

Rigel and Spica are familiar B stars.

B stars are somewhat cooler, 10,000 K to 30,000 K. The hydrogen lines are stronger than they are in the O stars, and lines of neutral helium instead of ionized helium are present.

Sirius, Deneb, and Vega are among the brightest A stars in the sky.

Cooler still are A stars, 7500 K to 10,000 K. By definition, the lines of hydrogen are strongest in this spectral type. Lines of singly ionized elements like magnesium and calcium begin to appear. (All elements other than hydrogen or helium are known as **metals** for this purpose.) O, B, and A stars are all bluish white.

Canopus, the second brightest star in the sky (not visible from the latitudes of most of the United States), is a prominent F star, as is Polaris, the north star.

F stars have temperatures of 6000 K to 7500 K. Hydrogen lines are weaker in F stars than they are in A stars, but the lines of singly ionized calcium are stronger. Singly ionized calcium has a pair of lines that are particularly conspicuous; they are easy to recognize in the spectrum. These lines are called H and K, from their alphabetical order in an extended version of Fraunhofer's original list (which we saw as Figure 4–23).

Besides the sun, alpha Centauri (actually the brightest of the three stars that together make up the bright point in the sky that we call alpha Centauri) is also a G star.

G stars, the spectral type of our sun, are 5000 K to 6000 K. They are yellowish, since the peak in their spectra falls in the yellow/green part of the spectrum. Hydrogen lines are visible, but the H and K lines are the strongest lines in the spectrum. The H and K lines are stronger in G stars than in any other spectral type.

Arcturus and Aldebaran, both visible as reddish points in the sky, are K stars.

K stars are relatively cool, only 3500 K to 5000 K. Their spectra show many lines, in contrast to the spectra of the hottest stars, which show few spectral lines. The lines in K stars mostly come from neutral metals. Note that the letter "K" is used in many astronomical contexts; see Box 24.5.

Betelgeuse is an example of such a reddish star of spectral class M.

M stars are cooler yet, with temperatures less than 3500 K, so they look reddish. Their atmospheres are so cool that molecules can exist without being torn apart, and their spectra show many molecular lines. Lines from the molecule titanium oxide are particularly strong and numerous. Look at the set of spectra shown and notice all the molecular lines in the spectrum of M stars, and that such molecular lines do not appear in the spectra of hotter stars.

For the very coolest normal stars, there are alternative spectral types to type M that reflect differences in the relative abundances of various elements. For example, stars that are relatively rich in carbon are called **carbon stars.** (These were originally called R and N stars.) S stars show absorption bands of the molecule zirconium oxide instead of the titanium oxide that dominates the spectrum of M stars. The zirconium and other heavy elements were formed deep inside the stars by nuclear fusion and have circulated up to the surface.

How do you tell the temperature of a star? You can look at the continuum, and use Planck's law. Or you can look at the spectral lines to give you the spectral types. In Figure 24–15, we have enlarged and brightened the ultraviolet end of the visible spectrum to show you how to tell temperature fairly easily. Notice how the K line of ionized calcium (Ca II) gets darker as you go from spectral type A5 down to spectral type M. In contrast, the H$\varepsilon$ and the Ca II H line are both at almost exactly the same wavelength, so as one fades out (which you can trace with H$\delta$ or H8), the other darkens, making the total darkness at that wavelength relatively constant between A5 and M. Thus the ratio of the darkness of the spectral line at the wavelength of K to that at the wavelength of H/H$\varepsilon$ tells you the spectral type.

## IN REVIEW

### BOX 24.5 Do Not Confuse: K

K stars: a spectral class
K: a kelvin, a unit of temperature
K: a strong line in many stellar spectra
K: the element potassium
K: an infrared filter band
K corona: the continuous spectrum of the solar corona
k: kilo-, a metric prefix

***Discussion Question***

Choose two other letters of the alphabet and for each give several examples in which the letter is used as a symbol.

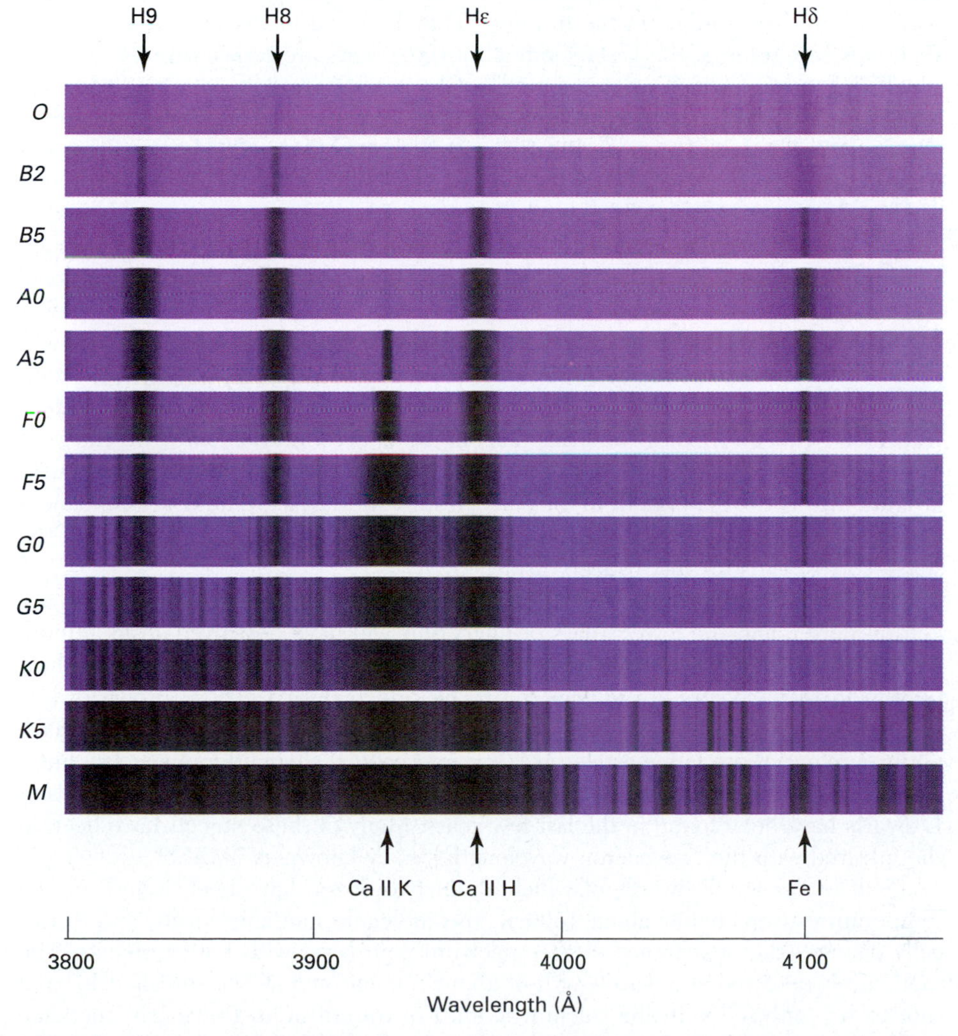


**FIGURE 24–15** The violet and blue part of the previous image, displayed to show more clearly the change in visibility of the H and K lines of ionized calcium (Ca II) versus the H-epsilon line of hydrogen as spectral types go from hotter to cooler.

WHAT'S IN A NAME?

## BOX 24.6 The New Spectral Types L and T

Scientists from Caltech and the University of Massachusetts, who obtained and reduced the data from the 2MASS infrared survey that discovered many stars cooler than the existing system of spectral types allowed, wanted to choose the best letters for the new spectral types that they thought necessary. C, J, N, R, and S had already been used for carbon stars and other cool stars. Some letters were omitted because they might lead to confusion: D could be confused with white-dwarf spectral types known as DA, DB, etc.; E could be confused with types of elliptical galaxies known as E0, etc.; I could be confused with 1 (one), so I0 could be misread as 10 or even Jupiter's moon Io; P could be misread to indicate a planet; Q could be misread to indicate a quasar; U could be misread to imply an ultraviolet source; V0 could be misread for the vanadium oxide molecule VO; W could be confused with Wolf-Rayet types; and X could be misread to imply an x-ray source. Along with OBAFGKM, that didn't leave a whole lot of letters. L was the closest available letter to M, the current cool end of the sequence, so they chose it. They continued in alphabetical order in available letters to get to T.

Still cooler spectral types have been named in the late 1990s. L dwarfs are cooler than M dwarfs, and T dwarfs are cooler yet. Stars of types L and T are intrinsically faint and are best studied in the infrared. They look red to the human eye. As we shall now see, some of the L stars and all of the T stars are brown dwarfs.

Brown dwarfs, which we mentioned in Chapter 18, are stars with too little mass to sustain the fusion processes inside that fuel normal stars, which we will discuss in Chapter 27. The hottest brown dwarf cannot be hotter than mid-M type. Since the coolest ordinary star cannot be cooler than mid-L, all the T dwarfs are brown dwarfs. That spectral type starts well below the minimum temperature of ordinary stars.

L dwarfs have temperatures between 1400 and 2200 K. The coolest ordinary stars are L3 or L4. The stars from type L5 and cooler, all brown dwarfs, are so cool not only outside but also inside that lithium remains in their spectrum if it was ever there, instead of being consumed and changed into other elements. (The nuclear processes that destroy lithium do not take place at the relatively low internal temperatures of these brown dwarfs, which have masses less than 60 times that of Jupiter.) But it isn't enough to know what happens in an object's interior, since we can't see there. Also very important to realize is that the gas in all relatively cool stars churns around, mixing the material from the core through the surface. Thus the abundances we see—the light we detect obviously comes only from the outer atmosphere of the star—corresponds to what is deep inside.

All low-mass objects cool off at their surfaces as they age. But their cores are getting hotter, and the lowest mass ordinary star will have destroyed all its lithium by the time it has cooled to M7. Thus, any objects cooler than M7 that still show lithium in their spectra are guaranteed to be brown dwarfs. Those brown dwarfs more massive than 60 times Jupiter's mass (those in the lower subtypes of M) have no lithium and those less massive (M4, say, and cooler) show lithium spectral lines. In the cooler L dwarfs, lithium-rich molecules are apparently forming. Hundreds of L dwarfs have been found in the last few years. Many of these objects have been in the infrared with the two-micron-wavelength survey known as 2MASS.

Still cooler are T dwarfs, of which about two dozen have been found. As the temperature drops below about 1500 K, the molecule methane forms, and drastically changes the appearance of the spectrum compared with hotter objects. The methane is detected at a wavelength of about 2 micrometers, the so-called K-band (not to be confused with the calcium K line) in the infrared. Previously, methane had been seen only in planetary atmospheres; it is not found in any true star. The brown dwarf known as Gl 229B, shown in Figure 18–11, is a T dwarf. We do not yet know what the coolest T dwarf is; perhaps the scale will wind up extending to include Jupiter. The definition of brown dwarf vs. planet is under much discussion these days, as we saw in Chapter 18.

We have listed the ordinary spectral types of stars, each of whose visible radiation shows an absorption spectrum, that is, a continuum crossed with dark lines. There are many unusual stellar spectra, though. **Wolf-Rayet stars** are O stars that not only show emission lines but also have the emission particularly broad in wavelength. The emission comes from shells of material that the star had ejected into the space surrounding it (Fig. 24–16) and results from strong outflowing "stellar winds."

The amount of **mass loss,** matter flowing out of stars, varies from spectral type to spectral type, but it can be considerable (Fig. 24–17). Much of the matter in interstellar space is the result of stellar mass loss.

In one of the first applications of astrophysics to the study of stars, in 1925, Cecilia Payne-Gaposchkin (then Cecilia Payne), long-term Harvard astronomer, graphed the strength of certain absorption lines versus spectral type. She used the graph together with her theoretical calculations to determine the temperature for each spectral type. It was she who first figured out that stars were almost entirely made of hydrogen, a conclusion so shocking that it took some time before it was accepted. It took a few years, further analysis by Donald H. Menzel, and evaluation by Henry Norris Russell before astronomers in general knew that the stars are made mainly of hydrogen.

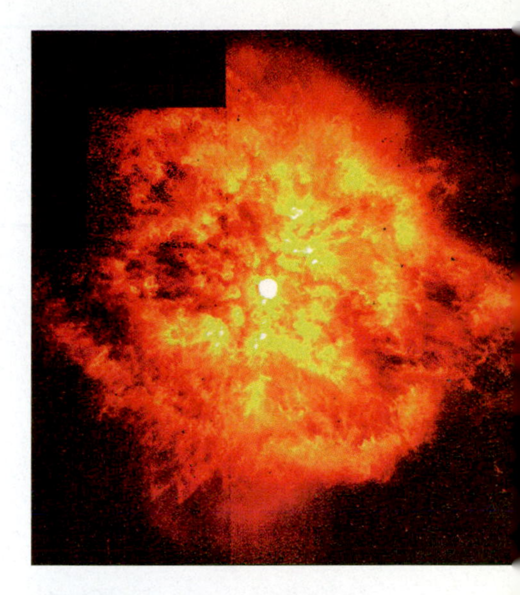

**FIGURE 24–16** A Wolf-Rayet star (*center*) surrounded by hot clumps of gas that are being ejected into space at tremendously high speeds. This Hubble Space Telescope image also shows chaotic structures extending 100 billion km into space around this hot star.

## 24.5b Spectra Observed from Space*

In the ultraviolet and x-ray regions of the spectrum, we often see radiation from above the star's surface. (This "surface" is not a sharp edge but is defined as is the apparent edge of the sun as seen in visible light on an ordinary day.) These higher levels are hotter than the surface, as was originally learned from studies of the sun. For example, we discovered from an x-ray telescope in space that stars of all spectral types give off x-rays. Though it had been expected that the coolest stars, the M stars, would be too cool to be so bright in x-rays, they are actually as much as 10 per cent

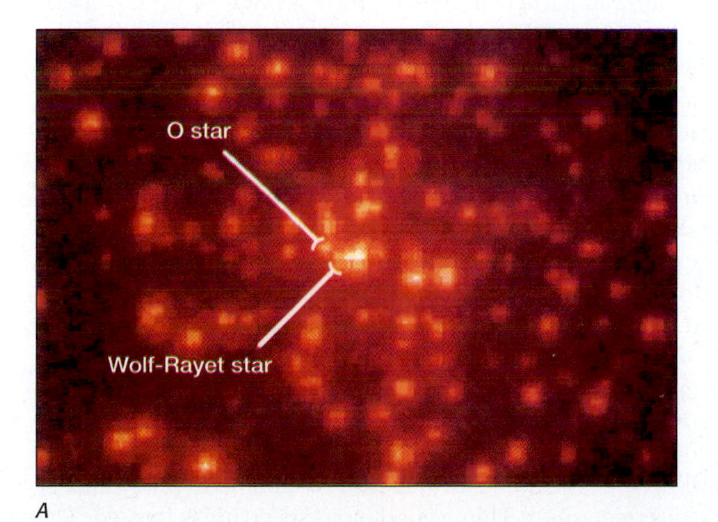


A

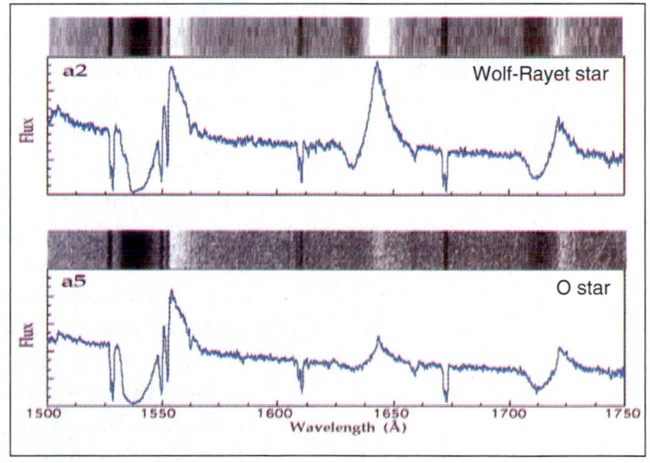


B

**FIGURE 24–17** (*A*) In this part of the star cluster R136a, we see a Wolf-Rayet star, near the end of its lifetime, and an O-type star. In this Hubble image, the stars are separated by only 0.2 arc sec, and they are blurred together when observed from the ground. (*B*) These Hubble Space Telescope spectra taken in the ultraviolet, with a tracing below the equivalent to the photographic view for each star, reveal that the helium spectral line that is emitted at 1645 Å is much stronger in the spectrum of the Wolf-Rayet star compared with its strength in the spectrum of the O-star. We deduce that the helium in the Wolf-Rayet star is flowing out at a much greater rate. The W-R star is losing mass so fast that it may disappear instead of eventually becoming a black hole.

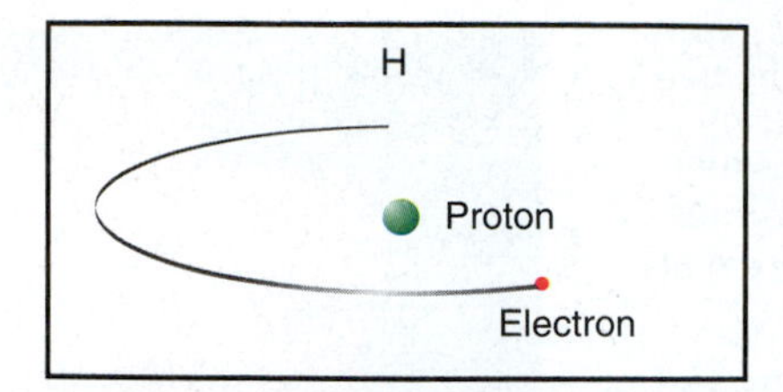


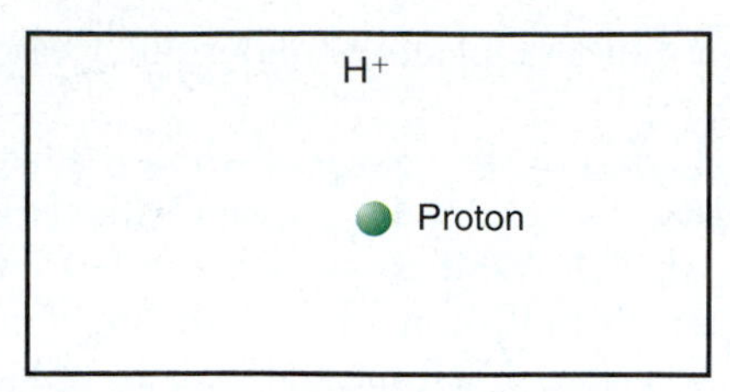


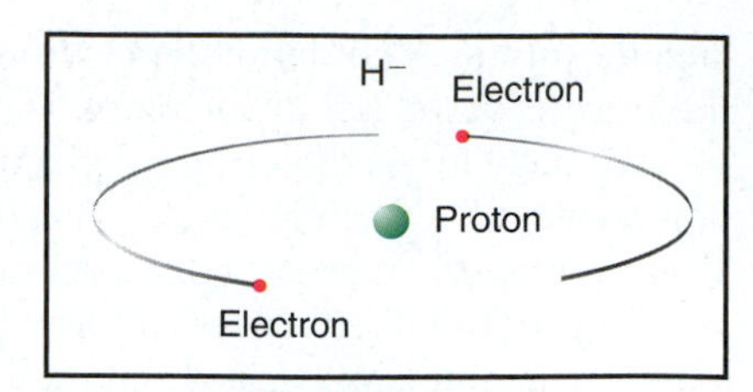


**FIGURE 24–18** Neutral hydrogen, the hydrogen ion $H^+$ (sometimes, though rarely, called the positive hydrogen ion), and the negative hydrogen ion $H^-$.

as strong in the x-ray spectrum as they are in the visible. But this doesn't say that we were wrong about the temperature of these stars' surfaces. The x-rays come from a hot **stellar corona** surrounding the stars at a temperature of over a million kelvins. As is the case for the solar corona, the energy to heat these stellar coronas presumably comes from magnetic waves rather than from the temperature of lower layers. The Chandra X-ray Observatory is observing the coronas of many cool stars.

Ultraviolet spectra of even cool stars taken from space show spectral lines from gas at temperatures of up to about 250,000 K. These spectral lines are formed in a **stellar chromosphere** between the star's ordinary surface and its corona, or in a **transition region** between the chromosphere and corona. Only since the launch of spacecraft could we study chromospheres and coronas in stars other than the sun.

### 24.5c The Continuous Spectrum*

The spectra we have discussed, with electrons jumping from energy level to energy level, give sets of spectral lines. For there to be an absorption line, there must be some continuous radiation to be absorbed. Where does the continuous spectrum come from? The mechanism that causes continuous emission is different for different spectral types. We are not discussing the fundamental cause of the energy we see in the continuum; that comes from nuclear fusion at the center of the sun. We are, instead, asking "What are the particular atomic processes that cause the actual photons we are seeing to be emitted?"

In the sun and other stars of similar spectral types, the continuous emission is caused by a strange type of ion: the **negative hydrogen ion** (Fig. 24–18). An ordinary ion has one or more electrons missing, and so it has a positive charge and could be called a positive ion. A negative ion, on the other hand, has one or more extra electrons. Only a tiny fraction of the hydrogen atoms are in such a state, with one extra electron, but the sun is so overwhelmingly (90 per cent) hydrogen that enough negative hydrogen atoms are present to be important.

Spectral lines arise from transition of electrons between energy states. Since a hydrogen atom has only one electron to begin with, when it is ionized it has no electrons at all and thus has no spectral lines.

When a neutral hydrogen atom takes up an extra electron, the second electron may temporarily become part of the atom, residing in an energy level. But the energy *difference* between the extra electron's original energy and its energy as part of the atom is not limited to discrete values, because the electron could have had any amount of energy before it joined the atom (Fig. 24–19). Alternatively, the extra electron starts free of the atom and is affected by the atom (changing velocity or direction), but it winds up still free of the atom. In this case too, the amount of energy involved is not limited to discrete values. Thus a continuous spectrum is formed. The case is thus different from the one in which an electron in an atom jumps between two energy levels, because in that case the electron has a fixed amount of energy in each level, and a fixed number subtracted from another fixed number gives a fixed number.

The processes of continuous emission and spectral-line absorption take place together throughout the outer layers of the stars rather than in separate layers. Supercomputers are able to handle complicated sets of equations that describe the transfer of radiation from the outer layers of a star into space.

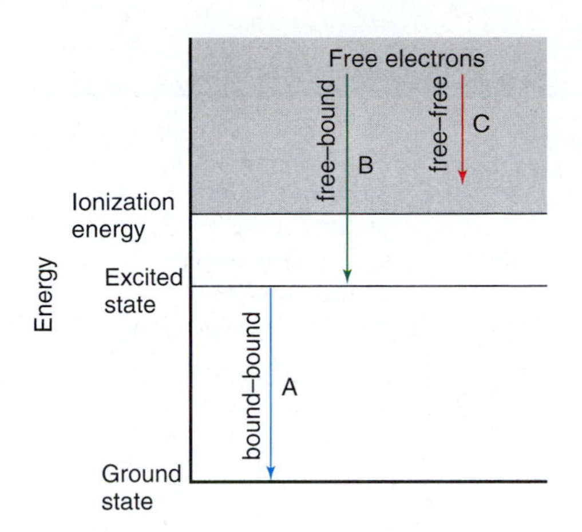


**FIGURE 24–19** A bound-bound transition (*A*) would go from one discrete level to another (that is, from a bound level to another bound level), analogous to jumping on a staircase from one step to another. A free-bound transition (*B*) would be analogous to jumping from some arbitrary height above the staircase (as from the sky with a parachute) to some particular step (that is, from being free to a bound level). In a free-free transition (*C*) the electron never joins the atom. The colors are drawn from bluer to redder as the arrows grow shorter, to indicate the bluer emission from more energetic photons.

## CORRECTING MISCONCEPTIONS

| ✖ *Incorrect* | ✔ *Correct* |
|---|---|
| All stars are the same color. | The colors of stars show their temperatures. |
| All red things in the sky are cool. | Red stars are cool, since we are seeing continuous radiation, but emission line sources that show red $H\alpha$ can be hot. |
| Stars shine because of spectral lines. | Continuous radiation comes mostly from the negative hydrogen ion. |
| The Balmer series is a hydrogen atom with electrons jumping to and from energy level 2. | The Balmer series is the set of spectral lines from hydrogen observed in the visible. (The misconception actually describes the Bohr atom.) |
| The Bohr atom has 5 levels. | The Bohr atom has an infinite number of levels. |

## SUMMARY AND OUTLINE

Stars (Sections 24.1 and 24.2)

Balls of gas held together by gravitation and generating energy by nuclear fusion; as gas gets hotter, the peak of the radiation moves to shorter wavelengths.

**Wien's displacement law:** $\lambda_{\max}T$ = constant.

As gas gets hotter, the total energy radiated grows rapidly, with the fourth power of the temperature.

**Stefan-Boltzmann law:** $E = \sigma T^4$.

Planck's law is an equation for the amount of radiation at each wavelength, given the temperature. It includes the other two laws.

> Wien's displacement law: $\lambda_{\max}T$ = constant
> Stefan-Boltzmann law: $E = \sigma T^4$
> Planck's law relates $\lambda$, $I_\lambda$, and $T$, where $I_\lambda$ is the intensity at wavelength $\lambda$.

Black body: matter whose radiation follows Planck's law precisely

The structure of an atom (Section 24.3)

Protons and neutrons with electrons orbiting the nucleus

Spectral lines (Sections 24.3 and 24.4)

Atoms can exist only in discrete states of energy (quantum theory).

When an atom drops from a higher to a lower energy state, it emits the difference in energy as a photon of a particular wavelength $\lambda$ and energy $hc/\lambda$. Such photons make an emission line. When an atom takes up enough energy to raise itself from a lower to a higher state, an absorption line forms in radiation that has been hitting a group of such atoms. Hydrogen's spectral lines fall in a pattern across the visible: the Balmer series. The Bohr atom, with each energy level having a different size, is a good way to visualize the hydrogen atom. The Balmer series corresponds to transitions between level 2 and higher levels: transitions up for absorption lines and down for emission lines. The Lyman series, which falls in the ultraviolet, involves transitions to the most basic level, the first level, also called the ground state. Other elements have different sets of energy levels from those of hydrogen, so they have different spectra.

Spectral types (Section 24.5)

Hottest to coolest: O B A F G K M L T

Hydrogen spectra are strongest in type A.

Calcium H and K lines are strongest in type G.

Molecular spectra begin to appear in type M.

New spectral types for cooler stars, including brown dwarfs, are L and T.

All types have absorption spectra. Only a few kinds of stars with emission lines exist.

From space we can observe stellar chromospheres and coronas. Many stars undergo mass loss.

A star's continuous spectrum comes from the negative hydrogen ion ($H^-$).

## KEY WORDS

astrophysics
Wien's displacement law
Stefan-Boltzmann law
Planck's law
black body
nucleus
electrons
proton
neutron
quarks
quantum mechanics
allowed states
energy levels
quantum
photon
ground state
first excited state
ionized
ion
positive ion
Balmer series
Bohr atom
principal quantum number
ground level
spectral types
stellar atmospheres
metals
carbon stars
Wolf-Rayet stars
mass loss
stellar corona*
stellar chromosphere*
transition region*
negative hydrogen ion*

*This term appears in an optional section.

## QUESTIONS

1. What are two differences between a star and a planet?
†2. (a) By how many times does the energy given off by a bit of gas increase when the gas is heated from 300 K (room temperature) to 600 K? (b) To 6000 K?
†3. The sun's spectrum peaks at 5600 Å. Would the spectrum of a star whose temperature is twice that of the sun peak at a longer or a shorter wavelength? How much more energy than a square centimeter of the sun would a square centimeter of the star give off?
†4. The sun's temperature is about 5800 K, and its spectrum peaks at 5600 Å. An O star's temperature may be 40,000 K. At what wavelength does its spectrum peak?
5. For the O star of Question 4, in what part of the spectrum does its spectrum peak? Can the peak be observed with the Keck telescope? Explain.
†6. One black body peaks at 2000 Å. Another, of the same size, peaks at 10,000 Å. Which gives out more radiation at 2000 Å? Which gives out more radiation at 10,000 Å? What is the ratio of the total radiation given off by the two bodies?
†7. A coolish B star has twice the temperature of the sun in kelvins. (a) How many times more than the sun does each square centimeter of the B star's surface radiate? (b) The B star's diameter is five times the sun's. How many times more than the sun does the whole star radiate, given that the surface area of a sphere increases as the radius squared ($A = 4\pi r^2$)?
8. Which contains more information, Wien's displacement law or Planck's law? Explain.
†9. What is the ratio of energy output for a bit of surface of an average O star and a same-sized bit of surface of the sun?
10. Star A appears to have the same brightness through a red and a blue filter. Star B appears brighter in the red than in the blue. Star C appears brighter in the blue than in the red. Rank these stars in order of increasing temperature.
11. (a) From looking at Figure 24–11, draw the Lyman series and the Balmer series on the same wavelength axis, that is, on a horizontal straight line labelled in angstroms every 100 Å from 0 to 10,000 Å. (b) Why is the Balmer series the most-observed spectral series of atomic hydrogen?
12. What is the difference between the continuum and an absorption line? The continuum and an emission line? Draw a continuum with absorption lines. Can you draw absorption lines without a continuum? Can you draw emission lines without a continuum? Explain.
13. Consider a hypothetical atom in which the energy levels are equally spaced from each other, that is, the energy of level $n$ is $n$. (a) Draw the energy level diagram for the first five levels. (b) Indicate on the diagram the transition from level 4 to level 2 and from level 4 to level 3. (c) What is the ratio of energies of these two transitions? (d) If the $4 \rightarrow 2$ transition has a wavelength of 4000 Å, what is the wavelength of the $3 \rightarrow 2$ transition? (*Hint:* Do not confuse the energy levels in this hypothetical atom with the distinct pattern of energy levels for the hydrogen atom.)
14. Does the spectrum of the solar surface show emission or absorption lines? Compare and/or distinguish between the sun's spectrum and the spectrum of a B star.
15. (a) Why are singly ionized helium lines detectable only in O stars? (b) Why aren't neutral helium lines prominent in the solar spectrum?
16. Compare Planck curves for stars of spectral types O, G, and M.
17. Why don't we see strong molecular lines in the sun?
18. Describe the spectral differences between M stars and L stars.
†19. Using Balmer's formula ($1/\lambda$ = constant $\times$ $(1/n^2 - 1/m^2)$), where $n$ is the principal quantum number of one level and $m$ is the principal quantum number of the other level) and the fact that the wavelength of H$\alpha$ is 6563 Å, calculate the wavelength of H$\beta$. Show your work.
20. What factors determine spectral type? What instrument would you use to determine the spectral type of a star?
21. List spectral types of stars in approximate order of strength, from strongest to weakest, of (a) hydrogen lines, (b) ionized calcium lines, and (c) titanium oxide lines.
22. If we are examining a stellar spectrum, explain how and why comparing the relative intensity of the absorption line at

about 3970 Å with the intensity of the calcium K line allows us to judge the star's spectral type.

23. If we take two stones, one twice the diameter of the other, and put them in an oven until they are heated to the same temperature and begin glowing, what will be the relationship between the total energy in the light given off by the stones? (The surface area of a sphere is proportional to the square of the radius.)

24. Why does the negative hydrogen atom have a continuous rather than a line spectrum?

†This question requires a numerical solution.

## PROJECT

1. Make up your own mnemonic for OBAFGKM. Also try one for the whole extended series OBAFGKMLT. Send in the best tries (perhaps as a class project) through this book's Web site.

## USING TECHNOLOGY

### W World Wide Web

1. Find images of Wolf-Rayet stars at the Anglo-Australian Observatory site, http://www.aao.gov.au/images.html.
2. Check out the Space Telescope Science Institute's images/captions about stars and recent press releases on stellar topics.
3. Learn about brown dwarfs from Prof. Gibor Basri at http://www.sciam.com/2000/0400issue/0400basri.html or at http://www.ency-astro.com.

### *REDSHIFT*

1. Consider Science of Astronomy: *Principia Mathematica:* The Stefan-Boltzmann Relation, Wien's Law, and Parallax.
2. Study Astronomy Lab: Blackbody, Wien's Law, Stefan-Boltzmann relation I and II.
3. Use the spectral-type filters in the main *RedShift* engine to show stars of each of the spectral types, in turn, in the region of the sky visible to the east tonight.

Finding the distances to the stars is not straightforward. Here we see an open cluster of stars in the constellation Crux. It is known as the Jewel Box. Studying clusters of stars has helped us establish methods of finding stellar distances.

# Chapter 25

# Stellar Distances and Motions

We have seen in Chapter 24 how we can analyze the spectrum of a star to tell us about the outer layer of that star. For example, we can tell how hot a star is from either its continuous spectrum or from its spectral lines. In this chapter, we see how information about the spectra of many stars can be put together to tell us about the overall properties of stars.

We begin by describing the scale in which astronomers give brightness. We also see how we can fairly directly measure the distances to the nearest stars and how these distances, together with classification by spectral type, can give us distances to farther stars. We see how graphing the temperatures and brightnesses of stars gives us an important tool: the Hertzsprung-Russell diagram. Then we study how astronomers tell the speed and direction in which stars are moving.

**AIMS:** To describe the absolute and apparent magnitude scales, the method of trigonometric parallax and how it tells us the distances to the stars, the Hertzsprung-Russell diagram and what it tells us about types of stars and their distances, and how we learn about stellar motions

## 25.1 LIGHT FROM THE STARS

The most obvious thing we notice about the stars is that they have different brightnesses (Fig. 25–1). Over 2000 years ago, the Greek astronomer Hipparchus (who lived up until about 127 B.C.) made a catalogue of stars and divided the stars into classes of brightness, using terms like "bright" and "small." Little else is known of his catalogue. By the early 1st century A.D., a system of six classes of stellar brightness was well known. In the *Almagest* of Ptolemy (written in Alexandria in about A.D. 140), the brightest stars were said to be of the first magnitude, a reasonable idea. Somewhat fainter stars were said to be of the second magnitude, and so on down to the sixth magnitude, which represented the faintest stars that could be seen with the naked eye. Ptolemy often quoted Hipparchus, and it is often not known which ideas belonged to each; most of what we know of Hipparchus comes from Ptolemy's *Almagest*.

### 25.1a Apparent Magnitude

In the 19th century, when astronomers became able to make quantitative measurements of the brightnesses of stars, the magnitude scale was placed on an accurate basis. It was discovered that the brightest stars that could be seen with the naked eye were about 100 times brighter than the faintest stars. This ratio corresponded to a difference of 5 magnitudes between the old 1st magnitude and 6th magnitude, a number that was used to set up the present magnitude system. Since this type of magnitude tells us how bright a star *appears,* it is called **apparent magnitude.**

**FIGURE 25–1** The individual stars we see in the sky are all in our Milky Way Galaxy and are at very different distances from us. Here we see a view toward the center of our galaxy.

The new system is based on the definition of stars that, on the magnitude scale, show a *difference* of 5 magnitudes are different by a *factor* of exactly 100 times. Thus a star of 1st magnitude is exactly 100 times brighter than a star of 6th magnitude.

| | |
|---|---|
| ½ mag = | 1.585 times |
| 1 mag = | 2.512 times |
| 2 mag = | 6.310 times |
| 3 mag = | 15.85 times |
| 4 mag = | 39.81 times |
| 5 mag = | 100 times |
| 6 mag = | 251 times |
| 7 mag = | 631 times |
| 8 mag = | 1585 times |
| 9 mag = | 3981 times |
| 10 mag = | $10^4$ times |
| 15 mag = | $10^6$ times |
| 20 mag = | $10^8$ times |

The brightnesses of the stars that had been classified by the Greeks were placed on this new magnitude scale (Fig. 25–2). Many of the stars that had been of the first magnitude in the old system were indeed "first magnitude stars" in the new system. But a few stars were much brighter and thus corresponded to "zeroth" magnitude, that is, a number less than one. A couple, like Sirius, the brightest star in the night sky, were brighter still. The numerical magnitude scale was easily extended to negative numbers. The new scale, being numerical, admits fractional magnitudes. Sirius is magnitude −1.5, for example. Physiologically, this type of measurement makes sense because the human eye happens to perceive equal *factors* of brightness (such a factor is the energy coming from one object divided by the energy coming from another) as roughly equal *intervals* (additive steps) of brightness, which is just how the magnitude scale is set up.

The magnitude scale was extended not only to stars brighter than 1st magnitude, but also to stars fainter than 6th magnitude. Since a difference of five magnitudes corresponds to a factor of a hundred times in brightness, 11th magnitude is exactly 100 times fainter than 6th magnitude (that is, 1/100 times as bright); 16th magnitude, in turn, is exactly 100 times fainter than 11th magnitude. The faintest stars that can be photographed from the ground are fainter than 24th magnitude.

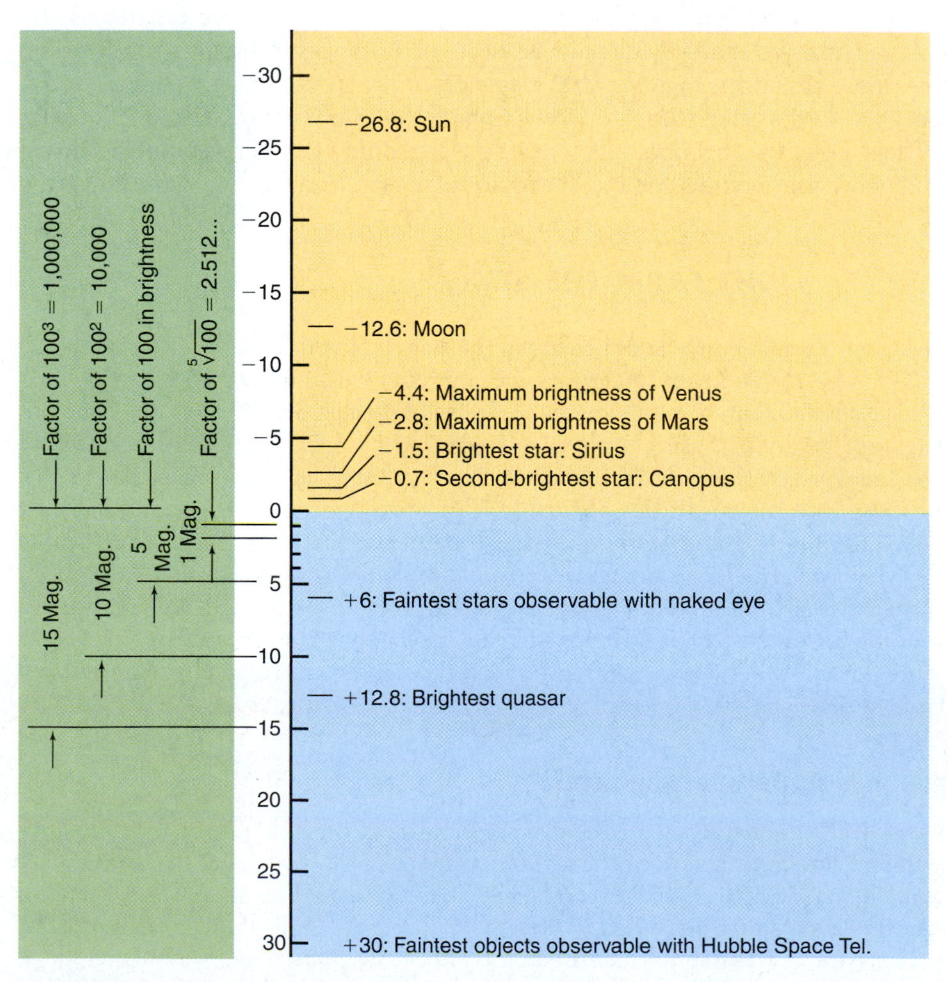


**FIGURE 25–2** The apparent magnitude scale is shown on the vertical axis. At the left, sample intervals of 1, 5, 10, and 15 magnitudes are marked and translated into multiplicative factors.

### 25.1b Working with the Magnitude Scale*

We have seen how a difference of 5 magnitudes corresponds to a factor of 100 in brightness. What does that indicate about a difference of 1 magnitude? Since an increase of 1 in the magnitude scale corresponds to a decrease in brightness by a certain factor, we need a number that, when 5 of them are multiplied together, will equal 100. This number is just $\sqrt[5]{100}$. Its value is 2.512 . . . , with the dots representing an infinite string of other digits. For many practical purposes, it is sufficient to know that it is approximately 2.5.

Thus a second magnitude star is about 2.5 times fainter than a first magnitude star. A third magnitude star is about 2.5 times fainter than a second magnitude star. Thus a third magnitude star is approximately $(2.5)^2$ (which is about 6) times fainter than a first magnitude star. (We could easily give more decimal places, writing 6.3 or 6.31, but the additional figures would not be meaningful. We started with only one digit being significant when we wrote down "2nd magnitude," so our data were given to only one **significant figure.** Our result is thus accurate only to one significant figure. Thus because 2.5 is accurate to only two significant figures (the 2 and the 5), the result of squaring it can also be accurate to only two significant figures. If we had given an original value to three significant figures, such as 2.51, then the result would be accurate to three significant figures, 6.31.)

We must be careful not to fool ourselves that we can gain in accuracy in an arithmetical process; the lowest number of significant figures we put in is the number of significant figures in the result.

In similar fashion, using a factor of 2.5 or 2.512 for the single magnitudes and a factor of 100 for each group of 5 magnitudes, we can easily find the ratio of intensities of stars of any brightness.

#### EXAMPLE 25.1 Magnitude Differences

*Question:* By what factor do the brightnesses of stars of magnitudes −1 and 6 differ (that is, what is the brightness ratio between the brightest and faintest stars, respectively, that are seen with the naked eye)?

*Answer:* Since the stars differ by 7 magnitudes, we have 100 from the first 5 magnitudes and 2.5 from each of the next two magnitudes, so the final factor is (100)(2.5)(2.5), which is about 600. The star of −1st magnitude is 600 times brighter than the star of 6th magnitude.

One unfortunate thing about the magnitude scale is that it operates in the opposite sense from a direct measure of brightness. Thus the brighter the star, the lower the magnitude (a second magnitude star is brighter than a third magnitude star), which is sometimes a confusing convention. This kind of problem comes up occasionally in an old science like astronomy, for at each stage astronomers have made their new definitions so as to have continuity with the past.

## 25.2 STELLAR DISTANCES

We can tell a lot by looking at a star or by examining its radiation through a spectrograph, but such observations do not tell us directly how far away the star is. Since the stars are but points of light in the sky even when observed through the biggest telescopes (with only a few exceptions), we have no reference scale to give us their distances.

The best way to find the distance to a star is to use the principle that was used in rangefinders, which used to be used to focus cameras. Sight toward the star from different locations, and see how the direction toward the star changes with respect to a more distant background of stars. You can see a similar effect by holding out your thumb at arm's length. Examine it first with one eye closed and then with the other eye closed. Your thumb seems to change in position as projected against a distant background; it appears to move across the background by a certain angle, about 3°. This is because your eyes are a few centimeters apart from each other, so each eye has a different point of view.

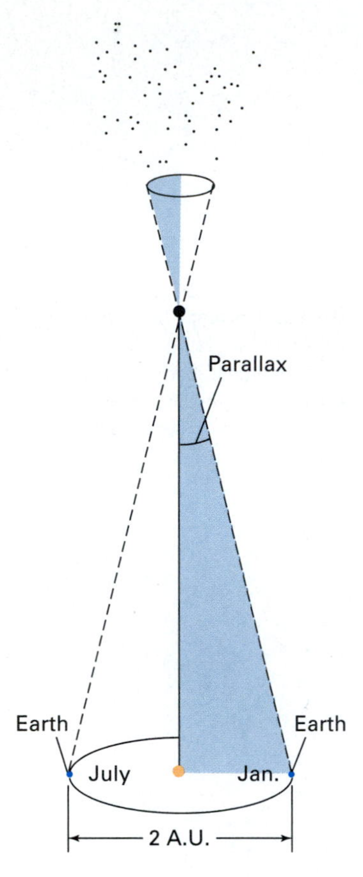


**FIGURE 25–3** The nearer stars seem to be slightly displaced with respect to the farther stars when viewed from different locations in the earth's orbit.

In close-up photography with a range finder or point-and-shoot camera, "parallax error" can cause you to mistakenly leave someone's head out of the picture, if you don't take account of the parallax caused by the fact that the lens and your viewfinder are in different places.

The angle to which Hipparcos pinpointed stars is about the same as the amount you would see my hair grow each second if you were standing one meter from me. It is equivalent to the angular size from the earth of a person standing on the edge of the moon.

Now hold your thumb up closer to your face, just a few centimeters away. Note that the angle your thumb seems to jump across the background as you look through first one eye and then through the other is greater than it was before. Exactly the same effect can be used for finding the distances to nearby stars. The nearer the star is to us, the farther it will appear to move across the background of distant stars, which are so far away that they do not appear to move with respect to each other.

To maximize this effect, we want to observe from two places that are separated from each other as much as possible. For us on Earth, that turns out to be the position of the earth at intervals of six months. In that six-month period, the earth moves to the opposite side of the sun. In principle, we observe the position of a star in the sky (photographically), with respect to the background stars, and several months later, when the earth has moved part way around the sun, we repeat the observation.

The straight line joining the points in space from which we observe is called the **baseline.** The distance from the earth to the sun is called an **Astronomical Unit** (A.U.). Our maximum baseline from Earth is 2 A.U. in length. The angle across the sky that a star seems to move (with respect to the background of other stars) between two observations made from the ends of a baseline of 1 A.U. (half the maximum possible baseline) is called the **parallax** of the star (Fig. 25–3). From the parallax, we can use simple trigonometry to calculate the distance to the star. The process is thus called the method of **trigonometric parallax.**

The basic limitation to the method of trigonometric parallax is that the farther away the star is, the smaller is its parallax. Thus, from the earth, this method can be used for only about ten thousand of the stars closest to us, extending only about 1 per cent of the way to the center of our galaxy; the other stars have parallaxes too small for us to measure. (These other stars, with negligible parallaxes, are the distant stars that provide the unmoving background against which we compare.) Even Proxima Centauri, the star nearest to us beyond the sun, has a parallax of only ¾ arc sec, the angle "subtended" (taken up across our vision) by a dime at a distance of 5 km! Clearly we must find other methods to measure the distances to most of the stars and certainly to the galaxies, which are more distant. But for the cases in which it can be used, the method of trigonometric parallaxes gives the most accurate answers.

A European Space Agency satellite devoted entirely to measuring positions was launched in 1989 and operated for about 3 years. It was named Hipparcos (*Hi*gh *P*recision *Par*allax *Co*llecting *S*atellite), after the name of the Greek astronomer whose name we usually spell Hipparchus. By compiling high-quality data on over a hundred thousand stars, Hipparcos has led to a tremendous improvement in **astrometry,** the study of the positions of stars. It also studied the stars' apparent motion across the sky compared with the overall celestial sphere, so-called "proper motions." It measured parallaxes to 1 to 2 milliarcseconds, giving distances to better than 10 per cent for all observed stars within 300 light-years of the sun. Its Hipparcos catalogue, released in 1997, contains distances for 118,000 stars to that precision. A secondary list, the Tycho-2 catalogue, contains less-accurate distances for a million stars, but contains accurate measurements of proper motions for a total of two and a half million stars. These catalogues greatly improve our fundamental knowledge of the distances of nearby stars. Moreover, our understanding of distant stars and even galaxies is often based on the distances to these closer stars, so the improvement is important for much of astronomical research.

Astrometric measurements are necessary not only for finding distances but also for providing accurate positions for faint optical objects so they can be identified with emitters in other parts of the spectrum, such as those identified in radio astronomy or with x-ray satellites.

Scientists are using the very high resolution of the Hubble Space Telescope to pinpoint the positions of a small number of stars even more accurately than was possible with Hipparcos.

Even with these impressive telescope satellites, stars not yet known to be nearby are undoubtedly lurking. One such star was discovered in 2000. At only about 13 light-years away, we have had to add it somewhere around 25th to the list of the 50 nearest known stars based on Hipparcos data that appears as Appendix 7. It was discovered in an infrared survey carried out at one of the European Southern Observatory's telescopes in Chile. A spectrum taken with the giant Keck telescope showed that it is an M9 dwarf, which means that it is intrinsically faint. This object's relatively large proper motion confirms that it is a nearby star, rather than a spectral class M9 giant at a larger distance. Since perhaps 70 per cent of all stars are M dwarfs, there may be over 100 such faint systems missing from the list in the Appendix.

To improve on the Hipparcos observations, NASA was to launch the U.S. Naval Observatory's Full-sky Astrometric Explorer (FAME), now cancelled. It was to give accurate positions of 40 million stars out to about 6000 light-years away, about a quarter of the way from Earth to the center of our galaxy. The Europeans don't want to give up their lead in astrometry, and will launch first a small mission and then, in 2012, a major one. The smaller mission, called DIVA for the German initials for German Interferometer for Multichannel Photometry and Astrometry (and also, in English, Double Interferometer for Visual Astrometry), will measure the positions and motions of about 35 million stars in our galaxy. Following its prospective 2003 launch, it will observe 300 times more objects than Hipparcos with a precision 5 times more accurate. The larger mission, GAIA (which actually is not an acronym; Gaia in Greek mythology was the goddess of the Earth), should eventually measure 30 times as many objects as DIVA with 20 times its precision, following its launch in 2012. So while the Hipparcos observations are limited to our local region of our galaxy, less than 1 per cent its diameter, within a decade or so we should have directly measured distances (that is, parallaxes) for objects in much of our galaxy.

## 25.2a Parsecs

Since parallax measures have such an important place in the history of astronomy, a unit of distance was defined in terms that relate to these measurements. If we were outside the solar system and looked back at the earth and the sun, they would appear to us to be separated in the sky. We could measure the angle by which they are separated. From twice as far away as Pluto, for example, when the earth and the sun were separated by the maximum amount (1 A.U.), they would be approximately 1° apart. As we go farther and farther away from the solar system, 1 A.U. subtends a smaller and smaller angle. When we go about 60 times farther away, 1 A.U. subtends only 1 arc min (60 arc min = 1°). From 60 times still farther, 1 A.U. subtends 1 arc sec (60 arc sec = 1 arc min). Note that the *angle* subtended by the astronomical unit is a measure of the *distance* we have gone (Fig. 25–4), similar to the way that we measure an object's parallax.

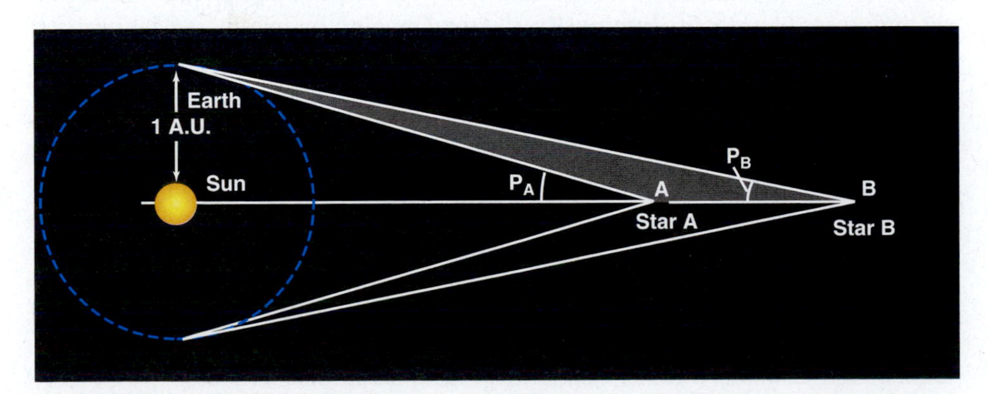


**FIGURE 25–4** Star B is farther from the sun than star A and thus has a smaller parallax. The parallax angles are marked $P_A$ and $P_B$. To **subtend** is to take up an angle; as seen from point A, 1 A.U. subtends an angle $P_A$. As seen from star B, the same 1 A.U. subtends an angle $P_B$.

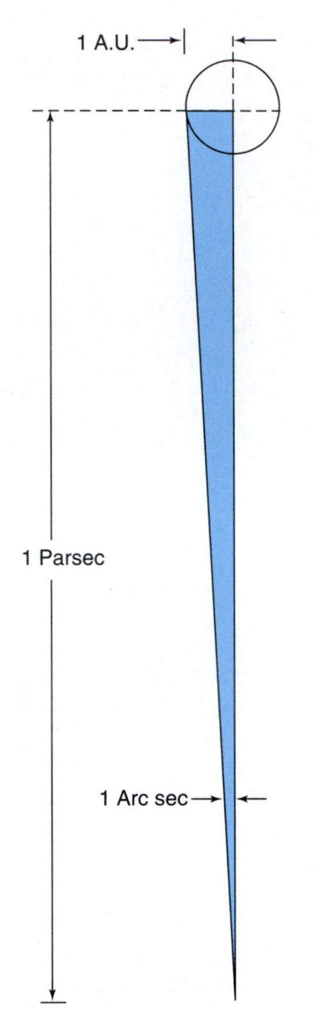


**FIGURE 25–5** A **parsec** is the distance at which 1 A.U. subtends an angle of 1 arc sec. The drawing is not to scale. An angle of 1 arc sec is actually tiny, the width of a dime at a distance of 4 km.

The distance at which 1 A.U. subtends only 1 arc sec we call **1 parsec** (Fig. 25–5). A star that is 1 parsec from the sun has a **par**allax of one arc **sec.** One parsec is a long distance. When talking about stellar distances, most astronomers tend to use parsecs instead of light-years, the distance that light travels in a year ($9.5 \times 10^{12}$ km). It takes light about 3.26 years to travel 1 parsec; thus 1 parsec = 3.26 light-years. Note that parsecs and light-years are both *distances,* just like kilometers or miles, even though their names contain references to their definitions in terms of angles or time. Similarly, if you travel at 65 miles per hour on the Interstate in your car for 3 hours, you have travelled 3 car-hours, which is equivalent to 195 miles. Note that a "car-hour" (given a constant speed) is a distance, just as a "light-year" is a distance (since the speed of light is a constant).

### 25.2b Computing with Parsecs*

The advantage of using parsecs instead of light-years, kilometers, or any other distance unit is that the distance in parsecs is equal to the inverse of the parallax angle in seconds of arc:

$$d(\text{parsecs}) = \frac{1}{p} \text{ (seconds of arc)}.$$

For example, a star with a parallax angle of 0.5 arc sec is 2 parsecs away ($1/0.5 = 2$), and it would thus be one of the nearest stars to the sun. A star with a parallax of 0.01 is $1/0.01 = 100$ parsecs away. Since the parallax angle is the quantity that is measured directly at the telescope, it was convenient to choose a distance unit very closely related to the parallax angle.

## 25.3 ABSOLUTE MAGNITUDES

We have thus far defined only the apparent magnitudes of stars, how bright the stars *appear* to us. However, stars can appear relatively bright or faint for either of two reasons: they could be intrinsically bright or faint, or they could be relatively close or far away (Fig. 25–6). We can remove the distance effect, giving us a measure of how bright a star actually is, by choosing a standard distance and considering how bright all stars would appear if they were at that standard distance. The standard distance that astronomers choose is 10 parsecs (about 33 light-years). We define the **absolute magnitude** of a star to be the magnitude that the star would appear to have if it were at a distance of 10 parsecs. Absolute magnitude describes the intrinsic brightness of a star, its **luminosity,** the total amount of energy the star gives off each second.

We normally write absolute magnitude with a capital $M$, and apparent magnitude with a small $m$. Sometimes a subscript signifies that we have observed through a special filter: $M_V$, for example, is the absolute magnitude through a special "visual" filter (which is approximately yellow).

If a star happens to be exactly 10 parsecs away from us, its absolute magnitude is exactly the same as its apparent magnitude. If we were to take a star that is farther than 10 parsecs away from us and somehow move it to be 10 parsecs away, then it would appear brighter to us than it does at its actual position (since it has been moved closer). Since the star would be brighter, its absolute magnitude would be a lower number (for example, 2 rather than 6) than its apparent magnitude.

On the other hand, if we were to take a star that is closer to us than 10 parsecs, and move it to 10 parsecs away, it would then be farther away and therefore fainter. Its absolute magnitude would be higher (more positive, for example, 10 instead of 6) than its apparent magnitude.

To assess just how much brighter or fainter that star would appear at 10 parsecs than at its real position, we must realize that the intensity of light from a star follows the **inverse-square law** (Fig. 25–7). That is, the intensity of a star varies inversely with the square of the distance of the star from us. (This law holds for all point sources of radiation, that is, sources that appear as points without length or breadth.) If we could move a star 2 times as far away, it would grow 4 times fainter. If we could move a star 9 times as far away, it would grow 81 times fainter, since 81 is 9 squared. Of course, we are not physically moving stars (we would burn our shoulders while pushing), but merely considering how they would appear at different distances. In all this, we are assuming that there is no matter in space between the stars to absorb light. Unfortunately, this is not always a good assumption (as we shall see in Chapter 33), but we make it anyway.

*A*

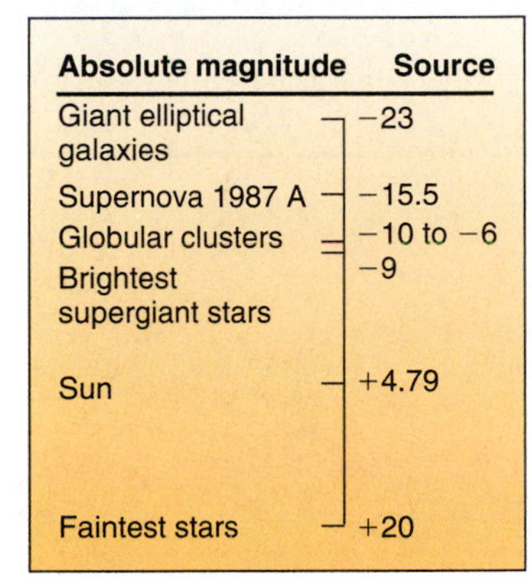

| Absolute magnitude | Source |
|---|---|
| Giant elliptical galaxies | −23 |
| Supernova 1987 A | −15.5 |
| Globular clusters | −10 to −6 |
| Brightest supergiant stars | −9 |
| Sun | +4.79 |
| Faintest stars | +20 |

*B*

**FIGURE 25–6** Some stars look bright because they are both intrinsically bright and close to us. A faint star could be intrinsically bright yet far away. A photograph of the sky shows stars at a wide range of distances and of intrinsic brightnesses. (*A*) Stars of different brightness—some are near, while others are far. The photo also shows nebulosity between the stars in the constellation Sagittarius. (*B*) The absolute magnitudes of some objects.

## EXAMPLE 25.2 Working from Distance and Apparent Magnitude

***Question:*** A star is 20 parsecs away from us, and its apparent magnitude is +4. What is its absolute magnitude?

***Answer:*** If the star were moved to the standard distance of 10 parsecs away, it would be twice as close as it was at 20 parsecs and, therefore, by the inverse-square law, would appear four times brighter. Since 2.5 times is 1 magnitude, and $(2.5)^2 = 6.25$ is 2 magnitudes, it would be approximately 1½ magnitudes brighter. Since its actual apparent magnitude is 4, its absolute magnitude would be $4 - 1\frac{1}{2} = 2\frac{1}{2}$, equivalent to the apparent magnitude it would have if it were 10 parsecs from us. (We see that we must subtract the magnitude difference, since the star would be closer and therefore brighter at 10 parsecs.)

As we just saw, if you know one type of magnitude and the distance, and want to find the other type of magnitude: (1) Find the ratio of the distances (the distance of the star and the 10-parsec standard distance—make sure the star's distance is also in parsecs so that you can properly take the ratio); (2) square to find the factor by which the brightness is changed; (3) convert this factor to a difference in magnitudes (using each 1 magnitude = a factor of 2.5; each 5 magnitudes = a factor of 100); (4) add or subtract the magnitude difference to change between absolute and apparent magnitude.

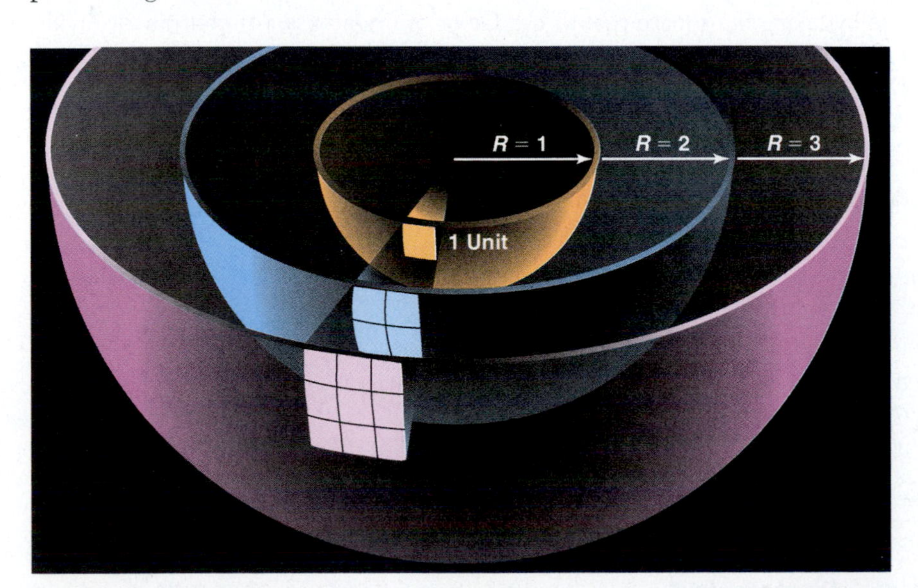


**FIGURE 25–7** The inverse-square law. Radiation passing through a sphere twice as far away as another sphere has spread out so that it covers $2^2 = 4$ times the area; $n$ times farther away it covers $n^2$ times the area.

A DEEPER DISCUSSION

## BOX 25.1 Relating Absolute and Apparent Magnitudes

The main method by which astronomers tell the distance to faraway objects is by comparing their apparent magnitudes (easily measured at the telescope) with their absolute magnitudes. The absolute magnitude is an intrinsic quantity, and is thus constant, while the apparent magnitude gets fainter the farther the object is away. We have just given examples of how to reason logically about this relation. The relation can be expressed mathematically:

$$m - M = 5 \log_{10}\left(\frac{d}{D}\right),$$

where the actual distance $d$ is in parsecs, as is the standard distance $D = 10$ pc. But don't let mere plugging into the formula allow you to lose track of what is going on.

To solve Example 25.2 using the formula, we are given $d = 20$ pc and $m = +4$. Thus $4 - M = 5 \log_{10}(20/10)$. The logarithm of 2 is about 0.3. Thus $M = 4 - (5 \times 0.3) = 4 - 1.5 = 2.5$.

Whenever you do a calculation, on a calculator or whatever, you should always check that the answer is reasonable. So for magnitude and distance calculations, you should be carrying out estimates with the methods given in the numbered examples, even if you use the formula given in this box.

### EXAMPLE 25.3 Deriving Distance from Absolute and Apparent Magnitude

***Question:*** A star has apparent magnitude 10 and absolute magnitude 5. How far away is it?

***Answer:*** The star would grow brighter by 5 magnitudes were it to be moved to the standard distance of 10 parsecs. It would thus brighten by 100 times. By the inverse-square law, it would do so if it were 10 times closer than its real position. Its distance must therefore be 10 times farther away than the standard distance of 10 parsecs, and 10 $\times$ 10 parsecs is 100 parsecs. Note that to find the distance in light-years, we must multiply by 3.26 parsecs/light-year. The distance here is thus 326 light-years.

As we just saw, if you know both types of magnitudes and want to find the distance: (1) Find the difference in magnitudes, by subtracting; (2) convert this difference in magnitudes into a factor by which the brightness is changed (using a factor of 2.5 = 1 magnitude; a factor of 100 = 5 magnitudes); (3) find the change in distance by taking the square root of the factor of brightness; (4) multiply or divide the standard distance by this factor of distance, multiplying if the star's apparent magnitude is fainter than its absolute magnitude (and, if necessary, converting from parsecs to light-years).

### EXAMPLE 25.4 Deriving Distance from Absolute and Apparent Magnitude

***Question:*** A star has apparent magnitude 10 and absolute magnitude 3. How far away is it?

(This is an example in which the numbers don't work out quite as smoothly.)

***Answer:*** The star would grow brighter by 7 magnitudes if we were to move it to the standard distance, making a factor of brightness of $(2.5)^2 \times 100 = 600$ (approximately). By the inverse-square law, it is thus growing closer by the square root of 600, which is about 25. (After all, $20^2 = 400$, and $30^2 = 900$, so the answer must be in between 20 and 30.) The star must thus actually be 25 $\times$ 10 parsecs, or 250 parsecs away.

## 25.3a Photometry*

The actual value of either absolute or apparent magnitude can depend on the wavelength region in which we are observing. Let us consider a blue star and a red star,

# A DEEPER DISCUSSION

## BOX 25.2 Linking Apparent Magnitude, Absolute Magnitude, and Distance

Astronomers have a formula that allows you to calculate one of the terms apparent magnitude, absolute magnitude, or distance, if you know the other two. But the formula merely does numerically what we have just carried out logically. The formula is $m - M = 5 \log_{10}(r/10)$, where $m$ is the apparent magnitude, $M$ is the absolute magnitude, and $r$ is the distance in parsecs. Note that if $r = 10$, then $r/10 = 1$, $\log 1 = 0$, and $m = M$. If $r = 20$, as in the first example above, $m - M = 5 \log 2 = 5 \times 0.3 = 1.5$ magnitudes. Be careful not to get carried away using the formula without understanding the point of the manipulations.

### EXAMPLE 25.5 Using Absolute and Apparent Magnitudes to Find Distance

*Question:* We see a G2 star of apparent magnitude +8 and are able to determine that it has absolute magnitude +5. How far away is it?

*Answer:* The absolute and apparent magnitudes differ by 3, which means that we raise 2.5 to the 3rd power to find the ratio of brightness. The star appears fainter than it would be if it were at 10 parsecs. It is approximately $(2.5)^3 = 6 \times 2.5 = 15$ times fainter. (More accuracy in carrying out this calculation would not be helpful because the original data were not more accurate.) By the inverse-square law, this means that it is approximately 4 times farther away (exactly 4 times would be a factor of 42 = 16 in brightness). Thus the star is $4 \times 10$ parsecs, or 40 parsecs, away. In Section 25.4, we will discuss the H-R diagram, which will allow us in most cases to find the absolute magnitude of a star whose spectral type we know.

#### Discussion Question

What are two possible disadvantages of using the formula given here without being able to manipulate the distances as shown in the text?

each of which gives off the same amount of energy. The blue star can seem much brighter than the red star if we observe them both through a blue filter, while the red star can seem much brighter than the blue if we observe them both through a red filter. Astronomers often measure a color index (Section 25.3b) by subtracting the magnitude in one spectral range from the magnitude in another. Since the color of a star gives its temperature, color index is a measure of temperature.

A standard set of filters (Fig. 25–8) has been defined for use in photoelectric systems, with one filter in the ultraviolet (called U), one filter in the blue (called B), and one filter in the yellow to more or less match the eye (called V for visual). Sets of equivalent filters exist at observatories all over the world. Hundreds of thousands

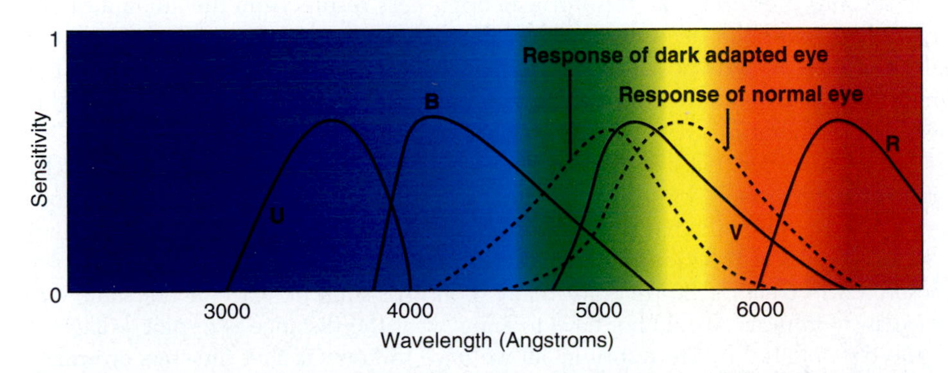


**FIGURE 25–8** The U, B, V, and R curves (ultraviolet, blue, visual = yellow/green, and red) represent a standard set of filters used by many astronomers. The response of the eye under normal conditions and the response of the dark-adapted eye are also shown. Filters often used in the visible and infrared and their central wavelengths in $\mu$m are U (0.36), B (0.44), V (0.55), R (0.70), I (0.90), J (1.25), H (1.6), K (2.2), L (3.4), M (5.0), and N (10.0), with further infrared extensions used by infrared satellites, soon to include the Space Infrared Telescope Facility (SIRTF).

of stars have had their colors measured with this **UBV** set of filters; we call the process **three-color photometry.** The name "photometry" is historic, since single-point devices called photometers were used, but now photometry is normally carried out with CCDs, which make images of the light passing through the selected filters.

A standard extension of this filter set into the red and infrared has been made. The bands correspond to windows of transparency in the earth's atmosphere. The bands are R, I, J, H, K, L, and so on. (Yes, "J, H, K" is the correct order, even though it isn't alphabetical.) The 2MASS infrared sky survey is being made in the J, H, and K bands, extending from 1.25 micrometers to 2.2 micrometers.

A "bolometer," the instrument used to measure the total amount of energy arriving in all spectral regions, derives its name from "boli," Greek for "beam of light."

The quantity of fundamental importance is not the magnitude as observed through any given filter (which is what we measure) but rather the magnitude that corresponds to the total amount of energy given off by the star over all spectral ranges. This is called the **bolometric magnitude,** $M_{bol}$.

### 25.3b Color Index

A measure of temperature often used in astronomy is the difference between the apparent magnitudes measured in two spectral regions, for example, B − V, blue minus visual. The standard system of filters was defined in the previous subsection.

The difference B − V is called the **color index** (Fig. 25–9). The blue magnitude, B, measures bluer radiation than the visual magnitude, V. A very hot star is brighter in the blue than in the visible; thus V is fainter, that is, a higher number, than B. Therefore B − V is negative for hot stars.

The B − V color index is zero for an A star of about 10,000 K. It falls in the range from −0.3 for the hottest stars to about +2.0 for the coolest. One can also compute a color index U − B for the U and B (ultraviolet and blue) magnitudes, or indeed a color index for magnitudes measured in any two spectral regions.

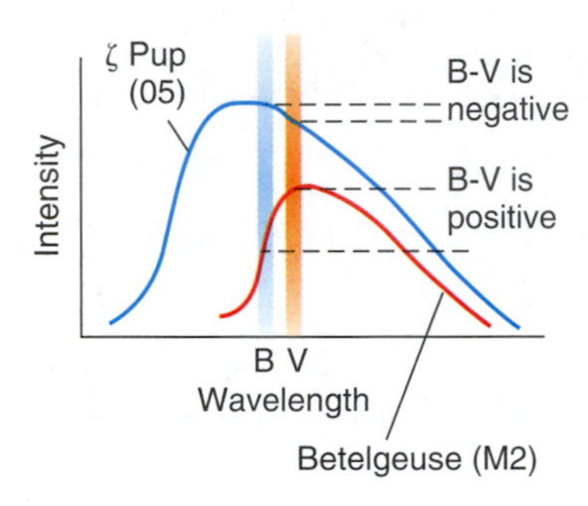


**FIGURE 25–9** Hotter stars have negative color indices while cooler stars have positive color indices. For example, the spectral-type O5 star is brighter in the blue B filter than in the "visual" V filter; that is, B is a lower magnitude than V, so B − V is less than 0. (For example, B of 2.0 less V of 2.3 gives B − V = −0.3.) Contrastingly, the spectral-type M2 star has B − V greater than 0. (B of 2.4 less V of 0.5 gives B − V = 1.9.)

## 25.4 THE HERTZSPRUNG-RUSSELL DIAGRAM

In about 1910, Ejnar Hertzsprung (Fig. 25–10) in Denmark and Henry Norris Russell (Fig. 25–11) at Princeton University in the United States independently plotted a new kind of graph (Fig. 25–12). On the horizontal axis, the $x$-axis, each graphed a quantity that measured the temperature of stars. On the vertical axis, the $y$-axis, each graphed a quantity that measured the intrinsic brightness of stars, the star's luminosity (which we defined earlier in this chapter as the amount of energy given off each second). For each star, this intrinsic brightness results from the amount of energy being generated deep inside. They found that all the points that they plotted fell in limited regions of the graph rather than being widely distributed over the graph. Thus stars tend to possess a limited set of the possible combinations of brightness and temperature.

But remember that a star can appear bright either by really being intrinsically bright or, alternatively, by being very close to us. One way to get around this problem is to plot only stars that are at the same distance away from us. How do we find a group of stars at the same distance? Luckily, there are clusters of stars in the sky (described in Chapter 26) that are really groups of stars in space at the same distance away from us. We do not have to know what the distance is to plot a diagram of the type plotted by Hertzsprung; all we have to know is that the stars are really clustering in space.

**FIGURE 25–10** Ejnar Hertzsprung in the 1930s.

If we somehow knew the absolute magnitudes of the stars, that would also be a good thing to plot. When we know the distances (for relatively close stars, we can get this from trigonometric parallax measurements, which is what Russell originally did), we can calculate the absolute magnitudes, which are actually what is plotted in Figure 25–12.

Such a plot of temperature versus brightness is known as a **Hertzsprung-Russell diagram,** or simply as an **H-R diagram.** Note that since H-R diagrams were sometimes originally plotted by spectral type, from O to M, the hottest stars are on the left side of the graph. Thus temperature increases from right to left. Also, since the brightest stars are on the top, magnitude decreases toward the top. In some sense, thus, both axes are plotted backwards from the way a reasonable person might now choose to do it if there were no historical reasons for doing it otherwise. When a color index is used on the horizontal axis to show temperature, we have a **color-magnitude diagram.**

When plotted on a Hertzsprung-Russell diagram, the stars lie mainly on a diagonal band from upper left to lower right (Fig. 25–13). Thus the hottest stars are normally brighter than the cooler stars. Most stars fall very close to this band, which is called the **main sequence.** Stars on the main sequence are called **dwarf stars,** or **dwarfs.** There is nothing strange about dwarfs; they are the normal kind of stars.

**FIGURE 25–11** Henry Norris Russell and his family, circa 1917. Prof. Russell was the subject of a major biography published in 2000.

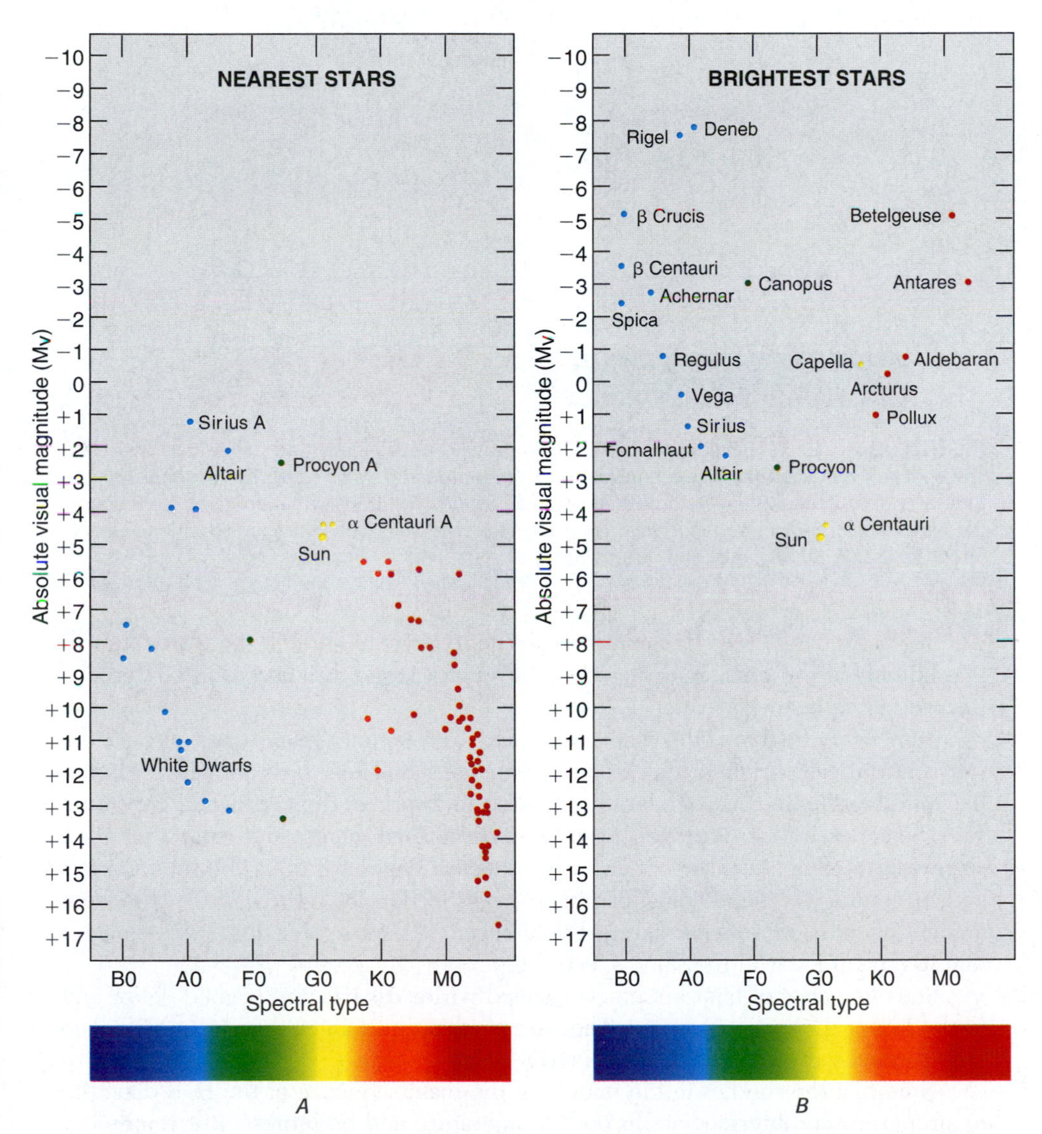


**FIGURE 25–12** Hertzsprung-Russell diagrams (*A*) for the nearest stars in the sky and (*B*) for the brightest stars in the sky. The brightness scale is given in **absolute magnitude.** Because the effect of distance has been removed, the intrinsic properties of the stars can be compared directly on such a diagram. Note that none of the nearest stars is intrinsically very bright. Also, the brightest stars in the sky are, for the most part, intrinsically very luminous, even though they are not usually the very closest to us. The color bars show the overall color of the star.

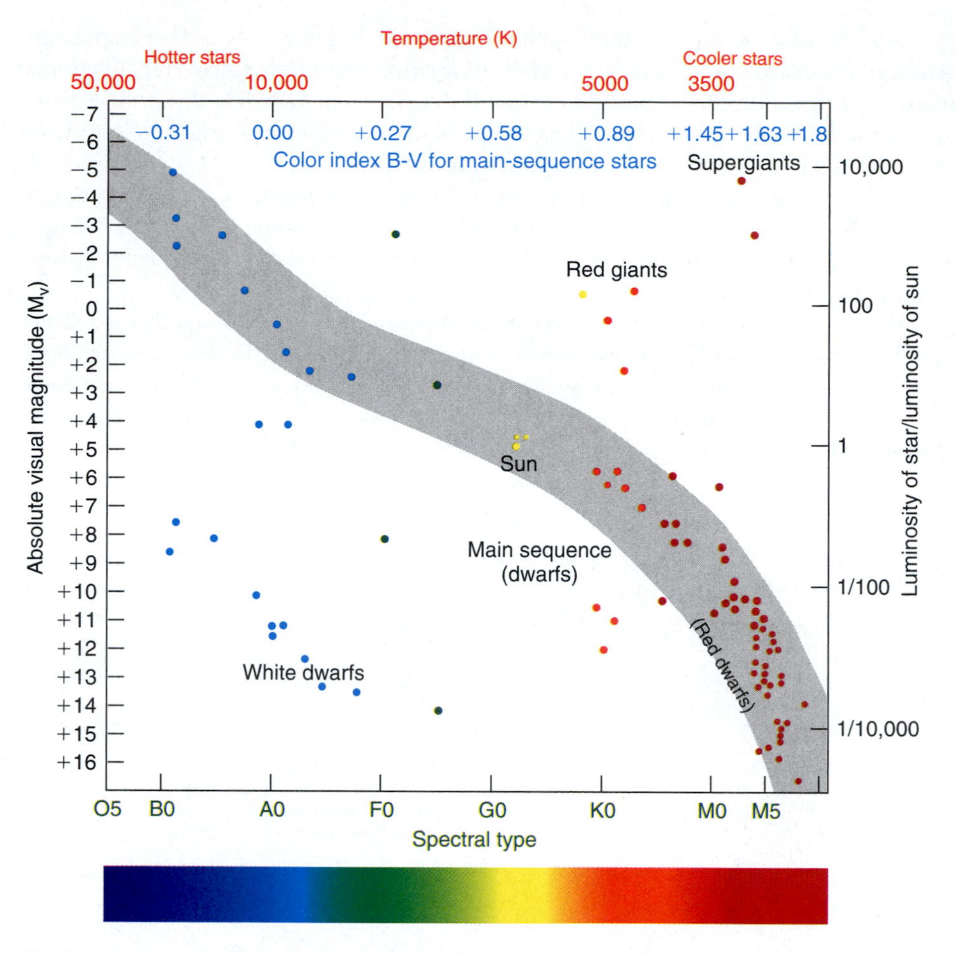


**FIGURE 25-13** A Hertzsprung-Russell diagram with both nearby and bright stars included. The spectral-type axis (*bottom*) is equivalent to the temperature axis (*top*). The absolute magnitude axis (*left*) is equivalent to the luminosity axis (*right*). The transformation given from spectral type to color index (B − V) is valid for main-sequence stars only; the correspondence is different for giants, supergiants, and other classes of stars.

The sun is a type G dwarf. Some dwarfs are quite large and bright; the word "dwarf" is used only in the sense that these stars are not a larger, brighter kind of star that we will define below as giant.

Some stars lie above the main sequence. That is, for a given spectral type the star is intrinsically brighter than a main-sequence star. These stars are called **giants,** because their luminosities are large compared to dwarfs of their spectral type. Some stars, like Betelgeuse, are even brighter than normal giants, and they are called **supergiants.** Since two stars of the same spectral type have the same temperature and, according to the Stefan-Boltzmann law (Section 24.1, Box 24.2), the same amount of emission from each area of their surfaces, the brighter star must be bigger than the fainter star of the same spectral type.

Stars in a class of faint hot objects, called **white dwarfs,** are located below and to the left of the main sequence. They are smaller and fainter than main-sequence stars (ordinary dwarfs) of the same spectral type.

Note that the sun lies in the middle of the main sequence of the H-R diagram; its properties are intermediate in both temperature and brightness. But remember that there are many more cool, faint stars than hot, bright ones. Of the 6000 stars we can see in the sky with unaided eye, only about 30 are less luminous than the sun—½ per cent. But if we consider a volume of space, of the 700 or so stars within 10 parsecs of us, over 96 per cent are less luminous than the sun. We must always be careful not to bias our conclusions by forgetting this point.

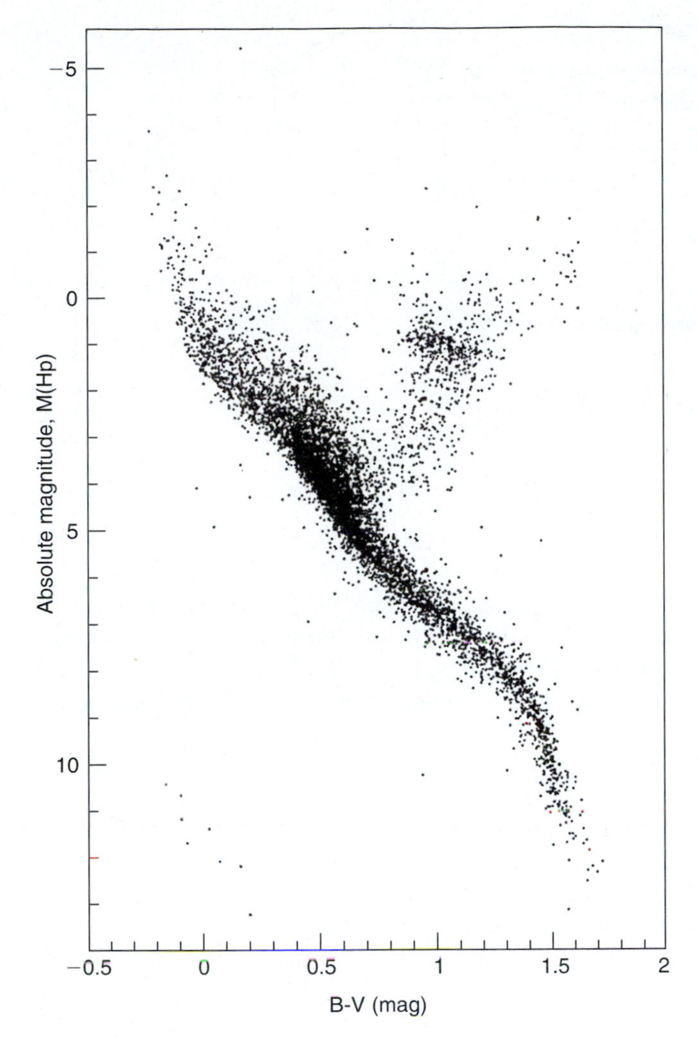


**FIGURE 25–14** The color-magnitude diagram of 8784 stars whose parallaxes are determined from the Hipparcos satellite to better than 10%. Six white dwarfs have been added at lower left.

The use of the Hertzsprung-Russell diagram to link the spectrum of a star with its brightness is a very important tool for stellar astronomers. The H-R diagram provides, for example, another way of measuring the distances to stars, as we shall see below.

H-R diagrams from the Hipparcos satellite are of unprecedented accuracy (Fig. 25–14). The Hipparcos and Tycho catalogues, compiled from Hipparcos observations, give millions of stellar distances, which allow the stars' luminosities to be calculated more accurately than ever before. We now await further spacecraft to carry the accurate distance measurements even farther out into our galaxy.

## 25.5 SPECTROSCOPIC PARALLAX

When we take a group of stars whose distances we can measure directly, by some method like that of trigonometric parallax, we can plot a Hertzsprung-Russell diagram. From this standard diagram we can read off the absolute magnitude that corresponds to any star on the main sequence.

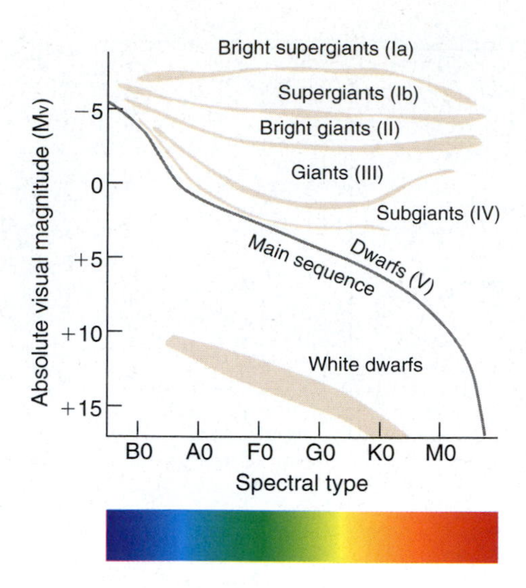


**FIGURE 25–15** One can tell whether a star is a dwarf, giant, or supergiant by looking closely at its spectrum, because slight differences exist for a given spectral type. For example, the absorption lines can be slightly broader. This procedure is called "luminosity classification." Once we know a star's luminosity class, we can place the star on the H-R diagram and find its spectroscopic parallax. Luminosity classes are specified by numbers (shown in parentheses) or names.

If we know that the star is on the main sequence, we can apply the standard Hertzsprung-Russell diagram to find the distance to any star whose spectrum we can observe, no matter how faint. Examining the spectrum in detail reveals to trained eyes whether or not a star is on the main sequence (Fig. 25–15). This "luminosity classification," with stars of the same spectral type having different brightnesses, was first spotted by subtle differences in sharpness of line in their spectra by Antonia Maury at the Harvard College Observatory about a hundred years ago.

For a main-sequence star, we first find the absolute magnitude that corresponds to its spectral type. We must also know the apparent magnitude of the star, but we can get that easily and directly simply by observing it. Once we know both the apparent magnitude and the absolute magnitude, we have merely to figure out how far the star has to be from the standard distance of 10 parsecs to account for the difference $m - M$ (using the inverse-square law).

We are measuring a distance, and not actually a parallax, but by analogy with the method of trigonometric parallax for finding distance, the method using the H-R diagram is called finding the **spectroscopic parallax.**

## 25.6 THE DOPPLER EFFECT

Astronomers use the Doppler effect (Section 4.8) to determine a star's **radial velocity,** its speed toward or away from us (on the radius of an imaginary sphere centered where we stand). The Doppler effect in sound is familiar to most of us, and its analogue in electromagnetic radiation, including light, is very similar.

Recall that when we say that an object is approaching us, its radiation is **blueshifted.** Conversely, when the emitting object is receding from us, the radiation is said to be **redshifted** (Fig. 25–16).

If we take the spectrum of the sun, with a spectrograph slit along the sun's equator, the result shows that the sun is rotating. One side is redshifted and the other is blueshifted.

### 25.6a Working with Doppler Shifts*

The wavelength when the emitter is at rest is called the **rest wavelength.** Let us consider a moving emitter. The fraction of the rest wavelength that the wavelength of light is shifted is the same as the fraction of the speed of light at which the body is travelling toward or away (or, for a sound wave, the fraction of the speed of sound).

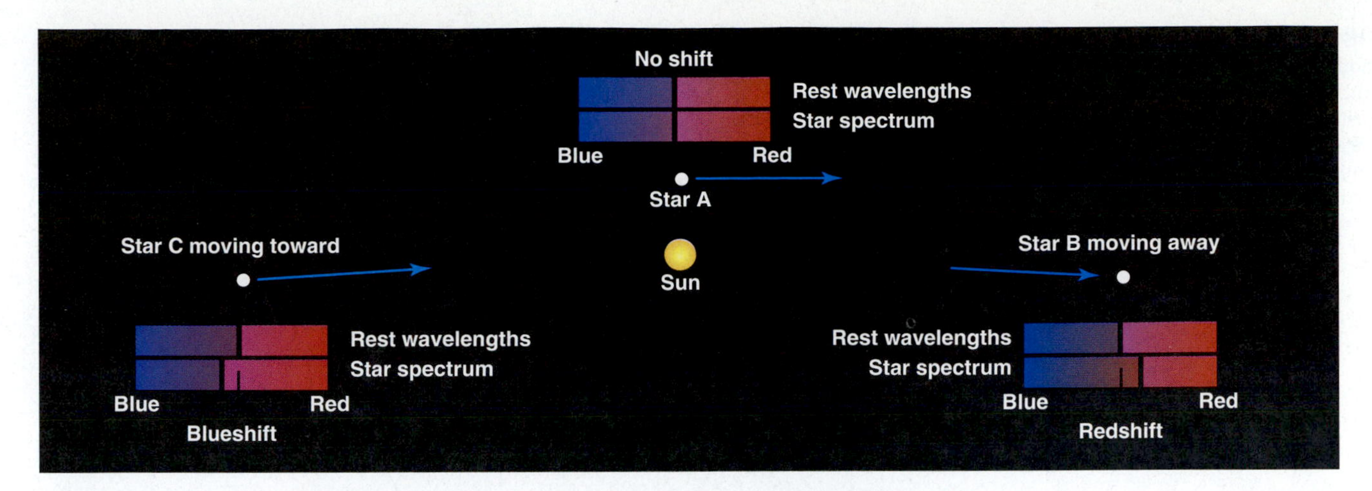


**FIGURE 25–16** The Doppler effect in stellar spectra as it appears for stars moving with respect to the sun and its nearby planet Earth. In each pair of spectra, the position of the spectral line in the laboratory is shown on top and its position observed in the spectrum of the star is shown below it. Lines from approaching stars appear blueshifted, lines from receding stars appear redshifted, and lines from stars that are moving transverse to us are not shifted because the star has no velocity toward or away from us. A short vertical line marks the unshifted position on the lower of each pair of spectra.

We can write

$$\frac{\text{change in wavelength}}{\text{original wavelength}} = \frac{\text{speed of emitter}}{\text{speed of light}}$$

or,

$$\frac{\Delta\lambda}{\lambda_0} = \frac{v}{c},$$

where $\Delta\lambda$ is the change in wavelength (the Greek capital delta, $\Delta$, stands for "the change in"), $\lambda_0$ is the original (rest) wavelength, $v$ is the speed of the emitting body along a radius linking us to it, and $c$ is the velocity of light ($= 3 \times 10^5$ km/sec). (We use the symbol $v$ for velocity instead of a symbol for speed since velocity is a term that signifies both speed and direction, and here the direction is known to be along a radius.) We define positive speeds as recession (redshifts) and negative speeds as approach (blueshifts). The new wavelength, $\lambda$, is equal to the old wavelength plus the change in wavelength, $\lambda_0 + \Delta\lambda$.

## EXAMPLE 25.6 Doppler Shift

***Question:*** A star is approaching at 30 km/sec (that is, $v = -30$ km/sec). At what wavelength do we see a spectral line that was at 6000 Å (which is in the orange part of the spectrum) when the radiation left the star?

$$\frac{\Delta\lambda}{\lambda_0} = \frac{v}{c} = \frac{-30 \text{ km/sec}}{3 \times 10^5 \text{ km/sec}} = \frac{-3 \times 10^3 \text{ km/sec}}{3 \times 10^5 \text{ km/sec}} = -10^{-4}$$

***Answer:*** Since $\Delta\lambda/\lambda_0 = -10^{-4}$, we have $\Delta\lambda = -10^{-4}\,\lambda_0 = -10^{-4} \times 6000.0$ Å $=$ $-0.6$ Å.

(Since the star is approaching, this change of wavelength is a blueshift, and the new wavelength is slightly shorter than the original wavelength.) The new wavelength, $\lambda$, is thus $\lambda_0 + \Delta\lambda = 6000.0$ Å $- 0.6$ Å $= 5999.4$ Å. It is still in the orange part of the spectrum. The change would be measurable using a sufficiently large spectrograph but would not be apparent to the eye.

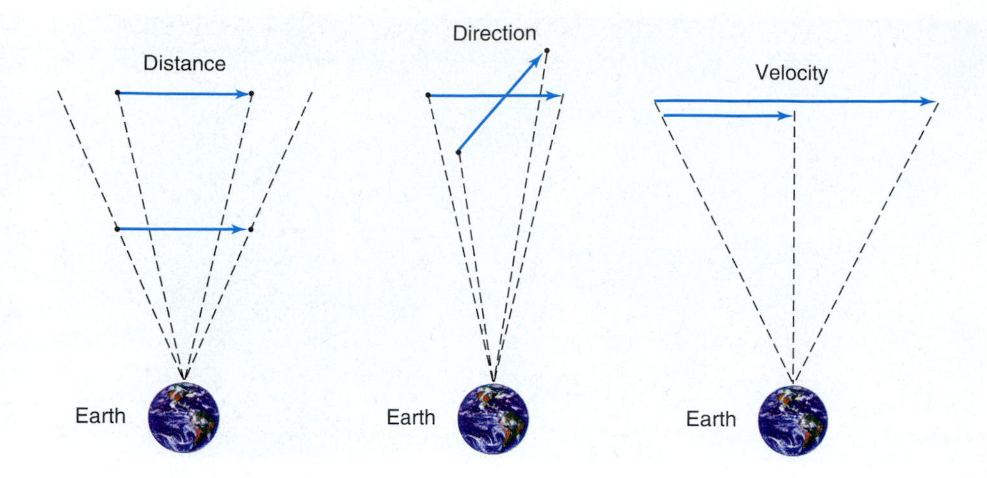


**FIGURE 25–17** The proper motion of a star depends on its distance from us (*left,* where equal velocities give different angles), the angle in which it travels (*middle,* where equal speeds in different directions give different angles), and its speed in space (*right*).

Note that since there are proportions on both sides of the equation, if we take care to use $v$ and $c$ in the same units (for example, km/sec, cm/sec, or whatever), the units of $v$ and $c$ will cancel out, and $\Delta\lambda$ will be in the same units that $\lambda$ is in, no matter whether that is angstroms, centimeters, or whatever.

The 30 km/sec given in the example is typical of the random speeds that stars have with respect to each other. These speeds are small on a universal scale, much too small to change the overall color that the eye perceives, but large compared to terrestrial speeds (30 km/sec = 30 km/sec × 3600 sec/hr = 108,000 km/hr).

## 25.7 STELLAR MOTIONS

On the whole, the network of stars in the sky is fixed. But radial velocities—velocities directly toward or away from us—can be measured for all stars using the Doppler effect. The Doppler shift allows us to measure only velocities toward or away from us and not to one side or another. If the object is moving at some angle to the radius of a circle having us at the center, then the Doppler shift measures only the part of the velocity (technically, the component of the velocity) in the radial direction (Fig. 25–17).

Further, some of the stars are seen to move slightly across the sky (that is, apparently from side to side) with respect to the more distant stars.

We usually deal separately with the part of the velocity of a star that is toward or away from us (the radial velocity), and the **angular velocity** of the star across the sky (how fast the object is moving across the sky in units of angle). The angular velocity is called the **proper motion** (Fig. 25–18), as we saw earlier in this chapter in the discussion of measurements from the Hipparcos satellite. Radial velocity (measured from the Doppler shift) and proper motion are perpendicular to each other. To make accurate measurements of star positions, such as those made by the Hipparcos spacecraft, astronomers must also measure the proper motion. Otherwise the movement from this cause could distort the parallax measurements.

The star with the largest proper motion was discovered in 1916 by E. E. Barnard and is known as Barnard's star. It moves across the sky by $10^1/4$ arc sec per year. Since the moon appears half a degree across, Barnard's star moves across the sky by the equivalent of the diameter of the moon in only 180 years.

Note that the Doppler effect gives you motion in only one dimension and proper motion gives you motion in only two dimensions. The actual velocity of a star in three-dimensional space, with respect to the sun, is called its **space velocity,** but when we detect a star's change in position in the sky, we know only through what angle it moved. We cannot tell how far it moved in linear units (like kilometers or light-years) unless we also happen to know the distance to the star.

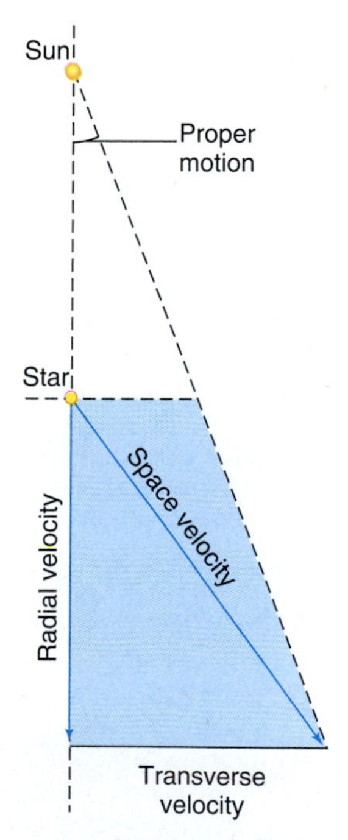


**FIGURE 25–18** If we know a star's proper motion and its distance, we can compute its linear velocity through space in the direction across our field of view. The Doppler effect gives us its linear velocity toward or away from us. These two velocities can be combined to tell us the star's actual velocity through space, its **space velocity.**

## CORRECTING MISCONCEPTIONS

**✖ Incorrect**

A bright star is close to us.

A dim star is far from us.

5 magnitudes is 100 times brighter, so 10 magnitudes is 200 times.

Stars are fixed in the sky.

**✔ Correct**

It can be intrinsically very dim but close.

It can be intrinsically bright but far.

Multiply factors, so 10 magnitudes is 100 × 100 = 10,000 times.

Many stars have measurable proper motions.

## SUMMARY AND OUTLINE

Magnitudes (Section 25.1)
- Lower (or negative) numbers correspond to the brightest objects.
- Five magnitudes' difference = 100 times in brightness; adding or subtracting magnitudes corresponds to dividing or multiplying brightness.
- Brightest star is magnitude −1.5; faintest stars visible to the naked eye are about 6th magnitude.

Trigonometric parallax (Section 25.2)
- We triangulate, using the earth's orbit as a baseline. We find the angle through which a nearby star appears to shift; the more the shift, the nearer the star.
- This method works only for the nearest stars; Space Telescope and Hipparcos have improved our capabilities.
- A *parsec* is the distance of a star from the earth when the radius of the earth's orbit *subtends* 1 arc sec when viewed from the star.

Absolute magnitude (Section 25.3)
- If we know how bright a star *appears,* and how far away it is, we can calculate how intrinsically bright it is. This can be placed on the same scale as that of *apparent magnitudes,* and is known as the *absolute magnitude.*
- The absolute magnitude is defined as the magnitude a star would appear to have if it were at a distance of 10 parsecs.
- The apparent brightness of a point object follows the *inverse-square law:* brightness decreases with the square of the distance.
- We can use the inverse-square law to relate apparent magnitude, absolute magnitude, and distance.

The Hertzsprung-Russell diagram (Section 25.4)
- A plot of brightness versus temperature
- Stars fall in only limited regions of the graph. The *main sequence* contains most of the stars; these stars are called *dwarfs.*
- *Giants* are brighter (and bigger) than dwarfs; *supergiants* are brighter and bigger still. *White dwarfs* are fainter than dwarfs, and so they fall below the main sequence.

Method of *spectroscopic parallax* (Section 25.5)
- Observing the spectrum of a star tells us where it falls on an H-R diagram; this tells us its absolute magnitude. Since we can easily observe its apparent magnitude, we can derive its distance.

The Doppler effect (Section 25.6)
- Radiation from objects that are receding is shifted to longer wavelengths: *redshifted.*
- Radiation from objects that are approaching is shifted to shorter wavelengths: *blueshifted.*

Stellar motions (Section 25.7)
- The *radial velocity* is measured from the Doppler effect.
- The *angular velocity* from side to side is measured by observing the *proper motion*—the motion of the star across the sky.
- This can be observed only for the nearest stars.

## KEY WORDS

apparent magnitude
significant figure
baseline
Astronomical Unit
parallax
trigonometric parallax
astrometry
subtend
parsec
absolute magnitude
luminosity
inverse-square law
color index*
UBV*
three-color photometry*
bolometric magnitude*
Hertzsprung-Russell diagram
H-R diagram
color-magnitude diagrams
main sequence
dwarf stars
dwarfs
giants
supergiants
white dwarfs
spectroscopic parallax
radial velocity
blueshifted
redshifted
rest wavelength
angular velocity
proper motion
space velocity

*This term appears in an optional section.

## QUESTIONS

1. Which star is visible to the naked eye: one of 4th magnitude or one of 8th magnitude?
2. Which is brighter: Mars when it is magnitude +0.5 or Spica, a star whose magnitude is +0.9 (Appendix 5)?
†‡3. Venus can reach magnitude −4.4, while the brightest star, Sirius, is magnitude −1.4. How many times brighter is Venus at its maximum than is Sirius?
†4. Venus can be brighter than magnitude −4. Antares is a first-magnitude star ($m = +1$). How many times brighter is Venus at magnitude −4 than Antares?
†5. Pluto is about 14th magnitude at most. How many times fainter is it than Venus? ($m_{\text{Venus}} = -4$, approximately.)
†6. If a variable star brightens by a factor of 15, by how many magnitudes does it change?
†7. If a variable star starts at 5th magnitude and brightens by a factor of 60, at what magnitude does it appear?
†8. The variable star Mira ranges between magnitudes 9 at minimum and 3 at maximum. How many times brighter is it at maximum than at minimum?
†9. Star A has magnitude +11. Star B appears 10,000 times brighter. What is the magnitude of star B? Star C appears 10,000 times fainter than star A. What is its magnitude?
†10. Star A has magnitude +10. The magnitude of star B is +5 and of star C is +3. How much brighter does star B appear than star A? How much brighter does C appear than B?
†11. How much brighter is a 0th-magnitude star than a +3rd-magnitude star?
12. You are driving a car, and the speedometer shows that you are going 80 km/hr. The person next to you asks why you are going only 75 km/hr. Explain why you each saw different values on the speedometer.
13. Would the parallax of a nearby star be larger or smaller than the parallax of a more distant star? Explain.
†14. A star has an observed parallax of 0.2 arc sec. Another star has a parallax of 0.02 arc sec. (a) Which star is farther away? (b) How much farther away is it?
†15. (a) What is the distance in parsecs to a star whose parallax is 0.05 arc sec? (b) What is the distance in light-years?
†16. Estimate the farthest distance for which you can detect parallax by alternately blinking your eyes and looking at objects at different distances. (You may find it useful to look through a window to outdoor objects, so that you can see them silhouetted against a very distant background.) To what angle does this correspond?
†17. Vega is about 8 parsecs away from us. What is its parallax?
18. What are the two fundamental quantities that are being plotted on the axes of a Hertzsprung-Russell diagram?
19. What is the significance of the existence of the main sequence?
20. What is the observational difference between a dwarf and a white dwarf?
21. Two stars have the same apparent magnitude and are the same spectral type. One is twice as far away as the other. What is the relative size of the two stars?
†22. Two stars have the same absolute magnitude. One is ten times farther away than the other. What is the difference in apparent magnitudes?
23. A star has apparent magnitude of +5 and is 100 parsecs away from the sun. If it is a main-sequence star, what is its spectral type? (*Hint:* Refer to Fig. 25–13.)
†24. A star is 30 parsecs from the sun and has apparent magnitude +2. What is its absolute magnitude?
†‡25. A star has apparent magnitude +9 and absolute magnitude +4. How far away is it?
†26. The brightest star in the nearest star system, alpha Centauri, has apparent magnitude 0 and absolute magnitude +4.4. How far away is it in parsecs? In light-years?
†27. Betelgeuse is apparent magnitude 0.4 and absolute magnitude −5.6. How far away is it in parsecs? In light-years?
†28. The first quasar to be discovered, 3C 273 (Chapter 36), was identified with what appeared to be a star of magnitude 13. It turned out to be 1 billion parsecs away. When the discoverer did the calculation, what absolute magnitude did it turn out to be? How many times brighter is it than the sun, whose absolute magnitude is about +5? The result astonished scientists.
†29. A jet plane travels 800 km/hr. Convert this speed to km/sec, and compare it with the speed of light, $3 \times 10^5$ km/sec. A red LED on a dial on the plane emits at 6600 Å. At what wavelength would we see it if we could detect it from the ground as the plane recedes from us?
†‡30. A star is receding from us at 1000 km/sec. The H$\alpha$ line ordinarily appears at 6563 Å. At what wavelength would it appear in the star's spectrum as seen from Earth?
†31. The 5250-Å spectral line from iron, used for measuring magnetic fields in the Sun, appears at 5252 Å in the spectrum of a star. At what velocity is the star moving with respect to us?
32. If a star is moving away from the earth at very high speed, will the star have a continuous spectrum that appears hotter or cooler than it would if the star were at rest? Explain.
33. (a) What does the proper motion of Barnard's star indicate about its distance from us? (b) What would Barnard's star's proper motion be if it were twice as far away from us as it is?
34. A star has a proper motion of 10 arc sec per century. Can you tell how fast it is moving in space in km/sec? If so, describe how.
35. Two stars have the same space velocity in the same direction, but star A is 10 parsecs from the earth, and star B is 30 parsecs from the earth. (a) Which has the larger radial velocity? (b) Which has the larger proper motion?

†This question requires a numerical solution.
‡Answers to selected questions: #3: $(2.512)^3 = 15$ times; #25: 100 parsecs; #30: 6585 Å

## USING TECHNOLOGY

### W World Wide Web

1. Look on the Space Telescope Science Institute's Web site for images of various stars, and note their magnitudes and how they compare with what we can see with the naked eye.
2. Find out how observing Doppler shifts has led to the discovery of planets around distant stars.

### REDSHIFT

1. Find Sirius, the brightest star in the sky, when it rises and sets today. Try to observe it.
2. Check the brightnesses of each of the stars in the Big Dipper and the Little Dipper, and observe them, noting how they appear. Compare the distances to each of those stars and comment on the relation to their brightnesses.
3. Find Proxima Centauri in *RedShift,* and comment on whether you can observe it from your location and why. Zoom in on its star system.
4. Check the absolute magnitudes, apparent magnitudes, and distances for the stars shown in the Hertzsprung-Russell diagram given in this chapter, and consider the differences between those for the nearest stars and the brightest stars.
5. Use the object filter for stars to observe the sky with stars up to the naked-eye limit. Next set the limit to 3rd magnitude, to approximate moderately light-polluted skies. Compare.
6. To learn about triangulation, look at Astronomy Lab: Ancient Astronomers: Parallax.
7. To learn about the Doppler shift, look at Principia Mathematica: Doppler Shift and Astronomy Lab: Galaxies: Doppler shift.

M15, a globular cluster in the constellation Pegasus, is visible to the naked eye as a hazy patch. It contains about 100,000 stars.

# Chapter 26

# Doubles, Variables, and Clusters

We often think of stars as individual objects that shine steadily, but many stars vary in brightness and most stars actually have companions close by. Also, many stars appear as members of groupings called clusters. By studying the effects that the stars have on each other, or by studying the nature of all the stars in a star cluster, we can learn much that we could not discover by studying the stars one at a time. This information even leads us to a general understanding of the life history of the stars.

**AIMS:** To discuss double stars, variable stars, and stellar clusters, and to draw conclusions about how to find their distances, their masses, and their ages

## 26.1 BINARY STARS

Most of the objects in the sky that we see as single "stars" really contain two or more component stars. Sometimes a star appears double merely because two stars that are located at different distances from the sun appear in the same line of sight. Such systems are called **optical doubles,** and they will not concern us here. We are more interested in stars that are physically associated with each other. We will use the terms **double star** and **binary star** interchangeably to mean two or more stars held together by the gravity they have between each other. (Even systems with three or more stars are usually informally called "double stars.") These double stars may form as separate components coalescing from a primordial stellar nebula, though some theoreticians think they can also form as a disk of gas and dust around a star gives birth to one or more companions.

The easiest way to tell that more than one star is present is by looking through a telescope of sufficiently large aperture. Stars that appear double when observed directly are called **visual binaries** (Fig. 26–1). The resolution of small telescopes may not be sufficient to allow the components of a double star to be "separated" from each other; larger telescopes can thus distinguish more double stars. When the components of a double star have different colors, they are particularly beautiful to observe, even with small telescopes. Five to ten per cent of the stars in the sky are visual binaries. Sometimes the components of visual binaries are separated so far that their period of revolution is 100 years or more.

**FIGURE 26–1** Albireo ($\beta$ Cygni) contains a B star and a K star, which make a particularly beautiful pair because of their different colors. Albireo is high overhead on summer evenings.

Astronomers can detect that stars are double in some cases even when they are not visual binaries—even those that only the Hubble Space Telescope can resolve. Even if a star appears as a single object through a telescope, if we can detect the presence of two sets of spectral lines from stars of different spectral types—of, say, one hot star and one cool star—then we say that the object has a **composite spectrum.**

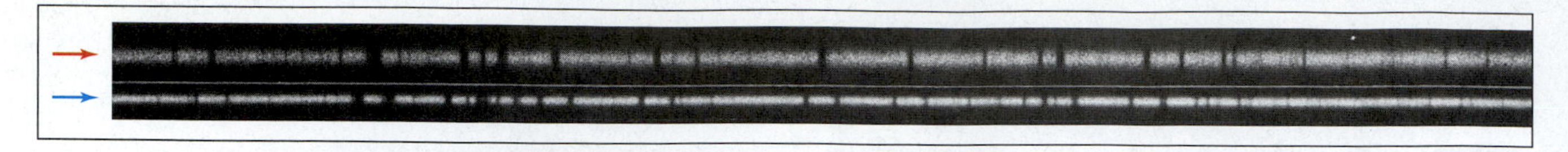

**FIGURE 26–2** Two spectra of $\kappa$ Arietis taken several days apart show that it is a spectroscopic binary. The lines of both stars are superimposed in the upper stellar absorption spectrum (*red arrow*) but are separated in the lower spectrum (*blue arrow*).

We can tell that a second star is present, even when an image and its spectrum appear single, if the spectral lines we observe change in wavelength with time. Such wavelength changes can occur only as the result of variable Doppler shifts, indicating that the speed of an object along a line linking it with us is changing. We deduce that two or more stars are present and are revolving around each other. Such an object is called a **spectroscopic binary** (Fig. 26–2). Each star in the system spends half of its orbit approaching us and the other half receding from us, relative to its average space motion. The variations of velocity in spectroscopic binaries are periodic. Note that even if the spectrum of one of the stars is too faint to be seen (making the system a "single-line spectroscopic binary," since only one set of spectral lines is observed), we can still tell that the star is a spectroscopic binary if the radial velocity of the visible component varies (Fig. 26–3). This technique has led to the discovery of planets around several stars, as we discussed in Chapter 18.

Careful spectroscopic studies have shown that two-thirds of all solar-type stars have stellar companions. Though the presence of companions of stars of other spectral types has not been studied in such detail, it seems that about 85 per cent of all stars are members of double-star systems. Few stars are single, an idea that is verified by noticing that many of the nearest stars to our sun (the latest list, based on Hipparcos satellite observations, appears as Appendix 7), stars we can study relatively well, are double. Often, mass can flow from one member of a binary system to the other, changing the evolution of each.

We can detect visual binaries most easily when the stars are relatively far apart. Then the period of the orbit is relatively long, over 100,000 years in most cases. So we know little about the motions in such systems because of our short human lifetime, even though most double stars discovered are visual binaries. It is harder to detect spectroscopic binaries, so we know fewer of them. But for this group, most of whose periods are between one day and one year, we are able to determine such important details of the orbit as size and period.

Sometimes the components of a double star pass in front of each other, as seen from our viewpoint on the earth. The "double star" then changes in brightness periodically, as one star cuts off the light from the other. Such a pair of stars is called an **eclipsing binary** (Fig. 26–4). The easiest to observe is Algol, $\beta$ Persei (beta of Perseus), in which the eclipses take place every 69 hours, changing the total brightness of the system within an hour from magnitude 2.3 to 3.5 and back. (A third star

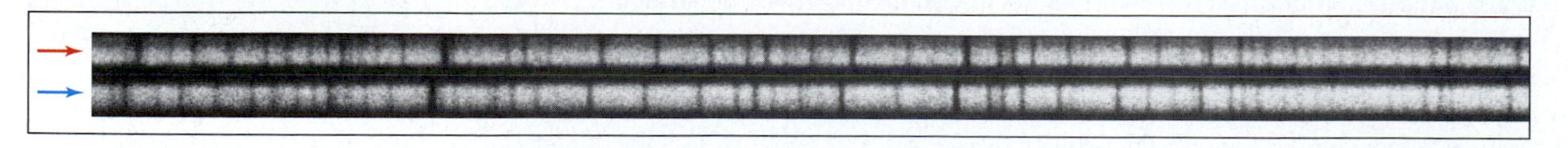

**FIGURE 26–3** Two spectra of Castor B ($\alpha$ Geminorum B) taken at different times show a Doppler shift. Thus the star is a spectroscopic binary, even though lines from only one of the components can be seen.

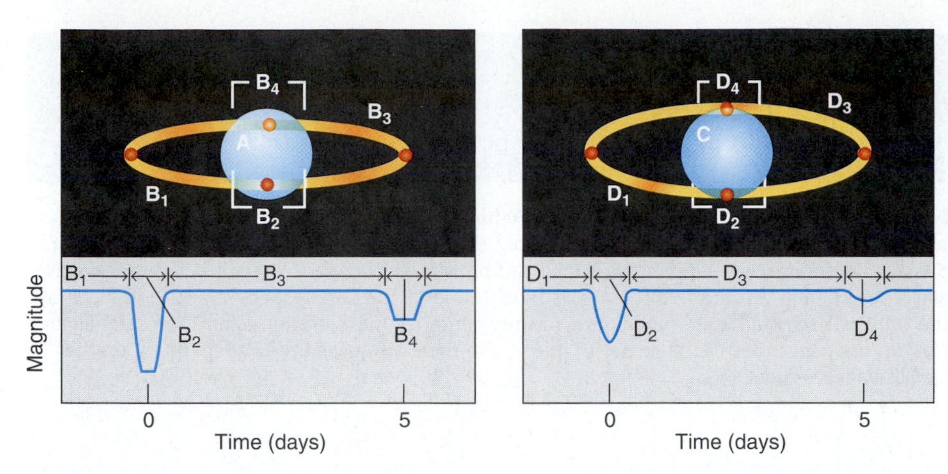


**FIGURE 26–4** Plots of the brightness of a star versus time are called **light curves.** (The graphs are usually in terms of magnitude, which decreases as brightness increases—upward on the graphs.) As the diagrams show, the shape of the light curve of an eclipsing binary depends on the sizes of the components and the angle from which we view them. At lower left, we see the light curve that would result for star B orbiting star A, as pictured at upper left. When star B is in the positions shown with subscripts at top, the regions of the light curve marked with the same subscripts result. The eclipse at $B_4$ is total. At right, we see the appearance of the orbit and the light curve for star D orbiting star C, with the orbit inclined at a greater angle than at left. The eclipse at $D_4$ is partial. From Earth, we observe only the light curves, and we use them to determine what the binary system is really like, including the inclination of the orbit and the sizes of the objects.

is present in the Algol system. It orbits the other two every 1.86 years and does not participate in the eclipses.)

Note that the way a binary star appears to us depends on the orientation of the two stars not only with respect to each other but also with respect to the earth (Fig. 26–5). If we are looking down at the plane of their mutual orbit, then we might

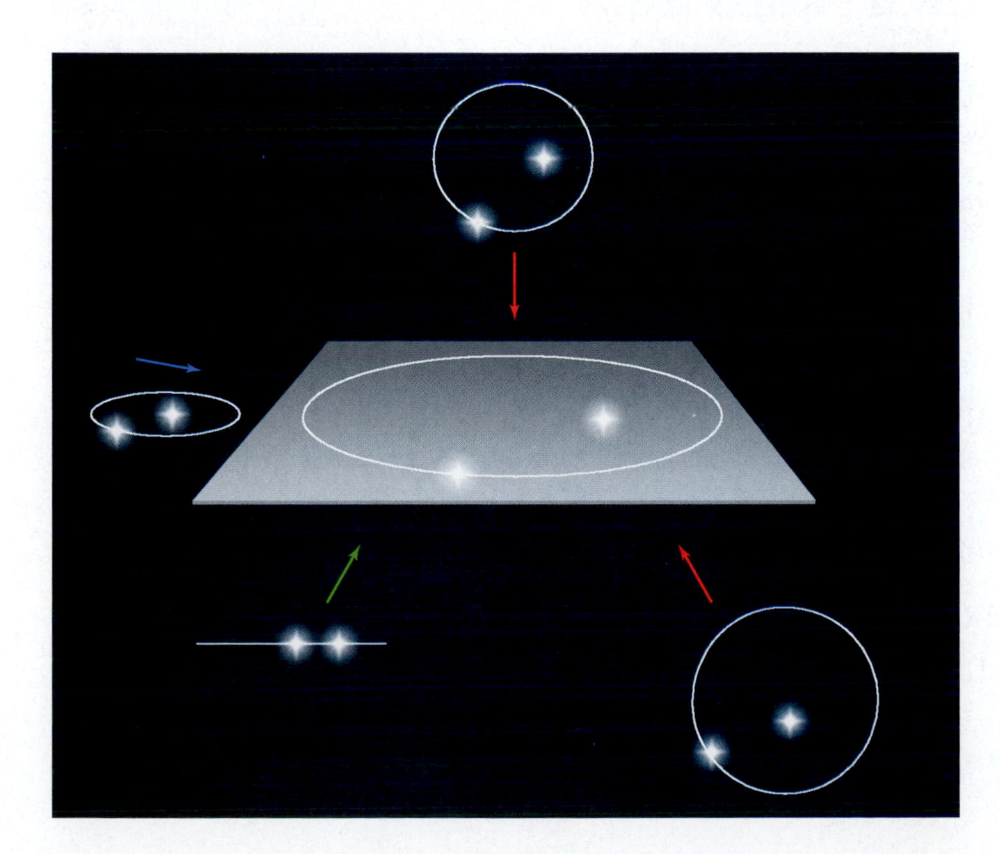

**FIGURE 26–5** The appearance and the Doppler shift of the spectrum of a binary star depend on the angle from which we view the binary. From far above or below the plane of the orbit, we might see a visual binary, as shown on the top and at the lower right of the diagram. From close to but not exactly in the plane of the orbit, we might see only a spectroscopic binary, as shown at left. (The stars appear closer together, so they might not be visible as a visual binary.) From exactly in the plane of the orbit, we would see an eclipsing binary, as shown at lower left. It could also be a visual binary.

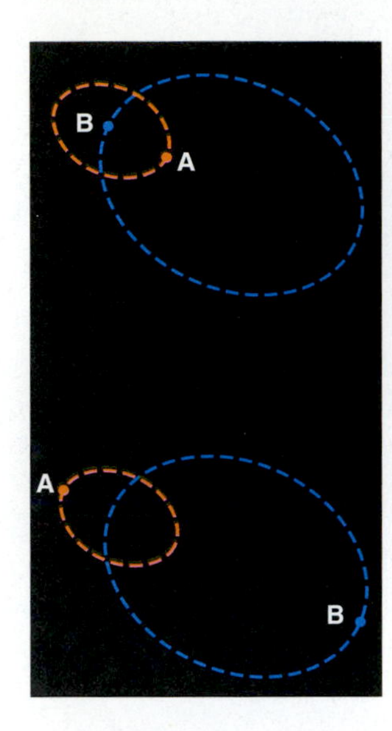


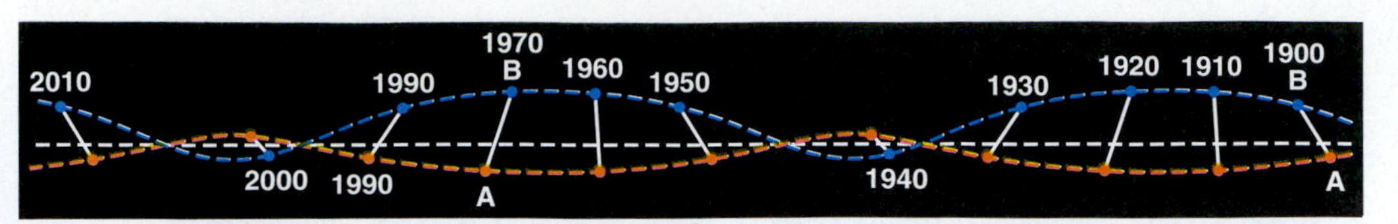


**FIGURE 26–6** Sirius A and B, an astrometric binary. Each is shown by a dot. From studying the motion of Sirius A (often called, simply, Sirius) astronomers deduced the presence of Sirius B before it was seen directly. Sirius B's orbit is larger because Sirius A is a more massive star. They contain 2.14 and 1.05 times as much mass as the sun, respectively. At one focus of the orbit of each star is the point known as the center of mass of the system. The stars always stay on opposite sides of the center of mass, with their distances from it dependent on the stars' relative masses.

see a composite spectrum; under the most favorable conditions the star might be seen as a visual binary. But in this orientation we would never be able to see the stars eclipse. We would also not be able to see the Doppler shifts typical of spectroscopic binaries, since only radial velocities contribute to the Doppler shift.

It is possible that a star could be a "double," but still not be detectable by any of the above methods. Sometimes, the existence of a double star shows up only as a deviation from a straight line in the proper motion of the "star" across the sky. Such stars are called **astrometric binaries** (Fig. 26–6).

A double star may fall into more than one category. For example, two stars that eclipse each other must be a spectroscopic binary, too.

Many of the celestial sources of x-rays that have been observed in recent years from orbiting telescopes turn out to be binary systems (Section 30.8). Matter from one member of the pair falls upon the other member, heats up, and radiates x-rays.

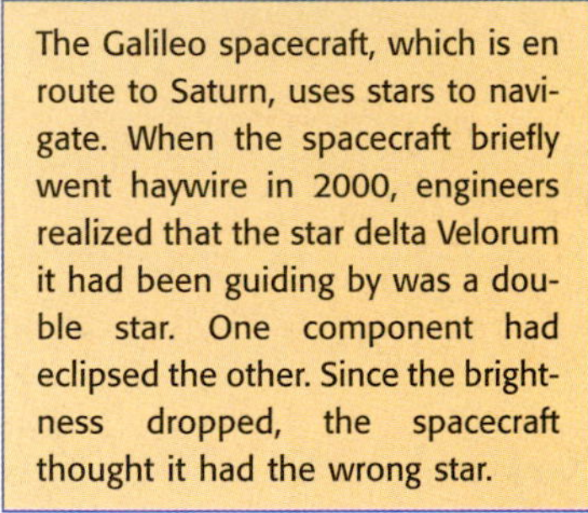
The Galileo spacecraft, which is en route to Saturn, uses stars to navigate. When the spacecraft briefly went haywire in 2000, engineers realized that the star delta Velorum it had been guiding by was a double star. One component had eclipsed the other. Since the brightness dropped, the spacecraft thought it had the wrong star.

## 26.2 STELLAR MASSES

The study of binary stars is of fundamental importance in astronomy because it allows us to determine stellar masses. If we can determine the orbits of the stars around each other, we can calculate theoretically the masses of the stars necessary to produce the gravitational effects that lead to those orbits. For a star that is a visual binary with a sufficiently short period (only 20 or even 100 years, for example), we are able to determine the masses of both of the components.

The method used is Newton's generalization of Kepler's third law, which we saw in Chapter 3. If you look back at that formula, you see that it includes a term with the sum of the masses of the two objects. While for planets, the mass of the planet is much less than that of the star, and so can essentially be ignored, in the case of binary stars, both objects can be close or even identical in mass. As we saw in Chapter 18 on the new planets discovered around other stars, only if we can determine the angle at which we are viewing the system do we know the actual mass.

We are more limited if a star is only a spectroscopic binary, even one with spectral lines of both stars present. If the star is not also a visual binary, we do not know the angle at which we are viewing the system, so can find only the lower limits for the masses—that is, we can say that the masses must be larger than certain values. Only if the spectroscopic binary is also an eclipsing binary do we know the angle of inclination, the angle at which the plane of the orbits of the member stars is inclined to our view, since the eclipses would not take place unless this angle were close to zero. Only if we know this angle can we find the individual masses.

From studies of the several dozen binaries for which we can accurately tell the masses, astronomers have graphed the luminosities (intrinsic brightnesses) of the

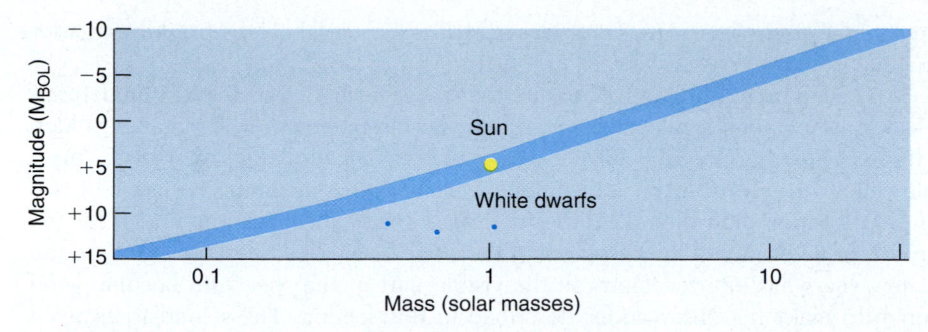


**FIGURE 26–7** The mass-luminosity relation, measured from binary stars. Most of the stars fit within the shaded band. The relation is defined by straight lines of different slopes for stars brighter than and fainter than magnitude +7.5. We do not know if there is a real difference between the brighter and fainter stars that causes this or whether it is an effect introduced when the data are reduced, at which stage we try to take account of the energy that the stars radiate outside the visible part of the spectrum. Note that white dwarfs, which are relatively faint for their masses, lie below the mass-luminosity relation. Red giants lie above it. Thus we must realize that the mass-luminosity relation holds only for main-sequence stars.

stars on one axis and the masses on the other axis. Most of the stars turn out to lie on a narrow band, the **mass-luminosity relation** (Fig. 26–7).

The mass-luminosity relation is valid only for stars on the main sequence. The more massive a main-sequence star is, the brighter it is. The mass, in fact, is the prime characteristic that determines where on the main sequence a star will settle down to live its lifetime. That is, the mass of a given star determines the temperature and luminosity the star has when it reaches the main sequence.

The most massive stars we know are about 50 times more massive than the sun, and the least massive stars have about 7 per cent the mass of the sun. From the theory of how stars reach different parts of the main sequence of the Hertzsprung-Russell diagram, we know that the most massive stars are hot stars, O and B stars and, to a lesser degree, A stars. Since these stars are bluish in color, we know that the most bluish stars we see in the sky are hot and massive. Sirius, Procyon, and Rigel, for example, which are early type A (that is, types A1 or A0) or late type B stars (that is, type B8), are all very massive.

## 26.3 STELLAR SIZES*

Stars appear as points to the naked eye and as small, fuzzy disks through large telescopes. Distortion by the earth's atmosphere blurs the stars' images into disks and hides the actual size and structure of the surface of the stars. The sun is the only star whose angular diameter we can measure easily and directly. The Hubble Space Telescope, with its high-resolution capability, has been able to image some giant and supergiant stars, including Betelgeuse (as we shall see in Figure 29–2).

Astronomers use an indirect method to find the sizes of most stars. If we know the absolute magnitude of a star (from some type of parallax measurement) and the temperature of the surface of the star (from measuring its spectrum), then we can tell the amount of surface area the star must have. The extent of surface area, of course, depends on the radius, and luminosity depends on surface area. So a star of twice the radius but the same temperature as another star has $2^2$, or four, times the surface area and thus four times the luminosity.

One type of direct measurement works only for eclipsing binary stars. As the more distant star is hidden behind the nearer star, one can follow the rate at which the intensity of radiation from the farther star declines. From this information together with Doppler measurements of the velocities of the components, one can calculate the sizes of the stars. One can sometimes even tell how the brightness varies across a star's disk.

It is much more difficult to measure the size of a single star directly. Sometimes we can tell how long it takes for a star's light to disappear when the star is hidden by the moon. Other methods for measuring the diameters of single stars use a principle called **interferometry,** a technique that measures incoming radiation at two different locations and then combines the two signals. This gives the effect, for the purpose of determining the resolution, of a single very large telescope. Only in the last few years has interferometry in the visible part of the spectrum become good enough to make the observations described in this section. The scientists involved can correct their data in real time blurring by the earth's atmosphere by using a technique similar to that of active optics. Several modern projects are building giant interferometers that give stellar images several times sharper than those from Hubble. Such an array of telescopes in Cambridge, England, separates the components of the star Capella, which are only $\frac{1}{20}$ arc sec apart. The two stars are separated by no more than 6 light-minutes, meaning that they are slightly closer together than the Earth and sun, and orbit each other every 104 days. The Palomar Testbed Interferometer has measured the diameters of many giant stars (Fig. 26–8).

Georgia State University's Center for High Angular Resolution Astronomy (CHARA) has built a stellar interferometer on Mt. Wilson in California. The array consists of five 1-m telescopes in a Y-shape that have a maximum separation between telescopes of 350 m. Since that makes it equivalent, for purposes of resolution, to a single telescope 350-m across, it provides images of 0.2 milliarcsec in the visible part of the spectrum, which has already allowed the scientists working with it to measure the brightness changes from the center to the limb of several giant stars. The array will operate in the near infrared as well.

Direct and indirect measurements show that the diameters of main-sequence stars decrease as we go from hotter to cooler; that is, O dwarfs are relatively large and F dwarfs are relatively small. Techniques have recently advanced enough to allow direct measurements of lower mass stars, which are smaller still, showing that this trend continues for K and M stars. As for stars that are not on the main sequence, red giants are indeed giant in size as well as in brightness.

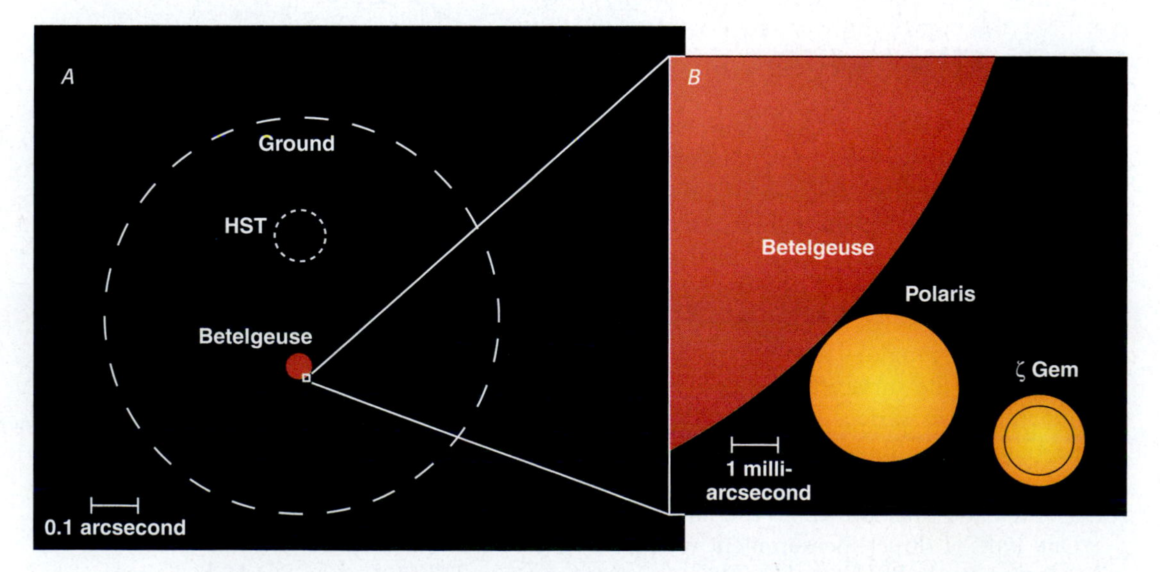


**FIGURE 26–8** The sizes of several giant stars measured with interferometry from the Palomar Testbed Interferometer. One of the stars is variable in size (as shown by the dashed circle marking its minimum size) as well as in brightness, as we will discuss in the following section. Even the smallest of the stars shown is nearly 100 times larger than our sun.

## 26.4 VARIABLE STARS

Some stars vary in brightness with respect to time. The most basic property of a star's variation is the **period.** The period of a variable is the time it takes for a star to go through its entire cycle of variation.

One way for a star to vary in brightness, as we have seen, is for the star to be an eclipsing binary. (For eclipsing binaries, the period of their variation coincides with the period with which they eclipse. Do not confuse this "period" with the period of revolution for each star, around the system's center of mass.) But individual stars can vary in brightness all by themselves. We say they are "intrinsic variables." Thousands of such stars are known in the sky. The periods of the variations can be seconds for some types of stars and years for others.

Besides ordinary variables of various types, some stars occasionally flare up briefly. These stellar flares often make a bigger difference in parts of the spectrum other than the visible (Fig. 26–9).

We will limit our discussion to three types of variables, one of which is especially numerous. The other two have provided important information about the scale of distance in the universe and thus are fundamental for cosmology.

### 26.4a Mira Variables

A type of variable star is often named after its best-known or brightest example. A red giant star named Mira in the constellation Cetus (the star is also known as o Ceti, omicron of the Whale) fluctuates in brightness with a long period (Fig. 26–10A), about a year. Red giant stars that share this characteristic are called **Mira variables.** The period of a given Mira-type star can be from three months up to about two years. The period of an individual star is not strictly regular; it can vary from the average period. The surfaces of these stars pulsate, moving in and out.

Mira itself is sometimes of apparent magnitude 9 and is then invisible to the naked eye. However, it brightens fairly regularly by about six magnitudes, a factor of 250, with a period of about 11 months. At maximum brightness it is quite noticeable in the sky. As it brightens, its spectral type changes from M9 to M5. Thus real changes are taking place at the surface of the star that result in a change of temperature. Mira is 400 light-years away.

Using the Hubble Space Telescope, scientists have been able to measure the diameter of several Mira variables. These stars turn out to be egg-shaped rather than round. We do not yet know whether there are actual irregularities in shape or if such irregularities show unresolved bright spots on the stars' surfaces.

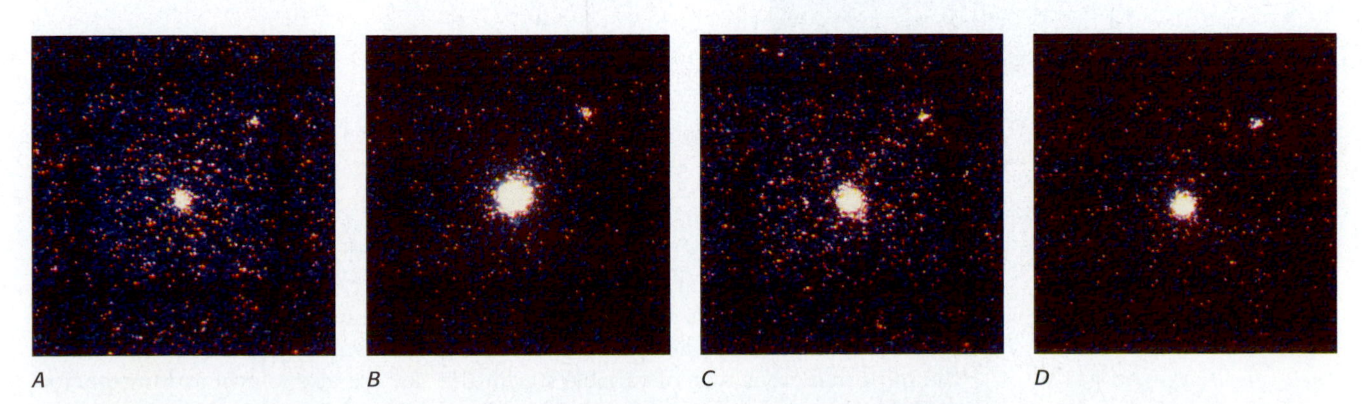


**FIGURE 26–9** An x-ray flare on Proxima Centauri, the nearest star, was observed with the Einstein Observatory in this series of 35-minute exposures taken on August 20, 1980. (*A*) Before the flare, (*B*) near the flare's maximum, and (*C*) and (*D*) the flare's decay. The steady x-ray flux probably comes from the star's corona. The flare's x-ray flux was about that of a bright solar flare. A second x-ray source appears at upper right.

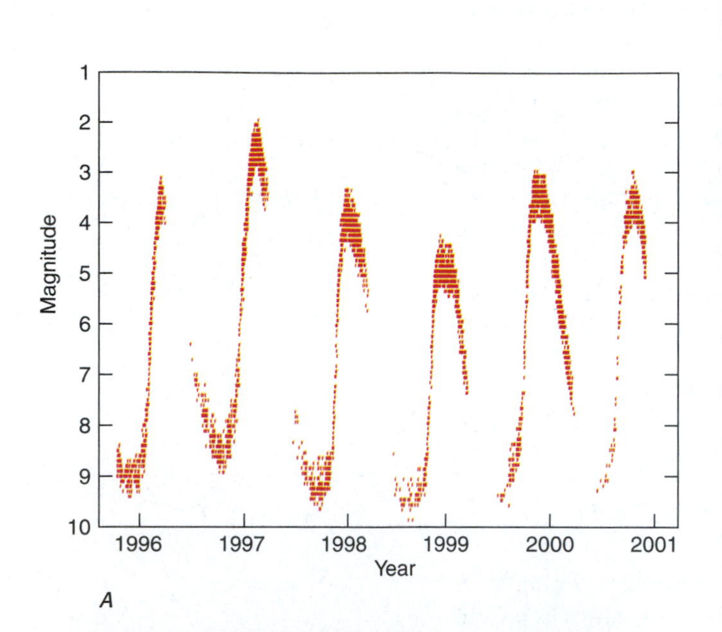


A

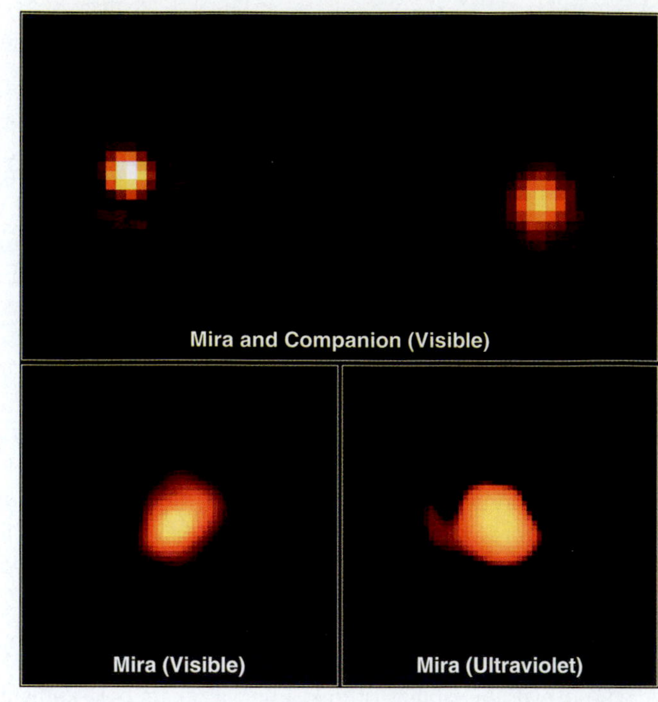


B

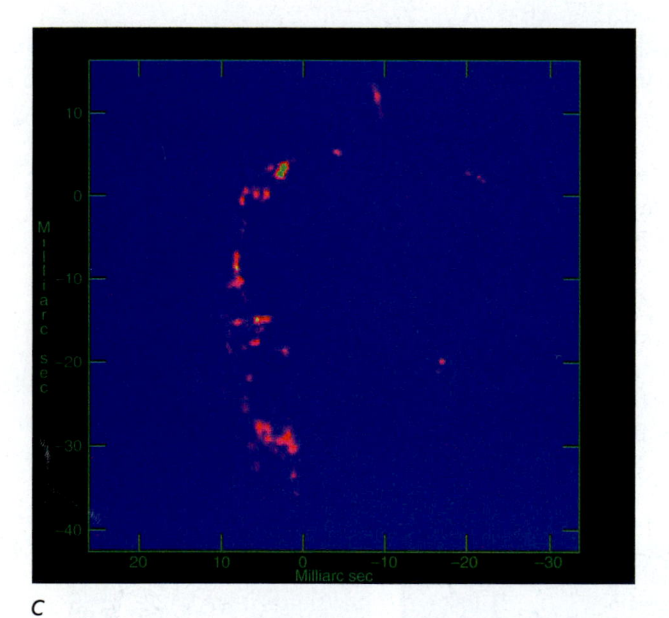


C

**FIGURE 26–10** Mira, the prototype of the class of long-period variables. (*A*) The light curve. The sun may be a Mira star in its distant future, when it is a red giant. (*B*) Hubble Space Telescope images of Mira and its white-dwarf companion. They are separated by only 70 A.U., corresponding at the 400 light-year distance to 0.6 arc sec. (*C*) Clouds of gas being ejected from the Mira-type variable star TX Camelopardalis over a 16-month interval. This ejection stage is thought to last only 10,000 years, a blink of the eye for a star at least 3 billion years old. The images, taken with the Very Long Baseline Array of radio telescopes that spans the United States, show that the gas is not ejected symmetrically. Similarly, the sun ejects gas asymmetrically in solar storms, so perhaps the star's magnetic field plays a role here too.

Mira stars are giants of spectral type M and have about 700 times the diameter of the sun. If such a star were in the center of our solar system, it would extend beyond Mars. Mira stars emit most of their radiation in the infrared. They are also the source of strong radio spectral lines from water vapor. These stars, which are the most numerous type of variable star in the sky, are also known as **long-period variables.**

Mira itself is one of the few long-period variables that has a close companion (Fig. 26–10*B*). The companion, a white dwarf, itself varies, and is known as VZ Ceti. When Mira itself is at its faint limit and VZ Ceti is at its brightest, the two

## WHAT'S IN A NAME?

### BOX 26.1 Naming Variable Stars

A naming system has been adopted for variable stars (often simply called "variables") that helps us recognize them on any list of stars. The first variable to be discovered in a constellation is named R, followed by the genitival form ("of . . . ") of the Latin name of the constellation (see Appendix 9), e.g., R Coronae Borealis (R of Corona Borealis), abbreviated R CrB. One continues with S, T, U, V, W, X, Y, Z, then RR, RS, etc., up to RZ, then SS (not SR) up to SZ, and so on up to ZZ. Then the system starts over with AA up to AZ, BB (not BA) and so on up to QZ. The letter J was omitted when the method was set up to avoid confusion with I in the German script of that time. This system covers the first 334 variables; after that, one numbers the stars beginning with V for variable (V335 Cygni, and so on). Variable stars with more commonly known names, such as Polaris and δ Cephei, retain their common names instead of being included in the lettering system.

***Discussion Question***

What number variable star in Cygnus is RS Cygni? What number would it be if the letter J were included?

stars are comparable in brightness. Hubble ultraviolet images show material extending from Mira toward the companion. It could be material pulled by the white dwarf's gravity or else material in Mira's upper atmosphere heated by the companion's radiation.

A network of radio telescopes known as the Very Long Baseline Array has been able to image clouds of gas being flung out from a long-period variable known as TX Camelopardalis (Fig. 26–10*C*). This Mira-type variable is about 1,000 light-years from us. Optical telescopes don't detect the clouds of gas, but the radio telescopes can pick up the emissions from silicon-monoxide in the clouds. When TX Cam and other Mira variables contract, they draw in nearby gas, and when they expand, they blow out even larger amounts of gas. TX Cam and other 3-billion-year-old Mira variables expel an amount of gas each year equivalent to the mass of the Earth.

### 26.4b Cepheid Variables

The most important variable stars in astronomy are the **Cepheid variables** (cef′e-id). The prototype is δ Cephei (delta of Cepheus). Cepheid variables have very regular periods that, for individual Cepheids (as they are called), can be from 1 to 100 days. δ Cephei itself varies with a period of 5.4 days (Fig. 26–11). Astronomers can tell Cepheid variables apart from other variables by the shape and regularity of their light curves. The shape of Cepheids' light curves is distinctive; δ Cephei itself, for example, rises sharply to maximum in about a day and a half and then falls more slowly, over about 4 days. The shape of the light curve has been said to resemble a shark's fin.

Cepheid variables are unstable stars. Once a star leaves equilibrium to become a Cepheid, gas pressure pushes it outward when it is too small, and gravity pulls it inward when it is too large. The star thus oscillates in size, surface temperature, and brightness.

Cepheids vary in brightness because their surfaces move in and out. We can follow the surface's motion from Doppler shifts of spectral lines. A Cepheid changes in luminosity mostly because its surface temperature changes and to a lesser degree because its size changes by perhaps 10 per cent. Cepheids are brightest as they pass through their average sizes while expanding; when they reach their maximum size, gravity pulls their surfaces in. When they get too small, pressure rises and pushes their surfaces outward. The surface overshoots, and the cycle continues.

Cepheids are important because of the relation that links the periods of their light changes, which are simple to measure, with their absolute magnitudes

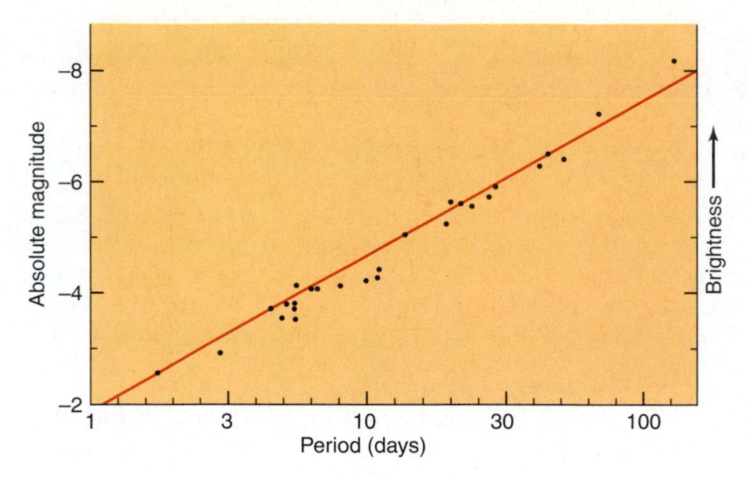


**FIGURE 26–12** The period-luminosity relation for Cepheid variables. The stars' actual average brightnesses are shown on the vertical axis.

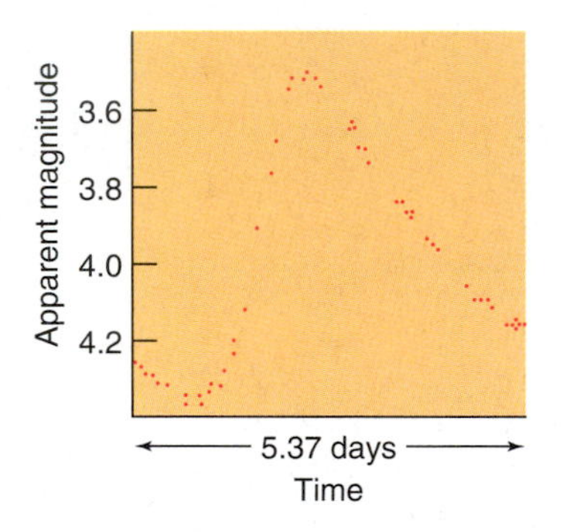


**FIGURE 26–11** The light curve for delta (δ) Cephei, the prototype of the class of Cepheid variables. Note that the distinctive shape of the curve reveals Cepheid variables; the light curves of other types of variables have other shapes.

(Fig. 26–12). For example, if we measure that the period of a Cepheid is 10 days, we need only look at this "period-luminosity relation" to see that the star is of absolute magnitude −4. We can then compare its absolute magnitude to its apparent magnitude, which gives us (by the inverse-square law of brightness) its distance from us.

The study of Cepheids is the key to our current understanding of the distance scale of the universe and has allowed us to determine that the objects in the sky that we now call galaxies (discussed in Chapter 34) are giant systems comparable to that of our own galaxy (discussed in Chapter 32). Cepheids are thus the most important type of star for revealing the universe.

The story started about 90 years ago at the Harvard College Observatory with Henrietta Leavitt (Fig. 26–13), who was studying the light curves of variable stars in the southern sky. In particular, she was studying the variables in the Large and Small Magellanic Clouds (Fig. 26–14), two hazy areas in the sky that were first reported back to Europe by the surviving crew of Magellan's expedition when they returned home following their sail around the world at far southern latitudes. Whatever the Magellanic Clouds were—we now know them to be galaxies, but Leavitt did not know this—they looked like concentrated clouds of material. It thus seemed clear that, for each cloud, all its stars were at approximately the same distance from the earth. Thus even though she could plot only the apparent magnitudes, the relation of the absolute magnitudes to each other was exactly the same as the relation of the apparent magnitudes.

By 1912, Leavitt had established the light curves and determined the periods for two dozen stars in the Small Magellanic Cloud. She plotted the magnitude of the stars (actually a median brightness, a value between the maximum and minimum brightness) against the period. She realized that there was a fairly strict relation between the two quantities, and that the Cepheids with longer periods were brighter than the Cepheids with shorter periods. By simply measuring the periods, she could determine the magnitude of one star relative to another; each period uniquely corresponds to a magnitude.

**FIGURE 26–13** Henrietta S. Leavitt, who discovered that Cepheids have a period-luminosity relation, at her desk at the Harvard College Observatory in 1916. Four years earlier, she had published an article reporting her discovery.

Henrietta Leavitt could measure *apparent* magnitude, but she did not know the distance to the Magellanic Clouds and so could not determine the *absolute* magnitude (intrinsic brightness) of the Cepheids.

To find the distance to the Magellanic Clouds, we first had to be able to find the distance to any Cepheid—even one not in the Magellanic Clouds—to tell its absolute magnitude. A Cepheid of the same period but located in the Magellanic Clouds would presumably have the same absolute magnitude.

To find the absolute magnitude of a Cepheid, it would seem easiest to start with the one nearest to us. Unfortunately, not a single Cepheid was close enough to the sun to allow its distance to be determined by the method of trigonometric parallax. (This situation has only recently changed, with the Hipparcos satellite. Hipparcos data show that Cepheids are more luminous and distant than previously thought. The result implies that the universe is about 10 per cent larger.) More complex, statistical methods had to be used to study the relationships between stellar motions and distances and so get the distance to a nearby Cepheid. Once we have the distance to a Cepheid, we can calculate its absolute magnitude from its apparent magnitude. Then we know the absolute magnitude of all Cepheids of that same period in the Magellanic Clouds, since all Cepheids of the same period have the same absolute magnitude.

From that point on it is easy to tell the absolute magnitude of Cepheids of *any* period in the Magellanic Clouds. After all, if a Cepheid has an *apparent* magnitude that is, say, 2 magnitudes brighter than the apparent magnitude of our Cepheid of known intrinsic brightness, then its *absolute* magnitude is also 2 magnitudes brighter. After this process, we have the period-luminosity relation in a more useful form—period versus absolute magnitude. We call this process "the calibration" of the period-luminosity relation.

Thus Cepheids can be employed as indicators of distance: First we identify the star as a Cepheid (by the shape of its light curve). Then we measure its period. Third, the period-luminosity relation gives us the absolute magnitude of the Cepheid. And last, we calculate (with the inverse-square law) how far a star of that absolute magnitude would have to be moved from the standard distance of 10 parsecs to appear as a star of the apparent magnitude that we observe.

**FIGURE 26–14** From the southern hemisphere, the Magellanic Clouds are high in the sky. They are not quite this obvious to the naked eye. Since all stars in a given Magellanic Cloud are essentially the same distance from Earth, their absolute magnitudes are in the same relation to each other as their apparent magnitudes.

When the calibration of the period-luminosity relation was worked out quantitatively by the American astronomer Harlow Shapley (the first syllable rhymes with "map," not "cape") in 1917, the distance to the Magellanic Clouds could be calculated. They were very far away, a distance that we now know means that they are not even in our galaxy! Instead, they are galaxies by themselves, two small irregular galaxies that are companions of our own, larger galaxy. The distances to nearby galaxies, such as M31 in Andromeda, as measured by Cepheids, were the key that allowed Edwin Hubble to show that M31 was outside our galaxy, with the resultant proof that our galaxy is an isolated building block of the universe.

Cepheids are bright stars (giants and supergiants, in fact) and can be observed in distant galaxies. The use of Cepheids is the prime method of establishing the distance to all the nearer galaxies, and one of the three "Key Projects" for the Hubble Space Telescope was to extend these measurements to more distant galaxies. Hundreds of Cepheids have been discovered with the Hubble Space Telescope in each of several distant galaxies. The ability of the Hubble Space Telescope to observe at intervals chosen for the stars' cycles rather than being limited by the day/night cycle and by weather, together with its ability to detect such faint stars (Fig. 26–15) led to the plotting of a period-luminosity relation for these galaxies. We shall discuss the important consequences of such Cepheid measurements in Chapter 35, where we give the final result of the Hubble Space Telescope's Key Project on the cosmic distance scale, as released in 2001. Cepheids are fundamental to how we determine the scale of distances in the universe.

For this cosmological reason, it is vital to know how far Cepheid variables are away from us, and a debate remains on the last few per cent of accuracy, even with the Hubble results. Only a handful were within reach of the Hipparcos satellite's distance determinations, and those were at the far end of the range and so not as accurate as we would like. The FAME spacecraft, cancelled in 2002, was expected to be able to map accurate, independent distances for 200 Cepheids. A European Space Agency project, GAIA, to be launched in 2012, is to be even more sensitive, detecting a billion objects including even more Cepheids.

The development of optical interferometry is allowing astronomers to measure the distance to at least one Cepheid in a way independent of parallax or other traditional

**FIGURE 26–15** Cepheids in a distant galaxy imaged with the Hubble Space Telescope. Note in the three insets how the star singled out changes over time.

methods. An article published in 2000 reported that the Cepheid variable known as zeta Geminorum had been pinned down by the Palomar Testbed Interferometer, whose telescopes are spaced over a region 110 meters wide. This star is one of the four brightest Cepheids visible in the northern hemisphere. Though the star is only about 1.5 milliarcsecond across (about 100th the size of a pixel on the Hubble Space Telescope), and changes in diameter over its 10-day cycle by only $\frac{1}{10}$ of that, the interferometer succeeded in measuring both, as was marked in Figure 26–9. Scientists assume that the star expands at the same rate in all directions. By equating the measured change in size over time (that is, side to side as we look at the sky) with the rate of the star's contraction and expansion in the line-of-sight measured from the Doppler effect (that is, in and out of the plane of the sky), they can calculate how far away the star is. They calculate that it is 1100 light-years from Earth. Their results are still uncertain by about 10 per cent, so aren't yet improving the accuracy of the distance-scale determination, but the method has great potential and accuracy will improve. The CHARA interferometer, which we discussed in Section 26.3, has a longer baseline and so should be able to measure more Cepheid distances in the next few years.

The north star, Polaris, is a Cepheid variable, though one that has varied by only about 0.1 magnitude. Like all Cepheids, it is a giant star, though it is a lot smaller than the supergiant star Betelgeuse. Its diameter has been measured by the same apparatus as in the previous paragraph. Polaris's fluctuations in magnitude have been decreasing over the years, until they are now only hundredths of a magnitude, and may soon cease entirely.

A Princeton/Warsaw project known as OGLE (*O*ptical *G*ravitational *L*ensing *E*xperiment) measures the brightnesses of millions of stars as often as possible, to see if some of the brightnesses of some distant stars are changing because of focusing by the gravity of a star that passes in front of them. As a side benefit, they discover hundreds of Cepheid variables. We can plot a period-luminosity relation for these Cepheids (Fig. 26–16), distinguishable, as always, from the shapes of their light

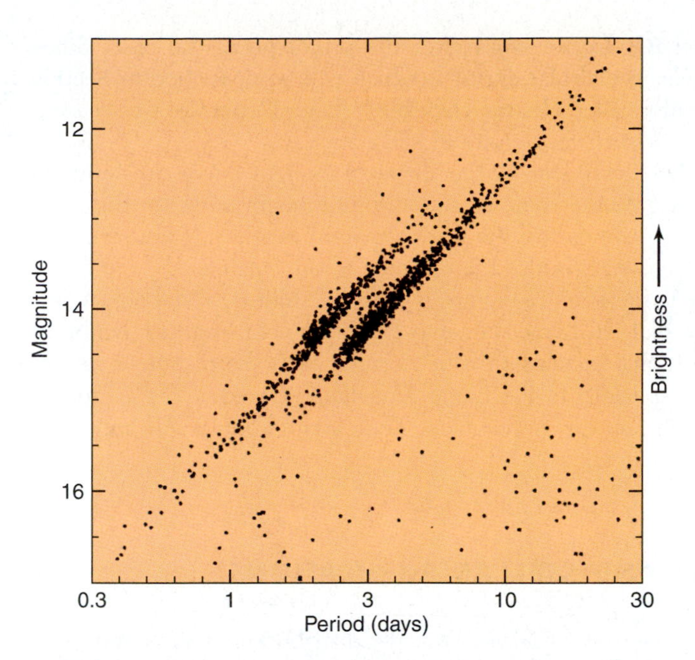


**FIGURE 26–16** The period-luminosity relation for Cepheids, revealed by the periods and magnitudes of hundreds of Cepheids discovered in the Large Magellanic Cloud by the OGLE team.

curves. All these stars are in the Large Magellanic Cloud, a neighboring galaxy to ours, and are all therefore at essentially the same distance from us. Thus though it measures a kind of apparent magnitude, the diagram shows the actual relation of the stars' brightnesses to each other. The diagram shows, somewhat surprisingly, that the Cepheids have two separate bands of period-luminosity relation. The band that extends to upper right is the traditional period-luminosity relation. The band to the left of it corresponds to stars pulsating not in the fundamental mode but in the first overtone. These musical terms correspond to notes from an organ pipe: the fundamental mode is when there is one wave in a pipe and the first overtone is when there are two waves in that pipe, each half as long and thus corresponding to a note whose pitch is twice as high.

Polaris is on the upper-left band. The diameter measured with the interferometer mentioned above allows us to calculate how bright it should appear. This calculation confirms that Polaris pulsates with the first overtone. Perhaps such stars are more likely than the other Cepheids to have their pulsations die out.

Fortunately, we are not much confused by the period-luminosity diagrams for the two kinds of pulsators. Note that all the stars in the upper-left band have periods of less than 6 days. Most of the Cepheids plotted in Figure 26–12 have periods of longer than that. So as long as we restrict ourselves to Cepheids with periods longer than 6 days, we do not have this ambiguity and can use the period-luminosity relation to measure distances reliably.

In the 1950s, when overlapping methods of finding distances were applied to some relatively nearby stars and clusters of stars, it was realized that there was a second type of Cepheid whose light curves looked the same but which were about 1.5 magnitudes fainter at each given period than the original type. The distances to many galaxies had to be recalculated because this distinction between ordinary Cepheids and Type II Cepheids had previously not been made. Our estimates of the distances to many distant galaxies doubled.

## 26.4c RR Lyrae Variables

Many stars are known to have short regular periods, less than one day in duration. Certain of these stars, no matter what their periods, have light curves of a specific distinctive shape (Fig. 26–17). All these stars have the same average absolute magnitude.

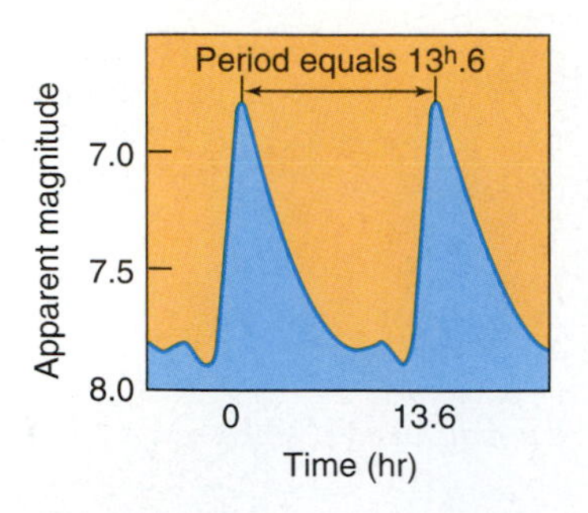


**FIGURE 26–17** The light curve of RR Lyrae, the prototype of cluster variables.

Such stars are called **RR Lyrae stars** after the prototype of the class. Since many of these stars appear in globular clusters (which will be described in Section 26.5), RR Lyrae stars are also called **cluster variables.** As for Cepheids, we see their brightnesses change as their surfaces pulsate in and out.

Once we detect an RR Lyrae star by the shape of its light curve, we immediately know its absolute magnitude, since all the absolute magnitudes are the same (about 0.6). Just as before, we measure the star's apparent magnitude and can thus easily calculate its distance, which is also the distance to the cluster.

Just how bright RR Lyrae stars really are is still the subject of debate. A few lines of evidence indicate that RR Lyrae stars may be 10 per cent brighter than most people have concluded. If that conclusion were correct, then the distances we deduce to nearby galaxies, especially to the Large Magellanic Cloud, and thus to farther galaxies, may be wrong by this percentage. The debate about such important details continues.

## 26.5 CLUSTERS AND STELLAR POPULATIONS

Even aside from the hazy band of the Milky Way, the distribution of stars in our sky is not uniform. There are certain areas where the number of stars is very much higher than the number in adjacent areas. Such sections of the sky are called **star clusters.** All the stars in a given cluster are essentially the same distance away from us and were also formed at about the same time.

One type of star cluster appears only as an increase in the number of stars in that limited area of sky. Such clusters are called **open clusters,** or **galactic clusters** (Figs. 26–18 and 26–19). The most familiar example of a galactic cluster is the Pleiades (Plee′a-dees), a group of stars visible in the evening sky in the winter. The unaided eye sees at least six stars very close together. With binoculars or the small-

**FIGURE 26–18** An open cluster of stars, NGC 3293 in Carina.

A DEEPER DISCUSSION

## BOX 26.2 Comparing Types of Star Clusters

| Galactic Clusters | Globular Clusters |
|---|---|
| No regular shape (also called "open clusters") | Shaped like a ball, stars more closely packed toward center |
| Many contain young stars | All have only old stars |
| H-R diagrams have long main sequences | H-R diagrams have short main sequences |
| Where stars leave the main sequence tells the cluster's age | All clusters have the same H-R diagram and thus the same age |
| Stars have similar composition to sun, Population I | Stars have lower abundances of heavy elements than sun, Population II |
| Hundreds of stars per cluster | 10,000–1,000,000 stars per cluster |
| Found in galactic plane | Found in galactic halo |

***Discussion Question:*** Which do you find prettier, galactic clusters or globular clusters? Why?

**FIGURE 26–19** An open cluster of stars, NGC 2244, surrounded by the Rosette Nebula.

The Pleiades are often known as the Seven Sisters, after the seven daughters of Atlas who were pursued by Orion and who were given refuge in the sky. That one is missing from the number in the myth—the Lost Pleiad—has long been noticed. Of course, a seventh star is present (and hundreds of others as well), although too faint to be plainly seen with the naked eye. The Pleiades seem to be riding on the back of Taurus.

est telescopes, dozens more can be seen. A larger telescope reveals hundreds of stars. Another galactic cluster is the Hyades, which forms the "V" that outlines the face of Taurus, the bull, a constellation best visible in the winter sky. More than a thousand such clusters are known, most of them too faint to be seen except with telescopes.

All the stars in a galactic cluster are packed into a volume not more than 10 parsecs across. Stars in galactic clusters seem to be representative of stars in the spiral arms of our galaxy and of other galaxies. When the spectra of stars in galactic clusters are analyzed to find the relative abundances of the chemical elements in their atmospheres, we find that over 90 per cent of the atoms are hydrogen, most of the rest are helium, and less than 1 per cent are elements heavier than helium. This is similar to the composition of the sun. Such stars are said to belong to stellar **Population I.**

Astronomers have long known that the easy-to-find open cluster the Pleiades contains dust that reflects starlight toward us (Fig. 26–20). Only recently, though, have astronomers realized that the dust is present because the stars have made a chance encounter with a cloud of interstellar dust. An image from the IRAS spacecraft shows the wake of the cluster. The hole is about 40 light-years long. The Pleiades apparently generates a shock wave that deflects gas and dust. An older idea—that the dust remains from the interstellar cloud out of which the stars were born—is erroneous.

The second major type of star cluster appears in a small telescope as a small, hazy area in the sky. Observing with larger telescopes distinguishes individual stars and reveals that these clusters are really composed of many thousands of stars packed together in a very limited space. The clusters are spherical, and they are known as **globular clusters** (Fig. 26–21).

*A*

*B*

FIGURE 26–20 (*A*) The wake of the Pleiades open star cluster through the dust and gas of interstellar space, in an infrared image from space. The stars themselves are in the bright emission to the right of center and give the energy that makes it shine. The material is much more extensive than the dust that shows in the visible reflecting the starlight. In this IRAS image, blue shows 12-$\mu$m, green shows 60-$\mu$m, and red shows 100-$\mu$m radiation. The wake is the dark region to the lower right (*B*) A Hubble image shows details in the dust from the star Merope, as this and the other Pleiades stars wade through a dust cloud.

**FIGURE 26–21** The globular cluster Omega Centauri, which contains hundreds of thousands of stars. We know of 154 globular clusters in our galaxy. The Hubble Space Telescope, because of its high spatial resolution, can detect individual stars in the cores of globular clusters. In the core of the globular cluster M15, for example, it found hot blue stars that may be stellar cores after the outer layers were stripped off by the gravity of nearby passing stars.

Globular clusters can contain 10,000 to one million stars, in contrast to the 20 to several hundred stars in a galactic cluster. A globular cluster can fill a volume up to 30 parsecs across. The stars are more closely packed toward the center than toward the periphery.

Most of the known globular clusters form a roughly spherical "**halo**" that goes far above and below our galaxy's disk. (We see few globular clusters in the disk because they are hidden by interstellar dust.) The abundances of the elements heavier than helium in stars in globular clusters are much lower, by a factor of 10 to 300, than their abundances in the sun. Since the abundances of heavy elements grow over time as the elements are formed in stars and spread through space, the globular-cluster stars have not experienced such stellar recycling and hence were burning steadily before all stellar explosions. They are thus older than Population I stars. Such stars, with their relatively low abundances of heavy elements, are **Population II.**

A typical halo globular cluster passes through the plane of the galaxy every 300 million years, as it orbits the galaxy's center of mass. These passages sweep the space between the stars in globular clusters free of gas and dust.

The distinction between Population I and Population II is oversimplified, and recent research has enlarged upon the idea, but the fundamental distinction remains useful.

## 26.5a H-R Diagrams and the Ages of Galactic Clusters

The Hertzsprung-Russell diagram for several galactic clusters is shown in Figure 26–22. For the purposes of this discussion, it is most important to recall that the horizontal axis is a measure of temperature and the vertical axis is a measure of brightness. The main part of the figure is actually a set of many individual diagrams like the two on the right, laid on top of each other.

Since all the stars in a cluster were formed at the same time out of the same gas, we can presume that they have similar chemical compositions and differ only in mass. The difference between one cluster and another is principally that the two clusters were formed at different times. Scientists using the newish 3.5-m telescope on Kitt Peak operated jointly by the University of Wisconsin, Indiana, Yale, and the National Optical Astronomy Observatories (WIYN) are using its relatively wide field to

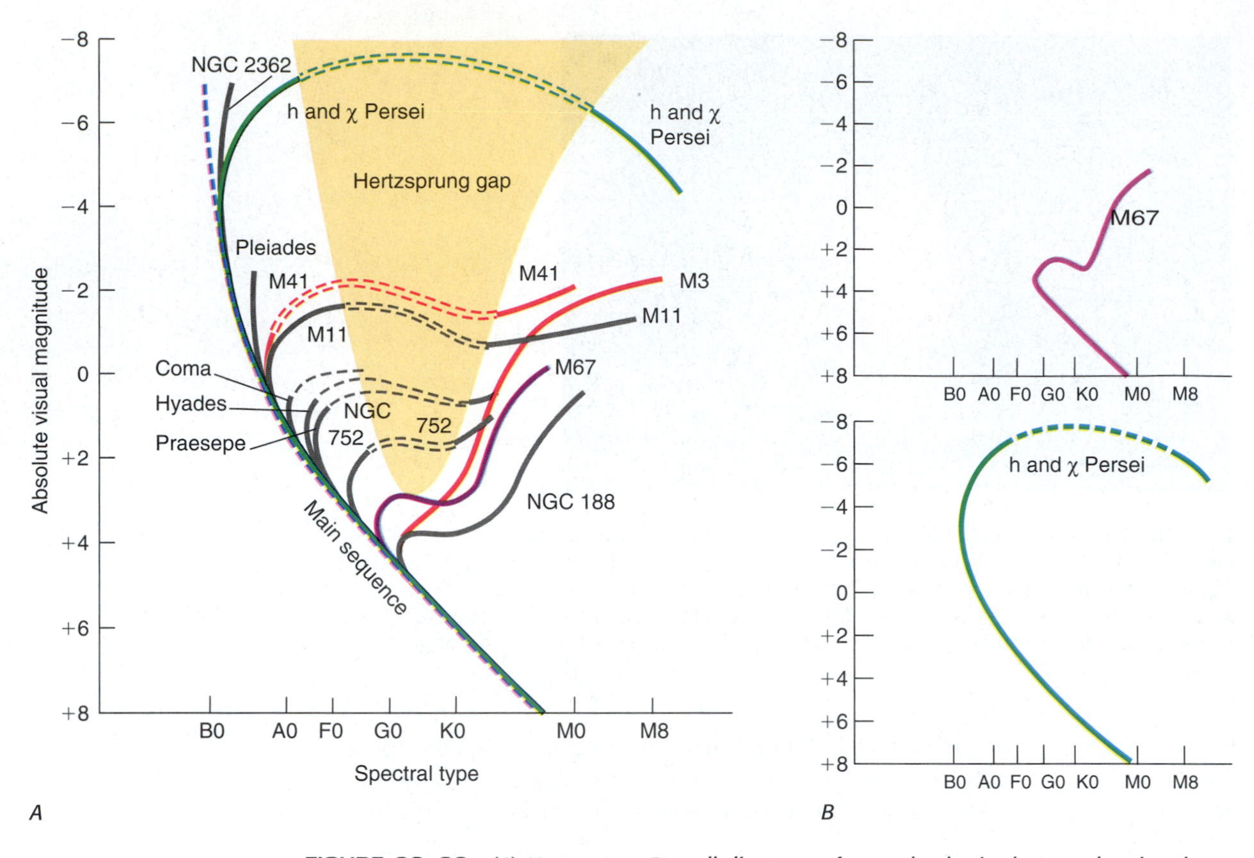


**FIGURE 26–22** (*A*) Hertzsprung-Russell diagrams of several galactic clusters, showing the overlay of several individual diagrams. By observing the point where a cluster turns off the main sequence, we deduce the length of time since its stars were formed (that is, the age of the cluster). The presence of the **Hertzsprung gap** (*shaded beige*), a region in which few stars are found, indicates that stars evolve rapidly through this part of the diagram. Part of the H-R diagram for the globular cluster M3 is included for comparison. An object's M number is its number in the Messier catalogue (Appendix 8), as we shall discuss in Chapter 32. (*B*) Two of the H-R diagrams of individual galactic clusters are shown separately here to illustrate how several of these are put together to make the composite diagram shown in *A*.

measure the abundances of carbon, nitrogen, oxygen, and other elements to study the evolution of stars in open clusters. The giant Hobby–Eberly Telescope constructed in Texas to do spectroscopy is enabling U. Texas, Penn State, and other collaborating astronomers to enlarge and continue the abundance work they have long carried out.

Studying H-R diagrams of star clusters has shown us that stars change with age—evolve, a theme to which we will return.

The H-R diagrams for different galactic clusters appear to be similar over the lower part of the main sequence but diverge at the upper part. Comparison of the diagrams for the different individual clusters has told us that stars evolve and given us a picture of how.

We see on the figure that the H-R diagrams for some clusters lie almost entirely along the main sequence, while others have only their lowest portions on the main sequence. Since stars spend most of their lifetimes on the main sequence, we can deduce that the cluster whose stars lie mostly on the main sequence must be the youngest. They would not have had time for many stars to have evolved enough to move off the main sequence. Thus NGC 2362 and the pair of clusters known as h and $\chi$ (chi) Persei (Fig. 26–23) are the youngest clusters in the composite diagram. (The clusters h and $\chi$ Persei have star-like names rather than ordinary cluster names because they were once confused with stars.) When a star finishes its main-sequence lifetime, its surface becomes larger and cooler. Thus the star moves upward and to the right on the H-R diagram.

The stars that finish their main-sequence lifetimes most rapidly (at which point they move off the main sequence) are the most massive ones. The most massive stars

A DEEPER DISCUSSION

## BOX 26.3 The Oldest Stars

It has generally been thought that the oldest stars were found in globular clusters. The ages astronomers derived for them depended on fitting the Hertzsprung-Russell diagrams for the clusters, comparing observations with theory. Stars in globular clusters, and various other stars, have low abundances of heavy elements because they were formed before these heavy elements were formed. In 2001, a direct measurement of the age of a metal-poor star was made for the first time. A Michigan State University astronomer and colleagues from Europe and Brazil used the very efficient high-dispersion spectrograph at the Very Large Telescope to detect an absorption line from singly ionized uranium in the spectrum of an isolated metal-poor star, one not in a cluster. Uranium is rare in old stars, since it was formed in supernovae and few supernovae had had time to go off in the early era.

They studied an isotope of uranium, U-238, that decays with a half-life of 4.5 billion years. Since the abundances of stable elements in the star is 12 per cent those in the sun, the uranium abundance presumably started at that level. Its abundance compared with other elements, given its rate of decay, indicated an age for the star of 12.5 billion years. The uncertainty in the measurement is plus or minus 3 billion years. Most of the uncertainty comes from inadequate knowledge of the nuclear properties, which can be refined in terrestrial laboratories. Better spectra of this star and spectra of similar stars would also help. Obviously, the universe is at least as old as the oldest stars in it.

are the O and B stars on the extreme upper left of the H-R diagram. They are more luminous and use up their nuclear fuel at a faster rate than the cooler, more numerous, ordinary stars like the sun.

In NGC 2362 and h and $\chi$ Persei, only the most massive stars have lived long enough to die and move off the main sequence. Theoretical calculations tell us that stars of the spectral type of the point where these two clusters leave the main sequence (about B0) have masses such that their main-sequence lifetimes are $10^7$ years. Thus these two clusters must be about $10^7$ years old. The stars that live for less time than $10^7$ years have died and moved off to the right. The stars that live for longer than $10^7$ years are still on the main sequence.

The Pleiades, on the other hand, must be older than $10^7$ years, since the stars with lifetimes of $10^7$ years have already died. Calculations tell us that stars of the spectral type where the Pleiades' graph turns off the main sequence (about B5) live $10^8$ years. Thus the Pleiades must be about $10^8$ years old. Similarly, the Hyades must be older than the Pleiades, since still more stars have had time to live their main-

**FIGURE 26–23** The double cluster in Perseus, h and $\chi$ Persei, a pair of galactic clusters that are readily visible in a small telescope and close enough together that they appear in the same field of view. Study of the H-R diagram reveals that they are relatively young. Perseus is a northern constellation that is most prominent in the winter sky. In Greek mythology, Perseus slew the Gorgon Medusa and saved Andromeda from a sea monster.

sequence lifetimes and die. Calculations tell us that stars at the turnoff point for the Hyades (about A0) have main-sequence lifetimes of $10^9$ years, so the Hyades must have been formed $10^9$ years ago.

We can thus read a cluster's H-R diagram like a clock, telling how long the cluster has lived by which of its stars (observationally) has turned off the H-R diagram and how old (theoretically) such a star is. Be sure to keep in mind that all the stars in the cluster formed at the same time, so we are telling the age of the cluster and all the stars in it.

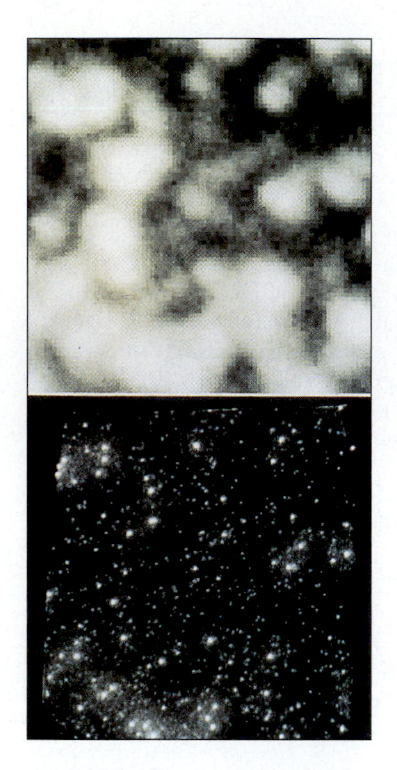

**FIGURE 26–24** The inner region of the globular cluster M14, seen both from ground (*top*) and with higher resolution from the Hubble Space Telescope (*bottom*).

## 26.5b H-R Diagrams for Globular Clusters

The H-R diagrams for all globular clusters look essentially identical. They all have stubby main sequences. The Hubble Space Telescope is allowing us to see the individual stars in globular clusters more clearly (Fig. 26–24).

From the fact that all globular clusters have very similar H-R diagrams (Fig. 26–25), we can conclude that they are all about the same age. From the fact that they all have H-R diagrams with short main sequences, we can conclude that the globular clusters must be very old. The latest detailed studies assign an age of about 13 billion years with an uncertainty of 1.5 billion years. If we assume that the globular clusters formed when or soon after our galaxy formed, then our galaxy must be about 14 billion years old, with an uncertainty of about 3 billion years. Just how different in age from each other individual clusters can be is a matter

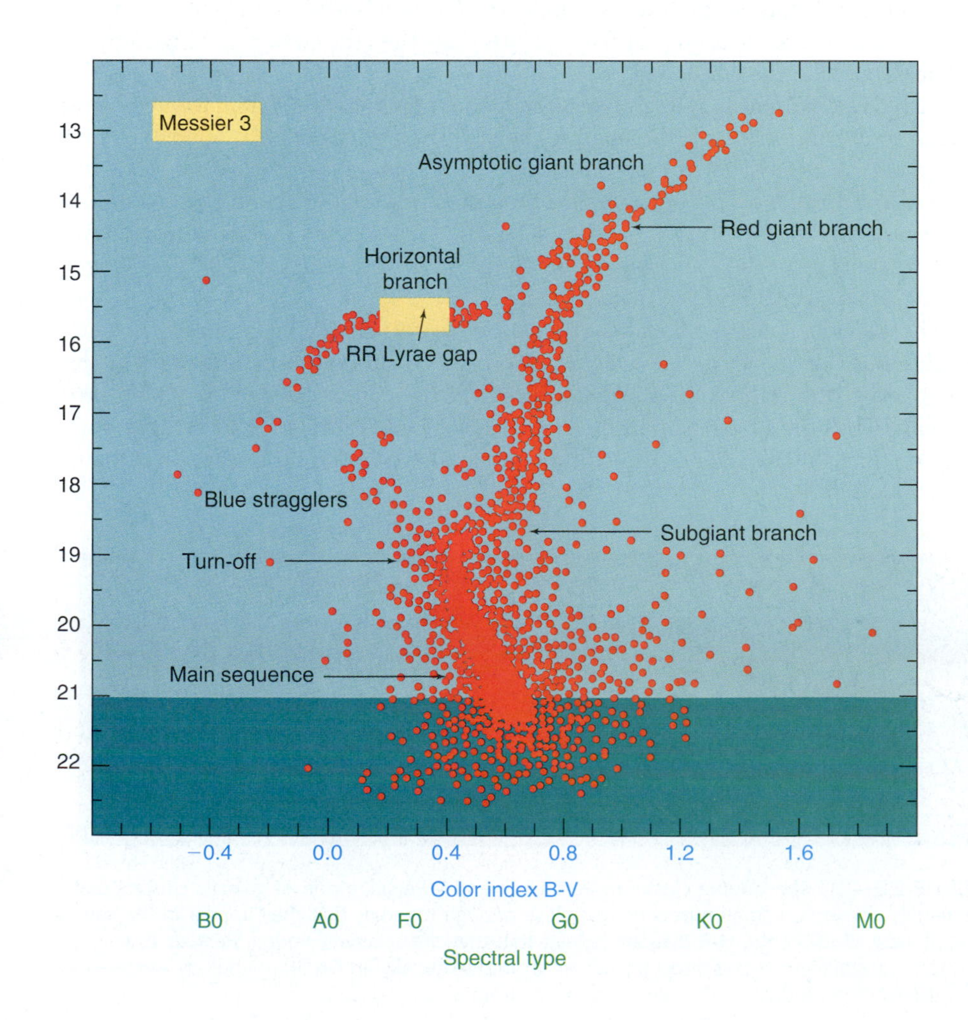


**FIGURE 26–25** The Hertzsprung-Russell diagram for the globular cluster M3, which was included schematically in Figure 26–22*A*. We can observe only the members of the cluster that are brighter than a certain limit. The stars in the more heavily shaded region at the bottom are newly observable because of the use of CCDs. The positions of stars on the horizontal axis were actually determined from measurements of the colors of the stars made at the telescope by comparing their brightnesses in the blue and in the yellow (which gives the color index B−V). The transformation to spectral type is approximate. Blue stragglers, found to the upper-left side of the turn-off, are stars that haven't turned off the main sequence even though they are leftward (blueward) of the turn-off point. They are thus straggling behind other stars that were at those points on the main sequence. Hubble Space Telescope observations have isolated some blue stragglers in the cores of globular clusters and shown that interactions between stars are responsible for this behavior.

**FIGURE 26–26** A globular cluster in the Andromeda Galaxy, M31. It is hundreds of times farther away than globular clusters in our own galaxy, but the Hubble Space Telescope is able to show it in such detail. The cluster has a higher abundance of heavy elements than globular clusters in our Milky Way Galaxy, which is leading us to model how the formation of stars and clusters differed in the two galaxies. Hubble is observing some two dozen globular clusters in M31.

of debate; a maximum difference of at least 2 billion years and perhaps even 5 billion years appears likely. The Hubble Space Telescope's spatial resolution is so good that it has photographed globular clusters in some distant galaxies as clearly as we have been formerly able to view globular clusters in our own galaxy (Fig. 26–26).

The H-R diagrams for globular clusters have prominent **horizontal branches.** The horizontal branch goes leftward from the stars on the right side of the diagram that have long since turned off the main sequence. It represents very old stars that have evolved past their giant or supergiant phases and are returning leftward. No open cluster (galactic cluster) has stars that old.

## CORRECTING MISCONCEPTIONS

| ✖ *Incorrect* | ✔ *Correct* |
|---|---|
| Most stars are single. | Most stars are multiple. |
| All stars have the same mass. | For main-sequence stars, stars of greater luminosities have greater masses. |
| Stars are steady in brightness. | Many stars vary in brightness. |
| Stars are too close to bother about if we want to learn about cosmology. | Important cosmological measurements depend on accurate knowledge of stars, such as Cepheid variables. |
| Stars are all made of the same stuff. | The proportions of the elements are quite different in younger and older stars. |

## SUMMARY AND OUTLINE

Binary stars (Section 26.1)
- Optical doubles, visual binaries, binaries with composite spectra, spectroscopic binaries, eclipsing binaries, astrometric binaries

Determination of stellar masses (Section 26.2)
- Mass-luminosity relation

Determination of stellar sizes (Section 26.3)
- Indirect: calculate from absolute magnitude or measure in eclipsing binary system
- Direct: interferometry

Variable stars (Section 26.4)
- Mira variables (Section 26.4a)
- Cepheid variables (Section 26.4b)
  - Period-luminosity relation
  - Uses for determining distances
- RR Lyrae variables (Section 26.4c)
  - All of approximately the same absolute magnitude
  - Used for determining distances

Clusters and stellar populations (Section 26.5)
- Galactic (open) clusters (Section 26.5a)
  - Population I: relatively high abundance of elements heavier than helium
  - Representative of spiral arms
  - 20 to several hundred members
  - Turn-off point on H-R diagram gives age
- Globular clusters (Section 26.5b)
  - Population II: relatively low abundance of metals
  - Representative of galactic halo
  - $10^4$ to $10^6$ members
  - Old enough to have H-R diagrams with horizontal branches

## KEY WORDS

optical doubles
double star
binary star
visual binaries
composite spectrum
spectroscopic binary
light curves
eclipsing binaries
astrometric binaries
mass-luminosity relation
interferometry
period
Mira variables
long-period variables
Cepheid variables
RR Lyrae stars
cluster variables
star clusters
open clusters
galactic clusters
Population I
globular clusters
halo
Population II
Hertzsprung gap
horizontal branches

## QUESTIONS

1. Sketch the orbit of a double star that is simultaneously a visual, an eclipsing, and a spectroscopic binary.
2. Define briefly and contrast an astrometric binary and an eclipsing binary.
3. (a)Assume that an eclipsing binary contains two identical stars. Sketch the intensity of light received as a function of time. (b) Sketch to the same scale another curve to show the result if both stars were much larger while the orbit stayed the same.
4. †How much brighter than the sun is a main-sequence star whose mass is ten times that of the sun?
5. †What does the mass-luminosity relationship tell us about the mass of the star 51 Pegasi (spectral type G2), around which planets have been reported?
6. †A main-sequence star is 3 times the mass of the sun. What is its luminosity relative to that of the sun?
7. †(a)Use the mass-luminosity relation to determine about how many times brighter than the sun are the most massive main-sequence stars we know. (b) How many times fainter are the least massive main-sequence stars?
8. When we look at the Andromeda Galaxy, from what mass range of stars does most of the light come? What mass range of stars provides most of the mass of the galaxy?
9. When we consider the gravitational effects of a distant galaxy on its neighbors, are we measuring the effects of mostly low-mass or high-mass stars? Explain.
10. A certain Cepheid variable has a period of 30 days. What is its absolute magnitude?
11. What is the ratio of apparent brightness of two Cepheid variables in the Large Magellanic Cloud, one with a 10-day period and the other with a 30-day period?
12. An RR Lyrae star has an apparent magnitude of 6. How far is it from the sun?
13. A Cepheid variable with a period of 10 days has an apparent magnitude of 8. How bright would an RR Lyrae star be if it were in a globular cluster near the Cepheid?
14. An astronomer observes a galaxy and notices that a star in it brightens and dims every 11 days. Sketch the light curve, label the axes, and explain how to find the distance to the galaxy.

†This question requires a numerical solution.

15. What is the absolute magnitude of an RR Lyrae star with an 18-hour period?
16. Briefly distinguish Population I from Population II stars.
17. Cluster X has a higher fraction of main-sequence stars than cluster Y. Which cluster is probably older?
18. What is the advantage of studying the H-R diagram of a cluster, compared to that of the stars in the general field?
19. Which galactic cluster shown in Figure 26–19 is about 8 billion years old?
20. From the position of the RR Lyrae gap and your knowledge of RR Lyrae stars, how far away is M3, the globular cluster whose H-R diagram is shown in Figure 26–22?

## USING TECHNOLOGY

### W World Wide Web

1. Look at the Anglo-Australian Observatory's photos of globular clusters at http://www.aao.gov.au/images.html.
2. Look at Hubble views of open and globular clusters and consider which are similar to each other and which look different. See http://oposite.stsci.edu/pubinfo/Subject.html.

### REDSHIFT

1. To learn about binary stars, look at Story of the Universe: Double Stars; and Guided Tours: Stars: Binary Star.
2. To learn about variable stars, look at Guided Tours: Stars: Variable Star; Principia Mathematica: Cepheid Variable and the Doppler Shift; Astronomy Lab: Stars: Cepheid Variable; Astronomy Lab: Galaxies: Cepheid Variable.
3. Turn on the deep sky objects, and look around to find globular and open clusters.
4. To learn about star clusters, look at Guided Tours: Galaxies: Open cluster, Globular cluster; Photos: The Galaxy: Open Clusters, Globular Clusters
5. Find the double cluster, h and $\chi$ Persei, zoom in, and examine various members and their spectral types.
6. Find out when and if you can observe globular clusters like M13 and Omega Centauri from (a) your home and (b) from the Cerro Tololo Observatory in Chile.

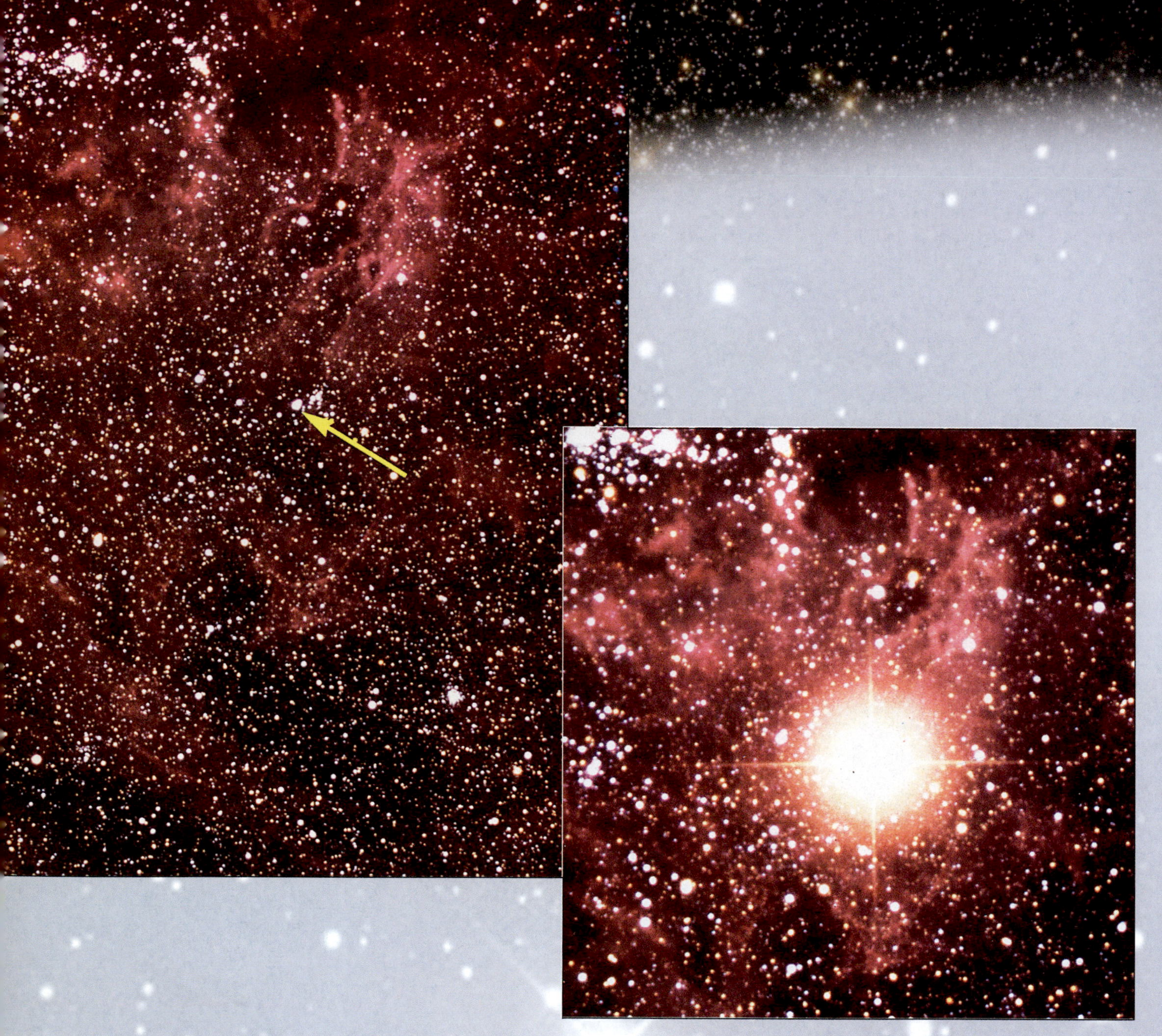

The supernova whose light reached us in 1987 is the closest and brightest supernova we have detected in nearly 400 years. It is located in the Large Magellanic Cloud. The arrow shows the star before it exploded. We then see the bright, overexposed supernova, SN 1987A, with spikes added to the image by the telescope.

# PART 5

# Stellar Evolution

We have seen how the Hertzsprung-Russell diagram for a cluster of stars allows us to deduce the age of the cluster and the ages of the stars themselves (Section 26.5). Our human lifetimes are very short compared to the billions of years that a typical star takes to form, live its life, and die. Thus our hope of understanding the life history of an individual star depends on studying large numbers of stars, for presumably we will see them at different stages of their lives.

Though we can't follow an individual star from cradle to grave, studying many different stars shows us enough stages of development to allow us to write out a stellar biography. Chapters 27 through 31 are devoted to such life stories. Similarly, we could study the stages of human life not by watching some individual's aging, but rather by studying people of all ages who are present in a city on a given day.

In Chapter 27, we shall study the birth of stars and observe some places in the sky where stars are being born now. We shall then consider the properties of stars during the long, stable phase in which they spend most of their histories.

In the chapters that follow, we go on to consider the ways in which stars end their lives. Stars like the sun sometimes eject shells of gas that glow beautifully; the part of such a star that is left behind then contracts until it is as small as the earth (see Chapter 28). Sometimes when a star is newly visible in a location where no star was previously known to exist (a "nova"), we are seeing the interaction of a dead solar-type star with a companion.

Occasionally, newly visible stars rival whole galaxies in brightness. Then we may be seeing the spectacular death throes of a star (see Chapter 29). What happens to massive stars after they explode is a topic of tremendous interest to many astronomers. The light from the explosion of a massive star reached Earth in 1987, in what may have been the most exciting thing to happen in astronomy in over 300 years. Some of the massive stars become pulsars (discussed in Chapter 30). Other massive stars, we think, may even wind up as black holes, objects that are invisible and therefore difficult—though not impossible—to detect (see Chapter 31).

When we write a stellar biography from the clues that we get from observation, we are acting like detectives in a novel. As new methods of observation become available, we are able to make better deductions, so the extension of our senses throughout the electromagnetic spectrum has led directly to a better understanding of stellar evolution. Our work in the x-ray part of the spectrum, for example, has told us more about tremendously hot gases like those that result from an exploding star. X-ray astronomy, which boasts of new telescopes in space, is intimately connected with our current exciting search for a black hole.

In these chapters, we shall see how important computers are for astronomy. Calculations that would have taken years or centuries to carry out—and indeed that never would have been carried out because of the time involved or the probability of error—are now routinely made in minutes. As computers grow faster and cheaper, as they do every year, our capabilities grow.

The importance of this part of the book is based on a theorem that states that all stars of the same mass and of the same chemical composition evolve in the same way, if we neglect the effects of rotation and magnetic field. Since the chemical differences among stars are usually not overwhelming, it is largely the mass that is the chief determinant of stellar evolution. Because of this fact, we need only study a few groups of stars to gain pictures of the evolution of most stars.

We can set up divisions based on the masses of stars. Collapsing gas that is less than about 8 per cent as massive as the sun never reaches stardom, and it becomes a brown dwarf. Stars more massive than that but containing less than about 4 solar masses eventually lose some of their mass and become white dwarfs; we call these **low-mass stars. High-mass stars,** which contain more than 8 solar masses, explode as one of the types of supernovae. After becoming supernovae, some high-mass stars wind up as neutron stars, which we may detect as pulsars or in x-ray binaries. Other high-mass stars wind up withdrawing from the universe in the form of black holes. The fate of **middle-mass stars,** between 4 and 8 solar masses, is less clear, though it is generally thought they wind up as low-mass stars. Note that in all cases, we are discussing mass (which is an intrinsic property) rather than weight (which is the force of gravity on a mass).

Let us start in the next chapter with the formation and the main lifetime of stars. The following four chapters will then discuss the death of stars.

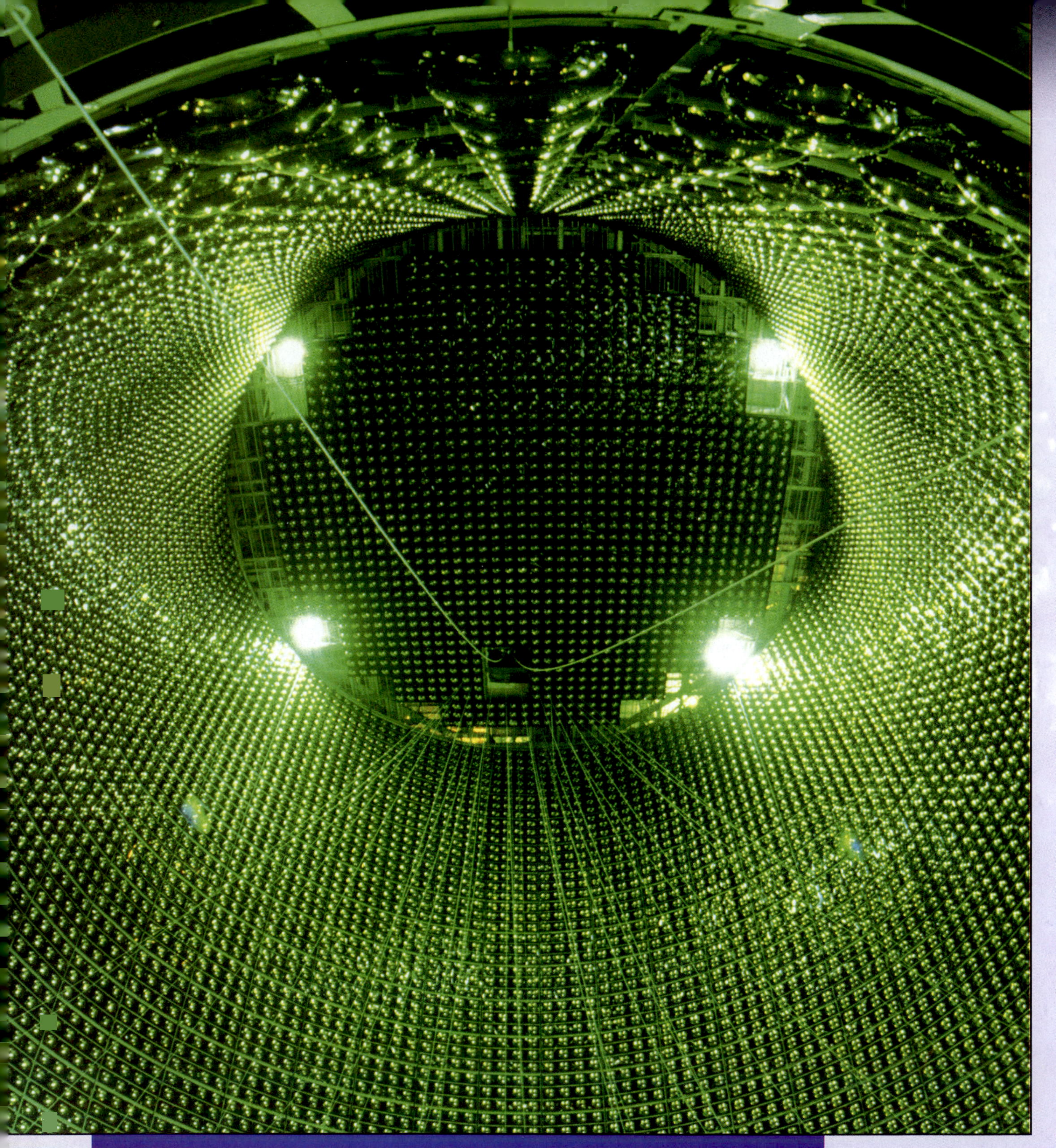

The Super-Kamiokande detector in Japan detects neutrinos from the sun and is providing a test not only of how the stars shine but also of whether our basic understanding of fundamental physics is correct. Here we see thousands of light-sensitive "photomultiplier" tubes that are detecting flashes of light from interactions between incoming neutrons and nuclear particles in the water that now fills this huge cavity.

# Chapter 27

# The Birth, Youth, and Middle Age of Stars

Even though individual stars shine for a relatively long time, they are not eternal. Stars are born out of gas and dust that may exist within a galaxy (Fig. 27–1); they then begin to shine brightly on their own. We now observe the dust and young stars especially well in the infrared (Fig. 27–2), not only with new types of sensors on ground-based telescopes but also from space with the Space Infrared Telescope Facility. Though we can observe only the outer layers of stars, we can deduce that the temperatures at their centers must be millions of kelvins (that is, millions of degrees Celsius above absolute zero). We can even deduce what it is deep down inside that makes the stars shine.

**AIMS:** To study the formation of stars, their main-sequence lifetimes, and how they generate their energy

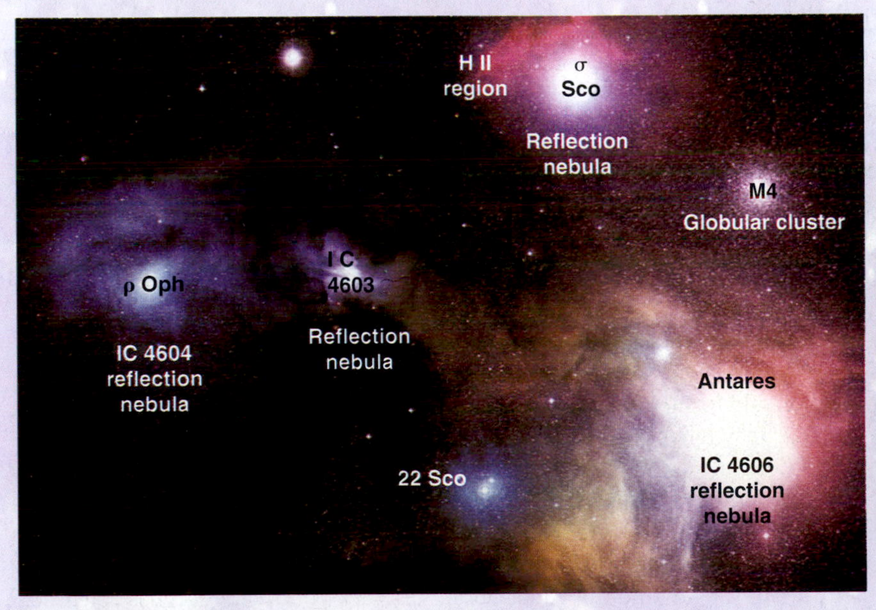


*A*

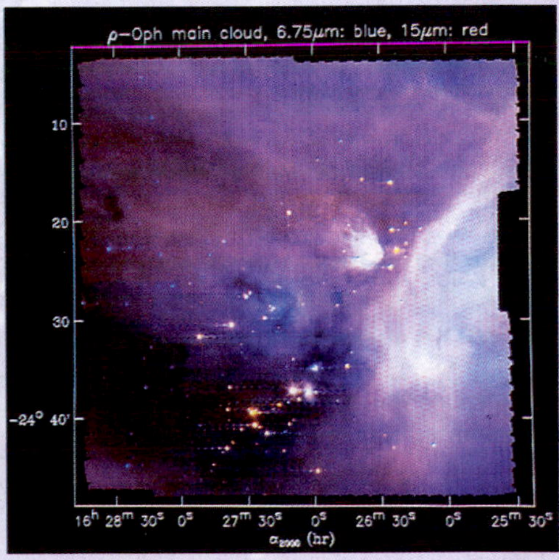


*B*

**FIGURE 27–1** (*A*) The clouds of gas and dust around $\rho$ (rho) Ophiuchi (in the blue reflection nebula) in visible light. The bright star Antares is surrounded by yellow and red nebulosity. The complex is 500 light-years away. M4, the nearest globular cluster to us, is also seen. (*B*) An infrared view of the region directly around $\rho$ Ophiuchi, covering only about 10% of the field of view of the optical image. In this ISOCAM view from the Infrared Space Observatory, infrared at 7 $\mu$m is reproduced as blue, and infrared at 15 $\mu$m is reproduced as red. The brightest regions are placental clouds glowing in the infrared because their dust has absorbed the ultraviolet light from newly born, massive stars.

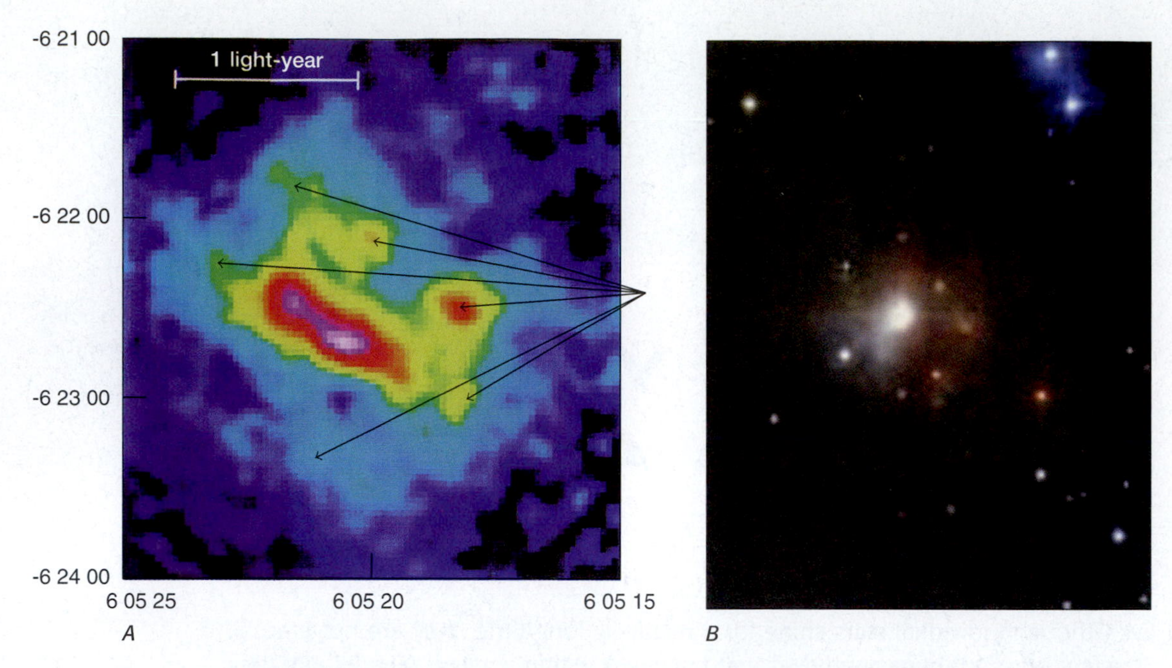


**FIGURE 27–2** (*A*) This view of a region in Monoceros reveals star formation. The image was taken with a camera that works at the long infrared wavelength of 350 $\mu$m with the Caltech Submillimeter Observatory on Mauna Kea. Hot gas surrounds two young, massive stars; arrows mark cool dust clumps that are likely to contain smaller, sunlike protostars. (*B*) The region of the star R Coronae Australis, imaged with the 2MASS infrared survey at a wavelength of 2 micrometers. The star is an example of an Ae/Be star, where the "e" means that emission lines are visible in addition to the ordinary A-star/B-star absorption lines. Such stars have 2 to 8 times the mass of the sun. The Corona Australis complex, one of the nearest star-forming regions, is only about 425 light-years away. The densest gas hides what is behind it, reducing its brightness by 35 magnitudes in the visible part of the spectrum. Over a dozen young stellar objects (YSOs) are visible. They are only about 3 million years old.

In this chapter we will discuss the birth of stars, then consider the processes that go on in a stellar interior during a star's life on the main sequence, and then begin the story of the evolution of stars when they finish this stage of their lives. The next four chapters will continue the story of what is called **stellar evolution.**

## 27.1 STARS IN FORMATION

Note that a given star does not move along the main sequence; it stays for a long time at essentially the same place on the H-R diagram. The main sequence shows up only when we plot the properties of a lot of stars.

The process of star formation starts with a region of gas and dust. The dust—tiny solid particles—may have been given off from the outer atmospheres of giant stars. A region may become slightly denser than its surroundings, perhaps from a random fluctuation in density or because a "density wave" in our galaxy (Section 32.5c) compressed it. Or a star may explode nearby—a "supernova"—sending out a shock wave that compresses gas and dust. In any case, once a minor density enhancement occurs, gravity keeps the gas and dust contracting. As they contract, energy is released; it turns out that half of that energy heats the matter (a result known as the "virial theorem"), causing it to give off an appreciable amount of electromagnetic radiation. Such a not-quite-yet-formed star is called a "protostar" (from the prefix of Greek origin meaning "first").

We can plot the position of a star on an H-R diagram for any particular instant of its life. Each of the pairs of luminosity and temperature corresponds to a point on the H-R diagram. Connecting the points representing the entire lifetime of the star gives an **evolutionary track.** A particular star, of course, is only at one point of its

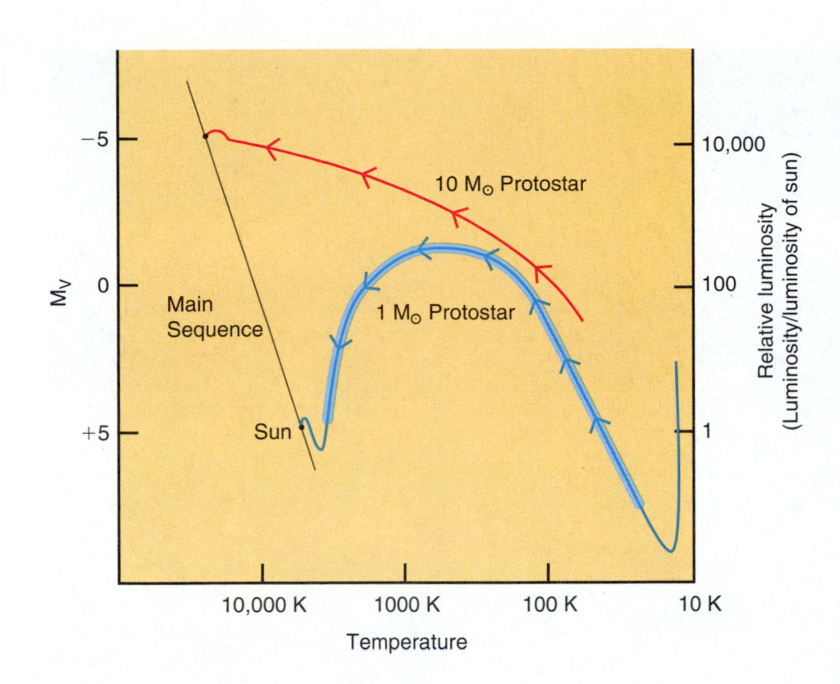


**FIGURE 27–3** Evolutionary tracks for two protostars, one of 1 solar mass and the other of 10 solar masses. ($M_\odot$ is the symbol for the mass of the sun. By saying "a star of 10 solar masses," we mean simply a star that has 10 times the mass of the sun.) The evolution of the shaded portion of the track for the 1-solar-mass star is described in the text. These stages of its life last for about 50 million years. More massive stars whip through their protostar stages more rapidly; the corresponding stages for a 10-solar-mass star may last only 200,000 years. These very massive stars wind up being very luminous. Note that stars evolve from the right side of the diagram toward the left side of the diagram along the paths shown.

Absolute visual magnitude is plotted at the left vertical axis. An alternative scale showing the star's luminosity relative to that of the sun is plotted on the right-hand vertical axis.

track at any given time. And note that an evolutionary track shows how the properties of a star evolve, not how or whether the star is physically moving in space.

The evolutionary track of a protostar depends on its mass (Fig. 27–3). At first, a protostar brightens, and the track bends upward and toward the left on the diagram—though actually the star may be hidden by the dust in which it is embedded. While the protostar's surface is brightening, the central part of the protostar continues to contract, and the temperature rises.

By this time, the dust has vaporized, allowing us to see the star itself. The star's gas has become opaque, so energy emitted from its central region does not escape directly. The outer layers continue to contract. Since the surface area is decreasing, the luminosity decreases, and the track of the protostar begins to move downward on the H-R diagram. As the protostar continues to heat up, it also continues to move toward the left on the H-R diagram (that is, it gets hotter).

The higher temperature results in a higher pressure, which pushes outward more and more strongly. Eventually a point is reached where this outward force balances the inward force of gravity for the central region. This is the time when the star reaches the main sequence of the Hertzsprung-Russell diagram and when we say that the star is born. We discuss this next phase of a star's life in the following section; let us now continue with the pre-main-sequence stars.

The time scale of the gravitational contraction that precedes the main-sequence lifetime depends on the mass of gas in the protostar. Very massive stars contract to approximately the size of our solar system in only 10,000 years or so. These massive objects become O and B stars and are located at the top of the main sequence (Fig. 27–4). They are sometimes found in groups in space, which are called **O and B associations.**

Less massive stars contract much more leisurely before they reach the main sequence. A star of the same mass as the sun may take tens of millions of years to contract, and a less massive star may take hundreds of millions of years to pass through this stage.

Theoretical analysis shows that the dust surrounding the stellar embryo we call a protostar should absorb much of the radiation that the protostar emits. The radiation from the protostars should heat the dust to temperatures that produce primarily infrared radiation. Infrared astronomers have found many objects that are especially bright in the infrared but that have no known optical counterparts. With infrared-

FIGURE 27–4 Massive, newly formed stars are visible in the infrared (*bottom*) but not in the visible (*top*) image. Both were taken with the Hubble Space Telescope, the top view with the Wide Field and Planetary Camera 2 and the bottom view with the NICMOS infrared camera. The view is of part of the 30 Doradus nebula in the Large Magellanic Cloud. The pictures show that the powerful radiation and high-speed material given off by the massive stars present here are triggering a new burst of star birth.

*A*

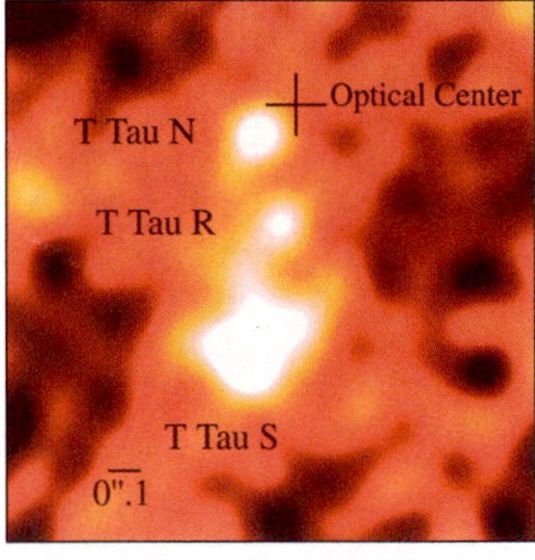


*B*

FIGURE 27–5 T Tauri. (*A*) Ground-based view in the near infrared from the 2MASS survey. Hind's Nebula (NGC 1555) wraps around T Tauri. (*B*) A radio map of T Tauri from the MERLIN interferometer. A third source is also revealed in addition to an outflow from the southern star. The position and size of the plus sign show the best estimate of the optical star's position and how uncertain that measurement is, as measured from the ground; Hipparcos gave a position of the optical star that even more convincingly overlaps T Tauri North.

sensitive telescopes on satellites and with the 2MASS full-sky mapping, we have discovered so many of these that we now think that about one star forms each year in our Milky Way Galaxy. These objects seem to be located in regions where the presence of a lot of dust and gas and other young stars indicates that star formation might be going on.

In the visible part of the spectrum, several classes of stars that vary erratically in their brightness are found. One of these classes, called **T Tauri stars,** includes stars of spectral types G, K, and M. These spectral types have relatively low masses. T Tauri stars are thought to be less than a few million years old and not to have reached the main sequence. Their visible radiation can vary by as much as several magnitudes. Astronomers work in the infrared to study the dust grains that surround T Tauri stars. Presumably, T Tauri stars have not quite settled down to a steady and reliable existence. They fall slightly above (to the right of) the main sequence of the H-R diagram, which is consistent with their being very young, as young stars have yet to move to the left onto the main sequence. Low-mass stars like these are now known to form more or less continuously in dark clouds of gas and dust. High-mass stars, on the other hand, form only in the densest parts of hotter clouds. Further, the high-mass stars form only when shock waves make the clouds collapse.

T Tauri stars are found in close proximity to each other; these groupings are known as **T associations.** The T Tauri stars are embedded in dark dust clouds. Also in these clouds are bright nebulous regions of gas and dust called **Herbig-Haro objects.** T Tauri itself (Fig. 27–5), the prototype of its class of variable stars, is near a typical Herbig-Haro object. It has an infrared companion that has recently ejected particles in opposite directions. Radio observations show that the particles are guided by a strong magnetic field.

We see Herbig-Haro (H-H) objects (Fig. 27–6) as a result of gas ejected from very young T Tauri stars. The supersonic jet-like gas flows—millions of times more powerful than the solar wind—interact with surrounding material to produce flowing shock waves. Jets of gas are sometimes seen; they probably come from repeated eruptions out the stars' poles. The heated regions, containing about one Earth mass of gas and dust, appear as Herbig-Haro objects. Some H-H objects reflect light from nearby stars. The stars that provide the energy to H-H objects sometimes are detectable only in the infrared.

Radio maps of T Tauri show that T Tauri's stellar wind—so strong that we might call it a gale—is not spherical. Observations of many T Tauri stars show that

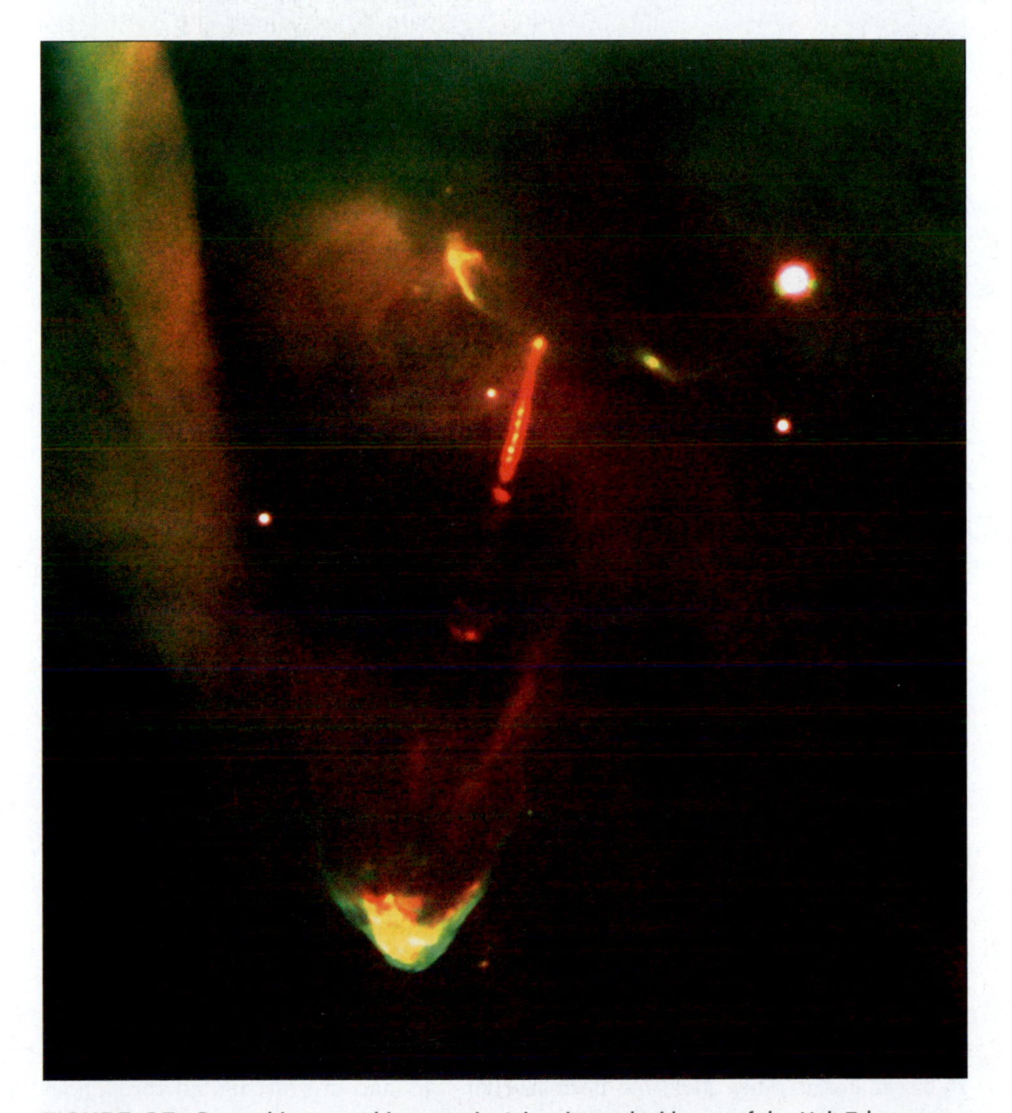

**FIGURE 27–6** Herbig-Haro Object #34 in Orion, imaged with one of the Unit Telescopes of the Very Large Telescope. We see two opposite jets ejected from a newborn star and ramming into the surrounding interstellar matter. The jets of gas were formed as the young star contracted under the force of its own gravity. Because a thick disk of cool gas and dust surrounds the star, the gas squirts outward along the star's axis of rotation at velocities of perhaps 1 million km/hr. The structure we see is produced by a machine-like blast of "bullets" of dense gas ejected from the star at high velocity. The appearance suggests that the star has episodic outbursts in which chunks of matter fall onto the star from a surrounding disk. The smallest features resolved are about the size of our solar system. The star itself is hidden by a dusty cocoon of gas. Herbig-Haro objects are named after George Herbig, formerly of the Lick Observatory and now of the University of Hawaii, and Guillermo Haro of the Mexican National Observatory.

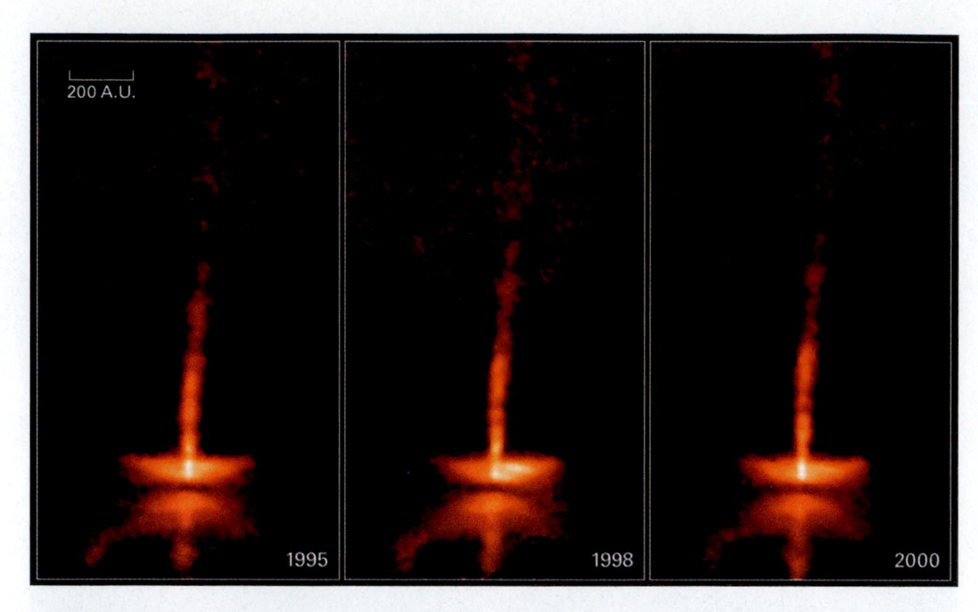


**FIGURE 27–7** The changes in HH-30 over a five-year period, as imaged with the Hubble Space Telescope. Our sun and planets probably formed from a disk of dust and gas like the one shown, about 450 A.U. across, which is blocking the star itself from our view. The star is only about a half-million years old. We see the star's light reflected by dust above and below the central plane of the disk. The star's light is also reflected by a dust cloud at top, which reveals changes in the star's brightness. The jets, held in place by the star's magnetic field, erupt from the star's poles. As in the previous image, knots of gas are emitted sporadically and travel at speeds up to about 1 million km/hr.

> The gas and dust from which stars are forming are best observed in the radio and infrared regions of the spectrum, respectively. We shall have more to say about these regions when we discuss interstellar matter in Chapter 33. In particular, in recent years we have learned of the importance of "giant molecular clouds" in star formation.

material is sent out in oppositely directed beams. This "bipolar ejection" (Fig. 27–7) of gas seems common and may imply that a disk of material orbits these stars, blocking outward flow in the equatorial direction and channeling flow at the poles. The source of energy driving the strong winds is not known, though a leading model uses magnetic-field–related waves to bring energy to the base of the stellar gales. Probably the result of the bipolar ejection, the H-H objects are often aligned as though they were ejected to opposite sides of the T Tauri star.

Classical (that is, ordinary) T Tauri stars show strong optical emission lines in addition to the absorption lines that all stars show (Fig. 27–8). These T Tauri stars also show excess radiation in the infrared and ultraviolet. The emission lines and excess radiation come from circumstellar (that is, "around the star") material—leaving intact the concept that the photospheres of stars show absorption spectra. When the absorption spectra from the photosphere travels through the circumstellar material, the emission lines are added. The T Tauri stars' radiation across the spectrum from x-ray through ultraviolet and visible varies strongly. Some "weak-line T Tauri stars" have been found from an x-ray survey. They are also low-mass stars not quite yet on the main sequence but, unlike the classical T Tauris, do not show signs of circumstellar material. They look like normal, cool stars and are distributed throughout

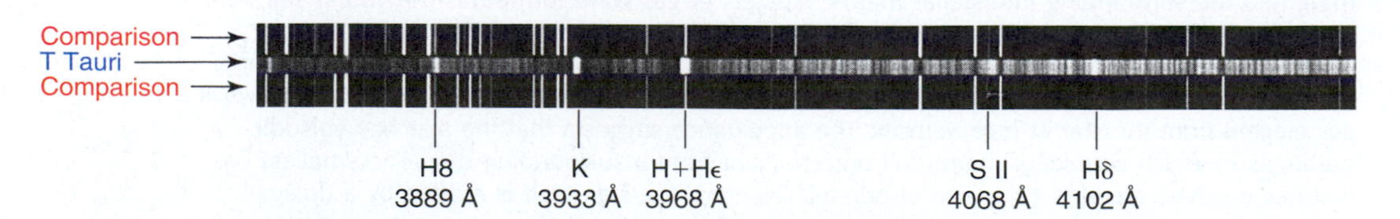


**FIGURE 27–8** T Tauri's spectrum shows strong, broad, optical emission lines. The star's spectrum is surrounded top and bottom by a comparison spectrum of a terrestrial source.

regions of star formation. We may come to understand the pre-main-sequence evolution of low-mass stars better from studying these weak-line T Tauri stars, because we can study the underlying stars directly without the interference of surrounding matter.

Solar-system–sized clouds of dust have been detected around several young stars, including HL Tauri (Fig. 27–9). The clouds are elongated, as though they are disk-shaped. The disks are best studied in the infrared and in submillimeter and millimeter waves. Studies of millimeter-wavelength emission from the disks show cool gas following Kepler's laws, confirming that the gas is orbiting the stars. The masses, sizes, and angular momentum of the disks are similar to those we think our solar nebula had when planets started to form. It seems reasonable that planetary systems are also forming around these young stars, some only 100,000 years of age. We discussed more about possible planets formed from dust around other stars, and about planets formed differently, near a pulsar, in Section 18.2. Apparently, the T Tauri stage plays a very important role in determining the masses of planets that are formed around these stars.

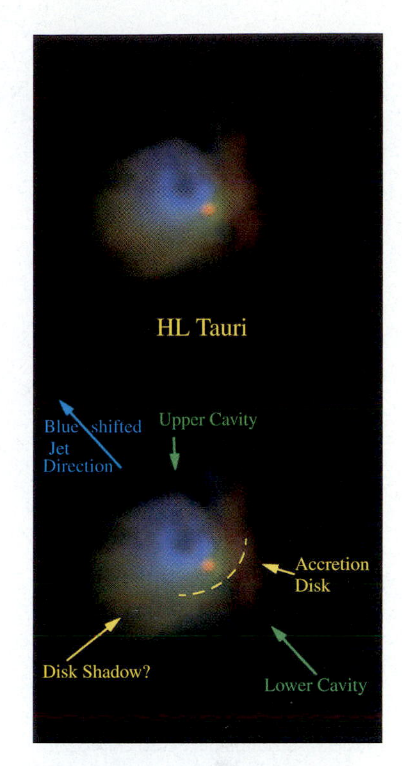


**FIGURE 27–9** The very young star HL Tauri is still embedded in the Taurus dark cloud, the molecular cloud from which it was born. This image shows the region of HL Tau in three colors in the near infrared (0.8, 1.2, and 1.6 micrometers), reproduced in false color as blue, green, and red. The image shows an accretion disk 150 A.U. in radius, perhaps like a protoplanetary disk. This last idea is endorsed by the measurement from radio observations that the gas revolves following Kepler's laws. We see a blue cavity (*above the disk*) and a fainter, more obscured red cavity (*below the disk*). These cavities are expanding at 30 km/sec as a strong 300 km/sec jet clears away the dust perpendicular to the disk. The disk contains about 40 Jupiters of mass. This mass corresponds with ideas about the formation of solar systems. HL Tau is young, having been collapsing for only 100,000 years.

## 27.2 NUCLEAR ENERGY IN STARS

All the heat energy in stars that are still contracting toward the main sequence results from the gravitational contraction itself. If this were the only source of energy, though, stars would not shine for very long on an astronomical timescale—only about 30 million years. Yet we know that even rocks on Earth are older than that, since rocks over 4 billion years old have been found. Some other source of energy holds the atmospheres of a star up against the star's own gravitational pull.

The gas in the protostar will continue to heat up until the central portions become hot enough for **nuclear fusion** to take place. Using this process, which we will soon discuss in detail, the star can generate enough energy inside itself to support it during its entire lifetime on the main sequence. The energy makes the particles in the star move around rapidly. For short, we say that the particles have a "high temperature." The particles exert a **thermal pressure** pushing outward, providing a force that balances gravity's inward pull. As we said above, it is this balance that stars on the main sequence have.

The basic fusion processes in most stars fuse four hydrogen nuclei into one helium nucleus, just as hydrogen atoms are combined into helium in a hydrogen bomb here on Earth. In the process, tremendous amounts of energy are released. We would like to be able to control fusion on Earth as well as it is controlled in stars. One major experimental approach to fusion on Earth uses magnetic fields to hold the plasma together at a high enough density and for a long enough time for fusion to occur. This "containment" of the plasma is difficult on Earth; containment is not a problem for the sun and stars, whose strong gravity keep the plasma dense at their cores. Containment is also not a long-term problem in making a bomb.

A hydrogen nucleus is but a single proton. A helium nucleus is more complex. It consists of two protons and two neutrons (Fig. 27–10). The mass of the helium nucleus that is the final product of the fusion process is slightly less than the sum of the masses of the four hydrogen nuclei that went into it. The mass of the resulting helium is 0.7 per cent less than the mass of the four hydrogen nuclei. Has the mass disappeared in the process?

The mass does not really simply disappear but is rather converted into energy according to Albert Einstein's famous formula $\boldsymbol{E = mc^2}$. Now, $c$, the speed of light, is a large number, and $c^2$ is even larger. Thus even though $m$ is only a small fraction of the original mass, the amount of energy released is enormous. The loss of only 0.007 (James Bond's number, equivalent to 0.7 per cent) of the mass of the central part of the sun, for example, is enough to allow the sun to radiate at its present rate for a period of at least ten billion ($10^{10}$) years. This fact, not realized until 1920 and

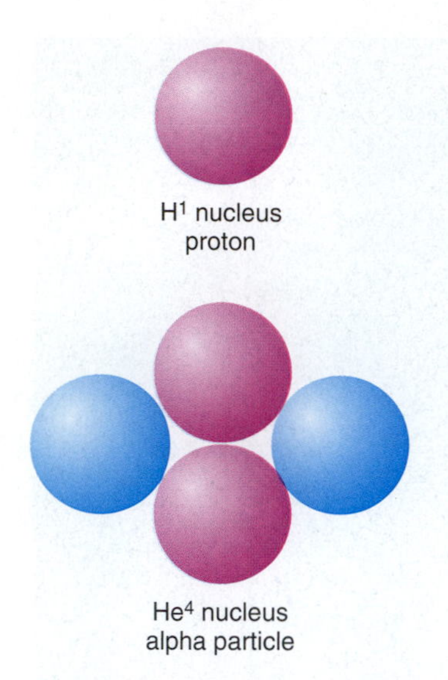


**FIGURE 27–10** The nucleus of hydrogen's most common form is a single proton, while the nucleus of helium's most common form consists of two protons and two neutrons. Protons and neutrons are made of smaller particles called quarks, which are discussed in Chapter 38.

In the interiors of stars, we are dealing with nuclei instead of the atoms we have discussed in stellar atmospheres, because the high temperature strips the electrons off the nuclei. The nuclei have positive electric charge. The electrons are mixed in with the nuclei through the center of the star; there can be no large imbalance of positive and negative charges, or else strong repulsive forces would arise inside the star. Such an ionized gas is known as a "plasma."

worked out in more detail in the 1930s, solved the long-standing problem of where the sun and the other stars got their energy.

All the main-sequence stars are approximately 90 per cent hydrogen (that is, 90 per cent of the atoms are hydrogen), so there is lots of raw material to stoke the nuclear "fires." We speak colloquially of "nuclear burning," although, of course, the processes are quite different from the chemical processes that are involved in the "burning" of logs or in cooking. In order to be able to discuss these processes, we must first discuss the general structure of nuclei and atoms.

## 27.3 ATOMS

An atom consists of a small **nucleus** surrounded by **electrons.** Most of the mass of the atom is in the nucleus, which takes up a very small volume in the center of the atom. The effective size of the atom, the chemical interactions of atoms to form molecules, and the nature of spectra are determined by the electrons.

The nuclear particles with which we need be most familiar are the **proton** and **neutron.** Both these particles have nearly the same mass, 1836 times greater than the mass of an electron, though still tiny (Appendix 2). The neutron has no electric charge, and the proton has one unit of positive electric charge. The electrons, which surround the nucleus, have one unit each of negative electric charge. When an atom loses an electron, it has a net positive charge of 1 unit for each electron lost. The atom is now a form of **ion** (Fig. 27–11). Astronomers use Roman numerals to show the stage of ionization: I for the electrically neutral state, II for the state in which one electron is lost, etc. Thus Fe I is neutral iron, Fe II is the state of $Fe^{+}$ ions, Fe III is the state of $Fe^{+2}$ ions, etc. (*Note:* The Roman numeral is one higher than the number of electrons removed.)

The number of protons in the nucleus determines the quota of electrons (and hence the chemistry) that the neutral state of the atom must have, since this number determines the charge of the nucleus. Each **element** (sometimes called "chemical element") is defined by the number of protons in its nucleus. The element with one proton is called hydrogen, that with two protons is called helium, that with three protons is called lithium, and so on.

Though a given element always has the same number of protons in a nucleus, it can have several different numbers of neutrons. (The number is always somewhere between 1 and 2 times the number of protons. Hydrogen, which need have no neutrons, and helium are the only exceptions to this rule.) The possible forms of an element having different numbers of neutrons are called **isotopes.**

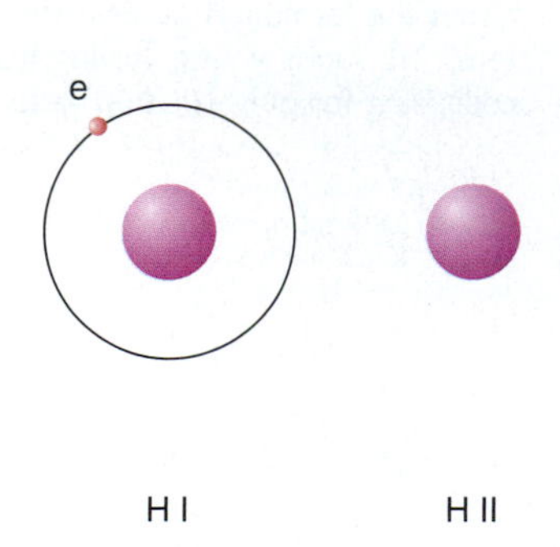


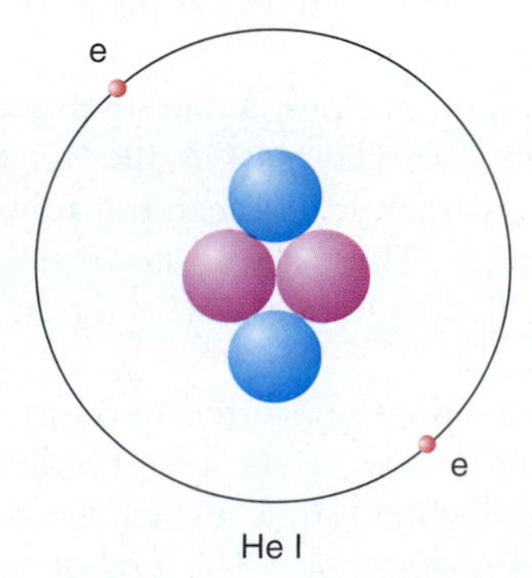


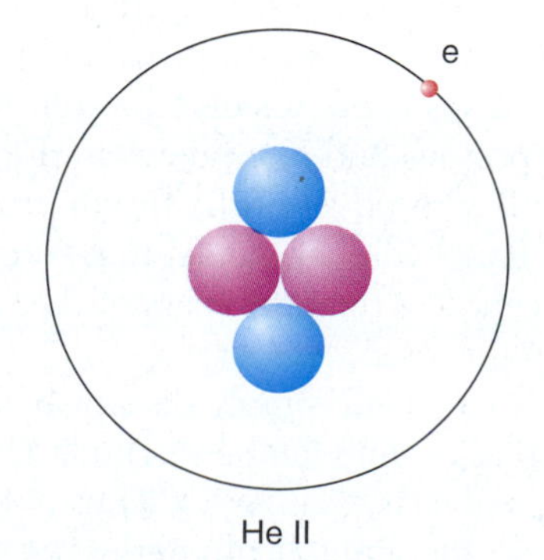

**FIGURE 27–11** Hydrogen and helium ions. The sizes of the nuclei are greatly exaggerated with respect to the sizes of the orbits of the electrons.

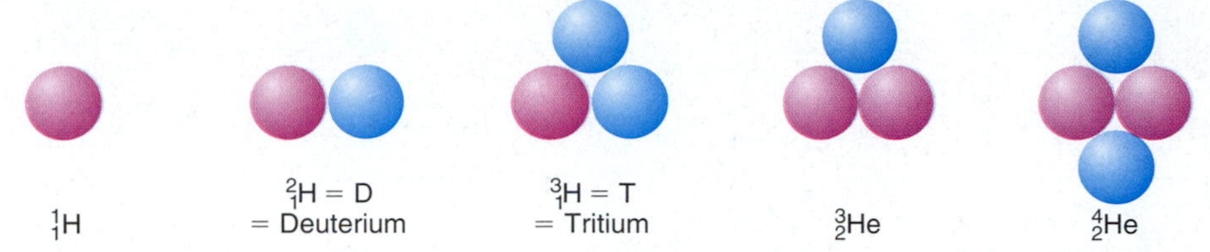


**FIGURE 27–12** Isotopes of hydrogen and helium. $^2_1H$ (deuterium) and $^3_1H$ (tritium) are much rarer than the normal isotope, $^1_1H$. Similarly, $^3_2He$ is much rarer than $^4_2He$.

For example, the nucleus of ordinary hydrogen contains one proton and no neutrons. An isotope of hydrogen (Fig. 27–12) called deuterium (and sometimes "heavy hydrogen") has one proton and one neutron. Another isotope of hydrogen called tritium has one proton and two neutrons.

Most isotopes do not have specific names, and we keep track of the numbers of protons and neutrons with a system of superscripts and subscripts. The subscript before the symbol denoting the element is the number of protons (called the **atomic number**), and a superscript is the total number of protons and neutrons together (called the **mass number**). Though when written out, both superscripts and subscripts are to the left of the element's symbol, when spoken, the superscript is read out last. Thus, $^4_2H$ is spoken as "2 helium 4," or just "helium 4," since helium automatically has mass number 2. For example, $^2_1H$ is deuterium in our notation, since deuterium has one proton, which gives the subscript, and a mass number of 2, which gives the superscript. Deuterium has atomic number = 1 and mass number = 2. Similarly, $^{238}_{92}U$ is an isotope of uranium with 92 protons (atomic number = 92) and mass number of 238, which is divided into 92 protons and 238 − 92 = 146 neutrons.

Each element has only certain isotopes. For example, most naturally occurring helium is in the form $^4_2He$, with a much lesser amount as $^3_2He$. (We sometimes informally read and write "helium-4" and "helium-3.") Sometimes an isotope is not stable, in that after a time it will spontaneously change into another isotope or element; we say that such an isotope is **radioactive.** Marie Curie and Pierre Curie did much of the basic work on radioactivity and gave the phenomenon its name shortly after it was discovered by Antoine Becquerel just over a hundred years ago. The three shared one of the first Nobel Prizes for their work.

In Box 26.3, we saw how radioactive decay of a uranium isotope has measured the age of the universe.

During certain types of radioactive decay, a particle called a **neutrino** is given off. A neutrino is a neutral particle (its name comes from the Italian for "little neutral one"). It has long been thought that neutrinos have no "rest mass," the mass they would have if they were at rest. An experiment in the late 1990s indicates that neutrinos in fact have some small amount of rest mass, though the experiment didn't show how much. (Einstein's special theory of relativity describes the relation of mass and velocity.) Still, observations we shall describe in Chapter 29 set low limits on the amount of mass a neutrino can have. If neutrinos have mass, then a substantial percentage of the mass of the universe could be in the form of neutrinos (see Section 38.2a on "dark matter").

## 27.4 ENERGY CYCLES IN STARS

Several sequences of reactions have been proposed to account for the fusion of four hydrogen atoms into a single helium atom. Hans Bethe, at Cornell University, suggested some of these processes during the 1930s. (He received the Nobel Prize in Physics for it in 1967.) The different possible sequences that have been proposed are important at different temperatures, so sequences that are dominant in the centers of very hot stars are different sequences from the ones that are dominant in the centers of cooler stars.

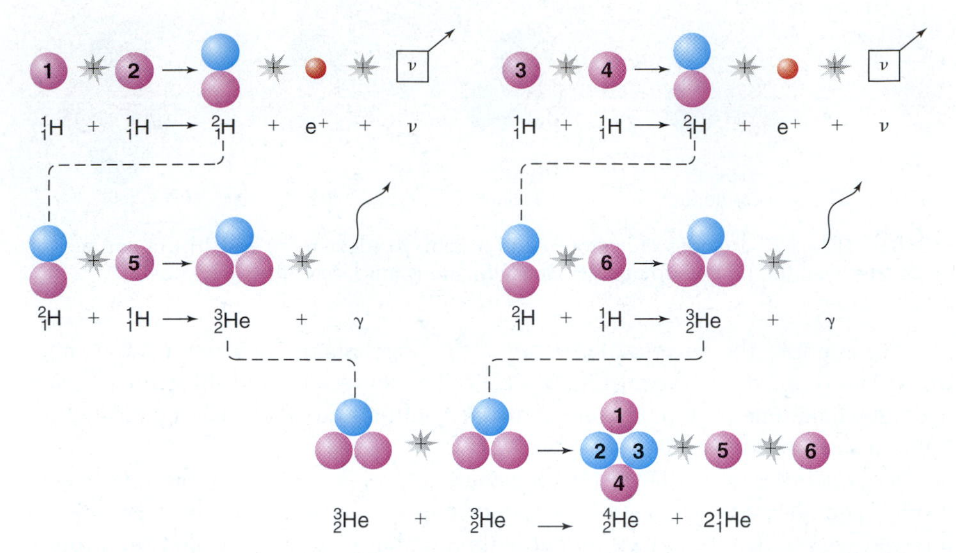


**FIGURE 27–13** The proton-proton chain; $e^+$ stands for a positron, $\nu$ (nu) is a neutrino, and $\gamma$ (gamma) is electromagnetic radiation at a very short wavelength. In the first stage, two nuclei of ordinary hydrogen fuse to become a deuterium (heavy hydrogen) nucleus, a positron (the equivalent of an electron, but with a positive charge), and a neutrino. (This stage is shown twice—at left and at right—to show the four original protons, which are numbered.) The neutrino immediately escapes from the star, but the positron soon collides with an electron. In a process not shown in the diagram, they annihilate each other, forming gamma rays. (A positron is an antielectron, an example of antimatter; whenever a particle and its antiparticle meet, they annihilate each other.) Next, the deuterium nucleus fuses with yet another nucleus of ordinary hydrogen to become an isotope of helium with two protons and one neutron. More gamma rays are released. (This stage is also shown twice, with the protons numbered.) Finally, two of these helium isotopes (one from each of the columns, as shown with dashed lines) fuse to make one nucleus of ordinary helium plus two nuclei of ordinary hydrogen.

When the center of a star is at a temperature less than 15 million K, the **proton-proton chain** (Fig. 27–13) dominates. Our sun makes most of its energy in this way. In the proton-proton chain, we put in six hydrogens one at a time and wind up with one helium plus two hydrogens, a net transformation of four hydrogens into one helium. But the six protons contained more mass than do the final single helium plus two protons. The small fraction of mass that disappears in the process is converted into an amount of energy that we can calculate with the formula $E = mc^2$.

For stellar interiors hotter than that of the sun, the **carbon-nitrogen-oxygen (CNO) cycle** (Fig. 27–14) dominates. The CNO cycle begins with the fusion of a hydrogen nucleus with a carbon nucleus. After many steps, and the insertion of four hydrogen nuclei, we are left with one helium nucleus plus a carbon nucleus. Thus as much carbon remains at the end as there was at the beginning, and the carbon can start the cycle again. Again, four hydrogens have been converted into one helium, 0.007 (0.7 per cent) of the mass has been transformed, and an equivalent amount of energy has been released according to $E = mc^2$. It has also been found that alternative cycles involving different oxygen isotopes sometimes occur.

Stars with even higher interior temperatures, above $10^8$ K, fuse helium nuclei to make carbon nuclei. The nucleus of a helium atom is called an "alpha particle" for historical reasons. Since three helium nuclei ($^4_2He$) go into making a single carbon nucleus ($^{12}_6C$), the procedure is known as the **triple-alpha process** (Fig. 27–15). In the triple-alpha process, two alpha particles first interact temporarily to form beryllium-8 ($^8_4Be$). The beryllium-8 is not very stable, and, depending on the temperature and density, it lasts only long enough for a level of one beryllium-8 atom per billion alpha particles to exist. However, this level is sufficiently high for a third alpha particle to interact with the beryllium-8 nucleus, forming carbon-12. A series of other processes can build still heavier elements inside stars.

The Greek letters derive from a former confusion of radiation and particles. We know now that $\alpha$ particles, formerly called $\alpha$ rays, are helium nuclei; $\beta$ particles, formerly called $\beta$ rays, are electrons; and $\gamma$ rays, as we have seen, are electromagnetic radiation of short wavelength.

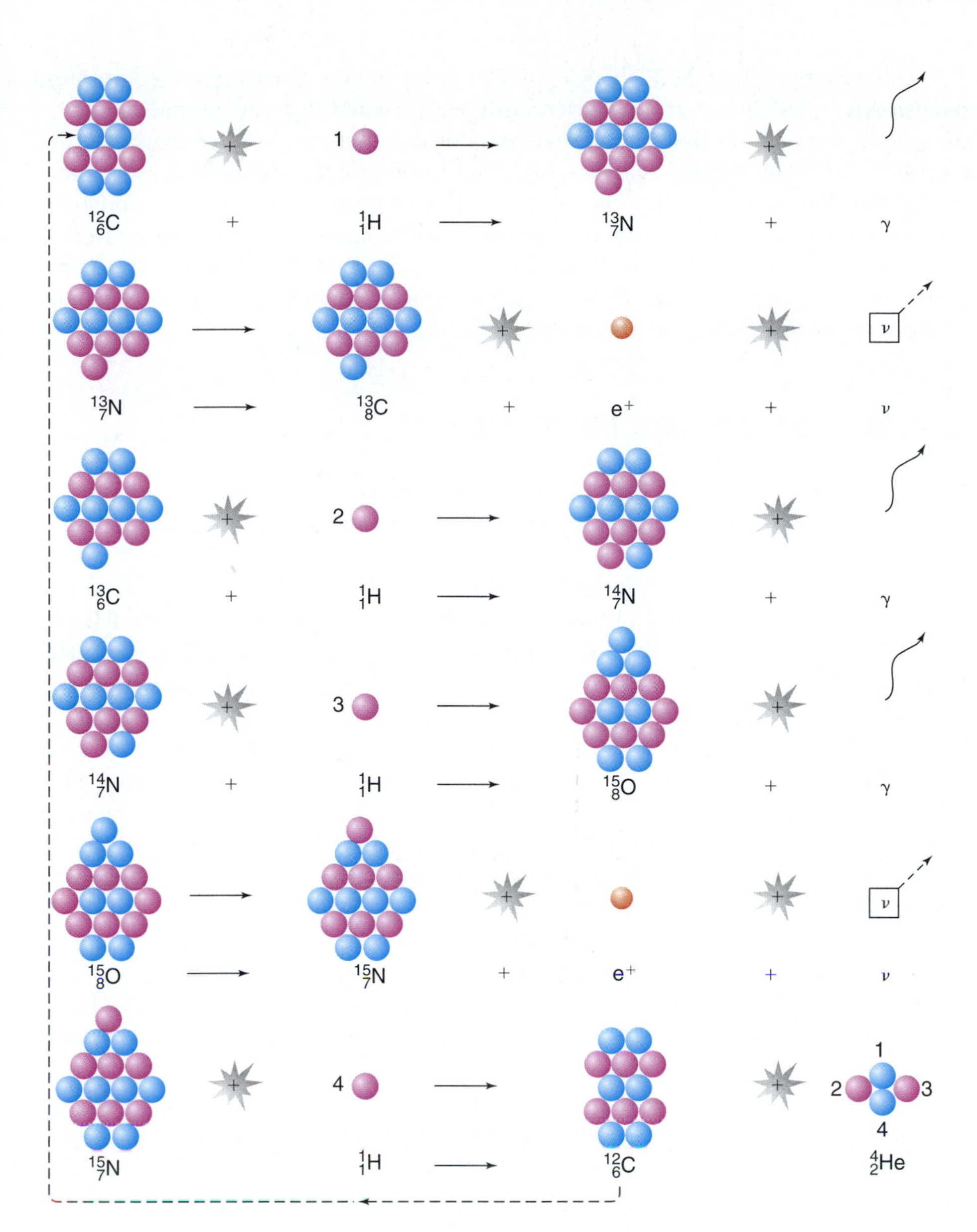


**FIGURE 27-14** The carbon-nitrogen-oxygen cycle, also called the carbon cycle. The four hydrogen atoms are numbered. Note that the carbon is left over at the end, ready to enter into another cycle.

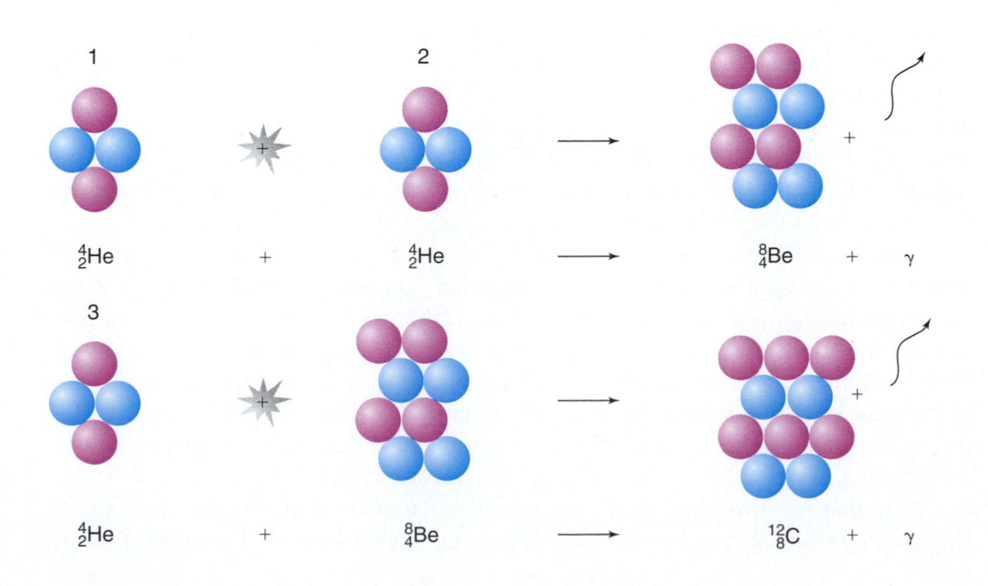


**FIGURE 27-15** The triple-alpha process, which takes place only at temperatures above about $10^8$ K. Beryllium, $^{8}_{4}Be$, is but an intermediate step, normally quickly decaying unless it fuses with a helium nucleus to become an excited state of carbon. This carbon nucleus soon emits the gamma ray shown.

The processes that build heavier nuclei from lighter ones are called **nucleosynthesis.** The theory of nucleosynthesis can account for the abundances we observe in stars and in the gas between stars of the elements heavier than helium. Currently, we think that the synthesis of isotopes of hydrogen and helium took place in the first few minutes after the origin of the universe (Section 38.1) and that the heavier elements were formed, along with additional helium, in stars or in supernova explosions (Section 29.2). William A. Fowler of Caltech shared the 1983 Nobel Prize in Physics for his work on nucleosynthesis, including the measurements of the rates of the nuclear reactions that make the stars shine.

## 27.5 THE STELLAR PRIME OF LIFE

Now that we have discussed basic nuclear processes, let us return to an astronomical situation. We last discussed a protostar in a collapsing phase, with its internal temperature rapidly rising.

One of the most common definitions of temperature describes the temperature as a measure of the velocities of individual atoms or other particles. Since this type of temperature depends on kinetics (motions) of particles, it is called the **kinetic temperature.** A higher kinetic temperature (we will call it simply "temperature" from now on) corresponds to higher particle velocities.

For a collapsing protostar, the energy from the gravitational collapse goes into giving the individual particles greater velocities; that is, the temperature rises. For nuclear fusion to begin, atomic nuclei must get close enough to each other so that the force that holds nuclei together, the **strong force** (Section 38.3), can play its part. But all nuclei have positive charges, because they are composed of protons (which bear positive charges) and neutrons (which are neutral). The positive charges on any two nuclei cause an electrical repulsion between them, and this force tends to prevent fusion from taking place.

However, at the high temperatures typical of a stellar interior, some nuclei have enough energy to overcome this electrical repulsion and to come sufficiently close to each other for the strong nuclear force to take over. The electrical repulsion that must be overcome is the reason why hydrogen nuclei, which have net positive charges of 1, will fuse at lower temperatures than will helium nuclei, which have net positive charges of 2.

Once nuclear fusion begins, the thermal pressure of the star's gas (the pressure resulting from temperature) provides a force that pushes outward strongly enough to balance gravity's inward pull. In the center of a star, the fusion process is self-regulating. If the nuclear energy production rate increases, then an excess pressure is generated that would tend to make the star grow larger. However, this expansion in turn would cool down the gas and slow down the rate of nuclear fusion. Thus the star finds a temperature and size at which it can remain stable for a very long time. This balance between thermal pressure pushing out and gravity pushing in characterizes the main-sequence phase of a stellar lifetime. These fusion processes are well regulated in stars. When we learn how to control fusion in power-generating stations on Earth, which currently seems decades off, our energy crisis may be over.

Stars more massive than 8 per cent of our sun's mass can sustain the fusion of protons, so we say that this limit represents the lowest-mass stars. Of great contemporary interest is the type of nuclear fusion that "burns" deuterium nuclei (1 proton + 1 neutron), instead of ordinary hydrogen nuclei (1 proton). The lower limit for deuterium "burning" is 1.3 per cent of our sun's mass. Deuterium-burning objects are known as brown dwarfs, and we have already seen (in Chapter 24) that the spectral-type sequence has been extended to L and M in large part to accommodate them.

Now that we have discussed stars much less massive than the sun, let us consider stars more massive than the sun. The more mass a star has, the hotter its core becomes before it generates enough pressure to counteract gravity. The hotter core leads to a higher surface luminosity, explaining the mass-luminosity relation (Section 26.2). Thus more massive stars use their nuclear fuel at a much higher rate than less massive stars, and even though the more massive stars have more fuel to burn, they go through it relatively quickly. The next three chapters continue the story of stellar evolution by discussing the fate of stars when they have used up their hydrogen.

## 27.6 THE SOLAR NEUTRINO EXPERIMENT

Astronomers can apply the equations that govern matter and energy in a star, and they can model the star's interior and evolution with a computer. Though the resulting model can look quite nice, nonetheless it would be good to confirm it observationally. Happily, the models are consistent with the conclusions on stellar evolution that one can make from studying different types of Hertzsprung-Russell diagrams. Still, it would be nice to observe a stellar interior directly. The attempts to do so have led to such exciting discoveries that they may turn out to affect our understanding of the fundamental nature of the particles and forces in our universe. Indeed, if I had to think of one part of astronomy with the best potential for leading to a revolution in physics it would be the topic we will now consider, the solar neutrino experiment.

Since the stellar interior lies under opaque layers of gases, we cannot observe directly any electromagnetic radiation it might emit. Only neutrinos escape directly from a stellar interior. Neutrinos interact so weakly with matter that they are hardly affected by the presence of the rest of the solar mass. Once formed deep in the sun's core, they zip out into space at near the speed of light. Raymond Davis, Jr., retired from the Brookhaven National Laboratory and working with Ken Lande at the University of Pennsylvania, has spent many years studying the neutrinos formed in the solar interior. John Bahcall of the Institute for Advanced Study at Princeton has been a major force behind the theoretical aspect of the program.

Neutrinos formed in the sun's normal proton-proton chain do not have enough energy for Davis's apparatus to detect them. So Davis's experiment can detect only the neutrinos that occur in branches of the chain that are followed less than 1 per cent of the time. Thus theoretical calculations of how often these branches are followed are an important part of the study.

FIGURE 27–16 The original neutrino telescope, deep underground in the Homestake Gold Mine in Lead, South Dakota, consists mainly of a tank containing 400,000 liters of perchloroethylene. The tank is now surrounded by water to shield it from unwanted cosmic particles.

How do we detect the neutrinos? Neutrinos, after all, pass through the earth and sun, barely affected by their mass. At this instant, neutrinos are passing through your body. Davis makes use of the fact that very occasionally a neutrino will interact with the nucleus of an atom of chlorine and transform it into an isotope of the gas argon. Argon interacts rarely with other atoms (and so is called a "noble gas"); this property makes it relatively easy to separate it from the rest of the matter in a tank of fluid so that the argon can be studied.

The transformation of chlorine into argon takes place very rarely, so Davis needs a large number of chlorine atoms. He found it best to do this by filling a large tank with liquid cleaning fluid, $C_2Cl_4$, where the subscripts represent the number of atoms of carbon and chlorine in the molecule. Davis has a large tank containing 400,000 liters (100,000 gallons) of this cleaning fluid. His neutrino telescope (Fig. 27–16) is set up 1.5 km underground (in a room in an active gold mine in Lead, South Dakota) to shield it from other particles, none of which pass through that much earth.

The experiments have continued, with increasing sophistication, for over 30 years. To everyone's surprise, his result is far below theoretical predictions.

In the last few years, other methods of detecting solar neutrinos have been worked out. One method works with large quantities of gallium. The metallic element gallium is sensitive to neutrinos of much lower energy than those with which chlorine interacts. It detects essentially all solar neutrinos, instead of only those with the high energy that the chlorine experiment can detect: gallium-71 interacts with an electron neutrino, and the results are germanium-71 plus an electron. The germanium-71 then decays with a half-life of 11.4 days, and the decay can be measured. A European collaborative, the Gallex consortium—with primarily scientists from Germany, France, and Italy, and with additional participation by scientists from the United States and from Israel—prepared 30 tons of gallium in the form of a water solution. Their tank is underground in the Gran Sasso automobile tunnel in Italy. A Russian experiment with U.S. participation, SAGE (Soviet-American Gallium Experiment; the word "Soviet" has not been changed to "Russian"), is using 56 tons of gallium in metallic form. (The gallium is so valuable that the Russian government is trying to get it back.) The values obtained by SAGE and Gallex are high enough to verify that the proton-proton chain fuels the sun but are low enough to indicate that some "new physics" beyond the standard model of the neutrino may be required.

Gallex is being expanded to a 100-ton Gallium Neutrino Observatory.

Another method uses purified water. It is sensitive to only very high-energy neutrinos that come from a minor branch of the proton-proton chain, but it has some other advantages, such as sensing the direction from which the neutrinos come. The first version of this experiment, called Kamiokande since it is in Japan's Kamioka zinc mine (the "nde" stands for "neutrino detection experiment"), was superseded by a much bigger Super-Kamiokande (Fig. 27–17). Super-Kamiokande's rate of 30 neutrinos a day is about 50 times higher than the rate in the chlorine experiment. A Hyper-Kamiokande, much bigger yet, is being planned with a cubic water tank 100 m on a side.

All the detection methods give results that are lower than the theoretical predictions (Fig. 27–18). But why? Is it the sun? Bahcall and most other theoreticians think that their knowledge of the solar interior is sufficient for the purpose, though they do keep improving their models. And now that we have detected solar neutri-

**FIGURE 27–17** The Super-Kamiokande neutrino detector in Japan, which was shown also in the photograph opening this chapter. We see the empty stainless steel vessel, a cube 40 m across, and some of the 13,000 photomultiplier tubes, each 50 cm across, surrounding it. Neutrinos interacting within molecules of the 50,000 tons of highly purified water (note the boat, photographed as the detector was partly filled) lead to the emission of blue flashes of light that are being detected.

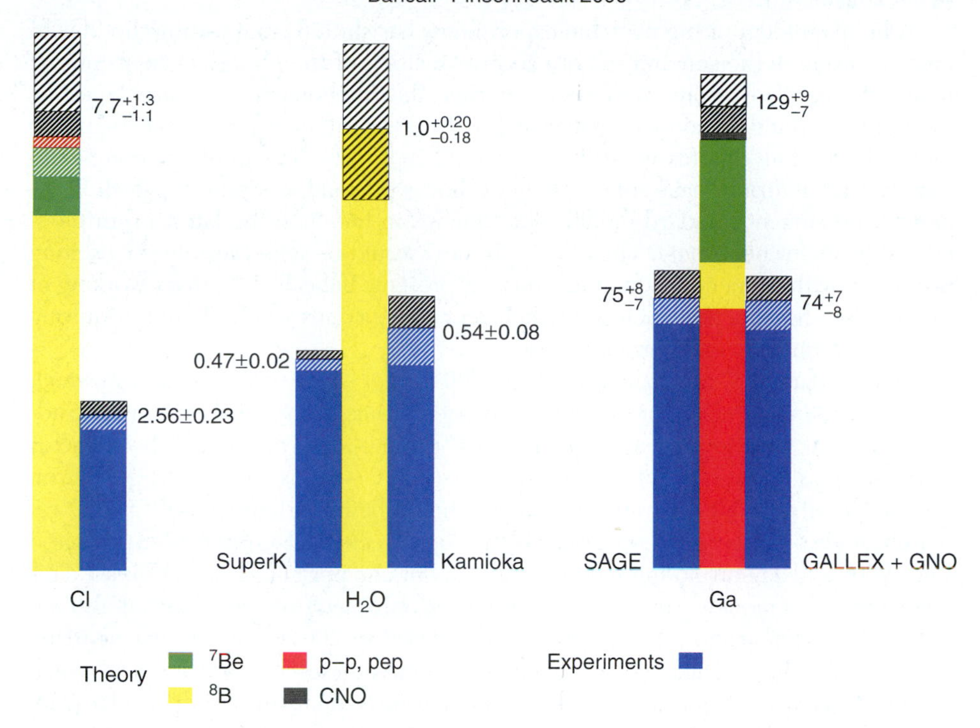


**FIGURE 27–18** The comparison of the results for gallium, chlorine, and water neutrino experiments. The gallium is the most sensitive, detecting neutrinos from proton-proton reactions in the solar core. The chlorine can detect only certain higher-energy and rarer events. The Super-Kamiokande detects only very high energy neutrinos that arise from the decay of boron-8, but its sensitivity and its ability to detect the direction of incoming particles makes it very valuable. Colors show the theoretical predictions for individual processes in the sun that make the detected neutrinos. Red shows expectations from the basic proton-proton chain and an alternative cycle involving an electron (e), green shows expectations from electron capture by beryllium-7, yellow shows expectations from decay of boron-8, and black shows expectations from the carbon-nitrogen-oxygen cycle, which is significant only in stars hotter than the sun. Blue shows the measurements. Note that the deficits are substantially larger than the uncertainties of the measurements or calculations, which are shown for each bar.

nos in various ways, no adjustment of the solar model can account for the relative deficiencies measured in all these different types of experiments.

Next, perhaps we do not understand the basic physics of nuclear reactions or of neutrinos as well as we thought. Davis's apparatus can detect only the form of neutrinos that the sun emits—"electron neutrinos." Perhaps a neutrino changes in type (there are three "flavors" of neutrinos: electron neutrinos, muon neutrinos, and tau neutrinos) during the 8 minutes the neutrino takes to travel at (or nearly at) the speed of light from the sun to the earth. Kamiokande is the source of the experimental evidence for this change in type, since it found a decrease in the number of electron neutrinos. But the details of the change are not yet known. Indeed, the tau neutrino itself was detected experimentally only in 2000.

In 1986, two Soviet scientists used a method worked out by an American scientist to point out that electron neutrinos are significantly scattered through the weak interaction as they travel through the very dense regions near the center of the sun. Since the Davis apparatus can detect only electron neutrinos, and since all neutrinos produced in the sun are electron neutrinos, such a transformation would make the Davis apparatus detect a smaller number of neutrinos. So the neutrino problem may be solved by this method, known after the initials of the scientists as

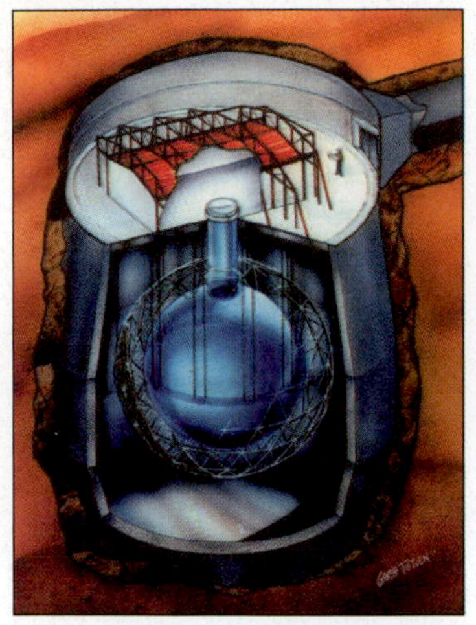

A

B

**FIGURE 27–19** The Sudbury Neutrino Observatory. (*A*) An artist's conception, showing the tank of deuterated water ("heavy water") that is the heart of the Sudbury Neutrino Observatory. (*B*) A photograph of the outside of the geodesic structure that surrounds the water tank to hold the photomultipliers that detect the signals. Note, for a sense of scale, the figures standing under the tank.

the MSW effect. Some other solutions involving nuclear-particle physics are also being considered.

The main focus of the neutrino experiments has shifted from testing the idea of nuclear fusion in the sun and stars to an investigation of the physics of neutrinos. If neutrinos "oscillate" from one type to another, this phenomenon is not covered by the current "standard model" that explains the elementary particles that make up matter in the universe (as we will discuss in Chapter 38). The phenomenon would require that neutrinos have some rest mass, however small, a condition which is not part of the current standard model (but which is in line with the latest experimental results about neutrinos). Thus "new physics" would be necessary, a very exciting possibility with tremendous ramifications. At present, Bahcall and others working in the field are hopeful that such exotic behavior of neutrinos will be found, requiring revisions in fundamental particle physics.

A Canadian/U.S. experiment is using 1000 tons of "heavy water" (water in which deuterium atoms replace the ordinary hydrogen atoms) in a nickel mine near Sudbury, Ontario. It is so well shielded that only three cosmic rays reach the detector per hour. This Sudbury Neutrino Observatory (Fig. 27–19) is looking for light given off when neutrinos hit the water. And there should be important spin-offs of the research, since other types of particles will also be detected. Data have been collected since 2000 and a neutrino-interaction rate of about one per hour is being found. This Observatory is uniquely capable of detecting not only electron neutrinos but also the muon and tau neutrinos. The first results, released in 2001, indicate that neutrino oscillations are indeed taking place, since this preliminary set of measurements (which detect electron neutrinos only) show fewer neutrinos compared with results from Super-Kamiokande, which include a small component of muon and tau neutrinos. Later SNO results will detect all neutrino types.

Still another type of detector for neutrinos at energies usefully lower than the water-based detectors has been built by German, Italian, Russian, and American scientists in Italy. This Borexino detector uses an organic liquid that gives off light when hit by electrons that are recoiling from their collisions with neutrinos. Both Sudbury and Borexino are sensitive to all flavors of neutrinos, though most sensitive to electron neutrinos, so they will be able to distinguish between problems in neutrino physics and problems in the solar model. Each of the detectors in this new generation collect more solar neutrinos in two months than all previous neutrino detectors had.

The ICARUS (Imaging Cosmic And Rare Underground Signals) detector uses liquid argon to detect high-energy solar neutrinos emitted in the chain that includes boron-8. The neutrinos leave visible tracks in the argon. It is at the same site in Italy as the Gallium Neutrino Observatory and Borexino. A module with 600 metric tons of liquid argon is in place; several modules will eventually total 5000 metric tons.

Several experiments are being set up in which neutrino detectors such as Super-Kamiokande, Gallium Neutrino Observatory, and ICARUS are used to detect beams of neutrinos generated at particle accelerators hundreds of kilometers away. The hope is to detect or rule out neutrino oscillations definitively.

The solar neutrino experiments, started as a test of fundamental stellar astrophysics, have turned into fundamental tests of our understanding of the universe and its constituents.

## 27.7 DYING STARS

We shall devote the next chapters to the various end stages of stellar evolution. The mass of the star determines its fate (Fig. 27–20). First we shall discuss the less massive stars, such as the sun. In Chapter 28, we shall see how such stars swell in size to become giants, possibly become planetary nebulae, and then end their lives as

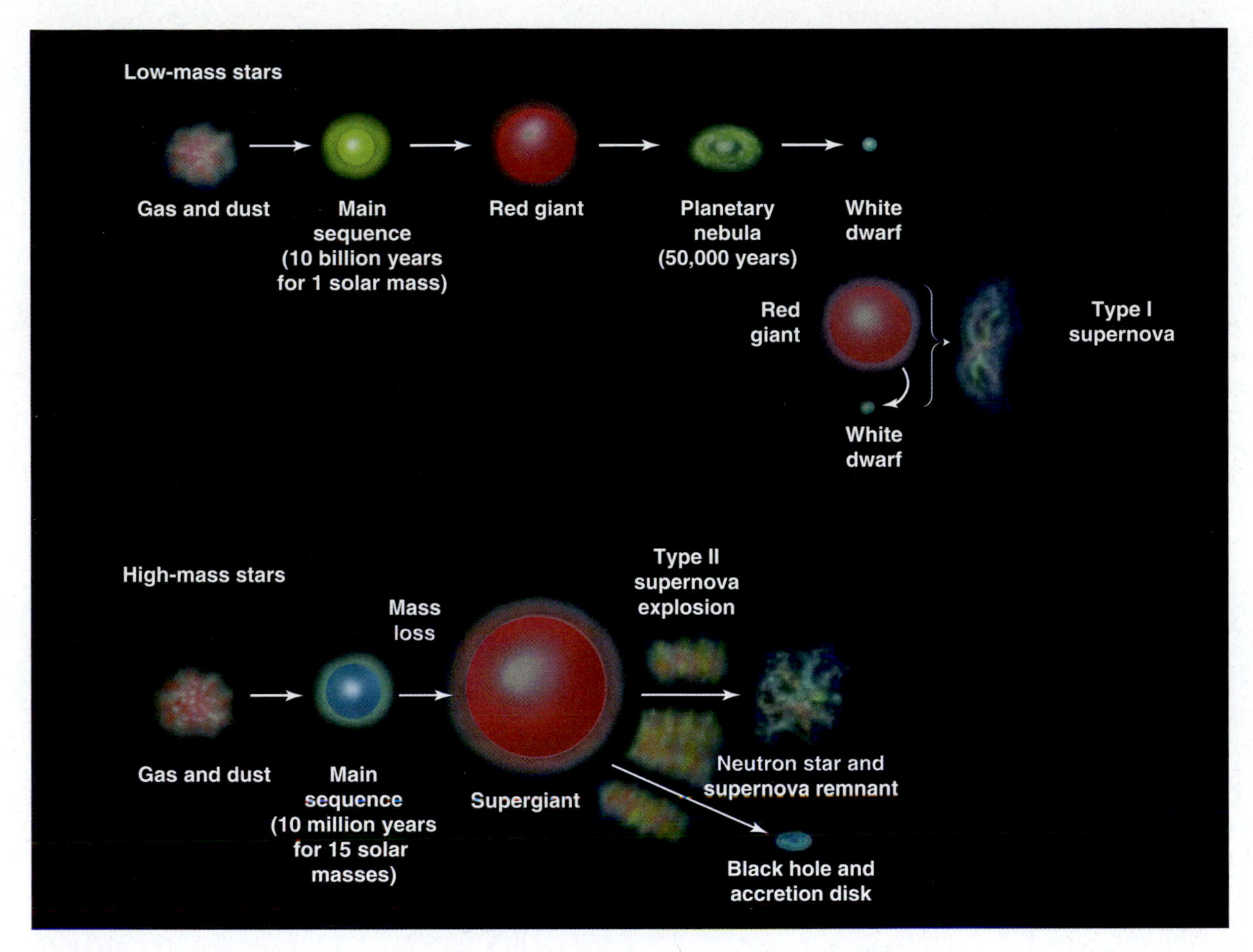


FIGURE 27–20 A summary of the stages of stellar evolution for stars of different masses; these stages will be described in the following four chapters.

white dwarfs. Some of these white dwarfs have companion stars and become either novae or supernovae.

More massive stars more regularly come to explosive ends. In Chapter 29, we shall see how these stars are blown to smithereens, and in Chapter 30 we will discuss how strange objects called "pulsars" are sometimes the remnants. Other remnants appear to be part of x-ray-emitting binary systems.

In Chapter 31, we shall encounter the strangest kind of stellar death of all. The most massive stars may become "black holes" and effectively disappear from view.

## CORRECTING MISCONCEPTIONS

| ✖ Incorrect | ✔ Correct |
|---|---|
| Stars don't change and live forever. | Stars evolve from birth to death. |
| Stars follow the same evolutionary path. | Evolutionary tracks depend on mass. |
| Radioactivity has always been known. | Radioactivity was discovered just over 100 years ago and was named by Marie Curie. |
| Stars shine from the carbon cycle. | Stars like the sun shine from the proton-proton chain. |
| Fe III is three-times-ionized iron, etc. | Fe III is twice-ionized iron, etc. |
| We can always feel high temperatures. | High temperatures mean fast-moving particles, which we feel only if there are a lot of them. |

## SUMMARY AND OUTLINE

Protostars (Section 27.1)
- Evolutionary tracks on H-R diagram: evolve from right to left
- Form when a cloud of gas and dust is compressed
- Observed in the infrared

T Tauri stars (Section 27.1)
- Irregular variations in magnitude
- May not have reached the main sequence
- Near Herbig-Haro objects
- Herbig-Haro objects show jets forming shock waves in the interstellar medium.

Stellar energy generation (Sections 27.2, 27.3, and 27.4)
- Nuclear fusion provides the energy for stars to shine.
- Proton-proton chain dominates in cooler stars, including the sun.
- Carbon-nitrogen-oxygen cycle is for hotter stars.
- Triple-alpha process takes place at high temperatures.
- Stellar nucleosynthesis accounts for most of the elements.

The main sequence: a balance of pressure and gravity (Section 27.5)

A test of the theory: the neutrino experiment (Section 27.6)
- Observations lower than theoretical predictions
- Possible explanations, including neutrino oscillations
- Is "new physics" necessary?
- New generation of experiments comes on line

Mass determines stellar evolution (Section 27.7).

## KEY WORDS

low-mass stars
high-mass stars
middle-mass stars
stellar evolution
evolutionary track
O and B associations
T Tauri stars
T associations
Herbig-Haro objects
nuclear fusion
thermal pressure
$E = mc^2$
nucleus
electrons
proton
neutron
ion
element
isotopes
atomic number
mass number
radioactive
neutrino
proton-proton chain
carbon-nitrogen-oxygen cycle
triple-alpha process
nucleosynthesis
kinetic temperature
strong force

## QUESTIONS

1. Since individual stars can live for billions of years, how can observations taken at the current time tell us about stellar evolution?
2. What is the source of energy in a protostar? At what point does a protostar become a star?
3. What is the evolutionary track of a star?
4. Describe the evolutionary track of the sun during the 10 billion years it is on the main sequence.
5. Arrange the following in order of development: O and B associations; T Tauri stars; dark clouds; sun; pulsars.
6. [†] Give the number of protons, the number of neutrons, and the number of electrons in: ordinary hydrogen ($^1_1$H), lithium ($^6_3$Li), and iron ($^{56}_{26}$Fe).
7. (a) If you remove one neutron from helium, the remainder is what element? (b) Now remove one proton. What is left? (c) Why is He IV not observed?
8. (a) Explain why nuclear fusion takes place only in the centers of stars rather than on their surfaces as well. (b) What is the major fusion process that takes place in the sun?
9. If you didn't know about nuclear energy, what is one possible energy source you might suggest for stars? What is wrong with this alternative explanation?
10. What forces are in balance for a star to be on the main sequence?
11. Use the speed of light and distance between the earth and sun given in Appendixes 2 and 3 to verify that it takes neutrinos 8 minutes to reach the earth from the sun.
12. What does it mean for the temperature of a gas to be higher?
13. (a) If all the hydrogen in the sun were converted to helium, what fraction of the solar mass would be lost? (b) How many times the mass of the earth would that be? (Consult Appendix 2.)
14. (a) How does the temperature in a stellar core determine which nuclear reactions will take place? (b) Why do more massive stars have shorter main-sequence lifetimes?
15. In what form is energy carried away in the proton-proton chain?

[†] This question requires a numerical solution.

16. In the proton-proton chain, the products of some reactions serve as input for the next and therefore don't show up in the final result. Identify these intermediate products. What are the net input and output of the proton-proton chain?
17. What do you think would happen if nuclear reactions in the sun stopped? How long would it be before we noticed?
18. Why do neutrinos give us different information about the sun than does light?
19. Why are the results of the solar neutrino experiment so important?
20. Why is the gallium experiment for solar neutrinos superior to the chlorine experiment?

## USING TECHNOLOGY

### W World Wide Web

1. What did the 2MASS sky survey find about young stars (http://www.ipac.caltech.edu/2mass/gallery)?
2. What has Hubble found about stars in formation (http://oposite.stsci.edu/pubinfo)?
3. Look at John Bahcall's Web page (http://www.sns.ias.edu/~jnb) and its links to find out the latest about the solar neutrino experiments.
4. Learn about the life stories of stars from NASA's educational site (http://imagine.gsfc.nasa.gov/docs/teachers/lifecycles).

### REDSHIFT

1. Find T Tauri when it rises and sets tonight. Observe it.
2. Use *RedShift*'s filters to observe the locations of O and B stars compared with the locations of G stars like the sun. What is the brightest O star you can see tonight?
3. Ponder Story of the Universe: Lives of the Stars; Science of Astronomy: Lifestyles of the Stars; and Science of Astronomy: Birth and Death of Stars.
4. Learn about nuclear fusion in stars from Science of Astronomy: Stars: What Is a Star? and from Astronomy Lab: Stars: Nuclear Fusion.

The Helix Nebula, NGC 7293, the nearest planetary nebula to us at a distance of 400 light-years. Long exposures show that its diameter in the sky is about the same as the moon's. The shell glows mainly H$\alpha$ and so is reddish. The violet and green lines of once- and twice-ionized oxygen are superimposed, giving a pinkish cast. The central part glows more with the emission of ions like oxygen and multiply-ionized neon; the ions are formed by the central star's ultraviolet radiation and radiate in the ultraviolet and blue. The human eye is not very sensitive to the H$\alpha$ that makes this photograph appear so red, and the enhanced green sensitivity of the eye's rods brings out the green oxygen radiation, so planetary nebulae appear faintly greenish.

# Chapter 28

# The Aging and Death of Stars Like the Sun

A star is a continual battleground between gravity pulling inward and pressure pushing outward. In the gas making up main-sequence stars, thermal pressure resulting from energy provided by nuclear fusion balances gravity. Here we will see what happens when fusion stops, and so no longer provides that pressure.

The sun is just an average star, in that it falls in the middle of the main sequence. But the majority of stars have less mass than the sun. So when we discuss the end of the main-sequence lifetimes of low-mass stars, including not only the sun but also all stars containing up to a few times its mass, we are discussing the future of most of the stars in the universe. In this chapter we will discuss the late (post–main-sequence) stages of evolution of stars that, when they are on the main sequence, contain less than about 8 solar masses. Recall that we term them **low-mass stars.** Let us consider a star like the sun, in particular, remembering that all low-mass stars go through similar stages but at different rates. We shall see their planetary nebula and white dwarf stages. The increased detail we can now see with the Hubble Space Telescope is giving us new insights (Fig. 28–1).

**AIMS:** To understand what happens to stars of up to about 4 solar masses when they have finished their time on the main sequence of the H-R diagram, including the planetary nebula and white dwarf stages

## 28.1 RED GIANTS

During the main-sequence phase of low-mass stars, as with all stars, hydrogen in the core gradually fuses into helium. This core takes up about the inner 10 per cent of the star. On the Hertzsprung-Russell diagram, the star stays in essentially the same place during this time, drifting upward only slightly. (This slight variation may make the sun enough hotter in about a billion years to evaporate the Earth's oceans.)

By about 10 billion ($10^{10}$) years after a one-solar-mass star first reaches the main sequence, no hydrogen is left in its core, which is thus composed almost entirely of helium. Hydrogen is still undergoing fusion in a shell around the core (Fig. 28–2). The sun is about halfway through this lifetime, so we have another 5 billion years or so to go.

Since no fusion is then taking place in the core, no ongoing nuclear process is replacing the heat that flows out of this hot central region. The core no longer has enough pressure to hold up both itself and the overlying layers against the inward force of gravity. As a result, the core then begins to contract under the force of gravity (we say it contracts **gravitationally**). This gravitational contraction not only replaces the heat lost by the core but also, in fact, heats the core up further. Thus paradoxically, soon after the hydrogen burning in the core stops, the core becomes hotter than it was before (because of the gravitational contraction).

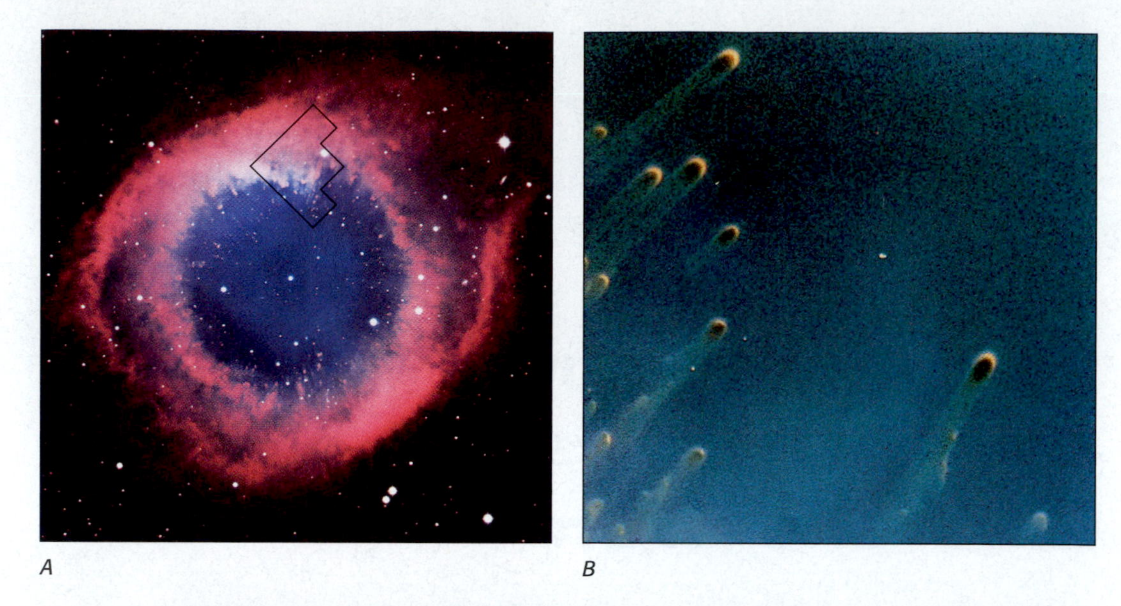


**FIGURE 28–1** This Hubble Space Telescope close-up of the Helix Nebula shows cometary-shaped images formed by the outward flow of stellar wind from the central star pushing on nebular knots or even breaking up the shell of gas. (*A*) The extent of the full HST field of view is outlined. (*B*) The heads of the knots are 100 A.U., roughly the size of the Oort Cloud around our solar system, and the tails perhaps 1000 A.U. The knots are relatively cool gas, un-ionized, embedded in the ionized gas, and are evaporating.

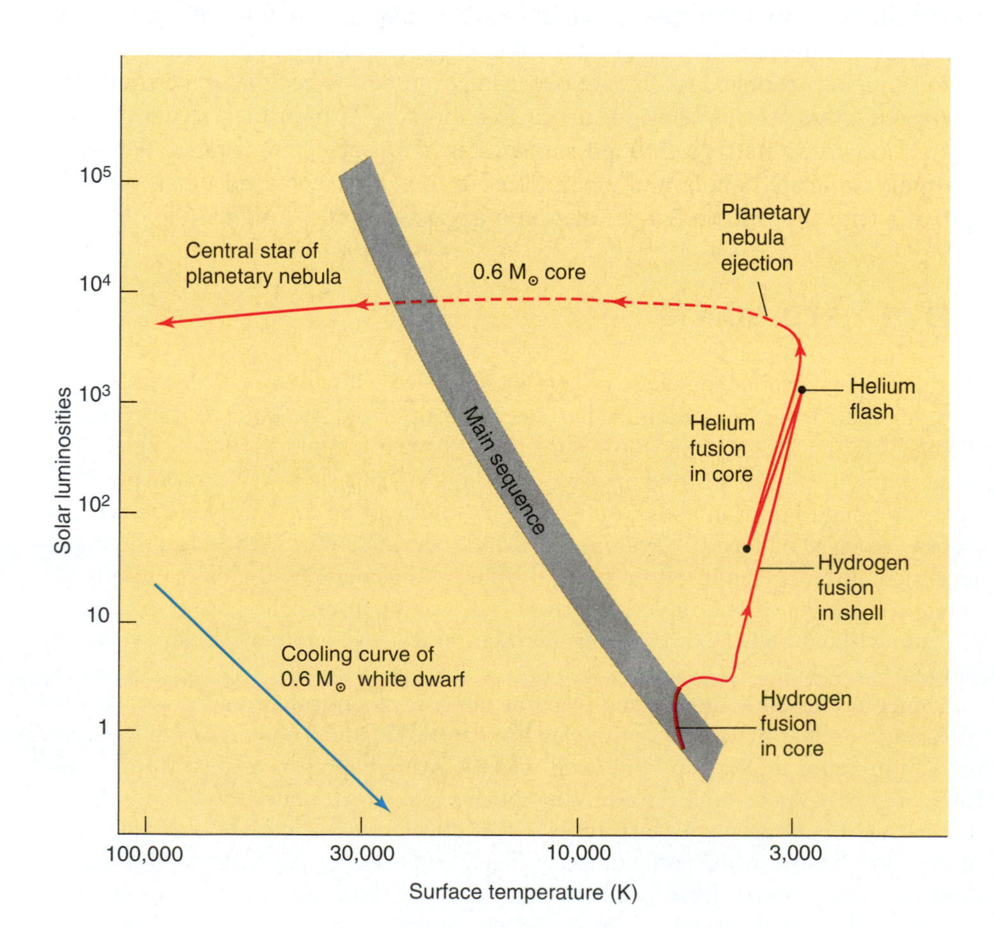


**FIGURE 28–2** A theoretical H-R diagram showing the evolutionary track of a 1-solar-mass star. We will soon describe its evolution after its red-giant stage. The vertical axis is the luminosity of the star, in units of the sun's luminosity; the horizontal axis is the temperature of the star's surface, increasing to the left.

Half the energy from gravitational contraction always goes into kinetic motion in the interior. This is an important theorem, the virial theorem, whose effects we met before in Section 27.1 when discussing the contraction of protostars. An example of such increased kinetic motion is the rise in temperature that we associate with something getting hotter. Most of the rest of the energy is radiated away.

As the core becomes hotter, the hydrogen-burning shell around the core becomes hotter too, and the nuclear reactions proceed at a higher rate. A very important phenomenon then occurs: part of the increasing energy production goes into heating and expanding the outer parts of the star. The total amount of energy is spread over the expanded surface area, and the surface temperature of the star decreases.

In sum, detailed calculations show that as soon as a main-sequence star forms a substantial core of helium with a shell of fusing hydrogen around it, the outer layers grow slowly larger and redder (because they are cooler). The star winds up giving off more energy (because the core temperature and nuclear reaction rates are increasing). The star's total luminosity—the emission per unit of surface area times the total area—is increased. The total luminosity increases even though each bit of the surface is less luminous than before (as a result of its decreasing temperature) because the total surface area that is emitting increases rapidly.

The process proceeds at an ever-accelerating pace, simply because the hotter the core gets, the faster heat flows from it and the more rapidly it contracts and heats up further. The hydrogen shell gives off more and more energy. The layers outside this hydrogen-burning shell continue to expand.

Our star, the sun, is still at the earlier, main-sequence stage, where changes are stately and slow. But in a few billion years, when the hydrogen in its core (about 10 per cent of the hydrogen in the whole sun) is exhausted, the time will come for the sun to brighten and redden faster and faster until it eventually swells and engulfs the current orbits of Mercury and Venus. (Mercury and Venus themselves may have receded, if the sun has lost enough mass.) At this point, the sun will no longer be a dwarf but will rather be what is called a **red giant.** (The region of the H-R diagram where these stars are found is called the "red-giant branch.") The red giant's surface will be so close to Earth (Fig. 28–3) that it will sear and char whatever is then here. A very strong stellar wind may form.

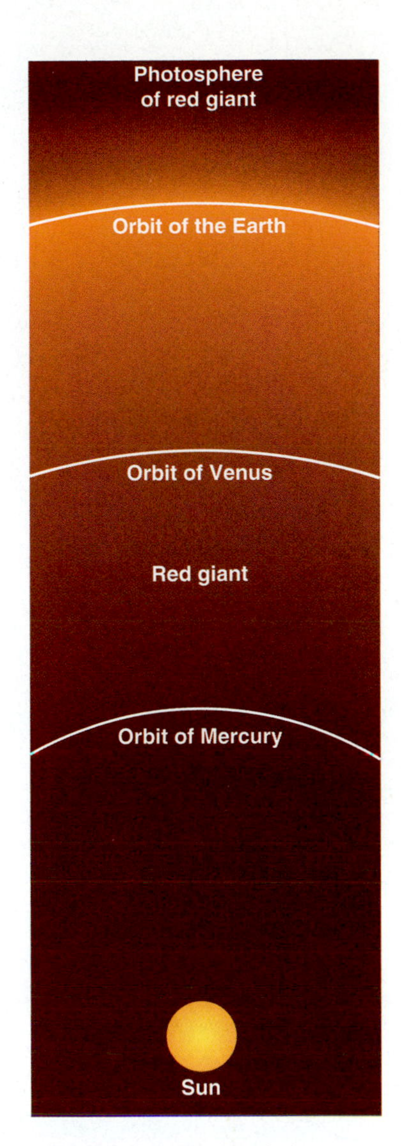


**FIGURE 28–3** A red giant swells so much that it can be the diameter of the earth's orbit.

Though they may be devastating to the Earth because they will come close, the outer layers of red giants are relatively cool for stars. (That's why they look reddish, after all.) Dust forms in them, and may eventually wind up in interstellar space. (We will see photographs in Chapter 31 of "absorption nebulae" made of such dust.) Regions farther from the star are sometimes found to contain molecules.

While the outer layers are expanding, the inner layers continue to contract and heat up, eventually reaching $10^8$ K. At this point, the triple-alpha process (Section 27.4) begins for all but the least massive low-mass stars, as groups of three helium nuclei (alpha particles) fuse into single carbon nuclei. For stars about the mass of the sun, the onset of the triple-alpha process happens rapidly. The development of this **helium flash** produces a very large amount of energy in the core, but only for a few years. This input of energy at the center reverses the evolution that took place just prior to the helium flash: now the core expands while the outer portion of the star contracts (because it is no longer receiving as much energy from the core as it was before the helium flash). As the star adjusts to this situation, it becomes smaller, hotter, and less luminous, moving back down and to the left on the H-R diagram to the "horizontal branch"; we see such a horizontal branch in the H-R diagram of a globular cluster (Section 26.5b). Helium now burns steadily in the core and during the next hundred million years is converted mostly to carbon. Some oxygen (which is just another helium nucleus more massive than carbon) is also formed.

## 28.2 PLANETARY NEBULAE

When the carbon core forms in a red giant, it contracts and heats up as did the helium core before it. This contraction drives up the rate at which hydrogen is being converted to helium and helium to carbon in shells surrounding the core. The star's parameters then are again at a point at the upper right on the H-R diagram. The star is now said to be on the second red giant branch, also called the "asymptotic giant branch." (It is called asymptotic because in a cluster H-R diagram, these stars form a sequence that nearly parallels and converges with the red-giant branch, and such nearly parallel and converging lines are called asymptotes.) We see some of these stars as Mira-type variables.

The hydrogen-burning and helium-burning shells produce energy unstably. These instabilities cause the star to shed mass in a strong stellar wind, and they also produce stellar pulsations in radius. This stellar wind is about a billion times stronger than our current-day solar wind, though it is in our future. We can measure the speed of the stellar winds by study of spectral features in the ultraviolet, whose Doppler shifts reveal velocities of thousands of kilometers/second.

As the pulsations grow, the star's outer layers become unstable. They lose a small fraction of their material with each pulse or with the irregularly blowing stellar wind; astronomers don't yet know which. As this gas cools, it forms dust. As the outer layers thin and the core is bared so that it can emit freely outward, the wind speed increases; the inner, faster-moving material then piles up against the outer, slower-moving material. Dense shells of gas are formed in this way. The shells sometimes are roundish, but often they extend farther in the plane of the star's equator as a result of denser winds there.

In a very short time (perhaps 1000 years) an entire envelope has been ejected. On this timescale, the ejected gas has moved several hundred astronomical units away from the star and has spread out enough to be transparent. A typical temperature for such gas is 10,000 K. We see it shining because it is ionized by ultraviolet radiation from the hot exposed core of the original star, now the "central star" of a **planetary nebula.** The light we see is the result of electrons rejoining the ions, forming excited atoms. The excited atoms often give off some of their energy as light when they fall to lower energy levels within the atoms. The most famous of these glowing planetary nebulae are the Ring Nebula (Fig. 28–4) and the Dumbbell Nebula (Fig. 28–5).

We know of about 1500 such objects in our galaxy that can be explained by this model, and there may be 10,000 in all. They were named "planetary nebulae" over two hundred years ago (in 1785) by William Herschel. When he discovered the first one in his small telescope, it looked similar to him to the planet Uranus (which he had also discovered). Both the planet and "planetary nebulae" appeared as small, greenish disks. But planets and planetary nebulae have only this semantic similarity.

The nebula's color is caused by the presence of certain strong emission lines of multiply (pronounced "mul-ti-plee'") ionized oxygen (that is, oxygen that has lost more than one electron) and other elements. We can determine the chemical composition of the nebula by studying these emission lines.

Planetary nebulae are exceedingly beautiful objects. They are actually semi-transparent shells of gas. When we look at their edges, we are looking obliquely through the shells of gas, and there is enough gas along our line of sight to be visible. The shapes of the shells show the interactions of the stellar winds that have blown. When we look through the centers of the shells, there is less gas along our line of sight, and the nebulae appear transparent. In the middle of most planetary nebulae, we see the **central star** from which the nebula was ejected. Central stars are very hot, some as hot as 200,000 K, so they appear high up on the left side of the H-R diagram. They are the hottest stars known. They are very luminous, in that they give off a lot of energy per second, but they are so hot that most of the energy they give off is in the ultraviolet. They therefore appear faint to the human eye.

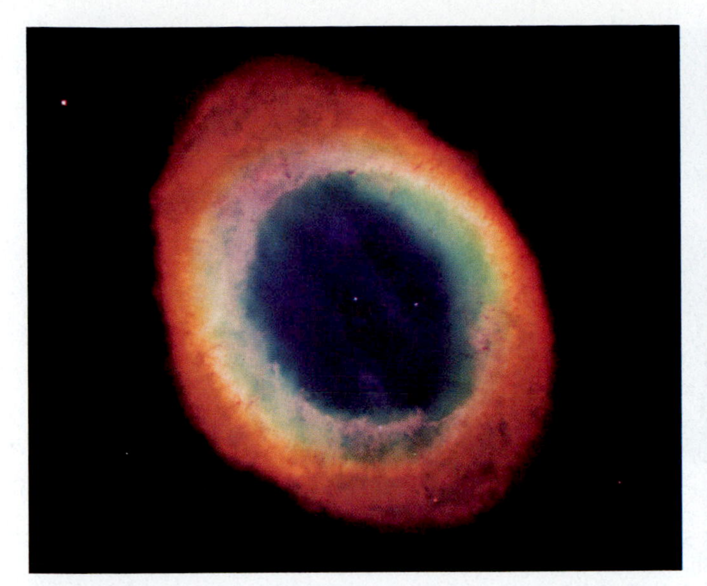

**FIGURE 28–4** The Ring Nebula in Lyra, M57, in a Hubble Heritage view. Rather than a ring, it is actually two lobes in a peanut shape and seen end-on. Bruce Balick, of the University of Washington, and British colleagues observed a faint outer halo and an inner halo of the lobes channeled by a ring formed by the original ejection of gas. The central star that ejected the gas nebula is easily visible.

**FIGURE 28–5** The Dumbbell Nebula, M27, a planetary nebula in the constellation Vulpecula, in a ground-based view. Its diameter in the sky is over one-fourth that of the moon. Radiation from the hot blue central star, visible in the image at the center of the planetary nebula, provides the energy for the nebula to shine.

Though all central stars have low mass, at the upper end of their mass range they can reach over 200,000 K, while at the lower end of the mass range, central stars never become hotter than about 80,000 K, still much hotter than our sun.

Astronomers are particularly interested in planetary nebulae because they want to study the various means by which stars can eject mass into interstellar space. The earliest stage of a planetary nebula is known as a protoplanetary nebula. The protoplanetary Egg Nebula (Fig. 28–6) shows us clearly the various shells that were emitted in sequence. As the progenitors of planetary nebulae evolve, they get smaller and the surface gravity of their central stars increases, since the surfaces are closer to the centers of mass. As a result, less and less mass is given off. At the same time, the temperature of the central star's surface increases, so the speed of the winds increase.

Many planetary nebulae do not have a simple shell structure. Constraining doughnuts of gas force some of the escaping stellar winds into bipolar shapes (Fig. 28–7). Sometimes, the shapes are like hourglasses, with gas limited to regions aligned with the poles (Fig. 28–8). We do not know whether some of the non-round planetary nebulae are shaped by the presence of a companion star or even of a planet. Alternatively, the magnetic field could play a role.

The many elliptical planetary nebulae have some similarities. In many of them, we can see the fast wind catch up with the slower wind emitted earlier (Fig. 28–9).

Sometimes the shapes revealed by the 10-times-improved resolution of Hubble force us to think of complex explanations (Fig. 28–10). This planetary nebula, NGC 6543, was the first one to be studied with a spectroscope. Its spectrum shows a blue H-beta line as well as a pair of green emission lines from ionized oxygen, a set of lines first known as "nebulium," when they were discovered in 1864. Only decades later, in 1928, were the "nebulium" lines identified. Much more recently, infrared satellites have added to our observing capabilities by showing us the cool dust (Fig. 28–11). The Chandra X-ray Observatory has revealed the superhot gas surrounding this planetary nebula, imaging for the first time the central bubble of hot

*(text continues on page 530)*

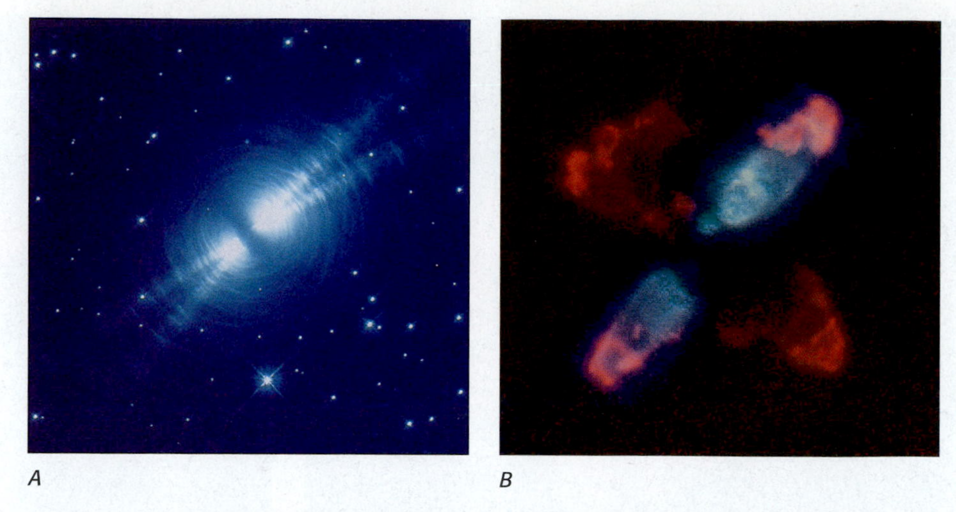


**FIGURE 28–6** Hubble Space Telescope views of a protoplanetary nebula, the Egg Nebula (CRL 2688). (*A*) In this visible-light view from WFPC2, the central star is hidden by a dense lane of dust. We are seeing light from the central star scattered toward us by dust farther away. Some of the light escapes through relatively clear places, so we see beams coming out of the polar regions. The circular arcs we see are presumably the shells of gas and dust that were irregularly ejected from the central star. (*B*) The false-color view from Hubble's Near Infrared Camera/Multi Object Spectrograph (NICMOS) showing starlight reflected by dust particles (*blue*) and radiation from hot molecular hydrogen (*red*). The collision between material ejected rapidly along a preferred axis and the slower, outflowing shells causes the molecular hydrogen to glow.

**FIGURE 28–7** Mz 3, a bipolar planetary nebula, imaged with the Hubble Space Telescope. Its name is from the 1922 catalogue of objects discovered by the theoretical astrophysicist Donald Menzel when he was doing an observational post-doctoral stint at the Lick Observatory.

FIGURE 28–8 The Etched Hourglass Nebula (MyCn18), a planetary nebula seen in one of the most fantastic images that the Hubble Space Telescope has taken. We see nitrogen (*red*), hydrogen (*green*), and oxygen (*blue*) radiation. This bipolar planetary nebula has two lobes of hot, expanding gas that are pushing outward from a constraining equatorial torus of gas. The lobes were emitted irregularly, hence the structure. The Egg Nebula may evolve into this shape one day. Note that the central star is off-center, probably explained by the presence of a faint companion.

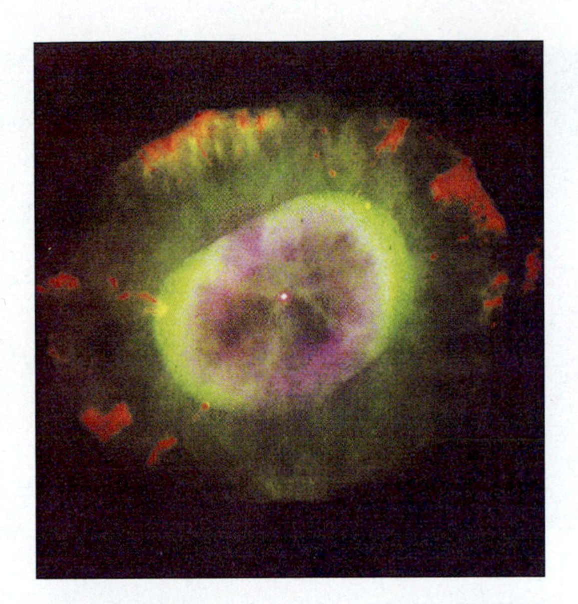

FIGURE 28–9 A Hubble Space Telescope view of the planetary nebula NGC 7662 shows the basic parts of elliptical planetary nebulae. We see the central cavity and shell caused by the fast wind and, around it, the material given off earlier. Colors show degrees of ionization and thus the energy of photons: singly ionized (*red*), doubly ionized (*green*), and triply or more ionized (*blue*).

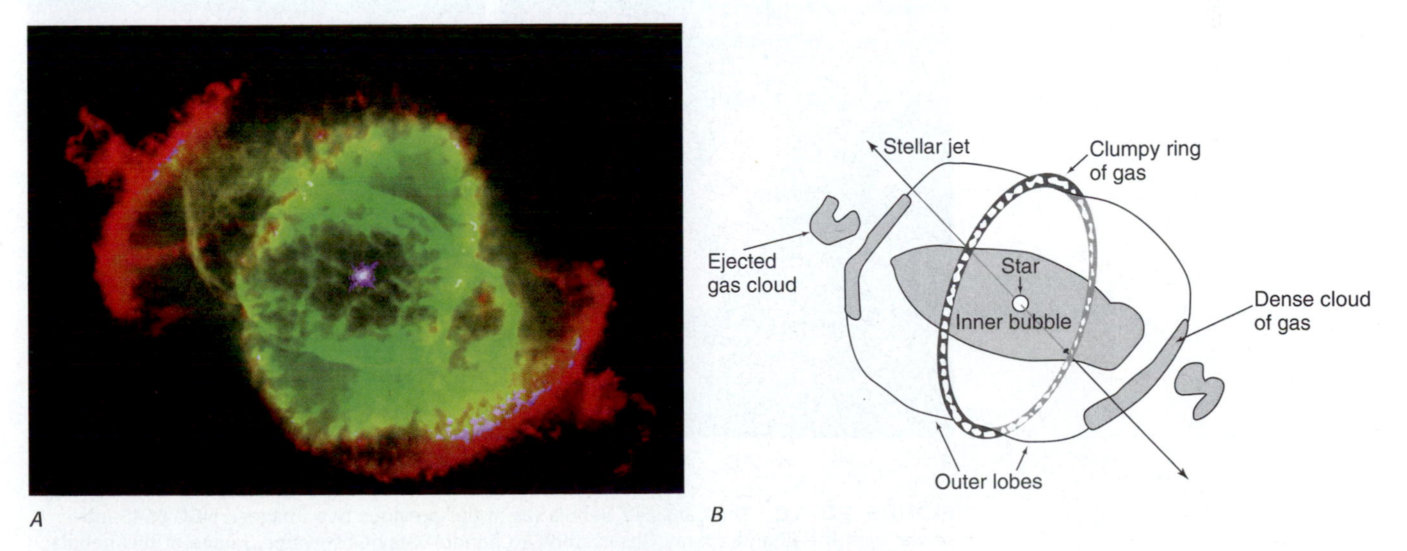


FIGURE 28–10 (*A*) Detail shown in some planetary nebula by the Hubble Space Telescope threatens to overwhelm our power of explanation. The Cat's Eye Nebula (NGC 6543) shows material ejected around the constraint of a disk or torus of gas. Neutral gas (*blue*), singly ionized gas (*red*), and doubly ionized gas (*green*) show the energy in available photons. (*B*) An artist's conception of how the structure arises. This young planetary nebula is at a temperature of 47,000 K.

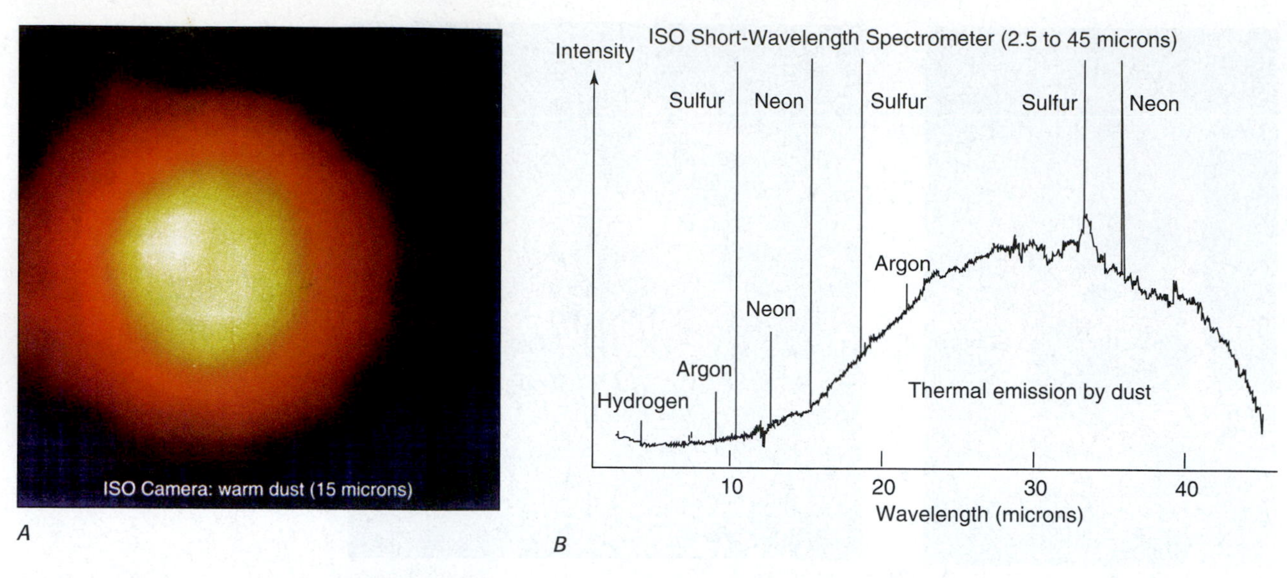


**FIGURE 28–11** (*A*) The Cat's Eye Nebula (as in the previous image), NGC 6543, observed with the Infrared Space Observatory (ISO) at the 15-micrometer infrared wavelength, which shows warm dust. Arms of this dust protrude in various directions, as the ejected matter interacts with material released from the star at earlier stages. (*B*) ISO's infrared spectrum of the Cat's Eye Nebula, showing both emission lines and thermal emission from the dust.

gas inside the expanding shell (Fig. 28–12). By comparing the x-ray and optical images, You-Hua Chu of the University of Illinois and colleagues conclude that the hot gas in the central bubble is causing the expansion of the nebula we see optically.

Planetary nebulae indicate something about the origin of interstellar matter and how it is enriched with heavy elements, since some planetary-nebula shells include material that underwent nuclear processing inside the progenitor star. Each plane-

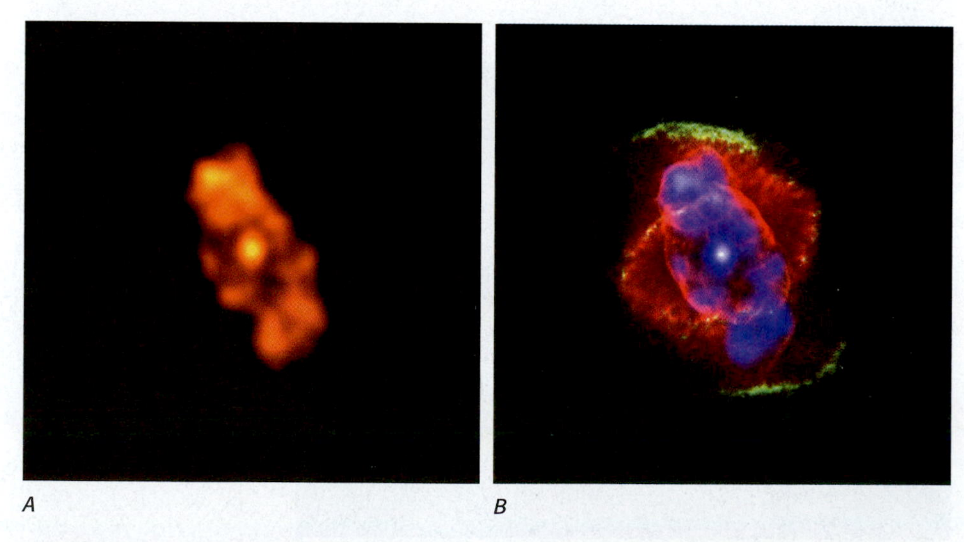


**FIGURE 28–12** The Cat's Eye Nebula (as in the previous two images), NGC 6543, observed with the Chandra X-ray Observatory. A Chandra X-ray Observatory image of this nebula shows the bright central star surrounded by a cloud of gas that is over 1 million kelvins. The brightness of the orange coloring of this false-color image corresponds to x-ray brightness. Scientists did not expect to find x-radiation so strong from the central star. (*B*) A false-color overlay of Chandra and Hubble data. The x-ray intensity is shown in purple, while the cooler material shown by Hubble appears in red and green. The gas, while hot, was not as hot as Chandra scientists expected.

tary nebula represents the ejection of 10 to 20 per cent of a solar mass, which is only a small fraction of the mass of the star. However, statistically, about one planetary nebula forms in our galaxy per year; hence within just one century, 10 to 20 solar masses of processed material enrich the interstellar medium. The carbon and nitrogen that are released in this way find themselves incorporated into stars that form later; the carbon in organic compounds on Earth may have formed in this way.

Note that forming heavy elements like carbon inside the core of a star does not explain how it gets to the surface so that it can be ejected in the planetary nebula. The gas in the star must circulate, bringing the elements created deep inside up to surface layers. This process is known as "dredge-up." In particular, while a low-mass star is rising on the H-R diagram to be a red giant, its convective outer layers penetrate so deeply that they reach the shell in which hydrogen is fusing into helium. Stars of higher mass also have later episodes of dredge-up. Helium, carbon, and nitrogen are the most important of the elements that are dredged up and therefore can be enriched in a planetary nebula. We often find twice as much helium as in stars and ten times as much nitrogen and carbon. Most of the nitrogen in our air was at one time formed in a distant star's core, dredged up, and released into space in a planetary nebula.

We can measure the ages of some planetary nebulae by tracing back the shells at their current rate of expansion and calculating when they would have been ejected from the star. Ages derived in this way have large uncertainties, since many effects could cause the expansion to speed up or slow down. Estimates for ages also come from observations of the central stars, interpreted by comparing their measured positions on the H-R diagram with theoretical calculations of how these positions vary with age. Most of the planetary nebulae are less than 50,000 years old. The data fit with our picture: after a longer time, the nebulae will have expanded so much that the gas will be invisible, and the central star may have cooled off enough that it is now unable to ionize gas and so cause the nebula to glow. Then only the central, contracted hot star is left to see. We identify it as a white-dwarf star. This path of stellar evolution on an H-R diagram is particularly interesting, since our sun may take it in 5 billion years or so.

Planetary nebulae in distant galaxies are now proving to be important for determining the distance scale of the universe (Section 35.5).

## 28.3 WHITE DWARFS

We have seen a low-mass star transform all the hydrogen in its core to helium and then all the helium in its core into carbon. The shell is ejected as a planetary nebula; now let us consider the core. Because it does not have enough mass, this core does not heat up sufficiently to allow the carbon to fuse into still heavier elements. Eventually, nuclear reactions no longer generate the energy needed to maintain the internal pressure that balances the force of gravity.

> Do not confuse the term "white dwarf" with the term "dwarf." The former refers to the dead hulks of stars in the lower left of the H-R diagram, while the latter refers to normal stars on the main sequence.

At this stage in their lives, all stars that contain less than 1.4 solar masses have the same fate. (We think that all stars that started with up to 8 solar masses lose enough mass to bring them below this 1.4-solar-mass limit.) Many or all stars in this low-mass group come from the red giant phase and may pass through the phase of being central stars of planetary nebulae. Other low-mass stars apparently do not go through a planetary-nebula stage, though we do not know why. Those that do not go through a planetary nebula stage lose parts of their mass in some other way, perhaps during the giant stage. In any case, the low-mass stars lose large fractions of their original masses.

When their nuclear fires die out for good, the stars with less than 1.4 solar masses remaining shrink in size and reach a stable condition that we shall describe below. As they shrink, they grow very faint (in the opposite manner to that in which red giants grew brighter as they grew bigger). Whatever their actual color, all these stars

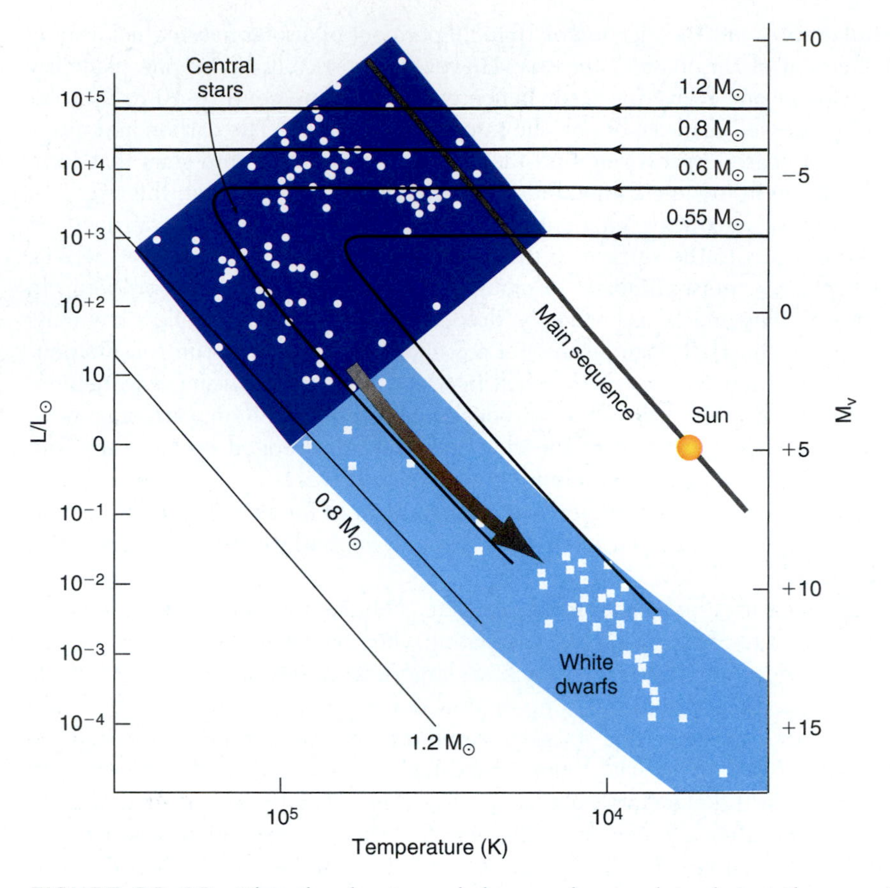


**FIGURE 28–13** When the planetary nebula expands around a red giant, the core of the star becomes visible at the center of the nebula. Since this **central star** has a high temperature, it appears to the left of the main sequence. These stars eventually shrink and cool, following the paths marked, to become white dwarfs, as shown on this H-R diagram. The paths the central stars take depend on their masses.

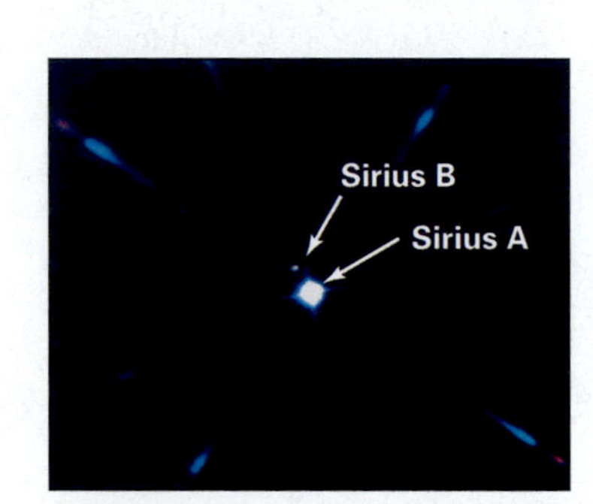


*A*

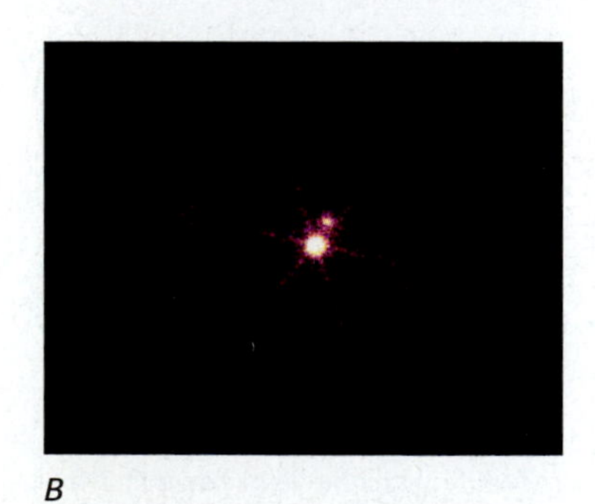

*B*

**FIGURE 28–14** Sirius A, the brightest star in the sky, has a white-dwarf companion known as Sirius B. Sirius B is about 10 magnitudes (10,000 times) fainter than Sirius A. (*A*) This optical image shows spectra of Sirius A going out from it in several directions. (*B*) Because Sirius B is so hot, about 25,000 K, it is bright in x-rays, as this Chandra X-ray Observatory image shows. The brighter of the two objects shown is the white dwarf. Sirius A, which is brighter in visible light, emits little light in x-rays, and the spot we see from it may be some of its ultraviolet radiation leaking through the detector's filter.

are called **white dwarfs.** The white dwarfs occupy a region of the Hertzsprung-Russell diagram that is below and to the left of the main sequence (Fig. 28–13). The best known white dwarf is Sirius B, the companion star of the brightest star in the sky (Fig. 28–14).

White dwarfs represent a stable phase in which stars of less than 1.4 solar masses live out their old age. The value of 1.4 solar masses is known as the **Chandrasekhar limit** after the astronomer S. Chandrasekhar (known far and wide as "Chandra"; this nickname later became the official name for the Chandra X-ray Observatory). Chandra calculated this value in 1930 when he was 19 years old, while on a boat from India to England to go to college; he later joined the faculty of the University of Chicago. He shared in the 1983 Nobel Prize in Physics for the discovery.

Chandrasekhar reasoned that something must be holding up the material in the white dwarfs against the force of gravity; nuclear reactions generating thermal pressure no longer take place in their interiors. The property that holds up the white dwarfs is a condition called **electron degeneracy;** we thus speak of "degenerate white dwarfs."

Electron degeneracy is a condition that arises in accordance with certain laws of quantum mechanics (and is not something that is intuitively obvious). As the star contracts and the electrons get closer together, there is a continued increase in their resistance to being pushed even closer. This shows up as a pressure. At very great densities, the pressure generated in this way exceeds the normal thermal pressure.

When this pressure from the degenerate electrons is sufficiently great, it balances the force of gravity, and the star stops contracting.

Thus the effect of the degenerate electron pressure is to stop the white dwarf from contracting; the gas is then in a very compressed state. In a white dwarf, a mass approximately that of the sun is compressed into a volume only the size of the earth (Fig. 28–15). (This great collapse is possible because atoms are mostly empty space: the nucleus takes up only a very small part of an atom.) A single teaspoonful of a white dwarf weighs 10 tons; it would collapse a table if you somehow tried to put some there. A white dwarf contains matter so dense that it is in a truly incredible state.

What will happen to the white dwarfs with the passage of time? The pressure from degenerate electrons doesn't depend on temperature, so the stars are stable even though no more energy is ever generated within them. Because of their electron degeneracy, they can never contract further. Still, they have some energy stored, and that energy will be radiated away over the next billions of years. Then the star will be a burned-out hulk called a **black dwarf,** though it is probable that no white dwarfs have yet lived long enough to reach that final stage. It will be billions of years before the sun becomes a white dwarf and then many billions more before it reaches the black dwarf stage.

**FIGURE 28–15** The sizes of the white dwarfs are not very different from that of the earth. A white dwarf contains about 300,000 times more mass than the earth, however.

## 28.4 OBSERVING WHITE DWARFS

White dwarfs are very faint and thus are difficult to detect. We find them by looking for bluish (hot) stars with high proper motions or, in binary systems, by noting their gravitational effect on the companion stars.

In the former case, by studying proper motions we find the stars that are close to the sun. Some are fainter than main-sequence stars would be at their distances; they must be white dwarfs. (White dwarfs at greater distances would be too faint for us to see.) To understand the latter case, we must realize that for any system of two masses orbiting each other, we can define an imaginary point, called the **center of mass,** which moves in a straight line across the sky. The individual bodies move around the center of mass, however, so the path in the sky of any of the individual bodies appears wavy. We have already discussed such astrometric binaries in Section 26.1.

At least three of the 45 stars within 5 parsecs of the sun (Appendix 7)—Sirius, 40 Eridani, and Procyon—have white dwarf companions. Another nearby object, known as van Maanen's star, is a white dwarf, although not in a multiple system. Hundreds of white dwarfs are known. So even though we are not able to detect white dwarfs at great distances from the sun, there seem to be a great number of them.

The Extreme Ultraviolet Explorer spacecraft, EUVE, was launched by NASA in 1992 to study the spectral region where many white dwarfs put out most of their radiation. Over 400 sources were catalogued. Radiation from the hot disks of gas around the white-dwarf member of a binary system is especially bright. White dwarfs are also observed with the Hubble Space Telescope (Fig. 28–16).

White dwarfs have been observed with temperatures as low as 4000 K and as high as 200,000 K (Fig. 28–17). Since white dwarfs are the cores of stars, revealed as the outer layers were lost, they are made of helium, carbon, and heavier elements. Some have atmospheres of hydrogen, but others have atmospheres composed entirely of helium. Their evolution is not always as straightforward as the path described above. Later in this chapter, we will discuss interchange of matter between members of binary systems. Further, even solitary white dwarfs are affected by mass loss.

Observations with a worldwide network of telescopes (the "Whole Earth Telescope") recently revealed that the outermost hydrogen layer in most white dwarf stars is thicker than had been thought. This thick layer acts as an insulator, and the star cools more slowly than had been realized.

**FIGURE 28–16** This Hubble Space Telescope view (*bottom*) of a small region only ⅔ light-year across in the M4 star cluster (*top*) revealed seven white dwarf stars (*inside blue circles*). The cluster may contain 40,000.

**FIGURE 28–17** The white-dwarf star shown here in the center of the planetary nebula known as NGC 2440 has not yet completely cast off the cocoon that surrounds it.

**FIGURE 28–18** Nova Herculis 1934 (DQ Her), showing its rapid fading from 3rd magnitude on March 10, 1935, to below 12th magnitude on May 6, 1935. The size of the star has not changed; the upper image merely appears larger because it is so much brighter, causing overexposure.

## 28.5 NOVAE

Although the stars were generally thought to be unchanging on a human timescale, occasionally a "new star," a **nova** (from the Latin for "new"; plural: **novae**), became visible. Such occurrences have been noted for thousands of years; ancient Oriental chronicles report many such events. Only in recent years have we found out that white dwarfs are at the center of the nova phenomenon. (A few of the events historically known as "novae" are now realized to be a grander type of event we call "supernovae"; they will be discussed in the next chapter.)

A nova is a newly visible star rather than actually a new star (Fig. 28–18). A nova is, in fact, a brightening of a star by 5 to 15 magnitudes or more, making it visible to the eye or to the telescope.

A nova may brighten within a few days or weeks. The nova ordinarily fades drastically within months and then continues to fade gradually over the years. Besides these "classical" novae, "dwarf novae" brighten at intervals of months by smaller factors.

Novae occur in binary systems in which one member has evolved into a white dwarf while another member is, usually, on the path toward becoming a red giant. The two are separated by about the distance between the earth and the moon. We have mentioned that the outer layers of a red giant are not held very strongly by the star's gravity. If a white dwarf is nearby, some of the matter originally from the red giant goes into orbit around the white dwarf rather than falling straight onto its surface. The orbiting gas forms a disk around the white dwarf. The disk gives off ultraviolet and x-rays that we detect from satellites, and irregularities in it cause the light from the system to flicker rapidly, on a timescale of seconds.

In a "classical" nova, some of this orbiting material falls onto the white dwarf's surface. After a time, enough builds up and enough energy is deposited to trigger

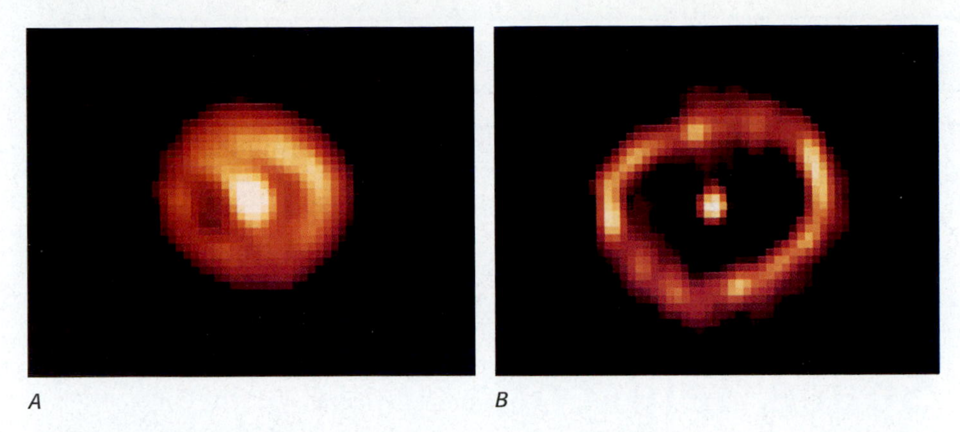


**FIGURE 28–19** The Hubble image of Nova Cygni 1992 (*A*) from May 1993 shows a shell of material expelled in a nova eruption. (*B*) A Hubble image after the repair shows that the shell is more elliptical. It also grew from 46 billion km to 60 billion km, and the bar that appeared earlier has disappeared.

nuclear reactions on the star's surface for a brief time. This runaway hydrogen burning causes the brightening we see as a nova; it lasts until all the hydrogen is consumed. Years after the outburst of light, the shell of gas blown off sometimes becomes detectable through optical telescopes and is even better seen with Hubble (Fig. 28–19).

Some nova are recurrent, happening over and over again. Hubble views of one such recurrent nova, T Pyxidis (Fig. 28–20) show that debris comes out in blobs, instead of making a smooth shell as had been thought. The recent ejecta are apparently plowing into older, fossil material from earlier explosions. When the newer material hits the older material, it brightens but slows. T Pyx is thought to be relatively massive for a white dwarf, so close to the upper limit for mass that even a tiny amount of extra material is enough to trigger an eruption. Scientists have permission to turn Hubble to this object within days after the next eruption.

*(text continues on page 538)*

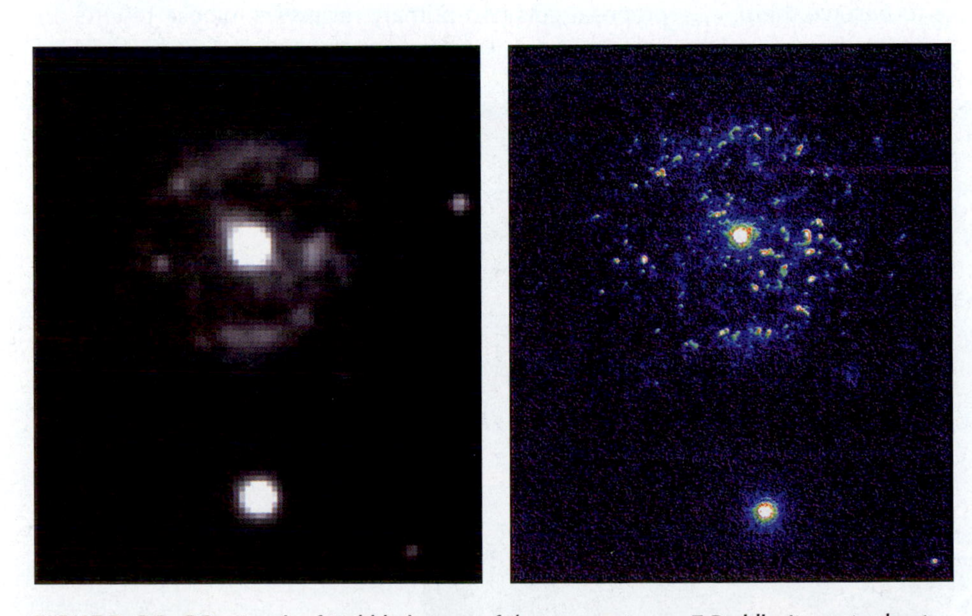

**FIGURE 28–20** A pair of Hubble images of the recurrent nova T Pyxidis. It erupts about every 20 years. Under Hubble's high resolution, the apparently smooth shell of gas surrounding the nova in ground-based images is seen to really be thousands of tiny blobs. They brighten, as one blob plows into another, and then fade. They take up a region about 1 light-year across. Other nova shells may well be similar.

ASTRONOMER'S NOTEBOOK

# Observing with the Hubble Space Telescope

Hubble has viewed over a hundred thousand celestial objects since its launch in 1990. Any astronomer can apply to be a Guest Observer. About once each year, an announcement is sent out from the Space Telescope Science Institute setting an application deadline for the next cycle. The notice gives about three months' warning for the application date. In that time, you must put together a substantial proposal. Of course, the heart of the proposal is the science itself that you propose to do. You must propose to do something that can be done uniquely on the Hubble Space Telescope. So your proposal will involve high-resolution imaging or ultraviolet spectroscopy, for example, techniques at which the Hubble Space Telescope excels.

The proposal must describe the details of your proposed observation with the Hubble Space Telescope (HST). You must provide a list of sources that you will observe—a set of white dwarfs, for example—and a calculation of how much time it will take to observe them. To compute the time, you must show how bright the objects are and use measurements of the detailed capabilities of HST for the type of observation intended. You might have to change the detail to which you can obtain a spectrum, for example, if it would take too long in the preferred observing mode. Often, you will consult with some of the scientists on the staff of the Space Telescope Science Institute about the details of the proposal.

At the deadline, roughly one thousand proposals are received, typically totalling six times more observing time than is available. After all, HST is only one telescope, and not a very large one at that by the standards of ground-based observing. When the applications are received at STScI (the lower case "c" in the acronym emphasizes that it is a science institute, rather than one of the instruments, which may have an acronym entirely in capital letters), Institute scientists check for technical feasibility. The proposals then go to one of 10 external panels. Each proposal gets two primary reviewers, whose reports are

A cutaway view of the Hubble Space Telescope and its scientific instruments.

evaluated by the appropriate panel, which assigns observing time for the highest-rated proposals. The full Telescope Allocation Committee (TAC) considers some of the next tier of proposals as well as the most extensive proposals, those for more than 100 orbits. Eventually, a final set of observing projects is decided upon and is forwarded to the Director for approval.

After acceptance and in the months before you arrive, your plans are considered by various committees and individual scientists at the Space Telescope Science Institute. One important detail is the selection of the guide stars that must be in the field of view for each observation. Major surveys of the sky were undertaken and then digitized to provide the information for choosing the vital guide star.

Eventually your plans are meshed with those of other successful applicants, and the time of the telescope is scheduled second by second. The general ideas that you have submitted must be turned into commands such as "turn telescope 17° west, open shutter, start science tape recorder, shut shutter, dump data to science tape recorder, turn tape recorder off," and so on.

When your observing time actually arrives, you are usually home, since you cannot change anything on the telescope in real time. More rarely, you may choose to travel to STScI, which is on the Homewood Campus of the Johns Hopkins University in Baltimore. There, staff scientists and technicians will help you, on the Internet or in person, prepare to interpret the data.

After your observation, the data are relayed to earth via intermediate satellites and then taken out of the stream of engineering and other data being beamed down by HST. They are then sent to you on computer tapes or over the Internet.

The archive of data is so large that perhaps someone has already taken observations of interest to you. Previous studies may have looked at only aspects of the data that don't concern you, leaving fresh Hubble data for you. Occasionally, some Hubble project was taken specifically to be open to "the community" of astronomers. The Hubble Deep Fields, whose importance we discuss later on, were such projects. They were taken in the 10 per cent of observing time whose assignment is at the discretion of the Director of the Space Telescope Science Institute.

As months pass, you will study your data, often together with your students. In addition to giving papers at meetings of the American Astronomical Society, you may speak at workshops and colloquia held at STScI to disseminate the results. Eventually, as you understand the data and their implications well enough, you write a formal article for a scientific journal. The *Astrophysical Journal* and the *Astronomical Journal* (now even available on-line) are common choices for American astronomers dealing with stars and galaxies. Meetings of the American Astronomical Society's Planetary Sciences Division and the Lunar and Planetary Science Conference are important events with the planetary scientists, who overlap less and less with other astronomers and who often publish their papers in the journals *Icarus* and *Journal of Geophysical Research*. Some of the more important and timely articles may find their way into journals that cover a wide range of sciences, principally *Nature* and *Science*. In all these cases, with public talks and publications your discoveries find their way to the astronomical community and to the world at large.

The sun reflects off the Hubble Space Telescope, with a crescent of Earth in the background, during the 1999 repair mission. We await the installation of the Advanced Camera for Surveys in 2001/2002 and the Cosmic Origins Spectrograph in 2003.

## 28.6 SUPERNOVAE OF TYPE Ia

In a classical nova, we have just seen how nuclear reactions are triggered on the surface of a white dwarf when material from a companion falls on it. In some cases, so much material falls onto the white dwarf's surface that the white dwarf is pushed over the Chandrasekhar limit. Then the electron degeneracy that has been supporting the white dwarf is overwhelmed and the star collapses. Within seconds, it is incinerated in a giant explosion known as a Type Ia supernova. We discuss the various types of supernovae in the following chapter.

## CORRECTING MISCONCEPTIONS

| ✖ ***Incorrect*** | ✔ ***Correct*** |
|---|---|
| Planetary nebulae have to do with solar systems. | Planetary nebulae are gaseous and have nothing to do with planets. |
| White dwarfs and dwarfs are the same. | A white dwarf is the dead hulk of a burned-out star and is found in the lower left of the H-R diagram, while "dwarf" refers to a normal star on the main sequence. |
| Brighter stars are bigger. | Almost all stars are mere points on images; brighter stars may appear bigger on film because of effects of overexposure on the film. |

## SUMMARY AND OUTLINE

Red giants (Section 28.1)
- They follow the end of hydrogen burning in the core.
- The core contracts gravitationally; the core and the hydrogen-burning shell become hotter.
- The star is red, so each bit of surface has relatively low luminosity; but since the surface area is very greatly increased, the total luminosity is greatly increased.
- Helium flash: rapid onset of triple-alpha process

Planetary nebulae (Section 28.2)
- Ages: less than 50,000 years old
- Mass loss: 0.1 or 0.2 solar mass
- Shapes depend on interactions of stellar winds

White dwarfs (Sections 28.3 and 28.4)
- End result of low-mass stars: less than 1.4 solar masses remaining
- Supported by electron degeneracy

Novae (Section 28.5)
- Interaction of a red giant and a white dwarf

## KEY WORDS

| | | | |
|---|---|---|---|
| low-mass stars | helium flash | white dwarfs | black dwarf |
| gravitationally | planetary nebula | Chandrasekhar limit | center of mass |
| red giant | central star | electron degeneracy | nova (novae) |

## QUESTIONS

1. What event signals the end of the main-sequence life of a star?
2. When hydrogen burning in the core stops, the core contracts and heats up again. Why doesn't hydrogen burning start again?
3. When the core first starts contracting, what halts the collapse?
4. How can a star reach its red-giant stage twice?
5. Why, physically, is there a Chandrasekhar limit?
6. [†] If you are outside a spherical mass, the force of gravity varies inversely as the square of the distance from the center. What is the ratio of the force of gravity at the surface of the sun to what it will be when the sun has a radius of 1 astronomical unit? (Use data in Appendix 2.)
7. Why is helium "flash" an appropriate name?
8. If you compare a photograph of a nearby planetary nebula taken 80 years ago with one taken now, how would you expect them to differ?
9. What sculpts the shapes of planetary nebulae?
10. Why is the surface of a star hotter after the star sheds a planetary nebula?
11. What keeps a white dwarf from collapsing further?
12. What are the differences between the sun and a one-solar-mass white dwarf?
13. When the sun becomes a white dwarf, approximately how much mass will it have? Where will the rest of the mass have gone?
14. Which has a higher surface temperature, the sun or a white dwarf?
15. Compare the surface temperatures of the hottest white dwarf and an O star. What spectral type of normal dwarf has the same surface temperature as the coolest white dwarf?
16. Sketch an H-R diagram indicating the main sequence. Show the evolution of the sun starting at the time it leaves the main sequence.
17. When the proton-proton chain starts at the center of a star, it continues for billions of years. When it starts at the surface (as in a nova), it lasts only a few weeks. How can you explain the difference?

[†] This question requires a numerical solution.

## USING TECHNOLOGY

### W World Wide Web

1. Look at images of planetary nebulae taken with the Hubble Space Telescope and with the Anglo-Australian Observatory. Consider their shapes and how they may have gotten that way. See http://oposite.stsci.edu/pubinfo/Subject.html and http://www.aao.gov.au/images.html.
2. Check out planetary nebula Web sites.

### REDSHIFT

1. Look at Story of the Universe: Lives of the Stars; Science of Astronomy: Lifestyles of the stars; Science of Astronomy: Birth and death of stars.
2. Look at the Guided Tour: Stars: Planetary Nebulae.
3. Find out when several bright planetary nebulae are visible. Your list could include the planetary nebulae listed in this chapter.
4. Look at images of planetary nebulae from the Photo Gallery. Can you explain their shapes?

A set of inner and outer rings, seen in the center of this Hubble Heritage image surrounds the supernova that went off in the Large Magellanic Cloud in 1987. Swirls of gas fill the picture. A dozen bright blue stars seen are hot objects, six times as massive as the sun and about 12 million years old, the same age as the massive star that exploded to make Supernova 1987A. An additional 500 stars of about the same mass as the sun are also in the region.

# Chapter 29

# Supernovae

We have seen how the run-of-the-mill, low-mass stars end: not with a bang but a whimper. Still, when the white dwarfs that result are in binary systems, huge explosions called supernovae may yet result.

High-mass stars, which contain more than about 8 solar masses when they are on the main sequence, put on a dazzling display. These high-mass stars cook the elements in the lower middle of the periodic table (say, carbon up to iron) deep inside. They then blow themselves almost to bits as supernovae, forming still more heavy elements in the process. The matter that is left behind in the stellar core settles down to even stranger states of existence than that of a white dwarf.

In this chapter we shall discuss supernovae in general, and then the appearance, starting in 1987, of the first supernova to be visible to the naked eye in almost 400 years. In the next chapter we shall see that some of the massive stars, after their supernova stage, become neutron stars, which we detect as pulsars and x-ray binaries. In the following chapter, we shall discuss the death of the most massive stars, which become black holes.

**AIMS:** To see how stars become supernovae, some as exploding white dwarfs and others as very massive stars whose cores sometimes collapse to become neutron stars, and to learn about the nearby supernova whose light and neutrinos we received in 1987

## 29.1 RED SUPERGIANTS

Stars that are much more massive than the sun whip through their main-sequence lifetimes at a rapid pace. These prodigal stars use up their store of hydrogen very quickly. A star of 15 solar masses may take only 10 million years from the time it first reaches the main sequence until the time when it has exhausted all the hydrogen in its core. This is a lifetime a thousand times shorter than that of the sun. When the star exhausts the hydrogen in its core, the outer layers expand, and the star becomes a red giant.

For these massive stars, the core can then gradually heat up to 100 million degrees, and the triple-alpha process begins to transform helium into carbon. After its ignition, the helium then burns steadily, unlike the helium flash of less massive stars.

By the time helium burning is concluded, the outer layers have expanded even further, and the star has become much brighter than even a red giant. We call it a **red supergiant** (Fig. 29–1); Betelgeuse, the star that marks the shoulder of Orion, is the best-known example (Fig. 29–2). Supergiants are inherently very luminous stars, with absolute magnitudes of up to −10, one million times brighter than the sun. A supergiant's mass is spread out over such a tremendous volume, though, that its average density is less than one-millionth that of the sun.

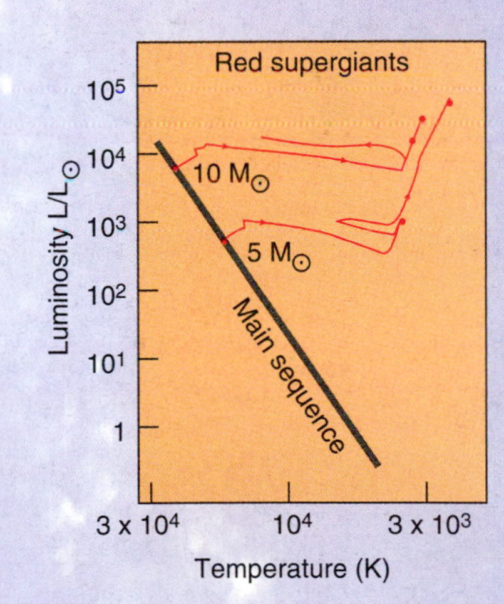


**FIGURE 29–1** An H-R diagram (as was Fig. 28–2 for less massive stars) showing the evolutionary tracks of 5- and 10-solar-mass stars as they evolve from the main sequence to become red supergiants. For each star, the first dot represents the point where hydrogen burning starts, the second dot the point where helium burning starts, and the third dot the point where carbon burning starts.

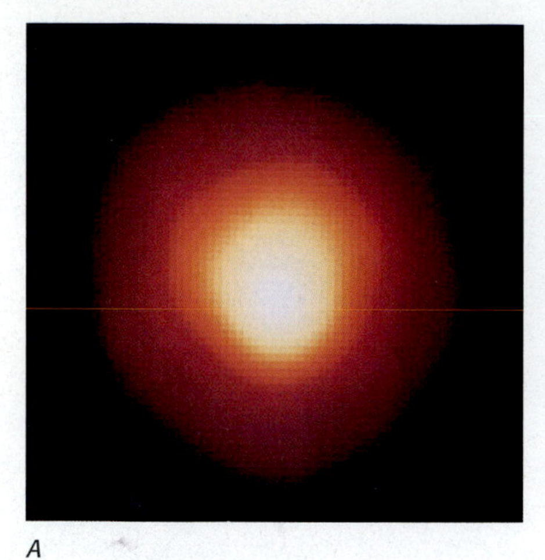

A

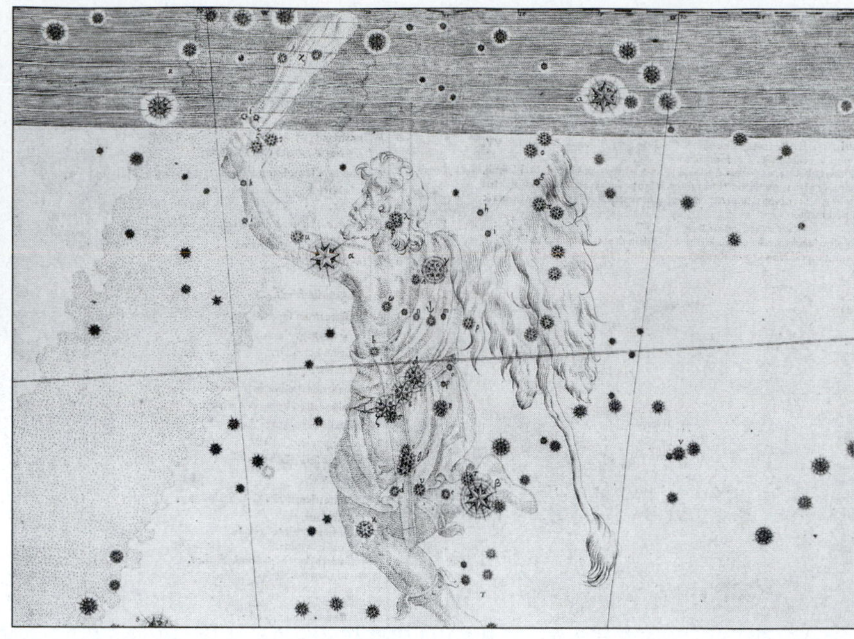

B

**FIGURE 29–2** (*A*) The red-giant star Betelgeuse, alpha Orionis, is revealed by the Hubble Space Telescope to have an ultraviolet atmosphere the size of Jupiter's orbit. The cause of the huge hot, bright spot, ten times the diameter of Earth, is unknown. It is 2000 K hotter than surrounding gas. (*B*) Betelgeuse is the star labelled $\alpha$ in the shoulder of Orion, the Hunter, shown here in Johann Bayer's *Uranometria,* first published in 1603. Betelgeuse, visible in the winter sky, appears reddish to the eye.

Astronomers have recently found two close companion stars to Betelgeuse. One is on an elliptical orbit that brings it within 1 A.U. of Betelgeuse, which may disrupt Betelgeuse enough to account for its variations in brightness.

The carbon core of a supergiant contracts, heats up, and begins fusing into still heavier elements. Eventually, even iron, a fairly heavy element, builds up. The iron core is surrounded by layers of elements of different mass, with the lightest toward the periphery. Also, mass escapes from the star and surrounds it. We will see in the next section that such stars are destined to explode, making spectacular displays.

## 29.2 TYPES OF SUPERNOVAE

A dying star can explode in a glorious burst (Fig. 29–3) called a **supernova** (plural: **supernovae**). Two quite different situations apparently lead to supernovae (Table 29–1). The best current models indicate that **Type Ia supernovae** represent the incineration of a white dwarf, which results from a low-mass star. On the other hand, **Type II supernovae** take place in high-mass stars and show that stellar evolution can run away with itself and go out of control. (To help remember which is which, recall that the less massive star leads to the lower Roman number—I instead of II.)

One type of supernova—Type II—is found only in spiral galaxies, galaxies that have arms of stars, gas, and dust in a flat disk. Our own Milky Way Galaxy is of this type. Spiral arms contain young, massive stars, which are not found elsewhere. Another type of supernova appears not only in spiral galaxies but also in galaxies that appear elliptical in shape. We think elliptical galaxies contain fewer young stars than spiral galaxies. We return to the study of galaxies in Chapter 34.

The two types can be distinguished by their spectra and by the rate at which their brightnesses change. Rudolph Minkowski of Caltech pointed out the distinction in the 1940s based on the observation that some supernovae—Type II—have prominent hydrogen lines when they are at maximum light, while hydrogen lines are absent in the others—Type I. Though all Type I supernovae lack hydrogen lines, Type Ia supernovae have a strong spectral line of ionized silicon. Type Ib has strong helium lines and Type Ic has weak helium lines or no visible helium lines. More re-

**FIGURE 29–3** (*A*) The brightest supernova visible from Earth's northern hemisphere in 55 years erupted in 1993, and is known as SN1993J. It is in the spiral galaxy M81, the second-nearest spiral from our own. M81, located only 8 million light-years from us, is always a popular galaxy to observe, so prediscovery images (*top*) existed. The supernova reached 10th magnitude (*bottom*), too faint for the naked eye but easily in reach of a wide variety of telescopes. (*B*) Supernova 1993J showed clearly in this x-ray view from ROSAT. (This German "Roentgen satellite," now defunct, was named after Wilhelm Roentgen, the discoverer of x-rays.)

*A*

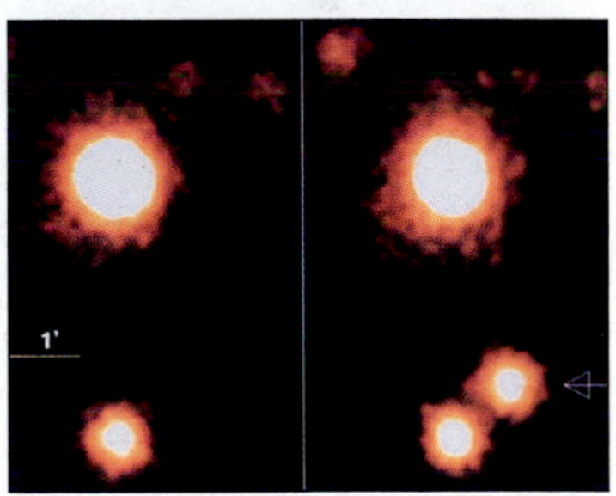


*B*

cently, it was learned that Type Ia supernovae come from white dwarfs while Types Ib and Ic supernovae, which lack the key spectral line from silicon, come from massive stars that have lost their outer layer of hydrogen before they exploded. Aside from the absence of their outer atmospheres, the collapse proceeds as for Type II supernovae. Let us ignore Type Ib and Ic supernovae, leaving the clear distinction between Types Ia and II.

Type Ia (again, the type from the incineration of white dwarfs) are more often seen than Type II because they are perhaps ten times brighter than Type II supernovae and take place in all types of galaxies. Type II supernovae are seen only in spiral galaxies. Indeed, they are usually found near the spiral arms, which endorses the idea that they come from massive stars. (Since the massive stars are short-lived, they are found in the spiral arms near where they were born.) In both cases, Doppler-shifted absorption lines show that gas is expanding rapidly: about 10,000 km/sec for Type Ia's and half that for Type II's.

Type II supernovae arise after a substantial core of heavy elements has been formed in a massive star (perhaps 8–12 $M_\odot$) and begins to shrink and heat up. The heavy elements represent the ashes of the previous stages of nuclear burning (Fig. 29–4). The stages of oxygen and magnesium "burning" (actually, undergoing fusion) to form heavier elements take less than 1000 years. Finally, silicon and sulfur "burn" to iron in only a few days. The conditions in the center of the star change so quickly that supercomputers are being used to keep up with the rapidly changing conditions in the models. Both because of this factor and because scientists keep thinking of new physical ideas that have to be included, we cannot know whether the following model ultimately will prove correct. In this model, the temperature becomes high enough for iron to form and then to undergo nuclear reactions. The stage is now set for disaster because iron nuclei have a fundamentally different property

**TABLE 29–1** Types of Supernovae

| | Type Ia | Type II |
|---|---|---|
| **Source** | White dwarf in binary | Massive star |
| **Spectrum** | No hydrogen lines | Hydrogen lines |
| **Peak** | 1.5 magnitudes brighter than type II | |
| | Sharper | Broader when graphed vs. time |
| **Light curve** | Rapid rise | |
| | Decay with several-week half-life | |
| | All have same magnitude | Different magnitudes |
| **Location** | All type galaxies | Spiral galaxies only |
| **Expansion** | 10,000 km/sec | 5,000 km/sec |
| **Radio radiation** | Absent | Present |

$\odot$ is the symbol for the sun. 8 $M_\odot$, read "eight solar masses," means "eight times as much mass as the sun has."

## WHAT'S IN A NAME?

### BOX 29.1 Supernova Types

**Observations:**

| Type I<br>No hydrogen lines | | | Type II<br>Hydrogen Balmer spectrum |
|---|---|---|---|
| Ia<br>Silicon line | Ib<br>No silicon line<br>helium lines | Ic<br>No silicon line<br>no/weak helium lines | |
| **Theory for precursor star:** | | | |
| Low-mass binary | High-mass lost envelope | High-mass lost envelope | High mass with hydrogen envelope |

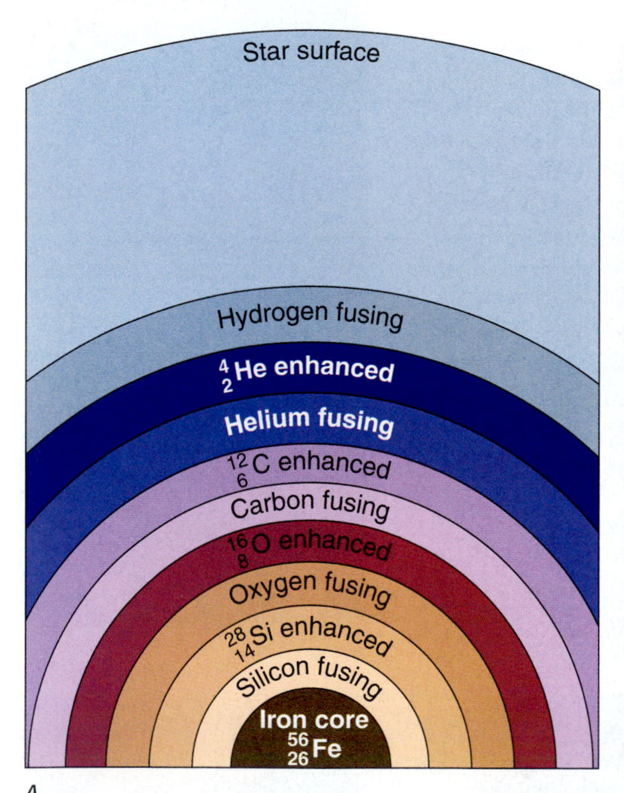


*A*

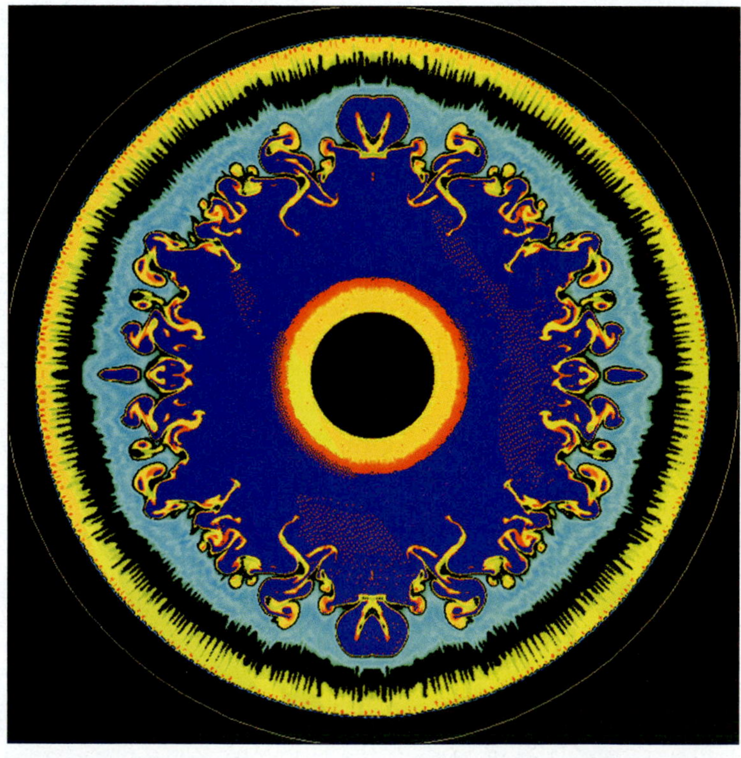

*B*

**FIGURE 29-4** (*A*) A massive star just before it explodes as a Type II supernova. The "onion-skin" shells, in which different elements are fusing in sequence. The oxygen found here is the major source of oxygen in the universe. (*B*) A supercomputer calculation showing the pre-supernova boiling phase of the protoneutron star. We see the inner part of the core; the scale is 200 km on a side. The supernova exploded 20 milliseconds later, and its subsequent evolution is depicted in the following figure. This figure shows instabilities that lead to gas mixing; on Earth, salad dressing mixes when shaken and flags flap for similar reasons. This and the following calculations were made by Adam Burrows, Evonne Marietta, and Bruce Fryxell at the San Diego Supercomputer Center.

from other nuclei when undergoing nuclear reactions. When other nuclei undergo fission or fusion, energy is given off. But when iron nuclei undergo fission or fusion, there is a net loss of energy. So processes that release energy no longer can build up heavier nuclei as the core shrinks this time.

As the core is compressed, it reaches the Chandrasekhar mass of 1.4 times the mass of the sun. Basically, a white dwarf forms inside and then has mass added to it to take it above the Chandrasekhar limit. We describe now the model worked out by Adam Burrows of the University of Arizona.

As the temperature climbs beyond billions of kelvins, the iron is broken up into alpha particles (helium nuclei), protons, and neutrons by the high-energy photons of radiation that are generated. Since these processes absorb energy, they diminish the pressure at the center of the star, helping the core collapse. The process goes out of control. Within seconds—a fantastically short time for a star that has lived for perhaps 10 million years—the inner core collapses and heats up catastrophically. The core drops from the size of Earth to the size of a city, and its density becomes greater than that of an atomic nucleus.

In a leading model for what happens next, as the outer core falls in upon the collapsing inner core, electrons and protons combine to make neutrons and neutrinos (Sections 25.3 and 25.6). The core collapses and (calculations show) bounces outward, since its gas has a natural barrier at nuclear densities against being compressed too far (Fig. 29–5).

The collision of the rebounding inner core with the supersonically collapsing outer core sends off shock waves that cause heavy elements to form and that throw off the outer layers. The shock wave leaves the core. A relatively old model, that the shock wave starts the explosion in less than 0.01 second (making the so-called "prompt explosion"), is no longer favored. It doesn't seem to work because, if the iron core is sufficiently large, the expanding shock wave loses too much energy to the infalling outer core. The favored model is as follows: the shock wave starts expanding outward, and then stalls about 10 milliseconds later because of all the energy it has given off to form neutrinos and to break apart nuclei in the gas. Then it is re-energized by neutrinos. The neutrinos are so numerous and energetic that enough of them interact with the outer layers of the star, which have continued to fall inward, and play an important role in blowing off these outer layers. The effect takes longer, several seconds, and so it is called the "delayed explosion." The energy given off in neutrinos in this burst equals, for that brief time, all the light given off by every object in the entire universe! In any case, the star is destroyed. Only the core may be left behind, as we will discuss in the following chapter.

An alternative model, advanced by University of Texas astronomers in 2000, is based on evidence that the collapse of the massive star may not be symmetric. Some observational evidence for this point of view comes from study of the bright,

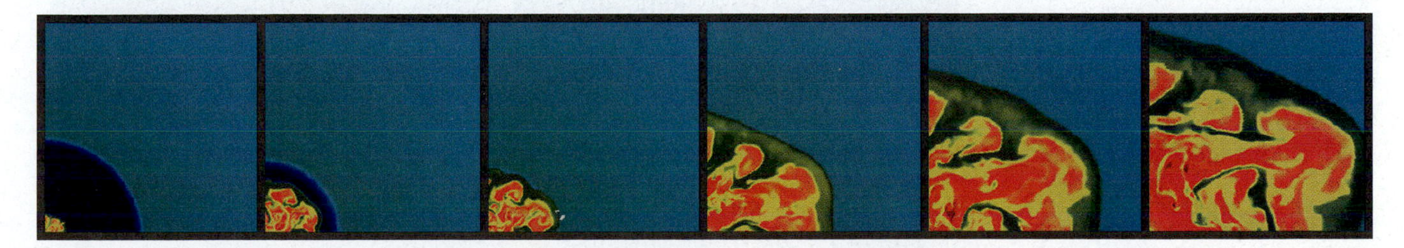

**FIGURE 29–5** A supercomputer simulation of the evolution of the supernova explosion in the innermost 2500 km of a massive star. Depicted are six snapshots of the density pattern of the ejecta for various consecutive times within about 100 milliseconds. Red is low density and green is high density. The transition from dark blue to light blue depicts the silicon/oxygen interface in the progenitor star. As is clear from the figures, the debris is quite clumpy.

relatively nearby supernova that went off in 1987, which we will discuss in detail later in this chapter. In the new model, the collapse leads to the formation of a dense neutron star (a type of object that we will discuss in the next chapter), and then the emergence of a jet from that neutron star. You have already seen in the photograph opening this chapter that the supernova has a set of rings. The jet would be nearly perpendicular to these rings. The jet could transfer energy from the rotation of the neutron star, providing enough energy to blow off the original star's outer layer as a supernova. Even if valid, this alternative model probably applies to a minority of supernovae.

Whatever model explains why the massive star's outer layers are blown off, the heavy elements formed either near the center of the star or in the supernova explosion are spread out into space. There, they enrich the interstellar gas. Though most Type II supernovae come from stars containing between 8 and 12 solar masses, these Type II supernovae don't supply as much of the heavy elements as those from still more massive stars do. (The upper limit for mass in stars is thought to be 60 or 100 solar masses, depending on how much they contained of the heavy elements when the stars were born.) In any case, when a star forms out of gas enriched by supernovae of both types, the heavy elements are present. Our understanding of element formation in Type II supernovae is circumstantial, since the outburst itself is hidden from our view by the outer layers of the massive star; we can only study the transformed material later on, when it has spread out considerably.

Spectroscopic studies of the star $\eta$ (eta) Carinae (Fig. 29–6) show that a gas condensation spewed out 160 years ago contains a relatively large abundance of nitrogen. This indicates that the star has processed material in the carbon-nitrogen cycle, brought it to the surface as the star's material mixed up, and ejected it. This processing/ejection cycle is expected from a massive star that will become a supernova soon—though "soon" could be next year or in 10,000 years.

Type I supernovae, the type that show no hydrogen lines, come from a variety of sources. We will consider the major source and limit our discussion to Type Ia. In

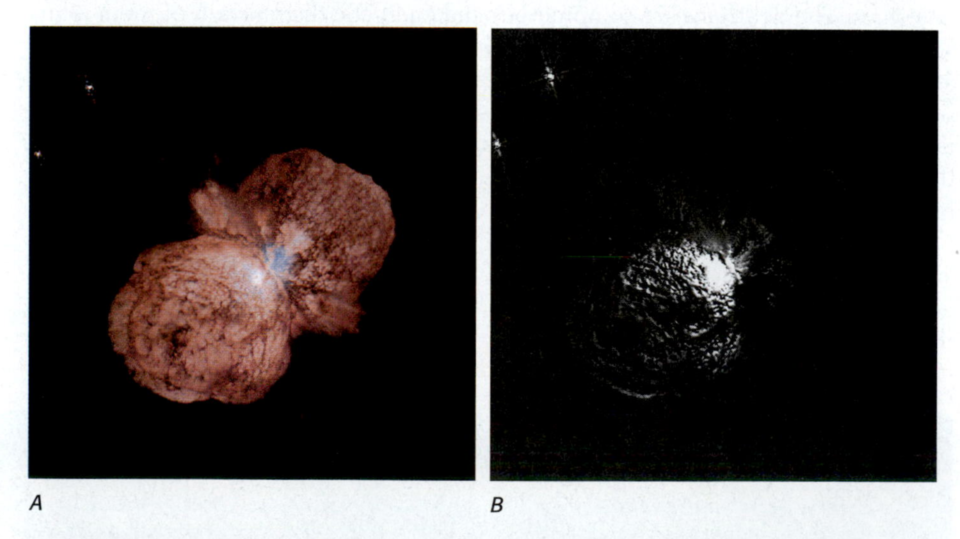


**FIGURE 29–6** (*A*) A Hubble Space Telescope view of the center of the Eta Carinae Nebula, a nebula around a massive star that may be a supernova in not too long. Eta Carinae is 9000 light-years away. A supernova there would be too far away to affect us, but it would appear brighter than Venus. The star $\eta$ Carinae exploded in 1843 and became the second-brightest star in the sky. The nebulae around it are expanding outwards from the explosion. We see two lobes of newer dusty ejecta that reflect starlight. The small spots on the lobes are the size of our solar system. The lobe at lower left is approaching us, and the lobe at upper right is receding. (*B*) A difference view, the subtraction of two images taken 17 months apart. Black shows where the material was located in the older image and white in the newer image. The difference image shows that the material closer to the star is moving more quickly than material farther away. Few astronomical objects show such drastic changes.

**FIGURE 29–7** A supercomputer simulation of the impact of a Type Ia supernova explosion on a main-sequence star companion. The white dwarf star that exploded just a few hours previously was situated just to the right of the picture. We see the collision and stripping of the supernova's stellar companion, which had earlier supplied the mass to induce it to explode. The companion is severely traumatized but not destroyed.

contrast to Type II supernovae, a Type Ia supernova apparently takes place in a binary system containing a white dwarf when mass from the second star (perhaps also a white dwarf) falls onto the white dwarf. Type Ia supernovae thus take place in the late stages of low-mass stars. The extra mass puts the white dwarf up to the 1.4-solar-mass Chandrasekhar limit, which leads to the star's thermonuclear explosion. Matter blasts into space in all directions, including past the binary companion (Fig. 29–7). After the sudden brightening, the light fades drastically for a month and then begins a slower decline. The energy comes (according to theoretical models) from radioactive decay of material produced in the supernova explosion. In particular, nickel-56, which was formed, decays into cobalt-56 and then iron-56, releasing energy. The optical spectrum of the long, slow decline can be explained as a composite of overlapping lines of ionized iron. NASA's Compton Gamma Ray Observatory verified this model by observing Type Ia supernovae in galaxies some distance away from us, detecting the gamma rays emitted as the nickel, cobalt, and then iron were formed. Since Type Ia supernovae expand so rapidly, they become transparent to gamma rays earlier than Type IIs, so their light curves eventually fall below those of Type IIs.

The star is apparently entirely disrupted, spreading heavy elements through the galaxy but often leaving no core behind. The fact that Type Ia supernovae don't show hydrogen in their spectra fits with these models, since the white dwarf may have shed its entire outer atmosphere before its incineration. Another point in favor of distinguishing models for Type Ia and Type II supernovae (Fig. 29–8) is that the former take place in elliptical galaxies, which lack young stars but which should have white dwarfs, while the latter occur in spiral galaxies, where we expect massive stars to be forming even now. (We don't always get a supernova when a white dwarf gains extra mass: theorists think that in some cases, two white dwarfs may merge, leaving an even denser core, a "neutron star," a type of star to be discussed in Section 30.1.)

Since supernovae, of which we still see remnants (Fig. 29–9), are the source of the heaviest elements in the universe, they provide many of the heavy elements that are necessary for life to arise. Heavy atoms in each of us have been through such a supernova explosion.

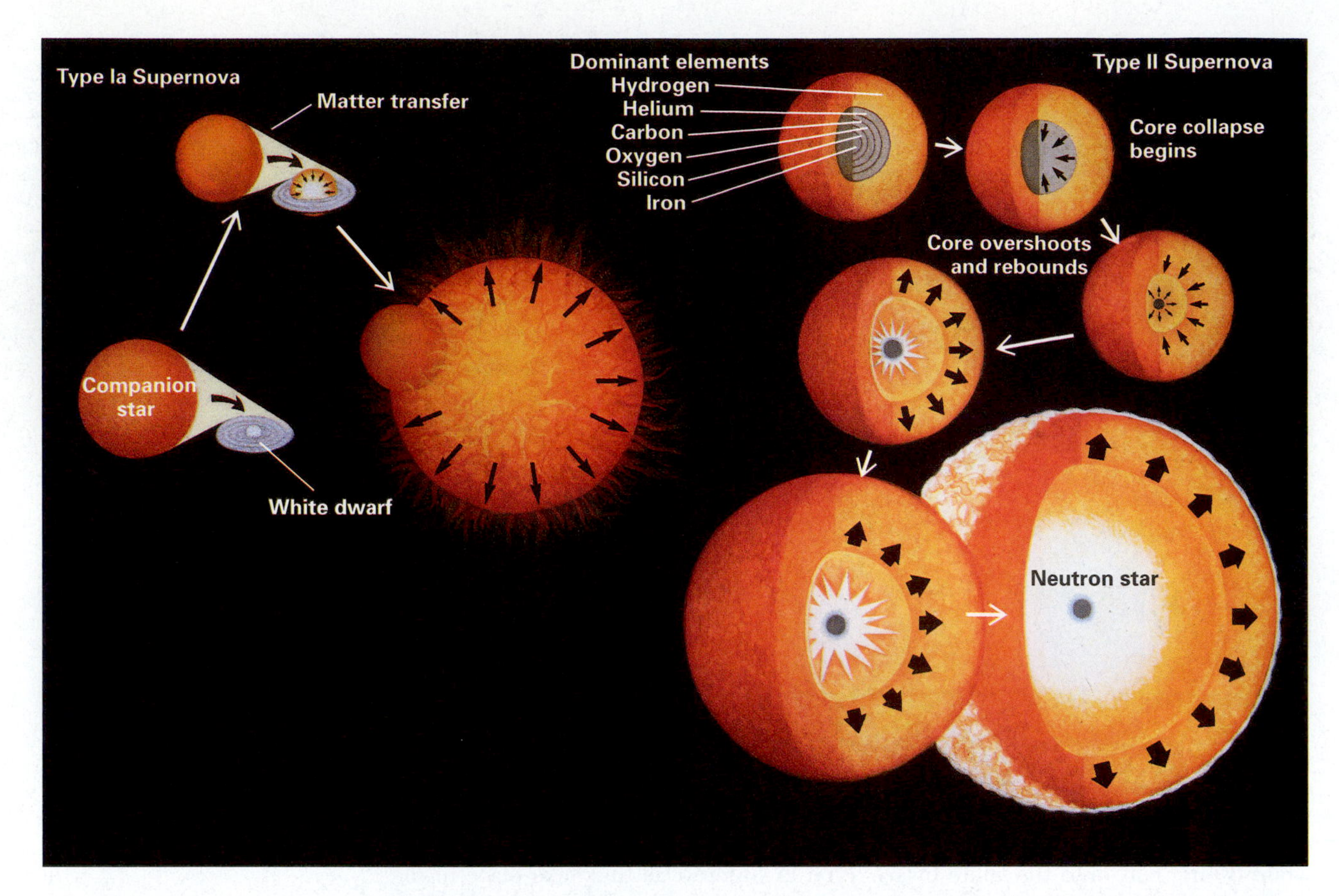


**FIGURE 29–8** Type Ia supernovae (*left*) come from the incineration of a white dwarf that is accreting matter from a neighboring giant. Type II supernovae (*right*) are the explosions of massive stars, usually from the supergiant phase. When iron forms at the center of the onion-like layers of heavy elements, the star collapses. In this model of the collapse, the core overshoots its final density and rebounds. The shock wave that results blasts off the star's outer layers.

**FIGURE 29–9** A small part of the Vela supernova remnant, which covers a region over 10 times larger across than the full moon. It came from a supernova that exploded 12,000 years ago. Studies of how the interstellar dust reddens stars, blocking the bluishness on the right side of the image, show that the nebula is about 1500 light-years away.

Among the heavy elements that come from supernova explosions, whether Type I or Type II, is a radioactive isotope, nickel-56. It decays with a half-life of 6.1 days to cobalt-56. The cobalt, in turn, is also radioactive and decays, with a half-life of 77 days (over 3 months), to iron-56, which is stable. These decays power the optical brightness and explain the fading of these supernovae.

The name "supernova" persists from a time when these events were thought to be merely unusually bright novae. The supernova in the Andromeda Galaxy in 1885 (known by the variable-star name of S Andromedae), since it was assumed to have the same brightness as a nova, gave a value for the distance to that galaxy that misled astronomers for years. They thought the Great Nebula in Andromeda was too close to be an independent galaxy. But in the 1930s, Walter Baade and Fritz Zwicky realized the distinction between novae and supernovae. A supernova explosion represents the death of a star and the scattering of most of its material, while a nova uses up only a small fraction of a stellar mass and can recur. One new similarity does exist: both novae and Type Ia supernovae are phenomena of binary systems in which one member is a collapsed star.

Though both involve white dwarfs, astronomers have no trouble distinguishing between a nova and a Type I supernova. The peak value of the light curve and the rate of fall-off of intensity are very different. The addition of a small amount of mass, which makes a nova, leaves the star basically unchanged, while the addition of enough mass to make a supernova disrupts the entire star.

## 29.3 DETECTING SUPERNOVAE

Whether Type I or Type II, in the weeks following the explosion, the amount of radiation emitted by the supernova can equal that emitted by the rest of its entire galaxy. It may brighten by over 20 magnitudes, a factor of $10^8$ in luminosity. Doppler-shift measurements show high velocities, about 5,000 to 10,000 km/sec, that prove that an explosion has taken place.

In our galaxy, the only supernovae thoroughly reported were in A.D. 1572 and 1604. Tycho's supernova was observed by the great astronomer (Section 2.6) in 1572. Kepler's supernova, observed by that great astronomer in 1604, was slightly farther away from Earth than Tycho's, though in another direction. Though these two supernovae were seen only 32 years apart, one actually exploded about 10,000 years after the other. The light curves measured then were accurate enough for us to tell that both were probably Type I. Only faint traces of them remain.

In A.D. 1054, chronicles in China, Japan, and Korea recorded the appearance of a "guest star" in the sky that was sufficiently bright that it could be seen in the daytime. No one is certain why no Western European drawings of the supernova were made, for the Bayeux tapestry from about that time illustrates a comet and thus shows that even in the Middle Ages people were aware of celestial events. A reference to a sighting of the supernova in Constantinople has been discovered, so perhaps it was not completely ignored outside the Far East. A 1999 analysis by George Collins and colleagues at Case Western Reserve University suggests that the supernova occurred several weeks earlier than had been thought, matching some European reports, and that later European reports may have been suppressed for reasons of religious politics.

A rock painting made by Native Americans in the American southwest (Figure 29–10) was thought for decades to show this supernova, though it now seems that it may well show Venus near the moon instead. The light curve was not well enough observed in that early time to allow us to determine whether it was a Type I or a Type II supernova. (It may even be an example of some rare additional type.) We shall discuss the remains, the Crab Nebula, in the next section.

The sightings of A.D. 393, 1006, and 1181 (Fig. 29–11) are now also thought to have probably been supernovae. A sighting in A.D. 185 has been thought to be a supernova, but recent work indicates that it might rather have been a regular nova enlarged in visible duration by being merged in people's minds by a comet three years later; the pitfalls of historical astronomy are revealed in this problem. The 1006 supernova, in the southern constellation Lupus, was apparently the brightest ever, perhaps apparent magnitude −26 (about as bright in our sky as our daytime sun). It remained visible, even in northern locations, for at least two years.

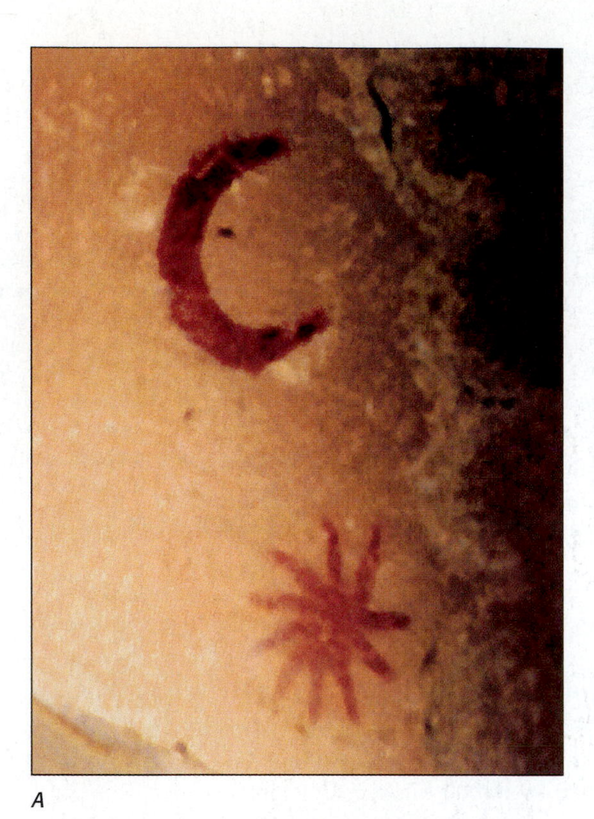

A

B

**FIGURE 29–10** *(A)* A Native American cave painting discovered in northern Arizona that may depict the supernova explosion 900 years ago that led to the Crab Nebula, though many people are skeptical of this explanation. It had been wondered why only astronomers in the Far East reported the supernova, so searches have been made in other parts of the world for additional observations. *(B)* A photograph showing a conjunction of Venus and the Moon. The cave painting could easily have shown such an event instead of a supernova explosion, showing how hard it is to guess what people were thinking long ago when they left no written descriptions.

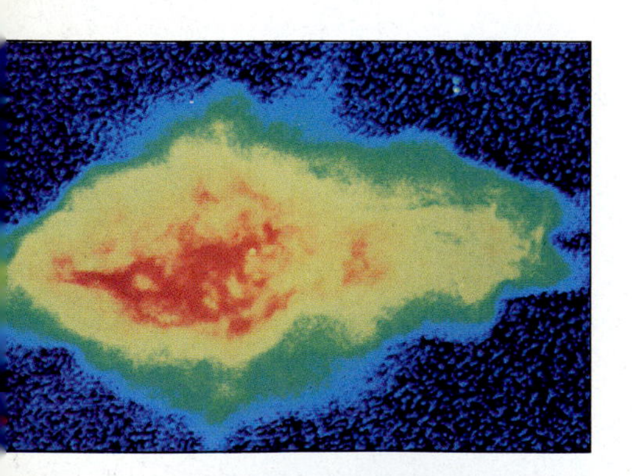

**FIGURE 29–11** The gas that expanded from the supernova of A.D. 1181, a radio source called 3C58, imaged with the Very Large Array (VLA) in the radio spectrum.

Through x-ray observations, we find a supernova remnant, RX J0852-4642, at a distance of only 800 light-years, a tenth the distance to the next nearest supernova in our galaxy. Tracing its expansion backward shows that its light reached us in about 1300. We don't know why the explosion remained unnoticed, though the fact that it is visible only from the southern hemisphere made it less likely that we would have a report.

Later in this chapter, we will show images of the supernova remnant Cassiopeia A, from a supernova thought to have gone off in 1680. Ambiguous notes taken by the English Astronomer Royal may have recorded this event. Though it was in the northern sky, and was thought to have become 100 times brighter than the brightest star, there are no clear records of it.

Type I supernova are very rare, perhaps occurring only every 300 years in our galaxy. Though we think there is one supernova every 30 to 50 years in our galaxy, mostly Type II supernovae, not a single supernova has been immediately noticed in our galaxy since the invention of the telescope. Dust has hidden most of them from us. In Section 29.6, we will discuss the supernova in the Large Magellanic Cloud, a satellite galaxy to our own visible only from southern latitudes; the light from this event reached us in 1987. We would obviously like to detect a supernova as soon as it goes off, so that we can bring to bear all our modern telescopes in various parts of the spectrum. Several new search programs, making use of electronic detectors and computer image-processing, are finding supernovae quickly and by the dozen. In Chapter 35 on the cosmic distance scale and Chapter 38 on cosmology, we will see how studies of the most distant supernova are revealing fundamentals about our universe.

**TABLE 29-2** Supernovae in our Galaxy and in the Large Magellanic Cloud

*These are all the supernovae known to have gone off in our galaxy and its satellite galaxies within the last millennium. The center of our galaxy is about 8.5 kpc from us.*

| Supernova | Year (AD) | Distance (kpc) | Peak Magnitude | Type |
|---|---|---|---|---|
| SN1006 | 1006 | 2.0 | –9 | Ia |
| Crab | 1054 | 2.2 | –4 | II |
| SN1181 | 1181 | 8.0 | ? | ? |
| RX J0852-4642 | ~1300 | ~0.2 | ? | ? |
| Tycho's | 1572 | 7.0 | –5 | II* |
| Kepler's | 1604 | 10.0 | –3 | II |
| Cassiopeia A | ~1680 | 3.4 | ~6? | II |
| *SN1987A* | *1987* | *50* | *3* | *II* |

*Reclassified from Type Ia to Type II in 2001.

## 29.4 SUPERNOVAE AND THEIR REMNANTS

The stellar shreds that are left behind are known as **supernova remnants** (Fig. 29–12). Optical astronomers have photographed two dozen of them in our galaxy alone, and others in nearby galaxies. Supernova remnants are prominent sources of radio waves and x-rays. Over 100 radio supernova remnants are known.

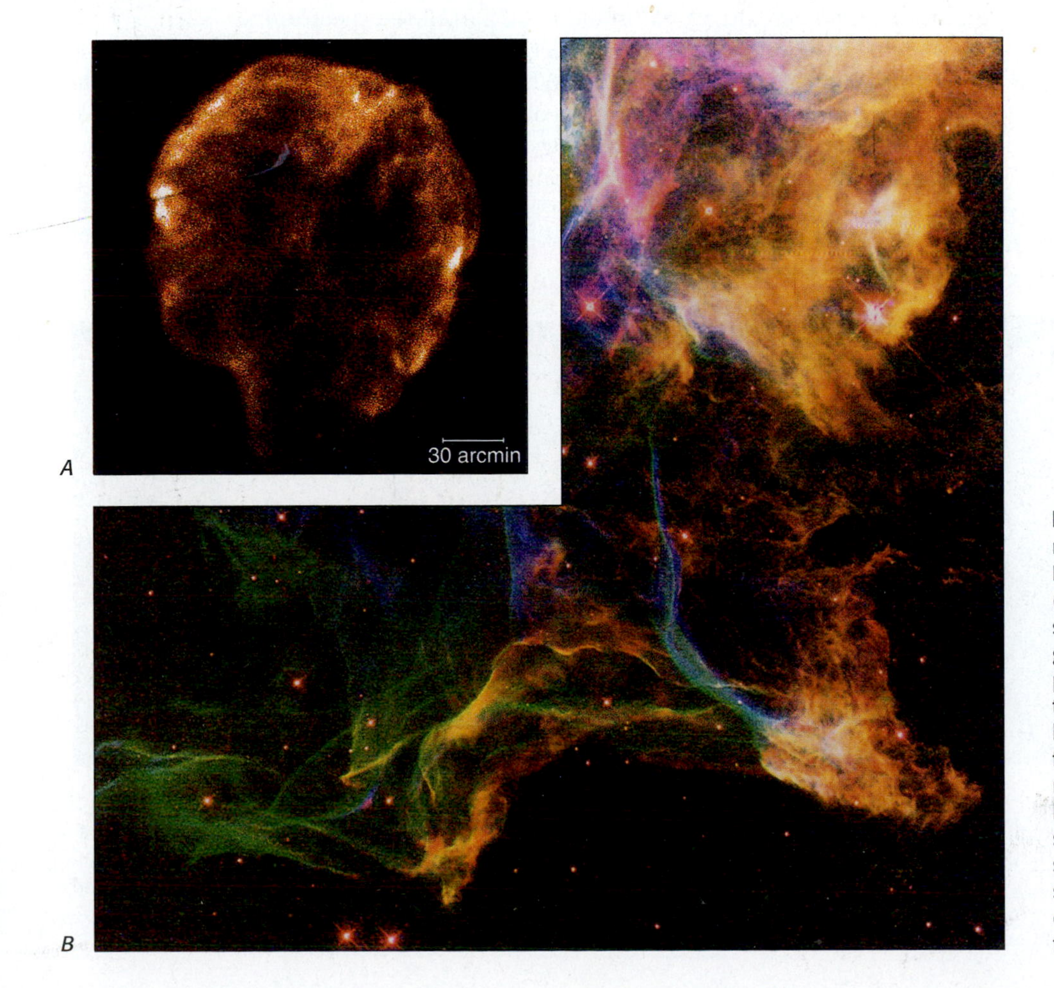


**FIGURE 29–12** The supernova remnant known as the Cygnus Loop, itself part of the Veil Nebula. (*A*) A wide-angle ROSAT x-ray view, showing the extent of the exploded gas. (*B*) A Hubble image of a small part of the Cygnus Loop. We see the effect of the shock wave. The hydrogen radiation (*green*) comes from a thin zone, only a few astronomical units across, immediately behind the shock wave. The emission from singly ionized sulfur (*red*) shows gas that has cooled after the shock wave passed. Doubly ionized oxygen (*blue*) was also formed by the shock front.

**FIGURE 29–13** The Crab Nebula, in a color presentation made especially for my books by David Malin of the Anglo-Australian Observatory from three black-and-white plates taken through visible-light filters at the Palomar Observatory.

The comparison of optical, infrared, radio, and x-ray observations of supernova remnants tells us about the supernovae themselves and about their interaction with interstellar matter. Few supernova remnants are known in the visible; most supernova remnants have been discovered in the radio part of the spectrum.

We find the most famous remnant in the location where Chinese astronomers reported their "guest star" in A.D. 1054. When we look at the reported position in the sky, in the constellation Taurus, we see an object—the Crab Nebula—that clearly looks as though it is a star torn to shreds (Fig. 29–13). New abilities to image at higher resolution than previously, as with the European Southern Observatory's Very Large Telescope (Fig. 29–14) and the Hubble Space Telescope (Fig. 29–15), reveal details

**FIGURE 29–14** The center of the Crab Nebula, imaged with the European Southern Observatory's Very Large Telescope.

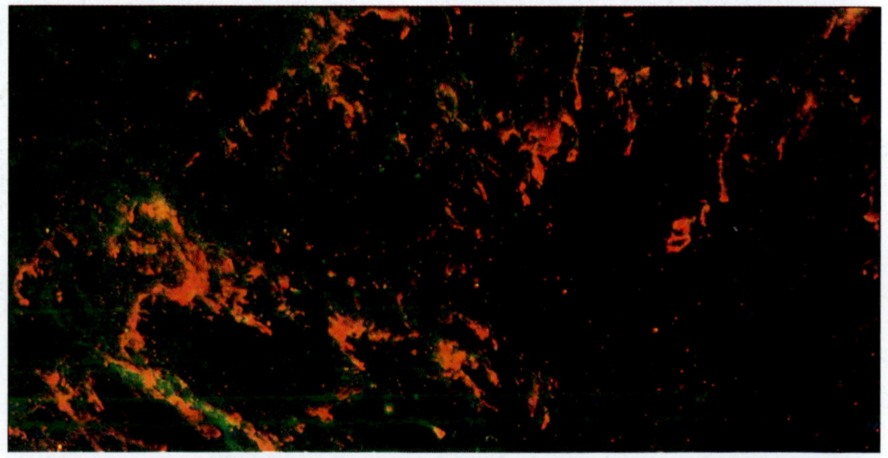

**FIGURE 29–15** Part of the Crab Nebula, imaged with the Wide Field Planetary Camera 2 on the repaired Hubble Space Telescope. In this false-color image, cool gas (neutral oxygen) is red, hot gas (doubly ionized oxygen) is green, and gas at intermediate temperature (neutral sulfur) is blue. "Fingers" of gas arise as the filaments interact with the cloud of gas and magnetic fields powered by the central pulsar.

in the shreds. The distance to the Crab is about ⅕ the distance to the center of our galaxy. We find the Crab's age by noting the rate at which its filaments are expanding. Tracing the filaments back in time shows that they were at a single point approximately 900 years ago. The Crab's age, location, and current appearance leave us little doubt that it is the remnant of the supernova whose radiation reached Earth 12 years before William the Conqueror invaded England. We will have more to say about the Crab when we discuss pulsars.

From a study of the rate at which supernovae appear in distant galaxies, we estimate that supernovae should appear in our galaxy about once every 30 years. Interstellar matter obscures much of the galaxy from our optical view. Still, we wonder why we haven't seen one in so long. It's about time. Radiation from hundreds of distant extragalactic supernovae is on its way to us now. Recently, we have detected supernovae in some of the nearest spiral galaxies. The light from a supernova in our galaxy may be on its way too. Maybe it will get here tonight.

## 29.5 X-RAY OBSERVATIONS OF SUPERNOVAE

NASA's Chandra X-ray Observatory and the European Space Agency's XMM-Newton mission, both launched in 1999, have each devoted a lot of their observing time to supernova remnants. The x-ray images show the distribution of million-degree-hot gas, and can be compared with images made in other parts of the spectrum (Fig. 29–16).

Supernova remnants are studied not only by imaging but also by taking x-ray spectra. The young supernova remnant E0102-72 (Fig. 29–17) in the Small Magellanic Cloud, for example, has revealed motions of 1000 km/sec and a shock front moving inward. The XMM-Newton's spectrometer (Fig. 29–18) detected x-ray spectral lines of highly ionized carbon and nitrogen, the first time they were detected in any supernova remmant.

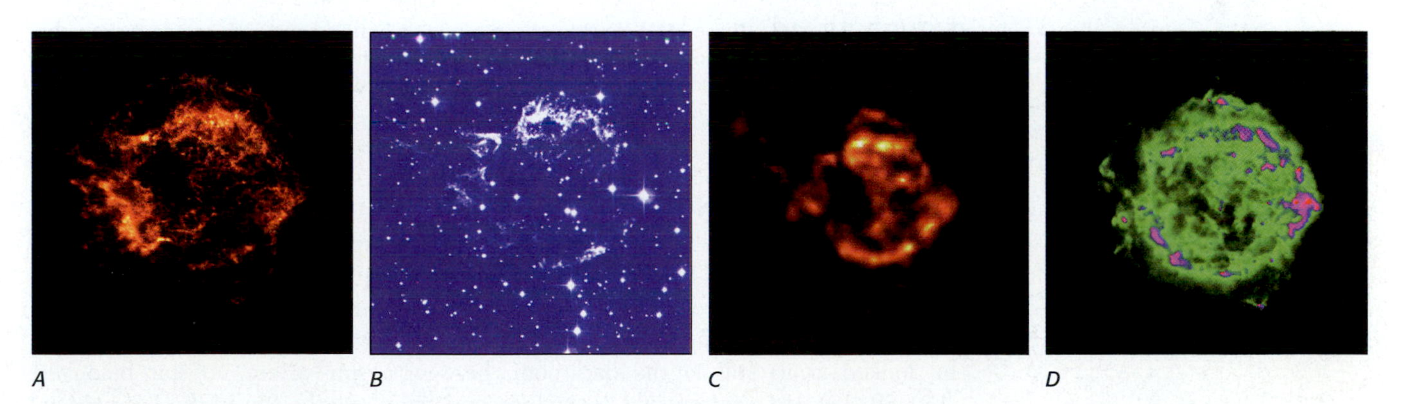


**FIGURE 29–16** The supernova remnant Cassiopeia A, imaged within various parts of the spectrum. The supernova went off in the 17th century. (*A*) This Chandra view shows the hottest gas. A fast outer shock wave and a slower inner shock wave are visible. The inner shock wave heated the gas to ten million kelvins. This image was the first to be released from CXO. (*B*) The optical image shows wisps of matter with a temperature of about 10 thousand kelvins. Spectra of the wisps show high abundances of heavy elements, so the wisps must be clumps of ejected material. (*C*) This infrared image from the Infrared Space Observatory (wavelengths 10.7 to 12.0 micrometers) shows dust grains that have been swept up and heated to hundreds of degrees by the expanding hot gas. (*D*) This radio image, made with the National Radio Astronomy Observatory's Very Large Array in New Mexico, shows emission from high-energy electrons spiralling around magnetic lines of force.

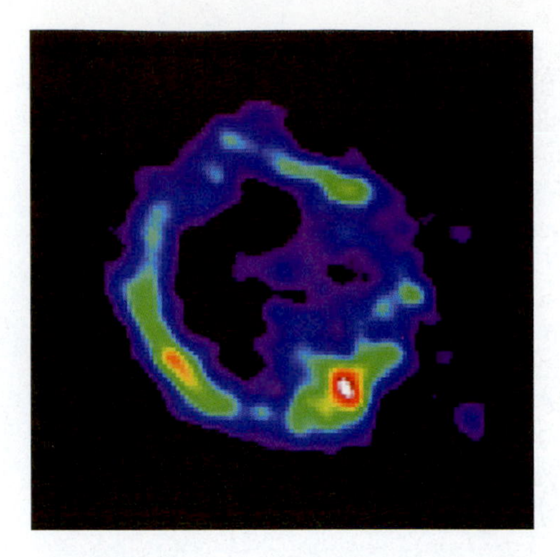

FIGURE 29–17 The young supernova remnant E0102-72 in the Small Magellanic Cloud. We see an overlay of observations that show the remnant's appearance about 1000 years after the supernova explosion. The Chandra x-ray image shows highly ionized multimillion-degree oxygen gas that has been heated by a shock wave that is rebounding back toward the star after material from the initial explosion hit surrounding gas. A compound image, available at http://chandra.harvard.edu/photo/cycle1/0015multi/, superimposes a Hubble Space Telescope image that shows dense clumps of oxygen gas at a temperature of 30,000 K. It also superimposes a radio image, made with the Australia Telescope Compact Array, that shows emission from high-energy electrons spiralling around magnetic lines of force. XMM-Newton spectra reveal internal motions, and show x-ray spectral lines of highly ionized carbon and nitrogen.

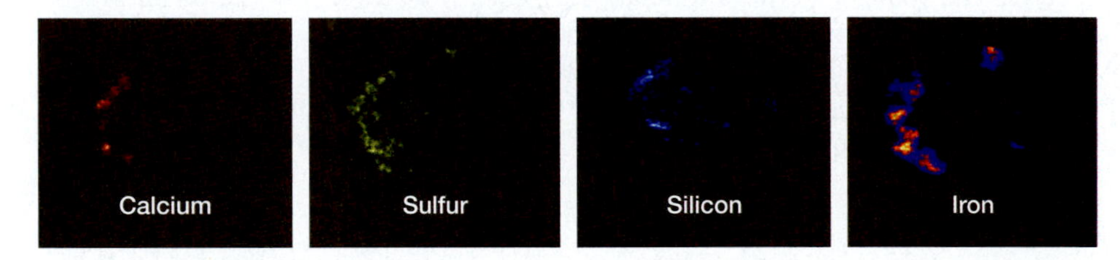


FIGURE 29–18 These XMM-Newton observations of Tycho's supernova remnant show the distribution of the abundances of several elements. Because the spacecraft could separate these spectral bands, we see that the elements form in different parts of the exploding star.

## 29.6 SUPERNOVA 1987A

On February 24, 1987, Ian Shelton was photographing the Large Magellanic Cloud. He was working for the University of Toronto at their telescope on Las Campanas in Chile. Fortunately, he chose to develop his photographic plate that night. When he looked at it, still in the darkroom, he saw a star where no star belonged (Fig. 29–19). He went outside, looked up, and again saw the star in the Large Magellanic Cloud, this time with his naked eye. He had discovered a supernova, the nearest supernova seen since Kepler saw one in 1604, five years before the telescope was invented. By the next night, the news was all over the world, and all the telescopes that could see the supernova were trained on it (Fig. 29–20). The event was perhaps the most significant to occur in astronomy in centuries. It received its official name from the International Astronomical Union's Central Bureau for Astronomical Telegrams: Supernova 1987A.

The "A" in Supernova 1987A means it was the first supernova to be discovered that year. Two years earlier, the IAU had certified the use of capital letters for supernovae in place of the lower-case letters that had been used informally.

Observations made independently by amateur astronomers just before and just after the explosion indicate that the supernova brightened within 3 hours, much faster than the few days of brightening for most other supernovae.

**FIGURE 29–19** Ian Shelton with Supernova 1987A just above the Tarantula Nebula, seen from the Las Campanas Observatory.

Though the supernova is not in our own galaxy, it is nearly so. The Large Magellanic Cloud is only about 50 thousand parsecs away, which the newspapers often gave with false accuracy as 163,000 light-years (multiplying the round number 50,000 by a precise conversion factor with 3 significant digits). Some parts of our own galaxy are farther away. Perhaps it is lucky that the supernova was there instead of closer, for a closer supernova would have been too bright for many telescopes. As it was, detectors are now so sensitive that many astronomers had to observe with most of their telescope mirrors masked off or through partially obscuring filters. The only unfortunate thing is that the supernova could not be seen from most of the northern hemisphere; the Large Magellanic Cloud is far enough south in the sky that it can be studied only with telescopes near the equator or in the southern hemisphere. One advantage of the supernova's location is that the Large Magellanic Cloud is circumpolar for southern-hemisphere observatories, so there were no long gaps in coverage.

The supernova, discovered at 5th magnitude, later reached slightly brighter than 3rd magnitude (Fig. 29–21), 20 or more times fainter than would be expected for a supernova at its distance, easily visible to the naked eye. Within a few nights it was realized that hydrogen lines were indeed present. That made it a Type II supernova, one that came from a massive star.

We even knew which star had blown up (Fig. 29–22)! It had been star 202 in the −69° declination band of the catalogue of Sanduleak, so it is generally known as Sk −69°202. Before its eruption, it had been seen to be a star of spectral type B3, a type that was not thought to go supernova.

Subsequent models put some of the facts together into a coherent picture. Spectral type B3 supergiants could go supernova after all. Their smaller radius compared with red supergiants made their outer layers denser and more likely to capture the energy produced by the shock wave. Thus a supernova from a B3 supergiant should be less luminous than normal supernovae, as was seen with Supernova 1987A. The energy emitted matched theoretical expectations of the explosion of such a 20-solar-mass star with a 6-solar-mass helium core. The supergiant had already lost perhaps 4 solar masses' worth of its outer atmosphere, which is why the supernova did not become as bright as expected. The outer atmosphere had had a high hydrogen content, so its loss explained why the hydrogen spectral lines were not at first detected. The fact that the supernova did not become as bright as most Type II supernovae may indicate why we have not had more examples of similar light curves—the objects with those light curves were too faint, so we missed them. International Ultraviolet Explorer (IUE) observations found an enriched nitrogen shell, perhaps the

*A*

*B*

**FIGURE 29–20** The supernova (*A*) before and (*B*) after February 24, 1987.

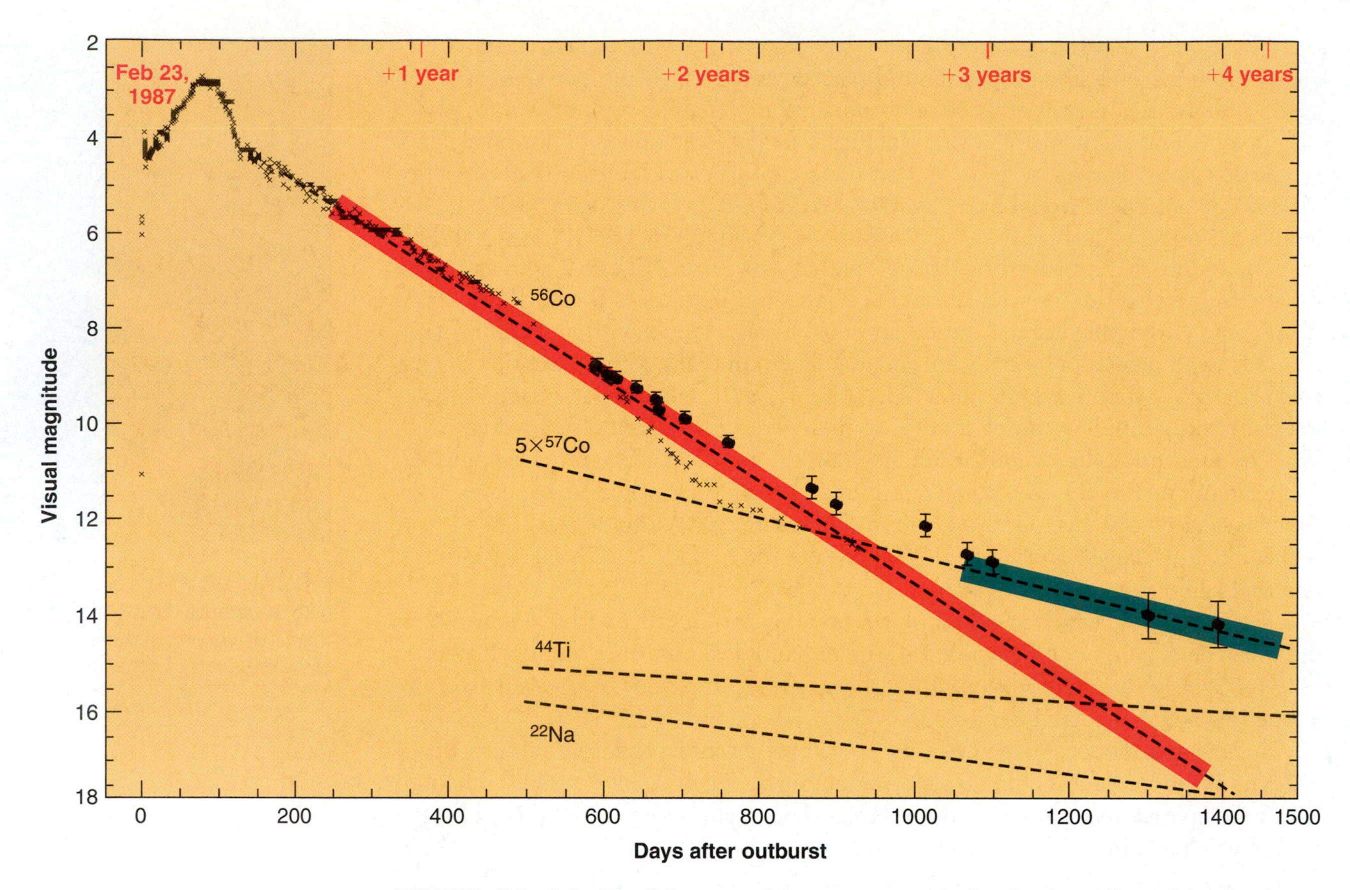


**FIGURE 29–21** The light curve of Supernova 1987A, showing its peak 80 days after its discovery. After day 120, the decay matches the prediction that the supernova light comes from radioactivity. A quick decay of the 0.07 solar masses of radioactive nickel-56 (half-life 6.1 days) to cobalt-56 is then followed by a slower decay (half-life 77 days). After several years, the radiation should be dominantly from titanium-44, whose half-life is 47 years.

result of a hundred thousand years of stellar wind removing the outer layers and changing the star from a red supergiant into a blue supergiant.

The supernova grew brighter in optical radiation for almost three months, an indication that the star that exploded had a substantial hydrogen envelope remaining even though it was a blue supergiant (Fig. 29–23). The envelope presumably was the reason why gamma rays and x-rays did not appear sooner than they did and did not become brighter. It may also have been significant that the object is in the Large Magellanic Cloud, which has a lower abundance of elements heavier than helium ("metals") than does our own galaxy.

Analysis of SN 1987A has shown that two mechanisms of explosion occur (Fig. 29–24). The abrupt expansion of the outer layers of the star comes in part from the shock wave formed as collapsing gases bounce off the core (known as the prompt explosion); this is then added to by the flood of neutrinos formed in the core (known as the delayed explosion). Scientists wondered whether a neutron star formed in the explosion. But no pulsar has appeared—as yet. (We discuss neutron stars and pulsars in Chapter 30.)

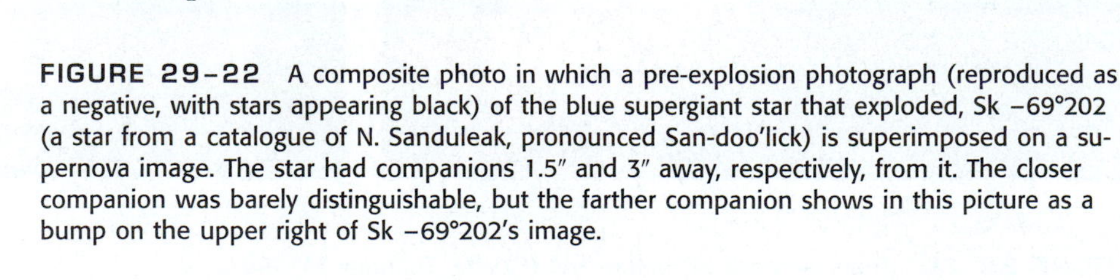

**FIGURE 29–22** A composite photo in which a pre-explosion photograph (reproduced as a negative, with stars appearing black) of the blue supergiant star that exploded, Sk −69°202 (a star from a catalogue of N. Sanduleak, pronounced San-doo'lick) is superimposed on a supernova image. The star had companions 1.5″ and 3″ away, respectively, from it. The closer companion was barely distinguishable, but the farther companion shows in this picture as a bump on the upper right of Sk −69°202's image.

**FIGURE 29–23** The Large Magellanic Cloud with the supernova near the reddish Tarantula Nebula.

### 29.6a Ring Around a Supernova

The Hubble Space Telescope was able to image not only the supernova itself but also the gas around it, showing two rings, as was shown in the photograph opening this chapter. The inner ring's radius is less than 1 arc second, comparable to the best ground-based seeing and so was not imaged clearly from the ground. Though Hubble wasn't up at the time when light from the explosion reached Earth, another spacecraft, the International Ultraviolet Explorer, saw the inner ring brighten 0.66 years later. Therefore, the inner ring is 0.66 light-years in radius. (After all, light travels at a speed of 1 light-year/year.) From the lack of emission across the ring, scientists conclude that we see an actual ring inclined at 43° rather than the edges of a sphere. The Space Telescope Imaging Spectrograph (STIS) has been able to take spectra that show the elements in the ring and their temperatures (Fig. 29–25).

Comparing the size of the ring in light-years with its angular size as seen in the sky has given its distance from Earth as 168,000 light-years, plus or minus only 5 per cent. Thus the whole Large Magellanic Cloud is at that distance. Measuring the whole cosmic distance scale is, in large part, based on the distance to the Large Magellanic Cloud (Chapter 35), so the supernova is playing an important role in cosmology.

Apparently, gas ejected in the supernova explosion has caught up with the gas that the star had given off as a stellar wind in the hundreds of thousands of years prior to the explosion. The ring's enrichment in nitrogen, measured from spectra, endorses the idea that the rings were ejected when the star was still a red supergiant. The ring's velocity of expansion shows that matter had been flowing outward for about 20,000 years. This period is presumably how long it took the star

**FIGURE 29–24** The current model of Supernova 1987A's explosion, in an artist's conception (*top*). The core of a massive star collapses under the force of gravity when nuclear fusion reactions shut down. This sends out a shock wave that plows into upper layers of gas (*center*). The outer layers of the star contract under the force of gravity. However, deep inside, a flood of neutrinos released from the imploding core boosts temperatures to 18 billion degrees (*bottom*). This hot, ballooning "bubble" of subatomic particles accelerates the shock wave until it slams into the infalling gas. The star blows apart, and a supernova is born.

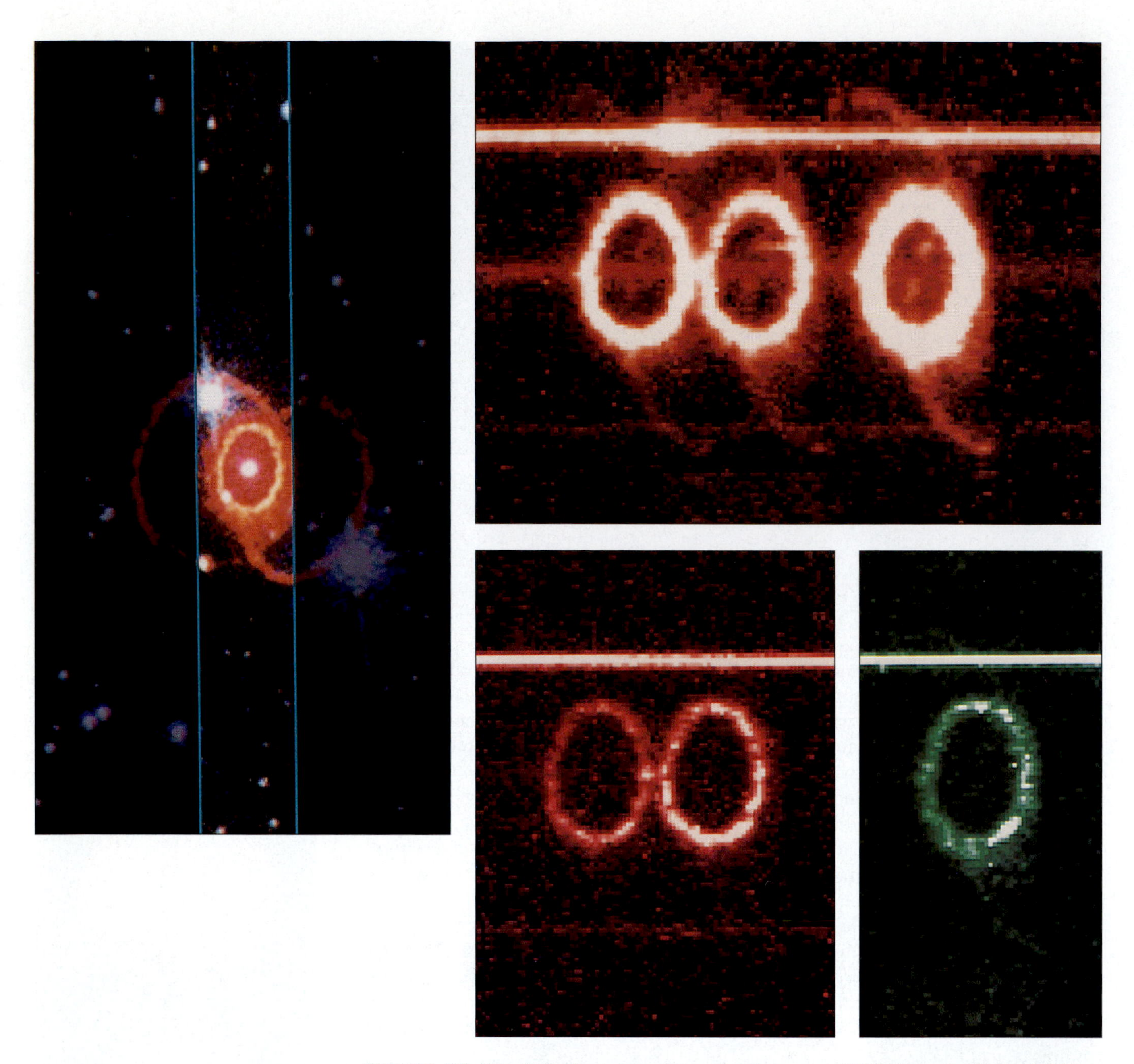

**FIGURE 29–25** (*left*) Wispy rings surrounding Supernova 1987A, imaged with the Hubble Space Telescope. The inner ring had been known (Fig. 29–26), but the symmetric outer rings were a surprise. These thin, sharp rings could be part of a huge shell shaped like an hourglass and illuminated from within by jets of particles or radiation coming from a rotating neutron star. In any case, the shape results in part from gas given off long before the supernova exploded. (*right*) We see parts of the spectrum of Supernova 1987A, taken with the Space Telescope Imaging Spectrograph (STIS). This image shows a "slitless spectrum," in which the bright rings of the supernova provide their own images at different wavelengths, without using a narrow spectrographic slit. Vertical lines define the wide slot; we see the slitless spectrum of everything within that slot. We see emission by two singly ionized nitrogen lines surrounding hydrogen—$\alpha$ (triple orange rings), singly ionized sulfur (double red rings), and doubly ionized oxygen (single green ring). The ratios of the rings' brightness in different spectral lines from the same element show the abundances. The ratios from different elements show temperatures.

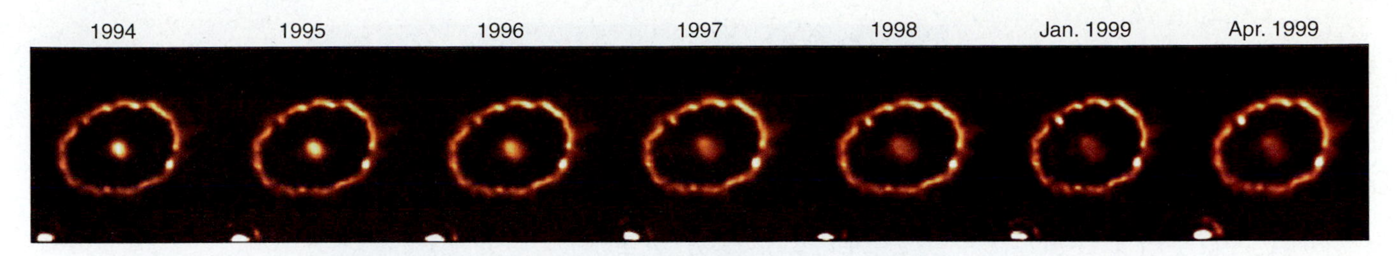


**FIGURE 29–26** The ring is made of gas thrown off by the star prior to its explosion as Supernova 1987A. Debris from the explosion is reaching the ring, making spots on it brighten. The ring is so small that it is a prime object for the Hubble Space Telescope.

to change from a red supergiant to a blue supergiant. The rings are now expanding outward at "only" about 100,000 km/hour, a speed slow in comparison with the supernova material in the center, which is expanding at 100 to 2000 times that speed.

Parts of the rings started brightening in 1994 (Fig. 29–26). The main part of the supernova's fireball is now hitting the inner ring. We expect major changes in its brightness and shape. As of 2001, 10 or more spots on the ring are brightening (Fig. 29–27).

The Chandra X-ray Observatory has observed the hot gas in the ring (Fig. 29–28).

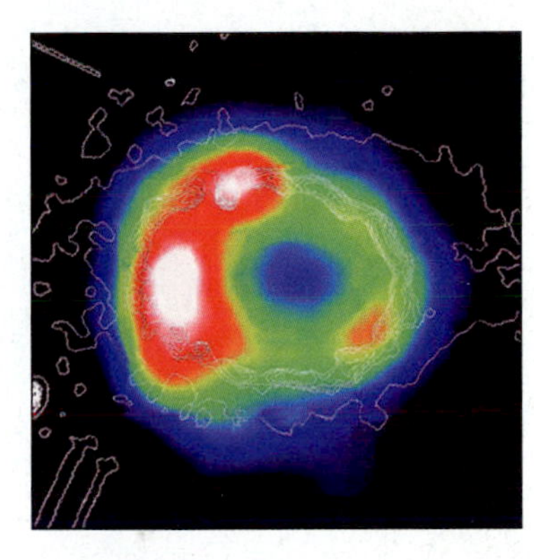

**FIGURE 29–28** The Imaging Spectrometer on the Chandra X-ray Observatory shows the expanding 10-million-kelvin shell of gas heated by the impact of the shock wave caused by the supernova explosion.

## 29.6b Neutrinos from the Supernova

The solar neutrino experiment, with a large tank of cleaning fluid in a mine, was not sensitive enough to the energy range of neutrinos emitted by the supernova. But at least two other experiments were. They had been set up to study the possible decay of the proton (which we discuss in Chapter 38), but they were fortunately operating for this event of the century. Both experiments contained large volumes of extremely pure water surrounded by sensitive phototubes to measure any light given off as a result of interactions in the water.

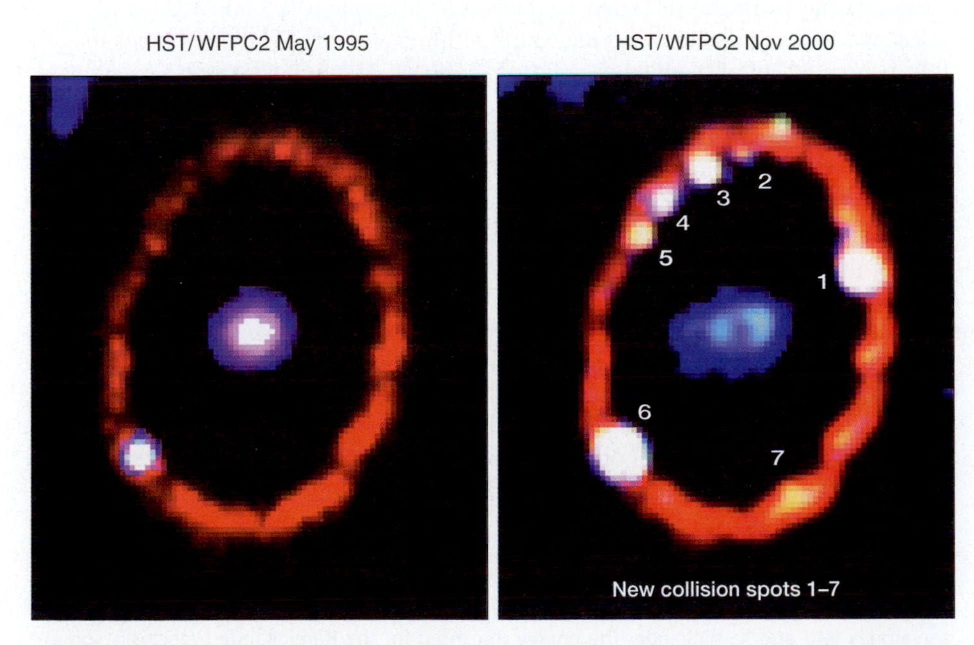


**FIGURE 29–27** These true-color views taken with the Hubble Space Telescope show that at least 10 spots have brightened.

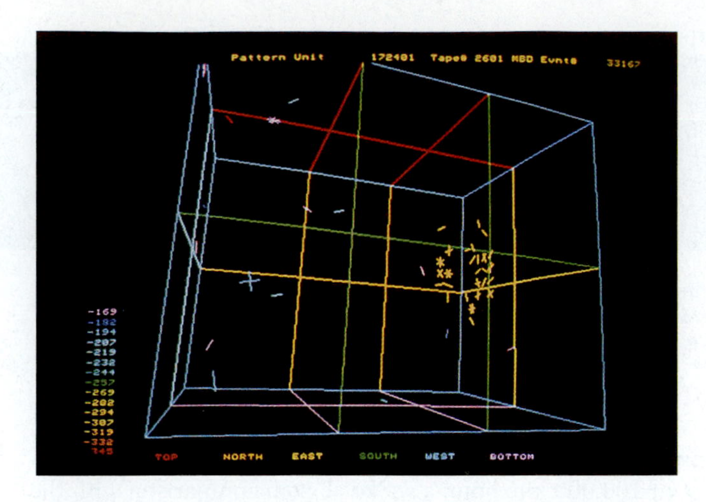


FIGURE 29–29 A video screen showing one of the neutrinos from the supernova interacting in the tank of water of the U. Cal/Irvine–U. Michigan–Brookhaven neutrino observatory (IMB). The yellow marks show the point of origin of the cone of photons generated by a neutrino interaction in the tank; tracing the path backward shows that the neutrino came from the supernova. Colors show arrival time, with intervals shown in nanoseconds. The neutrino supernova generated photons detected by the photomultipliers within a few nanoseconds. A few random interactions, shown in different colors, occurred at other times.

The IMB (University of California at Irvine–University of Michigan–Brookhaven National Laboratory) experiment in a salt mine in Ohio reported that 8 neutrinos had arrived and interacted within a 6-second period on February 23 (Fig. 29–29), three hours before the optical burst was first photographed and a day before it was noticed, whereas otherwise a single neutrino interacted about once every 5 days. This number indicated that $3 \times 10^{16}$ neutrinos from the supernova had passed through the 5000 tons of water in the detector. At the same time, the Kamiokande II detector in a zinc mine in Japan detected a burst of 9 neutrinos in two seconds followed by 3 neutrinos 9 to 13 seconds later. They could even tell that at least some of them were arriving from the direction of the Large Magellanic Cloud. A third neutrino telescope in Russia, detected 5 neutrinos within 5 seconds. The discoveries marked the beginning of a new observational field of astronomy: extrasolar neutrino astronomy.

The neutrino bursts preceded the optical brightening. The sequence matches the theoretical idea that the neutrinos are released as the collapsing core deposits energy, starting nuclear reactions. Also, the amount of energy carried in neutrinos closely matched theoretical predictions. Our basic idea of what happens in a supernova appeared verified.

Scientists studied the arrival times of the neutrinos (Fig. 29–30). If neutrinos had mass, they would have spread out in space, since neutrinos could then travel at different speeds. The fact that all the neutrinos arrived on Earth so close together is consistent with neutrinos being massless and gives a way of figuring out an upper limit on how much mass they may have. The upper limit is very small; scientists often use "electron-volts," a unit of energy, to express the amount of mass, since mass and energy are equivalent, as Einstein showed. In these energy units, the supernova ob-

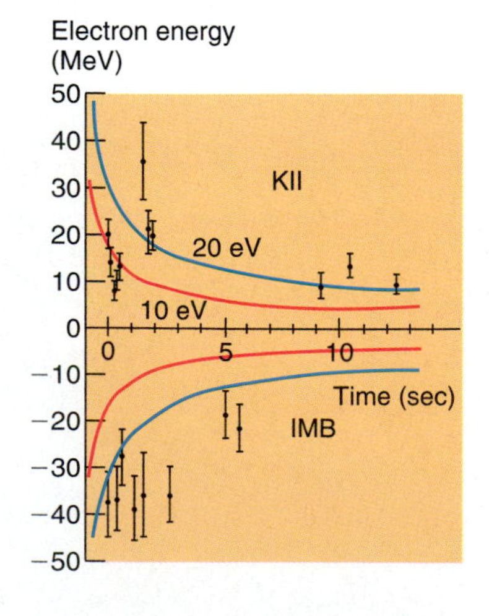


FIGURE 29–30 The arrival times and energies of the neutrino events at the neutrino observatories IMB and Kamiokande. The energy threshold for the Kamiokande detector is somewhat lower, 7 MeV, compared with 20 MeV for the IMB detector. (The Kamiokande times were uncertain by about 1 minute and were adjusted to match the IMB times.) The red lines outline the expected positions if neutrinos have mass 10 eV; the blue curves correspond to 20 eV.

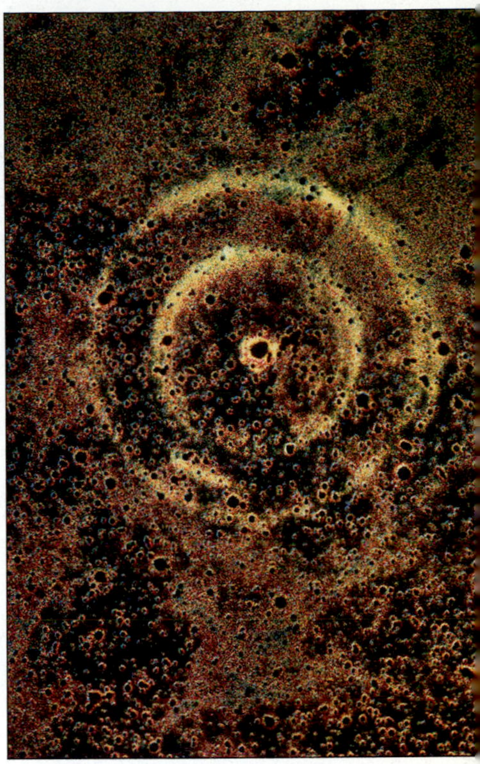

**FIGURE 29–31** A double light echo from Supernova 1987A, reflections in interstellar clouds in the Large Magellanic Cloud of the light from the bright supernova explosion. The light echoes became visible the following year. The two clouds making the echoes are 400 and 1000 light-years, respectively, from the supernova toward us. The photo shows two concentric rings around the overexposed image of the supernova itself, which was dimmed by a small obscuring disk that shows at the center. Continuing photos show the ring expanding. The image was made by subtracting photos from a pre-explosion photo, leaving only the light echo.

servations have shown that the neutrino mass must be less than about 16 electron volts (much less than, say, the 511-million-electron-volt rest mass of an electron). So studies of the supernova have even provided important information for basic particle physics (Chapter 38).

John Bahcall of the Institute for Advanced Study at Princeton has raised the possibility that many supernova explosions produce almost all their energy in neutrinos. These "neutrino bombs" may never become very bright but would show up in continued neutrino observations. If significant numbers are detected, then we have been underestimating the rate of stellar collapse in our galaxy by a misplaced reliance on optical observations. Super-Kamiokande, with 50,000 tons of ultra-pure water and 11,200 photomultipliers (Section 27.6) should be able to detect supernovae out to 650,000 light-years as well as solar neutrinos, and Hyper-Kamiokande should be able to detect supernovae even farther away. Here we see a major difference between astronomy and physics. We astronomers now have to wait for a suitable supernova to go off; we can't do the experiment on our own schedule.

### 29.6c Additional Studies of the Supernova

Supernova 1987A will remain visible to large telescopes for the foreseeable future. The supernova shell is turning transparent, allowing us to view the inner layers where the heavy elements were formed. We first observed x-rays that probably were formed as gamma rays hit the supernova debris. Somewhat later, we observed the gamma rays directly.

Most of the light we get from the supernova comes to us directly. But some is light that originally went in a slightly different direction and was later reflected toward us (Fig. 29–31). This reflected light tells us about interstellar space near the supernova.

What will we continue to learn about the formation of the heavy elements in the supernova, as we see to deeper and deeper levels? How will x-ray and radio emission vary? How much will the ring and the surrounding nebulosity brighten, as more of the debris hits the inner ring? And what even more exciting surprises may be in store? We must wait to see.

## 29.7 COSMIC RAYS

Most of the information we have discussed thus far in this book has been gleaned from the study of electromagnetic radiation, which can be thought of as sometimes having the properties of waves and sometimes having the properties of particles called photons. The neutrino studies of Supernova 1987A have already been an exception in which we have studied particles from space. Certain high-energy particles of matter travel through space in addition to photons and neutrinos. These particles are mostly protons or nuclei of atoms heavier than hydrogen moving at tremendous velocities; about 3 per cent are electrons. Together, they are called **cosmic rays.** Ordinary CCD exposures of astronomical objects show cosmic-ray "hits" (Fig. 29–32), which astronomers work to edit out of their images.

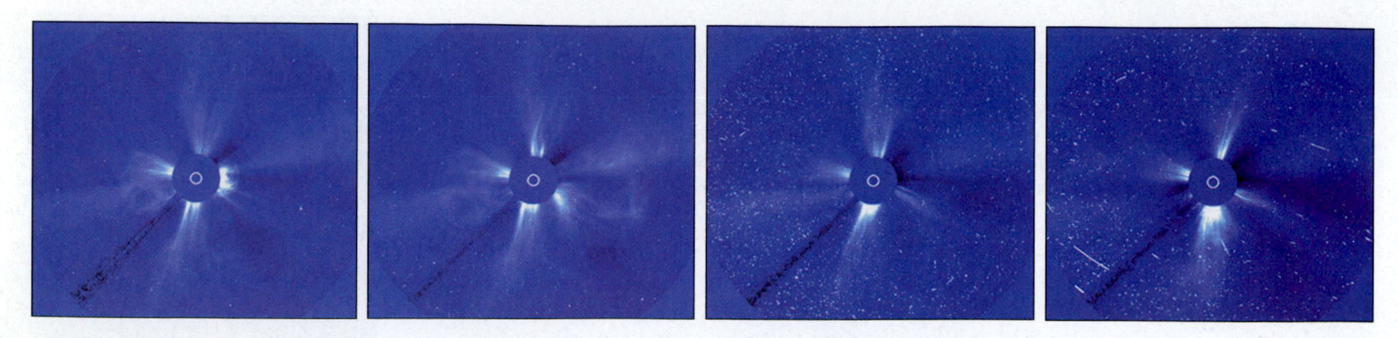

**FIGURE 29–32** Astronomers deal with cosmic rays on a daily basis, since they have to be eliminated from CCD images. Note how the solar mass ejection seen in the first frame soon led to solar cosmic rays reaching the SOHO spacecraft's coronagraph.

Some of the weaker cosmic rays come from the sun, and are increasingly known as "solar energetic particles" instead of "solar cosmic rays," but most cosmic rays come from farther away. This latter group is known as "galactic cosmic rays," though their ultimate sources are both within and beyond our galaxy. We receive fewer cosmic rays on Earth at the maximum of the sunspot cycle, because the stronger solar wind at that time sweeps cosmic-ray particles away and the stronger magnetic field at that time bends cosmic rays away from us. A graph of the cosmic-ray flux at Earth looks like a sunspot-number graph upside-down. Some of the cosmic rays are trapped by the Earth's magnetic field to make the Van Allen belts that we discussed in Chapter 8.

Cosmic rays provide about 30 percent of the radiation environment of the Earth's surface and of the people on it, and a higher percentage at greater altitude. It used to be thought that almost all the radiation we are exposed to comes from cosmic rays, from naturally occurring radioactive elements in the earth or in our bodies, or from medical x-rays. More recently, radon gas—emitted naturally from the Earth and sometimes found in houses, particularly when their ventilation is poor—has proved to provide about half the exposure for some people, though the rate of radon exposure depends greatly on location.

The origin of the non-solar "primary cosmic rays," the ones that actually hit the top of the Earth's atmosphere, as opposed to the "secondary cosmic rays" that hit the Earth's surface, has long been under debate. (The secondary cosmic rays are formed by collisions of the primary cosmic rays with nitrogen and oxygen atoms in molecules in our atmosphere. Neutrons and other subatomic particles result.) Because cosmic rays are charged particles—mostly protons and also some nuclei of atoms heavier than hydrogen as well as electrons—our galaxy's tangled magnetic field bends them. Thus we cannot trace back the paths of cosmic rays we detect to find out their origin. For example, some of the cosmic-ray particles are thought to come from colliding galaxies, but we are unable to trace them back to specific ones.

Historically, cosmic-ray scientists give energies in units of electron-volts (eV), based on how the earliest cosmic rays were detected. A 1 MeV cosmic-ray proton travels at 5 per cent of the speed of light; the rarer cosmic-ray electron travels at 94 per cent of the speed of light. A 1 GeV (giga-electron-volt, or billions of electron-volts) proton is travelling at 88 per cent of the speed of light. The corresponding 1 GeV electron is going at 99.999 98 per cent of the speed of light. These energy ranges are low for cosmic rays; see the set of metric-system prefixes in Appendix 1.

It is most likely that most middle-energy cosmic rays were accelerated to their high velocities in supernova remnants as shock waves from these objects reach into the interstellar gas. When you plot a graph that shows the density of particles vs. their energies, the curve has a "knee" at an energy of $10^{15}$ electron-volts, which is a million billion electron-volts (or 1 peta-electron-volt, 1 PeV). Thus many of the cosmic rays more energetic than that probably have a different origin, perhaps from pulsars or intergalactic shock waves, the latter possibly caused by colliding galaxies. These highest cosmic-ray particles may be iron nuclei. The "knee" itself may be from a nearby supernova remnant.

Above this "knee" in the cosmic-ray spectrum, events that reach Earth are too rare to be picked up with detectors aboard balloons or satellites. Thus systems spread out over large areas of ground are necessary. The high-altitude Chacaltaya Cosmic Ray Research Laboratory, at an altitude of 5220 feet (18,100 feet) in Bolivia, is above so much of the Earth's atmosphere that it is ideal for such research. (The atmospheric pressure there is only about half that at sea level.) Some of these highest energy particles are detected by the secondary cosmic rays they generate, which are detected on Earth as "extensive air showers." A "Fly's Eye" detector in Utah, whose multifaceted mirror looked like a fly's eye, collected a lot of light to look for blue flashes that are given off in the high-altitude interactions of cosmic rays with nitrogen atoms in our atmosphere. The Whipple Telescope, a 10-m telescope on Mt. Hopkins in Arizona, works similarly.

It has been suggested that a significant fraction of the high-energy cosmic rays in our galaxy are apparently from a single source—Cygnus X-3. This object was a discovery of space research. Its name means that it was the third x-ray source to be detected in the constellation Cygnus. Studies across the spectrum have shown that Cygnus X-3 gives off more energy than all but a couple of other sources in our galaxy, though dust obscures it in visible light. Current evidence indicates that Cygnus X-3 is a neutron star (Section 30.1) in orbit with a larger companion on the edge of our galaxy (Fig. 29–33).

Our atmosphere filters out almost all of the primary cosmic rays. To capture most primary cosmic rays directly, one must travel above most of the Earth's atmosphere in an airplane, balloon, rocket, or satellite. In 1984, a space shuttle carried cosmic-ray detectors aloft to remain in space for what was supposed to be a year on NASA's *L*ong-*D*uration-*E*xposure *F*acility (LDEF). Because of the explosion of the space shuttle Challenger in 1986, LDEF could not be retrieved until 1990 (Fig. 29–34). NASA's Compton Gamma-Ray Observatory was an important platform for studying gamma rays, which are intimately connected with the charged cosmic rays, until NASA removed it from orbit for safety reasons in 2000, fearing that parts of it would cause damage or injury on Earth if it was left to an uncontrolled descent. NASA launched the *A*dvanced *C*osmic *E*xplorer (ACE) in 1997. One of its instruments has detected tens of thousands of cosmic ray particles and identified the individual isotopes that have hit the spacecraft (Fig. 29–35). Some of the particles are cobalt and nickel, which are even heavier than iron. These particles are so rare that they must come to us straight from their sources, since it is very unlikely that they could be the result of collisions of still more massive and rarer particles.

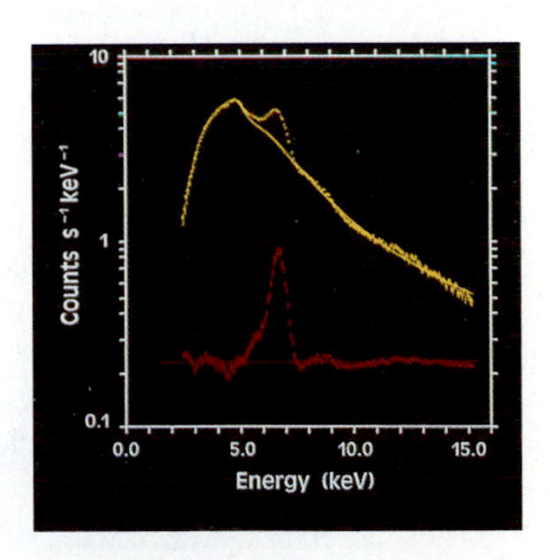


**FIGURE 29–33** The spectral line from highly ionized iron, observed from EXOSAT, is exceptionally strong in Cygnus X-3. The lower curve (*red*) shows the subtraction of one of the upper (*yellow*) curves from the other. The yellow curves show spectra at different locations in Cyg X-3, one in a place where the iron is ionized.

Though most of the cosmic rays come from our own galaxy, some, at least, of the highest-energy cosmic rays seem to come from the direction of the nearest large cluster of galaxies. Scientists suspect that a peculiar galaxy there, M87, may be giving off these particles, perhaps from a giant black hole. These cosmic rays, though individually energetic, are too sparse to be detectable from the spacecraft yet aloft or planned.

Several projects tried or are trying to study a variety of the highest-energy cosmic-ray particles, and neutrinos as well, by using clear ocean or lake water or clear ice. The technique is similar to that used in the neutrino experiments discussed in Section

FIGURE 29–34 The Long-Duration-Exposure Facility (LDEF) was launched by a space shuttle in 1984, carrying plastic and other materials aloft to be affected by the space environment. LDEF was picked up by a space shuttle and returned to Earth in 1989. It was examined to see the effects of the space environment on its surface and contents. Plastic sheets were scanned for cosmic-ray tracks. The photo shows many micrometeoroid impacts.

27.6, though with more collecting volume. Photomultipliers will detect flashes of light from secondary cosmic-ray particles and from neutrinos generated by supernovae. The flashes arise as the cosmic-ray particles slow down when they pass through water (in liquid form or as ice) or the neutrinos interact with protons in the water. Project DUMAND (*D*eep *U*nderwater *M*uon *a*nd *N*eutrino *D*etection) set up chains of giant photomultipliers 4.7 km deep in the clear ocean water near the island of Hawaii. But mundane technical problems with leaks and broken connections led to its loss of Department of Energy funding. Similar projects under water in the

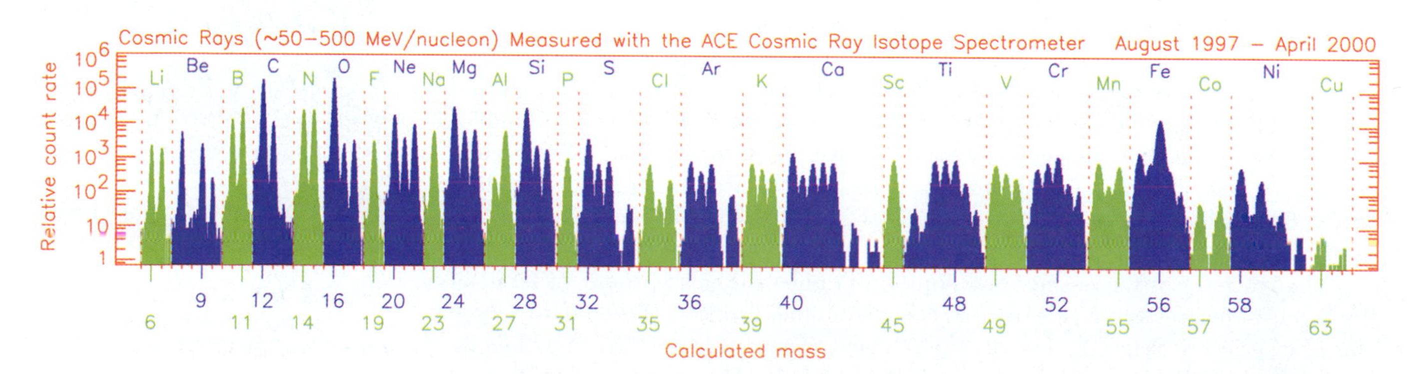


FIGURE 29–35 The mass spectrometer on the Advanced Composition Explorer (ACE), now in orbit, can distinguish individual isotopes that reach Earth as cosmic rays. Its successor will be ACCESS (Advanced Cosmic-ray Composition Experiment for the Space Station) for the International Space Station. In the diagram, isotopes for each of 27 elements are shown separately, with color alternating from element to element. Elements with even mass numbers are shown in blue and odd ones in green.

Mediterranean off Greece (Project Nestor, for *Ne*utrinos from *S*upernova and *T*eravolt sources), ANTARES off the coast of Marseilles, France, and in Lake Baikal, Siberia, are still underway. The last of these, already containing about 100 photomultipliers, has detected its first neutrinos.

A cubic kilometer of the clear ice deep underneath the ice cap at the South Pole is being instrumented as the *A*ntarctic *M*uon *a*nd *N*eutrino *D*etector *A*rray (AMANDA). Light can be seen almost 100 m away in the clear ice, about twice as far as in clear, deep ocean water. All these liquid water and ice projects may be on line in a few years, with substantial information returned around 2010.

Project Milagro, in New Mexico, is detecting high-energy cosmic rays up to 1 trillion electron volts, that is, up to the tera-electron-volt (TeV), range; such cosmic rays are expected to come from violent sources, perhaps even exploding mini black holes (which are described in Chapter 31). When gamma rays from space hit nuclei in our atmosphere, the nuclei give off subatomic particles just as they do for cosmic rays. Milagro contains a football-field-size tank of water a third of a meter deep, with photomultipliers to sense radiation from the atmospheric particles that result from cosmic rays or gamma rays. A smaller-size version, Milagrito, was on line in 1997–98, and Milagro itself is evolving in capabilities. These TeV detectors may open a new window on the universe, and history shows that such new windows often lead to major discoveries.

Only about 1 per week of the most energetic cosmic rays, those with energies 1 million times greater than 1 TeV (that is, 1 exa-eV, 1 EeV, to use the symbol shown in Appendix 1), reach each square kilometer of the ground. Therefore, to detect even rarer events at least 10 times more powerful than that, the Pierre Auger Observatory is covering a huge region, about 50 km on a side, with 1600 widely-spaced detectors—tanks of water with photomultipliers—that look for flashes of light from cosmic rays. Each tank contains 3000 gallons of water and will be spaced 1.5 km apart. At night, the scientists will use two different detectors to also look for simultaneous flashes of light in our atmosphere, from fluorescence, caused by the cosmic rays. (Auger was the French scientist who, in 1938, first observed extensive air showers of secondary cosmic rays.) One such region will be in Utah and another in Argentina; scientists from 20 countries are participating. Part of the ultimate Argentina detector, with 40 tanks and 1 fluorescence detector, went on line in 2001. Construction of the rest is to start in 2002 and be completed in 2004. The Utah system is awaiting funding. Even with the final system, the scientists expect to detect only 60 of these most powerful cosmic rays per year. Using accurate times from the Global Positioning System of satellites, they should be able to locate the direction from which the cosmic rays are coming to within about 2°.

Where do the highest-energy cosmic rays come from? Nobody knows. Scientists are considering colliding galaxies or even strange kinds of particles in our galaxy's outer regions—the dark matter we will discuss in following chapters—as sources. An individual cosmic ray carries enough energy to match the energy of a pitched baseball! Getting hit with one of these particles would be like being beaned by a major-league pitcher. The ICECUBE detector in Antarctica uses a huge volume of ice to try to detect such high-energy cosmic rays. It has been suggested that such cosmic rays would cause radio waves to be emitted on impact. If so, radio telescopes pointed at the moon might detect 250 ultra-high-energy cosmic rays per year, given the moon's large surface.

Whether the sun and the Earth's weather are connected has long been debated. Correlations over the years between sunspot number and weather on Earth have never panned out. Scientists are currently investigating a correlation between cloud cover over the Earth's oceans and the intensity of cosmic rays, which itself is correlated with sunspot number. We cannot predict whether this effort will pan out.

## CORRECTING MISCONCEPTIONS

| ✖ Incorrect | ✔ Correct |
|---|---|
| Stars are placid. | Some stars explode violently. |
| Supernovae are exploding single stars. | Type I supernovae explode because they are in binary systems, and matter from a companion flows onto a white dwarf. |
| We get only radiation from stars. | Sometimes we get particles: neutrinos or cosmic rays. |

## SUMMARY AND OUTLINE

Red giants and supergiants (Section 29.1)
- Stars of more than 8 solar masses that exhaust their hydrogen in the core become giants.
- Later, heavier elements are built up in the core, and the stars become supergiants.

Supernovae (Sections 29.2, 29.3, and 29.4)
- After cores of heavy elements form, massive stars explode as Type II supernovae.
- Type Ia supernovae come from less massive stars and may be the incineration of a white dwarf.
- Only a few optical supernovae have been seen in our galaxy.
- Supernova remnants can best be studied with radio and x-ray astronomy.

X-ray observations of supernova remnants (Section 29.5)
- Chandra and XMM-Newton study the hot gas.

Supernova 1987A (Section 29.6)
- The nearest, brightest supernova to be seen since 1604
- Erupted in 1987 in the Large Magellanic Cloud
- Decay of radioactive material formed
- Detection of neutrino bursts from the supernova

Cosmic rays (Section 29.7)
- High-energy particles in space
- Most probably come from supernovae
- The highest-energy cosmic rays have other sources.

## KEY WORDS

red supergiant
supernova
Type I supernovae
Type II supernovae
supernova remnants
cosmic rays (primary, secondary)

## QUESTIONS

1. Why are red supergiants so bright?
2. What are the basic differences between a nova and a supernova?
3. If we see a massive main-sequence star (a heavyweight star), what can we assume about its age, relative to most stars? Why?
4. What do we know about the core of a star when it leaves the main sequence?
5. What is special about iron in the core of a star?
†6. In a supernova explosion of a 20-solar-mass star, about how much material is blown away?
†7. A supernova can brighten by 20 magnitudes. By what factor of brightness is this? Show your calculations.
8. How do we distinguish observationally between Type I and Type II supernovae?
9. How do the physical models of Type I and Type II supernovae differ?
†10. A typical galaxy has a luminosity $10^{11}$ times that of the sun. If a supernova equals this luminosity, and if it began as a B star, by what factor did it brighten? By how many magnitudes?
11. How does $\eta$ Carinae endorse our model of supergiants?
12. Would you expect the appearance of the Crab Nebula to change in the next 500 years? How?
13. The light from Supernova 1987A came from the Large Magellanic Cloud, which is 50 kiloparsecs (kpc) away. When did the star actually explode?
14. What type of supernova was Supernova 1987A? Give two reasons why we think so.
15. From what event in Supernova 1987A did the neutrino bursts come?
16. Why was the neutrino burst significant for astronomy? For particle physics?
17. What are cosmic rays?
18. Where are secondary cosmic rays formed?

†This question requires a numerical solution.

## USING TECHNOLOGY

### W World Wide Web

1. Sign on to the International Astronomical Union Circulars to find out about current supernovae.
2. Look at homepages of ground-based observatories for supernova observations.
3. Look at Hubble's images of supernovae and supernova remnants and get links to supernova Web sites from the site for Supernova 1987A at http://heritage.stsci.edu.
4. Follow cosmic-ray observations at http://www.auger.org.

### REDSHIFT

1. Look at Science of Astronomy: Birth and Death of Stars; Story of the Universe: Lives of the Stars.
2. Find the position of Supernova 1987A in the Large Magellanic Cloud (LMC). In the Photo Gallery, examine the LMC and the supernova.

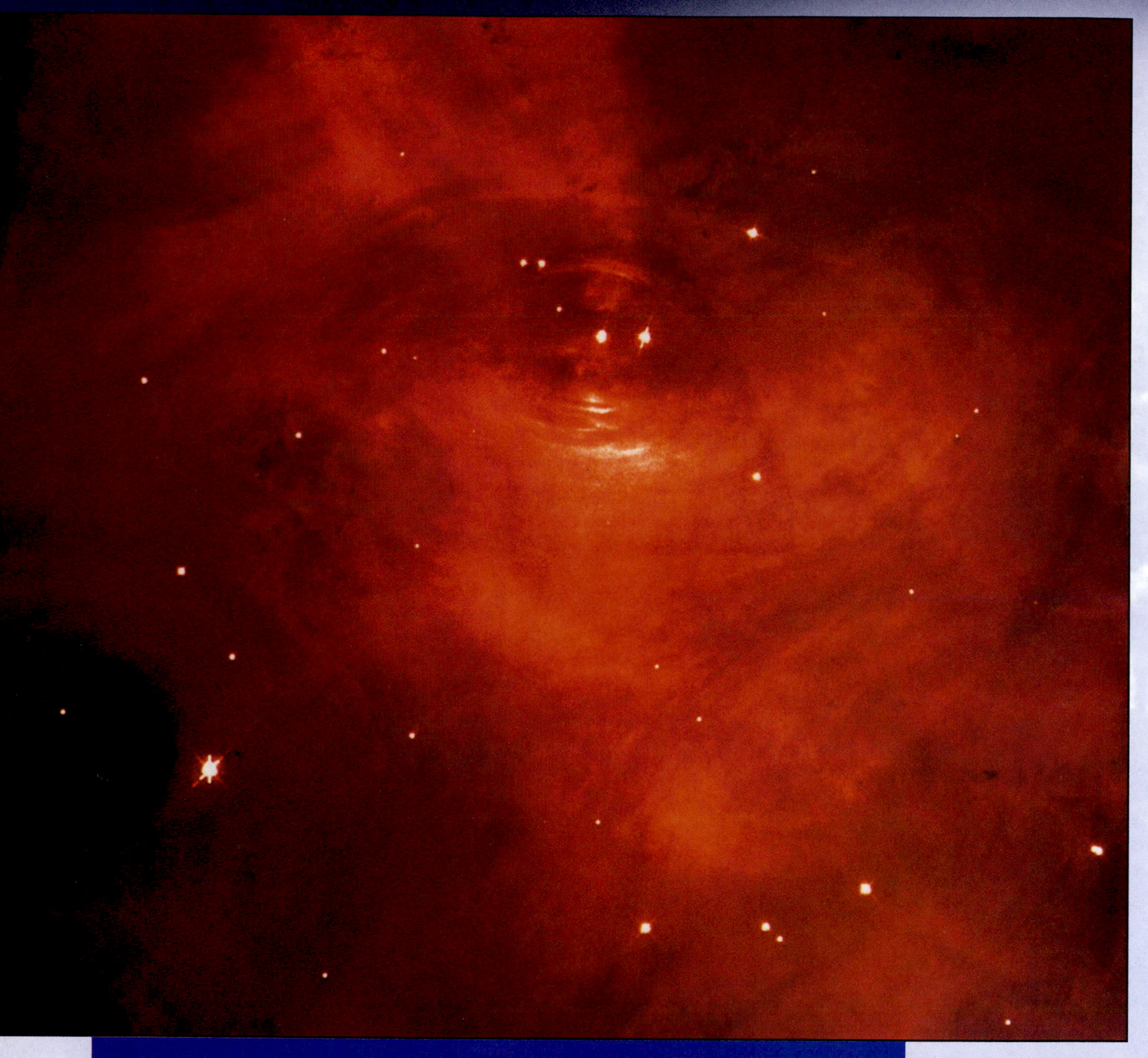

The center of the Crab Nebula, the remnant of a supernova explosion that became visible on Earth in A.D. 1054. It is viewed in a high-resolution yellow-light image (reproduced in red false color) taken with the Hubble Space Telescope. The pulsar, the first one to be detected blinking on and off in the visible part of the spectrum, is the left star of the pair at the center of the swirling gas. The details of the gas, moving at half the speed of light, change from day to day. In the long exposure necessary to take this photograph, the star turns on and off so many times that it appears to be a normal star. Only after its radio pulsation had been detected was it observed to be blinking in optical light. It had long been suspected, however, of being the leftover core of the supernova because of its unusual spectrum, which has only a continuum and no spectral lines, and its unusual blue color.

# Chapter 30

# Pulsars and Neutron Stars

In this chapter we study those massive stars that become neutron stars after their supernova stages. We can detect those objects as pulsars and x-ray binaries.

**AIMS:** To study how the cores of stars more massive than 8 solar masses sometimes collapse to become neutron stars, and to learn how we observe neutron stars as pulsars and as x-ray binaries

## 30.1 NEUTRON STARS

We have discussed the fate of the outer layers of a massive star that explodes as a Type II supernova. Now let us discuss the fate of the star's core.

As iron fills the core of a massive star, the temperatures are so high that the iron nuclei begin to break apart into smaller units like helium nuclei. Energy is carried off, the pressure drops, and the forces holding the star stable are overwhelmed by the force of gravity. The core collapses.

As the density increases, the electrons are squeezed into the nuclei and react with the protons there to produce neutrons and neutrinos. The neutrinos escape, thus stealing still more energy from the core, and may also help eject the outer layers if enough neutrinos interact as they pass through. In any case, a gas composed mainly of neutrons is left behind in the dense core as the outer layers explode as a supernova.

Following the explosion, the core may contain as little as a few tenths of a solar mass or possibly as much as two or three solar masses. This remainder is at an even higher density than that at which electron degeneracy holds up a white dwarf. At this density an analogous condition called **neutron degeneracy,** in which the neutrons cannot be packed any more tightly, appears. The pressure caused by neutron degeneracy balances the gravitational force that tends to collapse the core, and as a result the core reaches equilibrium as a **neutron star.**

FIGURE 30–1 A neutron star may be the size of a city, even though it may contain a solar mass or more. Here we see the ghost of a neutron star superimposed on a photograph of New York City. A neutron star might have a solid, crystalline crust about a hundred meters thick. Above these outer layers, its atmosphere probably takes up only another few *centimeters.* Since the crust is crystalline, there may be irregular structures like mountains, which would only poke up a few centimeters through the atmosphere.

Whereas a white dwarf packs the mass of the sun into a volume the size of the Earth, the density of a neutron star is even more extreme. A neutron star may be only 20 kilometers or so across (Fig. 30–1), in which space it may contain the mass of about two suns. In its high density, it is like a single, giant nucleus. A teaspoonful of a neutron star could weigh a billion tons.

Before it collapses, the core (like the sun's) has only a weak magnetic field. But as the core collapses, the magnetic field is concentrated. It becomes much stronger than any we can produce on Earth.

When neutron stars were discussed in theoretical analyses in the 1930s, there seemed to be no hope of actually observing one. Nobody had a good idea of how to look. But there should be a billion of them or so in our galaxy, based on how many supernovae had to go off to account for the abundances of heavy elements we find

If more than about three solar masses remain, then the force of gravity will overwhelm even neutron degeneracy. We shall discuss that case in Chapter 31.

in our galaxy. Let us jump to consider events of 1967, which will later prove to be related to the search for neutron stars.

## 30.2 THE DISCOVERY OF PULSARS

By 1967, radio astronomy had become a flourishing science. But one problem had thus far prevented a major discovery from being made: like radio receivers in our living rooms, radio telescopes are subject to static—rapid variations in the strength of the signal. It is difficult to measure the average intensity of a signal if the signal strength is jumping up and down many times a second, so radio astronomers usually adjusted their instruments so that they did not record any variation in signal shorter than a second or so. But in 1967 the only rapid variations that astronomers expected besides terrestrial static were those that correspond to the twinkling of stars.

An astronomer at Cambridge University in England, Antony Hewish, wanted to study this "twinkling" of radio sources, which is called **scintillation.** The light from stars twinkles because of effects of our Earth's atmosphere. The radio waves from radio sources scintillate not because of terrestrial effects but because radio signals are affected by clouds of electrons in the solar wind. Hewish therefore built a radio telescope uniquely able to detect faint, rapidly varying signals.

One day in 1967, Jocelyn Bell (Fig. 30–2), now Jocelyn Bell Burnell, then a graduate student working on the project and now a professor at the Open University in England, noticed that a set of especially strong variations in the signal appeared in the middle of the night, when scintillations caused by the solar wind are usually weak. In her own words, there was "a bit of scruff" on the tracing of the signal. After a month of observation, it became clear to her that the position of the source of the signals remained fixed with respect to the stars rather than constant in terrestrial time, a sure sign that the object was not terrestrial or solar.

Detailed examination of the signal showed that, surprisingly, it was a rapid set of pulses, with one pulse every 1.3373011 seconds. The pulses were very regularly spaced (Figs. 30–3 and 30–4). Soon, Burnell located three other sources, pulsing with regular periods of 0.253065, 1.187911, and 1.2737635 seconds, respectively.

FIGURE 30–2 Jocelyn Bell Burnell, the discoverer of pulsars. She used a radio telescope—actually a field of aerials—at Cambridge, England. The total collecting area was large so that it could detect faint sources, and the electronics were set so that rapid variations could be observed.

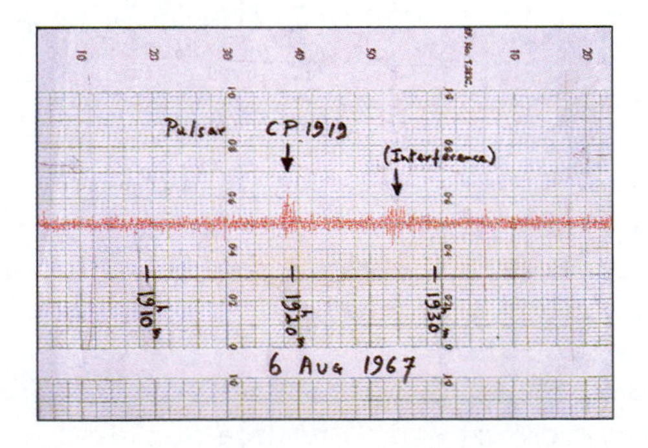


FIGURE 30–3 The up-and-down variations on this chart led Jocelyn Bell to suspect that something interesting was going on.

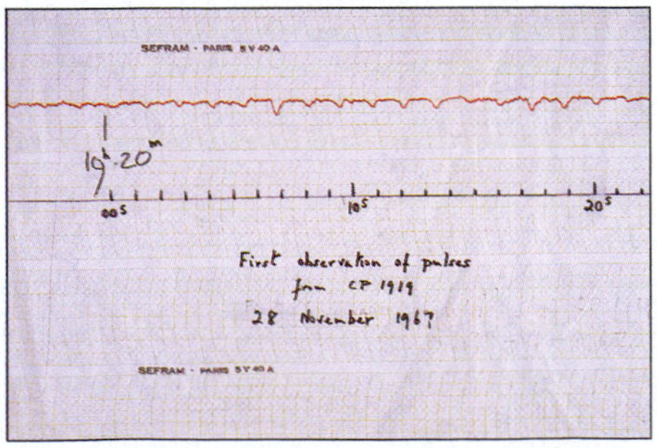

FIGURE 30–4 The chart record showing the discovery of the individual pulses from CP 1919.

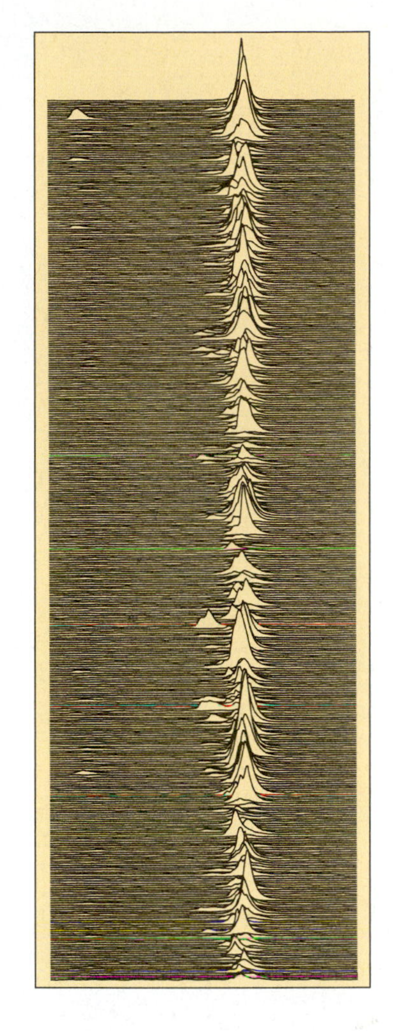

**FIGURE 30–5** A series of 400 consecutive pulses from pulsar PSR 0950+08. Each line from bottom to top represents the next 0.253 seconds. We can see structure within the individual pulses and an "interpulse" about halfway between some consecutive pulses.

One immediate thought was that the signal represented an interstellar beacon sent out by extraterrestrial life on another star. For a time the source was called an LGM, for Little Green Men, and the possibility led the Cambridge group to withhold the announcement for a time. The sources were briefly called LGM 1, LGM 2, LGM 3, and LGM 4, but soon it was obvious that the signals had not been sent out by extraterrestrial life. It seemed unlikely that there would be four such beacons at widely spaced locations in our galaxy. Besides, any beings would probably be on a planet orbiting a star, and no effect of a Doppler shift from any orbital motion was detected.

When the discovery was announced to an astonished astronomical community in 1968, it was immediately apparent that the discovery of these pulsating radio sources, called **pulsars,** was one of the most important astronomical discoveries of the decade. But what were they?

Other observatories turned their radio telescopes to search the heavens for other pulsars. Dozens of new pulsars were found. The pulse itself lasts only a small fraction of the period, the time from one pulse to the next. The periods of known pulsars range from thousandths of a second to four seconds. Pulses have different shapes, and even the pulses in a series from a single pulsar are not identical to each other (Fig. 30–5).

When the positions of all the known pulsars are plotted on a chart of the heavens, it can easily be seen that they are concentrated along the plane of our galaxy (Fig. 30–6). Thus they are clearly objects in our galaxy, for if they were located outside our galaxy, we would expect them to be distributed uniformly (since the Milky Way does not obscure the universe beyond at radio wavelengths).

There may be over 100,000 pulsars now active in our galaxy, of which we detect only a tiny fraction. A major mapping program has recently been under way in Australia. It has detected many more pulsars near the plane of our galaxy, which is best visible from the southern hemisphere.

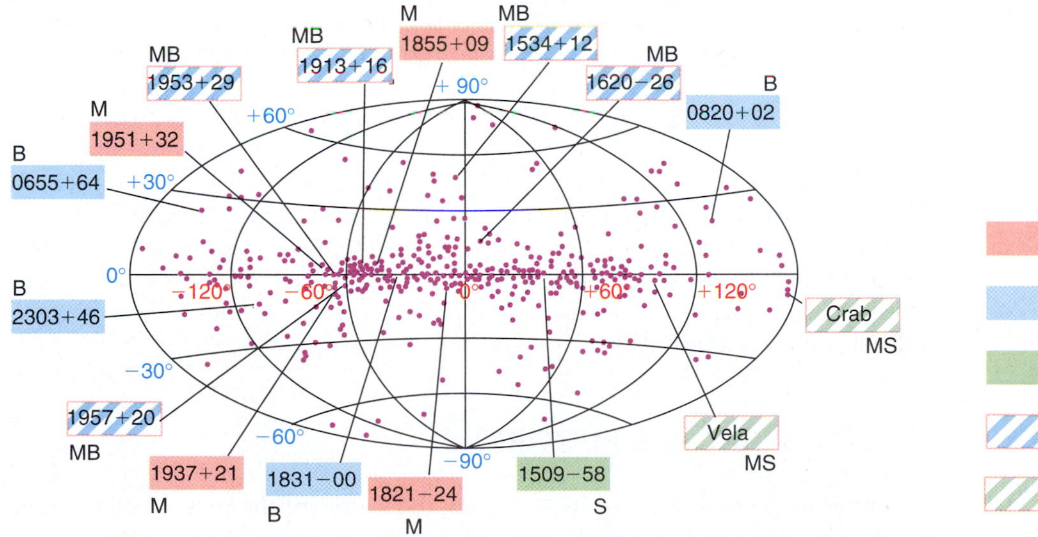


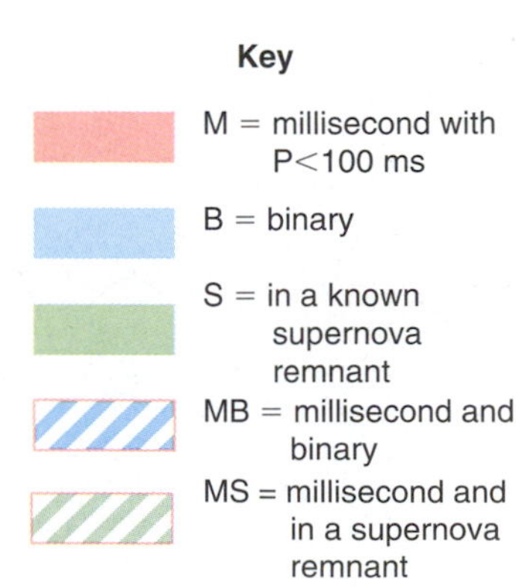


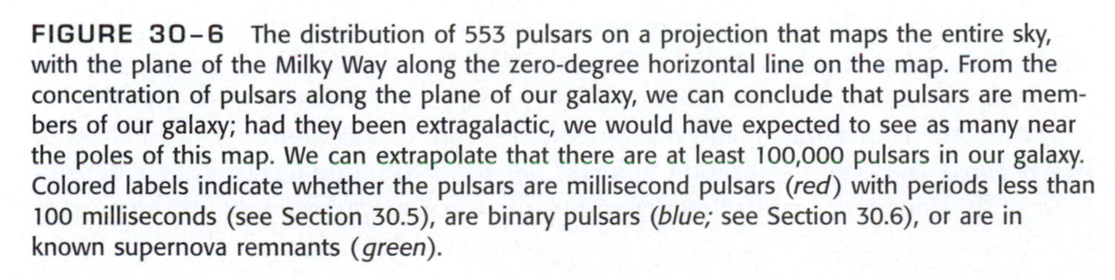

**FIGURE 30–6** The distribution of 553 pulsars on a projection that maps the entire sky, with the plane of the Milky Way along the zero-degree horizontal line on the map. From the concentration of pulsars along the plane of our galaxy, we can conclude that pulsars are members of our galaxy; had they been extragalactic, we would have expected to see as many near the poles of this map. We can extrapolate that there are at least 100,000 pulsars in our galaxy. Colored labels indicate whether the pulsars are millisecond pulsars (*red*) with periods less than 100 milliseconds (see Section 30.5), are binary pulsars (*blue;* see Section 30.6), or are in known supernova remnants (*green*).

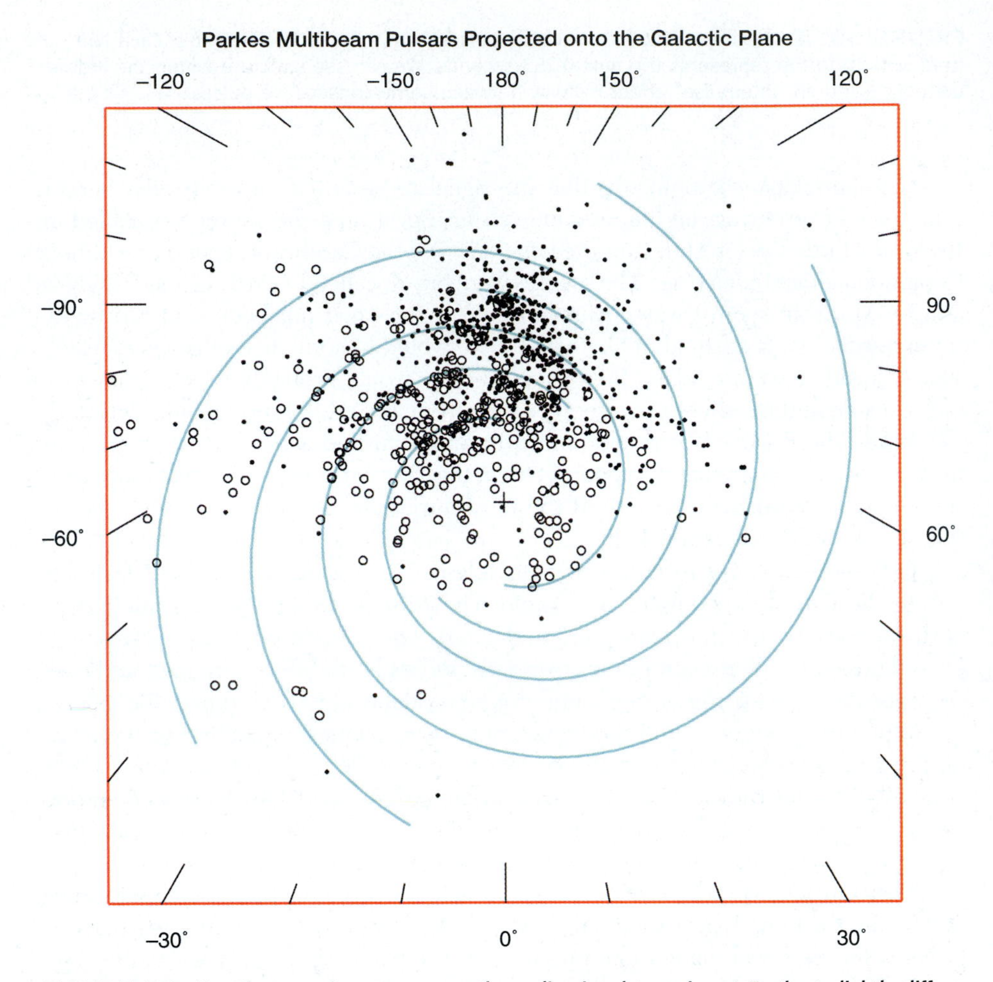


**FIGURE 30–7** Electrons in space cause the radio signal to arrive at Earth at slightly different times at different frequencies. From these time differences, given our understanding of the distribution of electrons in space, we can find the approximate distances to pulsars. Here we see the distribution of pulsars as though we were looking downward from high above our galaxy. The dispersion measure, shown as contour lines, is the average number of electrons per cubic centimeter times the distance to the pulsar; higher dispersion measures give longer time delays from frequency to frequency.

When the radio signals from pulsars pass through electrons in the gas between the stars, they travel at slightly different speeds. So pulses arrive at slightly different times at different frequencies. From the length of the delay, astronomers can figure out how many electrons have been in the way. Given an idea of the average density of electrons in space, they can use this information to get an idea of how far away individual pulsars are (Fig. 30–7).

## 30.3 WHAT ARE PULSARS?

Theoreticians went to work to try to explain the source of the pulsars' signals. The problem had essentially two parts. The first part was to explain what supplies the energy of the signal. The second part was to explain what causes the signal to be regularly timed, that is, what the "clock" mechanism is.

From studies of the second part of the problem, the choices were quickly narrowed down. The signals could not be coming from a pulsation of a normal dwarf star because the rate of pulsation of a star is known to be linked to a star's density. Stars of the density of main-sequence stars pulse too slowly to account for the pulsars.

They are too stiff. Also a main-sequence star would be too big. If the sun, for example, were to turn off at one instant, we on Earth would not see it go dark at once. The point nearest to us would disappear first, and then the darkening would spread farther back around the side of the sun (Fig. 30–8). The sun's radius is 700,000 km, so it would take 5 seconds to go dark. Pulsars pulse too rapidly to be the size of a normal dwarf star.

That left white dwarfs and neutron stars as candidates. Remember, at that time, neutron stars were merely objects that had been predicted theoretically but had never been detected observationally. Could the pulsars be the more ordinary objects, namely, special kinds of white dwarfs?

Astronomers could conceive of two basic mechanisms as possibilities for the clock. The pulses might be coming from a star that was actually oscillating in size and brightness, or they might be from a star that was rotating. A third alternative was that the emitting mechanism involved two stars orbiting around each other in a binary system. However, it was soon shown that systems of orbiting stars would give off energy in the form of gravitational waves (Section 30.7), making the period change. This was not observed, so orbiting stars were ruled out. This left the rotation or oscillation of white dwarfs or neutron stars.

Theoreticians have found that the denser a star, the more rapidly it oscillates. A white dwarf oscillates in less than a minute. However, it could not oscillate as rapidly as once every second, as would be required if pulsars were oscillating white dwarfs. Neutron stars, however, would indeed oscillate more rapidly, but they would oscillate once every 1/1000 second or so, too rapidly to account for pulsars. Thus, oscillating neutron stars were ruled out as well.

What about rotation? Consider a lighthouse whose beacon casts its powerful beam many miles out to sea. As the beacon goes around, the beam sweeps past any ship very quickly, and returns again to illuminate that ship after it has made a complete rotation. Perhaps pulsars do the same thing: they emit a beam of radio waves that sweeps out a path in space. We see a pulse each time the beam passes the Earth (Fig. 30–9).

But what type of star could rotate at the speed required to account for the pulsations? We expect a collapsed star to be rotating relatively rapidly, just as ice skaters rotate faster when they pull in their arms to bring all parts of their bodies closer to their axes of rotation (as we saw in Figure 7–21). If the sun, which rotates once a month, were to shrink by a factor that made it the size of a neutron star, its angular speed (speed measured in degrees of rotation per second, for example) would increase greatly. (Angular speed increases with the square of the radius in order to keep an object's angular momentum constant as its distribution of mass changes.) Rotation periods of 1 second are in the range we would expect for collapsed stars.

Could a white dwarf be rotating fast enough? Recall that a white dwarf is approximately the size of the Earth. If an object that size were to rotate once every 1/2 second, the inertial tendency for matter to continue moving straight (which in a rotating body is often loosely called "centrifugal force") would overcome even the immense inward gravitational force of a white dwarf. The outer layers would begin to be torn off. Thus pulsars were probably not white dwarfs; if a pulsar with a period even shorter than 1/2 second were to be found, then the white dwarf model would be completely ruled out.

Neutron stars, on the other hand, are much smaller, so the centrifugal force would be weaker, and the gravitational force would be stronger. As a result, neutron stars can indeed rotate four times a second. There is nothing to rule out their identification with pulsars. Since no other reasonable possibility has been found, astronomers accept the idea that pulsars are in fact neutron stars that are rotating. This idea is called the **lighthouse model.** Thus by discovering pulsars, we have also discovered neutron stars.

Note that we argued from elimination. For such an argument to be complete and convincing, we must be certain that we started with a complete list of possibilities (Table 30–1).

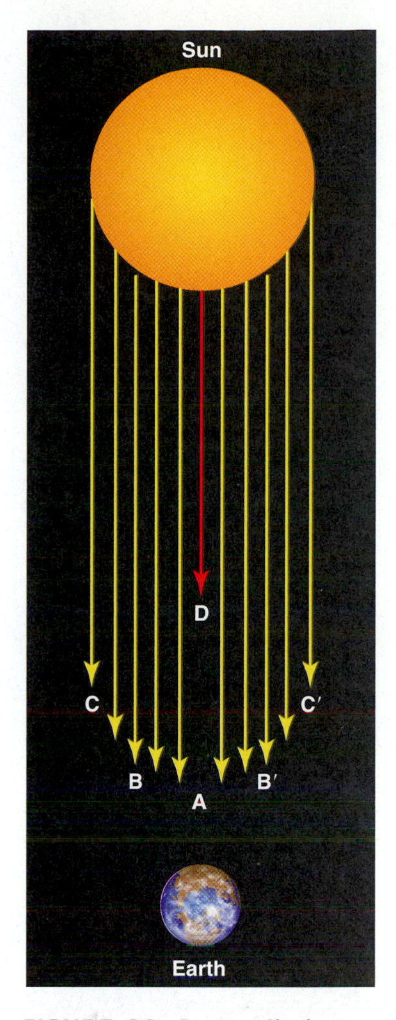


**FIGURE 30–8** Even if a large star were to turn off all at one time, the size of the star is such that it would appear to us to darken over a measurable period of time because radiation travels at the finite speed of light. Normal dwarf stars are too large and not dense enough to account for the rapid pulses from pulsars.

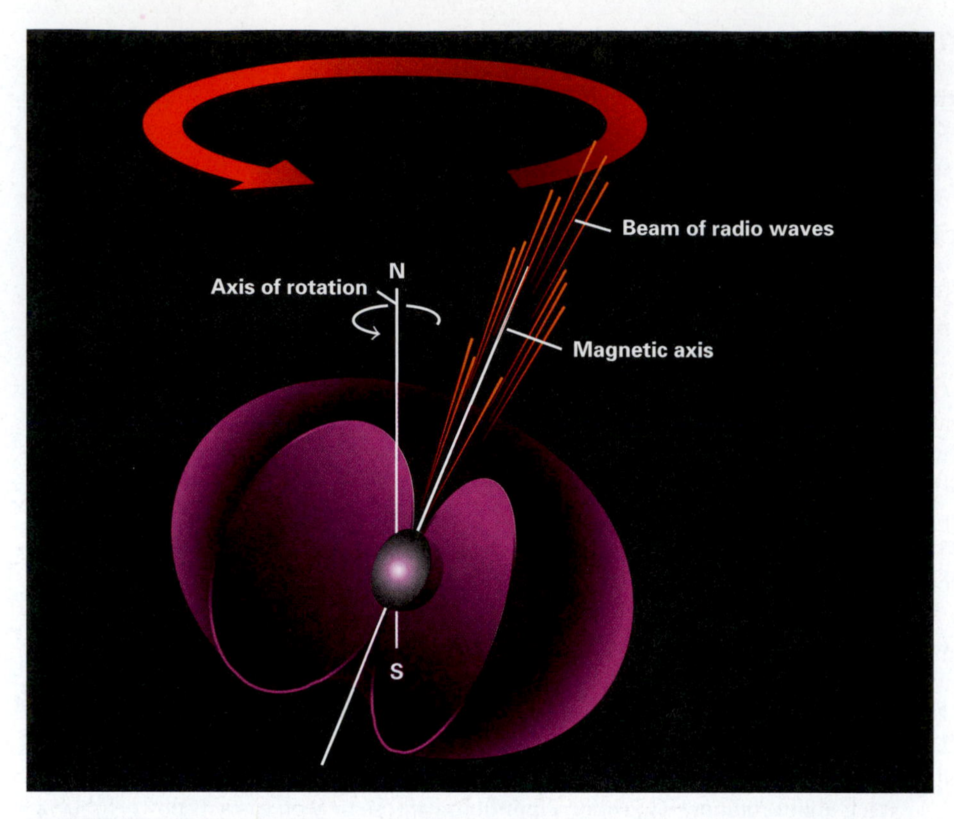


**FIGURE 30–9** In the lighthouse model for pulsars, which is now commonly accepted, a beam of radiation flashes by us once each pulsar period, just as a lighthouse beam appears to flash by a ship at sea. It is believed that the generation of a pulsar beam is related to the neutron star's magnetic axis's being aligned in a different direction than the neutron star's axis of rotation. The mechanism by which the beam is generated is not currently understood.

" . . . when you have excluded the impossible, whatever remains, however improbable, must be the truth." Spoken by Sherlock Holmes in "The Adventure of the Beryl Coronet" by Sir Arthur Conan Doyle.

Over 1500 pulsars have thus far been discovered. But the lighthouse model implies that there are many more, for we can see only those pulsars that are both close enough to us and whose beams happen to strike the Earth. For pulsars spinning at most orientations, the beams miss us, and we do not know that they are there. Except for a few in the Large Magellanic Cloud, all the pulsars we know are in our own galaxy. No known pulsar is strong enough to be detected from the distance of farther galaxies.

We have dealt above with only the second part of the explanation of the emission from pulsars: the clock mechanism. We have much less understanding of the mechanism by which the radiation is actually emitted in a beam. So we don't know, in spite of years of trying, just why pulsars shine! We know in general that we are seeing the effects of relativistic electrons (that is, electrons travelling at such high speeds that Einstein's special theory of relativity must be used) moving in spirals around magnetic lines of force. Presumably the beaming has something to do with the extremely powerful magnetic field of the neutron star. The radio waves may be

**TABLE 30–1** Possible Explanations of Pulsars

| Hypothesis | Probability |
|---|---|
| Regular dwarf or giant | Ruled out |
| Systems of orbiting white dwarfs | Ruled out |
| Systems of oriting neutron stars | Ruled out |
| Oscillating white dwarf | Ruled out |
| Oscillating neutron star | Ruled out |
| Rotating white dwarf | Ruled out |
| Rotating neutron dwarf | Most likely |

generated by charged particles that have escaped from the neutron star near its magnetic poles. These magnetic poles may not coincide with the neutron star's poles of rotation. After all, the Earth's north magnetic pole is not at the north pole (where the Earth's axis of spin meets the surface), but is rather near Hudson Bay, Canada. The pulsar beam seems to be a hollow cone, a cone with a central region containing gas in different conditions than gas at the edge of the cone. At higher frequencies, we detect mainly the edge of the cone.

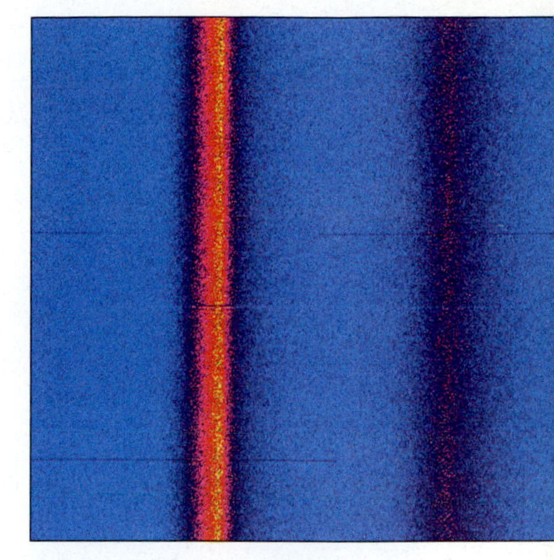


**FIGURE 30–10** A display of Hubble Space Telescope photometry of the pulsar in the Crab Nebula. Color shows intensity. Time goes from left to right; each horizontal line represents the sum of 100 consecutive periods, which together take 3.3 seconds. From left to right, we see first the main pulse and then the interpulse.

## 30.4 THE CRAB, PULSARS, AND SUPERNOVAE

Several months after the first pulsars had been discovered, strong bursts of radio energy were discovered coming from the direction of the Crab Nebula. The bursts were sporadic rather than periodic, but astronomers hoped that they were seeing only the strongest pulses and that a periodicity could be found.

Within two weeks of feverish activity at several radio observatories, it was discovered that the pulsar in the Crab Nebula had a period of 0.033 second, the shortest period by far of all the known pulsars. The fastest pulsar previously known pulsed four times a second, while the Crab pulsar pulsed at the very rapid rate of 30 times a second. No white dwarf could possibly rotate that fast.

Furthermore, the Crab Nebula is a supernova remnant, and theory predicts that neutron stars should exist at the centers of at least some supernova remnants. Thus the discovery of a pulsar there, exactly where a neutron star would be expected, clinched the identification of pulsars with neutron stars.

Soon after the pulsar was discovered in the Crab Nebula at radio wavelengths, three astronomers at the University of Arizona decided to examine its position to look for an optical pulsar, even though earlier optical searches had failed to detect a pulsar at other sites. They turned their telescope toward a faint star in the midst of the Crab Nebula, and it very soon became apparent that the star was pulsing! They had found an optical pulsar (Fig. 30–10).

X-ray observations from the Einstein Observatory revealed the structure of hot gas in the Crab Nebula and clearly showed the pulsar blinking on and off. The Crab supernova remnant was also seen by ROSAT (Fig. 30–11). The repaired Hubble Space Telescope obtained excellent views of the Crab Nebula and its pulsar (Fig. 30–12). With the increased resolution, wisps that had been previously detected were seen to take the shape of a whirlpool, presumably centered on a polar jet. This halo is 10,000 A.U. across. Within the halo we may have a polar wind, while outside the halo we may have an equatorial wind that powers a doughnut of emission. Near the pulsar, a knot of gas became visible; it had been lost in the glare in ground-based observations. It is only about 1500 A.U. from the pulsar. The knot and the pulsar line up with the direction of a jet of x-ray emission. The knot may be a "shock" in the jet, where the wind streaming away from one of the pulsar's poles piles up.

The Crab pulses across the whole spectrum, from radio waves down to short wavelengths. The Compton Gamma Ray Observatory even detected pulsed gamma rays from the Crab. Only the Vela pulsar is detectable over such a wide frequency range.

The Crab Pulsar, whose pulses were recently reimaged by the Very Large Telescope (Fig. 30–13), isn't the youngest known pulsar any more. A pulsar at a distance of 60,000 light-years on the far side of the Milky Way has been measured to be only 700 years old, an even quicker blink of an eye back in time. This pulsar is slowing even more drastically than the Crab Pulsar, and is rotating only three times per

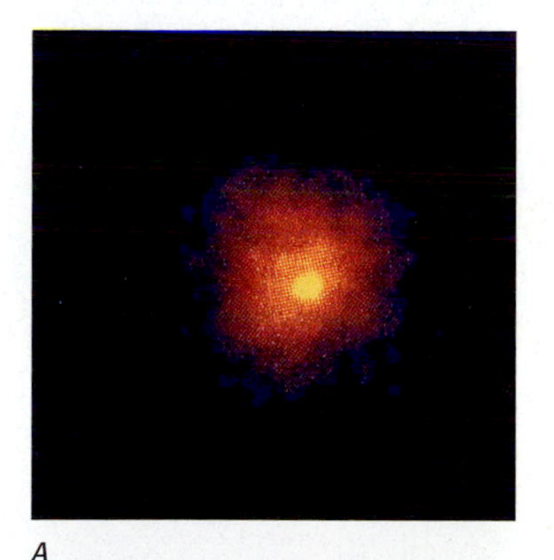

*A*

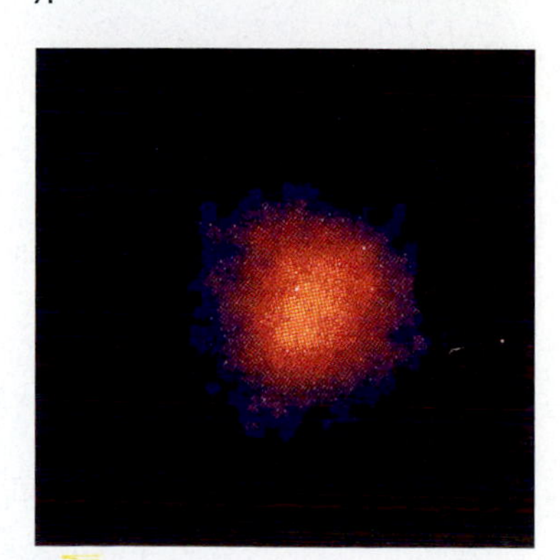

*B*

**FIGURE 30–11** ROSAT views of the pulsar in the Crab Nebula. (*A*) An x-ray view of the main pulse collected only when the pulsar is on, that is, when the beam of radiation is sweeping by the Earth. (*B*) The off phase of the pulsar. The satellite images x-rays, as can be told from its name, Roentgen Satellite, named after Wilhelm Roentgen, the discoverer of x-rays.

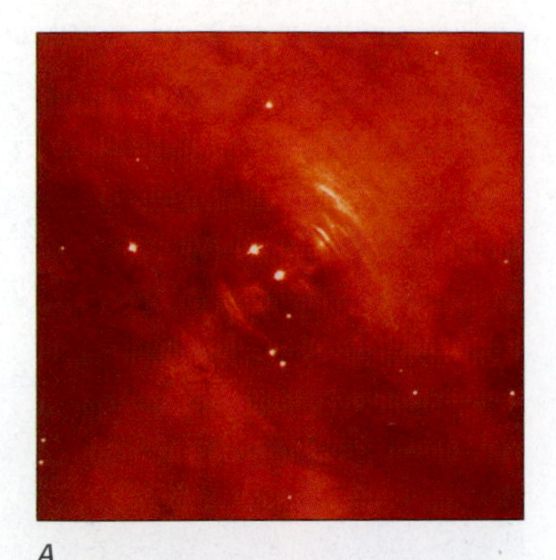

*A*

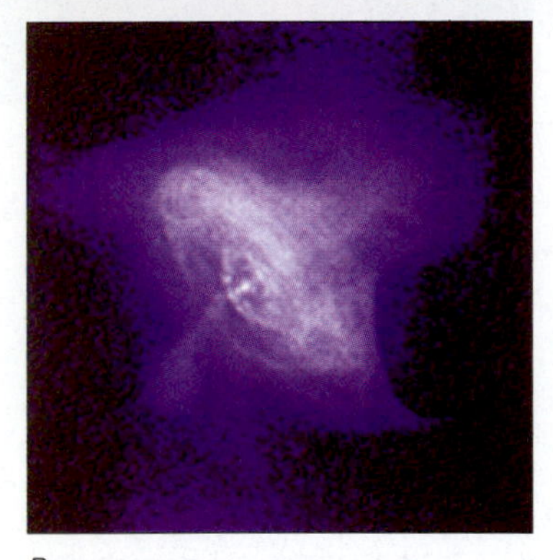

*B*

**FIGURE 30–12** The central region of the Crab Nebula. (*A*) An image by the Hubble Space Telescope, part of the image opening this chapter. Wisps of gas take on an apparent whirlpool shape. The pulsar, is in its "on" phase. An HST movie, available on the Web, shows waves of gas rippling outward along the pulsar's equatorial plane. (*B*) An image with the Chandra X-ray Observatory, showing outer and inner rings. The inner ring is 1 light-year across and the outer ring is 3 light-years across.

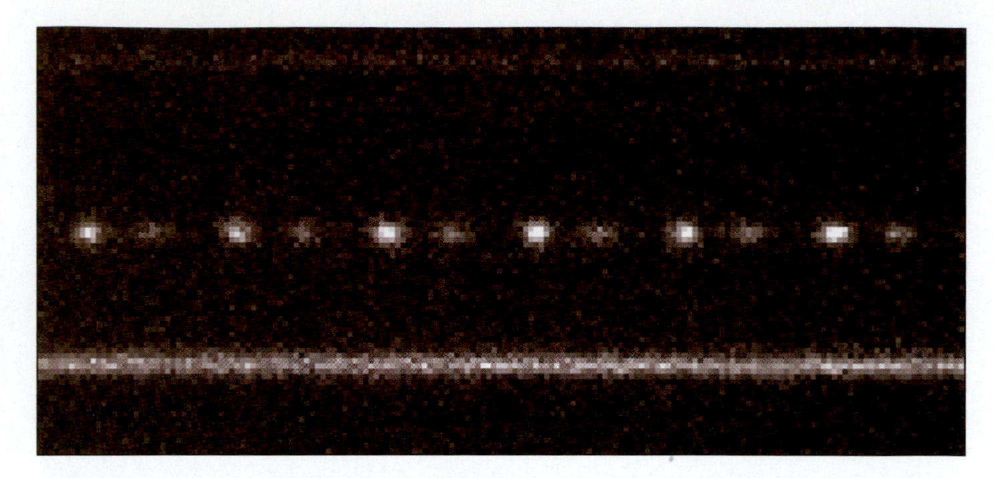

**FIGURE 30–13** Time goes from left to right, revealing that the star images at top and bottom are steady while the light from the Crab pulsar blinks on and off 30 times a second. The images were taken with the Kueyen 8-m telescope of the Very Large Telescope.

second now. It is slowing so much because its magnetic field is 10 times greater than that of the Crab pulsar. Some pulsars have been found with magnetic fields another 10 times greater, and are being called "magnetars." We will discuss this youngest pulsar and magnetars in Section 30.7.

Although 30 years have now passed, only three other objects have been found that pulse in the visible spectrum. One is the pulsar in the supernova remnant that covers much of the constellation Vela (Fig. 30–14). Several other objects are known that pulse x-rays or gamma rays, but the mechanism by which these pulses are generated is different from the run-of-the-mill radio-pulsing pulsar. When you say "pulsar," you mean that the star pulses radio waves.

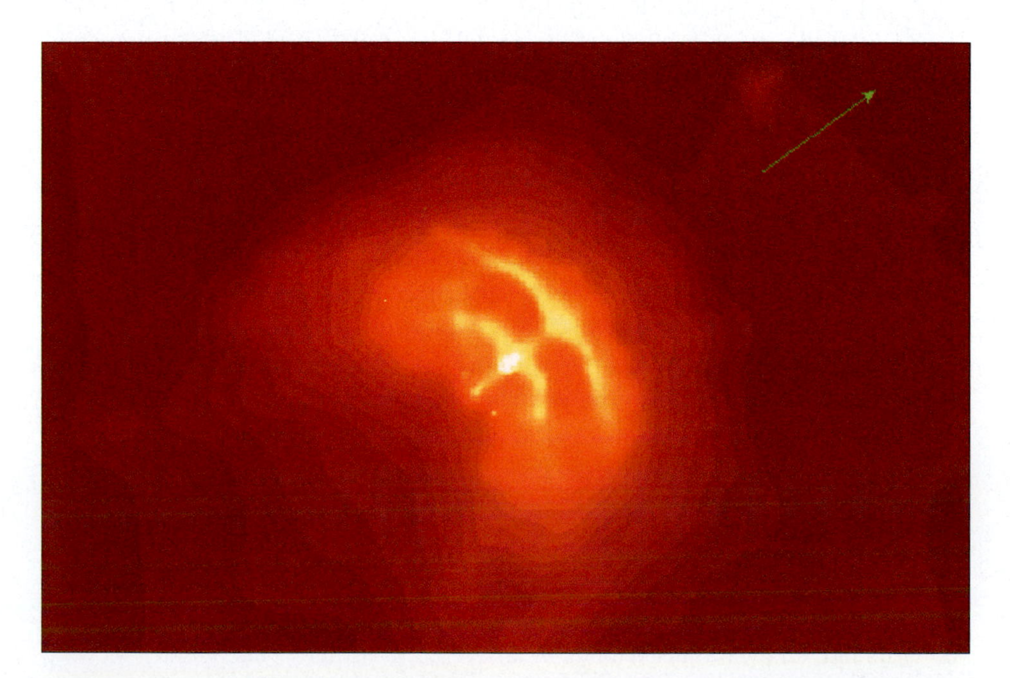

**FIGURE 30–14** This Chandra image shows a compact nebula around the Vela pulsar. We see a bow-like structure at the leading edge of the nebula, deep inside the large Vela 10,000-year-old supernova remnant. The two bows are probably the edges of tilted rings, shock waves excited by high-energy particles emitted by the neutron star. Jets from the poles of this neutron star are perpendicular to the rings. The green arrow shows the direction in which the pulsar is moving, which matches the direction of the jets and the swept-back look of the nebula. The image—3.5 arc min on a side, which corresponds to 0.2 light-years at that 800-light-year distance—was taken with Pennsylvania State University's instrument on Chandra.

## 30.5 SLOWING PULSARS AND THE FAST PULSAR

When pulsars were first discovered, their most prominent feature was the extreme regularity of the pulses. It was hoped that they could be used as precise timekeepers, perhaps even more exact than atomic clocks on Earth. After they had been observed for some time, however, it was noticed that the pulsars were slowing down very slightly. As a pulsar radiates and accelerates particles to high energy, it gives up some of its rotational energy and slows down. This slowing can be accounted for, allowing pulsars to be used as long-range clocks. Also, many pulsars slow down at some steady rate, and then a "glitch" suddenly occurs, with the pulsar immediately spinning faster. After this "glitch," the pulsar will resume its slowing. Astronomers have concluded that most of the glitches come from a cracking of the crust of the neutron star. The larger glitches may come from a difference between the neutron stars' crusts and their interiors.

The filaments in the Crab Nebula still glow brightly, even though over 900 years have passed since the explosion. Furthermore, the Crab Nebula gives off tremendous amounts of energy across the spectrum from x-rays to radio waves. It had long been wondered where the Crab got this energy. Theoretical calculations show that the amount of rotational energy lost by the Crab's neutron star as its rotation slows down is just the right amount to provide the energy radiated by the entire nebula. Thus the discovery of the Crab pulsar solved a long-standing problem in astrophysics: where the energy originates that keeps the Crab Nebula shining. The energy released by the slowdown of this single, tiny neutron star produces close to a million times more power than the sun's radiation.

The Crab pulsar, pulsing 30 times a second, seemed to be spinning incredibly fast. And since it is a young pulsar (only 900 years old) and is slowing down more rapidly than other pulsars, astronomers assumed that the shorter the period, the younger the pulsar and the greater the slow-down rate. But a pulsar was discovered spinning 20 times faster—642 times per second—and is thought to be old, not young, despite its great rotation speed. Even a neutron star rotating at this speed is on the verge of being torn apart. This "fast pulsar" was also known as the "millisecond pulsar," since its period is 1.56 milliseconds (0.00156 sec).

This pulsar's period remains very constant. Since astronomers think it is the strong magnetic field that slows a pulsar, the magnetic field of a millisecond pulsar must be relatively small for a pulsar. So it must be an old pulsar, over a million years old, in spite of its short period. More than four dozen other "millisecond pulsars" since discovered, ranging in period up to about 300 msec, share its properties (Fig. 30–15). Roughly half are now in binary systems. Theoretical models can explain the periods of most of the rest if they were once in binaries. We think they were typically formed over a hundred million years ago (compared with the 5- to 10-million-year age of most pulsars) and their rotation had decayed to very slow rates. Then mass flowing from the companion onto the compact neutron star speeds up the neutron star by supplying additional angular momentum, just as ice skaters speed up by pulling in their arms. So these are "rejuvenated pulsars."

Over 40 of the millisecond pulsars we have detected are in globular clusters. Many are in binary systems there, while so many stars are packed together in globular clusters that a companion star has been stripped off in some of the cases. Supercomputers are used to study the radio observations in searches for periodic signals

**FIGURE 30–15** Many of the millisecond pulsars have periods fast enough to hear as musical notes when we listen to a signal at the frequency of their pulse rate. The display, of millisecond pulsars discovered through 1999, is in order of right ascension (celestial longitude).

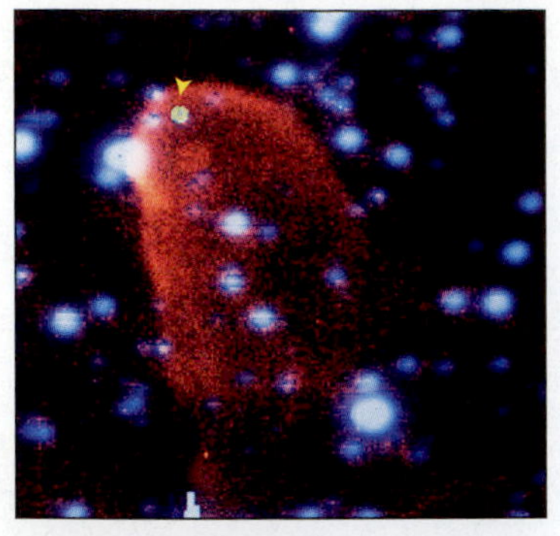

A

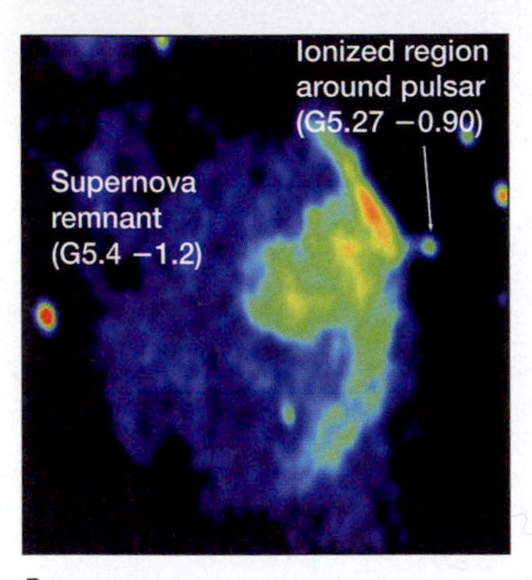


B

**FIGURE 30–16** (*A*) Hydrogen emission from the bow-shock wave around the eclipsing millisecond pulsar PSR 1957+20. The steady decrease in the period first found was thought to indicate that the companion was disappearing. But measurements through 1994 show that the period increased since. Thus the mass loss may result from the companion's rotation and magnetic cycles. (*B*) A VLA radio image of the expanding supernova remnant G5.4−1.2 and its pulsar, PSR B1757−24. The measured rate of expansion gives a different age than the spin-down rate of the pulsar.

from millisecond pulsars, and the rate of discovery is increasing. The dense globular cluster M15 has several pulsars in its core, with perhaps more waiting to be discovered.

The rotation of millisecond pulsars is so predictable that we may be able to use them to tell time more accurately over long periods than with atomic clocks. They give us time to within a millionth of a second per year. Having a few allows us to compare them, and thus to assess their accuracy as timekeepers.

Signals from one millisecond pulsar, the second fastest known, disappear abruptly for 50 minutes every 9 hours, indicating that the pulsar is eclipsed by its companion star or from gas coming from it. The astronomers who discovered the pulsar—Dan Stinebring, Andrew Fruchter, and Joseph Taylor—used the eclipse to deduce that the companion star is over 1.5 times the size of the sun, even though it has only 2 per cent of the sun's mass; energy from the pulsar has presumably bloated the companion. The companion may disappear in less than a billion years—it has been called the "Black Widow pulsar," in analogy with the black widow spider, which eats its mate. Later, evidence indicated that the companion may be losing mass by itself instead of being blown off by the pulsar. Optical observations have revealed (Fig. 30–16*A*) that the Black Widow pulsar and its companion are speeding through the galaxy, creating a teardrop-shaped shock wave in the interstellar medium.

Doubt was raised in 2000 about how accurate some of the ages are that scientists deduce by the rate that pulsars are slowing down. One pulsar has been found near a supernova remnant that has given a way to check (Fig. 30–16*B*). Comparing radio images taken in 1993 and 1999 with the Very Large Array, they find the supernova remnant is expanding at a rate that makes it at least 39,000 years old and perhaps 170,000 years old. Assuming that this pulsar resulted from the same event that formed this remnant, the pulsar must therefore be at least 39,000 years old and perhaps even 170,000 years old. Even the lower of those values is much older than the 16,000 years computed by the rate at which the pulsar is spinning down. More examples are needed before we can make definite conclusions as to the level of inaccuracy of the many pulsar ages we have deduced from spin-down rates.

A pulsar that is at the location of the A.D. 386 supernova remnant also gives information relevant to pulsar dating. X-ray observations, first from a Japanese satellite and then from Chandra, pin down the pulsar in the supernova right in the middle of the supernova remnant (Fig. 30–17), and show x-ray pulsations 14 times per second. If it is, in fact, a pulsar resulting from the supernova, it and the Crab

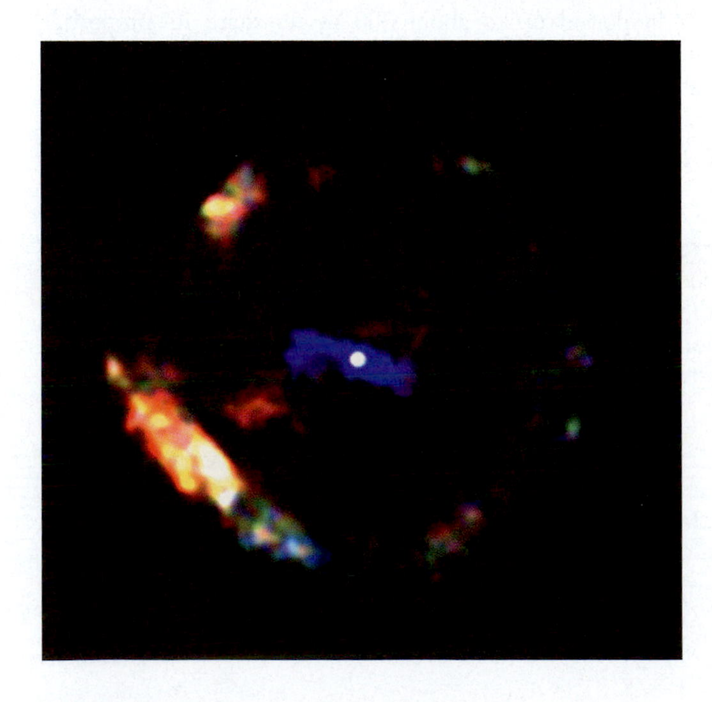

**FIGURE 30–17** A Chandra image of the supernova remnant from A.D. 386. The pulsar is visible in the center. Colors indicate the energy range of the x-rays, with red, green, and blue indicating increasing energy.

are the only two pulsars whose exact ages are known. It is interesting to have direct evidence at the rate at which a newly born pulsar spins. The disparity between the rates of the two young pulsars, this one and the Crab, is more than had been anticipated. Since ages of pulsars are often calculated from the current period, the rate at which the period is changing, and the period at birth, it is significant that we now have a better knowledge of the third of those parameters. The "characteristic age" calculated in this way is not necessarily the actual age for young pulsars. We can verify that this pulsar is young by the fact that its magnetic field is high. Magnetic fields of the older, rejuvenated pulsars, are much lower.

## 30.6 THE BINARY PULSAR AND PULSAR PLANETS

All pulsars are odd, but some pulsars are odder than others. In 1974, Joseph Taylor and Russell Hulse, then both at the University of Massachusetts at Amherst and now both at Princeton, found a pulsar whose period of pulsation (approximately 0.059 second, then one of the shortest) did not seem very regular. They finally realized that its variation in pulse period could be explained if the pulsar were orbiting another star about every 8 hours. If so, when the pulsar was approaching us, the pulses would be jammed together a bit, and when it was receding the pulses would be spread apart in time in a type of Doppler effect. The object is a **binary pulsar,** and this particular case is so famous that it is "*the* binary pulsar."

Interpreting the binary pulsar system on the basis of many years of pulse arrival-time data has told us much. The pulses arrive 3 seconds earlier at some times than at others, indicating that the pulsar's orbit is 3 light-seconds (1 million km) across, approximately the diameter of the sun. Thus each of the objects has to be smaller than a normal dwarf. The companion object is probably a neutron star, too. No pulses have been detected from it, but if it is giving off a radio beacon, it could simply be oriented at an unfavorable angle.

The binary pulsar has allowed us to derive the components' masses, which fall in the range we expected for neutron stars, about 1 to 3 solar masses. But the main interest in the binary pulsar has come about because it provides some of the most powerful tests of the general theory of relativity. In Section 31.1, we will discuss the excess advance of the perihelion of Mercury, which is only 43 seconds of arc per *century*. The binary pulsar, which is moving in a more elliptical orbit and in a much stronger gravitational field than is Mercury, should have its periastron advance by 4° per *year* (Fig. 30–18). (Periastron for a star corresponds to "perihelion," the closest approach of a planet's orbit to the sun.) Thus the periastron of the binary pulsar moves farther in a day than Mercury's periastron moves in a century. Observations of the binary pulsar on a continuing basis show that its periastron indeed advances 4° per year, and so they provide a strong confirmation of Einstein's general theory of relativity. As a result, Taylor and Hulse received the 1993 Nobel Prize in Physics for their discovery.

Still further important tests of Einstein's theory have been provided by the binary pulsar. In Section 30.8, we will see how long-term monitoring of its period reveals the existence of gravitational waves, as predicted. Further, the warping of space caused by the concentrated mass of the neutron star slows up signals coming from behind it. We will say more about this "Shapiro" effect and how it is caused by the warping of space in Section 31.1. At the end of that section, we will make some comments on how these pulsar tests of the general theory of relativity are even more important than the traditional tests made with the sun, since they test the theory in the vicinity of much stronger gravitational field. The three tests are known as "post-Keplerian," since they involve changes in the pulsar's pulses we receive in addition to those that we would expect (like the Doppler shift) from the ordinary orbit that follows Kepler's laws.

Studies of binary pulsars and the related theory continue to advance. A second pulsar, discovered in 1990, has even stronger and sharper pulses, enabling the system to be monitored more accurately. Its orbit is also closer to edge-on, making the

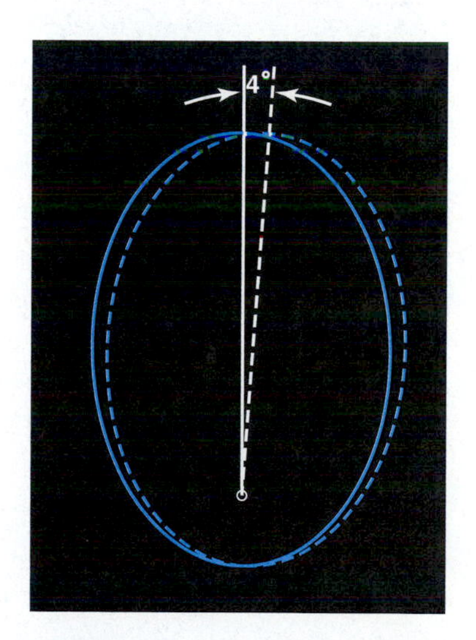


**FIGURE 30–18** The periastron of the binary pulsar PSR 1913+16 advances by 4° per year, providing a strong endorsement of Einstein's general theory of relativity. The angle is shown for the apastron, the farthest point of the pulsar from the other object. As we will see, the pulsar's change in period and Shapiro effect further test relativity. Further, the pulsar PSR 1534+12, discovered in 1990, provides additional useful data on these effects.

effects come closer to their potential. Another dozen tests of relativity can be carried out with these systems and analysis is under way.

The stars in this system are spiralling together. In a billion years or so, they will collide and coalesce. A black hole will probably result. In Chapter 32, we will discuss the type of gamma-ray burst that may also result from such an event.

Timing of another pulsar, PSR B1257+12, has revealed at least two planets around it. These were the first accepted planets outside our solar system (see Section 18.2).

## 30.7 MAGNETARS

Pulsars may seem strange enough, with magnetic fields over a billion billion times the magnetic field of the sun. But some neutron stars have magnetic fields a hundred or a thousand times even greater. They are "magnetars," honoring their huge magnetic fields.

Magnetars have properties that are different from pulsars. In regular pulsars, the magnetic field accelerates electrons away from the neutron star's surface. Gamma rays result, and some of them become electron-positron pairs of particles. Then the electrons and positrons annihilate each other, resulting in radiation over a broad band of wavelengths.

In magnetars, the magnetic field is so strong that the gamma-ray photons don't become electron-positron pairs. Rather, they split and become x-ray photons, still very energetic but less so than gamma-ray photons. Magnetars emit only x-rays.

An object has been found that is close to the dividing line between pulsars and magnetars. It is right in the middle of the supernova remnant known as Kes 75 and was located by the Rossi X-ray Timing Explorer. Astronomers had been searching in that remnant for over a decade to find signs of the neutron star that resulted from the explosion. The pulsar and the supernova remnant are about 700 years old, and the pulsar spins every 0.3 second. It is the youngest pulsar known, about 300 years younger than the pulsar in the Crab Nebula. Its magnetic field is 10 times greater than that of the Crab, and the pulsar is slowing down 10 times faster.

## 30.8 GRAVITATIONAL WAVES

Almost all astronomical data come from studies of the electromagnetic spectrum, with a small additional contribution of cosmic rays and neutrinos. But Einstein's general theory of relativity (which we will summarize in Section 31.1) predicts the existence of still another type of signal: **gravitational waves.** Any accelerating mass should, theoretically, cause these waves, which are ripples in the curvature of space-time.

Whereas electromagnetic waves affect only charged particles, gravitational waves should affect all matter. But gravitational waves would be so weak that we could hope to detect them only in situations where large masses were being accelerated rapidly. Two neutron stars revolving in close orbits around each other, as in the binary pulsar, is one possible example. Another would be a supernova or the collapse of a massive star to become a black hole.

Joseph Weber of the University of Maryland was the first to build sensitive apparatus to search for gravitational waves. The detector was a large metal cylinder (Fig. 30–19), delicately suspended from wires and protected from local motion. When a gravitational wave hit the cylinder, it should have begun vibrating; the vibrations could be detected with sensitive apparatus. His apparatus started working in 1969; though it sometimes started vibrating, the vibrations seem to have been from causes other than gravitational waves. However, his work encouraged other scientists to build more sensitive gravitational-wave apparatus; no gravitational waves have been detected yet. The International Gravitational Event Collaboration involves a network of five such bars, each cooled to close to absolute zero, to reduce thermal vibrations that could mask the effect being sought. Two of the bars are in Italy and one each

A

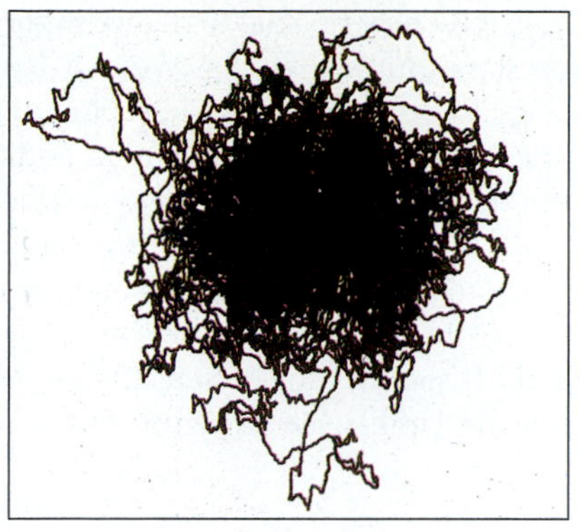

B

**FIGURE 30–19** (*A*) One of Joseph Weber's original gravitational-wave detectors from the 1960s, a large aluminum cylinder weighing 4 tons, delicately suspended so that gravitational waves would set it vibrating 1660 times per second. (*B*) The signal from a cooled bar, worked on by J. A. Tyson of Bell Labs, Lucent Technologies. We see the slowly changing readout over many weeks; a gravitational wave would cause a jump.

is in the United States, Switzerland, and Australia. In 2001, the team has reported that they detected no gravitational waves in a two-year period. From this information, they calculated that there couldn't be more than four merging neutron stars or black holes per year at the distance of the center of our galaxy. They also showed, valuably, that the network isn't bothered with false alarms.

In the meantime, the binary pulsar turned out to be a quicker way to show that gravitational waves exist. The gravitational waves emitted by the system carry away energy. This makes the orbits shrink, which results in a slight decrease of the period of revolution. By 1978, Taylor and colleagues had detected such a slight speedup (by 3 milliseconds per revolution!). The speedup has continued at precisely the rate predicted by the general theory of relativity. The pulsar now reaches periastron more than two seconds earlier than it would if its period had remained constant since it was discovered. The agreement of observations with predictions provides strong, though indirect, evidence for the existence of gravitational waves. The result contributed to the decision to award the 1993 Nobel Prize in Physics to Taylor and Hulse.

Current terrestrial gravitational-wave apparatus use laser interferometers to search for gravity waves. A Caltech–MIT consortium has built systems in which laser beams go up and back down long arms, so that interference properties of light can be used to measure minute changes in the lengths of the arms. The prototype had two perpendicular arms each 40 m long; the light was reflected back and forth thousands of times between mirrors in each arm. It was scaled up by a factor of 100, to 4-km arms (Fig. 30–20), to become LIGO (*L*aser *I*nterferometer *G*ravitational Wave *O*bservatory). The two LIGO stations are in Hanford, Washington, and Livingston, Louisiana, and are just going on line (2001). European, Japanese, and Australian scientists are building similar stations.

A

B

**FIGURE 30–20** (*A*) View of LIGO/Hanford. (*B*) Inside LIGO/Livingston.

LIGO may be able to detect supernova explosions or the mergers of pairs of neutron stars. But maybe not. We may have to wait for a second generation of this technology to be sensitive enough. Still, if and when it works, the capability will be fantastic. Gravitational waves are an additional window on the universe, quite different from the electromagnetic waves (plus cosmic rays) that we have been using up to now.

In the second decade of this millennium, the European Space Agency plans to launch an upgrade of LIGO. This Laser Interferometry Space Antenna (LISA) should be free of some of the noise sources from gravity and temperature that are found on Earth. It should thus be sensitive to the low-frequency gravitational waves that result from the large-scale structure of the universe.

## 30.9 X-RAY BINARIES*

Telescopes in orbit that are sensitive to x-rays have detected a number of strong x-ray sources, some of which are pulsating. Most of these objects pulse only in the x-ray region of the spectrum. One of the most interesting is Hercules X-1 (the first x-ray source to be discovered in the constellation Hercules). It pulsates with a 1.24-second period in x-rays and in visible light.

Hercules X-1 and the other pulsating x-ray sources are apparently examples of evolving binary systems. Theoreticians think that the x-ray sources are radiating because mass from the companion is being funneled toward the poles of the neutron star by the neutron star's strong magnetic field (Fig. 30–21). Unlike the slowing down of the pulse rate of pulsars (which give off pulses in the radio region of the spectrum), the pulse rate of the binary x-ray sources usually speeds up. The period of Hercules X-1, for example, is growing shorter.

The *R*ossi *X*-ray *T*iming *E*xplorer (RXTE) and the huge collecting area of the XMM-Newton missions continue to sensitively measure rapid variations in x-ray intensities. RXTE has observed the eruption of a neutron star 20,000 light-years away in the constellation Sagittarius. X-ray bursts of this star occur almost every day and last for about 10 seconds. They are probably fueled by helium gas that falls onto the neutron star from its white-dwarf companion. Carbon "ash" piles up on the neutron star's surface. But the big explosion lasted 3 hours and produced a thousand times

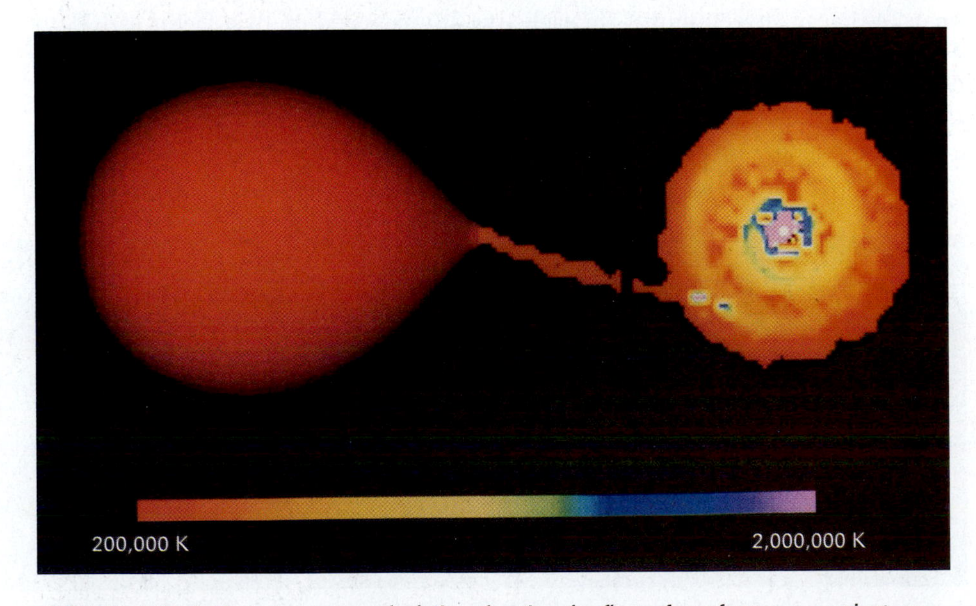


**FIGURE 30–21** A computer calculation showing the flow of gas from a supergiant companion to an accretion disk around a neutron star. We see the gas heated as it spirals into the center of the disk.

more energy. The scientist who made the observations suggests that these huge eruptions come from fusion of the carbon ash.

In Chapter 32, we will discuss a neutron star that pulses gamma rays.

## 30.10 SS433: THE STAR THAT IS COMING AND GOING*

SS433, one of the most exciting single objects in astronomy, contains spectral lines that change in wavelength by a tremendous amount. Its shifts of hundreds of angstroms correspond to Doppler shifts from velocities up to 50,000 km/sec. Such large shifts, about 15 per cent of the speed of light, are entirely unprecedented for an object in our galaxy. Furthermore, spectral lines appear simultaneously that are shifted to the red and to the blue by that same tremendous amount. Something seems to be coming and going at the same time.

"SS" indicates that it is from a catalogue of stars with hydrogen emission lines compiled by C. Bruce Stephenson and Nicholas Sanduleak of Case Western Reserve University.

SS433 was identified as a potentially interesting object because its optical spectrum showed emission lines. Then it turned out to be in the midst of an interstellar cloud thought to be a radio supernova remnant; it also seemed to correspond to an x-ray source. Following these discoveries, Bruce Margon and colleagues observed that a set of SS433's optical spectral lines apparently changes in wavelength from night to night. It took a lengthy series of observations before they realized that the wavelengths varied in a regular fashion (Fig. 30–22), though large changes take place from night to night. The wavelengths have a period of 164 days.

It was later discovered that SS433 is in orbit with a 13-day period around another object. It is thus an x-ray binary, though its x-ray emission is relatively weak for a binary x-ray source. The object is apparently a massive star, probably of spectral type O or B.

The best models indicate that SS433 is a collapsed object—a neutron star or a black hole—surrounded by a disk of material gathered from its companion. Such a disk of accreted material is known as an **accretion disk;** astronomers have recently been finding signs of accretion disks in many types of objects. In the leading model (Fig. 30–23), material is being ejected from SS433's accretion disk along a line. Because of the precession of the accretion disk, the line traces out a cone. (Precession is the wobbling motion, as of a child's top, that we have already met for the Earth's rotation in Section 6.3.) If we assume that SS433 is oriented with its spin axis more-or-less but not completely perpendicular to the direction in which we are looking, we see radiation given off by the material ejected upward as redshifted and radiation given off by the material ejected downward as blueshifted. The redshift and blueshift vary as the line traces out the cone. The extended series of observations give the angles and show that the material is ejected at 26 per cent the speed of light. High-resolution radio observations (Fig. 30–24) and x-ray observations (Fig. 30–25) show the jets, confirming the model.

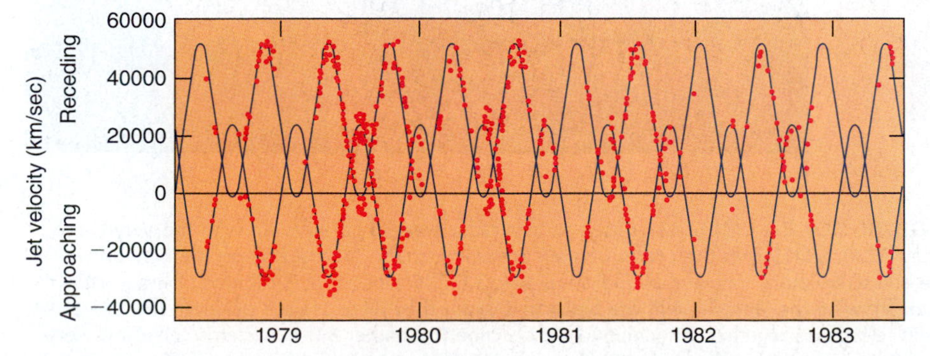


**FIGURE 30–22** The redshifts and blueshifts of spectral lines in SS433 vary in a regular fashion with a 164-day period. The red points are well fit by the two opposed blue curves.

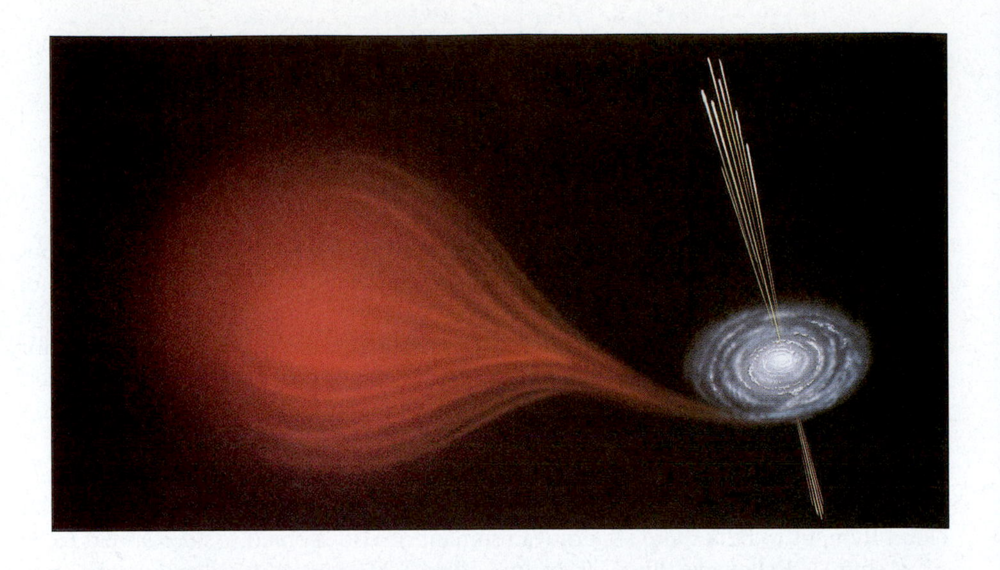

**FIGURE 30–23** A model of SS433 in which the radiation emanates from two narrow beams of matter that are given off by the accretion disk. The model follows ideas of Mordechai Milgrom, Bruce Margon, Jonathan Katz, and George Abell. The velocity of gas emitted is 80,000 km/sec in each direction, though we see a somewhat lower velocity because we do not see the jets head on. In some models, variations in pressure and density of the disk can cause the emission.

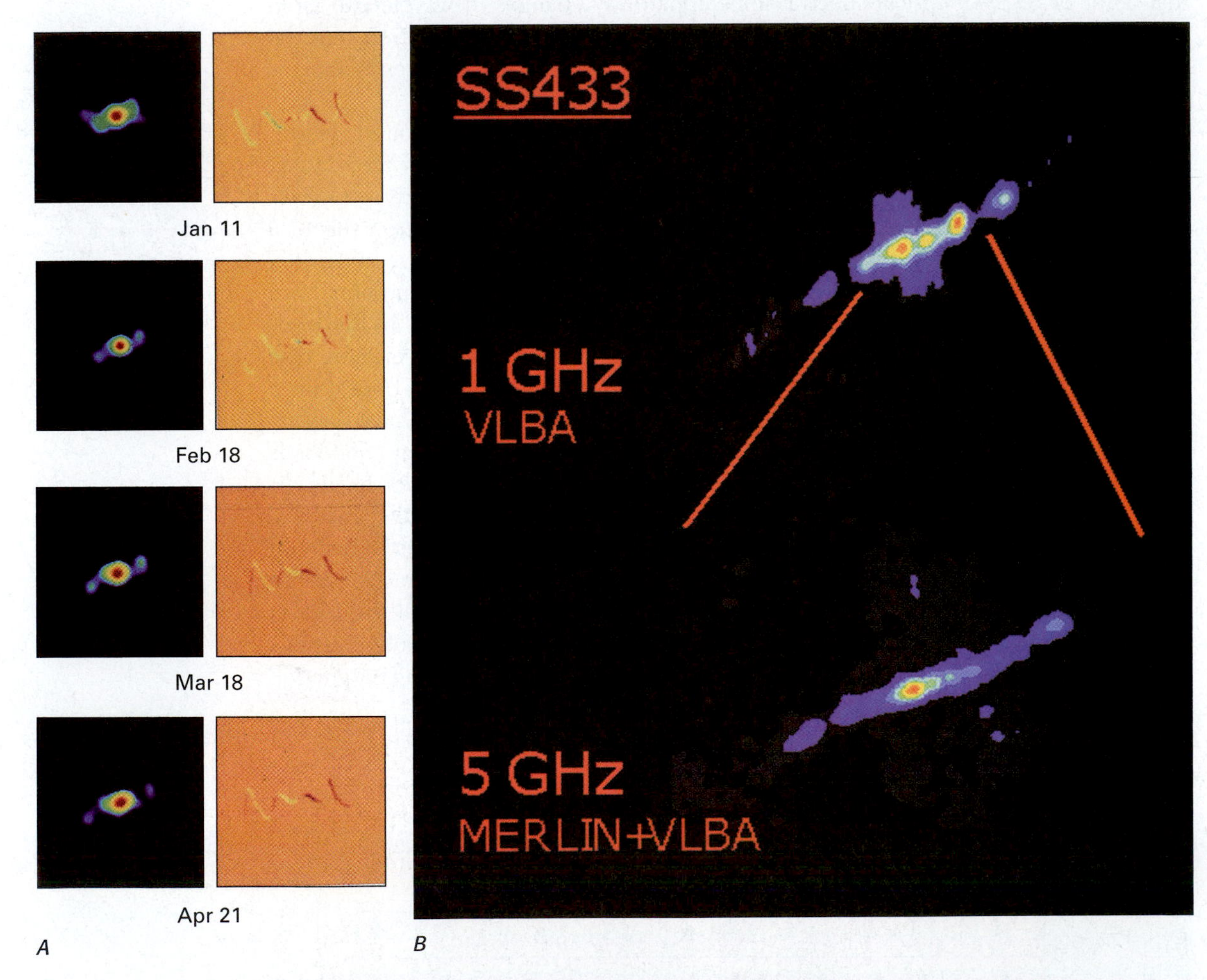


**FIGURE 30–24** (*A*) SS433 observed at a radio wavelength of 6 cm with the Very Large Array (VLA). SS433 is in the center of a supernova remnant. The column on the right shows the corkscrew paths followed by individual ejecta blobs over about 3 months of time, with color showing the Doppler shift the blobs had when they were at each position. (*B*) Use of 43 radio telescopes together (a technique we describe in Chapter 34) in systems called the Very-Long Baseline Array in the United States and MERLIN in the United Kingdom makes images of unprecedented detail. Their images of SS433 show the jets emerging from the central black hole or neutron star. Perpendicular to the jets the images show a collar of gas, something never before seen. Perhaps the emission from the collar will tell us about the reservoir of gas that feeds the jets.

Some of the high-resolution radio observations have actually shown knots of emission moving away from the core. Putting these angular measurements together with the velocities in km/sec from the optical spectra, we find that SS433 is about 16,000 light-years from us, about halfway to the center of our galaxy. A similar source in our galaxy with still greater velocities (92 per cent of the speed of light) was discovered in 1994. These objects are nearby examples of jets and other phenomena that exist on larger and more powerful scales in distant galaxies.

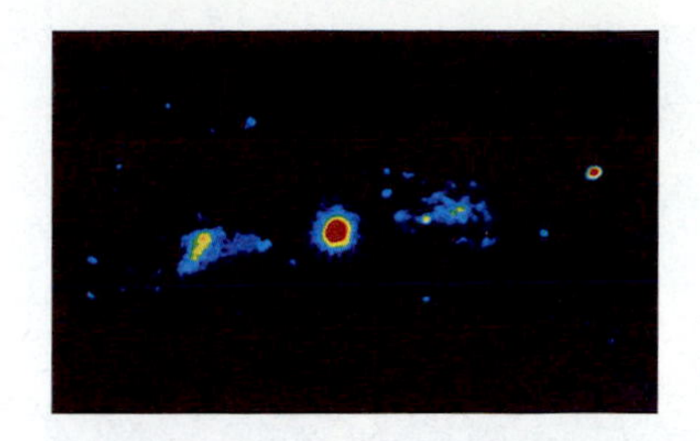

**FIGURE 30–25** An x-ray image of SS433 from the Einstein Observatory. The tunnels carved by the jets extend to the sides of the bright central image. SS433 is not an especially strong source; some of the other binary x-ray sources are 100 times stronger.

## 30.11 LONE NEUTRON STARS

It is now decades that people have been studying neutron stars through pulses of radio waves, optical light, gamma rays, or x-rays that come from those systems. But nobody found an isolated, non-pulsating neutron star until Fred Walter of the State University of New York at Stony Brook found one in 1996, based on x-ray ROSAT observations from 1992. It is 11 km in radius, matching theory. His discovery, made in x-rays, was confirmed by optical data from the Hubble Space Telescope (Fig. 30–26). It is very faint, only 26th magnitude, 100 million times fainter than the faintest star visible to the unaided eye. Because of its strong x-radiation, and because it is blue, we know it is very hot—about 600,000 K.

Later, Walter succeeded in measuring the object's parallax. He thus refined the distance measurement to 200 light-years. It is the closest neutron star to Earth.

The observations show that the neutron star is just about streaking across the sky, since it is moving ⅓ arc sec per year, only 30 times slower than Barnard's star. In only 5,400 years, it would travel the angular diameter of the moon. The star is speeding away from the group of stars in which it was born. Very Large Telescope observations turned up a bow-shock wave in front of the neutron star (Fig. 30–27).

A bright O star, zeta Ophiuchi, is streaking across the sky in another direction. Tracing its path back, puts it in the same place as the neutron star we are discussing about 1 million years ago. The two stars may have once made a binary system, with the current neutron star's predecessor acting as the primary member.

It was probably ejected about a million years ago to reach its position in the sky at its current speed. It should come as close as 170 light-years from Earth in about 300,000 years. Since this object is isolated, we know that it is hot because it is still

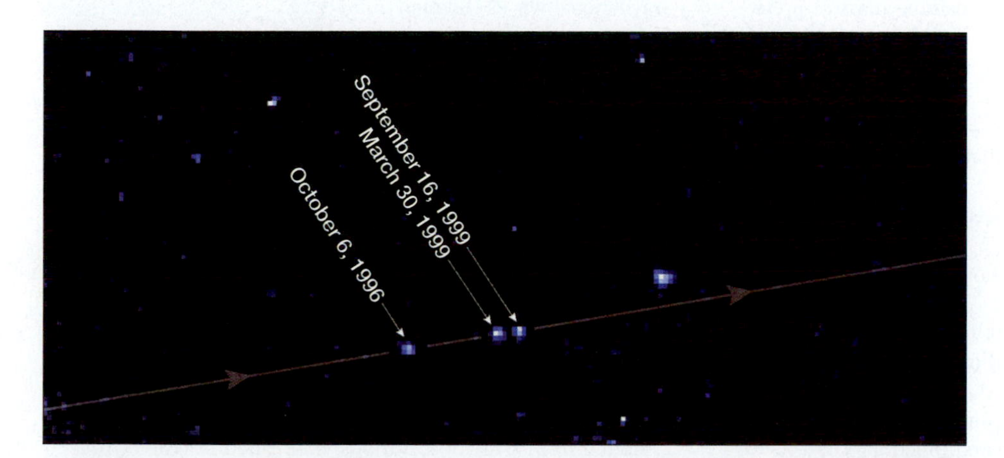


**FIGURE 30–26** The isolated neutron star RX J185635-3754 as it plows through space. (The J before the position denotes the system of time measurement used for specifying the star's position.) Three Hubble Space Telescope images are overlapped, showing a field of view only 8.8 arc sec across. All the stars line up except for the neutron star, which appears at the three different locations marked. Its parallax, caused by Earth's orbit around the sun, is 0.016 arc sec, too small to show on the image but corresponding to a distance of only 200 light-years.

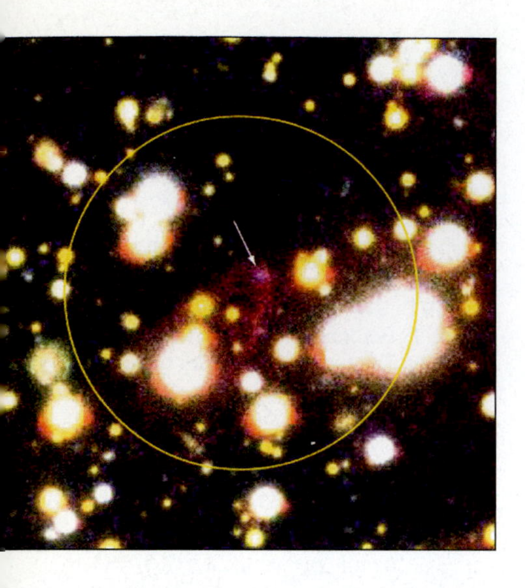

**FIGURE 30–27** The cone-shaped bow-shock wave formed by the isolated neutron star RX J185635-3754 as it plows through interstellar space.

young and cooling off rather than because it is picking up additional gas. Knowing its age, we can measure the speed at which neutron stars cool. The value deduced roughly matches that expected for neutron stars about a million years old.

Knowing its brightness and temperature allows us to deduce how big this neutron star is. The best value is 11 ± 3 km, a good test of our theory of neutron stars.

## CORRECTING MISCONCEPTIONS

| ✖ *Incorrect* | ✔ *Correct* |
|---|---|
| Pulsars are stars that pulsate. | Although pulsation was briefly considered, we now know that pulsars rotate rather than pulsate. |
| Pulsars have constant pulse rates. | Radio pulsars slow down; x-ray pulsars speed up. |
| Neutron stars are detected only as pulsars. | At least one non-pulsing neutron star has been found. |

## SUMMARY AND OUTLINE

After a star's fusion stops (Section 30.1)
- Neutron degeneracy can support a remaining mass of up to 2 or 3 solar masses.
- A neutron star may be only 10 km in radius and so is fantastically dense.

Pulsars (Section 30.2)
- Discovered with a radio telescope that did not mask rapid time variations
- Very regular signals; slowing down very slightly
- Periods 1.6 milliseconds to 4 sec

Pulsars are rotating neutron stars (Section 30.3)
- Process of elimination
- A pulsar is observed in the center of the Crab Nebula, a supernova remnant, where we would expect to observe a neutron star.

Pulsars emit mainly in the radio spectrum (Section 30.4).

Millisecond pulsars have been rejuvenated (Section 30.5).

Binary pulsars (Section 30.6)
- The periastron of the Hulse-Taylor binary advances 4°/year, confirming general relativity.

Magnetars (Section 30.7)

Gravitational waves (Section 30.8)
- LIGO and LISA experiment to search directly
- The binary pulsar is slowing down by just the predicted amount if gravitational waves are indeed being emitted from that system, confirming Einstein's theory.

X-ray sources (Section 30.9)
- Many caused by the infall of material from its companion onto a neutron-star member of a binary system

SS433 (Section 30.10)
- Spectral lines that shift periodically to the red and to the blue by about 15 per cent of the speed of light
- The Doppler-shifted radiation may come from beams ejected from an accretion disk about the neutron star in an x-ray binary.

Lone neutron stars (Section 30.11)
- Found from x-ray observations; confirmed in visible from Hubble
- Rapid motion across sky gives age; age gives cooling rate.

## KEY WORDS

neutron degeneracy
neutron star
scintillation
pulsar
lighthouse model
binary pulsar
gravitational wave
accretion disk*

*This term appears in an optional section.

## QUESTIONS

1. In your own words, replace the second column of Table 30–1 with the reasons why each of the explanations for pulsars involving single stars was ruled out except for the case of rotating neutron stars.
2. In view of current theories about supernovae and pulsars, list the pieces of evidence that indicate that the Crab pulsar is a young one.
†3. A pulsar has a period of 1 second, and a pulse lasts 5 per cent of the cycle. What is the pulse's width in seconds? What might that say about the size of the object emitting the pulse on (a) the oscillating model and (b) the rotating model?
4. Why did it seem strange that the fast pulsar is slowing down so slowly? Explain.
5. Explain two important consequences of the discovery of the binary pulsar.
6. Why is the discovery of gravitational waves a test of the general theory of relativity?
7. Sketch a model for an x-ray binary and explain how it leads to x-ray emission.
†8. A shift of the H$\alpha$ line in SS433 of 40,000 km/sec corresponds to a shift of how many Å in wavelength? (Ignore any effects of relativity.)
9. Describe why the precession of an accretion disk in SS433 can cause varying Doppler shifts.
10. Explain why the true velocities of the jets in SS433 may be even greater than the velocities we measure directly from Doppler shifts.
11. Why is it useful to find lone neutron stars?
†12. From the lone neutron star's parallax described in the chapter, show how you derive its distance.

†This question requires a numerical solution.

## USING TECHNOLOGY

### W World Wide Web

1. Explore the history of pulsars at the Lick Observatory's site: http://pulsar.ucolick.org/cog/pulsars/intro.html.
2. Listen to pulsar signals at the Princeton Pulsar Lab's site: http://pulsar.princeton.edu.
3. Find out the range of pulsar periods in the data at http://www.mpifr-bonn.mpg.de/div/pulsar.
4. Find out about the pulsar planets at the discoverer's site: http://www.astro.psu.edu/users/pspm/arecibo/planets/planets.html.
5. Learn from NASA's pulsar Web site: http://imagine.gsfc.nasa.gov/docs/science/know_l1/pulsars.html.

### REDSHIFT

1. Look at Story of the Universe: Lives of the stars; Science of Astronomy: Lifestyles of the stars; Science of Astronomy: Birth and death of stars.
2. Find the locations of some of the objects given in the diagram of special pulsars, and consider when and whether they are visible from your location at the time of your calculations.
3. Locate the Crab Nebula; examine its photos; observe it with a telescope.

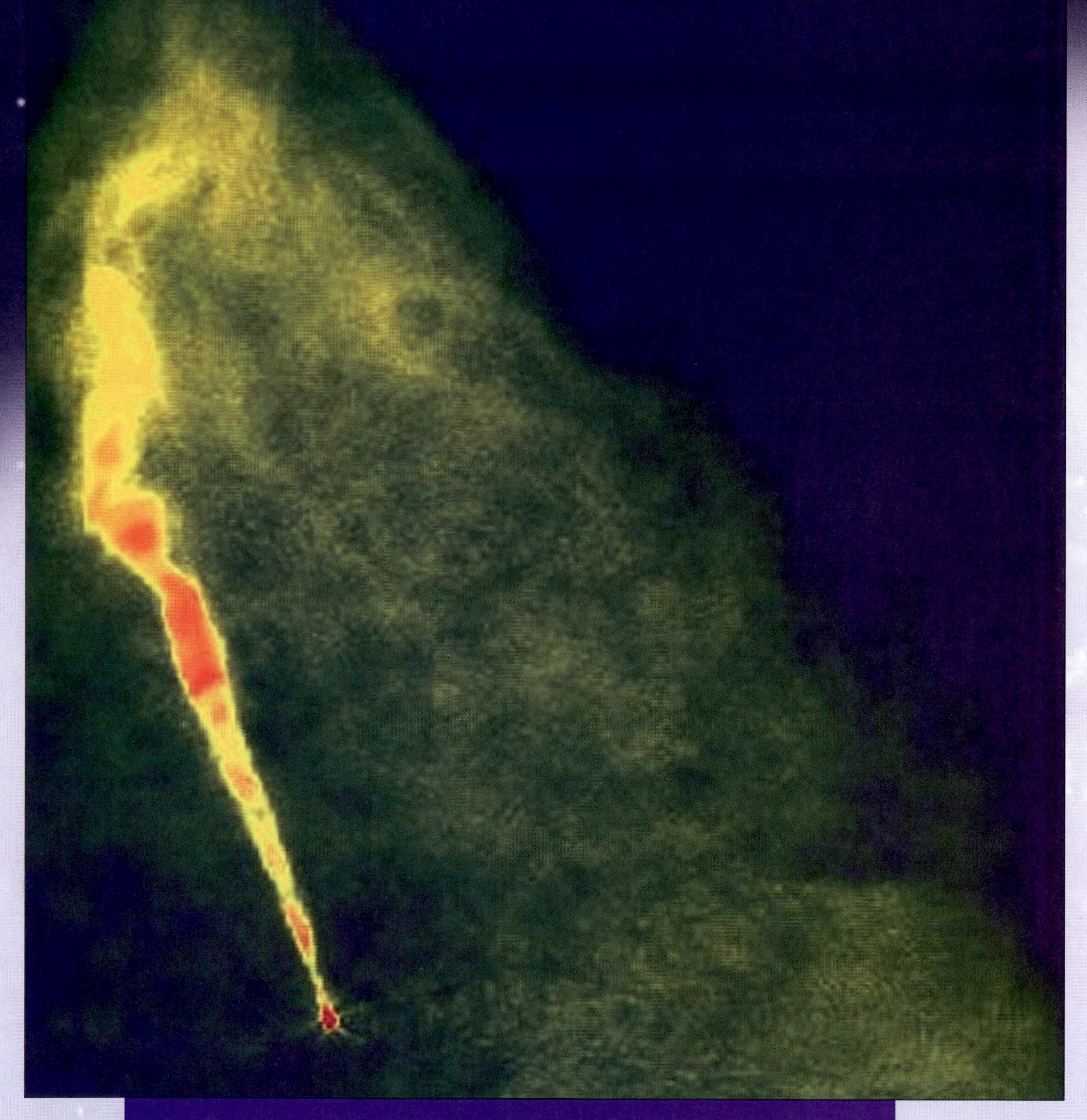

A radio image at high resolution of the center of the galaxy M87 in the Virgo Cluster. Using the National Radio Astronomy Observatory's VLA, we see a jet that, scientists have concluded, emanates from a supermassive black hole near the galaxy's core. In this chapter, we will discuss the evidence that leads to this conclusion.

# Chapter 31

# Black Holes

The strange forces of electron and neutron degeneracy support dying low-mass, middle-mass, and some high-mass stars against gravity. The strangest case of all occurs at the death of the most massive stars, which contained much more than 8 and up to about 60 solar masses when they were on the main sequence. They generally undergo supernova explosions like less massive heavyweight stars, but some of these most massive stars may retain cores of over 2 or 3 solar masses. Nothing in the universe is strong enough to hold up the remaining mass against the force of gravity. The remaining mass collapses, and continues to collapse forever. The result of such a collapse is one type of black hole, in which the matter disappears from contact with the rest of the universe. Later, we shall discuss the formation of black holes in processes other than those that result from the collapse of a star.

Though much about black holes can be understood in terms of the physics of Newton, they are actually constructs that follow from Einstein's theory of gravity, which is known as the general theory of relativity. We will therefore first make some general comments about Einstein and his work on related subjects.

**AIMS:** To understand the black holes that form from stars and how they are detected, and to comment on black holes much more massive or much less massive than the sun and stars

## 31.1 EINSTEIN'S GENERAL THEORY OF RELATIVITY

The intuitive notion we have of gravity corresponds to the theory of gravity advanced by Isaac Newton in 1687. We now know, however, that Newton's theory and our intuitive ideas are not sufficient to explain the universe in detail. Theories advanced by Albert Einstein in the first decades of the 20th century now provide us with a more accurate understanding.

Einstein's theory of gravitation is known as the general theory of relativity. (His earlier "special theory of relativity" describes motion but omits the effects of gravity.) The general theory, which Einstein advanced in final form in 1916, made three predictions that depended on the presence of a large mass like the sun for experimental verification. These predictions involved (1) the gravitational deflection of light, (2) the advance of the perihelion of Mercury, and (3) the gravitational redshift. Comparing the predictions with observations provided basic tests of the theory.

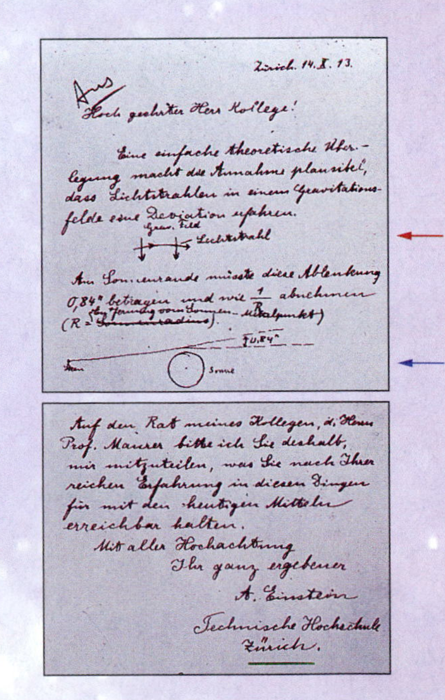

Zürich. 14. X. 13.

Hoch geehrter Herr Kollege!

Eine einfache theoretische Überlegung macht die Annahme plausibel, dass Lichtstrahlen in einem Gravitationsfelde eine Deviation erfahren.



Am Sonnenrande müsste diese Ablenkung 0,84" betragen und wie $\frac{1}{R}$ abnehmen (R = Entfernung vom Sonnen-Mittelpunkt).



Auf den Rat meines Kollegen, d. Herrn Prof. Maurer bitte ich Sie deshalb, mir mitzuteilen, was Sie nach Ihrer reichen Erfahrung in diesen Dingen für mit den heutigen Mitteln erreichbar halten.

Mit aller Hochachtung
Ihr ganz ergebener
A. Einstein

Technische Hochschule
Zürich.

FIGURE 31–1 The prediction in Einstein's own handwriting of the deflection of starlight by the sun. "Lichtstrahl" is "light ray" (*red arrow*). Here, Einstein asks Hale at Mt. Wilson if the effect could be measured without an eclipse, to which Hale replied negatively. In this early version of his theory, Einstein predicted (*blue arrow*) half the value he later calculated. The number he gives is slightly further wrong, because of an arithmetic error.

A DEEPER DISCUSSION

## BOX 31.1 The Special and General Theories of Relativity

In 1905, Einstein advanced a theory of relative motion that is called the **special theory of relativity.** A basic postulate is that, strangely, the speed of light is the same, no matter how the source of light and the observer are moving. For example, the speed of light is the same even if we are moving toward or away from the object that is emitting the light that we are observing. Ordinary objects, like raindrops, come at us faster if we are moving rapidly toward them.

Einstein's theory shows that the values that we measure for length, mass, and the rate at which time advances depend on how fast we are moving relative to the object we are observing. One consequence of the special theory is that mass and energy are equivalent and that they can be transformed into each other, following a relation $E = mc^2$. The consequences of the special theory of relativity have been tested experimentally in many ways, and the theory has long been well established.

"Special relativity" was limited in that it did not take the effect of gravity into account. Einstein proceeded to work on a more general theory that would explain gravity. Isaac Newton in 1687 had shown the equations that describe motion caused by the force of gravity, but nobody had ever said how gravity actually worked. Special relativity had linked three dimensions of space and one dimension of time to describe a four-dimensional space-time. Einstein's **general theory of relativity** holds that space-time is curved. It explains gravity as the effect we detect when we observe objects moving in curved space. Imagine the two-dimensional analogy to curved space of a golf ball rolling on a warped green. The ball would tend to curve one way or the other. The effect would be the same if the green were flat but there were objects with the equivalent of gravity spaced around it. In four-dimensional space-time, masses warp the space near them, and objects or light change their path as a result, an effect that we call gravity. Thus Einstein's theory of general relativity explains gravity as an observational artifact of the curvature of space.

In Einstein's own words, "If you will not take the answer too seriously, and consider it only as a kind of joke, then I can explain it as follows. It was formerly believed that if all material things disappeared out of the universe, time and space would be left. According to the relativity theory, however, time and space disappear together with the things."

### *Discussion Question*

Compare the general theory of relativity and a bed with you in it.

Einstein's theory predicts that the light from a star would act as though it were bent toward the sun by a very small amount (Fig. 31–1). We on Earth, looking back, would see the star from which the light was emitted as though it were shifted slightly away from the sun (Fig. 31–2). Only a star whose radiation grazed the edge of the sun would seem to undergo the full deflection; the effect diminishes as one considers stars farther away from the solar limb. To see the effect, one has to look near the sun at a time when the stars are visible, and this could be done only at a total solar eclipse (Fig. 31–3).

The British astronomer Arthur Eddington and other scientists observed the total eclipse of 1919 from sites in Africa and South America. The effect they were looking for was tiny, and it was not enough merely to observe the stars at the moment of eclipse. One had to know what their positions were when the sun was not present in their midst, so the astronomers had already made photographs of the same field of stars six months earlier when the same stars were in the nighttime sky. The duration of totality was especially long, and the sun was in a rich field of stars, the Hyades, making it a particularly desirable eclipse for this experiment. Even though his observations had been limited by clouds, Eddington detected that light was deflected by an amount that agreed with Einstein's revised predictions. Scientists hailed this confirmation of Einstein's theory; from the moment of its official announcement, Einstein was recognized by scientists and the general public alike as the world's greatest scientist (Figs. 31–4 through 31–6).

The experiment has been repeated, but it is a very difficult one. Though the data agree with Einstein's prediction, they are not accurate enough to distinguish between

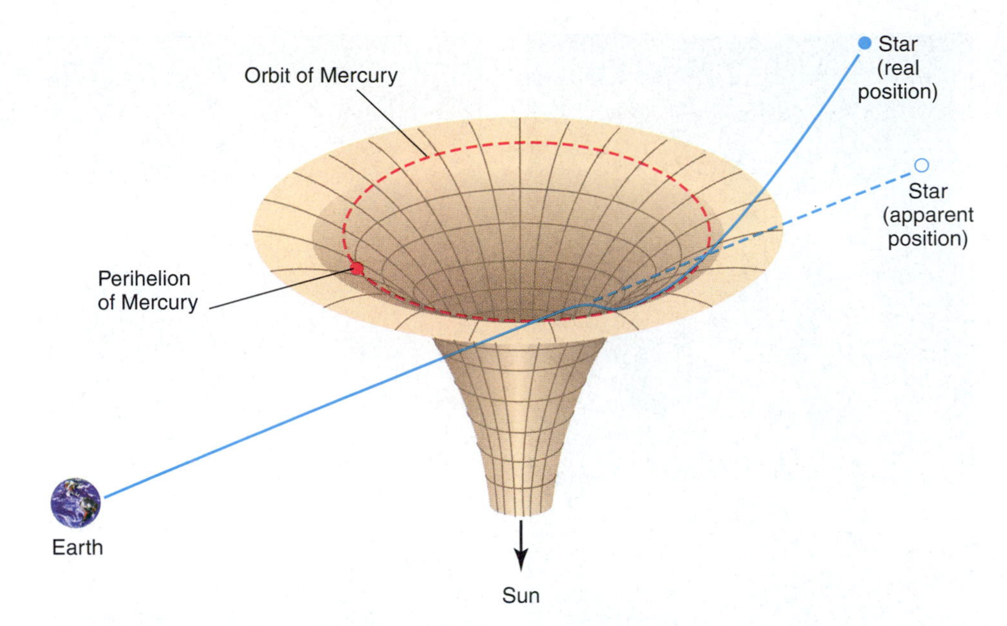


**FIGURE 31–2** Under the general theory of relativity, the presence of a massive body essentially warps the space nearby. This can account for both the bending of light near the sun and the advance of the perihelion point of Mercury by 43 arc sec per century, more than would otherwise be expected. The diagram shows how a two-dimensional surface warped into three dimensions can change the direction of a "straight" line that is constrained to its surface; the warping of space is analogous, although with a greater number of dimensions to consider.

Einstein's theory and newer, more complicated, rival theories of gravitation. Fortunately, the effect of gravitational deflection is constant throughout the electromagnetic spectrum, and the test can now be performed more accurately by observing how the sun bends radiation from radio sources, especially quasars. The results agree with Einstein's theory to within 1 per cent, enough to make the competing theories very unlikely.

A related test involves not deflection but a delay in time of signals passing near the sun. It is the same Shapiro effect since tested in some pulsars and discussed in Chapter 30. This test can now be performed with signals from interplanetary spacecraft, most recently with a spacecraft near Mars, and these data also agree with Einstein's theory.

A

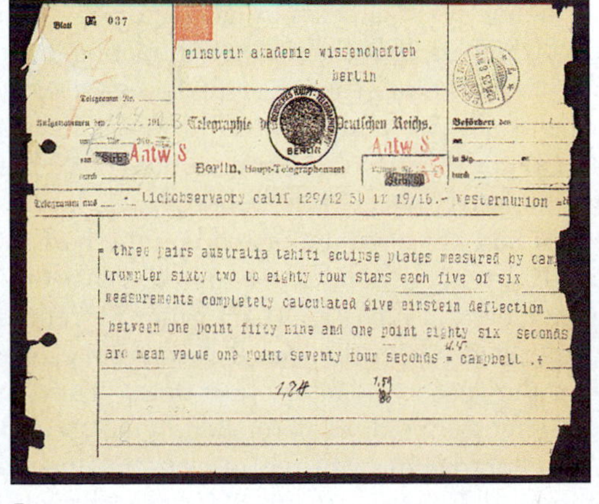

Blatt 037

einstein akademie wissenchaften
berlin

Telegraphie des Deutschen Reichs.
Antw S
Berlin, Haupt-Telegraphenamt

Telegramm aus – lickobservaory calif 129/12 50 1r 19/16.- westernunion

= three pairs australia tahiti eclipse plates measured by cam[bell]
trumpler sixty two to eighty four stars each five of six
measurements completely calculated give einstein deflection
between one point fifty nine and one point eighty six seconds
are mean value one point seventy four seconds = campbell .+

B

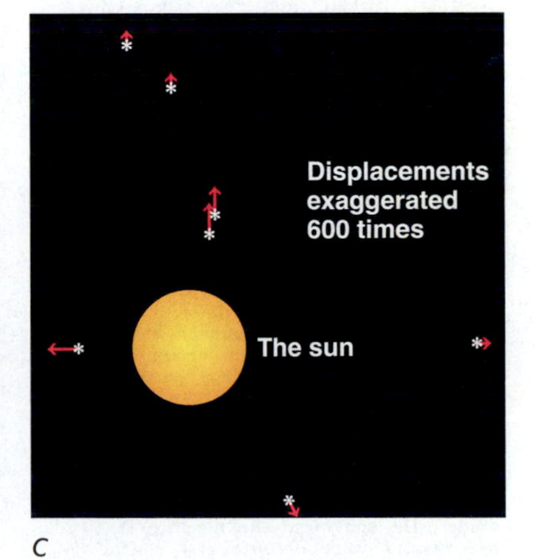


C

**FIGURE 31–3** (*A*) One of the photographic plates from the 1922 eclipse expedition to test Einstein's theory. The circles mark the positions of the stars used in the data reduction; the stars themselves are too faint to see in the reproduction. (*B*) The confirming results first found in 1919 were verified. (*C*) The field of view of the 1919 eclipse observations, showing (arrows) the deflection of the stars magnified by 600 times.

# ECLIPSE SHOWED GRAVITY VARIATION

## Diversion of Light Rays Accepted as Affecting Newton's Principles.

## HAILED AS EPOCHMAKING

### British Scientist Calls the Discovery One of the Greatest of Human Achievements.

Copyright, 1919, by The New York Times Company.
Special Cable to THE NEW YORK TIMES.

LONDON, Nov. 8.—What Sir Joseph Thomson, President of the Royal Society, declared was "one of the greatest—perhaps the greatest—of achievements in the history of human thought" was discussed at a joint meeting of the Royal Society and the Royal Astronomical Society in London yesterday, when the results of the British observations of the total solar eclipse of May 29 were made known.

There was a large attendance of astronomers and physicists, and it was generally accepted that the observations were decisive in verifying the prediction of Dr. Einstein, Professor of Physics in the University of Prague, that rays of light from stars, passing close to the sun on their way to the earth, would suffer twice the deflection for which the principles enunciated by Sir Isaac Newton accounted. But there was a difference of opinion as to whether science had to face merely a new and unexplained fact or to reckon with a theory that would completely revolutionize the accepted fundamentals of physics.

The discussion was opened by the Astronomer Royal, Sir Frank Dyson, who described the work of the expeditions sent respectively to Sobral, in Northern Brazil, and the Island of Principe, off the west coast of Africa. At each of these places, if the weather were propitious on the day of the eclipse, it would be possible to take during totality a set of photographs of the obscured sun and a number of bright stars which happened to be in its immediate vicinity.

The desired object was to ascertain whether the light from these stars as it passed by the sun came as directly toward the earth as if the sun were not there, or if there was a deflection due to its presence. And if the deflection did occur the stars would appear on the photographic plates at measurable distances from their theoretical positions. Sir Frank explained in detail the apparatus that had been employed, the corrections that had to be made for various disturbing factors, and the methods by which comparison between the theoretical and observed positions had been made. He convinced the meeting that the results were definite and conclusive, that deflection did take place, and

FIGURE 31–5 The first report of the events that made Einstein world famous. From *The New York Times* of November 9, 1919.

A B

FIGURE 31–4 (*A*) Einstein visiting California in the 1930s. (*B*) A postcard from Einstein to his mother, reporting to her that the English expedition had found the gravitational deflection that he had predicted.

Another of the triumphs of Einstein's theory, even in its earliest versions, was that it explained the "advance of the perihelion of Mercury." The orbit of Mercury, like the orbits of all the planets, is elliptical; the point at which the orbit comes closest to the sun is called the **perihelion.** The elliptical orbit is pulled around the sun over the years, mostly by the gravitational attraction on Mercury by the other planets, so that the perihelion point is at a different orientation in space. Each century (!) the perihelion point appears to move around the sun by approximately 5600 seconds of arc (which is less than 2°). Subtracting the effects of precession (the changing orientation of the planet's axis in space, described in Section 6.3) still leaves 574 seconds of arc per century. Most of that shift is accounted for by the gravitational attraction of the planets, as was worked out a hundred years ago. Still, subtracting the gravitational effects of the other planets leaves 43 seconds of arc per century whose origin had not been understood on the basis of Newton's theory of gravitation.

Einstein's general theory predicts that the presence of the sun's mass warps the space near the sun (Fig. 31–7). Since Mercury is sometimes closer to and sometimes farther away from the sun, it is sometimes travelling in space that is warped more than the space it is in at other times. This curved space should have the effect of changing the point of perihelion by 43 seconds of arc per century. The agreement of this prediction with the measured value was an important observational confirmation of Einstein's theory. More recent refinements of solar system observations have shown that the perihelions of Venus and of the Earth also advance by even smaller amounts predicted by the theory of relativity. And the verification has since been completely established by studying a pulsar in a binary system (Section 30.6).

Historically, in explaining the perihelion advance, general relativity was explaining an observation that had been known. On the other hand, in the bending of light it actually predicted a previously unobserved phenomenon.

As a general rule, scientists try to find theories that not only explain the data that are at hand but also make predictions that can be tested. This testing of predictions is an important part of the **scientific method.** Because bending of electromagnetic radiation by a certain amount was a prediction of the general theory of relativity that had not been anticipated, the verification of the prediction is traditionally said to be a more convincing proof of the theory's validity than the theory's ability to explain the perihelion advance.

Stephen Brush, a historian of science at the University of Maryland, showed in 1989 that this traditional view was not the actual case. The light-bending result indeed brought Einstein's work to public attention and brought praise from scientists. However, it was apparently too new a result to be completely convincing; many scientists preferred to see if other theories would prove to explain the result. Indeed, the result stimulated a search for other explanations. It was only ten years or so later that it became clear that no other explanation could be found. The advance of the perihelion of Mercury's orbit, on the other hand, was a well-established result that no other theory had succeeded in explaining. Also, the precession result tested a more fundamental part of Einstein's theory than did the light bending. Only with hindsight do scientists really consider the eclipse bending prediction to be of as much importance as the explanation of the precession of Mercury.

Thus far, we have discussed tests of general relativity in the solar system, where gravity is relatively weak. It could conceivably be possible that Einstein's general theory of relativity works in weak gravitational fields but not in very strong ones. It has therefore been important to test the theory in strong gravitational fields. The tests in involving binary pulsars we discussed in Section 30.6 provide just such tests in strong gravitational fields. The theory has been further verified for strong fields by the discovery of "gravitational lenses," which are detected as giant arcs that appear near some galaxies (Section 34.6) or as multiple images of distant quasars (Section 36.10). Weak field or strong field—Einstein's general theory of relativity has passed all the tests.

**LIGHTS ALL ASKEW IN THE HEAVENS**

**Men of Science More or Less Agog Over Results of Eclipse Observations.**

**EINSTEIN THEORY TRIUMPHS**

**Stars Not Where They Seemed or Were Calculated to be, but Nobody Need Worry.**

**A BOOK FOR 12 WISE MEN**

**No More in All the World Could Comprehend It, Said Einstein When His Daring Publishers Accepted It.**

Special Cable to THE NEW YORK TIMES.

LONDON, Nov. 9.—Efforts made to put in words intelligible to the non-scientific public the Einstein theory of light proved by the eclipse expedition so far have not been very successful. The new theory was discussed at a recent meeting of the Royal Society and Royal Astronomical Society, Sir Joseph Thomson, President of the Royal Society, declares it is not possible to put Einstein's theory into really intelligible words, yet at the same time Thomson adds:

"The results of the eclipse expedition demonstrating that the rays of light from the stars are bent or deflected from their normal course by other aerial bodies acting upon them and consequently the inference that light has weight form a most important contribution to the laws of gravity given us since Newton laid down his principles."

Thompson states that the difference between theories of Newton and those of Einstein are infinitesimal in a popular sense, and as they are purely mathematical and can only be expressed in strictly scientific terms it is useless to endeavor to detail them for the man in the street.

**FIGURE 31-6** These reports captured the public's fancy. From *The New York Times* of November 10, 1919.

## 31.2 THE FORMATION OF A STELLAR BLACK HOLE

When nuclear fusion ends in the core of a star, gravity causes the star to contract. We have seen in Chapter 28 that a star that has a final mass of less than 1.4 solar

**FIGURE 31-7** A golf ball rolling on a curving green appears to us to curve, even though it is rolling as straight as possible. Here Tiger Woods putts. This warping of a 2-D surface is similar to the warping of 3-D space by a gravitational mass.

IN REVIEW

## BOX 31.2 Updike on Relativity

"Welcome to the twentieth century, Mr. Wilmot; we all have some catching up to do. Time is the fourth dimension, it turns out, and slows down when the observer speeds up. The other three dimensions don't form a rigid grid; space is more like a net that sags when you put something in it, and that sagging is what we call gravity. Also, light isn't an indivisible static presence; it has a speed, and it comes in packets—irreducible amounts called quanta. Quanta aren't merely the limit of our measurement; they appear to be the fact. Energy is grainy! All these strainings of our common sense are fact."

(from *In the Beauty of the Lilies* by John Updike)

***Discussion Question***

Which phrases refer to Einstein's theory of (a) special relativity; (b) general relativity; (c) electromagnetic radiation?

masses will end its life as a white dwarf. In Chapter 29 we saw that a more massive star as well as some white dwarfs in binary systems will explode as supernovae, and Chapter 30 showed that if the remaining mass of the core is less than 2 or 3 solar masses, it will wind up as a neutron star. (Even if it is less than 1.4 solar masses, the force of the explosion may turn it into a neutron star instead of a white dwarf; on the high end of the range, our calculations are not accurate enough to know if 2 solar masses or 3 solar masses is the actual limit.) We are now able to observe both white dwarfs and neutron stars and so can study their properties directly.

It now seems reasonable that in some cases more than 2 or 3 solar masses remain after the supernova explosion. The star collapses through the neutron star stage, and we know of no force that can stop the collapse. In some cases, the matter may even have become so dense as the star collapsed that no explosion took place and no supernova resulted.

The value for the maximum mass that a neutron star can have is a result of theoretical calculation. We must always be aware of the limits of accuracy of any calculation, particularly of a calculation that deals with matter in a state that is very different from states that we have been able to study experimentally. Still, the best modern calculations show that the limit of a neutron star mass is 2 or 3 solar masses.

We may then ask what happens to a 5- or 10- or 50-solar-mass star as it collapses, if it retains more than 3 solar masses. It must keep collapsing, getting denser and denser. Another prediction of Einstein's general theory of relativity is that a strong gravitational field redshifts radiation. This prediction was verified both for the sun and for white dwarfs long before it was verified on Earth. Theoretically, such redshifts would be seen in a collapsing massive star. Also, radiation will be bent by a gravitational field, or at least appear to us on Earth as though it were bent. Further, the general theory of relativity indicates that pressure as well as mass produces gravity. This effect compounds the gravitational contraction of a very massive star.

As the mass contracts and the star's surface gravity increases, radiation is continuously redshifted more and more, and radiation leaving the star other than perpendicularly to the surface is bent more and more. Eventually, when the mass has been compressed to a certain size, radiation from the star can no longer escape into space. The star has withdrawn from our observable universe, in that we can no longer receive radiation from it. We say that the star has become a **black hole.**

Astronomers long assumed that the most massive stars would somehow lose enough mass to wind up as white dwarfs. When the discovery of pulsars ended that prejudice, it seemed more reasonable that black holes could exist.

Why do we call it a black hole? We think of a black surface as a surface that reflects none of the light that hits it. Similarly, any radiation that hits the surface of a black hole continues into the black hole and is not reflected. In this sense, the object is perfectly black.

## 31.3 THE PHOTON SPHERE

Let us consider what happens to radiation emitted by the surface of a star as it contracts. Although what we will discuss affects radiation of all wavelengths, let us simply visualize standing on the surface of the collapsing star while holding a flashlight.

On the surface of a supergiant star, we would note only very small effects of gravity on the light from our flashlight. If we shine the beam at any angle, it seems to go straight out into space.

As the star collapses, two effects begin to occur. Although we on the surface of the star cannot notice them ourselves, a friend on a planet revolving around the star could detect the effects and radio back information to us about them. For one thing, our friend could see that our flashlight beam is redshifted. Second, our flashlight beam would be bent by the gravitational field of the star (Fig. 31–8). If we shined the beam straight up, it would continue to go straight up. But if we shined it away from the vertical, the beam would be bent even farther away from the vertical. When the star reached a certain size, a horizontal beam of light would not escape (Fig. 31–9).

From this time on, only if the flashlight is pointed within a certain angle of the vertical does the light continue outward. This angle forms a cone, with its apex at the flashlight, and is called the **exit cone** (Fig. 31–10). As the star grows smaller yet, we find that the flashlight has to be pointed more directly upward in order for its light to escape. The exit cone grows smaller as the star shrinks.

When we shine our flashlight upward in the exit cone, the light escapes. When we shine our flashlight in a direction outside the exit cone, the light is bent sufficiently that it falls back to the surface of the star. When we shine our flashlight exactly along the side of the exit cone, the light goes into orbit around the star, neither escaping nor falling onto the surface.

The sphere around the star in which the light can orbit is called the **photon sphere.** Its size can be calculated theoretically. Its radius is 13.5 km for a star of 3 solar masses and varies linearly with the mass; that is, for a star of 6 solar masses it is 27 km in radius, and so on.

As the star continues to contract, the theory shows that the exit cone gets narrower and narrower. Light emitted within the exit cone still escapes. Note that the

The modern name "black hole" was suggested by John Wheeler in 1968 based on the general theory of relativity. A similar idea based on Newtonian gravitation had been suggested by John Michell in 1783 and Pierre Laplace in 1794.

**FIGURE 31–8** As the star contracts, a light beam emitted other than radially outward will be bent.

**FIGURE 31–9** Light can be bent so that it falls back onto the star.

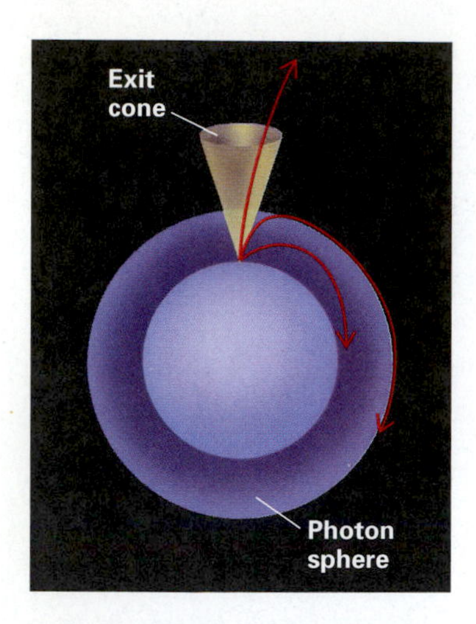


**FIGURE 31–10** When the star has contracted enough (the inner sphere), only light emitted within the exit cone escapes. Light emitted on the exit cone goes into the photon sphere. The further the star contracts within the photon sphere, the narrower the exit cone becomes.

| Escape velocities (km/sec): | |
|---|---|
| the moon | 2.4 |
| the Earth | 11.2 |
| Jupiter | 60 |
| the sun | 620 |
| white dwarf | 7600 |
| neutron star | 160,000 |
| 5-solar-mass black hole: | 300,000 |

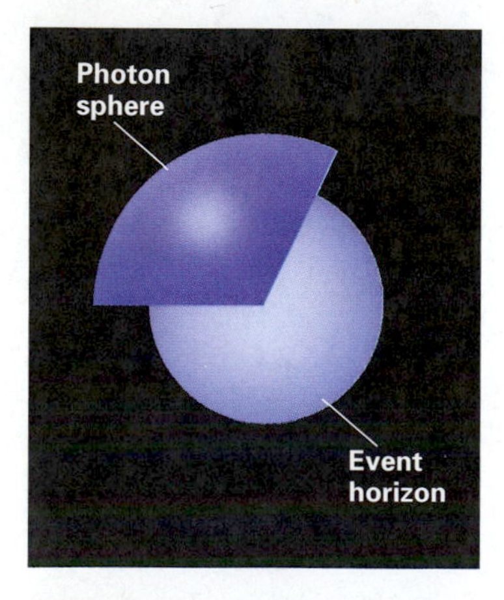


**FIGURE 31–11** When the star becomes smaller than its Schwarzschild radius, we can no longer observe it. We say that it has passed its **event horizon,** by analogy to the statement that we cannot see an object on Earth once it has passed our horizon.

photon sphere is unaffected by this further contraction. The photon sphere remains at the same height even though the matter inside it has contracted further, since the total amount of matter (that is, the mass) within has not changed.

## 31.4 THE EVENT HORIZON

If we were to depend on our intuition, we might think that the exit cone would simply continue to get narrower. But when we apply the general theory of relativity, we find that when the star contracts beyond a certain size the cone vanishes. Light no longer can escape into space, even when it is travelling straight up.

The solutions to Einstein's equations that predict this were worked out by Karl Schwarzschild in 1916, shortly after Einstein advanced his general theory. The radius of the star at the time at which light can no longer escape is called the **Schwarzschild radius** or the **gravitational radius,** and the spherical surface at that radius is called the **event horizon** (Fig. 31–11). The event horizon is not a physical surface. The radius of the photon sphere is exactly $3/2$ times this Schwarzschild radius.

We can visualize the event horizon in another way, by considering a classical picture based on the Newtonian theory of gravitation. The picture is essentially that conceived in 1796 by Laplace, the French astronomer and mathematician. You must have a certain velocity, called the **escape velocity,** to escape from the gravitational pull of another body. For example, we have to launch rockets at 11 km/sec (40,000 km/hr) in order for them to escape from the Earth's gravity. For a more massive body of the same size, the escape velocity would be higher. Now imagine that this body contracts. We are drawn closer to the center of the mass. As this happens, the escape velocity rises. When all the mass of the body reaches its Schwarzschild radius, the escape velocity becomes equal to the speed of light. Thus within that radius, even light cannot escape. If we begin to apply the special theory of relativity, we might then reason that since nothing can go faster than the speed of light, nothing can escape. Now let us return to the picture according to the general theory of relativity.

The size of the Schwarzschild radius depends linearly on the amount of mass that is collapsing. A star of 3 solar masses, for example, would have a Schwarzschild radius of 9 km. A star of 6 solar masses would have a Schwarzschild radius of twice 9 km, or 18 km. One can calculate the Schwarzschild radii for less massive stars as well, although the less massive stars would be held up in the white-dwarf or neutron-star stages and not collapse to their Schwarzschild radii (although conceivably some objects could be sufficiently compressed in a supernova). Since one couldn't, theoretically, put a tape measure into and out of a black hole to measure its radius, we would actually measure its circumference and divide by $2\pi$.

Note that anyone or anything on the surface of a star as it passed its event horizon would not be able to survive. A traveller would be torn apart by the tremendous difference in gravity between head and foot. (This difference is a tidal force, since this kind of difference in gravity also causes the tides on Earth, as we described in Section 8.3.) Since the traveller would be pulled out into long strands, the effect has even been called "spaghettification." If the tidal force could be ignored, though, the observer on the surface of the star would not notice anything particularly wrong as the star passed its event horizon. If we chose to stay with the observer at that time, we could still be pointing our flashlights up into space trying to signal our friends on the Earth. But once we passed the event horizon, no answer would ever come because our signal would never get out. (See Fig. 31–12.)

Once the star passes inside its event horizon, we lose contact with it. Theories indicate that it continues to contract and that nothing can ever stop its contraction. In fact, the mathematical theories predict that it will contract to zero radius, a situation that seems impossible to conceive of. The point at which it will have zero radius (it has infinite density there) is called a **singularity.** Strange as it seems, theories predict that a black hole contains a singularity.

**FIGURE 31–12** This drawing by Charles Addams is from cartoonbank.com. All Rights Reserved. © *The New Yorker Collection 1974 Charles Addams.*

Even though the mass that causes the black hole has contracted further, the event horizon doesn't change. It remains at the same radius forever, as long as the amount of mass inside doesn't change.

Chandra X-ray Observatory observations have apparently confirmed the reality of the event horizon of black holes. A team of scientists from the Harvard-Smithsonian Center for Astrophysics compared x-ray emissions in 6 x-ray binaries that contain neutron stars with x-ray emissions from 6 x-ray binaries that contain black-hole candidates. In 2001, they reported that the binaries with black-hole candidates emitted only 1 per cent as much energy as neutron stars. Their idea is that gas from the companion stars that impacts the neutron stars results in high temperatures, which are detected from Earth as x-rays. On the other hand, gas flowing from the companion stars toward black holes is simply swallowed by the event horizons (Fig. 31–13). After years of event horizons being mere theoretical constructs, we may now have finally detected their effects observationally.

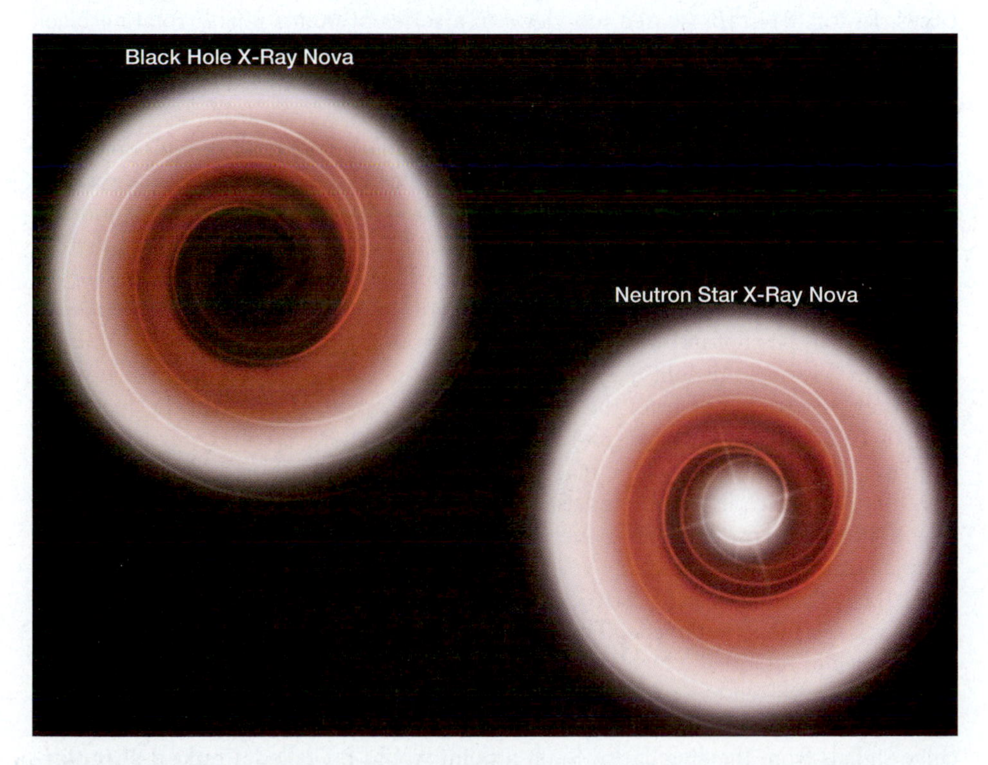


**FIGURE 31–13** The method by which evidence for the existence of a black hole's event horizon has been found, in an artist's conception. Chandra X-ray Observatory observations show that the neutron star system is much brighter in x-rays than the suspected black-hole system. If the gas in black-hole systems went into an event horizon, that would account for the difference.

## 31.5 ROTATING BLACK HOLES

Once matter is inside a black hole, it loses its identity in the sense that from outside a black hole, all we can tell are the mass of the black hole, the rate at which it is spinning, and what total electric charge it has. These three quantities are sufficient to describe the black hole completely. Thus, in a sense, black holes are simple objects to describe physically, because we only have to know three numbers to characterize each one. (By contrast, for the Earth we have to know shapes, sizes, densities, motions, and other parameters for the interior, the surface, and the atmosphere.) The theorem that describes the simplicity of black holes is often colloquially stated by astronomers active in the field as "a black hole has no hair." All other properties, such as size, can be derived from the three basic properties.

Most of the theoretical calculations about black holes, and the Schwarzschild solutions to general relativity, in particular, are based on the assumption that black holes do not rotate. But this assumption is only a convenience; we think, in fact, that the rotation of a black hole is one of its important properties. It was not until 1963 that Roy P. Kerr solved Einstein's equations for a situation that was later interpreted in terms of the notion that a black hole is rotating. In this more general case, an additional special boundary—the **stationary limit**—appears, with somewhat different properties from the original event horizon. At the stationary limit, space-time is flowing at the speed of light, so particles that would otherwise be moving at the speed of light remain stationary. Within the stationary limit, no particles can remain at rest even though they are outside the event horizon.

Let us consider a series of non-moving ("static") points at distances close to a black hole and visualize what happens to a sphere representing expanding wavefronts of light emitted from these points. In Fig. 31–14*A*, we see the wavefronts a standard time (say, 1 second) after they are emitted from each dot. For a point inside the event horizon, all the emitted waves go inward instead of outward, and no light escapes. In Fig. 31–14*B*, we see top views of a series of points near a rotating black hole. The rotation carries the wavefront to the side, and for a point at the stationary limit, some of the wavefront is outside the limit. These light waves can escape. Only for points within the event horizon do all wavefronts move inward.

The equator of a stationary limit of a *rotating* black hole has the same diameter as the event horizon of a *non-rotating* black hole of the same mass. But a rotating black hole's stationary limit is squashed. The event horizon touches the stationary limit at the poles. Since the event horizon remains a sphere, it is smaller than the event horizon of a non-rotating black hole (Fig. 31–14*C*).

The region between the stationary limit and the event horizon is the **ergosphere.** An object can be shot into the ergosphere of a rotating black hole at such an angle that, after it is made to split in two, one part continues into the event horizon while the other part is ejected from the ergosphere with more energy than the incoming particle had. This energy must have come from somewhere; it could only have come from the rotational energy of the black hole. Thus by judicious use of the ergosphere, we could, in principle, tap the energy of the black hole. We are always looking for new energy sources, and this is the most efficient known, although it is obviously very far from being practical.

A black hole can rotate up to the speed at which a point on the event horizon's equator is travelling at the speed of light. The event horizon's radius is then half the Schwarzschild radius. If a black hole rotated faster than this, its event horizon would vanish. Unlike the case of a non-rotating black hole, for which the singularity is always unreachably hidden within the event horizon, in this case distant observers could receive signals from the singularity. Such a point would be called a **naked singularity** and, if one exists, we might have no warning—no photon sphere or orbiting matter, for example—before we ran into it. Most theoreticians assume the existence of a law of "cosmic censorship," which requires all singularities to be "clothed" in event horizons, that is, not naked. In front-page news in *The New York Times* in 1997,

An oscillation 450 times per second in a small percentage of the x-radiation from a source near the center of our galaxy may be the proof that black holes rotate. The oscillation, measured in the "microquasar" GRO J1655-40 with the Rossi X-ray Timing Explorer, is thought to come from blobs in the black hole's accretion disk as they revolve at that rate. But theory shows that blobs couldn't revolve more often than 300 times per second if the black hole weren't rotating. For rotating black holes, the accretion disk extends farther down, where it can rotate faster. In "microquasars" like this one, about half the gas falls into the black hole and the other half streams outwards from the poles in jets, small-scale versions of the quasars we will discuss in Chapter 36.

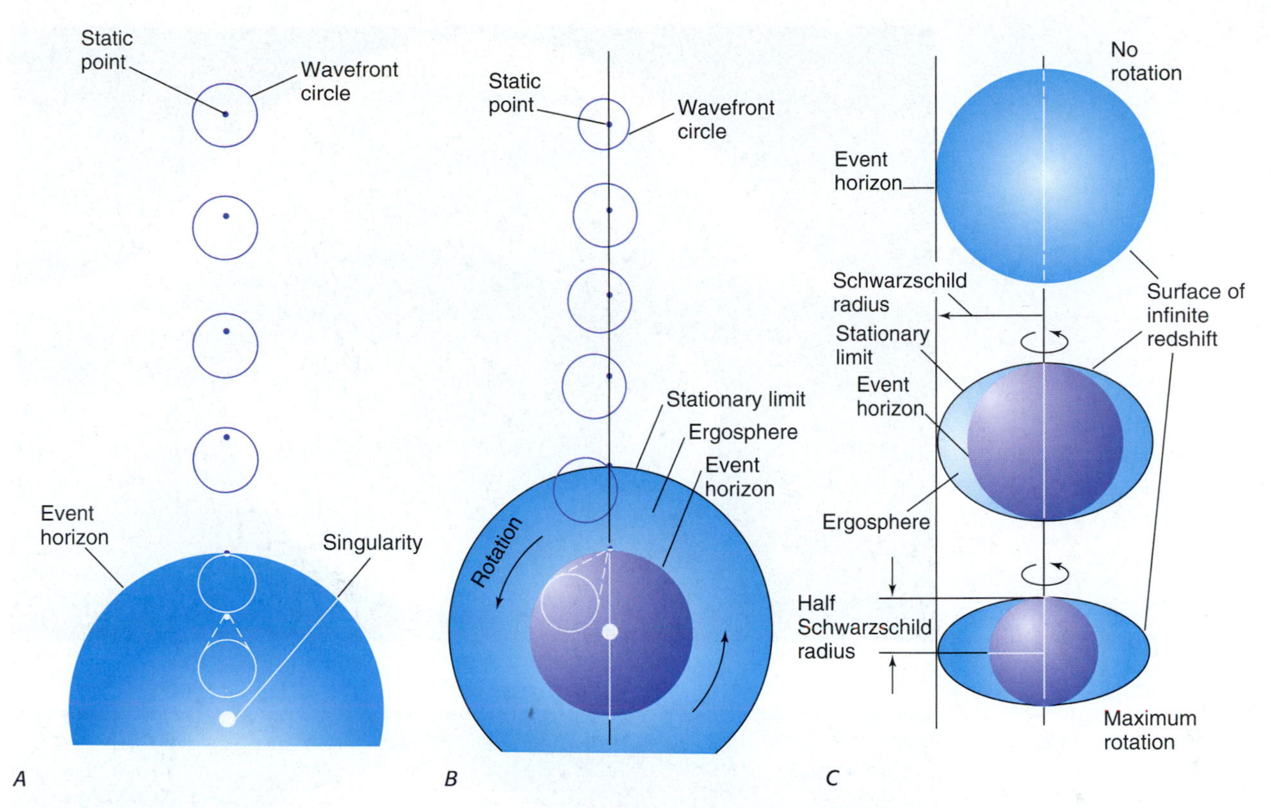


**FIGURE 31–14** (*A*) A static point that emits a pulse of light in all directions, and the circle showing the wavefront a standard time later for locations near a non-rotating black hole. The circle is dragged more toward the black hole the closer the static point is to the black hole. For a static point at the event horizon, the wave circles are entirely inside and no light escapes. (*B*) Near a rotating black hole, the wave circles in this top view are displaced not only inward but also in the direction of rotation. For a static point on the stationary limit, part of the circle is outside the stationary limit, so some light escapes. Some light can escape even from inside the stationary limit. (*C*) Side views of black holes of the same mass and different rotation speeds. The region between the stationary limit and the event horizon of a rotating black hole is called the **ergosphere** (from the Greek word *ergon,* meaning "work") because, in principle, work can be extracted from it.

theoretical physicist Stephen Hawking paid off a bet he had made with colleagues that naked singularities could not exist. A supercomputer model had shown their feasibility—not that they were probable.

## 31.6 DETECTING A BLACK HOLE

What if we were watching a faraway star collapse? Photons that were just barely inside the exit cone when emitted would leak very slowly away from the black hole. These few photons would be spread out in time, and the image would become very faint. As the star approached its event horizon, the photons that reach us would be extremely redshifted, so the star would get very red and then disappear, because it is too faint to see. Further, even if the star's surface would remain bright enough, it would seem to us to take an infinite time for a star to collapse enough to reach its event horizon.

But all hope is not lost for detecting a black hole, even though we can't hope to see the photons from the moments of collapse. The black hole disappears, but it leaves its gravity behind. It is a bit like the Cheshire Cat from *Alice's Adventures in Wonderland,* which fades away, leaving only its grin behind (Fig. 31–15).

**FIGURE 31–15** Lewis Carroll's Cheshire Cat, from *Alice's Adventures in Wonderland,* shown here in John Tenniel's drawing, is analogous to a black hole in that it left its grin behind when it disappeared, while a black hole leaves its gravity behind when its mass disappears. Alice thought that the Cheshire Cat's persisting grin was "the most curious thing I ever saw in my life!" We might say the same about the black hole and its persisting gravity.

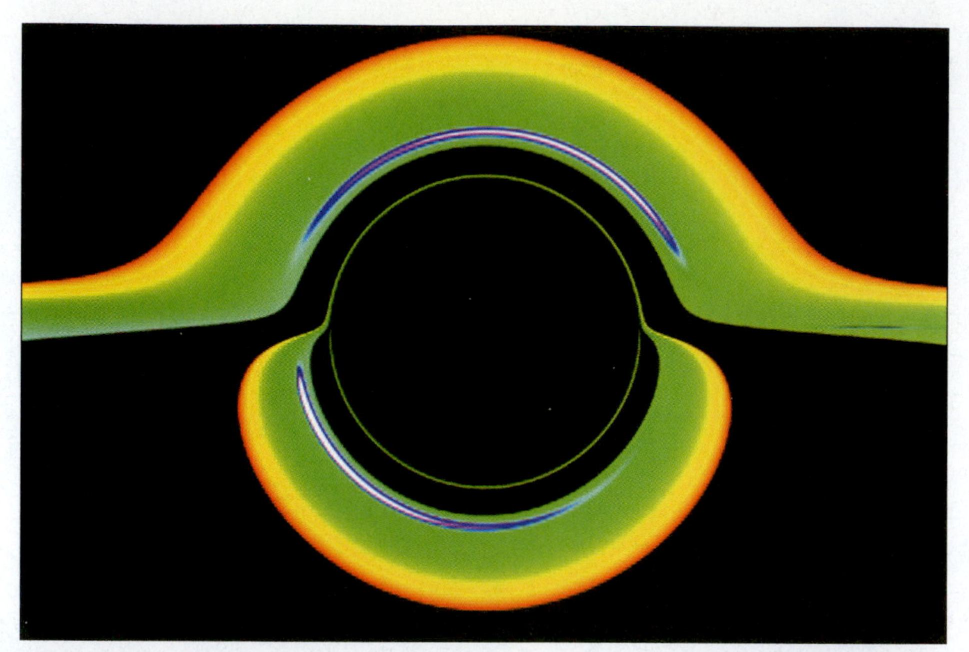

A

B

**FIGURE 31–16** The optical appearance of a black hole surrounded by a thin, flat accretion disk. The views are those from outside the black hole's event horizon. The rotation of the black hole causes the asymmetry we see. The apparent temperature of the accretion disk's gas is shown in false color, from coolest (red) through intermediate (green) to hottest (blue/white). (*A*) The case for a *non-rotating black hole.* We are hovering just above the inner edge of the accretion disk, which occurs at about 3 times the Schwarzschild radius. The top of the image shows the top of the disk, but the black hole distorts space so much that the bottom of the image shows the underside of the disk. A series of rings exists, with diminishing widths, that converge toward the photon sphere, which is 1.5 times the Schwarzschild radius. We see the distorted shape of the widest of these rings. (*B*) The case for a *rotating black hole.* Now we see the accretion disk of a rapidly rotating black hole, one spinning with 99.8% of the maximum possible speed. We are well outside the event horizon and slightly above the accretion disk. The disk material rotates counterclockwise around the black hole. The Doppler redshifting and blueshifting of the matter being accreted shows clearly. The accretion disk is really flat, but looks warped because of an optical illusion. Apparently, the light from the far side of the disk is bent into your line of sight by the gravitational lensing of the black hole.

The black hole attracts matter, and the matter accelerates toward it. Some of the matter will be pulled directly into the black hole, never to be seen again. But other matter will go into orbit around the black hole, and will orbit at a high velocity. This accreted matter forms an **accretion disk** (Fig. 31–16). The action of matter in the accretion disk can be modeled with supercomputers (Fig. 31–17).

Calculations show that the gas in orbit will be heated. Theoretical calculations of the friction that will take place between adjacent filaments of gas in orbit around the black hole show that the heating will be so great that the gas will radiate strongly in the x-ray region of the spectrum. The inner 200 km should reach tens of millions of kelvins. Thus though we cannot observe the black hole itself, we can hope to observe x-rays from its surrounding gas.

In fact, a large number of x-ray sources are known in the sky. They have been surveyed from a variety of x-ray satellites, most recently Chandra and XMM-Newton. Some of the x-ray sources have been identified with galaxies or quasars. Others pulse regularly, like Hercules X-1, and are undoubtedly neutron stars (Section 30.8). Some of the rest, which pulse sporadically, may be related to black holes.

It is not enough to find an x-ray source that gives off sporadic pulses, for one can think of other mechanisms besides matter revolving around a black hole that can lead to such pulses. Also, while regular pulses indicate that the compact object is a neutron star, one cannot conclude that a compact object with no detectable pulses is a black hole. One would like to show that a collapsed star of greater than 3 solar masses is present. (The value of 3 solar masses is theoretical; no neutron star has yet been found with more than 2 solar masses.) If more than that limit is present, we deduce that a neutron star would have collapsed, and the only possibility is that the object is a black hole.

We can determine masses only for certain binary stars. When we search the position of the x-ray sources, we look for a spectroscopic binary (that is, a star whose spectrum shows a Doppler shift that indicates the presence of an invisible companion). Then, if we can show that the companion is too faint to be a normal, main-sequence star, it must be a collapsed star. If, further, the mass of the unobservable companion is greater than 3 solar masses, it must be a black hole.

The main problem in proving that the companion's mass is high enough is that we cannot measure this mass by itself. By measuring the radial-velocity variations of

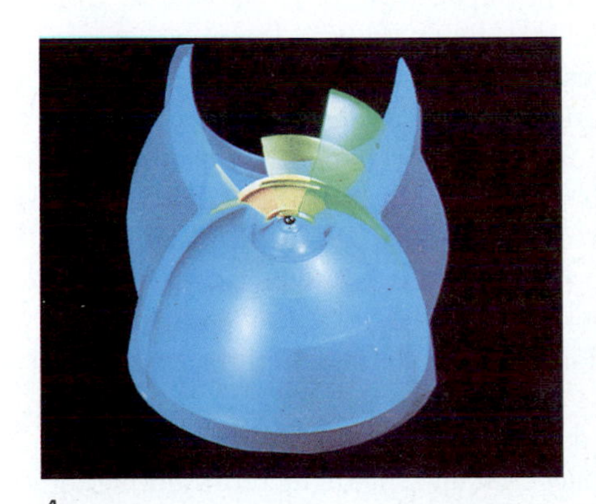

A

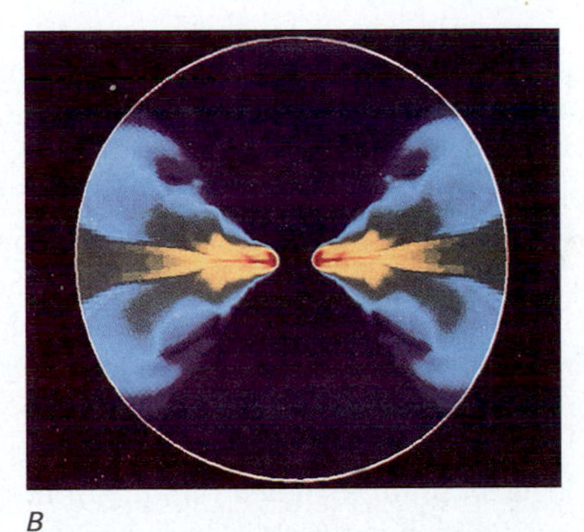

B

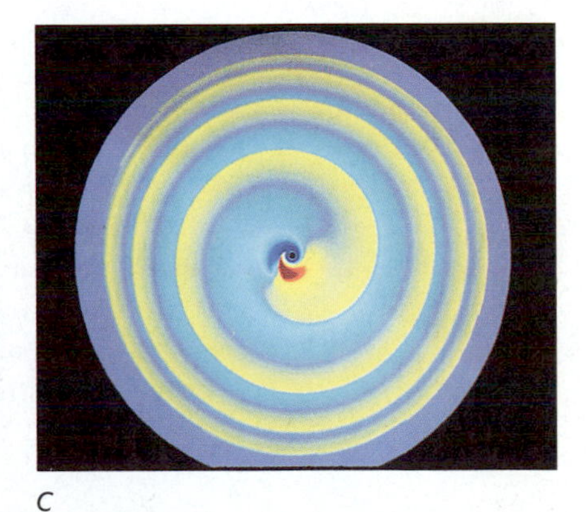

C

**FIGURE 31–17** Supercomputer simulations showing how a black hole's gravity affects the orbiting gas in the accretion disk that surrounds it. (*A*) A 3-dimensional view showing that when gas flows toward a black hole with high angular momentum, it does not flow into the hole. It is thrown outward, and as it splashes back, it opens an empty region that resembles a funnel. (*B*) Effects close to the event horizon show well in this side cross section of density; the calculations showed that the gas sometimes bounces back out. (*C*) This view from above shows gas density, with red being the densest gas. Instabilities cause a ring of gas to be drawn out into an orbiting blob. Narrow disks like this one are too unstable to be important in nature. Thicker disks are less unstable and form spiral waves.

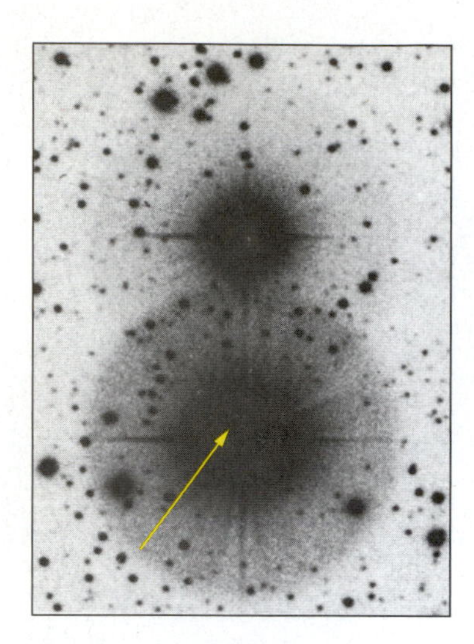

**FIGURE 31–18** The blue supergiant star HDE 226868. The black hole Cygnus X-1 that is thought to be orbiting the supergiant star is not visible. The image of the supergiant appears so large because it is overexposed on the film. The image does not represent the actual angular diameter subtended by the star, which is too small to resolve from the Earth.

the secondary object (from Doppler shifts), we can determine only a "mass function." (The situation is the same as that for "single-lined spectroscopic binaries," discussed in Chapter 26.) This mass function is the mass of the compact companion multiplied by the cube of the sine of the angle at which their mutual orbits are inclined to the plane of the sky (the plane perpendicular to our line of sight) and divided by a term that depends only on the ratio of the two masses. For any individual pair of stars, we usually have little idea what that angle is; from the fact that the stars can eclipse each other only if their orbit is at a high inclination we get some information. (Only when the stars do eclipse each other do we have a good value for the inclination.)

Because of the uncertainty in the angle of inclination and the uncertainty in the mass of the visible object, the mass function that we measure does not fully determine the mass of the unseen compact companion. However, the value of the mass function is an absolute lower limit to the mass of the compact star.

The longest known case of a black-hole candidate, though no longer the most persuasive, is Cygnus X-1, an x-ray source that sometimes varies in intensity on a time scale of milliseconds. In 1971, radio radiation was found to come from the same direction. A 9th magnitude star previously catalogued as HDE 226868 was found at its location (Fig. 31–18).

HDE 226868 has the spectrum of a blue supergiant (type O), and thus it has a mass of about 30 times that of the sun if it is a normal star about 2 kiloparsecs away. Its spectrum is observed to vary in radial velocity with a period of 5.6 days, indicating that the supergiant and the invisible companion are orbiting each other with that period. From the orbit, it is deduced that the invisible companion must certainly have a mass greater than 7 solar masses; the best estimate is 16 solar masses. Because this mass is so much greater than the limit of 3 solar masses above which neutron stars cannot exist, the object must be a black hole. It seems that too much mass is present to allow the matter to stop collapsing as a neutron star. Thus many, and probably most, astronomers believe that a black hole has been found in Cygnus X-1. But, recall, we can measure only the mass function. If the visible star's mass turns out to be extremely low for its spectral type, then the compact object could barely be a neutron star of 2.5 solar masses, so the case for a black hole is not proven.

Another promising candidate is LMC X-3 (the third x-ray source to be found in the nearby galaxy known as the Large Magellanic Cloud). Its 17th-magnitude optical counterpart orbits an unseen companion every 1.7 days. Since we know the distance to the Large Magellanic Cloud (especially from Supernova 1987A as well as from new measurements of variable stars in it) and thus to anything in it much more accurately than we know the distance to Cygnus X-1, our calculations for the invisible object's mass may be more accurate. Assuming that the visible companion is a normal star of the spectral type it appears, the invisible object seems to have at least 7 solar masses of matter, and so it should be a black hole. But if the visible companion is abnormal and has an extremely low mass, the companion's mass could possibly be slightly less than 3 solar masses. So this source is very likely a black hole, but not definitely so. (LMC X-1 is also a black-hole candidate, with a companion of perhaps 20 solar masses. But the mass function is very uncertain, and the case is not strong.)

A fourth promising black-hole candidate is A0620-00, the A standing for the x-ray catalogue of sources studied with the Ariel spacecraft and the digits representing the source's position in the sky, which is in the constellation Monoceros. The source was even brighter in x-rays than Cyg X-1 during an outburst in 1975 that lasted about a year, but no x-rays have been detected from it since. However, a star that appears like a K dwarf (V616 Mon) has been detected at its position and is about 1 kiloparsec away. Analysis of the star's 7.75-hour period indicates that the invisible companion's mass is surely greater than 2.7 solar masses and is probably greater than 4 solar masses, well above the apparent neutron-star limit. So again, the case for a black hole is very likely but not proven.

Other candidates for black holes include Nova Muscae 1991, which resembles A0620-00. Its mass is definitely greater than 2.6 solar masses and is probably greater than 5 solar masses, which is black-hole territory. Nova Ophiuchi 1977a and a handful of other objects have similar characteristics.

Perhaps the best current candidate for a black hole is in V404 Cygni, a recurrent x-ray nova. The name denotes the 404th variable star discovered in Cygnus, which is at a measured distance of 3.5 kiloparsecs, and the black hole would be its invisible companion. An intense outburst of x-rays that lasted about a year was discovered with the Ginga spacecraft in 1989. When the x-rays subsided, the companion star was then monitored. The compact star, based on the mass function, contains at least 6 solar masses, and probably 12 solar masses.

Separate teams from the University of Astronomy and from the Harvard–Smithsonian Center for Astrophysics found an unusual black-hole candidate in 2001, using the new 6.5-m MMT Observatory in Arizona. It is at least 6 and probably 8 times as massive as the sun. It is especially interesting because it is above the disk of our galaxy (as we will discuss in the next chapter), which makes our view of it unimpeded. The other black-hole candidates are in our galaxy's disk, which means that we have to look through the murk of interstellar matter to see them. Thus the new system can be studied in especial detail.

About a dozen such binary black-hole candidates are known (Table 31–1). Very-high-resolution radio images show changes on a daily basis in the material the systems eject (Fig. 31–19). Sometimes major changes in x-rays are detected, making the systems "x-ray novae" (Fig. 31–20). These objects eject material at high speed, resembling quasars (Chapter 36) but at a smaller scale. They are therefore called "microquasars."

Another twenty potential x-ray systems are suspected of harboring black holes, but are in the plane of the galaxy and are too faint to study through its bright stars and interstellar matter. The distribution in mass of x-ray binaries with suspected neutron stars and black holes as their compact objects can be plotted (Fig. 31–21).

Matter in the inner part of the accretion disk around a black hole (Fig. 31–22) would orbit very quickly. If a hot spot developed somewhere on the disk, it might beam a cone of radiation into space, similar to the lighthouse model of pulsars. If

*(text continues on page 602)*

**TABLE 31–1** Stellar-Mass Black Holes (The current list of compact stars more massive than 3 solar masses)

| X-ray Novae | Other Name |
|---|---|
| GRO J0422+32 | |
| A0620-00 | X-ray Nova Mon 1975 |
| GRS 1009-45 | |
| GRS 1124-683 | X-ray Nova Mus 1991 |
| 4U1543-47 | |
| GRO J1655-40 | X-ray Nova Sco 1994 |
| H1705-250 | X-ray Nova Oph 1977 |
| SAX J1819.3-225 | V4641 Sgr |
| GS2000+25 | |
| GS2023+334 | V404 Cyg |
| halo object XTE J1118+480 | |
| **Two-State X-ray Sources** | |
| Cyg X-1 | |
| LMC X-3 | |

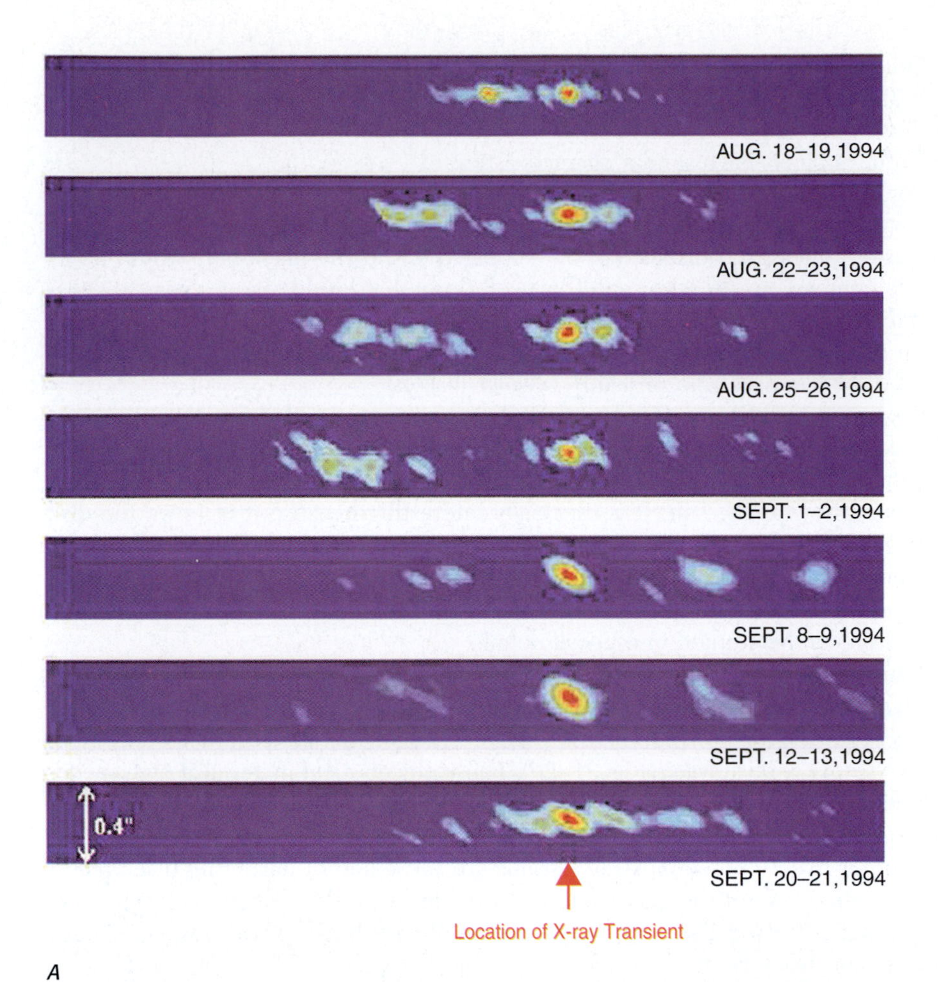


*A*

*B*

**FIGURE 31–19** (*A*) The ejecta from the x-ray binary GRO J1655-40 move and change from day to day. They are apparently given off by a black hole of 7 solar masses. In 2000, an eruption led to this object being pinpointed by the Rossi X-ray Timing Explorer. In the following months, joint observations of its ultraviolet radiation from Hubble and Extreme Ultraviolet Explorer coupled with highest-energy x-ray observations from Rossi and lower-energy x-ray observations from Chandra pinpointed the innermost edge of the object's accretion disk. It turned out to be about 100 km, much larger than the expected 40 km. Perhaps the accretion erupts into a hot bubble of gas, becoming invisible to these spacecraft, before it plunges into the black hole. (*B*) The nearest known black hole to us, accompanying the normal variable star V4641 Sgr. It is only 1600 light-years from Earth. In a 1999 flareup, the intensity of x-rays rose by a factor of more than 1000 in only 7 hours. This radio image, taken with the Very Large Array, shows jets larger than our solar system being ejected at 90% the speed of light. Perhaps there are many objects like this remaining undiscovered in our galaxy.

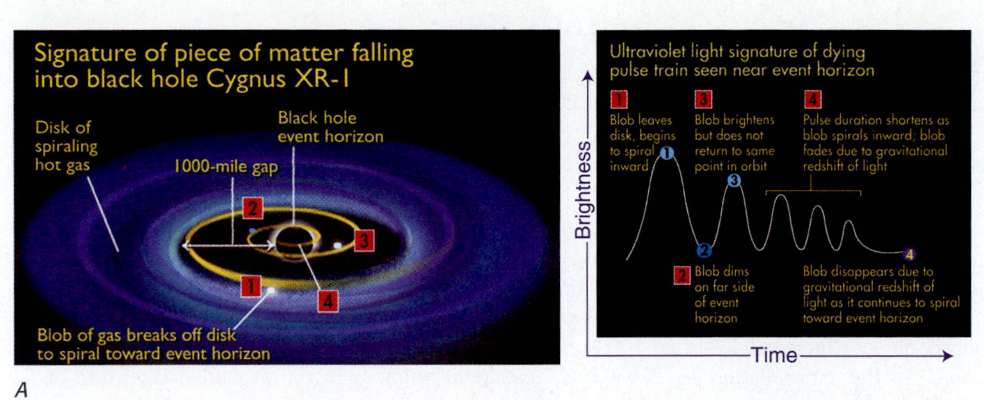


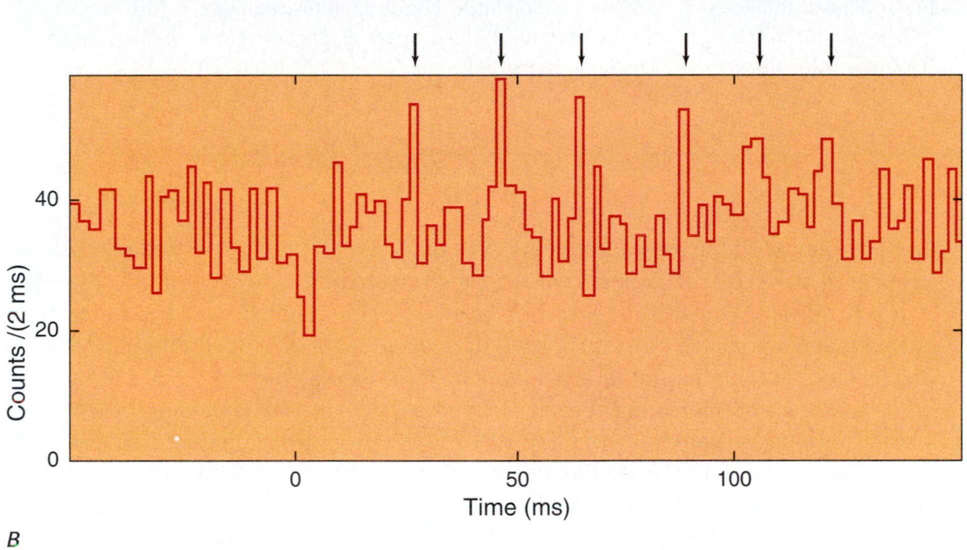


FIGURE 31-20 (*A*) An artist's sketch of the pulse trains for which Joseph Dolan is looking. (*B*) One of Dolan's pulse trains, with the individual pulses marked with arrows. A 200-millisecond interval is shown.

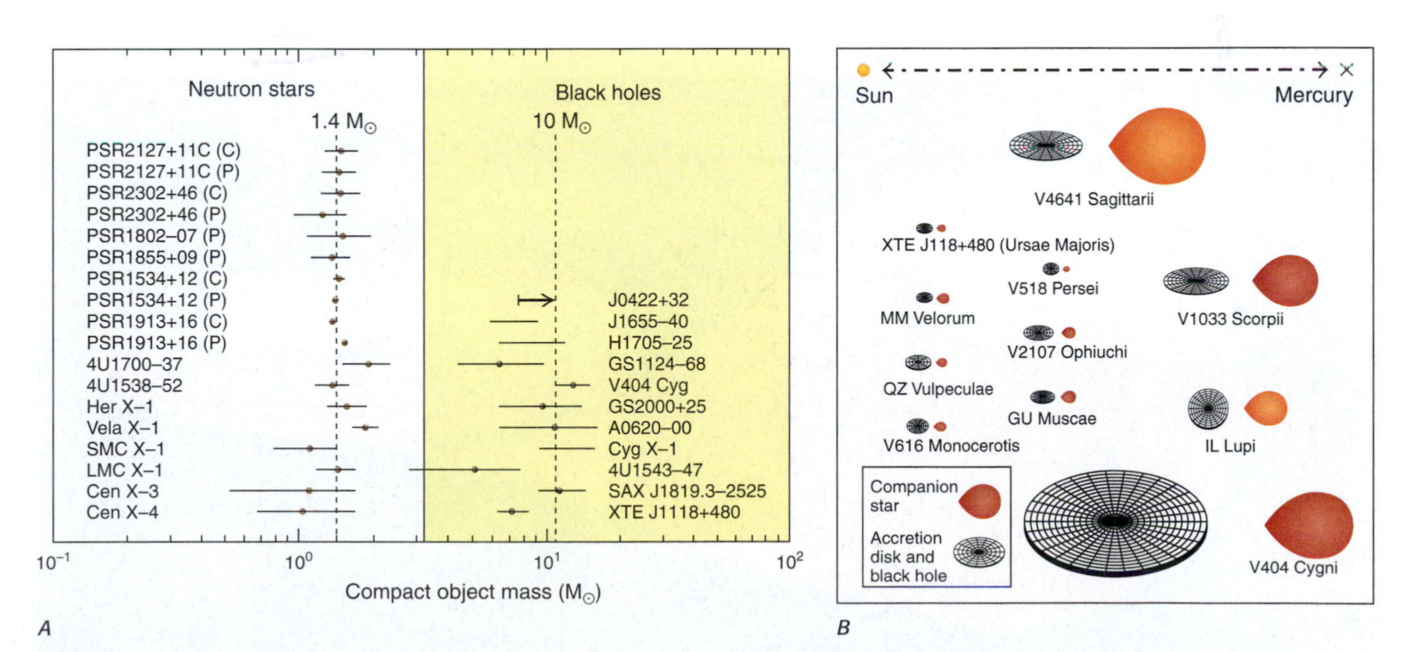


FIGURE 31-21 (*A*) The masses derived for neutron stars and candidate black holes in x-ray binaries. "P" is the primary and "C" is the companion object. (*B*) The sizes of the accretion disks, companions, and their separations for several black-hole candidates.

## WHAT'S IN A NAME?

### BOX 31.3 Black-Hole Candidates

Novae are merely named after the constellation they were in. Variable stars get the preface V and a number, after the variable-star lettering scheme discussed in Box 26.1 is exhausted. The first x-ray sources ever found, back a few decades when one could count the known x-ray sources on your fingers, have simple "X-" numbers plus the name of the constellation (Cygnus) or galaxy (Large Magellanic Cloud). The other names signify Galactic Radio Source for GRS or the satellite used to discover them: GRO for the Compton Gamma-Ray Observatory; A for the Ariel spacecraft; U for the Uhuru x-ray satellite, one of the first to study that part of the spectrum; H for High-Energy Astronomy Observatory-1; SAX for BeppoSaX; GS for the Japanese x-ray satellite Ginga; and XTE for the Rossi X-ray Timing Explorer.

The star name corresponding to Cygnus X-1, HDE 226868, is a catalogue number in the expanded extension of the most widely used Henry Draper catalogue, a 20th-century list of stars and their spectral classifications, hence *H*enry *D*raper—*E*xtended.

the hot spot lasted for several rotations, we could detect a pulse every time the cone swept past the Earth. The period of the pulses would be extremely short, less than a second. Joseph Dolan from NASA's Goddard Space Flight Center has found tentative evidence for such pulses. He used data from the High Speed Photometer that was in Hubble's original set of instruments. He looked at runs of 10,000 ultraviolet brightness measurements per second of Cygnus X-1. He was searching for a sign of a blob dropping off the inside of Cygnus X-1's accretion disk, an effect similar to water flushing down a drain. (Within about 3 times the Schwarzschild radius, no stable orbits exist, so the accretion disk has an inner edge.) As the blob orbits the black hole, we receive pulses as the ultraviolet light it emits brightens and fades. (When the blob is on the far side of the black hole from Earth, most of the light directed toward us winds up going into the black hole and not reaching us. The blob therefore looks fainter at those times.) The pulses should speed up and grow dimmer (because

**FIGURE 31–22** An artist's conception of the disk of swirling gas that would develop around a black hole like Cygnus X-1 (*right*) as its gravity pulled matter off the companion supergiant (*left*). The x-radiation would arise in the disk.

of the gravitational redshift) at a rate that corresponds to the shrinkage of the blob's orbit. It would disappear from our view before reaching the event horizon. He has two candidate sets, one with 6 pulses and the other with 7 pulses, each of which has an average period of 18 milliseconds. He thinks they show blobs breaking off the inner edge of the accretion disk and spiralling down toward the black hole. The period indicates that Cygnus X-1 would have a mass of 35 solar masses, several times higher than previously suspected.

Searches for the short-period pulses have been made with x-ray satellites. Cygnus X-1 and other black-hole candidates do indeed flicker in x-rays. But this method of detecting black holes is no longer as useful for deciding whether something is a black hole or a neutron star, since we now know that accretion disks around neutron stars can also flicker.

Some apparent black holes are ejecting jets of gas that can be followed with radio telescopes (Fig. 31–23). The *R*ossi *X*-ray *T*iming *E*xplorer (RXTE), with its huge

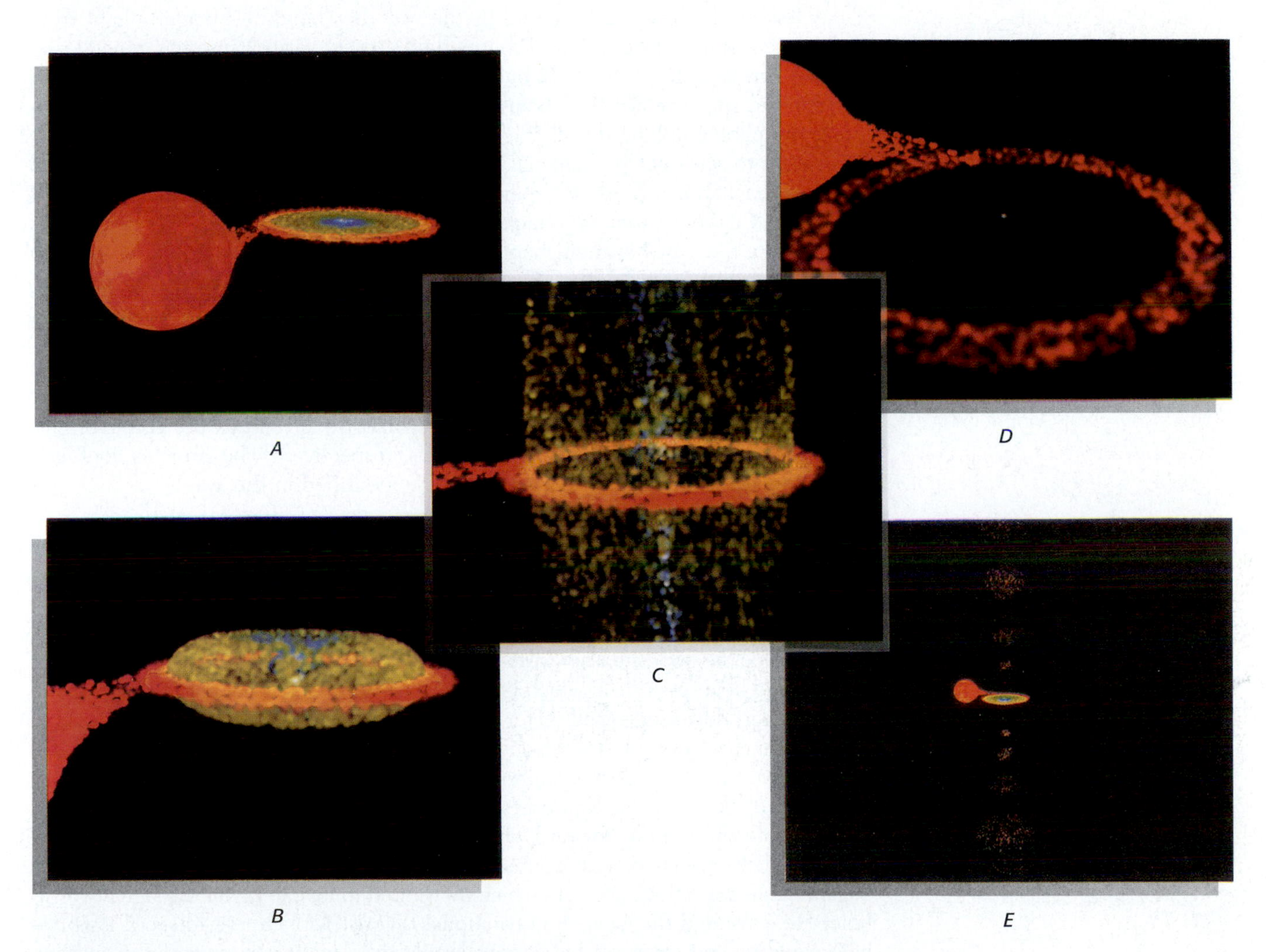


**FIGURE 31–23** A computer set of views showing the occasional disruption of the accretion disk surrounding a black hole. The black hole is orbiting a massive companion star, the red sphere at the left. The gas falling into the black hole is compressed and heated to millions of degrees. Different temperatures in these drawings are shown as different colors. (*A*) The hottest material, shown as blue/white in the center of the disk, is closest to the black hole. The Chandra X-ray Observatory measures radiation from such sources. (*B*) A disruption passes through the accretion disk's gas. (*C*) The gas in the disk accelerates to close to the speed of light and is ejected in jets. (*D*) The ejection of gas empties the disk center, and the black hole (smaller than the dot that marks its position in the center) attracts more gas. (*E*) The ejection repeats every half hour or so, forming the long jets that are seen from a distance.

collecting area and very high time resolution, is a satellite now observing a wide variety of other black-hole candidates in our galaxy. Shuang-Nan Zhang of NASA's Marshall Space Flight Center has exploited the idea that there is an innermost edge to the accretion disk. How far out this last stable orbit is depends on the black hole's spin. It is much closer in for a rotating disk than it is for a non-rotating disk. The closer in the orbit is, the more energy that can be extracted from gravity. Thus the closer the orbit, the higher the temperature. The temperature, in turn, shows up as the brightness of the x-rays emitted. So by studying the temperature of the x-rays with RXTE, Zhang has been able to find out the size of the innermost edge of the accretion disk. He finds that several objects known to have jets show high spins, close to the maximum allowed in Kerr's equations. He also finds that several objects that don't show jets have much lower spins. He speculates that the formation of jets, therefore, is connected with how fast black holes spin. So in the last few years, we have moved from wondering whether stellar black holes exist to being able to find out details of their properties.

A few black holes can be found by the way they bend and magnify light from more distant stars, an extension of the method used to verify Einstein's general theory of relativity in 1919 and 1922 but with a much greater effect. These black holes are alone in space rather than being in binary systems. Several observational programs are photographing the sky night after night and using computers to compare the images to see what is changing. We have already seen the result of the OGLE (*O*ptical *G*ravitational *L*ensing *E*xperiment, mentioned in Section 26.4b) project for measuring Cepheid variables, which they pick up as a side benefit. They are succeeding in a few cases in their main goal, which is to pick up brightenings of stars caused by gravitational lensing. These stellar-mass gravitational lenses are presumably black holes. (In Section 38.2a we will see how gravitational lenses on a larger scale are used to map "dark matter.") The *Ma*ssive *C*ompact *H*alo *O*bject (MACHO) collaboration also discovered some similar events: so far, two events out of the tens of millions of stars they routinely survey. Either the motion of the black hole past the farther star is slow or the black hole is very massive, given that the visible brightening they saw lasted 500 days and 800 days, respectively. The projects continue, and we expect more isolated black holes to be identified in this way.

## 31.7 WORMHOLES AND WHITE HOLES

As we have seen, nothing can escape from within the Schwarzschild radius of a black hole. What if the opposite also exists—a white hole, which some scientists have proposed could be a spout from which matter flows? Some scientists, and some science-fiction writers, have suggested that a black hole might be connected to a white hole, with matter that goes into the black hole coming out of the white hole. Though we don't know if any such exist, we already have a name for it: "wormhole."

What of the idea that you are stretched and crushed when you enter a black hole? If time always flows forward, once you enter the black hole you must be pulled inward, tidally stretched, and eventually crushed. In the 1980s, the astronomer Carl Sagan, wrote a novel, *Contact,* in which he wanted to invoke travel through a wormhole. He consulted the Caltech gravitational theorist Kip Thorne to see if a wormhole could indeed transport his heroine from one place in the universe to another. (Jodie Foster played the role in the movie; the character was based on Jill Tarter, now of the SETI Institute.) Calculations had shown for over two decades that the throat connecting the two parts of a wormhole would close up so quickly, in a tiny fraction of a second, so that nothing could pass through. Now Thorne and a student, following up on Sagan's question, still could not allow wormhole travel with the universe's constituents as we know them, but they invoked an otherwise unknown new kind of matter, which they called "exotic matter," to keep the throat open. Maybe new laws of physics will be discovered one day that permit this exotic matter, they said, though there is a limit on how scientific one can be while making such assertions. In this case, exotic matter could perhaps exist in situations when matter is very compressed, and when we would need a quantum theory to explain gravity. In Chapter 38, we will discuss some current attempts to link quantum theory and gravity.

As of the year 2000, Thorne states, "There is growing evidence, but nothing at all firm, that the laws of quantum field theory in curved spacetime may prevent the existence of the kind of stress-energy required to hold open a macroscopic wormhole." If this is really the case, wormholes can't exist.

## 31.8 MINI BLACK HOLES

We have discussed how black holes can form by the collapse of massive stars. But theoretically a black hole should result if a mass of any amount is sufficiently compressed. No object containing less than 2 or 3 solar masses will contract sufficiently under the force of its own gravity in the course of stellar evolution. But the density of matter was so high at the time of the origin of the universe (see Chapter 37) that smaller masses may have been sufficiently compressed to form what are called mini black holes.

**FIGURE 31–24** Stephen Hawking, who holds the professorial Chair at Cambridge University once held by Isaac Newton. Among Hawking's many theoretical ideas were black-hole radiation and mini black holes.

Stephen Hawking (Fig. 31–24), an English astrophysicist who now holds the professorial Chair at Cambridge University that was once held by Isaac Newton, in 1974 suggested the existence of mini black holes, the size of pinheads (and thus with masses equivalent to those of asteroids). There is no observational evidence for a mini hole, but they are theoretically plausible. Hawking has deduced that small black holes can seem to emit energy in the form of elementary particles (neutrinos and so forth). The mini holes would thus evaporate and disappear. This may seem to be a contradiction to the concept that mass can't escape from a black hole. But when we consider effects of quantum mechanics, the simple picture of a black hole that we have discussed up to this point is not sufficient. Hawking suggests that a black hole so affects space near it that a pair of particles—a nuclear particle and its antiparticle—can form simultaneously. The antiparticle disappears into the black hole, and the remaining particle reaches us. Photons, which are their own antiparticles, appear too.

Emission from a black hole is significant only for the smallest mini black holes, for the amount of radiation increases sharply as we consider less and less massive black holes. Only mini black holes up to the mass of an asteroid—far short of stellar masses—would have had time to disappear since the origin of the universe. Black holes a few times the size of the sun would take $10^{66}$ years to evaporate, which is far, far, far longer than the $10^{10}$-year age of the universe. Hawking's ideas set a lower limit on the size of black holes now in existence, since we think the mini black holes were formed only in the first second after the origin of the universe by the tremendous pressures that existed then.

In 1996, two other theoreticians were able to derive the formula for Hawking radiation by using a new theory, called "superstring theory," popular in the highest level of theoretical physics. (We will say more about superstring theory in Section 38.4.) They used a kind of object from this theory called a "brane." Since a "membrane" is a two-dimensional surface, it is an example of a "2-brane," and one can conceive of "3-branes," and so on into higher dimensions. The theoreticians constructed a mini black hole out of branes. Deriving Hawking's formula in this independent way endorses both Hawking's formula and superstring theory.

Hawking's original work on mini black holes followed on the suggestion by the Israeli physicist Jacob Bekenstein that black holes have a quantity associated with them that can be used to measure "information." "Information" is a measure of how messy a system is. Bekenstein has since shown that any system of a given energy and size has a maximum amount of this "information." This statement corresponds to a maximum rate at which a computer can calculate. It is remarkable that black holes can tell us something about computers! It shows that one never knows when it is worth studying something that seems obscure.

## 31.9 SUPERMASSIVE BLACK HOLES

On the other extreme of mass, we can consider what a black hole would be like if it contained a very large number, that is, thousands or millions, of solar masses. Thus far, we have considered only black holes the mass of a star or smaller. Such black holes form after a stage of high density. But the more mass involved, the lower the density needed for a black hole to form. For a very massive black hole, one containing hundreds of millions or billions of solar masses, the density would be fairly low when the event horizon formed, approaching the density of water. For even higher masses, the density would be lower yet. We have measured such high masses in the centers of various active galaxies and quasars (Fig. 31–25). The high resolution of the Hubble Space Telescope and the Chandra X-ray Observatory, including their imaging but especially their spectrographs, is giving increasing evidence of such extreme concentrations of mass moving in a way that shows that they must be black holes. In recent years, we have been finding supermassive black holes even in more normal galaxies, ones like our own (Fig. 31–26). The center of our galaxy seems to contain a black hole of over 2 million solar masses (Section 32.3). Though we cannot observe radiation from the black hole itself, the gamma rays, x-rays, and infrared radiation we detect would be coming from the gas surrounding the black hole. In the year 2000, scientists reported the discovery of a class of black hole intermediate between stellar size and supermassive, with several hundred times the mass of the sun.

If we were travelling through the universe in a spaceship, we couldn't count on detecting all the black holes by noticing volumes of high density. Far differently from the case of a black hole resulting from a collapsed star, which we discussed earlier, for a high-mass black hole we could pass through its event horizon without even noticing its tidal force. We would never be able to get out, but it might be hours on our watches before we would notice from the gravitational pull that we were being drawn into the center at an accelerating rate.

**FIGURE 31–25** The galaxy NGC 4261, whose core is imaged here with the Hubble Space Telescope, is thought to have at its center a black hole containing millions of solar masses of matter. The disk of dust we see is 800 light-years wide and contains 100,000 solar masses. It fuels a black hole with 1.2 billion times the mass of our sun that is concentrated into a volume about the size of our solar system.

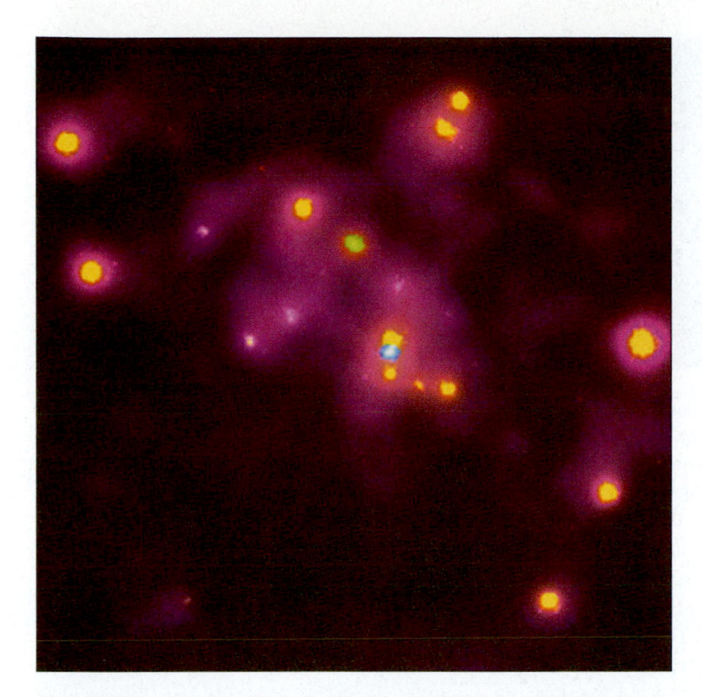

**FIGURE 31–26** A Chandra view of a tiny region at the core of the Andromeda galaxy, M31. The relatively cool object (blue in this color coding, which shows the brightness in different energy bands) is at the very center of the galaxy and is deduced to be a black hole that is millions of times more massive than the sun.

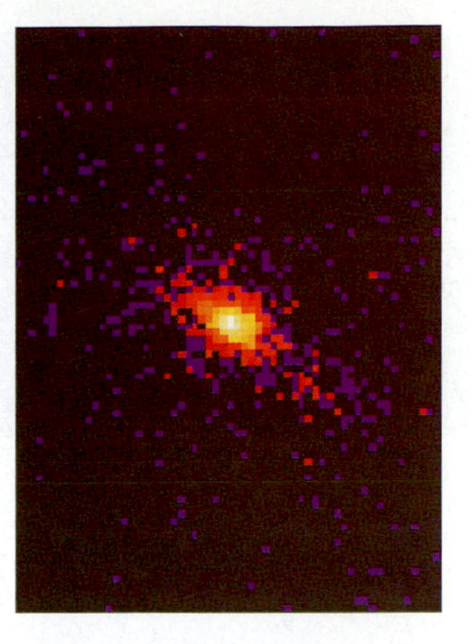

**FIGURE 31–27** An elongated cloud of gas, 3000 light-years across, in the center of the galaxy NGC 4151, imaged with the Chandra X-ray Observatory. A collimated jet of x-rays emitted from the vicinity of the supermassive black hole would account for the shape of the x-ray emission. The colors represent x-ray intensities, with white being the most intense.

The Chandra X-ray Observatory, since its launch in 1999, has found x-rays from dozens of supermassive black holes that are lurking in the centers of galaxies. We will discuss several of them later in this book, when we discuss galaxies and quasars. Let us just, for the moment, show one Chandra black-hole image, of a cloud of hot gas in the center of the galaxy NGC 4151. The gas is heated by x-rays given off from the vicinity of the black hole, which contains millions of times the mass of the sun. These x-rays must form a jet, since we see the gas cloud appears elongated (Fig. 31–27).

We are on the lookout for black holes everywhere (Fig. 31–28). Indeed, the whole universe is in a black hole, if certain ideas about the overall structure of the universe are true. There may be other universes outside our own, a position recently espoused by Sir Martin Rees, the Astronomer Royal (a British position of high honor), a noted theoretician.

**FIGURE 31–28** Directions to a small, deep cove on the shore of the Bay of Fundy. It got its name (probably a century or two ago) because it appears very dark as seen from the sea. Unlike in the stellar case, the cove's darkness results from its dark basaltic cliffs and its deep, narrow shape. Nobody lives there.

## 31.10 LIGO AND BLACK HOLES

Scientists expect black holes to be among the very strongest sources of gravitational radiation. Though gravitational waves have not yet been detected directly, supercomputer calculations show how they should act (Fig. 31–29). Scientists can calculate how they affect the space they are passing through. Maybe the laser observatories being built to look for gravitational waves, such as LIGO, will find them (Chapter 29). Calculations show that LIGO may be able to find gravitational waves from the last seconds of collisions and mergers of members of black-hole binaries or from massive stars collapsing to become Type II supernovae. The periods of the waves would range from a second or so down to a millisecond. To detect the more common black-hole binaries that are still a year away from merging or the formation of black holes from merging galaxies, sensitivity to gravitational waves of lower frequencies will be necessary. This sensitivity will have to wait for the Laser Interferometry Space

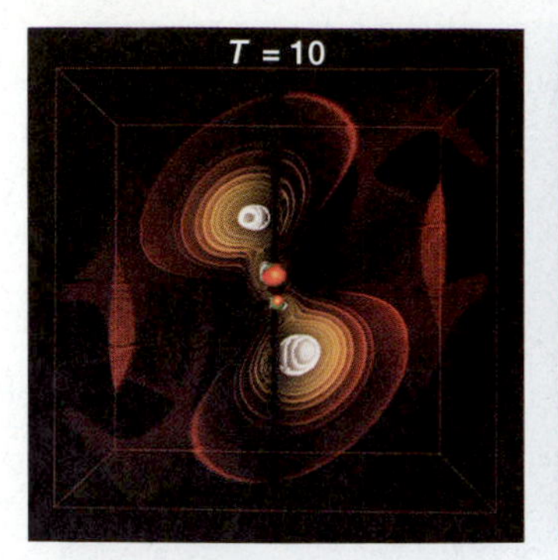

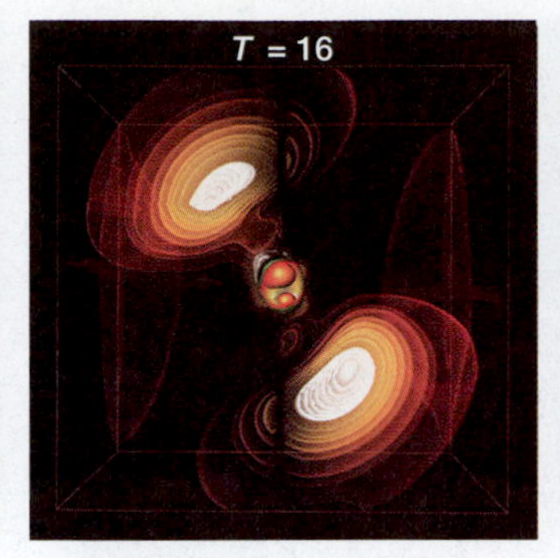


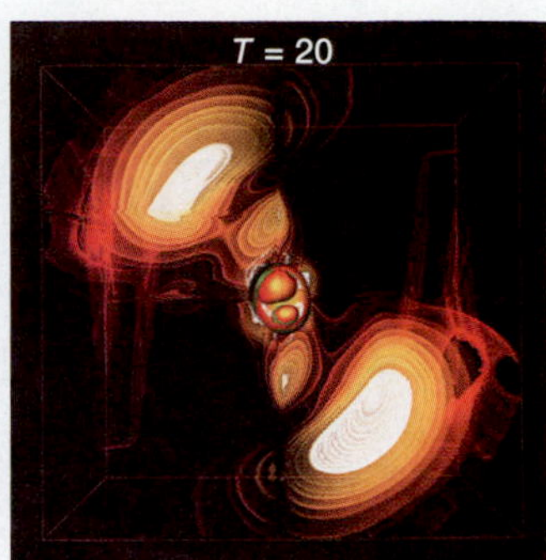


**FIGURE 31–29** Supercomputer calculations of the gravity waves that would result from an oscillating black hole. Two of the three spatial dimensions of the curved space that makes up the universe are shown projected into the kind of flat space that is more familiar to us. The white ring represents the location of the black hole's surface (a sphere in 3D), with its exterior above the ring. Colors show the amplitude of the gravitational wave.

Antenna (LISA) system of three orbiting satellites that has been proposed. LISA would be sensitive to periods that range from minutes to hours (Fig. 31–30). This joint NASA/ESA project might happen about 2010 or 2015. Kip Thorne hopes that after 2020 we might have successor projects that could detect every collision of black-hole pairs or neutron-star pairs in the whole universe. He calculates that we would detect several of these events every day.

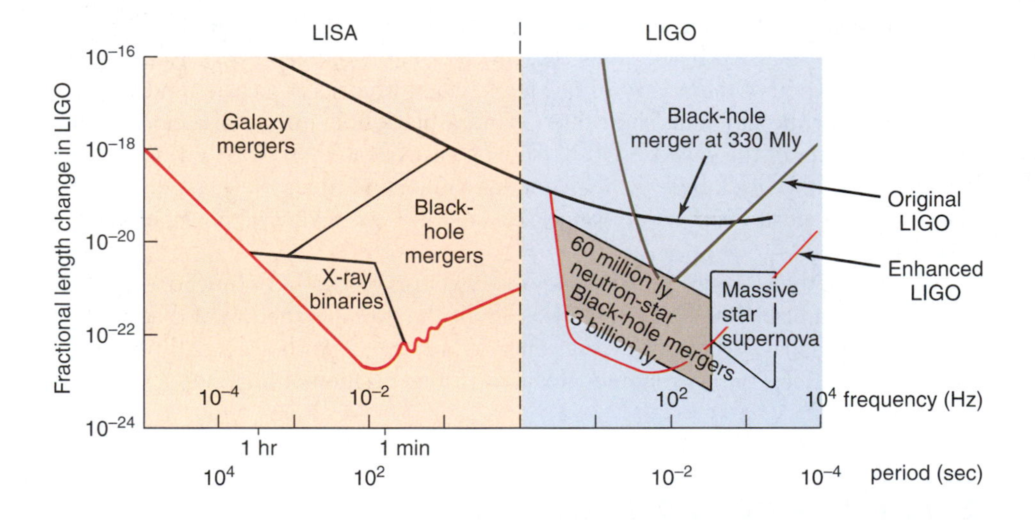


**FIGURE 31–30** The sensitivities of LIGO, just going on line; enhanced LIGO, expected in a few years; and LISA, the space version, hoped for in 2010–2015. Merger events are rare, especially at the limits in distance set by LIGO's current sensitivity. Many scientists think that we will have to wait for the later versions before we detect any gravitational waves.

## CORRECTING MISCONCEPTIONS

| ✖ *Incorrect* | ✔ *Correct* |
| --- | --- |
| Black holes can't be seen or detected. | We can see x-rays and other radiation emitted just outside the black hole and also detect gravitational effects. |
| Black holes suck in matter as do vacuum cleaners. | Black holes of millions of solar masses don't have strong gravity obvious outside their event horizons. |
| Einstein's general theory of relativity set the speed of light as a limit. | Einstein's general theory is a theory of gravity; the speed of light as a limit was set by the special theory 10 years earlier. |
| Black holes and black bodies are the same. | A black body is any radiating body whose emission follows Planck's law. Black holes are the result of strong gravity, as described here. |

## SUMMARY AND OUTLINE

General theory of relativity (Section 31.1)
- A theory of gravity; many verifications

Gravitational collapse to a black hole (Section 31.2)
- Occurs if more than 2 or 3 solar masses remain
- Critical radii of a black hole

Photon sphere: exit cones form (Section 31.3).

Event horizon: exit cones close (Section 31.4).
- Schwarzschild radius (gravitational radius) defines the limit of the black hole, the event horizon.
- Singularity

Rotating black holes (Section 31.5)
- Can get energy out of the ergosphere

Detecting a black hole (Section 31.6)
- Flickering x-radiation expected from accretion disks
- Detection in a spectroscopic binary, as an invisible high-mass companion
- Cygnus X-1, LMC X-3, A0620-00, and V404 Cygni: black holes?

White holes and wormholes (Section 31.7)

Non-stellar black holes (Section 31.8)
- Mini black holes could have formed in the big bang.
- Very massive black holes do not have high densities.

## KEY WORDS

| | | | |
|---|---|---|---|
| black hole | scientific method | gravitational radius | stationary limit |
| special theory of relativity | exit cone | event horizon | ergosphere |
| general theory of relativity | photon sphere | escape velocity | naked singularity |
| perihelion | Schwarzschild radius | singularity | accretion disk |

## QUESTIONS

1. Why doesn't electron or neutron degeneracy prevent a star from becoming a black hole?
2. Why is a black hole blacker than a black piece of paper?
3. Explain the bending of light as a property of a warping of space.
4. (a) How does the escape velocity of the moon compare with the escape velocity of the Earth? Would a larger rocket engine be necessary to escape from the gravity of the Earth or from the gravity of the moon? (b) How will the velocity of escape from the surface of the sun change when the sun becomes a red giant? A white dwarf? Explain.
5. [†] (a) What is the Schwarzschild radius for a 10-solar-mass star? (b) What is your Schwarzschild radius?
6. [†] What are radii of the photon sphere and the event horizon of a non-rotating black hole of 18 solar masses?
7. [†] What is the size of the stationary limit and of the event horizon of an 18-solar-mass star that is rotating at the maximum possible rate?
8. What is the relation in size of the photon sphere and the event horizon? If you were an astronaut in space, could you escape from within the photon sphere of a rotating black hole? From within its ergosphere? From within its event horizon?
9. (a) How could the mass of a black hole that results from a collapsed star increase? (b) How could mini black holes, if they exist, lose mass?
10. Would we always notice when we reached a black hole by its high density? Explain.
11. Could we detect a black hole that was not part of a binary system?
12. Under what circumstances does the presence of an x-ray source associated with a spectroscopic binary suggest to astronomers the presence of a black hole? In particular, discuss one deduction from optical observations and one special property of the x-rays.

[†] This question requires a numerical solution.

## USING TECHNOLOGY

### W World Wide Web

1. On the Space Telescope Science Institute Web site, http://oposite.stsci.edu/pubinfo/Pictures.html, find out about the galaxies imaged by Hubble as hosts for giant black holes.
2. View the black-hole simulation movies from the National Supercomputing Center at http://www.sdsc.edu/MetaScience/bin/meta_browse.cgi?123+120.
3. See the video that corresponds to the black-hole simulations of Figure 31–16 at http://www.pha.jhu.edu/~rauch/ViewsBHs/.

### REDSHIFT

1. Look at Story of the Universe: Lives of the stars; Science of Astronomy: Lifestyles of the stars; Science of Astronomy: Birth and death of stars.
2. Work with Astronomy Lab: Stars: Black Hole

*The Origin of the Milky Way* by Tintoretto, circa 1578. (*The National Gallery, London*)

# PART 6
# The Milky Way Galaxy

On the clearest nights, when we are far from city lights, we can see a hazy band of light stretched across the sky. This band is the **Milky Way**—that part visible with the naked eye from Earth of the aggregation of dust, gas, and stars that makes up the galaxy in which the sun is located.

Don't be confused by the terminology: the Milky Way itself is the band of light that we can see from the Earth, and the Milky Way Galaxy (sometimes called simply "the Galaxy") is the whole galaxy in which we live. Like other galaxies, our Milky Way Galaxy is composed of perhaps a trillion stars plus many different types of gas, dust, planets, etc. The Milky Way is that part of the Milky Way Galaxy that we can see with the naked eye in our nighttime sky.

The Milky Way appears very irregular in form when we see it stretched across the sky—there are spurs of luminous material that stick out in one direction or another, and there are dark lanes or patches in which nothing can be seen. This patchiness is simply due to the splotchy distribution of dust, stars, and gas.

Here on Earth, we are inside our galaxy together with all of the matter we see as the Milky Way. Because of our position, we see a lot of matter when we look in the plane of our galaxy. On the other hand, when we look "upwards" or "downwards" out of this plane, our view is not obscured by matter, and we can see past the confines of our galaxy. We are able to see distant galaxies only by looking in parts of our sky that are away from the Milky Way, that is, by looking out of the plane of the galaxy.

The gas in our galaxy is more-or-less transparent to visible light, but the small solid particles that we call "dust" are opaque. So the distance we can see through our galaxy depends mainly on the amount of dust that is present. This is not surprising: we can see great distances through our gaseous air on Earth, but if a small amount of particulate material is introduced in the form of smoke or dust thrown up from a road, we find that we can no longer see very far. Similarly, the dust between the stars in our galaxy dims the starlight by absorbing it or by scattering it in different directions.

The abundance of dust in the plane of the Milky Way Galaxy actually prevents us from seeing very far toward its center—with visible light, we can easily observe only $^1/_{10}$ of the way toward the Galactic Center itself. The dark lanes across the Milky Way are just areas of dust, obscuring any emitting gas or stars. The net effect is that we can see just about the same distance in any direction we look in the plane of the Milky Way. These direct optical observations fooled scientists at the turn of the century into thinking that the Earth was near the center of the universe.

We shall see in the next chapter how the American astronomer Harlow Shapley in the 1920s realized that the Milky Way is part of an isolated galaxy and that the sun is not in its center. This fundamental idea took humanity one step further away from thinking that we were at the center of the universe. Copernicus in 1543, by effectively removing the Earth from the center of the solar system, had already taken the first step toward removing the sun from the center of the universe.

Some of the dust and gas in our galaxy takes a pretty shape and may glow, reflect light, or be visible in silhouette. These shapes are the **nebulae.** Another class of objects visible in our sky was once known as "spiral nebulae," since they looked like glowing gas with "arms" spiralling away from their centers. (We still speak of the Great Nebula in Andromeda, Figure 34–2.) In the first decades of the 20th century, astronomers debated whether these objects were part of our own galaxy or were independent "island universes." We shall see, in Chapter 34 and in the introduction to Part 7 of this book, how we learned that the island-universe idea was correct. Our own Milky Way turned out to be part of a galaxy equal in stature to these other objects, which are now known to be spiral galaxies in their own right. Indeed, our Milky Way Galaxy is among the largest, brightest, and most massive galaxies that exist.

In recent years astronomers have been able to use wavelengths other than optical ones to study the Milky Way Galaxy. In the 1950s and 1960s especially, radio astronomy gave us a new picture of our galaxy. In the 1980s, we also benefited from infrared observations. Infrared and radio radiation can pass through the galaxy's dust and allow us to see our galactic center and beyond. In the 1990s, we had the launches of the Infrared Space Observatory and the NICMOS (*N*ear *I*nfrared *C*amera/*M*ulti-*O*bject *S*pectrometer) instrument on the Hubble Space Telescope giving us new infrared observational possibilities. We await NASA's major *S*pace *I*nfrared *T*elescope *F*acility (SIRTF) in 2002 and the instrumented airplane SOFIA (*S*tratospheric *O*bservatory *f*or *I*nfrared *A*stronomy) shortly thereafter, which will carry a telescope almost four times larger in diameter than SIRTF's.

In this part of the book, we shall first discuss the types of objects that we find in the Milky Way Galaxy. Chapter 32 is devoted to the general structure of the galaxy and its major parts. In Chapter 33, we describe the matter between the stars and what studying this matter has told us about how stars form.

A nebula in our Milky Way, NGC 3603, that shows various stages of the life cycle of stars. A blue supergiant star lights up a ring of gas around it. Stellar winds from a cluster of stars at the center hollow out space. Dark spots at upper right contain newly forming stars. This view is with the Hubble Space Telescope.

# Chapter 32

# The Structure of the Milky Way Galaxy

We have now described the stars, which are important constituents of any galaxy, and how they live and die. And we have seen that many or most stars exist in pairs and clusters. In this chapter, we describe nebulae: gas and dust that accompany the stars and that we named from our observations in the optical part of the spectrum. We also discuss the overall structure of the Milky Way Galaxy and how, from our location inside it, we detect this structure.

The Galaxy has about 200 billion times the mass of the sun in the form of visible matter. Of that visible matter (by which we mean matter detectable through electromagnetic radiation of any kind), about 96 per cent is in the stars and about 4 per cent is interstellar gas. The dust that shows up so well makes up only about 1 per cent of the mass of the gas, which makes it only 0.0004 the mass of the visible matter.

**AIMS:** To describe the basic components of the Milky Way Galaxy, and to understand why spiral structure forms

Nebula is Latin for "fog" or "mist." The plural is usually nebulae rather than nebulas.

## 32.1 NEBULAE

A **nebula** is a cloud of interstellar gas and dust that we see in visible light. (The word "nebula" comes from the Latin for cloud or mist.) When we see the gas actually glowing in the visible part of the spectrum, we call it an **emission nebula.** Sometimes we see a cloud of dust that obscures our vision in some direction in the sky. When we see the dust appear as a dark silhouette against other glowing material (Fig. 32–1), we call the object a **dark nebula** (or, sometimes, an **absorption nebula**).

The clouds of dust in the Milky Way or surrounding some of the stars in the open cluster known as the Pleiades (Fig. 32–2) are examples of **reflection nebulae**—they merely reflect the starlight toward us without emitting visible radiation of their own. (As we saw in Section 26.5, open clusters, also known as galactic clusters, are shapeless groupings of stars and are often younger than globular clusters.) Reflection nebulae usually look bluish because they are reflecting the bluish light from young, hot stars and because dust reflects blue light more efficiently than it does longer wavelengths. Whereas an emission nebula has its own spectrum, as does a neon sign on Earth, a reflection nebula shows the spectrum of the nearby star or stars whose light is being reflected.

In the heart of the Milky Way, we see the great **star clouds** of the Galactic Center, the bright regions where the stars appear too close together to tell them apart without a powerful telescope. The direction of the center of the Milky Way is readily apparent in the sky (Fig. 32–3).

**FIGURE 32–1** A dark nebula in Orion, NGC 1999, from the Hubble Heritage Program. Surrounding it, dust is reflecting starlight toward us, making a reflection nebula.

**FIGURE 32–2** The Pleiades, M45, an open cluster in the constellation Taurus, the Bull. Reflection nebulae are visible around the brightest stars. The spikes are artifacts caused in the telescope.

**FIGURE 32–3** A visible-light mosaic of the Milky Way, which wraps 360° around the sky.

**FIGURE 32–4** The central region of the Orion Nebula, imaged in the near infrared with the Very Large Telescope. The Trapezium stars show.

**FIGURE 32–5** The Horsehead Nebula, IC 434 in Orion, is dark dust superimposed on glowing gas. The bright star above the nebulosity at top is $\zeta$ Orionis, the leftmost star in Orion's belt.

The Great Nebula in Orion (see Section 33.9a) is an emission nebula. In the winter sky, we can readily observe it through even a small telescope, but only with long photographic exposures or large telescopes can we study its structure in detail. In the center of the nebula are four closely grouped bright stars called the Trapezium, which provide the energy to make the nebula glow (Fig. 32–4). In this region of the Orion Nebula we think stars are being born this very minute. Calculations released in 2000 show that in about 100 million years, the Orion Nebula will have evolved so much that it will look about like the Pleiades.

The Horsehead Nebula (Fig. 32–5) is another example of an object that is both an emission and an absorption nebula simultaneously. It also looks red because the hydrogen-alpha line is so strong. A bit of absorbing dust intrudes onto emitting gas, outlining the shape of a horse's head. We can see in the pictures that the horsehead is a continuation of a dark area in which very few stars are visible. Infrared observations show the glowing dust better (Fig. 32–6).

A

B

**FIGURE 32–6** (*A*) The Horsehead Nebula shows much less well in the infrared in the 2MASS survey, which shows glowing dust, than in the visible. (*B*) The optical image is from the Canada–France–Hawaii Telescope.

A

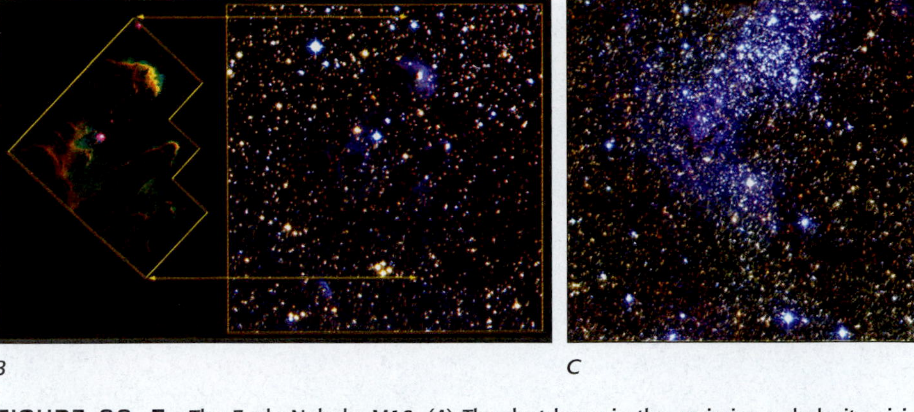

B C

FIGURE 32–7 The Eagle Nebula, M16. (*A*) The dust lanes in the emission nebulosity, visible in the visible. (*B*) Hubble Space Telescope observations through filters emphasizing the "elephant trunks" of absorption, paired with ground-based infrared observations. Stars are forming in the cocoon of protection formed by the dust that we see in silhouette as the "elephant trunks." (*C*) An infrared view of the Eagle Nebula from 2MASS. Note (*center*) the dark fingers that were the subject of the Hubble Space Telescope image.

We can show only a few of the beautiful nebulae here (Figs. 32–7 and 32–8). We have already discussed some of the most beautiful nebulae in the sky, composed of gas thrown off in the late stages of stellar evolution. They include planetary nebulae and supernova remnants.

FIGURE 32–8 An image of the Chameleon I region made with the Antu unit telescope of the Very Large Telescope. The three colors used were visual (yellow), red, and infrared. The infrared color penetrates the dust especially well.

## 32.2 THE SUBDIVISION OF OUR GALAXY

It was not until the 1920s that the American astronomer Harlow Shapley realized we were not in the center of the galaxy. He was studying the distribution of globular clusters, the spherical group of stars we discussed in Section 26.5. Shapley noticed that they were largely in the same general area of the sky as seen from Earth. They mostly appear above or below the galactic plane and thus are not obscured by the dust. When he plotted their distances and directions (using the methods described in Chapter 25), he saw that they formed a spherical halo around a point thousands of light-years away from us (Fig. 32–9). Shapley's touch of genius was to realize that this point must be the center of the galaxy.

We classify four components of our galaxy: (1) the nucleus, (2) the nuclear bulge, (3) the disk, and (4) the halo that contains the globular clusters.

**1.** *The nucleus:* About 8500 parsecs (25,000 light-years) from us, we find the center of our galaxy. It is hidden from our view by dust, but infrared and radio waves penetrate the dust. As we will discuss further later in this chapter, astronomers have concluded that a giant black hole, with millions of times as much mass as the sun, resides at the galactic center. The region within some tens of parsecs (perhaps 100 light-years) of the galactic center is considered the **nucleus.**

Within the nucleus, the central black hole is surrounded by a star cluster and—1.5 parsecs (5 light-years) out—a ring of molecular hydrogen and other gas.

**2.** *The nuclear bulge:* Our galaxy has the general shape of a pancake with a bulge at its center. This **nuclear bulge** is about 3000 parsecs (10,000 light-years) in radius, with the galactic nucleus at its midst. The nuclear bulge has the shape of a flattened sphere and does not show spiral structure. From Earth, we see it as the broadening of the Milky Way in the direction of the constellation Sagittarius, the direction in which the Galactic Center lies. The nuclear bulge contains mainly densely packed old stars as well as interstellar dust and gas.

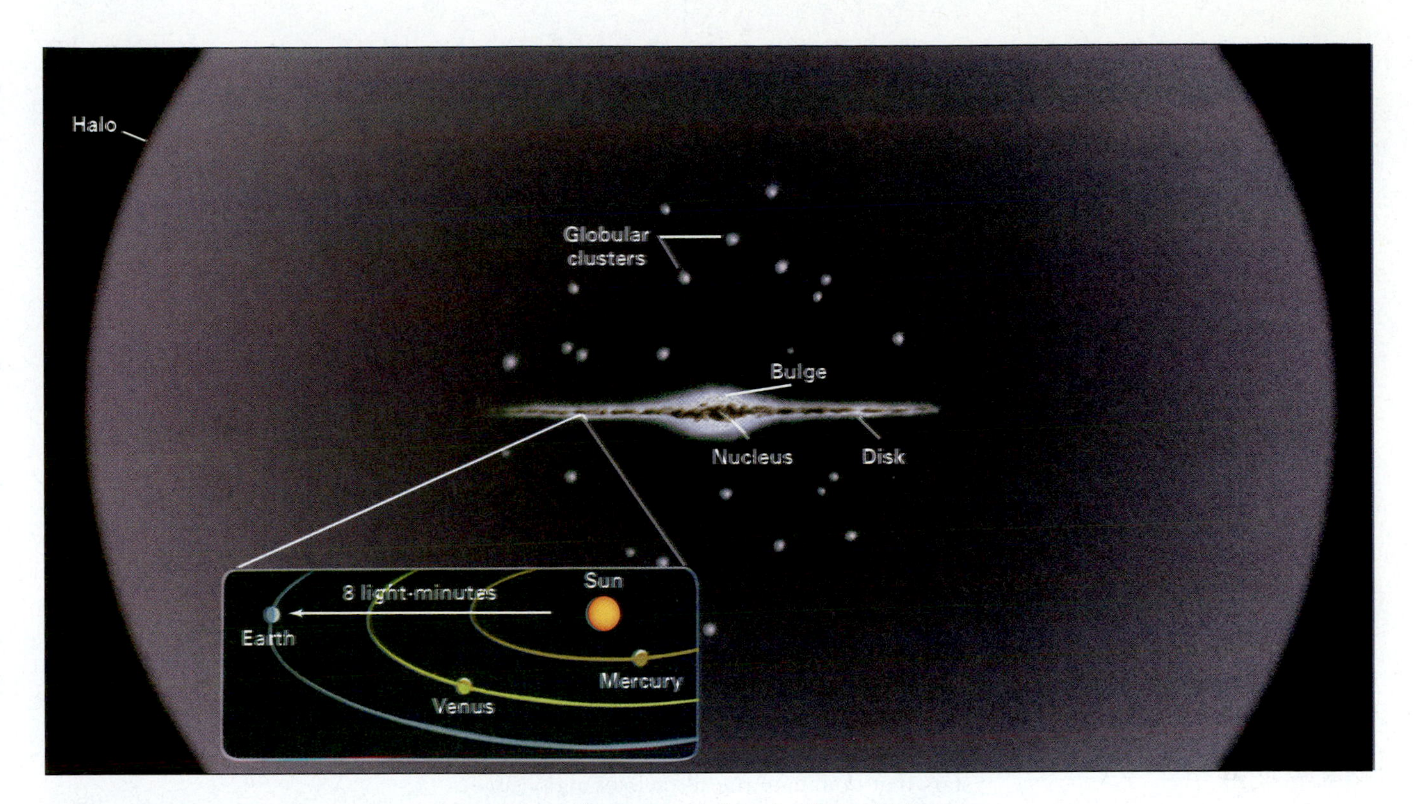


**FIGURE 32–9** The drawing shows the **nuclear bulge** surrounded by the **disk,** which contains the spiral arms. The globular clusters are part of the **halo,** which extends above and below the disk. From the fact that almost all the globular clusters appear in less than half of our sky, Shapley deduced that the Galactic Center is in the direction indicated.

Observations in the infrared have shown that the bulge is elongated in a direction toward the sun, and is actually a bar twice as long as it is wide. The conclusion means that our galaxy is a barred spiral rather than an ordinary spiral, a distinction we shall discuss in Chapter 34.

**3.** *The disk:* The part of the pancake outside the bulge is called the galactic **disk.** The stars in the disk extend out 15,000 parsecs (15 kiloparsecs or 50,000 light-years) or so from the center of the galaxy. The gas in the disk extends out about twice as far.

The sun is located about halfway out, about 8500 parsecs (8.5 kpc), about 25,000 light-years, from the nucleus. (The actual distance is still debated, with some astronomers holding out for a value as small as 7 kpc.) The disk contains all the young stars (Population I, as we discussed in Section 26.5) as well as interstellar gas and dust. It is slightly warped at its ends, perhaps by interaction with our satellite galaxies, the Magellanic Clouds. So, in the words of Leo Blitz of the University of California at Berkeley, Michel Fich of the University of Waterloo, Ontario, and Anthony Stark of Bell Labs, our galaxy looks a bit like a "fedora hat" with a turned-down brim.

It is very difficult for us to tell how the material is arranged in our galaxy's disk, just as it would be difficult to tell how the streets of a city were laid out if we could only stand still on one street corner without moving. Still, other galaxies have similar properties to our own, and their disks are filled with great **spiral arms,** regions of dust, gas, and stars in the shape of a pinwheel. So we assume the disk of our galaxy has spiral arms too, though in Section 33.4 we will see that the evidence that the arms that are detected are actually spiral is ambiguous, at least at distances closer to the center than our sun is.

The disk of the galaxy is, classically, very thin, only about 2 per cent of its width, like a CD. The height astronomers assign to it is 325 parsecs (1000 light-years). (Technically, it has a "scale height" of 1000 light-years, which is to say that for every 1000 light-years up from the plane of the galaxy, the number of stars declines by a factor of about 3.)

The disk looks different when viewed in different parts of the spectrum (Fig. 32–10). Infrared and radio waves penetrate the dust that blocks our view in visible light, while gamma rays and x-rays show the hot objects best.

In recent years, we have learned a lot more about our galaxy, not least by using automatic devices to count the numbers of stars of different brightnesses in different regions of the sky. The disk of stars extends about 350 pc (1000 light-years) above and below the plane of the galaxy, about four times more than the gas. It now seems that in addition to this classical "thin disk" of stars and gas, the galaxy also has a "thick disk" component. The number of stars in this "thick disk" decreases above and below the plane of the galaxy with a height of 1300 parsecs (4000 light-years), that is, four times the height of the "thin disk." The sun and 96 per cent of the stars near us are in the thin disk. Sirius, Vega, and Betelgeuse are examples. The rest of the stars near us, about 4 per cent, we consider to be part of the thick disk, as we can tell from how they are moving. Arcturus is an example.

SIRTF's observations are expected to penetrate the dust and show post-main-sequence giant stars. Studying them should give us a more accurate view of the total mass in the stars in our galaxy than the younger, rarer, individually more massive stars that dominate the ultraviolet and visible.

**4.** *The halo:* Older stars (including the globular clusters) and a very small amount of interstellar matter form a galactic **halo** around the disk. This halo is at least as large across as the disk, perhaps as much as 40 kpc (130,000 light-years) in radius, the distance from the center of the most distant globular cluster known. On its inside, the halo gradually merges with the "thick disk." On its outside, the halo extends far above and below the plane of our galaxy in the shape of a flattened spheroid.

Spectra from the *I*nternational *U*ltraviolet *E*xplorer (IUE) spacecraft showed that a modest amount of the gas in the halo is hot, 100,000 K. IUE discovered the spectra of this hot gas when pointing at more distant objects. The gas in the halo contains only about 2 per cent of the mass of the gas in the disk.

The halo contains relatively old stars of Population II (Section 26.5). These stars are rare near us—only one nearby star in a thousand. The older the stars, the lower the abundance of metals (elements heavier than helium, in astronomer talk). Thus halo stars have lower metallicities (percentages of metals) than stars in the thick disk which, in turn, have lower metallicities than stars in the thin disk. We shall see later on why we think that stars retain their original metallicity, and that later generations of stars form out of gas that has had heavier elements added to it.

How did the halo arise? The traditional view was that it all formed at one time, but more and more evidence is accumulating that the halo formed out of the debris of many colliding small galaxies.

**5.** *Dark matter:* We shall see in the next chapter (Section 33.6) how studies of the rotation of material in the outer parts of galaxies tell us how much mass is present. These studies have told us of the existence of a lot of mass we had overlooked before because we couldn't see it. This mass extends out 60 or 100 kpc (200,000 or 300,000 light-years). Believe it or not, the outer parts of our galaxy contain 5 or 10 times as much mass as the nucleus, disk, and halo together. And it makes our galaxy about 5 times larger across than we had thought. We now usually say that this dark matter is part of the halo.

The existence of this extra material had been suspected from older gravitational studies, which had led to the "missing-mass problem" (Section 38.2a). But only re-

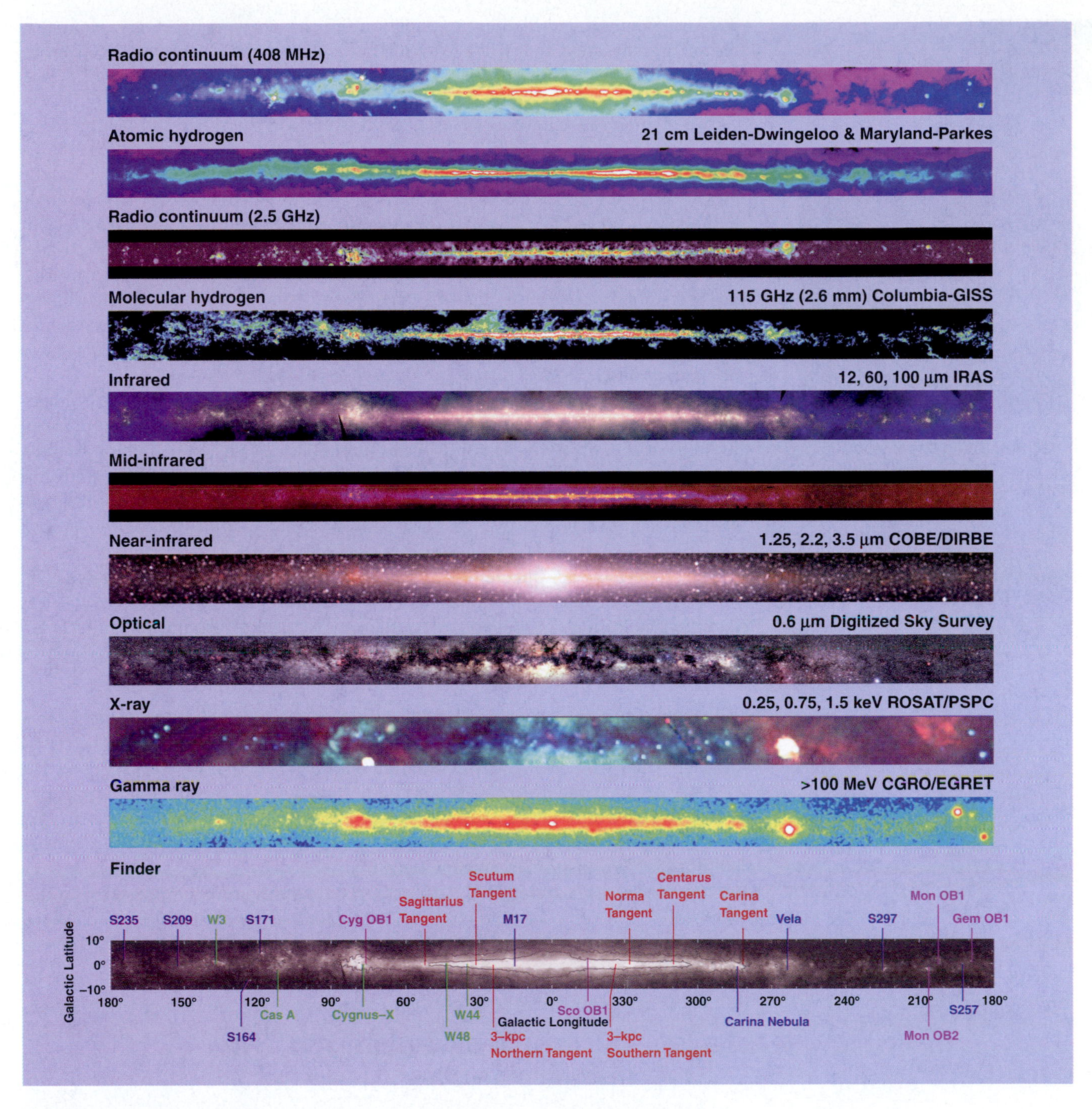


**FIGURE 32–10** Views of the Milky Way at a wide variety of wavelengths, from longer (*top*) to shorter (*bottom*). In all cases, we see the concentration toward the disk of the galaxy. In some wavelength bands, such as the long infrared wavelengths, the radiation comes all the way from the galactic center.

cently have we accepted the actual presence of so much non-luminous matter (that is, matter that isn't shining) in our own galaxy. If this non-luminous material were of an ordinary type, we would have seen it directly—if not in visible light, then in radio waves, x-rays, infrared, etc. But we don't see it at all! We detect only its gravitational properties.

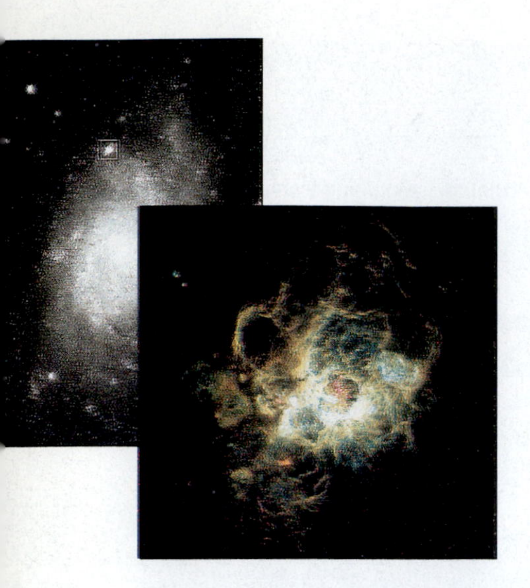

**FIGURE 32–11** A nebula in the galaxy M33, imaged with the Hubble Space Telescope. M33 (shown in the background in a wider view) is over 2 million light-years from us, over 20 times farther away than anything in our own Milky Way Galaxy.

The Sloan Digital Sky Survey (Section 5.5) has located two huge, elongated clumps of stars near the Milky Way in the galactic halo. They have been interpreted as streamers, stretched-out remnants of objects captured by the Milky Way Galaxy billions of years ago. So our galaxy may have been a cannibal, building itself up by merging with other galaxies or by pulling in their material. As the Sloan survey covers more of the sky, we will learn how these apparent streamers are distributed and whether there are indeed dozens or hundreds of them. We have long known of the Magellanic Stream, a narrow band of gas that covers a wide angle of the sky. It comes from the Magellanic Clouds, two companion galaxies of our own that are no farther away from us than the other side of our galactic disk. A third nearby companion to our galaxy, the Sagittarius Dwarf Galaxy, wasn't discovered until 1994, even though it is the nearest galaxy. It is only 24 kpc (80,000 light-years) away, but is beyond the galactic center (which is in the direction of the constellation Sagittarius) and only slightly above it, so our angle of view is blocked by dust.

What is the dark matter in the halo made of? We just don't know. A tremendous number of very faint stars seems unlikely, extrapolating from the numbers of the faintest stars that we can study. A large number of small black holes—the primordial kind rather than those formed from dying stars—has been suggested. Another possibility is a huge number of neutrinos, the subatomic particles of the type we discussed in Section 27.6 on the solar neutrino problem, if neutrinos have enough mass. (The current experimental status is that neutrinos have some mass, but that we can't determine how much.) From the neutrinos received from Supernova 1987A, scientists have been able to deduce very low limits for neutrino mass, making it less likely that neutrinos account for the unseen mass. Lots of planet-sized objects remains a possibility. Many scientists favor as-yet-undiscovered kinds of nuclear particles, as we will discuss in Section 38.2.

If the preceding paragraphs—stating that perhaps 95 per cent of our galaxy's mass is in some unknown form—seem unsatisfactory to you, you may feel better by knowing that astronomers find the situation unsatisfactory too. We can't rely on what the poet W. H. Auden described as "our faith in the naïve observation of our senses." But all we astronomers can do is go out and carry out our research, and try to find out more. We just don't know the answers . . . yet. We shall say more about the problem of this "dark matter" in Section 38.2.

Even if we haven't detected dark matter, our senses are being extended in distance. The Hubble Space Telescope works so fabulously well that we can even see nebulae in distant galaxies as well as ground-based telescopes show us nebulae in our own (Fig. 32–11).

## 32.3 THE CENTER OF OUR GALAXY AND INFRARED STUDIES

We cannot see the center of our galaxy in the visible part of the spectrum because our view is blocked by interstellar dust. We did not even know that the center of our galaxy lies in the direction of the constellation Sagittarius as seen from the Earth until Shapley deduced the fact from his study of the distribution of globular clusters. In recent years, observations in other parts of the electromagnetic spectrum have become increasingly important for the study of the Milky Way Galaxy.

Astronomers working in the infrared usually use the unit of microns for wavelength. One micron (1 micrometer in SI units, 1 $\mu$m) is 1 millionth of a meter (1/1,000,000 m), and is 10,000 Å.

### 32.3a Infrared and Radio Observations

For many years the sky has been intensively studied at optical wavelengths up to about 8500 Å (0.85 micron) and to a lesser extent up to 1.1 microns, which is in the near infrared. The lack of film sensitivity to the infrared, whose photons contain relatively low energy, is a major reason for this limitation. Only recently have especially sensitive electronic imaging devices been available in the infrared (Fig. 32–12). Atmospheric limitations, caused by the presence of only a few windows of transparency, have been another major factor. Since one can observe better in the infrared from locations where there is little water vapor overhead, the placing of telescopes at high-altitude sites like Mauna Kea in Hawaii has improved the situation considerably for ground-based infrared astronomy.

**FIGURE 32–12** A cluster of stars near the center of our galaxy, imaged with the infrared camera aboard the Hubble Space Telescope.

Another difficulty is the fact that the Earth's atmosphere radiates conspicuously in the infrared; the radiation coming into a telescope includes infrared radiation from the atmosphere, from the source being observed, and from the telescope itself. To limit the telescopic contribution, equipment is usually bathed in liquid nitrogen, or even in liquid helium, which is colder and more expensive.

A 1969 sky survey in the 2.2-micron window revealed 20,000 infrared sources. IRAS, the NASA/Netherlands/UK *I*nfrared *A*stronomical *S*atellite, with a 0.6-m telescope, mapped the sky at much longer wavelengths in 1983. IRAS sent back data for 10 months, until its liquid helium was exhausted. It discovered 245,389 sources.

Two-thirds of the infrared sources—158,000 objects—are stars within our galaxy. Another 65,000 objects are interstellar objects of the kinds we discuss in the following chapter. The rest, 9 per cent (22,000 objects), are apparently galaxies. Many of these galaxies are much brighter in the infrared than in the visible; long exposures showed their visible images.

IRAS mapped the entire Milky Way Galaxy; its view of our galaxy's disk penetrated to the Galactic Center. Another of IRAS's discoveries was that the sky is covered with infrared-emitting material, probably outside our solar system but in our galaxy. Since its shape resembles terrestrial cirrus clouds, the material is being called "infrared cirrus." (See Figure 5–25.) Subsequent ground-based observations have shown that the infrared cirrus corresponds to the distribution of neutral interstellar hydrogen. It also correlates with faint dust features barely detectable on optical photographs. Thus the cirrus is apparently dust grains embedded within the hydrogen gas.

The European Space Agency's *I*nfrared *S*pace *O*bservatory (ISO) worked from 1995 until its coolant ran out in 1998. It carried a 0.6-m telescope. Its ISOCAM was a camera that took images rather than mapping the sky widely; ISO also carried spectrographs. The University of Massachusetts's Two *M*icron All *S*ky *S*urvey (2MASS) has mapped the whole sky with much more sensitivity and resolution than the earlier maps (Fig. 32–13).

Infrared studies are continuing with major efforts. NASA's 0.85-m infrared telescope, the Space *I*nfrared *T*elescope *F*acility (SIRTF), will be the fourth and last in NASA's series of "Great Observatories," also including Hubble in the visible, ultraviolet, and the shorter end of the infrared; the Chandra X-ray Observatory in the x-ray; and the defunct Compton Gamma Ray Observatory. In 2002, the Japanese will launch their IRIS infrared satellite. NASA's SOFIA airplane, with a 2.5-m telescope, should start observations in 2004. Later in the decade, in 2007, the European Space Agency is to launch the Herschel Space Observatory. The Next Generation Space Telescope will be optimized for infrared studies.

## 32.3b The Galactic Nucleus

One of the brightest infrared sources in our sky is located at the position of the radio source Sagittarius A. Within that broad source is a much smaller radio source, Sgr A*, which marks the center of our galaxy; the asterisk in its name distinguishes this point source, pronounced "Saj A star," from Sagittarius A, which we now know is a whole complex of sources. Since the amount of scattering by dust varies with wavelength, we can see further through interstellar space in the infrared than we can in the visible.

**FIGURE 32–13** The 2MASS image of the central region of our galaxy.

The VLA and other radio interferometers have been used to survey the radio sky in the direction of the galactic center (Fig. 32–14). The compact radio source Sagittarius A* is the galactic center itself. The wide spectral lines in nearby infrared objects show that hydrogen and helium atoms there have high temperatures, indicating that a strong wind is being given off by the stars present near the center of our galaxy. A lot of energy is being emitted in the infrared: as much energy as if there were 20 million suns radiating. This energy apparently comes from dust grains that have been heated by ultraviolet radiation from the stars.

The position of Sgr A* has been pinpointed by simultaneously using a set of radio telescopes scattered around the world. Measurements with the whole-Earth-scale Very Long Baseline Array of radio telescopes reveal a source only 27 milliarcsec across, only 3.3 astronomical units at the distance of the Galactic Center.

The most precise knowledge of the black hole near the galactic center comes from tracking of the proper motions of 3 stars within 0.005 to 0.013 parsecs of

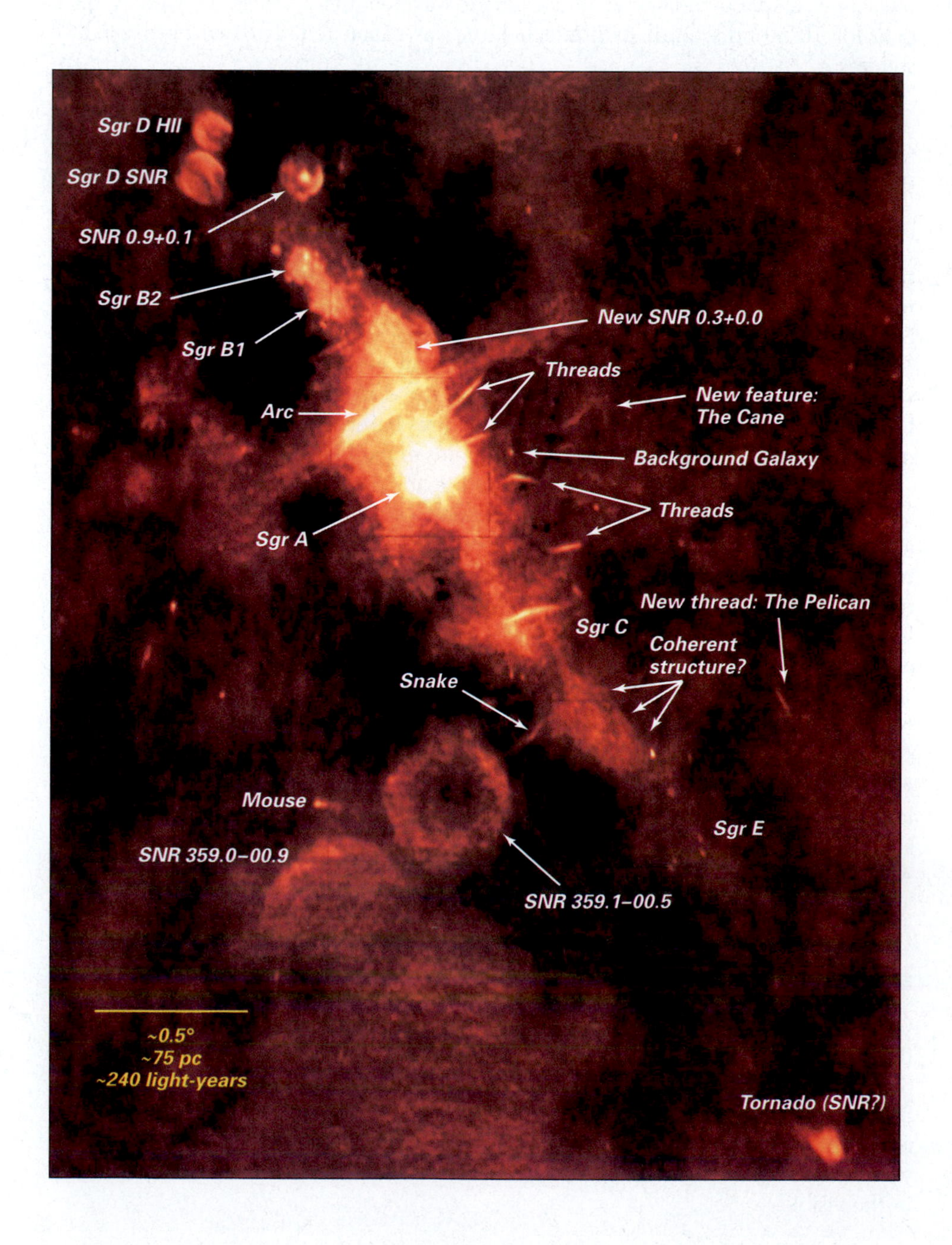


**FIGURE 32–14** A wide-field view of the center of our galaxy, observed at the VLA at a wavelength of 90 cm.

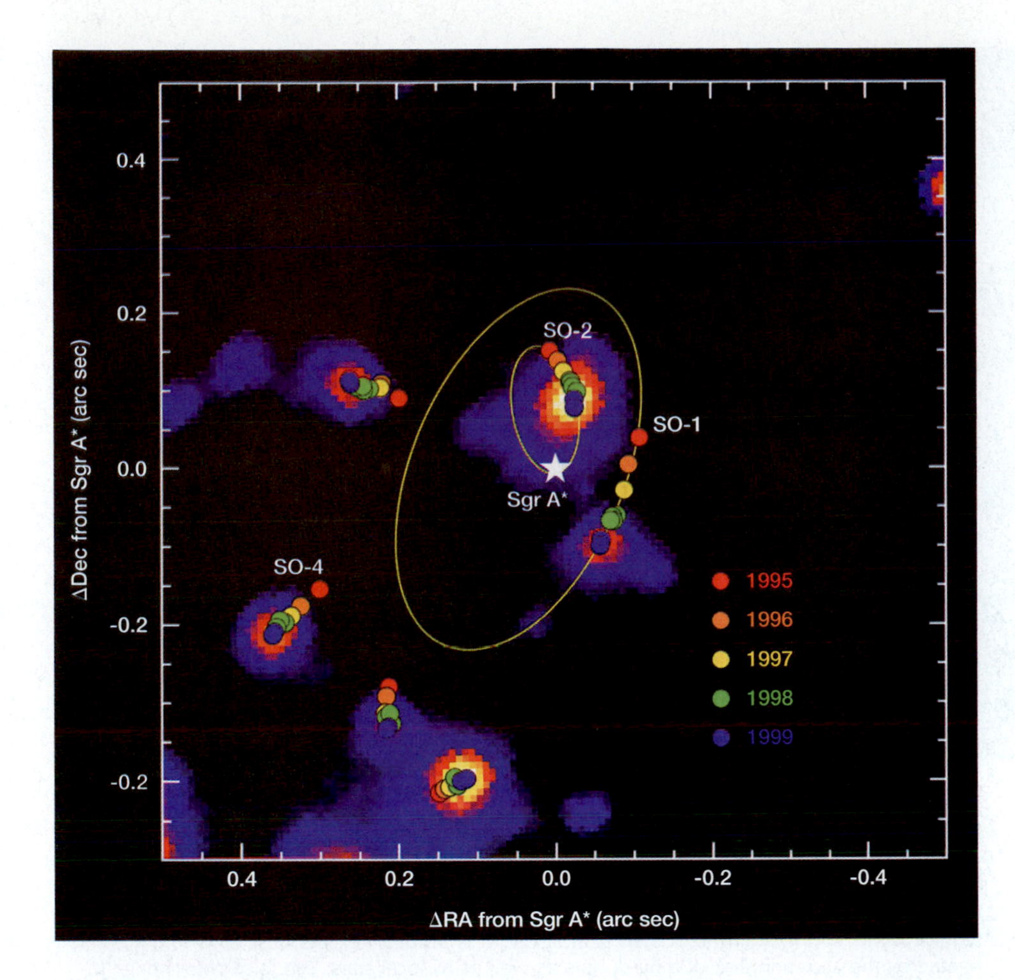


**FIGURE 32–15** The compact, non-thermal radio source Sgr A* is within 1 arc sec of the dynamical and gravitational center of our galaxy. These near-infrared observations, made with the Keck telescope, were used to follow proper motions of stars even at the 25,000-light-year distance from us of the center of our galaxy. At that distance, 1 arc sec = 0.1 light-years.

the core. Andrea Ghez of UCLA and colleagues took series of near-infrared observations over a 5-year period with the Keck I telescope (Fig. 32–15). They took many short exposures, about 1/10 second, each of which freezes the jitter in the image caused by turbulence in the Earth's atmosphere. Each of the stars accelerates about 3 to 5 millimeters per second, a similar acceleration to that the Earth experiences as it orbits the sun. The acceleration is toward Sgr A*, strengthening the case that Sgr A* marks the black hole whose existence is inferred. The velocities measured show the presence of a black hole of 2.6 million solar masses at Sgr A*. Since the results imply a density greater than 200 billion solar masses per cubic light-year, only a black hole fits these observations. The black hole would be only the size of the orbit of Mars around the sun. VLBI observations at wavelengths of 1.4 and 3 mm confirm that the radio source is smaller than 15 Schwarzschild radii. Our vision is not quite sharp enough to see the event horizon, but we are getting closer.

Though these observations strengthen the identification of Sgr A* with our galaxy's central black hole, we do not know why Sgr A* is so dark. Shouldn't we see more radiation from gas falling into the accretion disk? The evidence that a black hole is present is still indirect. Scientists are still working to rule out non-black-hole explanations completely. One of the stars reported above is so close to Sgr A* that it orbits in only about 15 years, so we have hope that observations over the next decade will give us the orbit in such detail as to rule out alternative explanations.

High-resolution infrared observations are improving, allowing an object that may correspond to Sgr A* to be identified (Fig. 32–16). It is still not clear which objects in different parts of the spectrum correspond to each other.

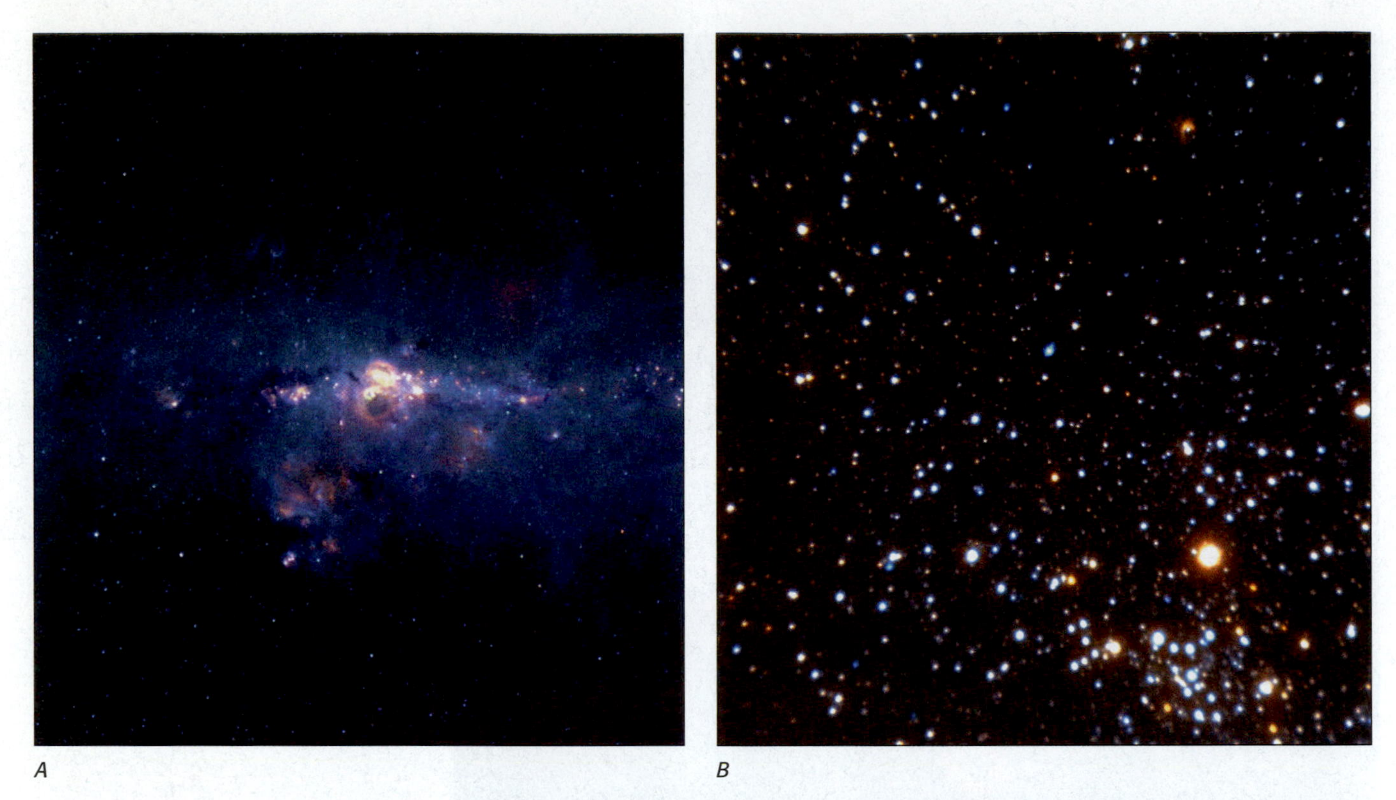


**FIGURE 32–16** (*A*) The center of our galaxy in the mid-infrared. A region 1.5° × 1.5°, including the galactic center, shows in this image from MSX (Midcourse Space Experiment), which made infrared observations from Earth orbit in 1996–7. We see largely dust and molecular clouds, which we discuss in the next chapter. The galactic center is the bright spot near the center. If it is indeed physically associated with Sgr A*, it is probably excess dust emission, possibly heated by the accretion disk of a black hole as well as by local stars. The false color image includes 6–11 microns (*blue*), 11–16 microns (*green*), and 18–26 microns (*red*). (*B*) Observations of the central 40 arc seconds (4 light-years) around the Galactic Center from the Gemini North Observatory. Adaptive optics were used for these near-infrared observations at the infrared filter bands known as H (1.6 $\mu$m, reproduced as blue) and K′ (2.15 $\mu$m, reproduced as red). The field of view is more than 100 times smaller than the observations in part A.

The observations seem to lead to the model that the central parsec of our galaxy contains a sphere of stars about 0.3 light-year in radius, a few small associations or clusters of newly formed B stars within that core, and an accreting black hole of perhaps 1,000,000 solar masses.

Not only radio and infrared but also gamma rays have been detected from the region of the center of our galaxy, though not at Sgr A*. Presumably, matter and antimatter have been annihilating each other there. In addition to an extended, steady source concentrated toward the Galactic Center, there is also a variable source called the Great Annihilator. The High Energy Astronomical Observatory-3's gamma-ray spectrometer confirmed the detection of the gamma-ray spectral line that results from electrons and positrons annihilating each other. The region emitting the Great Annihilator's gamma-ray spectral line is too small for the gamma rays to arise from cosmic rays interacting with the interstellar medium or from a collection of many supernovae, novae, or pulsars. The whole region has most recently been mapped and analyzed with the Compton Gamma Ray Observatory.

One kind of matter-antimatter annihilation occurs when an electron meets its antiparticle, a **positron.** The two annihilate each other completely, releasing energy in the amount $E = 2mc^2$, where $m$ is the mass of each of the two particles.

The high-resolution radio maps of our galactic center, now made with the Very Large Array, show a small bright spot that could well be the region of the central giant black hole. But though it appears on the images that a spiral a few parsecs across is present there (Fig. 32–17*A*), this view turns out to be an optical illusion; the "arms" are only apparently superimposed on each other. Doppler shifts seem to show that the streamers that make the apparent "arms" are falling into rather than outward from the nucleus. The conclusion is not universally accepted. Fortunately, the streamers are moving sufficiently fast that we have hope that we can actually see their proper motion across the sky in a few years, using high-resolution radio techniques. Then we will have a much better picture of their motion.

Work with the VLA has also revealed parallel filaments stretched perpendicularly to the plane of the galaxy (Fig. 32–17*B*). Because this "Arc" resembles a solar quiescent prominence, the structure seems to indicate that a strong magnetic field is present.

VLA observations have also shown a thin streamer of gas that may link the shell around the central black hole with the nearest gas cloud. The observations may explain how the central black hole continues to be fed. Chandra observations have shown the hot gas in the central supernova remnant known as Sgr A East, which is apparently near the galactic supermassive central black hole, known as Sgr A* (Fig. 32–18). The observations showed that the hot gas within the shell is highly enriched with heavy elements. These measurements support the idea that Sgr A East is a single supernova remnant, left over from an explosion 10,000 years

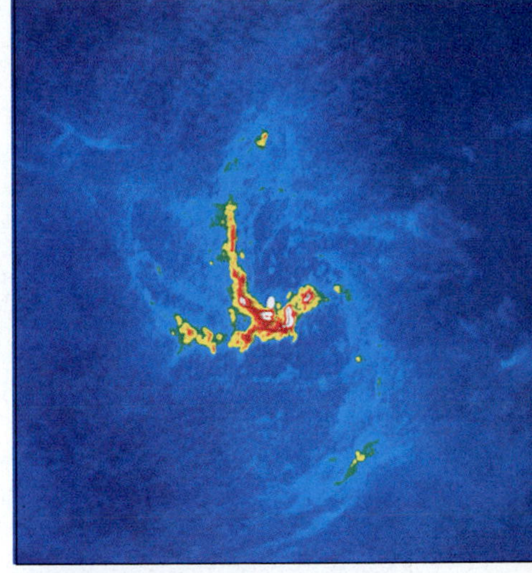

*A*

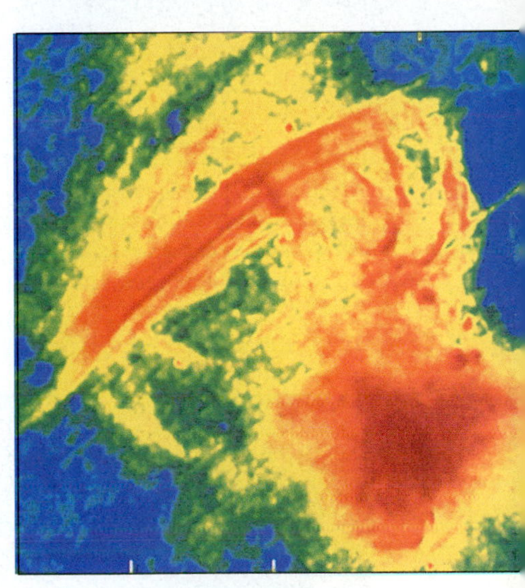

*B*

**FIGURE 32–17** (*A*) Subtracting the compact non-thermal radio source shows a spiral of gas within the central 10 light-years of our galaxy. The pattern is formed by ionized gas falling toward or orbiting around the center of our galaxy. Sgr A*, presumably the supermassive black hole at the center of our galaxy, shows as the bright white dot in this false-color image. The wavelength was 6 cm, the resolution was 1 arc sec, and the field of view was 3 arc min. (*B*) The Arc, parallel filaments that stretch over 130 light-years perpendicularly to the plane of the galaxy, which runs from upper left to lower right. The field of view in this VLA image is 25 arc min across; the wavelength used was in the continuous emission near 21 cm.

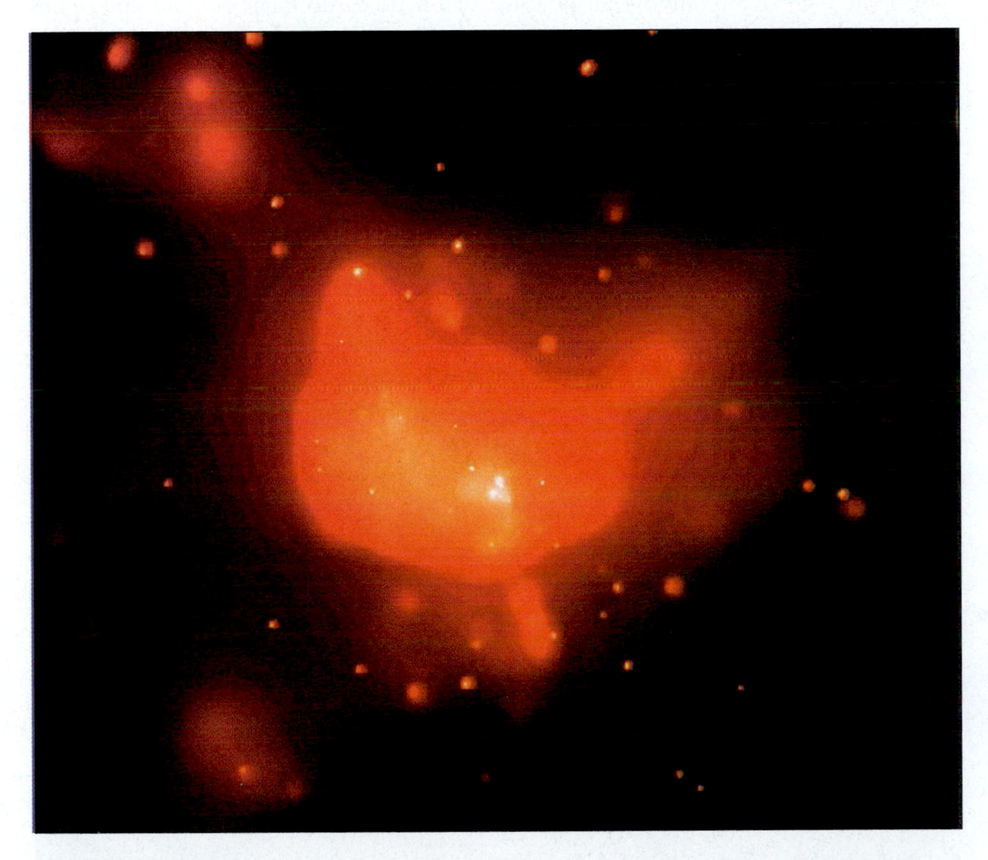

**FIGURE 32–18** The relationship between the supermassive black hole Sagittarius A* and the supernova remnant Sagittarius A East. This Chandra X-ray Observatory image has high enough resolution to separate the supernova remnant from other structures. The supernova remnant's emission shows in yellow and orange. Though long suspected of being a supernova remnant, the enhanced abundances of heavy elements like calcium and iron over their abundances in the sun endorse the idea. Sgr A*, the black hole at the center of the Milky Way Galaxy, is close to the white dots in the lower-right portion of the central object.

ago. Perhaps gas from this supernova fed the black hole. Since the black hole source is not now a bright x-ray source, the gas must have passed over it, though it could have occurred as recently as a few hundred years ago. These observations were made with the Penn State/MIT instrument on Chandra. Perhaps the regions of giant black holes in other galaxies also vary in brightness depending on nearby activity.

The center of our galaxy is a fascinating place under increasing scrutiny.

## 32.4 ALL-SKY MAPS OF OUR GALAXY

The study of our galaxy provides us with a wide range of types of sources. Many of these have been known for many years from optical studies (Fig. 32–19). The infrared sky looks quite different (Fig. 32–20) and was being entirely mapped by the 2MASS Project. The radio sky provides still a different picture. Technological advances have enabled us to study sources in our galaxy in the x-ray and gamma-ray regions of the spectrum as well.

## 32.5 GAMMA-RAY ASTRONOMY

A "fountain" of continuously erupting gamma rays seems to spew up out of one side of the Galactic Center, rising 3500 light-years.

Non-solar x-ray astronomy began in 1962, when Riccardo Giacconi and colleagues discovered x-rays from a source in the constellation Scorpius (and named it Scorpius X-1). Herbert Friedman and colleagues soon carried out additional x-ray work. The 1960s research was carried out with rockets rather than with satellites. A few dozen sources, including Scorpius X-1, the Crab Nebula, and the Virgo cluster of galaxies, were found.

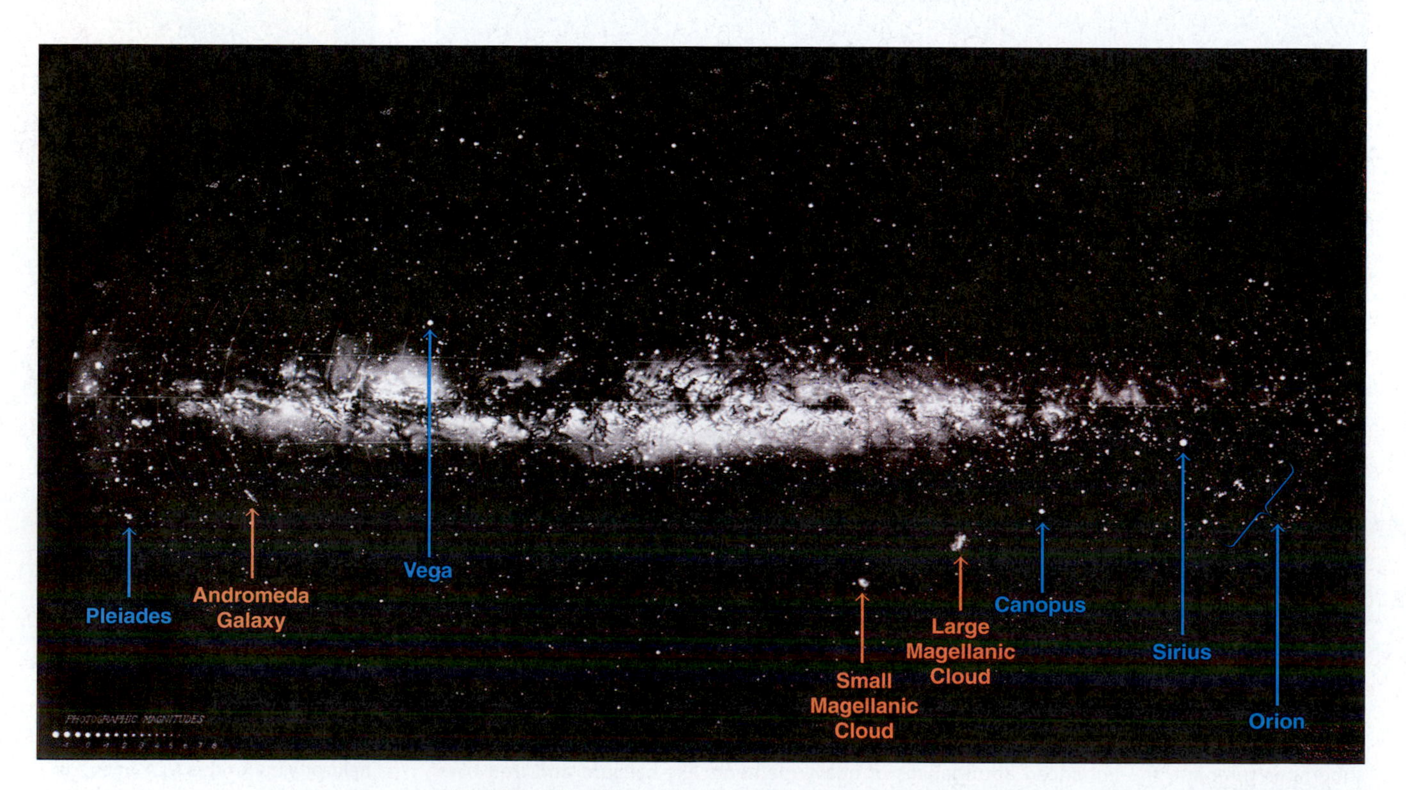


**FIGURE 32–19** A drawing of the Milky Way, made under the supervision of Knut Lundmark at the Lund Observatory in Sweden. Seven thousand stars plus the Milky Way are shown in this panorama, which is in coordinates (known as "galactic coordinates") such that the Milky Way falls along the equator.

U.S. satellites in the Vela series, whose prime purpose was to detect nuclear explosions, and HEAO-3 made early gamma-ray observations. Some of the strangest observations were gamma-ray bursts that were detected from time to time. The strongest for a long time, a burst observed on March 5, 1979, lasted $\frac{1}{5000}$ second and was followed for a few minutes by pulses with an 8-second period. During the short interval of this burst, the object gave off energy at a rate greater than that of the entire Milky Way Galaxy, even assuming that it was an object in our galaxy. Unfortunately, the spacecraft couldn't identify the direction from which the bursts came.

The gamma-ray situation changed dramatically in 1991 with NASA's launch of the Compton Gamma Ray Observatory, the second of NASA's series of Great Observatories, which mapped the sky (Fig. 32–21). It was named after Arthur Holly Compton, the American physicist who won the Nobel Prize for his studies of the interactions of photons and matter. In Section 35.7, we will discuss gamma-ray bursts, of which it discovered thousands. We will save our discussion for then since the sources of gamma-ray bursts turn out to be far beyond our galaxy.

One of the 271 gamma-ray sources the EGRET (*E*nergetic *G*amma-*R*ay *E*xperiment *T*elescope) experiment on the Compton Gamma Ray Observatory mapped is close to or coinciding with an x-ray binary system. This system is "only" 7000 to 10,000 light-years away, less than halfway to the center of our galaxy. Radio astronomers have found a jet in this system; perhaps it resembles SS433 (Section 30.9). Maybe other as yet unidentified EGRET sources may also be systems in our galaxy with similarly fast-moving jets. Since the object isn't especially bright in other parts of the spectrum, perhaps there are a lot of systems like it in our galaxy. They may even provide a high proportion of the high-energy particles and gamma rays in our galaxy.

The most studied gamma-ray object is Geminga (Fig. 32–22), a very strong gamma-ray-emitting region in our galaxy, located in the direction of the constellation Gemini. X-ray pulsations were discovered in 1992 using data from ROSAT, and these pulsations were confirmed in gamma rays with data from the Compton Gamma Ray Observatory. Thus, Geminga is a close cousin to the Crab and Vela Nebulae,

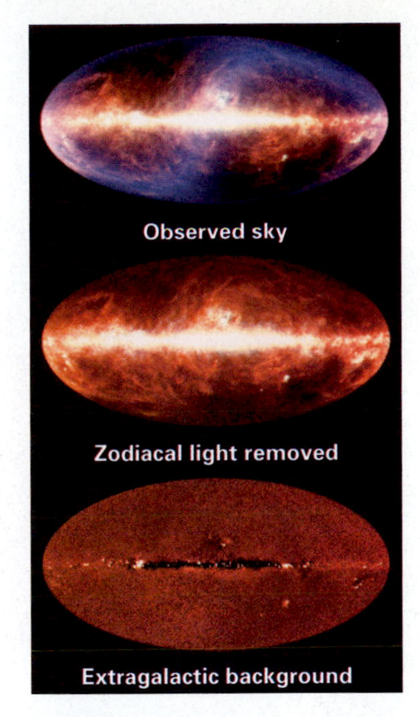


**FIGURE 32–20** Infrared maps of the sky from the Cosmic Background Explorer (COBE) satellite. The zodiacal light is sunlight scattered from dust in the plane of our galaxy. Subtracting it and subtracting the Milky Way as well as possible reveals a background of infrared radiation from beyond our galaxy, that is, extragalactic.

The Italian scientists studying the source chose its name, which is pronounced with a hard second "g," for **Gemini ga**mma-ray source; they found the name suitable since the gamma-ray source was unidentified for so long and since "geminga" means "it does not exist" in the Milanese dialect.

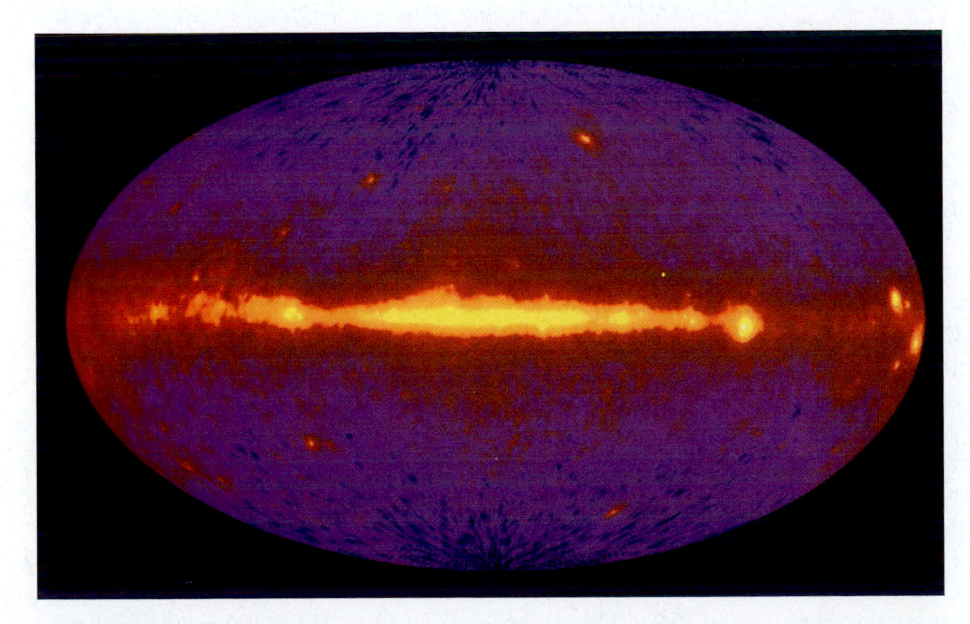

**FIGURE 32–21** This gamma-ray map made by the EGRET (Energetic Gamma-Ray Experiment Telescope) aboard the Compton Gamma Ray Observatory shows the plane of the Milky Way and several individual gamma-ray sources. EGRET found 271 gamma-ray sources. Most are extragalactic; a few are pulsars.

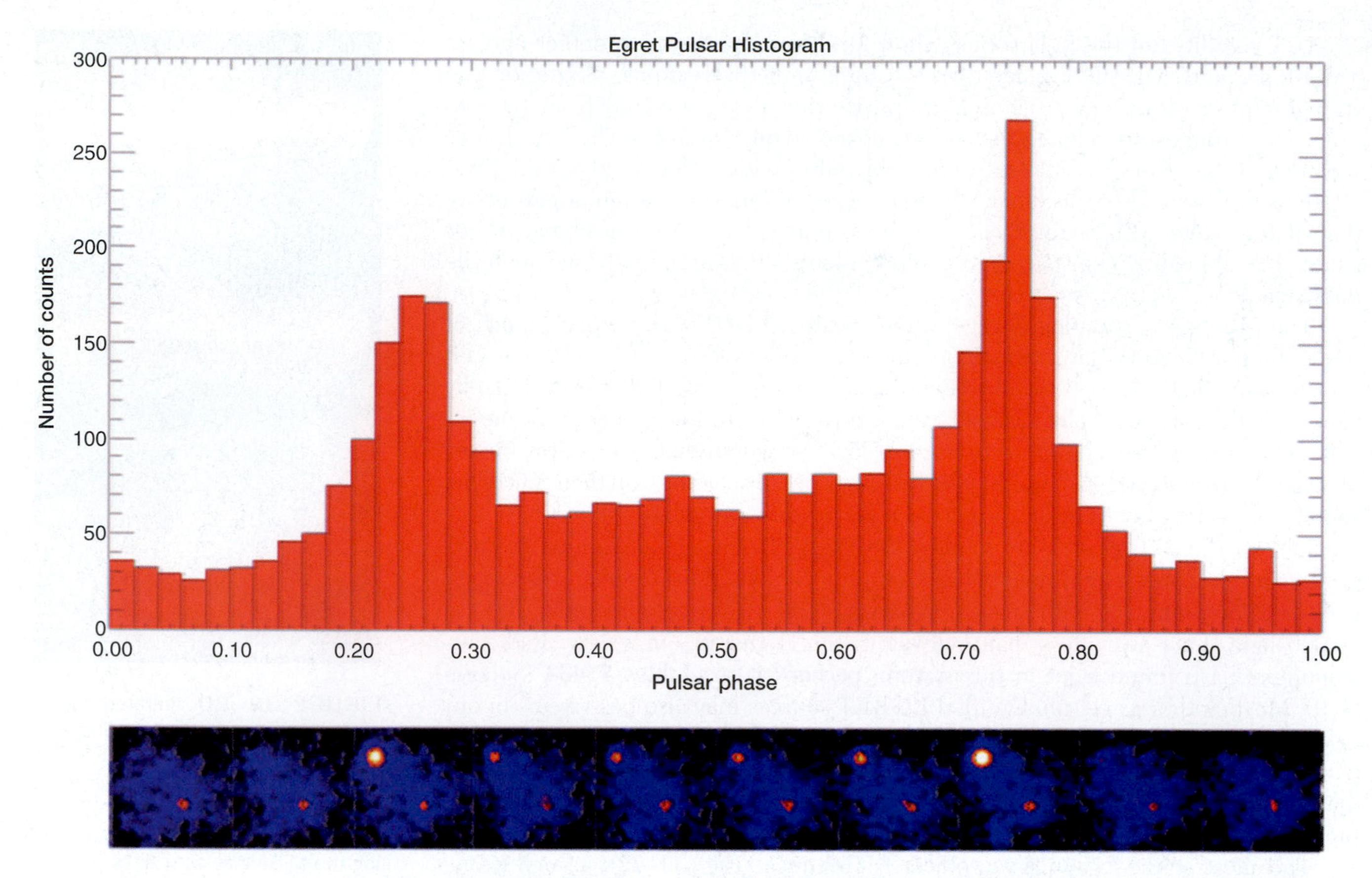


**FIGURE 32–22** Geminga (*upper left*) and the Crab Nebula, observed with the EGRET on the Compton Gamma Ray Observatory. The field of view is large: 60° square. We see Geminga pulsing; each image is exposed only at Geminga's pulse period of 0.237 second. Geminga is now the only pulsar not detected in the radio region of the spectrum. The Crab pulsar (*slightly below center and to the right*) appears steady because it pulses at a different period from the one used for this series.

which have pulsating neutron stars at their cores. Geminga and potentially the other gamma-ray sources in the Milky Way are therefore apparently powered by rotating neutron stars. HST has measured that Geminga is 500 light-years from us. Maybe we will find more exotic objects pulsing in gamma rays and yet remaining quiet in the radio spectrum (Fig. 32–23).

High-energy gamma rays make "extensive air showers" of rapidly moving subatomic particles when they hit the top of our atmosphere, as we discussed in Section 29.7. These particles emit light that can be detected by telescopes on the ground. A few projects are under way to search the sky for the blue flashes that would result from incoming gamma rays. These telescopes need not have high-quality imaging, and so can be very large. Project Milagro at the Los Alamos Scientific Laboratory in New Mexico is under way to measure such bursts. A preliminary project, Milagrito, already took some data. The *H*igh *E*nergy *G*amma *R*ay *A*stronomy (HEGRA) telescope is in the Canary Islands, and the *M*ajor *A*tmospheric *G*amma-ray *I*maging *C*erenkov Telescope (MAGIC), with a 17-m mirror, is to start observing at the same site in 2001. The American VERITAS (*V*ery *E*nergetic *R*adiation *I*maging *T*elescope *A*rray *S*ystem, which means "truth" in Latin, an acronym chosen to give the motto of Harvard, one of several participants), has seven 10.4-m mirrors.

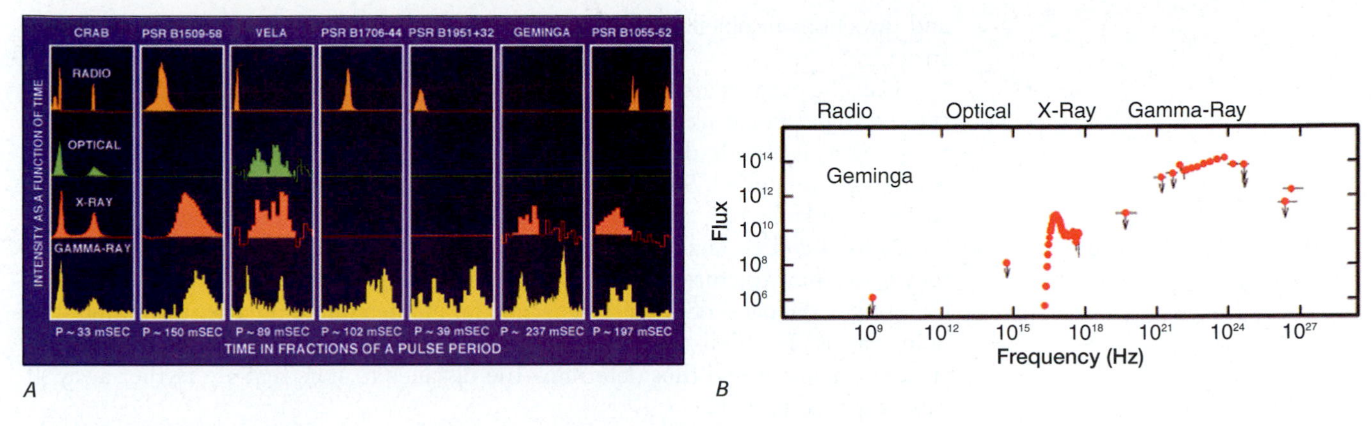


**FIGURE 32–23** Geminga is the only object that pulses gamma-rays but not detectable radio waves. *(B)* The strength of Geminga's radiation across the whole spectrum, from gamma-rays to radio.

It is to go on line near Mount Hopkins in Arizona during or after 2004, expanding the current, single Whipple Telescope (other American participants include Boston University, Caltech, Grunnel, DePauw, Iowa State, Kansas State, Northwestern, Purdue, UCLA, UC-Riverside, Chicago, Utah, and Washington University). A *H*igh *E*nergy *S*tereoscopic *S*ystem (HESS) is under construction in Namibia to cover the southern sky.

NASA's next gamma-ray orbiting observatory is to be GLAST (*G*amma-ray *L*arge *A*rea *S*pace *T*elescope). After its launch in 2006, GLAST is expected to observe at least ten times as many stellar-mass black holes as previously detected in gamma rays. It will also study pulsars and supernova remnants, gamma-ray bursts, diffuse backgrounds of high-energy gamma-rays that originate inside our galaxy and outside, and solar flares. GLAST will pinpoint objects ten times better than EGRET could, so it should be able to associate the still unidentified EGRET sources with known celestial objects.

One object, Markarian 501, has flared from being a puny gamma-ray source to become 10 times brighter than the Crab Nebula in spite of being 50,000 times farther away. This source is a blazar, a type of active galaxy thought to be powered by a giant black hole. So understanding the progenitors of the high-energy gamma-ray bursts will tell us about astonishingly violent events.

A

B

**FIGURE 32–24** The view from the level of a maze *(A)* makes it hard to figure out the pattern, which is readily visible *(B)* from above.

## 32.6 THE SPIRAL STRUCTURE OF THE GALAXY

### 32.6a Bright Tracers of the Spiral Structure

When we look out past the boundaries of the Milky Way Galaxy, usually by observing in directions above or below the plane of the Milky Way, we can see a number of galaxies with arms that appear to spiral outward from near their centers. In this section we shall discuss some of the evidence that our own Milky Way Galaxy also has spiral structure.

It is always difficult to tell the shape of a system from a position inside it (Fig. 32–24). Think, for example, of being somewhere inside a maze of tall hedges. We would find it difficult to trace out the pattern. If we could fly overhead in a helicopter, though, the pattern would become very easy to see. Similarly, we have difficulty tracing out the spiral pattern in our own galaxy. Still, by noting the distances and directions to objects of various types, we can tell about the Milky Way's spiral structure.

Galactic clusters are good objects to use for this purpose, for they are always located in the spiral arms. We have found the distance to galactic clusters by studying their color-magnitude diagrams. Also, we think that spiral arms are regions where young stars are found. Some of the young stars are the O and B stars; their lives are so short we know they can't be old. But since our methods of determining the distances to O and B stars from their spectra and colors are uncertain to 10 per cent, they give a fuzzy picture of the distant parts of our galaxy. Parallaxes measured from the Hipparcos spacecraft do not penetrate far enough into our galaxy to substantially help map it. The *F*ull-sky *A*strometric *M*apping *E*xplorer (FAME) was to observe smaller parallaxes and thus determine the distance to stars that are farther away. It was cancelled in 2002.

Other signs of young stars are the presence of regions of ionized hydrogen known as H II regions (pronounced "H two" regions). We know from studies of other galaxies that H II regions are preferentially located in spiral arms. In studying the locations of the H II regions, we are really again studying the locations of the O stars and the hotter B stars, since it is ultraviolet radiation from these hot stars that provides the energy for the H II regions to glow.

Scientists have long plotted the positions of the open clusters, the O and B stars, and the H II regions around the plane of the galaxy, and added the distances they have deduced. The results seem to show bits of three spiral arms in our part of our galaxy, and are unable to give information about possible spiral arms that are farther away.

### 32.6b Differential Rotation*

The latest calculations indicate that the sun is approximately 8.5 kiloparsecs from the center of our galaxy. From spectroscopic observations of the Doppler shifts of globular clusters (which do not participate in the galactic rotation) or of distant galaxies, we can tell that the sun is revolving around the center of our galaxy at a speed of approximately 200 kilometers per second. At this velocity, it would take the sun about 250 million years to travel once around the center; this period is called the **galactic year.** But not all stars revolve around the Galactic Center in the same period of time. The central part of the galaxy rotates like a solid body. Beyond the central part, the stars that are farther out have longer galactic years than stars closer in.

If this system of **differential rotation,** with differing rotation speeds at different distances from the center, has persisted since the origin of the galaxy, we may wonder why there are still only a few spiral arms in our galaxy and in the other galaxies we observe. An older star at the position of the sun could have made fifty revolutions during the lifetime of the galaxy, but points closer to the center would have made many more revolutions. Why haven't the arms wound up very tightly?

### 32.6c Why Our Galaxy Has Spiral Arms*

The leading current solution to this conundrum is a theory first suggested by the Swedish astronomer B. Lindblad and elaborated mathematically by the American astronomers C. C. Lin at MIT and Frank Shu, now at Berkeley. They say, in effect, that the spiral arms we now see are not the same spiral arms that were previously visible. In their model, the spiral-arm pattern is caused by a spiral **density wave,** a wave of increased density that moves through the stars and gas in the galaxy. This density wave is a wave of compression, not of matter being transported. It rotates more slowly than the actual material and causes the density of material to build up as it passes. A shock wave—like a sonic boom from an airplane—builds up and compresses the gas. In some galaxies, the compressed gas collapses to form stars. In other galaxies, the increased density leads to an increased rate of collisions of the giant molecular clouds we will discuss in the next chapter. These collisions may lead to the formation of massive stars.

So, in the density-wave model, the spiral arms we see at any given time do not represent the actual motion of individual stars in orbit around the Galactic Center (Fig. 32–25). We can think of the analogy of a crew of workers painting a white line down the center of a busy highway. A bottleneck occurs at the location of the painters. If we were observers in an airplane, we would see an increase in the number of cars at that place. As the line painters continued slowly down the road, we would seem to see the place of increased density move down the road at that slow speed. We would see the bottleneck move along even if our vision were not clear enough to see the individual cars, which could still speed down the highway, slow down briefly as they cross the region of the bottleneck, and then resume their high speed.

Similarly, we might be viewing only some galactic bottleneck at the spiral arms. Gas can pile up to make spiral arms that are visible without being permanent. New studies with an interferometer operating at millimeter wavelengths have shown that the newly formed stars are slightly downstream from the molecular complexes that the density wave forms from individual molecular clouds. In molecular-rich galaxies, this mechanism provides the luminous stars that define the arms. These stars heat the interstellar gas so that it becomes visible; we see young, hot stars and glowing gas outline the spiral arms (Fig. 32–26).

Some astronomers do not accept the density-wave theory; one of their major objections is the problem of explaining why the density wave forms. One alternative, very different theory says that stars are produced by a chain reaction and are then spread out into spiral arms by the differential rotation of the galaxy. The chain reaction begins when high-mass stars become supernovae. The expanding shells from the supernovae trigger the formation of stars in nearby regions. Some of these new stars become massive stars, which become supernovae, and so on.

Computer modelling of the **supernova-chain-reaction model** gives values that seem to agree with the observed features of some kinds of spiral galaxies. Each galaxy has a differential velocity, where "differential" means that regions at different distances from the center rotate at different speeds. At some distance from the center, the speed is a maximum. The degree of winding depends on the value for this maximum velocity.

It may be that the density-wave model and the supernova-chain-reaction model can be important to different degrees in different galaxies. The latter nicely reproduces the structure for galaxies with short spiral arms (known as "flocculent spirals"). ("Flocculent," in a standard dictionary, means having a fluffy or woolly appearance.) It does not do well for the beautiful, long, two-arm spirals that are often called "grand-design" spirals.

Debra Elmegreen of Vassar College and Bruce Elmegreen of IBM have shown that grand-design galaxies show the same spiral structure in the blue and the infrared while flocculent spirals (Fig. 32–27) look very different in those two parts of the spectrum. The effect can be explained by a density wave, which passes all parts of a galaxy, and so which leads to both old and young stars at the same locations. The supernova-chain-reaction model produces structure only in the young, short-lived component. Since that component shows up mainly in the blue, images of galaxies whose spiral structure formed in this way look different in different parts of the spectrum.

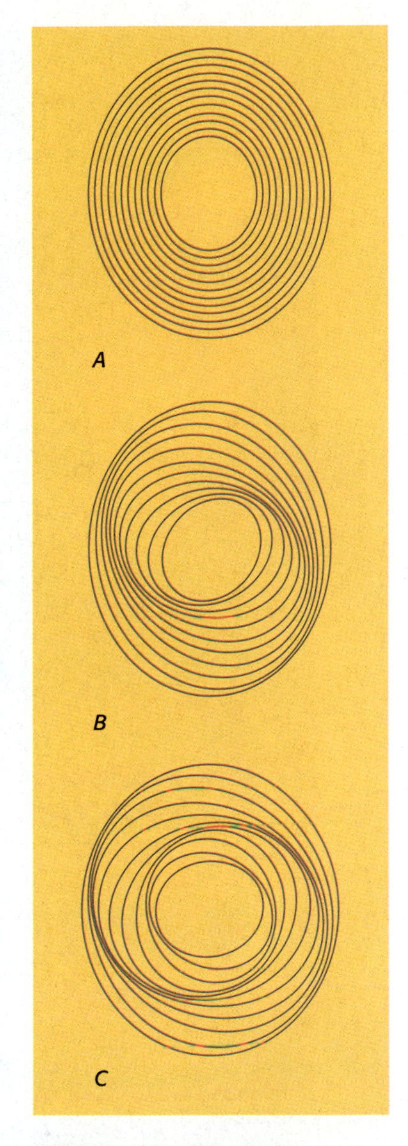


**FIGURE 32–25** Each part of the figure includes the same set of ellipses; the only difference is the relative alignment of their axes. Consider that the axes are rotating slowly and at different rates. The compression of their orbits takes a spiral form, even though no actual spiral exists. The spiral structure of a galaxy may arise from an analogous effect.

Note that we are not saying that stars' elliptical orbits cause the apparent spiral structure, since the structure we actually see comes from shining stars formed as a result of the rotation of a spiral wave of density enhancement.

**FIGURE 32–26** Studies of star formation in spiral arms of the grand design galaxy M51. The molecular clouds (Section 33.5) are traced by their emission of carbon monoxide (CO). The velocity of the carbon-monoxide gas shifts perpendicularly to the spiral arms. The shifts match predictions of density-wave models. The CO arm is offset slightly toward the nucleus. If the CO resulted from heating, then it would coincide with the visible arm of OB stars that heat the gas and cause the H$\alpha$ emission. Since it is displaced, we are seeing the location of maximum molecular density. The result matches the predictions of density-wave theory. The observations that the contrast is high between spiral arm and interarm regions, and that some clouds nevertheless appear between arms, are potentially important for interpreting carbon-monoxide observations of our own galaxy. The observations should lead to improved understanding of the process of star formation.

**FIGURE 32–27** The flocculent galaxy NGC 5055. Its appearance can be explained with the supernova-chain-reaction model. Local effects can, thus, provide wide-scale structure.

A Hubble image of the galaxy 2903 revealed clusters of stars that were born 5 to 10 million years ago (Fig. 3–28). It also showed spots that are H II regions where young stars are forming today. Since the two kinds of spots are far apart from each other, the scientists conclude that the star-forming activity shifts from place to place. This idea matches the supernova-chain-reaction model.

Some galaxies show signs of both types of processes.

**FIGURE 32–28** A Hubble image of the galaxy NGC 2903 in Leo, similar to our own galaxy but 25 million light-years away. The image is dominated by the bar, looking reddish because of the dust it contains, lying within the bluish spiral arms. The bar may be funneling material into the galaxy's center, causing stars to be born. The star formation regions are distributed in a 2000-light-year-wide ring, a feature seen in many other galaxies. The high Hubble resolution reveals that clusters of stars 5 to 10 million years old are separate from spots where stars are forming now. The conclusion is that star formation jumps around. The view is a false-color one constructed from exposures taken with Hubble's NICMOS instrument through filters at infrared wavelengths.

## CORRECTING MISCONCEPTIONS

| ✖ *Incorrect* | ✔ *Correct* |
|---|---|
| We can see the center of our galaxy. | The center of our galaxy is hidden by dust from visible light. |
| We can measure directly how far away the center of our galaxy is. | We must use indirect methods to detect the size of our galaxy since we can only measure parallaxes for less than 1000 light-years. |
| Galaxies have spiral arms because they spin and draw out material. | The spiral arms we see are probably illusions caused by density waves. |

## SUMMARY AND OUTLINE

Nebulae (Section 32.1)
- Emission; dark (absorption); reflection; planetary; supernova remnants
- The reddish glow results from H$\alpha$ emission.

Our galaxy has a nuclear bulge surrounding its nucleus, a disk, and a halo (Section 32.2).

Infrared observations (Section 32.3)
- Limited by atmospheric transparency and technology
- IRAS mapped the entire sky at relatively long wavelengths; ISO aloft
- Radiation from the galactic nucleus indicates a lot of energy is generated in a small volume there; a very massive black hole may be present.

High-energy astronomy (Section 32.4)
- ROSAT, Compton Gamma Ray Observatory, BeppoSAX, Chandra

Gamma-ray astronomy (Section 32.5)
- Gamma-ray bursters are very distant.

Spiral structure of the galaxy (Section 32.6)
- Bright tracers of the spiral structure (Section 32.6a)
  - Difficult to determine shape of our galaxy because of obscuring gas and dust and because of our vantage point
  - Galactic clusters, O and B stars, and H II regions are used to determine the locations of spiral arms.
- Differential rotation (Section 32.6b)
  - Sun is approximately 8.5 kpc from center of galaxy.
  - Central part of galaxy rotates as a solid body.
  - Outer part, including the spiral arms, is in differential rotation.
- Theories for spiral structure (Section 32.6c)
- Density wave: we see the effect of a wave of compression; the distribution of mass itself is not spiral in structure.
- Chain of supernovae

## KEY WORDS

Milky Way
nebula
emission nebula
dark nebula
absorption nebula
reflection nebula
star clouds
nucleus
nuclear bulge
disk
spiral arms
halo
positron
galactic year*
differential rotation*
density wave*
supernova-chain-reaction model*

*This term appears in an optional section.

## QUESTIONS

1. Why do we think our galaxy is a spiral?
2. How would the Milky Way appear if the sun were closer to the edge of the galaxy?
3. Sketch (a) a side view and (b) a top view of the Milky Way Galaxy, showing the shapes and relative sizes of the nuclear bulge, the disk, and the halo. Mark the position of the sun. Indicate the location of the galactic corona.
4. With a sketch, show why the globular clusters do not appear evenly distributed around us, duplicating Harlow Shapley's argument.
5. Compare (a) absorption (dark) nebulae, (b) reflection nebulae, and (c) emission nebulae.
6. How can something be both an emission and an absorption nebula? Explain and give an example.
7. †Emission nebulae shine because their atoms are ionized. Hydrogen atoms in their lowest energy state can be ionized by radiation of 912 Å and shorter. This wavelength is known as the Lyman limit because the Lyman series of hydrogen spectral lines extends up to it. By comparing with the Sun, what is the approximate temperature of a star that has a spectrum that peaks at the Lyman limit, and so emits significant energy short of that wavelength? What is its spectral type?
8. Sketch the Horsehead Nebula, indicating on the sketch where you see H$\alpha$ radiation and where you see light reflected off dust.
9. If you see a red nebula surrounding a blue star, is it an emission or a reflection nebula? Explain.
10. How do we know that the galactic corona isn't made of ordinary stars like the sun?
11. †Given the approximate size of our galaxy, and the average density of 1 atom per cubic centimeter, estimate how many hydrogen atoms our galaxy contains. Using the mass of a hydrogen atom from Appendix 2, what is the mass of the visible matter in our galaxy?
12. Why may some infrared observations be made from mountain observatories, while all x-ray observations must be made from space?
13. What radio source corresponds to the center of our galaxy?
14. Illustrate, by sketching the appropriate Planck radiation curves, why cooling an infrared telescope from room temperature to 4 K greatly reduces the background noise.
15. Describe infrared and radio results about the center of our galaxy.
16. What are three tracers that we use for the spiral structure of our galaxy? What are two reasons why we expect them to trace spiral structure?
17. Since Einstein's result was that $E = mc^2$, explain how, if $m_e$ is the mass of an electron, $2m_ec^2$ of energy results when an electron and a positron annihilate each other.
18. Why will x-rays expose a photographic plate, while infrared radiation will not?

†This question requires a numerical solution.

**19.** What types of objects give off strong infrared radiation?

**20.** What types of objects give off x-rays?

**21.** Discuss how observations from space have added to our knowledge of our galaxy.

**22.** Comment on our understanding of the gamma-ray bursts.

**23.** Why does the density-wave theory lead to the formation of stars?

†**24.** If a spacecraft could travel at 10 per cent of the speed of light, how long would it have to travel to get far enough out to be able to take a photograph showing the spiral structure of our galaxy?

†This question requires a numerical solution.

## USING TECHNOLOGY

### W World Wide Web

1. Look for Milky Way and nebular images in such sites as the Anglo-Australian Observatory (http://www.aao. gov. au/images.html), the Space Telescope Science Institute (http://oposite.stsci.edu/pubinfo/Pictures.html), and the European Southern Observatory at http://www.eso. org/outreach/info-events/ut1fl/astroimages.html.
2. Look at the 2MASS homepage to find out about its observations at http://www.ipac.caltech.edu/2mass and the Space Infrared Telescope Facility site at http://sirtf.caltech.edu.
3. Follow the motion of stars around the central black hole in our galaxy on the Web page of Andreas Eckhart and Reinhard Genzel, http://www.mpe-garching.mpg.de/www_ir/GC/prop.html. Check the Galactic Center Web site at http://www.mpifr-bonn.mpg.de/gcnews.
4. Follow gamma-ray astronomy and its future at http://cossc.gsfc.nasa.gov/cossc for the Compton Gamma-Ray Observatory and the Gamma Ray Large Area Space Telescope (GLAST) at http://www-glast.stanford.edu.

### REDSHIFT

1. Look at Guided Tours: Galaxies: Milky Way; Guided Tours: Stars: Emission Nebulae; and Science of Astronomy: The Milky Way.
2. Find when the center of the Milky Way in Sagittarius is up, and follow the angle of the Milky Way to the horizon through the night.
3. In *RedShift*'s sky, locate some of the nebulae mentioned in this chapter, and determine if you can see them at this season. When do they rise and set?
4. Classify some of the nebulae in the Photo Gallery as emission, absorption, or reflection. Find examples of each.
5. On Disk 2, study Story of the Universe: The Milky Way Galaxy.

The interstellar star-forming region of our Galaxy known as S106, imaged with the Subaru Telescope on Mauna Kea. A massive star, IRS4, is at the center of this nebula. About 20 times more massive than the sun, it is surrounded by a disk of gas and dust, which constricts the outflow of material. This false color infrared image is made at wavelengths of 1.25 $\mu$m (blue), 1.65 $\mu$m (green), and 2.15 $\mu$m (red). Since this nebula radiates continuum emission more strongly at 1.25 $\mu$m than stars that have the same flux that it has at 1.65 $\mu$m and 2.15 $\mu$m, it is fitting to portray it with the shortest wavelength of the false colors used, that is, as blue.

# Chapter 33

# The Interstellar Medium

The mixture of gas and the dust between the stars is known as the **interstellar medium.** The nebulae (Fig. 33–1) represent regions of the interstellar medium in which the density of gas and dust is higher than average. The interstellar medium contains the elements in what we call their **cosmic abundances,** that is, the overall fractions of all matter they make up in most of the universe (the cosmos).

In this chapter we shall see that the studies of the interstellar medium are vital for understanding the structure of our Galaxy and how stars are formed. We shall also discuss the fairly recent realization that units of the interstellar medium called "giant molecular clouds" are basic building blocks of our Galaxy.

**AIMS:** To discuss the interstellar medium, to describe the techniques (especially radio astronomy) used to study it, and to understand the relation of interstellar clouds to star formation

## 33.1 H I AND H II REGIONS

For many purposes, we may consider interstellar space as being filled with hydrogen at an average density of about 1 atom per cubic centimeter, although individual regions may have densities departing greatly from this average. Regions of higher density in which the atoms of hydrogen are predominantly neutral are called **H I regions** (pronounced "H one regions"; the Roman numeral "I" refers to the first, or basic, state, with hydrogen atoms that are neither ionized nor excited). Where the density of an H I region is high enough, pairs of hydrogen atoms combine to form molecules ($H_2$). The densest part of the gas associated with the Orion Nebula, for example, might have a million or more hydrogen molecules per cubic centimeter. So hydrogen molecules ($H_2$) are often found in H I regions. Since hydrogen atoms are almost entirely on their ground levels in H I regions, no Balmer spectral lines are emitted. These regions are best observed in the radio, as we discuss in Section 33.4. Hydrogen molecules also do not emit in the visible. They are studied largely from spacecraft in the ultraviolet.

FIGURE 33–1 The Lagoon Nebula, M8, an H II region in Sagittarius, glowing red since it emits primarily hydrogen (H$\alpha$) radiation.

A region of ionized hydrogen—that is, hydrogen with one electron missing—is known as an **H II region** (from "H two," the second state). Since hydrogen, which makes up the overwhelming proportion of interstellar gas, contains only one proton and one electron, a gas of ionized hydrogen contains individual protons and electrons. Wherever a hot star provides enough energy to ionize hydrogen, an H II region results. Emission nebulae are such H II regions (Fig. 33–2). They glow because the gas is heated. Emission lines appear after the electron and proton "recombine" to form an excited state of hydrogen.

Studying the optical and radio spectra of H II regions and planetary nebulae tells us the abundances of several of the chemical elements (especially helium,

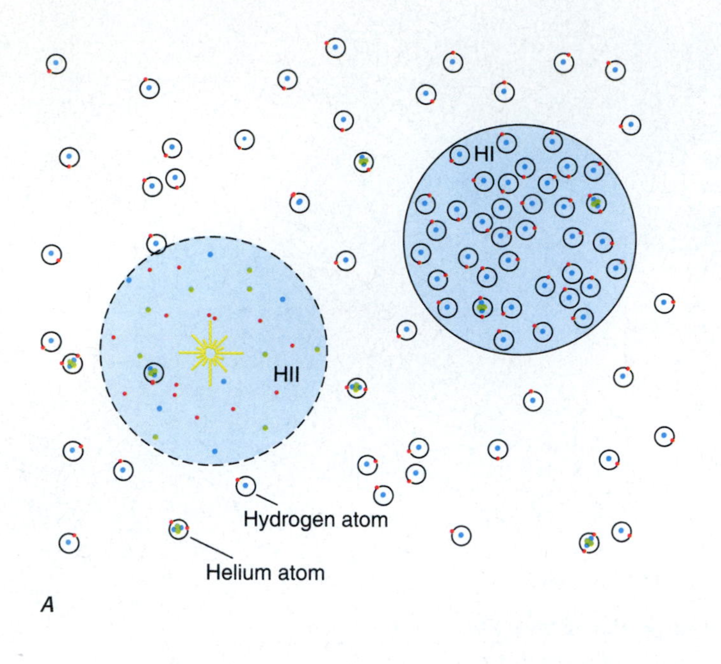


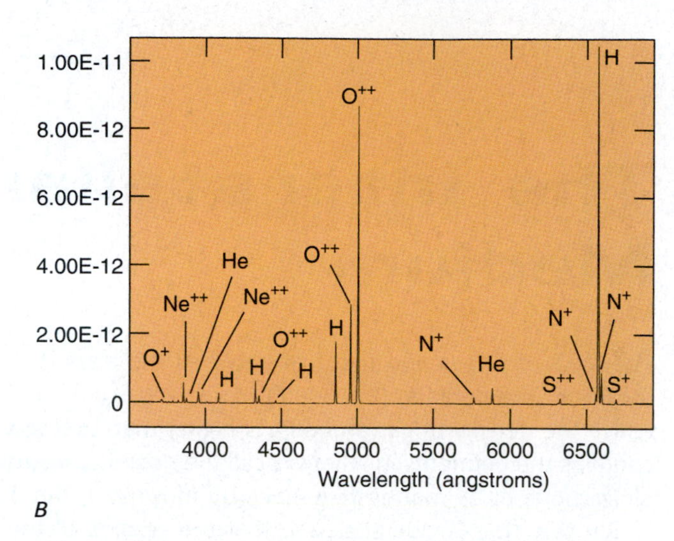


**FIGURE 33–2** (*A*) **H I regions** are regions of neutral hydrogen of higher density than average, and **H II regions** are regions of ionized hydrogen. The protons and electrons that result from the ionization of hydrogen by ultraviolet radiation from a hot star, and the neutral hydrogen atoms, are shown schematically. The larger dots represent protons or neutrons, and the smaller dots represent electrons. The star that provides the energy for the H II region is shown. Though all hydrogen is ionized in an H II region, only some of the helium is; heavier elements have often lost 2 or 3 electrons. (*B*) The emission lines in an H II region, including some of the members of hydrogen's Balmer series, helium, and "forbidden lines" of ionized oxygen and neon. (These "forbidden lines" are actually transitions that happen relatively rarely—they aren't completely forbidden from occurring by the rules of quantum mechanics.)

nitrogen, and oxygen). How these abundances vary from place to place in our Galaxy and in other galaxies helps us choose between models of element formation and of galaxy formation. The variation of the relative abundances from the center of a galaxy to its outer regions—the gradient of abundances—differs for different galaxies. The abundances of the elements known as metals (which, to an astronomer, are everything heavier than helium) are greatest in the locations where the most star formation took place; this variation indicates that these elements were indeed produced in stars.

## 33.2 INTERSTELLAR DUST

Distant stars in the plane of our Galaxy are obscured from our vision. In addition, many stars that are still close enough to be visible are partially obscured (Fig. 33–3). The amount of obscuration varies with the wavelength at which we observe. The blue light is **scattered** by dust in space more efficiently than the redder light is; for a given distance through the dust, more of the blue light has been bounced around in every direction. Thus, less of the blue light comes through to us than the red light, and the stars look redder—we say that the stars are **reddened.** This reddening is thus a consequence of the scattering properties of the dust. It has nothing to do with "redshifts," since the spectral lines are not shifted in wavelength by reddening.

A

B

**FIGURE 33–3** (*A*) A dark cloud, known as a Bok globule. Dust hides the stars beyond from our view. (*B*) The Hubble Space Telescope has imaged deep in the heart of the Lagoon Nebula, M8. This nebula is 5000 light-years away from Earth in the direction of the constellation Sagittarius. The image reveals a pair of one-half light-year-long interstellar "twisters." The nebula is an example of a stellar nursery, with stars being born from giant molecular clouds, which are in cocoons of interstellar dust. The images show emission from ionized sulfur atoms (*red*), from doubly ionized oxygen atoms (*blue*), and from the recombination of ionized hydrogen (*green*). An optical view of this emission nebula appeared as Figure 33–1.

The amount of scattering together with the amount of actual absorption of visible radiation by dust is known as the **extinction.** In the blue part of the spectrum, the total extinction in the 8.5 kpc between the sun and the center of the Galaxy is at least 25 magnitudes. (Remember that each 5 magnitudes is a factor of 100 times in brightness, so 10 magnitudes is $100 \times 100 = 10{,}000$ times, and 25 magnitudes is $5 + 5 + 5 + 5 + 5$ magnitudes, equivalent to a factor of $100 \times 100 \times 100 \times 100 \times 100 = 10^{10}$ times in brightness.) Most of this extinction takes place far from the sun,

## A DEEPER DISCUSSION

### BOX 33.1 Why Is the Sky Blue?

Just as interstellar dust scatters blue light more than it scatters red light, scattering by the electrons in air molecules in the terrestrial atmosphere sends blue light preferentially around in all directions. Since we are close to the scattering electrons, our sky appears blue. Further, when the sun is near the horizon, we have to look diagonally through the Earth's layer of air. Our line of sight through the air is then longer than a line of sight straight up, and most of the blue light is scattered out before it reaches us. Relatively more red light reaches us, accounting for the reddish color of sunsets.

#### *Discussion Question*

The scattering varies with one over the fourth power of the wavelength. The wavelength of red light is about 1.5 times the wavelength of blue light. By how much more is blue light scattered?

Astronomers have long wondered about the cause of the "diffuse interstellar bands," regions of spectrum that show absorption but are much wider (in wavelength) than normal spectral lines. It has long been suspected that complex carbon compounds could be the cause; buckminsterfullerene (carbon-60) is a strong possibility, now that its spectrum has been measured and seems to match many of the diffuse interstellar bands. A still newer theory is that these diffuse interstellar bands come from hydrogen in a complicated process in which two electrons are involved.

in regions with a high dust content. Even a tremendously bright object located near the center of our Galaxy would be dimmed too much—25 magnitudes—to be seen from the Earth in the visible part of the spectrum.

What is the dust made of? We know the particles are tiny—smaller than the wavelength of light—or else they would not scatter light this way, with the shorter wavelengths scattered more efficiently than longer wavelengths. From spectral studies, we know that at least some of the particles are carbon in the form of graphite. Other particles may be silicates or ices.

Interstellar dust grains form out of chemical elements that result from fusion in stars and supernovae. They can be made of carbon (in the form of graphite), silicon, magnesium, and iron in different combinations and can have different shapes. A typical grain has a core made of magnesium, silicon, or iron that was ejected from a dwarf star of spectral type M or from a red giant or supergiant and has picked up a surrounding mantle made of compounds of oxygen, carbon, and nitrogen. The larger grains may be the $10^{-5}$-cm size of smoke particles; others, 10 times smaller, include graphite particles. Other grains can be much smaller. We have long studied the dust grains by the spectrum of their extinction of starlight, and by laboratory simulation. We can now study them from their infrared radiation. And some interstellar grains have been identified in meteorites. Their relative abundances of isotopes indicate that these grains come from outside our solar system. They were in the cloud of gas and dust that collapsed to form the solar nebula. So, with meteorites in hand, we have some interstellar dust on Earth for us to examine and perhaps to give us clues about the formation of our solar system.

Interstellar dust is heated a bit by the radiation from stars that passes through space, thus causing the dust to radiate. Because the energy received is balanced by

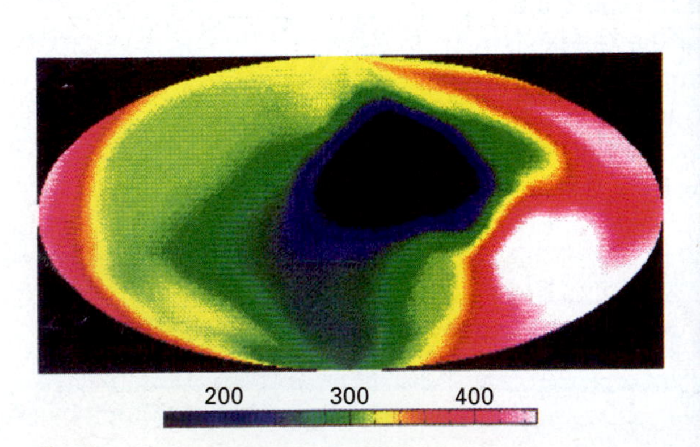


A

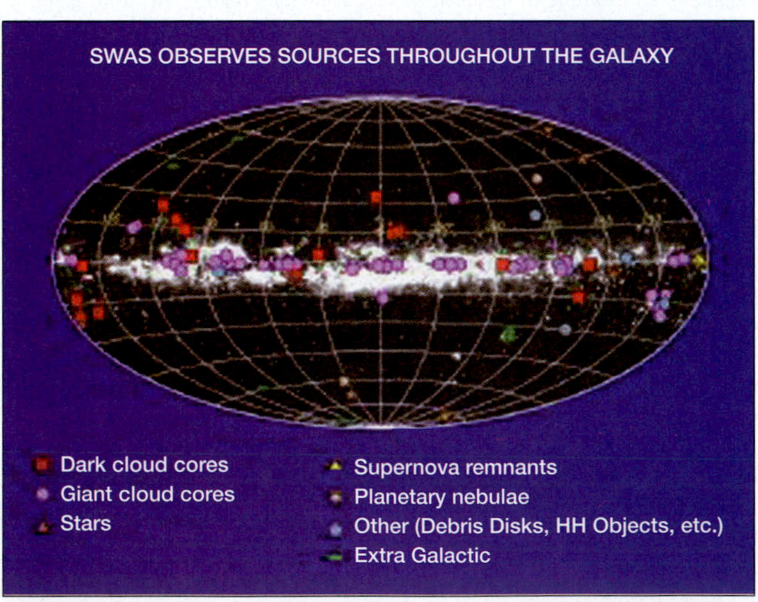


B

**FIGURE 33–4** (*A*) Our solar system lies in a region of space where the density of the interstellar gas is much lower than average. This all-sky display is from the University of Leicester's extreme ultraviolet (EUV) camera on ROSAT (most of whose instruments operated at shorter, x-ray, wavelengths). The scientists have measured the intensity of radiation received from white dwarfs, which emit strongly in the EUV, to measure the amount of absorption by hydrogen and helium in the interstellar medium. This map (the plane of our Galaxy is horizontal across the center) shows the distances, in light-years, to the edge of the boundary of the local bubble of low density. The edge of the bubble is nearest us roughly in the direction of the center of our Galaxy. The edge is farthest from us (*white*) in the direction of the star Beta Canis Majoris, where tunnel-like structures permeate the bubble wall. (*B*) A variety of sources observed from the Submillimeter Wave Astronomy Satellite (SWAS).

the amount even low-temperature dust radiates, the dust never gets very hot. Thus radiation from interstellar dust peaks in the infrared. In the infrared, we also detect the radiation coming from clouds of dust that surround stars. We find infrared radiation from so many stars forming in our Galaxy that we think that about one star forms in our Galaxy each year. Our abilities to observe in the infrared are increasingly rapidly. The 2MASS map of the sky is complete, and the Space Infrared Telescope Facility and other spacecraft will continue to improve our infrared abilities.

Similarly, since the interstellar gas is "invisible" in the visible part of the spectrum (except at the wavelengths of certain weak spectral lines), special techniques are needed to observe the gas in addition to observing the dust. A camera sensitive to extreme ultraviolet radiation has mapped our local region. We are in a "local bubble" of relatively low density (Fig. 33–4).

The Submillimeter Wave Astronomy Satellite (SWAS), launched in 1998, has surveyed the sky and observed some dozens of objects in detail. At wavelengths of about half a millimeter (that is, submillimeter), it observes water, oxygen molecules, neutral carbon, and forms of carbon monoxide and water with less common isotopes of carbon and oxygen included.

## 33.3 RADIO OBSERVATIONS OF OUR GALAXY

The rest of the electromagnetic spectrum carries more information in it than do the few thousand angstroms that we call visible light. We will first discuss some basic techniques of radio astronomy (see also Section 5.6b) and then go on to see how radio astronomy joins with other observing methods to investigate interstellar space. Fortunately, radio waves pass through not only the gas but also the dust in our Galaxy, allowing us to see farther in some directions than we can see in visible light.

**FIGURE 33–5** A 45° region of radio sky at a wavelength of 6.2 cm, with a photo of radio telescopes at the National Radio Astronomy Observatory in Green Bank, West Virginia, superimposed. The spherical objects are supernova remnants. The California Nebula, hydrogen ionized by hot stars, is at lower center. The point sources in this image are radio galaxies and quasars, hot stars as they would be in an optical picture. A new survey is being carried out with the Very Large Array and should show millions of radio sources.

The "G" in Galaxy or Galactic is commonly capitalized when referring to our own Milky Way Galaxy.

### 33.3a Continuum Radio Astronomy

All radio-astronomy studies in its early days were of the continuum. In radio astronomy, as in optical astronomy, studying the continuum means that we consider the average intensity of radiation at a given frequency without regard for variations in intensity over small frequency ranges. In short, we ignore any spectral lines. It was immediately apparent that the brightest objects in the radio sky are not identical with the brightest objects in the optical sky. The radio objects were named with letters and with the names of their constellations. Thus Taurus A is the brightest radio object in the constellation Taurus; we now know it to be the Crab Nebula. Sagittarius A is the center of our Galaxy (as we saw in the previous chapter); Sagittarius B is another radio source nearby, whose emission is caused by clouds of gas near the Galactic center. A continuum map of the sky at radio wavelengths looks different from optical maps (Fig. 33–5).

### 33.3b Synchrotron Radiation*

Continuum radio radiation can be generated by several processes. One of the most important is **synchrotron emission** (Fig. 33–6), the process that produces the radiation from Taurus A, a supernova remnant. Lines of magnetic field extend throughout the visible Crab Nebula and beyond. Electrons, which are electrically charged, tend to spiral around magnetic lines of force. Electrons of high energy spiral very rapidly, at speeds close to the speed of light. We say that they move at "relativistic speeds," since the theory of relativity must be used for calculations when the electrons are going that fast. Under these conditions, the electrons radiate very efficiently. (This is the same process that generates radiation in electron synchrotrons in some kinds of physics laboratories on Earth; hence the name synchrotron radiation.)

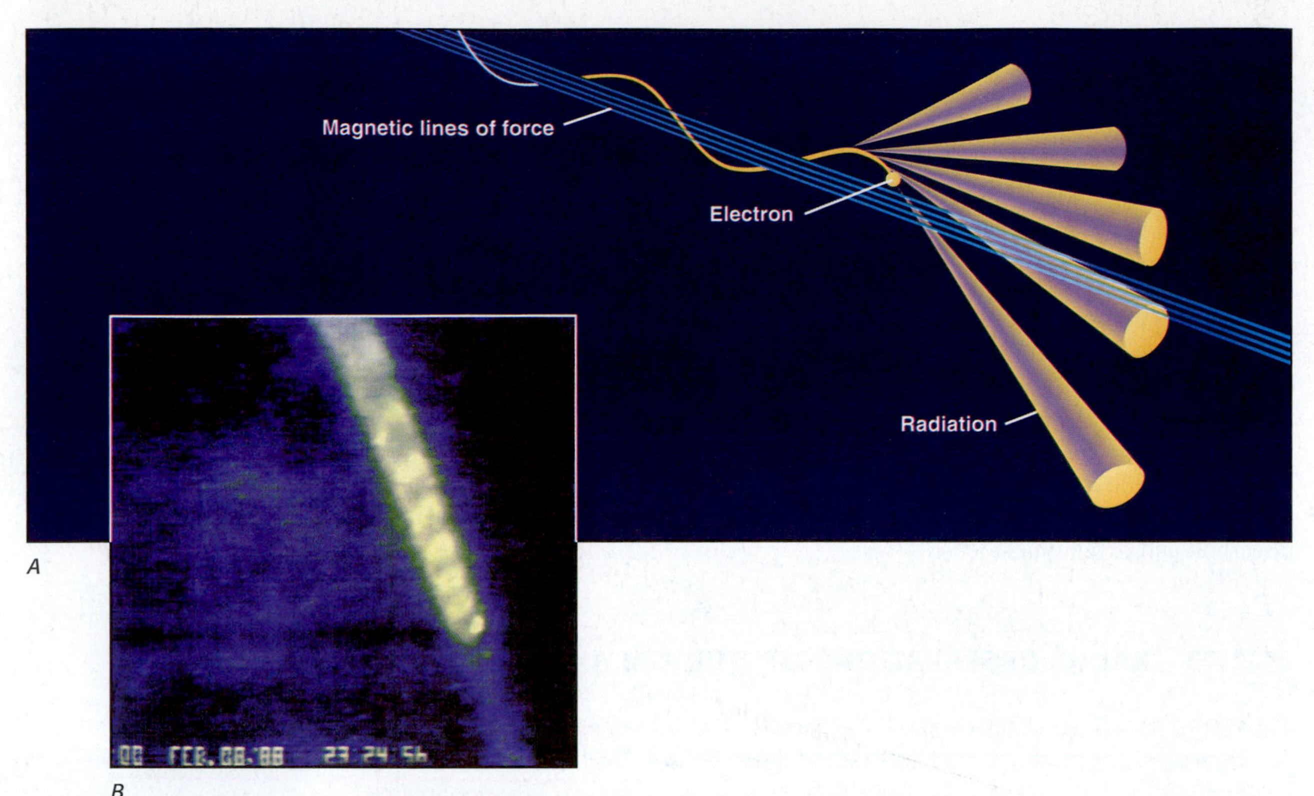


**FIGURE 33–6** (*A*) Electrons spiralling around magnetic lines of force at velocities near the speed of light (we say "at relativistic velocities") emit radiation in a narrow cone. This radiation, which is continuous and highly polarized, is called synchrotron radiation. Synchrotron radiation has been observed in both optical and radio regions of the spectrum. (*B*) Synchrotron radiation viewed in the Earth's atmosphere from the Echo-7 rocket. A beam was injected across the Earth's magnetic field and viewed with a TV camera.

Synchrotron radiation is highly polarized; that is, all the radio waves vibrate in the same plane—for example, all the electric waves go up and down rather than a mixture of up and down, side to side, and other angular orientations. Radiation given off because objects are heated is not polarized. The discovery that the optical radiation from the Crab Nebula is highly polarized revealed that it radiates by synchrotron radiation (Fig. 33–7). (Compare the straightforward optical view in Figure

**FIGURE 33–7** The polarization of the continuous radiation from the Crab Nebula (Figures 29–13, 29–14, and 29–15). This color display, made especially for my books by David Malin of the Anglo-Australian Observatory, assigns a color to a photograph taken at each of four angles of polarization: 0° (*blue*), 45° (*black*), 90° (*red*), and 135° (*green*). The plates were taken with the 5-m Palomar telescope. We see that the radiation is highly polarized and that different regions are polarized in different directions. The discovery of the polarization verified the prediction that synchrotron radiation was being emitted.

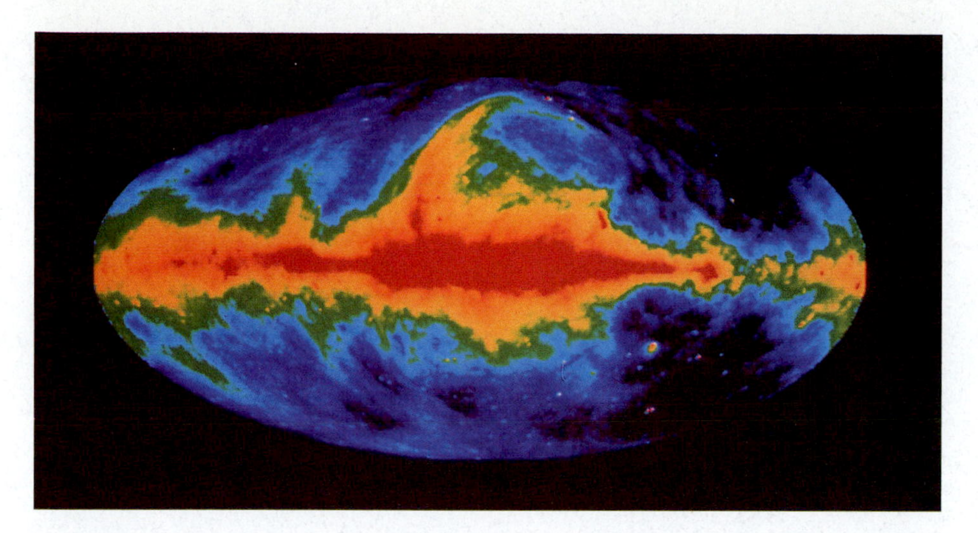

**FIGURE 33–8** Synchrotron radiation dominates the view of the radio sky in this image at 408 MHz (73-cm wavelength). On this map in Galactic coordinates, the plane of the Milky Way appears horizontally. The Galaxy's magnetic field bends the electrons to cause the synchrotron radiation, which appears both in the plane of the Galaxy and as the "spurs" that extend above the Galactic plane. The North Polar Spur, at top center, is about 500 light-years away and may have resulted from a supernova 300,000 years ago. Some of the sources on this map, compiled by several observatories together, are background galaxies.

29–13.) The radio radiation from the Crab is also highly polarized. (You may want to play with some polarizing sunglasses, which pass only light waves that vibrate in the same plane. A discussion question elaborates on this point.)

The intensity of synchrotron radiation is related not to the temperature of the astronomical body that is emitting the radiation, but rather to the strength of its magnetic field and to the number and energy distribution of the electrons caught in that field (Fig. 33–8). Since the temperature of the object cannot be derived from knowledge of the intensity of the radiation, we call this radiation **nonthermal radiation.** Synchrotron radiation is but one example of such nonthermal processes. The synchrotron process can work so efficiently that a relatively cool astronomical body can give off a tremendous amount of such radiation, perhaps so much that it would have to be heated to a few million degrees before it would radiate as much **thermal radiation** at a given frequency.

By **thermal radiation** we mean continuous radiation whose spectrum is directly related to the temperature of the gas. Thermal radiation follows a black-body (that is, Planck) curve. Anything that doesn't follow a black-body curve is a non-thermal source.

## 33.4 THE RADIO SPECTRAL LINE FROM INTERSTELLAR HYDROGEN

About 1950, though radio astronomers were very busy with continuum work, there was still a hope that a radio spectral line might be discovered. This discovery would allow Doppler-shift measurements to be made.

What is a radio spectral line? Remember that an optical spectral line corresponds to a wavelength (or frequency) in the optical spectrum that is distinctly more (for an emission line) or less (for an absorption line) intense than neighboring wavelengths or frequencies. Note that the higher the wavelength, the lower the frequency (Fig. 33–9). Similarly, a radio spectral line corresponds to a frequency (or wavelength) at which the radio noise is slightly more, or slightly less, intense. As we saw in Section 24.3, a spectral line corresponds to an atom's energy change, as an electron changes from one of an atom's energy levels to another. We saw there the size of a jump from one energy level of hydrogen to another that corresponds to spectral lines in the visible part of the spectrum. We also noted that as we made bigger jumps in energy (H-beta instead of H-alpha, for example), the wavelength got shorter.

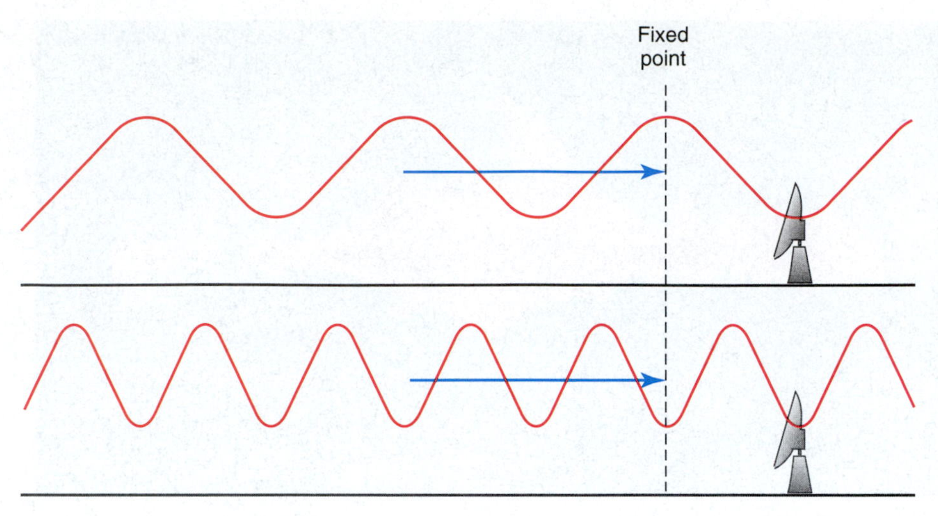


FIGURE 33–9 Radio astronomers often speak in terms of frequency instead of wavelength. Since all electromagnetic radiation travels at the same speed (the speed of light) in a vacuum, fewer waves of longer wavelength (*top*) pass an observer in a given time interval than do waves of a shorter wavelength (*bottom*). If the wavelength is half as long, twice as many waves pass; that is, the frequency is twice as high. The wavelength $\lambda$ times the frequency $\nu$ is constant, with the constant being the speed of light $c$: $\lambda\nu = c$. Frequency is given in hertz (Hz), formerly cycles per second. (Though the name of the unit, "hertz," is not capitalized, the symbol Hz is capitalized to show it is derived from someone's name.) The 21-cm line is at 1420 MHz.

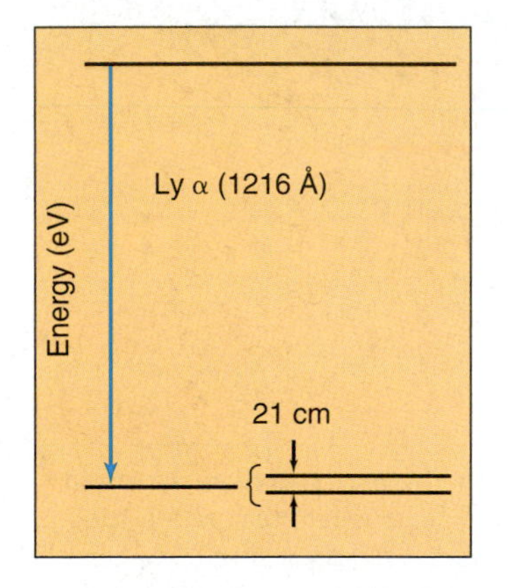


FIGURE 33–10 21-cm radiation results from an energy difference between two sublevels in the lowest principal energy state of hydrogen. The energy difference is much smaller than the energy difference that leads to Lyman $\alpha$. So (because $E = h\nu = hc/\lambda$) the wavelength is much longer.

So larger jumps in energy correspond to shorter wavelengths. The radio spectrum, on the other hand, has much longer wavelengths, so must correspond to smaller jumps in energy. None of the possible jumps in the energy-level diagrams in Section 24.3 are small enough to emit radio waves. We had to find an energy-level jump that was much smaller than anything then known.

## 33.4a The Hydrogen Spin-Flip

The most likely candidate for a radio spectral line that might be discovered was a line arising in interstellar hydrogen atoms. This line was predicted to be at a wavelength of 21 cm. Since hydrogen is by far the most abundant element in the universe, it seems reasonable that it should produce a strong spectral line. Furthermore, since most of the interstellar hydrogen has not been heated by stars or otherwise, it is most likely that this hydrogen is in its state of lowest possible energy.

This spectral line at 21 cm comes not from a transition to the ground state from one of the higher states, nor even from level 2 to level 1 as does Lyman alpha, but rather from a transition between the two sublevels into which the ground state of hydrogen is divided (Fig. 33–10).

For this astronomical discussion, it is sufficient to think of a hydrogen atom as an electron orbiting a proton. Both the electron and the proton have the property of spin; each one has angular momentum (Section 7.4) as if it were spinning on its axis.

The spin of the electron can be either in the same direction as the spin of the proton or in the opposite direction. The rules of quantum mechanics prohibit intermediate orientations. If the spins are in opposite directions, the energy state of the atom is very slightly lower than the energy state occurring if the spins are in the same direction. The energy difference between the two states is equal to a photon of 21-cm radiation.

If an atom is sitting alone in space in the upper of these two energy states, with its electron and proton spins aligned in the same direction, there is a certain small

If we were to watch any particular group of hydrogen atoms, we would find that it would take 11 million years before half of the electrons had undergone spin-flips; we say that the **half-life** (the lifetime before half of something has changed) is 11 million years for this transition. But even though the probability of a transition taking place in a given atom in a given interval of time is very low, there are so many hydrogen atoms in space that enough 21-cm radiation is given off to be detected.

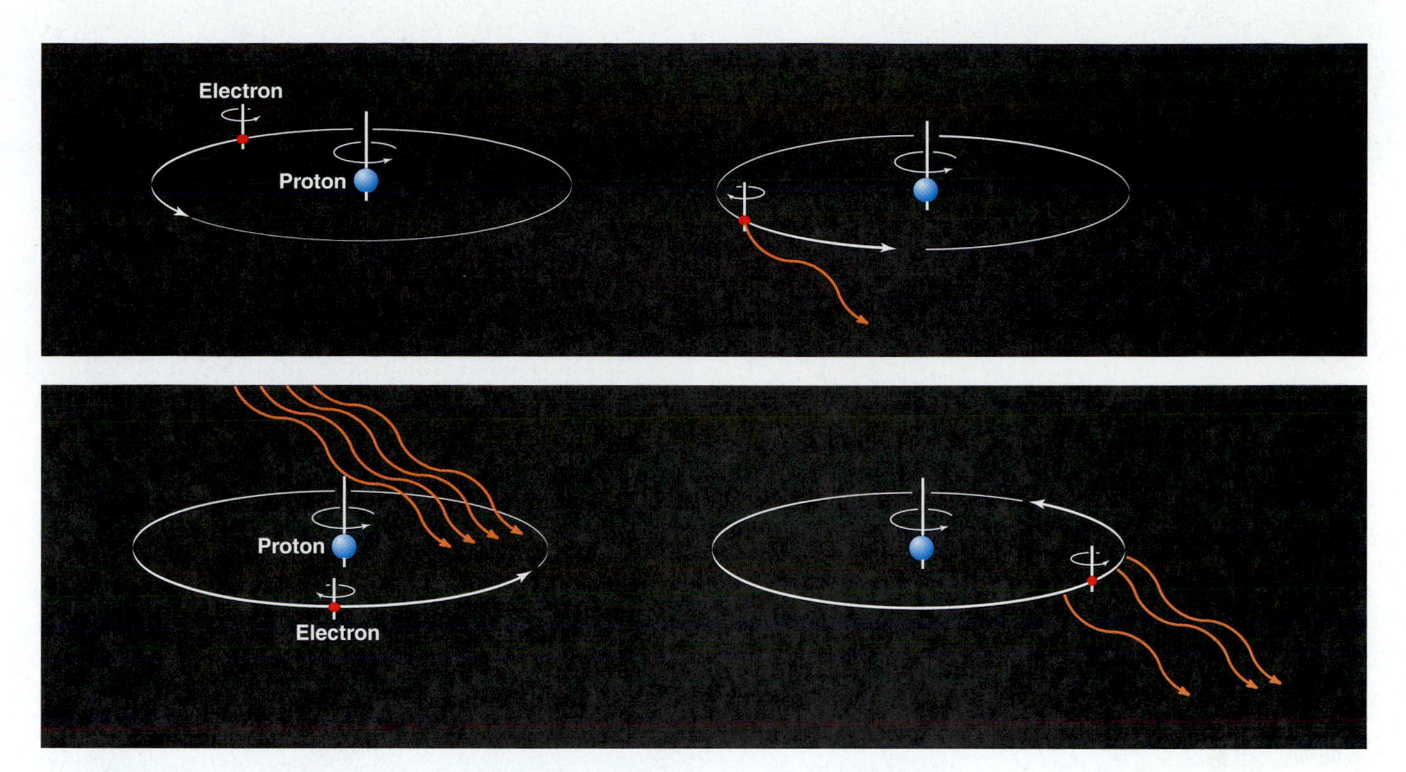


FIGURE 33–11 When the electron in a hydrogen atom flips over so that it is spinning in the opposite direction from the spin of the proton (*top*), an emission line at a wavelength of 21 cm results. When an electron takes energy from a passing beam of radiation, causing it to flip from spinning in the opposite direction from the proton to spinning in the same direction (*bottom*), then a 21-cm line in absorption results.

probability that the spinning electron will spontaneously flip over to the lower energy state and emit a photon. We thus call this a **spin-flip** transition (Fig. 33–11, *top*). The photon of hydrogen's spin-flip corresponds to radiation at a wavelength of 21 cm—the **21-cm line.**

We have just described how an emission line can arise at 21 cm. But what happens when continuous radiation passes through neutral hydrogen gas? In this case, some of the electrons in atoms in the lower state will absorb a 21-cm photon and flip over, putting the atom into the higher state. Then the radiation that emerges from the gas will have a deficiency of such photons and will show the 21-cm line in absorption (Fig. 33–11, *bottom*).

## 33.4b Mapping Our Galaxy

In fact, 21-cm hydrogen radiation has proved to be a very important tool for studying our Galaxy because this radiation passes unimpeded through the dust that prevents optical observations very far into the plane of the Galaxy (Fig. 33–12). Using 21-cm observations, astronomers can study the distribution of gas in the spiral arms (Fig. 33–13). We can detect this radiation from gas located anywhere in our Galaxy, even on the far side, whereas light waves penetrate the dust clouds in the Galactic plane only about 10 per cent of the way to the Galactic center.

But here again we come to the question that bedevils much of astronomy: how do we measure the distances? Given that we detect the 21-cm radiation from a gas cloud (since the cloud contains neutral hydrogen, it is an H I region), how do we know how far away the cloud is from us?

The answer can be found by using a model of rotation for the Galaxy, that is, a description of how each part of the Galaxy rotates. As we have already learned, the

*(text continues on page 652)*

# PHOTO GALLERY

## New Views of Nebulae

▶ The Bubble Nebula, NGC 7635, imaged with the Hubble Space Telescope. It is in Cassiopeia and is 7100 light-years from us. The bubble-shape is formed by an intense stellar wind flowing outward from a massive star, one containing 40 times the mass of the sun. The wind slows as it plows into the denser surrounding interstellar matter. Since this interstellar matter varies in density, the bubble's surface is not smooth. The star is off-center because the wind flows faster into less dense material.

The fingers of molecular gas at top are similar to the columns seen in the Eagle nebula. They are being illuminated by ultraviolet from the hot star of the Bubble Nebula, even though the outflow of mass from the star hasn't yet reached them.

▲ The black cloud Barnard 68, when imaged in the infrared, becomes somewhat transparent. Thus though no stars are seen through it at left, in the image taken with blue, visible, and near-infrared filters, the addition of a farther infrared band at 2.2 $\mu$m allows stars to shine through. In this false-color image, that "K-band" is translated to red, so the stars appear red at right.

Dark clouds like these, known as "globules," are thought to be the precursors of low-mass solar-like stars. This "Bok globule" was already shown in Figure 33–3.

The Horsehead Nebula in Orion looks very different when seen in mid-infrared wavelengths. In these views from the Infrared Space Observatory (ISO) at 7 and 15 $\mu$m, the main nebula is larger because of the glowing carbon compounds. The horse's head glows because of its own carbon compounds as well as dust. Newly formed stars show as bright dots, and are emphasized in the insets.

The Lagoon Nebula, M8, in the infrared from the 2MASS survey. See also Figures 33–1 and 33–3*B*.

**FIGURE 33–12** The sky at 21 cm, with black and dark blue indicating the smallest and red and white the largest hydrogen densities. The neutral hydrogen is obviously closer to the Galactic plane than the sources of synchrotron radiation.

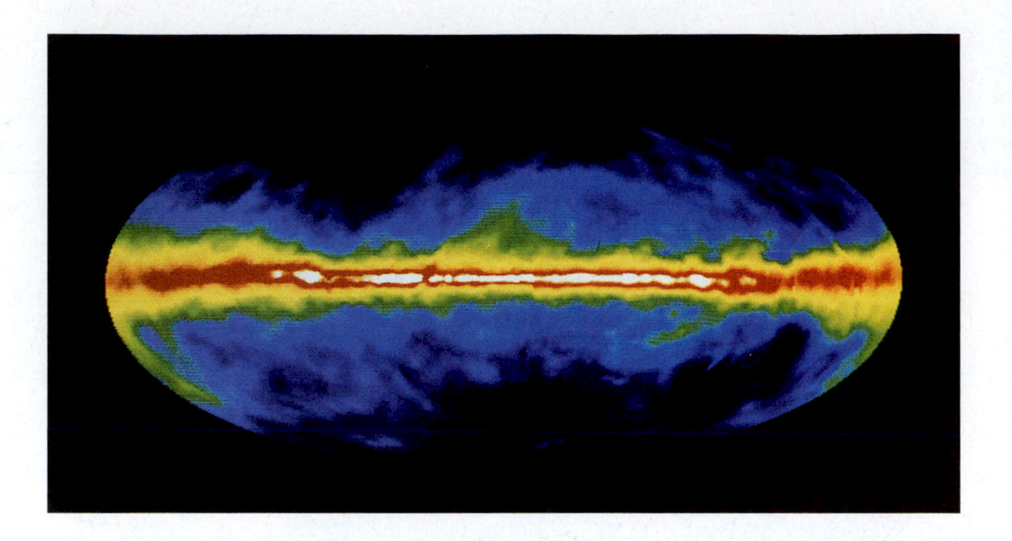

outer regions of galaxies rotate differentially; that is, the gas nearer the center rotates faster than the gas farther away from the center.

Figure 33–14 shows a simplified version of differential rotation. Because of the differential rotation, the distance between us and point A is decreasing. Therefore, from our vantage point at the sun, point A has a net velocity toward us. Thus its 21-cm line is Doppler shifted toward shorter wavelengths. If we were talking about light, this shift would be in the blue direction; even though we are discussing radio waves, we say "blueshifted" anyway. Blueshifted now simply means shifted to a shorter wavelength (higher frequency). If we look from our vantage point toward gas cloud C, we see a redshifted 21-cm line (shifted to a longer wavelength = lower frequency), because its higher speed of rotation is carrying C away from us. But if we look straight toward the center, clouds $B_1$ and $B_2$ are both passing across our line of sight in a path parallel to that of our own orbit. They have no net velocity toward or away from us. Thus this method of distance determination does not work when we look in the direction of the center, nor indeed in the opposite direction.

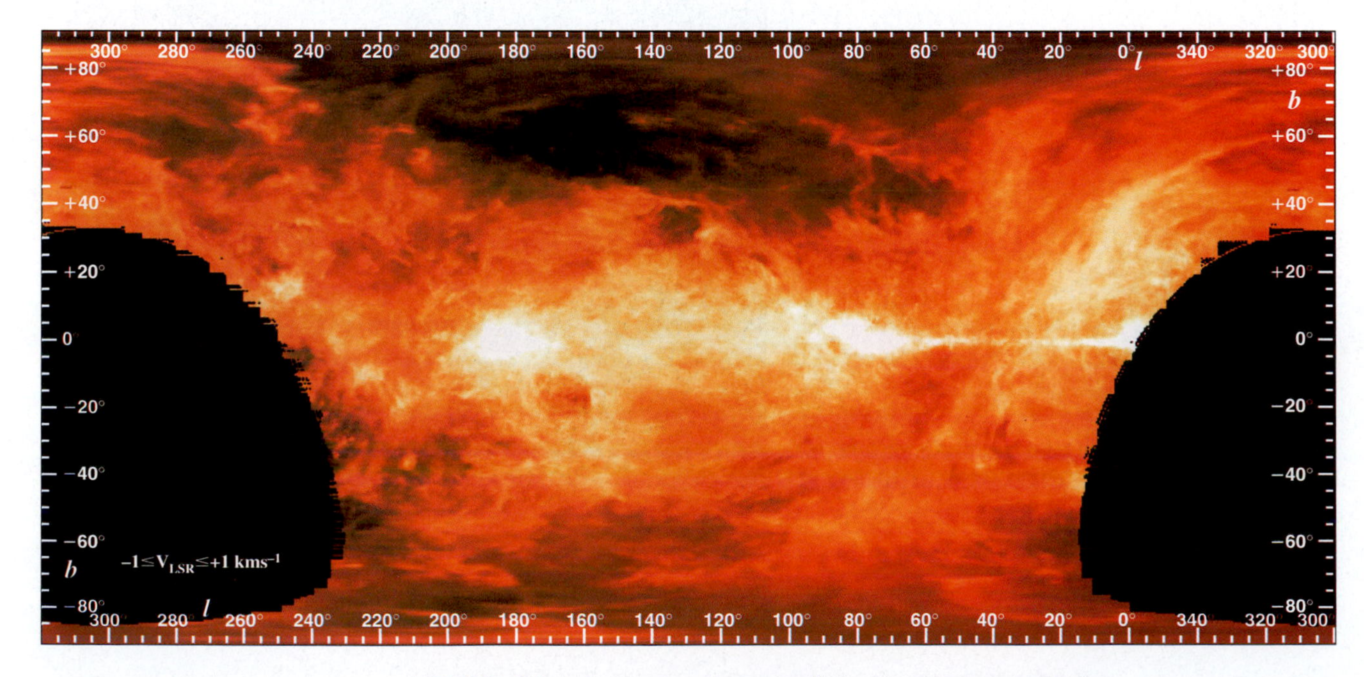


**FIGURE 33–13** A map of 21-cm radiation in a Mercator projection.

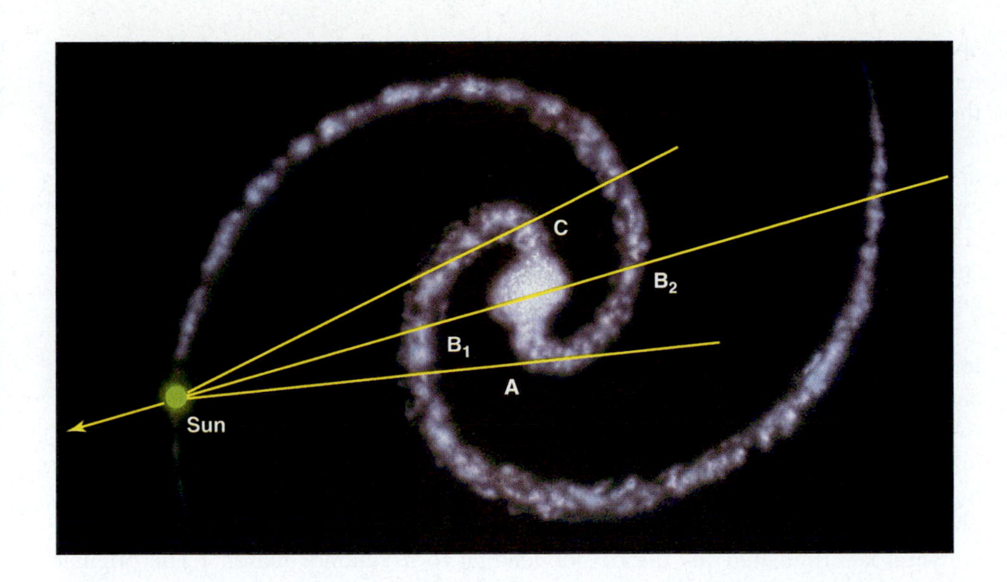


**FIGURE 33–14** Because of the differential rotation, the cloud of gas at point A appears to be approaching the sun, and the cloud of gas at point C appears to be receding. Objects at point $B_1$ or $B_2$ have no net velocity with respect to the sun and therefore show no Doppler shifts in their spectra. The closer in toward the Galactic center a cloud is, the faster it rotates, and so the larger its Doppler shift along any line of sight (other than that toward the center) when observed from the sun. (The difference between the Earth's and the sun's velocity is known and can be accounted for.)

Clouds farther from the Galactic center take longer to revolve than do clouds closer to the center, which corresponds to Kepler's third law, discussed in Section 3.1c and Box 3.4. Along any line of sight, the cloud with the highest velocity must be closest to the center.

Once we work out the law of differential rotation for the whole Galaxy by looking along various lines of sight, then we can apply our knowledge to any particular cloud we observe. From its observed velocity, we can tell how far it is from the center of the Galaxy: we figure out where along our line of sight in the direction we are looking we meet the ring of gas having the proper velocity to match the observations. By observing in different directions, we can build up a picture of the spiral arms.

### 33.4c 21-cm Radiation in External Galaxies

The 21-cm line of hydrogen has been fundamental to mapping the structure and rotation of many galaxies, not just our own. Brent Tully of the University of Hawaii and Richard Fisher of the National Radio Astronomy Observatory found some years ago that the speed of a galaxy's rotation is correlated with its brightness. This Tully-Fisher relation has been very useful for finding the distances to many galaxies and so has been important in cosmology.

The Square Kilometer Array (a possible version of which we saw in Figure 5–34) is to have 100 times more collecting area than today's biggest radio imaging telescopes. Radio astronomers' desire to be able to detect the faint 21-cm emission from structures formed early in the universe and in the galaxies that formed from them was a major reason for conceiving of the telescope. In fact, the choice of 1 kilometer for the diameter was linked to the ability of a telescope that size to image hydrogen gas in distant galaxies with at least 0.1 arc sec resolution, which is the resolution of the Hubble Space Telescope. The hydrogen 21-cm radiation, as always, penetrates even clouds of dust that might otherwise obscure the Galaxy.

## 33.5 RADIO SPECTRAL LINES FROM MOLECULES

For several years after the 1951 discovery of 21-cm radiation (it had previously been predicted), spectral-line radio astronomy continued with just the one spectral line. Astronomers tried to find others. One prime candidate was OH, hydroxyl, a molecule that should be relatively abundant because it is a combination of the most abundant element, hydrogen, with one of the most abundant of the remaining elements, oxygen. OH has four lines close together at about 18 cm in wavelength, and the relative intensities expected for the four lines had been calculated.

It wasn't until 1963 that other radio spectral lines were discovered, and four new emission lines were indeed found at 18 cm in 1965. But the intensity ratios were all wrong to be OH, according to the predicted values, and for a time we spoke of the "mysterium" lines. Mysterium turned out to be OH after all, but with the process of "masering" affecting the excitation of the energy levels of OH and thus amplifying certain of the lines at the expense of others, which are weakened.

The idea of discovering a new element was not unprecedented—after all, unknown lines at the solar eclipse in 1868 had been assigned to an unknown element, "helium," because they occurred only (as far as was known at that time) on the sun. But the periodic table of elements has been filled in during the last 100 years, so we can't expect to find new elements in that way anymore.

**Masers** and **lasers** are of great practical use on the Earth. (Maser is an acronym for *m*icrowave *a*mplification by *s*timulated *e*mission of *r*adiation, and lasers are their analogue using *l*ight instead of *m*icrowaves.) For example, masers are used as sensitive amplifiers. Masers were "invented" on Earth not long before they were found in space. For maser action to occur, a large number of molecules are pushed into a higher energy state in which they tend to stay. When "triggered" by a photon of the proper frequency, the molecules jump down together to a lower energy level. (Their emission is "stimulated" by the trigger, leading to the name "maser.") The original radiation is thus amplified, since there are now many photons at that wavelength instead of just one.

Since the interstellar abundance of OH seemed very much lower than that of isolated hydrogen or oxygen atoms (only one OH molecule for every billion H atoms), it seemed quite unlikely that the quantities of any molecules composed of three or more atoms would be great enough to be detected. The chance of three atoms' getting together in the same place should be very small.

In 1968, however, Berkeley's Charles Townes, Jack Welch, and colleagues observed the radio frequencies that were predicted to be the frequencies of water ($H_2O$) and ammonia ($NH_3$). The spectral lines of these molecules proved surprisingly strong and were easily detected.

Soon afterwards, another group of radio astronomers used a telescope of the National Radio Astronomy Observatory to discover interstellar formaldehyde ($H_2CO$) at a wavelength of 6 cm. This discovery (by Ben Zuckerman, now of UCLA; Patrick Palmer, now of the University of Chicago; David Buhl, now of NASA; and Lewis Snyder, now of the University of Illinois) was the first molecule detected that contained more than two atoms and that contained two "heavy" atoms, that is, two atoms other than hydrogen.

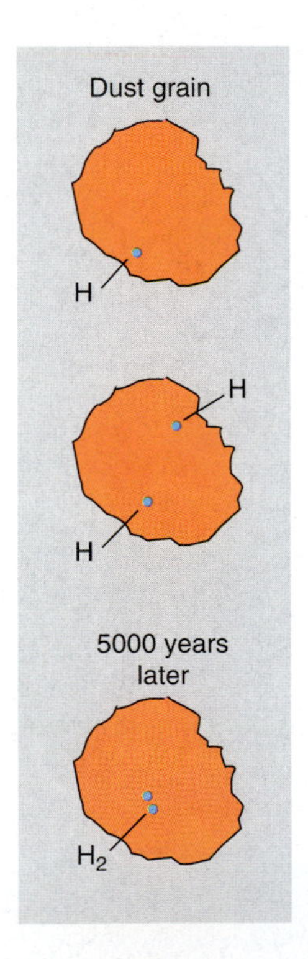


FIGURE 33–15 Hydrogen molecules are formed in space with the aid of dust grains at an intermediate stage.

Certain molecules form in the gas phase via ion-molecule reactions, but other molecules must form on grain surfaces. This latter category contains certain large molecules of biological interest, such as formic acid, acetic acid, and probably urea and glycine.

By this time, it was apparent that the earlier notion that it would be difficult to form large molecules in space was wrong. There has been much research on this topic, and finally the mechanisms by which molecules are formed is largely understood. For some molecules, including molecular hydrogen, it seems that the presence of dust grains is necessary. In this scenario, one atom hits a dust grain and sticks to it (Fig. 33–15). It may be thousands of years before a second atom hits the same dust grain, and even longer before still more atoms hit. But these atoms may stick to the dust grain rather than bouncing off, which gives them time to join together. Complex reactions may take place on the surface of the dust grain. Then, somehow, the molecule must get off the dust grain, thus being released into space as part of a gas. Perhaps either incident ultraviolet radiation or the energy released in the formation of the molecule allows the molecule to escape from the grain surface.

Though hydrogen molecules form on dust grains, a strong body of opinion holds that most of the other molecules are formed in the interstellar gas without need for grains. Recent theoretical and laboratory studies indicate that reactions between neutral molecules and ionized molecules may be particularly important. Many of these chains start with molecular hydrogen, formed on grains, being ionized by cosmic rays. There are still many gaps in the theory. It is likely that no single process forms all the interstellar molecules.

The list of molecules discovered expanded gradually from three-atom molecules like ammonia and water and four-atom molecules like formaldehyde, to even more complex molecules. Over 120 molecules have now been discovered in interstellar space—including, amusingly, even methyl and methyl alcohol (Table 33–1). One of the more complex is $HC_{11}N$, which has 12 "heavy" atoms and only 1 hydrogen. Even acetic acid, close to an amino acid, was discovered in interstellar space by astronomers at the University of Illinois at Urbana using the Berkeley–Illinois–Maryland Array.

**TABLE 33–1** Interstellar Molecules

| | | | |
|---|---|---|---|
| $H_2$ | hydrogen | CO | carbon monoxide |
| CSi | carbon monosilicide | CP | carbon monophosphide |
| CS | carbon monosulfide | NO | nitric oxide |
| NS | nitrogen monosulfide | SO | sulfur monoxide |
| HCl | hydrogen chloride | NaCl | sodium chloride |
| KCl | potassium chloride | AlCl | aluminum monochloride |
| AlF | aluminum monofluoride | PN | phosphorus mononitride |
| SiN | silicon mononitride | SiO | silicon monoxide |
| SiS | silicon monosulfide | NH | imidyl radical |
| OH | hydroxyl radical | $C_2$ | diatomic carbon |
| CN | cyanide radical | HF | hydrogen fluoride |
| $CO^+$ | | $SO^+$ | |
| CH | | $CH^+$ | |
| SH | | | |
| **Molecules with 3 atoms** | | | |
| $H_2O$ | water | $H_2S$ | hydrogen sulfide |
| HCN | hydrogen cyanide | HNC | hydrogen isocyanide |
| $CO_2$ | carbon dioxide | $SO_2$ | sulfur dioxide |
| MgCN | magnesium cyanide | MgNC | magnesium isocyanide |
| NaCN | sodium cyanide | $N_2O$ | nitrous oxide |
| $NH_2$ | amidyl radical | OCS | carbonyl sulfide |
| HCO | formyl radical | $C_3$ | triatomic carbon |
| $C_2H$ | | $HCO^+$ | |
| $HOC^+$ | | $N_2H^+$ | |
| HNO | | $HCS^+$ | |
| $H_3{}^+$ | | $C_2O$ | |
| $C_2S$ | | $SiC_2$ | |
| $H_2D^+$? | | $CH_2$ | |
| **Molecules with 4 atoms** | | | |
| $NH_3$ | ammonia | $H_2CO$(?) | formaldehyde |
| $H_2CS$ | thioformaldehyde | $C_2H_2$ | acetylene |
| HNCO | isocyanic acid | HNCS | thioisocyanic acid |
| $H_3O^+$ | hydronium ion | $HOCO^+$ | |
| $C_3S$ | | $H_2CN$ | |
| cyclic-$C_3H$ | | linear-$C_3H$ | |
| HCCN | | $H_2CO^+$ | |
| $C_2CN$ | | $C_3O$ | |
| $HCNH^+$ | | $CH_2D^+$ | |
| $CH_3$ | methyl radical | $SiC_3$ | |
| **Molecules with 5 atoms** | | | |
| $CH_4$ | methane | $SiH_4$ | |
| $CH_2NH$ | methyleneimine | $NH_2CN$ | cyanamide |
| $CH_2CO$ | ketene | HCOOH(?) | formic acid |
| $HC_2CN$ | cyanoacetylene | HCCNC | isocyanoacetylene |
| cyclic-$C_3H_2$ | | linear-$C_3H_2$ | |
| $CH_2CN$ | | $C_4H$ | |
| $C_4Si$ | | $C_5$ | |
| HNCCC | | $H_2COH^+$ | |

*(table continued next page)*

**TABLE 33–1** *continued*

| | | | |
|---|---|---|---|
| **Molecules with 6 atoms** | | | |
| $CH_3OH$ | methanol | $CH_3SH$ | methanethiol |
| $C_2H_4$ | ethylene | $CH_3CN$ | methylcyanide |
| $CH_3NC$ | methylisocyanide | $HC_2CHO$ | propynal |
| $NH_2CHO$ | formamid | $C_4H_2$ | butadiyne |
| $C_5H$ | | $HC_3NH^+$ | |
| $C_5O$ | | $C_5N$ | |
| **Molecules with 7 atoms** | | | |
| $CH_3C_2H$ | methylacetylene | $CH_3CHO$ | acetaldehyde |
| $CH_3NH_2$ | methylamine | $CH_2CHCN$ | acrylonitrile |
| $HC_4CN$ | cyanobutadiyne | $C_6H$ | |
| cylic-$C_2H_4O$ | | | |
| **Molecules with 8 atoms** | | | |
| $CH_3COOH$ | acetic acid | $HCOOCH_3$ | methyl formate |
| $CH_3C_2CN$ | cyanomethylacetylene | $C_7H$ | |
| $H_2C_6$ | hexapentaenylidene | | |
| $HOCH_2C(=O)H$ | glycoaldehyde | | |
| **Molecules with 9 atoms** | | | |
| $(CH_3)_2O$ | dimethyl ether | $C_2H_5OH$ | ethanol |
| $C_2H_5CN$ | ethylcyanide | $CH_3C_4H$ | methylbutadiyne |
| $HC_6CN$ | cyanohexatriyne | $C_8H$ | |
| **Molecules with 10 atoms** | | | |
| $(CH_3)_2CO$ | acetone | $CH_3C_5$ | cyanomethylbutadiyne |
| $NH_2CH_2COOH$? | aminoacetic acid | | |
| **Molecule with 11 atoms** | | | |
| $HC_8CN$ | cyanooctatetrayne | | |
| **Molecule with 12 atoms** | | | |
| $C_6H_6$ | benzene | | |
| **Molecule with 13 atoms** | | | |
| $HC_{10}CN$ | cyanodecapentayne | | |

*Note:* Molecules in blue have been identified in the ice phase as well as the gas phase. "?" indicates an uncertain detection.

Hundreds of spectral lines, some of which are undoubtedly from still other molecules, remain unidentified. Scientists have long thought that the ion of the molecule of three hydrogen atoms, $H_3^+$, could help many more complex molecules form. This molecular ion wasn't observed until the late 1990s, when it was found by its faint infrared spectral line.

Radio studies of molecular spectral lines have been used together with 21-cm observations to improve the maps of the spiral structure of our Galaxy. Observations of carbon monoxide (CO) in particular have provided better information about the parts of our Galaxy farther out than the distance of the sun from the Galaxy's center (Fig. 33–16*A*). CO is a tracer of the more abundant hydrogen molecules, which are harder to study. But there are still differences to be resolved between the different spirals that have been suggested. There are signs, probably real, of short partial arms or spurs, in addition to the four major arms. Our Galaxy has been called a "messy spiral."

The capability of the infrared space telescopes to study molecules is improving our understanding of the distribution of such molecules in space. For example, the Infrared Space Observatory observed water vapor in regions excited by shock waves given off as part of the stellar winds from old stars. It also observed the water vapor

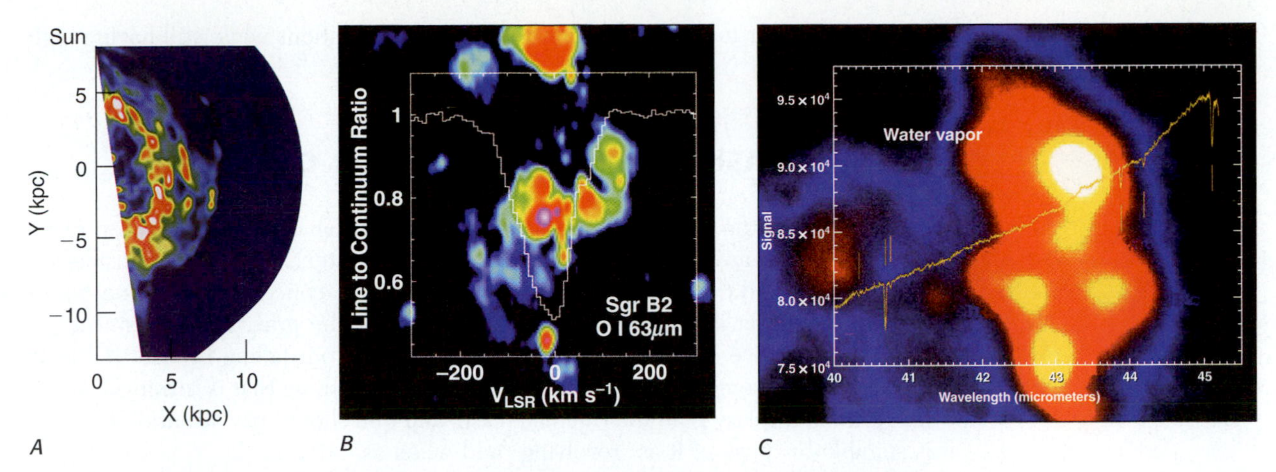


**FIGURE 33–16** (*A*) A map based on CO data does not show clear spiral arms. The point at (0, 0) on the graph marks the center of our Galaxy. The regions between 4–6 kpc and near 7 kpc are more ringlike than spiral. Studies of other galaxies are showing that clear spiral arms are common in the outer regions of galaxies but not necessarily in the inner regions, which may be the case with our Galaxy as well. (*B*) An ISO spectrum showing an absorption line from neutral oxygen overlain on a Very Large Array radio image of Sgr B2. Hundreds of spectral lines from molecules are detected in this source. (*C*) An ISO infrared spectrum (from its Short Wavelength Spectrometer, SWS) showing several lines from water vapor, overlain on an ISO image of Orion IRc2, an infrared source we discuss at the end of the chapter.

that cools the clouds of gas and dust that are collapsing to form stars in the first place. It has spectra and images showing atoms and molecules from the molecular cloud Sagittarius B2, located not far from the center of our Galaxy (Fig. 33–16*B*). More interstellar molecules have been detected in Sgr B2 than anywhere else.

The Submillimeter Wave Astronomy Satellite (SWAS) has detected water vapor at many locations in our Galaxy since its launch in 1998. Part of NASA's series of "small Explorers," it is run from the Harvard-Smithsonian Center for Astrophysics, with team members from MIT, U. Mass.-Amherst, Johns Hopkins, and elsewhere. In the coldest locations, it found only a few parts per billion of water vapor, less than had been anticipated. Perhaps the water vapor has frozen onto the surfaces of dust grains. In its observations of warmer gas clouds, where stars are being born, the temperature is thousands of kelvins and they found more than a few parts per million of water vapor. Formation of the water vapor cools the interstellar gas and may trigger star birth. SWAS can also observe at the frequency of spectral bands from oxygen molecules, though it has detected hardly any. It is unknown why the oxygen that must be there doesn't show.

Water vapor was also observed from the Infrared Space Observatory (Fig. 33–16*C*). The observations match those of water-vapor masers in showing that the water is in an expanding shell around the central object.

Studying the spectral lines from molecules in various parts of the spectrum provides information about physical conditions—temperature, densities, and motion, for example—in the gas clouds that emit the lines. For example, formaldehyde radiates only when the density is roughly ten times that of the gas at which carbon monoxide radiates. The clouds that emit molecular lines are usually so dense that hydrogen atoms have combined into hydrogen molecules and very little 21-cm radiation is emitted.

The largest millimeter telescope in the world is now in Nobeyama, Japan. At the end of the decade, we hope to have ALMA, the Atacama Large Millimeter Array, starting operation in the high, dry desert of northern Chile. It is a joint American/European project. It is to have 64 antennae, each 12-m across. It will be an interferometer in order to make very-high-resolution observations while still having high sensitivity.

## 33.6 MEASURING THE MASS OF OUR GALAXY

We can measure the mass of our Galaxy by studying the velocity of rotation of its gas clouds (which are revolving around the Galaxy's center). All the mass inside the radius of the cloud's orbit acts about as though it were concentrated at one point; the more mass that is present (and thus the stronger the gravity), the faster a gas cloud has to revolve around the center to keep itself from falling inward. (Objects "revolve" around something but "rotate" as part of something that is turning around; a gas cloud in a galaxy is an intermediate case, and one should pay attention to when it is suitable to speak of it as "revolving" and when as "rotating.")

The gravitational effect of the mass outside the cloud's orbit turns out to balance out (assuming spherical symmetry). Thus this exterior mass, however large, does not affect the cloud's velocity. Conversely, the cloud's velocity gives us no information about this mass, so we can measure the mass of our Galaxy only as far out as we can detect orbiting clouds.

From the velocity with which a gas cloud is orbiting, we can calculate how much mass is inside the orbit of the cloud. We use Newton's form of Kepler's third law (Box 3.4). Our own sun revolves around the center of our Galaxy in about 250 million Earth years, a period known as a "Galactic year." From this period and our measured distance from the Galactic center, we can calculate the mass internal to us.

It is thus important to measure the velocity of rotation of gas clouds in our Galaxy's disk at radii from the center as great as possible. Velocities have been

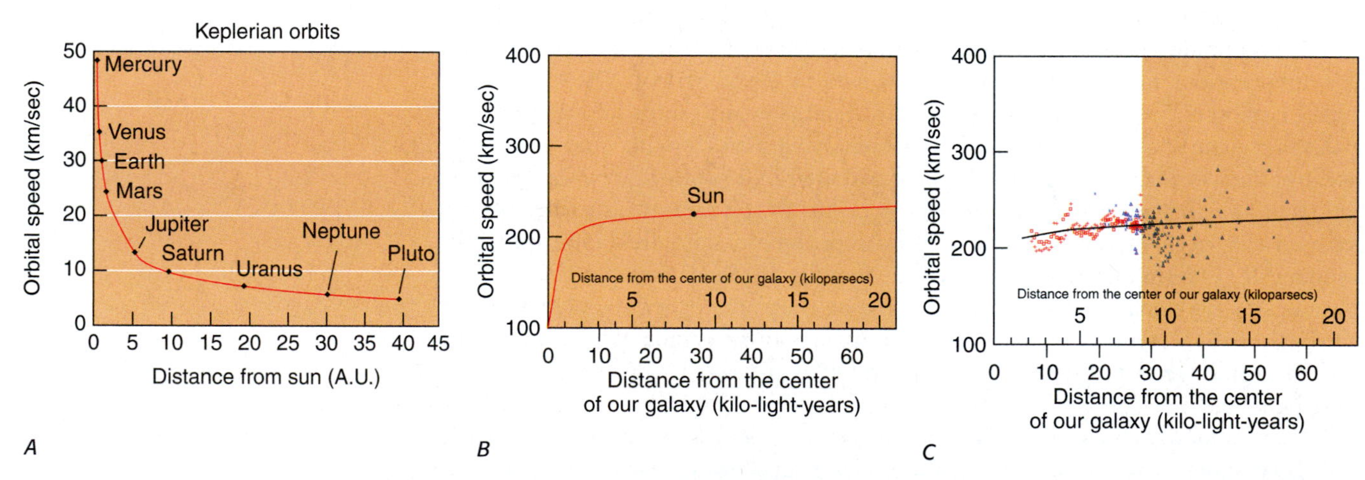


**FIGURE 33–17** (*A*) The rotation curve for our solar system, showing the Keplerian falloff of orbital speed. (*B*) A sketch of the rotation curve for our Galaxy. (*C*) The actual rotation curve of our Galaxy; the solid line is fit to the data points. The sun is 8.5 kiloparsecs from the Galactic center; the region beyond the sun is shaded. The inner part of the curve is based on 21-cm hydrogen observations (*red*). It had been anticipated that beyond the sun's distance from the Galactic center, the velocity would decline following Kepler's third law, because there would be little mass added as we go farther out. But by observing carbon monoxide (*blue*) we are able to make direct measurements of these outer regions (*yellow shading*). The observations (red pluses for northern-hemisphere 21-cm hydrogen observations, red squares for southern-hemisphere 21-cm hydrogen, blue crosses for northern-hemisphere CO, and blue triangles for southern-hemisphere CO) show that the velocity of rotation doesn't decline. This lack of decline of the curve with increasing distance from the sun must mean that there is more mass in the outer regions of the Galaxy than we had anticipated.

measured inside the sun's orbit from measurements of the 21-cm line, using the method discussed in Section 33.4. Leo Blitz, now at Berkeley, extended these measurements to much greater radii than was possible with 21-cm radiation by using radiation from carbon monoxide instead. The carbon monoxide, though present only in trace amounts in interstellar clouds, shares in the clouds' motions.

A graph of the velocity of rotation vs. distance from the center is called a **rotation curve** (Fig. 33–17). It had been expected up until about 1980 that our Galaxy's rotation curve would stop increasing beyond the sun's distance from the Galactic center and begin to decrease, since the Galaxy essentially ended. Then little more mass would be included inside as we went to larger radii. But the curve does not stop increasing and begin to decrease, which indicates that the Galaxy is larger and contains more mass than we had thought. It now seems that our Galaxy is twice as massive as the Andromeda Galaxy, and contains perhaps $10^{12}$ solar masses; this is at least twice as massive as had previously been calculated. If we assume that an average star has 1 solar mass, then our Galaxy contains about a trillion stars.

What form does this additional mass take? We don't really know, but we know that it isn't ordinary matter. We thus call it "dark matter," indicating that it is matter that doesn't give off light or any other radiation (not merely meaning that it doesn't give off visible light). We discuss dark matter further in Section 38.2. Scientists calculate, on the basis of various observations, that ordinary matter makes up only about 3 per cent of the overall stuff of the universe, and that dark matter makes up another 30 per cent. Thus ordinary and dark matter together make up about ⅓ of the universe.

## 33.7 MOLECULAR HYDROGEN

At the low temperature of interstellar space, only the lowest energy levels of molecular hydrogen are excited, and the lines linking these levels fall in the far ultraviolet. Because our atmosphere prevents these lines from reaching us on Earth, we had to wait to observe from space in order to observe lines from $H_2$. $H_2$ was first observed from a rocket in 1970.

In 1972, a 90-cm telescope was carried into orbit aboard NASA's third Orbiting Astronomical Observatory, named "Copernicus" in honor of that astronomer's 500th birthday (which occurred a year later). The telescope was largely devoted to observing interstellar material. The observers could point the telescope at a star and look for absorption lines caused by gas in interstellar space as the light from the star passed through the gas en route to us. Since it is easiest to pick up interstellar absorption lines if the star itself has no lines of its own, the scientists observed in the direction of B stars, which have few lines and are very bright.

Since the hydrogen molecule is easily torn apart by ultraviolet radiation, the observers did not find much molecular hydrogen in most directions. But whenever they looked in the direction of highly reddened stars, they found a very high fraction of hydrogen in molecular form: more than 50 per cent. Presumably, in these regions some of the dust that causes the reddening shields the hydrogen from being torn apart by ultraviolet radiation. There are also theoretical grounds for believing that the molecular hydrogen is formed on dust grains, so it seems reasonable that the high fraction of $H_2$ is found in the regions with more dust grains.

In the near infrared, molecular hydrogen emission has been used to map regions of star formation. Further, the hydrogen molecule emits faintly at a wavelength of 17 $\mu$m, emission resulting from the rotation of the molecule instead of the vibrations that cause the ultraviolet lines. The Infrared Space Observatory (ISO) was able to study this hydrogen emission not only in our own Galaxy but also in distant galaxies (Fig. 33–18A).

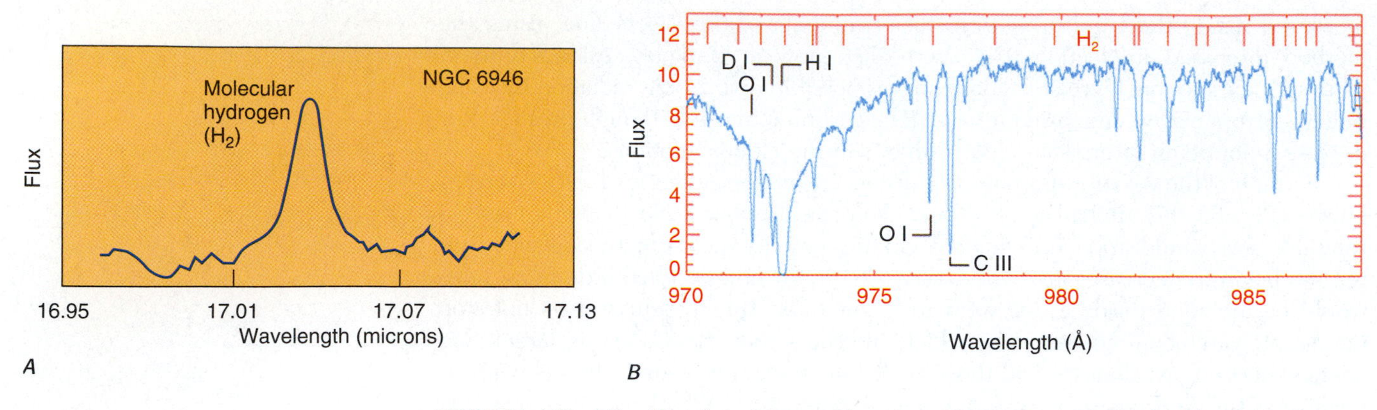


**FIGURE 33–18** (*A*) Infrared Space Observatory (ISO) observations of molecular hydrogen at a wavelength of 17 microns in the galaxy NGC 6946. (*B*) A *F*ar *U*ltraviolet *S*pectroscopic *E*xplorer (FUSE) spectrum showing molecular hydrogen. The positions of lines from the hydrogen molecule are marked near the top.

NASA's *F*ar *U*ltraviolet *S*pectroscopic *E*xplorer (FUSE) mission went into orbit in 1999. It observes, among other things, absorption bands of molecular hydrogen (Fig. 33–18*B*). It also observes this molecule with a deuterium atom substituted for an ordinary hydrogen atom. In Chapter 37, we will discuss the significance of this substitution.

## 33.8 THE FORMATION OF STARS

Most radio spectral lines seem to come only from a very limited number of places in the sky—molecular clouds (Fig. 33–19). (Carbon monoxide is the major exception, for it is widely distributed across the sky.) Infrared and radio observations together have provided us with an understanding of how stars are formed from these dense regions of gas and dust.

**Giant molecular clouds** are 50 to 100 parsecs across. There are a few thousand of them in our Galaxy (Fig. 33–20). The largest giant molecular clouds are about 100 parsecs across and contain about 100,000 to 1 million times the mass of the sun. Their internal densities are about 100 times that of the interstellar medium around them. Since giant molecular clouds break up to form stars, they only last 10 million to 100 million years.

Carbon-monoxide observations reveal the giant molecular clouds, but it is molecular hydrogen ($H_2$) rather than carbon monoxide that is significant in terms of mass. It is difficult to detect the molecular hydrogen directly, though there is over 100 times more molecular hydrogen than dust. So we must satisfy ourselves with observing the tracer, carbon monoxide, whose spectral lines are excited by collisions with hydrogen molecules. The carbon monoxide thus shows us where the hydrogen molecules are.

The Great Rift in the Milky Way, the extensive dark area that can be seen in the midst of the Milky Way, is made up of many overlapping giant molecular clouds.

A giant molecular cloud is fragmented into many denser bits 1 parsec in size. These bits must become much smaller and denser yet to form a star or, in many cases, several stars.

Sometimes we see smaller objects, called globules, or **Bok globules.** (They are named after Bart Bok, who extensively studied these objects.) Some of the smaller globules are visible in silhouette against H II regions. Others are larger and appear isolated against the stellar background. Two hundred large globules are known to be within 500 parsecs of the sun (5 per cent of the way to the center of the Galaxy), and there may be 25,000 in our Galaxy. It has been suggested that the globules may

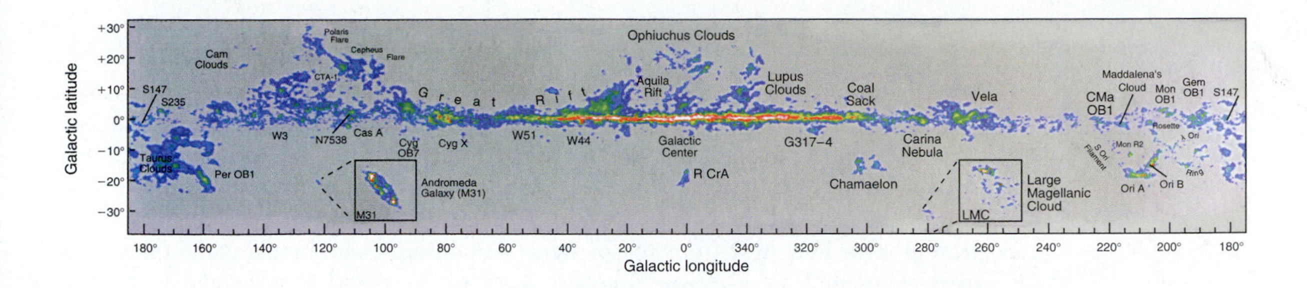


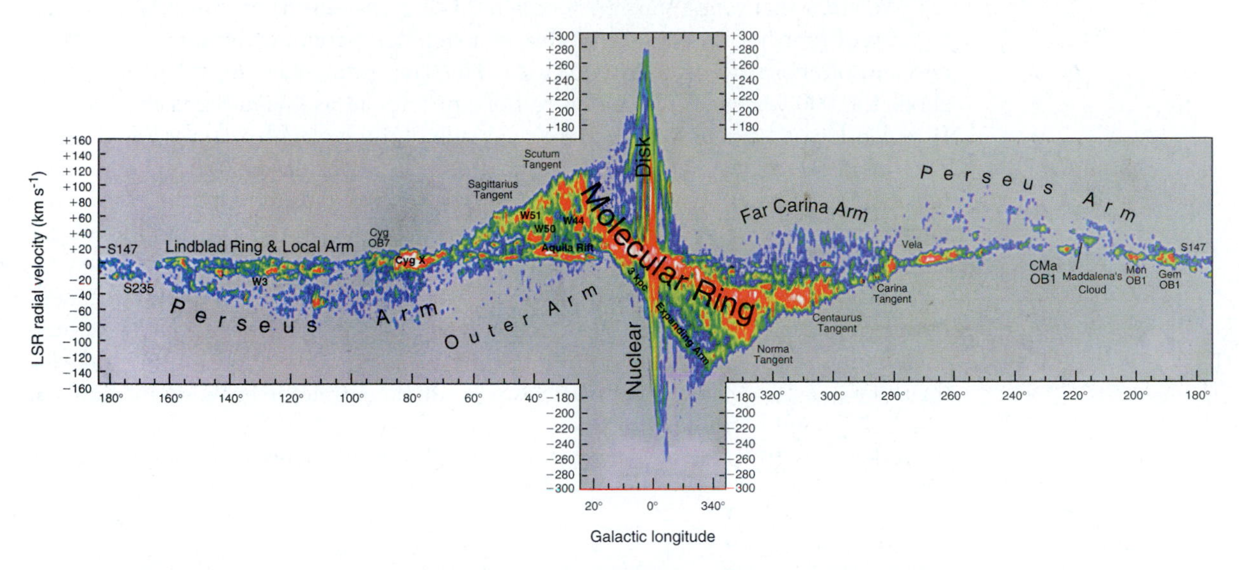


**FIGURE 33–19** Molecular clouds, mapped with a millimeter-wavelength telescope atop the Harvard-Smithsonian Center for Astrophysics from the middle of Cambridge, Massachusetts. The Galactic equator is spread out from left to right. The top graph shows the positions of the clouds. At the bottom, the vertical axis is radial velocity, so we can see which clouds are coming toward us and which are away. Note the huge velocities of clouds near the center of our Galaxy. "LSR" means the local standard of rest; measurements are taken with respect to this frame, which signifies no motion with respect to the average velocity of nearby stars.

be on their way to becoming stars. Molecular-line observations of the larger globules indicate that they may contain sufficient mass for gravitational collapse to be taking place. And though there had been some doubt whether stars were forming in smaller globules, IRAS discovered infrared radiation from at least some that indicates the presence of star formation inside. For example, a few stars, each about the mass of the sun and each only a few hundred thousand years old, were discovered in the globule called Barnard 5.

Observations of interstellar molecules have shown that gas flows rapidly outward from some newly formed stars in two opposing streams. These high-velocity **bipolar flows** typically have carried more than a solar mass of material outward and transport a lot of energy. They are thus clearly very important for the evolution of young stars. The gas flow extends only about 1 light-year, which can be traversed by gas at the high velocities measured in only about 10,000 years. Herbig-Haro objects (Section 27.1) are sometimes located at the end of one of the jets; the motion of the H-H objects can be projected back to the same location at which the jets originate. In this and in most cases, an infrared source (with often no image in the visible part of the spectrum) is located at the center. We have already met a jet of gas in the model of SS433, and we will learn about other jets in radio galaxies and in quasars. Astrophysicists are finding that signs of violent activity like jets are more common than had been suspected.

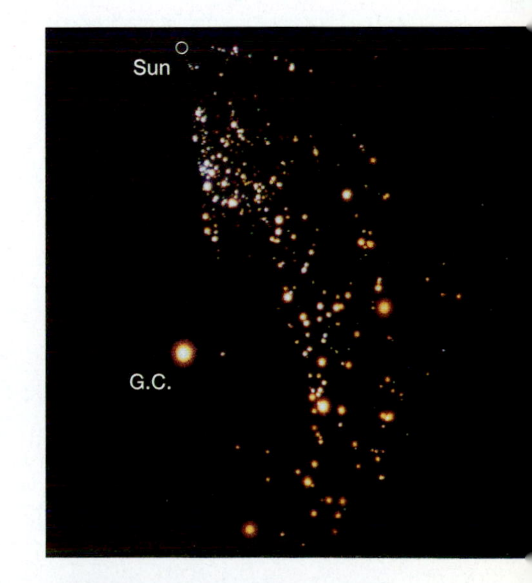


**FIGURE 33–20** A map of the Milky Way as it would appear from above, showing the molecular clouds in one part of our Galaxy. Two spiral arms show.

## 33.9 A CASE STUDY: THE ORION MOLECULAR CLOUD

Many radio spectral lines have been detected only in a particular cloud of gas located 500 parsecs (1600 light-years) from us in the direction of the constellation Orion, not very far from the main Orion Nebula (Fig. 33–21). This **Orion Molecular Cloud** is relatively accessible to our study because it is so close to us. Even though less than 1 per cent of the Cloud's mass is dust, that is still a sufficient amount of dust to prevent ultraviolet light from nearby stars from entering and breaking the molecules apart. Thus molecules can accumulate. Most of the cloud is molecular hydrogen, though the cloud is best studied by trace amounts of the molecule carbon dioxide.

We know that young stars are found in this region—the Trapezium (Fig. 33–22), a group of four hot stars readily visible in a small telescope, is the source of ionization and of energy for the Orion Nebula. The Trapezium stars are relatively young, about 100,000 years old. The Orion Nebula, prominent as it is in the visible, is an H II region located as a blister along the near side of the molecular cloud (Fig. 33–23). The nebula contains much less mass than does the molecular cloud. The ultraviolet radiation from the H II region causes a sharp front of ionization at the edge of the molecular cloud. A shock wave forms and concentrates gas and dust in the molecular cloud to make protostars.

The properties of the molecular cloud can be deduced by comparing the radiation from its various molecules and by studying the radiation from each type of molecule individually. The main Orion Molecular Cloud is 100,000 times more massive than the sun. Its density, 1000 particles per cubic centimeter in the outer limits at which the cloud is visible to us, increases toward the center. The cloud may actually be as dense as $10^6$ particles/cm$^3$ at its center. This is still billions of times less dense

*(text continues on page 665)*

**FIGURE 33–21** (*A*) A visible-light image of the Orion Nebula, an H II region on the side of a molecular cloud. (*B*) A view of Orion taken with an infrared array. Three infrared wavelengths known as J, H, and K (1.25 $\mu$m, 1.6 $\mu$m, and 2.2 $\mu$m, respectively) were used to make this false-color view.

A

B

**FIGURE 33–22** The center of the Orion Nebula, a region 2.5 light-years wide and 1500 light-years from us. (*A*) A view from the Hubble Space Telescope. The image is a mosaic composited from 15 separate fields of view, though it still covers an area only 5% that of the full moon. In the center of the image, we see the Trapezium, the four bright, hot stars whose ultraviolet light makes the Orion Nebula glow. Seven hundred other young stars are also present. We see shock waves caused by jets of hot gas given off by some of these stars. One major shockwave is seen as the diagonal red bar at lower left. We also see some of the protoplanetary disks ("proplyds") that seem to be solar systems in formation. In the image, light emitted by oxygen is shown as blue, hydrogen emission is shown as green, and nitrogen emission is shown as red. The result corresponds closely to what would be seen by an observer close up. About 2% of the area of the image has been filled in with ground-based images. (*B*) A wider view of the Orion Nebula, imaged with the Anglo-Australian Telescope.

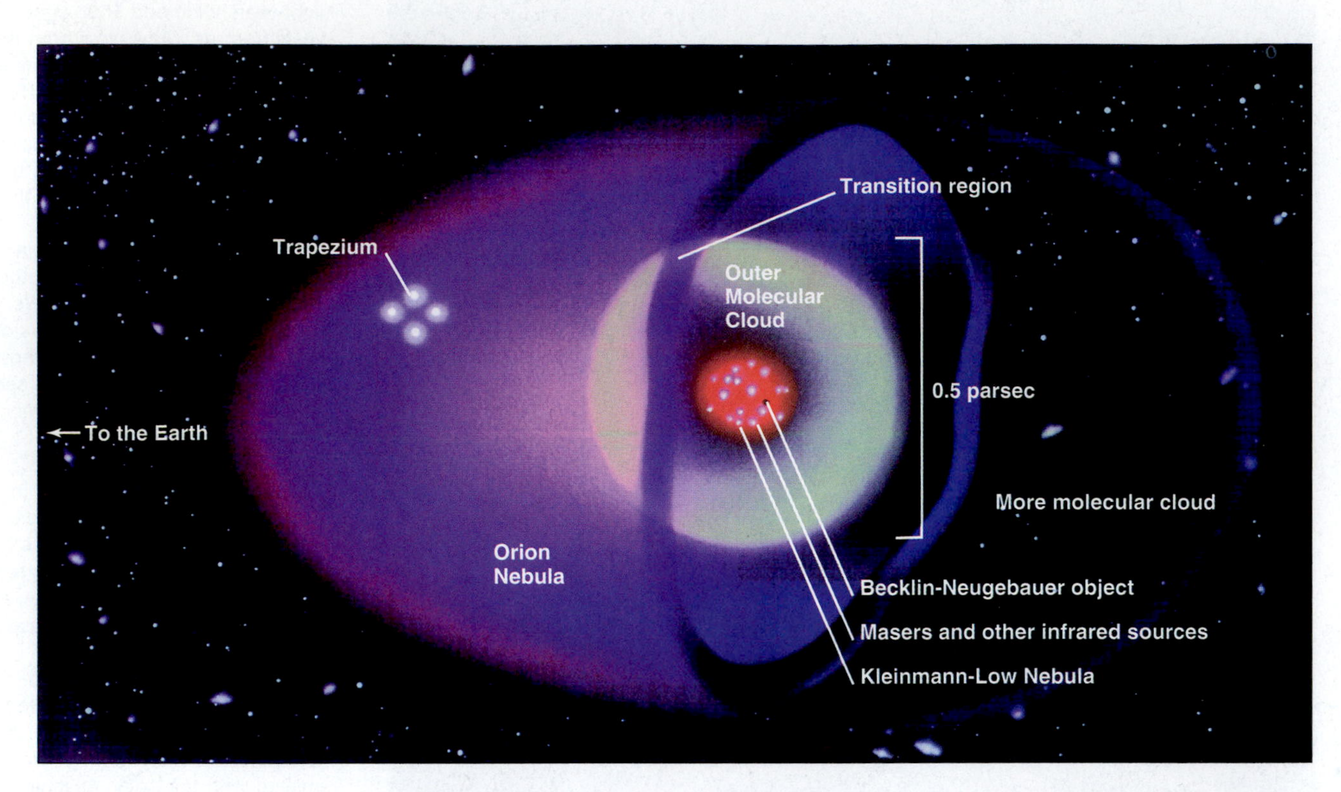


**FIGURE 33-23** The structure of the Orion Nebula and the Orion Molecular Cloud, proposed by Ben Zuckerman, now of UCLA. We are looking at this object from the left as you face the page.

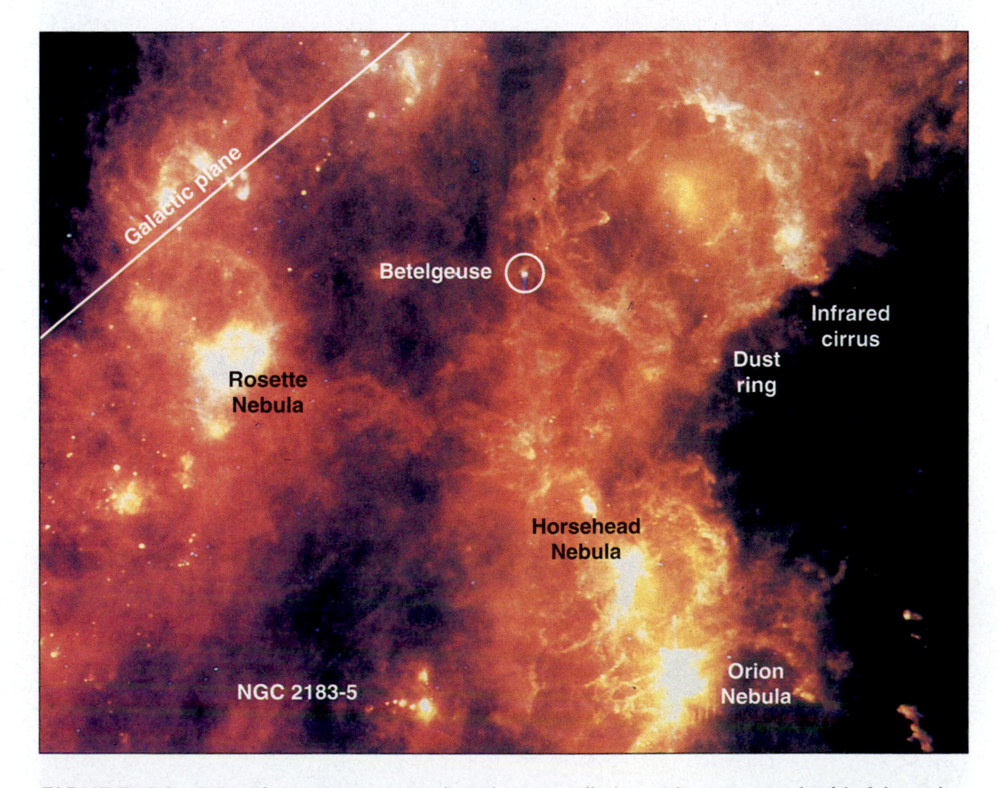


**FIGURE 33-24** The region surrounding the constellation Orion, as seen in this false-color infrared image constructed from data collected by the *I*nfrared *A*stronomical *S*atellite (IRAS), includes signs of different temperatures. The large circular feature at top can be seen at visible wavelengths but appears different in intensity and size here in the infrared. It corresponds to hot hydrogen gas and dust heated by the star Bellatrix, $\gamma$ Orionis, at the center of the ring. The brightest regions correspond to regions where stars are forming.

**FIGURE 33–25** Optical and infrared views of the heart of the Orion Molecular Cloud. The left image is from the visible-light WFPC2, and the right image is from the infrared NICMOS on Hubble. Most of the stars revealed by NICMOS were hidden from optical view. NICMOS reveals a chaotic, active region of star birth. Stars and dust that have been heated by the intense starlight show as yellow-orange. Emission from excited hydrogen molecules shows as blue. The brightest region is the Becklin-Neugebauer object, a massive, young star near the infrared source IRc2. Outflowing gas from BN causes the bow shock above it. The view is about 0.4 light-years across the diagonal; details as small as our solar system show.

than our Earth's atmosphere, and $10^{16}$ times less dense than the star that may eventually form, though it is substantially denser than the average interstellar density of about 1 particle/cm$^3$. This molecular cloud and one in the northern portion of Orion have together caused the Orion OB association—a group of hot stars—to form. The ultraviolet radiation and strong stellar winds from these stars of types O and B, as well as the supernova explosions from the ones that have already died, have made an expanding bubble perhaps 500 light-years across in the interstellar medium. One part of this bubble causes a large circular optical feature known as Barnard's Loop. The region can be studied especially well in the infrared (Fig. 33–24); the Infrared Space Observatory and the *N*ear *I*nfrared *C*amera and *M*ulti-*O*bject *S*pectrometer (NICMOS) on the Hubble Space Telescope peered through the dust (Fig. 33–25). The major new mountaintop observatories, such as the Subaru Observatory on Mauna Kea (Fig. 33–26) and the Very Large Telescope (Fig. 33–27), can take advantage of their positions above most of the water vapor in the Earth's atmosphere to make infrared views in windows of transparency.

One of the brightest of all the infrared sources in the sky, an object in the Orion Nebula that was discovered by Eric Becklin, now at UCLA, and Gerry Neugebauer of Caltech, is right in the midst of the Orion Molecular Cloud, in the Kleinmann-Low Nebula. This Becklin-Neugebauer object, also known for short as the BN object, is about 200 astronomical units across. Its temperature is 600 K. The BN object is a prime candidate for a star about to be born.

Other infrared sources are present near the BN object, and they probably also contain young stars or stars in formation. Small intense sources of maser radiation from various molecules also exist nearby. Such interstellar masers can only exist for

A

◀ **FIGURE 33–26** Infrared views made with the Subaru Telescope on Mauna Kea. Wavelengths between 1.2 and 2.1 $\mu$m were used. (*A*) A wide angle view. Hot stars such as the Trapezium appear as blue. Cooler sources and those obscured by dust appear yellow, orange, and red. In particular, the Kleinmann-Low Nebula is within the butterfly at upper right, which surrounds the infrared source IRc2. This object is a massive star, with 30 times the sun's mass, forming; its winds caused the butterfly. The Kleinmann-Low Nebula is a cluster of infrared sources just below the Becklin-Neugebauer object. These sources are on the side nearest us of a molecular cloud only a few light-years deep, and they may form an H II region like the Orion Nebula within a few thousand years. The source IRc2 is perhaps 10 times more luminous than the BN object but is hidden by even more dust. Most of the stars that show are members of the Trapezium OB association and less than a million years old; still more are not visible optically. The image covers about 5 × 5 arc min, which is equivalent to 2.1 × 2.1 light-years at the 1500-light-year distance of the Orion Nebula; thus everything in this image would fit well between the sun and our nearest star. (*B*) An enlargement showing the region of the Kleinmann-Low Nebula.

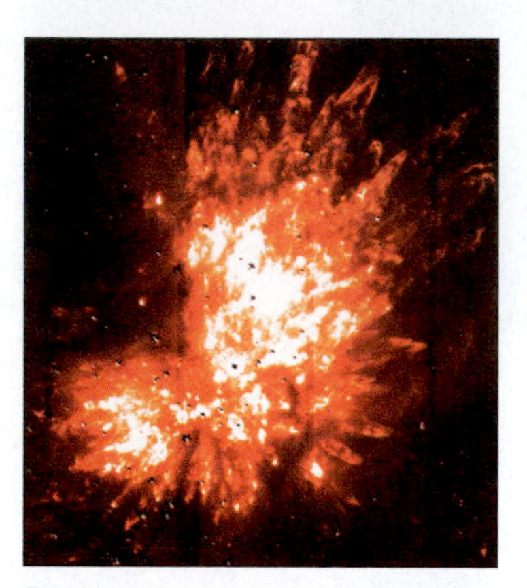

B

small sources, about the size of our solar system, and are also signs of stars in formation. The Orion region, with its active ongoing star formation, is an exciting observing direction for all technologies.

The Chandra X-ray Observatory, from above the Earth's atmosphere, can detect x-rays from the hot, young stars at the center of the Orion Nebula (Fig. 33–28).

▼ **FIGURE 33–27** (*A*) The yellow spot at the middle is the Becklin-Neugebauer Object surrounded by the Kleinmann-Low Nebula, deeply imbedded in the Orion Nebula. The image is with the ANTU unit telescope of the Very Large Telescope. (*B*) Infrared views of the Orion Nebula taken with the ANTU telescope of the Very Large Telescope. The orientation view at left is 3 arc min square, and is taken at wavelengths of 1 and 2 micrometers, which do not penetrate the dust all the way to the Becklin-Neugebauer Object. The view at right is at a wavelength of 20 micrometers, which does penetrate the dust (*C*) A Very Large Telescope infrared false-color composite of the center of the Orion Nebula. The Trapezium stars and the million-year-old associated cluster, which contains about 1000 stars, show.

A

B

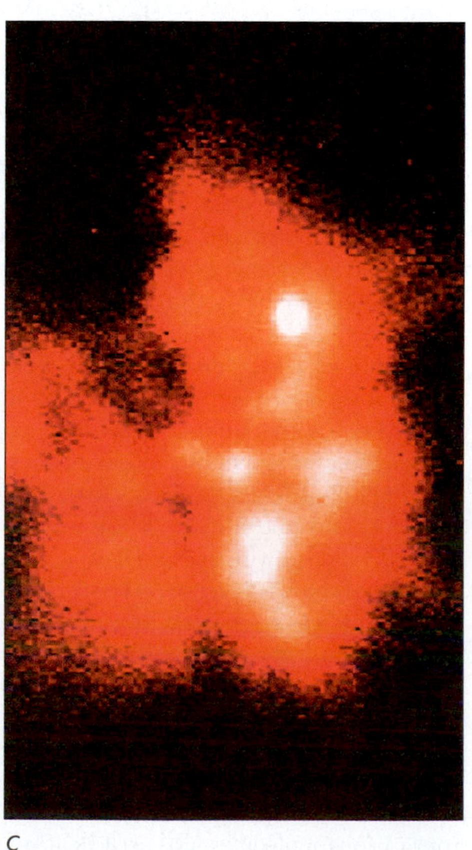

C

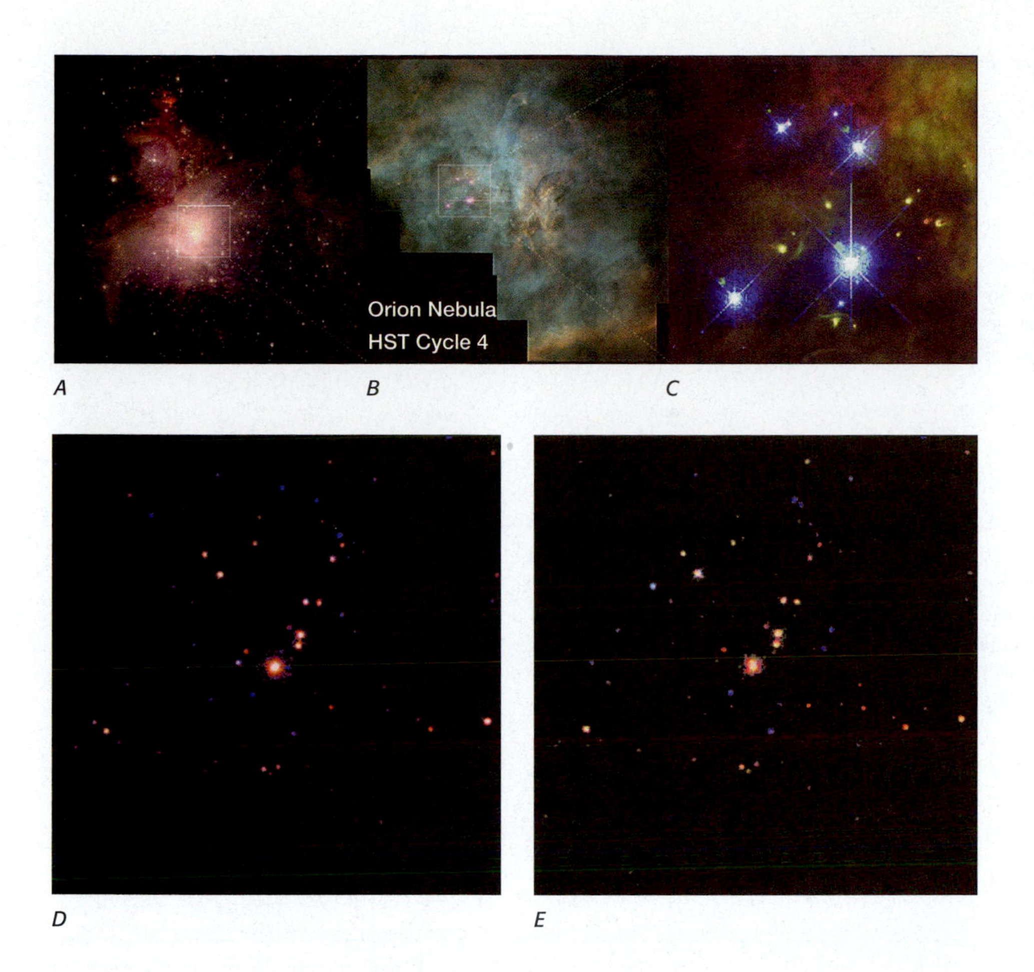


**FIGURE 33–28** (*A, B, C*) Hubble orientation images of the Trapezium. (*D, E*) Chandra's X-ray view of the Trapezium region at the core of the Orion Nebula shows not only stars but also protoplanetary disks. The protoplanetary disks (known as "proplyds") were discovered with the Hubble Space Telescope. Chandra scientists were surprised to find stars emitting not only at temperatures of 5 to 10 million kelvins, typical of stellar coronas, but also at temperatures of 60 million kelvins. The stars constantly change in x-ray brightness, sometimes within hours.

## CORRECTING MISCONCEPTIONS

| ✖ *Incorrect* | ✔ *Correct* |
|---|---|
| Interstellar space is invisible. | Gas and dust can prevent us from seeing far through interstellar space. |
| The Balmer series is the strongest set of spectral lines of hydrogen. | 21-cm radiation, from a split in the lowest energy level of hydrogen, arises from a more fundamental state of hydrogen than the Balmer series, and the Lyman series lines are very strong though cannot be seen through our atmosphere. |
| Hydrogen molecules in space are easy to detect. | Only from above the atmosphere have we been able to study hydrogen molecules. |

## SUMMARY AND OUTLINE

H I and H II regions (Section 33.1)
Interstellar reddening and extinction (Section 33.2)
Radio observations (Section 33.3)
- Continuum radio astronomy
- Spectra are measured over a broad frequency range.
- Radiation generated by synchrotron emission (electrons spiralling rapidly in a magnetic field) is found in many nonthermal sources.

Radio spectral line from interstellar hydrogen (Section 33.4)
- 21-cm spectral line from neutral hydrogen was discovered in 1951.
- Line occurs at 21 cm through a spin-flip transition, corresponding to a change between energy subdivisions of the ground level of hydrogen.
- Both emission and absorption have been detected.
- 21-cm radiation used to map the Galaxy

Distances measured using differential rotation and the Doppler effect

Radio spectral lines from molecules (Section 33.5)

"Mysterium" lines discovered at 18 cm—later identified as OH affected by masering process

Masers first developed artificially on Earth, but later discovered to exist in space.

Dozens of molecules have been discovered in space.

Analysis of molecular lines tells us physical conditions (e.g., temperature, density, and motions).

Measuring mass of our Galaxy (Section 33.6)

Revised measurements have doubled our value for the Galaxy's size.

Molecular hydrogen (Section 33.7)

Interstellar ultraviolet absorption lines from $H_2$, with abundances up to 50 per cent

The formation of stars (Section 33.8)

Molecular clouds contain gas collapsing to become stars.

Molecules are associated with dark clouds, such as the Orion Molecular Cloud, where the molecules are shielded by dust from being torn apart by ultraviolet radiation.

## KEY WORDS

interstellar medium
cosmic abundances
H I region
H II region
scattered
reddened
extinction
synchrotron emission*
nonthermal radiation*
thermal radiation*
spin-flip
half-life
21-cm line
masers
lasers
rotation curve
giant molecular cloud
Bok globule
bipolar flows
Orion Molecular Cloud
Becklin-Neugebauer object (BN object)

*This term appears in an optional section.

## QUESTIONS

1. List two relative advantages and disadvantages of radio astronomy compared with optical astronomy.
2. What is the ratio of the cosmic abundance of hydrogen to the cosmic abundance of helium?
3. Though cool gas does not give off much continuum radiation, we can detect 21-cm radiation from cool interstellar gas. Explain.
4. Describe the relation of hot stars to H I and H II regions.
5. Briefly define and distinguish between the redshift of a gas cloud and the reddening of that cloud.
6. What determines whether the 21-cm lines will be observed in emission or absorption?
7. Is 21-cm radiation thermal or non-thermal? Explain.
8. If our Galaxy rotated like a rigid body, with each point rotating with the same period no matter what the distance from the center, would we be able to use the 21-cm line to determine distances to H I regions? Explain.
9. Describe how a spin-flip transition can lead to a spectral line, using hydrogen as an example. Could deuterium also have a spin-flip line? If so, describe the likely process.
10. Why did it take so long to discover interstellar hydrogen molecules?
11. What was "mysterium"? Why was it thought to be strange?
12. How does a maser work?
13. Why did the abundance of heavy molecules predicted from observations of hydroxyl give too low a value?
14. Why are dust grains important for the formation of interstellar molecules?
15. Compare the roles of carbon monoxide and molecular hydrogen in giant molecular clouds.
16. Which molecule is observed in the most locations in interstellar space?
17. †Given the length of a Galactic year and our distance from the center of the Milky Way Galaxy, compute the mass inside the sun's orbit in terms of the sun's mass.
18. Explain what property of young stars and/or the region of space nearby was observed with infrared telescopes aboard spacecraft.
19. Describe the relation of the Orion Nebula and the Orion Molecular Cloud.
20. How do we detect star formation in another galaxy? Describe an example.
21. Optical astronomers can observe only at night. In what time period can radio astronomers observe? Why?
22. Discuss the choice of false colors for the different wavelength bands in Figure 33–21*B*.
23. Discuss the advantages of infrared arrays over prior abilities to observe in the infrared.
24. List three sciences besides astronomy that have applications to the study of interstellar space, and briefly describe one of the connections for each.
25. †Extinction of 25 magnitudes (Section 33.2) is what factor of dimming?

†This question requires a numerical solution.

## TOPICS FOR DISCUSSION

1. Find five terrestrial uses for lasers, and describe the properties of lasers that make them special. Consider the atomic properties of masers and lasers that lead to these terrestrial advantages.
2. Find out about the discovery of buckminsterfullerene on Earth and about some of its uses.
3. On a radio, mark the wavelengths that correspond to each frequency labelled. Make pieces of string of those sizes.
4. The blue sky is polarized because the reflection (scattering) of sunlight off air molecules reflects some angles of light waves better than others. Look through some polarizing sunglasses. ("Polaroid" is a brand name for sheets of polarizing material.) Rotating the glasses changes the plane that passes through the sunglasses, as you can see by looking at the daytime sky or at a reflection off a store window or car.

## USING TECHNOLOGY

### W World Wide Web

1. Check out the homepage for the Infrared Space Observatory (http://www.iso.vilspa.esa.es/science) and the Space Infrared Telescope Facility (http://sirtf.caltech.edu), and find out about the interstellar medium.
2. Find out what the National Radio Astronomy Observatory is discovering about the interstellar medium (http://www.nrao.edu).
3. Go to the 2MASS homepage, and compare its various images from its Picture of the Week at http://www.ipac.caltech.edu/2mass/gallery.

### REDSHIFT

1. Find pictures of the nebulae discussed here in the picture gallery.
2. Investigate the Orion region and the objects in it. Use the object filters to see stars of different spectral types.

Interacting galaxies. Many galaxies are in pairs or clusters.

# PART 7

# Galaxies and Cosmology

The individual stars that we see with the naked eye are all part of the Milky Way Galaxy, discussed in the preceding two chapters. But we cannot be so categorical about the conglomerations of gas and stars that can be seen through telescopes. Once they were all called "nebulae," but we now restrict the meaning of this word to clouds of gas and dust that are smaller than building blocks of the universe called "galaxies."

Some of the objects that were originally classified as nebulae turned out to be huge collections of gas, dust, and stars located far from our Milky Way Galaxy and of a scale comparable to that of our Galaxy. These objects are galaxies in their own right, and they are both fundamental units of the universe and the stepping stones that we use to extend our knowledge to tremendous distances.

In the 1770s, the French astronomer Charles Messier was interested in discovering comets. To do so, he had to be able to recognize whenever a new fuzzy object appeared in the sky. He thus compiled a list of about 100 diffuse objects that could always be seen (Appendix 8).

To this day, these objects are commonly known by their **Messier numbers.** Messier's list contains the majority of the most beautiful objects in the sky, including nebulae, star clusters, and galaxies. (In the 1990s, Patrick Caldwell Moore provided a list (Appendix 9) of 110 additional beautiful objects that are accessible to amateur observers.)

Soon after Messier's work, William Herschel, aided by his sister, Caroline, in England, compiled a list of 1000 nebulae and clusters. He expanded it in subsequent years to include 2500 objects. Herschel's son John continued the work, incorporating observations made in the southern hemisphere. In 1864, he published the *General Catalogue of Nebulae.* In 1888, J. L. E. Dreyer published a still more extensive catalogue, *A New General Catalogue of Nebulae and Clusters of Stars,* the NGC, and later published two supplementary Index Catalogues, ICs. The 100-odd nonstellar objects that have Messier numbers are known by them, and sometimes also by their numbers in Dreyer's catalogues. Thus what was long called the Great Nebula in Andromeda is very often called M31 and is less often called NGC 224. It is now often called simply the Andromeda Galaxy. The Crab Nebula = M1 = NGC 1952. Objects without M numbers are known by their NGC or IC numbers, if they have them.

Lord Rosse in Ireland in about 1850 made the largest telescope in the world at that time, a reflector with a metal mirror. When he looked through his telescope at the Messier objects, some of them showed traces of spiral structure, like pinwheels. They were called "spiral nebulae." But where were they located? Were they close by or relatively far away?

When such telescopes as the 0.9-m reflector at the Lick Observatory in 1898, and later the 1.5-m and 2.5-m reflectors on Mount Wilson, all in California, began to photograph the "spiral nebulae," they revealed many more of them. The shapes and motions of these "nebulae" were carefully studied. Some scientists thought that they were merely in our own Galaxy, while others thought that they were very far away, "island universes" in their own right, so far away that the individual stars appeared blurred together. (The name "island universes" had originated with the philosopher Immanuel Kant in 1755.)

The debate raged, and an actual debate on the scale of our Galaxy and the nature of the "spiral nebulae" was held on April 16, 1920, as an after-dinner event of the National Academy of Sciences. Harlow Shapley (pronounced to rhyme with "map lee") argued that the Milky Way Galaxy was larger than had been thought and thus implied that it could contain the spiral nebulae. Heber Curtis argued for the independence of the "spiral nebulae" from our Galaxy. This famous **Shapley-Curtis debate** is an interesting example of the scientific process at work.

Shapley's research on globular clusters had led him to correctly assess our own Galaxy's large size. But he also argued that the "spiral nebulae" were close by because proper motion had been detected in some of them by another astronomer. These observations were subsequently shown to be incorrect. He also reasoned that an apparent nova in the Andromeda Galaxy, S Andromedae, had to be close or else it couldn't have been as bright; nobody knew about supernovae then. Curtis's conclusion that the "spiral nebulae" were external to our galaxy was based in large part on an incorrect notion of our Galaxy's size. He treated S Andromedae as an anomaly and considered only "normal" novae.

So Curtis's conclusion that the "spiral nebulae" were comparable to our own galaxy was correct, but for the wrong reasons. Shapley, on the other hand, came to the wrong conclusion but followed a proper line of argument that was unfortunately based on incorrect and inadequate data.

The matter was settled in 1924, when observations made at the Mount Wilson Observatory by Edwin Hubble provided distances to some "spiral nebulae" and showed they were so far away they must be galaxies. Thus there were indeed other galaxies in the universe beside our own. In fact, we think of galaxies and clusters of galaxies as fundamental units in the universe. The galaxies are among the most distant objects we can study. Many quasars are even farther away and turn out to be certain types of galaxies seen at special times.

Galaxies and quasars can be studied in most parts of the spectrum. Radio astronomy and x-ray astronomy, in particular, have long proved fruitful methods of study. The study across the spectrum of galaxies and quasars provides tests of physical laws at the extremes of their applications and links us to cosmological consideration of the universe on the largest scale.

NGC 2997 in Antlia, a type Sb spiral galaxy whose structure is very similar to that of our own Milky Way Galaxy. This ground-based view shows the whole Galaxy. Note the bluish arms where stars are being formed and the red regions of ionized hydrogen.

# Chapter 34

# Types of Galaxies

**AIMS:** To discuss the different types of galaxies, to see that galaxies are fundamental units of the universe, and to consider methods used by radio astronomy to see fine details in distant galaxies

Though spiral shapes were seen in the sky by the Earl of Rosse and other observers at his telescope (Fig. 34–1), it took decades before people agreed on what they are. The question of the distance to these "spiral nebulae" was settled only in 1924 by Edwin Hubble. He used the Mount Wilson telescopes to observe Cepheid variables in two of the "spiral nebulae" and in another object. Following the line of argument that we described in Section 26.4b, he found variable stars in these objects and noticed that the shapes of their light curves showed that they were Cepheids. Then, from their periods, he found their luminosities. Comparing their luminosities with their observed brightnesses (that is, comparing their absolute magnitudes with their apparent magnitudes), he calculated how far they were. His measurements showed that the "spiral nebulae" and the other object were outside our own galaxy. From their observed angular sizes and their distances from us, it followed that these objects are not overwhelmingly different from the Milky Way Galaxy in size.

Since Hubble's work, there has been no doubt that the spiral forms we observe in the sky are galaxies like our own. For the rest of the book we shall strictly use the term **spiral galaxies;** the currently incorrect, historical term "spiral nebula" often hangs on in certain contexts, chiefly when we discuss the "Great Nebula in Andromeda," which is actually a spiral galaxy. As we shall see, galaxies can take other shapes as well.

Many of these shapes are represented in the Messier Catalogue and the Caldwell Catalogue (Appendixes 8 and 9). The Andromeda Galaxy (M31) at 2.4 million light-years and M33 at 2.6 million light-years away from Earth are the farthest objects you can see with your unaided eye. They appear as fuzzy blobs in the sky if you know where to look. M31 is shown on the star maps at the end of the book. It takes a telescope to see the spiral shape.

## 34.1 TYPES OF GALAXIES

As we saw with stars, it is often useful to have a classification scheme that has only a few categories. With stars, we limited ourselves to OBAFGKM and recently added L and T. Similarly, with a terrestrial map, we are used to simplification. For example,

A

B

**FIGURE 34–1** (*A*) The Earl of Rosse's telescope from about 1850, as newly restored at its site in the middle of Ireland by the current Earl. With a mirror 6-feet across, it was the largest telescope in the world from then through 1918. (*B*) Birr Castle, the telescope's site.

A

B

FIGURE 34–2 (*A*) M87 (NGC 4486), a galaxy of Hubble type E0(pec) in the constellation Virgo. Globular clusters can be seen in the outer regions. Two more distant galaxies appear to lower right. (*B*) The giant elliptical galaxy NGC 1316.

to create a map of the United States, we do not simply trace the United States onto a piece of paper thousands of kilometers long and wide. We simplify our categories of cities and roads. With galaxies, too, we choose a few major categories, and pigeonhole our varied examples into them.

Spiral galaxies are only one common type of galaxy. Many other galaxies have elliptical shapes, while still others are irregular or abnormal in appearance. In 1925, Hubble set up a system of classification of galaxies that we still use today; we normally describe a galaxy by its **Hubble type,** as we shall discuss.

### 34.1a Elliptical Galaxies

About one-third of galaxies are elliptical in shape (Fig. 34–2). The largest of these **elliptical galaxies** contain $10^{13}$ solar masses and are $10^5$ parsecs across (300,000 light-years, approximately the diameter of our own Galaxy); these **giant ellipticals** are rare. Much more common are **dwarf ellipticals,** which contain "only" a few million solar masses and are only 2000 parsecs (6000 light-years) across.

Elliptical galaxies range from nearly circular in shape, which Hubble called type **E0,** to very elongated, which Hubble called type **E7.** The spiral Andromeda Galaxy, M31 (Fig. 34–3), is accompanied by two elliptical companions of types E2 (for the galaxy closer to M31) and E5, respectively. It is obvious on the photograph that the companions are much smaller than M31 itself. We see the disk of M31 inclined at an angle of only 13° from edge-on.

We assign types based on the optical appearance of a galaxy rather than on how elliptical it actually is. After all, we can't change our point of view for such a far-off object. But even a very elliptical galaxy will appear round when seen end on. (Just picture looking straight at the end of an egg or a cigar or at the face of a hockey puck or other disk; they look round.) So the ellipticals are, in actuality, at least as elliptical as their Hubble type (which is based only on their appearance) shows; they may actually be more elliptical than they appear. Most seem to be describable by giving the dimensions of only two axes; some, though, apparently have different diameters in all three dimensions.

### 34.1b Spiral Galaxies

Spiral galaxies, with arms unwinding gracefully from the central regions, are a large fraction of all the bright galaxies in the universe. They form a majority in certain groups of galaxies.

Sometimes the arms are tightly wound around the nucleus; Hubble called this type **Sa,** the S standing for "spiral." Spirals with their arms less and less tightly wound (that is, looser and looser) are called type **Sb** (as in the previous image of the Andromeda Galaxy) and type **Sc** (Fig. 34–4). (Some classifications add a type Sd.) The nuclear bulge as seen from edge on (Fig. 34–5) or at a slight angle (Fig. 34–6)

FIGURE 34–3 The Andromeda Galaxy, also known as M31 and NGC 224, the nearest spiral galaxy to the Milky Way. It is type Sb and is accompanied by two elliptical galaxies, NGC 205, type S0/E5(pec) (*lower left*), and M32, type E2 (*right middle*). These galaxies are only 2.2 million light-years from Earth. The older red and yellow stars give its central regions a yellowish cast, in contrast to the blueness of the spiral arms from the younger stars there.

**FIGURE 34–4** (*A*) The Whirlpool Galaxy, M51, in Canes Venatici, a type Sc spiral galaxy. At the end of one of its arms, a companion galaxy, NGC 5195, appears. (*B*) M51 in the infrared imaged with the Infrared Space Observatory. Infrared images show the dust content. We saw the spiral arms of M51 as traced out from its carbon-monoxide spectral-line radiation in Figure 32-26. (*C*) A spiral arm of M51 as traced out from its carbon-monoxide spectral-line radiation, observed with the European IRAM 30-m radio telescope in the millimeter spectral region.

is less and less prominent as we go from Sa to Sc. On the other hand, the dust lane—obscuring dust in the disk of the galaxy—becomes more prominent.

Spectroscopic measurements from Doppler shifts indicate that galaxies rotate in the sense that the arms trail (where "sense" means whether the rotation is clockwise or counterclockwise as seen from some location).

Spiral galaxies can be 25,000 to 800,000 parsecs (80,000 to 250,000 light-years) across. They contain $10^9$ to over $10^{12}$ (a billion to a trillion) solar masses. Since most stars are of less than 1 solar mass, this means that spirals contain over $10^9$ to over

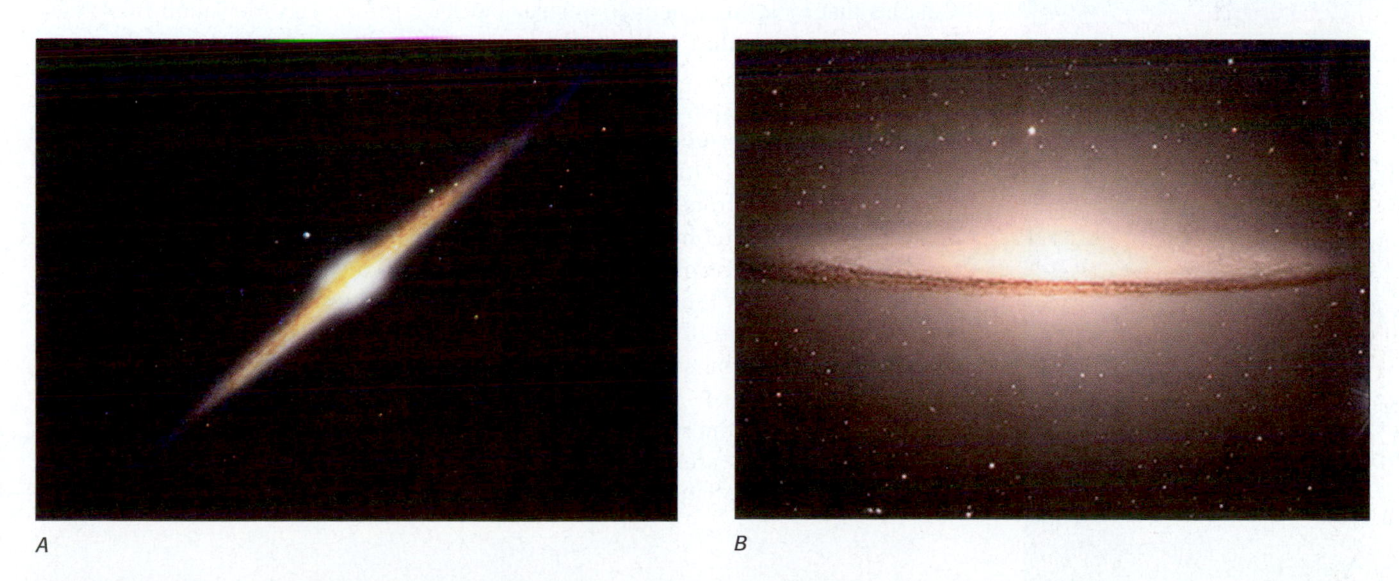


**FIGURE 34–5** (*A*) An edge-on view of NGC 4565, a type Sb spiral galaxy in Coma Berenices. Note its box-shaped central bulge. This view uses a blue image (0.4 $\mu$m) from a CCD and two infrared images (1.2 and 2.1 $\mu$m) from the U. Hawaii's infrared array, reproduced as blue, green, and red, respectively. The dust lane in the plane of the galaxy shows clearly. (*B*) The Sombrero Galaxy (M104) in the constellation Virgo. The dust lane again shows clearly as we look in the plane of the galaxy.

A

B

**FIGURE 34–6** NGC 253 in Sculptor, a type Sc galaxy, viewed from a low angle above the plane of its disk. Note that its central bulge is minimal. (*A*) In this visible-light image, the galaxy's light-absorbing dust lanes show clearly at this angle. (*B*) In this infrared image taken with 2MASS, the dust's emission lessens the contrast at this angle.

$10^{12}$ stars—we now think our own Galaxy has perhaps $10^{12}$. In recent years, we have learned a lot about galaxies by studying them in different parts of the spectrum.

In about one third of the spirals, the arms unwind not from the nucleus but rather from a straight **bar** of stars, gas, and dust that extends to both sides of the nucleus (Fig. 34–7). These are similarly classified in the Hubble scheme from a to c in order of increasing openness of the arms, but with a B for "barred" inserted: **SBa, SBb,** and **SBc.**

Evidence mounts that our Milky Way Galaxy is actually a barred spiral, though the bar is relatively small.

### 34.1c Irregular Galaxies

Of the galaxies we see, a few per cent show no regularity. The Magellanic Clouds, for example, are basically irregular galaxies (Fig. 34–8), though the Large Magellanic Cloud may actually be a sloppy barred galaxy. Irregular galaxies are classified as **Irr.** Since galaxies like the Magellanic irregulars are fairly faint, we don't see them very far away. There is evidence that they are the most common type of galaxy in the universe, which we find when we look at specific nearby volumes of space.

Irregular galaxies like the Magellanic Clouds have enough gas to form stars but relatively little dust, so star formation can be seen even in the visible. The ratios of abundances of elements in the interstellar gas in the Magellanic Clouds are somewhat different from the ratio in our own Galaxy, and the resulting differences are interesting to contemplate.

**FIGURE 34–7** A barred spiral galaxy, NGC 1365.

### 34.1d Peculiar Galaxies and Starburst Galaxies

Galaxies that look fundamentally regular but have some major deviation from regularity are called **peculiar galaxies.** In some cases, as in M82 (Fig. 34–9), it appeared at first look as though an explosion had taken place in what might have been a regular galaxy. But later, large clouds of molecular hydrogen and many supernova remnants were discovered at the center of the galaxy. Observations with the 45-m radio telescope of the Nobeyama Radio Observatory showed that this molecular gas is flowing outwards from M82's nucleus. This rapid wind of particles heats the gas and excites the neutral hydrogen that is present. We see the hydrogen in this outward flow from its H-alpha radiation, which is reproduced in red in this false-color image. Scientists now think that the outflow is caused by a "starburst," the almost simultaneous formation of many massive stars. M82 and its like are known as **starburst galaxies.** Starbursts could result from material funneled inward by a galaxy's bar, from collisions (see Section 34.4), or from gas flowing along the galaxy's magnetic field. In M82, the starburst could have been going on for 50 million years.

Peculiar galaxies are classified as the corresponding Hubble type followed by "(pec)": for example, Sa(pec).

## 34.2 THE HUBBLE CLASSIFICATION

Hubble drew out his scheme of classification in a **tuning-fork diagram** (Fig. 34–10). The transition from ellipticals to spirals is represented by type **S0.** S0 galaxies are called **lenticular.** Galaxies of this transitional type resemble spirals in having a bright

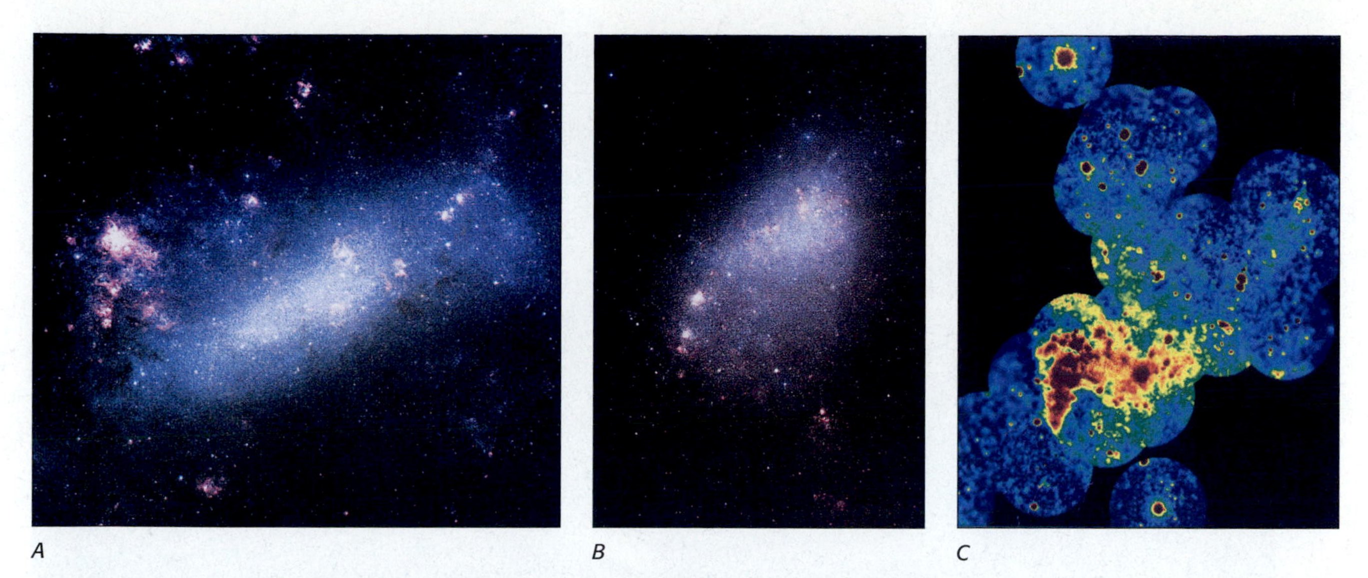

**FIGURE 34–8** The Large Magellanic Cloud (*A*) and the Small Magellanic Cloud (*B*). The LMC is about 50 kpc (about 170,000 light-years) and the SMC about 60 kpc (about 195,000 light-years) from us; they are about 20 kpc apart from each other. A few globular clusters, considered part of our Milky Way Galaxy, and some dwarf elliptical galaxies are also that far away. The LMC contains about 1/100 and the SMC about 1/1000 the mass of our Galaxy. The Clouds were first reported to Europe by 16th-century Portuguese navigators who had travelled to the Cape of Good Hope at the southern tip of Africa. They were named a few decades later in honor of Magellan, who was then circumnavigating the world by that route. (*C*) An x-ray mosaic of the Large Magellanic Cloud, made with ROSAT. Each of the circles shows the satellite's nearly 2° field of view. In this false-color image, dark blue and purple indicate low intensity and red indicates high intensity. The bright bar with a hook-shaped end is brightest near the optical object 30 Doradus, an association of hot, bright stars with surrounding nebulosity. The point source LMC X-3 appears at the top.

**FIGURE 34–9** (*A*) An orientation image of the spiral galaxy M81 (*lower left*) and M82 (*upper right*). (*B*) M82, a starburst galaxy that is a powerful source of radio radiation. It is 12 million light-years away. Stars form in starburst galaxies at rates tens or hundreds of times that in normal galaxies like ours. This detailed image was taken with the Japanese national Subaru telescope on Mauna Kea. M82 was once thought to be exploding, but then it was decided that gentler processes caused its form and non-thermal radiation. It now seems most likely that it is the site of strong star formation, mostly hidden by dust. The eruption of many supernovae may have blown a gigantic bubble in the interstellar gas and dust. We are apparently seeing, in H-alpha, gas and dust as it is blown out of that galaxy's disk in the form of filaments over 10,000 light-years long.

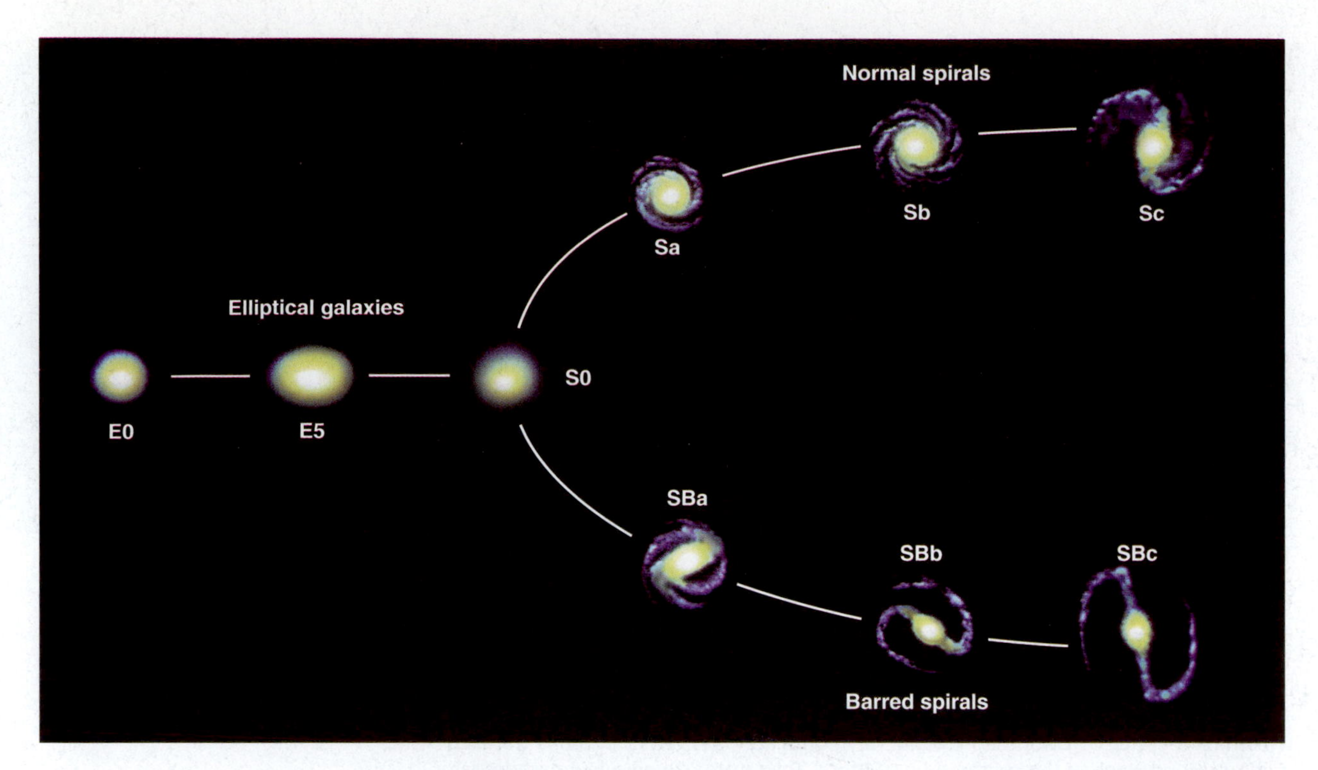


**FIGURE 34–10** The Hubble tuning-fork classification of galaxies. But there are intermediate types between the arms of the tuning fork.

nucleus and the shape of a disk but do not have spiral arms. All E6 and E7 galaxies are now thought to also be S0, but not vice versa.

It has since been shown, from optical observations and from studies of the 21-cm hydrogen line, that the amount of gas between the stars in galaxies is different in different types of galaxies. Elliptical galaxies have been thought to have essentially no gas or dust; the discovery with millimeter-wave radio telescopes of carbon-monoxide spectral-line radiation in several elliptical galaxies has led to some reevaluation of this point. Some optical and 21-cm radio observations have also detected gas and dust in ellipticals. In any case, spiral galaxies are known to have a lot of gas and dust. The relative amount of gas increases from types Sa (or SBa) to Sc (or SBc). The gas in the interstellar medium in an irregular galaxy is usually even denser.

The major sky-mapping projects cover all types of galaxies, so one can put together tuning-fork diagrams in visible light from the Sloan Digital Sky Survey (Fig. 34–11) and from the 2MASS infrared survey (Fig. 34–12).

Though no star formation is found in elliptical or S0 galaxies, the amount of star formation increases along the Hubble tuning-fork diagram toward type Sc. Only in the galaxies with a substantial gas and dust content—mostly the Sc, SBc, and Irr galaxies—are the O and B stars to be found. Since these stars have short lifetimes on a stellar scale, they must have been formed comparatively recently, within the last several million years. Star formation is well observed in the infrared (Fig. 34–13) and radio regions of the spectrum (Fig. 34–14).

## 34.3 THE CENTERS OF GALAXIES AND BLACK HOLES

We have seen the likelihood that a black hole of over a million solar masses is present in the center of our own Galaxy. Studies of nearby galaxies also indicate that giant black holes are present there, too. In particular, studies have been carried out of the motions of stars close to the centers of galaxies (Fig. 34–15). The stars' motions in

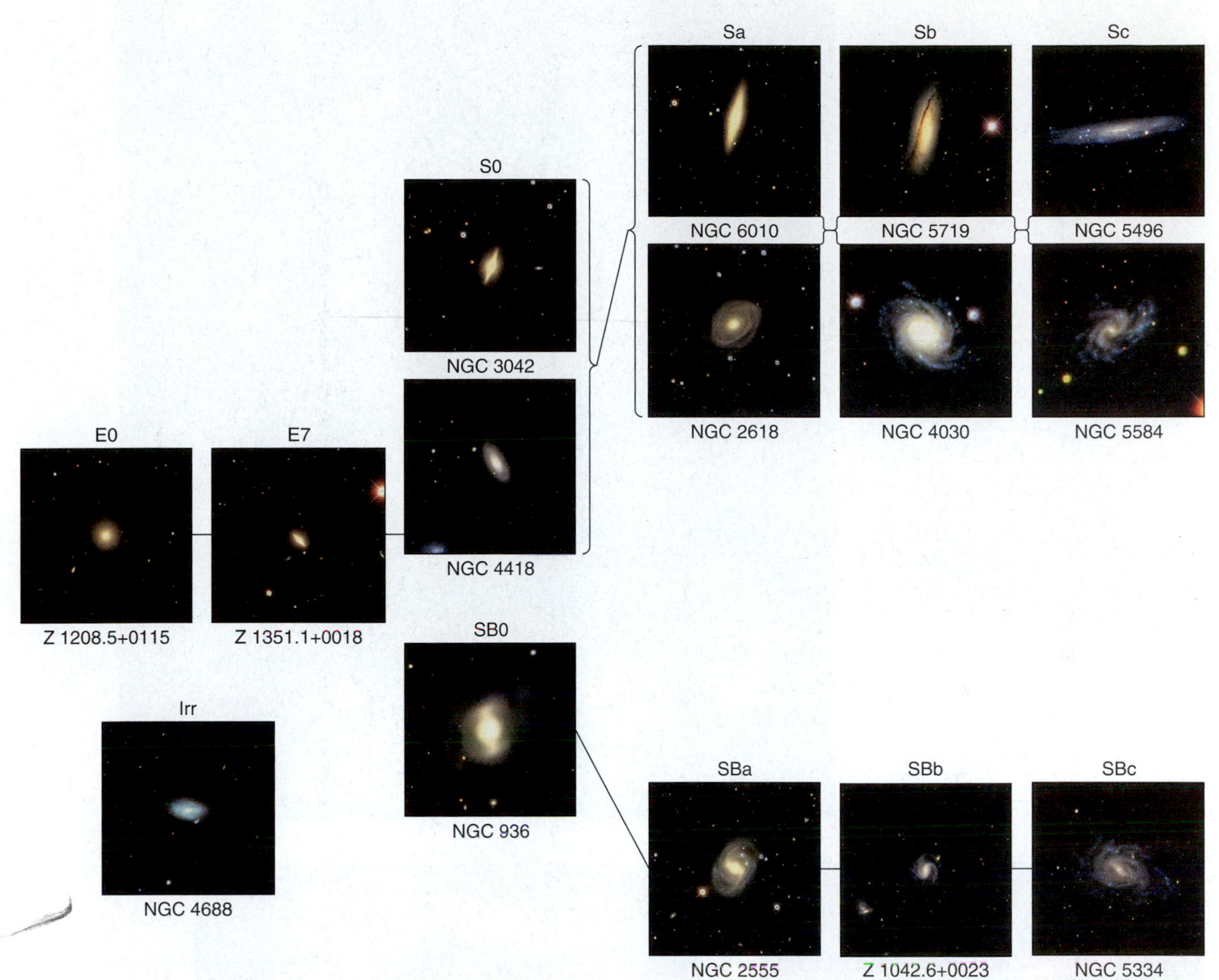


**FIGURE 34–11** Visible-light images of all the Hubble types of galaxies taken by the Sloan Digital Sky Survey.

a galaxy reveal the amount of mass that is present between them and the galaxy's center. If enough mass is present in a small enough volume, then it is in the form of a black hole. Black holes containing millions of solar masses of material seem to be present both in the Andromeda Galaxy (M31) and in its elliptical companion M32, and in other nearby galaxies as well. Studies with the Hubble Space Telescope are allowing us to observe closer to the nucleus of a galaxy than ever before (Fig. 34–16). The black holes have been determined to have hundreds of millions or billions of times the mass of the sun.

In Chapter 31, we distinguished among three classes of black holes: mini black holes, stellar-mass black holes, and supermassive black holes. In 2000, it was determined that there is a medium-mass category as well, partway between stellar-mass black holes and supermassive black holes. These black holes may have "only" a few hundred times the mass of the sun, nevertheless much too great a figure to have come from the collapse of a massive star. One such object, observed with the Chandra X-ray Observatory in the starburst galaxy M82 (which we saw as Figure 34–9), had its mass determined by the brightness of the x-rays, which are 100 times more

*(text continues on page 682)*

▶ **FIGURE 34–12** Contours of all the Hubble types of galaxies, as seen in the infrared with 2MASS.

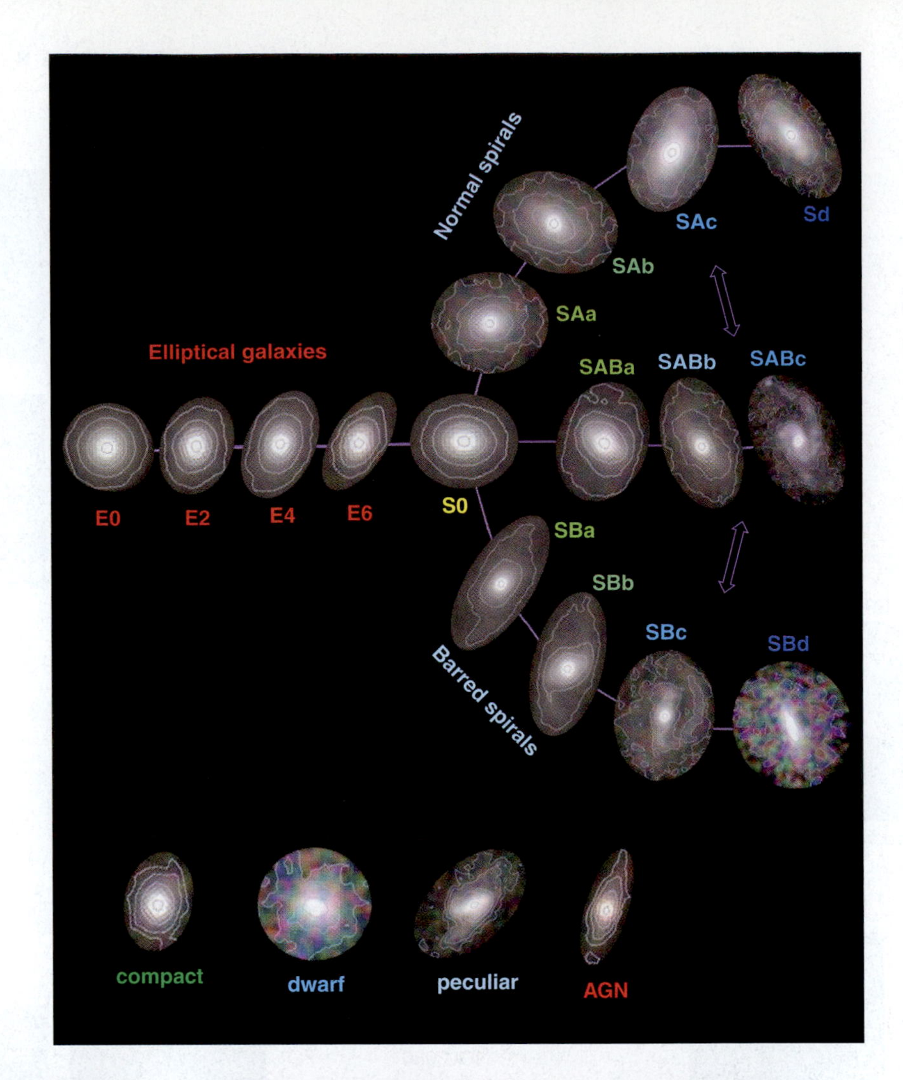


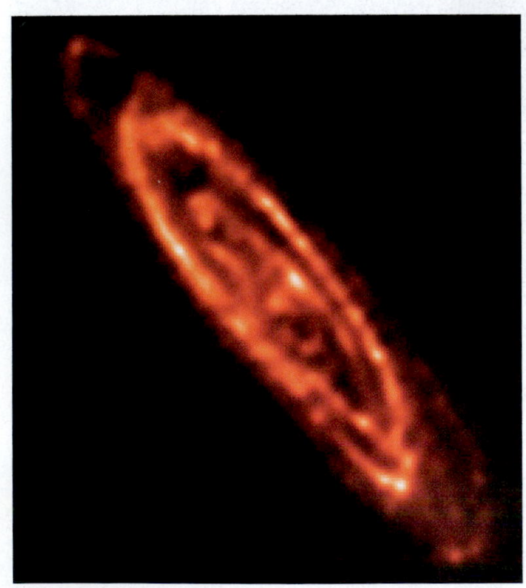

*A*

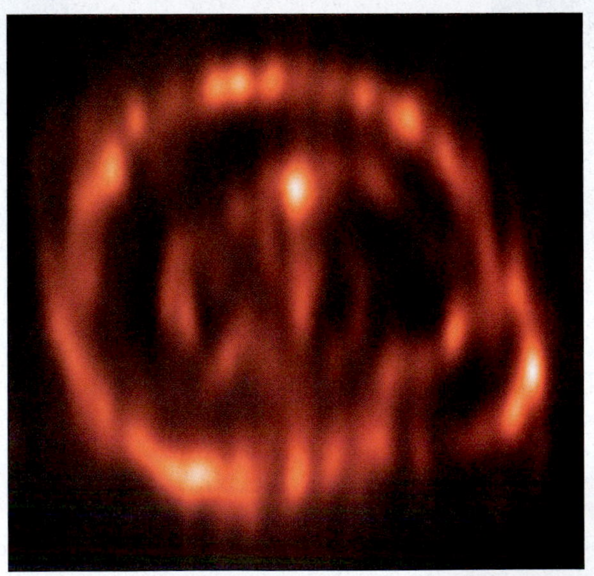

*B*

▲ **FIGURE 34–13** A long-infrared-wavelength view of the Andromeda Galaxy, M31. This Infrared Space Observatory (ISO) view is at a wavelength of 175 $\mu$m. (*A*) We see a ring in which a lot of star formation is occurring. (*B*) The ring shows better when the effect of the galaxy's tilt is taken out through calculations.

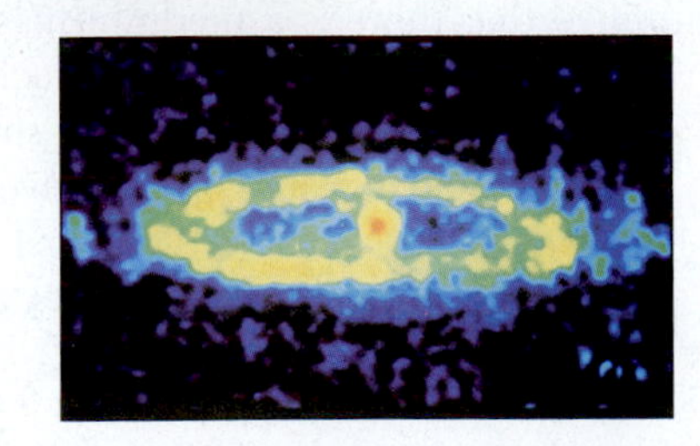

▶ **FIGURE 34–14** A radio map of the Andromeda Galaxy, M31, at 11 cm. As in the infrared view, the strongest emission is in a ring.

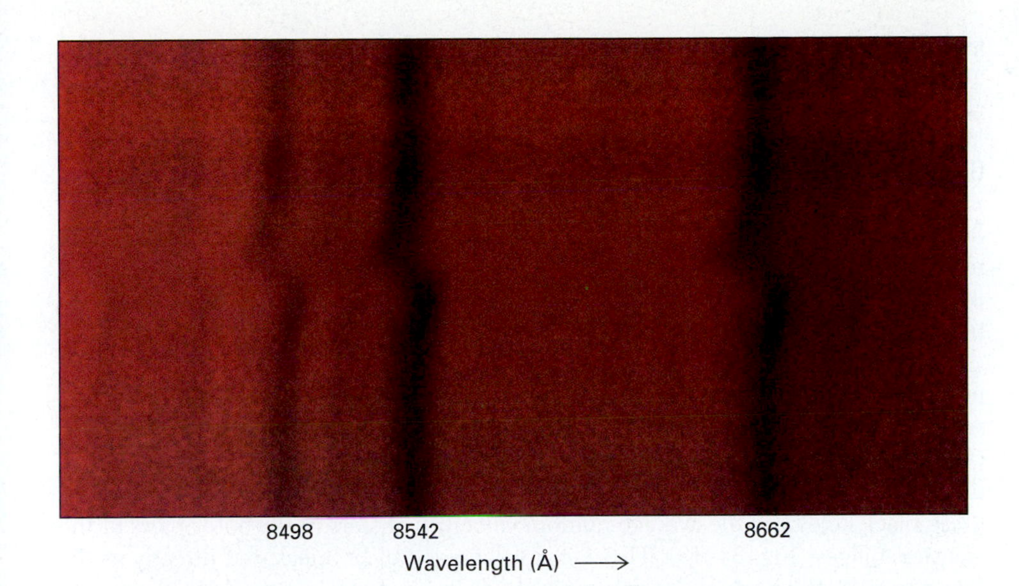


**FIGURE 34-15** This spectrum of M31, the Andromeda Galaxy, had the spectrograph's slit laid across the center of the galaxy. Note the abrupt change in wavelength as the slit crossed M31's center, showing an abrupt change in the direction of the velocities there. John Kormendy (now at U. Texas at Austin) has interpreted his spectrum to indicate the presence of a black hole with about 10 million times the mass of the sun.

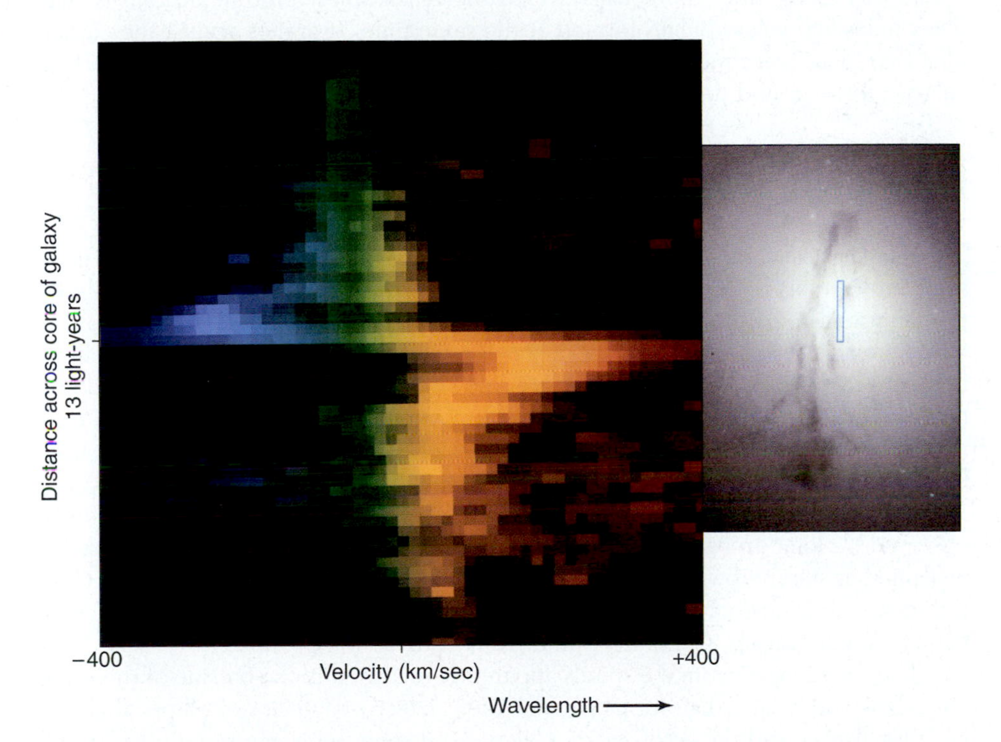


**FIGURE 34-16** (*right*) The nucleus of the elliptical galaxy M84. (*left*) The color image shows the spectrum of the region outlined in blue in the visible-light image of the center of the galaxy. The spectrum was taken with the Space Telescope Imaging Spectrograph (STIS), installed on the Hubble Space Telescope in 1997, and the visible-light image was taken with Hubble's WFPC2 camera. Wavelength goes from left to right; it is translated into velocities by the Doppler shift. The highest velocities of approach are at the left (colored blue, since they are blueshifts) and the highest velocities of recession (colored red, since they are redshifts) are at the right. Position on a line across the galaxy (shown in blue in the black-and-white photograph at extreme right) goes from top to bottom, as in the ground-based spectrum shown in the previous image. The resolution of the Hubble spectrum is much greater than that in the ground-based spectrum. The velocity reaches 400 km/s (880,000 miles per hour) within 26 light-years (that is, two pixels) of the galaxy's center. The increased resolution shows that the abrupt change in velocity at the galaxy's center, from the high green pixels just left of center to the low yellow pixels just right of center, happens over such a short distance (from top to bottom) that a black hole of 300 million times the mass of the sun must be present to provide gravity strong enough to keep the gas in its tight orbit. Each STIS pixel is only 13 light-years across.

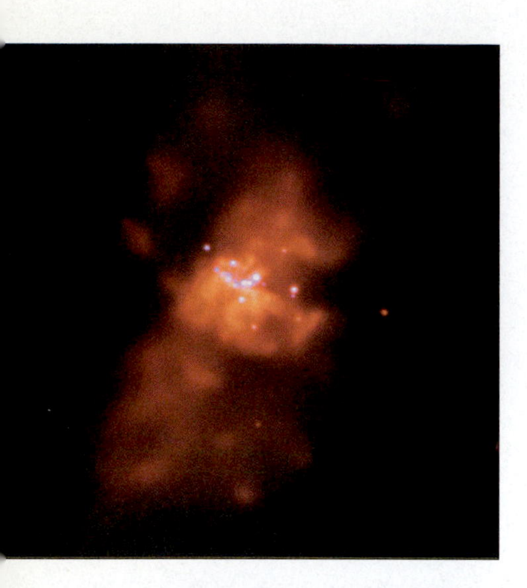

**FIGURE 34–17** The Chandra X-ray Observatory reveals many point x-ray sources distributed throughout M82. The field of view is 5 arc min on a side. The next figure shows a field of view about 10 times smaller on a side.

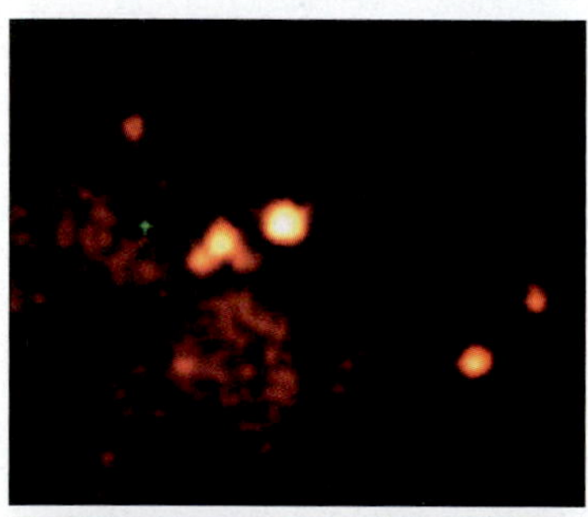

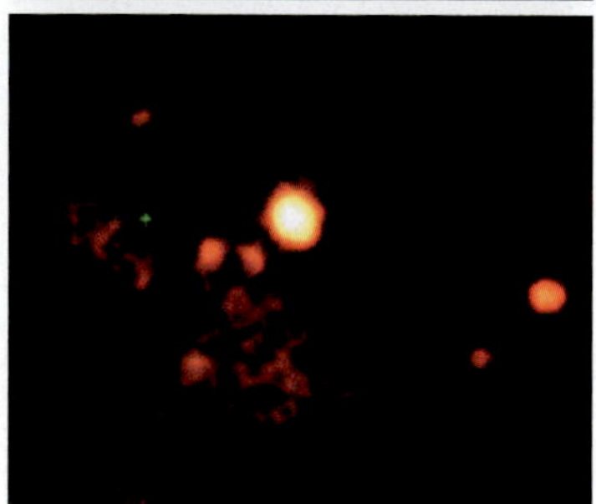

**FIGURE 34–18** These Chandra images show the effect of a mid-mass black hole 600 light-years from the center of the galaxy M82, the starburst galaxy that was shown in Figure 34-9. We see a region 35 arc sec across. This black hole contains at least 500 times the mass of the sun in a region about the size of the moon. It is about 600 light-years from the Galaxy's center (marked with a green cross) and varied in brightness over the two-month interval between the two images.

intense than the most powerful stellar-mass black hole in our own Galaxy. The x-rays are formed as accreted material falls onto the black hole and vary over time (Figs. 34–17 and 34–18). The brightness sets the lower limit to mass at about 500 solar masses. Calculating how quickly a much more massive body would have fallen into the center of the galaxy sets an upper limit of 1 million solar masses. Such objects may have formed in an early star-formation era of the galaxy, which led to such a high density of stars that a number of them collided and merged. Its location off-center may have resulted from a collision of M82 with M81 some 300 million years ago. This collision would have caused the starburst. Perhaps such mid-mass black holes eventually wind up as part of more massive black holes at the centers of galaxies.

Just a few years ago, we knew of no black holes in the center of galaxies, but now we know of more than 50. That number allows us to begin to do statistical analysis. John Kormendy of the University of Texas at Austin and colleagues have compared the masses of black holes at the centers of galaxies with various properties of the galaxies. He determined that a linear relationship existed between the masses of these black holes and the average random velocities of stars in the outer parts of the galaxies' bulges (Fig. 34–19). These star velocities can be measured directly at the telescope. Whether or not the galaxies have disks in addition to the bulges seems to be irrelevant. Kormendy suggests that the bulges form early in violent collapses. Starbursts that result from the collapses feed the black holes. The implication is that black holes and galaxy bulges formed at the same time. It argues against the earlier idea that black holes might have been born earlier in the universe's history and had galaxies form around them.

## 34.4 INTERACTING GALAXIES

Though galaxies were once thought to be isolated, we now realize that we must consider interactions between pairs of galaxies. For example, ring galaxies (Fig. 34–20) probably resulted from the passage of one galaxy through another. Sometimes the interactions lead to a very high rate of star formation; as we saw earlier in this chapter, the objects in which this occurs are known as starburst galaxies. This type of galaxy, now under much investigation, was discovered during an infrared survey carried out in the 1980s from a satellite. Over 90 per cent of the energy they emit may be in the infrared.

Theoretical work has been undertaken to explain the huge, faint shells recently observed on long exposures to surround some elliptical galaxies. The shells can be explained as stars thrown out by collisions of the ellipticals with spiral galaxies. Thus the idea is now current that some elliptical galaxies can result from the merger of pairs of spiral galaxies. Computer simulations endorse the possibility.

The gas stripped from the spirals apparently is carried to the outside of the remnant. It would then radiate in x-rays; we often detect x-ray halos of elliptical galaxies. Our Galaxy and the Andromeda Galaxy even seem on a line to interact in billions of years, so our sun may wind up in an elliptical galaxy someday.

In Section 32.5a, we saw how spiral structure can be maintained in a single galaxy by density waves or by a chain of supernovae. In some cases, spiral-like structure can arise from a gravitational interaction of two galaxies. Some scientists have used supercomputers to follow the evolution of a system over time, using the computers to follow the gravitational interaction between many particles, with each particle interacting by gravity with the center of mass of the other galaxy. Long arms, often called "tails," are drawn out by tidal forces (Fig. 34–21). The spin of the original galaxies contributes to the graceful curvature. The shapes can be made to match known galaxies (Fig. 34–22). In particular, The Antennae are the best-known pair of interacting galaxies.

ROSAT observations, now superceded by Chandra observations, of The Antennae revealed that x-rays are being emitted over the whole region of the merged nu-

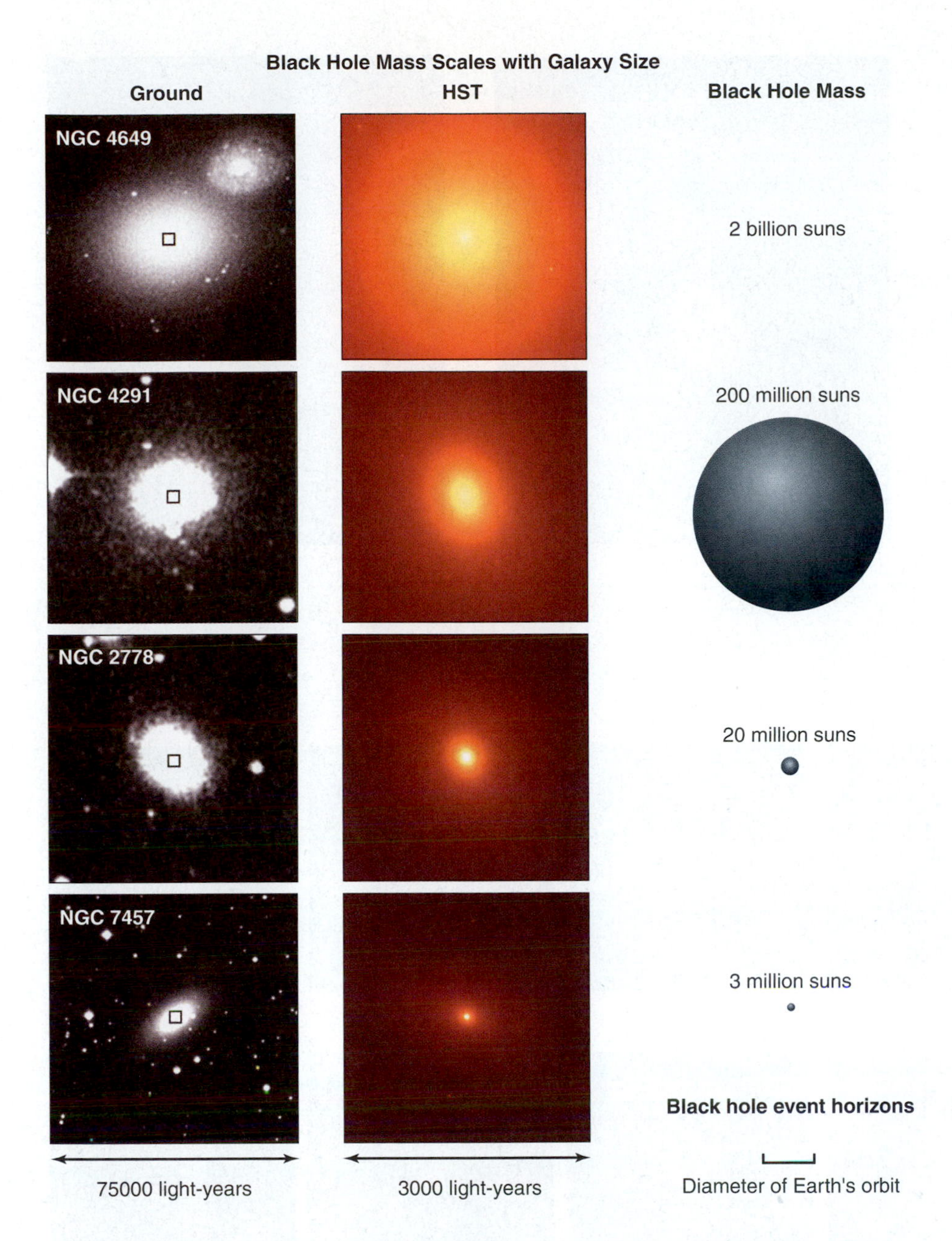


**FIGURE 34–19** The greater the star velocities measured in the midst of a galaxy, the more massive its central black hole. The mass of the central black hole has been correlated with galaxy mass.

clei. The x-rays probably come from the supernovae and x-ray binary stars that resulted from the rapid burst of star formation triggered by the collision. Infrared Space Observatory observations show the location of dust.

The Antennae involve collisions of galaxies of equal mass. If one galaxy is much more massive than the other, the structure that results from an interaction can resemble that of an ordinary spiral galaxy. Note that the Milky Way Galaxy has close companions—the Magellanic Clouds—and the Andromeda Galaxy has companions as well, so it is possible that there was a gravitational interaction to the spiral structure of our own Galaxy and other galaxies. Many examples of interacting galaxies are now known (Fig. 34–23).

Many astronomers believe that many elliptical galaxies may be the result of gravitational interactions of formerly spiral galaxies. The jury is still out on that one.

A

B

**FIGURE 34-20** (*A*) A Hubble Space Telescope view of the Cartwheel Galaxy, a ring galaxy caused by one of its satellite galaxies' passing through it. The Cartwheel, in Sculptor, is 500 million light-years away. (*B*) The polar-ring galaxy NGC 4650A. After a collision perhaps a billion years ago, the remnant of one galaxy has become the rotating inner disk of old reddish stars in the center. Little gas or dust was left there. The gas from another, smaller galaxy was probably stripped off and captured by the larger galaxy to form the ring of dust, gas, and stars. This ring is perpendicular to the old disk; we see both edge-on. The filters used for this false-color image were blue, green, and near-infrared.

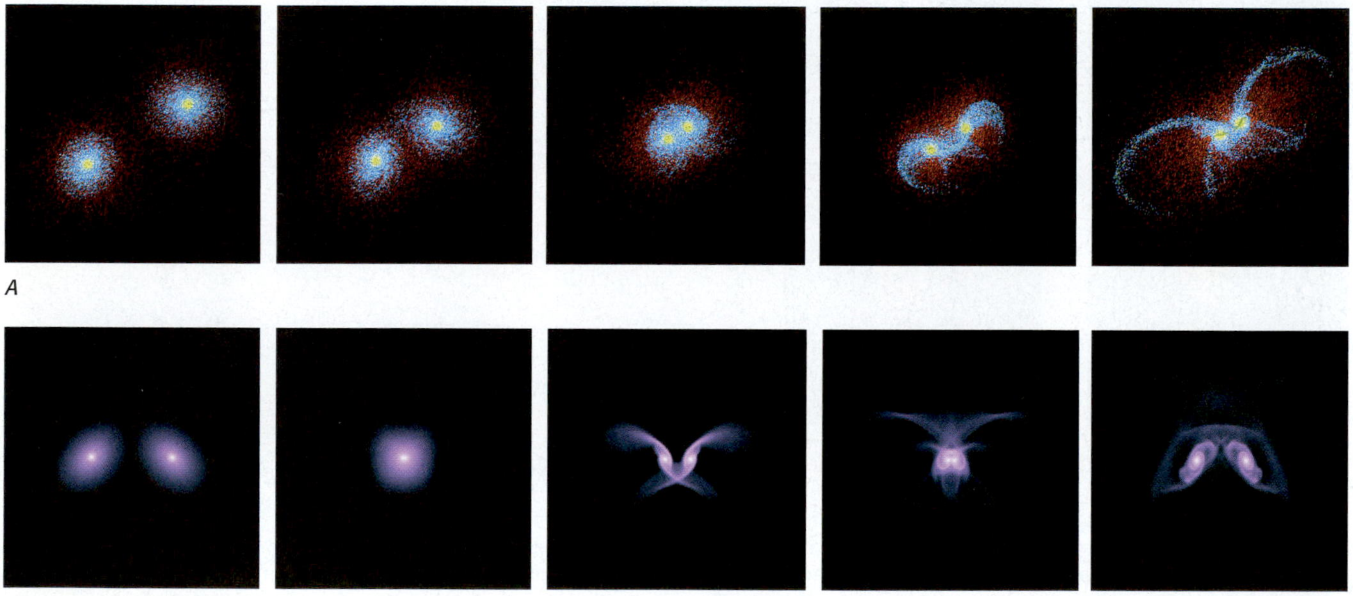

A

B

**FIGURE 34-21** Calculations made with supercomputers, showing a time sequence of the very close encounter of two identical model galaxies. (*A*) About 10,000 particles represent each galaxy. Time intervals are given in galactic years, which are each 250 million Earth years in duration. The supercomputer enabled Joshua Barnes (University of Hawaii) and Lars Hernquist (Harvard-Smithsonian Center for Astrophysics), both then of the Institute for Advanced Study in Princeton, to consider each galaxy to be made of a bulge, a disk, and a massive dark halo. (*B*) John Dubinski of the University of Toronto, in these more recent calculations, found general agreement with the earlier calculations of Barnes and Hernquist and of Alar and Juri Toomre. In the new simulation, the similarity with the arms of The Antennae is not as striking, since the faint, outer parts show better. In the simulation, each galaxy has a disk, bulge, and halo with a mass distribution similar to the Milky Way—the rotation curve is flat with a velocity of 220 km/sec like the Milky Way. The simulation contains 6 million particles—2 million particles in each disk, 500,000 in each bulge, and 500,000 in each halo. The two galaxies begin separated by a distance of 120 kpc and then swing by each other within 20 kpc at closest approach then receding to about the same distance by the end of the simulation. The time elapsed during the entire sequence is about 700 Myr. The size of each frame is 120 kpc on a side.

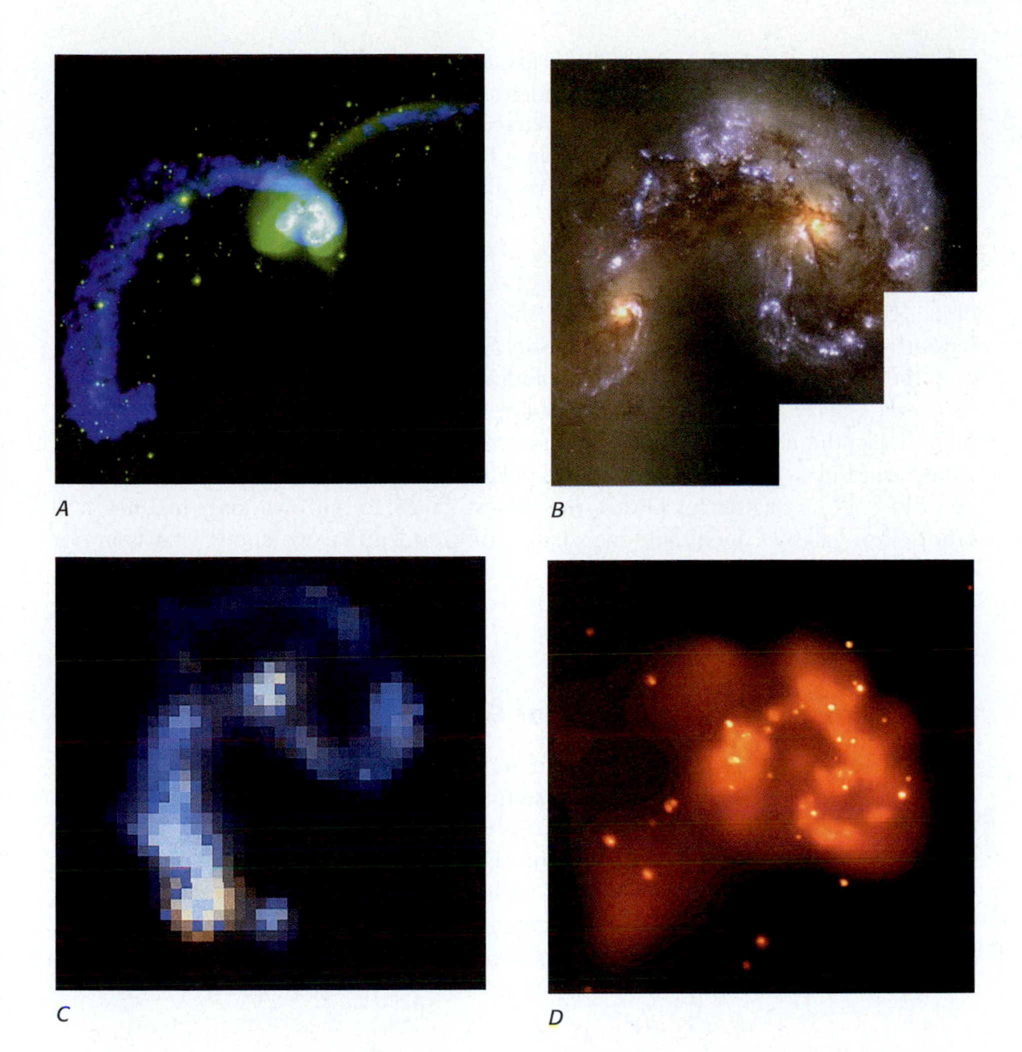


**FIGURE 34–22** The Antennae, the peculiar, interacting spiral galaxies NGC 4038 and NGC 4039. It is 60 million light-years from us, in the constellation Corvus. (*A*) False color (optical: green and white; radio: blue) highlights the tidal tails. (*B*) A Hubble Space Telescope image of the central region, showing details in the structure of the gas and dust. (*C*) The Infrared Space Observatory view. (*D*) The Chandra X-ray Observatory view. The dozens of bright point-like sources seen in x-rays are neutron stars or black holes pulling gas off nearby stars. The bright fuzzy patches are superbubbles, thousands of light-years in diameter, produced by the accumulated power of thousands of supernovae. The remaining glow of x-rays could result from many faint x-ray sources blurred together or from clouds of hot gas.

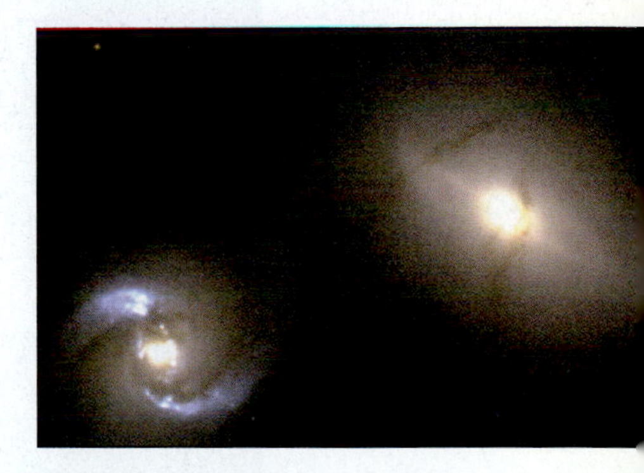

**FIGURE 34–23** An intergalactic "pipeline" of material flowing between two galaxies that interacted about 100 million years ago. The pipeline is the dark stream of matter that we see covering over 20,000 light-years of intergalactic space. It leaves NGC 1410 at left and wraps around NGC 1409 at right. The galaxies, in Taurus, were imaged with the Hubble Space Telescope.

## 34.5 CLUSTERS OF GALAXIES

Careful study of the positions of galaxies and their distances from us has revealed that most galaxies are part of groups or clusters. Groups have just a handful of members, while **clusters of galaxies** may have hundreds or thousands.

### 34.5a The Local Group

The three dozen or so galaxies nearest us form the **Local Group.** The Local Group contains a typical distribution of types of galaxies and extends over a volume 1 megaparsec (over 3 million light-years) in diameter. It contains three spiral galaxies, each at least 30 to 50 kiloparsecs (150 to 500 thousand light-years) across—the Milky Way, Andromeda (M31), and M33. These spiral galaxies contain about 95 per cent of the mass. Several of the forty or so other galaxies in the Local Group are ellipticals, including four regular ellipticals, two of which are companions to the Andromeda Galaxy. Others are dwarf ellipticals. The Local Group also contains at least four irregular galaxies, each 3 to 10 kiloparsecs (10 to 30 thousand light-years) across, including the Large and Small Magellanic Clouds. Besides these full-size irregular galaxies, at least a dozen other dwarf irregulars are known (Fig. 34–24).

Since our Milky Way Galaxy is one of the most massive members of the Local Group, it is located near the group's center, where it formed or fell under the groups' mutual gravity. The small galaxies in the Local Group are thought to be the type of object that would have first formed in the universe. Our Milky Way Galaxy is apparently the result of many mergers of such galaxies and protogalaxies. Indeed, even now the Sagittarius Dwarf Galaxy is colliding with the Milky Way Galaxy. The Andromeda Galaxy (M31) and our Galaxy are approaching each other at a rate that might make them collide in under 1 billion years. Such a collision, in current thought, would wind up with a merged elliptical galaxy as the result.

Finding the Sagittarius Dwarf, the closest galaxy to our own, only in 1994, and other close galaxies since then gives pause to the idea that we know what is in our neighborhood in the universe. Possibly, other nearby small galaxies will be found, also difficult to discover even though they are so close because they lie near the plane of our Galaxy and are thus hidden from our view by dust.

## 34.5b More Distant Clusters of Galaxies

In the vicinity of the Local Group, there are apparently other small groups of galaxies, each containing only a dozen or so members. The nearest cluster of many galaxies (a **rich cluster,** as opposed to a **poor cluster**) can be observed in the constellation Virgo and surrounding regions of the sky. It is called the Virgo Cluster (Fig. 34–25), and is about 15 Mpc (50 million light-years) away. It covers a region in the sky over 6° in radius, 12 times greater than the angular diameter of the moon. The Local Group is really an outlying part of the Virgo Cluster.

**FIGURE 34–24** (*A*) NGC 6822, a dwarf irregular galaxy. (*B*) A dwarf galaxy, a member of the Local Group, in the constellation Antlia. The individual stars are easily resolved.

**FIGURE 34–25** M84 (*center*) and M86 (*right*), bright elliptical galaxies, are the most prominent in this visible-light view of the center of the Virgo Cluster. NGC 4438 is the distorted galaxy at left, with NGC 4435 above it.

Other rich clusters are known at greater distances, including the Coma Cluster in the constellation Coma Berenices (Berenice's Hair). The Coma Cluster has spherical symmetry; its galaxies are concentrated toward its center, not unlike the distribution of stars in a globular cluster (which is a cluster of *stars*, and is therefore on a much smaller scale). Thus the Coma Cluster is a **regular cluster** as opposed to an **irregular cluster.** Dust in the clusters has been imaged from the Infrared Space Observatory (Fig. 34–26).

Rich clusters of galaxies are generally x-ray sources. Studies with x-ray telescopes revealed a hot intergalactic gas containing as much mass as is in the galaxies themselves. The temperature of the gas is 10 million to 100 million K. The gas is clumped in some clusters, while in others it is spread out more smoothly with a concentration near the center (Fig. 34–27). This difference may be an evolutionary effect, with gas being ejected from individual galaxies in younger clusters and spreading out as the clusters age. The high spatial resolution of the Chandra X-ray Observatory has revealed much structure that had been blurred out in earlier observations.

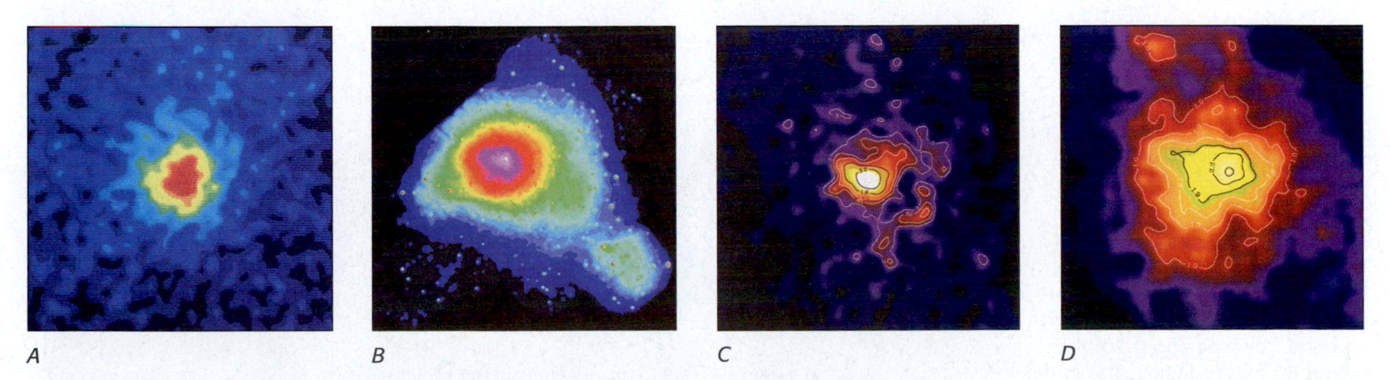


**FIGURE 34–26** The centers of the Virgo and Coma Clusters, displayed in false color. (*A*) The Virgo Cluster's x-ray emission, in an EXOSAT view, is sharply peaked around the central peculiar elliptical galaxy M87. (*B*) The Coma Cluster's x-ray emission, in a view from the XMM-Newton spacecraft. Chandra is now providing views of higher resolution of clusters of galaxies. (*C*) An Infrared Space Observatory (ISO) view of the Virgo Cluster. (*D*) An ISO view of the Coma Cluster.

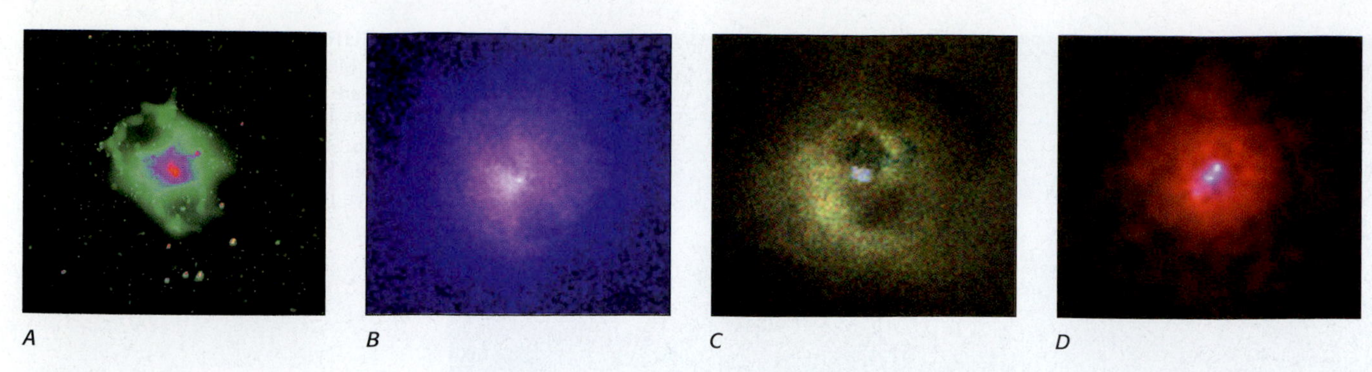


**FIGURE 34–27** X-ray images of clusters of galaxies taken with the Chandra X-ray Observatory and displayed in false color. We detect more irregularities than had been expected from earlier observations of clusters of galaxies, which didn't show such faint detail. It appears that clusters have been built up from smaller elements like the constituent galaxies or small groups of galaxies. (*A*) The central region of a compact galaxy group, HCG 62, which is smaller than a galaxy cluster. Green to purple to red indicate increasing x-ray intensity and therefore temperature. Such groups may fall together to become clusters of galaxies. (*B*) Hydra A, at a distance of 840 million light years. (*C*) The core of Perseus A, showing hot gas in and around the central supergiant galaxy. Colors indicate x-ray energies and therefore temperatures, with energy increasing from red to blue. The small dark patch at the 2 o'clock angle from the center results from the absorption of x-rays by gas in a smaller galaxy that is falling into the central galaxy. The bright blue spot in the center shows x-rays from gas around the central supermassive black hole. The twin dark cavities are thought to be bubbles of high-energy particles released from the vicinity of the black hole. The bright rims around the dark cavities show 30-million-kelvin gas that piled up as gas of twice that temperature fell from farther out. (*D*) 3C 295, a distant cluster 5 billion light-years from us.

X-ray spectra show that the intergalactic gas contains iron and other heavy elements with abundances that approximate those found in the sun. These abundances imply that the gas was ejected from galaxies, in which nucleosynthesis in stars formed the iron. The abundances also indicate that matter apparently flows out of most of the galaxies in a cluster. The matter is pulled by gravity toward the center of the cluster. A giant elliptical galaxy, like M87 at the center of the Virgo Cluster, may be so large because it has gobbled up gas and other galactic debris.

There may be as many as 10,000 galaxies in a very rich cluster, and the density of galaxies near the center of such a rich cluster may be higher than that near the Milky Way by a factor of one thousand to one million. Thousands of clusters of galaxies are known (Fig. 34–28).

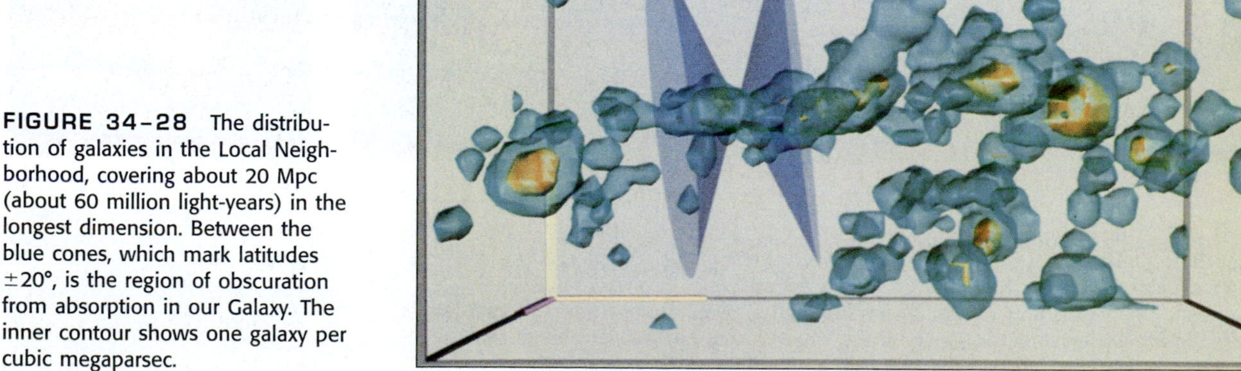

**FIGURE 34–28** The distribution of galaxies in the Local Neighborhood, covering about 20 Mpc (about 60 million light-years) in the longest dimension. Between the blue cones, which mark latitudes ±20°, is the region of obscuration from absorption in our Galaxy. The inner contour shows one galaxy per cubic megaparsec.

Every nearby very rich cluster is located in a cluster of clusters, a **supercluster.** The Local Group, the several similar groupings nearby, and the Virgo Cluster form the **Local Supercluster.** This cluster of clusters contains 100 member clusters roughly in the shape of a pancake that is 100 million light-years across and 10 million light-years thick. Superclusters are apparently separated by giant voids. In the next chapter, we will discuss more about these superclusters, and how they are revealed by contemporary large-scale mapping.

A University of Hawaii study of x-ray images has revealed over 100 massive galaxy clusters that are over 5 billion light-years away. Some of the clusters contain thousands of galaxies. The scientists revealed in 2000 that there are more distant clusters than had been expected. Chandra has even imaged one 10 billion light-years away (Fig. 34–29). As we will see in Chapter 38, the density of matter in the universe is an important parameter for cosmology, since it determines the overall strength of gravity permeating the universe. The density derived from these clusters of galaxies, reported in 2000, is in line with the latest value of density measured in other ways, which finds only about 30 per cent of the density needed to make the universe eventually collapse.

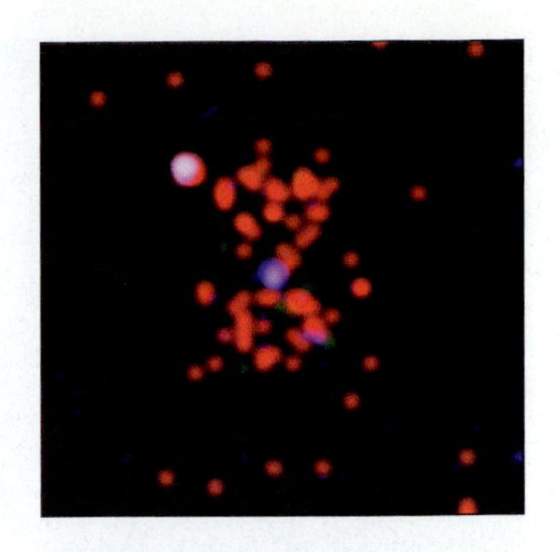

**FIGURE 34–29** 3C 294, an extremely distant x-ray cluster of galaxies, much farther away than any previously known cluster. The x-rays have been travelling from it to us for 10 billion years. This Chandra image shows how the cluster looked when it was only 20% of its current age. It shows an hourglass-shaped region of x-ray emission centered on a central radio source that was already known. The x-ray emission is over 600,000 light-years across. It had been known as the radio emission from a single galaxy, hence the 3C 294 name, but that galaxy is now revealed to be in the midst of a cluster. Clusters like this one are the largest systems in the universe that are held together by gravity.

## 34.6 GALACTIC ARCS AND GRAVITATIONAL LENSING

Strangely symmetrical giant arcs have been detected around galaxies, starting in 1987. At first it was thought that they might be gas in a cluster of galaxies, but now it seems that they are a result of a very strange phenomenon: gravitational lensing (Fig. 34–30). Gravitational lensing is explained with Einstein's general theory of relativity and results from the bending of light in a strong gravitational field.

The identification of a spectral line in one of the arcs showed that the gas that it is emitting is twice as far away as the galaxy around which the arc appears. Such measurements verify the relative distance to the arc and the cluster.

The gravitational-lensing model can explain why the arc is so regular. An object imaged straight ahead around a lensing object would appear as a ring, but any slight offset turns the ring into arcs. Further, the model places the object so far away that we are seeing the many hot young stars of a distant galaxy, explaining why the arcs appear so blue. Indeed, some arcs appear narrower when imaged in the red, which would result because a galaxy's red stars are more concentrated toward the galaxy's nucleus.

We shall meet other examples of gravitational lensing when we discuss the double quasars (Section 36.10) and MACHOs (Section 38.2a).

## 34.7 ACTIVE GALAXIES

Most of the objects that we detect in the radio sky turn out not to be located in our Galaxy. The study of these **extragalactic radio sources** is a major subject of this section. Chapter 36, on quasars, will further the discussion. Though we now think of quasars as a type of active galaxy, historically, they were discovered independently and so deserve a chapter of their own.

The core of our Galaxy, the radio source we call Sagittarius A, is one of the strongest radio sources that we can observe in our Galaxy. But if the Milky Way Galaxy were at the distance of other galaxies, its radio emission would be very weak.

Some galaxies emit quite a lot of radio radiation, many orders of magnitude (that is, many powers of ten) more than "normal" galaxies. We shall use the term **radio galaxy** to mean these relatively powerful radio sources. They often appear optically as peculiar giant elliptical galaxies. Radio galaxies, and galaxies that similarly radiate much more strongly in x-rays than normal galaxies, are called **active galaxies.**

Active galaxies are actually rare compared with ordinary galaxies. Still, just as newspapers tend to discuss crime instead of everyday happy occurrences, we will spend a fair amount of space in discussing these especially interesting galaxies.

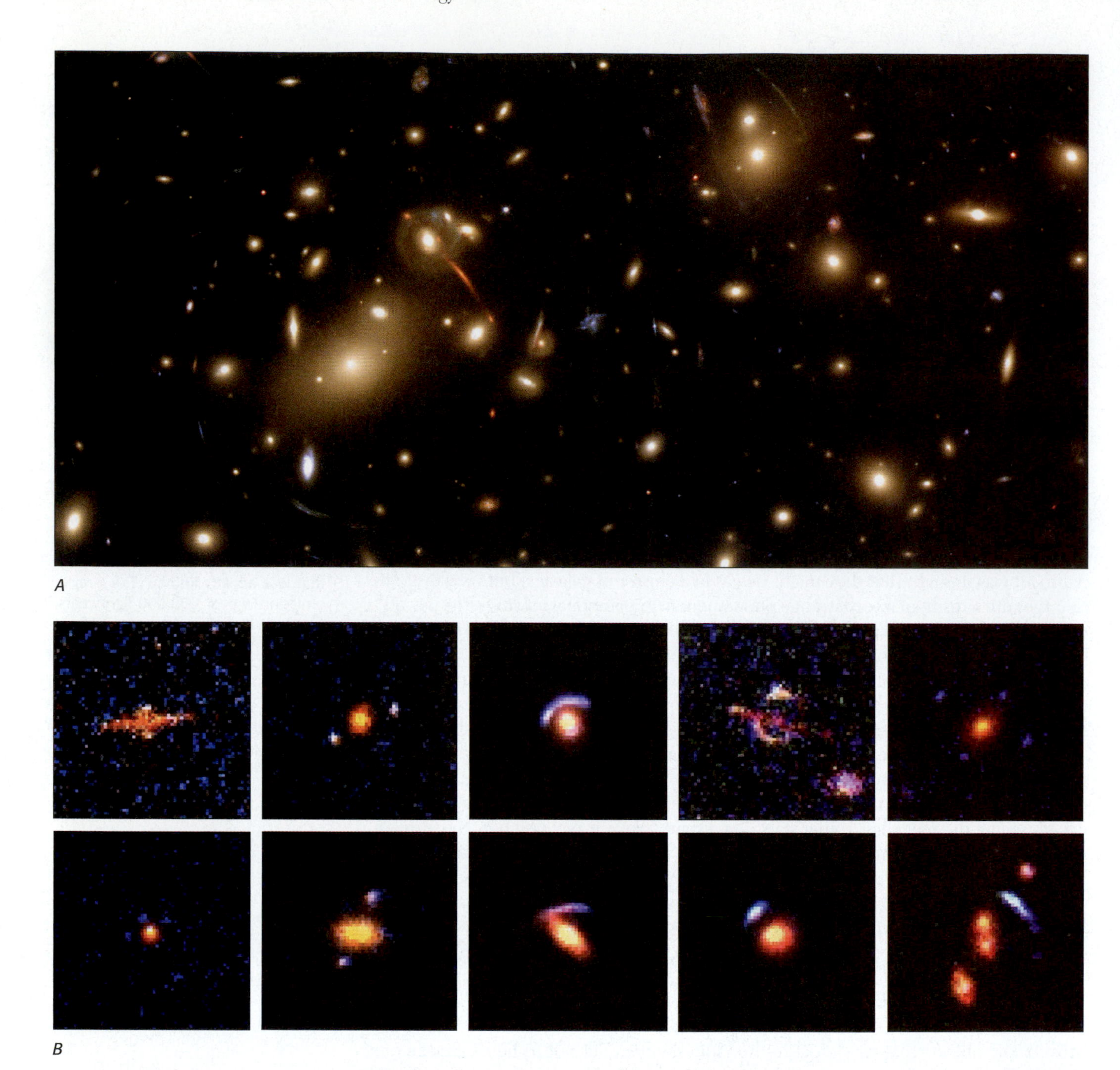


FIGURE 34–30 Astonishingly smooth arcs have been discovered in a few galaxy clusters. They represent gravitational lensing of a very distant galaxy as its light passes through a nearer but still distant cluster of galaxies. (*A*) Arcs formed by gravitational lensing by the cluster Abell 2218 (so-called from its number in a catalogue of clusters of galaxies compiled by George Abell). Since the lensing process does not change the color of light, parts of the arcs that have the same color come from the same background source. (*B*) A set of rings, arcs, and other gravitionally lensed patterns detected by Hubble in various sky directions.

### 34.7a Active Galactic Nuclei

Many of these galaxies have nuclei that are exceptionally bright. These sources are known as **AGNs** (from **Active Galactic Nuclei**); the whole source and not only the nucleus is often called an AGN. The most plausible and widely accepted source of energy for an AGN is matter falling onto an accretion disk around a massive black hole. In Chapter 36 we shall discuss extreme examples of AGNs called quasars.

One extreme example of an active galaxy has been discovered to give off extremely powerful gamma rays. This galaxy, known as Markarian 421, was already known to vary on a day-to-day basis. It became the strongest source in the sky at extremely short gamma-ray wavelengths when it gave off two bursts of photons, each lasting on the order of an hour. Each of these cosmic-ray photons carried an energy of approximately a trillion electron volts (1 tera-electron-volt, 1 TeV). The short duration indicates that the emitting region in the center of the active galaxy is not much larger than our solar system, pushing the existing theoretical models to their limits.

## 34.7b Radio Galaxies

The first radio galaxy to be detected, Cygnus A (Fig. 34–31), radiates about a million times more energy in the radio region of the spectrum than does the Milky Way Galaxy. Cygnus A and dozens of other radio galaxies emit radio radiation mostly from two zones, called **lobes,** located far to either side of the optical object. Such **double-lobed structure** is typical of many radio galaxies. (The images we show use the techniques we will discuss in the next section in order to provide high-resolution images.) Cygnus A's lobes are only 3 million years old, compared with perhaps 10 billion years old for stars in the central object.

The optical object that corresponds to Cygnus A—apparently a fuzzy, divided blob—has been the subject of much analysis, but its makeup is not yet understood. Only recently, and especially with new Hubble images, can we be sure that we see a single object partly obscured by dust. A whole scientific meeting has recently been devoted to this object alone.

The radio source Centaurus A has an optical counterpart (Fig. 34–32) that shows unusual wrapping by dust. Both x-ray and optical jets, aligned with the radio lobes, have been discovered. The jets, and the prodigious energy radiated, are thought to be caused by a spinning black hole of perhaps 100 billion times the mass of the sun.

Often the optical images that correspond to radio sources show peculiarities. For example, short visible-light exposures of M87, which corresponds to the powerful radio galaxy Virgo A, show a jet of gas 8000 light-years long. Light from the jet is polarized, confirming that the synchrotron process is at work here. High-resolution radio observations have shown that a very small source is present at M87's center. Studies of matter circling this small nucleus with the Hubble Space Telescope show that 3 billion solar masses are present in this small volume, indicating that it must be a black hole. In the next section, we will see how extremely high resolution observations of such jets are made with radio telescopes.

Magnetic fields are generated within a spinning accretion disk around the black hole. These magnetic fields spiral around the jet and confine it to a long, narrow tube. High-speed electrons and protons are accelerated near the black hole and race along the tube at almost the speed of light. The bright knot midway along the jet is where the jet becomes more chaotic. Near the bottom of the image, the jet runs into a wall of gas it has brought along ahead of itself.

As our observational abilities in radio astronomy have increased, especially with the techniques described in the next section, lobes and jets aligned with them have become commonly known. The best current model is that a giant rotating black hole in the center of a radio galaxy is accreting matter. Twin jets carrying matter at a high velocity are given off almost continuously along the poles of rotation. These jets carry energy into the lobes. (We may see only one, depending on the alignment and Doppler shifts.)

Not all radio galaxies have the double-lobed structure we have described. Others have a rounder structure—a core surrounded by a halo.

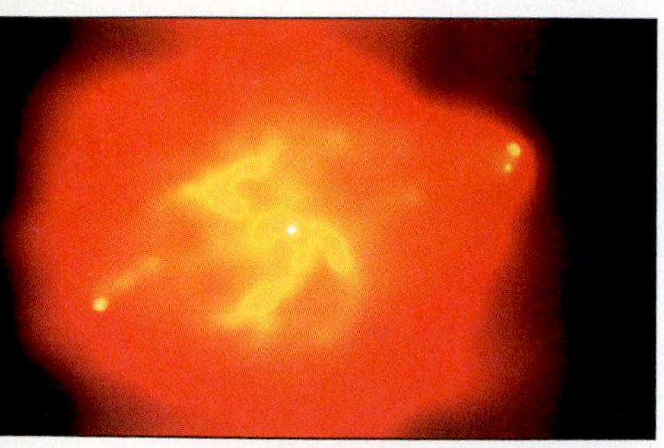

A

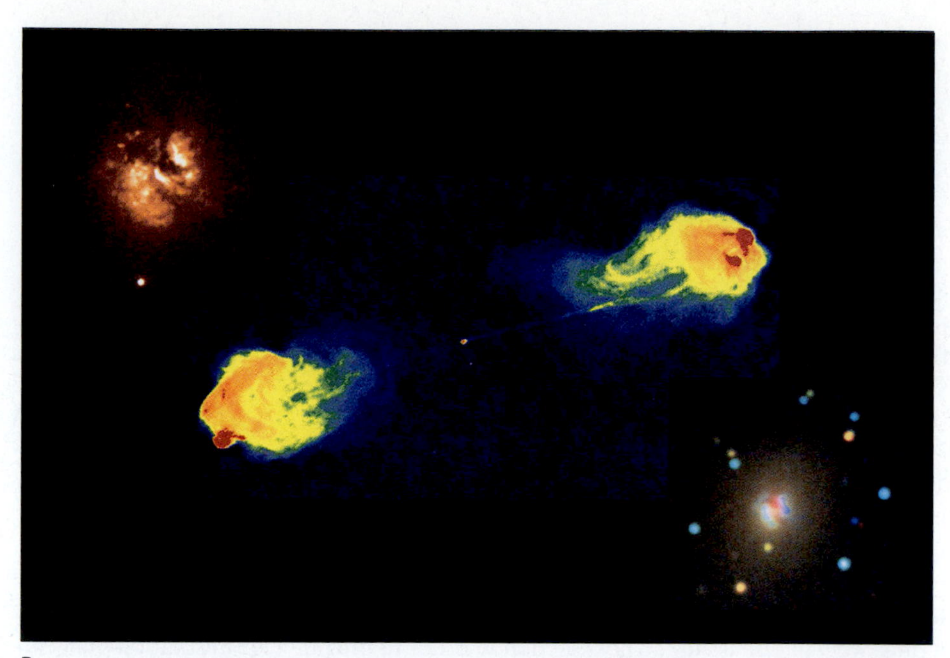

B

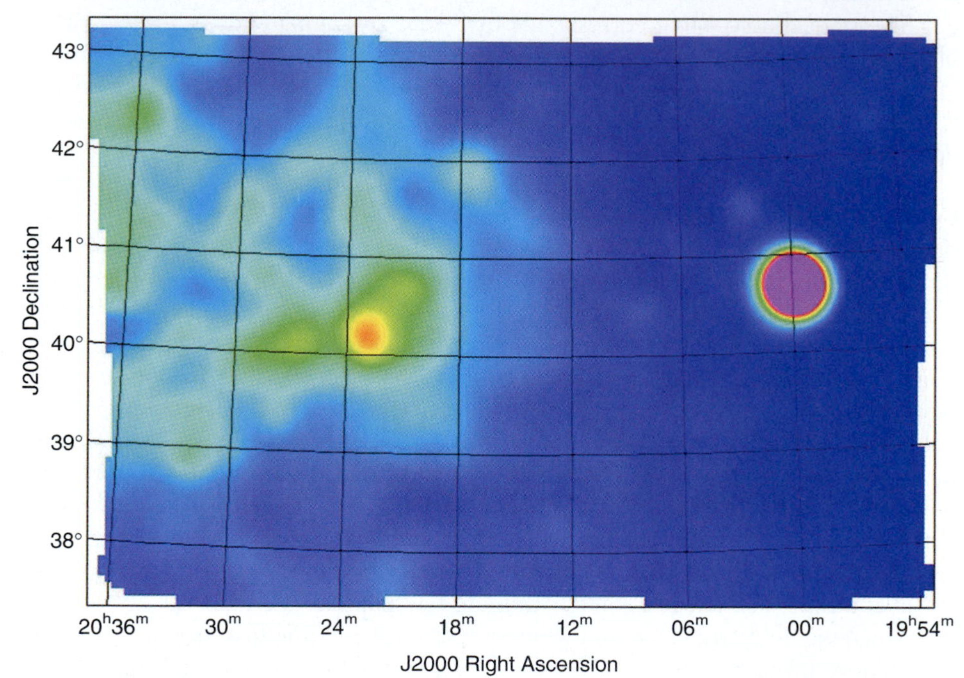


C

**FIGURE 34–31** (*A*) Cygnus A from Chandra. (*B*) A high-resolution radio map of Cygnus A, with shading and contours indicating the intensity of the radio emission. Electronic images of the faint optical object or objects that are observable, coinciding with the center of the radio image, are superimposed. (*Lower inset*) This optical ground-based view reveals a third, less prominent feature between the two main structures. This central region contains the central radio source. The resolution was improved over past images by using a computer-controlled monitor of the seeing and telescope guiding to form an image with 2/3-arc-sec resolution on a CCD at Mauna Kea. (*Upper inset*) A Hubble Space Telescope WFPC2 image. (*C*) The region including Cygnus A (which is the object that appears round, at right) was the first imaged with the Robert C. Byrd Green Bank Telescope, which opened in 2001. Red indicates the strongest radio emission. But the resolution was not sufficiently fine with this single dish, which was used at 800 MHz, to resolve Cygnus A (the circle on the right) in order to show the lobes of emission.

## 34.8 RADIO INTERFEROMETRY*

The resolution of single-dish radio telescopes is very low because of the long wavelength of radio radiation. Single radio telescopes may be able to resolve structure only a few minutes of arc or even a degree or so across. The techniques of interferometry are now used in radio astronomy to detect fine detail. Arrays of radio telescopes can now map the sky with resolutions far higher than the 1 arc sec or so that we can get with optical telescopes. Let us first describe how these radio interferometers work, and then discuss some of the high-resolution results.

**FIGURE 34–32** (*A*) Centaurus A, NGC 5128, looks like an elliptical or S0 galaxy surrounded by an extensive dust lane. The radio lobes of Centaurus A are superimposed on the optical image. (*B*) A Chandra Observatory x-ray view of the jet in Centaurus A.

## 34.8a Radio Interferometers

The resolution of a single-dish radio telescope at a given frequency depends on the diameter of the telescope. (A single reflecting surface of a radio telescope is known as a "dish.") If we could somehow retain only the outer zone of the dish (Fig. 34–33), the resolution would remain the same. (The collecting area would be decreased, though, so we would have to collect the signal for a longer time to get the same energy.)

Let us picture radiation from a distant source as coming in wavefronts, with the peaks of the waves in step (Fig. 34–34); we say that the waves are "coherent." If we can maintain our knowledge of the relative arrival times of the wavefront at each of two dishes, we can retain the same resolution as if we had one large dish whose diameter is equal to the spacing of the two small dishes shown. For a single dish, the maximum spacing of the two most distant points from which we can detect radiation is the "diameter." For a two-dish interferometer, we call it the **baseline.**

We study the signals by adding together the signals from the two dishes; we say the signals "interfere," hence the device is an "interferometer." Since the delay in arrival time of a wavefront at the two dishes depends on the angular position of an object in the sky with respect to the baseline, by studying the time delay one can figure out angular information about the object.

*(text continues on page 694)*

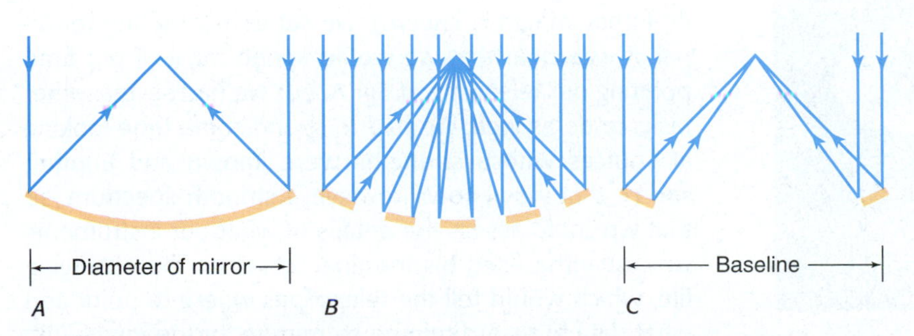


**FIGURE 34–33** A large single mirror (*A*) can be thought of as a set of smaller mirrors (*B*). Since the resolution for radiation of a certain wavelength depends only on the telescope's aperture, retaining only the outermost segments (*C*) matches the resolution of a full-aperture mirror. We can use a property of radiation called **interference** to analyze the incoming radiation. The device is then called an **interferometer.**

# ASTRONOMER'S NOTEBOOK

## Observing with the VLA

The Very Large Array of the National Radio Astronomy Observatory is an impressive instrument. Its 27 antennas march across the landscape, and it is hard to say whether it is more impressive when the telescopes are in their most compact status—with all the antennas within 1 km of each other—or when they are most spread out, with antennas disappearing over the horizon.

Observing time at the VLA is hard to get. Scientists work up detailed proposals and apply many months in advance. A committee at the VLA assesses all the proposals and gives out as much time as they can for the following quarter year.

My own project deals with the study of deuterium, a form of hydrogen, and what it tells us about the origin and future of the universe. In Chapter 38, I will discuss its significance for the study of cosmology. Here, let me describe what it was like to work at the VLA. I had been waiting years for a more capable telescope at the frequency I needed for the project. (It is beyond the FM radio band, which covers frequencies 88 to 108 MHz.) I joined Donald Lubowich of Hofstra University and the American Institute of Physics and K. Anantharamaiah of India's Raman Institute (then of the VLA's staff) to make deuterium observations.

*My own project deals with the study of deuterium, a form of hydrogen, and what it tells us about the origin and future of the universe.*

We were searching for deuterium in gas that lies between us on Earth and the bright source that is at the center of the Milky Way Galaxy. Deuterium, if it were present in measurable quantities, would make an absorption line. The Galactic center is marked by the radio source Sagittarius A (Sgr A for short, usually pronounced "Saj A"). From previous observations, we knew only whether or not deuterium appeared in the field of view of our telescope, which was $1\frac{1}{2}^\circ$ across, three times the angular diameter of the moon in the sky. The VLA would give us individual readings with a much smaller scale. We could know about each 2 arc min, less than $\frac{1}{10}$ the angular diameter of the moon.

Finally, the time for our observing came. At the VLA, we checked into our rooms and looked over the schedule. (Now observers ordinarily observe from headquarters in Socorro rather than coming to the telescope site.) It was mid-afternoon, and we were to observe that night from midnight to 8 a.m. The Galactic center was to be in view for $5\frac{1}{2}$ hours out of that time. With Anantha (as Anantharamaiah is known), we sat at a computer terminal. It was clear that we would spend most of our time pointing our telescope at Sgr A, but we had several other tasks to do as well. We had to spend some time looking at sources whose strengths were known and another source that was known to have a smooth spectrum, so that we could assess the details of what our instruments saw. Anantha used his terminal to set up the observing file, which would tell the telescopes where to point and what data to record minute by minute for the long night. We were to start with one of our calibration sources and then point to the Galactic center when it rose. Every 15 minutes we would briefly observe another of our cali-

▲ The smallest of the four possible configurations in which the VLA is arranged; here the telescopes are all within a 1-km circle. This arrangement provides lower resolution but a larger field of view than the most spread out configuration.

bration sources. When the Galactic center set, we would do more calibrations.

After dinner, we checked over our observing file. Midnight approached, the time I had been waiting years for. I was eager to have the observations. The telescope operator started on our observing file, and the 27 telescopes swung in unison to our first source. And what did we do next? The three scientists went to bed!

To bed? We had little choice. The telescope is so complex, and the types of data that come so involved, that only the most preliminary monitoring can take place while the observing run is going. The rest must be done by computer study, and we had the computer reserved for the morning. If we were to be fresh and awake for the computer work, we had to get some sleep.

I woke up at 5 a.m. and went into the control room. A computer screen blinked updates on the data taking, but there was nothing for me to do there. Everything seemed to be working. I walked out among the telescopes at dawn. They were silhouetted pristinely in the clear sky.

It took days of computer work to see what we had. The first stage of our data reduction was on the "pipeline" computer, which was optimized for handling the raw VLA data. After hours of work, we came up with a series of 64 maps, each map showing the strength of the signal in two dimensions. Together, the set is known as a "cube," and we could cut the cube with the computer in any dimension. For example, we could make a display of frequency versus position. Or we could sum point for point.

What did our data mean? We knew that we wouldn't solve the deuterium problem in this one observing run, but our observing had been a success. Even this single 5-hour run on the source Sgr A had told us something. We plotted graphs of the spectrum. If deuterium formed in small regions, it should show dips. None were seen, so we could place a limit on local formation of deuterium. The VLA allows us to pinpoint such localized areas of the sky.

If deuterium is not formed locally, then whatever is found was formed cosmically, in the first thousand seconds after the big bang. Our data fit in with this picture. We delivered a report at a meeting of the American Astronomical Society and published a scientific paper in the *Astrophysical Journal.*

Subsequently, Lubowich and I, working with colleagues and students, searched for deuterated molecules—molecules with deuterium substituting for ordinary hydrogen—in gas clouds near the center of our Galaxy. We used a radio telescope of the National Radio Astronomy Observatory on Kitt Peak to observe millimeter waves. We got a strong signal of DCN in Sgr A*.

▲ The VLA in its most compact configuration.

Our analysis, carried out in collaboration with astrochemists from England, showed that we detected 350,000 times more deuterium in the Galactic center than was expected. Our paper, published in the general scientific journal *Nature,* concluded that continuous streams of primordial gas from outside our Galaxy may be carrying deuterium-laden gas inward. We ruled out the idea that the extra deuterium was produced in the Galactic center by gamma rays or cosmic rays breaking up nuclei there because lithium, which would also result from such processes, is not enhanced there. We also interpret our observations to mean that our galaxy was never a quasar, which would have provided such cosmic rays or gamma rays.

We continued our attempts to observe deuterium in interstellar molecules with the 45-m Nobeyama radio

*(box continues next page)*

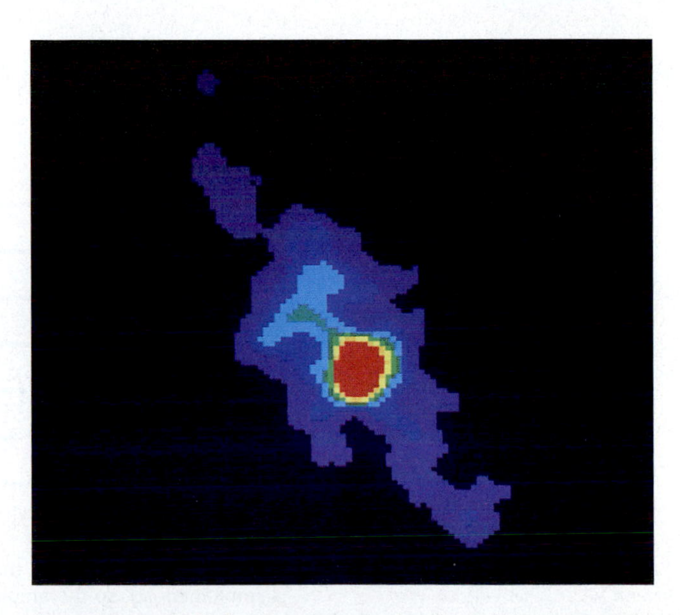

▲ The sum of all the maps in our cube showed the Arc near the galactic center.

telescope in Japan in 2001. We looked in the oldest (located at the edge of the Galaxy) and the youngest (located near the Galactic center) molecular clouds. The new Byrd Green Bank Telescope, 100-m across, and the eventual Atacama Large Millimeter Array (ALMA) should give excellent observations of deuterated molecules in the decades to come.

The best value for deuterium in stars in the nearest few hundred light years from the sun has been set by other astronomers from ultraviolet measurements with the Hubble Space Telescope and with the Far Ultraviolet Spectroscopic Explorer (FUSE). Since deuterium has been consumed to different extents at different places in our Galaxy in the billions of years since the big bang, it is important to survey the deuterium abundance across our Galaxy. Deuterium has even been discovered far in space, out toward a distant quasar. We discuss deuterium and its cosmological consequences further in Section 38.2b.

If the source were made of two points close together, the wavefronts from the two sources would be at slight angles to each other, and the time interval between the source reaching the two dishes would be slightly different. Thus an interferometer can tell if an object is double, even if it is unresolved by each of the dishes used alone.

### 34.8b Aperture-Synthesis Techniques

By suitably arranging a set of radio telescopes across a landscape, one can simultaneously make measurements over a variety of baselines, because each pair of telescopes in the set has a different baseline from each other pair and because various pairs are aligned in different directions. With such an arrangement one can more rapidly map a radio source than one can with two-dish interferometers. Also, using several dishes instead of just two gives that much more collecting area.

This interferometric technique is known as **aperture synthesis;** it was used to make the radio image of Cygnus A earlier in this chapter. The technique won a share of the Nobel Prize in Physics for Sir Martin Ryle, whose array at Cambridge, England, was the first of the modern configurations.

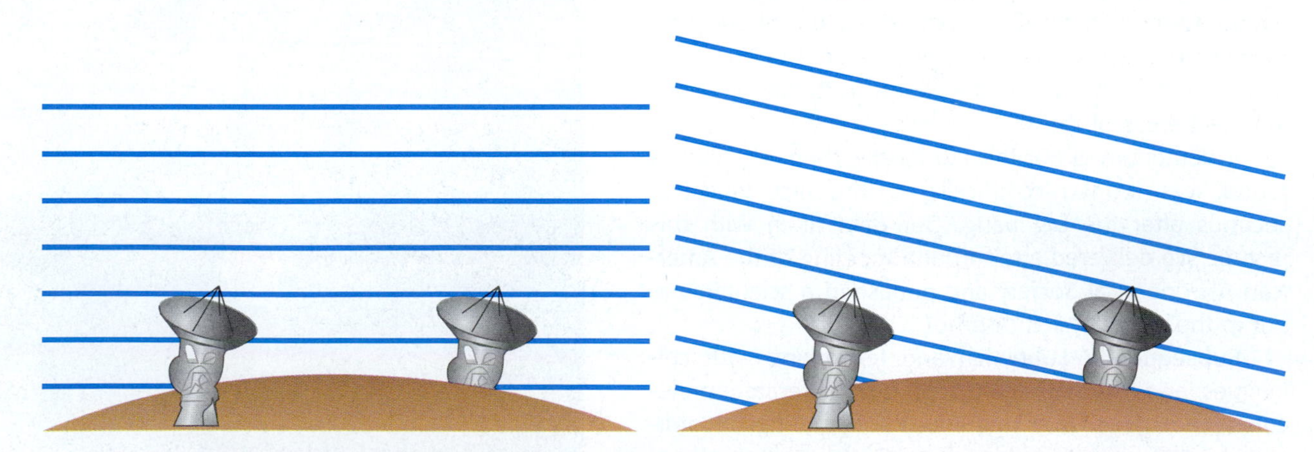

**FIGURE 34–34** In the left half of the figure, a given wave peak reaches both dishes simultaneously, so the amplitudes (heights) of the waves add. In the right half, the wave peak reaches one dish while a minimum of the wave reaches the other; the amplitudes subtract, and zero total intensity results. Thus "interference" results.

**FIGURE 34–35** The Very Large Array (VLA), near Socorro, New Mexico, with its dishes in its most compact configuration.

A fantastic aperture-synthesis radio telescope has been constructed in New Mexico by the National Radio Astronomy Observatory. It is composed of 27 dishes, each 26 m in diameter, arranged in the shape of a "Y" over a flat area that can be as much as 27 km in diameter (Fig. 34–35). The "Y" is delineated by railroad tracks, on which the telescopes can be transported to 72 possible observing sites; after an incident in which a storm came up on the site while a telescope was in motion, current rules allow moving telescopes only during the daytime and when they can be monitored visually. Sometimes, lower (though still high) resolution over a larger region of sky is needed; then the Y spreads out only over 1 km. The control room at the center of the "Y" contains powerful computers to analyze the signals, and the whole array is run from headquarters at Socorro, New Mexico, about 100 km away. The system is prosaically called the *Very Large Array* (VLA). It can operate at several wavelengths between 1.3 cm and 92 cm.

The VLA can make pictures of a field of view a few minutes of arc across, with resolutions comparable to the 1 arc sec of optical observations from large telescopes, in about 10 hours. Other arrays are also in use. The Berkeley–Illinois–Maryland Array (BIMA), in California, works at shorter radio wavelengths. An interferometer patterned after the VLA in size and scale was opened in India in 1997. This Giant Metrewave Radio Telescope contains 30 dishes each 45 m in diameter and is sensitive from 21-centimeter radiation up to wavelengths as long as 8 m.

A major new international project is placing an array of short-radio-wavelenth telescopes in a high, dry desert in Chile. This Atacama Large Millimeter Array (ALMA) is sponsored by the U.S. National Science Foundation, the European Southern Observatory, and institutions in France, Germany, the Netherlands, the United Kingdom, and Sweden. It is set to begin operations in 2009.

## 34.8c Aperture-Synthesis Observations of Galaxies

Aperture-synthesis observations show many giant double radio sources, much larger than any of the double-lobed sources previously known. Some are hundreds of times larger than our own Galaxy.

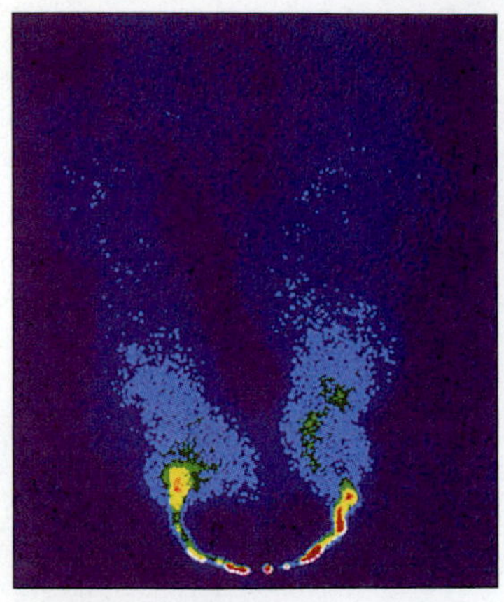

A

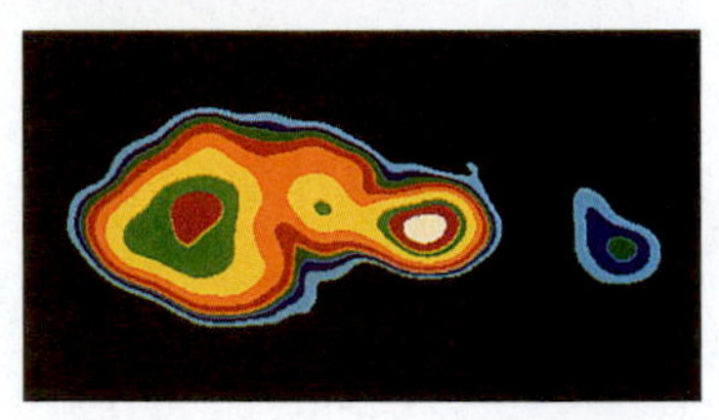

B

FIGURE 34–36 (*A*) The head-tail radio source NGC 1265, in a VLA view. The front end of the head of the radio galaxy corresponds to the position of an optical galaxy. (*B*) Very-Long-Baseline Interferometry (VLBI) images (see the following section) show a smaller field of view than VLA images. This VLBI image of NGC 1275 shows only 0.2 arc sec with a resolution of about 1 milliarcsec. We see a jet.

Interferometer observations have revealed the existence of a rare class of galaxies with "tails." They are called **head-tail galaxies** and resemble tadpoles in appearance. These galaxies expel the clouds of gas that we see as tails. The objects are double-lobed radio sources with the lobes bent back as the objects move through intergalactic space. High-resolution observations of one such galaxy with the VLA (Fig. 34–36) show that the source at the nucleus is less than 0.1 arc sec across, corresponding at the distance of this galaxy to a diameter of only 0.01 parsec (0.03 light-years or about 2000 A.U.). A narrow, continuous stream of emission leads away from the nucleus and into the tail. Such observations are being used to understand both the galaxy itself and the intergalactic medium, and they tie in with the x-ray observations of clusters of galaxies. Depending on the velocity of the galaxy and the density of the intergalactic medium, the lobes can be bent back by different amounts, so there is a range from lobes opposite each other to lobes slightly bent back to lobes bent back enough to make head-tail galaxies. Thus it makes sense that most head-tail galaxies are found in rich clusters of galaxies.

The discrete blobs that we can see in the tails indicate that the galaxies give off puffs of ionized gas every few million years as they chug through intergalactic space. Perhaps by studying these puffs, we can learn about the main galaxies themselves as they were at earlier stages in their lives. Head-tail galaxies seem to be a common although hitherto unknown type.

### 34.8d Very-Long-Baseline Interferometry

The first radio interferometers were separated by only hundreds of meters. The signals were sent over wires to a central collecting location, where the signals were combined. Telescopes were freed from the tyranny of wires with the invention of atomic clocks, which drift only three hundred-billionths of a second in a year. A time signal from an atomic clock can be recorded by a tape recorder on one channel of the tape, while the celestial radio signal is recorded on an adjacent tape channel. Since a wide band of frequencies must be recorded, videotape recorders are used, and their development was another necessary event. The radio signal recorded can be compared at any later time with the signal from the other dish, synchronized accurately through comparison of the clock signals.

No longer did dishes have to be near each other to make interference measurements. Now all that is necessary is that the two telescopes observe the same object at the same period of time; the signals can be compared in a computer weeks later. With this ability, astronomers can make up an interferometer of two or more dishes very far apart, even thousands of kilometers. This technique is called *V*ery-*L*ong-*B*aseline *I*nterferometry (VLBI).

VLBI can provide resolutions as small as 0.0002 arc sec, far better than resolutions using optical telescopes and even with aperture-synthesis radio telescopes (Fig. 34–37). But VLBI techniques are difficult and time-consuming. Also, we sample only a small area of sky at any time when we work at high resolutions, so it takes longer to study a region at high resolution than it does at low resolution. Therefore, VLBI techniques can be applied only to very small areas of sky. But for those few areas, chosen for their special interest, our knowledge of the structure of radio sources has been fantastically improved.

The next stage in VLBI work was to set up permanent networks of radio telescopes devoted to this purpose. Britain's MERLIN (*M*ulti-*E*lement *R*adio-*L*inked *I*nterferometer *N*etwork) with a 64-km-maximum baseline was first. It is being extended. The Australia Telescope, which began scientific observations in 1990, uses six 22-m telescopes on a 3-km baseline and can be linked. A larger American network—the *V*ery *L*ong *B*aseline *A*rray (VLBA)—is based at the Socorro, New Mexico, headquarters of the Very Large Array. The VLBA consists of 10 telescopes,

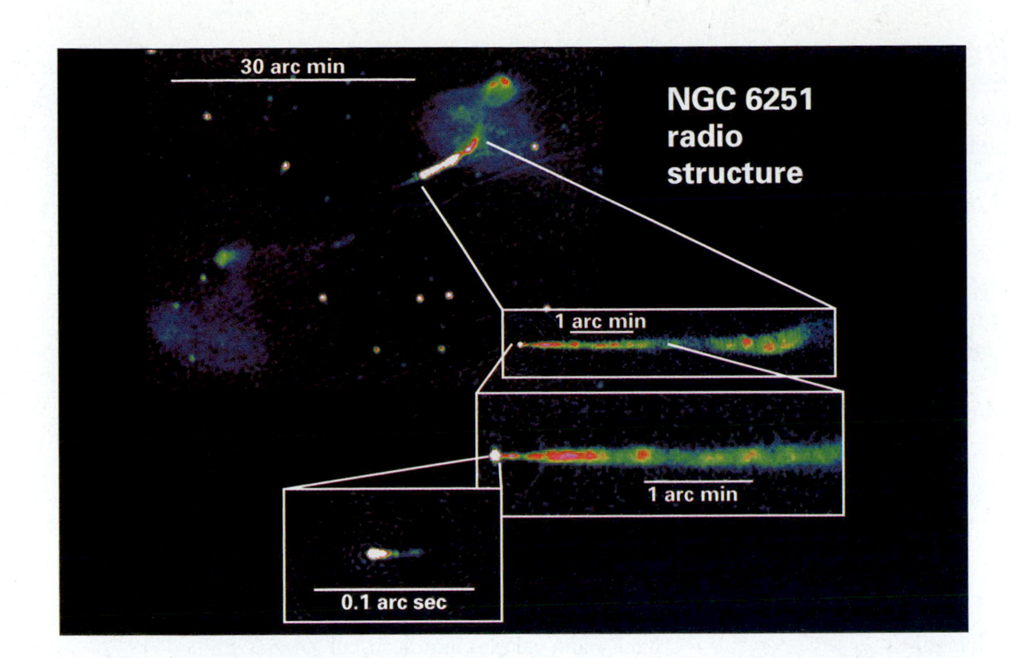


**FIGURE 34–37** Radio maps of the active galaxy NGC 6251 shown on four scales. One of the radio jets is pointing roughly in our direction, and thus appears much brighter than the other jet, which points away from us. The increasing resolution is achieved by going from the VLA to VLBI techniques.

each 25 m (about 80 ft) across, spread from Puerto Rico north to Massachusetts and west to Hawaii (which we saw in Figure 5–30). Its effective diameter is 8000 km. At a later date will come the addition of an antenna in space, increasing the baseline 4 times farther. In 1997, the Japanese launched a satellite, MUSES-B, with the first permanently orbiting antenna for VLBI work. The antenna is 8 m in diameter and is used in collaboration with antennas on Earth. It has been showing that the difficult task of integrating such a distant, orbiting antenna with ground-based antennas can be accomplished.

The Very Large Array itself is being expanded, with 8 additional antennas at distances up to 300 km and other improvements. This *E*xpanded *V*ery *L*arge *A*rray (EVLA) is to be ready by about 2010. It will have increased wavelength coverage, increased sensitivity, and increased resolution.

## CORRECTING MISCONCEPTIONS

| ✖ *Incorrect* | ✔ *Correct* |
| --- | --- |
| Spiral nebulae are in our Galaxy. | The "spiral nebulae" turn out to be galaxies comparable to our own. |
| Galaxies evolve along the Hubble tuning-fork diagram. | Once formed, in the absence of interaction, galaxies retain their Hubble type. |
| Ellipticals and spirals formed at the same time. | It seems that many ellipticals formed from collisions of spirals. |
| Galaxies are individual in space. | Most galaxies are in clusters. |
| We are limited in size of telescopes by the need for the parts to be connected. | Interferometers can have parts scattered over the world or in space. |

## SUMMARY AND OUTLINE

Observations and catalogues of nonstellar objects
- Lord Rosse's early observations of spiral forms
- Messier's catalogue, General Catalogues by the Herschels, *New General Catalogue* (NGC) and *Index Catalogues* (IC) by Dreyer
- Galaxies as "island universes"
- When Hubble observed Cepheids in galaxies, he proved that galaxies were outside our own Milky Way Galaxy.

Hubble classification (Section 34.1 and 34.2)
- Elliptical galaxies (E0–E7)
- Spiral galaxies (Sa–Sc) and barred-spiral galaxies (SBa–SBc)
- Irregular galaxies (Irr)
- Peculiar galaxies E(pec), S(pec)
- Amount of gas, and of star formation, increases toward Sc.

Centers of galaxies (Section 34.3)
- Giant black holes

Origin of galactic structure (Section 34.4)
- Ring galaxies from collisions
- Ellipticals may come from collisions of spirals.
- Gravitational interactions pulling out "tails"

Clusters of galaxies (Section 34.5)
- Local Group includes our Galaxy, 2 other spirals, and about 2 dozen other galaxies.
- Rich clusters, such as Virgo and Coma, are x-ray sources, containing hot intergalactic gas.
- Chandra has shown detail in distant clusters of galaxies.

Arcs in the sky result from gravitational lensing (Section 34.6)

Active Galaxies (Section 34.7)
- Some objects are powerful sources of radiation in the radio, x-ray, and infrared spectral regions.
- Double-lobed shape is typical of radio galaxies; sometimes a peculiar optical object is present at the center.

Radio interferometry (Section 34.8)
- Interferometers give resolution higher than optical observations.
- Aperture-synthesis arrays provide quicker maps.
- Current interferometers include the VLA (Very Large Array) in U.S.
- Giant radio galaxies and head-tail galaxies studied
- VLBI techniques leading to VLBA

## KEY WORDS

Messier numbers
Shapley-Curtis debate
spiral galaxies
Hubble type
elliptical galaxies
giant ellipticals
dwarf ellipticals
E0, E7, Sa, Sb, Sc, bar, SBa, SBb, SBc, Irr
peculiar galaxies
starburst galaxy
tuning-fork diagram
type S0
lenticular galaxies
clusters of galaxies
Local Group
rich cluster
poor cluster
regular cluster
irregular cluster
supercluster
Local Supercluster
extragalactic radio sources
radio galaxy
active galaxy
AGNs (Active Galactic Nuclei)
lobes
double-lobed structure
interference*
interferometer*
baseline*
aperture synthesis*
head-tail galaxies*

*This term appears in an optional section.

## QUESTIONS

1. What shape do most galaxies have?
2. Since we see only a two-dimensional outline of an elliptical galaxy's shape, what relation does this outline have to the galaxy's actual three-dimensional shape?
3. Sketch and compare the shapes of types Sb and SBb.
4. Draw side views of types E3, S0, Sa, Sb, and Sc, showing the extent of any nuclear bulge.
5. The sense of rotation of galaxies is determined spectroscopically. How might this be done?
6. Which classes of galaxies are the most likely to have new stars forming? What evidence supports this?
7. Describe three pieces of evidence that galaxies collide.
8. How does a ring galaxy arise?
9. How is it possible that galaxies could exist close to our own yet not have been discovered before?
10. What do infrared observations of galaxies show better than optical observations?
11. Describe what measurements we can make that reveal the mass of the central black hole of a galaxy. What information makes the link?
12. Discuss what the Chandra X-ray Observatory is showing us about x-ray emission from clusters of galaxies.
13. Compare the resolution and sensitivity of an interferometer with a single-dish antenna that scales the same outer limits of size.
14. Explain why an interferometer is sensitive to spatial details.
15. †For the same wavelength, roughly compare the resolution of VLBI spanning the United States with that of the VLA. Show your analysis.

†This question requires a numerical solution.

## INVESTIGATIONS

1. Stir tea leaves and make spirals. Notice what happens as your stirring makes the liquid speed up or slow down.
2. Play with a pinwheel. Which way does it rotate when blown on? Is it in differential rotation? Explain.
3. Try to see the galaxies M31 and M33 in the night sky with an unaided eye. Try again with binoculars. Sketch what you see in each case, including the surrounding constellations.

## USING TECHNOLOGY

### W World Wide Web

1. Check the Web site for the Great Debate on the origin of gamma-ray bursts, with its background of the Great Debate between Shapley and Curtis at http://antwrp.gsfc.nasa.gov/debate/debate.html.
2. Note a wide variety of galaxy types in the picture galleries of the Anglo-Australian Observatory (http://www.aao.gov.au/images.html) and of the European Southern Observatory (http://www.eso.org/outreach/info-events/ut1fl/astroim-galaxy.html).
3. Inspect the Hubble Space Telescope's high-resolution galaxy images at http://www.stsci.edu/pubinfo and search for ones showing galaxies with giant black holes.
4. Follow new measurements of x-rays from galaxies and clusters of galaxies at the Chandra X-ray Observatory site, http://chandra.harvard.edu.
5. Follow the Very Large Array's discoveries at http://info.aoc.nrao.edu/vla/html/VLAintro.shtml.

### REDSHIFT

1. Find M31, the Great Galaxy in Andromeda, and when it rises and sets today. How about three months from now?
2. Compare the variety of images of M31.
3. Compare images of various types of galaxies in the Photo Gallery.
4. Displaying deep-sky objects and constellation boundaries, locate the Virgo Cluster of galaxies and identify individual galaxies shown in this book.
5. Varying the magnitude limits for deep-sky objects, find out what are the brightest half-dozen galaxies in your sky this evening.
6. Look at the movies showing the mergers of galaxies, calculated in a supercomputer.
7. Look at the movie displaying the jet from the galaxy M87.
8. Look at the movie displaying the principles of gravitational lensing.
9. Study Science of Astronomy: A night on the mountain: The Messier marathon; Science of Astronomy: Galaxies: Islands in the sky; and Science of Astronomy: A night on the mountain: Galaxies: A comparative anatomy.
10. On Disk 2, study Story of the Universe: From Big Bang to galaxies.

The foreground galaxies shown are in the Coma Cluster, one of the nearer clusters of galaxies but still far enough away that local motions of its constituents are small compared with its overall participation in the expansion of the universe. Measurements in the Coma Cluster are thus particularly useful in determining the overall expansion velocity and, as we shall see, the age of the universe. Though our detailed studies of galaxies used to concentrate on the nearest cluster, the Virgo Cluster, Hubble Space Telescope views like this one are enabling us to make similar studies farther away. The Coma Cluster, in particular, is about five times farther from us than the Virgo Cluster. The brightest object seen here is NGC 4881. It is centered in the Planetary Camera, the small quadrant of Hubble's Wide Field and Planetary Camera 2. The image was taken to study the use of the globular clusters surrounding this galaxy as indicators of distance; these globular clusters are barely visible points as reproduced here. We know from our Milky Way that globular clusters are most numerous between absolute magnitudes −7 and −8. The investigation of NGC 4881 is to find the apparent magnitude at which its globular clusters are most numerous. If we assume that they are most numerous at the same absolute magnitude there as in our own galaxy, we can compare this absolute magnitude with the apparent magnitude to deduce the distance to NGC 4881. Only the elliptical galaxy in the Planetary Camera and a spiral at the right are in the Coma Cluster. Besides a few foreground Milky Way stars, all the other objects are galaxies far more distant than the Coma Cluster. One pair of galaxies even seems to be merging.

# Chapter 35

# Galaxies and the Expanding Universe

Over eighty-five years ago, in the decade before the problem of the location of the "spiral nebulae"—inside or outside our galaxy—was settled, Vesto Slipher of the Lowell Observatory took many spectra that indicated that the spirals had large redshifts. This work was to lead to a profound generalization. In these days when we can now simultaneously take a hundred spectra of galaxies in a half-hour, it is hard to realize that astronomers at that time took all night or even longer for a single spectrum.

**AIMS:** To learn how we measure distances to remote objects in the universe, how we study the expansion of the universe, why we think the expansion of the universe is accelerating, and how we map the universe

## 35.1 HUBBLE'S LAW

In 1929, Hubble announced that distant galaxies in all directions are moving away from us, and that the distance of a galaxy from us is directly proportional to its velocity. That is, he said that when the velocity we observe is greater by a certain factor, the distance is greater by the same factor (Fig. 35–1*A*). The diagram reveals how poor the data actually were by modern standards, and so shows how much a leap of genius it was for Hubble to advance that there was a straight-line relation between velocity and distance.

The relation is especially valuable because the velocity is directly determined from an observed quantity: the redshift. For all the redshifts then accessible to him, the relation between velocity and distance is the sample one we showed earlier in our discussion of the Doppler effect (Section 25.6). We can easily measure the redshift once we have made an image of the spectrum of a galaxy. It is the change in wavelength divided by the original wavelength, $\Delta\lambda/\lambda_0$. The redshift is often written simply as the letter $z$. Thus Hubble found the redshift $z$ was larger for galaxies that are farther away than for nearer ones. That is, the change in wavelength of a given spectral line was larger in more distant galaxies.

The proportionality between redshift and distance, or velocity and distance, is known as **Hubble's law.** Hubble, in collaboration with Milton Humason at Mt. Wilson, went on during the 1930s to establish the relation more fully (Fig. 35–1*B*). Note how much more obvious it is on this diagram that there is a straight-line relation between velocity and distance.

The graph would look about the same if we directly plotted redshift vs. distance instead of velocity vs. distance. The equation for Hubble's law is usually stated in terms of the velocity that corresponds to the measured wavelength, rather than in terms of the redshift itself. We can use the formula for the Doppler effect to link

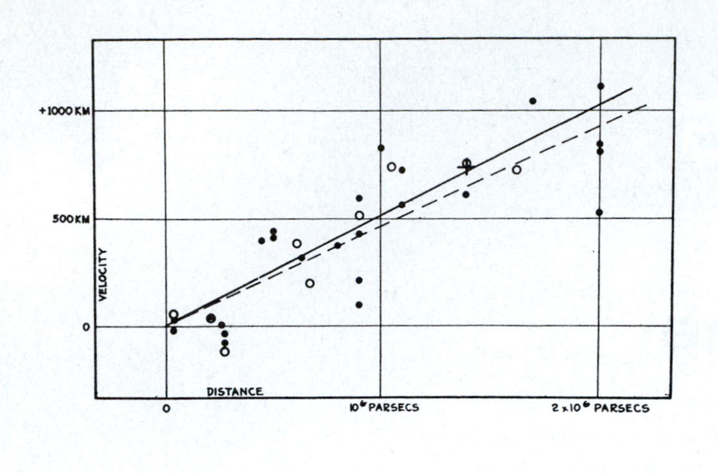


A

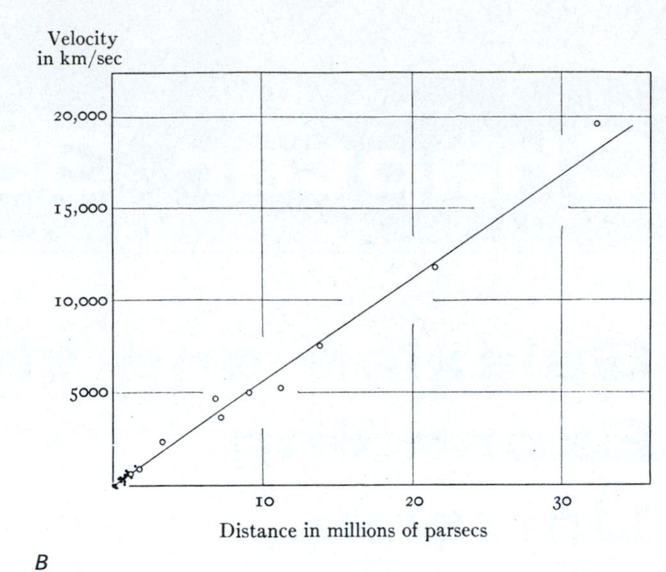


B

**FIGURE 35–1** (*A*) The velocity-distance relation, in Hubble's original diagram from 1929. Dots are individual galaxies; open circles are from groups of galaxies. The scatter to one side of the line or the other is substantial. (*B*) By 1931, Hubble and Humason had extended the measurements to greater distances, and Hubble's law was well established. All the points shown in the 1929 work appear bunched near the origin of this graph. $v = H_0 d$ represents a straight line of slope $H_0$. These graphs use older distance measurements than we now use, and so they give different values for $H_0$ than we now derive. Note that the straight line passes through "500 km" (it should say "500 km/sec") at left and 1 megaparsec at bottom, making Hubble's constant 500 km/sec/Mpc, far higher than the 75 km/sec/Mpc we now measure, using the latest values for distance. Current measurements show that all the distances shown are really about seven times greater than Hubble and Humason thought. Hubble's distances, given at the bottoms of the graphs in parsecs in the original handwriting and printing from the original published articles, are translated into light-years at the tops.

velocity and redshift. As we saw in Section 25.6a, reading first the left-hand side of the equation,

$$\frac{\text{change in wavelength}}{\text{original wavelength}} = \frac{\text{speed of recession}}{\text{speed of light}}$$

or, $\Delta\lambda/\lambda_0 = v/c$. This formula is valid for velocities much less than the speed of light. For velocities close to the speed of light, we must use a different formula, but we still say that the redshift $z = \Delta\lambda/\lambda_0$, the change in wavelength divided by the original wavelength.

Hubble's law states that the velocity of recession of a galaxy is proportional to its distance. It is written

$$v = H_0 d,$$

where $v$ is the velocity, $d$ is the distance, and $H_0$ is the present-day value of the constant of proportionality (simply, the constant factor by which you multiply $d$ to get $v$), which is known as **Hubble's constant.** We get the velocity from the redshift that we measure.

As we shall see in detail, there has long been a scientific fight over what Hubble's constant is. Allan Sandage of the Observatories of the Carnegie Institution of Washington is Hubble's intellectual heir. He has long used the world's largest telescopes to study the distances and redshifts of the farthest galaxies. Working over many years, Sandage and Gustav Tammann of the University of Basel in Switzerland

Astronomers always use km/sec/Mpc for the units of Hubble's constant, but in various scales of units, Hubble's constant is:
75 km/sec/Megaparsec
23 km/sec/million light-years
0.05 miles/hour/light-year
0.0005 inch/year/100 miles.

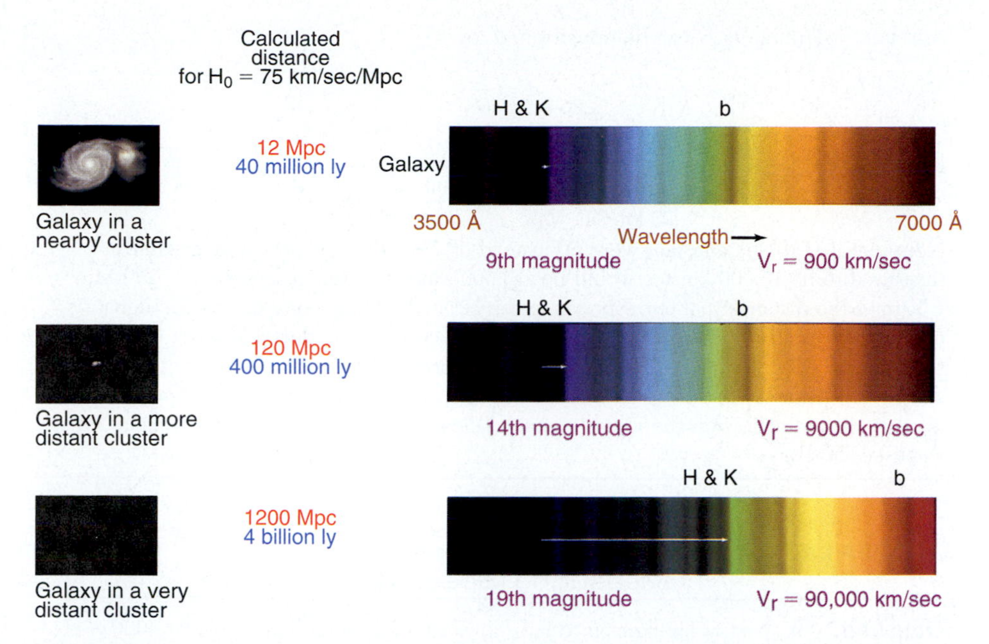


**FIGURE 35–2** Spectra are shown at right for the galaxies at left, all reproduced to the same scale. Distances are based on Hubble's constant = 75 km/sec/Mpc. Notice how the farther away a galaxy is, the smaller it looks. The arrow on each horizontal spectrum shows how far the H and K lines of ionized calcium are redshifted. The left side of the arrow marks the position of this pair of lines from a laboratory source on Earth, which therefore has no redshift. The right side of the arrow marks the position of this pair of lines as observed in a galaxy's spectrum. Their pattern—a close pair of dark Fraunhofer lines—is distinctive and so it is often easy to spot, though it doesn't show well in this color simulation. A spectral line known as "b" could also be used.

have derived a value of 55 km/sec/Mpc for $H_0$, as we shall describe in Section 35.2. This value is about 10 times lower than the one that Hubble originally announced, but Sandage and Tammann have used new techniques for finding the distance to far-off galaxies, incorporating also earlier corrections to the distance scale.

In recent years, other scientists have made similar sets of observations, and many have derived larger values for $H_0$. These values were first close to 100 km/sec/Mpc, so there was a major difference between one camp with values close to 50 and another camp with values closer to 100. (Astronomers mostly don't bother giving the units; "km/sec/megaparsec" is just understood.) The camp with the high value is now finding 75, in the middle of the old value of the two warring camps. In particular, we will see how Wendy Freedman and her colleagues, using the Hubble Space Telescope, found their final value to be close to 75. We will use 75 for the Hubble constant in the rest of this book.

Hubble's constant is given in units that may appear strange, but they merely state that for each megaparsec (3.3 million [$= 3.3 \times 10^6$] light-years) of distance from the sun, the velocity increases by 75 km/sec (Fig. 35–2). (Thus the units are km/sec per Mpc, spoken "kilometers per second per megaparsec.") From Hubble's law, we see that a galaxy at 10 Mpc would have a redshift corresponding to 750 km/sec, since (75 km/sec/Mpc) × (10 Mpc) = 750 km/sec. At 20 Mpc, the redshift of a galaxy would correspond to 1500 km/sec; and so on. The velocity deduced from the redshift for a given galaxy is the same no matter in which part of the spectrum we observe.

## EXAMPLE 35.1 Determining Distance from the Redshift

***Question:*** For $H_0 = 75$ km/sec/Mpc, how far away is a galaxy for which we measure a redshift that corresponds to 15,000 km/sec?

***Answer:*** Hubble's law can be transformed to

$$d = \frac{v}{H_0}$$

Thus

$$d = \frac{15{,}000\ \cancel{\text{km/sec}}}{75\ \cancel{\text{km/sec}}/\text{Mpc}} = \frac{200}{1/\text{Mpc}} = 200\ \text{Mpc}.$$

Note that if Hubble's constant were 50 instead of 75, then a galaxy whose redshift is measured to be 15,000 km/sec would be (15,000 km/sec)/(50 km/sec/Mpc) = 300 Mpc, 1.5 times the distance that corresponds to the larger Hubble's constant. So the debate over the size of Hubble's constant has a broad effect on our understanding of the size of the universe. The debate is often heated, and sessions of scientific meetings at which the subject is discussed have long been well attended. Almost all astronomers hope that the measurements made with the Hubble Space Telescope have settled the controversy, as we shall see shortly.

### EXAMPLE 35.2 Determining Distance by Measuring Spectra

***Question:*** A spectral line known to have a wavelength in the laboratory of 4000 Å is observed at a wavelength of 4100 Å in a galaxy. (The H line of calcium, for example, has a wavelength of about 4000 Å, and is readily identifiable in a spectrum.) How far away is the galaxy?

***Answer:*** The redshift

$$z = \frac{\Delta\lambda}{\lambda_0} = \frac{v}{c}$$

so

$$\frac{(4100\ \text{Å} - 4000\ \text{Å})}{(4000\ \text{Å})} \times c = v.$$

Thus

$$\left(\frac{100}{4000}\right)c = \left(\frac{1}{40}\right)c = v$$

Since

$$c = 300{,}000\ \text{km/sec}, \qquad v = \frac{(300{,}000\ \text{km/sec})}{40} = 7500\ \text{km/sec}.$$

Now, using Hubble's law, $d = v/H_0$,

$$d = \frac{7500\ \text{km/sec}}{75\ \text{km/sec/Mpc}} = 100\ \text{Mpc}.$$

Since 1 million parsecs is about 3.26 million light-years, the galaxy is about 326 million light-years away, which we should round to about 330 million light-years. (The original question was phrased with only two decimal places of accuracy, so our answer can't have more than two decimal places of accuracy.)

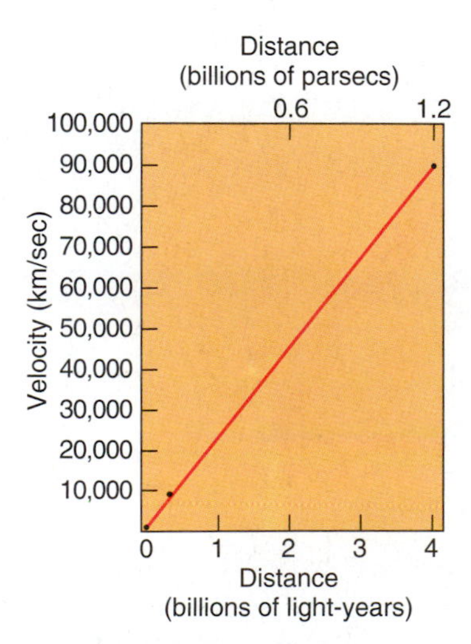


**FIGURE 35–3** The Hubble diagram for the galaxies shown in Figure 35–2.

Another issue being tested is whether Hubble's law follows the straight line very far out into the universe. It certainly is a straight-line relation between distance and velocity for billions of light-years (Fig. 35–3). Data in such a straight-line relation are called "linearly proportional." But does the relation curve 10 billion light-years away? In other words, is there a deviation from Hubble's law at the farthest reaches of the universe?

One technical point about which some scientists feel strongly: The redshift in Hubble's law is a consequence of the overall expansion of the universe rather than of objects merely speeding away from us in a static universe. We can make our calculations by thinking of the redshift as a Doppler effect, though theoretically it is slightly different.

## 35.2 THE EXPANDING UNIVERSE

The major import of Hubble's law is that since all but the closest galaxies in all directions are moving away from us at a rate proportional to distance, the universe is expanding. More precisely, it is expanding uniformly, as we will see below. Since the time when Copernicus moved the Earth out of the center of the universe (and the time when Shapley moved the Earth and sun out of even the center of the Milky Way Galaxy), we have not liked to think that we could be at the center of the universe. Fortunately, Hubble's law can be accounted for without our having to be at any such favored location, as we see below.

Imagine a raisin cake (Fig. 35–4) about to go into the oven. The raisins are spaced a certain distance away from each other. While the cake is rising and expanding, the raisins are moving apart from each other (though they are not themselves changing in size). If we were able to sit on any one of those raisins, we would see our neighboring raisins move away from us at a certain speed. It is important to realize that raisins farther from us would be moving away faster, because there is more cake between them and us to expand in a given unit of time. No matter in what direction we looked, the raisins would be receding from us, with the velocity of recession proportional to the distance.

The next important point to realize is that it doesn't matter which raisin we sit on; all the other raisins would always seem to be receding. Of course, any real raisin cake is finite in size; the universe may have no limit, so we would never see an edge. The fact that all the galaxies appear to be receding from us does not put us in a unique spot in the universe; there is no center to the universe. Each observer at each location would observe the same effect.

Note that in our analogy the size of the raisins themselves is not changing; only the separations are changing. In the universe, the galaxies themselves and the clusters of galaxies are not expanding; only the distances between the clusters (or, perhaps, the superclusters) are increasing.

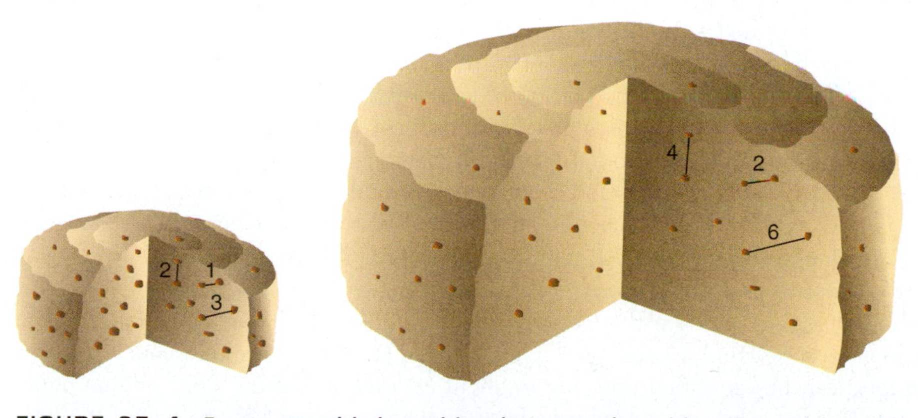


**FIGURE 35–4** From every raisin in a raisin cake, every other raisin seems to be moving away from you at a speed that depends on its distance from you. (Note that the raisins don't change in size.) This observational fact leads to a relation like the Hubble law between the velocity and the distance. The farther the raisin at first, the faster it is receding. Note also that each raisin would be at the center of the expansion measured from its own position, yet the cake is expanding uniformly. For a better analogy with the universe, consider an infinite cake; unlike the finite analogy pictured, there is then no center to its expansion. The numbers show the distance between pairs of raisins, and show the effect of doubling in size. Let us say that the cake doubles in size in an hour and that the distances are marked in centimeters. A distance of 1 at the left doubles to be 2 at the right, giving a rate of recession between the two raisins of 1 cm per hour. A distance of 2 cm at the left doubles to be 4 cm at the right, a change in distance of $4 - 2 = 2$ cm, making the rate of recession to be 2 cm per hour. Similarly, a distance of 4 cm at the left doubles to be 8 cm at the right, a change in distance of 4 cm, making the rate of recession to be 4 cm per hour. So in the raisin cake, as in the universe, the rate of recession is larger for objects that are farther away.

**FIGURE 35–5** M83, a spiral galaxy in Centaurus, is sufficiently close to Earth (3.7 Mpc = 12 million light-years) that its individual velocity makes it deviate from the "Hubble flow" of expansion. But farther galaxies and clusters of galaxies follow Hubble's law.

Following the style of the late George Gamow, using objects or locations of larger and larger scales in the universe, I can write my address as:
Jay M. Pasachoff
Williamstown
Berkshire County
Massachusetts
United States of America
North America
Earth
Solar System
Milky Way Galaxy
Local Group
Virgo Cluster
Local Supercluster
Universe

Note also that individual stars in our galaxy can appear to have small redshifts or blueshifts, caused either by their peculiar velocities or by the differential galactic rotation. Similarly, some of the nearer galaxies (such as M31 in Andromeda) have random velocities of sufficient size, or velocities less than our rotational velocity in our galaxy, so that they are approaching us. But all galaxies in distant clusters are receding. Still, the velocities of the nearer of those distant galaxies (Fig. 35–5) may be substantially affected by concentrations of mass so as to give misleading answers for measurements of the overall "Hubble flow" that leads to the determination of Hubble's constant.

## 35.3 THE DISTANCE SCALE OF THE UNIVERSE

The major problem for setting the Hubble law on the firmest footing is finding the distances to the galaxies for which redshifts are measured. We can't measure trigonometric parallaxes beyond the nearest region of our galaxy. We look for good "standard candles," objects of known intrinsic brightness, so that we can measure their distances by comparing their intrinsic brightness with the brightness we see. It is like telling how far away a streetlight is on a dark night by how bright it appears, since we know about how bright a streetlight should look when we are some known distance from it.

Only for the nearest galaxies can we detect **primary distance indicators.** The major example is Cepheid variable stars (Section 26.4b), for which we derive the distance directly by comparing absolute magnitude (from the period-luminosity relation) with the observed apparent magnitude. (Since the period-luminosity relation has been determined by Cepheids in the Large Magellanic Cloud, whose distance we measure in other ways, this type of distance determination is thought of as being "primary," like trigonometric parallax.)

Though prior to 1990, a handful of Cepheids were known in the nearest galaxies, the Hubble Space Telescope has shown many more Cepheids farther out (Fig. 35–6). Wendy Freedman of The Observatories of the Carnegie Institution of Washington and colleagues carried out one of the Hubble Space Telescope's three "Key Projects" using this method. (The other two Key Projects were a study of Quasar Absorption Lines and a Medium-Deep Survey of Galaxies.) They discovered Cepheids in 18 spiral galaxies and used additional Cepheid observations from 13 galaxies. For example, they found 50 new Cepheids in the spiral galaxy M81, many more than had been known. Observing over a period of months, they were able to plot light curves for these Cepheids. Their period-luminosity relation gave a distance of 11 million light-years to M81, with an uncertainty of only 10 per cent. They went on to compute period-luminosity relations for many other galaxies, and plotted a Hubble diagram based on the Cepheid measurements (Fig. 35–7A). From it, they could determine Hubble's constant. Even with the Hubble Space Telescope, Cepheid variables can be identified out only to about 30 million parsecs (100 million light-years). This is far by some standards, but less than 1 per cent of the size of the universe, which is over 10 billion light-years across, as we will see in Chapters 37 and

April 23

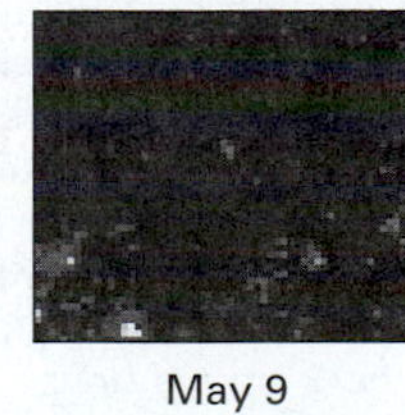
May 4

May 9

May 16

May 20

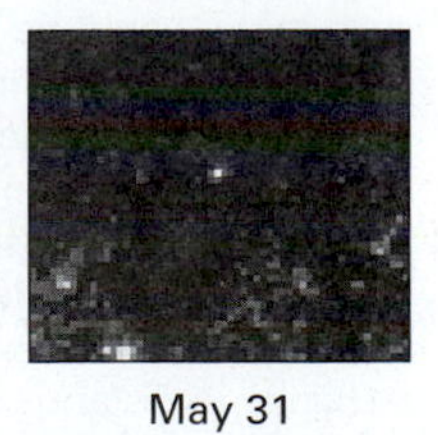
May 31

**FIGURE 35–6** Cepheids in the galaxy M100 observed with the Hubble Space Telescope. They were used to calibrate the cosmic distance scale.

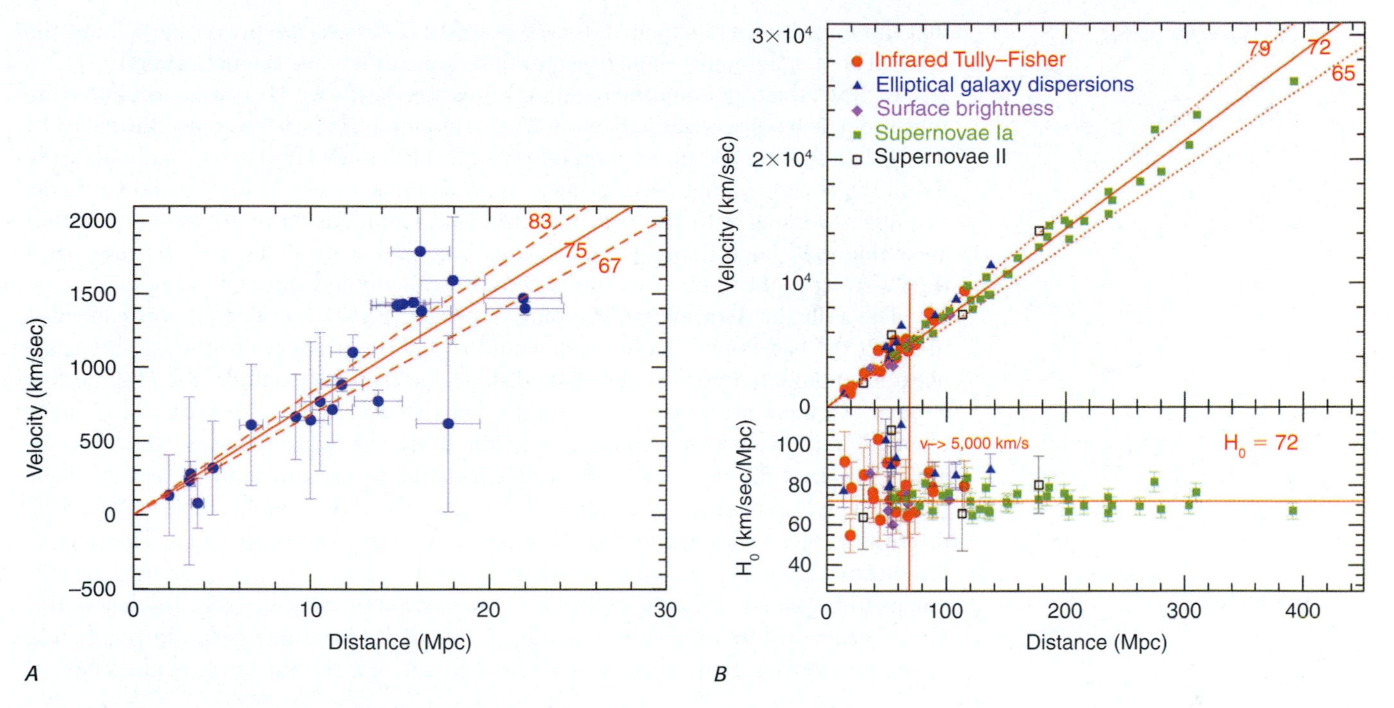


**FIGURE 35–7** (*A*) Hubble's law as determined from Cepheid variable stars observed with the Hubble Space Telescope out to a distance of about 20 Mpc (about 65 million light-years), our "local" region. The points at about 22 Mpc are averages for several clusters. This graph shows the final result for Cepheid variables of the Hubble Key Project on the Cosmic Distance Scale. (*B*) The Key Project's Hubble diagram made by evaluating and averaging several secondary distance indicators, including supernovae. Tying this result into the distance measured from Cepheid variables gives a Hubble diagram much farther out into space, beyond 400 Mpc (about 1300 million light-years). At the bottom, we see deviations of the individual values for Hubble's constant from the average.

38. They went on, though, to use their Cepheid measurements to calibrate methods of measuring distances even farther out. In the following section, we will say more about the final results of this Hubble Key Project.

Interestingly, the advantage of the Hubble Space Telescope was not only that it makes sufficiently fine images of stars so that the Cepheids could be picked out at the great distance of M81. The ability of a telescope in space to make observations at regular intervals, free of worries of weather, day/night, the moon, and seeing differences was also very important. Further, they could observe in the near infrared, cutting through dust to determine how much extinction by dust misled prior astronomers in deducing distances by making galaxies look fainter than they actually are. Other problems still remain in Cepheid distances, especially the question of how much the percentages of heavier elements in the stars affect their period-luminosity relationships. These percentages are known as "metallicities," so astronomers talk a lot about the "problem with metallicities." Though less than 1 per cent, the abundances of these elements (anything heavier than helium, to astronomers, is a "metal") affect the evolution of stars and thus their brightnesses. The Hubble Key Project took account of the effect of metallicity on Cepheid distances as best they could.

Beyond the nearest galaxies, in which we can observe Cepheids, we use **secondary distance indicators.** (They are "secondary" because they are not direct but are calibrated by the primary distance indicators.) Such a series of steps to measuring distances farther and farther out is known as the "cosmic distance ladder." The newer methods we discuss below are better than the traditional secondary methods, which are, therefore, little used anymore. Such older methods assumed, for example, that the magnitudes of supergiants or sizes of H II regions are everywhere about the same. These statements, though useful first approximations, are not accurate.

In 1997, the trigonometric parallaxes measured with the Hipparcos satellite were released. (As we discussed in Section 25.2, trigonometric parallaxes give distances by triangulation, the most direct method possible.) Even for Hipparcos, Cepheids were not in the distance range for which the most accurate measurements could be made. Scientists working with the data reported that Cepheids were 10 per cent farther away than had been thought. That means that they were really a bit brighter than they were thought to be. This fundamental result rippled through cosmology.

The *F*ull-sky *A*strometric *M*apping *E*xplorer (FAME) spacecraft was cancelled in 2002. We had hoped that its data would reveal the distances to the Cepheids accurately enough to improve the whole distance ladder. For example, the Hubble Key Project distances are linked to a rounding of the distance to the Large Magellanic Cloud (fondly known by astronomers as the LMC) of 50 kiloparsecs. Many people have measured the distance to the LMC over the years, and there is scatter in the results. The Hubble Key Project was able to observe 105 Cepheids in the LMC. Still, different projects have given slightly different answers for this distance. The period-luminosity relation's zero-point is set by it, so it is crucial for determining the distances to Cepheids. When the distance is measured more accurately, the whole distance ladder will be on a firmer footing. Though we have one accurate point from Supernova 1987A, that value cannot take into account the thickness of the LMC.

## 35.4 FINDING HUBBLE'S CONSTANT

We can determine Hubble's constant by measuring the velocity to a distant galaxy and dividing that measured velocity by the distance to a galaxy. But, as we shall now see, finding that distance is controversial. As we described, the "standard" value for Hubble's constant had been 55 km/sec/megaparsec, determined by Allan Sandage and Gustav Tammann, Hubble's scientific heirs. (Again, astronomers don't usually bother to give the units for Hubble's constant; they are always km/sec/Mpc—that is, kilometers per second per megaparsec.) But now the Hubble $H_0$ Key Project is completed, and its value, 72 km/sec/megaparsec, is widely accepted. We will discuss its result at the end of this section.

### 35.4a Secondary Distance Indicators and the Hubble Key Project

Several secondary distance indicators were worked out in the 1990s. For example, it has been shown that the distribution in brightness of planetary nebulae in, mostly, elliptical galaxies is a useful distance indicator between 5 million and 80 million light-years. The method is "secondary" because it uses a calibration based on the Cepheid distance scale. Distances from this method led to a Hubble constant near 80. George Jacoby of the Kitt Peak National Observatory, Robin Ciardullo of Penn State, and colleagues have championed this method.

Another method looks for the fluctuations of brightness in the image of a galaxy. The farther the galaxy, the more stars within each pixel and the smaller the statistical fluctuations from pixel to pixel. John Tonry of the University of Hawaii is credited with this method. It can be used out to about 70 Mpc (230 million light-years). The Hubble Key Project's analysis of this method gives a value of 70 with a statistical uncertainty of 5 (that is, $\pm 5$) and a systematic uncertainty of 6.

For the few galaxies in which a Cepheid distance is known and in which a Type Ia supernova was observed, the supernova method is also considered "secondary" (Fig. 35–7*B*). (Otherwise, the supernova method gives "tertiary" distances.) We assume that all Type Ia supernovae reach the same maximum brightness. Because we think they result when white dwarfs in binary systems barely go over their limit of 1.4 solar masses, it makes sense for them all to be the same brightness, a conclusion arrived at through study of the Hubble diagram for galaxies with such supernovae. But there are so many intermediate assumptions that the Sandage group is still finding a value close to 55 while the Hubble Key Project is finding a value close to 70.

The Tully-Fisher relation, a powerful device first exploited by Brent Tully of the University of Hawaii and J. Richard Fisher of the National Radio Astronomy Observatory, is often used. The relation is based on a link between how rapidly galaxies rotate and how bright they are (absolute magnitude). The speed of rotation is measured from their 21-cm hydrogen emission lines. The best relation is obtained when magnitudes are measured in the infrared. It can be used out to about 150 Mpc (500 million light-years). It gives values of about 70 for Hubble's constant, with a random error of about 3 and an allowance for systematic errors of about 7.

Another secondary distance indicator calculates the dispersion in velocity measured for stars in elliptical galaxies. This dispersion is correlated with the intrinsic luminosity of the elliptical galaxies similarly to the way that the rotation velocities are linked with luminosity for spirals. Sandra Faber and her graduate student Robert Jackson of the Lick Observatory originated this idea. The Hubble Key Project's analysis with this "Faber-Jackson method" gave a Hubble constant of 82, with statistical uncertainties of 6 and systematic uncertainties of 9, giving a total uncertainty of about 20 per cent.

One new way of measuring distance is a direct method independent of the chain of indicators. Championed by Robert Kirshner and Brian Schmidt of the Harvard-Smithsonian Center for Astrophysics, Ron Eastman of the University of California at Santa Cruz, and colleagues, it takes advantage of Type II supernovae, the explosions of massive stars. One such was detected in 1979 in the galaxy M100 (Fig. 35–8), in which we can also get distances from Cepheids. By 1985, radio waves from the supernova—probably from its interaction with gas surrounding it—had become detectable using the techniques of very-long-baseline interferometry. The scientists use

**FIGURE 35–8** A Hubble Space Telescope image of the galaxy M100 in the Virgo Cluster. Earlier, we saw Cepheids in it that were used to find its distances. A supernova occurred in it in 1979, allowing the galaxy to be used as a check on the accuracy of the supernova method for finding distance. The Key Project looked at dozens of galaxies like this one.

**FIGURE 35–9** A Hubble Space Telescope image, with the galaxy NGC 4881 appearing in the Planetary Camera segment which contains pixels with twice the resolution of the flanking squares. The galaxy is in the Coma Cluster, more than five times more distant than the Virgo Cluster and far enough away that it is much less affected by deviations from the smooth expansion. Enough globular clusters are visible around NGC 4881 to study their distribution in brightness, assessing a method of distance determination. The full Hubble frame appears as the photograph opening this chapter.

both optical photometry (measurements of brightness at given wavelengths) and spectra. The photometry gives the supernova's temperature and the number of photons reaching the Earth. This, in turn, depends on the radius of the supernova's shell with respect to the distance to the supernova. The spectra give the supernova's velocity of expansion and therefore, given its age, its size. Using our understanding of the laws of black-body radiation and deviations from it, the astronomers can compare the two types of measurements to find the object's distance from us. There are about two dozen such measurements. And what is this current result? These supernova values for $H_0$ are about 73 ± 6 from random errors and 7 from systematic errors for Hubble's constant.

High-resolution images from the Hubble Space Telescope have been taken to explore the possibility of using globular star clusters as distance indicators. The image (Fig. 35–9) shows a galaxy in the Coma Cluster of galaxies, far enough away that it should be in a region of relatively smooth expansion.

At still greater distances, we must resort to **tertiary distance indicators.** For the most part, Type Ia supernovae—the supernovae from white dwarfs in binary systems that are pushed over their limit in mass—are considered "tertiary." Though we know that they don't all reach exactly the same brightness, we can correct for the differences through careful study of their light curves. Because supernovae are a million times brighter than Cepheids, we can see them a thousand times farther away (because of the inverse-square relationship between distance and brightness), extending the distance scale. Automatic telescopes are now used to search the sky, finding many more supernovae with which to work. Studies of these distant supernovae indicate that the expansion of the universe is speeding up rather than slowing down. This surprising result is known as the "accelerating universe," and we discuss it in the next section.

Another tertiary method is to assume that the brightest member in a cluster of galaxies has the same absolute magnitude as the brightest members of other clusters. Sometimes, when one or two members are exceptionally bright, we consider instead the third-brightest member of a cluster, to lessen the possibility of one odd object being the brightest. This traditional method has been superseded by the more accurate Type Ia supernova measurements.

What is the best value for Hubble's constant? For a while, there have been two camps—those deriving 50 or 55 and those deriving 100. (Remember, the units are km/sec/megaparsec.) New studies have led these values to evolve over time (Fig. 35–10). Subtle effects can transform the same set of measurements from one conclusion to the other.

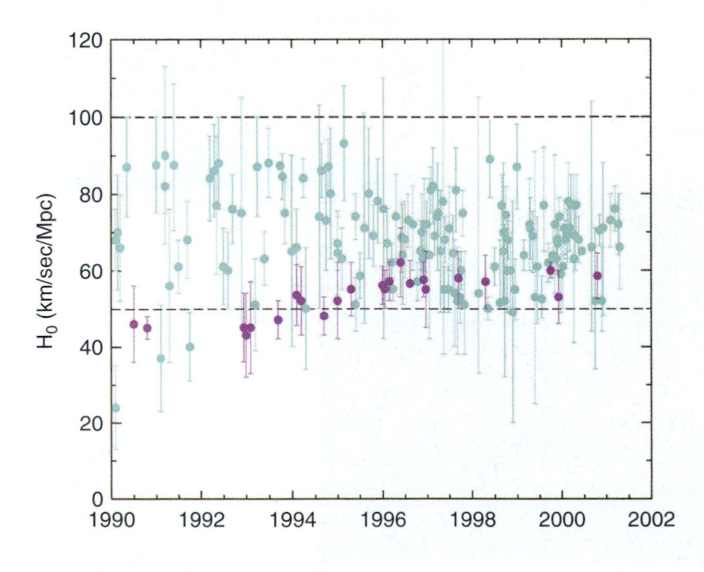


**FIGURE 35–10** Values of Hubble's constant reported by various groups of scientists as a function of time. The post-1990 values of the Sandage group, often lower than those of other groups, are shown in purple. Note at the right side of the graph that most of the recent values converged to 75.

The Hubble $H_0$ Key Project group of Wendy Freedman, Robert Kennicutt, Barry Madore, and colleagues announced their final result in 2001. Astronomers the world over have been waiting for their observations and their detailed analysis. The paper was signed by 15 astronomers, from the Carnegie Institution of Washington, Caltech, Rutgers, the National Optical Astronomy Observatories, the University of Arizona, Johns Hopkins, the Harvard-Smithsonian Center for Astrophysics, the Lick Observatory, the Dominion Astrophysical Observatory in British Columbia, and institutions in Australia and England. They were able to use the distances to galaxies they measured with Cepheid variables to provide distance scales for several other methods (Fig. 35–11). These methods included Type Ia supernovae, the Tully-Fisher relation in spiral galaxies, surface-brightness fluctuations, velocity dispersions in elliptical galaxies, and Type II supernovae. But in science it isn't enough to simply determine a value. One has to assess the limits of accuracy of that value. A lot of the work of the Hubble Key Project was to provide uncertainty limits on their measurements.

A good trick was to get beyond the distance at which local velocities are important. These to-and-fro ("peculiar") motions can be 300 km/sec randomly. In Section 35.6 we will see how large-scale motions of whole regions of the universe can distort the value we measure. So we want to look on as large a scale as possible, measuring beyond the regions that would distort our value. For velocities of recession of 3000 km/sec (divide by 75 to get a distance of 40 Mpc), the peculiar motions cause deviations of 10 per cent. If we go as far as 30,000 km/sec through studies of Type Ia supernovae, the effect of these peculiar motions is only 1 per cent.

The largest systematic errors the scientists allow for in the value for uncertainty they quote come from the large-scale flows, the uncertain distance to the Large Magellanic Cloud, the effect of metallicity on the Cepheid variable period-luminosity relation, and the calibration of the Wide Field and Planetary Camera 2 on the Hubble Space Telescope. They also consider lesser uncertainties from the reddening of Cepheid variables by interstellar dust, a bias in measuring Cepheids by not finding those of short periods, and the effect of stars not being seen as single on their images because many stars are crowded together.

The Key Project graph gives the Hubble constant by finding the slope of the straight line fit through the points. Averaging the results from five secondary

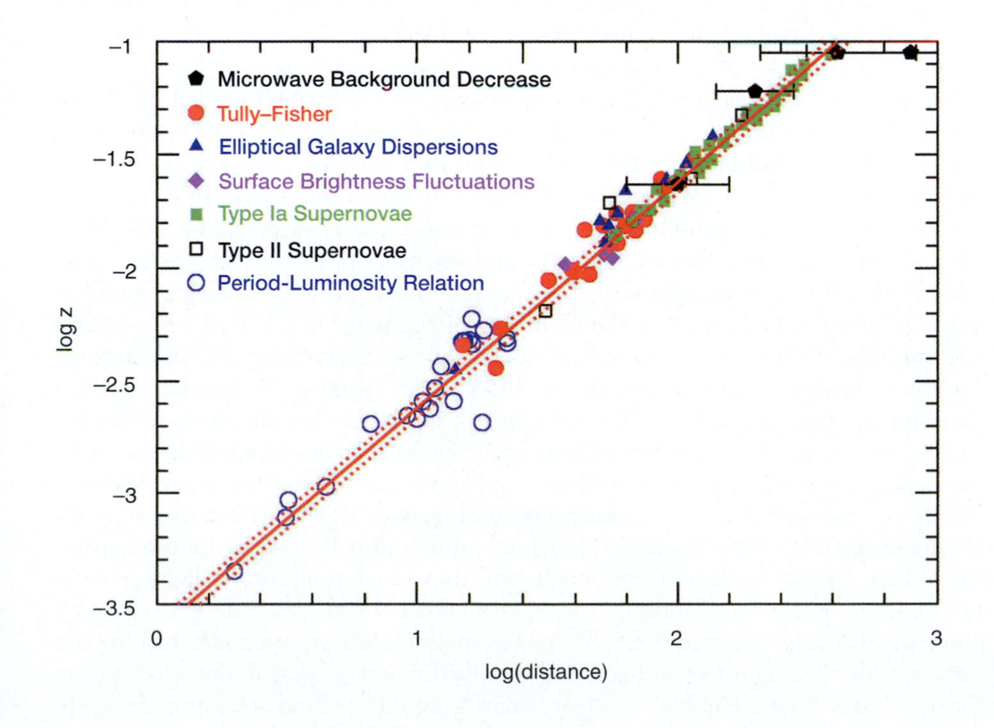


**FIGURE 35–11** The final Hubble diagram from the $H_0$ Key Project of the Hubble Space Telescope. The solid red line corresponds to $H_0 = 72$ km/sec/Mpc and the dash lines represent 10% uncertainties.

**TABLE 35-1** Hubble's Constant

The final publication of the $H_0$ Key Project of the Hubble Space Telescope (2001) gives (with first ± random and then ± systematic errors):

| | |
|---|---|
| Cepheid variables | 75 ± 10 |
| and from methods calibrated with Cepheids: | |
| 36 Type Ia supernovae | 71 ± 2 ± 6 |
| 21 Tully-Fisher clusters | 71 ± 3 ± 7 |
| 82 elliptical galaxy dispersions | 82 ± 6 ± 9 |
| 6 surface-brightness-fluctuations clusters | 70 ± 5 ± 6 |
| 4 Type II supernovae | 72 ± 9 ± 7 |
| **average** | **72 ± 3 (random)** |
| **final value for $H_0$** | **72 ± 8** |

methods that were calibrated with Cepheids, it gives a value of 72 ± 3 for random uncertainties and ±7 for systematic uncertainties, in units of km/sec/Mpc. The team finds a final value for Hubble's constant of 72, with an uncertainty of 8.

### 35.4b Future Hubble-Constant Measurements

Some uncertainties in Hubble's constant remain, especially the uncertainty in the distance to the Large Magellanic Cloud. If the LMC is at the distance shown by another study, Hubble's constant could be 79. A promising new technique for measuring distance without reference to the Large Magellanic Cloud involves masers. The galaxy NGC 4258 has several masers in its disk. The motions of the masers can be determined by very-long-baseline interferometry from side to side, and compared with the Doppler motions toward and away from us. If we assume those motions are the same, on the average, then we have the distance to the galaxy. As of 2001, the distance to this galaxy measured by Cepheids is almost 10 per cent farther than the distance measured from the masers. More such galaxies must be observed before we know what to make of the apparent discrepancy.

Scientists hope for still other methods independent of Cepheid variables and of the distance to the Large Magellanic Cloud. One of those methods uses the cosmic microwave background that we will discuss in Chapter 37. It provides a uniform background of the shortest radio waves. When those waves pass through a cluster of galaxies to us, they change slightly in frequency as they scatter off the hot gas in the cluster. (We see this gas with x-ray satellites like the Chandra X-ray Observatory.) We determine the distance to the cluster from its x-ray luminosity, which depends in part on the cluster's apparent size. Further, the diminishment of the microwave background doesn't depend on the cluster's distance but does depend on its thickness. If we assume that the apparent size and thickness of the cluster are the same (that is, that the cluster is spherical), we can measure the distance. This method has many potential pitfalls, such as the clumping of the gas that Chandra reveals, and the question of whether the clusters are really spherical. The most recent values for Hubble's constant determined by this method are about 60 ± 10. Many efforts are continuing, obtaining data for different clusters and higher-resolution observations.

Another hopeful method for measuring Hubble's constant at great distances independent of Cepheid and Large Magellanic Cloud problems involves gravitational lenses, which we discussed in the previous chapter, and quasars, which we discuss in the following chapter. If we see a quasar through two different paths because of the gravitational lens, the two paths differ in length and thus travel time for the light or radio waves. If the quasar varies abruptly, we see that variation with a time delay in one path compared with the other. Accurate use of the method depends on understanding the details of the lensing effect and knowing the distribution of mass in the lensing cluster of galaxies. So far, Hubble's constant values from this method are approaching 65.

IN REVIEW

### BOX 35.1 The Hubble Constant

The $H_0$ Key project of the Hubble Space Telescope revealed its final conclusions on Hubble's constant in 2001. Based on an average of the measurements by several methods, especially using the period-luminosity relation of Cepheid variables to calibrate other methods, they find:

$$H_0 = 72 \pm 8 \text{ km/sec/Mpc}.$$

### 35.4c Implications of Hubble's Constant

As we shall see in the following section and in Chapter 37, values for the Hubble constant of 72 indicate that the universe is about 13 billion years old. (This value assumes that gravitating matter makes up 30 per cent of the universe's total density and that, in line with current thinking, something else makes up another 70 per cent.) The value matches determinations of the ages of the oldest stars. Through some of the 1990s, calculations had globular clusters older than 12 billion years, which was a contradiction, but newer research has lowered globular cluster ages enough to eliminate the problem. In 2001, the age of an old star was revealed through study of radioactive decay in a uranium isotope in it, and also came out at about 12 billion years. There is no current conflict in ages between the age from Hubble's law and the age from the oldest objects.

An older model in which the expansion of the universe will eventually reverse, gives an age of the universe of only 8 billion years. The ages of stars we observe are in contradiction to that older model.

In sum, observations of Cepheids and of supernovae with the Hubble Space Telescope have determined the Hubble constant to about 10 per cent. There will still be much work to lessen the uncertainties, and some groups will still find other values, but the Hubble Key Project value will now set the standard for some time to come.

## 35.5 THE ACCELERATING UNIVERSE

It has been a struggle for over 70 years to observe galaxies as far away as possible to test Hubble's law. The brightest objects used are supernovae, so they give distances as far as possible. But finding supernovae used to be an unusual event, one depending even on amateur astronomers scanning the sky with binoculars or small telescopes. Within the last decade, those times have changed.

The ability to make telescope/detector combinations that are at the same time extremely sensitive and cover wide fields of view, and the computing that makes it possible to handle all those data, led to a revolution in finding supernovae. The method was worked out by the Supernova Cosmology Project at the Lawrence Berkeley Laboratory. They were able to use automatic methods to compare images of many galaxies taken at different times, noticing changes between images of members of a pair. The technique led to finding a dozen supernovae at each telescope run, and to finding them while their brightness had not yet reached its peak. (They had about two runs a year, after convincing skeptical telescope time-allocation committees to take a flier on their project.) The supernovae were identified so quickly that they could be followed with spectra and other observations while they were still on the rise. As a result, they could determine the peak of Type Ia supernovae and so better determine the distance scale of the universe. Formerly, most supernovae were located only after they had reached their peak brightnesses and were declining. That timing limited the accuracy of determining their peak brightness, which for Type Ia (white-dwarf-in-a-binary) supernovae limited the accuracy of finding the distances measured.

Two major projects are now using these methods to find huge numbers of supernovae. We discussed the Supernova Cosmology Project, headed by Saul Perlmutter of the Lawrence Berkeley Laboratory. The High-Z Supernova Search is headed by Brian Schmidt, formerly at Harvard and now in Australia at the Mount Stromlo and Siding Spring Observatory. Each has many collaborators. Alex Filippenko of the University of California, Berkeley, is an unusual individual by being on both teams. By "high-$z$" the scientists mean that they are studying supernovae with redshifts (the change in wavelength divided by the wavelength) from 0.2 to greater than 1. At the larger redshifts they are now surveying, the supernova cosmology project is finding a half dozen supernovae each observing run at a redshift of 1.2.

It is better to think of the redshift as resulting from an expansion of space than as resulting from the Doppler effect. (The latter is not quite the right explanation in the general theory of relativity.) As space expands, the wavelengths expand, and become redder. Still, it is sometimes helpful to think in terms of Doppler shifts. For those large redshifts, the transformation from redshift to velocity can no longer use the simple Doppler formula we gave. Redshifts we are used to on Earth are trivially small compared to this range. For example, a car going at 120 km/hr is going at 120 km/hr $\times$ (1 hr/3600 sec), which is about 1/30 km/sec. Since the speed of light is 300,000 km/sec, a speeding car is going at only (1/30)/(300,000) = 1/9,000,000 the speed of light. This value, about one ten-millionth the speed of light, is $z = 0.000\ 000\ 1$.

An original goal of both teams was to see what happens with the straight-line relation we know as Hubble's law when we get to the greatest possible distances. Hubble's law, as we learned, is a straight-line relation between velocity and distance. But it seemed that if the universe had enough gravity, the galaxies would eventually be pulled back and would slow down. Thus, it was confidently expected that when we looked out far enough, we would see the galaxies less far away than they should be by Hubble's law. We expected to use this method to find out how much matter there is in the universe by measuring its gravity.

So both teams went to telescopes, and found dozens of supernovas by comparing images from a month apart (Fig. 35–12). Then they calculated from the brightnesses of the peak intensity of the Type Ia supernova explosions how far the supernovae were away. At different telescopes, often one of the giant Keck telescopes, team members took spectra for these galaxies. From its spectrum, they could both determine whether the supernova was Type I or Type II and how fast it was moving. The results were a huge surprise.

From how fast a galaxy is moving (measured from the redshift), you can calculate how far it is away (from Hubble's law). Since you know the luminosity of a Type Ia supernova at its peak (they all have the same peak luminosity), you can easily figure out how bright you expect that supernova to look. But they were wrong! The supernovae looked fainter than expected.

The result, made public in 1998, was dazzling. The farthest parts of the universe seem to be speeding up, not slowing down. The expansion of the universe seems not to be uniform. Rather, the expansion of the universe is accelerating. That is, the rate of the expansion, previously thought to be decreasing over time (due to gravity), is actually increasing with time.

What is acceleration compared with uniform expansion? For the universe, the situation is complicated by the fact that when you look out in space, you are seeing how things were at an earlier time. For the rest of this paragraph, let us ignore this effect, and consider only a given time throughout the universe. We gave an example

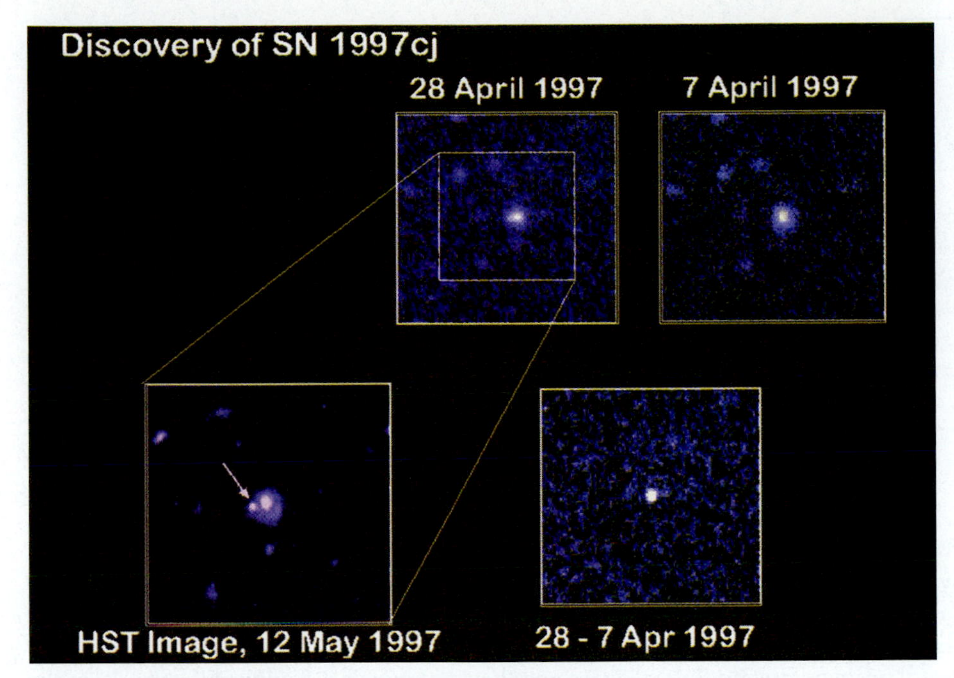


A

B

**FIGURE 35–12** The discovery of supernova 1997cj at a redshift of 0.5. The top two images show small regions cut out of large CCD images. Subtraction of the two images revealed, as shown at the bottom (28 April–7 April), a supernova. Spectra taken shortly thereafter with one of the Keck telescopes showed that it was a Type Ia supernova. The Hubble Space Telescope's high resolution showed the supernova clearly separated from the bright central region of its galaxy.

earlier, in the raisin-cake drawing, of distances of 2 units doubling to 4, distances of 4 units doubling to 8, and so on. Distances of 100 units would double to 200, according to that uniform rate of expansion. What the supernova people found, though, was that the smaller distances doubled on cue, but the larger distance of 100 changed to 205, not just 200. The distant objects were farther away and therefore fainter than they would be with a mere uniform expansion.

The result is so surprising that the teams have invested enormous efforts in testing possible reasons why they might be wrong. They have taken into account possible differences in the peak brightnesses of Type Ia supernovae, and made corrections by using information about their light curves. They have investigated as well as they could the possibility that intergalactic dust is dimming the distant supernovae instead of an acceleration of the universe. They checked the spectra of supernovae at different distances, near and far, to see if the supernovae were different at different times in the universe's evolution. As time has gone on for the last few years, they have not succeeded in finding fault with their data. They have become more secure with this surprising result that the expansion of the universe is accelerating. In 2001 they reported confirmation with the discovery of a Type Ia supernova very far back in time (10 billion light-years) but that was observed to be so bright that it could not have been dimmed by dust or evolution even though it was so far away. The scientists involved conclude that dust or evolution didn't affect their early observations of closer supernovae either endorsing the idea that the expansion is accelerating.

Normal gravity does not account for the acceleration of the expansion of the universe. In Chapter 37 on cosmology, we will discuss the current thinking of what in the universe could cause this acceleration. A leading idea is the "cosmological

IN REVIEW

**BOX 35.2 The Accelerating Universe**

Comparing images of galaxies taken about a month apart reveals supernovae soon after they exploded. Spectra then taken reveal which are Type Ia, which all have the same peak luminosity. The peak brightness we observe, compared with the peak luminosity, gives us the distance. Scientists have been surprised in the last few years to find that the most distant supernovae are slightly dimmer than predicted by Hubble's law. Therefore the distances to distant supernovae are slightly greater than predicted by Hubble's law. The universe's expansion is thus accelerating over its expected uniform rate. The phenomenon is known informally as "the accelerating universe."

constant," originally an idea of Albert Einstein's. The age we calculate for the universe depends on whether it is contracting, expanding uniformly, or accelerating. The amount of the cosmological constant now favored by observationalists and theorists is 0.7 of the density that would be needed to reverse the universe's expansion if it were ordinary gravity. Table 35–2 shows how the age of the universe depends on the value for Hubble's constant (now measured to be close to 75) and on the value for the cosmological constant.

To find still more supernovae, the proposed SNAP satellite (*S*uper*n*ova/*A*cceleration *P*robe) would carry a 2-m telescope aloft that would repeatedly image regions of the sky. It would carry visible-light and infrared imagers as well as a spectrograph. It should discover and accurately measure 2000 type Ia supernovae a year.

## 35.6 MOTIONS, GRAVITATIONAL PULLS, EVOLUTION, AND HUBBLE'S CONSTANT

Measuring distances is not our only problem: where we can measure distance well, the velocity may not be that of the cosmic expansion. And where we are sure the velocity we measure is the Hubble velocity (that is, very far out), the distances are hard

**TABLE 35–2 Ages of the Universe**

| For universes with no cosmological constant | |
|---|---|
| $H_0$ | Age (Gyr) |
| 55 | 11.9 |
| 65 | 10.0 |
| 75 | 8.7 |
| 85 | 7.7 |
| **For universes with cosmological constant of 0.7** | |
| $H_0$ | Age (Gyr) |
| 55 | 17.1 |
| 65 | 14.5 |
| 75 | 12.6 |
| 85 | 11.1 |

to measure. Our straightforward measurements of the distance to galaxies in the Virgo Cluster may not give us the true Hubble constant because our own galaxy is moving with respect to the Virgo Cluster as a whole, and, to a lesser degree, the Virgo Cluster is moving with respect to the average expansion of the universe. The expansion of the universe is thus not uniform from place to place except on the largest scale. The presence of a lot of mass in one place tends to slow down the expansion in its vicinity.

In particular, a large collection of galaxies known as the "Great Attractor" seems to be pulling the Local Group, the Virgo Cluster, and even the Hydra-Centaurus supercluster (Fig. 35–13) toward it. The idea was advanced by a group of seven astronomers known as the Seven Samurai—Alan Dressler of the Observatories of the Carnegie Institution of Washington, Sandra Faber of the Lick Observatory, David Burstein of Arizona State U., Gary Wegner of Dartmouth, and three other colleagues. They were investigating elliptical galaxies, making a uniform set of observations of hundreds of them when they discovered that the distances derived by Hubble's law varied in different parts of the regions they could investigate.

The Great Attractor includes tens of thousands of galaxies. The Great Attractor's gravity makes the surrounding region of the universe expand less rapidly. Not everyone accepts the reality of the Great Attractor, though much evidence does indicate that it is real (Fig. 35–14). Other giant superclusters, clumps of matter like the Great Attractor, appear elsewhere in the sky. The Great Attractor is probably merely the nearest such clump.

The methods we have to use grow less precise as we get farther from the sun. In particular, at the very farthest distances we are seeing galaxies that emitted their light very long ago. A galaxy may well have then had a very different brightness from the galaxies nearer to us with which we are comparing it. For example, there has been a belief that the brightest galaxies in clusters may be devouring other galaxies

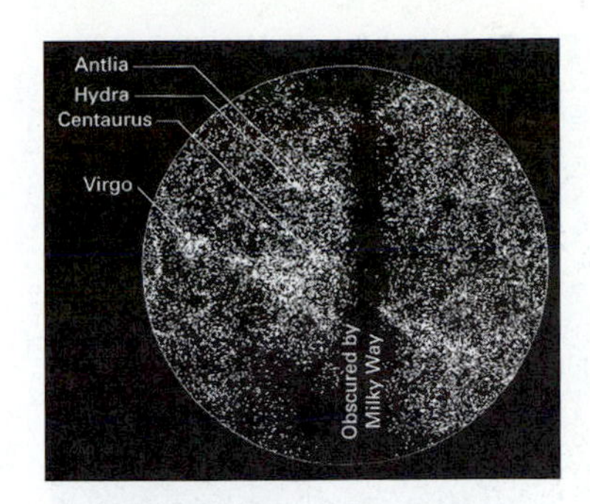


**FIGURE 35–13** The distribution of galaxies over half the sky. The graph is centered on the direction of the average motion of nearby elliptical galaxies relative to the cosmic background radiation (Section 37.4). Since redshift measurements show that the Virgo Cluster and the Hydra-Centaurus supercluster share in the motion, there must be a Great Attractor beyond, near the center of this image, exerting tremendous gravitational force. The dark strip across the center shows where the Milky Way prevents distant galaxies from being seen; it unfortunately cuts across the Centaurus band of clusters of galaxies that includes the Great Attractor.

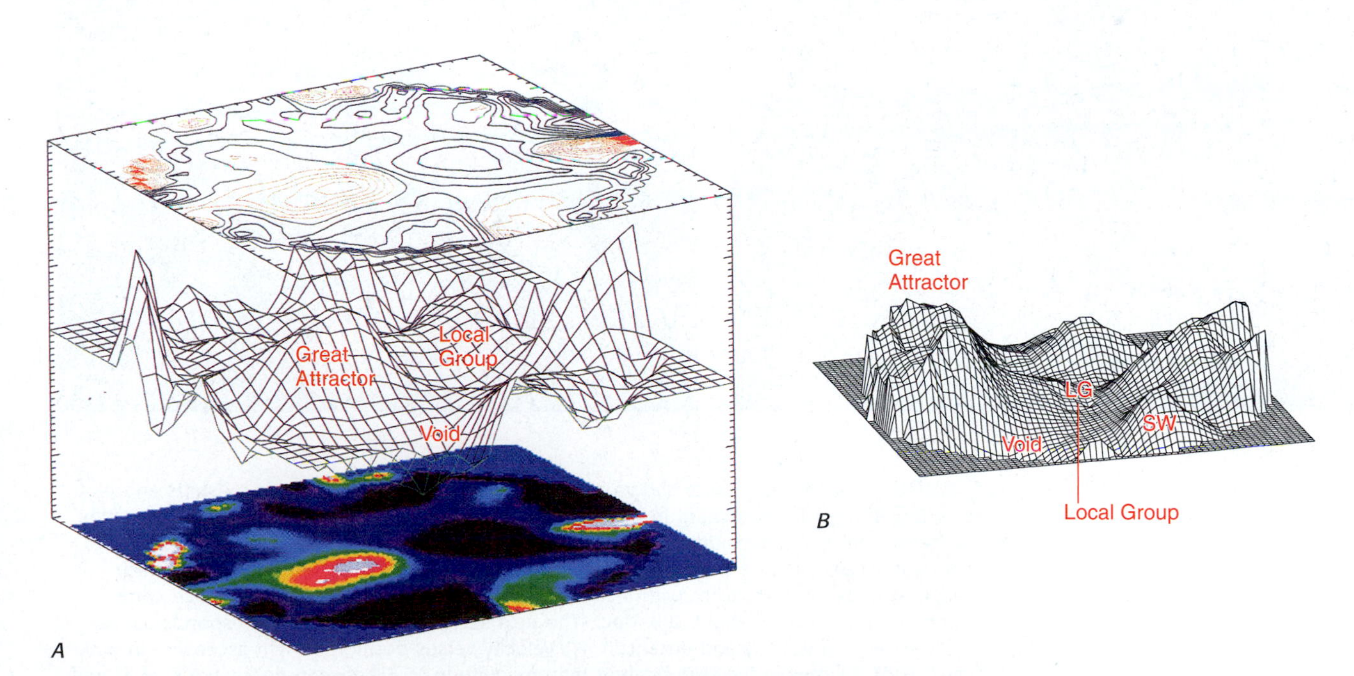


**FIGURE 35–14** Two sets of measurements in the direction of the Great Attractor, each of which reveals the distribution of mass on all sides of it. The results differ slightly but agree on the presence of the Great Attractor. We are in the center of this plot of mass density, which is shown on the z-axis (upward in the third dimension).

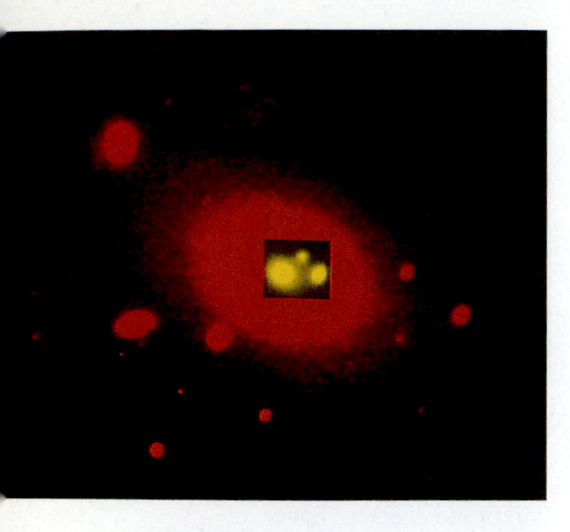

**FIGURE 35–15** Galactic cannibalism. Computer processing displays the central region of the large galaxy in a yellow box; we see the remains of two victims. The galaxy, NGC 6166, is the central galaxy in the rich cluster A2199.

and so growing brighter; this process is called **galactic cannibalism** (Fig. 35–15). The galaxies grow brighter right after merging, but they then soon fade again as stellar evolution takes over. If so, when we look to clusters at different distances, we do not know that the brightest galaxies we see are the same brightness.

## 35.7 SUPERCLUSTERS AND THE GREAT WALL

Stephen A. Gregory of the University of New Mexico and Laird Thompson of the University of Illinois concluded a while back that every nearby very rich cluster of galaxies is located in a cluster of clusters, a **supercluster.** A recent catalogue of superclusters includes 16 within 2 billion light-years of us, each of which contains two or more rich clusters. The Local Group, the several similar groupings nearby, and the Virgo Cluster form the **Local Supercluster.** This cluster of clusters contains 100 member clusters roughly in a pancake shape, on the order of 100 million light-years across and 10 million light-years thick. Superclusters are apparently separated by giant voids.

### 35.7a Slices of the Universe

To get a better overall picture of the universe, we need three-dimensional maps (Fig. 35–16). Two dimensions on the sky we get merely by noting the direction in which we are observing; the third dimension comes from Hubble's law. Margaret Geller, John Huchra, and their collaborators at the Harvard-Smithsonian Center for Astrophysics were among the first to provide such maps of the distribution. They observed the spectra of tens of thousands of galaxies to get the redshifts that gave them

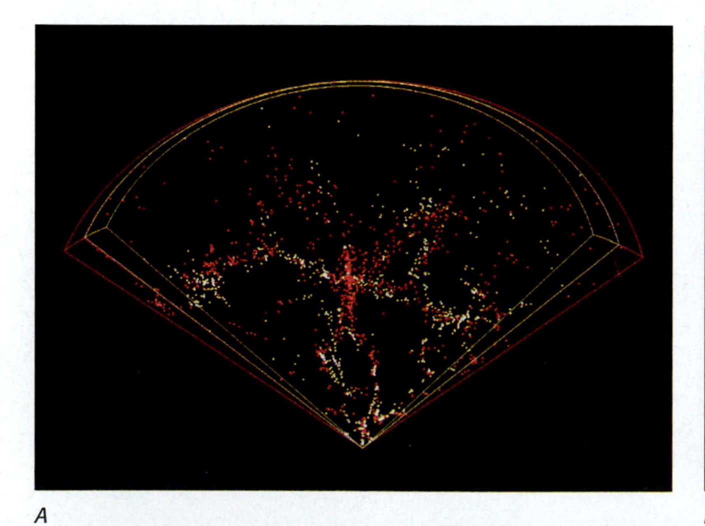

*A*

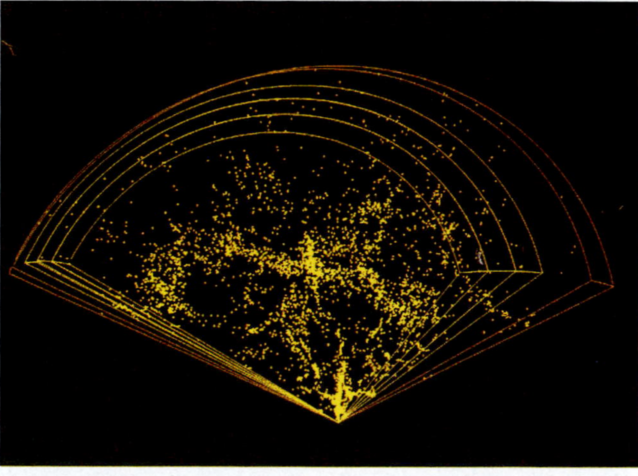

*B*

**FIGURE 35–16** Each slice shows distance from Earth (measured from redshifts and plotted along straight lines outward from the Earth, which is at the bottom point) versus position on the sky (in right ascension) for all galaxies in a strip of declination 6° in size. (Remember that right ascension corresponds to longitude and declination to latitude in the sky.) Each wedge extends out to galaxies with a velocity of recession of 15,000 km/sec, which corresponds to a distance of about 300 Mpc. (The next section describes the correspondence between measured velocity and distance.) (*A*) Velocity versus position in right ascension in pink for all 1065 galaxies in the strip brighter than magnitude 15.5 between declinations 26.5° and 32.5° and in white for the 702 galaxies in the next 6° strip of declination, between 32.5° and 38.5°. (*B*) Data for four slices. Note how the structures apparent in one slice continue to the next, indicating that the galaxies are on the edges of giant bubbles, like the suds in a kitchen sink, or perhaps sponge-like structures in space. The Coma Cluster of galaxies is in the center, apparently at the intersection of several bubbles.

distances. In the figure, we see the distance dimension plotted against one sky axis for several wedges—"slices"—of the universe.

The galaxies appear to be on vast sheets that form the surface of bubble-like structures. The structures are elongated with typical sizes of 50 megaparsecs by 30 Mpc by 5 Mpc (roughly 150 million light-years by 100 Mly by 15 Mly). The largest—at least 250 million light-years long—has been called the Great Wall. Calculations based on the recently popular theoretical idea that much of the universe is in the form of invisible particles known as "cold dark matter" (which we will discuss further) have been successful in explaining the formation of ordinary clusters of galaxies. But even this theory cannot account for structures of the size and shape of the Great Wall. As for the voids, they are not completely empty, but the density of galaxies in them is only 20 per cent of average.

Collaborations have extended the "slices of the universe" project to the southern hemisphere. And other astronomers have also made observations of thousands of galactic spectra and are compiling their own maps of the universe. Some such maps, instead of covering broad areas, use "pencil beams"—that is, narrow cones—and look farther out into space. They also find signs of voids surrounded by sheets of galaxies.

## 35.7b 2dF, Sloan, and 3-Dimensional Structure

The 4-m Anglo-Australian Telescope in Siding Spring, Australia, has been largely devoted in recent years to simultaneous measurements of galaxy spectra (and thus redshifts) over a relatively wide field per individual exposure. This 2dF project (from "2 degree field") has revealed that structures like the Great Wall exist in many locations (Section 5.5). It has shown that the features, such as voids and walls, found in the earlier "slices of the universe" project are typical, and that there are not still larger features forming a hierarchy of structures. The 2dF project has mapped the universe to great distances (Fig. 35–17). It is making 3-D wedges of different parts of the sky (Fig. 35–18).

Remember, though, the universe isn't expanding on scales smaller than tens of millions of light-years. Gravity dominates at smaller scales.

Note that the third dimension for 2dF and Sloan comes from measuring redshifts, which give distances through application of Hubble's law.

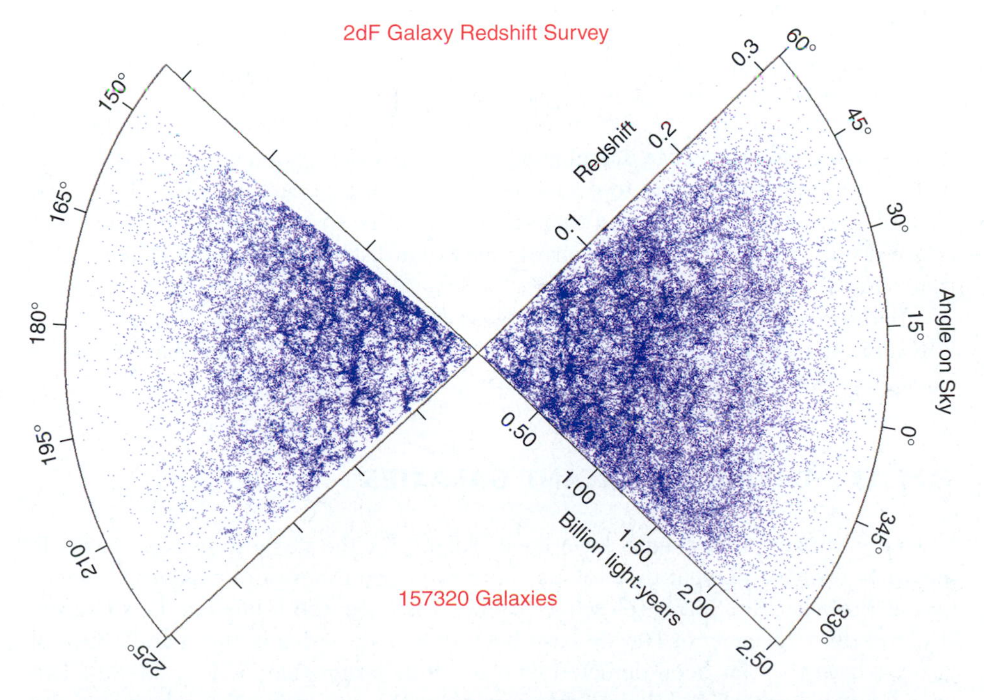


**FIGURE 35–17** A map of space, made by the 2dF project, showing a slice whose vertex is at the Earth, as in the previous illustration. The distances come from the redshifts they measure. We see tens of thousands of galaxies, and the structure they reveal in the universe.

**FIGURE 35–18** Regions of the celestial sphere that can be mapped by 2dF from its location in the Earth's southern hemisphere. It can nonetheless map some regions of our northern sky.

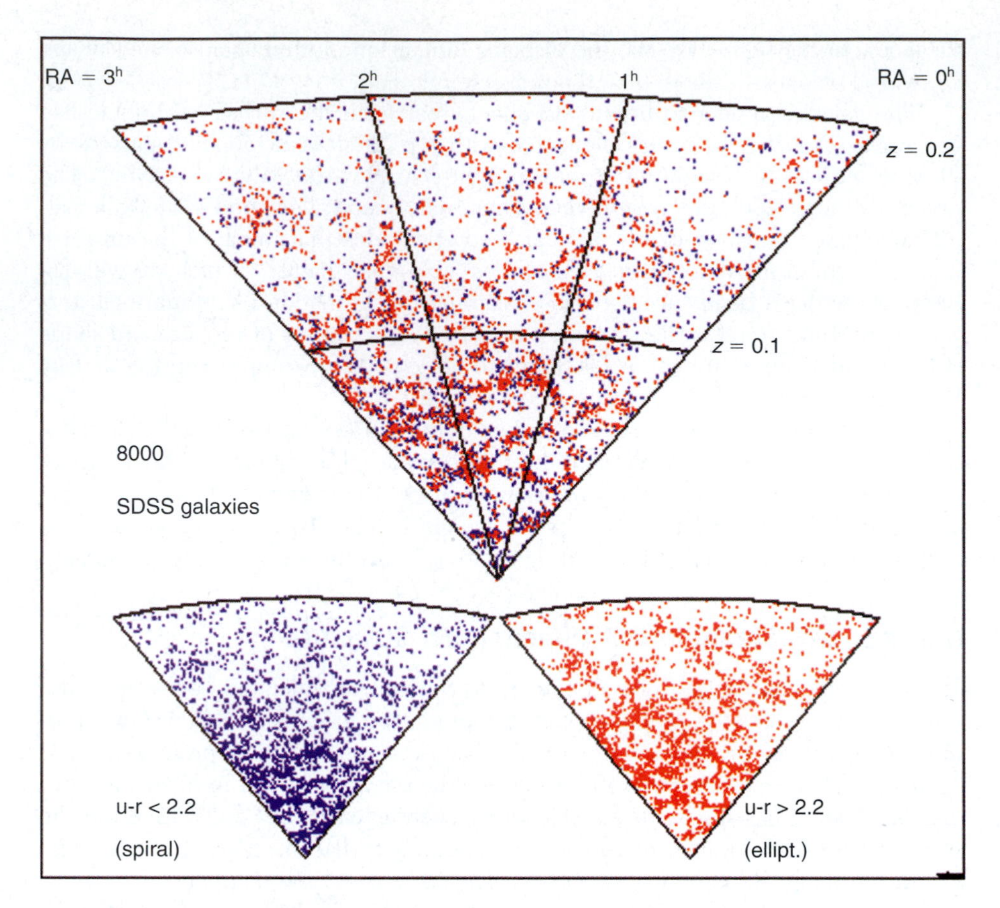


**FIGURE 35–19** A map of space, made by the Sloan Digital Sky Survey, showing a slice whose vertex is at the Earth, as in the previous two illustrations. We again see tens of thousands of galaxies, with their distances measured from their redshifts, and the structure revealed in space by the resulting 3-dimensional map.

**FIGURE 35–20** Just a few years ago, it would have been amazing to find a single galaxy with a redshift as large as 5. It would mean that the Lyman-alpha spectral line that originates at 1216 Å is redshifted by 5 × 1216 Å = 6080 Å, thus winding up at 1216 Å + 6080 Å = 7292 Å. This new value is beyond the red, in the near-infrared, though the spectral line started out in the ultraviolet! But through the searches of so many galaxies made by Sloan and 2dF, such distant galaxies are turning up in reasonable numbers.

The Sloan Digital Sky Survey (Section 5.5) is using a new telescope to image the 100 million brightest objects we can observe from Earth's northern hemisphere and to determine the distances to the million brightest galaxies by obtaining spectra. Their images are available on-line to everyone, student and researcher alike. They, too, have plotted wedges with tens of thousands of galaxies so far, showing distances out to a redshift of about 0.5, which corresponds to a distance of about 3 billion light-years (Fig. 35–19).

In the next chapter, we will see how observations of the redshifts of quasars from 2dF and Sloan have extended the 3-D mapping four times farther than it can be carried out with ordinary galaxies.

## 35.8 THE MOST DISTANT GALAXIES

Several groups of astronomers have been looking for the farthest galaxies, to study how they evolve and what they tell us about the early times of our universe. Scientists at Berkeley, the Space Telescope Science Institute, the Johns Hopkins University, and the University of Hawaii have been especially active in the search. Several galaxies have thus far been detected with redshifts larger than 5 (Fig. 35–20). The one shown is perhaps 13 billion light-years away. We are seeing the galaxy as it was 13 billion years ago. We say that the "look-back time" for this galaxy is 13 billion years. New observational techniques, like the use of CCDs, have reduced drastically the time necessary to take the spectra needed for this research. The current record holder was examined spectroscopically because of the steep slope of its radio spectrum, a property that is apparently linked to high luminosity.

Note that redshifts greater than 1 merely mean that the spectral lines are shifted by greater than their original wavelengths. At such great redshifts, the redshifts no longer correspond to $v/c$. A formula that is part of Einstein's general theory of relativity must be used.

The Hubble Space Telescope is regularly used to image very faint galaxies. It has discovered a population of faint blue galaxies that are very common between 3 and 8 billion light-years away (Fig. 35–21). Thus these galaxies must have been abundant when the universe was much younger, even though we no longer find galaxies like them today.

Hubble observations have provided enough of a sample of very distant galaxies that we can draw some general conclusions. It has shown that at least some elliptical galaxies formed very early, within a billion years after the big bang. Spiral galaxies did not start out as early in the universe.

**FIGURE 35–21** A long exposure (one full day) with the Hubble Space Telescope has revealed all these faint blue galaxies down to 30th magnitude in a tiny section of sky in the constellation Hercules. The region shown is one-tenth the diameter of the full moon.

## 35.9 GAMMA-RAY BURSTS

Some of the most distant objects in the universe turn out to be of a type that is not yet understood: the sources of gamma-ray bursts. We discussed early gamma-ray astronomy and the Compton Gamma Ray Observatory in Section 32.4.

The Compton Gamma Ray Observatory was able to map thousands of gamma-ray bursts during its 9 years of lifetime before its demise in 2000—about one a day (Fig. 35–22). They do not concentrate along the Milky Way, and so they are either very near or very far from us. The gamma rays come from all directions equally (an isotropic distribution), but there are fewer weak gamma-ray bursts than were expected, as though we are looking out past the edge of the distribution. The consensus had been that they are events in distant clusters of galaxies, but until recently there were some astronomers concluding that these gamma-ray bursts come from objects like neutron stars in our own galaxy.

Breakthrough observations were made in 1997, with the unprecedentedly accurate pinpointing of gamma-ray bursts made from BeppoSAX, an Italian-Dutch satellite launched the previous year. (Beppo was the nickname of an important Italian particle physicist and SAX is from the initials for the Italian equivalent of satellite for x-ray astronomy.) BeppoSAX also saw x-ray bursts from the same location as the gamma-ray bursts, and the positions measured allowed optical observations of the "afterglow" to be made from ground-based telescopes. A spectrum taken with one of the Keck telescopes showed absorption lines with a redshift from 7 billion light-years away, proving that the object giving off the gamma-ray burst was even farther than that, putting it among the most distant objects in the universe. Their observed brightness at their huge distances means that gamma-ray bursts are the most powerful explosions we know of in the universe. In the few seconds they last, they generate more energy than our sun does in its entire 10-billion-year lifetime. (The energy we compute assumes that they give off radiation equally in all directions. If their energy is largely beamed in one or more directions, then the total energy they give off would be lower.)

Only about a half-dozen optical afterglows have been detected each year. The events are simply named after their dates: GRB000301 was the Gamma-Ray Burst detected on [20]00 March 1. The redshifts for the first two dozen afterglows mostly ranged from 0.4 to 2, with one redshift as small as 0.01 and one as large as 4.5, as determined from the host galaxy (Fig. 35–23). The latter source is 11 billion light-years away.

What could cause such tremendous energy to be released? A collision of two neutron stars in a distant galaxy became the favored model for a while. But further observations have been made of some two dozen afterglows, which are seen in

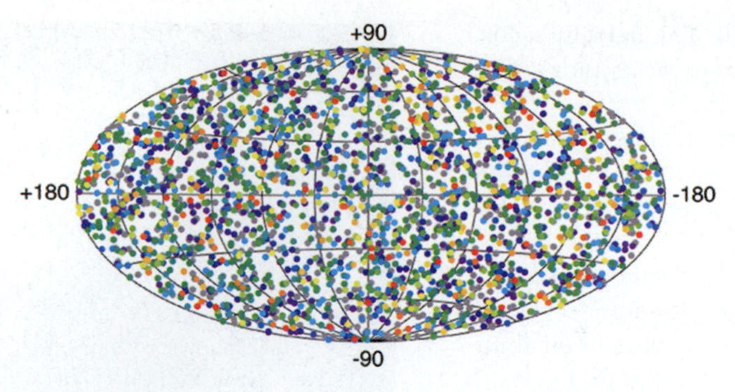


FIGURE 35–22 The distribution of the 2704 gamma-ray bursts recorded with the Burst and Transient Source Experiment on the Compton Gamma-Ray Observatory before the spacecraft's demise. The energy received from the burst, totalled over the duration of the burst, is color coded. Long duration, bright bursts appear in red while short duration, weak bursts appear in purple. Bursts with insufficient information appear in gray.

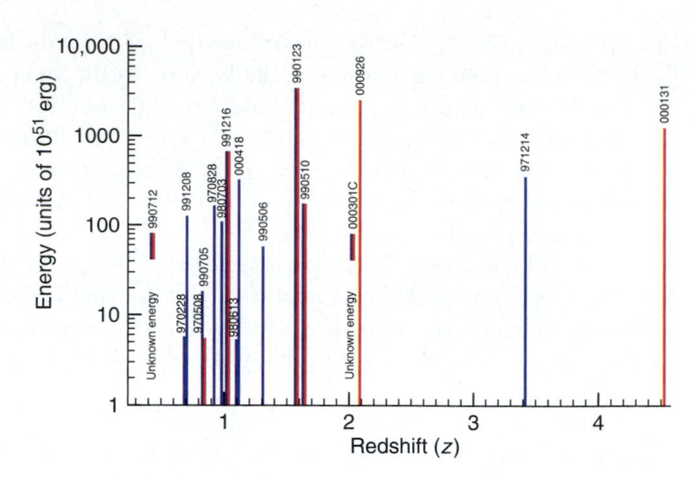


FIGURE 35–23 The energies emitted in gamma rays by gamma-ray bursts at different redshifts. The units on the vertical axis should be multiplied by $10^{51}$ to give the energies in ergs or by $10^{44}$ to give the energies in joules. (The number $10^{51}$ is significant since it is the energy given off by an ordinary supernova.) The distances were found from spectral lines in the galaxies that were the apparent hosts. Redshifts found from emission lines are shown in blue and redshifts found from absorption lines are shown in red.

optical, radio, or x-ray parts of the spectrum. They indicate that the bursts originate in regions of star formation (Fig. 35–24). There, hot, bright stars form, and they explode relatively quickly since their lifetimes are short. The current model for gamma-ray bursts is therefore extreme supernova explosions of massive stars, those from 20 to 300 times as massive as the sun. But only some of those supernovas would be bright enough, perhaps because they spin rapidly and emit jets of matter along their axes as a black hole forms inside. They are known as "hypernovas." Then, after the original supernova or hypernova, a fireball forms and then shock waves result and heat up both the ejecta and the surrounding material. Still, different gamma-ray bursts have different characteristics, and perhaps several different scenarios apply. Most obviously, many of the bursts don't have detectable afterglows, though perhaps the afterglows are merely too faint for our current telescopes. Observations of gravitational waves with LIGO would probably help distinguish between models invoking mergers of compact objects and models of hypernovae.

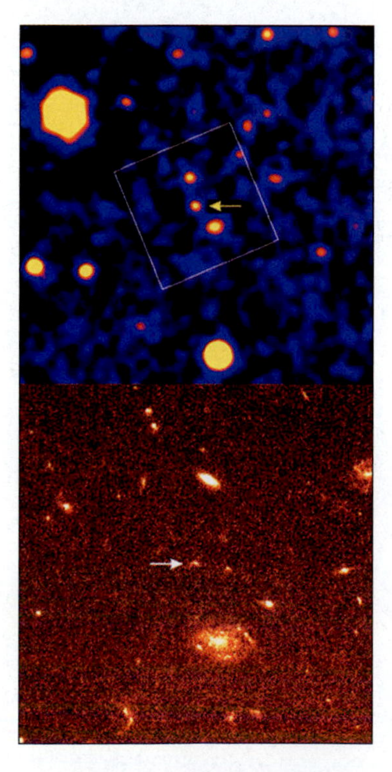

FIGURE 35–24 A gamma-ray afterglow, seen in the visible with the Keck telescope (*top*) two days after the burst and the Hubble Space Telescope (*bottom*) four months later.

Some of the afterglows are seen in x-rays with Chandra (Fig. 35–25). In one case, Chandra could be reoriented to make observations within two days of the discovery of the burst from the Compton Gamma Ray Observatory. Chandra's spectrographs were used to observe emission lines from iron. The redshift of the lines, which matches the redshift from an absorption line detected in the optical afterglow, gives a distance of 8 billion light-years for the source. They determined that material at least $1/10$ the sun's mass had been ejected and was moving quickly away from the object at 10 per cent the speed of light. These observations match the hypernova model better than they do the model that the bursts come from collisions of two compact objects. Chandra scientists think they are seeing the iron emission expected if the radiation from the gamma-ray burst and its afterglow ionize the surrounding region and the electrons then recombine with the iron atoms.

A gamma-ray burst in 2000 was the farthest yet measured. The spectrum of its afterglow showed that its source is 11 billion light-years away. When the burst was emitted, the universe was only about a 10th as old as it is now.

The High-Energy Transient Explorer (HETE-2) is a NASA mission launched in 2000 to study gamma-ray bursts. For the first time, it can locate hundreds of bursts

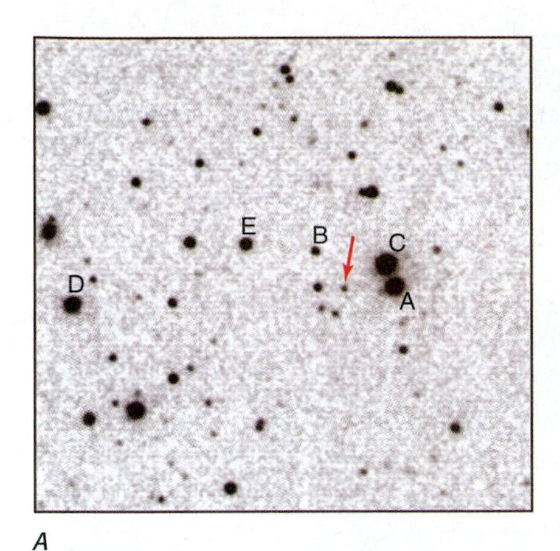


A

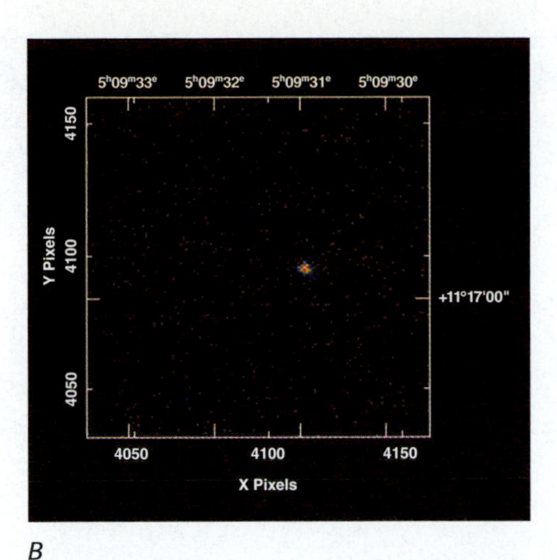


B

**FIGURE 35–25** The x-ray afterglow of GRB991216, observed with Chandra's High Energy Grating Spectrometer and Advanced CCD Imaging Spectrometer, showed both images and an emission-line spectrum from iron. (*A*) An optical image, with the apparent afterglow shown with an arrow. (*B*) An x-ray Chandra image with the Advanced CCD Imaging Spectrometer (ACIS). (*C*) An ACIS spectrum. ACIS can not only make images but also can determine the energy of each incoming x-ray. (*D*) A Chandra spectrum with its High Energy Grating Spectrometer for a higher-resolution spectrum.

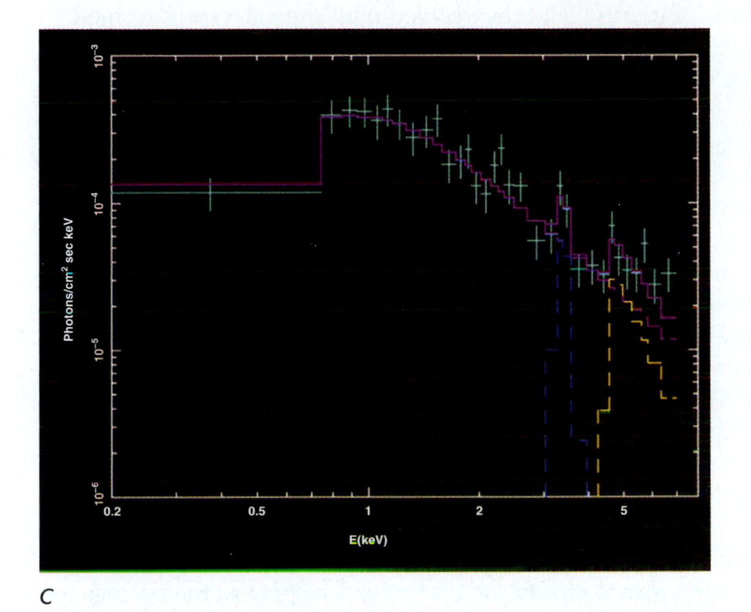


C

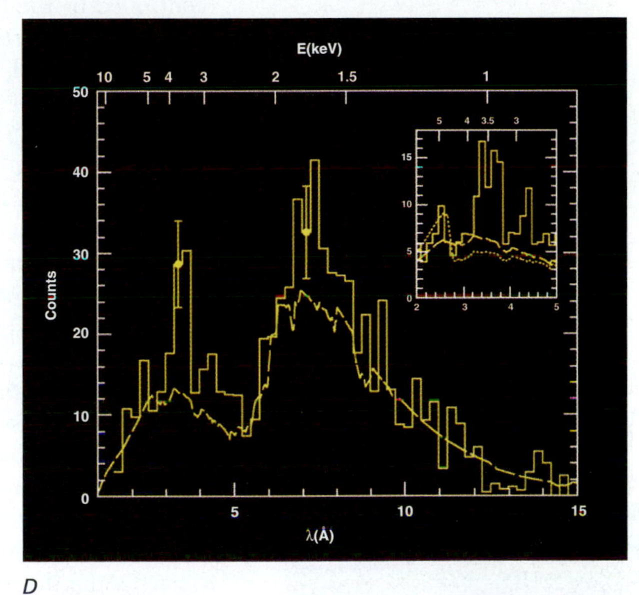


D

with pinpoint accuracy, and relay the accurate locations to the ground within seconds so that other observatories on the ground and space can be notified within 20 seconds so that they can observe the same events. It is expected to work for at least 4 years. BeppoSAX's lifetime has been extended; HETE-2 and BeppoSAX aloft at the same time provide better sky coverage than of either alone, as well as providing a mutual calibration. They hope to discover a burst with $z$ greater than 6, to take us back to the epoch at which the first stars were born.

The Swift spacecraft, to observe gamma rays and their afterglows at several wavelengths, is to follow in 2003. It should be able to send the bursts' coordinates to the ground within 50 seconds of their detection in gamma rays and to start taking x-ray and optical data within that time. It should detect about one burst per day.

## 35.10 THE ORIGIN OF GALAXIES

From the time that it was realized that galaxies were separate units in space, it has generally been believed that galaxies originated through some sort of **gravitational**

## WHAT'S IN A NAME?

### BOX 35.3 Swift

Unusually for a spacecraft, Swift is not an acronym; it is an adjective. The NASA spacecraft, from its middle-complexity Explorer series, is named after a species of bird that feeds while flying. The birder Roger Tory Peterson, famous for his *Field Guide to the Birds* (and originator of the series in which my *Field Guide to the Stars and Planets* appears), described this type of bird as "Flight very rapid, 'twinkling,' sailing between spurts."

**instability.** In this theory, a fluctuation in density either developed or pre-existed in the gas from which the galaxy was to form. This fluctuation grew in mass, collapsed (controlled by the force of gravity), and then cooled until the galaxy was formed.

Current theories of galaxy formation have to take into account the fact that the universe is expanding, which we discussed in this chapter, and the fact that the universe is filled with a glow called the background radiation, which we will discuss in Chapter 37. This background radiation was much stronger in the era when galaxies were formed than it is now. Also, we understand little about the magnetic fields of galaxies, how they formed, and the role they have played.

Further, we have learned in recent years, that the matter that we detect directly is only a minor fraction of the constituents of the universe. We learned about dark matter from the rotation curve of our galaxy.

To understand the giant voids discovered observationally, theoreticians must assume that galaxies form only in rare, especially high peaks in density. Where galaxies don't quite form we should see relative **voids** (Fig. 35–26), though the dwarf galaxies one would expect haven't been found. In this case, voids aren't empty of matter; it is just that the matter in them doesn't shine.

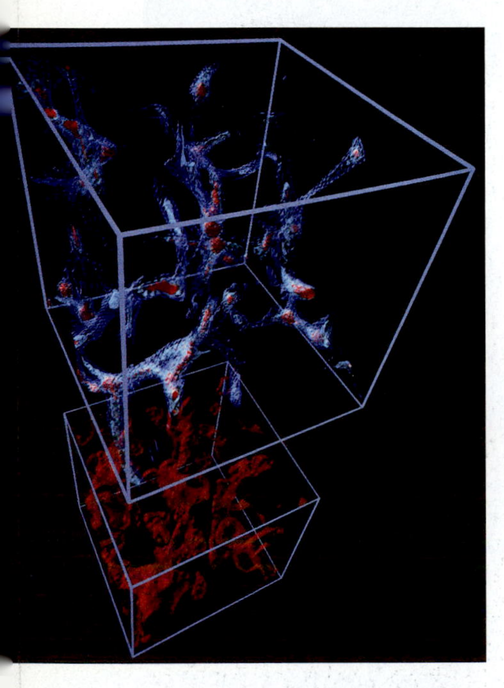

**FIGURE 35–26** These supercomputer calculations of the formation of clusters of galaxies show the existence of filaments and voids.

In the most widely accepted current theory, the dark matter is composed of exotic types of subatomic particles, none of which has yet been discovered. (We discuss dark matter further in Section 38.2a.) This dark matter is termed "cold" since it has (by the theory) little or no velocity to-and-fro with respect to the average expansion of the universe that follows Hubble's law. (Neutrinos, on the other hand, are considered **hot dark matter,** since they would have a large variation of velocity if the experimental result that they have mass is confirmed.) Since the **cold dark matter** does not interact well with photons, it could have started collapsing under gravity and forming large structures at early times in the universe when the matter we can see was still being spread out evenly by collisions with photons. Thus cold dark matter has been invoked to explain the formation of galaxies (Fig. 35–27). The theory, though, does not explain the most massive clusters of galaxies, and it seems far from explaining huge structures like the Great Wall. Further, the smoothness of the universe early on (which we will discuss in Section 37.4d) as revealed by spacecraft has made it even more difficult to understand how the universe could have gone quickly enough from an extremely smooth early state to the large-scale clumpy state we know it assumed within a billion years or so. The cold dark matter theory is considered to be integral to the solution.

There are still many unanswered questions regarding the theory that galaxies arose from gravitational instabilities. Theoreticians can't even agree whether individual galaxies formed first and later combined into clusters, or whether clusters of galaxies formed first, with individual galaxies condensing out of the larger bodies. Three-dimensional maps like those from the Sloan Digital Sky Survey and 2dF (remember: the redshift supplies the third dimention) seem to imply that structure of all sizes forms at the same time.

The theories that galaxies condense out of larger conglomerations follow from ideas of Yakov Zeldovich of the Institute of Applied Mathematics in Moscow. He showed that early in the universe flat sheets of matter could have formed, each taking approximately the shape of a pancake. In this **pancake model,** the pancake may have fragmented later on into clusters of galaxies and, in turn, into galaxies. In between the pancakes and filaments of galaxies are the giant voids, making a void the opposite of a supercluster. There is now observational evidence for such giant voids. More data and more analysis are needed. Faster supercomputers help in the understanding.

The Hubble Space Telescope has apparently detected some of the building blocks of galaxies. In a single long exposure, it found dozens of objects that are all the same distance, about 11 billion light-years away (Fig. 35–28). Each of these objects may be a cluster of a billion stars, and the current idea is that they are close enough together to merge to form galaxies. Further, we don't know of any such objects nearer to us in the universe, so perhaps we have found the "missing link" of galaxy formation. The possibility of "cosmic string" (Section 38.6) as an even earlier type of object around which galaxies formed is also being considered.

The formation of galaxies is intimately connected with whatever else was going on in the early years of the universe, and thus it is connected with the theories of cosmology we will discuss in Chapters 37 and 38. The theory that the universe inflated very rapidly in size in one of the first fractions of a second of time implies that any pre-existing structure or turbulence would have been smoothed out. Theoreticians are now working to explain how galaxies might have formed in this "inflationary universe" (Section 38.5). Further, the inflationary-universe model was worked out to explain an ordinary expanding universe, of the type we have considered for decades. The result of the last few years that the expansion of the universe is accelerating has required even further changes in our scientific models of space and time. In particular, if a cosmological constant indeed exists, it would make up about ⅔ of everything in the universe. It would obviously have a major effect on the formation of galaxies. In Chapter 37, we will learn about the detection of ripples in space set free about 300,000 years after the big bang. Clusters of galaxies probably had their origins in these ripples.

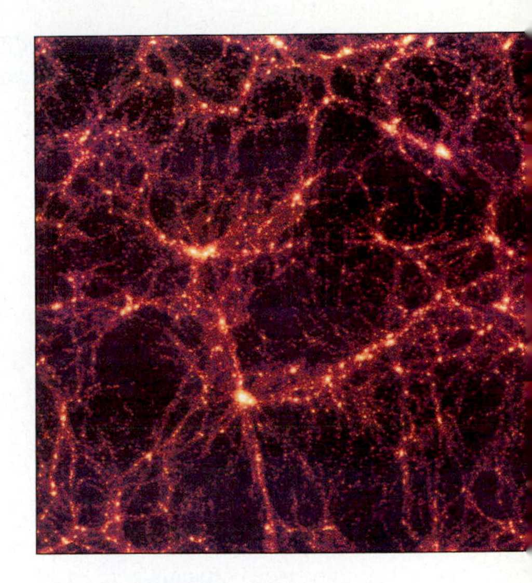

**FIGURE 35–27** Supercomputer calculations for rival mixtures of cold dark matter and cold dark matter + hot dark matter, calculated for a three-dimensional grid of 512 × 512 × 512 points. An inflationary universe (Chapter 38) is assumed, another idea that can be tested.

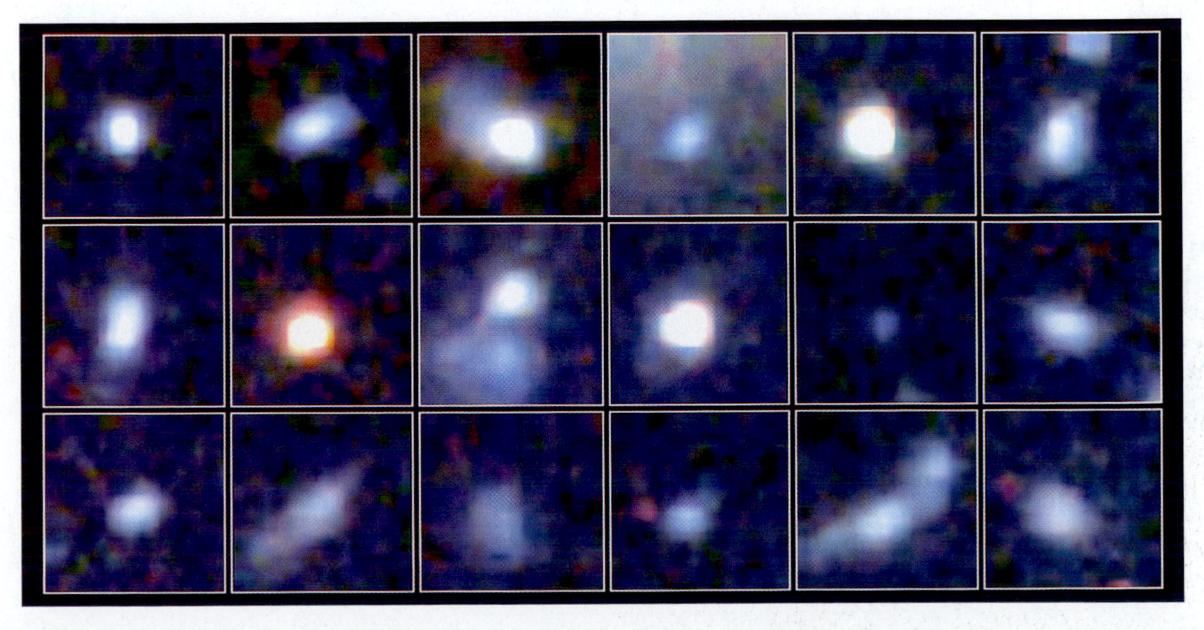

**FIGURE 35–28** Primordial building blocks of galaxies, imaged with the Hubble Space Telescope. Dozens of objects close to each other all have the same distance: 11 billion light-years.

## CORRECTING MISCONCEPTIONS

| ✖ Incorrect | ✔ Correct |
|---|---|
| The universe is static. | The universe is expanding. |
| The universe expands unpredictably. | There is a simple relation between velocity and distance. |
| The universe expands smoothly. | There are slight deviations from uniform expansion caused by mass concentrations. |
| We can readily measure distances. | It is hard to measure distances. |
| The universe is mostly ordinary matter. | The universe may be mostly dark matter, which is invisible. |
| An accelerating universe is the same as an expanding universe. | In an accelerating universe, distant objects are receding more rapidly than in a uniformly expanding universe. |

## SUMMARY AND OUTLINE

The universe is expanding. (Sections 35.1 and 35.2)
- Hubble's law, $v = H_0 d$, expresses how the velocity of expansion increases with increasing distance.
- Best current values for Hubble's constant are between 55 and 80 km/sec/Mpc.
- The expansion is universal and has no center.

The cosmic distance scale (Section 35.3)
- Cepheids are a primary distance indicator.

Finding Hubble's constant (Section 35.4)
- Still controversy over distances

Deviations from smooth expansion (Section 35.5)
- Great Attractor is a huge mass that distorts the Hubble flow.

Superclusters and the Great Wall (Section 35.6)
- Mapping slices of the universe shows huge structures.
- 2dF and Sloan projects mapping far into space by measuring redshifts

The most distant galaxies (Section 35.7)
- Galaxies detected almost back to the origin of the universe.

Gamma-ray bursts (Section 35.8)
- Found to be distant because they are beyond distant galaxies
- HETE and Swift continue Compton Gamma Ray Observatory's discoveries

Origin of galaxies (35.9)
- Gravitational instability theory: A density fluctuation grew, collapsed, and cooled.
- Opposite possibility: Clusters of galaxies formed first, and individual galaxies condensed out of them, as in the pancake model.
- Possibly endorsed by giant voids
- Supercomputers calculate models based on dark matter.
- Inflationary universe and ripples in background radiation must be taken into account.

## KEY WORDS

Hubble's law
Hubble's constant
primary (secondary, tertiary) distance indicators
galactic cannibalism
supercluster
Local Supercluster
gravitational instability
voids
hot dark matter
cold dark matter
pancake model

## QUESTIONS

**1.** To measure the Hubble constant, you must have a means (other than the redshift) to determine the distances to galaxies. What are three methods that are used?

**2.** Does $\alpha$ Centauri, the nearest set of stars to us, show a redshift that follows Hubble's law? Explain.

[‡]**3.** (a) At what velocity is a galaxy 3 million parsecs from us receding? (b) One at 3 million light-years? (*Note:* Your answer is valid to only one decimal place.)

[‡]**4.** A galaxy is receding from us at a velocity of 1500 km/sec. (a) If you could travel at this rate, how long would it take you to travel from New York to California? (b) How far away is the galaxy from us?

[†]**5.** (a) At what velocity in km/sec is a galaxy 10 million parsecs away from us receding? (b) Express this velocity in km/sec, miles/sec, and miles/hr.

[†]**6.** At what velocity in km/sec is a galaxy 10 million light-years away receding from us?

[†]**7.** How far away (in Mpc) is a galaxy with a redshift of 0.2? In km?

---

[†]This question requires a numerical solution.

[‡](For $H_0$ = 75 km/sec/Mpc) **3.** 225 km/sec; 70 km/sec **4.** (a) 3 sec (b) 20 Mpc.

†8. From Figure 35–1, compute the values that Hubble and, later, Hubble and Humason measured for Hubble's constant. Compare with the currently accepted value.

†9. The Hubble Space Telescope can observe stars of 30th magnitude, as opposed to 25th magnitude for ground-based telescopes. This extra ability allows us to observe Cepheids in galaxies farther out in the Virgo Cluster than we could before. How many times greater than before is the volume of space in which we can find the distances to galaxies in this way?

†10. If a galaxy's peculiar radial velocity is 20 per cent of its expansion velocity at the Virgo Cluster (16 Mpc), what percentage of its expansion velocity is its radial velocity at the Ursa Major Cluster (20 Mpc)? The Hydra Cluster (58 Mpc)? Comment on the importance of these percentages.

11. Why is it important to measure velocities not only at but also past the Great Attractor?

12. Why are we now able to measure slices of the universe, while Edwin Hubble was not?

13. How do we know that gamma-ray bursts are at huge distances?

14. What is the effect of dark matter on the formation of galaxies?

15. Why do we think that the Hubble Space Telescope has detected building blocks of galaxies?

## INVESTIGATIONS

1. Find out about the life and work of Edwin Hubble. Read a biography.
2. Draw dots on a balloon, measure the distances between several pairs of them, and measure again after you have blown up the balloon farther. Construct an analogy to Hubble's law.

## USING TECHNOLOGY

### W World Wide Web

1. Check the Web site for the Great Debate on the Cosmic Distance Scale, as staged at the National Academy of Sciences in Washington in 1996.
2. Follow the mapping of the universe from the 2dF Galaxy RedShift Survey and from the Sloan Digital Sky Survey at http://msowww.anu.edu.au/2dFGRS and http://www.sdss.org, respectively.
3. Follow the distance determinations from the High-Z Supernova Project and the Supernova Cosmology Project at http://cfa–www.harvard.edu/cfa/oir/Research/supernova/public.html and http://www.supernova.lbl.gov, respectively.
4. Study objects at different distances in the cosmic distance scale at http://heasarc.gsfc.nasa.gov/docs/cosmic/cosmic.html.
5. Follow improved knowledge of gamma-ray bursts from the Compton Gamma-Ray Observatory at http://gammaray.msfc.nasa.gov/batse, from the High-Energy Transient Explorer 2 at http://space.mit.edu/HETE, and from Swift at http://swift.sonoma.edu.

### REDSHIFT

1. Displaying deep-sky objects and constellation boundaries, and adjusting the limiting magnitude displayed, compare the overall dimensions of the clusters for which redshifts were measured in the set of four spectra shown in this chapter.
2. Check the data boxes for these clusters of galaxies, and compare the distances given with the dimensions. Do you detect a correlation?
3. Look at the movie showing the large-scale structure of the universe.

A Hubble Space Telescope image of a quasar, PKS 2349–014, whose spectral lines are redshifted by 17.3 per cent—that is, $z = 0.173$. The loops of gas around this quasar suggest that it is being fueled by gas from the merging of two galaxies. The galaxies above the quasar are probably interacting with it.

# Chapter 36

# Quasars

Quasars are enigmatic objects that appear almost like stars: points of light in the sky. But unlike the stars that we see, quasars occur in the farthest reaches of the universe and thus may be a key to our understanding the history and structure of space. Although quasars are relatively faint in visible light, many are among the strongest radio sources in our sky and therefore must radiate prodigious amounts of energy. Quasars turn out to be small (on a galactic scale), so ordinary methods of generating energy are not sufficient. Similarly, we would be surprised if we got a tremendous explosion out of a tiny firecracker. Consequently, the most exotic and efficient method of generating energy—matter being gobbled by a giant black hole—has been invoked to explain the energy of quasars.

**AIMS:** To describe the discovery of quasars, their significance as probes of the early stages of the universe, their connections with galaxies, and our knowledge of the giant black holes powering them

The word **quasar** originated from QSR, a contraction of "quasi-stellar radio source" ("quasi-," meaning "as if," "seemingly"). After all, they look almost, but not quite, like ordinary stars (Fig. 36–1). However, the most important characteristic of quasars is not that they are emitting radio radiation but that they are travelling away from Earth at tremendous speeds. After quasars were named, radio-quiet "quasi-stellar objects" (QSOs) were discovered and were also called quasars. So some quasars are radio-quiet, and others are radio-loud. In both cases, astronomers deduce distances by observing the quasar spectra, measuring the redshifts, and applying Hubble's law (Section 35.1). Since quasars have among the largest redshifts known, Hubble's law tells us that they are among the most distant objects that we see. We shall discuss the evidence on this point; only a few holdouts don't accept it these days.

**FIGURE 36–1** A ground-based optical image of 3C 273. The object looks like any 13th-magnitude star ("stellar") except for the faint jet that is visible out to about 20 seconds of arc (and therefore only "quasi-stellar"). The faint perpendicular spikes are artifacts resulting from the telescope and overexposure of the central image to bring out the jet.

## 36.1 THE DISCOVERY OF QUASARS

Quasars are a discovery resulting from the interaction of optical astronomy and radio astronomy. When maps of the radio sky turned out to be very different in appearance from maps of the optical sky, many astronomers in the 1950s set out to correlate the radio objects with visible ones.

Single-dish radio telescopes did not give sufficiently accurate positions of objects in the sky to allow identifications to be made, so interferometers—then just being developed—had to be used. At least one of the strong radio sources, 3C 48 (the *48*th source in the *3*rd *C*ambridge Catalogue, made at Cambridge, England), seemed suspiciously near the position at which optical observations showed a faint (16th-magnitude) bluish star. At that time, in 1960, no stars had been found to emit radio waves, with the sole exception of the sun, whose radio radiation we can detect only because of the sun's proximity to Earth.

In Australia, a large radio telescope was used to observe the passage of the moon across the position in the sky of another bright radio source, 3C 273. We know the position of the moon in the sky very accurately. When it occults—hides—a radio source, then we know that at the moment the signal strength decreases, the source must have passed behind the advancing limb of the moon, which is a curved line. Later on, the instant the radio source emerges, the position of the lunar limb marks another set of possible positions, another curved line. The source must be at one of the two points where these two curved lines meet. Three lunar occultations of 3C 273 occurred within a few months, and from the data, a very accurate position for the source and a map of its structure were derived. The optical object (already shown as Figure 36–1) is not completely star-like in appearance, for a luminous jet appears to be connected to the point nucleus. It is thus "quasi-stellar."

The radio emission from 3C 273 has two components. The discovery that one coincides with the jet (Fig. 36–2), and the other coincides with the bluish stellar object, clinched the identification of the optical object with the radio object. Maarten Schmidt photographed the spectrum of this "quasi-stellar radio source" with the 5-m Hale telescope on Palomar Mountain. The spectra of 3C 273 and 3C 48, both bluish quasi-stellar objects, showed emission lines, but the lines did not agree in wavelength with the spectral lines of any of the elements. The lines had the general appearance of spectral lines emitted by a gas of medium temperature, though.

The breakthrough came in 1963. At that time, Schmidt noted that the spectral emission lines of 3C 273 (Fig. 36–3) seemed to have the same pattern as lines of hydrogen under normal terrestrial conditions. Schmidt then made a major scientific discovery: he asked himself whether he could simply be observing a hydrogen spectrum that had been greatly shifted in wavelength by the Doppler effect. The Doppler shift required would be huge: each wavelength would have to be shifted by about 16 per cent toward the red to account for the spectrum of 3C 273. This interpretation would mean that 3C 273 is receding from us at approximately 16 per cent of the speed of light. Immediately, Schmidt's colleague Jesse Greenstein recognized that the spectrum of 3C 48 could be similarly explained. All the lines in the spectrum of

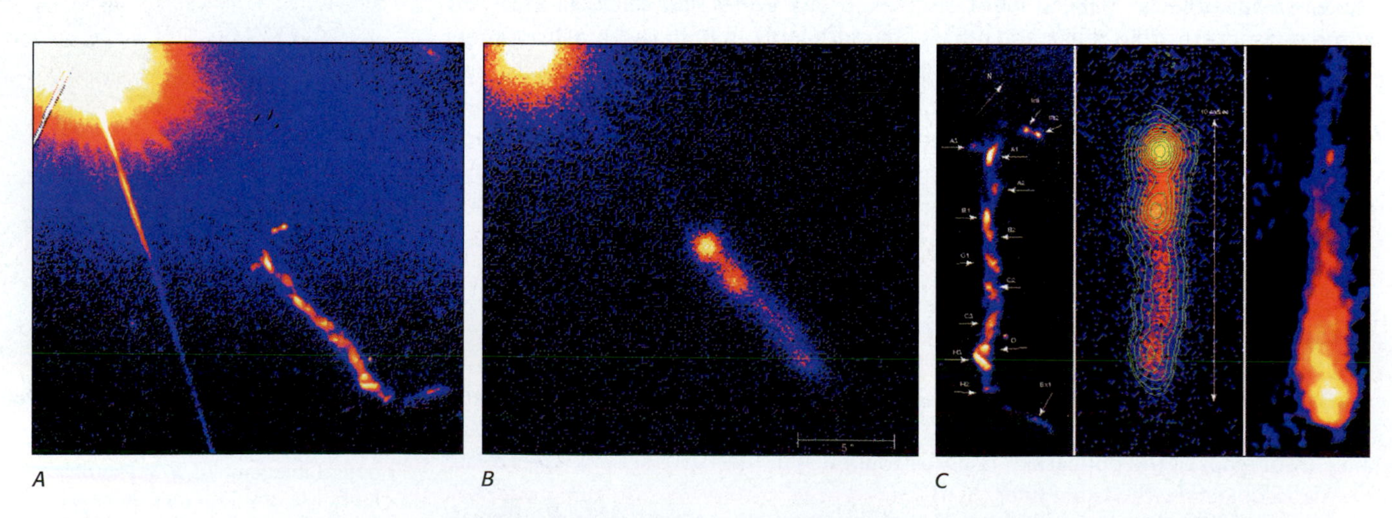


**FIGURE 36–2** (*A*) The quasi-stellar radio source 3C 273 (*at upper left*), taken with the Hubble Space Telescope, and the jet that extends 150,000 light-years from it. (*B*) 3C 273 from Chandra. A faint but definite stream of energy is seen to flow outward toward the lumpy part of the jet. The lumps, clouds of high-energy electrons giving off the x-rays Chandra detects, may be shock waves, cosmic traffic collisions on a large scale. (*C*) The jet of the quasar 3C 273 seen (*left to right*) in visible light from Hubble, in x-rays from Chandra, and in radio waves from the MERLIN interferometer.

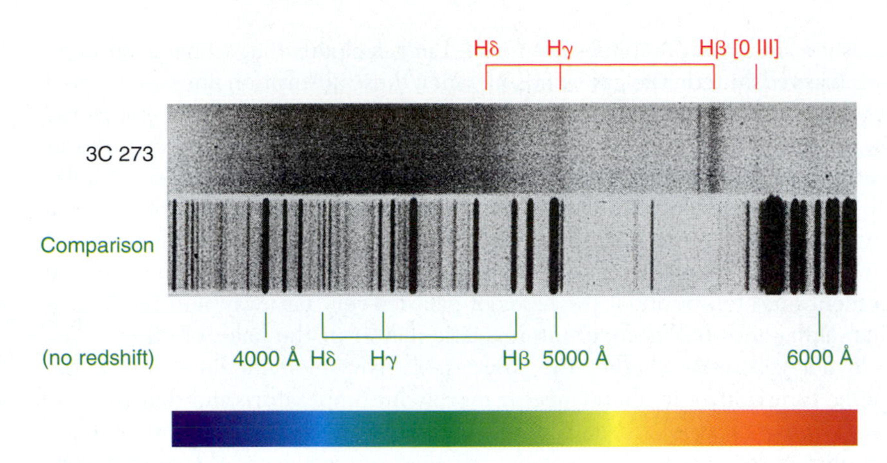


**FIGURE 36–3** Spectrum of the quasar 3C 273. The lower spectrum is a source on Earth; this spectrum consists of hydrogen and helium lines. This "comparison spectrum" establishes the scale of wavelength. The upper part is the spectrum of the quasar, an object of 13th magnitude. The Balmer lines H$\beta$, H$\gamma$, and H$\delta$ in the quasar spectrum are at longer wavelengths *(labels in red)* than in the comparison spectrum *(labels in green)*. The redshift of 16% corresponds, according to Hubble's law ($H_0 = 75$), to a distance of 4 billion light-years. Note that the comparison spectrum represents hydrogen and helium sources on Earth (more specifically, located inside the Palomar dome).

3C 48 were shifted by 37 per cent, a still more astounding redshift. These redshifts, now routinely observed, are often identified by the hydrogen emission spectrum (Fig. 36–4), which we discussed in Section 24.4.

The discovery was astounding because it implied that these objects were incredibly luminous. The luminosity was derived by combining the observed brightness of the quasar with the distance derived with Hubble's law. 3C 273, in particular, came out to be 100 times more luminous than our whole Milky Way Galaxy!

Later, absorption lines were discovered in quasars, in addition to the emission lines already known. Many quasars have several systems of absorption lines of differing redshifts, in addition to their set of broad emission lines. Some of these absorption lines seem to be formed in clouds of gas of differing velocities surrounding the quasars. Others are formed farther from the quasar as the light travels from the quasar to us. Ultraviolet spectra taken from space showed that our galaxy and the Large and Small Magellanic Clouds have extensive halos of gas that cause absorption lines detectable in the ultraviolet. This discovery indicates that the absorption lines of ionized carbon, silicon, magnesium, and other "heavy elements" in quasars might be formed in the halos of otherwise unseen galaxies between the quasars and us. Since each distant galaxy has a different redshift, we can see the same line at different wavelengths. In the previous chapter, we discussed the Hubble Key Project on the cosmic distance scale, which involved measuring lots of Cepheid variables, finding their distances, and then using the distances to the galaxies they are in to calibrate methods of measuring distances to galaxies that are still farther away. A second Hubble Key Project—there are only three—is to study absorption spectra of the nearest of the quasars. Understanding the spectra of these quasars should enable us to understand better the less-detailed spectra we measure from the most distant quasars.

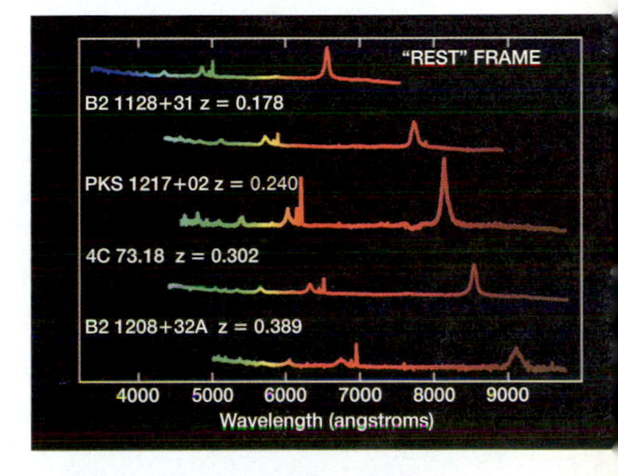


**FIGURE 36–4** The spectra of four quasars, showing how redshifts ranging from 17.8% to 35.9% shift the hydrogen emission spectrum. The rest wavelengths of the lines are 6563 Å for H$\alpha$, 4860 Å for H$\beta$, and 4340 Å for H$\gamma$.

The basic hydrogen line Lyman alpha in the spectrum of a quasar often appears in absorption at many different redshifts—presumably one for each gas cloud the light is passing through. (Remember that this spectral line, when observed from a source at rest, appears deep in the ultraviolet, far beyond the spectral region that comes through our atmosphere.) Thus, at a wide range of wavelengths shorter than that of the redshifted emission line of Lyman alpha, we see a "forest" of lines—the **Lyman-alpha forest.** The emission line is formed at the quasar itself and so has the largest redshift; the absorption lines are formed in gas clouds that aren't as far away and so aren't as redshifted. The gas clouds in which these absorption lines are formed seem to be made almost entirely of hydrogen, another line of evidence indicating that they are located out near the quasars where we are seeing to earlier times in the universe when few heavier elements existed. The nearby quasar 3C 273 shows few such lines, while high-redshift quasars show many, indicating that the neutral hydrogen clouds were more numerous early in the universe's history (Fig. 36–5). Some of the lines tell us about gas surrounding the quasars themselves, most of the heavy-element lines tell us about the halos of galaxies very far away, and the hydrogen Lyman-alpha lines tell us about intergalactic matter or the halos of other galaxies. We do not yet know whether the sources of heavy-element lines and of the Lyman-alpha forest differ in abundance or merely in temperature and density.

Though quasars had long been found by searching photographic plates for objects with ultraviolet excesses, the recent farthest quasars were discovered with CCDs. Candidate objects that have a high probability of being quasars can be found, as we described, by looking for star-like objects that seem unusually strong in some colors compared with others. Plots are made of the brightness of the object in one color versus its brightness in another color. Most of the hundreds of thousands of objects scanned fall in the same region of these "color-color diagrams," but a few stand out. (Originally, the quasar candidates appeared especially strong in the blue and in the ultraviolet, but now that we are seeing farther out, the color discrepancies have moved into redder parts of the spectrum.) Taking spectra of the few that stand out reveals distant quasars when the candidate objects turn out to have large redshifts.

The Sloan Digital Sky Survey in the United States (Fig. 36–6) and the 2dF survey in Australia are finding tens of thousands of quasars (Fig. 36–7). Sloan will eventually have a million quasars detected with redshifts (and thus distances) for 100,000

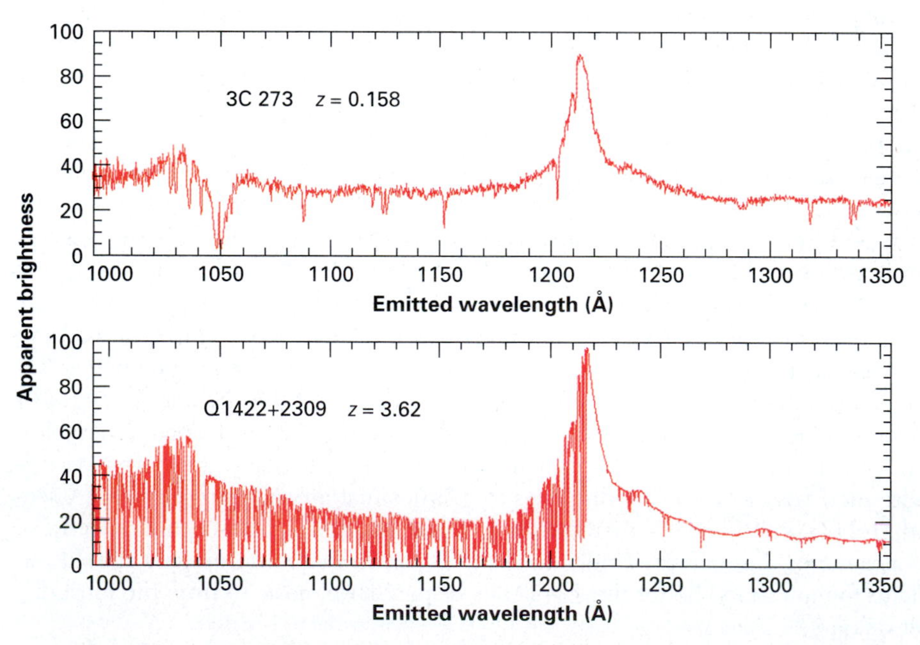


**FIGURE 36–5** The contrast between spectra of one of the closer quasars (*top*) and of a much more distant quasar (*bottom*), with only the latter showing the Lyman-$\alpha$ forest. Most of the absorption lines to the left of the highest peak in the bottom spectrum are produced by the hydrogen Lyman-$\alpha$ transition produced by intergalactic clouds at many different wavelengths from us. The upper spectra from the first quasar to be discovered (3C 273) show few such absorption lines because the quasar is closer than the intergalactic clouds and because there are fewer such intergalactic clouds at this epoch of the universe.

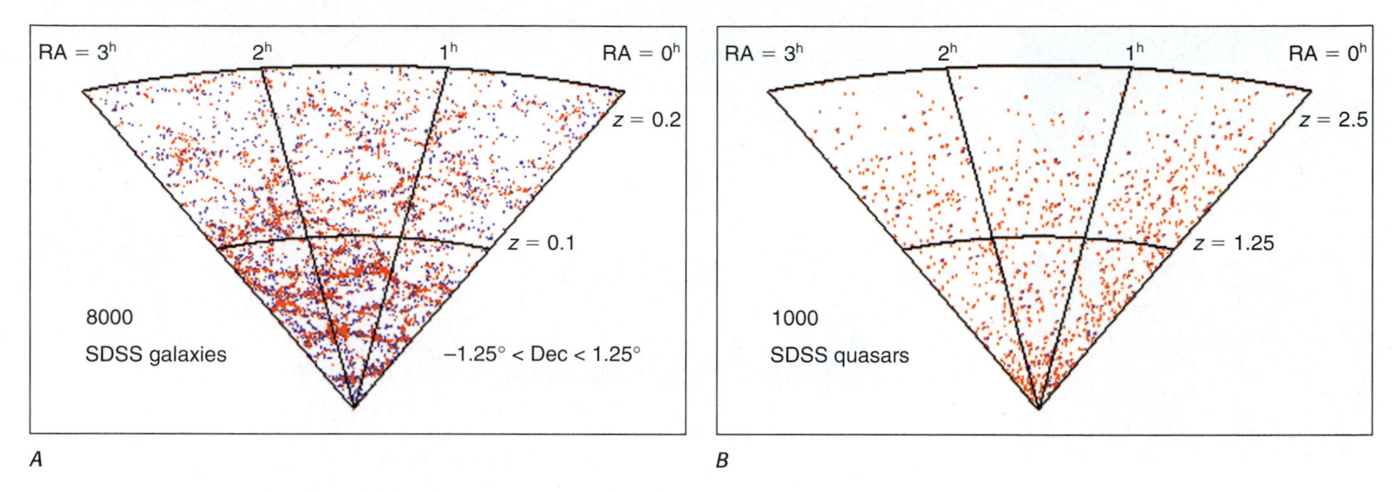


**FIGURE 36-6** Wedges of (*A*) galaxy and (*B*) quasar observations made by the Sloan Digital Sky Survey. For the galaxies, spirals are in blue and ellipticals in red. Remember that a strip across the sky in right ascension, containing a small region of declination, provides the angle of the wedge, since we are looking outward from Earth at these objects. And remember, most importantly, that the redshifts measured give the third dimension: the distance. Notice how much farther out in redshift $z$ the quasar observations extend.

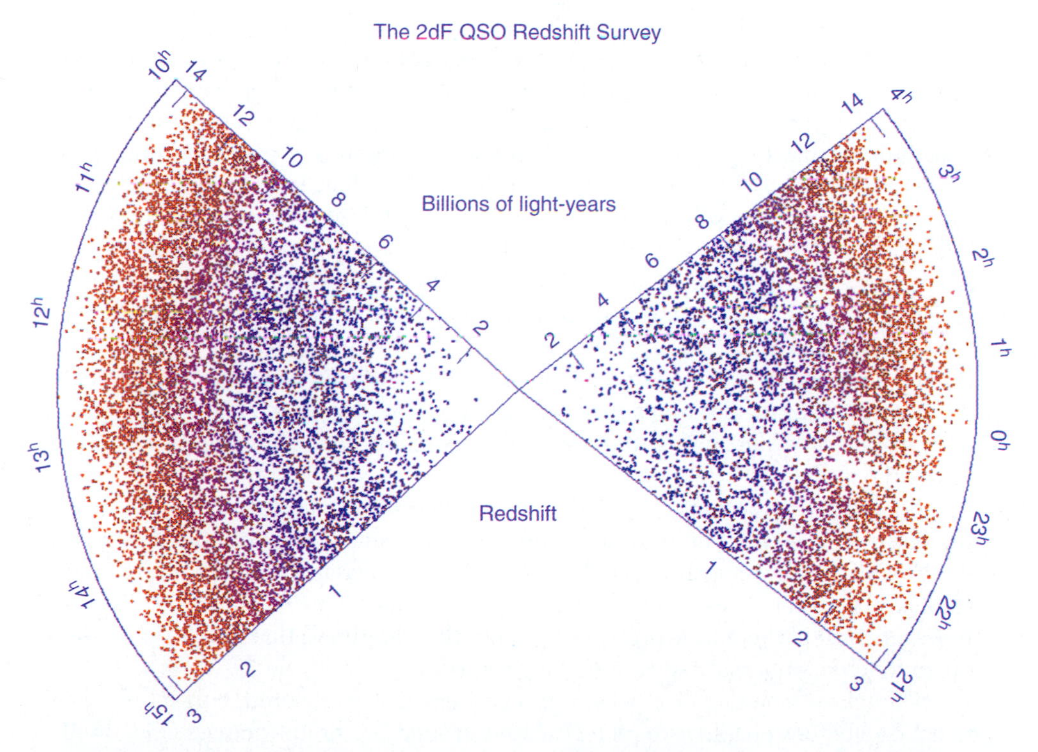


**FIGURE 36-7** Wedges showing 11,000 quasar observations made by the 2dF quasar survey. As in the previous diagram, the third dimension—outward on the wedge, is provided by redshift measurements. Fiber optics allow hundreds of such spectral measurements to be made at the same time. The positions of the objects from $10^h$ to $15^h$ of right ascension are shown in the wedge at left and from $21^h$ to $4^h$ are shown at right. A structure a billion light-years across shows and seems constant over the overall field of view, which is 10 times larger than that. The structure is much lumpier at larger scales than predicted. Since the lumpiness that is beginning to show up in the distribution originally arose in the epoch immediately after the big bang (see the next two chapters), this survey may help us understand conditions back then. The color of the dots has no significance other than expressing increasing redshift.

A

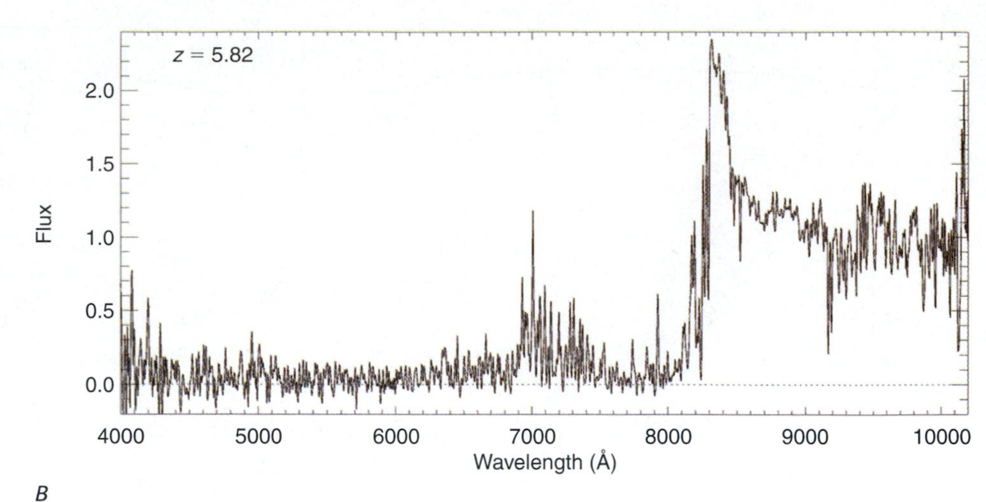


B

**FIGURE 36–8** (*A*) An extremely distant quasar, with its lines shifted by 5.8 times their original wavelengths. We see the detection in x-rays from the XMM-Newton spacecraft. The object is known from its Sloan Digital Sky Survey discovery number: SDSS 1044–0125. (*B*) The spectrum of this quasar has all its lines shifted by wavelengths 5.8 times their original wavelengths. The highest redshifts shift extreme ultraviolet lines like hydrogen's Lyman $\alpha$, at 1216 Å when measured in a laboratory on Earth, into the red and infrared regions of the spectrum. At these redshifts, we are looking back to within 2 billion years of the big bang (Chapter 37), when the scale of size of the universe was only 15% of its current scale.

of the brightest. 2dF expects 25,000 redshifts. (These surveys were discussed in Section 5.5, and the maps of galaxies and galaxy distances made with them were discussed in Section 35.7b.) One quasar that turned up in 2dF has a redshift of 5.8, making it the most distant object we then knew in the universe and still one of the farthest (Fig. 36–8). Its central black hole has been calculated to contain 3 billion times the mass of the sun. Examining x-ray sources has also proved fruitful to find quasar candidates. The $z = 5.8$ quasar has even been detected by the XMM-Newton spacecraft, in spite of its huge distance. (Recall that XMM-Newton has more collecting area than Chandra, and so is more sensitive even though it doesn't have the same spatial resolution.) Only 32 x-ray photons were detected in the 8-hour observation, about 10 times fewer than expected. Perhaps local gas near it is absorbing x-rays, or its black hole may be dragging x-rays inside it. (The x-rays bounce outward at some rate, but if the gas in the accretion disk is moving inward even faster, the x-rays don't get out.)

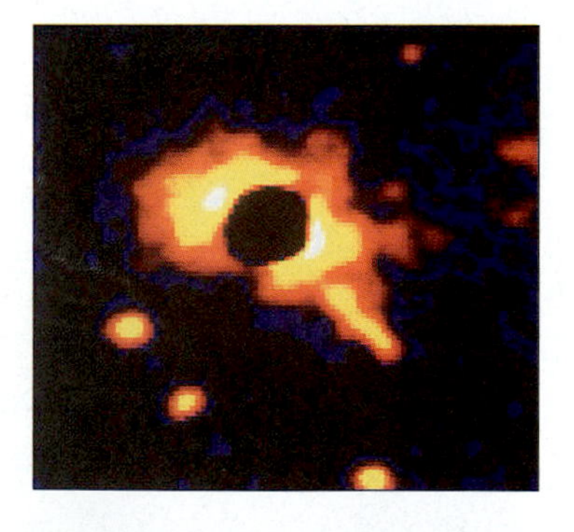

**FIGURE 36–9** Hiding the bright quasar 3C 273 with a dark disk, as was done here, has revealed nebulosity that resembles an elliptical galaxy. Quasars are now thought to be bright events in the centers of galaxies.

But however you find a candidate object, one must take spectra to prove that the object is a quasar. The procedure was very time-consuming until the production line Sloan and 2dF projects began discovering distant quasars routinely. Astronomers estimate that there have been perhaps a million quasars in the universe. By now, however, most of them have probably lived out their lifetimes; that is, they have given off so much energy that they are no longer quasars.

Though we couldn't do it when quasars were first discovered, current observational capabilities enable us to detect matter around the bright quasars (Fig. 36–9). Since this matter has the spectra of stars, such as those in distant galaxies, we now think that quasars are unusual events in the centers of galaxies—very unusual events indeed.

## 36.2 THE REDSHIFT IN QUASARS

That the redshift arises from the relative motion is accepted by almost all astronomers. Most objects in the universe show some redshift with respect to the Earth, again from the expansion of the universe. For velocities that are small compared to the velocity of light, the amount of shift in the spectrum is written simply, for rest wavelength $\lambda$, velocity $v$, and speed of light $c$, as the Doppler-shift formula

$$\frac{\Delta\lambda}{\lambda} = \frac{v}{c} \text{ (as discussed in Section 25.6).}$$

Astronomers often use the symbol $z$ to stand for $\Delta\lambda/\lambda$, the fraction by which the spectrum is redshifted. Recall that $\Delta\lambda$ is the change in the wavelength.

The redshifts of quasars are much greater than those of most galaxies. Even for 3C 273, the brightest quasar, $z = 0.16$ (read "a redshift of 16 per cent"). Thus Hubble's law implies that 3C 273 is as far away from us as many distant galaxies. To find the velocity of recession for quasars with redshifts larger than about 0.4, we must use a formula based on Einstein's general theory of relativity. Some quasars are receding at over 90 per cent of the speed of light. Hubble's law, for a suitable choice of Hubble's constant, then tells us they are over 10 billion light-years away.

## 36.2a Working with Non-Relativistic Doppler Shifts

Let us consider a redshift of 0.2, to pick a round number, and calculate its effect on a spectrum (Fig. 36–10). $z = 0.2$ implies that $v/c = 0.2$, and therefore

$$v = 0.2c = 0.2 \times (3 \times 10^5 \text{ km/sec}) = 6 \times 10^4 \text{ km/sec.}$$

The technique in the rest of this section is valid for all redshifts.

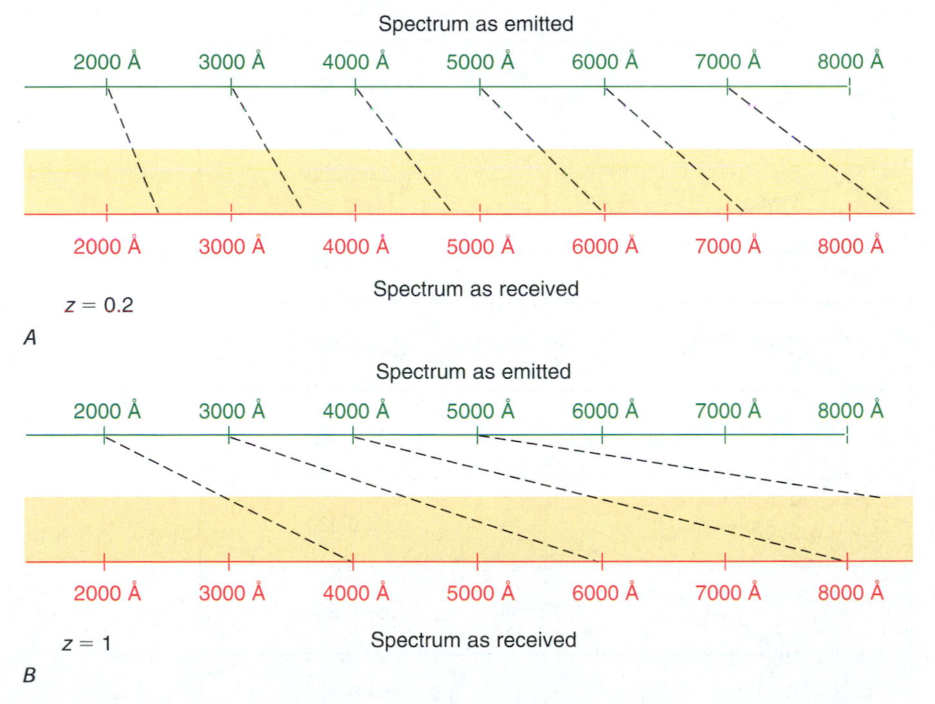


**FIGURE 36–10** *(A)* The wavelengths of the visible part of the spectrum are shown in green at top, and the wavelengths at which spectral lines would appear after undergoing a redshift of 0.2 ($\Delta\lambda/\lambda = 0.2$) are shown in red at bottom. The lines are shifted by 0.2 times their original wavelengths and wind up at 1.2 times their original wavelengths. The non-relativistic formula $\Delta\lambda/\lambda = v/c$ can be applied to give the approximate velocity; at still smaller redshifts this formula is even more accurate. *(B)* For a redshift of 1 ($\Delta\lambda/\lambda = 1$), as shown here, the lines are shifted *(bottom)* by an amount equal to their original wavelengths *(top)* and wind up at twice their original wavelengths. The velocity is a significant fraction of the speed of light, so the simple non-relativistic formula for the Doppler shift cannot be applied. The relativistic formula allows $z$ to be greater than 1 without the velocity exceeding the speed of light.

If a spectral line were emitted at 4000 Å in the quasar, at what wavelength would we record it on Earth on our CCD (or, over a decade ago, on our film)?

$\Delta\lambda/\lambda = \Delta\lambda/4000 = 0.2$. Therefore, the change in wavelength $= \Delta\lambda = 0.2 \times 4000$ Å $= 800$ Å. Note that we have calculated only $\Delta\lambda$, the shift in wavelength. The new wavelength is equal to the old wavelength plus the shift in wavelength,

$$\lambda_{new} = \lambda_{original} + \Delta\lambda = 4000 \text{ Å} + 800 \text{ Å} = 4800 \text{ Å}.$$

Thus the line that was emitted at 4000 Å in the quasar would be recorded at 4800 Å on Earth.

Similarly, a line that was emitted at 5000 Å is also shifted by 0.2 times its wavelength, and 0.2 of 5000 Å is 1000 Å. Then $\lambda + \Delta\lambda = 5000$ Å $+ 1000$ Å $= 6000$ Å. The spectrum is thus not merely displaced by a constant number of angstroms, but is also stretched more and more toward the higher wavelengths.

### 36.2b Working with Relativistic Doppler Shifts

The simple Doppler formula above is valid only for velocities much less than $c$, the speed of light. For speeds closer to the speed of light, we must use a formula from the special theory of relativity. Considering only radial velocity (and ignoring any velocity perpendicular to the line of sight, or assuming that the expansion is uniform),

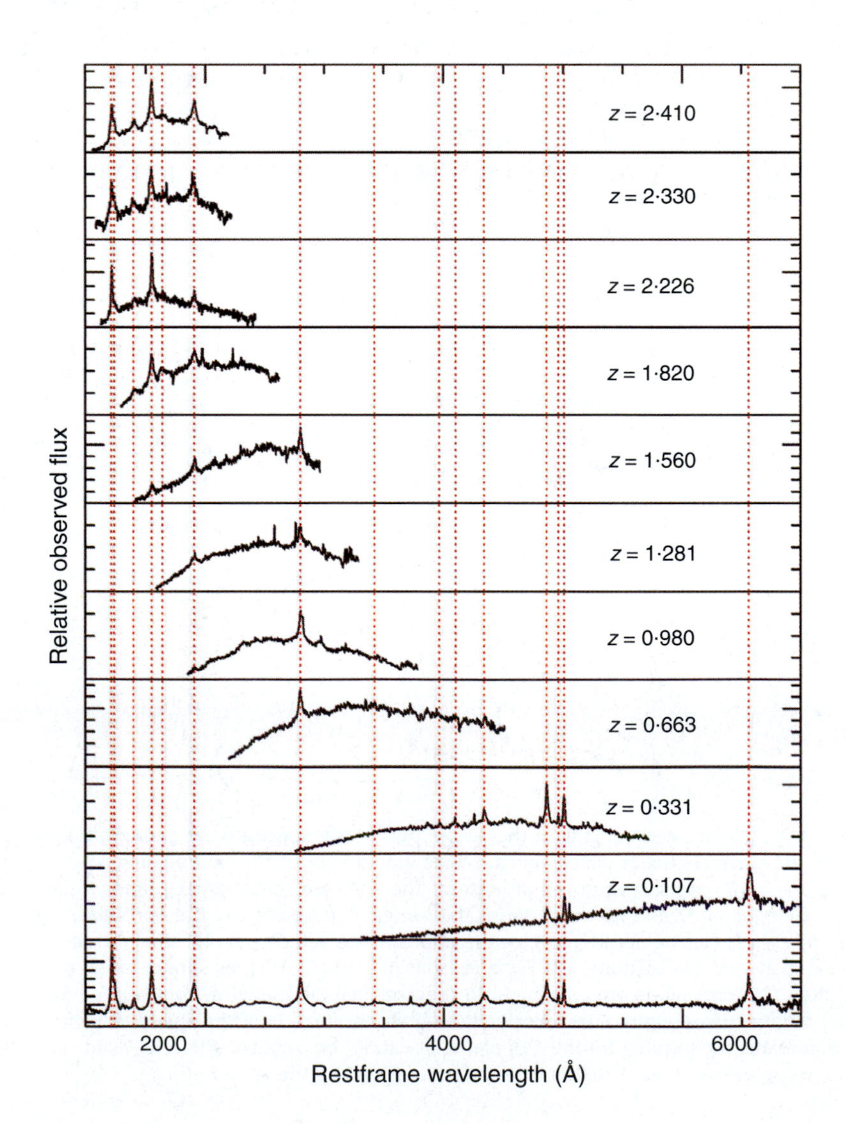


**FIGURE 36–11** The spectra of quasars of different redshifts, viewed in the visible but transformed into the wavelengths at which the spectral lines were emitted. A composite spectrum showing the whole range, an average of 10,000 quasar spectra, appears at the bottom. Notice how the part of the original spectrum we observe is different for quasars of different redshifts.

$$\frac{\Delta\lambda}{\lambda} = \sqrt{\frac{1 + v/c}{1 - v/c}} - 1.$$

Positive values of $v$ correspond to receding objects. When $v$ is much less than $c$, then the relativistic formula approximates the non-relativistic formula. (Note that if you substitute $v = 0$ in the relativistic formula, the redshift derived is indeed 0.) But when $v$ is close to $c$, $\Delta\lambda/\lambda$ is greater than 1 even though $v$ is still less than $c$. We still use the letter $z$ to stand for $\Delta\lambda/\lambda$.

For example, if $v$ = 90 per cent of $c$, $v/c = 0.9$.

$$\frac{\Delta\lambda}{\lambda} = \sqrt{\frac{1 + v/c}{1 - v/c}} - 1 = \sqrt{\frac{1 + 0.9}{1 - 0.9}} - 1 = \sqrt{19} - 1 = 4.4 - 1 = 3.4$$

There is no physical significance for $z = 1$; we merely have the shift, $\Delta\lambda$, equalling the original wavelength, $\lambda$. The quasar shown earlier (Fig. 36–8) with the very large redshift has $z = 5.8$, which means that its wavelengths are shifted by 580 per cent. This makes the new wavelengths $5.8 + 1 = 6.8$ times the original wavelengths. But its velocity of recession is not much over 90 per cent of the speed of light. The velocities of recession of all quasars are still less than the speed of light, as the special theory of relativity tells us they must always be.

It is instructive to look at the spectra of quasars of different redshifts (Fig. 36–11). Different portions of their ultraviolet spectra are shifted into the visible. A whole quasar spectrum shows emission lines of several different elements in different ionization states (Fig. 36–12).

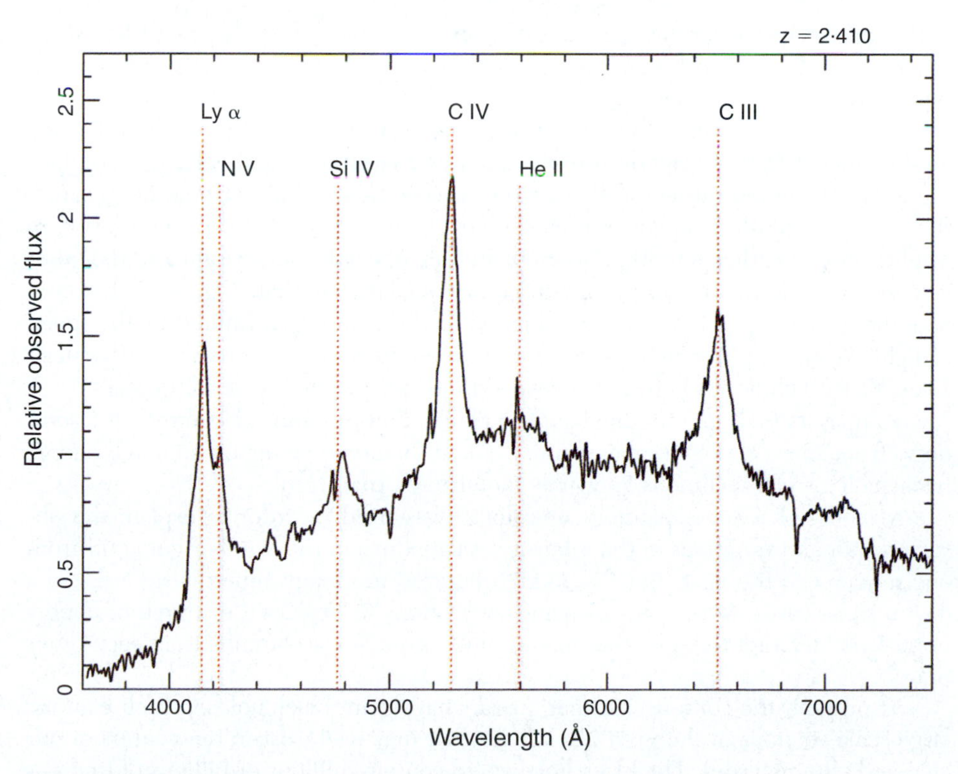


**FIGURE 36–12** The spectrum of the highest-redshift quasar from the previous figure, showing the elements whose spectral lines are detected in emission. As is usual, the Roman numerals indicate ionization. For example, C III is carbon-III, twice-ionized carbon, while C IV is three-times-ionized carbon.

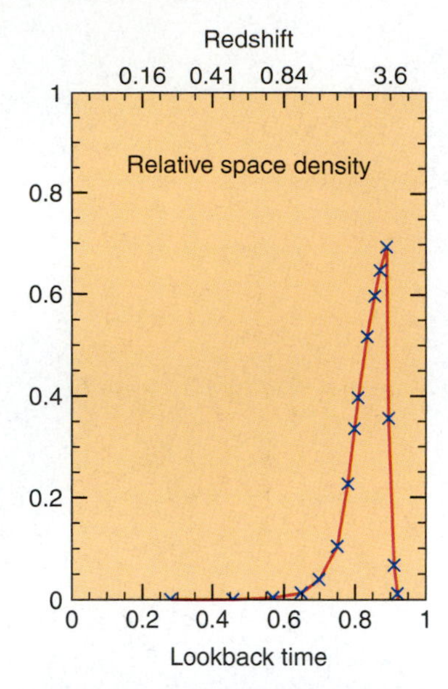


**FIGURE 36–13** The conclusion that there was a bright, spectacular era of quasars billions of years ago is shown in this distribution curve measured in a survey of quasars by Patrick Osmer of Ohio State and colleagues. The Sloan Digital Sky Survey and 2dF Quasar Survey are extending these results, with many more quasars at high redshift, in particular.

## 36.3 THE IMPORTANCE OF QUASARS

If we accept that the redshifts of quasars are caused by the Doppler effect, and we apply Hubble's law, we realize that the quasars with the largest redshifts are the farthest known objects in the universe. Thus we would like to use quasars to test Hubble's law, since deviations from the law would no doubt show up in the farthest objects.

If the quasars were found not to satisfy Hubble's law, on the other hand, then doubt would be cast on all distances derived by Hubble's law. In that case, when we were observing an object, we would never know whether the object satisfied the law or not. We would never be able to trust a distance derived from Hubble's law. At present, the evidence seems overwhelming that quasars do follow Hubble's law, that is, that their redshifts and distances are proportional with the same Hubble constant that we find for other galaxies.

If we accept the quasars as the farthest objects, then they are billions of light-years away, and their light has taken billions of years to reach us. Thus we are looking back in time when we observe the quasars, and we hope that they will help us understand the early phases of our universe. For example, a survey of the entire sky to look for quasars has shown that the number of quasars per volume of space increases as you go outward. Thus there were more quasars in the universe long ago. Among other things, this change shows that the universe has evolved.

Maarten Schmidt, the discoverer of quasars, dazzled the astronomical world by concluding (Fig. 36–13) that there was an abrupt epoch of quasar formation, an exciting time in the universe when the quasars were lit up like brilliant candles. The statistics being provided by Sloan and 2dF should pinpoint the turn-on time.

## 36.4 FEEDING THE MONSTER

If the quasars are as far away as the orthodox view holds, then they must be intrinsically very luminous to appear to us at their observed intensities. They are more luminous than entire ordinary galaxies.

But at the same time, the quasars must be very small because of the following argument: The optical brightness of quasars was measured on old collections of photographic plates and turned out to vary on a time scale of weeks or months. If something is, say, a tenth of a light-year across, you would expect that it could not vary in brightness in less than a tenth of a year. After all, one side cannot signal to the other side, so to speak, to join in the variation more quickly than that (Fig. 36–14). Somehow the whole object has to be coordinated, and that ability is limited by the speed of light. So the rapid variations in intensity mean that the quasars are fairly small. Even VLBI techniques have not succeeded in resolving the nuclei of quasars.

So quasars had not only to give off a tremendous amount of energy but also to do so from an exceedingly small volume. The difficulty of giving off so much energy from such a small volume is known as the **energy problem.**

A model of many exploding supernovae was tried in order to explain the observed sporadic variations in the intensity as variations in the number going off from one moment to the next. But this model required too many supernovae, a dozen a day in some cases. Other unconvincing suggestions to explain the source of energy in quasars included the idea that matter and antimatter are annihilating each other there.

At present, the consensus is that quasars have giant black holes in their centers, large-scale versions of the maxi black holes that may well exist in the centers of our own and other galaxies. The black hole would contain millions or billions of times as much mass as the sun (Fig. 36–15). The event horizon of a black hole of a billion solar masses is roughly the size of our solar system. As mass falls into the black hole, energy is given off. (The gravitational potential energy of the mass is converted into

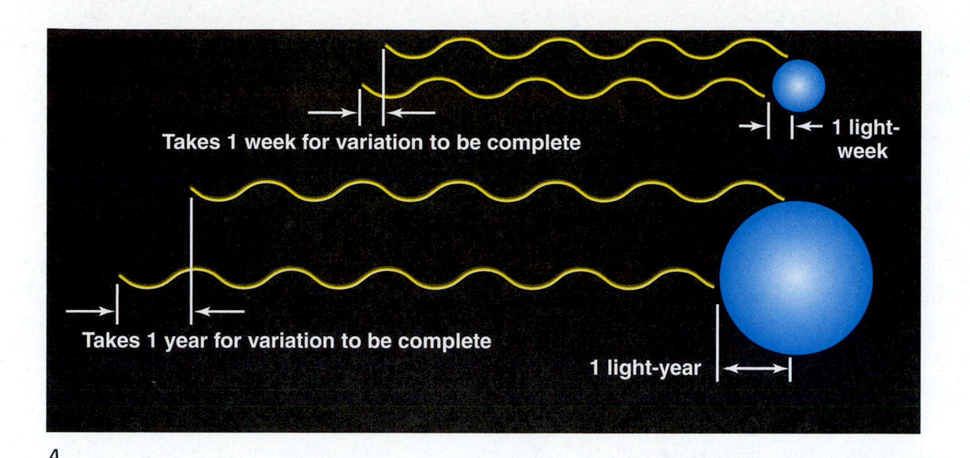


A

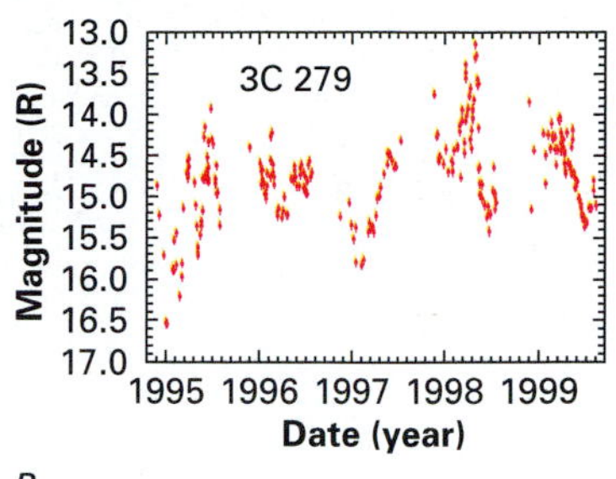


B

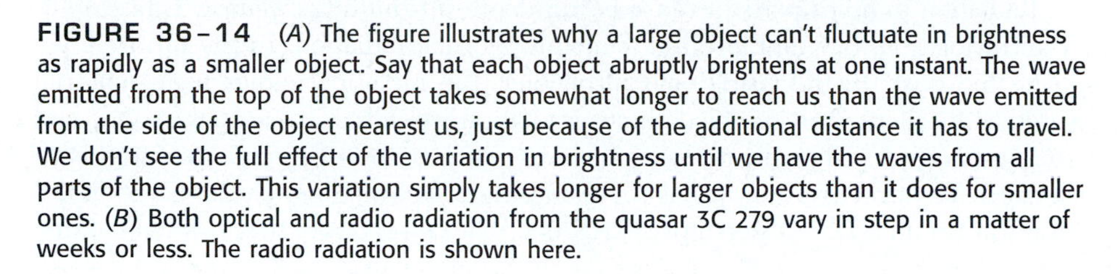

**FIGURE 36–14** (*A*) The figure illustrates why a large object can't fluctuate in brightness as rapidly as a smaller object. Say that each object abruptly brightens at one instant. The wave emitted from the top of the object takes somewhat longer to reach us than the wave emitted from the side of the object nearest us, just because of the additional distance it has to travel. We don't see the full effect of the variation in brightness until we have the waves from all parts of the object. This variation simply takes longer for larger objects than it does for smaller ones. (*B*) Both optical and radio radiation from the quasar 3C 279 vary in step in a matter of weeks or less. The radio radiation is shown here.

kinetic energy and radiation.) Black holes resulting from collapsed stars give off energy in a similar way, but on a much smaller scale. The central black hole in a quasar is known as "the engine" that produces the energy. The gas and dust around the engine is "the fuel." Together, the engine and the fuel are called "the machine."

Theoretical models show how matter falling into such a giant black hole would form an accretion disk. Data across the spectrum have supported this picture. The strength of ultraviolet radiation, for example, matches the predictions that the inner part of the accretion disk—where the friction of the gas is especially high—would reach 35,000 K. The matter in the accretion disk would eventually fall into the central black hole, so the process is called "feeding the monster." We will come back to the question of where quasars get their fuel.

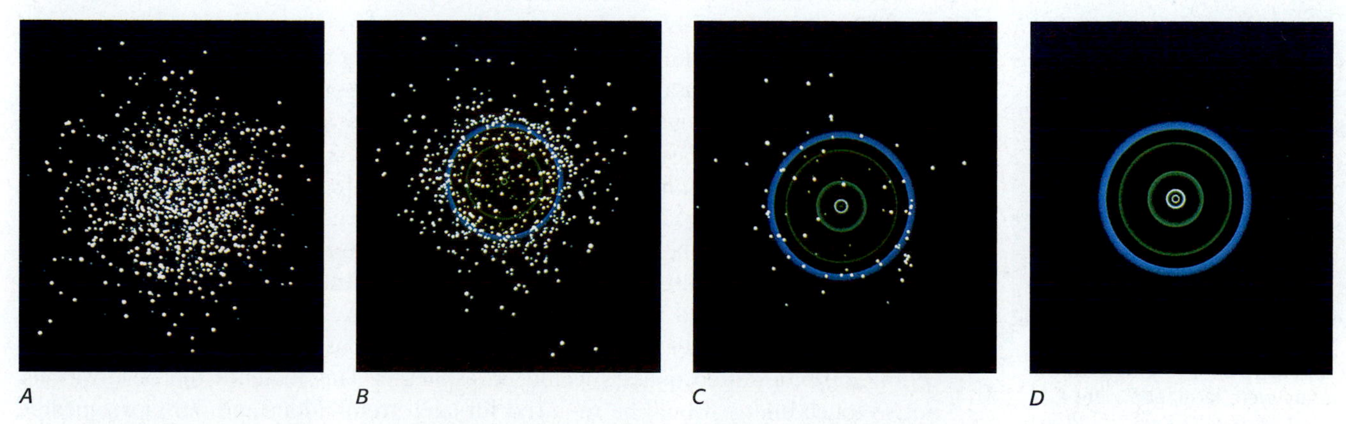

A B C D

**FIGURE 36–15** The formation of a quasar from a collapsing star cluster, as shown by supercomputer calculations based on Einstein's general theory of relativity. (*A*) The cluster, whose stars are moving at speeds close to the speed of light, is unstable and beginning to collapse. (*B*) During the collapse, a black hole forms at the center of the cluster. (*C*) The concentric circles represent spheres of light that had been sent out from the center of the cluster. (*D*) The light is trapped forever in the supermassive black hole that is at the center of the quasar. More and more of the mass in the central regions of the original cluster is consumed by the black hole.

The theory that black holes are at the centers of quasars fits with the observations that quasars have broad emission lines. These lines are so broad in wavelength because Doppler shifts of material going to and fro change the wavelengths at which we receive the emission lines. The motions that cause the Doppler shifts arise partly from infall of material toward the central black hole, partly from the material's rotation, and perhaps partly from the ejection of material.

Why do quasars have jets? Apparently, gas falling toward the supermassive central black hole is redirected by strong electromagnetic fields. The details of the process of jet formation are not understood.

Where does the radio emission of quasars come from? The electrons near the center of the quasar are accelerated to very high speeds, speeds near the speed of light. They spiral around lines of magnetic field, emitting synchrotron radiation, which we discussed in Section 33.3b. We verify that quasar radio waves are from this process by detecting their high polarization.

Radiation in other parts of the spectrum comes from other causes. Emission in the ultraviolet, visible, and infrared is mainly thermal radiation. In the infrared, we see heated dust. In the visible and ultraviolet, the radiation comes from material heated as it falls toward the black hole. Radiation we detect from quasars in x-rays and gamma rays is thought to come from the interaction of quasar radiation with electrons. The process is known as the Compton effect (and hence the name of the Compton Gamma Ray Observatory, NASA's Great Observatory that studied gamma rays).

## 36.5 THE ORIGIN OF THE REDSHIFTS

The most obvious explanation of quasar redshifts is the shift caused by the expansion of the universe, a "cosmological redshift" akin to a Doppler effect. But the distances implied are so large, and the energy problem at first seemed so difficult, that other explanations have been investigated very thoroughly.

### 36.5a Non-Doppler Methods

If the quasars are close, which would make the energy problem less serious (because their actual brightnesses would not have to be as great), we must think of some other way of accounting for their great redshifts. There are at least three other ways. They are all unattractive to most astronomers because they challenge Hubble's law and thus cast doubt on our knowledge of the whole scale of the universe. If Hubble's law were to fail for the quasars, we would never know when we could trust it.

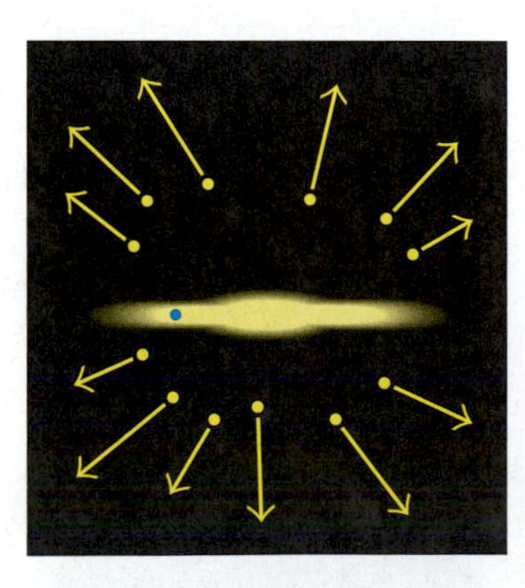

**FIGURE 36–16** The idea that quasars were local, and were ejected from our galaxy, could explain why they all have high redshifts. But a tremendous amount of energy would be necessary to make the ejection. So we might not find ourselves any closer to a solution of the energy problem—where all this energy comes from. The blue dot shows the position of the sun.

A first non-cosmological way also relies on velocity and assumes that quasars are close but are going away from us very rapidly. Quasars could be relatively close to us yet show high redshifts if they had been ejected from the center of a galaxy at very high velocities (Fig. 36–16). If such quasars occurred in another galaxy, we would expect to see some of them going away from us and some coming toward us. But all the quasars have redshifts; none of them has a blueshift, so this idea is unlikely.

If quasars exploded from the center of our galaxy or long enough ago from a nearby galaxy, all the quasars would have moved past us and would appear to be receding; no blueshifted quasars would be expected. This matches the observations, but so much energy would be required for such tremendous and perhaps repeated explosions that the energy problem would not be resolved.

A second non-cosmological possibility is the gravitational redshift predicted by Einstein's general theory of relativity and verified for the sun and for white dwarfs. But even before this method was ruled out by observations, scientists had not been able to work out such a system for quasars that would be free of other, unobserved consequences. And eventually, measurements were made of redshifts from the faint material that surrounds some quasars. The gravitational redshift theory predicts that the redshift would vary with distance from the quasar, but the observations contradict this. Scientists have thus discarded this disproved theory.

A third possibility is that quasars act on principles that we do not yet understand—some new kind of physics. But the basic philosophy of science forbids us from inventing new physical laws as long as we can satisfactorily use the existing ones to explain all our data. We discuss this idea at the beginning of the next chapter.

Now that we realize that giant black holes form a reasonable "engine," making the energy problem all but disappear, there is little need to invoke these non-cosmological methods for redshifts.

### 36.5b Evidence for the Doppler Effect

Some of the strongest evidence that the quasars are at their **cosmological distances** (their distances according to Hubble's law) instead of being **local** comes from a survey of the number of quasars. The quasars increase in number with distance from us in a manner that is difficult to account for on a local model but occurs naturally if quasars are at their distances according to Hubble's law.

Some of the strongest evidence that the redshift is a valid estimator of distance for quasars comes from a strange object named BL Lacertae. It was long thought to be merely a variable star (as the BL in its name shows), but later it was suspected of being stranger than that, partly because of its rapidly varying radio emission. Its spectrum showed only a featureless continuum, with no absorption or emission lines, so little could be learned about it.

Two Caltech astronomers had the idea of blocking out the bright central part of BL Lacertae with a disk held up in the focus of their telescope, thereby obtaining the spectrum of the haze of gas that seems to surround the apparent star. Its spectrum turned out to be a faint continuum with absorption lines typical of the stars in a galaxy. Thus BL Lacertae is apparently a galaxy with a bright central core. Further, the lines are redshifted by 7 per cent. Nobody doubts that a redshift of this amount arises from the expansion of the universe. Yet BL Lacertae, with its pointlike appearance and its rapidly varying brightness, appears almost like an extremely nearby quasar! About 100 similar objects (now called Lacertids or BL Lacs) have since been found to have redshifts of this order. Some, at least, appear to be embedded in nebulosity whose spectrum resembles that of an elliptical galaxy. Observations of a Lacertid with the Extreme Ultraviolet Explorer (EUVE) have revealed spectral features that have told us the composition and temperature of the gas. (Interestingly, emission lines appeared in BL Lac after its 1997 outburst, so at present BL Lac itself isn't even a Lacertid any more.)

The Lacertids appear to be a missing link between galaxies and quasars. The properties of quasars (assuming that they are at their Hubble's law distances)—such as the brightness of the core relative to that of the other gas that may be present—are more extreme than, but a continuation of, the properties of Lacertids. Since few doubt that the Lacertids are at their Hubble's law distances, the comparison with quasars strengthens the notion that quasars are also at their own Hubble's law distances.

Indeed, many astronomers now think that what we call some of these bright sources depends on the angle from which we view them. Both BL Lacs and quasars, in this model, are active galactic nuclei (AGNs), which we discussed in Section 34.7a. Together, they are known as "blazars." In the blazars, we are looking directly into the jets of gas they are emitting. When the jets are pointed away from us, the objects are ordinary radio galaxies.

Even before the launch of the Hubble Space Telescope, astronomers using ground-based telescopes succeeded in finding matter around some bright quasar images, and in taking the spectrum of that matter. The studies show that the quasars are embedded in host galaxies (Fig. 36–17) and are active phases in the lives of galaxies. We now know of many intermediate objects between ordinary galaxies and

Flare-ups of a BL Lacertae object in 1972, 1983, and late 1993 have led to the idea that the source contains a black hole of 20 million solar masses in an elliptical orbit around a black hole of 5 billion solar masses with a period of 9 years. When the objects are closest, material in orbit around each is stirred up; the 9-year period is stretched to 11.6 years by the expansion of the universe.

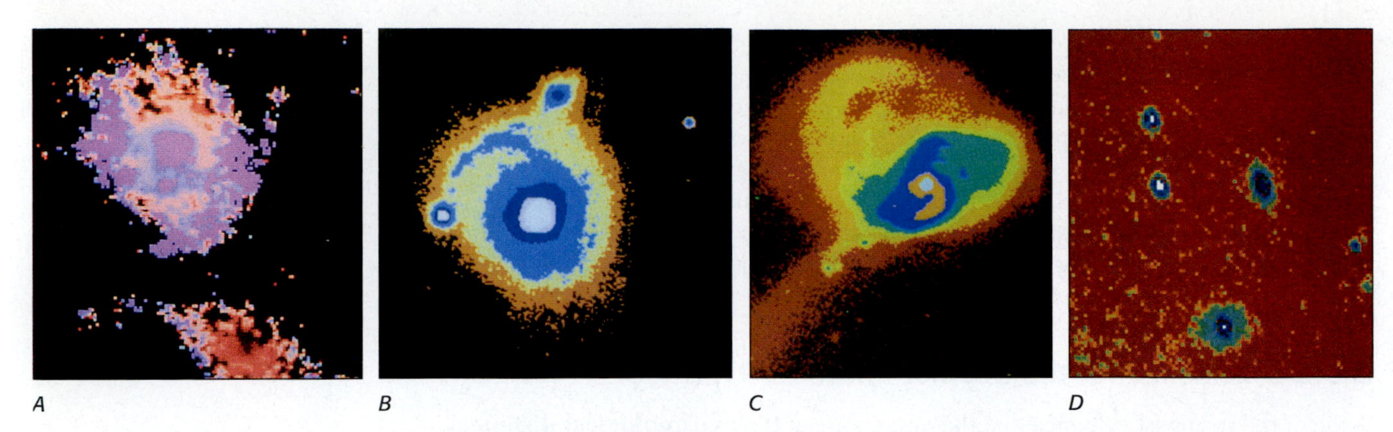


**FIGURE 36–17** (*A*) Quasar 0844+34 in its host galaxy. The system has a redshift of 0.06, which makes the nucleus either a very luminous Seyfert galaxy or a low-luminosity quasar. The companion is red and does not show much evidence of interaction, while the quasar host is blue, which indicates that star formation is going on. (*B*) The nearby quasar 0050+124 (also known as I Zw 1 from the Zwicky catalogue). The spiral arms of the host galaxy are clearly seen. (*C*) NGC 1614, another merging galaxy with infrared luminosity close to the luminosity of quasars. H$\alpha$ studies show that the outer curved loops are warped in the sky by tidal effects. (*D*) Note the fuzz around the quasar IE 2344+18, compared with the sharpness of the stellar images. All these views are ground-based.

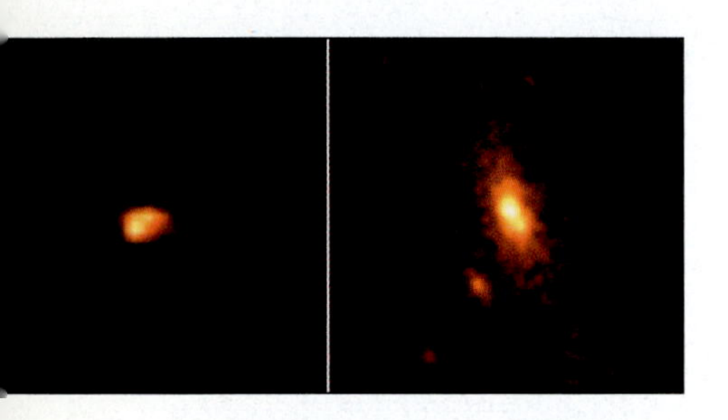

**FIGURE 36–18** A Type 2 quasar, a quasar in a tilted dust ring, viewed (*left*) from its x-rays by the Chandra X-ray Observatory and (*right*) from the Hubble Space Telescope. The comparison of the images reveals that the x-ray source is at the center of the optical galaxy. The observation that there are relatively few x-rays at low x-ray energy is consistent with the idea that these x-rays are being absorbed by a thick cloud of gas.

quasars. Thus quasars are now thought to be extreme cases of active galactic nuclei (AGNs).

Scientists using the Chandra X-ray Observatory have reported the discovery of a kind of quasar in which a surrounding ring of dust hides most of its radiation from us. Such a quasar is known as a Type 2 quasar (Fig. 36–18). Type 1 quasars, on the other hand, do not have their central black holes and accretion disks blocked from our view because they are tilted more favorably for us.

## 36.6 ASSOCIATIONS OF GALAXIES WITH QUASARS

A principal attack on the theory that quasars are at great distances from us has come from Halton Arp, now at the Max-Planck Institute for Astrophysics in Germany. On the basis of his observations, he concludes that quasars might be physically linked to galaxies, both the peculiar and the ordinary types. Arp contends that he has found many examples in which a galaxy and two or more quasars lie on a straight line, with at least one of the quasars on exactly the opposite side of the galaxy from another. He argues that the quasars and the galaxy, which have different redshifts, must therefore be linked. We know the distances to the galaxies from Hubble's law; if the quasars and galaxies are physically linked, then the quasars must be at the same distance from us as the galaxies. These distances are much smaller than the quasars' redshifts indicate.

If we were to find any cases for which Hubble distances cannot be trusted, then we could not rely on Hubble distances for quasars (or indeed for galaxies). One set of observations that has been used to argue such a case applies to a group of galaxies rather than to a mixture of galaxies and quasars. The redshift of one of the galaxies in Stephan's Quintet (Fig. 36–19), five apparently linked galaxies, is different from the redshift of the others. Such cases can be explained without rejecting Hubble's law if the object with a discordant redshift only accidentally appears in almost the same line of sight as do the other objects. Further, associations of a quasar with a distant group of galaxies with redshifts that do agree with each other have been found.

*A*

*B*

**FIGURE 36–19** Stephan's Quintet. Though NGC 7320 has a redshift much lower than that of the other galaxies, the Hubble Space Telescope image at right, with its WFPC 2 instrument, reveals that it really is closer to us by resolving individual stars in it. Part A is a ground-based photo used to label and orient the Hubble view shown in (*B*).

## A DEEPER DISCUSSION

### BOX 36.1 Stephan's Quintet

Stephan's Quintet was the first group of galaxies to be discovered. It was found by the French astronomer Edouard Stephan in 1877. That direction in the sky, in the constellation Pegasus, apparently shows a group of five galaxies. At first look, the individual galaxies seem to be about the same distance away from us, largely since they are in the same direction. But in 1961, Margaret Burbidge and Geoffrey Burbidge took spectra. Their spectra showed that four of the galaxies have about the same velocity of recession, about 6000 km/sec. That would put them, by Hubble's law, about 260 million light-years away. But one of the galaxies has a much lower redshift, only 800 km/sec. Is it closer, at its Hubble-law distance of 35 million light-years, or does Hubble's law hold for it? The stakes are high, because if Hubble's law doesn't always hold, we would never know when we could trust any measurements made with it.

The Hubble Space Telescope image of the Quintet reveals that the discordant galaxy, at lower left, is actually closer. We can make that conclusion because Hubble resolves individual stars in it but not in the other galaxies.

One of the other galaxies in the quintet, as well as a sixth galaxy off to the left in the ground-based image, turn out to be just passing through. Their somewhat different redshifts had long been noted, though they weren't very different. These galaxies brought energy into the system, which has prevented the system from collapsing. Bursts of star formation resulted from the galaxies' passage, and can still be seen. One galaxy is even in the process of spinning off a compact dwarf galaxy.

So Stephan's Quintet turns out really to be a gravitationally-bound trio of galaxies, with two more galaxies not too far away and with one foreground galaxy. And Hubble's law holds.

In some cases, it seems that a galaxy and a quasar with different redshifts are actually linked by a bridge of material. But this too can be a projection effect. Even in the "best" examples of galaxy-quasar bridges, different scientists do not agree whether the connection is real. The controversy will eventually be resolved with the Hubble Space Telescope or with active optics on ground-based telescopes.

In some cases, it seems that a galaxy and a quasar with different redshifts are actually linked by a bridge of material. But this too can be a projection effect. Even in the "best" examples of galaxy-quasar bridges, different scientists do not agree whether the connection is real. The controversy will eventually be resolved with the Hubble Space Telescope or with adaptive optics on ground-based telescopes.

### 36.6a Statistics and Non-Hubble Redshifts

Arp's data are subject to several objections, which are mainly statistical. Arp argues that the probability of finding quasars and galaxies so close together in the sky by chance is very small. But standard statistical methods are valid only before rather than after you know what is actually present. For example, if I flip a coin, show you that it has come up "heads," and then ask you what is the chance that it is "heads," you may answer "50 per cent." But the correct answer is 100 per cent. Once the deed is done, the odds are determined; after all, I showed you that the coin had come up heads. At a racetrack, you can't place your bets after the race is over. Similarly, once we look at a quasar and a galaxy that are apparently linked together, the odds are now 100 per cent that they are apparently linked. We can then no longer apply the simple statistical argument that the odds of finding two objects so close together by accident are so small that they must be physically linked.

The associations of galaxies with a quasar of a different redshift would be harder to explain as a projection effect if we found a cluster of galaxies in which not just one but two discrepant redshifts were found. No such cluster is known. So far, most astronomers do not accept Arp's arguments and believe the evidence shows that quasars are in fact very far away.

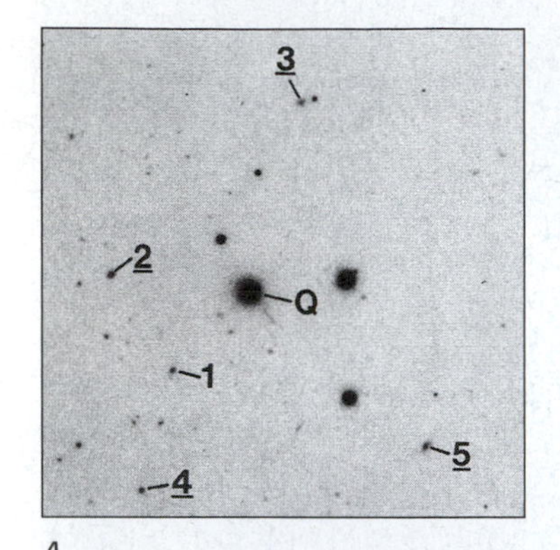

A

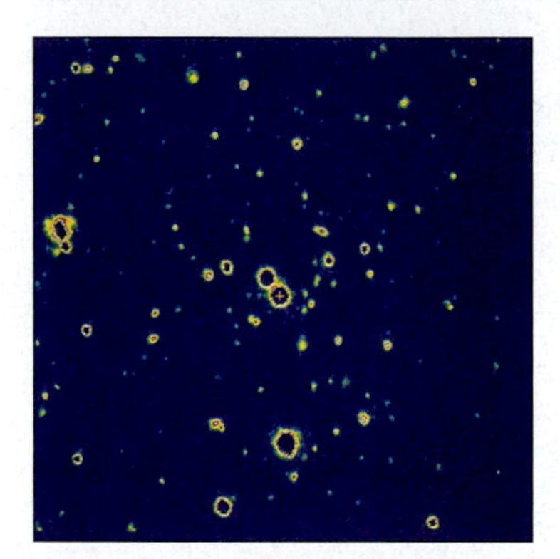
B

**FIGURE 36–20** (*A*) In this photograph taken by Alan Stockton on Mauna Kea, the central object is the famous quasar 3C 273, and the numbered objects are galaxies. The redshifts of the galaxies and the quasars agree. (*B*) This image of the binary quasar PKS 1145–071, taken with the first Keck Telescope in visible light, revealed a faint cluster of galaxies at the same large redshift.

Alan Stockton of the University of Hawaii carried out an important statistical test of the association of quasars with galaxies. He selected 27 relatively nearby quasars that were so luminous that the energy problem would be particularly severe. He carefully examined the regions around the quasars, and for 17 of the quasars found one or more galaxies apparently associated with them, as shown by their small angular separations.

For eight of the quasars, at least one of the galaxies nearby on the image had a redshift identical to that of the quasar (Fig. 36–20*A*). Since Stockton had chosen his sample of quasars before knowing what he would find, he was able to apply standard statistical tests to assess the probability that he could randomly find galaxies so apparently close in the sky to quasars. The probability of a projection effect making galaxies appear so close to a high fraction of quasars is less than one in a million. This indicates that some of the quasars and their associated galaxies are physically close together in space. Since the quasar and galaxy redshifts agree, the results strongly endorse the idea that the distances to quasars are those we derive from Hubble's law.

In several cases, a quasar was found in a cluster of galaxies (Fig. 36–20*B*). The quasars' redshifts agree with those of the galaxies. This conclusion has been endorsed by recent observations with the Very Large Telescope, finding objects nearby the quasars that are at the same redshift (Fig. 36–21).

### 36.6b Quasars and Galaxy Cores

In recent years, evidence has been mounting that quasars are extreme cases of galaxies rather than truly different phenomena. Some spiral galaxies have especially bright nuclei. A **Seyfert galaxy,** a type of active galaxy discovered by Carl Seyfert of the Mount Wilson Observatory in 1943, has a very bright nucleus indeed compared with its spiral arms (Fig. 36–22). Another type of galaxy, an **N galaxy,** also has an especially bright core. Can the quasars that we see be only the bright nuclei of galaxies? After all, although we can see both the nuclei and the spiral arms of nearby Seyferts, only the nuclei would be visible if such galaxies were very far away.

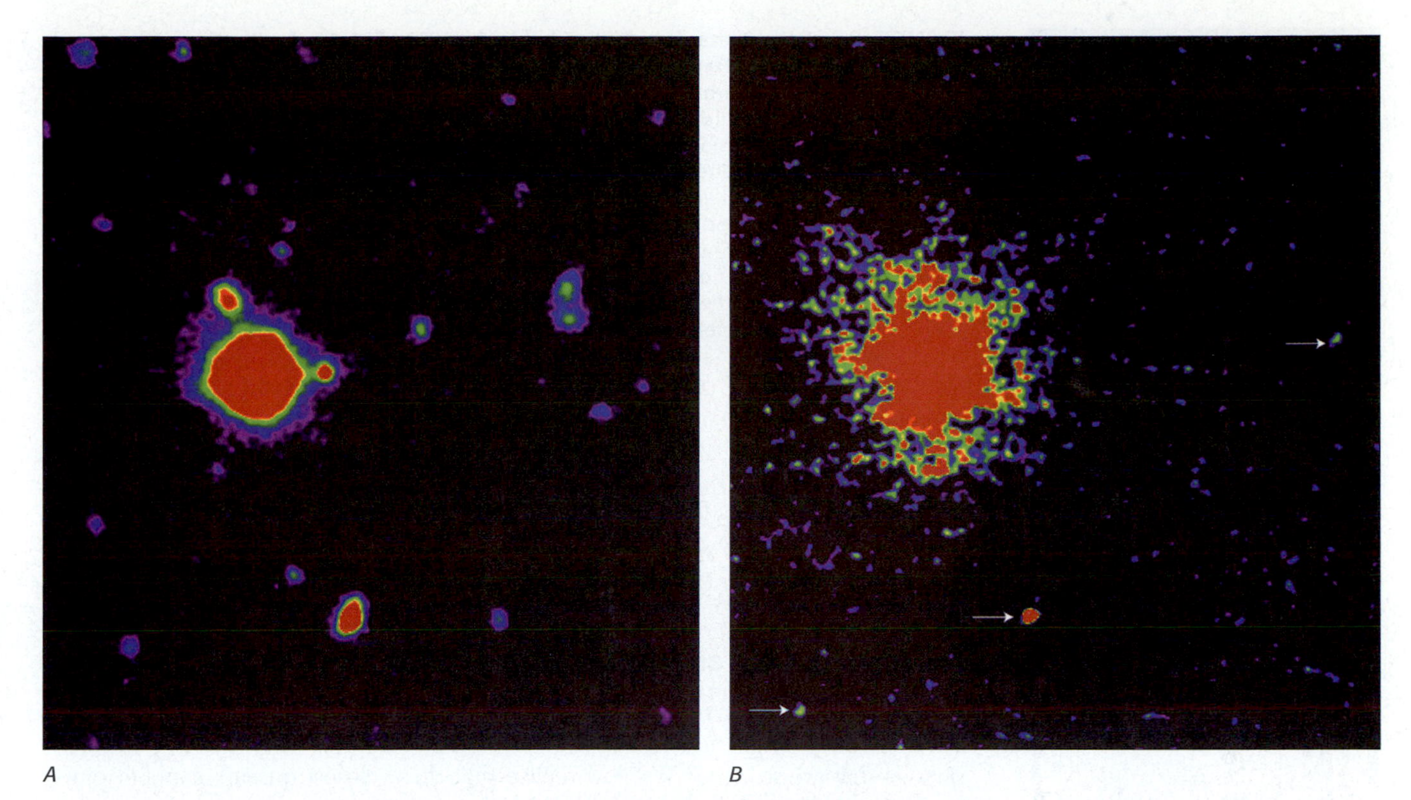

FIGURE 36–21 The world's largest telescopes, including the European Southern Observatory's Very Large Telescope (VLT), study quasars. We see observations of the radio-weak quasar J2233-606 in the Hubble Deep Field South. Its redshift $z$ is 2.2. (*A*) The quasar shown at a wavelength near to but not directly at that of redshifted Lyman-$\alpha$. We see tidal interaction with the nearby companion at its upper left. (*B*) The VLT view right at the Lyman-$\alpha$ wavelength. Not only does some hydrogen gas extending about 500,000 light-years around the quasar become visible but also some associated objects at the same redshift become visible from their Lyman-$\alpha$ radiation. These objects may be in the same cluster of galaxies as the quasar; they are about a million light-years away from it.

Seyfert galaxies not only have relatively bright nuclei but also have broad emission lines from various stages of ionization in their spectra, signs that hot gas is present. The exceptional breadth in wavelength of the lines could be caused by rapidly moving matter in the galaxies' cores, a sign of violent activity there. Two to five per cent of galaxies are Seyferts. Seyferts are quite bright in the infrared.

The fuzzy structure detectable around some quasars is technically known as **fuzz.** Its study has been revolutionized by the Hubble Space Telescope (Fig. 36–23). The results indicate that quasars are somehow related to the cores of galaxies. Perhaps galaxies go through a quasar stage during which their nuclei are very bright, or quasars are an extreme case of explosive events in galaxy cores.

The relation between quasars and the cores of active or peculiar galaxies has been consistently strengthened as we have been able to observe finer detail to greater distances. This linkage makes the quasars seem somewhat less strange, but at the same time it causes galaxies to seem more exotic. If we had known of BL Lacertae's redshift earlier on, quasars would not have looked as strange to us when they were discovered. And if collapsed objects like neutron stars and black holes had been in our minds when quasars were discovered, we would have had an obvious way to produce a lot of energy in a small space. In that case, we would not have said that "the energy problem" for quasars existed.

FIGURE 36–22 A Seyfert galaxy, NGC 7742, a galaxy with a very bright center.

An important link between quasars and galaxy cores has been discovered in the active galaxy Cygnus A, the second-brightest radio source in the sky (Section 34.7b). Though we can't see into the core of the optical object directly, given the band of

A

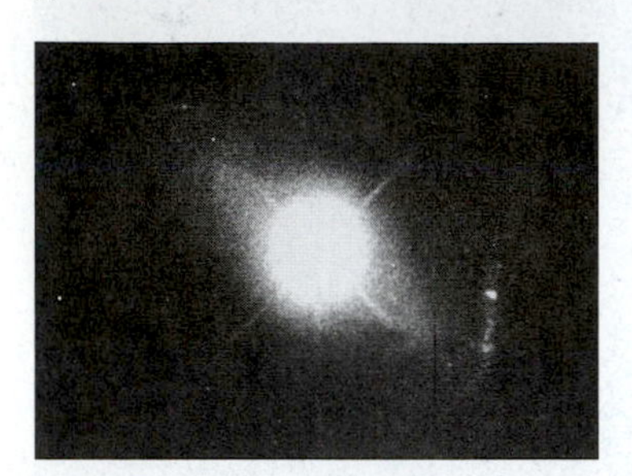

B

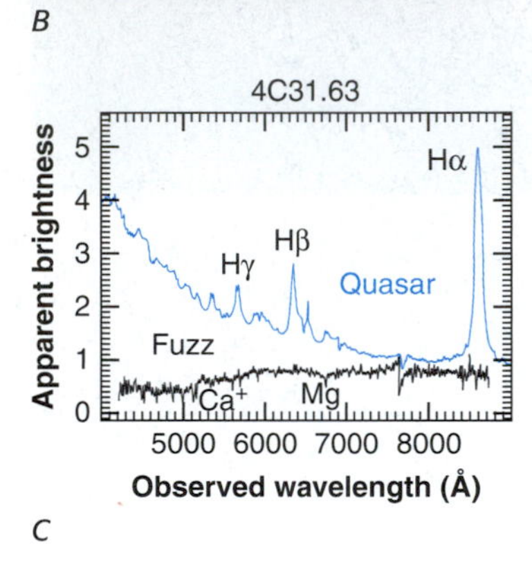


C

**FIGURE 36–23** Quasar fuzz, in the object QSO 1229+204. (*A*) A ground-based image made with the Canada–France–Hawaii Telescope on Mauna Kea. This image shows that the quasar is in the core of a barred spiral galaxy that is colliding with a dwarf galaxy. The resolution is about 0.5 arc sec. (*B*) The high resolution of the Hubble Space Telescope reveals structure in the fuzz. On one side of the galaxy we see a string of knots, which are probably massive young star clusters. They may have been formed as a result of the collision. The resolution is about 0.1 arc sec. (*C*) Keck telescope spectra of (*top*) the quasar 4C 31.63 ($z$ = 0.296) and (*below*) its "fuzz." The spectrum of the fuzz shows absorption lines typical of those that are produced by the relatively cool outer parts of normal stars. Thus the fuzz seen around nearby quasars is really starlight from galaxies surrounding the bright central object. In this double graph, the brightness of the quasar was reduced substantially to fit it on the same part of the vertical axis as the fuzz spectrum.

dust surrounding it, scientists using the Faint Object Spectrograph on the Hubble Space Telescope were able to study the ultraviolet light reflected off dust above and below the nucleus of the object. The spectrum reveals the typical emission lines of a quasar. The emission lines are very broad, probably because the gas in the nucleus is swirling at very high speeds. Maybe, as we saw, all the powerful radio galaxies harbor quasars hidden by dust because of the angle at which we are looking.

### 36.6c Quasars and Interacting Galaxies

Since almost all quasars are so far away, it seems that whatever provided the fuel to make them so bright must have soon been exhausted. So why can we see a few quasars that are so close? We now realize that these closest quasars may be rejuvenated, having received a new supply of fuel.

Some evidence that galaxies interact gravitationally has been known for a while. For example, we discussed (Section 34.4) the tails that have been explained as gravitational interactions. The occurrence of Seyfert galaxies, with their bright nuclei, is more common in galaxy pairs than in individual galaxies.

High-resolution studies of quasars have revealed that some apparent bumps on the quasar images are actually independent objects. In other cases, independent objects were already known from the work of Arp and others. These objects almost always have the same redshifts as the nearby quasars, and statistics now seem to show that the objects and the quasars are actually associated. These could be the objects that have been stripped of their outer layers to provide fuel for the quasar, rejuvenating the engine. Close encounters of galaxies with quasars in a way to provide the fuel for the black holes are rare, which could explain why nearby quasars are few in number. The members of one pair of quasars are so close to each other that they may be interacting. We may be seeing the exchange of gas that turns quasars on. Light has been travelling to us from these quasars for 12 billion years. Probably the collision rate was much higher 12 billion years ago, since the universe was less spread out, which would explain why most quasars have such high redshifts.

The quasars that are farthest away, those with the largest redshifts, are using up an original store of fuel. We see these quasars far enough back in time—to when rich clusters were forming. The quasars may have consumed some of the gas present in

**TABLE 36–1** Energies of Galaxies and Quasars

| | Relative Luminosity | | |
|---|---|---|---|
| | X-Ray | Optical | Radio |
| Milky Way | 1 | 1 | 1 |
| Radio galaxy | 100–5000 | 2 | 2000–2,000,000 |
| Seyfert galaxy/N galaxy | 300–70,000 | 2 | 20–2,000,000 |
| Quasar: 3C 273 | 2,500,000 | 250 | 6,000,000 |

these clusters. The closer quasars might be having a second youth, with new gas being introduced from encounters with other galaxies. Or, since it takes longer for the small clusters or groups of galaxies (where we see these quasars) to be stripped of their gas, these close quasars could be getting fuel for their black holes for the first time. Observationally, at least 30 per cent of quasars with redshifts up to 0.6—a fair way out into space—appear to be interacting, so the interaction model can be commonly applied.

According to this interaction model, Seyfert galaxies may simply have smaller black holes or less fuel than quasars. Starburst galaxies (galaxies with major amounts of star formation going on), Seyfert galaxies, and quasars may be part of an evolutionary sequence. Radio galaxies may be older quasars or objects that never received much fuel. (Some quasars resemble radio galaxies by also being double-lobed radio sources.)

## 36.7 QUASARS WITH THE HUBBLE SPACE TELESCOPE

Quasars and their surrounding fuzz and galaxies are being carefully studied with the Hubble Space Telescope. At first, observations seemed not to show galaxies associated with Hubble images of quasars; this fact seemed, to John Bahcall, "a giant step backward in our understanding of quasars." But as more Hubble quasar observations were made, the expected association between galaxies and quasars at the same redshift was found (Fig. 36–24). So Hubble views continue to endorse the idea that quasars are events that take place in the cores of galaxies, fed in at least some cases by interactions with nearby galaxies. Presumably, giant black holes are at the centers of the quasars.

To the surprise of many, Hubble images have shown that quasars can occur in the centers of both elliptical and spiral galaxies (Fig. 36–25). Only some of these host galaxies are interacting; others appear normal and single. So perhaps a variety of mechanisms turn on quasars. Radio-quiet quasars are found in elliptical or spiral galaxies. Radio-loud quasars are mostly found in elliptical or interacting galaxies.

We still don't know whether quasars are relatively short-lived stages in galaxies, perhaps lasting only 100 million years or less out of galaxy lifetimes that are 100 times longer. If so, our own Milky Way Galaxy could be a burned-out quasar. Or quasars

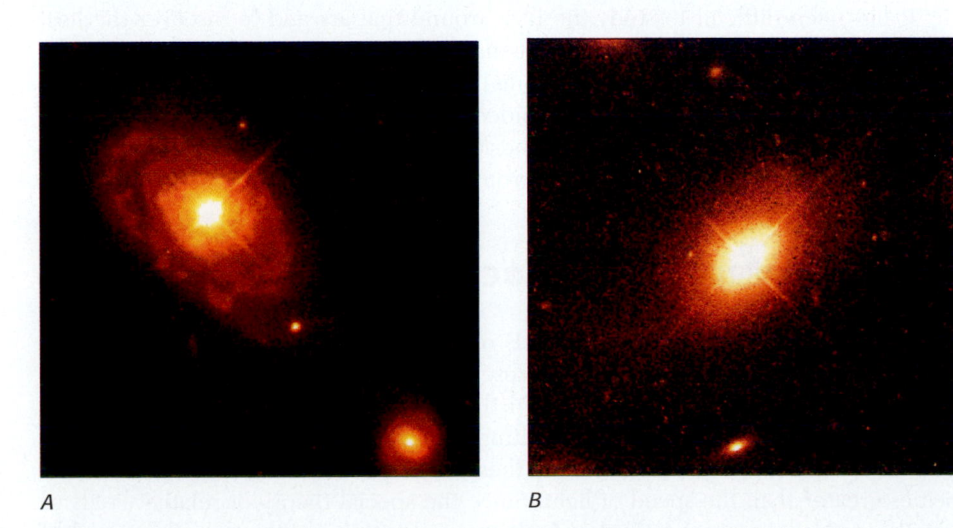

*A* *B*

**FIGURE 36–24** Quasars embedded in galaxies, as observed with the Hubble Space Telescope. (*A*) Quasar PG 0052+251, at redshift $z = 0.155$. It is in the center of an apparently normal spiral galaxy. (*B*) The quasar PHL 909, at redshift $z = 0.171$. It is at the center of an apparently normal elliptical galaxy.

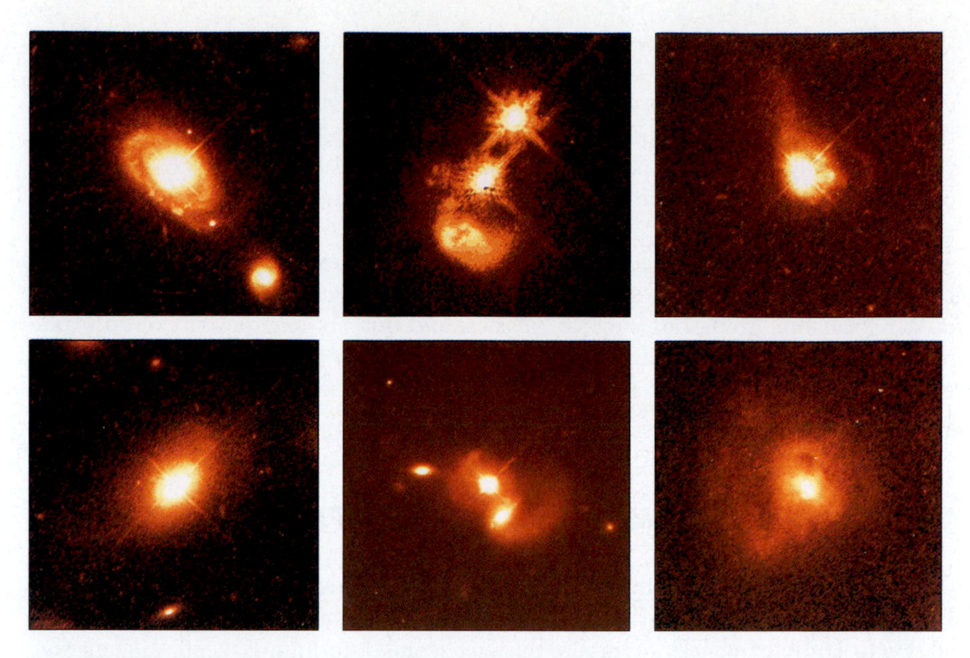

**FIGURE 36–25** Quasars in a variety of galaxy types, imaged with the Hubble Space Telescope. All these quasars are billions of light-years from us. Normal galaxies are in the left column; colliding galaxies are in the center column; and peculiar galaxies are in the right column. (*left top*) Quasar PG 0052+251 at the core of a normal spiral galaxy. (*bottom left*) Quasar PHL 909 at the core of a normal elliptical galaxy. (*top middle*) Debris from a collision between two galaxies fueling a quasar. A ring galaxy left by the collision (*bottom*) is fueling the quasar (*middle*), now 1/7 the diameter of our galaxy away from it. A foreground star is at top. (*bottom middle*) A quasar merging with the bright galaxy that appears just below it. The swirling wisps of dust and gas surrounding them indicate that an interaction is taking place. (*right top*) A tidal tail below a quasar, perhaps drawn out by a galaxy no longer there. (*right bottom*) A pair of merged galaxies have left loops of gas around this quasar.

could be long lived, which would mean that there were only a few very massive black holes around which they formed. Both the NICMOS experiment added to the Hubble Space Telescope in 1997 and the Advanced Camera scheduled for 2001 have coronagraph devices that can block out a quasar's glare to allow us to get good images close to the nucleus.

It has been difficult to study the fuzz around quasars and to discover the host galaxies because we are looking at faint material next to a very bright object, the quasar. An analogy is looking at a car at night in a snowstorm, and trying to see into the car even though you are all but blinded by the headlights. The Hubble Space Telescope at least turns on our own windshield wipers, wiping away the figurative snow near to us that had been impeding our view.

## 36.8 SUPERLUMINAL VELOCITIES*

Very-long-baseline interferometry (VLBI) observations with extremely high resolution (0.001 arc sec) have revealed the presence of a few small components in radio images of jets in a few of the quasars. Further, the observations have shown that in some cases these components are separating at angular velocities across the sky that seem to correspond (at the distances of these objects based on Hubble's law) to velocities greater than the speed of light. Since the special theory of relativity tells us that the apparent **superluminal velocities**—velocities greater than the speed of light—cannot be real, other theoretical explanations of the data have been sought.

The jets are presumably forced out of the quasar centers above the quasars' poles. The compression that formed the accretion disks would have made such high magnetic fields in the disks that particles couldn't cross them, leaving the poles as the best way out. The material ejected in this way presumably forms the lobes detectable with radio telescopes. The jets may be a larger-scale version of the jets of SS 433 (Section 30.9), a source in our galaxy. The calculated velocity at which some of the blobs in this source separate is $12c$ (12 times the speed of light), while other sources show velocities ranging up to $45c$.

A "Christmas tree" model, in which components are flashing on and off, had seemed possible at first. In this model, no rapid velocities need be implied, since the sources we see this year were not necessarily the same ones we saw last year. But years of observations (Fig. 36–26) showed that the components continue to separate from each other rather than flash on and off.

The current model involves the special theory of relativity, in that it depends on the fact that light travels at a finite speed. Picture a jet of gas that is moving rapidly almost directly toward us. The jet almost (but not quite) keeps up with the light it emits (Fig. 36–27).

Under these circumstances, the apparent separation of the objects is not the actual separation at any one instant of time. For certain angles of view, the apparent rate of separation can look much greater than the real rate of separation. For a correct understanding, we must keep careful account of exactly when light was emitted.

The "superluminal motions" just might give an important key to the understanding of quasars. The model we have just discussed requires that the knot of gas is moving almost right at us (within about 10°), which may seem statistically unlikely. But theory predicts that because of their motion toward us, the radiation from the rapidly moving components may appear to us to be concentrated in beams. We might detect only those quasars that are beaming radiation toward us (which would also explain why we sometimes see only one jet). If quasars concentrate energy in beams this way, then their total energy emission would not be as high as we had calculated on the assumption that they were emitting radiation in all directions at an equal rate. This effect would make quasar luminosities similar to luminosities of galaxies. But current thinking is that only a few of the quasars are beaming radiation at us, thus enhancing our measurements of their power.

The apparent superluminal expansion has been detected in several quasars and galaxies, and so it may be a fairly common phenomenon. There presumably would be a second jet of gas oriented away from us, to make the situation symmetric, but this jet would be moving away so fast it would be redshifted too much for us to detect it.

## 36.9 QUASARS AT SHORT WAVELENGTHS*

The Earth's atmosphere prevents the ultraviolet and x-ray region of the spectrum from reaching us, but our ability to launch telescopes in spacecraft has limited this handicap. Since many of the spectral lines we observe in the visible spectra of quasars were originally emitted by the quasar in the ultraviolet and have since been redshifted into the visible, it is particularly important to study the ultraviolet spectra of relatively bright nearby quasars. We use these ultraviolet spectra to understand the spectra of the more distant objects that we don't see as well because they are so faint, even though we can observe interesting parts of the spectrum in the visible.

Studying nearby quasars in the ultraviolet is one of the "key projects" for the Hubble Space Telescope. Both spectra and images are being obtained. Many of the absorption lines observed in the ultraviolet are the Lyman-alpha line of hydrogen at different wavelengths (the Lyman-alpha forest). Some of the Hubble images are in the "Snapshot Survey"; these are relatively quick views taken whenever there is a gap in other observing programs and thus without taking the time for the fine-guidance sensors to acquire the objects.

The sensitivity of the Einstein Observatory enabled astronomers to observe dozens of x-ray emitting quasars. ROSAT found still more quasars at x-ray and other

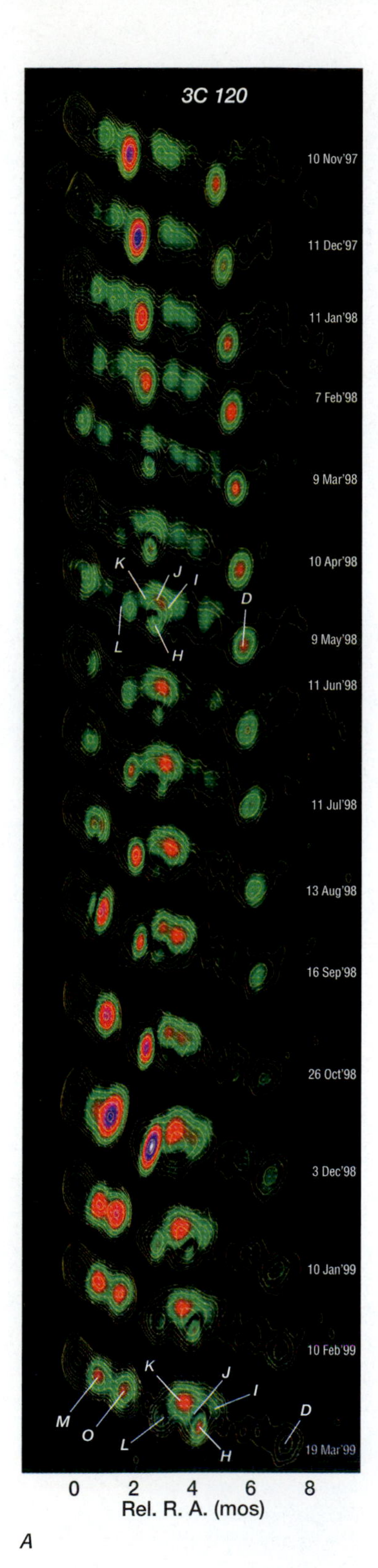




**FIGURE 36–26** (*A*) Superluminal velocities in the radio galaxy 3C 120 reach 6*c* relative to the core. See the animation of these VLBA images at http://www.sciencemag.org/feature/data/1052657.shl. (*B*) A series of views of the quasar 3C 279 with radio interferometry at a wavelength of 1.3 cm (22 GHz). The apparent velocity translates (for $H_0$ = 75 km/sec/Mpc) to a velocity of 8*c*, though the apparent superluminal velocities can be explained in conventional terms.

short wavelengths, and Chandra and XMM-Newton are doing better yet. The Compton Gamma Ray Observatory observed quasars at still shorter wavelengths, gamma-ray quasars (Fig. 36–28).

ROSAT has picked up an x-ray background; all or most of that background was generally thought to come from unresolved quasars, that is, quasars too far away or too close together to be detectable as individual sources. Chandra proved that this idea was correct. Its high resolution revealed that the apparent haze of x-rays detected with earlier spacecraft actually comes from blurred out individual objects. The discovery is considered one of Chandra's major triumphs.

## 36.10 DOUBLE QUASARS*

The astronomical world was agog in 1979 at the discovery of a pair of quasars so close to each other that they might be two images of a single object. The story began when inspection of the Palomar Sky Survey charts revealed a close pair of 17th-magnitude objects at the position corresponding to a radio source. When the spectra of the objects were taken, both objects turned out to be quasars, and their spectra looked identical. Even stranger, their redshifts were essentially identical.

It seemed improbable that two independent quasars should be so similar in both spectral lines and redshifts, so the scientists who took the first spectra suggested that both images showed the same object! They suggested that a gravitational bending of the quasar's radiation was taking place. We have discussed such gravitational bending as a consequence of Einstein's general theory of relativity and have seen that it has been verified for the sun (Section 31.1); in 1987, it was also used to explain galactic arcs (Section 34.6). For the quasars, a massive object between the quasar and us is acting as a gravitational lens, bending the radiation from the quasar one way on one side and the other way on the other side. As a result, we see the image of the quasar in at least two places. Sometimes, we can detect the galaxy that is doing the lensing (Fig. 36–29). In the double-quasar case, the lensing object seems to be a giant elliptical galaxy with a redshift about one-fourth that of the quasar. Theoretical studies have calculated from the observed separation of objects that the lensing galaxy must contain at least $10^{12}$ solar masses, which is possible for that

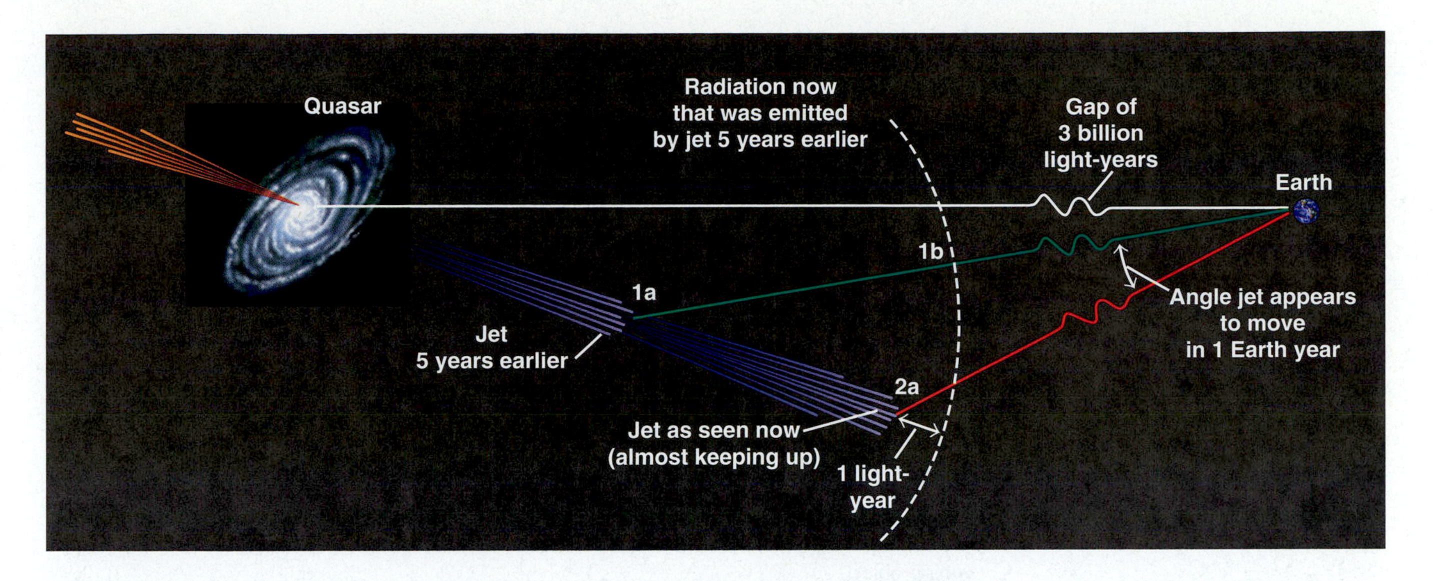


**FIGURE 36–27** The leading model for explaining how a jet of gas emitted from a quasar, shown at left with its accretion disk surrounding a central black hole, can seem to be travelling at greater than the speed of light when seen from Earth (*shown at right*). 1a marks the position of the jet 5 years earlier than the jet we are seeing now. (It takes another 3 billion years for the radiation from 3C 273 to reach us, an extra duration that we can ignore for the purposes of this example.) After an interval of 5 years, the light emitted from point 1a has reached point 1b, while the jet itself has reached point 2a. The jet has been moving so fast that points 2a and 1b are separated by only 1 light-year in this example. The light emitted then from the jet at point 2a is thus only 1 year behind the light emitted 5 years earlier at point 1a. So billions of years later, with a 1-year interval we receive light from two positions that the jet has taken 5 years to go between. The jet actually therefore had 5 years to move across the sky to make the angular change in position (that is, proper motion) we see over a 1-year interval. We thus think that the jet is moving faster than it actually is.

type of galaxy, though the "dark matter" that helps explain the radio images probably also contributes.

Detailed radio maps of the region have been made with several interferometers, most notably with the VLA (Fig. 36–30). The two images detected optically

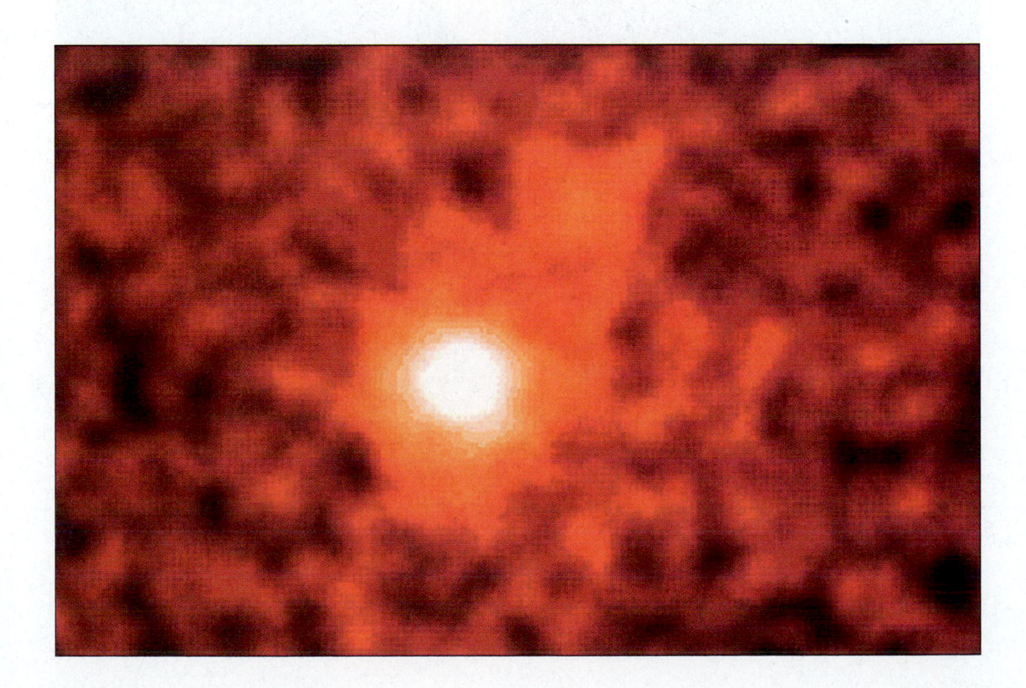

**FIGURE 36–28** The quasar 3C 279 appeared as the bright spot in this image made with the Compton Gamma Ray Observatory, but then faded below gamma-ray visibility. The quasar 3C 273 is the faint image above and to the right of the brighter image. Recall that gamma-ray directionality is fuzzy. Both these quasars are in the constellation Virgo but are separated by about 10°.

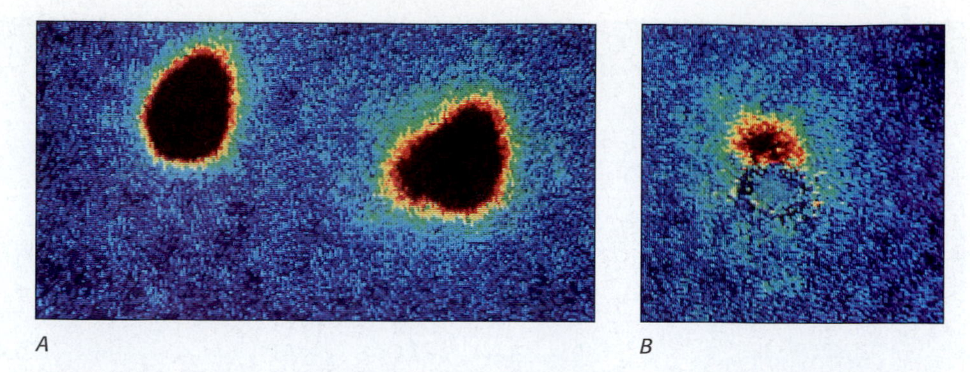


**FIGURE 36–29** This series of optical false-color views of the double quasar, 0957+561 A and B, reveals the intervening galaxy. (*A*) The actual images. The 17th-magnitude quasar images are bluish and are separated by only 5.7 arc sec. Their redshifts are identical to the third decimal place: $z = 1.4136 \pm 0.0015$. The intervening galaxy is only 1 second of arc from one of the quasar images. Its redshift is about 0.37. The objects are in the constellation Ursa Major. (*B*) The left image has been subtracted from the right image to reveal (*bottom*) just the intervening object.

coincide with the two sharpest radio peaks. All the bright regions observable on the radio map cannot be explained—perhaps "dark matter" (Section 38.2a) does the additional lensing. Sometimes observing in the infrared allows us to see the lensing object better (Fig. 36–31).

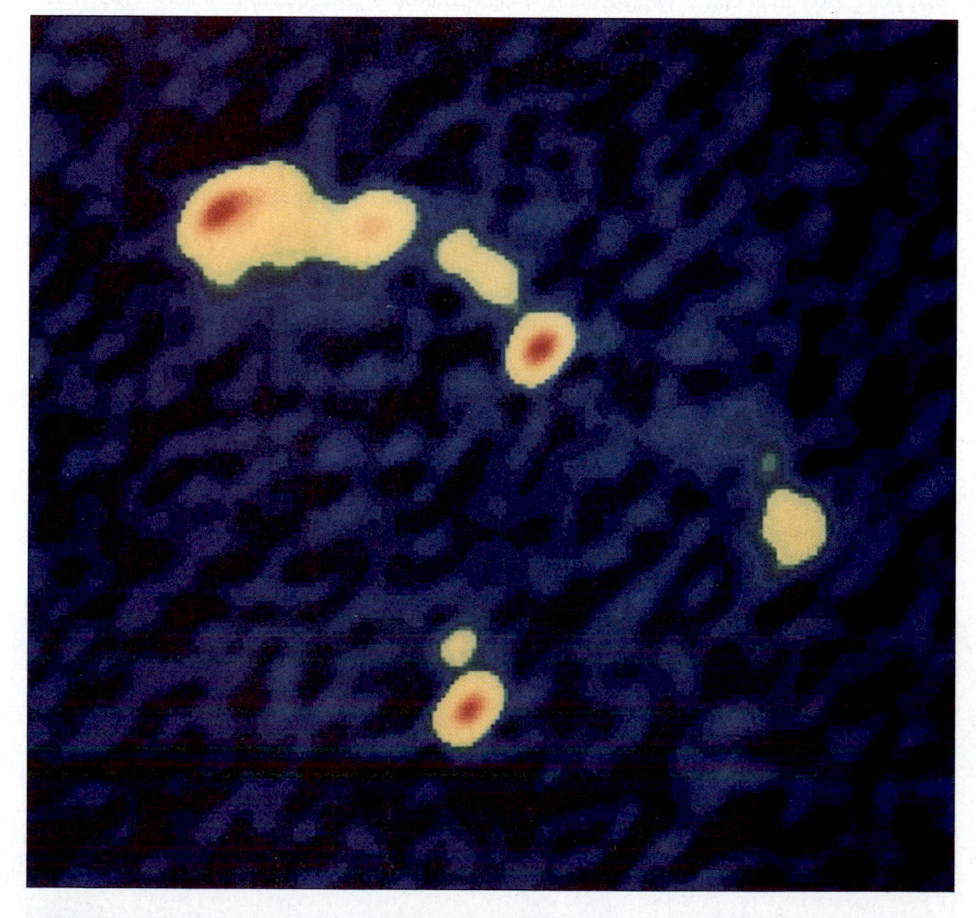

**FIGURE 36–30** A radio map of the double quasar, Q 0957+561 A and B, made with the VLA. The elliptical sources correspond to the optical objects. Additional images are also seen, which could result from an off-center gravitational lens.

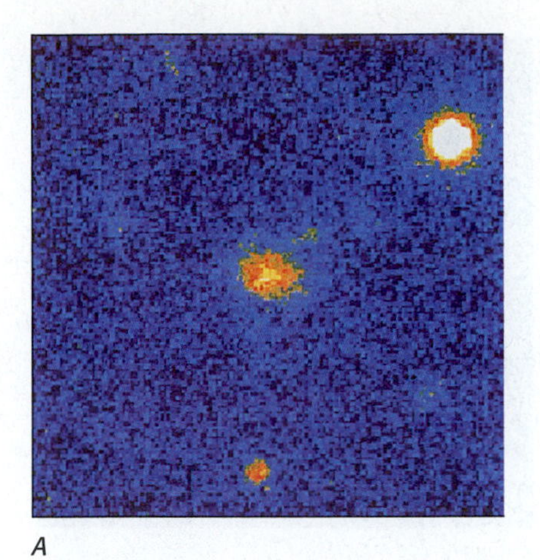

A

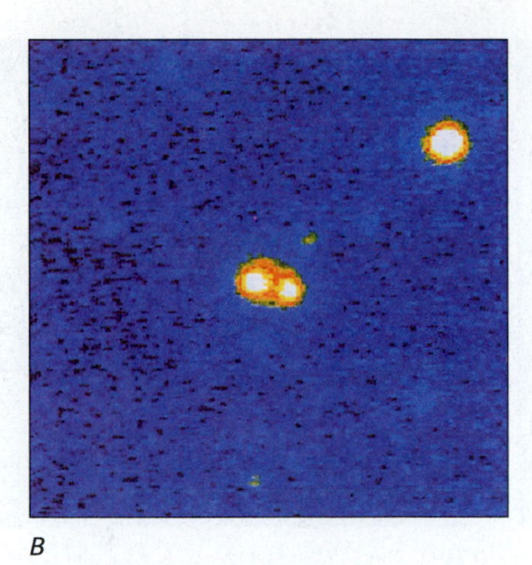

B

**FIGURE 36–31** Gravity lensing, imaged with one of the 10-m Keck Telescopes in Hawaii. (*A*) At the near-infrared wavelength of 1.2 $\mu$m, we see the lensing galaxy. (*B*) At a longer infrared wavelength of 2.2 $\mu$m, the background quasar is visible. Its light has been bent around the foreground lensing galaxy, so we see a double image. This system is known as MG 1131+0456.

One test of whether a gravitational lens is present is to see if the brightness fluctuations of the two optical images are similar. Since the light from the two images travels along different paths through space, though, the fluctuations could be displaced in time. There are indications that the delay in fluctuations from one member of the double quasar 0957+561 to the other is 415 days. Further observation and interpretation should give the mass of the lensing galaxy. Comparison of the brightness expected for this mass with the observed brightness may eventually give us an independent measure of the Hubble constant. In another object, an x-ray flare-up was detected in one member of a set of lensed images (Fig. 36–32), giving hope

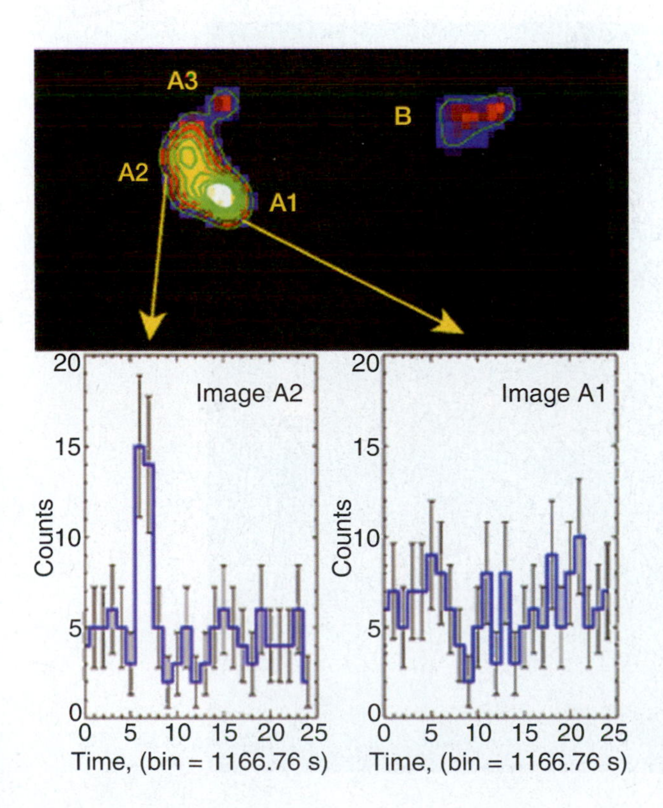


**FIGURE 36–32** Of the four lensed images of a single, distant quasar, RXJ 0911.4+0551, one showed an abrupt x-ray flare over a period of less than an hour. We see this brightening in Chandra observations. If another of the images brightens, we could have the desired time delay from which to calculate the quasar's distance and hence Hubble's constant.

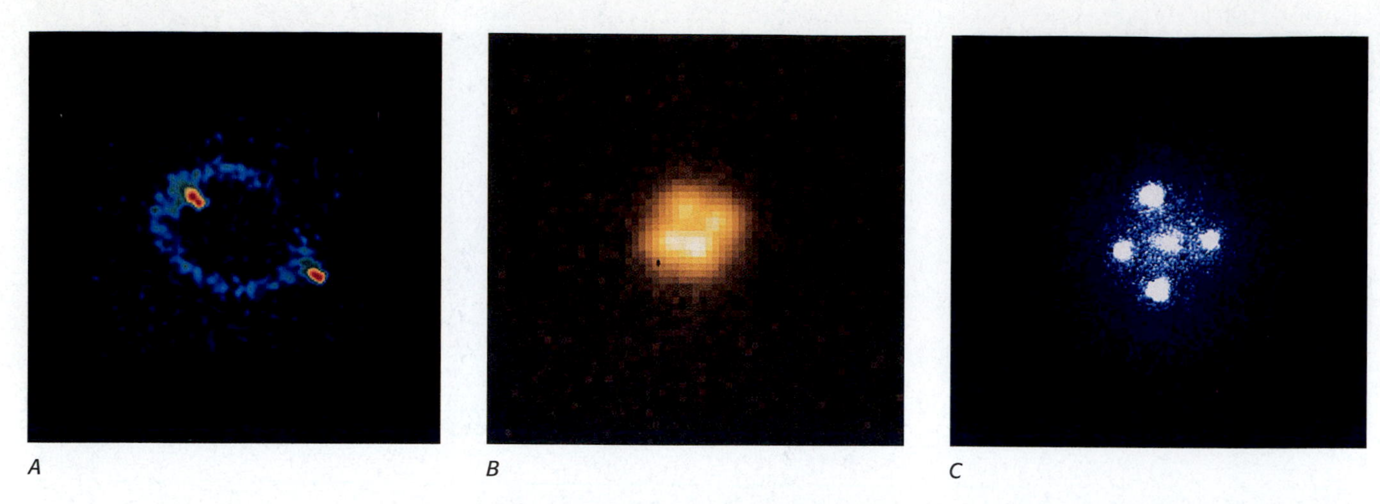


**FIGURE 36–33** (*A*) A quasar was so exactly aligned with the intervening gravitational lens that its radiation was spread out into an "Einstein ring." A second object is less well-aligned; its image appears slightly inside the ring on one side and slightly outside the ring on the other on this VLA image. (*B*) The "clover leaf," the quadruply lensed quasar H1413+117. The four images of comparable brightness are only 1 arc sec apart, and are barely resolved in this ground-based photograph. (*C*) Hubble Space Telescope image of the "Einstein Cross" (Q2203+10305), a quasar ($z$ = 1.70) that is gravitationally lensed into four images by a relatively nearby galaxy ($z$ = 0.039). The nucleus of the galaxy is visible in the center of the image. The four images of comparable brightness are separated by less than 2 seconds of arc and have essentially identical spectra. The rare configuration and identical spectra show that we are indeed seeing gravitational lensing rather than a cluster of quasars.

that such flare-ups will soon be detected in two or more members. Such a detection would allow the measurement of the time delay, and thus the determination of the distance to the quasar. This would be a powerful direct measurement independent of Hubble's law.

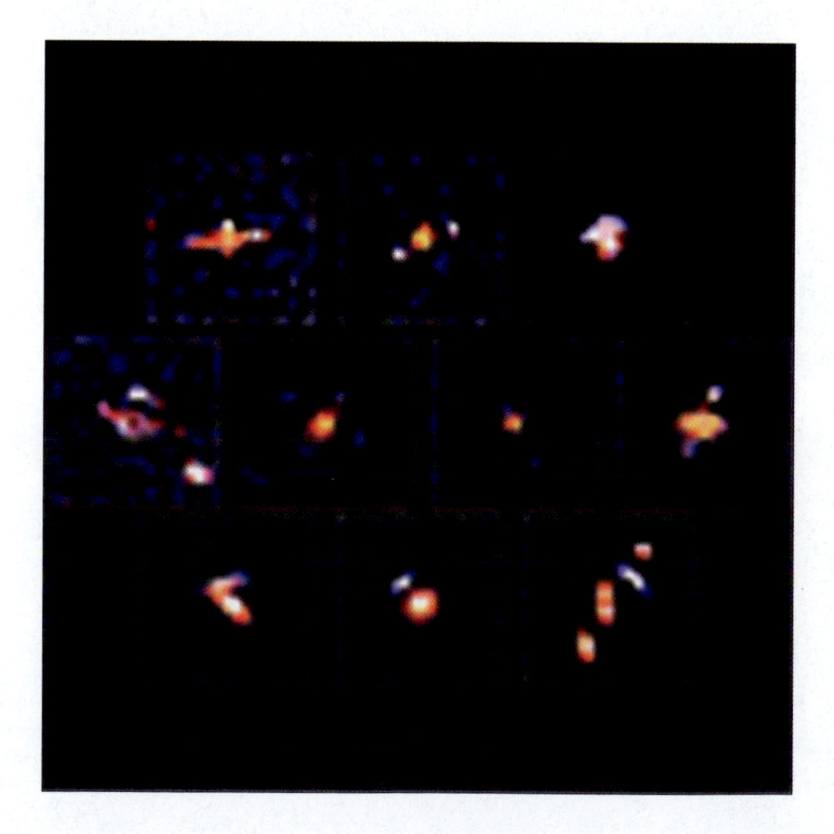

**FIGURE 36–34** A variety of gravitational lens systems, imaged with the Hubble Space Telescope.

The double quasar was the first known example of a gravitational lens. Over a dozen other multiple quasars have since been discovered (Fig. 36–33).

Of the 500 quasars imaged by the Hubble Space Telescope in its Snapshot Survey, only a half dozen new gravitational lenses showed up, in line with the predictions of certain cosmological models. These objects and the galactic arcs are exciting and valuable verification of a prediction of Einstein's general theory of relativity.

We can get a number of extra images because the lensing object is not a point and because the object being lensed is not exactly in line with it and us. For a point lens and an exactly aligned object, we can get an image that is a ring centered on the lensing object. Such a case is known as an "Einstein ring," and a very few are known. The discovery of an Einstein ring or of a quadruply lensed system would be so improbable without the lensing explanation that the discovery of gravitational lenses is generally accepted (Fig. 36–34).

## CORRECTING MISCONCEPTIONS

| ✖ *Incorrect* | ✔ *Correct* |
|---|---|
| Galaxies are the farthest things in the universe. | Certain quasars are among the farthest things. |
| Quasars are farther than galaxies. | We now know of some galaxies as far as many quasars. |
| Quasars are like pulsars. | Quasars are events in the centers of galaxies; pulsars are stellar. |
| Sloan and 2dF map only the positions of objects in the sky. | Sloan and 2dF measure redshifts, giving distances, in addition to positions. |
| Objects at the same distance can have different redshifts. | Essentially all astronomers think that Hubble's law holds for both quasars and galaxies. |
| Apparent velocities of parts of quasars indicate motions faster than the speed of light, violating Einstein's special theory of relativity. | The apparent superluminal velocities can be explained as projection effects in matter moving close to but less than the speed of light. |

## SUMMARY AND OUTLINE

Discovery of quasars (Section 36.1)
- QSR—quasi-stellar radio source
- Huge Doppler shifts in spectra
- Radio-quiet objects also included as quasars
- Sloan and 2dF are mapping tens of thousands of quasars and their redshifts.

Doppler shift in quasars (Section 36.2)
- Large redshifts mean large velocities of recession and, by Hubble's law, great distances.
- Relativistic formula used for redshifts close to or greater than 1

Importance of quasars (Section 36.3)
- Test of Hubble's law, which is thus a test of the accuracy of the distance scale
- Great distance, which means we view most of them as they were in an early stage of the universe

Energy problem (Section 36.4)
- Too much energy required from a small volume to be accounted for by ordinary processes
- Generally accepted: giant black hole is the engine.

Non-Doppler explanations of redshifts (Section 36.5)
- Quasars ejected from cores of galaxies at high velocities; no evidence of blueshifted quasars, however.
- Gravitational redshifts; conflict with observations
- Quasars operating under new physical laws?
- All these alternative explanations challenge Hubble's law, but they are not now needed since quasars can be understood as being powered by black holes.
- BL Lacertae and nearby quasars show that quasars are indeed at their Hubble-law distances.

Galaxies and quasars (Section 36.6)
- Question of physical link between galaxies and quasars

Stephan's Quintet and other sources with discrepant redshifts might be explained as chance alignments.
Other observations and statistical analysis show that quasars are indeed at their cosmological distances.
Seyfert and N galaxies also have bright nuclei; BL Lacertae and other nearby objects known to have distances corresponding to Hubble's law have properties in common with quasars.
Quasars seem to be events in cores of galaxies at some evolutionary stage; many quasars have surrounding fuzz that may be the underlying galaxies; the fuel may come from interactions with other galaxies.
Hubble Space Telescope observations find interactions.

Quasars imaged with Hubble (Section 36.7)
Quasars reside in all types of galaxy hosts.

Superluminal velocities of quasars (Section 36.8)
Components detected in several quasars seem to be moving apart at speeds greater than the speed of light.
Current explanation involves special theory of relativity and jets of gas pointing almost at us moving at velocities close to but less than the speed of light.

Short-wavelength observations of quasars (Section 36.9)
X-ray and gamma-ray observations of quasars

Gravitational lenses (Section 36.10)
Two apparent quasars with identical properties located very close together are apparently two images of one quasar formed by a gravitational lens.
Over a dozen gravitational lenses now known
Gravitational lenses are explained with the general theory of relativity as the result of a warping of space.

## KEY WORDS

quasar
Lyman-alpha forest
energy problem
cosmological distances
local
Seyfert galaxy
N galaxy
fuzz
superluminal velocities*

*This term appears in an optional section.

## QUESTIONS

1. Why is it useful to find the optical objects that correspond in position with radio sources?
†2. (a) You are heading toward a red traffic light so fast that it appears green. How fast are you going? (b) At $1 per mph over the speed limit of 55 mph, what would your fine be in court?
3. A quasar is receding at 1/10 the speed of light. (a) If its distance is given by the Hubble relation, how far away is it? (b) At what wavelength does the 21-cm line appear?
†4. A quasar has $z = 0.3$. What is its velocity of recession in km/sec, using the non-relativistic formula?
†5. We observe a quasar with a spectral line whose rest wavelength is 3000 Å, but that is observed at 4000 Å. (a) How fast is the quasar receding, using the nonrelativistic formula? (b) How far away is it if its distance is given by the Hubble relation? Specify the value you are using for Hubble's constant.
†6. (a) Consider a quasar at a redshift of 2. Calculate the wavelength at which the Lyman-$\alpha$ line, whose wavelength is 1216 Å from a source at rest, appears. (b) One of the farthest known quasars has $z = 5.8$. Show how you calculate the wavelength at which Lyman-$\alpha$ appears.
†7. A quasar is receding with a velocity 85 per cent of the speed of light. At what wavelength would a spectral line appear if it appears at 5000 Å when we observe it emitted from a gas in a laboratory on Earth? What part of the spectrum is it in when it is emitted, and in what part of the spectrum do we observe it from the quasar?
†8. A quasar is receding with a velocity 95 per cent of the speed of light. At what wavelength would the Lyman-$\alpha$ spectral line appear? (It is emitted at 1216 Å.) What part of the spectrum is it in when it is emitted, and in what part of the spectrum do we observe it from the quasar?
†9. A quasar has $z = 2$. At what velocity is it receding from us, and what is its distance from us?
10. What do quasi-stellar radio sources and radio-quiet quasars have in common?
11. Why does the rapid time variation in some quasars make "the energy problem" even more difficult to solve?
12. What are three differences between quasars and pulsars?
13. Briefly list the objections to each of the following "local" explanations of quasars: (a) They are local objects flying around at large velocities. (b) The redshift is gravitational.

†This question requires a numerical solution.

†7. 17,500 Å; emitted in visible (blue); observed in infrared.

14. If quasars were proved to be local objects, would this help solve the "energy problem"? Explain.
15. Explain how parts of a quasar could appear to be moving at greater than the speed of light, without violating the special theory of relativity.
16. Describe the implications of the observations of quasars in the x-ray spectrum. What new ability allowed these observations to be made?
17. What features of some quasars suggest that quasars may be closely related to galaxies?
18. Describe two ways in which Hubble Space Telescope observations of quasars are giving us new insights.
19. Explain how the clover-leaf quasar image arises.
20. How would the presence or absence of dark matter affect quasar lensing?

## USING TECHNOLOGY

### W World Wide Web

1. Find images of quasars from the Hubble Space Telescope.
2. Check on images of quasars from large ground-based telescopes around the world.
3. Report on the latest of the quasar surveys from the Sloan Digital Sky Survey (http://www.sdss.org) and the 2dF quasar survey (http://www.2dfquasar.org).

### REDSHIFT

1. Find the quasar 3C 273. When is it visible from your location today? How many times fainter than the faintest visible star is it?
2. Locate two dozen other quasars and BL Lacertae objects in *RedShift*'s sky. For the former, experiment with different values of the Deep Sky limiting magnitude.
3. For a dozen quasars from the previous question, compare the quasars' brightness and their redshifts, and comment on whether they follow Hubble's law.

The equivalent of the opposite of Olbers's paradox: Why is the sky dark at night? If we look far enough in any direction, we should see the surface of a star. Why then isn't the universe uniformly bright? This painting by the Belgian surrealist René Magritte shows the opposite, since in the painting our line of sight does not uniformly stop on trees in the forest or on the horse, but rather goes, impossibly, through them.

# Chapter 37

# Cosmology

In Armagh, Ireland, in the mid-seventeenth century, Bishop Ussher calculated that the universe was created at midday on Sunday, October 23rd, in the year 4004 B.C. Nowadays we are less certain of the date of our origin (though we do set out a split-second agenda for the first few minutes of time).

We study the origin of the universe as part of the study of the universe as a whole, **cosmology.** Even more than in other parts of astronomy, in order to study cosmological problems we use simultaneously both theoretical calculations and all our abilities to observe a wide variety of celestial objects.

The study of where we have come from and of what the universe is like leads us to consider where we are going. Is the universe now in its infancy, in its prime of life, or in its old age? Will it die? It is difficult for us who, after all, spend a lot of time thinking of topics like "What shall I watch on TV tonight?" or "What's for dinner?" to realize that we can think seriously about the structure of space around us. It is awesome to realize that we can conclude what the future of the universe will be. One must take a little time every day, as Alice was told when she was in Wonderland, to think of "impossible things." By and by, we become accustomed to concepts that may seem overwhelming at first. You must sit back and ponder when studying cosmology; only in time will many of the ideas that we shall discuss take shape and form in your mind.

**AIMS:** To study the origin of the universe, to learn the evidence that it follows our ordinary geometry, to study the evidence that matter makes up only a small fraction of the universe, and to understand how the background radiation tells us about one of the universe's early stages

## 37.1 OLBERS'S PARADOX

Many of the deepest questions of cosmology can be very simply phrased. Why is the sky dark at night? Analysis of this simple observational question leads to profound conclusions about the universe.

We can easily see that the night sky is basically dark, with light from stars and planets scattered about on a dark background. But a bit of analysis shows that if stars are distributed uniformly in space, then the sky shouldn't be dark anywhere. If we look in any direction at all we will eventually see a star (Fig. 37–1), so the sky should appear uniformly bright. We can make the analogy to our standing in a forest. There, we would see some trees that are closer to us and some trees that are farther away. But if the forest is big enough, our line of vision will always eventually stop at the surface of a tree. If all the trees were painted white, we would see a white expanse all around us (Fig. 37–2). The white on the trees that are farther away is of the same brightness as the white on the trees that are closer. Similarly, when looking up at the night sky we would expect the sky to have the uniform brightness of the surface of a star.

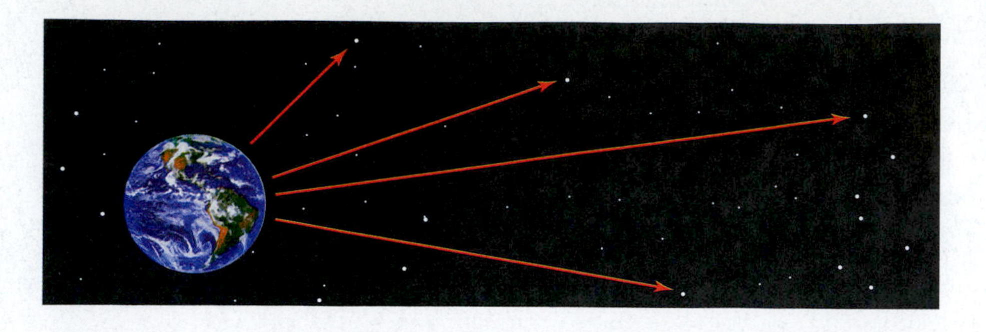

**FIGURE 37–1** If we look far enough in any direction in an infinite universe, our line of sight should hit the surface of a star. This leads to Olbers's paradox.

The name "Olbers" has an "s" on its end before we make it possessive, so we write of "Olbers's paradox" (or, alternatively, "Olbers' paradox"); "Olber's paradox," with an apostrophe before a sole "s," is incorrect, since his name wasn't "Olber."

The fact that this argument implies that the sky *is* uniformly bright, while observation shows that the sky is dark at night, is called **Olbers's paradox.** (A paradox occurs when you reach two contradictory conclusions, both apparently correct.) Wilhelm Olbers phrased it in 1823, although the question had been discussed at least a hundred years earlier. Solving Olbers's paradox leads us into considering the basic structure of the universe. The solutions could not have been advanced in Olbers's time because they depend on more recent astronomical discoveries.

We have phrased Olbers's paradox in terms of stars, though we know that the stars are actually grouped into galaxies. But we can carry on the same argument with galaxies and deduce that we must see the average surface brightness of galaxies everywhere. This is patently not what we see.

One might think that the easiest way out of this paradox, as was realized in Olbers's time, is simply to say that some material (we now know of interstellar dust) is absorbing the light from the distant stars and galaxies. But this doesn't solve the problem: because the sky is generally dark in all directions, the dust would have to be everywhere. This widely distributed dust would soon absorb so much energy that it would heat up and begin glowing. Given a long enough time, all the matter in the universe would begin glowing with the same brightness, and we would have our paradox all over again.

**FIGURE 37–2** The lower halves of trees are sometimes painted white, providing an example of a similar phenomenon to seeing a uniform expanse of starlight. Note that the surface brightness of a tree is the same whether the tree is close to us or far away.

One solution to Olbers's paradox lies in part in the existence of the redshift, and thus in the expansion of the universe (Section 35.2). This does not mean that the answer to the paradox is simply that visible light from distant galaxies is redshifted out of the visible, for at the same time ultraviolet light is continually being redshifted into the visible. The point is rather that each quantum of light, each photon, undergoes a real diminution of energy as it is redshifted. The energy emitted at the surface of a faraway star or galaxy is diminished by this redshift effect before it reaches us.

But the redshift is not the whole solution of Olbers's paradox. Indeed, calculations indicate that it would dim the background by only a factor of two. As we look out into space, we are looking back in time, because the light we see has taken a finite amount of time to travel to us. If we could see out far enough, we would possibly see back to a time before the stars were formed. E. R. Harrison of the University of Massachusetts has pointed out that this is the real way out of the dilemma. Harrison calculates that this explanation is a more important contribution to the solution of the paradox than is the existence of the redshift. In most directions, it can be calculated that we would not expect to see the surface of a star for $10^{24}$ light-years. We would thus have to be able to see stars $10^{24}$ years back in time for the sky to appear uniformly bright. However, the stars don't burn that long, and the universe is simply not that old. On the basis of Hubble's law, an age of somewhat over $10^{10}$ years seems to be the maximum. Not enough energy has been given off in the age of the universe for the sky to be uniformly bright. Even the cosmic background radiation that we shall discuss later in this chapter does not contain enough energy to affect this conclusion.

The fact that we have to know about the expansion and the age of the universe to answer Olbers's question—why is the sky dark at night?—shows how the most straightforward questions in astronomy can lead to important conclusions. In this case, we find out about the expansion of the universe or about the lifetimes of the universe and the stars in it.

**FIGURE 37–3** The central part of the cluster of galaxies that lies far beyond the constellation Fornax. When we look far out in all directions, we see clusters of galaxies. The question is whether we can hope to look far enough out that we find isotropy.

## 37.2 THE BIG-BANG THEORY

Where and when did the universe begin? Even this question is not too large for us to study. To get the overall picture, we need to make some simplifications. Astronomers looking out into space have assumed that on a sufficiently large scale the universe looks about the same in all directions (Fig. 37–3). That is, ignoring the presence of local effects such as our being in the plane of a particular galaxy and thus seeing a Milky Way across the sky, the universe has no direction that is special. Further, it is generally assumed that there is no change with distance either, except insofar as time and distance are linked.

These notions have been codified as the **cosmological principle:** *the universe is homogeneous and isotropic throughout space*. The assumption of **homogeneity** says that the distribution of matter doesn't vary with position (that is, with distance from the sun), and the assumption of **isotropy** (pronounced eye-sot'roh-pee) says that the universe looks about the same no matter in which direction we look. Actually, of course, only in certain directions can we see out of our galaxy without having our vision ended by the interstellar dust, but remember that we are ignoring inhomogeneities or lack of isotropy on this small scale. The discovery of giant voids and large structures like the Great Wall and the possibility of the Great Attractor (Section 35.7) are evidence that the universe may not, in fact, be homogeneous and isotropic even on a large scale. The large-scale mapping extending far into space by the Sloan Digital Sky Survey, 2dF, and related projects are showing us the structure of the universe on the largest possible scales and giving us information about how much clustering of individual objects goes on.

The explanations of the universe that essentially all astronomers now accept are known as the **big-bang theory.** Basically, this version of cosmology holds that once upon a time there was a great big bang that began the universe. From that instant on, the universe expanded, and as the galaxies formed they shared in the expansion. The big-bang theories satisfy the cosmological principle.

*Sky & Telescope* magazine had a contest in the 1990s to get a more dignified or otherwise better name than "big bang," but the original "big bang" name came out the winner. The term "big bang" was originally assigned derisively by Fred Hoyle, but it stuck and has become honorable. In 2001, one of the questions on television's *The Weakest Link* asked for the name of the theory of the universe suggested by Fred Hoyle. The contestant got it right! So you might read your textbook carefully in case you find yourself on a quiz show.

Many students ask whether the fact that there was a big bang means that there was a center of the universe from which everything expanded. The answer is no. First of all, the big bang may have been the creation of space itself. Furthermore, the matter of this primordial cosmic egg was everywhere at once. There may be an infinite amount of matter in the universe, so it is possible that at the big bang an infinite amount of matter was compressed to an infinite density while taking up all space. We would have to imagine our expanding raisin cake (Figure 35–4) extending infinitely in all directions, with no edge.

Consider a two-dimensional analogy to a universal expansion: the surface of a rubber balloon covered with polka dots (especially a non-expanding kind of polka dot, such as paper dots stuck on with adhesive). Though the balloon is three-dimensional, its surface has only two dimensions, and we consider only its surface.

Let us consider the view if we are sitting on one dot. As the balloon is blown up, all the other dots seem to recede from us. No matter which dot we are on, all the other dots seem to recede. The *surface* of the balloon is two-dimensional yet has no edge. Even if we go infinitely far in any direction, we never reach a boundary. (Remember, Columbus did not fall off the edge of the Earth.) There is no center to the surface from which all dots are actually expanding. It is much more difficult for most of us to visualize a three-dimensional situation like an infinite raisin cake (or, including time as a fourth dimension, a four-dimensional space-time). Yet the above analogy is valid, because our universe can expand uniformly yet have no center to the expansion.

Remember that the galaxies themselves are not expanding; the stars in a galaxy don't tend to move away from each other. It is space that is expanding, enlarging the distances among galaxies.

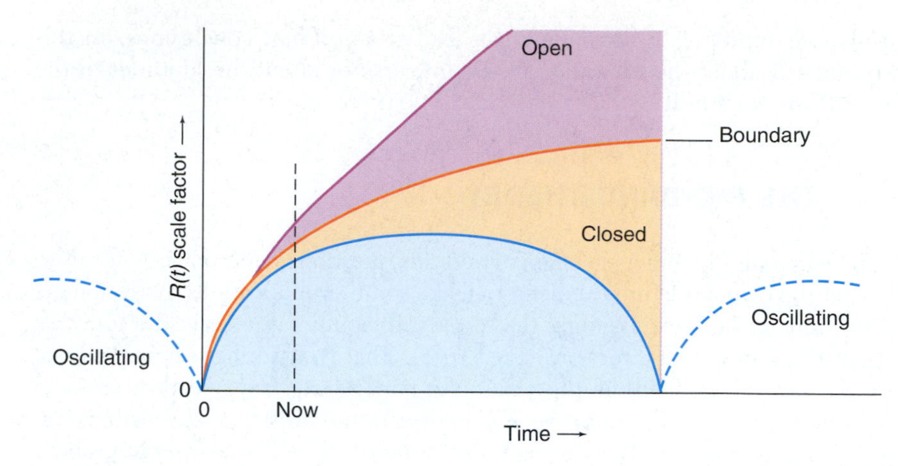


**FIGURE 37–4** To trace back the growth of our universe, we would like to know the rate at which its rate of expansion is changing. Big-bang models of the universe are shown; the vertical axis represents a scale factor, $R$, that represents some measure of distances and how they change as a function of time, $t$. The universe could be open and expand forever, or be closed and begin to contract again. If it is closed, we do not know whether it would oscillate or whether we are in the only cycle of expansion and contraction it will ever undergo. The boundary is a flat universe. We will see later that current evidence favors the flat model.

What will happen in the future? One possibility is that the universe will continue to expand forever. This case is called an **open universe.** It corresponds to the case in which the universe is infinite. The opposite possibility is that at some time in the future the universe will stop expanding and will begin to contract. This case is called a **closed universe** (Fig. 37–4), and it corresponds to the case in which the universe is finite (though it may still have no boundary, just as we can continue straight ahead forever on the surface of a balloon). In a closed universe, after the universe collapses enough we would eventually reach a situation that we might call a "big crunch," when we go back to high temperatures and high densities. The situation dividing open and closed is known as a **flat universe.** A flat universe forever expands at a slowing rate that would make it cease expanding only after an infinite time.

The open and closed universe can be considered in an analogy from geometry (Fig. 37–5). Einstein's general theory of relativity predicts that space itself could show properties of being curved, just as the surface of a saddle or the surface of a sphere

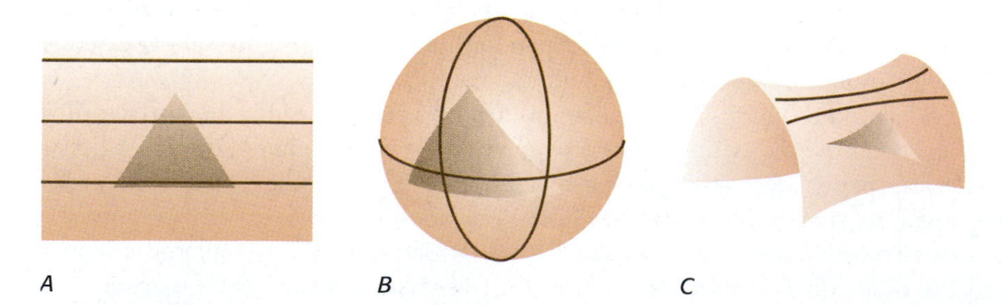


**FIGURE 37–5** Two-dimensional analogues to three-dimensional or four-dimensional space. (*A*) A flat universe is like an infinite sheet of paper (though in more dimensions); the laws of Euclidean geometry are obeyed. For example, the sum of the interior angles of any triangle is equal to 180°. In a flat, Euclidian geometry, one and only one line parallel to another line can be drawn through a point not on that other line. (*B*) A positively curved universe is like the surface of a sphere; the sum of the interior angles of any triangle is larger than 180°. In a positively curved space, no lines can be drawn through a point that are parallel to another line. (*C*) A negatively curved universe is like the surface of a saddle or potato chip in some respects; the sum of the interior angles of any triangle is smaller than 180°. In a negatively curved space, many lines parallel to another line can be drawn through a point.

A DEEPER DISCUSSION

## BOX 37.1 Einstein's Principle of Equivalence

Our notion of the structure of the universe is based on calculations using the general theory of relativity Albert Einstein advanced in 1916. At the basis of the theory is the idea Einstein formulated showing that gravity and accelerated motion are fundamentally equivalent, a notion known as the **principle of equivalence.**

Einstein explained the idea with a "thought experiment," though we can now have real examples for the situations (Figure A). Consider people in a closed elevator, "Einstein's elevator," who cannot see out. On the Earth's surface, they feel a downward pull caused by the Earth's gravity. But even if they were out in space in a location where there is no gravity, there would still be an identical force in one direction as the wall or floor pushes on them if their spacecraft accelerated forward in the opposite direction at the proper rate. Einstein's principle of equivalence holds that there is no way that people in the elevator, without contact with the outside, could tell whether such a pull came from gravity or from an acceleration.

In the vicinity of a massive object, there would be a force toward the object; the force would be in a different direction depending on where we were located with respect to the massive object. We could, following the principle of equivalence, consider ourselves to be in a curved space accelerating down a slope toward the massive object instead of being subject to the object's gravity. Thus we can explain gravity as a curvature of space. The observational tests of general relativity we considered in Section 31.1 are actually tests of only the principle of equivalence.

### *Discussion Question*

If you were standing on a scale on an elevator, how would your weight change when the elevator started to move up? Down? How would gravity have to change to account for these apparent scale readings, if you were not in an elevator?

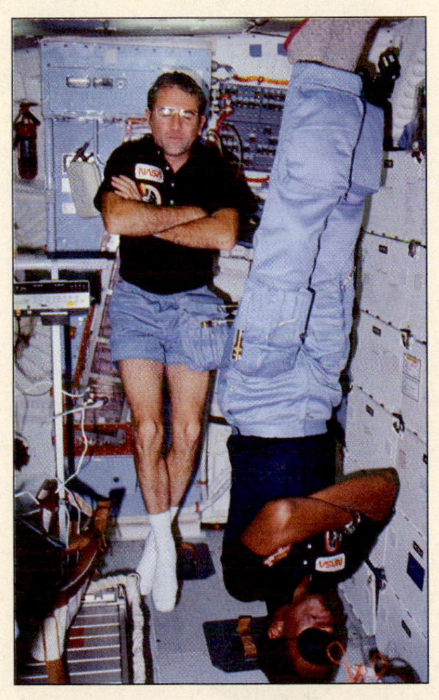

*A*

*B*

In "free fall," the spacecraft is accelerating toward Earth with the same acceleration that the astronauts have; they thus do not sense any gravity, since they are not pressed against anything. (*A*) Space-shuttle astronauts Richard H. Truly (*left*) and Guy Bluford (*right*) resting. (*B*) The team of astronauts floating inside the space shuttle, supporting a space lab mission, included Mark Lee, Robert Gibson, Jan Davis, Curtis Brown, Mamoru Mohri, and Mae Jemision (not shown).

is fundamentally curved. On neither of these curved surfaces, for example, can we draw straight lines, through a point near another line, that remain parallel out to infinity. That is, the two lines never cross and always remain the same distance from each other. Is space in the universe curved positively like a sphere, curved negatively

> If we assume that space is flat, the formula for the velocity is
>
> $$v = 2c\left(1 - \frac{1}{\sqrt{1+z}}\right).$$
>
> For $H_0 = 75$, reasonable models for closed, flat, and open universes give ages of about 13 billion years for distant quasars at $z$ of about 6. Closed universes give ages a few percent shorter and open universes give ages a few percent greater.
>
> Once we derive $v$, we can use Hubble's law, $v = H_0 d$, to find the distance to the object today, including the space created while the light travelled to us.

like a saddle, or flat? The closed universe would be like the surface of a sphere, which bends inward, while the open universe would be like the surface of a saddle, which bends outward. Evidence we will learn about in the next chapter strongly gives the answer that the universe is flat. That is, it follows the laws of geometry first laid down by the Greek mathematician Euclid in the third century B.C.

A closed universe is like the surface of a sphere in several ways, including the fact that "great circles" (lines circling a sphere that pass through opposite ends of diameters of the sphere) always intersect and cannot be parallel. We say that this universe has "positive curvature." An open universe is like the surface of a saddle in several ways, including the fact that an infinite number of lines parallel to a given line can be drawn through the same point. We say that this universe has "negative curvature." A flat universe is in between. A flat universe is considered "Euclidian," while curved universes are "non-Euclidian."

In principle, one could decide whether the universe is open or closed by counting the number of radio sources or of other galaxies in shells of constant thickness going outward from our galaxy, since the different geometries imply different changes in the number of sources (Fig. 37–6). But the fact that radio sources and other galaxies may have been brighter or fainter far back in time complicates matters too much.

**FIGURE 37–6** *(top row)* Consider dots uniformly distributed on each two-dimensional surface. If we count the total number of dots within circles of larger and larger radius $r$ in flat space, we find that the total number of dots grows exactly in proportion to the area, which is $A = \pi r^2$. But in positively curved space there are fewer dots within a given radius, as we will explain below. In negatively curved space there are more dots within a given radius. *(second row)* It is easiest to visualize these relationships by trying to flatten out the curvature in each case. The positively curved surface, when flattened, gives rise to empty sections, while the flattened negatively curved surface has folds with extra material. *(bottom row)* The resulting distribution of dots in the flattened versions shows a relative deficit of dots at large radii in the positively curved surface and a relative excess of dots in the negatively curved surface. Though counts of radio sources at different distances from us were made for a while to see if the effect could be measured, we began to realize that we see so far out into space that such arguments tell us more about how radio sources evolve over time than they do about cosmology.

Such studies really tell us mostly about the evolution of radio sources and of other galaxies.

What was present before the big bang? There is no real way to answer this question. For one thing, we can say that time began at the big bang, and that it is meaningless to talk about "before" the big bang because time didn't exist. We do not now think that the universe can remain in a static condition, so it seems unlikely that the universe was always just sitting there in an infinitely compressed state, whatever "always" means.

Of course, these possibilities don't answer the question of why the big bang happened. It is possible that there had been a prior big bang and then a recollapse, and that our current big bang was one in an infinite series. This version of a closed-universe theory is called the **oscillating universe.** But the oscillating-universe theory doesn't really tell us anything about the origin of the universe because in this case there was no origin.

Before we return to the study of the big-bang theories, let us proceed historically with a discussion of another model that was once seriously considered.

## 37.3 THE STEADY-STATE THEORY*

The cosmological principle, that the universe is homogeneous and isotropic, is very general in scope, but starting in the late 1940s, three British scientists began investigating a principle that is even more general. Hermann Bondi, Thomas Gold, and Fred Hoyle considered what they called the **perfect cosmological principle:** the universe is not only homogeneous and isotropic in space but also *unchanging in time*. A criterion for science is that we must accept the simplest theory that agrees with all the observations, but it is a matter of personal preference whether the cosmological principle or the perfect cosmological principle is simpler.

The theory that follows from the perfect cosmological principle is called the **steady-state theory** (Fig. 37–7); it has certain philosophical differences from the big-bang cosmologies. For one thing, according to the steady-state theory, the universe never had a beginning and will never have an end. It always looked just about the way it does now and always will look that way.

The steady-state theory must be squared with the fact that the universe is expanding. How can the universe expand continually but not change in its overall appearance? For the density of matter to remain constant, new matter must be created at the same rate that the expansion would decrease the density. Only in this way can the density remain the same.

The rate at which new matter would have to be created in the steady-state theory works out to be only one hydrogen atom per cubic centimeter of space every $10^{15}$ years. This rate is far too small for us to be able to measure. The "law of conservation of mass-energy" would thus not be testable at this level, so we cannot invoke it to disprove the steady-state theory.

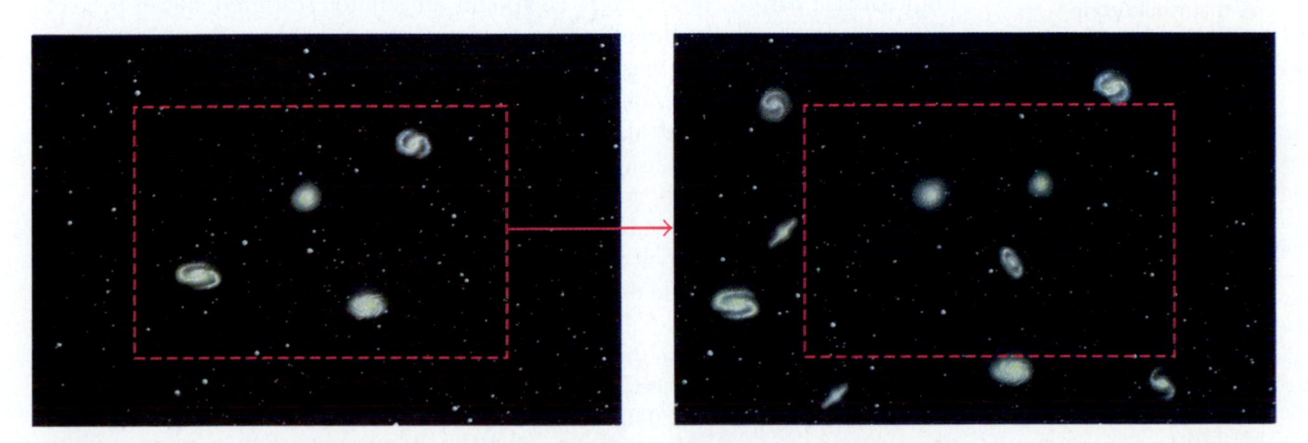

FIGURE 37–7 In the steady-state theory, as the dotted box at left expands to fill the full box at right, new matter is created to keep the density constant. In the picture, the four galaxies shown at left can all still be seen at right, but new galaxies have been added so that the number of galaxies inside the dotted box is about the same at it was before.

The matter created in the steady-state theory is not simply matter that is being converted from energy by $E = mc^2$. No, this is *matter that is appearing out of nothing* and is thus equivalent to energy appearing out of nothing.

For many years a debate raged between proponents of the big-bang theories and proponents of the steady-state theory. The evidence that came in—usually seeming to indicate that distant objects were somehow different from closer ones, which would show that the universe was evolving—seemed to favor the big-bang cosmologies over the steady-state theory. But none of this evidence was conclusive because alternative explanations for the data could be proposed or the steady-state theory itself could be modified (sometimes extensively) to be consistent with the discoveries.

The discovery of quasars provided some of the strongest evidence against the steady-state theory. The quasars are for the most part located far away in space, and so they were more numerous at an earlier time. Thus something has been changing in the universe, and change is not acceptable in the steady-state theory. As the evidence that the quasars were indeed at these distances grew, the status of the steady-state theory diminished.

In Section 37.5 we shall discuss the still stronger evidence that provided the crushing blow against the steady-state theory. This new evidence, more importantly, provides a major piece of the framework within which we now view cosmology.

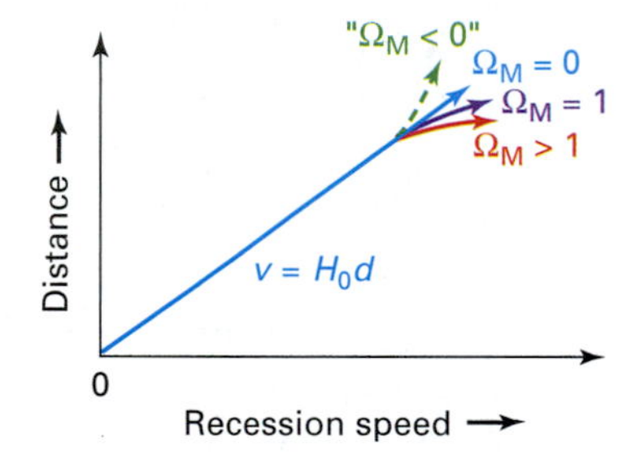


**FIGURE 37–8** A plot of the measured distances of objects against their recession speeds as given by their redshifts. At large recession speeds, there are differences that depend on how rapidly the universe's expansion rate is decelerating and hence on $\Omega_M$, the ratio of the average density of matter to the critical density. We can even plot the curve that would correspond to the density of matter being negative (that is $\Omega_M < 0$).

The oldest stars in our galaxy had been found to be 13 to 15 billion years old. This age was in conflict with some values measured for the Hubble constant, since the universe can't be younger than the stars in it. But the Hipparcos satellite's recalibration of the scale based on the distance to Cepheid variable stars has lowered this age to perhaps 11 billion years and slightly raised the age computed for the universe. These results eliminated the problem.

## 37.4 THE BIG BANG AS OF TODAY

Open or closed, the universe will last at least another 40 billion years, so we have nothing to worry about for the immediate future. Nevertheless, the study of the future of the universe is an exceptionally interesting investigation. Some of the methods of tackling this question involve measuring the amount of mass in the universe and thus the amount of gravity (as we will discuss in Section 38.2). Other methods of determining our destiny involve looking at the most distant detectable objects to see if any deviation from Hubble's law can be determined. This investigation of distant bodies has been carried out many times, but a deviation from the straight-line relation between velocity (measured for the redshift) and distance known as Hubble's law ($v = H_0 d$) has been found only in the last few years. Observations of supernovae made with the Hubble Space Telescope are now improving the distance scale and enabling an improved Hubble diagram to be plotted (Chapter 35). The discovery that the absolute brightness of a supernova at its maximum intensity is linked to the timescale over which it declines in brightness has tightened the points in the distant part of Hubble's law. The points are still too scattered, however, to enable us to determine the **deceleration parameter,** a term that gives the slowdown rate of the expansion and thus the curvature of the universe.

Another way of describing whether the universe will reverse its expansion because of gravity is to calculate the **critical density,** the density of matter barely sufficient to stop the expansion. The value of this critical density depends on the Hubble constant squared, so any errors in measuring the Hubble constant are compounded. (It also depends on the gravitational constant $G$ and on the speed of light $c$, which are constant, at least in our universe.) The critical density is shown as the blue line in the Hubble diagram in Figure 37–8.

If we consider Hubble's law with the current value for Hubble's constant, we can extrapolate backward in time (Fig. 37–9). We simply calculate when the big bang would have had to take place for the universe to have reached its current state at its current rate of expansion. This calculation indicates that for a Hubble constant of 75, the universe is somewhere between 10 billion and 14 billion years old. The Hubble Key Project on the distance scale, by measuring Cepheid variables in many galaxies and using the result to calibrate other methods of measuring distances, has refined Hubble's constant to ±10 per cent.

## A DEEPER DISCUSSION

### BOX 37.2 Omega: The Makeup of the Critical Density*

As we saw, the "critical density" is the density barely necessary to halt expansion of the universe. The ratio of the density of matter in the universe to the critical density is designated by the Greek capital letter Omega: $\Omega$. Omega is equal to (actual density)/(critical density). If this ratio is 1 ($\Omega = 1$), then the universe will just barely stop its expansion at an infinite time in the future. $\Omega$ less than 1 corresponds to an open universe, $\Omega = 1$ is a flat universe, and $\Omega$ greater than 1 corresponds to a closed universe. So determining the past and future expansion of the universe (Fig. 37-8) is essentially the same as determining $\Omega$. What we actually measure when we add up all the visible matter (stars, galaxies, etc., that are giving off light) is the ratio of the density of matter to the critical density, $\Omega_M$, where the subscript M stands for "mass." There turns out to be astonishingly little of this ordinary matter, only about 0.5 per cent ($\Omega_{VM} = 0.005$), where the subscript VM stands for "visible matter." If you add in all the matter whose radiation we discover in other ways (molecular hydrogen we detect in the ultraviolet, intergalactic gas we detect in the x-ray, and so on), we come up to about 3 per cent of the critical density ($\Omega_M = 0.03$), where the subscript M stands for "matter." In the next chapter, we will see how the dark matter we already discussed in Section 33.6 gives an additional 30 per cent of the critical density ($\Omega_{DM} = 0.30$).

But in the currently accepted model of the universe, that of "inflation" (which we discuss in the following chapter), the universe is right at the critical density. The rest of the critical density is provided by the cosmological constant. Thus rounded off, $\Lambda = 0.7$ (that is, the cosmological constant provides 70 per cent of the critical density). If $\Omega_\Lambda$ is the ratio of the effect of the cosmological constant to the critical density, then $\Omega_\Lambda = 0.7$.

Astronomers also use a quantity called the **deceleration parameter,** which is given the symbol $q_0$ (pronounced kew-naught, where "naught" stands for "zero") to describe how fast the expansion is slowing down. $q_0 = 1/2$ marks the dividing line between an open universe ($q_0$ less than 1/2, including 0) and a closed universe ($q_0$ greater than 1/2, including 1). In the flat case—the boundary between the open and closed—the universe would become infinitely large in an infinitely long time, with its rate of expansion approaching zero, and thus never quite collapse. Omega is twice $q$-naught. At present, with the discussion of the cosmological constant, the argument is more often framed in terms of Omega instead of in terms of $q_0$.

We now realize that the farthest galaxies, from which we had hoped to best determine a deviation from Hubble's law, may well have been very different long ago when they emitted the light we are now receiving. This point was made especially by Beatrice Tinsley in the 1970s, though it took the Hubble Space Telescope to show it clearly (Fig. 37–10). Thus we cannot assume that even such basic properties as their size and brightness were similar then to their size and brightness now. Since determining the distance to these distant galaxies depends on understanding their properties, it now seems that such considerations of galaxy evolution are too uncertain to allow us to determine in this way whether the universe is open or closed.

Big-bang theories actually arise as solutions to a set of equations that Einstein advanced as part of his general theory of relativity. Early solutions by Einstein

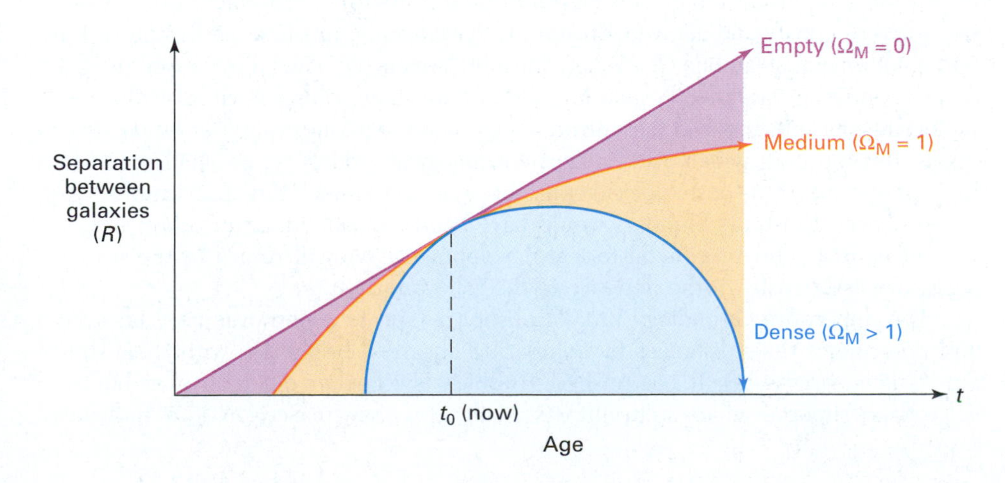


**FIGURE 37–9** If we could ignore the effect of gravity, then we could trace back in time very simply; the Hubble time corresponds to the inverse of the Hubble constant ($1/H_0$). Actually, gravity has been slowing down the expansion. The vertical axis again represents some scale factor.

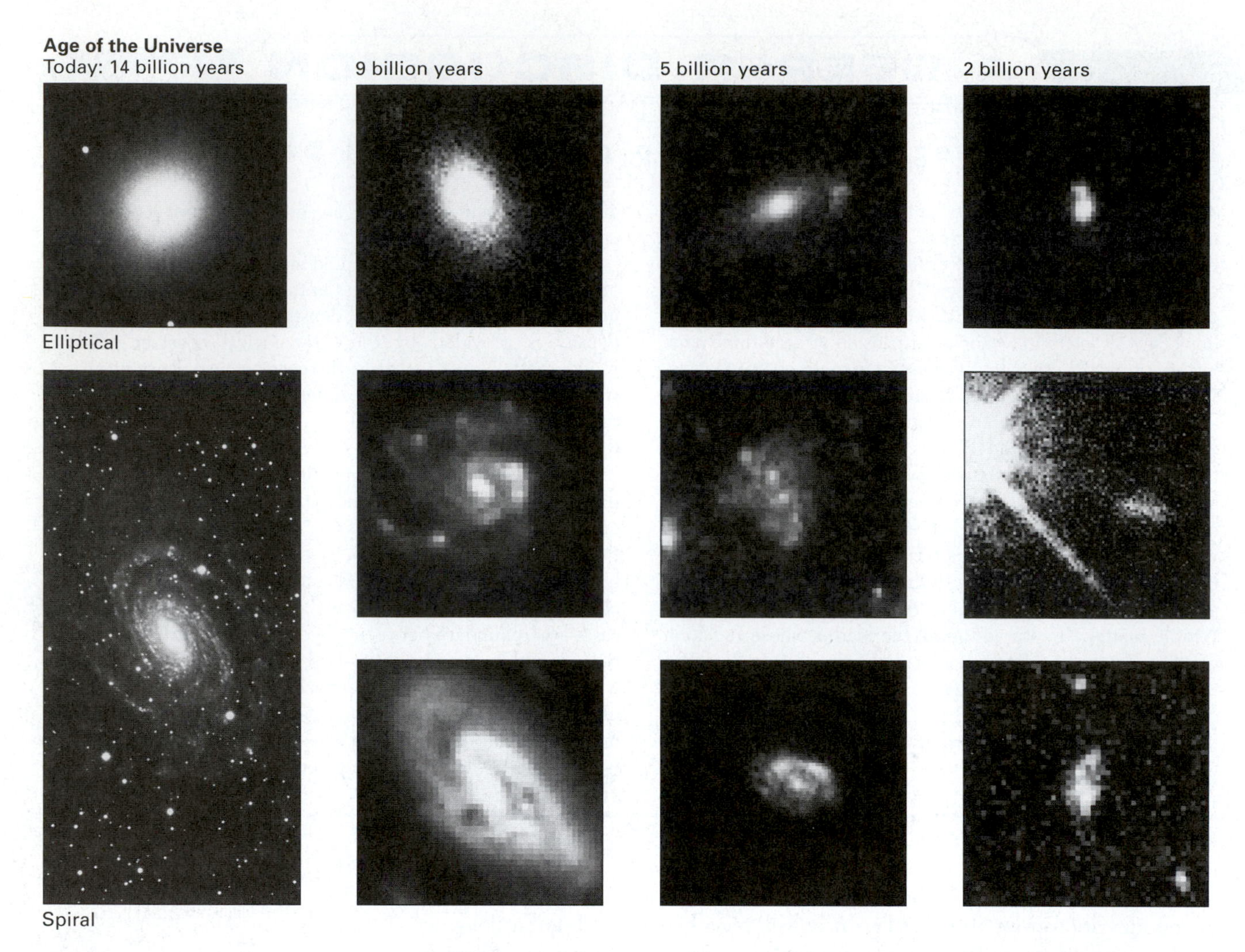


**FIGURE 37–10** Hubble Space Telescope views of some of the farthest galaxies that it is possible to see reveal that they have not yet taken on the regular shapes of galaxies in the current universe. The farther back in time we see, the less regular the shapes look.

himself (Fig. 37–11) and others were not valid. We now use solutions worked out in the early 1920s by the Russian mathematician Alexander Friedmann. The Belgian abbé Georges Lemaître discovered similar solutions slightly earlier.

**FIGURE 37–11** The statue of Albert Einstein at the National Academy of Sciences, Washington, D.C.

Einstein, for his own solutions, introduced arbitrarily a term into his equations to provide a repulsion force. This **cosmological constant** counteracted the attractive force of gravity and allowed Einstein to have a static universe. It was given the Greek letter capital lambda ($\Lambda$). When Einstein became convinced by Hubble's work that the universe was indeed expanding rather than static, Einstein withdrew his cosmological constant, and said that introducing it was the greatest blunder he had ever made. But the most widely used recent cosmological models, as we shall soon see, have reintroduced the cosmological constant in order to match the observations. Apparently even Einstein's blunders would have been fantastic ideas for other people. But let us first return to our historical development. We will return to the overall curvature and density of the universe in the following chapter.

The Supernova Cosmology Project observes Type Ia supernovae very far away and determines their distances by comparing observed brightness with their standard known brightness. It has pushed Hubble's law farther out than ever before, since these supernovae are so bright. Its result, as we saw in Section 35.5, indicates that the universe is accelerating (Fig. 37–12). As of 2001, this astounding result seems to get more and more backing. In the next chapter, we will see theoretical reasons why the universe should be just at the critical density. In the next section, we see observational results that also indicate that the universe is at the critical density, which corresponds to a flat universe. The best explanation of the supernova observations then builds on the measurements that the mass density is about 3 per cent of the critical density and that dark matter makes up about another 30 per cent. (The dark matter has gravity that we detect from the rotation curves of galaxies or by calculating the mass that has to exist in clusters of galaxies to keep them bound together.) The rest of the critical density, about ⅔ of it, then comes from the cosmological constant or something similar. By analogy with the term "dark matter" that completes the first third of the critical density, astronomers are using the term **dark energy**

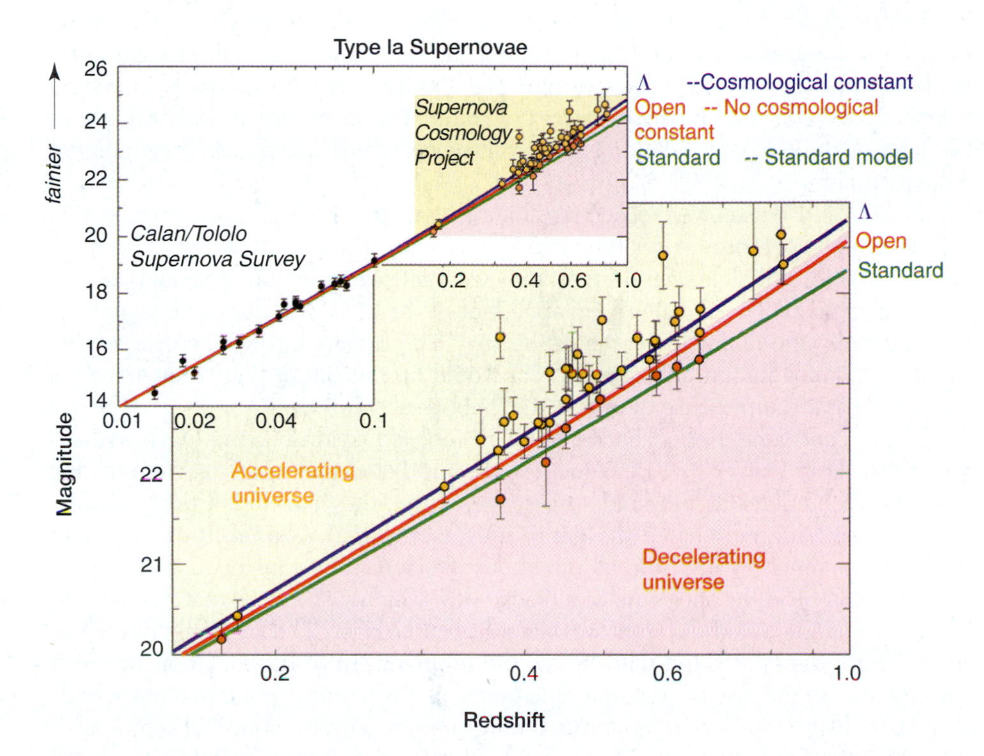


**FIGURE 37–12** This observational Hubble diagram is the result of observations of many distant Type Ia supernovae (the kind that come from incinerated white dwarfs in binary systems), as observed with the Supernova Cosmology Project. The lower-redshift part of the diagram is from a ground-based survey. The Supernova Cosmology Project's part of the upper-left diagram, for $z = 0.2$ to 1.0, is magnified in the lower right part of the diagram. Magnitude, shown on the $y$-axis, measures distance, since galaxies that are farther away are fainter. By convention, this diagram is plotted with magnitude (distance) on the $y$-axis and redshift on the $x$-axis, unlike our earlier Hubble diagrams, which had the axes interchanged. In principle, we can determine the future of the universe from the slight deviations of the measured curves from the straight-line Hubble law, especially at high redshift. On the curve for the "Standard" model, the universe is positively curved and is closed. It is finite, and would eventually begin to contract. For the curve for the "Open" model, the universe would be open, is infinite, and will expand forever. The surprising result of recent years is that the observations show that the distant supernovae are, in the range shown, somewhat fainter than expected (and thus higher on the graph). The best fit to the observations thus indicates that the expansion of the universe is accelerating, not decelerating after all. The observations are fit best in models with a cosmological constant $\Lambda$ of 70%, that is, $\Lambda = 0.7$.

On the graph, the "lambda" curve corresponds to the currently best accepted model: lambda of 0.7 and mass contribution of 0.3; the "open" curve has lambda of 0 (that is, no contribution from a cosmological constant) and mass contribution of 0.3; and the "standard" curve has lambda at 0 and mass contribution of 1, that is, mass density at the critical density.

to indicate the other two thirds that come from the cosmological constant or from whatever alternative source there is. In the next chapter (especially in Section 38.2), we will say more about the matter, dark matter, and dark energy that make up the universe.

The concepts discussed in the paragraphs above, and the idea that the matter you and I are made of is only a minor constituent of the universe, are not easy to digest. They may take hours, years, or a lifetime to come to terms with.

The importance of the discovery of the background radiation cannot be overstressed. Dennis Sciama, the British cosmologist, put it succinctly by saying that up to 1965 we carried out all our calculations knowing just one fact:

1. that the universe expanded according to Hubble's law.

After 1965, he said, we had a second fact:

2. the existence of the background radiation.

That may be a bit oversimplified, but it is essentially true. We now also credit

3. the abundances of light isotopes and elements, especially deuterium, helium, and lithium (Section 38.2b); and
4. the anisotropies in the background radiation, soon to be discussed.

## 37.5 THE COSMIC BACKGROUND RADIATION

In 1965, a discovery of the greatest importance was made: radiation was detected that is most readily explained as a remnant of the big bang itself.

That much is easy to state, and with this discovery of radiation from the big bang itself, then clearly the steady-state theory was discredited. The discovery was made by Arno A. Penzias and Robert W. Wilson of the then Bell Telephone Laboratories in New Jersey (Fig. 37–13). They were testing a radio telescope and receiver system to try to track down all possible sources of static. The discovery, in this way, parallels Jansky's discovery of radio emission from space, which marked the beginning of radio astronomy.

**FIGURE 37–13** Arno Penzias (*left*) and Robert W. Wilson (*right*) with their horn-shaped antenna in the background. Penzias and Wilson found more radio noise than they expected at the wavelength of 7 cm at which they were observing. After they removed all possible sources of noise (by fixing faulty connections and loose antenna joints, and by removing "sticky white contributions" from nesting pigeons), a certain amount of radiation remained. It was the 3° background radiation. Penzias and Wilson won the 1978 Nobel Prize in Physics for their discovery.

Penzias and Wilson were observing at a wavelength of several centimeters, which is in the radio spectrum. After they had subtracted, from the static they observed, the contributions of all known sources, they were left with a residual signal that they could not explain. The signal was independent of the direction they looked and did not vary with time of day or season of the year. The remaining signal corresponded to the very small amount of radiation that would be put out at that frequency by a black body at a temperature of only 3 K, 3° above absolute zero.

At the same time, Robert Dicke, P. J. E. Peebles, David Roll, and David Wilkinson at Princeton University had, coincidentally, predicted that radiation from the big bang should be detectable by radio telescopes. First, they concluded that radiation from the big bang permeated the entire universe, so that its present-day remnant should be coming equally from all directions. Second, they concluded that this radiation would have the spectrum of a black body, which just means that the amount of energy coming out at different wavelengths can be described by giving a temperature. Third, they predicted that though the temperature of the radiation was high at some time in the distant past, the radiation would now correspond to a black body at a particular very low temperature, only a few degrees above absolute zero (Fig. 37–14). If radiation could be found that came equally from all directions and had the low-temperature black-body spectrum that matched the prediction, then the theory would be confirmed.

The Princeton group continued the process of building their own receiver to observe at a different radio wavelength. They were soon able to measure the intensity at this wavelength, and they, too, found that it corresponded to that of a black body at 3°. This tended to confirm the idea that the radiation was indeed from a black body, and thus that it resulted from the big bang.

Actually, the Princeton group was not the first to predict that such radiation might be present. Many years earlier, Ralph Alpher and Robert Herman (in 1948) and George Gamow (in 1953) had made similar predictions, but this earlier work was at first overlooked.

Because the big bang took place simultaneously everywhere in the universe, radiation from the big bang filled the whole universe. The radiation thus has the property of being isotropic to a very high degree; that is, it is the same in any direction that we observe. It was generated all through the universe at the same time, so its remnant must seem to come from all around us now. The fact that the observed radiation was highly isotropic was thus strong evidence that it came from the big bang. It also shows that the early universe was very homogeneous.

### 37.5a The Origin of the Background Radiation

The leading models of the big bang consider a hot big bang. The temperature was billions upon billions of degrees in the fractions of a second following the beginning of time.

In the millennia right after the big bang, the universe was opaque. Photons did not travel very far before they were scattered by electrons. This process results in black-body radiation that corresponds to the temperature of the matter.

Gradually the universe cooled. After about 300,000 years, when the temperature of the universe reached 3000 K, the temperature and density were sufficiently low for the hydrogen ions to combine with electrons to become hydrogen atoms. This **recombination** took place suddenly. (It is called "recombination" even though there was no previous "combination.") Since hydrogen has mainly a spectrum of lines rather than a continuous spectrum and since no electrons remained free, recombination meant that the gas suddenly lost its ability to absorb photons except at a few wavelengths. Thus from this time on, most photons could travel all across the universe without being absorbed by matter. The universe had become transparent.

Since matter rarely interacted with the radiation from that time to the present, it no longer continually recycled the radiation. We were thus left with the radiation set free at the instant when the universe became transparent. This radiation is travelling through space forever, cooling all the time as its wavelengths expand. Observing it now is like studying a fossil. As the universe continues to expand, the radiation retains the shape of a Planck curve, but the curve corresponds to cooler and cooler temperatures.

Since the radiation is present in all directions, no matter what we observe in the foreground, it is known as the **cosmic background radiation.** In reference to its origin, it is also called the **primordial background radiation.**

In the above scenario, the universe has changed from a hot, opaque place to its current cold, transparent state. Such a change is completely inconsistent with the steady-state theory, which is no longer bothered with by almost any astronomers.

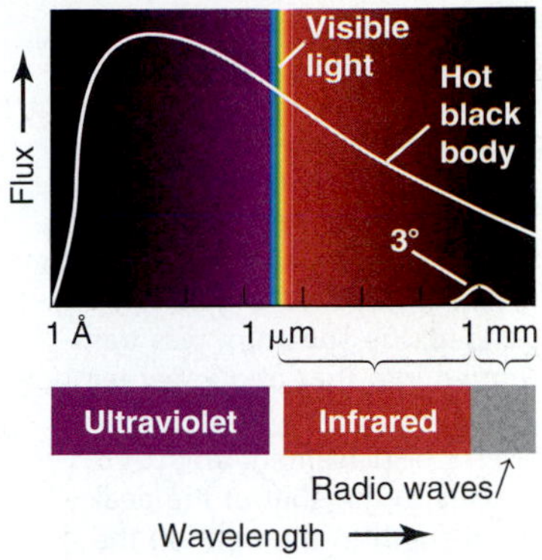

**FIGURE 37–14** Planck curves for black bodies at different temperatures. Radiation from a 3-K black body (*the inner curve at lower right*) peaks at very long wavelengths. On the other hand, radiation from a very hot black body (*the upper curve*) peaks at very short wavelengths.

### 37.5b The Temperature of the Background Radiation

Planck curves corresponding to cooler temperatures have lower intensities of radiation and have the peaks of their radiation shifted toward longer wavelengths. We have seen that radiation from the sun, which is 6000 K, peaks in the yellow-green and that radiation from a cool star of 3000 K peaks in the infrared. The universe, much cooler yet, peaks at still longer wavelengths: the peak of the black-body spectrum is at the dividing wavelength between infrared and radio waves. In a moment, we shall see just how cool the universe is.

Now, we recall that Penzias and Wilson measured a particular flux at the wavelength measured by their equipment. This flux corresponds to what a black body at a temperature of only 3 K would emit (Fig. 37–15). Thus we speak of the universe's **3° background radiation.** Following the original measurement at Bell Labs and at Princeton, other groups soon measured values at other radio wavelengths. These other values lay along the slope of one side of the black-body curve, and so they showed that the value was indeed about 3°. (In the Système International of units, this emission would be called 3 K radiation, but we shall join the astronomical community in continuing to call it 3° radiation.)

Unfortunately, though, for many years we were able to observe only at wavelengths longer than that of the peak, so only the right-hand half of the curve was measured. All the points corresponded to a temperature of approximately 3 K, but still, a black-body curve does have a peak. It would have been more satisfying if some of the points measured lay on the left side of the peak. Points lying on the left side of the peak would prove conclusively that the radiation followed a black-body curve

We have discussed black bodies in Section 24.2. Basically, the emission from a black body follows Planck's law of radiation (Fig. 24–1) in that for a given temperature there is an equation that tells us the intensity of radiation at each wavelength. The key fact to remember about Planck's law is that specifying just one number—the temperature—is enough to define the whole Planck curve.

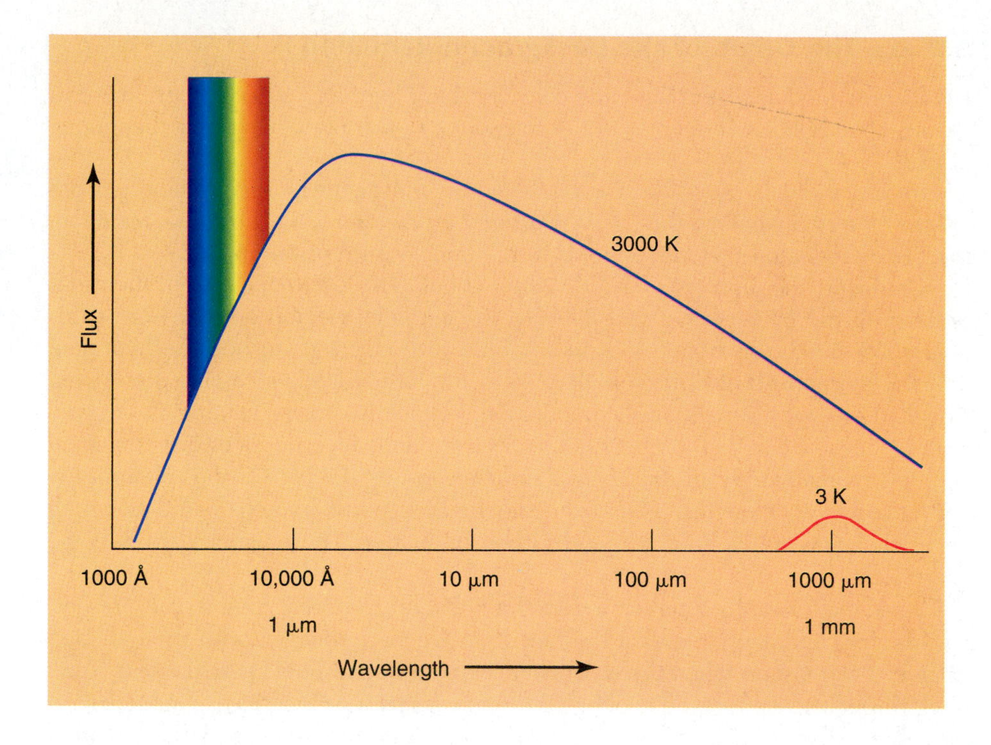


**FIGURE 37–15** The background radiation was set free at a temperature of 3000 K (*blue curve*) when the universe became transparent at the time of recombination, a million years after the big bang. Its peak wavelength was then in the infrared. As the universe expanded, the spectrum was transformed into that of a lower temperature (*red curve*). Notice how Wien's displacement law corresponds to the shift of the peak wavelength to the right on the graph. The Stefan-Boltzmann law holds that the temperature of 3 K is 1000 times less than that of 3000 K, so the energy involved is $(1000)^4 = 10^{12}$ or a billion billion times less.

and was not caused by some other mechanism that could produce a straight line, or some other form that happened to mimic a 3 K black-body curve in the centimeter region of the spectrum.

Though it would have been desirable to measure a few points on the short-wavelength side of the black-body curve, astronomers were faced with a formidable opponent that frustrated their attempts to measure these points: the Earth's atmosphere. Our atmosphere absorbs most radiation from the long infrared wavelengths.

Infrared observations from balloons seem to have proved unequivocally that the radiation follows a black-body curve. One point from a rocket experiment seemed to give a higher flux than expected from the black-body curve, and theoreticians worked out models that would explain this seeming extra infrared flux. For example, an otherwise undetected generation of stars before the ones we know now might have provided the flux. But the high point could have been the result of an imperfect correction of the results for the effects of the Earth's atmosphere. Was this point real (Fig. 37–16)? In Section 37.5d, we shall discuss the dramatic conclusion of this episode.

So at present, astronomers consider it settled that radiation has been detected that could only have been produced in a big bang. Accepting this interpretation clearly rules out the steady-state theory. It means that some version of the big-bang theories must hold, though it doesn't settle the question of whether the universe will expand forever or will eventually contract.

### 37.5c The Background Radiation as a Tool

Now that the spectrum has been measured so precisely on both sides of the peak of intensity, we have moved beyond the point of confirming that the radiation is black body. We can now study its deviations from isotropy.

A very slight difference in temperature—about one part in one thousand—has been measured from one particular direction in space to the opposite direction. Such a difference is called an **anisotropy** (pronounced an-eye-sot'roh-pee)—a deviation from isotropy (Fig. 37–17). To limit the effect of the Earth's atmosphere on the measurements, they were first made from low-water-vapor regions like the South Pole,

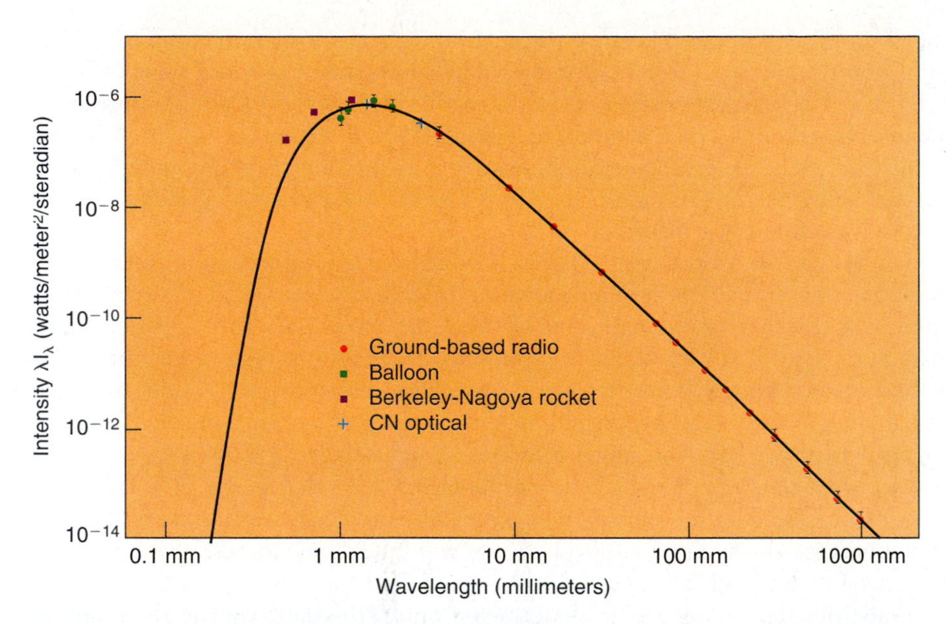


**FIGURE 37–16** Observations of the cosmic background radiation as of the time just before the launch of COBE (the Cosmic Background Explorer). The results of a rocket flight that showed an apparent deviation from a black-body curve in the submillimeter region of the spectrum, and thus caused worries that the spectrum was not that of a black body, were not verified by the satellite observations. The results from optical studies of the molecule CN, which can be interpreted to show how radio waves affect the molecules and thus give the strength of the background radiation at those wavelengths, are also shown.

from high-flying aircraft, and from balloons. The anisotropy measured is what would result from the Doppler effect if our sun were moving away from the region colored red and toward the region colored blue in the figure at about 600 km/sec with respect to the background radiation. (The blue stripe of variable intensity that runs obliquely around the globe is emission from our own galaxy.) Since we know the velocity at which our sun is moving as it orbits the center of our galaxy, we can remove the effect of our sun's orbital motion. We deduce that our galaxy is moving about 500 km/sec relative to the background about 40° from the direction of the Virgo Cluster. We do not know any particular reason why our galaxy should have such a velocity. A small yearly effect from the Earth's motion around the sun has also been detected. Both effects give the direction we are heading a slightly higher temperature than the direction behind us. The wavelengths are blueshifted, shifting the Planck curve toward the blue; a bluer Planck curve corresponds to a higher temperature. The overall velocity measured in these studies is similar to that deduced from the anisotropy in the background radiation.

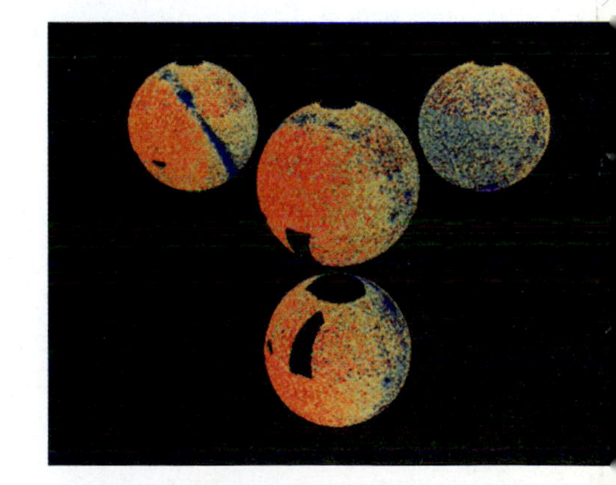

**FIGURE 37–17** The anisotropy of the background radiation is shown by the range of colors in this all-sky map made from four balloon flights. The total range is from +3 millikelvins (*reddish*) at upper left to −3 millikelvins (*bluish*). The center globe is a map of the sky brightness at 1.5 cm (19 GHz), at which the Earth's atmosphere contributes relatively little. Three reflecting mirrors show the parts of the globe that are hidden from direct view. No data are available for regions shown in black.

Measurements of the anisotropy of the background radiation are now being very effectively made from an observing station at the South Pole. The high altitude and exceedingly dry atmosphere allow the background radiation to be observed at shorter wavelengths than are otherwise normally accessible from the Earth's surface.

### 37.5d The Cosmic Background Explorer

COBE (Fig. 37–18), the Cosmic Background Explorer (pronounced "koh'bee"), was a NASA spacecraft meant to study the background radiation. It was launched in 1989, carrying enough liquid helium to cool its instruments for about a year; some of its instruments didn't need such cooling and lasted four years.

One of COBE's experiments was to study the spectrum over a wide wavelength range to compare with a black body. Indeed, it carried an experimental black body

**FIGURE 37–18** The COBE (Cosmic Background Explorer) spacecraft, remade smaller to fit on a Delta rocket after the explosion of the space shuttle Challenger. COBE was launched in 1989 and sent back data for four years. It fantastically improved the accuracy of measurements of the background radiation.

aloft to be placed in the telescope's beam from time to time, for comparison. This set of observations has turned out to be wildly successful. The set of points it measured (Fig. 37–19) agrees extremely closely with a black body at a temperature since updated, based on the final data set, to 2.725 ±0.002 K, where the uncertainty represents the scientists' estimate of systematic errors that might be cropping up. The analysis shows that the existence of an infrared excess radiation as suggested by earlier rocket observations was incorrect.

COBE also measured (Fig. 37–20) how isotropic—uniform from direction to direction—the cosmic background radiation is at three different microwave wavelengths (at the short-wavelength end of the radio spectrum). The COBE Science Team compared the observations at different wavelengths to determine which part of the radiation comes from our own galaxy. The anisotropy that results from the sun's motion in space—a Doppler shift—shows clearly as a change in color from upper right to lower left. The range is from +3 millikelvins (+0.003 K = +3 mK), shown in orange at upper right, to −3 millikelvins (−3 mK), shown in purple. The direction measured for the sun's motion agrees with previous measurements to within a few degrees. This type of anisotropy is known as a "dipole anisotropy" because it concerns only two ("di-") opposite directions ("poles").

Once this dipole anisotropy is subtracted out of the data, we can then look to see if there are any fluctuations left. These remaining fluctuations would be caused far out in the universe (far back in time), rather than by local effects such as our own motion in space. Various theories of cosmology, including the cold dark matter theory, predict that some small fluctuations should be detectable. In the next subsection, we will discuss the results of the COBE analysis of these data.

A third, infrared COBE experiment searched at various infrared wavelengths for radiation from the first stars and galaxies, the predecessors of our current stars and galaxies. It showed the Milky Way (Fig. 37–21) and other galactic sources. This experiment extended the observations of the Infrared Astronomical Satellite (IRAS) spacecraft.

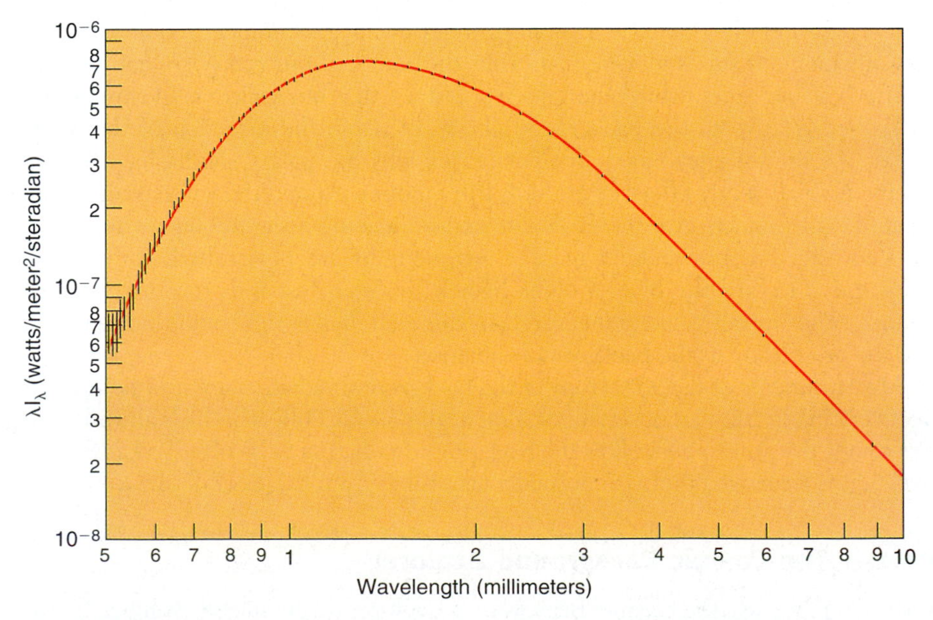


**FIGURE 37–19** This first spectrum based on COBE observations drew an ovation from astronomers when it was first shown at an American Astronomical Society meeting. It represents a proof at astonishingly high accuracy that the cosmic background radiation is a black body. The data points and their error bars (*black*) were fit precisely by a black-body curve (*red*) for 2.735 ±0.06 K, a value that has since been slightly improved, based on the final data set, to 2.725 ±0.002 K. The error bars are thinner than the thickness of the curve drawn!

FIGURE 37–20 Maps of the sky from all four years of COBE data at radio wavelengths of (*A*) 3.3, (*B*) 5.7, and (*C*) 9.6 mm. At the longest wavelength, we see a galactic source in Cygnus (*green dot at middle left*) and a sign of the galactic plane (*horizontal red at right*). The rest of the signal is from the cosmic background radiation. The asymmetry from bottom left to top right shows the anisotropy from the sun's motion. The range shown is ±3 millikelvins.

*A* 3.3 mm

*B* 5.7 mm

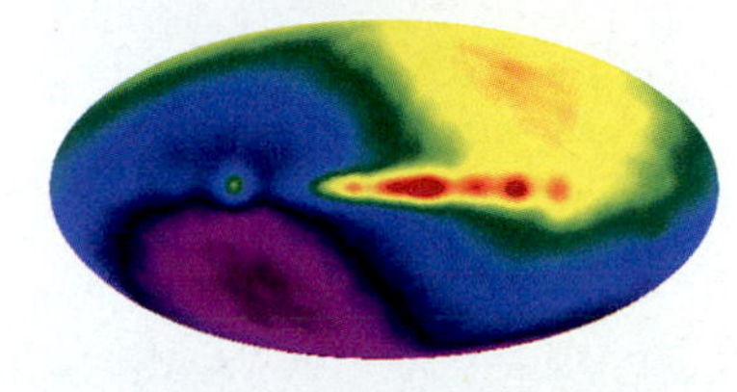

*C* 9.6 mm

## 37.5e COBE's Map

COBE's mapping experiment looked for (and found) anisotropy. With spatial resolution of 7° in angle, COBE's microwave experiment provided the best data available at the time. (By "spatial resolution of 7°," or "on a scale of 7°," we mean that the resolution is 7°, so that no features smaller than 7° are distinguishable.) On what angular scales can we find fluctuations? Probably on all scales, and scientists are using a variety of types of apparatus to look for such fluctuations.

Even from only the first months of data, fluctuations were detected. These fluctuations were 30 microkelvins (30 $\mu$K), one hundred times smaller than the 3 millikelvins (3 mK) dipole anisotropy we had already seen. Though at first the fluctuations showed only statistically—there were more fluctuations than expected on the assumption of random noise—analysis of all four years of data has provided much higher accuracy. On these final graphs (Fig. 37–22), the fluctuations we see are actual fluctuations in the background radiation.

## 37.5f Balloon and Ground-Based Observations

The existence of fluctuations has since been confirmed on smaller angular scales from experiments on balloons, from the ground in Antarctica, and from the ground in Saskatoon, Saskatchewan, Canada. (The last of these results was already shown as an inset in Figure 37–22). Are these the missing links that bring us back to how galaxies formed? Are the fluctuations shown by COBE the seeds from which galaxies grew? We shall return to the question in Section 38.7.

Two major balloon projects gave results in 2000. The BOOMERanG project (Fig. 37–23) and the MAXIMA project (Fig. 37–24) mapped regions of the sky at much higher spatial resolution than COBE's. They show detail in the fluctuations, which are only by 100 millionths of a degree in temperature. BOOMERanG stands for *B*alloon *O*bservations *o*f *M*illimetric *E*xtragalactic *R*adiation; it observed 3 per cent of the sky. MAXIMA stands for *M*illimeter *A*nisotropy E*x*periment *Im*aging *A*rray; it observed 3/10 per cent of the sky but at a resolution of 1/6 degree, higher than BOOMERanG's. Scientists analyze the data in very technical ways, looking among other things for the typical scales of size on which they see fluctuations (Fig. 37–25). Theory predicts that a flat universe will have fluctuations that are about 1° across, which is what the observations show. There are detailed predictions on the basis of theory regarding what other fluctuation scales should also be found. The predictions gave the height of a series of peaks on graphs showing the strengths of fluctuations at different scales (Fig. 37–26). You can think of the first peak as the fundamental note from an organ pipe and the second and other peaks as overtones. The data from these two spacecraft showed only the first peak clearly. They did not agree with the predictions for the second peak, which was predicted to be stronger than has been observed.

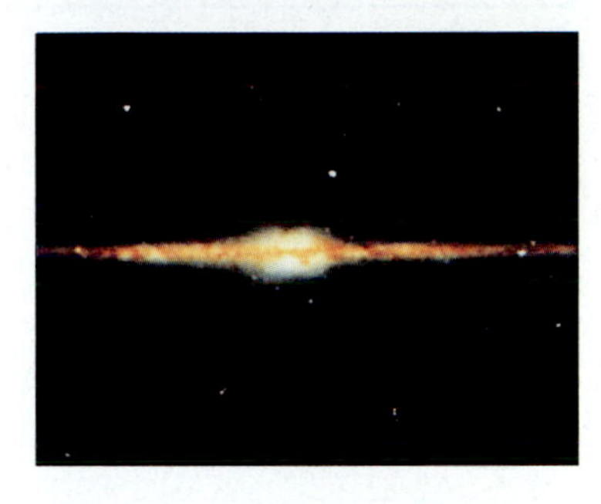

FIGURE 37–21 A near-infrared image of the Milky Way Galaxy obtained by COBE, at wavelengths 1.2 (*blue*), 2.2 (*green*), and 3.4 (*red*) microns. The image is presented in galactic coordinates, with the plane of the Milky Way horizontal across the center.

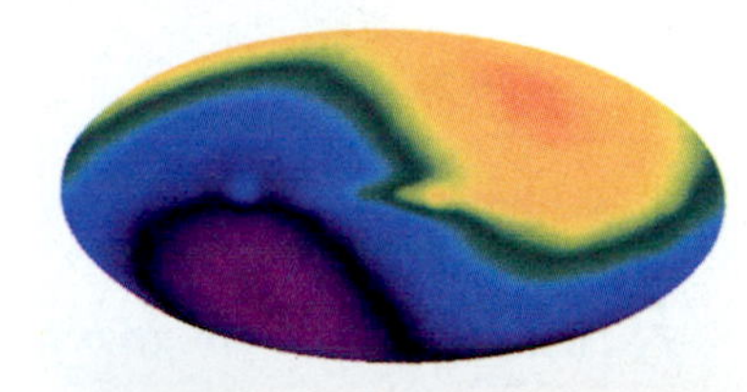

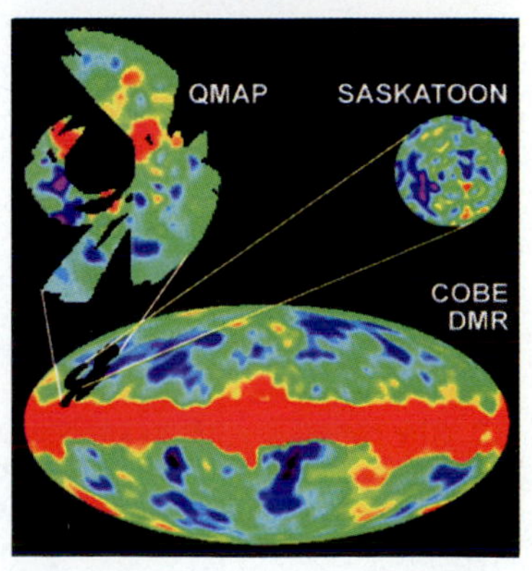


FIGURE 37–22 The full-sky map of variations in 3 K background radiation as analyzed from all four years of COBE data. Figure 37-20 showed variations of 3 millikelvins (0.12%). Here we see the result when the dipole anisotropy is subtracted. Variations of 30 microkelvins (0.01%) then show. The red band across the center of the all-sky map, which comes from COBE, results from the microwave emission of the Milky Way Galaxy. The variations above and below the Milky Way are fluctuations in the background radiation. Earlier data reduction could not pinpoint the fluctuations, but these final maps have enough statistical accuracy to do so. The results are also shown for a ground-based experiment carried out in Saskatoon, Saskatchewan, Canada. It observes only a small area around the north celestial pole but obtains higher spatial resolution within that region. In addition, the result from the QMAP balloon for approximately the same region is shown.

IN REVIEW

## BOX 37.3 Milestones in Understanding the Cosmic Background Radiation

- 1948: Background radiation at a few degrees above absolute zero predicted by Alpher and Herman; their prediction was forgotten.
- 1965: Background radiation discovered at 3 K by Penzias and Wilson; background radiation predicted by Princeton group, who were scooped.
- 1978: Penzias and Wilson awarded the Nobel Prize in Physics.
- 1989–1993: NASA's Cosmic Background Explorer aloft.
- 1997: COBE science team reports complete agreement with 2.7 K black body (Planck) curve and ripples—early seeds of structure—at or smaller than COBE's 7° angular resolution once the dipole anisotropy, caused by the sun's motion, is removed.
- 2000: BOOMERanG results reported that the universe is flat from the position of the first peak in the fluctuation spectrum, verifying the inflationary theory.
- 2001: BOOMERanG, MAXIMA, and DASI results show the second peak, further verifying the inflationary theory and measuring the baryon density.
- 2dF finds a fluctuation pattern representing the cosmic-background fluctuations at a later stage.
- NASA's Microwave Anisotropy Probe (MAP)
- 2007 European Space Agency's Planck mission.

Other experiments are resolving the problem. The TopHat balloon circled the South Pole in Antarctica for two weeks in 2001 and scientists are awaiting results at twice the resolution of BOOMERanG and MAXIMA. Its telescope is above, rather than below, its balloon to get an unobstructed view of the whole sky and to minimize contamination from the balloon itself. Caltech's Cosmic Background Imager in Chile and the Jet Propulsion Laboratory's Degree Angular Scale Interferometer (DASI) at the South Pole have also contributed to the discussion. Results from DASI, an array consisting of 13 individual telescopes, were released in 2001 (Fig. 37–27), as were further BOOMERanG and MAXIMA results. They clearly show not only the second peak but also the third peak (Fig. 37–28). The height of the peaks is in line with the predictions of the inflationary theory of cosmology we will discuss in the next chapter. The first peak indicates that the universe is indeed flat. The second peak gives the balance of matter in the form of baryons vs. non-baryonic matter.

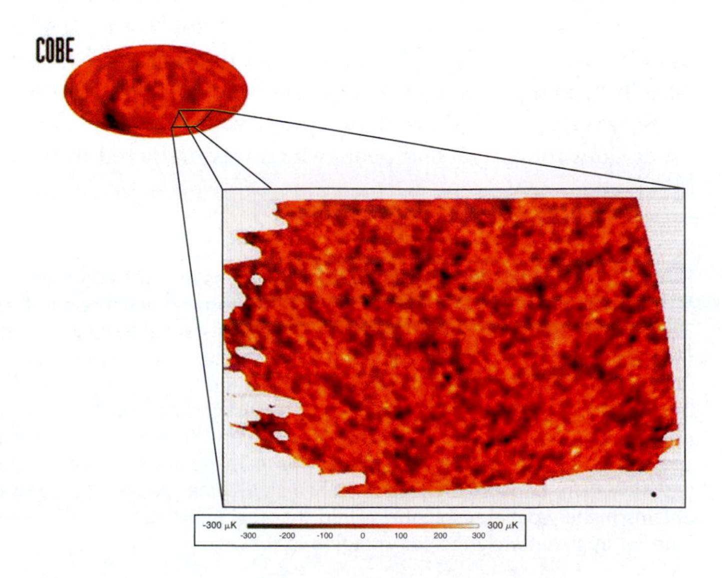


**FIGURE 37–23** The BOOMERanG project has mapped part of the sky at much higher detail than COBE (shown in an inset) or previous balloon missions could. We see variations in the 3 K cosmic background radiation over a part of the sky about 90° by 30°. (The size of the full moon is marked at lower right for comparison.) Note that the temperature differences, marked by the scale bar at top, varies from the red spots to the blue spots by only about 0.6 millikelvins, that is, 0.0006 kelvins.

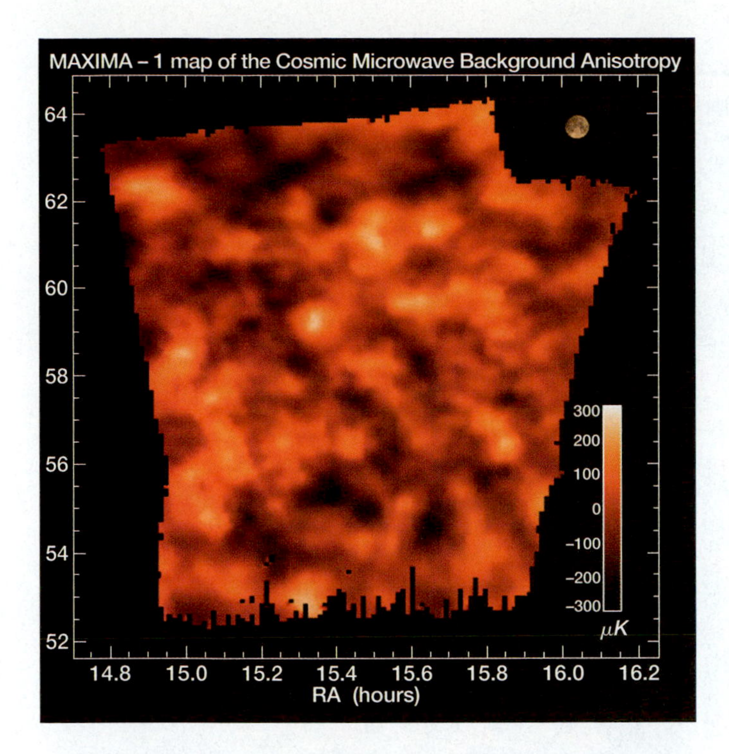


**FIGURE 37–24** The MAXIMA project was another balloon-borne instrument to map the microwave background at much higher detail than COBE. You can see that the scale of the fluctuations is approximately twice that of the diameter of the moon, which is ½° across. Fewer moon-size fluctuations were detected than had been expected.

The Very Large Telescope, on its high mountain in Chile, has also contributed to the study of the background radiation. It has long been known that certain spectral lines from carbon atoms in space result from the cosmic background radiation exciting carbon energy levels. Measurements decades ago placed a limit of a few kelvins, and were superseded by direct measurements that showed 3 K. Now one of the VLT telescopes has detected these carbon spectral lines as well as some spectral lines from molecular hydrogen while observing toward a quasar. It sees the spectral lines at a redshift of 2.34, which corresponds to a time when the universe was less than one-fifth its current age. At that time, the universe was hotter; after all, it has been cooling down since the time of recombination, when the background radiation was released. These spectral lines can arise only if the carbon and hydrogen is excited by something, presumably the background radiation. The results, indeed, show that the temperature where these lines arose was hotter than 6 K and cooler than 14 K. They agree with the predictions that the background radiation was 9 K that far back in time, which we obtain by observing clouds of gas that many light-years away.

### 37.5g Future Cosmic-Background Spacecraft

NASA has launched a satellite to build on the COBE results, and ESA (The European Space Agency) will launch an even better one in a few years. Both are to measure cosmic background radiation over the whole sky, looking for anisotropies. NASA's Microwave Anisotropy Probe (MAP) was launched on June 30, 2001. It has resolution of ⅓° of sky instead of COBE's 7°. It should find variations as small as 20 millionths of a kelvin. MAP is in the opposite Lagrangian point to the solar

*(text continues on page 782)*

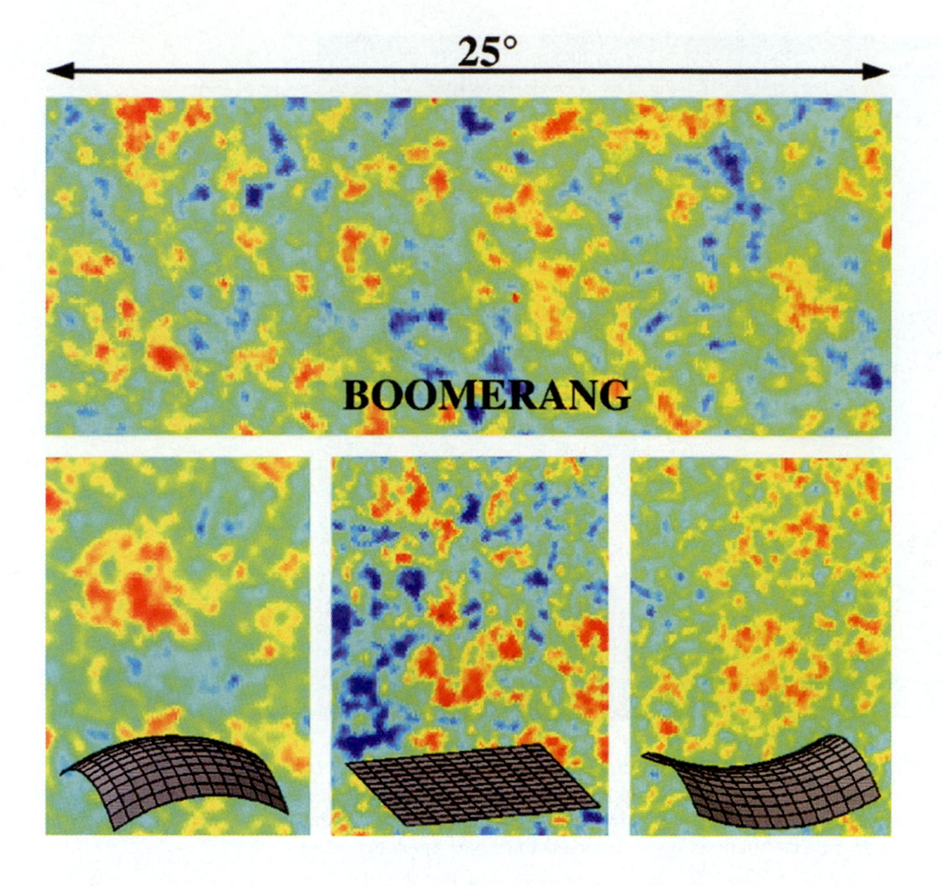


**FIGURE 37–25** The results on the size of fluctuations show that the universe is flat, neither open nor closed. The observations (*top*) agree much better with the calculated model for flat (*bottom middle*) than those for curved universes (*bottom left and bottom right*).

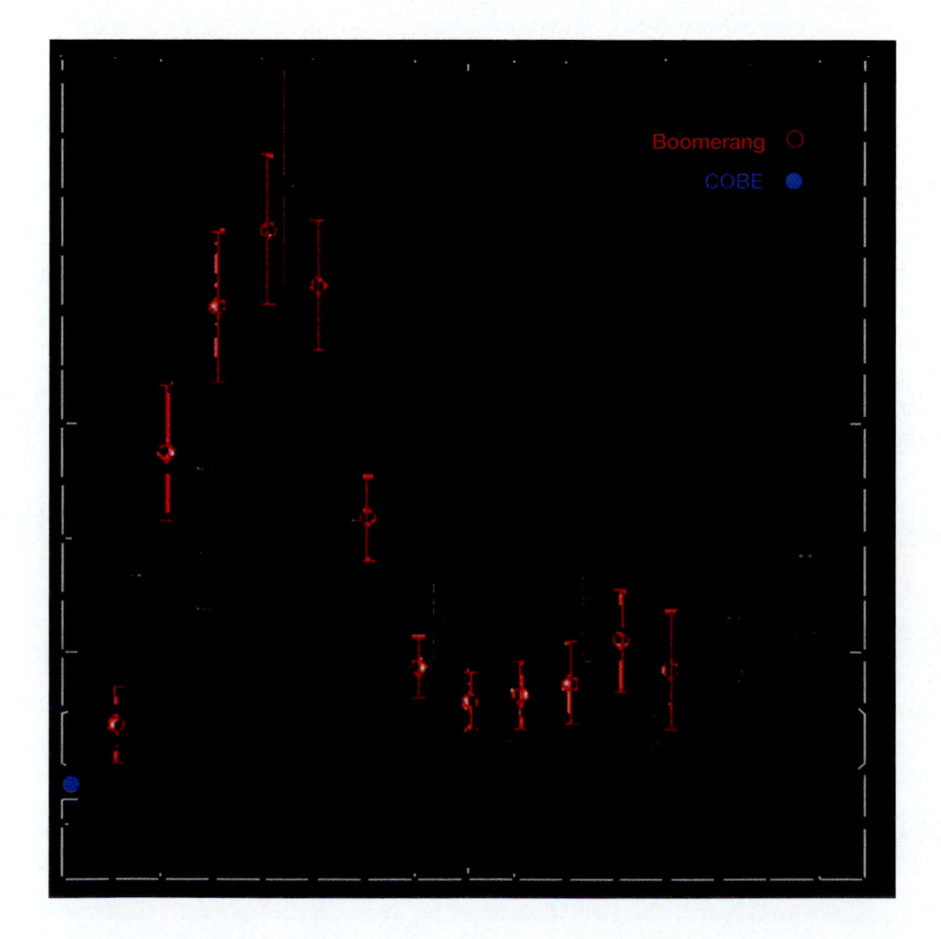


**FIGURE 37–26** A graph from 2000 showing the size of ripples on the *x*-axis and the relative number of ripples on the *y*-axis. Though some variations appear over a wide range of scales, the most common variations appear to be about 1° in diameter, or twice the diameter of the full moon. These variations indicate that the universe is flat. BOOMERanG and MAXIMA results largely agree. A major outstanding question, to be resolved by further observations, was the strength of the "second peak," a peak located to the right of the high peak that shows here, and even details of a possible third peak. Figure 37–28 will continue the story. Space data from NASA's MAP spacecraft should be even better.

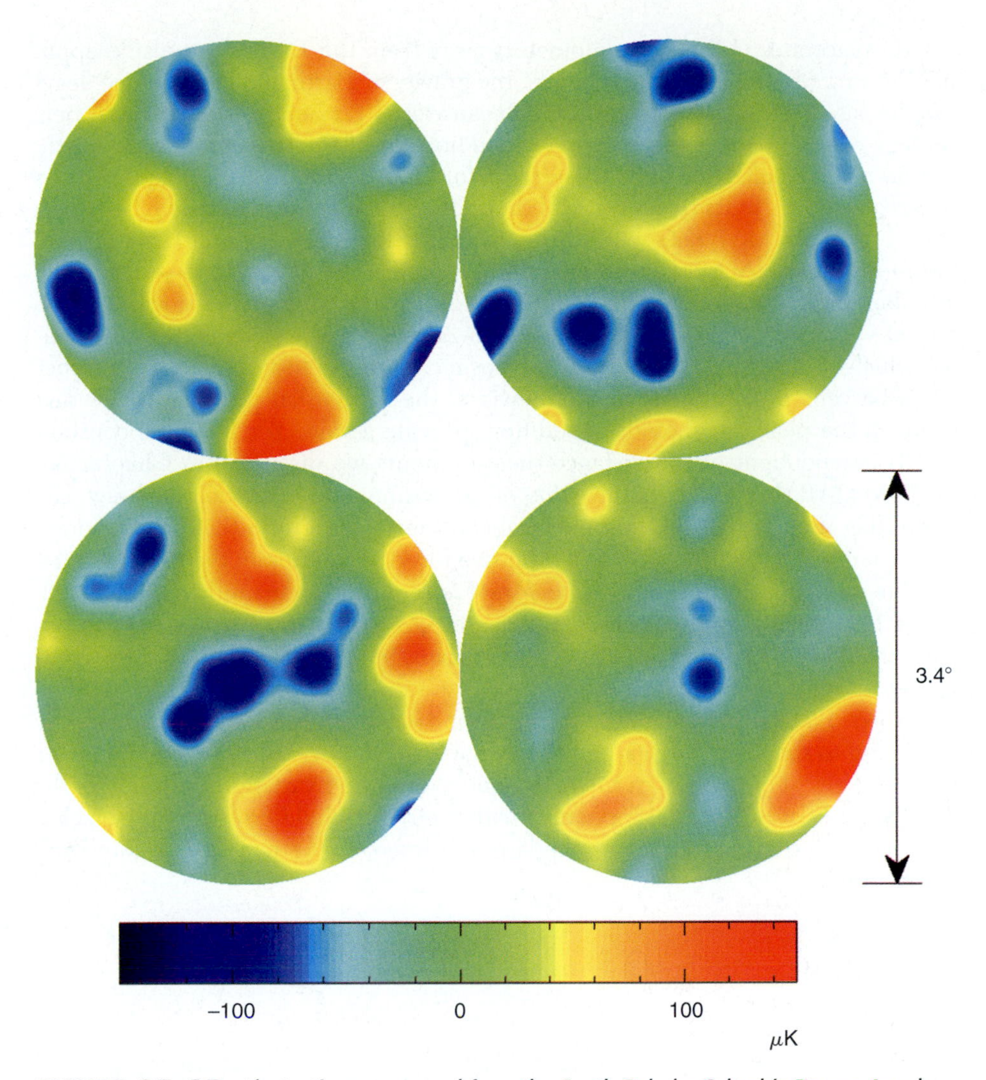


**FIGURE 37–27** Fluctuations measured from the South Pole by Caltech's Degree Angular Scale Interferometer (DASI) and reported in 2001.

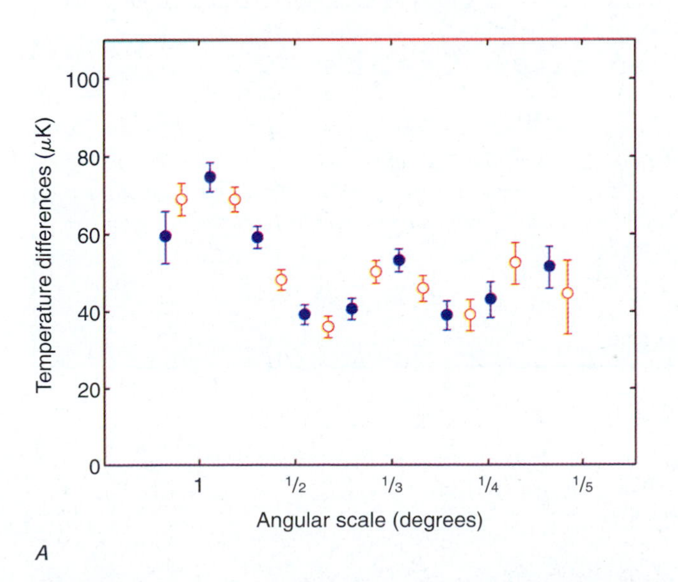


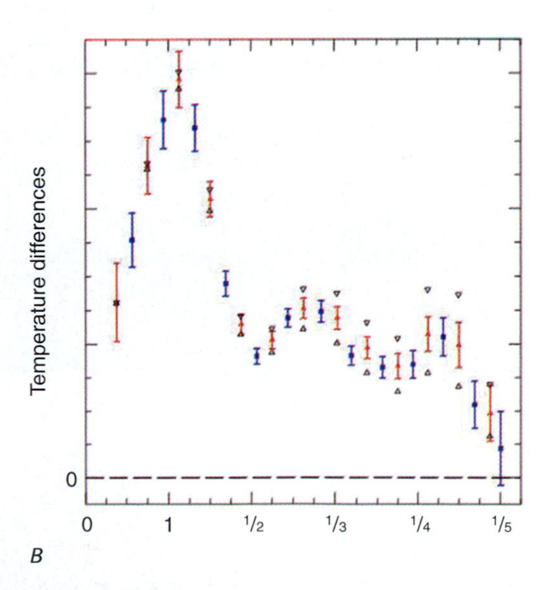


**FIGURE 37–28** The fluctuations measured by DASI, shown in the previous figure, when plotted by their sizes show not only the first peak that corresponds to a flat universe but also a second and even a third peak, which are predicted by the inflationary-universe theory.

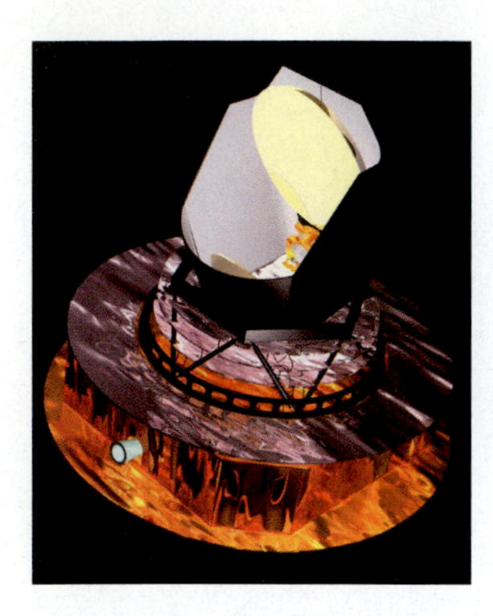

**FIGURE 37–29** An artist's conception of the European Space Agency's Planck spacecraft. It is to be launched in 2007 to study the anisotropy of the cosmic microwave background radiation on a small scale over the whole sky. Its detectors, some of them cooled to 1/10 K, will be sensitive to wavelengths between 1/3 mm and 1 cm.

SOHO spacecraft: 1.5 million kilometers away from the Earth on the side opposite the sun. (At the Lagrangian points, the gravity of Earth and sun balance, leaving the satellite in a stable position.) Because it is farther from Earth, the percentage of sky filled by the Earth, and the heating of the spacecraft from Earth, will be much smaller for MAP than it was for COBE. The maps it makes over its two-year lifetime in its final orbit will therefore be much more detailed. Theorists believe that the MAP data will be able to tell us the amount of dark matter and the form it takes. MAP will improve the measurements of overtone peaks in the distribution of fluctuations. It should also tell us a parameter that measures the rate at which the outward expansion is slowing down, the deceleration parameter, which is the same as specifying the cosmic matter density $\Omega_M$. It does so because baryons provide most of the inertia in the primordial pressure waves that result in the peaks. It should, in addition, provide a value for Hubble's constant that is independent of the distance measurements we discussed in Chapter 35. Further, MAP should tell us the value of the cosmological constant. We saw earlier in this chapter that this cosmological constant, $\Lambda$, is a term that can be added to Einstein's equations to provide a repulsive force to counteract the force of gravity. The idea that the cosmological constant is required is currently controversial, so the MAP results will be eagerly anticipated. As a scientist, I find it hard to believe that this single project will settle as many important cosmological controversies as is claimed in advance.

ESA's Planck project, with a 1.5-m mirror, is scheduled for launch by 2007, also to the Langrangian point known as $L_2$ (Fig. 37–29). Its resolution over the whole sky should be even finer, 1/6°, and should find variations of 6 millionths of a kelvin. ESA is pairing it for launch with their Herschel Observatory. Herschel is to carry a 3.5-meter telescope, the largest telescope ever in space, for studying the infrared.

## IN REVIEW

### BOX 37.4 Cosmic Background Radiation Graphs

The Planck curve of the cosmic background radiation graphs intensity vs. wavelength, as all Planck curves do, and it has a single peak, as all Planck curves do.

The fluctuation curve of the cosmic background radiation graphs the strength of fluctuations vs. the size of fluctuations. It shows at least two peaks. Do not confuse this graph with a Planck curve; both axes are different.

*Note that the cosmic background radiation was set free in the era of recombination, about 300,000 years after the big bang. Do not confuse this era with the era of nucleosynthesis, which was much earlier, only about 1 to 1000 seconds after the big bang.*

## CORRECTING MISCONCEPTIONS

| ✖ *Incorrect* | ✔ *Correct* |
|---|---|
| Galaxies and clusters of galaxies expand as the universe expands. | Galaxies and clusters of galaxies remain the same size, though they grow farther away from each other. |
| It is obvious that matter cannot be created out of nothing. | We have been able to prove that the steady-state theory is ruled out, but the rate invoked for the creation of matter is too low to rule it out straightforwardly. |
| The big-bang theory has a significant number of doubters among astronomers. | Essentially all astronomers have concluded that the big-bang theory is valid. Only a handful of astronomers are not convinced. |
| We can see back to the big bang. | The universe was dense and opaque about 300,000 years after the big bang, and we cannot see further back in any part of the spectrum. |
| The cosmic background radiation is uniform. | The CBR has a dipole anisotropy, which shows on the largest scale, and when that is removed shows small-scale anisotropies. |
| The CBR's Planck curve shows two or three peaks. | All Planck curves have only one peak; the two or three peaks associated with the CBR have to do with the graph of fluctuation size. |

## SUMMARY AND OUTLINE

Olbers's paradox and its solutions (Section 37.1)
- The universe's expansion diminishes energy in quanta.
- We are looking back in time to before the stars formed.

Big-bang theory (Section 37.2)
- Cosmological principle: universe is homogeneous and isotropic
- Hubble's constant and the "age" of the universe

Steady-state theory (Section 37.3)
- Perfect cosmological principle: universe is homogeneous, isotropic, and unchanging in time
- Continuous creation of matter is implied.
- Demise of the steady-state theory with discovery of quasars and 3K black-body radiation

Today's big-bang theory (Section 37.4)
- Supernovae have shown an accelerating universe.
- The universe is flat, since it is at the critical density.
- The cosmological constant is back in favor, and is thought to be 0.7.

Primordial background radiation (Section 37.5)
- Provides strong support for the big-bang theory
- Direct radio measurements gave the temperature.
- Anisotropy indicates motion of the sun.
- COBE data fit black body of $2.725 \pm 0.002$ K.
- COBE found fluctuations that may be seeds from which galaxies grew.
- BOOMERanG, MAXIMA, TopHat, and DASI studied smaller fluctuations.
- The scale of the fluctuations show that the universe is flat.
- NASA's MAP (2001) is to study smaller fluctuations.
- ESA's Planck (2007) will follow with even finer resolution.

## KEY WORDS

cosmology
Olbers's paradox
cosmological principle
homogeneity
isotropy
big-bang theory
open universe
closed universe
flat universe
oscillating universe
perfect cosmological principle*
steady-state theory*
deceleration parameter
critical density
cosmological constant
principle of equivalence
recombination
cosmic background radiation
primordial background radiation
3° background radiation
anisotropy

*This term appears in an optional section.

## QUESTIONS

1. Hindsight has allowed solutions to be found to Olbers's paradox. For example, knowing that the universe is expanding, we can come up with a solution. However, on the basis of the reasoning in this chapter, do you think it is possible that scientists might have used Olbers's paradox to reach the conclusion that the universe is expanding before it was determined observationally? Explain.
†2. Our universe is about $10^{10}$ years old. If we were riding on a light beam emitted from Earth now, how long would we have to wait on the average to arrive on a star, given that the average line of sight ends $10^{24}$ years after the big bang? (Assume for the calculation that the stars are still there and that the universe is not expanding.) What percentage of the $10^{24}$ years is the current age of our universe? The question demonstrates how long $10^{24}$ years is.
†3. Using the formula for the energy of a photon, how many times less energy does a photon corresponding to a wavelength of 1 mm (the peak of the cosmic background curve) have compared with a photon corresponding to 1 Å in wavelength (which would have been emitted by the background in the early universe)?
4. We say in the chapter that the universe is often assumed to be homogeneous on a large scale. Referring to the discussions in Chapter 34, what is the largest scale on which the universe does not seem to appear homogeneous?
5. List observational evidence in favor of and against each of the following: (a) the big-bang theory, and (b) the steady-state theory.
6. What is the relation of Einstein's general theory of relativity to the big bang?
7. What is "perfect" about the perfect cosmological principle?
8. If galaxies and radio sources increase in luminosity as they age, then when we look to great distances, we are seeing them when they were less intense than they are now. We thus cannot compare them directly to similar nearby galaxies or radio sources. If this increase in luminosity with time is a valid assumption, does it tend to make the universe seem to decelerate at a greater or lesser rate than the actual rate of deceleration? Explain.
†9. For a Hubble constant of 75 km/sec/Mpc, show how you calculate the Hubble time, the age of the universe ignoring the effect of gravity. (*Hint:* Take $1/H_0$, and simplify units so that only units of time are left.) Give your final answer in billions of years.
†10. What could the Hubble time be if instead of the value in the previous question the Hubble constant is 60 km/sec/Mpc? Compare with the answer from Question 9.
11. Comment on the additional effect that gravity would have on the answer to the previous two questions.
12. In actuality, if current interpretations are correct, the 3 K background radiation is only indirectly the remnant of the big bang but is directly the remnant of an "event" in the early universe. What event was that?
13. Apply Wien's displacement law to the solar photospheric temperature and spectral peak in order to show where the spectrum of a 3 K black body peaks.
14. What does the anisotropy of the background radiation tell us?
†15. By how much would the H$\alpha$ line (whose rest wavelength is 6563 Å) be shifted by the sun's velocity of 600 km/sec with respect to the background radiation?
16. Discuss the relation of the Planck curves for the primordial background radiation and for the sun in terms of Wien's law and the Stefan-Boltzmann law.
17. In terms of Planck's law, how many points do you have to measure to determine the cosmic background temperature? Explain.
18. Compare the magnitude of the ripples in the cosmic background radiation with the temperature itself.
19. Discuss the significance of the "second peak" in the fluctuation spectrum for cosmic ripples.
20. What advantages should MAP and Planck have over COBE?

†This question requires a numerical solution.

## TOPICS FOR DISCUSSION

1. Which is more appealing to you: the cosmological principle or the perfect cosmological principle? Discuss why we should adopt one or the other as the basis of our cosmological theory.
2. Who do you think deserves the Nobel Prize: the scientists who first observed the background radiation without having a theory, the scientists who had made rough predictions decades earlier, or the scientists who made theoretical models but hadn't yet carried out their planned observations? Note that the Nobel Prize rules allow the award to go to no more than three individuals. Compare with the actual prize award to Penzias and Wilson.

## USING TECHNOLOGY

### W World Wide Web

1. Look at the COBE homepage (http://space.gsfc.nasa.gov/astro/cobe) and then the BOOMERanG (http://www.physics.ucsb.edu/~boomerang), MAXIMA (http://cfpa.berkeley.edu/group/cmb and http://www.lbl.gov/Science-Articles/Archive/maxima-results.html), TopHat (http://topweb.gsfc.nasa.gov/tophat_overview.html), DASI (http://astro.uchica-go.edu/dasi/), and Cosmic Background Imager (http://www.astro.caltech.edu/~tjp/CBI/) homepages to see current results about the cosmic background radiation. A list of related sites is maintained at http://www. tac.dk/~hivon/cmb/.
2. Look at the homepages for MAP (http://map.gsfc.nasa.gov) and Planck (http://sci.esa.int) about the background radiation to see how these projects are coming along.
3. See if any of your questions on cosmology are in Prof. Ned Wright's "Frequently asked questions in cosmology" (http://www.astro.ucla.edu/~wright/cosmology_faq.html).
4. Learn about the Lagrangian points, to which spacecraft are sometimes launched and in which some moons of Jupiter are located, through the Microwave Anisotropy Probe's education program at http://map.gsfc.nasa.gov/m_mm/ob_techorbit1.html.

### REDSHIFT

1. Ponder the Story of the Universe: From Big Bang to Galaxies.
2. Look at Science of Astronomy: Beginnings and Endings: The Big Bang.

One quadrant of the Hubble Deep Field, the image that resulted when the Hubble Space Telescope observed one "normal" unremarkable region of space for a solid week. The images of some of the objects that resulted are the faintest ever detected. Almost every object you see on this image is a distant galaxy.

# Chapter 38

# The Past and Future of the Universe

How did the tremendous explosion we call the big bang result in the universe we now know, with galaxies and stars and planets and people and flowers? Obviously, many complex stages of formation have taken place, and what was torn asunder at the beginning of time has now taken the form of an organized system.

We can trace the expansion backward in time and calculate how long ago all the matter we see would have been so collapsed to a point. The result of this calculation and of other determinations of the universe's age is that the big bang took place some 10 to 14 billion years ago. It is difficult to comprehend that we can meaningfully talk about the first fraction of a *second* of a time so long ago. But we can indeed set up sets of equations that satisfy the physical laws we have derived, and we can make computer simulations and calculations that we think tell us a lot about what happened right after the origin of the universe. In this chapter we will go back in time beyond the point where the background radiation was set free to travel through the universe. This phase of time is known as the **early universe.**

Our knowledge of the structure of the universe and of the nuclear and other particles in it seems to be able to take us back to $10^{-43}$ second after the big bang. This moment in the history of the universe is known as the "Planck time." Before $10^{-43}$ second, the universe was so compressed that not only the laws of general relativity but also those of quantum mechanics have to be taken into account, and we are not at present able to do both simultaneously. So we can't even say $10^{-43}$ second after what, since we don't really know that there was a zero of time. But we measure time from the instant at which everything would have been together, extrapolating the measurable part of the expansion backwards.

In this chapter, we will first discuss how the chemical elements were formed. Then we will go on to discuss how our studies of the elements enable us to make predictions using the big-bang theories that have been current for the last few decades. Next, we will describe our knowledge of the basic physical forces that govern the universe and discuss what we learn about the universe's first second of time. Then we will see how our understanding of this early time has given us a new picture of the universe's evolution and has led to a different understanding of our planet's fate. Many of the ideas described in this chapter about the earliest times are much more speculative than other topics we have discussed. Still, in the last decade, cosmology has moved from a phase in which speculation was dominant to a science in which detailed measurements are driving the conclusions.

**AIMS:** To study the first second of time in the universe, how the lightest of the elements were formed, what we know about "dark matter" and "dark energy," and how the era of inflation explains many observations about the overall structure of the universe.

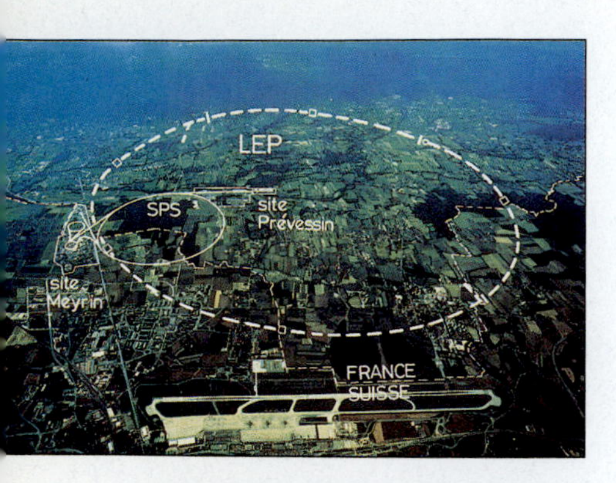


**FIGURE 38–1** The 27-km-circumference tunnel of the Large Electron-Positron accelerator (LEP) at the CERN laboratory is drawn onto this aerial photograph of a region outside Geneva. The tunnel is actually about 100 m underground. A more powerful accelerator is now being built in the same tunnel by installing more powerful magnets to bend the beams of particles.

## 38.1 THE CREATION OF THE ELEMENTS

Modern cosmology is merging with **particle physics**—the studies of the particles inside atoms, which are often carried out with giant atom smashers in which subatomic particles circle around a huge ring at speeds close to that of light and smash into other particles (Fig. 38–1). As we push further back toward the beginning of time in our understanding of the universe, it has become necessary to understand what particles were around and how they interacted with each other.

One striking but basic discovery, known for many decades, was that for every subatomic particle, there is a corresponding **antiparticle.** (A few particles are their own antiparticles, so no annihilation occurs. The photon is one such.) This antiparticle has the same mass as its particle but is opposite in all other properties. For example, an antiproton has the same mass as a proton, but it has negative charge instead of positive charge. Some particles have a property called spin; you can think of a top spinning. The corresponding antiparticles—an antiproton instead of a proton, for example—spin in the opposite direction. Antiparticles together make up **antimatter.** If a particle and its antiparticle meet each other, they annihilate each other; their total mass is 100 per cent transformed into energy in the amount $E = mc^2$.

We know that our universe, up to at least the scale of a cluster of galaxies, is made of matter rather than antimatter. If there were a substantial amount of antimatter anywhere, it would meet and annihilate some matter, and we would see the resulting energy as gamma rays. We do not detect enough gamma rays for this annihilation to be taking place commonly. And there is enough matter between the planets, between the stars, and between the galaxies to provide a continuous chain of matter between us and the Local Group of galaxies, showing that the Local Group is made of matter rather than antimatter. The chain probably extends to the Local Supercluster as well.

**Baryons** are the kind of subatomic particle that make up nuclei; protons and neutrons are the most familiar examples of baryons. Baryons are made of still more fundamental particles called "quarks," which we will define and discuss in Section 38.3. Up to about $10^{-35}$ second after the big bang, there was a balance between baryons and their antiparticles on the one hand and photons (which can be thought of as particles of light) on the other. Up to this time, the number of particles and antiparticles was the same, and there was a continual annihilation of the two; each time a particle and its antiparticle met, their mass was transformed into energy. Similarly, particle/antiparticle pairs (including protons and antiprotons) kept forming out of the energy that was available.

Since theory indicates that matter and antimatter should have been formed in equal quantities, we must explain why our universe is now made almost entirely out of matter. Though few nuclear reactions produce different amounts of matter and antimatter, enough of these imbalanced reactions could have taken place at a sufficient rate in the early universe to provide a slight imbalance: 100,000,001 particles for each 100,000,000 antiparticles. Then all the antimatter annihilated an equal amount of matter. The relatively small residuum of matter left over is the matter that we find in our universe today! The annihilations created 100,000,000 photons (light particles) for each baryon (nuclear particles like protons, neutrons, etc., as we said in the previous paragraph), a ratio that—as we can measure from the background radiation—holds true today. The 1980 Nobel Prize in Physics went to the scientists (James Cronin and Val Fitch) who first detected experimentally that particles can occasionally decay asymmetrically into matter and antimatter (Fig. 38–2). Major experiments are now going on to produce huge quantities of particles that decay asymmetrically into matter and antimatter. These "B-factories," so named because they are producing millions (rather than just a few) of particles known as "B mesons," are pinning down the statistics of these ideas. B mesons and their antiparticles, anti-B mesons, decay into other particles at slightly different rates.

Following the annihilation of most of the protons and essentially all of the antiprotons, the universe had expanded enough so that less energy was available in any given place. Since $E = mc^2$, there was no longer enough energy available to form

Great discoveries sometimes have humble beginnings, especially when the discoverers are not aware that they are making a great discovery. Because of this fact we do not have an attractive picture of the neutral kaon apparatus. I am sorry about this.

Sincerely,

James W Cronin

James W. Cronin

**FIGURE 38–2** James Cronin and Val Fitch received the 1980 Nobel Prize in Physics for their experiment that showed an asymmetry in the decay of a certain kind of elementary particle. This result may point the way to the explanation of why there is now more matter than antimatter in the universe.

quarks, and therefore to form protons and antiprotons, in particular. But until about a second had passed, there was still enough energy to form lighter particles like electrons and positrons (anti-electrons). Then essentially all the electrons and positrons annihilated each other, leaving a surplus of electrons. We were left with a sea of hot radiation, which dominated the universe, and which we now detect as the primordial background radiation. At that time, unlike the present, the photons each had so much energy that most of the universe's energy was in the form of these photons. Since neutrinos interact so rarely, they may not have been annihilated and many presumably remain from this era.

The results of computer-aided calculations of what happened after the first few seconds of the universe are less speculative than the reason why we wound up with a universe made of matter. According to these calculations, the universe's fantastically high density and temperature continued to diminish as the universe expanded. The conditions of high temperature and density of this stage are now being reproduced in huge physics experiments on particle accelerators—commonly known as "atom smashers"—on Earth (Fig. 38–3). The experiment whose result is shown reached densities over 20 times that in atomic nuclei. Such experiments are meant to spill the quarks and the particles that hold them together, called "gluons," into a sort of liquid known as a quark-gluon plasma.

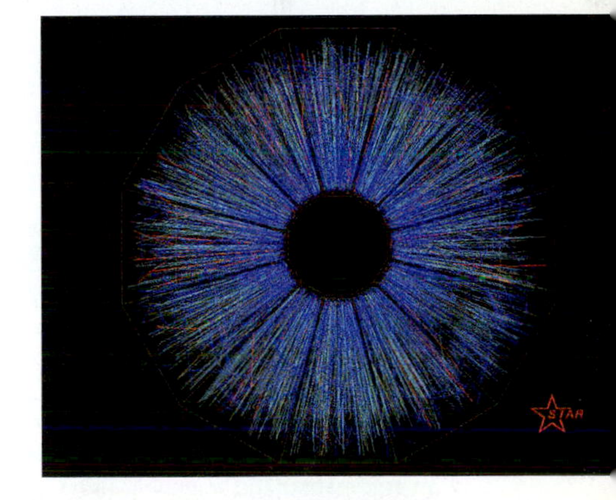


**FIGURE 38–3** Conditions in atom smashers in terrestrial laboratories can reach temperatures and densities typical of the early era of the universe. Here we see the straight-on view of a collision of two gold nuclei, each travelling at 99.99% of the speed of light, at the Relativistic Heavy Ion Collider at the Brookhaven National Laboratory, which is on Long Island in New York. The various colored lines show elementary particles being ejected in all directions. The data are being analyzed to find out the conditions when so much energy is poured into such a small space. The results hint that a quark-gluon plasma, a state of matter that hasn't existed since right after the big bang, resulted. This plasma may be slowing down particles shooting out the sides, which would explain why fewer high-energy particles are seen spraying out than otherwise expected.

After about 5 seconds, the universe had cooled to a few billion degrees. Only simple kinds of matter—protons, neutrons, electrons, neutrinos, and photons—were present at this time. The number of protons and neutrons was relatively small, but the number grew so that these particles had a relatively larger share of the universe's energy as the temperature continued to drop.

After about a hundred seconds, the temperature dropped to a billion degrees (which is low enough for a deuterium nucleus to hold together). The protons and neutrons began to combine into heavier assemblages—the nuclei of the heavier isotopes of hydrogen and elements like helium and lithium. The formation of the elements is called **nucleosynthesis.**

The first nuclear amalgam to form was simply a proton and neutron together. We call this a **deuteron;** it is the nucleus of deuterium, an isotope of hydrogen. Then two protons and a neutron could combine to form the nucleus of a helium isotope, and then another neutron could join to form ordinary helium. Within minutes, the temperature dropped to 100 million degrees, too low for most nuclear reactions to continue. Nucleosynthesis stopped, with about 25 per cent of the mass of the universe in the form of helium. Nearly all the rest was and is hydrogen.

To test this theory, astronomers have studied the distribution of helium in the universe to see if it tends to be approximately this percentage everywhere. But helium is very difficult to observe; it has few convenient spectral lines to study. Furthermore, even when we can observe it in stars we are observing only the surface layers, which do not necessarily have the same abundances of the elements as the

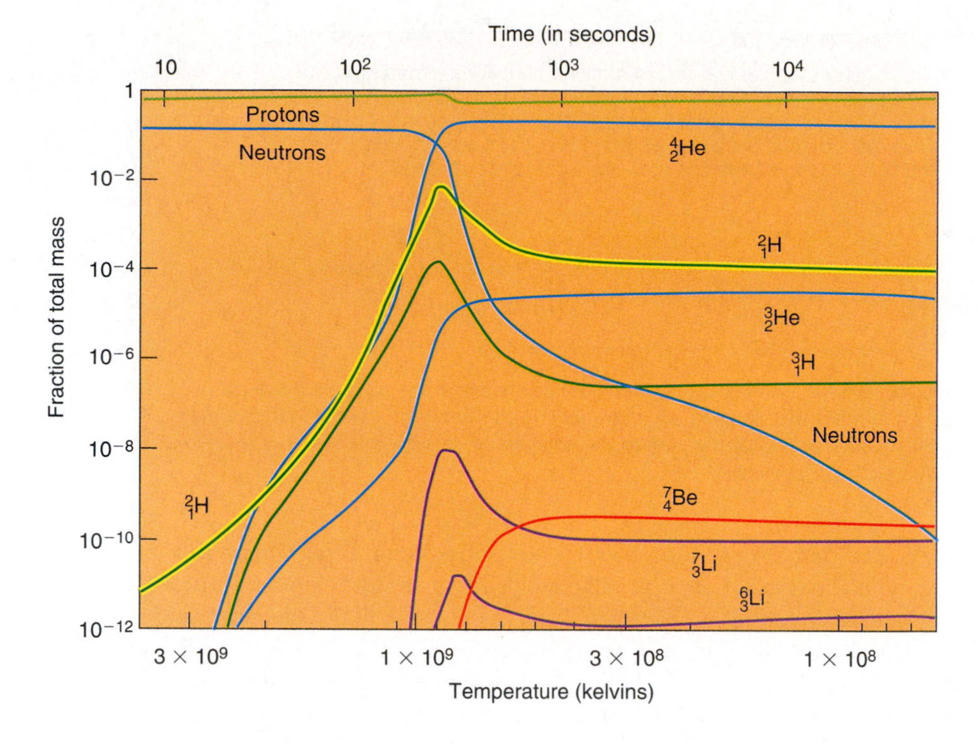


**FIGURE 38–4** The "standard model" of nucleosynthesis is a theoretical model that shows the changing relative abundances in the first minutes after the big bang. Time is shown on the top axis, and the corresponding temperature is shown on the bottom axis. The standard model was for years that of William A. Fowler of Caltech, Fred Hoyle of the University of Cambridge, and Robert V. Wagoner of Stanford, updated by Fowler and Wagoner. It has been updated further, especially by scientists at the University of Chicago and Fermilab, to include advances in our understanding of the relation of nucleosynthesis to elementary-particle physics.

interiors of the stars or the "cosmic" abundance distributed throughout the interstellar medium. "The helium problem"—whether the helium abundance is constant throughout the universe—has been extensively studied. Though the investigation is difficult, the helium is apparently uniformly distributed.

Only the lightest elements were formed in the early stages of the universe (Fig. 38–4). These lightest elements are hydrogen (1 proton), helium (2 protons), and smaller amounts of lithium (3 protons), beryllium (4 protons), and boron (5 protons). We get almost entirely only hydrogen and helium for the following reason: The nuclear state of mass 5 (the lithium isotope with three protons and two neutrons, or else the helium isotope with two protons and three neutrons), which can be reached by adding a proton to a helium-4 nucleus, is not stable; it doesn't hold together long enough to be built upon by the addition of another proton to form still heavier nuclei. Even if this gap is bridged, we get to another extremely unstable nuclear state—mass 8. As a result, heavier nuclei are not formed in the early stages of the universe. Only hydrogen (and deuterium) and helium (and perhaps traces of low-mass isotopes such as those of lithium with nuclear mass of 6 or 7) would be formed in the period soon after the big bang.

Some of the earliest quantitative work on nucleosynthesis in the big bang was described in an article published under the names of Ralph Alpher, Hans Bethe, and George Gamow in 1948. Actually, Alpher and Gamow did the work, and just for fun they included Bethe's name in the list of authors so that the names would sound like the first three letters of the Greek alphabet: alpha, beta, gamma. These letters seemed particularly appropriate for an article about the beginning of the universe.

The heavier elements must, then, have been formed long after the big bang. E. Margaret Burbidge, Geoffrey Burbidge, William A. Fowler, and Fred Hoyle showed in 1957 how both heavier elements and additional amounts of lighter elements can be synthesized in stars. In a stellar interior, processes such as the triple-alpha process (Section 27.4) can get past these mass gaps. Elements are also synthesized in supernovae (Section 29.2), as has been confirmed with observations of Supernova 1987A (Fig. 38–5) to study the gamma-ray spectral lines that result.

**FIGURE 38–5** Supernova 1987A (*lower right*), the site of formation of heavy elements. We see also the Tarantula Nebula (*upper left*). Both are in the Large Magellanic Cloud.

Astronomers thus believe that element formation took place in two major stages. First, light elements were formed soon after the big bang. Later, the heavier elements were formed in stars or in stellar explosions.

The origin of elements 4 and 5, beryllium and boron, was never quite satisfactorily understood until the interpretation of spectra from the Hubble Space Telescope. They are too massive to have been formed in the big bang but too fragile to have survived in a star. The spectra revealed more boron in relatively old stars than had been expected. There shouldn't have been enough heavy nuclei around in those early days to be able to be split to make the boron. The new interpretation involves cosmic rays of oxygen, carbon, and nitrogen that were sent out from supernova shock waves. In the model, these cosmic rays rammed into the hydrogen and helium formed in the big bang, with beryllium and boron splitting off.

The origin of element 3, lithium, was elucidated in 2001 by Caty Pilachowski of the National Optical Astronomy Observatories. She found that the largest and most massive of the red supergiant stars produced most of the lithium we have in the universe. Only about 10 per cent of the lithium came from the nucleosynthesis era. Deuterium, on the other hand, didn't form at all after the nucleosynthesis era. Since primordial lithium is polluted by lithium more recently formed, and similarly with beryllium and boron, only deuterium gives us a direct window into this early time.

The theoretical calculations of the formation of the elements use values measured in the laboratory for the rate at which particles and nuclei interact with each other. For his role in making these measurements and in the theoretical work on nucleosynthesis, William Fowler shared in the 1983 Nobel Prize in Physics.

## 38.2 THE FUTURE OF THE UNIVERSE

The question of the future of the universe has exciting new answers, with the discovery of the acceleration of the universe (Section 35.5). The question has long been: How can we predict how the universe will evolve in the distant future? We have seen that the steady-state theory has been discredited, so let us discuss the alternatives that are predicted by different versions of standard big-bang cosmologies.

The basic question used to be whether the mass in the universe had enough gravity to overcome the expansion. If gravity was strong enough, then the expansion would gradually stop, and a contraction would begin. If gravity was not strong enough, then the rate of expansion might slow, but the universe would continue to expand forever, just as a rocket sent up from Cape Canaveral will never fall back to Earth if it is launched with a high enough velocity. But at present, as we have seen, we realize that the amount of gravity from ordinary matter is far below that needed to reach the critical density. Based on the inflationary theory, to be discussed later in this chapter, the critical density is reached by adding the effect of the cosmological constant. Still, traditional methods are used to assess the mass in the universe, even if mass makes up only a minor fraction of what there is.

To assess the amount of gravity, we must determine the average mass in a given volume of space, that is, the density. It might seem that to find the mass in a given volume we need only count up the objects in that volume: one hundred billion stars plus umpteen billion atoms of hydrogen plus so much interstellar dust, and so on. But this method has severe limitations, because many kinds of mass are invisible to us (if, indeed, mass exists in such forms). How much matter is in black holes or in brown dwarfs or in the form of neutrinos, for example? Until a few years ago, we couldn't measure the molecular hydrogen in space, and until a few decades ago, we couldn't measure the atomic hydrogen in space either. X-ray and infrared observations now rule out a massive intergalactic medium. But adding all these types of matter, invisible to our eyes but not to our varied types of telescopes, increases the mass density from half a per cent to about 3 per cent.

We must turn to methods that assess the amount of mass by effects that don't depend on the visibility of the mass in any part of the spectrum. All mass has gravity, for example.

### 38.2a Dark Matter and Dark Energy

One place to investigate large-scale gravitational attractions is in clusters of galaxies. Clusters of galaxies appear to be large-scale stable configurations that have lasted a long time. But since galaxies have random velocities in various directions with respect to the center of the mass of the cluster, why don't the galaxies escape?

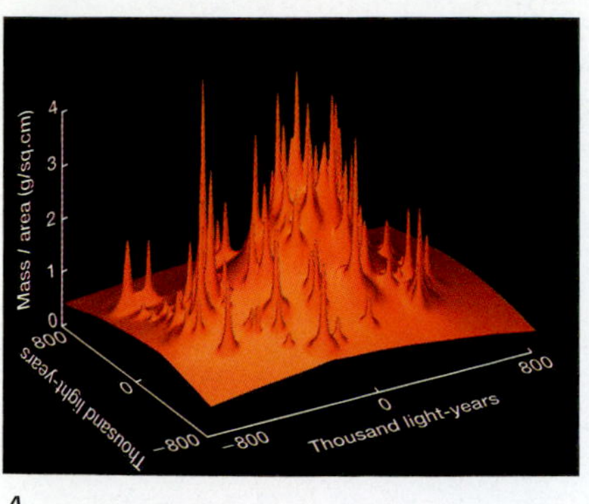


A

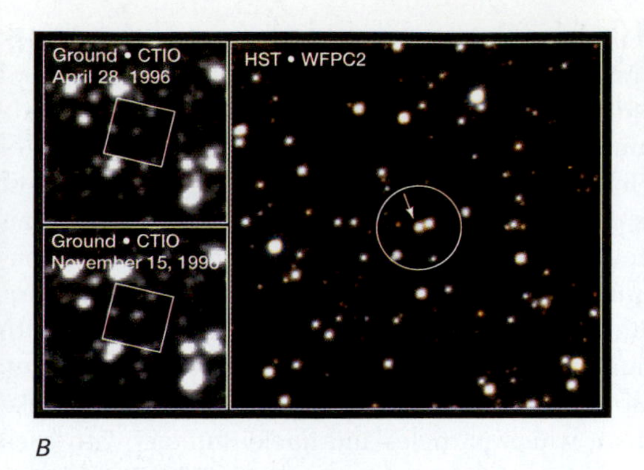


B

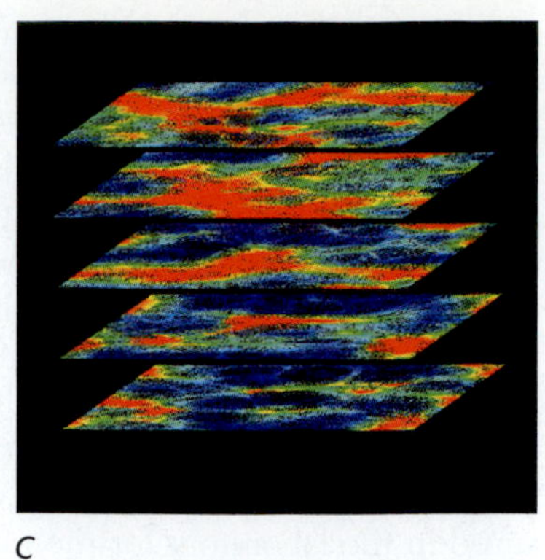

C

**FIGURE 38–6** (*A*) A calculation showing the distribution of dark matter in the cluster of galaxies 0024+16, based on observations of gravitational lensing. (*B*) A microlensing event, as a MACHO passed in front of a star. The star's light is bent by an angle more than 100 times smaller than Hubble's spatial resolution, so the separate images of the background star do not show. Instead, we see an apparent brightening of the background star, since the warping of space from the black hole magnifies the background star's image. The event was discovered, by the Massive Compact Halo Object (MACHO) collaboration, using a telescope in Australia, while the magnification was still increasing. The prompt announcement allowed telescopes in Chile and Australia to follow the event. The Hubble Space Telescope image shown was made after the micro-lensing event was over. It revealed that the lensed star was blended, in the blurrier ground-based observations, with other stars. Only with the Hubble observation could the mass of the black hole be determined. (*C*) The results of a stimulation involving 3 million particles of cold dark matter. Structure is smoothed at a scale of 10 Mpc. The results range from almost 10 times underdense (blue) to over 10 times overdense (red).

The missing-mass problem is the discrepancy found when the mass derived from consideration of the motions of galaxies in clusters is compared with the mass that we can observe. It is really a "missing-light problem"; the mass must be there.

If we assume that the clusters of galaxies are stable in that the individual galaxies don't disperse, we can calculate the amount of gravity that must be present to keep the galaxies bound. Knowing the amount of gravity in turn allows us to calculate the mass that is causing this gravity. When this supposedly simple calculation is carried out for the Virgo Cluster, it turns out that there should be 50 times more mass present than is observed; 97 per cent of the mass expected is not found. This deficiency has long been called the **missing-mass problem.** Since we cannot see it, the additional mass is **dark matter,** which we discussed briefly in Sections 33.6, 35.7, and 35.10. It shows up in gravitational lensing, so images of clusters of galaxies showing lensing can be interpreted to figure out what the distribution of dark matter must be (Fig. 38–6A).

The missing-mass problem also showed up in studies of the rate at which galaxies rotate. We have seen how our own galaxy's mass can be deduced by observing the rate at which objects in it revolve around the galactic center (Section 33.6). Vera Rubin of the Carnegie Institution of Washington and colleagues found that similar methods applied to distant galaxies also show that the revolution rate in each galaxy does not decline with distance from the galaxy's core as expected. Thus her pioneering studies of rotation curves showed the missing-mass problem.

More recently, scientists interpreting the data from the 2dF project (Section 35.7b), in which redshifts for thousands of galaxies give a third dimension to a map, concluded that baryons make up about 5 per cent and dark matter makes up about 30 per cent of the critical density. They made careful measurements of the amounts of clumping of galaxies in three-dimensional space, forming superclusters. One of their lines of argument compared their measurements of the clumping with the fluctuations in the cosmic microwave background. They also calculated how much gravity must be present to match the observed motions of galaxies. In both cases, they find a mass density of about ⅓ the critical density, requiring the presence of about 30 per cent dark matter.

In what form might the missing mass be present? The idea of a few years ago that neutrinos have mass (Section 38.2c) is in favor, since one experiment with Super-Kamiokande has verified the suggestion. Because they move at high speeds, very close to the speed of light, and because particles in a gas at high temperature have high speeds, neutrinos with mass would be an example of "hot dark matter." But neutrinos are thought to make up only about 2 per cent of the critical density, roughly as much as the ordinary matter.

As another type of constituent of the missing mass, there is much talk about hypothetical particles called "axions." These axions would be an example of "cold dark matter," since they would move at low speeds. Other as yet undiscovered elementary particles that physicists think might exist include some that are relatively massive for elementary particles but that don't interact with the normal ("strong") force between nuclear particles. They are thus known as *W*eakly *I*nteracting *M*assive *P*articles, or WIMPs. WIMPs are another potential kind of cold dark matter, and theory has it that they would have been formed in the first second of time after the big bang. The work of the cosmologists increasingly joins the work of physicists studying elementary particles. At present, the astronomical observations are constraining the properties and numbers of the hypothetical new particles. Other forms of dark matter are also being invoked to explain what may make up 97 per cent or so of the universe.

"Lots of things are invisible but we don't know how many because we can't see them."—Dennis the Menace

Another form the dark matter could take is objects like brown dwarfs or dim stars that are too faint for us to see. Such objects are called MACHOs, standing for *M*assive [*A*strophysical] *C*ompact *H*alo *O*bjects. (The name was chosen to counterbalance Weakly Interacting Massive Particles, WIMPs.) Several programs are under way to see if such objects occasionally go in front of stars, causing them to brighten for a few weeks by gravitational lensing (Section 34.6). Several hundred such "microlensing" brightenings have been reported in the surveys, which are using CCDs to observe tens of millions of stars looking toward the center of our galaxy or toward the Large Magellanic Cloud. Evidence that these changes represent microlensing instead of the brightening of a variable star is the fact that the events are the same in different colors and are symmetric over time. An effect akin to parallax for star distances sometimes enables scientists to distinguish between the presence of a massive lens and a slow relative motion between the lensing star and the lensed star.

The lensed objects in the image shown (Fig. 38–6B), discovered in 1996, as well as a second object discovered two years later, turn out to have six times the mass of the sun. Ordinary stars that massive would shine brightly enough for us to see them on their own. So we deduce that the objects are compact, and the only compact objects of that mass are black holes. It is nice to be able to detect black holes that, unlike the discoveries reported in the black-hole chapter, are single objects rather than objects that are members of binary systems.

The current understanding is that matter that has gravity makes up 30% of the contents of the universe. That is divided among 3% baryons, 2% neutrinos, and 25% cold dark matter. **Dark energy,** in the form of the cosmological constant or in some other form, provides the remaining 70%.

These surveys should eventually give us an idea of the fraction of the dark matter that MACHOs make up. Some indications are that the MACHOs make up 20 per cent of the mass of the Milky Way Galaxy's halo.

## 38.2b Deuterium and Cosmology

Perhaps the major method now being used to assess the density of baryons (ordinary particles like protons and neutrons, as we saw in Section 38.1) in the universe (the "baryon density") concerns itself with the abundance of the light elements. The light elements include hydrogen, helium, lithium, beryllium, and boron. Since these light elements were formed soon after the big bang (as we saw in Section 38.1), they tell us about conditions at the time of their formation. If we can find what the baryon density was then, we can use our knowledge of the rate that the universe has been expanding to determine what the baryon density is now.

But somehow we must distinguish between the amount of these elements that was formed in the big bang and the amount subsequently formed in stars. This consideration complicates the calculations for helium, for example, because helium formed in stellar interiors and then spewed out in supernovae has been added to the primordial helium. We have also seen how lithium, beryllium, and boron are sometimes seen subsequent to the era of nucleosynthesis.

Fortunately, the deuterium isotope of hydrogen is free of this complication. We do not think that any detectable deuterium is formed in stars (the deuterium formed as an intermediate stage in stellar fusion almost immediately fuses into helium), so all the deuterium now in existence was formed at the time of the big bang. Much of that original deuterium has been destroyed since it was formed, by being used up in stars.

Deuterium has a second property that makes it an important probe of the conditions that existed in the first fifteen minutes after the big bang, when the deuterium was formed. The amount of deuterium that is formed is particularly sensitive to the density of matter at the time of formation. A slight variation in the primordial density makes a larger change in the deuterium abundance than it does in the abundance of other isotopes. We measure the amount of deuterium relative to the amount of hydrogen.

The success of matching theory and observation for the abundances of deuterium and the light elements makes the study of nucleosynthesis into one of the pillars of modern cosmology. It joins Hubble's law (and, subsequently, the existence of background radiation and the ripples in it) as fundamental facts that any theory must explain.

Why is the ratio of deuterium to (ordinary) hydrogen sensitive to density? Deuterium very easily combines with an additional neutron. This combination of one proton and two neutrons is another hydrogen isotope (tritium). The second neutron quickly decays into a proton, leaving us with a combination of two protons and one neutron, which is an isotope of helium. If the universe was very dense in its first few minutes, then it was easy for the deuterium to meet up with neutrons, and almost all the deuterium "cooked" into helium. If, on the other hand, the density of the universe was low, then most of the deuterium that was formed still survives.

Theoretical calculations give the relation of the amount of surviving deuterium and the density of the universe, for a uniform universe (Fig. 38–7). This graph can be used to find the density that would have been present soon after the big bang, if the ratio of deuterium to hydrogen can be determined observationally. Most of the energy in the universe was in the form of radiation instead of matter in this early

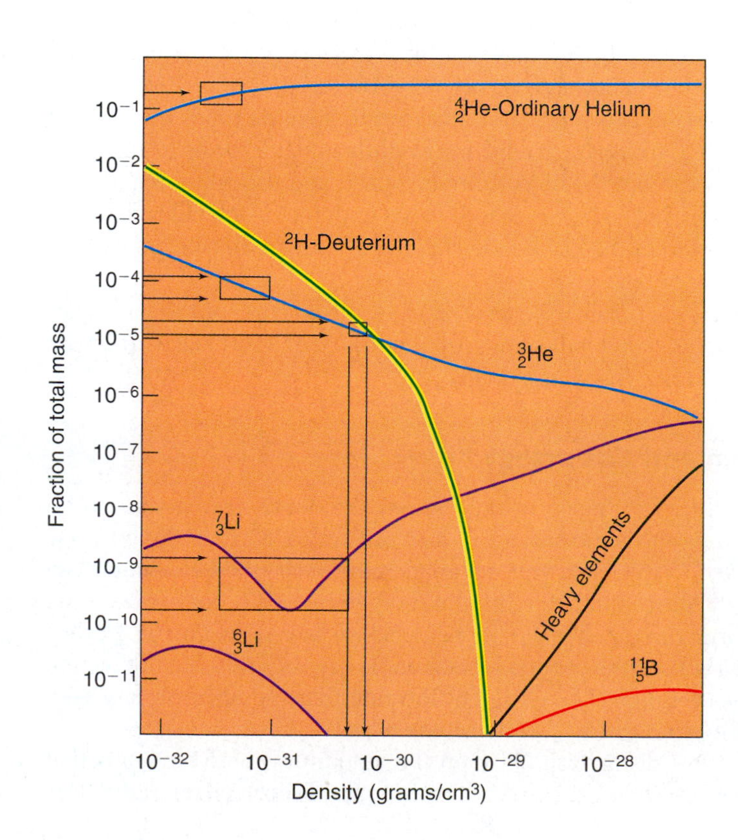


**FIGURE 38–7** The horizontal axis shows the current cosmic density of matter. From our knowledge of the approximate rate of expansion of the universe, we can deduce what the density was long ago. The abundance of deuterium is particularly sensitive to the time when the deuterium was formed. Thus present-day observations of the deuterium abundance tell us what the cosmic density is, if we follow the arrow on the graph. The measured abundances of helium-3 and of lithium-7 are also useful, though helium-4 is relatively insensitive to nucleosynthesis conditions.

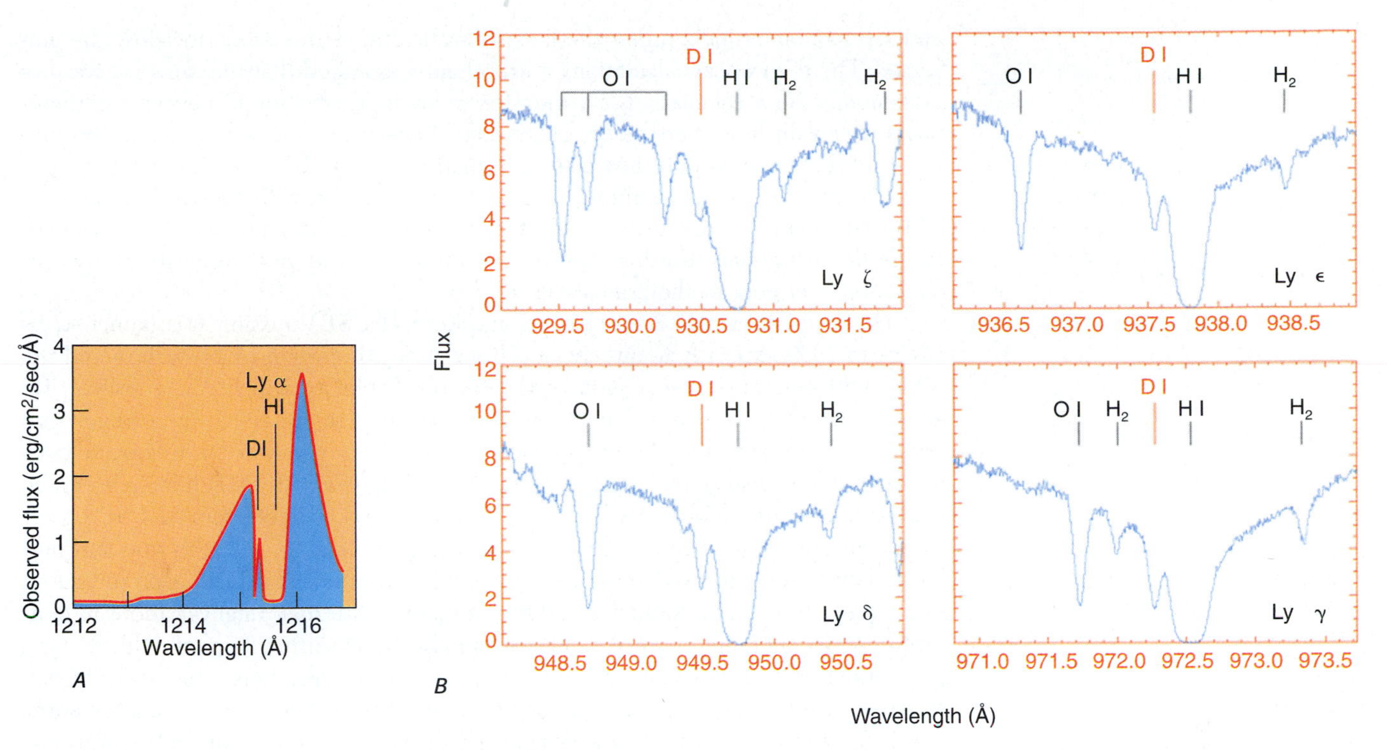


**FIGURE 38–8** (*A*) Observations from the Hubble Space Telescope of the Lyman-$\alpha$ lines of interstellar deuterium and ordinary hydrogen. The broad absorption that takes up the whole graph is from neutral hydrogen in the star Capella. The interstellar deuterium line is the narrow dip marked D I. The interstellar hydrogen line is marked H I. They are narrower than the stellar hydrogen line. (*B*) A spectrum of the star BD+28°4211 showing the Lyman-delta lines of interstellar hydrogen and deuterium from the FUSE spacecraft. The wavelength separation between the lines of hydrogen and deuterium is equivalent to a Doppler shift of 81 km/sec, and the x-axis is often labelled in velocity instead of in wavelength in such comparisons.

era, so the rate of the universe's expansion depended on the ratio of the number of photons to the number of baryons. Thus the deuterium measurements we are describing here give the "baryon density."

Unfortunately, deuterium is very difficult to observe. It has no spectral lines accessible to optical observation. Deuterium makes up one part in 6600 of the hydrogen in ordinary seawater, but it was not at first known how this ratio related to the cosmic abundance of deuterium.

Major uncertainties remain, but the detections of deuterium that have been made so far (Fig. 38–8) agree that the amount of deuterium is such that there is not, and was not, enough density in the form of ordinary matter to ever reverse the expansion of the universe. The results indicate, thus, that the universe will *not* fall back

IN REVIEW

## BOX 38.1 Pillars of Cosmology

At present, our understanding of the universe as a whole rests on four basic types of observations:

1. Hubble's law
2. The primordial background radiation
3. The distribution of fluctuations in the background radiation
4. The cosmic abundance of deuterium

on itself in a big crunch, unless there exists sufficient "dark matter" to "close the universe." The theoretical calculations until recently assumed that the universe was homogeneous. New calculations allowing for some concentration of matter into dense pockets explain how there could be enough deuterium to detect yet have the universe be flat, even without invoking dark matter.

The discovery of deuterium in a few distant quasars, with the Keck Telescope, adds to the evidence that ordinary matter does not "close the universe." The strength of the deuterium spectral line observed is interpreted to give a density of baryons only a few per cent of the critical density.

The Far Ultraviolet Spectroscopic Explorer (FUSE) mission was launched by NASA in 1999 explicitly to answer questions about the origin of deuterium and the other light elements. The region of the far ultraviolet to which it is sensitive, for about a 300 Å span of wavelength starting with the Lyman-alpha line of hydrogen and going longward, includes not only the Lyman series of atomic hydrogen and deuterium but also absorption bands of molecular hydrogen and the HD molecule (which contains one atom of ordinary hydrogen and one atom of deuterium). FUSE observations are pinning down the deuterium abundances in stars near the sun, and may have enough accuracy to determine whether the variations that have been measured in the past are from inaccuracies in measurement or are real. Real variations in deuterium abundances in nearby stars would complicate our efforts to model the changes over time in the abundances. The FUSE spacecraft has also been able to study redshifted hydrogen in the Lyman-alpha forest seen in absorption toward distant quasars, helping to improve our understanding of the distribution of baryons in the universe. FUSE is also sensitive to some moderately highly ionized ions, like five-times-ionized iron, and the amount of this O VI it has detected has shown that there is a substantial amount of hot, metal-enriched gas in intergalactic space. The enrichment has had to take place in stars, so this gas is not primordial.

### 38.2c Neutrinos

Neutrinos have the very interesting property of travelling very rapidly; indeed, we have long thought that they always travel at the speed of light. Now, matter cannot travel at the speed of light according to Einstein's special theory of relativity, because its mass gets larger and larger and approaches infinity as its speed approaches the speed of light. Relativity theory shows that the mass of an object is equal to a constant quantity called the **rest mass** divided by a quantity that gets smaller (approaching zero) as the object goes faster in approaching the speed of light. So the mass of an object gets larger as the object goes faster, and at speeds close to the speed of light is so large that it takes a tremendous amount of energy to accelerate it a little more.

Since the burst of neutrinos (Fig. 38–9) we received on February 23, 1987, from Supernova 1987A presumably left the star in the Large Magellanic Cloud all at the same instant, the spread in arrival times at Earth gives us a limit on how much mass neutrinos have, as we saw in Section 29.6b. (They would all travel at the same speed—the speed of light—if neutrinos were massless.) The limit is very small, so small that neutrinos probably cannot provide enough mass to close the universe. Experiments in terrestrial laboratories to detect neutrino mass have not succeeded, in spite of the discovery that neutrinos have some mass. In any case, neutrinos move so fast that they are considered "hot dark matter."

Observations from the Super-Kamiokande experiment, which we discussed in the context of its solar-neutrino observations (Section 27.6), have indicated that neutrinos oscillate from one type to the other. They can do so, theoretically, only if they have mass. So for the last few years physicists have accepted that neutrinos have some mass, even though the Super-Kamiokande experiment does not tell how much mass they each have. The mass is small, however. There are so many neutrinos that even with a small mass per neutrino, the total mass can be considerable. But we

**FIGURE 38–9** This cavity—deep underground in a salt mine near Cleveland—was filled with 10,000 tons of water to make the Irvine-Michigan-Brookhaven (IMB) detector. Twenty-four hundred photomultipliers were installed to detect the flashes of light that result as neutrinos or other elementary particles pass through. Its detection of a few neutrinos from Supernova 1987A gave us our best constraint on the mass that neutrinos can have. The limit is so low that it seems unlikely that the missing mass is in the form of neutrinos.

know, still, that the total mass in the form of neutrinos is far below that to have the universe reach the critical density, and is perhaps only about 2 per cent of the critical density.

## 38.3 FORCES IN THE UNIVERSE

The study of cosmology in many ways is becoming more and more dependent on the study of the smallest particles in the universe, known as particle physics. Our increasing knowledge of particle physics is being incorporated in our models of how the universe's earliest evolution went. Thus to understand modern trends in studying cosmology, we must become acquainted with some of the fundamental ideas of how particles and forces form and interact.

There are four known types of forces in the universe:

1. The **strong force,** also known as the **nuclear force,** is the strongest. It is the force that binds particles together into atomic nuclei. Although it is very strong at close range, it grows weaker rapidly with distance.

Only some particles "feel" the strong force. These particles, which include protons and neutrons, are composed of six kinds of particles called **quarks** (Fig. 38–10). (James Joyce used the word "quark" in *Finnegan's Wake,* and scientists have appropriated it.) The six kinds (flavors) of quarks are called "up," "down," "strange," "charmed," "top," and "bottom." ("Top" and "bottom" were sometimes less prosaically called "truth" and "beauty," respectively.) Each kind of quark can have one of three properties called **colors,** often called red, blue, and green. (These names are whimsical and do not

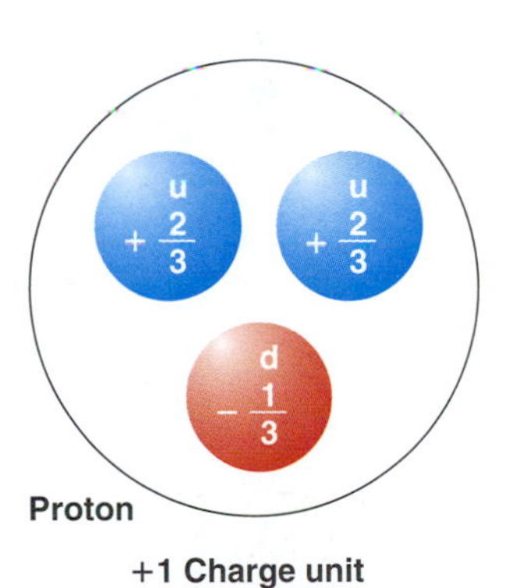


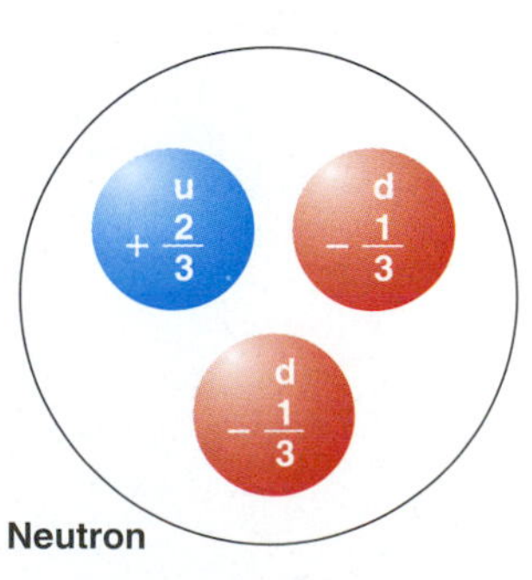


**FIGURE 38–10** The most common quarks are *up* (*u*) and *down* (*d*); ordinary matter in our world is made of them. The up quark has an electric charge of $+^2/_3$, and the down quark has a charge of $-^1/_3$. This fractional charge is one of the unusual things about quarks; prior to their invention (discovery?), it had been thought that all electric charges came in whole numbers. Note how the charge of the proton and of the neutron is the sum of the charges of their respective quarks. In more elaborate current models, notably the one championed by the physicist Richard Feynmann, the quarks themselves are pointlike and the spheres shown that make up protons and other particles are filled with gluons and pairs containing quarks and their antiquarks.

A DEEPER DISCUSSION

### BOX 38.2 Families of Matter

The strong force is carried between quarks by a particle; since this particle provides the "glue" that holds nuclear particles together, it is known (believe it or not) as a "gluon." Studies with atomic accelerators have led to the discovery of evidence for the existence of all six quarks and of the gluons. The sixth quark ("truth" or "top")—the last to be discovered—wasn't detected until 1995 because it is so heavy and is therefore very hard to make in an accelerator (because $E = mc^2$). It is 40 times more massive than the bottom quark, which is itself hundreds of times more massive than the "up" and "down" quarks. By itself, the top quark has about as much mass as a silver atom!

Theoretical work indicates that the observed cosmic abundances of helium and of deuterium could not arise if there were more than three pairs of quark types, each of which would come with its own type of electron and its own type of neutrino. We currently observe three pairs of quark types (up–down, strange–charmed, truth–beauty). The electron has two more massive variations, known as the muon ("Who ordered that?" said the physicist I. I. Rabi when the muon was discovered, since it appeared to be an electron but with a much higher mass) and the tau particle. To go with each type of electron is a type of neutrino (electron neutrino, muon neutrino, and tau neutrino). Measurements have since confirmed that there are only three types of neutrino. The tau neutrino wasn't isolated until it was detected at Fermilab in 2000.

We thus also conclude that there are only the three pairs of quarks that we already know. They make up the hadrons. These three families of quarks are joined with three families of leptons (electrons and electron neutrinos, muons and muon neutrinos, and taus and tau neutrinos). Together, quarks and leptons are types of particles known as fermions, since they follow a form of statistics worked out by Enrico Fermi. Gluons, on the other hand, which carry the strong force, follow a different kind of statistics. Such other particles, all of which carry forces between fermions, are called "bosons." Other types of bosons are photons, which carry the electromagnetic force, and the W and Z bosons, which carry the weak force. We shall see below that most particle physicists expect to find a Higgs field, as yet only a theoretical prediction, and a Higgs particle to go with that field. The Higgs field is thought to be the mechanism that gives particles their masses.

Together, all these particles make up "the Standard Model" of particle physics. It does a wonderful job of explaining the structure of matter. But some scientists are hoping for signs of "new physics" beyond the standard model, perhaps through the solar neutrino experiment or in atom smashers.

**The Standard Model**

| Fermions | | Bosons |
|---|---|---|
| *quarks* | *leptons* | photons |
| up | electron | gluons |
| down | electron neutrino | W and Z |
| charmed | muon | Higgs? |
| strange | muon neutrino | |
| truth (top) | tau | |
| beauty (bottom) | tau neutrino | |

have the same meaning that the words have in general speech, though the "strange" quark got its name because it had a property called "strangeness" that made it different from more ordinary particles.) Since each of the six flavors comes in three colors, there are 18 subtypes of quarks; since an antiquark corresponds to each quark, there are really 36. This number is so large that even quarks may not be truly basic.

2. The **electromagnetic force,** $\frac{1}{137}$ the strength of the strong force, leads to electromagnetic radiation in the form of photons. So most of the evidence we have discussed in this book, since it was carried by light, x-rays, and so on, was carried by the electromagnetic force. It is also the force involved in chemical reactions.
3. The **weak force,** important only in the decay of certain elementary particles, is currently being carefully studied. It is very weak, only $10^{-13}$ the strength of the strong force, and also has a very short range.
4. The **gravitational force** is the weakest of all over short distances, only $10^{-39}$ the strength of the strong force. But the effect from the masses of all the particles is cumulative—it adds up—so that on the scale of the universe gravity dominates the other forces.

Theoretical physicists studying the theory of elementary particles have in the past years made progress in unifying the theory of electromagnetism and the theory of the weak force into a single theory. Thus just as the forces known separately as electricity and magnetism were unified a 100 years ago into the electromagnetic force, the force of electromagnetism and the weak force have been unified into the **electroweak force.** (The 1979 Nobel Prize in Physics was awarded for this work.) The forces would be indistinguishable from each other at the extremely high temperatures, above $10^{26}$ K, that may have existed in the earliest moments of the universe.

The electroweak theory predicted that certain new particles could be discovered if a sufficiently powerful particle accelerator (informally called an "atom smasher") could be built. The new particles, known as W and Z, were predicted to be much more massive than the proton, so they could be created out of energy only if a lot of energy were available. (Remember that $E = mc^2$, so the more energy is available, the more massive the particles created can be.) The European atom smasher on the Swiss-French border at CERN (formerly the acronym for the European Center for Nuclear Research, but now known as the European Center for Particle Physics, to avoid the "dreaded" word "nuclear") was upgraded for the purpose. It reached such high energies in 1982 and 1983 that the W and the Z particles were indeed created and detected, a great triumph for the electroweak theory and a discovery that led to a Nobel Prize (Fig. 38–11). In 1989, an atom smasher at CERN, the Large Electron-Positron accelerator, started churning out thousands of Zs each month. Studies of the Zs determined various fundamental nuclear properties and quickly showed that there cannot be more than the three already known types of quark pairs—up and down, strange and charmed, and truth and beauty. A major increase in the energy available to create particles in LEP took place in 1996, and pairs of W particles—$W^+$ and $W^-$—could then be routinely created (Fig. 38–12).

To create particles of still higher energy, particle physicists need accelerators that can bring particles to higher velocities. The Superconducting Super Collider (Fig. 38–13) was the major project until it was cancelled in 1993. It was to use powerful superconducting magnets to bend protons and other particles of a related type, called "hadrons," in an oval the size of the Beltway around Washington, D.C.

**FIGURE 38–11** Carlo Rubbia and Simon van der Meer with the apparatus at CERN with which they and their colleagues discovered the $W^+$, $W^-$, and $Z^0$ particles. These discoveries verified predictions of the electroweak theory. Rubbia and van der Meer received the 1984 Nobel Prize in Physics as a result.

Particles made of quarks are hadrons; they and only they are subject to the strong force. Hadrons made of three quarks, like protons, are called baryons; mesons are hadrons made of two quarks.

**FIGURE 38–12** By studying the trails left in the electronic detector at CERN known as Opal by a collision of an electron and a positron, scientists conclude that a pair of W particles is formed here. Since there is barely enough energy to create the pair, they do not have much momentum and hardly travel at all before they decay, each into a muon and a neutrino. We see the tracks of the muons as measured in different parts of the detector; the neutrinos are invisible.

**FIGURE 38–13** Some of the tunnel dug in Texas for the Superconducting Super Collider, cancelled by Congress for financial reasons in 1993. The 87-km-circumference (54-mile) oval tunnel was to have carried the world's most powerful particle accelerator. Instead, American particle physicists are now participating in the Large Hadron Collider accelerator being built at CERN, which is basically a European collaboration.

**FIGURE 38–14** A second circular ring (*background ring*) has been dug at Fermilab, outside Chicago, to make the Tevatron, allowing for particles in it to collide with particles circulating in the original ring (*foreground*). Such "particle colliders" allow higher energies to be reached than with atom-smashers that smash atoms into stationary targets. The hint of the Higgs particle at CERN is at 114 times the mass of a proton, or 114 GeV (giga-electron-volts, that is, billions of electron volts). Collisions in Fermilab's Tevatron reach a total energy of 2 tera-electron-volts (2 TeV; a thousand billion electron volts). When divided among all the particles created at one time, the 2 TeV total energy barely allows particles of about 114 GeV to be created. CERN's Large Hadron Collider, to be ready in 2005, will be more powerful. It should reach 14 TeV overall.

Scientists had hope of creating the Higgs particle, a heavy particle thought to play an important role in assigning masses to the particles we know. Results in 2000 from the LEP experiment at CERN gave tantalizing indications of the Higgs particle. A very difficult decision then arose—whether to postpone the construction of a new, bigger, better accelerator, or whether to use the LEP to get better statistics on the events, in order to see whether the Higgs particle had really shown up. The new accelerator is to use the same 27-km-long circular tunnel as LEP (the tunnel drawn in on Figure 38–1), so the old one had to be shut to work on the new one. Finally, it was decided to stop the current experiments so that the newer accelerator wouldn't be delayed. That leaves a window of the next five years in which the new colliding-beam accelerator at Fermilab near Chicago (Fig. 38–14) will potentially be the only one powerful enough to find the Higgs particle.

The Higgs, if it exists, and the many particles being studied at the huge atom smashers are created from energy, following Einstein's $E = mc^2$ formula. The energy is made available when other particles are smashed together, the process that gives these huge particle accelerators the informal name "atom smashers." Once that amount of energy is available, particles whose total mass is equivalent to that amount of energy according to Einstein's formula can form. When LEP was almost finished with its lifetime, the scientists were allowed to run it at an energy slightly higher than its previous limit—something that could have led to its demise but it was being closed anyway. It is this extra bit of energy that may have brought the total energy high enough to allow Higgs particles to be created. It is the mass, or the equivalent amount of energy, that the Higgs particles have that scientists want to discover, so that theoreticians can incorporate the information in their theories of matter. The CERN candidates have masses of 114 GeV (that is, 114 giga-electron-volts), measured in energy units. But the number of probable sightings wasn't high enough to rule out, to an acceptable degree, the possibility that the sightings were of other phenomena that only by chance looked like a Higgs.

How would the Higgs particles work? Theory holds that there are so many of them that quarks and electrons have to essentially swim through them. The Higgs particles' drag on the quarks and electrons make the particles act as though they had inertia, which corresponds to mass. One explanation of how the Higgs particles give mass won a prize: Imagine the President of the United States coming into a room. People cluster around him as he makes his way across the room. The people may impede his progress. In the same way, the Higgs particles impede the progress of the subatomic particles. They seem to have inertia, which in our normal world arises from mass. At present, physicists can measure the masses of protons, neutrons, and other types of elementary particles, but don't know why they have the masses that they do. The Higgs particle, if it exists, could provide the answer.

Physicists are studying a family of theories that unify the electroweak and the strong forces. Such theories are necessary to explain the early universe. Though **grand-unified theories** (GUTs) have been discussed in recent years, other examples of "unified field theories" are currently more in favor. These unified field theories imply, surprisingly, that protons are not stable—they should decay with a half-life greater than $10^{30}$ years. So unified field theories imply that protons decay. Though this is a much longer time than the lifetime of the universe (which is "only" about $10^{10}$ years), given a sufficiently large quantity of protons, such a decay should be detectable. After all, 1000 tons of matter contain about $10^{32}$ protons, so perhaps 10 or 100 could decay each year in that volume.

We have already met (Section 27.6) some pieces of apparatus that have been used in the search for proton decay, since they have also been used to detect neutrinos from the sun and from Supernova 1987A. These detectors contain large volumes of water. After all, water, $H_2O$, is mostly protons, since an H nucleus is a proton and an oxygen nucleus contains 8 protons. A detector, for example, might contain about $2.5 \times 10^{33}$ neutrons and protons bound in nuclei. If any of them were to decay, the resulting particles would give off flashes of light.

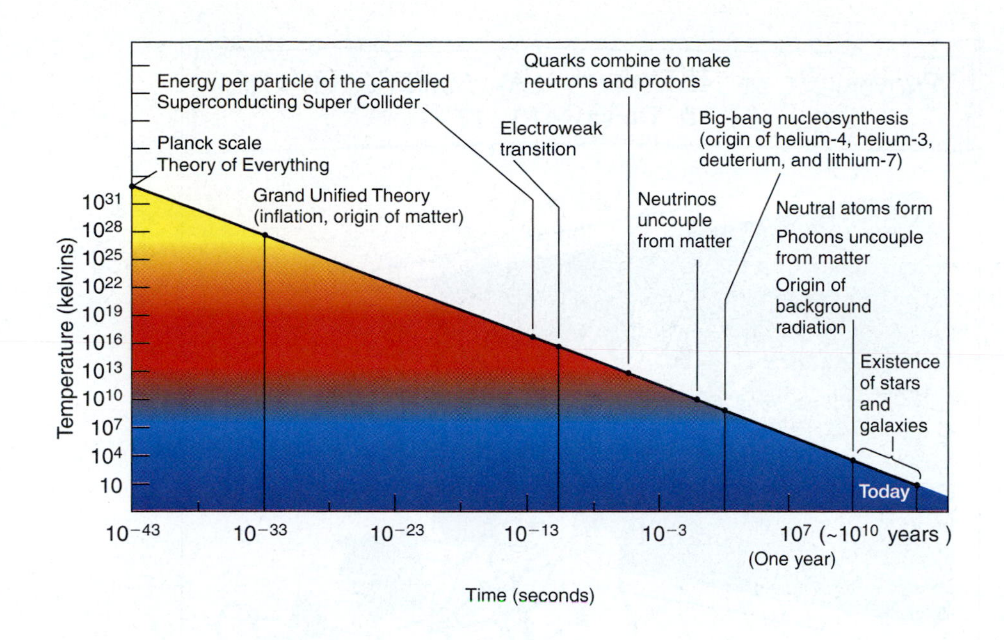


**FIGURE 38–15** As the temperature of the universe cooled, the forces became distinguishable from each other, and the universe evolved. Note that most of the horizontal axis shows the first year of the universe after the big bang. *(Adapted from* "Particle Accelerators Test Cosmological Theory" *by David N. Schramm and Gary Steigman. ©1988 by Scientific American, Inc. All rights reserved.)*

It is just such a change in the total number of baryons in the universe that is necessary to account for the current excess of matter over antimatter. Many scientists had expected to detect signs of proton decay by now, and a few candidate events are being studied, but the detectors have not observed such decay to a limit of $10^{31}$ years. Thus the simplest unified field theories, but not more complex ones, are ruled out experimentally.

Many theoreticians are working on theories that incorporate gravity with the other forces; "string theory" (discussed in the next section) is an example. String theory, thus, goes beyond "unified field theories" to include gravity, as well.

Basically, scientists believe that all the forces were unified at the extremely high energies that were present early in the universe. As the universe expanded and cooled (Fig. 38–15), basic symmetries among the forces were broken, and individual forces became apparent.

At early times in the universe, the exotic particles now being observed by physicists in giant accelerators had been present as well. Even isolated quarks may have been present, though they would have combined to form baryons (protons, neutrons, etc.) by the time 1 microsecond went by. The mini black holes described in Section 31.6 may have been formed in this era.

## 38.4 SUPERSTRINGS AND SHADOW MATTER

One of our major limitations in studying the universe is that we do not have a unified theory of gravity with the other fundamental forces. Many scientists are searching for this link. It has become an important branch of theoretical physics.

A major line of difficult and speculative theoretical research uses the concept of **superstring.** Superstring is a one-dimensional analogy to the zero-dimensional points that are often used in theories. Quarks, electrons, photons, and other elementary particles would be made out of one-dimensional elongated objects—"strings"—about $10^{-35}$ meter long that make loops so small that they seemed to be points.

We have dealt, in discussing Einstein's general theory of relativity (Section 31.1), with curvature of three spatial dimensions into a fourth dimension. **String theory** invokes vibrating 10-dimensional loops that are a billion trillion times smaller than a proton. Below, we will see that string theory is now considered as a reduced version of M-theory, which is framed in terms of an 11-dimensional space, with the 11th

**FIGURE 38–16** Theorists calculate that superstrings require an 11-dimensional space; we are more familiar with only a four-dimensional space, with three spatial dimensions and one time dimension.

**A Brief History of Gravity**
*(A Limerick) by Bruce Elliot*
It filled Galileo with mirth
To watch his two rocks fall to Earth.
He gladly proclaimed,
"Their rates are the same,
And quite independent of girth!"
Then Newton announced in due course
His own law of gravity's force:
"It goes, I declare,
As the inverted square
Of the distance from object to source."
But remarkably, Einstein's equation
Succeeds to describe gravitation
As spacetime that's curved,
And it's this that will serve
As the planets' unique motivation.
Yet the end of the story's not written;
By a new way of thinking we're smitten.
We twist and we turn,
Attempting to learn
The Superstring Theory of Witten!

dimension very small. In both string theory and M-theory, all but four of the dimensions have such strong curvature—are curled up so tightly (within $10^{-35}$ meter)—that we cannot detect them (Fig. 38–16). An analogy to the curling up of dimensions might be a garden hose on the other side of your lawn—it has three dimensions but at a distance looks like a one-dimensional line.

Though string theory is studied by many physicists, their progress stalled for several years as the mathematical problems became too difficult to solve. Indeed, some physicists criticize string theory as being only mathematical and not part of science, since it has not been able to make predictions that are testable. One of the prime proponents of string theory, Edward Witten of the Institute for Advanced Study (a research organization in Princeton, N.J., originally famous because Einstein worked there for many years), indeed was given the Field Medal for his work, the mathematicians' highest award (given that there is no Nobel Prize in Mathematics). Scientists today seem to speak of Witten with the tone of awe they once used for Einstein.

String theory is on the move again, with scientists making several breakthroughs in the 1990s. The scientists showed that several versions of string theory that were around are identical, and they produced some mathematical ways of advancing.

A theory is now developing that includes multidimensional surfaces, known as membranes, in addition to superstrings. A string, which has one dimension (length), is a one-brane; a soap bubble, whose surface has two dimensions, is a two-brane, and so on. Elementary particles are viewed, in string theory, as vibrations of these multidimensional membranes. They vibrate in the way that violin strings vibrate to make different notes. The resulting general theory is known as M-theory, where M stands for "membrane," or perhaps even "mother of all strings," or "mystery." Riffing on the temporary name change of the musician Prince, it has been called "the theory formerly known as string." M-theory exists in 11 dimensions. It reduces to the theory of strings when the size of the 11th dimension is very small.

To many (but not all) particle physicists, M-theory seems a promising way to get around theoretical difficulties in the extremely small scale where a quantum theory

A DEEPER DISCUSSION

**BOX 38.3 Parallels Between Planetary Orbital Theory and Cosmological Theory**

The new models of the universe, with parameters for not only ordinary matter but also dark matter and the cosmological constant, may seem complicated. But Sir Martin Rees of the University of Cambridge has pointed out that perhaps these complications are necessary, and parallel the complications added to theories of planetary motion in the 17th century. Here are some parallels in the development:

| Theory of Planetary Orbits | Theory of Cosmology |
|---|---|
| *Simple, unique theory (100 B.C.–A.D. 1609):* Orbits are exact circles | *Simple unique theory (1980–1995):* All matter, closed universe ($\Omega = 1, \Lambda = 0$) |
| *Newer theory, Kepler (1609):* Orbits are ellipses | *Newer theory (2000):* Baryons, dark matter, cosmological constant ($\Omega_{\text{baryon}}, \Omega_{\text{dark matter}}, \Lambda \neq 0$) |
| *Deeper underlying theory:* Newton's theory of universal gravitation (1687) | *Deeper underlying theory:* M-theory? |

of gravity is needed. (Such a quantum theory of gravity does not exist.) Thus M-theory (and previously string theory) is considered a prime candidate for a TOE—"theory of everything"—that simultaneously explains all four of the fundamental forces. We hope that with this theory one day we will be able to predict the complete set of elementary particles we observe experimentally, though we are not there yet.

String theory implies the existence of "shadow matter," able to interact with normal matter only through gravity. (Shadow matter may appear in simpler theories as well.) Since shadow matter would not be subject to the electromagnetic force, it would not generate light. It thus could be the dark matter that would solve the missing-mass problem. Will string theory and M-theory become, as some of its proponents have it, the route to the solution of many cosmological problems? Are they pieces of 22nd-century mathematics plunked down in the late 20th and early 21st centuries, merely a bit too soon for us to use? Only time and further research will tell whether string theory and M-theory become the basis for so much of astronomy in the 21st century the way that Einstein's theory of relativity was fundamental for cosmologists in the 20th century.

## 38.5 THE INFLATIONARY UNIVERSE

The application to cosmology of the exciting new theoretical and experimental results of particle physics has led to some big surprises in the last two decades. The results from particle physics have allowed scientists to push our understanding of the universe much further back in time than previously. In particular, now we can go back beyond the era of element formation.

The unified theories of the strong, weak, and gravitational fields got scientists thinking about the early universe and how it evolved from a hot, dense beginning. But is it actually true that the universe was hot from the very beginning? This assumption led to the realization that there were several problems that were extremely hard to resolve. A major theory, presented in the 1980s, holds that in a tiny fraction of the first second of time, the scale of the existing universe increased rapidly. The rate of this "inflation" of the universe was astounding. During this fraction of a second, the volume of the universe grew perhaps $10^{100}$ times larger than it would have grown according to the "standard" big-bang theory. The theory is called the inflationary cosmology. We thus live in an **inflationary universe.** As we shall see below, the inflationary-universe theory solves a number of outstanding problems of cosmology.

The basic concept, as just described, was thought of by Alan Guth of M.I.T. As the universe inflated, matter and energy were created. Thus inflation provided, in his quip, "The ultimate free lunch." His original model had the inflation starting at the first $10^{-35}$ second of time and lasting another $10^{-32}$ second. This first model had the major flaw that the inflation never stopped, which is contrary to our current observations that the universe now expands at the more modest rate defined by Hubble's law with the measured Hubble constant. This flaw was removed in models known as "new inflation" developed by Andrei Linde, first of the Lebedev Institute in Moscow and now at Stanford University; Andreas Albrecht, now of Imperial College, London; and Paul Steinhardt, formerly of the University of Pennsylvania and now at Princeton University.

The inflationary model came from ideas in particle physics. But now the inflationary theory stands on its own, to be tested independently of specific models of particle physics. It has been surviving the tests, most recently with the measurements of the small-scale anisotropies of the cosmic background radiation. Headlines in 2001 said that the measurements of the peaks in the graph showing the scales of fluctuations of the background radiation, measured from high-altitude balloons and from the South Pole (Section 37.5), back the idea of inflation. These results will provide new limits on particle physics. So in the end, particle physics may wind up benefitting from the inflationary theory rather than the relation being only the other way around.

### 38.5a Explanation of the Inflationary Universe

Though several variations of inflationary theories (the informal term used for theories of the universe using the idea that there was an era of inflation) have existed, let us concentrate on the currently most widely accepted model. It is based on an idea of Andrei Linde known as "chaotic inflation." Unlike earlier inflationary models, it uses only concepts already present in the grand-unified theories. Explaining it requires the notion of the physicists' type of "field," which we are most familiar with as the magnetic field that surrounds a magnet. Putting something in a field of any kind means that the inserted object can react with the field. In particular, putting a magnetic object in a magnetic field leads to a force being generated. Similarly, fields can be associated with the strong, weak, gravitational, and electromagnetic interactions.

The simplest version of chaotic inflation begins right after the Planck time—that is, at $10^{-43}$ second, and lasts about $10^{-35}$ second. The universe might even expand by ten to the billion-billion-power times, in any case much more than the $10^{30}$ times necessary for the theory to work. This model describes the universe as filled with a particular kind of field, similar in a general way to the fields used in the unified theory of weak, strong, and electromagnetic interactions. The differences among the strong, weak, and electromagnetic fields appear only when the whole universe becomes filled by this specific field, which physicists call a "scalar field." Unlike the electromagnetic field, whose consequences we can see when we turn on a light, the scalar field's consequences are invisible. However, it changes the properties of particles and their interactions.

It can be shown that if the scalar field was originally sufficiently large and homogeneous, it could change its value only very slowly. Therefore, for a long time its energy density remained almost constant. According to Einstein's general theory of relativity, a universe filled with matter that has a constant energy density should expand extremely rapidly, which corresponded to the stage we call "inflation."

Why is this version called "chaotic"? Because there is no reason to expect that the universe was uniform from the very beginning. Some regions, randomly, could be more dense than others, and they could contain a larger scalar field. Linde's model shows that only regions with a sufficiently large scalar field will inflate. Thus the inflation is nonuniform, and initially it seems chaotic. But in the end, each inflating region grows extremely large and becomes very homogeneous.

Of course, to be a useful cosmological picture, the rapid expansion not only had to happen but also had to stop, because our universe is not inflating today. One of the natural consequences of the chaotic-inflation model is that the inflationary era lasted only a short time. The universe then began expanding at a much more modest rate (Fig. 38–17). The universe has ever since continued to expand, but at the slower rate we observe today, just as in the earlier big-bang theories. The elements form, the background radiation is set free, and so on, on the time scale long believed.

All versions of inflation imply that the universe is very much larger than we had thought, in that the parts with which we could have been in touch have expanded far beyond our sight. The boundary of this volume, our past "light cone," is our "horizon," since—like a horizon on Earth (or in a black hole, as we saw in Chapter 31 with the event horizon)—we can't see beyond it. It is defined by signals travelling at the speed of light (Fig. 38–18), since relativity theory tells us that it is impossible for signals to travel faster.

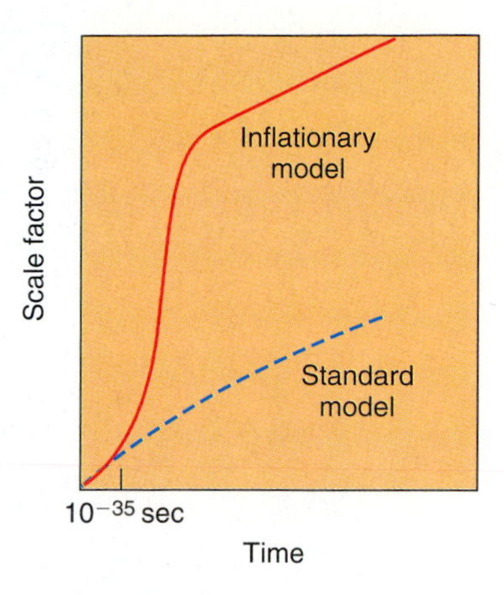


**FIGURE 38–17** The rate at which the universe expands, for both inflationary and non-inflationary models.

## 38.5b Major Questions Solved by Inflation

The inflationary theory has gained favor because it naturally explains some large-scale observational aspects of the universe.

### 1. Homogeneity

One major question that had been open in cosmology is how the universe became so homogeneous. According to the hot-big-bang model, our universe was supposed to have always been so large that neither light, nor any information about conditions at any given location, had had time since the big bang to travel across it. Thus matter far to one side of the observed universe has always been out of contact with matter on the other side. It has been a mystery how such widely different regions could have reached identical temperatures and densities. After all, we know that the temperature and density were nearly identical everywhere at the time the cosmic background radiation was emitted, since this background radiation is so isotropic.

In the inflationary-universe picture, our observed universe was much tinier than in the hot-big-bang picture, small enough that signals travelling at the speed of light had time to travel across it. The universe could thus be smoothed out. The whole part of the universe that we can potentially observe—the visible universe—was, before the inflation, much smaller than a single proton. It could contain no particles at all, but essentially all of the approximately $10^{86}$ particles in the observed universe today were produced from the energy released by the decaying scalar field at the end of inflation. The inflation then accounts for how such a tiny region could grow to be the size of our observed universe. It thus explains why the universe is so homogeneous.

At the distance to the fluctuations in the cosmic microwave background that we see, the universe's horizon should be only about 2° across. So by looking on a 7° scale with COBE, we were seeing units of the universe that are separate from each other. With the inflationary-universe model, we can best understand how such separated parts of the universe can be alike.

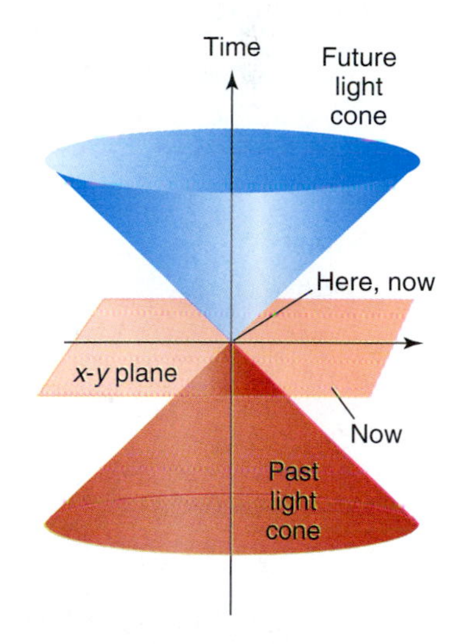


**FIGURE 38–18** If we plot a three-dimensional graph with two spatial dimensions (omitting the third dimension) and a time axis, our location can receive signals only from regions close enough for signals to reach us at the speed of light or less, our "past light cone." We can never be in touch with regions outside our light cone; we say that we are "not causally connected" with those regions. In the inflationary universe, the volume of space with which we are causally connected—where something here could cause something there, or vice versa—is much greater than the observable universe. It is easy to explain how a volume that is causally connected can be homogeneous—collisions, for example, can smooth things out.

### 2. Flatness

Another previously unresolved question is why the density of matter in the universe is so close to the critical amount that marks the difference between the universe's being open or closed—why the universe is so flat. Even from the existence of matter alone, we know that the density of matter is within one or two factors of 10 of the critical density. And the studies of the background radiation (Section 37.5) bring us within a factor of 2 or so of the critical density. Being even within one or two factors of 10 of the critical value is rather surprising, since there is no reason we couldn't be off by factors of, say, $10^{100}$. It can be shown that such a situation is highly unstable, like balancing a pencil on its point! A slight deviation of the early universe from flatness, or a slight deviation of the pencil from vertical, leads to a major change later on, such as the pencil falling over. Why are we in such an unstable situation? The inflationary-universe theory explains that during the inflation, space-time became very flat (Section 37.2), just as a small bit of the surface of a balloon flattens out as the balloon grows larger. After all, the Earth is so large that the ground around us seems flat, even though the Earth's surface is curved.

We may thus have been asking the wrong question all these years when we asked whether the universe is open or closed, wondering what kinds of studies would tell us the answer. The inflationary-universe theory implies that the universe is incredibly close to the borderline case between open or closed. Without a cosmological constant, an open universe will expand forever while a closed universe will eventually collapse. In this case, a universe on the borderline would expand forever at an ever-decreasing rate; the longer it expands, the less its rate of expansion will be. With a cosmological constant, though, even a closed universe will never collapse. Instead, it will expand forever with a constant acceleration. If the inflationary theory is right—and it is by far the dominant theory held by astronomers today—the universe is so close to the borderline that it would be impossible for any observations to tell us on which side it lies. This theoretical result fits with the observational results that the universe is flat.

### 3. Magnetic Monopoles

The inflationary universe also provides satisfactory answers for another question plaguing some theoreticians: why don't we see magnetic monopoles in the universe? A magnetic monopole is a point source of magnetic force; it is analogous to an electron, which is a single particle that has an electric force. But bar magnets always have two opposite poles (Fig. 38–19), and a single-pole magnet—a monopole—has never been discovered. Although scientists have toyed with the notion of monopoles for over a hundred years, it turns out that unified theories of the strong, weak, and gravitational fields predict that there would be monopoles. Further, these monopoles should be very massive and should have been produced in enormous numbers in the early universe. The monopoles never decay and rarely meet up with antimonopoles (in which case they would annihilate each other), so they should be around today. Yet none are seen. It is crucial to find some explanation of how the universe might have gotten rid of the high density of monopoles. The answer proposed by inflationary theory is very simple: inflation spread the magnetic monopoles out so much that there are at most only a handful in the part of the universe observable by us.

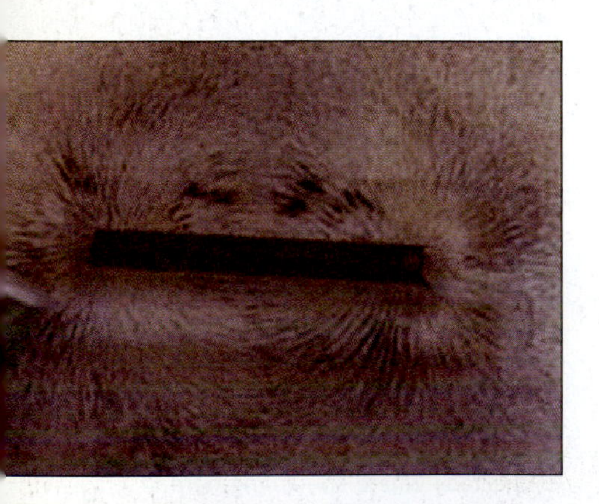

**FIGURE 38–19** A bar magnet has two poles. We know of no single-pole magnets, that is, monopoles. Here we see the effect of iron filings (literally, bits of iron filed off an iron nail) sprinkled onto a paper that covers a bar magnet.

### 4. The Existence of Matter and Radiation

A startling feature of the inflationary-universe theory is that it accounts for the existence of all the ordinary matter and radiation in the universe. In the theory, part of the energy existing in the early universe is transformed, after the period of tremendous expansion, into the matter and radiation that we detect today. Almost all the energy we observe in the universe today was produced by the transformation of energy that occurred at the end of the inflation. Since matter and energy can be transformed into each other, the model also shows why there is so much matter in the universe.

#### *5. Fluctuations in the Distribution of Matter*

Another prediction of inflation is the topic of much current discussion. The theory predicts a specific spatial pattern of variations of energy and matter after inflation. The variations arise first as subatomic fluctuations predicted by quantum theory on a microscopic scale, and they are inflated to become the large-scale structure we see today. This pattern should act as seeds from which galaxies eventually form. We discussed in Section 37.5 how the COBE spacecraft first observed these seeds and how they are now observed at higher spatial resolutions.

In certain cases, subatomic fluctuations amplified during inflation may become extremely large. The probability of such events is very small, but those rare parts of the universe where it happens begin expanding with a much greater speed. This expansion creates a lot of new space where inflation may occur, and where large quantum fluctuations become possible. As a result, the universe enters a stationary regime of self-reproduction: The inflationary universe permanently produces new inflationary domains, which, in their turn, produce still newer inflationary domains. Such a universe looks not like a single expanding fireball created in a big bang, but like a huge self-reproducing fractal consisting of inflationary domains of all possible types. (Fractals are mathematical constructs that are the same no matter on what scale you look at them. For example, if you smooth out the coastline from New York to Boston, you may travel 200 miles from one to the other. If you were to travel around every bay on the coast, the distance would be longer. If you were to travel around every tiny inlet, the distance would be longer still. On whatever scale you look, there are irregularities.)

Even physical conditions like the strength of gravity could be different within different exponentially large inflationary domains. Sir Martin Rees of the University of Cambridge, the Astronomer Royal of Britain, has called a collection of all these regions a "multiverse." Our universe may be only a single tiny piece of a "multiverse." (Remember, after all, that "uni-" means "one," while "multi-" means "many.")

### 38.5c Inflation Today

Over the last two decades, the inflationary theory has advanced through various versions. The theory, in general, has had triumph after triumph, most recently in predicting the "second peak" and "third peak" in the scale of fluctuations of the cosmic microwave background (Section 37.5e). The relation between physicists studying subnuclear particles and scientists studying cosmology has become much closer. We trust that our knowledge of the creation and evolution of the universe (Fig. 38–20) grows more and more accurate.

## 38.6 COSMIC STRING

The fluctuations we detect in the cosmic microwave background are thought to have evolved into the galaxies and the clusters of galaxies we see today. An alternative explanation for the origin of galaxies is long, very thin structures called "cosmic string," which could permeate the universe. Cosmic strings—if they exist—would be long, thin tubes in which matter and energy from an earlier, high-energy phase of the universe is trapped (Fig. 38–21). Such a tube could be infinitely long or it could form a closed loop. A single such loop could contain $10^{15}$ times the mass of the sun, equivalent to the mass of 10,000 galaxies. Fortunately, there clearly aren't many of them around now; they decay as time goes on. The smaller loops would have radiated away their energy in the form of gravitational waves, leaving only a few of the longest loops remaining today.

Theoretical calculation indicates that cosmic strings would be only $10^{-30}$ centimeters thick. Each centimeter of their length would have a mass of $10^{22}$ grams, a

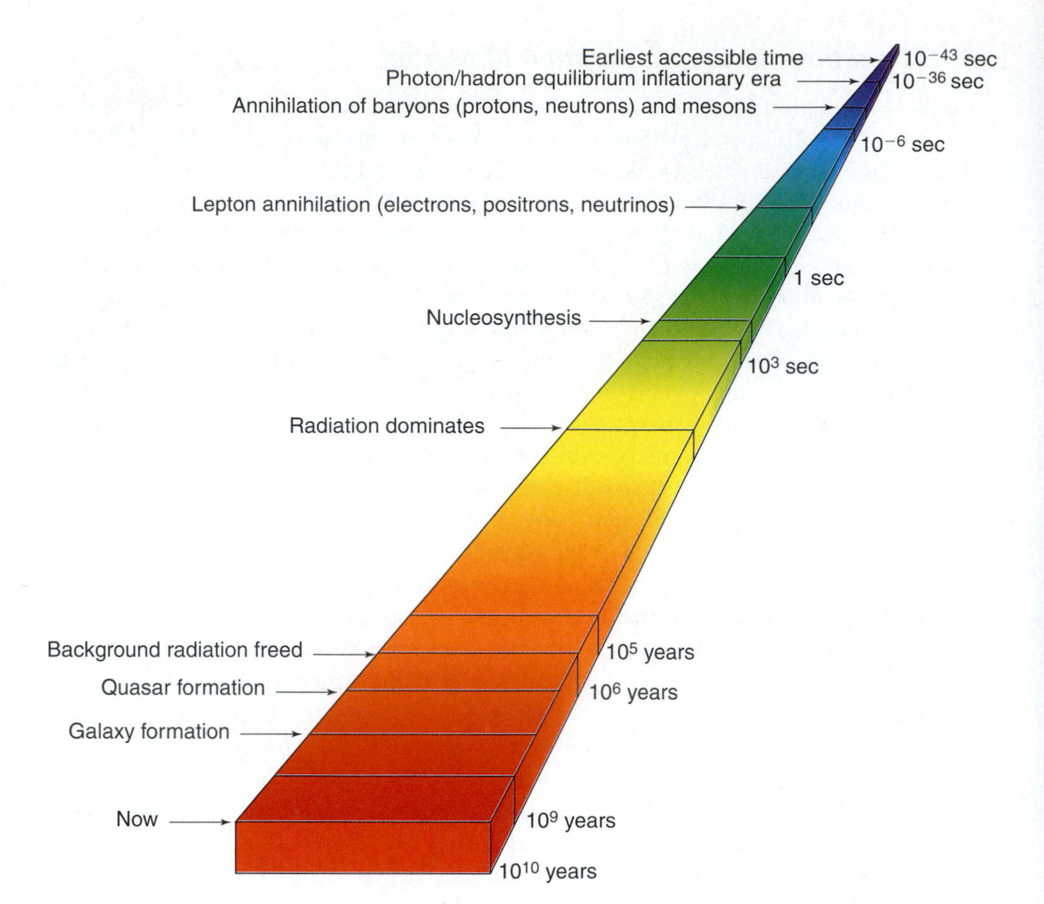


**FIGURE 38–20** Major events in the history of the universe.

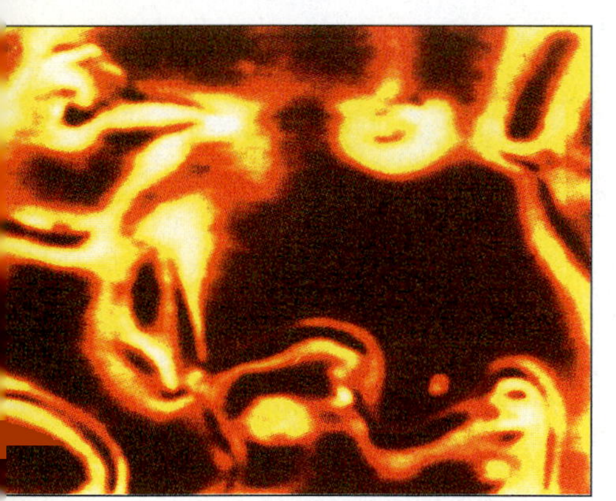

**FIGURE 38–21** A simulation of cosmic string formation in a liquid crystal (similar to those used in calculator displays) in a terrestrial laboratory. Superfluid helium can also be used to simulate cosmic string.

huge value equivalent to a hundred billion billion tons in a fingertip. The idea was that cosmic strings might have triggered huge explosions in the early years after the big bang. The wake of the strings might have been enhanced in density, providing the seeds for galaxies to form (Fig. 38–22). These galaxy-forming cosmic strings, though, would give off gravitational radiation. Studies of the steady period of one of Joe Taylor's millisecond pulsars (Sections 30.6 and 30.7) ruled out the simplest form of cosmic string.

Cosmic strings would have enough mass to make a gravitational-lens effect. If they exist, we may one day see an image of a galaxy cut off by a cosmic string. Such an effect could be seen on images taken for other purposes.

Theoretically, the cosmic strings would have been formed as defects in the change of state of matter (states of matter include solid, liquid, and gaseous) in the early universe. An analogy is the change of state observed in laboratories on Earth of helium–3, the light isotope of helium. Helium–3 becomes "superfluid" at a temperature of only 2/1000 kelvin, very close to absolute zero. The Nobel Prize in Physics for 1996 was given for the discovery of that phase change. "It helps us understand the models for order and the state of the universe after the Big Bang," a member of the Nobel Prize committee said.

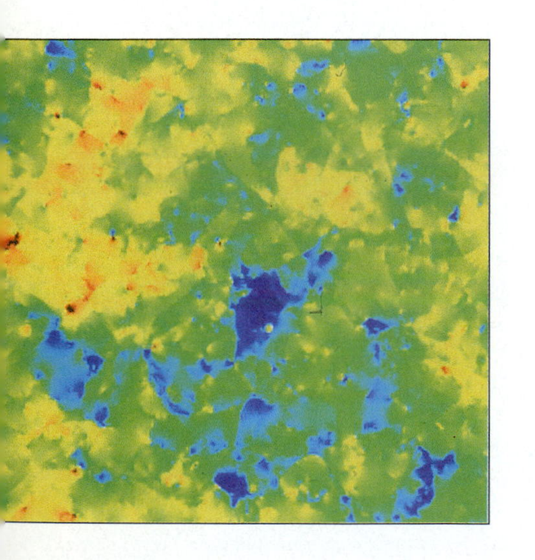

**FIGURE 38–22** If cosmic strings exist, they should generate a certain pattern of temperature contrasts in the cosmic background radiation. They would do so by acting as gravitational lenses. This computer simulation shows such a pattern over a field of view a few degrees across. Blue indicates gas that is hotter than average, and red indicates gas that is cooler than average. The study addresses the question of whether cosmic strings acted as seeds to begin galaxy formation. But strings seem to have been ruled out by the discovery of several peaks in the fluctuations of the background radiation.

IN REVIEW

## BOX 38.4 Keeping Strings Straight

Do not confuse:

A **string** of galaxies: galaxies one after the other in space, using the word "string" in its normal sense.

**Superstring:** the tiniest objects in the universe, only $10^{-35}$ m long, used in a theory (Section 38.4) that may explain all the known interactions of elementary particles and all the basic forces; in the theory, elementary particles are different vibrational modes of superstrings in an 11-dimensional universe.

**Cosmic string:** loops or giant strands that would stretch across the whole universe and could result from phase transitions in the early universe.

The ripples in the cosmic background radiation do not correspond to calculations with models of cosmic string, or of an alternative version of recent years known as "texture," and these models are on the ropes. The models invoking cosmic string in the universe's earliest epoch do not predict any peaks in the spectrum of the cosmic background radiation, so they apparently now have been ruled out. Strings could, instead, be produced after inflation. So they could make a minor contribution to some features of galaxy formation and the cosmic background radiation that we observe. But these minor effects were not what the backers of cosmic string theory had in mind.

## 38.7 THE HUBBLE DEEP FIELDS

A nondescript small region of the sky became the most heavily observed. At Christmastime 1995, the Director of the Space Science Telescope Institute devoted 10 days of observing under his personal control to round-the-clock imaging of one direction in space, using filters in both the visible and the near infrared. The result, after 100 hours of observation, was images so faint that it seemed likely that the objects were as far out in space as possible—that is, deep. This Hubble Deep Field found thousands of objects in its small region of sky, which is only the angular size of a grain of sand held at arm's length. The task of astronomers now is to understand each object. Since the direction chosen for the observation was far from the plane of our galaxy (where our viewing would be impeded by galactic dust), we assume that we are looking out as far in space as possible (Fig. 38–23).

Most of the objects in the Hubble Deep Field are apparently galaxies, some 3000 of them. They have a wide variety of sizes and shapes, and some are interacting. To prove that they indeed are very far away, we need redshifts. These redshifts are now being measured using the Keck and others of the world's largest ground-based telescopes (Fig. 38–24). Many of the faintest objects proved to be very far out, with redshifts of $z = 1$ to $z = 4$, which means that we are seeing them as they were in the early universe. Table 38–1 shows the ages of the universe at which we are seeing these high-redshift objects.

Many of the objects are too faint for even the largest telescopes to take spectra. Images of the Hubble Deep Field region through filters in various parts of the spectrum are being used to measure overall "colors," as in color indexes, and thus to give some idea of the redshift even though individual spectral lines aren't observable.

The Hubble Deep Field project was so successful that a Hubble Deep Field−South was taken about 3 years later. It resembles the Hubble Deep Field very much (Fig. 38–25). One of the most distant quasars known today, with $z = 5.8$, was discovered in it.

**FIGURE 38–23** The whole Hubble Deep Field; a section of it appeared as the opening photograph to this chapter.

**FIGURE 38–24** The Hubble Deep Field with redshifts measured with the Keck telescope shown for many of the objects.

**TABLE 38-1 Redshift and Lookback Time**

The relation of redshift and the time that light has taken to travel depends on the curvature of the universe. Here the calculations correspond to the current model: Hubble constant of 75, mass density of 30% and cosmological constant of 70%.

| Redshift ($z$) | Lookback Time (billions of years) |
|---|---|
| 0 | 0 |
| 0.2 | 2.46 |
| 0.4 | 4.27 |
| 0.6 | 5.63 |
| 0.8 | 6.67 |
| 1.0 | 7.49 |
| 1.5 | 8.89 |
| 2.0 | 9.80 |
| 2.5 | 10.33 |
| 3.0 | 10.73 |
| 4.0 | 11.25 |
| 5.0 | 11.57 |
| 6.0 | 11.77 |
| infinity | 12.57 |

One result from the Hubble Deep Fields is that there was an epoch of the early universe when stars formed at ten times the rate of today. This peak in star formation matches the idea from quasar counts of such an active period of the universe's youth.

The Chandra X-ray Observatory pointed for a million seconds (about 10 days) at each of two regions that contained the Hubble Deep Fields (Fig. 38–26). The field of view of Chandra's Advanced CCD Imaging Spectrometer is about three times in diameter, and so ten times in area, that of Hubble's camera. Many fewer x-ray objects are detected than visible objects. About half the objects apparently result from accretion onto supermassive black holes, while the other objects are nearer. For these nearer objects, the x-rays Chandra detects are probably the sum of x-rays

A

B

**FIGURE 38–25** The Hubble Deep Fields. (*A*) A portion of the original Hubble Deep Field. (*B*) A similarly-sized portion of the Hubble Deep Field–South.

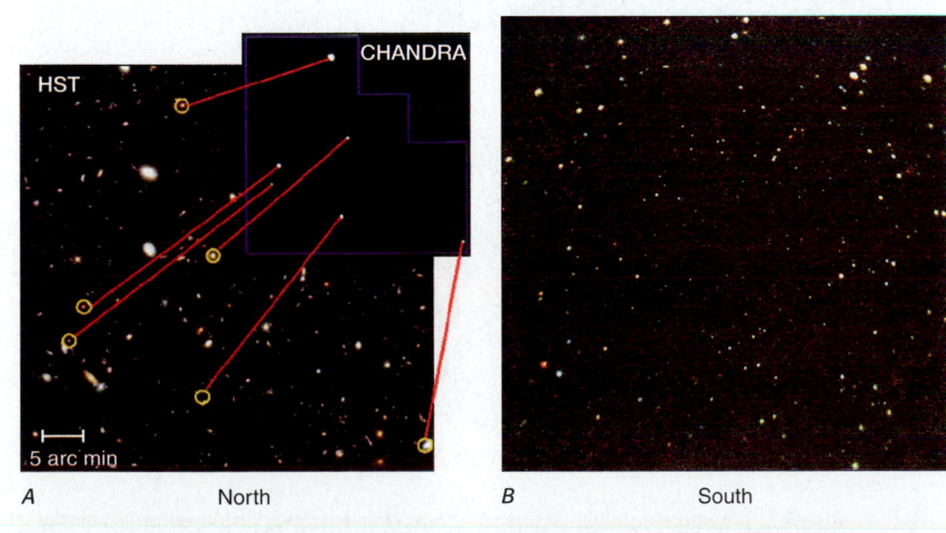


**FIGURE 38–26** The Chandra Deep Fields. (*A*) The portion of the Chandra Deep Field–North that coincides with the Hubble Deep Field, with Chandra's x-ray sources circled and overlain on the whole Hubble image. (*B*) The whole Chandra Deep Field–South, covering an area about ten times that of the portion shown for the northern field.

from stellar-mass black holes in binary systems, hot intergalactic gas, and supernova remnants. X-ray emission has now been detected from all the galaxies in the Hubble Deep Field with redshift $z$ less than 0.15. By comparing these nearer galaxies with the farther galaxies, Chandra scientists can now study the x-ray evolution of "normal" galaxies much like our own over the past 5 billion years.

## 38.8 COSMOLOGY TODAY

Today's theories of the creation and evolution of the universe are newly based on detailed observations. The fact that there are fluctuations when we look back to 300,000 years after the big bang (Fig. 38–27) seems to endorse our basic ideas about how this structure formed in the universe. The structure that we now see as clusters of galaxies and assemblages of clusters of galaxies have been analyzed to show the distribution in their sizes. The results, as we have seen, agree with the predictions of our current theory of inflation. We think we have a good general understanding of the evolution of the universe (Fig. 38–28).

Radical new ideas remain to be conceived. After all, though the ripples in the background radiation trace back to 300,000 years after the big bang, the epoch of nucleosynthesis traces back to 1 second after the big bang, and the Planck time was $10^{-43}$ sec after the big bang, these ideas still don't tell us how the universe formed. In Section 38.4, we discussed M-theory, which explains space as membranes in an 11-dimensional volume. A radical idea advanced in 2001 suggests that the universe formed as the result of the collision of two 4-dimensional membranes floating in

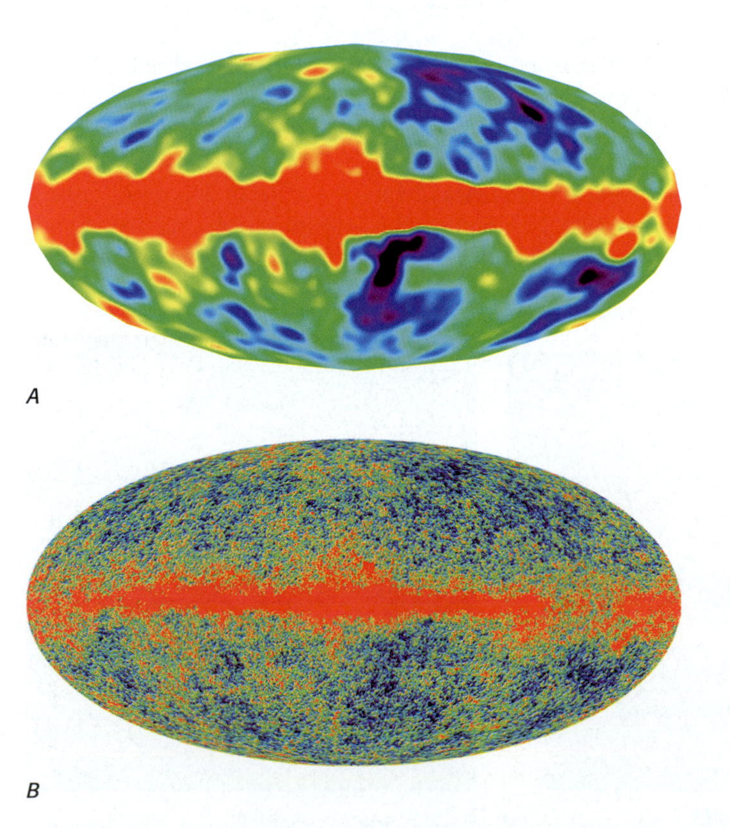


**FIGURE 38–27** Temperature fluctuations on the microkelvin level displayed on an all-sky map. (*A*) The reduction of all four years of COBE data. These fluctuations, about 5 parts per million, show the seeds from which structure in the universe evolved nearly 15 billion years ago. The detail in the image is real. (*B*) A simulation of the higher simulation resolution of similar data from the Microwave Anisotropy Probe. The improvement expected is fantastic.

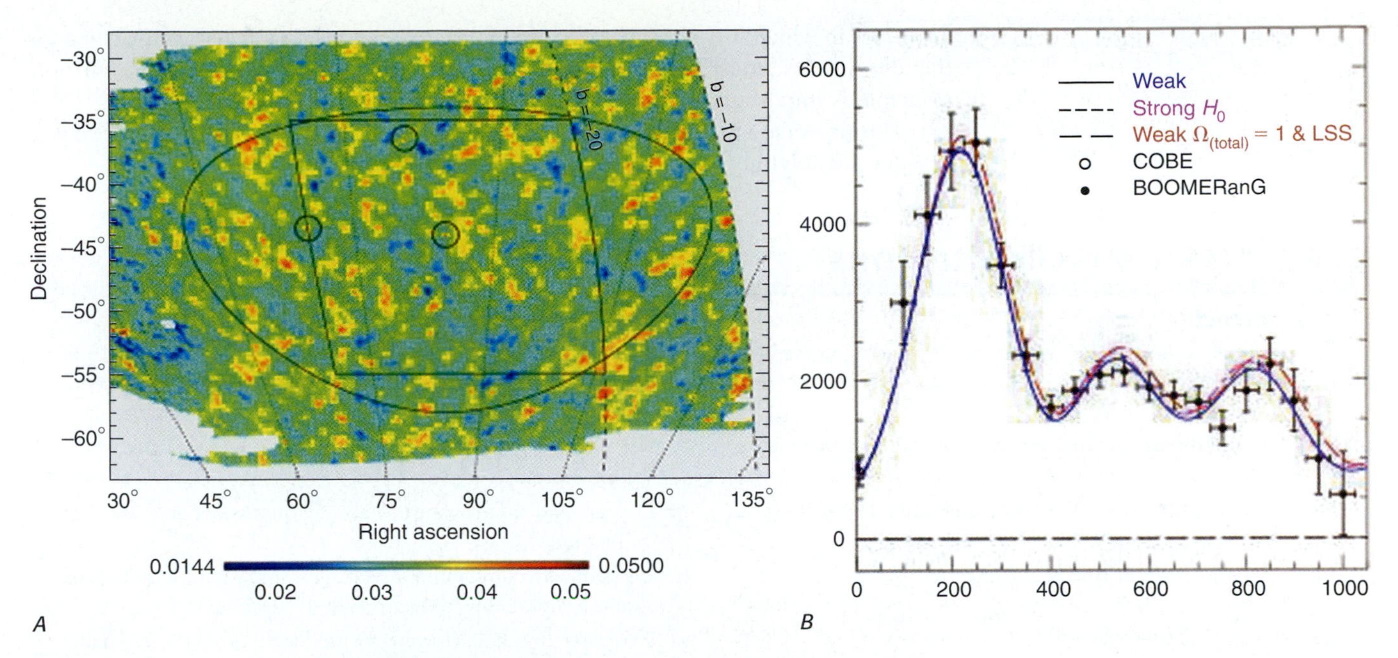


**FIGURE 38–28** (*A*) The BOOMERanG data released in 2001, an intensity map covering a larger region (the ellipse) than that used for the 2000 data reduction (the parallelogram). The reduction of the larger region showed three peaks in the fluctuations, as shown in Figure 37–28(*A*), indicating that the universe is flat and allowing the percentage of baryons in the universe to be calculated. Some point sources, quasars, that were omitted from the data reduction are circled. (*B*) The reduction of the data from the larger region showed three peaks in the graph of the sizes of the fluctuations versus their relative numbers (compare with Figures 37–26 and 37–28. The size and position of the peaks in the graph indicate that the universe is flat and allows the percentage of baryons in the universe to be calculated. (Do not confuse such diagrams of fluctuation peaks with Planck curves.) DMR was a COBE instrument and B98 refers to the 1998 BOOMERanG flight. LSS stands for "large scale structure."

5-dimensional space. It leads to a universe that is flat and isotropic without invoking inflation. This new and very exciting speculative possibility will require a lot of new work (Fig. 38–29).

If there are indeed various other parts of the multiverse, with other physical constants, why is our universe hospitable to our being in it? Perhaps, says the "anthropic principle," we are here rather than in a different part of the multiverse because our universe has conditions favorable for us to exist. After all, for example, if the gravitational constant were stronger, the universe would have collapsed before there was time for humans to evolve, or there wouldn't have been enough time for stars to cook up the heavy elements we are made of. So there is no possibility of life existing in

**FIGURE 38–29** The evolution of the universe, including the era 300,000 years after the big bang in which the universe turned transparent and the cosmic background radiation was thus set free.

universes in which it is not favorable for life to exist, and the fact that we exist means that our universe must have parameters within certain ranges. Whether the anthropic principle is important or merely circular is a topic of current debate.

The universe is a curious place, and the current research on cosmology makes it seem possibly more curious still.

## CORRECTING MISCONCEPTIONS

| ✖ *Incorrect* | ✔ *Correct* |
| --- | --- |
| The vertical axis of the graph of a Planck curve is temperature. | The vertical axis is intensity, or energy given off, while the horizontal axis is wavelength. Each temperature corresponds to a different Planck curve on these axes. |
| We know the full list of particles that make up our universe. | We are discovering new particles all the time in particle accelerators. |
| Protons and neutrons are the most fundamental particles. | Protons and neutrons are made of quarks, as are all other "baryons." |
| Electrons are made of quarks. | Electrons are fundamental particles, and along with muons and tau particles, they are called "leptons." |
| The elements always existed. | Hydrogen and helium were formed right after the big bang, and other elements were formed later in stars or supernovae. |
| The universe has always been expanding at about its current rate. | Most cosmologists think that the universe underwent a period of inflation. |

## SUMMARY AND OUTLINE

Creation of the elements (Section 38.1)
- Pairs of particles and antiparticles created out of energy in the early universe
- Light elements formed in the first minutes after the big bang; heavier elements formed in interiors of stars and in supernovae.

Future of the universe (Section 38.2)
- The missing-mass problem: discrepancy between mass derived by studying motions of galaxies in clusters of galaxies and the visible mass
- Deuterium-to-hydrogen ratio depends on cosmic density: all deuterium was formed in first minutes after the big bang; current evidence from deuterium is that universe is open.

Forces in the universe (Section 38.3)
- Four forces of nature, in declining order of strength: strong, electromagnetic, weak, gravitational
- Particles like protons and neutrons are made of quarks.
- Theory of electroweak force unifies electromagnetic and weak forces and led to discovery of W and Z particles, which are now being intensively studied.
- Grand-unified theories (GUTs) unify electroweak and strong forces and predict proton decay.

Shadow matter (Section 38.4)
- Superstring theory implies existence of shadow matter.
- Shadow matter would have gravity but no light.

The inflationary universe (Section 38.5)
- Universe grew rapidly larger in first $10^{-35}$ second.
- Explains homogeneity of background radiation, implies that universe is flat (on boundary between being open and closed), and explains why we don't detect magnetic monopoles

Cosmic string (Section 38.6)
- Defect in space-time; ruled out by cosmic microwave background observations

The Hubble Deep Fields (Section 38.7)
- Very long exposures of typical regions of space
- Very distant galaxies and quasars discovered.
- Regions also observed with other spacecraft and telescopes.

## KEY WORDS

early universe
particle physics
antiparticle
antimatter
baryons
nucleosynthesis
deuteron
missing-mass problem
dark matter
dark energy
rest mass
strong force
nuclear force
quarks
colors
electromagnetic force
weak force
gravitational force
electroweak force
grand-unified theories
superstring
string theory
inflationary universe

## QUESTIONS

1. Why can't we find out about the earliest universe before the Planck time?
2. An electron has one unit of negative charge. What is the charge of a positron, which is an antielectron? What is its mass compared with that of an electron?
†3. The mass of an electron is given in Appendix 2. What is the energy released in the annihilation of an electron and a positron? (Use the erg for the unit of energy, with 1 erg = 1 g $cm^2/sec^2$.) Compare your result with the energy being given off by the sun each second, also given in Appendix 2.
†4. Referring to the graph given in Fig. 38-4, what were the relative abundances of several light elements or isotopes 1 hour after the big bang? Compare these abundances to the abundance of protons.
5. What is the advantage of studying deuterium over studying helium, lithium, beryllium, and boron for assessing the density of the universe?
6. In your own words, explain why the abundance of deuterium is linked with the baryon density of the universe.
7. How did the discovery that the universe's expansion is accelerating affect our idea of whether the universe is open or closed?
8. How much energy would you have to put in to accelerate a proton until it was travelling at the speed of light? Explain.
9. Why do we feel gravity, given that the gravitational force is so weak relative to the other fundamental forces?
10. What are three advantages of the inflationary-universe theory?
11. What are the amounts of ordinary matter and dark matter relative to the critical density?
12. What would it mean for the universe to be a fractal?
†13. Compare the density of cosmic string with that of a cubic centimeter of water.
14. What is the relation between cosmic string and superfluid helium?
15. What is special about the Hubble Deep Fields? Discuss how they were taken and what they show.

†This question requires a numerical solution.

## TOPICS FOR DISCUSSION

1. What would be the effect on the conduct of your life or on your thoughts if you found out that the universe will expand forever, will begin to contract into a final black hole, or will contract and then oscillate with cycles extending into infinite time?
2. If one considers time as an arrow, with its direction fixed, would the direction of the arrow of time reverse if the universe were to start to contract?
3. Discuss whether astronomy or physics is more fundamental than the other, given the increasing overlap.

## USING TECHNOLOGY

### W World Wide Web

1. Look at the CERN information at http://public.web.cern.ch/Public, the Fermilab description at http://www.fnal.gov/pub/inquiring/matter/index.html, and the description at http://www.cpepweb.org for basic material as well as for the latest information about elementary particles.
2. The National Research Council's Committee on the physics of the universe was asked by the Department of Energy, NASA, and the National Science Foundation to identify science opportunities at the intersection of physics and astronomy. Learn about their *Connecting Quarks with the Cosmos* report, subtitled "Questions for the New Century," at http://www.nas.edu/bpa/reports/cpu/index.html.
3. Find a popular article about inflation at http://hbar.stanford.edu/linde.
4. Learn about inflation, cosmic string, quantum gravity, and other topics at http://www.damtp.cam.ac.uk/user/gr/public.
5. See NASA's site about the structure and evolution of the universe at http://universe.gsfc.nasa.gov.

### REDSHIFT

1. Look at Science of Astronomy: Beginnings and Endings: Yo, yo, So-so, or It's been nice knowing you.

# Epilogue

An artist's conception of the Planck and Herschel telescopes, to be launched into space together by the European Space Agency in 2007. Planck is to study the ripples in the cosmic microwave background at high resolution over the whole sky. Herschel carries a 3.5-m telescope for infrared studies of star formation and of distant, redshifted galaxies.

We have had our tour of the universe. We have seen the stars and planets, the matter between the stars, and the distant objects in our universe like quasars and clusters of galaxies. We have learned about explorations of planets and even of an asteroid by robotic spacecraft and seen fantastic close-up views. We have basked in the glory of a stellar explosion whose light and neutrinos have just reached Earth. We have learned how our universe is expanding and that it is bathed in a glow of radio waves. We have seen how tiny ripples in the glow are being studied from the ground, from balloons, and from space to reveal how the large-scale structure in our universe arose.

Further, we have seen the vitality of contemporary science in general and astronomy in particular. The individual scientists who call themselves astronomers are engaged in fascinating studies, often pushing new technologies to their limits. New telescopes on the ground and in space, new computer capabilities for studying data and carrying out calculations, and new theoretical ideas are linked in research about the universe.

Our knowledge of the universe changes so rapidly that within a few years much of what you read here will be revised. So your study of astronomy shouldn't end here; I hope that, over the years, you will keep up by following astronomical articles and stories in newspapers, magazines, and books (some are listed in the bibliography), on television, and, of course, on the World Wide Web. I hope that you will always consider the role of scientific research and realize its vitality. And I hope you will remember the methods of science—the mixture of logic and standards of proof by which scientists operate—that you have seen illustrated in this book.

# Measurement Systems



## Système International Units

| | SI units | SI Abbrev. | Other Abbrev. |
|---|---|---|---|
| length | meter | m | |
| volume | liter | L | ℓ |
| mass | kilogram | kg | kgm |
| time | second | s | sec |
| temperature | kelvin | K | °K |

## Other Metric Units

| | | |
|---|---|---|
| 1 micron ($\mu$) = 1 micrometer ($\mu$m) = $10^{-6}$ meter | $\mu$m | $\mu$ |
| 1 angstrom (Å or A) = $10^{-10}$ meter = $10^{-8}$ cm | 0.1 nm | Å |

## Prefixes for Use with Basic Units of Metric System

| Prefix | Symbol | Power | | | Equivalent |
|---|---|---|---|---|---|
| yotta | Y | $10^{24}$ | = | 1,000,000,000,000,000,000,000,000 | |
| zetta | Z | $10^{21}$ | = | 1,000,000,000,000,000,000,000 | |
| exa | E | $10^{18}$ | = | 1,000,000,000,000,000,000 | |
| peta | P | $10^{15}$ | = | 1,000,000,000,000,000 | |
| tera | T | $10^{12}$ | = | 1,000,000,000,000 | Trillion |
| giga | G | $10^{9}$ | = | 1,000,000,000 | Billion |
| mega | M | $10^{6}$ | = | 1,000,000 | Million |
| kilo | k | $10^{3}$ | = | 1,000 | Thousand |
| hecto | h | $10^{2}$ | = | 100 | Hundred |
| deca | da | $10^{1}$ | = | 10 | Ten |
| — | — | $10^{0}$ | = | 1 | One |
| deci | d | $10^{-1}$ | = | 0.1 | Tenth |
| centi | c | $10^{-2}$ | = | 0.01 | Hundredth |
| milli | m | $10^{-3}$ | = | 0.001 | Thousandth |
| micro | $\mu$ | $10^{-6}$ | = | 0.000001 | Millionth |
| nano | n | $10^{-9}$ | = | 0.000000001 | Billionth |
| pico | p | $10^{-12}$ | = | 0.000000000001 | Trillionth |
| femto | f | $10^{-15}$ | = | 0.000000000000001 | |
| atto | a | $10^{-18}$ | = | 0.000000000000000001 | |
| zepto | z | $10^{-21}$ | = | 0.000000000000000000001 | |
| yocto | y | $10^{-24}$ | = | 0.000000000000000000000001 | |

Examples: 1000 meters = 1 kilometer = 1 km
$10^{6}$ hertz = 1 megahertz = 1 MHz
$10^{-3}$ sec = 1 millisecond = 1 msec

*(continued)*

## Conversion Factors

1 erg = $10^{-7}$ joule = $10^{-7}$ kg · m$^2$/s$^2$
1 joule = $6.2419 \times 10^{18}$ eV
1 in = 25.4 mm = 2.54 cm
1 yd = 0.9144 m
1 mi = 1.6093 km $\simeq$ 8/5 km
1 oz = 28.3 g
1 lb = 0.4536 kg

## Binary Prefixes

| | |
|---|---|
| kibi | $2^{10}$ = 1,024 |
| mebi | $2^{20}$ = 1,048,576 |
| gibi | $2^{30}$ = $1.0737 \times 10^9$ |

Mars, on June 26, 2001, imaged with the Hubble Space Telescope when Mars was as close to Earth as it has been in a dozen years. Dust storms show at the north polar cap and at lower right. *[Credit: NASA and the Hubble Heritage Team, STScI/NASA; J. Bell (Cornell U.), P. James (U. Toledo), M. Wolff (Space Science Institute), A. Lubenow (STScI), J. Neubert (MIT/Cornell)]*

# Basic Constants



## Physical Constants (1998 CODATA Recommended Values)

| | | |
|---|---|---|
| Speed of light* | c | = 299 792 458 m/sec (exactly) |
| Constant of gravitation | G** | $= (6.67390 \pm 0.00001) \times 10^{-11}$ m$^3$/kg · sec$^2$ |
| Planck's constant | h | $= (6.626\ 068 \pm 0.000\ 0052) \times 10^{-34}$ J · s |
| Boltzmann's constant | k | $= (1.380\ 650\ 3 \pm 0.000\ 002\ 4) \times 10^{-23}$ J/K |
| Stefan-Boltzmann constant | $\sigma$ | $= (5.670\ 400 \pm 0.000\ 040) \times 10^{-8}$ W/m$^2$ · K$^4$ |
| Wien displacement constant | $\lambda_{max}T$ | $= (2.897\ 768\ 6 \pm 0.000\ 005) \times 10^{7}$ A · K |
| Mass of neutron | $m_n$ | $= (1.674\ 92716 \pm 0.000\ 00013) \times 10^{-27}$ kg |
| Mass of proton | $m_p$ | $= (1.672\ 621\ 58 \pm 0.000000\ 13) \times 10^{-27}$ kg |
| Mass of electron | $m_e$ | $= (9.109\ 381\ 88 \pm 0.000000\ 72) \times 10^{-31}$ kg |
| Rydberg constant | R | $= (10\ 973\ 731.568\ 549 \pm 0.000\ 083)$ m$^{-1}$ |

See http://physics.nist.gov/constants

## Mathematical Constants (to 100 places just for fun)

$\pi$ = 3.141 592 653 589 793 238 462 643 383 279 502 884 197 169 399 375 105 820 974 944 592 307 816 406 286 208 998 628 034 825 342 117 067 . . .

e = 2.718 281 828 459 045 235 360 287 471 352 662 497 757 247 093 699 959 574 966 967 627 724 076 630 353 547 594 571 382 178 525 166 427 . . .

## Astronomical Constants

| | | |
|---|---|---|
| Astronomical Unit* | 1 A.U. | $= 1.495\ 978\ 70 \times 10^{11}$ m |
| Solar parallax* | $\pi_\odot$ | = 8.794 148 arc sec |
| Parsec | 1 pc | $= 3.086 \times 10^{16}$ m |
| | | = 206 264.806 A.U. |
| | | = 3.261 633 ly |
| Light-year | 1 ly | $= (9.460\ 530) \times 10^{15}$ m |
| | | $= 6.324 \times 10^{4}$ A.U. |
| Tropical year (1900)* — (equinox to equinox) | | = 365.242 198 78 ephemeris days |
| Julian century* | | = 36 525 days |
| Day* | | = 86 400 sec |
| Sidereal year | | = 365.256 366 ephemeris days |
| | | $= (3.155\ 815) \times 10^{7}$ sec |
| Mass of sun | $M_\odot$** | $= (1.988\ 843 \pm 0.00003) \times 10^{30}$ kg |
| Radius of sun | $R_\odot$ | = 696 000 km |
| Luminosity of sun | $L_\odot$** | $= 3.827 \times 10^{26}$ J/sec |
| Mass of Earth* | $M_E$** | $= (5.972\ 23 \pm 0.00008) \times 10^{24}$ kg |
| Equatorial radius of Earth* | $R_E$ | = 6 378.140 km |
| Center of Earth to center of moon (mean) | | = 384 403 km |
| Radius of moon* | $R_M$ | = 1 738 km |
| Mass of moon* | $M_M$ | $= 7.35 \times 10^{22}$ kg |
| Solar constant | S | = 1 368 W/m$^2$ |
| Direction of galactic center (2000.0 precession) | $\alpha$ | $= 17^h 45.6^m$ |
| | $\delta$ | $= -28°56'$ |

*Adopted as "IAU (1976) system of astronomical constants" at the General Assembly of the International Astronomical Union that year. The meter was redefined in 1983 to be the distance travelled by light in a vacuum in 1/299,792,458 second.

**2000 values

## APPENDIX 3A Our Solar System: Intrinsic and Rotational Properties

| | Equatorial Radius | | | | | | | | |
|---|---|---|---|---|---|---|---|---|---|
| Name | km | ÷ Earth's | Mass ÷ Earth's | Mean Density ($g/cm^3$) | Oblateness | Surface Gravity (Earth = 1) | Sidereal Rotation Period | Inclination of Equator to Orbit | Apparent Magnitude During 2001 |
| Mercury | 2,439.7 | 0.3824 | 0.0553 | 5.43 | 0 | 0.378 | $58.646^d$ | 0.0° | −2.2 to +5.3 |
| Venus | 6,051.8 | 0.9489 | 0.8150 | 5.24 | 0 | 0.894 | $243.02^d$R | 177.3° | −4.6 to −3.9 |
| Earth | 6,378.14 | 1 | 1 | 5.515 | 0.0034 | 1 | $23^h56^m04.1^s$ | 23.45° | — |
| Mars | 3,397 | 0.5326 | 0.1074 | 3.94 | 0.006 | 0.379 | $24^h37^m22.662^s$ | 25.19° | −2.2 to +1.4 |
| Jupiter | 71,492 | 11.194 | 317.896 | 1.33 | 0.065 | 2.54 | $9^h50^m$ to $> 9^h55^m$ | 3.12° | −2.7 to −1.9 |
| Saturn | 60,268 | 9.41 | 95.185 | 0.70 | 0.098 | 1.07 | $10^h39.9^m$ | 26.73° | +0.2 to +0.2 |
| Uranus | 25,559 | 4.0 | 14.537 | 1.30 | 0.022 | 0.8 | $17^h14^m$R | 97.86° | +5.7 to +5.9 |
| Neptune | 24,764 | 3.9 | 17.151 | 1.76 | 0.017 | 1.2 | $16^h7^m$ | 29.56° | +7.8 to +8.0 |
| Pluto | 1,195 | 0.2 | 0.0025 | 2.1 | 0 | 0.01 | $6^d9^h17^m$R | 120° | +13.8 to +13.9 |

R signifies retrograde rotation.

The masses and radii for Mercury, Venus, earth, and Mars are the values recommended by the International Astronomical Union in 1976. The radii are from *The Astronomical Almanac 2001*. Surface gravities were calculated from these values. The length of the martian day is from G. de Vaucouleurs (1979). Most densities, oblatenesses, inclinations, and magnitudes are from *The Astronomical Almanac 2001*. Neptune data from *Science,* December 15, 1989 and August 9, 1991. Values for the masses of the giant planets are based on Voyager data for the mass of the sun divided by the mass of the planet (E. Myles Standish, Jr., *Astronomical Journal* **105,** 2000, 1993); Jupiter: 1047.3486; Saturn: 3497.898; Uranus: 22902.94; Neptune: 19412.24.

## APPENDIX 3B Our Solar System: Orbital Properties

| | Semimajor Axis | | Sidereal Period | | | | |
|---|---|---|---|---|---|---|---|
| Name | A.U. | $10^6$ km | Years | Days | Synodic Period (Days) | Eccentricity | Inclination to Ecliptic |
| Mercury | 0.387 099 | 57.909 | 0.240 84 | 87.96 | 115.9 | 0.205 63 | 7.004 87° |
| Venus | 0.723 332 | 108.209 | 0.615 18 | 224.68 | 583.9 | 0.006 77 | 3.394 71° |
| Earth | 1 | 149.598 | 0.999 98 | 365.25 | — | 0.016 71 | 0.000 05° |
| Mars | 1.523 662 | 227.939 | 1.880 7 | 686.95 | 779.9 | 0.093 41 | 1.850 61° |
| Jupiter | 5.203 363 | 778.298 | 11.857 | 4,337 | 398.9 | 0.048 39 | 1.305 30° |
| Saturn | 9.537 070 | 1429.394 | 29.424 | 10,760 | 378.1 | 0.054 15 | 2.484 46° |
| Uranus | 19.191 264 | 2875.039 | 83.75 | 30,700 | 369.7 | 0.047 168 | 0.769 86° |
| Neptune | 30.068 963 | 4504.450 | 163.72 | 60,200 | 367.5 | 0.008 59 | 0.769 17° |
| Pluto | 39.481 687 | 5915.799 | 248.02 | 90,780 | 366.7 | 0.248 81 | 17.141 75° |

Mean elements of planetary orbits for 2000, referred to the mean ecliptic and equinox of J2000 (E. M. Standish, X. X. Newhall, J. G. Williams, and D. K. Yeomans, *Explanatory Supplement to the Astronomical Almanac,* P. K. Seidelmann, ed., 1992). Periods are calculated from them.

## APPENDIX 3C Masses and Orbital Characteristics of Extrasolar Planets

| Star Name | M sin *i* ($M_{jup}$) | Period (d) | Semimajor Axis (AU) | Eccentricity |
|---|---|---|---|---|
| 1 HD83443 | 0.34 | 2.986 | 0.038 | 0.08 |
| 2 HD46375 | 0.25 | 3.024 | 0.041 | 0.02 |
| 3 HD179949 | 0.93 | 3.092 | 0.045 | 0.00 |
| 4 HD187123 | 0.54 | 3.097 | 0.042 | 0.01 |
| 5 Tau Boo | 4.14 | 3.313 | 0.047 | 0.02 |
| 6 BD−103166 | 0.48 | 3.487 | 0.046 | 0.05 |
| 7 HD75289 | 0.46 | 3.508 | 0.048 | 0.00 |
| 8 HD209458 | 0.63 | 3.524 | 0.046 | 0.02 |
| 9 51 Peg | 0.46 | 4.231 | 0.052 | 0.01 |
| 10 Upsilon And b | 0.68 | 4.617 | 0.059 | 0.02 |
| 11 HD168746 | 0.24 | 6.400 | 0.066 | 0.00 |
| 12 HD217107 | 1.29 | 7.130 | 0.072 | 0.14 |
| 13 HD162020 | 13.73 | 8.420 | 0.072 | 0.28 |
| 14 HD130322 | 1.15 | 10.72 | 0.092 | 0.05 |
| 15 HD108147 | 0.35 | 10.88 | 0.098 | 0.56 |
| 16 HD38529 | 0.77 | 14.31 | 0.129 | 0.27 |
| 17 55 Cnc | 0.93 | 14.66 | 0.118 | 0.03 |
| 18 HD13445=GJ86 | 4.23 | 15.80 | 0.117 | 0.04 |
| 19 HD195019 | 3.55 | 18.20 | 0.136 | 0.01 |
| 20 HD6434 | 0.48 | 22.09 | 0.154 | 0.30 |
| 21 HD192263 | 0.81 | 24.35 | 0.152 | 0.22 |
| 22 HD83443c | 0.17 | 29.83 | 0.174 | 0.42 |
| 23 GJ876c | 0.56 | 30.12 | 0.130 | 0.27 |
| 24 Rho CrB | 0.99 | 39.81 | 0.224 | 0.07 |
| 25 HD74156b | 1.55 | 51.61 | 0.276 | 0.65 |
| 26 HD168443b | 7.64 | 58.10 | 0.295 | 0.53 |
| 27 GJ876b | 1.89 | 61.02 | 0.207 | 0.10 |
| 28 HD121504 | 0.89 | 64.62 | 0.317 | 0.13 |
| 29 HD178911B | 6.46 | 71.50 | 0.326 | 0.14 |
| 30 HD16141 | 0.22 | 75.80 | 0.351 | 0.28 |
| 31 HD114762 | 10.96 | 84.03 | 0.351 | 0.33 |
| 32 HD80606 | 3.43 | 111.8 | 0.438 | 0.93 |
| 33 70 Vir | 7.42 | 116.7 | 0.482 | 0.40 |
| 34 HD52265 | 1.14 | 119.0 | 0.493 | 0.29 |
| 35 HD1237 | 3.45 | 133.8 | 0.505 | 0.51 |
| 36 HD37124 | 1.13 | 154.8 | 0.547 | 0.31 |
| 37 HD82943c | 0.88 | 221.6 | 0.728 | 0.54 |
| 38 HD8574 | 2.23 | 228.8 | 0.756 | 0.40 |
| 39 HD169830 | 2.95 | 230.4 | 0.823 | 0.34 |
| 40 Upsilon And c | 2.05 | 241.3 | 0.828 | 0.24 |
| 41 HD12661 | 2.84 | 250.5 | 0.795 | 0.19 |
| 42 HD89744 | 7.17 | 256.0 | 0.883 | 0.70 |
| 43 HD202206 | 14.68 | 258.9 | 0.768 | 0.42 |
| 44 HD134987 | 1.58 | 260.0 | 0.810 | 0.24 |
| 45 HD17051=iota Hor | 2.98 | 320.0 | 0.970 | 0.16 |
| 46 HD92788 | 3.88 | 337.0 | 0.969 | 0.28 |
| 47 HD28185 | 5.59 | 385.0 | 1.000 | 0.06 |
| 48 HD177830 | 1.24 | 391.0 | 1.10 | 0.40 |
| 49 HD27442 | 1.42 | 426.0 | 1.18 | 0.02 |
| 50 HD210277 | 1.29 | 436.6 | 1.12 | 0.45 |
| 51 HD82943b | 1.63 | 444.6 | 1.16 | 0.41 |
| 52 HD19994 | 1.83 | 454.2 | 1.26 | 0.20 |
| 53 HD222582 | 5.18 | 576.0 | 1.35 | 0.71 |

*continued*

## APPENDIX 3C *continued*

| Star Name | M sin *i* ($M_{jup}$) | Period (d) | Semimajor Axis (AU) | Eccentricity |
|---|---|---|---|---|
| 54 HD141937 | 9.69 | 658.8 | 1.48 | 0.40 |
| 55 HD160691 | 1.99 | 743.0 | 1.65 | 0.62 |
| 56 HD213240 | 3.75 | 759.0 | 1.60 | 0.31 |
| 57 16 Cyg B | 1.68 | 796.7 | 1.69 | 0.68 |
| 58 HD10697 | 6.08 | 1074.0 | 2.12 | 0.11 |
| 59 47 UMa | 2.60 | 1084.0 | 2.09 | 0.13 |
| 60 HD190228 | 5.01 | 1127.0 | 2.25 | 0.43 |
| 61 HD50554 | 4.91 | 1279.0 | 2.38 | 0.42 |
| 62 Upsilon And d | 4.29 | 1308.5 | 2.56 | 0.31 |
| 63 HD106252 | 6.81 | 1500.0 | 2.61 | 0.54 |
| 64 HD168443c | 16.96 | 1770.0 | 2.87 | 0.20 |
| 65 14 Her | 4.05 | 2000.0 | 3.17 | 0.45 |
| 66 HD74156c | 7.46 | 2300.0 | 3.47 | 0.40 |
| 67 Epsilon Eri | 0.88 | 2518.0 | 3.36 | 0.60 |

From exoplanets.org as of May 2001

# Planetary Satellites



| Satellite | Semimajor Axis of Orbit (km) | Sidereal Revolution Period (d | h | m) | Orbital Eccentricity | Orbital Inclination (°) | Radius (km) | Mass ÷ Mass of Planet | Mean Density (g/cm$^3$) | Discoverer | Visible Magnitude at Mean Opposition Distance |
|---|---|---|---|---|---|---|---|---|---|---|---|
| **Satellite of the Earth** | | | | | | | | | | | |
| The Moon | 384,400 | 27 | 07 | 43 | 0.055 | 18–29 | 1738 | 0.01230002 | 3.34 | — | −12.7 |
| **Satellites of Mars** | | | | | | | | | | | |
| 1 Phobos | 9,378 | 0 | 07 | 39 | 0.015 | 1.0 | 13 × 11 × 9 | $1.5 \times 10^{-8}$ | 1.95 | Hall (1877) | 11.8 |
| 2 Deimos | 23,459 | 1 | 06 | 18 | 0.0005 | 0.9–2.7 | 8 × 6 × 5 | $3 \times 10^{-9}$ | 2 | Hall (1877) | 12.9 |
| **Satellites of Jupiter** | | | | | | | | | | | |
| XVI Metis | 128,000 | 0 | 07 | 04 | | | 20 | $0.5 \times 10^{-10}$ | | Synnott/Voyager 2 (1980) | 17.5 |
| XV Adrastea | 129,000 | 0 | 07 | 06 | | | 13 × 10 × 8 | $0.1 \times 10^{-10}$ | | Jewett/Danielson (1979) | 19.1 |
| V Amalthea | 181,000 | 0 | 11 | 57 | 0.003 | 0.4 | 131 × 73 × 67 | $38 \times 10^{-10}$ | | Barnard (1892) | 14.1 |
| XIV Thebe | 222,000 | 0 | 16 | 11 | 0.015 | 0.8 | 55 × 45 | $4 \times 10^{-10}$ | | Synnott/Voyager 1 (1980) | 15.6 |
| I Io | 422,000 | 1 | 18 | 28 | 0.004 | 0.0 | 1830 × 1819 × 1813 | $4.68 \times 10^{-5}$ | 3.5 | Galileo (1610) | 5.0 |
| II Europa | 671,000 | 3 | 13 | 14 | 0.009 | 0.5 | 1565 | $2.52 \times 10^{-5}$ | 3.0 | Galileo (1610) | 5.3 |
| III Ganymede | 1,070,000 | 7 | 03 | 43 | 0.002 | 0.2 | 2634 | $7.80 \times 10^{-5}$ | 1.9 | Galileo (1610) | 4.6 |
| IV Callisto | 1,883,000 | 16 | 16 | 32 | 0.007 | 0.5 | 2403 | $5.66 \times 10^{-5}$ | 1.8 | Galileo (1610) | 5.7 |
| S 1975 J1 | 7,507,000 | 130.02 | | | 0.242 | 43.07 | 5 | | | Kowal (1975)/ Sheppard et al. (2000) | 21 |
| XIII Leda | 11,094,000 | 240 | | | 0.148 | 26.1 | 5 | $0.03 \times 10^{-10}$ | | Kowal (1974) | 20 |
| VI Himalia | 11,480,000 | 251 | | | 0.158 | 27.6 | 85 | $50 \times 10^{-10}$ | | Perrine (1904) | 14.8 |
| X Lysithea | 11,720,000 | 260 | | | 0.107 | 29.0 | 12 | $0.4 \times 10^{-10}$ | | Nicholson (1938) | 18.4 |
| VII Elara | 11,737,000 | 260 | | | 0.207 | 24.8 | 40 | $4 \times 10^{-10}$ | | Perrine (1905) | 16.8 |
| S/2000 J11 | 12,557,000 | 286.95 | | | 0.248 | 28.23 | 2 | | | Sheppard, Jewitt, Fernandez, Magnier | 22.4 |
| S/2000 J10 | 20,174,000 | 587.62R | | | 0.135 | 165.85 | 2 | | | Sheppard, Jewitt, Fernandez, Magnier | 22.5 |
| S/2000 J3 | 20,210,000 | 585.17R | | | 0.218 | 149.69 | 3 | | | Sheppard, Jewitt, Fernandez, Magnier | 21.8 |
| S/2000 J7 | 21,010,000 | 619.19R | | | 0.225 | 148.84 | 4 | | | Sheppard, Jewitt, Fernandez, Magnier | 21.2 |
| XII Ananke | 21,200,000 | 671R | | | 0.169 | 147 | 10 | $0.2 \times 10^{-10}$ | | Nicholson (1951) | 18.9 |
| S/2000 J5 | 21,336,000 | 633.68R | | | 0.235 | 148.96 | 2 | | | Sheppard, Jewitt, Fernandez, Magnier | 22.2 |
| S/2000 J9 | 22,304,000 | 682.68R | | | 0.259 | 164.79 | 3 | | | Sheppard, Jewitt, Fernandez, Magnier | 21.9 |
| XI Carme | 22,600,000 | 692R | | | 0.207 | 164 | 15 | $0.5 \times 10^{-10}$ | | Nicholson (1938) | 18.0 |
| S/2000 J4 | 22,972,000 | 713.52R | | | 0.28 | 164.99 | 2 | | | Sheppard, Jewitt, Fernandez, Magnier | 22.8 |
| S/2000 J6 | 23,074,000 | 718.71R | | | 0.261 | 164.83 | 2 | | | Sheppard, Jewitt, Fernandez, Magnier | 22.5 |
| VIII Pasiphae | 23,500,000 | 735R | | | 0.378 | 145 | 18 | $1 \times 10^{-10}$ | | Melotte (1908) | 17.0 |
| S/2000 J8 | 23,618,000 | 743.13R | | | 0.407 | 153.09 | 3 | | | Sheppard, Jewitt, Fernandez, Magnier | 21.7 |
| IX Sinope | 23,700,000 | 758R | | | 0.275 | 153 | 14 | $0.4 \times 10^{-10}$ | | Nicholson (1914) | 18.3 |
| | | | | | | | | | | Kowal (1975) | 20 |
| S/2000 J2 | 23,746,000 | 750.81R | | | 0.243 | 165.21 | 3 | | | Sheppard, Jewitt, Fernandez, Magnier | 21.8 |
| S/1999 J1 | 24,103,000 | 758.76 | | | 0.282 | 147.13 | 4 | | | Spacewatch, Minor Comet Center | 20.2 |
| **Satellites of Saturn** | | | | | | | | | | | |
| 18 Pan | 133,583 | | 13 | 48 | | | 10 | $8 \times 10^{-12}$ | | Showalter/Voyager 2 (1990) | |
| 15 Atlas | 137,670 | | 14 | 27 | 0.000 | 0.3 | 20 × 10 | | | Terrile/Voyager 1 | 18 |
| 16 Prometheus | 139,353 | | 14 | 43 | 0.003 | 0.0 | 70 × 50 × 40 | | | Collins/Voyager 1 | 16 |
| 17 Pandora | 141,700 | | 15 | 05 | 0.004 | 0.0 | 55 × 45 × 35 | | | Collins/Voyager 1 | 16 |
| 11 Epimetheus | 151,422 | | 16 | 40 | 0.009 | 0.3 | 70 × 60 × 50 | | | Fountain, Larson/ Reitsema/Smith | 15 |
| 10 Janus | 151,472 | | 16 | 40 | 0.007 | 0.1 | 110 × 100 × 80 | | | Dollfus (1966/80) | 14 |
| 1 Mimas | 185,520 | | 22 | 37 | 0.0202 | 1.5 | 209 × 196 × 191 | $8 \times 10^{-8}$ | 1.2 | W. Herschel (1789) | 12.9 |
| 2 Enceladus | 238,020 | 1 | 08 | 32 | 0.00452 | 0.0 | 256 × 247 × 245 | $1.3 \times 10^{-7}$ | 1.1 | W. Herschel (1789) | 11.7 |
| 3 Tethys | 294,660 | 1 | 22 | 15 | 0.00000 | 1.9 | 528 × 526 | $1.3 \times 10^{-6}$ | 1.0 | Cassini (1684) | 10.2 |
| 13 Telesto | 294,660 | 1 | 22 | 15 | | | 15 × 13 × 8 | | | Smith, Reitsema, Larson, Fountain (1980) | 19 |
| 14 Calypso | 294,660 | 1 | 22 | 15 | | | 15 × 8 × 8 | | | Pascu, Seidelmann, Baum, Currie (1980) | 19 |
| 4 Dione | 377,400 | 2 | 17 | 36 | 0.002230 | 0.02 | 560 | $1.96 \times 10^{-6}$ | 1.4 | Cassini (1684) | 10.4 |

*(continued)*

| Satellite | Semimajor Axis of Orbit (km) | Sidereal Revolution Period (d | h | m) | Orbital Eccentricity | Orbital Inclination (°) | Radius (km) | Mass ÷ Mass of Planet | Mean Density (g/cm³) | Discoverer | Visible Magnitude at Mean Opposition Distance |
|---|---|---|---|---|---|---|---|---|---|---|---|
| 12 Helene | 377,400 | 2 | 17 | 45 | 0.005 | 0.0 | 18 × 16 × 15 | | | Laques and Lecacheux (1980) | 18 |
| 5 Rhea | 527,040 | 4 | 12 | 16 | 0.00100 | 0.4 | 764 | $4.4 \times 10^{-6}$ | 1.3 | Cassini (1672) | 9.7 |
| 6 Titan | 1,221,830 | 15 | 21 | 51 | 0.029192 | 0.3 | 2575 | $2.4 \times 10^{-4}$ | 1.9 | Huygens (1665) | 8.3 |
| 7 Hyperion | 1,481,100 | 21 | 06 | 45 | 0.104 | 0.4 | 180 × 140 × 113 | $3 \times 10^{-8}$ | 1.9 | Bond, Bond, Lassell (1848) | 14.2 |
| 8 Iapetus | 3,561,300 | 79 | 03 | 43 | 0.02828 | 14.7 | 718 | $3.3 \times 10^{-6}$ | 1.2 | Cassini (1671) | 11.1 |
| 18 S/2000 S5 | 11,339,000 | 447.77 | | | 0.334 | 46.2 | 8 | | | Gladman | 22 |
| 19 S/2000 S6 | 11,465,000 | 453.05 | | | 0.319 | 46.64 | 5 | | | Kavelaars, Gladman | 22.6 |
| 9 Phoebe | 12,952,000 | 549 | 03 | 33R | 0.16326 | 177 | 110 | $7 \times 10^{-10}$ | | W. Pickering (1898) | 16.5 |
| 21 S/2000 S2 | 15,172,000 | 685.89 | | | 0.363 | 45.15 | 10 | | | Gladman | 21.3 |
| 22 S/2000 S8 (R) | 15,676,000 | 730.84 | | | 0.27 | 153.01 | 3 | | | Kavelaars, Gladman | 23.6 |
| 23 S/2000 S3 | 17,251,000 | 826.02 | | | 0.266 | 45.45 | 16 | | | Gladman,Kavelaars | 20.1 |
| 24 S/2000 S10 | 17,452,000 | 860.03 | | | 0.469 | 34.74 | 4 | | | Kavelaars, Gladman | 23 |
| 25 S/2000 S11 | 17,874,000 | 888.54 | | | 0.376 | 33.08 | 13 | | | Holman, Spahr | 20.5 |
| 26 S/2000 S4 | 18,231,000 | 924.58 | | | 0.536 | 33.54 | 7 | | | Kavelaars, Gladman | 22.1 |
| 27 S/2000 S9 (R) | 18,486,000 | 939.9 | | | 0.221 | 167.42 | 3 | | | Gladman, Kavelaars | 23.8 |
| 28 S/2000 S12 (R) | 19,747,000 | 1038.11 | | | 0.12 | 175.83 | 3 | | | Gladman, Kavelaars | 23.9 |
| 29 S/2000 S7 (R) | 20,144,000 | 1068.06 | | | 0.446 | 175.86 | 3 | | | Gladman, Kavelaars | 23.9 |
| 30 S/2000 S1 (R) | 23,076,000 | 1310.6 | | | 0.337 | 173.12 | 8 | | | Gladman | 21.7 |
| **Satellites of Uranus** | | | | | | | | | | | |
| 6 Cordelia | 49,770 | | 08 | 02 | < 0.001 | 0.1 | 25 × 18 | | | Terrile/Voyager 2 (1986) | |
| 7 Ophelia | 53,790 | | 09 | 02 | 0.010 | 0.1 | 27 × 19 | | | Terrile/Voyager 2 (1986) | |
| 8 Bianca | 59,170 | | 10 | 25 | < 0.001 | 0.2 | 32 × 23 | | | Voyager 2 (1986) | |
| 9 Cressida | 61,780 | | 11 | 07 | < 0.001 | 0.0 | 46 × 37 | | | Synnott/Voyager 2 (1986) | |
| 10 Desdemona | 62,680 | | 11 | 22 | < 0.001 | 0.2 | 45 × 27 | | | Synnott/Voyager 2 (1986) | |
| 11 Juliet | 64,350 | | 11 | 50 | < 0.001 | 0.1 | 75 × 37 | | | Synnott/Voyager 2 (1986) | |
| 12 Portia | 66,090 | | 12 | 19 | < 0.001 | 0.1 | 39 × 32 | | | Synnott/Voyager 2 (1986) | |
| 13 Rosalind | 69,940 | | 11 | 54 | < 0.001 | 0.3 | 18 | | | Synnott/Voyager 2 (1986) | |
| 14 Belinda | 75,260 | | 14 | 57 | < 0.001 | 0.0 | 32 × 16 | | | Synnott/Voyager 2 (1986) | |
| S/1986 U10 | 76,417 | | 15 | | 0.00215 | 81 | 64 × 32 | | | Karkoschka (2001) | |
| 15 Puck | 86,010 | | 18 | 17 | < 0.001 | 0.3 | 42 | | | Synnott/Voyager 2 (1985) | |
| 5 Miranda | 129,390 | 1 | 09 | 56 | 0.0027 | 4.2 | 240 × 234 | $0.2 \times 10^{-5}$ | 1.26 | Kuiper (1948) | 16.3 |
| 1 Ariel | 191,020 | 2 | 12 | 29 | 0.0034 | 0.3 | 581 × 578 | $1.6 \times 10^{-5}$ | 1.65 | Lassell (1851) | 14.2 |
| 2 Umbriel | 266,300 | 4 | 03 | 27 | 0.0050 | 0.4 | 585 | $1.4 \times 10^{-5}$ | 1.44 | Lassell (1851) | 14.8 |
| 3 Titania | 433,910 | 8 | 16 | 56 | 0.0022 | 0.1 | 789 | $4.1 \times 10^{-5}$ | 1.59 | W. Herschel (1787) | 13.7 |
| 4 Oberon | 583,520 | 13 | 11 | 07 | 0.0008 | 0.1 | 761 | $3.5 \times 10^{-5}$ | 1.50 | W. Herschel (1787) | 13.9 |
| 16 Caliban | 7,169,000 | 580R | | | 0.082 | 139.7 | 30 | | | Gladman | 22.4 |
| 17 Sycorax | 12,214,000 | 1,284R | | | 140.509 | 152.7 | 60 | | | Nicholson, Burns, Kavelaars (1997) | 20.09 |

Stephano (7,979,000 km orbital radius), Prospero (16,665,000 km), and Selebos (17,879,000 km) discovered in 1999 by Holman, Gladman, and Kavelaars.

| Satellite | Semimajor Axis of Orbit (km) | d | h | m | Orbital Eccentricity | Orbital Inclination (°) | Radius (km) | Mass ÷ Mass of Planet | Mean Density (g/cm³) | Discoverer | Visible Magnitude at Mean Opposition Distance |
|---|---|---|---|---|---|---|---|---|---|---|---|
| **Satellites of Neptune** | | | | | | | | | | | |
| 3 Naiad | 48,230 | | 7 | | < 0.001 | 4.74 | 29 | | | Voyager 2 (1989) | 24.7 |
| 4 Thalassa | 50,070 | | 7 | 30 | < 0.001 | 0.21 | 40 | | | Synnott/Voyager 2 (1989) | 23.0 |
| 5 Despina | 52,530 | | 8 | | < 0.001 | 0.07 | 74 | | | Synnott/Voyager 2 (1989) | 22.6 |
| 6 Galatea | 61,950 | | 10 | | < 0.001 | 0.05 | 79 | | | Synnott/Voyager 2 (1989) | 22.3 |
| 7 Larissa | 73,550 | | 13 | | < 0.0014 | 0.20 | 104 × 89 | | | Reitsema, Hubbard, Lebofsky, Tholen/ Voyager 2 (1989) | 22.0 |
| 8 Proteus | 117,650 | | 27 | | 0.0004 | 0.05 | 218 × 208 × 201 | | | Synnott/Voyager 2 (1989) | 20.03 |
| 1 Triton | 354,760 | 5 | 21 | 03 R | 0.000016 | 157,345 | 1353 | $2.09 \times 10^{-4}$ | 2.07 | Lassell (1846) | 13.5 |
| 2 Nereid | 5,513,400 | 360 | 5 | | 0.75 | 27.6 | 170 | $2 \times 10^{-7}$ | 2.03 | Kuiper (1949) | 18.7 |

+4 reported discoveries awaiting confirmation

**Rings and Ring Arcs of Neptune**

| Ring | Semimajor Axis of Orbit (km) |
|---|---|
| Galle | 41,900 |
| Leverrier | 53,200 |
| Lassell | 53,200–57,500 |
| Arago | 57,500 |
| Adams | 62,900 |
| Courage | 62,900 |
| Liberté | 62,900 |
| Egalité | 62,900 |
| Fraternité | 62,900 |

**Satellite of Pluto**

| Satellite | Semimajor Axis of Orbit (km) | d | h | m | Orbital Eccentricity | Orbital Inclination (°) | Radius (km) | Mass ÷ Mass of Planet | Mean Density (g/cm³) | Discoverer | Visible Magnitude at Mean Opposition Distance |
|---|---|---|---|---|---|---|---|---|---|---|---|
| Charon | 19,682 | 6 | 9 | 17 R | < 0.001 | 99 | 593 | 0.124 | | Christy (1978) | 16.8 |

Based on a table by Joseph Veverka in the *Observer's Handbook of the Royal Astronomical Society of Canada* and on the *Astronomical Almanac 2001*. Many of Veverka's values are from J. Burns. Density is average of Pluto and Charon. Inclinations greater than 90° are retrograde; R signifies retrograde revolution. Discoverers from Burns and Matthews, *Satellites* (1986), updated with correspondence with D. Pascu, R. Synnott, H. Reitsema, and R. Terrile (1994). Discoveries in 2000 and 2001 from Brett Gladman, Observatoire de la Cote d'Azur, Nice; David Jewitt, U. Hawaii; European Southern Observatory; Erich Karkoschka, U. Arizona; and Jet Propulsion Laboratory/Caltech. See http://ssd.jpl.nasa.gov/sat_props.html and http://ssd.jpl.nasa.gov/sat_elem.html.

# The Brightest Stars — APPENDIX 5

| | | Position (2000.0) | | | | | | Proper Motion | | |
|---|---|---|---|---|---|---|---|---|---|---|
| Star | Name | R.A. h m s | Decl. ° ′ ″ | Apparent Magnitude ($V$) | Spectral Type | Absolute Magnitude ($M_v$) | Distance $D$ (ly) | R.A. ″/yr | Dec ° | Radial Vel. (km/sec) |
| 1. $\alpha$ CMa A | Sirius | 06 45 09 | −16 42 58 | −1.46 | A1 V | +1.5 | 9 | 1.324 | 204 | −8 |
| 2. $\alpha$ Car | Canopus | 06 23 57 | −52 41 44 | −0.72 | A9 Ib | −5.4 | 313 | 0.034 | 50 | +21 |
| 3. $\alpha$ Boo | Arcturus | 14 15 31 | +19 10 57 | −0.04 | K2 IIIp | −0.6 | 37 | 2.281 | 209 | −5 |
| 4. $\alpha$ Cen A | Rigil Kentaurus | 14 39 37 | −60 50 02 | 0.00 | G2 V | +4.2 | 4 | 3.678 | 28 | −25 |
| 5. $\alpha$ Lyr | Vega | 18 36 56 | +38 47 01 | 0.03 | A0 V | +0.6 | 25 | 0.348 | 35 | −14 |
| 6. $\alpha$ Aur | Capella | 05 16 41 | +45 59 53 | 0.08 | G6 + G2 | −0.8 | 42 | 0.430 | 169 | +30 |
| 7. $\beta$ Ori A | Rigel | 05 14 32 | −08 12 06 | 0.12 | B8 Ia | −6.6 | 773 | 0.004 | 236 | +21 |
| 8. $\alpha$ CMi A | Procyon | 07 39 18 | +05 13 30 | 0.38 | F2 IV-V | +2.8 | 11 | 1.248 | 214 | −3 |
| 9. $\alpha$ Eri | Achernar | 01 37 43 | −57 14 12 | 0.46 | B3 V | −2.9 | 144 | 0.108 | 105 | +19 |
| 10. $\alpha$ Ori | Betelgeuse | 05 55 10 | +07 24 36 | 0.50 | M2 Iab | −5.0 | 522 | 0.028 | 68 | +21 |
| 11. $\beta$ Cen AB | Hadar | 14 03 49 | −60 22 22 | 0.61 | B1 III | −5.5 | 526 | 0.030 | 221 | −12 |
| 12. $\alpha$ Aql | Altair | 19 50 47 | +08 52 06 | 0.77 | A7 IV-V | +2.1 | 17 | 0.662 | 54 | −26 |
| 13. $\alpha$ Tau A | Aldebaran | 04 35 55 | +16 30 33 | 0.85 | K5 III | −0.8 | 65 | 0.200 | 161 | +54 |
| 14. $\alpha$ Sco A | Antares | 16 29 24 | −26 25 55 | 0.96 | M1.5 Iab | −5.8 | 604 | 0.024 | 197 | −3 |
| 15. $\alpha$ Vir | Spica | 13 25 12 | −11 09 41 | 0.98 | B1 V | −3.6 | 262 | 0.054 | 232 | +1 |
| 16. $\beta$ Gem | Pollux | 07 45 19 | +28 01 34 | 1.14 | K0 IIIb | +1.1 | 34 | 0.629 | 265 | +3 |
| 17. $\alpha$ Psc A | Formalhaut | 22 57 39 | −29 37 20 | 1.16 | A3 V | +1.6 | 25 | 0.373 | 116 | +7 |
| 18. $\alpha$ Cyg | Deneb | 20 41 26 | +45 16 49 | 1.25 | A2 Ia | −7.5 | 1467 | 0.005 | 11 | −5 |
| 19. $\beta$ Crucis | | 12 47 43 | −59 41 19 | 1.25 | B0.5 III | −4.0 | 352 | 0.042 | 246 | +20 |
| 20. $\alpha$ Leo A | Regulus | 10 08 22 | +11 58 02 | 1.35 | B7 V | −0.6 | 77 | 0.264 | 271 | +14 |
| 21. $\alpha$ Cru A | | 12 26 35 | −63 05 56 | 1.41 | B0.5 IV | −4.0 | 321 | 0.030 | 236 | −11 |
| 22. $\epsilon$ CMa A | Adara | 06 58 38 | −28 59 20 | 1.50 | B2 II | −4.1 | 431 | 0.002 | 27 | +27 |
| 23. $\lambda$ Sco | Shaula | 17 33 36 | −37 06 14 | 1.63 | B1.5 IV | −3.6 | 359 | 0.029 | 178 | 0 |
| 24. $\gamma$ Ori | Bellatrix | 05 25 08 | +06 20 59 | 1.64 | B2 III | −2.8 | 243 | 0.018 | 221 | +18 |
| 25. $\beta$ Tau | Alnath | 05 26 18 | +28 36 27 | 1.65 | B7 III | −1.3 | 131 | 0.178 | 172 | +8 |

Based on a table by Robert Garrison in the *Observer's Handbook 2001 of the Royal Astronomical Society of Canada,* with the permission of the RASC.

# APPENDIX 6 — Greek Alphabet

| Upper Case | Lower Case | | Upper Case | Lower Case | |
|---|---|---|---|---|---|
| A | $\alpha$ | alpha | N | $\nu$ | nu |
| B | $\beta$ | beta | Ξ | $\xi$ | xi |
| Γ | $\gamma$ | gamma | O | $o$ | omicron |
| Δ | $\delta$ | delta | Π | $\pi$ | pi |
| E | $\epsilon$ | epsilon | P | $\rho$ | rho |
| Z | $\zeta$ | zeta | Σ | $\sigma$ | sigma |
| H | $\eta$ | eta | T | $\tau$ | tau |
| Θ | $\theta$ | theta | Y | $\upsilon$ | upsilon |
| I | $\iota$ | iota | Φ | $\phi$ | phi |
| K | $\kappa$ | kappa | X | $\chi$ | chi |
| Λ | $\lambda$ | lambda | Ψ | $\psi$ | psi |
| M | $\mu$ | mu | Ω | $\omega$ | omega |

We see Saturn and its ring at different angles in this series of Hubble Space Telescope views from 1996 to 2000. *[Credit: NASA and the Hubble Heritage Team, STScI/NASA; Richard French (Wellesley College), J. Cuzzi (NASA's Ames Research Center), L. Dones (Southwest Research Institute), and J. Lissauer (NASA's Ames Research Center)]*

# Nearest Stars

## APPENDIX 7

| | Name | R.A. (2000.0) h | R.A. (2000.0) m | Dec. (2000.0) ° | Dec. (2000.0) ′ | Parallax $\pi$ ″ | Distance light-years | Proper Motion $\mu$ ″/yr | Proper Motion $\theta$ ° | Radial Velocity km/sec | Spectral Type | $V$ | B-V | $M_V$ | Luminosity ($L_{\odot} = 1$) |
|---|---|---|---|---|---|---|---|---|---|---|---|---|---|---|---|
| 1 | Sun | | | | | | | | | | G2 V | −26.75 | 0.65 | 4.82 | 1. |
| 2 | Proxima Cen | 14 | 29.7 | −62 | 41 | 0.772 | 4.24 | 3.86 | 282 | −22 | M5.5e | 11.05 | 1.90 | 15.49 | 0.00005 |
| | $\alpha$ Cen A | 14 | 39.6 | −60 | 50 | .742 | 4.40 | 3.71 | 278 | −22 | G2 V | .02 | 0.65 | 4.37 | 1.51 |
| | $\alpha$ Cen B | | | | | | | 3.69 | 281 | −18 | K0 V | 1.36 | 0.85 | 5.71 | 0.44 |
| 3 | Barnard's star | 17 | 57.8 | +04 | 42 | .549 | 5.98 | 10.37 | 356 | −111 | M4 V | 9.54 | 1.74 | 13.24 | 0.0004 |
| 4 | Wolf 359 (CN Leo) | 10 | 56.5 | +07 | 1 | .419 | 7.78 | 4.69 | 235 | +13 | M6 V | 13.45 | 2.0 | 16.56 | 0.00002 |
| 5 | BD+36°2147 HD95735 | 11 | 03.4 | +35 | 58 | .392 | 8.31 | 4.81 | 187 | −85 | M2 V | 7.49 | 1.51 | 10.46 | 0.006 |
| | (Lalande 21185) | | | | | | | | | | | | | | |
| 6 | Sirius A | 6 | 45.1 | −16 | 43 | .379 | 8.60 | 1.34 | 204 | −8 | A1 V | −1.45 | 0.00 | 1.44 | 22.49 |
| | Sirius B | | | | | | | 1.34 | 204 | | DA 2 | 8.44 | −0.03 | 11.33 | 0.0025 |
| 7 | L 726−8, BL Cet = A | 1 | 39.0 | −17 | 57 | .374 | 8.73 | 3.37 | 81 | +29 | M5.5 V | 12.41 | 1.87 | 15.27 | 0.00007 |
| | UV Cet = B | | | | | | | 3.37 | 81 | +28 | M6 V | 13.25 | | 16.11 | 0.00003 |
| 8 | Ross 154 (V1216 Sgr) | 18 | 49.8 | −23 | 50 | .337 | 9.69 | 0.67 | 107 | −12 | M3.5 V | 10.45 | 1.76 | 13.08 | 0.0005 |
| 9 | Ross 248 (HH And) | 23 | 41.9 | +44 | 10 | .316 | 10.32 | 1.62 | 177 | −78 | M5.5 V | 12.29 | 1.91 | 14.79 | 0.0001 |
| 10 | $\epsilon$ Eri | 3 | 32.9 | −09 | 27 | .311 | 10.50 | 0.98 | 271 | +16 | K2 V | 3.72 | 0.88 | 6.18 | 0.286 |
| 11 | CD−36°15693 HD217987 | 23 | 05.9 | −35 | 51 | .304 | 10.73 | 6.90 | 79 | +10 | M2 V | 7.35 | 1.49 | 9.76 | 0.01 |
| | (Lacaille 9352) | | | | | | | | | | | | | | |
| 12 | Ross 128 (FI Vir) | 11 | 47.7 | +00 | 48 | .300 | 10.89 | 1.36 | 154 | −31 | M4 V | 11.12 | 1.76 | 13.50 | 0.00034 |
| 13 | L 789−6 (EZ Aqr) = A | 22 | 38.5 | −15 | 18 | .290 | 11.27 | 3.25 | 47 | −60 | M5 V | 12.69 | 1.99 | 15.00 | 0.00008 |
| | = B | | | | | | | 3.25 | 47 | | | 13.6 | | 15.9 | 0.00004 |
| 14 | 61 Cyg A | 21 | 06.9 | +38 | 45 | .286 | 11.41 | 5.28 | 52 | −65 | K5 V | 5.22 | 1.17 | 7.51 | 0.084 |
| | 61 Cyg B | | | | | | | 5.17 | 53 | −64 | K7 V | 6.04 | 1.36 | 8.32 | 0.0398 |
| 15 | Procyon A | 7 | 39.3 | +05 | 14 | .286 | 11.41 | 1.26 | 215 | −4 | F5 IV-V | 0.36 | 0.42 | 2.64 | 7.45 |
| | Procyon B | | | | | | | 1.26 | 215 | | DA | 10.75 | | 13.03 | 0.0005 |
| 16 | BD+43° 44 A (GX And) | 0 | 18.4 | +44 | 1 | .280 | 11.60 | 2.92 | 82 | +12 | M1.5 V | 8.08 | 1.56 | 10.32 | 0.006 |
| | BD+43° 44 B (GQ And) | 0 | 18.4 | +44 | 2 | .280 | 11.64 | 2.92 | 82 | +11 | M3.5 V | 11.05 | 1.81 | 13.29 | 0.0004 |
| 17 | BD+59 1915 A | 18 | 42.7 | +59 | 38 | .280 | 11.64 | 2.24 | 324 | −1 | M3 V | 8.90 | 1.53 | 11.14 | 0.00 |
| | BD+59 1915 B | 18 | 42.8 | +59 | 38 | .280 | 11.64 | 2.27 | 323 | +2 | M3.5 V | 9.69 | 1.59 | 11.93 | 0.0014 |
| | Struve 2398AB = ADS 11632AB | | | | | | | | | | | | | | |
| 18 | $\epsilon$ Ind | 22 | 03.4 | −56 | 47 | .276 | 11.84 | 4.71 | 123 | −40 | K5 Ve | 4.69 | 1.05 | 6.89 | 0.148 |
| 19 | G 51−15 (DX Cnc) | 8 | 29.8 | +26 | 47 | .276 | 11.83 | 1.29 | 243 | +25 | M6.5 V | 14.79 | 2.07 | 16.99 | 0.00001 |
| 20 | $\tau$ Ceti | 1 | 44.1 | −15 | 56 | .274 | 11.90 | 1.92 | 296 | −17 | G8 Vp | 3.49 | 0.72 | 5.68 | 0.45 |
| 21 | L 372−58 = LHS 1565 | 3 | 36.0 | −44 | 31 | .270 | 12.06 | 0.84 | 119 | −20 | M5.5 V | 13.01 | 1.90 | 15.17 | 0.00007 |
| 22 | L 725−32 (YZ Cet) | 1 | 12.5 | −17 | 0 | .269 | 12.13 | 1.37 | 62 | +28 | M4.5 V | 12.05 | 1.83 | 14.20 | 0.0002 |
| 23 | BD+5°1668 | 7 | 27.4 | +05 | 14 | .263 | 12.39 | 8.74 | 171 | +18 | M3.5 V | 9.85 | 1.57 | 11.95 | 0.0014 |
| 24 | Kapteyn's star | 5 | 11.7 | −45 | 1 | .255 | 12.78 | 8.66 | 131 | +246 | M1 pV | 8.85 | 1.56 | 10.89 | 0.004 |
| 25 | CD−39°14192 (AX Mic) | 21 | 17.3 | −38 | 52 | .253 | 12.87 | 3.45 | 251 | +26 | M0.5 V | 6.68 | 1.41 | 8.70 | 0.028 |
| | (Lacaille 8760) | | | | | | | | | | | | | | |
| 26 | BD+56°2783 A | 22 | 28.0 | +57 | 42 | .250 | 13.18 | 0.99 | 242 | −33 | M3 V | 9.79 | 1.65 | 11.78 | 0.002 |
| | BD+56°2783 B (DO Cep) | | | | | | | 0.99 | 242 | −32 | M4 V | 11.46 | 1.8 | 13.45 | 0.0004 |
| | Krüger 60 AB | | | | | | | | | | | | | | |
| 27 | Ross 614 A | 6 | 29.4 | −02 | 49 | .243 | 13.43 | 0.93 | 132 | +17 | M4.5 V | 11.14 | 1.72 | 13.07 | 0.0005 |
| | Ross 614 B (V577 Mon) | | | | | | | 0.93 | 132 | | M5.5 V | 14.47 | | 16.40 | 0.00002 |
| 28 | BD−12°4523 | 16 | 30.3 | −12 | 40 | .235 | 13.91 | 1.19 | 185 | −22 | M3 V | 10.08 | 1.59 | 11.93 | 0.0014 |
| 29 | van Maanen's star | 0 | 49.2 | +05 | 23 | .233 | 14.02 | 2.98 | 156 | +43 | DZ7 | 12.39 | 0.55 | 14.22 | 0.0002 |
| 30 | CD−37°15492 | 0 | 05.4 | −37 | 21 | .229 | 14.22 | 6.10 | 113 | +23 | M3 V | 8.55 | 1.45 | 10.35 | 0.006 |
| 31 | Wolf 424 A | 12 | 33.3 | +09 | 1 | .228 | | 1.81 | 278 | −2 | M5.5 V | 13.04 | 1.84 | 14.83 | 0.0001 |
| | Wolf 424 B (FL Vir) | | | | | | | 1.81 | 278 | | M7 | 13.3 | | 15.1 | 0.00008 |
| 32 | L 1159−16 (TZ Ari) | 2 | 00.2 | +13 | 3 | .225 | 14.51 | 2.10 | 148 | −29 | M4.5 V | 12.27 | 1.81 | 14.03 | 0.0002 |
| 33 | L 143−23 = LHS 288 | 10 | 44.5 | −61 | 12 | .223 | 14.66 | 1.66 | 348 | | M5.5 | 13.87 | 1.83 | 15.61 | 0.00005 |

*(continued)*

| | Name | R.A. (2000.0) h | m | Dec. ° | ′ | Parallax $\pi$ ″ | Distance light-years | Proper Motion $\mu$ ″/yr | $\theta$ ° | Radial Velocity km/sec | Spectral Type | $V$ | B-V | $M_V$ | Luminosity ($L_{\odot}$ = 1) |
|---|---|---|---|---|---|---|---|---|---|---|---|---|---|---|---|
| 34 | BD+68°946 | 17 | 36.4 | +68 | 20 | .221 | 14.77 | 1.31 | 194 | −27 | M3 V | 9.17 | 1.49 | 10.89 | 0.0037 |
| 35 | CD−46°11540 | 17 | 28.7 | −46 | 54 | .220 | 14.80 | 1.05 | 147 | −16 | M3 V | 9.38 | 1.55 | 11.10 | 0.0031 |
| 36 | LP 731−58 = LHS 292 | 10 | 48.2 | −11 | 20 | .220 | 14.81 | 1.64 | 159 | −2 | M6.5 V | 15.60 | 2.10 | 17.32 | 0.00001 |
| 37 | G 208−44 = A | 19 | 53.9 | +44 | 25 | .220 | 14.81 | 0.74 | 143 | +11 | M5.5 V | 13.48 | 1.92 | 15.19 | 0.00007 |
| | G 208−45 = B | | | | | | | 0.74 | 143 | +6 | M6 V | 14.01 | 1.97 | 15.72 | 0.00004 |
| | G 208−44 = C | | | | | | | 0.74 | 143 | | | 16.66 | | 18.37 | 0.000004 |
| 38 | L 145−141 | 11 | 45.7 | −64 | 50 | .216 | 15.07 | 2.69 | 97 | | DQ6 | 11.50 | 0.19 | 13.18 | 0.00045 |
| 39 | G 158−27 | 0 | 06.7 | −07 | 32 | .213 | 15.31 | 2.04 | 204 | −41 | M5.5 V | 13.76 | 1.97 | 15.40 | 0.00006 |
| 40 | BD−15°6290 | 22 | 53.3 | −14 | 16 | .213 | 15.33 | 1.17 | 125 | −2 | M3.5 V | 10.16 | 1.59 | 11.80 | 0.0016 |
| 41 | BD+44°2051 A | 11 | 05.5 | +43 | 32 | .207 | 15.76 | 4.51 | 282 | +69 | M1 V | 8.76 | 1.55 | 10.34 | 0.006 |
| | BD+44°2051 B (WX UMa) | 11 | 05.5 | +43 | 31 | .207 | 15.75 | 4.51 | 282 | +68 | M5.5 V | 14.42 | 2.02 | 16.00 | 0.00003 |
| 42 | BD+50°1725 | 10 | 11.4 | +49 | 27 | .205 | 15.89 | 1.45 | 250 | −26 | K7 V | 6.59 | 1.36 | 8.15 | 0.0466 |
| 43 | BD+20°2465 (AD Leo) | 10 | 19.6 | +19 | 52 | .205 | 15.94 | 0.50 | 265 | +12 | M3 V | 9.41 | 1.54 | 10.96 | 0.0035 |
| 44 | CD−49°13515 HD204961 | 21 | 33.6 | −49 | 1 | .203 | 16.10 | 0.82 | 183 | +4 | M1 V | 8.67 | 1.48 | 10.20 | 0.007 |
| 45 | LP 944−20 | 3 | 39.6 | −35 | 26 | .201 | 16.19 | 0.44 | 49 | +7 | ≥ M9 V | | | | |
| 46 | CD−44°11909 | 17 | 37.0 | −44 | 19 | .198 | 16.45 | 1.18 | 217 | −41 | M4.5 V | 10.95 | 1.66 | 12.44 | 0.0009 |
| 47 | 40 Eri A | 4 | 15.3 | −07 | 39 | .198 | 16.45 | 4.09 | 213 | −43 | K1 Ve | 4.42 | 0.82 | 5.91 | 0.366 |
| | 40 Eri B | 4 | 15.4 | −7 | 39 | .198 | 16.45 | 4.09 | 213 | −17 | DA4 | 9.51 | 0.04 | 11.00 | 0.0034 |
| | 40 Eri C (DY Eri) | | | | | | | 4.09 | 213 | −46 | M4.5 V | 11.21 | 1.64 | 12.70 | 0.0007 |
| 48 | BD+43°4305 (EV Lac) | 22 | 46.8 | +44 | 20 | .198 | 16.47 | 0.84 | 237 | +1 | M3.5 V | 10.23 | 1.61 | 11.71 | 0.0018 |
| 49 | BD+02°3482 A | 18 | 05.5 | +02 | 30 | .197 | 16.59 | 0.97 | 173 | −7 | K0 Ve | 4.24 | 0.78 | 5.71 | 0.44 |
| | BD+02°3482 B | | | | | | | 0.97 | 173 | −10 | K5 Ve | 6.01 | | 7.48 | 0.086 |
| 50 | Altair | 19 | 50.8 | +08 | 52 | .194 | 16.77 | 0.66 | 54 | −26 | A7 IV-V | 0.77 | 0.22 | 2.21 | 12.25 |

Parallaxes and distances are the new ones from the Hipparcos satellite (1997), courtesy of Hartmut Jahreiss, updated in 2000. Parallaxes for the stars not observed by Hipparcos (usually because they are too faint) are from *The General Catalogue of Trigonometric Stellar Parallaxes,* 4th ed., by W. F. van Altena, J. T.-L. Lee, and E. D. Hoffleit (1995). Radial velocities, spectral types, and photometry are from Jahreiss's *Catalogue of Nearby Stars.*

# Messier Catalogue



| | | α | | δ | | | |
|---|---|---|---|---|---|---|---|
| M | NGC | h | m (2000.0) | ° | ′ | $m_v$ | Description |
| 1 | 1952 | 5 | 34.5 | +22 | 01 | 8.4 | Crab Nebula (Tau) |
| 2 | 7089 | 21 | 33.5 | −00 | 49 | 6.5 | Globular cluster (Aqr) |
| 3 | 5272 | 13 | 42.2 | +28 | 23 | 6.4 | Glob. cluster (CVn) |
| 4 | 6121 | 16 | 23.6 | −26 | 32 | 5.9 | Glob. cluster (Sco) |
| 5 | 5904 | 15 | 18.6 | +02 | 05 | 5.8 | Glob. cluster (Ser) |
| 6 | 6405 | 17 | 40.1 | −32 | 13 | 4.2 | Open cluster (Sco) |
| 7 | 6475 | 17 | 53.9 | −34 | 49 | 3.3 | Open cluster (Sco) |
| 8 | 6523 | 18 | 03.8 | −24 | 23 | 5.8 | Lagoon Nebula (Sgr) |
| 9 | 6333 | 17 | 19.2 | −18 | 31 | 7.9 | Glob. cluster (Oph) |
| 10 | 6254 | 16 | 57.1 | −04 | 06 | 6.6 | Glob. cluster (Oph) |
| 11 | 6705 | 18 | 51.1 | −05 | 16 | 5.8 | Open cluster (Scu) |
| 12 | 6218 | 16 | 47.2 | −01 | 57 | 6.6 | Glob. cluster (Oph) |
| 13 | 6205 | 16 | 41.7 | +36 | 28 | 5.9 | Glob. cluster (Her) |
| 14 | 6402 | 17 | 37.6 | −03 | 15 | 7.6 | Glob. cluster (Oph) |
| 15 | 7078 | 21 | 30.0 | +12 | 10 | 6.4 | Glob. cluster (Peg) |
| 16 | 6611 | 18 | 18.8 | −13 | 47 | 6.0 | Open cl. & nebula (Ser) |
| 17 | 6618 | 18 | 20.8 | −16 | 11 | 7 | Omega nebula (Sgr) |
| 18 | 6613 | 18 | 19.9 | −17 | 08 | 6.9 | Open cluster (Sgr) |
| 19 | 6273 | 17 | 02.6 | −26 | 16 | 7.2 | Glob. cluster (Oph) |
| 20 | 6514 | 18 | 02.6 | −23 | 02 | 8.5 | Trifid Nebula (Sgr) |
| 21 | 6531 | 18 | 04.6 | −22 | 30 | 5.9 | Open cluster (Sgr) |
| 22 | 6656 | 18 | 36.4 | −23 | 54 | 5.1 | Glob. cluster (Sgr) |
| 23 | 6494 | 17 | 56.8 | −19 | 01 | 5.5 | Open cluster (Sgr) |
| 24 | 6603 | 18 | 16.9 | −18 | 29 | 4.5 | Open cluster (Sgr) |
| 25 | IC4725 | 18 | 31.6 | −19 | 15 | 4.6 | Open cluster (Sgr) |
| 26 | 6694 | 18 | 45.2 | −09 | 24 | 8.0 | Open cluster (Scu) |
| 27 | 6853 | 19 | 59.6 | +22 | 43 | 8.1 | Dumbbell N., PN (Vul) |
| 28 | 6626 | 18 | 24.5 | −24 | 52 | 6.9 | Glob. cluster (Sgr) |
| 29 | 6913 | 20 | 23.9 | +38 | 32 | 6.6 | Open cluster (Cyg) |
| 30 | 7099 | 21 | 40.4 | −23 | 11 | 7.5 | Glob. cluster (Cap) |
| 31 | 224 | 0 | 42.7 | +41 | 16 | 3.4 | Andromeda Galaxy (Sb) |
| 32 | 221 | 0 | 42.7 | +40 | 52 | 8.2 | Elliptical galaxy (And) |
| 33 | 598 | 1 | 33.9 | +30 | 39 | 5.7 | Spiral galaxy (Sc) (Tri) |
| 34 | 1039 | 2 | 42.0 | +42 | 47 | 5.2 | Open cluster (Per) |
| 35 | 2168 | 6 | 08.9 | +24 | 20 | 5.1 | Open cluster (Gem) |
| 36 | 1960 | 5 | 36.1 | +34 | 08 | 6.0 | Open cluster (Aur) |
| 37 | 2099 | 5 | 52.4 | +32 | 33 | 5.6 | Open cluster (Aur) |
| 38 | 1912 | 5 | 28.7 | +35 | 50 | 6.4 | Open cluster (Aur) |
| 39 | 7092 | 21 | 32.2 | +48 | 26 | 4.6 | Open cluster (Cyg) |
| 40 | | 12 | 22.4 | +58 | 05 | 8 | Double star (UMa) |
| 41 | 2287 | 6 | 46.0 | −20 | 44 | 4.5 | Open cluster (CMa) |
| 42 | 1976 | 5 | 35.4 | −5 | 27 | 4 | Orion Nebula (Ori) |
| 43 | 1982 | 5 | 35.6 | −5 | 16 | 9 | Orion Nebula; smaller (Ori) |
| 44 | 2632 | 8 | 40.1 | +19 | 46 | 3.1 | Praesepe; open cl. (Can) |
| 45 | | 3 | 47.0 | +24 | 07 | 1.2 | Pleiades; open cl. (Tau) |
| 46 | 2437 | 7 | 41.8 | −14 | 49 | 6.1 | Open cluster (Pup) |
| 47 | 2422 | 7 | 36.6 | −14 | 30 | 4.4 | Open cluster (Pup) |
| 48 | 2548 | 8 | 13.8 | −5 | 48 | 5.8 | Open cluster (Hyd) |
| 49 | 4472 | 12 | 29.8 | +8 | 00 | 8.4 | Elliptical galaxy (Vir) |
| 50 | 2323 | 7 | 02.8 | −8 | 20 | 5.9 | Open cluster (Mon) |
| 51 | 5194 | 13 | 29.9 | +47 | 12 | 8.1 | Whirlpool Galaxy (Sc) (CVn) |
| 52 | 7654 | 23 | 24.2 | +61 | 35 | 6.9 | Open cluster (Cas) |
| 53 | 5024 | 13 | 12.9 | +18 | 10 | 7.7 | Glob. cluster (Com) |
| 54 | 6715 | 18 | 55.1 | −30 | 29 | 7.7 | Glob. cluster (Sgr) |
| 55 | 6809 | 19 | 40.0 | −30 | 58 | 7.0 | Glob. cluster (Sgr) |
| 56 | 6779 | 19 | 16.6 | +30 | 11 | 8.2 | Glob. cluster (Lyr) |
| 57 | 6720 | 18 | 53.6 | +33 | 02 | 9.0 | Ring N; planetary (Lyr) |
| 58 | 4579 | 12 | 37.7 | +11 | 49 | 9.8 | Spiral galaxy (SBb) (Vir) |
| 59 | 4621 | 12 | 42.0 | +11 | 39 | 9.8 | Elliptical galaxy (Vir) |
| 60 | 4649 | 12 | 43.7 | +11 | 33 | 8.8 | Elliptical galaxy (Vir) |
| 61 | 4303 | 12 | 21.9 | +4 | 28 | 9.7 | Spiral galaxy (Sc) (Vir) |
| 62 | 6266 | 17 | 01.2 | −30 | 07 | 6.6 | Glob. cluster (Sco) |
| 63 | 5055 | 13 | 15.8 | +42 | 02 | 8.6 | Spiral galaxy (Sb) (CVn) |
| 64 | 4826 | 12 | 56.7 | +21 | 41 | 8.5 | Spiral galaxy (Sb) (Com) |
| 65 | 3623 | 11 | 18.9 | +13 | 05 | 9.3 | Spiral galaxy (Sa) (Leo) |
| 66 | 3627 | 11 | 20.2 | +12 | 59 | 9.0 | Spiral galaxy (Sb) (Leo) |
| 67 | 2682 | 8 | 51.4 | +11 | 49 | 6.9 | Open cluster (Can) |
| 68 | 4590 | 12 | 39.5 | −26 | 45 | 8.2 | Glob. cluster (Hyd) |
| 69 | 6637 | 18 | 31.4 | −32 | 21 | 7.7 | Glob. cluster (Sgr) |
| 70 | 6681 | 18 | 43.2 | −32 | 18 | 8.1 | Glob. cluster (Sgr) |
| 71 | 6838 | 19 | 53.8 | +18 | 47 | 8.3 | Glob. cluster (Sge) |
| 72 | 6981 | 20 | 53.5 | −12 | 32 | 9.4 | Glob. cluster (Aqr) |
| 73 | 6994 | 20 | 58.9 | −12 | 38 | | Glob. cluster (Aqr) |
| 74 | 628 | 1 | 36.7 | +15 | 47 | 9.2 | Spiral galaxy (Sc) in Pisces |
| 75 | 6864 | 20 | 06.1 | −21 | 55 | 8.6 | Glob. cluster (Sgr) |
| 76 | 650-1 | 1 | 42.4 | +51 | 34 | 11.5 | Planetary nebula (Per) |
| 77 | 1068 | 2 | 42.7 | −0 | 01 | 8.8 | Spiral galaxy (Sb) (Cet) |
| 78 | 2068 | 5 | 46.7 | +0 | 03 | 8 | Small emission nebula (Ori) |
| 79 | 1904 | 5 | 24.5 | −24 | 33 | 8.0 | Glob. cluster (Lep) |
| 80 | 6093 | 16 | 17.0 | −22 | 59 | 7.2 | Glob. cluster (Sco) |
| 81 | 3031 | 9 | 55.6 | +69 | 04 | 6.8 | Spiral galaxy (Sb) (UMa) |
| 82 | 3034 | 9 | 55.8 | +69 | 41 | 8.4 | Irregular galaxy (UMa) |
| 83 | 5236 | 13 | 37.0 | −29 | 52 | 7.6 | Spiral galaxy (Sc) (Hyd) |
| 84 | 4374 | 12 | 25.1 | +12 | 53 | 9.3 | Elliptical galaxy (Vir) |
| 85 | 4382 | 12 | 25.4 | +18 | 11 | 9.2 | S0 galaxy (Com) |
| 86 | 4406 | 12 | 26.2 | +12 | 57 | 9.2 | Elliptical galaxy (Vir) |
| 87 | 4486 | 12 | 30.8 | +12 | 24 | 8.6 | Elliptical galaxy (Ep) (Vir) |
| 88 | 4501 | 12 | 32.0 | +14 | 25 | 9.5 | Spiral galaxy (Sb) (Com) |
| 89 | 4552 | 12 | 35.7 | +12 | 33 | 9.8 | Elliptical galaxy (Vir) |
| 90 | 4569 | 12 | 36.8 | +13 | 10 | 9.5 | Spiral galaxy (SBb) (Vir) |
| 91 | 4548 | 12 | 35.4 | +14 | 30 | 10.2 | M58? (Vir) |
| 92 | 6341 | 17 | 17.1 | +43 | 08 | 6.5 | Glob. cluster (Her) |
| 93 | 2447 | 7 | 44.6 | −23 | 52 | 6.2 | Open cluster (Pup) |
| 94 | 4736 | 12 | 50.9 | +41 | 07 | 8.1 | Spiral galaxy (Sb) (CVn) |
| 95 | 3351 | 10 | 44.0 | +11 | 42 | 9.7 | Barred spiral g. (SBb) (Leo) |
| 96 | 3368 | 10 | 46.8 | +11 | 49 | 9.2 | Spiral galaxy (Sa) (Leo) |
| 97 | 3587 | 11 | 14.8 | +55 | 01 | 11.2 | Owl Nebula; planetary (UMa) |
| 98 | 4192 | 12 | 13.8 | +14 | 54 | 10.1 | Spiral galaxy (Sb) (Com) |
| 99 | 4254 | 12 | 18.8 | +14 | 25 | 9.8 | Spiral galaxy (Sc) (Com) |
| 100 | 4321 | 12 | 22.9 | +15 | 49 | 9.4 | Spiral galaxy (Sc) (Com) |
| 101 | 5457 | 14 | 03.2 | +54 | 21 | 7.7 | Spiral galaxy (Sc) (UMa) |
| 102 | | | | | | | M101; duplication (UMa) |
| 103 | 581 | 1 | 33.2 | +60 | 42 | 7.4 | Open cluster (Cas) |
| 104 | 4594 | 12 | 40.0 | −11 | 37 | 8.3 | Sombrero N.; spiral (Sa) (Vir) |
| 105 | 3379 | 10 | 47.8 | +12 | 35 | 9.3 | Elliptical galaxy (Leo) |
| 106 | 4258 | 12 | 19.0 | +47 | 18 | 8.3 | Spiral galaxy (Sb) (CVn) |
| 107 | 6171 | 16 | 32.5 | −13 | 03 | 8.1 | Glob. cluster (Oph) |
| 108 | 3556 | 11 | 11.5 | +55 | 40 | 10.0 | Spiral galaxy (Sb) (UMa) |
| 109 | 3992 | 11 | 57.6 | +53 | 23 | 9.8 | Barred spiral g. (SBc) (UMa) |
| 110 | 205 | 0 | 40.4 | +41 | 41 | 8.0 | Elliptical galaxy (And) |

From Alan Hirshfeld and Roger W. Sinnott, *Sky Catalogue 2000.0,* vol. 2, courtesy of Sky Publishing Corp. and Cambridge Univ. Press.

# Caldwell Catalogue

| | NGC/IC | Constellation | Type | R. A. (2000.0) | Dec. (2000.0) | Mag. | Size (′) | Notes |
|---|---|---|---|---|---|---|---|---|
| 1 | 188 | Cepheus | Open cluster | $00^h 44.4^m$ | +85° 20′ | 8.1 | 14 | |
| 2 | 40 | Cepheus | Planetary nebula | $00^h 13.0^m$ | +72° 32′ | 12.4 | 0.8 | |
| 3 | 4236 | Draco | Sb galaxy | $12^h 16.7^m$ | +69° 28′ | 9.7 | 19 × 7 | |
| 4 | 7023 | Cepheus | Nebula | $21^h 01.8^m$ | +68° 12′ | — | 18 × 18 | Bright reflection nebula |
| 5 | IC 342 | Camelopardalis | SBc galaxy | $03^h 46.8^m$ | +63° 06′ | 9.2 | 18 × 17 | |
| 6 | 6543 | Draco | Planetary nebula | $17^h 58.6^m$ | +66° 38′ | 8.1 | 0.3/5.6 | Cat's Eye Nebula |
| 7 | 2403 | Camelopardalis | S0 galaxy | $07^h 36.9^m$ | +65° 36′ | 8.4 | 18 × 10 | |
| 8 | 559 | Cassiopeia | Open cluster | $01^h 29.5^m$ | +83° 18′ | 9.5 | 4 | |
| 9 | 8h2-158 | Cepheus | Bright nebula | $22^h 56.8^m$ | +62° 37′ | — | 50 × 10 | Cave Nebula |
| 10 | 663 | Cassiopeia | Open cluster | $01^h 46.0^m$ | +61° 15′ | 7.1 | 16 | |
| 11 | 7635 | Cassiopeia | Bright nebula | $23^h 20.7^m$ | +61° 12′ | — | 15 × 8 | Bubble Nebula |
| 12 | 6946 | Cepheus | Sc galaxy | $20^h 34.8^m$ | +60° 09′ | 8.9 | 11 × 10 | |
| 13 | 457 | Cassiopeia | Open cluster | $01^h 19.1^m$ | +58° 20′ | 6.4 | 13 | Phi Cas Cluster |
| 14 | 669/884 | Perseus | Double cluster | $02^h 20.0^m$ | +57° 08′ | 4.3 | 30 and 30 | Sword Hand region |
| 15 | 6826 | Cygnus | Planetary nebula | $19^h 44.8^m$ | +50° 31′ | 8.8 | 0.5/2.3 | Blinking Nebula |
| 16 | 7243 | Lacerta | Open cluster | $22^h 15.3^m$ | +49° 53′ | 6.4 | 21 | |
| 17 | 147 | Cassiopeia | dE4 galaxy | $00^h 33.2^m$ | +48° 30′ | 9.3 | 13 × 8 | |
| 18 | 185 | Cassiopeia | dE0 galaxy | $00^h 39.0^m$ | +48° 20′ | 9.2 | 12 × 10 | |
| 19 | IC 6148 | Cygnus | Bright nebula | $21^h 53.5^m$ | +47° 16′ | — | 12 × 12 | Cocoon Nebula |
| 20 | 7000 | Cygnus | Bright nebula | $20^h 58.8^m$ | +44° 20′ | — | 120 × 100 | North America Nebula |
| 21 | 4449 | Canes Venatici | Irregular galaxy | $12^h 28.2^m$ | +44° 06′ | 9.4 | 5 × 4 | |
| 22 | 7662 | Andromeda | Planetary nebula | $23^h 25.9^m$ | +42° 33′ | 8.3 | 0.3/2.2 | |
| 23 | 891 | Andromeda | Sb galaxy | $02^h 22.6^m$ | +42° 21′ | 9.9 | 14 × 3 | |
| 24 | 1275 | Perseus | Seyfert galaxy | $03^h 19.8^m$ | +41° 31′ | 11.6 | 2.6 × 2 | Perseus A radio source |
| 25 | 2419 | Lynx | Globular cluster | $07^h 38.1^m$ | +38° 53′ | 10.4 | 4.1 | |
| 26 | 4244 | Canes Venatici | S galaxy | $12^h 17.5^m$ | +37° 49′ | 10.2 | 16 × 2.5 | |
| 27 | 8888 | Cygnus | Bright nebula | $20^h 12.0^m$ | +38° 21′ | — | 20 × 10 | Crescent Nebula |
| 28 | 752 | Andromeda | Open cluster | $01^h 57.8^m$ | +37° 41′ | 5.7 | 50 | |
| 29 | 6005 | Canes Venatici | Sb galaxy | $13^h 10.9^m$ | +37° 03′ | 9.8 | 5.4 × 2 | |
| 30 | 7331 | Pegasus | Sb galaxy | $22^h 37.1^m$ | +34° 25′ | 9.5 | 11 × 4 | |
| 31 | IC 405 | Auriga | Bright nebula | $05^h 16.2^m$ | +34° 16′ | — | 30 × 19 | Flaming Star Nebula |
| 32 | 4631 | Canes Venatici | S0 galaxy | $12^h 42.1^m$ | +32° 32′ | 9.3 | 15 × 3 | |
| 33 | 6992/5 | Cygnus | SN remnant | $20^h 58.4^m$ | +31° 43′ | — | 60 × 6 | Eastern Veil Nebula |
| 34 | 6960 | Cygnus | SN remnant | $20^h 45.7^m$ | +30° 43′ | — | 70 × 6 | Western Veil Nebula |
| 35 | 4889 | Coma Berenices | E4 galaxy | $13^h 00.1^m$ | +27° 59′ | 11.4 | 3 × 2 | Brightest in Coma Cluster |
| 36 | 4559 | Coma Berenices | S0 galaxy | $12^h 36.0^m$ | +27° 58′ | 9.8 | 10 × 5 | |
| 37 | 6885 | Vulpecula | Open cluster | $20^h 12.0^m$ | +26° 29′ | 5.9 | 7 | |
| 38 | 4585 | Coma Berenices | Sb galaxy | $12^h 36.3^m$ | +25° 59′ | 9.6 | 16 × 3 | |
| 39 | 2392 | Gemini | Planetary nebula | $07^h 29.2^m$ | +20° 55′ | 9.2 | 0.2/0.7 | Eskimo Nebula |
| 40 | 3626 | Leo | Sb galaxy | $11^h 20.1^m$ | +18° 21′ | 10.9 | 3 × 2 | |
| 41 | — | Taurus | Open cluster | $04^h 27^m$ | +16° | 0.5 | 330 | Hyades |
| 42 | 7006 | Delphinus | Globular cluster | $21^h 01.5^m$ | +16° 11′ | 10.6 | 2.8 | Very distant globular |
| 43 | 7814 | Pegasus | Sb galaxy | $00^h 03.3^m$ | +16° 09′ | 10.5 | 6 × 2 | |
| 44 | 7479 | Pegasus | SBb galaxy | $23^h 04.9^m$ | +12° 19′ | 11.0 | 4 × 3 | |
| 45 | 5248 | Boötes | Sc galaxy | $13^h 37.5^m$ | +08° 53′ | 10.2 | 6 × 4 | |
| 46 | 2261 | Monoceros | Bright nebula | $08^h 39.2^m$ | +08° 44′ | — | 2 × 1 | Hubble's Variable Nebula |
| 47 | 6934 | Delphinus | Globular cluster | $20^h 34.2^m$ | +07° 24′ | 5.9 | 6.0 | |
| 48 | 2775 | Cancer | Sa galaxy | $09^h 10.3^m$ | +07° 02′ | 10.3 | 4.5 × 3 | |
| 49 | 2237.9 | Monoceros | Bright nebula | $06^h 32.3^m$ | +05° 03′ | — | 80 × 60 | Rosette Nebula |
| 50 | 2244 | Monoceros | Open cluster | $05^h 32.4^m$ | +04° 52′ | 4.6 | 24 | |
| 51 | IO 1613 | Cetus | Irregular galaxy | $01^h 04.8^m$ | +02° 07′ | 9.3 | 12 × 11 | |
| 52 | 4697 | Virgo | E4 galaxy | $12^h 48.8^m$ | −05° 48′ | 9.3 | 6 × 4 | |
| 53 | 3115 | Sextans | E6 galaxy | $10^h 05.2^m$ | −07° 43′ | 9.1 | 6 × 3 | Spindle Galaxy |
| 54 | 2508 | Monoceros | Open cluster | $08^h 00.2^m$ | −10° 47′ | 7.6 | 7 | |
| 55 | 7009 | Aquarius | Planetary nebula | $21^h 04.2^m$ | −11° 22′ | 8.0 | 0.4/1.6 | Saturn Nebula |
| 56 | 248 | Cetus | Planetary nebula | $00^h 47.0^m$ | −11° 53′ | 10.9 | 3.8 | |
| 57 | 6822 | Sagittarius | Irregular galaxy | $19^h 44.9^m$ | −14° 48′ | 8.8 | 10 × 9 | Barnard's Galaxy |
| 58 | 2360 | Canis Major | Open cluster | $07^h 17.8^m$ | −15° 37′ | 7.2 | 13 | |
| 59 | 3242 | Hydra | Planetary nebula | $10^h 24.8^m$ | −18° 38′ | 7.8 | 0.3/2.1 | Ghost of Jupiter |
| 60 | 4038 | Corvus | S0 galaxy | $12^h 01.9^m$ | −18° 52′ | 10.7 | 2.8 × 2 | |
| 61 | 4039 | Corvus | Sp galaxy | $12^h 01.9^m$ | −18° 53′ | 10.7 | 3 × 2 | |

*(continued)*

| | NGC/IC | Constellation | Type | R. A. (2000.0) | Dec. (2000.0) | Mag. | Size (′) | Notes |
|---|---|---|---|---|---|---|---|---|
| 62 | 247 | Cetus | S galaxy | $00^h\ 47.1^m$ | −20° 46′ | 9.1 | 20 × 7 | |
| 63 | 7283 | Aquarius | Planetary nebula | $22^h\ 29.6^m$ | −20° 48′ | 7.3 | 13 | Helix Nebula |
| 64 | 2362 | Canis Major | Open cluster | $07^h\ 18.8^m$ | −24° 57′ | 4.1 | 6 | Tau CMa Cluster |
| 65 | 283 | Sculptor | Scp galaxy | $00^h\ 47.8^m$ | −25° 17′ | 7.1 | 25 × 7 | Sculptor Galaxy |
| 66 | 5694 | Hydra | Globular cluster | $14^h\ 39.6^m$ | −26° 32′ | 10.2 | 3.6 | |
| 67 | 1097 | Fornax | SBb galaxy | $02^h\ 46.3^m$ | −30° 17′ | 9.2 | 9 × 7 | |
| 68 | 6729 | Corona Australis | Bright nebula | $19^h\ 01.9^m$ | −36° 57′ | — | 1.0 | R CrA Nebula |
| 69 | 6302 | Scorpius | Planetary nebula | $17^h\ 13.7^m$ | −37° 06′ | 9.6 | 0.8 | Bug Nebula |
| 70 | 300 | Sculptor | Sd galaxy | $00^h\ 54.9^m$ | −37° 41′ | 8.7 | 20 × 13 | |
| 71 | 2477 | Puppis | Open cluster | $07^h\ 52.3^m$ | −38° 33′ | 5.8 | 27 | |
| 72 | 65 | Sculptor | SB galaxy | $00^h\ 14.9^m$ | −39° 11′ | 7.9 | 32 × 6 | Brightest in Sculptor |
| 73 | 1851 | Columba | Globular cluster | $05^h\ 14.1^m$ | −40° 03′ | 7.3 | 11 | |
| 74 | 3132 | Vela | Planetary nebula | $10^h\ 07.7^m$ | −40° 26′ | 9.4 | 0.8 | |
| 75 | 6124 | Scorpius | Open cluster | $16^h\ 25.6^m$ | −40° 40′ | 5.8 | 29 | |
| 76 | 6231 | Scorpius | Open cluster | $16^h\ 54.0^m$ | −41° 48′ | 2.6 | 15 | |
| 77 | 6128 | Centaurus | Peculiar radio galaxy | $13^h\ 25.5^m$ | −43° 01′ | 7.0 | 18 × 14 | Cen A radio source |
| 78 | 6541 | Corona Australis | Globular cluster | $18^h\ 08.0^m$ | −43° 42′ | 6.6 | 13 | |
| 79 | 3201 | Vela | Globular cluster | $10^h\ 17.6^m$ | −46° 25′ | 6.7 | 18 | |
| 80 | 5139 | Centaurus | Globular cluster | $13^h\ 26.8^m$ | −47° 29′ | 3.8 | 36 | Omega Centauri |
| 81 | 8352 | Ara | Globular cluster | $17^h\ 25.5^m$ | −48° 25′ | 6.1 | 7 | |
| 82 | 6193 | Ara | Open cluster | $16^h\ 41.3^m$ | −48° 46′ | 5.2 | 15 | |
| 83 | 4945 | Centaurus | SBc galaxy | $13^h\ 05.4^m$ | −49° 28′ | 8.7 | 20 × 4 | |
| 84 | 5286 | Centaurus | Globular cluster | $13^h\ 46.4^m$ | −51° 22′ | 7.6 | 9 | |
| 85 | IC 2391 | Vela | Open cluster | $08^h\ 40.2^m$ | −53° 04′ | 2.5 | 50 | *o* Vel Cluster |
| 86 | 8397 | Ara | Globular cluster | $17^h\ 40.7^m$ | −53° 40′ | 5.6 | 26 | |
| 87 | 1281 | Horologium | Globular cluster | $03^h\ 12.3^m$ | −55° 13′ | 8.4 | 7 | |
| 88 | 5823 | Circinus | Open cluster | $15^h\ 05.7^m$ | −55° 36′ | 7.9 | 10 | |
| 89 | 6087 | Norma | Open cluster | $16^h\ 18.9^m$ | −57° 54′ | 5.4 | 12 | S Nor Cluster |
| 90 | 2887 | Carina | Planetary nebula | $09^h\ 21.4^m$ | −58° 19′ | — | 0.2 | |
| 91 | 3632 | Carina | Open cluster | $11^h\ 06.4^m$ | −58° 40′ | 3.0 | 55 | |
| 92 | 3372 | Carina | Bright nebula | $10^h\ 43.8^m$ | −59° 52′ | — | 120 × 120 | Eta Carinae Nebula |
| 93 | 8752 | Pavo | Globular cluster | $19^h\ 10.9^m$ | −59° 59′ | 5.4 | 20 | |
| 94 | 4755 | Crux | Open cluster | $12^h\ 53.6^m$ | −60° 20′ | 4.2 | 10 | Jewel Box Cluster |
| 95 | 6025 | Triangulum Aus. | Open cluster | $16^h\ 03.7^m$ | −60° 30′ | 5.1 | 12 | |
| 96 | 2516 | Carina | Open cluster | $07^h\ 58.3^m$ | −60° 52′ | 3.8 | 30 | |
| 97 | 3768 | Centaurus | Open cluster | $11^h\ 36.1^m$ | −61° 37′ | 5.3 | 12 | |
| 98 | 4609 | Crux | Open cluster | $12^h\ 42.3^m$ | −62° 58′ | 8.9 | 5 | |
| 99 | — | Crux | Dark nebula | $12^h\ 53^m$ | −63° | — | 400 × 300 | Coalsack |
| 100 | [IC 2944] | Centaurus | Open cluster | $11^h\ 36.6^m$ | −63° 02′ | 4.5 | 15 | Lambda Cen Cluster |
| 101 | 6744 | Pavo | SBb galaxy | $19^h\ 09.8^m$ | −63° 51′ | 8.3 | 16 × 10 | |
| 102 | IC 2602 | Carina | Open cluster | $10^h\ 43.2^m$ | −64° 24′ | 1.9 | 50 | Theta Cas Cluster |
| 103 | 2070 | Dorado | Bright nebula | $05^h\ 38.7^m$ | −69° 06′ | — | 40 × 25 | Tarantula Nebula |
| 104 | 382 | Tucana | Globular cluster | $01^h\ 03.2^m$ | −70° 51′ | 6.6 | 13 | |
| 105 | 4833 | Musca | Globular cluster | $12^h\ 59.6^m$ | −70° 53′ | 7.3 | 14 | |
| 106 | 104 | Tucana | Globular cluster | $00^h\ 24.1^m$ | −72° 05′ | 4.0 | 31 | 47 Tucanae |
| 107 | 6101 | Apus | Globular cluster | $16^h\ 25.8^m$ | −72° 12′ | 9.3 | 11 | |
| 108 | 4372 | Musca | Globular cluster | $12^h\ 25.8^m$ | −72° 40′ | 7.8 | 19 | |
| 109 | 3195 | Chamaeleon | Planetary nebula | $10^h\ 09.5^m$ | −80° 52′ | — | 0.6 | |

# The Constellations

| Latin Name | Genitive | Abbreviation | Translation |
|---|---|---|---|
| Andromeda | Andromedae | And | Andromeda* |
| Antlia | Antliae | Ant | Pump |
| Apus | Apodis | Aps | Bird of Paradise |
| Aquarius | Aquarii | Aqr | Water Bearer |
| Aquila | Aquilae | Aql | Eagle |
| Ara | Arae | Ara | Altar |
| Aries | Arietis | Ari | Ram |
| Auriga | Aurigae | Aur | Charioteer |
| Boötes | Boötis | Boo | Herdsman |
| Caelum | Caeli | Cae | Chisel |
| Camelopardalis | Camelopardalis | Cam | Giraffe |
| Cancer | Cancri | Cnc | Crab |
| Canes Venatici | Canum Venaticorum | CVn | Hunting Dogs |
| Canis Major | Canis Majoris | CMa | Big Dog |
| Canis Minor | Canis Minoris | CMi | Little Dog |
| Capricornus | Capricorni | Cap | Goat |
| Carina | Carinae | Car | Ship's Keel** |
| Cassiopeia | Cassiopeiae | Cas | Cassiopeia* |
| Centaurus | Centauri | Cen | Centaur* |
| Cepheus | Cephei | Cep | Cepheus* |
| Cetus | Ceti | Cet | Whale |
| Chamaeleon | Chamaeleonis | Cha | Chameleon |
| Circinus | Circini | Cir | Compass |
| Columba | Columbae | Col | Dove |
| Coma Berenices | Comae Berencies | Com | Berenice's Hair* |
| Corona Australis | Coronae Australis | CrA | Southern Crown |
| Corona Borealis | Coronae Borealis | CrB | Northern Crown |
| Corvus | Corvi | Crv | Crow |
| Crater | Crateris | Crt | Cup |
| Crux | Crucis | Cru | Southern Cross |
| Cygnus | Cygni | Cyg | Swan |
| Delphinus | Delphini | Del | Dolphin |
| Dorado | Doradus | Dor | Swordfish |
| Draco | Draconis | Dra | Dragon |
| Equuleus | Equulei | Equ | Little Horse |
| Eridanus | Eridani | Eri | River Eridanus* |
| Fornax | Fornacis | For | Furnace |
| Gemini | Geminorum | Gen | Twins |
| Grus | Gruis | Gru | Crane |
| Hercules | Herculis | Her | Hercules* |
| Horologium | Horologii | Hor | Clock |
| Hydra | Hydrae | Hya | Hydra* (water monster) |
| Hydrus | Hydri | Hyi | Sea serpent |
| Indus | Indi | Ind | Indian |
| Lacerta | Lacertae | Lac | Lizard |
| Leo | Leonis | Leo | Lion |
| Leo Minor | Leonis Minoris | LMi | Little Lion |
| Lepus | Leporis | Lep | Hare |
| Libra | Librae | Lib | Scales |
| Lupus | Lupi | Lup | Wolf |
| Lynx | Lyncis | Lyn | Lynx |
| Lyra | Lyrae | Lyr | Harp |
| Mensa | Mensae | Men | Table (mountain) |
| Microscopium | Microscopii | Mic | Microscope |
| Monoceros | Monocerotis | Mon | Unicorn |
| Musca | Muscae | Mus | Fly |
| Norma | Normae | Nor | Level (square) |
| Octans | Octantis | Oct | Octant |
| Ophiuchus | Ophiuchi | Oph | Ophiuchus* (serpent bearer) |
| Orion | Orionis | Ori | Orion* |
| Pavo | Pavonis | Pav | Peacock |
| Pegasus | Pegasi | Peg | Pegasus* (winged horse) |
| Perseus | Persei | Per | Perseus* |
| Phoenix | Phoenicis | Phe | Phoenix |
| Pictor | Pictoris | Pic | Easel |
| Pisces | Piscium | Psc | Fish |
| Piscis Austrinus | Piscis Austrini | PsA | Southern Fish |
| Puppis | Puppis | Pup | Ship's Stern** |
| Pyxis | Pyxidis | Pyx | Ship's Compass** |
| Reticulum | Reticuli | Ret | Net |
| Sagitta | Sagittae | Sge | Arrow |
| Sagittarius | Sagittarii | Sgr | Archer |
| Scorpius | Scorpii | Sco | Scorpion |
| Sculptor | Sculptoris | Scl | Sculptor |
| Scutum | Scuti | Sct | Shield |
| Serpens | Serpentis | Ser | Serpent |
| Sextans | Sextantis | Sex | Sextant |
| Taurus | Tauri | Tau | Bull |
| Telescopium | Telescopii | Tel | Telescope |
| Triangulum | Trianguli | Tri | Triangle |
| Triangulum Australe | Trianguli Australis | TrA | Southern Triangle |
| Tucana | Tucanae | Tuc | Toucan |
| Ursa Major | Ursae Majoris | UMa | Big Bear |
| Ursa Minor | Ursae Minoris | UMi | Little Bear |
| Vela | Velorum | Vel | Ship's Sails** |
| Virgo | Virginis | Vir | Virgin |
| Volans | Volantis | Vol | Flying Fish |
| Vulpecula | Vulpeculae | Vul | Little Fox |

*Proper names

**Formerly formed the constellation Argo Navis, the Argonauts' ship.

# Elements and Solar System Abundances APPENDIX 11

| | | Name | Atomic Weight | Abundance | | | Name | Atomic Weight | Abundance |
|---|---|---|---|---|---|---|---|---|---|
| 1 | H | hydrogen | 1.01 | 12.00 | 57 | La | lanthanum | 138.91 | 1.20 |
| 2 | He | helium | 4.00 | 10.99 | 58 | Ce | cerium | 140.12 | 1.61 |
| 3 | Li | lithium | 6.94 | 1.16 | 59 | Pr | praseodymium | 140.91 | 0.78 |
| 4 | Be | beryllium | 9.01 | 1.15 | 60 | Nd | neodymium | 144.24 | 1.47 |
| 5 | B | boron | 10.81 | 2.6 | 61 | Pm | promethium | (145) | |
| 6 | C | carbon | 12.01 | 8.55 | 62 | Sm | samarium | 150.36 | 0.97 |
| 7 | N | nitrogen | 14.01 | 7.97 | 63 | Eu | europium | 151.97 | 0.54 |
| 8 | O | oxygen | 16.00 | 8.87 | 64 | Gd | gadolinium | 157.25 | 1.07 |
| 9 | F | fluorine | 19.00 | 4.56 | 65 | Tb | terbium | 158.93 | 0.33 |
| 10 | Ne | neon | 20.18 | 8.08 | 66 | Dy | dysprosium | 162.50 | 1.15 |
| 11 | Na | sodium | 22.99 | 6.31 | 67 | Ho | holmium | 164.93 | 0.50 |
| 12 | Mg | magnesium | 24.31 | 7.58 | 68 | Er | erbium | 167.26 | 0.95 |
| 13 | Al | aluminum | 26.98 | 6.48 | 69 | Tm | thulium | 168.93 | 0.13 |
| 14 | Si | silicon | 28.09 | 7.55 | 70 | Yb | ytterbium | 173.04 | 0.95 |
| 15 | P | phosphorus | 30.97 | 5.57 | 71 | Lu | lutetium | 174.97 | 0.12 |
| 16 | S | sulfur | 32.07 | 7.27 | 72 | Hf | hafnium | 178.49 | 0.73 |
| 17 | Cl | chlorine | 35.45 | 5.27 | 73 | Ta | tantalum | 180.95 | 0.13 |
| 18 | Ar | argon | 39.94 | — | 74 | W | tungsten | 183.85 | 0.68 |
| 19 | K | potassium | 39.10 | 5.13 | 75 | Re | rhenium | 186.21 | 0.27 |
| 20 | Ca | calcium | 40.08 | 6.34 | 76 | Os | osmium | 190.2 | 1.38 |
| 21 | Sc | scandium | 44.96 | 3.09 | 77 | Ir | iridium | 192.22 | 1.37 |
| 22 | Ti | titanium | 47.88 | 4.93 | 78 | Pt | platinum | 195.08 | 1.68 |
| 23 | V | vanadium | 50.94 | 4.02 | 79 | Au | gold | 196.97 | 0.83 |
| 24 | Cr | chromium | 52.00 | 5.68 | 80 | Hg | mercury | 200.59 | 1.09 |
| 25 | Mn | manganese | 54.94 | 5.53 | 81 | Tl | thallium | 204.38 | 0.82 |
| 26 | Fe | iron | 55.85 | 7.51 | 82 | Pb | lead | 207.2 | 2.05 |
| 27 | Co | cobalt | 58.93 | 4.91 | 83 | Bi | bismuth | 208.98 | 0.71 |
| 28 | Ni | nickel | 58.69 | 6.25 | 84 | Po | polonium | (209) | |
| 29 | Cu | copper | 63.55 | 4.27 | 85 | At | astatine | (210) | |
| 30 | Zn | zinc | 65.39 | 4.65 | 86 | Rn | radon | (222) | |
| 31 | Ga | gallium | 69.72 | 3.13 | 87 | Fr | fancium | (223) | |
| 32 | Ge | germanium | 72.61 | 3.63 | 88 | Ra | radium | (226) | |
| 33 | As | arsenic | 74.92 | 2.37 | 89 | Ac | actinium | (227) | |
| 34 | Se | selenium | 78.96 | 3.35 | 90 | Th | thorium | 232.04 | 0.08 |
| 35 | Br | bromine | 79.90 | 2.63 | 91 | Pa | protactinium | 231.04 | |
| 36 | Kr | krypton | 83.80 | 3.23 | 92 | U | uranium | 238.03 | –0.49 |
| 37 | Rb | rubidium | 85.47 | 2.40 | 93 | Np | neptunium | (237) | |
| 38 | Sr | strontium | 87.62 | 2.93 | 94 | Pu | plutonium | (244) | |
| 39 | Y | yttrium | 88.91 | 2.22 | 95 | Am | americium | (243) | |
| 40 | Zr | zirconium | 91.22 | 2.61 | 96 | Cm | curium | (247) | |
| 41 | Nb | niobium | 92.91 | 1.40 | 97 | Bk | berkelium | (247) | |
| 42 | Mo | molybdenum | 95.94 | 1.96 | 98 | Cf | californium | (251) | |
| 43 | Tc | technetium | (98) | | 99 | Es | einsteinium | (252) | |
| 44 | Ru | ruthenium | 101.07 | 1.82 | 100 | Fm | fermium | (257) | |
| 45 | Rh | rhodium | 102.91 | 1.09 | 101 | Md | mendelevium | (260) | |
| 46 | Pd | palladium | 106.42 | 1.70 | 102 | No | nobelium | (259) | |
| 47 | Ag | silver | 107.87 | 1.24 | 103 | Lr | lawrencium | (262) | |
| 48 | Cd | cadmium | 112.41 | 1.76 | 104 | Rf | rutherfordium | (261) | |
| 49 | In | indium | 114.82 | 0.82 | 105 | Db | dubnium | (262) | |
| 50 | Sn | tin | 118.71 | 2.14 | 106 | Sg | seaborgium | (266) | |
| 51 | Sb | antimony | 121.75 | 1.04 | 107 | Bh | bohrium | (264) | |
| 52 | Te | tellurium | 127.60 | 2.24 | 108 | Hs | hassium | (267) | |
| 53 | I | iodine | 126.90 | 1.51 | 109 | Mt | meitnerium | (268) | |
| 54 | Xe | xenon | 131.29 | 2.23 | 110 | | | (269) | |
| 55 | Cs | cesium | 132.91 | 1.12 | 111 | | | (272) | |
| 56 | B | barium | 137.33 | 2.21 | 112 | | | (277) | |
| | | | | | 114 | | | (289) | |
| | | | | | 116 | | | (289) | |
| | | | | | 118 | | | (293) | |

Abundances given are logarithmic, based on hydrogen = 12.00.

The abundances given are solar photospheric for elements 3–10 and meteoritic for other elements; when both are available, differences are almost always slight. The abundance data were supplied by Nicolas Grevesse in 1994, and represent an update from Edward Anders and Nicolas Grevesse, *Geochimica et Cosmochima Acta,* **53,** 197–214 (1989).

Atomic weights are from the 1988 IUPAC (International Union of Pure and Applied Chemistry) report, plus the longest halflife for the heaviest elements.

*In 1997, a committee of the International Union of Pure and Applied Chemistry recommended the names given for elements 101 through 109.*

I thank Peter Armbruster of Gesellschaft für Schwerionen forschung mbH Darmstad for information about the heaviest elements, many of which he discovered.

For information about new (and old) elements, see http://www.webelements.com.

Standard Model of

# FUNDAMENTAL PARTICLES AND INTERACTIONS

The Standard Model summarizes the current knowledge in Particle Physics. It is the quantum theory that includes the theory of strong interactions (quantum chromodynamics or QCD) and the unified theory of weak and electromagnetic interactions (electroweak). Gravity is included on this chart because it is one of the fundamental interactions even though not part of the "Standard Model."

## FERMIONS

matter constituents
spin = 1/2, 3/2, 5/2, ...

**Leptons** spin = 1/2

| Flavor | Mass GeV/$c^2$ | Electric charge |
|---|---|---|
| $\nu_e$ electron neutrino | $<1\times10^{-8}$ | 0 |
| e electron | 0.000511 | −1 |
| $\nu_\mu$ muon neutrino | <0.0002 | 0 |
| $\mu$ muon | 0.106 | −1 |
| $\nu_\tau$ tau neutrino | <0.02 | 0 |
| $\tau$ tau | 1.7771 | −1 |

**Quarks** spin = 1/2

| Flavor | Approx. Mass GeV/$c^2$ | Electric charge |
|---|---|---|
| u up | 0.003 | 2/3 |
| d down | 0.006 | −1/3 |
| c charm | 1.3 | 2/3 |
| s strange | 0.1 | −1/3 |
| t top | 175 | 2/3 |
| b bottom | 4.3 | −1/3 |

**Spin** is the intrinsic angular momentum of particles. Spin is given in units of $\hbar$, which is the quantum unit of angular momentum, where $\hbar = h/2\pi = 6.58\times10^{-25}$ GeV s $= 1.05\times10^{-34}$ J s.

**Electric charges** are given in units of the proton's charge. In SI units the electric charge of the proton is $1.60\times10^{-19}$ coulombs.

The **energy** unit of particle physics is the electronvolt (eV), the energy gained by one electron in crossing a potential difference of one volt. **Masses** are given in GeV/$c^2$ (remember $E = mc^2$), where 1 GeV = $10^9$ eV = $1.60\times10^{-10}$ joule. The mass of the proton is 0.938 GeV/$c^2$ = $1.67\times10^{-27}$ kg.

### Structure within the Atom

Quark Size < $10^{-19}$ m

Nucleus Size ≈ $10^{-14}$ m

Electron Size < $10^{-18}$ m

Neutron and Proton Size ≈ $10^{-15}$ m

Atom Size ≈ $10^{-10}$ m

If the protons and neutrons in this picture were 10 cm across, then the quarks and electrons would be less than 0.1 mm in size and the entire atom would be about 10 km across.

## BOSONS

force carriers
spin = 0, 1, 2, ...

**Unified Electroweak** spin = 1

| Name | Mass GeV/$c^2$ | Electric charge |
|---|---|---|
| $\gamma$ photon | 0 | 0 |
| $W^-$ | 80.4 | −1 |
| $W^+$ | 80.4 | +1 |
| $Z^0$ | 91.187 | 0 |

**Strong (color)** spin = 1

| Name | Mass GeV/$c^2$ | Electric charge |
|---|---|---|
| g gluon | 0 | 0 |

**Color Charge**
Each quark carries one of three types of "strong charge," also called "color charge." These charges have nothing to do with the colors of visible light. There are eight possible types of color charge for gluons. Just as electrically-charged particles interact by exchanging photons, in strong interactions color-charged particles interact by exchanging gluons. Leptons, photons, and **W** and **Z** bosons have no strong interactions and hence no color charge.

**Quarks Confined in Mesons and Baryons**
One cannot isolate quarks and gluons; they are confined in color-neutral particles called **hadrons**. This confinement (binding) results from multiple exchanges of gluons among the color-charged constituents. As color-charged particles (quarks and gluons) move apart, the energy in the color-force field between them increases. This energy eventually is converted into additional quark-antiquark pairs (see figure below). The quarks and antiquarks then combine into hadrons; these are the particles seen to emerge. Two types of hadrons have been observed in nature: **mesons** $q\bar{q}$ and **baryons** $qqq$.

**Residual Strong Interaction**
The strong binding of color-neutral protons and neutrons to form nuclei is due to residual strong interactions between their color-charged constituents. It is similar to the residual electrical interaction that binds electrically neutral atoms to form molecules. It can also be viewed as the exchange of mesons between the hadrons.

## PROPERTIES OF THE INTERACTIONS

| Property \ Interaction | Gravitational | Weak (Electroweak) | Electromagnetic (Electroweak) | Strong: Fundamental | Strong: Residual |
|---|---|---|---|---|---|
| Acts on: | Mass – Energy | Flavor | Electric Charge | Color Charge | See Residual Strong Interaction Note |
| Particles experiencing: | All | Quarks, Leptons | Electrically charged | Quarks, Gluons | Hadrons |
| Particles mediating: | Graviton (not yet observed) | $W^+$ $W^-$ $Z^0$ | $\gamma$ | Gluons | Mesons |
| Strength relative to electromag for two u quarks at: $10^{-18}$ m | $10^{-41}$ | 0.8 | 1 | 25 | Not applicable to quarks |
| Strength relative to electromag for two u quarks at: $3\times10^{-17}$ m | $10^{-41}$ | $10^{-4}$ | 1 | 60 | Not applicable to quarks |
| for two protons in nucleus | $10^{-36}$ | $10^{-7}$ | 1 | Not applicable to hadrons | 20 |

**Baryons qqq and Antibaryons $\bar{q}\bar{q}\bar{q}$**
Baryons are fermionic hadrons.
There are about 120 types of baryons.

| Symbol | Name | Quark content | Electric charge | Mass GeV/$c^2$ | Spin |
|---|---|---|---|---|---|
| p | proton | uud | 1 | 0.938 | 1/2 |
| $\bar{p}$ | anti-proton | $\bar{u}\bar{u}\bar{d}$ | −1 | 0.938 | 1/2 |
| n | neutron | udd | 0 | 0.940 | 1/2 |
| $\Lambda$ | lambda | uds | 0 | 1.116 | 1/2 |
| $\Omega^-$ | omega | sss | −1 | 1.672 | 3/2 |

**Mesons $q\bar{q}$**
Mesons are bosonic hadrons.
There are about 140 types of mesons.

| Symbol | Name | Quark content | Electric charge | Mass GeV/$c^2$ | Spin |
|---|---|---|---|---|---|
| $\pi^+$ | pion | $u\bar{d}$ | +1 | 0.140 | 0 |
| $K^-$ | kaon | $s\bar{u}$ | −1 | 0.494 | 0 |
| $\rho^+$ | rho | $u\bar{d}$ | +1 | 0.770 | 1 |
| $B^0$ | B-zero | $d\bar{b}$ | 0 | 5.279 | 0 |
| $\eta_c$ | eta-c | $c\bar{c}$ | 0 | 2.980 | 0 |

$n \rightarrow p\, e^-\, \bar{\nu}_e$

A neutron decays to a proton, an electron, and an antineutrino via a virtual (mediating) W boson. This is neutron β decay.

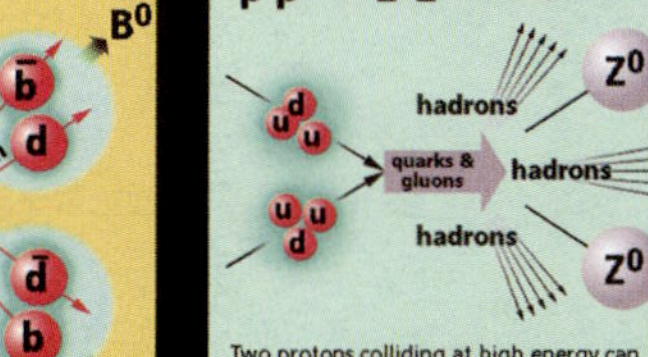


An electron and positron (antielectron) colliding at high energy can annihilate to produce $B^0$ and $\bar{B}^0$ mesons via a virtual Z boson or a virtual photon.

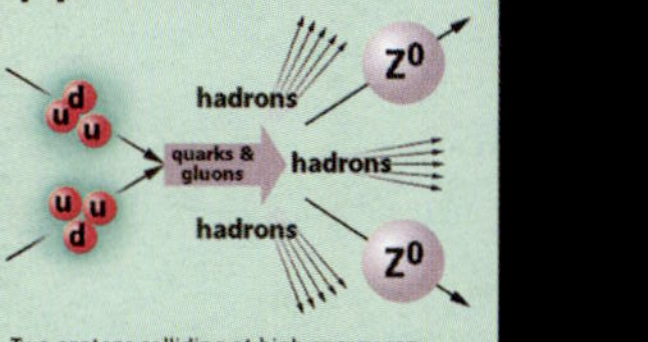


Two protons colliding at high energy can produce various hadrons plus very high mass particles such as Z bosons. Events such as this one are rare but can yield vital clues to the structure of matter.

# SELECTED READINGS

**See other books by the author at www.solarcorona.com.**

## Monthly Non-Technical Magazines in Astronomy

*Sky and Telescope,* P.O. Box 9111, Belmont, MA 02138, 800 253 0245, http://www.skypub.com.

*Astronomy,* 21027 Crossroads Circle, P.O. Box 1612, Waukesha, WI 53187, 800 533 6644, http://astronomy.com.

*Mercury,* Astronomical Society of the Pacific, 390 Ashton Ave., San Francisco, CA 94112, 800 962 3412, http://www.aspsky.org.

*StarDate,* 2609 University, Rm. 3.118, University of Texas, Austin, TX 78712, 800 STARDATE, http://stardate.utexas.edu.

*The Griffith Observer,* 2800 East Observatory Road, Los Angeles, CA 90027, http://www.griffithobs.org.

## Magazines and Annuals Carrying Articles on Astronomy

*Science News,* P.O. Box 1925, Marion, OH 43305, 800 552 4412, http://www.sciserv.org. Published weekly.

*Scientific American,* P.O. Box 3186, Harlan, IA 51593-2377, 800 333 1199.

*National Geographic,* P.O. Box 96583, Washington, D.C. 20078-9973, 800 NGC LINE.

*Natural History,* P.O. Box 5000, Harlan, IA 51593-5000, 800 234 5252.

*New Scientist,* 151 Wardour St., London W1V 4BN, U.K., 888 822 4342, rbi.subscriptions@rbi.co.uk; http://www.newscientist.com.

*Physics Today,* American Institute of Physics, 2 Huntingdon Quadrangle, Melville, NY 11747, 800 344 6902, http://www.aip.org/pt; http://www.physicstoday.org.

*Science Year,* World Book Encyclopedia, Inc., P. O. Box 11207, Des Moines, IA 50340-1207, 800 504 4425. The World Book Science Annual.

*Smithsonian,* P.O. Box 420311, Palm Coast, FL 32142-0311, Washington, D.C. 20560, 800 766 2149; www.smithsonianmag.si.edu.

*Discover,* P.O. Box 42105, Palm Coast, FL 34142-0105, 800 829 9132.

*The Planetary Report,* The Planetary Society, 65 North Catalina Avenue, Pasadena, CA 91106-2301, 818 793 5100, http://planetary.org/tps

## Observing Reference Books

Jay M. Pasachoff, *A Field Guide to the Stars and Planets,* 4th ed. (Boston: Houghton Mifflin Co., 2000). All kinds of observing information, including monthly maps and the 2000.0 sky atlas by Wil Tirion, and Graphic Timetables to locate planets and special objects like clusters and galaxies. See http://www.williams.edu/astronomy/fieldguide.

Jay M. Pasachoff, *Peterson's First Guide to Astronomy* (Boston: Houghton Mifflin Co., 1997). A brief, beautifully illustrated introduction to observing the sky. Tirion monthly maps.

Jay M. Pasachoff, *Peterson's First Guide to the Solar System* (Boston: Houghton Mifflin Co., 1998). Color illustrations and simple descriptions mark this elementary introduction. Tirion maps of Mars, Jupiter, and Saturn's positions through 2010.

H. J. P. Arnold, P. Doherty, and P. Moore, *The Photographic Atlas of the Stars* (Bristol, UK: Institute of Physics, 1997).

*The Astronomical Almanac* (yearly) (Washington, D.C.: U.S. Government Printing Office).

Michael A. Covington, *Astrophotography for the Amateur,* 2nd ed. (New York: Cambridge University Press, 1999).

Terence Dickinson, *Nightwatch: A Practical Guide to Viewing the Universe,* 3rd ed. (Willowdale, Ontario: Firefly Books, 1998).

Terence Dickinson, *The Universe and Beyond* (Willowdale, Ontario: Firefly Books, 1999).

Philip S. Harrington, *The Deep Sky: An Introduction* (Cambridge, MA: Sky Publishing Co., 1997).

Alan Hirshfeld, Roger W. Sinnott, and François Ochsenbein, *Sky Catalogue 2000.0,* 2nd ed., vol. 1 (Cambridge, MA: Sky Publishing Co., 1991); Alan Hirshfeld and Roger W. Sinnott, *Sky Catalogue 2000.0,* 2nd ed., vol. 2 (Cambridge, MA: Sky Publishing Co., 1985). Vol. 1 is stars and Vol. 2 is full of tables of all other objects.

Walter Scott Houston with commentary by Stephen James O'Meara, *Deep Sky Wonders* (Cambridge, MA: Sky Publishing Corp., 1999). From the master of observing.

Chris Kitchen and Robert W. Forrest, *Seeing Stars: The Night Sky Through Small Telescopes* (New York: Springer-Verlag, 1998).

David H. Levy, *Observing Variable Stars: A Guide for the Beginner* (New York: Cambridge University Press, 1998).

Jim Mullaney, *Celestial Harvest: 3000 Plus Showpieces of the Heavens for Telescope Viewing & Contemplation* (self-published, P.O. Box 1146, Exton, PA 19341, 1998); jimullaneysm@email.msn.com.

*Observer's Handbook* (yearly) (Toronto: Royal Astronomical Society of Canada, 136 Dupont Street, Toronto, Ontario M5R 1V2 Canada).

Stephen James O'Meara and David H. Levy, *The Messier Objects* (New York: Cambridge University Press, 1998).

Guy Ottewell, *Astronomical Calendar* (yearly) and *The Astronomical Companion,* Department of Physics, Furman University, Greenville, SC 29613, 864 294 2208; guyverno@aol.com; www.kalend.com.

Ian Ridpath, ed., *Norton's Star Atlas and Reference Handbook (Epoch 2000.0),* 19th ed. (Longmans, 1998). The old standard, updated.

Ian Ridpath, *Eyewitness Handbooks: Stars and Planets.* (New York, NY: DK Publishing, 1998).

Ian Ridpath, illustrated by Wil Tirion, *Stars & Planets* (Princeton, NJ: Princeton University Press, 2001).

Roger W. Sinnott, ed., *NGC 2000.0: The Complete New General Catalogue and Index Catalogues of Nebulae and Star Clusters* (Cambridge, MA: Sky Publishing Co. and New York: Cambridge University Press, 1998). A centennial reissue of Dreyer's work with updated data.

Roger W. Sinnott and Michael A. C. Perryman, *Millennium Star Atlas* (Cambridge, MA: Sky Publishing Co., 1997). Uses the Hipparcos data.

Wil Tirion, *Cambridge Star Atlas,* 2nd ed. (New York: Cambridge University Press, 1996). A naked eye star atlas in full color. A moon map, 24 monthly sky maps, 20 detailed star charts, and 6 all-sky maps.

Wil Tirion and Roger W. Sinnott, *Sky Atlas 2000.0,* 2nd ed., (Cambridge, MA: Sky Publishing Co. and New York: Cambridge University Press, 1998). Large-scale star charts.

Wil Tirion, Barry Rappaport, and George Lovi, *Uranometria 2000.0* (Richmond, VA: Willmann-Bell, 1986). Vol. 1 covers +90° to −6°. Vol. 2 covers +6° to −90°. Star maps to magnitude 9.5 and an essay on historical atlases.

Willmann-Bell catalogue, *Astronomy Books* (P.O. Box 35025, Richmond, VA 23235), 800 825 7827, http://www.willbell.com. They publish many observing guides and distribute all kinds of astronomy books.

## For Information About Amateur Societies

American Association of Variable Star Observers (AAVSO), 25 Birch St., Cambridge, MA 02138, http://www.aavso.org.

American Meteor Society, Dept. of Physics and Astronomy, SUNY, Geneseo, NY 14454, http://www.amsmeteors.org.

Astronomical League, the umbrella group of amateur societies. For their newsletter, *The Reflector,* write The Astronomical League, Executive Secretary, c/o Science Service Building, 1719 N St., N.W., Washington, D.C. 20030, http://www.astroleague.org.

Astronomical Society of the Pacific, 390 Ashton Ave., San Francisco, CA 94112, http://www.aspsky.org.

International Dark Sky Association, c/o David Crawford, 3545 N. Stewart, Tucson, AZ 85716, 877 600 5888, http://www.darksky.org.

The Planetary Society, 65 North Catalina Ave., Pasadena, CA 91106-2301, 818 793 5100, http://www.planetary.org/tps.

British Astronomical Association, Burlington House, Piccadilly, London W1V 0NL, U.K., http://www.ast.cam.ac.uk/~baa.

Royal Astronomical Society of Canada, 124 Merton St., Toronto, Ontario M4S 2Z2, Canada, http://www.rasc.ca.

## Careers in Astronomy

American Astronomical Society Education Office, 2000 Florida Ave., NW, Suite 400, Washington, D.C. 20009, fax 202 234 2560, aased@aas.org, http://www.aas.org. A free booklet, *A Career in Astronomy,* is available on request.

## Teaching

Jay M. Pasachoff and John R. Percy, *The Teaching of Astronomy* (New York: Cambridge University Press, 1990). The proceedings of International Astronomical Union Colloquium #105, held in Williamstown in 1988.

John R. Percy, *Astronomy Education: Current Developments, Future Coordination* (San Francisco: Astronomical Society of the Pacific, 1996). An ASP conference, held in College Park, MD, in 1995.

Lucienne Gouguenheim, Derek McNally, and John R. Percy, *New Trends in Astronomy Teaching* (New York: Cambridge University Press, 1997). The proceedings of International Astronomical Union Colloquium #162, held in London in 1996.

## General Reading and Reference

Paul Murdin, ed., and many others, *Encyclopedia of Astronomy and Astrophysics* (New York, NY: Macmillan and Institute of Physics, 2000), http://www.ency-astro.com.

Sybil P. Parker and Jay M. Pasachoff, *McGraw-Hill Encyclopedia of Astronomy* (New York, NY: McGraw-Hill, 1993). See updates at http://www.accessscience.com.

Jacqueline Mitton, *Cambridge Dictionary of Astronomy* (New York: Cambridge University Press, 2001).

Jean Audouze and Guy Israel, eds. *The Cambridge Atlas of Astronomy,* 3rd ed. (Cambridge: Cambridge University Press, 1994). Coffee-table size with fantastic photos and authoritative text.

Ronald Greeley and Raymond Batson, *The NASA Atlas of the Solar System* (New York: Cambridge University Press, 1997). Coffee-table size with fantastic photos and authoritative text.

Svend Lausten, Claus Madsen, and Richard M. West, *Exploring the Southern Sky: A Pictorial Atlas from the European Southern Observatory* (New York: Springer-Verlag, 1987). Beautiful and impressive.

David Malin, *The Invisible Universe* (Boston: Bulfinch Press, 1999). Fantastic color photographs and interesting historical and scientific information.

Stephen P. Maran, ed., *The Astronomy and Astrophysics Encyclopedia* (New York: Van Nostrand, 1992). Essays on all aspects of astronomy.

Stephen P. Maran, *Astronomy for Dummies* (Boston, MA: IDG Books, 1999). Straightforward.

Jacqueline Mitton, *The Penguin Dictionary of Astronomy* (New York: Penguin, 1993). Short entries on all aspects of astronomy.

Philip Morrison and Phylis Morrison, *Powers of 10* (New York: Scientific American Library, 1982). A classic progression through different size scales in the universe. A video version, a CD, and a DVD are available.

James Trefil, *Other Worlds: Images of the Cosmos from Earth and Space* (Washington, D.C.: National Geographic Society, 1999)

Neil de Grasse Tyson, Charles Liu, and Robert Irion, *The Universe: At Home in the Cosmos* (Washington, D.C.: Joseph Henry Press, 2000). Clearly written and lavishly illustrated.

## History

Michael Hoskin, ed., *Cambridge Illustrated History: Astronomy* (New York: Cambridge University Press, 1997).

I. Bernard Cohen, *The Newtonian Revolution* (New York: Cambridge University Press, 1981).

David H. DeVorkin, *Henry Norris Russell: Dean of American Astronomers* (Princeton, N.J.: Princeton University Press, 2000).

Owen Gingerich, *The Eye of Heaven: Ptolemy, Copernicus and Kepler* (New York: American Institute of Physics, 1993).

Rocky Kolb, *Blind Watchers of the Sky* (Reading, MA: Addison-Wesley, 1996). The evolution of world views, beginning with Tycho Brahe.

David Leverington, *A History of Astronomy: From 1890 to the Present* (New York: Springer-Verlag, 1995).

James Reston, *Galileo: A Life* (New York: HarperCollins,1994).

James Voelkel, *Kepler* (New York: Oxford University Press, 1999). A short biography in the Oxford Portraits series.

## Art and Astronomy

Roberta J. M. Olson and Jay M. Pasachoff, *Fire in the Sky: Comets and Meteors, the Decisive Centuries, in British Art and Science* (New York: Cambridge University Press, 1998, 1999). From the time of Newton and Halley to the present.

Jean Clair, ed., *The Cosmos: From Romanticism to the Avant-Garde* (New York, NY: Prestel, 1999). A mixture of objects, paintings, photographs, and so on, from a museum exhibition, with accompanying essays.

Jay Belloli, ed., *The Universe: A Convergence of Art, Music, and Science* (Pasadena, CA: Armory Center for the Arts, 2001). Exhibitions in Pasadena.

## Solar System

Walter Alvarez, *T.-Rex and the Crater of Doom* (Princeton: Princeton University Press, 1997). A personal account of the quest to understand the extinction of the dinosaurs.

J. Kelly Beatty, Caroline Collins Petersen, and Andrew Chaikin, *The New Solar System,* 4th ed. (Cambridge, MA: Sky Publishing Co. and New York: Cambridge University Press, 1999). Each chapter written by a different expert.

Reta Beebe, *Jupiter: The Giant Planet,* 2nd ed. (Washington, D.C.: Smithsonian Institution Press, 1996).

Eugene Cernan and Don David, *The Last Man on the Moon* (New York: St. Martin's Press, 1999). A first-person account.

Andrew Chaikin, *A Man on the Moon: The Voyages of the Apollo Astronauts* (New York: Viking, 1994). The story of the missions and the people on them.

Clark Chapman and David Morrison, *Cosmic Catastrophes* (New York: Plenum, 1989). An interesting account of the many ways in which life on Earth is threatened.

Gary W. Kronk, *Cometography: A Catalog of Comets,* vol. 1 (New York: Cambridge University Press, 1999). One by one.

Ken Croswell, *Planet Quest: The Epic Discovery of Alien Solar Systems* (New York: Free Press, 1997). Excellent early account of the search for extrasolar planets.

James Elliot and Richard Kerr, *Rings* (Cambridge, MA: MIT Press, 1984). Includes first-person and other stories of the discoveries.

Fred Espenak, *Fifty Year Canon of Lunar Eclipses: 1968–2035* (NASA Ref. Pub. 1216).

Donald Goldsmith and Tobias Owen, *The Search for Life in the Universe* (Reading, MA: Addison-Wesley, 1992). Describes the conditions thought to be necessary for life as we know it.

Donald Goldsmith, *The Hunt for Life on Mars* (Dutton, 1997).

Donald W. Goldsmith, *Worlds Unnumbered: The Quest to Discover Other Solar Systems* (Sausalito, CA: University Science Books, 1997).

David M. Harland, *Exploring the Moon: The Apollo Expeditions* (New York: Springer-Verlag, 1999).

Albert E. Hunt and Patrick Moore, *Atlas of Neptune* (New York: Cambridge University Press, 1994). Maps and tables.

Michael Lemonick, *Other Worlds: The Search for Life in the Universe* (New York: Simon & Schuster, 1998).

Eli Maor, *June 8, 2004—Venus in Transit* (Princeton: Princeton University Press, 2000).

Michael Maunder and Patrick Moore, *Transit: When Planets Cross the Sun* (New York: Springer-Verlag, 2000).

David McNab and James Younger, *The Planets* (New Haven, CT: Yale University Press, 1999). To accompany a TV series on planetary exploration.

David Morrison, *Exploring Planetary Worlds* (New York: Scientific American Library, 1993).

David Morrison and Tobias Owen, *The Planetary System,* 2nd ed. (Reading, MA: Addison-Wesley, 1996). All about the solar system.

J. H. Rogers, *The Giant Planet Jupiter* (New York: Cambridge University Press, 1995).

Carl Sagan, *Pale Blue Dot* (New York: Random House, 1994). One of astronomy's most eloquent spokeperson's last works.

Lutz D. Schmadel, *Dictionary of Minor Planet Names,* 4th ed. (New York: Springer-Verlag, 1999).

S. Alan Stern, *Our Worlds: The Magnetism and Thrill of Planetary Exploration As Described by Leading Planetary Scientists* (New York: Cambridge University Press, 1999).

S. Alan Stern and Jacqueline Mitton, *Pluto & Charon: Ice Worlds on the Ragged Edge of the Solar System* (New York: Wiley, 1997).

Peter Douglas Ward and Donald Brownlee. *Rare Earth: Why Complex Life is Uncommon in the Universe* (New York: Copernicus Books, 2000). Argues that intelligent, technologically advanced life is very rare in the cosmos.

Don E. Wilhelms, *To a Rocky Moon: A Geologist's History of Lunar Exploration* (Tucson: University of Arizona Press, 1993).

Ben Zuckerman and Michael Hart, eds., *Extraterrestrials—Where Are They?* 2nd ed. (New York: Cambridge University Press, 1995).

## The Sun

Leon Golub and Jay M. Pasachoff, *Nearest Star: The Exciting Science of Our Sun* (Harvard University Press, 2001). A non-technical, illustrated trade book. http://www.williams.edu/astronomy/neareststar.

Leon Golub and Jay M. Pasachoff, *The Solar Corona* (New York: Cambridge University Press, 1997). An advanced text.

Fred Espenak, *Fifty Year Canon of Solar Eclipses* (NASA Ref. Pub. 1178, Rev. 1987). Maps and tables.

Kenneth R. Lang, *The Sun from Space* (New York: Springer-Verlag, 2000).

Sten F. Odenwald, *The 23rd Cycle: Learning to Live with a Stormy Star* (New York: Columbia University Press, 2001).

Kenneth J. H. Phillips, *Guide to the Sun* (New York: Cambridge University Press, 1992).

Peter O. Taylor and Nancy L. Hendrickson, *Beginner's Guide to the Sun* (Waukesha, WI: Kalmach Books, 1995).

Jack Zirker, *Journey From the Center of the Sun* (Princeton: Princeton University Press, 2001). A non-technical trade book.

## Stars and the Milky Way Galaxy

Lawrence H. Aller, *Atoms, Stars, and Nebulae,* 3rd ed. (New York: Cambridge University Press, 1991). A classic, recently revised.

Ken Croswell, *The Alchemy of the Heavens* (New York: Anchor, 1995). An excellent account of our Milky Way Galaxy.

Ken Croswell, *Magnificent Universe* (New York: Simon & Schuster, 1999).

Nigel Henbest and Heather Couper, *The Guide to the Galaxy* (New York: Cambridge University Press, 1994). Exciting, contemporary results; profusely illustrated.

James B. Kaler, *Cosmic Clouds: Birth, Death, and Recycling in the Galaxy* (New York: W. H. Freeman, 1997).

James B. Kaler, *Stars and Their Spectra* (New York: Cambridge University Press, 1989). OBAFGKM.

Alfred Mann, *Shadow of a Star* (New York: W. H. Freeman, 1997). The story of the neutrinos from SN 1987A.

Laurence Marschall, *The Supernova Story.* (Princeton: Princeton University Press, 1994). Supernovae in general plus SN 1987A.

D. Mook and T. Vargish, *Inside Relativity* (Princeton: Princeton University Press, 1987). An excellent introduction to special relativity.

Kip Thorne, *Black Holes and Time Warps: Einstein's Outrageous Legacy* (New York: Norton, 1994). An excellent account of general relativity.

Wallace H. Tucker and Karen Tucker, *Revealing the Universe: The Making of the Chandra X-ray Observatory* (Cambridge, MA: Harvard University Press, 2001).

J. Craig Wheeler, *Cosmic Catastrophes: Supernovae, Gamma-ray Bursts, and Adventures in Hyperspace* (New York: Cambridge University Press, 2000).

## Galaxies and Cosmology

Fred Adams and Greg Laughlin, *The Five Ages of the Universe* (New York: Free Press, 1999). Tells the story of what will happen as the universe expands forever.

Bernd Aschenbach, Hermann-Michael Hahn, and J. Truemper, *The Invisible Sky: ROSAT and the Age of X-Ray Astronomy* (New York: Copernicus, 1998)

Mitchell Begelman and Martin Rees, *Gravity's Fatal Attraction: Black Holes in the Universe.* (New York: Scientific American Library, 1996)

Richard Berendzen, Richard Hart, and Daniel Seeley, *Man Discovers the Galaxies* (New York: Neale Watson Academic Publications, 1976). A historical review.

Dennis Danielson, *The Book of the Cosmos* (Boulder, CO: Helix Books, 2000). A compendium from millennia of writings about the cosmos, with useful commentaries.

Alan Dressler, *Voyage to the Great Attractor* (New York: Knopf, 1994). Personal and scientific stories.

Timothy Ferris, *The Whole Shebang: A State of the Universe(s) Report* (New York: Simon & Schuster, 1997, 1998). Contemporary cosmology from a gifted journalist.

Michael W. Friedlander, *A Thin Cosmic Rain: Particles from Outer Space* (Cambridge, MA: Harvard University Press, 2000).

Donald Goldsmith, *The Astronomers* (New York: St. Martin's Press, 1991). The companion to a PBS television series.

Don Goldsmith, *The Runaway Universe* (Cambridge, MA: Perseus Books, 2000). Describes the discovery of a nonzero cosmological constant.

Brian Greene, *The Elegant Universe* (New York: Norton, 1999). Superstrings and the universe; Phi Beta Kappa book award winner.

Alan H. Guth, *The Inflationary Universe: The Quest for a New Theory of Cosmic Origins* (New York: Helix Books/Addison-Wesley, 1997). By the originator of the inflationary theory.

Stephen W. Hawking, *A Brief History of Time, updated and expanded* (New York: Bantam Books, 1998). A best-selling discussion of fundamental topics. *The Illustrated Brief History of Time* (1997) is also available.

Paul W. Hodge, *Galaxies* (Cambridge, MA: Harvard University Press, 1986). A non-technical study of galaxies.

Craig. J. Hogan, *The Little Book of the Big Bang: A Cosmic Primer* (New York: Copernicus Books, 1998).

Michio Kaku, *Hyperspace* (New York: Doubleday, 1994). A very readable introduction to string theory and other dimensions.

Leon M. Lederman and David N. Schramm, *From Quarks to the Cosmos,* 2nd ed. (New York: Scientific American Library, 1995).

Mario Livio, *The Accelerating Universe: Infinite Expansion, the Cosmological Constant, and the Beauty of the Cosmos* (New York: John Wiley & Sons, 2000). Contemporary cosmology by an expert at the Space Telescope Science Institute.

Michael Lemonick, *The Edge of the Universe* (New York: Villard, 1993). The story of the universe and the people who study it.

Alan Lightman and Roberta Brawer, *Origins: The Lives and Worlds of Modern Cosmologists* (Cambridge, MA: Harvard University Press, 1990). Interviews with 27 cosmologists.

David Malin, *The Invisible Universe* (Boston: Bulfinch Press, 1999). By the master of ground-based astronomical color photography.

Dennis Overbye, *Lonely Hearts of the Cosmos: The Scientific Quest for the Secrets of the Universe* (New York: HarperCollins, 1991). Humanizing the cosmologists.

Jay M. Pasachoff, Hyron Spinrad, Patrick Osmer, and Edward S. Cheng, *The Farthest Things in the Universe* (New York: Cambridge University Press, 1995). The cosmic background radiation, quasars, and distant galaxies.

Martin Rees, *Before the Beginning: Our Universe and Others* (Boulder, CO: Helix Books/Addison-Wesley, 1997). By the Astronomer Royal, a major researcher in the field.

Martin Rees, *Just Six Numbers: The Deep Forces that Shape the Universe* (New York: Basic Books, 2000).

Michael Rowan-Robinson, *The Nine Numbers of the Cosmos* (Oxford: Oxford University Press, 1999).

Vera Rubin, *Bright Galaxies, Dark Matters* (Woodbury, NY: American Institute of Physics, 1997). Science and biography.

Allan Sandage and John Bedke, *The Carnegie Atlas of Galaxies* (Washington, D.C.: Carnegie Institution of Washington, 1994). Images to stare at and pore over.

Joseph Silk, *A Short History of the Universe* (New York: Scientific American Library, 1994).

Lee Smolin, *The Life of the Cosmos* (Oxford: Oxford University Press, 1997). Suggests that the properties of our universe resulted from natural selection.

George Smoot and Keay Davidson, *Wrinkles in Time* (New York: Morrow, 1993). A personal account of the discovery of ripples in the cosmic background radiation.

S. Alan Stern, ed., *Our Universe: The Thrill of Extragalactic Exploration as Told by Leading Experts* (New York: Cambridge University Press, 2001).

Steven Weinberg, *The First Three Minutes,* 2nd ed. (New York: Basic Books, 1993). A readable discussion of the first minutes after the big bang, including a discussion of the cosmic background radiation.

Ben Zuckerman and Matthew A. Malkan, *The Origin and Evolution of the Universe* (Boston: Jones and Bartlett, 1996).

# GLOSSARY

**absolute magnitude** The magnitude that a star would appear to have if it were at a distance of ten parsecs from us.
**absorption line** Wavelengths at which the intensity of radiation is less than it is at neighboring wavelengths.
**absorption nebula** Gas and dust seen in silhouette.
**accelerating universe** The model for the universe based on observations late in the 1990s that the expansion of the universe is speeding up over time, rather than slowing down in the way that gravity alone would modify its expansion.
**accretion disk** Matter that an object has taken up and that has formed a disk around the object.
**achondrite** A type of stony meteorite without chondrules. (See *chondrite.*)
**active galactic nucleus (AGN)** A galaxy with an exceptionally bright nucleus in some part of the spectrum; includes radio galaxies, Seyfert galaxies, quasars, and QSOs.
**active galaxy** A galaxy radiating much more than average in some part of the non-optical spectrum, revealing high-energy processes; radio galaxies and Seyfert galaxies are examples.
**active regions** Regions on the sun where sunspots, plages, flares, etc., are found.
**active sun** The group of solar phenomena that vary with time, such as active regions and their phenomena.
**aesthenosphere** A partly melted zone of the earth's mantle, under the lithosphere.
**AGN** Active galactic nucleus.
**albedo** The fraction of light reflected by a body.
**allowed states** The energy values that atoms can have by the laws of quantum mechanics.
**alpha particles** A helium nucleus; consists of two protons and two neutrons.
**alt-azimuth** A two-axis telescope mounting in which motion around one of the axes, which is vertical, provides motion in azimuth, and motion around the perpendicular axis provides up-and-down (altitude) motion.
**altitude** (a) Height above the surface of a planet, or (b) for a telescope mounting, elevation in angular measure above the horizon.
**amino acid** A type of molecule containing the group $NH_2$ (the amino group). Amino acids are fundamental building blocks of life.
**Amor asteroids** A group of asteroids with semimajor axes greater than Earth's and perihelion distances between 1.017 A.U. and 1.3 A.U.; about half cross the Earth's orbit.
**angstrom** A unit of length equal to $10^{-8}$ cm.
**angular momentum** An intrinsic property of a system corresponding to the amount of its revolution or spin. The amount of angular momentum of a body orbiting around a point is the mass of the orbiting body times its (linear) velocity of revolution times its distance from the point. The amount of angular momentum of a spinning sphere is the moment of inertia, an intrinsic property of the distribution of mass, times the angular velocity of spin. The conservation of angular momentum is a law that states that the total amount of angular momentum remains constant in a system that is undisturbed from outside itself.
**angular velocity** The rate at which a body rotates or revolves expressed as the angle covered in a given time (for example, in degrees per hour).
**anisotropy** Deviation from isotropy; changing with direction. Particularly important for the cosmic background radiation.
**annular eclipse** A solar eclipse at which the moon is too small in angular size to cover the solar photosphere, leaving a ring (annulus) of bright photosphere showing. Annular eclipses are visible at about the same rate as total solar eclipses: about every 18 months. The corona is not visible at annular eclipses, because too much bright photosphere remains.
**anorthosite** A type of rock resulting from cooled lava, common in the lunar highlands though rare on Earth.
**anthropic principle** The idea that since we exist, the universe must have certain properties or it would not have evolved so that life would have formed and humans would have evolved.
**antimatter** A type of matter in which each particle (antiproton, antineutron, etc.) is opposite in charge and certain other properties to a corresponding particle (proton, neutron, etc.) of the same mass of the ordinary type of matter from which the solar system is made.
**antiparticle** The antimatter corresponding to a given particle.
**aperture** The diameter of the lens or mirror that defines the amount of light focused by an optical system.
**aperture synthesis** The use of several smaller telescopes together to give some of the properties, such as resolution, of a single larger aperture.
**aphelion** For an orbit around the sun, the farthest point from the sun.
**Apollo asteroids** A group of asteroids, with semimajor axes greater than Earth's and perihelion distances less than 1.017 A.U.
**apparent magnitude** The brightness of a star as seen by an observer, given in a specific system in which a difference of five magnitudes corresponds to a brightness ratio of one hundred times; the scale is fixed by correspondence with a historical background.
**apsides (singular: apsis)** The points at the ends of the major axis of an elliptical orbit. The *line of apsides* is the line that coincides with the major axis of the orbit.
**archaea** A primitive type of organism, perhaps surviving from billions of years ago. Archaea are one of three major types of life, along with eucarya (which include plants and animals as well as smaller objects such as ciliates and slime molds) and bacteria.
**archaeoastronomy** The study of astronomy using archaeological techniques, used when written records are not available, as for Stonehenge or Native American sites, or when pictograms need interpretation, as for Mayan sites.

**association** A physical grouping of stars; in particular, we talk of O and B associations or T associations.
**asterism** A special apparent grouping of stars; part of a constellation.
**asteroid** A "minor planet," a non-luminous chunk of rock smaller than planet-size but larger than a meteoroid, in orbit around a star.
**asteroid belt** A region of the solar system, between the orbits of Mars and Jupiter, in which most of the asteroids orbit.
**astrobiology** The study of life anywhere in the universe, including past, present, and future life on Earth.
**astrometric binary** A system of two stars in which the existence of one star can be deduced by study of its gravitational effect on the proper motion of the other star.
**astrometry** The branch of astronomy that involves the detailed measurement of the positions and motions of stars and other celestial bodies.
**Astronomical Unit** The average distance from the Earth to the sun.
**astrophysics** The science, now essentially identical with astronomy, applying the laws of physics to the universe.
**Aten asteroids** A group of asteroids with semimajor axes smaller than 1 A.U. and aphelion distances greater than 0.983 A.U.
**atom** The smallest possible unit of a chemical element. When an atom is subdivided, the parts no longer have properties of any chemical element.
**atomic clock** A system that uses atomic properties to provide a measure of time.
**atomic number** The number of protons in an atom.
**atomic weight** The number of protons and neutrons in an atom, averaged over the abundances of the different isotopes.
**A.U.** Astronomical Unit.
**aurora** Glowing lights visible in the sky, resulting from processes in the Earth's upper atmosphere and linked with the Earth's magnetic field.
**aurora australis** The southern aurora.
**aurora borealis** The northern aurora.
**autumnal equinox** Of the two locations in the sky where the ecliptic crosses the celestial equator, the one that the sun passes each year when moving from northern to southern declinations.
**azimuth** The angular distance around the horizon from the northern direction, usually expressed in angular measure from 0° for an object in the northern direction, to 180° for an object in the southern direction, around to 360°.

**background radiation** See *primordial background radiation.*
**Baily's beads** Beads of light visible around the rim of the moon at the beginning and end of a total solar eclipse. They result from the solar photosphere shining through valleys at the edge of the moon.
**Balmer series** The set of spectral absorption or emission lines resulting from a transition down to or up from the second energy level (first excited level) of hydrogen.
**bar** The straight structure across the center of some spiral galaxies, from which the arms unwind.
**baryons** Nuclear particles (protons, neutrons, etc.) subject to the strong nuclear force; made of quarks.
**basalt** A type of rock resulting from the cooling of lava.
**baseline** The distance between points of observation when it determines the accuracy of some measurement.
**beam** The cone within which a radio telescope is sensitive to radiation.
**Becklin-Neugebauer object** An object visible only in the infrared in the Orion Molecular Cloud, apparently a very young star.
**belts** Dark bands around certain planets, notably Jupiter.
**beta particle** An electron or a positron outside an atom.
**big-bang theory** A cosmological model, based on Einstein's general theory of relativity, in which the universe was once compressed to infinite density and has been expanding ever since.
**binary pulsar** A pulsar in a binary system.
**binary star** Two stars revolving around each other.
**bipolar flow** A phenomenon in young or forming stars in which streams of matter are ejected from the poles.
**black body** A hypothetical object that, if it existed, would absorb all radiation that hit it and would emit radiation that exactly followed Planck's law ("thermal radiation").
**black-body curve** A graph of brightness vs. wavelength that follows Planck's law; each such curve corresponds to a given temperature. Also called a Planck curve.
**black-body radiation** Radiation whose distribution in wavelength follows Planck's law, the black-body curve.
**black dwarf** A non-radiating ball of gas that results either when a white dwarf radiates all its energy or when gas contracts gravitationally but contains too little mass to begin nuclear fusion.
**black hole** A region of space from which, according to the general theory of relativity, neither radiation nor matter can escape.
**black hole era** The future era, following the degenerate era, when the only objects (besides photons and sub-atomic particles) the universe will contain will be black holes.
**blueshift** A shift of optical wavelengths toward the blue or in all cases towards shorter wavelengths; when the shift is caused by motion, from a velocity of approach.
**B-N object** See *Becklin-Neugebauer object.*
**Bohr atom** Niels Bohr's model of the hydrogen atom, in which the energy levels are depicted as concentric circles of radii that increase as (level number)$^2$.
**Bok globule** A type of round compact absorption nebula.
**bolide** A very bright meteor from which a sound is heard.
**bolometer** A device for measuring the total amount of radiation from an object.
**bolometric magnitude** The magnitude of a celestial object corrected to take account of the radiation in parts of the spectrum other than the visible.
**BOOMERanG** *B*alloon *O*bservations *o*f *M*illimetric *E*xtragalactic *R*adiation; it observed the cosmic background radiation at high resolution over 3 per cent of the sky, most notably in a flight in 1998.
**bound-free transition** An atomic transition in which an electron starts bound to the atom and winds up free from it.
**breccia** A type of rock made up of fragments of several types of rocks. Breccias are common on the moon.
**brown dwarf** A self-gravitating, self-luminous object, not a satellite and insufficiently massive for nuclear fusion to begin (less than 0.07 solar mass).

**butterfly diagram** A diagram showing the locations of sunspots in latitude as a function of time. First drawn by E. Walter Maunder.

**calorie** A measure of energy in the form of heat, originally corresponding to the amount of heat required to raise the temperature of 1 gram of water by 1°C.

**canali** Name given years ago to apparent lines on Mars.

**capture** An outdated model in which a moon was formed elsewhere and then was captured gravitationally by its planet.

**carbonaceous** Containing a lot of carbon.

**carbonaceous chondrites** A type of carbon-rich stony meteor, thought to be primitive material from the early era of the solar system. Chondrules are embedded in carbon-rich minerals.

**carbon cycle** A chain of nuclear reactions, involving carbon as a catalyst at some of its intermediate stages, that transforms four hydrogen atoms into one helium atom with a resulting release in energy. The carbon cycle is important only in stars hotter than the sun.

**carbon-nitrogen cycle** The carbon cycle, acknowledging that nitrogen also plays an intermediary role.

**carbon-nitrogen-oxygen tri-cycle** A set of variations of the carbon cycle including nitrogen and oxygen isotopes as intermediaries.

**carbon stars** A spectral type of cool stars whose spectra show a lot of carbon; formerly types R and N.

**Cassegrain (telescope)** A type of reflecting telescope in which the light focused by the primary mirror is intercepted short of its focal point and refocused and reflected by a secondary mirror through a hole in the center of the primary mirror.

**Cassini** (a) The scientist, born Italian and later director of the Paris Observatory. (b) The NASA spacecraft en route to Saturn, carrying the Huygens probe to its moon Titan.

**Cassini's division** The major division in the rings of Saturn.

**catalyst** A substance that participates in a reaction but that is left over in its original form at the end.

**catastrophe theories** Theories of solar-system formation involving a collision with another star.

**CCD** Charge-coupled device, a solid-state imaging device.

**celestial equator** The intersection of the celestial sphere with the planet that passes through the Earth's equator.

**celestial poles** The intersection of the celestial sphere with the axis of rotation of the Earth.

**celestial sphere** The hypothetical sphere centered at the center of the Earth to which it appears that the stars are affixed.

**centaur** A solar-system object, with 2060 Chiron as the first example, intermediate between comets and icy planets or satellites and orbiting the sun between the orbits of Jupiter and Neptune.

**center of mass** The "average" location of mass; the point in a body or system of bodies at which we may consider all the mass to be located for the purpose of calculating the gravitational effect of that mass or its mean motion when a force is applied.

**central star** The hot object at the center of a planetary nebula, which is the remaining core of the original star.

**Cepheid variable** A type of supergiant star that oscillates in brightness in a manner similar to the star $\delta$ Cephei. The periods of Cepheid variables, which are between 1 and 100 days, are linked to the absolute magnitude of the stars by known relationships; this allows the distances to Cepheids to be found. Type II Cepheids (also known as W Virginis stars) are fainter at each period.

**Chandrasekhar limit** The limit in mass, about 1.4 solar masses, above which electron degeneracy cannot support a star, and so the limit above which white dwarfs cannot exist.

**Chandra X-ray Observatory** The major NASA x-ray observatory, the former Advanced X-ray Astrophysics Facility, launched in 1999 to make high resolution observations in the x-ray part of the spectrum.

**chaos** The condition in which very slight differences in initial conditions typically lead to very different results. Also, specifically, examples of orbits or rotation directions that change after apparently random intervals of time because of such sensitivity to initial conditions.

**charm** An arbitrary name that corresponds to a property that distinguishes certain elementary particles, including types of quarks, from each other.

**chondrite** A type of stony meteorite that contains small crystalline spherical particles called chondrules.

**chromatic aberration** A defect of lens systems in which different colors are focused at different points.

**chromosphere** The part of the atmosphere of the sun (or another star) between the photosphere and the corona. It is probably entirely composed of spicules and probably roughly corresponds to the region in which mechanical energy is deposited.

**chromospheric network** An apparent network of lines on the solar surface corresponding to the higher magnetic fields at the boundries of supergranules; visible especially in the calcium H and K lines.

**circle** A conic section formed by cutting a cone perpendicularly to its axis.

**circumpolar stars** For a given observing location, stars that are close enough to the celestial pole that they never set.

**classical** When discussing atoms, not taking account of quantum mechanical effects.

**closed universe** A big-bang universe with positive curvature; it has finite volume and will eventually contract.

**cluster** (a) Of stars, a physical grouping of many stars; (b) of galaxies, a physical grouping of at least a few galaxies.

**cluster variable** A star that varies in brightness with a period of 0.1 to one day, similar to the star RR Lyrae. They are found in globular clusters. All cluster variables have approximately the same brightness, which allows their distance to be readily found.

**CNO tri-cycle** See *carbon-nitrogen-oxygen tri-cycle.*

**COBE (Cosmic Background Explorer)** A spacecraft launched in 1989 that measured the cosmic background radiation for four years.

**coherent radiation** Radiation in which the phases of waves at different locations in a cross-section of radiation have a definite relation to each other; in noncoherent radiation, the phases are random. Only coherent radiation shows interference.

**cold dark matter** Non-luminous matter with no velocity dispersion compared with the expansion of the universe, such as mini black holes and exotic nuclear particles.

**color** (a) Of an object, a visual property that depends on wavelength; (b) an arbitrary name assigned to a property that distinguishes three kinds of quarks.
**color index** The difference, expressed in magnitudes, of the brightness of a star or other celestial object measured at two different wavelengths. The color index is a measure of temperature.
**color-magnitude diagram** A Hertzsprung-Russell diagram in which the temperature on the horizontal axis is expressed in terms of color index and the vertical axis is in magnitudes.
**coma** (a) Of a comet, the region surrounding the head; (b) of an optical system, an off-axis aberration in which the images of points appear with comet-like asymmetries.
**comet** A type of object orbiting the sun, often in a very elongated orbit, that when relatively near to the sun shows a coma and may show a tail.
**comparative planetology** Studying the properties of solar-system bodies by comparing them.
**comparison spectrum** A spectrum of known elements on Earth usually photographed on the same photographic plate as a stellar spectrum in order to provide a known set of wavelengths for zero Doppler shift.
**composite spectrum** The spectrum that reveals a star is a binary system, since spectra of more than one object are apparent.
**condensation** A region of unusually high mass or brightness.
**conic sections** Geometrical shapes that can be obtained by slicing a cone; they include ellipses (and therefore circles), hyperbolas, and parabolas.
**conjunction** Where two celestial objects reach the same celestial longitude; approximately corresponds to their closest apparent approach in the sky. When only one body is named, it is understood that the second body is the sun.
**conservation law** A statement that the total amount of some property (angular momentum, energy, etc.) of a body or set of bodies does not change.
**constellation** One of 88 areas into which the sky has been divided for convenience in referring to the stars or other objects therein.
**continental drift** The slow motion of the continents across the Earth's surface, explained in the theory of plate tectonics as a set of shifting regions called plates.
**continuous spectrum** A spectrum with radiation at all wavelengths but with neither absorption nor emission lines.
**continuum (plural: continua)** The continuous spectrum that we would measure from a body if no spectral lines were present.
**convection** The method of energy transport in which the rising motion of masses from below carries energy upward in a gravitational field. Boiling is an example.
**convection zone** The subsurface zone in certain types of stars in which convection dominates energy transfer.
**convergent point** The point in the sky toward which the members of a star cluster appear to be converging or from which they appear to be diverging. Referred to in the moving cluster method of determining distances.
**coordinate systems** Methods of assigning positions with respect to suitable axes.
**core** The central region of a star or planet.
**Coriolis force** The effect of an object's apparent swerving as seen from a rotating body's surface, as the object moves north or south of a body's equator. Though the object is proceeding straight, following Newton's laws of motion, its apparent swerving is attributed to a fictitious force, the Coriolis force, which isn't really a force at all.
**corona, galactic** The outermost region of our current model of the galaxy, containing most of the mass in some unknown way.
**coronagraph** A type of telescope with which the corona (and, sometimes, merely the chromosphere) can be seen in visible light at times other than that of a total solar eclipse by occulting (hiding) the bright photosphere. Also, any telescope in which a bright central object is occulted to reveal fainter things.
**coronal holes** Relatively dark regions of the corona having low density; they result from open field lines.
**coronal mass ejection** Globs of coronal plasma ejected approximately daily from the sun, realized in recent years to be a major cause of "space weather" affecting the Earth. Once thought to be caused by a flare but now realized usually to be a separate phenomenon.
**corona, solar or stellar** The outermost region of the sun (or of other stars), characterized by temperatures of millions of kelvins.
**correcting plate** A thin lens of complicated shape at the front of a Schmidt camera that compensates for spherical aberration.
**cosmic abundances** The overall abundances of elements in the universe.
**cosmic background radiation** The isotropic glow of 3° radiation.
**cosmic rays** Nuclear particles or nuclei travelling through space at high velocity.
**cosmic turbulence** A theory of galaxy formation in which turbulence originally existed.
**cosmogony** The study of the origin of the universe, usually applied in particular to the origin of the solar system.
**cosmological constant** A constant arbitrarily added by Einstein to an equation in his general theory of relativity in order to provide a solution in which the universe did not expand. It was only subsequently discovered that the universe is expanding after all. The cosmological constant now explains how the expansion can be accelerating.
**cosmological distances** Distances assigned by Hubble's law.
**cosmological principle** The principle that on the whole the universe looks the same in all directions and in all regions.
**cosmology** The study of the universe as a whole, including its origin and evolution.
**coudé focus** A focal point of large telescopes in which the light is reflected by a series of mirrors so that it comes to a point at the end of a polar axis. The image does not move even when the telescope moves, which permits the mounting of heavy equipment.
**critical density** The density at which the universe is on the dividing line between open and closed; inflation models drive the universe to this critical density. Current models are 0.03 visible matter, 0.3 dark matter, and 0.67 cosmological constant, to total the 1.0 critical density called for by inflation.
**crust** The outermost solid layer of some objects, including neutron stars and some planets.

**C-type stars** See *carbon stars.*
**cytherean** Venusian.

**dark energy** The term given to the cosmological constant or other cause of the universe's apparent acceleration in its expansion, thought as of this writing to provide 70% of the critical density.
**dark era** The final era of the universe when only low-energy photons, neutrinos, and some elementary particles (the ones that did not find a partner to annihilate) remain.
**dark matter** Non-luminous matter. (See *hot dark matter* and *cold dark matter.*)
**dark nebula** Dust and gas seen in silhouette.
**DASI** (pronounced "daisy") Degree Angular Scale Interferometer, which mapped the cosmic background radiation at high resolution from the Earth's South Pole.
**daughter molecules** Relatively simple molecules in comets resulting from the breakup of more complex molecules.
**deceleration parameter ($q_0$)** A particular measure of the rate at which the expansion of the universe is slowing down. For a universe of baryons, $q_0 < ½$ corresponds to an open universe and $q_0 > ½$ corresponds to a closed universe.
**declination** Celestial latitude, measured in degrees north or south of the celestial equator.
**deferent** In the Ptolemaic system of the universe, the larger circle, centered at the Earth, on which the centers of the epicycles revolve.
**degenerate era** The future era when the universe will be filled with cold brown dwarfs, white dwarfs, and neutron stars.
**degenerate matter** Matter whose properties are controlled, and that is prevented from further contraction, by quantum mechanical laws.
**density** Mass divided by volume.
**density wave** A circulating region of relatively high density, important, for example, in models of spiral arms.
**density-wave theory** The explanation of spiral structure of galaxies as the effect of a wave of compression that rotates around the center of the galaxy and causes the formation of stars in the compressed region.
**deuterium** An isotope of hydrogen that contains 1 proton and 1 neutron.
**deuteron** A deuterium nucleus, containing 1 proton and 1 neutron.
**diamond-ring effect** The last Baily's bead glowing brightly at the beginning of the total phase of a solar eclipse, or its counterpart at the end of totality.
**differential forces** A net force resulting from the difference of two other forces; a tidal force.
**differential rotation** Rotation of a body in which different parts have different angular velocities (and thus different periods of rotation).
**differentiation** For a planet, the existence of layers of different structure or composition.
**diffraction** A phenomenon affecting light as it passes any obstacle, spreading it out in a complicated fashion.
**diffraction grating** A very closely ruled series of lines that, through diffraction of light, provides a spectrum of radiation that falls on it.
**dipole** A pattern that is symmetric from one direction to the opposite direction. The cosmic background radiation has an overall dipole pattern, with one side of the sky warmer than average by one part in 100,000 than the opposite side, and varying from the directions of the dipole with a cosine function.
**dirty snowball** A theory explaining comets as amalgams of ices, dust, and rocks.
**discrete** Separated; isolated.
**disk** (a) Of a galaxy, the disk-like flat portion, as opposed to the nucleus or the halo; (b) of a star or planet, the two-dimensional projection of its surface.
**dispersion** (a) Of light, the effect that different colors are bent by different amounts when passing from one substance to another; (b) of the pulses of a pulsar, the effect that a given pulse, which leaves the pulsar at one instant, arrives at the Earth at different times depending on the different wavelength or frequency at which it is observed. Both of these effects arise because light of different wavelengths travels at different speeds except in a vacuum.
**D lines** A pair of lines from sodium that appear in the yellow part of the spectrum.
**DNA** Deoxyribonucleic acid, a long chain of molecules that contains the genetic information of life.
**Dobsonian** An inexpensive type of large-aperture amateur telescope characterized by a thin mirror, composition tube, and Teflon bearings on an alt-azimuth mount.
**Doppler effect** A change in wavelength that results when a source of waves and the observer are moving relative to each other.
**Doppler shift** The change in wavelength that arises from the Doppler effect, caused by relative motion toward or away from the observer.
**double-lobed structure** An object in which radio emission comes from a pair of regions on opposite sides.
**double star** A binary star; two or more stars orbiting each other.
**Drake equation** An equation advanced by Frank Drake and popularized also by Carl Sagan that attempts to calculate the number of civilizations by breaking the calculation down into a series of steps that can be assessed individually, such as rate of star formation, the fraction of stars with planets, and the average lifetime of a civilization.
**dust tail** The dust left behind a comet, reflecting sunlight.
**dwarf ellipticals** Small, low-mass elliptical galaxies.
**dwarfs** Dwarf stars.
**dwarf stars** Main-sequence stars.
**dynamo** A device that generates electricity through the effect of motion in the presence of a magnetic field.
**dynamo theories** Explaining sunspots and the solar activity cycle through an interaction of motion and magnetic fields.

**$E = mc^2$** Einstein's formula (special theory of relativity) for the equivalence of mass and energy.
**early universe** The universe during its first minutes.
**earthshine** Sunlight illuminating the moon after having been reflected by the Earth.
**eccentric** Deviating from a circle. In Ptolemaic astronomy, a deferent not centered at the Earth; short for "eccentric circle."
**eccentricity** A measure of the flatness of an ellipse, defined as half the distance between the foci divided by the semi-major axis.

**eclipse** The passage of all or part of one astronomical body into the shadow of another.

**eclipsing binary** A binary star in which one member periodically hides the other.

**ecliptic** The path followed by the sun across the celestial sphere in the course of a year.

**ecliptic plane** The plane of the Earth's orbit around the sun.

**ecosphere** The region around a star in which conditions are suitable for life, normally a spherical shell.

**electric field** A force field set up by an electric charge.

**electroglow** An ultraviolet emission on the sunlighted side of Uranus thought to result from sunlight splitting hydrogen molecules into protons and electrons, with the electrons accelerated perhaps by the magnetic field.

**electromagnetic force** One of the four fundamental forces of nature, giving rise to electromagnetic radiation.

**electromagnetic radiation** Radiation resulting from changing electric and magnetic fields.

**electromagnetic spectrum** Energy in the form of electromagnetic waves, in order of wavelength.

**electromagnetic waves** Waves of changing electric and magnetic fields, travelling through space at the speed of light.

**electromagnetism** The combined force of electricity and magnetism, which follows the formulae unified by Maxwell.

**electron** A particle of one negative charge, $1/1830$ the mass of a proton, that is not affected by the strong force. It is a lepton.

**electron degeneracy** The state in which, following rules of quantum mechanics, the further compression of electrons generates a high pressure that balances gravity, as in white dwarfs.

**electron volt (eV)** The energy necessary to raise an electron through a potential of one volt.

**electroweak force** The unified electromagnetic and weak forces, according to a recent theory.

**element** A kind of atom, characterized by a certain number of protons in its nucleus. All atoms of a given element have similar chemical properties.

**elementary particle** One of the constituents of an atom.

**ellipse** A curve with the property that the sum of the distances from any point on the curve to two given points, called the foci, is constant.

**elliptical galaxy** A type of galaxy characterized by elliptical appearance.

**emission line** Wavelengths (or frequencies) at which the intensity of radiation is greater than it is at neighboring wavelengths (or frequencies).

**emission nebula** A glowing cloud of interstellar gas.

**emulsion** A coating whose sensitivity to light allows photographic recording of incident radiation.

**energy** A fundamental quantity usually defined in terms of the ability of a system to do something that is technically called "work," that is, the ability to move an object by application of force, where the work is the force times the displacement.

**energy, law of conservation of** Energy is neither created nor destroyed, but may be changed in form.

**energy level** A state corresponding to an amount of energy that an atom is allowed to have by the laws of quantum mechanics.

**energy problem, the** For quasars, how to produce so much energy in such a small emitting volume.

**ephemeris** A listing of astronomical positions and other data that change with time. From the same root as *ephemeral.*

**epicycle** In the Ptolemaic theory, a small circle, riding on a larger circle called the deferent, on which a planet moves. The epicycle is used to account for retrograde motion.

**equal areas, law of** Kepler's second law.

**equant** In Ptolemaic theory, the point equally distant from the center of the deferent as the Earth but on the opposite side, around which the epicycle moves at a uniform angular rate.

**equator** (a) Of the Earth, a great circle on the Earth, midway between the poles; (b) celestial, the projection of the Earth's equator onto the celestial sphere; (c) galactic, the plane of the disk as projected onto a map.

**equatorial mount** A type of telescope mounting in which one axis, called the polar axis, points toward the celestial pole and the other axis is perpendicular. Motion around only the polar axis is sufficient to completely counterbalance the effect of the Earth's rotation.

**equinox** An intersection of the ecliptic and the celestial equator. The center of the sun is geometrically above and below the horizon for equal lengths of time on the two days of the year when the sun passes the equinoxes; if the sun were a point and atmospheric refraction were absent, then day and night would be of equal length on those days.

**erg** A unit of energy in the metric system, corresponding to the work done by a force of one dyne (the force that is required to accelerate one gram by one $cm/sec^2$) producing a displacement of one centimeter.

**ergosphere** A region surrounding a rotating black hole (or other system satisfying Kerr's solution) from which work can be extracted.

**escape velocity** The velocity that an object must have to escape the gravitational pull of a mass.

**event horizon** The sphere around a black hole from within which nothing can escape; the place at which the exit cones close.

**evolutionary track** The set of points on an H-R diagram showing the changes of a star's temperature and luminosity with time.

**excitation** The raising of atoms to higher energy states than the lowest possible.

**excited level** An energy level of an atom above the ground level.

**exclusion principle** The quantum mechanical rule that certain types of elementary particles cannot exist in completely identical states.

**exit cone** The cone that, for each point within the photon sphere of a black hole, defines the directions of rays of radiation that escape.

**exobiology** The study of life located elsewhere than Earth.

**exponent** The "power" representing the number of times a number is multiplied by itself.

**exponential notation** The writing of numbers as a power of 10 times a number with one digit before the decimal point.

**extended objects** Objects with detectable angular size.

**extinction** The dimming of starlight by scattering and absorption as the light traverses interstellar space.

**extragalactic** Exterior to the Milky Way Galaxy.
**extragalactic radio sources** Radio sources outside our galaxy.
**extremophiles** Archaea, bacteria, and other forms of life discovered to exist at temperatures or pressures more extreme than had been thought sufficiently hospitable.
**eyepiece** The small combination of lenses at the eye end of a telescope, used to examine the image formed by the objective.

**FAME** The U.S. Naval Observatory's *F*ull-sky *A*stro*m*etric *E*xplorer, to be launched in 2004.
**featherweight (stars)** A failed "star," contains less than 0.07 solar mass; becomes a brown dwarf.
**field of view** The angular expanse viewable.
**filament** A feature of the solar surface seen in H$\alpha$ as a dark wavy line; a prominence projected on the solar disk.
**fireball** An exceptionally bright meteor.
**first excited state** In Bohr's model of the hydrogen atom, the condition of the atom in which the electron is on the second lowest energy level, the first level above the ground level.
**fission, nuclear** The splitting of an atomic nucleus.
**flare** An extremely rapid brightening of a small area of the surface of the sun, usually observed in H$\alpha$ and other strong spectral lines and accompanied by x-ray and radio emission.
**flash spectrum** The solar chromospheric spectrum seen in the few seconds before or after totality at a solar eclipse.
**flat (critical) Universe** A universe in which Euclid's parallel postulate holds; one that is barely expanding forever. It has infinite volume and its age is ⅔ the Hubble time.
**flatness problem** One of the problems solved by the inflationary theory, that the universe is exceedingly close to being flat for no obvious reason.
**flavors** A way of distinguishing quarks: up, down, strange, charmed, truth (top), beauty (bottom).
**fluorescence** The transformation of photons of relatively high energy to photons of lower energy through interactions with atoms, and the resulting radiation.
**flux** The amount of something (such as energy) passing through a surface per unit time.
**flux tube** A torus through which particles circulate.
**focal length** The distance from a lens or mirror to the point to which rays from an object at infinity are focused.
**focus (plural: foci)** The point to which light is brought by a lens (optical or even gravitational) and at which an image is formed.
**force** In physics, something that can or does cause change of momentum, measured by the rate of change of momentum with time.
**Foucault pendulum** A heavy pendulum on a long wire that oscillates back and forth in a plane constant with the universe while the Earth rotates around it; the direct proof that the Earth rotates.
**Fraunhofer lines** The absorption lines of a solar or other stellar spectrum.
**frequency** The rate at which waves pass a given point.
**full moon** The phase of the moon when the side facing the Earth is fully illuminated by sunlight.
**fusion** The amalgamation of nuclei into heavier nuclei.
**fuzz** The faint light detectable around nearby quasars.

**galactic cannibalism** The incorporation of one galaxy into another.
**galactic cluster** An asymmetric type of collection of stars that shared a common origin.
**galactic corona** The outermost part of our galaxy.
**galactic year** The length of time the sun takes to complete an orbit of our galactic center.
**Galilean satellites** The four brightest satellites of Jupiter.
**Galileo** (a) The Italian scientist Galileo Galilei (1564–1642); (b) the NASA spacecraft in orbit around the planet Jupiter that sent a probe into Jupiter's clouds on December 7, 1995.
**gamma-ray burst** A brief burst of gamma rays, studied especially by the Compton Gamma-ray Observatory during 1991–2000, which found thousands of such bursts uniformly distributed in the sky; discovered to come, at least in many cases, from exceedingly distant and thus extremely powerful sources of energy.
**gamma rays** Electromagnetic radiation with wavelengths shorter than approximately 0.1 Å.
**gas tail** The puffs of ionized gas trailing a comet.
**general theory of relativity** Einstein's 1916 theory of gravity.
**geocentric** Earth-centered.
**geology** The study of the Earth, or of other solid bodies.
**geothermal energy** Energy from under the Earth's surface.
**giant** A star that is larger and brighter than main-sequence stars of the same color.
**giant ellipticals** Elliptical galaxies that are very large.
**giant molecular cloud** A basic building block of our galaxy, containing dust, which shields the molecules present.
**giant planets** Jupiter, Saturn, Uranus, and Neptune; or even larger extrasolar planets.
**giant star** A star more luminous than a main-sequence star of its spectral type; a late stage in stellar evolution.
**gibbous moon** The phases between half moon and full moon.
**GLAST** *G*amma-ray *L*arge *A*rea *S*pace *T*elescope, NASA's planned 2005 successor to the Compton Gamma-ray Observatory.
**globular cluster** A spherically symmetric type of collection of stars that shared a common origin.
**gluon** The particle that carries the color force (and thus the strong nuclear force).
**gnomon** The vertical stick or triangle used to cast the sun's shadow in a sundial.
**grand unified theories (GUTs)** Theories unifying the electroweak force and the strong force.
**granulation** Convection cells on the sun about 1 arc sec across.
**grating** A surface ruled with closely spaced lines that, through diffraction, breaks up light into its spectrum.
**gravitational force** One of the four fundamental forces of nature, the force by which two masses attract each other.
**gravitational instability** A situation that tends to break up under the force of gravity.
**gravitational interlock** One body controlling the orbit or rotation of another by gravitational attraction.

**gravitational lens** In the gravitational lens phenomenon, a massive body changes the path of electromagnetic radiation passing near it so as to make more than one image of an object. The double quasar was the first example to be discovered.

**gravitationally** Controlled by the force of gravity.

**gravitational radius** The radius that, according to Schwarzschild's solutions to Einstein's equations of the general theory of relativity, corresponds to the event horizon of a black hole.

**gravitational redshift** A redshift of light caused by the presence of mass, according to the general theory of relativity.

**gravitational waves** Waves that many scientists consider to be a consequence, according to the general theory of relativity, of changing distributions of mass.

**gravity** The tendency for all masses to attract each other; described in a formula by Newton and more recently described by Einstein as a result of a warping of space and time by the presence of a mass.

**gravity assist** Using the gravity of one celestial body to change a spacecraft's energy.

**grazing incidence** Striking at a low angle.

**great circle** The intersection of a plane that passes through the center of a sphere with the surface of that sphere; the largest possible circle that can be drawn on the surface of a sphere.

**Great Dark Spot** A giant circulating region seen by Voyager 2 on Neptune; it has since disappeared.

**Great Red Spot** A giant circulating region on Jupiter.

**greenhouse effect** The effect by which the atmosphere of a planet heats up above its equilibrium temperature because it is transparent to incoming visible radiation but opaque to the infrared radiation that is emitted by the surface of the planet.

**Gregorian (telescope)** A type of reflecting telescope in which the light focused by the primary mirror passes its prime focus and is then reflected by a secondary mirror through a hole in the center of the primary mirror.

**Gregorian calendar** The calendar in current use, with normal years that are 365 days long, with leap years every 4th year except for years that are divisible by 100 but not by 400.

**ground level** An atom's lowest possible energy level.

**ground state** See *ground level.*

**GUTs** See *grand unified theories.*

**$H_0$** The Hubble constant.

**$H\alpha$** The first line of the Balmer series of hydrogen, at 6563 Å.

**H I region** An interstellar region of neutral hydrogen.

**H II region** An interstellar region of ionized hydrogen.

**half-life** The length of time for half a set of particles to decay through radioactivity or instability.

**halo** Of a galaxy, the region of the galaxy that extends far above and below the plane of the galaxy, containing globular clusters.

**head** Of a comet, the nucleus and coma together.

**head-tail galaxies** Double-lobed radio galaxies whose lobes are so bent that they look like a tadpole.

**heat flow** The flow of energy from one location to another.

**heavyweight stars** Stars of more than about 8 solar masses.

**heliacal rising** The first time in a year that an astronomical body rises sufficiently far ahead of the sun that it can be seen in the morning sky.

**heliocentric** Sun-centered; using the sun rather than the Earth as the point to which we refer. A heliocentric measurement, for example, omits the effect of the Doppler shift caused by the earth's orbital motion.

**heliopause** The outer limit of the sun's influence on space; where the solar wind becomes weaker than interstellar matter.

**helioseismology** The study of the sun's interior through studies of waves measured with observations of surface variations.

**helium flash** The rapid onset of fusion of helium into carbon through the triple-alpha process that takes place in most red-giant stars.

**Herbig-Haro objects** Blobs of gas ejected in star formations.

**Herschel Infrared Observatory** The European Space Agency's infrared telescope, planned for 2007 and named after William Herschel, who discovered infrared radiation (after discovering Uranus and many other objects).

**hertz** The measure of frequency, with units of /sec (per second); formerly called cycles per second.

**Hertzsprung-Russell diagram** A graph of temperature (or equivalent) vs. luminosity (or equivalent) for a group of stars.

**high-energy astrophysics** The study of x-rays, gamma rays, and cosmic rays, and of the processes that make them.

**highlands** Regions on the moon or elsewhere that are above the level that may have been smoothed by flowing lava.

**H line** The spectral line of ionized calcium at 3968 Å.

**homogeneous** Uniform throughout.

**horizon problem** One of the problems of cosmology solved by the inflationary theory: why the universe is identical in all directions, even though widely separated regions could never have been in thermal equilibrium with each other since they are beyond each other's horizon.

**horizontal branch** A part of the Hertzsprung-Russell diagram of a globular cluster, corresponding to stars that are all at approximately zero absolute magnitude and have evolved past the red-giant stage and are moving leftward on the diagram.

**hot dark matter** Non-luminous matter with a large velocity dispersion, like neutrinos.

**hour angle** Of a celestial object as seen from a particular location, the difference between the local sidereal time and the right ascension (H.A. = L.S.T. − R.A.).

**hour circle** The great circles passing through the celestial poles.

**H-R diagram** Hertzsprung-Russell diagram.

**Hubble constant ($H_0$)** The constant of proportionality in Hubble's law linking the velocity of recession of a distant object and its distance from us.

**Hubble Deep Field** A small part of the sky in Ursa Major extensively studied by the Hubble Space Telescope in December 1995; the long exposures allowed astronomers to see far back in time. See http://www.stsci.edu/ftp/science/hdfsouth. Hubble Deep Field—South was observed subsequently.

**Hubble flow** The assumed uniform expansion of the universe, on which any peculiar motions of galaxies or clusters of galaxies is superimposed.

**Hubble's law** The linear relation between the velocity of recession of a distant object and its distance from us, $V = H_0 d$.
**Hubble time** The amount of time since the big bang, assuming a constant speed for any given galaxy; the Hubble time is calculated by tracing the universe backward in time using the current Hubble constant.
**Hubble type** Hubble's galaxy classification scheme: E0, E7, Sa, SBa, etc.
**hyperfine level** A subdivision of an energy level caused by such relatively minor effects as changes resulting from the interactions among spinning particles in an atom or molecule.
**hypothesis** The first step in the traditional formulation of the scientific method; a tentative explanation of a set of facts that is to be tested experimentally or observationally.

**IC** *Index Catalogue,* one of the supplements to Dreyer's *New General Catalogue.*
**igneous** Rock cooled from lava.
**image tube** An electronic device that receives incident radiation and intensifies it or converts it to a wavelength at which photographic plates are sensitive.
**inclination** Of an orbit, the angle of the plane of the orbit with respect to the ecliptic plane.
**inclined** Tilted with respect to some other body, usually describing the axis of rotation or the plane of an orbit.
**Index Catalogue** See *IC.*
**inertia** A property of mass used by Isaac Newton in his first law of motion to keep a body at rest if it was at rest or at motion with the same direction and speed unless acted on by a force.
**inferior conjunction** An inferior planet's reaching the same celestial longitude as the sun's.
**inferior planet** A planet whose orbit around the sun is within the Earth's, namely, Mercury and Venus.
**inflation** The theory that the universe expanded extremely fast, by perhaps 10 to the 100th power, in the first fraction of a second after the big bang. The concept of inflation solves several problems in cosmology, such as the horizon problem.
**inflationary universe** A model of the expanding universe involving a brief period of extremely rapid expansion.
**infrared** Radiation beyond the red, about 7000 Å to 1 mm.
**interference** The property of radiation, explainable by the wave theory, in which waves in phase can add (constructive interference) and waves out of phase can subtract (destructive interference); for light, this gives alternate light and dark bands.
**interferometer** A device that uses the property of interference to measure such properties of objects as their positions or structure.
**interferometry** Observations using an interferometer.
**intergalactic medium** Material between galaxies in a cluster.
**interior** The inside of an object.
**international date line** A crooked imaginary line on the Earth's surface, roughly corresponding to 180° longitude, at which, when crossed from east to west, the date jumps forward by one day.
**interplanetary medium** Gas and dust between the planets.
**interstellar medium** Gas and dust between the stars.
**interstellar reddening** The relatively greater extinction of blue light by interstellar matter than of red light.
**in transit** One body passing in front of another, as seen by the observer. Venus will be "in transit" when it passes in front of the sun in 2004, the first such event since 1882.
**inverse-square law** Decreasing with the square of increasing distance.
**ion** An atom that has lost one or more electrons. See also *negative hydrogen ion.*
**ionized** Having lost one or more electrons.
**ionosphere** The highest region of the earth's atmosphere.
**ion tail** See *gas tail.*
**IRAS** The Infrared Astronomical Satellite (1983).
**iron meteorites** Meteorites with a high iron content (about 90%); most of the rest is nickel.
**irons** Iron meteorites.
**irregular cluster** A cluster of galaxies showing no symmetry.
**irregular galaxy** A type of galaxy showing no shape or symmetry.
**ISO (Infrared Space Observatory)** A European Space Agency project, 1995–1998.
**isotope** A form of chemical element with a specific number of neutrons.
**isotropic** Being the same in all directions.

**joule** The SI unit of energy, 1 kg·m$^2$/sec$^2$.
**Jovian planet** Same as giant planet.
**JPL** The Jet Propulsion Laboratory in Pasadena, California, funded by NASA and administered by Caltech; a major space contractor.
**Julian calendar** The calendar with 365-day years and leap years every fourth year without exception; the predecessor to the Gregorian calendar.
**Julian day** The number of days since noon on January 1, 713 B.C. Used for keeping track of variable stars or other astronomical events. January 1, 2000, noon, began Julian day 2,451,545.

**Keplerian** Following Kepler's law.
**kinetic temperature** A measure of the average random speeds of particles in a gas.
**K line** The spectral line of ionized calcium at 3933 Å.
**Kuiper belt** A reservoir of perhaps hundreds of thousands of solar system objects, each hundreds of kilometers in diameter, orbiting the sun outside the orbit of Neptune. It is the source of short-period comets.
**Kuiper belt object** An object in the Kuiper belt; many comets and perhaps even the planet Pluto are examples.

**large-scale structure** The network of filaments and voids or other shapes distinguished when studying the universe on the largest scales of distance.
**laser** An acronym for "light amplification by stimulated emission of radiation," a device by which certain energy levels are populated by more electrons than normal, resulting in an especially intense emission of light at a certain frequency when the electrons drop to a lower energy level.

**latitude** Number of degrees north or south of the equator measured from the center of a coordinate system.
**law of equal areas** Kepler's second law.
**law of inertia** Newton's first law of motion, which states that a body at rest tends to remain at rest while a body in motion continues in its motion unless acted on by a force.
**law of universal gravitation** Newton's law of gravitation, operating throughout the universe, that the force of gravitation between two objects is equal to a constant (written $G$) times the product of the masses of the two objects and divided by the square of their mutual distance.
**leap year** A year in which a 366th day is added.
**lens** A device that focuses waves by refraction.
**lenticular** A galaxy of type S0.
**libration** The effect by which we can see slightly more than half the lunar surface even though the moon basically has one half that always faces us; the back-and-forth motion of an object around one of the stable points in the three-body gravitational problem.
**light** Electromagnetic radiation between about 3000 and 7000 Å.
**light cone** The cone on a space-time diagram representing regions that can be in contact, given that nothing can travel faster than the speed of light.
**light curve** The graph of the magnitude of an object vs. time.
**lighthouse model** The explanation of a pulsar as a spinning neutron star whose beam we see as it comes around.
**light pollution** Excess light in the sky.
**lightweight stars** Stars between about 0.07 and 4 solar masses.
**light-year** The distance that light travels in a year.
**limb** The edge of a star or planet.
**limb darkening** The decreasing brightness of the disk of the sun or another star as one looks from the center of the disk closer and closer to the limb.
**line profile** The graph of the intensity of radiation vs. wavelength for a spectral line.
**lithosphere** The crust and upper mantle of a planet.
**lobes** Of a radio source, the regions to the sides of the center from which high-energy particles are radiating.
**local** In our region of the universe.
**Local Group** The two dozen or so galaxies, including the Milky Way Galaxy, that form a subcluster.
**local standard of rest** The system in which the average velocity of nearby stars is zero. We usually refer measurements of velocity of distant objects to the local standard of rest.
**Local Supercluster** The supercluster of galaxies in which the Virgo Cluster, the Local Group, and other clusters reside.
**logarithmic** A scale in which equal intervals stand for multiplying by ten or some other base, as opposed to linear, in which increases are additive.
**longitude** The angular distance around a body measured along the equator from some particular point; for a point not on the equator, it is the angular distance along the equator to a great circle that passes through the poles and through the point.
**long-period variables** Mira variables.
**look back time** The duration over which light from an object has been travelling to reach us.
**luminosity** The total amount of energy given off by an object per unit time.
**luminosity class** Different regions of the H-R diagram separating objects of the same spectral type: supergiants (I), bright giants (II), giants (III), subgiants (IV), dwarfs (V).
**lunar eclipse** The passage of the moon into the earth's shadow.
**lunar occultation** An occultation by the moon.
**lunar soils** Dust and other small fragments on the lunar surface.
**Lyman alpha** The spectral line (1216 Å) that corresponds to a transition between the two lowest major energy levels of a hydrogen atom.
**Lyman-alpha forest** The many Lyman-alpha lines, each differently Doppler-shifted, visible in the spectra of some quasars.
**Lyman lines** The spectral lines that correspond to transitions to or from the lowest major energy level of a hydrogen atom.

**$M_{\odot}$** Solar mass; the mass of the sun, used as a unit of measurement. For example, a star with 5 $M_{\odot}$ has 5 times the mass of the sun.
**MACHOs** *M*assive *C*ompact *H*alo *O*bjects—like dim stars, brown dwarfs, or black holes—that could account for some of the dark matter. The name was chosen to contrast with WIMPs.
**Magellanic Clouds** Two small irregular galaxies, satellites of the Milky Way Galaxy, visible in the southern sky; the Small Magellanic Cloud may be split.
**magnetic field lines** Directions mapping out the direction of the force between magnetic poles; the packing of the lines shows the strength of the force.
**magnetic lines of force** See *magnetic field lines.*
**magnetic mirror** A situation in which magnetic lines of force meet in a way that they reflect charged particles.
**magnetic monopole** A single magnetic charge of only one polarity; may or may not exist.
**magnetosphere** A region of magnetic field around a planet.
**magnification** An apparent increase in angular size.
**magnitude** A factor of $\sqrt[5]{100}$ = 2.511886. . . in brightness. See *absolute magnitude* and *apparent magnitude.* An *order of magnitude* is a power of ten.
**magnitude scale** The scale of apparent magnitudes and absolute magnitudes used by astronomers, in which each factor of 100 in brightness corresponds to a difference of 5 magnitudes.
**main sequence** A band on a Hertzsprung-Russell diagram in which stars fall during the main, hydrogen-burning phase of their lifetimes.
**major axis** The longest diameter of an ellipse; the line from one side of an ellipse to the other that passes through the foci. Also, the length of that line.
**mantle** The shell of rock separating the core of a differentiated planet from its thin surface crust.
**MAP** *M*icrowave *A*nisotropy *P*robe, NASA's 2001 spacecraft for mapping the anisotropy of the cosmic background radiation.
**mare (plural: maria)** One of the smooth areas on the moon or on some of the other planets.

**mascon** A concentration of mass under the surface of the moon, discovered from its gravitational effect on spacecraft orbiting the moon.

**maser** An acronym for "microwave amplification by stimulated emission of radiation," a device by which certain energy levels are more populated than normal, resulting in an especially intense emission of radio radiation at a certain frequency when the system drops to a lower energy level.

**mass** A measure of the inherent amount of matter in a body.

**mass loss** The flow of mass out of stars as they evolve brings matter to interstellar space.

**mass-luminosity relation** A well-defined relation between the mass and luminosity for main-sequence stars.

**mass number** The total number of protons and neutrons in a nucleus.

**Maunder minimum** The period 1645–1715, when there were very few sunspots, and no periodicity, visible.

**MAXIMA** *M*illimeter *A*nisotropy E*x*periment *Im*aging *A*rray, an experiment that mapped the cosmic background radiation at high resolution from a balloon over the United States.

**mean solar day** A solar day for the "mean sun," which moves at a constant rate during the year.

**meridian** The great circle on the celestial sphere that passes through the celestial poles and the observer's zenith.

**mesosphere** A middle layer on the earth's atmosphere, where the temperature again rises above the stratosphere's decline.

**Messier numbers** Numbers of non-stellar objects in the 18th-century list of Charles Messier.

**metal** (a) For stellar abundances, any element higher in atomic number than 2, that is, heavier than helium. (b) In general, neutral matter that is a good conductor of electricity.

**meteor** A track of light in the sky from rock or dust burning up as it falls through the earth's atmosphere.

**meteorite** An interplanetary chunk of rock after it impacts on a planet or moon, especially on the Earth.

**meteoroid** An interplanetary chunk of rock smaller than an asteroid.

**meteor shower** The occurrence at yearly intervals of meteors at a higher-than-average rate as the Earth goes through a comet's orbit and the dust left behind by that comet.

**micrometeorite** A tiny meteorite. The micrometeorites that hit the Earth's surface are sufficiently slowed down that they can reach the ground without being vaporized.

**mid-Atlantic ridge** The spreading of the sea floor in the middle of the Atlantic Ocean as upwelling material forces the plates to move apart.

**middleweight stars** Stars between about 4 and 8 solar masses.

**midnight sun** The sun seen around the clock from locations sufficiently far north or south at the suitable season.

**Milky Way** The band of light across the sky from the stars and gas in the plane of the Milky Way Galaxy.

**Milky Way Galaxy** The collection of gas, dust, and perhaps 400 billion stars in which we live; the Milky Way is our view of the plane of the Milky Way Galaxy.

**mini black hole** A black hole the size of a pinhead (and thus the mass of an asteroid) or less, thought to be left over from the big bang; Stephen Hawking deduced that such mini black holes should seem to emit radiation at a rate that allows a temperature to be assigned to it.

**minor axis** The shortest diameter of an ellipse; the line from one side of an ellipse to the other that passes midway between the foci and is perpendicular to the major axis. Also, the length of that line.

**minor planets** Asteroids.

**Mira variable** A long-period variable star similar to Mira (omicron Ceti).

**missing-mass problem** The discrepancy between the mass visible and the mass derived from calculating the gravity acting on members of clusters of galaxies.

**molecule** Bound atoms that make the smallest collection that exhibits a certain set of chemical properties.

**momentum** A measure of the tendency that a moving body has to keep moving. The momentum in a given direction (the "linear momentum") is equal to the mass of the body times its component of velocity in that direction. See also *angular momentum.*

**mountain ranges** Sets of mountains on the Earth, moon, etc.

**multiverse** The set of parallel universes that may exist, with our observable universe as only one part.

**naked singularity** A black hole's singularity that is not clothed by an event horizon, a situation that may never exist.

**Nasmyth focus** A focus of an alt-azimuth reflecting telescope in which a tertiary mirror reflects light from the secondary mirror out to the side through the mounting axis of the telescope. Since the Nasmyth foci (one on each side) remain stationary as the telescope moves, heavy equipment can relatively easily be mounted there.

**neap tides** The tides when the gravitational pulls of sun and moon are perpendicular, making them relatively low.

**Near-Earth object** A comet or asteroid whose orbit is sufficiently close to Earth's that a collision is possible or even likely in the long term. See *Amor asteroids, Apollo asteroids.*

**nebula** (plural: **nebulae**) Interstellar regions of dust or gas.

**nebular hypothesis** The particular nebular theory for the formation of the solar system advanced by Laplace.

**nebular theories** The theories that the sun and the planets formed out of a cloud of gas and dust.

**negative hydrogen ion** A hydrogen atom with an extra electron.

**neutrino** A spinning, neutral elementary particle with little or no rest mass, formed in certain radioactive decays.

**neutron** A massive, neutral elementary particle, one of the fundamental constituents of an atom.

**neutron degeneracy** A state in which, following rules of quantum mechanics, the further compression of neutrons generates a high pressure that balances gravity.

**neutron star** A star that has collapsed to the point where it is supported against gravity by neutron degeneracy.

**New General Catalogue** *A New General Catalogue of Nebulae and Clusters of Stars* by J. L. E. Dreyer, 1888.

**new moon** The phase when the side of the moon facing the Earth is the side that is not illuminated by sunlight.

**Newtonian (telescope)** A reflecting telescope where the beam from the primary mirror is reflected by a flat secondary mirror to the side.
**N galaxy** A galaxy (probably elliptical) with a blue nucleus that dominates the galaxy's radiation. The emission lines are generally broader than those from Seyferts.
**NGC** See *New General Catalogue*.
**nonthermal radiation** Radiation that cannot be characterized by a single number (the temperature). Normally, we derive this number from Planck's law, so that radiation that does not follow Planck's law is called nonthermal.
**nova (plural: novae)** A star that suddenly increases in brightness; an event in a binary system when matter from the giant component falls on the white dwarf component.
**nuclear bulge** The central region of spiral galaxies.
**nuclear burning** Nuclear fusion.
**nuclear force** The strong force, one of the fundamental forces.
**nuclear fusion** The amalgamation of lighter nuclei into heavier ones.
**nucleosynthesis** The formation of the elements.
**nucleus (plural: nuclei)** (a) Of an atom, the core, which has a positive charge, contains most of the mass, and takes up only a small part of the volume; (b) of a comet, the chunks of matter, no more than a few km across, at the center of the head; (c) of a galaxy, the innermost region.

**O and B association** A group of O and B stars close together.
**objective** The principal lens or mirror of an optical system.
**oblate** With equatorial diameter greater than the polar diameter.
**Occam's razor** The principle of simplicity, from the medieval philosopher William of Occam (approximately A.D. 1285–1349): the simplest explanation for all the facts will be accepted.
**occultation** The hiding of one astronomical body by another.
**Olbers's paradox** The observation that the sky is dark at night contrasted to a simple argument that shows that the sky should be uniformly bright.
**one atmosphere** The air pressure at the Earth's surface.
**one year** The length of time the Earth takes to orbit the sun.
**Oort comet cloud** The trillions of incipient comets surrounding the solar system in a 50,000 A.U. sphere.
**open cluster** A galactic cluster, a type of star cluster.
**open universe** A big-bang cosmology in which the universe has infinite volume and will expand forever.
**opposition** An object's having a celestial longitude 180° from that of the sun.
**optical** In the visible part of the spectrum, 3900–6600 Å, or having to do with reflecting or refracting that radiation.
**optical double** A pair of stars that appear extremely close together in the sky even though they are at different distances from us and are not physically linked.
**organic** Containing carbon in its molecular structure.
**Orion Molecular Cloud** The giant molecular cloud in Orion behind the Orion nebula, containing many young objects.
**oscillating universe** The big-bang model in which expansion is followed by collapse to a big crunch and then by expansion again in a continual series of big bangs.
**Ozma** One of two projects that searched nearby stars for radio signals from extraterrestrial civilizations.
**ozone layer** A region in the earth's upper stratosphere and lower mesosphere where $O_3$ absorbs solar ultraviolet radiation.

**pancake model** A model of galaxy formation in which large flat structures exist and become clusters of galaxies.
**paraboloid** A three-dimensional surface formed by revolving a parabola around its axis.
**parallax** (a) Trigonometric parallax, half the angle through which a star appears to be displaced when the earth moves from one side of the sun to the other (2 A.U.); it is inversely proportional to the distance; (b) other ways of measuring distance, as in spectroscopic parallax.
**parallel light** Light that is neither converging nor diverging.
**parent molecules** The more complex molecules from which observed molecules were formed, as in a comet for which we see simple molecules that presumably came from a parent that contained more atoms.
**parsec** The distance from which 1 A.U. subtends one second of arc (approximately 3.26 light-years).
**particle physics** The study of elementary nuclear particles.
**Pauli exclusion principle** The quantum-mechanical rule that certain types of elementary particles cannot exist in completely identical states.
**peculiar galaxies** A galaxy with some marked deviation from the normal shapes of the Hubble classification.
**peculiar motion** The motion of a star with respect to the local standard of rest.
**penumbra** (a) For an eclipse, the part of the shadow from which the sun is only partially occulted; (b) of a sunspot, the outer region, not as dark as the umbra.
**perfect cosmological principle** The assumption that on a large scale the universe is homogeneous and isotropic in space and unchanging in time.
**periastron** The near point of the orbit of a body to the star around which it is orbiting.
**perihelion** The near point to the sun of the orbit of a body orbiting the sun.
**period** The interval over which something repeats.
**phase** (a) Of a planet, the varying shape of the lighted part of a planet or moon as seen from some vantage point; (b) the relation of the variations of a set of waves.
**phase transitions** Changes from one state of matter to another, as from solid to liquid or liquid to gas; phase transitions in the early universe marked the separation of the fundamental forces.
**photometry** The electronic measurement of the amount of light.
**photomultiplier** An electronic device that through a series of internal stages multiplies a small current that is given off when light is incident on it; a large current results.
**photon** A packet of energy that can be thought of as a particle travelling at the speed of light.
**photon sphere** The sphere around a black hole, $3/2$ the size of the event horizon, within which exit cones open and in which light can orbit.

**photosphere** The region of a star from which most of its light is radiated.
**pixel** Picture element, the smallest individual spot on a detector such as a CCD on which an image is formed.
**plage** The part of a solar active region that appears bright when viewed in H$\alpha$.
**Planck** The European Space Agency's spacecraft, planned for 2007, to map the anisotropy of the cosmic background radiation, improving on earlier observations with MAP.
**Planck's constant** The constant of proportionality between the frequency of an electromagnetic wave and the energy of an equivalent photon. $E = h\nu = hc/\lambda$.
**Planck's law** The formula that predicts, for gas at a certain temperature, how much radiation there is at every wavelength.
**Planck time** The time very close to the big bang, $10^{-43}$ seconds, before which a quantum theory of gravity would be necessary to explain the universe and which is therefore thus currently inaccessible to our computations.
**planet** A celestial body of substantial size (more than about 1000 km across), basically non-radiating and of insufficient mass for nuclear reactions ever to begin, ordinarily in orbit around a star.
**planetary nebulae** Shells of matter ejected by low-mass stars after their main-sequence lifetime, ionized by ultraviolet radiation from the star's remaining core.
**planetesimal** One of the small bodies into which the primeval solar nebula condensed and from which the planets formed.
**plasma** An electrically neutral gas composed of approximately equal numbers of ions and electrons.
**plates** Large flat structures making up a planet's crust.
**plate tectonics** The theory of the Earth's crust, explaining it as plates moving because of processes beneath.
**plumes** Thin structures in the solar corona near the poles.
**point objects** Objects in which no size is distinguishable.
**polar axis** The axis of an equatorial telescope mounting that is parallel to the Earth's axis of rotation.
**pole star** A star approximately at a celestial pole; Polaris is now the pole star; there is no south pole star.
**poor cluster** A cluster of stars or galaxies with few members.
**Population I** The class of stars set up by Walter Baade to describe the younger stars typical for the spiral arms. These stars have relatively high abundances of metals.
**Population II** The class of stars set up by Walter Baade to describe the older stars typical of the galactic halo. These stars have very low abundances of metals.
**positive ion** An atom that has lost one or more electrons.
**positron** An electron's antiparticle (charge of +1).
**precession** The slowly changing position of stars in the sky resulting from variations in the orientation of the Earth's axis.
**precession of the equinoxes** The slow variation of the position of the equinoxes (intersections of the ecliptic and celestial equator) resulting from variations in the orientation of the Earth's axis.
**precession of the solstice** The slow variation of the position of the solstices. (See *precession of the equinoxes.*)
**pre-main-sequence star** A ball of gas in the process of slowly contracting to become a star as it heats up before beginning nuclear fusion.
**pressure** Force per unit area.
**principle of equivalence** The idea from Einstein's general theory of relativity (1916) that gravity and accelerated motion are fundamentally equivalent and that their effects, therefore, cannot be told apart.
**primary cosmic rays** The cosmic rays arriving at the top of the Earth's atmosphere.
**primary distance indicators** Ways of measuring distance directly, as in trigonometric parallax.
**prime focus** The location at which the main lens or mirror of a telescope focuses an image without being reflected or refocused by another mirror or other optical element.
**primeval solar nebula** The early stage of the solar system in nebular theories.
**primordial background radiation** Isotropic millimeter and submillimeter radiation following a black-body curve for about 3 K; interpreted as a remnant of the big bang.
**primordial nucleosynthesis** The formation of the nuclei of isotopes of hydrogen (such as deuterium), helium, and lithium in the first 10 minutes of the universe.
**primum mobile** In Ptolemaic and Aristotelian theory, the outermost sphere around the earth, which gave its natural motion to inner spheres.
**principal quantum number** The integer $n$ that determines the main energy levels in an atom.
**prograde motion** The apparent motion of the planets when they appear to move forward (from west to east) with respect to the stars. (See also *retrograde motion.*)
**prolate** Having the diameter along the axis of rotation longer than the equatorial diameter.
**prominence** Solar gas protruding over the limb, visible to the naked eye only at eclipses but also observed outside the eclipses by its emission-line spectrum.
**proper motion** Angular motion across the sky with respect to a framework of galaxies or fixed stars.
**proteins** Long chains of amino acids; fundamental components of all cells of life as we know it.
**proton** Elementary particle with positive charge of 1; one of the fundamental constituents of an atom.
**proton-proton chain** A set of nuclear reactions by which four hydrogen nuclei combine one after the other to form one helium nucleus, with a resulting release of energy.
**protoplanets** The loose collections of particles from which the planets formed.
**protosun** The sun in formation.
**pulsar** A celestial object that gives off pulses of radio waves.

**$q_0$** The deceleration parameter, a cosmological parameter that describes the rate at which the expansion of the universe is slowing up.
**QSO** Quasi-stellar object; a radio-quiet version of a quasar.
**quantized** Divided into discrete parts.
**quantum** A bundle of energy.
**quantum fluctuation** The spontaneous formation and disappearance of virtual particles; it does not violate the law of conservation of energy because of Heisenberg's Uncertainty Principle.
**quantum mechanics** The branch of 20th-century physics that describes atoms and radiation.

**quantum theory** The set of theories that evolved in the early 20th century to explain atoms and radiation as limited to discrete energy levels or quanta of energy, with quantum mechanics emerging as a particular version.

**quark** One of the subatomic particles of which modern theoreticians believe such elementary particles as protons and neutrons are composed. The various kinds of quarks have positive or negative charges of $\frac{1}{3}$ or $\frac{2}{3}$.

**quasar** One of the very-large-redshift objects that are almost stellar (point-like) in appearance.

**quiescent prominence** A long-lived and relatively stationary prominence.

**quiet sun** The collection of solar phenomena that do not vary with the solar activity cycle.

**radar** The acronym for "radio detection and ranging," an active rather than passive radio technique in which radio signals are transmitted and their reflections received and studied.

**radial velocity** The velocity of an object along a line (the radius) joining the object and the observer; the component of velocity toward or away from the observer.

**radiant** The point in the sky from which all the meteors in a meteor shower appear to be coming.

**radiation** Electromagnetic radiation. Sometimes also particles such as alpha (helium nuclei) or beta (electrons).

**radiation belts** Belts of charged particles surrounding planets.

**radioactive** Having the property of spontaneously changing into another isotope or element.

**radio galaxy** A galaxy that emits radio radiation orders of magnitude stronger than that from normal galaxies.

**radio telescope** An antenna or set of antennas, often together with a focusing reflecting dish, that is used to detect radio radiation from space.

**radio waves** Electromagnetic radiation with wavelengths longer than about one millimeter.

**recombination** A free electron joining or rejoining an ion to make a neutral atom or a lesser ion. The era of recombination, when hydrogen atoms formed some 300,000 years after the big bang, set free the cosmic background radiation.

**reddened** See *reddening.*

**reddening** The phenomenon by which the extinction of blue light by interstellar matter is greater than the extinction of red light so that the redder part of the continuous spectrum is relatively enhanced.

**red giant** A post-main-sequence stage of the lifetime of a star; the star becomes relatively bright and cool.

**redshifted** A shift of optical wavelengths toward the red or in all cases toward longer wavelengths.

**red supergiant** Extremely bright, cool, and large stars; a post-main-sequence phase of evolution of stars of more than about 4 solar masses.

**reflecting telescope** A type of telescope that uses a mirror or mirrors to form the primary image.

**reflection** The bouncing off a surface by an object or wave.

**reflection nebula** Interstellar gas and dust that we see because it is reflecting light from a nearby star.

**refracting telescope** A type of telescope in which the primary image is formed by a lens or lenses.

**refraction** The bending of electromagnetic radiation as it passes from one medium to another or between parts of a medium that has varying properties.

**refractory** Having a high melting point.

**regolith** A planet's or moon's surface rock disintegrating into smaller particles.

**regular cluster** A cluster of galaxies with spherical symmetry and a central concentration.

**relativistic** Having a velocity that is such a large fraction of the speed of light that the special theory of relativity must be applied.

**resolution** The ability of an optical system to distinguish detail.

**rest mass** The mass an object would have if it were not moving with respect to the observer.

**rest wavelength** The wavelength radiation would have if its emitter were not moving with respect to the observer.

**retrograde motion** The apparent motion of the planets when they appear to move backwards (westward) with respect to the stars from the direction that they move ordinarily.

**retrograde rotation** The rotation of a moon or planet opposite to the dominant direction in which the sun rotates and the planets orbit and rotate.

**revolution** The orbiting of one body around another.

**revolve** To move in an orbit around another body.

**rich cluster** A cluster of many galaxies.

**ridges** Raised surface features on the moon, apparently volcanic, perhaps having developed on the crust of lava lakes, as volcanic vents, or from faulting; Mercury also has ridges.

**right ascension** Celestial longitude, measured eastward along the celestial equator in hours of time from the vernal equinox.

**rilles** Sinuous depressions on the lunar surface, apparently volcanic, perhaps huge collapsed lava tubes or lava channels.

**rims** The raised edges of craters.

**Roche limit** The sphere for each mass inside of which blobs of gas cannot agglomerate by gravitational interaction without being torn apart by tidal forces; normally about $2\frac{1}{2}$ times the radius of a planet.

**rotate** To spin on one's own axis.

**rotation** Spin on an axis.

**rotation curve** A graph of the speed of rotation vs. distance from the center of a rotating object like a galaxy.

**RR Lyrae variable** A short-period "cluster" variable. All RR Lyrae stars have approximately equal absolute magnitude and so are used to determine distances.

**S0** A transition type of galaxy between ellipticals and spirals; has a disk but no arms; all E7s are now known to be S0s.

**scarps** Lines of cliffs; found on Mercury, Earth, the moon, and Mars.

**scattered** Light absorbed and then re-emitted in all directions.

**Schmidt telescope** ***(Schmidt camera)*** A telescope that uses a spherical mirror and a thin lens to provide photographs of a wide field.

**Schwarzschild radius** The radius that, according to Schwarzschild's solutions to Einstein's equations of the general theory of relativity, corresponds to the event horizon of a black hole.

**scientific method** No easy definition is possible, but it has to do with a way of testing and verifying hypotheses.

**scientific notation** Exponential notation.

**scintillation** A flickering of electromagnetic radiation caused by moving volumes of intermediary gas.

**secondary cosmic rays** High-energy particles generated in the Earth's atmosphere by primary cosmic rays.

**secondary distance indicators** Ways of measuring distances that are calibrated by primary distance indicators.

**sector** Part of a circle bounded by an arc and two radii.

**sedimentary** Formed from settling in a liquid or from deposited material, as for a type of rock.

**seeing** The steadiness of the Earth's atmosphere as it affects the resolution that can be obtained in astronomical observations.

**seismic waves** Waves travelling through a solid planetary body from an earthquake or impact.

**seismology** The study of waves propagating through a body and the resulting deduction of the internal properties of the body. "Seismo-" comes from the Greek for earthquake.

**semimajor axis** Half the major axis, that is, for an ellipse, half the longest diameter.

**semiminor axis** Half the minor axis, that is, for an ellipse, half the shortest diameter.

**Seyfert galaxy** A type of spiral galaxy that has a bright nucleus and whose spectrum shows broad emission lines that cover a wide range of ionization stages.

**shadow bands** A phenomenon seen just before and just after total phases of solar eclipses in which low-contrast bands of light and dark race across the landscape. Shadow bands are caused by the Earth's atmospheric turbulence affecting light from the thin solar crescent.

**Shapley-Curtis debate** The 1920 debate (and its written version) on the scale of our galaxy and of "spiral nebulae."

**shear wave** A type of twisting seismic wave.

**shock wave** A front marked by an abrupt change in pressure caused by an object moving faster than the speed of sound in the medium through which the object is travelling.

**shooting stars** Meteors.

**showers** A time of many meteors from a common cause.

**sidereal** With respect to the stars.

**sidereal day** A day with respect to the stars.

**sidereal rotation period** A rotation with respect to the stars.

**sidereal time** The hour angle of the vernal equinox; equal to the right ascension of objects on your meridian.

**sidereal year** A circuit of the sun with respect to the stars.

**significant figure** A digit in a number that is meaningful (within the accuracy of the data).

**singularity** A point in space where quantities become exactly zero or infinitely large; one is present in a black hole.

**SIRTF** NASA's *S*pace *I*nfrared *T*elescope *F*acility (2002).

**slit** A long, thin gap through which light is allowed to pass.

**SOFIA** NASA's *S*tratospheric *O*bservatory *f*or *I*nfrared *A*stronomy, an airplane instrumented with a 2.5-m telescope, scheduled to start flying in 2004.

**SOHO** The *S*olar and *H*eliospheric *O*bservatory, a joint NASA and European Space Agency mission to study the sun. It is 1 million kilometers toward the sun from Earth, a location from which it can view the sun constantly with its dozen instruments, which include coronagraphs, cameras with filters that show the corona, and particle detectors.

**solar-activity cycle** The set of solar phenomena including sunspots, flares, prominences, coronal shape, and so on that depend on the sun's magnetic field and so vary with the same period.

**solar atmosphere** The photosphere, chromosphere, and corona.

**solar constant** The total amount of energy that would hit each square centimeter of the top of the Earth's atmosphere at the Earth's average distance from the sun.

**solar day** A full rotation with respect to the sun.

**solar dynamo** The generation of sunspots by the interaction of convection, turbulence, differential rotation, and magnetic field.

**solar flares** An explosive release of energy on the sun.

**solar rotation period** The time for a complete rotation with respect to the sun.

**solar seismology** (also, **helioseismology**) The study of the sun's interior through studies of waves measured with observations of surface variations.

**solar time** A system of timekeeping with respect to the sun such that the sun is overhead of a given location at noon.

**solar wind** An outflow of particles from the sun representing the expansion of the corona.

**solar year** (also **tropical year**) An object's complete circuit of the sun; a tropical year in between vernal equinoxes.

**solstice** The point on the celestial sphere of northernmost or southernmost declination of the sun in the course of a year; colloquially, the time when the sun reaches that point.

**space velocity** The velocity of a star with respect to the sun.

**space weather** Solar activity affecting interplanetary space and, in particular, the Earth and objects orbiting it.

**spallation** The break-up of heavy nuclei that undergo nuclear collisions.

**special theory of relativity** Einstein's 1905 theory of relative motion.

**speckle interferometry** A method obtaining higher resolution of an image by analysis of a rapid series of exposures that freeze atmospheric blurring.

**spectra** See *spectrum.*

**spectral classes** Spectral types.

**spectral lines** Wavelengths at which the intensity is abruptly different from intensity at neighboring wavelengths.

**spectral type** One of the categories O, B, A, F, G, K, M, C, or S into which stars can be classified from study of their spectral lines, or extensions of this system. The sequence of spectral types corresponds to a sequence of temperature.

**spectrograph** A device to make and record a spectrum.

**spectrometer** A device to make and electronically measure a spectrum.

**spectrophotometer** A device to measure intensity at given wavelength bands.

**spectroscope** A device to make and look at a spectrum.

**spectroscopic binary** A type of binary star that is known to have more than one component because of the changing Doppler shifts of the spectral lines that are observed.

**spectroscopic parallax** The distance to a star derived by comparing its apparent magnitude with its absolute magnitude deduced from study of its position on an H-R diagram (determined by observing its spectrum—spectral type and luminosity class).

**spectroscopy** The use of spectrum analysis.

**spectrum (plural: spectra)** A display of electromagnetic radiation spread out by wavelength or frequency.

**speed of light** By Einstein's special theory of relativity, the velocity at which all electromagnetic radiation travels, and the largest possible velocity of an object.

**spherical aberration** For an optical system, a deviation from perfect focusing by having a shape that is too close to that of a sphere.

**spicule** A small jet of gas at the edge of the quiet sun, approximately 1000 km in diameter and 10,000 km high, with a lifetime of about 15 minutes.

**spin-flip** A transition in the relative orientation of the spins of an electron and the nucleus it is orbiting.

**spiral arms** Bright regions looking like a pinwheel.

**spiral density wave** A wave that travels around a galaxy in the form of a spiral, compressing gas and dust to relatively high density and thus beginning star formation at those compressions.

**spiral galaxy** A class of galaxy characterized by arms that appear as though they are unwinding like a pinwheel.

**spiral nebula** The old name for a shape in the sky, seen with telescopes, with arms spiraling outward from the center; proved in the 1920s actually to be a galaxy.

**sporadic** Not regularly.

**sporadic meteor** A meteor not associated with a shower.

**spring tides** The tides at their highest, when the Earth, moon, and sun are in a line (from "to spring up").

**stable** Tending to remain in the same condition.

**star** A self-luminous ball of gas that shines or has shone because of nuclear reaction in its interior.

**starburst galaxy** A galaxy in which a region of high star-forming activity is seen, often detected through strong infrared radiation.

**star clouds** The regions of the Milky Way where the stars are so densely packed that they cannot be seen as separate.

**star clusters** Groupings of stars of common origin. The main types are globular clusters, which have spherical shape, and galactic clusters (also known as open clusters), which don't.

**stationary limit** In a rotating black hole, the location where space-time is flowing at the speed of light, making stationary particles that would be travelling at that speed.

**steady-state theory** The cosmological theory based on the perfect cosmological principle, in which the universe is unchanging over time.

**Stefan-Boltzmann law** The radiation law that states that the energy emitted by a black body varies with the 4th power of the temperature.

**stellar atmosphere** The outer layers of stars not completely hidden from our view.

**stellar chromosphere** The region above a photosphere that shows an increase in temperature.

**stellar corona** The outermost region of a star characterized by temperatures of $10^6$ K and high ionization.

**stellar evolution** The changes of a star's properties with time.

**stelliferous era** The current era, with lots of stars in the universe.

**stones** A stony type of meteorite, including the chondrites.

**stony-iron meteorite** A type of meteorite including properties of both stoney meteorites (that is, silicates) and iron-nickel meteorites. Pallasites are the dominant example.

**stratosphere** An upper layer of a planet's atmosphere, above the weather, where the temperature begins to increase. The Earth's stratosphere is at 20–50 km.

**streamers** Coronal structures at low solar latitudes.

**string theories** See *superstring theories.*

**strong force** The nuclear force, the strongest of the four fundamental forces of nature.

**strong nuclear force** The strong force.

**S-type star** A red giant showing strong ZrO instead of TiO spectral lines.

**subtend** The angle that an object appears to take up in your field of view; for example, the full moon subtends ½°.

**sunspot** A region of the solar surface that is dark and relatively cool; it has an extremely high magnetic field.

**sunspot cycle** The 11-year cycle of variation of the number of sunspots visible on the sun.

**superbolt** Giant lightning.

**supercluster** A cluster of clusters of galaxies.

**supercooled** The condition in which a substance is cooled below the point at which it would normally make a phase change; for the universe, the point in the early universe at which it cooled below a certain temperature without breaking its symmetry; the strong and electroweak forces remained unified.

**supergiant** A post-main-sequence phase of evolution of stars of more than about 4 solar masses. They fall in the upper right of the H-R diagram; luminosity class I.

**supergranulation** Convection cells on the solar surface about 20,000 km across and vaguely polygonal in shape.

**supergravity** A theory unifying the four fundamental forces.

**superluminal speed** An apparent speed greater than that of light.

**supermassive black hole** A black hole of millions or billions of times the sun's mass, as is found in the centers of galaxies and quasars.

**supermassive star** A stellar body of more than about 100 $M_\odot$.

**supernova (plural: supernovae)** The explosion of a star with the resulting release of tremendous amounts of radiation.

**supernova-chain-reaction model** The explanation of spiral structure in terms of a chain of supernova explosions, each one leading to more than one more.

**supernova remnants** The gaseous remainder of the star destroyed in a supernova.

**superstring theories** A possible unification of quantum theory and general relativity in which fundamental particles are really different vibrating forms of a tiny, one-dimensional "string" instead of being localized at single points.

**symmetric** A correspondence of shape so that rotating or reflecting an object gives you back an identical form; forces acting identically that now act differently as the four fundamental forces: gravity, electromagnetism, weak nuclear force, and strong nuclear force.

**synchronous orbit** An orbit of the same period; a satellite in geosynchronous orbit has the same period as the Earth's rotation and so appears to hover.

**synchronous rotation** A rotation of the same period as an orbiting body.

**synchronous satellites** Satellites in synchronous orbits, such as those appearing to hover over the equator of the Earth and used for broadcasting and other purposes.

**synchrotron emission** Polarized emission given off as an electron spirals at relativistic velocities around magnetic field lines. Supernova remnants, such as the Crab Nebula, and radio galaxies are powerful sources of synchrotron emission.

**synchrotron radiation** Nonthermal radiation emitted by electrons spiralling at relativistic velocities in a magnetic field.

**synodic** Measured with respect to an alignment of three astronomical bodies.

**synodic month** The month determined by the lunar phases, 29½ days, which depends on the relative positions of Earth, sun, and moon.

**synodic revolution period** The period of revolution as determined by the return to the same relative positions by three astronomical bodies.

**synthetic-aperture radar** A radar mounted on a moving object, used with analysis taking the changing perspective into account to give the effect of a larger dish.

**syzygy** An alignment of three celestial bodies.

**tail** Gas and dust left behind as a comet orbits sufficiently close to the sun, illuminated by sunlight.

**T association** A grouping of several T Tauri stars, presumably formed out of the same interstellar cloud.

**tektites** Small glassy objects found scattered around the southern part of the southern hemisphere of the Earth.

**temperature-magnitude diagram** A diagram of a group of stars with temperatures on the horizontal axis and intrinsic brightness on the vertical axis; also called a Hertzsprung-Russell diagram or a temperature-luminosity diagram.

**terminator** The line between night and day on a moon or planet; the edge of the part that is lighted by the sun.

**terrestrial planets** Mercury, Venus, Earth, and Mars.

**tertiary distance indicators** Ways to measure distance that are calibrated by secondary distance indicators.

**theory** A later stage of the traditional form of the scientific method, in which a hypothesis has passed enough of its tests that it is generally accepted.

**thermal pressure** Pressure generated by the motion of particles that can be characterized by a temperature.

**thermal radiation** Radiation whose distribution of intensity over wavelength can be characterized by a single number (the temperature). Black-body radiation, which follows Planck's law, is thermal radiation.

**thermosphere** The uppermost layer of the atmosphere of the Earth and some other planets, the ionosphere, where absorption of high-energy radiation heats the gas.

**three-color photometry** Measurements through U, B, V filters.

**3° background radiation** The isotropic black-body radiation at 3 K, thought to be a remnant of the big bang.

**tidal lock** When the rotation of a body is linked to its period of revolution because the tidal force keeps one face of the body in some simple relation (1:1, 1:2, 1:3, etc.) with a more massive body.

**tidal theory** An outdated explanation of solar system formation in terms of matter being tidally drawn out of the sun by a passing star.

**tide** A periodic variation of the force of gravity on a body, based on the difference in the strength of gravitational pull from one place on the body to another (the tidal force); on Earth, the tide shows most obviously where the ocean meets the shore as a period variation of sea level.

**TRACE** The *T*ransition *R*egion *a*nd *C*oronal *E*xplorer, a solar satellite studying the solar corona and the transition region between the chromosphere and corona at high spatial resolution.

**transit** The passage of one celestial body in front of another celestial body. When a planet is *in transit,* we understand that it is passing in front of the sun. Also, *transit* is the moment when a celestial body crosses an observer's meridian, or the special type of telescope used to study such events.

**transition zone** The thin region between a chromosphere and a corona.

**Trans-Neptunian Object (TNO)** One of the sub-planetary objects in the Kuiper belt; it was proposed but not accepted that Pluto should be considered as one.

**transparency** Clarity of the sky.

**transverse velocity** Velocity along the plane of the sky.

**trigonometric parallax** See *parallax.*

**triple-alpha process** A chain of fusion processes by which three helium nuclei (alpha particles) combine to form a carbon nucleus.

**Trojan asteroids** A group of asteroids that precede or follow Jupiter in its orbit by 60°.

**tropical year** The length of time between two successive vernal equinoxes.

**troposphere** The lowest level of the atmosphere of the Earth and some other planets, in which all weather takes place.

**T Tauri star** A type of irregularly varying star, like T Tauri, whose spectrum shows broad and very intense emission lines. T Tauri stars have presumably not yet reached the main sequence and are thus very young.

**tuning-fork diagram** Hubble's arrangement of types of elliptical, spiral, and barred spiral galaxies.

**21-cm line** The 1420-MHz line from neutral hydrogen's spin-flip.

**twinkle** A scintillation—rapid changing in brightness—and slight changing in position of stars as their light passes through the Earth's atmosphere.

**Type Ia supernova** A supernova whose distribution in all types of galaxies, and the lack of hydrogen in its spectrum, make us think that it is an event in low-mass stars, probably resulting from the collapse and incineration of a white dwarf in a binary system.

**Type II supernova** A supernova associated with spiral arms, and that has hydrogen in its spectrum, making us think that it is the explosion of a massive star.

**types, Hubble** Shapes of galaxies as classified by Hubble, including E0 . . . E7, S0, Sa, Sb, Sc, SBa, SBb, SBc, Irr.

**types, spectral** Types of stellar spectra used for classification, including OBAFGKMLT.

**UBV system** A system of photometry that uses three standard filters to define wavelength regions in the ultraviolet, blue, and green-yellow (visual) regions of the spectrum.

**ultraviolet** The region of the spectrum 100–4000 Å, also used in the restricted sense of ultraviolet radiation that reaches the ground, namely, 3000–4000 Å.

**umbra (plural: umbrae)** (a) Of a sunspot, the dark central region; (b) of an eclipse shadow, the part from which the sun cannot be seen at all.

**uncertainty principle** Heisenberg's statement that the product of uncertainties of position and momentum is equal to Planck's constant. Consequently, both position and momentum cannot be known to infinite accuracy.

**universal gravitation constant** The constant $G$ of Newton's law of gravity: force = $Gm_1m_2/r^2$.

**uvby** A system of photometry that uses four standard filters to define wavelength regions in the ultraviolet, violet, blue, and yellow regions of the spectrum.

**vacuum** Space that may seem empty but is actually teaming with virtual particles and quantum fields that can supply a vacuum energy, perhaps in the form of a cosmological constant.

**valleys** Depressions in the landscapes of solid objects.

**Van Allen belts** Regions of high-energy particles trapped by the magnetic field of the Earth.

**variable star** A star whose brightness changes over time.

**vernal equinox** The equinox crossed by the sun as it moves to northern declinations.

**Very Large Array** The National Radio Astronomy Observatory's set of radio telescopes in New Mexico, used together for interferometry.

**Very Large Telescope** The set of instruments, including four 8.2-m reflectors and several smaller telescopes, erected by the European Southern Observatory on a mountaintop in Chile through 2001.

**Very-Long-Baseline Array** The National Radio Astronomy Observatory's set of radio telescopes dedicated to very-long-baseline interferometry and spread over an 8000-km baseline across the United States.

**very-long-baseline interferometry** The technique using simultaneous measurements made with radio telescopes at widely separated locations to obtain extremely high resolution.

**virial theorem** In the limited sense here, that half the gravitational energy of contraction goes into heating.

**visible light** Light to which the eye is sensitive, 3900–6600 Å.

**visual binary** A binary star that can be seen through a telescope to be double.

**VLA** See *Very Large Array.*

**VLBA** See *Very-Long-Baseline Array.*

**VLBI** See *very-long-baseline interferometry.*

**VLT** See *Very Large Telescope.*

**void** A giant region of the universe in which no galaxies are found.

**volatile** Evaporating (changing to a gas) readily.

**wave front** A plane in which parallel waves are in step.

**wavelength** The distance over which a wave goes through a complete oscillation.

**weak nuclear force** One of the four fundamental forces of nature, weaker than the strong force and the electromagnetic force. It is important only in the decay of certain elementary particles.

**weight** The force of the gravitational pull on a mass.

**white dwarf** The final stage of the evolution of a star of between 0.07 and 1.4 solar masses; a star supported by electron degeneracy. White dwarfs are found to the lower left of the main sequence of the H-R diagram.

**white light** All the light of the visible spectrum together.

**Wien's displacement law** The expression of the inverse relationship of the temperature of a black body and the wavelength of the peak of its emission.

**WIMPs** *W*eakly *I*nteracting *M*assive *P*articles, an as yet undiscovered massive particle, interacting with other elementary particles only through the weak nuclear force, that may be a major type of cold dark matter.

**winter solstice** For northern-hemisphere observers, the southernmost declination of the sun and its date.

**Wolf-Rayet star** A type of O star whose spectrum shows very broad emission lines.

**W Virginis star** A Type II Cepheid, a fainter class of Cepheid variables characteristic of globular clusters.

**x-rays** Electromagnetic radiation between 0.1 and 100 Å.

**year** The period of revolution of a planet around its central star; more particularly, the Earth's period of revolution around the sun.

**Zeeman effect** The splitting of certain spectral lines in the presence of a magnetic field.

**zenith** The point in the sky directly overhead an observer. Note that it is not related to celestial coordinates or to particular stars; it is merely at altitude 90° for any given person.

**zero-age main sequence** The curve on an H-R diagram determined by the locations of stars at the time they begin nuclear fusion.

**zodiac** The band of constellations through which the sun, moon, and planets move in the course of the year.

**zodiacal light** A glow in the nighttime sky near the ecliptic from sunlight reflected by interplanetary dust.

**zones** Bright bands in the clouds of a planet, notably Jupiter's.

# ILLUSTRATION ACKNOWLEDGMENTS

**Preface and Table of Contents—p. v** Roger Bell, University of Maryland, and Michael Briley, University of Wisconsin at Oshkosh; **p. vi** Next Generation Space Telescope: NASA/Goddard Space Flight Center; **p. vii** Jeff Hester and Paul Scowen (Arizona State University); **p. viii** Jay M. Pasachoff; **p. ix** NASA's Johnson Space Center; **pp. x and xi** European Southern Observatory; **p. xiii** NASA, ESA, and The Hubble Heritage Team (STScI/AURA), image by R. Sahai (Jet Propulsion Lab, Caltech) and B. Balick (U. Washington); **p. xiv** Canada-France-Hawaii Telescope/CCD from Gerard Luppino, Institute for Astronomy, University of Canada-France-Hawaii Telescope, and MIT Lincoln Labs. Courtesy of Jean-Charles Cuillandre; **p. xv** NASA and The Hubble Heritage Team (AURA/STScI), data collected by Nolan R. Walborn (STScI), Rodolfo H. Barbá (La Plata Observatory, Argentina), and Adeline Caulet (France); **p. xvi** NASA and the Hubble Heritage Team (AURA/STScI)/Raghvendra Sahai (JPL) and Arsen R. Hajian (U.S. Naval Observatory); **p. xvii** NASA and The Hubble Heritage Team (AURA/STScI)/Debra Meloy Elmegreen (Vassar College), Bruce G. Elmegreen (IBM Research Division), Michele Kaufman (Ohio State University), Elias Brinks (Universidad de Guanajuato, Mexico), Curt Struck (Iowa State University), Magnus Thomasson (Onsala Space Obs., Sweden), Maria Sundin (Göteborg U., Sweden), and Mario Klaric (Columbia, SC); **p. xviii** (*top*) photo © Andrew Perala; **p. xviii** (*bottom*) NASA and The Hubble Heritage Team (AURA/STScI), including images from William Keel and Ray White III (U. Alabama); **p. xix** (*top*) NASA/JPL/University of Arizona; **p. xix** (*bottom*) NASA/JPL/Caltech and Malin Space Science Systems; **p. xx** (*top*) Jay M. Pasachoff; **p. xx** (*bottom*) Canada-France-Hawaii Telescope/CCD from Gerard Luppino, Institute for Astronomy, University of Hawaii, Canada-France-Hawaii Telescope, and MIT Lincoln Labs; **p. xxi** © 1999 Jay M. Pasachoff and Wendy Carlos; **p. xxii** © 1999–2001 Subaru Telescope, National Astronomical Observatory of Japan. All rights reserved; **p. xxiii** (*top*) The Hubble Heritage Team (NASA/AURA/STScI) using data collected by A. Hajian (U.S. Naval Observatory), B. Balick (University of Washington), H. Bond and N. Panagia (STScI), and Y. Terzian (Cornell University); **p. xxiii** (*bottom*) CISCO, Subaru 8.3-m Telescope, NAOJ; **p. xxiv** (*top*) John N. Bahcall (Institute for Advanced Study, Princeton), Mike Disney (University of Wales), and NASA; **p. xxiv** (*bottom*) BOOMERanG Consortium; **p. xxv** N.A. Sharp, Vanessa Harvey/REU Program/AURA/NOAO/NSF.

**Part 1—Opener** This image was produced by the Hubble Heritage Team (NASA/AURA/STScI) using data collected by the Hubble Heritage Team with the assistance of N. Walborn (STScI), R. Barba (La Plata Obs., Argentina), and A. Caulet (France); **Part 1—Figure 1** 1 NASA/JPL/University of Arizona.

**Chapter 1—Opener** ESO; **Fig. 1–1** Jay M. Pasachoff; **Fig. 1–3** © 1995 David Scharf; **Fig. 1–4** Jay M. Pasachoff; **Fig. 1–5** Neil Leifer for Sports Illustrated; © AOL Time Warner, Inc.; **Fig. 1–6** Joseph Distefano, City of Boston, and Abrams Aerial Survey Corp.; **Figs. 1–7 and 8** NASA; **Fig. 1–9** NASA/JPL/Caltech; **Fig. 1–10** with NASA/JPL/Caltech insets; **Fig. 1–11** NASA/JPL/Caltech; **Fig. 1–14** Axel Mellinger, Universität Potsdam; **Figs. 1–15 and 16** Hubble Heritage Team (AURA/STScI/NASA); **Fig. 1–17** Margaret Geller and John Huchra, Harvard-Smithsonian Center for Astrophysics; **Fig. 1–18** NASA COBE Science Team; **Fig. 1–19** The "Changing Sun" mosaics were created by Greg Slater and Charlie Little of Lockheed Martin Solar and Astrophysics Lab (LMSAL) with images from the Yohkoh mission of ISAS, Japan. The X-ray telescope was prepared by the Lockheed-Martin Solar and Astrophysics Laboratory, the National Astronomical Observatory of Japan, and the University of Tokyo with the support of NASA and ISAS.

**Chapter 2—Opener** Jay M. Pasachoff; **Fig. 2–1** British Library; **Fig. 2–2** Geoff Chester (Einstein Planetarium, Smithsonian Institution), now at U.S. Naval Observatory; **Fig. 2–3** Photo Vatican Museums; **Fig. 2–4** Courtesy of the Houghton Library, Harvard University; **Fig. 2–5** Biblioteca Nazionale Marciana, Venice; **Fig. 2–7A** Bodleian Library, Oxford, Ms. Marsh 144, p. 326; **Fig. 2–7B** Judaica Collection, The Royal Library, Copenhagen; **Fig. 2–7C** Courtesy of the Library of the Jewish Theological Seminary of America; **Fig. 2–8** Catherine Meisel; **Fig. 2–9** Jay M. Pasachoff; **Fig. 2–10** Torun Museum, courtesy of Marek Demianski; **Fig. 2–11** Photograph by Charles Eames, reproduced courtesy of Eames Office and Owen Gingerich, © 1990, 1998 Lucia Eames dba Eames Office; **Fig. 2–12A** Jay M. Pasachoff; **Fig. 2–12B** Huntington Library, San Marino, CA; **Figs. 2–13, 2–15, p. 26** (*left*) Jay M. Pasachoff; **Fig. p. 26** (*right*) courtesy of Adler Planetarium & Astronomy Museum, Chicago, Illinois; **Figs. 2–16A, 2–17** to **2–20** Jay M. Pasachoff; **Fig. 2–16B** Friends of Carhenge; **Fig. 2–22** U.S. Forest Service; **Fig. 2–23** © The Solstice Project, photograph by Philip Johnson Tulawetstiwa.

**Chapter 3—Opener** Jay M. Pasachoff; p. 34 From *Atlas Coelestis* by J. Doppelmaier (1742), provided by Jay M. Pasachoff; **Figs. 3–1, 3–2** and **3–3** Jay M. Pasachoff; **Fig. 3–8** Jay M. Pasachoff; **Fig. 3–10** William C. Livingston, National Solar Observatory, AURA; **Fig. 3–11** National Maritime Museum, London; **Fig. 3–12AB** Jay M. Pasachoff, after Ewen Whitaker, Lunar and Planetary Laboratory, U. Arizona; **Fig. 3–13A** Williams College-Hopkins Observatory, photo by Stephan Martin; **Fig. 3–13B** Jay M. Pasachoff; **Fig. 3–14** U. Michigan Library, Dept. of Rare Books and Special Collections, translation by Stillman Drake, reprinted courtesy of *Scientific American;* **Fig. 3–15** Jay M. Pasachoff; **Fig. 3–16** Akira Fujii; **Fig. 3–17** drawn from a plot made with *Visible Universe;* **Fig. 3–19** National Portrait Gallery, London; **Figs. 3–20 to 3–24** Jay M. Pasachoff

**Chapter 4—Opener** NASA and the Hubble Heritage Team (STScI/AURA); **Fig. 4–5** Chris Jones, Union College; **Fig. 4–6** American Museum of Natural History; Jay M. Pasachoff photo; **Figs. 4–10, 4–12,** and **4–15** Jay M. Pasachoff; **Fig. 4–18** Institute for Astronomy, U. Hawaii; **Fig. 4–19** Reproduced from U.S. Air Force Defense Meteorological Satellite Program (DMSP); **Fig. 4–20** Jay M. Pasachoff; **Figs. 4–21AB** Dennis di Cicco; **Fig. 4–21C** Akira Fujii; **Fig. 4–23** Jay M. Pasachoff; **Fig. 4–27** NASA/IoA/A. Fabian *et al.;* **Fig. 4–28** courtesy of Andrew Lange and Eric Hivon, Caltech; **Fig. 4–29A** Debra Elmegreen, Vassar College; **Fig. 4–29** inset National Radio Astronomy Observatory; **Fig. 4–29B** Robert W. Wilson.

**Chapter 5—Opener** NASA/Chandra X-ray Observatory Center/SAO, courtesy of Martin Zombeck; **Fig. 5–1** Jay M. Pasachoff; **Fig. 5–2** NOAO; **Fig. 5–3** © 1999 Subaru Telescope, National Astronomical Observatory of Japan; **Fig. 5–4** photo © Andrew Perala; **Fig. 5–5A** Xinetics, Inc.; **Fig. 5–5B** Richard Dekany (JPL), Thomas Hayward, Bernhard Brandl, and Don Banfield (Cornell); the AO system was developed by a JPL team led by Richard Dekany: imaged at the Palomar Observatory; **Figs. 5–6** and **5–8** European Southern Observatory; **Fig. 5–7** Photo courtesy of Gemini Observatory; **Fig. 5–10** © 1987 and 1979, ROE/AAO; **Fig. 5–11** Raytheon Optical Systems; **Figs. 5–12** and **5–14** NASA's Johnson Space Center; **Fig. 5–13** STScI; **Fig. 5–15** Lockheed Martin; **Fig. 5–16** Digital Instruments; **Fig. 5–17** Gerard Luppino, Institute for Astronomy, University of Hawaii; CCD from University of Hawaii, Canada-France-Hawaii Telescope, and MIT Lincoln Labs; **Fig. 5–18** NASA and The Hubble Heritage Team (STScI/AURA); Dr. Raghvendra Sahai (JPL) and Dr. Arsen R. Hajian (USNO); **Fig. 5–20** University of Chicago

News Office; **Fig. 5–21** Fermilab Visual Media Services; **Fig. 5–22** Sloan Digital Sky Survey; **Fig. 5–23** FUSE project/Johns Hopkins U., NASA; **Fig. 5–AN-A** © Richard J. Wainscoat; **Fig. 5–AN-B** Jay M. Pasachoff; **Fig. 5–24B** Eastman Kodak; **Fig. 5–25** Atlas Image obtained as part of the Two Micron All Sky Survey (2MASS), a joint project of the University of Massachusetts and the Infrared Processing and Analysis Center/California Institute of Technology, funded by the National Aeronautics and Space Administration and the National Science Foundation; **Fig. 5–26** NASA/JPL/Caltech; **Fig. 5–28** National Radio Astronomy Observatory; **Figs. 5–29 and 5–30B; 5–30A** NRAO; Jay M. Pasachoff; **Fig. 5–30A** NRAO; **Fig. 5–31** Courtesy of Leo Blitz, U. California at Berkeley; **Fig. 5–32** Dennis Downes, IRAM; **Fig. 5–33** Jay M. Pasachoff; **Fig. 5–34** courtesy of Australian National Telescope Facility, CSIRO; image by Ben Simons, Sydney VisLab, The University of Sydney.

**Chapter 6—Opener** Raymond Pojman, Carbondale, CO; **Fig. 6–2A** Jay M. Pasachoff; **Fig. 6–2B** Von Del Chamberlain, *When Stars Came Down to Earth: Cosmology of the Skidi Pawnee Indians of North America,* Los Altos, CA, and College Park, MD, Ballena Press, and Center for Archaeoastronomy, 1983; **Fig. 6–3** Pekka Parviainen; **Fig. 6–4** Jay M. Pasachoff; **Figs. 6–10AB** Peanuts reprinted by permission of United Feature Syndicate, Inc.; **Fig. 6–11** Phil Hudson; **Fig. 6–12** © 1997 Richard J. Wainscoat; **Fig. 6–14** Emil Schulthess, Black Star; **Fig. 6–16** Map Creation Ltd.; **Fig. 6–18** Mt. Vernon Ladies' Association of the Union.

**Part 2—Opener** LASCO Team, Naval Research Laboratories; and NASA.

**Chapter 7—Opener** NASA/JPL/Caltech; **Fig. 7–1** William C. Livingston, National Solar Observatory/AURA; **Fig. 7–2** © Jay M. Pasachoff and Wendy Carlos; **Fig. 7–4** Williams College—Hopkins Observatory, by Kevin Reardon, now at Capodimonte Observatory, Naples; **Fig. 7–6** NASA/JPL/Caltech; **Fig. 7–9** Akira Fujii; **Figs. 7–12AB** and **7–13** Jay M. Pasachoff; **Fig. 7–14** After Fred Espenak, NASA's Goddard Space Flight Center; **Figs. 7–15AB, 7–16** and **7–17** Williams College Eclipse Expedition and NASA's Goddard Space Flight Center's EIT Team/composited by Daniel B. Seaton, Williams College and Harvard–Smithsonian Center for Astrophysics; **Fig. 7–19** NASA/TRACE; **Fig. 7–21** ©1992 Brooks Kraft/Sygma; **Figs. 7–23ABCD** Department of Defense Visual Information Center; **Fig. 7–24** Alan Boss.

**Chapter 8—Opener** Total Ozone Mapping Spectrometer (TOMS), NASA's Goddard Space Flight Center; **Fig. 8–1** Naval Research Laboratory; **Fig. 8–2** Adapted from Raymond Siever, "The Earth," *Scientific American,* September 1975, and The Solar System (New York: W. H. Freeman and Co., 1975), © 1975 by Scientific American, Inc. All rights reserved; Drawn by Tom Prentiss; **Fig. 8–4** Italian Space Agency; **Fig. 8–5** U.S. Geological Survey, photo by R. E. Wallace; **Fig. 8–6** Lowell Whiteside, National Environmental Satellite, Data, and Information Service, National Geophysical Data Center, National Oceanic and Atmospheric Administration; **Fig. 8–7A** Courtesy of Stanley N. Williams, Arizona State U., from *Science Magazine,* 13 June 1980, © by the American Association for the Advancement of Science; **Fig. 8–7B** Photograph by James Zollweg; **Fig. 8–7C** Jay M. Pasachoff; **Fig. 8–8** M. F. Corrin, L. M. Gahagan, and L. A. Lawver, PLATES Project, University of Texas Institute for Geophysics; **Fig. 8–9** Jay M. Pasachoff; **Fig. 8–12** U.S. Naval Research Laboratory; **Fig. 8–13** Michael E. Mann, Department of Environmental Sciences, University of Virginia; from Mann, M.E., Bradley, R.S. and Hughes, M.K., "Northern Hemisphere Temperatures During the Past Millennium: Inferences, Uncertainties, and Limitations," *Geophysical Research Letters* **26,** 759–762, 1999; **Fig. 8–14** Pieter Tans, Climate Monitoring and Diagnostics Laboratory, National Oceanic and Atmospheric Administration, Carbon Cycle Group, Boulder, CO; and Charles D. Keeling, Scripps Institute of Oceanography, La Jolla, CA; **Fig. 8–15** TOMS, NASA's Goddard Space Flight Center; **Fig. 8–16** TOMS, NASA's Goddard Space Flight Center; **Fig. 8–17** NASA; **Fig. 8–18A** George E. Parks, U. Washington/NASA's Marshall Space Flight Center; **Fig. 8–18B** Courtesy of Stephen B. Mende, Space Sciences Laboratory, University of California at Berkeley; **Fig. 8–19** Christopher L. Grohusko, El Paso, Texas; **Fig. 8–20** Steele Hill, NASA.

**Chapter 9—Opener** U.S. Naval Research Laboratory; **Figs. 9–1** and **9–2** Daniel Good; **Figs. 9–4, 9–5,** and **9–6A** NASA; **Fig. 9–6B** Drawing by Alan Dunn; © The New Yorker Collection 1971 Alan Dunn, from cartoonbank.com. All Rights Reserved; **Fig. 9–7** U.S. Naval Research Laboratory; **Figs. 9–8, 9–9,** and **9–10** NASA's Johnson Space Center; **Figs. 9–11** and **9–12** R. Wobus, Williams College; **Fig. 9–13** NASA; **Fig. 9–14** U.S. Naval Research Laboratory; **Fig. 9–15** Gerald Wasserburg, Caltech; **Fig. 9–16** Drawings by Donald E. Davis under the guidance of Don E. Wilhelms of the U.S. Geological Survey; reproduced with the assistance of the Hansen Planetarium; **Fig. 9–17** © Edgerton Foundation, 1997, courtesy of Palm Press, Inc.; **Fig. 9–18** NASA; **Fig. 9–19** U.S. Naval Observatory; **Fig. 9–21** NASA; **Fig. 9–22** McDonald Observatory, U. Texas; **Fig. 9–23** Stephen Haggerty, U. Mass.; **Fig. 9–24** A. G. W. Cameron, Harvard-Smithsonian Center for Astrophysics; **Fig. 9–25** Michael Mendillo, and Jeffrey Baumgardner, Boston U., *Nature,* vol. 377, 405, 1995, courtesy of Macmillan Publishing Co.; **Fig. 9–26** NASA's Johnson Space Center; **Fig. 9–27** NASA/JPL courtesy of J. W. Head, Brown U.; **Fig. 9–28** Naval Research Laboratory; **Fig. 9–29** NASA/Ames.

**Chapter 10—Opener** Courtesy of Robert G. Strom, from his *Mercury: The Elusive Planet* (Smithsonian Institution Press), artwork by Karen Denomy; **Fig. 10–2** University of Southern California and Mt. Wilson Observatory, courtesy of Andy Grubb; **Figs. 10–5** and **10–12** Robert G. Strom, Lunar and Planetary Laboratory; **Fig. 10–6** JPL; **Figs. 10–7** ABCD NASA; **Fig. 10–8** NASA/JPL/Caltech; **Fig. 10–9** NASA/JPL/Caltech; **Fig. 10–10** NASA/JPL/Caltech; **Fig. 10–11** NASA, © Calvin J. Hamilton; **Fig. 10–13** Andrew E. Potter, Jr., and Thomas H. Morgan; **Fig. 10–14** David Mitchell, U. California, Berkeley; **Fig. 10–15** Mark Robinson, Northwestern U.; **Fig. 10–16** D. O. Muhleman and B. J. Butler, Caltech, and M. A. Slade, JPL; **Fig. 10–17** John Harmon, National Astronomy and Ionosphere Center; **Fig. 10–18** Ron Dantowitz and Marek Kozubal (Museum of Science, Boston) and Scott W. Teare (U. Illinois), with the permission of the *Astronomical Journal*; similar processing was carried out by Jeffrey Baumgardner, Michael Mendillo, and Jody K. Wilson (Boston U.).

**Chapter 11—Opener** NASA/JPL; **Fig. 11–1A** NASA's Johnson Space Center; **Fig. 11–1B** Jay M. Pasachoff; **Fig. 11–3** Vassar College; **Fig. 11–4** William Sinton and Klaus Hodapp (Institute for Astronomy, U. Hawaii), David Crisp (Caltech), Boris Ragent (NASA/Ames), and David Allen (Anglo-Australian Obs.); **Fig. 11–7** Courtesy of Donald B. Campbell, Arecibo Observatory; **Fig. 11–8** Courtesy of Valeriy Barsukov and Yuri Surkov; **Figs. 11–9** and **11–13** NASA/Ames Research Center; **Figs. 11–10, 11–14, 11–16, 11–17, 11–18, 11–19** and **11–21** NASA/JPL/Caltech; **Fig. 11–11** L. Esposito (U. Colorado, Boulder), and NASA; **Fig. 11–12** U.S. Geological Survey, Branch of Astrogeology, Flagstaff; **Fig. 11–14** NASA's Johnson Space Center; **Fig. 11–15** NASA/JPL/Caltech; **Fig. 11–16** Courtesy of Donald B. Campbell, Arecibo Observatory; **Fig. 11–20** David Anderson, © 1993. Reprinted with permission of *Discover* Magazine.

**Chapter 12—Opener** NASA/JPL/Caltech and Malin Space Science Systems; **Fig. 12–1** Lowell Observatory; **Fig. 12–2** NASA/JPL/Caltech and Malin Space Science Systems; **Fig. 12–3** USGS, Branch of Astrogeology, Flagstaff; **Figs. 12–5** © Caltech/NASA/JPL; **Fig. 12–6** NASA/JPL/Caltech and Malin Space Science Systems; **Fig. 12–7** NASA/MOLA Science Team; **Fig. 12–8** NASA/Mars Global Surveyor/Orbiter Laser Altimeter; **Fig. 12–9** Jay M. Pasachoff; **Fig. 12–10** Mosaic by Jody Swann, USGS, Branch of Astrogeology, Flagstaff; **Figs. 12–11** to **12–14** NASA/JPL/Caltech and Malin Space Science Systems; **Fig. 12–15** Digital mosaic by Tammy Becker, USGS, Branch of Astrogeology; **Fig. 12–16** Image processing by Alfred S. McEwen, USGS, Branch of Astrogeology; **Figs. 12–17** to **12–24** © NASA/JPL/Caltech; **Fig. 12–25** Dan Britt, U. Tennessee, and NASA; **Figs. 12–26** to **12–28** NASA/JPL/Caltech and Malin Space Science Systems; **Fig. 12–29** Steve Lee (U. Colorado), Jim Bell (Cornell Univ.), Mike Wolff (STScI), and NASA; **Fig. 12–30** B. Weiss; from *Science Magazine,* **290,** 791, 27 Oct 2000. © 2000 American Association for the Advancement of Science.

**Chapter 13—Opener** NASA/JPL/Caltech; **Fig. 13–2** Lunar and Planetary Laboratory, U. Arizona; **Fig. 13–4** Imke de Pater, U. California, Berkeley; **Fig. 13–5** Jay M. Pasachoff; **Fig. 13–6**

NASA/JPL/University of Arizona; **Figs. 13–7, 13–8** and **13–9** NASA/JPL; **Fig. 13–10A** Philip S. Marcus, U. California, Berkeley; **Fig. 13–10B** J. Someria, S. Meyers, H. L. Swinney, U. Texas at Austin, based on a computer model by Philip S. Marcus; **Fig. 13–11A** NASA/IRTF and John Spencer, Lowell Obs.; **Fig. 13–11B** Heidi Hammel (Space Science Institute, then at MIT), and NASA; **Fig. 13–12** NASA/JPL/Caltech; **Box 13.3 Fig. AB** H. Hammel, Space Science Institute, and STScI/NASA; **Box 13.3 Fig. C** W. M. Keck Observatory/Imke de Pater, James Graham, Garrett Jernigan, UC-Berkeley; **Fig. 13–14** NASA/JPL/U. Arizona; **Fig. 13–15** J. Clarke (U. Michigan) and NASA; **Fig. 13–16** NASA/JPL/Caltech; **Fig. 13–17** NASA/JPL/Caltech; **Fig. 13–18** NASA/JPL/Caltech; **Fig. 13–19** John Spencer (Lowell Observatory) and NASA; **Photo Gallery p. 242** (*top right*) NASA's Johnson Space Center; **p. 242** (*bottom right*) NASA/JPL/Caltech and Cornell University; **p. 242** (*left*) and **p. 243** (*top and middle left*) NASA/JPL/U. Arizona; **p. 242** (*middle right and bottom*) NASA/JPL/DLR (German Aerospace Center); **Fig. 13–20** U.S. Geological Survey, processing by A. McEwen, T. Rock, and L. Soderblom; **Fig. 13–21** Doyle T. Hall and Warren Moos, Johns Hopkins U.; **Fig. 13–22** NASA/JPL/Caltech; **Figs. 13–23, 13–24,** and **13–25** NASA/JPL/Caltech; **Fig. 13–26** NASA/JPL/Caltech and Arizona State University; **Figs. 13–27** and **13–29** NASA/JPL/Caltech; **Fig. 13–28** NASA/JPL/Caltech and Cornell U.; **Fig. 13–30** Mark Showalter, NASA/Ames; **Fig. 13–31** H. A. Weaver and T. E. Smith (STScI), J. T. Trauger and R. W. Evans (JPL), and NASA; **Fig. 13–32** David Jewitt and Scott Sheppard.

**Chapter 14—Opener** NASA/JPL/Caltech; **Fig. 14–2** Reta Beebe (New Mexico State University), D. Gilmore, L. Gergeron (STScI), and NASA; **Fig. 14–3** Erich Karkoschka, Lunar and Planetary Laboratory, U. Arizona; **Fig. 14–4** IRTF/NSFCAM, Keith S. Noll, Diane Gilmore, and David Soderblom; **Figs. 14–5** to **14–12** NASA/JPL/Caltech; **Fig. 14–13** redrawn from Augustin Sanchez-Lavega, J. F. Rojas, and P. V. Sada; **Fig. 14–14** J. P. Trauger (JPL) and NASA; **Fig. 14–15** NASA/JPL/Caltech; **Fig. 14–16A** Peter H. Smith and Mark T. Lemmon (U. Arizona) and NASA; **Fig. 14–16B** Claire Max, Bruce Macintosh, and Seran Gibbard, Lawrence Livermore National Laboratory, and W. M. Keck Observatory; **Fig. 14–16C** Canada-France-Hawaii Telescope/A. Coustenis/1998; **Figs. 14–17** to **14–30** and **14–32** NASA/JPL/Caltech; **Fig. 14–31** Dave Seal, NASA/JPL/Caltech.

**Chapter 15—Opener** NASA/JPL/Caltech; **Fig. 15–1** David Wittman, U. Arizona; **Fig. 15–2** Infrared Processing and Analysis Center/Caltech and University of Massachusetts; **Fig. 15–4** James L. Elliot, MIT; **Fig. 15–5** Erich Karkoschka (University of Arizona) and NASA; **Figs. 15–6** to **15–10, 15–12** to **15–18** NASA/JPL; **Fig. 15–11** Information from Norman Ness; style from *Sky & Telescope;* **Fig. 15–20** Erich Karkoschka, Lunar and Planetary Laboratory, U. Arizona and NASA.

**Chapter 16—Opener and Fig. 16–1** NASA/JPL; **Fig. 16–2** Master and Fellows of St. Johns College of Cambridge University; **Fig. 16–3** Charles Kowal, Space Telescope Science Institute; **Fig. 16–4** Tobias Owen, Institute for Astronomy, U. Hawaii, after R. Danhy; **Fig. 16–5A** Heidi Hammel (Space Science Institute, then at MIT) and NASA; **Fig. 16–5B** Infrared Processing and Analysis Center/Caltech and University of Massachusetts; **Fig. 16–6** W. B. Hubbard, A. Brahic, B. Sicardy, E.-R. Elicer, F. Roques, and F. Vilas, *Nature* 319, 636–640, 1986, courtesy of Macmillan Publishing Co.; **Figs. 16–7** to **16–11** NASA/JPL; **Fig. 16–12** NASA/JPL/Caltech, reprocessed by Caroline Porco, U. Arizona; **Figs. 16–13** to **16–20** NASA/JPL; **Fig. 16–21A** Richard Dekany (JPL), Thomas Hayward, Bernhard Brandl, and Don Banfield (Cornell); the AO system was developed by a JPL team led by Richard Dekany: imaged at the Palomar Observatory; **Fig. 16–21B** W. M. Keck Observatory; **Fig. 16–22** Image courtesy of C. Roddier, F. Roddier, J. E. Graves, M. Northcott and O. Guyon, Institute for Astronomy, University of Hawaii; **Fig. 16–23** Christophe Dumas and Richard J. Terrile (JPL), and the NICMOS team.

**Chapter 17—Opener** S. Alan Stern, Southwest Research Institute, and Marc Buie, Lowell Observatory, and NASA; **Fig. 17–1** © 1974 Charles Capen, Hansen Planetarium; **Fig. 17–2** Lowell Observatory photographs; **Fig. 17–3** U.S. Naval Observatory/James W. Christy, U.S. Navy photograph; **Fig. 17–4A** Keith S. Noll, STScI, and NASA; **Fig. 17–4B** Dr. R. Albrecht, ESA/ESO Space Telescope European Coordinating Facility, NASA; **Fig. 17–5** David J. Tholen, Mauna Kea Obs., Institute for Astronomy, U. Hawaii; **Fig. 17–6** Marc W. Buie, David J. Tholen, and Keith Horne; **Fig. 17–7** © 1998 Calvin J. Hamilton; **Fig. 17–8** Marc Buie, Lowell Observatory, and Alan Stern, Southwest Research Institute, NASA, ESA; **Figs. 17–9** and **17–10** James L. Elliot, MIT; **Fig. 17–11** David Jewitt, Institute for Astronomy, U. Hawaii; **Fig. 17–12** David Jewitt (IfA, U. Hawaii), Jane Luu (Leiden Obs.), and Jun Chen (IfA); **Fig. 17–13** David J. Eicher; **Fig. 17–14** Stephen C. Brewster, NASA/JPL.

**Chapter 18—Opener** courtesy of Geoff Marcy, University of California at Berkeley; **Fig. 18–1** © Gerry Gropp; **Fig. 18–2** Jack Schmidling; **Figs. 18–3 and 18–4** after a Cornell University visualization, imaging by Chris Hildreth and Wayne Lytle; courtesy of Alex Wolszczan, Pennsylvania State U.; **Fig. 18–5** after Geoffrey W. Marcy and R. Paul Butler, San Francisco State U. and University of California at Berkeley; **Fig. 18–6A** Michel Mayor and Didier Queloz, Geneva Observatory, University of Geneva; **Figs. 18–6B, 18–7A** Geoff Marcy, University of California at Berkeley, and Paul Butler, Carnegie Institution of Washington; **Fig. 18–7B** Geoff Marcy, Paul Butler, Debra A. Fisher, Steven S. Vogt; **Fig. 18–8** Geoff Marcy, San Francisco State U. and University of California at Berkeley; **Fig. 18–9B** Arto Oksanen, Nyrölä Observatory, Astronomical Association Jyväskylän Sirius, Finland, from *Sky & Telescope*, January 2001; **Fig. 18–10AB** NASA/JPL/Caltech; **Fig. 18–11** (*A*) Palomar Observatory, Caltech; (*B*) T. Nakajima and S. Kulkarni (Caltech), S. Durrance and F. Golimowski (John Hopkins U.), and NASA; **Fig. 18–12** D. A. Golimowski, S. T. Durrance, and M. Clampin, The Johns Hopkins U. and STScI; **Fig. 18–13** Al Schultz (Computer Sciences Corp. and STScI), Sally Heap (NASA's Goddard Space Flight Center), and NASA; **Fig. 18–14** C. R. O'Dell/Rice U. and Z. Wen, Rice U./NASA/STScI; **Fig. 18–15** Jane Greaves, J. S. Greaves; W. S. Holland; G. Moriarty-Schieven; T. Jenness; W. R. F. Dent; B. Zuckerman; C. McCarthy; R. A. Webb; H. M. Butner; W. K. Gear; and H. J. Walker, *Astrophy. J.* 506, L133–137, 1998 October 20; **Fig. 18–16** Helen Walker.

**Chapter 19—Opener** © Tony and Daphne Hallas/Hallas Digital Services; **Fig. 19–1** Jay M. Pasachoff; **Fig. 19–2** Akira Fujii; **Fig. 19–3** Naval Research Laboratory; SOHO is a joint ESA-NASA program; **Fig. 19–4A** H. A. Weaver (Applied Research Corp.), the HST Comet Science Team, and NASA; **Fig. 19–4B** H. A. Weaver (Applied Research Corp.), P. D. Feldman (The Johns Hopkins U.), and NASA; **Fig. 19–5** J. L. Bertaux, Aeronomie, Verrières-le-Buisson, France; **Fig. 19–6A** Akira Fujii; **Fig. 19–6B** Paul Feldman, The Johns Hopkins University; **Fig. 19–7** H. A. Levison: from Alan Stern and Humberto Campins; **Fig. 19–8A** C. Lisse, M. Mumma, NASA/GSFC; K. Dennerl, J. Schmitt and J. Englhausser, MPE; **Fig. 19–8B** NASA/CXC/SAO; **Fig. 19–9** Naval Research Laboratory, LASCO on the Solar and Heliospheric Observatory: SOHO is a NASA and ESA collaboration; **Fig. 19–10** National Portrait Gallery, London; by R. Phillips, c. 1720; **Fig. 19–11** The Bayeux Tapestry—11th Century; By special permission of the City of Bayeux; **Fig. 19–13** Royal Observatory, Greenwich, National Maritime Museum; **Fig. 19–14** James B. Kaler and Karen B. Kwitter; **Fig. 19–15A** photography by B. W. Hadley, © Royal Observatory Edinburgh; **Fig. 19–15B** James De Buizer and James Radomski, copyright 1996 The University of Florida Rosemary Hill Observatory; **Fig. 19–16** Soviet VEGA team; **Fig. 19–17** © Max-Planck-Institut für Aeronomie, Lindau/Hartz, FRG; photographed with the Halley Multicolour Camera aboard the European Space Agency's Giotto spacecraft, courtesy of H. U. Keller, additional processing by Harold Reitsema, Ball Aerospace; **Fig. 19–18** Jay M. Pasachoff and Steven Souza; **Fig. 19–19** NASA/Naval Research Laboratory; **Fig. 19–20** Soviet VEGA team; **Fig. 19–21** Susan Wyckoff, Arizona State U., Tempe; **Fig. 19–22** Jay M. Pasachoff; **p. 347** European Southern Observatory; **Fig. 19–23** Jane Luu (Leiden Obs.) and David Jewitt (Institute for Astronomy, U. Hawaii); **Fig. 19–24** Hal Weaver and T. Ed Smith (STScI), and NASA; **Fig. 19–25** Jay M. Pasachoff.

**Chapter 20—Opener** © 1998 Tony and Daphne Hallas, Hallas Digital Services; **Fig. 20–1** © James M. Baker; **Fig. 20–2** and **20–3** Jay M. Pasachoff; **Fig. 20–4** Allan E. Morton, courtesy of Meteor Crater,

Northern Arizona, USA; **Fig. 20–5A** Rick Wirth, courtesy of Paul Weiblen; **Fig. 20–5B** © 2000 Creators Syndicate, Inc. "Wizard of Id" cartoon by Brant Parker and Johnny Hart; By permission of Johnny Hart and Creators Syndicate, Inc.; **Fig. 20–6** Ghislaine Crozaz, courtesy of Ursula Marvin; **Fig. 20–7** NASA/JSC; **Fig. 20–8** Lawrence Grossman; **Fig. 20–9A** Jay M. Pasachoff; **Fig. 20–9B** Juan Carlos Casado; **Fig. 20–10** David Jewitt (Institute for Astronomy, U. Hawaii) and Jane Luu (Leiden Obs.); **Fig. 20–11AB** Harvard-Smithsonian Center for Astrophysics/International Astronomical Union Minor Planet Center and Central Bureau for Astronomical Telegrams; **Fig. 20–12** Thierry Pauwels, Koninklijke Sterrenwacht van Belgie (Observatoire Royal de Belgique-Royal Observatory of Belgium); **Fig. 20–13A** Anna M. Nobili, Università di Pisa, data from the *Ephemerides of Minor Planets,* 1987, from *Asteroids II,* eds. Richard P. Binzel, Tom Gehrels, and Mildred Shapley Matthews (U. Arizona Press, 1989); **Fig. 20–14** Andrew Chaikin, space.com; **Fig. 20–15** Laird Close (European Southern Observatory, Munich, Germany), Bill Merline (Southwest Research Institute, Boulder, CO USA); **Fig. 20–16** Lucy McFadden; **Fig. 20–17** Illustration from *The Little Prince* by Antoine de Saint-Exupéry, copyright 1943 and renewed 1971 by Harcourt, Inc., reproduced by permission of the publisher; **Box 20.2 Fig. A** © 2002 Universal City Studios, Inc. Courtesy of Universal Studios Publishing Rights. All rights reserved; **Box 20.2 Fig. B** Virgil Sharpton, Lunar and Planetary Institute; **Fig. 20–18** Alan B. Chamberlin, JPL; **Fig. 20–19A** Steven J. Ostro, JPL and R. Scott Hudson, Washington State U.; **Fig. 20–19B** Steven J. Ostro, JPL, and R. Scott Hudson, Washington State U., *Science Magazine,* **263,** 940–943, © 1994 American Association for the Advancement of Science; **Fig. 20–20A** U.S. Geological Survey; **Fig. 20–20B** NASA/JPL/Caltech; **Figs. 20–21, 22, 23** and **Photo Gallery** Johns Hopkins University Applied Physics Laboratory.

**Chapter 21—Opener** From "Star Wars: Episode I—The Phantom Menace," © Lucasfilm Ltd. & Courtesy of Lucasfilm Ltd. All rights reserved. Used under authorization; **Fig. 21–1** Drawing by Frank Modell, © The New Yorker Collection 1971 Frank Modell, from cartoonbank.com. All Rights Reserved; **Fig. 21–2** NASA/Ames; **Fig. 21–3A** © 1976 National Geographic Society, photo by Bruce Dale; **Fig. 21–3B** Photograph by Andrew Clegg/Cornell University; **Fig. 21–4AB** NASA's Johnson Space Center; **Fig. 21–5A** © E. Imre Friedmann; **Fig. 21–5B** Todd Stevens, photo from Pacific Northwest National Laboratory, operated by Battelle Memorial Institute; **Fig. 21–5C** Jim Fredrickson, Pacific Northwest National Laboratory; **Fig. 21–6A** NASA; **Fig. 21–6B** Jay M. Pasachoff; **Fig. 21–7** photo courtesy of Tony Acevedo, Arecibo Observatory; **Fig. 21–9** © Bill and Sally Fletcher; **Fig. 21–10** From *Glinda of Oz* by L. Frank Baum, illustrated by John R. Neill, copyright 1920; **Figs. 21–11** and **21–12** photo by Seth Shostak/SETI Institute; **Fig. 21–13** SETI@home, University of California at Berkeley; **Fig. 21–14** SETI Institute; **Fig. 21–15** *The Far Side®* by Gary Larson © 1982 FarWorks, Inc. All rights reserved. Used by Permission; **Fig. 21–16** NASA's Johnson Space Center.

**Chapter 22—Opener** Alan Title, Lockheed Martin Advanced Technology Center, and Leon Golub, Harvard-Smithsonian Center for Astrophysics/NASA/ISAS; **Fig. 22–1** Chris Malicki; **Fig. 22–2** William C. Livingston, National Solar Observatory/AURA; **Fig. 22–4** Optical Science Laboratory, University College, London, England; **Fig. 22–5** NOAO/National Solar Observatory; **Fig. 22–6** GONG Project, National Optical Astronomy Observatories, courtesy of John Leibacher; **Fig. 22–7** Solto (ESA & NASA) Michaelson Doppler Imager/Solar Oscillations Investigation (MDI/SOI) and VIRGO data imaged by A. Kosovichev, Stanford University; **Fig. 22–8** Jay M. Pasachoff; **Fig. 22–9** David H. Hathaway, NASA/Marshall Space Flight Center; **Fig. 22–10** Kanzelhöhe Solar Observatory of the Karl Frazens-University Graz; **Fig. 22–11** New Jersey Institute of Technology/Big Bear Solar Observatory; **Fig. 22–12** Robert A. E. Fosbury, European Southern Observatory; **Fig. 22–13** Eclipse image © 1999 Jay M. Pasachoff and Wendy Carlos, merged with a SOHO/EIT image from NASA's Goddard Space Flight Center, courtesy of Joe Gurman; **Fig. 22–14A** NOAO/NSO/Sacramento Peak Observatory; **Fig. 22–14B** Naval Research Laboratory; **Fig. 22–15** Naval Research Laboratory; **Fig. 22–16** NASA's Goddard Space Flight Center, courtesy of Joe Gurman. SOHO is a joint project of NASA and the European Space Agency; **Fig. 22–17A** Yohkoh image courtesy of Loren Acton, Montana State U.; **Fig. 22–17B** Stanford Lockheed Institute for Space Research from SOHO SOI/MDI; **Fig. 22–18** Smithsonian Astrophysical Observatory and IBM, courtesy of Leon Golub, Harvard-Smithsonian Center for Astrophysics; **Fig. 22–19** Alan Title, Lockheed Martin Advanced Technology Center, and Leon Golub, Harvard-Smithsonian Center for Astrophysics.

**Chapter 23—Opener** Williams College Expedition, directed by Jay M. Pasachoff; CCD images supervised by Stephan Martin and composited by Daniel B. Seaton with a LASCO image from the U.S. Naval Observatory and an EIT image from NASA's Goddard Space Flight Center; **Fig. 23–1** 7″ Starfire refractor, Wolfgang Lille/Baade Planetarium, Germany; **Fig. 23–2** Jay M. Pasachoff; **Fig. 23–3** and **23–5** NOAO/National Solar Observatory; **Fig. 23–4** The "Changing Sun" mosaics were created by Greg Slater and Charlie Little of Lockheed Martin Solar and Astrophysics Lab (LMSAL) with images from the National Solar Observatory, NOAO, courtesy of Jack Harvey (the figure is coordinated with Fig. 1–19); **Fig. 23–6** Data from Sunspot Index Data Center, Royal Observatory of Belgium; **Fig. 23–9** David Hathaway, NASA's Marshall Space Flight Center; **Fig. 23–10** © 1998 Jay M. Pasachoff and Wendy Carlos; **Fig. 23–10B** © 1999 Jay M. Pasachoff and Wendy Carlos; **Fig. 23–11** NOAO/National Solar Observatory and Yohkoh/NASA/ISAS; **Photo Gallery p. 418** SOHO EIT Team, NASA's Goddard Space Flight Center; **p. 419** SOHO LASCO Team, U.S. Naval Research Laboratory; **p. 420** Alan Title, Lockheed Martin Advanced Technology Center, and Leon Golub, Harvard-Smithsonian Center for Astrophysics/NASA/ISAS; **Fig. 23–12** (A) Nariaki V. Nitta, Lockheed Martin/NASA/ISAS; (B) Leon Golub/SAO/NASA/ISAS and TRACE Team; **Fig. 23–13** R. J. Poole/DayStar Filter Corp.; **Fig. 23–14** Jay M. Pasachoff; **Fig. 23–15** NASA and ESA; **Fig. 23–16** NASA's Marshall Space Flight Center; **Fig. 23–17** updated from John A. Eddy, then University Corporation for Atmosphere Research; **Fig. 23–18A** Pete Riley, Solar Physics Group, SAIC, San Diego; **Fig. 23–18B** NASA; **Fig. 23–19** NASA; image processing by D. G. Mitchell and P. C. Brandt; **Fig. 23–20** Claus Fröhlich, World Radiation Center, Physikalisch-Meteorologisches Observatorium Davos, Switzerland; and Judith Lean, U.S. Naval Research Laboratory.

**Part 4—Opener** © 1987 Royal Observatory Edinburgh/Anglo-Australian Observatory, photography by David Malin from U.K. Schmidt plates.

**Chapter 24—Opener** J. Hester and P. Scowen, Arizona State U., and NASA; **Fig. 24–1** (*left*) Jeff Hester and Paul Scowen, Arizona State U., and NASA/STScI/ESA; (*right*) Mark McCaughrean with the Calar Alto 3.5-m telescope; **Fig. 24–3** Palomar Observatory—National Geographic Society Sky Survey. Reproduced by permission from the California Institute of Technology; **Fig. 24–4** Bundesarchiv Koblenz; **Fig. 24–10A** American Institute of Physics, Niels Bohr Library, Uhlenbeck Collection; **Fig. 24–12** and **24–13** Harvard College Observatory; **Fig. 24–14** and **24–15** Roger Bell, University of Maryland and Michael Briley, U. Wisconsin at Oshkosh; **Fig. 24–16** Yves Grosdidier, Université de Montréal and Observatoire de Strasbourg; Antony Moffat, Université de Montréal; Gilles Joncas, Université Laval; Agnes Acker, Observatoire de Strasbourg; and NASA; **Fig. 24–17** NASA's Goddard Space Flight Center.

**Chapter 25—Opener** © 1981 NOAO/CTIO, photo by Gabriel Martin; **Fig. 25–6** © 1979 Royal Observatory Edinburgh/Anglo-Australian Observatory, from original U.K. Schmidt plates; **Fig. 25–10** Dorrit Hoffleit—Yale U. Observatory; courtesy of American Institute of Physics, Niels Bohr Library; **Fig. 25–11** American Institute of Physics, Niels Bohr Library, Margaret Russell Edmondson Collection; **Fig. 25–14** The Hipparcos Project and the European Space Agency, courtesy of M. A. C. Perryman.

**Chapter 26—Opener** Canada-France-Hawaii Telescope Corp., © Regents of the University of Hawaii, photo by Laird Thompson, then Institute for Astronomy, U. Hawaii, now U. Illinois; **Fig. 26–1** Williams College—Hopkins Observatory; **Figs. 26–2** and **26–3** Copyright U.C. Regents, UCO/Lick Observatory image; **Fig. 26–7** based on data from D. L. Harris III, K. Aa. Strand, and C. E. Worley, in *Basic Astronomical Data,* K. Aa. Strand, ed., U. Chicago Press, © 1963 by the U. Chicago; **Fig. 26–8** Palomar Test bed Interferometer, Caltech, courtesy of Ben Lane; **Fig. 26–9** Bernhard M. Haisch, Lockheed Palo Alto Research

Laboratory; **Fig. 26–10A** AAVSO; **Fig. 26–10B** Margarita Karovska (Harvard-Smithsonian Center for Astrophysics) and NASA; **Fig. 26–10C** NRAO/VLBA, A. J. Kemball, P. J. Diamond, National Radio Astronomy Observatory/Associated Universities, Inc.; **Fig. 26–11** observations obtained for Jay M. Pasachoff with the Automated Photoelectric Telescope; **Fig. 26–12** J. D. Fernie and R. McGonegal, reprinted from *Astrophys. J.* 275, p. 735, with permission of the U. Chicago Press; **Fig. 26–13** Harvard College Observatory; **Fig. 26–14** Akira Fujii; **Fig. 26–15** Wendy L. Freedman, Observatories of the Carnegie Institution of Washington, and NASA; **Fig. 26–16** OGLE Consortium from B. Paczýnski, from A. Udalski, I. Soszynski, M. Szymanski, et al., 1999, *Acta Astronomica,* **49,** 223; **Figs. 26–17 and 26–22** Reprinted by permission of the publisher from *The Milky Way* by B. J. Bok and P. Bok, Cambridge, Mass.: Harvard University Press, Copyright © 1941, 1945, 1957 by the President and Fellows of Harvard College; **Fig. 26–18** © 1977 Anglo-Australian Observatory, photograph by David Malin; **Fig. 26–19** © 1984 Anglo-Australian Observatory, photograph by David Malin; **Fig. 26–20A** Richard E. White, Smith College; **Fig. 26–20B** Terence Dickinson, Digitized Sky Survey, Chuck Vaughn, Hubble Heritage Team; **Fig. 26–21** Daniel Good; **Fig. 26–23** Shigeto Hirabayashi; **Fig. 26–24** NASA/ESA/STScI; **Fig. 26–26** Michael Rich, Kenneth Mighell, and James D. Neill; (Columbia U.), and Wendy Freedman (Carnegie Observatories), and NASA.

**Part 5—Opener** © 1987 Anglo-Australian Observatory, photography by David Malin from original U.K. Schmidt plates.

**Chapter 27—Opener** Y. Totsuka, U. Tokyo, Inst. for Cosmic Ray Research; **Fig. 27–1A** © 1979 Royal Observatory Edinburgh/Anglo-Australian Observatory, photography by David Malin from original U.K. Schmidt plates; **Fig. 27–1B** Alain Abergel, Institut d'Astrophysique Spatiale, Orsay, with the Infrared Space Observatory; **Fig. 27–2A** Todd R. Hunter, Caltech Submillimeter Obs.; **Fig. 27–2B** Infrared Processing and Analysis Center/Caltech and University of Massachusetts (IPAC/ U. Mass); **Fig. 27–4** NICMOS image: NASA/Nolan Walborn (Space Telescope Science Institute) and Rodolfo Barbà (La Plata Observatory, La Plata, Argentina); WFPC2 image: NASA/John Trauger (Jet Propulsion Laboratory) and James Westphal (California Institute of Technology); **Fig. 27–5A** IPAC/U. Mass; **Fig. 27–5B** T. P. Ray, T. W. B. Muxlow, D. J. Axon, A. Brown, D. Corcoran, J. Dyson, and R. Mundt, *Nature* 385, 415–417, 30 January 1997; **Fig. 27–6** European Southern Observatory; **Fig. 27–7** NASA, Alan Watson (Universidad Nacional Autónoma de México), Karl Stapefeldt (Jet Propulsion Laboratory), John Krist and Chris Burrows (European Space Agency/Space Telescope Science Institute; **Fig. 27–8** George H. Herbig, then Lick Observatory, now U. Hawaii; **Fig. 27–9** Laird Close, Institute for Astronomy, U. Hawaii; **Fig. 27–16** Brookhaven National Laboratory; **Fig. 27–17** Y. Totsuka, U. Tokyo, Inst. for Cosmic Ray Research; **Fig. 27–18** John N. Bahcall, Institute for Advanced Study, and Marc H. Pinsonneault, Ohio State U.; **Fig. 27–19AB** Lawrence Berkeley National Laboratory and Sudbury Neutrino Observatory.

**Chapter 28—Opener and 28–1A** © 1979 Royal Observatory Edinburgh/Anglo-Australian Observatory, photography by David Malin from original U.K. Schmidt plates. **Fig. 28–1B** C. Robert O'Dell and Kerry P. Handron (Rice University), and NASA; **Fig. 28–4** Hubble Heritage Team (AURA/STScI/NASA); **Fig. 28–5** Canada-France-Hawaii Telescope Corp.; **Fig. 28–6A** and **28–8** Raghvendra Sahai and John Trauger (JPL), the WFPC2 Science Team, and NASA; **Fig. 28–6B** Rodger Thompson, Marcia Rieke, Glenn Schneider, Dean Hines (University of Arizona); Raghvendra Sahai (JPL); NICMOS Instrument Definition Team, and NASA; **Fig. 28–7** NASA, ESA and The Hubble Heritage Team (STScI/AURA)/R. Sahai (Jet Propulsion Lab) and B. Balick (University of Washington); **Fig. 28–9** and **28–10A** U. Washington—U.S. Naval Observatory—Cornell—Arcetri Obs. (Florence, Italy) collaboration, courtesy of Bruce Balick, U. Washington; **Fig. 28–11** ESA/ISO, CEA Saclay and ISOCAM Consortium (*image*), SWS Consortium (*spectrum*). The SWS Consortium is led by the SWS PI, Th. de Graauw, Lab for Space Research, Groningen, NL; **Fig. 28–12** X-ray: NASA/University of Illinois at Urbana/You-Hua Chu et al., optical: NASA/HST; **Fig. 28–13** After C. Robert O'Dell in I.A.U. Symposium No. 34, *Planetary Nebulae,* eds. D. E. Osterbrock and C. R. O'Dell. Reprinted by permission of the International Astronomical Union; **Fig. 28–14A** McDonald Observatory, U. Texas; **Fig. 28–14B** Chandra X-ray Observatory HRC/LETG observation; **Fig. 28–16** Harvey Richer (U. British Columbia) and NASA; **Fig. 28–17** Hubble Heritage Team (NASA/AURA/STScI) using data collected by Howard E. Bond (STScI) and Robin Ciardullo (Penn State U.); **Fig. 28–18** Copyright U.C. Regents UCO/Lick Observatory image; **Fig. 28–19** F. Paresce and R. Jedrzejewski (STScI), NASA/ESA; **Fig. 28–20** Michael Shara (now at Rose Center for Earth and Space, American Museum of Natural History, New York), Robert Williams, (STScI), R. Gilmozzi (ESO), NASA; **Fig. AN–1** Dana Berry, STScI; **Fig. AN–2** NASA's Johnson Space Center.

**Chapter 29—Opener** Hubble Heritage Team using data collected by Robert Kirshner (Harvard/CfA), Nino Panagia (STScI), Martino Romaniello (ESO), and collaborators; **Fig. 29–2A** Andrea Dupree (Harvard-Smithsonian CfA), Ronald Gilliland (STScI), NASA and ESA; **Fig. 29–2B** Jay M. Pasachoff; **Fig. 29–3A** Jack Newton; **Fig. 29–3B** Joachim Trümper, Max-Planck-Institut für Extraterrestrische Physik; **Figs. 29–4B** and **29–5** Adam Burrows, U. Arizona; **Fig. 29–6A** J. Hester (Arizona State U.)/NASA; **Fig. 29–6B** J. Morse (STScI), Kris Davidson (U. Minnesota)/NASA; **Fig. 29–7** Adam Burrows, U. Arizona; **Fig. 29–8** Courtesy Encyclopaedia Britannica, Inc.; from the 1989 *Britannica Yearbook of Science and the Future;* illustration by Jane Meredith; **Fig. 29–9** © 1979 Royal Observatory Edinburgh/ Anglo-Australian Observatory, photography by David Malin from original U.K. Schmidt plates; **Fig. 29–10** photographs by Nile Root; **Fig. 29–11** Stephen P. Reynolds, N.C. State U./Hugh Aller, U. Michigan, using the VLA of NRAO; **Fig. 29–12A** Joachim Trümper, Max-Planck-Institut für Extraterrestrische Physik; **Fig. 29–12B** J. Hester, Arizona State U., and NASA; **Fig. 29–13** © David F. Malin, Anglo-Australian Telescope Observatory, and Jay M. Pasachoff, from Palomar Observatory plates made available by the California Institute of Technology, courtesy of Robert Brucato; **Fig. 29–14** European Southern Observatory; **Fig. 29–15** J. Hester and P. Scowen, Arizona State U., NASA/STScI; **Fig. 29–16A** NASA/CXC/SAO/Rutgers/J. Hughes; **Fig. 29–16B** MDM Obs.; **Fig. 29–16C** ISO; **Fig. 29–16D** VLA/NRAO/AUI; **Fig. 29–17** X-ray (NASA/CXC/SAO); Optical (NASA/HST); Radio (CSIRO/ATNF/ATCA); **Fig. 29–18** Bernd Aschenbach and colleagues, Max-Planck Institut für Extraterrestrische Physik, Garching b. Buenchen, Germany; **Fig. 29–19** © Roger Ressmeyer—Corbis; **Fig. 29–20** © 1987 Anglo-Australian Observatory, photograph by David Malin; **Fig. 29–21** Contributions to IAU Circulars, compiled by Daniel W. E. Green, Smithsonian Astrophysical Obs.; **Fig. 29–22** © 1988 Anglo-Australian Observatory, photograph by David Malin; **Fig. 29–23** © 1987 William Liller; **Fig. 29–24** Dana Berry, STScI; **Fig. 29–25** Christopher Burrows, ESA/STScI and NASA; STIS images—George Sonneborn (GSFC) and NASA; **Fig. 29–26** Peter Garnavich, Notre Dame; **Fig. 29–27** Peter Challis, Harvard-Smithsonian Center for Astrophysics; **Fig. 29–28** NASA/CXC/ SAO/ Pennsylvania State U./ D. Burrows et al.; **Fig. 29–29** John Learned, U. Hawaii High Energy Physics Group; **Fig. 29–30** diagram from Adam Burrows; reproduced by R. McCray and H. W. Li; **Fig. 29–31** © 1989 Anglo-Australian Observatory, photography by David Malin; **Fig. 29–32** LASCO Team, U.S. Naval Observatory; **Fig. 29–33** European Space Agency; **Fig. 29–34** NASA's Johnson Space Flight Center; **Fig. 29–35** Mass histogram from the ACE Cosmic-Ray Isotope Spectrometer provided by the CRIS science team at the California Institute of Technology, the Jet Propulsion Laboratory, Washington University, and NASA's Goddard Space Flight Center; **Table 29–2** Adam Burrows, University of Arizona, updated.

**Chapter 30—Opener** Jeff Hester and Paul Scowen (Arizona State U.), and NASA; **Fig. 30–1** background: NASA-Corbis; **Fig. 30–2** Jay M. Pasachoff; **Figs. 30–3 and 30–4** Master and Fellows of Churchill College Cambridge; **Fig. 30–5** James Cordes, Arecibo Obs. (operated by Cornell U. under contract with NSF); **Fig. 30–6** Joseph H. Taylor, Jr., at the Princeton Pulsar Physics Laboratory; **Fig. 30–7** Parkes Multibeam Survey; **Fig. 30–10** Jeff Percival and R. C. Bless, U. Wisconsin/NASA; **Fig. 30–11** courtesy of Joachim Trümper, Max-Planck-Institut für Extraterrestrische Physik; **Fig. 30–12A** Jeff Hester and Paul Scowen (Arizona State U.), and NASA; **Fig. 30–12B** NASA/Chandra X-ray Observatory Center/SAO; **Fig. 30–13** European Southern Observatory; **Fig. 30–14** Chandra X-ray Observatory ACIS Image, NASA/PSU/G. Pavlov, et al.; **Fig. 30–15** Walter Brisken, at the Princeton Pulsar Physics Laboratory; **Fig.**

**30–16A** Jeff Hester and Shrinivas Kulkarni, Palomar Observatory/California Institute of Technology, with the 5-m telescope, and NASA; **Fig. 30–16B** B. M. Gaensler/D. F. Frail/MIT/NRAO; **Fig. 30–17** NASA/McGill/Victoria Kaspi, Mallory Roberts (McGill U.), Gautum Vasisht (JPL), Eric Gotthelf (Columbia U.), Michael Pivovaroff (Therma-Wave, Inc.), and Nobuyuki Kawai (Institute of Physical and Chemical Research, Japan); **Fig. 30–19A** Jay M. Pasachoff; **Fig. 30–19B** Tony Tyson, Lucent Technologies/Bell Labs Innovations; **Fig. 30–20** Photos courtesy of Caltech/LIGO; thanks to Dave Beckett and Jonathan Kern; **Fig. 30–21** Hanns Ruder/Universität Tübingen; **Fig. 30–22** Bruce Margon, U. Washington, now at STScI; **Fig. 30–24A** NRAO, operated by Associated Universities, Inc., under contract with the NSF, courtesy of R. Hjellming; **Fig. 30–24B** Katherine Blundell (Oxford U.), Amy Mioduszewski (NRAO), Tom Muxlow (Jodrell Bank, U. Manchester) and Michael Rupen (NRAO); **Fig. 30–25** courtesy of M. Watson, R. Willingale, Jonathan E. Grindlay, and Frederick D. Seward, see *Astrophysical J.* 273, 688 (1983), courtesy of U. Chicago Press; **Fig. 30–26** NASA and F. Walter (State University of New York at Stony Brook); **Fig. 30–27** European Southern Observatory.

**Chapter 31—Opener** VLA of NRAO/AUI; **Fig. 31–1** Huntington Library, San Marino, CA, and the Albert Einstein Archives, courtesy of the Hebrew Univerity of Jerusalem, Jewish National University and Library; **Fig. 31–3A** Copyright U.C. Regents UCO/Lick Observatory image; **Figs. 31–3B** and **31–4B** Albert Einstein Archives, courtesy of the Hebrew University of Jerusalem, Jewish National University and Library; **Fig. 31–4A** Courtesy of the Archives, California Institute of Technology; **Figs. 31–5** and **31–6** © 1919 The New York Times Co. Reprinted by permission; **Fig. 31–7** AP/Wide World photos, Eric Risberg; **Fig. 31–12** © The New Yorker Collection 1974 Charles Addams from cartoonbank.com. All Rights Reserved; **Fig. 31–13** CXC/M. Weiss; **Fig. 31–14** after E. H. Harrison, U. Mass.-Amherst; **Fig. 31–15** Chapin Library, Williams College; **Fig. 31–16AB** Kevin Rauch, Johns Hopkins U.; **Fig. 31–17** John F. Hawley, Leander McCormick Observatory, University of Virginia; **Fig. 31–18** Palomar Observatory/California Institute of Technology photograph by Jerome Kristian; **Fig. 31–19A** Robert M. Hjellming and Michael P. Rupen, VLBA of NRAO/AUI; **Fig. 31–19B** Robert M. Hjellming et al., VLA of NRAO/AUI; **Fig. 31–20A** Ann Feild (STScI); **Fig. 31–20B** Joseph Dolan, NASA's Goddard Space Flight Center; **Fig. 31–21A** Phil Charles, U. of Southampton, England; **Fig. 31–21B** Jerome A. Orosz, Utrecht U., Netherlands; **Fig. 31–23** from *Image the Universe,* High Energy Astrophysics Science Archive Research Center, NASA's Goddard Space Flight Center; **Fig. 31–24** © Julian Calder/Woodfin Camp; **Fig. 31–25** L. Ferrarese (Johns Hopkins U.) and NASA; **Fig. 31–26** NASA/Chandra X-ray Observatory Center/SAO; **Fig. 31–27** Chandra X-ray Observatory ACIS/HETG Image; **Fig. 31–28** Roy Bishop, Acadia University; **Figs. 31–29ABCD** Max-Planck-Institute for Gravitational Physics (AEI) and Konrod-Zuse Center for Information Techniques, Berlin; courtesy of Werner Benger, Max-Planck-Institut für Gravitationalphysik, Albert-Einstein-Institut, Potsdam, Germany; International Numerical Relativity Group (NCSA/Laboratory for Computational Astrophysics-Potsdam-Washington U.; **Fig. 31–30** Extended from a merger of diagrams from *Physics Today* and from Visual Science, Inc., as adapted in *Sky & Telescope.*

**Part 6—Opener** The National Gallery, London.

**Chapter 32—Opener** Wolfgang Brandner (JPL/IPAC), Eva K. Grebel (U. Washington), You-Hua Chu (U. Illinois at Urbana-Champaign), and NASA; **Fig. 32–1** NASA and The Hubble Heritage Team (STScI); **Fig. 32–2** © 1985 Royal Observatory Edinburgh/Anglo-Australian Observatory, from original U.K. Schmidt Plates; **Fig. 32–3** Axel Mellinger, Universität Potsdam; **Fig. 32–4** Mark McCaughrean (Astrophysikalisches Institut Potsdam), and collaborators; European Southern Observatory; **Fig. 32–5** © 1979 Royal Observatory Edinburgh/Anglo-Australian Observatory, from original U.K. Schmidt Plates; **Fig. 32–6A** Image courtesy of Dr. Jean-Charles Cuillandre, CFHT; **Figs. 32–6B and C** Infrared Processing and Analysis Center/Caltech and University of Massachusetts (IPAC/U. Mass); **Fig. 32–7A** © 1986 Anglo-Australian Observatory, photograph by David F. Malin; **Fig. 32–7B** (left) Jeff Hester and Paul Scowen (Arizona State U.), and NASA/STScI/ESA; (right) Mark McCaughrean (Astrophysikalisches Institut Potsdam), with the Calar Alto 3.5-m telescope; **Fig. 32–8** European Southern Observatory; **Fig. 32–10** NASA's Goddard Space Flight Center Astrophysics Data Facility; **Fig. 32–11** Hui Yang (U. Illinois), Jeff J. Hester (U. Arizona), and NASA; **Fig. 32–12** Don Figer (STScI) and NASA; **Fig. 32–13** Infrared Processing and Analysis Center/Caltech and University of Massachusetts (IPAC/U. Mass); **Fig. 32–14** DoD High Performance Computing Resources; produced by N. E. Kassim, D. S. Briggs, T. J. W. Lazio, T. N. LaRosa, J. Imamura, and S. D. Hyman; original data from NRAO Very Large Array, courtesy of A. Pedlar, K. Anantharamiah, M. Goss, and R. Ekers; **Fig. 32–15** Andrea Ghez (UCLA); from *Nature* 407, 349–351, 2000, courtesy of Macmillan Publishing Co.; **Fig. 32–16A** Midcourse Space Experiment: Spatial Infrared Imaging Telescope (SPIRIT III) S. D. Price, M. P. Egan, R. F. Shipman (Air Force Research Lab./VSBC), E. F. Tedesco (Mission Research Corp.), M. Cohen (U. California Radioastronomy Lab.), R. G. Walker (Vanguard Research, Inc.), M. Moshir (JPL); **Fig. 32–16B** Photo courtesy of Gemini Observatory, National Science Foundation and the University of Hawaii Adaptive Optics Group; **Fig. 32–17A** Neil E. Killeen (CSIRO, ATNF, Australia) and K. Y. Lo, U. Illinois at Urbana-Champaign, with the VLA of NRAO; **Fig. 32–17B** Mark Morris (UCLA), Farhad Yusef-Zadeh (Northwestern U.), and Don Chance (STScI) with the VLA of NRAO; **Fig. 32–18** NASA/Penn State/G. Garmire et al.; **Fig. 32–19** Lund Observatory, Sweden; **Fig. 32–20** M. Hauser (STScI) and NASA; **Fig. 32–21** EGRET Science Team/NASA, courtesy of Carl Fichtel, Goddard Space Flight Center; **Fig. 32–22** P. Sreekumar (NASA/GSFC); **Fig. 32–23AB** NASA's Goddard Space Flight Center; **Fig. 32–24** Jay M. Pasachoff; **Fig. 32–25** after Agris Kalnajs, Mount Stromlo and Siding Spring Observatories; **Fig. 32–26** Stuart N. Vogel, U. Maryland; **Fig. 32–27** Zsolt Frei, Institute of Physics, Eötvös University, Budapest; **Fig. 32–28** Torsten Boeker, Space Telescope Science Institute and NASA.

**Chapter 33—Opener** © 1999–2001 Subaru Telescope, NAOJ. All rights reserved; **Fig. 33–1** © 1981 Royal Observatory Edinburgh/Anglo-Australian Observatory from original U.K. Schmidt plates; **Fig. 33–2B** Karen B. Kwitter, Williams College; **Fig. 33–3A** European Southern Observatory; **Fig. 33–3B** A. Caulet (ST-ECF, ESA) and NASA; **Fig. 33–4A** C. R. Barber and R. S. Warwick, University of Leicester; **Fig. 33–4B** Harvard-Smithsonian Center for Astrophysics, courtesy of Gary Melnick and René Plume; **Fig. 33–5** NRAO/AUI, James J. Condon, John J. Broderick, and George A. Seielstad; **Fig. 33–6B** © 1988, Regents of the University of Minnesota; collaboration of University of Minnesota, Air Force Geophysics Laboratory, and NASA, John R. Winckler, Perry Malcolm, and colleagues; TV camera experiment by Robert Franz; **Fig. 33–7** © David F. Malin, Anglo-Australian Observatory, and Jay M. Pasachoff, from Palomar Observatory plates made available by the California Institute of Technology, courtesy of Robert Brucato; **Fig. 33–8** Jodrell Bank, Max-Planck-Institut, Australian National Radio Astronomy Observatory at Parkes; combined at Max-Planck-Institut für Radioastronomie and Computing Center of Bonn University; **Photo Gallery p. 650** *(top)* NASA and D. Walter (South Carolina State University); **p. 650** *(bottom)* European Southern Observatory; **p. 651** *(top)* ESA/ISO/ISOCAM and L. Nordh (Stockholm Observatory) et al.; **p. 651** *(bottom)* Infrared Processing and Analysis Center/Caltech and University of Massachusetts; **Fig. 33–12** Image assembled by C. Jones and W. Forman from observations by A. A. Stark et al., M. N. Cleary and C. Heiles, and F. Kerr et al., See *Astronomy and Astrophysics Supplement* 36, p. 95; **Fig. 33–13** Dap Hartmann, Harvard-Smithsonian Center for Astrophysics, and W. B. Burton at Leiden Obs., see their *Atlas of Galactic Neutral Hydrogen,* Cambridge U. Press, 1997; **Fig. 33–16A** Dan P. Clemens (U. Arizona), D. Sanders (U. Hawaii) and N. Scoville (Caltech); **Fig. 33–16B** J. Keene, ESA/ISO and LWS Consortium; **Fig. 33–16C** ESA/ISO, SWS, C. M. Wright et al.; **Fig. 33–17C** J. Brand and L. Blitz, *Astronomy and Astrophysics* **275,** 1993, Fig. 3, p. 76; **Fig. 33–18A** Edwin A. Valentijn, Paul P. van der Werf, Th. de Graauw, and T. de Jong with the Short Wavelength Spectrometer on ISO of ESA, courtesy of Diego Cesarsky; **Fig. 33–18B** Courtesy of Scott D. Friedman (Johns Hopkins University) and George Sonneborn (NASA's Goddard Space Flight Center); **Fig. 33–19** Thomas M. Dame, Dap Hartmann, and Patrick Thaddeus, Harvard-Smithsonian Center for Astrophysics; **Fig. 33–20** Philip Solomon and Arthur Rivolo based on Stony Brook catalogue of giant molecular clouds, *Astrophys. J.* **319, 730**, 1987; **Fig. 33–21** Ian Gatley,

NOAO; **Fig. 33–22A** C. R. O'Dell (Rice U.) and NASA; **Fig. 33–22B** © Anglo-Australian Observatory, photography David F. Malin; **Fig. 33–23** After Ben Zuckerman, UCLA; **Fig. 33–24** IPAC, Caltech/JPL; **Fig. 33–25** WFPC2 image: C. Robert O'Dell, Shui Kwan Wong (Rice U.) and NASA; NICMOS image: Rodger Thompson, Marcia Rieke, Glenn Schneider, Susan Stolovy (U. Arizona), Edwin Erickson (SETI Institute/Ames Research Center), David Axon (STScI), and NASA; **Fig. 33–26AB** Subaru Telescope, National Astronomical Observatory of Japan; **Fig. 33–27** European Southern Observatory; **Figs. 33–28ABCDE** Orion studies by Norbert S. Schulz, Claude R. Canizares, Dave Huenemoerder, Sara-anne Taylor (MIT) and Joel Kastner (Rochester Institute of Technology); ABC: 2MASS Collaboration/John Bally, CASA Colorado and underlying image; DE: ACIS on Chandra from Penn State U.; **Table 33–1** from Marla H. Moore, NASA's Goddard Space Flight Center.

**Part 7—Opener** -© NASA and W. Keel (University of Alabama).

**Chapter 34—Opener** © 1980 Anglo-Australian Observatory, photograph by David F. Malin; **Fig. 34–1** Jay M. Pasachoff; **Fig. 34–2A** © 1987 Anglo-Australian Observatory, photograph by David F. Malin; **Fig. 34–2B** European Southern Observatory; **Fig. 34–3** Kazuo Shiota; **Fig. 34–4A** NOAO/N. A. Sharp; **Fig. 34–4B** M. Sauvage, J. Blommaert, F. Boulanger, et al., with ISOCAM of ISO, courtesy of Diego Cesarsky; **Fig. 34–4C** S. García-Burillo, M. Guélin and J. Cernicharo; **Fig. 34–5A** © 1991 Richard J. Wainscoat/Institute for Astronomy, U. Hawaii; **Fig. 34–5B** European Southern Observatory; **Fig. 34–6A** © 1980 Anglo-Australian Observatory, photograph by David F. Malin; **Fig. 34–6B** Infrared processing and analysis Center/Caltech and University of Massachusetts (IPAC/U. Mass); **Fig. 34–7** European Southern Observatory; **Fig. 34–8AB** © 1984 Royal Observatory Edinburgh/Anglo-Australian Observatory; **Fig. 34–8C** Steven L. Snowden, USRA/LHEA/GSFC; **Fig. 34–9A** Jack Newton; **Fig. 34–9B** © National Astronomical Observatory of Japan, Subaru Telescope. All rights reserved; **Fig. 34–11** Željko Ivezić and Robert Lupton for the Sloan Digital Sky Survey Collaboration; **Fig. 34–12** Adapted from T. H. Jarrett 2000, *Pub. Astron. Soc. Pacific* **112,** 1008; **Fig. 34–13** ESA/ISO/ISOPHOT and M. Haas, D. Lemke, M. Stickel, H. Hippelein, et al.; **Fig. 34–14** R. Wielebinski, R. Beck, E. Berkhuijsen; **Fig. 34–15** John Kormendy (U. Texas at Austin); **Fig. 34–16** G. Bower and Richard Green (NOAO), the STIS Instrument Definition Team, and NASA; **Fig. 34–17** NASA/SAO/CXC/PSU/CMU; **Fig. 34–18** NASA/SAO/CXC; **Fig. 34–19** NASA and K. Gebhardt (Lick Observatory); **Fig. 34–20A** Curt Struck and Philip Appleton (Iowa State U.), Kirk Borne (Hughes STX Corp.), and Ray Lucas (STScI), and NASA; **Fig. 34–20B** Hubble Heritage Team (AURA/STScI/NASA); **Fig. 34–21A** Joshua Barnes, Institute for Astronomy, U. Hawaii, and Lars Hernquist, Harvard-Smithsonian Center for Astrophysics, then both at Institute for Advanced Study; **Fig. 34–21B** David Dubinski, CITA, U. Toronto; **Fig. 34–22A** John Hibbard (NRAO) at the VLA of NRAO; **Fig. 34–22B** Brad Whitmore (STScI) and NASA; **Fig. 34–22C** ISOCAM image courtesy of Catherine Cesarsky, CEA/Saclay; L. Vigroux, F. Mirabel, B. Altiéri, F. Boulanger, C. Cesarsky, D. Cesarky, A. Claret, C. Fransson, P. Gallais, D. Levine, S. Madden, K. Okumura, and D. Tran; **Fig. 34–22D** NASA/SAO/CXC/G. Fabbiano et al.; **Fig. 34–23** NASA, William C. Keel (University of Alabama, Tuscaloosa); **Fig. 34–24A** © 1981 Anglo-Australian Observatory; **Fig. 34–24B** European Southern Observatory; **Fig. 34–25** © 1987 Royal Observatory Edinburgh/Anglo-Australian Observatory, from original U.K. Schmidt plates; **Fig. 34–26A** Exosat observations courtesy of Michiel van der Klis, astronomical institute "anton pannekoek," Amsterdam, The Netherlands; **Fig. 34–26B** Courtesy of Ulrich Briel, MPE Garching; **Figs. 34–26CD** M. P. Sirk, ESA/ISO; **Fig. 34–27A** NASA/CfA/J. Vrtilek et al.; **Fig. 34–27B** NASA/CXC/SAO; **Fig. 34–27C** NASA/IoA/Andrew Fabian et al., U. Cambridge; **Fig. 34–27D** NASA/CXC/SAO; **Fig. 34–28** Brent Tully, Institute for Astronomy, U. Hawaii; **Fig. 34–29** Andrew Fabian, Carolin Crawford, Stefano Ettori and Jeremy Sanders of the Institute of Astronomy, University of Cambridge; **Fig. 34–30A** NASA, A. Fruchter and the ERO team (STScI, ST-EFC); **Fig. 34–30B** Kavan Ratnatunga (Carnegie Mellon Univ.) and NASA; **Fig. 34–31A** NASA/UMD/A. Wilson et al.; **Fig. 34–31B** Radio sky: NRAO/AUI, observers James J. Condon, John J. Proderick, and George A. Seielstad; optical image from Laird Thompson, U. Illinois; *Astrophys. J. 279,* L47 (1984), courtesy of U. Chicago Press; (*lower inset*) Alan Stockton, U. Hawaii; (*upper inset*) Fernando Cabrera Guerra and NASA/ESA; **Fig. 34–31C** NRAO/Robert C. Byrd Green Bank Telescope; **Fig. 34–32A** © 1980 Anglo-Australian Observatory; **Fig. 34–32B** NASA/Chandra X-ray Observatory Center/Smithsonian Astrophysical Observatory; **Fig. 34–35** Riccardo Giovanelli and Martha Haynes, Cornell U.; **Fig. 34–36A** VLA of NRAO; **Fig. 34–36B** © NRAO/AUI, observers: R. C. Walker, J. D. Romney, and J. M. Benson; **Fig. 34–37** William C. Keel) (U. Alabama), Karl-Heinz Mack, Dayton Jones (JPL), and Rick Perley (NRAO); **Fig. AN–1** Jay M. Pasachoff; **Fig. AN–2** VLA of NRAOl; **Fig. AN–3** Jay M. Pasachoff, Donald A. Lubowich, and K. Anantharamaiah with the VLA of NRAO.

**Chapter 35—Opener** Hubble Space Telescope WFPC Team and NASA, courtesy of William A. Baum, U. Washington; **Fig. 35–1A** National Academy of Sciences; **Fig. 35–1B** From E. Hubble and M. L. Humason, *Astrophysical Journal 74,* 77, 1931, courtesy of U. Chicago Press; **Fig. 35–2** image: N. A. Sharp/NOAO; spectra: Ned Wright, UCLA, from B. Zuckerman and M. Malkan: *The Origin and Evolution of the Universe,* © 1996: Jones & Bartlett Publishers, Sudbury M.A www.jbpub.com. Reprinted with permission; **Fig. 35–5** © 1986 Anglo-Australian Observatory; **Figs. 35–6, 35–7A, 35–7B** Wendy L. Freedman, Observatories of the Carnegie Institution of Washington, and NASA; **Fig. 35–8** J. Trauger (JPL) and NASA; **Fig. 35–9** Hubble Space Telescope WFPC Team; **Figs. 35–10** and **35–11** Wendy L. Freedman, Observatories of the Carnegie Institution of Washington, and NASA; **Fig. 35–12AB** Brian P. Schmidt (Australian National U.) and the High-Z Supernova Search Team, and NASA; **Fig. 35–13** David Burstein, Sandra Faber, Roger Davies, Alan Dressler, D. Lynden-Bell, Roberto Terlevich, and Gary Wegner; image by Ofer Lahav, Institute of Astronomy, Cambridge U.; **Fig. 35–14A** Luiz N. da Costa, Wolfram Freudling, Gary Wegner, Riccardo Giovanelli, Martha P. Haynes, and John Salzer; **Fig. 35–14B** Avishai Dekel, U. Cal./Berkeley, and Hebrew University, Jerusalem; **Fig. 35–15** Rudolph Schild, Smithsonian Astrophysical Observatory; **Fig. 35–16A** John Huchra, Margaret Geller, and V. de Lapparent; © 1989 Smithsonian Astrophysical Observatory; **Fig. 35–16B** Margaret Geller and John Huchra; © 1993 Smithsonian Astrophysical Observatory; **Fig. 35–17** 2dF Galaxy Redshift Survey; **Fig. 35–18** Robert Smith, Astrophysics Research Institute, Liverpool John Moores University/2dF Galaxy Redshift Survey; **Fig. 35–19** Sloan Digital Sky Survey; **Fig. 35–20** Esther M. Hu, Richard G. McMahon, & Lennox L. Cowie, *Astrophys. J. Letters* **522,** L9 (September 1, 1999); **Fig. 35–21** Rogier Windhorst and Simon Driver (Arizona State U.); **Fig. 35–22** BATSE, Compton Gamma Ray Observatory Science Support Center, Goddard Space Flight Center; **Fig. 35–23** J. S. Bloom and the Caltech GRB group; *Science Magazine,* 5 Jan 2001, vol. **291**, p. 82, © American Association for the Advancement of Science; **Fig. 35–24** S. G. Djorgovski and S. R. Kulkarni (Caltech), the Caltech GRB Team, W. M. Keck Observatory and NASA; **Fig. 35–25A** S. Jha et al. (Harvard-Smithsonian Center for Astrophysics); **Figs. 35–25BCD** NASA/CNR/L. Piro et al.; **Fig. 35–26** Joan Centrella, Drexel U.;
**Fig. 35–27** Virgo Consortium; U. of Durham; **Fig. 35–28** Rogier Windhorst and Sam Pascarelle (Arizona State University) and NASA.

**Chapter 36—Opener** John N. Bahcall and Sofia Kirhakos (Institute for Advanced Study), and Donald P. Schneider (Pennsylvania State U.), and NASA; **Fig. 36–1** Herman-Josef Roeser (ESO NTT); Herman-Josef Roeser and NASA (HST); and Palomar Observatory, Caltech (POSS); **Fig. 36–2A** NASA/STScI; **Fig. 36–2B** NASA/CXO/SAC/H. Marshall et al.; **Fig. 36–2C** Optical: NASA/STScI; X-ray: NASA/CXO; Radio: MERLIN; **Fig. 36–3** Maarten Schmidt/Palomar Observatory/California Institute of Technology photograph; **Fig. 36–4** C. Pilachowski, M. Corbin/AURA/NOAO/NSF; **Fig. 36–5** William C. Keel (U. Alabama), Michael Rauch (Caltech), and NASA; **Figs. 36–6AB** Sloan Digital Sky Survey; **Fig. 36–7** Robert J. Smith, Liverpool John Moores University; 2dF QSO Redshift Survey; **Fig. 36–8A** XMM-Newton/European Space Agency/EPIC-pn image 0.5–2.0 keV; Niel Brandt, Shai Kaspi, and Donald Schneider (Penn State U.), Xiaohui Fan (Institute for Advanced Study), and Michael Strauss and James Gunn (Princeton University), with XMM-Newton scientists Matteo Guainazzi and Jean Clavel (VILSPA Science Operations Centre, Villafranca, Spain); **Fig. 36–8B** Richard L. White, SDSS Consortium; **Fig. 36–9** Tony Tyson (Bell Labs, Lucent Technologies), W. A. Baum

(Lowell Obs.), and Tobias J. N. Kreidl (Northern Arizona U. and Lowell Obs.); **Figs. 36–11** and **36–12** 2dF QSO Redshift Survey; **Fig. 36–13** Warren S., Hewett P. and Osmer P. 1994 *Astrophys. J.* **421** 412; **Fig. 36–14B** T. J. Balonek et al., Colgate, Foggy Bottom Observatory; **Fig. 36–15** Stuart L. Shapiro and Saul Teukolsky; **Fig. 36–17A** Richard Green (NOAO) and Howard K. C. Yee (U. Toronto), on the Steward Obs. 2.3-m tel.; **Figs. 36–17BC** John B. Hutchings (Dominion Astrophysical Obs.); **Fig. 36–17D** Matthew A. Malkan (UCLA) **Fig. 36–18** X-ray: NASA/IOA/Fabian et al., Optical: NASA/U. Durham/Smail et al.; **Fig. 36–19** Pierre-Alain Duc (Service d'Astrophysique Saclay, France) and Mariano Moles Villamate (Instituto de Matematicas y Fisica Fundamental Madrid, Spain); courtesy of Lars Lindberg Christensen, European Space Agency; **Fig. 36–20A** Alan Stockton (Institute for Astronomy, U. Hawaii); **Fig. 36–20B** Stanislav Djorgovski—Caltech/W. M. Keck Observatory, courtesy of Andy Perala; **Figs. 36–21AB** European Southern Observatory; **Fig. 36–22** Hubble Heritage Team (AURA/STScI/NASA); **Fig. 36–23AB** John B. Hutchings (Dominion Astrophysical Obs.)/CFHT and NASA/STScI; **Fig. 36–23C** Joe Miller and Andy Sheinis, University of California at Santa Cruz/Lick Observatory; **Fig. 36–24** John N. Bahcall and Sofia Kirhakos (Institute for Advanced Study), and Donald P. Schneider (Pennsylvania State U.), and NASA; **Fig. 36–25** John N. Bahcall (Institute for Advanced Study) and Mike Disney (U. of Wales), and NASA; **Fig. 36–26A** Stephen C. Unwin, Caltech; **Fig. 36–26B** José-Luis Gómez, Alan P. Marscher, Antonio Alberdi, Svetlana G. Jorstad, and Cristina Garcia-Miró, from *Science Magazine,* 29 September 2000, **289,** pp. 2317–2320, © American Association for the Advancement of Science; **Fig. 36–28** EGRET team, Compton Observatory, NASA; **Fig. 36–29** Alan Stockton (Institute for Astronomy, U. Hawaii); **Fig. 36–30** B. F. Burke, P. E. Greenfield, D. H. Roberts, with the VLA of NRAO; **Fig. 36–31** W. M. Keck Obs., picture by Keith Matthews and James Larkin (Caltech); **Fig. 36–32** George Chartas (Penn State U) and Marshall W. Bautz (MIT); CXC/ASO/NASA; **Fig. 36–33A** Jacqueline N. Hewitt (MIT), et al.; **Fig. 36–33B** Pierre Magain, et al., observations at ESO; **Fig. 36–33C** NASA and ESA; Kavan Ratnatunga (Carnegie Mellon Univ.), and NASA; **Fig. 36–34** Kavan Ratnatunga (Carnegie Mellon Univ.) and NASA.

**Chapter 37—Opener** René Magritte, *The Blank Signature* (1965). From the collection of Mr. and Mrs. Paul Mellon, © Board of Trustees, National Gallery of Art, Washington, D.C.; **Fig. 37–2** Jay M. Pasachoff; **Fig. 37–3** © 1984 Royal Observatory Edinburgh/Anglo-Australian Observatory, photograph from original U.K. Schmidt plates by David Malin; **Fig. 37–10** A. Dressler (Carnegie Institution of Washington), M. Dickinson (STScI), D. Macchetto (ESA/STScI), M. Giavalisco (STScI), and NASA; **Fig. 37–11** Jay M. Pasachoff; **Fig. 37–12** Supernova Cosmology Project, courtesy of Saul Perlmutter, Lawrence Berkeley National Laboratory; **Fig. Box 37.1A** NASA's Johnson Space Center, courtesy of Mike Gentry; **Fig. Box 37.1B** Jay M. Pasachoff; **Fig. 37–13** Bell Laboratories, Lucent Technologies; **Fig. 37–16** S. P. Boughn and R. B. Partridge, Haverford College; **Fig. 37–17** Map: E. S. Cheng (NASA's Goddard Space Flight Center), D. A. Cottingham (U. Cal.-Berkeley), S. Boughn (Haverford College), D. T. Wilkinson (Princeton U.), and D. J. Fixsen (GSFC/USRA), Graphics concept and execution: D. Hon, STX Corp.), supported by a grant from NASA, Astrophysics Division; **Fig. 37–18** NASA COBE Science Team, courtesy of Nancy Boggess; **Fig. 37–19** NASA COBE Science Team, special graphing courtesy of E. S. Cheng; **Figs. 37–20** and **37–21** NASA COBE Science Team; **Fig. 37–22** Updated in 2001 from Max Tegmark, Angelia de Oliveira-Costa, Marc Devlin, Barth Netterfield, Lyman Page, and Ed Wollack, in *Astrophys. J.* **474,** L77–80, 1996; **Fig. 37–23** BOOMERanG Consortium; **Fig. 37–24** MAXIMA Consortium/Paul R. Richards (University of California at Berkeley) et al.; **Fig. 37–25** BOOMERanG Consortium; **Fig. 37–26** BOOMERanG and MAXIMA Consortia, from the MAXIMA press release; **Figs. 37–27** and **37–28** Graphic courtesy of the DASI Collaboration; **Fig. 37–29** European Space Agency.

**Chapter 38—Opener** Robert Williams and the Hubble Deep Field Team (STScI) and NASA; **Fig. 38–1** Photo CERN; **Fig. 38–2** James W. Cronin (U. Chicago); **Fig. 38–3** Brookhaven National Laboratory, Department of Energy/STAR detector at the Relativistic Heavy Ion Collider; **Fig. 38–4** Robert V. Wagoner (Stanford U.); **Fig. 38–5** Photography by B. W. Hadley, © 1987 Royal Observatory Edinburgh; **Fig. 38–6A** Tony Tyson, Bell Labs, Lucent Technologies; **Fig. 38–6B** Hubble Space Telescope image: NASA and Dave Bennett (U. Notre Dame); event discovered by the Massive Compact Halo Object (MACHO) collaboration; ground-based images: Cerro Tololo Inter-American Observatory/NOAO/AURA; **Fig. 38–6C** N-Body Shop, University of Washington, Seattle; **Fig. 38–7** Robert V. Wagoner (Stanford U.); **Fig. 38–8A** Jeffrey L. Linsky, Alexander Brown, Ken Gayley, Athanassios Diplas, Blair D. Savage, Thomas R. Ayres, Wayne Landsman, Steven N. Shore, and Sara R. Heap, *Astrophysical J.* **402,** 694, 1993; **Fig. 38–8B** Courtesy of Scott D. Friedman (Johns Hopkins University) and George Sonneborn (NASA's Goddard Space Flight Center); **Fig. 38–9** Joe Stancampiano and Karl Luttrell/U. Michigan, © National Geographic Society; **Fig. 38–11** Philip Crane; **Fig. 38–12** Photo CERN, courtesy of Gail Hanson; **Fig. 38–13** Superconducting Super Collider Laboratory (R.I.P.); **Fig. 38–14** Fermi National Accelerator Laboratory; **Fig. 38–15** "Particle Accelerators Test Cosmological Theory," by David N. Schramm and Gary Steigman, copyright © 1988 by Scientific American, Inc. All rights reserved; Drawn by Andrew Christie; **Fig. 38–16** © 1989 Sidney Harris—*Physics Today;* **Fig. 38–19** Jay M. Pasachoff; **Fig. 38–21** Mark Bowick (Syracuse U.); **Fig. 38–22** David Bennett, François Boucher, and Albert Stebbins; **Fig. 38–23** Robert Williams and the Hubble Deep Field Team (STScI), and NASA; **Fig. 38–24** Robert Williams and the Hubble Deep Field Team (STScI), and NASA; with redshifts measured at the Keck Telescope compiled by M. Dickinson and Z. Levay; **Fig. 38–25A** Robert Williams and the Hubble Deep Field Team (STScI), and NASA; **Fig. 38–25B** Robert Williams and the Hubble Deep Field—South Team (STScI), and NASA; **Fig. 38–26A** NASA/JHU/AUI/R. Giacconi et al.; **Fig. 38–26B** optical: NASA/HST; X-ray NASA/PSU; **Fig. 38–27A** NASA's MAP Science Team; **Fig. 38–27B** NASA's MAP Science Team.

**Epilogue** European Space Agency.

# INDEX

References to illustrations, either photographs or drawings, are italicized. References to tables are followed by t. References to boxes are followed by B. References to Appendices are prefaced by A.

Significant initial numbers followed by letters are alphabetized under their spellings; for example, 21 cm is alphabetized as twenty-one. Less imporant initial numbers are ignored in alphabetizing. For example, 3C 273 appears at the beginning of the Cs. M1 appears at the beginning of the Ms.

## A

## B

## D



## E

## F

## G

## H

## N

## O

# T

## U



## V

## W



## X



## Y



## Z

# SPRING SKY

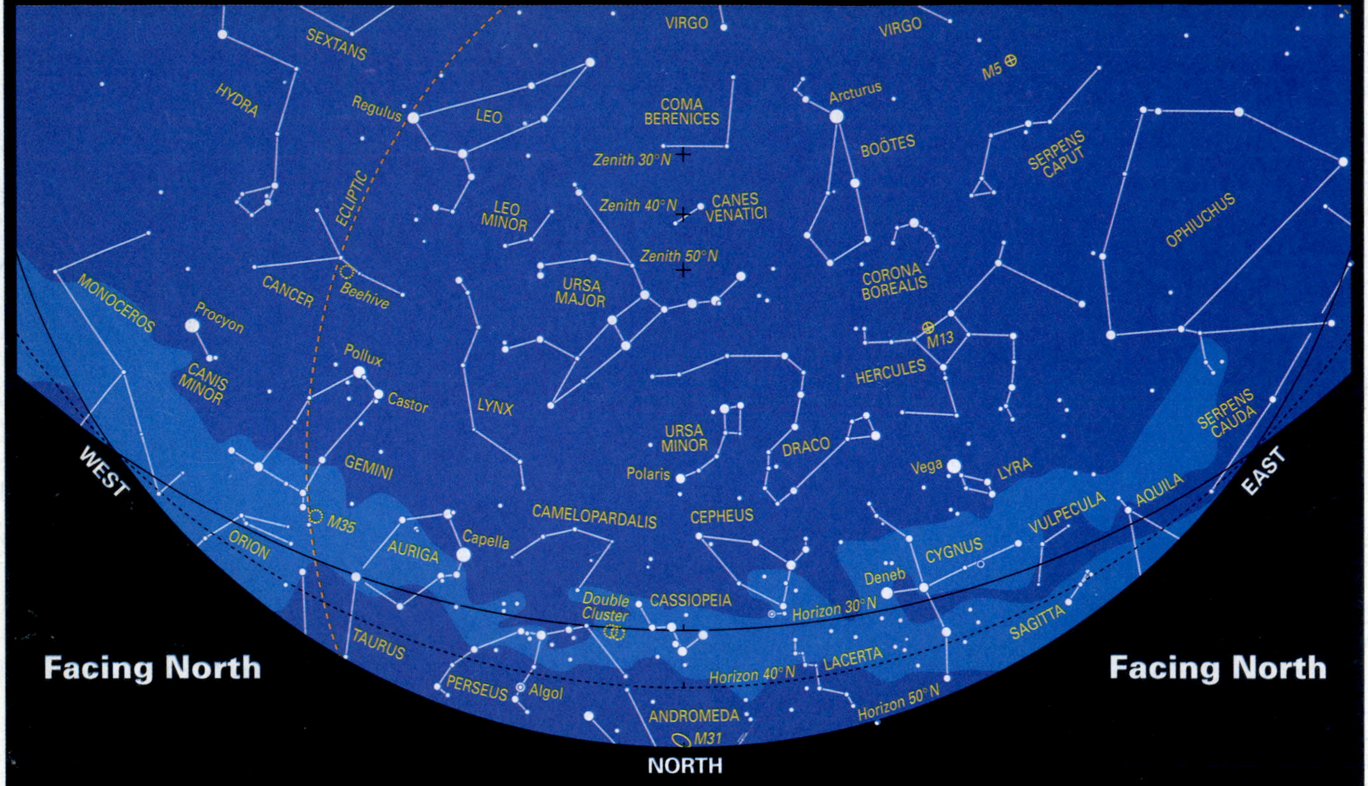

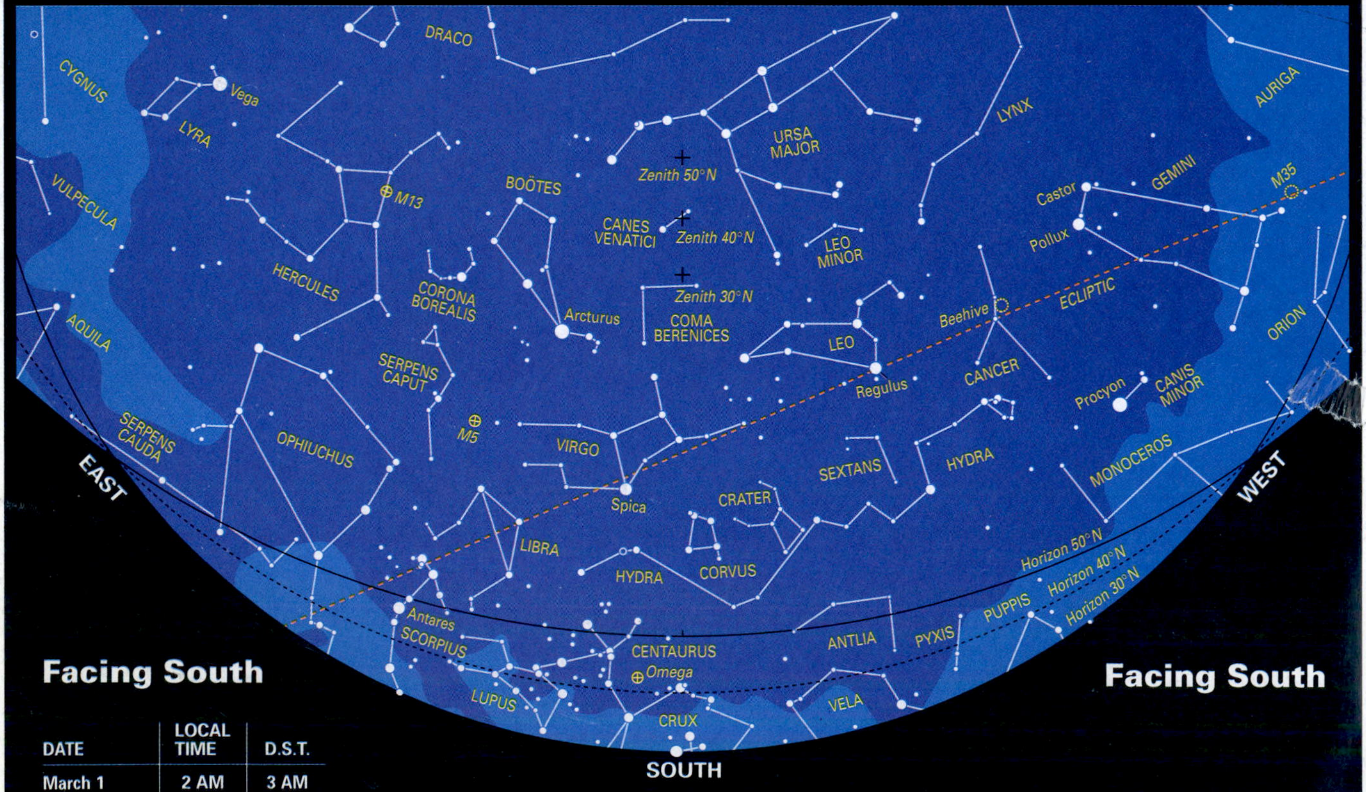

WEST — NORTH — EAST

**Facing North** **Facing North**

EAST — SOUTH — WEST

**Facing South** **Facing South**

| DATE | LOCAL TIME | D.S.T. |
|---|---|---|
| March 1 | 2 AM | 3 AM |
| March 15 | 1 AM | 2 AM |
| April 1 | Midnight | 1 AM |
| April 15 | 11 PM | Midnight |
| May 1 | 10 PM | 11 PM |
| May 15 | 9 PM | 10 PM |
| June 1 | 8 PM | 9 PM |
| June 15 | 7 PM | 8 PM |

**MAGNITUDES**

-1 · 0 · 1 · 2 · 3 · 4 · 5

Variable Stars
Open Cluster
Globular Cluster
Nebula
Galaxy

MAP BY WIL TIRION; FOR JAY M. PASACHOFF